An Ottoman Cosmography

An Ottoman Cosmography

Translation of Cihānnümā

By

Kātib Çelebi

Edited by

Gottfried Hagen
Robert Dankoff

Translated by

Ferenc Csirkés
John Curry
Gary Leiser

BRILL

LEIDEN | BOSTON

Originally published in hardback in 2021 as Volume 142 in the series Handbook of Oriental Studies – Handbuch der Orientalistik Section, Section 1, The Near and Middle East.

Cover illustration: Map of Gilolo, Molucca, Mindanao, R 49a; cf. M 133/134, 135/136 (Appendix, fig. 58–62) (Topkapı Palace Library, Revan 1624, with permission).

The Library of Congress has cataloged the hardcover edition as follows:

Names: Kâtip Çelebi, 1609-1657, author. | Hagen, Gottfried, 1963- editor. | Dankoff, Robert, editor.
Title: An Ottoman cosmography : translation of Cihānnümā / by Kātib Çelebi ; edited by Gottfried Hagen, Robert Dankoff ; translated by Ferenc Csirkés, John Curry, Gary Leiser.
Other titles: Cihānnümā. English
Description: Leiden ; Boston : Brill, [2021] | Series: Handbook of oriental studies. Section 1, The Near and Middle East, 0169-9423 ; vol. 142 | Includes index. | Translated from Turkish.
Identifiers: LCCN 2021023566 (print) | LCCN 2021023567 (ebook) | ISBN 9789004441323 (hardback) | ISBN 9789004441330 (ebook)
Subjects: LCSH: Geography–Early works to 1800. | World maps–Early works to 1800. | Turkey–Geography–Early works to 1800. | Cartography–Early works to 1800. | Spherical astronomy–Early works to 1800. | Cosmography –Early works to 1800.
Classification: LCC G114 .K3813 2021 (print) | LCC G114 (ebook) | DDC 910–dc23
LC record available at https://lccn.loc.gov/2021023566
LC ebook record available at https://lccn.loc.gov/2021023567

Typeface for the Latin, Greek, and Cyrillic scripts: "Brill". See and download: brill.com/brill-typeface.

ISBN 978-90-04-72962-9 (paperback, 2025)
ISBN 978-90-04-44132-3 (hardback)
ISBN 978-90-04-44133-0 (e-book)

Contents

Preface VII
References and Abbreviations IX
Outline of *Cihānnümā* X

Introduction to *Cihānnümā* 1
Gottfried Hagen

Introduction to the Translation 25
Robert Dankoff

Translation of Cihānnümā

Part 1, by Kātib Çelebi 35
Ferenc Csirkés, John Curry and Gary Leiser
Introduction 39
1 On the Earth 85
2 On the Seas 88
3 On the Major Parts of the Earth 104
4 On the Continent of Europe 105
5 On (the Continent of) Africa 107
6 On the Continent of Asia 110
7 On (the Continent of) America, i.e., the New World 112
8 On 90 Latitude 119
9 On the South Pole 122
10 On the Relation of Countries and Directions, according to the Triplicities of the Constellations and Planets 122
11 On the Inhabitants of the World with Respect to Religions 125
12 Island of Yāpūniā (Iaponia, Japan) 127
13 Map of (New) Guinea 132
14 Islands of Chīn u Hind (East Indies) 132
15 Islands of Gilolo and Molucca 134
16 Mindanao, etc. 138
17 Sumatra 143
18 Borneo 145
19 Island of Ceylon 147
20 Clime of Chīn u Māchīn (China) 152
21 Clime of Khitāy (Cathay) 161
22 Clime of India 181
23 Clime of Sind 206
24 Country of Makran 209
25 Clime of Zābulistān, Khwāst and Ashnāghar 211
26 Country of Kashmir and Tibet 215
27 Country of Sijistan 220
28 Country of Badakhshān, Ghūr and Ṭukhāristān 223
29 Clime of Kirmān, Hormuz and Lār 227
30 Clime of Fārs 231

31 Country of Khūzistān and Lūristān 246
32 Clime of Jabal (i.e., Jibāl) or Persian Iraq 251
33 Clime of Khurāsān and Quhistān 265
34 Country of Qūmis, Ṭabaristān and Māzandarān 284
35 Clime of Khwārazm, Daylam and Gīlān 287
36 Clime of Transoxiana 291
37 Clime of Turkistan and the Steppe 303
38 Clime of Azerbaijan 312
39 Clime of Arrān, Mūqān and Shirwān 320
40 Country of Dagestan, Georgia, etc. 328
41 Clime of Arminiyya 334

Part 2, by Ebū Bekr b. Behrām ed-Dimaşḳī 343
Gary Leiser
42 Clime of al-Jazīra (Upper Mesopotamia) 350
43 Clime of ʿIrāq al-ʿArab (Lower Mesopotamia) 363
44 Description of the Arabian Peninsula 383
45 Region of Shām (Syria) 430
46 Description of (Province of) Anaṭolı 483

Appendix: Maps and Diagrams
About Maps and Diagrams 531
Maps and Diagrams 533

Index of *Cihānnümā* 644

Preface

Cihānnümā is the summa of Ottoman geography and one of the axial texts of Islamic intellectual history. Kātib Çelebi (d. 1657), the leading Ottoman intellectual, sought to combine the Islamic geographical tradition with the new European discoveries, atlases and surveys. His own cosmography included a comprehensive description of the regions of the world, extending westward from Japan. He reached as far as the eastern Ottoman provinces before his death. Ebū Bekr b. Behrām ed-Dimaşḳī (d. 1691) continued the project with a survey of the Arab countries and the remaining Ottoman provinces of Anatolia. İbrāhīm Müteferriḳa combined the two texts, with additional notes and maps of his own, in one of the earliest Ottoman printed books, known as *Kitāb-ı Cihānnümā* (1732).

The English translation of *Cihānnümā* was sponsored by MEDAM (Medeniyet Araştırmaları Merkezi / Civilization Studies Center) of Bahçeşehir University. We are grateful for their funding and support during the initial stages of the project in the years 2009–2011. MEDAM simultaneously sponsored a Turkish translation, which was published in 2010. While we have benefitted from that effort, the translation offered here is based on a more rigorous scholarly approach. While our aim was an accurate translation of the entire book, we have carefully distinguished the parts of the three authors. Based on Kātib Çelebi's original manuscript, we have made hundreds of corrections, large and small, to Müteferriḳa's text. To a limited extent, we have also taken into account Kātib Çelebi's sources (Arabic, Persian and Turkish; Latin and Italian). So, at least for Part 1 by Kātib Çelebi, comprising two-thirds of Müteferriḳa's text, we offer an "edition in translation."

Kātib Çelebi inserted numerous diagrams and maps into his manuscript; these are included here in the proper location. Müteferrika adopted many of them into his edition and added others. These are gathered at the end of this book, together with translations and transliterations.

Gary Leiser translated Müteferriḳa's introduction and Kātib Çelebi's cosmography (M 1–70), also ed-Dimaşḳī's portion of the text, designated here as Part 2 (M 422–698). Ferenc Csirkés translated chapters 1–2 and 28–41 of Kātib Çelebi's world survey, designated here as Part 1 (M 70–96, 252–422). John Curry translated chapters 3–27 (M 96–252). The labels on the maps and diagrams were translated by Ethan Menchinger. Gottfried Hagen wishes to thank his students A. Barış Ekiz, Mackenzie Vaillancourt, Millie Wright, and Yikuan Chen for work on portions of this project, and his colleague Rudi Paul Lindner for invaluable help with aspects of the translation. The translation was supervised by Robert Dankoff and subjected to a final series of edits by Gottfried Hagen.

We thank the following institutions for making their holdings available to us and giving us permission to reproduce them in this volume:

Topkapı Palace Library, Istanbul, for the autograph copy of Cihānnümā, Revan K. 1624

Bayerische Staatsbibliothek München, for the copy of the printed edition, Istanbul 1732, Res/2 A.or. 371

University of Michigan Library, Special Collections, for Cluverius, *Introductionis in totam geographiam, tam veteram quam novam libri VI*, Z 232 .E5 1641eb

University of Michigan, Clements Library, for Gerardus Mercator, *Atlas Minor*, Atl2 1621 Me

Barry Lawrence Ruderman, Antique Maps Inc., La Jolla, CA, for the maps by Giacomo Gastaldi

Map of Gilolo, Molucca, Mindanao, R 49a; cf. M 133/134, 135/136 (Appendix, fig. 58–62)

References and Abbreviations

M	Müteferriḳa edition: *Kitāb-ı Cihānnümā*. Istanbul, 1145/1732
R	Kātib Çelebi's autograph ms. of *Cihānnümā*: Revan 1624, fol. 1–159a (Topkapı Sarayı Müzesi Kütüphanesi)
T	Turkish translation of M: *Cihânnümâ—Kâtip Çelebi*. Istanbul, 2010
Atlas Minor	*Atlas Minor Gerardi Mercatoris, J. Hondio plurimis aeneis auctus et illustratus*. Amsterdam, 1633
Baḥriye	*Kitâb-ı Bahriye Pirî Reis*. Istanbul, 1988
Disegno	Jacobo (Giacomo) Gastaldi, *Il disegno della terza parte dell'Asia*. Venice, 1580
Fabrica	Giovanni Lorenzo d'Anania, *L'universale Fabrica del Mondo, ovvero Cosmographia*. Venice, 1582
Introductio	Philippus Cluverius, *Introdvctionis in Vniversam Geographiam, tam veterem quam novam, Libri VI. Accessit P. Bertij Breviarium Orbis Terrarum*. Lugd. Bat.: Elzevir, 1641
Levāmiʿu n-nūr	Kātib Çelebi's autograph ms. of *Levāmiʿu n-nūr fī ẓulmeti Aṭlas Mīnūr*: Nuruosmaniye 2998
Lingua Franca	Henry and Renée Kahane—Andreas Tietze, *The Lingua Franca in the Levant*. Urbana, Ill., 1958
Nuzha	*The Geographical Part of the Nuzhat al-Qulūb composed by Ḥamd-Allāh Mustawfī of Qazwīn in 740 (1340)*, ed. G. Le Strange. Leiden and London, 1915

Outline of *Cihānnümā*

		{Müteferriḳa's Supplements}
		[Diagrams and Maps]

Part 1, by Kātib Çelebi

M	R	
		[Armillary sphere, M 0/1]
1	1b	Prefatory remarks
4	—	{Principles of geographic science}
		[Geometric figures, M 7/8]
8	1b	1. Sources used for this work; *Atlas Minor*
11	3a	Other sources
14	4b	2. Meaning of "geography"
16	4b	3. Goal and utility of geography
17	5a	4. Earth and heavenly bodies
		[Diagrams showing earth's sphericity, M 19/20]
22	6b	5. Celestial spheres and the sublunar world
		[Cosmological schemes, M 21/22]
		[Cosmological scheme, R 7a; M 25/26]
24	—	{Ancient astronomy, school of Aristotle and Ptolemy}
		[Star chart—planisphere, M 27/28]
		[Diagrams for theories of Aristotle and Ptolemy, M 33/34]
34	—	{Modern astronomy, school of Pythagoras and Copernicus}
46	—	{School of Tycho Brahe}
		[Diagrams for theories of Copernicus and Brahe, M 47/48]
48	7a	6. Circles and poles
50	8a	7. Division of circles and the earth
		[Zones, R 8b; M 49/50]
51	8b	8. Circles of latitude and climes
		[Climes according to the ancients, M 51/52]
		[Table of climes and latitudes, R 10a; M 51/52]
53	10b	9. Division of earth by longitude and latitude
		[Circles of distances, R 11a; M 49/50]
		[Circles of latitude and longitude, R 11b; M 49/50]
56	12a	10. Distances and surveying instruments
		[Measuring circumference of earth, R 12a; M 57/58]
		[Table of distances, R 12b; M 57/58]
59	14a	11. The four directions and the winds
		[Diagrams of winds, R 13a–14a; M 59/60]
62	15a	12. Rules of drawing maps
		[Compass, M 65/66]
66	15b	13. Difficulty of geography and need for correction
67	16b	14. Table of contents
70	19a	Chapter 1: The Earth
		[Scheme of earth with sun at apogee and perigee, R 19b]
		[World map, R 21a; M 71/72]
73	21b	Chapter 2: The Seas
		[Map of Mediterranean and Black Seas, M 75/76]
		[Maps of Adriatic and Aegean Seas, M 77/78]
92	29a	Monsoons

96	31a	Chapter 3: Major Parts of the Earth
		[World map, M 95/96]
96	31a	Chapter 4: Europe
		[Map of Europe, R 32a; M 99/100]
100	33a	Chapter 5: Africa
		[Map of Africa, R 34a; M 101/102]
102	34a	Chapter 6: Asia
		[Map of Asia, R 35a; M 103/104]
104	35a	Chapter 7: America
		[Map of America, R 37a; M 113/114]
108	36b	Journeys to America by Columbus, etc.
115	40a	Chapter 8: North Pole, Arctic Region
		[Map of Arctic Region, R 41a; M 119/120]
119	42a	Chapter 9: South Pole, Antarctic Region
		[Map of Antarctic Region, R 42a; M 119/120]
119	42b	Chapter 10: Relation of Countries and Directions …
		[Diagram of triplicities, R 42b]
122	44a	Chapter 11: Inhabitants of the World with Respect to Religions
		[World map: R 44a (sketch)]
124	44b	Chapter 12: Japan
		[Map of Japan, R 45a; M 125/126]
130	46b	Chapter 13: New Guinea
		[Map of New Guinea, R 46b; M 129/130]
130	47a	Chapter 14: East Indies
		[Map of East Indies, R 47b; M 131/132]
132	47b	Chapter 15: Gilolo and Molucca (Indonesia)
		[Map of Gilolo and Molucca, R 49a; M 133/134]
		[Map of Molucca, M 135/136]
136	49a	Chapter 16: Mindanao, etc. (Philippines)
		[Map of Celebes, Java, etc. R 50a; M 135/136]
144	51b	Chapter 17: Sumatra
		[Map of Sumatra, R 52a; M 143/144]
146	52b	Chapter 18: Borneo
		[Map of Borneo R 53a; M 145/146]
148	54a	Chapter 19: Ceylon (Sri Lanka)
		[Map of Ceylon, R 55a]
154	56b	Chapter 20: Chīn u Māchīn (Song, S. China)
		[Map of South China, R 58a; M 153/154]
165	61a	Chapter 21: Cathay (Khitan State, Liao, N. China)
		[Map of North China, R 69a; M 165/166]
193	71b	Chapter 22: India
		[Map of India, R 75a; M 193/194]
231	85a	Chapter 23: Sind
		[Map of Sind, R 87a]
235	86b	Chapter 24: Makran
		[Map of Makran, R 88a]
237	88a	Chapter 25: Zābulistān, Khwāst and Ashnāghar
		[Map of Zābulistān etc. R 89b]
241	90a	Chapter 26: Kashmir and Tibet
		[Map of Kashmir and Tibet, R 93a]
248	93a	Chapter 27: Sijistān
		[Map of Sijistān, R 95a]

252	95a	Chapter 28: Badakhshān, Ghūr and Ṭukhāristān
		[Map of Badakhshān etc., R 96b]
256	96b	Chapter 29: Kirmān, Hormuz and Lār
		[Map of Kirmān etc., R 98a]
262	99a	Chapter 30: Fārs
		[Map of Fārs and Persian Gulf, R 104a]
282	106a	Chapter 31: Khūzistān and Lūristān
		[Map of Khūzistān and Lūristān, R 108a]
		[Map of Iran, M 289–290]
289	109a	Chapter 32. Jabal (Jibāl) or Persian Iraq
		[Map of Jibāl, R 116a]
309	116a	Chapter 33. Khurāsān and Quhistān
		[(space for map) R 126a]
338	127b	Chapter 34: Qūmis, Ṭabaristān and Māzandarān
		[(space for map) R 129a]
343	129b	Chapter 35: Khwārazm, Daylam and Gīlān
		[(space for map) R 131a]
348	132a	Chapter 36: Transoxiana
		[(space for map) R 140a]
		[Map of Transoxiana, M 347/348]
366	139a	Chapter 37: Turkistan and the Steppe
		[(space for map) R 143a]
379	145a	Chapter 38: Azerbaijan
		[Map of Azerbaijan, R 146a; M 389/390]
391	150b	Chapter 39: Arrān, Mūqān and Shirwān
		[(space for map) R 152a]
		[Sketch map of Caspian: R 154a]
401	154b	Chapter 40: Dagestan, Georgia, etc.
		[(space for map) R 156a]
		[Map of Caucasus, M 431/432]
410	156a	Chapter 41: Arminiyya
411	156b	Province of Van
		[Map of Van province, R 159a]

Part 2, by Ebū Bekr b. Behrām ed-Dimaşḳī

M	
422	Province of Erzurum
429	Province of Trabzon
432	Chapter 42: al-Jazīra (Upper Mesopotamia)
433	Province of Mosul
433	Province of Mosul
436	Province of Diyarbekir
443	Province of Raqqa (Province of Urfa)
445	Province of Şehr-i Zūl (Shahrazūr)
449	Province of Kurdistan
451	Chapter 43: ʿIrāq al-ʿArab (Lower Mesopotamia)
451	Province of Basra
483	Chapter 44: Arabian Peninsula
	[Map of Arabian Peninsula, M 483/484]
498	Ḥijāz
523	Coastal Plain (*Tihāma*) of Ḥijāz

527	Najd-i Ḥijāz
527	Najd al-ʿĀriḍ ("Middle Najd")
532	Arabian Peninsula
552	Chapter 45: Syria
568	Jordan
571	Damascus
592	Qinnasrīn
593	Aleppo
595	Antakya
598	Provinces of Marʿash and Adana
610	Province of İçel
611	Sanjak of Alanya
612	Cyprus
614	Province of Ḳaraman
615	Konya
617	Sanjak of Nigde
618	Sanjak of Beyşehri
619	Sanjak of Akşehir
620	Sanjak of Aḳsaray
620	Sanjak of Kayseri
620	Sanjak of Ḳırşehri
622	Province of Sivas (Province of Rūm)
622	Sivas
623	Sanjak of Canik
624	Sanjak of Arapgir
624	Sanjak of Divrigi
625	Sanjak of Çorum
625	Sanjak of Amasya
626	Sanjak of Bozoḳ
	[Map of İçel, Ḳaraman, Anaṭolı and Sivas, M 629/630]
630	Chapter 46: Province of Anaṭolı
631	Sanjak of Germiyan
634	Sanjak of Ṣaruḫan
636	Sanjak of Aydın
638	Sanjak of Menteşe
638	Sanjak of Teke
639	Sanjak of Ḥamīd
641	Sanjak of Ḳaraḥiṣār-ı Ṣāḥib
641	Sanjak of Sulṭānöñi
643	Sanjak of Ankara
645	Sanjak of Çankırı
648	Sanjak of Ḳasṭamonı
651	Sanjak of Bolu
656	Sanjak of Ḫüdāvendigār
661	Sanjak of Ḳaresi
661	Sanjak of Ḳocaeli
667	Sanjak of Biga
669	Sanjak of Ṣıġla
	[Map of the Bosphorus, M 671/672]
671	Routes, Stages, and Distances of Anaṭolı and Ḳaraman
674	Kings and rulers who have conquered the province of Anaṭolı

Introduction to *Cihānnümā*

Gottfried Hagen

This book contains the first full translation of the world description by the Ottoman polymath Kātib Çelebi (1609–1657)[1] into a western language. Its title, *Cihānnümā*, meaning "Showing the World," refers to the legendary mirror which Alexander the Great—İskender in Islamic lore—had installed on top of the Lighthouse of Alexandria, enabling him to see everything in the world, so that he could defend his realm against his enemies and govern it justly. Iranian lore also tells us about the cup of King Jamshīd which had the same power and was known as *Jām-i Gītī-nümā*, meaning "World-showing Cup."[2] Since its first circulation in manuscript in the 17th century, Kātib Çelebi's *Cihānnümā* promised its readers comprehensive, practical, valid information about the entire world. To the modern historically-minded reader it presents insights into the way the world was understood in Islamic cultures. It is situated at a turning point, being both a summa of the Islamic geographical tradition and the first Muslim adaptation of the early modern atlas as the scientific representation of the world. In this essay I will first examine the intellectual, political and social context of the work, then discuss its complicated textual history and the editorial decisions that guided the present translation.

1 World, Empire, Geography

1.1 *Beginnings of the Islamic Geographical Tradition*

All civilizations form their own image of the world, based on their available knowledge and in accordance with their cultural traditions. The first global Islamic Empire, the Abbasid caliphate, was no different. While the Qur'an provides only rudimentary hints at an underlying cosmology, speaking of seven earths and seven heavens above them (for instance, Q 65:12), or a heaven like a canopy above a (flat?) earth (Q 21:32), it was under the Abbasids in the eighth and ninth centuries that Indian and especially Greek astronomy and mathematics were adopted in Baghdad. The *Geography* of Ptolemy (c. 90–c. 168) was translated into Arabic as *Ṣūrat al-arḍ* ("Picture of the World"). Its idea of a spherical earth, on which the location of cities could be defined by measuring latitude (from the equator) and longitude (from the far western boundary of the inhabited world), remains the basis of mathematical geography to this day. The division of the world into seven parallel "climes" or zones of equal latitude, and the drawing of maps using a grid of coordinates, proved to be a legacy that was continued for almost a millennium in the Muslim world.[3] An authoritative mapping of the world based on Ptolemaic parameters, commissioned by the caliph al-Ma'mūn (r. 813–833), appears to have been lost.[4] In a continuation of the same tradition, in the mid-12th century al-Idrīsī set out to compile all available locational data, using a map as an instrument of control. The silver planisphere on which his findings were engraved, on the order of his patron Roger II of Sicily, is also lost, but his text and the seventy partial maps, at which he arrived by dividing the seven latitudinal climes of the Ptolemaic tradition into portions of equal longitude, have come down to us as one of the most detailed mappings of the medieval world.[5]

At roughly the same time as the adoption of Ptolemy's work, another largely unrelated strand of geography emerged out of the administrative needs of the caliphate. Frequently denoted by the generic title of *Routes and Realms* (*al-masālik wa'l-mamālik*), this approach is most prominently preserved in the work of Ibn Khurradādhbih (two versions completed in 846 and 885, respectively) and Qudāma b. Ja'far (10th century).[6] Closely related, but with more literary ambition, are the descriptions of the Islamic world produced by al-Balkhī and his successors, al-Iṣṭakhrī, Ibn Ḥawqal, and al-Muqaddasī, all of which appeared during the 10th century. Uninterested in mathematical geography, these authors divided the world into either 14 or 21 historical or cultural zones which they described, often based on their own personal experience, with much attention to history and ethnographic detail. They also produced their own maps, one for each clime, that often show routes and borders in addition to cities, mountains and seas in highly stylized, geometrical forms.[7] Both Ptolemy's successors and the geographers of the

1 All dates in this chapter refer to the common era.
2 Berlekamp 2011, 91–92.
3 *Encyclopedia of Islam*, 2nd edition, art. Djughrāfiyā; art. Kharīṭa.
4 Tibbetts 1992, "The Beginnings of a Cartographic Tradition," 95f. Given the technicalities of mapmaking in the manuscript age, it seems a stretch to identify this with a map found in a 14th-century work, Ibn Faḍlallāh al-ʿUmarī's *Masālik al-abṣār fī mamālik al-amṣār*, as proposed by Fuat Sezgin.
5 A recent partial translation is that of Bresc and Nef (al-Idrīsī 1999).
6 See on this literature Antrim 2012 and especially Heck 2002.
7 Tibbetts 1992 "The Balkhī School of Geographers."

 DOI:10.1163/9789004441330_002

Abbasid tradition take the city as the essential unit and show little concern with nature where it does not relate to human geography.[8]

Nowadays we tend to link geographical literature to accounts of travel and exploration, as furnishing exciting new knowledge; and there is no doubt that information from travelers, sailors, merchants, ambassadors and others was incorporated into the literature discussed above.[9] As literary genres by themselves, however, reports of embassies, narratives of travels in exotic places, etc. are surprisingly scarce and remain largely unconnected to the learned traditions.[10] The greatest of the medieval Muslim travelers, Ibn Baṭṭūṭa (d. 1368/9 or 1377) was barely recognized by the geographers, who were never able to shed their doubts about the veracity of his account.[11]

1.2 *Compilations, Cosmographies, Encyclopedias*

The Ptolemaic and the Atlas of Islam traditions, which I have described above as two discrete strands, did not remain separate for long. Geographical knowledge turned out to be of interest to nearly everyone. Travelers and merchants wanted to know about routes and goods, administrators about taxes, military men about terrain, mathematician-astronomers about coordinates of cities, historians and philologists about the location of places and the orthography of toponyms. In short, geography became part of a canon of knowledge, a staple of general erudition (*adab*) required for the elite of the Abbasid caliphate and subsequent states. While geographical knowledge grew at a modest pace, enriched by travelers, traders, diplomats and literati, geographical literature grew much faster through ever new compendia. As we shall see, *Cihānnümā* can be considered the culmination of this tradition of compendia, while using many of its predecessors as sources.

The process of compilation was facilitated by the additive character of geographical knowledge, which allows discrete descriptions and anecdotes relating to different places to be juxtaposed at will. Moreover, much of this knowledge consists of lists, such as itineraries and catalogues of places, events, products, customs and other local features. It was easy for copyists and later readers and authors to add to such lists or to abridge them, since there was no narrative or analytical discourse, such as a travel account, that held them together. *Dictionary of Countries* (*Muʿjam al-buldān*), by the learned former slave Yāqūt al-Rūmī (d. 1229), was an early masterpiece of geographical compilation. In the introduction, Yāqūt meticulously juxtaposed the various descriptions of the cosmos and its divisions, assembling the knowledge of the time in an alphabetical sequence of articles on cities that are unsurpassed in their wealth of geographical, historical, biographical and literary information.[12]

As the political cohesion of the Islamic empire waned and ultimately vanished with the Mongol conquest of Baghdad in 1258, compilations of world geography became a new means of maintaining a cultural sense of unity. Zakariyā al-Qazwīnī (d. 1283), using much of Yāqūt's material, created a classic in a new genre: his two-part cosmography presented the wealth and diversity of the world as a testimony to God's omnipotence for the pious to contemplate.[13] In the geographical part, entitled *Monuments of Countries and Reports of their Inhabitants* (*Āthār al-bilād wa-akhbār al-ʿibād*), the material is arranged according to Ptolemy's climes, different latitudes being correlated with celestial bodies, thus accounting for differences of climate, temperament and culture. Characteristic for al-Qazwīnī's work is the overarching religious-philosophical agenda, starting from an Islamic cosmology based on Hadith. Over many centuries the pious had collected sayings attributed to the Prophet Muḥammad that related to natural phenomena—the movement of the stars, thunder and lightning, ocean tides, etc.—resulting in a strictly religious pattern of world explanation.[14] Al-Qazwīnī went about it in a more philosophical way in his work, *Marvels of Creation and Wonders of Existing Things* (*ʿAjāʾib al-makhlūqāt wa gharāʾib al-mawjūdāt*), explaining the title as follows:

> "Marvel" refers to the astonishment that befalls man because of his incapacity to understand the cause of something or how it works. For instance, if he sees a beehive and has not seen one before, he is struck with astonishment because he does not know who made it; and if he knew that it is the work of bees, he would be astonished that a weak animal could make such regular hexagons that a skilled engineer is incapable of making with compass and ruler. … If he happens to see a strange animal or rare plant or extraordinary fact, he

8 Miquel 1973.

9 A famous example is Sallām al-Tarjumān, who was sent by the caliph al-Wāthiq (r. 842–847) to the apocalyptic peoples of Gog and Magog and narrated his experience to the geographer Ibn Khurradādhbih, after whom many others repeated the tale (Krachkovskii 1957, 138–141).

10 Bonner and Hagen 2010.

11 Euben 2006, 46–89.

12 Yāqūt, tr. Jwaideh 1959.

13 al-Qazwīnī, Zakariyā b. Muḥammad, ed. Wüstenfeld 1848–1849. See Hees 2002; Berlekamp 2011.

14 See Heinen 1982, Introduction.

bursts out saying "Praise to God!" Throughout his life he sees things about that astonish the minds of the wise and perplex the souls of the intellegent.

Thus man is encouraged to study the cosmos and nature in order to recognize God's omnipotence:

> He should carefully observe the heavenly bodies, their sublime expanse and solidity, and how they are preserved from change and destruction until the day announced by scripture. In comparison to them, earth and air and seas are like a ring thrown into the desert. God said: "And heaven—We built it with might and extended it wide" (Q 51:47).

"Wonder," on the other hand, refers to something quite different:

> The word wonder applies to every marvelous thing that occurs rarely and contradicts common customs and familiar observation. This happens either through the effect of strong souls or the powers of celestial or elementary bodies, all this occurring through divine power and will.

Wonder, then, applies to miraculous events performed by prophets, saints and soothsayers, as well as inexplicable events in nature. This is the essence of the created world in its entirety:

> Everything that exists, except the Necessarily Existent [i.e., God], is created and made by Him; and every particular thing, be it substance or accident, whether characteristic or characterized, encompasses marvels and wonders in which God's wisdom and power and glory and greatness become manifest.[15]

Because of this universal scope, cosmography could theoretically comprise everything in the world and function as a true encyclopedia. This ultimately theological interest in the observation of the physical world, paired with a keen philosophical mind that distinguished regularities—what we would call the laws of nature—from the utterly miraculous, is displayed by al-Qazwīnī in exemplary form. But this interest was not limited to the genre of cosmography, since every geographical text can be subjected to a theological reading.[16]

Works on the marvels of creation subsequently became highly popular, in both senses of the word, as widely disseminated, and as appealing to non-academic interests. Later compilations, like the one usually attributed to Abū Ḥafṣ ʿUmar ibn al-Wardī (d. 1419), revel in sensationalism by including detailed descriptions of monsters and exotic creatures, purportedly in the service of edification, while lacking al-Qazwīnī's discerning philosophical gaze. Cosmographies were among the most widely read geographical texts in the Muslim world well into the 18th century, not only in Arabic, but in Persian and Turkish translations as well.[17]

The Mongol domination of Iran, Iraq, Anatolia and Central Asia created new political realities, and with them, new geographical perspectives as well. Both the Mongol Ilkhans in Iran and their Mamluk rivals in Egypt and Syria cultivated an elite class of administrators for whom the essential knowledge required—ranging from history to diplomatics to law to geography—was collected in massive encyclopedias. Those by Ibn Faḍlallāh al-ʿUmarī (d. 1349) and Shihābaddīn al-Qalqashandī (d. 1418) are among the most famous on the Mamluk side.[18] Noteworthy for us, because of its reception by Kātib Çelebi, is *Nuzhatu'l-qulūb*, the work of the Ilkhanid administrator Ḥamd Allāh Qazwīnī, known as Mustawfī ("Auditor-general," d. after 1339). It includes a detailed and often original description of the world, the sections on Īlkhānid territory reflecting information and practices in which the author was involved.[19]

Another compiler of geographical knowledge, half a century after Zakariyā al-Qazwīnī, was the prince of Ḥamā in Syria, Abū l-Fidāʾ (d. 1331), who was more interested in continuing the mathematical and descriptive tradition than providing edification or telling stories. His *Tables of Countries* (*Taqwīm al-buldān*) make for rather dry reading. The book consists of tables of coordinates of cities, arranged according to 28 regions, with pithy bits of historical information.[20] Evidently it met a demand for precise and easily accessible geographical information, since it was copied over and over for centuries. In the late 16th century an Ottoman compiler named Sipāhīzāde (d. 1589) simplified it further by arranging all the articles about cities in one alphabetic sequence. While he added some

15 al-Qazwīnī, Zakariyā b. Muḥammad 1848, 5–12; translation by GH.

16 See Hagen forthcoming.

17 The authorship of the Pseudo-Ibn al-Wardī is discussed in Sellheim 1976, 176–186. See also Coşkun 2007; Anonymous, ed. Kut 2012, Introduction.

18 Krachkovskii 1957, 399–430.

19 al-Qazwīnī, Ḥamdallāh Mustawfī, ed. Le Strange 1915; Krawulsky 1978.

20 Abū l-Fidāʾ al-Ḥamawī, ed. Reinaud and de Slane, 1840; tr. Reinaud and Guyard, 1848.

material from other geographical-philological dictionaries as well as cosmographies, he scanted the mathematical aspect and abbreviated the locational information, but made it easy to find what needed to be known about individual cities.[21] Throughout there is a pronounced adherence to classical models and sources, however outdated their information may have become over the centuries.

At the end of the 16th century another Ottoman scholar undertook a major attempt to combine the essential strains of this tradition. Within the framework of a cosmography that followed in the footsteps of al-Qazwīnī, Meḥmed ʿĀşıḳ (d. after 1598) organized his *Views of the World* (*Menāẓıru l-ʿavālim*) according to a combination of Ptolemaic climes and the historical regions laid out by Abū l-Fidāʾ. Information derived from the latter and from Ḥamdallāh Mustawfī and many other classics was supplemented with Meḥmed ʿĀşıḳ's own impressions during many years of travel. This represented the first acknowledgment, in the context of cosmography, that times had changed and that the geographical classics no longer sufficiently represented the world.[22] Despite his desire to revitalize the canon of classical geographical knowledge, Meḥmed ʿĀşıḳ continued to restrict himself to the traditional genres of geography and cosmography. He ignored newer writings rich with geographical information such as the maritime handbook and nautical charts of the famous sea captain Pīrī Reʾīs (d. 1553) that fascinated an Ottoman elite audience of the time.[23] Cosmography in particular had little interest in mapping other than highly schematic world maps in the visual tradition of the Balkhī—Iṣṭakhrī—Ibn Ḥawqal tradition.[24] Meḥmed ʿĀşıḳ also ignored the few isolated travelogues of exotic countries that had emerged in Ottoman letters, such as the adventures of the admiral and litterateur Seydī ʿAlī Reʾīs on his journey from India back to the Ottoman domains, or ʿAlī Ekber's early 16th-century account of China. Most starkly, his work never mentions the existence of a new world, even though it was shown on Pīrī Reʾīs' maps; moreover, the first Muslim book on the Americas, entitled *History of the West Indies*, or *Fresh News* (*Tārīkh-i Hind-i Gharbī*, or *Ḥadīs̱-i nev*) had been written in Ottoman Turkish around 1580.[25]

The work in which the boundaries between the genres were finally torn down, leading to a new and unified image of the world presented with the authority of learned tradition and experiential knowledge, was *Cihānnümā*. Its introduction includes a bibliography in which the author lists all the classical sources used in his work, including most of those discussed above. The bibliography serves to assert Kātib Çelebi's authority in a field that had never been taught as a subject of its own in Islamic institutions of higher education (madrasa), and hence could not provide him with an academic pedigree of teachers in the way a legal scholar would demonstrate his competence. Instead he had to list the books he used, or the books quoted in the books he used.[26] Kātib Çelebi tended to authenticate contentious information in particular by naming his sources and authorities.[27] Thus *Cihānnümā* not only takes a special place in the sequence of Islamic geographical literature, subsuming a good deal of it and positioning itself as part of a living tradition. But it also transcends the tradition by adding new material and, most importantly, by assigning geographical knowledge a new place in state and society, appropriate for its context in the 17th-century Ottoman Empire, to which we now turn.

1.3 *The Ottoman Empire and Its System of Governance*

The political situation of the Muslim world had changed completely since the Mongols and the Mamluks, who provided the context for much of the compilation and canonization of classical literature. The Mongols had established a suzerainty over much of Asia Minor, with only small portions ruled by the Byzantine Empire after the Fourth Crusade. As Mongol Ilkhanid control of the area decreased, several smaller principalities sprang up, ruled by local dynasties of Turkmen origin. Turkish nomads had migrated to Asia Minor since the eleventh century. Now their descendants—nomadic warriors on horseback wedged between the remnants of the Byzantine Empire and the more stable regions of Mamluk Syria and Ilkhanid Iran—assumed power in the cities of Asia Minor with their Persianate high culture. Many of these local enterprises in this frontier region were ephemeral and hardly left a historical record, while others persisted for generations, rising to regional importance. One of them, favored by luck and geographical position—or, in their own view, by divine blessing—expanded into a world empire: the Ottoman principality and empire-in-the-making.[28]

21 Sipāhīzāde wrote the book in Arabic first, then produced a Turkish abridged recension. See Koraev 2015; Hagen 2006.

22 ʿĀşıḳ, Meḥmed 2007. See also Hagen 1998.

23 Soucek 1992; Soucek 1996.

24 Pinto 2011.

25 For a full translation see Goodrich 1990.

26 The "genealogical trees" of his references are drawn in Hagen 2003, 530–531.

27 See for example his treatment of the Amū Daryā, which has, confusingly, changed course several times in medieval history (M 347–348, cf. Hagen 2003, 305–306).

28 A readable account of Ottoman history is Finkel 2006. The following account of structures and institutions relies heavily on Imber 2009.

Its semi-legendary eponym ʿOs̱mān probably had not been much more than the leader of a neo-tribal group of Turkmen raiders in the area of northwestern Asia Minor. It was his son Orkhān (d. 1324?), who first conquered an important city in the area (Bursa), and should be considered the actual founder of the Ottoman enterprise. The pace of the subsequent expansion of the Ottoman domains, at the expense of the little principality's Christian and Muslim neighbors, still leaves the historian struggling for an explanation. Only two generations after Orkhān, the Ottomans felt powerful enough to challenge the Mamluks and the new Turko-Mongol superpower, Tamerlane's empire. Their defeat at the hands of the latter in 1402 turned out to be only a temporary setback. Soon afterward the Ottomans again set their sights on the remnants of the Byzantine Empire and finally succeeded in taking the imperial city of Constantinople in 1453. It is from this point onward that we can truly speak of an Ottoman Empire. While at that time its center of gravity, in terms of economic resources and military engagement, lay in southeastern Europe, in the latter half of the 15th century it began shifting to the east, where the Ottomans helped crush the Turkmen empires of the Aqqoyunlu and Qaraqoyunlu in what is today eastern Turkey, Iraq and western Iran. Conquering Syria and Egypt from the Mamluks in 1517, it sought to stabilize its eastern frontier against the emerging power of Safavid Iran. The reign of Sultan Süleymān (r. 1520–1566), known in Europe as 'The Magnificent,' has often been described as the Golden Age, in which the Ottomans ruled over territory from Egypt and the Arabian Peninsula to the gates of Vienna, and in which Ottoman culture reached its greatest splendor and refinement. The last section of *Cihānnümā* gives an account of this Ottoman rise to glory, complete with the signs of divine approval that heralded it.[29]

The modern secular historian will prefer to attribute the success of the Ottomans in large part to the flexibility and pragmatism with which the ruling dynasty adapted to new situations. Typical of the empires of the time, the Ottoman state essentially functioned as an extended royal household in which administrative and military positions were held by the ruler's servants. As tax collection and military strength were the two interdependent main concerns of the pre-modern state, the Ottomans gradually replaced the military contingents of vassals and allies with a tightly regulated feudal cavalry along with a standing elite infantry of military slaves recruited from the Christian subjects of the empire (the devshirme or child levy).[30] The upper ranks of the military, who doubled as provincial administrators, were recruited through the same process. The fiscal administration centrally controlled the distribution of fiefs and taxation and kept meticulous records.[31] The imperial chancery communicated the imperial claims and policies to subjects and to other dynasties. We will see where our author fits into this system shortly. The second essential base of the Ottoman system of governance was a tightly regulated hierarchy of institutions of religious-legal education (madrasa), the graduates of which were either employed in those same schools or dispatched to serve as judges and jurisconsults throughout the empire. Like service in the imperial household, this career offered opportunity to rise to the highest ranks and govern the entire system. As with the imperial household, the top level of this hierarchy was based in Constantinople. A third pillar of culture and learning among the Ottoman elite was the dervish or Sufi orders, the institutionalized and systematized form of Islamic mysticism that had emerged since the 12th and 13th centuries. While some orders predated the Ottoman state, gradually extending their networks under their rule and patronage, others formed exclusively under Ottoman dominion.

Thus three locales were central for the Ottoman elite and many of its members were part of more than one. Just as training in the palace service qualified men for military assignments and governorships, men with madrasa training were able to enter the sultan's service or the chancery, while members from all groups might meet in the dervish lodges. The culture developed by those elites also provided a model for the many regional, ethnic and religious subcultures of the empire. Here we will focus on the elite, whose discourses shaped the ideas of men like Kātib Çelebi and who were in turn the primary addressees of his works.

The ethnic background of the elite was diverse: while the devshirme recruited Christian boys, to be converted to Islam, the madrasa career was the primary venue for free-born Muslims.[32] Besides their shared allegiance to the House of ʿOs̱mān, this elite was held together by a shared culture and language. The Ottomans used Turkish

29 Bacqué-Grammont 2009.

30 It should be noted that military slavery has a long tradition in the Middle East.

31 The use of the term feudalism is disputed; I am using it here to denote a contractual relationship of military service in return for the right to collect taxes and dues from the subjects in a specific area. Different from the European model, the Ottomans did not allow subinfeudation and the amount of taxes to be collected was, at least in theory, strictly regulated.

32 Fleischer 1986 gives an exemplary account of the early career of an Ottoman intellectual in the madrasa.

as their official language from their beginnings, and by the 16th century, with significant help from the imperial chancery, a new literary idiom had formed that was distinct from the vernacular spoken Turkish. Its structure was still Turkic, but its vocabulary and to some degree syntax incorporated vast amounts of Arabic and Persian which were the older transregional and transnational languages of the Muslim world. This new rich and sophisticated language, combining multiple traditions to form a distinct new whole, is symbolic of the Ottoman capacity for creative syncretism. We observe the same syncretism in classical Ottoman architecture, whose elements can be traced to different origins, yet the masterpieces clearly integrate them into an organic new whole. Or take Ottoman poetry, which is unthinkable without the rich heritage of Persian classical poetry, cultivated in the dervish lodges, yet emerges with its own style and import.[33]

1.4 *Ottoman Geographical Literature and State Patronage*

The various legitimating strategies of the Ottoman sultans furnish another example of cultural adaptiveness and integrative capacity. Pre-Islamic Turkic traditions saw the entire ruling house as imbued with royal charisma. In its Islamicized form, this charisma was now presented as the zeal of the warrior against the infidels. Ottoman sultans saw themselves as heirs of the Roman-Byzantine emperors while at the same time endorsing legal arguments that they were rightful successors to the Islamic caliphate, which had largely ceased to exist as a political institution in 1258. According to a primarily Persianate tradition of statecraft, Ottoman sultans highlighted the justice of their reign. Some even claimed to be the *mahdī*, the messianic figure whose appearance ushers in the end of time, thereby aligning themselves with analogous early modern European ideas.[34] Ottoman historical consciousness awoke in the contestation of these different claims. A rich historiography flourished since the late 15th century as a primary vehicle to formulate and challenge Ottoman imperial self-image and ideology.[35]

It has been argued that Ottoman geographical writing, which we outlined above as part of the Islamic geographical tradition, should also be analyzed in the same way, as formulating an imperial ideology. "[T]here was a heightened sensitivity to geographical knowledge, or in other words, scholarly and intellectual engagement in understanding and knowing the world's geography in the sixteenth-century Ottoman Empire. ... This new imperial vision underscored the Ottoman desire to become a world empire, and positioned the empire in relation to its rivals in both East and West."[36] Such a function of geography and cartography is amply attested in Renaissance and early modern Europe. The Western age of discoveries is unimaginable without systematic collection and codification of geographical knowledge. Maps showing the newly discovered territories were collected and guarded as state secrets in Spain and Portugal. The measuring and mapping of territories was an essential ingredient in the rise of the nation state.[37] In the medieval European context, on the other hand, maps had been put on display in symbolic function, whether in churches and monasteries as visual synopses of salvation history (e.g., the Ebstorf and Hereford maps) or as public show of the extent of the power of kings and princes.[38] To what degree, then, might the Ottoman production and use of maps and geographical literature be compared to that of Christian, or more precisely Latin Europe? Neither the state-sponsored practical use of maps and atlases, nor the public display in contexts of religion and government are attested in the Ottoman empire.[39] Yet it is indisputable that most geographical literature and many maps were produced in expectation of patronage, either from the sultans or from leading servants of the state. This is true for travelogues to exotic places such as ʿAlī Ekber's work on China (1520), for the maritime charts and world maps of Pīrī Reʾīs (drawn up between 1513 and 1528), for Sipāhīzāde's late 16th-century rewritings of Abū l-Fidāʾ's *Tables* and for numerous cosmographical works. The vast majority of Ottoman maps known today are in the collection of the imperial palace.[40]

Rarely do the authors of these works allow us a glimpse at their intentions. ʿAlī Ekber penned his description of China as a memorandum about administration and imperial justice.[41] Pīrī Reʾīs wished to demonstrate the usefulness of his nautical skills.[42] Meḥmed ʿĀşıḳ wanted to entertain and edify Meḥmed III whom he had known as a prince and provincial governor.[43] Seydī ʿAlī Reʾīs' narrative of his travels in India after a failed naval expedition and his arduous journey back to the Ottoman lands is first of all a tale of hardship overcome (in Arabic, *al-faraj baʿd al-shidda,*

33 Kuru 2013.
34 Imber 1987. For the last point see the groundbreaking studies of Cornell Fleischer, starting with Fleischer 1992.
35 Hagen and Menchinger 2014; Fetvacı 2013.
36 Emiralioğlu 2014, 5.
37 Konvitz 1987.
38 Birkholz 2004; Schulz 1987.
39 On the state gathering information see Agoston 2007.
40 Goodrich 1993; Karamustafa 1992.
41 Emiralioğlu 2012; Hemmat 2014.
42 Hagen forthcoming.
43 Hagen 2003, 109–110.

denoting a whole genre of anecdotes on this theme) and praise of the Ottoman homeland where he needed to obtain pardon for his military failure.[44] Sipāhīzāde's compilation "in its brilliant mediocrity" betrays a classicist literary and philological interest rather than a geopolitical agenda.[45] Yet this is only half the story, since the thoughts of Ottoman intellectuals in reading geographical accounts or viewing maps and cosmographical diagrams remain a matter of conjecture.

Geographical information can serve to guide a traveler from A to B, inspire the imagination to mental travel or to dreams of conquest, entertain the reader with colorful tales of the exotic or remind the pious of the omnipotence of the Creator. Did Süleymān I use Pīrī Re'īs' maritime atlas of the Mediterranean to plan campaigns against Italy, or did he read it as a reminder of the humility of human existence? Or did the detailed maps just make for the early modern equivalent of a beautiful coffee-table book?[46] Murād III (r. 1576–1595), Süleymān's reclusive, troubled grandson, Sufi and scholar more than military leader, had more geographical works dedicated to himself than any other sultan. Yet when he read the Turkish translation of ʿAlī Ekber's *Book of China* (*Khiṭāynāme*), was he interested in Chinese administration or in the reflected shortcomings of his own empire? Did he recognize in the *History of the West Indies*, presented to him in 1580, a reminder of the geopolitical realities of his day and a call to strategic action, or did he read it as an innovative addendum to al-Qazwīnī and pseudo-Ibn al-Wardī, as another set of *mirabilia* that attest to the greatness of the Creator?[47] Did he read Sipāhīzāde to learn about new horizons or about the correct spelling of toponyms in classical literature?[48] Were the pious tales in the more popular cosmographies inspiration of faith or stale repetitions of long-debunked legends? Almost any example of Islamic geographical literature lends itself to more than one of these readings. It remains significant, however, that the last great oeuvre of Ottoman geography before Kātib Çelebi was Meḥmed ʿĀşıḳ's great synthesis of classical Islamic cosmography, which, as indicated, ignored more recent genres like maritime charts and travelogues. Its practical value is minimal, but it was certainly considered entertaining and edifying. Yet by that time the shortcomings of the tradition had become apparent, as it had virtually nothing to say about vast areas under Ottoman control, prompting Meḥmed ʿĀşıḳ to insert his own experiences.

It is against this background—of multiple and ambiguous uses of geographical knowledge, in the form of literature as well as maps, in which the political (where it was present at all) was always intertwined with the pious and the aesthetic—that *Cihānnümā* stands out as an initially moderate but ultimately radical departure. But in order to understand its true significance, some information about the life of its author is in order.

2 Life and Times of Kātib Çelebi

2.1 *Youth: A Military and Bureaucratic Career*

The real name of our author was Muṣṭafā son of ʿAbdullāh, but he has always been better known under two nicknames that summarize his peculiar position in Istanbul society of his day: Kātib Çelebi and Ḥāccī Ḫalīfa.[49] He was born to a member of the Ottoman military class in 1609. His father belonged to one of the privileged cavalry contingents of the standing army (the so-called Six Divisions). At that time, men were no longer recruited to the standing cavalry or the janissary infantry through the devshirme, but came from other, usually Muslim backgrounds.[50] The former elite corps of the army was being transformed into a corporative enterprise with little interest in warfare.[51] Craftsmen and merchants of the bazaar found it increasingly advantageous to enroll in the janissary corps, which afforded them tax exemption and protection during unruly times. There is evidence that Muṣṭafā's mother came from a family of merchants, which would fit well with the overall picture of military and commercial connections. Members of the Six Divisions often held positions in the imperial administration; Muṣṭafā's father, for instance, worked in the fiscal chancery. As was common, the son (apparently, his only child to live to adulthood) was apprenticed in the same office, an indication of a modest middle class status at the time. Long training in a madrasa, that could have led to a scholarly career, was not in the cards for Muṣṭafā. His father did take care, however, to pro-

44 Vatin 1995.

45 Koraev 2015.

46 Hagen 1998.

47 For the different interpretations see Tezcan 2012; Casale 2010, 160–163; Hagen 1982; Hagen 1998; Hagen 2003, 99–103.

48 See Casale 2010, 187 and the objections in Soucek 2011; also the comprehensive assessment in Koraev 2015.

49 He mentions all these names in his autobiography. Three autobiographical passages in various works are the main source for his life; see Sarıcaoğlu 2002. An extensive account of his life is Hagen 2003, 7–78. A summary of his biography is available in Hagen 2007.

50 That a paternal uncle of Kātib Çelebi also served in similar positions would have been a violation of the rules of the *devshirme*.

51 The phenomenon described here is usually associated with the infantry, the janissary corps, but in the absence of specific studies on the cavalry I don't see a reason to assume that their situation was substantially different.

vide him a solid education, with tutors starting at an early age. Studying and then teaching others became his lifelong passion, explained by himself and his contemporaries by his being born under the ascendant of Mercury.[52]

As a member of the imperial chancery, the young man was also summoned to go on campaign with the imperial army. Since the late 16th century, the Anatolian provinces in particular had experienced dramatic social unrest, the ultimate cause of which is hard to pin down. Monetization of the economy led to both inflation and debasement of the currency. As a consequence, the standing army in particular saw the value of their salaries decline, which reduced their incentives to fight on behalf of the empire. Ottoman governors and provincial notables began hiring their own militias, contributing to the dissemination of cheap firearms, the raising of the tax burden, and the flight of peasants in conjunction with a rise in banditry. Climate change has been identified as a major cause of this Ottoman version of the global crisis of the 17th century.[53] Whatever the ultimate cause, all these factors came together to form a volatile mix of social and political tensions which reached a peak when Muṣṭafā was a teenaged apprentice.[54] Moreover, in departure from the Ottoman custom of succession, Sultan Aḥmed I was followed by his brother, not one of his sons. This brother, however, was deemed unfit to rule only three months later and was replaced by Aḥmed's son ʿOṣmān II, who was young and idealistic but no match for the entrenched interest groups in the capital. Rumors of radical plans to get rid of the unreliable janissaries incensed the latter and, in 1622, ʿOṣmān II was overthrown and executed by rebellious troops, the first Ottoman sultan ever to lose his life in an uprising.[55]

The campaigns in which Muṣṭafā participated were largely dealing with the aftermath of these traumatic events, against the backdrop of all the structural problems mentioned above. As the janissaries were blamed for the death of ʿOṣmān II, a governor in Anatolia in 1624 had used a popular spirit of revenge to rid himself of the local contingent in his province, and thus a means of sultanic control, and the imperial army was sent to bring him back into line. The next campaign sought to recapture Baghdad, whose Ottoman governor had used the opportunity to gain de facto autonomy under Iranian Safavid suzerainty in 1625. It took another campaign in 1628 to get the situation in Anatolia under temporary control, but a second attack of the imperial army on Baghdad over the course of 1629 and 1630 failed like the first. In 1633 the army again marched toward Baghdad; this time it wintered in Aleppo and, since the season was suitable, Muṣṭafā joined the pilgrimage caravan from Damascus to Mecca. After his return to the army he found that the campaign was being redirected against Safavid Iran, resulting in the conquest of Erivan in 1635, which would be lost again soon afterward.

Given the social disruptions, the trauma of regicide, and the obvious weakness of the military in establishing control in the interior and in winning a decisive victory, it is not surprising that the period was widely perceived as a crisis threatening the very existence of the empire. Historians today widely dismiss the characterization of Ottoman history after c. 1580 as one of decline, and in particular the notion that this decline constituted an irreversible process that would progress until the collapse of the empire in the early 20th century. At the time, however, the signs seemed all too obvious. Later in his life, Muṣṭafā would make use of the theories of the great North African historian Ibn Khaldūn (d. 1406) to suggest that the Ottomans were subject to the same cycle of rise and decline as every other dynasty in history, and that the process of decline that he noticed would be almost impossible to reverse.[56]

2.2 *From Fighting to Learning*

The political crisis for Muṣṭafā was compounded by a personal one. Muṣṭafā's father died during the catastrophic retreat from Baghdad in 1626, as did an uncle who also served in the military and chancery. Deprived of the patronage that was indispensable for a steady career in the scribal service, Muṣṭafā had to start over and, despite the help of elite patrons later on, never rose to a significant position in the administration. His two nicknames reflect the incongruence between his moral and intellectual status on the one hand and his middling professional rank in the bureaucracy on the other. His learned friends referred to him as Kātib Çelebi—*kātib* denoting a scribe and *çelebi* an educated gentleman who was not a madrasa graduate or one of the ulema. His colleagues in the chancery called him Ḥāccī Ḫalīfa—the clerk (*ḫalīfa*) who had made the pilgrimage to Mecca (Hajj). Both names subsequently became popular and eclipsed his given name and patronymic.[57]

52 *Cihānnümā*, first version (Vienna Draft), 4r.

53 White 2011.

54 İnalcık 1980.

55 On the events and their reflection in historiography see Piterberg 2003 and Tezcan 2002.

56 Thomas 1972 has demonstrated Kātib Çelebi's use of Ibn Khaldūn as documented in *Düstūrü l-ʿamel*. See also Hagen and Menchinger 2014.

57 Both names are explained in the autobiographical section in *Cihānnümā*, first version (Vienna Draft), 4r. Today, the name Kātib Çelebi is typically used in Turkish contexts, while Arab scholars prefer calling him Ḥājjī Khalīfa.

Besides the loss of his protector and the stalling of his career, Muṣṭafā considered those years on campaign very much a loss. While he drew on his experiences when narrating the campaigns in his chronicle *Summary of Histories* (*Fez̲leke-i Tevārīkh*), which also shows him as openly siding with the imperial cavalry,[58] it is remarkable that his geographical accounts of these areas rely entirely on written sources and never mention his own journeys with the army (they are mostly preserved in manuscript and not part of the text translated here).

Thus, while the army resumed its attempts to regain Baghdad under the increasingly assertive Murād IV (1623–1640), and in fact succeeded in 1639, Muṣṭafā decided not to go on campaign again, preferring, as he put it, the greater jihad of learning and self-improvement over the lesser jihad of armed struggle.[59] He was aided, as it seems, by one of the most controversial figures of the first half of the 17th century, Ḳāḍīzāde Meḥmed Efendi (d. 1635). Ḳāḍīzāde, who had himself been affiliated with a dervish order, had made a career as a brilliant populist preacher, having been appointed to the pulpit of one of Istanbul's most prominent mosques, Hagia Sophia. His main source of inspiration was a widely-read treatise by Birgivī Meḥmed Efendi (d. 1573) which in rigorous and concise form gathered the main tenets of faith, ritual and moral behavior as the basis of a pious life. While Birgivī advocated the classical Sufi practice of "remembrance of God" (*dhikr*) as a means of self-improvement, he insisted on rigid self-control and discipline, and thus rejected the ecstatic aspect of many Sufi rituals. He also favored an interpretation of the Qur'anic concept of "enjoining right and forbidding wrong" that opened the door to aggressive activism, a kind of puritan movement, nourished by a general sense of moral crisis in society. Ḳāḍīzāde's campaign since the late 1620s was directed mainly at some of the practices and beliefs of the Sufi orders, in particular the Ḥalvetīs, who also had much support in the Ottoman scholarly and judicial elite.[60] Muṣṭafā started to study with Ḳāḍīzāde in 1629, reading Birgivī and especially the great theologian and mystic al-Ghazālī's (d. 1111) *Revival of the Religious Sciences* (an apt title as motto of a movement like Ḳāḍīzāde's), which Muṣṭafā continued to hold in highest esteem.[61] His direct interaction with Ḳāḍīzāde was limited to a few months between campaigns, but Ḳāḍīzāde's influence lasted throughout his life. Not only did Muṣṭafā much later credit him for his own beginnings as a scholar, he also seems to have shared much of Ḳāḍīzāde's rigid rationalist religiosity. Although he distanced himself from Ḳāḍīzāde's activism late in his life, his work continued to manifest his distance from Sufism.[62]

Now in his late twenties, Muṣṭafā had long mastered all that he needed for the job of a scribe. Without hope for a prominent career, he shifted his energies not necessarily to piety (which one cannot verify, despite his interest in Ḳāḍīzāde), but certainly to books and study. Many prominent scholars of the time, in addition to training students in the madrasa, held lessons open to the public in mosques or in their homes. Muṣṭafā turned out to be a voracious student and reader, of Arabic and Persian as well as Turkish, and the list of books and topics he engaged in comes very close to the full madrasa curriculum. Looking back at his formation, he put particular emphasis on mathematics and astronomy, topics that he later complained had been very much neglected in Ottoman institutions.[63]

Already during his stay in Aleppo, probably the most important city for trading books in the Muslim world, he had rummaged through the stores of the booksellers. Now settled in Istanbul, he established himself as a kind of independent scholar. He was helped by two inheritances, one from his mother (an indication that she must have come from a well-to-do family) and the other from a wealthy merchant who was a relative. Muṣṭafā spent the money on fixing his house and taking a wife, but after achieving that, he spent his wealth first and foremost on books, assembling what may well have been the largest private library in Istanbul at the time. Reading to exhaustion during both daylight hours and by the light of a candle at night was how Muṣṭafā described his lifestyle; indeed, some of his autographs, like that of *Cihānnümā* itself, at times look like they were written late at night by a very tired hand. He went to the office probably no more than twice a week to retain his salary. On the other days, he started to give private lessons in the mid-1640s, especially teaching works of law, followed by those of mathematics and science. Several of his works, unfortunately barely studied today, have the character of scientific or edify-

58 Murphey 1993, 293; Piterberg 2003.

59 *Mīzānü l-ḥaḳḳ fī ḫtiyāri l-aḥaḳḳ* 1306/1888, 133.

60 On this dervish order see Curry 2010. The theological and social conflicts are masterfully described in Kurz 2011. The context of Ḥanafī legal thought is provided by Cook 2000. On Kātib Çelebi's intervention, which is an essential source for this problem, see below.

61 *Kashf al-ẓunūn ʿan asāmī l-kutub wa l-funūn* 1941, 23.

62 For Kātib Çelebi's treatment of the dispute, see *Mīzānü l-ḥaḳḳ fī ḫtiyāri l-aḥaḳḳ*, tr. Lewis 1957. A rare contemporary testimony vehemently rejecting his criticism of the Ḥalvetīs is Türer 2005, 458f. The literature about Ḳāḍīzāde and his followers is extensive, starting with Zilfi 1986. Most recently, Sariyannis 2012 also highlights its social implications.

63 This criticism has been rejected by El-Rouayheb 2008.

ing anthologies, some of which may directly relate to his teaching activities.[64]

2.3 *As an Author and Public Intellectual*

Despite his lack of academic credentials in the form of a madrasa diploma, Muṣṭafā was part of the leading intellectual circles of his time. The uppermost echelon of the religious-judicial hierarchy was firmly in the hands of a few aristocratic families of ulema, several of which maintained lively literary and intellectual salons where current events, history, geography and many other topics were discussed.[65] How Kātib Çelebi was introduced to those circles we do not know, but one of his shorter treatises, on legal questions related to astronomy, directly reflects the atmosphere of informal but learned disputation in such a setting, as it takes Shaikh al-Islam Bahāʾī Efendi to task for his lack of astronomical knowledge.[66] Even so, several shaikh al-islams—the highest religious-juridical post in the Empire—appreciated his works and at times endorsed them publicly. One of them helped him obtain a promotion in the chancery and another provided him with a precious manuscript.[67] Importantly, these circles were not restricted to Turkish-speaking Muslims but, occasionally at least, included Europeans residing in Istanbul, diplomats, scholars, etc.[68] Quite likely it was through these connections that he acquired some of European books that he used and that he met the erudite French convert who served as his translator of them.[69] In these circles Kātib Çelebi appears to have found a role akin to that of public intellectual, which compelled him to intervene in a number of public debates and to comment on current events. A short treatise on financial reform penned in 1653, *Rule of Action for Repairing the Breach* (*Düstūru l-ʿamel fī ıṣlāḥi l-ḫalel*), was the result of an investigation into the state account books by the grand vizier, but he set it in the philosophical framework of statehood that derived from Persian statecraft, and it remained influential far into the 18th century.[70]

Two important political developments of the time are specifically reflected in his works. One is the war over Crete, the last large island bastion of the Venetian seaborne empire in the eastern Mediterranean. The Ottomans had invaded it in 1645, but failing to gain a decisive victory, they found themselves caught in a protracted war of naval campaigns around the Aegean and of guerilla and siege warfare on the island itself. Kātib Çelebi's treatise on the states of Christian Europe, written in 1655, and rather trivial in nature due to its limited sources, may be considered a reflection of this confrontation.[71] Matters became more urgent when Venetian ships managed to blockade the Dardanelles in 1656. Suddenly a direct naval attack on the Ottoman capital appeared possible and the city panicked. Kātib Çelebi responded with a naval history of the Ottoman Empire, *Gift to the Nobles Regarding Campaigns on the Seas* (*Tuḥfetü l-kibār fī esfāri l-biḥār*), in which he recalled the past glory of the navy and held it up as the example to be followed if the Ottomans wanted to defeat the Venetians.[72] *Gift to the Nobles* invokes history as "magistra vitae" in the most immediate sense, as providing lessons for the political decision-makers of the present. It was in the course of this crisis that Köprülü Meḥmed Pasha became grand vizier (1656–1661), and he was to implement a program of rigid state recentralization that brought a considerable measure of stability after the crises of the preceding half century. There is evidence that Kātib Çelebi sympathized with him politically, suggesting that *Gift to the Nobles* should also be read as a program of reform for the navy.

The second development reflected significantly in Kātib Çelebi's works is the resurgence of the Ḳāḍīzāde movement. The first wave of activism had won support from, or coincided with the agenda of, the energetic but reckless Murād IV, leading to the destruction of several dervish lodges, the closure of coffeehouses and the prohibition of smoking, among other things. This wave had calmed down soon after the death of Ḳāḍīzāde in 1635. As the Venetian threat against the capital mounted, however, religious activism resumed under the preacher Üstü-

64 An overview of his work in Hagen 2007.

65 The seminal study on these intellectual circles is Wurm 1971. The importance of salons as forums of intellectual exchange is highlighted in Pfeifer 2015.

66 Partially edited in Şehsuvaroğlu 1957. It is also inserted in *Cihānnümā*; see translation at M 21.

67 They include most prominently Zekeriyāzāde Yaḥyā Efendi (d. 1644), Ḫocazāde Ebū Saʿīd Efendi (d. 1662), Ḫocazāde Bahāʾī Efendi (d. 1653) and ʿAbdürraḥīm Efendi (d. 1656). Bahāʾī Efendi previously owned the manuscript of *History of the West Indies* used by Kātib Çelebi (Hagen 1982–1998). Endorsements of his work by leading scholars preface the presentation copy to the sultan of Kātib Çelebi's *Gift to Nobles*, see *Tuḥfetü l-kibār fī esfāri l-biḥār* ed. İdris Bostan 2008, 59 and 1a.

68 Unfortunately, they are much better documented only for a period after Kātib Çelebi's death (Wurm 1971).

69 See Hagen 2007 on these translations. Several of them were preserved and subsequently circulated in their raw form, although they were clearly intended only for future integration into other works. There is no evidence that Kātib Çelebi himself learned a European language.

70 Thomas 1972, 65–124 *passim*; *Düstūru l-ʿamel fī ıṣlāḥi l-ḫalel*, ed. 1863, 119–140, tr. Can 1982.

71 *Irshādu l-ḥayārā ilā tārīkhi l-Yūnān ve r-Rūm ve n-Naṣārā*, ed. Yurtoğlu 2012.

72 *Tuḥfetü l-kibār fī esfāri l-biḥār*, 2r–2v; tr. Mitchell and Soucek 2012.

vānī Efendi, a successor of Ḳāḍīzāde, who also sought to capitalize on the turmoil and insecurity of the moment. Although he may well have been sympathetic to their agenda, Köprülü prioritized political stability over ideology, dispersing the rebels and banning their leaders from Istanbul. Kātib Çelebi's *Balance of Truth* (*Mīzānü l-ḥaḳḳ*), his last and one of his most influential works, recapitulates the points of contention between the Ḳāḍīzāde followers and their opponents in 21 short chapters. While Kātib Çelebi often sympathizes with the legal reasoning of Ḳāḍīzāde and Üstüvānī, he regularly rejects their activism and their interference in matters of personal belief and practice. In this sense, he provides a learned justification of Köprülü's crackdown, which, however, is argued in political rather than legal concepts.

That Kātib Çelebi himself was directly affected or involved in the events is nowhere attested. In contrast to his military years, we now perceive him as a humble, reclusive scholar, uninterested in career and wealth, conversing with his learned friends, fond of his books, his regular cup of coffee, and his flowers.[73] His only son had pre-deceased him at a young age. Kātib Çelebi himself died, probably of a heart attack, at less than fifty years of age in October, 1657.[74]

2.4 *The Encyclopedic Project*

Kātib Çelebi's untimely death cut short his work on a set of books that would be his most enduring intellectual legacy, his "Encyclopedic Project"[75]—the term signifies that these works were meant to gather and make accessible all important knowledge about the world that had accumulated in the course of history. The sheer expanse and nature of the project made it unlikely that it would have ever been finished; in fact, all its parts were left incomplete at Kātib Çelebi's death. Several of these parts can be seen as continuations of classical Islamic genres, but others are stunningly innovative. Since Kātib Çelebi worked on all of them more or less simultaneously, I will describe them here in logical rather than chronological order.

The first encyclopedia was a biographical dictionary entitled *Steps of Arriving at the Classes of Excellent Men* (*Sullam al-wuṣūl ilā ṭabaqāt al-fuḥūl*), in which Kātib Çelebi compiled short biographies of kings, viziers, scholars, poets and other men of note in alphabetical order.[76] Biographical compilations have a long tradition in the Islamic world. Usually they were restricted to particular class of people, scholars in particular, but we also find works on poets, calligraphers, etc. The most prominent precursor for Kātib Çelebi's comprehensive approach is Ibn Khallikān's (d. 1282) *Passings of Notable Men* (*Wafayāt al-aʿyān*), which was widely popular for centuries. *Steps of Arriving*, preserved only in the autograph manuscript, remains but a fragment as many entries are left to be filled in.[77] Capturing political and intellectual history through biography derived from the belief that history is always ultimately the result of individual decisions and, as such, a moral affair. Politics is therefore not the projection of a better future, but a return to previous states of justice and morality.[78] *Steps of Arriving* also shows Kātib Çelebi's fascination with alphabetization as a means to organize information, prioritizing easy access over narrative coherence.

Kātib Çelebi also gathered what he considered the essentials of world history in two other works. The first, *Summary of the Words of the Good* (*Fadhlakat aqwāl al-akhyār*), is a world history organized according to dynastic succession. It was written in Arabic, following a pattern that had become popular at the end of the 16th century and remained a constitutive form of Ottoman historiography throughout the 17th.[79] Its brevity and organization did not allow for much analytical narrative but instead turned it into a convenient handbook to look up specific kings and dynasties. It underscores another fundamental aspect of Ottoman historical thinking shared by Kātib Çelebi—that the dynasty (the same term also taking on the meaning of "state") was the unit of history, rather than, as modern thinking would have it, a nation, a class, a city, a concept, or anything else. The second work, *Summary of Histories* (*Fez̲leke-i tevārīḫ*), written in Turkish, provided an account of Ottoman history on which we relied when narrating Kātib Çelebi's military career. It too is incomplete, as the last chapters show considerable lacunae.[80] In addition, Kātib Çelebi provided a kind of key to both volumes in his *Tables of Historical Dates* (*Taḳvīmü t-tevārīḫ*), a chronological table of world history from creation to the present. Despite its brevity, this kind of compendium, judging from

73 He is mentioned as cultivating flowers in a book on Ottoman florists (Gökyay 1957, 15). There are indications that he enjoyed not only coffee but some opium confections as well.

74 A detailed account of his death has been noted on the flyleaf of *Cihānnümā*, second version (Topkapı Draft). The circumstances are discussed in Hagen 2003, 76–78.

75 Hagen 2007.

76 *Sullam al-wuṣūl ilā ṭabaqāt al-fuḥūl*, ed. İhsanoğlu 2010.

77 He included an entry about himself, an important source for his life; see Sarıcaoğlu 1991.

78 Hagen and Menchinger 2014.

79 *Fadhlakat aqwāl al-akhyār fī ʿilm al-taʾrīkh wa l-akhbār*, ed. es-Seyyid 2009.

80 Nevertheless it is an important source for a later master historian, Naʿīmā (d. 1716).

the large number of manuscripts, was in high demand in Ottoman learned circles. There are also several continuations beyond the date of Kātib Çelebi's death. And later historians used the chronological structure to fill in a more detailed narrative.[81]

The most popular of Kātib Çelebi's encyclopedic works was his bibliographical dictionary, *The Remover of Doubts Regarding Titles of Books and Names of Authors* (*Kashf al-ẓunūn ʿan asāmī l-kutub wa l-funūn*), written in Arabic. This too had its precursors, notably Ibn al-Nadīm's (d. 995 or 998) *Index of Books* (*Fihrist*). But Kātib Çelebi's work is hard to imagine without Ṭaşköprüzāde Aḥmed's (d. 1561) encyclopedia of the sciences, *The Key of Felicity* (*Miftāh al-saʿāda*), and it also owes much to Ibn Khaldūn's overview of the sciences in his famous prolegomena to history, *al-Muqaddima*.[82] Still, it is an achievement of herculean dimensions—14,500 entries on all the books Kātib Çelebi had seen or knew of, arranged into a single alphabetic sequence. Each entry contains information about the author, a table of contents if one was available along with an incipit, and information on translations, commentaries, glosses and other related works.[83] Inserted into the alphabetic sequence are 300 entries on individual sciences, each briefly discussing the main topic and giving lists of prominent works—for instance, the entry under history alone comprises 1,300 references. While parts of the work were written out as a fine draft, Kātib Çelebi continued to make additions to it.[84] *Remover of Doubts* was quickly recognized as an extraordinarily useful work. It continues to be an important resource for research in Islamic literature, precisely because of the alphabetical organization that facilitates quick consultation and the many cross-references that guide the reader further.

It is clear that the parts of the Encyclopedic Project discussed so far complement each other, inasmuch as they are all concerned with the same universal heritage of history and culture—Islamic humanities, we could call it—but approach it from different angles: biographical, chronological, or bibliographical. Moreover, they implicitly refer to each other: whoever looks up a book in *Remover of Doubts* can then find its author in *Steps of Arriving*, and vice versa; while looking up the date of a book or its author in *Tables of Historical Dates* and *Summary of Histories* will allow for historical contextualization. Such a project—to assemble the essential knowledge of mankind and make it accessible for quick reference rather than extensive study—differs from the massive literary encyclopedias written for the Mamluk scribal class mentioned above. It rather corresponds to the encyclopedism of the European Enlightenment, which was so essential in disseminating knowledge to the emerging bourgeoisie, as we shall see shortly.

In our emphasis on the complementary character of the parts of the Encyclopedic Project, we have thus far omitted one final principle of organization that Kātib Çelebi employed, viz., spatiality. The spatial or geographical dimension to the project is represented by *Cihānnümā*.

3 Text and Ideas of *Cihānnümā*

3.1 *Spatial Encyclopedia: First Version*

Cihānnümā occupies a unique place among Kātib Çelebi's works, and indeed represents an unusual case within the wider context of Islamic literature. That is because its evolution can be traced from its first inception to its final recension, although due to Kātib Çelebi's untimely death it did not reach completion. Kātib Çelebi himself stated that it was the outbreak of the Cretan war in 1645 that aroused his interest in geographical matters. Such a direct link to current affairs connects *Cihānnümā*, at least initially, more to the advice treatises discussed above than to the other components of the Encyclopedic Project. The way he went about it, however, immediately demonstrated a much wider horizon; in his own words, he "viewed every particularity in its larger context."[85] In an entry in *Remover of Doubts*, Kātib Çelebi describes *Cihānnümā* as alphabetically arranged, just like *Steps of Arriving* and *Remover of Doubts*. An early version that fits this description was only recently discovered. Essentially it is the text of Sipāhīzāde's alphabetized excerpt from Abū l-Fidā', to which Kātib Çelebi has started to add more recent material, mostly from Meḥmed ʿĀşıḳ's *Views of the World*. The margins of some sections are covered with small maps of lakes and rivers, which can also tentatively be attributed to his own hand. Why he abandoned his plan of a geographical dictionary we don't know, but we do know that he continued to use this manuscript as a repository for infor-

81 On continuations and expansion see Gökyay 1957, 48–54. The most detailed expansion was produced by Şemʿdānīzāde (d. 1779) under the title of *Mür'iyü t-tevārīḫ*. This was one of the earliest Ottoman printed works. A facsimile of an early manuscript is found in *Taḳvīmü't-tevārīḫ*, ed. Nurdan 2009, vii, 70, 172.

82 The epistemological concepts and the dependence on Ibn Khaldūn are discussed in Yurtoğlu 2009.

83 Birnbaum 1994; Birnbaum 1997.

84 The Istanbul edition (*Kashf al-ẓunūn* 1942–1943) is based on this autograph manuscript.

85 G.L. Lewis translates: "My method was to enter every plurality by way of unity, and to master first principles by comprehending universals" (*Mīzānü l-ḥaḳḳ fī ḫtiyāri l-aḥaḳḳ*, tr. Lewis 1957, 137).

mation, and thus mentioned it as part of the bibliography informing the final recension translated here.[86]

Reworking this early draft, he produced what has been recognized since the groundbreaking research of Franz Taeschner in the 1920s as the first version or recension of *Cihānnümā*.[87] Conceived as a cosmography along the lines of al-Qazwīnī and Meḥmed ʿĀşıḳ, it was to contain an introduction on the structure of the cosmos, followed by separate sections on the four elements, among which the one on water would describe oceans, seas, lakes and rivers, and the one on earth would be describe countries and cities, now arranged according to the historical-cultural regions or climes defined by Abū l-Fidāʾ and starting at the western end of the inhabited earth. The existing manuscripts all begin in the middle of the chapter on lakes, with some more lacunae, and the geographical chapters only cover the Iberian Peninsula (al-Andalus), North Africa (Maghrib), and the European parts of the Ottoman Empire (Rumeli), with this last part in particular containing much original information.[88]

This version already demonstrates significant innovation in Kātib Çelebi's concept of geography. Unlike his immediate predecessors, Meḥmed ʿĀşıḳ and Sipāhīzāde, he consulted all kinds of sources that included geographical information, including nautical literature and historiography, even grammatical works. As long as valid factual information could be obtained, he was happy to neglect boundaries of genre which his predecessors had observed.[89] On the other hand, although he was a scribe in the Ottoman fiscal administration, which itself preserved a tremendous administrative spatial or locational knowledge, there is no trace of tax administration records as a source for *Cihānnümā*. Instead he relied on a literary work, a widely-read reform treatise, not particularly up to date.[90] Copies of the first version of *Cihānnümā*, like the original rough draft, often include miniature maps in the margins, although it is impossible to determine if any of them are inspired by Kātib Çelebi's own ideas. Some copies have highly abstract maps in small medallions; in others, river maps run along the margin of a full page; some evidently later manuscripts have full-page maps as well.[91] As a concession to the alphabetic ordering he had abandoned in the transition, he now planned to add alphabetic indices to individual chapters.[92]

3.2 *Spatial Encyclopedia: Second Version*

Kātib Çelebi later wrote that he gave up this project due to a lack of sources for what would have been the next region to cover, i.e., Christian Europe. Although he set the book aside to work on other projects, he did continue to fill the blank pages and margins of a manuscript (known since Taeschner's seminal study as the "Vienna Draft") with additional information.[93] Things changed when he made the acquaintance of a learned French convert to Islam, probably a former member of the Capuchin mission to the Ottomans. This man, known to us only under his Muslim name of Shaikh Meḥmed İḫlāṣī, translated a number of European sources for Kātib Çelebi.[94] Since the mid-16th century, and sparked by the rediscovery of Ptolemy's *Geography*, Renaissance Europe had developed a sustained interest in maps and world descriptions, starting with the custom-made map assemblies by Italian cartographers. Abraham Ortelius (1527–1598) and Gerard Mercator (1512–1594), arguably the most prominent cartographers north of the Alps at the time, were both highly skilled and prolific mapmakers; but their most lasting impact in the history of cartography was to make the modern atlas a successful commercial product,[95] an endeavor in which they were succeeded in the 17th century by men like Jodocus Hondius, Jan Janszoon, Willem Janszoon Blaeu and Joan Blaeu. Ortelius' *Theatrum Orbis Terrarum* can be considered the first modern atlas. Published first in 1570, it was followed by innumerable reprints, redactions, supplements and translations.[96] The *Theatrum* responded well to

86 Hagen 2006, M 11. See also the description in *Kashf al-ẓunūn* 1942–1943, 622–623. The manuscript is University of Michigan, Isl. Ms. 215: https://hdl.handle.net/2027/mdp.39015081446448.

87 Taeschner 1926.

88 It drew the attention of the prominent orientalist Hammer-Purgstall, who translated it in 1812. The original was first identified in Taeschner 1923.

89 His sources comprised almost all Ottoman geographical literature discussed above as well as most of the Islamic texts. For extensive information on all identifiable sources see Hagen 2003.

90 ʿAyn ʿAlī's (d. 1609) *Ḳavānīn-i Āl-i ʿO̱smān* (see Hagen 1998).

91 The first type includes the Vienna Draft as well as Berlin, Staatsbibliothek, Ms. or. fol. 4056 and Vienna, Österreichische Nationalbibliothek, H.O. 191. Prominent examples of the second type are Vienna, Österreichische Nationalbibliothek, mxt. 762 and Beyazıt, Veliyüddin Efendi no. 2336. The third type is represented by Topkapı, Revan 1629. Not all manuscripts have been surveyed regarding the types of maps.

92 Hagen 2003, 180.

93 Taeschner 1923. The manuscript bears a (probably original) title *Draft of Cihānnümā on the science of geography* / *Müsvedde-i Cihānnümā fī fenn-i cuġrāfiyā*.

94 That he was Catholic is suggested by the character of some of the sources he provided and translated, but he has not yet been identified in the record of the order. İbrāhīm Müteferriḳa's preface to his edition of *Cihānnümā* is the only source, but is marred with hagiographical topoi (see translation at M 10).

95 Koeman et al. 2007, 1318.

96 Publishers frequently bought and sold the plates of famous atlases. Joan Batista Vrients acquired the plates from Ortelius as

the market demands of the time, defined by an emerging bourgeoisie that needed to plan for commercial travel, to keep track of political and military events, and to localize classical and biblical history. Yet it also served as item of prestige and representation.[97] We know of an order from a translator at the Ottoman court to obtain two copies of it from Vienna in 1573.[98]

Mercator, besides having a special type of map projection to his name, was following a more ambitious cartographic and scholarly agenda, designing his *Cosmographia* as a "synthesis of knowledge seeking to reconcile the observational sciences and biblical knowledge"—an increasingly precarious endeavor.[99] Its encyclopedic plan included the creation of the world, a description of the universe, a description of the oceans and land masses (divided into a contemporary atlas, a set of maps of Ptolemy, and maps of antique geography), genealogy and political history, and a world chronology. Only parts were published during his lifetime, and then reassembled in the *Atlas sive cosmographicae meditationes de fabrica mundi et fabricati figura*, published in 1595 by Mercator's sons and collaborators. A limited commercial success, Mercator's plates ended up in the possession of Hondius, who essentially held a monopoly in printed atlases after Vrients stopped publishing Ortelius' *Theatrum* in 1612. Hondius published several editions of Mercator's *Atlas* between 1606 and 1623; also a more affordable popular edition with 152 maps in a reduced format entitled *Atlas Minor*, which went through multiple editions between 1607 and 1621, at which date Hondius in turn sold the plates.[100]

Despite the very different commercial, technical and academic conditions under which European atlases were produced, Kātib Çelebi took to them quickly and thoroughly. Mercator's plan for his *Cosmography* was strikingly similar to Kātib Çelebi's Encyclopedic Project, although it is not likely that it was an inspiration since he encountered Mercator only in the abridged form of *Atlas Minor* which Meḥmed İḫlāṣī translated for him. Still, it is significant that he chose this work as the basis for his new effort in geography, *Sparks of Light in the Darkness of Atlas Minor* (*Levāmiʿu n-nūr fī ẓulmeti Aṭlas Mīnūr*). Although strictly a working translation, the Turkish rendering of *Atlas Minor* later circulated as a separate work, but its real significance is its function as the inspiration and main source for the rewriting of *Cihānnümā*, resulting in the second version. Ortelius' *Theatrum* was another source, but seems to have been used only selectively along with several less famous European geographical works, all of which are referenced in the bibliography of *Cihānnümā*.[101] In addition to these European sources he continued to gather Islamic texts, some of which he ended up using extensively. These include Amīn Aḥmad Rāzī's *Seven Climes* (*Haft iqlīm*), basically a Persian poetic anthology geographically arranged, which furnished much information on India, and Ḥamd Allāh Mustawfī's *Nuzhat al-qulūb*, which became one of his main sources for Iran.[102] On the basis of these new sources, Kātib Çelebi set out to rewrite *Cihānnümā*. This time, however, after a lengthy systematic introduction to geography, he chose to start in the east of the inhabited world, with Japan and China, and then work his way westward. As with the earlier recensions, he proceeded region by region, each chapter being accompanied by a half-page map usually copied from one of his European sources.

At the time of his death he had just finished the description of Asia up to the eastern borders of the Ottoman Empire. The autograph, on which he presumably kept working until the very end, and which I call the Topkapı Draft by analogy to the aforementioned Vienna Draft, is preserved in the library of the Topkapı Palace in Istanbul (TSMK, Revan 1624)—referred to as R in this book.

3.3 *The Idea of Geography*

Cihānnümā is not only the synthesis of an almost 1000-year-old Islamic tradition of geographical writing with European geographical and cartographic approaches, but also the result of a decade of the author's learning and thinking about geography. The textual history so richly documented in the sequence of autograph manuscripts allows us to appreciate Kātib Çelebi's originality in his

well as his competitor Gerard de Jode in 1601 and continued to publish the *Theatrum*, while he abandoned De Jode's *Speculum Orbis Terrarum* (first printed in 1578).

97 Koeman et al. 2007, 1319.

98 Ágoston 2007, 86–87. Earlier attempts to obtain European maps are documented in Arbel 2002.

99 Cosgrove 2007, 69.

100 Koeman et al. 2007, 1333.

101 These works are Philippus Cluverius, *Introductio in totam geographiam tam veteram quam novam*, first published Leiden 1624 (edition used: Paris 1635); Giovanni Lorenzo d'Anania, *L'universale fabrica del mondo*, Venice 1582 and 1596; *Commentarii Collegii Conimbricensis Societatis Iesv In Libros Meteororum Aristotelis Stagiritae*, Cologne 1596; a map of Asia by Giacomo (or Jacopo) Gastaldi; a world map by Iodocus Hondius. Passages in the chapter on Japan seem to go back to Giovanni Pietro Maffei, *Histoire des Indes*, Lyon 1603, but it is not clear if they are borrowed directly (information from Prof. Jean-Louis Bacqué-Grammont). Kātib Çelebi wrote down the table of contents of the third volume of Giovanni Ramusio's *Delle navigatione et viaggi*, Venice 1606, but does not seem to have used it in *Cihānnümā* (Goodrich 1990, 350 f.).

102 Kātib Çelebi also used sources directly later in his work that he had initially quoted indirectly from other texts.

approach to world description and to realize how this work is the culmination of the Islamic geographical tradition.

The first version of *Cihānnümā*, as indicated above, lacks an introduction. Possibly it was lost together with the first pages of the chapter on seas, lakes, and rivers. More likely it was never written, because Kātib Çelebi was dissatisfied with the frameworks existing at the time. The models available to him typically started with the structure of the universe and some key concepts like distances and coordinates.[103] Al-Qazwīnī's series of prefaces cited above, about man's understanding of the universe and its relation to knowing God, is already a unique case. Some cosmographies open with a brief creation account. The second *Cihānnümā*, by contrast, opens with a detailed explanation of the purpose and method of geography.

Thus, for the first time in the Islamic context, geography is not solely defined by identification with its object but as an intellectual pursuit with a special method. Classical cosmography proceeded on ever decreasing scales in describing the universe, from the celestial spheres to the earth to specific countries. In *Cihānnümā* Kātib Çelebi for the first time articulates these in terms of distinct scientific practices, ranging from cosmography as the description of the universe to geography dealing with the earth and chorography dealing with specific places or regions. It has been remarked that, writing more than a century after Copernicus, Kātib Çelebi still strictly adhered to a geocentric model of the universe, but there is nothing surprising about this since none of his European sources presented the heliocentric view. Other practices explained here in theoretical terms are mapmaking and navigation.[104]

Secondly, Kātib Çelebi becomes ever more assertive about the authority and unity of geographical knowledge. Geography for him was a book science, meaning that only books identified with known and verifiable authors could be trusted, and he remained highly skeptical about obscure authors (like ʿAlī Ekber Khiṭāyī), which explains why he hardly ever includes oral information from travelers or his own impressions from the days of campaigning. At the same time, *Cihānnümā* attests to his fascination with the empirical side of geography when he draws on nautical literature regarding navigation with maps, monsoons, etc.[105] For him, this knowledge needs to conform to observation and logical reasoning. Statements in scripture about natural phenomena like lunar eclipses that contradict observed realities cannot pose objections. Rather, where observational knowledge clashes with literal reading of scripture a metaphorical reading is warranted.[106] As he insists that geographical knowledge should be based on experience and observation, Kātib Çelebi categorically dismisses the vast lore of classical cosmography and makes an important step toward a unified world view in which all that occurs must be governed by the same quasi-natural laws.[107]

The third point worth underscoring takes us to the core of geographical thinking. Descriptive geography, as practiced in the footsteps of Ptolemy, was largely a synthetic procedure leading to extensive catalogs of seas, rivers, lakes, mountains, and especially cities. The division of these into "true climes" (7 in number) according to latitude meant organizing them according to astronomical and astrological qualities. "Conventional climes" (28 in number), as used by the Balkhī School and later by Abū l-Fidāʾ and Meḥmed ʿĀşıḳ, instead imposed a historical or cultural organization. In *Cihānnümā* each chapter is dedicated to a region as defined exclusively based on cultural and historical factors. Empiricism and pragmatism do away with the multiples of seven that made up both the "true" and "conventional" climes.

Kātib Çelebi designed an elaborate list of criteria he planned to discuss for every region, beginning with 1) its name, 2) boundaries, and 3) internal divisions. He then moved on to the political dimensions such as 4) government and politics, 5) penal law, and 6) religion. Then came cultural dimensions such as 7) sciences and learning, 8) crafts and trade, 9) war and courage, and 10) customs. The next sections mainly correspond to the categories familiar from Islamic geography: 11) cities, 12) buildings and monuments, 13) sympathetic qualities, 14) rivers, 15) mountains, 16) water and climate, 17) flora and agriculture, and 18) fauna. The last two items had already been included in the first version as innovations: 19) roads or itineraries, and 20) lists of kings.[108] This pattern, which Kātib Çelebi scrib-

103 One of the most elaborate texts is the introductory section of Yāqūt, *Muʿjam al-buldān* (tr. Jwaideh 1959).

104 The entire section is based primarily on Cluverius.

105 His main source is Seydī ʿAlī Reʾīs' *The Ocean* (*or: the Comprehensive*) / *al-Muḥīṭ*, which is dismissed by Tibbetts as a mostly literary rewriting of Arabic navigation manuals from the Indian Ocean (Tibbetts 1971, 44–46).

106 See translation at M 17 ff. The passage quotes al-Ghazālī's *Incoherence of the Philosophers*. For this argument in the context of al-Ghazālī's thought see Griffel 2009, 111–122.

107 Kātib Çelebi adopts another concept probably from al-Ghazālī to avoid determinism and reconcile regularity in natural phenomena with divine omnipotence; see discussion in Hagen and Menchinger 2014.

108 *Cihānnümā*, first version (Vienna Draft), flyleaf, no pagination; numbering in the original. The Turkish terms are: *taʿrīf, taḥdīd, taḳsīm, aḥvāl-ı salṭanat ve riyāset, aḥvāl-ı siyāset, aḥvāl-ı diyānet, aḥvāl-ı ʿilm ü maʿrifet, aḥvāl-ı ṣanʿat ve ticāret, aḫlāḳ ve ʿādāt, aḥvāl-ı ḥarb ve şecāʿat, bilād, aḥvāl-ı ebniye ve ās̱ār, ḫavāṣṣ,*

bled on the flyleaf of the Vienna Draft, is partly adapted from the categories in the country descriptions in *Atlas Minor*, but also bears evidence of Kātib Çelebi's thinking. It is applied quite precisely in some early chapters of *Cihānnümā*, with the exception of the sympathetic qualities which are a remnant of astrological geography.[109] In later chapters the pattern is barely recognizable and many of these items are not dealt with at all. Still, while short of geography as the science that explains how the world came into the state in which it currently exists, it is a fairly comprehensive plan of human geography.

The fourth point to consider is that this human geography was meant to be practical and useful. Kātib Çelebi enthusiastically embraced Mercator's argument that geography is a necessity for good government because, as he puts it, a conscientious home owner should not only know the rooms and passages in his own house but the situation of his property within the neighborhood and the city so that he can take appropriate measures in case of fire.[110] Given the constant danger of fire in Istanbul, and given the fact that the city's population was stricken by panic whenever the Venetian fleet threatened the straits, this argument must certainly have resonated with both the author and his later readers.

3.4 *The Maps of* Cihānnümā

Considering Kātib Çelebi's fascination with efficient ways of organizing and presenting information, it should come as no surprise that he considered maps an indispensable component of the project. I have argued elsewhere that his exposure to maps must have been limited, and that his first steps are very much experimental and unsatisfying.[111] His understanding of mapping evolved significantly over the course of his geographical and cartographical work. As we have noted above, the evidence of the maps in the Vienna Draft and other copies of the first *Cihānnümā* is inconclusive. We assume that the little marginal maps in the Vienna Draft are in his hand, but later copyists have taken all kinds of liberties and inserted their own ideas of maps in their products, including significant innovations. Kātib Çelebi's maps were sketches without consistent orientation or scale, more suited as illustrations of the text than as independent representations. On the other hand, there was space for a world map in two hemispheres, with a grid of coordinates, but not executed. The maps in *Levāmiʿu n-nūr*, the translation of *Atlas Minor*, are mostly free-hand copies of those in the original. While this technique requires significant practice and skill, there is no evidence that Kātib Çelebi had any technical devices to produce these copies. The frames of these maps often have little dots to indicate where the lines of latitude and longitude would be, but they are clearly an afterthought, and not the basis of the construction of the maps (which are, in fact, often left unfinished). The second version of *Cihānnümā*, however, is very much conceived as an atlas, a set of maps accompanied by textual explanation. Here again, it follows *Atlas Minor*, although the proportion of text with each map is much more various, because, as Kātib Çelebi explains, the absence of print technology makes copying an abundance of maps cumbersome and prone to errors (R 11b). Maps copied from Western models, Mercator or Gastaldi in particular, are almost uniformly oriented with north at the top, and often have latitude and longitude indicated on the frame.

Contrary to the popular assumption that the visual language of maps is quasi natural and universal, we see Kātib Çelebi experimenting with the symbols of mapping, starting with his depiction of countries or regions as non-contiguous rectangles in one of his earliest forays.[112] His later maps typically show coastlines, rivers, and cities, and in some cases mountain ranges and roads, with the grammar of symbols in part copied from the models, in part defined in an ad hoc fashion: sometimes the symbol for city seems to differ between smaller and larger cities (R 146a). A unique attempt of mapping confessions by placing crosses and crescents on the world map is described in the introduction of the second *Cihānnümā*, inspired again by Mercator/Hondius, but the map remained a sketch in the Istanbul Draft, and Müteferriḳa left it out, while reproducing the description.[113] Maps also raised other problems of translation: in his working translation of *Atlas Minor*, *Levāmiʿu n-nūr*, Kātib Çelebi left the scale in German miles, as he found it in the original, just as he did not bother to 'translate' labels into Turkish: in the aforementioned map of religions he initially transcribed 'Mahametānī', and added an interlinear translation 'Muḥammedīler' instead of translating 'Muslim' right away. But there is one instance, in his map of Fārs, where he made the step to 'translating' this element, as he replaced a scale (probably the Italian miles shown by Gastaldi, his likely model) with a scale in parasangs, 'according to the measure of the people of Asia' (R 104a).

enhār, cibāl, āb u hevā, nebātāt ve maḥṣūlāt, ḥayvānāt, mesālikü l-memālik, aḥvāl-ı mülūk. Cf. Hagen 2003, 220.

109 See e.g. Curry 2012.

110 See translation at M 17; and see Hagen forthcoming.

111 Hagen 2006.

112 Inserted into his copy of Sipāhīzāde, mentioned above.

113 M 123, R 44a–b. It is complete in *Levāmiʿu n-nūr* (Nuruosmaniye 2998, 421a), clearly inspired by Mercator/Hondius' Designatio Orbis Christiani (679).

We see Kātib Çelebi struggling to provide more detail especially to the west of India: here his European models were insufficient, instead he had to plot new maps based on the textual information from his sources. These are the rather rough maps from Sind to Jibāl in the Istanbul draft, in addition to several more which were left unexecuted; it is telling that Müteferriḳa did not include any of them in the printed edition, and instead had one of his collaborators draw up two new maps, one of Iran (M 289/290) and one of Transoxiana (M 347/348). The mathematical basis of drawing maps does not seem to have occupied Kātib Çelebi much. It is evident from many examples that show grid lines that he was handling ruler and compass with ease, but he only drew circles or straight lines, thus simplifying the problems of cartographic projection, which Ptolemy had already addressed. In fact, there is no indication in *Cihānnümā* that he considered the distortion resulting from projection problematic. The earliest instance that directly addresses such problems is found in an early eighteenth-century copy of his *Tuḥfetü l-kibār*, in the margin of a map of the old world and the seven climes.[114] Mapmaking, for Kātib Çelebi, remained very much an experimental process, open to new ideas and developments as new challenges arose, but remained substantially different from the mathematically informed cartography of a Mercator.

3.5 *Reception*

The reception history of *Cihānnümā* suggests that this resonance extended well beyond Kātib Çelebi's lifetime, that he had produced what the zeitgeist demanded. Despite its fragmentary character, the first version of the text must have circulated widely in fairly short order. There are a significant number of manuscripts from the second half of the 17th century, while there were no copies made from the expanded Vienna Draft which certainly remained in the author's hands until the end of his life. This suggests that he had permitted copies to be made of the earliest version, despite its being incomplete. The fates of the Vienna Draft and of the second version of *Cihānnümā* (= R) are intertwined with the intellectual history of the second half of the 17th century in very particular ways. Both manuscripts, together with the Sipāhīzāde manuscript and, incidentally, Kātib Çelebi's copy of the *History of the West Indies*, soon after his death passed into the possession of one of his acquaintances and patrons, Vişnezāde Meḥmed ʿİzzetī, who was at the time the center of an intellectual circle in the Ottoman capital.[115] While little is recorded of those circles or salons, we know some of the prominent men of the following generation who surrounded Vişnezāde, including another historian and polymath, Ḥüseyn Hezārfenn (d. 1691) and a geographer from Damascus, Ebū Bekr b. Behrām ed-Dimaşḳī (d. 1691). Another important person for us is the Dutch resident, Lewinus Warner (d. 1665), also a prominent orientalist. While there is no conclusive evidence that Kātib Çelebi met any of these men in person, they turned out to be crucial for the reception history of his geographical work. From their literary output, it seems fair to say that they shared Kātib Çelebi's idea that knowledge about the world, including history and geography as well as natural philosophy, was an essential prerequisite for a society to flourish and for a dynasty to rule successfully and maintain an empire. Moreover, Vişnezāde's circle overlapped in significant ways with the intellectuals who gathered around the new men in power at the time—Köprülü Meḥmed Pasha and his son Köprülüzāde Fāẓıl Aḥmed Pasha (d. 1676)—whose offspring and clients would dominate Ottoman politics well into the 18th century. While they prioritized political stability and administrative efficiency over everything else, the Köprülüs were sympathetic to the rigid religiosity of the Ḳāḍīzāde type to which Kātib Çelebi also adhered. Kātib Çelebi had also made a thinly veiled plea for Köprülü patronage in his *Guideline of Action* (*Düstūrü l-ʿamel*), and his maritime history *Gift of the Nobles* can be read as advice for the new vizier in the face of the Venetian threat.[116]

The politics of the Köprülü household, I suggest, also underwrote one of the most ambitious geographical projects of the time, another testimony to the changing place of geographical knowledge in Ottoman political thinking. By mid-century, Johannes Janssonius and Joan Blaeu, the two leading cartographic entrepreneurs in the Netherlands, had entered a kind of arms race for the most splendid and most expensive atlas publication. The winner ultimately was Joan Blaeu's *Atlas Maior*, which appeared in several languages between 1662 and 1672, an item appreciated not so much for its scientific value as for its aesthetic appeal and prestige, expressed also through its astronomical price, especially for the version with colored prints. Obviously, such works would also make impressive diplomatic gifts: thus, in 1668, Warner's successor as Dutch envoy in Istanbul, Justus Colier, presented Sultan Meḥmed IV with an 11-volume copy of the Latin edition of *Atlas Maior* bound in purple velvet. In 1675 the sultan ordered Ebū Bekr b. Behrām ed-Dimaşḳī to pro-

114 Mihrişah Sultan Kitaplığı 304, shown in the introduction to İdris Bostan's edition of *Tuḥfetü l-kibār*, 55.

115 Wurm 1971, 65–71.

116 Detailed argument in Hagen 2003, 62–76.

duce a Turkish translation of this work, which was completed in 1685. The exact translation process is unknown, but several different versions in both Arabic and Turkish exist today which are attributed to Ebū Bekr, the largest of them comprising nine volumes which are awaiting in-depth investigation.[117] Interestingly, despite its being a complete translation and based on much more recent sources than *Cihānnümā*, it was the latter that would become the canonical text of Ottoman geography.

This outcome was quite unlikely at the time. Vişnezāde seems to have passed the whole set of Kātib Çelebi's geographical manuscripts, including the two *Cihānnümā* autographs, on to Ebū Bekr, as we can see from many notes in his hand in their margins. Only one copy had been made from the second version until then, for Lewinus Warner, and it had been sent to Leiden with the rest of his manuscript collection after his death in 1665.[118] Otherwise, Ebū Bekr seems to have held on to Kātib Çelebi's autographs because he intended to use the material to supplement his translation of *Atlas Maior*, but he did not consider it as a work in its own right. In fact, the second version apparently began to circulate only after Ebū Bekr's death; at least, there is no manuscript other than Warner's that is datable to Ebū Bekr's lifetime. Then, however, the large number of copies indicates that it was in considerable demand, and frequently the first and second versions were copied together. Its canonization was then sealed in the early 18th century with the printed edition which is the basis of this translation, produced by the pivotal figure and Hungarian convert to Islam, İbrāhīm Müteferriḳa (d. 1745)—referred to as M in this book.

Printing in moveable type had long been practiced in the Ottoman domains in Hebrew, Greek and Armenian languages, but the Arabic script, with its ligatures between letters and their changing shapes, posed special technical challenges for print. As a result, early prints in Arabic type were quite expensive, while manuscript copying was still affordable, reliable, fast and aesthetically much more appealing. In other words, as opposed to the case of Western Europe, printing technology did not automatically offer the same advantages for this society, which explains why it was adopted so much later.[119] Müteferriḳa, most likely a Unitarian convert to Islam and certainly a remarkable political entrepreneur, rose in Ottoman service under the patronage of grand vizier Damad İbrāhīm Pasha (1718–1730), a towering political and intellectual figure, and made his imprint on Ottoman society by obtaining permission to print books in Arabic movable type. Between 1729 and 1745 he printed a total of seventeen books, mainly dedicated to history and geography, and arguably representing an enlightenment agenda of disseminating information that would be most relevant for statesmen and administrators.[120] If this strongly resembles Kātib Çelebi's program 80 years earlier, it is hardly surprising that no other author was represented as frequently among the books printed by Müteferriḳa: *Gift to Nobles*, *Tables of History*, and, as the most ambitious, *Cihānnümā* in 1732. Müteferriḳa had acquired several of the manuscripts that had been in Ebū Bekr's possession, among them both the Vienna Draft and the Topkapı Draft (= R). He started his edition with the latter but amended it in several instances, inserting longer or shorter digressions of his own. The most famous of these is a passage of around 20 pages in the introductory section of the work expounding Tycho Brahe's astronomy as well as the Copernican heliocentric system as superseding Ptolemy's geocentric system.[121] Where Kātib Çelebi's own text broke off in the Topkapı Draft, Müteferriḳa first continued with the notes Ebū Bekr had jotted down in its margins, then added an excerpt from another work of Ebū Bekr of almost 300 pages that covered the Asian provinces of the Ottoman Empire that Kātib Çelebi had not completed by the time of his death.[122] All these interventions are carefully identified by Müteferriḳa as "supplements by the publisher." The result was a massive volume of over 700 pages, including 45 pages of maps and diagrams executed as copper plate engravings by members of Müteferriḳa's workshop. A second volume was planned, to be based on the

117 See Wurm 1971, 39–47 on aspects of the translation. Brentjes 2005 and 2012 has done important work on the maps. Ebū Bekr complained about a lack of recognition for this work when he produced a Turkish abridgment about the Ottoman Empire (ed-Dimaşḳī, *El-fetḥu l-raḥmānī fī ṭarz-i devleti l-ʿOs̱mānī*, ed. Dorogi and Hazai 2011–2014).

118 Leiden, cod. Warner 1109.

119 Ottoman attitudes to printing are discussed in Sabev 2006 and Lindner 1998.

120 On books published, see Gencer 2010. The argument for an Ottoman enlightenment has been made most recently by Erginbaş 2014. The impact has been viewed with increasing skepticism; see Sabev 2014.

121 İhsanoğlu 1992, Morrison 2003. The new systems were not presented here for the first time in Ottoman literature, but this was the text that became most authoritative.

122 The immediate source for this supplement is not clear, as there is no known work by Ebū Bekr that contains this specific text. A manuscript in London (British Library, or. 1038) containing almost exactly this text was considered by Sarıcaoğlu 1991 as a supplement by Ebū Bekr; but it is anonymous and could just be an excerpt from his work by someone else. Given his ambitions it seems unlikely that Ebū Bekr himself would write such a supplement; and the collation made for the current translation reveals that it cannot be Müteferriḳa's immediate source.

Vienna Draft which also was in Müteferriḳa's hands, but it never came into being.

It was the printed edition that gave *Cihānnümā* its reputation as the key text of Ottoman geography, so characterized by Taeschner.[123] Although not really a commercial success, it was disseminated far more widely than any Ottoman manuscript text on geography.[124] It became available in provincial libraries and became the primary reference also for Europeans seeking information on the geography at least of the eastern parts of the Ottoman Empire. Despite a modern popular perception of a fundamental difference between the Ottoman Empire and early modern Europe, the history of *Cihānnümā* also reveals fundamental commonalities between Kātib Çelebi and many contemporary Europeans in the quest for knowledge. As we have seen, his idea of gathering all the knowledge about the world was not so different from Mercator's, and information gleaned from *Atlas Minor* blends seamlessly with data from Abū l-Fidā' and *Nuzhatu l-qulūb*. The quest for knowledge about the other was also, in many ways, mutual, although epistemologies certainly differed, as did access to sources. But just as Kātib Çelebi was seeking information about European states and dynasties, which he gathered for instance in his little treatise *Guidance of the Confused* (*İrşādu l-ḥayārā*), European scholars were eager to get their hands on Islamic manuscripts, frequently through the mediation of diplomats on mission to the Ottomans.[125] Lewinus Warner's acquisition of an early copy of *Cihānnümā* was only the beginning. Antoine Galland (1646–1715), who served in the French embassy in Istanbul, brought the first manuscript of the Arabian Nights to Paris; his subsequent translation, the first volume of which was published in 1704, set off an orientalizing frenzy of the popular imagination in Europe. This could only happen, however, because the ground had been prepared by more than a century of oriental research promoted by absolutist royals and princes that culminated in the first compendium of European knowledge about the Muslim world, Barthélemy d'Herbelot's monumental *Bibliothèque Orientale*, published in 1697. This massive encyclopedia—incidentally a testimony to the early French Enlightenment's infatuation with alphabetization of knowledge, later made most manifest in the *Encyclopédie* of Diderot and D'Alembert—rests to a large degree on none other than Kātib Çelebi's *Remover of Doubts*. Most likely, the two manuscripts of the latter that d'Herbelot had access to had been acquired by Galland, who during his stay in Istanbul in the 1670s had frequented the same circles of Vişnezāde, Hezārfenn and Ebū Bekr ed-Dimaşḳī to which Kātib Çelebi had belonged a generation before. A French translation was also prepared by another scholar from the same background, but was never published.[126] The learned Venetian priest Giambattista Toderini (d. 1799), who wrote a three-volume history of Turkish literature in the late 18th century, very naturally quotes Kātib Çelebi together with his European translators and counterparts, Galland and d'Herbelot.[127]

A similar interplay of Muslim and Western scholarship characterizes the reception of *Cihānnümā*, which promised access to new and valuable information on remote parts of the world. The Dutch scholar and diplomat Lewinus Warner not only acquired the first known copy made (albeit one without maps), but also produced a draft Latin translation of the entire descriptive part, obviously in the expectation of acquiring information unavailable from Western sources.[128] The popularity of the work among Western scholars as their primary reference for the geography of the Muslim world is attested by a number of references proudly given by Toderini, leading up to the Latin translation by the Swedish orientalist Matthias Norberg.[129] Many printed copies were acquired by Europeans and ended up in European libraries. While we should not disregard significant differences between Kātib Çelebi's thinking and that of early modern Europeans, the compatibility of geographical knowledge across cultures and the cross-cultural networks that produced this knowledge are important reminders how porous political, cultural and religious boundaries were in the 17th and 18th centuries. It is also worth noting the degree to which Kātib Çelebi's concept of knowledge as a basis for human action in the world, and his explanation of the world as governed by divinely instated and maintained regularities, was accessible and meaningful for many generations after him. As the discussion about modernity and modernization continues, as people in Turkey and elsewhere keep searching for an authentic genealogy of their modern identities, figures like Kātib Çelebi and books like *Cihānnümā* will take on increasing relevance as a primary source of evidence.

123 Taeschner 1935.
124 The sales for each work printed are collected in Sabev 2006.
125 *Irshādu l-ḥayārā ilā tārīkhi l-Yūnān ve r-Rūm ve n-Naṣārā*.
126 Dew 2009, 179. On the entire history of the Bibliothèque Orientale see ibid., 168–204, where the introduction of alphabetization is also discussed. See also Leiser 2010 on the comparison of Kātib Çelebi and d'Herbelot.
127 See Toderini 1787, especially vol. II, 117–121 on *Remover of Doubts*.
128 Leiden, University Library, or. 1130 (Schmidt 2012, vol. 1, 459 ff.).
129 See especially Toderini 1787, vol. III, 114–131. The translation of Norberg 1818 includes all descriptive parts but, aside from its many mistakes, omits the systematic introduction.

4 Principles of the Present Translation

The translation presented here is, pace Norberg, the first complete and critical translation of *Cihānnümā* to appear in print, aside from a modern Turkish version published under the auspices of Bahçeşehir University.[130] The first recension of the work was partially translated into German in the early 19th century by the pioneer orientalist Joseph von Hammer-Purgstall, an effort that, like Norberg's, drew much criticism for its linguistic and factual errors. Partial translations of the second *Cihānnümā* include Papazyan's rendering of the section on the Caucasus into Russian and a series of articles by Bacqué-Grammont that are exemplary in their philological apparatus and also, where possible, incorporate for comparison the unpublished partial translation by Armain made in the 18th century.[131] Still, taken together, these important efforts add up to only a small fraction of the total text.

The present translation was prepared by a team consisting of Ferenc Csirkés, John Curry and Gary Leiser, supervised by Robert Dankoff and subjected to a final series of edits by Gottfried Hagen. It is based on the printed edition (= M) which, as we have seen, was essential to the fame and prestige of the work. An 18th-century printed edition however, even an *editio princeps*, is not by itself a reliable basis for a critical translation. But we are in the fortunate position of having Kātib Çelebi's autograph text, the Topkapı Draft discussed above (= R), which was the very manuscript used by Müteferriḳa in preparing the printed edition. The translators have systematically collated the printed text with its urtext, the Topkapı Draft, and given precedence to the latter wherever there is divergence. With few exceptions, we have taken into account passages in the margin of the Topkapı Draft only when Müteferriḳa included them in the printed version. As noted above, Müteferriḳa supplemented this text with material from the work of Ebū Bekr ed-Dimaşḳī, which starts on page 422 of the printed edition and continues to the end. No manuscript has been identified as the immediate source for this part of the work, although the aforementioned London manuscript (British Library, or. 1038) closely parallels it and has been used to clarify a few problematic passages. The aim of this collective effort was to produce a readable and reliable text not weighed down by a massive text-critical apparatus or extensive commentary on the content but sufficiently scholarly for the reader to recognize the quality of what is one of the axial texts of Islamic intellectual history.

130 *Cihānnümā*, second version (Topkapı Draft), tr. Koyunoğlu et al., 2010.

131 Bacqué-Grammont 1996, 1997, 2006, 2012.

Bibliography

Works of Kātib Çelebi

Cihānnümā, first version (Vienna Draft, Österreichische Nationalbibliothek, ms. mxt. 389)

Translated by Joseph Hammer-Purgstall as *Rumeli und Bosna; geographisch beschrieben von Mustafa Ben Abdalla Hadschi Chalfa*. Wien: Im Verlage des Kunst- und Industrie-Comptoirs, 1812.

Cihānnümā, second version (Topkapı Draft, TSMK, ms. Revan 1624)

Edited by İbrāhīm Müteferriḳa as *Kitāb-ı Cihānnümā*. Istanbul, 1145/1732.

Translated by Matthias Norberg as *Geographia orientalis Gihan numa ex Turcico in Latinum versa*. Londini Gothorum: Literis Berlingianis, 1818.

Translated by Hüsnü Koyunoğlu et al. as *Cihânnümâ Kâtip Çelebi*. İstanbul Büyükşehir Belediyesi Kültür A.Ş., 2010.

Düstūru l-ʿamel fī ıṣlāḥi l-ḫalel

Edited as appendix to ʿAyn ʿAlī's *Ḳavānīn-i Āl-i ʿOs̱mān*. Istanbul, 1863, 119–140.

Translated by Ali Can as *Kâtip Çelebi: Bozuklukların Düzeltilmesine Tutulacak Yollar (Düsturu l-ʿamel fi islahi l-halel)*. Ankara: 1982.

Fadhlakat aqwāl al-akhyār fī ʿilm al-taʾrīkh wa l-akhbār

Edited by Muhammed es-Seyyid Mahmud as *Fezleketü akvâli'l-ahyâr fî ʿilmi't-târîh ve'l-ahbâr (Fezleketü't-tevârîh): Târîhu mülûki Âli ʿOsmân*. Ankara: Türk Tarih Kurumu, 2009.

Irshādu l-ḥayārā ilā tārīkhi l-Yūnān ve r-Rūm ve n-Naṣārā

Edited by Bilal Yurtoğlu as *Katip Çelebi'nin Yunan Roma ve Hristiyan Tarihi Hakkında Risalesi*. Istanbul: Atatürk Kültür Merkezi, 2012.

Kashf al-ẓunūn ʿan asāmī l-kutub wa l-funūn

Edited by Kilisli Rifat Bilge and Şerefettin Yaltkaya as *Keşf-el-zunūn*. Istanbul: Maarif Matbaası, 1941–1943.

Levāmiʿu n-nūr fī ẓulmeti Aṭlas Mīnūr

Edited by Ahmet Üstüner and H. Ahmet Arslantürk as *Levâmi'u'n-Nûr Fî Zulmet-i Atlas Minor*. İnceleme—tıpkıbasım. Ankara: TÜBA, 2017.

Mīzānü l-ḥaḳḳ fī ḫtiyāri l-aḥaḳḳ

Edition: Istanbul, Maṭbaʿa-i Ebū ẓ-Ẓiyā, 1306/1888.

Translated by Geoffrey L. Lewis as *The Balance of Truth*. London: Allen and Unwin, 1957.

Translated by Orhan Şaik Gökyay as *Mizanu'l Hakk fi İhtiyari'l-ahakk*. Istanbul, 1973.

Sullam al-wuṣūl ilā ṭabaqāt al-fuḥūl

Edited by Ekmeleddin İhsanoğlu as *Sullamu al-Wusûl ilâ Tabaqât al-Fuhûl / The Ladder of Elevation to the Lives of the Great and Famous by Generation*. 6 v. Istanbul: IRCICA, 2010.

Taḳvīmü't-tevārīḫ
Edited by Semiha Nurdan as *Takvîmü't-tevârîh: indeksli tıpkıbasım*. Ankara: Türk Tarih Kurumu, 2009.
Tuḥfetü l-kibār fī esfāri l-biḥār
Edited by İdris Bostan as *Tuhfetü'l-Kibâr fî Esfâri'l-Bihâr*. Ankara: T.C. Başbakanlık Denizcilik Müsteşarlığı, 2008.
Translated by James Mitchell and Svat Soucek as *The History of the Maritime Wars of the Turks*. Princeton: Markus Wiener, 2012.

Other Primary and Secondary Sources

Abū l-Fidā' al-Ḥamawī. *Taqwīm al-buldān*. Edited by J.T. Reinaud and MacGuckin de Slane as *Taqwīm al-buldān*. Paris, 1840.
Abū l-Fidā' al-Ḥamawī. *Taqwīm al-buldān*. Translated by J.T. Reinaud and Stanislas Guyard as *Géographie d'Aboulféda*. Paris, 1848.
Ágoston, Gábor. "Information, ideology, and the limits of imperial policy: Ottoman grand strategy in the context of Ottoman-Habsburg rivalry." In *The early modern Ottomans: Remapping the Empire*, edited by Virginia Aksan and Daniel Goffman, 75–103. Cambridge University Press, 2007.
Anonymous. *'Ajā'ibu l-makhlūqāt we gharā'ibu l-mawjūdāt (Turkish)*. Edited by Günay Kut as *'Acâyibü'l-mahlûkât ve garâyibü'l-mevcûdât (İnceleme—Tıpkıbasım). Süleymaniye Yazma Eser Kütüphanesi Nuri Arlasez Koleksiyonu, No. 128'deki Nüshanın Tıpkıbasımı*. Istanbul: Türkiye Yazma Eserler Kurumu Başkanlığı, 2012.
Antrim, Zayde. *Routes and Realms: The Power of Place in the Early Islamic World*. Oxford University Press, 2012.
Arbel, Benjamin. "Maps of the World for Ottoman Princes? Further Evidence and Questions Concerning 'The Mappamondo of Hajji Ahmed'." *Imago Mundi* 54 (2002): 19–29.
'Āşıḳ, Meḥmed. *Manāẓīru l-'avālim*. Edited by Mahmut Ak as Mehmed Âşık, *Menâzırü'l-avâlim*. Ankara: Türk Tarih Kurumu, 3 v., 2007.
Bacqué-Grammont, Jean-Louis. "Les routes d'Asie centrale d'après le Cihân-Nümâ de Kâtib Çelebî." *Cahier d'Asie centrale* 1–2 (1996): 311–322.
Bacqué-Grammont, Jean-Louis. "La description de Chypre dans le Cihân-Nümâ de Kâtib Çelebî." *Epetērida tou Kentrou Epistēmonikōn Erevnōn (Levkosia)* 23 (1997): 189–214.
Bacqué-Grammont, Jean-Louis. "Les merveilles botaniques du Cathay dans la Cosmographie de Kâtib Çelebî." In *Studies in Oriental Art and Culture in Honour of Professor Tadeusz Majda*, edited by Anna Parzymies, 107–116. Warsaw: Dialog. Academic Publishing House, 2006.
Bacqué-Grammont, Jean-Louis. "Kâtip Çelebi'nin Cihânnümâ'sında Osmanlı Devletinin Kuruluşu Hakkında Birkaç Not." In *Doğumunun 400. Yıl Dönümünde Kâtip Çelebi*, edited by Bekir Karlığa and Mustafa Kaçar, 251–260. Ankara, 2009.
Bacqué-Grammont, Jean-Louis. "L'Afrique dans la cosmographie de Kâtib Çelebi." *Osmanlı Araştırmaları* 40 (2012): 121–170.
Berlekamp, Persis. *Wonder, Image, and Cosmos in Medieval Islam*. New Haven: Yale University Press, 2011.
Birkholz, Daniel. *The King's Two Maps. Cartography and Culture in Thirteenth-Century England*. London: Routledge, 2004.
Birnbaum, Eleazar. "The Questing Mind: Katib Çelebi, 1609–1657. A Chapter in Ottoman Intellectual History." In *Corolla Torontonensis. Studies in honour of Ronald Morton Smith*, edited by Emmet Robbins and Stella Sandahl, 133–158. Toronto: 1994.
Birnbaum, Eleazar. "Kātib Çelebi (1609–1657) and Alphabetization: A Methodological Investigation of the Autographs of his *Kashf al-Ẓunūn* and *Sullam al-Wuṣūl*." In *Scribes et manuscrits du Moyen-Orient*, edited by François Déroche and Francis Richard, 235–263. Paris: Bibliothèque Nationale de France, 1997.
Bonner, Michael David and Gottfried Hagen. "Muslim Accounts of the Dār al-Ḥarb." In *The New Cambridge History of Islam. Vol. 4: Islamic cultures and societies to the end of the eighteenth century*, edited by Robert Irwin, 474–494. Cambridge University Press, 2010.
Brentjes, Sonja. "Mapmaking in Ottoman Istanbul between 1650 and 1750: A Domain of Painters, Calligrapher, or Cartographers?" In *Frontiers of Ottoman Studies: State, Province, and the West*, edited by Colin Imber, Keiko Kiyotaki, and Rhoads Murphy, 125–156. London and New York: I.B. Tauris, 2005.
Brentjes, Sonja. "On two manuscripts by Abū Bakr b. Bahrām al-Dimashqī (d. 1102/1691) related to W. and J. Blaeu's Atlas Maior." *Osmanlı Araştırmaları* 40 (2012): 1–22.
Casale, Giancarlo. *The Ottoman Age of Exploration*. Oxford University Press, 2010.
Cook, Michael A. *Commanding right and forbidding wrong in Islamic thought*. Cambridge University Press, 2000.
Cosgrove, Denis E. "Images of Renaissance Cosmography, 1450–1650." In *The History of Cartography, Volume Three (Part 1): Cartography in the European Renaissance*, edited by David Woodward, 55–98. University of Chicago Press, 2007.
Coşkun, Feray. "A Medieval Islamic Cosmography in an Ottoman Context: A Study of Mahmud El-Hatib's Translation of the *Kharidat Al-'aja'ib*." Unpublished MA thesis, Boğaziçi University, 2007.
Curry, John J. *The Transformation of Muslim Mystical Thought in the Ottoman Empire: The Rise of the Halveti Order, 1350–1650*. Edinburgh University Press, 2010.
Curry, John J. "An Ottoman geographer engages the early modern world: Katip Çelebi's vision of East Asia and the Pacific Rim in the *Cihânnümâ*." *Osmanlı Araştırmaları* 40 (2012): 221–257.
Dew, Nicholas. *Orientalism in Louis XIV's France*. Oxford University Press, 2009.

EI 2 = *Encyclopedia of Islam*, 2nd edition

ed-Dimaşḳī, Ebū Bekr b. Behrām, *El-fetḥu l-raḥmānī fī ṭarz-i devleti l-ʿOs̱mānī*. Edited by Ilona Dorogi and György Hazai as "Zum Werk von Ebū Bekr b. Bahram Dimişkī über die Geschichte und den Zustand des Osmanischen Reiches." *Archivum Ottomanicum* 28 (2011), 49–94; 29 (2012), 199–324; 30 (2013), 303–352; 31 (2014), 167–350.

El-Rouayheb, Khaled. "The Myth of the "Triumph of Fanaticism" in the Seventeenth-century Ottoman Empire." *Welt des Islams* 48 (2008): 196–221.

Emiralioğlu, M. Pınar. "Relocating the Center of the Universe: China and the Ottoman Imperial Project in the Sixteenth Century." *Osmanlı Araştırmaları* 39 (2012): 161–188.

Emiralioğlu, M. Pınar. *Geographical Knowledge and Imperial Culture in the Early Modern Ottoman Empire*. London: Ashgate, 2014.

Erginbaş, Vefa. "Enlightenment in the Ottoman Context: İbrahim Müteferrika and His Intellectual Landscape." In *Historical Aspects of Printing and Publishing in Languages of the Muslim World*, edited by Geoffrey Roper, 53–100. Leiden: Brill, 2014.

Euben, Roxanne L. *Journeys to the Other Shore*. Princeton University Press, 2006.

Fetvacı, Emine Fatma. *Picturing History at the Ottoman Court*. Indiana University Press, 2013.

Finkel, Caroline. *Osman's Dream: The History of the Ottoman Empire, 1300–1923*. New York: Basic Books, 2006.

Fleischer, Cornell H. "The Lawgiver as Messiah: The Making of the Imperial Image in the Reign of Süleymân." In *Soliman le magnifique et son temps*, edited by Gilles Veinstein, 159–177. Paris: La Documentation Française, 1992.

Fleischer, Cornell H. *Bureaucrat and Intellectual in the Ottoman Empire: The Historian Mustafa Ali (1541–1600)*. Princeton University Press, 1986.

Gencer, Yasemin. "İbrahim Müteferrika and the Age of the Printed Manuscript." In *The Islamic Manuscript Tradition*, edited by Christiane Gruber, 154–194. Indiana University Press, 2010.

Gökyay, Orhan Şaik. "Kâtip Çelebi. Hayatı—Şahsiyeti—Eserleri." In *Kâtip Çelebi. Hayatı ve Eserleri Hakkında İncelemeler*, 3–90. Ankara: Türk Tarih Kurumu, 1957.

Goodrich, Thomas D. *The Ottoman Turks and the New World: A Study of Tarih-i Hind-i garbi and Sixteenth-century Ottoman Americana*. Wiesbaden: Harrassowitz, 1990.

Goodrich, Thomas D. "Old Maps in the Library of Topkapi Palace in Istanbul." *Imago Mundi* 45, no. 1 (1993): 120–133.

Griffel, Frank. *Al-Ghazālī's Philosophical Theology*. Oxford University Press, 2009.

Hagen, Gottfried. "Kâtib Çelebi and Tarih-i Hind-i Garbî." *Güney-Doğu Avrupa Araştırmaları Dergisi* 12 (1982–1998; Prof. Dr. Cengiz Orhonlu Hatıra Sayısı): 101–115.

Hagen, Gottfried. "Kātib Čelebi's Darstellung der eyālets und sanǧaqs des Osmanischen Reiches." *Archivum Ottomanicum* 16 (1998): 101–123.

Hagen, Gottfried. "Some Considerations on the Study of Ottoman Geographical Writings." *Archivum Ottomanicum* 18 (1998): 183–193.

Hagen, Gottfried. "The Traveller Mehmed Aşık." In *Essays on Ottoman Civilization. Proceedings of the XIIth Congress of the Comité International d'Études Pré-Ottomanes et Ottomanes (CIÉPO), Praha, 1996*, 145–154. Prague: 1998.

Hagen, Gottfried. *Ein osmanischer Geograph bei der Arbeit. Entstehung und Gedankenwelt von Katib Celebis Ǧihannnüma*. Berlin: Klaus Schwarz Verlag, 2003.

Hagen, Gottfried. "Kâtib Çelebi and Sipahizade," In *Essays in honour of Ekmeleddin İhsanoğlu*, edited by Mustafa Kaçar and Zeynep Durukal, 525–542. Istanbul: IRCICA, 2006.

Hagen, Gottfried. "Kātib Çelebī." *Historians of the Ottoman Empire* (2007): https://ottomanhistorians.uchicago.edu/en/historian/katib-celebi.

Hagen, Gottfried. "Ptolemaeus Triumphans." Forthcoming.

Hagen, Gottfried and Ethan L. Menchinger. "Ottoman Historical Thought." In *A Companion to Global Historical Thought*, edited by Prasenjit Duara, Viren Murthy, and Andrew Sartori, 92–106. Oxford: Wiley Blackwell, 2014.

Heck, Paul L. *The Construction of Knowledge in Islamic Civilization: Qudāma b. Jaʿfar and his Kitāb al-kharāj wa-sināʿat al-kitāba*. Leiden: Brill, 2002.

Hees, Syrinx von. *Enzyklopädie als Spiegel des Weltbildes. Qazwinis Wunder der Schöpfung—eine Naturkunde des 13. Jahrhunderts*. Wiesbaden: Harrassowitz, 2002.

Heinen, Anton M. *Islamic Cosmology: A Study of al-Suyūṭī's "Al-hay'a as-sanīya fī l-hay'a as-sunnīya," with Critical Edition, Translation and Commentary*. Beirut: Orient-Institut der Deutschen Morgenländischen Gesellschaft; Wiesbaden: Franz Steiner, 1982.

Hemmat, Kaveh Louis. "A Chinese System for an Ottoman State: The Frontier, the Millennium, and Ming Bureaucracy in Khaṭāyī's Book of China," PhD diss., University of Chicago, 2014.

İhsanoğlu, Ekmeleddin. "Introduction of Western Science to the Ottoman World: A Case Study of Modern Astronomy (1660–1860)," In *Transfer of Modern Science & Technology to the Muslim world*, edited by Ekmeleddin İhsanoğlu, 67–120. Istanbul: IRCICA, 1992.

al-Idrīsī, Muḥammad. *Nuzhat al-mushtāq fī khtirāq al-āfāq*. Translated by Henri Bresc and Annliese Nef as *La première géographie de l'Occident*. Paris: Flammarion, 1999.

Imber, Colin. "The Ottoman Dynastic Myth." *Turcica* 19 (1987): 7–27.

Imber, Colin. *The Ottoman Empire, 1300–1650: The Structure of Power*. Basingstoke, UK: Palgrave Macmillan, 2009.

İnalcık, Halil. "Military and Fiscal Transformation in the Ottoman Empire." *Archivum Ottomanicum* 6 (1980): 283–337.

Karamustafa, Ahmet T. "Military, Administrative, and Scholarly Maps and Plans." In *The History of Cartography. Volume Two, Book One: Cartography in the Traditional Islamic and South Asian Societies*, edited by J.B. Harley and David Woodward, 209–227. University of Chicago Press, 1992.

Koeman, Cornelis, Günter Schilder, Marco van Egmond, and Peter van der Krogt. "Commercial Cartography and Map Production in the Low Countries, 1500–ca. 1672." In *The History of Cartography, Volume Three (Part 2): Cartography in the European Renaissance*, edited by David Woodward, 1296–383. University of Chicago Press, 2007.

Konvitz, Josef W. *Cartography in France, 1660–1848: Science, Engineering, and Statecraft*. University of Chicago Press, 1987.

Koraev, Timur K. "Un géographie arabe stambouliote: à la recherche du contexte intellectuel de l'*Awdah al-masālik ilā maʿrifat al-buldān wa-l-mamālik* de Sipāhīzāde." *Turcica* 46 (2015): 113–152.

Krachkovskii, I.Iu. *Arabskaia geograficheskaia literatura*. Moscow: Izdatel'stvo Akademii Nauk, 1957.

Krawulsky, Dorothea. *Īrān, das Reich der Īlḫāne: Eine topographisch-historische Studie*. Wiesbaden: Reichert, 1978.

Kuru, Selim S. "The literature of Rum: The Making of a Literary Tradition." In *The Cambridge History of Turkey. Volume 2: The Ottoman Empire as a World Power, 1453–1603*, edited by Suraiya Faroqhi and Kate Fleet, 548–592. Cambridge University Press, 2013.

Kurz, Marlene. *Ways to Heaven, Gates to Hell. Fażlīzāde ʿAlī's Struggle with the Diversity of Ottoman Islam*. Berlin: EB Verlag, 2011.

Leiser, Gary. "A figurative Meeting of Minds between Seventeenth-century Istanbul and Paris: The World Views of Kâtib Çelebî and Barthélemi D'Herbelot." *Journal of Oriental and African Studies* 19 (2010): 73–84.

Lindner, Rudi Paul. "Icons among Iconoclasts in the Renaissance." In *The Iconic Page in Manuscript, Print, and Digital Culture*, edited by George Bornstein and Teresa Tinkle, 89–107. Ann Arbor: University of Michigan Press, 1998.

Miquel, André. *La géographie humaine du monde musulman jusqu'au milieu du 11e siècle*. Paris: École des hautes études en sciences sociales, 1973.

Morrison, Robert. "The Response of Ottoman Religious Scholars to European Science." *Archivum Ottomanicum* 21 (2003): 187–195.

Murphey, Rhoads. "Ottoman Historical Writing in the Seventeenth-century: A Survey of the General Development of the Genre after the Reign of Sultan Ahmed I (1603–1617)." *Archivum Ottomanicum* 13 (1993/4): 277–311.

Pfeifer, Helen. "Encounter after the Conquest: Scholarly Gatherings in Sixteenth-century Damascus." *International Journal of Muslim World Studies* 47, no. 2 (2015): 219–239.

Pinto, Karen. "The Maps Are the Message: Mehmet II's Patronage of an 'Ottoman Cluster'." *Imago Mundi* 63, no. 2 (2011): 155–179.

Piterberg, Gabriel. *An Ottoman Tragedy: History and Historiography at Play*. Berkeley, CA: University of California Press, 2003.

al-Qazwīnī, Ḥamdallāh Mustawfī. *Nuzhat-al-Qulūb*. Edited by G. Le Strange as *The Geographical Part of the Nuzhat-al-Qulūb composed by Ḥamd-allāh Mustawfī of Qazwīn in 740 (1340) edited and translated*. London; Leyden: Luzac & Co.; E.J. Brill, 1915–1919.

al-Qazwīnī, Zakariyā b. Muḥammad. *ʿAjāʾib al-makhlūqāt wa gharāʾib al-mawjūdāt*. Edited by Ferdinand Wüstenfeld as *Kosmographie: I Die Wunder der Schöpfung, II Die Denkmäler der Länder*. Göttingen: Dieterich, 1848–1849.

Sabev, Orlin. *İbrahim Müteferrika ya da ilk Osmanlı matbaa serüveni (1726–1746). Yeniden değerlendirme*. Istanbul: Yeditepe Yayınevi, 2006.

Sabev, Orlin. "Waiting for Godot: The Formation of Ottoman Print Culture." In *Historical Aspects of Printing and Publishing in Languages of the Muslim World*, edited by Geoffrey Roper, 101–120. Leiden: Brill, 2014.

Sariyannis, Marinos. "The Kadizadeli Movement as a Social and Political Phenomenon: The Rise of a 'Mercantile Ethic'?," In *Halcyon Days in Crete VII: Political Initiatives 'From the Bottom Up' in the Ottoman Empire*, edited by Antonis Anastasopoulos, 263–289. Rethymno: Crete University Press, 2012.

Sarıcaoğlu, Fikret. "Cihânnümâ ve Ebûbekir b. Behrâm ed-Dımeşkî-İbrahim Müteferrika." In *Prof. Dr. Bekir Kütükoğlu'na Armağan*, 121–142. Istanbul: Edebiyat Fakültesi Basımevi, 1991.

Sarıcaoğlu, Fikret. "Kâtib Çelebi'nin Otobiyografileri." *İÜEF, Tarih Dergisi* 37, no. Prof. Dr. İsmet Miroğlu Hatıra Sayısı (2002): 297–319.

Schmidt, Jan. *Catalogue of Turkish Manuscripts in the Library of Leiden University and other Collections in the Netherlands*. Leiden: Brill, 2012.

Schulz, Juergen. "Maps as Metaphors: Mural Map Cycles of the Italian Renaissance." In *Art and Cartography: Six Historical Essays*, edited by David Woodward, 97–122. University of Chicago Press, 1987.

Şehsuvaroğlu, Bedi N. "İlham-al Mukaddes min-al Feyz-al Akdes risâlesi ve Kâtip Çelebi'nin ilmî zihniyeti hakkında birkaç söz." In *Kâtip Çelebi. Hayatı ve Eserleri Hakkında İncelemeler*, 141–176. Ankara: Türk Tarih Kurumu, 1957.

Sellheim, Rudolf. *Materialien zur arabischen Literaturgeschichte*. Wiesbaden: Franz Steiner, 1976.

Soucek, Svat. "Islamic Charting in the Mediterranean." In *The History of Cartography. Volume Two, Book One: Cartography in the Traditional Islamic and South Asian Societies*, edited by J.B. Harley and David Woodward, 263–292. University of Chicago Press, 1992.

Soucek, Svat. *Piri Reis & Turkish Mapmaking after Columbus: The Khalili Portolan Atlas*. London: Nour Foundation in association with Azimuth Editions and Oxford University Press, 1996.

Soucek, Svat. "About the Ottoman Age of Exploration." *Archivum Ottomanicum* 28 (2011): 313–342.

Taeschner, Franz. "Die Vorlage von Hammers 'Rumeli und Bosna'." *Mitteilungen zur osmanischen Geschichte* 2 (1923): 308–310.

Taeschner, Franz. "Zur Geschichte des Djihannuma." *Mitteilungen des Seminars für Orientalische Sprachen, 2. Abt.* 29 (1926): 99–110.

Taeschner, Franz. "Das Hauptwerk der geographischen Literatur der Osmanen, Kātib Čelebi's Ǧihānnümā." *Imago Mundi* 1 (1935): 44–47.

Tezcan, Baki. "The 1622 Military Rebellion: A Historiographical Journey." *International Journal of Turkish Studies* 8 (2002): 25–43.

Tezcan, Baki. "The Many Lives of the First Non-Western History of the Americas: From the New Report to the History of the West Indies." *Osmanlı Araştırmaları* 40 (2012): 1–38.

Thomas, Lewis V. *A Study of Naima*. Edited by Norman Itzkowitz. New York University Press, 1972.

Tibbetts, Gerald R. "The Balkhī School of Geographers." In *The History of Cartography. Vol. 2, Book 1: Cartography in the Traditional Islamic and South Asian Societies*, edited by J.B. Harley and David Woodward, 108–137. University of Chicago Press, 1992.

Tibbetts, Gerald R. "The Beginnings of a Cartographic Tradition." In *The History of Cartography. Vol. 2, Book 1: Cartography in the Traditional Islamic and South Asian Societies*, edited by J.B. Harley and David Woodward, 90–107. University of Chicago Press, 1992.

Tibbetts, Gerald Randall. *Arab Navigation in the Indian Ocean Before the Coming of the Portuguese: Being a Translation of Kitāb Al-Fawā'id Fī Uṣūl Al-baḥr Wa'l-qawā'id of Aḥmad B. Mājid Al-Najdī; Together with an Introduction on the History of Arab Navigation, Notes on the Navigational Techniques and on the Topography of the Indian Ocean and a Glossary of Navigational Terms*. London: Royal Asiatic Society of Great Britain and Ireland; Luzac, 1971.

Toderini, Giambattista. *Letteratura turchesca*. Venezia: Presso G. Storti, 1787.

Türer, Osman. *Mehmed Nazmi Efendi: Osmanlılarda Tasavvufî Hayat (Halvetilik Örneği). Hediyyetü l-İhvan*. Istanbul: İnsan Yayınları, 2005.

Vatin, Nicolas. "Pourquoi un Turc ottoman racontait-il son voyage? Note sur les relations de voyage chez les Ottomans des Vâkı'ât-ı Sultân Cem au Seyâhatnâme d'Evliyâ Çelebi." In *Études Turques et Ottomanes. Document de Travail no. 4 de l'URA du CNRS (décembre 1995)*, 3–15. 1995.

White, Sam. *The Climate of Rebellion in the Early Modern Ottoman Empire*. Cambridge University Press, 2011.

Wurm, Heidrun. *Der osmanische Historiker Ḥüseyn b. Ǧaʿfer, gen. Hezarfenn, und die Istanbuler Gesellschaft des 17. Jahrhunderts*. Freiburg: Klaus Schwarz, 1971.

Yāqūt ibn ʿAbd Allāh al-Ḥamawī, *Muʿjam al-buldān*. Translated by Wadie Jwaideh as *The Introductory Chapters of Yāqūt's Muʿjam al Buldān*. Leiden: Brill, 1959.

Yurtoğlu, Bilal. *Katip Çelebi*. Ankara: Atatürk Kültür Merkezi Yayınları, 2009.

Zilfi, Madeline. "The Kadızadelis: Discordant Revivalism in Seventeenth-century Istanbul." *Journal of Near Eastern Studies* 45, no. 4 (1986): 251–269.

Introduction to the Translation

Robert Dankoff

1 Conventions of Transcription and Terminology

For proper names derived from Islamic sources we use a standard transcription, with *c, ç, ş, ġ, ḳ* in Turkish names corresponding to *j, ch, sh, gh, q* in Arabic and Persian names. For names that are common in English we regularly substitute the common form—thus Borneo for Būrnūī, etc. Otherwise, for names derived from European sources the default is a transliteration, as in Ūssāqāyā (Osaka). In a few instances we have indicated (in brackets in the text or in footnotes) the spellings in the sources that Kātib Çelebi was copying, whether the Latin and Italian printed books and maps—*Atlas Minor, Fabrica, Disegno*, etc.; or the Arabic and Persian manuscripts—particularly of the *Nuzha*.

Parentheses () are used for anything implied by the text and all explanatory material provided by the translators; also to indicate emendations of the text. Brackets [] are confined to strictly textual matters; also to indicate spellings in sources that Kātib Çelebi was using, as signalled by a footnote. In Part 1, material not in Kātib Çelebi's main text (marginalia by himself and others, additions by Müteferriḳa) is marked by a vertical line to the left of the text. In the sections describing routes and stages, > marks the next stage, whether or not indicated in the text by *andan* or the like.

Titles and offices are given in the form common in English, thus: Sultan, Shah, Sharif, Ghazi, Agha, Bey (note that Bey substitutes for Beg); also Reʾīs (sea captain).

The following are either accepted as English words, and thus not italicized; or are left as technical terms, and thus italicized:

- Currency: dinar, dirham, toman (for currency; but note *tūmān* for administrative district); *aḳçe, ġuruş, para.*
- Weights and measures: oka, batman, donum; *mithqāl, irdabb, jarīb.*
- Religious terms: madrasa, hadith, shaikh, qadi, mufti, shaikh al-islam, muezzin, imam, ulema, Sharia, fatwa, dhimmi, qibla, mihrab; *molla* (chief qadi), *sayyid* (descendant of the Prophet), *minbar* (podium in a mosque), *maḥmal* (litter accompanying pilgrims from Cairo to Mecca).
- Administrative terms: padishah, vizier, emir, divan, sanjak (for *sancaḳ* and *livā*), sanjak bey, chavush, timar; *bey* (note that *bey* substitutes for *beg*), *beylerbey, emāret* (seat of an emir or sanjak bey), *ṣubaşı* (a kind of police officer—note M 571 *ṣubaşılıḳ* "precinct"), *sipāhī, defterdār, ḳılıç* (timar), *zeʿāmet* (timar providing an annual income of over 20,000 *aḳçe*), *ḳānūn* (sultanic law), *tekfur* (Byzantine governor or emperor), *tūmān* (administrative district in Iranian context; but note toman for currency).
- Culture terms: *qaṣīda* (a long poem, usually panegyric).

We have tried to be consistent in translating common terminology. In our translation, "district" generally has the sense of administrative district, translating *nāhiye*. The following list shows the English correspondents usually found:

ıḳlīm "clime"
memleket "country, territory, province"
ülke "country, region"
vilāyet "province, region, country"
eyālet "province" (*eyālet* is exclusively used for Ottoman province, especially in Part 2)
nāhiye "district, administrative district"
ʿamal, pl. *aʿmāl* "administrative unit"
ḳażā "qadi district"
jund "military district"
devlet "state, polity; dynasty"
melik, pl. *mülūk* "king, ruler"
ḥükūmet "government, prefecture"
ḥākim "magistrate, governor"
vālī, ʿāmil, mutaṣarrif "governor"
beg "lord, chieftain" (but note *bey* in Ottoman context)
emīr "emir", *imāret* "emirate"
reʾīs, baş "chief"
diyār "land, region"
belde, bilād "land, region, territory; city, town"
şehr "city"
ḳaṣaba "town, administrative center"
dār-i mülk "capital, administrative center"
rabaḍ "suburb"
muḍāfāt "annexes"
tevābiʿ "dependencies"
ulusāt "territories"
farsakh "parasang"
kārīz, qanāt "underground channel"
kharāj "tax, revenue"
jizya "poll tax"
bāj "toll, caravan toll"
reʿāyā "taxpaying subjects"

 | DOI:10.1163/9789004441330_003

TABLE 1

Marco Polo	*Disegno* (map)	R 50a (map)	R 51a (text)	M 135/136 (map)	M 142 (text)	T 212 (map)	T 218 (text)
Lambri	Lambri	Lānbrī	Lānbirī	Lānbrī	Lānbrī	Lanberi	Lanberi
Samara	Simara	Sīmārā	Sāmārā	Sīmārā	Sāmārā	Simara	Samarra
Fanfur	Farsur	Fārsūr	Fānfūr	Qārsūr	Qānfūr	Karsur	Kanfur
Dragoian	Dragoian	Drāghūyān	Drāghūyān	Drāghūyān	Drāghūyān	Dragoyan	Dragoyan
Felech	Felech	Felesh	Felek	Felesh	Felek	Feleş	Felek
Basma	Basma	Bāṣma	Bāżma	Bāṣma	Bāżma	Basma	Badma

waqf "endowment"
velī "saint, holy man"
rijāl, eren "saintly men"
khānqāh, zāviye, tekke "Sufi convent, dervish lodge"
khān "inn"
ribāṭ "hospice"
ṣoffa "terrace, platform"
ocaḳ "estate"
ocaḳlıḳ "hereditary grant"
arpalıḳ, iltizām, muḳāṭaʿa "tax farm, land grant"
ḫāṣ "imperial domain"
çiftlik "farm estate"

2 Problems of Text and Sources

Cihānnümā is a monster of a text, and translating it poses several serious problems. M (the Müteferriḳa edition) is a hybrid text in two parts. Part 1 consists of Müteferriḳa's printed transcription of R (Kātib Çelebi's autograph ms.), supplemented by his own extensive notes. Part 2 consists of Müteferriḳa's printed transcription of Ebū Bekr b. Behrām ed-Dimaşḳī's continuation of R, again supplemented by his own notes. Müteferriḳa also added numerous diagrams and maps, some of which are based on those in R, some original. He tried to smooth out and unify the two parts, but they are rather different. In content, Part 2 is almost completely devoted to the Ottoman provinces of the Middle East, while Part 1 has the rest of Eurasia. And crucially, we have the author's original for Part 1, including the maps, and we don't for Part 2. This means we are on much firmer ground in Part 1, insofar as what we are providing is a kind of "edition in translation," at least when it comes to the thousands of place names.

Regarding the place names especially, but for other aspects of the text as well, an edition would take into account the original information in the sources, European and Islamic, that Kātib Çelebi depended on. Consider the example in Table 1 above, the cities of "Java Minor"—a phantom conjured out of Ramusio's rendering of Marco Polo's travels.[1] Kātib Çelebi carefully copied these six place names from two different sources. On his map (R 50a) they exactly correspond to those of *Disegno*, which he frequently cites as "Giacomo's map." (The rendering of *mb* as *nb* is conventional in Ottoman Turkish spelling. Here he interpreted *ch* as standing for *š/ş/sh*.) In his text (R 51a) they correspond rather to those in Marco Polo's original, which he had access to according to a source as yet to be determined. (Note Sāmārā and Fānfūr, as opposed to Sīmārā and Fārsūr in the map which correspond to Gastaldi's errors. But he inadvertently made the ṣ of Bāṣma into a *ḍ/ż*. Here he interpreted *ch* as standing for *k*.) M simply copied Kātib Çelebi's map, confounding the errors by misreading Fārsūr/Fānfūr as Qārsūr/Qānfūr. This example shows the gamut of problems associated with the rendering of place names and dealing with the various layers of the text.

Since we took as our primary task the translation of M, we did not normally delve into these layers, with the exception that we gave priority to R in cases where M clearly got it wrong. Here are a few of many examples:

– Atlantic coast: R 22a line 19 *ve bu aralıḳda Britāniā'yla Ġāliā mā-beyninde oldıġından Britānīkūs ve Hiberniā ile Britāniā arasında olana Hibernīkūs derler ve Angliā'nuñ ḳısm-ı şimālīsi olan Sqūsiā ki selefde aña Qaledūniā derlerdi kenārına nisbetle Qaledūnīūs dediler* (The part between Britania and Gallia is called Britanicus, that between Hibernia and Britania is called Hibernicus. The part bordering on Scotia—known in olden times as Caledonia—which is the northern part of Anglia, was called Caledonius); M 75 line 18 *Ve Hiberniā ile* [one line skipped] *Sqūsiā ki selefde aña Qaledūniā der-*

1 A.E. Nordenskjöld, "The Influence of the 'Travels of Marco Polo' on Jacobo Gastaldi's Maps of Asia," *The Geographical Journal* 13 (1899), 403.

lerdi kenārına nisbetle Qaledūnīūs dediler (The part between Hibernia and Scotia—known in olden times as Caledonia—relative to its coast was called Caledonius) [thus T 130–131 *ve Hiberinya (İrlanda) ile Saksonya arasına—ki eskiler oraya Kaledonya (Caledonie) derlerdi—kıyılarına nispetle Kaledonyus derler*].

- Imagined creature responsible for tides: R 24a line 32 *bir ḥayvān olup menāḫiri oḳyānūs ḳaʿrında ola* (a living thing, whose nostrils are on the ocean floor); M 80 line 15 *bir ḥayvān olup müteaḫḫirīn oḳyānūs ḳaʿrında ola* (a living thing that, according to modern scholars, is on the ocean floor) [thus T 142 *bir canlıdır. (Bu canlı) son dönem bilginlerine göre, okyanusun içinde olmalıdır*].
- Cause of tides: R 25b line 5 from end *ʿUfūnet ḥarāret-i ġarībeden olur* (Putrefaction results from strange heat); M 84 line 5 *ʿUfūnet ḥarāret-i ʿazībeden olur* (Putrefaction results from the heat of sweet things) [thus T 146 *Kokuşma tatlı şeylerin ısınmasından meydana gelir*].
- Khālid b. al-Walīd's conquest of Hira: R 26b line 1 *Ehl-i Ḥīre'den ʿAbdülmesīḥ nām bir pīr görüp* (He met an old man from Ḥīra named ʿAbd al-Masīḥ); M 85 line 17 *Ehl-i ḫaberden ʿAbdülmesīḥ nām bir pīr görüp* (He met a knowledgable old man named ʿAbd al-Masīḥ) [thus T 147 *Abdülmesih adlı tecrübeli ve uzman bir ihtiyarı görüp*].
- America: R 27b line 25 *Amariḳa ʿarżı ḳavm-ı Yūnān'a ve Ejibsilere* (text: *Erbesilere*) *maʿlūm olduġı* (The land of America was known to the Greeks and the Egyptians); M 88 line 22 *Amariḳa ʿarżı ḳavm-ı Yūnān'a ve Erbesilere maʿlūm olduġı* (The land of America was known to the Greeks and the Arbesis) [thus T 151 *Amerika toprağının Yunanlılarca ve Arbesilerce (Araplar) bilindiği*].
- Cape of Good Hope: R 28a line 25 *Lākin Aṭlas'da ol yerlerüñ maʿmūr olduġı mesṭūrdur* (It is recorded in the *Atlas*, however, that those places are inhabited), M 89 line 3 from end *Lākin Aṭlas'da ol yerlerüñ maʿmūr olmaduġı mesṭūrdur* (It is recorded in the *Atlas*, however, that those places are uninhabited) [thus T 152 *Fakat Atlas'da o yerlerin mamur olmadığı yazılıdır*].
- Greeks and Romans' attributing divinity to objects in the world: R 44a line 20: *Cünūn vādīlerinüñ ḳanḳı semtine sālik olmadılar* (In what direction of the valleys of madness did they not stray!); M 123 line 2 *Cenūb vādīlerinüñ ḳanḳı semtine sālik olmadılar* (In what direction of the valleys in the south did they not stray!) [thus T 193 *Güneydeki vādīlerin hangi semtine yönelip gitmediler ki!*].
- Skill of the Chinese: R 58a line 9 *Ṣanʿati ṭabīʿat fiʿline müşābih olub* (Their crafts resemble the activity of nature); M 158 line 4 *Ṣanʿati fiʿline müşābih olub* (Their crafts resemble their activities) [Thus T 238 *Bunların sanatı, yaptıklarına benzer*].
- Length of the Safīdrūd: R 114a line 20 *yüz fersaḫ* (100 parasangs); M 304 line 25: *yüz yetmiş fersaḫ* (170 parasangs).
- Prostitutes in Cathay: R 66b line 11 *Dişileri resenbāzdır* (Their women are tightrope walkers); M 180 line 3 *Dişleri resenbāzdır* (Their teeth are loose) [thus T 263 *Dişleri döküktür*].
- Ambassadors in Cathay: R 67a line 9 from end *Elçiler iki ḳat oldılar lākin alınların yere ḳomadılar* (The Elchis bowed but did not place their foreheads on the ground); M 182 line 7 *Elçiler iki ḳat oldılar lākin ellerini yere ḳomadılar* (The Elchis bowed but did not place their hands on the ground) [thus T *Elçiler iki kat oldular. Fakat ellerini yere koymadılar*].
- Rise of the Mughal dynasty: R 80a line 10 from end *Ẓahīr al-Dīn Muḥammad Bābur b. Mīrzā ʿUmar Shaykh, Mā-varāʾaʾl-nahr'de Özbek devleti ġalebe edüb, mezbūr Kābul'a gelmiş* (When the Uzbeks took power in Transoxiana, Ẓahīr al-Dīn Muḥammad Bābur b. Mīrzā ʿUmar Shaykh came to Kabul); M 219 line 1 *Ẓahīr al-Dīn Muḥammad Bāyir b. Mīrzā ʿUmar Shaykh, Mā-varāʾaʾl-nahr'de evreng devleti ġalebe edüb, mezbūr Kābul'a gelmiş* (When Ẓahīr al-Dīn Muḥammad Bāyir b. Mīrzā ʿUmar Shaykh took power in Transoxiana he came to Kabul) [thus T 304 *Zâhirüddin Muhammed Bayır ibn Mirza Ömer Şeyh Mâverâünnehir'de devlet idaresini ele geçirip önce Kâbil'e gelmiş*].
- Brahmins: R 82a line 1 *Biri birlerine incinseler evlerini āteşe yaḳarlar ve babaların oġulların öldürürler* (If they are wronged they set fire to one another's houses and kill their fathers and their sons); M 222 line 2 from end *Biri birlerine incinseler evlerini āteşe yaḳarlar ve babaların oġulları öldürürler* (If they are wronged they set fire to one another's houses and their sons kill their fathers) [thus T 308 *birbirlerini incitseler evlerini ateşe verirler ve oğullar babalarını öldürürler*].
- Stages in Makran: R 86b line 9 *Bundan Pannūs iki merḥale şarḳa düşer* (Pannūs lies 2 stages east of here); M 236 line 12 *Bundan Patūs yigirmi ṭoḳuz buçuḳ merḥale şarḳa düşer* (Patūs lies 29½ stages east of here) [thus T 322 *Buradan Petus 29.5 merhale doğuya düşer*].
- Ghurid king Sayf al-Dīn Muḥammad: R 96a line 9 *İkinci sene Ġuz ṭāʾifesi cengine gidüb ḳavmi yedinde maḳtūl oldı* (In the second year of his reign he went off to do battle with the Oghuz tribe and was killed by his own men); M 255 line 7 *İkinci sene Ġazne ṭāʾifesi cengine gidüb ḳavmi beyninde maḳtūl oldı* (In the second year of his reign he went off to do battle with the Ghaznavids and was killed among his own men) [thus T 341 *Hüküm-*

darlığının ikinci senesinde Gazneliler ile savaş katıldı ve kendi halkı tarafından öldürüldü].

- Ghurid king Muʿizz al-Dīn Ḥusayn: R 96a line 5 from end *Mawlānā Saʿd al-Dīn Muṭawwal'ı anuñ nāmına te'līf etdi. Bu ṭabaḳanuñ kāmkārı idi. Yediyüz yetmiş birde vefāt edüb oğlı Ghiyāth al-Dīn Pīr ʿAlī yerine geçüb* (Mawlānā Saʿd al-Dīn dedicated his *Muṭawwal* to him. He was the fortunate monarch of this branch [of Ghūrids]. He died in 771 and was succeeded by his son Ghiyāth al-Dīn Pīr ʿAlī); M 256 line 2 *Mawlānā Saʿd al-Dīn* [one line skipped] *Pīr ʿAlī yerine geçüb* (Mawlānā Saʿd al-Dīn Pīr ʿAlī succeeded him) [thus T 342 *Yerine Mevlana Sadeddin Pir Ali geçti*].
- Bahrain: R 103b line 2 *Eyyām-ı ṣayfda Baḥreyn havāsı ġāyet ıssı iken bu ṣu cümleden laṭīf ü bāriddir, ol diyār ḫalḳı merecü'l-baḥreyn'den murād budur deyü Baḥreyn tesmiyesine sebeb ʿadd ederler* (In summer the weather in Bahrain is very hot and this water is the best and the coldest; the people of that region explain the expression *maraj al-baḥrayn* as referring to this and consider that to be the reason for calling it Bahrain); M 275 line 18 *Eyyām-ı ṣayfda Baḥreyn* [one line skipped] *tesmiyesine sebeb ʿadd ederler* (In summer they consider that to be the reason for calling it Bahrain) [thus (?) T 363 *Bahreyn asıl o sahillerin adıdır*].
- Countries bordering on the Black Sea: R 145a line 11 from end *Baḥr-ı Ḳırım sāḥilinde* (on the coast of the Crimean Sea); M 279 line 5 from end *Baḥr-ı Ḳum sāḥilinde* (on the coast of the Sea of Sand) [thus T 469 *Kum Denizi sahilinde*].
- Plain of Çaldıran: R 158b, right margin *Van ḳurbında* (near Van); M 417 line 9 from end *Kān ḳurbında* (near Kān) [thus T 506 *Kân yakınında*].
- Alexander's recovery of Bitlis: R 157b, upper margin *İskender daḫi ġayrı yoldan gidüp Bedlīs kefen der-gerden kilīd-i ḳalʿe ile varup* (so Alexander took a different route and Badlīs went to him with a winding-sheet around his neck and the key of the fortress in his hand) M 414 line 3 *İskender daḫi ġayrı yoldan gidüp ḳafādan Bedlīs kilīd-i ḳalʿe ile varup* (so Alexander took a different route and Badlīs went behind with the key of the fortress) [thus T 502 (with further misunderstandings) *İskender de başka bir yoldan gidip arkadan Bitlis'e Kilidkale yoluyla vardığı*].
- Bridges in Bitlis: R 157b, right margin line 9 *yigirmi bir köpri vardır* (there are 21 bridges); M 414 line 19 *yigirmi bir göz köpri vardır* (there is a bridge with 21 arches) [thus T 502 21 göz köprü vardır].

Regarding the *Nuzha*, one of Kātib Çelebi's major sources, we are fortunate to have the meticulous edition of Le Strange, who also consulted *Cihānnümā*. He writes:

> Ḥājī Khalfah repeatedly quotes the *Nuzhat*, and, as he evidently had excellent MSS. to work from, his readings of the place-names are valuable. Unfortunately, however, like all Oriental writers he is entirely uncritical. In the eastern regions, which lie beyond his personal ken, he inserts descriptions of Sirjān (in Kirmān), Zaranj (in Sīstān), and Arrajān (in Fārs), as though all these cities still existed in his day, when we know from history that, as a fact, the two former towns were destroyed by Timur, while Arrajān even before the time of Timur had been replaced by Bihbahān, which is the present existing town, of which place, however, Ḥājī Khalfah makes absolutely no mention. Then, again, with no mark of the borrowing, Ḥājī Khalfah frequently makes mention of towns, giving the sums of revenue due from each (e.g. Salam and ʿAyn in Armenia); but in most cases these appear to be simply paragraphs taken over bodily from the *Nuzhat*, and the sums for the taxes are those already given by Mustawfi, writing under the Īlkhānid administration three centuries before the time when the *Jihān Numā* was compiled.[2]

In the Preface to his edition, Le Strange sounded a despairing note regarding the correct spelling of place names: "My reading of unverifiable place-names are often as numerous as my MSS ..." (*Nuzha*, xvi). Based on the few examples below, Kātib Çelebi usually followed this source very closely, but sometimes misinterpreted it (discrepancies noted in bold; further misunderstandings in M occasionally noted):

- *Nuzha* 59 line 17 *Mardum* ***safīd-chahra*** *ve Shāfiʿī-madhhab and* (The people [of Abhar] are light skinned and of the Shāfiʿī rite).
  R 111a line 4 *Ḫalḳı ekṣer* ***siyāh-çerde*** *ve Şāfiʿī olurlar* (Most of the people are dark skinned and Shāfiʿī).
- *Nuzha* 72 line 18 *Ve ahl-i ānjā* ***safīd-chahra*** *and* (The people [of Asadābād] are light skinned).
  R 112b line 3 from end *Ḫalḳı* ***siyāh-çerdedir*** (Its people are dark skinned).
- *Nuzha* 131 lines 14 *Qilāʿ: eknūn shānzdah qalʿa-i maʿrūf u mashhūr ast va dar zamān-i mā-qabl* ***haftād*** *u chand qalʿa dar mulk-i Fārs būd* (There are now 16 well-known fortresses, but in the past there were some 70 fortresses in the land of Fars).
  R 102a line 9 from end: *Nüzhe'de Fārs'uñ selefde* ***altmış*** *ḳadar ḳalʿesi var idi, ṣoñra on altı ḳalʿe ḳaldı* (According

2 "Description of Persia and Mesopotamia in the year 1340 A.D. from the Nuzhat-al-Ḳulūb of Ḥamd-Allah Mustawfi, with a summary of the contents of that work," *Journal of the Royal Asiatic Society*, Jan., 1902, 49–74, p. 57.

to the *Nuzha*, in the past Fārs had some 60 fortresses, of which 16 remain).

M 271 line 2 from end: *Nüzhe'de Fārs'uñ selefde altmış ḳadar ḳalʿesi ḳaldı* (According to the *Nuzha*, in the past Fārs had some 60 fortresses that remained) [thus T 359 *Nüzhe*'de Fars'ın geçmiş devirlerden 60 kadar kalesinin kaldığı].

- *Nuzha* 154: *Ghūr vilāyatī ast va shahristān-i ān-rā Āhangarān khwānand.*

  R 95b, right margin *Nüzhe'de eydir: Ghūr bir vilāyetdir, şehrine Āhangarān dėrler* (According to the *Nuzha*, Ghūr is a province whose city is called Āhangarān).

- *Nuzha* 177 line 12 *Va-min Nīshāpūr ilā Harāt: Az Nīshāpūr tā* ***dīh-i Bād*** *haft farsang, az īncā rāhī ki bi-Sarakhs ravad yād karda shud, rāhī ki bi-Harī ravad tā Ribāṭ-i* ***Badīʿī*** [VARIANT: ***Badlaʿī***] *panj farsang ast, azū tā dīh-i* ***Farhādān*** *haft farsang, azū tā dīh-i Saʿīdābād haft farsang, azū tā dīh-i Khusrav panj farsang, azū tā shahr-i* ***Pūchkān*** [VARIANT: ***Būzjān***] *haft farsang, jumla bāshad az Nīshāpūr tā Pūchkān sī u hasht farsang, az īn maqām rāhī bi-Harī ravad va yakī bi-Qāyin va yakī bi-Bākharz va yakī bi-Sarakhs, bi-rāh-i Harī tā dīh-i Gulābād shash farsang, azū tā Kūshk-i Manṣūr dah farsang, azū tā shahr-i Fūshanj* [VARIANT: *Būshanj*] *shash farsang, azū tā shahr-i Harī haft farsang, jumla bāshad az Pūchkān tā Harī sī farsang va az Nīshāpūr shaṣt u hasht farsang, va az Dāmghān tā Harī bi-mūcab-i sharḥ-i mā-qabal ṣad u chihl u shash farsang va az Varāmīn tā Harī davīst farsang va az Sulṭāniyya davīst u panjāh u shash farsang.* (Nīshāpūr to Herāt.[3]—From Nīshāpūr in 7 leagues to Dīh-Bād, where the road to Sarakhs already given … goes off to the left. From Dīh-Bād it is 5 leagues to Rubāṭ Badīʿī, thence 7 to Farhādān village, thence 7 to Saʿīdābād village, thence 5 to Dīh Khusraw, thence 7 to the city of Pūchkān (or Būzjān). Total, from Nīshāpūr to Pūchkān it is 38 leagues; and from here roads go off to Herāt, Qāyin, Bākharz and Sarakhs. On the Herāt road it is, from Pūchkān, 6 leagues to Gulābād village, thence 10 to Kūshk Manṣūr, thence 6 to the city of Fūshanj (or Būshanj), and thence 8 leagues to Herāt. Total: from Pūchkān to Herāt it is 30 leagues, from Nīshāpūr 68, from Dāmghān to Herāt bu the road already detailed it is 146 leagues, from Varāmīn to Herāt 200, and from Sulṭāniyya 256 leagues.)

  R 124a *Nīshāpūr'dan Herāt'a* ***dīh-i Tāybād*** *Ḫ 7, bundan rāh-ı Sarakhs ayrılur ṣola gider, andan Ribāṭ-i* ***Badlaʿī*** *Ḫ 5, dīh-i* ***Farhāvān*** *Ḫ 7, dīh-i Saʿīdābād Ḫ 7, dīh-i Khusraw Ḫ 5, şehr-i* ***Lūḥ-kān*** *Ḫ 7, bundan ~~yol iki olur~~* [here Kātib Çelebi crossed out several words of his first translation attempt] *Qāyin va Sarakhs va Bākharz yolları ayrılur, dīh-i Gulābād Ḫ 6, Kūshk-i Manṣūr Ḫ 10, şehr-i Fūshanj Ḫ 6, Herāt Ḫ 7, cümle altmış sekiz fersaḫ olur, Dāmghān Herāt'dan yüz ḳırḳ beş fersaḫdır, Rāmīn yüz ṭoḳsan bir fersaḫ, Sulṭāniyya iki yüz elli bir fersaḫdır.* (From Nīshāpūr to Herat: village of Tāybād, 7 parasangs; the Sarakhs road goes to the left > Ribāṭ-i Badlaʿī, 5 parasangs > village of Farhāwān, 7 parasangs > village of Saʿīdābād, 7 parasangs > village of Khusraw, 5 parasangs > city of Lūḥ-kān, 7 parasangs; the roads to Qāyin, Sarakhs and Bākharz branch off here > village of Gulābād, 6 parasangs > Kūshk-i Manṣūr, 10 parasangs > city of Fūshanj, 6 parasangs > Herat, 8 parasangs; altogether 68 parasangs; Dāmghān is 145 parasangs from Herat, Rāmīn 191 parasangs, Sulṭāniyya 251 parasangs).

- *Nuzha* 192–192 *der sene-i iḥdā ʿashr va* ***sabʿmīʿa*** (in the year 711).

  R 113b19 ***altıyüz*** *onbirde* (in the year 611).

- *Nuzha* 220 line 15: *Āb-i Jāyjrūd az Kūh-i Damāwand bar mīkhīzad va bi-vilāyat-i Rayy mīrīzad ve dar ḥudūd-i Qūhad-i ʿulyā va Asān muqāsama mīkunad va qarīb-i chihil jūy azān bar dārand va akṣar-i vilāyat-i Rayy-rā azān ast va dar bahār harza-i ābash dar mafāza muntahī mīshavad.* (River Jāyij-Rūd.[4] This rises in Mount Damāwand, and flows through Ray province. In the districts of Upper Qūhad and Asān its waters are divided up and diverted into nearly forty channels, which serve to irrigate most of the Ray province. In spring time its flood waters make their way out into the desert.)

  R 114a line 24 *Jāyjrūd* [no dots under J's and Y; M 304 Khānkhūrūd, thus T 395 Hânhûrud] *Kūh-ı Demāvend'den çıḳub Ray vilāyetine dökülür, Qūhad ḥudūdından anı muḳāseme ėderler, ḳırḳ nehre bölünür,* [in margin] *ve bahārda fāżılı mefāzeye müntehī olur.* (Jāyjrūd: It rises in Mt. Damāwand and flows through Rayy province. In the borders of Qūhad it is divided into 40 channels. In spring its flood waters make their way out into the desert.)

Ebū Bekr b. Behrām ed-Dimaşḳī appears to have used the *Nuzha* as well in his continuation of *Cihānnümā*. In the following examples, it is hard to know whether to attribute the misreadings (noted in bold) to him or, which seems more likely, to Müteferriḳa:

- *Nuzha* 165 line 20 *Az Baghdād tā* ***dīh-i*** *Ṣarṣar du* [2] *farsang …* ***dīh-i*** *Farāsha 7 … Shaṭṭ-i Nīl 7 … Mashhad-i …ʿAlī… ki bar* ***sar****-i bayābān-i Najaf ast du farsang* (From Baghdad to village of Ṣarṣar, 2 parasangs … village of

3 English rendering: *The Geographical Part of the Nuzhat al-Qulūb composed by Ḥamd-Allāh Mustawfī of Qazwīn in 740 (1340)*, translated by G. Le Strange. Leiden and London, 1919, 170–171.

4 English rendering: ibid., 213.

Farāsha 7 ... Shaṭṭ-i Nīl 7 ... Mashhad-i ...ʿAlī ... which is at the edge of the desert of Najaf, 2 parasangs).
M 470 line 6 *Baghdād'dan Necef'e* ***depe-i*** *Ṣarṣar,* ***depe-i*** *Farāshar, Shaṭṭ-ı Nīl, Kūfa, Mashhad ʿAlī ki* ***sedd-i*** *beyābān-i Necef'dir* (From Baghdad to Najaf [the road goes through] mound of Ṣarṣar, mound of Farāshar, Shaṭṭ-ı Nīl, Kūfa, Mashhad ʿAlī which is the barrier of the desert of Najaf).

– *Nuzha* 171 line 2 *Az Baghdād tā Madāʾin shash* [6] *farsang ... Dayr ʿĀqūl 8 ... Jabal 7 ... Famm al-Ṣulḥ 10 ... Wāsiṭ 9 ...* ***Nahrābān*** *10 ... Fārūt 8 ... Dayr al-ʿAmmāl 5 ...* ***Ḥawānīt*** *7 ... Dijlat al-ʿAwrā 30 ... Nahr Maʿqil 10 ... Baṣra*
M 470 line 3 From Baghdad to Basra: > Madāʾin 6 parasangs > Dayr ʿĀqūl 8 > Jabal 7 > Famm Ṣulḥ 10 > Wāsiṭ 10 > **Shahribān** > Qārūth 8 > Dayr ʿAmmāl 5 **Ḥawāthit** 7—the road passes along the Shaṭṭ (Tigris) and through the Marshes, then crosses Nahr Asad and reaches Dijlat al-ʿAwrā > Nahr Maʿqil 10 parasangs > Basra.

– *Nuzha* 171 line 20 *Az Baghdād tā* ***Kangūr ... haftād u panj*** [*75*] *farsang ... Baydistān 5 ...* ***shahr****-i Nihāwand 3 ...* ***dīh-i*** *Farāmurz 4 ...* ***shahr****-i Burūjard 4 ... Ḥasanābād* [VARIANT: *Janābād*] *4 ... Mayān Rūdān 8 ... Manār* [VARIANT: *Miyār*] *3 ... shahr-i* ***Karaj*** *6 ... Dūn Sūn 4 ... Āsan 5 ... Sangān 6 ... jū-yi Murgh Kahtar 6 ... Isjqarān* [VARIANTS: *Ishgharān, Isfarāyin*] *7 ... Tīrān 7 ... jū-yi Kūshk* [VARIANT: *Mūy-i Kūshk*] *6 ... shahr-i Iṣfahān 4.*
M 470 line 7 From Baghdad to Iṣfahān: > **Kaylūr** 7 parasangs > Bandistān 5 > **nehr-i** Nihāwand 3 > **depe-i** Farāmurz 4 > **nehr-i** Rūjard 4 > Janābād 4 > Mayān Rūdān 8 > Miyār 3 > city of **Karkh** 6 > Dūn Sūn 4 > Sangān 6 > Murgh Kahtarā stream 6 > Isfarāna 7 > Bahrān 7 > Mūy-i Kūshk 6 > Iṣfahān 4.

Kātib Çelebi was equally exacting when translating from his European sources. Regarding the *Atlas Minor* (here noted as AM), it is instructive to compare the version in *Cihānnümā* with his earlier translation of the entire book, *Levāmiʿu n-nūr* (here noted as LN), as in the following examples.

– Place of publication: LN 4b line 7 *Felemenk'de Āznehembūm ʿamelinde basılmışdır.* R 2b line 16 *Felemenk'de Āznehembūm ʿamelinde* [M 10 line 4 from end *şehrinde*] *basılup* (It was printed in Holland, in the district of Arnheim, in the Christian year 1621).—This was based on the title page of the 1621 edition which shows Amsterdam as the place of publication but also has *apud Ioannem Iansonium Arnhemi.* We can conjecture that Kātib Çelebi's French intermediary, Meḥmed İḫlāṣī, explained *Arnhemi* as the genitive case, the nominative case being *Arnhemium*, and that this latter form lies behind *Āznehembūm*, miscopied from *Ārnehemīūm*.

– Ceylon: AM 633 *Incolae maritimi mahumetani magna ex parte sunt: mediterranei Gentiles:* (*Cingalas vocant*) *color illis albus, statura procera: abdomen prominens; ventrem enim curant. Imbelles sunt, exigui spiritus.* (Those who live on the coast are mainly Muslims. Those who live inland are heathens—they are known as Cingalas. They are white of skin, tall of stature, and with swollen bellies, since they indulge their stomachs. They are unwarlike and weak-spirited.) LN 406b line 5 from end: *Sevāḥilde sākin olan ehālīnüñ eks̱eri müselmānlardır, içerde olanlar put-perestlerdir. Bunlara Sīnghālas derler, anlaruñ levni aḳ, ḳāmeti uzun ve ḳarnı göbedir* (*göbekdir?*), *zīra çoḳ yerler, żaʿīflerdir ve* ~~*ʿaḳılları az*~~ *cānları azacıḳ cenge ḳādir degillerdir* (Those who live on the coast are mainly Muslims. Those in the interior are idol-worshippers—they are known as Cingalas. They are white of skin, tall of stature, and with swollen bellies, since they eat a great deal. They are weakly and ~~their intellects of little~~ their souls very little [or] not at all capable of war). R 54a line 23 *Bu cezīre sevāḥilinde sākin olanlar eks̱er müslimdir, içerde olanlar mecūs ve put-perestdir. ... Put-perestlere Sīnghālī derler, eks̱er anlar dirāz-ḳad ve beyāż ve gebes* (*gebeş?*) *ve ekūl olurlar, ol ecilden cebān ve żaʿīflerdir.* (Those who live on the coasts are mainly Muslims, those in the interior are Magians and idol-worshippers .... The idol-worshippers are known as Cingali. Most of the people are tall and white, with swollen bellies and inclined to gluttony. For that reason they are timid and weak.)

– Japan: AM 626 *Insula haec Marco Paullo Zipangri dicitur, olim Ghryse Magino, Mercatori Aurea Chersonesus: eam vulgo Iaponiam et Iapan vocant* (This island is called Zipangri by Marco Polo, Ghryse by Maginus, Aurea Chersonesus ["Golden Peninsula"] by Mercator, and Japoniam and Japan by the common people). LN 403b line 3 *Der cezīre-i Yāpūnīā yāḥūd Yāpān: Bu cezīre Pavlūs Marqūs ḳatında Rpānghrī ve Mājīnūs ḳatında selefde Khirīse ve Merqātūr ḳatında Avr ...(?) Kharsūzūs demekle maʿrūfdur* [in margin: *Altunlı Avrupa demekdir*]. R 44b line 10 from end *Der cezīre-i Yāpūnīā: cezīre-i mezbūra Çīn deryāsında ibtidāʾ-i cezāyir-i şarḳıyedir. İsminde Hāpān ve Cāpūn ve Yāpān* [below line: *ve Yāpen*] *lafẓları istiʿmāl olunur. Selefde aña Khirīse derlerdi. Merqātūr aña Altunlı Avrupa maʿnāsına Avrupa Khrsūnūs ismin ıṭlāḳ eder* (Island of Iaponia: It is the beginning of the eastern islands in the China Sea. The pronunciations Hāpūn, Jāpūn, Yāpān and Yāpen are also in use. Formerly it was called Khirīse. Mercator calls it Europa Khrusonus meaning "Golden Europe").—The confusion of Aurea and Avrupa, and Chersonesus and

Khrusonus, is perhaps to be blamed on Kātib Çelebi's French intermediary, Meḥmed İḫlāṣī.

- Osaka in Japan: AM 628 *Ossacaja civitas illustris, potens, libera, omnium, ut quidam autumant, totius Orientis opulentissima. Magna hic mercatorum undique concurrentium frequentia est, quorum qui mediocres facultates habent minimum possident xxx aureorum millia, qui majores, incredibilem summam.* (Osaka, an esteemed, powerful, free city—as is reported, the wealthiest of all in the entire East. It is a great confluence of merhants, of whom the middling sort possess a minimum of 30,000 goldpieces, and the greatest of whom possess an incredible sum.) LN 414a: *Ūssāqāyā nām şehr bir şerīf ü muʿteber şehrdir, ḫalḳı ġanī vü muʿāfdır. Baʿżılar żann üzre cümle şarḳuñ eñ māldārıdır. Eṭrāfdan aña bāzergānlar gelür ve anlaruñ vasaṭu'l-ḥāllarınuñ otuz biñ altun sermāyesi olur, büyük bāzergānlaruñ mālı bi-ḥisābdır.* (The city known as Osaka is a noble and esteemed city. Its people are wealthy and exempt from taxes—as some report, the wealthiest of the entire East. Merchants come there from all around. The middle class have 30,000 gold pieces and the great merchants have uncountable wealth.) R 46a *Ūssāqāyā* [M 128 Ūṣāqāyā, T 199 Oşakaya] *bir muʿteber ..... (?) ve muʿāf şehr olup ḫalḳı aġniyā ve māldār ve mecmaʿ-ı tüccārdur, anda olan bāzergānlaruñ* [in margin: *faḳīri biñ ġuruş sermāye ṣāḥibi ṣḥ*] *vasaṭu'l-ḥālı otuz biñ altun sermāye ṣāḥibi olup büyükleri mālı bī-ḥisābdur.* [in margin: *Üç biñ ḳadar ʿasker bu şehrüñ ṭaşrasında oturup dāyimā şehri muḥāfaẓa ėderler, lāzım geldikce şehre birer ikişer girüp gerü ṭaşra giderler. ṣḥ*] (Osaka: An esteemed, ... (?) and free city. Its people are wealthy and propertied, many of them merchants. For capital, [in margin: poor merchants have 1,000 *ġuruş*,] the middle class have 30,000 gold pieces, and the greatest have uncountable wealth. [in margin: Around 3,000 soldiers live outside the city as a standing guard. Whenever necessary, they enter the city one or two at a time and go back out.])—Clearly Kātib Çelebi translated the text in *Atlas Minor* first, then added in the margin two bits of information that he got from another source.
- East Indies: AM 625 *Gilolo ... quas vulgo nuncupant del Moro | Bandan parvae insulae numero circiter VII ... Bandan major caeteris, ac iisdem nomen communicans, Mira, Rosolargium, Ay, Rom, Neira, & Gunuape omnium minima.* (Gilolo ... which the common folk call del Moro .... Bandan consisting of the 7 small islands ... of which Bandan is the largest, the others being Mira, Rosolargium, Ay, Rom, Neira, and Gunuape which is the smallest.) LN 402b *Gilolo ... cumhūr dilince delmoro demekle maʿrūf | Bandan adaları yedi küçük adalardır ... Bunlaruñ esā[m]īsi Bandan ki sāyirden büyücek olmaġla bunuñ ismi sāyire iṭlāḳ olunur daḫi Mira Rosolargium Ay ve Rom ve Neira ve Gunyab ki cümleden eñ küçügi olup* (Gilolo ... which the common folk call *delmoro* | The Bandan islands are seven small islands ... named Bandan, which is the largest and so its name is applied to the others, also Mira, Rosolargium, Ay, Rom, Neira, and Gunyab which is the smallest). R 47b *Gilolo ... Aṭlas'da eydir ol diyār ḫalḳı delmor derler* | 50a *Cezīre-i Bandan: Aṭlas'da meṣṭūr olduġı üzre yedi cezīre-i ṣaġīredir ... Bandan nām cezīre sāyirinden büyücek olmaġla bunuñ ismi sāyire daḫi iṭlāḳ olunur, anlar Mira ve Solargium ve Ay ve Rom ve Neira ve Günyab nām aṭalardır ve bu Günyab cümleden küçük olup* (Gilolo ... It says in the *Atlas*: The people of that land call it *delmor* | Bandan island: As recorded in the *Atlas*, they are seven small islands ... The one named Bandan is the largest and so its name is applied to the others, which are Mira, Solargium, Ay, Rom, Neira, and Gunyab islands, of which Gunyab is the smallest).—In both instances, Kātib Çelebi mistranscribed what he had earlier written correctly: *delmoro* as *delmor* and Rosolargium as Solargium. [Note that Müteferriḳa further misread Gunyab as Gunyat (M 139 line 1 from end) and there are further misreadings in T 216: *Mira, Sularcinom, Ay, Zum, Tira ve Kunyat.*]

Lorenzo's *Fabrica* is another frequently cited source. Checking the original can resolve some obscurities in Kātib Çelebi's text, as in this example:

Japan: *Fabrica*, 275 *le danno nel suo Settentrione il golfo di Lonza, co'l paese d'Ania, & la terra de' Selvaggi; onde le nasce il Sole il mare Vermiglio nel nuovo mondo; & alla banda Australe l'oceano di Siam* (It is bordered to the north by the Gulf of Lonza with the country of Ania and the land of savages; to the east by the Red Sea of the New World; and to the south by the Sea of Siam). R 44b line 6 from end *ḥadd-i şarḳīsi yüz yetmiş altı derece ṭūlda* [in margin: *Berrāḳ Deryā maʿnāsına Māre Vermīlū kenārları*] *ve ḥadd-i cenūbīsi otuz bir derece ʿarżda* [in margin: *Sīām Ūqīānūsı*] *ve ḥadd-i şimālīsi otuz yedi derece ʿarżda* [above line: *Lūnrā körfezi ve Ānīā ülkesi .... (?)*] *olup* (Its eastern border is the coastlands of Māre Vermīlū meaning Clear Sea, at 176 degrees longitude. Its southern border is the Sea of Siam at 31 degrees (north) latitude. Its northern border is the region of the Gulf of Lūnrā and Ānīā .... (?) at 37 degrees (north) latitude.) Cf. M 124 line 7 *ḥadd-i şarḳīsi yüz yetmiş altı derece ṭūlda Berrāḳ Deryā maʿnāsına Nādre Vermīlū kenārları ve ḥadd-i cenūbīsi otuz bir derece ʿarżda Sīām Ūqīānūsı ve ḥadd-i şimālīsi otuz yedi derece ʿarżda Lūnrā Kūrūl ve Ānīā ülkesi kenārları olup.*

The interlinear scrawl translating *il golfo di Lonza, co'l paese d'Ania, & la terra de' Selvaggi* is hard to decipher, as we see from Müteferriḳa's unsuccessful attempt.

Checking Cluverius can also resolve difficulties in Kātib Çelebi's text, as in this example:

> Europe's borders: *Introductio* 44 Riphaeis montibus (in the Riphaean mountains). R 31b line 25 *Rīfa/Rīġa ṭaġlarında*; M 98 line 1 *Rīġa ṭaġlarında*.

The problem stems from Kātib Çelebi's ductus, in which *f* and *ġ* are sometimes indistinguishable. The correct reading in this case, *Rīfa* (for Riphae), is fixed by recourse to the original Latin text.

Another problem stems from the fact that Kātib Çelebi depended on his French informant for readings of Latin, and that informant tended to pronounce Latin names ending in *-thia* as *-tia* and those ending in *-tia* as *-sia*. An example of the former is R 21b line 6 from end and 34b line 1 *Sītīā* for Scythia.[5] Examples of the latter are R 22a lines 21–22 (M 75 lines 18–19) *Sqūsīā* for Scotia, *Jūsīā* for Jutia; R 32a (M 98) *Sārmāsīā* for Sarmatia, *Sqūsīā* for Scotia, *Qūrūāsīā* for Croatia, *Dālmāsīā* for Dalmatia. [Note that Müteferriḳa here misread *Sqūsīā* as *Sqūsnīā* (M 98 line 18) giving rise to T 163 Saksonya (Saxony instead of Scotland; again at T 130)]. In the translation we have simply restored the forms of the Latin original.

5 R 43b line 3 mistakenly has *Sfmūnīā* for Scythia, misread in M 121 line 11 as *Sqmūnīā*, thus T 191 Sakmonya.

Translation of Cihānnümā

Ferenc Csirkés, John Curry and Gary Leiser

Part 1, by Kātib Çelebi

[M 1] *Book of Cihānnümā* by Kātib Çelebi

In the name of God, the Compassionate the Merciful, to Whom we turn for help

[R 1b] The opening of the world-adorning pearls, and the title page of the world-displaying volumes (or, volumes of *Cihānnümā*); after having been adorned with praise and glory to the One most Holy and Exalted, and embellished with prayers for peace upon the Sultan of the Prophets:

It is not concealed from the minds of knowledgeable men that the science of geography, which is one of the branches of astronomy—I am referring in particular to the flat image of the earth depicted on maps—is so excellent a science and so desired a skill, that one who has experience of it and knows its subtleties can, while seated on the cushion of repose in the parlor of security and sociability, move around the world in an instant like world travelers who go to the ends of the earth. Such intellectual journeying enables one to acquire a level of knowledge that those who have spent a lifetime traveling are unable to reach. That is because, by continual recourse to the books of the geographers, every corner of the earth becomes engraved on the tablet of the heart; so that when it is time to recall it, it flashes instantly upon the imagination in general outline and one sees it in the mirror of the mind as though it is before one's eyes.

It is well known that no science is blameworthy qua science. As the saying goes, "A man is an enemy of what he does not know." Aside from that, while it may be temporarily blameworthy due to accidental circumstances, one cannot assume that one can always do without it. Rather, in the time of need, it may become the most necessary thing in the world.

Therefore, after I had traversed this valley somewhat and consulted "Routes of Countries" (*masālik al-mamālik*) and books of geography, and had drawn the shapes and figures of the lands and the seas many times, my determination to study it in detail was somewhat blunted. But prior travel to some islands of the sea now thrummed on the strings of the dulcimer of my mind. A new ardor was added to my previous inclination, and the desire arose to trace those figures on the pages of explication.

However, because it was clear that Islamic books were all inaccurate with respect to the countries of Europe and that Muslim writers fell short in describing most of the climes and countries, I have translated the abridgment of the book of *Atlas* (i.e., *Atlas Minor* of Hondius) which is the most recent of the geographical works written in Latin, and supplemented it with some useful information from the Islamic books. [M 2] From their combination I have extracted this book and I have set the table before those who wish to learn this science so that they can profit from it.

The combined book is in two parts, the first comprising several sections, the second arranged in chapters according to the names of the countries as found in the Table of Contents. However, before beginning these subjects, there is an Introduction. I have titled the book *Cihānnümā* ("Cosmorama") and I beseech those who will benefit from it to remember me for a blessing.

This book is of great value and one of the rare works of the age, the like of which has never existed in the Muslim era nor entered the treasury of any emperor in previous centuries. It has now been translated during the imperial reign of our patron, the Padishah who is the refuge of the world, the Sultan and son of the Sultan, Sultan Meḥmed Khan (IV) son of Sultan İbrāhīm Khan son of Sultan Aḥmed Khan—may God perpetuate his realm and sultanate and spread over the worlds his goodness and benefaction. When completed, it was presented to his imperial treasury, and thus the countries and kings of the world have been rendered comprehensible to all. At the beginning of the work, the chronogram was uttered: *Mirʾāt-ı dünyā ve mā fīhā oldı bu Cihānnümāy* ("This *Cihānnümā* became the mirror of the world and what is in it") 1058 (1648).

Supplement by the Publisher (İbrāhīm Müteferriḳa)

It should not be forgotten that the author of this delightful book, the late Kātib Çelebi, has described the arrangement of it as follows: "The combined book is in two parts, the first comprising several sections, the second arranged in chapters according to the names of the countries as found in the Table of Contents."

He also states, in the warning at the end of the eleventh chapter of the first part: "So in this book, the description of the countries of the world follows the movement of the ninth heavenly sphere (i.e., from east to west), with the motion of the whole earth following and transmitting the motion of the Primum Mobile, beginning with the eastern islands (of Asia) and ending with the western shores of America."

Thus it is a work of great renown in two volumes, comprised of numerous chapters and sections, of great size and a rarity of the age. However, the manuscripts

 | DOI:10.1163/9789004441330_004

that are in circulation were examined and the copy written in the author's own hand, which we succeeded in finding, is a draft. In the Contents Summary, which is in the fourteenth section of the Introduction, he sets out to explain what he intends to do. The chapters on the countries, climes and cities, whose names are given in the Table of Contents, were never completed in the aforesaid copies, which remain defective. Nevertheless, as this unrivaled work was written during the reign of the late Sultan Meḥmed Khan (IV), the annotated copy which was presented to the Imperial Treasury is in the late author's hand, as seen by the marginal note, "By the author" (*minhu*). This note was also inserted above.

It would be a great shame for such an excellent book and agreeable compilation not to be published and made available and thus be lost and destroyed. This being the case, since I have acquired the oldest manuscript and deem it worthy of attention by those interested in the subject, I have taken pains to have it published in multiple copies at the Imperial Publishing House. I diligently avoided making any changes in the original work, neither adding nor omitting anything. Only I considered it appropriate to insert in various places in the text, under the heading "Supplement by the Publisher," some useful information and clarifications of points of geometry and the natural sciences that I took from sources in my possession. I have also collected the views of all the ancient and modern philosophers on the number and nature of the upper and lower celestial spheres [M 3] and indicated them in summary fashion. In order to make the views of each school easier to understand, I have also inserted a variety of figures and provided an exposition of cosmology. And in the course of the account of countries and cities, I have inserted, here and there, useful information and maps taken from recent geographical works. Thus I have expanded the scope of the chapters and clarified the accounts of rulers, nations and genealogies.

This book begins its account of the countries and rulers of the world in the east and then moves west, ending in the clime of Armenia at the *eyalet* of Van, which is counted among the Ottoman domains. In accordance with the view of experts in this science and in conformity with the author's intention, I took pains to include all of the continent of Asia in the first volume, beginning with the islands of India and China in the east and extending to Anatolia as far as the beautiful city of Constantinople and the clime of the Arabian Peninsula.

Such a valuable and incomparable book has never appeared in the history of Islam or entered the treasury of any emperor, nor has its like been seen by the eye of Time.

This poor and humble one, dwelling in the convent of humility and encompassed by the circle of insignificance, İbrāhim the geographer, one of the *müteferriḳas* of the Sublime Porte, has been charged with publishing books at the Imperial Ottoman Printing House. With the hope that it will be a source of pride for this weak and insignificant slave, and with the boldness of sincerity to present it to the Imperial Treasury of our compassionate patron and generous lord, the present source of security and peace, caliph of the world and the age, refuge of the people of faith, epitome of the Ottomans, the Sultan and son of the Sultan, Sultan Maḥmūd Khan son of Sultan Muṣṭafā—may God perpetuate his realm and support his sultanate—I have cracked the whip of heart-stealing desire upon the hand of ardor and directed the rein of intention toward the desired goal.

This lowly slave has entreated God to facilitate his success in this matter and to bring to fruition his modest goal of providing a useful service to mankind. In like manner he entreats the Sultan to look favorably upon this work full of fault. There is no doubt that giving this valuable science (i.e., geography) the currency it deserves, and helping all Muslims and benefitting all believers by the spread and diffusion of the theoretical and the practical sciences, depend on the generous favor of the hero of the age and the incomparable attention of the Sultan who bears the mark of Solomon, his glorious majesty the king of kings.

I also request those who are knowledgeable and intelligent to meet the errors in this work, whether in expression and style or in typography and spelling, with forbearance and forgiveness and to cover our faults with grace and generosity. For the fame and currency of this wonderful science depend on the approval of the people of learning and the clemency of the people of nobility. May they examine this publication and find it worthy of favor, and may they become the means by which it is honored and circulated in the world. [M 4]

To begin:

It is well known among people of intellect and learning that the term World (*ʿālem*—"Cosmos, Universe") refers to everything in existence apart from the One whose existence is necessary (i.e., God). If the nature and condition of things are pondered and examined and if the status of existing things is studied and reviewed, it is certain that a sound mind would indicate and bear witness that all parts of the World are created.

Now the World consists of essences and accidents. Essences subsist by themselves; accidents do not.

Again, things are either composite, such as the body, or non-composite, such as substances.

This book is based on the science of geography, one of the branches of the science of philosophy and astronomy that studies the nature of existing things. Therefore, before beginning this subject, it is appropriate to review the basic principles that are the propaedeutic to cosmology and the axioms of geometry and of natural science that the expert geographers have elaborated upon in order to define, teach, describe, and write about geography; and to include a discussion of these matters in the introduction to this book.

Therefore it is the custom of most writers of books about geography to preface their works with a brief discussion of the essential features of the three dimensions. And in this regard, hoping that advanced students will acquire this science easily and that beginners will learn the basics of mathematics, and believing that they have the mental capacity to comprehend the subtle truths of this science, they have expatiated on the subject by providing the following definitions and examples.

In this world the existent physical bodies, whether solid or not solid, are beyond count. By the power of God the Eternal, all things differ from each other in shape and appearance according to their natural disposition. Among the solid bodies, we study one that has a specific shape, for example, one with six equal surfaces in the shape of a cube or what the people would call a backgammon die. First, we notice that it has three dimensions—length, breadth, and depth—and we call this die a SOLID. Second, we look at the six sides of this die and call these sides SURFACES. That is, looking only at the exterior surfaces of the die that strike the visual field, we see only length and breadth, not depth, because as soon as we add depth to our perspective it becomes a body. Third, we call the boundaries of the die—i.e., the edges of the surfaces that encompass the die—LINES. Here we only take account of length and do not give the line any share of the surface, because when the slightest breadth is added to a line it becomes a surface. Finally, we look at one of the lines and call the extremities POINTS. We do not give the point any share of the line but call the mere end of the line a point. Thus, the geometricians call any corporal entity that is perceptible but cannot be divided in any way, a point. [M 5] It is something that has a location but is not composed of parts. It is the extremity where a line stops at either end. Any entity that is perceptible and can only be divided in one direction is called a line. It has neither breadth nor depth, only length that ends at a point. Any entity that can be divided in two directions—i.e., into length and breadth, but not depth—is called a surface. This is something that has only length and breadth and ends at a line. And any entity that can be divided in three directions—i.e., has length, breadth, and depth—is called a solid. Here we mean by "solid" (*cism*, which also means "body") the solid that is taught in mathematics.

Lines are divided into two kinds, straight and curved. A STRAIGHT LINE is one in which all the imaginary points throughout its length are in the same alignment. In other words, the parts that extend between the extremities of the line are straight, none higher or lower, and when we bring one side into view, it prevents the other side and the middle from being seen. A CURVED LINE is the opposite of this: the parts throughout its length are laid out in bent form, i.e., some are higher and some are lower, and when we bring one side into view, the curve of the parts prevents the other side and the middle from being seen (i.e., one sees only half a curved line when viewed on the same plane).

Straight lines are either parallel or not parallel. PARALLEL LINES are two or more lines on a plane that are equidistant from each other throughout their length. Even if they extended to infinity in a straight line from both ends they would never intersect.

A surface, likewise, is either a plane or not. A PLANE SURFACE is one whose segments extending over the entire surface are on the same level. In other words, the parts of the imaginary lines extending over the surface from one end to the other in length and breadth are even with one another. Or if one connects by a straight line two imaginary points anywhere on the surface, this line does not separate from the surface in any way. In short, in all of the segments of the distance over which the surface extends, there is not the slightest raising or lowering. All of the segments are on a level, equivalent, and uniform.

A surface that is not a plane is the opposite of this. Some surfaces that are not planes are called SPHERICAL, such as the convex surface of a solid sphere and the convex and concave surfaces of a hollow sphere. Half of these spheres are called convex and concave hemispheres.

Plane surfaces that are parallel and not parallel are analogous to straight lines that are parallel and not parallel.

Every surface that is circumscribed with one or more lines is called a GEOMETRIC FIGURE. If it is circumscribed with three lines, it is a TRIANGLE. There are three kinds of triangle: 1) an equilateral triangle, of which all three sides are equal; 2) an isosceles triangle, of which only two sides are equal; 3) a scalene triangle, of which all three sides are different lengths. [M 6] If a surface is circumscribed by four lines it is a QUADRANGLE. If it is circumscribed by five lines it is a PENTAGON. The other figures are analogous to this.

A BORDER is a boundary and a FORM is a thing or an area that is encompassed by one or more borders.

A CIRCLE is a figure created by a curved line on a surface around a hypothetical point on that surface, all points on that line being equidistant from that central point. It is a flat surface that the curved line circumscribes. This circumscribed surface is the circle. The curved encompassing line is the CIRCUMFERENCE. The hypothetical point is the CENTER of the circle. Any straight line from that point to the edge is the RADIUS. A straight line that passes through the center and extends to both sides of the circumference is the DIAMETER, which is twice as long as the radius. The diameter divides the circle into two equal parts. The diameter and half the circumference encompass a semicircle.

A straight line that divides a circle in two but does not go through the center is a CHORD. It does not divide the circle into two equal parts but rather into a larger and a smaller part. Each of the two parts that are circumscribed by the chord and the circumference is a SEGMENT and each part of the circumference is an ARC. The midpoint of the chord is the SINE OF THE ARC. The perpendicular line that rises from the midpoint of the chord and meets the midpoint of the arc is the VERSED SINE OF THE ARC. Half the diameter of a circle is called ABSOLUTE SINE.

An ANGLE is a surface encompassed by two intersecting lines that do not merge at the point of intersection. There are two kinds of angles: 1) SUPERFICIAL ANGLE (i.e., angle formed on a surface), which is the convex area formed between two lines that intersect at a point without creating a straight line; 2) SOLID ANGLE, which appears in a body encompassed by one or more surfaces, such as the angle at the top of a cone or the angles of (the corners of) a house.

There are three kinds of angles formed on a surface: 1) RIGHT ANGLE, that is one of two equal angles formed when one straight line intersects another—the straight line that intersects the other is the PERPENDICULAR; 2) ACUTE ANGLE, that is smaller than a right angle; 3) OBTUSE ANGLE, that is larger than a right angle. It is not necessary for the lines of these two kinds of angles to be straight.

Triangles are categorized in the same way. One that has a right angle is a RIGHT TRIANGLE; one that has an obtuse angle is an OBTUSE TRIANGLE; and one that has neither a right angle nor an obtuse angle is an ACUTE TRIANGLE.

Similarly, if the four sides of a quadrilateral figure are equal and its four angles are right angles, it is a SQUARE. If its angles are right angles but its sides are not equal, it is a RECTANGLE. If its sides are equal and its angles are not right angles, it is a RHOMBUS. A quadrilateral whose sides are equal and whose angles are not right angles but whose opposing sides and angles are identical is a RHOMBOID. Otherwise (i.e., if one pair of sides is parallel) it is a TRAPEZOID.

It is also well known that [M 7] figures with more than four sides are called POLYGONS. And a solid that has one or more surfaces is called a SOLID FORM. If one supposes that there is a point in the middle of a solid form and all straight lines drawn from this point to the surface are equal, then it is a SPHERE. That surface is the SURFACE AREA of the sphere, the point is the CENTER of the sphere, and the lines are RADII of the sphere.

When a flat plane divides a sphere in two, it forms a circle. If it passes through the center of the sphere, it is a GREAT CIRCLE; otherwise it is a LITTLE CIRCLE. When a hypothetical point on the circumference of a rotating sphere completes its circuit it traces a circle; except the two opposite points, known as the POLES of the sphere or poles of motion, which are fixed. A diameter that goes between the two poles is the AXIS.

Of these circles, the one whose pole and whose center are the same as those of the sphere is called the ECLIPTIC. This is the largest of the circles. The other circles are parallel to it and get smaller and smaller; they are ORBITS around a hypothetical point. Any two circles (i.e., orbits) on the two sides of the ecliptic that are equidistant to the ecliptic are equal to each other; and the two poles of the sphere are also the poles of those orbits.

When two parallel planes cut a circle, the cross-section that results between them is the LATERAL CROSS-SECTION.

Natural Sciences

A NATURAL BODY is a substance in whose essence the corporeal dimensions—i.e., length, breadth, and depth—can be postulated on condition that they intersect at right angles. The body is either simple or compound. A SIMPLE BODY is one in which all of its parts are of the same nature. In other words, it cannot be divided into bodies of different forms and natures. Its nature is one (or, unvarying) such that anything that emerges from it does so in a single manner.

A COMPOUND BODY is the opposite of this. It can be divided into bodies having different natures and forms—such as minerals, plants, and animals, which are the three kingdoms of the natural world, whose fathers are the ethereal entities (the celestial bodies) and whose mothers are the elements (earth, water, air, fire).

It is established in natural philosophy that every simple body in its essential nature is in the shape of a sphere.

Simple bodies are divided into two types, celestial and elemental. CELESTIAL BODIES are the heavens (or celestial spheres) and everything related to them. They are called the ethereal bodies (i.e., those beyond the moon) and the upper world (i.e., the five superior spheres beyond that of the sun). ELEMENTAL BODIES are the four elements: earth, water, air, and fire. These and everything related to them are called the lower world (i.e., below the moon) or the world of being and dissolution.

Compound bodies are also of two kinds: complete and incomplete. A COMPLETE COMPOUND BODY is one that preserves its own compound form during a specific time, such as minerals, plants, and animals. An INCOMPLETE COMPOUND BODY is the opposite of this: it is not able to preserve its own compound form during a specific time, such as clouds, fog, smoke, rainbows, halos, meteors, and other atmospheric phenomena.

The motion of the celestial sphere is of two kinds, simple and varying. SIMPLE MOTION OF THE SPHERE, also called equable motion, is such that any hypothetical point on the surface or in the spherical shell that constitutes a sphere, in moving with that motion, marks off equal arcs in equal times from the circumference of that sphere, and creates equal angles in equal times in the circle at the center of that sphere.

[M FIGURES] [M 8]

For example, the ninth sphere, which is the largest sphere, moves from east to west around the center of the earth and completes its circuit in about one day and one night. A hypothetical point on the surface of this sphere marks off equal arcs in equal times with this motion, and creates equal angles in equal times around the center of the earth. In other words, a hypothetical circle on the circumference of this celestial sphere, e.g., the ecliptic, is divided into 360 equal degrees. As a hypothetical point on such an ecliptic moves with the motion of that sphere, it marks off 15 degrees every sidereal hour. The 15-degree arc marked off during the second hour is equal to that marked off during the first hour. Moving in this fashion, the angle created around the center of the earth during the first hour and that during the second hour are equal. The others are analogous to this. This motion is called equable motion around the center; otherwise it is not termed equable.

VARYING MOTION is different from this. Again it is divided into two types, elementary and compound: elementary is that which emerges from a single celestial sphere, compound that which emerges from more than one celestial sphere. Every elementary motion is simple and every varying motion is compound. But not every elementary and every compound motion is varying.

In order to more easily understand these phenomena, labeled figures and diagrams have been drawn (printed) on the previous page.

End of the publisher's supplement.

Introduction

Concerning the Principles on Which this Science is Based.

1 **On the books that were consulted and that served as sources for this work.** The European books that served as sources of the *Atlas* are also cited to some extent.

Āthār al-bilād wa akhbār al-ʿibād: a book arranged according to the seven climes, by Zakariyyā b. Muḥammad al-Qazwīnī, author of *ʿAjāʾib al-makhlūqāt*. It was composed in 674 (1275). The content of this book was included.

Aḥsan al-taqāsīm fī maʿrifat al-aqālīm by Shaikh Shams al-Dīn Muḥammad b. Aḥmad al-Maqdisī (or al-Muqaddasī), around 400 (1009). It mentions the countries and portrays the inhabited areas according to the conventional climes in the style of the Arabs. The content of this book was included.

Ishārāt ilā Maʿrifat al-Ziyārāt by the world-traveler Shaikh Abū'l-Ḥasan ʿAlī b. Abī Bakr al-Harawī, who died in Aleppo in 611 (1215). He travelled to most of the inhabited quarter and recorded the places he saw in the Muslim countries and the places of visitation. This book was read completely and a summary was included. [R 2a]

The *Atlas Major*, a recent geographical book of the Franks, a large and important work. It was begun by Gerardus Mercator (1512–1594) from Holland and completed by Ludovicus Hondius from the same country. The latter also summarized it under the title *Atlas Minor*. In Latin *major* means large and *minor* means small.

This humble one (Kātib Çelebi) translated *Atlas Minor* into Turkish and included the translation in this book. By this means I sought to fill in the gaps in the books that have been written about this science in Turkish and Arabic.

Preface to the translation

It is fitting that the caravans of praise and glory and the processions of sincere gratitude, which are manifest from the inhabited globe of the heart and from the city of the heart without guile and which proceed to the routes of the countries of the exalted assembly (the angels), halt at the threshold of sanctity and the final stage of intimacy, of the One who created heaven and earth, who brought into being the elements, who exists by necessity and pours forth goodness and generosity. [M 9] And it is fitting that the gracious tribes of pure blessings and excellent salutations, which emerge from the beautiful city (*belde-i ṭayyibe*, cf. Koran 34:15) of paradise and pass to the road of the tongue, enter the resplendent and delight-filled garden and pure and estimable domain of that pride of the world whose noble essence is the city of knowledge (i.e., Muḥammad). May greetings and peace to his family and noble companions ever be the gift of mankind, until the Day of Judgment.

Now, the writer of this book, the most insignificant creature, Muṣṭafā Ḫalīfe (Kātib Çelebi), ever since childhood, in the classroom of "Seek knowledge from the cradle", has extracted opinions and propositions by burning the midnight oil of perspicacity and intelligence in the lamp of thought and perception; and in the course of teaching and learning, has contemplated manners and matters thanks to the abundant blessings bestowed by God the Bounteous.

But I was not satisfied with what I had gained from the change of circumstances in traveling and staying put and the experience of the passing days and years. In order to acquire more knowledge and excellence, I studied books of history and prophetic biography. The Koranic verse "Do they not consider the kingdom of the heavens and the earth ..." (7:185), which was revealed about persons who stare like cattle at this world of bodies, led me to read books on astronomy; and the verses "Travel the world" (6:11) and "Look at the signs of God's mercy" (30:50) led me to study books on geography.

Whenever I saw the attention and skill with which the Christians investigated and presented this science, appropriated from the Greeks; and the foolishness and absurdity of the Muslims who ignored and neglected it; I could only sigh with regret and utter, "We belong to God and return to Him." Because all the books written in Arabic, Persian and Turkish about climes and regions, and about "Routes of Countries" (*masālik al-mamālik*), are erroneous and confused. Thus I decided that it was necessary to combine (the Christian and Muslim books) and to extract a book that I would entitle *Cihānnümā*.

After completing the introduction and first part of the work, I intended to write about the great islands in the Western Ocean—England, Hibernia (Ireland), and Iceland—but I was not able to record one part in a thousand of their information and description, even though they comprised many countries and kings. The shortcomings resulting from this failure blunted my resolve to finish the book; and because I was unable to obtain any source on these places, I was overcome with indolence. For I had previously examined the great book on geography written by Abraham Ortelius (i.e., *Theatrum Orbis Terrarum*) and also had copied the maps found in various places in the European book known as *Atlas Major* in order to incorporate them in *Cihānnümā*; and seeing the detailed information about those islands and other inhabited areas, I thought it would be better to translate those books rather than complete my own.

Around that time, Ḳara Çelebizāde Maḥmūd Efendi, who owned a copy of Abraham Ortelius's geography, suddenly died. While waiting to acquire this book from his estate, I succeeded in obtaining the book entitled *Atlas Minor*, which is the abridgment of *Atlas Major*, and the steed of ambition was spurred by my great eagerness to translate this book from Latin into Turkish. "When God desires something He prepares the intermediate causes."

Previously Shaikh Meḥmed Efendi had come to our city from France. He was a very capable man, familiar with the principles of geography [R 2b] and with an excellent knowledge of Latin. Because of his abilities he mastered Turkish in a short time. I had him read the book and expound it to me, and we reflected on its meaning, considering how best to convey the author's intention.

Thus I began to translate the *Atlas Minor* in the middle of Muḥarram in year 1064 of the *hijra* (December 1653). [M 10] I did not change the text, nor did I add or delete anything. Only when deemed necessary did I write a marginal note in order to clarify difficult names and explicate obscure places. And I made exact copies of the maps, tracing them from one page to another. I entitled it *Levāmiʿu'n-nūr fī Ẓulmet Atlas Minor* (Rays of Light in the Darkness of *Atlas Minor*), thinking this was appropriate for the subject of the book.

Because my purpose in all of this was to provide assistance to the people of Islam, I hope to receive forgiveness for my sins. "God wastes not the reward of him who does a good work" (Koran 18:30). For one cannot deny that the infidels, by making use of these sciences, have invaded most parts of the world and have resolutely and obsessively reduced the people of Islam to impotence.

Now begins the translation and the aims of the book laid out chapter by chapter. *Success comes from God, the Mighty, the Omniscient.*

End of the preface (to the translation of *Atlas Minor*).

Supplement of the Publisher (İbrāhīm Müteferriḳa)

That sincere friend, Shaikh Meḥmed Efendi, who was Kātib Çelebi's teacher in the science of geography and his master and professor in translating the *Atlas* and other Latin books, was famous and esteemed among the Christians of the European nations. He was the son of a very wealthy man. In his childhood he decided to follow the path of asceticism of the monks and he joined their caravan. Having by nature a general capacity for excellence, he devoted himself to the acquisition of the sciences. With his father's support and by means of his own maturity, native intelligence and innate aptitude, he acquired the various sciences in a short time and became distinguished among his peers.

Then, encouraged by his religious belief, he wished to gain fame for his earnest desire to expose the error of the clear Muhammadan religion, whose truth is manifest and bright as the shining sun at mid-day and the full moon over the horizons and the continents; and to expose as deficient the beliefs of Islam, whose foundation is preserved from obliteration. With this thought in mind, he came to Istanbul during the reign of Sultan Mehmed (IV) and in order to realize his desire read books of grammar and syntax, logic and semantics, in the customary fashion. He mastered Turkish and Arabic and, thanks to his ability and natural aptitude, familiarized himself with all the sciences. For a while he ranged in the valley of error, zealous to debase the coin of religion.

One day while reading books of Koranic exegesis, he happened to come across the noble verse, "O earth, swallow your water, and O heaven, cease your rain." (11:44) As he pondered this quintessential verse with an open mind, he became aware, by the grace of God, of both the letter and spirit of the miraculous eloquence contained in these profound words. The result was that he became the object of divine guidance and was honored with the nobility of Islam.

This is reported by reliable authority.

End of the publisher's supplement.

Translation of the passage that Hondius wrote giving his reasons for the composition of this work

First of all, this publication that is being translated was printed in Holland, in the district of Arnheim (text: Āznehembūm), in the Christian year 1621. Gerardus Mercator began to compose this book but he died before finishing it. His fellow citizen, Jodocus Hondius, completed it and had the maps engraved on copper plates.

After saying this, he praises this science and mentions those who write about it, including Abraham Ortelius, Daniel Cellarius, Antonius Maginus, Paulus Merula and Petrus Bertius. (He goes on to say:)

[M 11] "But the author of the *Atlas*, the excellent geometer Gerardus Mercator, my fellow-citizen, who is the most outstanding and skillful of them all, was not able to finish the book. Wishing to complete it, I assembled the relevant maps and completed their explications with the help of my kinsman Petrus Montanus and made a fair copy. Then those who are expert in this science and its scholars approved it and it was printed and published. However, because this book was highly detailed, those who were interested in it complained of its great size and high price and requested that an abridgment be made of it. I could not refuse their request and resolved to make an abridgment of Mercator's *Atlas*. In doing so, however, I was careful to convey the sense of that detailed book in the most succinct manner without sacrificing anything from the maps and explications of *Atlas Major* and without any loss of meaning. In order to make the work even more useful, I added the older plates after some of the figures. I completed this work in my college in the city of Amsterdam in the country of Holland on 17 March, 1607 of the Christian era."

This humble one (Kātib Çelebi) says: This *Atlas Minor*, which we translated, has 683 pages. Of these, 153 are devoted to the parts of the world, with one country depicted on each page along with its place names. On the other pages the author explicates these maps.

My original intention in translating that book (*Atlas Minor*) was to supplement this book (*Cihānnümā*). At the beginning of Ṣafar in the year 1065—corresponding to the beginning of December (1654)—I was about to translate the section on Bavaria, one of the countries of Germany, on page 438 of that book, [R 3a] with about one third of the work remaining. At that point I decided to rewrite *Cihānnümā* from the beginning and make a fair copy—I hope that God will vouchsafe me to complete it. Then the contents of the chapters of *Levāmi'u'n-nūr* (i.e., his Turkish translation of *Atlas Minor*) were extracted in their entirety and inserted in their proper place. Only matters concerning the lands of the infidels that are not suitable for extraction were left where they are, making the book somewhat less susceptible to the carping of fools.

[Marginal note by Kātib Çelebi] For if one were to make a full translation of the *Atlas* it would be necessary to mention repeatedly the monuments of the lands of the infidels and the names of their regiment commanders and their authors, and such things as wine and pigs and churches. Obscene passages of this kind were omitted.

End of the explanation of the book of *Atlas*.

Now we resume the account of the books of this science that serve as sources.

Avżaḥu'l-mesālik ilā maʿrifeti'l-büldān ve'l-memālik: Sipāhizāde Meḥmed Efendi, who was dismissed from the post of qadi of Istanbul and died in 997 (1589), arranged al-Malik al-Mu'ayyad (Abū l-Fidā')'s *Taqwīm al-buldān* in alphabetical order, then translated it into Turkish in somewhat abridged version and presented it to Koca Meḥmed Pasha. I added information, that I found over the years in the form of marginal notes. All of this has been used.

(*Kitāb-ı*) *Baḥriye*: Pīrī Re'īs, who was the brother (rather, nephew) of Kemāl Re'īs, one of the famous admirals of the Mediterranean, devoted his whole life to the science of navigation. In 960 (1553) he set out with the fleet from Suez and defeated the Portuguese at Hormuz. (Later) the governor of Basra, Ḳubād Pasha, accused him to the Porte of oppression and transgression and he was arrested in Egypt and executed. He prepared maps with detailed commentary on the islands and coasts of the Mediterranean, its harbors and shipping lanes, [M 12] and presented the resulting book to Sultan Süleymān Khan around the year 900 (actually 932/1526). I examined two versions of this work, one highly detailed and the other much shorter, and inserted appropriate parts (into my book). At the beginning of the larger version there are several hundred verses on seafaring, and many details (on this subject) scattered throughout, that are omitted in the smaller one. Both verse and prose are in the rough and ready language of sailors.

Tārīḫ-i Hind-i Ġarbī: a history of the countries of what the *Atlas* calls America and the common people call the New World. It is an abridged book describing the discovery of that continent and the marvels of its people. Based on a translation from some European folios, with supplements from *Murūj al-dhahab*, the *Kharīda*, and the commentary on the *Tadhkira*, it was composed around 990 (1582) and dedicated to Sultan Murād Khan (Murād III, r. 1574–1595). Because information on America is hard to find in other books, I have cited it along with the supplements to the translation of the *Atlas*.

Tuḥfetü'z-zemān ve ḫarīdetü'l-āvān: a book arranged according to the seven climes by Muṣṭafā b. ʿAlī, the timekeeper in the Mosque of Sultan Selīm. I extracted some information from this book.

Taqwīm al-buldān: This was composed by the ruler al-Malik al-Mu'ayyad ʿImād al-Dīn Ismāʿīl b. al-Malik al-Afḍal ʿAlī (better known as Abū'l-Fidā, d. 732/1331). He was a scion of the Ayyubid Kurds, the most distinguished ruler of Ḥamā, nay, the greatest scholar among caliphs and kings after al-Ma'mūn (Abbasid caliph, r. 813–833). In the introduction he states that he has not seen any book that treats this subject adequately. Ibn Ḥawqal, in his book, does not have detailed descriptions of the countries and fails to determine place names or give longitude and latitude. Sharīf al-Idrīsī in the *Nuzhat al-mushtāq* and Ibn Khurdād (i.e. Ibn Khurradādhbih) and others followed the same path as Ibn Ḥawqal. Astronomical tables and books of longitudes and latitudes do not determine place names or give descriptions of the regions and their conditions. Those books that are careful to determine place names—such as the *Ansāb* of al-Samʿānī, the *Mushtarik* of Yāqūt al-Ḥamawī, and the *Fayṣal* and *Muzīl al-irtiyāb* of Abū'l-Majd al-Mawṣilī, etc.—do not give longitudes and latitudes. Thus, when it was clear that none of these books alone would be able to explain this subject satisfactorily, he compiled this book around 730 (1330), giving the longitude and latitude of 623 cities according to 29 conventional climes in tabulated form in the manner of Ibn Jazla's *Taqwīm al-abdān*. [R 3b] In his apology for restricting his book to that number, he states:

"I do not claim to discuss all the cities comprehensively. In fact, it is impossible even to mention most of them, because very little information has come to us on the regions of the clime of China, which in area is almost a quarter of the inhabited world, and much of what has come to us is clearly erroneous. The same is true of the countries of India, which covers a vast area but is mostly unknown. We hardly know one percent of what there is to know about the lands in the north,—of the Turks, Tatars, Russians, Bulgars, Franks, Walachians, Poles and Kazaks (or Cossacks); or the area extending from Constantinople to the Western (Atlantic) Ocean; or the lands in the south—Nubia, Takrur (Sudan), Ethiopia and Zanzibar—which include the blacks and many other nations and tribes. For most of the books on "Routes of Countries" (*masālik al-mamālik*) only include the lands of Islam, and even these are not fully described. However, it is better to know these few than to be left completely ignorant."

The poor one (Kātib Çelebi) says: That book (*Taqwīm al-buldān*) is the best of the Islamic books on this science. Since it is as we have described, [M 13] one can judge the state of the other works by analogy to it and one can realize the superior quality of this *Cihānnümā* and of the translation of *Atlas Minor*.

The Geography (i.e., the *Almagest* of Ptolemy): The meaning of this term is explained in the chapter on the sphere of the earth. Ptolemaeus—which was Arabicized as Baṭlamyūs—was one of the philosophers of Alexandria in the year 900 of the calendar of Nebuchadnezzar and author of the great book known as *Almagest* based on his own observations. He wrote this book in order to describe the earth. To the extent that he was able, he described the borders of countries, 4,530 cities, mountains, rivers, and the conditions and descriptions of every region and its inhabitants. Later, during the reign of Caliph al-Ma'mūn, this book was translated into Arabic along with those of the other (ancient Greek) philosophers. It was translated by the distinguished al-Kindī. At present the original is extant and the translation, which no one has seen, is probably lost. However, the location of many places mentioned in this work is no longer known because, with the passage of the centuries, they have fallen into ruin or their names have changed.

When you are informed of places that you have found
They make you unhappy or happy as they do other people.

In the *Atlas* and other books, the cities that Ptolemy mentioned are identified by conjecture and deduction. However, because their features and names have changed, one cannot make use of the information in that book (i.e., the *Geographike Hyphegesis*). The method used in the transmission is that used in the transmission from the Arabicized version.

Kharīdat al-ʿajā'ib: This was the work of Shaikh Zayn al-Dīn ʿUmar b. Muẓaffar b. al-Wardī, who died in 758 (1357). Having no familiarity with this science, he followed the usual authorities and depicted the earth in Arab fashion, quite absurdly, then provided information about some climes and cities, revealing his ignorance to those qualified to judge. Subsequently, the uneducated were taken in by his depiction and copied his falsehoods into their own books, thus becoming accomplices in the perpetuation of error.

al-Rawḍ al-miʿṭār fī akhbār al-aqṭār: This was the work of Shaikh Abū ʿAbd Allāh Muḥammad b. ʿAbd al-Nūr al-Ḥimyarī who was from the Maghreb. He confined his work to the lands of Islam, providing information about a city or a place in proportion to its fame, and organizing it in alphabetical order. Where appropriate, I have extracted passages from this work. [R 4a]

ʿUmdat al-mahra: a book in Arabic in seven chapters by Sulaymān b. Aḥmad al-Mahrī. It describes how ships sail in the Eastern Sea (Indian Ocean).

In addition, I examined and made extracts from the same author's *Tuḥfat al-fuḥūl* and the commentary on it, as well as the Turkish translation of it by Ġalaṭalı Seydī ʿAlīzāde entitled *Muḥīṭ*.

[Marginal addition by Ebū Bekr b. Behrām ed-Dimaşḳī] *The book of Lorenzo*: the work of a writer named Lorenzo from the city of Anania in the country of Calabria in Italy. He discusses climes and cities in the manner of "Routes of Countries" (*masālik al-mamālik*), without maps. He entitled it *Fabrica del Mondo* which means "Buildings of the World." It was completed and printed in 1582 of the Christian era. It is about half the size of *Atlas Minor*. Places not found in other books were extracted from it and translated, including many useful passages.

Kashf al-mamālik: a book by Khalīl b. Shāhīn al-Ẓāhirī. It describes the clime of Egypt, its roads and conditions; and is arranged in 40 chapters. I examined it along with its abridgment and extracted suitable passages.

Masālik al-mamālik: the title ("Routes of Countries"—properly *Kitāb al-Masālik wa al-mamālik* "Book of Routes and Countries") of numerous books in Arabic and Persian, including Ṣāʿid b. ʿAlī al-Jurjānī's book in Persian and Aḥmad b. Sahl al-Balkhī's book (in Arabic). The latter was translated into Turkish by the late Şerīf Efendi, one of the *ulema* of Rūm, while he was a *müderris*, with the encouragement of Ġazanfer Agha, for the conqueror of Egri (Erlau) Sultan Meḥmed Khan (Meḥmed III, r. 1595–1603). He states that he was not familiar with this science. [M 14] Other books on climes and "Routes of Countries" are those of Abū ʿUbayd al-Bakrī, al-Masʿūdī, Ibn Ḥawqal and Ibn Saʿīd al-Maghribī; the *ʿAzīzī* of Ḥusayn b. Aḥmad al-Muhallabī, which he wrote for the Fatimid ruler al-ʿAzīz billāh (r. 365–386/975–96); the book on the roads of Andalus by Aḥmad b. ʿUmar al-ʿUdhrī; and the work of ʿAbd al-Raḥīm al-Mashhadī.

Muʿjam al-buldān: a book by Yāqūt al-Ḥamawī who died in 626 (1229). He subsequently abridged it and gave it the title *Marāṣid al-iṭṭilāʿ*. ʿAbd al-Mu'min Ibn ʿAbd al-Ḥaqq also abridged the original *Muʿjam*. In the introduction he states: "The purpose of writing this book was to specify the names of places. But al-Ḥamawī was not satisfied with this; he lengthened the book by adding etymologies offered in the lexicons, dubious information on latitudes and longitudes, and information on men discussed in books of genealogies. I shortened it by including only the names of places that occur in historical works." Jalāl al-Suyūṭī also abridged this book. I have included excerpts from the *Marāṣid*.

Introductio: a concise compilation by the writer Philippus Cluverius that serves as an introduction to ancient and modern books of geography. I translated this book also from Latin into Turkish and included excerpts from it occasionally in *Cihānnümā*. It was published in Paris in 1635 of the Christian era.

Menāzirü'l-ʿavālim: In 1005 (1596) the qadi Mehmed b. Ömer, with the byname ʿĀşıḳ, translated *Taqwīm al-buldān* into Turkish, plus supplements from *Ḥayāt al-ḥayawān*, *Mirʾāt al-zamān*, and other books, and additional information on cities that he had visited in Rumelia and Anatolia. I extracted from this book information that was reliable and useful and not found elsewhere. There is no clean copy of this work; it exists only in draft form.

Minhāj al-fākhir fī ʿilm al-baḥr al-zākhir: a work in seven chapters by Sulaymān b. Aḥmad al-Mahrī. It is one of the books about voyaging in the Eastern Sea (Indian Ocean).

Nuzhat al-mushtāq fī ikhtirāq al-āfāk: a work by Sharīf Muḥammad b. Muḥammad al-Idrīsī from Messina (Sicily). He gives an account of the seven climes in ten chapters each, describing "Routes of Countries" (*masālik al-mamālik*) and regions. Later some Maghrebis extracted sections from this book, omitting useful material, and copies of this printed nonsense were circulated. I extracted some useful information from this book.

> [Marginal note by Kātib Çelebi] At the beginning of *Künhü'l-aḫbār*, ʿĀlī commits a grave error in ascribing the *Nuzhat al-mushtāq* to the Prophet Idris. Its author lived around the year 700 (1300).

Haft Iqlīm: a Persian book arranged according to the seven climes by Amīn Muḥammad al-Rāzī. It was composed around 1010 (1601) and describes the poets and distinguished men of every region. I extracted information on the eastern region, places about which the author was knowledgeable.

Ḥamd Allāh Muṣṭawfī's *Nuzha* also includes discussions of the climes. I used it to correct information on the eastern regions. [R 4b]

2 On the meaning of "geography" which is the subject of this science.

Philippus states in his *Introductio*: "GEOGRAPHY is a science that deals with depictions (i.e., maps) of the entire globe, as far as our knowledge will allow, i.e., as far as our science comprehends."

> The poor one (İbrāhīm Müteferriḳa) says: This definition is circular. He should have said, "that deals with the conditions of the earth and water, as far as those accidental conditions are depicted on the globe as maps and borders and distances."

The term is a compound of two Greek words, *gaia* meaning "earth" and *graphia* which is derived from *grapho* meaning "to write." According to the rules of Greek grammar, when the two words combine, *gaia* becomes *geo* to form the word *geographia*.

> However—based on the principle that a widely accepted error is preferable (*Ġalaṭ-ı meşhūr evlā*), and also because it is somewhat easier to pronounce—[M 15] in this book we have followed the Arabicized spelling of this word and of other terms Arabicized from the Greek.

Just as the part is distinct from the whole but is included within it, geography is distinct from COSMOGRAPHY. In Greek, *kosmos* means "the world (or universe)." When combined with *graphia*, it signifies "depiction of the world".

The pertinent science is the science of ASTRONOMY, which deals with the depictions (or figures) of all the celestial spheres and sublunar world. Thus the science of geography is included in the science of astronomy and is a branch of it.

Similarly, HYDROGRAPHY, meaning "depiction of the sea," is also included in this subject.

Another branch of geography is CHOROGRAPHY. *Khoros* (in Greek) means "a certain district" and when combined with *graphia* it signifies "depiction of a district." It is a part of geography as a whole, and distinct from it.

Another distinct part of this whole is TOPOGRAPHY meaning "depiction of a certain city, town, or village." *Topos* means "a certain place;" in the compound it becomes topography.

Three of these four divisions are included in cosmography and are contained in each other as well, but they are also distinct from each other. Many books have been written about each of them. It would be possible to include all of these subjects in one book, but it would result in a lengthy work, so they wrote separate books on each subject. In this *Cihānnümā*, topography and chorography are included to a certain extent in geography.

> Supplement of the Publisher (İbrāhīm Müteferriḳa)
>
> After defining geography, Philippus expands on the topic in an addendum to his *Introductio*, thus:
>
> Our definition is a literal definition, not a true definition. Because those who originally pursued this science had no doubt or hesitation about (what they meant by) it, we need to provide a further explanation here.
>
> In accordance with the investigations of the ancients and the moderns who have studied the science of geography, this science is divided into two parts with respect to its aim: *mathematica*, i.e., mathematical geography; and *historica*, i.e., historical geography.

MATHEMATICAL GEOGRAPHY is a theoretical science that measures, specifies, and describes the size and position of the earth, its circles (parallels of latitude) and other divisions and parts according to the rules of geometry.

However, it should not be forgotten that the term "earth" here does not mean simply one of the four elements (earth). If one speaks in this science of "the earth" one means the globe that consists of the elements of earth, water, and air. The full subject matter of this [mathematical] part includes the land and the water, as well as the layer of air that surrounds them, for which the Greek philosophers used the scientific term *atmosphera*.

These elements are the material subject of this [mathematical] part. As for the formal subject, it consists of the quantifiable divisions and parts of the earth—size, position, circles (parallels of latitude), zones, climes, etc. The benefits derived from determining the divisions and parts in this way, and quantifying and knowing them, relate to the formal subject of this [mathematical] part.

Thus geography is a broader term that comprises several fields, and these are evident from the following categorization and definitions:

First, HYDROGRAPHY, which is the description of the ocean and other bodies of water.

Second, ATMOSPHERE, which is [M 16] the description of the air. This includes ANEMOGRAPHY, which is the scientific term for the description of the various winds that develop in the world, the seasons, the *ḥusūm* (the seven or eight days of cold weather expected about the week before the vernal equinox), floods, storms, and temperate weather.

Third, CHOROGRAPHY, which is the depiction of a given country, such as Spain, France, etc.

Fourth, TOPOGRAPHY, which is the depiction of a certain place—such as a city, village, fortress, or district and the orchards and gardens, fields and pastures, springs and rivers, mines, markets, forests, reed beds, buildings and lakes found therein—according to the rules of geometry and arithmetic.

So geography is the science that depicts and describes the structure of the world in its totality. Cosmography is the more general science that includes both geography and astronomy.

The second part of this science, HISTORICAL GEOGRAPHY, while providing various benefits and promoting the perpetuation of the human race, is a true science that depicts, specifies and describes the places on earth, one by one. In other words, it maps and knows, one by one, the lands and seas, islands and rivers, mountains and hills, gulfs and straits, valleys and deserts, cities and springs and other places that provide benefits of all sorts. It is, expressed in the terminology of this science, the knowledge and mapping of individual places. These are the material subject of the second part (of the science of geography).

The aim of this science is to facilitate travel and trade, communication, waging war and making peace and determining borders; and to provide the human race with the great benefits of civilization and society and other things. These pertain to the formal subject of this part (of the science of geography), as will appear in some detail in the texts below.

Some of the modern geographers speak of a third part of this science, which is termed *politica*, meaning "political." This part divides the world according to states (or dynasties), sultanates and governments—i.e., sovereigns, kings, sultans, and other rulers—and consists of depicting and describing and defining [the world] in this manner.

End of the publisher's supplement.

3 On the goal and utility of this science.

In addition to what was alluded to on this subject in the introduction of the book, the *Atlas* states the following:

"Geography is one of the extremely useful and beneficial sciences for the civilization and society of humankind. It is the most important and most necessary of the sciences for statesmen and government officials. Some philosophers have considered this science to be superior to the other rational sciences, because anyone ignorant of this science would be like a blind and deaf person when studying books on this subject. If a dispute arises over the borders of countries, this science resolves the difficulty. Even in minor matters, the utility of this science is clear. It is such a marvelous science that, because strange things that would never occur to one are recorded in it, it makes one both experienced in the vicissitudes of time and knowledgeable of the conditions of places. For it informs in detail about all the climes and countries and the unusual things found in them. Especially in our present time, when discoveries in every part of the world are being made by seafarers; lands and rivers, islands and seas, deserts and mountains, ponds and hillocks and forests are being mapped *in situ*; and longitudes and latitudes are being correctly determined by minutes and degrees. In addition, unusual events among people of every clime, in war and in peace, are being described as they occur. [M 17] Again, this science is more useful than all the others in political matters. Whoever knows and comprehends this science is commended and esteemed."

He goes on to mention several writers and then, in the course of his discussion of the globe, he says:

"There is a great need for this science in the following respect: the rising and setting of the stars, the change of day and night, the cause of summer and winter, and the location of events that occur in the world of becoming and dissolution can best be known by this science. The need for this science can be illustrated by the following analogy: If someone who owns a large mansion knows all the rooms within it—its stables and kitchens, storerooms and casements and other features—but does not know in which district and what quarter of the city it is located, [R 5a] then if a riot or a fire broke out in the city, he would not know if it was near or far from his house and would be remiss in taking the proper measures. So it is also important that he know the area around his house, his quarter and his neighbors, and even those farther afield and strangers."

To sum up: the sciences of cosmography and geography are among those things needed for the good order of civilization and human society. Whoever knows the maps and rules of geography, and can call them to mind, will have obtained more knowledge than those who travel a thousand years and endure thousands of labors and difficulties. The way to do this is to study books on this science with an open and critical mind and to discuss it with people who are well versed in it.

However, the following condition must be observed: first one should read and digest the general pages (of maps); then one should examine the detailed maps on the pages that follow; and on every page one should combine the detailed information with the general information. In other words, one should specify a given detail as pertaining to the general information previously given. Then all the aims of this science will be reached.

[Marginal note by Kātib Çelebi] By "the general pages" are meant the five maps at the beginning, those of the earth, of Europe, of Asia, of Africa, and of America.

4 On the roundness of the world of corporeal entities and the spherical shape of the earth.

Introduction

Leaving aside the proofs that have been offered concerning the sphericity of the celestial spheres and sublunar world, we must accept it here as an axiom, because all the principles of this science are based on it. There is no other possibility. If this philosophical view is thought to be contrary to the religious law, in order to dispel any concern about it and to ease the mind, I have translated verbatim and inserted here what Imam al-Ghazālī, who was the object of God's endless favor, wrote in his book called *Tahāfut al-falāsifa*. He says:

"It should be known that the difference between the philosophers and other groups is of three kinds:

"First, a dispute over a mere word, as when they call the Creator of the world a *substance*, explaining that term as a being that is not in a substratum, i.e., one that is free of place and is self-subsisting.

"Second, a view of theirs that does not conflict with any religious principle, so there is no need to dispute it in order to affirm the truth of the prophets—i.e., affirming the prophets does not require contradicting the philosophers. As when they say that a lunar eclipse occurs [M 18] when the earth comes between the sun and the moon so the light of the moon disappears, since it gets its light from the sun, and the earth being a globe completely surrounded by the sky, when the moon passes into the earth's shadow, it is cut off from the light of the sun; and a solar eclipse occurs when the moon passes between the observer and the sun, which happens when the position of the sun coincides with the ascending and descending nodes of the moon. We are not obligated to dispute such views, because it is not necessary to refute them. Anyone who thinks that trying to refute such views is a religious duty actually weakens the status of religion and commits an offense against it. [R 5b] For these matters are based on geometrical and arithmetical proofs and leave no room for doubt. If a man is informed about them and can confirm their proofs and can say why and when eclipses occur and how long they last, and you tell him this contradicts the religious law, it will not make him doubt his view on eclipses, since that is established with certainty, but will make him doubt the religious law; for how can that contradict what is certain? The harm of someone who supports religion by improper means is greater than the harm of someone who discredits it by proper means. As the saying goes, An intelligent enemy is better than an ignorant friend."

The Imam then relates the hadith about eclipses and continues:

"Someone may bring up the following objection: It is recorded that at the end of the hadith, the Prophet said, 'When God reveals Himself to something it bows down before Him,' and this shows that an eclipse is a prostration caused by divine revelation. To this objection we say: This addendum is not sound and we must reject it as spurious. The hadith is only recorded as we have given it. Even if the addendum were authentic, it would be simpler to interpret it metaphorically than to reject out of hand matters that are established with certainty. There are many plain texts that have been interpreted metaphorically and have not reached this degree of certainty—so that is the least one can do with regard to a text that is not based on a sound transmission. Nothing pleases heretics more than for someone who supports religion to assert that things like this are contrary to the religious law, [M 19] because that makes it easy for them to refute the religious law. That is because the real issue is whether the world was created or is eternal. Once its creation has been established, it makes no difference whether it is flat or round, and whether the celestial spheres and sublunar world are in 13 layers as the philosophers say or more or less. The important thing is that the world, however it is, was created by an act of God.

"Third, a dispute relating to one of the principles of religion, such as the creation of the world, the attributes of the Creator, and the resurrection of the body, (which are things that the philosophers deny). These are their views that must be refuted."

He expands on these topics in the introduction to the *Tahāfut*. I have cited and translated the Imam's words here so that simple-minded pietists should not think that the matters discussed below are subject to refutation by claiming that they are contrary to religion.

Confirmation

Following this Introduction, let us proceed to the geographers' confirmation of the roundness of the earth. I will first mention what is written on this subject in chapter 7 of the *Atlas* and then I will present the evidence.

According to the *Atlas*, in Latin the world is called *mundus* because it was created in such a perfect and beautiful arrangement.

Pliny says in the second article of his book on *Natural History* that all things in the world are connected to each other and the world itself consists of globes that encompass one another. Thus the natural shape of the world is round.

Aside from the works of sages who depict the earth as a globe, this claim is confirmed by much evidence based on the requirement of nature. [R 6a] Thus, wherever one looks, one sees a convex surface. Given that each section of the surface is an arc, as is known by the axioms of thought and by human experience, and because round shapes are the most voluminous (in relation to their circumference or surface), it is impossible for the other phenomena observed on earth and in the heavens to be any shape other than round. Those who believe that the earth is flat are the laughing-stock of the sages and paramount ignoramuses.

The tendency of heavy bodies to fall from anywhere toward the center of the earth was taken by Aristotle as conclusive proof that, although the lands and seas and mountains and rivers take different shapes, all together they form a globe.

According to other geographers, the fact that a lunar eclipse is caused by the shadow of the earth, and the fact that places have different latitudes and longitudes as travelers move across the face of the earth, are further proofs that the earth is round. The existence of *antaeci*, *periaeci* and *antipodes* stems from this fact, and *periscii*, *amphiscii* and *heteroscii* have also been established. The meaning of these terms is given in the discussion of the circles.

[Marginal note by Kātib Çelebi] *Amphoeci* are the inhabitants of the equatorial zone, *periaeci* are the inhabitants of the latitudinal zones on either side (of the equator). *Antipodes* are those who live foot to foot with each other. As *periaeci* live near the poles, and their shadows rotate around them. *Amphiscii* live near the equator, and their shadows fall to either side of it. *Heteroscii* live in the inclined parts, and their shadows always fall to the same side. All those are Latin words.

Here the account of the *Atlas* ends.

The following are some of the proofs mentioned in my astronomy books:

- The fixed stars revolve in parallel circles around the pole and those near the pole revolve in a smaller orbit that is always visible.
- The time during which the region extending from the fixed star, observed tangent to the circle of the horizon, to the edge of the zone is invisible increases according to the distance until it is visible and invisible for equal times, then the time that it is invisible increases and the time that it is visible gradually decreases, reaching a point near the other pole where it is always invisible.
- The rising star gradually ascends from the horizon to the zenith and then descends.

[M FIGURES] [M 20]

- The celestial bodies rise bit by bit and in like manner become full (like the moon).
- While a star is always the same size while circling in its orbit, it seems to be larger when rising above the horizon, because vapors that rise from the land and accumulate make anything that is behind them appear greater than it really is. For example, a grape seed appears bigger in water.
- Half the sky, or almost half, can always be seen from the surface of the earth.

- When a star rises, people in the east see it before those in the west. When it sets, the opposite occurs.
- As one travels north, the polar star and the northern stars get higher while the southern stars get lower in the sky. As one travels south, the opposite occurs.
- From the water, the convexity of the sea conceals the lower parts of the mountains, so one first sees the upper parts and then, as one approaches land, the lower parts.
- During the time that a star is visible, its ascent and descent are equal.
- At the equinox, when the sun is visible and invisible for an equal time, the shadow at the equator when the sun rises exactly corresponds to the shadow when it sets.

These examples and others like them are certain proofs that the earth and the heavens are round.

Supplement of the Publisher (İbrāhīm Müteferriḳa)

It was explained in the above text that a lunar eclipse occurs when the earth intervenes between the sun and the moon and the light of the moon is extinguished. It was also stated that at the time of a lunar eclipse, the earth casts a round shadow on the surface of the moon, and this is certain proof that the earth is spherical. If the earth were not a sphere but a different shape—e.g., a triangle, square, or hexagon—then the earth's shadow on the surface of the full moon during an eclipse would not be circular but would have to be triangular or square or hexagonal—which is counter to experience and observation.

It is also obvious that if a round solid body [such as the earth] blocks a round luminous body [such as the sun] and is the same size as that luminous body, the shadow cast by it will be cylindrical and the thickness of the shadow will be the same no matter how far it extends. If the solid body is larger than the luminous body, its shadow will be in the shape of a partial cone that widens as it extends. If the luminous body is larger than the solid body—as is the case with the sun and the earth—then its shadow will be in the shape of a cone that narrows as it extends until it disappears. All of these geometrical shapes are shown on the previous page.

Similarly, celestial events are observed (differently) from different places. Solar and lunar eclipses that occur at dawn in the east are not seen by people in the west. Those that occur at evening in the west are not seen by people in the east. Those that occur in the middle of the sky are not seen by people living at the antipodes on the other side of the earth. And those that occur in the middle of the sky on the other side of the earth are not seen by people living opposite, in this side of the earth. Actually a lunar eclipse that occurs in the middle of the sky in the upper half of the earth is seen by people in the west before those in the east. And the sun and the moon rise and set over those in the east before those in the west. These phenomena could not occur if the earth were not spherical.

Leaving all this aside, the possibility of traveling by sea to the East Indies and the West Indies [M 21] from the east or the west has been demonstrated. The fact that ships sailing from the west return from the east is the last of all the evidence to confirm the roundness of the earth and closes the door of dispute on it.

End of the publisher's supplement.

This science has some curious problems related to the principle that the earth is round. I will pose the problems and explain the paradoxical aspects.

1) A day (of the week) can be different for three persons. For example, from a given place, one person goes to the east, one to the west, and the other remains at that place. The two travelers set out in a straight line at the same speed. The one who goes east returns from the west and the one who goes west returns from the east. They meet on the same day at the place where the third person remained behind. If that day, according to the person who remained behind, is Friday, then according to the one who went west it is Thursday and according to the one who went east it is Saturday.
   The explanation is as follows: Assume the one going west travels seven days. Since he travels in the same direction as the sun, his sunset is later than the one going east—one-seventh of a revolution later. Thus in seven days the difference amounts to the loss of one day and night. According to him, he arrives on Thursday. Likewise, since the one going east travels in the opposite direction of the sun, [R 6b] the sun sets one-seventh of a revolution before the one going west. The total of these one-sevenths in seven days is a difference of one day and night. According to him, he arrives on Saturday.
2) At the earth's surface there is a deep well and above it a minaret. If one fills a bowl with water at the bottom of the well, that bowl will not contain the same amount of water at the top of the minaret, but will overflow.
   That is because, as the distance increases from the center of a circle, the convexity of the arc diminishes. The corporeal elements, wherever they are, being part of the sphere, the arc of the circle at the mouth of the bowl (i.e., the surface) is more bent at the bottom of the well because it is closer to the center of

the earth, while at the top of the minaret it is closer to a straight line and therefore the water will slightly overflow.

3) If two stones are dropped from two different places on the gallery of a minaret, the distance between the points where they land will be less (than the distance between the points where they were dropped). For example, if one side of the gallery is two cubits from the other, the places where the stones land will be less than two cubits from each other.
In like manner, the distance between the top and the base of two walls is not the same. That is because the starting point and end point of two plumb lines are not equal; as they approach the center of the earth, they gradually merge (and so are not parallel).

Note: We must qualify these paradoxes. The minaret and the wall that were mentioned must be very high, and the well very deep, for a difference to be detected.

Addendum: Some time ago, three similar paradoxes occurred to me. Because they pertained to jurisprudence, I asked Shaikh al-Islam Bahā'ī Efendi for his legal opinion, but no response was forthcoming. I then turned to divine favor and sought jurisprudential guidance with the blessings that descend from the spiritual world. I gave the title *İlhāmu'l-muḳaddes mine'l-feyżi'l-aḳdes* ("Hallowed Inspiration from Divine Favor") to the treatise I wrote on this subject.

1) Zeyd goes on an English ship to 90 degrees latitude, where there are six months of day and six months of night. How can he perform the five daily prayers and the fast of Ramadan?
2) Zeyd and ʿAmr discuss the matter of the sun rising from the west as one of the portents of the Day of Judgment. Zayd says that this is compatible with the laws of astronomy, and ʿAmr denies this. Zeyd explains that according to Taḳiyyüddīn the Astronomer, as the cincture of the constellations (zodiac) passes the ecliptic, it gradually, over a long period of time, coincides with the horizon of the equator, [M DIAGRAMS] [M 22] as do the cinctures of the other planets, at which point the sun rises from the west. If ʿAmr refuses to accept this and says, "If it is possible, may I be divorced from my wife," then is this divorce valid?
3) Zeyd says that there is a place other than Mecca where the qibla is in all four directions, and ʿAmr denies this. If each of them promises to free a slave if the other proves his point, which one of them will have to free a slave?

These three problems will not be answered here. It suffices for them to be presented for explication. Those who wish an explication can find it (in that treatise). Here I have only mentioned the matter in the briefest manner.

5 Summary of the status of the celestial spheres and the sublunar world

This subject has been referred to previously. The corporeal universe is composed of concentric spheres. The celestial spheres, above the sublunar world, are in layers like an onion. Thus the cosmos is in the shape of a ball, in the center of which is a point that is the center of the earth and of the universe, the lowest point (or true bottom, *taḥt-i ḥaḳīḳī*) from every direction.

The Greatest Sphere, which encompasses everything, is the highest point (or true top, *favḳ-ı ḥaḳīḳī*) and farthest limit of every direction. Being free of all celestial bodies, it is also called the sphere of Atlas. It makes one complete revolution from east to west in one day and night, turning the celestial spheres inside it. The daily rising and setting of the sun result from its motion. Its poles are the poles of the cosmos and its cincture is called the equinox.

Below it is the Sphere of the Constellations or Sphere of the Fixed Stars. There are twelve constellations along its cincture and all the fixed stars are set within it. Moving from west to east, it traverses one degree every 66 solar years. [R 7a] All the visible stars except for the seven planets are in this sphere.

Next is the Sphere of Saturn. Turning from west to east, it completes its revolution in 29 years, 5 months, and 6 days.

Next is the Sphere of Jupiter. Moving from west to east, it completes its revolution in 12 years.

Next is the Sphere of Mars. Turning in similar fashion, it completes a revolution in 1 year, 10 months, and 22 days.

Next is the Sphere of the Sun. Going from west to east, it completes its revolution in approximately 365¼ days. It is uncertain whether it is the fourth sphere.

Next is the Sphere of Venus. It completes its revolution in one solar year.

Next is the Sphere of Mercury. It also completes its revolution in one solar year.

The Sphere of the Moon. It completes its revolution in 28 days.

All of them move from west to east. There is detailed information on them in the astronomy books.

Now, the Sphere of the Moon is a hollow sphere and the four elements are contained within it. Earth and water are surrounded by the Sphere of Air (atmosphere), which in turn is surrounded by the Sphere of Fire. Because earth is

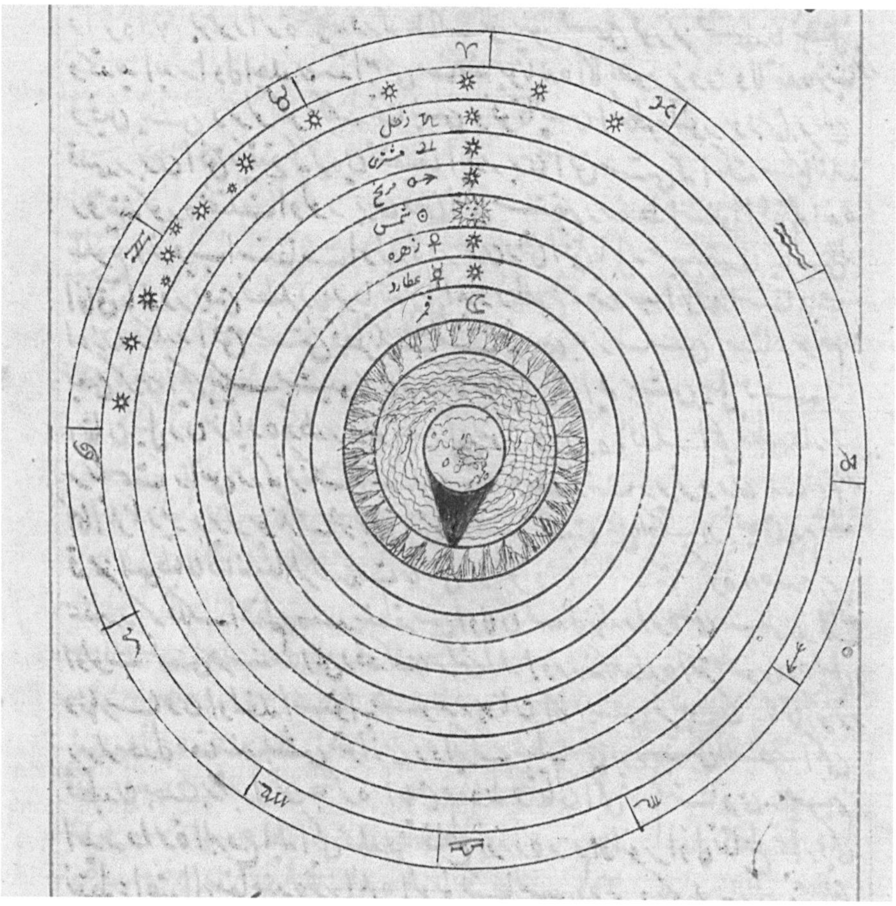

PLATE 1
Cosmological scheme, R 7a; cf. M 25/26 (Appendix, fig. 9–10)
TOPKAPI PALACE LIBRARY, REVAN 1624, WITH PERMISSION

the absolutely heavy element, it is at the lowest point (or true bottom, *taḥt-i ḥaḳīḳī*) that coincides with the center of the universe. As in the following scheme: [Plate 1]

Supplement of the Publisher (İbrāhīm Müteferriḳa)

By the permission and will of the Wise Creator, Originator of marvelous works, all the celestial spheres and sublunar bodies are in the form of a number of spheres set one inside the other, like the leaves of a cabbage. The larger encompasses the smaller in succession, tangential to and coinciding with each other from all directions. Together they are in the form of a single sphere. The physical world was laid out and established in this marvelous order and wise arrangement by the power of the Lord of absolute power. On this much, all the authorities on the science of astronomy and all the philosophers agree.

However, in order to explain the components of this marvelous structure—[M 23] i.e., to describe in detail the structure of the universe—they expand on the topic, enumerating the celestial bodies, describing how they are positioned and arranged, their motions, etc. In these matters they often disagree in their explanations and follow different paths.

From the time that the science of astronomy and philosophy appeared up to the present, the Pythagoreans and some modern philosophers who followed them, also the Peripatetics and most of the astronomers who followed them, have made calculations and presented proofs concerning the general structure of the universe and have drawn conclusions based on three types of forms and images. When the hidden reality behind the curtain of each of these forms and images was revealed by explaining and describing it in a satisfying manner, a group of people inevitably became beguiled by the heart-ravishing face adorned by the hairdresser of proofs, based on the laws of philosophy and astronomy, in the form and image of the universe that was manifest in the explanation and description of each type.

The commentaries and explanations of most of the philosophers, which differ in this way, have been reduced to three independent schools who have described, mapped, represented, and laid out the physical world, i.e., the sphere of the cosmos:

1. The school of Aristotle, who was the greatest philosopher, and his follower Ptolemy, who was the foremost authority in the science of astronomy.
2. The school of Pythagoras and Plato among the ancient philosophers, and the astronomer Copernicus among the moderns.
3. The school of the more recent astronomer, Tycho Brahe.

The Latin scholars call the view of the first school Ancient Astronomy and those of the second and third schools Modern Astronomy.

It should not be forgotten, however, that believing in such matters is not based on religious principle and doctrinal requirement. Whatever the configuration of the universe may be, whatever the composition of the celestial and sublunar bodies, and however the celestial sphere performs its revolution, it is impossible to deny the creation in time of the universe and all its parts, and that this wondrous structure could only have been made by the perfect Maker and sublime Lord; and this belief in creation *is* a religious requirement, as explained above.

Although the first school is preferable and universally accepted, while the second and third schools have been rejected; still, when the celestial bodies on high and the earthly bodies below—in short, the multifarious details of the universe—are taken into consideration and pondered in whatever configuration, so many subtle mysteries and truths are observed that the fingers of explanation fall short in enumerating them, and the tongue of explication is dumfounded in describing them.

Consequently, while there is not room here to fully explore the views of these three schools and to set down all of their proofs, nevertheless—in accordance with the title of this excellent book, *Cihānnümā* ("Cosmorama"), because knowing and taking into account the views of learned men in this arena, which is the vast space of the cosmos, is required for serious researchers—[M 24] a summary of the views of these three schools is presented in this chapter, and diagrams and maps that reflect their views have been added in order to facilitate each one's depiction of the world and representation of the cosmic sphere.

1 The school of the First Teacher Aristotle, his follower Ptolemy, and those who followed Ptolemy

Most of the Muslim philosophers as well have accepted Aristotle's view. According to this view, the ethereal bodies—which are the nine celestial spheres enumerated above—and the lower elements (the sublunar world) have, at their very center, an imaginary point which is the center of the world and is below the feet of all nations. The largest of these nine celestial spheres is the sphere of Atlas, which is the determiner of directions and the appointer of times. This sphere governs the other spheres and has them under its control. Once every 24 hours it turns all the luminous and fixed stars and the planets from east to west. The rising and setting that cause day and night are tied to the motion of this sphere.

The last of these nine spheres is the sphere of the moon, which encompasses from all directions the atmospheric beings and the world of becoming and dissolution. The spheres of the four elements are fixed within it according to their rank. In all of these, the side that encompasses is up and the side toward the center is down. Those creatures that move about on the surface of the earth always have their heads toward the sphere of the moon and their feet toward the center of the world.

While the authorities of this school are in agreement on these matters, they differ in other respects.

First, there is disagreement—spelled out in the astronomy books—on the positions and arrangement of the spheres, placing them in a different order, raising some and lowering others.

Second, there is disagreement on the number of spheres. Most of the ancients, including Aristotle, thought that there were only eight spheres and that the prime mover was the sphere of the Constellations (or zodiac), i.e., the sphere of the fixed stars. Others, detecting nine motions by astronomical observation, deduced that there were nine celestial spheres and assigned the sphere of Atlas as the prime mover.

The observations of most of the Muslim astronomers matched this latter view. They did not agree (with Aristotle) that there were less than nine, nor did they say there were more than nine. Later, at various times and places, observatories were built and new instruments were devised and more accurate observations were made. Nevertheless, they were unable to determine the ecliptic as it truly is. According to some adherents of this school, after its second motion the sphere of the Constellations (or zodiac) has a more erratic motion, now progressive and now retrograde, now from east to west and now from west to east, and also an odd trembling and shaking motion from north to south and from south to north, and these motions have caused the discrepancies in their observations concerning the ecliptic. It was subsequently determined that there were two spheres above the sphere of the Constellations that caused these two motions. As revealed by astronomical instruments, these spheres were luminous and transparent, and so pure that they were named crystalline, or icy, or watery. [M 25]

Assuming that the sphere of Atlas, which was the prime mover, was above them, these astronomers concluded that the total number of spheres was eleven. When the spiritually elect, who are the people of religion and faith, complete their sojourn in this inn of the transitory world and are removed to the eternal abode of the next world, there is prepared for them, by the grace of God, a magnificent place like paradise where they will receive their reward. The Christian theologians suppose that such a great sphere exists above the eleventh heaven and call it the sphere of the people of faith and the abode of the spiritually elect. Thus, according to them, the number of layers in the heavens is twelve.

THIRD, there is disagreement on emptiness (or void) and fullness (or plenum). While they agree that there is neither excess nor void throughout the heavens, in whole or in part—and in this they follow Ptolemy, who is the chief of the astronomers—they differ over what is beyond the heavens. Anaxagoras and some of the ancient philosophers say that the void is a real extension abstracted from matter. The Muslim theologians interpreted it as an imaginary (or postulated) extension. In their dispute, some tried to establish by philosophical proofs that plenum and void are impossible. Others preferred not to discuss this subject, saying that astronomical instruments cannot reach beyond the heavens and the human intellect does not have the ability to penetrate it; to inquire about something unknowable, or to try to characterize what is unknown—like inquiring about the extension in time of pre-eternity and eternity—only leads to presumption and deception.

As explained in detail above, the view of those who limit the greater spheres to nine is preferable, and it is the view accepted by the Muslim philosophers, while the other views have been rejected by the cognoscenti. But simply specifying the celestial spheres does not suffice to explain the motions of the heavens.

When the philosophers observed the behavior of the seven planets, they saw that their motion is sometimes moving forward, sometimes remaining stationary; sometimes fast, sometimes slow; and sometimes retrograde. Sometimes, as with the sun, their course is between the two points of rotation that constitute the ecliptic; and sometimes, as with the other planets, it goes well beyond the two points of rotation to the north and south. Sometimes their luminosity increases and sometimes it decreases. With all of these behaviors, sometimes they approach the earth and sometimes move away from it. Lunar and solar eclipses are also not uniform: sometimes they are total and sometimes they are partial.

As a result, contemplating this kind of variegated phenomena in the skies, as they (the astronomers) investigated and contemplated them, in explaining their causes, and in calculating times, and adjusting, regularizing, and equalizing [i.e., accounting for the variances in the planetary motions mentioned above], they had to postulate different kinds of epicycles inside the larger spheres, (that is, in the spherical shell between two parallel [inner and outer] concentric surfaces), some comprising the earth and others not, with the same or different centers and poles, and of equal or unequal thickness or thinness. They called these epicycles the lesser spheres and secondary spheres.

Because the Muslim philosophers too prefer the view of Ptolemy on this matter, I have briefly enumerated and defined the epicycles here in a manner consistent with the view of that philosopher, so that students of astronomy may visualize them easily. And in order to make the ideas tangible and the imagined things corporeal [M DIAGRAM] [M 26] and visible, I have drawn the figures of the epicycles of the greater spheres. The number and arrangement of the spheres correspond to the exposition in the text of the book, in descending order from highest to lowest.

THE HIGHEST SPHERE, WHICH IS THE NINTH—it is known as the sphere of spheres, the greatest sphere, the sphere of Atlas, etc.—as described in detail in the books of astronomy and as briefly touched upon above, is a spherical body enclosed by two parallel surfaces. Its center is the center of the earth and its pole is the pole of the earth. As it encompasses all the celestial bodies it is the boundary of the corporeal world. It is tangent to the convex [outer] surface of the sphere of the fixed stars that is within its concave [inner] surface. However, because it is assumed that void and plenum are impossible beyond the heavens, its own convex surface is not tangent to anything and no stars are visible on its surface. Given the breadth and scale of this sphere, as it moves from east to west around the center of the world, it completely turns the spheres inside it and partially turns the (sublunar spheres of) fire and air. It makes one rotation in 24 hours. There is no need to posit epicycles—i.e., lesser spheres—in the thickness of this sphere. Only it is postulated that the equator, which is one of the great circles, is located in the middle between the two poles encompassed by this sphere.

Next comes THE EIGHTH SPHERE, known as the sphere of the fixed stars and the sphere of the Constellations, which is fixed within the Greatest Sphere. Its center is the center of the earth and its pole is inclined to one side from the pole of the earth. It is a spherical body enclosed by two parallel surfaces. Its convex surface is tangent to the concave surface of the Greatest Sphere, which is above it, and its concave surface is tangent to the convex surface of the sphere of Saturn, which is within it. It is decorated and adorned with countless fixed stars and drawn and colored with the imaginary figures of the twelve constellations. In addition to moving from east to west around the center of the world and completing one rotation in 24 hours along with the Sphere of Spheres, which is the prime mover, it has its own motion with which it moves from west to east around a pole different from the pole of the world, over the cincture that inclines from the equinox, and as it slowly moves, it takes along the fixed stars that are embedded in its thickness, and all the spheres that are below it. It completes one circuit, according to some, in 40,000 solar years, according to Ptolemy, in 36,000 solar years. Indeed, some philosophers call the duration of this circuit the Ptolemaic Year and say that at the end of this period, the phenomena of the universe from end to end are repeated, as ordained by God who is eternal.

There is no need to posit epicycles—i.e., lesser spheres—in the thickness of this sphere. Only it is postulated that the circle of the cincture of the constellations (or zodiac), which is one of the greater circles, is encompassed by this sphere, and the images of the twelve constellations are delineated within this circle proper. In other words, the collective order (or pattern, *mecmūʿa-i heyʾet*) of the stars observed in each of the constellations, in each one-twelfth portion of that sphere, is seen to resemble a certain form and its name is given to that constellation. For example, the constellation of the Ram (Aries) is a portion in the area of this sphere whose stars, that were observed in that portion, if connected to each other by lines, display the form of a ram. The same for the other constellations.

Within the vast space of this sphere [M 27]—as described in astronomical tables and books of *Ṣuwar al-kawākib* (“*Figures of the Stars*”), and based on the observations of the ancients—48 images of animate and inanimate objects, comprising 1022 stars, were depicted. Twelve of these, comprising 346 observed stars, are the well known 12 constellations with their characteristic names. Of the rest, 21, comprising 360 stars, were determined in the northern part of this zone and 15, comprising 316 stars, were determined in the southern part.

Such was the enumeration, both of the constellations and the stars, according to the observations of the ancients in the area of this sphere. However, when the New World was discovered, and it became possible to go east by sailing west and go west by sailing east, and when ships went to India and China and back by sea south of the equator, some expert stargazers among the Western philosophers traveled to the eastern regions and made observations. They saw that in the polar region of Canopus there were many stars not named or described in the observations of the ancients or in the books of the images of the stars and not listed in the astronomical tables. They subsequently observed 90 new stars in that hemisphere, which they recorded and entered in their astronomical tables, and out of the figures that emerged from these new stars they depicted 12 new constellations. So now the number of observed stars, apart from the nebula, has reached 1112 and the number of constellations has reached 60. The celestial spheres manufactured at the present time include all of these images.

I give here a complete list of the figures of the stars (or constellations) that will serve as an introduction to the astronomical tables of the Muslim scholars.

- In the zone of the constellations (or zodiac): the Ram (Aries), the Bull (Taurus), the Twins (Gemini), the Crab (Cancer), Lion (Leo), the Spike of Grain, also called the Virgin (Virgo), the Scales (Libra), the Scorpion (Scorpio), the Bow, also called the Archer (Sagittarius), the Kid (Capricorn), the Bucket, also called the Water Pourer (Aquarius), the Fish, also called the Two Fish (Pisces).
- In the zone north of the zodiac: the Smaller Bear (Ursa Minor), the Larger Bear (Ursa Major), the Dragon (Draco), Cepheus, the Howling Dog, also called the Crier (Bootes), the Kneeler (Hercules), the Lyre (Lyra), the Bird, also called the Hen (Cygnus), the One with the Throne (Cassiopeia), Perseus who is carrying the head of the ghoul, the One Holding the Reins (Auriga), the Snake-charmer (Ophiuchus), the Serpent (Serpentarius), the Arrow (Sagitta), the Eagle, also called the Flying Eagle (Aquila), the Dolphin (Delphinus), the Segment of the Horse (Equuleus), the Second Horse (Pegasus), the Chained Woman (Andromeda), and the Triangle (Triangulum).
- In the zone south of the zodiac: Cetus, the Giant (Orion), the River (Eridanus), the Rabbit (Lepus), the Smaller Dog (Canis Minor), the Larger Dog (Canis Major), the Ship (Argo Navis), the Courageous One (Hydra), the Bowl (Crater), the Crow (Corvus), Centaurus, the Wolf (Lupus), the Censer (Ara), the Southern Crown (Corona Australis), and the Southern Fish (Piscis Austrinus).
- The twelve constellations discovered by the new observations: the Phoenix, the Braided Lock, the Indian Boy, the Indian Peacock, the Indian Bee, the Southern Triangle, the Shark, the Fly, the Winged

Fish, the Dragon, the Crow of Separation, and Noah's Pigeon.

Such is the list and brief description of the constellations comprised in the sphere of the fixed stars.

[M STAR MAP] [M 28]

I have depicted this sphere here, with the constellations and the figures of the stars, so that they will be recognized when seen.

I will now briefly describe the spheres of the seven planets:

THE SEVENTH SPHERE: It is the sphere of Saturn, the highest and largest of the three supersolar spheres, known as the Upper Spheres. It is titled Kaywān (in Persian). The astrologers call it the Greater Misfortune and say it has an Indian face; just as they call Jupiter, who is the bringer of joy, the Greater Fortune; and red Mars, who looks like an executioner, the Lesser Misfortune. The lady of heaven, shining Venus, and the messenger of heaven, beautiful-faced Mercury, are called the Lower (or subsolar) Spheres because they are below the sphere of the sun. The Upper and Lower together are called the Bewildered Five. In like manner, the luminous sun is called the Greater Luminary and the beautiful-faced moon is called the Lesser Luminary. Taken together they are known as the Seven Planets.

Before describing the structure of these spheres and drawing their shapes, I will first mention here some of the astronomical principles concerning the planets.

1) If a sphere is bounded by two parallel surfaces, i.e., if the distance between the two surfaces is equal everywhere, so that no part of the sphere is thicker or thinner than the other, then the center of both of these equidistant surfaces is the center of the sphere, and the center of the sphere is the center of the universe.
2) For any corporeal sphere that includes and encompasses the earth—such as the parecliptic, the deferent, and the oblique mentioned below—its two surfaces are parallel. The epicycles, which have been determined in the planetary spheres, or, according to the preferred school, in [all] the spheres except for the sun's, do not include the earth, and also have no concave surface, and therefore lie outside this definition.

Returning to our original subject, we say that for Saturn the astronomers have established three spheres: 1) The general sphere, which is called the parecliptic (*mümeṣṣel*, lit. "resembling") because it is equal to the sphere of the Constellations (Zodiac) with regard to center, cincture, poles, and rotation; 2) the eccentric sphere which is situated within the concentric and is called the deferent (*ḥāmil*, lit. "bearer") because it bears the epicycle; 3) the epicycle (*felek-i tedvīr*, lit. "sphere of turning"), called thus because Saturn is fixed in its body and turns along with it.

Therefore the greater sphere that is the parecliptic of Saturn is a spherical body bounded by two parallel surfaces. Its outer surface is tangent to the concave surface of the sphere of the fixed stars above it and its lower surface is tangent to the convex surface of the sphere of Jupiter below it. This sphere, like the other greater spheres enumerated above and below it, first of all follows the motion of the Greatest Sphere and moves with the Primum Mobile around the center of the world from east to west; second, it follows the essential motion of the eighth sphere and goes from west to east; and third, according to the rejected view mentioned above, it is said to be affected by the shaking and trembling motion of the eighth sphere along its width and goes shaking and trembling from north to south and from south to north. In any case, none of these motions is natural and essential; rather they are accidental motions resulting from the motions of the other spheres. [M 29]

However, every sphere also moves with a motion particular to itself, turning on its own axis and poles and completing its circuit. Thus the sphere of Saturn moves of its own accord from west to east and completes its circuit in a period known by calculation. Without doubt, the celestial spheres below Saturn are also characterized by essential and accidental motions, in an eastern and western direction. However, the fact that there is regularity in stasis, forward motion, retrograde motion, going slower and faster, being nearer and farther from the earth, and other observable phenomena that are relevant to all the planets, means that there must exist secondary spheres, i.e., lesser orbs (epicycles) that have different motions and rotations within the thicknesses (i.e., the spherical shells) of the greater spheres. Therefore, what one sees of their structures appears in different forms and they have been depicted here according to their conditions, as far as possible.

Accordingly, for the good condition and regularity of Saturn, it was hypothesized that there is a secondary sphere called the deferent inside the body of its greater sphere, i.e., within the thickness bounded by its two parallel (inner and outer) surfaces. However, it should not be forgotten that these lesser spheres are not perceived by the senses and are not known by observation. They are merely postulated in order to reconcile the phenomena observed in the spheres. They are like organs that serve for the well-ordered functioning of the general body of the sphere.

Now, the secondary sphere that has been identified for Saturn is a body enclosed by two parallel surfaces, which includes the earth, and its center is different from the center of the sphere, which is the center of the cosmos. The convex surface of this sphere and the convex surface of the primary sphere touch in one place, a point which can be indicated by the senses, but cannot be divided, which is shared between the two, and is called apogee. It is the furthest point relative to the center of the world. When the star arrives at this point, it has reached the extreme of distance and height from the center of the earth. In like manner, the concave surface of this sphere is tangent to the concave surface of the first sphere. The point where they converge, called the perigee, is the closest point relative to the center of the world. When the star reaches that point, it has reached the extreme closeness and lowness to the center of the earth. If, as described, the secondary sphere called the deferent is separated from the first sphere, and is so to speak removed, then necessarily two spheres of uneven thickness remain, one which would encompass the secondary sphere, and one encompassed by it. The encompassing sphere has its thinnest point in the apogee, and its thickest point in the perigee; with the encompassed sphere it is the opposite. Because these two spheres operate to supplement the celestial sphere, both are called supplementary.

What we have stated to this point does not suffice to regularize and harmonize the movements of the stars. They have restricted themselves to (postulating) an eccentric sphere (i.e., deferent) for the sun, but in the case of the other planets they have determined smaller spheres which do not include the earth, and which they have called epicycles. Thus, an epicycle for instance in the sphere of Saturn is a small sphere that does not include the earth, and is eccentric. It is fixed in the body of the sphere which carries it (i.e., deferent), and in a way embedded in it, so that the axis of the epicycle is parallel to the two surfaces of the deferent, that is, that the surface of the epicycle is tangent to the two surfaces of the deferent, so that it fits in exactly.

The star is also a spherical and solid body enclosed by one surface and is fixed in the epicycle [M 30] and in a manner embedded in it, so that the surface of the star is tangent to the surface of the epicycle in a point halfway between the two poles [i.e., on the equator of the epicycle]. In other words, the star is completely within the epicycle and its surface is tangent to the surface of the epicycle.

The Sphere of Saturn has been briefly described and the arrangement of its parts and the nature of its form and structure have been indicated. The Spheres of Jupiter, Mars and Venus are the same as that of Saturn in every respect. Each is a spherical body bounded by two parallel surfaces. The convex surface of each one is tangent to the concave surface of the celestial sphere above it and its concave surface is tangent to the convex surface of the heaven below it. Thus the center of the two surfaces is the center of the sphere, which is also the center of the cosmos. The parecliptic, deferent and epicycle were hypothesized, as was the case with the sphere of Saturn, within the (resulting) spherical shells. In this way, the disparate phenomena that were observed were reconciled. All of these spheres rotate from east to west in the first motion and from west to east in the second motion; and, according to one view, they also experience trembling and shaking as a result of an accidental motion. Aside from that, each one has a motion particular to itself, the amount, direction, and causes of which are described in the books of astronomy, while the period of rotation has also been briefly indicated in the text of this book. Since these four spheres share the same structure, with no differences among them, a description of one sufficed for them all.

Next is the Sphere of the Sun, which is the fourth sphere. The sun is the greater luminary, the most prominent and most luminous of all the stars. The nights, days, months, and years are determined in regular order by its motion. In the orderly progression of the seven planets, it occupies a place of repose in the middle, as though it were a lantern of light that radiated upward and downward. Its sphere is simpler than those of the other planets, all of its behavior being uniform and accounted for by the two spheres known as concentric and eccentric.

The center of the sphere of the Sun is absolutely the center of the cosmos, i.e., the Greatest Sphere. It is a spherical body enclosed by two parallel planes that includes the earth. Its convex surface is tangent to the concave surface of the sphere of Mars above it and its concave surface is tangent to the convex surface of the sphere of Venus below it. Because this sphere is similar to the sphere of the Constellations (Zodiac) with regard to its center, cincture, and poles, it is called parecliptic (*mümeṣ̱ṣel*, lit. "resembling"). However, because astronomers observed that in its motion the sun is sometimes fast and sometimes slow, and that its body is sometimes small and sometimes larger, a difference of opinion arose as to whether it is sometimes closer or farther from the center of the world. In order to resolve this difficulty, an eccentric sphere was hypothesized within the parecliptic. This secondary sphere is a spherical body enclosed by two parallel surfaces in the shell of the primary sphere. It includes the earth and its center is outside (or eccentric to) that of the cosmos. Its convex surface is tangent to the convex surface of the sphere of the parecliptic at a common point called the apogee that is the farthest point from the center of the world. When the sun meets this point it has reached the furthest distance from the center of the earth. The concave surface of this sphere is tangent to the concave surface of the primary sphere at a common point called the perigee, [M 31] which is the closest point relative to the center of the world. When the sun meets this point, it has reached the extreme closeness and lowness to the center of the earth. As recorded above, when the eccentric is separated and quasi removed from the body of the sphere of the concentric, there necessarily remain two spheres whose surfaces are not parallel and some of whose parts are thicker, some thinner. When these two spheres are added to the secondary sphere, the result is the complete first sphere; thus, because it is considered one single sphere, these [smaller spheres] are called supplementary. The sun, then, is a solid spherical body bounded by one surface. It is attached to, and quasi embedded, in the body of the eccentric halfway between the poles (i.e., at the equator), so that the axis of the sun is parallel to the (surfaces of) the spherical shell (that is) the eccentric, and the periphery of the sun is tangent to the surfaces of the eccentric in two points.

Next is the sphere OF VENUS. According to the most correct view, it is the third of the nine spheres. There is strong evidence that it is below the sphere of the Sun. Since the form and structure of this sphere are similar in every respect to those of the upper spheres (Mars, Jupiter, and Saturn), its features can be judged by analogy to them. It is similarly arranged in three spheres: 1) The parecliptic, which is similar to the sphere of the Zodiac and is the general sphere. It is a spherical body enclosed by two parallel surfaces that includes the earth and whose center is the center of the world. Its convex surface, according to the correct view, is tangent to the concave surface of the sphere of the Sun above it and its concave surface is tangent to the convex surface of the sphere of Mercury below it. 2) The deferent, which is situated within the thickness of the first sphere, i.e., the concentric, and whose center is hypothesized to be inclined away from the center of the world. 3) The epicycle, which is situated in the body of the (sphere of the) deferent and whose surface is tangent to those of the deferent. The star of Venus is fixed in the thickness of this sphere. The circumstances, position, and structure of this sphere are evident from those of its congeners, the upper spheres, so we may simply refer to them and skip the detailed information about Venus.

Next is the sphere OF MERCURY, the second of the spheres. Its position, structure, and parts (or epicycles) are somewhat different from the other spheres, so we will describe it in some detail.

The astronomers and philosophers have posited for Mercury three large spheres that include the earth and one small sphere that does not include the earth. The observed anomalies in its behavior are shown to be regular by the action of these four spheres. They are:

1) The parecliptic (*mümes̱s̱el*, lit. "resembling"), so called because with respect to pole, cincture, and motion it resembles the sphere of the Zodiac. Like the other parecliptics, it is a spherical body enclosed by two parallel surfaces that includes the earth and whose center is the center of the world. Its upper surface is tangent to the lower surface of the sphere of Venus above it; and its lower surface is tangent to the upper surface of the sphere of the moon below it.
2) The director is the first of the two eccentric spheres discussed below, and encloses the second one. It is the second of the four spheres of Mercury. It was called the director (*mudīr*, lit. "rotator") because it rotates the center of the second, enclosed eccentric. Like the other eccentric spheres that, as explained above, are located within the thickness of the parecliptic, [M 32] this sphere of the director is also located in the thickness of the parecliptic. Its convex surface is tangent to the convex surface of the parecliptic at a common point between them—perceptible but indivisible—called the apogee. Its concave surface is tangent to the concave surface of the parecliptic at a common point between them—perceptible but indivisible—called the perigee. The center of this sphere is outside the center of the world, as will be explained below.
3) Then, the second of the two eccenters, that is, the third sphere of Mercury, which is outside (or nonconcentric with) the first, the deferent of the center of the epicycle, and encompassed by the director. Just as the director is in the thickness of the parecliptic, this one is also fixed in the body of the director. The convex surface of the deferent is tangent to the convex surface of the director at a point called the apogee. The concave surface of the deferent is tangent to the concave surface of the director at a point called the perigee. The center of the deferent is eccentric to the center of the director.

The amount of its inclination and eccentricity was explained above.

4) The epicycle, the fourth sphere of Mercury, a lesser sphere that does not include the earth. It is embedded in the thickness of the deferent, i.e., between its two surfaces, such that the axis of the epicycle is parallel to the thickness of the deferent; and the surface of the epicycle is tangent to the surface of the deferent below and above.

The star of Mercury, which is a solid spherical body, is embedded in the thickness of the epicycle, as has been frequently illustrated and described. Accordingly, when the sphere of Mercury is shown in its total shape and structure with regard to this kind of position, arrangement, and division, [two] apogees and two perigees appear. Because one of them is like a part of the pareclip-tic, it is called the apogee of the parecliptic; and because one of them is like a part of the director, it is called the apogee of the director. The perigee of the pareclip-tic and the perigee of the director are analogous to this. In like manner, because the two eccentrics have four complementaries, two are said to be of the parecliptic for the director and two of the director for the deferent.

Next is the sphere OF THE MOON, the first of the spheres. Because its general sphere is divided into various parts (or epicycles), and its form and structure are rather different from the other planets, we will describe it in some detail.

Now, the sphere of the moon also comprises four spheres. The center of the first two is the center of the world. The center of the third is eccentric; it is the deferent (carrying) the epicycle. The fourth sphere is the epicycle.

1) The first of those that have the same center as the center of the world—it also encompasses the second—is known as the *jawzahr*, because the point of the *jawzahr* is on its periphery, and is known as the parecliptic because it resembles the sphere of the Zodiac. The convex surface of this sphere is tangent to the concave sphere of the sphere of Mercury above it, and its concave surface is tangent to the convex surface of the second of its own spheres.

2) The second of the first two spheres—the one that is encompassed by the first—is called the oblique because its cincture inclines in relation to the Zodiac and to the parecliptic circle. This oblique sphere is fixed within the *jawzahr*, i.e., within its lower surface. Its convex surface is tangent to the concave surface of the first sphere and its concave surface is tangent to the sphere of fire, which is one of the four elements.

3) The sphere of the deferent, which is in the thickness of the oblique sphere, as described above. [M 33] Its center is outside of (or eccentric to) the center of the world.

4) The epicycle, which is set in the body of the sphere of the deferent.

The moon is a solid spherical body. As described above with regard to the other planets, it is implanted in the thickness of the epicycle. Thus, the structure of the sphere of the moon has been drawn here so that it will be perceived as it is customarily observed, with its four spheres—the parecliptic, oblique, deferent, and epicycle—in the form of mathematical figures.

The astronomers who belong to the first school—the corporealists—when discussing the celestial spheres, abstract the problems from the evidence and, as described above, feel the need to discuss the spheres as bodies. So this kind of astronomy is called corporeal astronomy. As mentioned above, they have determined that there are 24 spheres, viz., the greatest sphere, the sphere of the fixed stars, two spheres of the sun, three spheres each for Jupiter, Saturn, Mars, and Venus, and four each for Mercury and the moon. Their figures have been depicted above.

The author of the *Almagest*, those who follow him, and most geometricians, when discussing the celestial spheres, are satisfied with circles and claim that this much is sufficient to provide proofs of celestial phenomena. So this science is called non-corporeal astronomy. For the eighth and ninth spheres they posit two intersecting circles that are the cinctures of those spheres.

For the sun they also posit two circles, the parecliptic and the eccentric, that are tangent at the point called the apogee. According to this view, there is no epicycle for the sun. But according to those who believe that it does have an epicycle, there are three circles: two deferents, one that is concentric and one that is eccentric, that intersect; and the epicycle, whose center coincides with the periphery of the deferent and whose periphery is tangent to the eccentric.

For each of the upper planets (Mars, Jupiter, and Saturn) and for Venus they posit five circles: the parecliptic; the oblique; the deferent; the equant, which intersects the deferent; the epicycle, whose center coincides with the periphery of the deferent.

For the moon they posit four circles: the parecliptic and the oblique, which intersect; the deferent, which is tangent to the oblique at the apogee; the epicycle, whose center coincides with the periphery of the deferent. For the moon, most astronomers do not posit a (separate circle) which carries the center of the deferent, because it is tantamount to the oblique. But some of them do mention it.

For Mercury they posit six circles: the parecliptic; the oblique; the deferent; the equant; the epicycle; the deferent of the center of the deferent. They do not posit a director, since the circle which carries the center of the deferent occupies its place; but those who do posit a director say that it is tangent to the oblique.

According to most geometricians, who only posit spheres in connection with circles, there are 34 spheres if there is no epicycle for the sun, otherwise 35. According to those who posit a deferent of the center of the deferent for the moon and a director for Mercury, the number is 37. For practitioners of corporeal astronomy, the number is 24 in either case—i.e., whether or not there is an epicycle for the sun. But in order to determine their motions, the practitioners of corporeal astronomy also need some extra spheres. [M DIAGRAM] [M 34] I have attached here the forms of the spheres according to the view of the corporeal astronomers; and in order to explain the spheres, I have added the figures drawn by the geometricians who only posit spheres in connection with circles.

2 (The school of Pythagoras and Plato among the ancient philosophers, and the astronomer Copernicus among the moderns)

As previously indicated, when the leading authorities in philosophy and astronomy, both ancient and modern, discovered and described the structure of the universe, consisting of the spheres of the ethereal and elemental bodies, and how they are arranged, their positions, and their motions and quiescences, they were divided into various schools and three groups emerged. The main outline of the authoritative view, preferred by the first school, was recorded above.

The philosophers who preferred the second school placed the luminous sun right in the middle of the cosmos, making it the center of everything. They thought that the earth revolved around the sun and that the heavens were quiescent. They sojourned in this valley and made inquiries in this path in order to support their claim and to arrange their ideas. The following is an exposition of their views concerning the upper and lower bodies contained in the sphere of the cosmos.

However, there is no need to reiterate the point that this is not a matter of religious belief or a requirement of nature. Indeed, while considering the views of this group to be erroneous and abhorrent, I have simply recorded them here in summary fashion and have attached the figures and forms that they have drawn in order to expound and illustrate their ideas concerning the nature of the celestial and earthly bodies. Thus, may the heroes of the endless field of virtue and learning and the experts in the vast plain of the sciences—the distinguished men of our age who are zealous in the pursuit of knowledge—provide the necessary rebuttals of their ideas after examining them and demolish their baseless views with clear proofs and strong evidence. In order to facilitate this desire, I have left ample margins outside the lines of these pages where they may write notes that will enhance the reputation and strength of the first school and will embellish this excellent book.

The Muslim philosophers have also ceaselessly rebutted this unworthy school, clearly setting forth their refutations in the books of astronomy. But no matter how baseless and false these views were, being the school of the ancient philosophers, they found supporters in every age and many books and treatises were written with the notion of defending them. However, because repeating the contents of these books here would needlessly extend our subject, I have not gone into detail in this matter and have summarized them.

In the hope of accomplishing the same purpose, the recent philosopher and author named (Edmond) Pourchot (1651–1734), one of the philosophers of the Christian nations, [M 35] in the third chapter of the third volume of the book he wrote discussing the views of the ancient and modern philosophers, summarized the views of the three schools on the nature of the corporeal world. First, he described who were the supporters of this baseless school and spokesmen of this false method, from ancient times to the present. Second, in order to clarify and illustrate their views on the arrangement, motion and quiescence, and nature of the celestial and worldly bodies, he drew pictures of the structure of the world according to their view. Third, in order to refute their arguments, he compiled the objections raised by their opponents, the evidence that they brought forward, and their questions and answers. I give here a translation of the third chapter of this book with the accompanying diagrams. The essence of what he stated is as follows:

In every period from ancient times to the present, there have always been people who were of the opinion that the earth moved. This was recorded and explained in various works. Indeed, Aristotle, the leading figure among the peripatetics, examined this matter in his books and stated that Pythagoras and his followers were the first to have this view. The world-illuminating sun, which is composed of fire and is the most perfect of all the elements and the center of all the celestial and worldly bodies, they placed at the exact center of the universe, encompassed and immobile, and made the earth move around it like one of the planets.

It is evident that this doctrine does not contradict the other doctrine attributed by Aristotle to Pythagoras and his followers, that the ethereal bodies move and with their motions create a song sung in tune by a single voice and harmony and order in their revolutions. This is because, according to Pythagoras and his followers, the earth is considered to be one of the planets and moves like them; and together with the other celestial spheres, with their particular motions, they maintain perfect order and harmonious arrangement. The harmonious song sung in tune in a single voice issuing from the spheres and the orderly harmony in their revolutions are merely allusions to the maintenance of perfect order and arrangement in their motion.

According to the report handed down by Archimedes, the philosopher Aristarchus agreed with Pythagoras on the motion of the earth. And according to Plutarch, a different Aristarchus claimed that this was also the view of Cleanthes of Samos who concluded from his astronomical observations that the celestial spheres were immobile and that the earth rotates on its own axis and also revolves on an oblique circle. Indeed, a group of Greeks denounced this philosopher as a heretic opposed to the Christian religion. Latin poets satirized him in their verses, likening the earth to a veiled virgin (Vesta) who staunchly maintained her chastity, did not associate with unseemly individuals, did not intend to accompany them in their journeying, and had no desire as yet for the embraces of lovers. [M 36]

Because their doctrine was contrary to the comprehension of the simple-minded and opposed to what they saw with their own eyes, the common people were completely unreceptive to the idea that the earth revolved around the sun in a circular path and treated its adherents with contempt, saying that those who defended such views had strayed from the faith and were outside of Christianity. Nevertheless, very many of the ancients adopted this view and it was maintained until the time of the recent astronomers. The historian Plutarch asserted that Plato also adopted this view toward the end of his life. Indeed, some of them constructed cells and houses in imitation of the structure of the cosmos, as they conceived it, and lit fires and lamps in the middle of their houses and cells in order to fully illuminate their buildings with light from the distant fire.

As observations continued to be made over the centuries and celestial phenomena were more accurately recorded, and more recently, with the refinement of the laws and instruments and the increase in experimentation and observation, as the understanding of celestial phenomena has become more orderly, this new view acquired so much currency that it won favor with Cardinal (Nicolaus) Cusanus (1401–1464), a great monk who was highly esteemed among the Christian nations and known for his great learning. While previously he had been determinedly and zealously opposed to it, when he was won over he became a fervent adherent.

Shortly thereafter, a Polish monk named Copernicus, who had an excellent knowledge of mathematics and was the most distinguished and learned scholar of his age, worked very hard on this matter. For thirty years, from 1500 to 1530 of the Christian era, using astronomical instruments and applying mathematical principles, he exerted great effort to confirm this (new) view. From his time until the present, this view has been named after him and is known as Copernican astronomy. Most recent scholars in the European countries have accepted this view. In particular Descartes, who was the most distinguished and learned of the modern philosophers, supported this view with some new and excellent principles that he invented in his new philosophy, and as a result it was named after him and became known as Cartesian astronomy.

This new astronomy, in outline, is as follows:

As mentioned above, the luminous sun is a fixed star located right at the center of the world and encompassed by something. Next to it is located the sphere of Mercury, or as they put it, the circle of Mercury, which is close to the sun and contains the body of the sun. The planet Mercury revolves around the sun, completing its circle in three months. Next is the circle of Venus, which surrounds the circle of Mercury; the planet of Venus completes its circle in eight months.

Next it has been established that there is a wide circle surrounding the circle of Venus. The earth, surrounded by the elements of air and water, is like a planet and [M 37] traverses that great circle in one year. Also located in this great circle is the circle of the moon, encompassing the body of the earth. The moon has the earth for its center and revolves around it, completing its circle in one month, similar to the position and motion of a star in the epicycle (in the corporeal system).

Next is the circle of Mars, which surrounds the great circle of the earth; the planet Mars completes the circuit of its own circle in about two years. Surrounding the circle of Mars is the circle of Jupiter; the planet Jupiter completes its own circle in 12 years. And surrounding the circle of Jupiter is the circle of Saturn; the planet Saturn completes its own circuit in 30 years.

In addition, just as the moon, as mentioned, within the great circle of the earth makes the earth its center and revolves around it, so four planets make Jupiter their center and revolve around it, and five planets make Saturn their center and revolve around it. These planets, observed by recent astronomers, have been newly named and the term used to describe them is "little moons".

After all of these comes the sphere—or, as they call it, the circle—of the fixed stars, which encompasses all of the other circles. The thickness of this circle is a vast space embellished with countless stars. Descartes imagined each of the fixed stars to be like a luminous sun of great mass. According to this new view, it is imagined that many stars (planets) revolve around the body of each of these fixed stars in the same way that planets revolve around the sun, which is the center of the world and illuminates it, as described above. I have drawn here the form in which the structure of the cosmos is imagined and presented according to this view.

According to this new view it is evident, first, that the earth moves from west to east on its own great circle, traversing the circle of the zodiac in a regular manner from day to day and completing its revolution in one year; and second, aside from its annual motion from west to east, it rotates around its own axis, completing this circuit in one day. The alternation of night and day results from this motion. Because of the earth's diurnal motion from west to east, it seems to us that the sun and all the stars move from east to west on a daily cycle. As an example of these kinds of motions by the earth, assume that a solid and perfectly round sphere (that is spinning around) is propelled along a perfectly flat plane: it will continue turning on its axis as it rolls the entire course of the plane. In like manner, the earth continuously turns on its own axis and around its own center while it moves in its own great circle from west to east, completely traversing the course of the circle of the zodiac.

Now, the earth is between the zodiac and the sun. Whenever it is in conjunction with one of the constellations of the zodiac, the sun appears in the constellation opposite that one. For example, when the earth is between Aries and the sun [M 38] and is in conjunction with Aries, the sun appears in Libra, which opposite Aries. When the earth is in Cancer—i.e., in conjunction with Cancer—the sun appears in Capricorn, which is opposite Cancer. In short, when the earth is in conjunction with one of the northern constellations, the sun appears in the southern constellation that is opposite it, and vice versa. As long as the sun remains in the northern constellations, earth is in the southern constellations. The reason is that as the earth moves through the southern constellations, it departs from its center (i.e., is farther from the sun) and widens its circle, and so it must necessarily tarry while moving in the south.

It is also evident that, according to this view, the earth's axis is located parallel to itself and to the axis of the equatorial plane, i.e., the axis of the cosmos. For, if it were parallel to the axis of the zodiac, night and day would always and everywhere be equal, and there would never and nowhere be an alternation of the four seasons. But because the earth's axis is parallel to the axis of the cosmos, it deviates 23½ degrees from the axis of the zodiac and inclines toward the plane of the zodiac at an angle of 66 degrees and 30 minutes. And because it always maintains its position parallel to the axis of the cosmos and rotates while continuously inclined toward a particular and perceptible side of the sphere, the seasons must necessarily alternate with the annual motion of the earth as it revolves around the sun. This can be observed on the globe manufactured according to the view of Copernicus, and can also be envisioned on this map projected on a plane.

Let us suppose that the fixed sphere (stars) consists of the circle BCDY and that the orbit that is traversed by the annual motion of the earth consists of the circle WKAH. Let us place the sun at the center of this circle. When, for example, the summer season arrives, i.e., when the sun is in Cancer and the earth is in conjunction with Capricorn opposite it, the earth is located at point K. The line S M, which is the earth's axis, is parallel to the axis of the cosmos. It deviates 23 degrees and 30 minutes from the axis of the zodiac and inclines toward the plane of the orbit of the earth 77½ degrees at the angle BKH. In this configuration, the sun's rays are perpendicular to the earth, i.e., the rays coming from the center of the sun to the center of the earth do not reach the surface of the earth at the equator; rather, they reach it 23 degrees and 30 minutes from the equator toward the North Pole at the Tropic of Cancer. When this occurs, they illuminate the northern hemisphere and abandon the southern hemisphere.

Later, at the beginning of autumn, the earth moves to point A. When this happens, the line SM, which is the earth's axis, remains parallel with itself and with the axis of the cosmos. [M 39] At this time the earth is in conjunction with Aries. When the sun is in Libra, its rays, which go out from the center of the sun toward the center of the earth, are perpendicular to the axis of the cosmos and reach the surface of the earth at the point of the equator considered as the first of Libra; so they extend equally over both hemispheres.

Then, at the beginning of winter, when the earth reaches point H, the axis SM remains parallel, exactly as it was before, while the rays of the sun now reach the earth perpendicularly at the Tropic of Capricorn, illuminating the earth's southern hemisphere and abandoning the northern hemisphere.

Later, at the beginning of spring, when the earth reaches point W in its annual motion—i.e., when it reaches the first of Libra—the sun appears in Aries. The rays of the sun, which go out from the center of the sun toward the center of the earth, reach the surface of the earth at the point of the equator that is the beginning of Aries. In this configuration, the rays of the sun again illuminate both hemispheres equally. In this case, however, since the illuminated side of the earth is facing the sun, it is not visible to us who are outside the diagram.

The alternation of day and night and their being longer and shorter, and the alternation of the four seasons are explained in this way in Copernican astronomy.

It it evident that a part of the circle of fixed stars—for example, part BC or part DY—that is parallel to the annual circle of the earth, appears like a tiny dot because of its extreme distance from us. At whatever point of its great circle the earth may be, its axis must always face the same point of the circle of the cosmos, i.e., the circle of the fixed stars, and the height and elevation of the pole of the earth must always appear in exactly the same direction and face the same star at the same magnitude over our heads. However close the earth actually is to this or that star, and whether further to the south or to the north, in its annual motion through the zodiac, i.e., in its own great circle, relative to the circle of the fixed stars which is in any case very far from us, the great circle of the earth that is traced by the annual motion of the earth will appear as a dot. While those who venture to place a limit on divine power might find this hard to believe, when carefully considered it is not, in fact, implausible.

Also, those who cannot accept the earth's motion, which is a concomitant of the Copernican view, must necessarily accept many other things that are far more difficult to believe, but are concomitants of the Ptolemaic view. One of these is the incredible speed of the Primum Mobile, i.e., the sphere of Atlas, which completes its circuit every day from east to west despite its huge extent and magnitude. Another is the motions of the celestial spheres and the speed of these motions. For each of them moves, in accordance with its own nature, from west to east, contrary to the sphere of Atlas within which they are all ensconced, while at the same time moving in conformity with the sphere of Atlas from east to west during its twenty-four hour circuit. The speed of these motions is 300,000 or 400,000 times faster than a bullet fired from a gun. [M 40] The Copernicans say that it is much easier for the earth—which is a small body and can move fast because of its spherical shape—to revolve around the sun than for the sphere of Atlas—which is a large body and whose convex plane is still of unknown shape—and the other spheres to fly around the earth which, compared to them, is a mere speck.

Another claim of the Copernican view is this: As the earth moves in its great annual circle, its axis always maintains the parallelism discussed above; but this does not mean that the earth's axis never and in no way changes its position. For the earth's axis moves so slowly that it traces a circle around the pole of the zodiac in 25,816 Coptic years (or "Egyptian years"—i.e., solar years of exactly 365 days). The radius of this circle is 23 degrees and 40 minutes. It necessarily results from this motion of the earth that the point of intersection between the circle of the zodiac and the equinoctial line, which is the point at which night and day are equal, must transit from east to west, contrary to the alternation of the constellations (or signs of the zodiac). It is for this reason that this motion is called the PRECESSION OF THE EQUINOX. It results from this motion of the earth that the motion of the fixed stars appears from west to east—i.e., the alternation of the constellations (or signs of the zodiac)—and that change distance from the point of the equinox toward the east becomes visible.

Furthermore, this motion of the earth's axis is not uniform but variable. For, from the time of Timocharis of Alexandria (ca. 320 BC–260 BC) to Ptolemy, the motions of the fixed stars were observed to vary about one degree every 100 years according to the order of the signs of the zodiac. From Ptolemy to Albatenius (al-Battānī, d. 317/929), the amount was determined to be one degree every 66 years. At the present time, according to the observations of Tycho Brahe, they vary one degree every 70 years. This being the case, they complete a circuit once every 25,816 Julian years. According to the observations of (Giovanni Battista) Ricciolo (1598–1671), it varies one degree every 72 years, and thus they complete a circuit once every 25,920 Julian years.

Now, while the fixed stars are essentially stationary and quiescent, in order to account for the apparent variance of their motion, Copernicus imagined the variance and inequality to this motion called the precession of the equinox and calculated that this variance and inequality was completed in 1717 Coptic years. However, the astronomers named Thebitius (Thabit b. Qurra, 826–901) and Alfonso (Alfonso X of Castile, 1221–1284) observed in the circle of the zodiac the variance and disturbance of the obliquity of the ecliptic and calculated that this variance and disturbance occurred in twice the time of the variance of the precession of the equinox, i.e., it completed in 3434 Coptic years. In order to resolve this problem, Copernicus explained the variance and disturbance by the motion of the earth's poles. [M 41]

We can envision this in the following way: The earth's axis moves only 24 minutes from north to south and again from south to north, and 2 degrees and 20 minutes from east to west and again from west to east. Thus the tip of the earth's axis traces a bent and twisted circle with this kind of bent and twisted motion. This occurs in twice the time of the variance in the precession of the equinox, and in the same period of time as the variance in the obliquity of the ecliptic that occurs in the circle of the zodiac, following the sequence of 1, 2, 3, 4, 5, 6, 7, 8, 9 as seen in the diagram. Thus, as the earth's axis moves and turns in the reverse order of the signs of the zodiac in 25,816 years, this circle of the pole, which is created by these seven, or perhaps more, segments, is fully traced around the circle of the zodiac.

Furthermore, in the Copernican view, the stasis and forward and retrograde motions that are observed in the planets can be easily understood and envisioned without recourse to the epicycles posited by Ptolemy. These are only features that appear so to us, not actual features of the external world. For we are on the surface of the earth, and when we look at the planets they sometimes seem to stand still or move forward or even move backward (relative to the fixed stars). But suppose we were on the sun, which is hypothetically the center: then this sort of illusion would never appear. That is because, as described above, the orbits of Mercury and Venus around the sun come before that of earth, and earth's orbit comes before those of the other three planets, Mars, Jupiter and Saturn. So sometimes Mercury and Venus are between the sun and earth and earth is between the sun and the other three planets.

Taking Mars as an example (of the upper planets), let us suppose that the sun is at point A, the earth's orbit is BHʿACTL, and the orbit of Mars is ṬDQRYP. By the time that Mars has traced an arc of this circle, earth has completed its orbit. Then let us suppose that the circle of the fixed stars is MFLaN. Thus, when earth is at point L and Mars at point Ṭ, then Mars appears at point M on the circle of the fixed stars. Now let earth transit from L to B and Mars from Ṭ to D. Then earth has transited to a point closer to being between Mars and the sun and Mars appears at point La on the sphere of the fixed stars. Since in this configuration the planet is perceived to hasten from point M to point La, following the order of the zodiac, this is called "forward motion." Then, when earth goes from B to H and Mars from D to Q, Mars is again perceived to be at point La; this is the "first stasis." When earth reaches ʿ and Mars reaches R, then Mars is perceived at point F [M 42] and thus appears to have gone backward, contrary to the order of the zodiac; this is called "retrograde motion." When earth reaches C and Mars reached Y, Mars is again seen at point F; this is the "second stasis." When earth transits to T and Mars to P, then Mars appears to be at point N and to have moved following the order of the zodiac; this is (again) called "forward motion." The stasis and forward and retrograde motions of the three upper planets occur in like manner.

As for the two lower planets—i.e., Mercury and Venus—because they are closer to the sun than the earth is, and so complete their orbits faster than the earth does, and because they come between the earth and the sun as they trace their orbits, they sometimes exhibit stasis or forward motion or even retrograde motion. Taking Mercury as an example, let us suppose that by the time earth traces the arc of its orbit represented by TBCDYF, Mercury has completely transited its orbit ṬNMLLa. Thus, when earth is at point T and Mercury is at point La, then Mercury must appear at point H on the orbit of the fixed stars. Later, when earth reaches B and Mercury reaches L, then Mercury appears at point P on the orbit of the fixed stars. Since it has moved following the order of the zodiac, it is called "forward motion." Then, when Mercury reaches M and earth reaches C, the planet again is perceived to be at point P; this is the "first stasis." When earth reaches D and Mercury reaches N, the planet is perceived at point Q, it goes backward counter to the order of the zodiac. Next when the earth reaches point Y and Mercury reaches point Ṭ, the planet will be perceived to be at point Q and thus appears to have gone backward, contrary to the order of the zodiac; (this is called "retrograde motion.") When earth arrives at point Y and Mercury at Ṭ, the planet still is perceived at point Q; this is the "second stasis." Finally, earth reaches point F and Mercury point La, so that the planet is now perceived at point ʿ and to have moved following the order of the zodiac; this is (again called) "forward motion." The description that we have given of Mercury applies to Venus as well, except that these alternations occur more slowly in the case of Venus since it takes longer to complete its orbit.

The explanations and examples given thus far suffice to illustrate this new—or rather, very old—view (i.e., Copernican or heliocentric astronomy). Our exposition of it being finished, we now turn to addressing the objections that have been directed against it.

FIRST, it was objected that this new view contradicts many of the pronouncements in scripture (lit., the heavenly books) [M 43] and any idea of this kind deserves no consideration whatsoever.

IN RESPONSE, they reject the major premise and argue as follows: If it is said that it is not worthy of consideration as statement of fact, as much as this will be conceded, but it will be of no use (i.e., besides the point). And if it is said that it is not worthy of consideration as the expression of a hypothesis, then this [conclusion] is not permitted.

Descartes rejected the minor premise as well, arguing as follows: Even according to this new doctrine, the earth in itself is not characterized by motion; what actually moves is the vortex that surrounds the earth and that is made up of that soft matter (ether). Furthermore, also according to certain laws that Descartes established, the earth must move more in accordance with the view of Ptolemy than that of Copernicus, since it is always surrounded by certain particles of soft matter that make up its vortex. Just as a man on board a ship remains motionless, so the earth is always motionless within certain particles of that soft matter. Since in Ptolemy's view the particles of air that surround the earth are continuously turning by being renewed and interchanged, the earth must also move according to Descartes' hypothesis.

Furthermore, there are many things in scripture based merely on what we see and having no connection to religious and ethical matters. For example, in the Torah the moon is called the great lamp (cf. Genesis 1:16, where the sun is called the great lamp and the moon the small lamp) while in fact the moon is not only smaller than the other planets but also receives its light from the sun. Another example from the Old Testament is the clear statement that the earth is always stationary (cf. Psalms 93:1, 96:10, 104:5). According to its meaning in context the statement is true, because it is preceded by the statement that one of creation goes and one comes; so the full statement is: "One of creation goes and one comes, but earth remains stationary." Evaluated in context, the statement that earth always remains stationary means that it always remains as it is, that in its totality it does not change no matter how much coming into being and passing away might occur on it. For this meaning one can compare the psalmist's (lit. David's) phrase "settled in its foundation" meaning that the earth is stationary or solidly constructed, as explained in the Hebrew exegesis, or that it is firmly established, as in the older translation. That is because the earth as a whole neither dissolves nor disappears and is not subject to any kind of corruption. Based on Job's statement about God (Job 9:6), "He makes the earth move from its place and its pillars shake and tremble," there are also those who deny that the earth is fixed in the middle of the world and is always motionless. [M 44]

SECOND, it was objected by resorting to the laws of astronomy that if the earth were far from the center of the cosmos and moved in its own great annual circle, the height of the north pole, for example, would not always be the same; and the stars directly overhead would not always be visible; and one half of the celestial sphere would remain invisible to us. In addition, the apogee and perigee would not be determinable. Proofs of the assertions that occur in these objections are known to those with even a slight understanding of optics and geometry.

IN RESPONSE, they say that if the earth moved as has been described, the height of the North Pole would always be the same; and the stars directly overhead would always be visible; and one half of the celestial sphere—i.e., the six visible constellations—would always be completely in our view. However, this would be contingent on our always remaining at a certain place on earth. That is because, as we have stated before, the sphere of the fixed stars is so far from us that next to it the earth's orbit appears as a dot, because the earth's axis is always parallel to the axis of the cosmos.

The reason we stipulated, with regard to these three matters, that we always be at a certain place on earth, is that the height of the North Pole, which is always in the same configuration, depends on our point of view and is constant as long as we are at a certain place on earth, i.e., as long as we do not change horizons and do not lose sight of or deviate from a certain point overhead. For example, if we move on the surface of the earth from north to south or from south to north, thus changing our position as well as our specific horizon and the specific point overhead, then another segment of the celestial sphere comes into view that we were not able to see before, and the segment we previously saw is now completely out of our view. In that case the height of the pole and the stars directly overhead are different.

As for the apogee and perigee not being determinable: During its orbit, when the earth comes to its farthest position from the sun it reaches its apogee and when it comes to its closest position to the sun it reaches its perigee. This is exactly like Ptolemy's view, except that according to Ptolemy, coming to a position near or far depends on the motion of the sun, while in this view it is the earth that moves.

Next, going back and rejecting the premise that is the basis of these responses, they say it is too implausible to believe that the sphere of the fixed stars is so far from us that next to it earth's orbit appears as a dot. In reply I say that implausibility is weak grounds for disbelief, and since the ordering of the universe, which is the intent of that minor premise, is achieved, that sort of implausibility and disbelief does little harm in this sort of science.

THIRD, it was objected based on the laws of physics that: [M 45]

- The lowest of all places is the center of the cosmos and therefore the earth, which is the heaviest of all bodies, should rest at the lowest of places.
- If the earth moved it would certainly be perceptible. Buildings and trees would be turned upside down. Heavy bodies would not fall perpendicularly to the ground, because the point that would be reached by falling straight down would move with the surface of the earth. While the birds were flying in the air, the earth would move their nests, so how would the birds be able to find their nests again if the earth took them with it.
- Of two bombshells fired at the same time, one toward the west and one toward the east, the first would travel much faster than the second, since the one going west is traveling in the opposite direction of the earth and thus will cover a greater distance than the one going east (in the same direction as the earth).

IN RESPONSE, taking the objections one by one:

- It has not yet been determined whether the earth is the lowest of all places. The astronomer Galileo refuted the proof of this hypothesis that Aristotle offered in his work *On the Heavens*.

Aristotle's argument was as follows: There is no doubt that heavy bodies are moved toward the center of the cosmos while light bodies are being moved away from the center of the cosmos. But heavy bodies seek the center of the earth while light bodies flee from the center of the earth. Therefore the center of the earth is also the center of the cosmos.

Galileo's refutation is as follows: The major premise in this argument either is null and void or is at least dubious. That is because, while we do observe that some heavy bodies move toward the center of the earth, it is neither obvious that they are brought to the center of the cosmos nor that Aristotle has proven this. In fact, if one reflects a bit, it is clear that this major premise begs the question.

Furthermore, it has not yet been determined whether the earth is heavier than the other heavenly bodies. How is it that the moon, for example, does not rest on the earth nor does the earth rest on the moon? Nor can it be said of these bodies that, of their own nature, some are higher and some lower; rather they are higher or lower in relation to us. While it is true that large stones and other such heavy bodies return to earth after being flung from it, that does not mean that the entire sphere of the earth must move from its position like a heavy body.

- Because we are ensconced together with the earth in that soft matter (ether) mentioned above, and move with the earth as in the current of a river, we cannot perceive the earth's motion. This also explains why buildings and trees are not overturned—indeed it proves why they must remain standing.

Then, whether the earth is stationary or whether it moves together with the soft matter, there is no obstacle or difficulty for a heavy body to fall to the ground perpendicularly, because apart from the motion of falling [M 46] it also certainly partakes of the motion of this soft matter. This is exactly like what happens when a stone is dropped from the spar on a ship: we know by experience that it falls to the bottom of the spar, whether or not the ship is moving. The reason is that, apart from the stone's falling motion, it also partakes of the motion of the ship.

Perhaps the true explanation is that neither the motion of the heavy body when it falls to the ground, nor of the stone when it falls on a ship, is ever in a straight line but rather traces a curved line, although it appears to us to fall perpendicularly. This is exactly like what happens when a ball is dropped on a ship. When observed from the deck, it seems to fall vertically. But when observed from outside the ship—i.e., from the shore—two motions are perceived: one is the vertical descent and the other is the joint motion with the ship. So the observer from outside imagines that the ball traces a curved line combining these two motions.

This being the case, just as the fish partake of the motion of the water in the sea, so birds partake of the motion of the air and therefore are not separated from their nests when they fly off.

- The bombshell fired toward the west does not move faster than the one toward the east, since it also partakes of the overall motion of the earth from west to east.

3 The view of Tycho Brahe and his followers

This philosopher's defective view of cosmology was recorded in summary fashion in the fourth chapter of the book mentioned above, and I have translated the gist of it here. He took a position contrary to both the Ptolemaic and Copernican schools and imagined that he had produced a third school more accurate than the previous two. It has become known after him as the Tycho Brahe astronomy. First he agreed with Ptolemy and placed the earth at the center of the cosmos. Second he agreed with

Copernicus and made the sun the center of the motions of all the planets except for the moon, which he said orbits the earth.

So according to this view, of bodies that orbit the earth and make the earth their center, first we have the moon, which is the closest to the earth and completes its orbit in one month; second we have the sun, which is at an intermediate distance from the earth and completes its orbit in one year; and third we have the sphere of the fixed stars, which is extremely distant from the earth and completes its very slow orbit in 25,000 years. Of bodies that orbit the sun, there are five of the planets, each traveling at its own speed: Mercury, which completes its orbit in three years; Venus, in eight years; Mars, in two years; and Jupiter [M 47] and Saturn, in well-known periods of time. In this system, as the sun traverses the circle of constellations (zodiac) in its annual motion, it takes along these five planets. Mercury and Venus, relative to their motion around the sun, do not include earth in their orbits, but Mars, Jupiter and Saturn do. Indeed, when Mars is at point B, it is closer to the earth than to the sun. All of this is illustrated in the diagram attached here.

Brahe said nothing at all about the reason for the observed daily motion of the heavenly bodies, and so his followers had to posit either a single mover, above the sphere of the fixed stars, that turns the entire cosmos in a daily motion from east to west; or else three movers—the moon, the sun, and the sphere of fixed stars—that slowly move from west to east traversing the signs of the zodiac, while at the same time they have a diurnal motion from east to west that is parallel to the circle of the equinoctial line. They imagined further that the five planets, aside from their own separate motions from west to east accompanying the sun as it traverses the zodiac, also have a diurnal motion from east to west with the movement of the sun—which in their view serves as the Primum Mobile—on a plane parallel to the circle of the solstice; and with all this, the axis of their orbits always keep parallel to themselves. While this system cannot be considered completely impossible, the difficulty of realizing it is obvious to anyone who thinks straight.

Regarding the physical nature of the heavenly bodies, there is difference of opinion among the cosmologists. According to the first school, that of Ptolemy, the ethereal bodies are solid bodies—i.e., they [can be described in terms of] volume, hardness, purity, and transparency. They do not expand or contract, do not grow looser or denser, and are not torn asunder or patched together. In their motions, they do not grow stronger or weaker, and do not reverse course or stop. And they never depart from their positions.

According to the second (Copernican) and third (Tychonic) schools, the heavenly bodies lack volume and hardness. They are fluid and soft, and can be torn asunder and patched together. Indeed, the founder of the third school, Tycho Brahe, around the year 1600 A.D., built an observatory in the lofty fortress of Uraniborg commanding the Öresund Strait, on the island of Hven in the Baltic Sea in the northern country of Denmark and declared that the heavenly bodies were soft and fluid; citing as proof of this that comets traveled through the planes of the spheres. He also defended the view that Mercury and Venus were sometimes above the sun and sometimes below it.

According to Tychonic astronomy, the heavens are only three in number: 1) the *empireum*, which surrounds everything and, in their belief, is the place of the blessed; 2) the sphere of the fixed stars; 3) the sphere of the planets, which they divide into orbits according to the different motions of the planets and the relative distances among them. Some posit a fourth sphere, above the sphere of the fixed stars, that serves as the Primum Mobile. [M 48] [M DIAGRAMS] In Copernican astronomy as well, the number of spheres is also limited to three: 1) the system described above, consisting of all the planets and elements; 2) the sphere of the fixed stars which we perceive; 3) another huge sphere that is beyond the sphere of the fixed stars and surrounds it—no matter how boundless in extent the thickness of that sphere is imagined to be—and is specified as the place of repose of the blessed.

After this, in the chapter mentioned above, he (Pourchot) explains the other matters according to the Tychonic system, including the reasons for the alternation of the seasons; the stasis and forward and retrograde motions that are observed in the planets; etc. We will omit these here for brevity's sake, leaving a representation of them to the diagrams; and in order not to prolong the subject, we will not go into the many refutations and responses that arise and will let what we have said suffice.

In translating this passage I have preferred the designations and spellings that are in current use.

Our discussion of cosmology ends here. We will not delve further into this subject, which is an endless sea and includes subtleties and mysteries that are the result of God's creation. Let me conclude with the eloquent words of the poet:

[Persian verse by Ḥāfiẓ]
Talk of the minstrel and of wine; seek less the secret
of Time,

For no one has solved, or ever will, through wisdom
that enigma.
[Verse in Turkish]
No one has solved the problem of the secret of the
celestial sphere,
If there is a solution, the knot of the Pleiades serves
as a question mark.

End of the publisher's supplement.

6 Description of the Circles and Poles

The geographers[1] have posited circles of various kinds and poles on a given sphere (or on the globe). These allow them to describe the earth and the celestial spheres. First, a sphere is assumed to have an AXIS. It consists of a line that begins on one side of the sphere or circle, goes through the center, and ends on the other side. This is also called the DIAMETER—*diametr* in Greek. The two ends of that line are the POLES, upon which the sphere turns—called in Greek *polos*, which derives from the word *poleo* meaning "to turn." On the celestial spheres and on earth, one of these is the NORTH POLE—called in Greek *ar[k]tikos*, which derives from *arktos*, meaning "bear," because the constellation of the Little Bear (Ursa Minor) is there. Opposite it is the SOUTH POLE—called in Greek *antar[k]tikos*, *anti* meaning "opposite" or "facing."

In Latin these have different names, mentioned in the *Atlas*. The North Pole is called *borealis* and *acolonaris*, because the wind called *boreas* and *acolo* blows from that direction; and also *septemtrionalis*, because the *triones* or Seven Stars of the Bear (i.e., the Plough or Big Dipper or Ursa Major) revolve around it. The South Pole is called *australis* and *notius*, because the wind called *auster* or *notus* blows from that direction; and also *meridionalis*, derived from *meridies* meaning "midday." [R 7b]

CIRCLE: a line or plane of which all lines drawn from the central point to the circumference are equal. That central point is called the CENTER of the circumference or center of the circle or center of the sphere. Then there are two kinds of circles: one that divides the sphere in half, called a GREATER CIRCLE; and one that does not divide it in half, called a LESSER CIRCLE. For the sphere of the earth eight circles have been specified, four greater and four lesser. The four greater circles are the equator, the meridian, the ecliptic, and the circle of the horizon. The four lesser circles are the two tropics (Tropics of Cancer and Capricorn) and the two polar circles (Arctic and Antarctic Circles). Five of these circles (the equator, the tropics, and the polar circles) are parallel, i.e., the distance between them is constant. The other three are oblique. [M 49] Two of them—the meridian and the circle of the horizon—are not depicted on the globe but are separate and mobile. The others are depicted on the globe and are fixed. Now we will describe them one by one.

CIRCLE OF THE HORIZON: a greater circle that separates and delimits the visible from the invisible part of the world. For this reason it is called *horizon* in Greek, which means "delimitor," and *terminator* and *finens* in Latin. The lower and upper parts of the globe are delineated by it. It has several categories:

- The true horizon, which divides the globe in half.
- The visible horizon, which varies according to the place of observation of the inhabitants.
- The straight horizon, which is the horizon at the equator where sunrise and sunset are in a straight line.
- The oblique horizon—a category that includes all the horizons except the straight horizon—i.e., one that inclines to one side from the poles and sunrise and sunset are generally not in a straight line.

On earth, the diameter of the visible horizon is 180 stadia, i.e., 22,500 paces, which has been estimated at not more than 5 5/8 German miles. The circle of the horizon has two poles, the *semt-i re's* ("direction of the head") which the European geometers have transformed to zenith and the *semt-i ḳadem* ("direction of the foot") which they call nadir. For the antipodes everything is the opposite, in other words, our nadir is the zenith of our counterparts (antipodes).

EQUATOR: A greater circle on the globe, located on the plane of the (celestial) equator which is the cincture of the greatest sphere (the ninth sphere or Primum Mobile). It is called the equator—*isi merinos* in Greek and *equator* in Latin—because night and day are equal in most of the inhabited quarter of the earth when the sun passes over it in its own motion twice a year. This is when the sun comes to the two equinoctial points—called *equinoctialis*—on the sphere (or globe), one at the beginning of Aries and one at the beginning of Libra.

The equator is the greater circle of the five parallel circles. It is equidistant from the two poles and divides the earth into two equal hemispheres, southern and northern. It traces a path that intersects the circle of the constellations (zodiac) at two places that are the two equinoctial points known as *equinoctialis*. Sailors call this circle simply *linea* meaning "the line".

1 Information in this section largely derives from Cluverius, *Introductio*, 3–15.

Ecliptic (circle of the constellations, zodiac): One of the greater circles traced on the globe. It is also called the circle of the constellations—*zodiakos* in Greek. It intersects the equator (forming two angles) opening towards the two solstitial points, and inclines to either side. This inclination, called the obliquity of the ecliptic, is about 23½ degrees to the south and north. Its poles tilt this much or toward each side of the pole of the cosmos.

The zodiac is divided into twelve sections known as the constellations, six in the north—the Ram (Aries), the Bull (Taurus), the Twins (Gemini), the Crab (Cancer), Lion (Leo), the Spike of Grain (Virgo); six in the south—the Scales (Libra), the Scorpion (Scorpio), the Bow (Sagittarius), the Kid (Capricorn), the Bucket (Aquarius), the Fish (Pisces). The reason is that these animals appear on this circle from the arrangement of the stars in the zodiac. Therefore the Greeks gave it this name, derived from *zodia* meaning "animal." The Latins called it *signifer*, [R 8a] meaning "sign bringer or signifier," because each of the constellations is a *signa*, i.e., a sign. They put the names of the constellations together in the following verse:

Sunt Aries Taurus Gemini Cancer Leo Virgo
Libra-que Scorpius Arcitenens Caper Amphora Pisces

[M 50] [diagrams]

The first hemistich contains the northern constellations, the second the southern constellations. *Que* is used as a conjunction meaning "and". In Latin Aries is the Ram, Taurus is the Bull, Gemini are the Twins, Cancer is the Crab, Leo is the Lion, Virgo is the Spike of Grain, Libra is the Balance, Scorpius is the Scorpion, Arcitenens is the Bow—also called Sagittarius, Caper is the Kid, Amphora is the Bucket—also called Aquarius, and Pisces is the Fish.

Each constellation is divided into 30 degrees, each degree into 60 minutes, and each minute into 60 seconds. Thus the circle is divided into 360 degrees. All the features of the sphere (or globe) are calculated according to this.

Circle of the meridian: A greater circle, mobile, that passes through the two poles of the cosmos and a specified zenith and intersects the equator, dividing the sphere of earth (or the globe) equally into eastern and western hemispheres. Midnight and midday are delineated by it. It is possible to consider each segment of the equator as a meridian. In Greek it is called *mezim* and *merinos* (i.e., *mesimerinos*, midday) while in Latin it is called *meridianus* meaning "the one that divides in half."

As for the four lesser circles that are parallel to the equator, they are the two tropics and the two polar circles.

The first two are the summer solstice or Tropic of Cancer and the winter solstice or Tropic of Capricorn. In Latin they are called *solstialis*, derived from *solstium* meaning "sun standing still," because when the sun comes to this spot, night and day have a slight pause and the sun seems to stand still. In Greek they are called *tropikos*, derived from *tropos* meaning "turning back" because the sun does not go beyond these circles and when it reaches one of them it turns back.

Circle of the polar tropic of the northern constellations: The Greeks call it *arktikos* because the feet of the Little Bear (Ursa Minor) seem to trace it; *arktos* means "bear." In Latin it is called *septemtrionalis* because Ursa Minor is composed of seven stars and this word means "the seven treaders of the pole." It is also called *borealis* and *acolonaris* which are names of that region.

Circle of the polar tropic of the southern constellations: It is called *antartikos* meaning "opposite *artikos*;" also *austrinus* or *australis*, because it is below the horizon and is always concealed from us, and *meridionalis*.

7 Division of these Circles and of the Earth

These four lesser circles—the two polar circles and the two tropics—divide the earth into five sections, called zones by the geographers. Each zone consists of the space between two lesser circles or in the middle of a circle.

One of them is the torrid zone, so called because it is below the tropic of the sun (i.e., between the Tropic of Cancer and the Tropic of Capricorn) and the ancients believed it was uninhabitable due to the extreme heat. It covers the space of 47 degrees between these two solstices and the equator passes through the middle of it.

Two of them are the frigid zones, so called because they are far from the course of the sun and the ancients believed they was uninhabitable due to the extreme cold. They cover the circular space of 23 degrees in the middle of the two polar tropics and are named after their own poles and circles.

The other two called the temperate zones, which are habitable because the heat and cold are moderate. The northern temperate zone lies between the Tropic of Cancer [M 51] and the Arctic Circle; the southern temperate zone lies between the Tropic of Capricorn and the Antarctic Circle. Both are at 43 degrees latitude.

Those who live in these zones are distinguished from one another with respect to shadow and position. With respect to shadow, those who live in the frigid zone are called *periscii* because their shadow turns on the surface of the horizon like a millstone; those who live in the temperate zone [R 8b] are called *heteroscii* because their shadow

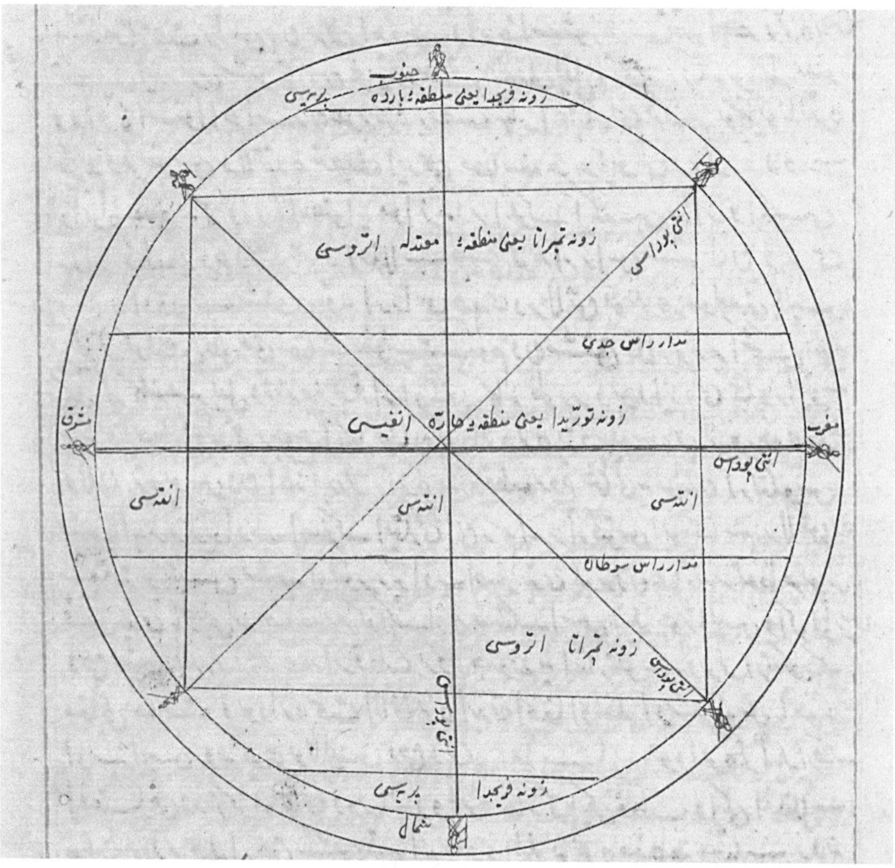

PLATE 2
Zones, R 8b
TOPKAPI PALACE LIBRARY, REVAN 1624, WITH PERMISSION

at noon falls to one side or the other (i.e., to the north in the northern zone and to the south in the southern zone); those who live in the torrid zone are called *amphiscii* because their shadow at noon falls sometimes to the south and sometimes to the north. With respect to position, the *periaeci* are those who live at one tropic and one latitude, from one end of a longitude to the other; the *antaeci* are those who live at the latitudes equidistant from both sides of the equator and at one longitude; the *antipodes* are those who live at equivalent latitudes on both sides of the equator, one being at the zenith and one at the nadir. These six divisions can be seen on the following circle: [Plate 2]

8 Concerning the Circles of Latitude and the Climes

The geographers divided the earth into several sections according to the different lengths of night and day and called each one a clime. The Islamic geographers distinguished two kinds of clime ("conventional" and "true"). The first is called the conventional clime—e.g., Anatolia and Syria. The author of *Taqwīm al-buldān* (al-Malik al-Mu'ayyad) speaks of 28 conventional climes. Other geographers, however, only spoke of true climes and made them subdivisions of three main zones, because most properly each section is divided into many and in each one are many subsections worthy of being called regions or climes. Under these circumstances, there is no neat and tidy arrangement. Some places that he considered as climes have so many divisions that, once they become known, it is determined that they each consist of several climes. Al-Malik al-Mu'ayyad's apology for not knowing them was given above (M 12 / R 3b). In this book, therefore, we have not taken the conventional climes into consideration. That is because they do not have objective boundaries. Not only are some round, some bow-shaped and some oblong, but also we would have to mention hundreds of climes. We have only referred to them where we describe the regions.

Here we shall discuss the category of true clime. The geographers have divided the climes as follows:

For those who live below the equator, [R 9a] night and day are always twelve hours each. Then the further a region is (from the equator), in the direction of the North Pole or South Pole, the greater the difference between night and day. So circles of latitude were posited parallel to the equator and the area between any two of them was called a clime. They were spaced such that the longest day (in any one at the summer solstice) is one half hour longer than the one closer to the equator. However, the word clime, which indicates the distance between the circles, is sometimes also used to mean the circle. With this division, one gains a general knowledge of a region's location and physical characteristics, and the difference between night and day, since regions at the same latitude have these things in common.

Within each clime three circles are posited—at the beginning, the middle, and the end—known in Greek as parallels. Ptolemy and his school specified 21 such parallels and seven climes in the northern hemisphere. All of the ancients limited the number of climes to seven. They posited climes only up to 50 degrees latitude, believing that the area beyond that was hardly habitable. The Islamic geographers followed suit—the extent of their knowledge [M 52] [M DIAGRAM] did not go beyond imitating the ancients. However, when later geographers learned that there were inhabited places beyond the seventh clime, they specified 24 climes going up to 67 degrees latitude. We have accepted their view in this book [see table after M 51].

THE FIRST CLIME: Some posit the beginning of this clime at the equator, but since those regions are mainly sea, most geographers posit its beginning at latitude 12 degrees and 40 minutes, where the longest day is 12 hours and 45 minutes. All agree that its middle is at latitude 16 degrees and 37 minutes, where the longest day is 13 hours, according to the Islamic geographers.

This, however, is contrary to the requirement that there be a half hour difference in each clime. Therefore we have rejected the other ridiculous matters that they mention along with this, because it would not be correct to enumerate rivers, mountains, and countries in places that are bounded by a hypothetical line. Supposing that one can account for such matters based on the rules of longitude and latitude stems from a lack of familiarity with the science (of geography). In this book, explication of the borders of the climes is consigned to tabular form and there is no need for a lengthy discussion of each one.

This clime extends to some of the islands in the Eastern Ocean and to the city of Meroe on the Nile.

THE SECOND CLIME: It extends from some coastal regions of India, passes through the Arabian Peninsula and southern Africa, and goes to the west.

THE THIRD CLIME: It includes some regions of China, India, Sind, and the northern coasts of Africa.

THE FOURTH CLIME: It extends from part of China to India, Tibet, Khurāsān, Jabal (al-Jibāl), Fārs, Rūm (Anatolia), part of France, and Spain.

THE FIFTH CLIME: It includes most of Transoxiana, Turkistan, Rūm, France, Italy, and Spain.

The fourth and fifth climes are the most temperate. Because of their fine climate, most of their inhabitants are well balanced in regard to external and internal (i.e., physical and mental) faculties and in regard to beauty.

THE SIXTH CLIME: It includes part of Spain, France, Germany, Rūm, the Black Sea, and Turkistan. This is also in the temperate zone.

THE SEVENTH CLIME: It passes through the regions of Poland, Russia, and northern Turkistan. The ancients determined that this clime ends at latitude 50 degrees and 20 minutes, where the longest day is 16 1/4 hours.

Some say that the end of the seventh clime is the place where human habitation ends, and north of this clime is the island of Thule, which they put at 63 degrees latitude, where the longest day is 20 hours. It is recorded in the *Almagest*, however, that a Slavic people lives at 64 [R 9b] degrees latitude, and in Ptolemy's *Geography* that human habitation reaches to 66 degrees latitude. The moderns record that even 70 degrees latitude is inhabited. In the *Atlas*, which I have translated, and elsewhere there are maps showing habitation even further than that, as will be detailed below.

According to the view of the moderns, THE EIGHTH CLIME passes through Wittenberg, THE NINTH CLIME through Rostock, THE TENTH CLIME through Hibernia (Ireland), THE ELEVENTH CLIME through Bohus, which is the fortress of Norway, THE TWELFTH CLIME through Gutia (Jutland), THE THIRTEENTH CLIME through the city of Bergen in Norway, THE FOURTEENTH CLIME through the city of Viborg in Finland, and THE FIFTEENTH CLIME through the city of Arotia in Sweden. Other climes pass through the cities of Norway, Sweden and Russia and other islands at equal distances (i.e., in equal widths of degrees of latitude). [M 53]

Thus it has been determined that there are 24 climes (extending) from the equator to where the longest day is 24 hours. It is not known whether there is a division of climes beyond that point and as far as the pole, because beyond that point it is impossible to observe the rule of a half hour difference between climes, since the longest day could be a week or a month or up to six months. The reason is that at the pole, half the celestial sphere is below the horizon and half is above the horizon, so the motion of the stars is horizontal.

If one wishes to know the longest day in a clime, one takes the number of the clime and adds half that number to the (length of the) day of the equator (i.e., to 12). For example, in the fifth clime, we add 2½ to 12 and get 14½. If the longest day is known, the number of the clime can be computed in the same fashion. For example, if the longest day of a city is 15 hours, we subtract 12 from 15 and get 3, then double it; the result is the sixth clime. If we multiply 3 times 4 (sic), we get the number of parallels. For example, 6 climes have 12 parallels.

According to the moderns, the nature of the climes as one goes south from the equator is exactly like that of the climes as one goes north: that side, which is the southern hemisphere, is also divided into 24 climes. But the geographers did not give special names to the southern climes; rather they give them names that correspond to the northern climes. For example, Antimeroe—*anti*, which means "opposite" in Greek, is prefixed to Meroe, so it means "Opposite Meroe".

Finally, the climes can be named after the places through which they pass. Thus, THE FIRST CLIME passes through the Mountain of the Moon and the source of the Nile; THE SECOND CLIME passes below the winter solstice (Tropic of Capricorn) and reaches Cape Corrientes (Cuba); etc.

The ancients gave no consideration to climes south of the equator because that region had few inhabitants and the cities were far from each other. In Ptolemy's *Geography* some cities are recorded as far as 10 degrees (south) latitude, but any beyond that are not mentioned. As a result of later discoveries, several countries were found at 30 degrees latitude. Afterwards, especially when South America was discovered, the southern hemisphere was also divided into climes.

The following table has been provided giving the latitude, longest day, and distance (width) of each clime, so that this information can be found easily when needed. The first column shows the number of the climes, the second shows the number of the parallels, the third shows the longest days, the fourth shows the latitude, and the fifth shows the distance. They have been indicated in degrees and minutes. [R 10a] The first parallel, which is the beginning of the first clime, is presumed to be at 4 degrees and 18 minutes, so the longest day shows a difference of 15 minutes. [Plate 3]

9 [R 10b] Concerning the Division of the Earth by Means of Longitude and Latitude

The geographers have posited and traced 18 meridians (lines of longitude) on the sphere of the earth (or the globe) and eight circles of latitude on each side (of the equator). As mentioned above, each circle is divided into 300 (sic; error for 360) degrees. The meridians specify the degree of longitude, and the nine parallel circles on each side of the equator, the degree of latitude. The distance between any two circles, in counting [squares like on] a chessboard and multiplying, is 10 degrees. [M 54] The start of latitude, by general agreement, is postulated to be from both sides of the equator: one is northern latitude and one is southern latitude. As for the start of longitude, there is disagreement. Ptolemy and the other Greeks determined that the starting point of longitude was the meridian of the Islands of Fortunata—called *Cezā'ir-i Ḫālidāt* (al-Jazāyir al-Khālidāt, "Immortal Isles") in our books and the Canary Islands in their own books—because that place was where the inhabited world, from west to east, began in their own regions at that time. Later, some recent geographers and Spanish sea captains started the beginning of longitude from the Azores archipelago that belongs to Holland. They stated that in those islands the needle of the compass points due north without leaning to one side or the other.

[Marginal note by Kātib Çelebi] In the region of Toledo also, which is in the middle of Spain, a pendulum faces due north, while elsewhere there is a deviation; this has been verified with observation instruments tonight.

Some considered the beginning of longitude to be from the western coasts. The distance between two lines of longitude is 10 degrees. The differences in the longitudes of regions are mainly based on these considerations.

The philosophers of the Orient established the beginning of longitude in their own regions at the end of the eastern coasts at the place called Gang Diz. This view does not conform to the one commonly accepted. The distance between the two starting points would be about 170 degrees, which is the distance covered by the inhabited quarter of the world.

In the middle of it (i.e., the longitudinal extension of the inhabited quarter) a place was posited over the equator called "the Dome of the Earth" (Qubbat al-Arḍ). This place is mentioned as Uzhayn in the books of the Islamic geographers; it is not recorded in the *Atlas* and other (European) books. Some said that it is hypothesized to be at 90 degrees longitude and 33 degrees latitude. Since this is not subject to proof, the geographers have tended to ignore it.

As a technical term in this science, THE LONGITUDE OF A PLACE (*ṭūl-ı beled*) is the arc of the equator between its intersection with the equinoctial line of the Canary Islands and its intersection with the equinoctial line of that place. THE LATITUDE OF A PLACE is an arc of the meridian of that place between its zenith and the equinoctial line. Whether in the southern or northern hemisphere, this latitude is equal to the elevation of the pole (above the horizon), or its angle (below the horizon) in the opposite hemisphere.

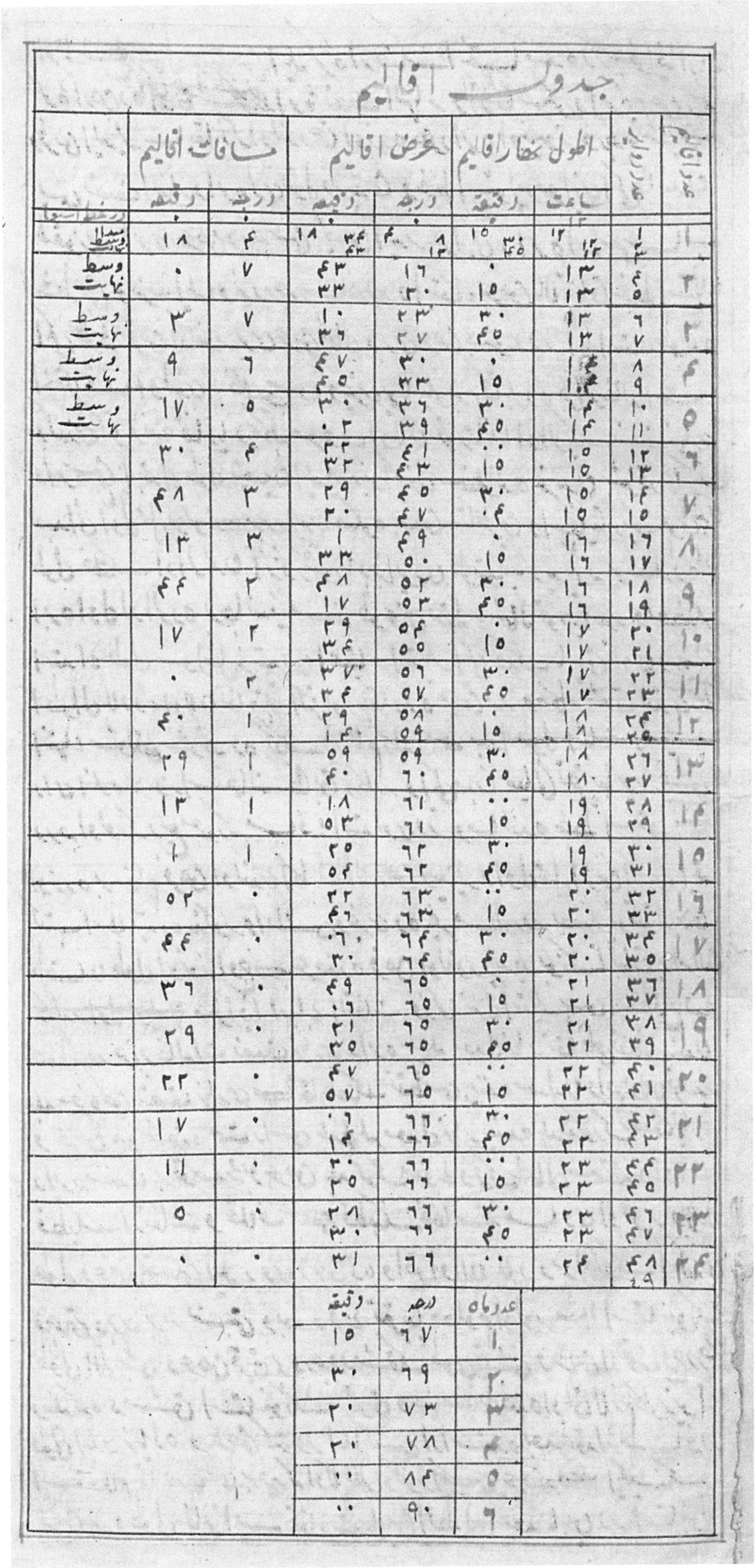

جدول اقاليم

عدد اقاليم	عدد ادوار	طول نهار اقاليم: ساعت	طول نهار اقاليم: دقیقه	عرض اقاليم: درجه	عرض اقاليم: دقیقه	مسافات اقاليم: درجه	مسافات اقاليم: دقیقه	
		۱۲	۰	۰	۰	۰	۰	درخط استوا
۱	۱ ۲ ۳	۱۲ ۱۲ ۱۳	۱۵ ۳۰ ۴۵	۴ ۸ ۱۲	۱۸ ۳۴ ۴۳	۴	۱۸	مبدا وسط نهایت
۲	۴ ۵	۱۳ ۱۳	۰ ۱۵	۱۶ ۲۰	۴۳ ۳۳	۷	۰	وسط نهایت
۳	۶ ۷	۱۳ ۱۳	۳۰ ۴۵	۲۳ ۲۷	۱۰ ۳۶	۷	۳	وسط نهایت
۴	۸ ۹	۱۴ ۱۴	۰ ۱۵	۳۰ ۳۳	۴۷ ۴۵	۶	۹	وسط نهایت
۵	۱۰ ۱۱	۱۴ ۱۴	۳۰ ۴۵	۳۶ ۳۹	۳۰ ۰۲	۵	۱۷	وسط نهایت
۶	۱۲ ۱۳	۱۵ ۱۵	۰ ۱۵	۴۱ ۴۳	۲۲ ۳۲	۴	۳۰	
۷	۱۴ ۱۵	۱۵ ۱۵	۳۰ ۴۰	۴۵ ۴۷	۲۹ ۳۰	۳	۴۸	
۸	۱۶ ۱۷	۱۶ ۱۶	۰ ۱۵	۴۹ ۵۰	۱ ۳۳	۳	۱۳	
۹	۱۸ ۱۹	۱۶ ۱۶	۳۰ ۴۵	۵۲ ۵۳	۴۸ ۱۷	۲	۴۴	
۱۰	۲۰ ۲۱	۱۷ ۱۷	۰ ۱۵	۵۴ ۵۵	۲۹ ۳۴	۲	۱۷	
۱۱	۲۲ ۲۳	۱۷ ۱۷	۳۰ ۴۵	۵۶ ۵۷	۳۷ ۳۴	۲	۰	
۱۲	۲۴ ۲۵	۱۸ ۱۸	۰ ۱۵	۵۸ ۵۹	۲۹ ۱۴	۱	۴۰	
۱۳	۲۶ ۲۷	۱۸ ۱۸	۳۰ ۴۵	۵۹ ۶۰	۵۹ ۴۰	۱	۲۹	
۱۴	۲۸ ۲۹	۱۹ ۱۹	۰۰ ۱۵	۶۱ ۶۱	۱۸ ۵۳	۱	۱۳	
۱۵	۳۰ ۳۱	۱۹ ۱۹	۳۰ ۴۵	۶۲ ۶۲	۲۵ ۵۴	۱	۱	
۱۶	۳۲ ۳۳	۲۰ ۲۰	۰۰ ۱۵	۶۳ ۶۳	۲۲ ۴۶	۰	۵۲	
۱۷	۳۴ ۳۵	۲۰ ۲۰	۳۰ ۴۵	۶۴ ۶۴	۰۶ ۳۰	۰	۴۴	
۱۸	۳۶ ۳۷	۲۱ ۲۱	۰ ۱۵	۶۵ ۶۵	۴۹ ۰۶	۰	۳۶	
۱۹	۳۸ ۳۹	۲۱ ۲۱	۳۰ ۴۵	۶۵ ۶۵	۲۱ ۳۵	۰	۲۹	
۲۰	۴۰ ۴۱	۲۲ ۲۲	۰۰ ۱۵	۶۵ ۶۵	۴۷ ۵۷	۰	۲۲	
۲۱	۴۲ ۴۳	۲۲ ۲۲	۳۰ ۴۵	۶۶ ۶۶	۰۶ ۱۴	۰	۱۷	
۲۲	۴۴ ۴۵	۲۳ ۲۳	۰۰ ۱۵	۶۶ ۶۶	۳۰ ۳۵	۰	۱۱	
۲۳	۴۶ ۴۷	۲۳ ۲۳	۳۰ ۴۵	۶۶ ۶۶	۲۸ ۳۰	۰	۵	
۲۴	۴۸ ۴۹	۲۴	۰۰	۶۶	۳۱	۰	۰	

عدد ماه	درجه	دقیقه
۱	۶۷	۱۵
۲	۶۹	۳۰
۳	۷۳	۲۰
۴	۷۸	۳۰
۵	۸۴	۰۰
۶	۹۰	۰۰

PLATE 3
Table of latitudes, R 10a; cf. M after p. 51 (Appendix, fig. 19–20)
TOPKAPI PALACE LIBRARY, REVAN 1624, WITH PERMISSION

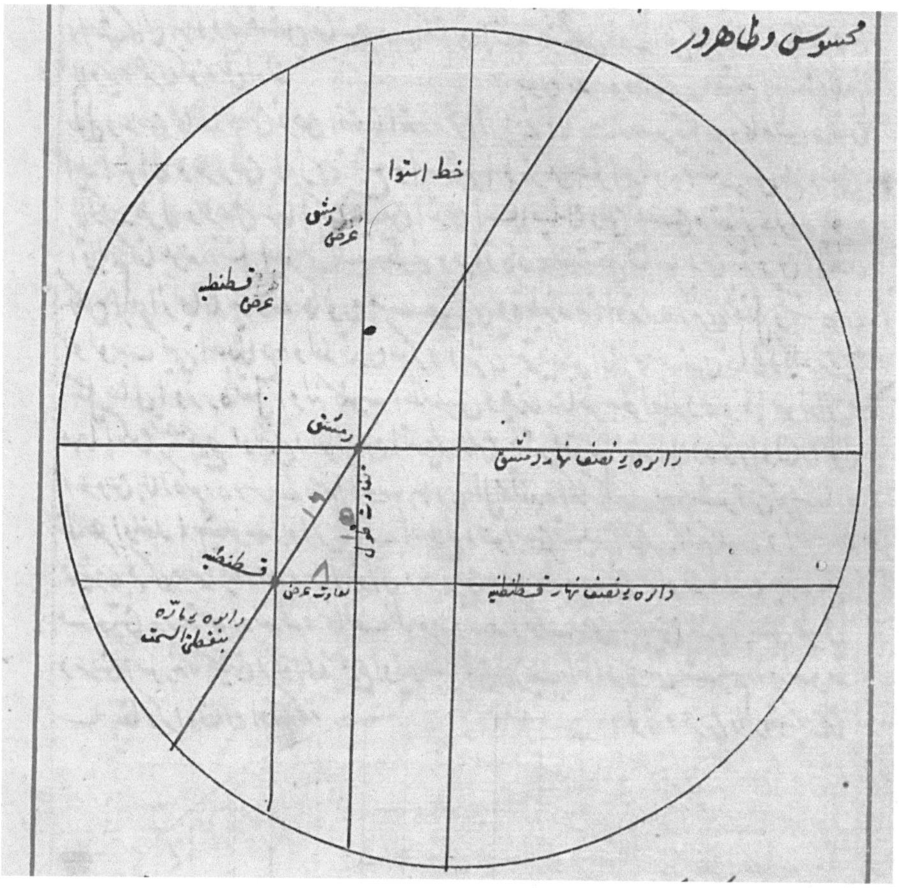

PLATE 4
Circle of distances R 11a; cf. M 49/50, middle diagram (Appendix, fig. 17–18)
TOPKAPI PALACE LIBRARY, REVAN 1624, WITH PERMISSION

With the determination of longitude and latitude, one knows approximately the locations of all cities and places on the face of the earth, their direction, their relation to each other, and the distances between them. For example, the longitude of Istanbul is 55 degrees and the latitude is 41 degrees and the longitude of Damascus is 70 degrees and the latitude is 33 degrees. This means that Damascus must be southeast of Istanbul, since its longitude is greater while its latitude is less.

The rule for determining the distance between two cities is as follows: If their latitude is the same but their longitudes are different, the difference between the longitudes is the distance between them, as is the case with Istanbul [R 11a] and İzmit. If their longitudes are the same but their latitudes are different, the difference in latitudes is the distance between them, as is the case with Istanbul and Bursa. If both their latitudes and longitudes are different, the distance between them is the hypotenuse of a right triangle. One side of the triangle is an arc of the meridian of the first city and the other side [M 55] is an arc of the circle of latitude of the second city. The hypotenuse is formed by an arc from the circle that passes through the point of the zenith of the two cities. We know the length of the first two sides: they are the differences of those latitudes and longitudes. One way to find the unknown side, which is the hypotenuse, is to add the squares of the two known sides and take the square root of the sum. The result is the number sought. For example, the difference in longitude between Damascus and Constantinople is 15 degrees and the difference in latitude is 8 degrees. The total of the square of the two sides is 289 and the square root of this is approximately 16. Thus the distance between the two cities is 16 degrees. Other distances can be found using the same method. This can be clearly seen on the following circle: [Plate 4]

SUPPLEMENTARY NOTES: The above account follows the arrangement of latitudes and longitudes given without maps in the books of the Islamic geographers. In the *Atlas* and other European geographical books, these matters are illustrated and tabulated on separate pages or on parts of pages, with the degrees of longitude and latitude given by number and section on the margins of the pages—known as *ḳarti* (map) and *tabula* (table)—that show the configurations of the countries. Usually the degrees of latitude are given on the left and right and the degrees of longitude are given at the top and bottom. They may be given in summary or detailed fashion, at intervals of six or ten minutes. The distances of these cities from one another in longitude and latitude are known from these tables with the help of a pair of compasses. In addition, each page has a scale line—known as *scala*—so that these distances can be calculated in miles. Generally one uses compasses to measure the extent of a place or the distance between two places according to that scale.

I have recorded all of this [R 11b] in detail in the translation of the *Atlas*. In this book, I did not feel compelled to copy all of the maps and have left some things out for conciseness' sake. For it would be a laborious task to copy all the maps from one manuscript to another. And since there is no printing in our country, it would be difficult to illustrate even a single page. So when a copy was made, there would be blank spaces left (where the illustrations would have gone) and the book would be defective. Therefore I was satisfied with including some of the general maps. The coordinates of the important cities recorded on these maps were included in the text; they were not consigned to the maps. So even if the maps had to be omitted, the book would not be defective but would stand on its own and adequately describe this subject.

The problem is that there are few scribes who can copy a text with all its illustrations properly in place—in our country there may be none—and the condition of those who can is well known. Nevertheless, it is to be hoped that those brethren who make copies of this esteemed book, or employ others to do so, will take pains to include the illustrations in their proper places, so the book will not turn into a boor stripped of his clothes or a bird with plucked tail and wings. That is because fine illustrations are one of the requisites of this science, which traditionally has been expounded by addressing both the mind and the senses. But what can we do about fools who think these illustrations are useless and cut them off when they copy the book? May God bring misfortune on their heads and cut off the days of their lives!

NOTICE ON COORDINATES: Since the rule of longitude and latitude must be used to determine the location of any place, the geographers have tried to establish coordinates using astronomical instruments, or at times by conjecture.

Latitude and longitude are determined by using quadrant and astrolabe, or by means of the times of an eclipse of the moon or sun. The method [M 56] is explained in treatises on those instruments and in astronomical tables. The sages have alluded to this by saying that on this subject, approximation stands in for calculation.

Ptolemy in the *Geography* established, as far as he could, the latitudes and longitudes of the cities that he recorded. The Islamic geographers also included them to some degree in the astronomical tables and books on the climes. Even separate books were written on this subject. However, as Abū al-Rayḥān al-Bīrūnī notes in *al-Qanūn al-Masʿūdī*, most of the coordinates given in those works are disorganized and confusing. He says that he copied some of them after correcting them, but that he usually copied them as they were, on the grounds that they were approximations. The authors of the *Atlas* and other recent books on geography, and the *Mappa Mundi*, claim that they too made corrections as far as possible.

However, when this humble person (Kātib Çelebi) drew a map according to the latitudes and longitudes of the cities mentioned in the *Taqwīm al-buldān*, it became clear that the places where the cities were cited to be were off by "the distance between two points of sunrise" (*buʿd al-mashriqayn*, Koran 43:38); so I could not accept Abū al-Rayḥān's claim that he corrected them. In the *Atlas* and the other European geographies as well, not only were the coordinates inconsistent, but also, when I made my translation, several well known places were obviously mistaken and I realized the true nature of these books.

[Marginal note by Kātib Çelebi] Thus, on most globes and in most geographies, the latitude of our city, Constantinople, is given as 43 or 45 degrees; whereas it has been proven from observations made with astronomical instruments that its latitude is approximately 41 degrees.

So it is not correct to claim that the precise latitude and longitude of a place can be given; rather, only approximations can be provided. The following is a sketch of the two times eight circles of latitude and longitude mentioned above: [Plate 5]

10 [R 12a] Concerning Distances and Surveying Instruments

According to the *Introductio* (of Philippus Cluverius), the Romans measured distances on earth in paces—one pace being approximately the length of a tailor's cubit—and they called a distance of 1000 paces a MILE. A stone was erected every mile and the distance between regions was determined with those stones. Later the word "mile" came to refer to the stone, so when they said "at the tenth mile" what they meant was "at the tenth mile-stone." The word *mīl* (meaning "an obelisk or mile-stone") derives from this.

Since every circle is divided into 360 degrees and every degree into 60 minutes, the ancients wished to determine for the sphere of the earth how many miles were in one degree. Ptolemy made some measurements in the Desert of Tadmur (Palmyra) and determined that one degree was 66 2/3 miles. He performed this measurement as follows: A marker was erected at a certain place in the desert and two

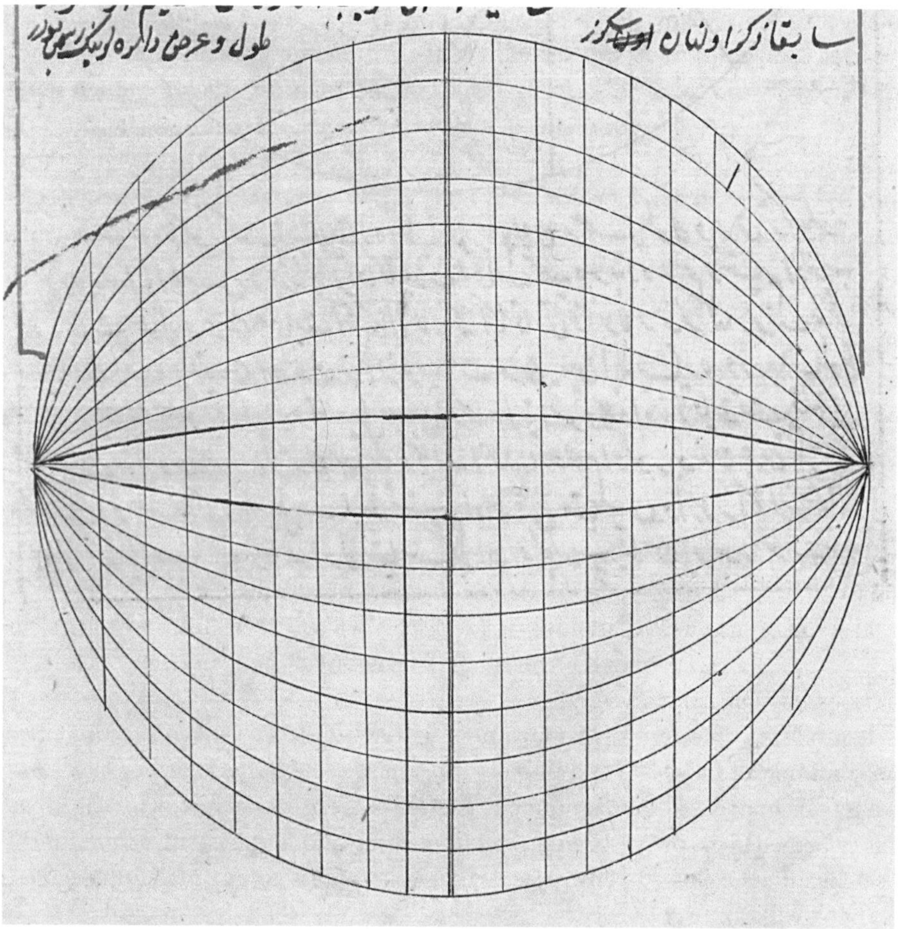

PLATE 5
Circles of latitude and longitude, R 11b; cf. M 49/50, lower right diagram (Appendix, fig. 17–18)
TOPKAPI PALACE LIBRARY, REVAN 1624, WITH PERMISSION

groups of people went in a straight line, one due north and one due south. At night they stopped and set up another marker at the place where the North Pole (i.e., the altitude of the pole star) was a difference of one degree from the altitude they had taken where the first marker was. Measuring the distance between the two markers in each direction, they found that one was longer than the other. As a precaution, they adopted the shorter measure—since in any case that distance would be included within the longer one—and it came to 66 2/3 miles. Multiplying that by 360, they learned the circumference, and from that the diameter and radius, and the size of the globe.

The Abbasid caliph al-Maʾmūn (r. 813–833) wanted to test the accuracy of this measurement. He charged ʿAlī b. ʿĪsā and another group, all of whom were among the learned of the age, with the task, which they performed in the desert of Sinjār (in northern Iraq) following the same procedure. In their measurement, one degree came to 56 2/3 miles. They ascribed the difference to a certain margin of error in the procedure. [M 57] Even today, the procedure is the same as that of the ancients. [Plate 6]

Again according to the *Introductio*, the Greeks measured distances in STADIA: one *stadium* was 125 paces, so there were eight *stadia* in one Roman mile. The Persians measure in PARASANGS: one parasang was 30 stadia. The Egyptians used the *üskeni* (i.e., Latin *schoenus*), about whose value there was disagreement: some said it was 60 stadia, some 40 and some 20. Today the Germans, Danes, Norwegians, Swedes, English, Scots, Poles, Czechs and Italians use the term MILE. The Spanish and French, however, use league (*levka*)—the Spanish pronounce it *legas* and the French *lieues*. The Russians and Muscovites use their own measure of distance called *verst*.

Furthermore, none of these peoples agree among themselves on the length of their own measure of distance. In Germany, for example, different miles are used in different regions, some long, some short, some medium. Similarly, in Spain and France and elsewhere there is disagreement about the league and the mile. Agreement should not be expected in this matter. [R 12b]

In terms of these measures, and taking the aforementioned disagreement into account, according to the *Introductio* and the *Atlas*, the length of ONE DEGREE is:

- 60,000 paces, each pace being four feet and each foot being 16 fingers.

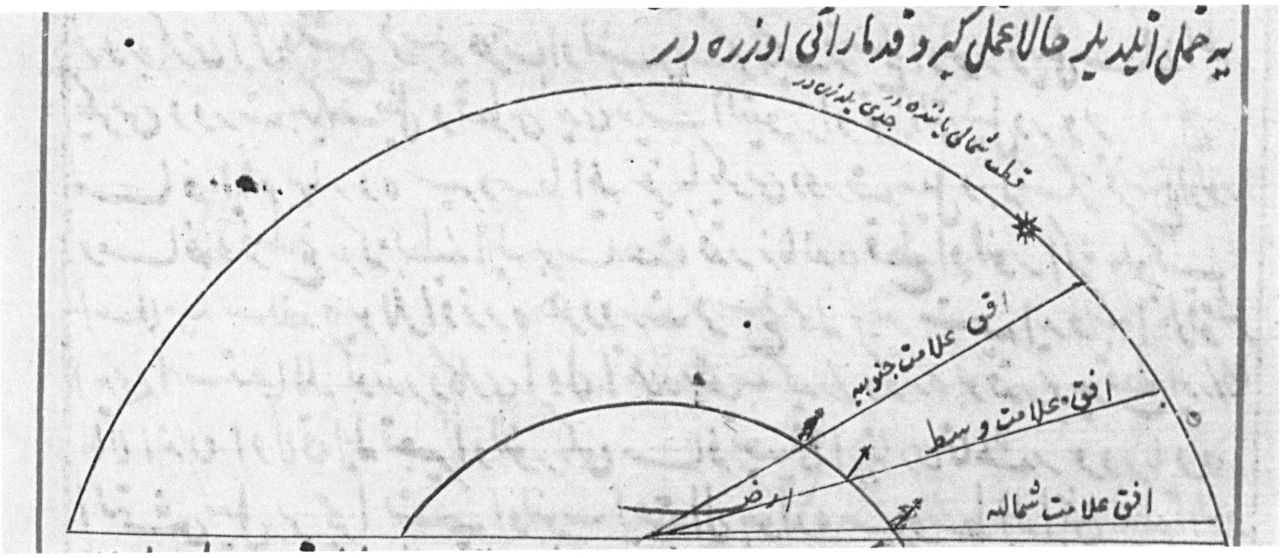

PLATE 6 Measuring circumference of earth, R 12a; cf. M 57/58, lower right diagram (Appendix, fig. 22–23)
TOPKAPI PALACE LIBRARY, REVAN 1624, WITH PERMISSION

- 480 stadia.
- In Germany, 10 long miles, 12 medium miles, and 15 short miles—this last being the one currently in use. The surveyors and geometricians of Holland and Sweden also use German miles.
- In terms of the league used in France, 25 miles or 20 long miles.
- 60 miles in Italian miles.
- 60 miles in English miles, but some say 50.
- 20 leagues in English leagues.
- 17½ leagues in Spanish leagues.
- The distance covered in 20 hours travel (by foot).

In the short mile of Germany the circumference of the earth is 5400 miles and its diameter is 1718 2/11 miles.

According to the Muslim surveyors and geometricians, the length of one degree, as mentioned above, is 66 2/3 miles—although according to the moderns this amount is 10 miles short—or 22 1/9 parasangs, each parasang being, by convention, 3 miles.

The MILE is 3000 cubits according to the cubit of the ancients or 4000 according to the cubit of the moderns. The CUBIT is 32 fingers according to the ancients or 24 according to the moderns—the cubit of the moderns being ¾ of the cubit of the ancients—and one mile is, by general agreement, 96,000 fingers. One FINGER is the length of six medium-sized barley grains placed side by side.

According to the moderns, when multiplied by 22 1/9 parasangs [per degree] the circumference of the earth is 8000 parasangs. Calculated in miles it is 24,000 miles and the diameter is 7636 miles.

The length of one STAGE is approximately 24 miles, or eight parasangs, walking at an average pace in the springtime. [M TABLES] [M 58] At a slow pace, one can cover the distance of one parasang in approximately one hour.

In the past, the Muslim rulers used the term *barīd* for a posting stage of four parasangs. They stationed horses along the roads at that interval which the couriers would mount and go speedily on their way, covering so many *barīds* per day. Nowadays the term (for such couriers in the Ottoman Empire) is *ulaḳ*. However, the distance covered is mostly the same.

At sea, 60 miles is termed one *mecrā*. It is the distance that can be covered (in one day) when sailing at a medium speed in mild weather. With a favorable wind there is no limit to how far one can go. All of this is presented in the following table: [Plate 7]

Addition by the Publisher (İbrāhīm Müteferriḳa)

Details on the Measurements of Land and Sea

As indicated in the text of this book, the people of every age and of every country have used different measurements in order to determine distances on land and sea. The elements of earth and water, taken as a whole, form a sphere, i.e., a round ball. In length and breadth—i.e., whether going from west to east or from north to south—the circle posited in the middle that encompasses it like a belt is divided in 360 equal degrees, by consensus of the astronomers. However, the length of one degree became subject to debate because of the different measurements, whether of parasang, mile, stage, hour, or other unit. According to the surveying of the

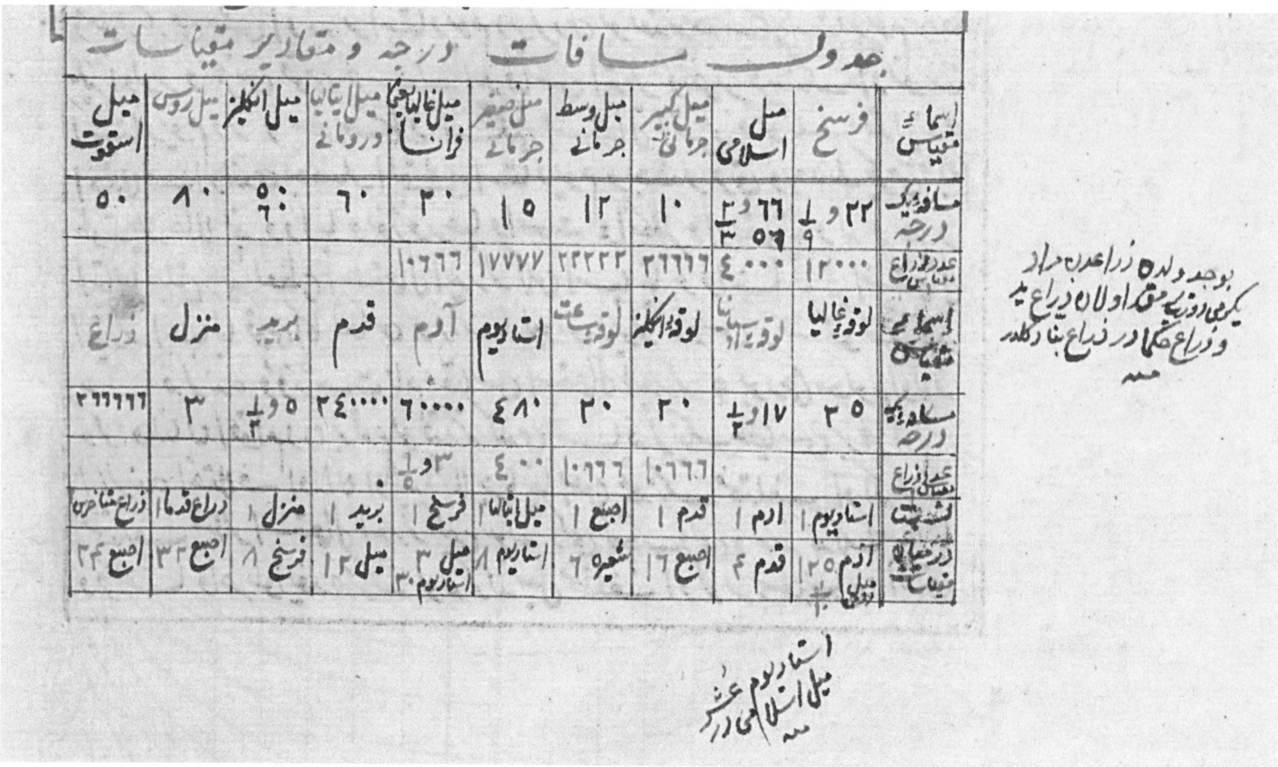
جدول مسافات درجه و مقادير مقياسات

PLATE 7 Table of distances, R 12b; based on *Introductio*; cf. M 57/58 (Appendix, fig. 22–24)
TOPKAPI PALACE LIBRARY, REVAN 1624, WITH PERMISSION

ancient mathematicians—Ptolemy and his followers—one degree on the earth's surface is 22 1/9 parasangs. One parasang is three miles, one mile is 3000 cubits, one cubit is 32 fingers, and one finger is six medium-sized grains placed side by side. Thus one degree on land is 66 2/3 miles.

One parasang is 9,000 cubits. One degree on land is approximately three stages at a slow pace. One stage iseight parasangs. One parasang is the distance covered in one hour at a slow pace. The distance covered in one day is approximately 24 miles. At sea, although with a favorable wind there is no limit to how far one can go, at a moderate pace one can only cover 60 miles in one day. Sailors call this one *mecrā*.

Up to this point we have mainly given the views of the ancients, but it is appropriate to give here the views of the moderns as well, so there will be no mistake.

According to modern surveys, one degree on the earth's surface is 18 8/9 parasangs. Thus one degree is 56 2/3 miles. One mile is 4000 cubits, one cubit is 24 fingers, one finger is six medium-sized barley grains placed side by side. There is a difference of three parasangs between the ancients and the moderns. In miles they differ by ten. This is not an actual difference but a verbal one. A mile has the same value for both; but a cubit is 32 fingers for the ancients and 24 for the moderns, a difference of eight; and a parasang is 9000 cubits for the first group and 12,000 for the second. In both views, one mile is a third of a parasang and one finger is the length of six medium-sized barley grains placed side by side.

There is also a difference with regard to halting places and stages. [M 59] If measured by a slow pace, like that of a caravan or an army, it is considered a moderate pace and one degree on land is three stages. An example is going from Istanbul to Büyükçekmece in one day. If one travels faster than this, it is considered a medium pace and one degree is two stages. An example is a horseman without baggage going from Istanbul to Silivri in one day. If one goes even faster, covering one degree of land in one day (it is considered a fast pace). An example is going from Istanbul to Çorlu in one day. In the first case the distance traveled is 1/3 degree; in the second, ½ degree; in the third, one degree.

Because the people of every age and country have used different standards of measurement, some people today in the Mediterranean and Black Sea region, even if they know nothing about calculating a degree, sometimes count that distance as 100 miles and sometimes as 60 miles. Astrologers and geographers consider one degree to be (the distance travelled in) 20 hours. The ancient Persians considered it to be 20 parasangs; the people of Morocco, 20 parasangs; the people of Hormuz,

100 miles; the people of Gujarat and Deccan in Eastern India, 30 miles; and the people of Eastern Kembayed (Cambodia), 100 miles.

The Christian nations also have differences with regard to measurements. The geographers generally discuss these differences and put various scale lines on their pages (i.e., maps) to guard against error.

It is clear to those who give thought to these principles that one can calculate the time needed to make a circuit of this globe of earth and water. Thus, if a person took Istanbul as the starting point and set out toward the west with the intention of traversing the globe lengthwise, he would pass through Europe and the New World and continue in the same direction as the sun until he reached India, then he would pass through India and Iran and arrive back in Istanbul. From our point of view, this person would have departed toward the west and returned from the east. In like manner, if he set out toward the north with the intention of traversing the globe breadthwise, he would pass through Muscovy and the newly discovered Nova Zembla, the most remote inhabited place in that direction; then he would go over the North Pole and travel through the place directly opposite us, arriving at the South Pole at the bottom of the earth; and thence via the clime of Africa and Ethiopia and the Mediterranean he would come back to Istanbul. This person would have departed toward the north and returned from the south. Walking the length and breadth of the globe in this manner at a moderate pace, he would traverse 4080 stages. Traveling at the speed of a horseman without baggage, he would cover 720 stages. And travelling by stage (*menzil ile*, i.e., one degree per day) and going around the earth in a straight line as required by the science of geography, it would be possible to make a complete circuit in 360 days.

(End of the publisher's supplement.)

11 [R 13a] **Concerning the Four Directions and the Winds**

The direction where the sun rises is called east or, in Latin, *oriens*; where it sets is called west or, in Latin, *occidens*; where according to those who live at the oblique horizons (i.e., not at the equator) the sun in its course leans to the other side, that direction is called south or, in Latin, *meridies*; the opposite side is called north or *septentrio*. Four more directions are inserted among these four.

[DIAGRAM] [M 60]

Between north and east is *oriens aestivus* meaning "summer east"; between south and east is *oriens hibernus* meaning "winter east"; between north and west is *occidens aestivus* meaning "summer west"; between south and west is *occidens hibernus* meaning "winter west".

Then they assign four major winds to the four directions and other winds between them. But the ancient and modern writers differ on this and the people of each region have their own division and terminology. We have illustrated these here in separate circles as far as we have been able to determine them. First, a circle showing the directions: [Plate 8]

The second circle shows the divisions of the ancient Romans and Greeks and Latins who sailed in the Inner Sea (Mediterranean). In Latin the general word for wind is *ventus*. Then the north wind is *aquilo*, the south wind is *auster*, the east wind is *subsolanus*, and the west wind is *favonius*. These are the primary winds of the four directions. Between them are subdivisions of two or three or seven winds, as shown in the following circles with the terms and usages of each people. [R 13b] The various groups divide this second circle into 12 sections and use them when they set out on the Mediterranean and the ocean beyond. However, the Europeans who sail on the ocean translate these terms into German and the other groups translate them into their own languages and adjust the division accordingly. The following is the division of the Latins: [Plate 9]

The third circle is the one used by the Romans and Italians. They divide the circle into 16 sections. [Plate 10]

[R 14a] The fourth circle has 32 sections according to the division of the Germans and other Europeans who sail on the Ocean Sea (Atlantic). In all they use four names in various permutations and combinations, as follows: [Plate 11]

The fifth circle shows the divisions of the Hindus and the various Muslim groups who sail on the Eastern Ocean. Their divisions are singular, i.e., there are 32 divisions and their names differ, as follows: [Plate 12]

[R 14b] As for the Muslims who sail on the Mediterranean, they denote the 32 divisions with ten different terms. They call the north *yıldız* (the star), the south *ḳıble* (the direction of Mecca), the east *günṭoġusı* (sunrise), the west *baṭı* (sunset), the northeast *poyraz*, the southeast *keşişleme*, the southwest *lodos*, the northwest *ḳarayel*; between every two is an *orta* and between every one and an *orta* is a *kerte* (half wind). They use the *kertes* in conjunction with the others to mark off the 32 divisions. They say, for example, *günṭoġusı'nuñ poyraz'dan yaña kertesi* (bearing east by northeast). The others are named in this same fashion. *Kerte* derives from the Greek word *ḳarta* meaning "quarter". That is why the pages that are drawn in quadrilateral shape are usually called *ḳarti*, of which *ḫarti* is a corruption. To designate a half *kerte* they say *meze kerte*.

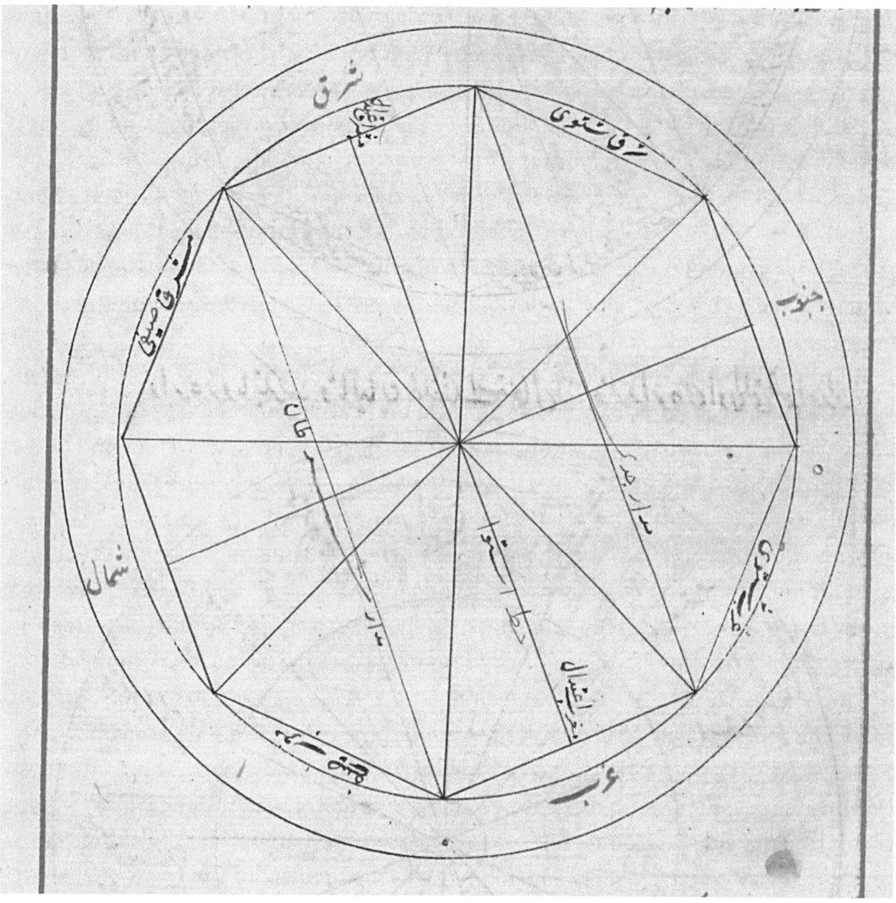

PLATE 8
Wind directions, R 13a; cf. M 59/60, upper right diagram (Appendix, fig. 25–26)
TOPKAPI PALACE LIBRARY, REVAN 1624, WITH PERMISSION

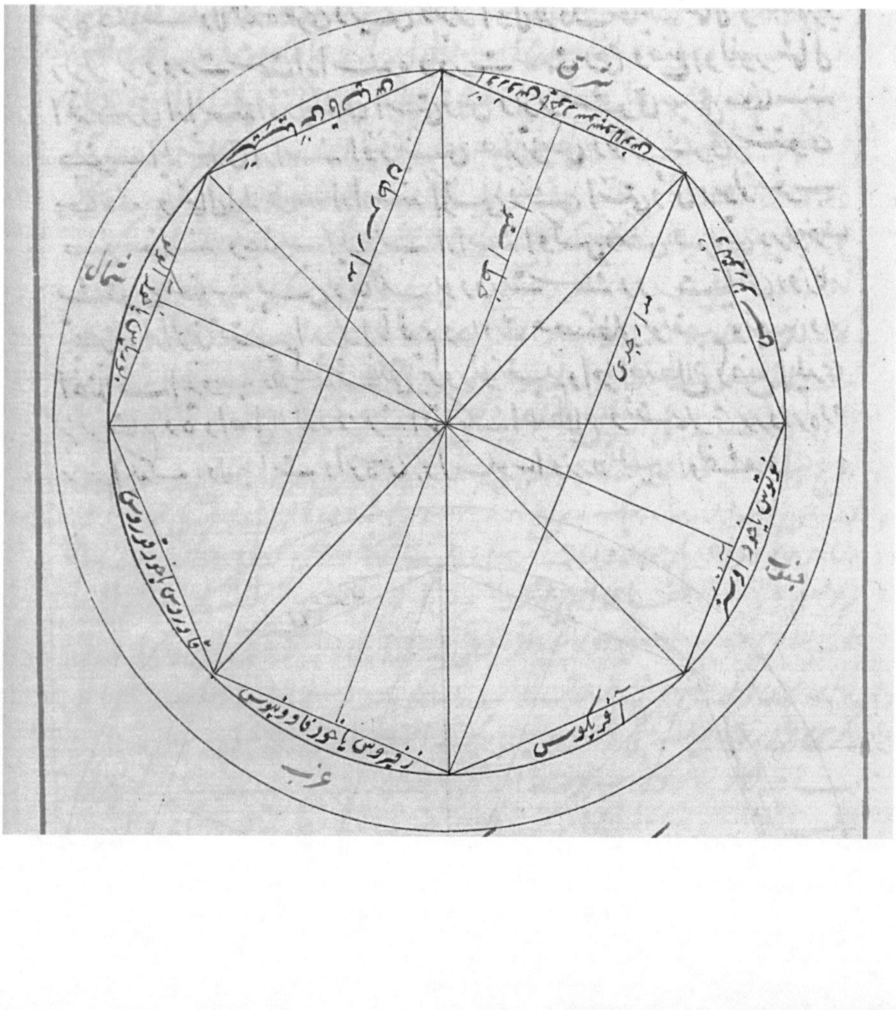

PLATE 9
R 13b, top; cf. M 59/60, diagram lower right (Appendix, fig. 25–27)
TOPKAPI PALACE LIBRARY, REVAN 1624, WITH PERMISSION

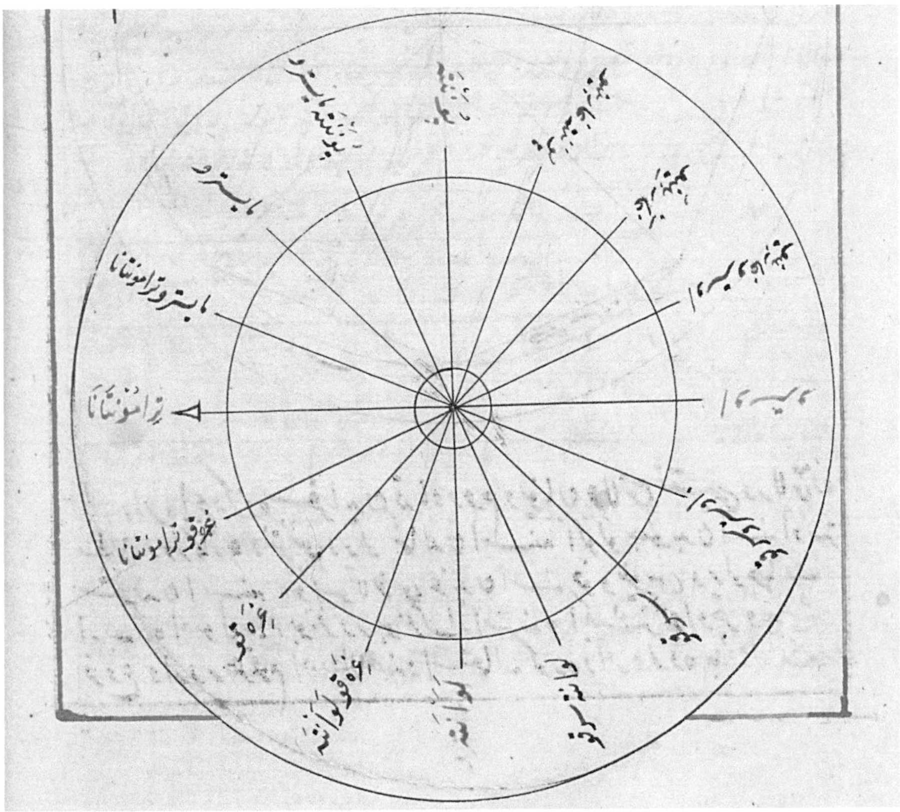

PLATE 10
R 13b, bottom; cf. M 59/60, diagram upper left (Appendix, fig. 25–27)
TOPKAPI PALACE LIBRARY, REVAN 1624, WITH PERMISSION

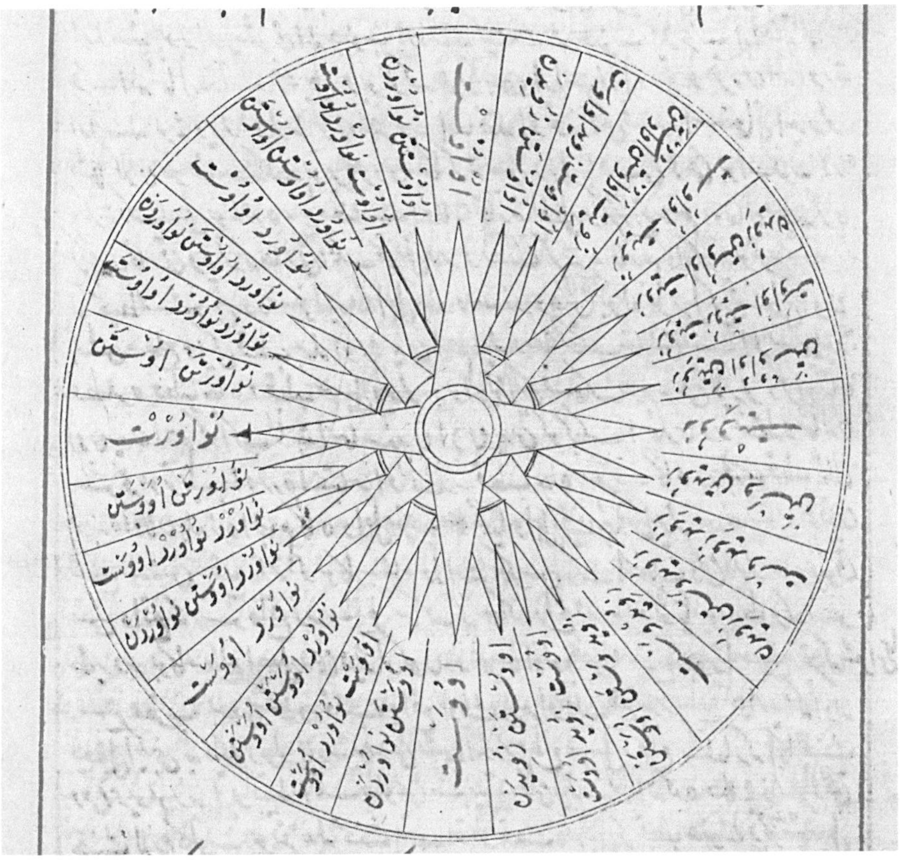

PLATE 11
R 14a, top; cf. M 59/60, diagram middle (Appendix, fig. 25–27)
TOPKAPI PALACE LIBRARY, REVAN 1624, WITH PERMISSION

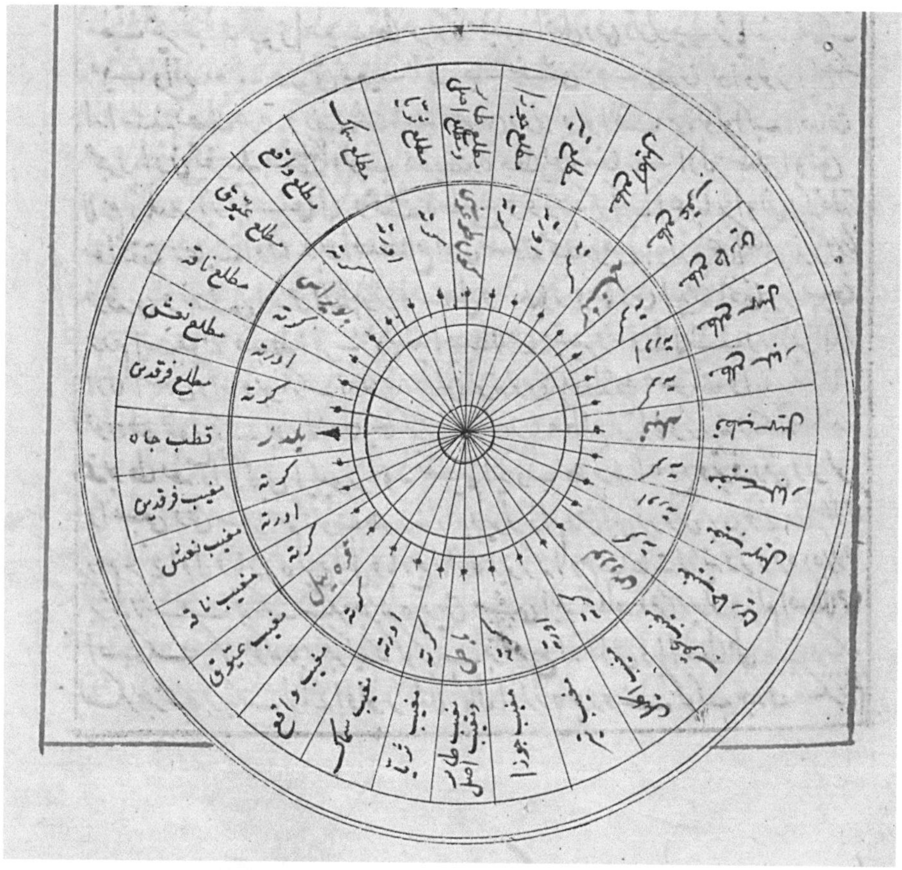

PLATE 12
R 14a, bottom; cf. M 59/60, diagram lower left (Appendix, fig. 25–27). These figures are based on *Introductio*, 28/30.

The people of India, Sind, China, and Persia who sail on the Eastern Ocean give 17 names to these 32 divisions, of which 15 are *maṭla*ʿs (rising positions, points above the eastern horizon) and two are poles. The opposite of the *maṭla*ʿs are called *maghīb* (setting position). The poles are Quṭb Jāh and Quṭb Suhayl and each of the 32 divisions is a *khan*. These 17 names are the names of 17 of the fixed stars, [M 61] arranged from the point of due north to the east as follows:

1. Quṭb Jāh (Pole Star): *jāh* is the Arabicized version of the word *gāh*. The Persians call the star that is near the North Pole Gāh, the Maghrebis call it Sumayyā, and astronomers call it Judā.
2. The *maṭla*ʿ of Farqadayn (Kochab and Pherkad, two stars of the Little Dipper).
3. The *maṭla*ʿ of Naʿsh (the four stars in the form of a quadrilateral in the constellations of Big Dipper).
4. The *maṭla*ʿ of Nāqa (Shedar), also called Sanām, Dhāt al-Kursī, and Kaff al-Khaḍīb (all names for Cassiopeia).
5. The *maṭla*ʿ of ʿAyyūq (Capella).
6. The *maṭla*ʿ of Wāqiʿ, also called Nasr-i Wāqiʿ (Vega).
7. The *maṭla*ʿ of Sammāk (the fisherman, Arcturus), also called Sammāk-i Rāmiḥ (Bootis).
8. The *maṭla*ʿ of Thurayyā (the Pleiades).
9. The *maṭla*ʿ of Ṭāʾir which is Nasr-i Ṭāʾir (Altair), also called *maṭla*ʿ *aṣlī* (the first *maṭla*ʿ) because it is seen at sunrise.
10. The *maṭla*ʿ of Jawzā (Orion).
11. The *maṭla*ʿ of Tīr (the arrow), also called Shiʿrā-i Yamānī (Sirius).
12. The *maṭla*ʿ of Iklīl (the Scorpion's Crown).
13. The *maṭla*ʿ of ʿAqrab (Scorpio, Antares)—in some books it occurs as Qalb meaning the heart of Scorpio (Scorpionus).
14. The *maṭla*ʿ of Ḥimārayn (two donkeys; alpha and beta Centauri) of which the first is called Maʿqil and the second Ẓalīm (Piscis Australis); they are also called Fārisayn (two horsemen).
15. The *maṭla*ʿ of Suhayl (Canopus in Argus).
16. The *maṭla*ʿ of Sulbār (Achernar), also called Muḥnith because a certain tribe believed that when the latter rose it was Canopus and referred to it as such. When this error was realized, they had to swear falsely (*ḥinth*) about it and so they called it Muḥnith
17. Quṭb Suhayl (the Pole of Canopus in Argus), which is the South Pole. Because the well-known star called Suhayl is near the South Pole, the word "pole" is attached to it.

In the western hemisphere, these stars are recited in the order of their *maghīb* (setting). After the Pole of Canopus come the *maghīb* of Ḥimārayn, the *maghīb* of ʿAqrab, the *maghīb* of Iklīl, the *maghīb* of Tīr, the *maghīb* of Jawzā, the *maghīb* of Ṭāʾir—also called *maghīb aṣlī*, the *maghīb* of Thurayyā, the *maghīb* of Sammāk, the *maghīb* of Wāqiʿ, the *maghīb* of ʿAyyūq, the *maghīb* of Nāqa, the *maghīb* of Naʿsh, and the *maghīb* of Farqadayn.

Each of these 32 *khans* is divided into a half *khan* and each half is divided into a quarter *khan*. By attaching these divisions to the names of these stars it is possible to designate them with very close approximation. It is not necessary in fact for the risings and settings to be exactly in those places, because in practice the risings and settings of the stars are not taken into account, such that one would have to check whether they correspond to those places.

The reason that a given term is reserved for a given place is that on this side (i.e., in the northern hemisphere) Farqadayn and Naʿsh are always visible while Canopus, Sulbār, and Ḥimārayn are always invisible. Then the Turkish method is more accurate for determining locations while the Indian terms are easier to pronounce. Although the former has only ten terms while the latter has 17, yet it is difficult to pronounce points with *kerte*, for example, *günṭoġusıʾnuñ poyrazʾdan yaña kertesi* ("the quarta / rhumb of east towards north-east")—it is easier to say "above the *maṭlaʿ* of Thurayyā" (for east by northeast).

Supplementary note: This group (the Indians) divides the circle into 210 sectors in determining distances, each one called *iṣbaʿ* (finger), also *tirfa*. Each *iṣbaʿ* has eight divisions called *zām*. Then there are two kinds of *zāms*: *ʿurfī* (conventional) and *iṣṭilāḥī* (technical). In that region, *zām-i ʿurfī* is used for a quarter of the day or night, which is a period of three hours; also for one-eighth of the distance that one can cover on land in one day and night by walking. The *zām-i iṣṭilāḥī* is one of the eight divisions of an *iṣbaʿ*, the *iṣbaʿ* of a star being the time required for it to move one-eighth the distance of its ascent or descent. For example, if it is moving northward, the constellation Capricorn gains a height of one *iṣbaʿ* in eight *zāms*. [R 15a] These terms are also used for a ship at sea that travels in this amount of time, whether in strong or light winds. In a moderate wind it travels one *zām-i ʿurfī* in a period of one *zām-i iṣṭilāḥī*, or about fifteen miles at the rate of five miles per hour. Thus one *iṣbaʿ* is 1 5/7 degrees, one degree is 7/12 of an *uṣbuʿ* or 4 2/3 *zāms*, and one *zām* is 14¼ miles. The entire circuit is 4680 *zāms*. [M 62] These are the calculations of the moderns. The ancients posited 224 *iṣbaʿs* to the circle and divided the *khans* accordingly. According to them, there were seven *uṣbuʿs* between every *khan*. According to the moderns there are 6 9/16 *iṣbaʿs*. This view was tested at the ascent of Capricorn and found to be correct, so the view of the ancients was discarded.

12 On the Rules of Drawing Maps

As mentioned above, in addition to the two branches of the science of geography for which maps are made, chorography and topography, there is another branch whose purpose is to depict the conditions of the sea, and that is hydrography.

So there are four branches (geography, chorography, topography, hydrography), each of which can be presented in general or in detail (for a total of eight):

1) General geographical maps. Most of the illustrations on European globes are in this fashion. In the *Atlas* and other European geography books, and in our translation, the globe is shown as two circles or one circle laid flat. The geographers have painted this circle on parchment or paper that is flattened out like a handkerchief. They call it *Mapa Munda* (Mappa Mundi), *Papa Munta* being a corruption of this. In Greek *mapa* means handkerchief (i.e., Latin *mappa* "a table-napkin") and *munda* means painted.

2) Detailed geographical maps. For geographical topics, general maps did not suffice and it was necessary to draw each region on a single page. The general maps on globes and at the beginning of books were broken up into sections and depicted in detail page by page; such as the 150 or so pages in the *Atlas* and the pages of this translation. The author of the *Atlas* sometimes calls these pages *ḳarti* and sometimes *tabula*. Since they are usually quadrilateral in shape, the Greek word meaning quarter was used for them.

3–6) The general and detailed styles of chorographical and topographical maps.

7–8) The general and detailed styles of sea charts. The term for these, *ḳarti*, has been corrupted to *ḫarti*. The common people also use *ḫarti* for the *Mapa Munda*. Longitude and latitude are not shown on sea charts, because it is not necessary to determine coordinates (for the Mediterranean). Only there is a wind rose showing the eight major winds in black, the *ortas* in green, and the *kertes* in red. Shoals are indicated by red dots, reefs by black dots, submerged rocks that are invisible by crosses, and streams by cracked lines to the shore. The names of famous cities on the shore are written in red while other places are written in black.

On charts of the Indian Ocean, east is at the top while north is to the left. In charts of the Western Sea (Mediterranean) west is at the top while north is to the right. Whichever region one sets out from, it is oriented accordingly.

Usually on sea charts a round piece of wood is affixed. The point indicating north is different from the others. In many of them a compass rose is drawn in which north is indicated in black while the other directions are distinguished from it.

THE COMPASS: The 32 winds are inscribed on a round pasteboard and it is placed in a box. Due north is indicated in black. The box is called the house of the needle. The tip of the needle is magnetised so that it points north. It is placed on an axle and covered with glass so it is not affected by wind. The northern point on that box is placed such that it points to the north on the chart. When the needle points to the north, the four directions and the winds are known by it.

It is also known [M 63] [R 15b] that the magnetised needle deviates seven degrees to the east from due north. Those who say that it points directly at the pole are mistaken. The *qibla* indicators coming from Europe were tested and all of them behaved this way. While this discrepancy is not noticed in the Mediterranean because of the short distances involved, it is exorbitant in the vastness of the ocean. Those who do not understand this believe the discrepancy results from the current.

On the charts are a number of lines on one or two sides, which are divided by dots, that indicate distance in miles. There are ten miles between every two dots and a marker is placed every 50 miles. In order to determine the distance in miles between two places, it can be measured with a pair of compasses. And to determine what wind to follow in order to go from one place to another, one puts a point of the compass on the wind parallel to the imaginary line that covers the two places and advances the compass; wherever it reaches, one follows that wind, setting the rudder accordingly. As they are sailing many days and nights in the open sea, with neither shore nor island in sight, and the ship is making a certain speed according to the wind (which will be known by experience), they take their bearings once or twice a day, and calculate the distance, and it is essential that current and drift be taken into account.

[Marginal note by Kātib Çelebi] The ancient seafarers used nine boards (wooden quadrant?) that differed according to *iṣbaʿ* (finger). The moderns currently use a board of approximately three or four spans. They call the route *deyre* (?). They have to know the seasons of the routes, the amount of cargo in the ships, the tides, and when storms occur.

A sea chart must have the winds and the islands properly located. Their accuracy can be checked by making sure that the interval between the centers of the 16 compass roses around them is equal, and the angles of the winds are also equal.

In addition, sea captains have instruments that are used in various ways and there are conditions and restrictions regarding sea routes. Providing detailed information on this would go beyond our subject, so this will suffice.

On this type of general chart, usually ten miles is covered by three dots. On this scale, it is not possible to show the nature of places on the shore that are less than ten miles, and to indicate many hazards. When expansion is needed, they illustrate the islands and coasts according to the eighth kind (of maps as explained above), region by region, with a page for each. These show details of harbors and other coastal features with the approximate mileage. It is possible to find this information in the *Baḥriye* of Pīrī Reʾīs and in some of the European books on the outer sea (i.e., the Atlantic and Indian Oceans). Those who sail on the Mediterranean usually consult the *Baḥriye*.

This entire subject is clearly marvelous—indeed, nearly miraculous!

[Marginal note by Kātib Çelebi, R 15a] When the needle of the compass is broken, it will not stop in any position. In that case the end that should point to magnetic north is struck several times with a knife and napped like the nap of broadcloth; then the tip pointing north is rubbed firmly, and it will work properly. The magnet is kept in a piece of red broadcloth from the odor of garlic and onion. If the magnet loses its power, it is soaked in fresh goat's blood or vinegar. If it is rubbed with olive oil, the iron (i.e., magnetism?) escapes; if it is soaked in goat's blood, it is drawn back in—this has been tested. While the magnet is in the bowels of the earth (i.e., not yet mined), the northern end points to the North Pole and the southern end points to the South Pole. When it is extracted, according to the soundest opinion, it maintains the same orientation.

[M 64] Supplement by the Publisher (İbrāhīm Müteferriḳa)

It should be known that one of the countless brilliant signs and abundant favors without end from God, and one of the limitless blessings that were bestowed simply for the benefit of His worshippers is the lodestone (magnet), that serves as *qibla* indicator for the people of Islam and shows the way to those who sail on the endless sea. All the ancient and modern philosophers have exerted great effort to rightly comprehend the essence of one piece of this stone and to discover and explain its nature, but until now they have been unable to do so.

According to some Latin books, this stone was first discovered in the mountains of Magnesia (*Maġnisa*, today Manisa) in Anatolia. The word *miqnāṭīs* (magnet) must come from the Arabicization of Magnesia or a corruption of it; or else it was the name of the person who discovered it. The property of this stone to attract iron, and hence to endow iron and steel with the same ability, had in fact been known to the ancients. However there is a third property that remained a hidden secret, inaccessible to the mind of man—viz., if a piece of lodestone is suspended in the air and left on its own, or if it is placed in a container and put in a pool or basin of water and left on its own floating on the water, it will always, because of the hidden property of its nature, turn parallel to the circle of the meridian, with its north pole—i.e., its side magnetised to face north—turned toward the North Pole of the earth, as can be observed in the compass needles currently in use.

The property of the magnet to point toward the pole was at first unknown to the people of the West. When they eventually learned about it, they were utterly amazed and imagined the following: the material substance of the lodestone must be a small sphere in the interior of the earth concealed by the soil, like the yolke inside an eggshell. That magnetic sphere, as they imagined it, must have poles that were congruent with or parallel to the poles of the sphere of the earth (or the globe), and must also have a meridian, an equinoctial line, and other circles by analogy to those on the earth.

Once this property was learned, in the Christian year 1402 a man appeared in the city of Amalfi, which belongs to the district of Naples in the country of Italy, who invented the compass and presented it to the owners of seagoing vessels, thus leaving behind a souvenir in this world.

Although 400 years ago people in the West had no knowledge that the poles of the magnet turned toward the poles of the earth, as described above; and so the invention of the compass and the knowledge that its needle could be magnetised to point north were delayed until that time; yet according to the reports of travelers who journeyed from west to east, the properties of the magnet had long been known to the people of China and instruments made with magnets were long in use among them. Those travelers reported that envoys were once sent by the Khaqan of Cochin to the Khaqan of China, and after the diplomatic ceremonies were completed and they were about to return, the Khaqan of China presented them [M 65] with various valuable gifts, including a jeweled *qibla* indicator and a compass inscribed with the four directions and the names of the winds. Since the country where the envoys were returning lay south of China, the Khaqan made the following witty remark: "This compass will show you how to go to your country; there is no need for a guide." They also reported that the people of China had been aware of the properties of the magnet for 2500 years.

As mentioned above, when the Western philosophers learned the properties of the magnet, they made a great effort to investigate why its needle pointed to the north. Some said it was the Polar Star that drew the magnet or the magnetised needle and directed it toward the North Pole. Others said it was a section of the heavens around the pole. Still others said the two poles of the earth consisted completely of lodestone. Some Sufis said this was a divine dispensation and that God's will brought it forth and intended it as a sign of His omnipotence and as a marvel for His worshippers to ponder and that He concealed its mystery in itself. Then some of the moderns, seeking a natural cause for such phenomena, said that it was an effect of the condition of the air. Others said that it was simply an emanation and influence of the earth and based on earthly causes, such as the hypothesized magnetic sphere that extended from south to north in the interior of the earth like veins and arteries. Most have adopted this view.

However, it should not be forgotten that the pointing of the needle towards due north is restricted to certain places. In some places it deviates somewhat to the west or east of due north. These deviations and motions of the needle are not uniform: it may move faster or slower, or go forward or backward or remain still. These phenomena have been tested and recorded over time. In Istanbul, where we live, as indicated in the text of the book, when a deviation to the east from due north is observed, it does not suffice in the construction of mosques and in the orientation of prayer-niches to simply look at the needle of a compass; one must also use a quadrants and astrolabe and other well-tried astronomical instruments in order to make accurate calculations. I have expanded on the topic here so that people will not be neglectful in this matter.

A marvelous story: In AH 1140 (1727) a dispute arose over the siting of the prayer-niche in the noble mosque built by the late Admiral Muṣṭafā Pasha in the park of Bebek near Rūmeli Ḥiṣār. The experts brought out several compasses and also a quadrant and astrolabe and other astronomical instruments. The orientation suggested by the compasses did not agree with those instruments, and a serious dispute arose. Eventually the compass readings were discarded and the prayer-niche was sited according to the elevation taken by those instruments. Some time later a few zealous individuals were stirred to action and began a thorough investigation. They made a *qibla* indicator with a large needle and when it was tested, the needle was observed to deviate 11½ degrees west of due north, which was contrary to received opinion. God knows best what is correct and recourse is to Him.

End of the publisher's supplement.

[M ILLUSTRATION] [M 66]

13 On the Difficulty of this Science and the Need for Correction and Assistance from Experts

It should be known that it is difficult to draw maps accurately, and making land maps is more difficult than making sea charts. That is because it is nearly impossible to copy all the locations in their proper places without making a mistake. Leaving aside the mess that Arab and Persian copyists make when they try to imitate the illustrations in Turkish, Persian, and Arabic books; even in European printed books, most of the cities and countries are wrong. Especially on the pages of the Muslim countries, Africa, and Asia there are a huge number of mistakes. This poor one (Kātib Çelebi), who travelled in Anatolia, Syria, Ḥijāz, Iraq, Jabal (Jibāl), and Azerbaijan and saw those places with my own eyes, observed that the maps they made of these places were highly unreliable. But despite so many errors, they are still preferable (to the Islamic ones). And they are excusable in this respect, as the author of the *Atlas* apologizes more than once, saying, "I collected the chorographies that I could find, but there were some I could not lay my hands on, and so had to copy what I found in old books. If the people of those regions would draw up chorographies of their own countries and attach them, it would be a great benefit." [R 16a]

As a result (of this problem), the Muslim geographers gave up trying to provide maps and illustrations and were satisfied with giving names and descriptions of places. Even so, they did not always hit the mark. Indeed, they committed many gross errors as a result of guesswork and conjecture. For example:

- Imam Masʿūdī, who was preeminent as a historian, states in the *Murūj al-dhahab* that some parts of the north shore of the Sea of Azov reach the North Pole; that the Golden Horn of Constantinople gushes forth from a sea; that near it is a city named Tolia; and that the sea known as the Sea of the Goths is the Varangian Sea (Baltic).
- Ibn al-Wardī is mistaken in his *Kharīda* when he says that a strait gushes forth from the Western Sea, passes behind the land of the Slavs, flows for a distance of two months, and empties into the Sea of Rūm (Marmara) before Constantinople.
- While the Jayḥūn River (Amu Darya, Oxus) actually flows into the Sea of Jurjān (Caspian), Ibn Ḥawqal says it flows into the Lake of Khwārazm (Aral Sea).
- While the source of the Nile is actually ten degrees south of the equator, al-Suyūṭī and others have written many impossible things about it, saying that it is a distance of three years away, and that it splits the sea.
- The author of *Menāẓirü'l-ʿavālim* (Mehmed ʿĀşıḳ) traveled throughout Turkey (*Rūm*) and when it came time to write about it asserted that the river of Edirne, the Tunca and Meriç, flows into the Black Sea [whereas it actually flows into the Aegean]; also that Çeşme is opposite Rhodes, while it is actually opposite Chios.
- ʿĀlī, the author of *Künhü'l-aḫbār*, attributes Sharīf al-Idrīsī's book entitled *Nuzhat al-mushtāq*, which belongs to the category of "Routes of Countries" (*masālik al-mamālik*), to the Prophet Idrīs.
- The author of the book on the New World (*Tārīḫ-i Hind-i Ġarbī*), writing about the Varangian Sea (the Baltic), interprets the phrase in the *Tuḥfa*, *ʿalā sāḥilih umma ṭiwāl kumāh*, which really means "On its coast is a tall and courageous people," to mean "The people on its coast are tall as mushrooms (*kam'ah*)."

Other objection could be raised to the errors in these books. The examples cited are testimony to their carelessness and inaccuracy; as in the verse, "A man has attained enough if he enumerates his faults."

ADDENDUM: That some historical information is mixed into this science (of geography) is because history [M 67] is known as the salt of the sciences, and geographers discuss matters of government and mores in the appropriate places of their works. The author of the *Atlas* expands on this topic in chapter 49, which is an introduction concerning Gallia (France), where he says:

"We did not come into this world for our own benefit alone; rather, God commands us to benefit others. Each person must strive to benefit others according to his ability, and my ability is to instruct people in the science of geography. Needless to say, geography is helpful and useful in preserving history. Moreover, its precedence over history comes from the fact that knowing it facilitates good government. That is because the science of geography has as its task is not only to record the status of countries but also to describe the rituals and principles of their people, the management of the affairs of state, and matters of government. Just as a painter falls short who is only preoccupied with the perfection of the limbs and pays no attention to gracious gestures and attitudes, so a geographer who only describes places and takes no interest in the condition of the people who live there and other unusual things is like a painter who depicts a naked corpse. Therefore I have devoted a few pages at the beginning of my discussion of every country to the organization of government and other remarkable matters, so that those interested in the conditions of civilization and public affairs will benefit from my book.

"It would hardly be surprising if there are deficiencies and errors in my book, since my sources do not discuss all the conditions of a country and its government. However, to the extent that I could find something on these subjects in the chronicles, I have included it. [R 16b] If someone notices an error concerning his country, he should bring it to my attention and point out the deficiency. I will then correct it and respectfully offer a prayer of gratitude. If everyone diligently corrected the information concerning his own country, this would be especially useful to students of geography."

The poor one (Kātib Çelebi) says: In my original translation of the *Atlas* I recorded the errors of this writer concerning the Ottoman Empire and its method of government just as they were written; but in this book I corrected them as far as possible. There are also many errors concerning the countries of Africa and Asia based on the works of their peoples. Hopefully it will be possible to correct these as well.

14 Table of Contents Showing the Chapters on the Countries, Climes, and Cities Mentioned in this Book

It is arranged alphabetically with the chapter numbers indicated. This is a summary table of contents. If one needs to know in which chapter of the book a country can be found, he can learn it from this. However, if one needs to know in which country a city or town is located, he should refer to the detailed table of contents (i.e., index) at the end of the book, where the names of countries and cities are arranged in alphabetical order.

[This Table of Contents has not been included in the current translation]

[R 19a] SECTION 1[2]

Section 1 of *Cihānnümā* is on the aims of this science (geography). It is divided into several chapters.

Chapter 1: On the Earth

Introduction: On the reason that land and sea are separated from each other

It is well known that water is one of the four elements. Because it is associated with a substance that is simple, cold, wet, and dense, its natural position is below the air and above the earth. While by nature it ought to completely cover the globe and engulf it, it does not cover it completely. Some astronomers attribute the fact that some places on the earth are exposed to divine favor, saying that God's will determined the creation of land animals and provided for their reproduction and nutrition, and that the natural cause for this is not apparent. Others, while accepting the first premise, dispute the second, saying that there is a cause for everything in the material world and that coming into being by natural causation is divine custom.

Now the reason the sea does not cover the earth completely is as follows: Because the sphere of the sun is eccentric (in Ptolemaic astronomy), it has an apogee and a perigee and its perigee now occurs at the end of Sagittarius which is a southern constellation. [M 71] When the sun in its course reaches the perigee, it comes close to the center of the earth and the temperature increases significantly. As long as it remains in the southern constellations, it is close to the earth; when it is in the northern constellations, it is at the apogee and far from the earth. Now the increase in temperature heats the element of water and causes it to move. For when water boils on one side of a large pan and starts to move, water elsewhere in the pan is drawn to it. Similarly, when the sea in the southern hemisphere [R 19b] starts to move due to the increased heat of the sun, the seas in other places are drawn to it and the northern parts of the world are left exposed. Thus: [Plate 13]

Because land and sea form a single globe, over time the blowing of winds and the flowing of floodwaters affected the places on earth that were exposed and hills and depressions appeared. When the waters started to flow, they rushed into low-lying places and flowed between the mountains and hills, and so seas formed here and there in the exposed land surface. However, the belief that this

2 This use of *bāb* ("section") is anomalous. There is no Section 2. Rather, the use of *faṣl* ("chapter") begins here and continues to the end.

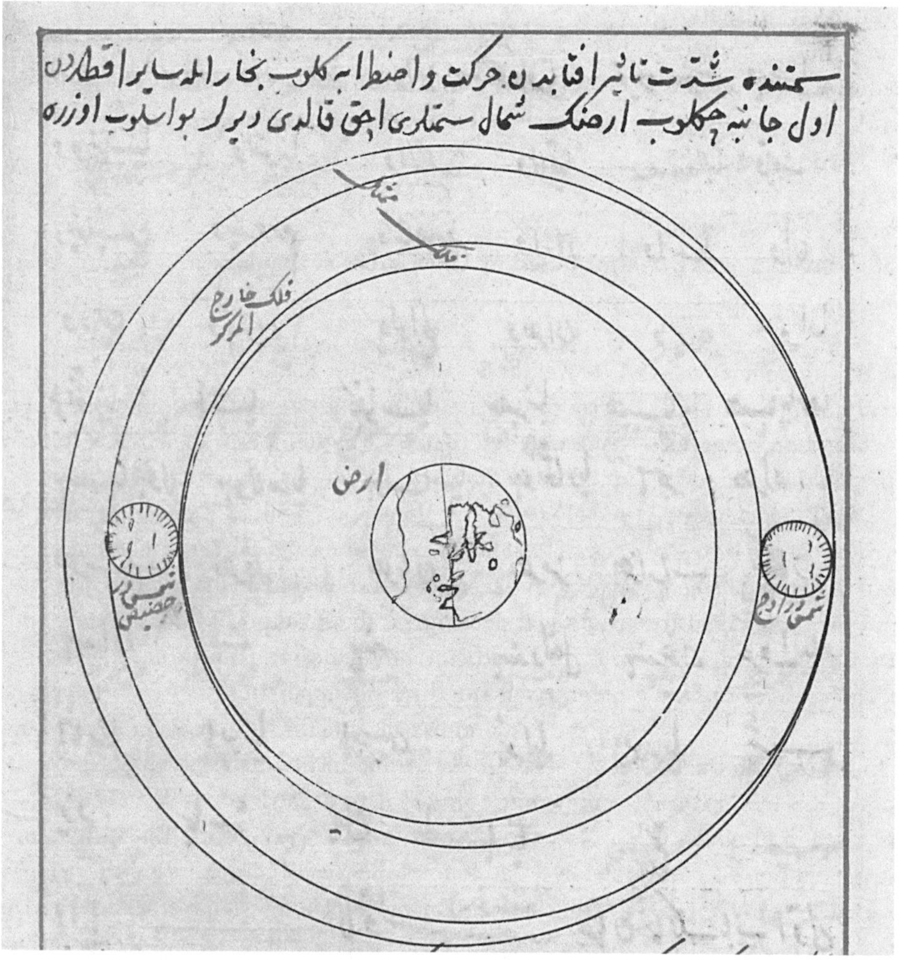

PLATE 13
Scheme showing earth with sun at apogee and perigee, R 19b (Appendix, fig. 30)
TOPKAPI PALACE LIBRARY, REVAN 1624, WITH PERMISSION

phenomenon is the cause of the exposure of land is dubious, since the exposed half of the earth is below the sun's perigee, and at the extreme point most if not all of it would be attracted to it.

> [Marginal note by Kātib Çelebi] Upon examination, it appears that the reason for the increase or decrease in temperature is not the closeness or farness of the sun. Rather, the increase comes from the perpendicular rays of the sun and the decrease from the oblique rays of the sun. When the rays are perpendicular, the reflection is intensified, just as with kindling mirrors.

To resume: The earth is a simple, cold, and dry body. Its natural position is that of the lowest of elements. Its coldness and dryness account for its being dense and held together, so that its exterior (or exposed surface—*zahr*) is the place where animals live and the interior (or concealed part—*batn*) is the source of plants and minerals.

The scholars, having proven with much evidence that the earth is round, have shown that no mountain on the face of the earth is higher than 2½ parasangs. So the highest mountain relative to the diameter of the earth is like one-seventh the width of a grain of barley relative to a cubit. The mountains do not prevent the earth from being round. Because of its great size it appears to be flat. The minds of those who have not acquired a share of the philosophical sciences cannot go beyond what they see with their own eyes, and because they see the place where they are as flat they think the whole world is flat. This opinion of theirs does not conform to reality.

At the center of the earth is an imaginary (or postulated) point that is the center of the cosmos and the true bottom. Heavy bodies incline toward it and gravitate toward it from all directions. If an obstacle arises, it proceeds and finds the point. Given that the distance from the earth to the sky is the same in all directions, heavy bodies remain in the middle of the sublunar world, whether by mutual repulsion or by attraction of the center. The fundamental element of earth is the pure element found at the center. It is colorless. [R 20a] Above it is the layer of mud, and above that is the layer of mixed materials where minerals and plants are formed. These layers surround each other concentrically.

Some believe that the total number of layers is seven, in conformity with the noble verse (Koran 65:12), "[God is He who created seven heavens,] and of the earth the like thereof." Others interpret this verse as referring to the seven climes. The tale of the ox and the fish, related by Ibn 'Abbās—if any credence can be given to it—is accounted for by the constellations of Taurus and Pisces. Such statements by the Companions of the Prophet probably derive from the fact that at the time of the rise of Islam, when the faith was not yet firmly established and people thought along the lines of the philosophers, the Companions did not want them to be deprived of grasping and relating the rules of the religion of Islam, so they responded to questions that were unrelated to the faith by recourse to the philosophical principle, "Speak to people according to their intellectual level." [M MAP] [M 72] Since the job of prophets and of the Companions of the Prophet is to teach the people matters of faith, not to explicate the reality of things, when they were asked about the phases of the moon, the verse (Koran 2:189), "Say: They are times appointed for men and for the pilgrimage" was revealed, so the people would know the kinds of questions to ask them and would not ask questions about matters outside the faith. Thus, with regard to fertilization and pollination, the Messenger of God said: "You know best the affairs of your world".

Later Naṣīr (al-Dīn) al-Ṭūsī and other astronomers asserted that the intersection of the solstice with the circle of the horizon results in four sections on the globe, the inhabited quarter being one of the two quarters in the north—which one is unspecified—while the nature of the remaining quarters is unknown and hidden, or else they are completely covered by the sea.

The poor one (Kātib Çelebi) says: The moderns have travelled the ocean and have determined the nature of the remaining quarters. Places unknown to the ancients have been discovered and are now known. Today all the exposed areas on the earth have been mapped out, and it is no longer possible to confine the inhabited areas to one quarter.

As drawn by the modern geographers, the land areas of the globe that are exposed from the element of water by the attraction and repulsion of the elements at the center of the sublunar spheres are the places on the two circles (maps of the eastern and western hemispheres) colored red and yellow. The yellow areas are landmasses and the small red areas are islands, while the remaining white areas are the seas. The line that divides the two circles in half is the equator. The red circles on either side of the equator are the Tropic of Cancer and the Tropic of Capricorn, which indicate the obliquity of the ecliptic, the extreme declination of the sun to the south and north. They extend to 23½ degrees latitude on each side of the equator. The sun follows its course for six months in the north, so there is summer on that side and winter on the other; then for six months it moves through the southern constellations, so winter is on the northern side. The small red circles near the poles indicate the polar tropics (Arctic and Antarctic Circles). The other circles, like the lines on a chessboard, are the circles of latitude and longitude, as previously mentioned. Distances of places near the poles are foreshortened, the closer they are to the circumference (of the circular map). At the other extreme, of places near the equator, distances are compressed as a concomitance of mapping the globe on a two-dimensional surface on these two circles. It is as though it (the globe) were cut in half along the (meridian which then appears as) circumference and squashed, making flat what was round, while the surrounding areas of necessity are slightly opened up and stretched out. Only in this way is it possible to indicate the borders and outlines on a hemisphere that is flattened out; it is not possible to show them as they really are. [R 20b] For this reason, when the geographers have to be precise, they refer to a standing globe (rather than a two-dimensional map).

Alternatively, one can take 36 arcs and spread them out on both sides, as though one were splitting a ball in half and then squashing it. However, because of the great distortion in this kind of map, modern geographers prefer to use two circles.

Now the two poles of the globe are at 90 degrees latitude, so they have six months of daylight and six months of night. The so-called "darkness of Alexander" refers to the six months of night at the North Pole; otherwise, no permanently dark region has been established by the geographers. [M 73] The rampart of Gog must also be in the regions near 90 degrees latitude. However, the geographers have not yet discovered it and determined its location. God willing, I will discuss this rampart in detail in the chapters on Cathay and Turkistan.

Measurements of the globe

According to Ptolemy's measurement, the circumference of the earth is 24,000 miles, the diameter is 7636 miles, and the surface area is approximately 20,363,636 parasangs. Based on this measurement, Jamshīd al-Kāshī calculated the distance to the heavenly bodies in his work entitled *Sullam al-samā'*. Using the radius of the earth and the cubit as units of measure, he determined that the distance from the center of the earth to the surface of the moon was 32 times the radius of the earth and 56 minutes and 59 seconds. He also determined other distances by geometrical means. When the distance of the earth's radius itself was deducted from this, the remainder was 41,664 parasangs and this was the distance between the earth and the sky. While this distance may seem implausible to those who do not understand mathematics, it is in fact correct and has been corroborated based on certain knowledge and scientific principles.

Finally, the difference between the measurement of the ancients and that made by al-Ma'mūn sometimes could not be reconciled with this calculation, but this measurement was attempted again after al-Ma'mūn. After the New World was discovered, three ships sailed west and about three years later returned from the east. It turned out that the calculation made by al-Ma'mūn was indeed correct.

It is written in some of the old books that the distance around the world is a 500 years' journey and the distance from the earth to the sky is also a 500 years' journey. There is also a hadith to this effect regarding the heavens. However, the intended meaning of this is to express its immensity, not to state a specific distance; for such numbers as 50, 70, 500, and 700 are used to express immensity. Just as in the noble verse (Koran 9:80), "Even if you ask forgiveness for them seventy times …" the number is not meant to be taken literally, so is the case in this hadith. Otherwise the calculation would be mistaken.

The places depicted on the second circle of this map of the world—which in extent are equal to and in population are equivalent to if not superior to the inhabited quarter (i.e., the "Old World")—are a New World whose wonders and marvels confirm the perfect power of the Creator and reveal the wisdom of His works, express the extent of His munificence and increase the certainty of those who take lessons from creation. This New World was unknown, however, to the ancient writers. It had not been seen or heard of since the time of Adam and had not been visited from the creation of the world until its discovery, when it became know by God's preordination and became famous under the name "New World." It is also known as the "West Indies" and the "New Indies," as discussed in the chapter on America. [Plate 14]

[R 21b] **Chapter 2: On the Seas**

Since the sphere of earth and the sphere of water have been briefly mentioned, in this chapter I will briefly discuss matters relating to the sphere of water, then I will begin to describe the regions of the earth, which is the aim of the science of geography. For this chapter I originally translated the eighth treatise of the book of *Meteora*—i.e., *Universe of Air and Elements*—a commentary on the philosophy of Aristotle made by the Academy of Coimbra belonging to Spain. I will summarize it here and cite the comments of the geographers on several topics.

1 On the formation of the sea

The ancient philosophers differed widely on the origin and formation of the seas [M 74] In the first part of this book Aristotle cites their views as follows:

Diogenes and others supposed that at first water completely covered the land. Sometime later, most of it evaporated due to the heat of the sun's rays. The remaining water was not able to evaporate and was concentrated in the sea, so some places on the earth were exposed.

Anaxagoras and others supposed that the sea, being salty like a pearl, was originally a kind of pearl of the land that emerged out of it as the earth warmed up from the heat of the sun. But this view is refuted by the fact that it (i.e., the sea water) was created together with the other elements. It is salty because it is mixed with particles of dry land heated by the sun and particles of sulfurous minerals that have dissolved and concentrated in the water. If the sea water were sweet, it would smell from being stagnant for a long time and the wind would spread the bad odor throughout the earth, putrifying the air and causing the destruction of the land animals.

2 On the position of the sea

Some ancient Roman and Greek authors supposed that the sea was higher than the land. In fact, the land is slightly higher than the sea. The proof of this offered by those ancient writers is recorded in the book mentioned above.

3 On the division of the seas

Authorities on hydrography—i.e., those who chart the seas—have divided all the seas into five: 1) Ocean, which is the term for the Circumambient Sea; 2) the Mediterranean; 3) the Red Sea; 4) the Persian Gulf; 5) the Caspian. Some moderns have divided them into seven, adding the following: 6) the German Sea; 7) the Black Sea. Some of the Islamic authors also follow this scheme. We will follow suit and provide here a summary description of the seven seas one by one.

1) Ocean: It is pronounced in various ways in Latin—Oqianuz, Osianuz, Woqthianus—all of which designate the sea. This term is the name of a man who in ancient times was believed to be a god. They named the sea after him because of his great fame. There is a detailed story about him in the books of the Greeks.

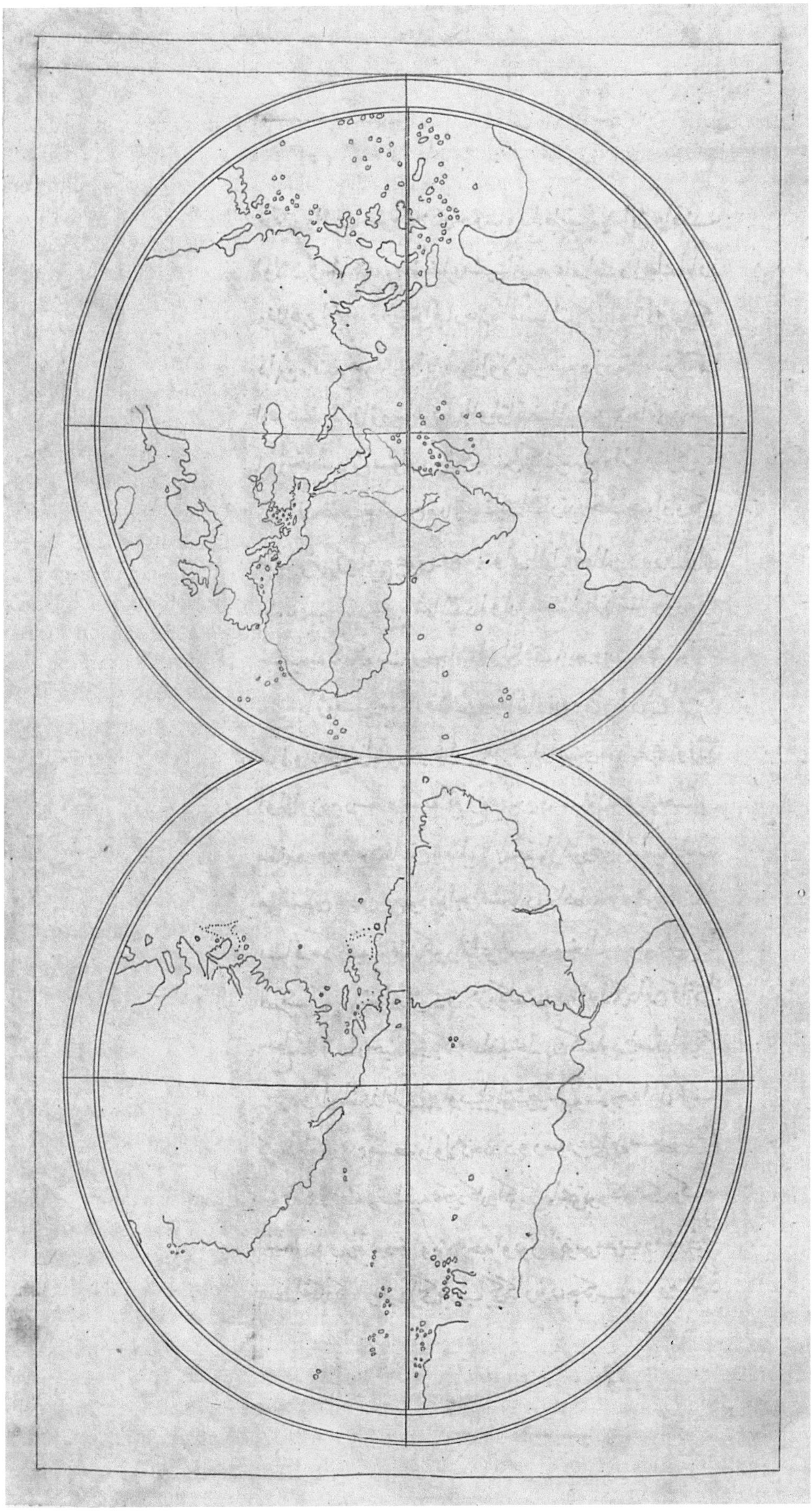

PLATE 14
World map, R 21a; cf. M 71/72 (Appendix, fig. 31–32)
TOPKAPI PALACE LIBRARY, REVAN 1624, WITH PERMISSION

The author of the *Introductio* (Philippus Cluverius) says that all the seas amidst the landmasses were called Ocean, then a division was made according to various places. The first division was according to the four directions of the world: the Eastern Ocean, the Western Ocean, the Southern Ocean, and the Northern Ocean. Then each was named after the region that bordered it.

First, the Northern Ocean was called Sarmaticus near the coast of Salmatia; Scythicus—i.e., the Tatar Sea—near the coast of Scythia; and Hyperboreus after the tribe dwelling on the northernmost coast. It was also called (Mare) Chronium after the cold planet Saturn (otherwise Chronos) which prevailed in those regions; and (Mare) Amalcium, meaning Frozen Sea. And because of the cold and darkness of that region, where the night is six months long, it was also called Mare Marusa, meaning Dead Sea.

The Eastern Ocean was at first called Sericus after the country of Serica (Cathay) on the shore of Qitāyā (Khitay, Cathay). Later it was called the China Sea or Sinnis Oceanus. Then the islands in the Eastern Ocean were called Hipparis Pelagus and Archipelagus. And it was named Lazari after the man who discovered it (i.e., Pizarro who discovered the Pacific in 1513). [R 22a]

The Southern Ocean was called Oceanus Hindicus where the Indian Ocean borders on India; [M 75] and Gangeticus where the river of India known as the Ganjes flows into the sea, viz., the Bay of Bengal. The Indian Ocean is also called Lāntkhīdāl (Lantchidol). The Persian Sea where it borders on Kirmān and Fārs is called Mare Persicum, meaning Sea of the Persians; and the sea off the coast of the Arabian Peninsula is called Arabicum. Then the sea off the coast of Comoros (*Jazīrat al-qamar*, also used for Madagascar) was called Mare Asperum ("Rough Sea") because of its many reefs; another name for it is Madagascar. So this half of the Southern Ocean—all of which is called the Sea of India—is distinguished as Gangeticus, Persicus, Arabicus, and Aspar. One part of the Southern Ocean begins at the famous cape known as Bona Ispiransa (Cape of Good Hope). Toward the west it is called Oceanus Ethiopicus, meaning Ethiopian Sea. It is known by this name as far as the mouth of the Niger River. The southern regions are sometimes called Mare del Sur, meaning Peaceful Sea (i.e., the Pacific Ocean) by some of the moderns. This designation is used for the ocean stretching from the east as far as the Magellan Strait.

The Western Ocean is first of all called Oceanus Atlanticus, meaning Sea of Atlas, after the Atlas Mountains on the west coast of Africa, also known as the Mountains of Mauritania, for the section from the Niger River as far as Cape Roca in Spain. From Cape Roca as far as Cape Lafour, belonging to the country of Celtica of the kingdom of Gallia (i.e., France), it is called Oceanus Galicus, meaning Sea of Gallia. And because it skirts Cantabria, belonging to Spain, it was also called Oceanus Cantabricus. The country of Gallia extends to Aquitane, where it is called Oceanus Aquitanicus. The part between Britania (Britain) and Gallia is called Britanicus, that between Hibernia (Ireland) and Britania is called Hibernicus. The part bordering on Scotia (Scotland)—known in olden times as Caledonia—which is the northern part of Anglia, was called Caledonius. Then the part from Britania to Jutia (Jutland), which is the country of Dania (Denmark), was called Oceanus Germanicus.

The ancients distinguished and divided the entire ocean using these names, most of which are still in use. ANOTHER DIVISION. Then the moderns went to distant lands and gave new names to the ocean that they discovered. In their view, the entire ocean has three parts: 1) the seas between Asia, America, and Magellanica (south polar region), called by the Spaniards Mar del Zur, meaning Southern Sea, and also Mare Pacificum because it is very peaceful; 2) the sea between America, Europe, and Africa extending to the equator, called Mar del Nort, meaning Northern Sea; 3) the sea extending south from the equator between America, Africa and Magellanica, called Mar d'Ethiopia; and the sea that flows east from between Africa, Asia and the south polar region, is called Mar d'India.

Division of the Seas: The ancients specified five large bodies of water within the inhabited quarter of the Old World. Some are bays and others are gulfs—i.e., bodies of water without straits: 1) The Sea of Sweden (Baltic Sea), recorded in Pliny's book as the Gulf of Codanus. 2) The Inland Sea, known as Mediteraneum. [M MAP] [M 76] 3) The Persian Gulf, known as Sinus Persicus; 4) the Red Sea, known as Mare Rubrum; 5) the Sea of Ṭabaristān and Jurjān, known as Caspium (Caspian Sea). We will discuss them one by one. [R 22b]

THE GERMAN SEA (BALTIC SEA): In our astronomy books and clime books is recorded as the Varangian Sea. The learned al-Shirāzī stated in his book *al-Tuḥfa*: *ʿalā sāḥilih umma ṭiwāl kumāh* ("On its coast is a tall and courageous people") by which he meant the Varangians and the Swedes. But the author of the history of the New World (*Tārīḫ-i Hind-i Ġarbī*) translated this as: "On the coast of this sea live a people who are as tall as *kam'ah*"—i.e., mushrooms. By failing to record that they were tall, courageous and heroic, he gave a mistranslation. Today in the language of its people it is known as the Baltic Sea. It is surrounded by the countries of Pomerania, Denmark, Sweden, Livonia and Prussia. Because the country of Germany is also near this sea, it is known in our country as the German Sea. Its shape and size can be seen on the map.

Sea of Rūm (Mediterranean): It extends from the Strait of Ceuta (Gibraltar) to the coast of Syria. Because it is enclosed between Europe and Africa it is called Mediterraneum, meaning Inland Sea. In our country it is known as Aḳ Deniz (White Sea). It has various sections named after the regions (or provinces, *vilāyet*) whose coasts it border on.

First of all, this sea is separated from the ocean and enters (the Mediterranean basin) from the Strait of Ceuta (Gibraltar). The ancients believed that this sea must have cut through the area between Calpe, at the Cape of Spain, and Mt. Abyla in Mauritania in Africa and poured (into the Mediterranean basin). The sea at the strait is called Gaditanum (Fretum) after the island of Ghādes (Cádiz), which is outside the strait. The strait is called the Strait of Hercules and the mountains on each side are called the Pillars of Hercules. Hercules was a highly esteemed ruler in ancient times and many stories are recorded about him in the books of the Greeks.

The southern coast of this sea borders Africa, the northern coast borders Europe, and the eastern coast borders Asia. We will discuss in order the European coastlines, giving the names part by part, moving from the strait to the interior.

Sea of Spain: The part at the southern coast of Spain is called Ibericum or Ispanicum Mare. The part in that vicinity around the islands of Yābisa (Ibiza), Minorca and Majorca, known as the Baleares, is called the Sea of Balearicum.

Sea of France: the part opposite the province of Narbonne is called Galicus Sinus and that opposite Genoa in the region of Ligorna (Livorno) is called (Sinus) Ligusticus.

Sea of Rome: The part opposite the Italian coast from Genoa to Sicily is called Tuscus (Sinus). The Greeks called it Tyrrhenum Mare and the Latins called it (Mare) Inferum. Corsica and Sardinia are in this vicinity and the waters around them are called Corsicum Mare and Sarduarum Mare. Then the sea extending from Sicily to Crete is called Siculum; from Crete to Cyprus, Creticum; and from Cyprus to the mainland of Asia, Cyprium. The sea lying between Sicily, Italy and Greece is called Ionium Mare, meaning Sea of the Greeks.

The Gulf of Venice (Adriatic Sea) is a part of the Ionian Sea, located between Italy and Illyricum—i.e., the coasts of Albania and Bosnia—and extending 700 miles to the northwest. [M 77] It is known as Adriaticum, named in ancient times after the city of Adria and still known by that name. The part that is west of Italy is called Inferum, meaning Lower Sea, and the part that is east of it is called Superum, meaning Upper Sea. This gulf also has sections. That off the coast of Albania and Bosnia is called Illyricum Mare, and this too is divided into two parts: The Sea of Dalmatia, which is the country of Albania, called Dalmaticum; and the lower part, named after Liburnia and called Liburnicum. The part between the Gulf of Venice and Sicily is called the Ausonium Sea because in ancient times this coast of Italy was known as Ausonia. Subsequently it was superseded by Adriaticum [R 23a] and has been known by that name ever since.

The Sea of Morea was called Corontiacus after Coron (the Peloponnesian port of Koron).[3]

The sea of the islands surrounded by Rumelia, Anatolia and Crete was called Aegeum Mare (Aegean). Subsequently it was called Archipelagos and the islands in it were called the Archipels. Various parts of this sea have different names after the names of islands.

The Straits: This sea between Asia and Europe passes through a strait and then opens out again. It is called Hellespontus. The Persian ruler Ardashīr (i.e., Xerxes, 486–465 BCE) made a bridge of boats on this strait and marched his army across it to Greece. Where the strait opens out is called Propontis (Sea of Marmara), which is the Sea of Constantinople. From here another strait goes off to the north, called Bosphorus Thracius, which is the Strait of Istanbul. Ardashīr's father Darius (d. 486 BCE) made a bridge here and marched his army across it. From this strait the waters open up again to the Black Sea, called Pontus Euxinus. The Sea of Azov, known as Lake Maiotis, is connected to the Black Sea by a strait currently called the Strait of Kerch and Taman; the ancient name was Bosphorus Cimmerius. The Tanais River, now called the Don, flows into this sea and separates Asia and Europe.

Next, the sea that extends off the coast of Asia from the Aegean to Caria is called Carium; that of Rhodes, Rhodium; that off the coast of Pamphylia, Pamphylium; that off the coast of Cilicia—i.e., Silifke—Cilicium; that off Syria—i.e., Shām—Syrium; that off Cyprus, Cyprium; that off Palestine, Palestinum; that off the coast of Egypt, Egyptiacum; that off Libya—i.e., Ifrīqiya—Lybicum or Africum; that off Numidia, Numidicum; and that off Mauritania—i.e., Algeria, etc.—Mauritanicum.

The length of the Mediterranean coastline, according to the measurements of Muslim sailors, is 13,057 miles. They say that coral is only found in this sea.

Sea of Qulzum (Red Sea): It is the sea of Mecca. It is called Sinus Arabicus (Arabian Gulf) after the Arabian Peninsula which is east of it. In the ancient books it was

3 This is apparently a misunderstanding of *Introductio*, 38–39: Inter Peloponnesum & Achaiam Corinthiacus est sinus (Between the Peloponnesus and Achaia is the Corinthian Gulf).

recorded as the Red Sea—Mare Rubrum and Mare Rosso. It is also called Mar di Mecca, meaning Sea of Mecca.

There are various opinions about the redness of this sea. Some said that it appears red from the reflection of the sun's rays; others that it is from the redness of the soil (i.e., the seabed). The Greeks called it Eritreum Mare, supposedly named after a ruler named Eritra, meaning red. [M MAPS] [M 78] As Solinus writes in the 45th chapter of his book entitled *Wonders of Creation*: "This sea was named Arutreus after one of the rulers of that region, Eritra, the son of Perseus and Andromeda." He also writes that there was a spring on the shore of this sea, and the fleece of any sheep that drank from it turned red. However, the Portuguese who sailed on this sea saw that it was red, but when they drew water from it they found that it looked like other water, so they realized it was from the sun and from the red sand at the bottom of the sea.

This sea begins opposite Socotra and ends at Suez. Its length is about 350 parasangs.

PERSIAN SEA: It is called Sinus Persicus, meaning Persian Gulf, after the land of Persia which is east of it; also Mare Persicum. Its length, from the south to the north and west, is about 280 parasangs.

SEA OF KHAZAZ [RECTE: KHAZAR] (CASPIAN): It is the Sea of Shirwān and Jurjān. It is called Caspius or Hircanus after two tribes who lived around it.

The ancients supposed that this sea was connected to the ocean in the north. Aristotle and others ultimately did not agree with this opinion, saying that it had no connection with the ocean. [R 23b] Ptolemy agreed, saying: "The Caspium Sea is surrounded on all sides by land. It is an island of water. Therefore it would be more proper to call it a lake." The moderns followed him in this matter.

The current name is Mare Sala, after a city on its shore. Some have claimed that there is an underground channel from this sea to the ocean; but it has not been proven to exist. Some said that Caspi and Hyrcani are mountains on the shore of this sea. And they recorded that it took twelve days to travel around it.

End of the description of the seas.

In the following pages, the remaining topics will be explicated in their proper place. For now let us continue with other matters concerning the seas.

4 On the various motions of the seas.

Because they are attached to a heavy element (i.e., water), by nature they move toward the center of the earth, just like the land. Besides this natural motion, they also move in some other ways. So the motions of the seas are:

1. The natural motion downward.
2. The motion impelled by the wind.
3. The motion from east to west. It has been tested and proven by sailors that, aside from the ebb and flow of the tides, there is a current in the ocean that flows from east to west. One can use it to sail westward from Spain to the New World in one month, while it takes three or four months to return. The Portuguese noticed this motion while going back and forth from the coast of Africa to the East Indies. For, after passing the Cape of Good Hope, no matter how favorable the winds, a ship sailing east went more slowly than one sailing west. The same phenomenon was observed in the Mediterranean: those travelling from Spain to the coast of Syria sailed more slowly outbound then they did on the return. The cause of this motion is thought to be the diurnal motion of the Greatest Sphere (or Primum Mobile), which has a hidden influence on the ocean. Just as it makes the other spheres turn, so it makes the seas turn as well, [M 79] for they are subject to its motion.
4. The motion from north to south. This motion has also been confirmed by the experience of sailors. Aristotle, in the first chapter of the second book of *De caelo* (*On the heavens*), says that the cause of this motion is that in the northern region the land is somewhat elevated and the sea flows from there to low places. There are several rivers in the north and they contribute to the flowing of the seas; thus the Don empties into the Sea of Azov and the Danube and other rivers empty into the Black Sea. Albertus (Magnus), who wrote a commentary on this book, says that because those regions are far from the sun and very cold, a great deal of water forms there and flows to the south. Some have objected to this, claiming that the southern region is colder than the northern. They cite the testimony on this matter of those who have travelled there and report that the cold is unbearable. The reason is that the stars in the southern region are few and small, and their light is not sufficient to dispel the intense cold and warm the area. Thus, a great deal of water also forms there and it must necessarily flow to the north. However, the commentator (Niphus), giving credence to the commentator Albertus, said that the cause of this current (in the seas) is not only the formation of the rivers but that the southern region is uneven. Some said that Aristotle did not assert this motion with regard to the ocean but only to the Mediterranean; for there it is quite evident that the Sea of Azov flows into the Black Sea and the Black Sea flows through

the Istanbul straits into the Mediterranean; but in the other seas it is not so evident. If one claims that all the seas are alike in this respect, then many problems arise. Therefore the commentators suggest that Aristotle only reported this claim; it does not represent his own opinion.

5. Rotary motion. This motion has reportedly been confirmed by experience especially in the Adriatic Sea. Its cause, however, is the same as that for the motion from east to west. For the coasts of the Mediterranean are narrow (i.e., the northern and southern coasts are close together) and furthermore they are broken up by many capes and gulfs. So the waters, while moving from east to west, strike the crooked coastlines and are forced back, and this creates rotary motion along the coasts.
6. [R 24a] The motion of the tides. The general term for this is *aestus*. With this motion the waters of the sea continually ebb and flow on the coasts. Its cause has been difficult for the philosophers to discover and verify. It has greatly perplexed them. I will summarize here what they have written about it.

5 On the reasons for the tides.

The tides differ according to time and place. First of all, in some regions of the sea the tide is either non-existent or barely perceptible. For example off the coast of Genoa and Italy, on the shores of France, off the coast of Barcelona, on the coast of Mexico in the New World, and on the islands near Cuba, the tide is so slight as to hardly exist. In other regions it is very great. For example, in the Indian Ocean at the mouth of the Indus River, in the Sea of the Goths (Baltic), in the sea of Flanders, Britain and Portugal, and in the Red Sea, there are very excessive tides. According to measurement, the tides of the ocean are greater than all others. And tidal motion at the seashore is noticeably greater than at a distance from the shore. Tidal motion also affects the rivers that flow into the sea differently. For example, during high tide the rivers of Portugal flow upstream. In Africa the rivers of the Atlas regions do not flow upstream. [M 80] The great Thames River in England flows upstream for about 50 miles, then returns. The Betis River (Guadalquivir) in Spain also flows upstream that far.

The duration of the tides as well is not everywhere the same. In the Portuguese Ocean (Atlantic) and elsewhere, high tide is generally six hours and low tide six. On the coast of Aquitaine, however, the ocean takes about seven hours to enter the Garonne River and about five hours to recede. Near Ethiopean Guinea, high tide is four hours and low tide eight. Along the Guinea coasts, high tide is so strong that a ship needs three anchors to be held firmly, otherwise it will be hurled against the land and destroyed. On the coast of Cambaia, where the mouth of the Indus River is located, the sea reaches places 30 parasangs away in two hours and then takes two hours to recede. It strikes with such speed that land animals that find themselves there cannot flee and save themselves. Nor can ships remain there; only in some places hollows have been dug where ships lay to during low tide.

It is also known that in some places the tides move slowly and in some quickly. In other places the tides do not begin at the same time nor end at the same time, but they begin about an hour later each day. For example, if they begin at noon on a given day at a given place, the next day they will begin an hour later.

6 On what activates the tides

The philosophers differed on this matter. They could not agree on a single cause for the tides and were divided into different schools of thought.

One group supposed that the entire visible world is a living thing, composed of the elements and possessing a mind and spirit, whose nostrils are on the ocean floor, and whenever it inhales and exhales it causes the tide to ebb and flow. Some said it was rather due to the inhaling and exhaling of the spirits in and around the ocean.

Plato thought there were caves at the bottom of the sea that were opened and closed by spirits. When they were opened, water entered and there was a low tide; when they were closed, there was a high tide.

Some philosophers supposed that the land moves according to an annual motion. Because the motion of the moon coincides with that of the land, a wind arises between the sea and the sky from the collision, which causes the sea to start moving.

The Commentator (of the Collegium Conimbricense) says that it is not clear what Aristotle thought about this matter. Eventually [R 24b] the view that based the motion of the tides on the sun was ascribed to him. In this view, just as the sun activates the motion of the winds, so it activates the motion of the seas. For as the winds get stronger the sea swells, and as they grow weaker the sea subsides.

Some said that the tides are caused by the lowness of certain coasts. For six hours at a time the sea strikes against a low coast and the water increases there while the opposite coast is exposed and left low; then it returns to the other side and in this fashion makes a continuous circuit.

Others supposed that the reason for the tides is the vapor rising from the water. They likened the sea to an animal afflicted with fever. When the vapor accumulates and breaks out, high tide occurs; when it stops, low tide occurs. Just as bad humors accumulate at intervals in the body of a feverish man and at a certain point he begins to shiver, then as they dissipates he stops shivering, so the cyclical buildup of vapors continuously causes the sea to move.

7 On the opinion of those who attribute the tides to the influence of the moon

While the philosophers differ in this matter, as mentioned above, most of them attribute the tides to the moon. [M 81] This is the prevalent view and is also the most plausible, as long as one takes into account the influence of the sun as well, since the moon receives its light from the sun. The opinion of those who attribute the tides to a cause other than the heavenly bodies is weak and commonplace, based on fantasy and legend. For, every day we experience and observe that the movement of the tides is connected with the diurnal motion of the moon. Thus:

First, just as the diurnal motion of the moon is divided into four quarters according to the four points of the celestial bodies, the sea also moves four times a day. There are two flood tides and two ebb tides. Thus, as soon as the moon appears on the eastern horizon, flood tide begins and lasts until it reaches the meridian. As soon as it passes the meridian, flood tide ends and ebb tide begins. It is high tide again from the time it appears on the western horizon until it reaches the nadir at midnight, then low tide again until it reappears on the eastern horizon. While this rule does not always hold true, if there is a discrepancy it does not nullify the general principle. For, as we have previously noted, there are differences among the tides. The moon does not always move below the horizon for twelve hours such that at all times and at all places the high and low tides would last six hours.

Second, the high tide can be great or small depending on the moon's relationship to the sun. Thus, when the moon is opposite the sun and is full, the high tide is great; when it is in conjunction with the sun and in crescent stage, the high tide is very weak. Then, as the moon waxes, the high tide increases. Therefore some authors have written that the ocean is the traveling companion of the moon, for it always shares in its mood and one observes that it is strong when the moon is strong and weak with the moon is weak.

Also, the time of the high tide is one hour later each day. Thus, if it begins at the first hour today, it will begin at the second hour tomorrow. For there are about 25 hours between moonrises.

Again, in some stages of the moon the high tide in winter is greater than that in summer. The reason is that in winter the sun visits the aqueous constellations, which strengthen the moon and increase its influence. The moon's influence on moist bodies is great, and for this reason the moon is said to be feminine by nature. Physicians attribute phlegmatic illnesses to the moon, and it is tried and tested that the brain and the bodily humors in which moistness is dominant are subject to the phases of the moon, and that the crisis periods of illnesses relating to humidity are coordinated with the moon's activity. [R 25a]

The sea is also affected by the moon's relationship with other planets. Thus, if the moon is oriented in favorable relation to Venus and is in a moist phase, the high tide will be very strong. If the moon is oriented toward Mars and is in a dry phase, the high tide will be weak. In this respect, one must also pay attention to the constellations in which the moon is found. For when the moon is in constellations that rise straight, the high tide is extended; when it is in constellations that rise obliquely, it is somewhat diminished; but on days when the moon does not appear, the duration of the high and low tides is the same.

Those who say that the moon causes the tides differ on the moon's influence. Averroes specified that the tides were affected by the motion and the rays of the moon. He said that because of these two factors, vapors arise in the sea, causing it to swell and produce high tide; when the vapors disappear, the tide ebbs. [M 82] However, Contarenus and others demonstrated that a third factor must be at work; otherwise a high tide would be impossible at the time of the new moon and when it is below the horizon, since the moon's rays would not reach the sea. So there must be another, invisible influence, other than motion and rays, so that at such times the moon can affect moist bodies and the tides occur. This invisible influence is described in detail in Aristotle's *De Caelo*.

Are the tides due to the expansion and contraction of the volume of water in the sea? There is a difference of opinion on this point. Contarenus argued pro, saying: "If the sea waters do not swell and expand, why do they suddenly arise (at high tide), and where do they go at low tide?" Most, however, argued con, saying: "Just as a magnet makes iron move, so the moon affects the sea and draws it by an invisible force, now lifting it up and drawing it to the shore, now drawing it away from the shore so the waters return to the low places by their own weight. The waters rise and fall depending on the various phases of the moon, its proximity to the earth, and the angle of its rays, whether obtuse, acute, or right angle. The excess water that appears during high tide comes from the vast expanse of the sea and during low tide returns there. There is no need for expansion and contraction." This opinion is closer to the truth.

It has also been established that some areas of the sea are greatly affected by the moon's light and have a greater high tide than other places. The reason for this is the reaction to the reflection of the rays, or there is some other invisible influence. Thus, when the moon is to the north there is a very high tide at the northern coasts, when it is to the south there is a very high tide at the southern coasts. But this situation is not constant everywhere.

8 On the doubts that have arisen about this doctrine and the replies that have been given to objections

First objection: Since the tides are caused by the influence of the moon, why is there no such influence on rivers and lakes?

Reply: Natural causes do not affect all objects, rather they affect those objects that are susceptible to such influence. Thus, a magnet does not attract every body to itself. In the same way, every body of water is not susceptible to the influence of the moon. Indeed, not every part of the sea is susceptible. There is much evidence for this. I will refer to it again below.

Second objection: During high tide, some of the rivers that flow into the sea recede while others do not. Why is this?

Reply: Some rivers do not recede because the high tide at such places is not strong or else the current of those rivers is so strong and fast that it overwhelms the tide.

Third objection: Why is the duration of the tides not the same everywhere?

Reply: In some places the duration of the tides is short, in some places it is long, and in some places high tide occurs quickly while low tide occurs slowly. The reason for this is as follows: Every place has a different topography. Some coasts are flat, some are high and some low. Many have straits or gulfs. All of these features either impede the movement of water or accelerate it. The blowing of the winds and the influence of heavenly bodies also play a role in this.

Fourth objection: In some seas the tides are slight and in some they do not occur at all. Why is this?

Reply: Some of the modern philosophers argue that in some parts of the Inland Sea (Mediterranean), [R 25b] such as the Sea of Azov and the Black Sea, there are no tides or they are imperceptible, because the impetus of the waters that flow from north to south prevents them; [M 83] and in certain regions of the ocean tides do not occur but the sea simply moves from east to west due to the volume and speed of the waters pouring in that direction. However, this reply is not convincing, for while it accounts for the lack of tides on the northern side of the Inland Sea, there still ought to be tides on the eastern and western sides; and since the ocean is subject to the motion of the moon, it ought to move everywhere from east to west.

The proper reply seems to be the following: Given the influence of the moon, its effect will vary depending on the nature of the places affected. Certain places are not very susceptible to tides, while other places are extremely susceptible; just as the sun's rays are everywhere effective in activating the vapors that provide the material of rain, but in some places this influence is strong and there is a great amount of vapor, so the rain is abundant, while in other places it is slight or there is no rain at all. The natures and capacities of the elements and compounds, and their susceptibility to the influence of the celestial bodies, are not uniform. The other planets also play a role, whether strengthening or weakening the influence of the sun and moon. Furthermore, the ocean moves in different directions depending on the different motions of the moon with respect to the earth, as discussed above.

Fifth objection: In some places there are wells with an underground passage to the sea, so the water fills up and recedes. But in some cases the timing of this event does not correspond with that of the tides: the water recedes during high tide and fills up during low tide, as is the case in Spain.

Reply: Wells that are filled during high tide at sea have a passageway that is extensive, and any obstacle has been removed while those that are filled during low tide have a passageway that is narrow and irregular, so by the time high tide reaches the well, low tide has begun and the situation is reversed.

Sixth objection: Under the bridge at the Euripus Strait, seven tides occur in a single day and night. Why is this?

Reply: Aristotle was very preoccupied with solving this problem. Indeed the Greek books record the improbable story that finally he was not able to comprehend the true nature of this phenomenon and out of dejection threw himself into the sea at that place, saying, "I have not been able to grasp you, so you grasp me!"

According to some of the modern geographers, this strait does not go back and forth seven times but rather is subject to the blowing winds and runs back and forth before the contrary winds. And if the tide does recur seven times, the reason must be that waves moving from the Aegean Sea become roiled among the islands and go backward, sometimes traveling in one direction and sometimes in another. Or else the position and situation of this strait cause this movement. There might be grottos in some of the depths and the rushing seawater fills them, then cannot find a way through and so comes back out. The different motions in the strait result from that, which is also the supposed reason for the circular motion in whirlpools.

In most matters relating to the tides, it is better to resort to natural causes, even if they are invisible and unknown. I have recorded what has been written on the subject, as far as the mind of man can comprehend. [M 84]

9 On the ultimate cause of the tides

Nature does nothing in vain, even in the most trivial thing, as Aristotle explained in paragraph 6 of chapter 8 of his *Meteorology*. The continuous motion of this element (water) is not without reason and ultimately has many benefits.

First benefit: Due to this motion, the sea does not putrefy, because motion repels putrefaction. The reason is that putrefaction results from strange heat. The exchange of air that results from motion prevents the influence of this heat. This is confirmed by experience by one who walks facing the sun: he is not affected by the sun's heat as much as one who sits facing the sun. To be sure, if he walks very quickly, the heat of his body will combine with the heat of the sun; the result will be more heat because of the strenuous movement and the exchange of air surrounding his body [R 26a] will have little effect in cooling it.

Second benefit: Due to this motion, the waters of the sea are cleansed, because stagnant water usually becomes turbid and polluted. In the moving sea, filth does not remain but is generally cast ashore. By this motion the sea expels carcasses that are thrown into it, and thus cleans and purifies itself. Fresh carcasses first sink to the bottom of the sea and then rise to the surface, as explained in Aristotle's book *De Caelo*.

Third benefit: The tides are very helpful to seafarers, because there are some ports that are impossible to enter without a high tide and ships can easily leave on the low tide.

10 On the exchange of land and sea

In chapter 14 of book 1 of this work (*De Caelo*), Aristotle says that over time the land undergoes great change because of water. This occurs in several ways.

1) Some places that are acrid and arid have their climate moderated by the sea. On the other hand, the land resembles animals in that, like them, it is sometimes young and sometimes old.
2) Some places that are exposed become submerged by the sea, while places that are under water become exposed. Because the sea's motion results from the power and influence of the celestial bodies, and the phases of the planets that sometimes activate storms and floods are conducive to this motion, the sea may rise beyond its limit and overflow the coasts, overwhelming a region and submerging it, thus subsuming it to itself; or else it may recede from a coast and expose the land, as though granting it to mankind as a gift. So it is reported that the islands of Delos and Rhodes emerged from beneath the sea. The islands of Anafi, Yana (?) between the Dardanelles and Lemnos, Alona (Alonissos), Thera (Santorini) which is one of the Cyclades, Therasia, and Yara (?) have emerged from the sea.
3) Over time, some places that were connected to the mainland became islands. Thus, Sicily broke away from Italy, Cyprus from Syria, the Euboea—i.e., Aġriboz—from Boeotia, and Besbycum from Bithynia. They were separated from the mainland by flooding. On the other hand, several islands have become joined to the mainland and attached to the nearby regions. Thus, Antissa became joined to Lesbos, Zephyrium to [Hali]carnassus (Bodrum), Athos to Mindos, Dromiscum and Yernes (?) to Miletus, and Narthacusa to Cape Parthenium.
4) As if in compensation for these lands that it gave, the sea has taken over some cities and islands, [M 85] such as the cities of Pira and Antsisa, which were around the Sea of Azov, and the cities of Alisas and Bor, which were in the Gulf of Morea. All of them were submerged and have remained under water.

Some believe that the Mediterranean basin on this side of the Strait of Ceuta (Gibraltar) was previously dry land. The ocean overflowed, opened the gap between the Pillars of Hercules, and submerged the area that is now this sea. Also there was an island off the Atlantic coast as large as Africa. At a certain point in history, the ocean overflowed and entirely submerged it. So it was that Christopholos (sic), who discovered America, while sounding the seas found the sea in that place to be marshy and grassy. This is evidence that water must have covered it at a later time. All things considered, the island submerged by the Atlantic and the flooding of the Mediterranean basin are baseless legends.

Supplement to Topic 10.

It is related in *Tārīḫ-i Hind-i Ġarbī* that Imam Rāzī stated the following in his work entitled *Mabāḥith mashriqiyya*: It is highly likely that in ancient times the inhabited quarter of the world was covered by the sea. This conjecture is confirmed by the fact that if we break open most rocks, parts of marine animals appear. The reason is that viscous mud exposed from under the water turns to stone in the heat of the sun.

The poor one (Kātib Çelebi) says: The fact that parts of marine animals appear does not prove that the earth was covered by the sea, since it is also possible that such animals are formed inside the earth.

Al-Masʿūdī (ca. 893–956) in his *Murūj* (*al-dhahab*) relates the following: Over time, the seas move and change position, but because of the huge volume of water and the vastness of the seas, this shifting is not known and it is assumed that they remain stationary. During the caliphate of Abū Bakr al-Ṣiddīq (11–13/632–634), [R 26b] Khālid b. al-Walīd set out to conquer Ḥīra (near the lower Euphrates) and went to Najaf. He met an old man from Ḥīra named ʿAbd al-Masīḥ from whom he heard marvelous things. One of them was the following: "I have lived long enough to have seen the Persian Gulf occupying the place where you have landed and its waves lapping at your feet. Ships bearing the goods of Sind and Hind used to come here and depart." Al-Masʿūdī says that today there is a distance of several stages between the sea and Ḥīra, as those well know who travel to Najaf.

Al-Masʿūdī says in another place that Aḥmad b. Ṭulūn met an old man in Cairo who informed him that in ancient times the area where Lake Damietta is located was free from sea water. From there to the sea was a day's journey. And there was a road between ʿArīsh in Egypt and Cyprus. The people of ʿArīsh used to go to Cyprus by land. And at the Strait of Ceuta (Gibraltar) there was a strong stone bridge twelve miles long. The people of al-Andalus (the Iberian Peninsula) crossed it to the Maghrib and the people of the Maghrib crossed it to al-Andalus. The water of the sea flowed under that bridge. Over time, the sea submerged the bridge and also covered its shores. Today when the sea is clear that bridge can be seen. When 251 years had elapsed since Emperor Diocletian ascended the throne (in 284 C.E.), the sea near Damietta advanced toward the area now known as the Lake of Tinnīs and Dimyāṭ (Tanis and Damietta) and engulfed low-lying villages and cemeteries. Subsequently the water gradually increased so that only one or two villages, which were built on high ground, remained.

Al-Masʿūdī also writes at the beginning of *Murūj al-dhahab* that the first ruler of India was the Great Brahmin [M 86] who initiated the science of philosophy and described the movements of the heavenly bodies. He expounded the influence of the stars on the sublunar world and established the existence of the First Cause. He wrote about all of this in a book entitled *Sind wa Hind*, where he says that the apogee of the sun sojourns in each constellation for 3,000 years and completes its circuit in 36,000 years. As the sun's apogee shifts to the southern constellations, the inhabited part of the world also shifts: prosperous places become desolate and desolate places become prosperous. He also claims that once in every circuit the world comes to an end and another world appears.

The poor one (Kātib Çelebi) says: The sun's pulling of the sea while it is at its perigee follows from this rule, because the transition from apogee to perigee is gradual. The prosperous becoming desolate and another world appearing also occur gradually, not all at once. The fact that land and sea exchange places, as previously discussed in detail, is not based solely on the movement of apogee and perigee, but this movement must play some role.

Naṣīr (al-Dīn) al-Ṭūsī says in the *Tadhkira*: Some say that as long as the perigee of the sun is in the southern constellations, the southern regions are hotter than the northern. Now heat attracts moisture, and so by necessity the element of water is drawn to the southern hemisphere. Assuming that the level of prosperity of the inhabited world is always dependent on the transition of the sun's apogee, when it enters the southern constellations, prosperity also goes in that direction. However, there is no clear proof and sound evidence for this claim. The seas that are in the north rather demonstrate the opposite of this.

The commentator on the *Tadhkira*, Mevlānā Niẓām, does not agree with the doctrine about the movement of the apogee and perigee and the shifting of the sea. He says that if the perigee, which results in heat, entered the northern constellations, the inhabited quarter of the world would certainly lose its prosperity and the places recently discovered in the south would become prosperous.

However, many questions have arisen about all these matters and many objections have been raised against. I am afraid that to discuss them all would require a long digression, so I will end here.

11 On partial and universal floods

In the books of the ancient prophets, among the favors of God that are enumerated are the barriers projecting from the land into the sea that prevent the seas from assaulting and engulfing the land. In chapter 38 of the Book of Job, God addresses Job as follows: "We have confined the sea with barriers and determined its limits and have said to it, 'You may not pass beyond here; here you will break your billowing surges.'" And in Psalm 104 is written: "O Lord, You have placed a limit to the sea beyond which it will not pass and cover the land." [R 27a] However, because of mankind's disobedience and other reasons, in this world of being and dissolution it sometimes happens that the sea spreads over the land and engulfs certain regions, as indeed occurred frequently in previous times.

First of all, the historian Genebrardus records in his history that before Noah's Flood there was a flood at the time of Enoch son of Seth in which one-third of the earth was destroyed. And several ancient authors record floods other than this one. As for the general flood that occurred at the time of Noah, it is mentioned in chapter 7 of the Book of Genesis in the Torah, which is one of the revealed books, and in several places in the Koran. It is also recorded in the works of ancient and modern authorities. And it has been accepted as a universal flood by all religious communities, although not by the Persians.

Now, some difficult problems arise with respect to this flood that need to be resolved.

1) Is it possible for the sea to flood this much from the force of nature and the effect of natural causes? [M 87] Avicenna (d. 428/1037) and some astrologers, following certain ancient authorities, were of the opinion that this was possible. Before them, Seneca (ca. 1 BCE–65 CE) wrote that sometimes a universal flood occurs in accordance with natural law; but it is a moot point whether the power of the sea alone is sufficient for this, or whether continuous rainfall and flooding rivers must contribute to it. His own opinion is that it is not necessary for it to come about from a single cause but it may occur from the convergence of several causes, including these contributing factors.

2) Is it possible for the universal flood to have occurred under the influence of natural events? One group affirmed the possibility while another group denied it. Aristotle too, in chapter 14 of book 1 of this book (*De Caelo*), was of the latter opinion, stating that God, who is the First Cause, does not give so much license to natural causes that, when it is necessary for this to occur, they should break the order of the world and destroy the habitation of mankind. The continuance of the order of the world was a concern of His exalted will, and so He balanced the elements with their opposites. Thus each element—characterized by coldness, hotness, dryness, and wetness—either repels or weakens the assault and overflow of the opposite element. So fire repels water with heat and water repels fire with cold. Furthermore, in order to counteract the violent influence of the celestial bodies, he gave the bodies in the world of elements (the sublunar world) certain natural characteristics by which they are strengthened, as though girded with a weapon, and can repel somewhat the influence of the stars, so that, while for the most part they are subject to that influence, in sum the result will be resistance. In the heavens as well he made the stars the opposite of one another so that the violent influence of one star is either completely repelled or is balanced by its opposite. In accordance with these principles, a universal flood does not result simply from the force of nature but occurs by the will and command of the primordial Necessary One (i.e., God).

3) Concerning the statement in this chapter of the Book of Genesis (cf. 7:11–12) that "The gates and the dams of heaven were opened and the rains fell on the land," commentators among the ancients supposed that above the sphere of the fixed stars there are actual waters which descend to earth when the gates of the heavens are opened. The moderns interpret this passage more correctly by saying that here the word "heaven" means "the air," (as in the phrase) "Everything that is above you is the heaven," and opening the gates refers to the release and contraction of the clouds.

4) After the universal flood, what happened to all the water, since it had risen far above the highest mountains? In reply to this they said that some of it entered porous places of the land and hollow caves; some of it, in miraculous fashion, turned into vapor and changed into air; while the present seas are the remnant of the flood. But these assertions have not been confirmed and cannot be regarded as true.

12 On other aspects of the seas

THE TASTE OF SEAWATER: The water of all the seas is salty (lit. bitter, *acı*). Nevertheless, the water of the ocean off the northern coast of Germany is sweet enough to drink. According to the book of *Atlas*, the reason for this is that a great deal of fresh water comes down from the mountains of Sarmatia and flows into this sea; and because the sun is distant from those regions, its effect is weak and it does not draw out the fine particles of the water. This explanation is in line with the view that the sun draws out the fine particles of water (leaving behind the dense particles which make the sea salty). However, [R 27b] the fact that in the north the waters of the Amalchium and the Chronium Seas (i.e., the North Sea and the Atlantic) are salty refutes this explanation. [M 88] The true reason is that the water off the northern coast of Germany consists solely of snowmelt that flows in the rivers from the mountains of Sarmatia.

SPECIAL FEATURE OF THE FRESHWATER SEA: It has been proven that ships in this sea can bear less weight of cargo than those in other seas. That is because salt water is essentially dense and therefore strong, so ships sailing on it can carry a heavier cargo, while with such a cargo on fresh water they would sink. Also, sailing in this freshwater sea is easier than sailing in salt water, since it is easy to move by cleaving through fine particles, while it is difficult to cleave through dense particles. Also, those who travel on this sea are more likely to get seasick and vomit because the waves are larger than those in other seas, sometimes billowing up to the sky, sometimes receding down to the sea floor.

Depth of the seas: It is nearly impossible to plumb the depth of the ocean off the Spanish and Atlas coast. A *būlīs* (?)—i.e., a plumbline—of 300–400 fathoms would not reach the bottom. The depth of the Sea of Germany is in most places only 60 cubits and nowhere more than 100 cubits. However, off some coasts of Norway the sea is so deep that the bottom is unknown.

Distances of the ocean: It is recorded in the *Tārīḫ-i Hind-i Ġarbī* that the distance from Spain to the region of Yucatan (in Mexico) is 5,600 miles. It takes eight or ten days to reach the Canary Islands from the Spanish port of San Luca, which is 1,000 miles away, located at 20 degrees latitude. It is 3,800 miles from the Canary Islands to the island of Hispaniola and it takes 30 days to reach sailing rapidly. From there to New Spain is 2,400 miles and to Yucatan is 1,400 miles. From Spain to the Straits of Magellan is 16,000 miles. It is 2,000 miles from the Canary Islands to Cape Verde and also from Cape Verde to the Cape of Santagostin. Departing the Canary Islands, they drop anchor at one of these two capes. It is 800 miles from Hispaniola to Yucatan. The other distances can be obtained, when necessary, with a pair of compasses and taking account of longitude and latitude.

13 On ocean voyages

According to the *Introductio* it is not easy to determine when the ancients began to travel on the ocean. The land of America was known to the Greeks and the Egyptians,[4] and the Tunisians (i.e., Carthaginians) sailed to that region, as will be discussed in the proper place. In recent times the entire globe has been circumnavigated several times. Previously the American continent, which is one of the primary continents of the earth, was unknown to the ancient Greeks and Romans. Now it has become known to some extent.

Pliny writes the following about the circumnavigation of the globe in chapter 67 of book 2: Beginning from Ghades at the Cape of Ceuta (Gibraltar) they examined the coasts of Spain and Gallia (France) and explored all of the West. Most of the Northern Ocean was explored during the reign of the Emperor Augustus. His fleet reached Cape Cimbria in the Sea of Germany and went as far as the edge of the frozen Sea of the Tatars.

Pliny says in the same chapter that the conquests of Alexander the Great—i.e., Alexander the Greek—reached the coasts of the Northern and Eastern Ocean and as far as the Arabian Sea where, [M 89] during the time of Julius Caesar, remains of sunken ships from Spain were found. It appears from this that the ancients must have sailed from the Western Ocean to the Eastern Ocean. When the Tunisians (i.e., Carthaginians) were at the height of their glory, a man named Hanno departed from Ghades at the Cape of Ceuta and, sailing around the coasts of Africa, reached the edge of the Arabian Sea. [R 28a] He wrote an account of this voyage. Also at that time a man named Hemilco was ordered to explore the seas of Europe.

Pliny also writes that a man named Eudoxus, fleeing from the Egyptian Pharaoh Latros (Lathyrus), came from the Strait of the Arabian Sea to Ghades. Some time before this, the author named Selīūs (Caelius Antipater) reports that he saw men who had gone from Spain to the Eastern Ocean and the region of Ethiopia.

The author named Cornelius (Tacitus) writes the following about voyages through the northern region: The king of Swabia once gave the Consul of Gallia a number of Indian captives as a gift. They had set sail from India on a trading voyage and were carried by a storm to the Sea of Germany. In recent times, however, the Dutch and the English sought a route through these regions, but they have not been able to find anything.

Thus far I have been citing the *Introductio*. Next let me recount the voyage of the Portuguese to the East Indies.

Voyage to the East Indies: Naṣīr (al-Dīn al-Ṭūsī) and other Islamic authors are uncertain whether the Western Ocean is connected to the Eastern Ocean. Some say that a connection from the north is doubtful but one from the south is certain. Masʿūdī in the *Tanbīh* cites Ptolemy and others regarding the southern connection.

Pīrī Reʾīs says at the beginning of his *Baḥriye*: The king of Portugal, who was gathering intelligence about the quarters of the earth, brought sailors and questioned them about India. They all provided information on the places they had seen. He gave them ships and sent them to various regions. Some of them reached the region of the blacks, but by the time they returned the king had died and, his successors being uninterested, the matter remained in obscurity. Forty years later, one of the captains who had gone to that region came to the king of Portugal and described his adventure, giving information on the route to India. He also asked for assistance to go there. The

4 For "Egyptians" the text has *Erbesiler*, an error for *Ejibsiler*.

king gave him the help that he requested and sent him forth. The captain set sail on a favorable wind, passed the equator and reached Cava Bona Esperansa meaning Cape of Good Hope. Previously they had searched for that cape mile after mile and found it after nine years. They sailed 1,000 miles every year and erected a marker at every stage, never leaving the coast. At the cape they placed a cross on a pillar. Now the captain landed there and found that marker. Many rivers flow into the sea at that place. On land there are no settlements as far as the source of the Nile. It is a lair of lions and desolate places. There are Negroes here and there on the coasts; their king is called Manikonko.

It is recorded in the *Atlas*, however, that those places are inhabited. Pīrī Reʾīs is to be excused because finding out about these matters is not the business of sailors.

Passing on from there (the Cape of Good Hope), they reached the waters off Comoros (*Jazīra-i qamar* ["Island of the Moon"], also used for Madagascar) and after exploring and confirming the routes, they returned to Portugal. Prior to this, [M 90] a fleet of seven ships had set out for India but did not find the way; six sank in a storm and only one returned. Subsequent to this, they determined the routes (to the Indies) with great care. They sailed there from the west and returned from the east. They occupied Malacca, Mogadishu, and the Antilles and from that time on made regular voyages.

Setting out from Spain in the spring, they sail 700 miles west by southwest to the island of Madeira. There they take on water and sail southward for 1,200 miles. 400 miles west from the point where Cape Verde is spied are eight islands where they go to take on water, then they sail southwest another 200 miles and reach the coast of the place called Terra Brazil meaning Land of Logwood. From there they set sail on the open sea, tacking southwest and south(east) until they reach 55 degrees latitude south of the equator. They cannot go beyond that because of fierce winds and storms.

Turning to the east, they sail 2,000 miles.[5] When the coast of Ethiopia appears on the left, they take the elevation (to determine the latitude) and, turning to the north, sail another 6,000 miles until they reach India. Until they pass below the equator [R 28b] there is no security from fear of the sea. They stop at Mogadishu and pass in front of Bāb al-Mandab.

The reason they take such a roundabout route is that those who sailed previously from Portugal to India with the westerly winds sometimes made landfall on the coast of the blacks and, with the winds calmed because of the heat (preventing them from sailing further), they suffered a great deal. Many of the sailors perished from the extreme heat. There would be no wind at all for 100 days. At night it would rain down filth and dew. Whirlwinds would arise and break the sails to pieces. Sometimes they encountered an easterly wind and it returned them to the west. Therefore they found the route on the open sea and began to sail with security. On the return journey they sail with the easterly wind, continuously following the Ethiopian coast.

Thus the distance from Portugal to India is 14,000 miles, sailing directly. Sometimes they set out to the south and made landfall on the Gold Coast. One time, in fact, a Portuguese ship was caught by a contrary wind and sailed to the south of the China Sea, reaching a coast in 25 days. This ship reported marvelous things, as will be discussed below.

End of the summary of Pīrī Reʾīs's discourse.

Supplement: The author of the *Hind-i Ġarbī* says that since the beginning of the 10th century (end of the 15th and beginning of the 16th century CE), Portuguese ships have sailed from the west to the Eastern Ocean. They have usually gone to the coasts of Sind and Hind. Through fortunate circumstance and excellent management they took control of the emporia of those regions and occupied them. If they are given free rein much longer, they will find the way to the Red Sea and seize the coasts of Ḥijāz and Yemen.

The poor one (Kātib Çelebi) says: For some time now Portuguese and Dutch ships have been entering the Red Sea and reaching the coasts of Yemen and Ḥijāz. They sail about in those regions and when they find the opportunity they capture merchant vessels and cut off shipping. However, now that a fortress has been built at the port of Jedda and guards have been stationed there, they (the Europeans) cannot disembark and engage in mischief, nor can they leave their ships and go one or two miles inland. But the evil and harm that they do at sea is beyond question. Their harm is more than that of enemies on land. In 1036 (1626–1627) some Flemish bertones seized merchant ships coming from India causing 600,000 *guruş* in losses, as was reported in a petition to the sultan by the governor of Yemen, Haydar Pasha, and the people of Yemen. Subsequently, seven bertones seized 14 merchant ships coming from India to Yemen, [M 91] plundered their goods and took those on board prisoner, then proceeded to the port of Mocha where they dropped anchor alongside British ships. Not a year goes by without several such incidents, but there is no way to stop them.

The author of the *Hind-i Ġarbī* goes on to say: It is a strange situation in which an ill-omened group act in such an audacious manner, sailing from the west to the east and bearing up against fierce winds and the calamities of

5 The text has "two miles," corrected according to *Baḥriye*, 123.

the seas, while the Ottomans, who are located half-way to those places relative to the Europeans, have no intention of conquering them, nor has any of the Ottoman sultans sought to subdue them. Yet countless benefits are to be had from launching a campaign in that direction, and there are countless reasons for wishing this to happen. If a fleet were assembled at Suez in Egypt and an army sent there, a wise commander could conquer those regions in a short time and the infidels would be driven out.

The poor one (Kātib Çelebi) says: Bahādur Shāh, one of the rulers of Gujarat, did send a letter of complaint to Sultan Süleymān asking for help. A fleet was assembled at Suez and the governor general of Egypt, Süleymān Pasha was sent to that area in 940 (1533). He arrived with more than 30 galleys, but the campaign produced no result but an expenditure of futile effort, and he returned. Again in 959 (1552) Pīrī Reʾīs, the author of the *Baḥriye*, was appointed admiral and sent to those regions with 25 galleys. He encountered some 70 infidel ships in the Strait of Hormuz, was victorious in battle [R 29a] and drove the infidels into their fortresses. He besieged them for a while and then went to Basra, seeking help from its governor, Ḳubād Pasha. Relations cooled between Pīrī Reʾīs and the Pasha because of the excesses of the sailors. When it appeared that Pīrī Reʾīs was going to be assassinated, he fled to Suez with three galleys. Ḳubād Pasha petitioned the Sultan, accusing him of treacherously taking money from the infidels and of plundering Basra. An order was given for his execution and he was beheaded in the state council of Egypt.

In 961 (1554) when Sultan Süleymān was on the Nakhjiwān campaign, he appointed Seydī ʿAlīzāde admiral and sent him overland, in order to bring the galleys remaining at Basra to Suez. The admiral went to Basra and set sail with those ships. While en route he encountered some infidel ships in the Gulf of Hormuz and a desperate battle ensued, but he broke away and continued on his way. Suddenly a contrary wind arose and scattered the ships. The infidels captured three of them and he entered the harbor of Bandar Surat with six. The infidels blockaded the entrance to the harbor and laid siege for four months. Finally, as it was impossible to return by sea, the admiral demolished the ships, put their cannons in the fortress, and returned overland. He wrote a book about the hardships and trials that he endured on this campaign. "Having the troubles of Seydī ʿAlī falling on one's head" became proverbial after that.

Now, the events that exercised the author of the *Hind-i Ġarbī* were experienced when this Exalted State (the Ottoman Empire) was strong. It is clear that experiencing such events again will be cause for regret, as the Crete campaigns in the present day amply testify. Since that book is very popular, [M 92] I feel quite justified in vigorously refuting some of the points mentioned in it, as I will occasionally do.

In addition to the voyages that have been mentioned, the voyage of Columbus to the West Indies, Magellan's circumnavigation of the globe, and the voyage to the subpolar region will be described in the pages of the supplement. For now the present account suffices.

14 On the seasons (or monsoons) of sea travel and its dangers

Although these topics have little to do with geography and properly belong to hydrography—i.e., the science of the sea—nevertheless, because this chapter is completely devoted to the sea, for completeness' sake I have decided to include them here, in summary fashion, along with other matters concerning the sea.

First of all, the Indians who sail in the Eastern Ocean use a dry measure known as *bahār*, while in Turkey we use *mud* and in Egypt, *irdabb*. To say that ships have come from India with *bahār* means that ships laden with cargo have arrived, not ships with hot spices. Anyone who sails the seas must know the seasons (or monsoons) of the routes, the cargo (*bahār*) of the ships, the speed of each ship according to its cargo (*yük*), the use of hawsers and cables, and the conditions and times of the tides. For example, low tide increases day by day from the beginning to the middle of each month, while high tide increases from the middle of the month to the end. In fact, until the end of the month it is not possible for seasonal ships to leave the Gujarati straits of Surat, Barwaj and Kenbaye (Cambaia) and head out to sea. The flow tide there is so strong that it will sweep away a ship at anchor in the face of a strong wind.

There are two kinds of distances at sea. One is by calculation and one is by experience—i.e., learned by the experience of sea travel. Differences in the latter are due to many factors, such as the strength of the wind, variation in the ship's movement, the weight of the cargo, whether the ship is moving fast or slow, the tides, and how salty the sea is—for ships move more slowly in salt water. Thus ships travel faster in the Black Sea than in the Mediterranean, because the Mediterranean is saltier than the Black Sea; and ships travel slower in the Indian Ocean and the Western Ocean than in the Mediterranean. The reason for this was discussed above. Because of all these factors, measurement by experience is erratic while measurement by calculation is exact.

One must also know the principles of the winds. Wind is air in motion. This is demonstrated by the fact that by moving the air with a fan one causes a wind. Whenever the seas surge and their movement is strong, the wind is also strong. One of the reasons for this [R 29b] is coldness. This is demonstrated by the fact that offshore winds begin at night, while onshore winds generally blow during the day and are still at night; the reason being that at night the shores are cold while the sea is warm. In summer this situation is reversed, because the land gets hot during the day because of the reflection of the sun's rays. In particular, sandy places are colder at night than mountainous places, and rainy places are colder than arid places. Winds are generated accordingly. Where there are clouds there is usually wind, especially when the clouds are in motion.

According to the Arabs there are four basic winds: northern, southern, western (*dabūr*), and eastern (*qubūl*). Those between the cardinal points are called *nakbā*. However, those who sail in the Indian Ocean call them by the names of stars, as was mentioned above. For those winds, the boundaries are determined in the sea where one is sailing, as first, middle, and last. They are called seasons (or monsoons), meaning the times when travel occurs. In the past they generally determined the beginning and ending [M 93] by the *nawrūz* (Persian new year, vernal equinox) of Yazdajird; but due to the leap year, they later substituted the Jalālī *nawrūz*. Any number with the preposition *fī* means that the first falls during that period. Thus if one says "*fī* 10 *nawrūz*" it means days one to ten, but if one says simply "15 *nawrūz*," it only means that day.

There are two kinds of monsoon. In contrast with the Mediterranean, where in open areas one must take eight or even sixteen winds into account, because the Indian Ocean is so vast there are only two, the east wind and the west wind, that blow six months each. Sailors call the west wind *rīḥ-i kaws* and the east wind *rīḥ-i azyab*, also *ṣabā*.

The first kind (which is the west wind)

It is in two parts. PART 1 OF THE FIRST KIND is called *ra's al-rīḥ* (head of the wind, i.e., beginning of *rīḥ-i kaws*), also *mawsim zaytūnī* (olive season / monsoon). They are:

- Monsoon of Aden to Gujarat, Kenken, and Manībār (Malabar): 360th of *nawrūz*—i.e., five days before the next *nawrūz*—if the *rīḥ-i azyab* is still, because Aden is the source of the *rīḥ-i azyab*. Sometimes, in fact, this wind is not still until nearly the 35th of *nawrūz*. If it starts on the 15th of *nawrūz* and one reaches Shiḥr (on the southern coast of Arabia) by the 45th, one can sail on to Gujarat and Kenken, but not to Malabar where there will be much rain and hazardous conditions.
- Monsoon of Shiḥr: It is at its height on the 15th of *nawrūz*. One can sail to Kenken on the 5th of *nawrūz* and to Malabar on the 360th of *nawrūz*.
- Monsoon of Ẓafār (on the southern coast of Arabia): 330th of *nawrūz*. Because Ẓafār is the source of the west wind, at that time one can sail to anywhere in India (or the Indies).
- Monsoon of the Swahili (East African) coast: 25th of *nawrūz* for Gujarat and the 35th of *nawrūz* for Muscat, Shiḥr, and Aden.

Monsoons of voyages *taḥt al-rīḥ* (beneath the wind). By *taḥt al-rīḥ* is meant the direction east of Indo-China (*Hind u Chīn*). Because the *ṣabā* wind blows from that direction, it is called "beneath it." For example:

- Monsoon of Gujarat to Malacca, Sumatra, Tanāṣara, Bengal, Marṭaban, and all the commercial ports of Indo-China: 360th of *nawrūz*. The height of this monsoon is 15th of *nawrūz*.
- Monsoon of Kenken: It runs from the 5th or the 1st of *nawrūz* to the 45th of *nawrūz*. The height of this monsoon is the 25th or the 15th.
- Monsoon of Malabar: 25th of *nawrūz*.
- Monsoon of Shiḥr to those regions: 340th of *nawrūz*. One goes with the *awlam* wind as far as Fartak (Ra's Fartak in Yemen) where one finds the west wind.
- Monsoon of Fartak and Ẓafār: the same.
- Monsoon of Muscat: 360th of *nawrūz*.
- Monsoon of Zaylaʿ (on the African coast of the Gulf of Aden), Aden, and Berbera (200 km east of Zaylaʿ) to Shiḥr and Mishqāṣ: The height is on the 85th of *nawrūz*; there is no use in going after that.
- Monsoon of Aden to Hormuz: 55th and 65th of *nawrūz*; there is no use in going after that.

PART 2 (OF THE FIRST KIND) is the end of (*rīḥ-i*) *kaws*:

- Monsoon of Jedda and Suakin to Malabar, Kenken, Gujarat and Hormuz: 145th of *nawrūz*.
- Monsoon of Zaylaʿ and Berbera: 155th of *nawrūz*. The monsoon of Aden is the same, or else five days later.
- Monsoon of Shiḥr: 165th of *nawrūz*.
- Monsoon of Mishqāṣ and Ẓafār: 166th of *nawrūz*.
- Monsoon of Fartak and Aden to Hormuz: [It starts on the] 155th of *nawrūz*.
- Monsoon of Qalhat (in Oman) and Muscat to those places: 166th of *nawrūz*. [R 30a] Voyages are possible to all the commercial ports of India until the 45th and 55th of *nawrūz*.

Monsoon of voyages *taḥt al-rīḥ*: At that time voyages can be made from Arabia and the ports of India to China and its islands. First of all:

- Monsoon of Aden, Shiḥr, and Mishqāṣ to Malacca, Sumatra, Tanāṣara, Martaban, Bengal, and the other Chinese ports: 145th of *nawrūz*. [M 94]
- Monsoon of Kenken: 170th.
- Monsoon of Malabar: 185th of *nawrūz*.

- Monsoon of the Swahili coast to Hormuz: 165th of *nawrūz*.
- Monsoon of Mogadishu and the Swahili coast to the Islands of Dhi'b: 185th of *nawrūz*; to Arabia: 190th of *nawrūz*.

THE SECOND KIND of monsoons, which is the east winds:

- Monsoon of Gujarat to all the Arabian islands and commercial ports: from 205th of *nawrūz* to the 5th or 10th; to Muscat, Qalhat, and Hormuz: until the 25th of *nawrūz*; there is no use in going after that.
- Monsoon of India: The best time is from 330th of *nawrūz* to 340th, for at that time there are no violent rainstorms.
- Monsoon of Kenken: from 215th of *nawrūz* until the 5th of the following *nawrūz*; there is no use in going after that, but one can sail to Hormuz until the 15th of *nawrūz*.
- Monsoon of Hormuz: from 205th of *nawrūz* to the 330th, if one sails along the coast; but if one sails across the sea it is possible from the 330th to the 365th.
- Monsoon of Gujarat to the Swahili coast: from the 1st of *rīḥ-i azyab* which is 205th of *nawrūz* until the 320th; there is no use in going after that.

Monsoons of *taḥt al-rīḥ*:

- Monsoon of Bengal to Aden, Mecca, Hormuz, Arabia and places nearby: from 280th of *nawrūz* to the 310th; but to Ceylon: until 330th of *nawrūz*.
- Monsoon of Malacca and Marṭaban to these places: from 280th of *nawrūz* until the 330th.
- Monsoon of Sumatra: from 260th of *nawrūz* to the 320th; to Bengal: from 330th of *nawrūz* to the 15th of the following *nawrūz*; there is no use in going after that.
- Monsoon of the Islands of Dhi'b: from 240th of *nawrūz* to the 350th; to Gujarat and Kenken: until the 60th.
- Monsoon of Dabūl-i Sind: from 240th of *nawrūz* to the 330th; there is no use in going after that. The same for Comoros (*Jazīrat al-qamar*, also used for Madagascar).
- Monsoon of Kanīwa to Sofala: from 240th of *nawrūz* to the 290th; but from Sofala to Kilwa: from 25th of *nawrūz* to the 55th.

Notice: This science of the sea is based on experience and reason. The sea routes and the monsoons are only learned from experience, while conceiving the status of the stars and the rules of calculation are purely intellectual. Determining measures and distances is based on combining the two. Learning about a place from its inhabitants is not like learning about it at second hand. However, it is better to learn about if from others, as long as they are experienced individuals, if the inhabitants have little knowledge. Knowledge obtained from verified sources is definitive.

Dangers: According to seafarers there are ten dangers involved with sailing the seas. These include:

- The appearance of mirages at certain places and times.
- In the Red Sea, dozing off and not paying attention to the coast.
- Failing to inspect the ship's hawsers when preparing for a voyage and especially when there are strong winds.
- Forgetting to reduce or slacken the sails at night when it is cloudy and rainy.
- Failing to observe the Circle of *Najm al-zawj* ("Star of the pair"). *Najm al-zawj* is an imaginary star. The Indians call it *Kawkab-i jūkī* ("Star of the Yoghi") and the Eastern astronomers call it *Şeker yıldızı* ("Star of sugar"). It appears now one day, now two days before the west wind, on the days of the months of TWRK (*nawrūz*?) when the sun and moon are in conjunction. It appears in the east on the 1st, the 11th and the 21st; in the southeast on 3–2 (i.e., the 2nd, the 12th and the 22nd); in the south on 3–3; in the southwest on 3–4; in the west on 3–5 [M 95]; in the northwest on 3–6; in the north on 3–7; in the northeast on 3–8; below the horizon on 3–9; above the horizon on 3–10 (i.e., the 10th, the 20th and the 30th). On whatever day, sailors do not sail in the direction that it appears. It is very famous and highly esteemed among them, like the Circle of *Rijāl al-ghayb* (men of the occult world, saints).

In addition, the year has inauspicious days [of travel]. When the moon is in Scorpio and is on the "Burning road" [R 30b]—i.e., from the 19th day of Libra to the fourth of Scorpio—it is forbidden to travel on land, and according to some on the sea as well; but it is recorded in the calendars of the Arab astrologers that sea travel during this time is auspicious. They even take into consideration the hours of the day and do not travel during the hours of the sun, Saturn, and Mars; but they do travel during the hours of Jupiter, Venus, Mercury, and the moon.

Five periods of heavy rain are experienced in the Eastern Ocean.

1. Rain of *dāmān*: In India, on the 175th of *nawrūz*, which is the beginning of *maṭar-i fīl* ("rain of the elephant").
2. Rain of *uḥaymir*: It comes from the vicinity of Madraka in the land of Aḥqāf and goes as far as Shiḥr and sometimes Aden. It begins on the 215th of *nawrūz*, or just before or just afterward. It begins with the appearance of Scorpio.
3. Rain of *arbaʿīn* (the forty days of midwinter beginning with the winter solstice): in the Sea of Hormuz, on the 280th of *nawrūz*.
4. Rain of *banāt* (the quadrilateral and the three outer stars in the Great and Little Bears): It is known as

the wind of winter. It comes between northeast and northwest. This is the time of the setting of *banāt*. It goes from Maṣīra to near Aden, i.e., to all of Arabia. It starts on the 280th of *nawrūz* and end on the 300th.

5. Rain of *tisʿīn* (ninety): In India and at sea, beginning on the 310th and 20th of *nawrūz*. It comes from the shore. This rain also occurs in the land of Aḥqāf. On the 55th of *nawrūz* it also occurs in *taḥt al-rīḥ*, i.e., in China. It is stronger than all others.

Concluding notes

It is well known among sailors regarding the stars that rise at the second dawn, that if any star appears large during its rise, any wind that occurs at that parallel will be strong.

According to Turkish sailors, if the beginning of well-known storms coincides with the beginning of the Arab month, that storm will be powerful. It has been confirmed by experience that the strength of storms does not usually last more than 24 hours but sometimes the storm calms down and sometimes it changes. It has also been confirmed by experience that if the moon is in a windy constellation there will be a big storm that day, if it is in a watery constellation it will rain, in an earthy constellation it will be moderate weather, in a fiery constellation it will be pleasant weather.

They take precautions at the beginning of storms that come from the east, northeast, north, and northwest; and at the end of those that come from the west, southwest, and south; because such storms are severe.

At the time of the rise of a star, they do not travel at sea but wait for it to set. If the wind shifts, they prepare to travel from the southwest to the south and turn from the northeast to the north and northwest. They do not trust clear winter nights and do not fear cloudy summer days. At the time of *bonaccia*[6]—i.e., *meltem* (off-shore breeze that blows daily for a period in summer)—if the western horizon is red at sunset, there will be a strong wind the following day. Gulf winds during the time of *meltem* turn to *imbat* (steady summer south wind) in the afternoon—i.e., they reverse themselves. Sometimes a cloud moves in the opposite direction of the wind and goes backward because of its strength and the wind does not cease. Hence the saying, "The cloud does not enter the sail."

End of the matter found in the *Muḥīṭ* and the *Baḥriye*. If one wants more detailed information, he should consult the original works.

6 A period of calm weather; see *Lingua Franca* #103. The text has *pūrūnatsa*, in error for *bonatsa*.

15 On other matters regarding the praise of the sea

According to the *Atlas*, the sea has countless benefits. In particular, the nature of this element (water) is much more skillful than the other elements at fashioning life. [M MAP] [M 96] For after generating and nourishing various kinds of animals in the earth (or on land) and in the air, it forms many marvelous animals in the water (or in the sea), the like of which are not found in the other elements. It not only imitates simple types (or species) of land animals, but also compounds, making a horseman the model for a sea horse, and even imitates a human with a mermaid and a sea-monk; thus showing forth different kinds and classes of persons in different kinds of fish. It also imitates plants, bringing forth vegetal gems such as coral and showing its skill by nourishing other gems such as ambergris and pearls. [R 31a] All of this is evidence for the omnipotence of a Wise Creator and is a means to understand the customary causes (i.e., the workings of nature).

End of the discourse concerning the sea. Next begins the discussion of matters relating to the earth, which is the subject of geography.

Chapter 3: On the Major Parts of the Earth

According to the *Atlas*, the ancients sometimes divided the globe into two continents, sometimes three. Before the discovery of the New World, the three-continent view was in favor. Since then, the moderns have made a fourth continent out of that clime.

Mercator at first divided the globe into three continents: 1) the inhabited quarter; 2) America, meaning the West Indies; 3) Magellanica, meaning the southern land. We however divided the globe into five continents: 1) Europe, 2) Asia, 3) Africa, 4) America, 5) Magellanica. But we also added the land under the North Pole at ninety degrees latitude known as *polus arcticus* thus making it six continents. We will explain each of these in a separate chapter.

According to the *Introductio*, the globe is divided into three continents separated by ocean. They are the ones that Mercator mentions, beginning with the inhabited quarter which is our dwelling place. Most of it has been known from ancient times. In what manner the ocean is bordered by the continents has been explained in the chapter on the sea; there is no need to repeat it. This continent was also divided into three continents in every age: Europe, Asia and Africa. This division has been in favor since the time of the sons of Noah. It is thought that Shem dwelt in Asia, Ham in Africa and Japheth in Europe. The borders of these three continents have shifted relative to the tribes that appeared with the passage of time, and the geographers demarcated them with various borders according to their own wish. Some of the compilers who came in the Middle Ages of the world made two continents out of the inhabited quarter and restricted it to Asia and Europe. They considered Africa as part of Europe and thought that the ocean had cleaved the mountains at the Strait of Ceuta (Gibraltar) and rushed toward Syria, thus separating Africa from Europe. Some considered Africa to be part of Asia. The recent compilers made four continents: Europe, Asia, Egypt and Africa. These divisions and their borders were expounded in the books of the old compilers. Here too they will be explained in summary fashion, with the borders of each one and the discrepancies among them, God willing.

Chapter 4: On the Continent of Europe

First we will explain the name of this continent, then its borders, then its component parts.

The cause of naming: There was dispute over how this continent got the name of Europe. The ancient historian Herodotus said the reason for naming it [M 97] is not known. Some transmitted from a historian named Isidorus that in ancient times, a ruler named Agenor, who and ruled the coasts of Syria and the Maghreb and made the city of Tyre his capital, had a beautiful daughter named Europa Tyria. According to their claim, their god named Jupiter fell in love with her and, assuming the form of a bull, brought the girl from Sidon to Cyprus or to Crete, and this continent was named after that girl.

In other versions, someone kidnapped the girl and placed her in a boat in the form of a bull and carried her off; or the picture of a bull was painted on the stern of that boat; or a man from Crete named Taurus, meaning "bull," conquered Tyre and took that girl prisoner along with its people and brought her to Europe, which got its name in that way; or a soldier named Europa, whose standard had a bull depicted on it, once came and conquered this clime, which got its name in that way; or they likened the delightful qualities of this clime to that girl's beauty called it Europa; or Europa [R 31b] was the name of a magnificent ruler over these lands and the clime became known by his name, since it is recorded in the ancient histories that a person named Europa governed these lands.

The historian Becanus refuted those who said that this name was Greek. He said that before the Greeks it was known as Sīmber (Cimbri, Cimmerians), which is the name of the people who lived in these lands. In their language *eur* means "good" and *rop* means "many people," so the compound phrase means "land whose people are very good." In this age it is called Rūm and Frengistān.

The borders of Europe: Ptolemy and the ancient historians defined this clime as extending from the eleventh tropic in the fourth clime to the twenty-first tropic at the end of the seventh clime, and from 36 to 54 latitude and 17 to 64 longitude. Later, when inhabited lands were discovered as far as 72 latitude, it was necessary to adjust the latitude of the border on that side. And if its longitude is reckoned from the Cape of Tesāwensā (S. Vincente, Cape Finisterre) in Spain to the Don River, it extends from 17 to 71 degrees longitude.

But when its borders are demarcated according to the four directions, one says that its southern border consists of the Strait of Ceuta (Gibraltar), the southern tip of the island of Sicily named Odyssia, and the Cape of Mania in the Morea Peninsula (Peloponnese). Its northern border is the Cape of Wardhuys, which is the last settlement at 71½ degrees. Its western border is the ocean, which also surrounds its northern side.

Its eastern border is disputed. According to most men of science, it is a hypothetical line drawn from the source of the Don River to the North Sea. But some of them say that this is a hypothetical border, not a determined border. Plato, Aristotle and their followers consider the demarcating border as the lands between the Volga River and the Black and Caspian Seas. Abraham Ortelius fixes it as the line drawn to the north from the Mediterranean through the Black Sea, the Sea of Azov and the Don River and its source.

The author of the *Atlas* says: We prefer the line from the source (of the Don) northward that Ptolemy determined, and we say that the demarcating border extends from the Sea of Azov through the Taman Peninsula, the Black Sea, the straits of Istanbul and the Mediterranean (Bosporus and Dardanelles) and through the Islands (Archipelago) to Ifrīqiya (Tunisia).

According to the *Introductio*, those who say that the Don River is the border [M 98] fixed its source in the Ripheaan Mountains near the North Sea, which is erroneous. Those who consider the border as the land between the two seas claimed that the Caspian Sea is connected to the North Sea, which is not so. In any case, the shores that extend to the mouth of the Ob River on the north side are outside of the border while being attached to Sarmatia, which is a part of Europe. Therefore, to fix a true border that unites these points, it is necessary to say that a line should be drawn through the straits of the Mediterranean Sea, Black Sea and Sea of Azov to the Don River, then the line which is drawn upstream to where the two

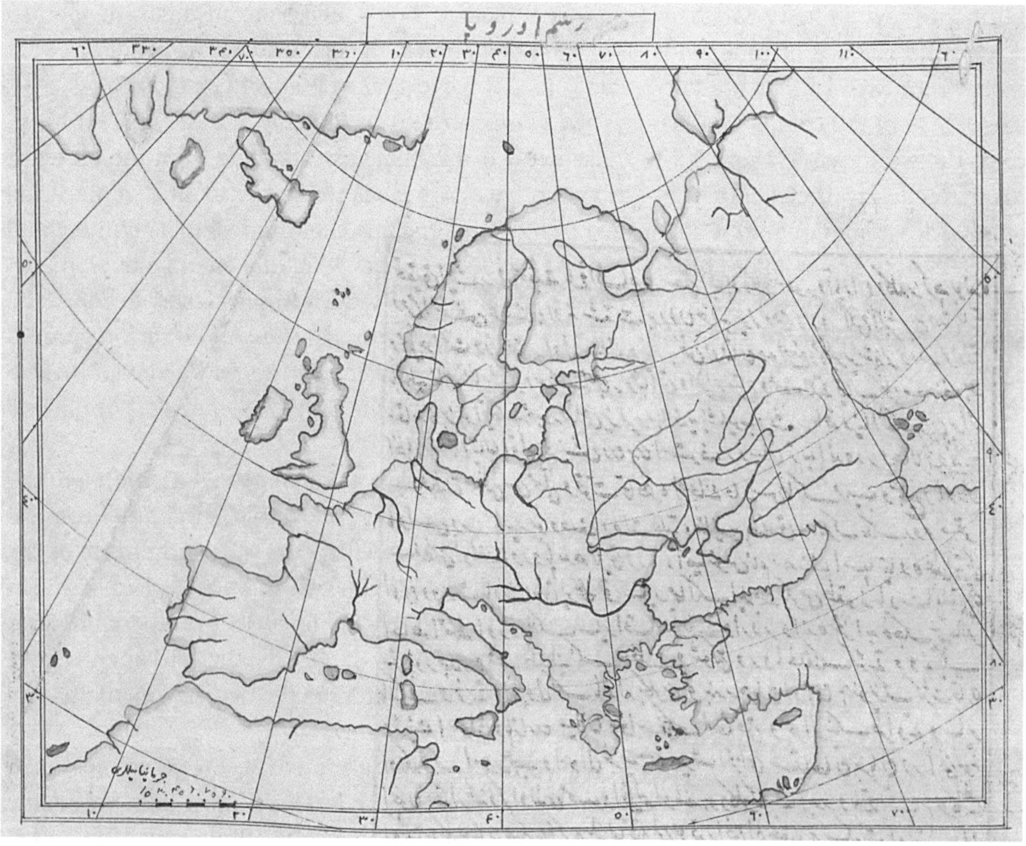

PLATE 15
Map of Europe, R 32a; cf. M 99/100 (Appendix, fig. 41–42)
TOPKAPI PALACE LIBRARY, REVAN 1624, WITH PERMISSION

rivers approach each other and above the Ob River as far as the North Sea is the demarcating border.

Distance: The length of Europe from east to west is approximately 900 German miles from Spain's Cape of Vīn-sen (S. Vincente, Cape Finisterre) to the mouth of the Ob River. The widest places are 500 miles with the aforementioned mile as soon as it comes from the Cape of Mania in the Morea to the Cape of Iscandia [Scritofinnia][7] known as Nort-Kent [Noortkyn].

Division: This clime has been divided into various provinces since ancient times, so their names are presently known as follows: Ispania [Hispania], Gallia, Germania, Vindelicia, Rhaetia, Italia, Noricum, [R 32a] Pannonia, Illyricum, Epirus, Graecia, Macedonia, Thracia, Moesia, Dacia, Sarmatia; among the large islands of the ocean, Britannia or Anglia, Hibernia, Thule or Islandia, Frislandia; among the islands of the Inland Sea (Mediterranean), the two Balearics—i.e., Majorca and Minorca, Sicilia, Sardinia, Corsica, Malta, Creta, Euboea—i.e., Ağrıboz, and other small islands. [Plate 15]

Presently, it is divided into many sovereignties and governments. The notable among them are these: Ispania [Hispania], Gallia or France, Anglia or England, Scotia (Scotland) which is a kingdom in England, Hibernia (Ireland), Norvegia (text: Norvergia; Norway), Suedia (Sweden), Dania (Denmark), Germania which is presently a kingdom of Rome, Polonia or Leh (Poland), Bohemia or Czech, Hungaria or Magyar, Slavonia (text: İsqlāvūnīā) or Arnavud (Albania), Croatia or Ḫırvat, Dalmatia or Bosnia, Servia or Sirf (Serbia), Bulgaria or Rumelia, Transylvania or Erdel, Lesser Tataria or Crimea or Perekop, Napoli (Naples), Sicilia (Sicily). There is but one arch-duchy, Austria, and three grand-duchies, Muscovia (Muscovy), Lithuania and Tuscia (Tuscany). All these are mentioned in the *Atlas* and the *Introductio* and well known among the men of the science. As for the countries and governments from Romania to Hungary that date from after the Muslim conquests, they will be discussed in due course, God willing.

Likening: The historian Strabo likened the shape of Europe to a woman, saying that its head is Spain, its neck France, its body Germany, its right arm Italy, its left arm Denmark and Sweden, its tail Romania.

In some of the Islamic books, the inhabited quarter is likened to a bird. The ancients said that its tail is Spain; and when they saw how thickly settled it was, they specified the bird as a peacock. But likening the inhabited quarter to a bird is very farfetched, and makes no sense at all.

Eulogy and praise: According to the *Atlas*, due to the excellent climate of the lands of Europe, its people are

7 Spellings in brackets according to *Introductio* 46.

mostly energetic and healthy. [M 99] Although it is the smallest of the continents, it is the most thickly settled and has the most moderate and pleasant climate, the most plentiful crops and most flavorful foods, the most splendid dwellings and the most beautiful and skillful people—as Pliny confirms. [R 32b] The truth is that the other continents, despite their great extent, are not as thickly settled and prosperous. Even in Europe there are uninhabited places above 60 latitude due to the extreme cold. But most of it is settled and intensely cultivated. Even places that in ancient times were thought to be barren do have cities and fortresses here and there and are not completely empty. And the trees and vegetation and goods and products have no equal.

The author of the *Atlas* begins his description of Europe on a hyperbolic note, saying, "We must sing the praises of this clime. Macedonia has produced a man like Alexander. Italy since the beginning of creation has nurtured the most noble of Romans. And today Rome is the seat of the greatest sovereignty and the refuge of mighty and valiant heroes who defend their country."

The poor one (Kātib Çelebi) says: This description holds good for the exalted Ottoman state, and since the territory of Rūm is the basis of their sultanate, Europe is indeed worthy of praise.

According to Damianus, Europe is superior to the other continents in that it encompasses 29 sovereign territories, 14 of which are in Spain. Among its many glories are the imperial majesty of Constantinople; the multitude of the people of Rome; the vastness of the wealth of Venice; the wealth of the lords of Naples; the great number of merchants and the beauty of the markets and bazaars of Genoa; the pleasant climate and fresh fields of Milan; and the marvels of the other regions, which excite the wonder of world-travellers. "I have only mentioned these few examples," says Damianus, "which are sufficient to show that Europe is the envy of the climes and that the other continents, let alone being its equal, are not even worth mentioning in comparison."

He goes on to say: "Aside from the great number of ancient monuments and mighty buildings of this clime, its corporal delights and spiritual pleasures, its war-making capacity and administrative genius, and its mastery of philosophy and the sciences have no equal. Europe nurtures and preserves the arts and the sciences to such a degree that by display of marvelous inventions and wonderous devices it has become the mother of science and art. In sum, one cannot exhaust the recounting of the beauty of this clime.

"Let us also mention some of its faults. The people of Franconia are stupid and violent and lack modesty. The Bavarians are spendthrift, gluttonous, quarrelsome and cruel. The Swedes are flighty, loquacious and boastful. The people of Thuringia are suspicious, dirty and quarrelsome. The Saxons are hypocritical, deceptive and stubborn. The Flemings are weak and cowardly, prissy and corpulent."

In saying they are corpulent, the author seems to praise his own people. But it is said that no corpulent infidels are so filthy and soft as the Flemings.

"The Italians are arrogant, spiteful and fantastical. The Spaniards are pompous, ill-fortuned and acquisitive. The French are prissy, impatient and excitable. The people of Sīmberī (Cimbri) [M MAP] [M 100] are crude and malicious, ill-favored and factious. The Poles, Russians and Tatars are gluttonous, thieving and arrogant. The Czechs are ruthless and plunderers. The Albanians and Bosnians are malicious, capricious and quarrelsome. The Magyars are heretical, vain and sour. The Romans are ill-starred and wretched.

"Every people has a certain characteristic. For example, one speaks of a Polish bridge, a Czech monk, an Austrian soldier, a Swedish nun; Italian piety, Russian religion; a German fast, a French compact—all of these are without benefit, and so they have become proverbial for whatever is worthless or useless."

The reason this author gives for the predominant morals of these groups will be explained in the chapter on trigonometrics (ch. 10). The author of the *Atlas* made a number of comments on each of the other continents; but in this book we have changed the organization. [R 33a]

Rivers and mountains: It says in the *Atlas* that there are medicinal waters and many lakes and great rivers in Europe, each serving as a kind of fortress for its land, and excellent sources of fish.

One of the great mountain ranges of this clime is the Pyrenees, which divide France and Spain. Another is the Alps, between France and Italy, permanently covered with snow. Here and there in these mountains are forests filled with all kinds of wild animals. But there are no beasts of prey such as lions and tigers. Most of these places are hunting grounds.

Chapter 5: On (the Continent of) Africa

On the naming of this clime, the author of the *Atlas* says that according to the ancient historian Niceas, it was named Afer after the famous commander Afrus who journeyed to this clime with Hercules. According to the historian of the Jews, Josephus, it is named after Afer who was a descendant of Abraham. According to the Latin historian Festus, the name was formed from the Greek word *friki* meaning "extreme cold" (i.e., Gk. *phrikē* "shuddering, shivering") with the negative prefix *a-* to indicate "there is no extreme cold" since it is between the two tropics. But the Arabs derived it from *farq* "separation", since this clime is separate from their own lands. They also relate that it was named after a ruler named Āfrīq. The Greeks call this clime Libya, relating that it was named after Libya the daughter of a ruler named Epaphus; or else that it is named after *libs*, which is the wind that blows from that direction (i.e., Gk. *lips* "SW wind, Lat. Africus"). In the books of the ancient prophets, it is recorded with the name Qamazyā. The Indians call it Bazqat.

Borders: The northern border of this clime from the Strait of Ceuta (Gibraltar) eastward is the Inland Sea, i.e., the sea of Rūm and Shām (Mediterranean). Therefore this sea is called (Mare) Africum and (Mare) Libicum. The eastern border is the land of the Arabs, i.e. the Arabian Peninsula; or else the Red Sea, known as the Sea of Qulzum (Gulf of Suez). Some among the ancients held the eastern border of Africa to be the Nile River, and included a portion of the territory of Egypt in Asia. But according to most of the schools, the Arabian Sea is the eastern border. The southern border is the Ethiopia Sea, known as the Sea of Ḥabesh. The western border is the Atlantic Ocean, known as the Western Ocean.

Distances: The length of Africa from the Strait of Ceuta (Gibraltar) to the Cape of Bona Esperansa (Cape of Good Hope) is 700 German miles. Its length, from the cape known as Capo Verde up to the Cape of Zaylaʿ near the mouth of the Arabian Sea, is 550 German miles. Its circumference is approximately 3,030 [M 101] miles.

The clime of Africa is connected to the clime of Asia by an isthmus 25 German miles long between Suez and ʿArīsh. Therefore it is called the Great Peninsula—"peninsula" meaning an island that is connected to land. The rest of it is surrounded by the sea. The equator nearly divides this clime in half, and the two tropics (of Cancer and Capricorn) cross it. The two parts extend beyond those tropics to the south and north more than ten degrees each. To be sure, the extension from east to west is less than Europe. But the extension from south to north is about twice that of Europe. Some of the continent has inlets from the sea, which conceal its extent. But not every place is populated as in Europe.

Divisions: It says in the *Introductio* that the ancients described the regions of this clime close to the Inland Sea and had no clear idea of its interior. They were not able to discover what lay beyond the Nile or anything about the Mountains of the Moon.

According to the *Atlas*, the Romans divided this clime into six parts: 1) Numidia; 2) Carthage, which is Tūnis; 3) Brāshīū (Bizacchium); 4) Tripoli, which is Ṭarabulūs; 5) Mauretania Caesariensis; 6) the other Mauretania.

Ptolemy, at the beginning of the fourth article, mentions 12 countries: 1) Mauretania Tingitana; 2) Mauretania Caesariensis; 3) Numidia; 4) Africa proper; 5) Cyrenaica; 6) Marmarica; 7) Libya proper; 8) Upper Egypt; 9) Lower Egypt; 10) Inner Libya; 11) Ethiopia, which is Ḥabesh; 12) Inner Ethiopia. [R 33b]

Leo Africanus, one of those who consider the eastern border of this clime to be from the Nile, divides it into four parts: 1) Berberia; 2) Numidia; 3) Libya; 4) Nigrita, meaning Land of the Blacks. But this scheme is not favored, since Egypt remains outside it.

The author of the *Introductio* says that in ancient times this clime was divided into these provinces:[8] 1) Aegyptus; 2) Cyrenaica; 3) Africa Minor; 4) Troglodytae; 5) Garamantes; 6) Numidia; 7) Mauritania; 8) Gaetulia; 9) Libya interior; 10) Arabia Troglodytica; 11) Aethiopia. At present, it is divided into seven provinces: 1) Aegyptus; 2) Berberia; 3) Biledulgerid (Bilād al-Jarīd); 4) Desert of Sarra; 5) Nigritae, meaning Blacks; 6) Aethiopia Interior, which is the kingdom of Abyssinia (Ḥabeshī Sultanate); 7) Aethiopia Exterior.

Description: The soil of most of Africa is sandy and brackish, and there are few water sources other than the various rivers. But the places that are inhabited are very productive, giving a return of one hundred to one. Especially the production of Mauretania is amazing, with clusters of grapes one cubit long.

Both grazing animals and beasts of prey are plentiful in this clime. Due to certain celestial phenomena, huge elephants and terrible dragons develop here that eat the other animals. There are lions and wild buffalo; a donkey with horns, according to Herodotus; an animal called *tores* that is a cross between a wolf and a jackal; dragons; wild rams, large hedgehogs, tigers, ostriches, and various snakes, especially the horned viper. By God's command, a small animal the size of a mouse is prevalent that all the other animals are wary of. Only, according to Heredos, there are no pigs and no deer in these lands.

There is said to be an animal called *ṣal* (basilisk) that can kill a man with a single glance. Some say that this is legend. But it is recorded in the age of Pope Leo (IV, 847–855) that a *ṣal* made its lair under a vault in the district of the Lucia Church in Rome (i.e., Sta. Lucia in Selci)

8 Spellings according to *Introductio*, 319, 332. Cluverius does not number them, and specifies that Troglodytae is the proper name of Africa Minor.

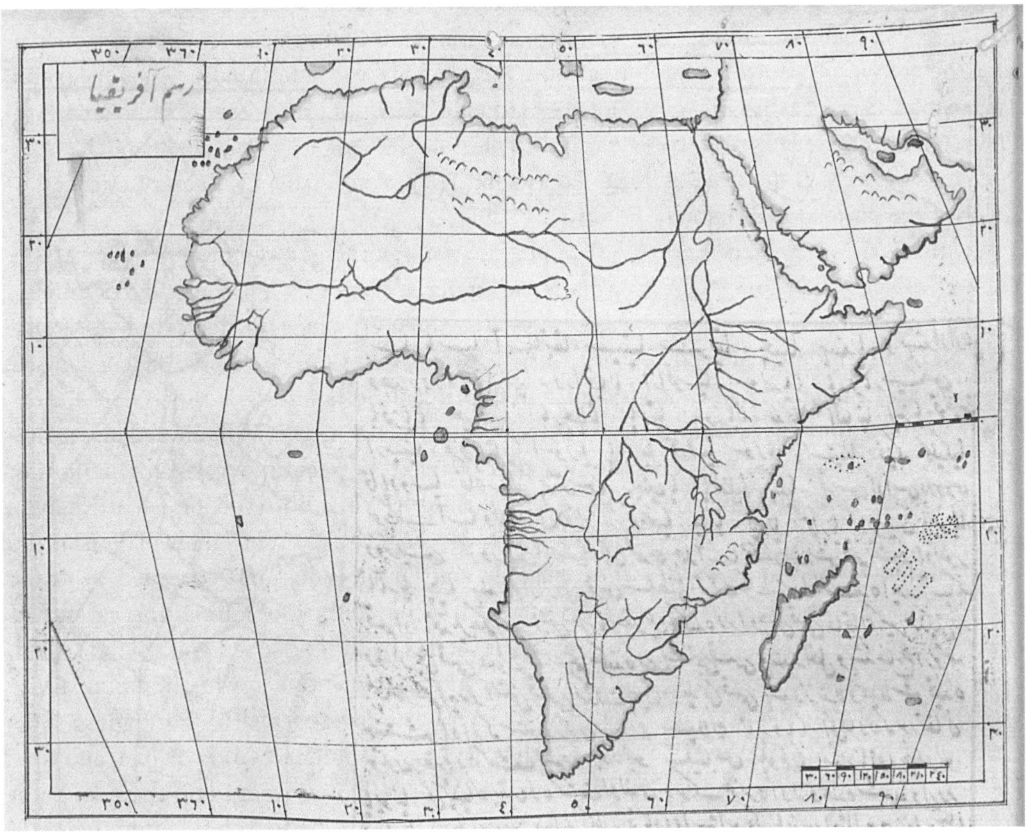

PLATE 16
Map of Africa, R 34a; cf. M 101/102 (Appendix, fig. 43–44)
TOPKAPI PALACE LIBRARY, REVAN 1624, WITH PERMISSION

[M MAP] [M 102] and a great plague arose in Rome due to the putridness of its breath.

They say that the reason why so many kinds of animals develop in this clime is the lack of water and the extreme heat, because heat and dryness compel frequent mating of animals and so various kinds of animals come into being.

Rivers and mountains: There are large lakes in Africa. The best-known is Lake Zenbere which is 50 miles in circumference. The Nile, Zaire and Cuama (Zambezi) Rivers all flow from this lake. Aside from these, the Black River (Niger), the Senaga[l], the Cambram and the Spiritus Sancti River all flood like the Nile and irrigate the lands that they reach.

There are large mountains here, the largest being the Atlas whose peaks in some places rise out of the sands and into the heavens. For that reason it was called the Pillar of Heaven. It starts at the ocean shore in the west and extends eastward in a zigzag pattern until the border of Egypt where it turns. Since most of this mountain is steep and rocky, there are few places where it can be crossed. Because of its great height, the upper half is always snow-covered and cold, while its lower half is forested and full of springs. Sometimes the north wind blows strongly on these mountains and breaks off great masses of snow which pour down in an avalanche from above, burying many large trees and animals below.

Near the sea is a mountain called Sierra Leone meaning Lion Mountain. It is always covered with clouds and passing ships hear loud frightening noises coming from it.

As for the Mountain of the Moon, which was also known among the ancients: it lies below the Winter Solstice (Tropic of Capricorn) and is very lofty and steep. Wild men dwell on it, and there are wide and deep valleys at its base.

There are also several mountains in the country of Vānqūla (Angola) which have plentiful silver mines. They are known as Qāntābaras.

Islands: The famous islands around Africa include, in the Atlantic, the island of the Blessed Harbor (Porto Santo), Madeira, Canaria (Canary Islands) and Perīdā (Cape Verde) meaning Green Cape. Islands in the other parts of the ocean are: Santa Anton, Santa Vīnensī, Santa Lucia, Santa Nicola, Saltpan Island (Tuzla Adası), Bula Vista, Santa Jacobi, Mājū, and Fire Island (Ateş Adası). In the Sea of Ḥabesh are Sultan Island, Santa Toma, Santa Lorensi, etc. [R 34a] [Plate 16]

Chapter 6: On the Continent of Asia

On the naming of this clime, the author of the *Atlas* says that according to Varro it was named after Asia, the wife of Japheth son of Noah. Some say that it is derived from Asius, the son of Lidus. In one version, this Asius in the time of idol-worship was the founder of the famous idol named Paladium in the city of Troy, which is now known as Eski İstanbulluḳ. At first that city was called by his name, then the entire clime was called Asia.

The name Asia originally referred to the provinces of Anatolia. Its sense gradually broadened and the name of the part was applied to the whole. So Anatolia became known as Asia Minor meaning Little Asia.

This clime is mentioned in the books of the ancient prophets under the name Semeyā.

Borders: The eastern border of this clime is the Eastern Sea. The northern border is the Northern Ocean, known as Scythicus Ocean, meaning the Tatar Sea. [M 103] The southern border is the India Sea. The western border is the Arabian Sea or the Sea of Qulzum (Red Sea), the Strait of Suez which is between it and the Inner Sea, the Sea of Rūm (Mediterranean), the Strait of Istanbul (Bosporus), the Black Sea, the Sea of Azov, the Don and Ob Rivers.

Distances: According to some, Asia extends from 52 to 196 degrees longitude. But on Mercator's map it extends from the borders of Asia Minor at 57 degrees longitude to 178 degrees.

It says in the *Introductio*[9] that the length of Asia from the Hellespont, meaning the Mediterranean Sea Strait (Dardanelles), to the Cape of Malacca, which is the furthest border of India, is 1300 German miles. Its width from the mouth of the Arabian Sea to Cape Tabin, which is at the Strait of Anian, is 1300 German miles. On the western side it is connected to Europe, or rather overlaps with it, because the Tatars, who are an Asian tribe, dwell on the edges of Europe and the Sarmatians, who are a European tribe, dwell on the edges of Asia.

According to Pliny and Strabo, among the ancients, all the eastern provinces of Egypt are included in Asia, because they and their followers hold the western border of Asia to be the Nile. But the opinion of most is contrary to this.

Divisions: It says in the *Introductio* that in ancient times Asia was divided into two parts: Asia Major and Asia Minor.

The provinces of Asia Major are these: [R 34b] Sarmatia, Asiatica, Scythia, Serica, China, India, the Indian islands, Gedrosia, Carmania, Drangiana, Aracusia, Sogdiana, Paropamisis (text: Pārūtārmīnes), Bactriana, Hyrcania, Margiana, Parthia, Susiana, Media, Albania, Iberia, Colchis, Armenia, Mesopotamia, Assyria, Babylonia, Arabia, Syria, Palestina, Phoenicia, Cilicia, Cappadocia, Galatia, Pontus, Bithynia, Pamphylia, Lycia and the island of Cyprus.

The provinces of Asia Minor are these: Phrygia, Mysia, Lydia, Caria, Aeolis, Ionia, Doris, and the island of Rhodes.

But now the whole of Asia is divided into five distinct parts: Tataria, China, India and its islands, the Persian Sultanate and the Turkish Sultanate.

It says in the *Atlas* that the Muhammadans govern the first part of Asia; the second is ruled by the lord of Muscovy; the third is the country of the Great Khaqan; the fourth is governed by the Sofi Shah of Persia; the fifth is Inner and Outer India, and governed by many lords; the sixth is the country of China; the seventh is the scattered islands of the India Sea and Pacific Ocean, of which the most famous are Ceylon, Taprobana, the two Javas recently discovered by the Portuguese, Borneo, Celebes, Palohan, Mindanao, Gilolo, Japan, and New Guinea that was recently discovered and there is doubt that it is an island because its southern extremity is still not known.

Authors like Strabo and Arian among the ancients divided Asia in various ways. And Ptolemy divided it into 47 parts and made a map of its provinces in the fifth, sixth and seventh books of his *Geography*.

The poor one (Kātib Çelebi) says: In this book there are separate chapters and commentary devoted to the parts of Asia according to the chapter divisions that are current among the Muslims and in their books.

Description: It says in the *Atlas* that some places in Asia have a moderate and pleasant climate. But the northern and southern regions, according to Marius, [M MAP] [M 104] are not moderate in regard to heat and cold, the people's humors are various and their morals bad.

The blessings and prosperity of these regions are so great as to be proverbial among their inhabitants, because the soil of Asia is very fertile and productive. There are plentiful foodstuffs, vast amounts of metals and merchandise and fine perfumes, spacious pastures, and many kinds of drugs and medicinal herbs throughout the provinces. The people have the utmost intelligence, wealth and bodily strength. The first man was created here and the prophets who came from his descendents grew up in this clime. The invention of arts and sciences and the laying down of sacred law occurred here. It was here that the great ones of mankind first gained authority over the small. There are various types of wild ani-

9 The information here derives from *Introductio*, 240.

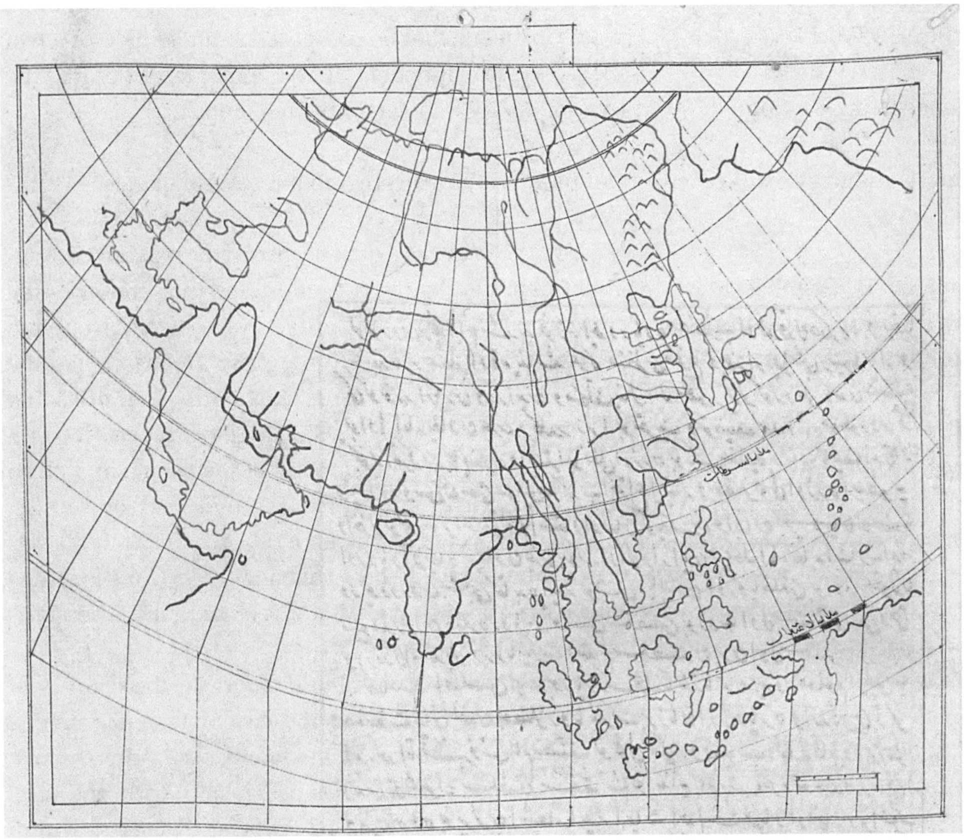

PLATE 17
Map of Asia, R 35a; cf. M 103/104 (Appendix, fig. 45–46)
TOPKAPI PALACE LIBRARY, REVAN 1624, WITH PERMISSION

mals here, camels and elephants being especially abundant.

Mountains and rivers: Asia has large mountains, including the Taurus range that divides Asia between east and west and is the home of various tribes. This mountain range passes through the east of India with one chain going north and one going west and south. It passes by the shores of the Black and Mediterranean Seas in a zigzag pattern and comes to the Riphean Mountains. The author of the *Atlas* says that these mountains have different names in different places and he describes them from beginning to end; but we have summarized his account because they are not well known. There are 14 names in total, including some designated as "Gate", such as Porta Armenia; Porta Caspia, meaning Iron Gate; Porta Silesia; etc. He goes on to say that in most places the width of these mountains is 3000 stadia and their length is 45,000 stadia which is the length of Asia. The range extends from the shore of Rhodes (i.e., the Anatolian shore opposite Rhodes) all the way to China.

There are large lakes in this clime, also a sea that is like a lake [R 35a] in that it has no connection to the ocean. Its famous rivers include the Tigris and Euphrates, mentioned in the Torah; the Jordan; the Jayḥūn (Oxus); the Indus; the Ganges, which is a large river on the borders of China; etc. [Plate 17]

Ancient monuments: Among the ancient monuments in this clime are marvelous buildings and wondrous temples. Three of the Seven Wonders of the World are here: 1) the Walls of Babylon; 2) the Temple of Diana; 3) the Tomb of Mausolus which is in Ayaṣluġ (Ephesus); 4) the Lighthouse of Alexandria; 5) the Colossus of Rhodes; 6) the Pyramids of Egypt; 7) the statue of Jupiter which is in Albania. All of these are in ruins except for the Pyramids of Egypt, which are mentioned in their proper place. The author of the *Atlas* says that the great Temple built by Solomon in Jerusalem is worthy of being added to these seven.

The poor one (Kātib Çelebi) says: There are also buildings in the Muslim books, such as the Arch of Chosroes and the Palace of Khawarnaq, that did not find fame among the Romans. They will also be mentioned in their proper places. This much is sufficient in summarizing the conditions of Asia.

Chapter 7: On (the Continent of) America, i.e., the New World

Several inquiries will be mentioned in this section.

1st inquiry: Whether this clime is known or unknown among the ancients

The author of the *Introductio* states: It is possible to affirm that this region was known by the people of Europe in ancient times, as transmitted from the teachings of Plato and Diodorus. It is recorded in Plato's book *Timaeus* that the Egyptian priests explained it to Solon the Athenian who lived 600 years before the birth of Jesus (peace be upon him) [M 105]. They said that there was a large island apart from Asia and Africa across from the Strait of Gibraltar. Its name was Atlantis. But afterward, by reason of a great earthquake, it was entirely submerged by water in one day and one night and the boats at that time could not pass out of the Strait of Gibraltar, because it was filled up with mud. Later on it was opened (again).

Diodorus writes in his fifth book that the Phoenician people, in ancient times, were traveling along the coastlines of Africa outside of (the Strait of) Gibraltar when a great wind rose up and took them to faraway regions of the ocean. [R 35b] After being blown about with a violent storm for several days, they landed at a large island to the west of Africa whose view was exhilarating, full of meadows and fruit trees and navigable rivers. They saw inhabited places with imposing buildings.

Now, in truth this island cannot be anything other than America, because everything that is described here is found there. It is implausible that Diodorus and Solon, who compiled a book about this according to the relation of the Egyptian priests and Strabo, should say this without reason and invent lies.

According to the *Geographia Minor*, which is attributed by some to Aristotle and by others to Theophrastus, they indicate America when they say there are large islands other than Asia, Africa and Europe. Diodorus states in the fifth book that the people of Tunis (i.e., the Carthaginians) used to block the people of Europe from entering this island. After that, the Romans defeated the rulers of Tunis, and the Tunisians also were unable to go in that direction. From the beginning of Roman rule, those lands remained unknown. But the author of the *Atlas* says it was not known at what time these lands became inhabited. The true state of affairs was not known. Some record that the Romans at some time came to these lands. But an author named Gasparus (Varrerius, ca. 1500–1574) firmly rejected this statement; he said it is a claim impossible to prove. Seneca, one of the ancient poets, mentioned these places in an ode that he composed in praise of a sorceress named Medea. But poems are not proof texts, because the basis of poetry is imagination and fantasy.

2nd inquiry: Why this region has several names

First, in the year 1492 of the Christian calendar, Christopher Columbus, by his own effort and the support of the emperor of Castile, went on a voyage of discovery and persevered in such a great enterprise. He discovered the island of Hispaniola and some neighboring territories, and called it the West Indies, because in some respects it corresponded to the East Indies. Most of the people in both go about naked, and their customs resemble each other. Also, this was discovered at the same time when East India was discovered in Asia. Others called it *Novus Orbis* meaning New World, because in size it was close to the other parts [of the world] together.

Five years after Columbus's report, in the country of Italy, a captain residing in the city of Florence named Amerigus from the Vespusius family (i.e., Amerigo Vespucci) set out on this business again with the support of Manuel the King of Portugal. He set sail from Cadiz, crossed the equator and discovered the countries of Paria and Brazil along the eastern coast of the New World [M 106]. On account of that, all of this region was named America after him. Subsequently, many people came from Spain, France and England and everyone who discovered a province named it with his own language and name. But the influence of the work of the captain Amerigus was the continent being named after him.

3rd inquiry: Concerning the borders of this region

These lands are circumscribed by the Ocean on every side, except for one side which is not known. Its eastern border is the Atlantic Ocean, which is known as *Mar Del Nort*—meaning Sea of the North—and Sea of Atlas. Its southern border is the Strait of Magellan at 50 degrees latitude; this strait divides the space between America and the Southern Land (Antarctica). Its western border is bounded by the Pacific Ocean, which is known as *Mar del Zur* meaning Peaceful Sea. Its northern border is unknown lands beyond 67 degrees latitude; they still have not been explored. The *Introductio* says it is thought to be bordered by the Frozen Sea (Arctic Ocean), because it is underneath the Pole, and Asia and Europe, which are situated opposite to it, are bordered by this sea. [R 36a]

4th inquiry: Concerning the extent of this land

It says in the *Atlas* that America extends from north to south and is in the form of two great islands that are linked to each other by a slender isthmus. The northern one is called America Septentrionalis and the southern one America Meridionalis. The length of its shores is 32,000 miles approximately by true reckoning. Its shores have all been traveled by ship, investigated and surveyed; only the northern side remains.

It says in the *Introductio* that the shores of this land (America) appear when 330 German miles are traversed northwest from the mouth of the Niger River in Africa. While it is not larger than the inhabited quarter of the world, it is equal to it. If the distance is considered in relation to the islands, it is closer to Europe than Africa, because it is 200 miles between Canada and Hibernia (Ireland). The greatest length is 2,400 German miles from the Strait of Anian to the Cape of Magellan. The greatest latitude is 1,300 miles from Cape Fortuna, which is close to the Strait of Anian, to Cape Breton in Nova Fransa (Nova Scotia).

5th inquiry: Concerning the divisions of this region

It says in the *Introductio* that all of America is divided into two parts: the northern, known as Mexicana; and the southern, known as Peruviana—the majority of this falls south of the equator.

The north, whose main city is Mexico, has these divisions: Nova España, Nicaragua, Yucatan, Florida, Apalachin, Norumbega, Nova Fransa, Terra Labradoris—meaning Land of Laborers—and Estotilandia. On its western side there are two countries along the Strait of Anian: Quivira and Anian after which the strait is named.

South America, known as Peruviana after the name of one country, has many countries, but not all of them have yet been conquered. The countries that have been conquered are: Castella Aurea, Bogota, Peruvia, Ploppaiana (Popajana, Popayán), Chile, Chica, Brazilia, Caribana, Guiana (text: Buiana). All of these are along the coasts of the ocean. Inland are found the Amazon country, Paguan, Picola, Moxos, Uram, Charcas.

[M 107] These provinces have two great cities that are almost without equal, Mexico in the north and Cuzco in the south. All of these will be explained in detail below, God willing. Pertaining to this region there are several islands, including Hispaniola, Cuba and Jamaica.

6th inquiry: Concerning the mountains and rivers of this region and some buildings and reports

The author of the *Atlas* states: Formerly, wheat and grapes did not use to grow in these places. Instead of wine, they drank the juice of a kind of fruit called maize (*mays*) [Mayz].[10] Agriculture was not done according to custom. They used to leave three or four seeds in a hole and cover it up. Several stalks would emerge, from every stalk three or four ears, and from every ear up to 100 kernels, which they would eat.

They made a kind of bread called cassava (*każābī*) [Cazabi] by beating a root the size of a turnip called yuca (*jūqa*) [jucca]. It is a type of plant filled up inside like a reed, that grows at the time of sowing. They cut it up in pieces and bury it in the ground; whenever they need some they pull it up and beat it and press out the juice, which is poisonous. They make the bread from the pulp. After two days it is like stone. They take out as much as they need and leave the remainder to rot.

There are two types of fruit there that resemble a cherry, one called potato (*baṭāṭās*) [Battatas], the other *hāyās* [Haias]. It is big and sweet. They plant it and six months later eat the fruit. There is an amount of sweetness in the flavor, but a lot cannot be eaten, it is not good for eating. The provisions of this land's people are meager, their food is tasteless [R 36b] and productive of wind.

The wild grape trees in this region are plentiful. Its seed is black, resembling that of a wild plum, but its husk is thick and has a seed. There are olive trees also, but the fruit smells and tastes bad, one does not eat it. There is a tree resembling a plane tree known as *hūvī* [Hovi]—it is *dulb* ("sycamore") in Arabic—and the pine and the trees called *māmey*, *ghavānnā* and *aney* [Platani, Pinea, Guiava, Mamei, Guanavana] are numerous. Among the types of plants, sugar cane, cotton and flax are plentiful. The other types of plants and fragrant extracts are limitless, as are diamonds and pearls among the types of gems, and silver and gold mines.

Animals: Originally there were no cattle, horses, donkeys, mules, goats or dogs in this region; they were brought afterward. The peoples of this country initially fled in terror when they saw a man on horseback; because they had never seen a horse previously.

According to the *Atlas*, they did not know what a mouse was in their houses or on the ground. A ship came from the city of Antwerp in Flanders and a mouse left it when it approached the shore while crossing the Strait of

10 Spellings in brackets according to *Atlas Minor*, 18–20.

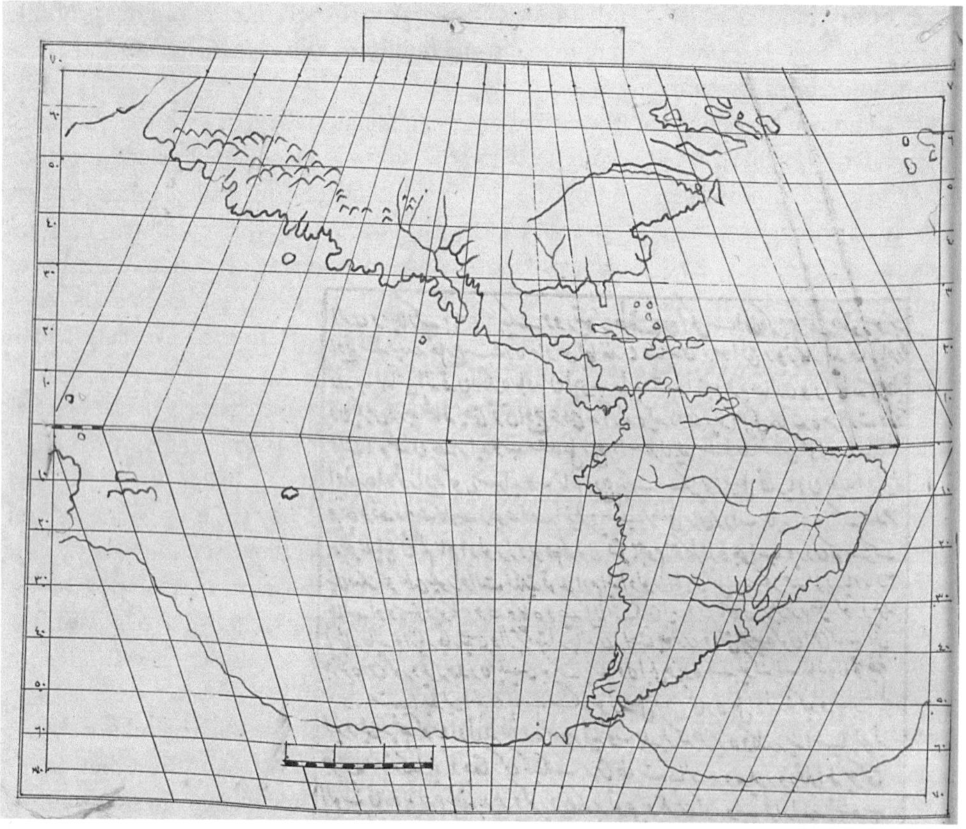

PLATE 18
Map of America, R 37a; cf. M 113/114 (Appendix, fig. 47–48)
TOPKAPI PALACE LIBRARY, REVAN 1624, WITH PERMISSION

Magellan. It procreated, and the mice multiplied so much that they took over the continent.

The poor one (Kātib Çelebi) says: It is established that mice propagate wherever there is putrefaction. As for their multiplying where there is no putrefaction, it cannot be contemplated; however, particular places are exempt from this.

In this land there are so many different animals that only a portion of them are known. One type of these, its nose and trunk resemble those of a fox. It has feet like a monkey and ears like a bat. On its belly it has a pouch like a bag. If it goes somewhere, it puts its young inside the pouch and takes it along.

There is another animal of that country known as *qāshqūy* [Cascuij] (i.e., capybara?). It is black and furry and shaped like a pig. Its skin is very firm, its eyes are small, its ears are like those of a donkey, it has cloven hooves and a small trunk like an elephant's trunk. Its voice is so frightful and vehement that if it bellows all of a sudden it makes a man deaf. But its meat is very tasty.

In addition to these, wild boars and mountain tigers are plentiful and very savage. Lions are also plentiful, but the lions of that place are very timid; if they to see a person, they flee. Peacock, partridge, jungle fowl and various [M 108] birds that do not resemble the birds of these lands (i.e., the Old World) are plentiful. They will be mentioned in part later on in the detailed maps of America.

Mountains: There are great mountains in this region. One of them is very high, fire comes out of its summit all the time, its flame can be seen from 100 miles away in every direction at night. A monk took an iron bucket with a chain and went there with four people thinking that it would have gold in it, but when they dangled the bucket into a fissure it immediately melted, they tried it again with a thick chain, but the fire melted that too.

Rivers: The rivers of this land are plentiful, but the most important are two rivers, the Maranium (Amazon?) and the Ordeliana (Orinoco?). Another is called Argenteus meaning Silver River; the Spanish call it Rio de (la) Plata.

Buildings: In this country there are great cities, excellent roads, elaborately ornamented palaces and villas, and royal gardens. There is also a royal treasury, described more fully below; it is sufficient merely to mention it here, since this section is a synoptic overview. For now, let us mention the voyages that the Christians made to these lands, so that this book shall also be sufficient on that subject.

7th inquiry: Concerning the journeys that the moderns—Columbus, Magellan and others—made to the Americas, related by abridgment from the *History of the New World*

Columbus's First Journey. He made countless voyages and obtained knowledge of surroundings and directions. He found fame by drawing maps and making surveys.

He went out beyond Ceuta (i.e., the Strait of Gibralter) with the intent of voyaging to the Indies, and took anchor at an island called Madeira. Suddenly a boat landed at that island whose crew had been killed by the violent surges of the sea. Only the captain and one or two people were left alive. Columbus took the captain home, [R 37a] [Plate 18] feasted him and made him relate his adventure. He said: "We had journeyed to the coasts of the Maghreb for commercial purposes. Suddenly, a contrary wind arose that took us westward into the open ocean. We acceded to fate and went where the wind took us. We passed by so many islands and coasts. Finally the wind was favorable again and turned our ship back to this direction. But the surges of the fearful sea sent most of the crew to destruction. Now you see that my own condition is not good." The captain lived for one or two days, then passed away.

This story stuck in Columbus's mind and he became obsessed with discovering those parts. But he was not able to outfit a ship, so he petitioned the Portuguese king and requested help. The king paid no heed, saying that there was no inhabited place in that direction and that Columbus's speech was on account of a vain desire and a baseless idea.

But the author of the *Atlas* does not concur in this opinion. He states that some envious people asserted that Columbus did not attempt this task on his own initiative, but that previously he got news about a Spanish sailor who had traveled to those lands, and then made his voyage. So, on account of the aforementioned historian saying that it is the speech of envious ones that he transmitted, he suggests that Columbus did voyage in that direction based on his own idea.

After that, he came to the ruler of England and sought help in the aforementioned aim. The English king also replied with a rejection. But Columbus never despaired, nor gave up. Finally he came to the ruler of Spain, entering his court at Aragon. He made many petitions and put forward his proposal. The Spanish king was about to campaign against the fortress of Granada [M 109]. He promised to help if he was vouchsafed victory. He seized the fortress, and nothing remained of the control of the rulers of the people of Islam in Andalus. When he came to a place near Granada called Santa Fe he met with Columbus and made good on his promise. He pledged 16,000 goldpieces and gave an order that Columbus should be helped wherever he went, with the proviso that however much money and wealth he acquired, Columbus would take one tenth and bring the remainder to him.

Columbus took the gold and went to the city of Palos. He outfitted three ships, placing forty men on each one. In fact, very few men were willing to go on this journey of their own accord, and he put mainly criminals on the boats whether willing or unwilling. He took provisions and an amount of goods for trade [R 37b] and set sail from the port of Cádiz, going outside of the Strait of Gibraltar, in the Christian year 1492 which is the Hijra year 903. After coming to the Canary Islands with a favorable wind and taking on provisions, he unfurled the sails and headed directly to the west. He always proceeded twenty degrees to the north from the Tropic of Cancer, meaning at 43 degrees latitude, thus avoiding the heat and cold on either side. He constantly observed the sunset point with his quadrant.

In this manner they traveled 33 days and were 3,800 miles distant from the Canary (Islands). Many times they regretted the voyage and aimed to go back. When the crew mutinied against Columbus, saying, "You drove us into this whirlpool of catastrophe; we will all be lost in this endless sea," he replied, "Your salvation lies with a man who knows navigation and uses astronomical instruments; if you kill me, all of you will perish at sea," and he calmed them alternately with promises and threats.

Suddenly they saw an uninhabited island with tall trees and flowing rivers. They rested a bit, then went six days more and saw six more islands. Two of those were larger than the others. They called the bigger of those two islands Hispaniola and the smaller one Gonave.

From there they went 800 miles more to the northwest and reached a coastline. They traveled around in those parts many days and realized that it was not an island. They returned to Hispaniola and, while disembarking, one of the boats struck a rock and sprung a leak. They took the cargo onto small boats and landed the men on shore. Observing that a group of people had seen them and fled, they caught up with them and seized a woman. Columbus showed her respect and gave her some gifts, but she was unable to understand his language and he explained by gestures that she should bring her people. So the woman came to her people, presented the gifts and precious items, and told them that they posed no danger.

A group from among them approached the boat bringing silver and gold, fruit and bread, and various birds and animals, and they began to buy and sell freely. They were especially attracted to the little things, such as needles and earrings. They bought each one for its weight in gold and were pleased. For several days they came and went in this fashion and did a lot of business. After that their ruler, who had the title cacique, came with an entourage and met with Columbus, bringing many gifts from the goods of the island. Although the two groups [M 110] did not understand each other's speech, they demonstrated their intentions with signs.

Then Columbus requested permission from the cacique to build a fortress on that island and leave in it a small number of men to learn their language. The cacique gave his assent and assisted them. When he finished, Columbus put thirty-eight people from his own group and loaded the goods of the island onto the two boats. Taking ten men from the people of the island he returned to Spain and came to the city of Palos in fifty days. A full year had passed since the beginning of the journey.

He went to the emperor of Spain in Barcelona, presented his gifts and explained the descriptions and reports of that island. Columbus's gifts were: thirty parrots of which some were completely red, some yellow, and some of various colors; several small rabbits whose ears and tails resembled those of a mouse, and their color that of a squirrel; a type of bird called *galipavos* (wild turkey) that resembled a peacock; a root from a certain plant that tasted like sugar; and an herb that tasted like cinnamon. The ruler of Spain was captivated by this account and showed great interest in Columbus, giving the government of that province over to him. He gave an order that Columbus should henceforth receive two-tenths of the wealth that he acquired and should go again with 1,500 men and seventeen ships.

Columbus's Second Journey: The aforementioned captain came to Seville, which became the West Indies port of trade, and saw to the provisioning. In the Christian year 1493 he departed from the port of Cádiz [R 38a] and came to the Canary Islands. He went 5,280 miles in a southwest direction with a favorable wind, and on the twenty-third day he came to the island of Dominica (Santo Domingo), one of the Caribbean Islands, whose inhabitants were cannibals. They turned from there and came to another island that was covered with flat lands, shady trees and flowing waters, but they did not see anyone; they gave it the name Galita (Marie-Galante). Continuing some ways they came to the largest of the Caribbean Islands. It had a village upon a flowing river and its inhabitants were cannibals who took fright and scattered. The crew attacked and seized ten individuals, plus thirty women and boys whom they were fattening up to be eaten. They caught a large number of parrots and headed back to Hispaniola. Along the way they saw and passed by innumerable islands.

When they came to Hispaniola they found the fort that they had built previously had been destroyed, and they did not see any of the people they had left behind. They asked the people of the island, who said: "There is a cacique on this island who is greater than the rest. He came with a group and destroyed the fort and killed its people." Columbus maintained good relations with them and said nothing bad. He found a good harbor on the island, disembarked and built a fort which he called Isabella after the wife of the Spanish ruler, because she had helped Columbus a great deal. He treated the people of that district well and acquired wealth without end. They loaded twelve of the seventeen ships with money and goods and sent them back to Spain; they returned to the [M 111] port of Cádiz in 1494.

Columbus built another tower in the middle of the island and gave it the name Santa Toma. Many gold nuggets were found in those districts which amounted to ten drams' weight. He placed his brother there with a number of men and descended to the coast again, equipped the five remaining ships, and left to explore the surrounding areas. They sailed all around Hispaniola, then headed westward for another eighty miles. They saw a large island and gave it the name Cuba. From there they journeyed south and saw another large island which they named Jamaica. They disembarked there and clashed with its people. After making a truce, they proceeded to the west, voyaging 880 miles in 70 days. In some places they encountered a formidable current which only allowed them to travel one mile per day. They passed innumerable islands and reached a coastline. There they saw a people and established good relations. They reported that the aforementioned coastline extended to an immense plain whose extent was not known. From there they returned to Hispaniola.

Columbus dwelt there for an extended period. He maintained friendly relations with some of the island's people, while others he subdued by force of arms. He amassed gold without end. He placed his brother in charge of the island and himself returned to Spain in 1495. He presented his gifts to the King, including gold nuggets amounting to 150 and 200 drams apiece; raw ambergris; brazilwood; gum lac; cloaks made of cotton; and felt carpets. And he reported on the islands and coasts to which he had traveled.

Columbus's Third Journey: Again in the Christian year 1498, Columbus departed from San Luca (Sanlucar) with eight ships and turned toward New India. Previously his brother had constructed one or two fortresses on Hispaniola and gone to the western half with one or two galleons and imposed a tribute on its cacique. Columbus now came to Madeira island and sent five ships to Hispaniola. He himself sailed with three ships above the equator and he came to the Hesperides which are thirteen desolate islands close to the corner of New India called Cape Verde, which means Green Promontory. From there they went 180 miles further toward the equator. [R 38b] when there remained five degrees to the equator, they fell upon calm seas (i.e., the Doldrums) and suffered from a great heat. The iron hoops on the casks shattered and the water spilled out. The wheat in the storage hold began to burn, so they dumped it into the sea. All of them nearly perished from the violent heat.

On the tenth day they found a bit of an easterly wind and turned west. After some time they saw great mountains on the coast. They entered a harbor which they called Dragon's Mouth on account of the great suffering they endured there. Disembarking, they saw pleasant places, well-proportioned trees, delightful streams, sweet-smelling flowers and sweet-singing birds. They called that Paria and stayed there for one or two days. Finding that both poles were absent (i.e., that the compass needle did not indicate north and south) they tested with astronomical instruments and found that the earth was not completely spherical. They depicted it in the shape of a pear, with Paria as the stem; and they recounted that the nights there were very bright.

The poor one (Kātib Çelebi) says: This claim is disputable. Supposing the earth to be in the shape of a cone, the top of the cone, corresponding to the stem of the pear, would be Paria which is near the equator. [M 112] Both poles could not be lost at that point; rather they must appear, because the perceptible horizon is situated beneath the actual horizon. If he had said that poles are found absent since Paria is a sunken place like the place of the flower of a pear, then it would have been correct. But it is not necessary that the earth have a conical shape because of this. What is the basis of this judgment? If one were to gaze from all of the sunken places that are below the equator, both poles would appear to be absent. In short, this claim is ignorant talk, uninformed about the principles of astronomy. The world's not being a perfect sphere does not preclude its being spherical; because the height of the largest mountains is negligible in proportion to the diameter of the earth. For it is established by clear proof that the seventh of the width of a barleycorn in relation to one cubit is equally imperceptible. And for nights to be very bright appears in many places by means of accidental circumstances.

Columbus proceeded from there and saw a great river that was 112 miles wide and 30 fathoms deep. In the place where it entered the sea there were great waves and dreadful noises. There also they suffered much. Arriving at a place between the west and south where the surface of the sea was completely filled with grass and sedge, they crossed it with a thousand hardships and turned to the north.

Returning that same year to Hispaniola, they saw that many of the Spaniards had perished. The reason was that they had forced that country's people into continuous servitude and made them dig mines; so they had abandoned agriculture and gone to other places, and the land had become barren. Also, the Indians had come from the surrounding areas and waged war, and many people had fallen on both sides.

After Columbus's departure, fifteen caciques allied to form an army and fought fiercely with the Spanish. In the end they were defeated, the reason being that they were naked and they had no weapons since there was no iron mine on the island; while most of the Spanish wore chain mail, carried guns and were swordsmen. The Spaniards put the majority of that nation to the sword and took their caciques prisoner.

After that Columbus clashed with his brother over a certain matter and the Spaniards split into two factions. They brought the case to the ruler of Spain who in turn dismissed both of them from office. Columbus remained out of office in Spain for up to three years until, by intercession of some notables, he was again appointed and sent off.

Columbus's Fourth Journey: In the Christian year 1502, Columbus once again journeyed to New India with four ships. When he arrived to the island of Hispaniola, the ruler did not cede the island to him and refused to obey him. Columbus, lacking the power to fight them, [R 39a] resorted to trickery. He said to the people of the island: "Because you have been cruel to us, your god has become angry at you. His sign is this, that tomorrow the sun will darken." An eclipse was due the next day as Columbus knew from the almanac. The people of the island became apprehensive and gathered in the morning to observe. When the eclipse occurred, they were completely terrified. They came to Columbus bringing many gifts and made a truce. Then, Columbus established a great port in a district of the aforementioned island, and gave it the name Santa [G]loria. Columbus stayed there for one year and came to Spain again. After that, he was not destined to travel again and in the year 1506 he drank the wine [M 113] of death.

The people of Hispaniola used to worship idols in the form of devils. And the unclean spirits that were associated with them gave news that "in the near future, a group shall come to this land, and shall destroy the majority of you and shall take this land from your possession." That group composed songs from this report and recited them in their gathering place and wept. When eighty years had passed, the hornets of catastrophe fell upon their heads. The Spaniards seized the majority of those places in twenty years and took more than 40,000 prisoner and made many more hundreds of thousands prey for the sword.

The journey of Captain Magellan setting out toward the west and returning from the east:

There was a person from the kingdoms of Portugal, a corsair named Ferdinand Magellan. He became informed by some philosophers that it was possible to reach the Moluccan Islands from the Western (Atlantic) Ocean. For this reason, he was always thinking about it. He made a petition to the Portuguese King and talked about this idea, requesting his assistance, but that king paid him no heed. When he came to the Spanish ruler, he offered assistance with five ships, 200 soldiers and other necessities.

Magellan departed from Seville in the Christian year 1519. Setting sail with a favorable wind, he voyaged toward the setting sun for many days and arrived at the Cape of Santagostin (Santo Agostinho). From there he traveled southwest, and was caught by a powerful winter storm. He rested for twenty days at an unknown location and established cautiously friendly relations with the people of that district, they being black-skinned negroes, thirteen spans in height, who wielded bows and arrows and javelins. They found one of them alone and eight people were not able to capture him. They invited one of them to the ship by means of a ruse and put him in irons; he refused to eat and perished.

Setting sail from there, when they had gone about forty miles, a great cyclone arose, pulled in one of the boats with its crew and heavy baggage, and dropped them at the top of a mountain. The boat shattered, but most of the crew survived. From there they went another 120 miles and came to the place known as Promontory of Women (ʿAvret Burnı). From there they came to a strait like the Strait of Ceuta (Gibraltar). It is now called the Strait of Magellan. It is 440 miles long and nine miles wide at the narrowest point.

Passing through this strait, they continued westward—that is, to the east, in one estimation—until they came to the island named Anbigana where they saw several small boats. Continuing on, they came to the island of Cebu. The people of the islands that were nearby came and established friendship with Magellan, but the people of the island of Mactan came and did not submit. For this reason Magellan went with a number of soldiers and made war on them. In the midst of the fighting he was killed with a number of his companions. The 115 people who remained loaded sugar, cloves and cinnamon from Cebu into two boats and having report of the Moluccan Islands, they journeyed for a long period and stopped at many islands until they came to a large island named Būrnāy (Borneo). Its people were all blacks. They met its king and gave gifts.

Leaving the Moluccas behind them, they came to the island of Cimbombom and found many pearls there, most of them as big as eggs. From there they came to Tidore, one of the many islands of the Moluccas, in the Christian year 1521. [R 39b] Its ruler was a person named Manṣūr from among the Arabs. He came aboard a ship and they placed gifts before him and presented the good will from the ruler of Spain. Manṣūr [M 114] regarded them well and gave them permission to conduct trade. The Spanish traded on that island for five months. They had established a price for a bag of cloves as thirty cubits of linen or ten cubits of red broadcloth or four cubits of yellow broadcloth. Many groups came from the other islands and opened up business also. They loaded up endless amounts of cloves, cinnamon, parrots and a type of honey that had a taste resembling sugar, and they left. One of the ships fell behind along the way and the other came to the port of San Luca (Sanlucar) in the year 1522 with a captain named Bastian (Sebastian). All of the period of the journeys fell fourteen days short of three years. In that period, they traversed a distance of 50,000 miles.

The Spanish King was quite delighted by this news. It is related by some Mudéjars that he built a dockyard for Magellan's ship and covered it with red broadcloth. They used to visit it, saying that he accomplished something that had not been done since the time of Adam, circumnavigating the entire globe. The King of Portugal was very despondent at this news and regretted that he had not believed what Magellan told him previously. He sent a man to the ship that had not yet arrived, ordering him to attack it and to hang its captain on the ship's boom. For this reason, strife and contention arose between the two factions (Spain and Portugal), they fought many times until a truce was made.

Mention of the voyage of Alexander the Great: It is recorded in some of the Islamic books that Alexander outfitted a ship in order to ascertain the shores of this sea. He ordered them to voyage up to a year, learn the conditions of the peoples and tribes along the coasts, and return. So with this aim, the crew journeyed many winds, but hey saw nothing save the surface of the sea and the dome of the sky, and they despaired of ever reaching the other side. At the point of turning back, they saw a boat and approached. It too was loaded with equipment and men. But the two groups were unable to understand each other's speech. The people of Alexander gave them a woman and took from them a man in exchange. They returned and informed Alexander of their adventure. They married the aforementioned person to a woman and when the two of them had a son, he used to understand the language of his mother and father. He translated and he said on behalf of his father: “The king who was the ruler in the direction from the seacoasts was greater than Alexander. Wondering whether there were lands capable of conquering the face of the earth and kings other than him, he outfitted a ship for exploration and sent us forth. We voyaged for a long period until we encountered the ship of Alexander.”

This story is transmitted from Samarqandī in the (book entitled) *Kharīdat al-ʿajāʾib* and it is related as taking place in the Sea of Shirvān (the Caspian). The historian of New India (i.e., the author of *Tārīḫ-i Hind-i Ġarbī*) objected, saying: “This report is not correct because that sea’s boundaries and distances are known. What is proper is that the aforementioned story be connected with the Sea of the Ocean, and that the aforementioned boat came from New India.” Some informed people also objected, saying: “It is incorrect and silly to say that it came from New India, because before the Spanish came to those lands there were no boats, only rafts were used to reach some nearby places along the coast. When the Spanish came with large ships and the people of the New World saw them, at first glance they thought that they were fish.”

The poor one (Kātib Çelebi) says: It is evident according to the Commentator on the *Hayʾat* Qāḍīzāde, that this report has no basis in fact but belongs to the category of fiction. [M 115] There are many aspects of the story that corroborate this point, evident to those with insight. Because if this Alexander is Dhū’l-Qarnayn who was the builder of the Wall—to whom the name of Alexander is attached by way of confusion and error—a long and distant time has passed, since his age was the age of the Prophet Abraham, and no histories or biographies remain outside of the Revealed Scriptures that give information about those times [R 40a] which could be interpreted and transmitted to the Islamic books. And if this Alexander is Alexander the Great who was the ruler of Macedonia and the student of Aristotle, he did not journey to the west. We have made a translation from the Greek history about his campaigns and his biography, transmission and movements. Nor is it written in the historical or geographical books that he sent a boat beyond the Mediterranean. This story was only written to make known that it is a false report from the spurious remarks of the ignorant.

The journeys on the remaining oceans will be explained in detailed sections, with the conditions of their countries which were discovered.

Chapter 8: On 90 Latitude

According to the arrangement of the book of *Atlas*, after the four complete maps, we will begin by recounting the regions at 90 latitude, which is the North Pole, and the surrounding regions; because it is appropriate that the conditions of the two poles follow those four maps so that the entire extent of the globe shall be made known.

The North Pole is called *polus ar*[*c*]*ticus* in Latin. The author of the *Atlas* says: “After the complete maps of the four continents, we follow the *Geographia* of Ptolemy, who is the sultan of the geometricians, and begin from below the North Pole and the surrounding regions. Thus we shall descend by degrees from highest to lowest and make a circuit of east and west, south and north. We hope for the grace and help of God to benefit us and all people in this work.” At this point he gives the explanation of pole, axis of rotation, and south and north that I have given above, and commences his account, saying: “The North Pole (i.e., the north star) is always visible to us and the south pole is not visible. The countries in the vicinity of the North Pole are these: Groenlandia, Frieslandia, Nova Zembla.”

Groenlandia: It was given this name in the Dutch language because of the verdure and freshness of the trees and plants. It is an island in that district that has not yet been fully expolored. It falls between the polar circle and the North Pole. Its southern boundary is 65 latitude; its northern shore extends to 77 degrees latitude.

According to Captain Nicola—he came to the vicinity of that island in the Christian year 1480 and was caught in great storms and tossed about in the sea a long time—winter here lasts nine months, during which it snows constantly and never rains. The snow that falls at the beginning of winter remains until the end and does not melt. Despite the great volume of snow, it does no harm to the pasturage and fodder. So the grass grows without stint and the milch-beasts graze in large flocks, digging away the snow and eating the grass. The people of the country take so much milk that [M 116] one sees mountains of butter and cheese heaped up here and there on the docks and along the seashore. People come from the surrounding regions and take it away for a cheap price.

There are two settlements known to us on this island, where the islanders dwell. One is called Alba, with the monastery of Santoma nearby. The other is also a monastery. The mountain nearby constantly spews fire, like Mt. (Aetna in) Sicily, and hot water flows continuously from a spring at its base. The cells of the monastery are like a bathhouse from the heat of this water. They use no other fire, and they cook their food in that water. This monastery is constructed of a stone called tuff that crumbles to sand if it is rubbed. It is the lava that erupts from the volcano. The water of the hot spring warms the areas around it and the gardens in its path [R 40b] causing flowers and plants to grow. That place is always smiling with various types of flowers. The sea near the volcano never freezes, so boats go around on it and fish swim in it. So many sea creatures flock here, fleeing from cold places on this island, that people come from all around to hunt them and store them for future use.

Mare Pi[g]rum, meaning "Lazy Sea," laps at—i.e., surrounds—Groenlandia. This sea is called (Mare) Glaciale and (Mare) Concretum meaning "Icy and Frozen Sea."

FRIESLANDIA: A large island that was unknown to the ancients. The air of this place is very bad and unwholesome. Plants do not grow, and its land does not accept agriculture. It is a dry land. Its people make a living entirely through fishing. Its chief town has the name of the island and is under the rule of the king of Norway. Its people catch all kinds of fish from the sea and pile them up on the shores. Boats come from all around and take them away in loads.

The sea around this island is called Itarbum (Icarium?), because the western side is so strewn with reefs that ship movement is impossible. Therefore its inhabitants call this island Icaria. This island was discovered in recent times by effort of the English.

NOVA ZEMBLA (NOVAYA ZEMLYA): an island at 76 degrees latitude. The air is extremely cold on account of blowing winds, and so it is uninhabited. The cold is unbearable. Plants and trees do not grow there. Nothing lives there except carnivorous animals like wolf, bear and fox—such animals are very plentiful in the northern regions. There is also a sea creature here known as walrus. Its mouth is bigger than an ox's and resembles that of a lion. Its skin is furry like that of a pig. It is four footed. Two large white tusks jut out of its upper jaw; they are like elephants' tusks and are equal in value.

There are three *fretum*—i.e., straits—on this island. One is the Weygats Strait that extends from west to east up to the corner known as "Crucifix Corner." An English captain named Guillaume (Willem) Barents came to the southern coast of this strait and found wild men known as Samoyed. He recounted that their clothes were made of animal furs, like the garments of men who live in ancient forests, but the men are not so wild. They are generally possessing intelligence, short of stature, broad-faced, small-eyed, [M 117] with short and crooked legs, quick and agile and strong. They hop about like fleas. They cover themselves with the skin of an animal called *zenjifiros* (?) They have carts that, hitched to one or two such animals can pull one or two men so fast that a horse of these lands cannot catch up at a gallop.

The second is the Frobisher Strait, named after Captain Martin Frobisher. In the Christian year 1577, while seeking a route through those regions toward China and Cathay (i.e., the Northwest Passage), came to this strait where he saw many islands. The inhabitants of those islands were men, but a people with the character of wild beasts. They eat raw the flesh of fish and other creatures, and wear the skins of the wild animals they hunt. They graze like cattle and are transhumant. They cover their tents with the skin of a fish called *balina* (i.e., whale). This fish is a large animal in the sea that spouts water from its mouth to the height of a spear.

In those places [R 41a] the weather is always extremely cold and the sky is always dark with clouds. The people yoke their dogs, which resemble the wolves of this country, and make them pull their loads over the ice. Their weapons are bows, arrows and slingshots. [Plate 19]

Trees do not grow in those places. The people neither plow nor sow seed, but eat whatever grows naturally. Their livelihood mainly comes from hunting. They hunt deer, which are plentiful, and they drink the warm blood of other animals. When they cannot find it, they take ice in their mouth and lick it. The waters freeze from the extreme cold and the ground too is frozen solid, so there is no running water; they use snow and ice instead.

This tribe is tough, sturdy and courageous. They do battle with wild beasts. They are especially skilled in hunting birds. They make a type of boat (kayak) from tree bark in which a single man sits and rows with a single oar, holding a snare in one hand to catch birds. They brought one of these boats to England as a model (for a single scull).

The third is the Davis Strait, named after the English captain John Davis. He discovered it in 1585 while travelling around the coastlines of America and Groenlandia and seeking a passage to China. He started from 53 degrees longitude and went as far as 75 degrees.

The four straits that are depicted on the page (i.e., the map at the top of R 41a, Plate 19) were taken from the travel account of Jacob from the city of Buscod (Buscoducum, s'Hertogenbosch). It is related from him that an English monk and surveyor from the city of Oxonia (Oxford) discovered lands near the pole and went to 90 latitude. Measuring the altitude with an astrolabe, he estimated where the polar star came to the zenith, and established it by this means.

Mercator relates from this Jacob that there is a large whirlpool in these four straits. The water flows from four directions toward the inside of the whirlpool to such a degree that ships that enter there have no possibility of turning back because of the force of the wind. Despite this, no wind blows in those places sufficient to turn a windmill. But this account contradicts that of a captain named Lucianus; and others who have travelled there have not reported seeing this kind of strait.

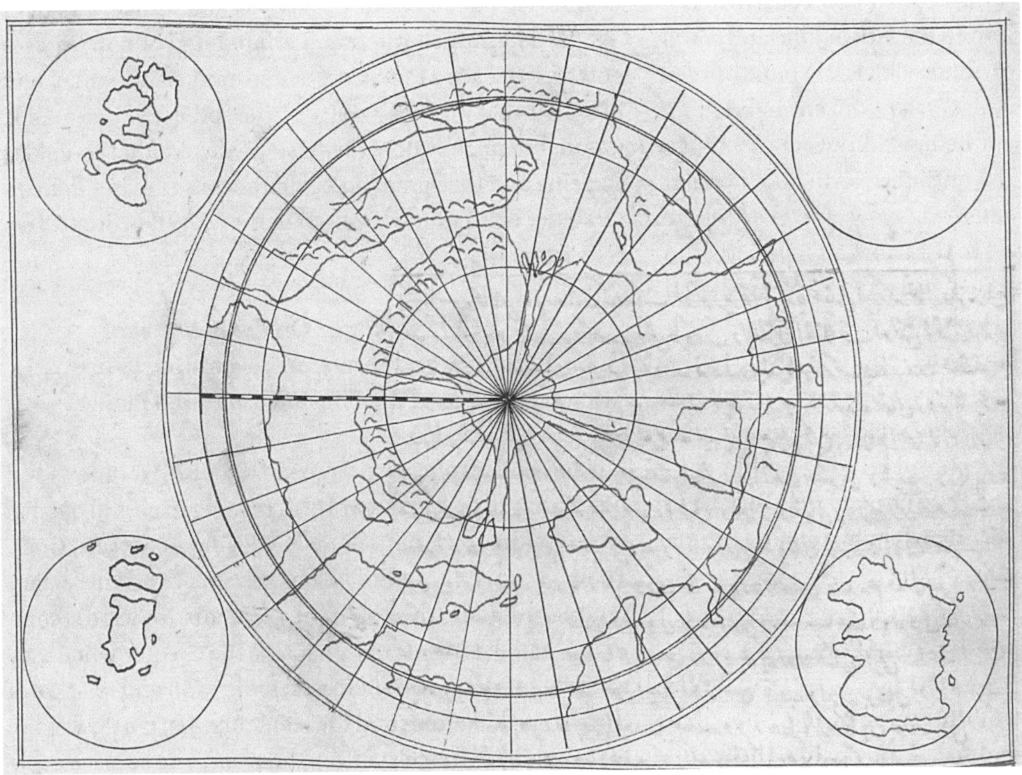

PLATE 19
Map of the north polar region, R 41a; cf. M 119/120, top (Appendix, fig. 49–50)
TOPKAPI PALACE LIBRARY, REVAN 1624, WITH PERMISSION

The Dutch explored those regions as far as 81 degrees latitude and saw that is was all sea. As for 90 latitude, which is below the pole, [M 118] [R 41b] an author named Julius [S]caliger investigated it in the 37th chapter of his book, as follows:

Is is possible to travel via the North Sea all the way to China, or is it not? The question must be debated according the proofs of the two sides. Some say it is possible. When one wishes to travel there, one emerges from the strait of the Dvina River and goes to the eastern corner of the Tatar country. On turning the corner, the north wind turns into the west wind and one proceeds eastward.

But those who say this do not know the viability of the winds of that sea and of its coasts. There the east and west winds blow so lightly that they are not distinguishable from each other, while the north wind is so strong that it completely dominates. In places where the boreas is dominant, the sea surface becomes frozen solid, like a stone pavement, from the extreme cold that lasts up to ten months. And there are many reefs in the sea. In summer, when the longest day is 24 hours, the coasts are shrouded in fog in the morning. It clears up after noon for two hours at most, after which the foggy conditions return. The other side of the strait is the same, as far as the Pole. Thus, the sea lanes in these regions are blocked and perilous. Sometimes, when the ice melts, ice floes the size of islands collide.

In 1594, a group from Holland travelled to that region and tried to make a passage from the North Sea to the east coast of the New World. But they suffered greatly from ice and from extreme cold at night. For this reason, the captain named Guillaume (Willem Barents) declared that it was impossible to travel via the Nassau Strait to the land of China. And he was speaking from experience, because after exploration he made a sound estimate and determined that the other side of this sea is not connected to the Eastern Ocean and that the coast goes around the Pole and turns westward, becoming the coast of the New World. He deduced from this that if its other side were connected to the sea, there would be ebb and flow of tide there also. But that is not the case; rather, it is a lagoon like a pool. Such was his definitive proof.

Nevertheless, some do not accept this and assert that there is strong hope that such a passage will be found on the northern side of Nova Zembla, because there are always new fleets being outfitted for travel to China, and one day the desired result will be achieved. So they urge people on. But so far no definitive result has been obtained for either side.

Mercator says that their ships came to 81 degrees longitude and could not pass beyond because of ice floes and because they were left in darkness. So they turned back and came to 76 degrees. On the 4th of November in the year 1597 the sun set and did not rise again until the 24th of January. During these two months of night they disembarked on the shore of Nova Zembla and constructed a tent. They remained inside until the 14th of June and could not move because of the extreme cold. For 13 months they wandered in such perilous places, far from any habitation, lacking means of making a living and devoid of human necessities. They proceeded until the North Star stood at 76 degrees. When they emerged, snow was everywhere and they remained trapped in the snow for a period of 10 months. During that time they were attacked more than once by large and ferocious bears and sea monsters [M 119] and had to fight them off and defend themselves. In the end, they abandoned their ships and managed to get out with rowboats. Before them no one had ever gone to that region. Their journey put the proverbial voyage of the Argonauts in the shade. [R 42a] They boasted that they were worthy of being called the Argonauts of Ultajia (Ultima Thule?).

Chapter 9: On the South Pole

The ninth chapter concerns the region of the globe that is the South Pole. In some of the Islamic books, the mainland in this region is dubbed the Land of Gold. According to the *Introductio*, the geographers have described this land and named it Magellanica and Terra Incognita of the South. But none of it has been explored until now except the coastlines. Its extent is nearly double that of any other continent. It is called Magellanica because it was Captain Magellan who first discovered and described its coasts on the New World side.

Lorenzo says that the country of Lāfāq (?), which is one of its coasts, was discovered at a cape located at 140 degrees longitude and 15 degrees south latitude. Its people are white, but so savage that they have no possibility of converse or commerce with foreigners. Boats sometimes come to it from Java, but they do not let them approach. A captain named Michael Lopes departed from the port of Nativita in New Spain in the Christian year 1565 and discovered this country. He reported that the wealth and gold there are greater than that of Peru.

Pīrī Re'īs, at the beginning of the *Baḥriye*, recounts that a Portuguese ship went south of the China Sea with a contrary wind and in 25 days came to a coast. They saw many naked men using weapons, shooting arrows with shafts made of reed and arrowheads of fish bone. There were people in weird forms and strange shapes, some with horns, some with half bodies and chirping like birds, some with one eye in their forehead, some with elephant ears, some with their head in their chest and a tail like a dog.

That is the story he recounts. But there is doubt about its truth. In reality, those who travel the sea and roam the world have no idea what this large island is like, and have not yet determined whether it is inhabited or not. No river entering the sea at its coast is depicted on the map, and it is plausible that the aridity of the soil is the reason it is uninhabited. They do write to some degree that its cold is extreme and the winter and cold there are greater than on this side. This point was noted in the Magellan Strait. This is its form: [Plate 20]

[R 42b] Chapter 10: On the Relation of Countries and Directions, according to the Triplicities of the Constellations and Planets

Muslim philosophers have proven that the conditions and traces that appear in the world of coming to be and passing away are the work of the Freely-chosing Actor (i.e., God). They say that the etherial bodies and heavenly occurrences influence the world of elements, by God's command, and that the primary cause is the Creator while they are secondary causes. There is a connection and relation of the natural kingdoms and the elements—in particular, the earthy globe—to the stars and heavens. The seven earthly climes are arranged in relation to the seven planets. Furthermore, there is an established connection of each of the countries and lands with the twelve constellations. This connection is the relation of the lands with the triplicities of the constellations. And the triplicities of the constellations are as follows:

The cincture of the eighth celestial sphere, which is the sphere of the zodiac, is divided into twelve parts, each of which is called a constellation; as was explained above in the introduction. [M 120] [M MAP] In determining the triplicities, from each of the three constellations one counts forward four in the order of the constellations. Thus:

Aries—Leo—Sagittarius
Taurus—Virgo—Capricorn
Gemini—Libra—Aquarius
Cancer—Scorpio—Pisces

From these one obtains four equilateral triangles:

The First Triplicity: The constellations are masculine and pertain to fire and the north. Of directions, what is between the north and west winds belongs to this. As these constellations contain the houses of the Sun, Jupiter and Mars, the governor of this triangle is the Sun by day and Jupiter by night.

The Second Triplicity: The constellations are feminine and pertain to earth and the south. What is between the south and east winds belongs to it.

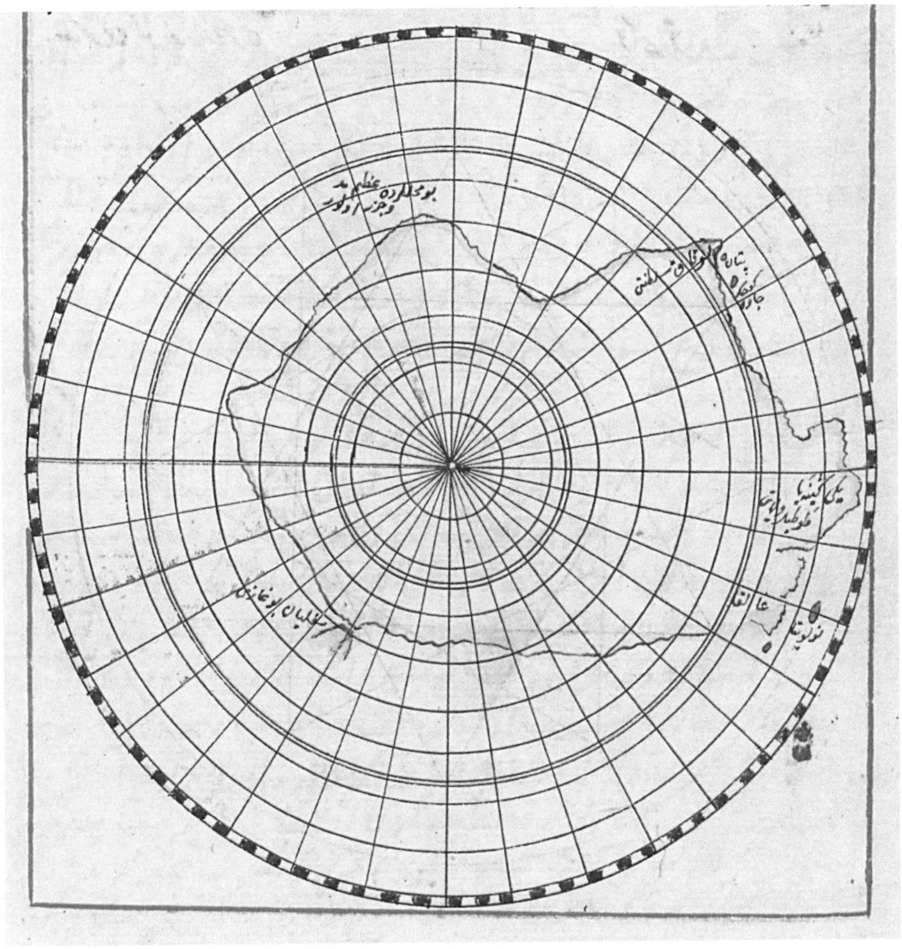

PLATE 20
Map of the south polar region, R 42a; cf. M 119/120, bottom (Appendix, fig. 49–50)
TOPKAPI PALACE LIBRARY, REVAN 1624, WITH PERMISSION

The Third Triplicity: The constellations are masculine and pertain to air. As they contain the houses of Saturn and Mercury, its governor is Saturn by day and Mercury by night. What is between the north and east winds belongs to it.

The Fourth Triplicity: The constellations are feminine and pertain to water. They contain the houses of Venus and Mercury. What is between the south and west winds belongs to it. Its governor is Venus by day and the Moon by night. [Plate 21]

[R 43a] Likewise, the Inhabited Quarter is considered as four continents resembling the triangles of the constellations.

The first continent: Europe, as it is between west and north, belongs to the first triplicity. The tribes and nations that dwell here are brave and unruly in most matters by reason of the domineering nature that is inherent in the first triplicity. In general they are given to the use of arms and punitive measures and capable of enduring fatigue and hardship, and are clean and elegant, since nighttime and Mars share in governance. The prior components of the triplicity are masculine, the later components are feminine.

The Europeans are for the most part careless in the matter of their women. They are not jealous. They have a greater desire for men than for women. Especially England and Germany are similar to Aries and Mars, and so their inhabitants are for the most part savage and impetuous, and their character resembles that of predatory beasts. The lands of Italy, i.e. the territories of Rome; and Gallia, i.e. the provinces of France; and the island of Sicily belongs to Aries and the sun, and so most of their inhabitants are people of punishment and domination. The land of Spain and Portugal is bound to Sagittarius and Jupiter, and so most of its inhabitants are ingenuous, clean and good-natured.

After these come the land of Macedonia, i.e. Rumelia; the land of Thrace, i.e. the vicinity of Istanbul; Crete, Cyprus and the coasts of Asia Minor, i.e. Anaṭolı; and the area comprised by the Black Sea and the Mediterranean. While they are included among the lands of the first triplicity, they are considered similar to the second circle and Venus and Saturn share in their governance. Therefore most of their inhabitants are moderate in form and conduct, close to one another, of sound temperament, people of punishment and domination, and given to violence and

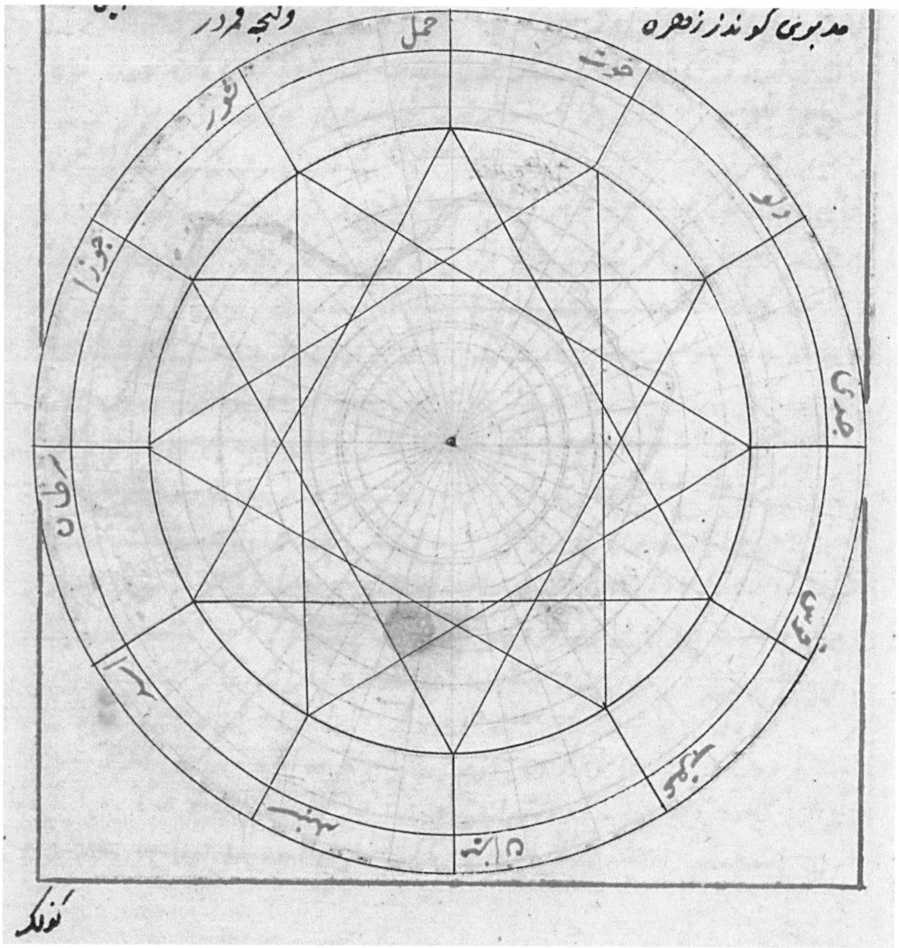

PLATE 21
Diagram of triplicities, R 42b
TOPKAPI PALACE LIBRARY, REVAN 1624, WITH PERMISSION

liberty. Since Mars and Mercury are the possessor of the triplicity, they engage in study and love struggle and music and are clean. Since its governor is Venus, they are hospitable. In particular, Macedonia and Constantinople are similar to Capricorn and Saturn, and so their grandees possess sovereignty and domination.

The second continent is Asia: Its lands are connected to the second triplicity. As the governor of this triplicity is Venus in daytime and Saturn (at night) its inhabitants greatly esteem these two planets. Because of their similarity to Venus, they have a great desire for dance, movement and sexual intercourse with women; shamelessness and impudence; no desire for men; love of ornamentation in dress and furnishing; over-indulgence in the training of their bodies; and in general a resemblance to women; and a predominance of heat in their natural humor. [M 121] But because Saturn shares in the governance, their carnal desires are influential and strong; they are violent in war; and people of lofty ambitions.

This is the rule of the overall resemblance of this clime to the triplicity. But the separate parts have individual tendencies as well. Thus, the land of Persia being related to Venus and Taurus, its people wear colorful clothing—indeed, even their shirts are not plain white. Babylon and al-Jazīra—i.e., the area between the Tigris and Euphrates and the region of Baghdad—being connected to Virgo and Mercury, have many people given to astronomical observation and learning. India and Sind, being like Capricorn and Jupiter, most of its people are immoral, ugly, dirty, ungrateful and stubborn.

The middle of Asia—the land of Idumea, i.e., the coastal region of Jerusalem; the land of Syria, i.e., the country of Damascus (Shām); and the Arabian Peninsula, including Ḥijāz and Yemen—resemble the first triplicity. Since the governor of these is Jupiter, Mars and Saturn, the majority of its inhabitants [R 43b] are people of domination and commerce. They glory in deceit and trickery, cowardice, and contempt for others. The settled places of Arabia, being related to Sagittarius and Jupiter, are mostly attracted to abundance and ease.

The third continent, known as Scythia (text: Sfmūnīā), is the northeast part (of the Inhabited Quarter). This continent is bound to the third triplicity. Most of the lands of Turkistan, Georgia, Daylam, Transoxiana, Cathay and Khotan fall into this portion. Since the governor of this is Saturn and Jupiter, most of its people are of a philosophical disposition, chaste and pure. In particular the land of Azerbaijan, being bound to Gemini and Mercury, the people are given to movement, violence and villainy. The districts of Transoxiana, being bound to Aquarius and Saturn, most of the people are savage and cruel. Others are also like that.

The fourth continent is Africa. The southwest part which is Libya, and the Egyptian lands which are Ethiopia, and the part known as Maghreb which is Mauritania, are bound to the fourth triplicity. And since Mars by night and Venus (by day) share in the governance of this triplicity, its people are not free from the interference of women in their rulers' affairs. In most matters men and women mingle together, and one woman may marry several men. The men go about in the dress and ornament of women. Most of them are soothsayers.

Especially the Mediterranean coast of Africa, being bound to Cancer and the moon, the people are mainly merchants and their lands are accustomed to luxury and ease. The westernmost districts of the Maghreb, being bound to Scorpio and Mars, the character of its people resembles that of predatory beasts. They are savages. They are rarely free of disputation and antagonism and cannot refrain from killing one another.

The countries of Upper Egypt and Ethiopia (*Ḥabeş*), being governed jointly by Saturn, Jupiter and Mercury, the people of those lands have various customs. They esteem their dead. They obey governors who come from abroad. They indulge in sex. They are a cowardly and base people possessing weak spirits. The cities of Alexandria and Cairo, being bound by Gemini and Mercury, the people are largely possessors of understanding and intelligence, and inclined to the extraction of secrets and to study. The middle of the land of Ethiopia, being bound by Aquarius and Saturn, the people are inclined to eating meat and fish. Their eating and drinking is like that of cattle.

Up to this point, it is the summary of Ptolemy's discourse in the Tetrabiblos (*Maqālāt-i Arbaʿa*). The individual cities that are bound to constellations were also excerpted from that work and from some esteemed books of astrology.

Lands of Aries: [M 122] Fārs, Azerbaijan, Palestine, the cities on the Black Sea coast, Samarqand, Jurjān, Kabul, Anṭākiya (Antakya, Antioch), Ṭūs, the island of England, Gallia (France), Germany, Sweden.

Lands of Taurus: Cyprus, the small islands (Cyclades?) in the Aegean, Kurdistan, Georgia, Hamadān, Ḥulwān, Diyārbekir, Gaza, Qaysāriya (Kayseri, Caesaria), Tarsus, the island of Messina (i.e., Sicily), Rūs (Russia), Leh (Poland).

Lands of Gemini: Egypt, Ṭarābulus Gharb (Tripoli in Libya), Barca, Ḥimṣ, Ḥarrān, Gīlān, Māzandarān, Qazwīn, Kirmān, Tustar, Marw, Sardinia, Flanders, Cordoba, and Greater Armenia.

Lands of Cancer: Ifrīqiyā (i.e., Tunisia), Fez, Algiers, Tunis, Jerusalem, China, Turkistan, Granada, Constantinople, Venice, Genoa.

Lands of Leo: Damascus (Shām), Malaṭya (Malatya, Melitene), Ruhā (Urfa, Edessa), Nuṣaybīn, Sind, Demir Kapu (Darband), Puglia, Rome, Qusṭanṭiniyya-i Gharb (Constantine in Algeria), Prague.

Lands of Virgo: Mosul, Jazīra, the country of Anatolia between the Euphrates and the Mediterranean, Babylon in the vicinity of Baghdad, the capital of Ethiopia, Ṣanʿāʾ, Sūs al-Aqṣā (in Morocco), Ṭanja, Jazīrat al-Khaḍrā (Algeciras in Spain), Crete.

Lands of Libra: Mecca and Medina, Balkh, Baalbek.

Lands of Scorpio: Ḥijāz, Qayrawān, Basra, Nahrawān, Tabri, Aleppo, Rayy, Zangistān, Jazāyir-i Gharb (Algeria), [R 44a] Trabzon, Fās (Fez), Amasya.

Lands of Sagittarius: Spain, Iṣfahān, Quhistān, Tyre, Raqqa, Sinjār, Baghdad, Bosnia, Budin (Buda), Mora (Morea, Peloponnesus).

Lands of Capricorn: Ethiopia, India, Sind, Qūmis, Ghūr, Kabul, Oman, Ahwāz, Makran, Macedonia, Bulgar, Muscovy.

Lands of Aquarius: Circassia, Rūs (Russia), Zangibār (Zanzibar), Kurdistan, the middle of Ethiopia, Farghāna, Ra's al-ʿAyn, Sulṭāniyya, Ghaznīn (Ghazni), Tatar, Nogai, the Arabian Desert.

Lands of Pisces: The provinces of Ṭabaristān, Simnān, Portugal, Yemen, Upper Egypt, Egypt, the vicinity of Alexandria, the Islands (or Archipelago, *cezīreler*).

Chapter 11: On the Inhabitants of the World with Respect to Religions

This chapter is the 156th chapter of the *Levāmiʿu'n-nūr*, which is the translation of the book of *Atlas*. I have included it here, since it is apropos. To be sure, Mercator in the *Atlas* planned out and divided (this chapter) according to his own religion. In this book I have written it according to the requirement of Islam. Mercator said:

It has become known to us from travels on land and sea that every group of mankind worships a deity according to its own opinion. There is no wild or savage tribe that does not have its deity. Leaving aside works written in ancient times, wherever we went in our travels, we saw its people worshipping some kind of deity. The ancient Egyptians worshipped the ox, horse, snake, crocodile, hawk, dog, wolf, cat and dung beetle, and considered such creatures as gods. The ancient Syrians worshipped a fish—it is mentioned in scripture as Dagon. The people of Tyre

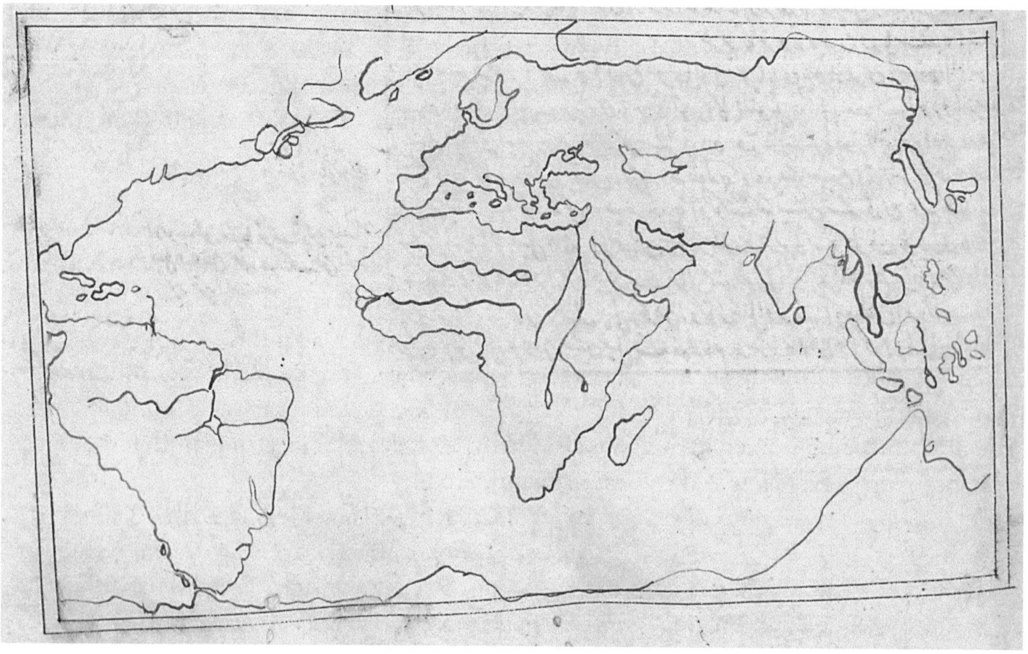

PLATE 22
Sketch of world map, R 44a
TOPKAPI PALACE LIBRARY, REVAN 1624, WITH PERMISSION

and Sidon worshipped an idol named Baal. The people of Damietta took onions and garlic as their deity. [M 123] The Persians worshipped the sun and fire.

As for the Greeks and Romans, regarding this topic: In what direction of the valleys of madness did they not stray! Whatever object they observed, whether natural or man-made, that they saw was beneficial or harmful for mankind, they immediately attributed divinity to it. In general, they built temples to all the unknown gods, out of concern lest any sort of deity be abandoned or neglected in their worship. And on the gate they wrote: "This temple was constructed for worship of the gods of Asia, Europe and Africa, and for worship of the unknown gods of other unknown places," and they worshipped there as well.

But all this is absurd. Just as God is one, so the path of worship is also one. This path is not known with the intellect; it is according to the will of God; because He himself knows His own essence and knows how be ought to be worshipped, and He teaches this through the prophets. [Plate 22]

[R 44b] He who worships in this way worships the True God.

Distinguishing worshippers among mankind according to rational possibilities, one says that a god is either real or not real. A god that is not real is other than God. Again, a true God is worshipped either by divine law or by intellect. Therefore worshippers must belong to one of three categories:

- Those who worship a true God by divine law. (On the map) a sign is placed on their dwelling in the shape of a crescent and a finial (i.e., they are the Muslims).
- Those who worship by intellect. They are the Christians. Their dwelling is indicated by a sign in the shape of a cross.
- Those who worship a false god. They are worshippers of idols and heavenly bodies and worshippers of spirits. A sign in the shape of an arrowhead is placed on their dwellings.

From these signs the dwelling places of the inhabitants of the earth are known in accordance with their religions. Because this is one of the general principles of the science (of geography), it was presented before the other chapters. After this, commentary on the remaining parts will commence.

Warning: It should be known that because the authors of the *Atlas* and the other books of geography are mainly from European lands, after the general principles they start with the continent of Europe, which they describe in the utmost detail. Then, at the end of the book, they deal with Asia and Africa in summary fashion. It is as though they favored writing about the conditions of their own lands and passed quickly over the remainder. Actually, because those lands are so far away, they cannot learn about them in any detail; and what they do write about them is not free of error.

Abraham Ortelius, after the general principles, did not separate the detailed description of America from the summary description after the general section of the continents, but brought the description of its regions before the regions of Europe. After completing the description of America he wrote about Europe and then Asia and Africa. The final third of his book was given over to maps of

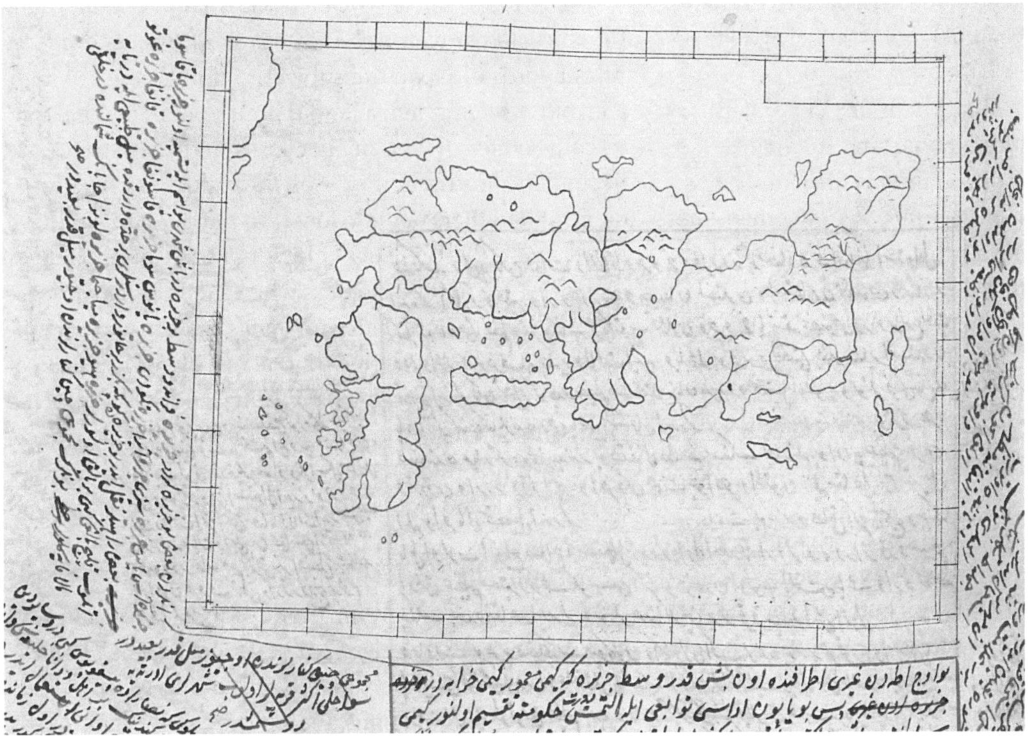

PLATE 23
Map of Japan, R 45a; cf. M 125/126 (Appendix, fig. 51–53)
TOPKAPI PALACE LIBRARY, REVAN 1624, WITH PERMISSION

the ancient geographies, from which he drew many benefits.

Mercator in the *Atlas*, after the general maps, began with the detailed description of Europe. After describing Africa and Asia, he commented on America in a few pages and ended the book by copying seven maps of the ancient geographies. In the translation, the *Levāmiʿu n-nūr*, I followed his organization.

But in this book I considered it better to abandon imitating him and chose a different way to organize the material. I begin with the eastern part of Asia. Then come detailed sections on Africa, Europe and America. To the maps of the countries of Asia I have added many useful things from the books of the Orientals. So in this book, the description of the countries of the world follows the movement of the ninth heavenly sphere (i.e., from east to west), [M 124] with the motion of the whole earth following and transmitting the motion of the Primum Mobile, beginning with the eastern islands (of Asia) and ending with the western shores of America. God willing, I will be granted success on this path by divine grace and will be vouchsafed to attain my goal.

Chapter 12: Island of Yāpūnīā (Iaponia, Japan)

It is the beginning of the eastern islands in the China Sea. The pronunciations Hāpūn, Jāpūn, Yāpān and Yāpen are also in use. Formerly it was called Khirīse. Mercator calls it Europa Khrusonus meaning "Golden Europe" (confusion with Aurea Chersonnesos meaning "Golden Peninsula").

This island is in the Pacific Ocean. Its western border is the Sea of China at 163 degrees longitude. Its eastern border is the coastlands of Māre Vermīlū [Mare Vermiglio][11] meaning Clear Sea, at 176 degrees longitude. Its southern border is the Sea of Siam at 31 degrees (north) latitude. Its northern border is the region of the Gulf of Lonza (text: Lūnrā) and Ania and .... (?) at 37 degrees (north) latitude. It is a large island stretching 150 Teutonic miles from east to west and approximately 60 miles wide. Lorenzo records its length as 1,600 miles and its breadth as about one third of its length.

Attached to the west side there is another island named Aqsīmen extending lengthwise from south to north within these borders. It is 60 miles long and 30 miles wide. Between the two is a third island named Aqsīqūqam which is 30 miles long and 5 miles wide. [R 45a] [Plate 23]

Aside from these three, there are around 15 mid-size islands in the vicinity, some of which are developed and some in ruins. They include: Ferāndū, a mid-sized island whose main city is called by this name; Yāqāsūmā; Ās-Yās; Sīzī; Āīrā; Tekūdī; Yānūs-Sūmā; Qāsūnghā; Qānfā; Ghūtū—some monks relate that there is a beast on this island resembling a dog which, when its teeth reach a certain extent, jumps into the sea by natural inclination and becomes a fish like a swordfish; Hīū; Mīājīmā; Lonza—it is

11 Spellings in brackets according to *Fabrica*, 275.

north of Japan and its people come to Malacca (for trade). These islands taken together are approximately 300 miles from the coasts of China.

This island of Japan along with its dependents is divided into 64 prefectures. Some are sultanates (i.e., sovereign entities). The main country is Miyako. The second island includes nine prefectures, the esteemed ones being Wūsūken and Fenāyūra that is a dependent of the Būjen sultanate. The third island has 4 prefectures.

Sovereignty

Japan's sovereignty is subject to three individuals living in the capital city of Miyako. The administration of public affairs is in their hands.

1. The highest official is called Zāzū [Zazzo].[12] He has authority over religious matters, like the Pope. He dismisses and appoints a subordinate body known as Tūnī [Toni] who are like the bishops among the Christians and are charged with other religious maters. And he has a qualified and learned deputy.

The religious practice of this Zāzū is as follows: He uses earthenware vessels. He keeps the fast every month from the crescent moon to the full moon. During the fast he is extremely abstinent, he only wears white and does not approach his wife. The wife of this official must be virtuous and nobly born. After the full moon he breaks the fast, wears red, resumes pleasure and enjoyment, and goes on the hunt. It is a great crime for this official to touch the ground with his bare foot; they will remove him because of this. This official has no property of his own. He has many dependents and many expenses, but the Japanese lords see to all his needs as far as they are able. He does not lie down at night [M 125] without commending his body to the keeping of a special idol. He thus has one idol for every day of the year and rotates them.

2. The second official is called Wūāū [Voo]. He sees to the ranking of the people's offices and degrees, and is charged with their appointment and dismissal. There is no nation in the world that has such a high esteem for rank as Japan. This official assigns everyone's rank according to his merit, and gives each one a badge that they use instead of a seal. All the Elchis (official emissaries) of the lords of the island attend on this official and bring copious gifts in order to be the object of his favors. He gets great wealth from these gifts, but his private salary is minimal. After he dies his son or another relative takes his place; the office is inherited.

3. The third official is called Kūhāqāmā [Cuhacama]. He is in charge of travel and residence, war and peace. All the lords of Japan are subject to him. There are fourteen autonomous and splendid lords possessing the right to coin money. Other smaller lords and magistrates are dependent on these. If one of them deviates from obedience, the others act in concert to punish him. Of these fourteen are the Ūjī [Euge] chief and the Ghūnj [Gunge] lords who see to public affairs when the sultanate is lacking in administrators. This Kūhāqāmā's pomp and magnificence are of such a degree that he resembles the kings of Europe. Most of his pages are the sons of lords. A specific group of soldiers is assigned to his service in his palace. Once, when a Kūhāqāmā converted to Christianity, the other lords killed him and also killed the Christians who took his place.

The rest of the people are divided into five classes: 1) Magistrates. They are in the position of overlord and lord. Their chief is designated as Tūnūs [Tonos][13] and is addressed by his subordinates as "So-and-so Tūnūs". 2) Adminstrators of religious affairs. They shave their heads and beards completely and pass their lives as monks. They are celibate and, like the lords of Malta (i.e. the Knights Templar), combine monkery and warfare. They are all called bonze. There are various sects and splendid academies. They mainly live in the city of Fīānūmā with their chief and they teach and study. Most of them follow the views of the Brahmins. The paintings in their temples also resemble Brahmin paintings. Their chief college is in a city named Bāndū. Most of the people of Japan gather there for education and gradually advance to the level of bonzes. This rank is highly desired and prized among them. They are very wealthy due to their great endowments. It is related concerning some of these bonzes that when they die, a wind takes their corpse up into the air and it disappears. Clearly this is a superstition; but as the Shaikh (Avicenna) says at the end of the *Shifā'*, "You will suffer no loss by leaving the extraordinary things that you hear about in the realm of possibility." 3) Urban notables and grandees. There are two ranks of these: one is the servitors and one is the notables of the land. Each of them is fixed in his position until he dies. However poor he becomes, he does not lose his rank, and however rich he becomes, he does not rise to higher rank. In times of war, the grandees assist in the fighting. 4) Merchants and craftsmen. 5) Peasants and cultivators. They place great importance on maintaining the holders of these ranks.

12 Spellings in brackets according to *Fabrica*, 276. Cf. *Atlas Minor*, 628: Zazo, Voo, Cubacama.

13 Spellings in brackets according to *Atlas Minor*, 628–629.

Punitive Regime

They banish or execute criminals in this land. There is no lesser [M MAP] [M 126] punishment than banishment. They parade a thief around in a cart and after making a spectacle of him they hang him on the outskirts of the city. As for other criminals, they strike off their heads without warning. Murderers are forced to take their own lives. They do not drag out court cases, but act quickly.

The Japanese settle guard soldiers outside of the main cities, just as the Tatars do, and they do not allow them inside the city.

Religion

Most of them are idol-worshippers. In otherworldly matters they bow to idols named Fūtūqī [Fotoquos]. In supplications for health, children and sustenance they have recourse to small idols called Qāmīs [Camis].

At the head of all the idols are two large idols known as Āmīdā and Zāqā [Amida & Xoca]. The bonzes say in their nonsensical fashion that this Zāqā was a person. When he was born, a great many snakes flew into the air and came and bowed at his feet. When he grew old, he gave them a book containing a number of commandments. Āmīdā they also claim to be the possessor of a revealed law.

The people believe in these idols so fervently that they will kill themselves for their sake in order to go to paradise. Sometimes they jump into the sea, sometimes they leap from a cliff; or they enter a cave and do not come out until they die from hunger.

Knowledge and skill

There are able and intelligent ones among the people of this land. All the literary and calligraphical arts are found among them. There are colleges in many places and a famous academy in the city of Bānūāūm [Banoum] whose chief instructors are bonzes. They maintain great pomp and ceremony and confer teaching positions.

In the city named Būnqūm there is an academy of monks known as Jesuits. There the Japanese people learn Portuguese and those who come from Europe study Japanese.

There is much divergence in the Japanese language. [R 45b] It is a single language, but the vocabulary and expressions and script vary widely. They use several expressions for one meaning. There are particular terms and usages for the elite and the poor, for men and women and children. In the written language as well, they have special pens and specific varieties of script for verse and prose and for books and letters. There is one variety of speech and a corresponding pen in which a single character indicates an entire sentence.

In sum, those who know Japanese prefer it over other languages. But learning it is difficult and requires a long period of study. They write its script from top to bottom, and are astonished at lines running right and left. They print Frankish books and also use printing in their own books.

But because the Japanese only see Chinese and Indians they divide the world into Japan, China and India and think there are no countries other than those.

The characters used in the script are symbolic, as with the ancient Egyptians. These characters do not indicate a word but rather a meaning. There are approximately 5,000 characters that are familiar to the scholars of Japan and of China. But the common people do not understand them, because they are difficult to learn; for them there is a syllabary. It does not resemble the script of any other nation.

Industry and commerce

The Japanese are marvelously skilled in every craft. The Portuguese make long journeys to Japan for trading purposes and stay for a long time. They obtain great wealth by exchanging their goods for pearls and other gems and for gold.

War and valor

The Japanese are for the most part [M 127] courageous and inclined to warfare, and use various kinds of weapons. And so it is greatly esteemed to be tall of stature and corpulent. The majority of the soldiery is of this type. They fight until the age of 60 and then retire. The Japanese clip their beards; they do not shave.

Other mores and customs

The Japanese grandees and notables leave a small lock of hair on the back of their necks. The middle classes pluck the hair of half their head. The children pluck the hair of their foreheads. It is a great shame and embarrassment to touch one another's forelock. All of them pluck their whiskers with tweezers.

The Japanese are white, well-kempt (?) and handsome. They furnish their houses with rough clean rush matting, like carpets, on which they sit and sleep. For a pillow they use a stone or piece of wood.

They have great patience and fortitude in the face of hunger, thirst, heat and sleeplessness. Even in extreme cold they plunge infants into the river to wash them. After weaning, they separate them from their mothers, raise them in harsh conditions, and accustom them to the hunt. But they consider nothing more loathsome than poverty. For this reason, the women would rather kill their sons than let them become poverty-striken and come into the service of the notables.

The women wear silk clothing and are covered from head to foot. Their shoes are artfully made out of wheat stalks. The men go around with great conceit (or, putting on airs), fully clothed and armed.

The Japanese, like the Chinese, are greatly concerned with cleanliness and propriety. They do not even let geese or chickens or other domesticated animals into their houses, thinking they would defile them, but keep them in the field. At meal times they sit on their knees and pick up their food with two fork-sticks (chopsticks) and do not soil their hands. They are careful to take off their shoes outside so as not to dirty their mats.

On the coasts and in the cities, those of humble state subsist on vegetables, rice and fish. The notables eat mostly game, but they are also partial to fish. They give splendid feasts where they change the table for each course. Everyone has a plate of juniper or pinewood in front of him, which is also changed for each course. They sprinkle gold dust on their food. They pile up food on their plates and set up cypress branches here and there for decoration. They stick gold leaf on the beaks and feet of cooked birds and put them on very valuable plates.

The Japanese do not know viticulture and the fermentation of wine. They use a kind of arrack or wine made from rice that is like millet beer to toast one another. They make strange gestures and in this way honor their guests.

The Japanese are greatly occupied with a certain hot beverage that preserves bodily health for a long time. They do not know wine other than this. Into boiling water they place the grounds of a certain herb that they call *kīā* or *khīām* (tea). Nobles and peasants alike prepare it with their own hands, and there is a small room in every home for that purpose. If a guest comes, they certainly give him a cup of it. Because of this, they make splendid utensils and ewers, cups and trays. They all take pride in these teahouse utensils according to their status.

Poor Japanese use whale oil instead of regular oil and in some places they light wood chips and straw branches instead of candles. [M 128]

Cities

This island has large, splendid and well-ordered cities. Most are in the interior. The seat of sovereignty is a city named Miyako. It is located on the northern shore of a lake at 30 degrees (north) latitude and 169 degrees longitude. Formerly it was a very large city, 21 miles in circumference. Most of it lies in ruins because of civil disturbances and wars; less than a third of it remains, but the remaining part is very densely settled. [R 46a] Presently its ruler lives there. His primary courthouse is there, and three people are appointed to administer justice.

Ūssāqāyā [Ossacaja][14] (Osaka): An esteemed, ... (?) and free city. Its people are wealthy and propertied, many of them merchants. For capital, poor merchants have 1,000 *ġuruş*, the middle class has 30,000 gold pieces, and the greatest have uncountable wealth. Around 3,000 soldiers live outside the city as a standing guard. Whenever necessary, they enter the city one or two at a time and go back out.

Būnghūm [Bungum]: A large city, the seat of sovereignty and chief city of a province, and situated in a suitable place. Christians are plentiful here. They study Japanese in its college and the Japanese people learn Portuguese and Latin. There is a good port on the coast.

Qūyā [Coja]: All of the lords and notables are buried there because it is a city devoted to the idol known as Qūnbūdāsī [Combodasi]. If one cannot be buried there, they at least extract a tooth and bring it there for burial from a faraway place, believing that if a small piece of someone is buried there, he shall be resurrected in the afterlife in a good condition. They claim that this Qūnbūdāsī was the inventor of the script they use. And they say that, like the expected Mahdi, he has not died; he had a grave dug, entered it and disappeared, and will re-emerge. A great temple is built atop that tomb, where crowds of bonzes strive in his service, go into seclusion and worship.

Fīūnghū [Fiongo]: A city 18 parasangs from Miyako. Part of it was swallowed up in the earthquake of the Christian year 1596, and the remainder was burned and devastated in civil disturbances, so little has remained.

Amānghūzākī [Amangasaqui] (Nagasaki): A great and esteemed city. It lies in the middle of the island.

Dūsīken [Vosuquin], Fūnālūm [Funajum] and Tūsā [Tosam]: They are famous and esteemed cities.

The remaining cities in the book of Lorenzo are:

Qūnghūsīmā: A city near the coast. The people of this place were the first to become Christians.

14 Spellings in brackets according to *Atlas Minor*, 628.

Qāqātā: A great entrepot.

Dūqūsūdā, Būnghū and nearby Zawāw, Mānfātī, Būnjen, Chīqūn, Yuwāmī, Tāīnghū, Dūkhī, Rīmā.

Fīānūyāmā: The bonzes of most of these provinces live here, and so it is a famous city.

Nārā (Nara): A city famous for three things: 1) There is a pagoda, meaning an idol, made from bronze and even larger than the Colossus of Rhodes. 2) Tame deer are plentiful and its people worship them. 3) There is a large lake at the edge of the city, with innumerable fish dedicated to the idols: they are fed with rice every day and it is forbidden to catch them.

Naghazū, Tūnūqātū, Ūwīād, Meynī.

Qāwāchī: Capital of the Tūsūn clan.

Bāndū: The main college of the bonzes is there.

Ākūtiya: A famous city and gathering-place of merchants. But its cold is extreme since it is on the north of the island.

Japan has numerous ports, including Ūkhīnūfāmānūs [Ochinofamanus] (Okinawa) [M 129], a famous port where many ships are always gathered.

Buildings

Japan has many splendid temples, separate monasteries for men and women, and magnificent palaces. According to the account of Paulus (Marco Polo), the roof of the ruler's palace is covered with strips of gold instead of lead. The ceilings of the houses of their chieftains are covered in gold plates, studded with jewels, and highly ornamented.

In these times, the Tāīqū [Taico], who is the ruler of those lands, built a marvelous and splended palace furnished with 1,000 precious and ornamented rush mats known as *tatami*, with fringes made of silk and gold. Each mat measured eight spans by four spans. The timbers were made from a very valuable tree with an extraordinary construction. The splendor and ornamentation of the interior cannot be expressed in words. In front of the pavilion of this palace was built an arena for various games and public spectacles, with towers on either side, each three or four storeys high. There are many other buildings of similar splendor; but because earthquakes are not infrequent on this island, most of the buildings are made of wood, although here and there one finds splendid stone buildings as well.

Rivers and mountains

Japan has several small rivers and in some places beneficial hot springs. A large fish-filled lake is located at the edge of the city of Miyako. Most of the area around it is a garden of the bonzes. There are numerous high mountains scattered around the country. Two of these are famous and very high, and one of them has fire continually spouting from the summit. Satan appears there in an illuminated cloud to some idol-worshipping ascetics. They come and make supplication before that mountain. The other one, which is called Qīje Nūyāmā, rises several parasangs above the clouds.

Climate, products, flora and fauna

The climate of this island is good and salubrious, but the cold is extreme, and the land is not very fertile or productive. They harvest rice in September. Most of the notables eat a flavorless bread made from barley and rice flour. In some places they harvest wheat in May and make a type of flaky pastry from it.

But while the ground is not very fertile, it does have many mines of various kinds. With gold and silver mines especially numerous, they draw the goods of the world to themselves. The fruit-bearing trees are like those in other lands. Finally, the giant juniper trees in places are extremely tall and massive; they are used to make ship masts and columns for large buildings.

Marvels

There is a certain fruit resembling a date whose tree has the strange property of not liking moisture. If there is too much moisture in the ground, it wilts up as though affected by poison. To ameliorate the situation, they take it out of the soil and dry it in the sun, then plant it in a hole in the ground filled with dry sand and crushed iron slag. It quickly resumes its freshness, sprouts leaves and regains its vigor. If a branch breaks off, [R 46b] they nail it back on with iron nails and it regenerates itself.

This island's animals, both domesticated and wild, all graze in flocks in the open fields. The animals include sheep, pigs chickens, geese, wolves, rabbits, deer, and various birds without limit, especially francolins, pigeons and quails.

Supplement

Formerly there was a ruler of this island named Dāīr. During a long period of security and freedom from care, he fell to carousing and lost touch with humanity. The lords did not submit to him and everyone [M 130] took control of his own province. Afterwards two chief magistrates known as Kūbī got rid of him. Then one of them killed the other and seizing royal authority (lit. the sultanate) became overbearing and autocratic. He is now called Tīqūsāmā.

Some Portuguese claim to have discovered Japan. But Antonius (Antonio Galvão, fl. ca. 1544) writes in his account of the discoverers of the New World that Antonius Mora, Franciscus and other sea captains, while journeying from the city of Dūdarā to China in the Christian year 1542, a violent wind sent them to the island of Yābūn (Japan). So it was they who discovered it.

An author named Mafeos (Jean-Pierre Maffei, d. 1603) wrote in detail about Japan. Ortelius and Mercator quoted from him, and we have translated it here from those two sources.

Chapter 13: Map of (New) Guinea

Some coasts and islands were discovered extending from the equator to 10 degrees south latitude and from 160 to 210 degrees longitude. They were given the name New Guinea, because they found a complete measure of resemblance in climate and coastlines to Guinea in Africa. It is still not known whether the south side of these coasts is connected to the land of Guinea or whether it is another island; they only explored up to here.

The conditions of this island are not given in the *Atlas*. It was copied here from the large circular world map drawn in Holland in the Christian year 1613 and revised by Jodocus (Hondius), who completed the book of *Atlas*, since this humble one (Kātib Çelebi) happened to have it in my library. In the book of *Atlas* and in some old maps of the globe, the south side is drawn connected to the land to the south according to conjecture, but it is not known for certain.

Lorenzo says:[15] Some call this New Guinea the Country of Parrots. Its people are all blacks and negroes and idol-worshippers. There is plentiful gold in their provinces which they exchange for iron. This country is approximately 2,000 miles from the opposite coast. The Spaniard Viglia Lopes (text: Vīlalūpes; i.e., Ruy López de Villalobos, d. 1544) was sent to explore the Molucca islands in the Christian year 1545 from New Spain (i.e., the Americas) and he discovered this country. Eight years previous to this, Ferrando Galleco set out from the country of Peru and discovered Terra Galleca, meaning Land of Galleca, to the east of this country. Its people call it Cailoco, meaning Great Coast. It is a vast country that appears at 15 degrees south latitude. Its air is temperate and its land is fertile. Gold, jewels and perfumes are found there in abundance. Because of that, the Spaniards say that the island from which Solomon brought gold is this one. They confirm that this country is connected to the Land of Fire (Terra del Fuego) in the south. Its people are savages and cannibals. They do not interact with anyone. They paint their bodies in motley colors and have thick skins. They go around naked. This country has islands named San Giacomo, San Petro and Noluita, and the Candelara reefs. These names were given by Portugal; they are known by other names among the natives. [Plate 24]

[R 47a] Chapter 14: Islands of Chīn u Hind (East Indies)

The fourteenth chapter is about the islands of the Pacific Ocean. There are twelve conspicuous islands in the book of *Atlas*, other than the elongated island of Korea near the coast of China and the above-mentioned islands of Japan and (New) Guinea. The full map with small islands around it [M 131] is copied from two sheets (of a larger map?) and is assembled on one sheet. It serves as an introduction to the following detailed description. Since there is not very much explanation in this chapter in the geographies of Mercator and Ortelius, and since their maps are limited to this map, we somewhat expanded the map of the third part of Asia that the compiler named Giacomo from Castille drew in the Christian year 1562,[16] and supplemented it from other Muslim books. If there is any discrepancy between the maps in this regard, it is not surprising, because it is difficult to make the detailed information conform to the summary and to make each place conform to the whole, especially as the maps are contradictory. In any case, whether it is a detailed exposition or a summary, it cannot be free of approximation, let alone conforming with external reality.

Lorenzo says: These eastern islands are plentiful on both sides of the equator in East and South Asia, beyond the Cape of Singapore. If they were brought together, it would be vaster than Europe. Although gold is found in this group (of islands), the merchants who travel to that district do not condescend to buy gold but load their ships with valuable perfumes and spices. A Portuguese captain named Antonius Debreyu (António de Abreu, d. ca. 1514) discovered these first from the east, then Ferdinandos Magellan went there from the west.

15 Spellings of proper names in this paragraph according to *Fabrica*, 272.

16 *Il Disegno Della Terza Parte Dell' Asia*, by Jacobo (Giacomo) Gastaldi.

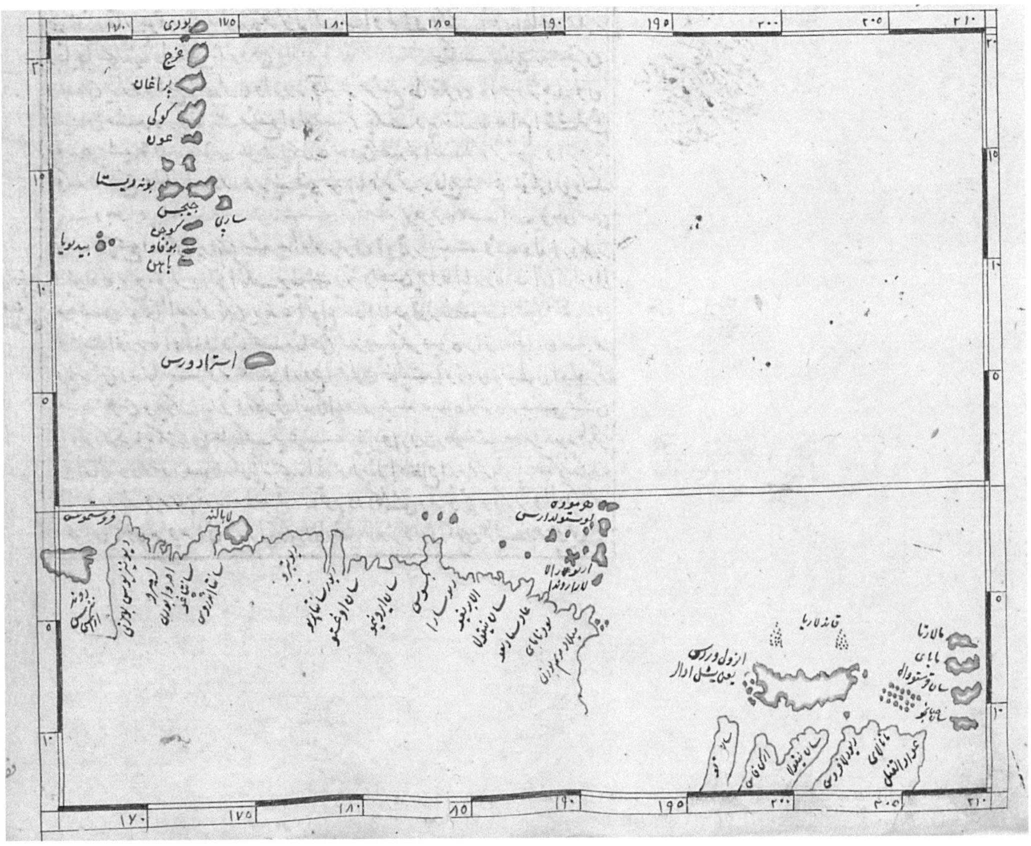

PLATE 24
Map of New Guinea, R 46b; cf. M 129/130 (Appendix, fig. 54–55)
TOPKAPI PALACE LIBRARY, REVAN 1624, WITH PERMISSION

Because these islands are under the equator and the orbit of the sun, the people are mostly weak of disposition, short of stature, and capricious of intellect and cannot endure hardship. Some of the islanders are black complexioned while others are white; some are refined and wear clothes while others are savage and go around naked; some are Muslim while others are Christian and the majority are idol-worshippers according to their ancestral rite. The subjects are very submissive to their lords—when they see them they shut their eyes and fall prostrate. They name them with the names of the sun, moon and other planets and believe in their divinity. It is a tribe of little intellect.

The twelve islands which were copied from the map of Mercator are:[17] 1) Meneses; 2) Luconia; 3) Calamianes; 4) Mindanao; 5) Ceiram; 6) Gilolo; 7) Celebes; 8) Borneo; 9) Timor; 10) Java; 11) Sumatra; 12) Ceylon.

Meneses: A square-shaped island close to 100 Teutonic miles in circumference at 170 degrees longitude and 2½ degrees south latitude. It is approximately 10 miles from the Cape of New Guinea, and since it was discovered by the lord of Meneses from Spain, it is named after him (i.e., the Portuguese explorer Jorge de Menezes who discovered New Guinea in 1526).

Luconia: A long island situated at 155 degrees longitude and extending from 10 to 20 degrees (north) latitude. It is around 60 Teutonic miles east of China. There are three cities and fortresses on this island.

Lorenzo says: These are called the Lekīne islands. They are many islands, but two of them are the main islands. They are known as Greater Lekūnīā and Lesser Lekūnīā. On Greater Lekūnīā there is abundant gold, choice fabrics and silk. The people take many valuable goods for sale to the coasts of India. These are a very orderly people and more humane than the Chinese. Their color is white. They defend their own territory and do not allow foreigners. The only known city of this island is that of Sīnk. It is an entrepot and a city on the coast. Previously, this island's people [M MAP] [M 132] used to come to Malacca every year to buy spices, but they have stopped coming since the Portuguese conquered it. Now they go to China and buy spices from the city of Qāntā (Canton) where the Portuguese bring them.

This island is recorded on the map of Giacomo with the name of Philippina, and it is written: "They also call the small islands on its eastern shore the Philippina islands."

At the end of the Hijri year 1010 (1601), the Spanish ruler of Mexico in Peru came to the coasts of China from the Pacific Ocean and became ruler over this island.

17 Spellings according to *Insulae Indiae orientalis Mercator-Hondius* 1608.

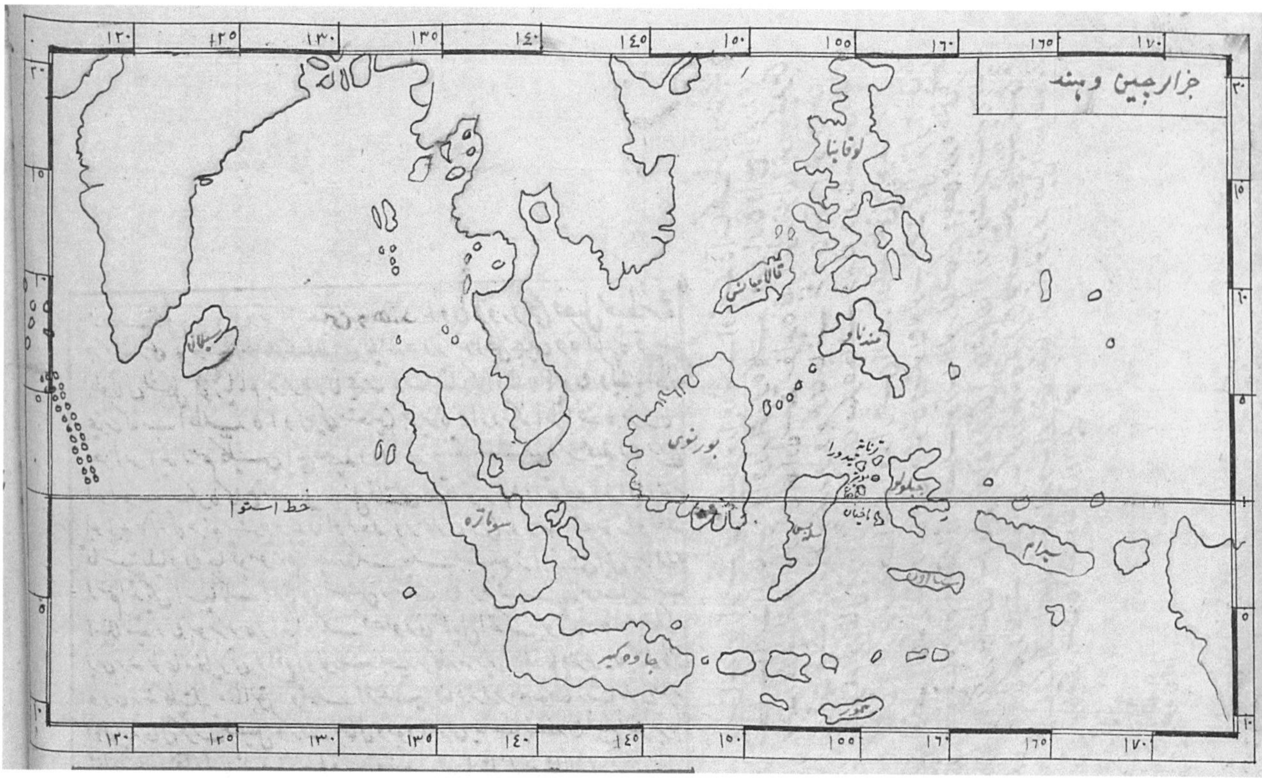

PLATE 25 Map of East Indies, R 47b; cf. M 131/132, 133/134 (Appendix, fig. 56–59)
TOPKAPI PALACE LIBRARY, REVAN 1624, WITH PERMISSION

Lorenzo says that these islands were discovered in the time of the Spanish ruler Philippos II and were called Philippina after him. It is now in the hands of the Spanish. Its people call this the island of Tendāy. Its cities are: Sīdāy, on the north side—Giacomo called it Cangu; Pāghāns, in the middle—Giacomo called it Ciabu; Mān, in the middle to the south—on Giacomo's map it is written as Polo.

Calamianes: An elongated island at 153 degrees longitude and 11 degrees (north) latitude. It is situated more than 50 miles east of Qāmpāyā in China. There is a vast triangular-shaped reef between this and the coast. Its length is 90 Teutonic miles and its breadth (lit. base—i.e., of the triangle) is 30 miles, and there are some routes (i.e., sealanes) amidst so much distance. A ship going from China to this island and the islands around it uses these routes.

This island is recorded on Giacomo's map as nearly square in shape with the name Palohan, and three cities are indicated: Pulo; Palahan, in the middle; and Carazan.[18] South of this island is Borneo, the distance between being 30 Teutonic miles.

Ceiram: An elongated island extending from east to west close to New Guinea at 165 degrees longitude and 5 degrees south latitude. It is written in the *Atlas* as Papuas Ceiram and was named after its discoverer. But on Giacomo's map no island is drawn in that place; rather, a triangular-shaped island is drawn to the west of Philippina with the name Papuas. Perhaps that one too was named after its discoverer, and he discovered this one afterwards. [Plate 25]

[R 47b] Chapter 15: Islands of Gilolo and Molucca

The fifteenth chapter is about the islands of Gilolo and Molucca, among the islands of the Pacific Ocean.

Gilolo, also called Zhilolo: The equator divides it in half at 160 degrees longitude. It is a large island nearly 200 Teutonic miles in circumference. To the east is Ceiram, to the north is Mindanao, to the west is Celebes, and to the south are the islands of Cenaon.

It says in the *Atlas*: The people of that land call it del Moro[19] (text: Delmūr). There are several islands and this is one of them. Its weather is very hot and cold and its chief product is rice which is very plentiful. They make a bread from the pith of a certain tree known as sago (the

18 Spellings according to *Il Disegno Della Terza Parte Dell' Asia*.

19 Spelling according to *Atlas Minor*, 625; cf. *Levāmiʿu n-nūr*, 402b line 11.

sago palm) and they drink the sap of that tree instead of wine. A type of wild chicken is plentiful here that does not resemble our chickens. On the seashores are turtles without limit. Their meat resembles veal in flavor.

This island's people are cannibals, a ruthless and uncivilized tribe. But Lorenzo says its people are Muslims.

It says in the *Tārīḫ-i Hind-i Cedīd*: Magellan's ship came to Gilolo and found a group from Spain there. They had come previously and built a strong fortress on this island. So the islanders gave them 20 quintals of cloves in addition to what was necessary for the people on the ship. [M 133]

In *Taqwīm al-buldān* this island is inscribed with the name of Sīlī at 170 [degrees] longitude, and is recorded in some books with the name of Sīlā. It says in *Nihāyat al-arab* that this Sīlā is six islands. A group of ʿAlawīs fled from the Umayyads and settled there. Its cities are Pūntekāl, Sūyā, Ghūntū, Mālīqī, Pākūr, Siyām and Maghākāris—six (!) cities in total. They are mentioned with conflicting longitude and latitude.

The Islands of Mālūqā (Molucca), also called Melūk and Mūlūqā and Melūka and Mālūka: Several small islands below the equator at 155 degrees longitude. They are near the western side of Gilolo, falling between it and Celebes.

It says in the *Atlas* that this name is applied specifically to five islands:[20] 1) Ternāta, also called Tārāntā and Ternīātī [Ternatae, Tarante]; it is the largest and head of this group. 2) Tīūdūr, also called Tīdūrī and Tīdūre [Tidor, Theodori]. 3) Mūtīr, also called Mūtīl, Mūtel and Mātīl [Motir, Muthil]. 4) Mākhīyān, also called Mākīān and Mākere [Machian, Mare]. 5) Bākhīān, also called Bākhīūn and Būqīān [Bachian, Bachianum]. All of them are within 25 miles' distance. The largest measures barely six miles around.

The esteemed among these are Tīdūrā and Ternāta. Ternāta has two good ports. When Magellan's ship came to Tīdūrī, its ruler was an Arab named Manṣūr. He came to the ship in a small boat wearing a shirt on his back, a belt girded on his waist, and a handkerchief wound on his head, and barefoot. He boarded the ship and sat down. They dressed him in a yellow velvet coat and placed before him several lengths of broadcloth, satin, and fine linen and several glass goblets, a rosary, a mirror, a knife and a scissors. They offered greetings from the King of Spain and requested permission to trade. Manṣūr exchanged pleasantries with them and gave them permission buy and sell. Magellan's people stayed there five months engaging in commerce. They set a price of 30 lengths of linen cloth or 10 lengths of red broadcloth or 4 lengths of yellow broadcloth for one bag of cloves. Many groups came from the vicinity and they conducted business.

Climate

The air of these (islands) is very bad and unwholesome, but on account of its abundant wind merchants come and go and stay. In the mornings the air is misty, and the heat is intense until noon. In the afternoon a wind comes up and blows until sunset.

Plants and products

The soil of these islands, being very dry, absorbs the unceasing rains like a sponge and does not let them flow into the sea. But they grow perfumes and spices in abundance, some of which are remedies for fever. These islands [R 48a] are widely famous for that reason. Nutmeg, mastic, aloeswood, sandalwood, saffron, pepper, cloves and cinnamon grow without limit and without cultivation. They are found on all of them, but cinnamon is most plentiful on Mātīl.

The cinnamon tree is shaped like the pomegranate tree, but its leaves resemble bay leaves and its fruit, bay berries. Its bark peels off with the heat of the sun. They gather it and put it in the sun and it dries out. They extract a water and oil from the leaves that for its pleasant smell and great benefit is more valuable than rosewater.

The clove tree is especially common on Ternāta and Mūtīr. It mainly grows on rocks. The tree is large. Its leaf resembles the bay leaf, and its bark is like that of an olive tree. In the fourth year it gives fruit. At first, its flowers are like orange blossoms. [M MAP] [M 134] After they (the flowers) fall, its fruit grows mostly in stages in the shape of its bulb and hangs in clusters like grapes. It is green at first, then turns white. When it is completely ripe it turns red, and when dried it turns black. The fruits are gathered, fortified by soaking in seawater, then stored in pits. They take two harvests in a year.

These clove trees grow so densely that the sun does not penetrate through the branches. The best kind is that which grows at the highest elevation. That which grows in flat places bears little fruit. Some cloves ripen in hard stony ground and, after falling, the rains wash them down and they are gathered there, because it is not possible to ascend (to the heights where they grow). The people of the islands have shared out these trees among themselves. They subsist on them, because the other necessities of life come from other lands and they trade with these cloves.

The aloeswood tree is the size of a pomegranate tree and its leaves are like those of sweet basil. The aloeswood

20 Spellings in brackets according to *Atlas Minor*, 622.

that grows in stony ground is moister and sweeter smelling than those grown elsewhere. The leaves of some are like those of a peach, of others like those of a pear.

The nutmeg tree is also large. If its fruit is extracted from the shell while fresh and moist, it is like gum mastic.

There is a type of honey on these islands whose flavor is close to sugar.

Animals

Parrots are plentiful on these islands.

Especially the Phoenix—also called *mānāqūdīātī* or *mānqūdiyātā* (manucode, bird-of-paradise) meaning Bird of God—sometimes falls from the sky. It is a lightweight bird with a slight body and little flesh, a long neck and beak, a flat head like a magpie, a long forked tail like a swallow, and no feet. It is always in the air and manages to sustain itself while flying. The feathers on its head are as thin as a hair and differentiated with difficulty. It is green in color like the wild duck, but light green and glittering like an emerald. It is light green under the throat, gradually turning to yellow under the belly. The feathers on top of the neck are light in weight and dense black in color. When approaching the tail, the feathers get larger and the color lighter. The wings are more than a span when opened and the color of boxwood underneath and speckled white and black on top. The pinions are not attached to each other but remain separate when opened.

This bird is desired and esteemed among the people. It is recorded in the *Atlas* in the manner described above.

According to *Tārīḫ-i Hind-i Ġarbī*, its feet are one span. It has no wings but it remains in the air and does come down to the ground as it lives. Sometimes it alights on clove and cinnamon trees and eats their leaves. Whenever the wind is still it alights and rests, when it becomes windy again it flies off. It has beautiful plumage and ornamented crests and its flesh is beneficial for some illnesses.

That is what is written (in *Tārīḫ-i Hind-i Ġarbī*). But if it has no wings it is not possible to fly and to always remain in the air. And his saying that it has feet is a mistake. This poor one (Kātib Çelebi) happens to have seen that bird, and it is as described in the book of *Atlas*.

Religion

The people of these islands are mostly idol-worshippers and sun-worshippers. Some worship the moon. They ascribe rulership to the sun in the daytime and to the moon at night. They say the moon is female and the sun is male, and the other planets came into being from those two. They believe that all the heavenly bodies are gods, but the sun and moon are the parents of the other planets, which are lesser divinities. [M 135] When the sun rises they greet it with certain verses and poems, and they offer reverence and love to the moon at night. They supplicate them for children, food and other necessities.

After these islands were discovered, the Portuguese king appointed a governor to Ternāta and had a splendid monastery constructed and supported by pious endowments. Monks lived there and misled the people and invited them to Christianity. Some converted to Christianity and others remained in their heathen religion.

Mores and customs

The people on these islands are inclined toward compassion and justice. They do not like war and they prefer peace. They mostly obtain their means of subsistence from merchants in exchange for cinnamon, cloves and the like. [R 48b] Because the islands are a source of spices, ships come from foreign lands as spice traders and bring various goods for sale. In particular, Spanish and Portuguese ships come and go every year.

These Melūka islanders are a capricious and warlike group possessing weak intellects. They do battle with each other for no reason. They have a great respect for divination. They know nothing of trade; their boats are all oar-driven and small. They do not sow crops and subsist on very little. Although their lands are fertile, they avoid the trouble of sowing crops.

Mention of the Discoverer of Molucca

The captains who undertook sea journeys were mentioned previously. Captain Magellan's expedition, when he went from the west and came from the east, was expounded in summary fashion in the *Tārīḫ-i Hind-i Cedīd*. Mercator, author of the *Atlas*, gave a detailed exposition, and it has been copied here since it is not without benefit.

It is recorded in the 147th chapter of *Levāmiʿu'n-nūr* that as the Portuguese and Spanish nations discovered the New World stage by stage, they endeavored to attach the newly-discovered countries to that of their own rulers, and were in his service for this reason. Because of that, conflict developed between them. Pope Alexander VI intervened and divided the New World between them, lest they fail to do what is right. He established conditions and laws and ruled that whatever was found to the east of the islands known as Hesperides would belong to the Portuguese, and whatever was discovered below there to the west would belong to the Spanish. Thereafter, the Spanish undertook journeys from those islands and explored America, while

PLATE 26
Map of Gilolo, Molucca, Mindanao, R 49a; cf. M 133/134, 135/136 (Appendix, fig. 58–62)

Portugal headed east from there and found and seized quite a few countries. With the passage of time, the Portuguese out of covetousness for gain and wealth began to extend their control to the sector assigned to the Spanish. Later, conflict developed once again on a certain issue. Ferdinand Magellan, who was a Portuguese captain but took the Spanish side, conquered and divided. He claimed the Molucca islands for Spain according to the division of the Pope.

The reason for this judgment was as follows: the Portuguese King Emmanuel had not given Magellan the office that he promised in exchange for his service. So Magellan came to the Spanish ruler, Emperor Carolus V in the Christian year 1519 and by his command went to explore the aforementioned islands from the western route. Initially, he sailed south from Spain as far as 52 degrees south latitude. From there he turned to the west, [M MAP] [M 136] crossed an endless sea in the southern hemisphere, and finally arrived at the Molucca islands that were the destination in the east. He himself fell there in battle. His companions passed the coasts of Asia and Africa in the northern hemisphere and returned to Spain. They circumnavigated the globe in a single journey with boldness and perseverance. No human being had dared to do this until that time. Of the ships, some were sunk and some were captured. Only one boat remained. It arrived in Spain safely, bringing abundant perfumes and pearls—some of the pearls were the size of a pigeon egg and some the size of a chicken and goose egg. So these sailors are worthy of eternal remembrance and praise, even more than the Argonauts. The Argonauts departed from Greece, went around the Black Sea and returned, thus becoming proverbial for a distant journey. But Magellan and his companions went around the entire world. Their ship is properly compared with the constellation of Argo Navis in the heavens. For that reason, this ship was given great honor and became a site of pilgrimage.

Thus did Magellan go from the west and arrived at these islands from the Pacific Ocean by way of the Straits of Magellan. And so Portugal withdrew from them. But Portuguese ships do not avoid coming and going, because the east is closer to them. [R 49a] [Plate 26]

Chapter 16: Mindanao, etc.

The sixteenth chapter explains Mindanao, Māt, Cebu, Senāvan, Bāndan, Sejetis and other islands of the eastern and southern Pacific Ocean.

Wandanāo, with *fatḥ* of the *wāw* (i.e., the first vowel is *a*): According to the *Atlas*, a large, triangular-shaped island at 151 degrees longitude and 10 degrees latitude, with circumference close to 200 miles in Teutonic miles. It is also called Mindanao. It has four cities: Mindanao; Būrān, also called Būrnū; Mālāqā; and Qānūlā, also called Qānābā. The people are of dark complexion and have long hair and long faces. They weave splendid matting from date leaves. Viewed from a distance, one would think it is gold. They make a cover for their pudenda from that matting; otherwise they go around naked and barefoot. Their women are beautiful. They have big boats, iron swords, cannons and gunpowder. And they shoot poisoned projectiles with blowguns, which kill whatever they touch. The people wear cotton robes and their lords wear a jeweled golden crown on their heads. There are many gold mines on these islands, and endless supplies of bread and rice.

Lorenzo says that the Spanish have some forts here and the people use artillery. Cinnamon is produced on this island, but it is a variety different from Ceylon cinnamon; the taste is sweet and resinous. In the Christian year 1527, under the prompting of a captain named Cortes, the Spanish ruler gave three ships to a person named Sāya Patrū [R 49b] and sent him to the Pacific Ocean so that he would explore it completely. The aforementioned also went 1,400 miles and came to the aforementioned island. He saw many islands, only two which were big, one being Mindanao and the other Visayas. He departed from there and came to the island named Nātī upon which the Portuguese had a strong fortress. The aforementioned island is at 145 degrees longitude and 5 degrees latitude. It is recorded in full on the map of Giacomo.

Sebūt (Cebu): An island at 159 degrees longitude and 9 degrees [M 137] latitude. Giacomo writes Zebūt. According to *Tārīḫ-i Hind-i Ġarbī*, Captain Magellan passed by the strait and went to explore the islands of the Pacific Ocean. He came first to the island of Anbaghana (Ambon) where they found a lot of white coral. And they saw several small boats whose owners had long hair and black and red teeth and clothes made of date leaves, and who said they came from the east. Passing this island by, they came to Cebu. Its emir was called Amāra. And there were many islands close by.

Cebu is a big island. Gold mines, sugar and ginger are on it. Its people make such fine white porcelain that if poisoned food is placed within it, it shatters instantly on account of its perfect delicacy. They cure the clay of the aforementioned porcelain more than 50 years. Because the kings desire this type of porcelain, it is also desired and valued among the other people. They do not even know why they desire it, but considering that any kind of Chinese cups, bowls and plates have value in and of themselves, they trade in these wares, doubling the price out of imitation. What makes them err in this respect is the property these fine vessels have of not withstanding poison. As for that type, they say it is forbidden to be exported (from China?) to these lands. These porcelain properties are detailed in the chapter on China.

The people of Cebu go about naked. Their ruler binds an apron around his waist, wears an embroidered cap on his head, and puts rings of precious stones on his fingers. They all anoint themselves with oil and stain their mouths and teeth red with a pear-shaped fruit called *arīqa*. In their assemblies they fill a bowl with water and drink it through long reeds.

Plants: Rice and millet are plentiful on this island, but they don't eat them. They make a bread from dates, dry it out and eat it. They make wine from the rice. They poke holes in date trees and every morning drink its sap.

In this place grows a fruit known as *qūqūs*, which is long and wide like a watermelon. Its leaves are in layers like an onion, and its shell is like a dry gourd. They burn it and make a medicament from its ashes. Exactly in its middle there is a thing like date-palm fiber, from which come fine cords. The tree of the aforementioned fruit is like the date palm. It gives produce like a bunch of grapes. They burn the roots of this tree and distill its water into bottles; it is a medicament for various illnesses.

Animals: There are fish in the vicinity of this island that sometimes fly like birds, fall onto the land, and go back into the sea.

The inhabitants of the nearby islands came and met together with Magellan, except for the inhabitants of Māghūtān Sīlāpū (Mactan), also called Māntān and Mātān, which is 16 miles distant from Cebu, who did not submit. Magellan came with a number of men and made war on them. He was killed in the middle of the fighting. The 115 men who were left loaded a quantity of spices from Cebu onto two ships and set out again to the west. They came to an island named Būrnāy.

Būqānūr: The people of this island are very occupied with the science of magic. As confirmed by the Portuguese, some of them can make themselves invisible. Their enemies are very afraid of them because they are invisible and can kill whomever they want, and no one can defeat them.

Because of this, the peoples of the islands call them *ūrānī sāngūy* meaning "devil-men."

There is a type of valuable tree on the island called *pālū dīmālūqū*, it is a medicament for various illnesses.

Nearby are islands named Qānghī, Jūkī, Jūghāmā, Būrū and Lūmātūlā. In olden times these did not have any specific [M 138] chieftain. Every one had a chief known as *sān badārī* who governed only his own tribe. Afterwards the Portuguese invaded and subjugated them. Nutmeg is plentiful on these islands.

Ūda, also called Ayūndan: A rectangular island at 153 degrees longitude and 5 degrees south latitude. Cinnamon is plentiful here.

Zūlūnāqī, also called Sūlāqā: It is located at 154 degrees longitude and 6 degrees south latitude.

Mālwa, also called Mālwā: It is close to the northern coast of Timor Island. The people of these two islands are a very savage tribe. In the Christian year 1581, they set sail from Peru and came to these two islands for exploration. Although they had much wealth and merchandise, they did not stop on account of its people being savage and immoral.

Antūkatū and Mūdā: These are also near Mālwā. Here they make bread from a plant called *sago* (the sago palm) according to the custom of the people of Sumatra. There are huge and fine chickens here that are unlike the chickens of Turkey or India. The eggs are huge and the meat is delicious, like that of a francolin. The people of this island decorate their faces with spots. Their languages are diverse and savage. Its major cities are Rāo, Sakūtiyā and Tūlūn. Among these the majority are Christian.

Sināon: A rectangular island at 160 degrees longitude and 5 degrees south latitude in the *Atlas*. Lorenzo names it Sīnpāghū. Nearby is the island of Sūbandī, also called Jūnbadīd.

Jūāghānā: It is at as far as 160 degrees longitude and 5 or 6 degrees south latitude.

Jelāqayū

Būtāhūr, also called Būtābūr.

Anbūna: It is at 155 degrees longitude and between 2 and 5 degrees south latitude. The people of these islands are all pirates.

Tiyūr, Qārā and Qāsābūn, also called Qāzābūn, are close to the southern coast of Gilolo.

Anbālā Ūdinbūna, Qālārīrī and Bādīā are small islands which fall from 154 to [1]58 degrees longitude and between 2 and 5 degrees south latitude.

Then, to the south of Wandanāo, come the islands of Chīpitīā, Mūnūrīā, Qāghāyān, Bātāchīnā, Ghārghūs also called Qārīūs, Dāghārāla also called Dāghūādā, Bānkīrā also called Sānkīr, Adāpūās also called Aqlāpūās.

Then come the small islands to the north of Gilolo. They are:

Selebe: Lorenzo named this island Celebi and said that the people engage in piracy. They are a savage people and cannibals. It is their custom that an unmarried man take a rod in his hand and possess one of the married women. If there is no rod in his hand, the husband may complain about the man. As long as the rod is present, if he complains he is killed.

Sāghūnīā and Zayūn, also called Zūlū: The people of these islands have become Christians.

(Then come) the islands of Selānū, Sārānghānī also called Sārānghām, Zūlū, Pāwīlūghān also called Pānīlūghūn, Būhāl also called Būhūl, Qāndīghārā, Chīpūqū also called Yapūqūā, and Būtūān. Besides this, there is a famous city known as Nadū. A few of its people [M 139] have become Christians.

Qālāghān, also called Qālūghān.

Chīmbūbūn, also called Himbūbūn: An island at 156 degrees longitude and 15 degrees latitude. Here there is a strange tree that after its leaves break off, it moves as if it possessed a soul. Lorenzo says this is not strange because some stones, such as *astirodita* and *trūqita*, if placed on marble moistened with juniper oil, one moves straight ahead and the other turns sideways.

(Then come) the islands of Zānbūtā, Chīānā also called Zhīānā, Qabāyū, Qamāqā, Kūīālū, Qabāy, Lapānā, Qāwī, Nūzā, Tīqūlū, Pūlū, Tarā, Jalā, Masārā also called Masānā, Qātīghān, Hīpūs also called Khīpat, Hūmūr also called Hūmānū, Lūsūn, and Īrā which means "Sultan Island" (Italian: Isola de Ré) whose people make a type of garment from wheat stalks that from a distance looks like gold.

Lebāse: It consists of low reefs (Italian: Le Basse).

Apriloko reefs: It means "Open your eyes!" (Italian: Apri l'occhio)

Ījārdīnī also called Lārdīnī: It means "Vegetable Gardens" (Italian: I Giardini)

Ledūāsūrellā: It means "Two Sisters" (Italian: Le Due Sorelle)

Qūrāllī: It means "Coral Reef" (Italian: Coralli)

Zāmāl.

Lādrūnī: It means "Thief Islands" (Italian: Ladroni).

The people of these islands go around naked. They engage in piracy. The men are ugly and the women are pretty. They have a type of chaika called *bārqa* (Spanish: barca) which they row front and back and they go about on the sea with great speed.

San Lazaro: One of the wonders of this island is the large sperm whales that cause great damage to boats. A type of small bird, called the *lāns*, is the enemy of this whale. It enters its nose (i.e., blowhole) just as it emerges from the

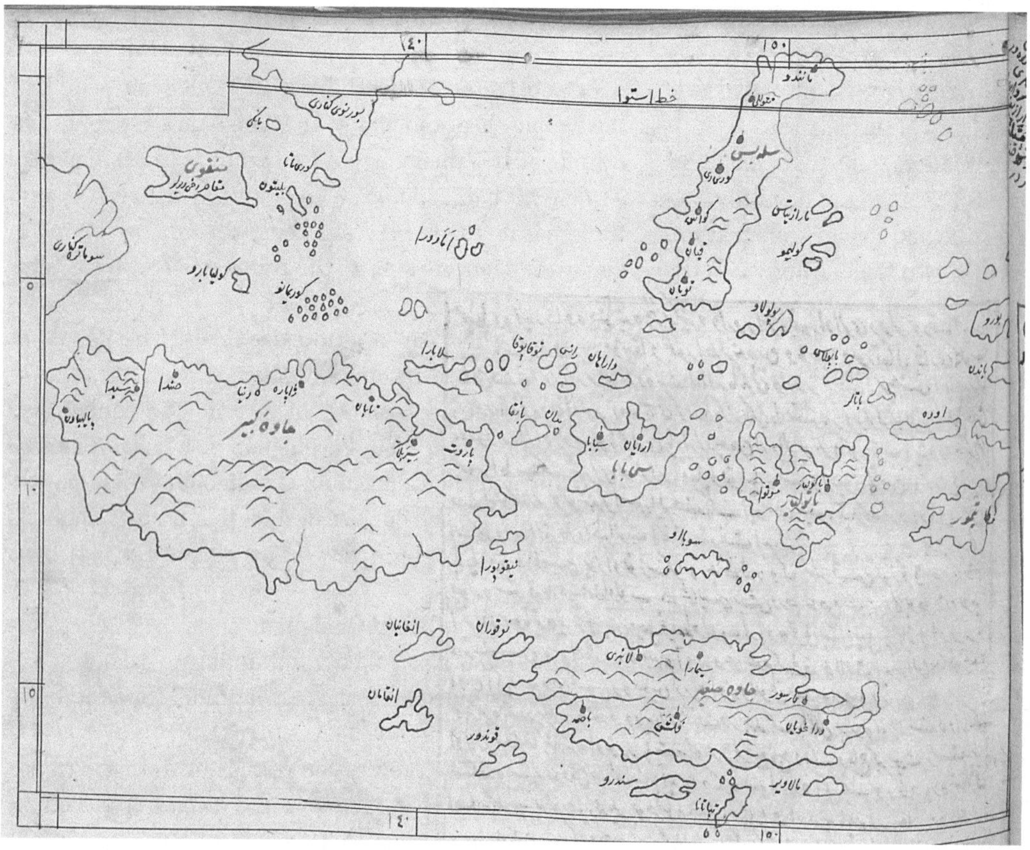

PLATE 27
Map of Moluccae (Java, etc.), R 50a; cf. M 135/136 (Appendix, fig. 60–62)
TOPKAPI PALACE LIBRARY, REVAN 1624, WITH PERMISSION

water and eats its heart and kills it. This creature does not show its head very much out of fear of this bird.

Luzon: An island famous for its abundant gold. It has a city named Manila, where the Spanish go.

Arāzīqā, also called Arāzrīqas: A triangular island. Next to it, east of Mindanao, is a string of islands. All of these islands are recorded in the book of Lorenzo, and most of them conform to the map.

The islands of Anbūn (Ambon) are recorded at 155 degrees longitude and 5 degrees south latitude. It says in the *Atlas* that it is an island with a circumference of 500 miles. The name of this island is also applied to the islands in its vicinity. The climate is very cold and brackish. The people are cannibals and pirates. [R 50a] [Plate 27]

Celebes (Sulawesi): An island that, according to the map of the *Atlas*, extends to 150 degrees longitude and as far as 6 degrees south latitude from the equator, with one side 2 degrees to the north. It has four major cities: Tetüllī, Kūrīnrī, Qiyān and Tūbān. The map of Giacomo has six cities and their names are written differently. The name of this island is also applied to the small islands around it.

Bandan: According to the *Atlas*, it consists of seven small islands located south of the Molucca islands, approximately 7 degrees, or between 4 and 5 degrees south latitude. Because Bandan is a little larger than the others, its name is also applied to them. They are:[21] Mira, Rosolargium (text: Solargium), Ay, Rom, Neira and Gunuape. Gunuape is the smallest and is desolate because of fire continually breaking out [M 140] from the ground. This fire is also called *volkan* (i.e., volcano). Another volcano appears on the map of Giacomo at 156 degrees longitude and 15 degrees latitude. On these (volcanic) islands there is always smoke during the day and fire at night.

Plants

On the Bandan islands brazil-wood grows. Sandalwood and aloes-wood and gum-mastic trees are plentiful. Nutmeg trees are widespread.

The nutmeg is a big, dense and shady tree, resembling a pine or oak. Its leaves resembles those of a peach tree, its flowers those of a dogrose. The nut has three layers of shell: the top one is thick; the middle one is reticulated—this is what we call *besbāse* ("mace of nutmegs") and the Spanish call *macis*; and the third one is thin, like a membrane. When it is planted it produces a fruit like a carob, but this is inedible.

21 Spellings according to *Atlas Minor*, 625; cf. *Levāmiʿu n-nūr*, 402b line 5 from end.

There are two types of ginger on these islands, one that is cultivated and one that grows wild. The cultivated variety is good. The herb is like saffron.

There is cinnamon on these islands as well. It is especially plentiful in the nests of the phoenix, because this bird, by reason of its hot bodily temperament, is attracted to cinnamon and collects it. This bird is fiery and lives on hot islands; it is not harmed by fire.

Timor: A medium-sized island at 155 degrees longitude and 10 degrees south longitude. Here white and yellow sandalwood is plentiful. Its people exchange for iron. [R 50b] It says in the *Muḥīṭ* that there are many islands to the south of Java called the Timor Islands. Lorenzo says that Chinese, Arab and Portuguese traders come to the islands of Visayas and Mūtīqā and the Timor Islands and buy white sandalwood, because there are sandalwood forests on these islands.

The sandalwood is a tree with azure flowers and fruit that resembles a cherry but is tasteless. They say that the people of this island recite long-established prayers to split open these trees and some spirits appear and give news about the future.

The cities of the Timor Islands are: Manāpūtūmīā, Srūājīū and Qāmen. They load up plentiful sandalwood from these. The best sandalwood is known by the name of these cities. When the boats come, the inhabitants learn which port they set out from and seek them out accordingly.

Greater Java: A large island extending east to west. The west coast is at 138 degrees longitude, the east coast at 147 degrees longitude, the north coast at 5 degrees south latitude. Turning westward from this side (the north coast) there is a strait 45 miles wide to Sumatra. The south coast is indefinite on the map of the *Atlas*, indicating that it is unexplored; but on the map of Giacomo it is fixed at 12 degrees south latitude. The island is 500 miles in length.

Government and religion

The lords of this island are numerous and their religions various. The chief of all of them lives in the city of Maghāpaghā, which is the capital, in the middle of the island. He wears a jeweled crown on his head and a brocade robe on his back. He is an idol-worshiper. In the past, the others were idol-worshipers also and subservient to him. Then two lords and the people of several cities on the coast converted to Islam and separated from him. The people of this island tend to believe in superstitions.

Crafts and warfare

These people are very clever and capable in every craft. It is as if every one was created for (his particular place in) the order of the world. In horseback riding, in fortifying cities, in the use weaponry, in the employment of artillery, they are superior to the people of the eastern realms. When making bows and arrows, their master craftsmen [M 141] take account of the stars and the wound inflicted by an arrow made at that time cannot be healed, it is definitely fatal. These people gird themselves with a long straight dagger, just like a skewer, which is called *qārīzī* (kris).

Mores and customs

The Javanese are given to shows of pomp and magnificence. They wear richly embroidered and ornamented silken garments. They shave their front locks and beards, leaving a little hair at the back of the neck. Most are haughty and proud; they bow before no one. If any falls captive, he first kills the master and then himself. They are a very distrustful and obstinate. The notables among them engage in the hunt. They board carriages and hunt with birds and dogs. They go about on foot or on horseback. Their wives are more beautiful and adorned than the women of the other islands. The coast dwellers are mainly pirates. Their boats are of the Chinese type (?) and they sail in the Indian fashion. They use an implement composed of two pieces of wood like the *bālisternā* (?) of the Franks, and they utilize maps. But they cannot discern the winds and degrees to that great a degree of exactitude, so their maps are crude.

Conditions of the lands

All the seaports of Java are toward the east and north. There is no port or harbor on the south side. It is not possible to lay anchor on that side due to the violent wind and storms that come from the pole.

This island is split down the middle by high mountains and the people on both sides are safe from each other. The mountain that lies on the north side is called Sunda. The province of Sundabārī is to its north.

The capital, Maghāpaghā, in the middle of the island, is a very green and pleasant city. Pālipān is a famous seaport and where the China and India boats dock; it is called the Port of Java. The special port of the island was Singapore; later the port of Malacca was opened and Singapore was demoted; but the Muslims are reviving its fortunes. Deymā, Kīndābū and Jāpārā are merchant centers and well-known cities. Timir is a Muslim city and a capital.

Sīrillā, Pārūt, Aghāchīn, Warbāla and Banden are among the well-known cities; each has its specific ruler. The south side of the island is called Shāndī.

Some of the seaports on the north side are Jershīk, Serbāy and Lāshim. In some texts it was cities named Pālmede, Bāmtem, Sūndaqalāba, Amāya, Ūdā, Aqūd, Sājda, Lipāyū, Betūm, Sedāra, Jīse, Perkūān and Perātqa. It is written according to this order as far as 9 degrees south latitude. The multiplication and variation in the names are an obstacle to their exact placement.

Plants, animals and products

Java is more productive than the other islands. Indian and other (merchants) mainly take such things as rice and meat. Gold, gems and various odoriferous substances are mined. Storax gum is associated with this place; it is called Javan frankincense. One of the famous medicaments is *seng-rīze* (?). They hunt an animal called *qābāl* that has value because of the special property of the bones that stop bleeding. There is also an animal called *ghāndā* whose horn is esteemed. They also hunt a beast called *pāsa* for a valuable stone like a jewel that comes out of its belly. [R 51a]

Most of their commerce is cinnamon, gold, and various silken and cotton stuffs. Especially their cloth of gold is a marvel, even finer than that of Phrygia, which was famed among the ancients. Previously, the ruler of Java [M 142] sent a large sheet of it to Alfonsez, the Portuguese commander, on which he had illustrated the famous (Javan) cities, rare animals, trees and rivers, also the campaigns that he had waged. All were illustrated so finely that even the Flemish, who were proficient in illustration, were not able to imitate it. Alfonsez sent the sheet to the Portuguese king who was very pleased with it.

Lesser Java, also called Anbābā: It is not mentioned in the *Atlas*. On the map of Giacomo it is a large island extending east to west from 145 degrees to 155 degrees longitude, and (north to south) from 13 degrees to 17 degrees south latitude. It is half the size of Greater Java, and it is situated one day's journey south and one day east of it. The cities of this island are: Lānbirī, Sāmārā, Fānqūr, Drāghūyān, Felek and Bāżma [Lambri, Simara, Farsur, Dragoian, Felech, Basma].[22] Much camphor is found on this island. The people of the island give the name of camphor to a tree, the shade of which is more beneficial than that of a plane tree.

22 Spellings in brackets here according to *Il Disegno Della Terza Parte Dell' Asia.*

Lorenzo says there are several islands to the south of Java that were discovered later. They are:

Anquman (Andaman): Portugal has not been able to explore all of it yet. According to the report of Auiadat, this island's people go about naked and are a very savage and cannibal tribe. On the map of Giacomo the name of this island is written as Anqāmān and it is shown at 140 degrees longitude and 15 degrees south latitude. In the *Muḥīṭ* it is written as Andamān and is said to consist of two islands, Greater Andaman and Lesser Andaman. The strait lying between the south side Greater Andaman and the north side Lesser Andaman is called Bīrūn Shīrū. East of it lies a shallow reef known as Kenākil, the area between them being five miles.

Mālāyūt: A very famous port. When the Javanese head south they stop here. On the map it is written as Mālāwīr and shown at 152 degrees longitude and 16 degrees south latitude, close to the coast of Lesser Java.

Betān, written as Ptātānā: A small, elongated island at 149 degrees longitude and 17 degrees south latitude. This also lies near the southern coast of Lesser Java.

Sūndūd, written as Senderū: At 147 degrees longitude and 17 degrees south latitude. Since the coastal waters are shallow, the sailors lift the rudder while passing it. These three islands are strung from west to east along the southern coast of Lesser Java.

Nūqūpar (Nicobar): A small island near the eastern coast of Greater Java at 144 degrees longitude and 10 degrees south latitude. It is written on the map of Giacomo as Nīqūpūrā [Nucopora].[23]

Nūserpānā, written as Nūqūrān: A small island at 144 degrees longitude and 13 degrees south latitude. It falls between the two Javas.

Then Lorenzo mentions many islands to the north and east of these Java islands. Some of these were mentioned above. The rest are as follows:

Sābāo, written as Subao on the map of Giacomo: A small island at 149 degrees longitude and 12 degrees south latitude.

Bātūlīār: A mid-sized island in the shape of a triangle at 151 degrees longitude and 10 degrees south latitude. On the map of Giacomo it has two cities: Bākūlīār and Mūnūā [Baculiar, Motua]. [M 143]

Sībābā (Simbawa), written as Ārām: A large island at 149 degrees longitude and 7 degrees south latitude. It has two cities: Arāyān and Sibābā, both on the northern coast.

23 Spellings in brackets here according to *Il Disegno Della Terza Parte Dell' Asia.*

This island is also called Zimbūyūn. It is between Borneo and Molucca. When Magellan's ship came here, they found many pearls. Most were as big as a pigeon egg, some were as big as a chicken egg. 11 okkas of flesh came out from some of the shells, and 20 okkas came out of one of them, but that one was devoid of pearls.

Kūnyāpā, written as Bānqā: At 144 degrees longitude and 7 degrees south latitude, between Greater Java and Arām.

Medānā, written as Medān: A small island between Bānqā and Arām.

Nūqāpūqā: A small island at 144 degrees longitude and 6 degrees south latitude.

Dāryāhāmā, written as Dārāhān: close to the northern coast of Arām Island at 146 degrees longitude and 6 degrees south latitude. [R 51b]

Sālāpārā: An island in the shape of a triangle close to the eastern and northern coast of Greater Java at 143 degrees longitude and 6 degrees south latitude. It is also called Salābārā.

Amādūrā: Three small islands at 144 degrees longitude and 4 degrees south latitude. It is to the north of Greater Java.

Qūrīmānū: As many as 15 small islands around 139 degrees longitude and 6 degrees south latitude.

Qūlīnābārū: A square-shaped island to the west of the Kūrīmānū Islands at 135 degrees longitude and 5 degrees south latitude. It is also called Kūlīābārū.

Beleytūnā: A small island at 136 degrees longitude and 4 degrees south latitude.

Kūrīmātā: A small island at 137 degrees longitude and 2½ degrees south latitude. There is a strait between this island and the south coast of Borneo.

Māqāzār (Makassar): A mid-sized island at 135 degrees longitude and 2½ degrees south latitude. It is also called Maqāṣar. It falls between Sumatra and Borneo. Lorenzo says its people are idol-worshippers and they do not let strangers onto the island. They are a cultured and wealthy people. Its conspicuous cities are Portillo, Mālāqādū and Sīderūm. Sīderūm is a populous city built on the shore of a large lake. Its well-known commerce is gold and a type of valuable wavy (?—*meneviş*) dye called *lāqā*. The chief ruler of the island lives there, the other lords obey him. Another populous city is Sūpānā, which has an independent ruler.

Mārīzūnet, written as Mārāzītās: Two small islands close to the east coast of Celebes Island at 151 degrees longitude and 3 degrees south latitude.

Pūlālāor: A small island close to the south coast of Celebes at 150 degrees longitude and 5 degrees latitude. It is also called Pūlūlāo.

Pāpūlās: A small island at 145 degrees longitude and 6 ½ degrees latitude.

Tenetūm: A small island in the vicinity of Pāpūlās. It is estimated at 145 degrees longitude.

Selāqū: A small island at 145 degrees longitude and 1½ degrees south latitude. It is also called Sūlāqū. [M Map] [M 144] These two islands are close to the Molucca islands. The rest have been discussed there (i.e., in the notice on the Moluccas).

Chapter 17: Sumatra

The seventeenth chapter explains the islands of the Pacific Ocean, Sumatra and the other small islands of India (i.e., the East Indies).

As for Sumatra, it is an island of great length, extending from 134 to 144 degrees longitude and from 5 degrees (north) latitude to 5 degrees south latitude. The equator nearly divides it in half. Lorenzo says that some of the recent geographers have fallen into error by claiming that it is the island known as Taprobana. But Ptolemy determined that island in another place. The ancients called this island Aurea Chersonnesos ("Golden Peninsula"), believing—from the mistaken account of merchants—that it was connected to the mainland. It was called it *aurea* meaning "golden" because of the abundance of gold. But this island is divided from the country of Siam by a strait, and because the strait is narrow they thought it was attached (to the mainland). That strait was perilous, and no one dared cross it. The Arabs who lived in India sometimes reached that strait, and several times their boats got stuck amongst tree roots and branches and they suffered a great deal.

This island is egg-shaped (or oval) and covers a great distance both in length and breadth. It is 600 miles long and more than 400 miles wide at its middle part. At its beginning, on the northern coast, is the mountain of Lāmrī. At its end, in the far south, it is called Nīkū-Termen.

Administrative division

This island is divided into six sultanates. 1) Qāmpār; 2) Pāzan; 3) Dākin, also called Pehyer/Peyir; 4) Orū; 5) Sunda; 6) Mānanqāo. [R 52a] [Plate 28]

On the map of Giacomo there is also a country named Bīrīā [Birea].[24]

24 Spellings in brackets according to *Il Disegno Della Terza Parte Dell'Asia*.

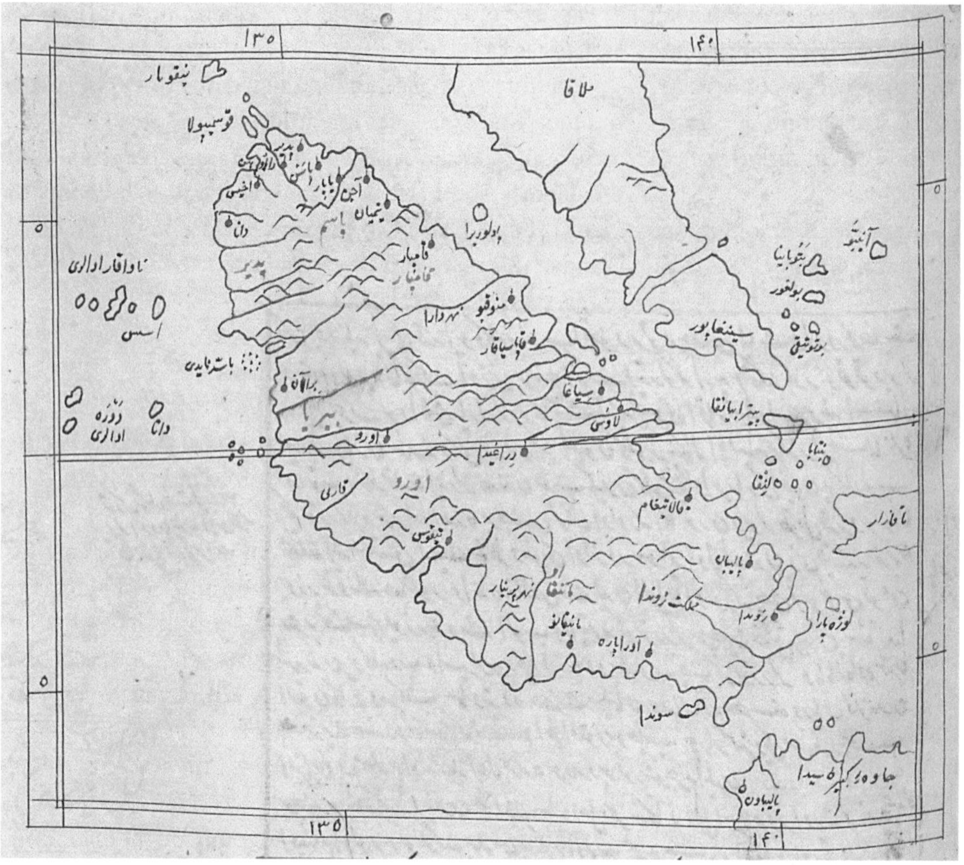

PLATE 28
Map of Sumatra,
R 52a; cf. M 143/144
(Appendix, fig. 63–64)
TOPKAPI PALACE LIBRARY,
REVAN 1624, WITH
PERMISSION

Cities

Gomo; Sepīlā; Dākin, also called Ājin [Achen] and Akām; Pāzan [Pacen]: A famous seaport and gathering place of merchants; numerous ships come here from China and Arabia. Tamīān [Timian], also called Tānīām; Qāmpār [Campar], also called Kūnpār, and across from it a small island named Pūlūdra [Poluerera]; Manūqbū [Manaucabo], also called Mākūnā; Qāpāsīāqār [Capasiacar], a small island in front of it named Qānādūs [Canados] which is fronted with reefs; Sīānā [Ciagua]; Sābān, written as Laus on the map of Giacomo. Qālātīghān [Calatigaon]; Dalīnghā [Delinga], written as Pālīān [Paliban] on the map; Sunda, also called Sīdā; this island is on a cape. All these cities are strung out in a line on the eastern coast.

On the western coast are: Andrāpārā; Mānānqo, also called Mān Qāno; Tīqūs [Tiquos]; Orū [Auru]; Brālān [Biraen]; Dānā [Data]; Akhīs [Achis]; Dāohen [Daohen]; Padīr [Pedir], also called Pādir; Pās; Yāpār—it is attached to the eastern coast.

There is also a city in the central part of the island named Drāghīdā.

According to the *Muḥīṭ* this island's seaports are the following: Lāmrī, on the north coast; Sumatra, i.e., Pāzān; Mandra, also a new and flourishing port on the east coast near Lāmrī whose ruler is famed for justice; Qayṣūr, i.e., Qāmpār; Qayṣūrī, an island that gives its name to camphor; Manqābūr, on the western coast; Flūnj, on the eastern coast.

Behind (west of?—*ẓahrinde*) Sumatra there are cannibals, known as Buṭīkh, very much feared by the people.

In sum, it has 22 cities along its length and breadth.

Rivers and Mountains

This island has numerous rivers. One is the Dārā, which has a reef at the mouth. Another is the Prīnār, which flows west and south. In addition to these [M 145] there are 19 irrigation canals.

It has numerous mountains here and there, the most famous of which is Lāmrī on the north coast.

Plants, Animals and Products

It says in the *Muḥīṭ* that this island is a habitat of aloewood, pepper, long pepper, and camphor. Silk is also produced. There are gold mines and, in some places, gold dust is found. Civet, i.e. *Galia moschata*, is plentiful. The camphor trees are so large that the shadow of every tree covers a great distance. When they tap the upper part of these trees, the sap of camphor flows; and if they scratch the lower part, the camphor comes out like gum. After the camphor is removed, the tree dries out.

Lorenzo says that this island is a very productive country. There is plentiful gold and lac. There is a very prized scent known as benzoin which is collected from a tree resembling the almond tree. A drug known as *kūkūba* is found widely, as are pepper and aloe-wood. But the aloe-wood found on Qāmpā[r] is better than this.

The people of this island collect silk from certain trees. [R 52b] But this land's silk is not so good.

There is a plant called *sango* (the sago palm) that is very plentiful. They make a kind of bread from its pith that resembles barley bread.

They split the branches of a tree known as *qūqū* and a very delicious sap flows which they use in place of wine.

Plentiful opium comes from Dillī (Dili in East Timor) to the city of Pāzan. Merchants from Bengal bring it and sell it to the Muslims with whom they engage in commerce. Ships also come here from Aden in the west, load up opium and various spices, and bring them back.

In the vicinity of this city is a bird known as *nūrī*. It is the size of a parrot and has feathers of different colors, a pleasant appearance and sings beautifully. So it has a high value, each one selling for 100 *ġuruş*.

Chinese and Persian merchants come here to buy large boards of pure sandalwood for 100 *ġuruş* apiece and take them to their countries. The Chinese merchants also go to Sunda to load up pepper and also to gather slaves and bring them to China.

The city of Māmānqā is a gold mine. It is recorded in the Holy Scriptures that every year Solomon sent his ship from the Red Sea to the island of Hind and brought back copious gold, scents and spices. Lorenzo determined according to contextual references that this refers to that island (i.e., Sumatra).

Religion and warfare

The people of this island have various religions. The seacoast dwellers are Muslim. Those in the middle of the island are idol-worshipers. All alike are warriors. The Muslims use poisoned bows and arrows. And they use bombs. But they—and others as well—learned the art of war from the Turks; because they requested help and brought them to that region on account of the Portuguese invasion and having been defeated many times. Among Portuguese battles, this victory is famous: that in the Christian year 1579 a Portuguese named Men Lopes, with 12 ships, sank approximately 100 galleys of this island's people.

Conclusion

There are a large number of islands in the vicinity of Sumatra: Dūra, one or two small islands below the equator at 130 degrees longitude. Dānā, a small island below the equator at 132½ degrees longitude. Āses, a small island at 134 degrees longitude and 3½ degrees [M 146] latitude. Nāwāqārā, a medium-sized island at 133 degrees longitude and 3½ degrees latitude; the coasts are shallow and the central part has a gradually rising mountain; it has a substantial city named Bārū, very productive in gold. Orū. Ḥam. Andrājīdā, most of whose people are cannibals. These islands are recorded from the book of Lorenzo.

But it is recorded in the *Muḥīṭ* that there are numerous islands to the west of the outer side of Sumatra known as Mīqāmārūs. Its people are cannibals and the men there are like beasts of prey. Then there are two islands with great mountains known as Andrasābūr. The distance between them and the outer side of Sumatra is 114 miles. Then there is an island named Matahārī on the south side. There are also many islands beyond these on the south side that are also called Mīqāmārūs. Two of them have large mountains. The small islands on the north side and the island reefs on the east side are known as Flūbānīq. Since it resembles Telājīā in Gujerat, the Indians also call it Telājīā. They pass by that place with great caution, tending along the Sumatra coast via the Fansūra route from Jāmas-Fala.

Chapter 18: Borneo

The eighteenth chapter is on Būrnūī, Pūlūhūn and the other islands among the islands of China.

Būrnūī (Borneo), also called Būrnīo and Barnīo. It is a large island extending from 145 to 151 degrees longitude and reaching from 2 degrees south latitude to 7 degrees north latitude. [R 53a] [Plate 29]

Mercator says that this island was recorded with the name Bona Fortuna in the geography of Ptolemy. Its south side passes through the equator and its north side lies in front of China. Some say its circumference is three months' distance and some 2,200 miles.

Cities

This island is divided into several governments. The seat of sovereignty is a city named Būrnīo. It is a large city whose houses are carefully constructed out of wood, in a marshy part of the sea like the city of Venice, on the northwest side of the island at 150 degrees longitude and five degrees latitude. Its population is around 25,000; the majority are Muslim.

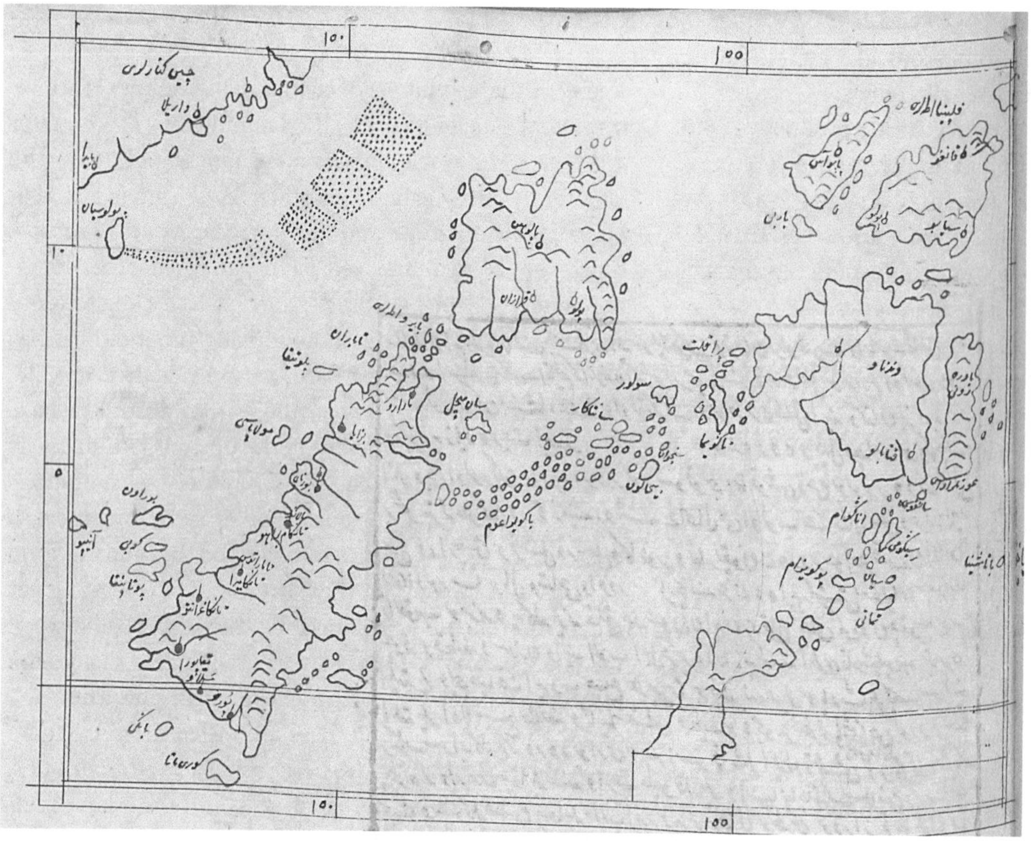

PLATE 29
Map of Borneo, R 53a; cf. M 145/146 (Appendix, fig. 65–66)
TOPKAPI PALACE LIBRARY, REVAN 1624, WITH PERMISSION

The remaining cities are: Qābūrā, also called Tīāpūrā, on the west side; Tāmīāo; Temārātūs; Mālāno, also called Lāo; Tāminker, also called Tānkāmīr; Būkāūrid; Pekāsīr, also called Bezāyād; Bārāhān, also called Tānkām Brāho; Qānserāo, also called Selāo. As recorded in the *Atlas* and others.

Conditions of sovereignty and religion

This island has several lords. Some are Muslim and some are idol worshippers.

The chief of all lives in the city of Būrna. He has great pomp and majesty. His followers mostly wear gold-embroided and jeweled garments. His palace servants are lovely girls. Magnificent spreads are laid out for him in silver and porcelain vessels. His army is always defending the city. If there is a war, the ruler undertakes it. No one is allowed to meet with him or speak with him—when it is required, an interpreter announces the requests of envoys through a reed. The ruler's decrees are written on a type of leaf.

The city of Mālānū has its own governor as well.

The majority of this island's people do not have a religion. They known nothing of Paradise or Hell, reward or punishment. [M 147]

Mores and customs

Some of the Būrnūīs go around clothed and some naked. They wear soft felt caps on their heads. There are many Arabs here. They practice circumcision. They are familiar with writing but do not use paper for writing. They are friendly with each other and do not like war and fighting. Their lords go out hunting, and they never speak to anyone.

Lorenzo states that their buying and selling is by action (i.e., barter), not by speech. They lay out each of their goods and increase the price as much as it can bear, then remove it when the price is accepted. Their transactions are highly trustworthy; they do not know tricks and do not practice treachery. [R 53b]

This humble one (Kātib Çelebi) says: Perhaps when the Portuguese and other merchants came, they had to conduct business by this means, since transactions were necessary and they did not know each other's language. But it is implausible that they would not speak with their own kind, and would not find an interpreter but stick with this way of doing business. In the clime-books of the Muslims it is recorded that cloves are sold on most of the islands of the Eastern Ocean in this fashion (i.e., by barter). This must be the source of the error.

The people of this island have a great desire for mirrors, linen cloth, iron and quicksilver.

Plants and animals

They have camels, elephants, water buffaloes, handsome horses and goats. But there are no donkeys whatsoever. According to Lorenzo, the bird-of-paradise is also found here. A tree like laurel grows here; whenever its leaf falls to the ground, it immediately comes to life and moves about like a snake. Rice, cinnamon, cloves, ginger, camphor, and *aghārīqūn* (?) are without limit. Of precious stones, there are diamonds. Other wealth and commodities are copious.

Supplement

In the *Tārīḫ-i Hind-i Ġarbī* it says that when Magellan's ship departed from Cebu, they came to this island and its people were all black. Several men left the ship with broadcloth, pins and mirrors. They were met by a large group of the islanders and conducted to the city. At night they made them a feast, and in the morning they mounted each one on an elephant and sent them to their ruler's palace.

First they saw an armed group, then men with striped velvet garments, jeweled daggers and valuable rings. Finally they came to a great palace whose four walls were covered with cloth, and there were latticed windows on one side. Entering the reception room, they knelt and gave the greeting of the ruler of Spain, offering friendship and saying, "He requests good relations with you." The interpreter conveyed this news to the vizier inside with a reed, he in turn to the ruler inside. Word came back to them via the interpreter with the reed: "You have come from a great distance. Whatever you wish (to see), see it, and whatever you want from this country, take it." He accepted the Spaniards' gifts, hung a gilded sash on each one's shoulders, and gave each a measure of sugar, chickpeas, cloves and cinnamon.

He brought them out of the palace, mounted each one on a horse, and sent them to the house of the steward. They brought various types of food on 12 big porcelain plates: various turned and roasted meats garnished with vinegar and lemon, stuffed pastries and soup with pasta, and fruits they did not recognize. At night they lit candles in silver candlesticks and illuminated the house roofs with lanterns. Many well dressed servants stood at attention. In the morning they again mounted them on elephants and showed them around the city. They gave them two elephant loads of food and drink, ginger and cloves, and dropped them off at the seacoast. They showed them the way to the Molucca islands and saying, "It lies behind you," sent them off toward the east once again.

Around Borneo [M 148] there are small islands. These are the ones that are known by name: Jūbī, also called Kūbī; Nūpālāsī, also called Mūpūlāsī; Būrā Ūn; Pūlūtīqā; Tābārān; Bāīrū, small islands on the north side; San Michel, also called San Michael; Santa Clara (?), to the east of Borneo; San Juan; Tārīūmā, also called Tākūīmā; Prāchel, also called Prāqlt; Sūlūr, near Mindanao, abounding in sulfur that is brought to China and Molucca and sold; Bitcāī, also called Bānkī; Tābūghū, which has productive iron mines and so many traders gather there, according to Lorenzo.

Some clime-books (i.e., the Muslim cosmographers) mention a big island south of China named Ghūr where is found the Ghūrī iron mine. Its ruler is always fighting with the people of China. They say there are long-horned oxen here. But the position of this on the map is not known. [R 54a]

Dona is a long island drawn on Giacomo's map with the name of Yādra Doni [ya delle dóne][25] meaning "Island of Women" (Italian: Isola delle donne). A narrow strait separates it from Wandanāo. The Island of Women (Jazīrat al-Nisā) found in some clime-books is perhaps an error based on this. That island is said to be in the China Sea and to be entirely populated by women. They conceive by eating a certain fruit that grows on the island, and they only give birth to girls. It is recorded that they do not allow a single man among them and kill any who go there. But this also seems highly implausible.

Chapter 19: Island of Ceylon

The nineteenth chapter is on Seylān (Ceylon/Sri Lanka) and the other islands from the extremity of the Indian peninsula. It is also called Sīn and Zāyla or Zeylān [Zeilan].[26] The Persians, Turks and Arabs call it Sarandīb and Sarandīl. The Indians call it Sangādīb and Habarnā. According to the *Atlas*, it is disputed which of the islands known as Tāprūbānāyā [Taprobana] or Pānī-Jerīs [Panigeneris] in Ptolemy's *Geography* it is; Ortelius prefers the first, Mercator the second.

Borders

This island is nearly square in shape, extending from 121 to [1]23 degrees longitude and from 6 to 10 degrees (north)

25 Spelling in brackets here according to *Il Disegno Della Terza Parte Dell' Asia*; presumably *ya* is an abbreviation of *isola*.

26 Spellings in brackets here according to *Atlas Minor*, 630–633.

latitude, and it is separated by a little distance from Cape Comori[n] at the south of India. Lorenzo says that the length of this island is 250 miles and its circumference is 1,000 miles. Mercator says that the circumference is 240 parasangs, and according to the statement of some of them, 900 or 700 miles. Its length is 240 miles or 250 miles and its breadth is 140 miles.

Division and conditions of authority

Mercator says that in ancient times, an independent ruler governed the entire island. Afterward he was assassinated by a ruse and his lords divided up the country amongst themselves. Presently it has 9 magistrates. The chief magistrate lives in the city known as Qūl-Mūkhī. The other lords send him gifts and taxes annually. Those nine lords are these with their titles: 1) Qūl-Mūkhī [Colmuchi] is their ruler; 2) Yānāsāpītān [Ianasapitan]; 3) Trīkī Nāmāl [Triquinamale]; 4) Bātī Qūlūn [Batecolon]; 5) Vīllāsen [Villassem]; 6) Tānānāqā [Tananaca]; 7) Lāūlā [Laula]; 8) Ghālle [Galle]; 9) Qānd [Cande]. When their chief lord is enthroned, he mounts an elephant and travels around the city, and he wears a crown encrusted with rubies whose like is rare.

Johannes (Jan Huygen van Linschoten) [M 149] relates in *Mesālik al-Memālik* that a surgeon named Rāyū [Raju] killed the great lord and, being a master of judgment and discernment, seized the country. The Sīnghālīs [Cingales] (Sinhalese) submitted out of fear. He besieged the fortress named Colombo which was in the hands of the Portuguese, but he could not take it because reinforcements arrived.

Conditions of religion, warfare and crafts

Those who live on the coasts are mainly Muslims, those in the interior are Magians and idol-worshippers. In recent times the Portuguese arrived and conquered it. They built a fortress near the capital and a few of its people became Christians. The idol-worshippers are called Sīnghālī [Cingalas] (Sinhalese). Most of the people are tall and white, with swollen bellies[27] and inclined to gluttony. For that reason they are timid and weak. Instead of guns and swords they fight each other with reeds, and that rarely. They are the most skilled of people in weaving. Indeed, they are skilled in every craft. They do extremely fine work in gold, silver, iron and ivory. They make marvelous iron gunstocks.

They are also skilled in painting. It is recorded in the *Atlas* that a humble painter painted a crucifixion that was a fathom in length and presented it to the chief monk. The priest put it in a box and sent it to the king of Spain. When they opened and saw the crucifixion, they agreed that no European painter was capable of such a perfect likeness. It became greatly esteemed and entered the treasury.

The people of Ceylon are famous for the art of entertainment. They tour India and China and stage marvelous performances.

Mores and customs

They only wear garments below the waist and do not cover the upper body. They cover the head with a thin cloth, using mainly silk and cotton fabrics. They adorn their ears with gold and jewelry and wear jewel-encrusted waistbands.

Climate

Although Ceylon is close to the equator, its air is better than that of the eastern lands. For that reason, many peoples have called it the Paradise of Adam. [R 54b] The summer heat and winter cold are in perfect moderation, the ground is always adorned with flowers, and the riverbanks are verdant meadows.

Plants, animals and products

The soil of Ceylon is very fertile and productive. Its trees are always bearing fruit. There are bitter oranges, lemons, pomegranates, citrons, dates and other fruits. They do not grow grapes, but make wine from dates. The fruit from this date is called *qūkūn* [*Coquen*].

They use the wood of this date palm to make boats and its leaves to make sails like rush matting. They twist its fiber to make ropes and tie the planks with cords made of date fiber. They do not use iron nails.

Finally, there is little rice cultivated here; they import it from the Malabar country. Of spices, there are cinnamon, cloves and pepper. They have coconuts, aloeswood, various types of perfume and musk. The cinnamon is fine and plentiful, the forests being full of this tree.

There are mines of gold, silver and other metals; but the kings protect them out of pride and do not allow them to be mined. Of jewels, there are ruby, turquoise, a hyacinth-colored jewel known as *seylānī*, another gold-colored jewel like amber, and diamonds. Pearls are plentiful in its seas. Flax, sulfur and other commodities are without limit.

27 Reading KBS as *gebeş*; cf. *Atlas Minor*, 633: *abdomen prominens*.

Of animals, deer, rabbit, boar and other game beasts abound. Elephants are especially plentiful here, and the elephants of Ceylon are better, [M 150] bigger and more esteemed than those of other lands.

The elephant is the mightiest of land beasts. Its feet are round, its eyes like those of a pig and ears like shields. For a nose it has a long hollow trunk that it uses for eating and drinking, so in that respect (i.e., raising food to its mouth) it is like a human. Because its skin is very thick, it is not bothered by the stings of other animals. Its color is mostly bluish-gray; white elephants are valuable and rare.

This beast is naturally wild. On Ceylon and in other places where it is found, the wild ones go around in herds and do not separate from each other. When captured, they are tamed in a short time with hunger.

The elephant accepts training to such a degree that it seems to have intelligence. It understands its trainer's signals and obeys them. The female goes around pregnant for two years, gives birth once in her entire life, and lives 200 years. Elephants are good for battle but are afraid of fire.

On Ceylon they measure the elephants and there is a fixed price per span, so the larger, the more valuable. It is also a tested fact that the elephants of other lands pay a certain respect to the elephants of this island.

The male elephant grows two tusks. When they get very big they are called *ʿāj* (ivory tusks) and are sold by weight. The female tusks are not so large.

This animal appears slow moving, but when it walks, a man on foot cannot keep up with it by running. It is so strong-willed that it sometimes gives up food and sleep to accomplish its goal.

A story is related in the *Atlas* attributed to a certain Christophorus who said that he observed the following in the city of Cochin: When the owner of an elephant did not feed it as usual, the elephant complained openly. Its owner showed that the bowl it ate food from had a hole and made a sign saying, "if you want your food, take this to the kettle-maker and let him patch it." The elephant took the bowl and came to the kettle-maker. The kettle-maker, as a joke, struck it once or twice with a hammer and gave it back. When the elephant brought it home, its owner was displeased, scolded it, and sent it back. Once again the kettle-make tricked the elephant and gave back the bowl. This time, the elephant submerged it in the water and saw that it had a hole in it. It complained to the people wordlessly, attacked the kettle-maker and was about to kill him. The people rushed in and saved him, and he patched the bowl and gave it back. The elephant tested it again with water, then returned home. If the story is true, it proves the intelligence of this animal.

There are giraffes in Ceylon. This is also a strange animal. Its front legs are tall and the back legs short, its neck is long, speckled and spotted. There is another beast known as *kūīl* that resembles a polecat. It is a great enemy of snakes and they fight. If bitten by a snake, it finds an herb called *yervā serīnterā* in Portuguese, and by eating the herb counters the venom.

The feathers of the birds of this island are speckled and striped with various colors and delightful to look at. They say this is due to the heat, but clearly it is an innate characteristic.

The capital city of Ceylon is Qūl-Mūkhī, also called Seylān. It is located in the north [R 55a] [Plate 30] on the strait facing India, and there is pearl diving in its vicinity. In the recent past Portugal made a fortress close to it. Lorenzo indicates it on the map by the name of Colombo. It is an important city, with a good port. Its people used to worship the tooth of a pagoda, meaning an idol. The Portuguese took it and its people offered considerable gold to recover it, but they refused to give it back and had it pulverized. From Lorenzo's observation that the highest-ranking ruler lives there and collects land tax from the six other lords, it appears that this Colombo is the Qūl-Mūkhī mentioned in the *Atlas*.

[M 151] The other cities are: Belī-Tūra. Banībā. Belīghān. Ghānāqūrā: The Nūqūrā Islands, a group of small islands, lie next to it. Jenūmāqūtān. Nāpātānā. Terīqālāmāt.

Mākhī: This may be the Qūl-Bākhī mentioned in the *Atlas*. Across from it is the island of Jene-Pātān. Its population is Christian. Portugal gathered them out of fear of the Narsinga ruler in India and settled them on this island.

The first cape of this island (Ceylon), in the north, is known as Merāshī; the cape of the south is known as Dennūr. Around it are Ṭūṭājām, the port of Ghālī, Bālījām; and after Dennūr, Mekātem, Rās-Kādī, Terkenāmelī, Merāshī, Shelem, Menāder, Kedermelī, Mendem, and Brawlī.

Among Ceylon's famous ports are: Colombo, large port on the west side. Brawlī, also a famous port in the west. Ghālī, better suited for ships than all the others; it is on the main island in the southwest. Belījām, a port on the south side of the island. There are numerous smaller ports as well which larger ships do not enter.

In some texts the cities of this island are recorded as follows: Qālītūr, Ghermālīa, Pesqārīa and Ḥānā, from 6 to 8 (degrees north) latitude.

Mountains

Most of the mountains in Ceylon are on the south side of the island.

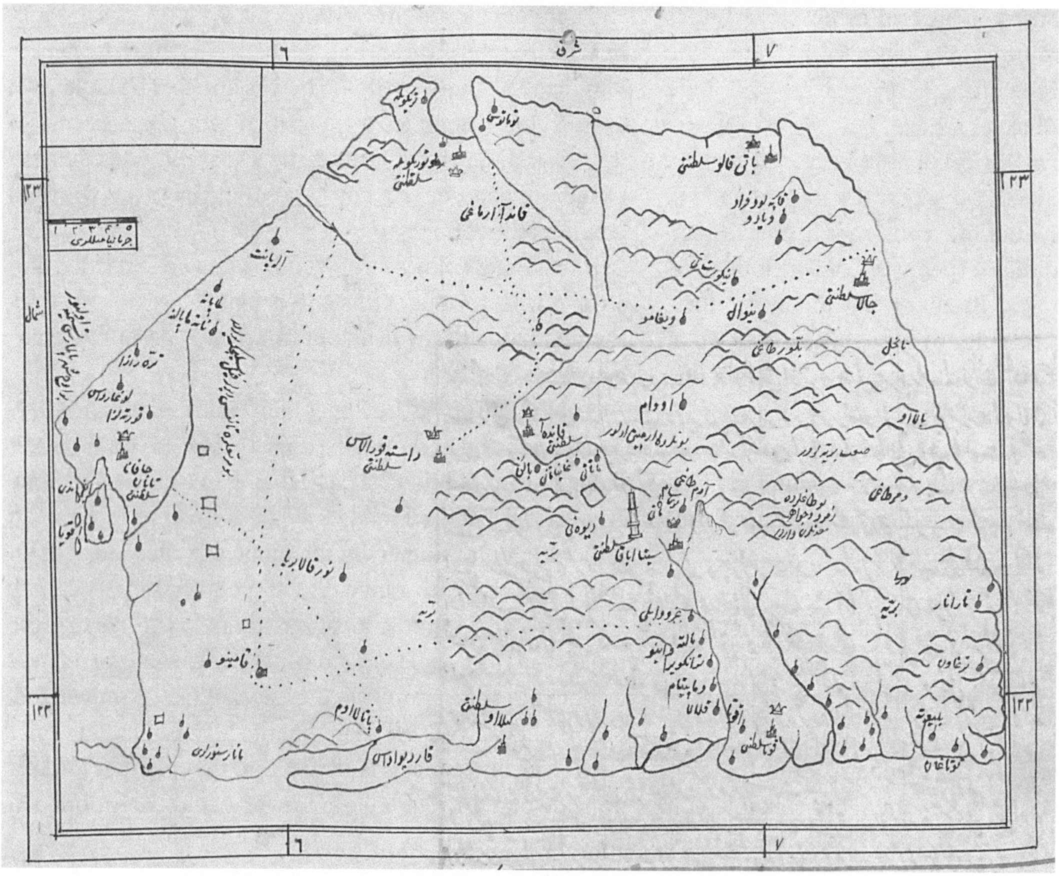

PLATE 30
Map of Ceylon, R 55a; cf. Atlas Minor, p. 647 (Appendix, fig. 67)
TOPKAPI PALACE LIBRARY, REVAN 1624, WITH PERMISSION

There is a very high mountain near the middle of the island known as Rāhūn. On account of Adam's Footprint on a rock there, people say that Adam descended upon this mountain and left his footprint, which is more than two spans in length. But the idol-worshippers say it is the footprint of someone named Sūrghūrmūyān who was a prophet, according to their assertion. He must be the one who taught the worship of idols to the Chinese and Japanese peoples (i.e., the Buddha). The site of this footprint is a great pilgrimage site. They come without end from every nation—from India, China and other lands—and gather there. They bathe in the lake at the foot of the mountain, according to their claim that everyone who bathes in this fashion and visits the footprint will have all his sins forgiven.

At the summit of the mountain there are two splendid tombs, believed to be the tombs of Adam and Eve. The Ceylonese enter into seclusion in a cave that is there. Some ascend to the summit with steps and chains. This mountain can be seen from several days' distance. In the belief of the Indians, [R 55b] this island is the paradise of Adam and he was created there.

The area around this mountain is a forest with large trees and is full of swamps and marshes with large snakes.

Lorenzo says that one of the amazing things on Ceylon, in the river known as Ārūtān, is a certain fish: whoever grasps it becomes feverish and the fever leaves him as soon as he lets go.

The Fāl Islands

These are three island chains west of Ceylon, at 115 degrees longitude and 7 degrees latitude from the equator, extending from southeast to northwest. In each chain there are known islands. The first chain, toward the open sea—i.e., on the northwest side—is known as Belī-Fenīn, meaning Greater Fāl. From there southward are the islands known as Sharyā Fenīn, meaning Lesser Fāl. Then come Beter, Ekti, Kel Fiter, Benjārem, Kel Fetī, Beqrā Fetī and Belī Fetī. In the middle chain are Shetlākem, Kenj, Mencelem, Emmīnī, Fennī and Kawar. In the chain toward the continent are Kelletī, Ender, Enderwā, Kefīnī and Melkī. Most of these [M 152] are reefs.

Other islands

The islands that were discovered to this point and had their locations determined have been recorded here in accordance with the Frankish geographies and globes.

Now I am appending some of the islands that are mentioned in the Islamic sources where the information is in agreement. Because most of these are unknown, I record their names here briefly lest it be thought that they were passed over in ignorance.

Ṣanf: In the *Taqwīm*, an island 200 miles long extending east to west from the islands of China. The best aloeswood comes from here. According to the *Taqwīm*, Qamār is to the west of this. What is meant by Qamār is Sumatra. It is also known as the country of Qampār. Ṣanf is also written as Ṣanjī. Perhaps it is Java.

Ṣandābūlāt: An island 200 miles long to the east of Ceylon. Teak is said to grow there, also triangular-shaped grapes.

Kele: In the *Taqwīm*, an island between China and India, 800 miles in length. When the *Taqwīm* states that Qayṣūr is among its cities, it is clear that it is Sumatra.

Lenjālūs, also written Lenkālūs: It is said to be in the Sea of India near the equator.

Mehrāj, also called Serīre: In the *Taqwīm*, a large island at 147 degrees longitude. Ibn Saʿīd claims they are many islands and named after the rulers of Mehrāj. The *Tārīḫ-i Hind-i Cedīd* states: "There is no end to the dominion of the aforementioned king, and no limit to his armies. If one were to travel for many years with a swift boat, it would not be possible to go around all of his islands." What nonsense is transmitted in that book! There is no doubt that such words are clear falsehood.

The Dīb islands: It says in the *Muḥīṭ* that the islands are close enough to be seen from each other, except there is a large distance between Hedmetī and Süweydū. Meḥel, Fūnīghār, Kemīl and Eddū are among them. But to the south of these there are too many to be counted. Some people allude to this by saying that there are 10,000.

The Wījī-Andrāwī islands: They are 15 islands. A strait runs between them and the most shallow water is three fathoms.

Krī-Nāk-Bārī.

Keshfelā-Andrāwī: The distance between these two and the previous one is around 4 miles.

Manjal-Fūla.

Sarjel.

Jāmas-Fala.

Andramendā

Zānj, in the *Taqwīm* written as Rānj. According to the *Taqwīm* it is a large island between the Indian Ocean and China. According to the *Tārīḫ-i Hind-i Cedīd* all the islands of India are called Zānjāt because coconuts, known as *zānj*, are mainly imported from these islands.

Rāmnī (Rāmīn): From the report of Ibn Saʿīd that this island is 800 parasangs in circumference, it has to be either Java, Sumatra or Borneo. Masʿūdī says there are around 1000 islands near Ceylon and are collectively known as Rāmīn.

Anqūja, also written as ʿAnqūja and Almūja: A long island encompassing 400 miles in the Sea of China. Perhaps it is Luconia.

Waqwāq: One of the islands of China. They write that it is behind Mt. Iṣṭīfūn and near the coast. They say that there is a certain tree here known as the Waqwāq tree. Its leaves resemble the fig leaf. In January [R 56a] it brings forth a flower like a date flower from which two human feet emerge. Gradually, by the end of April the complete human body appears—beautiful girls hanging by the hair like clusters. At the end of June they begin to fall, and all have fallen by mid-July. [M 153] As they fall they call out, saying "Wāq Wāq" two or three times, then expire. There is no bone in their bodies. This island is connected to that of Rāmīnī. There are mines and forests, but it is uninhabited. Sometimes a ship is tossed there with a violent wind. It is certain that the breasts and pudenda of these girls are those of women. Some men have sex with them and experience a sweet fragrance and great pleasure.

The poor one (Kātib Çelebi) says: This island has still not been discovered and its location is not known. Judging by some of the above description, there is no basis in truth; it should rather be considered as tales from the *Wāqwāqnāme*. To be sure, the emergence of such things from the actions of nature and the play of humors and circumstance is not impossible, because there are plants in human form and the genesis of animals from plants does occur. The three kingdoms of nature (animal, vegetable and mineral) can emerge from one another with some admixture. Those who grasp the sciences of alchemy and of agriculture and the foundation of wisdom do not reject these matters out of hand. It is possible that the origin of this tree is something composed of the matter of humans and plants. Aside from this, it is within the realm of possibility that someone knowledgeable about the properties of nature and the subtleties of the foundation of wisdom might combine the original matter of both types and produce a different type of being composed of the two.

Alqarūd: An island whose circumference is 600 miles. Ibn Athīr and Ibn Wardī relate that its population is entirely monkeys and there are no human beings among them.

Elbīnumān: Its people are Christians and valiant knights. As many heads that they cut off, so many wives do they take.

Jazīrat al-rukh (Island of the Rukh): A mythical bird with a 10,000-fathom wingspan is said to be here. In Persian dictionaries it is called *khutū*.

Berṭāyīl: Its people have faces like shields. At night they exchange heaps of cloves for heaps of goods. Because of the sound of a drum that constantly comes from a mountain, people suppose that the Antichrist must reside here.

Jazīra Muḥtariqa (Burning Island): Once every 30 years, when a planet comes to the zenith above this place, all the vegetation and the island itself burn.

Jazīrat al-ʿūd (Island of Aloeswood): A flock of cranes one cubit tall supposedly reside here. Once a year they are attacked by birds.

Sekser: Its people have dog heads and strap legs and ride on men's shoulders.

Ṣaydūn: This Ṣaydūn is supposedly a wizard king who built a castle on the island atop a marble pillar. Solomon made war on him and toppled it.

Selāhaṭ: There is supposedly a type of fruit here that makes the eater lose his mind.

Meskūne: Its people are supposedly invisible for 3 months and visible for 6 months.

Jazīrat al-qaṣr (Castle Island): Supposedly there is a castle of crystal here. Anyone who gazes on it becomes weak and falls, but if he eats some grass of the island he gets better.

Aṭūrān: Its people have the heads of cattle.

Jazīrat al-timsāḥ (Crocodile Island): Its people have dog tails.

Jazīrat al-saḥāb (Cloud Island): Supposedly a white cloud permanently rests over it.

Melātī: Its ruler has white elephants.

Khālūs: Its people are supposedly naked cannibals and its soil is a silver mountain.

Alqaṭrīa: Its people are Christians.

Elselāsī: Falcons and hawks abound here.

Yesfehān.

Jazīrat al-ḥayrān (Isle of Wonder): Its people are short of stature and long of beard.

Saʿālī: Its men and women are indeterminate and have their molars on the outside of the mouth.

Jazīra-i Tinnīn (Dragon Island): A dragon appeared here which Iskandar destroyed with a copper ox.

Ṣūṣā: It has a city made of white stone. [M MAP] [M 154] Its people's voices can be heard but they themselves are invisible.

Qalhāt: Its people are polecats.

Nūrya: Sea monsters as big as a mountain roam in its vicinity.

Lāqa: It was ruined by a multitude of snakes. The scent of its aloeswood does not dissipate as long as it is not removed from the island.

Most of these are lies and fables typical of the clime-books (i.e., the Muslim cosmographers). The authors of *ʿAjāʾib al-makhlūqāt* ("Wonders of Creation") mention these absurd matters simply in order to arouse wonder. Then they get transmitted by shallowminded authors who take them literally. I mention them here in order to inform readers of the truth of the matter. The subject won a place in this chapter, which has a detailed account of the discoveries in the islands of the Pacific Ocean and a summary account of matters that are unknown or dubious.

Some small islands on the edges of this region still remain. They will be put forth in the account of the other countries. For now, we begin with the large continent of Asia, [R 56b] starting from the clime of China.

Chapter 20: Clime of Chīn u Māchīn (China)

The twentieth chapter explains the clime of China. It is the first of the conventional climes on the shores of the Pacific Ocean. It is also Arabicized as Ṣīn. They relate that it was named after Chīn son of Yāfith (Japheth son of Noah). The ancient Greeks called it Sīnā, its indigenous people call it Tāmen, and some contemporaries call it Manji. This clime is the most excellent of the climes of Asia.

Borders

In the east China is surrounded by as much as 2,000 miles of ocean. In the west are the Dāmāzī Mountains behind which are Tibet and some of the countries of India. In the south are the ocean as far as the Pacific and the Siam country of India. In the north it is divided from Cathay and the land of the Tatars by large mountains. In the north is a wall 300 Teutonic miles in length that was built to repel Tatar attacks in the passes of the Bū-ūtūr and Qūrān mountains.

Distance: Lorenzo says that China is 32 degrees (north) latitude in its middle and 46 degrees at its end, and it extends from 160 to 166 degrees longitude. But the depictions in the Atlas and the circular world map (of Hondius) contradict this, extending approximately from 135 to 155 degrees longitude and from 20 to 54 degrees (north) latitude. Thus, its extent from north to south is 450 Teutonic miles and from east to west is 270 Teutonic miles. According to Philippus, it is 600 Teutonic miles long and 300 Teutonic miles wide. All these measurements are approximate. The shape of the clime of China is closer to a square than a triangle.

Regional division

The clime of China is divided into 13 or, according to some, into 15 prefectures. In olden times each of these had its own ruler. At present it is all a single sultanate under the rule of one emperor, according to Lorenzo, but he does not describe its divisions.

The Persian *Qānūnnāme*, which I think was written by a merchant and presented to Sultan Selīm II (r. 1566–1574), records that China and Cathay form a single country with a total of 12 divisions. But as drawn on the circular world map (of Hondius) and as recorded in the geography books, China and Cathay are independent sultanates and separate countries, each having defined borders. Therefore I have paid no heed to the words of that ignorant person (i.e., the author of the Persian *Qānūnnāme*), which is mostly idle talk, nor to the stories connected with the clime of Cathay that are recorded there, nor to the statement that the clime of China is without divisions. [M 155]

Conditions of authority

Lorenzo says[28] that the emperor of China is the greatest of the rulers of the east. He resides most of the time in a city named Quinsai (Hangzhou). He possesses many troops and much treasure. On land he mostly fights with the ruler of Cathay and at sea with the ruler of Japan, and he is wealthier than the other rulers. They bring a yearly income of 6,000 okas of silver from the country of Canta (Canton). The income derived from a tithe on silk and from the salt customs is sufficient for the army salaries. It is related that this emperor had a war lasting four months with the lords of Moin, Tipora and Barma, and that he spent 2,000,000 goldpieces during that period. He wears a crown and sits on a throne in the fashion of the ancient Roman emperors. His ornamentation and ceremony was of their manner. At present, in grandeur and pomp he surpasses them to a great degree. But this emperor does not appear before the people. Even his palace servants never see him. His only converse is with his wife who serves him in the harem along with eunuchs who are in control of several hundred concubines. They provide his food to his wife through a revolving cupboard (set into the wall between two rooms) and she herself performs the office of taster and then puts it before her husband. Despite this, nothing happens in his domain without his knowledge. [R 57a] He has spies everywhere. His viziers are four in number. The people of China call them Coloui and they resemble the viziers of the Ottomans. But they are required to be knowledgeable, experienced and advanced in years. The following words are inscribed on his imperial cipher: "The god of heaven gives peace to whoever wishes it. All that seek peace shall find it."

28 Spellings of proper names in this paragraph according to *Fabrica*, 263.

The engraving on the emperor's seal is in the form of a serpent. His palace is very splendid and its rooms are ornamented with gold, silver and jewels. He has a bodyguard of mountain troops, tall and fierce, who wear striped garments and bear long and broad swords that look like clubs. The Chinese call them *ālīamīnī*. Three hundred of them are men of strange dress resembling demons.

This ruler of China is of Moghul ethnicity. He has great power over his people and no one can defy his command. He pays the people's debts, frees the imprisoned, and gives great benefits to his troops. It is a custom from olden times to marry his sister, and he gives his own daughter in marriage to those famed for bravery.

His frontiers are guarded out of fear of the Tatars. None can enter the realm from abroad without a diploma of permission sealed with the emperor's seal. If an envoy arrives from abroad he sits at the frontier and the imperial magistrate informs the emperor. Until news comes, all his expenses are paid. When permission is granted, he is given a large retinue and brought to the capital via palanquin. He is greeted with gifts by many eminent people and conducted with great honor. A day is appointed for him to meet with the emperor. On that day he comes to the court and a master of ceremonies goes in front of him. They conduct him before a glass pane where the emperor is sitting and points to it, saying, "That place is the station of the Emperor." The envoy kowtows and raises his hands four times to the sky. The privy secretary comes and records his (i.e., the emperor's?) words and decisions. After the imperial desire is perfectly conveyed, the envoy rises and departs, backing out.

This Chinese emperor is very attracted to grandeur and false esteem. He titles himself "Son of God" and does not allow any independent magistrate except himself. [M 156] Only he has a deputy called Tutan who hears and decides all the people's suits and sees to other important matters. One day a year he holds a general assembly and divan. They put a wooden staff in a central location. Those who come and give greetings are ordered to strike that staff with an ax. Eventually that staff takes on a strange form that would not have occurred if they had carved it intentionally.

Punitive regime

The punishments and penalties of the Chinese are by torture. Among them there is no greater crime than theft. They mostly execute murderers and thieves by long imprisonment and starvation. They do not execute criminals condemned to death immediately but drag out their punishment so that most of them die in prison and in chains. Some are executed by flogging. They have a kind of leather whip five fingers broad which they moisten. The criminal has his hands tied and two public scourgers flog him with this whip. It draws blood with every blow and after fifty or sixty blows he expires. One or two thousand criminals are executed by flogging every year. And about 10,000 imprisoned criminals are found in every city.

Religion

[R 57b] Because this country of China is governed under the rites and laws of unbelievers, their emperor acts according to that.

It is recorded in the *Atlas* that in the belief of the people of China the creation and direction of lowly beings are dependent on superior beings and especially the heavens. They believe that Heaven is a god. They refer to it by the first letter in the alphabet and worship the sun, the moon and the other planets. They depict Satan in a hideous form and worship and bow down to him also. They say, "We remain on good terms with him lest his evil strike us."

Lorenzo says: The majority of these people are idol-worshippers. Among them there are worshippers of heaven, the stars and fire. Some worship their famous heroes. There are followers of different religions. Their rites are mainly that of the Zoroastrians. They come to their women during menstruation and do not perform ablutions.

Matters of knowledge and skill

The Chinese are very involved in the science of religious law and most of their magistrates are drawn from the scholars of religious law. They are called Lūteʾī. They are attracted to astrology, philosophy and geometry.

They do not measure or estimate the distance and remoteness of countries with degrees of the heavenly spheres or with the measures of other people. Rather, they call the distance that a man's voice can reach *li* and use this instead of the mile and the parasang.

Instead of letters, they have more than 5,000 symbols that represent meanings and are used by their scholars. They are known and commonly used among the scholars of Cochin-China and LKH (?) and Japan. Even though they do not know each other's languages, they exchange correspondence with these symbols, because they do not indicate sounds but meanings and do not depend on the knowledge of language.

They use clocks and printing. But they engrave a page and print it; they do not use type setting, so it is not as easy as European printing.

Craft and commerce

The majority of Chinese are tradesmen and craftsmen. Nothing of idleness or ignorance is found among them. For that reason, their country is so prosperous that there is no mountain or plain devoid of settlement and agriculture.

They are very able and competent in making various tools. Their manufactures are so strong and solid that whoever sees any of their works appreciates it and considers it a work of nature (rather than artifice). In the matter of craft, they consider the people of other lands to be blind and the Greeks to be one-eyed. Printing books and forging cannons are ancient crafts among the Chinese. [M 157] They do not know their beginning.

Commerce in this land is an esteemed and renowned profession. Merchants in perfume, silk, pepper, saffron and other spices come and go from all quarters of the earth. And they thoroughly investigate the crafts. They are especially skilled in fine work in painting, and this art is esteemed as religious service among them.

It is related that Mani the Painter came to China and found its climate pleasant. He liked the form and features of its people and saw their aptitude for painting. He subdued them with this art. He made a book (with illustrations) for them and they all became Manichaeans. They relate the following story from him: When he first gained fame as a painter, the Chinese masters tested him. They painted a fountain, placed an earthenware jug in his hand and sent him to fetch water. Mani came and looked, and when he realized the situation he took out his pen and painted a dead dog. He came and said, "The water of the fountain is pleasant but they left a dead dog there, so I could not fetch water." When they came and looked, they dropped their pens and fell at his feet saying, "You must be Mani." He smiled and said, "How did you know?" They said, "Each of us is without equal in the art of painting, but we cannot remove this dead dog from the fountain." Mani said, "I will take away that picture with my finger." He showed them his method and demonstrated his skill. After this, they all accepted and submitted. Afterwards, his work gained such currency that the Chinese acknowledged him as a prophet. The rite that he established is still esteemed and practiced among that people.

On the manufacture of porcelain

It is mentioned in the *Atlas* that the material of porcelain is mother-of-pearl, eggshells and some types of earth. They pound it and knead it and bury it in the ground. After lying there for 80 or 100 years, they take it out and work it. This clay is left as an inheritance for their descendants. Most of

PLATE 31
Map of China, R 58a; cf. M 153/154 and Atlas Minor 631 (Appendix, fig. 68–70)
TOPKAPI PALACE LIBRARY, REVAN 1624, WITH PERMISSION

what is used is the stock of the grandfathers that is worked in turn by the grandchildren; and according to their law, however much old clay one takes out, that much new clay must be put in.

According to the *Qānūnnāme*, the source of porcelain is a very fine white stone that they crush and pass through a sieve. There are three pools carved out of rock and connected to each other by passageways. Into the first pool they pour the crushed stone and dissolve it in water. They drain its water into the second pool, and after dissolving there for some time, they drain its water into the third pool. The water is drawn off, leaving the clay behind which they bury and use after a long time. The porcelain worked from the first pool is of low quality, that from the middle one is of middling quality, and that from the third one is of excellent quality. On the porcelain that is made in the winter they paint the flowers of that season. And in the other seasons as well, they paint whatever flowers are appropriate to that season.

The finest of the aforementioned vessels are made ten at a time and placed into cannisters of Cathay iron, the mouths of which are sealed. They are sold for 100,000 dirhams apiece. Whatever comes out, whether broken or sound, [R 58a] [Plate 31] is accepted by the customer as his luck. The workshop of the emperor makes the highest quality. This type has a sign on the bottom. There is some porcelain that is sold on the spot for 1,000 dirhams. Pieces like that are not exported abroad. And some types are worth their weight in gold.

There are three special qualities in porcelain that no jewel has except for jade. 1) Whatever (liquid) is placed inside it [M 158], its dregs settle to the bottom and it becomes pure. 2) It does not age. 3) Nothing can scratch it other than diamond; it is used as a touchstone for diamonds.

Eating food and drinking wine from porcelain increases the intellect and gladdens the heart. However thick it might be, if the painting made on it is not visible, the painting becomes visible if it is held to a fire or a torch, or to sunlight.

Lorenzo says: The people of China are very clever. Their crafts resemble the activity of nature, and in Germany and Flanders they seem marvelous.

Matters of war and bravery

The people of China are not that courageous. They prefer craft and commerce over courage and peace over war. But weapons, cannon and guns are in use among them. Their army is always at war with the Tatars and some of the people of the eastern islands. They also have ships on the sea and mount campaigns to India and the Molucca islands.

Mores and customs

The people of China for the most part are like those of Holland, tall of stature with wispy beards and small eyes. Their language resembles German. Being mostly of moderate temperament, they live long and do not age quickly.

According to the *Atlas*, the Chinese have mainly broad faces and flat noses. They shave their beards. Some among them are handsome. Those living near the equator are swarthy and those far from the equator are wheat-complexioned. They live very courteously according to their own traditions. The other people of the world appear uncultured and savage to them. Their women are beautiful and veil their faces. They do not cover their heads but attach combs made of ivory to their hair—there may be twenty combs in a single woman's hair. Their husbands are extremely jealous and do not let their wives go outside alone. A man marries as many women as he wishes, keeping each wife in a separate house. A married women who commits adultery is put to death. Prostitutes [R 58b] are not allowed in the city but are settled on the outside. Sodomy is not a sin among these people. Ephebes are common among them, like the fornicating women of the Indians.

Their men wear a headcover like a conical cap. They eat carrion and do not wipe themselves. Their food is mostly rice, and they make their wine also from rice.

Their weddings are usually at the beginning of March during the new moon. That is also their New Year. Their holidays are impressive, with feasting many days in a row. They bring dancers, musicians and storytellers to the gathering and give themselves over to eating and drinking.

They rarely leave their own country, and foreigners may not enter their provinces without permission.

The Chinese notables are a great and esteemed class. If their ranks and levels are written down they are distinguished by color: the highest in gold, the middling in silver, and the lowest in azure or other colors. They have many games in the manner of the ancestors.

When a man dies, they seat him on a throne and feast before him for several days. They burn before him several garments on which are depicted various animals, believing that those burnt animals will be resurrected along with him and serve him. Then all his relatives and friends gather and bid him farewell. They burn him with camphor, aloes and other fragrant woods.

The Chinese know the science of navigation and sail with big ships. But they make their ships from extremely thick boards—four or five spans thick—due to the violence of ocean storms and the multitude of whales.

There is little gold and silver in their own country, [M 159] but plenty comes from Japan, and it is valued by weighing.

It is recorded in some books that the Chinese are tribes and clans like the tribes of the Arabs, and they record their genealogies. A person can trace his ancestors to fifty generations. The tribal subdivisions do not intermarry lest there be a break in the lineage.

Both men and women wear silk and brocade. The poor and slaves go about wearing low-quality silk. Their houses are spacious and their sitting rooms are adorned with sculptures and vessels. They are an elegant people, who take pride in the cleanliness of their clothing and the refinement of their residences.

Territories and cities of China

According to the *Atlas*, China has 250 famous cities; the word *fū*, meaning "city," is written after the name of each one, e.g., Canton *fū*. As for villages, they are innumerable; the word *jū*, meaning "village," is written after the name. Most of the cities are on the banks of navigable rivers. Each one has a strong fortress with a wide and deep moat.

This clime, especially the seacoast, is so populous that there is only a distance of two or three miles between two cities.

Philippus says: China is divided into several prefectures. On the seacoast are Canton, Fūkūyam, Khaykūyām, Nānkūy, Qasāntūm and Pākīn; those in the interior are Qisyās, Qānsī, Sānqī, Sūkhnām, Hūānūm, Jūānā, Fūqām, Sūīnām, Kūīkhan, Kūyānqī, Cochin-China and Qūānsī. There is disagreement on the matter of its capital. Some say it is Pākīn, some Qsāntūm, and some KLNH.

Qāntā: A country sharing a border with the country of Pegu in the south of China. Its capital is a city named Qandā located at 145 degrees longitude and 22 degrees (north) latitude at the mouth of the Qāntā River. The Emperor of China used to be here in ancient times. It is the ancient capital, also called Namṭāy. Philippus says: Its circumference is three Teutonic miles.

Nāntū: A large city near Qāntā. Being a famous entrepot built on a river that is known by this name, 500 and sometimes as many as 1,000 ships come and gather here from the surrounding areas. They buy such commodities as copper, alum, cinnamon, silk, musk and rhubarb. Excellent swords are manufactured in this city.

Nearby along the coast are the cities of Nāntū, Birākī and Mātān.

Among the important cities in the interior is Sīngūmātū, a large entrepot and commercial zone. Many merchants from Japan and India come here but they do not enter the city but sit outside; such is the law of China. On the map of Giacomo this city is near the bank of the Biraqlān River.

The island of Īmān off the coast of this country is a pearl-hunting ground.

The island of Bīghāmā.

The island of Sānqūān—which some call Damāghān—is an island known for commerce. The Portuguese have taken up residence there and look for an opportunity to enter China, but the Chinese do not allow foreigners into their provinces. [R 59a]

Fukien: A country bordering Qāntā on the coast of China. Its capital is Fūkīnīā, also called Fūkīān. It is located on a large navigable river at 151 degrees longitude and 25 degrees (north) latitude. The streets and quarters of this city are laid out according to geometric principles, and perfect symmetry and proportion were followed in its markets and houses. The aforementioned river encompasses most of the city. Many ships come up it against the current [M 160] with loads of salt and spices.

In the *Qānūnnāme* this is a section dealing with Fūkensī. Fine fabrics of satin, velvet and linen and other clothing commodities attributed to China are there.

Nearby on the coast, in a place like an island, is the city of Chīn-Shīū. Portuguese go there every year and load up spices from China. The Emperor of China showed graciousness and assigned a residence for them in order to sell goods.

Nearby is the city of Chanīqū—written as Shīshū in the *Atlas*. It falls at 153 longitude and 25 (north) latitude.

Qānāla, also called Qūnālā: It is inland from the sea on the Mātān River.

Mahārī, written as Masūrū on the map of Giacomo: It is inland on the Fukien River.

Nīzārū, also called Nāghārī: It is on the river that goes to Qāntū.

In front of this country, Āghūdā and Būghātrā are small islands near the coast.

The country of Kekīān, also called Qūqīān; in the *Atlas* it is written as Shkūyām: It is the third country on the coast of China from south to north. Its capital, Qūqīān, is located on the seacoast at the mouth of the Qarāmārān River at 153 degrees longitude and 31 degrees (north) latitude. The cities nearby are the following:

Nīmpū: According to the map in the *Atlas*, an entrepot city on the island of Āshīū at the place where the aforementioned river empties into the sea. It is also called Līāmpū. They load up silk and porcelain goods from here. It is related in the *Atlas* from Joannes Barryos (i.e., the *Décadas de Ásia* of João Barros, d. 1570) that in this city, they loaded 166,000 *batmans* of silk onto boats in approximately three months.

North of this city, on the coast, are the cities of Āghūnārā and Tārtāhū; and in the interior, in the mountains, the cities of Sīguyā, Dūtīūn (or Dūītūn?) and Sājānfū—also called Sījānfū—it is a city in the mountains near Lake Kūyān.

Sabārsā—also called Sābārzā: It is on a river that flows into the Qūrūmarā(ī?)n River.

Isqābānā: It is on a river that flows into Khaqīān.

Kelīm, written in the *Atlas* as Kūlānīm: An imposing and important city, on the coast.

Pīlghū: A small island lying in front of Āghūnārā.

To the north of these is the country of Nānqī (Nanking), shown in the *Atlas* as Nām-Kūyī at 153 degrees longitude and 38 degrees (north) latitude. Its capital is a city built on a river known as Nānqī in the aforementioned place. It is also called Nānqīn. It is the largest and most splendid of the cities in the interior. The Emperor of China also has a palace there.

On the coast is Arjīs-Sārā, written on the map of Giacomo as Āsīsarā. It is on the gulf of the Nānqīn River. Along with this is the city of Ās-Pīsīā located on the coast. In these places there is a people called Petūrī, most of whom are dwarfish, wicked and tricky.

To the south of Nānqī are two cities on the Āghūnārā River: Pāgharfā, also called Prāgharāfā, and Qarāqāran, also called Qūrāqānū.

South of these is the city of Wānī, located at the foot of a mountain.

The island of Ispiyalū is a small island off the shore of this country.

The country of Zeytūn

It is recorded in the *Atlas* as Qsāmtūn and on the map of Giacomo as Kisāntūn, and is a well-known land on the coast north of Nānqīn. It is called the country of Zeytūn because olive (*zeytūn*) trees are without number here in the mountain and in the plain. Its capital is a city named Zeytūn, recorded in the *Atlas* as Sāmtūn, and located at 154 degrees longitude [M 161] and 42 degrees (north) latitude. It is an entrepot more famous than Alexandria. Many ships come from India and Cathay [R 59b] and load up silk and sugar. In this land they used to sell sugar in skins like honey without purifying it or congealing it; afterwards they learned how to do this.

This city of Zeytūn is one-half stage inland from the sea, with a fresh-water harbor in front of it where ships may enter. Its people drink from that water and from wells. Khanbaliq (Peking) is 30 stages' journey from here.

In some of the clime-books (i.e., the Muslim cosmographers) the river of this place is written as the Ḥamdān River.

The people of Zeytūn burn their dead with sandalwood or logwood according to their means, put their ashes in a pouch and throw it into the Ḥamdān River. Once a year a large number of people gather on the bank of this river to perform this rite.

There is a kind of tree in this land whose fruit has a varying taste. They say by way of exaggeration that it has 100 different flavors.

A marvel: Once a Khurāsāni came to Zeytūn to trade and saw in the countryside that the people were ploughing, even though the sun was not in Leo and it was not the season for sowing. "What are you sowing at this season?" he asked. "We're sowing iron," they replied. He thought it was a joke. The peasant explained: "When the sun is in its own house the air is hot. This ground is an iron mine. We make furrows and the heat of the sun causes bits of iron to leach into the bottom of the furrows and gather there. At night they freeze and solidify. When the sun enters Scorpio this stops and every furrow has a heap of iron. Then everyone goes to his own field and collects it."

The cities around Zeytūn are the following: Barmā, written Barman in the book of Lorenzo, and Sīnādyā: Two cities on the coast south of the Gulf of Zeytūn.

Cities in the interior: Qānjīū, on the Zeytūn River. Zenjīū, writen Sī-Shīū in the *Atlas*, on a river one stage inland from the coast north of Zeytūn. Tanzū, on a river two stages inland from the coast south of Zeytūn. Awtar, on the aforementioned river above Tanzū at the foot of the mountains. Ābrāghānā, in the mountains to the west of Zeytūn; the river of this place flows into the Zeytūn River.

Opposite Zeytūn and near the coast is the island of Zānjīā (Korea). On the map of Giacomo it is drawn as a small island. But on the recent globes and in the *Atlas* it is drawn as a long crescent-shaped island with the name Korea, extending from south to north, its north side almost connecting to land opposite Kense and its south side passing opposite Nānqīn. Its length is just over 15 degrees (of latitude). On the map of the *Atlas* this island is at 157 degrees longitude and extends from 34 to 50 degrees (north) latitude.

Some globes and world maps indicate one or two cities on this island, including Qūrī, Tūqsām and Shenshū.

In eastern China this island is known as Tenzū. The perfume known as *tenzū* of Cathay is said to come from here; but it is noted that Tenzū is on the mainland.

Philippus says: Korea is an important island opposite China. Its people call it Tūkā Qāūlī. It is 230 Teutonic miles long and 50 miles wide and has several prefectures.

The country of Quinsai (Hangzhou), also called Khinsai: It is a famous and large province in the north of China. Its capital [M 162] is a city named Kuense. On the map of the *Atlas* it is a large city and capital of the clime, located at 155 degrees longitude and 50 degrees (north) latitude. Its circumference is around 100 miles. It is located in a place where a large river named Pulisangu that comes from Cathay empties into the ocean. It is surrounded by water, like the city of Venice, and there are so many bridges that they cannot be counted. It is a delightful and splendid city. Because of that, the Chinese call it Quinsa meaning City of Heaven. All things that mankind require are found there. The splendid gardens and delightful springs, the innumerable waterfowl and fish in its lake defy description. The people of this city catch them with a kind of tame bird, like a crow, known as *lūtre*. [R 60a] They take these birds with them on a boat and as soon as they see a waterfowl or fish, let them loose; the birds dive into the water and immediately pull it out.

Because this is the capital city, the Chinese emperor usually resides here. Merchant ships come here from all quarters of the earth and conduct business.

Philippus says: All authors are in agreement that this city borders on the north with Tatar and Cathay. It is said to be the throne of both the Emperor of China and the Khan of Cathay. Its circumference is 100 Italian miles. It has several hundred stone bridges, constructed high enough that large ships with sails spread can pass underneath them. There is a large lake in the interior of the city, seven miles in circumference, with two islands, on each of which a great palace has been built for the emperor. It was related from Marcos (Marco Polo) that 30,000 soldiers are permanently stationed there to guard the city. But these descriptions do not accord with common sense. And some say that after Marcos this city was destroyed and no longer remains in this state.

According to the *Taqwīm* another name for Khansā is Khānghū. It is one-half stage inland from the sea and one stage's distance in length and breadth. Its market extends from one end of the city to the other. Its roads are paved with stone and its houses are lofty multi-storey wooden palaces.

Masʿūdī says in his *Murūj al-dhahab*: This city is located on a river greater than the Tigris, and its people are Muslim, Jewish, Christian, Zoroastrian and other mixed classes. Since there are six or seven stages between this city and the sea, boats enter the river and come before the city.

It is true that this report is corroborated by Abū'l-Rayḥān (al-Bīrūnī), who writes in the *Qānūn* that (Khansā) is at 160 degrees longitude and 14 degrees (north) latitude.

Ibn Saʿīd, however, corroborates the report of the corroborator, since he says that it is a great entrepot situated at the mouth of a river and that it has many fruits and sugarcane grows there.

This city has plenty of rice and vegetables and various cereals. Painting is the craft of the majority of the people.

In the clime-books (i.e., the Muslim cosmographers) generally, cities are described with the names of Khānjū, Khānkū and Khānqū, and many details are found arising from conjecture and fantasy. Since these reports are unreliable, I have not deemed it proper to transmit them.

One of the cities in the vicinity of Khansā is the city of Ūnkūīn. It is situated at the mouth of a large lake that is connected to the sea.

Another is Āūtek, which is also on the coast south of Kense. In its vicinity [M 163] sugarcane grows without limit.

Khūān, also called Chūān and written on the map of Giacomo as Khīnām. The gulf located above 50 degrees (north) latitude is named after this place. The Chinese station a large garrison there out of fear of the Tatars.

Janjīū: A city in the interior on the Kense River. It is also called Jānzū. The science of medicine has fame and currency there. The Chinese preoccupy themselves with that science there with great devotion.

Tinje Kūī: It is in the interior on the Zeylūn (= Zeytūn) River.

Sinkīfū: It is on the Ḥamdān River.

Qānījīū: It is also on the aforementioned river. It is also called Qānīzū. According to Lorenzo, it is possible to go by boat from one to another of these places, because when Kublai Khan from the line of Chingis Khan took control of this country, he constructed a great canal.

Tāpinzū: A city on the coast at the mouth of the Kense River.

Sīnzū: It is also a city on the coast, north of Kense.

Pānjī: A city on the frontier of China in the north. Above these the country of Āīnā, one of the countries of Cathay, is outside the borders of China.

To this point I have written about the coasts of China and the nearby cities. From here on I cannot describe the countries of the interior in detail. I content myself with transmitting their names as they appear on the map of the *Atlas*.

Lorenzo writes that among these are cities named Qūnghū, Kūīn-Chīn, Ūrnān, Sīkhīmā, Qādāslī, Shīānsū, Kūīāns and Sānsī [Confo, Quincin, Urnan, Sichima, Cadasli, Scianso, Quiansin; Sansi].[29] He says that each one has its own Tutan, which is the Chinese term for "magistrate". Lorenzo transmitted from the Chinese that long ago in Sansi a great flood occurred and fire rained from the sky, and the large lake near that city is the remnant of that flood.

On the map of the *Atlas* the countries in the interior are the following: Hūnāo, Sānsī, Jūnnā, Qūqūām, Qawānsī, Kūshū, Sūīnām (or Sūnīām) and Cochin-China. The names of their cities are recorded on the map.

Conditions of buildings

The Chinese fortify all their cities and fortresses with strong walls and moats and make canals flow around them. Especially they have sealed off the high mountain passes in the north with walls and towers stretching approximately 500 miles in order to block the Tatars of Cathay. It is recorded in the *Atlas* and noted by the Chinese historians [R 60b] that for 93 years prior to their writing the people of China saw endless suffering from the Tatars. Finally the Chinese emperor Terīn-Zū, after finding deliverance from the hand of the Tatars, had this wall built with great forethought and planning. He completely walled off the clime of China and appointed garrisons stationed at intervals along the wall.

The gates of Chinese cities are very well constructed. Their roads are laid out level and geometrically and are straight and broad. Fifteen horsemen can ride side by side on any road. On these roads there are bridges over the canals in various places.

It is related from the Portuguese that in the city of Fūkesh they saw a strong and splendid, well-constructed tower, built of marble and standing on forty columns. Each column was 40 spans tall and 12 spans wide. They say that it is superior to the monuments of Europe.

There are many splendid temples in the cities and villages of China. But the majority of their houses are constructed of wood.

Rivers and lakes

China has many navigable rivers and lakes, and so there are innumerable boats in this land. The number of sailors [M 164] can be estimated by the number of people in all of China. There are at least as many who live on boats as those who live on land. These rivers irrigate the cities and surrounding districts and greatly benefit the people. Their well-known rivers are the following:

- Qāntā: A large river. It emerges from a lake in the south of China and flows into the sea near the city of Qāntā. There are many islands in it and cities on it.

29 Spellings in brackets according to *Fabrica*, 265.

- Nāntū: A large river. It passes by close to the Qāntā River. There are many grassy islands in it. The geese and ducks of the people in the area graze there.
- Barqalān: A wide river. It emerges from Moin Lake and flows into the Qāntā River. The Sīnkūmātū River is also nearby.
- Qarāmār: It comes from the province of Cathay, passes by the large lake named Khīāo, and flows into the ocean at Kekīān. It is a large river. This river widens out to a lake near Nīmpū. It is also called Qūrūmarān.
- Zeytūn, also called Zeylūn; in Islamic books it is written as Ḥamdān. It comes from Cathay and divides into two branches at the city of Tāja. The larger branch flows east and empties into a large lake, emerging from the other end and becoming gradually wider until it empties into the sea at Zeytūn. It is the largest of the rivers of China. There are many cities on it.
- Pulisangu: It comes from Cathay and reaches a lake, emerging from there and flowing into the sea at the city of Kense. Its mouth is a very wide reedbed resembling a lake. The city of Kense is at its mouth.
- There are six lakes in China:
- Heylām: It is near Cochin-China.
- Chīn (?): It is connected to the sea. A river enters it and its two tributaries flow into the sea. The country of Moin is in its vicinity.
- Kūyān-shī: A large lake near the middle of the clime. In its middle there is a large island and city named Kūyān-shī. Three rivers enter this lake and its tributary flows into the sea from four places. Each of them is a large river.
- Qūqūām: It is west of Kūyān-shī. Its tributary is connected to Kūyān-shī Lake from two places. This lake also has an island in the middle of it.
- Shīn-fū: It is in the middle of the clime. Many small rivers also flow into this, and rivers emerge from two or three places.
- Sānsī: A large lake in the northwest of China. It resulted from a flood that occurred in the Christian year 1557. That flood submerged seven big cities in addition to many towns and villages, drowning all of their people. At present it is a large lake.

Mountains

In addition to the large mountains that surround China, its northern coast is a district of great mountains. There are also some small scattered mountains in the middle (of the clime), most of them forested [R 61a].

Climate

Most of China's rivers are pleasant and good tasting. The weather is good and temperate, the soil is fertile and productive. Because the air is salubrious, one rarely sees people who are blind or sick.

Plants, animals and products

Rice is sown in the well-watered places of this country and is harvested four times a year. The dry places are fields of other cereal grains. Its valleys and moutains are mostly pine trees, and its vegetable gardens are full of various kinds of fruit-bearing trees. Only in China and India grapes and figs are rare. Most of its mountains and plains have mulberry trees that (i.e., whose leaves) they dry for silk production. Flax is also plentiful. In its land there are also gold, [M 165] silver, copper and iron mines.

Among the commodities (for export) of this land are gems, pearls, musk, sugar and especially rhubarb. There is a tree here that the Chinese call *lām-pālām* and that we call chinaroot (*çöpçini*). It is a medicinal herb of benefit for various tumors and ailments. The root of this tree is white and heavy and dense.

China's important commodities are silk, cloth, rhubarb, musk and pearls. Its silk is very abundant. It is related by contemporary reports that 166,000 batmans of silk were exported in a period of three months from the city of Līāmpū alone. Its copper also goes by boat to India and Sind. It also has cloves, but they are not as good as the cloves of the Molucca Islands.

Riding animals and walking animals, hunted beasts, sable and ermine, and various types of birds are plentiful. There is an abundance of ducks there on account of it being a well-watered place. There are innumerable other waterfowl as well. As many as 12,000 are hunted in a day in the city of Qāntā alone.

In the summer the Chinese, like the Egyptians, bury 2000–3000 eggs in manure and a little while later the young emerge. In the winter they put them into reed baskets, light a uniform fire under them, and get chickens in this way also.

There are fish without limit in the rivers and lakes of China, and in the sea.

Rulers

I was unable to get any information on the conditions of the rulers of the clime of China in the Islamic books. Only in the chronicle named *Akhbār al-duwal* (by Qaramānī Aḥmad b. Yūsuf), six individuals are recorded up to the time of Anūshirwān. But I did not deem it suitable to transmit this information, since it manifestly belongs to the category of false reports without basis in fact. If I manage to obtain any sound information in the future, I shall add it in the margin. Nor is anything on this subject mentioned in the Frankish histories that we translated.

The conditions of this clime are for the most part what we in Turkey (*Rūm*) consider dream and imagination. In the following chapter we will summarize what is recorded in nonsensical fashion in the *Qānūnnāme*. Those who wish should view it there.

Supplement

One or two anecdotes are mentioned in some books about the wonders of China.

One of them is a wheel that turns and stops as much as it is desired without someone moving it.

In one of the villages of China there is a lake where the people come once a year and put a horse into it and keep it from coming out. As long as the horse is in the water, it rains, and when it has rained enough, they take it out, slaughter it and leave its flesh on the mountain for the birds to devour.

There is a spring in China that, when a sick person drinks from it, he either dies or lives.

And there is a mill of which the bottom stone turns and the top stone stays in place. It grinds flour without any grit.

Chapter 21: Clime of Khitāy (Cathay)

The twenty-first chapter explains the region of Cathay, which is among the regions of Asia.

Description

Cathay is also called Qitāy, Khiṭāy and Khitā. According to Lorenzo, this clime was given the name Kūl. Its old name was Serica. Some say it was corrupted from the word Kūtāy meaning "place of descent of divine lights" and it was given this name because ten tribes of the Israelites came there and settled after the exile.

Borders

In the east are the Eastern Ocean and China; in the south are India, the Ganges River, and the mountains and fortresses of China; in the north are Chorzā and the Frozen Sea; in the west are the countries of Ghālāsīā, i.e., Kashmir, and Khotan. The author of the *Atlas* put the western border as West Asia, included Turkistan and Transoxiana on this map, and [M MAP] [M 166] mentioned it in very summary fashion. But we put Cathay in this chapter and gave separate chapters to Turkistan and Transoxiana.

Distances

On the map of the *Atlas*, it is drawn from 120 [R 61b] to 160 degrees longitude and from 50 to 70 degrees (north) latitude. Philippus says that this Cathay country extends from the Nawrūsī Mountains on the uninhabited Tatar boundaries as far as Cape Ānīān. Its length is approximately 600 Teutonic miles.

Division

Philippus says this Cathay, which they call Serica, was the old Tataristan and was a great sultanate among the Tatar people. The ancient divisions of Tataristan are: Khitāy, Tangut, Tāyinghū, Tandūk, Qāmūl, Qīār, Qīām, Qarakhitay, and Tabar. These comprise Old Serica, most of Tataristan beyond Imaus Mountains (i.e., Caucasus), and a portion of Tataristan to the west of Imaus.

Sultanate of Cathay

In this section we transmit the summary of the translation of the *Qānūnnāme* and *Rūznāme-i Khitāy*. The *Qānūnnāme* is a Persian book in twenty chapters composed in the reign of the older Sultan Selīm (I) and later translated. The *Rūznāme* is a tract in which Shāhrukh Mīrzā (r. 1407–1449), son of Tīmūr, while Khaqan in Khurāsān in the year 822, sent Shādī Khwāja, one of the notables of his state, as envoy to the Khan of Cathay. He appointed Khwāja Ghiyāth al-dīn Naqqāsh from among the scholars as a traveling companion, and commanded him to record the affairs and events that he witnessed from his departure until his return to Herat. Khwāja Ghiyāth al-dīn Naqqāsh wrote it down as he was commanded and presented it when he came back after three years. What he presented was recorded in the *Ḥabīb al-siyar* as transmitted from the *Maṭlaʿ al-saʿdayn* and is mentioned here in translation.

Laws of Cathay

In Chapter 13 of the *Qānūnnāme* he writes that the reason for laying down the laws of Cathay was that in ancient times, out of necessity, a woman named Laozi of the lineage of Khaqans became Khan of Cathay and it being necessary to lay down laws in order to administer the territories. A learned sage by the name of Barjīn Kazīn, seeing the helplessness of the woman, presented himself. She appointed him vizier and he busied himself with the administration of affairs and laid down laws according to the principles of wisdom and intellect. He drew up a ledger and laid it down that whoever wished to control the wealth of Cathay must act according to the laws recorded in that ledger. Eventually those laws became precedent and they have been carried out for thousands of years, without change or substitution. Opposing them is considered a great crime and offense. Even the Khaqan himself can be removed according to the law. In their view, sin means abandoning the law, so that if he sins three times he forfeits the rulership and his sons also are guilty and lose the lineage of royal authority.

This humble one (Kātib Çelebi) says: later, the laws laid down by Chingis Khān, which are termed *yasa*, were generally accepted among the idol-worshippers dwelling in Tataristan and Cathay and most people follow them.

Law of the Khaqan

To begin with, he who is Khaqan must be the most knowledgeable of the royal dynasty. If someone dies or is deposed for violating the law, a suitable one among the sons takes his place. Based on the consideration that the sons and descendants of the deposed are ill-omened and that he would exercise his vengeance and spite against the pillars of the state, he is put with all his followers and retinue into a fortress [M 167] [R 62a] where all their needs are provided and no one is permitted to meet with them.

The capital of the Khaqan of Cathay is a large city named Khanbaliq. His palace and court are there.

The other territories of Cathay are 12 courts in total, each with an audience hall like that of the capital. Nine viziers sit as administrators in each, and whatever happens they present to the Khaqan. The Khaqan does not converse openly with the commanders and viziers; if a matter of importance arises, the eunuchs summarize the petitions and letters, and the sign of his acceptance is red script.

The time of the Khaqan is divided into three parts, one for eating and drinking, one is for sleep and relaxation, and one for writing and signing (official documents) for adminstering matters of state. When this (latter) time comes, there is no possibility of postponing it. If he abandons it three times in a month he is deposed.

Lorenzo says the Khan of Cathay is the greatest of the Tatar khans and he is called Ulugh Khan. The Muscovites refer to him as Qzār Qātāīskī meaning "Qaysar of Cathay." His subjects pay him so much regard that they take the water he washes his hands in and use it to wash their faces, believing that it is a blessing among them and their sins are thereby forgiven.

This Khaqan has a special language; he speaks with others through an interpreter.

When the Khaqan, who has succeeded by inheritance to his father's throne, dies, the grandees of the seven esteemed tribes gather in the compound of the vizier, all wearing white robes for mourning. After prayer and eulogy, they spread a black felt in the middle of the palace and bring the new Khaqan. He sits (on the felt) and they appoint him according to Chingisid tradition, pledging their allegiance as follows: "Look at the sun above and confess to the eternal God. You are His shadow. Conform your own administration to His wish, so that your position in paradise shall be greater than this. If you do the contrary, you will find your penalty in the world also like the others and nothing will remain for you save this black felt." Afterwards, everyone puts on red clothes and sticks a crest in his headgear. Then the *kadi* of the city places a crown on his head, and they also crown his wife and glorify and esteem her as much as they do the Khan. All the lords come into the presence of the Khaqan and kneel three times, pledging their allegiance and kissing his foot. Each of them brings great gifts, a total of nine each according to their custom because the Tatars view this number as auspicious. Chingis used to kneel to God nine times a day. They write his name in all the churches with gold letters on boards painted with vermillion, and they call him the Son of God. They place his word at the level of divine law and greatly esteem it. The Khaqan rarely appears to his subject and lords.

Description of the palace and audience hall

As described in the *Atlas*, this palace is square in shape, each side one mile long and the walls 12 cubits high. They whitewash the outside and paint it with red designs. There is a tower in each corner, resembling a great and lofty pavilion, and a similar tower in the middle of each wall, comprising eight pavilions in total, every one of them a storage magazine for cannons and guns, swords and bows, and other implements of war. In the middle of these four walls is built another palace like a citadel. It is the private quarters of the palace belonging to the Khaqan. This does not have an upper floor, but its base is raised ten feet above the ground. The ceiling of this palace is very high [M 168] and is decorated, painted and gilded with various and wondrous craftwork. Its walls are ornamented with silver and gold leaf and illustrated with battles of olden times and marvelous events depicted with astonishing skill and dazzling the eyes of the beholder.

According to the *Qānūnnāme*, the Khaqan's palace is seven fortresses inside of one another, each with orchards and gardens, idol temples and audience halls and picture galleries. The throne of the Khaqan is made of red gold and is surrounded by the image of a dragon. On one side of the 12 audience halls stand eunuchs, on the other side, slavegirls. They wear jeweled crowns, garments of gold brocade and waistbands studded with jade. One slavegirl and two eunuchs sit in the presence of every vizier, and officers and slavegirls arrayed in ranks stand behind him.

The eunuchs in each of these seven palaces are appointed to a specific duty. In the first, they look after the orchards and gardens. In the second, they bring in complaint reports and petitions and letters that have been received. In the third, they summarize those letters for the Khaqan. In the fourth, they see to the needs of the audience hall eunuchs. In the fifth, they keep the record-books of the wealth of Cathay and make collections and disbursements; the lion house is also there. In the sixth, there are 12,000 residences. In the seventh are the family and concubines of the Khaqan.

In this (seventh) palace there are no men other than the Khaqan. It has gates on all four sides, each with seven guardhouses and each appointed for a specific task. At one, the produce of the 12 territories arrives and is submitted. At another gate, responses to petitions are issued. Here has been placed the drum of justice. From another gate the Khaqan emerges once or twice a year with various types of adornment, inspects the audience halls and returns through the same gate. These gates are forged from Cathay steel.

The inner palace has wondrous gates with multicolored lanterns and banners. These lanterns are lit and at midnight all the courtiers are present, because sometimes the Khaqan comes out before the first light of dawn, and if an office holder is not present, he is dismissed from office and is deemed deserving of being bound and imprisoned.

When the Khaqan sits with an imposing air on the dragon-shaped throne, permission is granted to the commanders, notables, soldiers and envoys who proceed in groups through the gates and take up a position in their appointed places. Tibetans and Kalmyks, Indians and Chinese stand with assorted strange clothing and manners. On the right side of the Khaqan sit the viziers, commanders, scholars and other military officials. On the left sit the notables of the beaurocrats and chancery officials. Behind the Khaqan's throne stand the eunuchs and slavegirls of the palace. No one dares to gaze upon the Khaqan and those in his circle. The Cathay commanders who are seeking a position and the envoys from abroad stand and present themselves before the Khaqan. It sometimes occurs that the Khaqan places the throne at the third gate and the above-mentioned viziers, commanders and envoys stand there. If the Khaqan does not appear at court two times in a month, it is forgiven; if he fails three times, he becomes a sinner. At three sins he forfeits the rulership. Sin in their opinion is opposition to the law.

In Khanbalıq, all matters great and small are presented to the Khaqan. In other cities, the eunuchs who are in charge resolve the matter if it is small, but if it is great it must be presented to the Khaqan.

If the Khaqan wishes to go outside, the people of the city close all the shops [M 169] and conceal themselves.

Lorenzo says there are two courts for administration of sovereignty, each presided over by 12 intelligent and experienced viziers. One, called *tāy*, sees to matters of warfare; the other, called *sīnk*, sees to matters of punitive regime. Several great lords wearing bejeweled and gem-encrusted crowns are in constant attendance at these.

According to Ghiyāth al-dīn, the front of the gate of this court is paved with carved stone 700 feet in length, and five elephants are tied on each side. The Khaqan's throne, which is in the inner pavilion, is visible from the gate. The elevation of this throne is three cubits. Over it is placed a finely worked and painted canopy made of yellow satin, 60 cubits in length and 40 cubits in breadth, with columns elevated to eight cubits each. In front of these columns are three gates, the middle gate being larger than those on the left and right and reserved for the passage of the Khaqan. Other people come and go from the two sides. The imperial pavilion is ennobled by these gates and canopy. A great bell is hung over these gates, and on the court's public days several people come out and stand at attention. Several thousand men assigned to the court come at the time of morning prayer and draw up in rows. About 2,000 singers and lively musician and singing-groups also stand to one side, and about 2,000 of the soldiery line up with unsheathed swords and battle-axes displaying their might.

This court is spacious. It is surrounded with vaulted antechambers and hallways, and here and there arches atop great pillars. The roads are all paved with carved stone. In the public court they set up a triangular throne-base and place a golden throne on it. When the sun rises, two masters of the palace raise the curtain of the gate of the private quarters of the palace from two sides and the Khaqan emerges. The musicians standing above the pavilion sound the drums and bells and the three gates open. First the commanders of ten thousand, then of one thousand and of one hundred, run in groups and fall into rows left and right. Each has a staff in his hand, one cubit long and one-quarter cubit wide, which he gazes upon—it is a breach of etiquette to look anywhere else. In the *Qānūnnāme*, these staffs are made of elephant tusk and ornamented. Behind the commanders, armored men and spear-bearers stand silenly.

The Khaqan ascends the throne from a five-step silver stairway and sits. On either side of the throne stand two moon-faced slavegirls with their hair wound atop their heads, their faces and necks uncovered and adorned with large pearl earrings, and grasp pen and paper. Whatever the Khaqan commands, they write down and present it inside. If changing a judgment is deemed suitable, he sends a decree and the commanders act accordingly.

After the Khaqan sits on the throne, whatever matters are to be presented are presented. If there is an envoy, they bring him before him and make him stand 15 cubits away. One of the commanders kneels and reads to the Khaqan the document containing the conditions of the envoy. Then they instruct the envoys who kowtow three times and kiss the ground, take into their hands the letters they have brought in yellow satin purses [R 62b] and raise them up. It is a law of the people of Cathay that everything pertaining to the Khaqan is placed in something that is yellow. Some of the commanders take the letters and submit them to the master of the palace, who then hands them to the Khaqan. After opening them and considering the matter, the Khaqan gives them back to the master.

The Khaqan now descends from the throne and sits upon a chair. [M 170] As many as 3,000 garments and 2,000 robes are brought, with which the notables and men of state and children are dressed. The leader of the envoys is brought before the Khaqan who, after some questions and answers, dismisses him saying, "Go and dine." A chair and a table for each person are put in the open space of the audience hall and they feast. Afterwards, they are sent to the post station.

Feast of the Elchis (official emissaries)

It is transmitted from Khwāja Ghiyāth (al-dīn) that when they went to the palace for the feast, the crowd of people was greater than before. They passed through the first and second courtyards of the palace, the throne room of the Khaqan being in the second. They entered through a silver gate and reached an esplanade that was spacious and delightful.

In the middle of a pavement of carved stone, a splendid throne is set up beneath a high canopy around 60 cubits in length. On three sides are silver stairways, above a man's height, and two masters of the palace stand near the throne. A small throne is placed like a chair at the foot of that big throne. The corners and feet of the throne are very well crafted. On the left and right are posted lords standing armed. In one corner, a band of musicians take up rows beneath a large tent. The chief musician sits on a chair. Seven tents of seven colors are set up in front of the throne, surrounded by 2000 armed retainers. In front of the gate of the private quarters of the palace, on either side of the gate, stand two masters of the palace holding the strings of a splendid silk curtain. Just as the Khaqan comes, they raise the curtain and the musicians all at once set the bowstrings in motion. As soon as the Khaqan is seated on the throne, all fall silent.

The Khaqan of that time was Dāīmenk Khan. A group of hairs in the middle of his beard were very long and bound into three or four circles at the tip. Above his head was a yellow satin curtain raised about 10 cubits, on which four dragons were depicted attacking each other.

They came before him and kowtowed five times. They went to the first courtyard, where food was served and they dined. The Khaqan's kitchen had been arranged inside a yellow satin enclosure near the seven-colored tents. They put three plates apiece in front of the persons being honored. As the servers continually brought forward food and drink, the musicians and singers played and sang songs. Then pretty boys and girls began dancing and cavorting, crumbling food morsels in thousands of places on the open space of the dais and eating them up, and no one meddled with them. This gathering and feasting lasted from morning to evening. Finally the last note was sounded and the people dispersed group by group.

Law of the New Year and the festival

On the 27th day of Muḥarram, the Khaqan's troops were decked out and the following evening that is the start of the New Year, the people ornamented the houses, markets and shops with lanterns and torches, turning night into day. The Khaqan emerges from his palace and comes to his own compound in pomp and magnificence. No one wears white clothes, because among the people of Cathay, white is the color of mourning.

The new army camp

It was a tall building that had been completed in 19 years. From its gate to the end of the building it was 1925 feet, all of it constructed from carved stone and baked brick. Around 100,000 people come from Cathay, China, Kalmykia and Tibet and gather there. The Khaqan entertains the commanders and notables with great feasts. There at night they light the night-torch, which is a mountain-like tower made of wood and erected in the esplanade of the palace [M 171]. Its outside is completely covered with cypress branches and strung with myriads of lamps with interconnected wicks—when one is lit from below the flame immediately reaches all of them.

Khwāja Ghiyāth (al-dīn) says: That year the Cathay astrologers had made the judgment that the Khaqan [R 63a] would suffer harm from fire in this year, so of necessity they did not light the night-torch. But afterwards lightning struck the palace and all of it was destroyed.

The Khaqan stays in this army camp and carouses for seven days. He gives monetary gifts and frees slaves, and no criminal is punished in those days.

Their festival is also at the New Year. The reckoning of tax accounts is also completed at that time. For an entire month they are engaged day and night in music making and singing and carousing.

Law of public judgment

Whenever the dispatching of royal decisions to surrounding areas for a specific matter becomes necessary, the court comes into session and the throne is set up. When the Khaqan takes his seat, the people kowtow. They bring a chair and three people ascend it in front of the throne. Two of them take the Khaqan's decision and one of them reads it in a loud voice so that all the people can hear. Then they raise it aloft, drawing it up with silk tied to the top of a yellow tree. The people and musicians return to their stations, and they send copies of the decision from the post station to the surrounding districts.

Khwāja Ghiyāth (al-dīn) says: The decisions read that year occurred in the month of Ṣafar. On the tenth of this month, three years had passed since the royal night-torch and another night-torch season came when slaves, sinners and courtiers in arrears should be forgiven. It had been written that the Khaqan forgave all except those guilty of blood crimes.

Law of the hunt

Whenever the Khaqan goes on the hunt, the people of Cathay make earthen forts here and there 500 feet in length and breadth where they camp. They set up two tents in the middle of that enclosure, each 25 square cubits with four poles. The entourage of the Khaqan settle around them in tents and canopies of yellow satin with shot gold.

Khwāja Ghiyāth (al-dīn) says: During the hunt, the Khaqan fell from the horse we gave him. He got angry at us and we were brought before him. He mounted the black horse with white socks that Ulugh Beg had sent. He put his beard in a black satin pouch, wore a red garment of gold brocade, and covered himself with a gold brocade robe. On his left and right were the commanders of the hunt, behind him seven slavegirls in a small covered palanquin borne on the shoulders of servants. Another large palanquin was drawn on the shoulders of 70 individuals. The rest of the people go on the left and right at the distance of a bowshot.

Mevlānā Yūsuf Qāḍī, who was one of the twelve royal courtiers, was present when this happened. He interceded with the Khaqan on our behalf, then approached us and said, "God had mercy on you strangers and the Khaqan has forgiven your sin." When the Khaqan drew near, we kowtowed. He told us to mount and had Shādī Khwāja brought next to him.

"A rarity or gift sent to the Khaqan must be excellent so that it be a cause of increased good will," he said. "This horse that you sent is very old and unruly."

"This horse is a keepsake of his majesty, lord of the auspicious conjunction, Amīr Tīmūr," Shādī Hoca replied. "Sultan Shāhrukh sent it with careful consideration to honor you."

The Khaqan accepted this excuse and engaged in hunting cranes. Because excellent horses are valuable and rare in those lands, he appointed men, saying "Bring us excellent horses from Qara Yūsuf," and he returned to Khanbaliq. [M 172]

Law of worship

The Khaqan has a special temple in Khanbaliq. Once a year he goes to that temple in order to see those prisoners who are deserving of execution. There are no images there, only the walls are painted with calligraphy and designs. Having fasted the previous day, he sits in the royal palanquin borne by several thousand soldiers clad in iron and holding swords. When he arrives, the Khaqan descends from the palanquin barefoot and bare-headed. He kneels at the gate, enters with a greeting and standing on one foot, laments and entreats as follows:

"O God, You are all-knowing and all-seeing. You gave me the rulership. You assigned the execution of these people to my command. I have acted with caution as far as my own intellect allows. You know the remainder."

When evening comes he bows his head, leaves the temple, breaks his fast and goes to the palace. In front go elephant trains and golden carts, while musicians play surrounding the palanquin. Only two of the Muslim eunuchs mount horses; the remaining commanders all walk on foot. [R 63b]

Law of traveling

If the Khaqan travels anywhere, or mounts a horse or sits in the palanquin, he takes his concubines with him. A band of musicians surrounds him, performing as they go. As many as 4000 footsoldiers march along—so tall that their heads are in a line with their mounted chiefs—all wearing gilded armor and jeweled helmets. Of these, 1000 bear an unsheathed sword like a club and 1000 bear a heavy mace on their shoulders—the sword-bearers marching in front, the mace-bearers behind; 1000 on the right and 1000 on the left are fully armed, with painted banners on their shoulders and lanterns of various colors suspended from the tips of their spears. They march in formation with all sorts of pomp and majesty.

The Cathay commanders when they go on campaign, however high-ranking officers or generals they may be, at every stage all their comforts and luxuries are provided by the Khaqan. At no time do they mount horses, nor do they load their baggage on carts—not for lack of horses or mules, but because their law stipulates that they go inside a litter borne by rotating contingents of 50 strapping and powerful men. Some of the soldiers in the commander's train march slowly in front, with gilded and silver-plated maces. Some bear large silver- and gold-plated unsheathed swords on their shoulders. Another group suspends lanterns of various colors on multicolored standards, and when night falls they light the lanterns. With this pomp and majesty they advance the litter stage by stage. The rest of the baggage goes in front, in boxes suspended on rods, each box borne by two men. Bedclothes and varicolored satin curtains, supplies of food and drink and servants and retinue are provided at every stage according to the commander's rank. The notables of that city come out to greet him and offer hospitality. When the commander arrives at one stage, its people inform those of the next stage of his office and assignment, and they pay their respects according to his rank. When he comes in this fashion to within one parasang of Khanbaliq, that pomp goes away, all his supplies and baggage are loaded onto rented carts, he himself mounts a horse or mule at his own expense and proceeds to the city. According to law, he is mounted alone when he enters the city.

Law of the *pāy*

The *pāy* is a tablet adorned with various inscriptions. On it is written the rulings and commands of the eunuchs sent out from the palace on important assignments. Their *pāy* goes one day in advance, and on that day the people of the city tie down animals such as dogs, chickens and pigs, clean the streets, and decorate the shops. [M 173] The possessors of the *pāy* are majestic eunuchs with authority to control the country. The eunuchs assigned to them go out from the palace one day in advance, wearing gold brocade and colored satins, proceeding in formation with various arms as described above. At the palace gate the eunuch mounts a palanquin. His soldiery marches before him with pomp and ceremony, bearing up the palanquin. If anyone's dog or chicken is found outside, its owner gets 70 lashes and is put in prison. Whatever town or city this eunuch comes to, its notables come out to receive him and offer hospitality.

The commands and prohibitions of these eunuchs hold sway over the commanders and the territories. Their *pāy* goes two or three stages before them, so the people of the city are ready, market prices are regulated and malefactors flee. When he arrives in a city, the cry is made saying, "Let those who have suffered injustice come forth!" Anyone who fails to come is a sinner according to the law. Then he inspects the city and learns the events that occurred from the time of the previous inspector to the time of his arrival. If an injustice has occurred, he rights it. He has covert agents who spy out the conditions and inform him about them. He in turn records all of the matters in writing and informs the Khaqan. Thus, the territories of Cathay are populated and flourishing.

Most of those eunuchs are Muslims and are the Khaqan's privileged courtiers and favorites. They emerge from government service with great favor, re-enter the palace and manage their own wealth. They wear a belt studded with jade and are among the highest-ranking eunuchs, at the level of the Khaqan's son.

Law of ranks of the commanders and privileged courtiers

For every ten commanders [R 64a] there is a chief known as *shījen*. The commander of 10,000 soldiers is called *ḥūḥū*, of 20,000 *yān-semzen*, of 30,000 *semzen*, of 40,000 *yān-denben*, of 50,000, *denben*. On the day of battle, 50,000 soldiers have three great commanders of, the first known as *ṭay-ken*, the second as *dūtānk*, the third as *denben*.

The *ṭay-ken* is a eunuch who has emerged from the Khaqan's palace and is the absolute deputy of the Khaqan at the level of his son. The entire command of the city is entrusted to him. There is also one *ṭay-ken* for each of the other twelve courts.

The *dūtānk* is a magnificent commander in the court and foremost of all the commanders of the chancery. Control of the treasuries and account of income and expenditure are entrusted to him. These two ranks are held by eunuchs, having attained this rank by virtue of their reproductive organ being cut off voluntarily and by serving with integrity in the Khaqan's palace. When they die, since they have no children, their wealth reverts to the treasury. This *dūtā[nk]* must be proficient in accounting. There is a superintendent above him, again one of the eunuchs, but he is not a magistrate.

The *denben* is the officer over 50,000 soldiers. He has progeny, and when he dies, his ranks are given to his sons.

Such is the organization of the twelve courts of the territories of Cathay.

Laws of the soldiery

The insignias and ranks of the soldiers of Cathay are as follows: Every 10 individuals have one horsetail and one tent. Every 30 individuals have three horsetails, three tents and one standard. The commander of every 1,000 persons is assigned one red horsetail, two kettledrums, two reed pipes, two horns and two cymbals. 50,000 soldiers are given 5000 horsetails, 2000 standards, 1000 cannon and 50,000 muskets. They say that this organization of 50,000 soldiers is installed in each court. Aside from the eunuchs, these draw salaries and land grants from the Khaqan's treasury.

In the cities [M 174] garrison soldiers stand ready with armor and coats of mail, cannon and muskets and other implements of war. They each receive one *mudd* of rice, one *mudd* of wheat, and twenty dirhams of silver per month. They publicly tether their horses in the marketplace and the town quarters and get their horses' feed from the state treasury. If someone's horse dies he gets 100 lashes and is given another one.

In cities other than Khanbaliq, every morning all the salaried soldiers with their *lords*, clad in mail and outfitted with their implements of war, proceed to the court square of the city and take up positions in two battle-lines. Just as two enemies face each other and do battle, so these stand face to face making attacks, knocking their opponents from their horses and tying them up. Their goal however is not to kill but to vie for promotions and ranks by the display of swift riding and courageous action. After the battle is over, they remove their captives' bonds; but meanwhile they cannot avoid being wounded or injured. They consider battling in this fashion as a game. Those who are injured do not enter the lists until they recover.

The discipline of their commanders is such that they are able to muster 50,000 fully equipped and armed soldiers in one hour. When they come from the court, they remove their weapons and in plain clothes attend day and night at the gate of their lords in the service to which they are assigned. And each day at dawn they attend the Khaqan's court in Khanbaliq. They may not abandon this duty for a single day without excuse.

There is no city in the territories of Cathay without a garrison of 5000 or 10,000 or 20,000 or as many as 50,000 troops. According to their law, they are absolutely forbidden to engage in pleasure-outings or the hunt. It rarely occurs that the Cathay armies make war on the enemy. They do not leave their soldiers and commanders inactive, however, but always engage them in some task, whether repairing a city, a milestone or a fortification. At the very least, they dig moats around the cities or fortresses. They say that when soldiers are idle, it is an invitation to civil strife. [R 64b]

Lorenzo says that 12,000 Tatars are always present to protect the Khaqan. They are called *kūīstīān*. In other provinces, garrison troops live in tents outside of the capital cities.

Law of war

Their rule in fighting is that when they are one stage away from the enemy, they line up their carts around them and in one hour, dig a trench in front of them. Their combat is very severe. They draw the line (of carts) in front of the soldiers who stand ready with gun in hand. First they fire their cannon, then they fire the 600,000 or 700,000 guns. They only suffer defeat when they are making camp or breaking camp. If the enemy attacks them, and they are unable to draw their carts in front of them or to their side, he will achieve victory. In fact, a brave commander of the Kalmyks named Esen Tayshi with 60,000 troops attacked the Cathay Khaqan Chīn-Khūār in 854 (1450), broke them by this means and took him prisoner. Ever since this battle, the Khaqan has not gone to war in person but sends the army wherever necessary.

Among the Cathayans, fleeing combat and dying in combat are one and the same. Even if 100,000 soldiers flee, by law they must all have their heads cut off. They will not cede to the enemy even a rush mat's worth of ground from their territory, though the entire army be destroyed.

By nature, the Cathayans are not given to warfare. If an enemy appears, they pay them off and make peace. For that reason their territories are flourishing.

Law of salaries

All the people in the country of Cathay, be they military [M 175] or civilian, are registered by name and description; no one is omitted from the register. And everyone gets a ration according to his condition, by the month or by the year. Men are appointed to the service of the commanders and other high officers according to rank. If the lord of a city or the commander of 20,000–30,000 troops is removed from office, only four or five of his retainers stay at his side. Since all the people belong to the Khaqan, only a few men are specifically appointed to the lords and commanders, and therefore no one can rebel or raise an insurrection.

Law of coinage

Instead of silver and copper coins, the Cathayans issue square pieces of paper in that amount, strike a seal upon them like a die, and use them as currency. They call this *jāw-jengīzī*. Anyone who does not accept it is punished. They do not strike pure gold and silver coins, but cut them into pieces with a shears and use them as a medium of exchange by weight. They call 10 dirhams of silver one *sīr*. They give 70 small coins for one silver dirham, and buy a bowl of food for one coin, which is cheap. In that country, men and women alike are moneychangers and assayers. They all know the value and grades of the various currencies that are in use among them.

Law of grain supplies and provisions

Heaps of straw and firewood are piled up next to every city and marked with a symbol like a bird. They are visible from a distance, so that people come and take what they need. Once every six months when the heaps of wood and straw are depleted—the straw being used by soldiers as fodder for their horses—they are replenished and supplies are given once a week to other travelers.

Rice, wheat, barley, and other grains and fruits, fresh and dried, are collected in the imperial storehouse in every city. Soldiers are given grain once a month and fruit once every two days. Merchants that come with the title of Elchi are given a three years' ration of grain and various dried fruits. Grain stocks placed in the storehouses are depleted once every six months and are then replenished. There is no city or fortress that does not have these storehouses. These stored goods and provisions are obtained from the fines exacted from criminals.

Law of eunuchs of the palace

The eunuchs and slavegirls who dwell in the imperial palace advance rank by rank. The eunuchs number 7000. The virgin girls number 12,000 of whom 3000 serve the Khaqan in turns as concubines. Once in four or five years, the eunuchs go around the country and collect beautiful girls whom they bring and present to the Khaqan's mother or (closest female) relative. Those who do not receive acceptance are sent home with gifts and favors. Likewise, they also send off the women in the palace who are old with gifts according to their station.

In this palace, respect is paid to the eunuchs above all others, and therefore anyone in the country who has a fine-looking son has him castrated in infancy. First they go and inform their lord. After getting a license [R 65a] they cut off the boy's penis and place him on the ground in the presence of their lord. The boy is attended to and, if found in good health, is taken to the palace with the license and presented to the chief eunuch, from whom he gets permission to serve. His service is tested for some time by those in the outer palace; he then enters the inner palace. He traverses the ranks over a period of time and, if proven capable, assumes positions at the top of the hierarchy as they fall vacant.

Eunuchs of the first grade are responsible for the imperial orchards and gardens.

Eunuchs of the second grade bring inside the writs of complaint and letters and petitions coming from around the palace and take the replies back out again.

Eunuchs of the third grade sit on the inside with the slavegirls of the audience halls, [M 176] translate and summarize the writs and letters that arrive, send them to the Khaqan with the girls and attach the royal cipher when they come back. The Khaqan considers these documents. If he accepts their contents he draws a red line over them; if not, he does not draw a line. When they come back out they are put in a pouch and sealed and sent to whatever audience hall they originated from. The possessors of this grade are more esteemed than the second.

Eunuchs of the fourth grade are responsible for the register and number of the palace slavegirls and eunuchs. In each of the twelve audience halls, esteemed eunuchs stand on one side and auspicious concubines stand on the other.

Eunuchs of the fifth grade are responsible for keeping the register of jewels, silver and gold, fabrics and goods in the imperial treasury and guarding them.

Eunuchs of the sixth grade are responsible for the affairs of the twelve lords.

At the seventh grade are the favored concubines and family of the Khaqan.

Law of guards

There are 1000 guardhouses in the seven levels of the fortress of the outer palace, with ten men stationed in each one guarding it night and day. At night they go around with arms and torches and ring bells, which they put down here and there, striking them with a stick. There are also designated drums that are beat on top of the fortress. The guardhouses are arranged at the road entrances. Each one's bells have a distinctive sound and they are struck with a stick and rung in turns.

Law of drums and kettledrums

Above the fourth gate there is an esteemed bandleader with 50 men who play the kettledrums. A bell is suspended by a chain from the roof, which is like a vaulted ceiling, and 400 eunuchs are assigned to ring it. They skillfully pull back a large beam like a ship's mast a distance of 40 paces. When the drumming is completed, they release the beam from its place and strike the bell with such force that all the inhabitants of Khanbaligh hear its sound and know that the Khaqan has come out. When the Khaqan sits on that dragon-decorated throne of gold, it is played once again and the gates are opened all at once. Various flutes and stringed instruments are played around the Khaqan. Amidst those sounds an ugly drum is beat with a wailing noise that makes one forget the pleasure of hearing the other instruments. They intend by this to deflect vanity and as a warning. These instruments are exclusive to the Khaqan's palace.

And a kettledrum of justice is placed at each of those gates, with someone assigned to each one. Anyone who has suffered injustice kneels down at the first gate and strikes the kettledrum. When the guard at the second gate hears it, he strikes the kettledrum there. When the seventh kettledrum, which is in the presence of the Khaqan is struck, they have that oppressed person brought and inquire into his condition. The eunuchs write down his complaint and inform the Khaqan. No matter who brings the complaint, the penalty is exacted according to the law. No one gets any special consideration. When this drum of justice is struck, the commanders lose their wits and some die of fright; because it is struck whenever they witness an injustice on the part of one of the commanders, as if to say, "He did not see to my right according to justice."

Law of the princes

In most cities of Cathay, beautiful palaces are arranged for the sons of the Khaqan, their walls covered with red, white, yellow and green porcelain tiles and windows set with chrysolite. The income of the wine-shops in each city is appointed to that prince. There are some princes who [M 177] are so poor they must work as day laborers. Their father fell from the rulership because of a sin, and so sovereignty has been cut off from his children. If necessary, they bring the most knowledgeable of the princes who subsisted on the brothel revenues in the palaces [R 65b] and make him Khaqan. The son of the Khaqan who is heir to the throne is at the third gate of the palace, in his own audience hall. Three times a year the commanders and high officials attend his court and hold court like the court of his father.

Law of the seal of the provinces

In each of the twelve courts is a seal of white jade, as big as the palm of a hand, engraved with the picture of a dragon. It makes a stamp with a red color. Since their paper is a thin silk, they can seal 10 official orders with a single seal.

Lorenzo says that instead of patent letters of rank the Khaqan has sheets of silver and gold engraved with a lion, falcon or sun. If the person it is given to is an idolworshipper, the name of the idol Naghāy is inscribed; if he is Muslim or Jewish, the name of God; if he is Christian, the name of Jesus; and from this the condition and desire of the possessor are known.

Law of the court of the commanders

In every city an audience hall has been built by the Khaqan, according to the rank of its commander, where the notables gather every morning and conduct affairs of state. Each hall has three administrators: the governor of the city, the steward, and someone responsible for income and expenditures (i.e., treasurer). Each of them checks on and fears the others.

Each of the twelve regional courts has its own seal and script.

In the outer palace of the Khaqan there is a bureau for each region where a concubine official sits flanked by two eunuchs—the one on the right is the steward and the one on the left is the inspector of registers. They see to the petitions, letters and other business coming from the regional courts.

Law of viziers

Administration of the audience halls in the outer palace is in the hand of six viziers. One is the enforcer of the law of the provinces of Cathay. One is the supervisor of the silver, gold and silk stuffs of the provinces. One is in charge of rice, wheat, barley, dried fruits, firewood and straw. One is the supervisor of the military. One is in charge of fortresses and other buildings in the provinces. One is responsible for the conditions of those confined in prison. Once a year these viziers make a register of affairs under their charge and send it to the officials of the inner audience halls. They in turn make summary registers that they present to the Khaqan.

Law of jurisdiction and judicial opinion

It is imperative that the jurisconsults of Cathay be advanced in age and the most knowledgeable in the law. They reach the position of jurisconsult after working their way through the judicial hierarchy. They have a large and splendid audience hall with orchards and gardens near the prison. Here they sit in great majesty, trying cases of the imprisoned and giving judicial opinions according to the law. In their laws there is no more serious matter than confinement and execution. In these matters they give their full attention. They draw up writs of execution and other penalties and send them to the palace. The Khaqan is informed and the requisite law is carried out.

Law of detention and the imprisonment of criminals

There are two prisons in Khanbaliq, known as Shīnbū and Mekbū. Torture in the second prison is severe and it is rare for anyone to get out alive. Torture in the first prison is mild and most of the prisoners emerge with life and limb unimpaired. Each of the prisons has a separate place for women. There are prisons in other cities as well, each with an audience hall in front [M 178] where the criminal is tried. The conviction is recorded in writing and presented to the Khaqan. An order is then issued and the criminal is put into bonds and imprisoned.

The Khanbaliq prisons have specific places for the 12 court prisons with an audience hall built in front of them. Three majestic commanders sit and write the name and description of the sinners who come before them, their crime and punishment, the court that they came from, and their age and the current date. They imprison them on specific levels according to their crime. One of the three sits on a seat of precedence and listens, the other two sit next to him and record the proceedings.

Every morning, after seeing to their affairs at the Khaqan's court, the prison commanders go to their audience halls to try the criminals and mete out punishment according to their crime. In the law, prison matters take precedence over the other affairs of the country. The Khaqan is ever cognizant of this. The prisoners' food and drink is provided from state revenue. They are fed once a day and are given leave to answer the call of nature twice a day.

Law of punishment and penalty

Lorenzo says that the punishments and judgments of the Cathayans are severe. According to the opinion of the Stoic sages, all crimes are equal in their opinion. But it is recorded in the *Qānūnnāme* that some of the criminals are subjected to torture and some are beaten with a rod. Contrary to the punishments that occur in other *vilayets*, some have heavy leaden collars put on their necks, some have boards like a coffin attached to their throats and heavy leaden fetters attached to their feet. [R 66a] Some are hung by the hair with their hands placed in fire or in red-hot ceramic pots. Some are bound tightly to boards with chains. Some are crucified. Some—those who have committed patricide or matricide—are placed inside a narrow box.

Then those who survive the binding and torture are brought to the marketplace or other public thoroughfare and displayed in public, with heavy bonds and boards on their necks or fetters on their hands and feet. Finally they are given 100 strokes on the backside—if it is a woman, her backside is covered with breeches for the beating. The criminal is fined several *mudds* of rice or wheat or millet according to his crime. If poor, he is made to work without pay for several years as an elephant-keeper or night-guard or the like. When his sentence is complete, he shows his record to the magistrates and he is given another 100 strokes and a document of release. Thus, after paying their penalty, the criminals are taken out in groups and publicly displayed before the guardhouses in the marketplaces according to their crime. They are happy at this, it being a sign of their release.

Law of execution and retaliation

The Khaqan holds court once a year to inspect criminals condemned to be executed. They are brought before him, each held by ten executioners, and named and they confess their crimes. Being of the belief that the Khaqan knows all, they cannot deny what they have done. The Khaqan too is silent, thinking that this belief is most beneficial in the matter of punishment. Especially in the twelve law-courts, their confessions are presented to the Khaqan every month for three years, after which he orders their execution. Then several thousand murderers with red marks on their heads emerge in groups from the Khaqan's palace and the death penalty is carried out in the execution grounds.

The head of an executed man is put in a box with his name and description and the court order and kept for 30 years, lest one of his relatives come and claim that he was executed unjustly. [M 179] Then by order of the Khaqan the box is found and the claimant silenced by the court order. After 30 years they remove the head and throw it into the sea.

If two people—men or women—have a fight and one of them dies, the other one along with ten individuals from his people or tribe are clapped in chains and put in prison. After exacting retaliation on murderers, the Khaqan goes to his temple, as explained above.

Law of theft

According to the *Atlas*, if a small amount is stolen, the thief gets as many strokes as he can bear, from 7 to 100. If he has stolen a horse or a valuable item, he is executed by a sword thust into his belly. If he restores nine times the value of what he stole, he is freed. If he is not able to do so, he is put to death. And if he commits the crime a second time, he cannot escape death.

If a son opposes his father, there is a special judge for this. After a long period of imprisonment his face is branded and he is released.

Law of prostitutes

There are separate districts for prostitutes in every city of Cathay. The smallest such district is 500 houses; districts of 1000 houses also exist. Most of the prostitutes are daughters of lords and notables who were found guilty of a crime, imprisoned and executed. Their sons become soldiers at a low rank. Their wives and daughters are banished to the brothels and they never see each other again. Their lineage dies out. One such crime is to establish a relationship with the prostitutes in these brothels.

If there is a drought and the prayer for rain becomes necessary, these prostitutes do it. The amount of rain and snow that falls in every province and the amount of crops sown are presented to the Khaqan. Wherever lack of rain is reported, a decree comes ordering the prayer for rain. Then these women bid farewell to their retainers and go to the temple of the province where the prayer is to be performed. If rain does not fall after the prayer for rain, the governor of the province puts to death as many of these women as he wishes, on the grounds that they were previously condemned to death; because when they are banished to the brothels, it is on condition that they will be put to death if their prayer for rain is not accepted.

In the temple, the women are given a meatless diet from the pious endowment for a time, based on the belief that they are purified by this. Then the temple musicians and singers take their seats and strike up various songs and melodies. The temple dancers dance to exhaustion, then get up in groups and weep and lament before the idol; to such a degree that (God) Who Gives Without Cause has mercy on those miserable ones [R 66b] and responds to their entreaties. At times so much rain falls that some places are flooded.

It is for this reason that the Cathay sages frighten the prostitutes with the death penalty, so that when they cry and lament and pray for rain out of fear for their lives, (God) The Living One, the Helper will overlook their sin committed out of ignorance and will grant rain. The Muslim jurists as well have permitted the prayer for rain of unbelievers, based on the abundance of divine mercy. So the sages have laid down the law in this manner. And sometimes if it does not rain and drought continues, they sacrifice some of those prostitutes.

Schools have been built for the children of those women in the brothels. They teach the girls musical instruments and singing and the boys dancing and various amusements and games.

These women go about in the markets with wine glasses, playing music and singing and dancing. Those who draw a salary from the Khaqan may not converse or drink with them; if caught, [M 180] they are put to death. Most of those who go to them are craftsmen and merchants, and quite a few dissipate their wealth in those brothels. If they become impoverished and go begging, as a rebuke no one gives them a penny and they starve to death. And when one of that sort dies, they throw him into the sea.

These prostitutes have strange and marvelous amusements. Their women are tightrope walkers. They put on strange garb and perform marvelous deeds.

Law of the couriers and post-stations

Yām is a courier. In order to send news and letters quickly, they keep horses and mules in posting stations called *yām-khāna*. There are splendid *yām-khānas* for travelers in the country of Cathay. Merchants who sojourn in that land with the title of *Elchi* ("official emissary") are housed in those staging posts. There are separate places to unload their carts of their materials and goods. Each apartment of the *yām-khāna* has splendid satin and velvet furnishings and other provisions ready at hand, and around it are small rooms for servants. A party of ten people get a daily ration of one sheep, one goose, two chickens, two *batmans* of wheat, one bowl of rice, two plates of sweetmeats, one cannister of honey, various other vegetables, and several servants.

Law of travelers and merchants

Those who go by land to the kingdom of Cathay go in the status of Elchi. Otherwise they do not allow merchants and others to enter. When merchants approach the frontier, they are registered in parties of ten and a portion of their merchandise is sent as gifts to the Khaqan. The Khaqan invariably takes the horses. Their merchandise is recorded in a register and every party of ten is assigned a place. They send on ten or twenty people to Khanbaliq. The rest settle at the post-station of Kenjū, a city ten stages distant. Living expenses are provided by the Khaqan. Every day a party of five gets a ration of one sheep, a quarter bushel of rice and other provisions. Whatever of their merchandise is suitable to the Khaqan is purchased, and even the gifts are paid for in full. They are cared for as long as they stay, and when they leave they get favors from the Khaqan and all their belongings are returned, only forty dirhams of silver apiece are taken as tax.

If a merchant comes in the status of Elchi with a caravan of 100 or 200 men, carts, servants and other provisions are appointed for them. Because of this, it is related that every day 1000 carts of silk alone, irrespective of other goods, come from China to Khanbaliq. The merchants mainly bring diamonds, wool, scarlet broadcloth, jade, coral, lions, leopards, elephants, and packhorses.

For Arabian horses, men are assigned as escort out of respect from as much as 100 stages distant and they are sent to the Khaqan with their owner. Twelve servants are assigned to each horse, six to run alongside with colored lanterns attached to spears, three to run in front and three behind.

A lion is valued as much as ten horses. A lynx is worth half a lion. One who makes a gift of a lion is given 30 cases of goods. For satin, velvet and other textiles gifts are given according to this rate. Each of the men, in addition to the price of the gift, is given one load of silk stuffs and clothed with three robes of honor, one on top of each other, including footwear and boots.

Transgressions of travelers

There is no punishment for travelers other than imprisonment. If a traveler commits murder or another crime, (because) he owns nothing of the goods of this world they record it in their registers and put him in prison. [M 181] In their mind, travelers are considered to be desert dwellers and country bumpkins (who) unlike their own people have no desire for etiquette or honor and no respect for their laws.

Law of the Elchi

In Cathay every ethnic group has its own interpreters. Seven esteemed commanders are appointed specifically for the business of the Muslims. When it is time to license Elchis, the interpreter of every people comes before the Khaqan [R 67a] and presents himself.

[Continuation of the Persian envoy's report]

Khwāja Ghiyāth (al-dīn) says that when ten stages remained to the pass of Sekjū, which is the frontier of Cathay, a detachment of Cathayans received them and recorded their Elchis and their number. In a meadow where there was a raised platform they set up a canopy and gave a feast. They brought dried and fresh fruit on china plates. The accompanying merchants were recorded as the Elchi's servants. Then the governor brought them to his own yurt.

The yurt among the Cathayans is a group of tents set up tent-rope to tent-rope in the shape of a square with a path on each of the four sides. The middle is a wide, open space. In the center, a two-columned marquee is erected on a high platform, its skirts raised, and chairs are placed on the right and left.

The Elchis were seated on the left, because the Cathay commanders esteem the left more than the right. They placed food before everyone. Then musicians and dancers come and beautiful slave-boys with pearls in their ears stood like girls. Around the gate, armor-clad soldiers with spears in their hands drew up in rows.

Then the drinking began. The governor of the assembly went around with a winebowl in his hand and flower-sprigs inside a box. To whomever he gave the winebowl, he placed a sprig on his head. The dancers made strange figures out of cardboard and performed dances in the Cathay manner. Some slaveboys went around with plates of hazelnuts, grapes, peeled chestnuts, onions, cured garlics and sliced watermelons and muskmelons, and offered these bowls to whomever the winebowl was given. That day the drinking party lasted from morning to evening.

The next day they set out again and, cutting stages (i.e., traveling quickly, not stopping at staging-posts, etc.) in the desert, came to the fortress of Qarāwul, and from there to Sekjū where they settled in a post-station at the city gate. The Cathayans provided for their needs according to their number, bringing one silk nightgown and one servant for each one of them. They proceeded in this manner from here until the capital. At each post 150 horses and donkeys and carts are ready at hand for the Elchis. The servants who take care of the horses are called *qū*; those who look after the donkeys, *elūfū*; and the cart-drivers, *jīnfū*. Every cart has 12 attendants who haul it with ropes on their shoulders. Young boys wearing pearl earrings wind up their forelocks on their heads and bring the Elchis from one post-station to the next.

There are great feasts in every city and special feast-houses for the Elchis called *rūsūn*. The front of every *rūsūn* is oriented toward the capital. A throne is set up and a curtain hung. A person stands next to the throne and the Elchis sit on felts that are spread out in front of it. The person standing by the throne calls out three times in the Cathay language and the *eji*s, who are the frontier governors, kowtow. The Elchis also [M 182] are obliged to prostrate themselves. The gifts brought by the Elchis were sent to the city of Qamchū, as mentioned previously.

They arrived in Khanbaligh and came before the Khaqan's throne. At that time, 700 fugitive slaves and criminals had been brought, some with cangues (lit. two forked branches) on their necks, some with their heads passing through long pierced boards. The Khaqan sent some of these to prison and ordered a group of them to be executed.

Now one of the commanders knelt and read the sheet containing the information about the Elchis. Interpreters came and told them to kowtow to the Khaqan by putting their heads on the ground three times. The Elchis bowed but did not place their foreheads on the ground. Then they took their letters in their yellow satin purses and raised them up, and they were taken and given to the Khaqan. The Elchis were then brought to the former open space and a table and chair were placed before everyone. After dining they proceeded to the post-station.

When they were about to depart, the *shiqāwul* (minister of guests) came and summoned them, saying, "The Khaqan will give you gifts today." They proceeded to the capital and saw that the Khaqan was seated with *shīr*s—i.e., small tables or footed trays—in front of him. He made a sign and they put these before the Elchis. On each *shīr* was a little silver, 20 or 30 Cathayan satins, and tin (?—*kalāy ve cāv*) separately appointed for themselves and for their wives. They took these and departed. The gold, silver, and other things bestowed by the Khaqan are of the best quality.

Religion

The Cathayans say that Shakyamuni [R 67b] was a prophet who called the people to the truth. They relate many miraculous stories about him. They say he lived 3,000 years before the *hijra*. They believe, in accordance with what is written in their books, that the Khaqans of old were mainly worshippers of God and wise sages, and that idol-worship is a practice of the ignorant.

They call Muḥammad—peace be upon him—*shīn-chīn*, meaning "best of the people." They have an affection for Islam and for Muslims. The call to prayer is made five times a day in their lands. The Muslims wear turbans and perform the congregational prayer.

One day the commanders assembled and said: "Thousands of Muslim households are mixed with ours. Let us eradicate them. They do not give wealth such that they would be considered beneficial." The Khaqan said: "Our forefathers never acted in this way, so how can we? We judge them according to their manifest behavior. At present, there is no manifest denier among them, and what have we to do with their hidden thoughts?"

The Cathayans are broad-minded in matters of religion and know nothing of fanaticism and hostility. Whatever their religious belief may be, their effort in this world is to abide by the law. Thus some of the people are idol-worshippers, some worshippers of the sun or the moon or cows. Christians and Jews also came to those lands in olden times and many groups settled there. Houses of worship were constructed by the Khaqan for each group. In Khanbaligh alone there are four congregational mosques for Muslims.

Lorenzo says that the Khan of Cathay is a heterodox Christian. But the Muslims deny it and claim that he is an idol-worshipper. Most of the people are Christian and Nestorian. A group of them are idol-worshippers. They believe in two principles. One is known as Creator of Souls, whom they worship by burning incense [M 183] and from whom they seek intelligence and understanding. The other is known as *Nāgāy* or Creator of Bodies, whom they worship by making silver and gold statues and from whom they seek matters connected to wealth—whatever it is they want, they sacrifice something of that kind.

According to the *Atlas*, most of the Cathayans accepted the law of Muḥammad in the Christian year 1246. But according to the account in the *Qānūnnāme*, Dīn-Ṭāy Khan and the people of the palace converted to Islam around the year 900 (1495). All of the ignorant Cathayans worship the Khaqan, because it was rumored that he had become a Muslim, and so the majority of them converted to Islam also. The ignorant Cathayans believe that the Khaqan is divine. And they claim that there are 300 gods who do not appear to them; only one of them appears, and he is the Khaqan. They say that the one true deity created these 300. They believe that all crimes are known to the Khaqan and they cannot deny them. The Khaqan too is silent, thinking that this belief is most beneficial in the matter of punishment.

A great temple was built, dating from the time of Shakyamuni, where the Cathayans from all around perform great austerities and which they visit and circumambulate. Near the temple is a tall mountain, in the middle of which a large stake has been driven with a rope tied to it. A man whose austerities have reached perfection binds that rope to his waist and, holding onto it, climbs up the stake and sits on it. He unties the rope and lets it fall. If he is able to sit 40 days and nights without falling asleep or growing afraid, he acquires the power of flight from his great austerity. He flies to the top of the mountain and from there goes wherever he wishes. But if he falls asleep or his eyes grows dim and he falls, he dies.

Ascetics of this kind are reduced to skin and bones. Their food and drink consist of certain medicinal herbs which they boil in water and consume. They also learn to hold their breath—some of them breathe once in three days and three nights.

Knowledge and skill

Many Cathayans are of sophisticated mind. Astrology has a great importance among them. Four skilled practitioners of this science live with their families in separate quarters in the first grades of the Khaqan's palace. They enjoy great respect and do not come outside. All their provisions are provided by the Khaqan. Each one produces a calendar on New Year's Day and presents it to the Khaqan. He in turn gives the four calendars to the other scholars, ordering them to produce a common text based on those points that they have in agreement. The scribes then make copies, ten at a time, by placing ten pages of very thin silk paper on top of one another. With this method, however, each copy is only written on one side of the page. From 100 copyists they get 1,000 calendars. It is customary to give one calendar to those who govern 10 people. [R 68a] Thus the calendar spreads to the various climes. The Cathayans closely observe its times and hours; draw lessons from it for advice and public punishment; and find matter in it for poems, riddles and puzzles.

The Cathayans have a special writing system with 44 letters. Some of them are shared with the people of Japan and China. They use a special language in their worship.

Craftsmanship

The Cathayans are generally capable and well versed in every craft. Especially in painting they are equal to the Chinese. The ceilings and walls of the houses of the notables are covered with paintings.

They [M 184] make cloth with millstones. They are also skilled in the medical arts. They consider all people other than themselves to be blind.

Warfare

The Cathayans are adept in the business of war. From the frontiers to the capital there are mileposts and sentries at every stage. If an enemy appears, they signal with smoke by day and fire by night, thus notifying in a single day all who are within a one-month's journey. If the enemy is approaching from the east they signal with one fire; if from the north, two fires; from the south, three; and from the west, four. And they always have sentries guarding the mileposts and bells that are suspended and rung continuously. The ladders of those mileposts are made of rope and are pulled up if an enemy comes. Water and provisions and war material are at the ready. They shoot arrows rather than bullets, and the arrowheads are tipped with poison. All Cathayans great and small are skilled in the manufacture of gunpowder and fireworks are widespread. They all know how to shoot cannon and guns.

Mores and customs

The Cathayans never seek to harm others out of religious zealotry. They only strive to maintain control of their territories according to wisdom. Usually when they go to any country but their own they consider the rest of the world to be empty and a wilderness. That is because most of the people who come to the Khaqan under peace terms are desert dwellers. Any tribe that comes to their country to settle, their wealth is not confiscated.

In their laws, it is a great sin to drink to the point of losing consciousness. They sip one or two cups of wine during meals.

Everyone wears sables or other furs according to their rank. And they wear long clothes, because Cathay is part of the clime of Tataristan.

The people are mainly of middling height, with broad and fleshy faces, sunken eyes, and little hair on their bodies—other than beards, which most of them shave off. Their bodies are strong, their intellects ample. The women are tall and graceful and have ruddy cheeks and narcissus eyes. None of them is ever swarthy, nor are there blue-eyed blonds among them. The men bear up under hunger and fatigue. When thirsty, they pierce the veins of their mounts and drink the blood.

The common people are mostly dirty and uncouth, knowing nothing of God and divine law. They rely on their weapons and their strength. Most of them are desert-dwellers and nomadize with tents. Some of them live on carts, to which they attach sails and travel over the plains with great speed as on a ship, navigating by the northstar.

Coinage is rare among the Cathayans; they exchange goods by barter.

They eat foul foods, especially half-cooked meat and cheese, and drink mares' milk. From this milk they make a kind of intoxicant called *ḳımız* (koumiss), which is like white wine but very flavorless, and from millet they prepare *boza* (millet beer) They are a very dirty people and know nothing about washing hands and clothes.

In the cities of Cathay, playing ball games and manipulating puppets are esteemed skills. The prostitutes mostly learn acrobatic tricks.

This is their custom in laying out their dead: for the notables, there is a mountain where they are borne and placed into an assigned cave. Their wives also have a special cave, and their horse is let loose on that mountain to graze by itself and no one bothers it. That grotto is very spacious. Many concubines and palace servants dwell there with five years' worth of provisions; when their provisions run out, they also die.

This country of Cathay, being cold, has furnaces like those of Germany, [M 185] using red-hot stones.

The diet consists mainly of rice and millet. In times of drought, everyone by decree keeps a sufficient measure for himself from his stocks and sells the remainder to the people. If that measure is not adequate, a daily ration is allotted to the poor from government storehouses. When those stores too are depleted, orders go to the country that anyone who brings 100 *mudd* of grain be given command over ten men, and by this method unlimited grain comes from the surrounding districts.

If fire breaks out in a city, they have three measures to fight it: 1) Watchmen are at the ready day and night who put out a fire when they see one. 2) Every night a large amount of water is made ready in every house. Anyone who neglects this measure, and whose house is burning, is thrown into the fire along with his extended family. 3) To prevent fire from spreading, they build town quarters and markets separated from one another.

In their mills, the quern-stone is placed underneath, not on top. Anyone who places it on top is put to death for violating the law.

Cities

The *Qānūnnāma* and Khwāja Ghiyāth (al-dīn)'s *Rūznāme* give a twelve-fold division of the cities of Cathay—in the *Qānūn*[*nāme*] it includes the cities of China. Because this division is contrary to the rule of the science (of geography—i.e., has no geographical basis), I did not consider it appropriate to include here.

According to the divisions of Mercator and Philippus mentioned above, Tataristan includes Cathay and other regions. I will first mention the cities of the country of Cathay whose borders have been outlined here, and then [R 68b] will add the other countries, because the corresponding map of the *Atlas* includes them.

The country of Cathay mainly consists of flat plains. Mountains are few. There are lakes and rivers in various places.

The capital is a large city named Khanbaliq. This is a word in the Ghūrid (?) language that the Cathayans use to mean "capital city". The old name for it was Īddūm Sariqa. It is located at 154 degrees and 5 minutes longitude and 56½ latitude. They say that this city is two, one old and one new. The new one was built by Kublai Khan son of Tolui Khan of the Chingisid dynasty. According to the *Ḥabīb al-siyar*, when Kublai succeeded his brother Mengu Khan in Qaraqorum in 658 (1260) and became Khaqan he sent armies to China several times on raiding expeditions. After this he commanded the city of Khanbaliq to be built on the Pūlīs River near the city of Jengdū, which was the seat of authority of the khans of Cathay. It was laid out as a square, four parasangs on each side. He ordered a canal to be dug from Khanbaliq to the canal that flows to the city of Zeytūn, the entrepot of China and India, 40 stages away, so that ships could come from the sea of India and China to the middle of the city. Thus Khanbaliq became very prosperous in a short time and was made the capital of Cathay. Kublai died in 693 (1293). It is at present the capital of the Great Khaqan.

At the end of the *Ḥabīb al-siyar* it says that in Khanbaliq the abovementioned canal is 30 cubits wide. The road that goes along its bank extends to the capital of Māchīn and is completely paved with stone for 40 stages. Willows and other shade trees are planted on both sides, so travelers go in the shade. And on both sides there is a continuous row of villages, inns and idol temples. This road reaches the frontiers of China in this manner. [M 186]

The Khaqan's palace, called Kök-ṭāq ("Blue Vault"), is in the middle of the city. The Khaqan has a pleasure-dome there. The rivers of this place freeze in winter. All fruits are found here except grapes, which are scarce.

It is recorded in the *Qānūnnāme* that a river on the north side reaches the city. The Cathayans dug that river from its deepest part and excavated a channel that powers 30 mills. It goes around the seven fortresses that comprise the Khaqan's palace and reaches their houses. After leaving the city and irrigating the surrounding orchards and gardens, it flows on.

In this city they burn a type of black stone instead of wood.

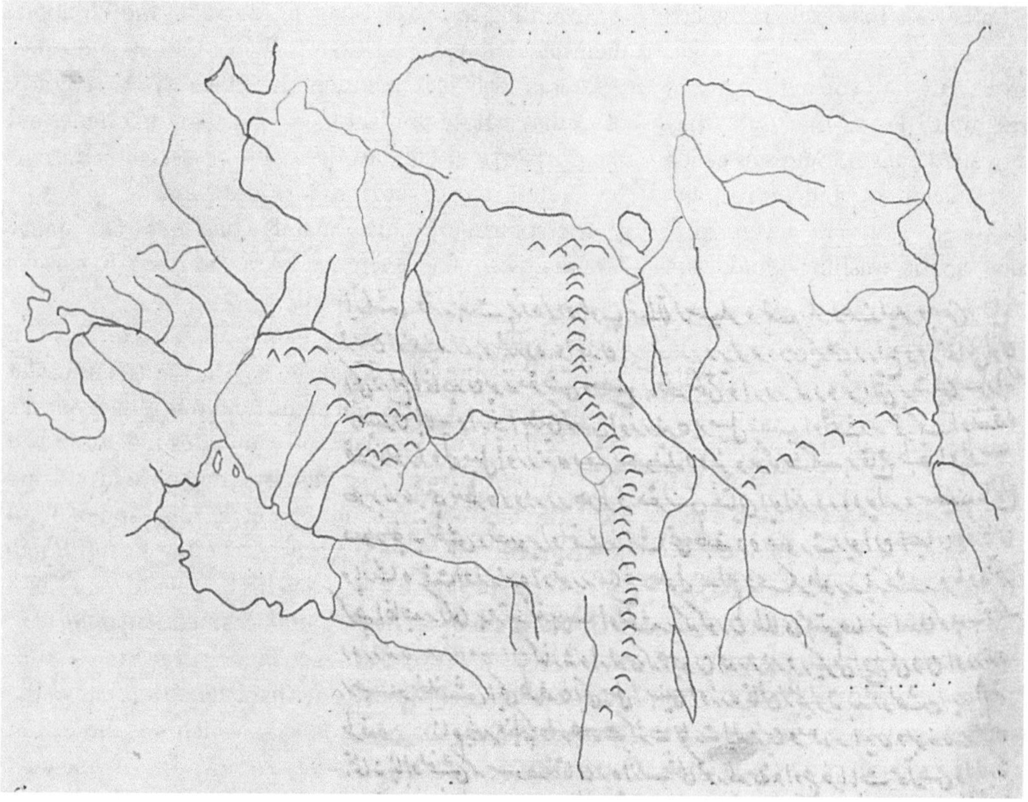

PLATE 32
Map of Cathay, R 69a; cf. M 165/166 (Appendix, fig. 71–72)
TOPKAPI PALACE LIBRARY, REVAN 1624, WITH PERMISSION

It says in the *Atlas* that the circumference of Khanbaliq, along the bank of the Pūlīs-Sānkūs River, is 24 Italian miles. Some books say 25 German miles. It has 12 gates, and there is a suburb in front of each of them. It is an entrepot for trade in jewels, gold and silver. Since that place is a transit point of the Kalmyks, the Cathayans made its buildings into markets. It was rebuilt in 840 (1436). Every ethnic group has its own town quarters and its own insignia. Four congregational mosques have been built for the Muslims alone.

It is related that various goods come to this city from the surrounding areas. Every day as many as 1000 loads of silk come from China alone. The Chingisids brought most of the wealth of Asia that they plundered to Khanbaliq. It piled up there and the city became very wealthy.

The city itself uplifts the spirit with its wide spaces and delightful climate. And it has a magnificent bridge that has no equal in the world.

Khwāja Ghiyāth (al-dīn) says that Khanbaliq is 99 *yām* (post-horse stages) from Sekjū, which is the frontier of Cathay. Each *yām* is close to a city. Between every two *yām* are several *qarghū*, *key*, and *fū*. *Qarghū* refers to a watchtower that is built to a height of 60 cubits and always has two sentries sitting at the top. The towers are just within sight of each other. If a crisis develops—e.g., the appearance of foreign troops—they immediately light a fire on the *qarghū* by night or make smoke signals by day and communicate to each other. So the news reaches the capital, a three-months journey away, in a single day and night. After the general situation has been made known in this way, the details are conveyed in letters by the people of the *key* and *fū*. *Fū* is a hamlet of several houses where a group of people lives. They are charged with conveying any letter or report that comes to them to the next *fū*. *Key* is similar. The Distance between one *key* or *fū* and the next is 10 MRH, 16 MRH constituting one parasang. The people of *key* and *fū* live in their hamlets and sow and harvest crops. Each one is like a village.

These are the cities mentioned in the report of Khwāja Ghiyāth (al-dīn):

They departed Samarqand on 10 Ṣafar (24 August) [R 69a] [Plate 32] and arrived in Chalgā-i Yulduz on 6 Jumādī al-ūlā (16 November) while the sun was in Cancer. In the wilderness the water had frozen to two finger-lengths. Rain and dew fell most of the time.

They reached the city of Ṭarqān in the last part of Jumādī al-ākhira. The people are mainly idol-worshippers. They have large temples. On a platform of one of them was placed a large idol that they said was a representation of Shakyamuni.

They reached Qarakhoja on 5 Rajab and halted in Qāmal on 21 Rajab. There was a lofty and splendid mosque of Fakhr al-Dīn in this city. A large temple was built nearby [M 187] with many large and small idols inside it and around it. At the gate was placed the image of two demons attacking each other. This city had a governor named Mengli Timur.

Departing from there, they went 25 more stages in the wilderness. In these plains were places with yaks.

Cutting stages, they reached Qarāwul, a very strong fortress in the mountains with only one road leading to it.

Then they came to the large city of Sekjū on the frontier of Cathay. It has a solid wall in the shape of a square. Inside are broad markets and roads 50 cubits wide that were watered and swept. Swineherds went about in most of the houses and mutton and pork were hung in the butcher shops and sold. There are four splendid marketplaces with, at the entrances, wooden vaults and gates decorated at the top with carved and painted ceilings.

In the moat around the city wall are covered towers, built every 20 feet, and gates at the four corners placed opposite one another. Every gate is a great distance from the middle of the city, but because of the straightness of the roads, they are all visible from each other. And atop each gate is a two-tiered pavilion.

There are numerous temples here, each one covering an area of 10 *jarīb*s. The idols are carved out of wood and the floors are paved with wooden boards. In those temples, handsome boys proclaim revelries.

According to the *Taqwīm*, Sūkjū (sic) is a city about the size of Ḥimṣ on a flat plain at 107 degrees longitude and 40 degrees latitude. It has streams that come from the surrounding mountains and vegetable gardens. There are four stages between this place and Khānjū.

At ten *yām* from Sekjū they came to the city of Qamjū, which is bigger than Sekjū. The highest-ranking of the frontier governors, known as *dānkji*, lives here. There is a temple here with an idol inside that is 50 cubits tall. Its foot is 9 cubits and the circumference of its head is 21 cubits. Small idols are arranged on top of its head. The ministers of the idols in the temple manipulate it with great skill so that it moves as if it were alive. Other images are inscribed on the temple walls. [R 69b] The temple was hung with gold brocade curtains and ornamented with gilded thrones and candlesticks and chairs and tiles.

One of the wonders of this city is the so-called Wheel of Heaven, a finely carved wooden octagonal pavilion, 20 cubits around and 12 cubits high, that rotates. At its base is a solidly built subterranean chamber. A pole going from the ground to the summit is fixed atop an iron base. As that pole is turned, the entire pavilion gently rotates with it. This pavilion has several stories, each with several balconies and finely carved Cathayan vaults open to every side. Walls and ceilings are covered with various and marvelous paintings. On the lower story is a depiction of the court of the notables and men of state, the emperor seated on his throne, eunuchs and soldiers standing to the right and left. While the pavilion is made of wood, it is gilded to such a degree that the observer thinks it is solid gold.

On the 4th of Shawwāl they reached the shore of the Qarāmūrān, [M 188] a river as large as the Jayḥūn. It was crossed by a pontoon bridge of 23 boats, fastened on either side by a thick chain attached to an iron stake. On the other side was a large city named ..... [*lacuna in text*], very prosperous and beautiful. A large idol-temple was built here whose like was not seen on the frontier of Cathay. It had three prostitute quarters with many beautiful concubines dwelling in each one.

Traveling on from there, they crossed several rivers and on the 20th of Dhū'l-qaʿda reached the city of Ṣadīn-fū. It was very magnificent. In its great temple they saw a large gilded idol of molten bronze, 50 cubits high, known as the Idol of a Thousand Hands. Every limb had a hand and in the palm of every hand was an eye. It is one of the famous idols of the people of Cathay. It is set on two stone platforms in a lofty multi-storey building of novel construction, with porches and balconies all around. The first storey comes up to the idol's heel, the second only to its knee, the third to just above the knee. Over its head is a splendidly carved vault. Visitors were strolling about the several stories of the temple, both inside and outside. It is well known that each of the idol's feet is ten cubits high, and that 100,000 donkey-loads of rice were expended for all of it. Aside from the great idol, there are numerous small idols arranged in rows in that temple, each one painted in many colors. On the walls are painted mountains and various types of trees and animals. The painters have attained a peak of perfection, creating works that leave the viewers wonderstruck. There was also a Wheel of Heaven in this city, even more magnificent than the pavilion in Qamjū.

The report of Khwāja Ghiyāth (al-dīn) ends here.

These are the cities of Cathay mentioned in the *Qānūn-nāme*:

- Dī-Teng: a city near Khanbaliq. It has many silver mines and abundant cloth and other goods. As in Khanbaliq, here too they burn a black stone instead of firewood.
- Sālār-fū: a city considered as belonging to Hīzā, which is the fourth part of Cathay. It has cloves, pepper and bedda nut, but salt is a rare and valuable commodity.
- Wānsī and Wāndūn: two great cities on the shore of the Eastern Sea in the twelfth section at the end of the earth. They are as big as Khanbaliq. Their main product is sugar which they sell at three dirhams per Iraqi *raṭl*.
- Genjān-fū: a large city. As many as 30,000 Muslims live there.

- Beshbaliq: one of the cities of Cathay.
- Almaliq: a city 20 stages from Tengī.
- Kenjek: a city one parasang from Tengī.
- Qāyendū: a famous city. Here they make an extremely thin veil from tree bark. Philippus writes that the country of Cathay was famous for this product in olden times, but Lorenzo writes that it is a cloth more delicate than silk.

He (Lorenzo) mentions the cities of Bayqūndil, Pāzānfū, Sīndīfū, Akhīlūfū and Qāchānfū in the country of Cathay as places with many perfumes. Then he lists cities named Rūsīn, Tāqān and Pānqūnīā. This Pānqūnīā is a hunting ground, so it is mainly the Khaqan who comes and stays there. Then Kuwudāghāy, Pāīn, Jīndū, Tāyfū, Kūnzāfū, and Ūghūnt, which is the main city of this country. Its women are beautiful because of its fine climate. Every year they collect beautiful girls from this country for the Khaqan. The best weapons are manufactured in the city of Sendīchīn. The workshop of the Khaqan's armory is here. [M 189]

- Tāīnfū: one of the countries of Cathay and an extensive and fertile province. Philippus says it is famous among the Tatars as a place where weapons are produced. Even Pliny, among the ancients, preferred the iron of this province over the iron of other places.
- Sīānghānūr: since it is a good hunting ground, the Khaqan lives there approximately two months of every year. From there in the northwest direction the city of Sāīndū is also a good hunting ground, but the cold in winter is extreme.
- Pālīsānghū, Lūrīn, Yūnes, Qalmanghū, Rfālqān, Bāydā, Arqānārā, and Bārāt: they are on the shore of a lake.
- Dānghū and Ghūnzā: at present orchards are planted there; in other places of Cathay, there are no orchards.
- Beyond Khanbaliq, on the shore of the ocean, is the city of Qāzārdīn. Many ships enter the river and come before it. The Khaqan mainly hunts here.
- North of Qāzārdīn are Setīn-kūy, Chānqū, Qārlī, Bāsqūl, Qāydū, and Chīūrzā, each of them the main city of a province.

This is the full amount recorded by those authors concerning the cities of Cathay.

Now begins the description of some of the countries of Tataristan in the vicinity of Cathay.

Qalmāq (Kalmyks): a country north of Cathay. Its people are a tribe of the infidel Tatars. They mostly wear sable furs. In winter they turn the fur to the inside, in summer they wear it on the outside. Their bedding and pillows are sheepskins. [R 70a] They also spin sheep's wool for their shirts. Among them, a needle is highly prized and valuable. They will buy one needle for one sheep. Their arrowheads are made of wood. Even their horseshoes and nails and cooking pots are made of wood, because there are no iron mines in their country. They put meat into a wooden pot and pour water over it, then put in red-hot stones. The meat cooks from the heat. That dish is called *cūş* in Turkish.

The number of horses, camels, sheep and other cattle among them is beyond reckoning. They mainly eat antelope and wild horse and camel. Since their land is very flat and consists of fine steppeland, they get 140 bushels of wheat for every bushel they plant. The Cathayans only covet the lands of these people. Every two or three years they clash with the Kalmyks, seize a part of their lands and, after a while, make cities and fortresses. The thirteenth part of the realm of Cathay is the lands they have seized from the Kalmyks.

Qaraqorum: a large city in the Kalmyk Steppe. It is under the rule of the Khaqan of Cathay. Most of his army lives here. Its people are wonderful craftsmen. They weave fine fabrics. There are many lovely boys and girls. But the people are star-worshippers. Mines of tin, silver and copper are plentiful near here. Rhubarb can be gotten in the nearby mountains at no cost. The climate is pleasant. It is an ancient city close to Cathay. There is a temple here whose walls are made of lodestone. It is recorded in some books that an iron idol stands in the middle of it.

Philippus says that this country extends from Cathay up to the Northern Ocean and the Strait of Anian and comprises the areas known as Old Tataristan. The name Tatar first appeared there. There are many countries and hordes in its steppe, but their names are not known with certainty.

Lorenzo says that one of the countries of Tataristan is Qarāzān. Its people live in the mountains and they do not obey the Khaqan. [M 190] They are a very violent and savage tribe. They always carry poison with them in order to not fall alive into the hands of the enemy, and they never accept peace terms.

Qarāyān is in the middle of a large salt lake. Their women gild their teeth, and when they give birth they force their menfolk to lie in the bed with them. Gold is plentiful in this country. The people are a dim-witted and savage tribe.

Tanduq: a country north of Cathay. It had its own ruler before the Tatars invaded it. Nestorian Christians are numerous among its people. Philippus says that the country of Tanduq was known among the ancients as the sultanate of Prester John. After that they mistakenly attached this name to the ruler of Ethiopia.

Tangut: Philippus records this as being a country also. He mentions that the craft of printing was invented here 1000 years ago.

Ania: according to Lorenzo, a large country northeast of Cathay. Between them is a strait. Its main city is named Anian. However, the contemporary author of the *Atlas* and others did not record this country. They also mention an island in that direction named Līmānqū. Its people are brave and savage cannibals. It has its own languages and its lands are rich with gold. Its main city is called Dūn. The Tatars tried to conquer it many times but were not able to. In this place there is a kind of stone known as *ghārūf*. Lorenzo recorded that it stems the flow of blood. However, there is no trace of this island on the new maps.

After these places, the Tatars have numerous hordes in the steppes, as many as there are in empty Tataristan. For a six-month distance as far as the Kipchak Steppe—i.e., the length and breadth of the desert of Tibet and Cathay—there are dwellings of Chingisids and other Tatar tribes [R 70b] who nomadize with tents. Since there are rivers, lakes and reedbeds here and there, they generally rest in one place during the summer. In winter all the rivers freeze and they go wherever they want. They have large dogs pull carts over the ice. And there are caravanserais on the roads from stage to stage. Merchants travel there to trade in falcons, sable and leather.

[Resumption of description of Cathay]

Buildings

There are many important buildings in the clime of Cathay, especially the large and splendid palace of the Khaqan in Khanbaliq; and in the city of Qīāndū (Shangdu) the splendid gold-embellished palace that Kublai Khan built out of marble and cut stone, over whose walls a bowshot cannot reach and whose gates are made of Cathay steel.

It is recorded in the *Qānūnnāme* that when the Cathayans plan to take over a place in the land of the Kalmyks, they load innumerable carts with pickaxes, shovels and other construction tools and depart with soldiers. When they come to the place they want, they build a city or fortress out of clay in two or three days; dig moats around it and put up gates; erect a marketplace, audience hall, idol-temple and posting station; and fill it with people.

If a military force comes to a staging post, as a rule they surround it with fortifications or trenches. At the very least they make a kind of wall by piling up stones. They do not leave a staging post unmarked. It is their job to make cities in forbidding valleys and to dig trenches in places as far away as one or two months' journey. Along the seacoast they have made a city or fortress at every staging post for a distance of a six months' journey.

Rivers and mountains

Most of the territory of Cathay is flat and well watered. On the west side, the Imaus Mountains extend from south to north. Other mountains [M 191] and forests are few. According to Lorenzo, the important rivers (and lakes) are the following:

- Qārūmūrān River: a broad and deep river, like a lake. It is too broad to have a bridge. It empties into the ocean.
- Qarāyān Lake: a large salt lake abounding in oysters and fish. It is also called Qarāīm Lake. It is in the country of Qarāīm. Its circumference is 100 miles.
- Qūyāntū River: a broad and deep river. Its breadth is as much as six, eight or ten miles. Its length is almost 100 stages. Fish are abundant.
- Pūlī-sānghīs: a great river. It passes by Khanbaliq and empties into the ocean. It is navigable by laden ships.
- Qānpīlū Lake: a large lake in the country of Qānpīlū. Pearls are very plentiful here, but hunting for them is forbidden except by permission of the Khaqan.

Climate

The weather of this clime is astoundingly immoderate. On summer days dreadful thunderclaps occur that cause many to die from fright. Sometimes a great heat wave strikes; then there is a cold snap and it snows. The wind blows with such vehemence that it stops men on horseback and knocks them over and uproots trees. In summer there are frequent rainstorms, but the ground hardly gets wet. It never rains in winter, but plague never strikes and foodstuffs are cheap.

Animals

In the clime of Cathay there are many cattle, goats, pigs and horses. The Khaqan alone has 10,000 milch mares whose milk they drink. They keep 10,000 hunting falcons and hunt francolin and crane. In the Khaqan's palace they keep lions, tigers, leopards, cheetahs, lynx, elephants and other beasts of prey; and especially the Tibetan dogs, as tough and fearsome as lions and let loose in battle.

Caged birds of strange colors include red-footed and red-billed Balkhi crows, ducks, parrots, turtledoves, and nightingales. Their cages are covered with brocade.

Yaks are abundant in the Cathay desert. The yak bull is so bulky and powerful that it sometimes snatches a horseman from the saddle and holds him for a time on its horn. The tail of this animal is esteemed. They attach it to their horse's neck. It is white and beautiful.

Lorenzo says that in Pānqūnīā there is a creature called *mārīn-kūzākh*. Its face is like that of a girl, its body that of a lion, its tail is like the two-forked tail of a scorpion. As long as it is left alone it is harmless.

It is recorded in the *Atlas* that in the city of Kuve-līnfū there is a type of chicken whose feathers are like the fur of a black cat. But it lays a huge number of eggs.

Plants and products

There are gold and silver mines in these regions. There is a mine for a type of black stone that they burn in place of wood. The land is fertile. Various grains and cereals are plentiful, especially rice. Abundant products include silk, ginger, cinnamon, pepper, cloves, rhubarb, sugar, musk, and pine resin. There is a root here called *khīāqbāy* [R 71a] that is a remedy for various ills. When the Cathay notables go anywhere they take it with them. It is forbidden to sell it to the people of foreign lands lest that cause a slump in the market for rhubarb and other drugs. In these lands, only grapes are rare, other fruits are plentiful.

The *Qānūnnāme* cites travelers' reports as follows: "We went a distance of 100 days into Cathay territory [M 192] and saw mountain and plain completely turned into arable land. They even brought soil on carts to stony and sandy places and sowed crops there as well. For this reason the animals here are kept in the houses. Trees had been planted in a continuous row on both sides of the roads. Three things in that country have no equal in the world: the jujube, which is as large as an apple and is delicious; and two kinds of flower, the poppy and the water lily, each of which is about the size of a small dish and has 100 petals and marvelous color."

The author of the *Atlas* reports another marvel:[30] in the country of the Zavolhen Tatars there is a kind of seed that resembles a melon seed but is smaller and more elongated. When it is sown, a plant grows called *boranetz* (text: *būrāns*) meaning "lamb." It looks exactly like a lamb in shape and size. It is three feet tall and its feet, head, ears and hooves are all those of a lamb. Only it grows a special kind of hair instead of horns. Its skin is very delicate and fine: the notables of that region make caps out of it that they send each other as gifts; they are highly esteemed. And they say that the inside of this plant is like crab meat: when it is cut, it exudes very delicious blood. If the root of the plant is extracted from the earth it reaches to the lamb's belly. This lamb grazes on the surrounding grass and grows fat and handsome. When the grass runs out it wilts and falls down and dies. The same if they intentionally pluck out the grass—it cannot then live. This lamb is attacked by wolves and other beasts.

Phenomena like this are deemed strange since they are marvels of the natural world. But the spontaneous generation of an animal from a plant is permissible and real.

Roads

This clime has three roads from the western direction: the Kashmir road, the Khotan road, and the Moghulistan road. The Khotan and Kashmir roads are thickly settled and water and fodder are plentiful. Only at the other end water and fodder are scarce for a 15-day stretch of road. But at every stage one can get water if one digs a well up to a man's height—in some places it is enough to dig just one cubit.

As for the Moghul and Chagatay road, it is broad and pleasant. Tīmūr intended to go via that road (to conquer Cathay in 1405). He ordered a fortress built at every staging post and thousands of troops to be garrisoned in each one so that they could sow and reap grain. But he was not vouchsafed to undertake that campaign, and died on the way.

From the frontier of Cathay to the bank of the Oxus is a three-month journey. A staging post is assigned for every day's journey as far as the Sekjū Pass. From there it is another three months to the eastern extremity of Cathay. At every stage there are numerous sentries and a posting station, as mentioned above. In the interior of the country travelers may spend the night on the plain. Every stage is a town or a fortress, and caravanserais have been built by the Khaqan that provide abundant comforts to those who sojourn there.

From Khanbaliq, which is the capital, to Khansā is a 40-days' journey whether by land or by sea. Qamjū is also forty stages.

Kings

Originally the clime of Cathay was ruled by many khans who were descendants of Türk son of Japheth and belonged to Tatar tribes. They continued until the appearance of Chingis Khan. Then the Chingisids invaded from Moghulistan to Cathay and seized control, extinguishing the dynasty of the old Khaqan. For a time, the Kalmyk khans were in power and the Cathayans submitted to them. The rule went back to the descendants of the original Khaqans [M 193] and the khanate and government were established among them. After Chingis, six of his descendants exercised the khanate in the country

30 Cf. Atlas Minor, 610, report about the Vegetable Lamb of Tartary (*Planta Tartarica Barometz / Boranetz*).

known as Qaraqorum and Ulugh Yurt. Their conditions are recorded in the *Fezleke*; the remaining khans are not mentioned here.

Chapter 22: Climе of India

The twenty-second chapter is about the clime of India.

In the *Atlas* it is written as India Orientalis meaning Eastern India. It mentions that it is a large part of Asia and an important country, and that it is named after its great river. That river and that clime are called Hind in all languages.

Borders

This clime is bounded on the west by the Hind (Indus) River, on the north by the Taurus Mountains, on the east by the Eastern Ocean and the Dāmāzī Mountains, and on the south by the Indian Ocean—it embraces this sea and its capes stick out like two horns.

Distance

This is an extensive clime. The geographers report from reliable sources that a forced march from the western to the eastern border takes at least 60 days, and that a ship with a favorable wind can hardly make the circuit of its coastline in that much time. Philippus says the length of this clime is approximately 600 German miles from the source of the Indus to the Cape of Malacca, and its greatest width is 450 German miles from the Dāmāzī Mountains to Cape Comorin. According to the map of the *Atlas* it extends from 110 to 145 degrees longitude, and ends at 35 degrees (north) latitude from the equator. Thus, they say that the land of India is more extensive than China, but China is more populous than India.

Division

Ptolemy and his followers divided this clime in two parts, east and west of the Ganges River. They called the western part Hind Intra Gangem, meaning Inner India, and the eastern part Hind Extra Gangem, meaning Outer India—i.e., the one on this side (of the Ganges) and the one on that side. What the Persians and other nations call Hindustān is Inner India. In some books of the Prophets, this part is written as Hūlā or Hūlāt and Ūlāt. The other part was known among the ancients as Seria. Marius wrote that the Indians call it Māken or Mājen, and Mercator and his followers related that China is also called Māchen. As this western part falls between the Indus and the Ganges, it was called Greater Mesopotamia in imitation of the Greeks who called the region between the Tigris and Euphrates "Mesopotamia" meaning "Land between two rivers."

Actually, this country between the ocean and the two rivers is nearly rhomboidal in shape. According to Lorenzo, it is approximately 900 miles in length from the sources of the two rivers in the north to Cape Comorin and 200 miles (in breadth) between the mouths of the rivers. The Ghats Mountain range, which is a branch of the mountains that contemporaries call the Caucasus, stretches from north to south and ends at Cape Comorin, thus dividing this country (i.e., India) into two sections down the middle. The territories of Cambaia (Khambat, Gujarat) and Deccan and those of Kanara and Malabar lie to the west of these mountains; those of Narsinga and Orissa lie to the east; those of Multan, Delhi, Qūsbatīr and Chitor are in the middle.

Philippus says: India has a new division. The comprehensive map of India was not known in earlier times. [M 194] [MAP] Travel there was impossible, because of the savagery of the nations between us and India, and the sea lanes were not yet open. So Europeans had little idea about this clime. They only knew the reports about it in the works of the ancient authors. Finally, a Portuguese known as Vasco da Gama set sail from Spain in the Christian year 1497, passed the Cape of Bona Esperansa (Cape of Good Hope) and, after going around all of Africa, reached the coast of India. He brought news to the King of Portugal. Later, Portugal sent ships and took the coast of India by force. They built many fortresses to defend it [R 72a] and appointed a deputy governor. For a long time they did not allow anyone (to trade or settle) there who was not Portuguese or Spanish. In recent times the Dutch also began to voyage in those regions and (they too) discovered India at that time.

So now its borders and territorial divisions are not as they used to be. In olden times it was divided into two parts and, reportedly, into 27 sultanates, some of them quite small. But contemporaries have divided all of India into 9 countries: 1) Cambaia, 2) Narsinga, 3) Malabar, 4) Orissa, 5) Bengal, 6) Siam (Thailand), 7) Pegu (Burma), 8) Cambodia, 9) North India, which is the third part of all of India.

Philippus recorded this division as given here and made summary mention of the cities of each one. Mercator recorded only two parts in the *Atlas* and mentioned only two or three cities. Lorenzo too recorded nine sultanates in two sections and went into some detail. He recorded the countries of Bengal, Aracan, Pegu, Siam and Cambodia east of the Ganges, and mentioned those of Qīāntāy, Barma, Māchīn, Moin, and Qāīm to the north.

The *Taqwīm al-buldān* cites some travelers to the effect that India is in three parts: in the west is Gujarat, also called Juzarat; in the east is Menībār; and in the north is Maʿbar. But this division is not found among the geographers.

The poor one (Kātib Çelebi) will summarize the reports of these three writers (Philippus, Mercator, Lorenzo) in nine sections, adding useful information from the Islamic books.

1 Cambaia, which is the sultanate of Gujarat

Lorenzo includes Cambaia in the Sultanate of Gujarat. Philippus records Gujarat as a country from the Cambaia Sultanate. He says that the country of Cambaia is on the right flank of India, i.e., on the western cape, at the mouth of the Indus. Its length is approximately 160 German miles and its width is nearly that.

Its main city is Cambaia, from which the country takes its name. The Indians call it Kenbāyet (Khambat). It is the choicest of all the cities of India in size and beauty, and for that reason it is called the Cairo of India. It is an entrepot and gathering place of merchants.

This place has a bay (or gulf) known as the Bay of Cambaia (or, Gulf of Cambay). It is a hazardous place because the tide there is so great. Sometimes the sea ebbs as much as three miles, great rocks emerge from the water and many ships are wrecked. For that reason, seafarers come there only after taking on skilled pilots from the city of Diu.

Cambaia is a pleasant city on a river, about three miles inland. The wealthy merchants there are not to be found in any other city. Copious spices and goods come to this city from the surrounding area [M 195]. So much ivory comes from the district of Rūfālā in particular that it cannot be described. Nevertheless, it is all expended in this city, because the people here are so pompous and vain that they decorate their rooms with ivory.

The poor one (Kātib Çelebi) says: Cambaia is located at 115 degrees longitude and 24 latitude. From here Aḥmadābād is three stages to the north and Barwaj three stages to the south. Its construction is baked brick and white marble. It has few vegetable gardens. The mules of this place are famous. Three languages are spoken in these provinces, Arabic and a special provincial language. Its people are also mixed.

Aḥmadābād: the capital of Gujarat. It lies three stages northeast of Kenbāyet. It is a city with a very populous and pleasant hinterland. Its markets are not narrow and dirty like those of other Indian cities; they are spacious and clean and the shops are two and three stories each, beautiful and cheerful. The people are mostly good-natured, with refined gestures and movements. They are a mixture of idol-worshippers and Muslims.

The author of the *Haft Iqlīm* cites the *Ṭabaqāt-i Akbarī* that a certain Gujarati king, Aḥmad Shah b. Sultan Muḥammad b. Muẓaffar, liked the climate of a town named Asāwal and consulted with one of the notables among the shaikhs, Shaikh Aḥmad Kahtūb, about building a city there. He deemed it reasonable and began it in Dhū'l-qaʿda 813 (March 1410). He laid out a great city-wall on the bank of the Indus comprising mosques and markets and town quarters. Later, during the reign of the ninth ruler of this dynasty, Sultan Maḥmūd II, a city named Muḥammadābād was built in a place 10 *kuroh* (an Indian measure akin to a mile) from Aḥmadābād, and bazaars and shops were erected on both sides (of the road) between the two cities. It became more and more built up and eventually the two became like one city. The women of these cities are mostly beautiful, white and amorous.

Surat: On the coast, five stages south of Aḥmadābād, the entrepot of Gujarat. [R 72b] It is an impregnable fortress and a strong city. According to the *Ṭabaqāt-i Akbarī*, one of the statesmen of the Gujarati Sultan Mahmūd, a certain Safar Agha, who was known by the title Khudāwand Khan, built the fortress in H 947 (1540) to repel the Portuguese invasion. The infidels came with ships and deployed their artillery. There was much fighting and they left defeated and broken. Eventually, by way of conceding, they offered a great deal of money to abandon the building of the fortress; but that was not possible.

The fortress is 15 cubits wide and 20 cubits tall. The construction is very solid: every stone was riveted to the one next to it with an iron pin. It is as though carved from a single block of stone. The walled city is now the entrepot of India. Merchant ships from the Red Sea and the Persian Gulf lay anchor there and buy the merchandise of India.

Barwaj (Bharuch, Broach): a famous port and entrepot and city on the eastern shore of the Gulf of Kenbāyet. It lies between Ṣūrat, two stages to the south, and Kenbāyet, three stages to the north.

Diu: an entrepot and fortress on an island at the mouth of the Gulf of Kenbāyet. According to the map of the *Atlas*, it is located at 111 degrees longitude and 21 degrees latitude.

Kenbāyet lies to the east of Diu and has commercial ports for the southern and eastern shores of the Gulf: Nawa-nagar (Bhavnagar?), two stages from Kenbāyet, on the eastern shore, and Gugli on the southern shore. [M 196]

The Portuguese built Diu by a subterfuge when they first came to India and established a garrison. The people of the region were unaware, learning about it after its completion. They waged many wars, but to no avail. It is now the Portuguese capital, an entrepot for ships coming and going. According to the *Taqwīm*, the water of this place is gotten from rainfall.

Sū-manāt (Somnath): a famous city on the coast, between Diu and Mangalore, one stage from each. According to the *Haft Iqlīm*, there was a great temple here with golden idols, and the large idol known as Manāt, which was removed from the Kaʿba with the appearance of the Prophet and brought to India. This temple was made for it and the city was given the name Sū-manāt for this reason. The Brahmins have much to say about this idol, most of it nonsense.

In 416 (1025) Sultan Maḥmūd of Ghazna crossed the desert of Hind and Sind with his army and conducted a religious war (*ġazā*) against this city. During the siege there were many battles and thousands of Hindus were put to the sword. Finally he was vouchsafed victory and the warriors headed toward the temple. The Hindus brought out the idol with much weeping and wailing and 50,000 infidels sacrificed themselves before the idol. Sultan Maḥmūd entered the temple and saw a great structure atop 56 tall and wide columns, each of which was encrusted with jewels. The idol was made of jewel-encrusted stone and was five cubits high. Sultan Maḥmūd broke it to pieces with a mace. He brought one piece to Ghazna and used it as a paving stone in his Friday mosque; it is still extant.

In the temple were 300 musicians and 500 dancing-girls, mostly the daughters of notables who had pledged themselves to the service of Manāt. Most of the villages in the vicinity were endowed property for the temple, and thousands of Brahmins engaged in austerities and worship around it. Their circumstances and false customs are recorded in some of the chronicles.

It is reported from one knowledgeable source that there is another Sū-manāt beside this one. It is considered part of Multan province and lies between Multan and Kashmir.

Guwwa (Goa), also spelled Ghuwwā and, on the map of the *Atlas*, Ghūghā (Guga): a large entrepot of the Portuguese near Diu on an island of the Indus. It has magnificent buildings and a strong fortress.

Daman, also called Damaon: a fort of the Portuguese on the mainland across from Diu. These forts are garrisoned by Spanish troops. Daman lies one stage south of Barwaj, on the seacoast.

Patan, also called Pachan: a famous city and fort, three stages from Aḥmadnagar. Most of the populace are master weavers. It is on the road from Kenbāyet to Lahore. [R 73a] Shaikh Farīd Shakarganj was from this city.

Dabāl-pur (Dibalpur): a city one stage northeast of Patan. It is also called Dapār-pur and Dibalpur. It is 43 miles southwest of Lahore on the road to Kenbāyet.

Chūhīr, also called Chahanī: a city 25 miles southwest of Lahore. A community of people from Bukhara lives there. Dibalpur is 18 miles from here.

Mangalore: a port city two stages west of Diu on the seacoast. Sū-manāt lies one stage east of here. In the *Ḥabīb al-siyar* [M 197] Mangalore is recorded as an entrepot on the frontier of Bījānagar. Two parasangs from this city is a square-shaped temple, 10 cubits on each side and 5 cubits in height. All the walls are covered with bronze sheathing. Inside are four platforms. On the platform in front of the door is a golden idol in the form of a man, with eyes made of rubies. It is a masterpiece of skilled craftsmanship.

Jā-mahr: a city and port two stages west of Mangalore. It is the first place west of Gujarat on the coast of the Indian Ocean. Below this, four stages to the west, one comes to Lāharī, the commercial port of Sind. Jā-mahr is now also called Yām-mahr and Zagad. Its province is mentioned as the land of Rāzbūt (Rajput). Zagad is a port near Jā-mahr. All of these are located at 105 degrees longitude and 20 degrees latitude.

Tirād: a city on the Lahore road four stages east of Aḥmadābād. It also has an administrative district (*nāhiye*).

Sarkhīj: a city 10 stages northeast of Aḥmadābād. Lorenzo says that it is the city of Nīzā (i.e., Nysa) built by the commander named Bacchus on top of Mt. Meron.

Nahlwāra: It was the capital of Gujarat in olden times. It is a famous city in a flat place three stages distant from the seashore. It is also called Nahr-wāra. Kenbāyet is the port of this place. The town quarters are scattered amidst streams and vegetable gardens.

Tāna-sar, also called Chānā and Tahānīsar: a city and entrepot two stages east of Aḥmadābād. Merchants gather there. The people are mainly infidels. According to the *Haft Iqlīm*, although it is a small city, it is very prosperous. Its weavers have made different kinds of goods very well.

There was previously a temple there that the Hindus called Jagarsūm where they worshipped a great idol. Sultan Maḥmūd of Ghazna conducted a religious war (*ġazā*) against it in the year 402 (1011), tore down the temple and used the idol as the threshold of the Friday mosque in Ghazna. (The poet) ʿUnṣurī recalls this story in one of his *qaṣīdas*.

In this city is a pool called Kurchehīt. It is a sacred place of the Brahmins. The Hindus come there from the surrounding districts to bathe on days of solar and lunar eclipse. They scatter silver, gold and jewels in it, in the belief that they will be compensated 70-fold.

Chalk is found in the plain here, and ebony and logwood are plentiful. Chalk is extracted by the following method: Reeds 10 spans long grow in the reedbed of the valley of this city. Every year, in the simoom season, they burn up from the heat of the sun and their roots turn to chalk and are gathered. Sometimes, whatever animals happen to be in that reedbed also burn up, bones and all, and their remains are also sold as chalk.

Sagwān: a city one stage east of Tānasar.

Jalore: a city midway between Nākor and Nahlwāra on the Lahore road eight stages from Aḥmadābād.

Qāmpā'il: According to Philippus and Lorenzo, it is also one of the great cities of Gujarat, worthy to become a capital. It is built in a high place defended by seven fortresses. However, there is no city by this name in other books [M 198]. There is Qāndābīl in Sind, discussed below.

In addition to these, some knowledgeable sources record the following as Gujarati cities: Chāpāner. Prūde. Rādhānpur. Krāy. Dhūnikā. Dhndūkā. Qaprūrenj. Nūsārī. Elī-lātī. Lāl-pur. Lāchhūrī.

Lorenzo also mentions the following: Kūsīānā. Kharwār. Qūrīnār. Mūdrāfāwā. Mūhā. Tālāyā. Ghūndīn. Mākīghān. Ghānder. Yārūq, at the mouth of the Bārbānd River. Tāpeten. Reyner. Nūsqārī. Ghāndūīn. Māīn. Kalmāīn. Bāzāīn, also under the control of the Portuguese: it lies on a strait and they obtain much wealth from it every year. Dawadan. Āsīr.

Sindāpūr, also called Sindān: a city at the end of Gujarat on the shore of a bay three stages from Tāna-sar. There are 15 parasangs between this place and Manṣūra, and it is a place of qat, henna and bamboo

Sufālat al-Hind, also called Sūfāra: a port five stages from Sindāpūr.

After these are the cities of Māhūda, Menderī, Kūlem and Maʿbar.

In the mountains of the interior are the Rajput tribe. He (Lorenzo) records their cities as Qurūdī, Wāmastā, Arā and Ārjentū.

Deccan

Philippus recorded this country as part of the country of Cambaia and said that in olden times the entire sultanate was called Deccan. Lorenzo considered it to be a separate sultanate and divided it into three parts according to the writings of the ancients. However he mentioned some of its cities and said that the meaning of the word Deccan is "bastard," because its people mingled with the tribes of Daylam who came to conquer these regions.

The extent of this country is 250 miles from the mouth of the Bāt River to the Alīghā River. The Ghāt Mountain being in the middle of this place, one part is on the mountain and the other two are on its two sides. All of it lies to the south of Gujarat. The author of the *Haft Iqlīm* says the country of Deccan comprises approximately 360 fortresses.

Aḥmadnagar, also called Āmadnagar: the capital of Deccan. It is located at 115 degrees longitude and 20 degrees latitude. [R 73b] It is exceptional among the cities of Deccan with regard to climate and mountains and plains. It has a strong fortress on the edge of the city that is impregnable. The city's water supply is completely supplied by several underground channels. On one side it has a delightful excursion spot, like the Garden of Iram, with an esplanade that has a large lake and a lofty building.

According to the *Ḥabīb al-siyar*, the capital city is Bījānagar; it is discussed below.

Dawatabad: a famous and large city, one stage west of Aḥmadnagar and three stages east of Barwaj. It produced fine paper and cloth. In olden times this city was called Dīwgīr. It was always the capital before Islam. The people are mainly weavers and have attained perfection in this craft. It is recorded in the *Haft Iqlīm* that the fortress of this city is among the major fortresses in the world.

Chayūl: a famous city on the seashore that is the emporium and port of Deccan. One reaches it in seven stages from Āmadnagar.

Jūnīr: a city with sweet water and delicious fruit. It also has a strong fortress.

Qadarābād: a city two stages east of Chayūl.

Piranda: a city northeast of Aḥmadnagar and three stages from Lahore.

Sūlāpūr: a city one stage east of Piranda. [M 199]

Ḥirj: a city one stage east of Qadarābād.

Rasūl-pur: a city one stage east of Piranda.

Darāsin: a city one stage north of Sūlāpūr.

Bidar, also called Bīdar: an ancient city 90 parasangs west of Aḥmadnagar. Patan lies two stages west of this place. Bidar is a city as big as Cairo, with orchards and gardens. Its water is gotten from wells. Lorenzo says that this city is the capital of the Deccan Sultanate. In olden times the rulers used to live there.

Māhūr: a fortress on a mountain.

Golī-konda (Golconda): a capital city three stages east of Bidar. It was the capital of Quṭb al-Mulk. It is smaller than Āmadnagar; it is 108 parasangs between them.

Rān-tamūr: a magnificent and strong fortress on a mountain. It lies eight stages west of Agra on the Agra Road from Aḥmadābād. The city of Hendūn, which is the frontier of the province of Rana Sanga, is three stages east of here.

Nīsā: a city between Aḥmadnagar and Dawatabad, one stage from each.

Bijapur: one of the cities of Deccan. It was the capital of ʿĀdil Khan (Yūsuf ʿĀdil Shah, 1459–1511).

Bījānagar (Vijayanagar): one of the cities of Deccan. It is five stages east of Goa, with mountains in between. According to the *Ḥabīb al-siyar*, this city has a seven-layer fortification, one surrounding the next. The outer layer consists of rectangular stones, half of them on the ground and half at a man's height, rising about fifty cubits from the ground. They are planted firmly and connected to each other on the exterior so that neither infantry nor cavalry can get through to the (inner) fortress.

These fortresses have strong and solid gates. The middle fortification is the palace of the ruler known as Kriās. Between the first and second and third are fields and orchards and several buildings. From the third to the seventh are countless shops and markets, and four bazaars are connected to the royal palace, each constructed with raised vaults and terraces and open places.

Bījānagar has fine and fragrant roses. There are many jewelry shops in the market, selling rubies, diamonds, pearls and emeralds. To the right of the palace is a very tall and extensive audience hall constructed atop forty pillars. The people of this city, even the shopkeepers, wear jewel earrings and jewel-encrusted necklaces and expensive rings on their fingers.

To the right of the audience hall mentioned before is a mint. Opposite that is a police station, and behind it is a brothel. This brothel is like a bazaar, 100 cubits long and 15 cubits wide. The rooms and halls are open in front and have splendid ceilings and columns. They are swept out and watered down very morning and chairs are placed in front of every room. Young and beautiful prostitutes, covered in gold and finery, pass by and sit down. Several slavegirls stand in front and invite passersby to pleasure. Anyone who wishes goes and enjoys himself.

Balgām: It was the capital of Asad Khan. An independent country is known by this name.

Īlaḥ-pūr: belongs to the country of ʿImād al-Mulk.

In addition to these, cities dependant on Deccan are Senghīl, Berār, Wāīn, Ūnā, Pūnājknā, Sūpūh, Serūl, Kaḥlī, Sendhūrī and Bālāpur; and cities in the country of ʿĀdil Khan are Sitāra, Chenden, Menden, Kīhej and Awazpur.

Some of the Deccan cities in the book of Lorenzo [M 200] are the following:

Qāol: an important city on the coast, now under Portuguese control.

Qātāpūrā and Qārāpātan: also is on the coast.

Goa: an emporium of this (text: *bir*, error for *bu*) country. On the map of the *Atlas* it is on the coast at 115 longitude and 16 latitude. Lorenzo says that the Portuguese captain Alfonsez (Afonso de Albuquerque, d. 1515) took it and established control. It is situated in the middle of a small island. The Portuguese deputy resides there because the air is pleasant and it is a place of commerce. All sorts of goods arrive from the Molucca Islands and other emporia and are shipped abroad. All the Indian lords have agents who reside there and apply to its governor for permission to embark on sea voyages. Whoever wishes to travel in the Indian Ocean must get a permit and a Portuguese standard from him, according to Lorenzo. We now hear, however, that Holland and England have gone to those regions and Portugal no longer has that degree of sovereignty.

Previously, when Spain conquered Portugal, the Portuguese deputy remained in his former position. Aside from him, there is an archbishop in this city; the other bishops in the east are subservient to him. [R 74a] And there are several colleges here where various languages and the philosophical and Christian (i.e., theological) sciences are taught.

The Portuguese fleet winters in Goa. Sometimes the galleons, galleys and other ships exceed 200. This is a very strong and well-fortified city, with beautiful and splendid buildings.

Sīntāqūrā: on the coast south of Goa, at the mouth of the Alīghā River. As there are five small islands in front of this place, it called Ārken-dīwā, meaning Five Islands in Hindi.

Dānājer: the choice city of this country. Its governor lives there on account of its delightful gardens.

Līspūr: There are some mines nearby where they extract diamonds once every five years.

He mentions many other cities in addition to these, but they are not well known.

2 Narsinga

Philippus says this country is on the western horn of India, and the eastern and western seas encompass it on both sides. Its length is approximately 150 German miles from the border of Cambaia and Orissa to Cape Comorin. At its widest it is 90 miles between the two seas.

This county has two capitals. One is Narsinga after which the country is named. It is located at 120 degrees longitude and 15 degrees latitude. The other, Bīsnāghār, is on the mainland in the middle of the two seas.

The capital of this country (i.e., Bīsnāghār) is a magnificent city four miles around, located on the Nāghawandī River. It is also called Bījānagar. It has splendid palaces and temples. Copious gems are extracted from its mountains and are sold from the state treasury. Countless merchants come and exchange goods with the jewel-dealers. Because horses are prized in this city, they do not take customs dues from anyone who brings horses. The prostitutes of this city are so wealthy that one of them can pay the salary of several thousand soldiers from her own money and equip them for battle. According to Lorenzo, they even participate in battles and encourage their lovers to fight.

In addition to these, Philippus records the following cities in this country: Awanūr. Bataqālā. Mangalore, which is under Portuguese control. Qolmāndel. Maliapur (Mylapore) [M 201].

The Christians of India dwell in this Maliapur and the tomb of Thomas the Apostle is there. Therefore, says Lorenzo, this city is also called Santo Mio (Portuguese: São Tomé). Previously, the Portuguese rebuilt this city with splendid palaces, according to their style, and garrisoned troops there. Once every three years they remove the old troops and bring in new ones. They give a monthly stipend of three *ġuruş* and an amount of pepper for each soldier. In Lisbon that pepper is worth more than 20 *ġuruş*.

The ancient writers have many stories about this city. For example, there was a mirror in the San Toma Church in which the things of the world appeared, just as in the Mirror of Troy. But when the Portuguese came they did not find this mirror, only a cross on which was written that Thomas was sent by the Messiah to this region to preach the Gospel and one day, while he was praying, a Brahmin came and killed him.

According to Lorenzo, the following cities, dependent on Narsinga, are on the seacoast and belong to the country of Kanara:

Awanūr: it is still under Portuguese control. Every year many rice ships land there.

Bātāqālā: a city famous for commerce. It pays taxes to Portugal. Rice and other products are distributed from there to the countries of India

Bāndoro. Brāshaloro. Bāqānor. Qārnāt. Mānjenā.

These are on the coast. And on the mainland:

The country of Ghārzūpān. It has its own governor, but he is subordinate to the lord of Narsinga.

Baghāpūr.

These are the cities. According to some knowledgable sources, Santo Mio is on the coast, two stages north of Neke-beten. As it has no harbor, ships anchor at a distant spot and transport their goods with small boats.

Pālīqāt: a fine city, seven parasangs north of Santo Mio, at 124½ longitude. It has skilled painters.

Pālīūn: a magnificent city two stages inland from the coast at 123½ longitude and 15 latitude. Its port is Pālīāqāt. Remarkable implements and tools are found here. Its painters are among the best. They weave Bayramī cloth, striped stuff and waistbands. There is a grass here called *chāya*, like a slender blue string, that is mixed with crimson dye and used for painting. It is a pleasant red color and never fades. But it is rare and valuable. It is not exported.

Penta-poli: a big city on the coast. It lies two stages north of Pālīāqāt. It is a fine place producing various goods. The *chāya* paint is found here as well.

Ghawāndī: a city that lies two stages northeast of Penta-poli at the mouth of the Nāghawandī River.

Banāket: It is located on the seashore at one of the mouths of the Nāghawandī River. From there along the coast one reaches Qālīghān, which is somewhat smaller than Banāket.

3 The country of Malabar

It is also called Manībār. It is in a corner of Cape Comorin at the end. Its length is 70 and its width 45 German miles. This country is a densely settled part of India, with pleasant climate and plentiful water sources and date palms. The people mainly subsist on dates. Other fruits are also good.

The capital of this country is a large city named Calicut. On the map of the *Atlas*, it is located on the seacoast at 116 degrees longitude and 13 degrees latitude. According to Philippus, the king of this place is known as Samori, meaning "Great Lord". [M 202] This city is vast, but its buildings are not that tall. Previously, many battles with the Portuguese occurred here, and for that reason it is in somewhat ruined condition. Idol-worshippers are many here. They depict Satan in gruesome form and worship him in order to be safe from his harm.

In the *Taqwīm* it is recorded with the name Kale. The people are a mixture of Hindus, Muslims and Persians. There are tin mines, camphor and bamboo.

It says in the book *Theatrum* (i.e., *Theatrum Orbis Terarum* of Abraham Ortelius) that in olden times this city was known as Sīmīlā. It is one of the important cities of India. It has no fortification and many places in the interior are empty. Its houses are made of stone, scattered and distant from each other. Their interiors are painted with strange images. The roofs are covered with date palm leaves. The doors are very large and well crafted. Each house has a wall that encloses it in the front. There are many lakes and streams where the people generally go to bathe. Their ruler is an idol-worshipper. His subjects show great submission. When he emerges from the palace, he enters a bejeweled and gem-encrusted palanquin that is carried on men's shoulders and is surrounded by musicians and dancers. The notables of state walk behind the palanquin, with a sword in one hand and a shield in the other, as is their custom; and a parasol is held above it. This city has many very wealthy merchants. Various perfumes, silk, gold brocade fabrics, and beaten vessels of copper and tin are plentiful there.

Qānūlūz: an important city not far from Calicut. It has many fine goods and rarities. The Portuguese have a strong fort there. Its people worship idols, the sun, the moon, and cows. It produces spices and ginger. It has its own governor, but he is ruled by the Portuguese.

On the coast is Qāla, which Philippus records as Qūlet. The Portuguese have a fort here.

Tāwūr: It has its own chieftain. He is Christian.

Bālīānqor and Ghrānghānūr, also called Qrānghānūr.

Cochin: one of the famous cities of India. Then the city of Calicut (i.e., its inhabitants?) was transferred to this place. Ships are loaded here with the spice cargoes of Malabar and the Portuguese take most of it. Across from this city is an island named Wāīpen. Since its governor is the chief of the idol-worshippers, the Hindus go there to make sacrifices. They greatly esteem this island.

Pūrqā: It has its own governor. The Portuguese have a strong fort there also.

Tranganor and Cape Comorin, which earlier was known as Qūrī: There are many Indian Christians there, but they are Nestorians and live in the mountains. Every year the Baghdad Patriarchate sends them priests.

In the interior are Waspūr, Qorqorān, Qūtūghāno, Rapalīno, Trānkālūr, Pārāpūrān and Bālūret. [R 74b]

Lorenzo says the Maldive Islands are across from this country. They are said to be several thousand islands. However, they are all small and low-lying. Two of the larger islands—Māldīwā and Qāndālūs—are under the control of the Portuguese. Copious fish are caught in the Maldives and exported abroad. They collect beads and mother-of-pearl and in many places use them in place of coinage. Plentiful ambergris is found on these coasts. The boats of these islands are made with wooden nails, because iron is scarce.

The poor one (Kātib Çelebi) says: These islands are called the Fāl Islands. They were discussed above. [M 203]

4 The country of Orīqsā (Orissa)

It is a country stretching along the shore of the Gulf of Ganges between the lands of Narsinga and Agra, on the one hand, and the country of Bengal on the other. Its main city is Orīksā, also called Orīshā. But its capital, known as Rāmānā, is on the bank of the Ganges River. Many merchants come there and buy ivory and jewels. In the *Atlas* it is recorded at 125 degrees longitude and 20 degrees latitude, and several other cities are also recorded.

In the book of Lorenzo this country is named Oristan, and he says that the coast of this country is very stormy. Since ports are few, it is not very built up, but its interior is very populous. On the seacoast are Penāqūt, Qālpaghān, Bāsānāpatan, Osānāpatan, Qālīnapatan, Nāshanjapatan, Pūlūrū, Penājīāt and Cape Sqūghūrā, also called Cape Khurmā on account of the copious dates there. A small gulf known as Orīsān begins from there and comes up to the mouth of the great river named Ganga (Ganges).

In the *Haft Iqlīm*, this country is recorded as Orissa. It produces three or four *leks* of *rūsa*. It is divided into two prefectures. There is much elephant hunting in this country. Its people write on tree leaves with an iron pen. At present, the Afghans have conquered these lands. Orissa is also called Gajapati. Among its dependencies is a place called Persūtem (Puri?) where there is a temple built on the seashore with a large idol known as Jagannath. There are many stories about this idol among the Hindus.

Mānīghābatan (Mangapatnam): a city and entrepot on the seashore eight stages east of Orissa. It produces coconuts, red pepper, sugar, lac and rice, which are exported abroad. To get to Bengal from here, one goes to the east and north. It is at 142 longitude and 20 latitude.

Pipili: a city and entrepot on the coast of the country of Orissa. It lies two stages northeast of Mānīghābatan. Whatever is found there is also found here.

Pūlārī: an entrepot of Orissa at 21 latitude. It lies two stages northeast of Pipili.

Māsūlī-patan (Masulipatam): a big city on the seashore at 120½ degrees longitude and 16½ latitude, the capital of Quṭb al-Mulk. Stuffs, goods and painted things are plentiful. Mānīghāpatan lies eight stages northeast of here. With the exception of this city, most of the city-dwellers of Orissa are infidels.

5 The Country of Bingāla (Bengal)

This country is named after the city of Bangāla. Situated at the mouth of the Ganges and other rivers, and in their vicinity, its length is 150 German miles and its width is 70 German miles.

As for the city of Bangāla, it is a famous entrepot built on an island of the Qūsmīn River at 135 degrees longitude and 23 degrees latitude. It is also called Banghālā. Its people are wealthy and well known as traders. Various types of cotton cloth are woven here, and good preserved fruits are in this country. Among the Hindus this city is known as Chātīgūn or Ṣātīghān (Chittagong). Some say that Chātīgūn is Greater Bangāla and Ṣātīghān is its small entrepot, the distance between the two being 100 parasangs.

Lorenzo says that Sātīghān is a city famed for trade at the mouth of the Ganges. The road by the seashore is difficult and rocky. Those who go there travel by sea; it is more arduous by land.

Pūlārī: a city on the seashore three stages from Ṣātīghān [M 204].

According to the *Haft Iqlīm*, the climate of Bengal is very temperate. Rice, sugar, silk, pepper and other fruit are plentiful. There is a fruit called *kūla* that resembles an orange and is more flavorful, and another called *lenken* that resembles a pomegranate. These areas get so much rain that the land is like the sea, and the main means of travel and transport is by boat. The Bengalis are masters in the craft of weaving. They weave such thin cotton cloths and muslins that 27 cubits fit into the palm of a hand.

The aforementioned (*Haft Iqlīm*) says that the entire country of Bengal is 300 *kuroh* in length and 260 *kuroh* in breadth, each *kuroh* being one mile. The whole is divided into 22 *tūmāns* (administrative districts).

Tūmān of Adīnar: Its tax revenue is 597,573 dinars. Silk thread is produced in some districts (*parganāt*).

Tūmān of Sharīf-ābād.

Tūmān of Madān: One of its dependencies is a diamond mine named Hīrbūr. And one of its industries is salt production: they burn the grass of this place while yet fresh and boil its ashes to produce salt. Sātīkānūn, one of the famous entrepots of India, is considered part of this *tūmān*. *Khāṣṣa*, *ṣaḥn* and other types of cloth are made there. *Ṣaḥn* is a fine fabric of the *chūtār* variety. This city is also called Chātigām (Chittagong). However, some knowledgeable people deem it among the dependencies of Delhi.

Tūmān of Sulaymān-ābād: The famous cloth known as *ṣaḥn-i Sulaymān-bādī* is made here. Maḥmūdābād is one of the appendages of this place. Long pepper and *maḥmūdī* elephants are plentiful here.

Sanārkānūn: This *tūmān* has tax revenue of 303,003 dinars. Its only agricultual product is rice, but among its famous cloth goods are *khāṣṣa*, *malmal* (muslin), *natk* and *gangajil* (?).

Tūmān of Srī-Hast: Its people are mainly merchants.

Tūmān of Jannat-ābād: *chūtār-i rāmūtī* is made here. And Gaur, the capital of Bengal in olden times, is within this *tūmān*. The Gaur citadel is one of the important fortresses of India. On its western side is the Ganges. On its eastern, northern and southern sides a seven-layered moat was cut and filled with water. The distance between each one (i.e., from one moat to another) is one-half *kuroh*, i.e., one-half mile; the width of each moat is approximately three tent-ropes; and the depth is such that an elephant cannot cross. This city is on the shore of the Ganges, eight stages north of Chātīgūn, which is known as Greater Bengal. [R 75a] [Plate 33]

The governor of Bengal lives there. He is submissive to the descendants of Jalāl al-dīn Akbar.

Kūra-kāt: the frontier of the country of Kawaj. Rice, silk and areca nuts are produced here.

Tūmān of Bārīk-ābād: The sugar, areca nuts and long peppers are good here. *Khāṣṣa*, *mu'minī*, and *ṣaḥn* are well known among its cloth goods; especially *khāṣṣa shahbārpūrī* is the *khāṣṣa* of this *tūmān*.

Lorenzo says that the country of Bengal is somewhat narrow on the sea side, but its interior is broad. It is better than the other regions of India in regard to sugar, rice and meat. One can buy an ox for six carlinos and a sack of rice for one carlino.

Qātīghān: A coastal city built on one of the mouths of the Ganges. The Portuguese have a *feitoria* (factory) there, i.e., a trading post.

Farther inland is Oro, a magnificent city built on the shore of a large lake stemming from the Ganges [M 205]. It has splendid palaces. The people are Muslims and idol-worshippers. The governor's palace is decorated with images of gold, silver and lapis lazuli.

After that are two cities famous for trade named Sarnūwan and Mārāzyā, also on the Ganges. Silver and gold and aloes-wood are brought to these places, and many merchants come from China.

The people of this country are also mainly merchants. The salamander bird is found here. The Hindus call it the *qūqnūs* (phoenix) and tell many stories about it. The ancients wrote that it is a bird. However, since all things in the natural world are perishable, reproduction is necessary, and a species of animal cannot be restricted to a single individual.

Bay of Bengal: Lorenzo says that the ancients did not know very much about this bay and did not describe it. We will provide a measure of detail. This bay is known among the Hindus as Dībānākūn, also as the Ganges Bay. It enters from the ocean between two landmasses and splits India into two sections. It is more than 800 miles distance between the two capes, then it narrows as it goes north and ends at the mouth of the Ganges, which is at 22 degrees latitude. That sea (the Bay of Bengal) is in the shape of a triangle, and that point is like its angle.

Many merchant ships sail on this bay. Aside from the Bengalese, Burmese and Siamese, there is no lack of Persians, Arabs, Chinese, Javanese and Portuguese. Abundant pearls are extracted from this sea. Sailors sometimes see there the fish that they call mermaid (*banāt al-baḥr*, lit. daughters of the sea); it has the face of a woman and the lower part is a fish.

There are few ports on the eastern shore of this bay. Most of the settled places are on the western shore. [R 75b]

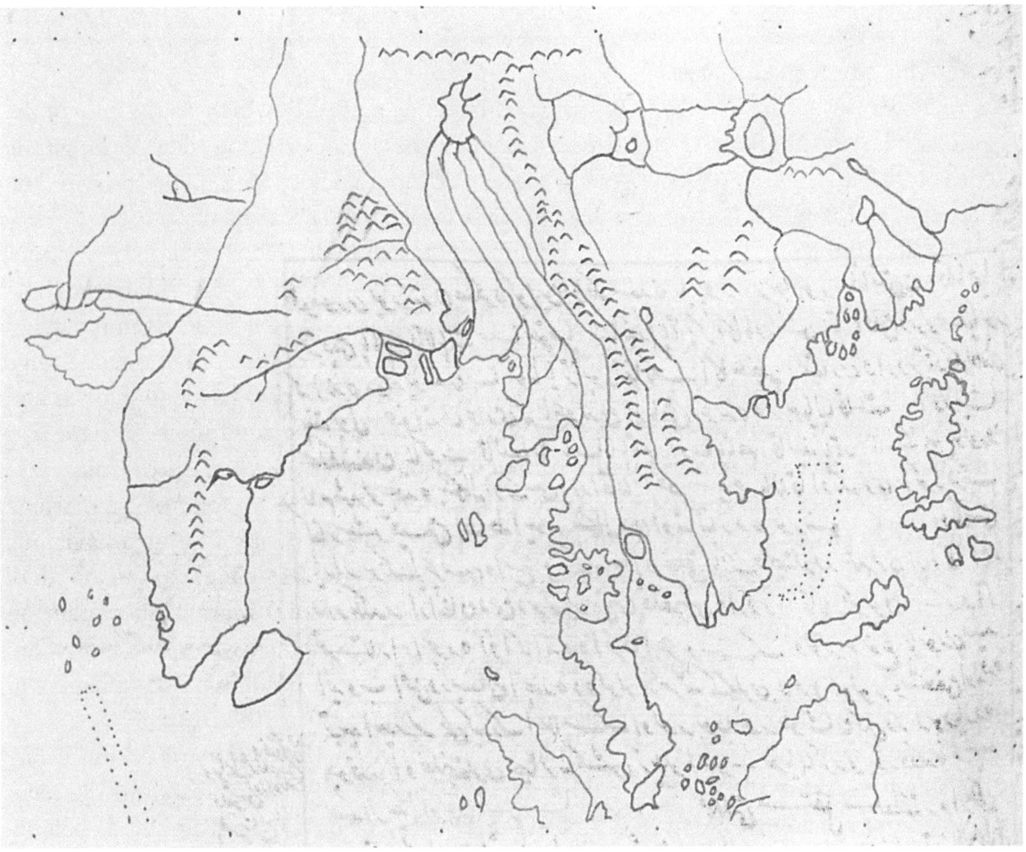

PLATE 33
Map of India, R 75a; cf. M 193/194 (Appendix, fig. 73–74)
TOPKAPI PALACE LIBRARY, REVAN 1624, WITH PERMISSION

After Cape Comorin, in the country of Kūlīn are the cities of Maipur; Trīqānbūr; Mānānqūrt; Qābāmrīā; Trīmīn-pātām; Qālāpāt; Sānqūmrīā; Sāndrā-pātān, also called Mādras-pātān; Maliapur (Mylapore), where the Apostle San Toma Pater is held in great respect by the Portuguese, as mentioned above; Penta-poli; Ghawāndī-wārī, at the mouth of the Nāghawandī River; Pālīāqāt; Banākūt in the country of Orissa; Qālīghān; Bāsānā-pātām, Awsā-pātām, Qālina-pātān, Māsanga (?), Pūlūrū. Then the country of Bengal and the countries of Aracan, Pegu, Siam and Malacca which encompass the eastern shore of this bay.

Lorenzo writes that the country of Aracan is on the further shore (i.e., across the bay) from Bengal. Its interior is vast, its produce plentiful, and its people elegant and friendly. Their cities are Aracan, also called Arākhān, which is the capital and has a strong fortress; Bāqālā; Kūbūdā; Sarūāa; Qūstārā; and, in the interior, Āwā where gems and musk are plentiful. They hunt a certain creature and pound the flesh into the skin so that even the bones are broken up. When it rots they cure it and its flesh gradually becomes the finest musk. This creature is like a rabbit and is abundant in this land. They hang silver bells on the genitals of their infants and they go around naked, producing strange sounds as they walk. This place also has many horses and elephants. Philippus records that the country of Āwā is attached to Pegu.

6 The country of Pegu (Burma)

It is an independent sultanate, named after its capital, the city of Pegu. On the map of the *Atlas* this city is built on an island at 136 degrees longitude and 17 degrees latitude. It has a large and well-constructed fortification, [M 206] delightful buildings and splendid palaces. Its houses are made of bamboo, with unusual craftsmanship. Previously, when the ruler of Barma conquered it, he had a palace built that had no equal in the world. The great events of the rulers of the East and many animals and trees were illustrated there, in a manner that those who saw them thought them to be alive. This country stretches from the Cape of Naghirās (or Nigharās?) 300 miles to the south and its eastern border is the Menān River. It is superior to the other parts of India in grain and rice. It has a ruby mine. The Bay of Pegu is along the coast. Several rivers empty into it from Fetāntāy (?) Lake, including the Pegu River and the Martaban River.

Martaban: a large and famous city and entrepot at the mouth of that river, belonging to the country of Barma. Precious and splendid plates and vessels are manufactured there and exported abroad. However, they are not pure and painted like the porcelain of China and Cathay; only they are strong and sturdy. This city is on the coast, south of Bangāla. The tides there are severe and cause shipwreck. Pegu is three stages southeast of Martaban. The river empties into the sea south of the city.

Barma: a city between Martaban and Bangāla. A river also comes close to it and empties into the sea.

The remaining cities of this country, according to Lorenzo, are the following: Wāghārūn; Trāghān; Tāwās; Qūsmīn; Sīryān; Dugūn, also called Ṭūghūn. It has a famous *wārīllā*, meaning a temple. It is located at such a height that from it most of the country of Pegu can be seen, just as the entire country of Egypt can be seen from the Pyramids. The people of Pegu greatly esteem this place and make pilgrimage to it from faraway places. Sāwulās is a city on the Bay of Pegu. Formerly it was called Sābārīqū.

7 The country of Siam

Both Siam and Sian are in use. According to Philippus, this is a large sultanate, 300 miles in length and 160 German miles in breadth. On the south it reaches the Cape of Malacca, which is India's eastern horn. Lorenzo says it is named after the Siam River. Its domains extend from near the mouth of the Ganges up to Singapore. Its capital is Sian, a great and magnificent city located at 140 degrees longitude and 15 degrees latitude at the source of the Menān River which empties into the Gulf of Sian from the aforementioned lake. The cities on the coast are the following: [R 76a]

Tanāṣarī: A large city well known for commerce. A river like the Nile flows through the middle of the city. Because there are places with too much rain, this city tends to suffer dearth.

Lūnghūra. Tūrrān. Pedān. Perā.

Malacca: a city at the end of the cape (i.e., the Malay Peninsula). Previously, Alfonsez (Alfonso de Albuquerque, d. 1515) of Portugal had taken this city. It is the main city of a large country, extending 260 miles along the coast. There are many islands around it. Being a crossroads and, as it were, the center of the East, merchant ships come to this city from all around bringing various goods. The coinage was originally tin; later gold and silver coins began to be minted. The city is on a peninsula known as the Island of Gold. It is an entrepot famous for perfume. It is four German miles in circumference.

Lorenzo records the city of Singapore on this cape (or peninsula).

8 The country of Kambuja (Cambodia), also called East Kenbāyet

It was named after its main city, which on the map of the *Atlas* is a magnificent city located at 143 degrees longitude and 13 degrees latitude. [M 207] Merchants come and do a massive trade in silver and gold and aloeswood. According to Philippus, its capital is Dīām; according to others, it is known as Ūdīā (Udong?). As for the Gulf of Kambuja (Gulf of Thailand), which is the angle between India and China to the east of the Cape of Malacca, it is not as large as the Bay of Bengal. However, it is has many storms and contrary winds. Sometimes the wind blows so hard that ships cannot withstand it and are lifted into the air.

The cities on the coast are as follows: Pān, which Portugal destroyed some time ago; Pūntī-qān; Qālāntā; Pātān, Lūghūr; Kūī; Peh-prīn; Bāmblāqūt, at the mouth of the Siam River; Sīrī; Zāqūbadrā; Tāwārnā; Pūlāqāndūr, an island near the Cape of Kambuja where the aforementioned wind blows violently; Bārdān; Bāydā, a large and prosperous city; Wārīllā, a city on the coast.

On the map of the *Atlas*, the country of Cochin-China is drawn east of this country of Cambaia, on the frontier of China. Lorenzo records it at this point as belonging to Cambaia and mentions several cities there. He says that the coast of this place is very dangerous and there is much shipwreck, but nevertheless merchants are not deterred from coming, because one journey there is more profitable than six journeys to China. He records several islands in front of this country, most of them located opposite the country of Kense which belongs to the clime of China.

9 The northern part of India

Philippus says they used to call the sultanate of this part the Great Moghul. This country is the third part of India and its capital is Delhi. It is situated on the borders of Cambaia and Narsinga. Its important cities are Mandaw, Sānghā, Multan and Kītūr.

The poor one (Kātib Çelebi) says: transmission from the translated European books ends at this point. Hereafter is additional material (from Islamic books).

This ninth part is the broad country of India. At present it is the seat of government and the kingdom of the Timurid rulers of India. The other kings of India are submissive to him. Its capital consists of several magnificent cities.

One of these, Lahore, is a large city and ancient seat of sovereignty at 123 degrees longitude and 31½ degrees latitude. It is located on a flat plain on the bank of the Indus and has a two-layer fortress with 12 gates. The inhabited part of the city lies mainly outside the city wall, while the sultan's palace is inside. It has many Friday mosques and bathhouses, but their roofs are not capped with lead. Most of the houses have earthen roofs; windowpanes are rare.

The palace is well constructed, with beautiful ornamentation. It has a splendid hall that is completely walled with mirrors, so that one's reflection is manifold.

The winds of this city are violent and its rains frequent.

In the *Taqwīm* the name of this place is recorded as Lawhūr. It is also called Lahāwar.

According to the *Haft Iqlīm*, the present prosperity of this city is such that it surpasses the cities of the inhabited quarter of the world in the multitude of goods and the necessities of life. Its people are masters of every craft and pursue every occupation with excellence. [R 76b] In fruit season there are so many grapes and watermelons that no one pays attention. Watermelons ripen twice a year. And snow and ice are plentiful even in summer.

Among the dependencies of Lahore is the mountain of Nagarkot with a lofty and impregnable fortress. In the foothills, in a domed building, [M 208] there is a plain piece of stone that is greatly venerated by the Hindus. Twice a year, thousands of people make pilgrimage to it and circumambulate it with bare head and bare feet. After circumambulating it, anyone in need cuts a piece of his own tongue and buries it at the threshold of that place, presenting his need and making supplication. On account of his complete devotion and sincerity, his need is met and his tongue is healed.

The area from Nagarkot to the country of Kawaj, which is the most distant *vilayet* of Bengal, is ruled by feudal lords, each controlling a distance of three or four stages. They resemble the people of India in religion and language.

Sīhrind, also called Sirhind and Chihnind: a city 120 miles east of Lahore. It lies on the Delhi road from Jālantar. Delhi is also 120 miles from this place. Previously it was part of the country of Sāmāna. In the year 753 (1352), Sultan Fīrūz Shah (Tughluq, Sultan of Delhi, 1351–1388) arrived there and made it a seat of government. He built a strong fortress there called Firozabad. It is still famous for its promenades and pleasure gardens. Among its inhabitants are many literati and skillful painters.

Samānā: a city three stages west of Sīhrind.

Panipat: a well-known city in those parts. Its people are mostly master knife makers.

Sulṭān-pur: a city 30 miles southeast of Lahore.

Jālantar: a city on the Delhi road, 42 miles from Lahore. It lies on the north side of the road. Sulṭān-pur is between this place and Lahore.

Sialkot: a magnificent city and fortress four stages north of Lahore. It lies on the east bank of the Chenab River.

Sūdra and Kūpra: one stage west of Sialkot on the bank of that river. It was two cities, but then Kūpra was destroyed and little more than a village remained. However, Sūdra is still a populated city.

Hazara: a city on the west bank of that river, one stage west of Sūdra. It is also called Hazārā and Hazārān. It lies on the Kabul Road four or five stages northwest of Lahore. The Bābā Ḥasan River comes from the northeast and passes alongside this place. This is not the Hazara in the country of Farḥāla.

Chendenot (Chiniot): a city northwest of Lahore on the bank of the Chenab. It is at the foot of a mountain on the Kandahar Road, two stages west of Hazara. It is also called Chīnot and Chendālpot.

Sīwa: on the bank of the Chenab, three stages west of Chendenot. Its original name is Shūr, but it is erroneously called Sīwa. The Wīhat River meets the Chenab two stages east of here, and the Rāwī River meets the Chenab one-half stage west. This Shūr is a town attached to Multan. It also has a fortress. Dates, pomegranates and bitter oranges are grown in its gardens.

Qashāw, also called Khashāb: a city 10 stages northwest of Lahore on the Jahīnkūt Road. Jahīnkūt is reached in three days from this place.

Bihrā: a city northwest of Lahore, on the Kabul Road between Lahore and Khashāb, on the bank of the Wīhat River.

Qaṣūr: a city near Sulṭān-pur, two stages south of Lahore.

The country of Dehlī (Delhi)

A large country to the east of Gujarat. Its capital is Dillī. On the map of the *Atlas*, it is a large capital city on a level plain located at 120 degrees longitude and 20 degrees latitude. In the *Taqwīm al-buldān*, it is specified as Dillī, with *kasra* on the *dāl* and a silent double-*lām*. [M 209] The three forms are variants.

The ground of this place is stony and sandy. A great river (the Yamuna), the size of the Euphrates, flows by one parasang away. Between Delhi and the sea is a great distance. It is 15 stages southeast of Lahore.

The Friday mosque of this city has a strange minaret (the Qutub Minar) constructed of red stone like the Lighthouse of Alexandria. [R 77a] Its lower part is broad and polygonal, and one ascends it with 360 stairs.

This city has a city wall built out of brick. Around it are a small number of vegetable gardens and vineyards. The people are mainly Muslim.

According to the *Haft Iqlīm*, Dehli surpasses all the cities of India in its delightful climate and pleasant orchards and fields.

It is recorded in the histories of India that in olden times this city was extremely prosperous. Later it somewhat fell into ruin. It was repopulated in the year 304 (916) and remained in the hands of the pagans of India for nearly 300 years. Sultan Quṭb al-Dīn Aybek conquered it in 588 (1192), and since that time it has been in the hands of the Muslims. Old Delhi is presently in ruins; only the aforementioned mosque and minaret are left of its former buildings. The Friday mosque is built on 1000 columns. The minaret is 80 paces in circuit and 130 cubits in height.

As for New Delhi, it was built one parasang from the old city on the bank of the Jawn (Jumna, Yamuna) River, by the command of Sultan Jalāl al-Dīn Khaljī in the year 688 (1289). It has an extremely pleasant climate. Around it are paradisical excursion spots, gardens and tall buildings that defy description. Among these, the tomb of the Emperor Humāyūn is a delightful and lofty building. Outside the city there is a structure known as Fīrūz Shah's Hunting Ground. In its middle is a great column, 30 cubits tall and 3 cubits in diameter, carved out of a single block of stone. It is said to be buried another 30 cubits in the ground. It seems impossible that such a huge stone could have been erected. There are many such marvels in this city.

Many learned individuals are from Delhi, such as Niẓām (al-Dīn) Awliyā and Mīr Khusraw.

Agra: a large city and capital 80 miles southeast of Delhi. It is related from the *Ṭabaqāt-i Akbarī* that in olden times, Agra was one of the dependencies of Bayāna (?). Sultan Iskandar (Sikandar Lodi, Sultan of Delhi, 1489–1517) sought to populate this place. After him, Shīr Khan and Salīm Khan also expended every effort and it became a prosperous and magnificent city. In the time of Jalāl al-Dīn Akbar (Mughal emperor, r. 1556–1605), Iram-like buildings and soul-enhancing orchards reached perfection and the city acquired splendor and commercial vitality. That is owing to the Jawn (Jumna, Yamuna) River that passes through the middle of the city. On both banks are lofty pavilions and palaces and gardens, like so many gardens of paradise.

This city has a fortress whose like has not been built or seen. By the time that lofty structure was complete, four years had passed and the wealth of Qārūn had been spent. Its two sides, extending to the river, are joined with hewn stones and iron clamps and rivets, so the fortress appears to be cut from a single stone. With this fortification, its ornamentation and beauty are to such a degree that one cannot get enough of looking at it.

Many learned individuals and excellent men have come from this city as well, including Abū'l-Faḍl b. Shaykh Mubārak, author of the *Tārīkh-i Akbarnāma*; [M 210] Shaykh Abū'l-Fayḍ, known by his pen name Fayḍī-i Hindī, who wrote a Koran commentary with undotted letters, also the *Mawārid al-kilam* and a collection of poetry; and Waḥshī, author of a collection of poetry.

Lucknow: a small city east of Agra. It is also called Laklaw. Its air is pleasant. The people are mainly boyers, and they make good bows.

Ḥiṣār: a large city northeast of Agra.

Hānsī: a strong fortress in the environs of Ḥiṣār. It was a famous city in olden times.

Barwaza: also a city near Ḥiṣār.

Tīra-mal: a city three stages northwest of Delhi. The idol-worshippers hold this place in great esteem. Like Jerusalem, many pious foundations have been appointed for its upkeep.

Kālbī: a seat of government in that area. Its city is on the bank of the Jawn (Jumna, Yamuna) River. The vegetation of that place is famous.

Other than these, among the cities in the vicinity of Delhi are Jāna-pur, Jājāmū, Sarmandal, Baliar, Kidārā, Shaqrīz, Pūrub, Chīlandhar, Chzūrlā, [R 77b] Dirwal, Nariat and Malnair.

Separate countries and regions of India

Now we will describe some of the countries and cities on the northern side and the interior of the clime of India that are not known to be included in any other part.

Rana Sanga (Mewar): a country of pagans located between Gujarat, Dawatabad and Lahore. Its ruler is known as Rana Sanga. Its capital is a city named Chūtpūr, also called Jūdapūr, Udaipur and Uda. It is west of and close to Agra. It is located on an arid mountain and has a fortress. Aḥmadābād is 10 stages from here. In the book of Lorenzo, this city is recorded as Chitor and the country of Sānghā.

Rantāpūr: a city in an arid place that draws water from deep wells. It rains only rarely. The rest of the country of Rana Sanga is also like this.

Chītawar (Chittor): a city and district four stages from Udaipur. One reaches Aḥmadābād is a three days' journey from here.

Dana-pur: also part of this country.

Sayyidpur: a city 10 stages east of Patan.

Kūlī: a city 10 stages north of Sayyidpur.

Tākwar: a city 10 stages north of Chūtpūr and four stages west of Lahore. Its people weave good broadcloth. The *saqarlāṭ* here is preferred over that in the other cities of India.

Jalore: a city three stages southwest of Tākwar.

Sirohi: a city ten stages south of Chūtpūr.

Qinnawj (Kannauj)—variants are Qinawj, Qinnūj, Qinūj: According to the *Taqwīm al-buldān*, a country and city in the extreme east of India. It is considered the Cairo of India. It has as many as 300 jewelers' shops. There is a gold mine. The Ganges River flows from a place 40 parasangs east of here. It is recorded at 140 degrees longitude and 28 degrees latitude. It is far from the sea; most of its cities are in the mountains. The people are generally idol-worshippers. Their ruler is called *nawda*. The raid (*ġazā*) of Sultan Maḥmūd ended at this place.

In the *Haft Iqlīm* this city is recorded as Kawaj. It lies to the northeast of Bengal. Its northern border ends at the frontier of Cathay, between which and here merchants are always coming and going. Another border is Kūra-kāt, which is called Shām on the frontier of Cathay. It is 20 stages from the frontier of Cathay to the country of Kawaj. The products of Kawaj are silk, pepper and horses. In India, this country is called Tānkehen. [M 211]

There is a cave in this country that the Hindus greatly venerate, thinking it is the dwelling of the demon named Āī. There is a gathering and festival there once a year. On that day they make sacrifices of every kind of animal. The *bhogis* also sacrifice themselves. They are a group that has pledged to sacrifice themselves on the path of Āī, saying, "Āī wants us." From that day until the same day the following year when he will be killed, he is permitted to do whatever he wants—or so he claims—and he may dally with any woman or girl whom he wishes, without hindrance.

Telangana: According to the *Haft Iqlīm*, a prosperous province, including villages and administrative districts, contiguous to the country of Deccan. In the past it was always customary for one of the esteemed rajas of India to rule in this country. Its capital is Golkonda, a city with many orchards and delightful marketplaces and buildings. It was the throne of the Bahmanids.

Probably the fortress of Kalinjar, also known as Mahra (?), which Sultan Maḥmūd conquered in his religious war (*ġazā*) against Kannauj, belonged to this country. It is also called Kālinjar. It is a city on a mountain, and a province. Shīr Shah used to live there. It is 15 stages northeast of Agra. Jalāl al-Dīn Akbar conducted religious war against it, took it and destroyed it. Shīr Shah was killed while firing a cannon. Afterward Jalāl al-Dīn Akbar repopulated the city.

Rūy-tāz: a city on a mountain in the vicinity of Agra.

Purna: Sultan Maḥmūd attacked it during his religious war against Kannauj and its lord converted to Islam. It is a city.

Manaḥ: a fortress of the Brahmins in those parts. Sultan Maḥmūd conquered it.

Mathura: a city containing a great idol temple. Sultan Maḥmūd seized it and destroyed the temple.

Gwalior: a fortress in the province of Nandā. Sultan Maḥmūd conquered it as well. [R 78a]

Lekīnhūt: a city in the vicinity of Bengal.

Pūtwāl, also called Pūtūhāl: a province. Its ruler's dynastic name is added to Kekkerān. It lies 6 or 7 stages northwest of Lahore, between Hind and Sind and Kashmir. Its rulers are named Kekkerān after the Moghul tribe. Its capital is Farḥāla, a city located at 122 degrees longitude and 33½ degrees latitude. It is also called Farhāla. It lies on the west bank of the Wīhat River. From here the journey to Lahore is seven days.

Rahtās: a city and fortress on the bank of the Ṣuwān River, 1½ stages south of Farhāla.

Sūkra: a city 3½ stages southwest of Farḥāla. It lies between Nīlāb and Farḥāla.

Rawalpindi: a city on the west bank of the Wīhat River, 4 stages southwest of Farḥāla. Nīlāb lies 3 stages to the east from here.

Damtūr: a small city 4 stages north of Farḥāla. The inhabitants are mainly Turkmen, and it is a place rich in mountain rice, wheat and honey. It lies between Nīlāb and Paklī.

Paklī: a famous city on a mountain northwest of Farḥāla, located at 110½ degrees longitude and 34½ degrees latitude. It lies between Damtūr and Kashmir. It has good rice and pomegranates. Its inhabitants are Turkmen. They wear gold necklaces. The Hazara River passes nearby, while the Sind River is to the east. Damtūr lies 5 stages east of here.

Hazara: a city that lies 4 stages southwest of Paklī, north of where the Bābā Ḥasan and the Sind Rivers meet. [M 212]

Mārglī: a city 2 stages east of Hazara and ½ stage from the Sind River.

Nīlāb: a mid-sized city at the foot of a great mountain 3 stages west of Rawalpindi. It is located at 110 degrees longitude and 32½ degrees latitude. The Sind River, also called the Mehran River, emerges from this mountain and flows south. From here both Lahore and Kabul are 10 stages distance via Kashmir.

In addition to these places, northwest of India are the countries of Multan, Kandahar, Kabul, Ashnāghar, Khwāst, Tibet and Kashmir. They lie between Hind and Sind and Kirmān and Khurāsān, beyond the border of India, and so a separate chapter will be devoted to each of them.

Now we will discuss the remaining features of India.

Authority

In the ancient polities of India the people were divided into seven categories: philosophers, farmers, shepherds, tradesmen, soldiers, judicial administrators (lit. qadis), and administrators of state affairs. Originally their laws were not recorded in writing; later they were collected in a volume.

Philippus says that there are three great sultanates in India: Cambaia, Narsinga, and Bengal.

Lorenzo says that the four groups that exist in the world are gathered in India: Jews, Christians, idol-worshippers and Muḥammadans. In olden times the latter were few, but by coming and going frequently for trade, they converted most of the idol-worshippers to their faith.

The padishah of these people belongs to a great dynasty. He rules over Cambaia, Deccan, Multan, Qūsbatīr and Delhi.

The rulers of Bengal in olden times were Ethiopians. As with the Circassians in Egypt, their slaves, who all come from Zaylaʿ, gradually become the rulers. They are Muslim.

Because meat is rare in his country (Bengal), merchants cannot trade unless they bring meat. The grandeur of this ruler is such that his vessels and other implements are studded with gold and silver and gems and his retinue have substantial salaries. His prayer-leader alone gets an annuity of 12,000 ġuruş. He has many great lords in his service known as *lāsqārī*. The ruler of Tipora (Tripura) pays tribute to him. Ambassadors of the rulers of Qāʾūr, Awrīqān and Aracan reside in his city. This ruler is very much drawn to music. In his court he employs musicians of Cambaia and Narsinga and of his own sultanate.

Aside from this, the kings of India generally do not give salaries to the military. When a campaign is afoot, they issue a summons and the soldiers come at their own expense.

The ruler of Gujarat is a powerful monarch, commanding up to 30,000 soldiers and elephants. On the sea side he is always doing battle with the Portuguese; but as the Rajputs keep harassing him on the land side, he has ceded most of the coast to the Portuguese and made peace with them. He has copious slaves and prisoners. As is the case with other rulers, most of his soldiers are of the slave type. [R 78b]

The country of Deccan consists of several parts, each having magnificent lords whose duty is to defend the province and to attend on the Deccan ruler once a year. Each of them has a palace in the city of Bīdar where his sons reside as hostages. When they come into the ruler's presence they place their hands on their knees, bow their head, greet him as if prostrating in prayer, [M 213] and withdraw. This they do twice. The ruler has little role in governing the country—that is in the hands of the lords—but they nonetheless fear him. His court is more splendid and glorious than theirs. His servants wear bejeweled and gem-studded garments. He selects the beautiful women of his own province and takes them to his palace.

There is a tribe in that country known as Wene-zārī, similar to the Tatars: they raid the province and give a share of the booty to the ruler.

It is related in the *Ḥabīb al-siyar* that Bijanagar is the capital of Deccan, and that its palace is the middle of a seven-layered fortress. A lofty and spacious audience hall is built atop 40 columns to the right of the sultan's pavilion. In front of it is a platform about the height of a man, 30 cubits in length and 6 cubits in breadth, upon which a book rests. It is the registry. The majordomo, known as *wanāhik*, sits amidst the 40 columns in an independent seat of government. The officers stand in line in front of him with their staffs. The supplicant approaches with a gift, places his face on the ground and makes a petition. He (the majordomo) then makes a ruling according to the law of justice and it is executed.

When the *wanāhik* gets up from the audience, he passes through several colored tents. On both side eulogists shower him with benedictions. When he goes before the ruler from the audience-hall, seven gatekeepers are sitting there, each with a group of assistants. They present their petitions. He enters alone through the seventh gate, submits the matters of sovereignty and finance to the ruler, and immediately departs.

On the left side of the palace is the mint, where three kinds of gold coin are struck: 1) an alloyed coin called *darha*, nearly 1 *mithqāl* in weight; 2) one called *partāb* that is half of that; 3) one called *fanam* that is one-tenth of a *darha*—this is mainly what is in circulation. They also make a coin of pure silver called *tār* that is one-sixth of a *fanam*—this is also in circulation; and a copper coin called *chatal* that is one-third of a *tār*. At a designated time they bring the gold pieces to the mint from all of the countries, and once every four months the soldiers are paid a stipend from the mint. No one's stipend is assigned from the province; (rather they are paid out of) treasuries that have been excavated in the ground and are full of gold and silver. Opposite the mint is a police station with 12,000 guards. Their daily salary, amounting to 12,000 *fanams*, is gotten from the income of the whorehouse that is behind the mint.

Kamāl al-Dīn ʿAbd al-Razzāq says in the *Maṭlaʿ al-saʿdayn*:

When we came to Bījānagar with a message from Khāqān Saʿīd in Muharram 848 (1444), the ruler was sitting in grandeur among the 40 columns with soldiers in jeweled garments standing on the left and right. The Raja wore a necklace made of large pearls and a cloak of olive-green satin. He was tall and slender, with a dignified expression. We stood on his left side. He said: "We are happy that the Great Emperor has sent an envoy to us." They brought a plate filled with silver and gold.

At that time, this ruler had 700 wives. No two wives were in the same house, and ten-year-old boys did not enter the harem. If word came of a beautiful girl somewhere in his country, she would be brought with her parents. He would take the girl into the palace and send off her parents with great favor.

Once a year the sultans of Bījānagar arrange an imperial feast [M 214] known as *masāwī*. All the commanders and governors under his rule attend. 1000 elephants are decked out with weapons and litters, and tumblers perched on top. For three successive days in the middle of the month of Rajab they set up camp in a broad plain, a beautiful place. Trellises are bound together, three or four levels each, decoratively painted with birds and animals and plants, and pretty slave-girls sitting on top. A nine-vaulted palace on 40 columns is erected for the ruler. In the middle is a vast square where instrumentalists with rose cheeks and girls beaming like the sun dance and sing, the tumblers display their skills, and trained elephants sway and move their trunks in time with the rhythms of the musicians.

The governor of Malabar is also the ruler of an independent country, with lords under his control. In his presence there is always a candlestick topped by a red ruby like a carbuncle that sheds light all around. His intention thereby is to proclaim that the holders of government require intellect that is of suchlike brilliance.

The king of Narsinga is also a great ruler. The idol-worshippers call him *Rākhū*. He has many great lords under his authority who receive an annual income of 100,000 or 200,000 gold pieces and stand at his gate. This ruler takes into the palace the most beautiful daughters of the notables in his country. Should he receive a complaint concerning any of the provincial governors, he brings him into a special room and gives him a beating with his own hands; afterwards he dresses him in a gold brocade robe of honor.

This ruler has a large number of Brahmin soldiers in his court who are highly esteemed. And he has a kind of soldier called *nāīrī*. At first they play with cudgels. Once they have demonstrated their skill, he inducts them into this society (of *nāīrī*) by girding each one with a sword reddened with blood. They take a vow before the ruler and swear to treat cows and Brahmins with kindness and to sacrifice themselves in battle. Their wives are held in common between themselves and the Brahmins. Their sons do not inherit from them; they are sons of the public at large; only their mothers are known and the ruler gives them a monthly stipend of one-half *guruş*. This group is skilled on the field of battle and in swordplay. They are 40,000 strong. They challenge one another and fight on account of their wives.

This ruler takes a heavy customs duty from merchants. He executes his lords for the most trifling crime and seizes their wealth. If their wealth becomes too great, he imprisons them.

The great sultanate in India is that of the Mughal emperor. The other Muhammadan rulers in India obey him. The rulers of Deccan, Cambaia, Mārdū and Sānghā used to be appointed by him. He mainly does battle with the Yashilbash ("Greenhead") Tatars, i.e. Uzbeks, and with the Sofis, i.e. the Qizilbash ("Redheads" = Safavids). He keeps a large number of horses, elephants and camels and possesses many troops and treasuries. He sometimes also clashes with the kings of Bengal and Badakhshan. His banner is white lines on a green ground. His soldiers speak Persian. They are warlike and wield various weapons. They are Afghans, gathered from every tribe, and employ cannons and muskets.

Lorenzo said that in his time the Mughals had little independence; in these times, however, they have overcome the other kings around them [M 215] and added their countries to the ninth part of India. He (the Great Mughal) has rendered the countries of Bengal, Malabar, Deccan, Gujarat, Kashmir and Burhanpur submissive. Only some of the pagan Indian kingdoms bordering with China, and the Portuguese on the coast, have not submitted.

Lorenzo also says the ruler of Siam is the wealthiest of all the kings of India. He has on land 3000 elephants and at sea a fleet of many ships, known in their language as *yerāy*, that resemble the Turkish galleys. He is at peace with the king of Portugal [R 79a] and they exchange gifts every year. Before the Muḥammadans took Malacca it was under his rule. Later, the Portuguese took it from them and made peace with him.

Kings

Nothing definitive is known about the sovereigns in these countries before Islam. Only the Raja of India is recorded at the time of Alexander the Greek who journeyed to that region, but about him as well nothing has been confirmed.

Pliny says the Indians are an ancient people who never once left their own countries. They say that their first ruler was Liber Baba (Liber Pater). Later the Persians invaded and seized the countries between the Indus and the Ganges. After the Persians, Alexander routed the Indian ruler Por (Porus), but he did not attack the ruler of the Gangaridai. After that, each country had its own ruler, and this continued for some time. From Liber to Alexander there were 153 rulers overs a period of 2403 years.

According to Strabo, no one was victorious over the Indians except Hercules and Alexander, and the aforementioned Liber. Cyrus and Semiramis also did battle with them, but were not victorious.

Masʿūdī writes in the *Murūj al-dhahab* that the first king of India was Albahar the Great—he is the aforementioned Liber; then there was Nāhūd; Dāmān; Fūr (Porus), who did battle with Alexander; Dābshalīm who commisioned the *Kalīla wa Dimna*; and Kūrash. He mentions these rulers' names, but has no definitive information about them.

It is recorded in some of the chronicles that in the year 94 (712), while Ḥajjāj (b. Yūsuf) was in Fārs, he made Muḥammad b. Qāsim al-Thaqafī commander and sent him to India, and that he conquered and raided some cities.

After the Islamic conquest some Muslim kings are mentioned in the chronicles; but their governors, as well as the idol-worshipping kings, go unmentioned, so one cannot say anything about them.

The first of the Muslim kings to conduct *gaza* into India was Nāṣir al-Dīn Sabuktagin. Then his son, Maḥmūd of Ghazna, marched into India in 392 (1001) and made many conquests. He mustered an army and entered the interior of India eight times, taking Kannauj and Somnath. After him, his son Sultan Masʿūd and his son Sultan Ibrāhīm and his son Bahrām Shah conducted campaigns and appointed deputies to the northern part of India. Later, the son of Bahrām Shah, Khusraw Shah, established an independent sultanate in Lahore for eight years. When he died in 555 (1160), his son Khusraw Malik Shah succeeded him, but because he was given to drinking and carousing, Shihāb al-Dīn Muḥammad of the Ghūrids, who had conquered Ghazna, marched against him and took India from his hands. The Ghaznavids came to an end in 579 (1183) [M 216] and control of India shifted to the Ghūrids.

This Shihāb al-Dīn was vouchsafed many conquests in India. In 590 (1194) he appointed one of his personal retinue, a slave-soldier named Quṭb al-Dīn Aybek, as governor in Lahore and himself went to Ghazna. By the time Quṭb (al-Dīn) died in 607 (1210) he had established an independent government in India and conducted several *gazas*. Because his young son Ārām Shah was incapable of governing the Iletmish (Iltutmish) Dynasty came into being. One of Quṭb (al-Dīn)'s emirs and statesmen, Shams al-Din Muḥammad, who had the title Iletmish, was made regent. He reigned as sultan in Delhi for 26 years and performed several ghazā; he died in 633 (1236). Because his son, Fīrūz Shah, busied himself with amusement and play, the emirs placed his sister Sultan Raḍiyya on the throne; she sat and governed in the garments of a man. Later they installed Bahrām Shah, son of Iletmish; he was put to death after ruling for two years. His successor Masʿūd, son of Fīrūz Shah, followed the path of justice; he was deposed after four years. Nāṣir al-Dīn Maḥmūd, son of Shams al-Dīn (Muḥammad), who succeeded him, was a just and intelligent padishah; the *Ṭabaqāt-i Nāṣirī* was compiled in his name and he reigned as sultan in Delhi for nearly 20 years.

Then arose the dynasty of Ulugh Khan, which was a branch of the Iletmishids. When Nāṣir al-Dīn (Maḥmūd) ruled in Delhi, he made Ghiyāth al-Dīn Balbān, who was his slave and son-in-law, his vizier and awarded him the title of Ulugh Khan. Having entrusted matters of governance to him, he (devoted himself to piety) writing two copies of the Koran in a year and spending what he earned on his own sustenance. When he died in 650 (1252), no heir to the sultanate remained and Ulugh Khan became independent sovereign. He was intelligent and just. His elder son, Sultan Muḥammad Khan, became his heir-apparent and governed Multan, while his other son, Nāṣir al-Dīn Bughra Khan, was governor of Lucknow. Aytemür, one of the Mughal emirs, marched against Muḥammad Khan with a massive army and wreaked destruction between Lahore and Dibalpur. As Muḥammad was martyred in battle, his father grew ill from grief and died after 20 years of rule.

The son of Bughra Khan, Muʿizz al-Dīn Kay-Qubād, succeeded his grandfather and busied himself with drinking and carousing. Getting wind of this, his father started from Lucknow and had a meeting with his son on the bank of the Āb-i Sarw—Mīr Khusraw wrote his poem *Qirān al-Saʿdayn* about this episode. After that [R 79b] he gave him much admonition and counsel, but it was no use: excess of drinking made him paralyzed and afflicted with palsy. The emirs installed his son Shams al-dīn in his place while still a child. After his father had ruled three years and himself two months, he was assassinated by Malik Fīrūz Khaljī.

Appearance of the Khaljī Dynasty: Some of the emirs, seeing the condition of Shams al-dīn, communicated with Jalāl al-Dīn Fīrūz Khaljī, one of the provincial deputies, and invited him. He came with the Khalj emirs in 689 (1290) and seized control of the country. He built New Delhi and reigned as sultan for seven years, making great conquests. Then his nephew and son-in-law, ʿAlā al-Dīn, [M 217] who was governor of Agra, rebelled and Fīrūz went to repulse him. In 695 (1296) he came to the bank of the Ganges and ʿAlā al-Dīn begged for forgiveness. During their meeting, Fīrūz was killed by Maḥmūd Sālim from the rabble of Sāmānā.

ʿAlā al-Dīn, usurping the throne, came to Delhi and blinded Fīrūz's young son Ibrāhīm with a stylus. He reigned as sultan for 20 years, carrying out great religious wars and taking many countries. He commited affairs of state to his deputy, Kāfūr Hazār Dīnārī, and appointed his son Khiḍr Khan as his heir. When he died in 725 (1324), Khiḍr Khan was deposed and imprisoned and his other son, Shihāb al-Dīn, succeeded while still a child. His followers blinded Khiḍr with a stylus—Mīr Khusraw wrote a book about this—and two months later brought his brother Quṭb al-Dīn Mubārakshāh to the throne.

Quṭb al-Dīn reigned four years and four months. He used to wear women's clothes and make fun of the emirs. He made Khusraw Khan, one of his personal slaves, governor through favorable treatment. This Khusraw asserted his independence, had Quṭb (al-Dīn) killed at the hands of his own faction, and succeeded him under the regnal title Sultan Nāṣir al-Dīn. He killed ʿAlā al-Dīn's son Farīd Khan and also Mengü Khan. Seeking revenge, Ghāzī Malik (Tughluq), one of the emirs, marched on Delhi, routed the troops, took Khusraw captive and had him torn to pieces. This was the end of the Khaljī dynasty.

Appearance of the Tughluq Shah Dynasty: This Tughluq Shah is the aforementioned Malik Ghāzī Mubārakshāh. The Turks called him Qutluq Shah; it became Tughluq through frequent usage. When he did away with Khusraw in 720 (1320), none of ʿAlā al-Dīn's children was left. Tughluq Shah assumed the throne in Delhi. He was an intelligent and just ruler. He conducted religious war against Lucknow and subdued the governor of Bengal. Having designated his son Ghiyāth al-Dīn Muḥammad as heir, he was crushed beneath the walls of a pleasure-palace that he had built near Tughluqābād when they collapsed as he was feasting. He ruled for four years and was succeeded by his son who assumed the title Sultan Muḥammad Tughluq Shah. His strange behavior and wasteful extravagance overburdened and alienated the people. He died in 752 (1351) after ruling for 27 years. According to his last will and testament he was succeeded by his nephew Fīrūz Shah, the son of his father's sister.

Fīrūz Shah built Firozabad and had several canals dug and cities rebuilt. In 787 (1385), having reached old age, he appointed his son Muḥammad Khan as plenipotentiary. He died (in 790/ 1388) after ruling for 38 years. He is counted among the greatest of sultans, with many monuments to his name. His son Muḥammad Khan, having been led astray by corrupt people, had rebelled against his father and fled. Ghiyāth al-Dīn Tughluq Shah b. Fatḥ Khan b. Fīrūz Shah, who was installed in his place, became sunk in drinking. His nephew Abū Bakr, son of his brother Ẓafar Khan, rebelled and gained the support of the emirs. Ghiyāth was put to death, his rule having lasted 53 days, and was succeeded by Sultan Abū Bakr. His uncle, Sultan Muḥammad b. Fīrūz, marched on Delhi from Nagarkot. Although he was defeated, and then was defeated again when he collected troops a second time, some of the emirs secretly sought him out. Abū Bakr fled after ruling 1½ years, and in 792 (1390) [M 218] Muḥammad Shah was seated on the throne in Delhi. Abū Bakr was put in prison and died there.

Muḥammad Shah ruled for 6½ years. At his death the emirs installed his son ʿAlā al-Dīn Iskandar Shah in his place, but he too died after one month and they declared Muḥammad Shah's youngest son, Maḥmūd Shah, as padishah. [R 80a] He gave each of the emirs a new title. Khwāja Jahān got the title Malik al-Sharq (King of the East), given to him when he went from Kannauj to the province of Bihār. He made Jaunpur the capital and took possession of the region. Several individuals from his line exercised sovereignty. After him, Muqarrab Khan became heir-apparent. The emirs in each province asserted independence. When three years had passed, Tīmūr appeared and marched on Firozabad. Maḥmūd was routed and fled to Gujarat. Delhi lay in ruins for two months. Tīmūr had appointed Khiḍr Khan to Multan. Sultan Maḥmūd went to Kannauj and died there in 804 (1401), having ruled for 20 years. That was the end of the Tughluqids.

Appearance of the dynasty of Khiḍr Khan: He served as governor in Multan for seven years and had the Friday sermon made and coins struck in the name of Tīmūr and his son Shāhrukh. He died in 824 (1421) and was succeeded by his son Sultan Mubārak Shah. During his reign the emirs and kings were active and many battles took place. He was assassinated at the instigation of his vizier Kamāl al-Dīn, in the city that he had built on the Jawn (Jumna, Yamuna) River, in 837 (1433), having ruled for 13 years. During the reign of his successor, his son Sultan Muḥammad, the kings of the vicinity continued to be disobedient and to raise the banner of independent rule. He had given sovereignty over Lahore and Dibalpur to one of his emirs, Malik Bahlūl. After reigning for two years he died in 847 (1443). His son and successor ʿAlā al-Dīn busied himself with drinking and died after a rule of seven years.

He was succeeded by Malik Bahlūl (Lodi), who followed the path of justice. He fought battles with the eastern kings Maḥmūd Shah, Muḥammad Shah and Sultan Ḥusayn, and emerged victorious. He ruled for 38 years and died and

894 (1489). His son Niẓām Khan took his place and was enthroned in Delhi with the regnal title of Sultan Iskandar. He clashed with Sultan Ḥusayn, ruler of Lucknow, and ʿAlā al-Dīn, ruler of Bengal, and emerged victorious. He reigned for 28 years and died in 924 (1518). Sultan Ibrāhīm, his eldest son and successor, gave sovereignty over Jaunpur to his brother Jalāl Khan. When the latter rebelled, he placed his five (other) brothers who were with him in prison. Jalāl Khan marched upon Agra with 30,000 horsemen; he was captured and executed. After Ibrāhīm had reigned as sultan for seven years, the emirs turned away from him. The governor of Lahore, Dawlat Khan, went to Kabul and incited Bābur Mīrzā, a descendant of Tīmūr, to advance on India. Dawlat Khan himself died on the way. The emperor Bābur clashed with Sultan Ibrāhīm in the vicinity of Panipat and routed him. Sultan Ibrāhīm fell along with a group of his emirs. The Afghan era came to an end there.

Appearance of the Timurid (Mughal) Dynasty: [M 219] Previously, when the Uzbeks took power in Transoxiana, Ẓahīr al-Dīn Muḥammad Bābur b. Mīrzā ʿUmar Shaykh came to Kabul and took Kandahar. In 930 (1523) he gained victory in India and seized many treasures and sources of wealth. He made his son Muḥammad Humāyūn Mīrzā governor over Agra.

Humāyūn clashed with Rana Sanga, mentioned above, and emerged victorious. He was famous for courage, and when Bābur Mīrzā died in 937 (1530), he succeeded him as padishah. The chronogram of his enthronement is *khayr al-mulūk* ("best of kings"). His brothers Mīrzā Kāmrān, Mīrzā ʿAskerī and Mīrzā Hindāl each became governor over a country. When the ruler of Gujarat, Bahādur, showed opposition, he marched against him and routed him and distributed his country among the emirs. And in 947 (1540) when the King of Bengal, Shīr Khan, rebelled, he marched against him, but was himself routed and fled. First he went to Kandahar, then to Herat where he met with Shah Ṭahmās. The Shah carried out the obligations of respect, gave him a military reinforcement and sent him to Kandahar. [R 80b] Humāyūn recovered the aforementioned capital from the hand of the usurper and made peace with Shīr Khan. He got his brothers out of the way and continued his reign until his death in 962 (1554). He was a scholar and knowledgeable ruler.

His eldest son, Jalāl al-Dīn Akbar Muḥammad, succeeded him. He overpowered the sons of Shīr Khan and conquered Gujarat. He also took Kashmir. He continued his reign until his death in 1012 (1603). His son Sultan Salīm (Jahāngīr) assumed the throne and ruled for 20 years. He had an intelligent vizier named Āṣaf Khan. Salīm Shah died in 1032 (1622) and was succeeded by his son Khurram Shah with the regnal title Shah Jahān, and he turned out to be intelligent and just. He conquered most of the countries of India and the Uzbeks. He is now sovereign in the clime of India.

The Shīrkhānid dynasty: This Shīr Khan, whose name was Farīd, was from the tribe of Sūr. His grandfather Ibrāhīm came to India and was in the service of the emirs. His father Ḥusayn held the land grants of Jaunpur. Of his 12 sons, Farīd was resentful toward his father. He studied Arabic and history in Jaunpur and entered the service of Dawlat Khan. When his father died, he took charge of the land grants and came into the service of Bahādur Khan who had the title Sultan Muḥammad. One day, while hunting, a lion threatened Sultan Muḥammad. Farīd slew the lion with a sword and was awarded the title Shīr ("Lion") Khan. Later, Sultan Muḥammad turned away from him and gave some of his land grants to his brothers. He then switched his allegiance to Sultan Junayd, one of Bābur's emirs, and marched upon Muḥammad Khan with the Mughal army. The latter went into the Rahtās mountains and Shīr seized his country. He attended the court of Bābur for several days, but they did not get along, so he returned to the service of Sultan Muḥammad and was made tutor to his son Jalāl Khan.

Muḥammad died at that point, and Jalāl succeeded. Shīr married his mother and became an independent ruler. He marched upon the ruler of Bengal, Ibrāhīm Khan, and defeated him in battle. Seizing his elephants and cannons, he invaded Bengal and conquered many fortresses. Finally, while besieging the fortress of Kālinjar, the gunpowder caught fire and his death occurred on the same day as the conquest of the fortress. He ruled 15 years as an emir and 5 years as a sultan.

In 955 (1548), the emirs made Jalāl Khan padishah and gave him the regnal title of Islām Shah. [M 220] Later he became known as Salīm Khan. His rule lasted nine years, and when he died his son Fīrūz Khan, who was still a child, succeeded. Shīr Khan's nephew Mubāriz Khan, the son of his brother Niẓām Khan, killed Fīrūz on the third day and took up residence in the province of Sarīr. He reigned for three years, with the regnal title Sultan Muḥammad ʿĀdil, doing battle with his opponents. Finally he was killed near Jinār, where he went to remove Bahādur Khan, and his dynasty came to an end.

The Sultans of Gujarat: When after the death of Fīrūz Shah the provincial emirs raised the banner of independent rule, Muẓaffar Shah was governor in Gujarat. In 773 (1371) he began to rule in his own name. He was succeeded by his son, Muḥammad Shah—or, according to another account, by his grandson Sultan Aḥmad who constructed Aḥmadābād in 813 (1410). His son Sultan Maḥmūd was next in line; then his son Muẓaffar; and after that his son Iskandar. The latter, after ruling for seven months, was deposed and executed by his vizier. Because his son Maḥmūd Shah, who succeeded, was very young, ʿImād al-Mulk acted as regent. The emirs summoned Bahādur Khan, the vizier of the governor of Delhi Ibrāhīm Khan, who came and killed Maḥmūd Shah and became ruler. The Portuguese cultivated good relations with Bahādur Khan, but eventually the captain of Goa killed him by means of a ruse (on board of his ship in 943/1537). After that, his son Shah Laṭīf became ruler. His vizier Daryā Khan brought Sultan Maḥmūd [R 81a] out of prison and took the oath of allegiance to him. He continued ruling until he was killed in 961 (1554). Then Aḥmad Shah, a member of his household, succeeded, but he died shortly thereafter.

After this, Cenābī writes that Shīr Khan was one of the sons of the abovementioned Sultan Maḥmūd. But the account given above is correct.

The Kings of Deccan and Gulbarga (Bahmanids): The one of the Delhi Sultans who conquered the country of Deccan is Ḥasan Kākūr (?) who had the regnal title Sultan ʿAlā al-Dīn Khaljī. He seized Dīwgīr (Devagiri, Deogiri), known as Dawlatabad, with its dependencies, which belonged to Sultan Muḥammad, in 448 (1056) and became independent ruler. His line continued as sultans in the cities of Gulbarga, Bīdar and Aḥmadnagar. Because Ḥasan's genealogy goes back to Bahman b. Isfandiyār, they are also called the Bahmanid Sultans. After ruling for 21 years he died and was succeeded by his son Sultan Muḥammad. He died after ruling for 18 years the oath of allegiance was given to his son Mujāhid Shah. A year later his cousin Dāwud Khan killed him and took his place, but he too was killed by his emirs after one month in office and the sultanate shifted to Maḥmūd (*recte* Muḥammad) b. Ḥasan Shah. He continued for 19 years and his son Ghiyāth al-Dīn became ruler after his death. In 790 (actually 799/1397) he was blinded with a stylus through the treachery of one of his slaves and Sultan Shams al-Dīn took his place. His brothers Fīrūz Khan and Aḥmad Khan marched against him and routed his forces. Fīrūz Khan ascended to the throne, captured Shams al-Dīn and put him to death, after he had ruled for 57 days.

Fīrūz Shah turned out to be intelligent and just. He took Vijayanagar and died after 25 years of rule. Because his son Ḥusayn was very young, [M 221] The Great Khan Aḥmad Khan succeeded. He defeated Sultan Hūshang, the ruler of Kahraka (?), and died after ruling for 12 years. His eldest son ʿAlā al-Dīn Aḥmad Shah, who succeeded him, followed the path of justice. He defeated Naṣīr Khan, the governor of Burhanpur, who had raised an army and marched on Deccan. Some injustices took place during his reign and several thousand men were executed. He died in 862 (1458) after ruling for nearly four years. Next in line was his son Humāyūn Padishah who died after ruling for three years.

One of his sons, named Niẓām Shah, who was eight years old, was proclaimed padishah. The unbelievers of Orissa took this opportunity to invade, but they were repulsed by the people of Deccan. When Sultan Maḥmūd Khaljī was informed about the sultanate of this child he marched to Deccan with an army and gained victory in battle. Niẓām died in 867 (1463) and his brother Muḥammad Shah succeeded with the regnal title Sultan Aḥmad Shukrī. The vizierate was given to Maḥmūd Gīlānī, who gained fame under the title Khwāja-i Jahān. He gradually became the object of the courtiers' envy. They forged a letter from him to the ruler of Orissa and, casting him under suspicion, had that intelligent and virtuous old man put to death. Then each one raised the banner of opposition. Three months later Sultan Muḥammad also died, having ruled for 19 years.

His son and successor, Sultan Maḥmūd, fell to drinking and revelry and the emirs claimed independence in every quarter, dividing the country of Deccan among themselves. Sultan Maḥmūd nominally remained sultan for 37 years and died in 924 (1518). During his reign, Ismāʿīl ʿĀdil Khan, Qutb al-Mulk, ʿImād al-Mulk b. ʿImād al-Dīn—who was Daryā Khan—and Malik Aḥmad Niẓām al-Mulk assumed control over the countries. After this, the Bahmanid dynasty died out and the kingdom of Deccan reached its terminus. A few times they tried to claim the name of sultan, but did not succeed.

The Niẓāmid dynasty: Among the aforementioned usurpers, Malik Aḥmad Niẓām al-Mulk was independent ruler in Dawatabad and elsewhere for 12 years or, in another account, 19 years. When he died, his son Niẓām al-Mulk Burhān Shah succeeded at seven years of age. [R 81b] A certain ʿAzīz al-Mulk took control. The emirs incited ʿImād al-Mulk against him. ʿAzīz defeated him, but then his eyes were put out with a stylus and Shah Ṭāhir became vizier. He converted Burhān to Twelver Shiism. Eventually Niẓām was defeated in battle against Bahādur, the ruler of Gujarat, near Burhanpur. He became ill and died after 55 years of rule.

His two sons, ʿAbd al-Qādir and Mīrānshāh Ḥusayn, began quarreling and the people split into two factions. Finally Shah Ḥusayn emerged victorious and the Deccan sultanate was fixed on him. He continued for 12 years, then became ill from an excess of drinking and died. His son Murtaẓā Niẓām Shah took his place. For seven years his mother Humāyūn Shah conducted affairs and fought many battles with the surrounding kings. Eventually Murtaẓa imprisoned her and committed affairs to Khwāja Mīrak, granting him the title Jangīz Khan. Jangīz [M 222] performed many services for him and put his state in order. Murtaẓā Niẓām continued for 24 years. In 996 (1588) his son Shah Ḥusayn usurped his position and put him to death. His father's vizier, Mīrzā Khan, pretending to pay him allegiance, entered the castle and imprisoned him. He wanted to make Shah Ḥusayn's cousin, Ismāʿīl b. Burhān, the padishah. The people of Deccan revolted and besieged the castle. Mīrzā Khan cut off Shah Ḥusayn's head and threw it down from the castle wall. The Deccanese proceeded to set fire to the castle and massacred the seditious ones inside. The reign of Shah Ḥusayn lasted 18 months. Many excellent men perished during this civil war. The sultanate was settled on Ismāʿīl Shah, who ruled for 1½ years. This event occurred in 999 (1590).

The Kings of Talingāna and the Quṭbid dynasty: With the disarray of the Bahmanids, Sulṭān Qulī Quṭb al-Mulk seized this country and became ruler. Being of a vicious character, he was assassinated at the hand of a Hamadānī. His elder son Jamshīd succeeded and his younger son Quṭb al-Mulk Ibrāhīm fled. Jamshīd ruled for 7 years, after which his brother Ibrāhīm ruled for 34 years, being on good terms with the Persian Shahs. When he died in 969 (1561) his son Muḥammad Qulī Quṭb Shah succeeded him.

The Kings of Khaljī in Oudh: They are a branch of the Ghūrids. Previously a group of the Khaljis of Ghūr had come to India with Sultan Shihāb al-Dīn. One of them, Muḥammad Bakhtiyār, a very strong and robust individual, was victorious in battle over the Hindus, routing their elephants and invading Oudh and Lankūtī. He had the Friday sermon made and coins struck in his own name. After that he marched on Tibet and was defeated. He died in 602 (1205), and one of his followers, Muḥammad Shīrān took his place. He too fell in battle with the Hindus. By order of Sultan Quṭb al-Dīn, ʿAlā al-Dīn Mehran was appointed. Being stupid and bloodthirsty, he was assassinated and Ḥusām al-Dīn ʿIwaḍ took his place. In 624 (1227) he too was killed in battle with Malik Nāṣir al-Dīn Maḥmūd, and their dynasty came to an end.

The particulars of the kings of India, as mentioned in the Islamic chronicles, are summarized to this extent.

Religious belief

All of the sects and religions are present in India—Jews, Christians, Zoroastrians, Muslims, idol-worshippers, star-worshippers, worshippers of plants and animals, worshippers of the four elements, worshippers of the heavens—90 different sorts, including 48 false sects. Those who dwell in the west and north of the country mainly pass as Muslims. Most of these believe in reincarnation. Some do not eat without a toothbrush and a ritual ablution, but they fail to perform the ablution when they are ritually impure. The majority of those who dwell in the east and south of the country are idol-worshippers and Brahmins, with many differences among them.

Muḥammad Yūsuf Harawī, who travelled around India and recorded the strange things that he witnessed, writes: When I reached the borders of Kālibī I saw a crowd whose heads were uncovered and their hair unkempt. They both travel the highway of Islam [R 82a] and at times worship idols. On their waist is a *zunnār* belt and in their hand is a rosary. If they are wronged, they set fire to one another's houses and kill their fathers and their sons.

Except for the Muslims, Christians and Jews, fornication is permitted among the Indians. One woman has several husbands. [M 223]

The religious affairs of the idol-worshippers are entirely in the hands of the Brahmins, who claim to be descendants of Abraham. Since they belong to many different sects, the astrologers consider that they are related to the moon. The differences of opinion among them are greater than the differences of the philosophers over matters of natural science. Some of them worship cows and elephants at night. Some worship the sun and the stars, saying, like the ancient Persians, that all things emerged from the sun. Others worship the elements and their own heroes of the past, like the ancient Greeks.

In Gujarat the most eminent of the polytheists follow the opinion of Pythagoras and venerate the Manichaeans. They are called Baniari. Killing is a great crime among them, even if it is only an ant. They fumigate with various types of incense and practice monogamy. They consider charging interest to be permissible. Most of them fast (during the day and) at night eat a piece of sugar and drink milk. They feed ants with sugar water. They will their wealth after they die to a place in need of water.

There is a group known as Abduti who abstain from everything for a period, then engage in all sorts of lewdness and debauchery. Some affirm the existence of the Creator and deny the Messengers. Some worship gods of various sects. Others do not weary their souls, deny all beliefs and pay no heed to worship.

The pagans of Pegu (Burma), like the Christians, worship a god whom they depict in three equal forms and whom they call *sān-rūī* meaning "one god in three." Their dervishes are called *dūlīs*. They have monasteries where bells are rung and they worship in a special language.

In the country of Orissa there is a temple in a place called Purusotam (Puri) known as Chaganat (Jagannath, Juggernaut). A large and revered idol has been placed there. When the Hindus get married, they first place their wives in the service of that idol for three days; if their virginity is taken away, the men rejoice, otherwise they deem it inauspicious. The Hindus rend their chests and flagellate their backs in that temple, then rub the wounded places on that idol and are healed. Anyone, whether a believer or unbeliever, who behaves improperly before that idol, is put to death.

It is recorded in the *Haft Iqlīm* that when Mawlānā Luṭfullāh Nīsābūrī, traveling on land and sea, came in a company to that temple, the Brahmins allowed them inside on condition of proper conduct. When one of them intentionally spat toward the idol, he immediately expired and died. The Mawlānā states: "We were very strongly affected by this event, wondering at the mystery that a false idol should have such an effect. In our sleep we were vouchsafed a dream that resolved the problem. Someone explained that because human souls had been directed toward it for so many years, a connection emerged in between. Since it is a conduit of souls, that effect manifests itself."

The sciences

India has no lack of scholars and sages. In their view, India is the source of wisdom. The first of the sages, the greatest brahmin, manifested wisdom and invented the implements of warfare. He recorded the figures of the heavens and constellations in a book named *Sind-hind* meaning "Age of ages" and established the prime origin (i.e., the beginning of the world). He said that the cycle of the revolution of the world divided into four periods. The first lasted 1,728,000 years, the second 1,296,000 years, the third 864,000 years. [M 224] In the fourth period the natural human lifespan is 100 years, in the third it was 1000 years, in the second 10,000 years and in the first 100,000 years.

The Brahmins and the Sophists who came after him were concerned with astrology, medicine and wisdom, and they had no equal in the science of arithmetic. The Greeks and Arabs took it from them. They established the science of magic, charms, spells and talismans. Their soothsayers are without equal in divination.

They write from left to right. They have two types of script. One of them is written on coconut palm leaves with an iron pen; these leaves are up to two cubits long, but the writing becomes effaced. The other is written on black paper with a kind of soft white stone that is whittled like a pen; the writing appears as white and cannot be erased; it is used by the scribes of Deccan. [R 82b] The letters resemble Arabic letters. They have various languages. They have great esteem for augury and fortune telling, and consider astrology and magic pre-eminent among the sciences. The science of medicine is also required, as stipulated by their lords; but they cannot rival the Greeks and the Arabs in medicine.

Punitive regime

Lawsuits and disputes occur rarely among the Indians. They do not require witnesses or evidence. They depend solely on oral testimony. If someone bears false witness, they cut one joint from the tip of each of his fingers. If someone injures another person's limb, he has his hand cut off and must also pay the indemnity. If he injures the eye or hand of an artisan, he is put to death.

Craft and commerce

Artisans are greatly esteemed among the Indians. All artisans are exempt from taxes and also are provided with grain supplies out of the state revenue. Trade is flourishing in many places of India. Especially fragrances, spices, gems, silk, cotton and other goods are exported abroad.

War and bravery

For the most part the Indians are courageous and ruthless warriors. They bring elephants into battle and fire cannons and muskets from atop them. Horsemen are rare. They sometimes drive 30,000 elephants into battle.

Mores and customs

Indians are mainly dark-complexioned, of middling height, slight of build, and have long beards. They do not know grief or anxiety. When the beard grows down below the belly, they shave it off. They let their hair grow long. The grandees deck themselves with gems and don various garments. The middle classes wear cotton and broadcloth. The lower classes wear animal and bird skins. Some go around naked, only covering their pudenda. The women wear only undergarments.

There is a great esteem for family and lineage among them. They marry as many wives as they wish, paying a dowry of two oxen for each one. Most of the women are harlots. They themselves are a nation of tricksters. They are not resolute in an opinion; but they are very obedient to their lords.

Their food is mainly rice. They slaughter animals by striking them on the head, not cutting the throat. Both men and women wear earrings and bracelets.

Among the pagans, one woman has several husbands, all of whom provide her needs, and they divide her time among themselves. Each of them has a token, which he attaches to the door when he visits her. If another one comes, he sees the token and passes by. Their children do not inherit from their fathers; instead the sister's son inherits, because the children are common to all.

Fornication is generally permissible among the pagan Indians, the only exception being those who mingle with Muslims. They may marry their own daughters and sisters, [M 225] and their aunts on their mother's and father's side.

In their eyes, there is no greater crime than killing and eating a cow. Anyone who kills or eats a cow is put to death. And they rub dried cow dung on their faces and eyes.

According to Masʿūdī, farting is not shameful among these people, but coughing is shameful. They do not consider it proper to hold in gas.

The Indians are assiduously occupied with bodily pleasure. They generally regard lewdness and debauchery as permissible. They nourish a desire for beverages and aphrodisiacs.

They have no shortcoming in hospitality. A traveler in their lands does not have to take along water or provender. At every stage of the journey he will find available all the necessities of life, and there is no hindrance to any pleasure that he may desire.

In the country of Orissa, before the Afghan conquest, there was a wealthy man named Makand. He had 400 wives, each in her own house, whose needs he provided and whom he established as a pious foundation for travelers. When a stranger arrived, a supervisor of this foundation, appointed and paid by Makand, had him washed and dressed and put into one of those houses where he stayed alone with one of the wives. In the morning, he gave him road-money and sent him off. Makand considered this arrangement as one of his pious acts that would assure his place in the afterlife.

The pagans of the Indians burn their dead; they know nothing of graves. Those neighboring on Sind burn and bury them in secret. In the country of Siam, when a man dies, his wife dresses and adorns herself and enters the funeral pyre while it is aflame to burn together with him. The Brahmins urge the women to do this, saying that one who burns in this fashion goes straight to paradise. Sometimes a man has several wives, all of whom burn together with him, and they fight over who gets to enter the fire first.

Buildings

The houses of India are generally constructed of wood and earth and covered with date-palm leaves. In olden times, however, the palaces of the grandees were constructed with some magnificence, and their ruins point to their being more splendid than those of ancient Egypt and Greece. [R 83a]

Climate

Gentle and violent winds blow very strong in India, especially easterly winds. Since it is a very large country, its weather is not the same in every place. Places close to the equator are hot, some places are thoroughly temperate, and the northern district is cold. In some places, there are two harvests and two summers in one year. Between the two, blowing of the wind known as etesian moderates the heat and there is a kind of winter. All of the weather conditions are harmless, and even one who lies down in the night frost suffers no discomfort. The pleasantness of its waters is boundless.

Rivers

India has vast rivers. Some of the geographers have determined that this clime has as many as 60 rivers, most of which flood in the summer season like the Nile and, by irrigating the land, are the cause of the country's prosperity. The best known are the following:

Indus: The ancient Indians called this river Ayandar and Sandus, so the part of the country to the west was called Sind, and Hind (India) was derived from Ayandar. The river is known to various groups by different names. In the Muslim books it is recorded as Mihrān (Mehran), or else as Duwīl and Qārshad.

It rises in a place close to the source of the Ganges River. It emerges from the foothills of Mt. Naugrocot and flows 900 miles from north to south. According to some knowledgeable people, its source is south of the Kashmir mountains at 109½ degrees longitude and 35 degrees latitude. It passes east of Ashnāghar and joins the Kabul River near Dūnbadī (text: Rūbandī). From there it goes southwest and is joined by the Hazara River. [M 226] Northwest of Nīlāb It turns to the southwest. In two stages it passes below the great mountain of Jahīnkūt and in two more stages comes to Chawbāra, then in two stages each to Pīlūtū, Dīra-i Ismāʿīl Khan and Dīra-i Fatḥī Khan, then in four to Sītpūr. From there it passes the juncture with the Chenab River and joins the Wiyāh River. In ten stages it comes to Kūfdī and Bāwlā, then in one stage to the fortress of Mītla, in five stages to Pakar, in five to Sahwān, and in five to Nagarchacha. Two stages further it empties in the sea from two great mouths. This is the course related by Shaykh ʿAlam al-Dīn al-Qamarī.

Some say that the Indus becomes three branches north of Nagartata. One of them passes west of Tata and enters the sea near the entrepot of Lāharī. Another empties into the sea one stage east of Lāharī near the town of Ranipur.

The Indus traverses a total of 42 stages. The greatest width is 50 stadia and the greatest depth is 15 paces. In it are many beasts of prey, crocodiles and lizards—this crocodile is attracted to human flesh and never stops growing until the end of its life. Its water spoils the color of animals that come from other lands. It has up to 20 tributaries.

Some knowledgeable people say that five rivers emerge from the mountains of Kashmir: 1) Wiyāh: It flows southeast, passes to the southeast of Lahore, and empties into the Indus near Uchcha. 2) Rāwī: It flows south toward Lahore, turns to the west, and empties into the Indus below Sīwar. 3) Chanhāw (Chenab): It flows southwest and empties into the Sind River near Multan. 4) Wīhat: It flows into the Sind River near Bahrā. 5) Rāwī (sic). The largest of these is the Chenab, also called the Chandarbāk. It comes from the mountains of Kashmir, (passes) northeast of Lahore, reaches the north of Sialkot, then Sūdra and Kūpra, and goes toward the west to Hazara and reaches Chandapūt. Above Sīwar it is joined by the Wīrta River and it enters the Sind River between Uchcha and Sītpur at 106 longitude and 29 latitude. It is very fast flowing.

Over the Chenab on the Lahore road from Kashmir are two bridges made of stout rope. One of them consists of four ropes, each as thick as one's arm, stretched three cubits above the surface of the water and bound securely to branches of very tall trees along the bank known as *yāhū*. Those cords are called *zanpa*. When the water recedes, they are stretched to stakes below; when the river is in spate, they are raised to the trees above. The width of the river at that point is 100 cubits. Above those four ropes are bound two other ropes [R 83b] which travelers grasp as they walk across on the four ropes whose width is a half-cubit. Their horses get across near the bridge by swimming. At no time can one wade across.

The other is on the Kampar Road. Ropes 70 or 80 cubits long, twisted out of a kind of grass growing in Kundafar, are bound tightly to trees at the river's edge and elevated about 15 spans abover the surface of the water. A rope 1½ cubits long and a quarter cubit wide is bound to the end of a forked tree trunk and passed over that stretched rope so it hangs like a ring. One who wishes to cross the river enters the fork and sits on top of that rope. Another long rope [M 227] is tied to that forked tree, one end on either side of the river, by which the person crossing is pulled to the other side and when he gets off it is pulled back. The rope for sitting and each of the ropes has a special name.

There is a similar rope and fork bridge in the village of Damkand on the Bartal Road of Kashmir. It is impossible for horses to cross there in the spring. In the summer, they attach ropes to the horse's neck and tail and pull it across from both sides.

Among the rivers that empty into the Chenab are the Wīhat: It emerges from the Kashmir mountains at 112½ longitude and 35 latitude and flows west to Farḥāla and Bahrā, then to the juncture with the Ṣuwān River and to Bahrāw. One stage east of Shūra and three stages above Multan it flows into the Chenab. Another is the Rāwī: It reaches Lahore from the northeast and continues northwest, passing north of Multan and flowing into the Chenab River below at 107½ longitude and 29½ latitude.

The Wiyāh River comes from east of Lahore and passes by its south. Continuing to the northwest, it flows into the Mehran River below Uchcha and above Mītla.

Lorenzo says that the Multan River is called Īdāsīas (Hydaspes). Seven rivers flow into it. One of these, known as Hapāziyas (Hyphasis), was the terminus of Alexander's campaign.

The Ṣuwān comes from 114 degrees longitude and 35 degrees latitude and flows into the Wīhat between Farḥāla and Rawalpindi.

The Ghānj (Ganges), in the east of India, is as large as the Indus. It is also called Gāng and Geng. The river recorded in scripture by the name Fīzūm [Phison][31] (i.e., Pishon) is thought to be this one. Mercator says that the Ganges is now called the Qāntāūm [Cantaon]. It emerges from the mountains of Tibet and Tataristan, flows to the south and empties into the sea from one or two mouths in the Bay of Bengal. Its narrowest place is two German miles and its widest up to five miles. Its greatest depth is 100 feet. So it is larger than the Indus.

Lorenzo, however, states that it is not as large as the ancients thought it was. Around 20 other rivers flow into the Ganges. It is nowhere wider than 12 miles and its greatest depth is eight fathoms. Large ships from the country of Bengal navigate beyond the city of Oro. The current of this river is gentle. Although it emerges with great clamor from its source and pours through steep ravines, when it comes to the plain it first collects in a lake and then emerges and flows gently.

The water of the Ganges is very pleasant and delicious, sweet-smelling and beneficial. It resembles the water of the Nile. On both sides are copious banana palms. The river has crocodiles and dolphins, which they call pigfish. According to Pliny, it has eels over 300 feet long that the Hindus revere, feeding them rice and saying they are

31 Spellings in brackets according to *Atlas Minor*, 620.

blessed fish. They also have a great faith in this river and come from faraway places to perform ablutions in it. Those who are unable to do this have its water brought to them. The ruler of Narsinga has water brought from it every week and washes with it. They claim that this ablution washes away sins. Government agents sit on the riverbank and collect great wealth from the Hindus. [R 84a]

Here and there are temples and shrines. The Hindus go on fasts and retreats there. Some worship the river, some sacrifice themselves and hurl themselves into the water and it carries them away. They say that if something impure is thrown into this river, the air darkens and wind and rain appear. ʿUtbī mentions that Sultan Maḥmūd tested this during the Kannauj campaign [M 228] and found it true. Kannauj is between two arms of this river. Fortresses and cities of Brahmins are linked to one another along this river.

The Ghānghā flows from north to south from a large river and empties into the Gulf of Orissa. The Brahmins venerate this river also, as much as the Ganges. Several rivers flow into this one as well—the Pāla, the Qarūsyār and the Banwar—after which it is known as the Ghānghā and gains esteem. These rivers all emerge from the Ghats Mountains and irrigate the countryside.

The Nāghawandī is a large river in the country of Narsinga. It emerges from the Ghats Mountains and flows southeast, gradually increasing in size and watering most of the country of Narsinga, before it reaches the cities of Bīsnāghār and empties into the sea.

In addition to these rivers, the Mandona, Chaberis, Campu and Mīqūn, also called Menon,[32] all emerge from a lake and empty into the sea in Pegu (Burma) and Siam (Thailand). Each of them is a large river.

Lakes

There are several large lakes in India. The largest is the Chiamay[33] which is 400 miles in circumference and 500 miles distant from the sea. The aforementioned rivers emerge from this lake. Another is the Cincuihai which is one mile in width and surrounded by dense forest. It is recorded in the history of Alexander that when his army was dying of thirst they found life in this place.

Mountains

The northern part of India consists of large mountains known to the Europeans as Paropamisus, Naugocrot, and Caucasus. There are also many mountains in the interior, none of them lacking in vegetation, called the Ghats. According to Solinus, the trees of the forests of India are so high that it is not possible to shoot an arrow over them. These mountains extend beyond India to Kabul, Badakhshān and Khurāsān, as far as the border of Cathay, and embrace Kashmir and Tibet. Most of the rivers of India flow from these mountains.

Plants and products

In India, various grains and cereals are plentiful, especially barley and rice; also sugar cane, as well as various kinds of fruit; plus pepper, frankincense, cloves and ginger; and in the southern part, cinnamon, hyacinth, indigo, myrobalan, bedda nut, coconut, marking nut, sandalwood and countless other valued drugs, harmful and beneficial herbs, roots and trees are endless.

According to Solinus, in the vegetable gardens of India is a fig tree that six men can hardly embrace and whose shadow covers two stadia. Its leaves are as broad as a shield and its fruit is very delicious. In their marshes grows a reed so huge that a segment cut between two nodes is used like a boat. As for the ebony tree, it is restricted to India. Coconut palms grow in the steep ravines.

In Gujarat there is genuine costus root known as *pūqū*. In the country of Delhi is a type of fruit that in its natural state is male or female; it is highly valued. Again in Gujarat is a tree known as *tārī*: they make an incision in the branch and hang jars on it that collect a kind of wine from its sap, which is drunk. In the country of Malabar is a type of cotton that is twisted instead of rope, since hemp does not grow there, and its leaves are used to roof their houses and to make vessels.

With regard to minerals, gold and silver mines are plentiful. However, a disagreement arose as to whether India had gold, copper, iron and lead mines. Diodorus was of the opinion that they existed, Pliny that they did not. As for precious stones, [M 229] pearls, diamonds, the gold-colored gemstone known as chrysoprase, and red rubies are plentiful. Silk is very plentiful. And there are different kinds of perfumes, ivory and ambergris. Ambergris is especially plentiful on the Malabar coast. There is much discussion about what this substance really is. Lorenzo writes that it is said to be a kind of gum that grows at the bottom of the sea, and this is plausible; but it also seems plausi-

32 Cf. *Atlas Minor*, 620: Mandona, Chaberis, Ava, Campumo, Menam, Menon; accurately reflected in *Levāmiʿu n-nūr*, 400a, lines 3–4 from end.

33 Spellings of lakes and mountains according to *Atlas Minor*, 620.

ble that it is a kind of beeswax. There are three varieties: white, gray, and black. The white one is the most valuable.

Animals

The wild beasts of India are larger than those of other lands. Elephants in particular are very plentiful. [R 84b] They go about in herds. In height they reach nine cubits, even ten or eleven in some places. The largest of them are desired by kings, who buy them at high prices, the price varying according to the size. Most of these elephants are domesticated and obey commands.

In the wilderness there are serpents as big as elephants that do battle with elephants. India has many snakes. The people make kebabs of them and eat them like eels. And there is a kind of giant ants that resemble the crabs of this place. These too are eaten, seasoned with pepper.

It says in some books that there is a sandy tract in the wilderness of Delhi where there are ants the size of dogs. There is gold dust there, but it is inaccessible because of those ants. Only when it is midday or midnight the ants enter their lairs and people come with containers and hastily fill them with the gold dust.

The dogs of India are large and ferocious. They fight with lions. In addition there are tigers, rhinoceros, white monkeys, and an animal called *būqalamūn* (chameleon) which, they say, take nourishment from the air. Types of birds include motley parrots and birds of various colors that have no equal in other lands. The peacock and other birds are without number.

One of the characteristics of India is that there are no pigs. In the country of Malabar there are no horses—they are brought there from surrounding areas. In the eastern part of Siam tigers (*bebr*) are plentiful in the mountain forests. And there is a kind of animal known as *mārīqī* with the face of a girl, the tail of a scorpion and the body of a lion. Tigers (*ḳaplan*) are also plentiful.

There are many rhinoceros in the plains and wilderness area of Gujarat. In the morning they come to the lakes to drink, and because there are many large snakes there, they use the horn on their forehead to stir up the water and remove its putridity and harm. That is a special feature of the rhinoceros horn. This beast is enemy to the elephant and friend to other creatures.

In the country of Mandu are many cattle with a large hump like a camel. They kneel like camels and carry loads. The Hindus, especially the Brahmins, most definitely revere cattle and consider it forbidden to slaughter or eat them. When a cow dies it is buried, on the grounds that it is a blessed animal.

In Delhi there is an animal called *pāsa*, the size of a yellow goat, from whose stomach emerges the stone known as *pānzahr* (bezoar). The special property of this stone is that it repels poison and putridity. This property pertains to the species and does not derive from the quality of the elements.

Solinus says there are sperm whales in the Indian Ocean whose length is more than four donums. And there is a fish called *physetera* whose bulk and length are beyond measure. It rises to the height of ship's mast, like a tall column, and blows so much water that it generally sinks the ship.

In Delhi there is a type of sheep that has six tails, one in the usual place, one on its chest, two on its shoulders, and two on its thighs.

In sum, there is no end to the wonders of India.

Routes of India [M 230]

From Aḥmadābād, which is the capital of Gujarat, Kenbāyet is 3 stages to the southwest, Barwaj is 6 stages to the south, Sarkhīj is 10 stages to the northeast.

From Patan, Lahore lies to the northeast. First, Dibālpur, 1 stage > Chūhbar, 1 stage > Lahore, 2 stages.

From Patan, Sayyidpur is 10 stages to the east and Sarkhīj is 10 stages to the south.

From Mangalore, Zakad is 1 stage to the west, Jāmhar is 2 stages, Lāharī is 4 stages. From Mangalore, Sūmanāt is 1 stage to the east.

From Aḥmadābād, Jālūr is 8 stages. From Jālūr, Lahore is 6 stages.

From Aḥmadnagar, the capital of Deccan, Kūlī is 18 stages and Dawlatabad is 3 stages to the west. From there, Barwaj is 3 stages to the west. From Aḥmadnagar, Bandarjūlī is 7 stages and Qadarābād is 5 stages, Aḥmadābād is 20 stages to the northeast, and Bījānagar is 10 stages to the southeast.

From Agra, Rāntamūr is 8 stages to the west > Hindūn, 3 stages to the east.

From Bīdar, Kūlīkanda is 3 stages to the east, Srūhī is 15 stages to the north. From Kūlīkanda, Sayyidpūr is 10 stages to the south.

From Ṣātīghān, Pūlārī is 3 stages, and Gaur is 8 stages to the north.

From Lahore, Kabul is 30 stages to the northwest, Delhi is 15 stages to the southeast, Nīlāb is 15 stages, Multan is 25 stages to the west. [R 85a] Sulṭānpur is 3 stages > Māchīwān, 1 stage > Sahīrand, 6 stages > Pāyīl, 2 stages > Tanīsar, 2 stages > Kītal, 2 stages > Panipat, 2 stages > Delhi, 3 stages.

From Sulṭānpur, Jālantar is 1 stage on the Delhi Road.

From Lahore, Chahnī is 2 stages to the southwest > Dibālpur 1½ stages > Patan, 1 stage. From Dibālpur, Multan is 8 stages

From Lahore, Khashāb is 10 stages to the northwest > Jahīnkūt is 3 stages. From Lahore, Hazara is 5 stages on the Kabul road. From Hazara, Khashāb is 3 stages.

From Delhi, Lahore is 15 stages to the northwest and Agra is 8 stages to the southeast.

From Delhi, Nākor is 15 stages to the west and Panipat is 4 stages > Tānīsar 3 stages > Kītal 4 stages > Mālīnar 5 stages > Samānā 5 stages > Tarhāda 4 stages > Sulṭānpur 1 stage > Lahore 2 stages.

From Jaunpur, Aḥmadābād is 10 stages, Chītawar is 4 stages, and Lālapūr is 7 stages to the west.

From Sayyidpūr, Kūlī is 10 stages to the north and Patan is 10 stages to the west.

From Farḥāla, Rawalpindi is 4 stages to the southwest, Damtūr is 3 stages to the northwest, Nīlāb is 5 stages, and Sūkra is 3½ stages > Nīlāb, 1½ stages.

From Rahtās, Rawalpindi is 3 stages to the northeast.

From Nīlāb, Rawalpindi is 3 stages to the east.

From Damtūr, Paklī is 5 stages to the east. From Paklī, Hazara is 4 stages to the southwest. From Hazara, Mārklī is 3 stages to the east. From Nīlāb, Mārklī is 1½ stages to the west. From Paklī, Swat is 7 stages to the northwest. From Ashnāghar, Swat is 4 stages to the west.

From Māsūlīpatan in Narsinga, Kandpūlī is 3 parasangs > Kūlī, 3 parasangs > Badar, 15 parasangs > Aḥmadnagar, 5 parasangs. From Badar, Aḥmadnagar is 5 stages > Barwaj is 1 stage to the west.

Supplemental note

In the *Aḥsan al-taqāsīm* the difference between India and China is noted as follows: [M 231] The Chinese are sodomites, the Indians are fornicators; the Chinese are mainly beardless or wispy-bearded, the Indians have long beards; the Indians do not approach women during menstruation, the Chinese do; the Indians use a toothbrush and perform ablution, the Chinese do not. The land of India is larger than China, but China is more populous (or prosperous—*maʿmūr*) than India. In both countries there is not a single date palm.

So much suffices for India. Now we shall discuss the countries on its northwest borders, God willing.

Chapter 23: Clime of Sind

The name derives from Sind son of Ham (son of Noah). It is recorded in some books, and is well known among the geographers, that the Indus River was given the name Sind, which then attached to the regions around the river.

Borders

To the west, the borders of Makran and the desert of Sijistān; to the south, the Indian Ocean; to the east, the land of Gujarāt; to the north, the countries of Tibet, Kashmir and Kabul. According to the *Taqwīm*, the lands of Makran, Ṭūrān and Nadha are included within the borders of Sind. In this chapter we will discuss the countries of Kalwa, Nadha and Sind. In the *Atlas* and other books that we have translated, this clime is passed over with one or two words in the India section, because they have no knowledge of these regions.

First of all, most of the cities of Sind lie to the west of the Mehran River. The well-known ones are the following:

Dābūl, also called Dabūl: In olden times it was a famous seaport, located at 101½ degrees longitude and 22½ degrees latitude. The Sind River lies on its eastern edge. At present it is a small ruined city situated on a spit of land at the mouth of the river, and is a place of extreme heat. This city has become completely ruined over time, and the name Dabūl was attached to the surrounding district.

Lāharī: The current seaport of Sind, two stages east of Dabūl, located at 102½ longitude and 22½ degrees latitude. [R 85b] It is a city and trading center to the east of where one branch of the Sind River empties into the sea. The other branch, which passes to the west of Tata, passes to the south of this place. The ebb and flow of the tide reaches as far as the city, so its water is sometimes acrid.

Manṣūra: A large city at 105½ longitude and 25½ latitude. An inlet from the Mehran River surrounds this city, which is like an island in its midst. It is also called Nahwāra. The reason it was named Manṣūra, according to the *ʿAzīzī*, is that it was founded during the reign of the Abbasid caliph Abū Jaʿfar al-Manṣūr, by ʿUmar b. Ḥafṣ al-Muhallabī, known as Hazārmard.

In the past, Manṣūra had its own king. He wore earrings in the manner of the emperor of India. Its circulating currency was the *qāhirī* which weighed 50 *akçe* apiece. It was conquered in the beginning of Islam and the majority of its people are Muslim.

There are also cities named Manṣūra in Egypt, Khwarazm, Ifriqīya (Tunis) and Yemen. They will be discussed in their proper place.

Multan: A famous city at 107½ longitude and 29½ latitude. The Chenab River passes by one hour south of here and flows west to Uchcha. This city has large surrounding districts. It extends to the border of Makran in the west and Manṣūra in the south, and Ghazna lies 160 parasangs to the north.

Lorenzo says that the women of this city are riders and warriors.

Some thought that Multan was part of India, but Abū Rayḥān (al-Bīrūnī) in the *Qānūn* considered it part of Sind. It has a solid fortress. Around it for half a parasang are orchards and gardens [M 232] and splendid pavilions and palaces where its kings reside.

Bārkhān: A town and fortress belonging to Multan, west of the Sind River on the Kandahar road, located at 114½ longitude and 30 degrees and 20 minutes latitude. Sītpur is seven stages from here across the desert. To the northwest it is nine stages of mountains and desert to the territory of the tribe of Malik-Tata. The Bārkhān tribe resides in those mountains.

Bīrūn: One of the port cities of Sind, located at 104 degrees longitude and 25 latitude. It lies between Dabīl and Manṣūra, which is 25 parasangs away.

Jahīnkūt, also called Chīnkūt and Chankūch: A large city on the bank of the Sind River, located at 109½ degrees longitude and 32 degrees latitude. It is also the name of its province. The city is situated on both sides of the river, which passes through the middle of it, and there are great mountains on both sides that extend as far as Nīlāb, five stages away. The province (*vilāyet*) is a country (*memleket*) with cheap foodstuffs. The city is a famous entrepot, forming a tripod along with Kabul and Nīlāb, Ja[hī]nkūt lying to the south of them. In these mountains of Jahīnkūt live the Afghans, also called Awān.

Chawbāra: An ancient city of the Afghans, located at the foot of a great mountain on the bank of the Sind River. It lies two stages below Jahīnkūt.

Khānapur: A city on the bank of the Sind River, two stages below Chawbāra.

Pīlūtū: A city on a great mountain, two stages below Khānapur, with many date groves and orchards. The Sind River passes nearby.

Dayra-i Ismāʿīl Khan: A city on a plain on the bank of the Sind two stages below Pīlūtū.

Dayra-i Fatḥī Khan: A city on a plain on the bank of the Sind, two stages further down.

Sītpur: A city on the bank of the Sind, located at 107 degrees longitude and 29½ degrees latitude. It lies three or four stages below Dayra-i Fatḥī. The river passes by the south side of the city.

Uchcha: A city opposite Sītpur, southeast of the Sind River, located at 107 degrees longitude and 30 degrees latitude. Multan lies three stages east of here. A half stage to the south, the Chenab and Rabāh Rivers join and then flow into the Sind.

Bāwalā, also called Bālā and Obāwalā: One of the cities of Multan, located at 106 degrees longitude and 28 degrees latitude. It lies between the Sind River and Multan and is three stages to Uchcha. [R 86a] The aforementioned river (i.e., the Sind?) reaches this place.

Matīla: A city one stage distant from Bāwalā on the west bank of the Sind River, located at 106½ degrees longitude and 28½ degrees latitude.

Pakar: A city and capital of a province atop a hill at 105½ degrees longitude and 28 degrees latitude. The kings of Sind live there. The Sind River divides in two and surrounds this mountain. On its southern edge is the city and fortress of Lawhirī, on its northern edge was built the city and fortress of Sakar.

Takar: A city four parasangs from Pakar.

Sīhwān: A city on the bank of the Sind River at 104½ longitude and 26½ latitude. The river passes it by on the south. It is five stages below Pakar. After the river passes this place, it flows to the west. The cities of Diu, Dabīl and Nagartata are each ten stages from here. [M 233]

Nagartata, also called Nagarchacha: A city like an island in a place where the Sind River divides in two, located at 102½ degrees longitude and 24½ latitude. The western branch goes one stage and empties into the sea near Lāharī. The eastern branch also goes that far. This city has a strong fortress. Some knowledgeable people say that the city's name is Tata. Langar is the city's entrepot, 1½ stages distant, on the seashore.

Azūr: A city on the bank of the Mehran River at 105½ degrees longitude and 28 degrees latitude. Manṣūra is estimated at 30 parasangs from here.

Ṣālihābād: A city on the bank of the Sind, seven stages below Sīhwān.

Rāhmī: A town on the bank of the Sind, one stage north of Nagartata.

Sadūsān: It city with cheap foodstuffs, located at 104 degrees longitude and 38 latitude. It lies west of the Mehran River.

Darabīla: A city located at the same longitude and at 26½ latitude.

Pūnāfād: A town on the Sind frontier, eight stages from Kīj, on the Dabūl road toward Makran.

Ataḥ: A city at 106½ degrees longitude and 29 latitude.

Gihindī, also called Gindī: A town one stage from Bālā. From here one reaches Pakar in four days.

Sarbal: A city at 103 degrees longitude and 25 latitude.

Ānrī: A city to the east of the Mehran, on the road going to Multan from Manṣūra, located at 105½ longitude and 27½ latitude. Manṣūra is 40 miles from here and the Mehran River is some distance away.

Qālarī: A city on the east of bank of the Mehran, on the Manṣūra and Multan road, located at 105½ longitude and 26 latitude. One of its districts lies west of the Mehran. The Mehran does not flow very much west of here; rather it divides in two, the larger branch flowing in the direction of Manṣūra to the west, the smaller one inclining north-east and rejoining the larger one 12 miles below Manṣūra.

Jāmrawar, also called Yāmrawar: It lies two stages to the east of Kandkūla. Paklū is three stages to the east on the shore of the Indian Ocean.

Kandkūla, also called Jāmmahar: It lies two stages to the east of Lāharī and is a port city and capital of the sultan of Mahar.

Chandawar: A small city one parasang east of Multan.

Basmand: Another small city nearby, with a two-level fortification.

Both of these are one parasang from the river, and their water comes from wells.

In addition to these, the cities of Bibipur, Mubārakpur, Harar, Lafkar, and Banjū are reckoned as cities of Sind.

Ṭūrān province: A district on the frontier of Sind. Its town, Qazuwār, is atop a hill that lies on a flat plain at 101 degrees longitude and 31½ latitude. From here, Multan is estimated at 20 stages and Bust at 80 parasangs. Ṭūrān itself is a small city of Zoroastrians and full of good things.

Nadha province: An extensive district on a level plain, located in the middle of Ṭūrān, Makran, Multan and Manṣūra. It is west of the Mehran River. The people are mainly nomads and desert-dwellers, with pastures for camels and sheep. The camels known as *bukhtī* (Bactrian) are raised here. The town of this district is Qandābīl, located at 95 longitude and 28 latitude. Its people are also desert-dwellers who enter Ṭūran to conduct trade. [R 86b] The people of this region are many tribes. [M 234] They have spread out along the bank of the Mehran River and constructed houses made of reeds.

Fāmhal, also called Tāmhal; and Sadūsān, Ṣīmūn, Kasāna: four cities whose people are Muslims and who perform the Friday prayer. In these places, rice is prevalent among the grains. Honey, coconuts and bananas are also abundant.

Kasāna is also called Tāniya. Between here and Fāmhal is desert. Ṣīmūn is a prosperous district whose people are a mixture of Muslims and Zoroastrians. They wear no clothing, only a waist-wrapper.

As for Arsābīl and Qandābīl, they are two large cities. There are two stages between them and the sea is one-half parasang away.

Kabarkākiyān: Another big city, on the edge of the desert.

Rāhūk and Kalwān: Two districts with little water.

Kalwa district: A province between Sind and Makran. It has flowing water. Their people are called Baluch. They are a corrupt tribe. Their rulers come from among themselves. Kalwa itself is a large city at 102½ degrees longitude and 29 latitude. It has a few orchards and gardens in its vicinity. Kīj is 10 stages from there, the Sind River is 10 stages, and Kandahar is 20 stages.

One town of this district is Parūm: a city on the bank of a river, with orchards and gardens, located at 101½ longitude and 28½ latitude. Its grain crop is good. Kalwa is ten stages away.

Another is Sīwa: a town at 100½ degrees longitude and 27½ latitude. It has few gardens. Kīj is five stages away, Kalwa is a four days' journey. The Kalwa River comes from the direction of Naghara and flows southwest. It reaches Kalwa, Parūm and Sīwa and empties into the Indian Ocean.

Characteristics of Sind

The Sindians, like the Indians, are a mixture: some are star worshippers, some are Zoroastrians, and some are Muslims. There are many wealthy men and merchants. Among them adultery and sodomy are not shameful; rather, they chide those who do not engage in it. Their soothsayers are numerous. Various kinds of medicinal herbs and drugs for fever are found. Fine fabrics and clothing and other products are plentiful.

According to Ibn Khurradādhbih, the product of Sind in the tax farm of ʿImrān b. Mūsā al-Barmakī was 10 *yük* (1 yük = 100,000—here probably of dirhams). After that it was sometimes more or less under the administration of the deputies of the sultans and caliphs. At present it is controlled by the emperor of India.

In Sind there are many marshlands and reedbeds. The people feed themselves mainly by hunting ducks and catching fish. They resemble the Arabs, just as the Makranis resemble the Kurds.

Routes

From Dabīl, Manṣūra is 6 stages, Bīrūn is 4 stages, the entrepot of Lahore (sic; error for Lāharī?) is 2 stages to the east, Būnāfād is 2 stages, and Sīhwān is 10 stages.

From Manṣūra, Multan is 12 stages, Ṭūrān is 15 stages, the border of Nadha on the other side of the Mehran River is 8 stages, and Bīrūn is 15 parasangs.

From Multan, Ghazna is 160 parasangs, Ṭūrān is 20 stages, Lahore is 11 stages, and Sītpur is 3 stages.

From Lahore, Sialkot is 4 stages, Tata is 1 stage.

From Ṭūrān, Bust is 80 parasangs.

From Nadha, Tīz is 15 stages.

From Sīhwān, Dabūl is 10 stages, Tata is 15 stages, Darabīla is 2 stages to the southwest, and Pakar is 3 stages to the northeast.

From Sialkot, Sūdra is 1 stage.

From Rāhmī, Chācha is 2 stages, Darabīla is 5 stages to the northeast, and Nagartata (text: Pakartata) is 10 stages to the south. [M 235]

From Jandanūt, Hazara is 2 stages.

From Jahīnkūt, Nīlāb is 5 stages and Kabul is 5 stages.

From Bālā, Gihindī is 1 stage and Matīla is 3 stages. From there Ataḥ is 2½ stages and Sītpur is 3 stages. From Ataḥ, Bibipur is 2½ stages and Sītpur is 2 stages.

From Diu, Mangalore is 2 stages, Jāmhar is 2 stages and Lāharī is 4 stages.

From Kīj, Būnāfād is 8 stages.

For the rivers and mountains of this clime, as well as its mores and other matters, it suffices to draw an analogy from the India chapter.

Chapter 24: Country of Makran

It is west of Sind.

Borders

To the west, Kirmān; [R 87a] [Plate 34] to the east, Sind; to the south, the Indian Ocean; to the north, Ashnāghar, Khwāst and Zābulistān.

Ibn Ḥawqal says that this is a very extensive district, most of it desert. Crops being few, it is never free of drought and scarcity. Its people resemble the Kurds. They speak Persian and wear long cloaks and turbans. There are many merchants among them.

It is related from some knowledgeable people that they are Ḥanafīs. Their ruler is a *sayyid* (descendant of the Prophet) of sound lineage from the line of ʿAbd al-Qādir-i Gīlānī. In recent times Sayyid Ḥusayn-i Qādirī raised the banner of (independent) government.

This province is a country comprising many cities and fortresses. Its cities and its capital are the following:

Kīh: A large city located at 96 degrees longitude and 27½ latitude. There are mountains to the north and south. One of the streams flowing from Sarbār reaches this place and passes it on the east. Hormuz lies ten stages west of here, Kīj lies ten stages to the east.

Kīj: A city and strong fortress, larger than Aleppo, located at 98½ degrees longitude and 27½ degrees latitude. The Nahang River passes by the fortress on one side and on another side is a mountain difficult to cross. North of the city are mountains and south of the city is desert all the way to the Indian Ocean, a ten days' journey.

Dīzak: A city at 97½ longitude and 29½ latitude. A river as big as the Orontes comes from the north and reaches this place. Kīh lies ten stages to the southwest and Jāl three stages to the east.

Jāl: A city and fortress at 98½ degrees longitude and 29½ latitude. A large river comes from the northwest, passes by Jāl to the north, and empties into the Indian Ocean north of Panjpūr. Kandahar lies eight stages to the south.

Pinpūr: A city at 96 degrees longitude and 27½ degrees latitude. A great river passes it to the north and empties into the Mākshīd River.

Panjpūr: A city at 98½ degrees longitude and 20½ degrees latitude. The river coming from Dīzak passes north of here and after five stages joins the Mākshīd River. Kīj is five stages from here.

Nāwak: A city at 99 longitude and 30 latitude. A river reaches here coming from the south and joins the Sarbār River.

Kushk: A fortress at 96 longitude and 28½ latitude. It lies between Dīzak and Pinpūr and is reckoned among the districts of Sarbār. A river comes from east of here, flows south and joins the Sūrnīkūr River.

Mand: A city with a fortress at 96 longitude and 26½ latitude. [M 236] It lies between Kīj and Kīh. The Nahang River passes next to it. It has an independent lord.

Pīshīn: A city at 96½ longitude and 27 latitude. Mand lies to the west. Kīj is five stages from here.

Qaṣrkand: A city between Pīshīn and Kīh, located at 96 longitude and 26½ latitude. The Kūrkas River passes between here and Kīh. [R 87b]

Pīrūzābād: A city at 96½ longitude and 27½ latitude. It lies between Qaṣrkand and Kīh. The Kūrkas River passes it to the east.

Sarbār: A district on a mountain to the north of Kīh and Kīj. It comprises villages and fortresses and has many rivers. It has one large river known as Mākshīd which flows in a southerly direction.

Ispaka: A city and fortress in the district of Sarbār, located at 96½ longitude and 28½ latitude. The lord of the district lives here. It has a weakly flowing river that comes from the north and sometimes runs dry. Kīh is four stages to the south of here.

Pannūs: A town in the district of Sarbār at 96 longitude and 28 latitude. It lies between Ispaka and Kīh.

Damshik: A city on a mountain at 95½ longitude and 28 latitude, and the frontier of the Qizilbash (i.e., Safavid Persia). Pannūs lies two stages east of here.

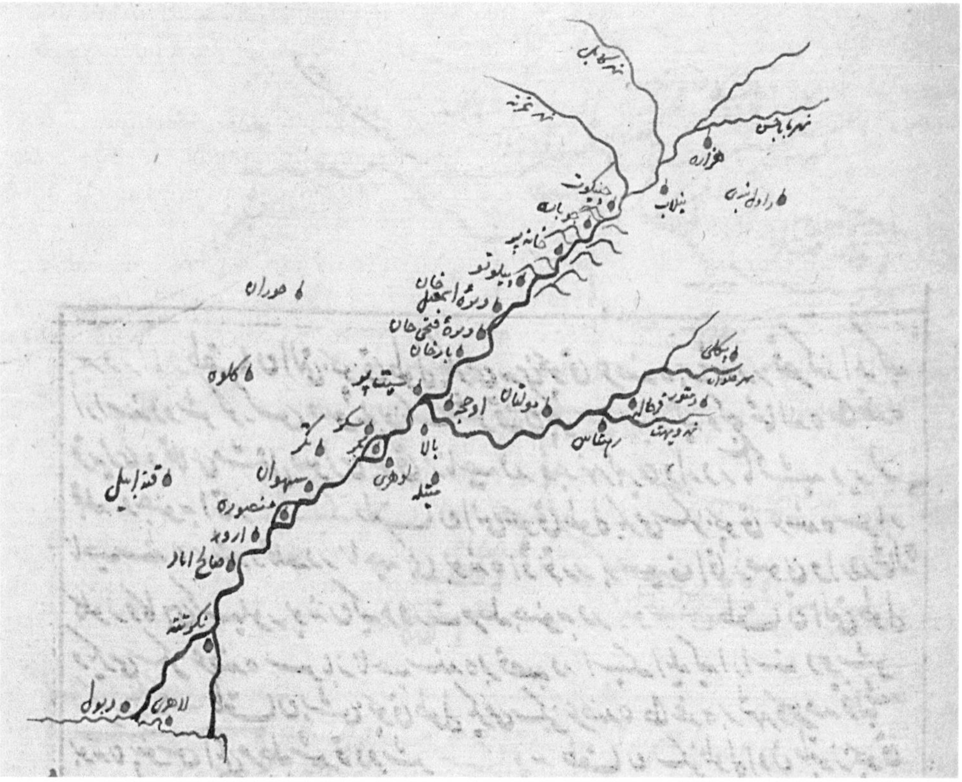

PLATE 34
Map of Sind, R 87a (Appendix, fig. 75)
TOPKAPI PALACE LIBRARY, REVAN 1624, WITH PERMISSION

Sipāwand: A city at 98 longitude and 19½ latitude. It is reckoned as part of Makran. The mountains that are the sources of the Sūrnīkūr River lie north of here.

Tīz: A port city eight stages by land from Hormuz and four stages by sea. The *Taqwīm* records it as the town (or administrative center—*ḳaṣaba*) of Makran and on the edge of the Mehran (River), but the locals say it is very far from the Mehran. The Kūrkas and Kūr[k]ang Rivers join near here and empty into the sea.

Aside from these places, on the road going from Kīj to Dabūl are BLB (?), Malān, Shīrābād, Khushkābād, Sirnigīn and Lallīn, towns separated from each other by one stage, and all of them attached to Kīj. They get their water from springs and wells and have no farming or garden districts.

Rivers of Makran

Nahang: A river as big as the Nile. It comes from the direction of Ghazna and Arkūb and from Badakhshān, passes Kīj to the southeast, reaches Mand south of Dāran, then turns to the south and empties into the sea two stages west of the port of Kawādar at a place called Dastyārī.

Kūrkang: It comes from the direction of Nāwak and passes to the east of Pīrūzābād, then to the west of Pīshīn and procedes to the southwest. In that place it is called the Sūrnīkūr. After flowing a considerable distance it unites with the Kūrkas River and empties into the sea next to Tīz. According to one report, the Kushk River also joins this river and it enters the Sea of Hormuz (Persian Gulf) between Ḥidār and Pīshīn.

Kūrkas: It comes from east of Sipāwand, passes Dīzak, Kushk, Pinpūr and Kīh, reaches the west side of Qaṣrkand, joins the Sūrnikūr and empties into the Sea of Hormuz near Tīz.

Mākshīd: It comes from the direction of Ghazna and reaches Nāwak, Jāl and Panjpūr, then passes to the east, continues to one stage west of Kīj and joins the Nahang River near Dāran.

Routes of Makran

From Kīj, Kīh is 10 stages to the west and Jāl is 8 stages to the north.

From Kīh, Hormuz is 10 stages to the west.

From Dīzak, Jāl is 3 stages to the east and Kīh is 10 stages to the southwest.

From Jāl, Kandahar is 10 stages to the northwest, Panjpūr is 4 stages to the east, and Sijistān is 8 stages to the north [M 237].

From Pinpūr, Dīzak is 10 stages to the southeast and Jāl is 4 stages to the west.

From Panjpūr, Nāwak is 4 stages > Jāl 5 stages.

From Dīzak, Kūshk is 3 stages > Pinpūr 4 stages.

From Kīj, Mand is 5 stages > Kīh 5 stages.

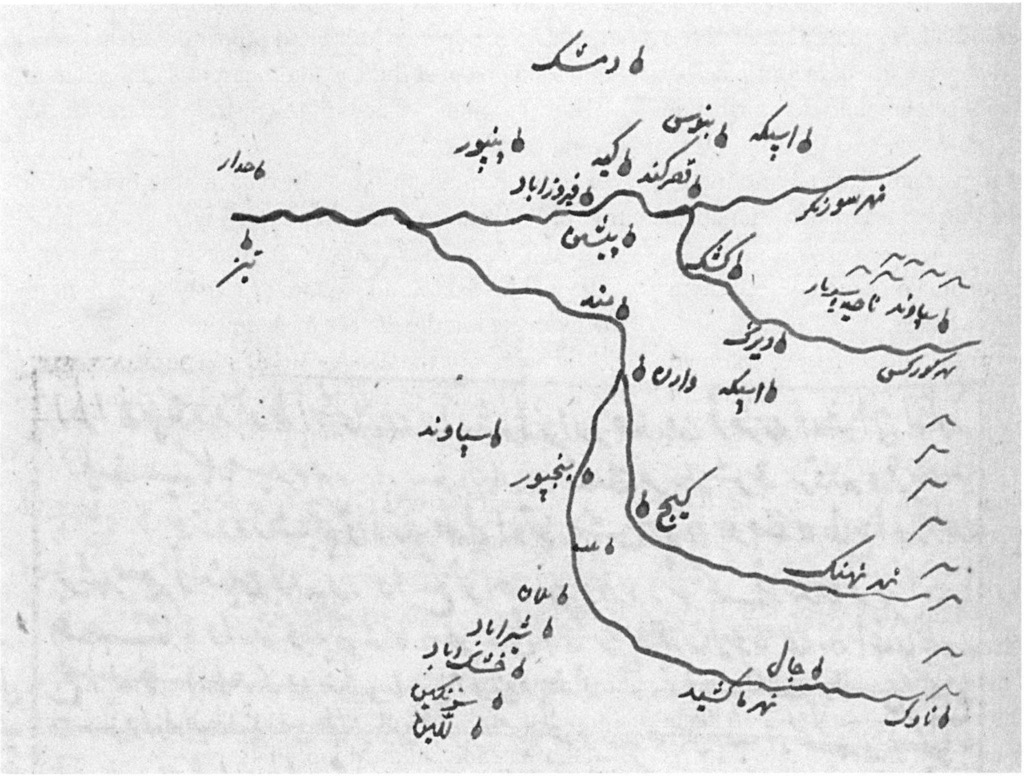

PLATE 35
Map of Makran, R 88a (Appendix, fig. 76)
TOPKAPI PALACE LIBRARY, REVAN 1624, WITH PERMISSION

From Kīj, Pīshīn is 5 stages.

From Pīshīn, Mand is 3 stages.

From Kīh, Qaṣrkand is 1 stage > Pīshīn 2 stages.

From Qaṣrkand, Firozabad is 5 stages to the north > Kīh 5 stages.

From [Qaṣr]kand, Ispaka is 4 stages to the north > Kīj 12 stages.

From Ispaka, Pannūs is 2 stages to the south > Kīh 2 stages.

From Pannūs, Damshik is 2 stages to the west > Hormuz 10 stages.

From Kīj, Dabūl is 7 stages.

The remaining characteristics are not known. [R 88a] [Plate 35]

Chapter 25: Clime of Zābulistān, Khwāst and Ashnāghar

Several countries are included in this chapter.

ZĀBULISTĀN: The Persians call it Bākhtarzamīn. It is an oblong-shaped country surrounded by mountains.

Borders

To the east, Barshawar and some regions of India; to the west, Kūhistān and Hazara; to the north, the country of Qunduz and Andaz where the boundary is located on this side of Hindu Kush; to the south, Qazmal and other places of Afghanistan.

Its capital is Kabul: A pleasant city with orchards and gardens on the bank of a river at 105½ degrees longitude and 33½ degrees latitude. It has a solid and strong fortress that can be ascended from only one place. Its people are a mixture of Muslims, Indian heathens and Jews. In olden times there was great esteem for this city among the Indians. Their ruler was enthroned here, and he would not become ruler unless he was crowned (here).

Because it is surrounded by snow covered mountains and its climate is both hot and cold, dates cannot be grown, but cotton and saffron grow instead. In the mountains are iron mines, medicinal herbs and aloeswood. However, the myrobalans of Kabul, on account of coming here first from India and being sold, are named after this place; otherwise they do not grow in this province. In olden times, the income of Kabul was 15 *yük* (1.5 million) dirhams and 90,000 more were acquired from the spice trade.

Around Kabul are many districts (*nāhiye*) and a string of villages and fortresses.

According to the *Haft Iqlīm*, Kabul consists of 14 *tūmāns*. The largest of them is:

Tūmān of Nīknār: rice, wheat, oranges, lemons, and sugarcane are good and plentiful.

Tūmān of ʿAlī Shang: A river passes through the middle of it. There are prosperous villages in its district. It has good rice. The people of this place use *ghayn* in place of *kāf*; thus they call this province Lamghān. Its north side reaches Hindu Kush.

Hindu Kush: A fortress atop a snow-covered mountain, six stages from Kabul. From there, one reaches Khanjān in three days.

Khanjān: A fruit-filled city on a mountain. From there, one crosses over to Ghūr in two days.

Tūmān of Daranūr: The wine of this place is well known.

Tūmān of Kaznūrkal: It is on the frontier of Kāfiristān. In this place, up to the frontier of Swat (text: Swād) and Bajūr (text: Najūr), the custom is as follows: If a woman is accused of a crime, they put her on a board and lift it up from four sides. If she is guilty, it will seem heavy; if not, it will seem light.

Tūmān of Baḥrāw: Pomegranate trees are plentiful in its mountains; also *jalghūzadūn* which the locals call *ḍaḥk*, a kind of tree that gives off light like a candle. And there is a kind of fox in the mountains that has wings of skin like a bat's wings between its front paws and chest and leaps as though flying. It is called the flying fox.

Tūmān of Ghūrband: [M 238] One crosses here through a mountain pass over to Ghūr, which is why it is called Ghūrband. There are silver and lapis lazuli mines, but they are unworked.

Between Ghūrband and Ābbārān are two meadows that in the spring season resemble paradise in beauty and freshness. There are fragrant tulips there called *lālagul*.

Aside from this, Istālif and Astarghanj are two delightful places in Kabul that have no equal. Mīrzā Ulugh Beg Abū Saʿīd used to call one of them Samarqand and the other Khurāsān.

One *parasang* from there (i.e., Ghūrband?) is a valley that is a delightful excursion spot and hunting grounds known as Khwāja Sihyārān. It has a spring [R 88b] surrounded by plane trees, oak trees and red and yellow Judas trees.

Kābul-i Ṣughrā ("Little Kabul"): A fortress and village one-half stage east of Kabul.

Bamiyan: A small city and strong fortress on a mountain at 104½ degrees longitude and 35½ latitude. It has extensive districts. It lies between Khurāsān and Ghūr, and is ten stages from Balkh. A river coming from Ghazna passes nearby and goes to Gharjistān where it joins the Jayḥūn (Oxus).

Bamiyan has no orchards or gardens; its fruit comes from the surrounding area.

According to the *Nuzha*, Bamiyan's climate is cold. At the time of the Mongol conquest, Bāmgān son of Chaghatay was killed there, and so Chingis completely destroyed it. He named it Mūsā Baliq and ordered that no one build a staging post there. After that it was repopulated.

There are many villages in the district of Bamiyan, also several towns.

Panjhīr: A medium-sized city on a mountain in the district of Bamiyan. It has a pleasant climate. Its produce is grain and a little fruit. It has silver mines in the mountains.

Baghchūr, Sakāwand, Alḥarā (?): They are among the towns located in the district of Bamiyan.

The author of the *Nuzha* reckoned Bamiyan as one of the dependencies of Balkh; however, it is rather distant from Balkh and closer to Ghazna and Kabul.

Country of Khwāst and Ghazna: A large province in the vicinity of Kabul.

Borders

To the south, the province of Ashnāghar; to the west, Kandahar and Sijistān; to the east, Kabul; to the north, Badakhshān and Balkh. Mountains separate this place and Ashnāghar.

Its capital is Ghazna: a famous city and prosperous seat of sovereignty, located at 104½ degrees longitude and 33 degrees latitude. It is also called Ghaznīn. A river passes along its edge and flows into the Kabul River.

In olden times it was a small city. Amir Sabuktegin and his son Sultan Maḥmūd revived and expanded it. After that, it was plundered and destroyed in the time of Chingis.

Because of the severe cold, it has no orchards and gardens. It is a mountainous city famous for its salubrious air and delightful water. Because of the excellent climate, the people mainly have long lives and sound constitutions. There are gold mines. Snakes, scorpions and other vermin do not appear. They say that the paucity of fruit in Ghazna is a proof of its excellence, because much fruit is usually associated with much illness.

During the reign of the Ghaznavids, the prosperity of this city was so great that there were 2000 mosques and madrasas and its bird-hunters sold 10,000 sparrows per day. This is reported in the *Tārīkh-i Mubārakshāhī*, citing the *ʿAjāyib-i ʿālam*. Bābur in his memoirs cites the poet Farrukhī to prove that at that time they called Ghaznīn and Kandahar "Zābulistān."

Many eminent men have appeared in this city, including Shaikh Abū'l-Majd Muḥammad b. Ādam, known as Ḥakīm Sanāʾī, author of the *Ḥadīqa*; [M 239] Shaikh Raḍī al-Dīn ʿAlī Lālā; and Yaʿqūb Charkhī, one of the great shaikhs.

Kalālkū: A town at 106 degrees longitude and 33½ latitude. It lies one stage to the southeast of Kabul.

Khūshī: A town three stages distant from Kabul.

Patan: A large city one stage distant from Khūshī. It lies to the west of Kabul.

Mīrsakāna: It lies one stage to the west of Patan.

Rabī: A town lying eight stages to the east of Kabul.

Mazdikhil, Pūṣūkhil, Sankhil, and Shābilkaza: Towns lying from west to east with one-half stage between them.

Nūraza: A town lying one stage to the east of Shābilkaza.

Malīza: A town one stage to the east of Nūraza.

Tārūza: A town one-half stage to the east of Malīza. It lies between Shābilkaza and Nūraza.

Gardīz: A large city on the bank of the Dalan River two stages to the south of Kabul. It is located at 105½ longitude and 32½ latitude.

Warduk: A town 5 stages to the west of Kabul.

Dīkān: A town one stage to the west of Warduk.

Khwāst, also called Khāst: A province between Andarābe and Ṭukhāristān, one of the dependencies of Balkh, located at 103 degrees longitude and 36½ latitude. The author of the *Taqwīm* mentions that when Qutayba came, the King of the Turks was using it as a place of refuge.

Naghar: At 106½ longitude [R 89a] and 32½ latitude, a large city lying five stages to the south of Kabul. It is located on the bank of the Shāmil River.

Pālītar: A city on the Dalan River, three stages to the west of Kabul.

Ūrgūn: A city two stages northwest of Naghar, located at 104½ longitude and 32½ latitude. It has a district and villages and is dependent on Ghazna, which is a three days' journey away. West of here is desert and wasteland for 12 stages, with no populated place, until one reaches Kandahar.

Fīrūzkūh: A strong fortress between Herat and Ghaznīn. It was the capital of the Ghaznavid rulers. Its weather inclines to cold and its water comes from springs.

Maymand: A small town near Ghazna. The vizier of Sultan Maḥmūd, Abū'l-Hasan ʿAlī b. Aḥmad al-Maymandī, came from here. It is a hot place with many date palms. Other fruit is also plentiful. Its water comes from a stream. It lies to the south of the Pentepel Mountains.

COUNTRY OF ASHNAGHĀR: A province to the north of Khwāst and Naghar.

Borders

To the south, Sind; to the north, Kashmir; to the east, the mountains of India; to the west, the districts of Kabul.

Towns and cities

Duwāwa, also called Miyānī Duwāw: A large city at the confluence of the Panjkūra River from the west and the river coming from the Kūbar mountains in the east, located at 108½ longitude and 34 latitude. The Kabul River also joins there and goes to Dūnbadī.

Ashnāghar: A large city at the confluence of the Hazara and Sind Rivers, located at 109 degrees longitude and 34 degrees latitude. It lies seven stages to the northwest of Nīlāb. This province was named after this place.

Bījāwar, also called Peshawar: A large city at 107½ degrees longitude and 34 latitude. It lies one stage to the west of Duwāwa. The river here comes from the west and empties into the Duwāwa River.

Swat: the name of the district of Ashnāghar. It is where the Hazara River empties into the Sind River. Ibrāhīm Khan, the son of Kajū Khan, raised the banner and established an emirate there in recent times. [M 240]

Kajū: A district of Ashnāghar. It lies two stages south of Swat, where the Kabul River—i.e., the Hazara River—joins the Sind River. Kabul is ten stages to the west of here.

Muqām: A district and villages one stage northwest of Ashnāghar, located at 108½ longitude and 34½ latitude.

Panjkūra: A city three stages from Ashnāghar, located at 109 degrees longitude and 34½ latitude. The river here flows to Bājūr.

Bājūr: A city three stages northeast of Peshawar, located at just over 109 longitude and 34½ latitude.

Nagarhā[r]: A city three stages to the east of Kabul, on the western slope of a high mountain, located at 106½ degrees longitude and 34 latitude. That mountain is called Kūh-i Safīd (White Mountain). Its rivers collect in a lake called ʿAbdī Khāṣ. People make pilgrimage to this lake, thinking it is blessed water.

Kunar: A province on a mountain 12 stages northwest of Ashnāghar. The Sind River comes from there. Its people are a mixture of Muslims and heathens.

Rivers

There are several rivers in these countries:

Kabul, also called Hazara since it has a thousand (*hazār*) towns and cities on it, a thousand being an expression for "very many." It flows from north to south. From Kabul it turns southeast and reaches Nagarhār in four stages, Peshawar in two, and Duwāw in two. South of Duwāw the Panjkūra and Swat Rivers join, then flow into the Kabul River. This flows for one more day and joins the Sind River next to Dūnbadī, southeast of Ashnāghar. All of these cities are on north of the river.

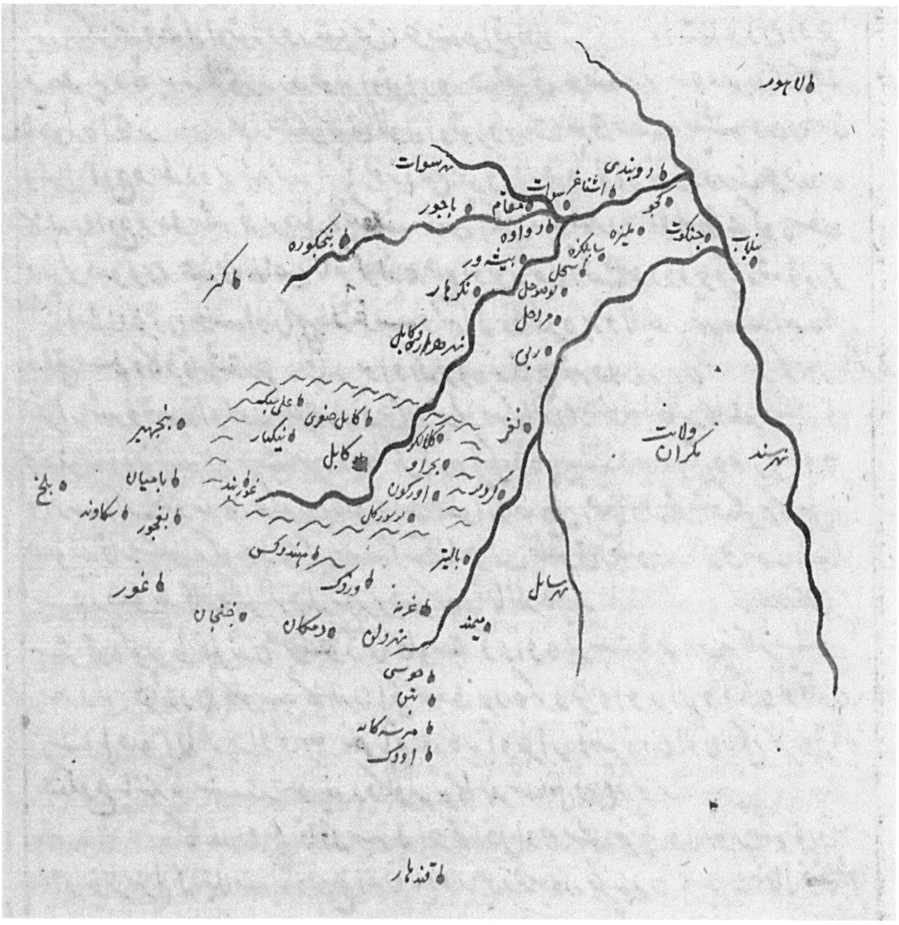

PLATE 36
Map of Zābulistān Khwāst Ashnāghar, R 89b (Appendix, fig. 77)
TOPKAPI PALACE LIBRARY, REVAN 1624, WITH PERMISSION

Panjkūra: It comes from the west and, together with the Swat River that comes from the Kūper Mountains in the east, joins the Kabul River near Duwāwa.

Dalan: It comes from Ghazna and reaches Gardīz in a place two stages from Kabul. Above Gardīz it joins the Sābil (?) River and reaches the city of Naghar. It empties into the Sind River 12 stages from Malīza near Jankūj (Jahīnkūt). This is also called the Sābil (?) River.

Routes of Kabul and Ghazna

From Kabul, Balkh is 15 stages of difficult stopping places by caravan journey, and Lahore is 30 stages to the southeast. Thus Nīlāb is 100 parasangs and from there Lahore is another 100 parasangs.

From Ghazna, Bamiyan is 45 parasangs to the north, [R 89b] Bust is 140 parasangs, and Kabul is 3 stages to the east. To the west of Ghazna, on the Kandahar road, Khūshī, Patan, Mīrsakāna and Awardak are 1 stage each, after which one enters the desert and reaches Kandahar in 15 days.

From Kabul, Kajū is 10 stages to the east; Hindu Kush is 6 stages > Khanjān 3 stages > Ghūr 2 stages > Maymand 3 stages of desert > Balkh 2 stages with villages in between.

From Kabul, Ghazna is 3 stages to the west; Kalālkū is 2 stages > Khūshī 1 stage.

From Kabul, Rabī is 8 stages to the east > Mazdikhil, Pūṣūkhil, Sankhil, Shābilkaza, ½ stage each to the east > Manūrza, Malaza, Taghaza, 1 stage each.

From Malīza, Jankūj (Jahīnkūt) is 12 stages > Tārūza, one-half stage to the east.

From Kabul, Gardīz is 2 stages to the south and Warduk is 5 stages to the west > Ghazna 2 stages.

From Warduk, Dīkān is 1 stage to the west.

From Kabul, Naghar is 5 stages to the south.

From Gardīz, Naghar is 3 stages to the east.

From Kabul, Pālītar is 3 stages.

From Naghar, Ūrgūn is 2 stages to the northwest.

From Ghazna, Ūrgūn is 3 stages to the west.

From Ashnāghar, Nīlāb is 7 stages to the southeast, and Swat is 2 stages.

From Duwāwa, Peshawar is 1 stage.

From Swat, Kajū is 2 stages to the south and Muqām is two stages to the south.

From Ashnāghar, Muqām is 1 stage to the northwest.

From Kajū, Kabul is 10 stages to the west.
From Ashnāghar, [M 241] Panjkūra is 3 stages.
From Muqām, Panjkūra is 1½ stages.
From Peshawar, Bājūr is 3 stages to the northwest.
From Duwāwa, Bājūr is 1 stage to the northeast.
From Panjkūra, Bājūr is 1 stage to the west.
From Ashnāghar, Bājūr is 3 stages.
From Kabul, Nagarhār is 3 stages to the east.
From Peshawar, Nagarhār is 2 stages to the west.
From Ashnāghar, Nagarhār is 2 stages. [Plate 36]

[R 90a] **Chapter 26: Country of Kashmir and Tibet**

It is pronounced Kishmīr and Qishmīr. Originally it was the name of a tribe deriving from India and Sind and the Turks; later the name was applied to the place where they lived. It is an oblong country surrounded by mountains like a fortress and belongs to the fourth clime.

Borders

To the south, Delhi and Lahore in India; to the north, Badakhshān and part of Khurāsān; to the east, the territories of Tibet; to the west, the halting-places of the Afghans. Its length from west to east is nearly 40 parasangs, its width from south to north is 20 parasangs.

It is all flat open plains and comprises 2000 villages and 32 districts (*pargana*). These plains are decked out with meadows and pools, flowers and rivers and trees, all as delightful as a garden of Paradise. Its pleasant climate is the cause that its people are proverbial for beauty of face and refinement of character.

It is bruited about that in the mountains and plains of this country are 100,000 farmed villages. All sorts of fruit are delicious and plentiful. To be sure, because of the cool climate it cannot produce as many oranges and lemons and dates as hot places, but still it produces more than nearby areas.

The capital of this clime and seat of sovereignty of that country is the famous city of Srinagar, located at 110½ degrees longitude and 37 latitude. A river as broad as the Tigris flows through the middle of it. In this city, only 2000 houses are engaged in weaving cashmere shawl, the skill of that art being limited to them alone on the face of the earth. For that reason the people of Kashmir mostly wear cashmere shawl, which is famous around the world.

Because of the delightful climate and the beauteous mountains and plains, the people of Kashmir are mainly given to music and singing, while they are indifferent in matters of food and dress. Their staple food is rice, and that usually left over from the night before. But that well-tuned region is entirely a place of wine and music, dancing and mirth, whose tulips betoken rosy cheeks and whose produce is such that 400 batmans of saffron are exported annually to India alone.

Lorenzo says the people of this country are mainly idol worshippers. They are much occupied with magic, both the practice of it and the instruction. The tribe known as Yoghi is originally from this country. They are devoted to chastity and worship and they wander about Hindustan in the garb of monks, entering caves and engaging in spiritual exercises. Some live 200 years on account of constant fasting.

The Kashmir River, as broad as the Tigris, emerges entirely from a spring at the foot of the mountain and flows through the middle of the city. There are fifty bridges over it, seven of them alone in the city of Nigar (Srinagar). After passing by there, it is named after the places that it reaches, such as Dandān and Jamad. Eventually it empties into the Chenab River above Multan.

Mountains

The mountains of Kashmir are like a city wall constructed around the country, and thus [M 242] its people have remained safe from the incursions of the Mongols and other raiding peoples. There are only three passages to Kashmir through these mountains: 1) The Khurāsān Pass, which is very narrow and difficult of access. It is impossible to get through with pack animals. If necessary, travelers on foot load up as much baggage as they can and carry it. 2) The India Pass, which is similarly difficult and narrow. 3) The Tibet Pass. While the road is somewhat easier, nothing is found along it for several stages except poisonous grasses, and riding animals or pack animals rarely get through it without mishap.

Routes

From Kashmir to Lahore there are 13 roads.

1. The Kihtwār Road. It has three branches, the first being 200 *kuroh*, the second 260 *kuroh*, and the third 258 *kuroh*. The Indians use *kuroh* for mile [R 90b]. We use numerals to indicate the length of each stage in *kuroh*.

The first branch, in the southwest direction: from Kashmir > village of Panpūr 7; it abounds in saffron > Wanatpūr 6; it is on the bank of the Dit River which goes to Nīlāb in the vicinity of Rahṭās > village of Wajiblārū 7; it is also on the bank of this river > village of Arkāmū 6 > village of Kuhan.

One *kuroh* east of here is a marvelous pool, seven cubits square and the depth of a bow, built of dressed stone. Every year, from the time the sun enters Gemini and for twenty days thereafter, water begins to emerge from the bottom of the pool at sunrise. By mid-day it is completely full and flows out with a force that turns two mills. After mid-day it begins to recede and by late afternoon not a trace of water remains. A half hour later it bubbles up again, so that by midnight it is completely full and flows out, then again recedes. In the morning it repeats the cycle and this continues for twenty days, when it stops and the pool stands empty of water until the following year. There are four holes at the bottom of the pool where the water comes out and goes back in. Each hole has a specific name. When the water is receding, if a person throws three or four clay pellets of different shapes one at a time while naming those holes, the water takes each pellet to the hole named and to no other.

Another pool, in a place named Sandabrārī, is recorded in the *Haft Iqlīm* as one of the marvels of Kashmir. It too is made of dressed stone, but it has no holes or fissures. When the sun enters Taurus, every day before sunrise water starts oozing out and gradually begins to flow. After five or six hours it begins to recede until none remains. So this must be the same as the previous pool, the only discrepancy being whether there are holes or not.

From the village of Kuhan > Kitilpāyī 5 > palace of Mular 8 > Kirnār 15; it is within Kihtwār, in a valley, and the road is all downhill > Chātrū 10, in the middle of the valley > village of Changalistān 12; the Ārū River comes here from Mular and irrigates it > Chandarpāk 8, on the bank of the river that the Indians call Chenab; there is a rope bridge there by which one crosses > city of Kihtwār 10, on the bank of the river; this place has a separate ruler; inns have been built in this stretch of the road > fortress of Wāwarīḥājī 7 > fortress of Pāybasanū 8, on a mountain top; the ruler of this place, known as *tkur*, is an enemy to the ruler of Kihtwār > village of Damānisandū 10; it has a separate ruler known as *anū* > [M 243] town of Badargāh 20; its ruler is a woman. The first half of the road beyond is prosperous and under the authority of the ruler of Basanū.

From there (Badargāh) the road divides in two, the first branch going to Bāndarwār. It is flat plains as far as the base of Mt. Bilāl 10 > the ascent of Mt. Bilāl 7 > the descent to Bāndarwār 8, a town with a separate ruler > fortress of Mānkūt 10, with a separate ruler > small town of Sāmbū 10 > Sangrū 15, a village within India > Gāwanīsarāy 10 > village of Kinpin 15 > Lahore 10. The second branch from Badargāh > village of Kalbān 10 > base of Mt. Balāl which its people call Zalāl 4 > summit of the mountain 10 > small town of Pachal 6, at the base of the mountain; it is recorded as attached to the government of Rāmchand. From here (Pachal) the road splits in two, one branch going east to the village of Lālar 7, attached to the government of Sarbal in the east > Karan 15, the town of Bāndarwār. The other branch from Pachal > village of Chandnī 10, also attached to the government of Rāmchand; the road is through jungle > town of Jāniknū 8; its ruler is one of the eunuchs of the emperor of India; the road is entirely within a river valley > village of Dansāl 10, attached to the government of Parsarām (?) > small town of Ban 5 > Sialkot 10 > Lahore 35.

2. The Kampar Road. It has four branches, the first being 184 *kuroh*, the second and third 182, the fourth 185.

From Wajiblārū > village of Lūkbaūn 4, a village on a hilltop > village of Dūrū 5 > village of Kampar 4, within Jākīr > village of Chūn 2, on the skirt of Mt. Tandūk > Rāndsang 5, at its summit. From this point the road divides in two, one branch on the right > village of Pīristān 15. This road, called Shālūt, is for horsemen but is blocked by snow in the winter. The other branch on the left > several caves known as Hūya 5, on one side of the aforementioned mountain. This road is for people on foot, difficult of access. At the midpoint of the route one ascends to the top of the mountain. Sometimes contrary winds blow and bring down snowdrifts the size of houses. It is dangerous and it is not a winter road. From there (Hūya) > Kakjādū 3, on a flat place > village of Pīristān 5, [R 91a] > village of Hāla 4, halfway up Mt. Sūnūn to the south, it is the place of the Qanjsiyah (?) sheep > village of Kanwit 8, south of the summit of Mt. Hampardan, its old name is village of Sapū and it has a separate ruler > village of Dandarhit 5, south of the summit of Mt. Dārā > village of Chūk 3, on the bank of the Chandarpāk River, its road is entirely mountains > village of Kundafar 2, attached to Rāmchand; one crosses the river here with a rope bridge, as previously mentioned > place known as Kāwkul 5, halfway up Mt. Dakkū.

Kāw is the name of a certain tree, while *kul* is the generic term for tree. The way it was named is as follows: Some time ago, Ādam Khan, the son of Sultan Iskandar, ruler of Kashmir, was vexed on account of his father. When he came to this place he had in his hand a stake from that tree and planted it in the ground. After some time it took hold and became a mighty tree.

> village of Batūt 4, on the upper part of Mt. Dakkū, attached to Rāmchand > village of Kalā 8, halfway up Mt. Dakkū, attached to Rāmchand > town of Chalī, also called Pachal 4. The remainder of the route was noted above.

3. The Katrū Road, also called the Pānhāl Road. The road divides in two from Lūkbūn. The first is Dūrū, to the west of which it goes to Katrū 3 > village of Manazmūh 1, on the skirt of Mt. Gāzmār, a place where *zakāt* is collected > Sūktel 4, at the summit of that mountain > village of Nūkām 4, on the southern skirt of the mountain.

One bowshot from there, on the India side, in a place called Rāsad, a round blue-colored stone one cubit tall and one and a half cubits in diameter is driven into the ground. It is called *shanar* meaning "truth". It is always [M 244] the custom in Kashmir that if two people quarrel over a matter of truth and falsehood, they come to this truth stone and stand on its south side. It is said that the one with the rightful claim can jump over the stone to the north side, but the one with the false claim cannot. The ruler of Kashmir, Mīrzā Ḥaydar, forbade the people from doing this, but the custom persisted.

> village of Charal 2 > village of Arzankūt 4; these two villages belong to Pānhāl > village of Chatalwās 2, attached to Pānhāl; in the middle of the road is a valley known as Dīwnār > village of Lūdū 4, halfway up Mt. Māchīl, to the south > village of Pāynatāl 4; on this road are two river valleys known as Nīlaārū and Kūdahlārū > village of Rāk 4.

These two villages are within Zulmattū and have separate rulers. In the middle of the road is a small mountain named Akarū and an idol temple built on top of it. From here it is 2 *kuroh* to the bank of the Chenab River, all downhill. After crossing the river and going along the bank another 2 *kuroh* one comes to the ascent of Mt. Gang. Horses and donkeys cannot make this crossing.

> village of Gang 2, on the top of the mountain > village of Kūtar 2 > village of Katrū 3, halfway up the mountain, attached to Takhtmal > village of Shadanū 5, at the base of Mt. Lūdarak, attached to Takhtmal, on the India side > village of Pūchī, also called Pūtī 3, the road is all downhill > village of Chulhan 4, on the upper part of Mt. Dakū; all of these are attached to Takhtmal > village of Mihr 2 > village of Narūrū 2, on the skirt of Mt. Kītīklū, on the India side > town of Kirmiz 3, the place of the government of Takhtmal > village of Mamhāl 3, within Jāniknū > village of Dānyāl 4, within the province of Jamū; in the middle of the road is a small mountain called Chabdak > town of Jamū 7; on its road is a mountain called Kājdār > village of Ban, having crossed the Tūhī River 4 > city of Sialkot 10. This road is 140 *kuroh*.

4.[34] The Barnal Road. From Wintpūr the road divides in two, one branch going to Wajiblārū, the other to Nāīnū on the bank of the Wat River 5 > village of Kadawnī 3, on the bank of the river, a grain depot > village of Hābilishī 30 > village of Jūkām 4 > village of Wāsangkand 3. Wāsang is a spring in the middle of this village whose water is very cold in summer and dries up in winter. This village is on the skirt of Mt. Sundar. From this village the road divides in two. The eastern branch, for travelers on foot, goes up Mt. Sundarbār.

> village of Mahū 10, on the south side of that mountain. The western branch, for travelers with mounts, goes in the direction of Mt. Wāndaspak > Barsarāy 6, at the summit of that mountain > village of Mahū 6. These are within the district (*pargana*) of Barnal > village of Warnga 5, in the middle of the Kaz Valley, with Mt. Sundarbār to the east and Mt. Butar to the west; these are high mountains; the Kaz River flows through the valley and is forded twice, also with horses > village of Chardanū 4; these two villages are within Jākīr. Mawḍiʿpanār 4, a halting place of caravans in the middle of the valley > village of Dīmandal 5 > town of Sumurkūt 6; this is the original Barnal, it has a fortress > village of Damkand 3, on the bank of the Chenab, where one crosses by a rope attached to a forked tree > village of Tankhar 3, on the bank of the river on the India side; it has good rice and pomegranates > village of Kartan 5, on top of Mt. Ḥāh; shawl is woven here > village of Pūtī, also called Pūchī 4. These villages are attached to Takhtmal. This road is 141 *kuroh*.

5. The Dānduwār Road, also called the Mankanār Road. It is suitable for horses. After Wajiblārū, there are two more roads aside from the two previously mentioned. The eastern branch comes to the village of Zundar 5 > Dīwasar 4.

[R 91b] In the village of Phāl belonging to this district (*pargana*) of Dīwasar [M 245] is a spring whose depth is the height of a man and whose length and breadth are 100 cubits. The water flows out with a force that turns one mill. If a person has a need, they put a measure of rice and water in a new earthen jar, seals its mouth tightly with a skin, and throw it into the water invoking the name of that person. It sinks and after some time comes out. If the rice has cooked properly, they draw a favorable omen and say that that person will prosper that year; if not, they believe the opposite.

> village of Dānduwār 4 > village of Mahū 7.

The western branch from Wajiblārū > village of Kadawnī 3 > village of Kalāhkām 5, attached to Dīwasar > village of Tākmū 5, on the skirt of Kutal-i Zarmark > village of Bārnahār 15; this road is completely mountainous and suitable for horses > village of Lah 5, in a valley > village of Shār 5 > village of Gayal 7; its road is not level; in its middle is the Tūsh River that flows into the Chenab; in summer one crosses with a rope and forked tree, in winter one wades across > village of Jān 4, on the bank of the Chenab > village of Chābkārī 4.

From here the road divides in two. In the eastern branch one crosses here by boat > village of Shalāl 4, on a hilltop > village of Anbār 3, all downhill; these are connected to Lāldīw > Dāmkar 7, a small town connected to Lāl; it has

34 R repeats "3" here and the numbering is off subsequently; corrected in M.

a fortress. In the western branch one goes along the bank of the Chenab > village of Kīlū 7, on the Kashmir side > Dāmkar 5, crossing by boat > village of Chaprār 7 > Sialkot 8. This road is 141 *kuroh*.

6. The Mayamundar Road. From Kashmir > Chānpūr 3.

Among the wondrous things here us a weeping willow tree that the people of Kashmir call *baran*. Its trunk is five cubits, and two persons together cannot reach around it. It is nearly four cubits in circumference and 30 cubits in height. Its branches bend down to less than two cubits from the ground. It has many branches at the top. Its expanse and circumference are nearly 50 cubits. They say that if someone shakes one of the branches, all the branches are set in motion. This tree is between two hills on the bank of the Wat River. Its branches are connected to one another and easily damaged. There are many trees like this in the district (*pargana*) of Dachinpāra. The more branches there are, the greater the motion, because the motion is assisted by the air between them.

> village of Rāmū 7 > village of Mayamundar 8 > village of Manazkām 5 > Hākuwās 5 > village of Bādrund 5 > village of Dūlū 2 > Gagarū 5 > village of Gayal 5. One of the roads is 121 *kuroh*, the other is 120.

7. The Sidaw Road. From the village of Rāmū, the road divides in two, one branch going to Mayamundar.

- Western branch: > village of Warwaḥ 7 > Jamālnagar 5, a ruined village > summit of Mt. Burzal 15 > village of Badal 12, on its skirt > village of Surāl 15 > village of Pawunī 15; on this road are villages some distance away and one passes a mountain named Dāramī > village of Kaltin 7. From here the road divides in two.
- Eastern branch: > town of Maynāwar 6. One crosses the Chenab by boat at Sikkapatan and enters India 4 > village of Chak 1 > Sialkot 4. In the western branch > Bahlūlpūr 7, a small town > Sialkot 7. In this region one crosses the Chenab by boat. Both branches are suitable for horses. One is 126 *kuroh*, the other is 125.

8. The Hīrpūr Road. From Jānpūr > village of Khānpūr 2 > village of Zawur 3 > village of Hīrpūr 4, near Mt. Badayāl.

On the skirt of the mountain is a stone house named Patmaran. From here the western road goes to the right, up the mountain. > Machlū, 1 *kuroh*. From there it is downhill along the middle of the mountain > Basātara, a stone house, 3 *kuroh*. From there one goes two *kuroh* [M 246] further down, called the Zaznār Descent. Then one comes to the middle of the mountain in one *kuroh* > place named Nārbarār 1; its road is very narrow; two people and a horse can pass, one leading it and the other driving it from behind > palace of Duwind 3 > Pīrpanjāl 2, at the beginning of the ascent > Kūkchī 13, at the end of the descent; it is not a winter road; one ties a sheepskin to one's bottom and slides down, but it is dangerous > village of Pushna 12 > Bīramkala 5, comprising several villages where they collect customs tax; it has a small river, 3 or 4 cubits wide and 2 cubits deep, but when it is in spate one fords it 18 times, 27 in summer > village of Kanawānī 4 > town of Tanū 3 > village of Barut 3 > Rājwīral 4. [R 92a]

The eastern road from Patmaran to Kāmlanūkūt 7, a ruined fortress on a high mountain > Mt. Hastawnaz 3 > village of Kandpār 3, a caravan halting place. From there the road divides in two.

- Eastern branch: > Baknazīl 15, a ruined fortress > Sīl which is the village of Kalānī 15 > Rājwīrī 8 > village of Sangpūr 7 > Kārnakūt 5 > Nawshahr 6 > village of Mūjawāl 5 > village of Badū 3 > Banbar 7 > village of Dirrikiw 7 > village of Kātahāl 15, on the bank of the Chenab > village of Pīrkānī 7 > fortress of Kakarjama 8 > village of Saynasar 6 > Shīrkar 7 > village of Patyāla 3 > palace of Kujaz 6.
- Western branch from Kandpār > Pīrpanjāl 3.

Elephants, horses and camels go by the [Ha]stawnaz route, not by the others.

9. The Charhārū Road. From Kashmir > village of Supuk 7 > village of Sūrish 5 > village of Yabhun 3, on the skirt of Mt. Charhārū > summer pasture of Yārmarak halfway up the mountain 5 > Kanzawal 5, a ruined castle at the summit of the mountain; from here it is downhill > Hawīn 3, a great cave that holds 1000 men > village of Kātrū 7, in the plain two *kuroh* distant from the mountain > village of Suhran 6 > Mabatpal 4, a stone as big as a house > Hastapālūt 7, in the middle of Mt. Balar > village of Barut 4 > Brājwīrī 7. This road is 155 *kuroh* in total. In summer it is suitable for horses; in winter it is impassable because of much snow and severe wind.

10. The Juyl Road. From the city (i.e., Srinagar) > village of Kalanahāmū 7 > village of Lūlapūr 5 > village of Pīrūzpūr 6 > a village on the skirt of Mt. Chuyl 1 > Barzilal 7, a stone house halfway up the mountain. One goes through the valley for three *kuroh* and goes up another four. From there > Sarbal 8, at the top of the mountain > Shādaj 15, halfway up the mountain > village of Kātrū 5. This road is 169 *kuroh*.

11. The Badsangar Road. From the city (i.e., Srinagar) > village of Patan 15 > village of Ushkaw 7 > village of Kachahāmū 8 > village of Kantamal 4, one side of it is mountain and one side the Bhat River > village of Māhūrū 7. This road is very narrow, since it goes through the mountain pass and for four *kuroh* two men go with a horse, one leading and the other driving it from behind > village of Shahkūt 5 > village of Ur 1; from here one goes to Mt. Badsangar > Hālan 15, on the southern skirt > village of Dīwkār 8 > village of Prūnaj 8.

From here the road divides in two. One goes to the left > Suhran 7. The other goes to the right > village of Kīdabarū 7 > town of Bāhūrū 15 > village of Pūhālan 15 > town of Sulṭānpur 15 > village of Jilum 7, on the bank of the Jilum (Jhelum) River; one crosses the water by boat > village of Khawāṣpur 8, on the bank of the Chenab > village of Kāthāl 7 > village of Pīrkānī 7. Horses, elephants and camels go as far as Prūnaj; only horses go beyond thhat. The first road is 194 *kuroh*, the second 174 *kuroh*.

12. The Paklī Road. From Patan > Bārīla 12, on the way one crosses the Wat River by a pontoon bridge > village of Tatmal 8 > village of Duwārkantal 6, flanked by the mountain on the right side and the Wat River on the left > village of Dīw 10, away from the road. [M 247] > village of Wazlū 8 > village of Dankūt 5; on this road one passes Mt. Kuwāram > Ranbal 8, a wooden gateway flanked by the mountain on one side and the Wat River on the other > village of Pūlyās 3 > town of Nakrī 5, its ruler is subordinate to Kashmir > village of Nūpūr 4 > village of Wār 5, on the bank of a great river known as Pīdarang; it is crossed by means of rafts called *shrāh* by the Kashmiris who tie ropes on both sides and haul horses and men across > village of Naynusak 15; Mt. Shankar is in the middle of this road and after that one crosses a river known as Kusna > town of Paklī 15 > village of Māngal, with a fortress 6 > town of Shāhrukh Sultan 8 > village of Gajargām 5 > village of Hazara 8 > palace of Mādūsang 7, known as the palace of Kharbuza (or "palace of watermelon") > town of Rawalpindi 6 > village of Hātapa 7 > Rahṭās 6 > village of Jilum 3. It is 204 *kuroh* in total and suitable for all animals.

13. The Karnū Road. From Patan > town of Suypūr 10 > village of Patrū 7 > village of Trāhkām 8 > village of Tabhandū 3, on the skirt of Mt. Zar > village of Tangdār 15 > village of Acharbul 10 > town of Hībang 12 > Paklī. This road is only for foot travelers; it is 216 *kuroh*.

Kings of Kashmir

According to the *Haft Iqlīm*, in olden times Kashmir was governed by the Indians. In A.H. 654 (1256), Sultan Nāṣir al-Dīn b. Shams al-Dīn, who was the ruler of Delhi, spread thc banner of Islam in this region. A certain Zayn Ḥasan became governor and ten of his descendants ruled independently for 160 years, reigning as sultans. [R 92b] One was Sultan Zayn al-ʿĀbidīn who ruled for 52 years (1423–1474). Kashmir was never more prosperous than during his time. He paid no regard to whether one was pagan or Muslim, but only esteemed knowledge and talent. He sought out men skilled in every craft from all the surrounding regions and brought them to his court. He settled them in that land and paid them due respect, whatever their religion might be, not harrassing them in the least. Therefore some said that he acted according to the religion of the Yoghis (*jūkiyān*). After his death, the country was governed by a group of local people until 990 (1582) when it fell under the sway of the Shah.

But Mīrzā Ḥaydar says in the *History of India* that Shams al-Dīn came to Kashmir in the guise of a Qalandar dervish. At that time a woman was governing the country and Shams al-Dīn entered her service. Eventually he achieved his desire, married the woman, made the people obedient to himself and assumed the throne. His son ʿAlā al-Dīn reigned after him, then in succession Sultan Quṭb al-Dīn and his son Sultan Iskandar who favored the religion of Islam and destroyed most of the idol-temples. His successor was Zayn al-ʿĀbidīn who made Kashmir prosperous. After him, several of the Yoghis of that country governed, and during the reign of one of them, Yūsuf Khan, it was conquered by the Persian Shah. Yūsuf Khan submitted and for many years acted as khan (i.e., Safavid governor) on his behalf. He was skilled in poetry and music, and several poets and excellent men entered his service.

The country of Tibet: It is pronounced Tubbat in the same pattern as *sukkar*, as mentioned in the *Taqwīm*; but Masʿūdī writes it as Thubbat. It is an extensive and pleasant country in the vicinity of Khurāsān.

Borders

To the east, the mountains of China; to the south, Kashmir and India; to the west, the lands of Khurāsān; to the north, the steppes of Cathay and Khotan.

Its famous city is Ghataḥ: According to Ibn Saʿīd, a famous and prosperous city atop a mountain at 130 degrees longitude and 40½ degrees latitude. [M 248] Its fortress has one gate.

Another is Qaṣr-i Ḍaḥak, located on a mountain at 125½ longitude and 38½ latitude. To the east of this mountain is a plain extending 6 stages from south to north, whose width is 3½ stages, surrounded as with a hoop by the Sunbul Mountains where musk deer graze on hyacinth (*sunbul*). Streams flow down from every side and collect in the middle of the plain, a delightful body of sweet water known as the Pool of Tibet, located at 126 degrees longitude and 41 latitude.

The Tibetans are flat-nosed and round-faced like the Chinese. They trade in silver, iron, gems, musk and tiger pelts. Masʿūdī says that the special qualities of Tibet are marvelous. Its climate is salubrious and delightful, its soil is productive, and its various types of fruit and flowers are unlimited. Its inhabitants know no grief. It is the special quality of this place that worry and sorrow do not afflict

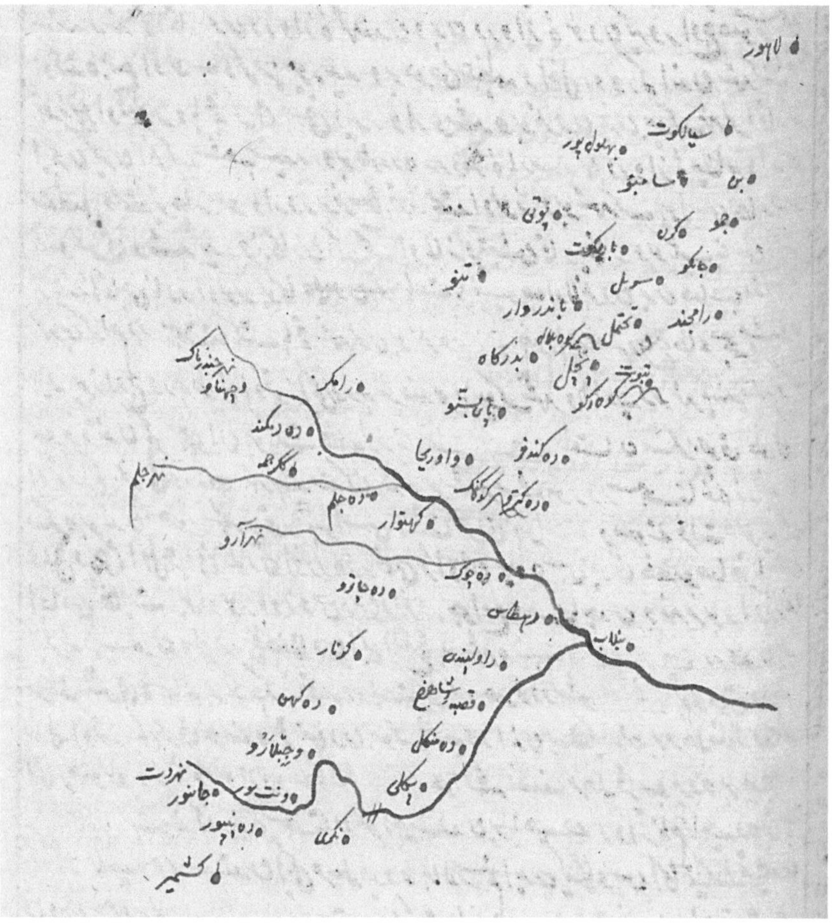

PLATE 37
Map of Kashmir, R 93a (Appendix, fig. 78)
TOPKAPI PALACE LIBRARY, REVAN 1624, WITH PERMISSION

them. All of them are laughing and cheerful, have a gentle temperament, and engage in various types of dance and amusement. They even carry the funeral biers with music and song. Generally in this country the sanguine humor is ascendant in all creatures, both speaking and mute, and affects their nature.

Lorenzo says that there are large reeds in this country and abundant forests with many beasts of prey. For that reason, the roads are dangerous. Nevertheless, every year merchants travel there to obtain coral, because coral is used there as currency instead of silver coins. They also buy musk and make a great profit from that.

The musk deer resembles a gazelle, with head like that of a pig and two fangs jutting from its mouth like an elephant's tusks. They are raised in flocks. [R 93a] Every year a swelling develops in their belly and gradually increases over the course of a month. When it reaches maturity, it becomes very irritated. The deer rub themselves on some rocks to pierce the swelling, and the blood that flows from that turns to musk. The best kind is said to be yellowish. These deer produce excellent musk because they consume Indian hyacinth. Therefore merchants travel here from faraway places.

Cinnamon and cloves are also obtained here, but they are not esteemed like the spices of Molucca and are not exported. The Tibetans season meat with spices and eat it raw. It is a crime for women to undo their own waistbands. The men do not grow beards but all of them shave or pluck them out.

The *Ḥabīb al-siyar* records a certain mountain in Tibet where, if a fire is lit, water comes out and extinguishes it. Some kings tested this by heaping up naptha and sulfur and when they set it on fire, water gushed out and extinguished it, so it was impossible to kindle the fire. [Plate 37]

Chapter 27: Country of Sijistan

Arabicized as Sīstān. An extensive province between Khurāsān and Kirmān and Sind.

Borders

To the east, Afghanistan, India and Makran; to the north, part of India, Ghazna and Ghūr; to the west, Kirmān and Khurāsān; to the south, the desert of Kirmān.

In olden times it was so prosperous that one *jarīb* of land, despite having little water, was sold for 1,000 dinars. Then the Ghūr and Isfirār entered this country; later it fell to ruin and they departed. [R 93b]

The capital of this province is the city of Zaranj, [M 249] located at 97 degrees longitude and 32½ latitude. It is a large city and the name of the province is also used for the city. It lies between Khurāsān, Makran, Kirmān and Sind. In olden times the seat of sovereignty was Rāmshahr; it fell to ruin and Zaranj was built in its place.

Zaranj has a fortification and a suburb. In its moat is a spring whose water fills up the moat together with that of other streams that collect in it. The suburb is also surrounded by a wall and a moat.

The fortress has five gates: Iron Gate and Old Gate both open to the Pārs road; Karkūna Gate opens to the Khurāsān road; the fourth gate, Panbakī, opens toward Bust; the fifth, Ṭaʿām Gate, is completely made of iron.

The suburb also has 13 gates: Mīnā Gate on the Pārs side, Karkān Gate, Ashrak Gate, Darsārā Gate, Shuʿayb Gate, Khawīk Gate, Darkār Gate, Darpanbakī Gate, Karkūna Gate, Aspar Gate, Ghanjūh Gate, Rastān Gate, and Zangabār Gate. All are made of clay. Neighborhood mosques and Friday mosques are plentiful. The markets are full and prosperous. Yaʿqūb b. Layth has a market here; he made it an endowment for Mecca and his Friday mosque.

Numerous rivers flow in this city. One of these enters at Old Gate and another at New Gate (same as Iron Gate above?). Three streams that come together in front of a mosque turn a waterwheel and flow into a large pool; from there the water is distributed to the houses of the city and irrigates the gardens. Aside from these, larger rivers flow in the suburb; the ones that enter the city are branches of them. From Pārs Gate to Mīnā Gate is a continuous and extended bazaar, extending as much as one parasang.

The weather of this city is warm, though not free of cold and harsh wind. It does not snow in the winter. The buildings are mainly sun-dried brick; wooden buildings are scarce.

Among those from Zaranj is the founder of the (Karrāmiyya) sect, Muḥammad b. Karrām (d. 255/869).

The rest of the cities of Sijistān are as follows:

Rukhkhaj: A large city and district located at 103 degrees longitude and 32 degrees latitude. The Helmand River passes along its borders. One of its associated towns is Baghjuwān.

Bust: A famous city on the bank of the Helmand River at 98½ longitude and 32½ latitude. There is no more prosperous city than this in Sijistan after Zaranj. It has a pleasant climate, the air resembling that of Iraq and the water that of the Euphrates. The people's manners are also like those of Iraq. It is rich in produce and fruit. Dates and plums are especially fine and grapes are also abundant. It has Friday mosques and neighborhood mosques, and a pontoon bridge on the river. Among the excellent men from here is Abū'l-Fatḥ (al-Bustī, d. 354/965).

Qarnayn: A small town one stage northeast of Zaranj. It lies to the left while going toward Bust.

Khwāsh: A large town at 100½ degrees longitude and 32½ latitude. It is bigger than Qarnayn and lies one stage to the left when going from there toward Bust. It has flowing water and date palms.

Ḥiṣn al-Ṭāq: A small city one-half stage from Khwāsh, with many orchards and pastures and villages. Produce and fruit, especially grapes, are disseminated from there throughout Sīstān. Its fortification is a strong castle, difficult of access, atop a high mountain and surrounded by a river. The kings of that region keep their treasury there. They say it is impregnable. [M 250]

Darghash: A town on the bank of the Helmand in the Duwār district of Sijistān.

Duwār district: It is two stages from Bust. It was the land of Rustam which Kay-Kāwus had given him to rule. It is also called Dāwar. It was the capital of Sūrī, the ancestor of the Ghūrid sultans. It is attached administratively to Kandahar.

Sarwān: A small city two stages from Bust. It has grapes and other fruit.

Kandahar (Qandahār): A city and fortress ten stages east of Zaranj at 100 degrees longitude and 33 degrees latitude. According to the *Taqwīm* it is one of the fortresses built by Alexander. A river comes from a mountain to the west of the city, passes it to the south and encircles the fortress. It is an extensive city with a large population. In the middle of the fortress is a deep well made of granite. Whenever they draw water from it, pieces of wood and grass come out, proving that it has a channel from the outside. As this fortress is the frontier between the Persian Shah and Emperor of India, it is seldom free of siege and conflict. [R 94a]

Farāh: A large town 7 stages northwest of Kandahar at 97½ degrees longitude and 34 degrees latitude. It has 60 villages and sown fields and date groves. Next to it flows a river over which a great bridge has been built. One parasang away is a mountain known as Mārandakī where water constantly drips from a vault carved out of stone. The people go there on pilgrimage to pray and say what they need. If the dripping increases somewhat they depart happy, thinking the need is granted; otherwise they are despondent. Among those from this city is Abū Naṣr, author of the *Niṣāb-i Ṣibyān*.

Nih-Bandān: A fortress and town in the mountains 6 stages west of Farāh. Its flowing water comes from Farāh.

Qaṣr: A city where the governor of the province resides.

Safkhāy: A city larger than Qaṣr.

Mābayn: A town in the district of Rukhkhaj.

Rūdān: A small town close to Farūrmand. It has flowing water and abundant farmland.

Khalaj: A place whose inhabitants are a people that came to this province in ancient times and settled between India and Sīstān. Their form and mannerisms resemble the Turkmen people and they speak the Turkish language.

Patta: A nomadic tribe of the Afghan people. Their great ones are called Malik-Patta. They are in the mountains and raise camels, sheep and cattle. Their settlements are 15 stages northwest of Sītpur and 7 stages southeast of Kandahar. They are obedient to the ruler of Sīstān and pay taxes to the Qizilbash (i.e., Safavids).

Characteristics of Sīstān

The tax revenue of Sijistān was originally 67 *yük* (6.7 million); after various calamities and flight of the population it dropped to 9 *yük* (900,000). It is entirely flat sandy tracts. Only there is a mountain near the city of Bām (?). It is 1/3 parasang in elevation and one of its sides is made of pure moving sand. There are pilgrimage sites in that region. The people come on Friday eve and say that they hear the sound of loud drums and kettledrums. This story is related from the *Tārīkh-i Mubārakshāhī*.

Cold and fierce winds blow in Sijistān, but it does not snow. The wind is always moving the sand from one place to another. Sometimes a place is buried in the sand. They block the sand from the cities by various devices, such as building something like a wall around it and putting vaults and holes in its base so the wind enters through those channels and it blocks the sand. [M 251] On the plains people make a living mainly by operating windmills. It is a country with strong winds.

The area between Sīstān and Kirmān is full of large buildings and ancient ruins. Sīstān is a province of cheap provisions. Its pomegranates, grapes and dates are plentiful. Its people mostly lead lives of wealth and ease. Asafetida is found in plenty in its plains. They raise hedgehogs like cats, because there are so many snakes.

HINDMAND (HELMAND) RIVER: It is the largest river of Sijistān. It comes from the east of Ghūr, reaches the borders of Rukhkhaj and Bust, passes them to the south, flows from east to west, twists back and passes to the south of Zaranj.

Some have claimed that the Helmand empties into Lake Zara. The truth is that when it comes to Bust, it separates into branches—the Ṭaʿām, Bāsarūz and Sitārūz Rivers—that irrigate many villages. It is the Sitārūz that goes to Zaranj; when it floods, one can go there from Bust by boat. The Malā and other rivers separate off from this.

The Helmand has violent currents. There are many cities and town on it. Many Indians practice austerities on the bank of this river. Many rivers separate off from it and many empty into it, so it neither increases nor decreases.

FARĀH RIVER: It flows from the Farāh Mountains and comes to the castle of Nih-Bandān, then turns south and empties into Lake Zara.

LAKE ZARA: A large lake in Sijistān. Its length is 30 parasangs and its breadth is approximately one stage. There is an island in the middle that is inhabited and cultivated. As the lake water is very fine, the fish that are caught are excellent, [R 94b] and there are many reedbeds. All around the lake are farmlands and villages, except for the side toward Sijistān which is ruins and wasteland. The Helmand and Farāh rivers empty into this lake.

Routes of Sīstān

From Zaranj to Herat: Karkūna 1 stage > Bastara 4 parasangs; one crosses a branch of the Helmand on this road > Jarīr 1 stage > bridge over the Farāh River 1 stage > Farāh 1 stage > Kūsān which is the frontier of Sīstān 1 stage > Khāsān 1 stage > Kārīz-sarī 1 stage > Siyāh Kūh 1 stage.

From Zaranj to Bust: > Rasūq > Sarūr > village of Ḥarūrī > Pashak River; over it is a bridge built out of brick > hospice of Dahak > hospice of Arsūra > hospice of Karūd > hospice of Hastān > hospice of ʿAbd Allāh > Bust—1 stage each.

From Bust to Ghazna: > hospice of Fīrūzmand > hospice of Maʿūn > hospice of Kru-Rukhkhaj > Kamīnābād > Khurāsān > hospice of Sarāb > Awnī > hospice of Jangalābād > village of ʿAwm > village of Khwāst > village of Ḥūm > Khābsār—1 stage each.

From Sīstān to Bālīn the road is through the desert. From Rukhkhaj: hospice of Sangīn > hospice of Jangī > hospice of Bam > Sinjān—a total of 14 stages.

From Sīstān to Kirmān and Fārs: > Khāwrān > hospice of Wārang > hospice of Qāṣī > hospice of Karāmjān > staging post of Sīnj, one of the districts Kirmān—a total of 8 stages.

Thus, from Zaranj, Herat is 6 stages. From Bust, Ghazna is 14 stages. From Farāh, Kandahar is 5 stages. From Kandahar, Pakar is 12 stages; Kīj is 18 stages to the south; Mashhad is 20 stages; Zaranj is ten stages to the west; Hazara is 12 stages; Nahalwāra is 5 stages. From Farāh, Nishapur is

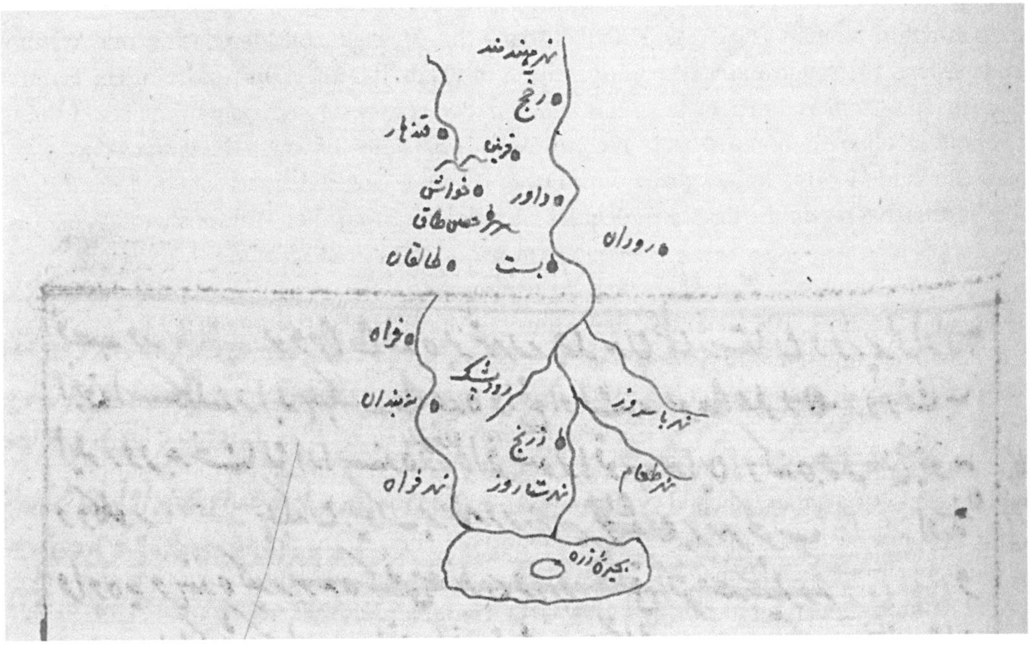

PLATE 38
Map of Sijistan, R 95a (Appendix, fig. 79)
TOPKAPI PALACE LIBRARY, REVAN 1624, WITH PERMISSION

10 stages; Herat is 5 stages. From Zaranj, Nih-Bandān is 4 stages to the north and from there Khabiṣ is 6 stages to the southwest.

Kings of Sīstān

This clime was conquered in the age of the Rightly-Guided Caliphs and was ruled by Umayyad and Abbasid governors. During the governorship of Dirham b. Ḥusayn, Layth b. Ṣaffār appeared [M 252] and, while he had been a bronze merchant, became a military commander under Dirham's patronage. After the death of Layth, his son Yaʿqūb took his place. In 237 (851), Dirham also died, and as his son Ṣāliḥ was incapable of governing, Yaʿqūb took charge of affairs, assumed the throne and became independent ruler. He seized Khurāsān from the Ṭāhirids and also conquered Kirmān and Fārs. He died in 265 (878) and his brother ʿAmr succeeded.

ʿAmr was obedient to the caliph and took the royal patent. He conquered Quhistān, Ghazna, and Māzandarān. Later he rebelled and he tried to take Baghdād. The emir Ismāʿīl-i Sāmānī attacked him on behalf of the caliph, routed him in the district of Balkh and took him captive. After two years of imprisonment he was executed in 289 (902). He reigned for 22 years.

Now his grandson Ẓāhir b. Muḥammad was enthroned in Sīstān and conquered Fārs. (The caliph) Muʿtaḍid sent the emir Badrī against him and defeated him. Layth b. ʿAlī b. Layth became ruler after Ẓāhir. He drew up an army in Fārs. He was succeeded by his brother Muʿaddal b. ʿAlī who ruled for a time. In 300 (912) ʿAmr b. Yaʿqūb became ruler. He was succeeded by Aḥmad b. Khalaf, who resided in Herat.

Khalaf b. Aḥmad succeeded his father. He turned out to be virtuous and intelligent. He ordered a Koranic commentary of 100 volumes to be compiled. He died in [3]99 (1008) and was succeeded by his son Ṭāhir. The Palace of Ṭāhir is named after him.

In 400 (1009) the Saffarids came to an end. After that, Sijistan was governed by the Ghaznavids and Seljuks. Finally it was conquered by the Persian Shahs and is now under the control of the Qizilbash khans (i.e., Safavid governors). They govern as far as Kandahar; the area beyond that is under the control of the (Mughal) Emperor of India. Kandahar is a bone of contention between them; sometimes the Indians take it and sometimes the Qizilbash. [R 95a] [Plate 38]

Chapter 28: Country of Badakhshān, Ghūr and Ṭukhāristān

BADAKHSHĀN: A country between the Jayḥūn (Oxus) and Murghāb Rivers. The name is also used for its capital city.

Borders

To the south, the Murghāb River; to the north, the Jayḥūn River; to the east, Turkistan; to the west, Balkh and Khurāsān.

The capital city is located at 111 degrees longitude and 37 degrees latitude. It has prosperous villages and a fortress. This fortress is said to have been built by Zubayda the daughter of Jaʿfar al-Dawānīqī. The Kharār River, which flows next to the city, empties into the Murghāb. The city is surrounded by orchards and gardens. In its mountains are many mines of ruby, turquoise, lapis lazuli and rock crystal; Badakhshān rubies and lapis lazuli are famous. Musk is also abundant. And there is a kind of stone known as *ḥajar-i fatīla* ("wick stone", asbestos) that resembles bezoar stones.

According to the *Haft Iqlīm*, the country of Badakhshān has many rivers, trees, and pastures. The people are mainly nomadic tribes. They raise fine horses. It is a province with excellent climate and verdant steppes and plains.

One of the shahs of Badakhshān in olden days, Shah Nāṣir-i Khusraw, had a bathhouse with a square-shaped dressing room, outfitted with 24 doors. Each door opened to a bath, each bath received light through a glass pane, and each bath had its special features. But it fell to ruin and only traces of it remain.

In the mountains is an animal with a tufted tail. [M 253]

The towns of Badakhshān are the following:

Jirm: A town seven stages east of Qundus and one of the dependencies of Badakhshān. The tomb of Shah Nāṣir-i Khusraw was here.

Qundus: A city of orchards and gardens seven stages distant from Badakhshān. It has flowing streams. It is located slightly off the road to the east, going from Balkh to Kabul, and two stages from Balkh, with a waterless desert between them.

Khinjān: A town between Kabul and Qundus. It has abundant fruit and produce. It is located in a valley in the mountains. A river flows through the middle of it. It is on the road half way between Kabul and Balkh, six stages from each.

Andarāb, also called Andarāba: A town two hours north of Khinjān, between Ghazna and Balkh, located at 103½ longitude and 35½degrees latitude. Mt. Panjhīr, which is nearby, has a silver mine. At present it is a large village with much produce, orchards and gardens. Its streams flow to Khinjān.

Ashkmish: A town with orchards and gardens, in a flat area one stage south of Qundus. Its springs and streams are excellent and cold. There is another mountain to the east of it.

Ghūrī: A town and district north of Khinjān. From here one crosses a desert in three days and reaches Maymana and from Maymana to Balkh is two stages with prosperous villages between them.

The other towns of Badakhshān are Rustāq, the fortress of Ẓafar, Baghlām.

Ṭukhāristān: A large district located to the west of Badakhshān. Ibn Ḥawqal considered this one of the appendages of Balkh. It borders on Turkistan. The Jayḥūn and Murghāb Rivers flow through here from Badakhshān. It comprises several cities, of which the famous one is:

Walwālij: A city at 102½ degrees longitude and 37½ degrees latitude. It is also called Walwālish and Walj. It is located on a mountain of black stone 20 parasangs from Balkh; the mountain is 8 parasangs in circumference. It is the country of the Hayāṭila (Hephthalites). From here to Ṭālaqān is six parasangs.

Ṭālaqān, also called Ṭāyaqān and Ṭāyakān: A town of Ṭukhāristān in the vicinity of Balkh, located at 98 degrees longitude and 37½ degrees latitude. It is in mountains and a river flows in front of it. It has orchards and gardens and, in the mountains, villages. The people are mainly weavers. It is a prosperous province with abundant fruit and grain.

Rāwān and Siminjān, also called Sihbān: A small city. On the eastern side are three town quarters, a strong fortress, and many orchards and streams. Grapes, figs, peaches, and pistachios are very plentiful here, as recorded in the *Nuzha*.

Iskalkand: A small town in the mountains. [R 95b]

Hulbuk: Also a city in the mountains. All of the cities of Ṭukhāristān are in the plain; only Iskalkand and Hulbuk are in the mountains. These mountains are between the Khashāb and Badakhshān Rivers. Numerous streams flow from here and in front of ..... (?—*medāyin öñinde*) they all enter the Jayḥūn.

Munk: A larger city than Hulbuk. A river from Panjhīr passes here, then flows through Chārbāna to Farāwa and India.

Dhuwālīn and Dharāb: Both cities have many streams and trees.

Wakhsh: Also recorded as one of the cities of Ṭukhāristān; however, I have placed it among the cities of Khuttalān in Transoxiana.

Country of Ghūr: [M 254] It is bordered in the east by Sijistān and on the other three sides by Khurāsān. It is a mountainous province. In the past it had large populous fortresses and cities, villages and arable fields. Subsequently it fell into decline. It has many inaccessible fortresses. It is surrounded by Muslim territory, but Ghūr itself is a land of heathens. They speak the language of Khurāsān. There are very few Muslims. The climate is excellent, the fruit incomparable. There are abundant arable fields, pastures, and silver mines. It extends in one direction to Gharjistān, in another to Herat, and in another to Ribāṭ-i Kardān. On the borders of Bamiyan, its mountains extend from Khurāsān as far as Panjhīr.

According to the *Nuzha*, Ghūr is a province whose city is called Āhangarān. It is located at 99 degrees longitude and 35 degrees latitude and includes 50 villages. It produces wheat and some fruit. The people have a reputation for stupidity.

According to some books, the city of this country, Ghūr, is one of the dependencies of Badakhshān. It is a delightful city located between Herat and Ghazna, and its district is called Ghūrī. Its capital, Fīrūzkūh, is an inaccessible and strong fortress and the seat of the ruler of Ghūr. Fīrūzkūh is the Arabicized form of Pīrūzkūh. Another inaccessible fortress in this region is that of Chinār; it would be difficult to find the like of it in the world. Another is Girdkūh.

Radaf: One of the towns of Ghūr, located at 99 degrees longitude and 32 degrees latitude.

Routes of Badakhshān and Ghūr

From Badakhshān, Balkh is 13 stages, Qundus is 7, Ṭālaqān is 7, Jirm is 7.

From Qundus, Khinjān is 6 stages, Balkh is 2 and there is a desert between them, Andarāb is 4, the village of Ashkmish is 1.

From Khinjān, Kabul is 6 stages, Balkh is 6 stages, Andarāb is 4 hours to the north.

From Ṭālaqān, Khuttalān is 4 stages, Jirm is 4, Qundus is 7 stages to the west.

Rulers

Shahs of Badakhshān: They claimed descent from Alexander. For many years government remained in the hands of this dynasty and no one caused trouble. The governorship of that province was surrendered to them in return for a trifling tax. Sultan Abū Saʿīd Guragan (Timurid ruler, r. 855–873/1451–1469) coveted the country of Badakhshān and intended to exterminate their shahs. He seized and executed Sultan Muḥammad, the last of them, and his sons (in 871/1466). Ultimately, he himself was put to death by Uzun Ḥasan (of the Aq Qoyunlu, r. 861–882/1457–1478). Subsequently, the government of Badakhshān passed into the hands of the deputies of the kings of Khurāsān and then into the hands of the Persian shahs (Safavids).

Kings of Ghūr: A certain Bisṭām, a descendant of Ḍaḥḥāk (a tyrant of Iranian mythology), had taken refuge in the mountains of Ghūr and settled there. During the caliphate of ʿAlī (35–40/656–661), one of his descendants, a certain Tansī, became a Muslim and received a patent of authority signed by ʿAlī himself. It remained in their hands until the time of Bahrām Shāh (Ghaznavid ruler, r. 511–545, 547–552/1117–1150, 1152–1157). When Abū Muslim (leader of the revolutionary Abbasid movement, d. 136/753–754) appeared on the scene, Fūlād, a member of this dynasty, was the ruler. Thaḥī was on the throne during the reign of Hārūn al-Rashīd (r. 170–193/786–809) and Sūrī during the time of the Saffarids. His son Muḥammad (r. 401–420s/1011–1030s) was the ruler during the reign of Sultan Maḥmūd (of Ghazna, r. 388–421/998–1030). He was succeeded by his son Abū ʿAlī and then his nephew Muḥammad, who fought against Ibrāhīm the Ghaznavid. [R 96a] Muḥammad was succeeded by his son Quṭb al-Dīn Ḥasan, from whom the Ghūrid sultans are descended. His son ʿIzz al-Dīn Ḥusayn had seven sons. They were divided into two branches.

The first branch of Ghūrid kings included Quṭb al-Dīn Muḥammad b. ʿIzz al-Dīn Ḥusayn, known as Malik-i Jibāl ("King of the Mountains"). He became son-in-law to the Ghaznavid Bahrām Shāh (r. 512–547/1118–1152) [M 255] and when he went (to Bahrām Shāh's court in Ghazna, Bahrām Shāh) poisoned him on suspicion. This was the source of the enmity between the Ghūrids and Ghaznavids. His brother Sūrī (i.e., Sūrī b. Ḥusayn, r. 540–544/1146–1149) succeeded him. He marched on Ghazna but was defeated. Bahrām Shāh captured and executed him in 544/1149. Bahāʾ al-Dīn Sām b. Ḥusayn succeeded him and set out to take revenge for the death of his brothers, but he died en route.

Then ʿAlāʾ al-Dīn Ḥusayn Jahān-sūz became ruler (r. 544–556/1149–1161). When he marched on Ghazna he defeated Bahrām Shāh and won great fame. He was the first of his dynasty to rule independently, seizing Ghazna and part of Khurāsān. He died in 556 (1161) and his son Sayf al-Dīn Muḥammad succeeded him. In the second year of his reign he went off to do battle with the Oghuz tribe and was killed by his own men.

His cousin, Ghiyāth al-Dīn Muḥammad, who now became ruler, seized Khurāsān and became a mighty emperor who carried out numerous religious wars. He died in 599 (1202) and was succeeded by his brother Shihāb al-Dīn Muḥammad who had been governor of Ghazna, India and Sind. After assuming the throne, he took control of Ghūr and Khurāsān as well. He fought several battles against Quṭb (al-Dīn) Muḥammad the Khwārazm Shah and was victorious. In 602 (1205) he was betrayed by the *fidāʾīs* (Ismāʿīlī commandoes) at Lahore and assassinated. Ghiyāth al-Dīn Maḥmūd b. Ghiyāth (al-Dīn) Muḥammad, who succeeded him, was killed in the seventh year of his reign by treachery of the Khwārazmians. His son Bahāʾ al-Dīn Sām succeeded him. Three months later, ʿAlāʾ al-Dīn Atsız the Khwārazm Shah sent an army to Ghūr and besieged Fīrūzkūh. He captured the city and sent Sām and his brother Muḥammad to Khwārazm.

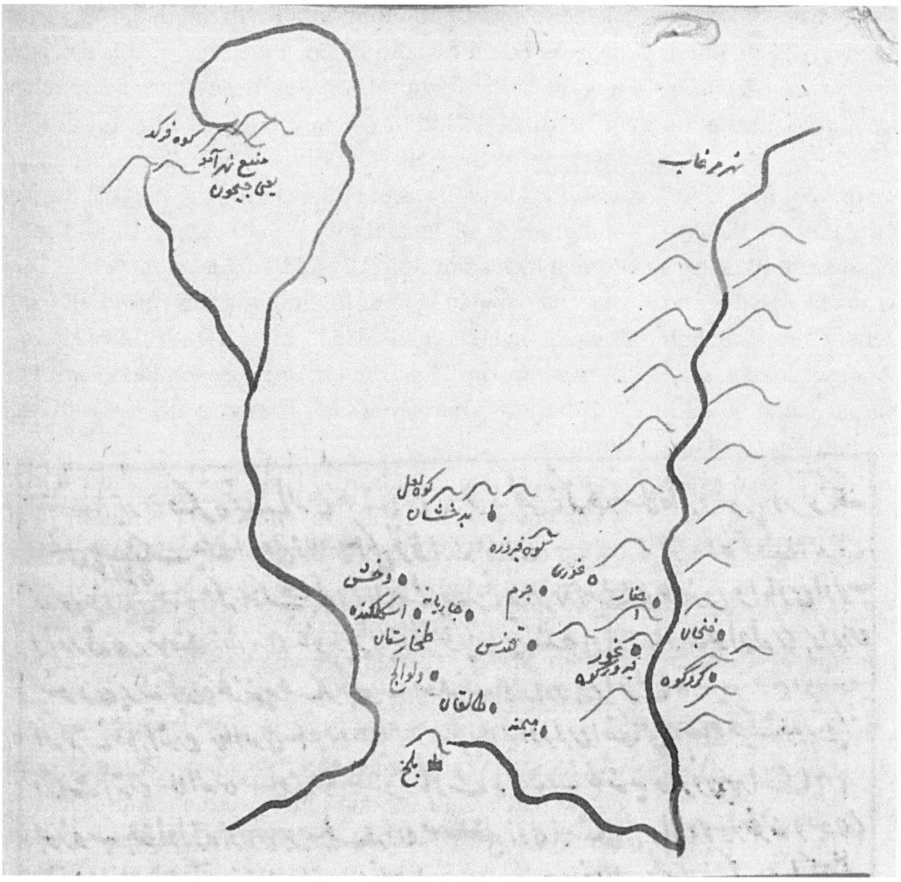

PLATE 39
Map of Badakhshān Ghūr Ṭukhāristān, R 96b (Appendix, fig. 80)
TOPKAPI PALACE LIBRARY, REVAN 1624, WITH PERMISSION

During the Mongol interregnum the two brothers were thrown into the Jayḥūn and this lineage came to an end.

The second branch of Ghūrid kings raised the banner of independent rule at Bamiyan. Ghiyāth al-Dīn (Muḥammad) had routed his nephews. One of them, Malik Masʿūd b. Fakhr al-Dīn (rather, Fakhr al-Dīn Masʿūd), succeeded his father on the throne (r. 540–558/1145–1163). He was succeeded by his brother Shams al-Dīn Muḥammad (r. 558–588/1163–1192), who was a virtuous ruler. Imam Fakhr (al-Dīn) al-Rāzī dedicated his *Risāla-i Bahā'iyya* to his son. When he died in 602/1205 (sic) his son Bahā' al-Dīn Sām succeeded him (r. 588–602/1192–1206). The new ruler marched on Ghazna and fought against Ilduz (the Turkish commander who had seized the city) but was defeated and returned to Bamiyan. His brother Jalāl al-Dīn deposed and succeeded him (r. 602–612/1206–1215). Ultimately Jalāl al-Dīn was captured and executed by the Khwārazmians and this branch of the Ghūrids came to an end.

Ghūrid Kings of Kart: They are a third branch of the Ghūrids. Their first king was Shams al-Dīn Muḥammad Kuhīn, grandson of Rukn al-Dīn Marghī who was the cousin of Ghiyāth al-Dīn Muḥammad b. Sām. When Chingis appeared, Shams al-Dīn was governor of Khurāsān. He submitted to the Mongols and was allowed to retain the government of Ghūr. He became independent ruler in 644 (1246) (rather, 643/1245). Being of an even temper and lofty ambition, he made many conquests. When his appointed hour of death arrived at age 76, he was succeeded by his son Muḥammad Kuhīn, who died in Jīsārghūr (or Jīsār of Ghūr?) in 705 (1305). His son and successor, Fakhr al-Dīn, reigned for more than one year. He built the fortress of Ikhtiyār al-Dīn. The poet Rabīʿī (of Būshanj, d. 702/1399–1400) dedicated the *Kartnāma* to him.

His brother and successor, Ghiyāth al-Dīn, ruled until his death in 729 (1329), having made many conquests in Khurāsān. Then his son Shams al-Dīn ruled one year and another son Ḥāfiẓ al-Dīn ruled two years before being put to death. During the latter's reign, the Ghūrids took control of affairs. [M 256] His brother Muʿizz al-Dīn Ḥusayn succeeded him and overcame the usurpers. He ruled for 39 years (732–772/1332–1370) and left behind many monuments. He was intelligent and just. Mawlānā Saʿd al-Dīn (al-Taftazānī) dedicated his *Muṭawwal* to him. He was the fortunate monarch of this branch (of Ghūrids). He died in 771 (1370) and was succeeded by his son Ghiyāth al-Dīn Pīr ʿAlī. Tīmūr appeared during his reign and invaded Khurāsān, seizing it from him. In 784 (1382), Tīmūr sent him and his sons to Transoxiana and imprisoned them. With their executions there, this branch of the dynasty came to an end. Subsequently Ghūr came under the control of the kings of Khurāsān and was taken over by the shahs (Safavids). [R 96b] [Plate 39]

Chapter 29: Clime of Kirmān, Hormuz and Lār

The 29th chapter is on the clime of Kirmān and the country of Hormuz and Lār.

KIRMĀN, also pronounced Karmān: According to Ibn al-Kalbī, it is an extensive and populous clime, named after Kirmān b. Falūj, a descendant of Yāfith (Japheth son of Noah).

Borders

To the south, the Persian Gulf; to the west, the cities of Fārs and the Sea of Hormuz; to the east, the land of Makran; to the north, the desert of Khurāsān. Thus it is recorded in the *Taqwīm*. But some geographers affirm that it is bordered to the west by Fārs and Lār, to the south by Hormuz, to the east by the desert of Makran, and to the north by the desert of Sīstān; for they consider Lār and Hormuz to be a separate province.

With regard to distance, it is estimated at 180 parasangs in length and breadth. It is a populous territory, but its cities are scattered. Its territory is mainly coastal and extremely hot. One fourth of the territory is summer pasture. From Sīrgān to the city of Bām, which is at the border of Pārs (sic), the climate is excellent, with abundant rivers and trees; plentiful fruit of all kinds, especially dates and figs; many arable fields and much livestock.

Kirmānī zinc is famous.

During the time of the Chosroes (Sasanians), 600 *yüks* (1 yük = 100,000) of revenue were collected. Afterwards the amount fell somewhat.

The capital of Kirmān is Gawāshīr, also called Bardshīr. It is a city located at 95 degrees longitude and 29½ degrees latitude. It is recorded that its fortress was built by Ardashīr b. Bābak, and the name (i.e., Bardshīr) is after him. Also Gushtāsb built a fire temple there.

It is recorded in the *Simṭ al-ʿulā* that al-Ḥajjāj (the Umayyad viceroy of Iraq) sent an army to conquer it but he was not able to do so. It was later conquered during the caliphate of ʿUmar b. ʿAbd al-ʿAzīz (r. 99–101/717–720). By order of ʿUmar they built the Old Mosque. The emir ʿAlī b. Ilyās built Bāgh-i Sīrjān (Sīrjān Garden) and Qalʿa-i Kūh (Mountain Castle). The Seljuk [ruler of Kirmān] Tūrān Shāh (r. 477–490/1085–1097) built the mosque known as Tabrīzī Mosque. Among the famous Sufis buried in this city are Shāh-i Shujāʿ Kirmānī and Niʿmat Allāh Walī (d. 834/1430–1431).

Bam: A city two stages east of Gawāshīr, located at 95½ degrees longitude and 30½ degrees latitude. Its climate is unpleasant. It has a strong fortress. Friday prayer is held at three places: the mosque near the bazaar, known as Masjid-i Khawārij (Mosque of the Kharijites); the Masjid-i Ahl-i Sunnat (Mosque of the Sunnis); and another mosque in the fortress where the alms of the Muslims are stored.

Jīruft: As for longitude and latitude, they are about the same as the previous cities. This is a city of cheap foodstuffs and a meeting place of merchants. It is also called Asfī. It is east of Sīrjān. Snow, walnuts, and bitter oranges are found here at the same time. All the pastures and arable fields are irrigated by rivers. When it was conquered by Islam, this area was an oak forest. The Muslim soldiers cut them down and built villages, naming each village after its inhabitants. The water here comes from Dīwrūd. As the climate is hot, dates and bitter oranges are abundant.

Sīrjān: The largest city of Kirmān. It is two stages west of Jīruft, located at 92½ degrees longitude and 29½ degrees latitude. [M 257] There are underground canals within the city walls. The buildings are domed and, because of the scarcity of wood, are constructed of masonry.

Sīrjān has a strong fortification, built during the time of Hārūn al-Rashīd. During the time of Muslim rule, the governors of Kirmān resided there. This fortress has eight gates, including Bahyābār Gate, which is the gate of the city square; Marḍī Gate; and Maymand Gate, which is the western gate and opens to the west. Outside this latter gate is a ruined pavilion known as the Chamberlain's Pavilion (Qaṣr-i Ḥājib).

This city has a council hall and splendid markets. Most of the people are wealthy merchants. Their women are very beautiful. The climate tends to be hot. Products include wheat, cotton, fruit and dates. Most of the people are transmitters of hadith. The villages and districts get their drinking water from wells.

Narmāshīr: A city on the edge of the desert, located at 93 degrees longitude and 31 degrees latitude. It is one stage west of Bam. It is recorded in the *Tārīkh-i Kirmān* that it was founded by Ardashīr b. Bābak. [R 97a]

Zarand: A city 29 parasangs from Sīrjān.

Bāfd: Formerly, a city with a hot climate, one stage south of Kirmān. It is now in ruins. It has abundant streams, arable fields and pastures, hunting grounds and meadows.

Khabiṣ: A city on the edge of the desert between Kirmān and Sijistān located at 93 degrees longitude and 31 degrees latitude. It has abundant orchards and gardens, dates and other fruit and grain. The climate is hot. A river rises from a mountain to the west and flows to this city, irrigating its orchards. Sijistān is a ten-day journey east of here. It is reported in the *Nihāyat al-arab* as a special feature of Khabiṣ that it never rains within the city walls.

Sapanj: A city on the frontier of Sijistān and one of the districts of Kirmān. It was founded by ʿAmr b. Layth. It is called Bridge of Kirmān (Qanṭara-i Kirmān). There is no bridge, however; it is simply known by this name.

Village of Sayf: A village of orchards and gardens, one stage north of Khabiṣ.

Hormuz, also called Hurmūz: It is the port of Kirmān, located at 92 degrees longitude and 20½ degrees latitude. It has many date palms and is a very hot place. Formerly it was on the coast; when the Tatars destroyed it, the people moved to the island off its shore and settled there. This island is called Jarūn and Zarūn. It is located west of Old Hormuz, twelve miles from the mainland, inside the straits of the Persian Gulf. Water there is a precious commodity. It is brought from the port of Kūmrū, because the water on the island is brackish.

The Island of Kīsh is west of here, three days sailing. It is a famous emporium, the bazaar of Hormuz and gathering place of merchants. The slave market for female slaves is much prized. In the Turkish (slave?) market, at the beginning of the marketplace, is a four-cubit high column made of fish vertebrae. This fish, which is like a turbot, was killed by a sea-monster and left on the shore. They say that the sea-monster had a pointed tail like a snake and that it went in one side of the fish and came out the other.

In the *Theatrum*, one of the books that I translated, Johannes says that with respect to pearls, there is no place as productive as Hormuz. It has little fresh water and few stores of grain; everything it needs is brought from the outside. The pearldivers get oysters within a distance of three stages around the island. The pearls extracted there are larger and finer and more valuable than those found elsewhere. [M 258]

In 1506 of the Christian era, the Portuguese captain Francisco de Albuquerque seized this city and levied a heavy tax on the people. Later, when the Persian shahs (Safavids) conquered Kirmān, they ruled Hormuz jointly with the Portuguese.

Hormuz has a large and strong fortress, as depicted in the *Theatrum*.

According to the *Haft Iqlīm*, Ardashīr b. Bābak originally built the city on the coast and named it Hormuz. Because it had no security from the assaults of thieves and malefactors, the governor of that region, Malik Quṭb al-Dīn, made the seaport of Jarūn his seat of residence and the rest of the people migrated there. But the *Majmūʿ al-ansāb* records that Ayāz, one of the slaves of the emir Maḥmūd (of Ghazna), made the port of Jirūn the capital. Later, during the reign of Sultan Shihāb al-Dīn b. Safar Shāh, the Franks (i.e., Portuguese) captured Hormuz.

The reason for their victory was the following: During his reign, Shihāb al-Dīn followed the path of cruelty and oppression. Darwīsh Nūr al-Dīn, one of the notables of the city, cautioned him about this, but he would not listen. So a delegation of the people went to Goa and invited the Franks to their city. The thankful Franks came with their ships, invaded the island and built a prominent fortress. The people of Hormuz fought the Franks, but to no effect. So they cut their ties with the island and went to the emporium of Kashmīr. At that time the Franks, obliged by the exigency of the time, sent an envoy to the ruler of Hormuz. They made peace on condition that one-fourth of the customs duties would go to the ruler of Hormuz and three-fourths would go to the Franks. This agreement remained in force for some time.

The other towns and cities in the vicinity of Hormuz and Kirmān—located at 91, 92, and 93 degrees longitude and 27 and 28 degrees latitude—are the following: Hurmuzak, Karmū, Kūristān, Shaḥnab, Dashtbir, Manūkān, Kūjard, Rūsān, Bandar Ibrāhīmī, the village of Kūh-i Mubārak, Kat, Jāshak, Tanka, Tays, Māskān, Chakīnaw, and Māhān.

Rūbār: A plain inhabited by Arabs, located at 93 degrees longitude and 29 degrees latitude. Sometimes they are obedient to the ruler of Kirmān and sometimes they are rebellious.

Daryāy (Shiʿb-i Daryāy): A populous ravine (*shiʿb*). It has villages, vegetable gardens, and places of excursion.

Mīnāb: A large city, located at 93 degrees longitude and 28½ degrees latitude. [R 97b] It is between Lār and Makran, ten stages from each.

COUNTRY OF LĀR: A country northwest of Hormuz and ten stages south of Shiraz. It has an independent king.

Borders

To the west, the coasts and commercial ports of the Persian Gulf; to the north, Fārs; to the east, part of Fārs and Kirmān; to the south, the desert of Hormuz and Kirmān.

Its capital is Lār: A large city ten stages northeast of Hormuz, located at 91 degrees longitude and 29 degrees latitude. As the climate of Lār is extremely hot, it does not suit the men but is very agreeable to the women. Lemons, bitter oranges, sugar cane and oranges are grown here. For the most part the people are thin and emaciated and bad-tempered. They are archers. But their women are hospitable, even-tempered and nice-looking. Guinea worm are widespread here, due to the fetid air, because there are few streams. There are wide and deep tanks that fill up during the rainy period, which lasts about forty days. The city folk drink this water for about six months. That is the reason for the illness of worm. [M 259] From here to the sea is a journey of seven or eight days.

Its dependencies are the following:

Darpaz, meaning "beautiful": A beautiful city located at 91 degrees longitude and 28½ degrees latitude. Lār is two stages to the southwest and Ramāykān is one stage to the east.

Ramāykān: A city near Darpaz and Tāram, two hours from each. Markh and Pūhān are located on either side of it.

Pūhān: A city located at near 92 degrees longitude and near 29 degrees latitude.

Tāram and Tazrak: Two cities whose longitude and latitude are each ten minutes more than Ramāykān and Pūhān. Tazrak is on the east and Tāram is on the west. They are half a stage apart. They are both east of Darpaz. Tāram is a large city north of Pūhān. They also call it Ṭāram. It has a fortress.

Tazrak Pakan, also called Pīnbīnī: A city between Pūhān and Tazrak, located one stage south of Ramāykān and north of Kupuhra.

Kupuhra: A city one stage southwest of Baymand.

Baymand: A city at near 90½ degrees longitude and 29 degrees latitude.

Nukār: A city at 92½ degrees longitude and 30 degrees latitude.

Dastkard: A city at one degree less than the longitude and latitude of Nukār.

Kalkā: city two stages from Lār, between Unj and Lār. It has a stream.

Khawr: A city east of Lār, between Dastbar and Bīnī, located at 92 degrees longitude and 28½ degrees latitude. It is on the road from Sīrjān to Hormuz.

Khūshanābād: A city on the frontier of Lār, approximately at the longitude and latitude of Khawr. It is between Tazrak and Tāram and Bīnī, in the direction of Kirmān.

Rūz: A city on the edge of the desert.

Aside from these, some authorities record that the country of Lār has a number of fortresses, including those of Marjān, Kūrz, Naw (or "Newcastle"—*Qalʿa-i Naw*) and Maymūn. Another is the city of Bābā Bayrām.

Island of Andarāwī. Lār has two islands in the Persian Gulf, just in front of Bandar Ray (Ray Shahr). One is this island, which is in the direction of Old Hormuz. According to the *Taqwīm*, its circumference is 12 parasangs; it has villages; there is pearl diving; and it is between Qays and Sīrāf. The other is a small island north of this one. Both islands are close to the shore.

Mountains

There are several mountains in Kirmān, Hormuz and Lār:

Mt. Qufṣ: The most famous and populous mountain of Kirmān. To the south is the sea; to the north are the borders of Jīruft, Rūdān and the mountainous area of Abū Ghānim; to the east are Ḥawās and the desert; to the west are the borders of Sīrjān and the land of Baluch.

The Baluch are a people and tribe. They dwell on the skirt of these mountains, from the border of Mughūn and Lāshkard to the district of Hormuz. There are abundant arable fields and date groves. They are desert dwellers. No matter how many dates are blown down by the wind, they consider it shameful to pick them up; rather, they leave them for the poor to glean. And they never do harm to anyone. But the people on Mt. Qufṣ are Kurds and evildoers. Every band of them has a chief. They come down the mountain on foot and roam as far as the borders of Fārs and Sīstān, committing highway robbery. They are of Arab origin.

Bāzar Mountains: The summer pasture of Kirmān. There is a silver mine here, and another one on Mt. Qufṣ. And so these mountains are called the mountains of mines. There are mines of gold, silver, Kirmānī iron, copper and brass (i.e., tin?). They are a string of mountains that extend for two stages from Jīruft toward Shiʿb-i Daryāy. There are many productive villages and places of excursion there.

Daryāy River: It comes from Hanjān with a very [M 260] turbulent and swift current. It turns about 20 mills and there is no place where it can be forded. Apart from it, there is no large river in Kirmām, and there is none that is navigable.

Routes of Kirmān and Lār

From Sīrgān (= Sīrjān) to Kāhū is 2 stages, from there it is 2 stages each to Ḥasanābād and Rustāq.

From Sīrgān [R 98a] [Plate 40] to Dūrdān and Hamad is 4 parasangs. From Hamad to Kurdkān is 2 parasangs. From Kurdkān to Amās is one stage.

From Sīrjān to Ribāṭ-i Mushrifān is 2 stages. From Amās to Rūdān is 1 stage.

From Sīrjān to Shāristān via the Bam road it is one stage each to the following halting places: > Sīmābrūd 1 > Bahār 1 > Ḥibān 1 > ʿAbīrā 1 > Karʿūn 1 > Rāsin 1 > Shāristān 1. From here the road turns to the left to Jīruft: > village of Kūz 1 > Jīruft 1.

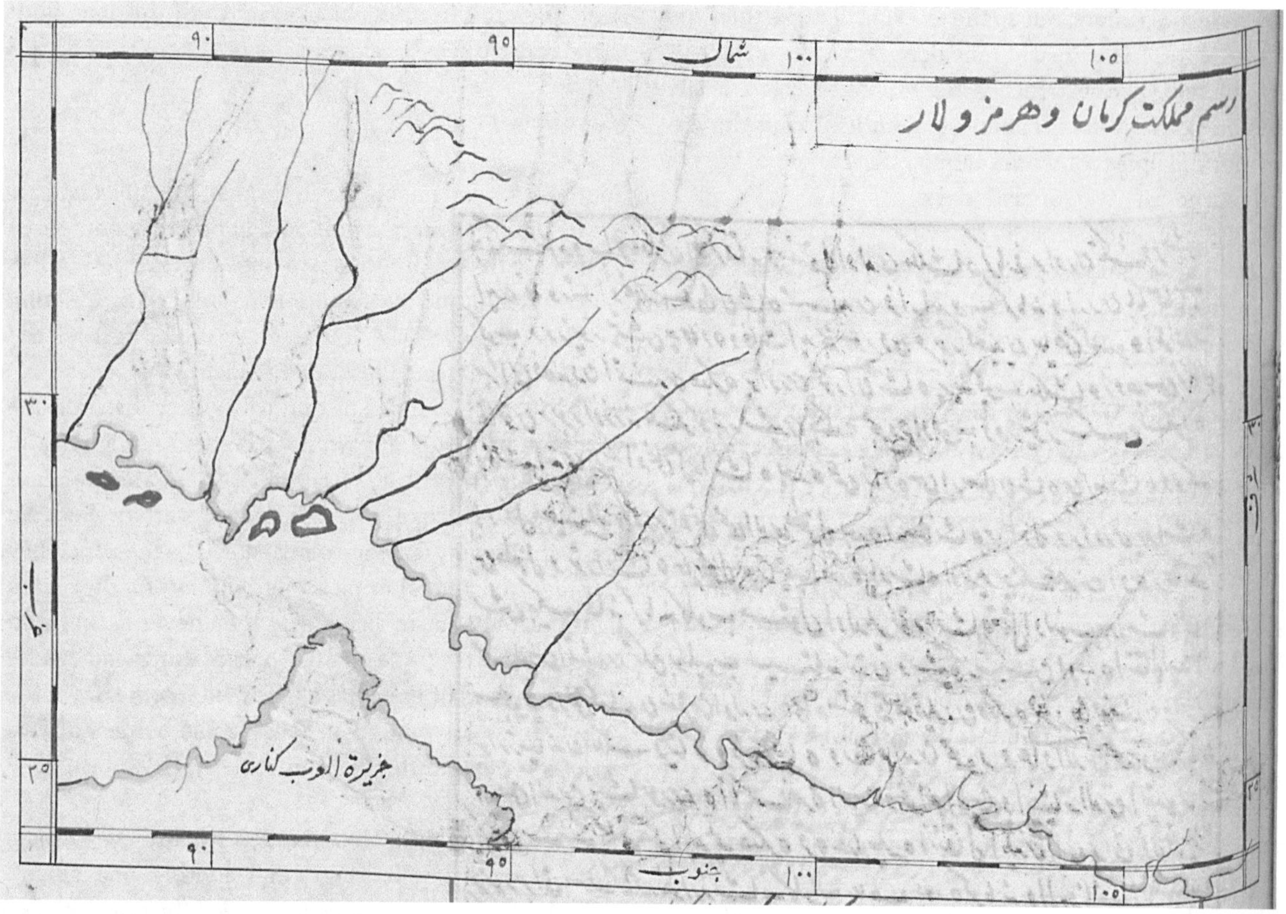

PLATE 40 Map of Kirmān Hormuz Lār, R 98a
TOPKAPI PALACE LIBRARY, REVAN 1624, WITH PERMISSION

From Sīrgān: > Tākht 2 > Jīr 1 > Mt. Naqra 1.

From Sīrgān: > Zarand 2 (?) > Bardshīr 2 > Khamrūd 1 > Zarand 1 (sic?). From here the desert is one KRY (?) stage.

From Sīrgān: > Narmāshīr 1 > Bahraj 1 > Jīruft 1—in the direction of Fārs; > Mūghūn 1 > Lāshkard 1 > Sūrqān 1 > Mūrqān 1 > Jarwān 1 parasang > Kasīnān 1 > Rūyīn 1 > Tāram 1.

From Jīruft to Hormuz: > Lāshkard 1 > Kūmīn 1—to the left; > Rigān River 1 > Maywajān 1 mile > Hormuz 1 > seashore 1 > the port 1.

From Sīrjān to Shiraz is 65 parasangs.

From Kūjard to Sīrjān: Shamīl 1 > 1 > Khawr 1 > Bīnī 1 > Sīrjān 1.

From Kūjard to Jīruft: > Ḥanāb 1 > Jakīnū 1—it belongs to Hormuz; > Manūkān 1 > Rūbār 1 mile > Jīruft 1.

From Kumrū, which is the port of Hormuz, to the city of Lār is 4–5 stages. It is in ruins; here and there are caravansaries and ponds.

From Lār to the seashore is 4 (?) stages: > village of Wahīj 1 > Nīmanda 1 > Kirih 1 > Kīl 1—these are villages; > Kāzarūn 1 > seashore 2.

From Hormuz to Dabūl-i Sind: > Bandar Ibrāhīm 1 > Kūh-i Mubārak 1 > Jāshak 1 > Tīz 2 > Nanak 2 > Pūzam 2 > Kuwādar 6 > Kalamat 4 > Bārū 6 > Dabūl 6. According to one account, from Kuwādar to Dabūl is 5 stages.

Rulers of Kirmān

During the caliphate of ʿUmar (r. 13–23/634–644), Mujāshiʿ b. Masʿūd al-Sulamī was appointed military commander and he took Sīrjān, which was the capital. Afterwards Rabīʿ b. Ziyād conquered the surrounding area. The people later violated the peace treaty and Jīruft was taken by sword. Most of the people fled by boat. Mujāshiʿ divided up the lands of Kirmān and gave them to the Arabs. Then Quṭn b. Qubayṣa became governor of Fārs and Kirmān. He ruled until the Umayyad period. In both the Umayyad and Abbasid periods, governors were in control. Sometimes the rulers of surrounding areas interfered in their affairs or were in conflict with them. Finally Kirmān passed into the hands of the Seljuks.

The Seljuk Dynasty: Qāwurt b. Jaʿfar b. Mīkāʾīl b. Saljūq, whose original name was Aḥmad, was a heroic figure.

He was also called Qara Aslan ("Black Lion"). In Ṣafar 433/October 1041 he became governor of Kirmān. After Alp Arslan died (in 465/1072), he raised the banner of sultanate and became independent and took Shiraz from the Daylamis. In 465 (1072) he fought against Sultan Malik Shāh near Hamadān; [R 98b] he was captured along with his sons and put to death.

Now Malik Shāh gave Kirmān to his brother Takash. Qāwurt's son Sultan Shāh escaped from prison and went to Kirmān. He succeeded his father and remained on the throne until his death in 477 (1085). Sultan Shāh covered the Kaʿba with a white cloth cover and furnished it with a silver waterspout. His brother Tūrān Shāh later succeeded him (r. 477–490/1085–1097). Malik Shāh's wife Turkān Khātūn sent an army against Tūrān Shāh. He was wounded in the ensuing battle [M 261] and died in 488 (1095).

His son Irān Shāh succeeded him, then, in turn, his nephew Arslan Shāh b. Kirmān Shāh, his son Mughīth al-Dīn Muḥammad, his son Muḥyī al-Dīn Tughrul, Arslan Shāh b. Tughrul, his brother Bahrām Shāh b. Tughrul, and Tūrān Shāh b. Tughrul. During the reign of the last, 5,000 tents of the Oghuz tribe invaded Kirmān and Tūrān Shāh was put to death. Bahrām Shāh's son Muḥammad ascended the throne, but was not able to hold his position and went to Sīstān. He died in 583 (1187), bringing the Seljuk dynasty of Kirmān to an end.

Various Rulers: After the Seljuks, from among the Oghuz tribe and the Khwārazmian emirs, were Malik Dīnār, Farukh Shāh, Niẓām al-Dīn Maḥmūd, ʿImād al-Dīn Muḥammad, Bughā Tekin, Raḍī al-Dīn-i Nishābūrī, Atabeg Nuṣrat al-Dīn, Qiwām al-Dīn, and Ikhtiyār al-Dīn. They ruled until 620 (1223), after which sovereignty passed to the Qara Khitāy.

Dynasty of Āl Barāq (the Qutlughkhanids): This Barāq came from the Qara Khitāy to the Khwārazm Shah, Sultan Muḥammad (r. 596–617/1200–1220), and became chamberlain, then governor of Kirmān. When Chingis Khan defeated Sultan Muḥammad, he declared his independence and took control of Kirmān. The subsequent rulers from this dynasty were: Quṭb al-Dīn Muḥammad, Mubārak Khwāja, Sultan Ḥajjāj, Pādishāh Khātūn, Jalāl al-Dīn Suyurghatmish, Muẓaffar al-Dīn Muḥammad, and Quṭb al-Dīn Shāh Jahān. Their rule continued until 704 (1304) and came to an end when the Mongols awarded the sultanate of Kirmān to someone else.

No independent dynasty arose in Kirmān after them. It fell under the control of deputies of the rulers of Iraq and Persia. It is currently ruled by the Persian shahs (Safavids).

Sultans of Hormuz: While in the past the government of Hormuz was attached to the rulers of Kirmān, during the Mongol interregnum the Āl Barāq were unable to maintain their independence and the Salghurids took over the government of Hormuz. In 647 (1249), Rukn al-Dīn Maḥmūd-i Qalhātī rose up and invaded Hormuz and the islands of the Persian Gulf. After his death, power was held by his son Nuṣrat, then by Masʿūd and his slave Ayāz. In 711 (1311) control of the dynasty passed to Kurdān Shāh and his sons. Ayāz died that year and Kurdān Shāh became sultan. Subsequently the rulers were, in turn, his son Quṭb al-Dīn Tahamtan, Tūrān Shāh, Muḥammad Shāh, Sayf al-Dīn Fīrūz Shāh, Tūrān Shāh, Shihāb al-Dīn, Salghur Shāh, Tūrān Shāh, and Salghur Shāh. The Portuguese invaded the island during the reign of the last-named Salghur Shāh in the year 916 of the Hijra (1510). Later Sultan Muḥammad became governor and signed a peace agreement with the Portuguese. His descendants nominally ruled the island until 1000 (1591), after which the Persian shah took over Hormuz as well and ruled the island jointly with the Portuguese.

Sultans of Lār: It is recorded that they were the descendants of Gurgīn Mīlād, one of the heroes of Kay Khusraw. The first of them to accept Islam was Jalāl al-Dīn Īraj b. Gurgīn. He did so during the reign of ʿUmar b. ʿAbd al-ʿAzīz. His descendant Quṭb al-Dīn Muʾayyad became ruler in 594 (1197). He was succeeded in turn by the following members of his lineage: Muẓaffar al-Dīn Kālinjār, Muʾayyad II, ʿAlāʾ al-Mulk, Kālinjār II, Ḥājī [M 262] Sayf al-Dīn, Ḥājī ʿAlāʾ al-Mulk, Ḥājī Mubāriz al-Dīn, Quṭb al-Dīn Mubshir, Jahān Shāh, ʿAlāʾ al-Mulk, Hārūn, Muḥammad, and Anūshirwān who was known as *shāh-i ʿādil* (the just shah). In 948 (1541) [R 99a] Anūshirwān was martyred by a madman and was succeeded by his cousin, Ibrāhīm Khān. The sultanate continued until 1000 (1591) when the government of Lār also passed to the Persian shahs.

Chapter 30: Clime of Fārs

The 30th chapter is on the region of Fārs, a famous country in Asia. This clime is reported to be named after Fāris b. Nāsūr b. Sām (Shem son of Noah). It is Arabicized from Pārs.

Borders

To the west, Khūzistān and some coastland of the Persian Sea (i.e. the Persian Gulf); to the south, the Persian Sea and Hormuz; to the north, the desert between Bilād-i Jabal (Jibāl) and Khurāsān and Fārs, which ends at the borders of Iṣfahān; to the east, Kirmān and Sijistān.

It is nearly a square in shape. As it has many cities and villages, the only ones recorded in this book are those where the congregational Friday prayer is performed and those designated as city (*şehr*) and town (*ḳaṣaba*).

In praising this clime, Ḥamd Allāh writes: Concerning the people of Fārs, the Prophet said: "God has chosen from amongst His people the Quraysh from the Arabs and the Persians (*Fāris*) from the Iranians (*al-ʿAjam*)." Therefore the people of that region are called the select of Fārs (*akhyār-i Fāris*).

And the following hadith is related in the *Muʿjam al-buldān*: "The farthest people from Islam are the Byzantians; but even if Islam were hung on the Pleiades, the Persians would reach it."

Division

The land of Fārs is in two parts, southern and northern. The northern part is flat; the southern part is mountainous.

In the *Nuzha* it is divided into two, an inland part and a part along the sea. In the past, the inland provinces were divided into five regions (*kūra*): 1. Ardashīr Khūra, 2. Iṣṭakhr, 3. Dārābjird, 4. Shābūr Khūra, 5. Qubād Khūra. Each of these comprised several provinces and cities.

In total, the breadth is 150 parasangs, the length from Yazd to Jūr is 320 parasangs.

Tribes of Fārs

In the terminology of that region, the word for "tribe" is *zam*. They include the following: Zam Jīlūna, also called Zam Mīḥān; Zam Aḥmad b. al-Layth, also called Zam Lawāyiḥān; Zam Aḥmad b. Ṣāliḥ; Zam Shahrī, also called Zam Bārīḥān; Zam Aḥmad b. Ḥusayn, also called Zam Kārmā; and Zam Ardashīr.

And there are many Kurdish clans in Fārs. There are said to be over 500,000 nomadic households who migrate to winter and summer pastures. Many of these households have over 200 followers and retainers.

Countries

As in the past the capital of Fārs was Iṣṭakhr, its region (*kūra*) took precedence over the others. However, since now the capital is Shiraz, Ḥamd Allāh mentioned its district first.

Region of Ardashīr Khūra: It is named after Ardashīr b. Bābak, the first Sasanian king. The cities in this district (*nāhiye*) are the following:

Shiraz: the capital of the clime of Fārs, an important city and famous emporium. It is a city founded in Muslim times, with an abundance of water and numerous orchards and gardens, located at 88½ degrees longitude and 29½ degrees latitude. It was built in 76 (695–696) by Muḥammad b. Qāsim b. ʿUqayl, the nephew of al-Ḥajjāj. Every house in it has a vegetable garden and flowing water. However, the roads are narrow and dirty.

During the reign of ʿAḍud al-Dawla this city was so flourishing that there was no room for his troops. So he built a town south of it named Fanā Khusraw Kard and his troops lived there. Revenue amounting to 20,000 dinars used to be obtained from this town; but it is now in ruins.

At that time Shiraz had no city wall. Ṣamṣām al-Dawla, son of ʿAḍud al-Dawla, [M 263] had a rampart built 2,500 paces in circumference. Later it was rebuilt by Saʿd b. Zangī. Around 620 (1223–1224), when it was falling into ruin, it was again rebuilt by Sharaf al-Dīn Maḥmūd Shāh. He built guardhouses out of brick over the towers.

Shiraz consists of 17 town quarters. It has nine gates: Iṣṭakhr Gate, Darrāk Mūsā Gate, Bayḍā Gate, Kārizūn Gate, Salam Gate, Fanā Gate, New Gate, Dawlat Gate (or Gate of Fortune) and Saʿādat Gate (or Gate of Felicity). The *Aḥsan al-taqāsīm* lists eight gates in the following order: Iṣṭakhr, Tustar, Band Āsitāna, Ghassān, Salam, Kawādir, MNDR and MHNDR. Iṣṭakhr Gate resembles the Minā Gate in Mecca. It is the most beautiful place of the city.

Shiraz has excellent air and fine water. Its markets rarely lack sweet basil. [R 99b] One of its solidly built underground channels is the channel of Ruknābād, constructed by Rukn al-Dawla Ḥasan b. Buwayh. Apart from this, there are several other channels, solidly built and never in need of repair. Most of them come from streams. In spring a flood comes from Mt. Darrāk, flows past the outskirt of the city and emptiues into a lake.

Grain in this city is of middling quality. Food prices are generally high. Of fruit, the *mithqālī* grapes are very fine and delicate. Cypress trees are prominent and sturdy.

Most of the inhabitants are of a swarthy complexion and slender build. They were formerly Sunnis of the Shāfiʿī rite, but now are mostly Shīʿīs. The wealthy inhabitants are mainly from abroad.

In the past many great men came from this city, and so it was called "bastion of saints" (*burc-ı evliyā*). In particular, it is said that saints have never been absent from the Old Mosque (Jāmiʿ-i ʿAtīq) which was built by ʿAmr b. Layth. The Saʿd b. Zangī and Sunghur and other Friday mosques, the dervish lodges and madrasas, the hospital of ʿAḍud al-Dawla, etc.—these comprise more than 500 buildings. Famous blessed tombs in the city are those of Sībawayh the grammarian; Muḥammad and Aḥmad b. Mūsā al-Kāẓim; Shaikh Abū ʿAbd Allāh al-Khafīf, built by the Atabeg Zangī; and Shaikh Rūzbihān.

Shiraz has several excursion grounds, such as the one between Jaʿfarābād and the *muṣallā* (open-air prayer ground) which is the tomb of Ḥāfiẓ of Shiraz; or that of Khaljān (?) which consists of orchards and gardens extending about 2 parasangs from the city as far as Mt. Darrāk.

Ten parasangs from Shiraz is a spring known as Āb-i Jāsht, the water of which has purgative and medicinal qualities. In the autumn season several thousand people make an excursion to that place and stay for three days drinking the water. While drinking, one expresses the intention to be purged above (through vomiting) or below (through emptying the bowels), and the purgation takes place accordingly. If one drinks without expressing intention, there is no purgative effect and it may cause bloating or even be fatal. It is quite a marvel.

According to the *Nuzha*, the state dues from Shiraz are 450,000 dinars and all of Fārs is administratively dependent on it. Its dependencies, as listed in the *Nuzha*, are the following:

Sayf: several hot-climate districts on the seacoast. Some are called Sayf of Abū Zubayr, some Sayf of ʿImāra. It is mostly populated by Arabs. The only products are grain and dates. It is also called Sayf al-Baḥr. Ḥawz al-Sayf, on the other hand, is a small town in that region below Sīrāf. It has a hot climate and is located at 87½ degrees longitude and 29 degrees latitude.

Būshkāmāt: A number of districts and villages in that vicinity. It is extremely hot. Grain and dates are abundant. [M 264]

Tūḥ: A city built of clay, now in ruins and populated by Arabs. In the past it was a great city. It is 12 parasangs from Jannāba. It is also called Tūh or Tūz. Dates are abundant. It has fine linen. A river flows by it.

Ḥīr: A mid-sized city with a moderate climate. The water is fine, the soil fertile. There are fruits that are coastal (hot climate) and others in the summer pastures (cold climate). It has a strong castle. Game animals of mountain and plain roam its uncultivated area.

Khabrar: A hot-climate district with an abundance of dates. The people make weapons.

Additional districts include Marzdān, Dādhan and Dawān. Some have a hot climate, others are in the mountains with a moderate climate. They produce grain, cotton, fruit and rice.

Sarwistān: A mid-sized town 3 stages from Shiraz, located at 87 degrees longitude and 29½ degrees latitude. It has streams and vegetable gardens. It is at a hot-climate place in the Kūhijān province. The climate is bad; dates and grain crops are abundant.

Sīrāf: the largest and best-known seaport of Fārs, an entrepot city located at 89 degrees longitude and 27 degrees latitude. It is 63 parasangs south of Shiraz. The southern desert ends here. It has neither sown fields nor livestock and is surrounded by wilderness. However, its houses are solidly built and splendid merchant houses. Some merchants spend over 30,000 dinars to build houses of teakwood brought from Zanzibar. Rain water is stored in cisterns. There are two or three fountains and three Friday mosques. The water of these fountains comes from Mt. Jam which overlooks the city. Fruit grows at the foot of this mountain. The grammarian Abū Saʿīd al-Sīrāfī came from this city.

Najīram: A town on the seacoast, dependent on Sīrāf. The distance between them is recorded as 13 parasangs.

Ṣanamgān: A delightful city. It has a marvelous feature: A river flows through it with a bridge across. From the bridge upward is a summer pasture, with such trees as walnut and plane. From the bridge downward is a hot-climate area with such trees as orange and bitter orange. Its wine is undrinkable unless mixed with two or three times more of water.

Fīrūzābād: A famous town near Shiraz, also called Jūr. Its old name was Jūr, which is the Arabicized form of Gūr. Whenever ʿAḍud al-Dawla went there to hunt, people would say, "ʿAḍud al-Dawla has gone to Gūr (or has gone after *gūr* meaning wild ass)." Since *gūr* also means grave, ʿAḍud al-Dawla thought this was not nice and changed the name to Fīrūzābād, meaning City of Victory. This is according to the *Taqwīm*. Other authorities say that it was restored by Fīrūz Shāh [R 100a] and so was named after him. In the past, the region (*kūra*) of Ardashīr was called the district of Fīrūzābād.

It is an ancient city with strong walls. The Barāra River rises from the mountains and flows by it. When Alexander came here he lay siege to the city but was unable to take it, so he dammed up the river on the mountain side and it became like a sea and submerged the city. Later Ardashīr opened the dam and let the river flow. He built a circular city.

It has an earthen fortress and four gates: Mihr Gate in the east, Bahrām Gate in the west, Hormuz Gate in the south and Ardashīr Gate in the north. Inside the walls there are several streams, and going out from each gate are orchards and gardens and picnic spots for a distance of one parasang. Its rose-water is fine and famous and is exported abroad. The distance between Fīrūzābād and Shiraz is [M 265] 24 parasangs. The air is hot and putrid.

Kārzīn: A small town close to the sea, located at 89½ degrees longitude and 28½ degrees latitude. It has an ancient fortress and extramural settlement. The climate is hot. It is on the bank of the Mukān River.

Nearby are the two small cities of Fīrūz and Īrad. The climate is hot; dates are abundant. Water is gotten from the Mukān River.

Kawār: A small and delightful city with a climate, however, that tends to be hot. It has many villages and fields producing grain crops and fruit. Most of Shiraz's needs are met from here. The pomegranates and almonds are excellent. Game is plentiful in the uncultivated area. Finally, its people are mainly thick-natured and crude.

Khānasitān: A desert and arid place on the seacoast, 30 parasangs in length and breadth. It comprises several villages and has very fertile soil. When it rains in spring or autumn, 1 *man* of seeds yields 1,000 *man* of crops. If it does not rain, it does not even yield seeds.

Maymand: A small city in this district. The climate is hot; grain crops, dates, grapes and all other kinds of fruit are plentiful. The people are mainly craftsmen. It is located south of Shiraz and two stages east of Jūr.

Tīr: A fortress on a lone mountain 3 parasangs southeast of Shiraz. There is one fountain inside the fortress and another at its base. Its surrounding area for a day's journey distance is populated and has pastures.

District of Iṣṭakhr: an ancient country. According to the *Nuzha*, before Iṣṭakhr there were no buildings in the clime of Fārs. The entire region lengthwise from Yazd to Hazārdirakht and breadthwise from Quhistān to Sard belongs to this district.

Iṣṭakhr city: an ancient town on a level plain, located at 88½ degrees longitude and 30 degrees latitude. As this was the capital of Fārs in the past, there are still traces of large buildings of the ancients. Ardashīr transferred his capital from here to Jūr. It is said that Gayūmarth founded it for his son named Iṣṭakhr, that Hūshang expanded it and Jamshīd completed it. It was 10 parasangs wide and 14 parasangs long. Inside the walls were several buildings, fields and villages; as well as three strong fortresses on three mountains: Iṣṭakhr, Shikasta and Shaykarān, known as the Three Towers.

It is related in the *Fārsnāma* that there are no buildings n Fārs older than this. It is in a kind of river-valley with an open plain on one side. Whenever it rained, water accumulated. ʿAḍud al-Dawla built a dam on that side and a large reservoir of 17 stairs (?—*pāye*) with columns and a roof. The water of this reservoir was enough for 1000 men for one year. In general, this city was proverbial for its fortification and gained great fame. Today, however, neither its fortress nor its extramural settlement remains, only fortress-like solid houses and gates.

At the foot of the mountain in Iṣṭakhr, Jamshīd laid out and constructed a square-shaped palace made of black granite rocks, with one side adjacent to the mountain and the other three sides facing the plain. It was 30 cubits high, and one ascended to it by ladders from two sides. In this building were circular and square-shaped pillars made of black stone, each one bearing a weight of 100,000 batmans. And each of the rooms was breathtaking, with splendid paintings of Burāq and Jamshīd. A hot spring that rose in the mountain flowed into the palace through a crevice in the masonry. On top of the mountain are huge caves called Zindānbād by the populace.

At the advent of Islam, [M 266] the people of Iṣṭakhr broke their allegience several times and the Muslims killed them and destroyed the city. During the reign of Ṣamṣām al-Dawla, Qutalmish brought an army [R 100b] and destroyed its remnants. Only a small town remained. Among the ruins, the building built by Jamshīd became like Indian tutty (i.e., like dust). The columns that have remained of that building are now known as the Forty Minarets. There are many stories about them. Some say they were the house of Humāy daughter of Bahman, others that they were the mosque of Solomon. Perhaps it was first a house and then a mosque.

The grain and grapes of Iṣṭakhr are good; the delicious apples are famous, half sweet, half sour. However, the air is bad and the water is dirty because it flows through rice fields.

Abarqūh, the Arabicized form of *bar kūh* (meaning) "on the mountain;" also called *dar kūh* ("in the mountain"): A town near Yazd, located at 89 degrees longitude and 31 degrees latitude. Although there are no trees, still it is blessed with many fine things. They say that no rain falls inside its walls, or at any rate, that it rains rarely and then mostly in the surrounding area. Originally the city was on top of the mountain, as the name indicates; later it was rebuilt on the plain below. Its water is gotten from underground channels. The climate is moderate. Rice, grain crops and cotton are fine.

No Jew is allowed to stay here for more than forty days.

The tomb of Ṭāwūs al-Ḥaramayn is a place of pilgrimage. According to Ḥamd Allāh, it cannot be roofed and even if it is covered only with a shade it collapses.

In Marāgha, which is one of its dependences, there are great cypress trees numbering more than those in Kashmir and Balkh.

The state dues from Abarqūh were 140,400 dinars.

The district of BRD is the largest district of the region (*kūra*) of Iṣṭakhr. Friday prayers are held there.

The two towns of Katha and Maybad are on the edge of the desert. Katha has much fruit, which is exported to Iṣfahān.

The town of Khūmāqān district is called Maskān.

Isfandān: A small city. It has a fortress. Nearby is Qumistān, a large village and summer pasture. The large and solid cave in its mountain is sometimes used as a refuge.

Iklīd, the Arabicized form of Kilīd: A small city with flowing water and a fortress. The climate is moderate. Fruit and grain crops are abundant.

Surmaq, also called Jurmaq; the Arabicized form of Surma: A town near Iqlīd but more cheerful and pleasant. It is on the edge of the desert. It has many estates and villages dependent on it.

Bayḍā: the largest city of the district of Iṣṭakhr, on a level plain 8 parasangs from Shiraz, located at 88 degrees longitude and 30 degrees latitude. The construction of its fortification is attributed to Gushtāsb son of Luhrāsb the Kayānid. It is called Bayḍā because it is built out of white (Ar. *bayḍā*) stone. According to the *Nuzha*, it is because its soil is white. Its old name is Shānk.

The climate is moderate and there is an abundance of flowing water, grain crops and fruit. It has a pasture 10 parasangs in length and breadth. 3 parasangs from the city are meadows with delightful streams.

According to the *Haft Iqlīm*, every grape seed yields 10 *mithqāl*; at present, however, it has gone to ruin.

Several eminent people came from this city, including the qadi Nāṣir al-Dīn ʿAbd ʿAllāh b. Muḥammad al-Mufassir and Ḥusayn b. Manṣūr al-Ḥallāj.

[M 267] Kharīz: A small city with abundant streams, grain crops and grapes, and a moderate climate. Its water comes from the Rūd-i Garm. The state dues are 25,500 dinars.

Ḥarmā: A small, pleasant city in that vicinity, with a moderate climate and abundant streams, fruit and grain. It has a strong fortress. Its water source is separated from the Āb-i Gard by a dam and irrigates the district. The state dues are 52,500 dinars. The fortess gets water from a cistern.

Ṣāhta: A small city the products of which are grain and fruit. There is an iron mine here.

Kāmfīrūz: A district along the river. It has a great forest with many lions. Its cold is severe.

Kamīn and Fārūn: two cities. They have flowing water. Fruit and grain are abundant. The climate is moderate. They have many dependencies. Within their borders there is game without limit.

Māyīn: A small city and summer-pasture with a moderate climate, in the midst of the mountains, on the Kūshk-i Zar road. Grain and fruit are abundant. Most of the inhabitants are thieves. At the foot of its mountain lies Ismāʿīl b. Mūsā al-Kāẓim. This city is a separate, independent district.

Yazdkhwar, also called Yazd-kura: A small town in the Iṣṭakhr district at the end of the desert, located at 89 degrees longitude and 32 degrees latitude.

The *Nuzha* records it as a village and also a summer pasture, and says that its only crops are walnuts and grain. It also considers the city of Yazd one of the *tūmāns* of Persian Iraq, located at the above-mentioned longitude and latitude. In ancient books this city was considered as belonging to the region (*kūra*) of Iṣṭakhr. Water is gotten from underground channels. the climate is moderate. Water channels pass through the middle of the city. The people have made them into pools and cisterns. Its products are grain, cotton, fruit and silk; the pomegranates are excellent. The inhabitants are mainly weak-minded, with few clever ones among them. Together with its dependencies, the state dues from here are 25 tomans and 1,000 dinars.

So much for the *Nuzha*. According to the *Taqwīm*, Yazdkhwar is the aforementioned city, but it is between Iṣfahān and Kirmān and is a separate province.

According to the *Haft Iqlīm*, the women of Yazd are beautiful and comely. There are many delightful orchards and elegant buildings. Of the excursion grounds of Yazd, Baft is especially fine. Many worthy people have compared Kāzargāh of Herat with Baft; however, each of them has a special quality not found in the other. Baft is 4 parasangs from Yazd between two high mountains. It includes delightful buildings and pleasant orchards and gardens. A river flows through it, but most of the time it is dry. If this river did not dry up, there would be no place on earth comparable to it. The two banks of this river are two regions (*maḥalle*), one known as *garmsīr* (hot region), the other as *sardsīr* (cold region). Their climates are so different that the crops of *garmsīr* ripen 20 days earlier than those of *sardsīr*.

The main goods traded in Yazd are textiles and beverages. Of eminent people, Sharaf al-Dīn ʿAlī Yazdī, the qadi Kamāl al-Dīn Mīr Ḥusayn Maybudī and Mawlānā ʿAbd Allāh, come from this city. [R 101a]

District of Dārābjird: It is named after Dārāb b. Bahman b. Isfandiyār the Kayānid. Today most of the province of Shabānkāra is considered as belonging to this region (*kūra*).

Shabānkāra: An extensive territory whose borders reach as far as Kirmān and the Persian Gulf. At the time of the Seljuks its state dues were 200 tomans. Later it declined to 26 tomans [M 268] and 600,000 dinars.

According to the *Haft Iqlīm*, in the past Fārs used to be five regions (*kūra*). Now it is 10 districts (*bölük*), of which Shabānkāra is the largest. Its capital is Dārābgird.

Dārābjird, the Arabicized form of Dārābgird: A town and fortress on a flat plain, located at 91 degrees longitude and 29½ degrees latitude. In the middle of the town is a dome-like hill and the round fortress was built on top of that hill. The city was round, as though marked off by a pair of compasses. The fortress had a huge moat. Later it fell to ruin and was rebuilt. It has four gates and its circumference is estimated to be one parasang.

This city has a covered market with two gates. The climate is hot. Grain, fruit and dates grow here. It is preferable to other cities in being surrounded by shade-giving trees and soul-refreshing rivers. In the mountains nearby is a seven-colored salt-mine, as well as mines of quicksilver and mummy. Shiraz is 50 parasangs distance from here. The tomb of Diḥyat al-Kalbī, who made great efforts to conquer Fārs, is here.

Mummy is produced in the form of drops from a crevice in the aforesaid mountain. It is a precious and fine medicine. Just as much as the Byzantian kings boast of their sealed clay (Lemnian earth), the kings of Fārs boast of this mummy. Only 20 *mithqāls* are produced in a year, and that belongs to the sultans.

According to the *Haft Iqlīm*, this mine was discovered during the reign of Farīdūn. A local hunter was chasing a wounded mountain ram when it entered a crevice and was concealed. As it happened, water was dropping from that crevice and the ram drank of it and was healed. The next morning the hunter caught it and brought it to Farīdūn, informing him of the water. Farīdūn broke the leg of a bird and had it drink that water, and it was healed. He seized control of the area, and it has been under state control ever since.

There is also artificial mummy. A red-haired boy is kept until he is 30 years old and then placed in a stone trough filled with honey. The opening of the trough is shut tight, and after a while the contents turns to mummy. This type is better then the one from the mine.

According to the *Aḥsan al-taqāsīm*, in the middle of Dārābjird is the Mummy Dome (*Ḳubbe-i mūmiyā*) on a hill with a Friday mosque and some markets. The gate of the Mummy Dome is made of iron and is guarded. In the month of Mihr, the governor of the city, the qadi and the professional witnesses come and open the gate. A naked man goes inside and brings out (the liquid) amounting to less than one *raṭl*, which they put in a vessel, seal it and send it to the governor of the province.

Jahrum: A town located at 89 degrees longitude and 28 degrees latitude. It was built by Bahman b. Isfandiyār. It has flowing water in underground channels. Grain, fruit and cotton grow here. The climate is hot.

Five parasangs from Jahram is the strong fortress of Khursha, situated on top of a high mountain. The climate there is moderate, tending to hot. It was built by Khursha, the governor of Jahrum, who was appointed by the brother of Ḥajjāj. Later he revolted, relying on this fortress.

Samīrān: A fortress in Jawīm Abī Aḥmad belonging to the province of Abrāhistān. The inhabitants bear arms and are highway robbers, waylaying on foot. The climate is hot; water is gotten from cisterns. The aforesaid province is a place of grain crops and dates. However, it is considered to belong to the Ardashīr-khūra district.

Jawīm Abī Aḥmad: an extensive area [M 269] covering ten parasangs, all of it date groves. It is surrounded by mountains and a small river flows on one side. It has a Friday mosque.

Fasā, also called Basā: The largest city of the Dārābjird district, located at 89 degrees longitude and 29 degrees latitude. It is a town as large as Shiraz. Most of the wood used in its buildings is cypress. The climate is moderately hot. Walnuts and bitter oranges are harvested when it snows. It is 27 parasangs from Shiraz. Its construction is attributed first to Ṭahmūrath, then Gushtāsb the Kayānid and Bahman. Initially it was triangular. Āzādmard, at the behest of Ḥajjāj, put it into a different shape. Then it fell to ruin and was rebuilt by Atabeg Jawlī. It has many dependencies and districts. Its water source is not a river but underground channels. The relational adjective of this place is Fasawī; the people of Fārs use Basāsīrī.

Īj: the Arabicized form of Īg. According to the *Nuzha*, it is the capital of Shabānkāra. [R 101b] It is a city 3 stages east of Shiraz, located at 91½ longitude and 29½ degrees latitude. It has incomparable bows. In the past this place was a village. In Seljuk times it was raised to the top of the mountain by the Ḥasnūya dynasty and made it into a kind of fortress. The learned qadi ʿAḍud al-Dīn (al-Ījī, d. 1355, religious scholar) comes from this town.

Zarkān: A town below Īj. It has a moderate climate, but the water is not good. Its grain, cotton and dates are good and plentiful.

Iṣṭahbānān: A city with a moderate climate and full of trees. Until Atabeg Jawlī of the Shabānkāraids appeared, the Seljuks had brought it to ruin; later it was rebuilt.

Parak: A city located at 90 degrees and 50 minutes longitude and 29½ degrees latitude, between Nīrīz and Dārābjird. Dārābjird is 3 parasangs away. It has a strong fortress. Grain crops and dates are in abundance. This is the frontier of Kirmān.

Nīrīz: A city and fortress as big as Aleppo, located at 90 degrees longitude and 29½ degrees latitude. It is 5 stages east of Shiraz and Yazd is 5 stages to its north. Its most important commodities are steel, various weapons and iron implements. The climate tends to be hot. Raisins are plentiful.

Ḥamd Allāh considers the Lār country as belonging to Shabānkāra. He gives a summary account. Since we have given a detailed account above, there is no need to return to it.

District of Shābūr Khūra: It is named after Shābūr b. Ardashīr. In the past its largest city was Shābūr; now it is Kāzarūn.

Kāzarūn: A large city with salubrious climate and fertile soil, located at 87½ degrees longitude and 29½ degrees latitude. It is on a flat plain. Most of the houses and the Friday mosque are on a high hill, while the markets and merchant villas are below. It is a two days journey to the sea. The buildings are made of stone and lime. Originally it was three villages, which were made into a city by Fīrūz b. Yazdgird. His son Qubād developed its construction further. Now the living quarters are scattered and spread out. However, it has substantial and solid, lofty villas and palaces, each one like a fortress.

The climate is hot. Water is gotten from three channels that belong to the three (original) villages. It has various coastal fruits, such as oranges, bitter oranges and lemons. There is one fruit like a date called *khatlān* (?) which has no match in the world. Cotton and linen fabrics are plentiful and are exported abroad. Therefore Kāzarūn is called the Damietta of the Persians. Several eminent individuals have their tombs here. It is a city with many districts and dependencies. [M 270]

Bishāwur: originally it was Bināʾ-i Shābūr ("Building of Shābūr"); with the passage of time it was simplified to Bishāwur. It is a city located at 87½ degrees longitude and 30 degrees latitude. It was built by Ṭahmūrath. At that time it was called Dayn Dulā (?). When Alexander the Greek conquered Fārs he destroyed it. Shābūr b. Ardashīr rebuilt it and it became known as Bināʾ-i Shābūr.

As it is blocked on the north and has a hot climate, the air is putrid. A river flows next to it. Its products, such as grain, rice, oranges, bitter oranges and lemons, have no local value; anyone traveling through can eat them (without cost). Fragrant flowers such as water lily, violet, jasmine and narcissus are abundant. Silk is also plentiful.

Outside the city is a large statue of a man carved out of black stone. Some say it is a talisman. Kings hold it in honor and make pilgrimage to it.

Shahristān: According to the *Aḥsan al-taqāsīm*, it was the administrative center (*kaṣaba*) of Shābūr and a large city in the past. Later it somewhat fell to ruin. However, it is a city with many fine qualities and unique aspects. There are flowing springs, abundant vegetable gardens, and endless kinds of fruit and grain. Bitter oranges, olives, dates and figs are gathered. The buildings are made of stone and lime. The Friday mosque is outside the city in the middle of gardens.

This city had four gates: Hormuz Gate, Mihr Gate, the Bahrām Gate, City Gate. A canal coming from the moat encompassed the city; it was crossed by a bridge. There was also a fortress next to the town, situated at the foot of a mountain known as Dhanab ("Tail"). In its two valleys it had gardens and villages. Outside of the city was a large bridge; the Kāzarūnīs destroyed it when they conquered Shahristān.

Jaruh, also called Garuh: A small city near Shiraz. The Amīr Dam (Band-i Amīr), one of the lofty buildings of the world, is between here and Shiraz.

Khanjān: A fortress in this region (*kūra*). It has no extramural settlement.

ʿAbd-i Jān, also called Dast-i Bārī: A small city with a hot climate. It has a spring and wells with bitter water. Most of the inhabitants are shoe-makers and weavers.

In the midst of the mountains are two more cities named Khasht and Kamārukh. They have streams. There is no fruit other than dates.

Nawbandjān, also called Nawbandgān: A city built on a flat plain, located at 87½ degrees longitude and 30½ degrees latitude. It was large in the past. [R 102a] It was destroyed at the time of the interregnum and rebuilt by Atabeg Jawlī. The climate is hot. There are various fruits and fragrant flowers. Water comes from the mountains nearby. It has several districts and dependencies, lowlands and highlands. Qalʿa-i sapīd ("White Fortress") is one parasang below this city.

Shiʿb-i Bawān: A famous valley and a most delightful excursion ground, 2 parasangs from Nawbandjān. It is named after Bawān b. Īrān b. Aswad b. Sām (Shem son of Noah). According to the *Nuzha*, this valley is between two mountains, three parasangs long and half a parasang wide. All of it is made up of intertwined gardens, streams and meadows. The climate is fine and the scenery delightful. It is adorned with fruit trees. A large river flows through the valley, with villages along it here and there; but they are invisible until one enters them because of the vegetation. One of these villages is known as Gurgān. Snow is rarely absent from the mountains on either side. [M 271] As this valley is forested from end to end, one never sees the sun.

The sages have said the most beautiful places of the world are four: 1) this valley, 2) the Ghuta of Damascus, 3) Soghd of Samarqand, 4) the Ubulla Canal—or else, in one version, the Meadow of Sindān. These are all mentioned at their appropriate place. When the famous poet al-

Mutanabbī came to this excursion ground (Shiʿb-i Bawān) he described it in his *qaṣīda* with the incipit: "The campsites of al-Shiʿb are as delightful among campsites as spring is among the seasons of the year." However, those sages never saw the Strait of Constantinople (the Bosphorus), otherwise they would have mentioned it before all of these places.

District of Qubād: It is named after Qubād b. Fīrūz, the father of Anūshirwān. Its cities are the following:

Arrajān, also pronounced Arajān or Arraghān: A large town at the extreme border of Fārs on the Khūzistān side, located at 85½ degrees longitude and 31 degrees latitude. It has lowland and highland and is landward and seaward—i.e., it is one stage from the sea and has date palms and olive trees. It is a city lost amidst date palms and fruit trees.

This city has seven (sic; error for six) gates: Ahwāz Gate, Ray Shahr Gate, Shiraz Gate, Ruṣāfa Gate, Maydān Gate and Kayyālīn Gate. Next to the fine Friday mosque, with its well-proportioned minaret, are the marketplace and the covered market with four gates. The wheat market is famous.

Arrajān is a mine of figs and olives, a treasury of soaps in Fārs. Its winter pasture is delightful, but in summer the weather is rather hot. Water is gotten from the Ṭāb River which flows through that province. There is a bridge across it consisting of one vault, 160 cubits (wide) and 50 cubits high. All kinds of fruit are abundant here; the *malīsi* pomegranate is especially fine. The people are mostly camel drivers.

On the borders of Arrajān is a fortress known as Ṭanbūr. It also has several districts. In one of them is a village called Ṭaryān where a well is recorded of which the water flows from its mouth with a force that would turn a water-mill and the depth of which has never been determined.

Ray Shahr (Bandar Ray): A city and entrepot, one of the seaports of Fārs, located at 87½ degrees longitude and 29½ degrees latitude. Its founding is attributed to Luhrāsb the Kayānid. Later it was rebuilt by Shābūr b. Ardashīr. It is a mid-sized city on the seacoast with a very hot climate. Dates and linen are abundant. The people are mostly maritime traders and generally voyage in the summer.

Mahrūyān, pronounced Māhī Rūyān by the people of Fārs: A city on the coast, the port of Arrajān, located at 86 degrees longitude and 30 degrees latitude. It has a flourishing marketplace and a small river. The climate is hot and putrid. Nevertheless it is a busy entrepot, because anyone going from Fārs to the sea by way of Khūzistān, or from Basra to Khūzistān, stops here. There is no fruit other than dates. The chief source of income is from ships. It also grows much flax. This port is the extreme border of Fārs on the sea. From here it is 170 parasangs along the seacoast to Ḥiṣn Ibn ʿUmāra, where the land of Fārs ends.

Ḥiṣn Ibn ʿUmāra: A strong fortress at 91½ degrees longitude and 29½ degrees latitude. It is in the direction of Kirmān. From Sīrāf it can be reached along the seacoast, passing through scattered mountains and deserts. It is also called Qalʿa-i Dānbān and Jalandī. Later it fell to ruin.

According to the *Nuzha*, in the past Fārs had some 60 fortresses, of which 16 remain. Some of them have already been mentioned. The rest are as follows:

Qalʿa-i aspīd (Qalʿa-i sapīd, "White Fortress"), also called Saʿīdābād: It is 1 parasang from Nawbandjān. [M 272] Ziyād b. Umayya resided here during the caliphate of ʿAlī when he was governor of Fārs, so it was also called Qalʿa-i Ziyād. According to the *Fārsnāma*, it is an ancient fortress. After it lay in ruins for many years it was restored by Abū Naṣr al-Dawwānī during the time of the Seljuks. It is on top of a mountain that is 20 parasangs in circumference and is adjacent to no other mountain. Only one road leads to the fortress. The top of the mountain is a flat meadow with streams and fruit trees. Water is at hand in wells. The mountain is surrounded by a vast plain.

Qalʿa-i Kāwiyān: A strong fortress built on top of Mt. Ṭīn.

Qalʿa-i Ābāda: A small fortress with a moderate climate and water gotten from cisterns.

Qalʿa-i Khawādān: A fortress similar to it (i.e., to Ābāda Fortress) in Fasā province.

Qalʿa-i Shahāda: A fortress 4 parasangs from Fīrūzābād on top of a great mountain with a pleasant climate and plenty of grain crops. It was built by the Masʿūdīs.

To this point is the text of Ḥamd Allāh, with a few additions. [R 102b]

In some manuscripts (of the *Nuzha*?) the attachments (*tetimmāt*) to Fārs are given as follows:

Jannāba: A town with cheap foodstuffs, a port of Fārs where the river Sīrīn empties into the sea, located at 86½ degrees longitude and 30 degrees latitude. The climate is hot. Today most of it is in ruins. From here Kāzarūn is 24 parasangs and Shiraz is 44 parasangs.

Shatar: A village 1 stage from Jannāba. The Ṭāb River empties into the sea near here.

Shīlān: A fortress and village on top of a mountain at the seacoast. Bīdahān is 2 stages hence.

Bīdahān: A town and one of the ports of Fārs, located at 88½ degrees longitude and 29 degrees latitude. It is 2 stages southwest of Kāzarūn, 3 stages east of Ray, and 2 stages west of Bih Dah, a port and village which is the frontier between Fārs and Hormuz.

Again according to the *Nuzha*, there are many meadowlands (*marghzār*) in Fārs. These are the most famous ones:

Marghzār-i Kūshk-i Zard: an excellent pasture 10 parasangs long and 5 parasangs wide. It has large villages and many springs. The climate is cold.

Marghzār-i Dasht-i Rūn: A delightful pasture 7 parasangs long and 5 parasangs wide. It has many streams but its climate is even colder than that of Kūshk-i Zard. The hospice of Ṣalāḥ al-Dīn and the desert of Shahriyār are in this plain.

Marghzār-i Dasht-i Arzan: It is 2 parasangs long, 1 parasang wide. It is on a lakeshore. On one side of it is a thicket with beasts of prey.

Marghzār-i Arashkān: A pasture 5 parasangs long and 3 parasangs wide between Guwār and Shīrdār. A stream flows through it. Near here as well is a thicket with lions.

Marghzār-i Sandān: A pleasant pasture, 10 parasangs in length and breadth, with many streams and flourishing surrounding areas. When figs ripen a pond appears in the middle and during the hot season it dries up again. This is also a famous excursion ground in Fārs, similar to Shiʿb-i Bawān.

Marghzār-i Fālī: A delightful meadow 3 parasangs long and 1 parasang wide on the bank of the Andarāb. Its grass is good for livestock in winter but bad in summer.

Marghzār-i Kālān: A delightful place 4 parasangs in length and breadth next to the tomb of the mother of the prophet Solomon. This tomb is square-shaped and built of stone.

Marghzār-i Kāmfīrūz: A delightful pasture on the bank of Rūd-i Kur. It is dangerous, however, because its thicket is a haunt for lions. [M 273]

Marghzār-i Nargis: A famous place near Khān Āzād in the vicinity of Kāzarūn, 3 parasangs in length and 2 parasangs in breadth, where the only plant is wild narcisssus (*nargis*) whose fragrance is intoxicating.

The remaining characteristics and affairs of Fārs

Some places in this clime are mountain pastures such that no trees or fruits grow there because of the severe cold. In other places—in the vicinity of Iṣṭakhr and Bayḍā and elsewhere—it is impossible to sleep during the day because of the heat. In the mid-regions of Fārs—places such as Nawbandjān, Kāzarūn, Sīrāf and Arrajān—the climate is moderate. In such districts as Shiraz and Shābūr—indeed, almost everywhere—one finds snow; it comes from places near and far (?). Trees can be found at most places. Fruit is excellent.

The people of Fārs are of several classes. There are more Zoroastrians than Jews; and there are some Christians as well. Kāzarūn is known for its one-eyed people and Shiraz for its paralytics. In the past there was a preponderance of hadith scholars; the populace was mainly Sunni of the Ḥanafī rite, with some Dāwūdīs; the jurists were Muʿtazilīs; and Shīʿīs predominated along the coast. Today, under the rule of the shahs (i.e., Safavids), most of the populace are Imami Shīʿīs. The people of this land follow Zoroastrian customs at Nīrūz (Nawrūz) and Mihrajān.

Men in the hot regions of Fārs are thin, with wheat complexions and sparse beards. Three languages are current among them: Persian, which is spoken in every city; Pahlawī, which is what they mostly use for writing, but not everybody understands it; and Arabic.

As to dress and appearance, the magistrates wear an inner garment of *qabā* (a close long gown worn by men) and an outer garment of full *durrāʿa* (an upper garment), a small turban, a swordbelt, and boots narrow at the shank. The qadis wear a sort of conical hat (*külāh*) with their turban sash hanging down and a thin shirt, but no *durrāʿa* or boots.

The people of Fārs have fine abodes and distinguished houses. They are hospitable and friendly to strangers. They are well-mannered and modest. Being mainly merchants, they are quite covetous and stingy.

Commodities

Arrajān is famous for soap, molasses, olive oil, figs and bathcloths; Mahrūyān for fish and dates; Sīrāf for pearls, bathcloths and linen; Dārābjird for beddings and curtains; Jahrum for carpets and felt; Shiraz for clothing, curtains, brocade and silk curtains, bedding, bathcloths, [R 103a] handkerchieves, and plums; Sābūr for ointments, flowers and fruit; Jūr for rosewater; Iṣṭakhr for rice and foodstuffs; Dārābjird for mummy, colored salt and boneless fish; the mountains of Nīrīz for balsam, emery, *maghnīsiyā* stone, iron mines and white lead.

In general, oils for all kinds of medicaments are the best in this clime, except for oil of *khīrī* and oil of violets which are better in the Kūfa region. Basā is famous for its fabrics and brocades. In Yazd and Abarqūh excellent stuffs are made from cotton and silk. From Sīrāf, being a port, are exported aloeswood, ambergris, camphor, precious stones, rattan, ivory, ebony, black pepper, sandal, and various perfumes and medicaments.

In the past, the tax from Fārs was 330 *yük* (1 yük = 100,000) of dirhams. Faḍl b. Marwān made it into a tax farm for 350.

Rivers and mountains

There are several rivers in Fārs. According to the *Aḥsan al-taqāsīm*, in the Sea of Sīrāf (the Persian Gulf) is a place where seamen dive in with waterskins and bring up sweet water. At the bottom of the sea [M 274] there is said to be a spring bubbling up. Another source like this is reported from Bahrain.

Ṭāb River: It rises in the mountains of Lūristān near Murj and is joined by the Mas River. Passing near Arrajān, it waters Pārs and the fields of the tribes in the middle of Khūzistān. This river separates the two provinces. It empties into the sea next to the village of Shatar. It is a great river with no fords. According to the *Āthār al-bilād*, in Arrajān there is a marvelous bridge with a single vault 80 spans across. If a camel with a lance on it passes under the bridge, the lance does not touch the vault.

Dakān River: It rises in Fārs from the mountain of the village of Shād-āfarīn and irrigates the district known as Rūstā-yi Shāh, then flows through the plains of Jamragān, Kūmār and Kāzrīn, watering some of the districts of Sīrāf. The streams from the surrounding mountains also join this river. It empties into the Persian Gulf between Najīram and Sīrāf. There is no river in this clime that is more beneficial or whose course is more populated. Dakān is a village on this river. The length of the watercourse has been estimated to be 55 parasangs.

Kur River: It rises in the Kalār mountains of Fārs. After joining the Shiʿb-i Bawān and Māyīn Rivers and the other small rivers of Fārs, it reaches the Kāmafrūz (Kāmfīrūz) district, waters Sakān and Ṭasūj, and empties into Lake Bakhtigān.

This river is stingy. It cannot be used for irrigation wherever there is no dam. The first dam on it is that of Rāmjard. It was an ancient construction, restored by Fakhr al-Dawla Jawlī the Seljukid, who named it Fakhristān. The second is the ʿAḍudī Dam, the like of which is rare in the world. It waters Upper Kirmān. (The third is) the Qaṣār Dam, which irrigates Lower Kirmān. This too had been destroyed and was repaired by Jawlī.

Fīrūzābād River: Rising from the Ḥasanāt (?) mountains, it irrigates Fīrūzābād, joins the Sitāragān River, and empties into the sea. Its length is 12 parasangs.

Nishāwur River: Rising from the Nishāwur mountains, it irrigates Nishāwur and empties into the sea between (Nishāwur and) Mābadastān. Its length is 9 parasangs.

Farawāt River: It rises in the region (*kūra*) of Juwayn at a place called Farawāt. In Iṣṭakhr it passes under the Khurāsān Bridge and reaches Rūd-i Nīrīz. That river rises from the district of Arrajān, waters Jīghān and Jawr and the valley of Ardashīr, and empties into the sea.

Dīwrūd: It rises on the borders of Kirmān, flows fast enough to turn 20 water mills, and empties into the sea.

Shīrīn River: rising in Mt. Dīnār, it empties into the sea at the borders of Jannāba. It is a great river, difficult to cross.

Aside from these, there are many small rivers in Fārs.

Mt. Sābūr: According to Ibn Ḥawqal, one of the wonders of Fārs is the mountain in the district of Sābūr where the kings of Persia and their march lords and the elders of the guardians of the fires (i.e., Zoroastrian priests) are depicted (Naqsh-e Rostam).

Mt. Darāk: It is 2 parasangs from Shiraz, which gets its snow from there.

Mt. Dārābjird: The mines of colored salt are here.

Lake Bakhtigān, also called Lake ʿAmr: an oblong salt lake some 20 parasangs from the edge of Nīrīz. The Kur River of Fārs empties into this lake. Shiraz is 2 parasangs from here. On one side it is adjacent to the borders of Kirmān. Salt is crystallized in its water.

Lake Dasht-i Arzan (Dasht-i Arzhan): A sweet water lake in the district of Sābūr, 10 parasangs in circumference. It sometimes dries up. Shiraz gets its fish from this lake. [M 275]

In the Shābūr district is another lake as well, near Kāzrīn, some 10 parasangs in circumference. It also has salt water, but fish are caught in plenty.

Aside from these, there are several small lakes in Fārs.

The Sea of Fārs (Persian Gulf): It extends east-west from the border of Sind to Oman and south-north from Oman to Basra. The territory of Fārs is to its east. From Basra to Oman is 300 parasangs and beyond Oman to Muscat and Ra's Jumjuma is 50 parasangs, which together gives the whole length of this sea. The west coast extends from Oman to Bahrain and Qaṭīf and ʿAbbādān which is the estuary of the Tigris. There it turns and the east coast extends through Mahrūyān, Sayf Baḥr, Sīrāf, Ḥisn Ibn ʿUmāra and the Strait of Hormuz where it turns east and reaches the shores of Makran.

In this sea are the following islands:

Qays: an ancient island. In the past its rulers were famous and prosperous. They even considered the kingdom of Fārs as subservient to them and called it their "palace" (*dawlatkhāna*). The state dues were 491,300 dinars. The other islands in the Persian Sea are also its dependents. Located 4 parasangs from the coast at 88 degrees longitude and 25½ degrees latitude, this island is 4 parasangs in length and breadth and has an abundance of grain crops and dates and a very hot climate. Water is gotten from the rain, stored in cisterns. It is surrounded by pearl-diving areas.

Bahrain: an island near the west coast, 10 parasangs in length and 5 parasangs in breadth. It has streams, orchards and gardens, and many villages. Sometimes it belongs to the region of Hijr, sometimes to Fārs. Originally Bahrain (Ar. *baḥrayn* "the two seas") was the name of those shores; the island took its name from that. [R 103b]

Seydī ʿAlīzāde says in the *Mirʾātüʾl-memālik*: On the coast of Bahrain seamen dive approximately 8 fathoms or more with skins in their hands and bring up sweet water. In summer the weather in Bahrain is very hot and this water is the best and the coldest. The people of that region explain the expression *maraj al-baḥrayn* ("meeting place of the two seas"—Koran 55:19) as referring to this and consider that to be the reason for calling it Bahrain. The real reason for the name will be explained in the chapter on the Arabian Peninsula in the section on Bahrain.

Khārg, also called Maḥram: A small island 1 parasang in length and breadth near the coast of Bahrain and Oman. Crops, dates and fruit are good. There are many places for pearl-diving. The pearls here are superior to those at other islands. In the past it belonged to the Qubād-khūra district. Kīsh is 25 parasangs from here, Basra 35. The tomb of Muḥammad b. Ḥanafiyya here is a place of pilgrimage.

There are some small islands aside from these, also one or two near Hormuz. Some have been mentioned, some omitted.

Routes of Fārs

From Shiraz, Iṣfahān is 10 stages or 72 parasangs to the northwest, Sīrāf 60 parasangs, Iṣṭakhr 12 parasangs, Kāzarūn 22 parasangs. The border of Khūzistān is 60 parasangs to the west, Fasā is 27 parasangs, Dārābjird 50 parasangs, Tustar 6 stages > Daspul, 2 stages > Bayāt fortress, 7 stages.

From Yazd, Kirmān is 10 stages to the southeast; Shiraz is 10 stages to the southwest; Kāshān is 5 or, in another report, 9 stages.

From Dārābjird to Lār is 4 stages to the south; Sīrjān is 3 stages to the east; Nīrīz is 4 stages to the west.

From Kāzarūn, [M 276] the entrepot of Kanbāwa is 4 stages to the west > entrepot of Ray Shahr, 1 stage to the east; the city of Dawraq is 3 stages to the west > Basra, 1 stage.

From Shiraz, Sarwistān is 3 stages > Jannāba, 2 stages; altogether 44 parasang.

From Sīrāf, Nāwband is 2 stages > Shīrdūn opposite Kīsh, 10 stages.

From Shiraz, Fīrūzābād is 24 parasangs > Kāzarūn, 16 parasangs.

From Shiraz, to the east, Tabriz is 5 stages.

From Jūr, Sīrāf is 24 parasangs, Kāzarūn is 16 parasangs.

From Shiraz, to the southwest, Dārkān is 34 parasangs; turning left from there, the city of Shabānkāra is 34[35] parasangs; going straight, Hormuz is 59 parasangs.

From Shiraz, Kirmān is 91 parasangs; Abarqūh is 39 parasangs > Yazd, 28 parasangs, making 68 parasang from Shiraz.

From Shiraz, Nawbandjān (sic; error for Nawbandjān?) is 25 parasangs > frontier of Khūzistān, 37 parasangs, making 62 parasangs from Shiraz.

From Shiraz to the frontier of Kirmān by way of Iṣṭākhr: > Iṣṭākhr, 12 parasangs > Ziyād-ābād, which is one of the dependants of Khūzistān, 8 parasangs > observatory of KLWDR, 8 parasangs > village of Jarmābān, 6 parasangs > ʿAbd al-Raḥmān and Būdī-jān, 6 parasangs > Ṣāhil al-Kubrā, 8 parasangs > hospice of Sarmaqān, which is the frontier of Kirmān, 8 parasangs > hospice of Basht-ham, 9 parasangs > Sīrjān of Kirmān, 9 parasangs.

From Basā, Kāzarūn is 10 parasangs; Jahrum, 10 parasangs.

From Shiraz, Bayḍā is 8 parasangs; Yazd, 74 parasangs; Tūḥ, 30 parasangs; Jahrum, 32 parasangs.

From Jūr, Kāzarūn is 16 parasangs.

From Sīrāf, Jahrum is 12 parasangs.

From Māhrūyān, Ḥiṣn Ibn ʿUmāra is 160 parasang, which is the length of Fārs.

From the border of Kirmān to the border of Iṣfahān: the boundary (*sınur*) of Rūdān > Anāra, 18 parasangs > Mahraj and Katha, 5 parasangs > Maybud, 10 parasangs > ʿAqd, 10 parasangs > Nāyīn, 15 parasangs > Iṣfahān, 45 parasangs.

The border on the side of Khurāsān is altogether 120 parasangs. On the side of Khūzistān, from Māhrūyān to the border of Iṣfahān as far as the region of Sābūr, is 60 parasangs.

Fire-temples of Fārs

In the past the inhabitants of this clime were Magian fire-worshippers, as shown by the fire-temples here and there which are held in esteem. The most famous ones are these: [R 104a] [Plate 41]

- Kārbān: A large building known as the Fire-temple of Fārs. An inscription in the Pahlawī language says that 30,000 dinars were spent on it.
- Gunbad-i Kalūsh: near Shābūr.
- Khīfa: in Kāzarūn.
- Masūbān: in Shiraz.

35 Text has *iç dört*, error for *otuz dört*; cf. *Nuzha*, 178.

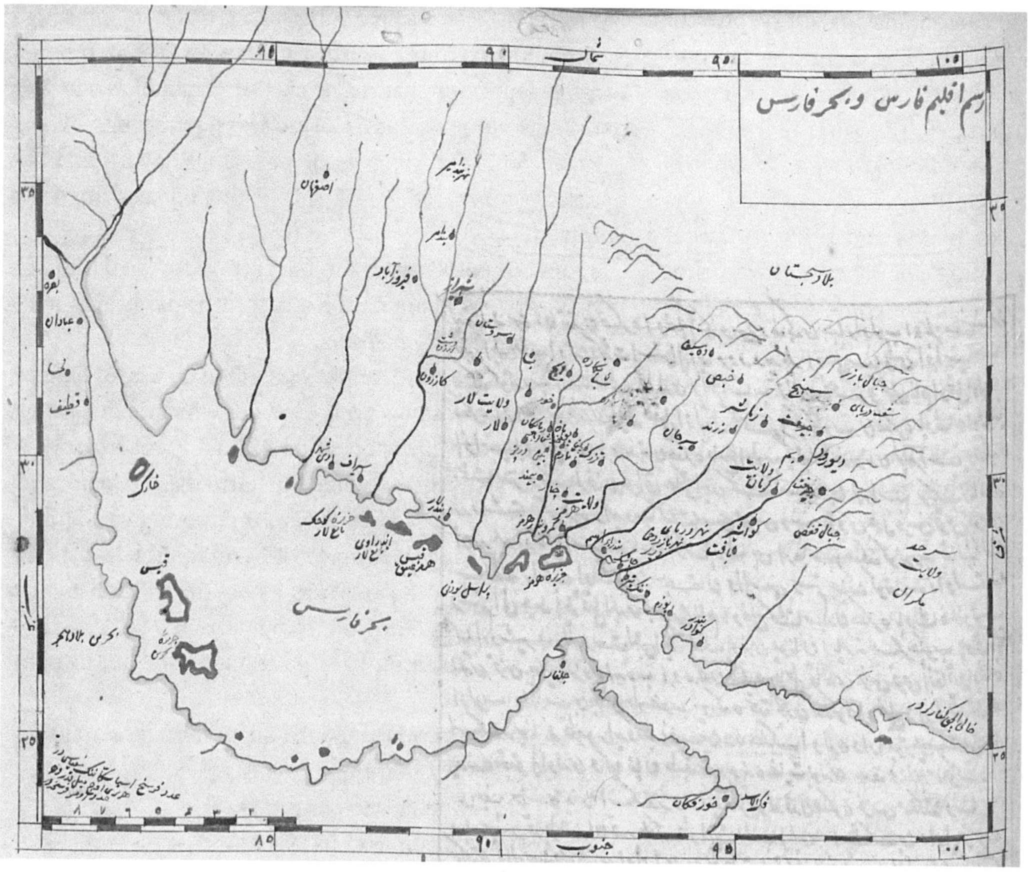

PLATE 41
Map of Fārs and Persian Gulf, R 104a (Appendix, fig. 81)
TOPKAPI PALACE LIBRARY, REVAN 1624, WITH PERMISSION

Kings

Since ancient times the country of Fārs has been the seat of sovereignty of the padishahs of Iran. The might and prestige of the empire of the Chosroes (Sasanians) reached such a degree that most kings of the inhabited quarter were subservient to them. They (i.e. the rulers of Fārs) were of four dynasties:

1st dynasty: Pīshdādīs
In the opinion of the people of Fars (or: the Persians), the Pīshdādīs are the most ancient kings of humankind. Due to their justice and liberality they became famous under the name Pīshdād ("Lawgiver").

Gayūmarth is the first of this dynasty. He was dubbed Gulshāh ("Rose King"). In the religion of the Persians he was the first man, i.e., Adam. The Islamic historians record him as a descendant of Shem son of Noah, because all nations and religions other than the Persians affirm the universal Flood, [M 277] while the Persians assert that it did not go beyond Babylon. And they assert that their lineage goes back to Adam.

After him in this dynasty came Hūshang, son of Siyāmak; Ṭahmūrath, grandson of Hūshang; Jamshīd, brother of Ṭahmūrath; Ḍaḥḥāk, son of Mirdās, also called Bīwarāsb; Farīdūn, grandson of Jamshīd, who succeeded Ḍaḥḥāk after Kāwa the blacksmith revolted against him and killed him; Manūchihr, grandson of Farīdūn, who fought against Afrāsiyāb son of Turk, but was defeated and Afrāsiyāb subdued Iran. Later Afrāsiyāb was driven back to the land of the Turks by Zav son of Ṭahmāsp, one of the grandchildren of Manūchihr, who took Iran from his hands. He was succeeded by his son Garshāsb, with whom this dynasty came to an end.

There is much dispute concerning the extent of this dynasty and the events that took place. It is mostly idle stories, not worth spilling ink over.

2nd dynasty: Kayānīs
They are called Kayānī after the first part of their names (i.e., Kay). The first was Kay-Qubād, son of Zāb, son of Zaw. In his time the Jayḥūn (Oxus) became the border between Iran and Turan. After the death of Zav, when Afrāsiyāb again had designs on Iran, Zāl raised Kay-Qubād, who was a descendant of Farīdūn, to the throne. Rustam, son of Zāl, was the champion of Kay-Qubād, and Afrāsiyāb made peace with his father, Bushang. [R 104b]

Kay-Qubād was succeeded by his son, Kay-Kāwus. He was a tyrant who assumed the title Namrud, meaning "immortal"; with frequent usage it came to be pronounced Namrūd. He sent his son Siyāwush to war against Afrāsiyāb, but Siyāwush made peace with him and remained at his court. He married one of Afrāsiyāb's daughters and had a son named Kay-Khusraw. Later Afrāsiyāb killed Siyāwush in an act of cruelty.

Kay-Khusraw became padishah after Kay-Kāwus and killed Afrāsiyāb. After some time he withdrew from the throne and was succeeded by Luhrāsb, a descendant of Kay-Qubād. Bukht Naṣr (Nebuchadnezzar) was governor of Iraq on his behalf.

(The next king was) Gushtāsb, son of Luhrāsb. In his time appeared Zardusht (Zoroaster) the sage, who converted Gushtāsb to the Magian religion. All the Persians became Magians. Gushtāsb sent his son Isfandiyār to war against Arjāsb, the padishah of Turan. When Isfandiyār was victorious he had designs on the throne, so Gushtāsb sent him to war against Rustam, the governor of Sijistān, who killed him by a ruse in Zāwulistān.

Gushtāsb was succeeded by his son Bahman, who assumed the titles Kayā Ardashīr and Dirāz-dast. He led an army against the land of the Greeks, conquering many provinces. He married his own daughter, Humāy. Humāy being pregnant when Bahman died, she became padishah by agreement of the Persians. When her son, Dārāb son of Bahman, was born and grew up, he assumed the throne. Later, during the reign of his son Dārā (Darius), Alexander the Greek appeared and they fought a battle at Niṣībīn. Dārā was routed and put to death by his own followers. The Kayānī dynasty ended with him.

3rd dynasty: Party Kings and Ashkānīs (*Parthians*)

After defeating Dārā, Alexander appointed to the country of the Persians and the other Eastern lands some 90 party kings. Ashk son of Dārā was made governor of the country of the Persians; hence this dynasty was named the Ashkānīs. His lineage ruled for the next 500 years, as follows: Ashk son of Ashk; Shābūr son of Ashk; Bahrām son of Shābūr; Balāsh son of Bahrām; Hurmuz son of Balāsh; Narsī; Balāshān, Jūd; Fīrūz; Khusraw; Ardawān the Elder; Ardawān [M 278] the Younger. There is dispute concerning their order, names and length of rule. The dynasty continued until the 512th year according to the Alexandrian calendar, coming to an end with Ardawān the Younger.

4th dynasty: Sasanians

They are also called Akāsira, because each one had the title Kisrā, which is the Arabicized form of Khusraw. Ardashīr son of Bābak, a descendant of Sāsān son of Bahman, defeated the Ardawānīs and took control of the country of Fārs and the other countries. He became a mighty padishah. He was succeeeded by his son Sābūr who made the Roman emperor pay tribute and had the books of the Greeks translated into Persian.

He was followed on the throne by his son, Hurmuz; Bahrām son of Hurmuz; Bahrām son of Bahrām; Bahrām III; Narsī son of Bahrām; Hurmuz son of Narsī.

Next was his son, Shābūr Dhū al-Aktāf. The painter Mānī and Manicheism appeared during his reign. He fought against the Roman emperor and was defeated. When Julius Caesar came to Ctesiphon he was killed, and so they made peace. Later Shābūr fell captive in Rome and escaped. The Vault of Kisrā was built by him.

The kings after him were as follows: his brother, Hurmuz; his son, Sābūr son of Sābūr; his other son, Bahrām son of Sābūr; Yazdajird Athīm son of Sābūr; and his son, Bahrām Gūr, who was valiant, just and given to drink; then Yazdajird son of Bahrām; Hurmuz son of Yazdajird; Fīrūz son of Yazdajird; Balāsh son of Fīrūz; Qubād son of Fīrūz; and Anūshirwān son of Qubād.

Anūshirwān had the title "the Just". He was padishah at the rise of Islam. He ruled for 48 years and died in the year 888 according to the Alexandrian calendar. He was succeeded by his son, Hurmuz, whose chief commandant was Bahrām Chūbīn.

(The next king was) Parwīz son of Hurmuz, who was the Khusraw to Shīrīn and who built the Qaṣr-i Shīrīn. He tore up the letter of the Prophet inviting him to accept Islam, and was killed by his own son, Shīrūya. When Ardashīr son of Shīrūya succeeded his father as king, one of the commanders, Shahriyār, seized power and put him to death. After that [R 105a] Būrān daughter of Khusraw Parwīz ascended the throne along with her sister Azarmī-dukht. After a short time they too were killed. Khusraw, one of the sons of Ardashīr, and Fīrūz and Farrukhzād, two sons of Anūshirwān, were each put on the throne and then put to death after six months.

Next to rule was Yazdajird son of Shahriyār. By that time the dynasty was weak and signs of decline were visible. During the caliphate of ʿUmar, the army of Islam came from Medina commanded by Saʿd b. Abī Waqqāṣ and clashed with Yazdajird and his commander Rustam. The Persian army was routed, Rustam was killed in battle, and Yazdajird fled and was put to death. After the battle of Qādisiyya in the 14th year of the Hijra (635/6), the Sasanian dynasty came to an end. Henceforth the Persian countries as well as Iraq came under the administration of Muslim governors.

Āl ʿUmāra, also called Āl Jalandī: an ancient lineage. They are a people with large territories and strong fortresses, extending from the coast of the Sea of Fārs (Persian Gulf) to the border of Kirmān. It is said that the king who seized the ship in the story of Khiḍr and Moses in the

Koran was one of them. They were powerful rulers with their own armies who, according to the *Masālik*, never submitted to the kings around them. Indeed, one of their rulers, ʿAbd Allāh b. Aḥmad al-Jalandī, waged war for two years against ʿAmr b. al-Layth, but could not defeat him. However, their dynastic and political affairs are not discussed in the chronicles.

After the Muslim conquest, the kingdom of Fārs was in the hands of Umayyad and Abbasid governors. [M 279] Sometimes dynasties appeared that ruled by conquest. The first of them was the Buwayhids.

Buwayhids (Buyids)

The Daylamite Abū Shujāʿ Buwayh was a poor man. His three sons, ʿAlī, Ḥasan and Aḥmad, were in the service of Mākān, the governor of Khurāsān. When Mardāwīj the Daylamite defeated Mākān, they went over to his side. His older brothers gave ʿAlī the governorship of Karḥ and the title ʿImād al-Dawla. Later, estrangement occurred between him and Mardāwīj. With some 900 soldiers he defeated the governor of Iṣfahān, Abū Bakr Muẓaffar b. Yāqūt, who commanded 10,000 troops. ʿImād al-Dawla took the city, achieving great renown. Then, under the encouragement of Zayd b. ʿAlī Nawbandjānī, he went to Arrajān and in 320 (932) conquered Shiraz. Gradually, he subdued the entire country of Fārs. In (3)22 (934) he sent his brother, Rukn al-Dawla Ḥasan, to Kāzarūn; he conquered the city.

In (3)26 (937–938) he sent his other brother, Muʿizz al-Dawla Aḥmad, to Ahwāz (Khūzistān); he seized Wāsiṭ and Basra. Before this, he had already taken Kirmān and Sijistān. Later Muʿizz al-Dawla conquered Baghdad and the Abbasids withdrew from state affairs. He died in 356 (967). Rukn al-Dawla Ḥasan, who was governor of Persian Iraq, died in 336 (947–948).

In (3)38 (949–950) ʿImād al-Dawla despaired of his life and made his nephew, ʿAḍud al-Dawla Fanā Khusraw b. Rukn al-Dawla, heir apparent. After the death of his father, ʿAḍud al-Dawla had become an independent, and powerful world-conqueror. He seized (Arab) Iraq as well and built the wall around Medina. He died in 372 (982–983), leaving behind several charitable works in Fārs.

He was succeeded by his son, Ṣamṣām al-Dawla Abū Kālinjār Marzubān, who ruled for four years. After him, his brother, Bahā al-Dawla Abū Naṣr Shahinshāh—along with his two sons, Sulṭān al-Dawla Abū Shujāʿ and Qawām al-Dawla Abū al-Fawāris, and ʿIzz al-Mulūk Abū Kālinjār Marzubān b. Sulṭān al-Dawla—raised the banner of rulership in Fārs and Kirmān. This ʿIzz al-Mulūk built a great wall with 12 gates around the city of Shiraz. He died in 440 (1048–1049), having governed in Fārs and Ahwāz (Khūzistān) for 25 years, and was succeeded by his son, al-Malik al-ʿAzīz Abū Manṣūr Fulāsutūn—which is the Arabicized form of Pūlādsutūn ("Steel Pillar"). Because of a quarrel between him and his brother, al-Malik al-Raḥīm Abū ʿAlī Kay-Khusraw, Amīr Faḍlūya Shabānkāra seized the opportunity and revolted. In 448 (1056–1057) he captured al-Malik al-ʿAzīz and imprisoned him in a castle. With this the Buwayhid dynasty came to an end, and the Shabānkārīs came forth.

Shabānkāra Dynasty

Their lineage goes back to Ardashīr. The father and ancestors of Amīr Faḍl b. Ḥasan, known as Faḍlūya, were military commanders of Fārs. In the same year (448/1056–1057) he imprisoned Abū Manṣūr and subdued the kingdom of Fārs. In 464 (1071–1072) Niẓām al-Mulk, the vizier of Alp Arslan, marched against him and did away with him. After this, Niẓām al-Dīn Maḥmūd b. Yaḥyā, known as Quhnūya, became governor of Shabānkāra. He built the governor's palace in Īj.

He was succeeded by his son, Amīr Mubāriz al-Dīn, who ruled for a year; then by the latter's son, Niẓām al-Dīn and then his other son, Malik Muẓaffar Muḥammad, who became governor in 624 (1227). He was a man of great erudition and a promoter of excellence. He was martyred in (6)58 (1260) [M 280] when Hulagu was besieging (Shiraz?) and the Mongols destroyed the governor's palace. Then his son Quṭb al-Dīn governed for one year; his nephew Niẓām al-Dīn for three years; the latter's brother, Nuṣrat al-Dīn Ibrāhīm, for two years; Malik Jalāl al-Dīn for 17 years; and Bahā al-Dīn Ismāʿīl for seven years. After them, while Ghiyāth al-Dīn, Niẓām al-Dīn and Ardashīr were governing in mutual antagonism, the Āl Muẓaffar did away with them and conquered the country.

Āl Muẓaffar

Amīr Muẓaffar's father was Ghiyāth al-Dīn Khurāsānī. He was upright and brave; therefore when he was in the service of the governor of Yazd, Atabeg Yūsuf Shāh, he obtained the office of Yasawul in the army of Arghūn Khan and Ghāzān Khan gave him the governorship of Hazara. He died in Shabānkāra in 713 (1313–1314).

His son, Amīr Mubāriz al-Dīn Muḥammad, served Abū Saʿīd and became governor of Yazd. In 733 (1332–1333) he took Kirmān and later took control of Fārs. In 760 (1359) he was blinded by his sons, Shah Shujāʿ and Maḥmūd. He was succeeded by Jalāl al-Dīn Shah Shujāʿ, who died in (7)86 (1384–1385) and whose son, Zayn al-ʿĀbidīn, became governor. It was during his tenure that Tīmūr conquered Fārs. Shah Yaḥyā was confirmed as governor by Tīmūr. After that his brother, Shah Manṣūr, revolted, taking Fārs from his hands. Tīmūr came back to Shiraz and Shah Manṣūr was killed in battle. Tīmūr massacred the rest of the Muẓaffarids, and their dynasty came to an end in 795 (1392–1393).

Īnjūids

During the time of the Chingisids the term *īnjū* was used for the ruler's personal estates. Sharaf al-Dīn Maḥmūd Shāh b. Muḥammad Shāh, who was from Fārs, was appointed to that office (of governing Fārs) during the reign of Sultan Abū Saʿīd and accrued money and estates without limit. He became the source of that dynasty and seized Fārs—both land and sea (i.e., both the territory of Fārs and the Persian Gulf)—and Kirmān and Shabānkāra. After the death of Abū Saʿīd in 736 (1335–1336), Arpa Khan suspected him of revolt and had him put to death.

His son Masʿūd Shāh, who had gone to Anatolia (*Rūm*), came to Shiraz after the death of Arpa Khan and conquered it. In 743 (1342) he was killed and succeeded by his brother, Amīr Shaikh Abū Isḥāq, who governed Fārs for some time. Then it was conquered by Mubāriz al-Dīn Muḥammad Muẓaffarī, who harried Amīr Shaikh, defeated him in battle several times, laid siege in 754 (1353–1354) and drove him out of Shiraz. Finally Amīr Shaikh was killed in (7)58 (1357) and the Īnjūids came to an end.

Kākūyids

Abū Jaʿfar Muḥammad Kākūya, maternal uncle of Sayyida Khātūn who was the mother of the Daylamite Majd (text: Maḥd) al-Dawla Rustam, was governor of Iṣfahān. In 443 (1051–1052) the Seljuk sultan Toghril removed his son, Abū Manṣūr Farāmurz, from Iṣfahān and gave him the governorship of Yazd. After that, the Kākūya dynasty was transferred from Persian Iraq to Fārs.

Abū Kālinjār Garshāsf b. Muḥammad Kākūya was governor of Hamadān. The Seljuks took it in 437 (1045–1046) and he went to Shiraz. The governor of Ahwāz died in (4)43 (1051–1052). Amīr ʿAlī b. Farāmurz became governor of Yazd after his father. In 536 (1141–1142) he fell in battle against the Qarakhitay. He had a special relationship with Sultan Sanjar. He gave the governorship of Yazd to one of his followers, Sām b. Wardān, by way of viceregency, giving him one of his daughters in marriage. This [M 281] Amīr ʿAlī had several good works. One of them is the wall he built in 515 (1121–1122) around the ʿAlī Riḍā shrine in Mashhad.

Atabegs of Yazd

When Sām b. Wardān became governor of Yazd by way of vicegerency through one of the daughters of Amīr ʿAlī, he built a lofty madrasa. As his was a mild character, he left state affairs to his brother, ʿIzz al-Dīn Lashkar before dying in 590 (1194). Lashkar was brave and valiant. At times he was governor of Shiraz or Iṣfahān on behalf of the Seljuks. He died in 604/1207 and was succeeded by his son, Wardān Zūr, who ruled for twelve years.

After him his brother, Manṣūr Sipahsālār, became governor; he was known as Sultan Quṭb al-Dīn Khaljī. Having withdrawn from the world, he died in 616 (1219–1220) and was succeeded by his son, Maḥmūd Shāh, who ruled for 13 years. After his death, his son, Salghur Shāh became governor. He built Salghurābād. Next came his son Ṭughī Shāh, who died in 670 (1271–1272). His son ʿAlā al-Dawla assumed the throne.

In 673 (1274) a great flood came and destroyed Yazd. The inhabitants of the city built houses on top of a hill, which became known as the Sar-i Jamʿ district and is now called Sarjam. One month later ʿAlā al-Dawla died and his brother Yūsuf Shāh became governor. He made efforts to repair the damage caused by the flood and restore the city walls; but he was unable to govern because he was given to drink. He sent troops against Sultan Ghāzān. When Yūsuf Shāh left towards Sīstān, the emirate of Yazd came to an end and control of Yazd was assumed by the Mongol divan.

Salghurids

They are also called the Atabegs of Fārs. They descend from Mawdūd b. Salghur who came with the Turkic tribe of Salghur to Khurāsān and settled there. With the advent of the Seljuks he entered their service, becoming a chamberlain. His tribe went to Fārs, summering and wintering at Mt. Kīlūya.

In 552 (1157–1158) [R 106a] the governor of Fārs, Yāwzāna, was killed when he revolted against Sultan Masʿūd. Sunqur b. Mawdūd, who was the Atabeg bringing up the younger son of Sultan Muḥammad, was appointed governor of Fārs. Having held the office for three years, he was imprisoned by a ruse of his brother, Takala. This Sunqur has a Friday congregation mosque and a hospice in Shiraz and Takala built an inn.

After Sunqur, his brother Muẓaffar Zangī became governor. He defeated the army of Alp Arslan. He died in 570/1174 and was succeeded by his son, Muẓaffar Takala, who died in (5)90 (1194). The merchants' inn in Shiraz was built by him. Sunqur's son Ṭughril quarreled with him and there were several battles between them, the country of Fārs becoming ruined during their strife. Finally Ṭughril fell captive and in (5)99 (1202–1203) was blinded.

Takala was succeeded by his brother, Muẓaffar Abū Shujāʿ Saʿd b. Zangī, who made the kingdom of Fārs flourish with his justice. He was an intelligent and just padishah, as mentioned and praised by Shaikh Saʿdī. Dying in 618 (1221–1222), he was succeeded by his son, Muẓaffar Abū Bakr, who followed the path of his father, and the fortune of this dynasty rose further. He died in 658/1260–1261, to be succeeded first by his son, Muẓaffar Saʿd, then by the latter's son, Muḥammad, and then by his cousin, Muḥammad Shāh, who each ruled for one or two years.

The Salgurids were eliminated by the Mongol onslaught in 66 (1263). The daughter of Saʿd b. Abī Bakr, [M 282] Anas Khātūn, was appointed for one year on behalf of the Mongols. When she too died, the dynasty came to an end and the country of Fārs came under Mongol control. Later it went into the hands of the Muẓaffarids, remaining under their rule until the advent of Tīmūr. Afterwards it was governed by the governors (*nüvvāb*) of the Timurids, and later came under control of the shahs (Safavids). It is still under the rule of the Persian shahs.

Chapter 31: Country of Khūzistān and Lūristān

Khūzistān

Khūz is the name of a people whose land is called Khūzistān and Ahwāz. It is a country with a flat plain and abundant rivers. Its borders (are as follows): in the west, Wāsiṭ in Arab Iraq; in the south, the seashore extending from ʿAbbādān to Mahrūyān (Mahrūbān) and Dawraq and the borders of Fārs with that side forming an arc; in the north, from Iraq to Ṣaymar, Karakh, Rūdbār, the mountains of Lūr to the borders of Jabal; in the east, the Tāb river, which flows between Fārs and Iṣfahān and, after some of the rivers of Khūzistān flow into it, empties into the sea at a place near Mahrūyān. This country falls between Fārs and Basra. It is surrounded by Arab Iraq, Kurdistan, Luristan, and Fārs. Within its borders, from Mahrūyān to opposite ʿAbbādān, there is no sea but the bay that enters from the Persian Gulf. And there are no mountains and sand dunes other than Tustar and Jundīsāpūr; most of its territory is a gentle, flat plain. There is no snow and ice except in the environs of Rāmhurmuz. Drinking water is not gotten from wells. The lands near the Tigris are well watered, whereas the lands distant from it are dry. The climate is hot and dates grow in most places. Many diseases are prevalent; but fruits and grains are abundant and sugarcane plantations are found everywhere.

Especially the people of ʿAskar Mukram speak both Arabic and Persian. The nobility use the Khūzī language. Their manners are like those of the people of Iraq. Most of them are greedy, bad-tempered, sallow-complexioned, skinny and have wispy beards. The *muʿtazila* school has currency there. Jews and Zoroastrians are many, Christians are few. The people of Ahwāz being mainly Shīʿīs and Ḥanafīs, they do not cover their private parts in the public bath. They eat rice bread and ride oxen.

According to the *Nuzha*, this country has twelve cities. In the time of the caliphs the state dues coming from here were more than 300 tomans; later they fell to 1½ toman.

Tustar: the capital of Ahwāz. This is the Arabicized form; the common people say Shūstar. It is located at 86½ degrees longitude and 31½ degrees latitude. It is an ancient city on the aforementioned river (i.e., the Karun). According to the *Qāmūs*, its city wall was the first to be built after the Flood. Because the city was on a high place, Shābūr built a large dam on the aforementioned river [R 106b] and raised its water one mile to the city. Therefore they called that dam Shādirwān ("Fountain").

According to Ḥamd Allāh, Hūshang built this city; when it was in ruins, Ardashīr restored it, giving it the form of a horse. Shābūr built a large dam for the water of the city and made the channel of Dashtābād flow from that dam, for the center of Tustar province is on that desert (*dasht*). According to the *Masālik al-mamālik*, there is no dam that is more solid; but the Band-i Amīr built by ʿAḍud al-Dawla in Fārs is even greater and more solid than this.

The city has four gates. Its climate is very hot and the simoom blows during most of the summer. As for its water, [M 283] it is very fine and digestive. So even when its people eat coarse foods, it does them no harm. Cotton, sugarcane and grains are abundant. Fine stuffs and brocades are woven here. There is a certain type of rice with a pleasant smell called *panj angusht* ("five fingers"). This rice even when it is scarce (and expensive) is preferable to the rice of Shiraz even when it is (abundant and) cheap.

The people are mostly of a black complexion and skinny. Few among them are affluent. The cemetery is in the middle of the city; the Friday mosque is in the middle of the bazaar.

The city has four hunting grounds: 1) Rakhshābād, which is 15 parasangs long and 12 parasangs broad; 2) Zawraq, which is 20 parasangs long and 10 parasangs broad; 3) Mashhad-i Kūfī, which is 10 by 6 parasangs; 4) Ḥawīza, which is 20 by 12 parasangs. All are good pasturages.

Since the heat is extreme, strangers are unable to tolerate it after the fortieth day of spring. Grain crops are reaped when the sun is in Taurus; if they are left in the ear until the sun reaches Gemini, they go to ruin.

Ahwāz: an ancient city and a large district, located at 85 degrees longitude and 32 degrees latitude. It is also called Hurmuz Shahr and Sūq Ahwāz. In former times it was the capital of Khūzistān. This city had two parts, the Persian side and the Iraqi side. It was like an island surrounded by the river (the Karun) and there was a bridge built of baked bricks and called Qanṭara-i Hinduvān that linked the two parts. Overlooking the bridge was an elegant mosque built by ʿAḍud al-Dawla. Using mills and water wheels they raised the water to the high ground of the reservoirs of the city. The water gathered behind a dam below and became like a lake. It was diverted to three places and irrigated many fields. As in Baghdad, the waters of Ahwāz were navigable from the low part of the river as far as Basra. There are many mills on the rivers. Iṣfahān is eighty parasangs from here, ʿAskar Mukram is one stage's distance.

One of the characteristics of this land is that any kind of perfume loses its scent. The air is thick and humid. Fever is constant and people's faces show no trace of redness.

Cities of Ahwāz

ʿAskar Mukram: It is 10 parasangs west of Ahwāz. It is a city with many small scorpions. Sugar and *fānīd-i ʿaskerī* (a kind of candy) are produced here. In ancient times it was a village. When Ḥajjāj sent Mukrim b. al-Ghurar with an army to fight against Khurdād b. Fāris, he set up camp at the aforementioned village and stayed there for a while, and the soldiers erected buildings there. Later it became a city. The author of the *Taqwīm* writes that it is the only city in Ahwāz that was founded (after Islam). But according to the *Nuzha* it is a large and ancient city, first built by Ṭahmūrath, then by Shābūr, and there is no city in Khūzistān with a more pleasant climate. In the *Haft Iqlīm* it is written that this city is now in ruins. Tustar is eight parasangs from here.

Ṭīb: A city between Wāsit and Ahwāz, located at 84 degrees longitude and 33 degrees latitude. A village by the name of Qurqūb is approximately seven parasangs (from here). Al-Ṭībī, the commentator on the *Kashshāf*, comes from this city.

Sūs: An ancient city 10 parasangs from Qurqūb. Its castle was built by Shābūr, who settled here the people he exiled from Rūm (i.e., Roman captives), and he called the place Shābūr Khūra. Oranges grow in its gardens. The tomb of Daniel is west of it. He is said to have resided there as a captive of Nebuchadnezzar. [M 284] At the time of the (Muslim) conquest a coffin was found there. The people revered it as Daniel's coffin and at times of draught they would take it out and ask for his intercession. Abū Mūsā al-Ashʿarī had a cellar made of stones and mortar on the bank of the river flowing by the city, placed (the coffin) inside it and caused the river Sūs to submerge it. Saying that the corpse of a prophet should not be accessible to people, out of reverence he made it invisible. This city is also now in ruins.

Jundīsābūr: Arabicized from Gund-i Shābūr. Located at 85½ degrees longitude and 31½ degrees latitude, it is the chief city of Khūzistān, strongly fortified and extensive. Founded by Shābūr b. Ardashīr, it was renovated and enlarged by Dhū 'l-Aktāf ("the man of the shoulder-blades," i.e. Shābūr II). Its climate is hot and unpleasant. There is much sugarcane and crops, and dates are also abundant. Tustar is eight, Sūs six parasangs from here.

Jubbā: A city with many date palms eight parasangs from ʿAskar Mukram. Sugar is abundant here. It has a large district, rivers and villages. Abū ʿAlī Jubbāʾī, the great *muʿtazilī* (theologian), comes from here. [R 107a] His name is with one *b* or two (Jubāʾī or Jubbāʾī).

Rāmhurmuz: One of the cities and districts of Ahwāz, built by Hurmuz b. Shābūr and located at 86 degrees longitude and 30½ degrees latitude. A place with a hot climate, an abundance of grain crops, cotton and sugarcane, it is approximately 19 parasangs from Ahwāz. Fine stuffs are made here. ʿAḍud al-Dawla built an elegant mosque in its bazaar. The date palms of this city are amidst the gardens; they cannot be seen. Salmān-i Fārisī and a group of noble people come from this city.

Rustāq al-Zuṭṭ, meaning, bazaar of Zūt: A district of Ahwāz 7 parasangs from Rāmhurmuz. The climate is very hot, but it is a prosperous place.

Hāyizān: A prosperous district of Ahwāz. Both Hawmat al-Batt and Hāyizān are districts of summer pasturage with a pleasant climate by a river.

Ḥawīza: A mid-sized town, built by Shābūr II. It is hot; however, as its climate is better than the other cities of Khūzistān, it has an abundance of grains, cotton and sugarcane. The *Nuzha* has a separate entry on this city, not under Ahwāz.

Dizfūl: It is located at 84 degrees longitude and 31½ degrees latitude. Shāpūr built it near Jundīsābūr on both banks of the river. He built a large, 42-arch bridge across the river, with a breadth of 15 cubits. It was called the Ābdīsak Bridge and the city was named after it. He built a 50 cubits high water wheel below the city, which distributed the water of the city. On the outskirts of the town there are trees called Zarrīn Dirakht ("golden tree") with yellow flowers. They have many blossoms but do not yield fruit.

Dastgir: It is written Dastwā in the *Taqwīm*. It is a city of Ahwāz, very hot and putrid. It is said to have been built by Hurmuz b. Shābūr. It has a strong fort made of clay.

Sūq-i Arbaʿā ("Wednesday Market"): A town built on both banks of one branch of the river that branches at Ahwāz. In the middle of it there was a lofty wooden bridge. Boats would pass under it. The Iraqi side being more prosperous than the Persian side, the Friday mosque was built on the Iraqi side.

Mashruqān: A mid-sized city on the Dashtābād River, located at 85 degrees longitude and 31 degrees latitude. A large district belongs to it. It is very hot.

Ṭarārak: A mid-sized city. [M 285] Sugarcane is more abundant here than elsewhere.

Sanlīl: one of the districts of Khūzistān, four parasangs from Arrajān.

Rujān: A town near Sūs. It is recorded in some books that Qubād b. Fīrūz founded the city by the name of Rujān between Fārs and Ahwāz and settled there the people whom he had driven out of Hamadan.

Ḥiṣn Mahdī: A town located at 84½ degrees longitude and 30½ degrees latitude, where the Tustar river flows into the Persian Gulf. Being a frontier and a port close to the sea, al-Mahdī (Abbasid caliph, r. 775–785) built its fortification. It has inns and a Friday mosque on the bank of the river and abundant date groves and fields. The rivers of Khūzistān come from Ahwāz and Dawraq, and, uniting by this city, they ebb and flow with the ebb and flow of the sea. Sūq-i Arbaʿā is 16 parasangs from here.

Dawraq: A small city and castle in Ahwāz province, located at 85 degrees longitude and 30½ degrees latitude. It is on the Iraqi side of the river. It has an extensive market. It is on the route of Hajj pilgrims from Kirmān and Fārs. ʿAskar Mukram is approximately four stages hence.

Bandar Mājūr: A city two stages from Dawraq, located on the Persian Gulf at 85½ longitude and 30½ altitude.

Bāsiyān: A mid-sized city on the river Tustar. It is also called Bāsān. Ḥiṣn Mahdī is two stages hence. Similar to Dawraq, it can be approached both by land and by the river.

Matūth: A town between Qarqūb and Ahwāz.

Karkha: A small town. A fair is held there on Sunday. It has a fortress and a garden.

Baṣnī: A town on the bank of the Dujayl river. It has two strong fortresses and an open-air prayer-grounds in the middle. Fine curtains and woolens are manufactured here.

Greater and Lesser Mabādir: Two prosperous districts where dates are abundant.

Īdaj: A small town in the Rāmhurmuz district. There is snow (in winter) which they take to Ahwāz. They get water from the river of Shiʿb-i Sulaymān. Melons are abundant. Since it is closed off on the northern side, it has a bad climate.

Lūr. According to the *Taqwīm*, it is a district (*rustāq*) of Khūzistān; but actually it is a district (*nāhiye*) and mountain range between Tustar and Iṣfahān, with a length and breadth of six stages. The people are Kurds. It has autonomous rulers. In the past it belonged to Khūzistān, but now it is separate and is called Lūristān. It has two parts, Greater and Lesser Lūr. Lesser Lūr is a substantial province. In the time of the Atabegs the state dues from it amounted to 100 tomans; those from Greater Lūr were close to that.

Lūrkān. A small town in Lūr. It has a bad climate and unwholesome water. Grapes are abundant.

Sūq-i Asal. One of the districts of Ahwāz. There is a mountain nearby. Fire continually breaks out of it, illuminating the surroundings by night, while during the day smoke can be seen reaching to the sky.

Aside from these, some texts mention cities by the names of Kalīvān, Iram-i Sulaymānān, Bāzār-Ḥāyizān, Ḥal-Banān as located in Ahwāz, but no details are known.

Rivers

Most of the rivers of Khūzistān are navigable. The largest is the Tustar (i.e., Karun). It rises in the Zard and Lūr Mountains, reaches Tustar in 30 parasangs, a short distance; hence it flows cold. Its water [R 107b] is pleasant and digestive. Shābūr built above it the Shādirwān ("Fountain"), referring to a dam, and distributed the water in three directions. And he sent the water all around Tustar, four channels from the west and two channels from the east. These two (i.e., the eastern and western channels?) unite at the border of [M 286] ʿAskar Mukram. Then the rivers Dizfūl and Karkha flow into it, after which, according to the *Nuzha*, it flows by Ahwāz, and into the Shaṭṭ al-ʿArab, i.e., the Tigris. It is 80 parasangs long. The report in the *Taqwīm* that the Tustar flows into the Persian Gulf near Ḥiṣn Mahdī is also true, for it is recorded in the *Aḥsan al-taqāsīm* that the Ahwāz (i.e., Karun) and Tigris flow into the sea. The two rivers are separated by desert. In the past people went by boat down the river from Ahwāz to the sea and up the Tigris to Ubulla. Since this route was dangerous, ʿAḍud al-Dawla made a large, 4-parasang long canal from the Ahwāz (i.e., Karun) to the Tigris. Boats now use this. Therefore this river has two branches, one that flows into the Tigris and another that flows into the sea.

Dizfūl: It rises in the mountains of Greater Lūr. After it flows by Jundīsābūr, Dizfūl and Mashruqān, it flows into the Tustar (i.e., Karun). It is 60 parasangs long.

Karkha: It is also called the Sūs. Rising from Mt. Arwand, it is joined by the Dīnawar, Kūlkū, Sīlākhūr and Khurramābād rivers, flows through Ḥawīza province, and after joining the Dizfūl and Tustar rivers, it flows into the Shaṭṭ al-ʿArab. Together with the Shaṭṭ al-ʿArab, it is 120 parasangs long.

Mashruqān: It is a branch of the Tustar. It stretches from Tustar to ʿAskar Mukram, where there is a large bridge over it. People take this river to Ahwāz. They go 6 parasangs on the river, after which the water peters out and only the riverbed is left, so they go two more miles on land. The river comes to an end watering sugar cane fields. It is said that this riverbed is the most fertile land in Khūzistān.

[Mountains]

Kūh-i Zarda ("Yellow Mountain") is in Lūristān. The Jū-yi Sard ("Cold Stream"), which is the source of the Zandarūd in Iṣfahān, and the Tustar (i.e., Karun) also rise here.

There is a mountain in Lesser Lūr called Hawīn. Ḥamd Allāh reports that there is a marcasite mine there.

Distances and roads in Khūzistān

There are two roads from Pārs to Iraq, one through Basra, one through Wāsiṭ.

The Basra road: Arrajān > Asal, two stages > a village called Vīrān, 1 stage > Dawraq, 1 stage > Bāsān, 1 stage > Ḥiṣn Mahdī, two stages, which can also be traversed by sea, since it is easy to go from Dawrāq to Bāsān by river > Sāb, two stages > the bank of the Tigris, which is the end of the border of Khūzistān and where the Tigris can be crossed.

The Wāsiṭ road: Arrajān > Bāzār-i Sanlīl, 1 stage > Jundīsābūr, 1 stage > Qurqūb, 1 stage > Ṭīb, 1 stage > district of Wāsiṭ.

ʿAskar > Ahwāz, 1 stage > Dawraq, three stages.

There is another road from ʿAskar to Wāsiṭ that passes through Tustar. There are four stages from ʿAskar to Īdaj, three stages from Ahwāz to Rāmhurmuz. These three form a triangle.

ʿAskar > Sūq Arbaʿā, 1 stage > Ḥiṣn Mahdī, 1 stage.

Sūs > Baṣnī or Bardūn, 1 stage each > Matūth, 1 stage.

Tustar > ʿAskar, 1 stage.

Ahwāz > Dawraq, four stages.

Ḥiṣn Mahdī > Ubulla, ten parasangs.

Dawraq > Basra, 1 stage.

Kings

As recorded in the *Haft Iqlīm*, Khūzistān was conquered during the reign of caliph ʿUmar (r. 634–644). Peace was made with its then ruler, Hurmuzān, and he went to Medina and converted to Islam. According to one account, some of these lands were conquered through the efforts of Abū Mūsā al-Ashʿarī.

[M 287] Afterwards Umayyad and Abbasid governors were ruling. During their tenure they seized the states that appeared in the countries of Fārs and Jabal. In particular, a few polities appeared in Khūzistān and Lūristān and became independent. One of them is the state of Greater Lūr, also called āl Faḍlūya or the Greater Atabegs, since the province of Lūristān was divided in two and given to two brothers in the year 300 (912–913). One part was called Greater Lūr, the other part Lesser Lūr.

[Lūristān]

Greater Lūr. Fatḥ al-Dīn Faḍlūya, who was from the progeny of Kay-Khusraw, lived in western Syria, and migrated with his following, his son ʿAlī and ʿAlī's son Ibrāhīm, to Mafārqīn and stayed there for 60 years. His son Muḥammad went to Azerbaijan and filled some offices for 45 years. After him, his son ʿAlī was commander of the army for 5 years and was killed in the year 584 (1188–1189). His son Muḥammad ruled for 10 years, (while) his brother Hazārāsp was in the caliphal seat (i.e., Baghdad). [R 108a] [Plate 42]

He was granted most lands of Lūristān. He came (to power) and ruled for 34 years. He was succeeded by ʿImād al-Dīn Abū Ṭāhir b. Muḥammad, who ruled until he died in 626 (1228–1229). After that his son, Nuṣrat al-Dīn Hazārāsp became ruler. He fought against Takla b. Zangī's army and won. After he died in 646 (1248–1249), his son Muẓaffar al-Dīn took Takla's place and defeated Saʿd b. Zangī. He also fought against the ruler of Lesser Lūr, Ḥusām al-Dīn Khalīl, and routed him. In 656 (1258–1259) Hulagu killed him. When his brother Arghūn became ruler, he remained on good terms with the Mongols. Lūristān flourishing in his time, he died having ruled for 15 years. His son Yūsuf Shāh became ruler. Since he was brave and valiant, Abaqa perpetuated his rule. After he passed away in 680 (1281–1282), his son Afrāsiyāb sat on the throne. Resisting the Mongols, he routed their army sent against him. Finally in 696 (1296–1297) he was captured and killed. Afterwards his brother Nuṣrat al-Dīn Aḥmad became ruler. He ruled wisely and remained on good terms (with the Mongols) until he died in 733 (1332–1333). He left the kingdom to his son, Yūsuf Shāh who, after a 23-year long rule, was captured and imprisoned by Muḥammad Muẓaffar. Since Yūsuf Shāh had no son, their lineage came to an end with him.

Lesser Lūr: It is also called Āl Khūrshīd or the Lesser Atabegs. They paid the state dues to the sultans of Iraq, governed in the name of the capital (i.e., Baghdad). In 580 (1184–1185) Shujāʿ al-Dīn Khurshīd became governor (on behalf of) the Chingisids. Having subjugated Lūristān with his good conduct, he passed away in 621 (1224–1225). His nephew Sayf al-Dīn Rustam became governor. After that his brother Abū Bakr killed him and ruled for some time. He was poisoned by his wife and was succeeded by his brother Garsāshf (Garshāsp). His cousin (of

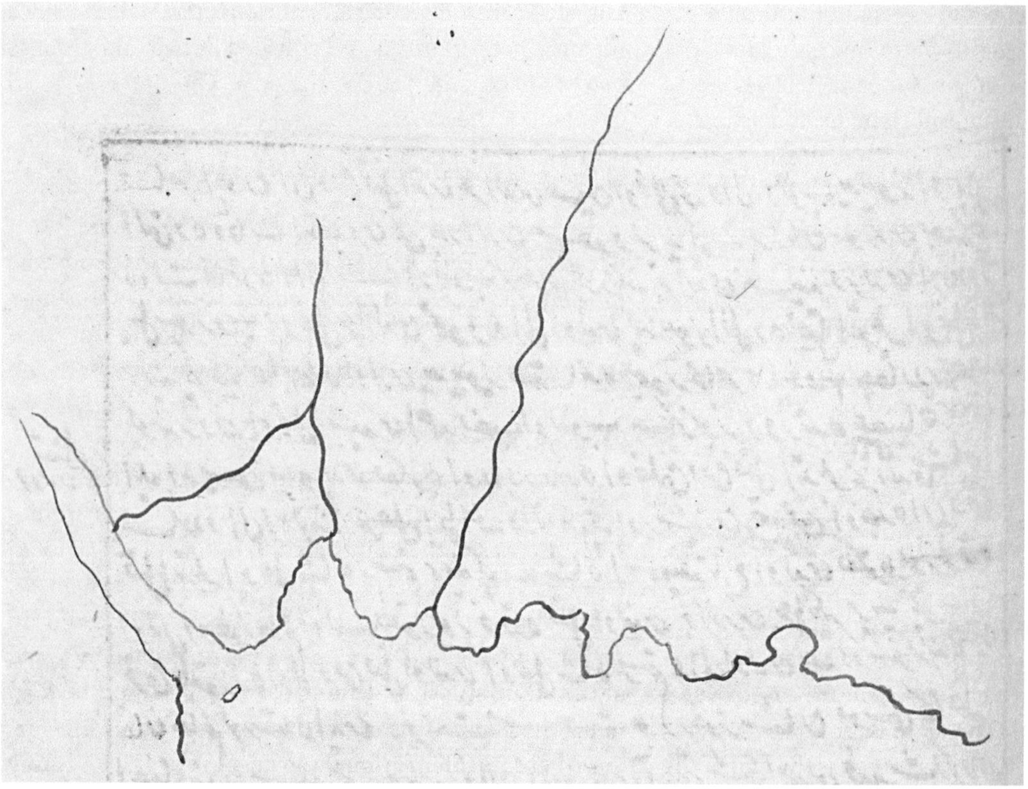

PLATE 42
Map of Khūzistān and Lūristān, R 108a
TOPKAPI PALACE LIBRARY, REVAN 1624, WITH PERMISSION

his paternal uncle) Khalīl b. Badr defeated him and conquered the country. He struggled against Sulaymān Shāh and was killed in 640 (1242–1243). His brother Masʿūd succeeded him, ruling for 18 years. He was a scholar and legal expert. After his death his nephew Shah Tāj al-Dīn became the governor, who was put to death by Abaqa in 677 (1278–1279). Then his two sons, ʿIzz al-Dīn Ḥusayn and Falak al-Dīn Ḥasan, ruled jointly. [M 288] After they died one after the other in 692 (1293–1294), his (i.e. Shah Tāj al-Dīn's) brother Jamāl al-Dīn Khiḍr was governor for a year. He was put to death (by Abaqa?) at a hunt and was succeeded by Ḥusām al-Dīn ʿUmar from the descendants of Badr, who was shortly dismissed. (The next governor,) Ṣamṣām al-Dīn had killed (a certain) Ilyās and was put to death by Ghāzān in retaliation. [R 108b] ʿIzz al-Dīn Ḥusayn's son Muḥammad became governor and died in 716 (1316–1317). He was succeeded by his cousin (of his paternal uncle) Badr al-Dīn Masʿūd who ruled for some time. After that his wife Dawlat Khātūn was appointed; then (the office) was given to the woman's brother Ḥusayn. After he ruled for 14 years, his son ʿIzz al-Dīn Maḥmūd became governor. He was killed in 750 (1349–1350). He was succeeded by his son ʿIzz al-Dīn, who reigned supreme. In 790 (1388–1389) Tīmūr besieged him, removed him and sent him to Samarqand. His son, Sīdī Aḥmad, was appointed in 804 (1401–1402), when his father was put to death on suspicion of rebellion. He was flayed. His son was also killed in [8]14 (1411–1412). ʿIzz al-Dīn Shāh Ḥusayn became governor, who laid his hands on neighboring lands. He was also killed in [8]73 (1468–1469). He was succeeded by his son Shah Rustam, who became a follower of Shah Ismāʿīl. Then his son Oghuz became ruler and was killed in 940 (1533–1534). He was succeeded by his brother Jahāngīr but he was also killed in [9]49 (1542–1543). His son Shah Rustam became governor. At the time of the Baghdad campaigns, Ḥusayn Khān was the ruler of Lūristān.

The Mushaʿshaʿ dynasty in Khūzistān. Sayyid Muḥammad b. Sayyid Falāḥ, a descendant of Mūsā Kāẓim, was from Baghdad, and became a disciple of Shaikh Aḥmad b. Fahd. The shaikh had a book on the occult sciences. When he (i.e. the shaikh) was dying, he commanded one of his servants to cast this book into the Euphrates. Sayyid Muḥammad got control of the book, and through occult means he made some ignoble Arabs his disciples. Their splendor gradually increased so much that people were flocking around them. They could bend a drawn sword by holding it against their belly. Becoming more and more famous by the day, he announced his mission. He took entire Khūzistān under his control. After him his son ʿAlī was the governor. He plundered Arab Iraq. He pretended that the spirit of ʿAlī b. Abī Ṭālib was incarnated in him, and his followers believed it. Finally, when he was killed in the vicinity of Kūh Gīlūya, he was succeeded by his brother Sayyid Muḥsin. During his reign the dynasty gained great currency and his followers thought him to be divine. They abolished the laws of Islam and took the path of heresy. After that his sons, Sayyid Ayyūb and Sayyid ʿAlī became governors. When Shah Ismāʿīl captured Baghdad, he took Khūzistān from their hands and killed both of them in 914 (1508–1509). After this Sayyid Falāḥ b. Muḥsin was appointed. He was succeeded by his son Sayyid Badrān, who submitted to the Shah. (But) when Sajjād b. Badrān became governor, he turned away (from the Safavids) again.

THE RAʿNĀSHIYYA DYNASTY. Raʿnāsh is one of the villages of Dizfūl. Qivām al-Dīn Raʿnāshī was the teacher of Sayyid Muḥsin's children. One of his sons, Shaikh Muḥammad, was the governor of Dizfūl, but was eventually killed by his nephew, Khalīl. He was succeeded by Khalīl, who quarreled with Sayyid Badrān. When he even showed contempt for the shah, he (the shah) sent against him the emirs of Mt. Gīlūya and Badrān. [M 289] Since the shah died at this time, Khalīl escaped and ruled until he died in 945 (1538–1539). His son ʿAlā al-Dawla became ruler. Considering him to be in opposition, Shah Ṭahmās came to Dizfūl in 949 (1542–1543) and ʿAlā al-Dawla fled to Baghdad. According to Ghifārī, the dynasty ended with him. Accordingly, all of Khūzistān and Lūristān is (now) under the rule of the Persian shahs (i.e., Safavids).

Chapter 32: Clime of Jabal (i.e., Jibāl) or Persian Iraq

Iraq either means a place with many trees or it is the Arabicized form of Īrān Shahr. If mentioned without qualification it signifies Arab Iraq, i.e., the region of Baghdad. Being its neighbor and adjacent to it, this country was named Persian Iraq. It is bordered in the west by some of Arab Iraq and Azerbaijan, [R 109a] in the south by Khūzistān and Lūristān, in the east by the desert of Khurāsān and Fārs, and in the north by Azerbaijan and Daylam. Its divisions, according to direction, are Zanjān in the north, Abhar and Iṣfahān in the south, Rayy and Qazwīn in the east, Hamadān in the middle. Originally the capital was Iṣfahān; then it became Hamadān; now it is Qazwīn. It is considered the throne of the Persian shahs (i.e., Safavids).

Ḥamd Allāh says that this clime has 40 cities. The climate of most of them is moderate; partly it is cold or warm. Its length is 160 parasangs from Safīdrūd to Yazd, its breadth 100 parasangs from Gīlānāt to Khūzistān.

In the past, four of its cities were much esteemed: Iṣfahān, Hamadān, Qum and Rayy. As reported from his grandfather Amīn al-Dīn Naṣr Mustawfī, during Seljuk rule the state dues were 2,520 tomans. But today there are only three great cities: Iṣfahān, Qazwīn, and Hamadān.

Iṣfahān: It is a great and ancient city among lowlying mountains, located at 85 degrees longitude and 33½ degrees latitude. Being a staging area of the Sasanian army, it was called Sipāhān ("Cavalry"). Later it was Arabicized as Iṣfahān. According to the *Nuzha*, Iṣfahān was originally four villages: Karrān, Kūshk, Jūbāra, and Dasht. These were villages with arable lands founded by Ṭahmūrath and Jamshīd. First Kayqubād I made Iṣfahān his capital. People came here and it gradually became a great city. When Nebuchadnezzar expelled the Jews (from Palestine) most of them came to this city and, pleased with its climate, settled here. They built houses at a place one mile away, which is now called Yahūdiyya. It is like a separate town.

Rukn al-Dawla Ḥasan b. Buwayh surrounded the old city with a wall of 21,000 cubits in circuit and twelve gates. The city has twelve districts, each of them resembling a separate town. Similar to Syria, the houses are made of clay, which, however, looks like stone. The author of the *Aḥsan al-taqāsīm* says that the Friday Mosque in the marketplace has a minaret 70 cubits high. It is entirely built of clay, and there is nowhere a breech in it. Some of the bazaars are covered, some are open aired. The weather is moderate in both summer and winter; rain, storms, earthquakes, lightning, vermin and scorpions are rare. When buried (i.e., stored underground?), grains remain unspoiled. Chronic diseases and plague are rare, and the soil preserves the dead.

The river Zandarūd comes from the *qibla* direction (i.e., south) and flows by the outskirts of the city. Some branches are diverted and flow into it. Aside from that, the wells are 4–5 cubits (deep). [M MAP] [M 290] The water in these wells springs up fresh and is hardly less than the flowing water.

Grain crops and other produce are always middling. Every seed brought from somewhere else and sown here yields no less than in its original place, except for pomegranates, which do not grow well. However, this is also indicative of how good Iṣfahān's climate is, as pomegranates are proven to yield well only in putrefactive climate. Fruit prices are always low here. Most fruits are fine and tasty; especially watermelons and apples are very sweet and good. They are exported to Turkey and India. Breeding livestock is also good, and there are large pasturage-lands and hunting grounds. A large meadow is Balāsān. There are fine parks (or orchards, *bāğ*) and gardens, especially the Naqsh-i Jahān Park, which is the most famous park of Iran.

There are many madrasas, dervish convents, charity buildings, mosques and Friday mosques. One is the madrasa of Sultan Muḥammad-i Seljuqī, the threshold of which is a large idol. The Hindus wanted to buy it for its weight in gold, but the aforementioned sultan did not give it to them and for the sake of religion he made it the threshold of the madrasa.

Most inhabitants of the city are of white complexion and courageous. They are rarely free from struggle, discord and civil strife. It is related in the *Rawḍ al-miʿṭār* that in the Mongol period, the Shāfiʿīs and Ḥanafīs having been fighting zealously against each other since ancient times, the Ḥanafīs brought in the Mongols and they massacred the Shāfiʿīs. The pleasantness of the climate is in contrast with the city's discord. Complaining of the city's inhabitants, Kamāl al-Dīn Ismāʿīl Iṣfahānī wrote the following verses [in Persian]:

As long as Jūbāra has been the gate of the desert
There has been no remedy for killing and strife.
O Lord of heaven and earth!
Send a bloodthirsty padishah,
Let him make the gate of the desert like the desert
And divert the stream of blood away from Jūbāra.
May its population increase;
May He smash them into a hundred pieces!

And in hadith, as reported by ʿAbd Allāh b. ʿAbbās: "The Antichrist comes from among the Jews of Iṣfahān, until he comes to Kūfa."

It is reported by those who have seen this city that it has two sections: Jahūdistān ("Jew town"), which is bigger than Hamadan; and Çulhalı ("Weavery"), which is as big as Baghdad. houses are located sparsely, each with a garden of 1–2 donums and flowing water. Since the soil is salt marsh, nothing grows without cultivation. However, there is no other city in the Jibāl with blessings as abundant as this. Iṣfahān being the emporium of Fārs, Kūhistān, Khurāsān and Khūzistān, merchants from these provinces gather here. Matchless cotton stuffs and other textiles are woven and exported abroad. Iṣfahān collyrium is also famous.

Allāh Verdi Khān built a large, 17-arch bridge over the river Zandarūd. In the middle of each arch he built a pavilion, and the bridge became the promenade of the city. When Shah ʿAbbās saw it and was pleased with it, the Khān gave it to him as a present. Shah Ismāʿīl made a minaret there out of animals' heads.

On the south side of the square of Iṣfahān, the shrine of Aaron is famous. Shah ʿAbbās erected a matchless elegant bath (there) and he is reported to have built several (other) lofty edifices as well.

The state dues from Iṣfahān used to be 35 tomans. Together with its districts it was 50 tomans. The city taxes alone amounted to 70 *yük* (1 yük = 100,000) of *akçe*. The city had 8 districts and 411 villages. [M 291] This district, 80 parasangs in length and breadth, has 17 markets and large villages (*rustāq*). Its districts are as follows:

- Chay: It is in (the vicinity of) the city and comprises 75 villages, including Ṭihrān, Mārbānān, Jāvān [R 109b] and Shahristān. This Shahristān is also called New Iṣfahān. It was founded by Alexander the Greek and restored by Fīrūz the Sasanian.
- Mārbīn: It comprises 58 villages. In the following verse a poet intimates that since in most places the gardens (or orchards, *bāġ*) adjoin one another, the entire district is like a garden [Persian verse]: "Mārbīn is a copy of (the garden of) Iram / The sun is like a silver coin in it." Ṭahmūrath built a castle, Bahman a fire temple, in this district.
- Karārij: It comprises 36 villages, the biggest of them being Ashkāvand and Qabrāwān. Gardens come after one another in this district also.
- Qahāb: It comprises 40 villages. People get water from underground channels; hence its name.[36]
- Parḥawār: It comprises 32 villages, the biggest of which is Dīh-i Ḥur. Here also drinking water is obtained from underground channels, while in all the other districts people drink the water of the Zandarūd. Bahman b. Isfandiyār had a fire-temple in Dīh-i Ḥur. The inhabitants of this district were fire-worshippers up to the reign of Shah ʿAbbās. In his time the Magians removed the fire and went to India. Jalāl al-Dīn took it from them and put it in the treasury.
- Alkhān: It comprises 20 villages, the largest of which were Kūpān, Diranjān and Kalīshād.
- Bara'ān: It comprises 80 villages, the largest of which were Ashkshān, Dabīristān, Fashārān and Kūhān.
- Rūdastar (Rūdasht): It comprises 60 villages. The biggest towns were Fārifā and Qūrṭān. The other villages also were like towns, each of them composed of 1000 house, markets, madrasas, dervish convents, mosques, and baths.

There are many eminent people from Iṣfahān. One is the very learned Shams al-Dīn Muḥammad b. Maḥmūd, who died in 688 (1289–1290). A commentator on the *Tajrīd*, he was an old Iṣfahāni. Another is the very learned commentator Shams al-Dīn Abū al-Thanā Maḥmūd b. ʿAbd al-Raḥmān al-Iṣfahānī, who died in 749 (1348–1349). The *Sharḥ ṭawāliʿ wa-tafsīr* is his work. A third is Kamāl al-Dīn Ismāʿīl b. Muḥammad, who died in 635 (1237–1238) and was a famous poet.

36 Presumably *qahāb* is a local variant of *qanāt*. Cf. *Nuzha*, 51, lines 1–2: *va āb-i kārīz mī-khwarand ba-dān sabab qahāb mī-khwānand.*

Rayy. Located at 87 degrees longitude and 35½ degrees latitude, Rayy was a great ancient city. Now it is mostly in ruins. It used to be called "mother of the cities of Iran" and "shaikh of cities." It is closed off (by mountains) in the north. The climate is hot, the air putrid. It is a place with unpalatable water and frequent plague. Its construction goes back to the ancient kings of Persia. When it was in ruins, Farīdūn restored it. It was conquered by Qarṭ b. Kaʿb al-Anṣārī or Nuʿaym b. Muqarrin during the caliphate of ʿUmar. Of the Abbasids, Mahdī lived here as governor during the caliphate of his father, Manṣūr, and restored it. Since his son Rashīd (Hārūn al-Rashīd, Abbasid caliph, r. 786–809) was born here, the city was called Mahdiyya. After Baghdad, this was the most flourishing city in the Mashriq.

When more than 100,000 people were killed during riots among the inhabitants, the city again fell to ruin. Under the Chingisids, it was completely destroyed. Shaikh Najm al-Dīn-i Dāya says the following in his *Mirṣād al-ʿibād*: "In the city of Rayy, which is my hometown, 700,000 people are said to have been martyred during that civil war." During the reign of Ghāzān Khān, Malik Fakhr al-Dīn somewhat rejuvenated the city, but it was not like it had been before.

[M 292] In the past its settled area was one and a half parasangs long. Two rivers flow through it from the direction of Daylam. One, called Mūsā, has canals (coming out of it). It is situated southwest of Mt. Damāwand on the edge of the desert. Since it has a great run-off, there are many trees and fruits, and its cotton produce is large.

Some of Rayy's well-known districts (*mahalle*) are Dahak-i Naw, Baṣrābād and Sārbānān. The famous gates are: Marsiyān Gate, which opens toward Qazwīn; Kūhak Gate, which opens toward Qum; Sāṭir Gate, which opens toward Kūhistān of Iraq; Khurāsān Gate; Hishām Gate, which opens toward Qarmas (?). It had many finely-built caravanserais and markets.

The author of the *Haft Iqlīm*, who was from Rayy, says that Rayy now has four sections (*bölük*), and the rest of the 11 sections that had belonged to it in the past have been incorporated into Rustamdār.

This land has been granted so many blessings that it is impossible to describe. During the forty years when the Persian shahs resided in Qazwīn, Rayy met most of their needs. Kāshān's grain and most of the people's needs came from Rayy. Rustamdār, Sāwa and Qum also benefitted from this city.

The different fruits are quite tasty and abundant here. Especially its watermelons, grapes, *malīsī* pomegranates, peaches, figs, apricots and pears have no match anywhere. Accordingly, even in the autumn season fever is rampant.

The tomb of the jurist Imām Muḥammad b. al-Ḥasan and the Koran-reader al-Kisāʾī are here. The *nisba* (relativizing adjective) of Rayy is Rāzī. Many eminent people came from this city, including: the unique of his age, Imām Fakhr al-Dīn Muḥammad b. ʿUmar al-Rāzī, who died in 606 (1209–1210), a Shāfiʿī scholar and author of a great Koran commentary; the other Fakhr al-Dīn, Aḥmad b. ʿAlī, who died in 360 (970–971), a Ḥanafī scholar who wrote the *Aḥkām al-Qurʾān*; Quṭb al-Dīn Muḥammad b. Muḥammad al-Taḥtānī, died in 672 (1273–1274), a commentator on the *Shamsiyya* and the *Maṭāliʿ*. It is also said that master thieves come from this city.

In the vicinity of Rayy, at the foot of a mountain north of the city, is Ṭabarak castle. The province of Qaṣrān is in front of that mountain. It is comprised of a few villages with no main town. The largest of the villages is called Qaṣrān. Another is Fīrūzrām, which was built by Fīrūz the Sasanian and is now called Fīrūzān. It has abundant grain and cotton. Usually there is plenty and low prices. Of fruits, pomegranates, pears, ʿAbbāsī grapes and peaches are plentiful. Their excessive consumption can lead to fever. Most of the inhabitants of this land are Shīʿī.

The state dues from Rayy province were 15 tomans and 1,500 dinars.

Ṭihrān (Tehran): an estimable town with a good climate. Rāmīn used to be a village and later became a town. According to Amīn Aḥmad al-Rāzī, at present one of the capitals of Rayy province is Rāmīn, the other is Ṭihrān.

Ṭihrān was built up by order of Shah Ṭahmāsp, its markets were decorated and adorned. Now, with its canals, shady trees and paradise-like gardens, it is superior to other cities and lands. The north side, known as Samīrān, is a mountain valley and a piece of the gardens of paradise. In the past it was called Shamʿ-i Īrān ("Candle of Iran"). In its suburbs the different kinds of fruits are delicious.

Two parasangs from here is a mountain village known as Kand-i Salafān, an envy-enticing garden of paradise with flowing streams, an abundance of trees and measureless fruits. [M 293] Pears and peaches are so good and tasty that whoever sees them wishes to keep them in his mouth like his tongue. Located at 86½ degrees longitude and 35½ degrees [R 110a] latitude, it has fine climate. Of its products, cotton, grain and fruit are better than in Rayy. The inhabitants are Shīʿī.

Sulṭāniyya: A new city, located at 86 degrees longitude and 39 degrees latitude. Formerly it was a dependency of Qazwīn. Later it was considered as the capital of Iran and independent *tūmān*, and ten cities were made dependent on it. According to the *Taqwīm*, this city belongs to Azerbaijan; but since Ḥamd Allāh writes in the *Nuzha* that it belongs to this region (i.e., Jibāl or Persian Iraq), his opinion was followed. He says that the fortification was begun by Arghun b. Abaqa. Since it was completed by his son, Öljeytü Sultan, the city became famous under his name.

The tomb of Öljeytü is there; it was completed in 700 (1300–1301). It is in a flat place, with water supplied from underground channels. The fortification is square. Each of the bastions is 500 cubits high and made of cut stone. The walls are wide enough for four horsemen to ride abreast on top of them. It is one station from the Daylam Mountains. There are few places in it with gardens and trees; but the climate is fine. Water can be found at 3–4 cubits depth. Fruits are brought here from the surrounding region. There are plenty of pasturages and an abundance of hunting grounds. After its construction, people came from every direction and the city became so prosperous that it was filled with people of every craft, religious community and sectarian division. Most of the populace speak Persian. Its state dues are 20–30 tomans.

The tomb of the aforesaid sultan (Öljeytü) is like a large Egyptian pyramid; it is 100 cubits in diameter and 120 in height. Despite such adornment and workmanship, it was completed in forty days. The padishah invited the leading scholars, *sayyids* and shaikhs of the (Mongol / Ilkhan) empire and gave a huge banquet under that lofty dome. Shaikh Ṣafī al-Dīn Ardabīlī sat on one side, Shaikh ʿAlā al-Dawla Simnānī on the other. ʿAlā al-Dawla partook of the meal, while Ṣafī al-Dīn refrained from eating. When the meal was finished, the padishah said: "There is no doubt about the greatness of the two holy men. What is the significance of one of them eating and the other abstaining? If the food is religiously permissible, why did one of them not eat? If it is religiously forbidden, why did the other eat?" Shaikh Ṣafī said: "ʿAlā al-Dawla is the sea. For the sea, nothing is impure. It accepts whatever comes into it." ʿAlā al-Dawla said: "Ṣafī al-Dīn is a falcon. The falcon does not alight upon every food." The padishah was pleased with their mutual praise and gave generous favors to both.

Qazwīn: Located at 85 degrees longitude and 37 degrees latitude, and now the capital of Persian Iraq, Qazwīn is a magnificent city. There are many hadiths related about its excellences. One such hadith is: "Respect Qazwīn, for it is one of the loftiest gates of Paradise," which is reported by one of the scholars of Qazwīn, Imam Rāfiʿī, in his chronicle entitled *Tadwīn*. Hence Qazwīn is also called "Gate of Paradise." As Ḥamd Allāh was from Qazwīn, he goes into great detail about this city in his chronicle entitled (*Tārīkh-i*) *guzīda*, while he mentions it in the *Nuzha* in summary fashion. The gist (of both accounts) is the following:

Shābūr b. Ardashīr escaped from the captivity of the Roman emperor and arrived in the region, he considered his joining his army to be well-omened and founded the city by the name of Shābūr between the river and Abhar, where there was a small garden and a monastery at the time. As is reported in the *Tadwīn*, the fort of Qazwīn is [M 294] a district (*mahalle*) in the middle of this city. It was built in year 463 of the Alexandrian calendar. The ruins of its walls still remain. At that time a unit from Shābūr's army were guarding it against the depredation of the Daylamites. Shābūr collected an army and marched against the Daylamites. He massacred them as far as the seashore. Hence a bitter enmity rose between the Daylamites and the Qazwīnīs. However, some say that while fighting against the Daylamites, one of the Sasanians saw a breech in their own ranks. He said to one of his followers: *ān kash wīn*, meaning, *ān kunjrā nigar wa lashkarrā rāst kun*, i.e., "See that corner and straighten the breech!" Having been vouchsafed victory from that direction, Shābūr founded a city there and named it Kashwīn. Arabicizing it, they called it Qazwīn.

During the caliphate of ʿUthmān, Walīd b. ʿUqba, who was the governor of the two Iraqs, sent Saʿd b. al-ʿĀṣ (sic; recte Saʿd b. Abī Waqqāṣ) to this province. They camped around it during the siege. People gradually became settled and the city grew. The Abbasid al-Hādī bi'llāh Mūsā founded another town in the vicinity which was called Madīna Mūsā; it is now called the district (*mahalle*) of Dhawj and Jawsaq. His slave Mubārak-i Turkānī added another town; it is called Mubārakābād.

During the caliphate of Hārūn al-Rashīd, the populace, in order to be safe from the depredation of the Daylamites, ordered the building of a wall that would encompass these three towns (Qazwīn, Dhawj and Jawsaq and Mubārakābād) and other districts, and a Friday mosque. At the time, the wall was called Rashīdābād. After Hārūn al-Rashīd died it was left unfinished. Later caliphs worked on it until al-Dāʿī ilā'l-Ḥaqq Ḥusayn b. Zayd invaded Qazwīn. Caliph Muʿtazz sent Mūsā b. Būqā to repel him. Mūsā was victorious and finished the wall in the year 254 (868). It became a large city.

Later it became somewhat ruined. The vizier of Fakhr al-Dawla, Ismāʿīl b. ʿAbbād, with the byname of Ṣāḥib, in 373 (983–984) [R 110b] saw the ḥadiths about the virtues of Qazwīn, restored (the wall) and had a house built for himself called Ṣāḥibābād in the district of Jawsaq. In 411 (1020–1021) the wall was somewhat damaged in a dispute between the Qazwīnīs and the Daylamite commander Ibrāhīm b. Marzubān. It was repaired by Amīr Abū ʿAlī al-Jaʿfarī. In 572 (1176–1177) Sultan Alp Arslan's vizier, Ṣadr al-Dīn Muḥammad Marāghī also renovated it and made bastions and battlements of brick. Afterwards the Mongol army destroyed this building. It measured 10,300 fathoms in circuit and had 230 towers and 7 gates.

The city of Qazwīn is comprised of nine districts (*maḥalle*), two in the center and seven where these gates are. The central ones are Shahristān and Surkh; the gates are Abhar, Ardāq, Rayy, Dāmghān, Dastjird, Dhawj and Jawsaq. Today there is no trace of the wall.

The climate of the city is moderate. It obtains water from underground channels. There are many gardens. Grapes, almonds and pistachios grow in abundance. One of the peculiarities is that the orchards here are irrigated only once a year. Shāhinī grapes are very fine. Melons and watermelons are also good. Grain crops and fruits are usually cheap. There are hunting grounds and good pasturages. Camels from Qazwīn are more highly valued than those from elsewhere.

There is a fountain 3 parasangs from the city that freezes in the hot summer weather. When the [M 295] ice in the city runs out, they get ice from there. Because the rivers of the city come from runoff of snow and rain, their water shrinks to a trickle in hot weather and some of it runs out before reaching the city. There are five rivers altogether. One of them, which is named after Daraj, flows through the middle of the city, at times causing fear there, for the flow is toward the houses. Most of the orchards of the city are watered from it. Aside from this, there are six underground channels. One is the Khumārtash branch which flowed through most of the districts of Qazwīn. Another is the Ḥājibī channel. According to the *Haft Iqlīm*, the best water is that of the Khumārtāsh, which was made by an ascetic of that name in 500 (1106–1107).

Shah Ṭahmās made the city his capital for thirty years, and there was much construction during his reign. Markets, public baths and other institutions were made everywhere in the city. Shah ʿAbbās brought water by adding another canal. Its water is still flowing in the city and the market.

The Qazwīnīs are excellent conversationalists and most of them are knowledgeable in the science of music. As one poet has put it [Persian verse]:

> The shah needs four classes of professionals from four cities
> If he wants to perpetually sit on the throne of lordship:
> Musicians from Khurāsān, boon companions from Qazwīn,
> Artisans from Iṣfahān, and soldiers from Tabrīz.

The cemetery is inside the city; previously it was outside, but when the wall was erected it was included inside.

Mosques. The great Friday mosque called Muṭṭalabī, belonged to the Shāfiʿīs; each of its corners was made by (a different) person. The Friday mosque of the Ḥanafīs was built by the ascetic Khumārtāsh in 548 (1153–1154). The Masjid-i Tūt, which had been an idol temple, was converted to a mosque by Muḥammad b. Ḥajjāj. The Shīʿīs do not like this mosque, saying that during the time of the Umayyads ʿAlī was cursed here.

The city was conquered by al-Barā b. ʿĀzib and Zayd b. Khalīl (al-Jabal) al-Ṭāʾī. The siege of the town built by Shābūr took place in the time of ʿUmar's caliphate. The people feigned to accept Islam in exchange for peace, (but) they reverted. ʿAbd al-Raḥmān al-Ḥārithī rallied again an army and subdued the city. This time they converted to Islam sincerely. Considered as people of received religions (*erbāb-ı meẕāhib*) some Jews were also settled there. When Hārūn al-Rashīd surrounded the city with the wall, he added to Qazwīn some of Hamadān, Abhar and Qāmrān, making it a region (*kūra*). Mūsā b. Būqā added to it districts such as Rāmina, Zahrā, Kharqānīn and Ṭālaqān. In the reign of the Jaʿfarīs, Zanjān, Ṭārumayn, Rūdbār (text: Rūdbād) and Raḥmatābād were also added. When during the time of the Mongols, the Iftikhārids added Sāwa and Juhar (Chūhar), it had 8 districts and 300 villages. As revenue 5 tomans came from the city and 5 tomans from the districts.

Some eminent people came from this city, such as the hadith scholar Ibn Māja; the Shāfiʿī imam, Imām Rāfiʿī; Najm al-Dīn ʿAlī ʿUmar al-Kātibī; and the author of the *ʿAjāʾib al-makhlūqāt*, Zakariyā b. Muḥammad.

Abhar: Located at 84½ degrees longitude and 37½ degrees latitude, Abhar was built by Kay-Khusraw 12 parasangs from Qazwīn. It had a mud-brick castle. Bahā al-Dīn Ḥaydar, one of the sons of Anūsh Tigin Atabeg, built another castle and named it Ḥaydariyya. The wall around the city [R 111a] was 5,500 fathoms.

The climate is cold. Water is obtained from a river with the same name as the city, which comes from the borders of Sulṭāniyya and flows toward Qazwīn. Grain crops and fruits are abundant and good, but bread is not so good. There is also some cotton. Of fruits, [M 296] the pears and apricots are fine.

Most of the people are dark skinned and Shāfiʿī. Abū Ḥāmid relates: "I reached Abhar in 524 (1130). Its qadi Abū'l-Yusr ʿAṭā recounted: In our place is a castle called Ardashān. On a nearby mountain, over a cave there is a hole as small as the mouth of a cup, whence a handful of branches grow downwards. If someone comes underneath the hole, he finds them. There are no more and no less than fifteen branches. It is unknown what type of tree it is, nor what keeps them tied together. If one is untied, no one knows how to tie it again. If it is torn out and removed from the cave, another grows in its stead. If it is removed a thousand times, another one will certainly be there. The qadi gave me one of its branches."

Abhar had 25 villages. Its state dues were one toman and 4,000 dinars. Of the eminent ones, Athīr al-Dīn Mufaḍḍal b. ʿUmar al-Abharī comes from this city.

Zanjān: Located at 84 degrees longitude and 37½ degrees latitude, it is a city with its northern border on Persian Iraq. Abhar is southwest of it. The total circuit of its wall is approximately 5,000 fathoms. Its foundation goes back to Ardashīr. It was destroyed in the Mongol invasion and is now a small town. A river comes from the border of Sulṭāniyya and flows by the city. The climate is moderate. In winter they store ice in the wells. Cotton is abundant and good. Of fruits, the figs are fine. The people are white skinned and are Shīʿī. The state dues from cityare 10,000 dinars and its 100 villages yield 20,000 dinars. Zanjān belongs to the Sāwa region (*bölük*). Eminent people who come from here include Shaikh Akhī Faraj; ʿIzz al-Dīn, who is the author of the *Taṣrīf*; and Ṣadr-i Jahān, etc.

Rūdbār: A vilayet 60 parasangs north of Qazwīn; it is crossed in the middle by the river Zandarūd.

There are as many as 50 fortified castles in this province. The greatest of them is Alamūt, which was the capital of the Ismāʿīlīs for 170 years. It is located at 84½ degrees longitude and 36½ degrees latitude. It was built in the year 246 (860–861) by Dāʿī ilā al-Ḥaqq Ḥasan b. Zayd al-Bāqirī, and was captured by Ḥasan-i Ṣabbāḥ in 483 (1090–1091). Before then it had been called Āluh Amūt (text: Ālah Alamūt) "Eagle's Nest". In 654 (1256) it was destroyed on the order of Hulagu Khan.

The Rūdbār province has summer pastures here and there and fertile fields. Grain, cotton and fruit grow in abundance. Most of the inhabitants are Bāṭinīs (i.e., Ismāʿīlīs). is recorded that the state dues amount to 8,000 dinars. Of the (famous) shaikhs, Abū ʿAlī al-Rūdbārī comes from this province.

Sāwa: at 85½ degrees longitude and 35 degrees latitude, Sāwa is a city 30 parasangs west of Rayy. It has a fine market and houses. Originally its site was occupied by a lake. According to Ḥamd Allāh (Qazwīnī), on the night of the birth of the Prophet its water was absorbed into the ground and it dried up. This was one of the harbingers (of the coming of the Prophet). Later a city was built on the site of the lake.

However, according to the *Haft Iqlīm*, it is an ancient city built by Ṭahmūrath and the lake was on the edge of the city. It dried up that night and its traces still visible.

The city wall fell to ruin and was restored by vizier Khwāja Ẓahīr al-Dīn ʿAlī who paved its roads with baked bricks. The circuit of that wall [M 297] was 8,200 cubits. Later his son, Shams al-Dīn Ṣāḥib, enclosed the land down to the river with a 4,000 cubits long wall.

The climate is rather hot, but salubrious. Water comes from underground channels and the Muzdaqān (text: Murdaqān) river. In winter they store ice in the wells. The inhabitants are mostly Shīʿī and Shāfiʿī. Grain, cotton and fruit are abundant, but the grass and barley are not good for animals.

The state dues are 2½ tomans. Four districts and 105 villages are registered as belonging to Sāwa. Another 4½ tomans are collected from the villages. Poets from this city include Salmān (Sāwajī) who has a divan, and Qāḍī Ṣadr al-Dīn who has a treatise on metrics.

Āba: A city 5 miles from Sāwa. It is also called Āwa. Qazwīn is 16 parasangs from here.

Sujās and Suhraward: Located at 82½ degrees longitude and 37 degrees latitude, they were originally two cities. Both were destroyed during the Mongol interregnum and are now villages. The province consists of mountain pastures. It produces grain crops and some fruit; 100 villages are registered as belonging to it. It is one station south of Sulṭāniyya. Arghun Khan's tomb was on Mt. Sujās, being hidden here according to Mongol custom. Later his daughter discovered it and erected a mausoleum and dervish convent over it. That place is called Anjarūd.

There is a big palace there, with a large pool in the courtyard. Its depth is not known. Water is constantly flowing into it from two channels, [R 111b] enough to drive one mill each. If the two channels are blocked, the water of the pool stops and does not rise; if they are unblocked, it flows as before. The water has never flowed too much or too little, but is always stable, which is quite a marvel. This palace was built by Abaqa Khan. It is surrounded by good pasturage.

The state dues from this province are recorded as 1½ toman. Shaikhs from this city include Abū al-Najīb ʿAbd al-Qādir; Shihāb al-Dīn ʿUmar b. Muḥammad; and Shihāb al-Dīn Yaḥyā b. Ḥabash, the philosopher who was put to death.

Sarjahān: It is a castle on a mountain opposite Ṭārumayn, 5 parasangs east of Sulṭāniyya. 50 villages belonged to it, and it was considered a district. The largest village is Quḥūd, called Ṣāyin Qalʿa by the Mongols. Destroyed during the Chingisid interregnum, it was restored when Sulṭāniyya was built. Since it is a place of summer pastures, its produce is grain. And because it is on the (public) road, it is exempt from state dues.

Ṭārumayn: A place one station north of Sulṭāniyya, located at 84 degrees longitude and 36½ degrees latitude, with a hot climate and much fruit and yielding an abundance of crops. Sulṭāniyya obtains most of its fruits from here. Especially its olives are unequaled. In the past there was a city here by the name of Fīrūzābād. Lower Ṭārum which used to be the capital is now in ruins, whereas in Upper Ṭārum there is a town to which 100 villages belong. This province is divided into five administrative units, including Upper Ṭārum; Lower Ṭārum, where the dependencies of the castle of Shamīrān are 50 villages; Nasāzar (text: Nasārū), a large village with 8 other villages as dependencies; and Lower Ābād consisting of 25 villages. The state dues of Ṭārumayn are 6 tomans and 4,000 dinars.

Ṭālaqān: It is a city and district located at 84½ degrees longitude and 37½ degrees latitude. It is between two great mountains. It is a cold place in mountainous country between Qazwīn and Abhar. [M 298] It has hardly any villages worth mentioning. It produces grains and some walnuts. The people are mainly Bāṭinīs (i.e., Ismāʿīlīs). Its state dues are one toman.

Kāghad Kunān: It was a mid-sized town called Khūnaj. Because of the fine paper produced there it gained fame as Kāghad Kunān ("Papermakers"). Its climate is cold, and it is a place with no produce other than grain. Water comes from the mountains and flows into the Safīdrūd. Only a village has remained of it by now. In the past, 35 villages belonged to it. It was destroyed during the Mongol interregnum. The Mongols later settled down and practiced agriculture, so it was called Mughūliyya.

Mazdaqān: A city in the Sāwa district. The wall is 3,000 fathoms in circuit. Its climate tends to be cold. It obtains water from a river that is named after this city and comes from the border of Sāmān. Grains and grapes are good. The state dues are 1 toman.

Qum: Located at 86½ degrees longitude and 35 degrees latitude, it is an ancient fortified city. The wall was 10,000 fathoms in circuit. It is one of four great cities composing its district which is 100 parasangs long and 100 parasangs wide. However, its wall being now in ruins, there is little left of the city itself. It has a temperate climate. Water is from a river that comes from Jarbādqān. The wells are brackish. In winter people store ice in the wells. Grain and fruit are produced. Pomegranates, pistachios, hazelnuts, pears, figs and watermelons grow abundantly, and there is much cotton. The place is good for cypress trees. The people are fanatical Imami Shīʿīs. State dues are 4 tomans.

According to the *Taqwīm*, after being destroyed in the year 83 (702) in the battle between the armies of ʿAbd al-Raḥmān b. Ashʿath and al-Ḥajjāj, the wall was rebuilt by ʿAbd Allāh b. Saʿdān. At that time Qum was comprised of seven villages close to one another. The defeated soldiers killed their commanders, occupied the city and rebuilt it. The seven villages became districts (*maḥalle*). One of them was Kumīdān, which was shortened and Arabicized as Qum.

In Qum light and delicate blue cups are made. Niẓāmī, the author of the *Khamsa*, and Khāja Masʿūd are from Qum.

South of Qum and Sāwa is a very high mountain which grows opoponax and other beneficial herbs.

Kāshān: Its Arabicized form is Qāshān. Sometimes it is pronounced as Bāshān. Located at 87 degrees longitude and 34½ degrees latitude, it is a town smaller than Qum and is attached to it. It was built by Zubayda, the wife of Hārūn al-Rashīd. It is on the edge of the desert. There is a mud brick castle outside it known as Fīn (text: Qīn). The climate is hot. Water is drawn by underground channels from a river that comes from the Qahrūd (text: Fīrūz). [R 112a] In winter there is much ice which they store in wells. The produce of this city is middling. Watermelons and grapes are excellent. The people are mostly Shīʿī and given to philosophizing. There are few among them who are ignorant or lazy. Of vermin, the scorpions here are numerous and deadly, (but) they do not generally harm travelers. The state dues from here were 11 tomans and 7,000 dinars. Eminent people from this city include ʿIzz al-Dīn Maḥmūd, the commentator on the *Tāʾiyya* (?); ʿImād al-Dīn Yaḥyā, the commentator on the *Miftāḥ*; ʿAbd al-Razzāq, etc.

Ardistān: Located at 87½ degrees longitude and 33½ degrees latitude, 18 parasangs from Iṣfahān, on the edge of the desert, it is a city and 50 villages. Bahman b. Isfandiyār built a fire temple there.

Because its ground is white sand, [M 299] it yields only little but has good fruits. Of celebrities, Mawlānā Muḥammad, who wrote the *Mahāratnāma* (Book of Proficiency) on astrology, geomancy and mathematics, comes from this city.

Jarbādiqān: It is the Arabicized form of Darbāygān or Gulbāygān. The city is located at 85½ degrees longitude and 34½ degrees latitude, between Karaj and Hamadān. It is said to have been founded by Humā, the daughter of Bahman the Kayānid. It has a moderate climate; water is gotten from a river with the same name. Its grain products are fine. There are 50 villages in the district (of Jarbādiqān). The state dues are 4 tomans and 10,000 dinars. Dalījān (text: Dalnjān), a mid-sized city in the past but now in ruins, is one of the villages of this district.

Farāhān: It is located at 84½ degrees longitude and 34½ degrees latitude. In the past it was a mid-sized city, but now is in ruins. It is said to have been founded by Ṭahmūrath. It has 20 villages. There is a lake with fine hunting grounds around it. The state dues from it are 7,000 dinars.

Karaj: Arabicized from Karah. It is located at the same longitude and latitude as Farāhān. Originally it consisted of scattered villages one parasang across. During the reign of Hārūn al-Rashīd, Abū Dulaf Qāsim b. ʿĪsā b. Idrīs al-ʿIjlī settled here with his following and built a palace and fortress. Widely known therefore as Karaj of Abū Dulaf, it became a large city. It is a place of grain cultivation and animal husbandry. Hamadān is 4 stages, Iṣfahān 60 parasangs hence. Due to the severity of the cold (in winter), its fruit is obtained from the surrounding regions. According to the *Nuzha*, north of Karaj is a mountain called Rāsmand, at the foot of which Khusraw made a big fountain. Karaj also has a pasture 6 parasangs long and wide. There is a strong castle called Farrazīn nearby. The state dues of Karaj were 1 toman and 1,000 dinars. The river Garmrūd, also called Āb-i Garm, is in this region.

Naṭanz: A mid-sized city 20 parasangs across in the vicinity of Iṣfahān. Nearby are approximately 30 villages. The state dues from it were 10 toman and 2,500 dinars. There is a castle in this province. Because it once had a governor by the name of Washaq, it has become associated with him.

Nawbahār: A town on the road from Rayy to Iṣfahān. There is another Nawbahār in Khurāsān as well.

Zawwāra: A city on the brink of the desert. It is said to have been built by the brother of Rustam. There are 30 villages dependent on it. The state dues from it are 8,000 dinars.

Burūjird: A city with numerous trees and rivers, 8 parasangs from Hamadān. It abounds in saffron. It has two Friday mosques, a new and an old one. The climate is moderate.

Ṣaymara: It used to be a small town. It has crops, trees and water sources. Located in the mountains, it was the only place with date palms. It is now in ruins.

Hamadān: A large city and ancient settlement located at 83½ degrees longitude and 36 degrees latitude, on the eastern slope of Mount Arwand. It has many flowing streams and vineyards, flower gardens and vegetable gardens. There are abundant fields and pastures. Foodstuffs are plentiful and cheap. According to the *Nuzha*, it used to have an earthwork fortress 12,000 fathoms in circumference, but it fell to ruin. As reported in the *Ṭabaqāt*, the settled area of this city was 2 parasangs across at the time. [R 112b] Having been destroyed by Nebuchadnezzar, it was restored by Dārā b. Dārā. In the Islamic era it was conquered by Badīl b. Waraqā [M 300] in the 23rd year (643–644) of the Hijra. It was sacked and ruined, and its inhabitants massacred, by the Daylamite Mardāwīj around 320 (932/3), and again by the Chingisids in 618 (1221–1222). Later it was restored. Ever since then it has been flourishing and been famous as the abode of kings and a mine of eminent people. However, the (winter) is severely cold. It is therefore that Ibn Khālawayh al-Hamadānī said about it: "A country in summer superior to the Garden (of Paradise), but in winter worse than the Fire (of Hell)." Someone else said [Arabic verse]:

> Hamadān is soul destroying with its cold and frost,
> But its warmth can also be relied upon.
> Its winter has overcome its summer and its autumn,
> As though its July were December.

The people were mostly Shīʿī and Muʿtazilī in the past. Badīʿ al-Zamān Hamadānī wrote the following [Arabic] verse about them:

> Hamadān is my city whose virtues I recount,
> But it is one of the ugliest of cities.
> In ugliness, its youths are like old men;
> In intellect, its old men are like youths.

However, its steppes and plains are verdant, and its springs and rivers betoken Paradise. Every grief-stricken one who comes to this city becomes happy and cheerful and his grief disappears. Therefore they say that the people of the region are taken with pleasure and amusements.

In the year 1039 (1629–1630) the author of these lines (Kātib Çelebi) was with Grand Vizier (Ġāzī) Ḫüsrev Pasha who conducted a raid on Hamadān and I toured the city. After making a foray of 2–3 stages into the desert, we went into the villages, with their interjoined gardens and vineyards, and alighted at the outskirts of the city. We saw that it was finer than described. Words cannot express its delightful climate, the beauty of its plains, its flowing streams, verdant pastures and splendid gardens. It is a mid-sized city in circumference, with 5000–6000 houses. Here and there are neighborhood mosques and Friday mosques, but they are verging on ruin because of neglect in the hands of the Rāfiḍīs (i.e., Safavids). The market is in the middle of the city where four roads intersect. It is covered over and white stuccoed, and the shops are clean. The houses are all one-storied earthen buildings, the interiors adorned to the utmost with pictures. Here and there are palaces with paintings in the Persian style. South of the city there is a polo arena with tall stakes (as goal-posts). Public baths are sparse; a small one has been constructed outside the city, with Kashan tiles and two turrets in the Persian style, but it is cramped. Most of the other houses and buildings also have small doors and are cramped and dark, resembling merchants' houses. We saw no trace of the magnificent buildings and ancient monuments. Its population was driven away and the city was empty. The Ottoman army stayed there three days, cutting down the trees and burning the city. They destroyed the other villages (of the province) as well and left for Darguzīn and Baghdad.

Ḥamd Allāh (Qazwīnī) says that the state dues of this city are 10½ tomans.

The province of Hamadān is comprised of five districts:

- Farīwār: It is in the vicinity of the city. It comprises 75 villages extending 2 parasangs. The most notable of these are Shahristān, Fakhrābād, Qāsimābād and Kūshk Bāgh. Especially Nāmshābrūd, which is the envy of the Garden of Paradise, is reckoned as belonging to this district. It consists of 10 villages and gardens joined to one another and extending 2 parasangs. Sunlight does not reach the ground due to the abundance of trees.
- Azyārdīn: 41 villages. Some of the most notable are Dāqābād, Sayfābād and Kurdābād.
- Sharāmīn: 40 villages.
- Aʿlam: 35 villages.
- Sardarūd [M 301] and Barahnarūd: 21 villages.

The state dues from these districts were 13 tomans and 6,000 dinars.

Some of the eminent people of Hamadān are ʿAyn al-Quḍāt, Khwāja Yūsuf, Sayyid ʿAlī b. Shihāb, Shaikh ʿAṭṭār, Abū al-Aʿlā al-Ḥāfiẓ, Abū al-Ḥusayn b. Jahḍam, etc. The tombs of some of them are places of visitation.

Asadābād: Located at 83 degrees longitude and 35 degrees latitude, it is a small city west of Mt. Alwand, 9 parasangs from Hamadān on the Iraq road. Its climate is moderate. Water is gotten through underground channels from Mt. Alwand. Cotton, grain, fruit and grapes are produced here. Its people are dark skinned. 35 villages are registered as its dependencies. The state dues from it are 1 toman and 5,500 dinars. We passed by this town on the way back from Hamadān. It was a town with earthen houses, situated on an elevation in a level plain. It is a place for crops. [R 113a]

Kharqān: A district comprising 40 villages. It is 12 parasangs north of Qazwīn. The climate tends to be cold; water comes from the mountains, and there are springs. A place of fruit and grain. Its most notable villages are Āb-i Arwān, Alīshār, Kalanjabīn, Tarak, Alwīr and Sayfābād. According to the *Nuzha*, the state dues of Kharqān are 10,000 dinars. Abū al-Ḥasan al-Kharqānī comes from this region. There is a hot spring here whose water is so hot that it can cook an egg. It is a proven cure for headache and scab. It is covered with a dome and has pools and washbasins. A river flowing nearby drives mills.

Darguzīn: Located at 85½ degrees longitude and 36½ degrees, it used to be a village 2 stages east of Hamadān, in the Aʿlam district. Later it became a town. It is an affluent place with flowing streams, abundant gardens and orchards, grain crops and cotton. Grapes and other fruit are good.

According to the *Nuzha*, the people are Sunnis of the Shāfiʿī rite, and are pure in their faith, being followers of Shaikh al-Islām Sharaf al-Dīn al-Dargazīnī. The state dues from it were 1 toman and 2,000 dinars. It is built on an elevation.

However, during the Hamadān campaign we had turned toward Qazwīn and reached this town in 3 stages. We saw a fine and pleasant town on a level plain, just as described above. A new Friday mosque was built in the middle of it. There was no lack of stones and baked bricks in the construction of its houses. The houses were dense and compact and surrounded by gardens and orchards. We laid waste to the city; but because it could not supply sufficient water for the army on the Qazwīn road, we returned from there to Baghdad.

Surkhāb is four stages east of Darguzīn.

Sāmān: A large village in the vicinity of Kharqān. The climate tends to be cold. Water comes from the mountains; after it reaches the Muzdaqān river, it goes toward Sāwa. Sāmān produces grain crops, grapes, and some other fruit. The state dues from it were 1,200 dinars.

Surkhāb: A district located at 87½ degrees longitude and 36½ degrees latitude. Qazwīn is 5, Simnān 1½ stages east of it.

Nihāwand: It is said to (derive from) Nūḥ Āwand. Located at 83½ degrees longitude and 35½ degrees latitude, 14 parasangs south of Hamadān. It is on a hill. The climate is moderate. The water is fine and comes from Mt. Alwand. It is a mid-sized city with many vineyards and an abundance of grain crops and fruit. There is little cotton. The inhabitants are Kurds. The villages in its district registered as amounting to 100. The state dues are [M 302] 3 toman and 7,000 dinars. There are many nomadic Kurds around it. 12,000 sheep are collected from them annually. The event of *Yā sāriyat al-jabal* is reported to have taken place here during the reign of ʿUmar b. al-Khaṭṭāb (referring to the Battle of Nihāvand in 639, 641 or 642). The tombs of a number of martyred Muslims and some monuments of the Persian kings can be found here.

Nuwayrī, the author of the *Nihāyat al-arab*, writes about a rock here: if a person wants to find out about something lost or stolen, or a runaway slave, he goes and sleeps underneath that rock, and the condition of the lost item is made known to him in his dream.

The author of the *Mukhtaṣar-i muʿjam* (i.e., *Marāṣid al-iṭṭilāʿ*) writes that chiretta (*qaṣab al-zarīra*) grows here. It has no smell until it ripens, but starts to give a smell when it withers.

Damāwand, sometimes pronounced Dunāwand or Dubāwand: An ancient city and a district 12 parasangs northeast of Rayy at the foot of Mt. Damāwand. It counts as the end of Persian Iraq in the direction of Ṭabaristān. This town is said to have been the capital of Ḍaḥḥāk and Jamshīd. It has a district comprising 44 villages and 3 castles, the names of which are Fīrūzkūh, Ustūnāwand and Gul-i Khandān. It takes four days to get to Āmul of Māzandarān. Simnān is south of it, Astarābād in the east.

Zarand: The Arabicized form of Zarang. Located at 83½ degrees longitude and 36½ degrees, it is a town in the Iṣfahān district. There is another Zarand in Kirmān.

Khān Lanjān: A in the vicinity of Iṣfahān.

Khwār: A town in the vicinity of Rayy. It is in the direction of Simnān. It has villages. Water comes from Mt. Damāwand.

Rūdrāwar: A small town with plenty of fruit and water sources in the vicinity of Hamadān. Saffron is abundant. Its district is referred to with this name.

Rūdgard: A small town half a parasang from Karaj, founded by one of the Abū Dulaf viziers. It is a field of saffron.

Qaṣr al-Luṣūṣ: A town 7 parasangs from Asadābād. It is also called Kingiwar. When the army of Islam first alighted here, their horses were stolen, so they called the place Qaṣr al-Luṣūṣ ("Castle of Thieves"). According to the *Taqwīm*, it is a strong fortification.

Dīnawar: Located at 82½ degrees longitude and 36 degrees latitude, it is a large town three stages northwest of Hamadān. Fruit and water sources are plentiful. There are many plains. Foodstuffs are plentiful and cheap. The inhabitants are Kurds. In the past, 38 *yük* (3,800,000) of dirhams were obtained from here as tax (*kharāj*). Between here and Hamadān there are mountains ranged north to south with a pass between. Ibn Qutayba and Abū Ḥanīfa al-Dīnawarī were from this city.

Qirmīsīn: The Arabicized form of Kirmānshāh. It is close to Dīnawar, 30 parasangs from Hamadān, in the mountains. It is a renowned city of Persian Iraq. Mainly saffron is grown here.

Qaṣr-i Shīrīn: A ruined city located at 82½ degrees longitude and 35½ degrees latitude [R 113b] near Qirmīsīn. It is named after Khusraw's Shīrīn. There are many magnificent buildings and ancient monuments still standing. Since it is on the road from Hamadān to Baghdad, we passed through it at night and saw these monuments.

Māspidān: A city in the Sīrawān district amidst mountain trails 2 stages from Ṣaymara. Streams flow through the middle of the city. It is recorded as located at 83 degrees longitude and 33½ degrees latitude.

Famous products

The bows of Qazwīn; the mineral collyrium, [M 303] locks, salted meat and dairy products of Iṣfahān; the watermelons and peaches of Rayy; the yoghurt of Dīnawar.

Mountains and rivers

Mountains are found here and there in this clime:

Arwand, also called Alwand: A large mountain 30 parasangs (in length), 1 parasang west of Hamadān. Snow is never absent from the summit. Numerous streams flow from it in every direction. This mountain is visible from a distance of 20 parasangs. Its environs are full of excursion grounds with amenities; wherefor one of the poets said [Persian verse]:

Spring season, and the skirt of Alwand, and we are gallants (*lawand*).
O winebibber, bring forth the goblet and laugh at the beard of ascetics.

One end of this mountain stretches to Azerbaijan, the other to Iraq, branching out into every direction. Finally the peak close to Hamadān is called Mt. Alwand.

Ashkahrān: A mountain in the region of Iṣfahān. According to Ḥamd Allāh (Qazwīnī), there are huge snakes there.

Altar: Originally it was Aʿlā Tar, now called Altar from frequent use. The author of the *Nuzha* does not share this view, saying that Altar is rather the original name.[37] It is a high mountain north of Qazwīn and connected with other mountains. There is snow on the summit on the Rūdbār side.

Bīsutūn: A very famous and very high mountain of black stone on the Baghdad road below Asadābād. Ḥamd Allāh (Qazwīnī) writes that the surrounding district is Kurdistan. Since it is situated on a plain, it is visible from a distance of 20 parasangs. Its perimeter is also 20 parasangs. The author of the *Nuzha* states that he measured the height of this mountain on the order of Öljeytü in 611 (error for 711/1311–1312) and it came to 4,008 tailor's cubits. Clouds usually hang over it.

The highway passes by the foot of this mountain. At that point, the south of the mountain is sheer, like a wall from top to the bottom, as though it had been carved out. Niẓāmī recounts in his poem entitled *Khusraw and Shīrīn* that one day Khusraw said to Farhād: "This mountain is

37 Cf. *Nuzha* 192, lines 9–10; the edition has Alnaz and Aʿlā Naz.

in our way; make a passage through it!" And Farhād cut a passage. Ḥamd Allāh rejects this, saying it is a foolish legend and that Niẓāmī never saw the mountain but wrote about it from hearsay. The author of these lines (Kātib Çelebi) has had the opportunity to see it, and his (i.e. Ḥamd Allāh's) rejection of Niẓāmī's story is appropriate.

West of that mountain there is a grotto known as Khusraw's Shabdīz (i.e. Ṭāq-i Bustān). It is a ledge carved out of stone, 7–8 cubits in length and breadth. The ceiling is arched. On the outside, high up on both sides, are two angels carved out of stone. On the inside is the statue of a horse carved into the wall with an armored man riding it who should represent Khusraw. His arms are broken, and the passage of time has damaged the horse as well. In front of the ledge is a vast plain. From the bottom of the ledge a great cold stream gushes out into the plain with a force enough to drive two watermills. The water irrigates the fields there, but it is heavy and one cannot drink a lot from it. This ledge is not very high up. I approached from the plain on horseback and was able to go on top of it by a stairway.

Damāwand, also called Danbāwand: A famous, very high mountain east of Rayy. Its summit can be seen from 50 parasangs away, like a dome. It is perpetually covered with snow. It is difficult to climb, for it is 3 parasangs high and hard going. On the top is a plain 100 *jarībs* across. Its sandy middle is like a pond. As there are no trees on the summit, this mountain is also called Kūh-i Aqraʿ (Bald Mountain). [M 304] Because it is like a lake and there are some 70 sulphurous wells at the summit, smoke is constantly rising from it and the air makes one lose consciousness. There are often avalanches on this mountain.

According to the Persian fabulists, Ḍaḥḥāk is imprisoned in this mountain. The story can have no basis in truth. He must have been killed in Makran.

This mountain is the easternmost extremity of the Jabal region (i.e., the clime of al-Jibāl, or else the mountainous region). Since Damāwand is an ancient town at the foot of this mountain, the mountain is named after it. [R 114a]

Rāsmand: North of the city of Karaj it rises to a point above the plain like Mt. Bīsutūn. It is 10 parasang in circuit. The Kītī Pasture, the most famous pasture land of Iraq, is 6 parasangs long and 3 parasangs wide, and lies north of this mountain. It is watered by a spring at the foot of the mountain, named after Khusraw.

Rāmand: A mountain west of Qazwīn and north of Kharqān. It is densely populated, with many villages and fields. It is not a very high mountain.

Karkas. It lies between Iṣfahān and Mt. Damāwand. It is surrounded by desert.

Fīrūzkūh: It lies between Rayy, Bisṭām, Dāmghān and Ṭabaristān. It is also called Mt. Alfār. Danbāwand is one branch of this mountain. It is surrounded by cultivated areas.

Sāwa: A high mountain one stage from (the town of) Sāwa in the direction of Kharqān. There is a portico-like cave there with various figures painted on it. At the far end of the cave water drips down from an upper basin. It is said to have medicinal properties.

Ṭabarak (in) Rayy: There is a silver-mine here, but the income from it does not meet the expense of production.

Namaklān: A high mountain consisting of clay between Āwa and Qum. It is not connected to any other mountain. As its soil is quite salty, snow does not stick to it and water does not flow (but gets absorbed in the ground). It is impossible to climb, since one's feet sink. Rayy is 3 parasangs away, and the mountain is visible from 3 parasangs away.

[Rivers]

Zandarūd: After emerging from the borders of the Jūy-i Sard at Mt. Zard, one of the mountains of Greater Lūr, it passes Rūdbār of Lūristān, arrives at Fīrūzān and Iṣfahān, and flows into the ground of Gāwkhānī swamp of the Rūydashtīn district. It is 60 parasangs long. This river has the following characteristic: even if it is dammed up completely at one place, as much water re-emerges from that point and becomes a river again. For that reason it is called Zāyandarūd ("Birthing River"). At sowing time it does not lie idle but irrigates the crops; so it is also called Zarrīnrūd ("Golden River").

Safīdrūd ("White River"): The Turks call it Hūlān. It rises in the Besh Barmaq Mountains; flows through Kurdistan; receives the Zanjānrūd, Bīstrūd and Miyānajrūd, also the rivers of Ṭawālish and Ṭārumayn; and after its confluence with the Shāhrūd in Barra province, one of the dependences of Ṭārumayn, it empties into the Caspian Sea in Gīlān. This river is 100 parasangs long. As it cannot be used for irrigation, it is not administered.

Shāhrūd: It is two branches in Rūdbār of Qazwīn. One rises in the Ṭālaqān mountains of Qazwīn, the other in Mt. Shīr. It flows through the provinces of Rūdbār Alamūt and empties into the Safīdrūd in the Barra district of Ṭārumayn. Up to that point the river is 35 parasangs long, and it flows another 50 parasangs in Azerbaijan. Here, just like the Safīdrūd, it is not used for irrigation and has no administration.

Jāyjrūd (Jājrūd): It rises in Mt. Damāwand and flows through Rayy province. In the borders of Qūhad it is divided into 40 channels. In spring its flood waters make their way out into the desert.

Garmrūd: It rises in the Ṭālaqān mountains of Qazwīn and disappears in the desert after 35 parasangs.

Qamarrūd: [M 305] It rises at Mt. Khānīsār in Jurbādqān (province) and passes Jarbādiqān and Qum. It is 35 parasangs long. It is absorbed in the desert.

Gāwmāsārūd: One branch rises from Mt. Arwand; another from Asadābād, Māmshārūd and the hill of Farīwār; another from Mt. Rāsmand, Karaj and the mountains of the Kīsū Pasture. It passes Hamadān and Sāwa. At a place near Āwa, Khwāja Shams al-Dīn (Muḥammad b. Muḥammad Juwaynī, d. 1284), the *ṣāḥib dīwān*, erected a dam and because of that it becomes a lake. In spring the excess (i.e., what is not used for irrigation?) disappears in the desert. In summer Sāwa and Āwa get most of their water from the reservoir created by this dam. The Gāwmāsārūd is some 40 parasangs in length. Similar to the Zandarūd, this is also a *zāyanda* ("birthing") river.

Zanjānrūd: Rising at the borders of Sulṭāniyya, it receives the streams coming from the Zanjān mountains. After it passes the Zanjān province it empties into the Safīdrūd. It is 20 parasangs long.

Abharrūd—God is most great (*Allāhu akbar*): It rises at the borders of Sulṭāniyya and Mt. Sar-i Asad, meaning "Mt. Lion Head" and flows toward Abhar and Qazwīn. In spring it unites with the rivers of Qazwīn and flows to the desert. It is 20 parasangs long. It is also a "birthing" river like the Zāyanda in Abhar and Qazwīn, meaning that if it is dammed up its flow increases.

Āb-i Ṭārum: Rising in the Ṭārum mountains, it empties into the Safīdrūd, [R 114b] irrigating the fields of Ṭārum.

Āb-i Kāshān: rises in the Qamṭar mountains and flows to (the city of) Kāshān. In spring its lower course goes to the desert. In that season sometimes there is a big flood, causing a lot of damage to Kāshān. However, in summer it does not even reach Kāshān but waters (only) the upper villages.

Āb-i Muzdaqān (text: Murdaqān): It comes from Kharqān and Hamadān, passes Muzdaqān (text: Murdaqān) and comes to Sāwa. In the past it emptied into the lake of Sāwa. At the birth of the Prophet this lake dried up and a road was made that led to the river. The river disappears in the desert. It is 25 parasangs long.

Āb-i Qazwīn: It comprises four streams. In spring they flow and are enough for the gardens. If the flow is not strong enough, some of the gardens remain dry. In summer it does not reach Qazwīn.

Kardānrūd: It rises in the Ṭālaqān mountains and passes through the district of Qazwīn. In spring it reaches the desert, but in summer, not even Qazwīn.

Āb-i Kharqān: It rises in the Kharqān mountains. In spring it gets to the Khushkrūd. Their waters mix and disappear in the desert. However, in summer it does not go beyond Kharqān. All these rivers that flow to the desert are absorbed in the ground and disappear.

Roads and Stages in Persian Iraq

From Sulṭāniyya: Sakirābād village, where the Khurāsān road branches off, 24 > Ribāṭ-i Ḥājib 6 > Ribāṭ-i Wāsiq 7 > Sāwa 5.

From Kāshān: Mahrūd village 8 > Asṭa village 6 > Ribāṭ-i Mūrcha-yi Khurd ("Hospice of the Little Ant") 6 > Sīn village 20 > Iṣfahān 4.

From Asṭa > Māyīn, 13, but there is no populated place on the way > Sīn.

Iṣfahān is 48 parasangs from Kāshān, 64 from Sāwa, 106 from Sulṭāniyya.

From Sulṭāniyya > Tabrīz, to the northeast, 32 > Astarābād 37 > Ṭārum 10 > Rayy 50 > Zanjān 50 > Sujās 5 > Qazwīn 19 > Qum 54 > Kāshān 64 > Hamadān 20 > Yazd 14.

Iṣfahān > Iṣfahānak village 3 > Mahyār village, which marks the border with Fārs, 5 > Qūmis city 6 > Rūzkān 5 > Yazdkhwar 7. This is altogether 26 parasangs. The road used in winter turns left at Band-i ʿAḍudī (the dam built by ʿAḍud al-Dawla); the road used in summer turns right towards Kūshk-i Zard.

From Hamadān to Iraq: > Asadābād 9 > Qaṣr-i Duzdān (= Qaṣr al-Luṣūṣ) 7 > Qanṭarat al-Nuʿmān 4 [M 306] > Bīsutūn 2 > Kirmānshāh 8 > Zubaydiyya 8 > Maraj (?) castle 9 > Ḥulwān 10.

From Hamadān > Sāwa 30 > Rayy, toward the east, 30. The route is level, there are no mountains.

From Hamadān > Bārsīn 10 > Ūd 8 > Qazwīn 14 > Abhar 12 > Zanjān 15.

From Hamadān > Mādarān Castle 12 > Ṣaḥna 4 > Dīnawar 4 > Shahrazūr 4 > Ḥulwān 4.

From Hamadān > Rāman 7 > Wardkard 10 > Karaj 10 > Burj 12 > Ḥūmanjān 10 > Iṣbahān 30. Through desert.

From Hamadān > Rūdraward 7 > Nihāwand 9 > Lāshtar 9 > Khwāst 12 > Lūr, through desert, 30 > Andāmīsh, a town with a bridge, 2 > Jundīsābūr, 2.

From Hamadān > Sāwa 30 > Zanjān 30.

From Qum > Sāwa 12.

From Qum > Kāshān 18.

From Rayy > Qazwīn 27.

From Dīnawar > Ṣaymara 35.

From Dīnawar > Sarwān 4 > Ṣaymara 7.

From Iṣfahān, Ardistān is 34; Īdaj of Greater Lūr 45; Yazdajird of Lesser Lūr 66; Jarbādiqān 31; Sāwa 64; Qazwīn 92; Qum 52; Kāshān, via Māyīn to the northwest, 32; Naṭanz

20; Māyīn 26; Nihāwand 84; Shiraz, seven stages to the southwest, 72.

The direct road from Iṣfahān to Rayy goes through the desert and is difficult. The road through cultivated areas: > Kāshān > Qum, to the northwest > Rayy, to the northeast.

From Qazwīn, Rayy is 30; Surkhāb 32; Ardabīl 56; Tabrīz 80; Sāwa 34.

From Rayy > Kāshān 24 > Qūmis 8 > Dāmghān 64.

From Sāwa > Qum 12.

From Hamadān, Zanjān is 30; Simnān 32; Qum 40, all by a level road; Rayy 40; Darguzīn 16.

From Qum, Rayy is 11; Hamadān 40; Qazwīn 80 to the northwest.

From Asadābād, Dīnawar is 17; Mosul 40; headwater of the Zāb River 10; Marāgha 60; Shahribān 48 > Baghdad 16.

From Kāshān > Yazd 73.

From Qaṣr-i Shīrīn, Qaṣr-i Luṣūṣ is 50; Khāniqīn 7; Ḥulwān 5; Shahrazūr 20; Baghdad 30; Hamadān 58.

From Damāwand, Rayy is 12; Āmul 33; Simnān 64 to the south; Astarābād 96 to the east.

The rest (of the distances) can be deduced from the intermediary stages.

Conditions of kings and rulers of Persian Iraq

Before Islam, these lands were governed by commanders (*sarhang*) of the Persian kings. They were gradually taken from the hands of their march lords (*marzubān*) and officials (*ʿāmil*) during the time of the Rightly Guided Caliphs. Most were conquered during the caliphate of ʿUmar and were governed by officials (*ʿāmil*) and governors (*vālī*) on behalf of Abū Mūsā al-Ashʿarī and Nuʿmān b. Muqarrin [R 115a] who were companions of the Prophet and commanders (*sardār*) of the East. During the Umayyads, clients of Ḥajjāj were appointed as provincial governors (*mutaṣarrif*). In the Abbasid era, it was sometimes governed by their own officials, sometimes dynasties arose and governed independently.

Rulers of Qazwīn. According to the *Tārīkh-i guzīda*, under the Sasanians the territory of Qazwīn was governed by their own lords. During the reign of ʿUmar the governors were Abū Dujāna al-Anṣārī and Kuthayyir al-Ḥārithī. Under the reign of ʿUthmān it was given to Walīd b. ʿUqba. During the time of ʿAlī the governors were Rabīʿ b. Ḥaytham (?) Kūfī, Abū al-ʿArīf al-Hamadānī, ʿUbayda b. ʿAmr al-Salmānī and Qurṭ b. Arṭāh. Afterwards, when Ḥajjāj became the governor of most of the kingdoms of Iran on behalf of the Umayyads, he gave the governorship to his son Muḥammad. After Ḥajjāj, the governors of Qazwīn were Yazīd b. Muhallab, Qutayba and Naṣr b. Sayyār.

In the Abbasid era Persian Iraq was given to the Barmakids; afterwards it was governed by ʿAlī b. ʿĪsā b. Māhān, then by the Ṭāhirids. In order to protect those lands from the Daylamites, Hārūn al-Rashīd removed Qazwīn from the control of the governors of Iraq and gave it to Fakhr al-Dawla Manṣūr al-Kūfī, who was Ḥamd Allāh Mustawfī's fourteenth great grandfather and who was from the progeny of Ḥurr b. Yazīd Ribāḥī (al-Riyāḥī), sending him with a troop to Qazwīn in 203 (818–819). His sons governed there for 22 years and bore the sobriquet Fakhr al-Dawla. When Ḥasan b. Zayd al-ʿAlawī asserted his authority and conquered most of Persian Iraq in 251 (865–866), they submitted to him for two years. On the order of the caliph, al-Muʿtazz, Mūsā b. Būqā deposed him and served as governor for 38 years. Later Qazwīn was occupied by Ilyās b. Aḥmad al-Sāmānī. In 294 (906–907) with caliphal decree the province was given to Ḥamd Allāh's twelfth (great grandfather) Fakhr al-Dīn Abū ʿAlī. The Daylamites took most of Iran in 321 (933) and governed it for a century. When Sultan Maḥmūd of Ghazna took Persian Iraq in 421 (1030) he appointed as governor Ḥamd Allāh's ninth great grandfather, Abū Naṣr b. Fakhr al-Dawla, who had not yet come of age, and he even gave him the office of *mustawfī* (treasurer). It is thus that the Mustawfīs got their *nisba*.

After that the province was assigned to the governor of Qum, Ḥamza b. Yasʿa, and to the emir Abū ʿAlī Muḥammad al-Jaʿfarī. His progeny, the Jaʿfarīs, governed for as long as 60 years; the last one, Fakhr al-Maʿālī Sharafshāh b. Aḥmad b. Muḥammad al-Jaʿfarī, died in 484 (1091–1092). Afterwards ʿImād al-Dawla Tūrān, a slaveboy of the Seljuk Malik Shāh, became governor and ruled with his children for 50 years. Because most of the time they served at the Sultan's capital, the affairs of the province of Qazwīn were administered by his slaveboy Zāhid Khumārtash, who therefore has many good works in both Qazwīn and in Mecca. When Zāhid died in 503 (1109–1110), the people of Qazwīn asked the caliph, al-Muqtafī, for a governor. The caliph sent his slaveboy Baznaqsh Bāzdār, who took office in 535 (1140–1141). The country remained with his progeny for 106 years, the last one of them being Nāṣir al-Dīn b. Muẓaffar b. Alparghun b. Baznaqsh Bāzdār.

In the Mongol era, by decree of Mengü Qā'ān (the emperor Möngke) in 651 (1253–1254), Iftikhār al-Dīn Muḥammad became governor; he and his brother Imām al-Dīn Yahyā governed for 27 years. In 677 (1278–1279) Ḥusām al-Dīn Amīr ʿUmar al-Shīrāzī became governor; after that Abaqa appointed Khwāja Fakhr al-Dīn al-Mustawfī as governor. Later the office was given back to the Iftikhārīs, and by the end of Öljeytü's reign it had mostly been held by them. This is how in the *Guzīda* Ḥamd

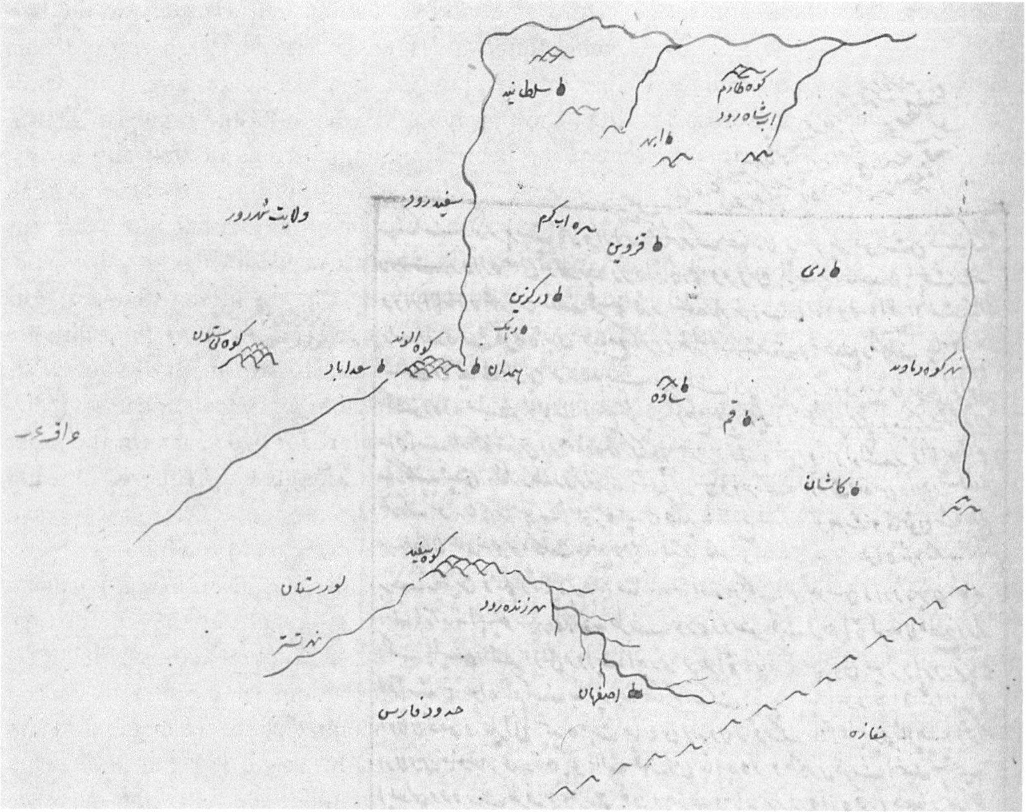

PLATE 43
Map of Jibāl or Persian Iraq, R 116a (Appendix, fig. 85)
TOPKAPI PALACE LIBRARY, REVAN 1624, WITH PERMISSION

Allāh reports down to his own time, as supplemented by other histories.

Dynasty of the Daylamites: They are also called Buwayhids (Buyids). Because they ruled mainly over the clime of Fārs, they were dealt with in that section. [R 115b] The members of this branch of the dynasty who ruled as independent governors over Persian Iraq are the following:

Rukn al-Dawla Ḥasan b. al-Buwayh made Iṣfahān his capital and took most of Persian Iraq under his control. He ruled for 30 years and was over 80 years old when he died in 366 (976). [M 308] He was succeeded by his son Mu'ayyid al-Dawla Abū Naṣr Buwayh, who ruled over Jurjān and Ṭabaristān as well. When he died in 373 (984), his brother Fakhr al-Dawla ʿAlī came from Khurāsān and took control over Persian Iraq. Ṣāḥib b. ʿAbbād was his vizier. In 387 (997) he too passed away in Rayy and was succeeded by his son Muḥammad al-Dawla Rustam. Because he was little, his mother Sayyida Khātūn was governing and when she died in 419 (1028–1029) the affairs of state became disordered. In [4]20 (1029–1030) Maḥmūd of Ghazna came and took over the realm. He seized Rustam together with his son Abū Dulaf and sent them to Khurāsān. This branch (of the dynasty) ended with this. Around 410 (1019–1020) his brother Shams al-Dawla became governor of Hamadān and gained independence. Ibn Sīnā was the vizier of this Shams al-Dawla.

Kākūyid Dynasty: *kākūya* is the word for "maternal uncle" in the Daylamite languge. Ḥusām al-Dīn ʿAlā al-Dawla Muḥammad b. Dūshman Ziyād, nicknamed Abū Jaʿfar Kākūya, was the brother of Majd al-Dawla Rustam's mother. In 398 (1007–1008) he acquired Iṣfahān as a grant (*iqṭāʿ*) and was governor for a number of years. He promoted excellence and was a good administrator. He passed away in 433 (1041–1042). When he was succeeded by his son Ẓahīr al-Dīn Abū Manṣūr Farāmurz, he was opposed by his brothers. Finally at their instigation Toghril the Seljuk besieged Iṣfahān and drove out Abū Manṣūr in [4]43 (1051–1052). He gave him the governorship of Yazd and Abarqūh in exchange; their rule was transferred to Yazd, as has been related in the Yazd section.

Dynasty of the Heretics (*malāḥida*): They are also called Ismāʿīlīs and Āl Ṣabbāḥ. The first member of the dynasty was Ḥasan b. ʿAlī, whose line goes back to one of the kings of Yemen, Yūsuf b. Ṣabbāḥ al-Ḥimyarī. After traveling a lot, he went to Mustanṣir al-ʿAlawī (the Fatimid caliph) in Egypt. He wanted to be his propagandist in the eastern lands. Receiving permission, he came to the fortress of Alamūt in one of the districts of Qazwīn. By some means he infiltrated the fortress and became master of it in 483 (1090–1091). He wrote books propagandizing for the Fatimids. As he displayed asceticism and abstinence, he attracted a following. Gradually he took DRKWL

(Dizkūh) near Iṣfahān, the fortresses of Khālanjān and Ṭabas, Girdkūh in Dāmghān, and other castles and his followers brought them under control.

Ḥasan knew the mathematical and occult sciences. When he died in 518 (1124), he designated Buzurg-Ummīd as legatee. After ruling nearly 14 years he also passed away in 523 (1129) and was succeeded by his son Kiyā Muḥammad. He died in 557 (1162) and was succeeded by his son Ḥasan who displayed corruption and was nicknamed Khwand. In 561 (1166), he too died and was succeeded by his son Muḥammad. He ruled until he died in 607 (1210). His son Kiyā Khudāvand Ḥasan, known as Naw Muslim ('Neo-Muslim'), succeeded him, and burnt the books of his forefathers and ancestors. Having espoused the outward interpretation of the Shariah, he passed to the other world in 618 (1221). He was succeeded by his son ʿAlā al-Dīn Muḥammad, who returned to the ways of his ancestors. His son Khāwar Shāh succeeded him after killing him in 653 (1255) while he was asleep. Not long afterwards Hulagu took their fortresses. In 655 (1257–1258), Naṣīr al-Dīn Ṭūsī [M 309] came out from the fortress of Maymūn through the mediation of Khāwar Shāh's vizier and (Khāwar Shāh) was killed at the hands of the Mongols. In this way Naṣīr al-Dīn Ṭūsī gained proximity to Hulagu. With the aforementioned (Khāwar Shāh), the Ṣabbāḥ dynasty came to an end. When they had appeared the surrounding kings and the Abbasid caliphs attacked and defeated them several times. Finally, the Chingisids exterminated them, destroying their fortresses. [R 116a] [Plate 43]

Chapter 33: Clime of Khurāsān and Quhistān

The 33rd chapter is on the lands (or cities, *bilād*) in the clime of Khurāsān and its desert and in Quhistān. Khurāsān means "Land of the Rising Sun". The ancient Persians said *khūr* for sun and *sān* for direction, and these were compounded to form Khurāsān.

As this clime lies in the middle of the climes, its climate is fine; the people are of a balanced humor, clever and wise, mighty and strong. Khurāsān is a noteworthy clime containing several great cities and some 500 towns. Since it is a place with many important holy men and shaikhs, quite a few accounts and works have been written about it.

Borders: To the east, the districts of Sijistān and some of the regions of India; to the north, the regions of Transoxiana and Turkistan; to the west, the desert between Persian Iraq, the regions of Jabal, Ṭabaristān and Khurāsān; to the south, the desert that separates Fārs from Kirmān and Khurāsān. It is roughly square in shape. Its extremities are like two arms extending into the desert, one between Herat and Ghūr to Ghazna, the other through Qūmis and the borders (?—*firāk*) of Qayṣarān.

According to Ibn Furāt, Khurāsān extends in length from Dāmghān to the Āmū Ṣuyı (Jayḥūn) and in breadth from Zaranj to Jurjān.

According to Ḥamd Allāh, formerly Quhistān, Qūmis and Māzandarān were within the borders of Khurāsān. Under the Mongols, Khurāsān, Quhistān, Qūmis, Māzandarān and Ṭabaristān were considered separate provinces. But (today only) Qūmis, Māzandarān and Ṭabaristān are considered as separate provinces.

The greatest cities of Khurāsān are Nishapur, Herat and Balkh. For this chapter we have excerpted Muʿīn Zamjī's *History of Khurāsān*, which we supplement with other descriptions and reports from the *Nuzha* and other works. In the past, the capital city of this great clime was Nishapur; later, Herat became the capital city.

Herat: A great city and ancient entrepot located at 95½ degrees longitude and 34½ degrees latitude. Situated in a flat plain, it is surrounded by gardens, orchards and streams. Canals have been made to flow even inside the city. Water comes from a barren mountain two parasangs distant on which there is a fire temple called Sirishk. [R 116b] As it approaches the city it divides into a number of branches, each of which irrigates a separate district. On that mountain they cut black millstones; otherwise it has no benefit.

The city has a fortress, a moat filled with water, and five gates. The Balkh Gate is made of iron, the others of wood. The governor's palace is in Khurāsānābād, separated from the city by half a parasang. Formerly it was an army camp.

The Grand Mosque, finely built and richly adorned, is on the outskirts of the city. A Kufic inscription on the portico of the ruler's private chamber attests that it was built by Sultan Ghiyāth al-Dīn Muḥammad (b.) Sām (r. 558–599/1163–1203) in 597 (1200–1201). It is a solidly constructed and blessed place (or shrine—*maḳām*), built in a curious shape and unique style. [M 310] To quote Mawlānā Saʿd al-Dīn al-Kāshgharī: "While the Koran contains verses on the Grand Mosque of Mecca, the bounty of this blessed place is no less than that." It has five gates and four terraces. Originally this was the only grand mosque in the city. Later, during the reign of Shāhrukh Sultan, Gawhar Shād Khātūn erected another richly adorned Friday mosque outside the city.

Ikhtiyār al-Dīn Castle is a strong and fortified castle in the northern part of Herat. Its towers and bastions are proverbial for their firmness. It has two gates, one that opens to the horse market in the north, the other to the city wall in the south. Quhandiz ("Old Castle") of Herat, today called Muṣarrakh, is said to have been built by Nebuchadnezzar.

There are many stories about the founders of Herat. In one version, it was founded by one of the emirs of Narīmān who was named Herat and was later renovated by Alexander.

In another version, it was founded by Shamīra, the daughter of Ḥamān son of Farīdūn. She was the queen of her people and built the Shamīrān fortress north of the city. The descendants of this people, who called their kings Khwasha, multiplied during the time of Manūchihr. They built Quhandiz, with four pavilions on its four sides and gates in the south and north, and with the fortress of Shamīrān inside. Quhandiz may have been enlarged to 12,163 cubits in the time of Bahman b. Isfandiyār.

In another version, first the town of Awba was built. Some time later, two brothers migrated from Kawāshān and settled in Khayabāt. Their descendants went to Shamīrān, the daughter of Bahman son of Isfandiyār, who was queen of Balkh. She granted them permission to build a fortress and they named it after her. Dārā (Darius) son of Dārā[b] built Quhandiz after that. Later Alexander came, and because the people of Quhandiz were complaining about the Turks, he built Herat. It was enlarged by Ashk son of Dārā.

In another version, Ḍaḥḥāk had a daughter (named) Herat. She first built Awba and then Herat.

Herat, when it was in ruins, was restored by the Apostles (of Jesus). There is a quatrain on this topic [Persian verse]:

Bihrāsp laid the foundation for Herat.
Gushtāsb placed another building inside it.
Bahman made another building after that.
Alexander of Rūm destroyed it all.

Shamīrān and Quhandiz, the citadels of Herat that are preserved today, are to the south. In the past they were surrounded by another citadel by Muʿizz al-Dīn Ḥusayn Kart—that is now in ruins. The castle of Ikhtiyār al-Dīn is inside the citadel, connected to the north of the city.

That citadel has five main gates: Malik Gate in the north, Iraq Gate in the west, Fīrūzābād Gate in the south, Khwash Gate in the east, and Qatjāq Gate in the northeast. Three smaller gates have been built to each large gate, except Malik Gate, which has but two smaller gates. Inside the citadel are markets named after the gates. Each gate has bazaars on both sides as far as the central market place, except Qatjāq Gate, which has none. The Malik bazaar extends from the foot of the fortress to the central marketplace. Aside from these, each of them has another bazaar parallel to it. Outside each of the gates is a parasang-long bazaar down to the end of the settled area of the city. In each quarter can be found a small bazaar comprised of various shops.

This city wall comprises two segments 10 cubits from each other. It has 149 towers, a circuit of 7,300 feet and a diameter [M 311] 1,900 feet in both length and breadth, from the Malik to the Fīrūzābād Gates and from the Khwash to the Iraq Gates. The moat of the city wall [R 117a] is 20 cubits wide.

The Grand Mosque is between the Khwash and Qatjāq Gates. It was confirmed by examination that the climate of that part of the city was finer than elsewhere. Otherwise it would have been appropriate that the Grand Mosque be in the middle of the city. The Ḥawīja Canal passes near the Grand Mosque; there is no other flowing water inside the city. The city itself has few gardens and orchards, but 3–4 parasangs hence there are plenty. Its *fakhrī* grapes and its watermelons are good.

The Sultan Ḥusayn Madrasa and Khānqāh (Sufi lodge) in Sarānjīl have no match. The building, built by Sultan Aḥmad Mīrzā on the ʿAlī Asad Tower between Qatjāq and Khwash Gates is famous for its loftiness and high station. The city wall built by Ḥusayn Kart was very extensive, one parasang in both length and breadth, stretching from the head of the Injīl Bridge to the Shaikh-i Ḥaram Ravine, and from the vicinity of Mulāsiyān to the Bridge of the Tentmakers. After capturing the city, Tīmūr considered that this wall was impossible to hold and defend, because of its size, and so he destroyed it.

Afterwards the settled area of the city increased so much that its breadth rose to two parasangs, extending from the Valley of Two Brothers to the Mālān Bridge. Indeed, buildings, orchards and villages came one after the other for 30 parasangs from Awba to Kūsūy[a], all of it considered as one city. Those who have seen it say that Cairo is but one quarter of Herat; Cairo, however, has more annexed areas.

The Herat River is known as Rūd-i Mālān. Flowing among the villages, it irrigates the fields and the villages on both sides. Some of these villages are as big as towns. One of them is the shrine, which is a gathering place for scholars, shaikhs and holy men. It has a market with 1,000 shops and all kinds of merchandise. Sultan Ḥusayn erected a Friday mosque there as well. Another is Siyāwushān, which in some years produces 3,000 loads of grapes. The other villages are comparable to these.

The dependencies of the city are divided into ten districts (*bölük*). 1. Tūrān and Tūniyān, 2. Qūrdān and Bāshtān, 3. Kambarāq, 4. Sabqar, 5. Khiyābān, 6. Kadāra, 7. Injīl, 8. Alanjān, 9. Adwān and Tīzān, 10. Jūn.

The Gardens, orchards and meadows around the city are without limit. In particular, the excursion ground of Kāzargāh, near the city, which extends from the head of the bridge over the Jakān stream to the bridge of Niʿmatābād, is full of pleasant gardens and orchards. This excursion ground has mountains on one side and open plains on the other three. Its site is higher in altitude than Herat. Sultan Ḥusayn built quite a few buildings there, making it extremely flourishing. The built up area extends from the Injīl River to the foot of Mt. Zinjīrgāh and joins the Valley of Two Brothers. Sultan Abū Saʿīd brought water from Rūd-i Pāshān, which flows from the middle of Mt. Zinjīrgāh, and started the Fountain of Māhiyān one parasang from Kāzarghāh.

Kāzargāh was originally Kārzārgāh ("Battlefield"). It was called this because when the Khārijites revolted in 206 (821–822), ʿAbd al-Raḥmān b. ʿAbd Allāh came from Nishapur to Herat, and there was a great battle near the plain where it lies. Later its pronunciation was lightened to Kāzargāh. Those on both sides who fell in this battle were buried in seven pits. People [M 312] came, built houses, a mosque and a hospice. Those buried there as a blessing eventually included Khwāja ʿAbd Allāh-i Anṣārī (Sufi saint, d. 1088), and so the place was considered blessed and became a place of visitation.

As for Khiyābān, it too is a blessed site, being exceptional on account of its excellent climate and fine constructions. The tombs of several eminent people as well as the open-air prayer grounds of the city can be found there. Commoners and aristocrats both consider the place to be noble and auspicious.

Concerning Herat, they relate a hadith from Ḥudhayfa b. al-Yamān that it is "the best of Khurāsān." The Heratis are tender hearted, swordsmen (?—*silāḥ-dūz*), intrepid, Sunni, mischievous (?—*ʿayyār-pīsha*), of good moral character, and praiseworthy. As Herat has always been a source of sciences and spring of scholars, in every century eminent people have issued from there. A certain poet said about this city [Persian verse]:

If someone asks you which is the best of cities,
In a true answer you would have to say Herat.
The world is like a sea, Khurāsān like mother-of-pearl,
And Herat is like a pearl in the middle of the shell.

One of the annexes of Herat is Awba, extending from the outskirts of the city to the river. These (annexes) are excellent places, vying with one another for pure water, delicious fruit, and fine climate. Here (in Awba) there is a hot spring at the foot of the mountain. Its water is so hot that it is impossible to enter it. They mix it with cold water found nearby, and people come from far and wide to bathe in it. It has medicinal properties. Sultan Abū Saʿīd erected a small building over it. Sultan Ḥusayn (Bayqara, Timurid ruler, r. 1469–1506) built it up with splendid buildings, gardens and orchards. In the mountain nearby there is a mine of white stone resembling marble and used to make columns, tablets, and *muʿābire* (?) chests. [R 117b] Many of the monuments in Herat are built from this stone, such as an obelisk with a tablet on the tomb of Khwāja ʿAbd Allāh Anṣārī, which is a wonderful tombstone.

(Another of the annexes is) rūd. Most fruit come to Herat from there. Its apples are especially delicious. It has villages, fields and excursion grounds with streams—a garden of Paradise and blessed place of visitation.

Another of the annexes is Kūhpāya-yi Shāfilān, which has a very extensive territory comprising fine fields and villages. One of the towns there is a small town by the name of Karūcha. Located on top of a mountain, it is a town with shops and markets. There are iron and lead mines in this district. Iron to Herat is supplied from here. Apples, apricots, pears and peaches are abundant. There is also a hot spring there called Chashma-yi Safīd Kūh (White Mountain Spring) with very beneficial water. Sultan Ḥusayn Bayqara built a lofty edifice here. People from all directions and regions come here for recreation, and the sick bathe in the water and are healed.

Isfizār: Since ancient times it was a flourishing province known as the Garden of Herat. Due to disasters and civil strife, it is now mostly in ruins. According to the *Taqwīm*, it is four towns next to each other with streams, gardens and orchards east of Herat, located at 97½ degrees longitude and 34 degrees latitude. According to the *Nuzha*, it is a mid-sized city with a few villages and abundant grapes and pomegranates. According to the *Haft Iqlīm*, it is now known as Shīzar. In the past it had a fortress called Ḥiṣār-i Muẓaffar, built on top of a high rock. [M 313] Muʿīn al-Dīn says that Rūd-i Isfizār passes by the foot of that mountain.

Isfizār had a Friday mosque, a public bath, and splendid buildings. In the middle of the fortress a fountain flowed with sweet water; later it dried up, and as there was no more water, the fortress was neglected and fell to ruin. Now only the walls of the fortification remain.

On a level with this fortress there is another ruined fortress called Shāristān on top of a mountain in the middle of the plain. The Rūd-i Isfizār flows between these two. Shāristān has an excellent climate, the envy of the Garden of Iram. It is a stronghold built on a circular rock. The ground in the middle of the fortress is soft; if someone digs one cubit deep, water rises from it. Although it is like that on the outside, too, it was a strong fortress. Some called it Shahristān-i Bilqīs (City of the Queen of

Sheba). When one of the eminent people of the country, Sharaf al-Dīn ʿAbd al-Qahhār, came here from Herat and saw the delightful shade and the delectable water, he composed the following hemistich: "The water and air of Isfizār betoken Paradise." Muʿīn al-Dīn Zamjī, a native of this place and author of the *Tārīkh-i mubārakshāhī*, added this hemistich: "Now because of your advent it boasts before Heaven."

In the town of Isfizār there were some 3,000 shops. Zāwal was one of the sections (*bölük*) of this district (*nāhiye*). Eighty underground channels flowed in its 3-parasang broad territory, each of them with enough force to turn a water mill. A flourishing village or a fortress lay at the head of each. Now some of these fortresses are intact, some are in ruins.

A fine pear called *khusrawānī* is produced here; it has no equal anywhere. There is also a strange kind of seedless jujube like a seedless grape. In the entire province, this jujube grows on a single tree in Barzīnābād village, one of the establishments of Barzīn Ḥakīm. The grape called *ṣāḥibī* is so delicate that if it falls on the ground while one is eating, it breaks apart.

The rivers and underground channels of the town of Isfizār all flow from west to east. In the districts of ʿAbqal, Khīrān and Firangān there are neighborhood mosques and Friday mosques.

Būshanj, also called Fūshanj and Pūshang: It is the Arabicized form of Pashang. It is a city with flowing waters, gardens and orchards, located on a flat place at 95 degrees longitude and 34 degrees latitude 7 parasangs northwest of Herat. The Herat River flows here and on to Sarakhs. Sometimes it peters out before getting there. The city has a fortress, a moat and three gates. The junipers growing here can be found nowhere else in Khurāsān. This town is the most ancient one of Khurāsān. Since the city was built by Pashang, son of Afrāsiyāb, it was also called Pashang city. Later it was Arabicized as Fashanj and gradually came to be known as Fūshanj. Some said that its founder was Hūshang. [R 118a]

It is a place with a large district and many fields and crops. Fruits, especially grapes, are good. They say 100 varieties of grapes grow here. Some of its towns are Khurdkard, Rūj and Kūsūya. The ancient hospice and the neighborhood mosque are said to have been built by Abraham. People make a pilgrimage to it every year. It is well known that if an official pays a visit to it he will certainly be dismissed from office. Therefore people in government service do not come here. There is a mountain near that hospice where there is a footprint on a rock that is said to belong to Abraham. A river flows by the foot of that mountain. [M 314] It is a very fine excursion ground, especially in the spring.

Kūsūya: A small town two stages northwest of Fūshanj. It had a fortress with four old walls in the middle of the plain. When Dawā Khān arrived there with 10,000 Mongol troops in 695 (1295–1296), the approximately 200 inhabitants got together and resisted them. Strange to say, after laying siege to the place for some time and fighting several battles, the Mongols retreated in misery and defeat. The sage Jāmāsb is said to be buried here.

Falbandān: one of the annexes of Herat, it is a spacious plain full of sown fields. Watermelons are so fine and abundant that Falbandān supplies all of Khurāsān with them. It is confirmed that they have grown one watermelon weighing 10 *mans*, and one muskmelon weighing 20 *mans*.

Bādghīs: A district 40 parasangs long and 30 parasangs wide bordering on Herat near Mālīn. Its town is called Bāsīn. In the past this was the capital of the Hayāṭila (Hephthalites). Being very windy, it was called Bād-khīz ("Raise the Wind") and was Arabicized as Bādghīs. It is located at 95½ degrees longitude and 35½ degrees latitude. Dahistān is one of its towns.

In this province there is a forest five parasangs in breadth and length, mostly pistachio trees. The inhabitants make a living by picking pistachios and taking them to Herat and other cities. It is tried and tested that if someone takes the pistachios gathered by another, his donkey will be eaten by boars, but if he commits no treachery, the donkey will suffer no harm.

This province has many water sources and sown fields. It flourished greatly at the time of Sultan Ḥusayn Bayqara.

Near Bādghīs is one of the most wonderful castles of the world, Nartūka. It is situated on top of a granite mountain, and of a solidity and inaccessibility that are rarely matched. The only path up to it is so narrow that only one man can go on it at a time. And after half a parasang the path stops and the rest of it, to the bottom of the castle, is rocks for 1000 cubits. No king or sultan has ever been able to conquer it by force. As it is very high up, roses bloom there in July.

Timber for the buildings of Herat, as well as firewood, comes from this province. However long this timber is in the building, it will neither rot nor become worm-eaten. Horses, sheep, and other livestock are very plentiful here and are taken to the surrounding places to be sold. Grain and other crops are likewise abundant. The state lands alone produce an annual 40,000 loads of grain, one load weighing 100 *mans*. If one *man* of seeds of any grain except rice is sown, it will produce 100 *mans* without irrigation and toil. Undoubtedly, this characteristic is due to the richness of the soil and the excellent climate.

The district (*nāhiye*) has three important institutions (*sarkār*):

1. The hostel (*langar*) of Amīr Ghiyāth, founded during the time of Tīmūr. It is now a flourishing town with some 300 shops, public buildings (or soup-kitchens, *ʿimārāt*) and gardens. There are fine grapes and a kind of watermelon known as *bābā shaykhī*. How the town was named is as follows: Mīr Ghiyāth was a lovely descendant of the Prophet who sold sweets. He had a shop in the Khwash Market. An ecstatic dervish by the name of Bābā Ākhī Maḥmūd-i Jāmī, who lived on the bank of the Injīl River, drew Mīr Ghiyāth to himself with his spiritual attractions and made him his disciple. He gave him permission to change his residence, saying, "Cast anchor (*langar*) wherever you hear the sound of a bell." Mīr went to the aforementioned place and alighted there, although it was empty. When his dervishes were digging a place in the ground for the hearth, they heard the sound of a bell coming from the ground. [M 315] He "cast anchor" there. The place whence the noise came is now said to be in that dervish hostel (*āstāna*). Gradually it became a town, the sultans raising buildings, and it is now a place of visitation.
2. A flourishing town called Chihil Dukhtarān ("Forty Maidens"). Mīr ʿAlīshīr has a splendid, lofty hospice there. There is a famous tomb and place of pilgrimage devoted to forty maidens. People resort to it, saying that their needs are met there. A stream flows next to the town, which has no ford in the spring. Sultan Ḥusayn Bayqara built a masonry bridge over it.
3. Jarlān, a great tomb and place of visitation.

As for Kārīz, a place in one of the administrative units of Bādghīs, Ḥukm b. Hishām (i.e., the impostor Hāshim, known as al-Muqannaʿ), possessor of the "moon of Nakhshab," is from there. He was the vizier of Abū Muslim. During the time of al-Mahdī (Abbasid caliph, r. 775–785), he revolted in Khurāsān, captured the fortresses of Kish and Nakhshab, and claimed divinity. There is a famous story of how in Nakhshab he used to produce a light like moonlight in a pit. [R 118b]

The summer pasture of Bābā Khākī: A fine meadow of Bādghīs, where most of the sultans and nobles of Herat go in the spring for an excursion. There is another such beautiful excursion ground called Takht-i Malik, a place with fine pastures and delightful meadows amidst snow-capped mountains. Another excursion ground is the summer pasture of Hazār Mīsh ("Thousand Rams"). In spring its tulip gardens betoken the rose gardens of Paradise.

Karkh, also called Karūkh: A district and town near Herat. It is located at 97½ degrees longitude and 35½ degrees latitude. It is a valley between mountains, 20 parasangs long and wide, completely interlaced with trees, gardens and orchards. It is a place with a pleasant climate and abundant grain and fruit crops.

Balkh, "the *qibla* of Islam": Located at 101 degrees longitude and 36½ degrees latitude and bordering on Ṭukhāristān, Badakhshān and Bamiyan, it is an ancient city and a magnificent settled area that comprises districts (*nāhiye*) and administrative units (*aʿmāl*). The urban area is on a broad and level plain, half a parasang long and wide, and is 4 parasangs from the closest mountains. A river called Dahās flows by its outskirts, turning watermills. The city is surrounded by gardens and orchards. Citrons, sugarcane, water lilies, dates, and other coastal fruits are abundant; grapes and watermelons are fine. Four melons make a single load. Other kinds of fruits are also plentiful. The Jayḥūn (Oxus) River passes on the eastern border.

Balkh has been the nursery of kings and the destination of mystics. In no period has there been an absence of eminent people. Ibrāhīm b. Adham, Shaqīq (al-Balkhī) and the Barmakids are from Balkh. It was a great city in pre-Islamic times as well. People came from far and wide to its fire temple called Bahār, which was respected and renowned among Zoroastrians. Traces of it are still extant. This fire temple was the endowment of the Barmakids' ancestors, who were the viziers of the Sasanians from the time of Ardashīr. From this lineage came Jaʿfar, Faḍl and Yaḥyā, who served as viziers under the Umayyads and ʿAbbasids. The world resounded with their renown for generosity and munificence.

Balkh was conquered for Islam by Aḥnaf b. Qays during the caliphate of ʿUthmān b. ʿAffān. It was enclosed by a city wall made of clay, in which six gates were placed: those of Nawbahār, Raḥba, Hindūyān, Jahūdiyān, Shaṣtband [M 316] and Yaḥyā. The Friday mosque is in the middle of the city. The Dahās River passes through the Nawbahār Gate and irrigates the surroundings. There are plenty of gardens and orchards in the neighborhood of each gate.

The Bactrian camel is found in Balkh. The city is said to have been founded by Gāyūmarth, completed by Ṭahmūrath and restored by Luhrāsb who surrounded it with a wall. According to a tract by Sultan Malik Shāh, most of the inhabitants are unzealous. According to the *Haft Iqlīm*, water was brought here by Kay-Kāwus, who made it flourish.

Since the fortress was destroyed by Aḥnaf, it was restored by Naṣr b. Sayyār. As most of his slaves were Indians, the fortress was called Qalʿa-i Hinduwān. At the time of Chingis Khan's invasion, Friday prayers were performed at altogether 1,200 places in the city and the villages. The welcoming and submission of the inhabitants proved useless; the entire city was destroyed.

[Persian verse]

He made all of Balkh like the palm of the hand,
He crushed all of its lofty buildings.

Later it gradually became flourishing again. Now it has a strong fortification with a moat full of water.

In the year 865 (1460–1461) when Mīrzā Baysonqur was governor of Balkh, a grave appeared in the village called Khwāja Khayrān, one of the annexes of Balkh, 3 parasangs from the city. On it was inscribed: "This is the grave of the Lion of God, ʿAlī b. Abī Ṭālib." A lofty building was soon erected above it, and it became a town with shops and a public bath. The river known as the Shāhī was made into an endowment for that grave, and the place gradually gained such renown that it became a place of circumambulation (i.e., a shrine) and a gathering place for the people of Balkh.

Shaburghān, also called Shabūrqān: A town and district 19 parasangs from Balkh, located at 100½ degrees longitude and 36 degrees latitude. It has plenty of sown fields and fruit crops, as well as streams and gardens. The watermelons are excellent; they are dried (for export). Its district is called Jūzjān or Jūzjānān. Fāryāb and Yahūda are also cities of this district.

Andkhūr (Andkhūy): one of the districts of Balkh. [R 119a] This was the dwelling-place of Tīmūr's spiritual mentor, Sayyid Jamāl al-Dīn Barka. Tīmūr had such a belief in him that he requested to be buried at the foot of his grave. For when he first came to the Jayḥūn (Oxus) region, the Sayyid offered him drum and banner (symbols of authority). Tīmūr regarded it as a good omen, and as his political fortune increased by the day, so increased his belief in and regard for the Sayyid.

Most of this province consists of endowments to the Holy Cities. There are extensive province and districts between Herat and Balkh, such as Sān, Jahār-yak, Jījaktū, Maymana, Qaysār, Murghāb, Marūjāq. All of them have flourishing villages, broad plains and meadows. Most of the inhabitants are nomads. They have flocks and belong to a tribe. The people of Herat greatly benefit from them, becoming rich from their flocks and livestock, grain crops and rice.

Marw-i Shāhijān: an ancient town on a level plain far from mountains, located at 97 degrees longitude and 37 degrees longitude. It is said to have been founded by Alexander. According to the *Nuzha*, it was founded by Ṭahmūrath and made into a capital by Alexander of Rūm. Later Abū Muslim raised there a Friday mosque and a lofty and lavishly decorated governor's palace. This was a portico 60 cubits in length and breadth, with a 30-cubit roof, a 50-cubit high dome, and gates on all four sides.

Caliph al-Ma'mūn lived in Marw while he was governor of Khurāsān. The Banī Layth, Chaghri Beg of the Seljuks, as well as his grandchildren Sultan Sanjar and Malik Shāh, lived here. He (?) surrounded the city by a wall 12,300 fathoms in circumference.

[M 317] The air is fetid, so the inhabitants are sickly, guinea worm being especially frequent. Water is gotten from the Marwrūd, which is also called the Murghāb River. As the soil is brackish, grain crops and silk are plentiful and profitable. A sown field yields a hundredfold already in the first year. Of fruits, grapes, watermelons and pears are good; they are dried and exported abroad.

Many eminent people came from here in the past, such as Burzōē the physician, Buzurgmihr the sage and Bāyazīd the minstrel in the Sasanian period; as well as Abū Muslim-i Khurāsānī, Burayda b. al-Khaṭīb who was a companion of the Prophet, Bishr-i Ḥāfī, ʿAbd Allāh b. al-Mubārak, and ʿAsjadī.

According to Muʿīn al-Dīn, Marw is comprised of two towns: Old Marw, now in ruins, built by Sultan Malik Shāh where his solidly constructed tomb is; and New Marw, built by Mīrzā Sanjar.

Prior to the construction of New Marw, the city was ruined to some degree during the rule of Sanjar the Seljuk. Later it was gradually restored. When Chingis Khan finished massacring and plundering in Balkh, he sent his son Tolui Khan with 60,000 troops to Marw, to besiege it. After 22 days of battle, the commander of the Marw garrison, Mujīr al-Mulk, who was one of the emirs of Sultan Muḥammad Khwārazm Shāh, came out with matchless presents and submitted. Tolui drove the people out of the city into the plain for four days. He gave 400 of them safe conduct and distributed the rest among the Mongol soldiers. Each one got 400, all of whom he massacred. That several hundred thousand souls were massacred, only four individuals remaining, is reported by the author of the *Ḥabīb al-siyar* from Sayyid ʿIzz al-Dīn Nassāba.

The city remained in ruins until the reign of Sultan Shāhrukh. In that period all the people were gathered. Mīrzā Sanjar made an effort to restore the city and founded New Marw. Now Old and New Marw are two separate towns. As is put by Mīr ʿAlī-shīr [Persian verse]:

Love makes the kingdom of the heart flourish for old and young.
You make Sanjar, the builder of Old Marw, enraptured (*sanjar*).

Marw has many dependencies, such as Mākhān and Talakhdān.

According to the *Masālik*, the old castle (*quhandiz*) of Old Marw is the work of Ṭahmūrath, the city itself that of Alexander. There are three Friday mosques: one built at the beginning of Islam; the Old Friday mosque; and the Māhān (= New) Friday mosque. The market was in front of the old Friday mosque. [R 119b]

Several canals flowed in the central marketplace: Rūd-i Hurmuz Farra (text: Qarra); Rūd-i Māhān, which flowed by the New Friday mosque as well as the governor's palace and the dungeon; Rūd-i Razīq, which flowed by the city gate and whose water was distributed in reservoirs—the Old Friday mosque and most of the districts of the city were on this canal.

The city had four gates: 1) Shāristān and Jāmiʿ; 2) Shaykhān; 3) Bar 4) Mashkān. The latter, situated towards Khurāsān, was where Ma'mūn's palace was located.

All of these canals came from a great river known as Marw al-Rūd, also Murghāb. At the village of Dīh-i Rawnaq it branched off in every direction into small streams.

The inhabitants of Balkh have a reputation for stinginess. There is a saying, "All the world's roosters call for the hen when they find a grain, except for the Marw rooster, which does not." From this, people have inferred that the people of Marw are stingy.

Sarakhs: A town surrounded by sand, located at 92 degrees longitude and 37¼ degrees latitude, in a flat place between Nishapur and Marw. It has no flowing water. Sometimes on winter days the overflow of the Herat and Fūshanj Rivers comes here. There are many pasturages in the vicinity. The main wealth of the people is camels.

[M 318] According to the *Nuzha*, the circuit of the city wall was 5,000 fathoms. It was a strong, earthen rampart. According to the *Haft Iqlīm*, it was one of the highly regarded fortresses of Khurāsān.

The climate is warm. There are grapes and watermelons. At the time when Muḥammad Khān Shaybānī conquered Khurāsān, 160,000 houses were registered in Sarakhs. When Shāh Ismāʿīl set out toward the region to do battle, Muḥammad Khān drove the populace to Transoxiana. Therefore Sarakhs was left in ruins for a time. It started to flourish again during the reign of Shāh Ṭahmās. Now its buildings are in good condition.

Of eminent people, the shrines of Saʿd al-Dīn al-Taftāzānī and Shaikh Abū'l-Faḍl Ḥasan are famous. The great Ḥanafī (jurist) Shams al-A'imma (al-Sarakhsī) comes from here.

Mihana, also called Mayhana: A town in Khābarān, one of the districts between Herat and Marw. It comes between Abīward and Sarakhs. The tombs of Asʿad-i Mayhanī and Abū Saʿīd-i Abū al-Khayr are important places of visitation.

Khābarān, also called Khāwarān: A flourishing district between Sarakhs and Abīward. Several eminent people were born here, such as Abū ʿAlī Shādān and the poet Anwarī, who said [Persian verse]:

> Be happy, O air and water of Khāwarān, for due to divine favor
> Like water of the sea or earth of the mine you are nourishing pearls.

Abīward, also called Bāward: A town located at 93 degrees longitude and 38 degrees latitude near Mayhana, 6 parasangs from Kūfan. It is 3 stages west of Marw and 1 stage east of Nasā. It has a bad climate. Of shaikhs, Fuḍayl-i ʿIyāḍ (d. 187/803), while of eminent people, Mawlānā Aḥmad Dānishmand comes from this city.

Kūfan: A town located 6 parasangs from Abīward at 93 degrees longitude and 36½ degrees latitude. It was founded by Muslims and built by ʿAbd Allāh b. Ṭāhir.

Nasā or Nisā, also called Jayghūl: A town located at 92¼ degrees longitude and 39 degrees latitude, one stage north of Abīward. It has abundant water sources and vegetable gardens and a large district. Between Nasā and Marw is desert. Nishapur is southwest of it. As reported in the *Taqwīm*, the climate is bad, hence the illness caused by guinea worms (?—*ʿarak-ı medīnī*) is prevalent. According to the *Haft Iqlīm*, however, it is a heart-refreshing place on account of its climate. It is famous for its 12,000 fountains, 12,000 plane trees and 12,000 holy men. Therefore Nasā is also called Little Damascus. It is said that next to the Sufi convent of Abū ʿAlī Daqqāq, there are the tombs of 401 eminent people. Imām Aḥmad (al-Nasāʾī, d. 303/915), the master of hadith and compiler of a hadith collection, comes from this place.

Ṭūs: located at 91 degrees longitude and 36 degrees latitude, it is a district comprising a number of towns. The main town is called Ṭābarān. Between this and another town called Nūqān is six parasangs. One of the Twelve Imams, Imam ʿAlī b. Mūsā Riḍā, died in Ṭūs in the year 203 (818–819) and was buried in the village of Sanābād, one of the dependencies of Ṭūs. It later came to be known as Mashhad and became very famous. Men of station and people of authority built tombs for themselves in the vicinity of the shrine and adorned it with splendid and lofty structures. It became a town. Especially under the Persian shahs (i.e., Safavids), [M 319] this shrine was full of golden and silver lamps and candlesticks. The name of Ṭūs was forgotten and the entire province became known as Mashhad.

Samʿānī relates that the district of Ṭūs comprises 1000 villages. It was conquered in the year 29 (649–650) in the time of caliph ʿUthmān b. ʿAffān. The tomb of Hārūn al-Rashīd is also in Sanābād.

In Ṭūs there is a kind of white stone used to make vessels. In Nūqān there are mines of serpentine, turquoise and malachite.

The tomb of Imām Ghazālī is in Ṭābarān. Naṣīr al-Dīn (al-Ṭūsī), Niẓām al-Mulk and Firdawsī come from this land (i.e., Ṭūs). According to Ḥamd Allāh, this land was first built up by Jamshīd; after falling to ruin it was restored by Ṭūs son of Nūdhar [R 120a] who named it after himself.

Of fruits, grapes and figs are abundant and sweet. In the vicinity of Ṭūs can be found the famous Rāyagān meadow, which is 12 parasangs long and 5 parasangs wide.

Chil-hazār: A stage between Ṭūs and Mashhad, located at 91 degrees 50 minutes longitude and 36½ degrees latitude. It can be reached in two days from Mashhad.

Khwāf: Located at 93 degrees longitude and 35½ degrees latitude, it is one of Nishapur's districts with many villages. Among its towns are Salāma, Sanjān, and Zawzan where King Zawzanī erected lofty buildings. Grapes, watermelons, pomegranates and figs are good; silk is abundant. In the *Taqwīm*, Zawzan is recorded as a large town in Quhistān between Herat and Nishapur, located at 93 degrees longitude and 35 degrees latitude. Of eminent people, al-Zawzanī, the commentator of the *Miftāḥ*, and of shaikhs, Zayn al-Dīn al-Khwāfī and Shāh Sanjān, come from Khwāf.

Jasht: A town with 50 villages connected to the Harīrūd. It is a place with an abundance of grain crops and fruits. The white apple is especially good.

Bākharz: A district between Nishapur and Herat. It has fields, gardens and orchards. Its town is also called Mālīn and Mālān. Located at 86 degrees longitude and 34½ degrees latitude, it has many villages. Its watermelons and halva are famous throughout Khurāsān. Most of the fruits of this district go to Herat. Its shaikhs are Sayf al-Dīn al-Bākharzī, Abū Saʿīd al-Mālīnī, and Zayn al-Dīn Abū Bakr al-Tāybādī who takes his *nisba* from one of the villages of this district; also al-Bākharzī (d. 467/1075) the author of the *Dumyat* (*al-qasr wa ʿuṣrat ahl al-ʿaṣr*).

Zāwa: The town of one of the provinces dependent on Herat. It has a strong earthen fortress and 50 villages. Some villages get water from underground channels, some from a small river. Its products are grain, cotton and silk. Fruit is abundant, grapes are without limit. Quṭb al-Dīn, the head of the Ḥaydarī order, is buried there. It is a place of visitation.

Jām: A famous town located at 93 degrees longitude and 36 degrees latitude, two stages southeast of Bākharz in the vicinity of Nishapur. 200 villages are registered as dependent on this city. Gardens and orchards are plentiful, fruits abundant. All its streams flow in underground channels. Mawlānā ʿAbd al-Raḥmān-i Jāmī came from this city. Of the great shaikhs, the tomb of Shaikh al-Islām Aḥmad-i Jāmī, nicknamed Zinda Fīl ("Living Elephant") is a famous place of visitation. The watermelon called Bābā Shaykhī is very fine. Near Jām, at the foot of the mountains, there is a fountain [M 320] whose water is warm in winter and freezing cold in summer.

Naysābūr (Nishapur): It was built by Shābūr (II, r. 309–379) Dhū'l-Aktāf. Its site was originally a reed bed, and since he had the reeds cut and a city built in its stead, it was named Nay-i Shābūr ("Reed of Shapur").

According to the *Nuzha*, it was founded by Ṭahmūrath, then fell to ruin. Ardashīr built the city of Nah between Kirmān and Sijistān. When his son Shābūr was governor of Khurāsān, he asked his father for that city, but he did not give it to him. Shābūr nevertheless set about restoring it and he named it Nah-i Shābūr. As recorded in the *Tārīkh-i Khurāsān*, the word *nah* means "city" in ancient Persian.

Located at 91 degrees longitude and 36½ degrees latitude, it is a magnificent and famous and flourishing city. The settled area is located on a flat place one parasang in length and breadth. In Persian they say Nishāwur. In the past it was the capital of the clime of Khurāsān.

Most of the water comes from underground channels and the air is salubrious. It is a noteworthy province, of which the vegetation is mainly sorrel, the soil is mainly bottomland, and the stones are mainly turquoise. As recorded in the *Nihāyat al-arab*, there is a kind of edible clay that is found nowhere else on the face of the earth; one *raṭl* of it is sometimes sold for a dinar. The turquoise mine is also a characteristic of Nishapur. When the Samanid Ismāʿīl entered the city, he said in admiration that if the underground streams only flowed above ground, and the rubbish and dungheaps that are above ground were only underground, this city would have no equal.

The inhabitants are faulted for lack of hospitality. As has been put by Murādī [Persian verse]:

Do not dismount in Nishapur in the evening,
For in this city strangers are despised.

According to the description of the city in ancient times, as given in the *Masālīk*, all the buildings were made of mud. Both parts of the city, the walled inner city and the suburb, became flourishing. The Friday mosques, a dungeon and the governor's palace were in the suburb. Between the dungeon and the palace was a quarter-parasang long military camp. The palace was built by ʿAmr b. Layth.

The walled city was a fortress with four gates: Wazpūl, Kūy-i Maʿqil, Quhandiz and Pūl-i Takīn. Quhandiz ("Old Castle") was built outside the walled city.

The suburb is surrounded with a wall and gates adjoining each other. The gates are: Jang Gate, which opens toward Balkh and Transoxiana; [R 120b] ʿUqāb Gate, which opens toward Iraq; and Sarbas Gate, which opens toward Quhistān and Fārs.

The markets of the city are in the suburb. There are flourishing and elongated marketplaces. One, called Bāzār-i Buzurg ("Grand Bazaar"), extends to the east and ends at the Ḥusayn Cemetery. To its left is a bazaar that extends as far as Sar-i Pul-i Ḥusayn. Its market is close to the small Ḥusayn Square. The governor's palace can be found there.

The streams of the city flow in underground channels by the markets, public bath and inns; then rise to the surface and irrigate the gardens and orchards. There is also a large canal that waters the fields of the villages.

Fine silks and delicate cotton fabrics are woven and exported. However, around 810 (1407–1408) there was a landslide and, according to common report, most areas of (the city) collapsed.

According to Ḥamd Allāh, the circuit of the city wall was 15 square fathoms, and the quarters were laid out like a chessboard, 8 segments long and 8 segments wide. Shābūr made great efforts in the construction, and under the Sasanians [M 321] this was the capital of Khurāsān.

Down to the age of the Ṭāhirids the governors of Khurāsān lived here. Later the Ṭāhirids moved to Balkh and Marw. The Banī Layth (Ṣaffārids) made this city the capital again. When it was destroyed in an earthquake in 605 (1208), another city called Shādyākh was built in the vicinity. Its walls were 6,900 fathoms in circumference, but it was also destroyed in an earthquake in 679 (1280–1281). In one of the corners yet another city was built, which is still extant. It is in front of a mountain to the south. Its walls are 15,000 fathoms in circumference. Water is gotten through underground channels. The water of the river of Nishapur comes from the mountain in the east. That mountain is very high and two parasangs east of the city. There are forty watermills along those two parasangs on the river, which flows forcefully. Five parasangs to the north, at the foot of the mountain there is a fine spring called Chashma-i Sabz whose water is green. Amīr Chūbān built a pavilion near that spring. On Friday nights a frightful voice can be heard from it.

Muʿīn al-Dīn says: In the past Nishapur was called Īrānshahr. It was the most magnificent city in Khurāsān. In 505 (1111–1112) it was destroyed by the cruelty of the Oghuz. The reason was that two servants had a dispute over a watermelon. They each went to their own lord. The two amirs blindly took sides and fought. One of them brought the Oghuz tribe from the Turks, the other one took the side of Sultan Sanjar. The two armies fought, Sanjar was defeated and the Oghuz destroyed the city. Later it was somewhat restored. At the time of Chingis Khan it was destroyed again. First came Chingis' son-in-law Ṭughājār, who was struck by an arrow and killed during the pillage of the city. By the decree of Sultan Jalāl al-Dīn, the governor of Nishapur was Sharaf al-Dīn, the Amīr-i Majlis. Knowing that Chingis would certainly attack, he made preparations. When Tolui Khan b. Chingis was finished with the destruction of Marw, he turned towards this city and arrived with a large army. In revenge for his brother-in-law, Chingis ordered the inhabitants of Nishapur massacred. The siege took place in the middle of the month of Rabīʿ II. Sharaf al-Dīn positioned 12,000 armed troops at each gate. The battle went on continuously for eight days. After many fell on both sides, the Amīr-i Majlis came out of the city with imams and city notables, going before Tolui Khan and pleading for safe conduct. Tolui did not give it, but had them arrested. Later they were taken in front of the city and killed. In an all-out assault the Mongol army came on like a flood. The fighting lasted until the next morning, [R 121a] when the Mongols entered the moat. Breaching the walls at several places, approximately 10,000 Mongols climbed up the ramparts. Desperate, the Muslims were determined to fight to the death for their goods and families. They struggled until they had no more strengh, and the sun of their lives set on Saturday. The Mongol army kept up the killing and pillaging for four days and four nights. Only four archers were left alive. (The Mongols) killed even the cats and dogs. They destroyed all the city walls and houses, razing them to the ground. [M 322] It took twelve days to assess that more than a million people were martyred. When the Muslims heard of this horrific event they renewed the mourning of Karbalā. In short, the city having been destroyed by earthquake three times, by the Oghuz occupation in 505 (1111–1112), and then by the Mongol destruction, was gradually restored.

Its eminent shaikhs were Abū Ḥafṣ Ḥaddād, Abū ʿUthmān Khayrī, Ḥamdūn Qaṣṣār, Abū ʿAlī Daqqāq, Abū al-Qāsim Naṣrābādī, Abū Muḥammad Murtaʿish, Abū Ḥamza, Abū Bakr Farā, Farīd al-Dīn Shaikh ʿAṭṭār. Its hadith scholars were Imām Muslim and Ḥākim. Other imams were Jawharī, Thaʿlabī, Wāḥidī, Thaʿālabī, Shaikh Riḍā, Niẓām Aʿraj, ʿUmar Khayyām, Kātibī the poet, Mīr Ḥusayn Muʿammāyī and Muḥammad Mu'min. These are some of the leading individuals of this city.

Sabzawār: A district and administrative center 20 parasangs from Nishapur, located at 88 degrees longitude and 36½ degrees latitude. Previously the administrative center of the district was Khusrūjird, which is the administrative center of Bayhaq. Then it became Sabzawār. Bayhaq is several villages in the aforesaid district. According to the *Taqwīm*, the largest was Khusrawjird, and the administrative center of Sabzawār was smaller than that. The *Nuzha* says the same.

It is a city with a moderate climate, and a solidly built and extensive market. Its products are wheat, some fruit and grapes. Forty villages belong to it. All the inhabitants are Imami Shīʿīs.

According to *Tārīkh-i mubārakshāhī*, in the time of Sultan Muḥammad Khwārazmshāh, the inhabitants of Sabzawār were accused of heresy (*rafḍ*, i.e., Shīʿīsm). They came before him and denied it. The ruler said to them: "If you are telling the truth, bring forth someone from among you who is named Abū Bakr." They tried hard but could not find anyone. Later they did find a weak and puny individual named Abū Bakr and brought him forth. Sultan Muḥammad said, "This man is neither dead nor alive." They said, "My padishah, pardon us. There is no Abū Bakr in our province better than this one." Mawlānā (Jalāl al-Dīn Rūmī) nicely versified this story in the *Masnavī*.

The main square of Sabzawār is such a delightful place that Paradise is said to be either under it or over it. Today the square in the middle of the city is known as Dīv-i Safīd ("White Demon").

Imām Bayhaqī, Yamīn al-Dīn Ṭughrāyī, Ibn Yamīn, and Amīr Shāhī are considered the great men of this region.

Tarshīz (Turshīz): According to Muʿīn al-Dīn, a pleasant city in Khurāsān. Ḥamd Allāh ascribes it to Quhistān. It is a small city and hot-climate region with a populace known for bravery and shedding of blood. There is a strong fortress built by Bahman b. Isfandiyār. Its water comes from underground channels. It has many villages. Grain and other crops are abundant. Fruits, especially figs, grapes and *malīsī* pomegranates are good. Silk is also among the products.

In the district of Turshīz there is a large village known as Kashmīr where it is said there are never earthquakes. Several strong castles are among the dependencies of Turshīz: Pardār, Minkāl, Mujāhid-ābād and Ātashgāh. Fruit and grain from this province are exported to Nishapur and Sabzawār.

Juwayn: The Arabicized form of Guwān. It is one of the districts of Nishapur. Its administrative center is called Āzādwār. It is a place with orchards and gardens and contiguous villages, three stages long and one mile wide, located at 92½ degrees longitude and 37 degrees latitude. [M 323] Its water is from underground channels. It has many vegetable gardens.

Coming from this district are the Shāfiʿī imams Imām Qushayrī and Imām al-Ḥaramayn (al-Juwaynī, ʿAbd al-Malik ibn ʿAbd Allāh, 1028–1085). Eminent men include Khwāja Shams al-Dīn Muḥammad *ṣāḥib-i dīwān*, ʿAṭā Malik (Juwaynī, 1226–1283) and Muʿīn al-Dīn.

Ḥamd Allāh says: Previously Juwayn was inside the borders of Bayhaq. Now it is an independent province.

Baḥrābād: A pleasant town of that district. Various fruits, especially grapes, are abundant. Shaikh Saʿd al-Dīn Ḥamawī and Shaikh Riḍā al-Dīn ʿAlī Lālā are buried here.

Isfarāyin: A town in the vicinity of Nishapur located at 91½ degrees longitude and 37½ degrees latitude. It is also called Mihrajān. Due to its many benefits, the pleasantness of the climate and the freshness of its meadows, Kisrā (Khusraw Anūshirwān) named it Mihrajān, saying that every day here is like the Day of Mihrajān in other countries [R 121b]—Mihrajān being Nawrūz and the spring season, the most pleasant time of the year.

Ḥamd Allāh says: Isfarāyin is a mid-sized city. In its mosque there is a large bronze bowl 12 *khayāṭī* cubits in circumference; no larger bowl is known to have ever been made. There is a strong fortification in the north of this town. A small river flows at the foot of this castle. Approximately 50 villages are dependent on it. Walnut trees, grain and grapes are abundant.

Imams from this city include Abū Isḥāq Isfarāyinī. Shaikhs include Nūr al-Dīn ʿAbd al-Raḥman Isfarāyinī; and among recent ones, ʿIṣām al-Dīn, Shaikh Azrī.

Amīn Aḥmad (Rāzī) says: Isfarāyin is a very extensive and pleasant province. In all the lands of Khurāsān, no place is superior to Isfarāyin and Qāyin, but in terms of pleasant climate, Isfarāyin even supersedes Qāyin. Its fruits are very fine. In particular, its pears are sent abroad as gifts. There are very old plane trees; most are said to have been planted in the time of Anūshirwān.

Jājurm (Jājarm): A mid-sized city near Isfarāyin, located at 91 degrees longitude and 37½ degrees latitude. It falls between Nishapur and Jurjān. As it is surrounded for a distance of two stages by poisonous plants, it was safe from the invasion of foreign armies, and it was secure and protected during times of interregnum. It has a fortification with its district around it. It produces grain and fruit.

Khabūshān: One of the important towns of Khurāsān in Ustuwā, which is a district of Nishapur. It is also called Khūjān. It was restored by Hulagu and further built up by Arghun Khan. It has a fine climate. Its grain and fruit are good. Of imams, Najm al-Dīn Abū al-Barakāt comes from here.

In the *Nuzha*, Ḥamd Allāh writes of a castle called Kalāt with a town called Ḥaram at its foot, as well as a small city called Marīnān. He says that they are administrative units of Nishapur, have flowing water, orchards and some villages dependent on them.

Addendum

According to the *Taqwīm* and other sources, the remaining cities of Khurāsān are the following:

Dandāniqān (Dandānaqān): A town located at 95½ degrees longitude and 37 degrees latitude two stages from Marw-i Shāhijān on the Sarakhs road. It has so much cotton and silk that it is proverbial for these products.

Sarmaqān—also called Jarmaqān: A town near Isfizār. It is six stages southeast of Tirmidh. There are 22 parasangs between it and Ṣaghāniyān.

[M 324] Qarīnayn, also called Bargudīn: A town in the valley of Marw al-Rūd, located at 97 degrees longitude and 38 degrees latitude. Marw al-Rūd is four stages from here.

Marw al-Rūd: A city located at 94 degrees longitude and 38 degrees latitude, 40 parasangs north of Marw-i Shāhijān. It has gardens and orchards and a fine climate. A river flows past it, irrigating the city's vegetable gardens and flower gardens. To the west of it, the road passes a desert for three parasangs and arrives at a mountain.

One of its dependencies, Qaṣr-i Aḥnaf is on the Balkh road one stage from here; it has flowing waters and orchards. One of its villages is Ṣāghān, which is the Arabicized form of Chāghān. To its west is a desert. Imām Ṣānī comes from this village.

Farāwa: A town in the direction of Khwārazm, located at 90 degrees longitude and 39 degrees latitude. It is also called Ribāṭ-i Farāwa. It was built by ʿAbd Allāh b. Ṭāhir during the caliphate of Maʾmūn as protection against the Oghuz. There is a single spring and a few vegetable gardens, separate from the villages. The hadith scholar Imām Farāwī comes from here.

Sīrzān: A town located at 92 degrees longitude and 36 degrees latitude.

Būzjān: A town located at 93 degrees longitude and 36 degrees latitude, four stages from the administrative units of Nishapur. It is between Herat and Nishapur.

Baghashūrā: A town located at 96½ degrees longitude and 36½ degrees latitude, between Marw and Herat. Its soil and air are salubrious. In the desert, water is gotten from wells. [R 122a] Its administrative center is called Kawn, its inhabitants are called Baghawī. Imām Baghawī comes from here.

Kushmayhan: One of the administrative units of Marw-i Shāhijān, it is a large village located at 97 degrees longitude and 36½ degrees latitude. It is located at the edge of the great desert, 5 parasangs from there (i.e., from Marw-i Shāhijān). It is famous for raisins, which are exported abroad.

Qubādhiyān (Qubādiyān), also called Qawādiyān: A district of Balkh with many vegetable gardens.

Fāryāb: The Arabicized form of Bāryāb. It is a town with many vegetable gardens, located at 89 degrees longitude and 37½ degrees latitude in the vicinity of Balkh, in the Jūrjān district. It is 22 parasangs from Balkh. In order to distinguish it from Fārāb in Transoxiana, people from here are called Faryābī or Fāryābī.

Zamm: A town located at 99 degrees longitude and 37½ degrees latitude on the bank of the Jayḥūn (Oxus), with the border of Khurāsān to its west. As foodstuffs are cheap, its people as well as the villagers are devoted to animal husbandry. Camels and sheep are plenty here. Āmul of the Stream (Āmul-i Shaṭ) is to the northwest.

Āmul: A town located at 97½ degrees longitude and 38½ degrees latitude, two miles along the west bank of the Jayḥūn in the direction of Bukhara. It is also called Āmū. Today the Jayḥūn joins it and is known as Āmū Ṣuyı (i.e., Amu Darya). It is also called Āmūl-i Shaṭ. The Jayḥūn flows northwest past here to Khwārazm.

Gharja: According to the *Nuzha*, it is a province located at 99 degrees longitude and 36½ degrees latitude and comprises 50 villages. He writes that it is the province known as Gharjistān. It has arduous and impenetrable mountains, many mountain passes and strong fortifications. Apples, pears, figs and pomegranates are good. The people are vehement of temper.

Kālif: A small town on the bank of the Jayḥūn. It reportedly has good climate and plentiful fruit. [M 325] Ḥamd Allāh mentions that the Jayḥūn is 3,000 fathoms wide here. He also mentions two small cities named Isfūrān and Yasāzān (Nasārān?) in the vicinity of Marw-i Shāhijān, but they are not found in the *Taqwīm*.

Country of Quhistān

It is the Arabicized form of Kūhistān. As recorded in the *Taqwīm*, it is a large district between Nishapur, Herat, Iṣfahān and Yazd. Its towns (i.e., administrative centers—*ḳaṣaba*) are far from each other, separated by large deserts. Water is gotten from underground channels; there is no flowing water above ground.

According to the *Nuzha*, it borders on the desert of Khurāsān, Transoxiana and Kabul. State dues from here are included in the country of Khurāsān. Its capital is Shahrastān. Tūn, Qāyin, Khūsaf and Janābdār are considered its major cities.

Shahrastān: A town located at 91½ degrees longitude and 39 degrees latitude between Khurāsān and Khwārazm on the border of Nishapur. It is called Ribāṭ-i Shahrastān. It was founded by ʿAbd Allāh b. Ṭāhir during the time of Caliph Maʾmūn. One of the eminent people from this town is Imām Muḥammad Shahrastānī.

Qāyin: A town of Quhistān, located on the edge of the desert at 94 degrees longitude and 33 degrees latitude. It is a city with scattered villages, water gotten from underground channels, and only a few vegetable gardens. It is between Iṣfahān and Nishapur. It is close to Ṭabas.

According to Ḥamd Allāh, it is a large city with a strong fortress. Most houses have underground reservoirs. It has a moderate climate. Its products are grain and fruit and there is plenty of saffron. [R 122b] Most of the inhabitants are mounted horsemen, wielding arms. Ploughing is done after forty days of the summer have passed. The yield is plenty. The grass is very suitable for animals. According to the *Taqwīm*, Quhistān mainly consists of this (i.e., grass).

Ṭabas: Two towns close to each other in the eastern part of the land of Quhistān. They are located at 97½ degrees longitude and 33 degrees latitude.

One is Ṭabas-i Mīnān, a small city with a warm climate and water gotten from underground channels. Grain crops are watered for seventy days. Around it is a tract seven days across; if someone eats but a single vetch that has grown from its soil he will die instantly. Nearby is a well with a lot of water flowing into it in winter. In summer it overflows and irrigates the fields. There is another well in which the shape of the moon can be seen.

The other is Ṭabas-i Gīlakī, a small city on the edge of the desert, one stage west of Tirmidh, between Nishapur and Iṣfahān. Both (Ṭabas-i Mīnān and Ṭabas-i Gīlakī) produce plenty of silk. Yazd is seven stages hence. As it has quite a warm climate, dates and sour oranges are plentiful. This is the only place in Khurāsān where these fruits grow. Water is gotten from a spring that drives two mills. There is a strong fortification with no grass around it. A few villages depend on it.

Tūn: A city and administrative center located at 92 degrees longitude and 35½ degrees latitude, near Qāyin, two stages south of Mashhad on the Kirmān road. Ḥamd Allāh says: It was a large city in the past and now is midsized. Its founder laid it out so that the fortification is in the middle, surrounded in concentric circles by the bazaar, then houses, then orchards, and then fields. Ditches and dams were dug around the fields. [M 326] They fill with rain water and dry out, then watermelon is planted. It is quite good. The city has a moderate climate. Water is from channels. Grain, fruit and silk are plentiful.

Tanja: An administrative center like Tūn, it is a small city also called Junābid or Gunābid. It has a strong fortification built on a sand hill, at the foot of which are orchards and villages. Water is gotten from channels coming from four parasangs distance. A castle has been built on both the northern and the southern ends of it, one called Ḥawāsar, the other called Darjān. This province produces grain and fruit. Silk is abundant.

Dasht-i Bayāḍ: A province the administrative center of which is called Fārs. It is a summer pasture in relation with Junābid and Tūn. Fruits of summer pasturage such as walnuts and almonds grow here.

Pīrchand: A town with plenty of saffron and little grain. It has a few villages. Fruits are grown there.

Khūsaf: A small city with a river passing by. A few villages belong to it; drinking water comes from underground channels.

Zīrkūh: A province with three towns: Asʿadan, Rāabr (?), Shāriḥat. It has grain crops, silk and cotton in abundance, and also much fruit. Water is gotten from underground channels.

Dara castle: A strong fortress near Qāyin. A spring rises above it. Jujube is abundant. It yields a lot of grain but not much fruit.

Muʾminābād: A province comprising several villages. It has a strong castle, built by the heretics (i.e., Nizārīs). It is a very strong fortress.

These are the 16 provinces recorded in the *Nuzha* as belonging to Quhistān. The *Taqwīm* registers Zawzan and the nearby town of Niyābud as being in Quhistān. Water there is said to be gotten from underground channels. However, we have followed Muʿīn al-Dīn, who reports Zawzan as belonging to Khwāf, because he is a native of that place.

Desert of Khurāsān

According to the *Masālik*, the Khurāsān desert is bordered in the east by Makran and part of Sīstān; in the south by Kirmān, Fārs and part of Iṣfahān; in the north by Khurāsān and part of [R 123a] Sīstān; and in the west by Qūmis, Rayy, Qum and Kāshān. Around this desert are the cities of Ḥabḍ, Rūd and Tarmāshīr in Kirmān; the two cities of Ṭabas and Qāyin in Quhistān; Dāmghān and Simnān in Qūmis; Māyīn and Yazd in Fārs; Iṣfahān, Qum, Kāshān and Rayy in Persian Iraq.

According to Ḥamd Allāh, this desert begins in Sūhqān, one of the villages of Qazwīn, whence it extends to Hormuz and the shore of the Indian Ocean, gradually becoming wider. In the south it passes by the borders of Sāwa, Qum, Kāshān, Zawāra, Māyīn, Yazd, Kirmān, Makran, ending in the sea. In the north it touches the borders of Rayy, Qūmis, Khurāsān, Quhistān, Zāwul and Sīstān and reaches the sea. It is in the third clime. Its length is 400 *parasangs*; its breadth starts with 1–2 parasangs and exceeds 200 parasangs at the seashore. Although the inhabitants of this desert are fewer than in any other desert in the land of Islam, there are more highwaymen and bandits.

According to the author of the *Masālik*, built houses are rare among these people, who are desert dwellers and nomads. Travel is impossible without a guide, because it borders on different climes and so is a haunt for evildoers. When they commit a crime or rob a caravan within the borders of one clime [M 327] they flee behind the borders of another. They have secret roads, unknown to outsiders. When a caravan sets out, it follows a familiar route that is known to have water. The robbers, however, have secret water sources and hidden gathering places where they hide stolen goods. One such place is the mountain known as Kūh-i Kargas, which is surrounded by two miles of desert. This mountain has a water source called Āb-i Bayda.

There are villages here and there in this desert, which has a pleasant climate. A small town called Sipanj on the Sīstān road is also counted as being of these (villages), but it belongs to Kirmān. On the Nishapur road from Iṣfahān there are three villages called Jarmaq, the Arabicized form being Jarma. It has a spring, date palms, camels and livestock.

According to Ḥamd Allāh, Ṭabas-i Gīlakī, Kuhbānān and the city built by Ardashīr, are considered as being the cities in this desert, aside from the ones on its edges.

Routes of the desert

There are a number of roads known in this desert, including those from Iṣfahān to Rayy, from Kirmān to Sīstān, and from Fārs and Kirmān to Khurāsān. There are eight roads altogether.

1. From Rayy to Iṣfahān, the shortest way: > Dara, 1 stage; it is a desert for 2 parasangs > a place called Barakchīn, 1 stage; there are wells here with briny water; rain water collects in a pond outside the monastery (?—*deyrden ṭaşra*) > Kāj, 1 stage; it is all desert > Qum, 2 parasang > village of Kabrān, 1 stage > Qāshān, 2 stages; a flourishing place; it can be reached following the edge of the desert > Ribāṭ-i Badr, 2 (stages); it has a fortification that encompasses 50 houses and is surrounded by fields and pasturages; some of the road goes through desert > Ribāṭ-i ʿAlī b. Rustam, 1 stage; it can be reached following the edge of the desert. The border of this desert joins Mt. Kargas, which is on its left, Siyāh Kūh ("Black Mountain") being on its right; the road goes between them, the distance between the two being 9 parasangs. In the aforementioned Ribāṭ the houses of the castle are guarded by sentries. Nearby are a village and a pond > Iṣfahān, 1 stage. Altogether 10 stages.
2. Māyīn to Khurāsān: > Mazraʿa, 1 stage; it has flowing water on the edge of the desert > Jarma, 4 stages; every two parasangs are one or two domes and a pond > Nawkhānī (?), 4 stages; every two or three parasangs there is a dome and a pond > Ribāṭ-i Ḥūrān, 1 stage > Bardasīr, 2 stages > Nishapur, 5 stages. There is a large village 3 parasangs from Ṭabas. Altogether 7 stages.
3. The saltmarsh road from Kirmān to Khurāsān: > Shūra: the name of a special water source and a stage in the desert on the border of Kirmān; 1 stage > Maʿzal spring, 1 stage. After this there are no populated places. > The stage of ʿUmar b. Sarḥ, 1 stage; there is a large pit here with red soil > stage, 1 stage; here several domes have been built [R 123b] and a well dug > 1 stage; there is a pool here that is filled by rain water. The Khurāsān road here turns right > Dirakhtistān, 1 parasang; it is said that amongst these trees are 71 kinds of tree with a human shape > Āb-i Shūra, 1 stage > pond, 4 parasangs; this is a large pond that fills with rain.
4. The Rāwan road from Kirmān to Sijistān: > Rāwan, 1 stage; it is on the border of Kirmān > Badjūy, 1 stage; it has a tiny stream > Sūd Dawwāra, 1 stage > ruined hospice, 1 stage; brigands are not absent here > Dayr-i Burqān, 1 stage; it is a hospice comprising 20 houses and a mill [M 328]; it has some meagerly flowing water; crops are sown at its foot and date palms grow. > domes, 2 parasangs; it has flowing water and date palms, but is not inhabited, being mostly a haunt for brigands > Tirshak, 1 stage; here at a two-parasang (long) place a pond and a dome have been made. Tīrshak is the name of a delightful water source > Chūr, 1 stage > Bast, 2 stages. Altogether 7 stages.
5. The Khabīṣ road from Kirmān to Quhistān: > Khabīṣ: one of the cities of Kirmān on the edge of the desert > Wāraq, 1 stage; it is in ruins and has no water > Shūra, 1 stage; a broad plain. Most of its water is (from) floods after rain > Mt. Arsal, 1 stage; a small mountain in the desert > Ḥawḍ, (or, a pool?) 1 stage. This road is dangerous. > Ribāṭ-i Chashma, 2 stages. At this stage there is flowing water. This place is protected by some 200 guards. It has some fields as well. If one goes two parasangs hence in the direction of Khurāsān one reaches a tract with small black stones. For four parasangs (the road) leads through this

stony tract. Aside from this, some places have small white stones, others small green stones. > Kawkūr, 1 stage; a flourishing village on the border of Kūhistān > Khūst, 2 stages. Altogether 9 stages.

6. The Yazd road, which goes to Khurāsān. Khayr (?), 1 stage; it has flowing water and a pond, but is uninhabited > Khwān, 1 stage; it is a desert, but there are desert dwellers who live in tents. They have livestock and fields. > Pul-i Siyāh and (Pul-i) Safīd, 1 stage; it is not inhabited; its pond is from rain water > Sabāʿīd, 1 stage; a village with 400 houses and flowing water > Yasht-i Bādām, 1 stage > Ribāṭ-i Muḥammad, 1 stage > Rīg-i Biyābān, 1 stage; there is a caravanserai and a pool that fills with rain water, but (the place) is not inhabited > Ribāṭ-i Kūrān, 1 stage; a solid, masonry building, accommodation for five-six persons. It has flowing water > Ribāṭ-i Kara, 1 stage > Zād-i Ākhirat, 1 stage. It has a caravanserai and a well, but it is uninhabited > Bustādārān, 1 stage; a village with 300 houses and flowing water. There are fields; water is gotten through underground channels > village of Dayr, 1 stage; a village with 500 houses. It has flowing water and fields > Dādūya, 1 stage; it has a caravanserai and a well, but is not inhabited > Ribāṭ-i Zangī, 1 stage; 3–4 people live here. It has flowing water > Āstalisht, 1 stage; it has a lake, a caravanserai and flowing water, but is uninhabited > Barsīr, 1 stage; this is on the border of Nishapur. 2 parasangs hence a pool has been built with an inn beside it. Altogether 16 stages.
7. A new road: From Narmāsīr > Rustān, 1 stage; a village with date palms at the edge of the desert. Beyond it there is no habitation. > Chashma-i Surāb, 1 stage; village of Salim, 4 stages; a village. These four stages are all desert. > Herat, 2 stages. Altogether 8 stages.
8. The Sīstān road: From Narmāsīr > Sipanj on the border of Kirmān 5 (stages) > Sīstān, 7 stages. Altogether 12 stages.

Rivers and mountains of Khurāsān

This clime has few rivers and mountains. These are the famous ones:

Murghāb River: It is also called Āb-i Zarbaq, because it branches at the village by the name of Zarbaq. It rises in the Murghāb and Bādghīs mountains and comes to Marw al-Rūd. It passes through part of Khurāsān and empties into the Caspian Sea.

Āb-i Saḥar: It rises in the Saḥar mountains and goes to Nishapur, irrigating some places on the way. It is 3 parasangs long.

Āb-i Shūrarūd: It passes through the middle of Nishapur province. The water of the mountains on the two sides also empties into it. [M 329] It irrigates the fields of the entire province.

Āb-i Darbād: It rises in Mt. Darbād; in spring its overflow empties into the Shūra.

Āb-i Kharw: Rising in the mountains at the borders of Nishapur, it empties into the Shūra.

Mt. Kanābid and Raband: Two mountains in Qūhistān province. Firdawsī mentions them.

Mt. Gulistān: [R 124a] It is in Ṭūs. There is a portico-like cave and vestibule (grotto?—*dehlīz*). When one advances in it for some time, one emerges into the light. It is like an enclosed graveyard. There is a spring whose water turns to stone.

Mt. Shatān: It is at the border of Jājurm. There is a crevice there from which water flows that drives two watermills.

Roads and stages

From Nīshāpūr to Herat: village of Tāybād, 7 parasangs; the Sarakhs road goes to the left > Ribāṭ-i Badlaʿī, 5 parasangs > village of Farhāwān, 7 parasangs > village of Saʿīdābād, 7 parasangs > village of Khusraw, 5 parasangs > city of Lūḥ-kān, 7 parasangs; the roads to Qāyin, Sarakhs and Bākharz branch off here > village of Gulābād, 6 parasangs > Kūshk-i Manṣūr, 10 parasangs > city of Fūshanj, 6 parasangs > Herat, 8 parasangs; altogether 68 parasangs; Dāmghān is 145 parasangs from Herat, Rāmīn 191 parasangs, Sulṭāniyya 251 parasangs.

From Nishapur to Turshīz in Quhistān: Ribāṭ-i Saddī, 5 parasangs > Ribāṭ-i Nūrkhā, 4 parasangs > Chāh-i siyāh, 3 parasangs > village of Dāna, 5 parasangs; at this stage there are seven flourishing villages and flowing streams > village of Bamrū, 4 parasangs > Turshīz, 7 parasangs. From Turshīz Tūn is 25 parasangs, Qāyin 31 parasangs, Lūḥ-kān, 36 parasangs.

From Herat to Marw: Sagābād, 5 parasangs > Bādghīs, 5 parasangs > Baʿra-shūr, 8 parasangs > Tūn, 5 parasangs > Marghzār-i Dara, 5 parasangs > Būsrūd, 5 parasangs > Marw al-Rūd, 3 parasangs.

From Herat to Marw al-Rūd is 37 parasangs > town of Aḥnaf, 5 parasangs > Ḥūrāb, 4 parasangs > Asadābād, 6 parasangs > Farshī, 7 parasangs > Ḥatābād, 5 parasangs > Mahdīābād, 7 parasangs > Fārs, 6 parasangs > city of Marw, 7 parasangs; from Marw al-Rūd to Marw is 48 parasangs.

From Sarakhs via Balkh to the Jayḥūn (Oxus) on the border of Iran: Ribāṭ-i Jaʿfarī, 10 parasangs > Mīl-i ʿUmarī, 7 parasangs > Ribāṭ-i Bū Naʿīmī, 7 parasangs > Āb-i Shūr, 5 parasangs; this is a sandy stage, a desert with no flowing water > Hindū, 5 parasangs > Marw al-Rūd 5; altogether 35 parasangs; > Ribāṭ Sulṭān, 7 parasangs > village of Kūjābād, 5 parasangs. On this road the city of Ṭālaqān on the right is 6 parasangs. From Kūjābād > Āb-i Garm, 7 parasangs > Kabūtarkhāna, 5 parasangs > Masjid-i Rāzān, 7 parasangs > Āsitāna 7 parasangs. The city of Fārāb is two parasangs from here to the right. From Āsitāna > Ribāṭ-i Kaʿb, 6 parasangs > city of Shubūrqān, 10 parasangs. There is no flowing water from Āsitāna to Shubūrqān > village of Salbāʾān (?), 2 parasangs > city of Balkh, 2 parasangs.

Balkh is altogether 62 parasangs from Marw al-Rūd, 107 parasangs from Sarakhs, 140 parasangs from Nishapur, 224 parasangs from Dāmghān, 278 parasangs from Rāmīn, 334 parasangs from Sulṭāniyya.

From Balkh: > Āsyākard (?), 6 parasangs > Tirmidh on the Jayḥūn, 6 parasangs.

From Nishapur: > Kurdān, a village near Asadābād on the border of Qūmis, 7 stages > Dāmghān, 5 stages.

From Nishapur: > Sarakhs, 6 stages > Marw, 6 stages > Āmul on the bank of the Jayḥūn, 6 stages.

From Nishapur: > Isfarāyin 5 stages.

From Nishapur: > Pūshang, 4 stages > Herat, 1 stage,

From Herat: > Isfarāyin, 3 stages > Dara, 2 stages.

[M 330] From Qāyin to Herat is 8 stages. From Marw to Herat is 12 stages. From Marw to Bāward is 6 stages, to Nasā is 4 stages. From Herat to Marw al-Rūd is 6 stages. From Herat to Sarakhs is 5 stages. From Balkh to the bank of the Jayḥūn is 2 stages. From Balkh to Andarāba is 9 stages, to Bamiyan is 10 stages, to Ghazna is 8 stages. From Balkh to Badakhshān is 13 stages, to Khwārazm is 21 stages.

From Nishapur: > Pūzkān, 4 stages > Mālin—not the Mālin in Herat—1 stage > Jām, 1 stage > Pashkān, 1 stage > Zawzan to the left, 1 stage > Qāyin, 3 stages > Kaydaram (?), a dependency of Nishapur, 1 stage > Sāyand, 2 stages.

From Nishapur to Khusrawgird (Khusrawjird) is 4 stages.

From Nishapur: > Khāwarān, 1 stage > Mihrajān, 2 stages > Isfarāyin, 2 stages.

From Balkh: > Khalam, 2 stages > Wālīn, 2 [R 124b] stages.

From Balkh: > Baghlān, 6 stage > Waka, 1 stage > Shubūrqān, 3 stages > Fārāb, 3 stages > Ṭālaqān 3 stages > Marw al-Rūd, 3 stages.

From Qāyin: > Zawzan, 3 stages > Ṭabasīn, 2 stages > Wachūn, 1 stage > Khūst, 1 parasang.

From Herat to Nishapur and Marw and Sijistān is 11 stages.

From Balkh to Farghān is 30 stages, to Rayy is 30 stages to the west, to Sijistān and Kirmān is 30 stages east to the south, to Khwārazm is 30 stages to the north.

Again from Balkh to Multan is 30 stages, to Kabul is 20 stages, to Bukhara is 12 stages. There is no city in between, only villages. Some say that the direct road from Balkh to Kabul is 5 stages, 3 of which are in the desert.

From Balkh to Shubūrqān is 2 stages, to Marw is 12 stages.

From Marw to Nishapur, Herat and Bukhara is 12 stages each.

From Sarakhs to Nasā is 28 parasangs to the north, to Nishapur is 5 stages.

Marw > Abīward, 3 stages to the west > Nasā, 1 stage to the north.

From Nasā to Khwārazm is 8 stages to the east. From Ṭūs to Nishapur and to Nasā are 2 stages each, to Jām is 3 stages to the northwest. From Marw al-Rūd to Bukhara is 7 stages to the east.

Rulers of Khurāsān

Muʿīn al-Dīn reports in his 8th chapter from the book of ʿAbd al-Raḥmān al-Fāmī that the governors of Khurāsān formerly resided in Herat and sent agents and deputies to the other cities. In the 29th year of the Hijra (649/650) Herat was taken by capitulation, and in the same year Marw also went into Muslim hands. A year afterwards Nishapur was taken, and when ʿAbd Allāh b. ʿĀmir came to Khurāsān he stayed there and subdued it by warfare. Most cities of Khurāsān were conquered by the aforesaid ʿAbd Allāh. After that Saʿīd b. ʿUthmān b. ʿAffān became governor. ʿAbd Allāh b. ʿĀmir returned to Khurāsān and governed there. After him, Ḥukm b. ʿAmr al-Ghaffārī, ʿAbd Allāh b. Ziyād, ʿAbd al-Raḥmān b. Ziyād, Sālim b. Ziyād, and ʿAbd Allāh b. Ḥāzim al-Sulamī were governors for ten years at the time of Qutayba b. Muslim. After him Umayya b. ʿAbd Allāh al-Qurashī governed for four years, (then) his son Yazīd and Yazīd's brother Faḍl b. Muhallab. Then Qutayba b. Muslim was governor for two years, but he was murdered in 96 (714–715). After that Wakīʿ b. Aswad governed for a month. And while Yazīd b. Muhallab was again governor of Iraq, he came to Khurāsān and stayed for a month, making his son Muhallab governor of Khurāsān. Six month later ʿUqba b. al-Ḥawāj al-Ḥārithī governed for one year and two months. [M 331] Leaving ʿAbd al-Raḥmān b. Naʿm al-Ayādī in his place, he himself went to ʿUmar b. ʿAbd al-ʿAzīz. After one year and seven months Saʿīd b. ʿAbd al-ʿAzīz al-Qurashī succeeded him, governing for one year. (Then) Asad b. ʿAbd Allāh al-Bajlī governed for three years, later leaving as successor Ḥakam b. ʿAwāna and going to Iraq. Six months later Asad came back and was governor for five months. Then Junayd b. ʿAbd al-Raḥmān was governor for three and a half years, appointing as successor ʿAmmāra for two months. ʿĀṣim b. ʿAbd Allāh al-ʿĀmirī governed for 3 months, Asad b. ʿAbd Allāh al-Bajlī for four years, Jaʿfar b. Ḥanṭala and Naṣr b. Sayyār al-Laythī for six years.

When Abū Muslim rebelled, rule was transferred to the Abbasid dynasty. The last Umayyad governor was Ibrāhīm b. al-Ḥārith al-Ḥanafī. He was succeeded by Nahār b. ʿAbd al-Raḥmān al-ʿĀmirī. The Umayyad governors governed for 72 years.

In 128 (745–746) Abū Muslim appointed ʿUthmān al-Kirmānī governor in Herat. He was succeeded by Abū Manṣūr b. Ṭalḥa al-Khuzāʿī and Dāwūd b. Kazzāz al-Bāhilī. In 146 (763–764) Laʿriyān (?) revolted. After the suppression of their revolt Mushamrakh al-Ḍabbī took Herat in 164 (780–781). After that the governors were Jinān b. Nuʿmān, Hārūn b. Ḥamīd, Qaṭn b. Muḥārib al-Dhuhalī, Yazīd b. Jarīr, Ismāʿīl b. Ghazwān. At the time of the latter, in the year 178 (794–795), a great terror arose. Mushamrakh became governor three times, and in 191 (806–807) Rāfiʿ b. Layth al-Kārī came to power. Hārūn al-Rashīd came to Khurāsān in order to remedy the situation [R 125a] but died in Ṭūs. Then Muḥammad b. Shāddād became governor for six months in 200 (815–816), succeeded for 37 months by Hārūn b. Ḥusayn. The Kārizgāh incident (?) took place in his time. He appointed Ibrāhīm b. Muḥammad as successor. Then Yaʿqūb b. ʿAbd Allāh Ḥājib became governor and Ilyās b. Asad al-Sāmānī, ʿAzīz b. Nūḥ and ʿAbd Allāh b. Muḥammad al-Maʿbadī took control of the administration. In his time there was a serious drought. When Ibrāhīm b. al-Ḥusayn was governor in 222 (837) a great comet appeared from the east, the like of which had not been seen before. ʿAbd Allāh b. Ṭāhir died in that year. The governor of Herat was Jibrīl b. ʿAbd Allāh. He was succeeded by ʿAbbās b. ʿAbd Allāh. After that his deputy Ibn Ḥusayn, ʿAzīz b. al-Sarī, Muḥammad b. Nūḥ and Ḥusayn b. ʿAbd Allāh were governors.

In 256 (870) Yaʿqūb b. al-Layth came out of Sīstān and gave the province to Dāwūd b. Manṣūr al-ʿĀdil. When Dāwūd died in the same year, he appointed his brother ʿAbbās to succeed him. Later ʿAzīz b. al-Sarī and Aḍram b. al-Sayf came. Then Ṭāhir b. Ḥifṣ occupied Herat. He went into battle against Maʿmar b. Muslim and was killed. Maʿmar b. Muslim conquered Herat, appointing as successor Hārūn al-ʿAbbāsī. Then Muḥammad became governor and in 260 (873–874) ʿAmr b. al-Layth, whose brother Yaʿqūb died in Ahwāz, ʿAmr becoming his deputy. He appointed as successor ʿAlī b. Ḥasan al-Dirhamī. When in 267 (880–881) ʿAmr went to Sīstān, after him the governors were Baḥr b. Aḥnaf, ʿAmr b. ʿAmmāra and Yūsuf b. Maʿbad. Unexpectedly, when Rāfiʿ b. Harthama came from Marw to Herat, he arrested Yūsuf and made Abū Jaʿfar b. Aḥmad his deputy. Later Shādān, was the deputy of ʿAmr b. al-Layth in Nishapur, became governor of Khurāsān. [M 332] In 287 (900) ʿAmr went to Balkh and was killed at the hands of Ismāʿīl b. Sāmānī.

The Samanid state, etc.: According to the aforesaid chronicle (of Muʿīn al-Dīn), the first governor sent by the Samanids to Herat was Abū ʿAlī Ḥasan b. ʿAlī Marw al-Rūdī, coming to office in 287 (900). He was succeeded by Sīmjūr and then by Ismāʿīl b. Muḥammad Dahistānī. The latter was followed by Muḥammad b. Harthama, the brother of Rāfiʿ and then Aḥmad b. Sahl, who was sent by Naṣr b. Aḥmad. He laid siege to Herat for twenty days, taking it by capitulation in 306 (918–919). He was succeeded by Abū al-Fawāris, who went to Nīshāpūr. The people of Herat elected ʿAbd al-Raḥmān b. Muḥammad as emir. He held office for fourteen months. Abū Manṣūr Mahānī governed for a year, to be succeeded by his brother Ḥasan. Then Sīmjūr became governor. He was succeeded by Abū ʿAlī Qumī, (but) the people of Herat elected Abū ʿAmr Saʿīd b. ʿAbd Allāh Ḍabbī as emir and Abū Bakr Ṭaghār became the governor's chamberlain.

In 318 (930–931) Abū Zakarīyā Yaḥyā b. Aḥmad and his two brothers, Manṣūr and Ibrāhīm, who were imprisoned in Bukhara, got out and sent Sāsī to Herat. Abū Bakr fled. Then Manṣūr b. ʿAlī was appointed (governor) by Naṣr b. Aḥmad. After that Sāsī returned and Amīr Abū Zakarīyā gave the province to Qaratigin. He captured and killed fourteen notables of the city and subdued the people. After Abū Zakarīyā departed, Naṣr b. Aḥmad came and gave the province to Sīmjūr, then went in pursuit of his brother. Abū Zakarīyā returned from Samarqand to Herat and reappointed Qaratigin. When he departed, Qaratigin went to Fūshanj and Abū Bakr Muḥammad b. Muẓaffar and the chamberlain Ṭaghār came and made Ibrāhīm b. Fārsī governor. In Ramaḍān of 320 (September–October 932) they dismissed him, giving the office to Manṣūr b. ʿAlī. After him the governors were Abū al-ʿAbbās b. al-Jarrāḥ, Mulakā (?) Tigin, Abū Jaʿfar ʿAzīz and Ibrāhīm Sīmjūr. (This last one) destroyed the city wall and went to Nishapur. In 338 (949–950) ʿAbd al-Raḥīm Māraqī tore down the southern vaults of the Mālān Bridge.

In (3)41 (952–953), when Ibn al-Jarrāḥ was governor, he went to the mountains (i.e., rebelled). The emir Abū Yaḥyā Asʿad b. Muḥammad Sāmānī came and closed off Quhandiz, battled with the garrison of Shamīrān [R 125b] and killed a faction of its people. In that year, Abū ʿAlī Tūlakī pillaged Ṭuḥāb and burnt its mosques. After that the province of Herat was given to Abū al-Ḥasan Muḥammad b. Ibrāhīm Sīmjūr and Abū Manṣūr b. ʿAbd al-Razzāq. In their time as well there were a number of battles and civil wars. After that the governors were Alp Tigin, once again Abū al-Ḥasan Sīmjūr, Abū al-Ḥasan Fāryābī and Ṭalḥa b.

Muḥammad Nasafī. In 354 (964) the Carmathian discord occurred and their chief Abū al-Ḥasan Dāwūdī was killed. After that Abū al-Ḥasan Mazanī became governor.

In 360 (970–971) the province came under the jurisdiction of Sultan Maḥmūd Sabuktigīnī (Ghaznavi) and Abū ʿAlī Sīmjūr came (as governor). He was followed by his son Abū al-Qāsim, then Abū al-Ḥasan in (3)71 (981–982), the chamberlain Tāsh in (3)73 (983–984), and once again Abū ʿAlī. In (3)81 (991–992) Lamangū was appointed governor. In (3)82 (992–993) Nūḥ b. Manṣūr and Sabuktigīn came and fought with Abū ʿAlī. Abū ʿAlī was defeated and Amīr Sabuktigīn occupied the *vilayet*. He was followed by Bektüzün in (3)88 (998). In (3)96 (1005–1006) the Khan of the Turks (?—*Türkān ḫānı*) came and caused a lot of damage. In (39)8 (1007–1008) [M 333] he fought a battle with Sultan Maḥmūd and was defeated. After this the province submitted to Sultan Maḥmūd, his son Masʿūd governing.

In 428 (1036–1037) the Seljuks came and took Herat by capitulation. Masʿūd again got the upper hand, but eventually he was defeated at Marw, went off to Ghazna and was killed in (4)32 (1040–1041). During this time Herat was left without a ruler. Shaikh Abū Muḥammad ʿIṣamī overcame and took control. Finally he fell in battle against the castle warden of Shamīrān and was succeeded by his brother Shaikh Rāfiʿ. The latter was subsequently killed by Manṣūr b. Ashʿath, who became governor through that victory. The Seljuks came and went but could not capture the city. Most of the people scattered due to famine. Quhandiz and the extramural settlement were destroyed at this time.

The Seljuk conquest of Khurāsān: After this the sultanate was established in the hands of Muʿizz al-Dawla Mūsā, who killed Manṣūr and took the city into his control. Then Abū Shujāʿ Alp Arslān the Seljukid came and subdued the city. His vizier, the famous Niẓām al-Mulk Ḥasan, governed the empire wisely and made it flourish. Alp Arslān sent his son Shams al-Dawla Ṭughānshāh to Herat. The next governor was Ẓahīr al-Mulk Abū Naṣr Saʿīd b. Muḥammad Naysābūrī. When Alp Arslān died and was succeeded by his son Malik Shāh, the province was again given to his brother Ṭughānshāh. This Ṭughān followed the path of oppression and rebellion; so Malik Shāh removed him and put him in the castle of Iṣfahān. Amīr Barsaq became governor. He was succeeded by Mu'ayyid al-Mulk Abū Bakr ʿAbd Allāh b. Niẓām al-Mulk who governed until Niẓām al-Mulk was martyred in Ramadan of 485 (1092–1093). When Malik Shāh also died and the affairs of state became disturbed, Malik Arslān came to Herat. After that Amīr Qul Sāriʿ, Yurt Alp Arslān and his vizier ʿImād al-Mulk Aḥmad b. Niẓām al-Mulk became governors. ʿImād fought a battle with his brother Malik Arghūn and was defeated. The province submitted to Arghūn. Some time later he was killed at the hands of his own slave, and the son of Malik Shāh, Barkyaruq came to Herat. He appointed governor Amīr Ḥabashī, who caused a lot of damage. Finally Sultan Sanjar b. Malik Shāh came and saved the land from his evil. In Shaʿbān 493 (June–July 1100) Ḥabashī the heretic attacked Sultan Sanjar between Herat and Isfizār with a group of scoundrels, but they were defeated and killed. Barkyāruk was also there, and he went to Iraq. When Sultan Ghiyāth al-Dīn Muḥammad died in 512 (1118–1119), the sultanate was taken over by his brother Sanjar.

Thus, from the emergence of Islam to the time of Sultan Sanjar, Herat, the capital of Khurāsān, was first governed by the Muʿāzids, scions of Muʿāz b. Muslim. After that, the Ṭāhirids, Samanids, Sīmjūrids, Maḥmūdids (Ghaznavids), [R 126a] [SPACE FOR MAP]

Seljuks and Khwārazmshāhs governed in this order. Now the Ghūrids and the Kart rulers invaded. The special importance attached by these two dynasties to Herat is greater than the others and their good works are still extant.

There were five Ghūrid sultans, who ruled for 64 years. We mentioned them in Chapter 28; but there is no objection to giving a summary acount of their governors of Herat here.

When Maḥmūd b. Ibrāhīm of the Ghaznavid dynasty became ruler, he gave the emirate of Ghūr to Ḥusayn b. Sām who achieved a high status. [M 334] When the Ghaznavid state came to an end, in 545 (1150–1151) ʿAlā al-Dīn Ḥasan b. Ḥusayn led an army and took Herat, making it his capital. He was an intelligent ruler. After governing for six years, he passed away, and was succeeded by his son, Sayf al-Dīn Muḥammad. After Sultan Sanjar died he took Balkh and gave it to his nephew, Muḥammad. He was attacked by the Iraqi army, and they fought a battle in 558 (1162–1163). Sayf al-Dīn fell in the battle and was succeeded by his nephew, Ghiyāth al-Dīn Muḥammad. He fought great battles against the Oghuz army. He appointed his brother, Shihāb al-Dīn Abū al-Muẓaffar, as his deputy in Herat and himself went to Ghazna where he established himself. Some time later Ghiyāth al-Dīn sent Shihāb al-Dīn against one of the kings of India by the name of Sangīn. Shihāb al-Dīn and his 30,000 men defeated 100,000 Indians. Sangīn fell in the battle, and his son Kawkār had to pay tribute. In the meantime the Oghuz tribe found an opportunity, coveting Khurāsān. Ghiyāth al-Dīn came with his brother and besieged and conquered Nishapur. Khurāsān was subjugated to them again. They appointed Pahlawān Muḥammad Ḥaramak as governor. When in 597 (1200–1201) Ghiyāth died and was succeeded in Ghazna by his brother Shihāb al-Dīn, Sultan Muḥammad Khwārazmshāh sent an army to Marw. After a huge battle Muḥammad Ḥaramak fell and Sultan Muḥammad took Marw. [R 126b] Shihāb al-Dīn rallied his forces and marched against him, but he was defeated and fled to Ghazna and India. After that arbitrators intervened and made peace, leaving Balkh and Herat to Shihāb al-Dīn, Marw and Nishapur to the Khwārazmshāh. After some time Shihāb fell martyr to the cruelty of (Nizārī) commandoes. He was succeeded by Maḥmūd b. Sām, who was not very brave, so the emirs ruled the country while he himself stayed in Herat. After governing for seven years, in 610 (1213–1214) he was found murdered in his home.

After this, the country was taken over by the Khwārazmians. Sultan Jalāl al-Dīn appointed as governor of Khurāsān Malik Shams al-Dīn Muḥammad Jūzjānī. When Jalāl al-Dīn's relations with Chingis Khan went foul, the Mongol army destroyed Transoxiana, then came over to Khurāsān led by Tolui the son of Chingis. Shams al-Dīn resisted, preparing for war and locking himself up in Herat, which at that time had 190,000 fighting men. When the two sides began battle, there were thirty thousand casualties in a single day. After seven days of continuous warfare the Mongols were victorious, Shams al-Dīn having fallen in the battle. The Heratis fell into dispute over the question of fight or capitulation. Finally the city capitulated. The Mongol infidels by order of Tolui divided the inhabitants into four groups, leading each group out of the city through a different gate and massacring them. On each side 600,000 people were killed, as is reported in the chronicle *Sirāj-i Minhāj*. That day until evening the bloodthirsty sword held sway, being lifted on the following day. Approximately 200,000 people were left alive, but even of these, [M 335] the followers of Jalāl al-Dīn were massacred. Tolui gave the governance of the city to Malik Abū Bakr and its military governance to a man named Mengetei, then turned back. This event occurred in 617 (1221).

After this the people of Herat did not stay quiet but killed these two, massacring their following and appointing Malik Mubāriz al-Dīn Sabzawārī as their ruler. Filled with anger, Chingis Khan dispatched against them Eljigidei at the head of a 60,000 strong Mongol army. He ordered that this time no one of the inhabitants of Herat be left alive. Eljigidei came in Shawwāl 618 (November–December 1221) and laid siege to the city. From the vicinity 50,000 infantry and cavalry came to the city, fully armed, making sorties and engaging in huge battles. The siege and battle went on until the middle of (6)19 (1221), the Mongols gradually becoming stronger and the Khurāsānīs weaker. The latter disputed among themselves whether to capitulate or fight. At this point Eljigidei found an opportunity and at the beginning of Jumādā I (June 1222) he took the city from the direction of the Earthen Tower. He ordered that everyone—great and small, young and old—be put to the sword. The Mongol infidels swarmed in and massacred the people, then for seven days tore down the houses and destroyed all the buildings. This time 300,000 souls were killed. The Mongols departed after eight days, but again sent 2000 horsemen to round up and kill anyone hiding or having fled. Only sixteen people were able to entrench themselves at some inaccessible place and thus be saved. They came back and for some time only they lived in the city. Gradually, in eleven years, the city was restored.

Invasion of the Kart rulers: The first of the dynasty was Malik Shams al-Dīn Muḥammad Kart. His father Abū Bakr Kart was commander-in-chief in the service of the Ghūrid sultans. He was a kinsman of Sultan Shihāb al-Dīn. When the Chingisid Mengü Qan (= Möngke Qaghan) became khan, he went and received the patent of rule for Herat and the province of Nīmrūz. Later he joined forces with Arghūn and became governor of the empire as far as the river of Sind (Indus). Acknowledged as striving for the good order and welfare of all, he established justice throughout the empire.

When Hulagu Khan conquered Khurāsān and Iraq, Malik Shams al-Dīn for some reason turned against him. In 658 (1259–1260) he fought a battle at the border of Sīstān with the Mongol army sent against him and defeated them. He returned to Herat and again fought against the Ilkhānid army and was victorious. After that through the dispatch of envoys and display of obedience he became the object of Ilkhānid favors. When Abaqa (Khan) learned about his bravery in the battle against Berke Khan at the borders of Baku he honored him greatly. He was wounded in ten places during that battle. In 675 (1276–1277) he went to Abaqa Khan's court where, due to the calumny of ill-wishers, the Khan's disposition toward him changed. He was detained and not permitted to return to Khurāsān. One day in the bath he was fed poisoned melon and he died in Tabrīz in Shaʿbān (6)76 (January 1278).

His son Rukn al-Dīn Kart, who was called Malik Shams al-Dīn-i Kihīn ("the Younger, Junior"), succeeded him in (6)77 (1278–1279) and came to Herat. In (6)80 (1281–1282) he led an army against Kandahar, taking much booty. In (6)82 (1283–1284) he appointed his son ʿAlā al-Dīn in his place and himself went to the castle of Khaysār. In (68)3 Arghun Khan came to Herat, bestowing favors on ʿAlā al-Dīn. They gradually became enemies, [M 336] and ʿAlā al-Dīn went to his father. The Khurāsānīs were greatly aggrieved and most of them departed. Tegüder came with an army of 10,000 Mongols, looting and destroying Herat. Only two people were left alive in each quarter of the city.

In (6)90/1291 Ghāzān Khan sent Amīr Nawrūz with 5,000 horsemen to Khurāsān. He granted two years of tax exemption and in a short time Herat was resettled. He sent a letter to Malik Shams al-Dīn, inviting him to his province. Shams al-Dīn apologized and asked for pardon, sending his son Fakhr al-Dīn with this request. Amīr Nawrūz showed him kindness and granted him the governance of Herat. Ghāzān Khan sent a robe of honor and a patent of rule. In (6)96 (1296–1297) when Nawrūz turned against Ghāzān, Fakhr al-Dīn was by his side. Ghāzān sent an army against them and there were several battles. After that Fakhr al-Dīn turned against Nawrūz, had him arrested and sent him to Qutluqshāh. He killed him and departed from Herat.

In 698 (1298–1299) Amīr Tegüder came from Iraq to Herat with 3,000 men. Fakhr al-Dīn settled them in Herat, but since they were a rapacious, bloodthirsty lot, and were violent with the inhabitants, complainants went to Ghāzān Khan, who sent his brother Khudābanda to them. Khudābanda set out from Māzandarān, sending people from Nishapur to Fakhr al-Dīn, calling for the Tegüderids and the Sanjarids. As they made excuses and pretexts, he (Ghāzān) came to Herat, and there were many battles and sieges. In the end he received 30,000 dinars and turned back and left. In 699 (1299–1300) Fakhr al-Dīn fortified the fort by repairing the bastions, the walls and the moat. [R 127a] He surrounded a field at the foot of the fortress with a wall to serve as a fairground for festivals. He also dug a moat on the inside and greatly fortified it. He built a dervish lodge and restored the mosques. At the foot of the fortress he built the famous market known as Bāzār-i Malik ("King's Market").

In 702/1302 Öljeytü Sultān b. Arghun succeeded his brother Ghāzān and summoned Fakhr al-Dīn, who gave a pretext and did not go. Khudābanda sent against him Amīr Dānishmand Bahādur with 10,000 horsemen, who arrived at Herat and laid siege to it. After Fakhr al-Dīn came out of the city and there were several battles, Bahādur sent a man with the message, "I cannot oppose Öljeytü's orders. It would be beneficial to both parties if, out of respect for the honor of the ruler, the King (Malik, i.e., Fakhr al-Dīn) would go to the mountains with a safe conduct and leave one of his sons in his place." Fakhr al-Dīn consented and went, leaving Jamāl al-Dīn Muḥammad Sām in defense of Ikhtiyār al-Dīn Castle. Bahādur entered Herat and beheld the fortifications. Wajīh al-Dīn had one of the gates torn down, saying, "This fortress is the cause of the rebellion of the people of Herat." After that Muḥammad Sām invited Bahādur into the fortress for a feast. Bahādur, however, left some of his men to wait in ambush and do away with Muḥammad. Jamāl al-Dīn found this out and when Bahādur arrived, he killed him together with his following. After that his (i.e. Bahādur's) sons came and besieged Herat. He fought with them as well and defeated them.

In 706 (1306–1307) Malik Fakhr al-Dīn died and was succeeded by his brother Ghiyāth al-Dīn Muḥammad. He went to Öljeytü's camp and submitted to him, then returned (to Herat) and governed independently from the river of Sind (Indus) to the Amu (Oxus). No ruler experienced such esteem and favor from the Chingisids as he.

In 710 (1310–1311) [M 337] Prince Yasūr came from Transoxiana and seized the territory from Shaburghān to Murghāb at Öljeytü's orders. Ghiyāth al-Dīn made excuses for not meeting with him (?) and he remained on good terms with him until his father Öljeytü's death. After Abū Saʿīd assumed the throne, in (7)18 (1318–1319) Yasūr became desirous of taking Khurāsān. Later he lived in Jām for a time. In 719 (1319–1320) he appointed Mubārakshāh as commander and sent him to lay siege to Herat. He himself also came and for eighteen days there were great battles in front of the Fīrūzābād Gate. When 40,000 of Yasūr's men had fallen and he still could not conquer the city, he destroyed and pillaged its environs and departed in despair and disappointment. He was finally killed in (7)20 (1320–1321). Ghiyāth (al-Dīn) restored the Friday mosque of Herat, and built a public bath, a pavilion, two caravanserais and numerous dervish lodges, establishing pious foundations. In 721 (1321–1322) he went on pilgrimage and his son Shams al-Dīn Muḥammad served in his place. After he came back he died in (7)29 (1328–1329) and Shams al-Dīn governed Herat. But he did not show great independence, and his brother Muʿizz al-Dīn Ḥusayn conquered him and took over Herat. After the death of Sultan Abū Saʿīd he took control of most of the country and became such a successful ruler that he surpassed all his peers. Provoked by some evildoers, Amīr Ghazghan, one of the Chingisid emirs, came and laid siege to Herat. After fighting for forty days, he finally made peace and departed.

In 771 (1369–1370) Muʿizz al-Dīn died, and his son Ghiyāth al-Dīn Pīr ʿAlī became governor. While he was quarreling with his brother Malik Muḥammad, Khwāja ʿAlī Mu'ayyad Sabzawārī revolted, taking Nishapur and Sabzawār. Ghiyāth al-Dīn repeatedly attacked him and they fought many battles.

Tīmūr's invasion of Khurāsān: In 781 (1379–1380), when Tīmūr's envoy arrived, Ghiyāth renewed the agreement because of his father Muʿizz's fidelity. In 782 (1380–1381) Amīr Tīmūr came to the summer pasture of Balkh and summoned Ghiyāth. Ghiyāth hesitated and in the end refused to go, so Tīmūr proceeded against Herat. Ghiyāth's brother Malik Muḥammad welcomed him and received favors. Tīmūr first came to Fūshanj and fought with the inhabitants who came out against him. After a week's resistance the city was taken and destroyed and its people massacred. Then he came to Herat and laid siege to it. While Malik Ghiyāth was occupied with drinking and carousing, ladders and ballistas were raised and war cries were sounded. After three days he demolished the city wall. Some 2000 garrison troops were led before Tīmūr, [R 127b] who pardoned them. Seeing this, the Heratis leaned to the side of Tīmūr. Intermediaries passed between the parties and the Heratis realized that they could not survive with fighting and obstinacy. Ghiyāth came out and surrendered the city in exchange for safe conduct. On Tīmūr's order the old city walls were torn down—such splendid towers and iron gates were razed to the ground and the moat was filled up. The gates were transferred to Shahr-i Sabz, the treasures and hoards were taken out, protection money was collected from the people, and some 200 of the notables and householders were transported to Shahr-i Sabz. This event took place in Muḥarram 783 (April 1381). In that year Ghiyāth remained as governor of Khurāsān while Tīmūr went to Samarqand. When Ghiyāth returned to his former opposition, Tīmūr came to Khurāsān for the second time, seized Ghiyāth and his sons and took them to Transoxiana. He gave the governance of Herat [M 338] to his son Mīrānshāh. In (7)85 (1383) he returned and exterminated the remainder of the Ghūrids in Khurāsān. Ghiyāth too, together with his sons, was killed while prisoners in Özkend.

In this period, the strife and discord caused by the Ghūrīs in Khurāsān, especially in Herat and Isfizār, was unprecedented. They got their comeuppance, but several thousand innocent Muslims who were with them also perished, and because of them Khurāsān was destroyed.

In 799 (1397) the governance of Herat was given to Shāhrukh b. Tīmūr, and during the rule of this prince (Mīrzā) it regained its former condition. When Sultan Shāhrukh died in 850 (1447) his grandson Bābur Mīrzā took possession of Khurāsān. He passed away in (8)61 (1456–1457), and while his sons Shāh Maḥmūd and Mīrzā Ibrāhīm were contending with each other, Mīrānshāh's grandson Abū Saʿīd seized Khurāsān from their hands and established himself in Herat. In (8)63 (1458–1459) he fought with Jahānshāh (Qaraqoyunlu) and afterward they made peace. In 872 (1467–1468) he coveted Azerbaijan and marched against Uzun Ḥasan (Aqqoyunlu), but was routed at Qarabāgh, taken captive and put to death.

(At this time) Sultan Ḥusayn Manṣūr b. Bayqara, who was in Khwārazm, came and seized Khurāsān. He defeated Uzun Ḥasan in battle at Chinārān and ruled as sultan until his death in 911 (1506). He was an intelligent and virtuous ruler. In Khurāsān, especially in Herat, Sultan Shāhrukh and Sultan Ḥusayn have many good works, such as madrasas, mosques, inns and baths. During their time this city came to flourish again as before.

After him Khurāsān was conquered by Muḥammad Khan Shaybānī the Uzbek, who seized it from the hands of Badīʿ al-Zamān and the other sons of Ḥusayn Bayqara. In (9)16 (1510–1511) Shāh Ismāʿīl Ṣafawī, after taking Fārs and Iraq, attacked Khurāsān and took it from Muḥammad Khan. The Uzbek armies later found an opportunity once or twice and took it back, but they were not able to keep it, and until now Khurāsān has remained in the hands of the Rāfiḍīs (i.e., Safavids).

Saintly men: Khurāsān is said to have as many saintly men as the other lands (together?), and as compared with other lands the great men coming from here are more numerous. The prominent men coming from other cities have already been mentioned. Of famous Heratis, many are religious scholars such as Shaikh al-Islām Ḥafīd Taftāzānī; Sufi shaikhs such as Shaikh al-Islām ʿAbd Allāh Anṣārī and those stemming from the Chishtī Sufis such as the Khwājagān (i.e., Naqshbandis); and poets such as Ḥakīm Azraqī, Mawlānā Saʿīd, Rukn Ṣāyin, Mawlānā Panāhī, Mīr Sayyid Muḥammad Jāmabāf, etc.

Chapter 34: Country of Qūmis, Ṭabaristān and Māzandarān

In this chapter (these) three provinces will be mentioned.

QŪMIS: The Arabicized form of Kūmis. It falls between Khurāsān and Persian Iraq. It is a province with several cities. Its capital is Dāmghān, located at 89 degrees longitude and 37 degrees latitude, a town with little water and cheap prices on the edge of the desert three stages south of Nishapur. According to the *Nuzha*, it was built by Hūshang. Its city wall was 8000 fathoms in circuit. Its climate tends to be hot. Its pears are very good.

In the vicinity of one of the villages called Chahār-dih, there is a spring with little and yellowish water in it. If someone pollutes it with feces, a huge [M 339] wind starts to blow in Dāmghān, breaking and tearing out trees. When someone comes and cleans the spring, the wind calms down. [R 128a] The great Ḥanafī chief qadi Abū ʿAbd Allāh al-Dāmghānī comes from this city.

Simnān: One of the cities of Qūmis, located at 87 degrees longitude and 36 degrees latitude. It is on the edge of the desert, and is smaller than Dāmghān and larger than Bisṭām. It falls between Rayy and Dāmghān. Water is gotten from a stream. The climate is moderate. Its pomegranates, pistachios and grapes are excellent. One of the great shaikhs, Shaikh ʿAlā al-Dawla al-Simnānī, comes from this city. To the west a town by the name of Samank adjoins it. From Simnān to Nishapur is three stages. One and a half stages south of Simnān is the town of Surkhāb.

Bisṭām: One of the towns of Qūmis. It is a small city located at 90 degrees longitude and 37 degrees latitude. It has a moderate climate and produces grain and fruit. Jurjān can be reached from here in two days. One of the great shaikhs, Bāyazīd al-Bisṭāmī, comes from this city. He is also buried there.

ṬABARISTĀN: The Arabicized form of Tabaristān. In the *Nuzha*, Ḥamd Allāh locates it within the same borders as Qūmis, saying that it adjoins Khurāsān, Persian Iraq, the desert and Māzandarān. Aside from the towns of Qūmis, he registers as within its borders the towns of Khwār, Girdkūh, Fīrūzkūh, Damāwand, Maryam and Kharraqān, while he considers the rest as belonging to Māzandarān province. The *Taqwīm al-buldān* cites the *Marāṣid* to the effect that Māzandarān is another name for Ṭabaristān. We have followed suit, and thus we do not have a separate section for Māzandarān.

The story of its name: as this land was mostly oak forest mountains, an army cut down the forests with axes when passing through, and so it was called Tabaristān (*tabar* being Persian for "ax"); or it indicates that the trees were cut down and cities were built in their stead, so it was named after the instrument (i.e., the ax).

Borders: to the east, Khwārazm; to the south, Khurāsān; to the north, the Sea of Shirwān (Caspian); to the west, Daylam and Gīlān. Again, according to the *Taqwīm*, Ṭabaristān is divided into two districts: Qūmis, which adjoins Khurāsān, and Māzandarān to the north of it. This division shows that what will be mentioned henceforth are the cities of Māzandarān. He (i.e., Abū al-Fidā) goes on to say that Ṭabaristān is a province surrounded by very steep mountains and the central part is flat country.

According to the *Masālik*, most people in Ṭabaristān have eyebrows joined together, i.e., widely spaced eyebrows; are very hairy; and speak rapidly. Their food is mainly rice and fish. They love garlic.

According to the *Nuzha*, MĀZANDARĀN with its appendages is seven *tūmans*: 1. Jurjān, the capital; 2. Mūrustāq; 3. Astarābād; 4. Āmul and Rustamdār; 5. Dihistān; 6. Rūʿad; 7. Sāstān.

Jurjān: The Arabicized form of Gurgān. Located at 90 degrees longitude and 39 degrees latitude, it is a famous region (*belde*) on a flat area near the mountains. As rain is never absent and a great river flows from the mountains, reaching the outskirts, the climate is warm and humid and plague is frequent. It is a place hostile to strangers; hence it is said, "Jurjān is the graveyard of Khurāsān." Finally, trees and plants are very hardy; a tree that takes ten years to grow elsewhere, here grows in a single year. Fruits of the plain and of mountain pastures can be found—it is a region combining features of land and sea, plains and mountains—such as dates, grapes and jujubes; as well as grain, cotton and silk. [M 340] In summer they bring the snow down from the mountains.

According to Ḥamd Allāh, the city wall was 7,000 fathoms in circuit. The city was conquered by Yazīd b. Muhallab during the reign of Sulaymān b. ʿAbd al-Malik. Until the Būyids, the city was populous and victorious, but during the reign of the Būyids they suffered defeats and were broken. In the time of the Mongols the inhabitants were massacred with no one left alive. Later it started to flourish again. Most of the people are Shīʿī and generous. When Fīrūz, the Sasanian king, was at war with the Turānians, he built a 50-parasang long wall in order to ward off their attacks and he blockaded this region.

Of ancient monuments, there are two millstones nearly 2 cubits thick and 20 cubits in diameter. The grave of Muḥammad b. Jaʿfar al-Ṣādiq is a place of visitation there. One of the mystic luminaries, the very learned Sharaf al-Dīn Sayyid Sharīf ʿAlī b. Muḥammad al-Jurjānī, who comes from here, died in 816 (1413–1414). This city of Jurjān has an independent history.

Ābaskūn: Located at 89 degrees longitude and 37½ degrees latitude, it is a town (*belde*) on the Caspian shore, the seaport for Ṭabaristān and Jurjān. [R 128b] It is three stages from Jurjān, whose river flows here.

Āmul: A mid-sized town located on the seashore at 87½ degrees longitude and 37½ degrees latitude. Sāriya can be reached in two days from here. The town with the same name to the east, at the end of the border of Ṭabaristān, is 5 (?) parasangs from here, Sālūs is 12 parasangs.

According to Ḥamd Allāh, it is a big city built by Ṭahmūrath. Its climate tends to be hot. Fruits of both mountain and seashore grow here, such as almonds, walnuts, grapes, dates, bitter oranges and lemons. Its muskmelons are excellent. Abū Jaʿfar Muḥammad b. Jarīr al-Ṭabarī comes from this city. There is another Āmul in Transoxiana; therefore this one is called Āmul of Ṭabaristān. Its pastures are very good for sheep.

Masʿūdī says in the *Murūj*: It is located where a large river coming from Turkistan empties into the sea. The river cuts the city into three sections. The middle one, like an island, is where the capital is. The other two are alongside it, and a pontoon bridge has been built to one of them.

Māmaṭīr: A town in the district of Āmul. It is six parasangs from Āmul. Sārī is also six parasangs from here.

Hīm: A town 5 parasangs east of Āmul at the end of the border of Ṭabaristān. It is the port of Sārī. It is also called ʿAyn al-Hīm.

Sārī: A mid-sized city built by Ṭaḥmūrath at 88 degrees longitude and 37½ degrees latitude. It is also called Sāriya. It is approximately 4000 fathoms in circuit. It produces fruits and grain crops. It has many districts. Āmul is 2 stages hence, Astarābād 4 stages. It is west of Simnān.

Astirābād or Istirābād (Astarābād): A town located at 89½ degrees longitude and 38 degrees latitude between Sāriya and Jurjān. According to the *Nuzha*, it is a mid-sized city near the sea named Astārābād, with moderate climate producing silk, grain and fruit. Jurjān is 2 stages hence, Sāriya 4 stages, Āmul 39 parasangs. According to the *Taqwīm*, Istir is the name its founder. However, it is apparent that it has this name because it is a place of mules (*astar* in Persian). The great grammarian Shaikh Raḍī comes from this city.

Dihistān: A town located at 91 degrees longitude and 38½ degrees latitude between Jurjān and Khwārazm, at the end of the border of Ṭabaristān. In the Islamic period it was built by ʿAbd Allāh b. Ṭāhir. In ancient times it belonged to Qubād b. Fīrūz. It has a warm climate. Its water is gotten from a stream. It yields little fruit. [M 341]

Jarbādqān: The Arabicized form of Darbāykān. A town located at 89½ degrees longitude and 38 degrees latitude between Jurjān and Astarābād. Another Jarbādqān is recorded in Persian Iraq; there are two cities with this name.

Rustamdār: According to the *Nuzha*, a province comprising 100 villages. Water is gotten mainly from the Shāhrūd river and the climate tends to be hot.

Rūʿad: A mid-sized city located at 88 degrees longitude and 37 degrees latitude.

Rūyān: A large city located at 86½ degrees longitude and 36 degrees latitude in the mountains of Ṭabaristān. It is also called Shāristān. It is built at the head of a steep mountain pass. It has many districs. The region (*bilād*) of Jabal and the border of Qazwīn are 16 parasangs hence. The great Shāfiʿī scholar Abū al-Walīd, author of *al-Baḥr*, comes from this city.

Nāqil: A town located at 87 degrees longitude and 36½ degrees latitude in the vicinity of Āmul. It has many streams and pastures. Āmul is 4 parasangs hence, Sālūs 1 stage.

Lārijān: A town located at 86½ degrees longitude and 36½ degrees latitude in Ṭabaristān between Āmul and Rayy, each of which is 15 parasangs from it.

Kizhih: A town located on the seashore at 90 degrees longitude and 40 degrees latitude, 4 stages north of Jurjān.

Farwa: A town located at 90 degrees longitude and 41 degrees latitude, five stages from Kizhih. In the *Taqwīm* it is written as Farāwa. It is southwest of Khwārazm. [R 129a]
[SPACE FOR MAP]

Ḥamd Allāh includes Damāwand at the border of Qūmis and Ṭabaristān, saying that its town is called Pīshān, it was built by Gayūmarth and it has a cold climate. However, we mentioned it in the section on Persian Iraq.

Wīma: A town located at 87½ degrees longitude and 36½ degrees latitude in the district of Dunbāwand between Rayy and Ṭabaristān. Its climate is cold; grain crops and water are plentiful.

Girdkūh: A village (*rustāq*) of Manṣūrābād 3 parasangs from Dāmghān. Sown crops and produce are plentiful.

Fīrūzkūh: A castle at the foot of (Mt.) Damāwand. Because of its cold climate, no trees grow there. The Khwār River (Āb-i Khwār) passes below the castle. Its grain crops are plentiful.

Kharraqān: A village with a fine climate and plenty of water, belonging to Bisṭām. The great shaikh Abū al-Ḥasan al-Kharraqānī is buried here.

Aside from these, the *Nuzha* registers the cities of Kabūd Jāma, Nīm Mardān and Shahrābād in Ṭabaristān: Kabūd Jāma is a broad vilayet with abundant grain and silk. Nīm-mardān is an island 3 parasangs from Astarābād. Shahrābād in the past was a town; now it is in ruins.

Sālūs: According to the *Taqwīm*, the last city of Ṭabaristān, on the seashore toward the west. It is difficult of access. The road from Ṭabaristān to Daylamān goes through there.

Saʿīdābād: According to the *Masālik*, a town in the vicinity of Rūyān, built by Saʿīd b. Daʿlaj, the deputy of the Abbasid caliph al-Mahdī.

Jurjān River: It rises in the mountains of Māzandarān, flows through the Yenişehir valley, passes by Maydān-i Sultan and Jurjān and irrigates the fields for some distance, then empties into the Caspian Sea before Ābaskūn. For most of its course it is very deep and still. Because its banks are full of gullies and swamps, it is extremely difficult to cross. Not a day passes without a man or beast being drowned. This river is approximately 50 parasangs long.

Rulers

Before Islam, this clime was ruled by the Persian kings (Sasanians). [M 342] Anūshirwān gave it to his elder brother Kayūs who governed it for seven years before he rebelled and was killed at the hands of his brother. After the dynasty of the Persian kings came to an end, in the year 45 of the Hijra, Anūshirwān's grandson Bāwand became governor at the request of the people of the country. The dynasty of his descendants is called Bāwandiyya. Fourteen of them were governors until 397 (1006–1007).

There were three classes of the kings of Māzandarān [R 129b] who ruled between the years 45 (665–666) and 750 (1349–1350). Occasionally ʿAlawīs or deputies of the caliph interfered, and that dynasty at times died out, at times reemerged. Because they were mainly in the mountains of Ṭabāristān, they were also called Kings of the Mountains. The first class ended with the rule of Qābūs at the above-mentioned date (397 (1006–1007)).

The second class appeared in 466 (1073–1074). A member of the Bāwandid clan, Iṣbahbad Ḥusām al-Dawla Shahriyār b. Qārin asserted his rule, taking over the inheritance of the kingdom. Eight of his descendants became governors down to Shams al-Mulūk Rustam, when in 606 (1209–1210) Abū al-Riḍā Ḥusayn b. Muḥammad al-ʿAlawī rebelled against him and killed him, this class ending with him.

The third class are the lineage of Ḥusām al-Dawla Ardashīr b. Kaydkhwār who was a descendant of the previous Ḥusām al-Dawla. He appeared in 635 (1237–1238), taking Ṭabaristān. Eight individuals governed until 750 (1349–1350), the last being Fakhr al-Dawla Ḥasan.

In that year Kiyā Ḥasan al-Jilāwī killed Fakhr al-Dawla Ḥasan and took over the country. Thus emerged the Jilāwiyya dynasty, of which six members governed until the year 909 (1503–1504) when the dynasty came to an end with the emergence of the Qizilbash.

During the time of the Bāwandids, the ʿAlawī dynasty of Ṭabaristān emerged in 250 (864–865) when the great propagandist (*dāʿī*), Ḥasan b. Zayd occupied Māzandarān. Seven members of this dynasty governed until 316 (928–929), when they came to an end.

In 315 (927–928) emerged the Ziyādid dynasty. Mardāwīj b. Ziyād, the commander of Asfār b. Shīrūya, occupied Ṭabaristān and Jurjān. These provinces were governed by nine of his descendents until the year 470 (1077–1078). One of them, Shams al-Maʿālī Qābūs, was a famous ruler.

The Dābūya kings, who were descended from Gāwpāra: Dābūya took control of Rustamdār in the year AH 40 (661–662). Five of his descendents governed until the year 140 (757–758), when the dynasty came to an end.

Bādūspān, another descendant of Gāwpāra, settled in Rūyān in AH 40 (660–661), the people of Rustamdār became subject to him. 35 of his descendants governed until the year 881 (1476–1477), at times sharing the rule with descendants of the Prophet. They were followed by the Gayūmarth dynasty, who also claimed descent from Gāwpāra. They took the castle of Nūr in Ṭālaqān and Kajūr in Rustamdār and are also known as the Kings of Kajūr. As they belonged to Imami Shīʿism, they did not come to an end with the emergence of the (Safavid) shahs.

The Qiwāmiyya dynasty, who stem from the ʿAlawīs, were descendants of Sayyid Qiwām al-Dīn al-Marʿashī. He emerged in the year 760 (1358) in Āmul in Ṭabaristān and seized the country. 23 of his descendants governed there until 979 (1571–1572), and because they submitted to the shahs, [M 343] their rule was extended for a while. They are called the *shurafā* (nobles, descendants of the Prophet) of Māzandarān. They sometimes shared rule with the Rūzafzūniyya dynasty, who are but a few individuals.

As all of these dynasties are detailed in the *Fezleke*; they have only been summarized here.

Chapter 35: Clime of Khwārazm, Daylam and Gīlān

The 35th chapter is on the countries of Daylam and Gīlān and Khwārazm.

Daylam is the name of a *jīl*, meaning a tribal group of the ancient nations. The name of that people came to stand for their settlements. According to the *Jāmiʿ al-uṣūl* by Ibn al-Athīr, a certain Bāsil b. Ḍubba of the Banī Muḍar went there and married and Daylam was born to him, so he was Abū Daylam ("Father of Daylam").

As for Gīlān, it is the plural of Gīl, which was Arabicized as Jīl, and is also called Jīlān. It is the name of a *ṣiqʿ*, meaning a region (*diyār*). It has several villages, but no big city, and it is adjacent to the land (*bilād*) of Daylam, so it is considered to be within the borders of Daylam.

Borders of Daylam and Gīlān

To the west, part of Azerbaijan; to the east, Rayy [R 130a] and Ṭabaristān; to the south, part of Azerbaijan and the land of Qazwīn; to the north, the Caspian Sea.

According to Ibn Ḥawkal, the land of Daylam has both plains and mountains. The plains are called Jīl and Jīlān. The northern slopes of the Daylam Mountains down to the seashore are long narrow valleys, a day's journey wide more or less, extending all along the coastline. In some places the sea beats against the mountain, with only a narrow path along the coast before it widens out again. At its widest it is two days' journey wide.

Most of the people of this region are puny, frivolous and unclean. They have a peculiar language, neither Persian nor Arabic. In the past they were Magians; later some of them converted to Islam.

Ḥamd Allāh says in the *Nuzha* that Gīlān is comprised of twelve cities. In length it extends along the coast of the

Caspian Sea from Safīdrūd to Mūghān, in breadth from the Daylamān Mountains to the sea. At the time of the Mongols the state dues from it were 2 tomans.

The largest city of Daylam is Lāhijān, located in a plain at the foot of the Daylam Mountains, two stages from the sea, at 84 degrees longitude and 37½ degrees latitude. It is a large city with orchards and gardens and streams. It is an administrative center (*dār-ı mülk*).

Langarū: A port and entrepot town on the seashore two stages east of Lāhijān. As many as 100 ships anchor in its harbor.

Iṣfahbad: A mid-sized city located at 85 degrees longitude and 38 degrees latitude. Its produce is grain and some fruit. There are approximately 100 villages in its district. State dues from it are 2 tomans and 9000 dinars.

Tūlam: A city located on the coast two stages from the mountains at approximately the previously mentioned longitude and latitude. Its products are grain, rice, oranges and bitter oranges. It is in a plain on the bank of the Nawkhāla River, which comes from Māsūla. It is also an administrative center, the residence of a lord.

Māsūla: A town at the foot of the mountains. The Nawkhāla and Khandkhāla rivers come (respectively) from its east and south, passing west of Tūlam and emptying into the sea in front of that city after joining the Māsūla River coming from the east.

Fūmin: the Arabicized form of Pūmin. It is a large city located approximately at the aforementioned longitude and latitude. Its products are grain, rice and silk. It has an extensive district. It is located between the sea and the mountains. Māsūla to its southwest, Tūlam to its northwest and Sālūs on the sea coast, are each 1 stage hence. The Khandkhāla and Nawkhāla rivers pass through it.

Kālār: A town located approximately at the aforementioned [M 344] longitude and latitude in the Jabal-i Khākhāl district, southwest of Lāhijān. It is attached administratively to Ardabīl.

Rasht: A town on a plain one stage east of Pūmin. It is two days' journey from the sea.

Shaft: A small city located at 85½ degrees longitude and 38 degrees latitude, in the mountains one stage east of Rasht. It is one day's journey from the sea.

The climate of these two cities is hot and fetid, and their inhabitants are uncouth and doltish. But they produce grain, rice and silk.

Kūtam: A mid-sized city and administrative center with orchards and gardens one stage east of Shaft. It is one stage from the sea. The governor of Gīlān resides here. It has a (permanent) bazaar.

Kūjistān: an entrepot and port on the seashore, one stage east of Kūtam, on the border between Gīlān and Lāhijān. It is a profitable place, with ships coming and going from Gurgān, Ṭabaristān and Shirwān.

Dūlāb: A large city and administrative center on the seashore, two stages west of Tūlam. There are 20,000 houses and many streams. Its district is called Kaskar. Ardabīl is 3 stages to the west, the Daylam mountains are 1 stage to the south, and Qizil Aghach five stages to the west.

Lashinshāh: A large city on a plain with orchards and gardens and streams between Lāhijān and Ḥasnigāh. It is one stage from the mountains and one stage from the sea. It has many plane trees. The inhabitants are mostly soldiers and merchants.

Pashpishā: A city east of Lashinshāh, abounding in orchards and gardens and water sources. It is between the mountains and the sea. [R 130b] It produces wheat and rice. Its streams come mainly from the east and its inhabitants are mostly soldiers.

Ḥasnigāh: A mid-sized city on the coast, one stage east of Langar, with orchards and gardens. A big river comes from the mountains east of it, passes west of it and empties into the sea. The mountains are three stages hence.

Tangālīn: also called Tamjān. A city on the coast with streams three stages east of Pashpishā. Most of the inhabitants are descendents of the Prophet.

Karinjiyān: also called Karjān or Kirjān: A town in the mountains two stages northeast of Tangālīn. According to the *Nuzha*, it was built by Ardashīr.

Rūdbār: according to the *Taqwīm*, citing Ibn Ḥawqal, it is a city in the mountains and the administrative center of Daylam. There are villages by the name of Rūdbār belonging to Baghdad, Marw, Ṭūs and Shāsh as well.

Kūka: A town on a plain one stage east of Kūchisbān. A river called Safīdrūd, bigger than the Euphrates, comes from north of Kūchisbā[n], passes by Kūka and empties into the sea. The mountains and the sea are each one stage from Kūka.

Kīsim: A city one stage east of Kūka on a plain at the foot of the mountains. It has orchards and gardens; water is gotten from springs. It is also considered an administrative center. It has a (permanent) bazaar. It is two stages from the sea.

Dilimān: A big city two stages east of Kīsim with gardens, orchards and a (permanent) bazaar. The seashore is 4 stages from here. Qazwīn is 4 stages. Lāhijān is 2 stages to the east.

Khushkrū: A town in the mountains one stage east of Karinjiyān. The sea is two days' journey from here.

Safīd Rūbār: A city between Kūka and Lāhijān, one stage from both. It has gardens, orchards and a (permanent) bazaar. The sea is two days' journey from here as well.

Some features of Gīlān

According to Khwāndamīr, Gīlān is surrounded by [M 345] steep rocky mountains and passes with oak forests, and plains with springs of abundant water. Rain is rarely absent. People eat mostly rice, fish and birds; mutton and greasy foods are quite detrimental there. Sometimes it rains for several days and nights and the people become desperate. If at night they hear the barking of a fox and afterwards that of a dog, they cheer up, taking it as a sign that the rain will stop and the next day will be clear. And Qazwīnī writes that this is based on experience.

The country of Gīlān has two parts: one is Lāhijān with its dependencies, the other is Rasht and Fūmin with their annexes. Each has an independent governor. Their dependencies and annexes are mixed together in the written sources, and I have not managed to disentangle them.

Country of Khwārazm

Regarding the pronunciation, the *Taqwīm*, citing the *Marāṣid*, says that the vowel after *Kh* is between *a* and *u*, and the *alif* (for *ā*) is not a true *alif* but a kind of furtive *alif*. Regarding how it got its name, Ibn al-Furāt says that it is a place of ignominy, because its people submit only if they are treated with disdain (*khwār*). However, in the chapter on the clime of Fārs it was recorded that the word *zam* means tribe.

The *Haft Iqlīm* reports that in olden days a certain king exiled a people to this desert, where they built houses and settled. Some time later that king sent someone to inspect the exiles. When he arrived he saw that they had built houses and collected firewood, and were catching fish in the river and cooking them. In their language *khwar* meant meat and *zam* meant firewood, making the compound term *Khwārazm* meaning "a place for meat and firewood". The king pardoned them and sent 400 Turkish women for the 400 exiled men. They were fruitful and multiplied and became the people of a clime.

They are brave men, for the most part, and can field 30,000 valiant horsemen.

This country is a small clime surrounded by deserts and cut off from Khurāsān and Transoxiana. It is oblong in shape. The breadth of its cultivated areas along the bank of the Jayḥūn (Oxus) as far as Gurgānj gradually increases to two or three parasangs. At its widest, there is a village at the foot of the mountains that is 5 parasangs (from the Jayḥūn).

Borders

To the south, Khurāsān; to the west, part of the land of the Turks; to the east, Transoxiana; to the north, the land of the Turks. The cultivated area ends with the two banks of the Jayḥūn; beyond them there is no populated area.

The climate of this clime is quite cold. Hands and feet are lost because of the extreme cold. The Jayḥūn freezes within the borders of Khwārazm.

The administrative center is Greater Gurgānj (Gurgānj-i Kubrā), also called Ūrgānj. It is a great city on the southwest bank of the Jayḥūn, located at 94 degrees longitude and 10 minutes to 43 degrees latitude. It was highly flourishing until the year 610 (1213–1214). It was destroyed by the Tatars, but was completely restored afterwards. Today it forms the administrative center and capital city of Khwārazm. Āmul is 12 stages hence, Buḥayra 6 stages.

Lesser Gurganj (Gurganj-i Ṣughrā): A mid-sized city on the west bank of the Jayḥūn, 10 miles from Greater Gurgānj, at the same longitude and latitude. It is also called Gurganj-i Fasawī and Jurjānīyya-i Khwārazm. It is one of the mother cities of the Jayḥūn. According to the *Haft Iqlīm*, many hadiths are related about its excellence.

Kāt: A city located at 94 degrees longitude and 41½ [R 131a] [SPACE FOR MAP]

degrees latitude. [M 346] In the past it was the chief city of Khwārazm. As it was on the east bank of the Jayḥūn, it was flooded and destroyed by the river and the inhabitants rebuilt it in an elevated place. In the past it had a citadel and a Friday mosque, near which was the Khwārazmshāh's palace. The small river called Jardūr flows through the city and into the Jayḥūn. The marketplace was on its two banks. Khwārazm has no other city or town on the east bank of the Jayḥūn. From here come Qiwām al-Dīn, the Ḥanafī jurist who wrote the *Miʿrāj al-Dirāya*; Ḥusām al-Dīn, the commentator on the *Isagoge*; Nāṣir al-Dīn al-Maṭarzī, the author of the *Maghrib*; Khwāja Abū al-Wafā; Pahlawān Maḥmūd Pūryā; Mawlānā Kamāl al-Dīn Ḥusayn; etc.

Zamakhshar: A large village located at approximately the same longitude and latitude (as Kāt). ʿAllāma Jār Allāh, author of the *al-Kashshāf*, was from this village; he died in the year 538 (1142–1143) in Gurgānj.

Hazārasp: an impregnable castle on the west bank of the Jayḥūn, located at 95 degrees longitude and 41 degrees latitude. It is six parasangs from Kāt. It is like an island located in the Jayḥūn River. It has only one built road. When the Khwārazmshāh Atsız locked himself up here, the Mongol army took the castle after a five-month siege and massacred the people.

Darghān: A town at the end of the border of Khwārazm, 24 parasangs from Hazārasp in the direction of Marw.

Khīwa: A flourishing town east of the Jayḥūn, one stage above Ūrgānj. It is on the west bank of the Jayḥūn. Kāt is 2 stages up from here.

Wazīrshahrī: A town opposite Khīwa. It is also called Khīwaq. The shrine of Shaikh Najm al-Dīn Kubrā is there.

Some features of Khwārazm

According to the *Masālik*, Khwārazm is a place of many blessings and plentiful fruits. Only there are no walnuts. Lovely linens, fine woolens and beautiful silks are woven here. The inhabitants are mostly merchants, practice chivalry and travel a lot. Their language is different from that spoken in Khurāsān and their ways and manners are also peculiar to themselves. As this clime is adjacent to the land of the Oghuz and the Turks, warfare and combat are rarely absent, and most of the inhabitants are skilled in battle and brave frontier warriors. However, the majority of them are said to be Muʿtazilites. Gold, silver and other sorts of metals are found in this region. In the past the tax from here was 489,000 dirhams.

Rivers

Most of the rivers in the clime of Khwārazm are tributaries of the Jayḥūn (Oxus). We will discuss the Jayḥūn in the next chapter, which is on Transoxiana.

According to the *Masālik*, the place called Ṭāhiriyya, at the extremity of Khwārazm on the Upper Jayḥūn, is south of the river. [R 131b] Six parasangs above there a tributary called Gāw-khwāra branches off the Jayḥūn. It is five cubits broad and as deep as the height of two men. It is navigable. Most of the villages of Khwārazm are on its banks. After two parasangs another tributary called Rūd-i Karba branches off that irrigates some of the villages. It is not very broad. There are many tributaries that branch off west of the Jayḥūn, such as the Rūd-i Hazār, Rūd-i Kard Aḥwās, Rūd-i Āl, Rūd-i Būh, Rūd-i Ḥayra. Some of them are navigable. Every village has its own canal that branches off the Jayḥūn and irrigates the fields.

Lake Khwārazm (the Aral Sea): the Tatars call it Ḳara Deñiz (Black Sea) and Öküz Ṣuyu (Ox Lake). According to Ḥamd Allāh, it is more than 100 parasangs in circumference. [M 347] Its water is brackish. It is six stages north of Gurgānj. The distance between this and the Caspian is nearly 100 parasangs. Some of the water of the Jayḥūn (Oxus) and the Shāsh River (Jaxartes) empties into this lake.

Masʿūdī says in the *Murūj al-dhahab* that there is no larger lake in the inhabited quarter (of the world). That is the opinion of the clime-books (i.e., the Muslim cosmographers). However, the European geography books do not have a map of this lake; and in the *Atlas*, which has been translated, there is neither denial nor confirmation of it. Apparently they have no knowledge of this clime.

Although a huge river like the Shāsh River (Jaxartes) empties into it, it neither changes in taste nor increases in size. For this reason, some of the common folk say that underneath it there is a passage to the Caspian. This is nonsense, for the fine particles of the river that empties into it are dissolved and transformed into their elements, so it is not necessary for the lake to overflow. Common folk know nothing of the laws of natural science. They make judgments according to what they see on the surface.

Talisman

There is a large mountain in the vicinity of Khwārazm, with castles and villages, and in that mountain, on top of a high hill, there was built a tomb with a dome and four gates. The hill is surrounded by a pond, and through the gate of the tomb one can see ingots of gold piled up. There is no way to cross the pond and reach the tomb. Those who have tried it with a boat or a raft have drowned. Any wood or light material placed into the pond immediately sinks and cannot be retrieved.

When Sultan Maḥmūd of Ghazna came to Khwārazm, he wanted to conquer this talisman. He mustered Turks and peasants to cut stones and trees and for some time they and his soldiers kept pouring them into the water, but they all disappeared without a trace, and he had to abandon the effort.

Abū Ḥāmid al-Andalusī says that he traveled to Khwārazm three times and saw this talisman.

Routes of Khwārazm

From Gurgānj to Samarqand is 10 stages and to Nisā is 10 stages. This road is through desert and wasteland, with water only at every second or third station.

Of the cities of Khurāsān, Marw is 124 parasangs, thus: Gurgānj > city of Sūrāwān, 6 parasangs > Andrānyān, 2 parasangs > Mār-i Ḥasmīn, 6 parasangs > village of Azraq, 7 parasangs > Hazārasp, 10 parasangs > Sandbūr, 10 parasangs > Ribāṭ-i Dahān-i Shīr, 4 parasangs—it is a narrow valley between two mountains through which the Jayḥūn flows with a roar that frightens whoever hears it; > city of Ḥafrband, 5 parasangs > Darghān,

7 parasangs—it is the end of the border of Khwārazm > Ribāṭ-i Būd, 10 parasangs > Ṭāhirī, 10 parasangs > Sangābād, 6 parasangs > Ribāṭ-i Nawshākir, 7 parasangs—here there is quicksand for 2000 fathoms; > Chāh-i Bīrūn, 7 parasangs > Ribāṭ-i Sūrān, 8 parasangs > Ābdān-i Ganj, 8 parasangs > village of Saqrī, 2 parasangs > Marw, 5 parasangs.

Rulers

The Būyid dynasty appeared in this land, which belonged to the country of Daylam, so they are also called Daylamites. In the beginning, Abū Shujāʿ Buwayh was a poor man from the people of Daylam. He made a living by fishing. His three sons—ʿAlī, Ḥasan and Aḥmad—were in the service of Mākān, one of the deputies of the kings of Iraq. When Mardāwīj-i Daylamī defeated Mākān, they left him and went over to Mardāwīj [M MAP] [M 348] and became mighty and powerful. One of them, ʿImād al-Dawla ʿAlī, in the year 320 (932) [R 132a] asserted his authority and took Iṣfahān. Gradually he conquered Fārs and Iraq, as was mentioned in the chapter on the clime of Fārs. Eighteen members of this dynasty governed Fārs, Iraq, Daylam and Ṭabaristān until the year 448 (1056). The governors of Daylam were generally appointed by them, while they resided in Fārs and Iraq.

Sultans of Gīlān: They are of two classes, the kings of Lāhijān and the kings of Rasht.

The first class, dubbed Kār Kiyā, are a clan of descendants of the Prophet, descended from ʿAlī Aṣghar. During Tīmūr's rise to power, one of them, Mahdī Kiyā, was appointed governor at the request of the people of Gīlān. Ten of his descendants governed down to the time of Shāh Ṭahmāsp. Lāhijān with its dependencies was in their hands.

The second class, who are descendants of Isḥaq Āwand, are the Amīra, also called the dynasty of Bāj. Payah-pas, Kaskar, Fūmin and Rasht were under their governance. ʿAlā al-Dīn, Ḥusām al-Dīn, Amīr Dawāj, and Muẓaffar Khān inherited the governance from their father in this order. Muẓaffar Khān came over to Sultan Süleymān at the conquest of Baghdad (in 1534). Because he was a Sunni, the Persian Shah drove him out of his territory and gave it to the sultans of Gīlān.

At first they were Shīʿīs, then they became Rāfiḍīs (i.e., fanatic pro-Safavids). But Shah ʿAbbās drove them out as well, took Gīlān and appointed governors and khans from his own men. Of this dynasty, Khān Aḥmad fled to the Porte during the time of Sultan Murād Khān and remained in Turkey. Shah ʿAbbās reiterated his demand for him in each of his letters, but it was not granted.

Dynasty of the Khwārazmshāhs: They are descendants of Ānūshtigin, one of the slaves of the Saljuq emirs. At the time when Dāwūd Beg came to Khwārazm, by order of the Seljuk Sultan Barkyaruq, the Turks had killed its governor. He (Dāwūd Beg or Sultan Barkyaruq) considered Anūshtigin's son Quṭb al-Dīn Muḥammad suitable and appointed him governor of Khwārazm in the year 440 (1048–1049) and dubbed him Khwārazmshāh. Later he became a favorite of Sanjar and received confirmation in his office from him. His reign and his majesty gradually increased. Eight of his descendants ruled over most of Iran and Turan up to the year 628 (1230–1231). They waged several wars against the surrounding kings. Eventually, when Chingis attacked them, they were defeated in battle against the Mongols, their state was destroyed and their dynasty came to an end.

Chapter 36: Clime of Transoxiana

The 36th chapter is on the clime of Tūrān, known in Islamic books as *Mā warāʾ al-nahr* ("What is beyond the river"). The river refers to the Jayḥūn (Oxus), east of which is Tūrān, and this side of which is Khurāsān and Iran.

The *Taqwīm*, citing the *Qāmūs*, says that Transoxiana is the land of the Hayāṭila (Hephthalites) and that Tūrān is the name of the whole of Transoxiana.

According to Kemālpaşazāde, Tūrān is the plural of Tūr, and because Tūr b. Farīdūn became governor of this clime, it got the name Tūrān. But in the *Nawādir-i Akhbār* it says that the Persians call the Turks Turkmān, and since Turkistan is east of Iran, the area to the east was called Tūrān, which is a corrupted form of Turkmān.

Borders

To the west, Khwārazm; to the east, the border of India; to the south, the Jayḥūn, [M 349] which extends from the border of Badakhshān to the border of Khwārazm, since it flows in a zigzag pattern from east to west; in the north, the borders of Turkistan.

This clime is located between the two rivers, the Jayḥūn (Oxus) and the Sayḥūn (Jaxartes). It is divided into seven *tūmans*, each of which fields 10,000 troops. Most of its districts are located on the bank of the Jayḥūn and extend from Bukhara to Soghd, Samarqand and Usrūshana. In the time of Īlak Khan (the Qarakhanids) the administrative center was Marghīnān; later it became Samarqand.

Samarqand: A great and famous city located at 99 degrees 16 minutes longitude and 39½ degrees latitude, [R 132b] south of the Soghd valley, on an elevation overlooking that valley. It has a great city wall and moat. A river flows through the middle of the city.

The fortress has four gates: Chīn Gate in the east, Nawbahār Gate in the west, Bukhara Gate in the north, and Kash Gate in the south. There is flowing water in the town quarters and in every house. The place called Sarṭāq is a flourishing bazaar and splendid market. If one ascends the citadel and looks at the city, one sees mainly verdant trees covering the buildings. The moat is filled with water. The suburbs have their own several gates, including those of ʿĪd-āward (ʿAbd-āward?), Anfusak, Sarakhs, Afshīna, Kūhak, Rawshanīn, Dīwrū, and Farkhunda. The city is built of timber and clay. The inhabitants are for the most part excellent and handsome.

The *Haft Iqlīm*, citing the *Āthār-i bilād*, says that this city was built by Kay-Kāwus b. Qubād and that later Alexander the Great built a strong wall around it.

In the *Rawḍat al-ṣafā*, toward the end, it says that in olden days Samarqand had a fortress 50,000 fathoms in circumference. It collapsed with the passage of time, but the great hero Garshāsp restored it, using the treasure he found there. Later Gushtāsb the Kayānid made the city flourish and built a wall between Transoxiana and Turkistan. When it was the turn of Alexander the Great, he also enlarged and renovated it.

Then Shamar, one of the Tubbaʿs (pagan kings) of Yemen, came to Tūrān and ruined the city. From that time on it came to be known as Shamarkand meaning "Shamar's Ruin". It was Arabicized as Samarqand. In another version, at that time Soghd was greatly flourishing. The aforementioned Shamar destroyed it and built this city, which was called Samarqand. In this case, *kand* means "village" and the word is a compound. In the first meaning, *kand* is interpreted as the Persian word meaning "to tear out".

Ḳınalızāde ʿAlī reports to have heard from certain scholars that *Semiz-kent* is a compound meaning "large village" and that it was Arabicized. He says that in his opinion, *kent* is more accurate than *kend*, because most of the place names of that region are compounded with *kent*, such as Tashkent.

According to the *Ḥabīb al-siyar*, during the reign of Walīd b. ʿAbd al-Malik, Qutayba b. Muslim was sent by Ḥajjāj to Samarqand. After five months' siege the governor surrendered the city in exchange for paying a fixed annual tribute. Qutayba built a mosque there and burnt the idols he found.

According to Bābur's memoirs, the people of Samarqand became Muslim during the caliphate of ʿUthmān. One of the second generation of Muslims, Qutham b. ʿAbbās took control of that province; his grave near Samarqand, next to the Iron Gate, [M 350] is known as the Tomb of the Shah.

During the time of Tīmūr Samarqand was so flourishing that it had no match either in Iran or Turan. In Ulugh Beg's time the city expanded further. That ruler built a lofty madrasa and dervish convent in the middle of the city, and an observatory in the outskirts. That is where he made the famous astronomical tables.

On the edge of this city is a small mountain known as Kūhak. The stones brought from there are used to pave all the city's streets. The river flows from the south, the Bunam and Chaghāniyān Mountains. There are charitable foundations with tax-exempt villages assigned to the water channels in the city. In Samarqand the river divides into an eastern and western branch with several channels, each of which irrigates a valley. So if you go around Samarqand a distance of seven or eight days' journey, you will see villages and towns with orchards and gardens, an abundance of fruit and streams. If one goes up a high place and looks around, he can see the verdancy of the forests. In summertime the water in the channels increases when the snow in the mountains of ʿArjistān, Usrūshana and Samarqand melts and there are torrents.

The districts of Samarqand are:

Barm: A district with no water. Yet it is flourishing, with plenty of cereals and livestock, and the inhabitants are wealthy.

Bārkant: It is close to Usrūshana. There is flowing water but no pasture. It is a narrow place.

Village of Qūzaʿ: It adjoins Bārkant.

Bājar: an extensive district, one day's journey long, at the border of Samarqand.

Village of Kayūr (or Kīwar?): this too is a large district and forest.

Village of Wadān: A district that includes plain and mountain in the vicinity of Kash. The inhabitants of both places claim descent from Bakr b. Wāʾil and are called Abbasids.

Samarqand has 12 villages (*rūstā*) altogether, 6 in one direction and 6 in another, each separated by a stage or half a stage.

Samarqand is surrounded by delightful pastures. One is Kān-i Gil, east of the city. The Qara Ṣu ("Black River"), also known as Āb-i Raḥmat ("Water of Mercy"), flows through the middle of it. Another is Yurt-i Khān. The Qara Ṣu flows past it and goes to Kān-i Gil. This place is entirely surrounded by water, which can only be forded at two or three places. Another is Ūlānak, which is situated on the shore of a lake.

Most fruits in Samarqand are good, especially apples, pears, pomegranates, grapes and watermelons, one being more delicious than the other.

Famous products are paper, parchment that is like paper, sal ammoniac and quicksilver.

Important individuals from this city include the Naqshbandī shaikh Khwāja Aḥrār ʿUbayd Allāh; the poet Niẓām al-Dīn Aḥmad b. ʿAlī ʿArūḍī; the Ḥanafī jurist Abū al-Layth Naṣr; the searcher for truth and author of the *Ṣaḥāʾif*, Shams al-Dīn Muḥammad; etc.

Damascus of Samarqand: A fine city built by Tīmūr half a stage west of Samarqand. It is related that Tīmūr erected a city wall like that of Damascus in Syria and settled people there.

Soghd: A famous valley near Samarqand and one of the four main excursion spots of the world. It is also written as Ṣoghd. Ibn Ḥawkal described this excursion spot in great detail, because the Soghd Valley is 8 stages long, extending from the border of Bukhara to that of Bunam. These are all places with interlacing trees, streams and meadows. As there is abundant water, the trees do not grow tall. In the middle of the valley are villages, orchards and gardens; with fields and plowlands on both sides and pastures behind. On the banks of the streams [M 351] are meadows, and in the gardens incomparable pavilions, pools, and gazebos. All along this 8-day long road are picnic places, orchards and vegetable gardens. Such spaciousness and such delight cannot be imagined in the Rūd-i Abla (?) or the Ghouta of Damascus. And the fruits of Soghd are more delicious than anywhere else.

The towns of Soghd, in the middle of the valley, are Baykand and, along the Soghd River, Būmaḥkath, Garmīna, Dabūsiya, Hajrūd, Kasāsa and Sapīdrūd whence Samarqand can be reached.

The beginning of Soghd valley is more than 20 parasangs from Samarqand. Wafī is in the middle of the valley. All the rivers come from the Bunam mountains. The river divides into branches, each of which irrigates a village. The great Ḥanafī scholar Shaikh al-Islām ʿAlī al-Soghdī is from one of these villages.

Farbar: A flourishing town with cheap foodstuffs on the bank of the Jayḥūn, located at 96½ degrees longitude and 39 degrees latitude. It is in the direction of Bukhara and a crossing point from Transoxiana [R 133a] to Khurāsān. The hadith scholar Imām Farbarī, one of the transmitters of the *Ṣaḥīḥ* of Bukhārī, is from this town.

Bukhara: A famous city and flourishing emporium located on a plain west of Soghd-Samarqand at 97½ degrees longitude and 39½ degrees latitude, with plentiful orchards and gardens and flourishing districts and villages. According to the *Ḥabīb al-siyar*, it derives from the word *bukhār* meaning "an assemblage of the sciences" in the language of the Magians. In the past its name was Maḥlath. If one climbs up the citadel and looks around, he sees meadows everywhere. It is a delightful place. The entire plain is spread out like a green cloth. Here and there in the midst of the meadows are elegant pavilions, lofty and exquisite buildings that display the mastery of the local carpenters.

Of Samanid construction are a great fortress, incomparable markets and Friday mosques, and the seven gates of the city wall. Most of the Bukhara districts are inside the city wall, which encompasses an area 12 parasangs in length and breadth. There are no mountains or deserts inside it, only villages, fields, orchards and gardens. These districts are Ṭawāwīs, Tawraq, Lower Farghāna, Būm, Rūstākā, Ḥastawān, Lower and Upper Farwān, and Arwān. And there are several districts outside the wall, each with canals to carry water from the river, including the village (*rūstā*) of Garmina, Manāḥis, Upper Farghāna, and the district (*rustāk*) of ʿIrqīd (or ʿArqiya?).

The channels of Soghd-Samarqand flow to Bukhara; the main channel flows inside the city, turning the mills and irrigating the fields. All the orchards and gardens are along the channels. After they pass Bukhara they flow to Baykand and into the Jayḥūn (Oxus). There are numerous other channels as well, such as Rūd-i Buzurg, [Rū]d-i Nawkand, Rūd-i Māgahān, Jūybār-i ʿĀriḍ and Rūd-i Nawbahār. Several channels branch off the main channel of the Rūd-i Soghd, flowing to the villages and irrigating the fields. One is Rūd-i Kāfirkām, which irrigates a large district as far as Ḥarmash. Rūd-i Nawkand also belongs to a district (*nāhiye*). Rūd-i Barja, Rūd-i Basta, Rūd-i Armīniya and Rūd-i Farwān-i Suflī (Lower Farwān) each bring water to one or two districts.

Firewood in Bukhara is not gotten from the mountains, but all of it is cut from the gardens. There is no better fruit in Transoxiana than that of Bukhara. Its soil is so fertile that a field of one or two donums can provide for an entire family. However, because the population is large, [M 352] the produce is not sufficient and grain must be brought from the surrounding areas.

The inhabitants of Bukhara speak Soghdian. Most of them are good and virtuous people. They wear caps and mantles. Aside from the markets there are temporary bazaars held at specific times. The citadel is famous for its auspiciousness. They say that no army mustered under a banner there can be routed. And none of its rulers was ever buried outside of the citadel. The inhabitants are greatly obedient to the kings and governors.

They say that Bukhara, Samarqand and Balkh form the angles of a triangle each side of which is 25 parasangs. Samarqand is northeast of Bukhara.

According to the *Haft Iqlīm*, Bukhara is called Fākhira ("Boastful") because on doomsday it will boast of many martyrs. So many saints and scholars come from this town that [R 133b] they are without limit. In a single century 4000 jurists forgathered there, each of whom was authorized to issue legal opinions and was expert in the fundamentals and subdivisions of the law. There are masters of hadith, such as Imām Abū ʿAbd Allāh Muḥammad b. Ismāʿīl, author of the *Ṣaḥīḥ* of Bukhārī; members of the Maḥbūbī dynasty, the progeny of ʿUmar b. Māzan, such as Ṣadr al-Sharīʿa I and II, Burhān al-Sharīʿa and Tāj al-Sharīʿa; Naqshbandī shaikhs such as Khwāja ʿAbd al-Khāliq, Khwāja Muḥammad Bābā, Sayyid Amīr Kulāl, Khwāja Bahā al-Dīn, Khwāja ʿAlā al-Dīn ʿAṭṭār, Khwāja Ḥasan ʿAṭṭār, Khwāja Muḥammad Pārsā and his son Abū Naṣr Pārsā, Sayyid Burhān al-Dīn Khwānd (text: Khāwand) Shāh, Amīr Khwānd Muḥammad and his son Khwānd Amīr; poets such as Nāṣir and Khwāja ʿIṣmat, Mawlānā Barandaq, Khayālī, Sayfī, and Hāshimī.

Baykand: A town one stage from Bukhara, located at 98 degrees longitude and 39½ degrees latitude. The Tatars destroyed its walls. It has an elegant Friday mosque. It is said to have had 1000 hospices in the past.

Yanghikand: A town on the bank of the Sayḥūn, located at 88 degrees longitude and 47 degrees latitude. It is 125 parasangs north of Bukhara. The Sayḥūn flows through it and after a distance of two stages empties into the Aral Sea.

Jand: A town close to Yanghikand, located at 97 degrees longitude and 47 degrees latitude on the bank of the Sayḥūn.

Ṭawāwīs: A town located at 97½ degrees longitude and 38 degrees latitude, one of the annexes of Bukhara and within the city wall. It is seven parasangs from Bukhara. In the past it was a flourishing town with many scholars. It had a citadel and a walled city (*shāristān*), but they have been destroyed. It has an annual fair. In spring (lit. the vegetable garden season) people from Transoxiana come here for an excursion. It has many orchards and gardens.

Garmīniya: A town between Bukhara and Samarqand. It is bigger than Ṭawāwīs and has many villages. It is seven parasangs from Ṭawāwīs.

Dabūsiya: A small town on the Khurāsān road south of Soghd Valley. It is between Bukhara and Samarqand, five parasangs from Garmīniya. It has no villages. The Ḥanafī jurist Abū Zayd ʿAbd Allāh b. Muḥammad al-Dabūsī comes from this town.

Nasaf: the Arabicized form of Nakhshab. A town with many streams, orchards and gardens, located on a flat, broad plain at 88 degrees longitude and 39½ degrees latitude. The mountains are 2 parasangs hence in the direction of Kash. The streams flowing from Kash join here into a large river that irrigates the villages and then empties into the Jayḥūn. The land between the Jayḥūn [M 353] and here is a desert.

Nasaf has a ruined citadel and a suburb with four gates, the Kash gate, the Bukhara Gate, the Samarqand Gate and the West gate. The emir's palace is on the bank of the large river at a place called Sar-i pul. The prayer-grounds for religious festivals are at the Bukhara Gate, whereas the bazaar is in the suburb. Nasaf has a number of districts and two administrative centers, Bazda and Gashta.

Among those who come from Nasaf are Ḥāfiẓ al-Dīn ʿAbd Allāh Muḥammad, a great Ḥanafī jurist and author of the *Madārik wa manār*; Abū Ḥafṣ ʿUmar b. Muḥammad, author of the text of the *ʿAqāʾid* and the *Taysīr*; and Abū al-Muʿīn, author of the *Tabṣira*.

Bazda: A castle and town six parasangs from Nasaf, located at 99½ degrees longitude and 38½ degrees latitude. It is reported to have been built by Afrāsiyāb. The jurist Shaikh al-Islām ʿAlī al-Bazdawī and his brother Abū al-Yasar, who was a scholar of legal fundamentals and authored a famous text, came from this town.

Kash / Kish: A large town near Nasaf, located at 99½ degrees latitude and 39½ degrees latitude. It is reported to have a length and breadth of 3 parasangs, [R 134a] but this must be together with its districts. It is a place with cheap foodstuffs and abundant fruit. Its fruit ripens earlier than elsewhere, but the air is bad. In the past it used to be called Shahr-i Sabz ("Green City"). Between here and Samarqand there is a mountain that can be crossed in one day.

Kash has a citadel, suburb and walled town. The bazaars are in the suburb. The buildings are made of clay and timber. There are many orchards and vegetable gardens and abundant cereal crops. The walled town has four gates: the Āhanīn ("Iron") Gate, the ʿAbd Allāh Gate, the Qaṣṣābān ("Butchers") Gate, and the Shāristān gate, also called the Turkistan Gate because there is a village called Turkistan outside the gate.

It has two rivers: the Qaṣṣārīn ("Fullers") River, which rises in Shahristān; and the Aswar River. Both flow through the gate. There are quite a few streams in the villages, too, such as Rūd-i Chāch, Rūd-i Ḥasak and Rūd-i Jawān. Rūd-i Chāch is one parasang from Samarqand. Rūd-i Ḥasak flows along the Balkh road, 8 parasangs from the city.

Ishtīkhan: A large village and district in Soghd-Samarqand, 7 parasangs from Samarqand. It is a place of cheap foodstuffs, with orchards and gardens. According to Ibn

Ḥawqal, it has a citadel, a bazaar, villages and streams. A group of scholars originated from here. It is north of Soghd Valley.

Kushānīya: A district and town, two stages in length and one stage in breadth, located at 98½ degrees longitude and 40 degrees latitude in the very center of the Soghd region. The town is larger than Ishtīkhan and has more villages. However, the territory of Ishtīkhan extends to 5 stages. Both are north of the Soghd Valley.

Arbījan: A town in the Soghd region, near Kushānīya. It is said to be going to ruin.

Zāmīn: A town 2 stages east of Samarqand, located at 99½ degrees longitude and 40 degrees latitude, on the Soghd-Farghāna road. Before it is a plain and behind it are the Usrūshana Mountains. Persian manna (*terencebīn*) is abundant. There are streams, orchards and gardens.

Shāsh (Tashkent): the Arabicized form of Chāch. It is located at 101½ degrees longitude and 42 degrees latitude. According to the *Taqwīm*, [M 354] it is a large town, dependent on Samarqand. It is in a plain beyond the Sayḥūn. The houses have running water. This is the finest and most salubrious city of Transoxiana. Its district has about 15 towns. Farghāna is 5 stages from here, Khujand 4.

According to the *Masālik*, the town of Shāsh is half the size of Binkath. It has a rampart and a suburb and many streams, orchards and gardens. The border of Chāch is along the Īlāq (river). The inhabitants of this vicinity are Ghuzz and Khalaj, Muslim tribes who engage in jihad. They mostly live in tents and are submissive to no one.

According to the *Haft Iqlīm*, Shāsh is an ancient city. It is also called Banākath, but today is known as Tashkent.

I have seen reports in some knowledgable sources that Tashkent is located at 101 degrees longitude and 41½ degrees latitude. It is a large city, 7 stages east of Samarqand on the Khitāy road. Andijān is also 7 stages.

According to the *Taqwīm*, citing Ibn Ḥawqal, Binkath is the town of Shāsh. It has a citadel and suburb, and outside this suburb is another suburb. There are streams everywhere in the city. The Friday mosque is adjacent to the citadel wall.

According to the *Masālik*, the citadel has three gates: the Abū'l-ʿAbbās Gate, the Kash gate and the Āhanīn ("Iron") Gate. The emir's residence and the Friday mosque are located on the wall of the citadel. Inside the walled town there is only one bazaar, the middle of which is a public square. A wall has been built, extending from the mountain known as Sābligh to the Chāch Valley. It is called The Wall of ʿAbd Allāh b. Ḥamīd.

As for Tashkent, it is not mentioned in the *Taqwīm*. Again, according to the *Haft Iqlīm*, with regard to buildings, markets [R 134b] and other features, this city is simple and not very imposing; on the other hand, its land is adorned with streams and gardens, and its meadows and orchards are dressed in roses, tulips and other flowers. The seven-colored tulip in particular can be found nowhere else. Tashkent is famous for its tulip gardens, just as Bukhara is famous for its red roses.

In this province is a well whose water is purgative. If some water is taken a short way from its source, it turns into blood, and if taken further afield, it turns to stone. If a piece of cloth is dipped into menstrual blood and then cast into the well, a thunderbolt descends that can destroy buildings. And there is a spring in this land whose water does not flow except when there are clouds in the sky.

From the Shāsh region come the great Shāfiʿī hadith scholar and jurist Imām Muḥammad b. ʿAlī b. Ismāʿīl, known as Qaffāl-i Shāshī; Abū Bakr Muḥammad b. Aḥmad al-Mustaẓharī; Fakhr al-Dīn Banākatī, who is the author of a chronicle; Badr-i Shāshī; Fāḍil-i Tashkentī; etc.

Tunkat: one of the cities of Shāsh, also called Tuna. In the *Taqwīm* it is reported from Ibn Ḥawqal that Tunkat is the town of Īlāq. It has a citadel, a suburb, an emir's palace and streams. The populated areas of Īlāq, its vegetable gardens and crop fields, reach as far as the Shāsh valley. There are many gold and silver mines in the mountains of Īlāq.

The author of the *Masālik* says that Tuna is a city with no orchards or vegetable gardens. It has a citadel, a walled town, and a Friday mosque. The people in general are fine looking [M 355] and chivalrous. Grain supplies mostly come in boats on the Rūd-i Chāch.

However, according to the *Taqwīm* citing Ibn Ḥawqal, it has many vegetable gardens. The wall extends from Mt. Sāblagh to the Shāsh valley, protecting the city from incursions by the Turks.

He also reports that there is another river here known as the Īlāq. But the above-mentioned wall is the one located in Banākath. What the sources agree on is that Īlāq is the name of the entire Shāsh region from the border of Nawbakht to Farghāna. Avicenna's disciple, Ḥakīm al-Īlāqī, comes from this region.

Isfījāb: A large town on the frontier between Transoxiana and Turkistan, located at 100 degrees longitude and 43½ degrees latitude. Ibn Ḥawqal says that the citadel is in ruins, but the city and the suburb are in good condition. The outer wall encircles the suburb for about a parasang and there are streams and vegetable gardens within the wall. It is in a plain. The *Masālik* gives the names of the four gates of the outer wall: the Kūchak ("Small") Gate, the Farjān Gate, the Shākirāna Gate, and the Bukhara Gate.

The great Ḥanafī scholar and commentator on the *Muḥtaṣar* of al-Ṭaḥāwī, Shaikh al-Islām ʿAlī al-Isbījābī, comes from this town.

Cenābī says in his chronicle that this town is also called Sayrām; but I have not seen this in other books. In that case the annotator of the *Muṭawwal*, ʿAlāʾ al-Dīn al-Sayrāmī, would come from this town.

Usrūshana, recorded in the *Marāṣid* as Ushrūsana: A large town in the Hayāṭila region of Transoxiana, located at 101 degrees longitude and 41½ degrees latitude, between the Sayḥūn and Samarqand. It is estimated to be 26 parasangs from Samarqand.

According to Ibn Ḥawqal as reported by the *Taqwīm*, Usrūshana is the name of a clime that is mostly surrounded by mountains. Bordering on the Farghāna region in the east, the borders of Samarqand in the west, Shāsh and part of Farghāna in the north, and part of the border of Kash and Ṣaghāniyān in the south, it is a *vilayet* that comprises some 400 fortresses and several cities.

According to the *Masālik*, Usrūshana is called Būmaḥkath in the language of its inhabitants. Its districts are: Arān, Banāmkath, Kawkab, ʿAraq, Sābāṭ, Zāmīn, Dīzak. It is a flourishing city. The buildings are all made of clay and timber. The inner walled town has a wall and the suburb is surrounded by another wall. The walled town has two gates, the Shāristān Gate and the Bālātīn Gate. [R 135a] The Friday mosque is in the walled town.

A river flows through the city that turns some ten watermills. The two riverbanks are lined with trees. The river flows into the moat surrounding the city wall, then spreads to the palaces and gardens, irrigates the fields, and flows on. The suburb has four gates: the Zāmin Gate, the Ibn Samand Gate, the Ibn Ḥikmat Gate and the Kuhliyān Gate.

Various minerals are in abundance here, especially vitriol and sal ammoniac. The sal ammoniac mine is a cave in the mountains, whence steam arises. It looks like smoke during the day and like fire at night. A roof is installed above it and a wall with a door. Since there is no vent, the sal ammoniac crystallizes out of the steam layer-by-layer and sticks to the roof. From time to time a person wearing wetted felt opens the door and quickly leaps aside, because the confined steam bursts out violently, killing anyone standing in the way. After the steam is vented, [M 356] they take the roof down and remove the sal ammoniac, then reinstall the roof. As long as the steam is not confined it is not harmful; but the crystals accumulate only if it is confined.

In Usrūshana there is a type of rose that remains blossoming until the end of summer.

Sābāṭ: one of the towns of Usrūshana, located at 101½ degrees longitude and 41½ degrees latitude on the Farghāna-Shāsh road, 3 parasangs southeast of Usrūshana. Sābāṭ is the Arabicized form of Balāshābād. There is another Sābāṭ in Iraq near Ctesiphon.

Usbānīkath: one of the dependencies of Isfījāb, it is a town on the Fārāb road one stage from it and 9 parasangs east of Usrūshana.

Shāwkath: one of the towns of Shāsh, located at 101½ degrees longitude and 41 degrees latitude.

Farghāna: the name of a province comprising several towns and villages, located at 102½ degrees longitude and 41½ degrees latitude. As recorded in the *Haft Iqlīm*, its borders are Kāshghar in the east, Samarqand in the west, Kūhistān and Badakhshān in the south, and Turkistan and Ṭarāz in the north plus other regions destroyed during the Tatar invasion. The Sayḥūn, which is known in that region as the Khujand River, comes from the northeast of this province and flows through its middle.

According to Bābur's memoirs, mandrake is said to grow in Farghāna.

There is no place in Transoxiana as extensive, with such large villages and such cheap foodstuffs as this. There are silver and gold mines in Akhsīkath, quicksilver in the Sūkh Mountains, a pitch mine in the Rustā-yi Zarrīn province, and a petroleum well, copper, lead and iron mines, and also a turquoise mine, in the Astar mountains.

Farghāna was founded by Anūshirwān, who brought people from every household and settled them here. He called it Arhūkhāna; later it came to be called Farghāna.

The author of *Manāhij al-ʿibād*, Shaikh Saʿīd al-Dīn, and the commentator on the *Ṭawāliʿ*, ʿAbd Allāh b. Muḥammad ʿUbaydī, come from this province.

Again according to the *Haft Iqlīm*, there are seven large and small cities in the region of Farghāna, five south of the river, two north of it. Those in the south are as follows:

Andijān: A city in the middle of Farghāna with very strong fortifications. There are several streams flowing inside the fortress. Grain and fruit are good, but the air is bad, and eye sickness is rampant. According to the *Taqwīm*, it is the Arabicized form of Andigān. It is on the Kāshghar road, two stages east of Marghīnān. Its river comes from the south, reaches the city and then empties into the Awsh River.

Awsh: A town and castle southeast of Andijān. It has three gates: the Kūh Gate, the Āb Gate and the Maʿkada Gate. The climate is fine. There are meadows everywhere around it. Its river comes from the southeast, passes through the city and goes toward Yasī. Sirāj al-Dīn, author of the *qaṣīda* with the incipit *yaqūlu l-ʿabd*, comes from this town.

Marghīnān: A town seven parasangs west of Andijān. It is in the district of Lower Nasā, half a stage east of the Shahrūqiyā River and southwest of Kāshghar. Its river comes from the southeast, flows south of Marghīnān and empties into the Shahrūqiyā River. [M 357] It has delicious pomegranates and apricots. The best known of those who come from this town are the Ḥanafī jurists Imām Ẓahīr al-Dīn ʿAlī b. ʿAbd al-ʿAzīz and Imām Burhān al-Dīn ʿAlī b. Abī Bakr, the excellent author of the *Hidāya*, who died in 592 (1195–1196).

Asfara: the Arabicized form of Aspara. It is a town in a mountainous region 9 parasangs southwest of Marghīnān on the Samarqand-Kāshghar road. It is a place with abundant orchards and gardens and streams. Its river comes from the south, winds around the town to the east and goes to the Shahrūqiyā. [R 135b]

Khujand: A large city on the bank of the Sayḥūn, located at 102 degrees longitude and 40 degrees latitude. The river flows past it and goes to Fārāb where it is called the Shahrūqiyā River. The city is on the south bank, on the Samarqand-Kāshghar road; Kāsghar can be reached in 35 days, and it is 10 stages from Samarqand.

According to the *Haft Iqlīm*, Khujand is a city with very strong fortifications, 5 parasangs west of Andijān. There is abundant fruit, especially pomegranates. However, eye disease is rampant—even sparrows' eyes are ill.

In the district of Khujand there is a place called Kand-i Bādām, which is a flat plain where the wind is constantly blowing. It is also called Dasht-i Darwīsh ("Dervish Plateau"), because once some dervishes were overtaken there by a strong wind and perished while saying to each other, "Hay dervish, hay dervish!"

One of the great poets, Shaikh Kamāl al-Dīn (Khujandī), comes from this city. As is reported in the *Naghamāt*, he hid his poetic gift under a veil. He has a divan. Of later poets, there are many Khujandīs in Mecca and Medina.

Akhsīkath: one of those seven cities (?), located at 101½ degrees longitude and 42 degrees latitude, on the north bank of the Khujand River, on a plain one parasang from the mountains. The *Haft Iqlīm* writes it only as Akhsī. In the territory of Farghāna, after you pass Andijān, there is no city as extensive as this. It is 10 parasangs from it (i.e., Andijān) and has a strong and fortified castle. It has a type of watermelon called Mīr Tīmūrī that is found nowhere else.

According to the *Masālik*, Akhsīkath is the town of Farghāna. It has a citadel, a walled town and a suburb. The emir's residence is in the citadel, the Friday mosque is in the city, and the public prayer-grounds are on the bank of the river. The city has five gates: the Zinjīr ("Chains") Gate, the Dafīna Gate, the Kāshān Gate, the Ādhīna Gate, and the Hābasār Gate. In the suburb are many streams and orchards and gardens. The Ḥanafī scholar Athīr al-Dīn al-Akhsīkathī, author of the *Uṣūl*, comes from this city.

Kāsān: A town and district in the region of Farghāna, located at 101½ degrees longitude and 42½ degrees latitude. It has many villages. As mentioned by Yaʿqūbī in agreement with other authorities, in the past it was the administrative center of Farghāna and one of the beauties of the world, but it was destroyed in the invasion of the Turks. There is another Kāsān in the clime of Persian Iraq, which is mentioned in its proper place.

Nasā: two towns and districts of Farghāna, adjacent to each other. One is called Upper Nasā, the other Lower Nasā. They are located on a grassy and well-watered plain. In Upper Nasā there are wells of pitch and a kind of black stone that burns like coal and the ashes of which are used as soap. Three okas of this stone are sold for one dirham. [M 358] Aside from these there is another Nasā in Khurāsān, which is the famous one.

Khwākand: A town in the vicinity of Farghāna. It is sometimes pronounced Khwāqand with a *q*.

Isbīdbalān: the *Taqwīm*, based on Ibn Ḥawqal's report, registers it as a city of Farghāna. There are gold and silver mines in the mountains.

Qubā: A town in one of the districts of Farghāna, located at 102 degrees longitude and 42½ degrees latitude. It is in the direction of Shāsh, close to Akhsīkath. Its citadel is in ruins, but the city and suburb are in good condition. The suburb is also surrounded by a wall. It has plentiful water sources and vegetable gardens.

Kand-i Bādām: located at 102½ degrees longitude and 40½ degrees latitude, on the south bank of the Sayḥūn and one stage north of Khujand.

Khīsa Kūz: A town one stage west of Kand-i Bādam, on the bank of the Sayḥūn. It is one stage north of Khujand, with a mountain between the two, and one stage east of Zāmīn. [R 136a] [SPACE FOR MAP]

Sūkh: A town located at 104 degrees longitude and 41½ degrees latitude. There are quicksilver mines in its mountains.

Tirmidh: A magnificent city on the bank of the Jayḥūn, located at 99 degrees longitude and 37 degrees latitude. All its houses and markets are paved with bricks. The closest mountain to it is one stage away. It has abundant villages and districts that are irrigated by the Ṣaghāniyān River. The Jayḥūn flows by south of the city. There is a distance of one stage between Balkh and this city. Therefore, some consider it to belong to Ṭukhāristān. However, as it is on the other side of the river, it is a famous entrepot in Transoxiana.

It has a citadel, a walled town and a suburb. The Friday mosque is inside the city. The buildings are made of clay. The people get their water from the Jayḥūn. Its fields are irrigated by the Rūd-i Chaghāniyān. The hadith transmitter and author of the *Sunan*, Imām Abū ʿĪsā Muḥammad, and the author of the *Nawādir al-uṣūl*, Ḥakīm al-Tirmidhī, come from this town.

Jūminjān: One of the districts of Tirmidh, also called Jūmingān.

Wāshjird: A town appended to Ṣaghāniyān, located at 102 degrees longitude and 38½ degrees latitude. It abounds in saffron. The castle of Rāsib is 6 parasangs hence.

Shibliya: one of the towns of Usrūshana. One of the great sheikhs, Shaikh Shiblī, comes from here.

Ṣaghāniyān: A town south of Usrūshana, located at 100½ degrees longitude and 38½ degrees latitude. It is also called Chaghāniyān. It has a large district and villages, a citadel, and abundant streams, orchards and gardens. Its streams flow toward Tirmidh. Imām Ṣaghānī, author of the *Mashāriq*, comes from this town.

Shūman: A town in the district of Ṣaghāniyān, located approximately at the same longitude and latitude.

Kharashkath: A town on the bank of the Sayḥūn.

Najānīkath: A town in the vicinity of Samarqand. Samʿānī thinks it is in the direction of Usrūshana; others consider it to be one of the administrative units of Usrūshana.

Wadhār: A town 4 parasangs from Samarqand. It has a fortress and a Friday mosque.

Ūzkand: one of the a towns of Farghāna, located at 104½ degrees longitude and 40½ degrees latitude. It is two stages east of Andijān and one stage west of Awsh. It is Arabicized as Ūzjand. It has a citadel and a walled town, and streams and orchards and vegetable gardens within the wall. [M 359] Its climate is warm. The famous jurist and author of *al-Fatāwā*, Imām Qāḍīkhān Ḥasan b. Maḥmūd al-Ūzjandī, comes from this town. [R 136b]

Māymargh: A large village on the Nasaf-Bukhara road and a town near Samarqand. A place on the bank of the Jayḥūn is also called Māymargh. The Ḥanafī jurist Fakhr al-Dīn Muḥammad comes from here.

Khadaysar: an administrative unit of Usrūshana on the border of Samarqand. It has a famous hospice.

Baskath: One of the towns of Shāsh.

Bārsinkath: also one of the towns of Shāsh, close to Baskath.

Budaḥkath: one of the towns of Shāsh. Some consider it as belonging to Isfījāb.

Butam: one of the districts of Transoxiana. It has great mountains and many villages, but the climate is very cold.

Shahrūqiyā: A city on the north bank of the Sayḥūn, located at 100 degrees longitude and 40½ degrees latitude. It is two stages south of Tashkent and five stages northwest of Khujand. The Sayḥūn is also called by the name of this city.

Khuttalān: according to the *Taqwīm*, a district of Transoxiana, east of Balkh. Ibn Ḥawqal says that Khuttal and Wakhsh are two districts adjoining each other. The town of Khuttal is comprised of Halāward and Lāwakand, falling between the Wakhshāb and the Badakhshān rivers. These towns have abundant water sources and trees, and are located on a plain with cheap foodstuffs.

Halāward: A town in the of Khuttal region, located at 100 degrees longitude and 37½ degrees latitude. This district is on a plain adjacent to Transoxiana with abundant rivers and forests and cheap foodstuffs.

Wakhsh: A town of Khuttalān at the foot of a mountain in the vicinity of Balkh. Its district is called Wakhshāb. There are many gold and silver mines here and gold dust is collected from the streams. Some consider Khuttalān and Wakhsh as towns of Ṭukhāristān, not Transoxiania, since they are south of the Jayḥūn. But most consider them to be in Transoxiana.

Rivers

Aside from the small ones, there are two great rivers in this clime: the Jayḥūn (Oxus) and the Sayḥūn (Jaxartes).

Jayḥūn: A great river, also called Balkh River, Amū River and Tabar River. It separates Iran from Turan. Its main tributary is the Kharnāt River which rises in the mountains of Badakhshān and flows to the northwest. Several tributaries flow together in Qubādābād. The Wakhsh River comes from Ṭukhāristān and empties into it at the borders of Balkh, above Tirmidh. After that it is called the Jayḥūn, a designation not used further up. In Tirmidh it is joined by the rivers of Chaghāniyān, whence it continues its way to Kālif, the town of Zumm, and Āmul Shaṭṭ, i.e. Āmū. It does not irrigate anywhere down to Zumm, but there it is used to a certain extent, and in Āmū it is used completely for irrigation. It mostly benefits the inhabitants of Khwārazm.

After this, some channels branch off in the districts of Balkh and Tirmidh. Then it contracts and passes through a narrow gorge between two mountains called Dahān-i Shīr ("Lion's Mouth")—it is not more than 100 fathoms wide—and reaches the village of Lūmīna, a dependency of Herat. Gurganj in Khwārazm is close to that gorge. After it exits that gorge there is a sandy tract for two parasangs where the water is absorbed. Beyond it [M 360] the river reemerges and continues to the province country of Khwārazm. It is impossible to cross that sandy tract, which is a huge marsh.

In Khwārazm the Jayḥūn divides into several large channels—including the Kākhwāra, the Hazarasp, the Gardānjūy, the Karba and the Ḥara—which irrigate the fields of many villages and provide water to the entire country of Khwārazm. All of these are navigable. Some of them empty into the Aral Sea. But the main branch of the Jayḥūn flows through Khwārazm and then the Ḥulam Pass, which the Turks call Karlāwa. It is a narrow, stony valley. The river flows through that pass with such a terrifying noise that it can be heard from two parasangs away.

According to Ḥamd Allāh, it empties into the Caspian six stages from Khwārazm at a place called Khalkhāl, where only fishermen live. But the *Masālik* and the *Taqwīm* both say that it empties into the Aral Sea. It is possible that one branch of it empties into the Aral Sea.

The entire course of the river is approximately 300 parasangs. In winter it is so frozen that a caravan can cross it; people bore through several fathoms of ice and draw water from it as from a well.

The tributaries that flow into the Jayḥūn are the Mākhash, the Būyān, the Fāraghī, the Andījār, and the Wakhshāb. They flow together near Qubādābād and empty into the Jayḥūn.

The Wakhshāb River rises in Turkistan and goes to Wakhsh territory. It flows under a bridge built at the foot of a mountain between Khuttalān and the border of WLSKR (Lashkar?), then it goes to the borders of Balkh and joins the Jayḥūn above Tirmidh. In Transoxiana the first regions (*kūra*) on the Jayḥūn are Khuttalān and Wakhsh. Afterwards the vicinities of Wajān and Saqīna are all lands of infidels, which produce fine slave boys and slave girls.

Sayḥūn (Jaxartes): also called Shāsh, Chāch, Khujand and Shahrūqiyā. The people of that region also call it Gul-i Zaryūn ("Golden Rose"). Its main tributary comes from the Munjatīn Mountains in Turkistan. Coming from the northeast, it reaches Ūzgand and Shāsh and flows through the middle of Farghāna province. Flowing from north to south, it receives a number of affluents such as the Ḥarsāb and Rūd-i Ūs (Awsh?), becoming as large as two thirds of the Jayḥūn. It flows by Akhsīkath, Khujand, Fārāb and Baykand, reaching Fanākath, which is also called Shahr-i Janna ("Paradise City"). After passing through that place, according to the *Haft Iqlīm*, it is absorbed in the sands of Turkistan; according to the *Taqwīm* and the *Nuzha*, two stages after Yanghikand it empties into the Aral Sea.

In regard to length and breadth, the Sayḥūn is like the Jayḥūn. It freezes in winter, and can be crossed by caravans. It irrigates several regions. After it passes the border of Ṣabrān, both of its banks are inhabited by the nomadic Oghuz tribe, most of whom are Muslims.

There are various reports on the discharge of the Jayḥūn and the Sayḥūn. The *Taqwīm*, citing Ibn Ḥawqal, says that both empty into the Aral Sea; and citing the *Rasm-i maʿmūr*, reports that one channel branches off the Jayḥūn at 91 longitude, passes near Khujand and empties into the Caspian Sea. The *Masālik* also says that both the Sayḥūn and the Jayḥūn empty into the Aral Sea; but the statement that there is a twelve-days' distance between the places where they reach it is contradictory, because the length of the Aral Sea does not allow for this. The *Nuzha*, which says that the Jayḥūn empties into the Caspian and the Sayḥūn into the Aral Sea, seems closest to the truth. The report of the *Haft Iqlīm*, [M 361] that the Sayḥūn disappears in the sand, is also possible, although it would be more credible if it referred to the Jayḥūn. [R 137a]

Būy River: according to the *Nuzha*, it rises in the mountains of Soghd and Ṣaghāniyān and forms a lake. When it emerges from that, several other large and unfordable streams flow into it and it irrigates the province. There is no ford on it at all. The main tributary flows through Soghd, Samarqand and Bukhara and these provinces flourish because of this river.

It should be noted that the Sayḥūn, the Jayḥūn, the Nile and the Euphrates originate from the rivers of Paradise and flow from beneath the Sidra (or Lote Tree) of the Boundary (Koran 53:14), as recorded in a sound hadith. Some commentators interpret these rivers as metaphors for faith and Islam; others interpret them literally.

Qāḍī ʿIyāḍ, in his commentary on (the *Ṣaḥīḥ* of) Muslim, prefers the former alternative, saying that since these are mainly rivers in the lands of Islam and most of those who drink from them are Muslims, and since the bodies of the Muslims nourished by them will mostly go to Paradise, it is as if they rose in Paradise or as if their source were the Sidra. In this case, doubts are removed and there is no need for further explanation.

Imām Nawawī, on the other hand, gives a literal explanation, saying that these four rivers originate from under the Sidra and descend to the bottom of the earth, although we do not know how they descend, and then each of them rises from a separate direction. This is what the external meaning of the hadith requires, and there is no rational or legal impediment against it.

Imām Nawawī's adversaries, however, do not grant that there is no rational impediment to this belief, because the rational explanation for the origin of rivers is something else entirely. This subject would require a lengthy discussion. Whoever so desires should open that door.

Mountains and mines

There are many mountains in Transoxiana, the most important ones being:

Mt. Sārwān: A populated mountain with a good climate south of Samarqand. There is a Christian temple there called Zardgard.

Mt. Asfara: A mountain at the extreme end of Shāsh. There is a petroleum well and an iron mine, and turquoise is also mined here.

Mt. Farghāna: A large mountain with mines of quicksilver, copper, lead, gold, petroleum, tar, sal ammoniac and pitch. And there is a kind of stone that is burned instead of firewood.

Mt. Warka: It extends from a place near Bukhara through Samarqand and Kash, adjoins Usrūshana, then turns back, reaches the border of Farghāna and then China. The Usrūshana Desert is contiguous with this mountain from the border of Farghāna and Īlāq as far as Khirkhīr. It contains various mines, also a lake whose shores are frozen in summer and warm in winter.

Plants and products

This clime has many minerals and plants. Fruit is so abundant that in Soghd, Usrūshana and Farghāna it is used to feed animals. Even if there is no crop for one or two years, it makes no difference in the prices. It is a place with an abundance of meadows and pastures, rivers and fields. Therefore it has excellent livestock, although sheep come mainly from Turkistan.

Famous products include silk, wool, and Samarqand paper. Furs of fox, sable and squirrel are plentiful. Musk comes here first from Tibet and then is exported to the surrounding areas. Fine fruit trees grow in the mountains between Farghāna and Turkistan, but no one owns them. There are many types of herbs and flowers. Pistachio is especially abundant in the mountains of Transoxiana.

In the past, the revenue from Bukhara was 11 *yük* (1 yük = 100,000) (and) 89,200 dirhams; [M 362] from Farghāna, 28,000; from Usrūshana, 5000; from Shāsh, 60 *yük* of *missīsiya* dirhams, which resemble copper; from Āmul, 293,400; from Kash, 111,500; from Fārāb, 55,000; from Tirmidh, 2000; from Khujand, 10,000 dirhams.

Mores

The inhabitants of Transoxiana are mostly righteous, chivalrous and hospitable. They fight over guests. The author of the *Masālik* says: "In Soghd I saw a hostel that was like a palace. The gate was open and nailed to the wall. They told me it was the hostel of so-and-so, that its gate had been open like this for more than 100 years and that inside provisions were ready for travelers. Probably they nailed the gate to the wall lest a traveler should come and, not finding the gate open, go somewhere else. For generations they have shown great respect to the travelers who sojourn at this hostel."

They generally spend their wealth on good works. There is no waystation in the desert without a hospice or hostel. It has been reported that in this clime there are almost 10,000 soup kitchens and lodges providing food to travelers and fodder to their animals. [R 137b]

Transoxianians for the most part are valiant and brave and skillful in battle. Since they are surrounded by non-Muslims and continuously engage in religious war (*ghazā*), they have tremendous facility with instruments of war and the use of weapons. In particular, there is no other province where the people have military provisions like the people of Chāch and Farghāna. Also, they greatly obey their rulers, and for this reason the past kings chose their soldiers from amongst the Turks.

Routes of Transoxiana

From Samarqand to Balkh: > Shahr-i Sabz > town of Shaikh Ṣādiq > town of Ḥiṣār > town of Dih-i Naw > town of Ṣafā > town of Būyā > Tirmidh > Balkh, 1 stage each.

From Samarqand to Usrūshana, 2 stages; to Farghāna, 35 parasangs.

The Farghāna road, starting from the Jayḥūn at Tirmidh: > Baykand > Ṭawāwīs > Garmīna > Dabūsiyya > Samarqand > Āthārkīt > Ribāṭ Saʿd—the road to Chāch goes off from here; > Marzghad > Zāmīn—the road to Banākath goes off from here; > Sābāṭ > Ūzkand > Sāwkat > Khujand > Kanda > Sūkh > village of Warīz > Amash > Qubā > Awsh > Ūzkand, 1 stage each.

The Chāch road as far as the frontier of Islam, beginning from Atārkīd: > Qaṭrān valley, 1 stage; and again, Atārkī > Ḥarfān > Dīzak > Shaqq Ḥusayn > Ustūrkat > Sakat > Ribāṭ Abū al-ʿAbbās, also called Ayqar > ʿArkaz > Isbījāb > Bāḥakath, 1 stage each; > Ṭarāz, 2 stages—there is no populated place in between.

When following the bank of the Jayḥūn, it is 22 stages to Ṭarāz. If one goes from Samarqand to Khwārazm through populated country, it is 12 stages.

From Āmū[l] > Wīza > Mardūyīn > Aspās > Mughān > Ṭāhiriyya > ʿĀna > Jangband > Sadw > Hazārasp, 1 stage each.

The Badakhshān road: From Manak > Kūbrī (Köpri "Bridge"), 6 stages; > Wakhshāb > Abarkand > Halāward > Halīk > Manak, 2 stages each. The village of Manak is the passage to Badakhshān.

Distances

From Tirmidh: > Ḥarmangān, 1 stage > Dār-i Zangī, 1 stage > Chaghāniyān, 2 stages > Janga, 2 stages > Shūmāna, 2 stages > Ayūnān, 1 stage > Yashkar, 1 stage > Pul-i Sangīn, 1 stage. Again from Ayūnān: > Īlāq, 1 stage > Darband, 1 stage > Gāwkand, 2 stages.

From Chaghāniyān: > Zaytūn > Kūrāb > Rīg in the vicinity of Rasht, 1 stage each.

From Tirmidh: > Qabāryān, 2 stages > Chaghāniyān, 3 stages.

Distances of Bukhara: From Būmaḥkath [M 363] to Baykand is 1 stage, to Ḥijāra is 3 parasangs. From the city to Maʿkān is 5; to Zubda, 4; to Ṭawāwīs, 4; to Badmāḥkath, 1 parasang.

Distances from Soghd and Samarqand: From Samarqand > Atārīkath, 4 > Dharʿas, 4 > Baḥīkath, 9. From Samarqand > Dawān, 2 > Kayūrmaḥkath, 2. From Samarqand > Isbījar, 7 > Kāsān, 5 > Zarqān, 3. From Kāsān > Arhajar, 2 parasangs.

Between Kash and Naḥshab are 3 stages. From Kash to Chaghāniyān is 6 stages, to Būkat is 5 stages, to Sūya is 2 stages. From Naḥshab to Gashta is 4 parasangs, to Bazda is 6 parasangs.

Distances of Usrūshana: From Ḥarkāna > Darak, 5 parasangs > Zāmīn, 9 > Sābāṭ, 3 parasangs. From Usrūshana > Daʿkath on the Khujand road, 3 parasangs > Khujand and ʿAraq, 6 parasangs.

Distances of Shāsh and Isbījāb: From Banākath > Ḥarsīkath, 1 parasang > Sūrkat, 3 parasangs > Dihqānkat, 2 > Bārankat, 1 > Bankath, 2 > Chagharkat, 2 > Farankat, 2 parasangs; > Abrūkat > Kuttāl > ʿAdwāl > Kafarna > Ajbūrīn, 1 stage each.

Between Rūd-i Īlāq and Rūd-i Chāch, on the west side of Būkbak, the channels of Balkh are 5 parasangs in length and breadth. From Ḥabābnijkath (?) > Chāch, 2 parasangs > Bīkat, 3 > borders of Sakana, 1 > Ḥābūrkat, 3 > Ḥūmānkat, 4 parasangs.

From Baykand > Isbijāb, 4 stages > Bīkat, 2 stages.

From Sābīkand to Fārāb is 2 stages. From Sāwghar to Ṣīrān is 1 stage. From Akhsīkath to Saylāb is 1 stage. From Akhsīkath to Kāsān is 5 parasangs, to the border is 9 parasangs. From Qubā > Isbiqān (?), 3 parasangs > [R 138a] Kanār-irūd, 7 parasangs. From Sūkh > Makāḥis, 5 parasangs > Awāl, 10 parasangs. From Qubā > the city of Mughān, 7 parasangs.

Kings and rulers

During the caliphate of Muʿāwiya, when Transoxiana was in the hands of the infidel Turks, Saʿīd b. ʿUthmān b. ʿAffān came with the Muslim army and took Bukhara, but he could not conquer Samarqand. Later, during the caliphate of Yazīd b. Muʿāwiya, Muslim b. Ziyād conquered it, but the inhabitants revolted. During the time of Walīd b. ʿAbd al-Malik, Qutayba b. Muslim al-Bāhilī came with an army and conquerted it again. After Qutayba there were several more governors appointed by the Umayyads.

During the early days of the Abbasids, governors were appointed according to the directions of Abū Muslim. When Maʾmūn became caliph and went from Khurāsān to Iraq, he appointed the Samanids as governors of the cities of Khurāsān and Transoxiana. Thus emerged the Samanid dynasty in the year 204 (819), and for 191 years nine members of the dynasty governed Khurāsān and Transoxiana. As expressed by their court poet ʿUnṣurī [Persian verse]:

Nine men of the Samanid dynasty are famous,
Who are mentioned as being emirs of Khurāsān:
One Ismāʿīl, one Aḥmad and one Naṣr,
Two Nūḥs, two ʿAbd al-Maliks and two Manṣūrs.

Their capital was Bukhara. The dynasty ceased to exist in 395 (1004–1005).

After them, Sultan Maḥmūd ruled Transoxiana and appointed governors there. In 604 (1207–1208), during the tenure of the Seljuk governors, the Khwārazmshāh ʿAlā al-Dīn Muḥammad took control and stopped naming the Abbasid caliph in the Friday sermon in Khurāsān and Transoxiana. Shaikh Shihāb al-Dīn [M 364] al-Suhrawardī came with a missive from Caliph Nāṣir and counselled them to resume, but to no avail.

The emergence of Chingis Khan: In 616 (1219–1220) some merchants were sent to Transoxiana by the khan of the Great Yurt, Chingis Khan. The governor of Bukhara on behalf of Khwārazmshāh ʿAlā al-Dīn, who was also his paternal uncle, cruelly put the merchants to death and confiscated their goods. Chingis Khan requested that the governor be turned over to him, but ʿAlā al-Dīn had the envoys killed and Chingis led an army to Transoxiana. After a three-day battle, ʿAlā al-Dīn was defeated and fled to Qazwīn. From there he went toward Gīlān and died in Ābaskūn.

This event was the reason for Chingis' assertion of authority (in the Khwārazmian empire). After the Khwārazmshāh's defeat, Chingis marched against Bukhara and laid siege to it for three days. The 50,000 troops who were holding the fortress could not withstand the Mongol onslaught and surrendered. After their weapons and goods were taken from them, Chingis ordered the entire population of Bukhara to be massacred. All were put to the sword except one man who went to Khurāsān and related what had happened. This event took place on 4 Muḥarram 617 (18 March 1220). Nothing was left alive in Bukhara. They even slaughtered the dogs and cats. And they burnt and destroyed the city.

Then the Mongols advanced on Samarqand. The inhabitants resisted, and when they sought terms, their weapons and goods were taken and they too were massacred like the people of Bukhara. The Mongols burnt and destroyed the city, and went to Khurāsān. Chingis's massacres in Turkistan, Iran and Turan during the six years until 621 (1224–1225) reached such a level that humankind came close to being exterminated in those lands. In that year Chingis returned (to Mongolia), leaving his sons in his stead.

The Chaghataid governors: After the emergence of Chingis, the countries of Turan went under the governance of his son Chaghatay. Twenty-nine Chaghataid khans ruled until 806 (1403–1404) when the last of them, Maḥmūd Khān, was killed by Tīmūr and his line came to an end.

The Timurid dynasty: Tīmūr was born in Shaʿbān 736 (March–April 1336) in one of the villages of Kash. In his twenty-fifth year he married the sister of Amīr Ḥusayn, the governor of Turkistan, [R 138b] and he became known as Küreken meaning "Brother-in-law". With Tīmūr's support Amīr Ḥusayn took Transoxiana. Later dischord broke out between Amīr Ḥusayn and Tīmūr, and when his sister died, their bond was broken. In 771 (1369–1370) Amīr Ḥusayn was killed at Tīmūr's hands along with his sons in the fortress of Hinduwāna, and the government of Transoxiana was left to Tīmūr. One of the great sayyids (descendants of the Prophet), Sayyid Baraka, gave him a drum and a banner, and he became padishah in that year. He made Samarqand his capital and the Friday sermon was recited in his name.

In (7)73 (1371–1372) he conquered Khwārazm. In (7)75 (1373–1374) he subjugated Moghulistan. In (7)82 (1380–1381) he undertook the conquest of Iran. In (78)4 (1382–1383) he took Māzandarān. In (78)5 (1383–1384) he defeated the Awghānīs (Afghans) of Sīstān. In (7)88 (1386) he set out on a three-year campaign in which he raided and subdued the countries of Azerbaijan, Shirwān, and the Kipchak Steppe. In (7)93 (1390–1391) he fought a battle with Toḳtamış, the khan of the Steppe, [M 365] the like of which had not been seen before, and Toḳtamış was defeated.

In (7)94 (1391–1392) Tīmūr set out on a five-year campaign in which he crossed the Jayḥūn and laid waste to the climes of Māzandarān, Khurāsān, Persian and Arab Iraq, Jazīra (Upper Mesopotamia) and Armenia. Finally, after conquering Ghazna and Kabul and India, he returned to Samarqand. Six months later, in Muḥarram 802 (September–October 1399), he set out on a seven-year campaign. He again crossed the Jayḥūn, went through Khurāsān and Iraq and reached Sivas, which he destroyed, then spent the winter in Karabagh. In 804 (1402) he defeated Yıldırım Bāyezīd Khan on the plain of Çubuḳ and spent the winter in Kütahya. Then he went to Baghdad and reconquered it. In 806 (1403–1404) he conquered Georgia. In 807 (1405) he crossed the Jayḥūn and returned to Samarqand. After a few days' stay he set out in the middle of winter on a campaign against Cathay. When he reached Utrār, destiny intervened and he died on 17 Shaʿbān of that year (18 February 1405). His bier was taken to Samarqand and he was buried in his tomb.

Twenty of his progeny ruled over Transoxiana and Khurāsān until 917 (1512), the last being Mīrzā Bābur. The most notable rulers among them were Sultan Shāhrukh, Sultan Ulugh Beg Muḥammad, Sultan Abū Saʿīd and Sultan Ḥusayn Bāyqarā. Especially notable was Ulugh Beg, who built an observatory in Samarqand and compiled a famous astronomical table under the supervision of the leading scholars, Ghiyāth al-Dīn Jamshīd al-Kāshī, Ḳāḍīzāde-i Rūmī and ʿAlī Ḳuşçı. With this beautiful work, he gained fame and took precedence among the Muslim rulers. Sultan Ḥusayn was also an intelligent and virtuous ruler who wrote a delightful book entitled *Majālis al-ʿushshāq*. He had such a perfect royal companion as Mulla Jāmī and such an excellent vizier as Mīr ʿAlī Shīr (Nawāʾī).

In that year (807 (1405)) the Shaybakī (Shaybanid) dynasty emerged. When the Uzbeks conquered Transoxiana, Mīrzā Bābur fled to Ghazna and Kabul, establishing himself there; his descendants conquered India, and the present ruler of India is from this lineage. The Shaybakīs, who are Uzbeks, are the descendants of Shaybān b. Jochi b. Chingis. There were Shaybakī khans residing in the Steppe down to Shaybak Khan b. Shāh Būdāgh in the sixth generation. In about 906 (1500–1501), this Shaybak conquered some portions of Turkistan and Transoxiana and defeated Mīrzā Bābur several times in battle. Eventually Bābur went to India, while Shaybak Khan [R 139a] took control of those lands and then crossed the Jayḥūn and took Khurāsān as well.

In the meantime, Shāh Ismāʿīl also emerged, conquering Azerbaijan and Iraq. His Qizilbash troops entered Khurāsān on several occasions, but Shaybak Khan routed them with his Uzbek forces. Thus Shāh Ismāʿīl became Shaybak Khan's archenemy. In 916 (1510) near Marw he killed him in a surprise attack, turned his skull into a drinking cup, and took Khurāsān from the Uzbeks. However, Shaybak's paternal uncle, Küchkünji Khan, remained in possession of Transoxiana. After that a number of khans such as ʿUbayd Allāh Khan, ʿAbd al-Muʾmin Khan and Iskandar Khan, have continued to rule down to the present age. [M 366] Today Transoxiana and Turkistan are in their hands, and they never cease to war with the Qizilbash.

Chapter 37: Clime of Turkistan and the Steppe

This is a long and broad clime.

Borders

To the east, the frontiers of Cathay; to the north, the Kipchak Steppe and the deserts of Muscovy; to the south, Badakhshān, Transoxiana, Kashmir and the borders of Tibet; to the west, Khwārazm, Dagestan and the Don River in the Kipchak Steppe.

Cities

Although its cities are few and scattered and it is mostly desert, it is nevertheless considered on a par with other countries because even the desert is cultivated and populated. Its main province is:

Kāshghar: A long and broad province in Turkistan, located 15 stages northeast of Andijān. It is bordered in the north by the mountains of Moghulistān. Many rivers descend from these mountains and flow to Kāshghar. In the south it is bordered by part of the country of Shāsh and part of the desert. Its western border is an oblong mountain branching off the aforesaid mountains. The rivers from this mountain flow to the east. All the provinces of Kāshghar and Khotan are at the foot of this mountain. Its eastern border runs from Turfan to the Kalmyk territory. Its eastern part and some of its southern part is a vast desert and scrubland and endless steppe.

Distance of Kāshghar: From Shāsh to Turfan is a three-day journey.

In ancient times there were populated cities in those deserts. Now only the names of two of them have survived, Thūb and Kanak. The rest have been buried in the sands. The steppe (i.e., Bactrian) camel is found in these deserts and is hunted.

The administrative center (*dār-i mülk*) of Kāshghar is at the foot of the western mountain. A number of streams rise from the mountain, bringing water to the fields. One of them is called Taman. In the past this river reached the city of Kāshghar, flowing through its middle. Mīrzā Abū Bakr destroyed that city and had it rebuilt on one side of that river, which now flows along its edge.

According to the *Taqwīm*, citing the *Qānūn* of Masʿūdī, Qāshghar—with Q—is a large city. The inhabitants are Muslim. It is also called Ordukand.

Shaikh Saʿd al-Dīn al-Kāshgharī comes from this region.

Yārkand: A town, the capital (*pāytaḫt*) of the province of Kāshghar, located at 112 degrees longitude and 42½ degrees latitude. In the past it was a large city. Later it fell to ruin and became a haunt for wild beasts. Mīrzā Abū Bakr found the climate suited his disposition and settled it, making it his capital (*dārü'l-mülk*). He channeled streams there and built lofty buildings. He surrounded it with a 30-cubit high fortification and created some 12,000 orchards around it. On account of the streams and trees and gardens full of flowers, there is no place in Kāshgar more pleasant than Yārkand.

The river there has the finest water. Its level decreases in spring and rises in July. The river bed is a source of jade. However, the air is not pure.

For the most part, the climate of Kāshghar is cool and salubrious and the people enjoy good health. Although fruit is plentiful, illness is rare [R 139b] and the yield of fruit is small.

The inhabitants are of four types: 1) taxpaying subjects; 2) Qawchīn, who are made up of cavalry troops; 3) Aymāq; 4) Shariah office holders and executors of pious foundations.

Lākhūf, [M 367] which is three stages from Yārkand, is full of rivers, trees and gardens. From there, Khotan is a 14 days' journey. Aside from waystations, there is no inhabited place on that road, it being entirely a desert.

Yañiḥiṣār: A town near Yārkand, located at 110½ degrees longitude and 42½ degrees latitude.

Khotan: according to the *Taqwīm*, a town at the far end of Turkistan beyond Yūzkand. It has many rivers. It is located at 116 degrees longitude and 42 degrees latitude.

According to the *Haft Iqlīm*, Khotan is a famous town, but now is on the verge of ruin. There are two rivers flowing in its region, one called Baqrā-tāsh, the other called Bāūrang-tāsh. Jade is found in their riverbeds, and the people of Khotan trade in it and profit from it. The chief commodities are cotton cloth and silk. Wheat is also abundant. Once a week on Friday some 20,000 people come from the surrounding areas and districts and there is a bazaar with a huge crowd.

Yasī: located at 101 degrees longitude and 43 degrees latitude, it is a famous city and the capital (*dār-i mülk*) of Turkistan. In the past it was the seat (*taḫt*) of the Uzbek khans. The Naqshbandī shaikh Khwāja Aḥmad (Yasawī) comes from this city.

Sabrān: A city one stage west of Yasī.

Otrar: one stage from Yasī in the direction of Tashkent. It is also called Ūtrār. The Ḥanafī jurist Qiwām al-Dīn comes from this town.

Turfan: A town on the road from Samarqand to Cathay. As it is 18 stages from Andijān in the middle of Moghulistan, some say it is between Kāshghar and Khotan. Cathay is 20 stages from here.

Tandū: A town located at 114 degrees longitude and 39 degrees latitude.

Barsājān: A town east of Kāshghar, located at 114½ degrees longitude and 41 degrees latitude. According to Ibn Saʿīd, the two administrative centers in Turkistan are Kāshghar and Barsājān. At the time of Faḍl b. Yaḥyā al-Barmakī its inhabitants were pagan Turks. Later they converted to Islam and submitted to the Seljuks. From here the lands of Islam reach Tibet.

Bughbūriya, spelled thus according to the *Taqwīm*: A large town in the Turkish lands, located at 118 degrees and 42 degrees latitude. The Yaʾjūj River flows to its east. The city of Azkashiya is to its southeast. Bughbūr was originally a title of the kings of China—so much is reported from Ibn Saʿīd. In that case, Bughbūr is the Arabicized form of Faghfūr.

Balāsāghūn: A town near Kāshghar, on the other side of the Sayḥūn, located at 101 degrees longitude and 47½ degrees latitude.

Chigil: A Turkish town near Ṭarāz. The inhabitants being people of beauty and charm, their beauty is proverbial among poets. They worship Canopus, Gemini and the outer three stars of the Little Dipper. Among them it is not forbidden to marry their own daughters or sisters.

Khazaz: according to the *Taqwīm*, a Turkish clime north of Bāb al-Abwāb (Darband). Its town is called Itil. The Khazaz (*recte* Khazar = Caspian) Sea, known as the Shirwān Sea, and a large river that empties into it (i.e., the Volga) are named after this town. The town is on both banks of the river, the river flowing in the middle. The people are mostly Muslims and Christians; some are idol worshippers. Aside from Turkish, they speak their own language as well. [M 368] Their physiognomy is also different. One group, called Qara Khazaz, is black and one group is white. The idol worshippers enslave each other's children and sell them.

The ruler of Itil resides on the western side (of the river). The inhabited part on that side is approximately one parasang in length. The buildings are mainly of stone, but most of the people live in tents. There are Friday mosques in three places. Bazaars and public baths are few. In a place away from the riverbank there is a large palace for the ruler made purely of tiles—in that country, tile buildings are restricted to him. The fortress has four gates, one of which opens on the river, while the western gate opens to the desert.

According to written sources, the ruler of this kingdom is a Jew and the leading statesmen are also Jews, while the other approximately 4000 (state) servants are from different nations. They prostrate themselves before those to whom they wish to show respect. They provide for some 12,000 regular troops; irregulars must provide for themselves. There are nine qadis; everybody is subject to one of them.

Of the aforementioned groups, the Muslims are the most numerous. No one lives in villages. The surroundings within 20 parasangs are cultivated. The Muslims and the merchants dwell in the eastern part of the city. Candles are brought from Rūs. The language of the Khazaz people resembles Turkish. No other group can understand it.

Ismīd: A city in Khazaz with many orchards and gardens and wooden buildings. Most of the inhabitants are Muslim and have built mosques. But their king is Jewish and depends on the patronage of the governor of Khazaz and Sarīr. [R 140a] [SPACE FOR MAP; CF. M 347/348 (MAP OF TRANSOXIANA)]

The frontier of Sarīr is 2 parasangs hence. Sarīr is what today is known as Dagestan. It is described in the chapter on Shirwān.

Burṭās: one of the districts of Khazaz on the bank of the Volga. The inhabitants are also called Burṭās. They are two tribes, one at the far end of Khazaz close to Bulgar, the other dwelling together with the Turks.

Fārāb: according to Ibn Ḥawqal, a Turkish region on the other side of the Sayḥūn. It is one or two stages in length and breadth. Its city is in a wooded valley with fields to its west. Both the district and the city are large.

Aṭrāz: also called Ṭarāz and Bāygī. It is located 100 degrees longitude and 44½ degrees latitude. The inhabitants belong to the Shāfiʿī legal school. Many scholars emerged here. The philosopher Abū Naṣr al-Fārābī and the author of the *Qānūn al-adab* come from this region. In the past it was a renowned city. It was destroyed by Uzbek troops. Only a nearby cave is known under this name.

Shalj: A town 4 parasangs from Ṭarāz. The people are Muslim, as is the case with Ṭarāz.

Qaraqum: It means Black Sand. According to the *Taqwīm*, it is a town, capital of the Moghul country at the extreme end of the land of the Turks. Ibn Saʿīd specifies it at 116½ degrees longitude and 30½ degrees latitude.

However, this name is a mistake for Qaraqarim (Qaraqorum), mentioned above in the chapter on Cathay, and is the dwelling place of the Kalmyk Tatars.

Khāwas, also spelled Khāwaṣ: A town between Kāshghar and Samarqand, located at 100 degrees longitude and 41½ degrees latitude. Khujand is 10 stages from here.

Akhsū: A town located 7 stages north of Yañiḥiṣār, at 110½ degrees longitude and 44 degrees latitude, administrative center of the ruler of Kāshghar and Yārkand.

Sinjū: A town 6 stages south of Yañiḥiṣār and 12 long stages west of Tibet. [M 369] It is the same distance east of Kāshghar, thus midway between the two. Kashmir is 15 stages to the south.

Ūrṭāgh, a famous mountain, is the summer pasture of the ancient Oghuz tribes. Kur Ṭagh is their winter pasture. They are two large mountains. In that region is a city called Īnānj and nearby is another ancient city known as Qari Ṣayram. The majority of the inhabitants are Muslim Turks.

Products and features of Turkistan

In most of its features it is like India. There are musk, squirrel, sable, ermine, black fox, and white rabbit. One of the features of Turkistan is the rain stone, called *sang-i yada* by the Persians and *jada tash* by the Turks.

Routes of Turkistan

In the past it took 14 days to get from Khotan to Cathay. The area between them was partly inhabited. The people had no need of a caravan, [R 140b] so one or two went on their own. At present this road is blocked due to the Kalmyk invasion. The route taken today is 100 stages.

According to the account of Khwāja Ghiyāth (al-Dīn), Ulugh Beg's envoy: from Samarqand > Chalgā-i Yulduz, two months > the city of Ṭarqān, by the end of the third month > the city of Qāmal, 20 stages > the pass of Sagjū, which is the frontier of Cathay, 25 stages. All the land in between is desert.

From Kāshghar, the capital of Turkistan, Samarqand is 30 stages to the west and Lahore is 24 stages to the south.

From Yasī, Samarqand is 17 stages, Bukhara is 25 stages and Tashkent is 9 or 10 stages.

From Tashkent, Samarqand is 6 stages to the west and Andijān is 7 stages.

From Andijān > Moghulistan, 8 stages > Ṭarqān, 10 stages.

Inhabitants of the desert of Turkistan

It should be known that according to the books of history, after the Flood, Türk son of Japheth settled in the eastern lands. His descendents broke up into tribes, each one taking a certain area for its dwelling place. Those areas were named after them. Most of them were nomadizing in the steppes of Turkistan, lodging in tents. Türk was succeeded by Alanja Khan. Two sons were born to him from the same mother, one called Moghul, the other Tatar. He divided the entire country between these two sons. At that time these two *oymaqs* came forth from the Turkish tribes, becoming known as the Moghul and Tatar *oymaqs*.

From the Tatar *oymaq* eight individuals ascended to rulership. Later their lineage came to an end, the state being left to the Moghul branch. Those eight individuals are: 1) Tatar son of Alanja, 2) Buqa, 3) Yalanja, 4) Esli, also called Ilkhan, 5) Atsiz, 6) Ordu, 7) Baydu, 8) Sunj. In Sunj's time the Moghuls defeated them and conquered the country. With Sunj the Tatar state came to an end.

The Moghul branch consists of nine individuals. For this reason the Turkish tribes greatly respect the number nine. They are recorded as follows:

Moghul son of Alanja. This designation (Moghul) became the generic name of many tribes and peoples. It was applied metaphorically and generalized to people who are not from them, and so Turks and Tatars are also called Moghuls. In fact, the Moghuls were a branch of the Turks. Later they multiplied and overcame the other branches. They conquered most of their countries and named them after their own clans. As the first of these were the Tatars, [M 370] all of them came to be called Tatar. Today the Moghul tribes in Arabia, India and Cathay are called Tatar.

This Moghul had four sons: Qara Khan, Oghuz Khan, Kur Khan and Or Khan. He was succeeded by Qara Khan. He lived in Qaraqum, spending the summer and winter at Aq Taq and Kur Taq. As he was pagan, his son Oghuz Khan revolted and overthrew him, taking his place. Since he was a believer and good adminstrator, he laid down the laws of the Turks and controlled and subjugated the tribes. He conquered the lands from Artlās and Ṣayram to Bukhara, and killed his opponents. This Oghuz is similar to Jamshīd among the Persian kings. His reign is recorded as lasting 67 years. It is he who invented and established the names of the peoples and tribes known as Uyghur, Qanqli, Qibchaq (Kipchak), Qarluq and Khalaj.

- Uyghur was his designation for a tribe that fled from their enemies and came over to him; it means "joining" (*ulaşmaḳ*).
- Qanqli was his designation for a group of men whom he assigned to supervise carts loaded with booty that he took in battle.
- Qibchaq: When he was defeated in Turkistan by one of the Turkish sultans, It Baraq, on the way back he saw that a woman whose husband had fallen in battle had given birth in the hollow of a tree. Out of compassion Oghuz Khan took the boy and adopted him. When the boy grew up and became chief of a tribe, it was designated by the term and place name Qibchaq, which is

the diminutive of *qobuq* meaning "hollow". The famous Steppe was added to it.

- Qarluq: [R 141a] A group of his army was left behind because of snow (*qar* / *ḳar* in Turkish) and rain and were designated with this name.
- Khalaj: He had issued a warning that none of the soldiers in his campaign of conquest should stay behind. It so happened that a woman gave birth on the way and stayed behind because of hunger. Oghuz took the child and designated his name as Qal-ach (Turkish *ḳal aç* meaning "stay hungry"). Later he became the chief of a tribe and the corrupted form of the name, Khalaj, became the name of that people.

The sons of Oghuz: Oghuz had six sons: Ay ("Moon"), Kün ("Sun"), Yulduz ("Star"), Kök (text: KWJK; "Sky"), Taq ("Mountain"), Tengiz ("Sea"). One day on the hunt they found a bow and three arrows. They came to their father and asked him to distribute these between them. Their father gave the bow to Ay, Kün and Yulduz who broke it into three and distributed the parts among themselves. Therefore Oghuz designated them as Bozuq ("Broken"). He gave the arrows to the other three and designated them as Üç Oq (Turkish *üç oḳ* meaning "three arrows"). They became the chiefs of several tribes. He made the Bozuq his heirs and appointed them the right wing of his army, subordinating to them the Üç Oq in the left wing. He designated Ala Taq as their dwelling place. Hence the Bozuq became superior to the other progeny of Oghuz and were respected. From them appeared 24 tribes. Some of these went to Transoxiana and Khurāsān and assumed a different form. The Chingisids and the Seljuks emerged from them, as is recorded in the *Rawḍat al-ṣafā*.

Oghuz was succeeded by Ay, Kün and Yulduz. Then Mengli son of Yulduz, Tengiz, and Ilkhan son of Tengiz became rulers in this order. Some chroniclers maintain that after Oghuz the countries of Turkistan remained in the hands of his progeny for 1000 years. Ilkhan was a contemporary of Tūr son of Farīdūn. Tūr occupied Turkistan and Ilkhan fell in battle. His son Qayān was saved and fled to the desert called Ergene Qon, [M 371] a place surrounded by great mountains like a fortress. He fortified himself there and made it his dwelling place. With the passage of time his children and followers multiplied and that place proved too narrow. One group of them went back to the old camping grounds and seized them from the Tatars. They were called the Qayat or Qayi people. Their chief was one of Qayan's sons, Yulduz son of Mengli.

Bozanjar Qa'an, son of Alanqo, daughter of Jūbīna, daughter of the aforementioned (Qayan?), was made ruler of Moghulistan. He was alive at the emergence of the Abbasid dynasty. He established the laws of the Moghul sultanate. He defeated the Khitayans in battle. He limited the position of khan over the territory and tribes of the Moghul people to Alanqo's progeny. They were convinced that the khanate of one who is not from that dynasty is not sound. They still believe that. Indeed, because Tīmūr was not of that lineage, he did not attain the rank of khan. What counts for them is the idea.

Then Bozanjar's progeny became khan in this order: Qaydu, Buqa, Baysunqur, Tumana, Qayli, Qajuli, Bartan, and Yisuga (Yesügey) Bahādur. Yisuga died in 502 (1108–1109) and was succeeded by his son, Chingiz. His capital was Qaraqum, also called Ulugh Yurt. Having subdued the Moghul khans, he went to Iran in 616 (1219–1220). After destroying Transoxiana and Khurāsān he returned and died 624/1227. He established the laws of the Yasaq, which are still observed among them. He divided the countries among his four sons, Tolui, Jochi, Chaghatay and Ögetey. Two years later two of them resigned the throne to Ögetey.

Chaghatay became ruler in Turan (Turkistan) and 29 of his progeny ruled over Kāshghar, Badakhshān, Balkh, part of Khwārazm and Ghazna. Their dynasty came to an end in 806 (1403–1404) with the emergence of Tīmūr.

Of Ögetey's progeny, according to the *Jahān-ārā*, 19 ruled until 830 (1426–1427). Sometimes they invaded Cathay and sometimes the Cathayans conquered them. They resided in Qaraqarim (Qaraqorum).

As for Tolui's progeny, after Hulagu's invasion, 17 of them ruled Iran until 754 (1353–1354). [R 141b]

Of Jochi's progeny, the sons of Berke Khan conquered the Kipchak Steppe and the region of Crimea and remained there—the dynasty of the Tatar khans is still in that region. Regarding this lineage, Ghaffārī records the khans of Aq Orda (White Horde) and Kök Orda (Blue Horde). Kök Orda is the khanate of the tribes to the right of Ulugh Yurt. Aq Orda is the polity of the *ulus* on the borders of Ala Ṭāgh; they were extant until 970 (1562–1563). The sultans of Khwārazm are also recorded as from this lineage, as detailed in the chronicles. My aim here is to give a summary account of where these tribes live.

Sharaf-i Yazdī says in the *Muqaddima* that the Moghuls live in the east, far from cultivated areas, in a large steppe area seven months in circuit. It is adjacent to the borders of Cathay in the east, the Uyghurs in the west, the Qaraqar lands in the north and the clime of Tibet in the south. Their food is mostly meat, their clothing, the skins of wild animals.

As recorded in some chronicles, the Turks are nine tribes, either steppe dwellers or others:

- Oghuz: the greatest tribe. Their original homeland was the region of Cathay. The Seljuks are from this tribe.
- Qayi: [M 372] they surpass the Oghuz in number. They came from the territory of Sārī and spread as far as the borders of Armenia.
- Khīrākhīr: they were given this name by a certain Maghsūn. Their settlements were close to the Bakhtāl and Kimek.
- Ḍarīḥ: they live on a high mountain known as Jabal Yūnus. They left Turkistan because of rebellion against the khan and settled there. They are divided into nine clans of which the Chigili and the Henli each comprise three.
- Kimek: They are steppe-dwellers and most are fire-worshippers. They burn their dead. They fast for one or two days a year. They have abundant flocks.
- Bakhtākin: They do not dwell in one place but have dwellings 30 stages apart. They are a nomadic tribe. On one side of them is the Kipchak Steppe, to the west the land of Khazaz and the Slavs. The area between the Bakhtākiyān and the land of Khazaz is a desert adjacent to large mountains. Two tribes of the Turks dwell there, the Ṭūlāsī and the Ghuz. They are in constant strife against the Bakhtākiyān.
- Maḥraqa: They nomadize in an area 100 parasangs in length and breadth between two great rivers, the Volga and the Don. The Slavs and Russians are in constant warfare with them. Usually the Maḥraqa are victorious and take captives, whom they bring to Turkey and sell there.
- Ṣaqlāb (Slavs): They are forest-dwellers, 15 stages from the Bakhtākiyān. Most of them are fire-worshippers. Their food crop is millet, their drink, honey (or mead). Their commanders are called *sarbatāw*, their viziers, *sūbakh*.
- Rūs (Russians): they are recorded as a tribe of the Turks, although this is evidently not the case.

In the European books the aforementioned place (i.e., Turkistan) is described in a different way. We give here their summary account:

Tataristan, written as Tataria. The Greeks and Latins call it Scythia, while the Hebrews call it Magog. It is on the Don River, which divides Asia from Europe. It was called Tataria because 300 years ago those regions were conquered by the Tatar tribe.

The modern geographers have delineated the boundaries of Tataria as follows: in the south, the Shirwān (Caspian) Sea, the Jayḥūn River and the mountains of Dalanguer and Naugrocot; in the west, the Don River and the Lake of Chetai (Aral Sea); in the east, the Sea of Cathay (?), which turns toward the north and the Land of Darkness, i.e., to the 90th latitude.

Armenian authors have called this region "Deep Asia". The Alti Ay (Altay) mountain runs through the middle of this area and divides it into two parts. This mountain is a branch of Mt. Imaus; several other branches go to the north and end at the ocean.

Division: the ancients divided this clime into two parts, this side and the other side of Mt. Imaus. But we divide it into three parts: 1) Sarmatia, which is Asian Sarmatia and empty deserts; 2) the country of Chagatay, which the ancient called Scythia, on this side of Mt. Imaus; it is the sultanate of Turkistan; 3) the country of Serica and Tangut, which is Scythia beyond the Imaus; Ptolemy records it as terra incognita.

Warning: Lorenzo and Philippus fixed the boundaries and divisions of these regions in the manner described, and Mercator followed them in the *Atlas*. However, in this book the division of Serica, which is the country of Cathay, has been described above. Now we will summarize in three sections whatever has been written (by the Europeans) aside from that (Serica) and aside from the cities of Turkistan recorded in this chapter.

Sarmatia: [M 373] it is empty Tataristan. There are fewer inhabited areas here than elsewhere. Its borders are the Shirwān (Caspian) Sea, the Don River and the Qitay Lake. It extends from the 60th to the 120th degree longitude, and from the 55th to the 67th latitude. Most of it is plain and steppe. The desert region known in other books as the Kipchak Steppe is this region. At present the Noghay Tatars nomadize on the edge of the steppe on this side.

According to Lorenzo, before the Tatars, some tribes by the name of Cumani and Polovzi (Polovtsy) lived within these boundaries. Remnants of them can still be found today. The rivers called Volga, Kasal and Sūr flow through this clime. The soil is quite fertile. However, the Tatars mainly attend to warfare and the hunt, and do not know agriculture. In some places they do sow millet, getting limitless produce, and they eat that. Camels, horses, small livestock are very abundant here, and are exported to Persia. The people [R 142a] nomadize with tents made of felt. Most of them dwell in carts which they drive on the plains with sails. Steering them like a boat, they can go wherever they want. There are *ulus* chieftains among them, most of whom are subject to the Duke of Muscovy. They are divided into tribes, each of which is called a horde (*ordu*). They also have fortresses here and there.

The Kazan Horde: They are held in high regard among them. There are three tribes belonging to this horde. Their people, as opposed to the other Tatars, go to war on foot.

They are excellent archers: whatever they aim at, they do not miss. Although they are ruled by Muscovy, they are not like her other subjects. Among them there is a group occupied with superstitions and magic. They can set in motion the wind, thunder and rain, and then quiet them down. 30,000 warriors can be fielded from this horde. They have a city on the bank of the Volga called Kazan. When the duke of Muscovy, Jovannes (Ivan), conquered it, he brought a large number of people from the land of Livonia and settled them there. The khan had no choice but to submit.

Astraqan (Astrakhan): It is what is now called Azhdarhān. According to Lorenzo, it is somewhat south of Kazan, near the Baku (Caspian) Sea. It was taken by the duke of Muscovy, Demetrius, and remained for a time under the rule of Muscovy. It is on the bank of the Volga River. This horde was also named after the city.

Beyond this, further inland, are the cities of Khāār, Hūlīzā and Qūndūnī; and some cities on the shore of the Baku (Caspian) Sea; and a river known as the Bāīqū, which rises in the country of Siberia.

There is a species of animal there that is like a sheep and has valuable horns. It likes the sound of drums: when it hears a drum, it comes frolicking to it. The Tatars hunt it by beating drums.

Near the mouth of the aforementioned river (Volga?) is a city by the name of Qāmīnazār where the khans of the Zāwlaq Tatars were buried. It was invaded and destroyed by the Tatars of Perekop (in Crimea).

Near the Kasal River there is the city of Qatābūzī. And in the interior on the bank of the Jānqū river is the city of Sawāīkh (Saraich), which was a famous and flourishing place and the capital of the Cumani before the Tatars invaded and destroyed it. It was restored by the Nogai Tatars, whose khans still reside there, and there are three hordes under their governance. Finally, they pay tribute to Muscovy, giving it every year many [M 374] horses and felts made of white wool and, when called upon, joining it on campaign with 4000 horsemen.

So writes Lorenzo. However, it is well known that the Nogai Tatars are subject to the Crimean khans. It is possible that, as Lorenzo suggests, they maintain good relations with Muscovy as well, since they share a border with them.

He goes on to say that a bit north of them is the Bulgar tribe. Their language and ceremonial are similar to those of the Rus.

The poor one (Kātib Çelebi), however, says that the Bulgars converted to Islam in the Abbasid era. Their habitat is between Khazaz and Rus. When the days are longest (at the summer solstice) there is no nighttime prayer for ten days, because dawn appears before evening twilight turns to night. For that reason they sent someone to Khwārazm, posing this question to Imam al-Baqqālī: "In our country it dawns before evening twilight turns to night. Are we required to make up the nighttime prayer that we miss?" Al-Baqqālī gave the legal opinion that the nighttime prayer did not have to be made up. But Shams al-A'imma al-Ḥulwānī disagreed and said that it had to be made up. Each of the two imams tried to prove his own claim and convince the other. Shams al-A'imma sent one of his assistants to Khwārazm, instructing him to put the following question to al-Baqqālī: "What do you say about someone who drops one of the five daily prayers? [R 142b] Is he not an unbeliever?" He came and posed the question to al-Baqqālī while he was teaching at the Friday mosque of Khwārazm. Imam al-Baqqālī in turn posed his own question, thus: "And what do you say about someone whose two feet have been cut off at the ankle? How many obligatory acts are there in his ablution?" The fellow answered, "Three, because the fourth (i.e., the obligation to wash one's feet) is impossible." Al-Baqqālī said, "The obligation of the five daily prayers is similar to this." The fellow was convinced and those present at the gathering gave their approval. When the responsum reached Shams al-A'imma he admitted that it was right. The rest of this problem is explained in our treatise *Üç mes'ele* ("Three Problems").

According to the *Rawḍ al-mi'ṭār*, the Bulgars' settlements are on the bank of the Volga. They number some 500 houses and they have mosques.

Again Lorenzo says: North of the Bulgar people is the country of Siberia. It is mainly forest. Some of the Nogai Tatars have settled there. They have a city by the name of Hakhā. It is very prosperous and famous for its trade. Some precious stones are sold there; also a kind of fruit called "fruit of life" which strengthens the heart and which they put into their foods. The Tatars of Cathay meet the Muscovites here for trade, bringing much merchandise from Qānbālū—i.e., Khanbaliq—which is 460 parasangs from here.

On the bank of the river Kasal live the Chabanlu and Kasallu Tatars. They strike square-shaped coins. Formerly they would pay out gold and silver by weight, like those living beyond the Imaus.

The poor one (Kātib Çelebi) says: They must be the followers of Shabān son of Jochi and Kashīlū Khan.

Again Lorenzo says: Again in the north, in the steppes not very far from the Qaspiyum (Caspian) Sea, live the Qazaq Tatars. They are the champion warriors of the tribes living in the northern lands. They are quite prone to magic and theft. They rob the Muscovite and Tatar merchants going to and coming from Cathay.

The poor one (Kātib Çelebi) says: These are not the Don Cossacks or the Cossacks of Özi (Ochakov). They are discussed in the chapter on Crimea.

Again Lorenzo says: To their north is the Qitay Lake. It is a very large lake, more like a sea. In it there are small islands [M 375] where Tatars live and have dealings with Muscovy. They have a type of cart called *isleten* (sled) which is pulled by tamed deer (i.e., reindeer) at a high speed over the ice. Abundant fish are caught in the lake; and there are a lot of birds. The middle of the lake is located at 65 degrees latitude. The Ob River takes its rise from there. Because the Ob is over 80 miles wide at its source, some think that the Caspian Sea is connected with it through an underground passage, or that it is linked with the Northern Ocean. Indeed, the prophet Solomon said that all rivers are connected to one another under the ground.

The poor one (Kātib Çelebi) says: This lake is recorded in the Islamic books as the Khwārazmian Lake (i.e., the Aral Sea). The *Taqwīm al-buldān* cites the *Rasm-i maʿmūr* to the effect that the middle of Lake Khwārazm is at 90 degrees longitude and 43 degrees latitude. The Jayḥūn, flowing from the east, empties into its southern part. It is 600 miles or 100 parasangs in circumference. The Sayḥūn also empties into it. Khwārazm is 6 stages from here, the Caspian, 20 stages. Masʿūdī has recorded the same thing in the *Murūj al-dhahab*. On the lakeshore there is a mountain known as Jaʿāʿar, which is covered by thick ice in winter. In the chapter on Transoxiana we have already mentioned the disagreement over where these rivers, especially the Jayḥūn, debouch. What Lorenzo calls the Ob River is that river (i.e., the Jayḥūn?). In sum, this lake [R 143a] [SPACE FOR MAP]

is recorded in the ancient works but is not mentioned in the modern books, the *Atlas* in particular.

Again Lorenzo says: On the shore of this lake is a city called Crustina, a large emporium where Tatar and Russian merchants come together.

Then there is the dwelling place of the Bashghurd (Bashkurt) Tatars and the Tūman Tatars, whose horde is subject to the khans of Cathay; just as the other Tatars beyond Mt. Imaus are all subject to it. The Muslims among them all have recourse to this khan when it comes to warfare and strife with Tatar pagans.

As for the Atil (Volga) River, also called Adil and Hedil: It is recorded in the books of the ancient Greeks and Latins under the name of Rha. As reported by Lorenzo, it originates from Lake Finou at the borders of Muscovy; according to others, it rises from Lake Volgou and flows southeast. Its source is located approximately at 64 degrees latitude and 70 degrees longitude. With 60 or so rivers emptying into it from the mountains of Russia and Bulgar, it becomes a huge river. And none of those rivers is easy to cross. This river is said to be the largest river in the world after the Nile, around 600 parasangs long. It winds past Itil, Kazan, and other cities; breaks up into several large delta branches; and empties into the Hyrcanian (Caspian) Sea, also called the Khazaz and the Shirwān Sea, near Astraqān (Astrakhan), also called Azhdarhān. It can be seen flowing even ten parasangs into the sea. Its color and taste do not change, and it is navigable by large vessels. However, this river is of little benefit, since no place is irrigated by it. In winter it freezes and can be crossed on foot. The Bulgar and Kimek tribes live on its banks. As it approaches the sea it comes near the Don River, but then it bends and flows east. The place where the river Don empties into the Sea of Azov is a great distance from Azhdarhān.

The story of connecting the two rivers: During the time of Sultan Selim II, the Kazan Tatars sent a letter to the Porte, [M 376] [R 143b] complaining about Muscovy and proposing the plan to subjugate the Muscovites by connecting the two rivers with a canal. Ḳoca Meḥmed Pasha (the grand vizier) approved the plan, saying that this would facilitate sending ships with grain supplies and military reinforcements to Demir Ḳapu (Darband) and Shirwān, and that ships would be able to go (from the Black Sea) to the coast of Gīlān and Ṭabaristān (i.e., the Caspian Sea). After consulting with some experts, he bestowed the province of Kefe (Kaffa in Crimea) on Çerkes Kāsim Bey, who at that time was director of the financial administration of Asia Minor, because he had knowledge of that region. Çerkes Kāsim Bey sent some trustworthy agents to survey that area and reported to the Porte that the distance between the two rivers (at the narrowest point) was six miles. The grand vizier mounted the expedition and, in the year 976 (1568–1569), dispatched the admiral of the fleet, Mü'eẕẕinzāde ʿAlī Paşa with pickaxes and shovels and all kinds of military equipment. He proceeded with excellent warships as far as Azov, and then with caiques to the staging area, while the Tatar khan came by land. After excavating for three months, only a third of the canal had been dug. A rumor appeared among the troops that the winter there would be severe. When this fear was aggravated by a food shortage, they dropped the pickaxes and shovels and other tools and departed.

Some say that the opening up of this route was not to the Tatar khan's liking and that the rumors that frightened the troops originated with him. Be that as it may, all these efforts went up in the air, because the agents had not enough authority for such an enormous undertaking. Every job has the right man to do it. Such a work can only be successful if rulers themselves supervise it. The excavation of the Sakarya River is a similar case, which will be explained below.

As for the second part, which is Tataristān beyond the Imaus mountains: According to Lorenzo, they are the Moghul Tatars who spread to most places in Asia, as has already been mentioned. They are the original Tatars who took their name from the Tatar River. They dwelt at the place known as Monghāl, which was a region of the north. From there they spread into the steppes, at times defeating their neighbors, at times being defeated by them.

Chingis emerged from them. He deceived the Tatars by claiming to have a secret connection with Noghay, the Tatars' idol, and made them submit to him. Then he overcame his neighbors and gradually destroyed most of Asia. After Chingis, his followers became divided and their prestige waned. The Pitorsi Tatars, who resided in the Albanian mountains, converted to Christianity, joining the sect of Muscovy. The majority of the Tatars living beyond the Imaus mountains are idolators. The rest of their affairs are mentioned in the chapter on Cathay.

Characteristics of the Tatars

Among humankind, this people are wielders of steel. Their nature is accustomed to movement, like the water and the wind. They are strong as Ahriman and descend from demons. Their horses want neither fodder nor provender. They eat whatever the saplings of their arrows sprout in the garden of the hunting grounds. They roam the desert as swift as the wind, without themselves tiring or their horses sweating. The steeds they mount need neither horseshoe nor nail. [R 144a] The meat they eat needs neither cauldron nor skewer. They swim in the sea and require neither sailor nor ship. They ride in the steppe and take with them neither horsecloth nor fodder. Although they wear sheepskins, they are wolves in the mountain country of battle. They do not shrink from either bird or beast and they eat them whether dead or alive. [M 377] They live to a ripe old age without feeling the pinch of time. Their teeth never fall from the mother-of-pearl of their mouths. Their faces never turn yellow like autumn leaves. The light of their eyes does not dim with the passage of the months like the moon and the sun. Their backs are not bent with the turning of the ages like the wheel of heaven. At age 100 their faces are smooth as sword blades and their eyes are sharp as a sharpened arrowhead. [Persian verse:]

> There is no beard around the dimple of his chin.
> How can peach-fuzz sprout from the surface of ice?

Their eyes are small as specks. Their faces are large as shields. Their foreheads are bare cliffs in the mountain of firmness. Their heads are tall mounds in the desert of awesomeness.

This discourse is corroborated by the hadith, "Leave the Turks alone as long as they leave you alone"—i.e., Do not attack them as long as they do not attack you.

Lorenzo says that they are mostly swarthy of complexion and iron of countenance, scanty-bearded, broad-faced, narrow-eyed, powerful and ruthless, a savage tribe. Their language is similar to Turkish, and their customs are also close to the Turks. They take their wives with them when they go to battle. They are excellent archers, able to hit a target while riding a horse. They can cross great rivers on horseback.

Although their horses are ungainly, they are tough and fast. The Tatars ride them all day long and then let them range in the steppe. In summer or winter they get by with whatever grass they find. And they use camels as horses. Their camels are small and black and have a strange shape; they do not resemble the camels in this country.

The Tatars usually mount raiding expeditions in Poland, Muscovy and Circassia; but sometimes they consider it advisable to make peace with them. They eat every animal except the pig and drink mare's milk. They extract the sap from a delicious root called *baltıraḳan* (asafetida) and dip their bread in it. They drink *müdevvene* (?) sherbet. When traveling, they sometimes cut the veins of their horse and suck the blood. They wear felt. In winter they cover themselves with some sheepskins.

Their punishments are according to the Turkish manner. They put adulterers to death; banish murderers from the horde; and usually kill thieves with severe torture. While they mostly pass as Muslims, crypto-idolaters are not absent from among them. Their qadis exact a large penalty when they are discovered, but nevertheless they are not forbidden (from practicing idol-worship) since it is an ancient custom. The rest of their affairs are mentioned in the chapter on Crimea.

Gog, Magog and the rampart of Dhū'l-Qarnayn

They are mentioned in the Koran. According to the commentators, Gog and Magog are a branch of the Turks who live at the extreme end of the inhabited world. They have short stature and big ears; they are like beasts in the form of wild men. They dwell in a separate region in the land of the Turks. They multiplied and spread in that region until it could no longer contain them. Then they began to spread in this direction to the borders of the middle of the inhabited world. With the corruption that was in their nature, they subverted the affairs of the people of those regions.

During the period of Abraham, when Dhū'l-Qarnayn the Himyarite—he was one of the Himyarite kings, i.e., the rulers of Yemen, whose travels to the east and west are mentioned in the Noble Verse (i.e., the Koran) and reported in notable chronicles—went to those regions, the people complained to him about oppression. Between them and that rebellious people stood a mountain, and it was through a gap in that mountain that Gog and Magog used to pass back and forth. Only by closing that gap with a mighty rampart would the hand of usurpation and invasion be repelled. And since that depended on the aspiration of such a powerful ruler as Dhū'l-Qarnayn, they all beseeched him to block the way before that people by building such an edifice. Thus he would save from their evil the inhabitants of so many lands, and would give them new life, and would leave a beautiful memory on the page of Time. [R 144b]

Dhū'l-Qarnayn [M 378] succeeded in carrying out the task. He built the famous rampart, as is recorded in chronicles and commentaries. Gog and Magog have remained on the other side of the rampart, which will stand as long as God wishes. There are signs mentioned in books on the holy law that at the end of time they will destroy the rampart and emerge.

The history books also report a number of nonsensical stories, such as that Dhū'l-Qarnayn was Alexander the Greek, or that a man went to the rampart of Gog during the reign of al-Wāthiq bi'llāh. These are not to be taken seriously, because it is mentioned in prophetic hadith and recorded in the *Ṣaḥīḥayn* that Dhū'l-Qarnayn lived at the time of Abraham, and that they met and embraced each other in Mecca. Alexander the Greek lived later, and it has been confirmed that there were 1958 years between the birth of Abraham and the victory of Alexander the Greek over Dārā (Darius), which marks the beginning of the Alexandrine calendar. In accordance with the [Arabic] verse,

> O you who have married the Pleiades to Canopus:
> May God lengthen your life, how can the two of them meet?

attribution does not allow for approximation (?—*ḥamle taḳrīb bulunmaz*). As Dhū'l-Qarnayn was one of the titles of the kings of Yemen, and Iskandar must be the Arabicized form of Alexander, which means "good man" in Greek, how could the Arabic and Himyarite title and the Greek name be connected? Moreover, Iskandar / Alexander is not mentioned in the Koran, and the one who appears in the Noble Verse is Dhū'l-Qarnayn. As surmised by the author of the *Tafsīr-i kabīr*, Alexander was famous abroad, and so it was conjectured that the description and epithet mentioned in the Koran referred to the object of that fame. They made the attribution (*ḥaml*) without investigating the basis for it. No chronicle has survived with reports on the kings and events of the time of Abraham, recourse to which would resolve the doubts. As with the passage of time there remained no source, they stuck to this conjecture. However, when one investigates the matter, one concludes that Alexander is not Dhū'l-Qarnayn.

Location of the rampart: In the book of Ibn Saʿīd it is recorded as being in the 10th part of the 5th clime. The first thing mentioned in that part is the fortress of Dhū'l-Qarnayn, which he built at 163 degrees longitude and 40 degrees latitude, in order to defend the rampart. The gate of the rampart is north of that fortress. The rampart adjoins the mountain that extends from north to south and surrounds the people of Gog and Magog. The source of the river of Gog and Magog is also in that part. It rises in the mountain at the head of the rampart and empties into the Circumambient Ocean. The country of that tribe is along this river, the rampart falling to their west. The city of Gog, which is their capital, is located at the outlet of this river, at 160 degrees longitude and 43½ degrees latitude, as recorded in the *Geography* of Ptolemy. To its southeast is the city of Magog, located to the north of that city (?) at 161 degrees longitude and 42 degrees latitude. To the east of these two cities is the Circumambient Ocean, there being no other inhabited area.

The poor one (Kātib Çelebi) says: the longitude and latitude mentioned here pertain to the lands of China and Cathay. For a number of reasons it is not accurate that the Rampart of Gog is in those regions. One is that the dwelling of that people must be far from moderation and close to the 90th latitude, because as latitude increases on the face of the earth, moderation decreases and gradually disappears. Indeed, as was pointed out at the beginning of the book, [R 145a] the inhabitants of places close to the 90th latitude [M 379] are like animals in human form, and the description of Gog and Magog fits them.

Now it is thought that that place is located in the northeast, close to the 90th latitude. However, aside from Ibn Saʿīd al-Maghribī, the geographers have not located it. His assertion that Ptolemy located it in his *Geographia* is not acceptable. There is no such thing in the Greek original of the *Geographia*. Perhaps it was added by the translators in Baghdad. Those who wrote geographical works in Latin and Greek were not obliged to determine its location, because those regions were not places they discovered and wrote about. For them the northeast is still unknown and, as is written in the *Atlas*, the hope remains that one day it will be possible to cross through that region to the east; but for now it must remain as terra incognita. The following story, given by the historians, may be nonsensical, yet it corroborates this claim (i.e., that it is located in the northeast).

Story: Ibn Khurdād, one of the Abbasid viziers, reports from Sallām the Interpreter that al-Wāthiq bi'llāh saw the rampart in a dream. In order to corroborate it, he sent Muḥammad b. Mūsā al-Khwārazmī, one of his astrologers, to Ṭarkhān, the king of Khazaz. Isḥāq b. Isrā'īl, the king of Armenia helped him get from Tiflis to the ruler of Sarīr, who sent him to the king of Lān, who in turn sent him on to Ṭarkhān. After that he went for 26 days and reached a fetid land. He went for 10 more days and came to cities that had been destroyed by the people of Gog. He went for 27 more days amidst those ruined cities and arrived at the mountain where the rampart was built and a nearby city.

His eyewitness report about that city and about the rampart was recorded (by Ibn Khurdād) as transmitted from Sallām, as follows: That valley is 150 cubits wide; the rampart is constructed in the shape of a 50-cubit high gate made of iron and brick, the gaps filled with brass; he saw the sidepost and key of this gate; and above the gate is built a 100 cubit-high wall. That is the report, with other such nonsense added in the margins.

Whether this story is true or not, the claim that it is located in Andalusia is untenable; it is certainly in the northeast.

Qāḍī al-Bayḍāwī in his commentary transmits the report that there are two ramparts, which are the mountains of Armenia and Azerbaijan. It is only given as hearsay, not as fact. It is possible to explain the report by claiming that those mountains adjoin the mountains of Gog and Magog.

Chapter 38: Clime of Azerbaijan

The most prevalent pronunciation is Adhrabījān; some say Ādharbayjān. It is the Arabicized form of Ādharbāygān.

The author of the *Taqwīm* states: "It is a famous clime. Azerbaijan, Arminiyya and Arrān are actually three separate climes; but because they intersect with one another, scholars have mentioned them in the same chapter. However, we have added to it the region in the north, on the northern shore of the Black Sea." Thus he extended its limits and, excusing himself due to lack of detailed knowledge, he only gave a summary of these regions as well.

As for this poor one (Kātib Çelebi), in this chapter I only discuss Azerbaijan, presenting the other climes in separate chapters.

Borders of Azerbaijan

To the west, Arminiyya and Diyarbekir; to the south, Persian Iraq and Shahr-i Zūr; to the east, the land of Daylam, Ṭabaristān and Māzandarān; to the north, Shirwān and the Caspian Sea.

Divisions

In the *Nuzha*, Ḥamd Allāh divides this clime [M 380] into ten *tūmans* and includes some other regions. It has 27 cities. The climate of most of them tends to be cold; some are moderate. It is surrounded by Iraq, Mūghān, Georgia, Arminiyya and Kurdistan. Its length from Mākū to Khalkhāl is 95 parasangs, its width from Mājarwān to Mt. Sībān is 55 parasangs. In the past its capital [R 145b] was Marāgha; now it is Tabriz.

Tabriz, written as Tibrīz in the *Lubāb*, colloquially pronounced as Tawrīz: A famous city located at 88 degrees longitude and 38½ degrees latitude at the end of an extensive plain, on a plateau near the western foot of Mt. Sahand, in the midst of Eden-like gardens. The Āb-i Talkh ("Bitter River"), also called the Surkhāb ("Red River"), flows by it.

Ḥamd Allāh says that this city was settled in 175 (791) by Zubayda, the wife of Hārūn al-Rashīd. It was destroyed by an earthquake in 244 (858–859), but the Abbasid caliph al-Mutawakkil had it restored. It was completely destroyed by an enormous earthquake on 14 Ṣafar 434 (9 October 1042).

Qāḍī Rukn al-Dīn al-Khūyī writes in his book, *Majmaʿ arbāb al-mamālik*, that an astrologer known as Abū Ṭāhir al-Shīrāzī was then in Tabriz and predicted that the city would be destroyed that night in an earthquake. Based on this, the magistrates evacuated the inhabitants from the city to the desert. The prediction came true and the city was completely destroyed, leaving 40,000 people under the rubble. Amīr Ruwwādī, who was the governor appointed by the caliph, started rebuilding it in 435 (1043–1044), during the sign of Cancer as designated by the aforementioned astrologer.

The astrologer declared that Tabriz would never again be destroyed by earthquake, although there was danger of flooding. According to Ḥamd Allāh, this prognostication has proven true for more than 300 years. Although there have been several earthquakes, none of them destroyed the city. Later, underground channels were built, and

PLATE 44
Map of Azerbaijan, R 146a; cf. M 389/390 (Appendix, fig. 88–89)
TOPKAPI PALACE LIBRARY, REVAN 1624, WITH PERMISSION

because there were many outlets, their vapors became weaker.

In the past, the city wall of Tabriz was 6000 fathoms and had ten gates. In the Mongol period, when Tabriz was the capital, people built houses outside the wall, enlarging the city. The areas outside of the gates became like small cities themselves. Ghāzān Khan erected another wall encompassing the newly built houses. Now all the buildings, orchards and town quarters as well as Mt. Walīyān and Mt. Sanjān, were inside that wall. This wall was 25,000 fathoms in circumference.

Ghāzān Khan added gates in six directions: Ūjān, Shirwān, Sardrūd, Shām, Sarār and Dawīr. To the west, outside the city wall at a place known as Shām, he built a small city in which was a lofty building that would serve as his tomb; it had no match throughout Iran. To the east, the vizier Saʿīd Khwāja Rashīd (al-Dīn) built a small city at a place called Mt. Walīyān, inside Ghāzān's wall; he named it Rashīdī and built lofty edifices in it; his son, the vizier Ghiyāth al-Dīn Muḥammad al-Rashīdī enlarged those buildings and erected several large buildings. To the southwest, in the district of Niyārmiyān, outside the wall, the vizier Khwāja Tāj al-Dīn ʿAlīshāh built a large Friday Mosque with a courtyard of 50 cubits [M 381] and a hall larger than the Arch of Chosroes.

Ḥamd Allāh in the *Nuzha* recorded the old conditions of the city in the manner given above. At present, however, it is not in that state. I arrived there in 1045 (1635–1636) with the late Sultan Murād (IV) Khān and stayed for three days while it was being sacked. Here is its description as I saw it:

Those city walls had been torn down without a trace. Only here and there were traces of some large buildings that were lying in ruins. One is a lofty arch that is presumably the hall of the Khwāja ʿAlīshāh Friday Mosque. The large building called Shām-i Qazān [R 146a] [Plate 44] is visible from a distance, like the Galata Tower, in the midst of gardens. It is bordered in the east by the aforesaid mountain (Mt. Walīyān); in the west, south and north, by an extensive plain that ends at Lake Tabriz (Lake Urmiya). The streams flow in that direction. Here and there are richly decorated houses and paradisiacal gardens, rose beds and poplars, streams and pavilions.

The quantity and extent of these gardens were such that it took the Ottoman army three days to cut down their trees and penetrate (the forest). That multitude of men disappeared (in the forest) and could not cut down one tenth of the trees. When the army decamped, the trees seemed as dense as before.

As for the vineyards, the grapes are seedless, juicy and delicious, and so plentiful that even the troops' animals ate them without apparent diminishment. As Ḥamd Allāh confirms, one can truly say that the Ghouta of Tabriz effaces the copy of Soghd of Samarqand and the Ghouta of Damascus and is the envy of Shiʿb-i Bawān and Māmshānrūd of Hamadān.

This famous city has richly adorned shops made of stone, a large open market and a covered market, Friday mosques loftier than the most illustrious buildings, and minarets and dervish lodges decorated with painted tiles.

One of them is the Sulṭān Ḥasan Friday Mosque, built by the Aqqoyunlu ruler Uzun Ḥasan. It is a solid, richly adorned building in the style of the sultanic mosques of Istanbul, made of dressed stones and covered with a lead roof. A huge piece of *balghamī* marble several cubits in length and width was installed at the edge of the hall of the mihrab, imparting splendor to the Friday Mosque. This stone is a rarity of the age, the like of which is seen in no other mosque. The south side of this Friday Mosque is a spacious square. [R 146b] Most of the town quarters and markets are located south and east of it.

East of the square and adjacent to this mosque is another beautiful Friday Mosque, which was built by Shah Ṭahmāsp and partially destroyed by the (Ottoman) army. Above the door of the Sultan Ḥasan Mosque the names of the Four Companions were cut out of the stone in large letters; the Qizilbash erased them except for the name of ʿAlī.

Aside from this, the Jahānshāh Friday Mosque inside the city is from the Qaraqoyunlu rulers. It is a splendid and delightful mosque, smaller than that of Sulṭān Ḥasan—it has a single dome and minaret, and covered with painted tiles inside and out, including the minaret. The window gratings are carved out of *balghamī* stone with marvelous art. Most of the houses of the city are scattered and made of earth.

According to Ḥamd Allāh, there are 24 large villages in the mountains and plains around the city, each like a town. Four of them are on the western side of the city. Also on that side, along the lakeshore and in a valley from three to twelve parasangs (in length) and five parasangs in breadth, is the source of most of Tabriz's grain crops. [M 382] It is a fruitful and fertile place, comprising 30 villages, most of which are like towns, such as Shabistar, Dāmghān, Kūzakunān and Ṣūfiyān. Another district is one parasang north of Mt. Surkhāb and four parasangs from the city. It has good crops and comprises 40 villages. Aside from this, the districts of Nīmrūd and Badūstān are to the north. The state dues from these districts are 27 tomans; together with the city tax on commerce (*tamgha*) they amount to 115 tomans.

Amīn Aḥmad says in the *Haft iqlīm*: Most fruit in Tabriz is good and tasty, especially apples, pears, apricots and grapes—however much one speaks of them, there is scope for more (praise). The Mihrānrūd, which comes from Mt. Sahand, is divided into some 100 channels that flow to the orchards and gardens of Tabriz, but it is not enough. The people mostly have white complexions and pleasant countenances. They are boastful, acquisitive, and tough. And for the most part they are unfaithful and untrustworthy, as this [Persian] verse attests:

The Tabrizi is never a friend by nature,
The world is the kernel, the Tabrizi the shell.
He whom you do not find loyal to a friend
Is either from Tabriz or has a Tabrizi character.

In reply, one of the Tabrizis wrote the following quatrain:

Tabriz is like Paradise, its people are of purity.
Like a mirror, they are clear of the rust of cruelty.
You have said they are not loyal in friendship,
But nothing appears in the mirror but your reflection.

According to the *Rawḍ al-miʿṭār*, the people of this land are known for being occupied with learning. As has been said about them by Ḥāfiẓ Abū Ṭāhir al-Salafī:

We think that the land of Azerbaijan in the east
In grammar and letters is like Andalusia in the west.
You will not find any of its people falling short
But all are seriously engaged in the search for knowledge.

Many eminent men come from this city, including the following:

- Shams al-Dīn Muḥammad b. ʿAlī b. Malikdād: he was a disciple of Mawlānā (Jalāl al-Dīn Rūmī) and is buried in Khūy. His shaikh was Abū Bakr Sallabāf Tabrīzī. Some say that he was a disciple of Shaikh Rukn al-Dīn Sajāsī; that he came to Konya in 642 (1244–1245) where he associated with Mawlānā; and that he died in (6)45 (1247–1248). Some say that he is buried in Konya.
- Shaikh Maḥmūd Shabistarī: the author of the *Gulshan-i rāz*. He died in 720 (1320–1321).
- Sayyid Qāsim-i Anwār: one of the deputies of Ṣafī al-Dīn Ardabīlī. He settled in the Sarāb district of Tabriz. He

also received mystical guidance from Shaikh Ṣafī's son, Ṣadr al-Dīn Mūsā. He died in the Kharjird district of Jām in 837 (1433). He has a famous divan of poetry.
- Khwāja Humām: he was expert in poetry and epistolography and has a commentary on al-Jaghmīnī (astronomer, d. 745/1344–1345). He was a contemporary of Shaikh Saʿdī with whom he exchanged a few witticisms.
- Quṭb al-Dīn ʿAtīqī and his son Jalāl al-Dīn: they were famous for virtue. [R 147a]
- Mawlānā Muḥammad ʿAṣṣār: author of the *Mihr [u] mushtarī*.
- Mawlānā Muḥammad Ḥanafī: he was famous for virtue and perfection. He was a commentator on the *Ādāb*.
- Dastūr Saʿīd Khwāja Rashīd and his son Khwāja Ghiyāth al-Dīn: they were eminent viziers.
- Shams al-Dīn ʿUbaydī: he has commentaries on the *Maṭāliʿ* and on Euclid.
- The qibla of calligraphers Mīr ʿAlī: a master of *taʿlīq*.
- Mawlānā Mīrak.
- Kamāl al-Dīn Chalabī (Çelebi) Bey: a disciple of Mīrzā Jān. [M 383]
- Of the ancients: Amīn al-Dīn Muḥammad b. Muẓaffar. He died in 621 (1224–1225).
- Of litterateurs: the commentator on the divan of al-Mutanabbī, Abū Dhakariyyā Yaḥyā b. ʿAlī, known as Khaṭīb-i Tabrīzī. He died in 502 (1108–1109).

All of these are from the city of Tabriz.

Ūjān: a town 8 parasangs from Tabriz, located at 81½ degrees longitude and 37½ degrees latitude. It has a market and streams. According to Ḥamd Allāh, in the old registers it is recorded as one of the dependencies of Mihrānrūd district; but that is not correct. The city wall, constructed originally of stone and lime, was built by Ghāzān Khan. It was 1000 fathoms in circumference. The climate is cold. Water comes from Mt. Sahand. Fruit and cotton do not grow; its products are wheat and vegetables. The inhabitants are Muslims and Christians. The state dues are 10,000 dinars. Ghāzān Khan assigned the yield of Ūjān to the endowment of his own tomb.

Ardabīl: located at 89 degrees longitude and 39 degrees latitude east of Mt. Saylān. It was built by Kay-Khusraw the Kayānid. The climate is cold. Water comes from that mountain, and since it is pure and delicious and digestive, the inhabitants are gluttonous. Most of them are disciples of Shaikh Ṣafī al-Dīn. 100 villages belong to this city. During the time of Kay-Khusraw it had a castle on top of Mt. Saylān; it is now in ruins. The state dues from Ardabīl were 65,000 dinars.

Abū Ḥāmid al-Andalusī says that he saw a black stone in Ardabīl that gave a sound like the sound of steel. It is shaped like a cow's kidney, is shiny, and weighs over two batmans. When it does not rain in Ardabīl, that stone is put on a cart and brought to the city. It starts to rain immediately and does not stop until it leaves the city.

This city has more mice than other places. And so the people purchase cats to catch them.

According to the *Haft iqlīm*, the markets of Ardabil are clean, the baths delightful, the trees and streams plentiful. Only fruit is sparse—nothing is grown but apples and pears—so fruit is imported from Ṭālish and Tabriz, which are nearby.

The Safavid dynasty: Ardabīl became famous for the shrine of Shaikh Ṣafī al-Dīn Isḥāq b. Jibrīl b. Ṣāliḥ b. Quṭb al-Dīn. After the emergence of the (Safavid) shahs, its prestige increased. It has now reached the point that a condemned criminal, no matter how great his crime, who takes refuge in that mausoleum will definitely be saved from the hard hand of the Shah.

Shaikh Ṣafī's lineage goes back 21 generations to Imam Mūsā al-Kāẓim. Ancestor by ancestor, his lineage consists of great Sufis. In Shiraz he himself was in the company of Shaikh Abū ʿAbd Allāh Khafīf and Shaikh Saʿdī. After completing his training on the mystical path under Zāhid Gīlānī, he passed to the prayer rug of deputyship in Ardabil and became famous. He died in 735 (1334–1335) and was buried in his dervish lodge.

Later the shahs who were his descendents were buried next to him. They embellished the lodge and the tomb with silver gates and golden lamps and it became a magnificent shrine. However, his descendents abandoned his path, taking the road of heresy (*rifẓ ve ilḥād*, i.e., Shiism). For the sake of worldly fortune, they left the path of shaikhdom and the Sunna and chose innovation (i.e., heresy—*bidʿat*). Now the clime is in their hands. [R 147b]

Famous men from Ardabīl include the Sufi shaikh Abū Dharʿa; and the eminent scholar Mawlānā Ḥusayn, a disciple of Jalāl al-Dawwānī, [M 384] who died in 955 (1548–1549) and has a commentary on the *Sharḥ-i jadīd*.

Khalkhāl: it is near Ardabīl, two stages in the direction of Gīlān. It was a mid-sized city; later it became a village. It has four districts, in the mountains: Asad, Khānanda-mīl, Sanjar and Zanjalābād. Their governors used to reside in Fīrūzābād. After that was destroyed, Khalkhāl became the capital. In that vicinity is a valley with a spring whose water freezes in summer, and near it another spring in whose water one can boil eggs. One parasang from Khalkhāl is a mountain that resembles a wall. Its upper part juts out approximately 15 cubits. From that projection a stream flows that turn two mills and comes to Khalkhāl, irrigating its fields and gardens. Those regions have delightful meadows; game is plentiful and fat. The state dues from Khalkhāl are 30,000 dinars. Mawlānā Ḥusayn Khalkhālī, who wrote commentaries on *Ithbāt wājib wa tahdhīb*, and also commentated on the *Talkhīṣ miftāḥ*, comes from this village.

Dārmarz: a district comprising 100 villages. Its state dues are 29,000 dinars.

Shabāhrūd: a district adjacent to Dārmarz. It comprises 30 villages. The climate tends to be moderately warm. Wheat crops are good, but fruit is sparse. Taxes are 10,000 dinars.

Mushkīn: a *tūmān* located at 82½ degrees longitude and 37½ degrees latitude. Seven districts and towns are recorded as dependent on it. As it is on the south side of Mt. Saylān, the climate tends to be putrid. Its streams come from that mountain. Wheat and fruit are abundant. The state dues are 5,200 dinars. It used to be called Darāwī. When it was ruled by the Mushkīn Kūrjī (?) it became known under that name.

Anād and Arjāq: two towns on the south side of Mt. Saylān. Anād was built by Fīrūz the Sasanian king, while Arjāq was built by his son Qubād. The climate is moderately cool. Its water flows from that mountain. Vineyards are plentiful. Grapes and walnuts grow in abundance.

Āhar: a small town in that vicinity. The climate is cold. Water is gotten from Mt. Ashknīr and from springs and underground channels. Its produce is wheat and some fruit. 20 villages belong to it. Its taxes are fixed at 15,000 dinars.

Khiyār: a town on the south of Mt. Saylān. The climate tends to be hot. Water comes from that mountain. Wheat is plentiful; orchards are rare. Most inhabitants are bootmakers. The state dues are 10,000 dinars.

Darāward: in the past it was a town; now it is a province and winter pasture. Wheat, cotton and rice are produced here.

Kalīz: a town in a the mountains, in a forest. It has a strong fortress. A river flows by it. The climate is moderate. It is a fertile area with wheat and fruits.

Gīlān-i Faṣlūn: a district comprising 50 Ṭālish villages. Its people have no share in humanity. Products are wheat, cotton and rice.

Murād-i Naʿīm: a district comprising 30 villages. It produces wheat, grapes and other fruits. At some places it borders on the Aras (Araxes). The state dues are 8,700 dinars. [R 148a]

Nawdiz: a ruined castle on top of a mountain. It has flowing water and a few villages belonging to it. Its governor resides in the village called Hūl. [M 385] The climate tends to be hot. There are many orchards and plenty of wheat, cotton and rice. Taxes are 15,000 dinars.

Yāft: a district comprising some 20 villages in the midst of an oak forest. Its products are wheat and some fruit. The climate is hot. Taxes taken from here are 4,000 dinars.

Khūy: a town on a plain located at 79½ degrees longitude and 39 degrees latitude. The climate tends to be hot. Water comes from the Salmās Mountains. It has many orchards along the Aras. A very fine pear is grown in its gardens that is not found elsewhere. The people are white of countenance and good-looking and of Cathay origins. Therefore Khūy is called the Turkistan of Iran. It has 60 villages and earthen forts. The state dues are recorded as 53,000 dinars. Marand is to the southeast, Salmās to the southwest from here. According to Ḥamd Allāh, the *tūmān* of Khūy is four cities: Khūy proper, Salmās, Urmiya and Ushnūya.

Salmās: a city west of Tabriz at the extreme border of Azerbaijan, located at 79 degrees longitude and 38 degrees latitude. Its city wall was built by Khwāja ʿAlīshāh al-Tabrīzī, but it was destroyed again. The climate tends to be cold. The river comes from the mountains of Kurdistan and empties into Lake Tabriz. There are many vineyards with good grapes. Wheat and other grains are also plentiful. The state dues are recorded at 39,000 dinars. East of this town is Lake Tabriz. On the west are the slopes known as Kuskun-kıran and the mountains of Van and Kurdistan. Those mountains are the border (between the Safavids and the Ottomans).

Urmiya: a town on the shore of Lake Tabriz south of Salmās, located at 79½ degrees longitude and 37½ degrees latitude. The climate is hot and putrid. Its water comes from the mountains. Orchards are abundant, with very fine grapes, pears and apricot. 120 villages belong to it. The state dues are recorded as 74,000 goldpieces. The eminent scholar Qāḍī Sirāj al-Dīn al-Urmawī comes from this city. The Shaikh of Rūmiya, Mahmūd Efendi, who was put to death by Sultan Murād (IV) Khan in Diyarbekir (in 1048/1639), came from this city, which is why he was called the Shaikh of Rūmiya. His father and ancestors were Naqshbandi shaikhs in this town.

Ushnūya: a town in the mountains one stage southwest from Urmiya. It has a better climate than Urmiya and its water is delicious. Wheat and other grains and grapes are plentiful. 60 villages belong to it, paying 19,000 dinars in taxes.

Sarāh, also called Sarāw: a town between Tabriz and Ardabīl in a wide plain southeast of Mt. Saylān. According to Ḥamd Allāh, it is considered a separate *tūmān*. The climate is cold. Produce is (mainly) cereals; grapes and (other) fruit are sparse. The Sarāh River rises in Mt. Saylān, passes by the town and flows into Lake Tabriz. The inhabitants of this place are white of countenance and gluttonous. Four districts and 100 villages are registered as belonging to this *tūmān*, paying a total of 81,000 dinars in taxes. In the Ardabīl campaign, the Tatar Khan fought on the plain of Sarāh and was defeated by the Redheads (*surḫ ser*, i.e., Qizilbash = Safavids). One of those four districts is:

Warzand, also called Barzand: it is 20 parasangs from Ardabīl.

Miyānaj: [M 386] the Arabicized form of Miyāna. It is two stages from Marāgha. It was formerly a town, but now is the size of a village. The climate is hot. [R 148b] It is a watery and stony place.

Garmrūd: a province with a pleasanter climate than Miyānaj, comprising 100 villages. There are wheat, grapes and (other) fruit, rice and cotton. Its water comes from the mountains and eventually flows into the Safīdrūd. The state dues are 25,800 dinars.

Marāgha: a city 7 parasangs southwest of Tabriz, located at 85 degrees longitude and 38½ degrees latitude. Originally it was a village. When Marwān b. Muḥammad the Umayyad arrived here at the head of the Muslim army he let his animals graze in the pastures there. As they were rolling in the grass (*tamarrugh*), and as he liked the place, the city that was built there was named Marāgha. Later its castle was destroyed by Ibn Abī'l-Sāj.

It is a clean and pure town with delightful orchards and gardens and streams. Outside the city is a hill where the great scholar Naṣīr al-Ṭūsī built an observatory in 657 (1259). The *Zīj-i Ilkhānī* (Ilkhanid Astronomical Table) was written in this observatory. They say that traces of it remain to this day.

According to Ḥamd Allāh, the climate of Marāgha is moderate and tends to be putrid, because Mt. Sahand blocks it from the north. There are many orchards and water sources. The Rūd-i Ṣāfī rises in (Mt.) Sahand, passes by this city and flows into Lake Tabriz. Its products are wheat, cotton and fruits. Grapes are usually found in abundance.

Six districts belong to Marāgha. It is considered a separate *tūman*. The state dues are recorded as 185,500 dinars. The inhabitants are white complexioned and look like Turks.

Naṣīr's observatory: Ṣafadī writes in his *Wāfī bi'l-wafāyāt* that when Khwāja Naṣīr wished to build an observatory in Marāgha, he submitted its expenses to Hulagu, who deemed them excessive, saying, "What is the use of astrology that so much money should be spent on it?" Naṣīr explained: "Let me show you. Order someone to take a large basin and drop it from that high place without warning." When the order was carried out and the basin was dropped, the noise caused a fright among the soldiers, who were caught by surprise, while Naṣīr and Hulagu knew the cause and so were not bothered by it. "This is the use of astrology," said Naṣīr. "Those who are aware of the circumstances do not suffer pain and anxiety as those who are unaware do, and they act accordingly." Hulagu agreed that the observatory should be built and he gave a great deal of money for the construction. By the time it was finished, the cost of maintaining the instruments alone amounted to 20,000 dinars.

Tasūy: a small town near Salmās. Its water comes from the mountains and flows into Lake Tabriz. Its products are wheat and some fruit. State taxes come to 25,000 dinars.

Daskhwāqān: a small town with a moderate climate. Its water comes from Mt. Sahand. It has many vineyards producing matchless grapes. It also produces wheat, cotton and fruit. It has 20 villages; the state dues are 23,300 dinars.

Nīlān: a small town. It has many vineyards and much wheat and cotton. Its water comes from the Rūd-i Juft and from springs. The people are Turks. It pays 10,000 dinars as state dues.

Marand: [M 387] a town between Khūy and Tabriz, located at 86½ degrees longitude and 39 degrees latitude. It is 14 parasangs northwest of Tabriz. The climate is moderate. Its water comes from the mountain on its western side; it is a delightful river. Its gardens and fruits are plentiful; the peaches and apricots are especially fine. It is fertile with wheat and other grains. Constituting a separate *tūmān*, it has 100 villages. [R 149a] The state dues are 24,000 dinars.

Darbāz: a district north of Tabriz with more than 50 villages. The climate is warm. Its water comes from the mountains and eventually flows into the Aras. Wheat, cotton and fruit are abundant and exported to surrounding regions. The state dues are 40,800 dinars.

Zanūz: a town with many orchards. Its products are wheat, fruit and grapes. Especially fine is the white apple known as *qablī*. The state dues are 3000 dinars.

Karkar: a town near the Aras River. It produces fruit, wheat and cotton. Ḍiyā al-Mulk al-Nakhjiwānī built a magnificent bridge over the Aras near Karkar, one of his great good works.

Akhbān: it is also called Kārkhāna ("Workshop") because of its copper mine.

Ordūbād: a delightful town on the bank of the Aras River. It has many orchards and gardens and abundant wheat and fruit. Its river rises in the Qabān Mountains and, after irrigating its gardens, empties into the Aras. Every house has flowing water. The people have a passion for marvelous buildings and build splendid houses. Hence the saying, "The Ordūbādis are house-worshippers, the Tabrizis are woman-worshippers, and the Nakhjiwānis are gold-worshippers." Ordūbād is famous for its pears, and especially the Sultani apples. When they ripen, which occurs at the rise of the Pleiades, they are covered with cutouts of suitable verses and flowers and the exposed areas are lined with red. They are sent as gifts to the surrounding regions.

Mākūya: a castle in a mountain gully. It has a village below it that is overshadowed by the mountain and does not see the sun until midday. According to the *Taqwīm*, it is a town near Shirwān (in) the Kharzān Pass, located at 74½ degrees longitude and 39½ degrees latitude.

Khūnaj: a town at the extreme southern border of Azerbaijan, located at 81½ degrees longitude and 37½ degrees latitude. It is 13 parasangs from Marāgha.

Warthān: a town 7 parasangs from Ganja, located at 83½ degrees longitude and 39½ degrees latitude. The Aras River flows 2 parasangs from here.

Mountains

Famous mountains in this clime include:

Mt. Sahand: it is surrounded by Tabriz, Marāgha and Ūjān. Its circumference is estimated to be 25 parasangs. Snow is never absent from its summit. There is a lake as well. The tomb of one of the noble Companions, Usāma b. Shurayk, is on this mountain.

Mt. Saylān: a very high mountain. It is visible from 50 parasangs away. Snow is never absent from its summit. Its circumference is estimated to be 30 parasangs. Ardabīl, Surāh, Mushkīn, Anār, Arjāq and Khiyār are located around the foot of this mountain.

Rivers and bodies of water

In this clime there is one large lake and several rivers.

Lake Urmiya: also called Lake Tabriz. In the *Taqwīm* it is written as Lake Talā. Its water is acrid. No animals live in it, although some say there are sharks in it. It is oblong in shape, extending 139 miles from west to east with some deflection to the south. [M 388] Its width is estimated to be half its length, i.e., 65 miles. Marāgha is 3 parasangs to the east, Salmās is to the west. Traveling on land, its length is nearly three stages; by boat it can be crossed in a day. Its circumference is six stages. It is 17 parasangs west of Tabriz.

In this lake there are a few islands with ruins. [R 149b] According to the *Taqwīm*, the one in the middle used to be called Talā. It is a mountain island. On the top there was a fortress carved out of the rock where the treasury of Hulagu was stored. It is now in ruins. Ibn Saʿīd locates that fortress at 72 degrees longitude and 39½ degrees latitude. Several rivers flow into this lake, but it has no outlet.

The rivers of this clime are the following:

Andarāb: It rises from Mt. Saylān, passes through Ardabīl where it is called the Ardabīl River, then it joins the Andarāb (!), flows underneath the Khwāja ʿAlīshāh Bridge and conflows with the Āhar River. Its length is approximately 25 parasangs.

Āhar: It rises in the mountains of Karīwa-i Armaniyān ("Ravine of the Armenians"), which in the Mongol language is Gökçe Nīl ("Blue Indigo"), and flows by Qalʿa-i Naw and Āhar. When it passes the village of Baylaqān, it unites with the Andarāb River and both empty into the Aras. Its length is approximately 20 parasangs.

Ṣāfī: It rises from Mt. Sahand, goes to Marāgha, then joins the Juft River. It is 20 parasangs long.

Naftū: it rises in the mountains on the borders of Karīwa-i STN'(?) ("Ravine of Satnā [?]") and unites with the Juft. It is approximately 15 parasangs long.

Juft: it rises in the mountains of Kurdistan at Siyāh Kūh, unites with Ṣāfī nearby Marāgha and flows into Lake Tabriz. It is approximately 20 parasangs long.

Sardrūd, also called Bāwīlrūd: It rises from Mt. Sahand, flows by the places called Sard and Bāwīl, and joins the Sarāwrūd. It is approximately 6 parasangs long.

Marwrūd: it rises at Mt. Sahand, flows near Marāgha, and joins the Juft River in Kāwdawān. It is 8 parasangs long.

Sajad and Kadīw: They rise as two rivers, then unite and empty into the Safīdrūd. They are 8 parasangs long.

Shālrūd: it rises in the Shāl mountains and empties into the Safīdrūd at Barbadaq Castle. It is also 8 parasangs long.

Garmrūd: it rises at Mt. Sarāw, passes the province of Garmrūd, and joins the Miyānaj River. Its length is approximately 10 parasangs.

Miyānaj: it rises in the Ūjān mountains, passes Ūjān and Miyānaj, and joins the Hashtrūd in the Miyānaj Plain. Its length is approximately 20 parasangs.

Hashtrūd: It rises in the mountains of Marāgha and Ūjān, unites with the Miyānaj River at the border of Miyānaj, after which it joins the Safīdrūd. Its length is approximately 20 parasangs. The Miyānaj Bridge, a magnificent structure with 32 arches, was built over this river by Shams al-Dīn Ṣāḥib-dīwān (Juwaynī, vizier under the Mongols, in office 1263–1284). As for the Safīdrūd, it rises in Persian Iraq in the Besh Barmaq Mountains and, after joining other similar-sized rivers that come from elsewhere, empties into the Caspian Sea in Gīlān. It was mentioned in the chapter on the clime of Jabal.

Andarūd: it rises from Mt. Sahand and ends in the orchards of Tabriz. When it is in spate, the overflow unites with the Sarāwrūd and empties into Lake Tabriz. Its length is 7 parasangs.

Ūjān: it rises from Mt. Sahand and flows by Ūjān, joining the Sarāwrūd in 7 parasangs.

Saraw, also called Surkhāb and Sarāwrūd: It rises from Mt. Saylān [M 389] in the Ashkamad Mountains and flows by Sarāw. At the borders of Kūlū it flows over salty and brackish ground and the water becomes salty. It also makes the Ūjān River salty. North of Tabriz it flows under the bridge and empties into the lake. Its entire length is approximately 40 parasangs.

Marand: It rises in the Marand Mountains and flows from west to east. In the spring it cannot be forded. Before arriving at the city of Marand it goes underground for around 4 parasangs and disappears, then it surfaces again and flows to Marand. What remains after Marand unites with the Khūy River and empties into the Aras. Its length is approximately 8 parasangs. As for the Aras River, it is discussed in the chapter on Arrān.

All this is written in Ḥamd Allāh's *Nuzha*.

Routes of Azerbaijan

From Tabriz westward to Erzurum: > Marand, 11 parasangs > Khūy, 12 > Sekmānābād, 6 > Band-i Māhī, 3 > Erciş, 8 > Malādhjird, 8 > Ḫınūs, 10 > Ravine of Aq Band, 15 > Pāsīn, 5 > Erzurum, 6. Altogether 79 parasangs.

From Tabriz northward to Qarabāgh: > village of Armaniyān, 8 parasangs—it is in the Ravine of Gökçe Nīl; [R 150a] in the past, Vizier Khwāja Saʿd al-Dīn al-Sāwajī had a hospice there, and the Amīr Niẓām al-Dīn Yaḥyā al-Sāwajī also built a hospice > city of Āhar, 6 parasangs > village of Baylaqān, 8 parasangs—on this road in the valley of Quruchāy there is a hospice built by Khwāja Tāj al-Dīn ʿAlīshāh > Ribāṭ-i Alīwān (Hospice of Alīwān), 8 parasangs > village of Barzand (text: Būzand), 6 parasangs > Qarabāgh, 4 parasangs. The bank of the Aras, which is the border of Qarabāgh, is 10 parasangs south of Qarabāgh.

From Tabriz to Sulṭāniyya: > Saʿīdābād, 4 parasangs > city of Ūjān, 4 parasangs > village of Sangalābād, 4 parasangs > village of Turkmān, 4 parasangs—this village used to be a city called Dayr-i Khwān > Miyāna, 6 parasangs > Sarjam, 6 parasangs—here is a hospice built by Vizier Ghiyāth al-Dīn Muḥammad and another one built by his brother Jalāl al-Dīn Muḥammad > Ribāṭ-i Sagbā (Hospice of Sagbā) built by Vizier Tāj al-Dīn ʿAlīshāh, 7 parasangs > Zanjān, 7 parasangs > Sulṭāniyya, 5 parasangs. Altogether 46 parasangs.

From Ardabīl to Sulṭāniyya: village of Tālish, 6 parasangs > village of Sanjīd-i Khalkhāl, 6 parasangs > city of Kaghadh-kunān ("Paper-makers"), 6 parasangs > village of Tūb (?) Suwāri, 7 parasangs > Zanjān, 7 parasangs > Sulṭāniya, 5 parasangs. Altogether 37 parasangs.

From Ardabīl to Qarabāgh: > Ribāṭ-i Arshad, 8 parasangs > village of DRLQ, 8 parasangs > Bājarwān, 4 parasangs > city of Barzand, which is a village; it is toward the south from this stage, on the right when going (from Ardabīl to Qarabāgh); > Maḥmūdābād, 8 parasangs. Altogether 40 parasangs.

Also, from Tabriz to Nakhjiwān is 12 parasangs; to Ardabīl, 25 parasangs; to Qazwīn, 7 stages—one can also get there in 7 days from Ardabīl.

From Ardabīl to Khūnaj is 27 parasangs; to Marāgha, 30 parasangs; then to Urmiya, 30 parasangs; to Dīnawar, 60 parasangs; to Khūnaj (!), 13 parasangs.

From Khūy to Nakhjiwān is 3 stages; to Arjīs, 6 stages; to Salmās, 7 parasangs; then to Urmiya, 16 parasangs.

Again, from Tabrīz to Marand is 2 stages; then to Khūy, 1 stage.

From Khūy to Nakhjiwān is 3 stages; then to Wabīl (?), 3 stages.

From Marāgha to Dīnawar is 30 parasangs. From Salmās to Urūmiya is 14 parasangs.

Ardabīl > Miyāna, 20 parasangs > Khūna, 8 parasangs > Lūsra, 10 parasangs > Marāgha, 10 parasangs > Dayr-i Kharqān, 2 stages > Tabriz, 2 stages.

Ardabīl > Sapīdrūd Bridge, 2 stages > Khūna, 2 stages > Zanjān, 2 stages.

Rulers of Azerbaijan

[M MAP] [M 390] Part of this clime was conquered during the time of Caliph ʿUmar, who gave its governance to ʿUtba b. Farqad. Then, during the time of Caliph ʿUthmān, Walīd b. ʿUqba conducted a religious war against Arminiyya and Azerbaijan and exacted 800,000 dirhams in tribute from them, as provided by the treaty of Ḥudhayfa b. al-Yamān made earlier. After that, Azerbaijan was administered by governors of the Umayyads and Abbasids. At times it was governed by the Seljuks or the kings of Khurāsān. 618 (1221–1222) it was taken and destroyed by the Chingisids.

The Chingisid dynasty (i.e., Ilkhanids): After Hulagu took Iraq, he made Azerbaijan the administrative center. His descendents—Abaqa, Aḥmad, Arghun, Gaikhatu, Baydu and Ghāzān Maḥmūd—sometimes resided in Khurāsān, sometimes in Azerbaijan. After Ghāzān, his brother Sultan Muḥammad Khudābanda built the city of Sulṭāniyya in 705 (1306–1307) and transferred his capital there from Tabriz. In 736 (1335–1336) his son Abū Saʿīd Bahādur Khan died, his place being taken by Arpa Khan. However, after Abū Saʿīd Bahādur the dynasty began to decline and seven of his descendents ruled some places like "party kings" (*mülūk-i ṭavāʾif*) until 754 (1353–1354).

The descendents of Ṭughā Tīmūr became extinct in 812 (1409–1410).

The Qaraqoyunlu of Mosul and Jazīra: In 777 (1375–1376) Qara Muḥammad and Qara Yūsuf, sons of Turkmān Bayram Khwāja, emerged to power and conquered Azerbaijan. It was governed by the son of Qara Yūsuf, Iskandar, then by Jahān Shāh who built a Friday Mosque in Tabriz. He was succeeded by his son, Ḥasan ʿAlī. In 873 (1468–1469) he was defeated in battle by Uzun Ḥasan and the dynasty came to an end.

The Aqqoyunlu Dynasty: Qara Aylūk (Yölük) ʿUthmān, from the Aqqoyunlu clan of the Turkmen tribe, was appointed governor of Diyarbekir by Tīmūr. After him his son Ḥamza, then his nephew Jahāngīr, and then Uzun Ḥasan were governors. They carried on continuous warfare against the Qaraqoyunlu rulers. Finally Uzun Ḥasan was victorious and took Azerbaijan from them. [R 150b] He ruled Khurāsān and Iraq as well until his death in 883 (1478–1479).

When the ruler of Iran, Abū Saʿīd b. Mīrānshāh b. Tīmūr, haughtily marched against him, he was routed and killed in Qarabāgh. After this Uzun Ḥasan stood in high repute. He was defeated in battle by Ebū'l-Fetḥ Sultan Meḥmed Khan (the Ottoman sultan, Fātiḥ Meḥmed II). He was succeeded by his son Khalīl, then by Yaʿqūb. When after the death of Sultan Yaʿqūb, his brothers Aḥmad, Murād, Alwand and Muḥammad fought against each other, Shāh Ismāʿīl, the descendent of Shaikh Ṣafī, saw the opportunity and conquered the country (in 914/1508).

The Safavid Dynasty: They are the Safavid family mentioned above in connection with Ardabīl, the descendants of Shaikh Ṣafī. Shaikh Junayd b. Ibrāhīm b. ʿAlī b. Mūsā b. Ṣafī al-Dīn Isḥaq, following his ancestors in turn by heredity, sat on the prayer rug of mystical guidance in the convent in Ardabīl. As he gained fame with so many disciples, Jahānshāh b. Qara Yūsuf, the ruler in Tabriz, banished him from Ardabīl. Junayd went to Diyarbekir and married Uzun Ḥasan's sister; from this union Shaikh Ḥaydar was born. The desire to revolt and ascend to the peak of rulership was fixed in Junayd's mind. He rallied his disciples and marched off, on the pretext of religious war (*ġazā*) against Georgia. When he reached that region, he confronted Khalīl the Shirwānshāh [M 391] and defeated him in battle.

His son Ḥaydar designed the twelve-gored red headgear (*tāj*) and made his disciples wear it. He too marched in that direction with the call to revolt in the name of Imami Shiism. The Shirwānshāh informed the ruler of Tabriz, Yaʿqūb, who sent his commander Amīr Sulaymān against him. Shaikh Ḥaydar was defeated and fell in the battle.

His son Shāh Ismāʿīl, being yet a child, was not put to death but was left in Iṣṭakhr castle. He remained there until Yaʿqūb's death in 896 (1490–1491), when he found the opportunity to leave. He revolted in 906 (1500–1501), first defeating the Shirwānshāh whom he put to death. He sat on the throne in Azerbaijan and proclaimed Imami Shiism, killing and eradicating Sunnis. In a short time he took Khurāsān and Iraq. He continued to sow corruption until his defeat at Çaldıran by Sultan Selīm.

His descendants—his son, Shāh Ṭahmāsp, then his son Ismāʿīl, then Khudābanda, Shāh ʿAbbās, Ṣafī, and ʿAbbās Mīrzā—have ruled those regions up to the present time and perpetrated the rites of heresy (*rafḍ*, i.e., Shiism). For approximately 20 years, Azerbaijan was controlled by the Ottomans, who appointed a *beylerbey*. But in 1012 (1603), because of the Hungarian campaigns and the Celālī revolt, Shāh ʿAbbās took it back through many wars and stratagems. It is now under the control of that abominable group.

Chapter 39: Clime of Arrān, Mūqān and Shirwān

The 39th chapter is on the clime adjacent to Azerbaijan in the north.

Borders

To the east, Azerbaijan, Gīlān and Daylam; to the south, Arminiyya, now called Van, and Kurdistan; to the west, the province (*eyālet*) of Erzurum and the mountains of Georgia; in the north, the Caspian Sea and the province (*vilāyet*) of Dagestan. These borders embrace three countries (*memleket*)—Arrān, called Rawān, Mūghān and Shirwān—which we discuss in one chapter because they interpenetrate each other.

Arrān

As reported by the *Taqwīm*, citing the *Tuḥfa*, it is named after Arrān son of Japheth. In the *Aḥsan al-taqāsīm* it is written as al-Rān. This indicates that its use without the definite article (*al-*) is a way to pronounce it as Rawān (or Revān). [R 151a] Originally it was the name of the province whose capital was Ganja. Later it became the designation of a well-known castle. According to Ḥamd Allāh, Arrān is the name of the province between two rivers, the Aras and the Kur. This being so, we will now list the cities that are found in this province.

Rawān (Revān, Erevan): located at 87 degrees longitude and 39 degrees latitude, it is the town and capital of the region known as Chuḳur Saʿd. A stream called Zangī Ṣuyı flows by the foot of its fortress on the west. It is an earthen fortress situated on a level place overlooking that stream. It has two gates, the Tabriz Gate to the south and the Shirwān Gate to the north. Surrounded by orchards and gardens, it is a place of wheat and fruit. Inside the fortress is a market, a Friday mosque and a palace that overlooks the river.

In 991 (1583–1584) Rawān was taken (for the Ottomans) by Ferhād Pasha, who left a garrison here. In 1014 (1605–1606) it was retaken by Shah ʿAbbās after a seven-month siege. Ten years later, Meḥmed Pasha mounted a siege but could not conquer it. In 1045 (1635–1636) Sultan Murād Khan (IV) marched there and took it in seven days; [M 392] but after he returned (to Istanbul) Shah Ṣafī reconquered it. It is now a fertile province in the hands of the Redheads (i.e., Safavids).

Mt. Arghī (Ağrı Dağı, Ararat): a great mountain one stage southwest of Rawān. It is visible from several stages away. It is a circular mountain whose top half is always covered with snow and whose circumference comprises several stages. The Aras River (Araxes) flows by it on the northeast. The plain (*ṣaḥrā*) of Eleşkird falls between the river and the mountain. Another side is the plain (*ova*) of Çaldıran; and so its northern side is called the province (*eyālet*) of Çıldır.

Ganja: a city of orchards and gardens situated on a level place four stages east of Rawān. As figs and other fruits are plentiful, the climate is unwholesome. Eating figs brings on fever.

Ḥamd Allāh says that Ganja was founded by Alexander the Greek and restored by (the Sassanian king) Qubād son of Fīrūz. In the past it was a very large and prosperous city. There are still traces of its lofty buildings. Its pomegranates, grapes and hazelnuts are excellent. Water is gotten from a stream known as the Barbar.

Bardaʿ: city built on an elevated place two stages northwest of Ganja. It is a Muslim city, having been founded in the year 39 (659–660). It has a summer pasture called Tarak, a delightful place with meadows and streams. People migrate there in summer.

Ibn Saʿīd says that Bardaʿ is encompassed by two canals that originate from the Kur River. Some have claimed that this city is Andarāba, but it is not.

Near Bardaʿ there is a hazelnut grove with an endless number of very fine hazelnuts, better than in Samarqand. The chestnuts here are also very good. And there is also a kind of fig found nowhere else. Silk production is high, because there are plenty of mulberry trees, and the silk is exported abroad. The mules here are of superior quality and are held in esteem. The Ḥanafī scholar Abū Saʿīd al-Bardaʿī comes from this city.

Nakhjiwān (Nakhchivan): the Arabicized form of Naqsh-i Jahān ("Ornament of the World"); it is also pronounced as Naqjawān. It is a town located at 88 degrees longitude and 38½ degrees latitude. Most of its buildings are of brick. Its products are wheat, cotton, grapes and (other) fruit. The inhabitants are light complexioned.

According to Ḥamd Allāh, the fortress of Alinjaq, the ford of Qabān and other places are dependent on Nakhjiwān. It is an independent *tūman* with state dues amounting to 115,000 dinars.

Baylaqān: formerly an ancient city founded by Qubād; now it is nothing more than a village. Its construction is of brick. The climate is hot. Its products are wheat and other grains, rice and cotton. Warthān is 6 parasangs hence.

Amīn Aḥmad, citing the *Ṣuwar-i aqālīm*, says that Baylaqān had a very solid fortress; Hulagu Khan conquered it, massacred the inhabitants and destroyed the fortress. The Ilkhanid army had laid siege to it for some time but was unable to take it for lack of stone missiles for the catapults. Instructed by Khwāja Naṣīr al-Ṭūsī, they made wooden balls filled with lead and shot them from the catapults instead of stones. So they were able to conquer the city. Later, when Amīr Tīmūr reached Baylaqān on his way back from Rūm, he wished to restore it. Mīrzā Shāhrukh set about the project. They attempted to divert the Barlās River, but ran into obstacles. Later the river was successfully diverted and it is still flowing.

Mūghān

It includes the region of Qarabāgh. It is also called Mūqān. According to the *Taqwīm*, it is a province named after Mūqān son of Japheth. It is on the south shore of the Caspian. [M 393] It is bounded by Arminiyya, Shirwān, Azerbaijan and the Caspian Sea. In the era of the Atabegs it provided over 300 tomans of revenue. Later this decreased to 30 tomans and 3000 dinars.

According to Ḥamd Allāh, in the area that extends from the Sang-bar-sang Pass to the bank of the Aras River, wherever Mt. Saylān disappears from sight the herbage is poisonous and lethal for livestock, especially in spring when hungry animals eat it; but wherever Mt. Saylān can be seen, this danger is not present.

According to other knowledgable sources, Mūghān lies between Shamākhī, Nawshahr and Maḥmūdābād, one stage from each—Nawshahr being a town near Mūghān—and the Kalla-khāl River flows through it.

Bājarwān: Formerly it was the (main) city of Mūghān; now it is only a village. The climate tends to be hot. Its source of water is a stream that descends from the neighboring mountain. Its only product is wheat. [R 151b]

Barzand: a midsized town. After it was destroyed, Afshīn, the slave of Caliph al-Muʿtaṣim, restored it. Later it was again destroyed and now remains as a village. The climate tends to be hot. Water is gotten from springs. The soil is rich and fertile.

Fīlsuwār: it was founded by one of the Būyid emirs, Pīlasuwār. It was a town; later only a village remained of it. Water is gotten from the Bājarwān River. Its product is wheat.

Maḥmūdābād: a town in the Qarabāgh district on the Gāwpārī Plain. It is also called Shal (?). It is on the seashore between where the Kur River and the Chapa-khāl empty into the sea. Plenty of fish are caught here. It was built by Sulṭān Maḥmūd Ghāzān.

Qarabāgh: a town and district of Mūghān, east of Shirwān. There are a number of towns in this district, including Butlān and Alqāsiya.

Hamshahra: it is located two parasangs from the coast. In the past it was called Abrshahr. Firdawsī refers to this city when he says in the *Shahnāma* [Persian verse],

If this Abr-shahra (sic?) were like Farhād the hero
In battle it would have taken away splendor from the world.

Shirwān

According to Ḥamd Allāh, Shirwān is the name of a province that extends from the bank of the Kur to Bāb al-Abwāb, i.e., Demir Ḳapu Derbendi ("Iron Gate Pass," Darband). In the era of the Khaqans its state dues were 100 tomans, later reduced to 11 tomans and 3000 dinars. Its capital is recorded at 87 degrees longitude and 40 degrees latitude.

According to the *Taqwīm*, it is pronounced Sharwān and is a town in the vicinity of Bāb al-Abwāb. Others have said that it is the name of a province whose town is Shamākhī. Today it is known as Darband-i Khazaz. It was founded by Anūshirwān; to lighten the pronunciation, they dropped Anū and called it simply Shirwān. There are other views as well that will be discussed below.

Amīn Aḥmad says in the *Haft iqlīm*: In the past, Shirwān was the name of a city. Today it is the name of a province that encompasses several cities, including Baku, Shamākhī, Aras, Bāb al-Abwāb, etc.

Shamākhī: According to the *Taqwīm*, citing the *Marāṣid*, it is the capital of the region of Shirwān, a flourishing town one stage west of the Caspian.

Ḥamd Allāh says that it was founded by Anūshirwān. Its climate tends to be hot. Moses's Rock (Koran 18:63) and the Fountain of Life are there.

The poor one (Kātib Çelebi) says: It is unlikely that the Confluence of the Two Seas (Koran 18:60) is in this region.

According to the *Haft iqlīm*, this city is small but quite prosperous. In one year, 20,000 bales of silk are traded. [M 394] Excellent fruits include pomegranates, apples, and watermelons.

Shamkūr: According to the *Taqwīm*, a village near Bardaʿa, one of the administrative districts of Arrān. There is a very tall minaret there.

According to Ibn Saʿīd, it is east of Tiflis. Some have said that Shamkūr is Shamākhī, but this judgment is based on the similarity of the names.

Qabala: According to Ḥamd Allāh, it is near Darband (i.e. Shirwān) and was built by Qubād son of Fīrūz. Silk and grain crops are good. The climate is pleasant. One of its dependencies is a valley, in the middle of which a very hot spring rises with steam burning like a flame. After flowing a few feet, it enters a hole and disappears under ground; then after a bowshot's distance, it re-emerges and continues to flow. It is a most delightful stream.

Fīrūz-Qubād: According to Ḥamd Allāh, citing the *Majmaʿ al-buldān*, it is a town with a pleasant climate near Darband.

Gushtāsfī: a large city and province on the bank of the river, built by Gushtāsb son of Luhrāsb. Canals branch off the Kur and Aras Rivers, irrigating this region. It has many productive villages. Wheat and rice are good; fruit is sparse. The inhabitants are light complexioned. They speak Pahlawī, the language of the people of Jīlān. The state dues are 117,500 dinars.

Tiflis (Tbilisi): a fortress and city located at 83 degrees longitude and 43 degrees latitude. It is on the Kur River, east of the mountains of Georgia. There is a mountain on one side and the river on the other. Wheat is fine, but fruit is sparse. This city has two fortresses and three gates, two of which face each other. The Kur River flows in the middle and the fortresses are on either side, on two huge jutting rocks with a narrow ravine between them, estimated at 10 cubits wide where the river passes.

In ancient [R 152a] [SPACE FOR MAP; CF. M 431/432 (MAP OF CAUCASUS, APPENDIX, FIG. 91–92)] books this city is recorded as the administrative center of Georgia. After being conquered by the Muslims, the Georgians recaptured it for a while, but it was conquered again. Now it is dependent on Shirwān.

Outside the city wall to the east are numerous hot springs with pleasant and moderately hot flowing water. Bathing areas with small stonework domes have been constructed over several of the springs. They are ancient buildings. This water has no smell of sulphur. Bathing in it is very beneficial. Amīn Aḥmad, citing the *ʿAjāʾib al-makhlūqāt*, says that if you put ten eggs in one of those springs, nine of them get cooked and one of them disappears.

Bāb al-Abwāb: the fortress known as Demir Ḳapu and, in written sources, as Bāb al-Ḥadīd ("Iron Gate" in Turkish and Arabic respectively). It is a solid fortress and famous entrepot, located at 81½ degrees longitude and 42½ degrees latitude, on the western shore of the Caspian. It was built to block the pass between the mountains of Georgia, known as Mt. Alburz, and the sea. It is the port of Khazaz, Sarīr, Gurgān, Ṭabaristān, Karj (Kerch) and Qaytāq. At present it is mostly in ruins and has remained but a town. Linen cloth and saffron are exported from here to the abovementioned places. There are saffron meadows on most of the seashore.

This fortress is 21 parasangs north of the place where the Kur River empties into the sea. It is hemmed in on the east by the sea and on the west by huge mountains, the space between being a flat plain two or three miles wide. The tribes—Khazaz (i.e., Khazars) and Turks—living north of it used to come through it in order to pillage the countries of Shirwān, Azerbaijan and Arrān. The Persian kings never ceased warring with them until the reign of Nūshirwān. [M 395] Nūshirwān made peace with them on condition that he build a fortress here. He raised a wall from the seashore to the mountain top, putting an iron gate every 3 miles, and settled guards inside the fortress, thus blocking the way of those tribes. Some say it was first built by Isfandiyār and Anūshirwān merely restored it.

It is situated on a level place, at the foot of a circular mountain. It has two gates, one on either side, and only two walls. A spring of fine water rises in the middle. The fortress is 10,500 cubits long and 550 cubits wide and has 70 towers. Its height, [R 152b] according to tradition, is the same as the land wall of Constantinople. The wall, inside and out, is completely made up of dressed stones, the ones on the inside being huge stones, each four cubits in length and width. To the east is the sea; to the south, a wide plain; to the west, the aforementioned mountain on which grow various species of flower.

Masʿūdī says in the *Murūj al-dhahab* that its wall reaches two miles into the sea. He describes how it was built as follows: They loaded inflated oxhides with stones, rendered a solid mass with iron and lead, and constructed the wall as they sank. When the hides reached the sea floor, divers dived in and pierced them, so the wall settled firmly in the ground.

The poor one (Kātib Çelebi) says: This report is not the sort that can be confirmed. If the place of the construction is not deep, according to the rule for building in water, such structures are built inside caulked chests. The deeper it is, the more difficult it is to do that. It is possible that the oxhides were laid out on both sides so that one could walk on them; otherwise it makes no sense to build on hides. That Imam Masʿūdī reports this as if he believed it shows that he did not know anything about the rational sciences.

Regarding the naming of Bāb al-Abwāb (lit. "Gate of Gates"), the *Rawḍ al-miʿṭār* states that the gates after which the city is named are the mouths of the tributaries that disappear in the desert nearby. Bāb Ṣūl, Bāb al-Lān, Bāb al-Sarīr, Bāb Fīlān Shāh, etc. are all the mouths of tributaries. This city was built on this side, as the gate of these gates, and so was called Bāb al-Abwāb.

Aras: a town built by Nūshirwān on the bank of the river, three stages northwest of Shamākhī. The river is named after it. The climate is quite hot. There are few buildings. The Gulistān Castle, which is the residence of the kings of Shirwān, is a strong fortress on a steep elevation in the vicinity of Aras.

Qizil Aghach: a city on the coast one stage east of where the Kur River empties into the sea. Ardabīl can be reached in two days from here.

Nawshahr: a town on the bank of the Kur River one stage south of Shamākhī.

Langarkunān: a town on the coast, two stages east of Qizil Aghach, where the Barāzrū River coming from Ardabīl empties into the sea.

Lori: a fortress on the Kur River on the border with Georgia.

Gori: this is also a famous fortress near Lori.

Baku: a town on the coast near Darband, northeast of Shamākhī, located at 84½ degrees longitude and 40½ latitude. It is one of the famous cities of Shirwān. Its climate tends to be hot. It is surrounded for nearly 10 parasangs by a stony tract with no trace of vegetation. However, if one digs to 10 spans deep, soil appears. Some of these stones [M 396] are burnt instead of firewood.

One of the dependencies in its vicinity is ʿAmaliyān, where there are nearly 500 wells from which one gets white and black petroleum. In that region the people dig a hole in the ground and put in a pot or a cauldron, and the heat of the earth cooks the food. All this is recorded in the *Haft iqlīm*.

Niyāzābād: a town on the coast between Baku and Darband.

Men

Famous individuals who came from this clime are:
- Khāqānī, the famous poet, who stems from the Great Khaqan, Manūchihr. He died in 582 (1186–1187).
- Falakī, the poet.
- Mawlānā Kamāl al-Dīn Masʿūd, who was unique of the age in theology, logic and philosophy. He died in 905 (1499–1500).

- Mawlānā Ṣadr al-Dīn, whose son came to Rūm.
- Mawlānā Rukn al-Dīn, who was very skilled in medicine. He dedicated his *Mir'at al-shifā* to Sultan Meḥmed the Conqueror.
- Mawlānā Yūsuf Qarabāghī, a disciple of Mīrzā Jān, who was expert in the rational sciences. He died not long ago.
- Abū al-ʿAlā-i Ganjī, a famous poet in Shirwān.
- Tiflisī, a Shāfiʿī jurist and author of the *Ṭabaqāt*.
- The author of the *Qānūn al-adab*.

Rivers

There are two great rivers that flow through this clime:

[1.] Kur: it rises in the Qālīqān Mountains and flows from west to east, passing to the south of Lori and Akhisqa.

[Marginal addition by Ebū Bekr ed-Dimaşḳī] According to some reliable sources, the source of the Kur is in the mountain between the fortress of Göle and Kars, closer to the former. Above the source is the stone image of an ox, positioned in such a way that the water gushes out of its mouth and nose. Its tributary pours onto the Göle Plain, is drawn close to Göle, and continues into the Göle Ravine. If not for that ravine, the plain, which is surrounded by mountains, would be like a lake and turn into a swamp. In ancient times they cut through that ravine for this reason and scraped out that rock, creating a cistern for that water. The river flows through this passageway carved out of the rock and through the ravine, reaches the houses of Ḳara Ardahan and the foot of its castle, and flows in front of Arjūka (?) and Khartūs Castle. The Akhalkalak River flows into the Aras at that point, Khartūs being located near the corner where they join. After this it passes before the castles of Azbūr and Khajrak, joins the Akhiskha River on the Biñlik meadow, then flows before Azghūr castle and through the Badra Ravine and becomes like a sea.

Now the Kur reaches Tiflis where it flows between two huge jutting rocks, the space between which is very narrow and the fortress being on top of them. [R 153a] Then the Qiniq River flows into it from the Zagam Mountains. It turns south and then northeast, serving as the border between Shirwān and Azerbaijan. In Arrān, according to Ḥamd Allāh, it divides into two branches, the smaller branch emptying into Lake Shamkūra, the bigger one joining the Aras River in Yurtbāzār and then emptying into the Caspian. It is a river approximately 200 parasangs long. This Yurtbāzār is 10 parasangs east of Shamākhī. It is also called Jawād. A reedbed can be found nearby the town.

Some knowledgeable sources relate that the Kur River divides into three branches, each of which empties into the sea at a different place: [M 397] 1) Rastkhāl: a large tributary of the Kur. It flows through a place one stage from Aras, reaches Shamākhī, flows north of it and empties into the sea at a place that is visible (from Shamākhī?). There is a fishing village there. 2) Kalla-khāl: It branches off the Kur after the Rastkhāl, reaches Mūghān, flows east and empties into the sea. 3) Chapa-khāl: It branches off the Kalla-khāl, flows east below Nawshahr, passes north of Maḥmūdābād and empties into the sea.

The Kur has a highly esteemed fish known as *zarqī*. It is not utilized for irrigation anywhere in its course, providing no benefit in that regard.

[2.] Aras: It rises on the summer pasture of Bingöl, flows to the border of Pāsīn and passes underneath the Choban Bridge. At Rawān it is joined by the Zangī River, which rises from Lake Gökçe (Lake Sevan) opposite Bāyezīd Castle in the north. Again near Rawān (Yerevan) it is joined by the Arpa Çayı coming from the Ḳara Ṣu and the Chuḳur Saʿd. After being joined by those streams, the Aras turns east at the foot of Mt. Arghī (Ararat), flows through Eleşkird before Nakhjiwān, passes underneath the Chūlhā Bridge, is joined by the Ganja and Qubān Rivers opposite Mt. Qaraja and the fortress of Qahqaha. It then flows under a bridge, is joined by one branch of the Kur coming from Maḥmūdābād, passes underneath the Jawād Bridge, and plunges down at a place known as Arasbār where a major highway passes below it. Near there it joins the Kur River and becomes like a lake before emptying into the Caspian at the borders of Gushtāsfī. This river is utilized for irrigation throughout its course, thus providing general benefit. Its length is approximately 150 parasangs.

The Dwellers of Rass mentioned in the Koran (50:12) were a people living on this river, according to one tradition.

It is mentioned in the *ʿAjāʾib al-makhlūqāt* that if someone crosses this river on foot with his body submerged from the waist down, and then presses a woman in difficult labor on the back, she will give birth easily. And some books record that if someone suffering from Guinea worm crosses it on foot, he will be cured of that illness.

The poor one (Kātib Çelebi) says: On the way to Tabrīz during the Rawān campaign (1635) we were crossing the Aras River with Sultan Murād Khan (IV). It was the season when rivers are low, and the water at the ford reached to our saddles. A strong trooper was swept away by the current. The felicitous Padishah caught him by the hand and he was barely saved from drowning. But to be saved from Guinea worm while crossing this river on foot is only possible if the afflicted one goes to the next world. Otherwise it cannot be crossed on foot anywhere below its source after it passes the Choban Bridge. Its curative property might only be apparent when crossing the river near its source.

The Aras is a beloved and famous river. It forms the eastern border of the clime of Arminiyya. Ḥāfiẓ praises this river thus [Persian verse]:

> O zephyr breeze, if you pass by the bank of the Aras,
> Kiss the soil of that riverbed and make your breath musky.

Samūr: it rises in the mountains of Georgia. The rivers of Dagestan also flow into it. It flows into the Caspian between Demir Ḳapu and the district of Niyāzābād. It is fast flowing and difficult to ford. In the riverbed there is a soapy clay and razor-sharp stones.

Gūsfand: also called Ḳoyun Ṣuyı ("Sheep River"). It is four stages northeast of Demir Ḳapu. It rises in the mountains of Shamkhāl and empties into the Caspian. This too is difficult to ford and fast flowing, but it can be crossed at many points.

Siwinj: it rises in western Dagestan [M 398] and empties into the Caspian. As its flow is slow, it is hardly visible. It is half a bowshot wide, as recorded by (Meḥmed) ʿĀşıḳ in the *Menāẓirü'l-ʿavālim*.

Barāzrū: it comes from the direction of Ardabīl and flows to the northeast, passes south of the city known as Langarkunān and empties into the sea to its east.

Mountains

The famous mountains of this clime are the following:

Alburz (Mt. Elbrus): it is west of Demir Ḳapu. It is part of a mountain chain that extends from Turkistan towards the Ḥijāz and is said to be more than 1000 parasangs long. For this reason, some people who spout nonsense suppose that it is Mt. Qāf.

The western side of this range is adjacent to Georgia. Those mountains are called Elegzi; then, after Erzurum, Qalīqalā (Cilicia); after reaching Antakya and Maṣṣīṣa, Mt. Likām. At those places it separates Shām (Syria) from Rūm (Anatolia). When it reaches between Ḥimṣ and Damascus, it is called Mt. Lebanon; and when it reaches between Mecca and Medinah, Mt. ʿArj. These [R 153b] are all discussed in their proper place. The eastern part of the Alburz range, adjacent to Arrān and Azerbaijan, is called Mt. Qaytaq; when it reaches the borders of Gīlān, it is called Mt. Daylam and Ṭabaristān; when it reaches the borders of Iraq, it is called Mt. Raqūda and Ṭarfak.

Many peoples live in these mountains. In the north of the Alburz is a people known as Qaytaq. Its southern part is like a (stone) wall dressed with an adze. The width of this mountain is as much as 10 stages. Some say that there is no passage to the other side except through Demir Ḳapu. Others have confirmed the existence of ten such passages, the largest being in the middle of the mountain where another solid wall was constructed known as Bāb al-Lān. The people of Shirwān call these mountains Darband; others call them the mountains of Demir Ḳapu.

The Shirwān (Caspian) Sea

In the *Taqwīm* it is written as the Khazaz Sea; in the *Atlas* as Mare Gaspium. Although it was discussed in summary fashion at the beginning of this book, since it adjoins this clime, some further details here are appropriate.

It is like an oblong lake extending from east to west and is not connected to any other sea. It is as large as the Black Sea. According to the map of the *Atlas*, its western border is located at 85 degrees longitude, its eastern border at 105 degrees longitude, its southern border at 42½ degrees latitude, its northern border at 47½ degrees latitude. So its total length is 20 degrees and its width is 5 degrees.

Such large rivers empty into it from every direction and are simply absorbed, that those who only see the surface of things say that at its bottom there is a passageway through which the water flows to the Black Sea. They have even tried to prove this preposterous claim with the false testimony that a beggar's bowl fell into the Caspian and showed up in the Black Sea. But to no avail. Once the basic principles of natural science are understood, there is no need for such nonsensical claims in order to dispel the fancies born of ignorance. Since we have explained these matters at the beginning of the book, we need not repeat them here.

Shores of the Caspian Sea: To the west, the country of Shirwān, including Demir Ḳapu, Niyāzābād and Baku Castle; to the south, the country of Mūghān, including Band-i Māhī, which is the mouth of the Kur and Aras Rivers at the borders of Qarabāgh, then eastward to the country of Gīlān, including Pūmin, Māsūla, the city of Rasht, Langarūn and Rūd-i Sar; to the east, the country of Ṭabaristān, including Rustamdār, Old Āmul, Māzandarān, [M 399] Ṭāhān and Sar-i Maydān; to the north, the desert and Siyāh Kūh, known as the Qaytaq Mountains.

The people of Khazaz (i.e., Khazars) dwells in these regions. From the mouth of the Murghāb to Hashtdar, which is the mouth of the Atil (Volga), and from Āmul to Samandar, is 7 stages in the desert. From Samandar to Shirwān is 4 stages with few inhabited places.

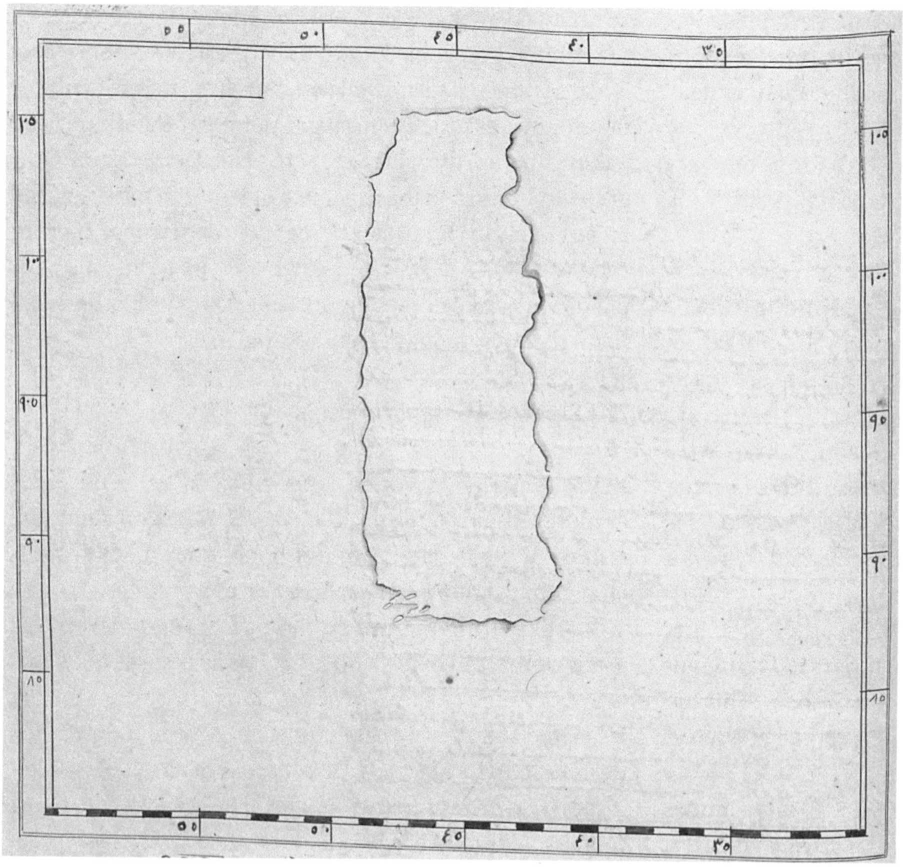

PLATE 45
Map of Caspian, R 154a
TOPKAPI PALACE LIBRARY, REVAN 1624, WITH PERMISSION

This sea has no tides. The water is muddy and dark, not clear like the Red Sea or the Sea of Oman. No white stones can be seen at its bottom, and no commodities like pearls and coral are gotten from it. But there are several species of sea animals, such as big fishes and other strange creatures.

There are six or seven small islands in it. Black and white petroleum can be found on most of them. Only one of them is inhabited. One is like a round mountain of clay with streams running down to the sea. Another has forests and meadows, but so many snakes that it is spoiled and uninhabitable. Another, known as Kūh-i Siyāh ("Black Mountain"), is very big with many streams and trees, but is in ruins. Another, near the western shore, that has grass and fresh water, is used for raising cattle.

Routes of countries of Shirwān, Arrān and the Caspian shores

From Mūqān to Tabriz is 10 stages south; to Bāb al-Abwāb, 2 stages; to the mouth of the Kur River, 16 parasangs northeast > Demir Ḳapu, 21 parasangs.

From Ganja: > Bardaʿ, 2 stages > Shamākhī, 14 parasangs > Tiflis, 43 parasangs.

From Bardaʿ to Warthān is 7 parasangs. [R 154a] [Plate 45]

From Ābaskūn to the vicinity of Khazaz is 300 parasangs > Dihistān, 6 stages.

From Āmul to Samandar is 8 stages; to the border of Barṭās, 20 stages; to the country of Sarīr, 3 stages.

From Āmul to Bulgar is one month's journey through the desert. It can be reached in two months if one goes by sea.

From Bardaʿ: Warthān, 7 parasangs > Baylaqān, 7 parasangs > Īlkhān, 7 parasangs > Barzand, 7 parasangs > Ardabīl, 15 parasangs.

From Bardaʿ: > Barzand, 17 parasangs > Shamākhī and Shirwān, 3 parasangs each, crossing the Kur River.

From Shirwān: > Lāyijān, 2 stages > Darband, 20 parasangs, crossing the Maymūn Bridge.

From Ganja: > Shamkūr, 4 parasangs > the city of Khiyār, 11 parasangs > the fortress of Abarkandmān, 10 parasangs > Tiflis, 12 parasangs > Maras, 12 parasangs > Dūmīsh, 12 parasangs > Gulgūn, 16 parasangs > Dibīl, 16 parasangs.

From Shamākhī to Arash is 3 stages to the northwest; Baku, 2 stages; to Rastkhāl River which branches off the Kur, 1 stage; to the sea, 1 stage to the east; Nawshahr, 1 stage.

From Demir Ḳapu to Ajdarkhān (Astrakhan) is 7 or 8 stages.

From Shamākhī to Mūghān is 1 stage. From Mūghān to Maḥmūdābād and Nawshahr is 1 stage each. From Nawshahr to Qizil Aghach, 2 stages by sea and 3 stages by land. From Qizil Aghach to Ardabīl is 2 stages.

Rulers of Shirwān and Arrān

As is recorded in the *Rawḍ al-miʿṭār*, Azerbaijan was conquered by the commander of the Muslims, Surāqa b. ʿAmr, during the caliphate of ʿUmar. He sent Bakīr b. ʿAbd Allāh to defend the region of Shirwān. Putting ʿAbd al-Raḥmān b. Rabīʿa in the vanguard, he too set out in that direction. At the time Shirwān was ruled by Shahriyār, a descendant of the Persian kings. [M 400] When ʿAbd al-Raḥmān and Bakīr arrived near Shirwān, Shahriyār came under a safe conduct and said: "I am on the frontier with several enemies, who belong to different religions, and I seek protection from you. Let me be a solid rampart, and let the governorship be yours." He submitted and agreed to pay the poll tax. When this was presented to Surāqa, he saw it as reasonable and presented it to Caliph ʿUmar, informing him about it, and he gave permission.

In the Umayyad period, Maslama b. ʿAbd al-Malik conducted religious war in these regions, and the people converted to Islam. Muḥammad b. Yazīd, a descendant of Bahrām Chūbīn, became governor of Shirwān and Demir Ḳapu, governing a range of one month's journey. In 180 (796–797), the people of Khazaz (Khazars) took Demir Ḳapu and 140,000 Muslims fell as martyrs in that battle, an event unprecedented in Islam.

Abū Ḥāmid says that after Maslama took Bāb al-Abwāb and Ṭabaristān, Hishām b. ʿAbd al-Malik went to that region. When he was about to depart, the governor of the country asked him to leave behind a token. Hishām gave his sword. They set it up on a stone like a prayer-niche and put it in an elevated place. For some time, people made pilgrimage to it. [R 154b] The strange thing is that in winter it did not matter what color the pilgrims wore, while in summer they had to wear white; if they wore a different color, there would be huge rainstorms that year, spoiling the fruit and causing dearth. This was their tried and true experience.

In the Abbasid period, these regions were ruled by the governors of either Arminiyya or Azerbaijan. In the Mongol period, the Chingisids took control.

Then the Shirwān sultans emerged. In the *Niẓām al-tawārīkh*, Qāḍī Bayḍāwī records them as descended from Bahrām Chūbīn and Ardashīr Bābakān. In the *Jahān-ārā*, Qāḍī Aḥmad Ghifārī makes them descendants of Nūshirwān, showing their genealogy as follows: Abū al-Muẓaffar Manūchihr b. Kisrān b. Kāwus b. Shahriyār b. Garshāsf b. Afrīdūn b. Farāmurz b. Sālār b. Zayd b. Jawn b. Marzbān b. Hurmuz b. Anūshirwān. After this Manūchihr, twelve of his descendants ruled in Shirwān, in the following order:

- Farrukhzād b. Manūchihr.
- Gushtāsb, after whom the place called Gushtāsbī is named.
- His son, Farāmurz.
- His son, Farrukhzād.
- One of his offspring, Kay Qubād, who was famous for his justice.
- His son, Kāwus, who died in 774 (1373–1374).
- His son, Hūshang, who ruled for ten years. After his death, the sultanate was transferred to Shaikh Ibrāhīm b. Sulṭān Muḥammad, who joined Tīmūr in battles and died in 820 (1417–1418).
- His son, Sulṭān Khalīl, whose name (considered as a chronogram) gives the date of his succession (820/1417–1418). His rule lasted until 867 (1462–1463), when he was succeeded by the Shirwānshāh, whose name also gives the date of his succession. In 906 (1500–1501) he was martyred by Shāh Ismāʿīl's tyranny.
- His son, Bahrām Beg, who ruled for one year.
- His brother Ghāzī Beg, who ruled for one year.
- His brother Shaikh Ibrāhīm, also called Shaikh Shāh, who had good relations with Shāh Ṭahmās and died in 939 (1532–1533).
- His son Sulṭān Khalīl, who was put to death in 945 (1538–1539) when Shāh Ṭahmās conquered Shirwān.

[M 401] With this the Shirwānid dynasty came to an end.

Later, during the reign of Kūr ("Blind") Khudābanda (Safavid shah, reg. 985–996/1578–1588) and until the year 1000 (1591–1592), such Ottoman commanders as Ḳara Muṣṭafā Pasha, Os̱mān Pasha and Ferhād Pasha fought several battles there against the Qizilbash and conquered Shirwān. Then, during the Celali revolts, Shāh ʿAbbās (I, r. 996–1038/1588–1629) found an opportunity and took Rawān, Azerbaijan and Shirwān around the year 1015 (1606–1607). Those regions now belong to that side (i.e., the Safavids).

As for the Atabegs of Ildegiz who governed Arrān, the first of them—Shams al-Dīn Ildegiz—was of Turkish origin, in the service of the Seljuk Sultan Masʿūd (reg. 529–547 (1134–1152)). He gradually became an illustrious emir and, in 540 (1145–1146), was made governor of Arrān. He performed religious war against Georgia and was victorious. He died in 568 (1172–1173). His son and successor, Muḥammad Jahān Pahlawān, took Tabriz and governed for 14 years. After his death, his place was taken by his brother, Qizil Arslān. Five years later he was killed at the hands of his followers and was succeeded by his nephew Abū Bakr. In 594 (1197–1198), he too was put to death and his brother Muẓaffar al-Dīn Özbek succeeded him. In 622

(1225–1226), while he was in the fortress of Alinjaq, Jalāl al-Dīn Khwārazmshāh invaded the country and captured him. With his death, the dynasty came to an end.

Chapter 40: Country of Dagestan, Georgia, etc.

The 40th chapter is on towns, deserts and mountains that remain of Asia north of the Caspian and northwest of Shirwān.

ṬAĠISTĀN (Dagestan): A province recorded as Sarīr al-Lān in the *Taqwīm al-buldān*. [R 155a]

Borders

To the west, the Circassian country; to the south, Georgia; to the east, the Shirwān (Caspian) Sea; to the north, the Khazaz (Khazar) lands.

It encompasses 10 stages in length and breadth. As most of its territory is mountainous, it is called Ṭaġistān (from Turkish *ṭaġ* "mountain"). Its capital is a town and fortress by the name of Quymuq (Kumyk). Here and there in the mountains there are flourishing villages and districts. The inhabitants are mainly Sunnis of the Shāfiʿī rite. They have independent governors amongst them.

In 1048 (1638–1639) Maḥmūd Khan, while going on the Haj, came to Istanbul and met with Sultan Murad Khan (IV). When his son Qorqmaz fell in battle against the Circassians in (10)58 (1648–1649), his brother Chalandar Khan became governor. Today the governance of this region is given over to him.

According to the *Taqwīm*, this region is called Sarīr because one of the Persian kings appointed one of his relatives as governor there and set up a throne (*sarīr*). He gave him complete independence and made him sit on the throne like a padishah so that he would overcome the surrounding peoples.

This country has a mixture of peoples from different nations. Its mountains are contiguous with the "mountain of languages" called Qaytaq.

[Marginal addition by Ebū Bekr b. Behrām ed-Dimaşḳī] Qaytaq: a country the capital of which is the city of Kubachi and the castle of Qarīsh. Kubachi is a large and flourishing and famous town with many religious scholars and pious individuals. The scholars are held in great esteem.

Bamkhū: an important city in the direction of Georgia. It is in the hands of the governors of Dagestan.

Enderay: one of the cities of Dagestan in the direction of Russia.

[M 402] Some places in Dagestan that are toward Shirwān and Ganja are the following.

- Nakhwad: a large village, like a town. Some of the governors of Dagestan are dwelling around here.
- Khajmās: Also a large village, toward Qabā from Shirwān, at the skirt of Dagestan. From Nakhwad one goes to Kichi Dehne and to Dehne, Khajmās, Qabā and Demir Ḳapu.
- Kichi Dehne: a well-known village one stage from Nakhwad on the Ganja road.
- Ulu Dehne: a village away from the Ganja road, two miles from Kichi Dehne. From Ulu-Dehne to Khajmās is 2 stages > Qabā, … stages > Demür Ḳapu, … stage. The road here reaches the shore of the Caspian Sea. One crosses the Samūr River and passes through a ruined place to the seashore.

Among the famous castles and places conquered by Tīmūr, during his seven-year campaign, from the tribes of infidels living (?) north of Mt. Alburz, the castle of Kūlā and the castle of Ṭāwus were in the hands of a tribe who dwelt in the Alburz. Especially the fortress of Ṭāwus was situated on the third level of a high mountain, which was higher than an arrow's shot. Although this fortress seemed impregnable, he enlisted the Mekrit tribe, who were part of his army and whose specialty was to operate in difficult terrain. They climbed to the first level using ladders, then pulled up the ladders and climbed one more level, and in this fashion climbed to the third level. The fortress was underneath this overhang (?—*kemer*). Then some of them descended to the fortress with the ladders, others swung down on tentropes with one end tied around their waists and the other fixed on top of the mountain. They stood their ground fighting with swords and arrows and conquered the fortress willy-nilly. The defenders of the castle killed many of them with stones and axes, but another regiment joined the fray and in this way it was conquered.

There is much in this region, because wild bees are plentiful in the mountains. So much honey could be found in those mountains that Tīmūr's army could be fed with it.

Appendix (by İbrāhīm Müteferriḳa)

Ṭarkhū: an important city in the direction of Georgia. It is in the hands of the governor of Dagestan.

Mawlānā 'Abd al-Raḥīm al-Turkashī reports: Inside the mosque of one of the Quymuq (Kumyk) villages is a pond whose water gushes out twice a year, once when the sun is in Aries, and once when it enters Libra. The strangest thing about it is that no one can ever see when the water rises. Many people waited to see it gush out, but they were not able to witness it.

Nearby, in the village of Qabā there is a pond whose water is lukewarm in spring and autumn and hot in winter, while in summer it becomes so cold that it freezes and the people of that region use the ice for fever and illnesses with burning heat.

Enderay: also a city of Dagestan. It is in the direction of Russia.

Terek: a fortress belonging to Muscovy beyond Bāb al-Abwāb on the Astrakhan road near the Caspian coast.

Ūj: a fortress on the coast, on a spit of land, between Ṭarkhū and Terek. It was built by the Tsar of Muscovy in 1135 (1722–1723).

Mt. Qaytaq: an extensive mountain range, covering 20 stages in both length and breadth. It has many streams [M 403] and forests. It extends southwest from Demir Ḳapu along the Caspian. There it is called Mt. Alburz, as mentioned above. The mountains recorded in the *Atlas* under the name Alpes are close to it etymologically. This range is called the "mountain of languages" because peoples and nations speaking various languages dwell here. One of their tribes is called Qaytaq, whose settlements are north of this mountain. The southern side of the mountain, which is mainly in the shape of tall cut rocks, is populated by a tribe called Lagzī (Lezgi).

Appendix—From the book of (Ebū) Bekr Efendi

In this region is a settlement and walled town of 40,000 houses known as Ṭabasarān. It has its own governor. There is a lake here with the grave of a saint on its shore. The water of this lake sometimes flows white as milk, then freezes and looks like clotted cream. In a nearby cave there is a sword, said by surrounding populace to be the sword of Maslama, who conquered the region for Islam. In spring they bring great donations and alms to that cave, paying a visitation to the sword and making a new sheath for it. On their way to the cave they also bring one *ṣāʿ* of barley, scattering it on the ground as they go. After putting the sword into the new sheath they retrace their steps and, strange to say, not a single seed of barley can be found on the ground where they scattered it.

Another appendix—On the country of Circassia

This province extends from the Taman Strait somewhat removed from the Black Sea coast and ends at Dagestan. The Abaza (Abkhaz) province to its south extends to the Black Sea coast and ends at the Makrīl (Mingrel) province. The Circassians have eleven *qabaq*s, meaning "thousand"; some of them are Muslim, some pagan.

This people have their own customs and rites. However, some of their customs agree with those of the Jews. Indeed, three of the Israelite tribes are said to have come to this region and settled here, and the Circassians and Abkhaz are said to have disseminated from those tribes.

The *qabaq*s of the Circassians are, first of all, Tamān, Tamrak, Jaghākā—these are pagan; then Jānā-kabīr, Jānā-ṣaghīr, Būndūq, Ḥātūqāy, Būluq-khāy, Bastanī, Qabarṭāy—these are Muslim.

This people have wheaten complexion (i.e., darkish), black eyebrows, moles (or beauty marks). Their fairness and beauty are apparent. They make intelligent slave-boys. They steal one another and sell one another to slave traders. Once the parents have sold their son, even if they see him at the slave trader's shop they cannot get him back. This is the tradition among them. The annual tax they pay to their beys is in captives: they steal one another and give the captives to their beys, who in turn sells them to the slave merchants.

The ascetics of this people are called *dāqūq*; they never eat chicken. They worship a tree known as *qūdash*. One of their tribes is called Akhūkh, who dwell on Mt. Akhūkh; another is called Mahasnī, who dwell near the Azov region.

Appendix: On the Abaza (Abkhaz) province

They live on the Black Sea coast, south of the Circassians. This people also have their own rites and traditions. Their border begins from the port of Ṣūcha (Sochi) and goes as far as the Qūdar River. That region is called Sukhūm. The Abaza lords are each named after a harbor—e.g., the lord of Achūyā; the lord of Soghuq Ṣu, where there is a very important church known as Māllūja; the lord of Pari-Jamtā; the lord of Kajī; also the lords of Qāghir-Darbandī, Būchandalar, Ariqlu-kachīlar, Qāmish, Ūshghān, Māmāy, Qaya-subashlari. [M 404]

Sukhūm (Sukhumi): It is on the eastern shore of the Black Sea. The people of this town are Muslims. It is located at the beginning of the border of Abaza (Abkhazia), three stages from Trabzon. To the north is the province of Tabantā (Tayantā?) and the tribes of Kāwbāy and Alan.

Anjās: Originally the name of a tribe, in our times known in the Islamic countries as Abaza, one of those who dwell in Qara-azhdahān (Astrakhan). This city is on the Black Sea coast, on a mountain that is above a bay. It is northeast of Sukhūm. Near Abaza is the tribe of Lāzkī (Lezgi), and near them is the tribe of Lawand, who have the fortress of Karam, and one of their important fortresses is that of Zagam.

Stages

From Kefe (Kaffa) to Demir Ḳapu: Kefe > the Strait of Kerch, 20 miles; here one crosses to Taman > Taman castle > the city of Kuban, 5 stages > the province of Makrīl (Mingrelia)—among this tribe, a measure of cloth and an old shirt are worth more than a purse of *akçes*; they are great thieves > the Heyhāt Plain, which stretches a distance of 20 days along the seashore, after which one journeys 5 more days drinking from ponds and streams > Beş Depe > the Terek River, 5 stages—on the banks of this river are forests with graceful trees reaching up to heaven > the place called Qabarṭāy, where the Terek and Aḳ Ṣu Rivers can be crossed by a bridge > Kanlu Sevinç River > Ḳoyun River > the province of Shamkhāl > İnceṣu, 4 stages > Demür Ḳapu, 1 stage.

Supplement by the Publisher (İbrahīm Müteferriḳa)

Some time ago, the people of Dagestan invaded Shirwān, took its capital, Shamākhī, and incited a rebellion against the Persian shahs. From the east appeared the Afghans, who reached as far as Isfahan, the capital of the Safavid shahs, and conquered it (in 1722). The building of the Safavid state was shaken by the cold wind of discord and was on the point of collapse. Meanwhile, the Tsar of Muscovy (i.e., Peter the Great of Russia) had been for some time on the lookout for an opportunity, and the numerous peoples and various tribes who dwelt on Mt. Alburz and its inner and outer (i.e., northern and southern?) slopes, and who had never been subject to anyone or conquered by any state, submitted to him.

Now for some time the Ottomans paid no attention to the distinction and division of borders in that region. Dagestan and Circassia were open areas with unclear boundaries, and that Tsar conceived the desire to conquer those lands. He especially intended to lay his hands on the Georgian territories and somehow or other to turn the hearts of the Georgians in his direction. He set out towards Shirwān and by a stratagem took Darband (in 1722). The terrible news of this event was a wakeup call to the Ottomans. The foresightful statesmen and far-seeing ministers realized that it was an utmost state necessity to block such a potentially disruptive move. Thus, hasty measures were taken by the Ottomans to seize Tiflis (Tbilisi), the capital of Georgia, [M 405] and appoint a governor to Shamākhī, the capital of Shirwān. The favor of the people of these regions thus gained, by good administration they were attracted to submit. Thus those provinces were adjoined to the Sublime State and the Muscovite Tsar's ill intentions in this region were curtailed.

After a survey of the conditions of the Black Sea, it was thought appropriate as a precaution to build forts at certain places on the coast. In particular, some experts reported that a fort should be built at Fāsh and road opened from there to Tiflis; then it would be easy to bring equipment to Fāsh with the fleet and transport it to Tiflis and to send reinforcements to Shirwān and Dagestan when necessary.

In 1135 (1722–1723) Hunbaracı Meḥmed Agha, one of the architects of the Sublime State, was appointed to investigate the region. The substance of the reports that he sent to the Porte is as follows:

The fortress of Sukhūm on the Abaza coast of the Black Sea, on territory separated from Georgian Mingrelia, is now under the control of Yorgi, the nephew of Rustam Bey. It has been examined and measured and a plan has been drawn up.

The wall of the fortress is 1210 cubits in circumference. On the sea side, 250 cubits of the wall have collapsed and are now under water. On the land side, 200 cubits of the wall are extant, but the moat is cracked, and roughly 760 cubits are level with the ground with mountain trees growing above it.

The western side of the fortress is a rushy swamp. On the southern side is a sweet-water lake. If it were dredged (?) and its mouth opened (i.e., a passage to the sea dug out?), frigates could winter there. The depth of the lake away from the shore is 1 fathom.

The fortress is exposed to the sea—there is no harbor—and the southeasterly, southerly and southwesterly winds buffet it. Ships can anchor there only in summer. It is not a harbor for the imperial fleet. Aside from those three winds, the Abaza coast is altogether safe.

There is a stream coming to the fortress from a quarter-hour's distance. It might be better if its mouth were blocked and it were channeled into the lake near the fortress.

The fortress is surrounded by level ground. Three quarters of an hour to the north are mountains where wood and timber are everywhere available. Other building materials—lime, stone and sand—can be found in the forts of Arslan Giray and Anapaya (Anapa), three hours from Sukhūm. The materials are drawn to the coast by carts, then transported to Sukhūm by ship.

In Abaza there are no carpenters, day laborers or carts at all. Axmen can be found in the Gönye sanjak of Trabzon province, while stonecutters and lime-workers can be found in the Hemnişin (Hemşin) district. When the castle of Anḳara (?) was being built not far from Sukhūm, Bīcān Dādyān, the khan of Mingrelia came and provided carts.

Description of Ṣūjaq (Sucuk, Sochiq, i.e. Sochi): the harbors of Sochi and Gelincik on the Black Sea were examined and measured and a plan has been drawn up as required.

The mouth of Sochi harbor is 13 miles to the end on the Crimean side and 19 miles to the end on the Circassian side. The strait of the harbor is 5 miles wide. At the cape on the Crimean side toward the southeast are 2 miles of shallows. The bottom of the harbor is pure sand.

In the mountains on all four sides are freely accessible fresh-water springs, [M 406] as well as timber. The soil and the climate are fine. There is no trace of its ancient castle. Nearby there is a commanding height. For five miles from the mouth of the harbor on the Crimean side there is level ground with no high ground above it; it is suitable for building.

This harbor has no match on the Black Sea. There is 9 miles between it and Gelincik. The mouth of Gelincik harbor is 2 miles wide, its far end is 5 miles. A small fleet can winter in it.

GEORGIA: A province (*vilāyet*) in the mountains, between the Caspian Sea and the Black Sea, 20 stages wide. Borders: in the east, Demir Ḳapu and Shirwān; in the south, the provinces (*eyālet*) of Çıldır, Kars and Erzurum; in the west, the Black Sea; in the north, the provinces (*vilāyet*) of Abaza and Dagestan.

Georgia is divided into several regions, each of which has its independent governor. All the inhabitants belong to the Georgian people and are of the Christian faith. The territories adjacent to the lands of Islam have a mixed population of various tribes known as Chinchāwāt, Achiq Bash, Dādyān, Makrīl (Mingrel), etc. All of them are called Georgians and each one has its own governor and region. Among them the Makrīl is a vile people living on the Black Sea coast. The Achiq Bash and Dādyān are considered pure Georgians.

Zagam: a region in the vicinity of Tiflis. Its former governor was Ṭahmūrath. Its current governor is Rustam, the greatest khan (i.e., Persian provincial governor) of Georgia.

Achiq Bash: a region in the middle (of Georgia), surrounded by Zagam, Demir Ḳapu, Dagestan and Dādyān.

Dādyān: a region near the Black Sea coast, bordering on Abaza. Its governor is in constant war with Abaza. On one side it reaches the Fāsha River.

Makrīl (Mingrelia): a district on the Black Sea coast where the Fāsha (Phasis) River empties into the sea. Its borders are adjacent to Gönye.

These provinces contain many fortresses and districts that are not detailed in the books of geography. I record here the gist of what is found in the chronicles.

When Sultan Süleymān Khan set out with Alqāṣ Mirza on an eastern campaign in 955 (1548–1549), he sent vizier Aḥmed Pasha to Georgia. He marched there with the army of Islam and conquered the following fortresses:

- Brākān: a strong fortress whose rampart reaches to heaven. Since its governor commanded a great army, he took the field but was defeated and his fortress [R 155b] was conquered.
- Kūrmak: a solid fortress.
- Pnāk: also a strong fortress. It is a pilgrimage place of the Georgians.
- Brnāk, Karmak, Ṣmāghār, Akhā: fortresses that were taken at that time.
- Njākh and Imrakhor: two fortresses near Tortum.
- Tortum: a small fortress 4 stages from Erzurum. It was taken in the Shaʿbān month of that year.
- Aqcha-qalʿa: a solid fortress in that vicinity. Built on an elevated rock, it is the key to Georgia. It was taken in the same year.
- Ashkṣūr, Kāmkhīṣ, Nakbar, Anzar: fortresses difficult of access. They were given up by their governors.
- Pertekrek: one of the great fortresses of Georgia. It was conquered at that time.
- Yavana Valley: A flourishing province of Georgia. A sanjak bey was appointed here; also in Tortum and Kāmkhīṣ.
- Fortress of Nūlāh: it is in an inaccessible place. It too took great effort to conquer.
- Riyāsta: [M 407] a place in the Alburz that has very steep roads.
- Masākin-i Qirim Shams: a district with several fortresses. Its roads are so steep that one cannot travel on them either on horse or on foot; at many places one must crawl up and tumble down. Tīmūr himself did so.
- Fortresses of Nākū, Mīkā and Direklü: they were conquered and razed to the ground.
- Ashkūja: a fortress whose inhabitants were pagans. It was conquered.
- Fortress of Narkas: it was conquered and razed to the ground. The inhabitants withdrew into the caves and fissures of some high mountain passes and built forts level with the caves. Battles were fought from the moats and those in the caves were defeated and all the forts conquered.

Guril: a country located in the northwest of Georgia. It has its own governor. In 997 (1588–1589) Ghāzī Giray Khan marched there and raided the following fortresses and towns: Brāslū; Vindosa; Keramine—this was the fortress of Pūpe; Būrleske, a voivodeship; Kenarpej, a duchy; Elwāw, the capital of Guril. Apart from these, the following cities were raided: Blachekū, Bsār, Bīqū, Ūlānū, Khalfek, Senpāū, Blāwche, Mejyor (?), Qusṭanṭīn, Kerāslū, Bāzālī, Kūpīn, Yānpūl, Shūnbār, Kuşkerman, Izbārāsh, Shlūznīche, Yāshlū, Krāmenz, Ḥūzhāūtse, Mnāchīn, Dlāchb/trche (?), Būlāskīrū, Chūzbūsūrū, Shrāwke, Hūrāwek, Ṣāntāū, Zenkū, Qāmets, Isqale, Trenpūl, Qālnūske, Kūlense, Būdyūrāūj, Bāzlāūdse, Snātīn, Ḥālīj, Tūrchīn, Rḥānīn, Berzen, Qāmenke, Zlūchū, Zlūske, Klī-nān, Būrske, Bāwārāū, Shāhūrde, Lūqātīn. Nothing is known about them.

Mt. Legzi: a mountain range extending east of the Black Sea and west of Mt. Alburz. It is called the Georgian Mountains. There are many nations living there, who are said to speak 72 languages—this is an expression of the multitude of various peoples. These mountains adjoin the mountains of Erzurum.

No more information has been obtained on Georgia.

Appendix: Some passages have here been included from Ebū Bekr b. Behrām ed-Dimaşḳī, the translator of the *Geography*.

Chapter on the province (*eyālet*) of Kars. Borders: To the southeast, Rawān and Çıldır; to the north, Akhiskha; to the west, Erzurum. This province consists of 6 sanjaks: Kars, the capital; Ardahan; Kūchak; Khūjawān; Zārūshān; Ḳağızman. Also Kachwān, which is an appendix to Ḳağızman. This province is made up of Georgians. After the Muslim conquest the Georgians reconquered it. Later, Tīmūr destroyed its strongest castle, which is Kars. After the Ottoman conquest of this region, Ḳara Muṣṭafā Pasha came in 988 (1580–1581) and revived it, providing it with a city wall, mosque, public bath and moat. Since olden times it has had a number of mosques and blessed tombs, including that of Shaikh Abū'l-Ḥasan al-Kharaqānī. At one point, when Tīmūr Leng was laying siege to Kars and was unable to take it, it was in the hands of a certain Fīrūz Bakht. Finally Tīmūr took it on terms of safe conduct for the defenders and destroyed it. Today it has a special garrison. It is located between two mountains. The Aras River flows nearby.

Chapter on Ardahan. It is located six stages northeast of Erzurum. It has a small city wall and fortress. It is surrounded by oak forests. The trees in Ardahan are mainly sloe; the berries are not big but have a special taste.

Chapter on Qāghizmān. [M 408] It is located at the foot of a mountain between Kars and Pāsīn, with another mountain opposite. Between these two mountains is the Qāghizmān ravine. The Aras River flows in the valley between the two mountains near Qāghizmān and the Choban Bridge crosses it west of the city.

Near Qāghizmān, at a place called Üç Kilise ("Three Churches"), is a well-known Armenian church with many monks. Once a year in spring they gather different kinds of flowers, which they put in a large cauldron filled with water. They cover the mouth of the cauldron with rugs and for forty days they continuously recite incantations upon the flowers. During those forty days the water boils from the heat of the church and becomes like fermented liquor. When it clarifies, they take the flower-water and sell it for one gold piece per *mithqāl*. It is rubbed on the face, eyes, mouth and nose of the dead. They believe that angels have come down from heaven and caused this water to boil. The only other place such flower-water is produced is the Akhtamar Church. These men are of two sects, one belonging to the Akhtamar Church, the other to Üç Kilise.

The highest mountain of this region is called Ṣoğanlu Yaylası. To the southeast of it is Mt. Ağrı (Mt. Ararat). Below it is a great forest. No one can climb to the peak of this mountain; it is so high that it is above the clouds.

Fords between Kars and Rawān: the Shaḥna Ford on the road to the village of Qara Khān; the Qalʿa Ford in front of Shūragil; the Qūshāwnīk Ford; the Gerekmez Ford; the Ağzı Açıḳ Ford; the Ḳañlu Ford; the Bekren Ford from Qarabāgh. All of these fords are along the Qara Ṣu. Also in that region are the Mālīn, Aqcha-qalʿa, Sürmelü, Merd, and other (fords?).

Chapter on Akhiskha, also called Akhisqa. It is within the borders of Georgia. In the past it was the Çıldır province. As Çıldır was destroyed in wars, it was added to Akhisqā and the two were made into an independent province.

Borders: to the east, Kars; to the south, Çıldır; to the west, the mountains of Georgia; to the north, Tiflis. Its sanjaks are as follows: Acara; Ardanuç; Greater Ardahan; Azadhan; Küçük Söke—it belongs to Oltu; Petekrek, a hereditary grant (*ocaḳlıḳ*); Penk; Postḫū; Tavüsker; Çıldır; Ḥacrek; Khartūs; Şavşad, a hereditary grant; Göle; Livāne, a hereditary grant, Mākhjīl, Nıṣf-ı Livāne, a hereditary grant; Mamravān; Akhalkalak; Terālet. This province includes 656 *ḳılıç* (timars) of which 97 are *zeʿāmet* (timar providing an annual income of over 20,000 *aḳçe*) and 559 are certified and uncertified timars. Together with their armed retainers they provide 1700 troops.

Description of Akhiskha (Akhaltsikhe): It is now the seat of the province. It was ruled by one of the governors of the Georgian tribes, Manūchahr son of Ghaza, whose possession of it was strengthened by way of inheritance. In 988 (1580–1581), when Vizier Muṣṭafā Pasha was ordered by Sultan Murād Khan (III) to conquer Shirwān, some of the nobles in this region who had converted to Islam were given this province as a hereditary grant. Later it was taken by the Qizilbash. Then in 1045 (1635–1636) Sultan Murād Khan (IV) son of Aḥmed Khan gave a number of troops to Vizier Kenān Pasha, who took the fortress by capitulation after a 20-day siege. He also conquered six other castles in the vicinity, [M 409] all of which were given as a hereditary grant to Sefer Pasha. They are now governed by that dynasty who have erected Friday mosques, public baths, madrasas and inns. In the vicinity are several ancient tombs said by the people of the province to belong to the Muslims of old.

The city of Akhiskha is located east of Kars, between Lori and Gori. Lori is between Kars and Akhiskha, somewhat to the east toward the interior. A large river flows north of Gori, reaches Tiflis and then the plain of Lori, whence it flows on the south edge of Akhiskha. Just to the southeast it touches the north of Lori Castle and goes on to Tiflis. To the south of Akhiskha and this river is the region of BRRQYA, which is located between this river and the TYHWY River.

Azghūr: a castle atop a lofty cliff on one bank of the Badra Ravine, at the mouth of the strait overlooking the Kur River. It belongs to the Ottomans, while Badra castle on the other side of the strait belongs to the Qizilbash.

Akhalkalak (Akhalkalaki): it means White Castle in the Georgian language. It is northeast of Erzurum and southwest of Gori. Between Gori and Akhalkalak are 5 stages. Akhalkalak is a fortress atop a great hill in a broad plain between two valleys—in one of these valleys flows a small stream—and behind it on the road to Tiflis. It used to be the seat of the province; as it was a frontier, it was destroyed. The climate is cold. Its produce is wheat and fruit.

Ardanuç: (a castle) carved out of the stone, atop a lofty stone mountain several minarets high. It has a single road, by which one can ascend half way with a loaded animal, after which one can only proceed on foot, and that with a thousand difficulties. Overlooking the castle is another mountain. Between the two mountains is a large valley with a low tower on one side. The natural wall on the side of the mountain contains a cistern, and in the middle of the castle a large cistern has been carved out of the rock. Thus, no harm can befall (the defenders). The town quarters and marketplace are located at the foot of the mountain, where there are a Friday mosque, madrasa and public bath built by Sefer Pasha.

Final remarks: The remaining inhabitants of the Northern Steppe

Baqrāj: as mentioned in the *Haft iqlīm*, it is originally a Turkic tribe. The men have no beard or moustache. Their land is one month's journey in extent. They believe in the divinity of ʿAlī—God forbid! They have a ruler descended from Yaḥyā b. Zayd, whose badge is a long beard. He taxes every item belonging to his subjects. There are no cattle in that land.

Kīmāk (Kimek): another Turkic people, also called Keymās and Kemyās. Their land is more than one month's journey in extent. Most of them wear animal skins. They are experts at the science of the rain-stone. Gold and silver are plentiful, and diamonds can also be found. These people have no ruler; whoever among them is over eighty years old becomes a spiritual leader and is worshiped. They do not eat meat.

In this land is a stone that is sure to produce rain if put into water. And there is a pit with water only a span deep, yet it never runs out even if a whole army drinks of it.

Tagharghar (*recte* Tughuzghuz, Toquz Oghuz): These are also a Turkic tribe. Their territory adjoins that of Khīrkhīr, Kīmāk, Kharluj and Bulgar.

Kharluj (*recte* Kharlukh, Qarluq): the Qālāj (Khalaj) tribe, who live midway between Tagharghar and Khīrkhīr in the north, behind the Saqlāb.

[M 410] Khīrkhīr (*recte* Khirkhiz, Qirgiz): a people. Their territory is between the Tagharghar and Kīmāk on the one hand and the Ocean and the land of the Kharluj on the other.

Saqlāb or Ṣaqlāb: a Turkic tribe. Their territory is a two months' journey in length and breadth.

Aside from these, the tribe of Rūs is usually recorded at this point. However, as their habitat is mainly within the boundaries of Europe, we have postponed their description until after the chapter on Muscovy. Since some of these tribes are recorded *en passant* in the chapter on Turkistan or the chapter on Cathay, there is no harm in reiterating somewhat.

To this point, the discussion has been about the northeastern part of Asia. Those regions have been dealt with in detail in several chapters. From now on we will speak about the borderlands of the Ottoman Empire adjoining the continents of Europe and Africa in the west. Then we will deal with the rest of the continent, devoting separate chapters to the provinces (*eyālet*) in those regions, viz.,

Van, Çıldır, Erzurum, Diyarbekir, Aleppo, Shām (Syria), Baghdad, Basra, Şehr-i Zūl, Raqqa, Marʿash, Sivas, Anaṭolı, Trabzon and Kars. [R 156a] [SPACE FOR MAP]

Chapter 41: Clime of Arminiyya

The 41st chapter is on Arminiyya, one of the conventional climes.

Borders

There is disagreement about the borders of this clime. While the *Taqwīm* renders the name as Irminiya, the *Lubāb* gives it as Arminiyya, then alligns its border with Azerbaijan, saying that Azerbaigan (sic), Arrān and Arminiyya are three large climes that interpenetrate one another. Because it would be difficult to depict and describe them separately, the geographers have treated all three together.

Sharīshī, the commentator of the *Maqāmāt*, says that Arminiyya has three parts: 1) Dabīl, Qālīqalā, Khalāṭ (Ahḷāṭ) and Shamīsāṭ; 2) Bardaʿa and Baylaqān; 3) Nakhjiwān.

According to Ḥamd Allāh, the province of Arman has two parts: Lesser and Greater. Lesser Arminiyya is not within Iran; Greater Arminiyya is east of it; north of it is Rūm; south of it is Syria; west of it is the Mediterranean. Its main regions are Sīs, Tarsus and Cyprus. Greater Arminiyya is within the borders of Iran and is known as the *tūmān* of Ahḷāṭ. It borders on Lesser Arminiyya, Rūm, Diyarbekir, Kurdistan, Azerbaijan and Arrān. It runs lengthwise from Arzan al-Rūm (Erzurum) to Salmās and in width from Arrān to the end of the province of Khalāṭ. Its capital is Khalāṭ. Its state dues used to be nearly 200 tomans, later reduced to 39.

The poor one (Kātib Çelebi) says: in this case, Greater Arminiyya presently consists of the provinces (*eyālet*) of Van, Kars and Erzurum; while Lesser Arminiyya consists of the provinces of Adana and Marʿaş.

According to some chronicles, in the past the capital of Arminiyya was Ahḷāṭ. When their dynasty came to an end, the people became tax-paying subjects and scattered to Tarsus and Maṣṣīṣa. At present their chief city is Sīs. [R 156b] Their ancestors derived from Arman b. Layṭī b. Yūnān. The tribes of Anjār, Karj, Sabāsta, Sādarya and Ṣabbār are from them. Later all of them converted to Christianity. Now the remnants of that people are all Christians and non-Muslim subjects.

In the *Taqwīm al-buldān* the following are given as cities of Arminiyya: Elbistan, Adana, Erciş, Azerbaijan (i.e., Tabriz), Bitlis, Bardaʿa, Baylaqān, Tiflis, [M 411] Ahḷāṭ, Dabīl, Dwīn, Sulṭāniyya, Sīs, Tarsus, Malatya, Muş, Van, Vesṭān, Muş (sic), Arzan al-Rūm (Erzurum), Malāzcird. As these cities are scattered throughout this province, we do not follow here al-Malik al-Mu'ayyad (the author of *Taqwīm al-buldān*), but are content with discussing the province of Van which, according to the author of the *Nuzha*, consists of Greater Arminiyya. Actually we have not followed either of those scholars, since we devote separate chapters to Marʿaş—called Lesser Arminiyya by the author of the *Nuzha*—and to Erzurum and Kars which are in Greater Arminiyya. The terminological conventions of our own day have obliged us to abbreviate in this fashion.

PROVINCE (*EYĀLET*) OF VAN

A frontier province (*memleket*) of the Ottoman Empire on the east. Borders: to the east, the country (*ülke*) of Tabriz, known as Azerbaijan; in the south, the districts (*nāhiye*) of Kurdistan and Sulṭāniyya; in the north, the provinces of Çıldır and Kars; in the west, the province of Diyarbekir.

Its sanjaks number 13, plus one governorship (ḥükūmet) administered by its governor as private property. They are as follows: 1) Van, the pasha sanjak and seat of authority; 2) Adilcevaz; 3) Erciş; 4) Muş; 5) Bargiri; 6) Kārkār; 7) Kesani; 8) Esbāberd (cf. Esyāber below); 9) Aġakis; 10) Ekrād-ı Benī Ḳoṭur; 11) Bāyezīd castle; 12) Bardaʿ; 13) Ovacık. This is the cadastre of ʿAyn ʿAlī Efendi. However, in his *Ṭabaqāt*, Ḳoca Nişāncı enumerates them as follows: Vesṭān, Ḥīzān, Ḥakkārī, Müküs, Sīrvī (cf. Şīrvī below), Altak, Selmās, Dunbuli, Ustūn. In this matter changes occur and disagreement arises with the passage of time. As it is a matter of convention, the division of countries is not fixed in every age and breaches are bound to occur in any division. The same is true for the divisions of the other provinces and the borders of sanjaks and regions (*ülke*). One must not expect a definitive judgment in this matter.

Now we will give the regions as they are recorded in the books of geography and describe as far as possible the sanjaks of this province.

Van: a famous town and strong fortress on a flat plain on the eastern shore of Lake Erciş (Lake Van), located at 83½ degrees longitude and 37 degrees latitude. In the year 940 (1533–1534) Sultan Süleymān Khan set out on the campaign against the two Iraqs, and while he was in Aleppo, Grand Vizier İbrāhīm Pasha took Van from the shah of Persia and its people submitted. The chronogram of the conquest is: *Aldı ḥiṣār-ı Van'ı Süleymān Şāhımız* ("Our Shah Süleymān took the fortress of Van"). The fortress was rebuilt and Van was made a separate province.

There are an inner and an outer fortress. The inner fortress is built upon high dressed stones on the western side of the city, with the city and the suburb on one side and the outside of this strong fort on the other. The eastern side, which is that of the city, is built of dressed stones like the fortress. As its external side is a slope descending to the plain, strong two-storeyed towers and walls have been built here and there.

The outer fortress, on the flat plain, is surrounded by a stone wall, as is customary. It has three gates. To the east is the Tabriz Gate, to the south is the Port Gate. To the southeast are orchards and gardens.

Both the city and the fortress are removed somewhat north of the lake, but there is a road from the fortress to the lake, and at times of siege [R 157a] boats still operate. All the houses and town quarters, the palace of the *beylerbey*, the marketplace, the public bath and the Friday mosque are within the city walls. Ḫüsrev Pasha built a fine Friday mosque, a madrasa and a shrine (*türbe*), assigning pious endowments for them. This building complex was completed and worship was performed in Rajab 975 (January–February 1568). Rüstem Pasha also built a public bath here in 958 (1548–1549). [M 412]

The castle warden, garrison troops and janissary guards are stationed in the inner fortress, while 1500 soldiers with their commander as well as the *beylerbey* and his attendants are in the outer fortress. Ships come to its port from Erciş, Aḫlāṭ and Adilcevaz. The surrounding mountains are half a parasang away.

[Marginal addition by Ebū Bekr ed-Dimaşḳī, R 156b] When Tīmūr took the fortress of Van—commemorated in the chronogram *Kaywān bigirift* ("Saturn conquered it")—he ordered one of his emirs to demolish it. After much effort, it proved impossible to tear it down completely. So they were content with destroying some recently constructed towers and ramparts, which did not do the slightest damage to the castle. From this one gets an idea how solidly built Van castle is.

Bargiri: a small town and sanjak north of Van and 8 parasangs east of Erciş. Its fortress was in ruined condition; Sultan Süleymān had it restored. Between here and Erciş is the village of Haydarbağı. There is a proverb related to it: "He goes for six months and reaches Haydarbağı."

Arjīsh (Erciş): a fortress and town and sanjak two stages northwest of Van on the northern shore of the lake. It is situated on a plain and surrounded by gardens and walnut trees. Ḥamd Allāh says that its castle was built by the vizier Khwāja Tāj al-Dīn ʿAlīshāh. Its products are grain crops and cotton. The state dues were 74,000 dinars. It has a strong fortress and a garrison.

Lake Erciş (Lake Van): an oblong-shaped lake extending from west to east and a bit to the south. From most places (on its shores) the other side is visible. It is 60 parasangs in circumference. It tastes somewhat bitter and briny. Once a year a kind of fish called *tırīḥ* appears, which is delicious and invigorating, and is exported abroad. Van is on the eastern shore of this lake. After Van, following the shore to the north, one reaches first Band-i Māhī, then Erciş, Adilcevaz and Aḫlāṭ.

Akhtamar: a fortress on a small island near the southern shore of Lake Erciş. In the past, the lake was named after this fortress. It is an arrow's shot distance to Vesṭān and three or four hours from Van.

Germābe: a hot spring near Erciş at a place called Diyādīn. The water gushes out from around the spring like a jet, with a force that would turn one water mill. There is another hot spring that boils up like a cauldron and flows on. Again, nearby, a delightful stream emerges like the Water of Life.

Adilcevaz: A strong fortress and town and sanjak on a hill of white earth on the lakeshore, two stages west of Erciş. The fortress slopes up the hill from the shore. Several of its turrets are now under water. The town is built on an elevation facing the lake and is completely surrounded by a wall. The lake is to the south of the fortress. I have seen in some compilations that the name Adilcevaz is a corruption of *Hāt al-jawz* ("Bring the walnuts!").

Aḫlāṭ (Ahlat): In the *Taqwīm* it is written as Akhilāṭ and the form with initial A is given (as opposed to Khalāṭ). Today it is pronounced with initial A. It is a town on a plain one stage west of Adilcevaz, with orchards and gardens, a moderate climate and abundant fruit. It has a fortress on a hill.

According to Mīr Sharaf (Sharaf Khān Badlīsī, author of *Sharaf-nāma*), this city is of ancient construction. At times the Armenian kings made it their capital. [M 413] During the time of Anūshirwān, his paternal uncle Jāmāsb lived here, governing his province.

The climate is quite pleasant. There are many orchards and the fruit is juicy. The apricots and apples are especially fine. The Aḫlāṭ apple, one of which weighs 100 dirhams, is famous in Azerbaijan.

In the past, Aḫlāṭ fell to ruin during periods of interregnum. First, in 626 (1229), Sultan Jalāl al-Dīn Khwārazmshāh took the city from the Seljuks by force of arms and laid it waste. Then it was destroyed by the Mongol army. In 644 (1246–1247) most of its buildings collapsed as a result of a huge earthquake. In 955 (1548–1549) it was taken by Shah Ṭahmās who razed the fortress to the ground. Sultan Süleymān abandoned the old city and built a fortification on the lakeshore; however, while the old fort is in ruins, the new one was has not been built up that much.

Streams flowing nearby empty into the lake. The lake is somewhat far from the town.

In the past, several eminent people came from this city, including the following:

- Muḥammad b. Mulkdād, the author of the *Talkhīṣ-i jāmiʿ*, which is a received text in jurisprudence.
- Sayyid Ḥusayn, a chief authority in the exoteric and esoteric sciences. With regard to the *Jifr-i jāmiʿ* ("The Encompassing Onomancy") in particular, he was one of the celebrities of the age. Guided by that science, he determined that Chingis Khan would arise, and so before that occurred he migrated together with some 12,000 households of his followers and dependents and settled in Cairo. His tomb is still there, in the quarter known as the quarter of the Aḫlāṭīs.
- Mawlānā Muḥyi'l-dīn, who was brought by Khwāja Naṣīr (al-dīn Ṭūsī) to the observatory in Marāgha.

Mt. Seybān: a large mountain to the south of Khalāṭ. It has many inhabitants. Snow is never absent from its peak, which can be seen from 50 parasangs away. It is 50 parasangs in circumference and has extensive pasturage. So much is recorded in the *Nuzha*. However, if what is meant by this mountain is Mt. Arghī (Ağrı, Ararat), that is in the other direction, near Rawān.

Tatvan: a large village one stage west of Aḫlāṭ. This is the endpoint of the lake.

Badlīs (Bitlis): a strong fortress and town in a river valley one stage west of Tatvan, located at 81½ degrees longitude and 37½ degrees latitude. A river flows through the town, with houses and town quarters on both banks, and crossed by a bridge in the middle. As the entrance and exit ways are narrow and steep, the houses are surrounded by the fortification.

In his chronicle (*Sharaf-nāma*), Mīr Sharaf Khān Badlīsī details the history of Bitlis, saying that it is one of the monuments of Alexander and is named after one his slaves. On his way from Iraq and Babylonia to Rūm, Alexander tested the climate of the places he visited and the water of the rivers that empty into the Tigris. When he reached the confluence of the Bitlis River, he found its water extremely pleasant and fine. Following along the bank, he came to the confluence of the Kusūr and Ribāṭ Rivers and found the water of the Kusūr more excellent. Proceeding to the source of the Kusūr, he saw that it was a verdant place and delightful mountainous area, and rested there for a few days. Today that place is called Khaymagāh-i Iskandar ("Alexander's Campground"). After that he ordered that a city and strong fortress be built at that passage, stipulating that no ruler like himself should be able to conquer it. His slave Badlīs built a solid fortress between the Kusūr and Ribāṭ Rivers, two parasangs from the abovementioned spring.

[Marginal addition by Ebū Bekr b. Behrām ed-Dimaşḳī] According to the chronicle of Sharaf Khan, Bijanoğlı Süleymān Bey was commanded by Sultan Ḥasan (Aqqoyunlu) to conquer Bitlis castle. He laid siege to it for three successive years from spring to the beginning of winter, each year wintering in Mardin and resuming the siege at the beginning of spring. The defenders were in dire straits. Most of them perished from dearth of supplies and the onset of disease. Only seven of them were left alive, but still it was impossible to take the castle by force. At that time the commander of the fortress was İbrāhīm Bey. Finally he surrendered by a truce.

[M 414] [R 157b] To this day the Bitlis fortress has stood as a mighty rampart in that passage. But because it is triangular in shape, it is always subject to vicissitudes and subversion.

Indeed, when Alexander was returning from Iran, Badlīs revolted and did not give him way, so Alexander had to take a different route. Badlīs then went to him, with a winding-sheet around his neck and the key of the fortress in his hand, and excused himself, saying: "I strengthened the fortification at your command, because you stipulated that no one like yourself should be able to conquer it." Alexander was pleased and gave him the governorship of the fortress.

It is related that, in the past, Bitlis was infested with snakes. A certain wise man made a talisman of carved stone in the shape of a snake. After it was placed in the forecourt of the castle, the snakes disappeared.

The climate of Bitlis is essentially very pleasant. To be sure, the winter is generally harsh, but the populace bears it without difficulty. Of the fruit of the summer pastures, apples and pears are excellent. Grapes and watermelons come from the surrounding area. In summer the inhabitants move to the orchards. They remain in the surrounding plains and meadows for six months and return in autumn.

Since ancient times this city has been exempt from noncanonical taxes in exchange for guarding the road. Muslim governors have erected several buildings as charity works, including 21 bridges of carved stone in the middle of the city, crossed by people from both sides, and four great Friday mosques: 1) The Old Friday Mosque: it is the work of an ancient sultan, according to the inscription in Kufic script. 2) The Red Mosque: it was originally an Armenian church and was converted to a mosque after the conquest. 3) The Emir Şemseddin Mosque: it is at the place known as Gök

Meydan; a dervish lodge nearby belongs to it. 4) The Sharaf Khan Mosque: he was the grandfather of the historian Mīr Sharaf. Next to it he built a madrasa and dervish lodge as well, known as the Sharafiyya. Aside from these there are five other madrasas—known as Shukriyya, Idrīsiyya, Khaṭībiyya, Ḥāji[bi]yya and Ikhlāṣiyya—with students living and studying in each. They had several eminent teachers, such as Muḥammad Sharānshī.

Among the famous emirs of Bitlis is Ḫüsrev Pasha, who built a double public bath of carved stone, also two inns, 100 shops, two olive presses and several granaries, making them a pious foundation for his dervish lodge in Rāhvā. Ḳāḍī Meḥmed Cān Efendi, one of the well-born grandees of the city, offered the following chronogram for this building: *Binā-yı ʿimāret-i Ḫüsrevāne* ("Building of the Khusraw-like edifice").

This town is a natural pass connecting Azerbaijan with Diyār Bakr, (Diyār) Rabīʿa and Arminiyya. [M 415] Those comimg from Turkistan, Iran and Khurāsān and going to Ḥijāz and Syria must pass through Delikli Ḳaya ("Pierced Rock") south of Bitlis. There is no other way; otherwise one must go around. It is one parasang south of Bitlis. In the past the road passed by it on one side. With the passage of time the water came up to the foot of the rock, blocking the road. One of the charitable ladies of the time had the rock pierced and the road cleared. She also had a mosque built in Bitlis and a great bridge in the middle of the town; they are known as Pül-i Ḫātūn ("Lady's Bridge") and Mescid-i Ḫātūn ("Lady's Mosque").

According to al-Wāqidī, Bitlis and Aḫlāṭ were conquered for Islam during the caliphate of ʿUmar in the 27th year of the Hijra with the efforts of ʿIyāḍ b. Ghanam. At that time the ruler of Aḫlāṭ was an unbeliever named Būstīnūs, famous in the region. The ruler of Bitlis, Barwand, surrendered the city peacefully.

The people of this town are mainly Armenian dhimmis. The Muslims are mostly Shāfiʿīs, with a few Ḥanafī. For the most part they are brave, generous and hospitable. There have always been eminent and talented people from here, including the following:

- Mawlānā ʿAbd al-Raḥīm. He wrote an excellent commentary on the *Maṭāliʿ* and has books on logic and rhetoric.
- Mawlānā Muḥammad Barqalaʿī. He excelled in jurisprudence and hadith. In grammar he wrote commentaries on Khabīṣī and Hindī under the penname Mīr Sharaf, which are held in high regard by elite and commoner as well.
- Leader of the mystics, Shaikh ʿImād Yāsir. He was the disciple of Abū'l-Najīb al-Suhrawardī and the shaikh of Najm al-Dīn al-Kubrā.
- Mawlānā Ḥusām al-Dīn. He was an eminent scholar and, in Sufism, the protegé of Shaikh ʿImād. He wrote a commentary on the language of Sufism.
- His son, Mawlānā Ḥakīm Idrīs. For a time he was the chancellor of the Aqqoyunlu sultans. Later he was the boon-companion to Sultan Selīm (I). On the order of Sultan Bāyezīd (II) he wrote a Persian history of the Ottoman dynasty entitled *Hasht Bihisht*. When Shah Ismāʿīl gave currency to the heretical sect (i.e., Shiism) and rose to power, Mawlānā Idrīs penned the following chronogram: *Madhhab-i nā-ḥaq* ("False sect" = 906/1500–1501). When Shah Ismāʿīl heard of this, he ordered his boon companion, Kamāl al-Dīn Ṭabīb-i Shīrāzī, to write a letter in an elegant style and ask Mawlānā about it. When the letter arrived Mawlānā Idrīs wrote a reply, saying: "I did compose that chronogram, but the words are Arabic. What I actually said is: *Madhhabunā ḥaq* ('Our sect is true')." Shah Ismāʿīl was pleased with the answer and desired his company, but Mawlānā rejected it, sending a versified apology.
- His son Abū'l-Faḍl. He was *defterdār* at the Porte and adorned with excellence (*faḍl*). He wrote an appendix to his father's chronicle.
- Shaikh Abū Ṭāhir-i Kurdī. He is mentioned in the *Nafaḥāt*, and is buried west of Bitlis in the town quarter of Kusūr.
- The poet Şükrī. He was one of the boon companions of Sultan Selīm (I) and composed the *Selīmnāme*.
- Mawlānā Mūsā. The Shukriyya madrasa was transferred to him from his grandfather, the 120-year old Shāh Ḥüseyn.

Towns and districts around Bitlis

Rāhvā: a place between the village of Tatvan and Bitlis. It is half a stage east of Bitlis. Ḫüsrev Pasha built two large caravanserais, one lofty dervish lodge, one public bath, ten shops and one mosque. He brought water there from a distance of approximately 12,000 cubits away [M 416] and re-settled some 30 households of taxpaying subjects. Food is provided to travelers.

This complex was built at a very appropriate place, because in the past there were numerous caravanserais between Bitlis and Tatvan, [R 158a] all of which were in ruins and travelers were in danger of perishing from the severe cold. Sultans and governors intended several times to restore that place. Many buildings were started but were abandoned when obstacles appeared. In 980 (1572–1573) this complex was complete and travelers began to come.

Muş: according to the *Taqwīm*, a small town at the mouth of a valley on a flat place at the foot of a mountain. The plain, known as Muş Ovası (Muş Valley), is very extensive, covering two stages' distance.

Mīr Sharaf says: "Muş is a district of Bitlis. It is a city of ancient construction, and traces of its ruined fortress and buildings are still extant. At the time of my grandfather, Sharaf Khan, a fortress was built on top of a mountain one parasang south of the city. Later Sultan Süleymān restored half of the ancient castle, the part situated on a high hill in the western side of the city. Today it is defended by some fifty troops."

Muş means "flock" (text: *deme*, error for *reme*) in Armenian. Because there is so much of that (i.e., because there are so many flocks of sheep and goats), there are few fruit trees in this province. But there are orchards in the mountains and the city. Grains are in abundance. The taxpaying subjects have countless livestock. Sheep and water buffalo are especially abundant, because it is a place abounding in meadows and pastures.

The Muş Valley is approximately 8 parasangs in length and 3 parasangs in breadth and very flat and level. Surrounding it are forested mountains and verdant and snow-covered mountain pastures. Delightful streams flow out in all directions. The Euphrates comes from the north of that plain, cutting off one-third of it and flowing toward the south. The Ḳara Ṣu comes from the east, flows in the middle and empties into the Euphrates. In autumn, matchless falcons are caught in the mountains and various birds are hunted and fish are caught in that plain. The villages are totally inhabited by Armenians.

According to Ḥamd Allāh, the state dues in the past, during the time of the Chingisids, were 69,000 dinars. Under the Ottoman sultans they amount to 15,000 dinars, according to the *Defter-i Ḫākānī*, and 4000 dhimmis have been registered.

Ḫınūs: Another district of Bitlis. According to Mīr Sharaf it belongs to Bitlis, but other local authorities consider it one of the sanjaks of Erzurum. It has vast mountain pastures. One of them, Bingöl pasture, is famous. Another is Ṣu Şehri, and another is Mt. Sharaf al-Dīn, where Mīr Sharaf's ancestors and the Kurdish tribes are settled.

Produce and crops are plentiful. There are two streams, one called Beyāż Ṭoz ("White Salt") and the other Namak-i Aḥmar ("Red Salt"), each of which produces 4000 gold-pieces in revenue every year. The state dues from Ḫınus are like those of Muş. Finally, there are few Armenians among the taxpaying subjects; most of its villages are timar or *ze'āmet* (timar providing an annual income of over 20,000 *aḳçe*). Some 4000 mounted troops can be fielded from here.

In this district are bred famous wind-swift horses. The only crops are grains.

One of the marvels here is Bulanıḳ Göl ("Turbid Lake"), approximately 1 parasang in circumference. Its water is always turbid and red, [M 417] and its tributary, a small river, in turn flows red and muddy. It never becomes clear. Another is Nāzük Göl ("Elegant Lake"), located between Aḫlāṭ and Bulanıḳ Göl. Its water is very sweet and transparent. In winter it freezes so hard that a caravan can cross it, and when it is about to melt it cracks so loud that it can be heard 3 parasangs away. After the ice melts, countless fish appear in it and go into the surrounding smaller rivers. The people of Ḫınus and Aḫlāṭ rush here to fish. One person can catch three loads in a day. Several times the government sent agents, imagining they would collect large sums in taxes; but it chanced that in those years there were no fish and nothing was gained.

Mt. Nemrūd: a large mountain north of Bitlis. Nemrūd (Nimrod) is said to have gone up to the summer pasture there and built several castles and lofty mansions. When he became the object of divine wrath, [R 158b] his constructions collapsed and water sprang from the site. Today there is a lake on top of the mountain, 2000 cubits high and some 3 parasangs in circumference. It is surrounded by sheer rock and can only be approached on three paths, one by foot alone, the other two by horse as well. Along the lakeshore are hot springs here and there. The stones around it are as if burnt.

On the northern side the mountain has subsided and a slag-like material is emerging from the ground and pouring down. It has become a large heap, 600 cubits long and 50 cubits high, which is gradually getting bigger.

The poor one (Kātib Çelebi) says: At that place there are mines of sulphur and petroleum that produce formations of this kind, as explained in books of natural science. There are many such places on the face of the earth. It seems marvelous to one who is not aware of the source, and has given rise to many speculations.

Ḳoṭur: a district and sanjak two stages east of Van. It has a small fortress in a valley. Sultan Murād (IV) stopped here on his way back from the Rawān campaign. Some Redheads (*surḫ-ser*, i.e., Qizilbash = Safavids) had locked themselves up in the fortress and would not submit. It was not vouchsafed to him to take it by force. Later, when it was taken by peaceful capitulation, it was decided that the fortress be demolished and that place on the border was left empty. Today it belongs to the Tabriz province and the fortress is reported to have been renovated.

Bāyezīd Castle: a castle and sanjak north of Ḳoṭur. As it is located on the frontier, it is constantly subject to attack.

[Marginal addition by Ebū Bekr ed-Dimşḳī] The plain of Çaldıran: an elongated and broad plain near Van. The two great rulers, the Ottoman Sultan Selīm (I) and the Safavid Shah Ismāʿīl, both of whom claimed the title *ṣāḥib-qirān* ("Lord of the Auspicious Conjunction"), fought a battle on this plain (in 1514). As Sultan Selīm came out victorious, the title *ṣāḥib-qirān* was granted to him. Shah Ismāʿīl was defeated and fled. Henceforth his star was fallen. This was a great and famous event.

Vesṭān: According to the *Taqwīm*, a town on a flat place on the lakeshore one stage southeast of Van. The plain ends there; beyond are the mountains of Kurdistan. At present it is the size of a village.

[Marginal additions by Ebū Bekr ed-Dimaşḳī. R 157b–158a] Kārjīkān: one of the districts of Bitlis. At times it has been an independent emirate.

Armūk: a castle on the shore of Lake Aḫlāṭ (Lake Van). It is very difficult of access. Its state dues are 13,800 dinars. Thus according to the *Nuzha*.

Türkeri: a small city. Reportedly it was a large city in the past. It is located on a hill. [M 418] A large river flows by it coming from Alaṭaġ. Orchards are plentiful. Various kinds of fruit grow there. Within the city on one side is a strong castle. The state dues are 25,000 dinars.

Meymān: a town with plentiful orchards and fruit. The state dues from it are 12,000 dinars.

Ḫarādīn: a small city. It used to be flourishing and large. Its state dues are 5,300 dinars.

Selem: a town. Its state dues are 7200 dinars.

ʿAyn: a mid-sized city. Its state dues are 4300 dinars.

Kusūr: a small city. Its state dues are 4300 dinars.

Üçkān: a district, one of the administrative units of Muş. Sometimes it is attached to Ḫınus and considered an independent emirate.

Dabīl: according to the *Taqwīm*, it was a flourishing and famous city of Arminiyya and its ancient capital. It had a rampart superior to and larger than that of Ardabīl, but later it was altered. Christians here are numerous. The church and the mosque have been built close to each other. The people make incomparable carpets. The crimson color here is very fine.

Sekmānābād: a district near Khūy. The Dunbulī tribe resides there.

[Addition by Ebū Bekr ed-Dimaşḳī] Sharaf Khan says that the lineage of the Dunbulī goes back to a certain ʿĪsā, one of the emirs of Syria. Originally they came from Syria, entering the service of the Persian kings, and they gave him Sekmānābād, one of the administrative units of Khūy, as a hereditary grant. After that the tribes and clans gathered day by day and became known as the Dunbulī tribe. At first they were Yazīdīs; later, when some emirs and chieftains became Sunni, they abandoned unbelief and heresy and most of that tribe found right guidance, while many remained in that aberration. According to one report, the Dunbulī were a clan of the Yaḥyā tribe who settled in Sekmānābād. Indeed, among the tribes and clans of Kurdistan, this clan is called Dunbul-i Yaḥyā and their emirs are called ʿĪsā Beylü.

Some of these beys govern in Khūy itself, and the Ḳoṭur valley, Abaḳa and the district of Ovacuḳ, which is one of the administrative units of Nakhjiwān, were under their control. Some of them also gained control of half of Abaḳa, the district of Süleymānsarāy, Dere-i Alakis, which is one of the administrative units of Nakhjiwān, and the emirate of Shurūr. When those regions were ruined by the marching back and forth of the Ottoman army, they were seized by the shahs. At times, the region of Abakay, Çaldıran, Süleymānsarāy and Sekmānābād were given to them by the Ottomans, according to ancient custom. In recent times, the Ottomans gave some of them the district of Çaldıran as a sanjak. Again, the chronicle of Sharaf Khan depicts the Dunbulī tribe as steppe-dwellers.

After Sultan Murād IV took Rawān on 10 Rabīʿü-l-evvel in the year 1045 (24 August 1635) he turned towards Tabriz and camped on the bank of the Aras River. The Dunbulī tribe came along with some 500 households, the people of the Shurūr district with some 300 households, and the Pesyān tribe also with some households, and petitioned for territory. They were resettled, the Dunbulī being sent to the qadi district of Erzincan, the people of Shurūr to the qadi district of Tercān, and the Pesyān to Pāsīn. Orders were issued to settle them in empty, ruinous places and abandoned villages. [M 419] Those with property were given a sanjak and a emirate, and the notables of the tribes were favored with a timar or *zeʿāmet* (timar providing an annual income of over 20,000 *aḳçe*).

Ḫoşāb: the hereditary grant and government seat of the Maḥmūdīs, who are among the beys of Kurdistan. According to the *Nuzha*, Ḫoşāb is a town whose state dues are 1000 dinars. As can be seen on some maps, Ḫoşāb is a castle, a town and a sanjak between Van, Vesṭān and Selmās, south of Lake Van, at a place like a corner where the spurs of two mountains meet. Its west side opens to an extensive desert opposite Vesṭān that extends as far as the border of Ḥakkārī province where there are again mountains. From those two mountains rise two streams that join below—i.e., to the west of—Ḫoşāb. The river then flows northwest and then north, passes under a stone bridge and, somewhat east of Akhtamar, empties into Lake Van from the south.

Sharaf Khan mentions this river when he speaks of the bank of the Ḫoşāb River known as Jam-i mīrāḥime (?). And he refers to the castle, which is the hereditary grant of the Maḥmūdīs, as the castle of Āshūb. He also writes that the Maḥmūdī beys administered Albāḳ, one of the districts of Ḥakkārī. He also mentions that emirs were sometimes appointed from them to the district of Kārjīkān; the castle of Tuwān, one of the administrative units of Ḥūl (?); Sīrūm, one of the administrative units of Marāgha (?); the castle of Mākū, one of the administrative units of Nakhjiwān; and the region of Ordūbād. He also writes that Aḳçaḳalʿe, Sekmenābād and Bargiri have their own emirates.

In Ḫoşāb is the madrasa of Ḥasan Bey, one of the Maḥmūdīs, and he himself is buried there. In the year 993/1585, when Özdemür[oġlı] ʿOs̱mān Pasha was general commander, he fell in battle against the Qizilbash in Saʿdābād; a year later he was brought to Ḥasan Bey's tomb in the ʿAṭāyī madrasa and buried there. Ḥasan Bey has several other mosques and madrasas. He was the first of the Maḥmūdīs to abandon the Yazīdī sect and become a Sunni.

The emirate of Ḥakkārī

It is east of the emirate of ʿİmādiye and west of the districts of Vesṭān in the province of Van. In the north it borders on the emirate of Bitlis and some of the sanjaks of Kurdistan in the province of Diyarbekir. In the south it borders on the province of Shahr-i Zūr (Shahrazūr) and the emirate of Sohrān—several castles and districts are reckoned as belonging to this emirate. In the east it borders on the Qizilbash (i.e., the Safavid realm), so they are mostly empty places.

This emirate is a dependency of the province of Van. It is the province of the Ḥakkārī clan of the Kurds. Originally its emirs and governors and the Ḥakkārī clan administered it from Van and Vesṭān. Later they took the territory under direct control. When the Safavid and Ottoman rulers made Van into a province, they gave them a part of their provinces. Now the seat of the emirate of the Ḥakkārī emirs is Cūlāmerg, a castle and town located southeast of Vesṭān.

Kūz: a district on a level plain in the mountains east of Cūlāmerg. A river rises from there, and another one comes from Cūlāmerg. After a curve to the west they join south of Cūlāmerg. The river flows in the mountains and passes Çil, a sanjak and castle belonging to Ḥakkārī, and the castles of Bīsutūn and Ustūn. In Zībārī, one of the districts of ʿİmādiye, [M 420] it flows under a stone bridge and joins the river coming from the castle of Kūz (text: Kūr), a bit to the southeast. Further down, near Ḥarīr, it joins a river coming from the sanjak of Bāyān. Close to where they empty into the Tigris from its south bank they join a large river coming from the sanjaks of Terliyān (?) and Kestāne and empty into the Tigris just north of Sāmira.

Cūlāmerg: a town in the mountains, nestled against a small mountain. It is on a level place, while its associated castle is at the top of a mountain to its west. The emirs of Ḥakkārī have several public buildings, Friday mosques and madrasas there. Nearby is the castle of Emīr Dāvūd and Birādūst; to the north is Müküs castle and emirate; to the west, as far as the city of ʿİmādiye, are again some castles belonging to the emirate of Ḥakkārī; to the east is the district of Kūz and, east of that, Ḫoşāb.

Albāḳ: a sanjak in the mountains just north of Selmās and opposite that town. It has usually been under the administration of the emirs of Ḥakkārī. It is directly east of Cūlāmerg.

Şaḳaḳ: a district and region on top of a mountain north of Cūlāmerg castle. A stream descends from that mountain and empties into Lake Van. Şaḳaḳ is between Müküs and Cūlāmerg, but closer to Cūlāmerg.

Zerīl: a castle in the corner of a mountain directly west of Cūlāmerg. North of this is the castle and territory of Şīrvī.

Bīsutūn: also a large castle on a mountain top, located between ʿİmādiye and Zerīl. The river coming from Cūlāmerg flows just to the south of here.

Çil castle: there are several castles and districts belonging to it. It is a separate emirate, though for a time it was considered a dependency of Ḥakkārī. It is a large castle south of and opposite Bīsutūn. It too is located in the corner of a great mountain. The Cūlāmerg River flows past between these two mountain corners.

Dīzī: another castle belonging to Ḥakkārī. It is behind the mountains south of Çil castle and is itself on top of a mountain, located near Mīr Dāmerd castle and just east of Ustūn castle. At the foot of this mountain rises a stream from a place like a small lake. Further down it becomes a large river that flows under a stone bridge just west of Mīr Dāmerd and another stone bridge just north of Mīr Nāṣır Biradūst castle, flows by that castle on the east, reaches Lake Urmia from the west and empties into it.

Ustūn: another castle on a mountain, near Rūmī castle, to the west of Van. Its territory is shared by the district of Zībādī (recte: Zībārī).

Hlor: one of the castles in the mountains northeast of ʿİmādiye and southwest of Bīsutūn. Its town is on a level place between three mountains, the castle being on top of a great mountain to the north. A stream rises from the two mountains southwest of it, from the direction of ʿİmādiye, and flows toward the ʿİmādiye plain. After it passes by ʿİmādiye and joins the Khābūr River, it empties into the Shaṭṭ (Tigris).

Bāyān: one of the strong castles of Kurdistan. Bijanoġlı Süleymān Bey, one of the great emirs of (Uzun) Ḥasan Bey Aqqoyunlu, laid siege to ʿİmādiye castle. [M 421] When winter came, he was obliged to lift the siege. He went to the district of Serdī to spend the winter. At that time the governor of Ḥakkārī was ʿİzzeddīn Şīr Bey. While most of the Ḥakkārī castles had fallen to the rule of Ḥasan Bey, ʿİzzeddīn Şīr Bey had locked himself up in Bāyān castle. Now Süleymān Bey sent him the following message: "As long as the castles of ʿİmādiye, Görgil, Seri Yazlulı and Bāyān are in the hands of the Şenbūner, we have nothing to fear. In the eyes of the Kurds, your tents on the plain are like water buffalo turds." The governors of Ḥakkārī are known by the title Şenbū, which is a corruption of the word *şīne*. Its origin is explained in detail in Sharaf Khan's chronicle.

Bāyān castle is near Ḥarīr, just to the southeast of it. It is a strong castle on the eastern shore of a small lake in front of a mountain pass known as Derbend-i Püşt. It is quite a distance from Cūlāmerg. Near the city, the outlet of this lake joins the two rivers coming from Balaban and the sanjak of Kestāne in the south. Then it flows west through the plains of the sanjaks of Dūrkāmī and Semāvkū and passes west of Irbil. At a place near the Shaṭṭ (Tigris) it joins the river flowing from the district of Zībārī and empties into the Shaṭṭ north of Sāmira.

Above Irbil this river forks into two branches, an island forming between them, then they rejoin. On that island is a village called the castle of Altun Köprisi (Golden Bridge), also called Esyāber (cf. Esbāberd above). For some time the castle and district of Aġakis used to belong to it; later it was made into a separate emirate.

Esbāber (text: Esyāber): sanjaks and castles located somewhat northwest of Vesṭān. The sanjak of Kārkār, the sanjak of Esyākerd and another town—these three sanjaks and castles are located on the foot of a mountain extending from southeast to northwest. Arranged in this order, they face the plain that ends in the mountain near Vesṭān.

Sanjak of Keysān: a castle and town at the foot of the corner of a mountain east of the aforementioned towns and castles. The town at the end, which is at the corner of the mountain, is opposite Keysān castle, there being a short distance between them.

Müküs: a magnificent city and castle on a plain west of the aforementioned mountain. Ḫīzān is somewhat to its northwest. A river flows in front of Müküs and, after passing through the territory of Ḫīzān, joins the Bitlis River just below Deliklü Ḳaya. This river rises at the aforementioned mountain and reaches Müküs in a short distance. The district and castle of Kārkār formerly belonged to Müküs; later it was made into a separate emirate.

Ḫīzān: a flourishing castle and town on top of a mountain in the northwest, south of Bitlis and west of the sanjak of Keysān. Streams issue from two places on this mountain and, after joining, flow into the Müküs River and then into the Bitlis River.

Sharaf Ḫān says that Ḫīzān is a lightened form of *saḥar-khīzān* ("early risers"), the people being early risers and diligent in prayer and piety.

Ḫīzān castle is a construction of Islamic times. According to the inhabitants, it was built by the one who founded Marāgha of Tabriz. However much this poor one (Ebū Bekr b. Behrām ed-Dimaşḳī) has studied chronicles, I have not come across the founder of Marāgha. Only recently I read that Hulagu (text: Hlū) Khan restored it (Marāgha) and made it his capital.

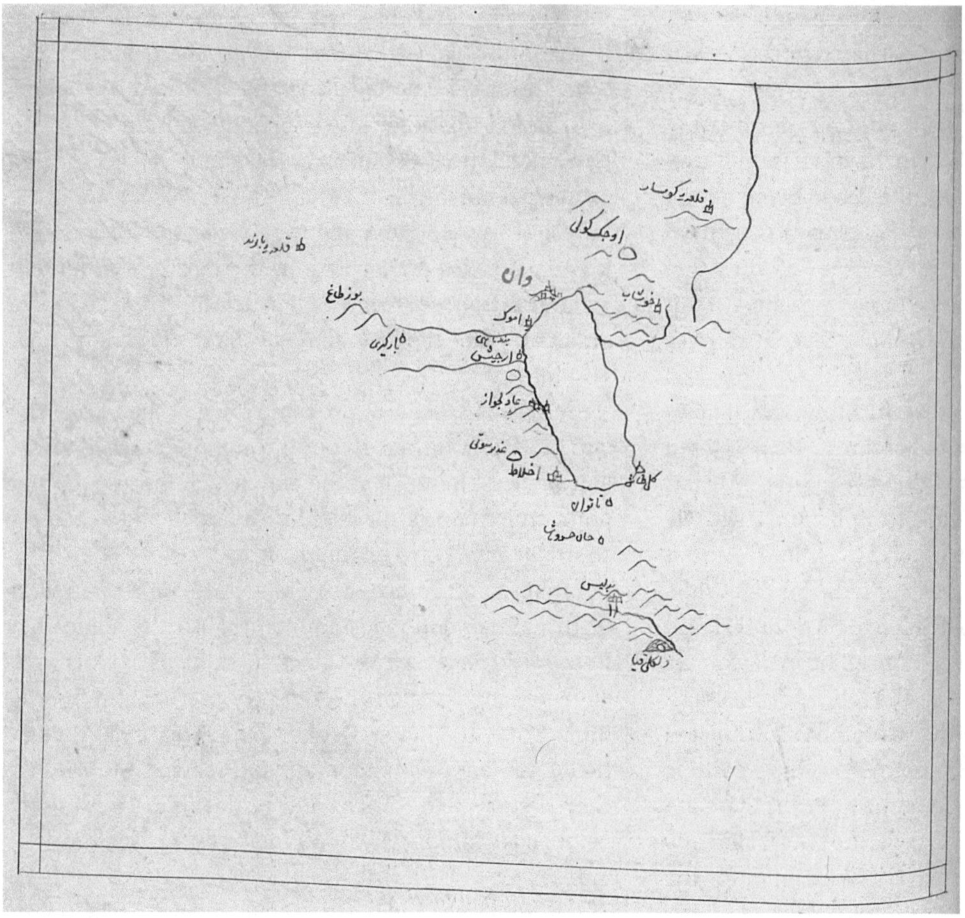

PLATE 46
Map of Van Province, R 159a (Appendix, fig. 90)
TOPKAPI PALACE LIBRARY, REVAN 1624, WITH PERMISSION

Part 2, by Ebū Bekr b. Behrām ed-Dimaşḳī

[M 422] Supplement of the Publisher (İbrāhīm Müteferriḳa): We should not overlook the fact that, as indicated at the beginning of the book, the esteemed author, the late Kātib Çelebi, promised that the account of the countries and kings of the world would extend from east to west. However, the text written in his own hand ends here; and the other available copies of his work also do not record anything beyond the borders of the province of Van. Nevertheless, according to the rather clear arrangement of his book, the many clear statements of the author from the beginning of the book to this point, and what he says where the text breaks off, he had planned to complete it. Indeed, he specifically says this in the 41st chapter describing the clime of Armenia, which we have printed. In the course of pointing out the different opinions regarding the delimitation of this clime, he says, "I will describe in separate chapters the provinces of Erzurum and Kars, which are included in Greater Armenia." It is obvious from this that he was going to write about the other countries and that he would have had to increase the size of the book by including many additional chapters. Probably the last part of the book is lost, or else his precious life must not have allowed him to complete it according to his ambition and promise.

This insignificant slave (Müteferriḳa), to the extent that he can deduce from the arrangement of the text the intention and goal of the late author, has made a great effort humbly to complete the work by following in the footsteps of the late author in describing and enumerating the (remaining) countries and kings. Because the province of Kars and Çıldır has previously been described in summary fashion in connection with the accounts of Georgia and Akhiskha, I begin here with the province of Erzurum. Consideration is also given here to the view of those who believe that the towns of this province belong to Greater Armenia. Afterwards, I try to describe systematically the remaining climes of the continent of Asia and, with the help of God, set out to complete the book.

In the course of describing the (remaining) countries and rulers, I have in some places selected passages from Ebū Bekr b. Behrām ed-Dimaşḳī, the translator of *Coġrafya-i Kebīr*, and summarized them; but in most places I have preferred to cite them exactly as found.

Province of Erzurum

This province is located immediately to the east of the province of Rūm. Some assert that it is part of Armenia.

Borders

To the east are the provinces of Rawān and Kars; to the north are the province of Trabzon and Georgia; to the west is the province of Sivas; and to the south are Diyarbekir and Van. This province consists of 11 (sic, error for 14?) sanjaks. Erzurum is the pasha sanjak; İspir, Pāsīn, Tortum, Ḫınūs, Ḳaraḥiṣār-ı Şarḳī, Ḳızuçan, Mamrevan, Kiğı, Mecingerd, Malazgird, Tekman, Eleşkird, and Bāyezīd. There are 15 fortresses and 5,157 timars with and without government certification. As prescribed by *ḳānūn* (sultanic law), there are 7,800 soldiers with their retainers. Those from Mecingerd and İspir are considered separately; those from Eleşkird and Bāyezīd are considered separately; and those from Ḳızuçan are considered separately. This province has a special kind of timar called a rotational timar: three or four people hold this timar jointly and go on campaign in rotation. Nothing like this exists in other Ottoman provinces.

Sanjak of Erzurum: The city of Erzurum is the capital of the province and has a post for a molla (chief qadi) who is paid a salary of 500 *akçes* per diem. Its qadi districts are Erzincan, Bayburt, Lower Tercan, Upper Tercan, Tortum, Urla (text: Deverler)—also known as Gümüşḫāne, Ṭorul, Ḳuruçay, Ḳızuçan, Ḳoray, Gercanis, Kelkīt, Kemaḫ, Gevanis, Merdcan, and Yaşyağmur Deresi. [M 423]

Description of Erzurum

It is a walled city. Inside is an old city and a new mosque. There are also public baths, markets, and a covered market. It has abundant arable lands. But it produces no fruit and has no trees. Lumber comes from a place two days' journey distant. Most people burn dry dung. It has many flowing springs. From one of them, called Cennet Bıñarı ("Spring of Paradise") there flows a stream. Inside the Tabriz Gate is a famous domed church. It measures 50 cubits on each side. Part of its vault collapsed on the birthday of the Prophet, and it was impossible to replace. The Muslims built a mosque the length and width of the Kaʿba across from it. It was called Nümūne-i Kaʿba (the mosque modeled on the Kaʿba).

 DOI:10.1163/9789004441330_005

The city has three gates. One, the Tabriz Gate, is on the north side of the city. Near this gate, but outside it, is a spring called Şāmḫāne-i Çene that spills into stone troughs. Another is the Georgia Gate and a third is the Erzincan Gate. There is a place for promenades, heading out from the Tabriz Gate. The flowing waters there are incomparable. The pilgrimage site of Shaikh ʿAbd al-Raḥmān is found there. At a place near Erzen is an image of a dragon that looks real. There are stone remains on this mountain that altogether, from a distance, look like a real dragon.

Comments: What they call Qālīqalā is Erzurum. Some of the emirs of Byzantium were dispersed (about Anatolia) and became like independent princes. One of them became the ruler of Armenia and later died there. After him one of his wives named Qālī became ruler and built the city of Qālī. This woman had her own image depicted on a gate. Many Persian merchants come to this city. Customs duties are collected from them and this produces considerable revenue. The grave of Abū Isḥāq is found in this city. It is a place of visitation. Its subordinate districts are Ḫınūs, İspir, Mürsḳulı, Boğaz, Şuşehir, and Tercan.

Kelkīd: It includes about 100 villages. It is the name of a qadi district and a small town a distance of three stages from Sivas in the direction of Istanbul. It has a mosque, public bath, and flat plains. Most of its houses are made of wood. It is near a summer pasture called Çemen Ṭağı. The Turkmen and Ulus tribes pass the summer there. Many creeks descend from this summer pasture and flow into the plains. It is a mountainous and populated region.

Kemaḫ: A prosperous town on the Euphrates River. It has a fortress. It is a distance of one stage from Erzincan. Tīmūrṭaş Pasha, one of Yıldırım Bāyezīd's commanders, conquered it in 798 (1395). A characteristic of this town is that small birds like sparrows land there. These birds, which are like the *salwā* (Koran 2:57, 7:160, 20:80) that descended upon the Israelites in the spring and resemble quail, descend in clouds upon the environs of this town. They are an important source of food for the people of the town. Their taste resembles that of halva. If their young are not taken (by the people) within a few days and remain where they are, they spread their wings and fly away. The cheese from this town is highly prized. Its districts are Ḳuruçay and Vādī-i Vank (text: Vādī-i Rank).

Urla: an esteemed qadi district. Because it is near a silver mine it is also called Gümüşḫāne ("House of Silver"). The mine, being close by, only a distance of one or two hours, led to the creation of a large and prosperous town. This mine is operated by the inspectors of the emirs and their agent. Its people speak Greek. This place is a tax farm which extracts silver, copper, and gold. There is a small river that flows from Gevanis. It is under the control of a *ṣubaşı* who is subordinate to the Pasha of Erzurum. [M 424]

Tercan, also called Dercan: A medium-sized city.

Erzincan: A city with expansive pastures 40 parasangs from Erzurum. There is a mountain here with a cave. A stream descends from the ceiling of this cave and becomes hard as stone. Because a number of earthquakes have occurred in this city, most of the buildings have fallen down. The Seljuk sultan ʿAlāʾ al-Dīn Kay-Qubād rebuilt its rampart. The air is excellent. The Euphrates flows through the outskirts of the town. Cereal grains, cotton, grapes, and fruit are abundant.

Ḳolor: A large village near the Senūr Valley and one stage to the south of Bayburt.

Mināreliköy: Ferruḫşād Bey, a former emir of Sultan Selīm, took possession of a number of villages, about 14, in this area in return for his service while on campaign. He built a mosque, an inn and a public bath in this one.

Bayburt: This town is two stages from Mināreliköy in the direction of Erzurum. It is on the road northwest of Erzurum and southeast of Trabzon. Bayburt is two stages from Erzurum and three stages from the Trabzon Plain. There is a mosque where one can perform the Friday prayer and three or four public baths. On one side of the town is a hill on which there is a fortress. The air is cold. The sown fields and cereal grains are abundant. The Cū-yı Rūḥ (Çoruḥ) or Çoraḳ River flows through the middle of the town. It is a large river that flows at the edge of town and east of the fortress. Lumber comes from a mountain two days' journey distant. They cut the trees there and put them in the river. When the trees reach the city, each person looks for those on which he has placed his mark and retrieves them. This river flows to Trabzon. This district is the imperial domain of the Pasha of Erzurum and is to the south of Trabzon.

Tekman: This town is very close to Erzurum. The town and its fortress, as well as a number of lakes, are between two mountains to the northeast of Erzurum. The pasturage tax is exacted from the Ulus tribe at the Bingöl summer pasture. The winters are long. Indeed, it snows while the crops are growing and they take in the harvest. When the summer comes, they do the threshing.

Gevanis Ḳalʿesi: This is the estate of Murād Khan. The Pasha of Erzurum places a *ṣubaşı* in charge of it. A small river flows on one side of it. This is in a mountainous and forested district.

Yağmur Deresi: Also a mountainous and forested district.

Sanjak of Ḳaraḥiṣār-ı Şarḳī: Its qadi districts are Ebū'l-ḫayr, Ezköy, Özḳar, İskefer, Aḳşar Ovası, Aḳköy, Ulubey, Aybaştı, Bāzārṣuyı, Şuşehir, Behrāmşāh, BWLM'N (Pülümür?), Bayrāmlu also known as ʾRDR (Ordu?), Çamās, Ḥīsmāne (Ḥapısmāne?), Zeġābe, Sīs, Orta, Şebḫāne, Şīrān, Ḳırıḳ, Ḳoyunlıḥiṣār, Mīlāş, and Yaʿḳūb Bey Derbendi also known as Pencşembe (Perşembe).

Description of Ḳaraḥiṣār-ı Şarḳī: A town between Erzurum and Tokat and it has a fortress. They call it Ḳaraḥiṣār-ı Şābīn.

Ḳoyunlıḥiṣār is east of Tokat. It is two military stages from Sivas on the Erzurum road. Between Sivas and Ḳoyunluḥiṣār are fine plains.

Niksar is to the east of Koyunluḥiṣār toward the north. There is a dirt hill on which there is a small fortress. It is reached by a narrow road. Below this hill there is a difficult foot path.

Şebḫāne: A place from which alum is extracted. It is a tax farm: the guards of Ḳaraḥiṣār-ı Şarḳī [M 425] obtain their pay from tax farming this place.

Sanjak of Pāsīn: A broad plain with a small walled city called Ḥasanḳalʿesi. It is located about half a stage to the east of Erzurum. The fortress is located on a plain. It has a Friday mosque and permanent markets. There are several districts: Ḳaşnı, Köni, Eynek and Çiçekrek. This Pāsīn consists of two fortresses: the Upper Fortress, which stretches from the edge of Ḥasanḳalʿesi to Choban Bridge and, beyond that, the fortress of Mecingerd which is the Lower Fortress. Kös Ṭaġı, which is in this region, is between Pāsīn and Ḳaġızman. This mountain extends from east to west. To the south and east is Ḳızılcaṭaġı. Between these two (mountains) is a wide plain. Near it is Aḳṭaġ, on one side of which flows the Aras River. A river that comes from the Büyükgöl summer pasture flows over the Pāsīn Plain into the Aras. South of Ḥasanḳalʿesi is Alaṭaġ where the Murād River rises.

Sanjak of Mecingerd, also called Lower Pāsīn: It has a strong fortress. It consists of three districts: Ḫorāsān, where the bey resides; Zevīn; and Ṣaġān. The fortress is located on a bank of the Aras River.

Sanjak of İspir. This city is two stages from Erzurum. To the northeast of the sanjak, surrounded by high mountains, is a valley with some flat areas. Here are a fortress and the city. There are vineyards and gardens on both sides of the river that flows through this valley. The river flows toward Bayburt and from there empties into the Black Sea. İspir is full of fruit. The villages located along the Black Sea are inhabited by Greek infidels. They are dhimmis. Every year they pay 600 quintals of beeswax as state tax. The area where they live is forested and full of *aġılu çiçek* (*azalea pontica*) and thus there is a lot of wild honey which they collect from the trees in the mountains. In İspir they press clotted cream together with honey. In small casks they spread one layer of cream and one layer of honey. It stands from summer to winter. One cannot tire of its taste. The honey is a preservative and it does not spoil. It becomes a dry (i.e., clotted) cream mixed with fine honey. In this district people hunt with falcons found nowhere else. The İspiri falcon gets its name from here.

Sanjak of Ḫınūs: It is located a distance of three stages southeast of Erzurum. It has several districts: Baçan, Bulanın, and Ḫındırıs. They are at the foot of Bingöl (mountains).

Description of Ḫınūs: Its fortress is a large natural rocky place. In the center is a depression and all around it large rocks rise up from the ground. Their height is 10 cubits. The ground around the top of the (natural) wall is flat. In the middle of the depression surrounded by the walls of Ḫınūs is a high hill. On this hill the town of Ḫınūs has a strong fortress. Inside are guards. On the east side of this fortress, at the foot of the hill, flows a small river. This river flows through the fortress from a fissure on the south side. This fissure is blocked with an iron grate. The outlet for the river is on the northwest side and comes from here. Most of the people are Kurds belonging to the Besyāniyān clan. There is a Friday mosque and market. There are excellent horses. More than 100,000 of the Ulus clan spend the summer in this district

Sanjak of Mamrevan, also called Namrevan: A mountainous and forested region with a fortress, northeast of Ḫınūs. [M 426] Beyond it one comes to Oltu.

Sanjak of Kiġı: A small city, at the southern extremity of Erzurum. One part of this region is a plain and the other part is mountainous. Its district is a sanjak. Here there is a mine (an iron mine) for *yuvalaḳ*, meaning cannon balls, that are found nowhere else. The people are Kurds belonging to the Besyān tribe.

Sanjak of Ḳızuçan: A very inaccessible, mountainous and forested place. The people are also Kurds belonging to the Besyān tribe. Sorrel grows abundantly there.

Sanjak of Malazgird: A small town built of black stone. It has springs but no trees. It is close to Aḫlāṭ. The Murād River flows by it. Near the town is a marvelous bridge. West of the town is the Malazgird River. Malazgird is two stages southeast of Erzurum. Its best-known district, Ḳatadi Ṣarıṣu, has an ancient fortress, a capital of the Armenians. Mt. Sübḥān is to the west of it, a very high mountain. The people are a clan of swarthy Kurds known as the Bātrek tribe.

Eleşkird: An old sanjak between the Çaldıran Plain and Rawān. It is no longer considered as a sanjak.

Bāyezīd Castle: A fortress and sanjak north of Ḳoṭur, on the frontier with the Persians. There are two other fortresses in this sanjak, Diyādīn and Ḫamur, both of which are subordinate to it (the fortress of Bāyezīd). Pehlūl

Bey rules this sanjak as a hereditary grant. The people who live here are also a clan of Kurds from the Besyān tribes. They are very brave clansmen and the Qizilbash (i.e., the Safavids) are very fearful of them. The Murād River flows through the Bāyezīd Plain. At one place it goes underground. It resurfaces and flows again at distance of four hours. Ağrı Ṭağı (Mt. Ararat) overlooks this sanjak. The Çaldıran Plain is located here.

Tortum: A small town in a small valley about two stages north of Erzurum. Nearby is a saltpeter mine. The streams that originate from the Tortum valley flow through the district of Aḳçaḳalʿe, then through part of Georgia, and then empty into the Black Sea. The district of Aḳçaḳalʿe is to the north of the district of Tortum. Aḳçaḳalʿe is a small fortress on a hill. It is garrisoned with guards. The air and streams, and the fruit including pears and apples are excellent.

Mountains

Alaṭaġ: Here there is a large summer camp ground and pasturage. There is much game. The Murād River rises from four or five sources and flows from springs. The Mongol Arghūn Khan (r. 1284–1291) had a palace built here.

Bozcaṭaġ: extends from east to west.

Mt. Kūs (Köseṭağı): It is near Ḳızılcaṭaġ.

Aside from these pine groves and forests abound.

Egerlü[ṭaġ]: It can be seen from Erzurum. It is high and extends towards Ḥasanḳalʿe.

Rivers

The Murād: It has two sources: 1) In Alaṭaġ it bubbles up from several places and becomes a large river that goes through a pass called Çarmur where it divides into four channels. A large stone bridge called Cüdāmen Şāh spans this river which flows under it and on the other side joins the Malazgird River. 2) From the Bingöl summer pasture it flows south and joins the other branch. Then it merges with the Ḳaraṣu on the Muş plain, flows past Genç, Çapaḳçur and Palu, and merges with the Euphrates at Reşvān.

The Euphrates: It reaches Erzurum from the Şuġni (Şuġti?) Valley in the Qālīqalā Mountains, [M 427] then flows through the territories of Erzincan, Kemaḫ, Ḳuruçay, Egin, and Reşvān. Here it merges with the Murād River. It then flows near Ḥekīm Ḫānı. The Ḳırḳgeçid River, which comes from the direction of Malatya, merges with the Euphrates at the Nevşar ford. Then below Şamīsāṭ (recte: Sumaysāṭ, Samsat), Rūmḳalʿe, Birecik, and Raqqa, it merges with the Rūhā (Urfa) River and then with the Khābūr River. West of the confluence are the towns of Dayr and Raḥba. Then it flows to Maqām-ı ʿAlī and Ḥamās. Here it cuts through the Ḥamur Mountains and flows to ʿĀna, Hayʾat and Ḥadītha. Then it goes past the outlet of the canal that Sultan Süleymān dug at Karbalāʾ and past the outlet of the ʿAqarqūq (recte: ʿAqarqūf) Canal, which is on the east side of the Euphrates. From there it flows to the east of Ḥilla, then past the outlets of the Shāhī, Rumāḥiyya and Samāwāt rivers. In the territory of the sanjak of Jawāzir it merges with the Shaṭṭ (Tigris). These two rivers become one large river like a sea, in the midst of which many islands appear. All streams and branches of the river meet near the fortress of Qurna. Then it flows east of Basra, passes by Qalʿa Jadīda ("New Castle"), and empties into the Persian Gulf. The source of the Euphrates is near Erzurum. One who bathes in its water in the spring will be immune from illness for a year.

Deşt-i Erzen Lake: A small lake, three parasangs in circumference. The water is sweet and many fish appear in it. The land around it is fertile and it is very productive. Ottoman troops may perform their duty there in the winter without hardship. Sometimes because of the severity of the winter no crops can be harvested because three months are needed to grow the crops and bring in the harvest. Crops are sown in the summer months and at the end of summer they have to store them in granaries.

Roads and Stages

Erzurum to Kars: Ilıca, which is near Erzurum > Menzilḫāne 3 hours > Ḥasanḳalʿe, also called Deveboynı 2 hours > Ḳaraḳovacalar 5 hours > Choban Bridge, Żiyāmīr (?) 5 hours > Ṭumāṭāmı, also called Ṣoġan Yaylası, 3½ hours > Kars 2 hours.

Kars to Rawān: Kars > Zaʿīm İsmāʿīl 7 hours. At this stage there is a difficult ford to the other side of the Kars River. The water flows with great force. At the bottom of the river are huge rocks > Şaḥne Geçidi 5 hours—it is the fording place of the Kars River > Şūregīlī 4 hours—the Arpa River flows by this stage > Beygöl 3 hours—here is the Ağır River which is created from melting snow from Mt. Eẕ-ẕikr and has incomparable trout > Şūrbḫāne 4 hours—a grassy flat plain > ʿAyārān 4 hours—with plentiful grass > ʿAbdallar, also called Ṭayılar, 4 hours. The arable lands of Rawān reach this stage (and extend) as far as the village of Gedik which is close to Rawān.

Erzurum to Sivas: > Aşḳāle > Tercan > Kelkīt > Şīrovası > Ḳaraḥiṣār > Sivas.

Erzurum to Diyarbekir: > village of Kāfir, a distant stage. > Aḳçaḳalʿe, where one crosses a large river > Ḫān-ı Geylān, distant and difficult to cross—one must pass over lofty mountains and cross wide rivers > vicinity of Palu, a distant stage, on the bank of the Euphrates > Ḳızlar Geçidi 4 hours > Şuʿābaşı 3 hours > Deyr-i Ḳıyme 5 hours > halting place of ʿAyār Çayı, 3 hours > Pīr Ḥüseyin Depesi 5 hours > Sulṭān Depesi, [M 428] also called Seyrān Depesi > Diyarbekir.

Another road from Erzurum to Diyarbekir: > pasture of Ḥaydargöliler > Aġagöli > Ḳurdyurdı > Ḳarġabāzārı > Köpriḫān > Geylān > Semāviye > Murād River > Ḳızılbıñar > Gölek Baba > Tekye Çayı > Ilıca > ʿAyārbaşı > Pīr Ḥüseyin Depesi > Ḳaṭırbel > Diyarbekir.

Another road from Erzurum to Diyarbekir: > Nerdibānlar > Mamaḫātūn—she was a daughter of the [Aq]qoyunlu prince ʿIzz al-Dīn > İki Aḫur ("Two Stables")—there is a small mosque in town and a public bath outside town, later repaired by Meḥmed Pasha (apparently the governor of Erzurum, d. 1647) > Tercan > Keşīşḫānı > village of Çemen > Erzincan—the road is easy > Mt. Kemaḫ—at this stage one goes through places along the Euphrates that are steep and dangerous > Boġrat Lake, also called Ḫoġīs—the road is difficult > Ḫüstū Ḫānı—the road is difficult > a mountain known as Ḳaracaṭaġ > Ḫüstū mountain pass > Çemişgezek > Pertek > crosses the Murād River > Ḫarpert (Harput) > village of Ḫābūse—half the road is difficult and half is easy > Başḫān > Ortaḫān > Serbeten > Şelīle > Diyarbekir.

From Erzurum to Erzincan: > Ilıca, which is a dependency of Tercan, 3 hours > Ḫānlūc, which is in Bādeklü > Erzincan.

From Erzincan to Sivas: > Ḫvāce Aḥmed > Ezrencik 3 hours > Sūrzāde > Hospice of Ḫvāce Aḥmed > Sivas.

From [Erzu]rum to Tabriz: > Malazgird 6 hours > Erciş 8 hours > Band-i Māhī 8 hours > Nevşehir 3 hours > Sekmānābād 5 hours > Fersenk 6 hours > Khūy 12 hours > Maraud > Ṣofyān > Tabriz.

Status of kings and rulers

After the Iranians, Armenians, and Romans (or Byzantines) the Muslims appeared. During the period of the caliphs, the first to conquer Erzurum was ʿIyāḍ b. Ghanam, by treaty.

When the Abbasid state declined, usurpers took over this region. From among them in 556 (1161) arose the dynasty, known as the sultanate, of the Saltuqids (text: Salīqiyya, error for Saltuqiyya) in Erzurum. In 559 (1164) the ruler of this dynasty (Emir Saltuq) and its leading figures were taken prisoner in a great battle that they fought with the Georgians. Sulaymān b. Ibrāhīm b. Sulaymān, who was known as Ruler of Aḫlāṭ and Shah of the Armenians, and was married to the sister of Emir Saltuq (text: Salīq), Shāh Bānuvān Khātūn, sent magnificent gifts to the Georgians and rescued him. Later Emir Saltuq's (text: Salīq) son Muḥammad became governor of Erzurum.

After that the Seljuk sultan Rukn al-Dīn Sulaymān Sanjar seized the province of Erzurum. In 598 (1202) Malik Shah (the son of the Saltuqid Muḥammad) and the aforesaid Sulaymān fought the Georgians and defeated them. Later Malik Shah came to Kakheti and was imprisoned. After Malik Muḥammad someone named Chāqdash became ruler of Erzurum.

In 598 (1202) the dynasty of the Mankūjiya (Mengüjekids), who were a branch of the Seljuks, took control of Erzincan and Kemaḫ. This province was given to Mankūj Ghāzī by Sultan Alp Arslan. He was succeeded by Fakhr al-Dīn Bahrām Shah b. Dāwūd b. Mankūjiya, who was succeeded by his son Malik Dāwūd.

In 728 (1327) the Chobanid dynasty arose. The famous Tīmūrtash was from this dynasty. He conquered it (Erzurum) on behalf of Abū Saʿīd who was a Chingisid (i.e., the Ilkhanid Abū Saʿīd, r. 1317–1335). After the death of Abū Saʿīd, Shaikh Ḥasan, the son of Tīmūrtash, became governor of Erzurum in 738 (1337). Among later rulers, Malik [M 429] Ashraf was oppressive and brutal. During his reign, a righteous man named Qadi Muḥyiddīn Bardaʿī fled from his oppression to Khānī Beg Khan and took refuge with him. Malik Ashraf marched on him with a powerful army. It is reported that when he reached Tabriz, God sent a great darkness and stormy winds. All of his troops and their horses became confused by this and Malik Ashraf was discomfited. Khānī Beg Khan then attacked and killed him.

In 809 (1406) the Aqqoyunlu, who are the Bayindiriya, arose and invaded this province. ʿUthmān (i.e., Qara Yülük ʿUthmān, r. 1378–1435?), appointed him (Khānī Beg Khan?) governor of Anatolia since he was a descendant of Tīmūr. Afterwards Iskandar b. Qara Yūsuf killed him and his son Yaʿqūb became ruler of Erzurum. Later the Ottoman Sultan Selīm went to war against Shah Ismāʿīl and was victorious and conquered these regions.

Province of Trabzon

The sanjaks of Trabzon, Gönye and Batum have been consolidated and made into a *beylerbeylik*. It contains 454 *ḳılıç* (timars) of which 56 are *zeʿāmet* (timar providing an annual income of over 20,000 *aḳçe*), and 398 are certified and uncertified timars. Exclusive of Gönye, there are according to *ḳānūn* (sultanic law) 700 armored troops. There are 14 fortresses.

Most of this province is along the seacoast. It is five to six stages long and three stages wide. It is difficult, rocky country with large mountains. The forests seem endless. On some of the main roads, there are bridges across the valleys making it possible to go from one side of the mountains to another. The bridges are bound together with rods of wild vine (*bryonia dioica*). From one bridge to another one goes up and down mountains as high as many minarets.

The regions toward Erzurum are also mountainous and it is mainly uphill as far as the borders of Erzurum. The qadi districts of this province are Atina, Ḳabayarġūl, Aḳçaḳalʿe, Arẖova, Of, Plaṭane Tirebolı, Rahvi (Arhavi), Rize, Sürmene, Ṭorul, Giresun, Keşāb, Kürtün, Gönye, Maçka, Māyer, Mayāvri, Yavabolı—also called Görele, and Liçe.

Description of Trabzon

This province is very beautiful. It has abundant walnuts, hazelnuts, apples, oleaster, and all kinds of mountain fruits (or berries). Trabzon, which is the major city of this province, is located near the eastern corner of the Black Sea. It is both a city and a fortress. Most of its inhabitants are Lezgi, whom the people call Laz. The Lezgi Mountains, to the southeast of Trabzon, are part of a chain including the Qaytaq, Alsin and Alburz Mountains extending as far as Bāb al-Abwāb (Darband) in Dagestan. These mountains go beyond Georgia and reach the banks of the Tigris and Euphrates.

Many peoples live here: Mingrelians, Georgians, Abkhazians, Circassians and Laz; both Muslims and non-Muslims. The Muslims are Shāfiʿīs. They call those who live near the extremity of Trabzon, Laz. Southwest of Trabzon, in the Çepni Mountains, live Turkish tribes who mix with the Laz. They speak both Turkish and Persian and are Rāfiḍīs (i.e., Shīʿīs) who worship the Shah of Iran.

In 865 (1460), when Fātiḥ Sultan Meḥmed conquered Sokhumi which is northeast of Trabzon and Ḳızıl Aḥmed Bey fled to Uzun Ḥasan, he also conquered Kasṭamonı, Trabzon, and Sinop and undertook religious war against Georgia. [M 430]

Trabzon is a city with strong firm walls. It has two fortifications, called the Lower Castle and the Middle Castle, and a strong citadel called the Tower, which has a garrison and a Friday mosque. The citadel has no gate that opens to the outside of Trabzon; there is a gate on the north wall that opens to the Middle Castle and a small gate on the south side that is kept locked and only opened when necessary.

The Middle Castle is rectangular and has four gates: 1) the New Friday Mosque Gate, on the east wall near the gate (of the citadel) that opens to the Middle Castle; 2) the Tannery Gate, at the end of the eastern wall, outside of which is a tannery and beyond which is a small river that flows through a broad valley and has a large bridge over it; 3) The Dungeon Gate, at the end of the western wall, over which is the Trabzon prison and outside of which is a small river that flows through a broad valley toward the Zaġanos Gate where it is crossed by a wood and masonry bridge three cubits in length that will be mentioned again below; 4) The Lower Castle Gate opens to the Lower Castle from the north wall. Before the conquest, this Middle Castle was a church in the hands of the Christians. After the conquest, Sultan Meḥmed transformed it into a Friday mosque. A madrasa whose rooms are connected to this mosque is the work of Sultan Meḥmed.

The fortification known as the Lower Castle is square in shape. Certain places along the northern wall are connected to the sea. This fortification also has four gates: 1) the Zaġanos Gate, which leads to the Zaġanos quarter; 2) the Sūtẖā Gate, outside of which is a Christian quarter on the seashore; 3) the Rubble Gate; 4) the Candle Market Gate, where beeswax is made.

Outside the walled city, in all directions except the sea on the north, Trabzon is surrounded by a large suburb where there are gardens and orchards. The western suburb has four Friday mosques while the eastern one has two. The mosques on the west side are: 1) The Ḫātūniye Mosque. When Sultan Selīm Khan's mother died she was buried at the western side of the fortification outside the Zaġanos Gate, and near her grave was built a fine mosque. There are also a madrasa, a guest house, a kitchen, an oven, and a stable for travelers. After the dawn and dusk prayers, meals are prepared for the poor and students. And there is a school for boys, who are also provided with food and an allowance. 2) The Süleymān Bey Mosque. There is an open space in this mosque called Ḳabaḳ Meydānı ("Pumpkin Square"). 3) The Ayasofya Mosque. It is on the shore and had been a church that was transformed into a mosque. It has large marble pillars and is decorated all around with olive trees. 4) The Erdoġdı Bey Mosque. This was also originally a church. There are other mosques as well.

There are all kinds of fruit in this town: grapes, cherries, pears, *bey* pears, *gülābī* pears, Sinop apples, fox grapes (*Solanum nigrum*), unrivaled purple figs, pomegranates, bitter oranges, and a fruit called *ḳara yemiş* similar to the black cherry in size, [M 431] which is a specialty of Trabzon and very sweet.

There is fishing in the sea of Trabzon. They fish for MRLH (?), *mezgit* (whiting, *Gadus euxinus*; text: *mezgir*) and turbot. During the *ẖamsīn* (Pentecost, the fifty days before the spring equinox) they catch a small fish that the people of Trabzon call *ẖabsi*, a mispronunciation of *ẖamsīn*. The town sophisticates make fun of them, for during the *ẖamsīn* the fishermen go out in small boats and when they make a catch a horn is blown to alert the people. The sound of this horn can be heard from a distance of two or three parasangs. When they hear it they say, "The *ḫabsi* have arrived" and rush out, falling all over each other. Nevertheless, these small fish are difficult to digest and are unwholesome. The (beverage) called *māhī-keş* ("fish-puller") is also unwholesome. During the winter this small fish has a bad odor that disturbs people in their homes.

Giresin (Giresun): A small town without a wall on the seashore. It has a mosque and a market. It is full of gardens and orchards. It is located between Samsun and Trabzon. From here Samsun is four stages to the west and Trabzon is three stages to the east. It has a ruined fortress on a mountain. It looks toward the sea. In this town near the seashore there is currently a small lake in which agates are found. They resemble Yemenite carnelians. Another stone that is found here resembles the *ʿaynülhirr* ("cat's eye").

Rize: It is between Trabzon and Gönye. It is fortified and has a garrison. It overlooks the sea. There is high quality cloth (*bez*) made here. It is famous as Rize cloth and is given as a gift. Rize is in a flat area. There is a market in town through which a river flows.

Sanjak of Gönye: It is on the frontier with Georgia. It consists of a fortress and a town on the seashore. Its qadi districts are Atina, Arḫavi, Sumla, and Vice. The Güril and Makrīl (Mingrelian) peoples live along the frontier. The Çoraḳ (Çoruḫ) river flows through here and empties into the sea. All kinds of abundant fruit are found in this district.

Sanjak of Batum: It is situated near the Güril.

Mountains

Mt. Yayla Mescidi: It is between Trabzon and Bayburt. Its waters are sweet. There is always snow on this mountain. There is a certain small white animal that lives here in the snow. It moves about in the ice, but one has to be attentive to spot it. It is a kind of worm. If one sips (the snowmelt from the snow worm) it has the taste of fresh water (*āb-ı zülāl*). There is an old mosque on this mountain called Yayla Mescidi ("Mosque of the Summer Pasture").

Ḫoşoğlan Mountains: A range of large and small mountains two parasangs south of Trabzon. At the foot of these mountains someone named Ḫoşoğlan built a lodge and his name was given to the mountains and to the river. The people of Trabzon call these mountains Ağacbaşı. Their length is 50 parasangs. They are in the middle of the highway that goes from Trabzon to Bayburt and there is one narrow passage; elsewhere they are impassable. The government has appointed some villagers to guard the bridge. If those coming from Trabzon do not show a pass from the city's governor they cannot pass through.

Rivers

The Ḫoşoğlan River flows down from the Ağacbaşı Mountains through Trabzon and empties into the sea. Other rivers are the Sürmene, Of, Rize and Atina, also called Ṣoğukṣu. These rivers flow into the sea near the aforesaid towns. The Faş is another large river that flows near Faş and then into the sea. All of these rivers rise in the mountains of Georgia.

Status of kings and rulers

Some Muslim groups conquered this province after it had been under Persian and Roman (or Byzantine) rule. [M Map of the Caucasus] [M 432] Later, Fātiḥ Sultan Meḥmed II conquered it and its previous ruler, Ḳızıl Aḥmed Bey, fled and took refuge with Uzun Ḥasan.

Supplement of the publisher (İbrāhīm Müteferriḳa). The 41st chapter of this rare book of the age began with a description of the clime of Armenia, one of the conventional climes, and ended with an account of the province of Van. The author was unable to complete the sections describing the remaining climes, as he had promised to do. The account of the clime of Armenia given in the aforesaid chapter conforms with the ideas of all the scholars, and what follows agrees with the views of the men of culture. In accordance with their expectations, the towns of the province of Erzurum, definitely being included in Armenia, were described in the aforesaid chapter, and the towns of the province of Van immediately followed. Because the province of Trabzon, as previously mentioned, is adjacent in some places to the territories of Erzurum and in others is connected to Georgia, they have all been depicted together on a single map (reference to the above map of the Caucasus) and an account of the above-mentioned province (i.e., Trabzon) has been given after that of Erzurum, in order to facilitate reference to it.

The description of the features of these two provinces was extracted exactly as found in the work of (Ebū) Bekr b. Behrām ed-Dimaşḳī. It was then inserted according to the arrangement of the chapters of *Kitāb-ı Cihānnümā*. The description of the clime of al-Jazīra, which is one of the conventional climes known in Greek and Latin sources as Mesopotamia, begins in chapter 42.

Chapter 42: Clime of al-Jazīra (Upper Mesopotamia)

Because this clime is encompassed by the Tigris and Euphrates rivers, and the area not encompassed by them is small, it is called al-Jazīra ("The Island"). Currently there are three provinces in al-Jazīra: Raqqa, Mosul, and Diyarbekir. Some places in other provinces are subject to them. After Noah, the capital of al-Jazīra was Nineveh, which today is called Mosul.

Borders

al-Jazīra's borders to the north, west, and south are with Bilād-i Rūm (Anatolia). To the north, Malatya, Sumaysāṭ (Samosata, Samsat; text regularly has Şamīsāṭ), and Qalʿat al-Rūm (Rūmḳalʿe) are on the Euphrates; to the west, veering south, is Bīra (Birecik); and to the south, Bālis, Raqqa, Raḥba, Hīt and Anbār. At Anbār the Euphrates remains outside the borders of al-Jazīra. Then the border of al-Jazīra curves from Anbār to Tikrīt and runs through Jazīrat Ibn ʿUmar and Āmid. Thus, part of Armenia and part of Anatolia are in western al-Jazīra; part of Syria and the (Syrian) desert are in southern al-Jazīra; ʿIrāq al-ʿArab is in eastern al-Jazīra; and another part of Armenia is in northern al-Jazīra.

This (clime of) al-Jazīra includes Diyār Rabīʿa, Diyār Muḍar, and part of Diyār Bakr (Diyarbekir). Diyarbekir belongs to Abū Bakr b. Wāʾil and Diyār Rabīʿa belongs to his father. They are Ottoman sanjaks. The rest of this region belongs to the bedouin Arabs: the Banī Nizār b. Maʿadd b. ʿAdnān. Diyār Muḍar belongs to Muḍar b. Nizār b. Maʿadd b. ʿAdnān. Both are brothers of Rabīʿa. As a designation of Muḍar, Muḍar al-Ḥumr and his brother, they are called Rabīʿat al-Faras ("Rabīʿa of the Mares"). The reason they were given this name is that during the distribution of their inheritance gold was given to Muḍar and mares to Rabīʿa. The entirety of al-Jazīra was conquered by ʿIyāḍ b. Ghanm at the time of (the caliph) ʿUmar. Much of al-Jazīra is in the hands of various Kurdish and Arab chiefs; [M 433] the rest of it consists of Ottoman sanjaks.

Province of Mosul

Borders

To the east it extends from Kurdistan to Shahrazūr; to the south, Sawād al-ʿIrāq (lower Mesopotamia); to the north, Diyarbekir; to the west, the province of Raqqa.

Division

It is divided into six sanjaks: Old Mosul, which is the pasha sanjak; Bājwānlu; Tikrīt; Harūyāna; Ḳaradāsni; Būdāsni, which is Yazīdī.

Description of Mosul

This city is the administrative center of al-Jazīra. The circuit of its fortification is 1,000 paces. It is on the west bank of the Tigris. Opposite Mosul, on the east bank, is the city of Nineveh, built by Nīnavī son of Bālūs, one of the kings of Assur, in the year 1073 after the Flood. Its circuit is 60 miles. Currently the moat around Mosul is deep and its environs are full of gardens. The weather is fine in the spring, very hot in the summer, feverish in the fall and severe in the winter. The town has fine buildings and there are summer houses along the Tigris.

The shrine of the Prophet Jirjīs (i.e., St. George) is in the middle of Mosul. He lived in the city of Ramla. (At that time) the ruler of Mosul was an oppressive infidel who worshiped an idol named Aflūn (Apollo), called upon the people to worship this idol, and put to death those who opposed him. During that time Jirjīs received the prophetic mission and demonstrated his prophethood with many miracles, but the ruler continually punished him and tormented him. They say that for seven years he was subjected to all kinds of trials and that he was killed 70 times, but each time he immediately returned to life. Finally, due to the wrath of God, a cloud of torments fell upon the infidels who were killed and destroyed. His tomb is today near Ramla.

Water is brought to this city via an underground channel from the Tigris. Water wheels have been set up on the Tigris, and in the middle of the river are mills on boats that are turned by the river.

Near the city of Assur, south of Mosul, the Lesser Zāb River meets the Tigris. Assur is now in ruins. It is mentioned in the Torah. It was the kings of Assur who destroyed Jerusalem.

Mosul has two fortifications, parts of which are in ruins, and a citadel.

The city of Nineveh: It is where God sent Jonah in year 149 after the death of Solomon. Jonah summoned the ruler of the city and its people to the path of God. When the people persisted in error, Jonah warned them, saying "If you remain in unbelief for three more days, the punishment of God will surely come upon you." And without taking leave, he turned and left the city. Soon a dark cloud appeared from God and enveloped the city. Under these frightening conditions, the people sought for Jonah and when they did not find him they bade farewell to and forgave each other and, with faith and sincere repentance, humbled themselves before God and prayed and repented. That sign (i.e., the dark cloud) was lifted and Friday the tenth of Muḥarram became the day of salvation for the people. When Jonah had left the city he had departed in anger. At the Mediterranean coast, he boarded a heavily laden ship. The ship was violently tossed at sea. The sailors could make no headway and said, "There is a fugitive among us." [M 434] So they cast lots. The lot fell upon Jonah three times and they threw him into the sea. By God's command, a great fish swallowed Jonah. The fish followed the ship and when its passengers went ashore the fish disgorged Jonah from its belly. Because of the great heat in the belly of the fish, Jonah's body was thoroughly cooked and he became like a newborn child. Because his movement had become weak, by the grace of God a fresh gourd vine sprouted (and covered him) so that flies would not alight on him. (The archangel) Gabriel appeared and massaged his body. Then his eyes opened and his beard etc. began to grow. Afterward Jonah went to Jerusalem and subsequently died in the village of Ḥalḥūl (text: Jaljūliyya) in year 815 after the death of Moses.

In sum, the people of Mosul are charitable, friendly, and gracious. In a village near Nineveh there is a tomb that the people of the region claim to be the grave of Jonah.

Mosul produces high quality *boğası* (a kind of twill woven in stripes). In the old days they also made high quality copper pots and pans.

There is a petroleum well opposite Mosul on the bank of the Tigris, and very near the city, just as one goes out in the direction of Baghdad, is a place as large as a public square where tar bubbles up to the surface and spreads out. They use this tar on ships and in public baths.

The black cloth (*bez*) of Mosul is extremely fine.

There is a Friday mosque in this city. It has a prayer niche exquisitely carved from stone—something that every visitor should see.

Most of the people of Mosul know four languages: Arabic, Persian, Turkish, and Kurdish.

At a place one stage south of Mosul in the desert near the Tigris there is a hot spring with a masonry dome. There is a single pool to which one descends by a stairway. A kind of pitch-black resin emerges here. It has a pleasant taste and odor.

East of Mosul there is a spring called Ra's al-Nāʿūra ("Head of the Waterwheel") where indigo grows in abundance.

Ḳarġa Çamı: A stage between the Bath of Imam ʿAlī and Ṭopraḳḳalʿe.

Tall al-Tawba ("Hill of Repentance"): A hill east of Mosul. When the people saw the torments of Jonah with their own eyes they went to the top of this hill. They gathered in prayer and repented their sins. God then put an end to his torments. There is a martyrium on this hill. The people visit this martyrium every Friday eve (Thursday night) and make vows.

ʿAqr al-Ḥamīdiyya: One of the administrative districts of Mosul. The Ḥamīdiyya are a clan of Kurds who live in this region.

Shūsh: An administrative district of Mosul. It is a famous fortress in Jibāl on the east bank of the Tigris. It gives its name to the Shūshī pomegranate seeds.

Ḥakkār: An administrative district of Mosul. It is a district and a town above Mosul on the side of a mountain.

Ḥadītha: A town on the Tigris belonging to Mosul. It is near the Greater Zāb River, east of the Tigris, and 14 parasangs from Mosul.

Sanjak of Tikrīt: This is the last of the cities of al-Jazīra. It is on the west bank of the Tigris, six stages from Mosul. The Isḥāqī Canal is south and east of Tikrīt. Isḥāq b. Ibrāhīm, who was chief of police during the reign of Caliph Mutawakkil (r. 847–861), excavated this canal, which marks the beginning of Sawād al-ʿIrāq (lower Mesopotamia). The river Dujayl branches off near Tikrīt and irrigates the agricultural lands (*sevād*) of Samarra all the way to Baghdad. Shāpūr b. Ardashīr Bābāk built the fortress of Tikrīt, but it is now in ruins. A source of naphtha is found there, in the Isḥāqī Canal. [M 435] The thicket begins here.

Sinn: A flourishing town above Tikrīt on the Tigris. The Lesser Zāb River meets the Tigris at Sinn. It is ten parasangs from Ḥadītha.

Balad: (A town) belonging to Diyār Rabīʿa. It is on the west bank of the Tigris. From Balad to Mosul is six parasangs. This place is also called Ribāṭ and Balad al-Khaṭīb ("Town of the Preacher"). Yūnus b. Mattā (Jonah son of Amittai, i.e., the prophet Jonah) lived here.

Barqaʿīd: It has a fortification and market. It is 17 parasangs from Mosul.

Khaḍir: It is in the desert opposite Tikrīt and is in ruins.

Tall Aʿfar: A very forested place between Sinjār and Mosul, six parasangs from Balad.

Kafr Tūthā: (A town) belonging to Diyār Rabīʿa. It has forests and rivers. It is a town on flat ground, 15 parasangs from Dārā.

Ḳızılcaḥān: It is on the Tigris in the direction of Mosul. Ṭopraḳḳalʿesi is between it and Mosul. It is in ruins.

Bawāzīj: A well-known town between Tikrīt and Irbil.

Qalʿa-i ʿAjūr ("Clay Fortress"): It is on a small hill on the west bank of the Tigris and is in ruins. Opposite it the Khābūr River meets the Tigris.

Sanjak of Bājwānlu: They call the people who live near Kirkuk the Bājwān (text: Bājlān) tribe. They are nomads and a Kurdish *ulus* (tribal confederation). Some of them cultivate arable fields. During the harvest they take in the crops and put them in pits. When needed they take them out and bring them to their *obas* (large nomadic tents). Their *yurt* (ancestral grazing grounds) is between the two Zāb rivers and their arable lands are at the foot of Ḳara-caṭaġ.

Bāna: A sanjak opposite Mosul where some Kurdish tribes live.

Harūr: An inaccessible castle north of Mosul. It is 30 parasangs from Mosul and 3 from ʿImādiyya and is included among the towns of Ḥakkārī as a sanjak.

Roads and stages

From Mosul to Tuz-khurmā (text: Tāze-ḫurmā): > Ḥarqal 2 hours > Karbīl and the Zāb River 3 hours > Baghdad 4 four hours > Kīrīl 3 hours > Zāb River next to Baghdad 4 hours > Band-i Shamāmak 3 hours > Bardāwūd 2 hours > İnceṣu 4 hours > the bridge of Altunṣuyu 2 hours. One crosses two more rivers on this road. > Göktepe 2 hours > head-waters of the Khāṣṣa River 2 hours—a very beautiful river > fortress of Kirkuk 2 hours. There are many canals along these roads as well as swamps and rocky areas. > Dāqūq 3 hours > Tuz-ḫurmā 3 hours.

From Mosul to Shahrazūr: Jallād Khānī (or Cellādḫānı, "Inn of the Executioner") 3 hours > village of Qūsh 4 hours > Ḥāriz River 3 hours > Zāb River 2 hours > Shamāmak 3 hours > Bardāwūd 2 hours > Çuḳurbostān River, 5 hours—it belongs to Khānzāde Khātūn > Şelfederesi Çayı 5 hours > Gümsi 4 hours > village of Merebek (Mere Bey?) 4 hours > Siperbārik 3 hours > Altunṣuyu 5 hours > Boġaz 1 hour > village of Aġaçlar, also called Elḳāş, 3 hours—it belongs to Kirkuk > Ḳaynarca, also called the village of Bāzyān near Derbend Pasha > swamp of ʿAlī Shāh 4 hours > Ṭāsānadası, also called Ḫāntepesi, 5 hours > Serçınar 3 hours—a beautiful place > Rāżiyān 2 hours > Hebān, also called Bencīn 4 hours > Chāġān, also called Güberhebān 3 hours > Seyyid Ṣādıḳ 6 hours > Chāġān River 4 hours > Shahrazūr.

Status of kings and rulers of al-Jazīra

The first dynasty (or state, *devlet*) to arise was that of Assur in the year 985 after the Flood. The first king was Bālūs who ruled for 62 years. After him his son Nīnavī reigned for 52 years and he built the city of Nineveh. After him his wife Sāmirā [M 436] ruled for 42 years. She built the city of Samarra. Then her son Rāmish became ruler followed by Aryūs and Arālyūs. The latter taught the people how to wear jewelry and the art of warfare. The last of these rulers, Qūlīras, was extremely timid and cowardly. When the Balūkh people (i.e., Babylonians), who belonged to the Kayāniyān (dynasty), besieged Nineveh, he fled the city unable to oppose them. In the end he threw his treasure and himself into the flames.

After the Balūkh people conquered Nineveh they went on campaign against the Israelites and began to collect an annual tax from them. Later they were conquered by the Ashkāniyān (Parthians). They in turn were conquered by the Sasanians, i.e., the Chosroes. Then the dynasty (*devlet*) of Islam appeared. In year 18 of the hijra (639 C.E.), during the reign of Caliph ʿUmar, ʿIyāḍ b. Ghanam conquered Mosul. Later the Abbasids conquered it. In 323 (935) the Ḥamdānid dynasty arose in Mosul. The first ruler of this dynasty was ʿAbd Allah b. Ḥamdān. After this dynasty that of the ʿUqaylids appeared. They were (originally) from Bahrain. When the Banī Taghlib defeated them they fled that region and came to al-Jazīra. Eventually they made their way to Mosul and occupied it. At that time, the first part of 380 (990), Abū'l-Rawād and then the dynasty of the Aqsunqur Atabegs of the Seljuks had arisen in Mosul. The ʿUqaylids took Mosul from them (in the same year). Later in 486 (1093), when Ibrāhīm the ruler of Mosul was killed, the Seljuk (ruler of Syria) Tutush finally put an end to the ʿUqaylid dynasty.

In 630 (1233) on the death of Maḥmūd b. ʿIzz al-Dīn, the dynasty of the Atabegs came to an end. Badr al-Dīn Lu'lu', supported by Hulagu, conquered Mosul. Then in 659 (1261) a branch of the Ayyubid dynasty conquered al-Jazīra. Then it was conquered by the Aqqoyunlu. Around that time Tīmūr arose, but they (the Aqqoyunlu) submitted to Tīmūr and so were left in possession. Later the Ottoman Sultan Selīm conquered it while on the way to Egypt.

Province of Diyarbekir

This province is located on both sides of the Tigris. Because the Arab tribe Bakr b. Wā'il b. Qāsiṭ conquered this region from the Iranians it was named after them (Diyār Bakr = Country of Bakr).

Borders

To the east is the province of Van, to the north Erzurum, to the west Sivas, and to the south Raqqa and Mosul.

Administrative subdivisions

There are 19 sanjaks and 5 *ḥükūmets*. Eight of the sanjaks are hereditary grants of Kurdish beys. They can neither be dismissed from, nor appointed to, office. The Ottoman sanjaks are the following: Erġani, Aḳçaḳalʿe, Āmid, Çemişgezek, Ḥıṣn Keyf, Ḫābūr, Ḫarput, Siʿirt, Sincan, Siverek, Mafārḳīn, Mazgird, and Nuṣaybīn. As for the sanjaks that are hereditary grants, they are Ataḳ, Pertek, Tercil, Çapaḳçur, Çermik, Ṣaġmān, Ḳulp, and Mehrānī. As for the *ḥükūmets*, they are Egil, Palu, Cizre, Ḫazo, and Genç. The troops raised from this province come from 730 (text: 7030) *ḳılıç* (timars), of which 42 are *zeʿāmets* and 688 certified and uncertified timars. Apart from the Kurdish beys there are 1800 armored troops.

Sanjak of Āmid: The pasha sanjak. It has a post for a molla (chief qadi) earning 500 *aḳçes* per diem. Its qadi districts are Çüngüş, [M 437] Ṣavur, Ṭulp, Mardin, Maġazgird—also called Mazgir, Magirmeydān, and Hini—also called Hani.

This city has immense walls made of black stone. All the houses and quarters are inside the walls. There are four gates: Mardin Gate, Mountain Gate, Tigris Gate, and Erzurum Gate. There is a citadel that encompasses many Friday mosques, madrasas, public baths, soup kitchens, and inns. It is located on a slope of the city. It is a strong fortress that includes the sultan's palace, some houses, and quarters. The city is in a flat place. The citadel, which is at one end, overlooks the plains on the other side of the Tigris. On the banks of the Tigris below the fortress are melon gardens. When the Tigris recedes, everyone goes to his plot of land and plants melons, household by household in the sandy and stony soil. They dig away the stones, mix a handfull of pigeon droppings with five or ten melon seeds and bury them in the sand among the stones. The melons are extremely sweet. In melon season the people go out to their gardens and erect reed huts. They adorn the huts with green and festive arches of ivy and bindweed and the like. They make sandy bands in the river and enjoy themselves for some days eating and drinking. This city serves as a winter quarters for troops. Rulers and military commanders have often spent the winter here. South of Āmid in the direction of Mosul, there is a stone bridge over the Tigris.

Mardin: It is southwest of Āmid and part of Diyār Rabīʿa. Its fortress is on a mountain. The distance from the base of the fortress to the top is two parasangs. Mardin is almost half way up the mountain. It is impossible to seize this fortress by storm. There are deadly snakes in Mardin. There is a rock crystal mine on this mountain. The suburb in front of the walled town is enormous and contains markets and madrasas. The houses of Mardin are like flights of steps, with one residence above another and every street overlooking the street below. There is little spring water; most of the drinking water comes from cisterns. Its plums are very famous.

Ṣādır: A town and castle near Mardin. It is located two stages to the south of Āmid. The Tigris flows near it. It has a Friday mosque, public bath, and a number of shops. Most of the gardens are devoted to growing plum trees. It is located in the eastern foothills of the Sulṭānyaylaġı Mountains. A small river that rises in the mountains traverses this town and flows into the Tigris. There is a ruined citadel.

Ḥīzān: It has many trees, and hazelnut trees are especially abundant. There is a lofty fortress among the mountains.

Mafārḳīn, also called Miyāfārkīn: It is the administrative center of Diyarbekir. As for its gardens and orchards, they are like those of Nuṣaybīn. The grave of Sayf al-Dawla b. Ḥamdān is there. The Mosul road from Miyāfārkīn in the direction of Ḥıṣn Kayfā is six days. It goes via Mardin. There is another road that takes eight days. There is a mountain to the north. Miyāfārkīn is at the foot of this mountain. There is a small river at a distance of a racecourse from Miyāfārkīn. It is called ʿAyn-i Ḥavz. The water gushes up and flows from a spring. It irrigates the gardens to the north of Miyāfārkīn.

Ḥıṣn Kayfā (or Ḥıṣn Keyf, Hasankeyf): A large city on the east bank of the Tigris. Its fortress is to the north. It is also on the riverbank and on a rocky mountain. There is a bridge between the city and the mountain with the fortress on it. This city has excellent grapes, known as *ḥısnī* (text: *ḥıssī*) grapes. [M 438] The first emir of Ḥıṣn Keyf, Merd-i Māhmere, built this city. He was from the Kurdish Ayyubid dynasty. Previously Ḥıṣn Keyf was called Ra's al-Ghūl and then with the construction of the fortress of Kayfā it became known as Ḥıṣn Kayfā ("Fortress of Kayfā").

According to one story, a man named Ḥasan, who was from a distinguished Arab family, was seized and imprisoned here. One day he sent word to the aforesaid emir asking him to give him his mare. The *emir* of the fortress granted his request. By some means he escaped from the prison, mounted the mare and made her gallop across the open square of the fortress. At the end of the square, he spurred on the horse and leapt over the wall into the Tigris. When he landed in the water he swam away and saved himself. For this reason this place is also called Ḥasan Kayfā.

Its districts: Beşirī, Beherdī, Ḫandaḳī, Ṭūrmaḥallī, Pesendīde.

Tribes of Ḥıṣn Kayfā: Āştī; Çelekī, forest dwellers; Greater Kürdeli; Lesser Kürdeli, forest dwellers, where the gall oak grows; Mehrānī, rice growers; Cangī, Cezbonī, also called Ceze or Buḫtī; Istorkī; Şaḳāḳī; Rişānī. There are also Yazīdī tribes: Ḫandaḳī; Behmere, mountain dwellers, where there are mountain sheep, mountain goats, lynx, and martens; Necbūmī, also called Ṭūrkāşkī; Sarhānī, also called Sahrānī; Beşīrī who live near Batman where they have excellent arable lands and where there are lions. There are also Arabs of the Maḥlamī clan who live in oak forests where manna falls from the trees and there are excellent grapes.

Sanjak of Nuṣaybīn: The city of Nuṣaybīn is the administrative center of Diyār Rabīʿa. To the north is a great mountain. The Hermas River rises from this mountain and flows near Nuṣaybīn. Nuṣaybīn is full of flower gardens; there are said to be 40,000 gardens along the river. White roses are peculiar to this city; there are no red roses at all. They call the mountain of Nuṣaybīn Mt. Jūdī. They say that Noah's Ark came to rest on this mountain after six months and eight days at sea. The scorpions are deadly. Nuṣaybīn is very unhealthy. Its people are disagreeable (*serd*; or, reading *sert*, unfriendly).

The Village of Dārā: This is the place where Alexander fought Darius (Dārā), and it was named after him. It is now a village.

Sanjak of Sinjār: It is part of Diyār Rabīʿa, south of Nuṣaybīn. The mountains here are called the Akhzāb range. Except for Sinjār no town grows dates in al-Jazīra. Sinjār is three stages from Mosul. The walls of Sinjār are at the foot of the mountains that are north of the city. There are many gardens and irrigation canals. It has excellent public baths. Its houses are spacious and paved with marquetry (*fusūs* for *fusūskāri*). The water pipes drain into octagonal basins made of stone. The ceilings are decorated with red, yellow, green, and white glass panes. There are rivers and forests, and plentiful bitter oranges and oranges.

The story is told about one of the slave girls of Sultan Malik Shāh according to which the astrologers said, when she was about to give birth in the land of Sinjār, that if she did not give birth on that day, her son would be a great king. So Malik Shāh ordered them to postpone her giving birth. They say that when Sultan Sanjar was born the city was named after him. He was a great king.

The palace of the governor of Egypt, ʿAbbās b. ʿAmr al-ʿAnawī, is near Sinjār. It is a marvelous construction, overlooking gardens and streams. After ʿAbbās the kings made it a residence.

Near Sinjār is a mountain called Çatalgedük. Behind the mountain is a lake called Khātūniyya. In the middle of the lake is an island on which there is a large village. [M 439] On the west side of the island is a hill called Hawātiyya. A large column has been erected on it. A little to the north of it is the cemetery of Şeyḫ Kendi ("Shaikh Village").

Sanjak of Siʿird: It is part of Diyār Rabīʿa. It is on a mountain but is surrounded by flat ground. The Tigris is nearby to the southwest. Miyāfārkīn is to the north of Siʿird; they are one and a half days' journey apart. Āmid is to the south of Siʿird; they are four days' journey apart. The drinking water of the people of Siʿird comes from springs. Siʿird is surrounded by mountains. Figs, pomegranates, and grapes are abundant. There is a famous grape here called the Shāfiʿī grape which is very sweet. Rain makes all the trees fruitful.

Siʿird is northeast of Mosul; there are five stages between them. Siʿird is on a flat plain and the Bitlis River flows to the south. There is another small river that flows into the Bitlis River.

The Fortress of Zirḳī: Also on flat ground. The Bitlis River rises above it. Kefender is a bit higher in the mountains above it.

Sanjak of Ḫarput: A fortress and town on a mountain, overlooking a small lake. It is composed of an *emāret* (read thus, for *ʿimāret*—i.e., it is the seat of an emir or sanjakbey) and a qadi district. It has a great plain that reaches all the way to the borders of Pertek and Çemişgezek. This place is also called Ḥiṣn Ziyād. It is two stages from Malatya. The district of Ulubad belongs to it. Sumaysāṭ (Samsat), which is near it, is part of Diyār Muḍar. It is between Āmid and Ḫarput.

Sanjak of Erġani: It is composed of an *emāret* and a qadi district. It is a fortress and city on a mountain between Ḫarput and Āmid. It has several districts. In the foothills there is a small lake known as Gökçek. It belongs to Ḫarput.

Sanjak of Siverek: It is composed of an *emāret* and a qadi district. It is beyond Ḳaracaṭaġ when coming from Āmid and is located on a plain between the Euphrates and the Tigris.

Sanjak of Çemişgezek: At the upper reaches of the Euphrates and is composed of an *emāret* and qadi district (*ḳażā*) encompassing many districts (*nāhiye*) on the Āmid side of the river. There is a fortress in the center of Çemişgezek. It is at a corner of the mountains. It is on the Euphrates and in front of the city is an open plain. It had previously been a large province (*vilāyet*) encompassing Kurdistan. Its ruler governed 32 fortresses and 15 districts. The province was surveyed during the reign of Sultan Süleymān. The town of Çemişgezek, the district of Ṣaġmān, Cezīre-i Küfre and the sheep tax were annexed to the imperial domain (*ḫāṣ*) and the rest of the province was divided into 2 *emārets* and 14 *zeʿāmets* and (other) timars. Later Sultan Süleymān removed the district of Ṣaġmān and bestowed it on some of the descendants of the rulers of Çemişgezek. The three remaining *emārets* are Mecingerd, Pertek, and Ṣaġmān. Pertek is a hereditary grant (*ocaḳlıḳ*). The part that is the plain is open and broad.

Çermik: It is near Siverek and between two mountains opposite the Nevşār ford on the Euphrates. It is a hereditary grant.

Çapaḳçur: It is on the Murād River on the Erzurum side. It is also a hereditary grant.

Tercil: This is on a mountain near the Murād River in the direction of Āmid going from Ḥānī. It is a hereditary grant. Panbuḳlı Stream rises from Mt. Tercil and joins the Tigris below the Seyyid Ḥasan River.

Ataḳ: It is near Tercil on a plain belonging to the fortresses of Muş, Felek, and Tercil. It is also a hereditary grant. Ṣalāt Stream rises from a corner of the nearby mountain and empties (into the Tigris) below Panbuḳlı Stream.

Mafārḳīn: A ruined town at the Altunṣuyu ("Golden River") bridge between two mountains opposite Āmid. It is a corruption of the name Miyāfārḳīn.

Jazīrat Ibn ʿUmar: [M 440] The fortress of the eighth Umayyad caliph, ʿUmar b. ʿAbd al-ʿAzīz, who in matters of justice was a second ʿUmar b. al-Khaṭṭāb. It is on the bank of the Tigris. When the river is in flood, it divides in two and surrounds the city and fortress. They had to build stone barriers to prevent the river from causing damage to the fortress. Because the people always had to go back and forth by bridge, this place was called Jazīra ("Island"). And because it was associated with ʿUmar they called it "Island of Ibn ʿUmar".

The districts and fortresses belonging to this Jazīrat (Ibn ʿUmar) are well known. They are:
- District of Görgil, opposite Old Mosul. This is where Mt. Jūdī is located, the place where Noah's Ark came to rest. The Kurdish tribes that inhabit this district are: Şehr-i Lūrī, Şehirli, Görgili, and İstorī—these are Muslims; Benvīd, Gāven, Porş, and Hīvrel—these are Yazīdīs.
- District of Berke: Abode of the Berke tribe. It has a fortress. It is also called Bersī.
- District of Erūḫ: Abode of the Erūḫ tribe. Its fortress is formidable.
- District of Pervez: Abode of the Pervez tribe. It has a fortress. This tribe has three branches: Cāstolānī, Bezm, and Kerāfān.
- District of Bādān: The Kāmrasī tribe is found here. It is also called the Dūrbādān.
- District of Ṭanzī: The fortress of this district is called Kāhūk (text: Kelhūk). The Kārisī tribe is found here.
- District of Feyek: There are four tribes here.
- District of Ṭoro.
- District of Heyşem: Most of the people settled here are tribute-paying Armenians. The Çelekī tribe is found here. It is a very productive district.
- District of Şāḫ: Armenians are found here. There is a fortress. The Şīlvī tribe also lives here. It has fine pomegranates.
- District of Teşetel castle and Ermeşāṭ castle: The Berāsī clan is found here. They are one of the largest branches of the Buḫtī tribe.
- District of Gīr castle, also called Ḳamīz: The Kārisī and Ḳarīşī clans found here are attached to the Ṭanzī tribe.
- District of Dīranda: It has the fortress of Larke and is one of the districts of the Tanzi tribe. Some of the people settled here are bedouin Arabs. They are the clans of Ṭahrī, Ṣaghānī, and the Banī ʿUbāda. There are Armenians in this region, most of whom speak Arabic. The Vālersāyī tribes are: Dunbulī, Nūkī, Maḥmūdī, Şeyḫ, Tīzīn—also called Biznī, Māsek, Reşkī—also called Reşī, Mursī—also called Alūtaşī, Miḫ, Nehrānī, Peygānī, Bilānī, Setūrī, Şīrūyān, Dūtūrānī and Erdānī who are Yazīdīs.
- District of Fenek: A flourishing town and fortress opposite Jazīrat (Ibn ʿUmar). The palace of the beys of Jazīra is here. This fortress is sometimes (?—*gā[hī]*) given to the beys of ʿImādiye. The tribes of Fenek are the Necbūy, Şaḳāḳī, Seyrānī, and Gūmiye.

Genc: A *ḥükūmet* in the mountains in the direction of Bitlis from Çapaḳçur. The Murād River flows through the Genc plain.

Egil: A town, qadi district and *ḥükūmet* in the mountains in the direction of Āmid from Genc. It is located in a valley between two mountains. The source of the Tigris is in a corner of these mountains.

Sanjak of Ḫazo: A well-manned fortress and prosperous city. There are mountains on two sides. Here is a broad open valley. Two streams come from the mountains, one from each side of the valley. They almost surround the city. Although the city is not near the streams, it is on flat ground between them. Immediately below the city the two streams merge and pass under a stone bridge. This river, which is known as the Erzen, meets the Tigris below Ḥıṣn Kayfā. Around this city [M 441] are the fortresses of Felek, Ṣāṣūn, and Kefender. Ṣāṣūn is a fortress on a mountain in the direction of Mafārkīn between Delikli Ḳaya and Ḫazo.

Mountains

Kāre: These are the highest mountains in Kurdistan. They are opposite the city of Jazīrat (Ibn ʿUmar). Mt. Jūdī is one of the branches of these mountains. The peak that overlooks Mt. Jūdī is never free of haze and fog, snow and rain. It seems to be hidden in darkness. Because it is so high, Noah's Ark came to rest on this mountain (Mt. Jūdī). In these mountains, there is a kind of wasp that makes honey combs in the ground, in holes like ant holes, and produces honey. It is fine honey, which people bring as a gift when they travel. The wax has an aroma like ambergris, and some grades of the honey are as potent as pastilles of ambergris. It is mostly found on Mt. Jūdī.

Mt. Jūdī: It is one or two hours east of Jazīrat Ibn ʿUmar. There are no trees at all on this mountain. Some herbs grow, such as thyme and lavender. The ground is rocky and brackish. In some valleys on the north side, and in sheltered places, there is snow winter and summer. The place where Noah's Ark came to rest at the summit of this mountain is well known. It is a place of pilgrimage and has a Friday mosque. There is a seasonal festival here once a year, attended by pilgrims and merchants.

At the foot of this mountain is the village of Ṣemānīn ("Eighty"), which is like a small town. The village has a Friday mosque, madrasa, and some charitable institutions. It is two hours to the east of Jazīrat Ibn ʿUmar. The passengers on Noah's Ark first disembarked at this place. This mountain can clearly be seen from Mosul.

In the summer season and in harvest time, manna falls on the gall oak trees (*mazi = Thuya orientalis*) in the mountains of Kurdistan. The people of that region benefit greatly from it. In the summer time what falls is dry and it falls on the leaves. They spread kilims under the trees and collect the manna by shaking the trees. The Kurds call manna *gezengūy*. In harvest season, what falls is moist and it falls on the oak galls themselves. They collect it fresh, mix it with water, put it in kettles and boil it until it thickens. The Kurds call it *cizik*.

Rivers

Ḳaracaṭaġ River: The mountain (Ḳaracaṭaġ) is behind and to the south of the city of Āmid. It is between the sanjak of Siverek and Āmid. The river flows from this mountain and below the Diyarbekir bridge it meets the Tigris. From another mountain at the end of Ḳaracaṭaġ rises another river called the Gökçeṣu. Its headwaters are two streams which, after a short distance, merge and become a large river. It passes under a stone bridge and joins the Tigris below Ḳaracaṭaġ River. This area of Āmid (i.e., the province of Diyarbekir) and the area west of it are open plains.

Languages

The inhabitants of this province speak Arabic, Turkish, Persian, Kurdish, and Armenian. The Muslims among them belong to the Ḥanafī and Shāfiʿī rites. The Kurds are divided into two groups: one is Muslim belonging to the Shāfiʿī rite and the other is infidel Yazīdī.

Roads and Stages

From Diyarbekir to Aleppo: > Canfeza meadow 2 hours > Ḳızıltepe 4 hours > Ḳocaṭaġ 3 hours > Elmalı 3 hours > Acıgöz 3 hours > ʿĀbidūn 6 hours—a rocky road but with cold water > Cülāb River 4 hours > Rūhā (Urfa) 5 hours > Üçpıñar 7 hours > Şakaf Han, also called Beştepe, 3 hours—there are good shallow-dug wells on the roads > Meşhedpıñarı 3 hours—excellent roads and cold [M 442] water > Birecik 5 hours—it is near the Euphrates > Nizib 2 hours > Tall Bashar 4 hours—ʿAyntāb is near this place > Tirişken 6 hours—the road is excellent, with two bridges, one called the Telkarn > Marj Dābıq 3 hours—the road is level and the ʿAynṭāb River flows through here. Sultan Selīm fought (the Mamluk Sultan) Ghawrī on this plain. At the stage opposite, beyond this place there is the shrine of the prophet David. It is on a mountain between Kilīs and ʿAzāz > bridge of Semuk 3 hours—here one crosses the river that comes from ʿAynṭāb. The river is deep and the road is rocky > Cebelān 2 hours > Aleppo 3 hours.

From Diyarbekir to Mosul: > Ḳaraköprü 3 hours. Near an inn half way on the road to Göksu, cold water gushes out of the ground in four or five places. The roads are excellent > Şühūdpıñarı 4 hours—as one approaches this stage, one passes through a rather broad valley > Shaikh Zevli 3 hours—here there is a mountain and valley > Ḫarzem 5 hours > Ḳaradere 5 hours > Nuṣaybīn 5 hours—the road is level and not rocky > Cerraḥ River 5 hours > Gedlekşemāḫī, also called Şemāḫī, 4 hours—one fords the river at three or four places > Dillikār 6 hours—the Ṣafvān River roads are rocky but there is water in the valleys > Ḫābūr 4 hours—the road is level > Kefr-i Zemān (Kafr Zamān) 6 hours—here one fords the Tigris to the stage opposite > Ebū Saʿīd, also called Ebū Şuʿbe, 3 hours > Delefseme 5 hours—the road is excellent and there is flowing water, but there are also swamps > Old Mosul 5 hours > Ḳaraseydi 4 hours > Ḥamāl Kendi (village of Ḥamāl) 3 hours—the road is excellent > opposite Mosul, 3 hours.

From Diyarbekir to Van: > Arpa Çayı 2 hours > Ḳaratepe 6 hours > Batman River 5 hours > Ḫato 8 hours > Üve[y]s el-Ḳaranī 6 hours > Eşek Meydanı 5 hours > Kefender 4 hours. At this stage the road passes through Deliklüṭaş ("Pierced Rock") to a great plain. In ancient times they made a tunnel (reading *naqb* for *naḥt*) measuring 30 cubits long, 10 cubits wide, and the height of three men, and travelled through it. > Bitlis 5 hours > Tatvan 5 hours > Kārmūḫ 5 hours > Savur 4 hours > Adilcevaz 6 hours > Kenzik 9 hours > Erciş 3 hours > Ḳaraköy 4 hours > Band-i Māhī 4 hours > Canıgel 5 hours > Ḳasımoğlı Çayı 5 hours > Van 2 hours.

From Diyarbekir to Malatya: > village of Şelbe 2 hours > Ḳaraköprü, also called Malan—here one crosses the pass known as Deveboynu ("Camel's Neck") > Erġani, 6 hours > Suçuk summer pasture 5 hours. At this stage via the pass at the summit it is 6 hours to Behremaz or else 5 hours to Gölbaşı. > village of Malkoç Efendi, also called Ḳara Baġnik, 6 hours—the road here is level > bank of the Euphrates 6 hours > Malatya 7 hours.

Another road from Diyarbekir to Malatya, via Aḳdegirmen: > Erġani plain 12 hours. One passes over the summit at Ḳızderesi > village of Ḫūḫ > Ḫān of Sinān Pasha > Malatya.

Status of kings and rulers

Diyarbekir was ruled by the Persians, the Romans, then the Persians again, then by the Arab tribe of Bakr Ibn Wā'il, which came from (the direction of) Persia, and then by the Arab kings of Kinda who arose from Diyarbekir. The first among the latter who took the city was Ḥujr Ākil al-Murād b. ʿUmar from the Kinda tribe. This occurred 200 years before the Hijra. Afterwards [M 443] it was ruled by his son ʿAmr and then by his son al-Ḥārith who was allied with Qubād (Kawad), the heretical king of Persia. During his reign the heretic Mazdak appeared. He claimed to be a prophet and ordered the people to share their wealth and women. Qubād joined the religion of this heretic. Ḥārith became subject to Qubād. al-Ḥārith was succeeded by his son Ḥujr as ruler. This Ḥujr was the father of the famous poet Imru'l-Qays. The Banī Asad killed Ḥujr and put an end to his dynasty. When Imru'l-Qays learned that his father had been killed, he obtained help from the Banī Bakr and Banī Thaʿlaba and went to war against the Banī Asad and defeated them. Eventually they fought another great battle. Although he obtained help from the Caesar of Engüriyye (Angora, Ankara), the enemy conspired to send a poisonous garment to his bath. They dressed him in it and he died.

Afterwards, in the reign of (caliph) ʿUmar, (the Muslims) conquered this region. Then the Abbasid caliphs took control of it. Then the Marwanid dynasty arose in Diyarbekir. The first Marwanid was Abū ʿAlī b. Marwān al-Kurdī in 392 (1001) and the last was Manṣūr b. Saʿīd in 478 (1085). Then came the Banī Artuq b. Eksük, who were a branch of the Seljuks. This Artuq was a commander of Malik Shāh, the son of Alp Arslan. Malik Shāh sent him with a great army to conquer Diyarbekir. Later Tīmūr arrived on the scene. He besieged and conquered Mardin and invaded the region all around it. Eventually the Artuqid Shihāb al-Dīn Ibn Aḥmad took possession of it. At that time the Aqqoyunlu clan of the Turkmen, who were called the Bayindiriya, appeared. The first of them was ʿAlā' al-Dīn Ṭūr ʿAlī (text: Ṭūrghul), then Fakhr al-Dīn Quṭlu[gh] Bey, then his son Qara Yülük (text: Eylük) ʿUthmān who allied himself with Tīmūr. Tīmūr appointed him over Āmid, Mardin, and their surrounding territories. Then in the year 908 (1502) Shah Ismāʿīl invaded these regions. Finally, they were conquered by the Ottoman Sultan Selīm.

Province of Raqqa

This is now a province called Urfa.

Borders. To the east it borders on Mosul, to the south on the plain of Sinjār, to the west on the Euphrates, to the north on Diyarbekir.

Administrative subdivisions. Banī Rabīʿa, Jamāsa, Dayr and Raḥba, Raqqa, Rūhā—also called Urfa, Sarūj, Khābūr, Ḥarrān, Julāb, Nīraj, Darā and Banī Qays. In this province there are 653 *ḳılıç* (timars) of which 28 are *zeʿāmet* and 616 certified and uncertified timars. It has 1,400 troops with their retainers. Rūhā is the pasha sanjak.

The story of Abraham and Nimrod, according to the people, took place in this city. But in the (holy) books this occurred in a village named Kūsha, which was one of the villages of Babylon.

The Urfa citadel is on raised ground and two high columns have been erected on its wall. At two places at the foot of the wall, abundant water gushes forth and spreads out like (two) lakes. The circuit of the walls is seven miles. One of the streams emerges from the foot of the exterior walls and flows from west to east. It creates a large pond-like area that extends for 100 cubits. On the north side of the city are gardens. On the south side are a Friday mosque, a dervish convent, and at the extremity, the villa of Shaikh ʿAlī, which overlooks the pond. It is an incomparable promenade, and the town notables come here to relax.

The source of that stream is inside the Friday mosque. It is a place of pilgrimage, the shrine of Abraham. It was built where the water gushes forth and people make pilgrimage to it. At the end of each of these two lakes (where the water flows out) there are two or three mills. The water passes through the city and then irrigates the gardens outside. [M 444] No one catches and eats the fish at this place of pilgrimage. It is well known that misfortune will strike those who catch the fish.

Near this city are hills, mountains, and large caves. The mountains are rocky and full of chasms. The pomegranates are unrivaled. The people of Urfa are very courageous.

Ḳoçḥiṣār: A town and fortress at the corner of a mountain between Urfa and Nuṣaybīn. A river rises in this mountain and meets another river that comes from the direction of Nuṣaybīn and also passes near Ḳoçḥiṣār. Their confluence is below Ḳoçḥiṣār. This river goes around the corner of the mountain and then reaches the plain of Nuṣaybīn.

Jamāsa: A fortress on a hill on the west bank of the Euphrates. Mt. Ḥamr extends from the desert of the Arabs and intersects with the Euphrates in this area. A little above here is the Shrine of ʿAlī, and a little above this is Raḥba. A little above that, on a hill, is Dīr which forms a sanjak with Raḥba. It is located opposite Raqqa and is a fortress on a hill. The river that comes from Urfa and merges with the river that flows by Koçḥiṣār and Raqqa, meets the Euphrates below this hill.

Sanjak of Khābūr. There are two fortresses on a mountain between Ra's al-ʿAyn and the Euphrates. This mountain extends from Ra's al-ʿAyn and from near Karak, which is the source of the Khābūr River, toward the Euphrates. The Khābūr River goes around this mountain and, at the lower end of it, flows into the Euphrates. The Arabs of the Banī Rīsha, who are also called the Mawālī, spend their summers in this region and their winters in Salamiyya.

Māksīn: A city on the Khābūr River. It is seven parasangs from Qarqīsā.

Sarūj: This is now a district of Urfa. It is in ruins. It is one stage from Ḥarrān. There is much water and abundant gardens and fruit. The pomegranates, pears, quinces, and plums are fine. It is also one stage from Bīra.

Ḥarrān: An old city belonging to Diyār Muḍar. It was built in the year 3323 (after the Creation?). It is very large, but now lies in ruins. It was built by the Canaanites. Large and marvelous buildings are still standing. There is a hill in the city on which there was a temple of the Sabians. This hill is associated with Abraham.

Raḥba: About two parasangs from Ḥarrān. There is a mountain here. Mālik b. Ṭawq al-Thaʿlabī built it (in the reign of Caliph al-Ma'mūn, 198–218/813–833). It is a famous city on the Euphrates between Raqqa and ʿĀna, but it is now in ruins. Later in 721 (1321) the ruler of Ḥimṣ, Shīrkūh b. Muḥammad, founded New Raḥba. It is a halting place for caravans coming from Iraq and Damascus.

Raqqa: This place is also called Rāfiqa. It is part of Diyār Muḍar and is northeast of the Euphrates. At an earlier time it had been the administrative center of Diyār Muḍar. It is now in ruins. Gardens and all kinds of fruits are plentiful.

Ra's [al-]ʿAyn: Also called ʿAyn Warda. It is part of Diyār Rabīʿa. It is situated on level ground where more than 300 clear springs emerge (Ra's al-ʿAyn = fountainhead). The Khābūr River originates from these sources.

Qarqīsā: A city near Raqqa on the Euphrates and Khābūr rivers. It is part of Diyār Muḍar. Jarīr b. ʿAbd Allah al-Bajalī (a companion of the Prophet and military leader) died here.

The fortress of Jaʿbar: Also called Dūsariyya, having been founded by Dūsar, who was a slave of (the Ghassanid ruler) Nuʿmān b. Mundhir. Later Jaʿbar Qushayrī took possession of it; and because he lived there a long time it was named after him. This fortress is on the east bank of the Euphrates. It is 25 parasangs from Bālis. Facing it on the west bank of the Euphrates is the territory of Ṣiffīn where the famous event took place. This fortress is seven parasangs from Raqqa.

Description of al-Zūr (Shahrazūr): Also called Būk. This Zūr stretches along both banks of the Euphrates from Bālis all the way to ʿĀna. [M 445] Here and there are mulberry trees, so dense that they are like an impenetrable forest. However, there are narrow roads from the Euphrates that allow passage to each grove. The entrance to each road is surrounded by a broad area of mulberry trees that encircle it like a castle. These places are inhabited by different clans of nomads with their tents (text: *obarlar*, error for *obalar*). These clans produce silk. Every year around Zūr they earn 300–400 purses (1 purse = 50,000 akçes). The clans who live in Zūr pay taxes to the Mawālī Arabs. The leading clan is the Banī Samak. In Zūr there are lions, lynx, wild boars (text: *hizebr* "lions", error for *ḫınzīr*) and other animals.

Supplement of the publisher (İbrāhīm Müteferriḳa): We should not overlook the fact that the territories of the province of Şehr-i Zūl were previously mentioned and described to some extent in the chapter on the clime of Jabal (Jibāl), i.e., ʿIrāq al-ʿAjam (Persian Iraq, Ch. 32). However, because this clime in some places borders on al-Jazīra and ʿIrāq al-ʿArab, and because its description relates also to the territories of Kurdistan that follow (below), and because it is considered an independent province outside of the laws and administration of the Ottoman state, it seemed appropriate to provide more details on it here. We have again extracted the information from Ebū Bekr b. Behrām ed-Dimaşḳī's geographical work exactly as found.

Province of Şehr-i Zūl (Shahrazūr)

This province (*vilāyet*) is a province (*eyālet*) on the border with Persia. It is considered part of the clime of Kurdistan. Most of it consists of provinces of the Gūrān clan. The Sohrān clan lives around the area of Ḥarīr. The rulers of Gūrān and the emirs of Ardalān had authority over (this province). Sometimes they were subject to the Ottomans and sometimes to the Qizilbash (i.e., Safavids). Consequently most of its territory was taken away from them and annexed to the province. They currently make the town of

Ḥasanābād their capital. It is a city near Hamadān on the Şehr-i Zūl—Hamadān road.

Borders: It meets the provinces of Baghdad and Mosul and the *ḥükūmet* of ʿImādiyya, Ḥakkārī, Azerbaijan, and ʿIrāq al-ʿAjam.

Administrative subdivisions: There are 32 sanjaks. They are the sanjaks of Erbīl, Şemāmek, Ḥarīr, Kūy, Ebrūmān, Ūştī, Bāf, Berend, Balḳās, Bīl, Evṭārī, Mt. Ḥamrīn, Cengūle, Dūrāmān—also called Dāverān, Dūlcevrān, Serūcek, Seyyid Būrencīn, Shahrabāzār, Shahrazūr, ʿAcūrḳalʿa, Ġāzī Kashān, Merkāve, Hezārmerd, Rūdīn, Mehrvān, Şemīrāh, Ḳaraṭaġ, Chāghān, Ḳızılca, sanjak of Behbeh, sanjak of Zenge, sanjak of Kirkuk, and Incīrān.

Description of the land of Shahrazūr: The ancients called it Nīmrāh ("Midway") since it is the half way point on the road between Madāʾin (Ctesiphon) and the fire temple of Azerbaijan. The Sasanian Qubād b. Fīrūz founded it, so they used to call it Şīr-i Fīrūz ("Lion of Fīrūz"); and because it was very difficult (*zūr*) to govern, it was called Şehr-i zūr (Shahrazūr), i.e., Şehr-i zūl. It is in the mountainous region of ʿIrāq al-ʿAjam. It is a small city. Its land borders on Marāgha; there are six stages between them. The people are unfriendly. It is eight stages from Baghdad. Near Shahrazūr is a ruined domed building (said to be) the tomb of Alexander. It must be empty because his mother took his coffin to Macedonia.

This city is twelve parasangs from Ḥulwān and five stages from Mosul. The castle of Gül-ʿanber is the capital of Shahrazūr and is on the Shahrazūr plain. This plain is broad. At the end there are mountains running south to north. Springs emerge at the base of a low hill at the foot of the mountains. A large river flows from the mountains through a valley and passes through the plain near this hill. Where it passes the side of the hill [M 446] the late Sultan Süleymān built a fortress. The citadel is on that hill and the exterior fortress is enclosed by the river. Previously the pashas of Shahrazūr lived there. Shah ʿAbbās invaded (this area) and destroyed it. Troops were drawn up in regiments and, on 1 Şaʿban 1039 (15 March 1630), its reconstruction began.

In the mountains at the end of the above-mentioned valley is a pass known as the abode of Azraq Jāzū ("Blue Witch") and the cave of Khalṭ-i Kalām ("Confused Speech"). At the entrance to the pass, a rampart and fortress were built known as the Fortress of Ẓālim ʿAlī ("ʿAlī the Cruel"). Halfway between this and Gül-ʿanber is another fortress known as the Fortress of Charkh-i ẓālim ("the Cruel Wheel"). There is a natural cave at the arch of a sheer and lofty cliff. Below the cave they have cut a stairway from stone and in various places they have placed windows to let in light. This cliff is a mountain that reaches to the skies. There is another mountain opposite the cave where there is a ruin called the Fortress of Yazdajird. A river rises from the tops of these two mountains and flows through the plain of Shahrazūr. The Shahrazūr River is at the entrance to this valley.

Erbil: It is in Bilād-i Jabal ("Land of the Mountain"—i.e., al-Jibāl) and is the administrative center of the land of Shahrazūr. It is a recent city (i.e., not an ancient city). It is between the Greater and Lesser Zāb Rivers. Erbil is two stages from Mosul. At one side of the city within the walls is a high hill on which there is a fortress. Erbil is on flat ground. It has many water channels. Two canals enter the city and flow to the Friday mosque and to the government building (*dāru's-salṭana*). Bilād-i Jabal is commonly known as Irāq al-ʿAjam.

Previously the Turkmen Abū Saʿīd Gökböri b. Abī al-Ḥasan ʿAlī, who had the titles Malik-i Muʿaẓẓam and Muẓaffar al-Dīn, possessed this city. His good works can still be seen. Such good works are unheard of in other places. It is reported that every day a large crowd of poor would assemble in Erbil and he would bestow on everyone according to his need. In winter he gave a special winter garment and in summer he gave a separate allowance. He built a separate *han* for the blind and assigned them an allowance. He also built a separate hospice for widows and assigned them an allowance. He built a hospital and designated what was to be given to the patients according to their need. He also built a hospice for foundlings, and he designated wetnurses for the infants. He also built a large inn for travelers from abroad and designated funds for morning and evening meals for the residents. He even gave each traveler, when he departed, a gift according to his needs. He built a richly adorned madrasa and appointed Ḥanafī and Shāfiʿī professors to it. Sometimes he came to the madrasa and held magnificent banquets for the students. He built a large lodge for Sufis. Twice a year he sent a considerable sum to the seaports for the ransom of prisoners from the infidels. Also every year he appointed a man to accompany the pilgrimage caravan (to Mecca) and while en route this man distributed food and water to the poor. Each year he sent 5,000 goldpieces to the holy cities of Mecca and Medina. He also built some buildings in Mecca. He was the first to bring water to Mt. Arafat and he constructed a number of reservoirs on that mountain. [M 447]

He took an inordinate interest in the birthday of the Prophet. It is reported that every year for this occasion people—including religious scholars, Sufis, preachers, Koran readers, poets, and others—came to Erbil from Baghdad, Jazīra, Mosul, Sinjār, Nuṣaybīn, the Persian cities, etc. About a month before the celebration, he erected about twenty wooden domed buildings, each several stories high. These cupolas were decorated with all kinds of ornaments and on each storey singers and performers entertained the people. The people of the city engaged in buying and selling. These cupolas extended from the fortress gate to the Sufi lodge gate. Every day following the afternoon prayer, Muẓaffar al-Dīn went down to these cupolas and strolled about to view the singers and shadow-play performers. He would spend the night at the lodge and have *semāʿ* (ecstatic dance) performed. (The next day) after the morning prayer, he would go hunting. He would do this every day during the festival.

When the time for the actual birthday celebration arrived, a large number of camels, cattle, and sheep were brought to the city square to the accompaniment of music and song. There they were slaughtered and all kinds of meals were prepared. On the night of the birthday, after evening prayer, they performed *semāʿ* in the fortress. Then, in a procession of large candles and lamps, he came down to the lodge. In the morning he sat on an elevated throne. All the distinguished men, notables, leading figures, and others assembled and two magnificent meals were brought out, one for the people in general and one for those in the lodge. Afterwards they performed *semāʿ* and the preachers preached. Then he gave robes of honor and gifts to the religious scholars, Sufis, etc. according to their rank. This was repeated every year.

It is reported that Muẓaffar al-Dīn was never defeated in battle. Rather, by the grace of God he was always victorious. He died on 18 Ramadan, 630 (27 June, 1233). While his corpse was being sent to Mecca, where he was to be buried according to his last will and testament, an obstacle arose, so he was buried near the Shrine (of ʿAlī) in Kufa. His wife Rābiʿa (text: Rabīʿa) Khatun died in 643 (1245) and was buried in the madrasa that she built at the foot of Mt. Qāsyūn in Damascus. She endowed this madrasa for the Ḥanbalīs.

Fortress of Keşāb: This is one of the fortresses in the districts of Bilād-i Jabal. It is a small fortress between the Zāb and the Tigris, just above the confluence of the two rivers. It is two stages west of Erbil.

Kirkuk: An earthen fortress on level ground among some hills two stages east of Mosul. The whole city is within the walls. The river passes at the base of the fortress. Currently the pashas of Shahrazūr live in this fortress. It is considered to be the ruler's residence (*dāru'l-mulk*).

Fortress of Mihribān: It is on the Hamadan road near the mountain pass, at the extremity of Yüksekyaz[ı]. Behind the fortress to the east, going toward the pass, it is mountainous. North towards the lake it is a flat plain, where the valley turns and goes to the pass. A small river flows from this valley and into the lake. This place is full of reed beds and swamps, but there is a ford.

Dūlcevrān: A desert that extends from Kirkuk to the regions of Baghdad and Shahrazūr. If one goes from Kirkuk to Shahrazūr one reaches the defile of İmām-ı Şāh at the end of the desert, beyond which is Jibāl. [M 448] This Dūlcevrān is a sanjak on level ground on the other side of the defile. The mountain to the right of the defile is also a sanjak, known as Ḳaracaṭaġ.

Hezārmerd: Also a sanjak. It is on the mountain at the extremity of the plain of the Dūlcevrān sanjak.

Shahrabāzār. A sanjak in the Shahrazūr mountains overlooking the nearby plain of Sharazūr.

Ġāzīḳıran: A sanjak and fortress in the mountains.

Ḳızılcaḳalʿe: A well-known and esteemed fortress on the other side of the Chāghān defile going from Shahrazūr towards Irāq al-ʿAjam.

Yelenkān: A fortress and town in the vicinity of Ḳızılca. The fortress is on a high mountain. It is the original residence of the emirs of Gūrān and is too high to be taken by storm.

Şehmerān: A town on level ground at the middle ridge of a mountain on the south side of the Shahrazūr plain. The Diyāla River flows below it. It is reached by a vine ladder.

Serūcek: A fortress in the mountains of Shahrazūr and a sanjak with many districts.

Merkūh: A sanjak is in a defile-like place to the left of the defile of Īmān-ı Şāh at the extremity of the Kirkuk desert.

Balaban Fortress: A great fortress on a lake near Ḥarīr. The outlet of this lake immediately joins the nearby AltınKöprü River.

Ūştī: A sanjak in the mountains on the other side of the province of Ḥarīr.

Samāḳlu: A sanjak on Mt. Samāḳlu near Ḥarīr. Below it on a plain is the sanjak of the [D]evrekānī clan.

Fortress of Geldim: It is built on the same plan as the Fortress of Ẓālim as is the Fortress of Orman. But the town is below the fortress on high ground. The following fortresses are also found in this region: Nevī, also called Nevīn, Meşīġale, Ḥasīd-i Meyve, Küre, Vilek, and Mabker.

Description of Ḥarīr: A town without a fortress on a flat plain near its mountain. It has new and old Friday mosques, public baths, and markets. The territory of Ḥarīr is entered via a difficult road, which is called Çārdīvār ("Four Walls"). It is where the valleys of three rocky mountains converge. Since the path was blocked, the segments (of the path) were made into a road with a broad wall. Altogether it is a province between two mountains. Athwart the valleys of those two mountains is another mountain known as Samāḳlu. This province includes many districts and fortresses.

Devīn: A province attached to Ḥarīr. There is a flat plain behind one of the two mountains (of Ḥarīr). It reaches as far as the mountain, and the area in between is called Devīn. There are villages and districts. These provinces are at the frontier of the Ottoman Empire with Azerbaijan.

Rūbīn: A well-known fortress in the mountains (or in Jibāl). It is on the frontier of Ḥarīr.

The districts of Ḥarīr are Avān[g], Yālkān, Bayān, Samāḳlu, and Şakābād.

The district of Avāng: It has a fortress surrounded by red stones. In olden times the first of the rulers of Sohrān laid siege to Avāng fortress in order to capture it. Because the besiegers fought among those red stones, the people call it Sohrān meaning "reddish" (*surḫāne*). It is a Kurdish expression.

Fortresses of Herbīl and Bayān: They are strong fortresses in Kurdistan. Şerefoğlı Süleymān Bey, one of the great emirs of the Aqqoyunlu Ḥasan Bey, laid siege to the fortress of ʿImādiyya, but when winter came he had to abandon the siege. He then went to the district of Sīrvī to spend the winter. At that time ʿİzzeddīn Şīr Bey was the ruler of Ḥakkārī. Most of the fortresses had come under Ḥasan Bey's authority, so he [M 449] had fortified himself in the fortress of Bayān. This being the situation, Süleymān Bey sent him a message saying "Since the fortresses of ʿImādiyya, Kūregīl, Sevī Bāzūkī and Bayān Şenbū are in our hands, we have no reason to fear you." The rulers of Ḥakkārī had the title Şenbū.

With regard to the fortress of Bayān: It is near Ḥarīr, a little to the southeast. It is a strong fortress at the eastern side of a small lake at the entrance to the Peşt defile. Near the city, the outlet of the lake meets the river that flows north from Balaban and the sanjak of Kestāne. It passes through the plains of the sanjak of Devrekani and the sanjak of Samāḳlu (text: Samāvḳū) to the west and flows west of Erbil. Near the Tigris it joins the river coming from the district of Zībār and north of Samarra flows into the Tigris. Above Erbil this river divides into two branches. They create an island between them and then they reunite. On this island there is a village called Altun Köprisi.

There are other fortresses in this region. They are Çınar, Ḫūşīr, Zencīre, Şerqabū, Bāskī, Verān, Pāre, Perted, Laʿlāb, and Bārīl. Other features of the towns of Kurdistan will be given later in the description of Van.

Province of Kurdistan

There are various accounts of the Kurds. According to some, the Kurds are a branch of the Arabs. According to others, Ḍaḥḥāk of the Serpents had pains in his shoulders and every day he cooked the brains of two men as medication. Since two men were killed every day, the people were greatly oppressed (because they had to slaughter the victims and provide the brains). Finally, the man responsible for carrying out this task, out of pity, killed one man each day and mixed his brains with those of some sheep and sent them to Ḍaḥḥāk. In this way he secretly saved one of those poor victims. The ones who were saved went to live in the foothills of Jibāl. Little by little they reproduced and became a large number. They were dubbed Kurds.

There are four (sic, error for three?) divisions of the Kurds that have emerged over time, so that the manners, customs, and dialects of each division are different. They are the Kurmānc, Kelehrān and Gūrān.

Originally the province of Kurdistan extended from Hurmuz as far as the borders of Malatya and Marʿash. To the north is the province of Arrān, to the south Mosul and ʿIrāq al-ʿArab. Most of these people are courageous, impulsive and proud. They take pride in being thought of as bandits and highwaymen.

The Kurdish clans are Sunnis belong to the Shāfiʿī rite. However, some of them belong to the Yazīdī rite, namely, the Sinnī, Tāsnī, and Ḫāluy tribes in the territories of Mosul and in Syria. They consider themselves to be disciples of Shaikh Hādī, who was one of the Marwāniyya spiritual leaders. They departed from the mystical path and fell into error, becoming heretics and apostates. Most of them are ignorant (of religion). Their shaikhs are renowned for wearing black turbans, thus they are known as the Ḳarabaş ("Black Head"). They never let their wives run away and they purchase places in paradise from them. They refrain from cursing Satan, Yazīd, indeed, everything. They say Satan is an angel brought close to God. They also have a corrupt belief according to which Shaikh Hādī took upon himself their obligation of prayer and fasting, and on Judgment Day they he will bring them into paradise without a reckoning. They bear great enmity toward those learned in the exoteric sciences.

While many learned men have appeared among the Kurds, they do not follow the path of the sciences. For example, they are destitute of the arts of calligraphy, poetry, and epistolography. [M 450] The word "Kurd" presumably means courageous, brave. The most famous and brave figures from early times were from this people, such as Rustam son of Zāl, Bahrām Chūpīn, Gurgīn-i Mīlād (text: Mīlār) and Farhād the champion of Shīrīn, who was from the Kelehrā[n] clan.

The Kurds never act in harmony and agreement with each other. After learning of the prophecy of Muḥammad, kings from all directions responded to his call. At that time the ruler of Turkistan was Oghuz Khan. He sent an ugly Kurd named Bağuz to the felicitous threshold of the lord of men and jinn (i.e., Muḥammad). When the best of men, Muḥammad saw the ugly envoy he was repelled by his appearance and inquired what his tribal origins were. Learning that he was a Kurd, the Prophet said: "May God—praised and lofty—never vouchsafe this people to live in harmony, because the world is bound to come to ruin at their hands." Thus, subsequently, this people never obtained sovereignty. Most of them are shedders of blood and unclean. The blood-price for killing a man is either one horse, one cow, or one or two sheep; or else the relatives of the killer come to a peaceful resolution by giving a girl in marriage to the relatives of the victim.

Most Kurdish men have four wives. Those among the rulers of Kurdistan who possess many tribes are known in reference to their tribes; those who possess fortresses—such as Ḥakkārī and Jazīra—are known by the name of their fortress. Most sultans have had no desire for the territories of this people, but have been content with gaining their submission. Some frontier zones of Kurdistan between the Ottoman and Persian realms have been designated *ocaḳlıḳ* ("hereditary grant") or *ḥükūmet* ("government"—i.e., quasi-independent principalities).

The core mountains of Kurdistan are an immense range stretching from the borders of Fars and Kirmān to the mountains of Van and Erzurum. These mountains never lack snow. In many places there are springs and large rivers and pleasure-grounds. There are 18 provinces (*vilāyet*) in Kurdistan. The weather is mild. During the reign of Süleymān Shah the (annual) government revenue was 200 tomans and 1,500 dinars.

Districts (*nāhiye*) of Kurdistan

Alanī: A town with excellent air and running water. It is a district with many places to hunt game.

Elbester: A medium-sized city. It has a fire temple called Arvakhsh.

Bahār: A fortress. It was the residence of Süleymān Shah during his reign.

Ḥaqshiyān: A sturdy fortress on the Zāb River.

Darband-i Tāj Khātūn: A city that is now in ruins.

Darband-i Zangī: It has a mild climate and abundant water, but the populace are filthy.

Darbīl: A city with fine water and air.

Dīnawar: It was described while discussing ʿIrāq al-ʿAjam. The same for Shahr-i Zūl (Shahrazūr) and Keykūr.

Sulṭān Jamjālābād, also called Sulṭān Yārmanjān: A town with excellent air at the foot of Mt. Bīsutūn. (The Mongol) Sultan Öljaytü (r. 1303–1316) founded it.

Kirmānshāhān, which is called Qarmāsīn in books: It was founded by (the Sasanian ruler) Bahrām b. Shābūr. Qubād b. Fīrūz rebuilt it and built a villa. His son Anūshirwān built a pleasure-dome here. It measured 100 by 100 cubits. The envoys of the Faghfur of China, the Khaqan of the Turks, the Raja of India and the Roman Caesars used to assemble here. It is now a village. Near it is Shabdīz which Khusraw Parwīz built. There is a garden in its plain measuring two by two parasangs. Here were fruits from high mountain and coastal plain. Outside of it were pastures and many animals.

Karand and Khoshān: two villages at the head of the ravine of Ḥulwān. Karand is in ruins, but Khoshān is prosperous. The air is mild and water comes from the mountains to the meadows [M 451] and vineyards.

Māndasht: A province that includes 50 small villages on a plain. The air is mild and water comes from the nearby mountains.

Marsīn: A fortress with a town next to it. There is abundant water and the air is mild.

Wasṭām: A large village opposite Shabdīz. The air is mild and its water comes from the Kevelkū River which rises from Mt. Bīsutūn.

Tuz-ḫurmā ("Salt-date"): A verdant and delightful village with date-palms. A stream of bitter water issues from below a cupola. Where the water issues forth they have made ditches like garden terraces. After they have been filled brimful of water, the stream is returned to its normal channel. This takes one or two days. The water stands in the ditches (and evaporates), leaving pure salt. This water emerges mixed with petroleum. They fill jugs with it and after they sit for a while, the petroleum rises to the surface. They collect the petroleum and pour out the water. This Tuz-ḫurmā is one or two stages beyond Kirkuk in the direction of Baghdad.

Also one or two hours from Kirkuk is a small hill called Gürgür Baba. The top is flat and the size of two threshing floors. Wherever one digs here fire erupts. For example, those who come on pleasure-outings dig a hole here, place cauldrons and pans over it, and cook their meals. From the moment they start to dig, flames appear. When they are finished, they throw some earth in the hole and the fire and flame disappear. A short distance to the west is a small stream and three petroleum wells. On the surface is petroleum and below it is water. Those who want any of it take it and use it. Sometimes in order to make a spectacle, they throw a piece of burning cotton or cloth into one of the wells. Inside the well there is a huge roar and flames shoot up to the sky like minarets. It is a fireworks spectacle for the discerning eye. After a flame shoots up, smoke continues for a while. When the well is empty of petroleum the fire goes out. Near here is a well of tar. It flows out of the ground and spreads over the plain. If a man or animal steps in it, he gets stuck and can't get out. They mix this tar with sand and use it for the pavement of public baths and on ships.

The other districts are: Kalūsh, Shakās, Hāvār, Seymān, Rāvdān, Taġsū, and Shamīrān.

Supplement of the publisher (İbrāhīm Müteferriḳa):

Chapter 43: Clime of ʿIrāq al-ʿArab (Lower Mesopotamia)

The 43rd chapter describes the clime of ʿIrāq al-ʿArab, which is one of the conventional climes. To the west it borders on al-Jazīra and the desert, to the south on the desert, the Persian Gulf, and Khūzistān, to the east on Bilād al-Jabal (i.e., Jibāl) as far as Ḥulwān, and to the north again on the territories of al-Jazīra. The length of the region of ʿIrāq al-ʿArab extends from Ḥadītha in the north to ʿAbbādān in the south along the banks of the Tigris, that is, to where the Tigris flows into the Persian Gulf. Its breadth, from east to west, stretches from Qādisiyya to Ḥulwān. Thus, this region constitutes approximately the provinces of Basra and Baghdad.

Province of Basra

From the time that the Ottomans took control of ʿIrāq al-ʿArab in 1079/1668, it was governed as a hereditary grant. Pashas were appointed to it from the Ottoman capital and it was divided into several sanjaks and districts.

Borders: To the east is the Persian Gulf and the frontier with the [M 452] Qizilbash (i.e., Safavids). To the north is the province of Baghdad, to the south is the region of Laḥsā, and to the west are the sanjaks of Kūfa and Najaf.

Sanjaks and fortresses: Abū ʿArna, Basra, Raḥmāniyya, Zakiyya, Qabān, Qaṭīf, Madīnat al-Qilāʿ, fortress of Bögürdelen, fortress of Suwayb, fortress of Saʿīd, fortress of Qūrna, Ḳalʿe-i Cedīde (al-Qalʿa al-Jadīda, "New Castle"), Kūt Dāwudiyya, Kūt Abū Manṣūr, fortress of Sharash, fortress of Birāgh Nawāda, fortress of ʿAntar Canal, fortress of Madīna, fortress of Ṣāliḥiyya, Kūt Abū Suwayd, fortress of Dād b. Saʿd, fortress of Kūt Baḥrān, fortress of Manṣūriyya, fortress of Fatḥiyya, Kūt Sūra, fortress of ʿAqārā, fortress of Shālūshiyya, Kūt Muʿammar, fortress of Arslāniyya, fortress of Dakhna.

Description of Basra: In the year 14 (635) ʿUtba b. Ghazwān laid out and built Basra on the order of Caliph ʿUmar to found a city next to the sea on the border of Iraq for the Banī Fāris (i.e., Persians?) and as a home for the Muslim Arabs. He built there a mosque of reeds and established himself with 800 men. ʿUtba b. Ghazwān recited the first Friday sermon there and it became famous. It was recorded in the *Ṣaḥīḥ* of Muslim.

The city is on level ground. It is called the "Dome of Islam." To the east are many canals. To the west, near the mouth of the Shaṭṭ is the Arabian peninsula. Thanks to the flooding of the tributary that comes from the mountains, Basra is the government center of the province, full of vineyards and gardens. Because large ships come here with their goods from India and other lands with the monsoon, it has become a large entrepôt. Southwest of Basra is a mountain called Sanām. South and west of the city is desert. To the east are many canals.

It is reported that at one time Basra had 8000 irrigation canals and 7000 mosques. The Ubulla Canal in Basra is 12 miles long, with continual villas and gardens all along its banks. All the canals that surround Basra on the eastern side flow into one another and most of them are tidal basins. Thus at flood tide, the canals overflow and irrigate the fields and gardens. As the tide ebbs, the canals resume their natural course.

Basra is a land of date palms and orchards the likes of which are not seen anywhere else. There is an enormous variety of dates. The dates of Basra are superior to those of Nakhīl, Shayāh, Ḥamāma and other regions.

Basra is 80 parasangs from Kūfa. On the way to Basra, the Tigris divides in two. One is the river proper and the other is the Maʿqil branch. Then they reunite and flow toward Basra. Then they divide into many branches. The water irrigates the district of ʿAbbādān and other places. Basra extends along the river. Some of its buildings are on the desert side of the city. On this side is a gate from which it is three miles to the river.

There are three fine Friday mosques, constructed on stone columns: One, next to the Bādiya (Desert) Gate, is the Old Mosque; another is toward the interior of the city; and there is another Friday mosque that ʿAbd Allāh Ibn ʿĀmir made of sun-dried brick and Imām (Caliph) ʿAlī ordered to be expanded.

There are three famous markets: One, Qaṭʿa al-Kalā, is next to the river; the others are the Great Market and the Bābil ("Babylon") Market. All the markets have excellent and unrivaled public baths.

South of the city is a valley called Wādī'l-Nisā' ("Valley of Women") because women go there to collect mushrooms. Mt. Sanām is a half stage from Basra. The water there is salty because the flood-tide pushes the sea water a distance of three days' north of Basra. [M 453] At a point parallel with Basra the water of the Shaṭṭ mixes with the sea water and becomes salty.

Basra has abundant cotton, milk, meat, and vegetables and there is thriving trade. The tombs of Ṭalḥa, Zubayr, and Anas b. Mālik are here. In addition there are many graves of members of the Aṣḥāb (Companions, contemporaries of the Prophet), Tābiʿīn (Followers, the second generation of Muslims), and great men (of early Islam). The gardens of Basra are on somewhat raised ground so at high tide when the level of the Shaṭṭ rises it irrigates them. Because of the flow and ebb of the tide, the lower land becomes saline.

The Basrans take great interest in keeping track of the pedigree of horses, sheep, and pigeons. For example, they record a horse's pedigree. If the parents are not known, no regard is paid to the foal. When a horse is bought or sold, witnesses confirm its descent on both sides saying it was the offspring of such and such. They also record the pedigree of sheep. They call a certain kind of sheep al-ʿAbdiyya—i.e., related to ʿAbd Qays. They say that when a delegation from ʿAbd al-Qays came to the Prophet, he prayed for them. One of them, who was named ʿUbāda, told the Prophet that he liked sheep. The Prophet took a ram by its ears, with his own blessed hands, and gave it to the man. Wherever the Prophet touched the ears of that sheep with his blessed fingers, he left a mark like a white ring. When that man went to the province of Bahrain, he introduced that ram to his own sheep as breeding stock and the sheep that were born were marked with this white ring on their ears. Those sheep were called ʿAbdiyya, and to this very day the sheep with this white mark on their ears are descended from that line. Having a great desire for them, the people of Basra brought this kind of sheep from Bahrain. Indeed, they report that such sheep have sold for as much as 50 goldpieces, and if the white ring appeared on the ears of a ram, the price was 400 goldpieces. They also pay such regard to pigeons that one sold for 700 goldpieces.

Mukhtār: A city on the Tigris one stage from Basra and near ʿAbbādān. There was an uprising in this town during the reign of the Abbasid caliph Muʿtamid. Caliph Muhtadī executed the participants.

Mashān: A small town above Basra. It is a beautiful place for excursions and known for its many dates.

Ubulla: At the mouth of the canal that flows to it from the Tigris. It is a small, pretty town. One border is the Ubulla Canal, which goes as far as Basra. Another is the Tigris, from which the Ubulla Canal branches off and turns toward Ubulla. The main arm of the Ubulla Canal reaches the sea from ʿAbbādān. The length of this canal is four parasangs from Basra to Ubulla. There are villas and gardens all along the canal, so it is like one long garden. There are channels through which the tide pushes and then through which the water of the Ubulla Canal returns.

Ubulla is to the east of Basra. During high tide, the salt water of the sea reaches the mouth of the Maʿqil Canal. It took its name from Maʿqil b. Yasār who had it dug here in the reign of Caliph ʿUmar. There is no salt water beyond the Maʿqil Canal.

Ubulla has two sides, east and west. Previously the east side was known as Shāṭ ʿUthmān, (built by) ʿUthmān b. Abān in 674 (1275). This side used to be prosperous. There are trees and canals here. This side is on the Tigris and its canals come from the Tigris. [M 454] There is a shrine here where ʿUmar b. al-Khaṭṭāb's comrades in arms are buried. There used to be a lotus tree at this shrine, every branch of which was (the size of) a date palm and every stalk was seven cubits. People used to fumigate the bark of this tree as a cure for fever. When Tābkīn was the governor of Basra, people persuaded him to cut down this tree for some reason and he was dismissed from Basra for cutting down the tree.

The west side of Ubulla is in ruins, but it has a shrine called Mashhad al-ʿAshāra. They say that prayers are answered here.

ʿAbbādān: It is near Khashabāt, where the Tigris flows into the sea. The mouth of the Tigris is southeast of ʿAbbādān. ʿAbbādān is on the shore and is four stages from Mahrūyān and one and a half stages from Basra. It is south east of Basra. This Khashabāt is a landmark for ships and boats at sea. ʿAbbādān has no agriculture and stock raising. The people of ʿAbbādān put their trust in God. Their livelihood comes from everywhere (thanks to international trade). There is a group of people here who refrain from worldly affairs and devote themselves to prayer.

The fortress of Zaynī: This place is opposite ʿAbbādān and very close to it. It is a round island at the mouth of the river.

The fortress of Fakhrī: It is to the west of the mouth of the river.

Qalʿa-i Jadīda ("New Castle"): It was apparently built to guard the mouth of the river. There are two more islands side by side just below the island at the mouth of the river, one called Fakhriyya and the other Shitābī. In shape and size, Qalʿa-i Jadīda resembles the island at the mouth of the river on which is the fortress of Fakhriyya. These three islands form a triangle.

On a level place next to Basra, on the Basra Canal, is a fortress called Kilid al-Baḥr ("Lock of the Sea"). Between this fortress and Basra is Qalʿa-i Sadda.

At the mouth of the Basra branch of the river and opposite Kilid al-Baḥr is Maqām Imām ʿAlī ("Shrine of Imām ʿAlī"), a large town with a Friday mosque and minaret. On one side of the shrine in the direction of Khūzistān is the fortress of Qabān. This is the furthest extent of the Ottoman border. There is a fortress on the bank of the Shaṭṭ near where it enters the Persian Gulf. In order to go, for example, from Qalʿa-i Jadīda to this fortress, one must cross to the opposite side of the Shaṭṭ. The area between them is desert. Above it are three more fortresses close to each other. They are in the hands of the Qizilbash (i.e., Safavids). Near these fortresses is the outlet of a lake. Its waters flow into the Persian Gulf below the bank of the Shaṭṭ (i.e., below where the Shaṭṭ enters the Persian Gulf). Between the outlet of this lake and the bank of the Shaṭṭ is the territory of Qabān. It is like an island.

Raḥmāniyya: This is near Basra. The area between it and the Maqām Imām ʿAlī is a desert. It is on the west bank of the Shaṭṭ beyond where it passes the shrine, that is, on the Kūfa and Baghdad side (or in the direction of Kūfa and Baghdad). In this area two or three canals start from the Shaṭṭ toward the desert. One of these is the Ṣadr-i Ṭawīl Canal. It is just above Raḥmāniyya and it belongs to Basra.

Jazāʾir al-Shaṭṭ: A branch of Shaṭṭ al-ʿArab curves like a semi-circle, then returns to the river proper. They call the semi-circular island, which is created by this branch of the Shaṭṭ, Jazāʾir ("Islands"). It belongs to Basra. The outer edge of Basra borders on the Ṣadr-i Ṭawīl Canal.

The Fortress of ʿUthmān Muḥammad: It is in the middle of an island in the Shaṭṭ al-ʿArab, just where the canal that they started flows into the western desert. It is located in the desert above the Jazāʾir al-Shaṭṭ. During the reign of Sultan Selīm, this ʿUthman Muḥammad revolted. Later, when he submitted, he was given this place as a sanjak.

Shaṭṭ al-ʿArab: [M 455] Sanjak of Jawāzir, which was mentioned above, is an island in the shape of a semi-circle at the place where the Euphrates and Tigris merge. Then the river divides into two branches that go around the island. At the extremity of the island the two rivers again merge. These two rivers then divide into three branches. Two of them become the western and northern branches. One can travel by galley the easternmost of the two western branches. The eastern branch, which is the third of these branches, separates from the other two branches and is somewhat below them and it surrounds this island. In the middle of the island is the sanjak of Madīna. At its extremity is the fortress of Qūrna.

Qūrna: A strong fortress opposite Raḥmāniyya. From here the main arm of the two rivers flows straight into the sea.

One or two canals from this branch of the river divert water to the eastern desert, which is opposite this island. The fortress of Fatḥa (Fatḥiyya?), which is on the bank of one of them, is between the river and the sea. Above on the river side is the sanjak of ʿArna. It is between the river and Baghdad. These areas are specifically called Jazāʾir-i Shaṭṭ al-ʿArab. (Here) the two rivers divide into many branches. At the *kert* (?) of the ʿAqāra Canal the area is flooded because it is a tidal basin and creates a marsh. Troops have to be careful when crossing it. There is a district in this region called Ṣadr Baḥrayn and a fortress called Zarnūk. Fortresses are also found at ʿAqāra. In 975 (1567) when there was an uprising of the ʿAlawīs (ʿAliyān), Selīm II appointed İskender Pasha, who was the pasha of Baghdad, to the position of commander-in-chief and sent boats and troops to that region and brought it back under control.

The fortress of Zakiyya: A strong fortress. At one time it was the capital of the Mushaʿshaʿ dynasty. Later the Ottomans took it and incorporated it into a sanjak with other places under their control. It is east of the Tigris; rather, it is some distance from the river and on the seashore. It is near the place where the aforesaid river meets the sea, opposite and a bit below the Shaṭṭ-i Raḥmāniyya.

Baṭāʾiḥ ("Marshes"): From one direction, at Zuqāq-i Qaṣaba, the Tigris enters one marsh. As soon as it emerges, it enters another great marsh. When it emerges from that, it enters a third marsh and then a fourth. They are like four lakes. The people who live there call a marsh *hawr*. Where the Tigris emerges from Baṭāʾiḥ it is called Dijlat al-ʿAwrā. Here the Basra Canals flow out.

Roads and Stages

From Basra to Baghdad: > east bank of the Tigris where one crosses the Shaṭṭ > fortress of Suwayb > fortress of Saʿīd > Samra Māryān > Sīḥ > Maqrūn > Ghadīr b. Hārūn > Zakiyya > ʿAbd Allāh b. ʿAlī > ʿUbaysī > Kūbīḥ > Abū Ṣadrī > ʿAbd Allāh b. ʿAṭā > Quʿayba > Ṣuwayḥ > Makshafa > Sāl b. Ṣadra > Kūt Ḥūr > Jawāzir > Baghdad.

Another road from Basra to Baghdad, west of the Euphrates: one crosses the ʿAshār Canal > Ribāṭ Azraq > Shams al-Dīn > Nukhaylāt Bākīr > Maʿqil Canal > Kūt Banī Manṣūr, opposite which is Qūrna > fortress of Sharash > Maṣabb Ṭabā > fortress of Bidāʿ Narāda > Ṣadr-i Āl-i Ṭawīl > ʿAntar Canal > Madīna > fortress of Ṣāliḥiyya > Kūt Dāwudiyya > Kūt Abū Suwayd > the fortress of Dād

b. Saʿd > Kūt Baḥrān, opposite which is Fatḥiyya > Bāṭnā > Manṣūriyya > Khayrī > ʿAqār > Thawr > Shālūshiyya > Kūt Muʿammar > fortress of ʿArcha > [M 456] the fortress of Nakīb > fortress of Karīm > Samāwāt, opposite which are Khālid and Kamīsha > Rūmāḥiyya, opposite which is Cem Sultan > Ḥilla, where one crosses the Euphrates > Ortaḥān > Baghdad.

The middle road from Basra to Baghdad: Qūrna > Hadīr > ʿAbra-i Amīr al-Muʾminīn > Āl-i Muʾminīn > Āl Ḥusayn > Fatḥiyya > al-Sabʿ Canal > Jadīda-i ʿAfrād > ʿAbd Warqā > Manṣūriyya > Sikandarī > Shaṭṭ al-Ḥimār > Qalʿa-i Jadīda > Daka > fortress of Qaṣr > Jawāzir > Ṣadr ʿAmmār > Luqmāniyya > Baghdad.

From Basra to Qaṭīf is six stages and from Basra to Wāsiṭ is eight stages.

From Basra to Bahrain: > ʿAbbādān, 12 parasangs > Khūshāb, 2 parasangs > Bahrain, by sea, 70 parasangs. There are two reefs on this route, known as Ghaws and Kasar. If a boat passes over the top of them, it will suffer damage. Otherwise on this route the depth of the water is 70–80 fathoms.

From Basra to the island of Qays: > island of Kharg, 50 parasangs > island of Alān, 80 parasangs > island of Abrūn, 7 parasangs > island of Chīn, 7 parasangs > Qays, 7 parasangs.

Status of kings and rulers

(This area was) first (conquered by) ʿUtba (text: ʿUqba) b. Ghazwān during the reign of Caliph ʿUmar. After the Ṣaḥāba (Companions of the Prophet), during the reign of the Umayyad caliph Muʿāwiya, Ziyād b. Samra (expelled) the Azāriqa (Kharijites), a band of heretics who arose in Basra and Ahwāz, in 69 (688). They were led by Nāfiʿ b. Azraq. This band cursed ʿUthmān, ʿAli, Muʿāwiya, Ṭalḥa, Zubayr, and ʿĀʾisha. Later, in 132 (729), the Abbasid dynasty appeared and it took control of this area. Their first governor was Sulaymān the uncle of Ṣaffāḥ. Then, in 334/945, the Buwayhid Muʿizz al-Dawla occupied it. Then the Seljuks took it, then the Mongols under Hulagu, then the Ilkhanids, then the Aqqoyunlu, then the Mushaʿshaʿ, and then the Ottoman sultan Süleymān took Baghdad. When he did so, the Mushaʿshaʿ who were the governors of V’LYK’ (?) submitted to him and gave him the keys to the fortresses. He left the area to them as a hereditary grant on condition that they mention the Ottomans in their Friday sermons and stamp the Ottoman sultans' names on their coinage.

When Shah ʿAbbās took Baghdad, a prominent and wealthy man named Rustam, the captain of a military contingent, was sent to Basra (by the Ottomans) at the time of the Qizilbash incident (i.e., with the Safavid takeover). The people there, who were Sunnis, fled and rallied to him. He took power and seized Basra, where he became independent, holding it as a hereditary grant from the Ottomans. His son ʿAlī Pasha succeeded him. When ʿAlī Pasha died, his son Ḥüseyin Pasha took his place and was granted the rank of vizier on condition that he retire. Instead, Ḥüseyin Pasha became independent, and behaved badly, so in 1077 (1666) the governor of Baghdad Firārī Muṣṭafā Pasha was appointed in his place and the abovementioned ʿAlī Pasha and his son fled to India. Subsequently, his deputy Yaḥyā Pasha took his place [M 457] and governed in Basra, which was a tax farm for him with an annual income of 700 purses. When he realized that he would not be able to earn that much money, he too fled to India. As a result Basra was subsequently governed by pashas sent from Istanbul.

Province of Baghdad

The city of Baghdad is the capital (seat of government) of ʿIrāq al-ʿArab. Originally, the abode of the (Abbasid) caliphate had been Kūfa, then Anbār, and then it was transferred to Baghdad. The early Muslims called ʿIrāq al-ʿArab (Lower Mesopotamia) "Iran".

Borders: The borders are delimited by the Najd Desert, Basra, the province of Khūzistān, Kurdistan, Mosul, Urfa, and the Syrian Desert. Its length from Tikrīt to ʿAbbādān is 125 parasangs. Its width from the pass of Ḥulwān to the Najd Desert, which is opposite Qādisiyya, is 80 parasangs. The area is 10,000 (square?) parasangs. At the time of Caliph ʿUmar, ʿIrāq al-ʿArab was established as an endowment for the Muslims and surveyed. It came to 360 times 100,000 plots (1 *jarīb*—a plot of 250 sq. ft.) of arable land. There were 40,000 plots in a parasang and 60 cubits in each plot. ʿUmar assessed a plot of wheat at four dirhams, a plot of barley at two, and a plot of date palms at eight. Each plot of date palms had 40 trees. Every plot of fruit was taxed at six dirhams. The dhimmis were counted and their number came to 500,000. They were divided into three levels. The highest level paid a poll tax of 88 dirhams, the middle level paid 24, and the lowest level paid 12. The total came to 1,280 *yük* (1 yük = 100,000) of dirhams, that is, 2,133 tomans. Later during the reign of Caliph Nāṣir, the total reached 3,000 tomans. At the time of the author of the *Nuzha* it was set at 300 tomans.

In 1048 (1638) after Sultan Murād IV wrested Baghdad from the rule of the Shah Safī, the Grand Vizier Ḳara Muṣṭafā Pasha made peace with the Persians. Their agreement established the border which has remained in place up to the present. Within the province of Baghdad, the Ottomans took possession of the regions (*maḥalle*) called

Jasān, Badra, Mandaljīn, Dartang, Darna, and Sarmīn as far as Sarmīl, the plains between Mandaljīn and Dartang, the Żiyāʾ al-Dīn and Hārūnī branches of the Jān tribe, the villages west of the fortress of Zanjīr, the fortresses at the head of the fortress of Ẓālim near Shahrazūr, the environs of Shahrazūr facing the mountains and the fortress of Ẓālim as far as Chāghān Pass, and Ḳızılcaḳalʿe and the places belonging to it. However, Akhiskha, Kars, Rawān, the villages of Shahrazūr east of the fortress of Zanjīr—namely Dayra and Zarwadī, also called Zardyā—and Mihribān and its subject areas would be in the possession of the Shah and would not be harrassed by this side (i.e., the Ottomans). The fortress of Zanjīr, which was on top of a mountain, the fortresses of Qūtūr and Mākū which were on the frontier of Rawān, and the fortress of Manārbard, which was in the district of Kars, were to be pulled down by both sides.

Administrative divisions: This province has 18 sanjaks. They are Āl-i Ṣaliḥ, Baghdad, Bayāt, Tartang (Dartang), Jangūla, Jawāzir, Ḥilla, Darna, Rūmāḥiyya, Bālā at Demirkapı, Zangābād, Samāwāt, ʿImādiyya, Qarāniyya, Ḳaraṭaġ, Karna, Gīlān, Wāsiṭ, *hükūmet* of ʿImādiyya. Currently these sanjaks are given as land grants. The bedouin tribes of Akhshāmāt and Ajmāsāt are very numerous. A detailed account of this is given in my book *Jawalān al-afkār*.

Districts (*nāhiye*): They are Nahr-i Aḥmadiyya, Nahr-i Sultan Sulaymān, Banī Mālik, Kīsha, [M 458] Bilād Rūzīn, Rūmāḥiyya, Zayd, Khāliṣ, Ṣadr ʿAmmāra, [A]jmāsāt Ḳara Ulus; sanjaks of Ḥilla, Jawāzir, Wāsiṭ, ʿArja with Ḥūr Hūyishla, Samawāt, Rūmāḥiyya with Ḥūr ʿUthmānī, ʿĀna, and Qaṣr-ı Shīrīn; districts of Zangābād, Khurāsān, Mahrūr, Rūzīn, Ḥaska, Khālid and Ḥūr Wayla, Kabsha and Ḥūr Ayriq, Ḥāliṣ (cf. Khāliṣ above?), and Banī Mālik; towns of Mandaljīn, Ḳızılribāṭ, Shahribān and the Hārūniyya Bridge, and Jasān; fortress of Niqāb; Citadel of Ḳaraşeker; Nahr Muḥāwīlayn, Wasaṭ al-Nīl, Nahr Qudus, Nahr Ṣalanba (Ṣaltaba?) wa Ḥūriyya, Nahr Shāhī Bāshiyya, Nahr Ṭahmāsiyya, Nahr Sīb, Nahr Musīb Wasaṭ al-Nīl-i ʿAtīq, Nahr ʿĪsā, Nahr Riḍwāniyya, Nahr Iskandariyya; the banks and district of Ḥūr-ı ʿAqarqūf; ships (*safīna*—i.e., districts) of Bahrūz, Sīb, Fallūja, Imām-ı Aʿẓam, Ṣaʿr Qāwushān; and an incalculable number of villages and farms; clans of Āl Yaḥyā, Awlād Ḥāzir, Awlād Mūnis, Zawāmil, Muḥāwira, ʿArnaʿāt, Awlād Nūr al-Dīn, Rabīʿāt, Awlād Qaws, Awlād Masʿūd, Awlād Ḥamadān, Āl Abū ʿArīf, ʿUqayl, Banī Ḥakīm, Zunayrāt, Razīd, Lawāṭīs, Āl Ḥawiyya, Āl Majīr, Āl Ḥasanayn, Āl Dabāb, Āl Amīra, Āl Badr; and many others.

Sanjak of Baghdad: The pasha sanjak and a *mevleviyet* (post for a molla or chief qadi). It is the seat of government of Iraq al-ʿArab. Its qadi districts are ʿArja, al-Imām al-Aʿẓam, Imām Ḥusayn, Imām ʿAlī, Imām Mūsā, Jasān, Jawāzir, Ḥaska, Ḥilla, Akhshāmāt, Khāliṣ belonging to Shahribān, Khurāsān, Dujayl, Dartang, Darna, Rūmāḥiyya, Samawāt, Shahribān, Ṣawlīna, Ḳızılribāṭ (?-text: qzrb't), Ḳara Ulus, Muḥāwir, Mashhadayn, Manhard, Mandaljīn, ʿĀna, Hīt belonging to ʿĀna, Ḥadītha, and Ḥabba.

Description of Baghdad: It has the titles Dār al-Salām ("Abode of Peace"), Burj al-Awliyāʾ ("Tower of Saints"), Zawrāʾ ("the Crooked or the Slanting") because the interior gates of Baghdad are concealed from the exterior gates, and Dār al-Khilāfa ("Abode of the Caliphate)". It was termed Dār al-Salām because the Tigris was called Wādī al-Salām ("Valley of Peace").

The Abbasid caliph Abū Jaʿfar al-Manṣūr built this city on the bank of the Tigris in 148 (765). He made it his capital and called it the "Abode of Peace." As is well known, it is located on the east bank of the Shaṭṭ (Tigris). The flowing river washes up against its castle walls. It is in the shape of a bow two miles long. It has a strong wall made of (baked) bricks and lime that goes all around the city. It has a deep moat. (The fortifications are) known as Aḳḳule ("White Tower") and Burj al-ʿAjam ("Tower of the Persians"). It has a deep moat, into which some water from the Shaṭṭ flows and so the city is surrounded by water.

This wall has four gates: Imām-ı Aʿẓam (Abū Ḥanīfa) Gate, Aḳ ("White") Gate, Ḳaranlıḳ ("Dark") Gate, and the Jisr ("Bridge") Gate. There are also postern gates that open to the Shaṭṭ from the palace. The Jisr Gate extends to the town called Kuşlar Ḳalʿesi ("Castle of Birds") and unites these two cities. From the Shaṭṭ Gate to the Imām-ı Aʿẓam Gate are 12 towers. The distance is 700 cubits. From there to the Aḳ Gate are 34 towers. The distance is 2,050 cubits. From there to Burj al-ʿAjam are 26 towers. The distance is 2,850 cubits. From there to Ḳaranlık Tower are 36 towers. The distance is 2,850 cubits. From there to the Shaṭṭ are four towers. The distance is 50 cubits. From there to the bridge [M 459] are 33 towers. The distance is 2,650 cubits. From here to the upper Shaṭṭ are 18 towers. The distance is 1,050 cubits. According to this calculation, there are a total of 163 towers and the distance is 12,400 cubits.

Inside the city are impressive markets and some squares. It is filled with monumental buildings and charitable institutions. There is a Mevlevi lodge in a delightful place overlooking the Shaṭṭ. Inside the outer walls is a strong citadel garrisoned with janissary troops that come every year from Istanbul. The citadel has its own moat and gate. Inside the citadel are barracks for the troops. One side of the citadel is next to the Shaṭṭ and extends to the Imām-ı Aʿẓam Gate. Above the citadel walls on the Shaṭṭ is a palace composed of lofty pavilions and numerous houses. This is where the governor of the province lives. Overlooking the Shaṭṭ are an audience hall and tiled vaults, all of which are very fine. A beautiful garden has also been laid out with bitter orange, lemon and cypress trees.

There are 12,000 salaried troops raised among the local inhabitants who reside in this city. The weather is extremely hot; even marble fractures in the heat. It is a great emporium. Merchants come by *ghurāb* from Basra with goods from India. They discharge their cargo and trade here. Baghdad has abundant fruit. From Basra come dates; from Baṭā'iḥ, sugarcane and rice; from Wāsiṭ, apples, *razākī* grapes and *ya'qūbī* lemons; and from Shahribān, pomegranates. The soil of Baghdad itself produces dates that are unrivaled, as well as lemons, bitter oranges, and rice.

It is reported that when the Caliph Manṣūr wanted to build this city he ordered a group of astrologers to select a propitious time to do so. On the recommendation of one of them, Nawbakht, they selected a phase of the constellation Sagittarius, when it was ascending and the sun was in that phase. They agreed that this ascendancy indicated that the city would have many buildings, it would last a long time, and it would have a large population. Manṣūr was delighted with this. Afterwards Nawbakht said to Manṣūr, "O Commander of the Faithful, there is additional good news. The death of the caliph and the ascendancy of the city (i.e., when its constellation is ascending) will never coincide, so no caliph will die in the city." Indeed, what Nawbakht said proved true. Manṣūr died while on pilgrimage to Mecca. Mahdī died in Nuṣaybīn. Hādī 'Īsā died in Bādir, Rashīd died in Ṭūs. Ma'mūn died in Tarsus. Mu'taṣim, Wāthiq, and Mutawakkil died in Sāmarrā'.

One of the buildings in Baghdad, built by al-Muqtadir bi'llāh (Abbasid caliph, r. 908–932), is the Dār al-Shajara ("Palace of the Tree"), one of the marvels of the city, comprising a spacious residence and extensive gardens. The audience hall was between two large pools and there were eighteen trees of gold and silver, each with many branches decorated with jewels in the shape of fruits. The trees to the right and left of the pool were decked out with figures of fifteen horsemen wearing silks and brocades, with swords and spears in their hands, advancing in single file.

Tombs (of holy men) and venerated martyria: West of Baghdad are the tombs of Imām Mūsā Kāẓim, Imām Muḥammad Jawwād Raḍī, Imām Abū Yūsuf, Imām Mūsā, Imām Aḥmad Ḥanbalī, Ibrāhīm (b. Adham), Junayd al-Baghdādī, Sarī al-Saqaṭī, Ma'rūf Karkhī, Shiblī, Ḥusayn Manṣūr al-Ḥallāj, Ḥārith [M 460] al-Muḥāsibī, Abū Ya'qūb al-Buwaiṭī, Abū Muḥammad Rī'ashī, and Abū'l-Ḥusayn Ḥaḍarī. East of Baghdad in the small town of Ruṣāfa are the tombs of Imām-i A'ẓam (Abū Ḥanīfa) and the Abbasid caliphs. In the center of Baghdad are the tombs of Shaikh 'Abd al-Qādir-i Gīlānī and Shaikh Shihāb al-Dīn Suhrawardī.

Town of Imām-i A'ẓam: A fortress and town a distance of one hour to the northeast of Baghdad. It is on the Baghdad side of the Tigris. There is a flat plain with date plantations between it and Baghdad. It has a special public bath and noble Friday mosque. There is a special place in the mosque where Imām-i A'ẓam (Abū Ḥanīfa) is buried—may God be pleased with him.

(Town of) Imām Mūsā Kāẓim: A prosperous town on the west side of the Tigris across from the town of Imām-i A'ẓam. The resting place of Imām Mūsā is to one side of the courtyard of the Great Mosque and is under a dome. Imām Abū Yūsuf is buried under a dome resting on four pillars in the courtyard of this mosque opposite the resting place of Imām-i A'ẓam.

Dujayl: Originally it was on Nahr Dujayl (the Dujayl Canal), one or two hours distant from the mouth of the canal. It was a prosperous town and qadi district on the main road with many farms and villages around it. Later, during the time of the Abbasid caliphs, the Dujayl canal was raised from the Tigris and one end of it connected to the Euphrates, a journey of two or three days to the west of the Tigris. Because of a lack of maintenance it fell into disrepair and this town and the villages around it were ruined. In 1061 (1651) the governor of Baghdad Murtaḍā Pasha took an interest in this canal. He cleaned it and brought water through it, returning it to its original state, and settled farmers along it. Today it has again begun to fall into disrepair because it requires a lot of money to keep it productive.

Ḳuşlar Ḳal'esi ("Castle of Birds"): A fortification on the west side of the Tigris opposite Baghdad. Special troops and guards are stationed there. It is a city with pubic baths, Friday mosques, and markets.

'Ukbarā: A town above Baghdad on the Tigris. It is ten parasangs from Baghdad.

Qaṭrabul: between Baghdad and 'Ukbarā. In the past it was the resort of caliphs and refuge of pleasure-seekers. The inhabitants of this town are Christians. It has many churches.

Bazdān: A town from which many religious scholars have come.

Bābil (Babylon): A ruined city on the left side of the road going from Ḥilla to Baghdad. The area around it has turned to forest. Its people say that the sorcery of the Well of Hārūt and Mārūt was performed in the place called Jabal Fulāna; today it has been covered. Bābil was the capital of the kings of the Suryān (Assyrians) and the Nabateans. They say the ruler Hūshang built Bābil 200 years after the Flood and this was the city where people first gathered and there was a confusion of tongues.

It is reported that at the time of Qāliʿ seventy-two giants gathered together in the territory of Bābil and consulted. Fearing lest another flood would soon appear because of their wicked behavior, each of them, deluded by his strength and power, built a strong castle and fortification for himself. No matter what Qāliʿ advised them to do, they ignored his words. Each of these seventy-two giants began to build a fortress for himself (one on top of another). The structure rose to 700 cubits. The upper section, which started 100 cubits above the base, was designed in the form of stories, the interiors of which were hollow. There were six stories. The wall of each story was 2000 cubits wide. According to their erroneous ideas, their building was one-half built; they were determined to go as high as the acme of heaven. [M 461] When Qāliʿ saw this he cursed it. It was fated that God's wrath would befall them. Storm winds arose and the four corners of the foundation began to crumble. The wind blew fiercely for three days and nights. In short, the people were bewildered and terrified. While in this plight, according to a fanciful report, their languages changed and they scattered. In short they were shown seventy-two roads and each went in a different direction where a different language was spoken.

It is related that this was supposedly the location of Paradise where Adam lived. Bābil was also the seat of Nimrod, an agent of Ḍaḥḥāk who had a fortress built there.

Kawnī: It has a market and Friday mosque. It is two parasangs from Nahr al-Malik (the Royal Canal) and six from Qaṣr Ibn Hubayra. Qaṣr Ibn Hubayra is a city near the main branch of the Euphrates.

Karbalāʾ: It is located in the desert west of Qaṣr Ibn Hubayra and attributed to Yazīd b. ʿUmar b. Hubayra al-Fazārī. This Yazīd was the governor of Iraq during the reign of the last Umayyad caliph Marwān al-Ḥimār. This Qaṣr is near Jisr Sūrān which is near Bābil.

Ṣarṣar: There are two such places. One is Upper Ṣarṣar which is a village on Nahr ʿĪsā in Baghdad. The other is Lower Ṣarṣar. It is to the right of the road that pilgrims from Baghdad take when they set out for Mecca. It is three parasangs from Baghdad and two from Nahr al-Malik, which branches from the Euphrates and irrigates the arable land of Iraq.

Madīnat Nahr al-Malik is on a branch of the Euphrates. One crosses this branch by a bridge.

Sukra: the village of Nahr al-Malik. It has a Friday mosque and ancient monuments.

Tall ʿAqarqūf: A hill constructed by Kay-Kāwus. It is to the east of the Euphrates. Here is the grave of Jumjuma Sultan. It is a little above Ḥilla on the east bank of the Euphrates and near the river. From there (the Euphrates) a canal was begun that irrigated the fields of ʿAqarqūf. It is now in disrepair.

There is a village in this area called Kūthī. The fire of Nimrod occurred there and he died there.

It is related that Abraham was born in this village. The astrologers informed Nimrod of his birth saying it was fated and the future of the child would be such and such. When Abraham grew up be began to attack the idols of the idol-worshippers, so Nimrod decided to throw Abraham into the fire. At that time he was 16 year old. When he threw Abraham in the fire he (Abraham) said *Ḥasbunā Allāh wa niʿma al-wakīl* (God is sufficient for us and is an excellent guardian, Koran, 3:173). Nimrod was watching in a castle. Then he saw Abraham sitting in a verdant garden. In short, Abraham remained there for 40 days. Nimrod's rage intensified and he wanted to subject Abraham to further tortures. But God sent gnats to Nimrod. One of them flew into his nose (and entered his brain) and killed him. Afterwards Abraham and his wife Sarah set out on his migration toward Jerusalem.

Qawsān: A small city two parasangs from Baghdad. It has 100 villages attached to it.

Muḥawwal: A small city two parasangs from Baghdad. It is on the bank of Nahr ʿĪsā west of Baghdad. Its orchards reach all the way to Baghdad. Here (Caliph) Muʿtaṣim [M 462] bi'llāh built a pavilion that was unrivaled. Some remains of it can still be seen.

Nīl: A district containing many villages. Its fields and orchards are very productive.

Dayr ʿĀfūl: It is near Baghdad and ten parasangs from Madāʾin.

Sābāṭ: It is near Madāʾin-i Kisrā. It took its name from a river (or canal?) that goes through the middle of Nahrawān. It is four parasangs from Baghdad. It has a number of districts but most are in ruins. This city is counted as one of the seven cities of Iraq. Today this district (sic) belongs to Jalūlā.

A town close to it is Kulwāz. Jalūlā is one of the districts of Sawād-i Baghdad and is on the road going from Baghdad to Khurāsān. It is seven parasangs from Khāniqīn.

Khāniqīn: This is one of the towns of the districts of the arable land of Baghdad. It is on the Hamadān road between Qaṣr-i Shīrīn and Ḥulwān.

Ḥulwān: It is five stages from Baghdad and is the last of (the seven) cities of Iraq. Ḥulwān is reached by ascending the mountains. Most of its fruit are figs and Mawṣif (?) dates. Snow always falls on the mountains.

Madāʾin: This city was the capital of the Chosroes (Sasanians). It is one stage below Baghdad. The Vault of Khusraw (Īwān-i Kisrā = Ṭāq-i Kisrā below) is a marvelous building. Its width is 36 paces and its length is 64 paces. Its height is 70 cubits and the height to the pinnacles is 150 cubits. At the time of the birth of the Prophet, the upper part accidentally broke away and fourteen pinnacles fell to the ground. The Persians call this place Rūmiyat al-Madāʾin. This Madāʾin was the greatest of the seven cities of Iraq. The others were Qādisiyya, Rūmiya, Ḥīra, Bābil, Ḥulwān, and Nahrawān. They are all in ruins. Ṭahmūrath began the construction of Madāʾin and Jamshīd completed it. He built a stone bridge here over the Tigris. The Persians destroyed it. Today to the west of it there is only a small town in the eastern quarter of which, opposite the Vault of Khusraw, is the tomb of Salmān Fārisī.

Rūmiya: After Anūshirwān captured Antioch (from the Byzantines, Rūm), he had a city built near Madāʾin that was modeled after it. When he finished construction he brought the people exiled from Antioch and settled them in Rūmiyya. When they entered the city everyone found his own home. Its resemblance to Antioch was so close that the people thought that they were actually entering that city.

Zāwān: A province between the two rivers. It belongs to Nahrawān. Its crops are unrivaled.

District of Khāliṣ: one of the districts of Baghdad. It is in the desert around Imām-i Aʿẓam. It has abundant villages and farms.

Khāliṣ is a district belonging to Nahrawān. It has thirty villages.

Qūbābād: It was built by a princess of the Chosroes named Qūbā. It is an important province on the Khurāsān road. Today they call it Yaqūbādān. A canal coming from the river of (or Nahr) Nahrawān brings water to the villages of this district. It is full of date groves and orchards. Oranges (*nārenc*) and bitter oranges (*turunc*) are especially abundant. There are eighty villages in this district.

Qaṣr-i Shīrīn: Khusraw Parwīz built it for his beloved Shīrīn. It is a great fortress whose circuit is 1000 paces. Khusraw built another fortress for himself west of here where he had his capital. He also built a large and lofty hospice. The Ḥulwān river flows through this area. The air is very bad; the simoom winds blow. The stream bed thought to be Shirin's milk channel is the water channel of the capital.

It is related that at the Ṭāq-i Wustān (Ṭāq-i Bustān) stage west of Mt. Bīsutūn, at a place that ends in a plain, a very cold stream emerges, (strong enough to) operate one or two [M 463] mills. It flows from above a ledge about a cubit high at a place carved out of the mountain. In the recess of the arch an image of Khusraw and Shīrīn has been carved from the stone and (below) Rustam is mounted on a horse and wearing armor. With the passage of time some of the carving has eroded. On each side of the celestial vault has been placed the image of an angel.

Wāsiṭ: A city in the desert on the Baghdad side of the Tigris between Baghdad and Basra. A canal was constructed to it starting from the Tigris. This is the last of the sanjaks of Baghdad in the direction of Basra. The reed pens of Wāsiṭ are famous. Half of it is on one side of the Tigris and half on the other. There is a bridge of boats between the two sides. The name of the city derives from the fact that it is half way (*wasaṭ* = middle) between Basra and Kūfa. Furthermore it is 50 parasangs in each direction from Baghdad, Ahwāz, and Kūfa. al-Hajjāj Ibn Yūsuf al-Thaqafī (the Umayyad governor) laid out the city. He began in 84 (703) and finished in 86 (705).

Shākhā: one of the districts of Wāsiṭ. A (religious?) community came into existence here and is named after it.

Jarjarāyā: A town near the Tigris, on the west bank, between Baghdad and Wāsiṭ. It is four parasangs from Dayr ʿĀqūl and nine from Jabal.

Jabal: A town between Baghdad and Wāsiṭ on the Shaṭṭ and Tigris (!). Many people trace their origin to it. One of them was Abū'l-Khaṭṭāb Jabalī. There was a dispute between him and Abū'l-ʿAlāʾ al-Maʿarrī.

Nuʿmāniyya: A well-known town between Baghdad and Wāsiṭ. It is a town in the district of the Greater Zāb.

Famm al-Ṣulḥ: on the west bank of the Tigris near Wāsiṭ and 12 parasangs from Jabal. The wedding of the daughter of Ḥasan b. Sahl, the vizier of (Caliph) Maʾmūn, took place here.

Shahribān: A large and prosperous town on the east bank of the Diyāla. It is three stages toward the northeast of Baghdad. It has a Friday mosque, public bath, and a small market. Its gardens are full of date trees, lemon and bitter orange trees, pomegranate and fig trees. The pomegranates are very large and unrivaled. Their vineyards also contain all kinds of grapes. The canal that rises from Diyāla flows through the middle of town.

Kūfa: on a western branch of the Euphrates. There is agreement that the word Kūfa indicates metaphorically that the promises and covenants of its people are not valid. In like manner there is a proverb that states, *al-Kūfī la yūfī* (The Kufan does not fulfill his promises). It was built during the caliphate of ʿUmar. Today it is in ruins. Saʿd (b. Abī) Waqqāṣ had rebuilt it. Its water comes from Nahr Nāhiya (or *nehr-i nāhiye*, the district canal?). It is full of date orchards. Its grain, cotton, and other crops are fine. Noah's Oven is in a mosque of this town—it is the mosque where ʿAlī was wounded, and today it is the Old Friday Mosque of Kūfa. Apart from ʿAlī's houses there is no other habitation. There are several blessed cemeteries.

ʿAlī's tomb is a distance of one hour from here in the territory of Najaf in the direction of the Ḥijāz. ʿAlī dug a well there and the water was extremely sweet. There is no other well with sweet water in Kūfa.

Imām ʿAlī: A town inside a large fortification on a level place in the territory of Najaf near Kūfa. It has a Friday mosque, public bath, and flourishing markets. ʿAlī's tomb is there in the shape of a square. There are cells all around it, on the ground floor and upper floor, constructed with Kashani tiles. [M 464] Shah Ṭahmāsp built the tomb and its courtyard. His name is written on the tiles in *jalī* script. The tiles are gilded. The entire structure is roofed. It is a separate enclosed mosque where six Korans have been placed. In two walls of the courtyard are six stories of rooms. The courtyard is next to the walls. Outside the courtyard is a Friday mosque. Imām ʿAlī is a one-hour journey from Kūfa.

Dhū'l-Kifl: the village where the tomb of the Prophet Dhū'l-Kifl is found. It is three hours distant from Kūfa. That place is also called Sarmalāḥa. There are some buildings in that district built by the Safavids. They are beautifully adorned. Collected here are votive offerings and gifts from treasuries that have come from everywhere. There is a special storage place where gold, silver, and jeweled candelabra are safeguarded. There is also a special library with a large endowment. ʿAḍud al-Dawla al-Daylamī established a lofty soup-kitchen on top of it. Ghāzān Khan built here a residence for descendants of the Prophet and a dervish lodge. The Mongol sultan Öljeytü built a mosque and minaret.

(Najaf:) a town near Kūfa founded by ʿAlī and completed by Manṣūr-i Dawāniqī. He also built the wall around it, and around Kūfa. Its circuit is 18,000 fathoms. "Moon stone" (*ḥajar al-qamrāʾī*) or Najaf stone is found at Najaf. It expands during the first fifteen days of the month and shrinks during the last fifteen days, following the phases of the moon.

Khawarnaq: A canal in the territory of Kūfa. There is an impressive villa here with that name. It is mentioned in poetry and in the annals of the Arabs.

Ḥīra, also called Ḥīra-i Bayḍāʾ ("White Hira"): In earlier times it was a glorious city with many canals. Before Islam it was the residence of the dynasty of Nuʿmān b. Mundhir. Mundhir b. Imruʾ al-Qays became a Christian in Ḥīra and he built great churches here. Ḥīra was built when the Tubbaʿ marched in this direction from Yemen. It is one parasang from Kūfa.

The noble place of the martyrium of Imām Ḥusayn: It is on the same plan as the Shrine of ʿAlī. Every year during the month of Muḥarram it has a fair to which pilgrims and merchants flock. The keepers of these two noble martyria are chosen from *sayyids* (descendants of the Prophet). Imām Ḥusayn is one stage distant from Ḥilla and in the Karbalāʾ plain toward the north. Kūfa is two stages west of Ḥilla. There is a prized date here like mastic called *mafthal*.

Ḥilla: between Baghdad and Kūfa. They call it the Ḥilla of the Banī Mazyad. The first to come here, lay out the residences of the city, and enlarge it was Sayf al-Dawla Ṣadaqa (b. Manṣūr) b. Dubays b. ʿAlī b. Mazyad al-Asadī. He arrived in 495 (1102). Ḥilla is mostly on a flat area west of Baghdad and west of the Euphrates. Its buildings are strung along the bank of the Euphrates. Its walls are in ruins. Most of its gardens and orchards are on the east bank of the Euphrates. Ḥilla is two parasangs from Baghdad. At Ḥilla there is a bridge over the Euphrates. In Ḥilla they make shawls, turbans and horse bits without rival, also fine celadons hardly different from genuine (i.e., Chinese or Korean?) celadons except with respect to weight.

Fallūja: on the Euphrates above Ḥilla one stage to the east. It is a famous port. Boats come here from Birecik. It is a village with inns. A river comes from it which is a branch of the Euphrates. It meets the Shaṭṭ between Imām Mūsā and Kuşlar Ḳalʿesi, at the point known as Minṭaqa. When the river overflows they make *keleks* (rafts of inflated skins). The governor of Baghdad [M 465] Ḥāṣekī Meḥmed Pasha built a dam at the place called Minṭaqa. This is because when the water was in flood it sometimes damaged the fortification of Baghdad.

Anbār: on the bank of the Euphrates. The shrine of the first Abbasid caliph Abū'l-ʿAbbās al-Saffāḥ ʿAbd Allāh is here.

Ḥadītha is one parasang from Anbār. This Ḥadītha is called Ḥadīthat al-Nūr ("Ḥadītha of Light") and is on the mid-Euphrates.

ʿĀna: an island in the Euphrates. Ḥadītha is below it and below that is Hīt. ʿĀna is at the extremity of the border of Baghdad. It has a very large number of olive trees. Olives are grown nowhere else in the province of Baghdad. It has many religious scholars, virtuous men, lovers of music, and people of refinement. This district is very prosperous. It is about one parasang from Anbār. Previously Nuṣayrīs lived in this area but today they are very few in number.

Hīt: It belongs to ʿĀna. It is above Anbār. The tomb of ʿAbd Allāh b. al-Mubārak is here. There are springs of tar and petroleum here. It is eight parasangs from Anbār.

Qādisiyya: A small city with many date palms and streams. Qādisiyya, Ḥīra, and Khawarnaq are all located between the desert and the sown of Iraq. Qādisiyya is to the east of the mountains and rivers. It is 15 parasangs from Kūfa and the road between them is on the pilgrimage route. This Qādisiyya is the place where Saʿd (b.) Abī Waqqāṣ fought the famous battle against the Persians.

Samāwāt: an extended plateau in the Wādī al-Samā near Kūfa and belongs to the tribes of the Banī Kalb. It is a place where there are no stones at all. It is a qadi district of many villages and nomads. The people are fanatical Sunnis.

Sanjak of Rūmāḥiyya: between Nahr Shāhī, which is below Ḥilla, and Nahr Rūmāḥiyya, which is below that. It is irrigated by its own canal. Below it is Nahr Samāwāt. It is an area with a Friday mosque, public bath, orchards and date groves. The town quarters are laid out amidst gardens.

Jawāzir: at the confluence of the Euphrates and Tigris. Two hundred Qizilbash (i.e., Safavid) *beys* are stationed here from year to year in rotation. They do not leave the fortress for a year. When one man's tour of duty is up they immediately replace him with another. They also supply their grain from outside. The fortress has a postern gate through which they are taken. All the lands of the sanjak are under the control of the Ottomans and an emir is placed in charge. Between this place and Jasān there is a desert.

Jasān: A low-lying qadi district between Jawāzir and Dartang, on the Iranian border. Its villages are inhabited by Iranians. Nahr Āftāb, which comes from Mt. Gīlān behind the Ḥīra mountains reaches as far as the base of the aforesaid fortress. It is in the middle between the fortress of Badra and the inn of Mughūlī. Badra takes its water from Nahr Āftāb.

Dartang: A fortress on the frontier with the Qizilbash (i.e., the Safavids), built on a mountain top, at a narrow pass on the main highway to Fars. It is a strong and well-built fortress that was built to close and guard the aforesaid highway. Dartang is known for its beautiful women and lovely boys whose beguiling quality is acknowledged by one and all. The fortress of Dartang is in a valley and the Diyāla River flows by it. The sanjaks of Gīlān and Kartan are very close to Dartang. Qaṣr-i Shīrīn is also in the vicinity of Dartang.

Zangābād: an independent district west of Khāniqīn.

Darna: similar to Dartang. It is comprised of a fortress in the mountainous area on the highway to Hamadān.

Jayl: A sanjak in the desert behind Ḥamra. In this area large canals were excavated from the Diyāla River which go to the surrounding villages. This is a prosperous area. Ḳızılribāṭ has been a sanjak in this area from of old. Badrāy, [M 466] Kashāb, and some towns belong to it. It has a stream whose water is brackish, but there is an underground channel one parasang distant whose water is sweet. In Badrāy there is excellent sugar cane. Its dates are also famous.

Jangūla: located in the area of Qalʿa-i ʿAjam in a mountainous region and is situated on a mountain. A river flows from this mountain.

Sanjak of Bayāt: near the district of Jangūla. It is a fortress located on level ground along a river in a valley where Mt. Jangūla and another mountain meet.

Dih-i Bālā: A sanjak on Mt. Īwān on the border. A river that rises from this mountain goes up to Mandalī. There is a wide plain between this mountain and Mt. Gilān.

Gīlān: A fortress dependent on Dartang. It is on a mountain toward Iran.

Ḳarand: A fortress at the foot of a mountain near Dartang. There is a plain between it and Gīlān. A river flows through this plain.

Ḳaraṭaġ: A sanjak on a mountain opposite Baghdad. It is at the extremity of the desert behind Mt. Ḥamra.

Old Baghdad: A place of ruins on the Baghdad side coming from Sāmarrāʾ. It is on the bank of the Tigris and opposite the fortress of Tikrīt (Tikrit is north of Sāmarrāʾ).

ʿĀshiq and Maʿshūq: Āshiq is not far from Baghdad. Maʿshūq is a little above Āshiq.

Sāmarra: Originally named *Surra man raʾā* ("He who sees it is delighted"), it was built by (Caliph) al-Muʿtaṣim. It is on the east bank of the Tigris twelve parasangs north of ʿUqbarā. It has sound air and soil. (The son of al-Muʿtaṣim Caliph) al-Wāthiq expanded it to the city of Hārūniyya. (His son Caliph) al-Mutawakkil expanded it to the city of Jaʿfariyya. The length of Sāmarrāʾ including these cities was seven parasangs and the width was one parasang. al-Muʿtaṣim made this place his capital.

Today it is a town nearly in ruins twelve hours from Baghdad. It has a place of pilgrimage: one passes through numerous streets and houses to a roofed-over place where there is a well. If one looks inside it, a full moon is clearly visible. People believe that this is a sign of the presence of the Mahdi. Those who currently live there are *sayyids* (descendants of the Prophet).

al-Muʿtaṣim supervised the construction of this large town. It has a Friday mosque with a tall minaret. al-Muʿtaṣim put a magnificent basin in the mosque. It is 22 *bāʿ* (1 *bāʿ* = 2 armlengths) in circumference. In front of the mosque are the tombs of the Infallible Imām ʿAlī al-Naqī (the tenth imam of the Twelver Shīʿa), Imām Mūsā (al-Kāẓim, Imām ʿAlī al-)Riḍā (the seventh imam of the Twelver Shīʿa, who is buried there), and Imām Ḥasan al-ʿAskarī (the eleventh imam of the Twelver Shīʿa).

Ṭuz-ḫurmā: one stage from Dāqūq in the direction of Baghdad. Between it and the village of Dāqūq is the city of Kirkūk.

Dāqūq: A medium-sized city. Its air is excellent. It has petroleum wells.

Ṣadrayn: This province has a great abundance of grain, dates, animals, and water.

Baṭā'iḥ (Marshes) of Wāsiṭ and Basra: This is where all the canals (or branches) of the Tigris converge. In flood stage the canals fill and turn into lakes. The resulting marshes are eighty parasangs in length and width. They originated at the beginning of Islam when the Persians and Arabs were fighting. Throughout them are towns and villages. The principal town is called Jāmda. The people of this town were Chaldeans and Sabians, descended from the people of Seth. Abū Isḥāq al-Ṣābī (astronomer, d. 335/946) was one of them.

Baṭā'iḥ (Marshes) of Kūfa: This is where the flood waters of the Euphrates collect. Bedouin Arabs nomadize around it. It is a refuge for highwaymen.

The *hükūmet* of ʿImādiyya. It is located opposite Jazīra and extends opposite Mosul. Most of its territory is mountainous. Baghdad and Shahrazūr are behind it (south of it). Most of its fortresses are in the direction of Shahrazūr. Its capital is the city of ʿImādiyya. It is surrounded by mountains. It is subject to Baghdad.

Description of ʿImādiyya: It has a well-built stone fortress laid out on a level place on the top of a mountain. It is three stages to the northeast of Mosul. [M 467] Below it are flowing streams and abundant gardens. Aqsunqur, one of the (commanders) of ʿImād al-Dīn Zangī, destroyed the fortress of Āsib, which was one of the largest fortresses of Ḥakkārī and built this one nearby in its place. Sharaf Khan relates that ʿImād al-Dīn Zangī built it. The fortress and city are round and on a broad plain. Some quarters and their environs are approximately 100 cubits (1 *gez* = 26 inches) high, 50 or 60, some 20. Inside the fortress are two wells. They were reached by digging through the rock. These wells provide water for the city's public baths, madrasas, and other institutions. They bring other water from outside by various means.

Their language is a mixture of Kurdish and Arabic. Most of the people are upright and charitable. The governors have built many madrasas and Friday mosques. There are many religious scholars and learned men. The most important tribe of ʿImādiyya is the Muzūrī and then the Zībārī. Zībār is the name of a river that flows through the province of ʿImādiyya. Because this clan lived along this river they were named after it. This river is also called the Junūn (Mad River) because it flows very swiftly. There is also another tribe, the Rādgānī, a name that the common people have altered to Raygān. The other tribes of ʿImādiyya are the Parwārī, Mamī, Sabābirūī, Tīlī, Bahlī—*baḥlū* meaning "valley" in the dialect of the people of ʿImādiyya.

The fortresses of ʿImādiyya are as follows: ʿAqr[a] is a prosperous town of 1,200 Muslim and Jewish homes. Wahūk and Dayr Maqlūb are where the Mīrzādegān and Banī Aʿmām govern (respectively?) as rulers of ʿImādiyya. Bībashar is under the control of the Arīkānī tribe. Qalāta. Shūsh. Aḥmarānī. Bāzīrānī is under the control of the Zībārī tribe. Karmalīs is a town on a mountain, having the Shaṭṭ on one side and Bāshqara and Barzān on the other.

Description of the district of Zakho: This is one of the important districts of ʿImādiyya. It is the home of the Sindī and Sulaymānī tribes. The people call Zakho the province of the Sindiyān. Most Kurdish religious scholars were raised in this province. In the past the province of Sindiyān had independent rulers. They controlled it as a hereditary grant. The governors of ʿImādiyya now occupy it.

Wahūk: in front of Zakho on the river that meets the Khābūr River. From Shūsh, ʿAqra, and Dahūk toward Shahrazūr there are fortresses near one another. Dayr Maqlūb is a fortress located in the region of the Shaṭṭ in the direction of Gilān. The district of Zībārī is located between Gilān and Hakkārī. These fortresses and districts form an arc in front of Jazīra and surround the city of ʿImādiyya in the direction of Shahrazūr. The fortress of Dayr Maqlūb, however, is in an area outside this ring. The district of Zakho is between the Khābūr and Hīzal rivers.

Rivers

The Tigris: A great river like the Euphrates. It emerges with an awful and grotesque sound from a cave at the base of a ruined fortress north of Diyarbekir. While flowing to Āmid (Diyarbekir) it merges with several rivers. It flows under a bridge east of Āmid. Then, by joining the Hīnī, Sayyid Ḥasan, Tarjīl, Atāq, Basharī, also called the Altın Köprü ("Golden Bridge"), all of which come from the (north)east, and then, below (i.e., southeast of) them, by joining the Erzen and Bitlis rivers, which come from Khazo, it becomes a great river. It flows under a bridge at Jazīra (Ibn ʿUmar) [M 468]. Then the Hīzal and Khābūr rivers merge with it from the east and at Mosul they flow as one under a bridge. The (Greater) Zāb River, which (enters the Tigris) below ʿAlī Ḥammāmı, comes from the mountains of ʿImādiyya. The Altın Köprü (Lesser Zāb?) River, which is in front of (on the Mesopotamian side of) Ṭopraḳḳalʿe, comes from Ḳaraṭaġ and Erbil and a separate branch comes from Hakkārī and Zībārī, which are in the vicinity of Van. They all flow (into the Tigris and) past Tikrit and Old Baghdad and then reach Baghdad. Below Baghdad the Diyāla River joins the Tigris. After passing the Vault of Chosroes (Ṭāq-i Kisrā), it joins the Euphrates at Jawāzir (i.e., where they form the great marsh). Going from Jawāzir to Jamāsa it divides into numerous branches around Wāsiṭ. Later they all join together again. They call the region encompassing all these different river branches the Jazā'ir (the islands of the) Shaṭṭ al-ʿArab. The Khurramābād River, which comes from Mt. Alwand, and the Sayr Ahwāz River which comes from Jamāsa and Khūzistān also join the Tigris. They all come together around Qūrna, then empty into the Persian Gulf near Basra. The length of this river from its source is 400 parasangs.

(Greater) Zāb: It descends from the mountains of ʿImādiyya east of Mosul. It meets the Tigris two stages below (south of) Mosul. Because it flows very swiftly it is also called the Junūn ("Mad") River. And the Lesser Zāb is called the Altun ("Golden") River. There is a bridge over it below Erbil on the road to Baghdad. The two Zābs take their name from Zawā b. Ṭahmāsp of the (legendary) Pishdadian dynasty (of Persia). One (the Greater) emerges from the mountains of Armenia and the other comes from Diyarbekir (province) and they empty into the Tigris at the borders of Ḥadītha. Below Wāsiṭ the Tigris splits into five main channels: the Qalā, Aʿrāf, Jaʿfar, Sīsān, and Shāhī. They are such that not enough water remains in the main channel for it to be navigable. Below Maṭārī there is deeper water. When the Euphrates emerges from the marshes it joins the rivers coming from Khūzistān. This (the combined waters) is called Shaṭṭ al-ʿArab. Below Basra it empties into the Persian Gulf. They (also) call this the Arwand River. It is larger than where the (Greater) Zāb empties into the Tigris. The Zāb empties into the Tigris from TN (?) JH Narrows below Kirkuk. There are extensive gorges and forests at the narrows. There are also reed beds and lion lairs.

Bitlis: It joins the Aswad ("Black") River in front of the fortress of Reşen. It rises from two different sources. The larger branch emerges from a large cave in a mountain near the fortress of Müküs. From the extremity of the cave to its mouth, this branch cuts its way through the rock and gushes out. The stones in the cave are shiny and there are places like balconies and ledges—it is a place of excursion. The water gushes out and joins the other branch at the foot of the mountain below Müküs near the fortress of Makhlaṭ. Then it flows below Jazā'ir Yūsuf, meets another river below QLHBRDN (?), continues under the bridge of Emir Süleymān and joins the Bitlis River in front of the fortress of Zeres near Siʿird. A stone emerges from the middle of the river near Reşen; it is called Ḳarıncaḳayası ("Ant Rock") because it is covered with ants.

Description of the Tigris: The Tigris has special boats called *ghurāb*. Most of them come from Basra. From Mosul to Baghdad one usually travels by *kelek* which is also used for travel on the Euphrates from Birecik to Ḥilla. The *kelek* is described as follows: [M 469] They nail boards together in a checkerboard pattern. The smallest have four and most have twelve to fourteen compartments to which are tied inflated skins. One cubit above these compartments they build a deck with wooden planks in the shape of a bench (or ledge, *ṣofa*). They load a lot of cargo onto them, and use them on the Euphrates and Tigris in this way, as they constantly go to the two rafts under the deck and inspect the inflated skins. Along some banks of the Shaṭṭ, for example, in front of Mosul and here and there below (south of) Mosul, there are petroleum springs. As it comes from the ground, petroleum floats to the surface of the water. Whatever does not flow away they collect in receptacles. Sometimes they set it on fire to make a spectacle; it can burn for some distance from the surface of the water.

Diyāla: It meets the Tigris at a place three hours below (south of) Baghdad. It is formed from all the waters around Shahrazūr. The Darna and Dartang rivers come from Tuzḫurmā near Ḳızılribāṭ. The Taşköprü River meets them. (The Diyāla) flows near Shahribān [and] Quryān and empties into the Tigris between Baghdad and Ṭāq-i Kisrā. As is the case with the Euphrates, many canals and channels branch off from the Tigris and irrigate the arable land of Iraq.

Mountains

Jabal Ḥamra ("Red Mountain"): A string of bare and low mountains stretching toward Arḍ-ı Surkh ("Red Land"). Coming from the Arabian Peninsula, it crosses the Euphrates near Jamāsa; continues across the desert of Jazīra-i ʿUmar; crosses the Tigris opposite ʿĀshiq, Maʿshūq, and Old Baghdad; skirts the desert of Baghdad; crosses the Diyāla River near Ḳızılribāṭ; goes on to the desert of Wāsiṭ; crosses the Daspul River between Zakiyya and Daspul Bridge; then passes again from desert to desert, ending at the Persian Gulf. In some areas of Shahrazūr and Mosul a mineral is found in these mountains that burns like beeswax and is black as tar. The people of the province call it *mūmyā-i maʿdanī* ("mineral tar"; bitumen). It is a black mineral like black rock-crystal. They melt it in the fire and pour it like wax. There is *mūmyā-i maʿdanī* in Mosul. Around Baghdad in places not covered by flood waters, where there is wilderness and briny soil, they obtain from the soil a white powder called *güherçile* (salt peter, potassium nitrate). If they boil this soil with water, the *güherçile* separates out and appears in a pure state.

Plants and Animals

The lands of ʿIrāq al-ʿArab are very productive. Its cotton and various kinds of fruit are of high quality. In ʿIrāq al-

ʿArab during the Abbasid caliphate, the peasants were forbidden to slaughter cattle because the cattle were needed for plowing and sowing. Lions abound in the deserts and in the thickets around the rivers and canals; antelope in the plains; wolves, leopards, bears, hyenas, cheetahs, jackals, foxes, and gazelles in the mountains of Kurdistan; and geese, cranes, partridges, francolins, and quail. Many kinds of fish are caught in the rivers and canals. In Baghdad there is a well-known type of fish called *shabūṭ*.

There are currently two categories of taxpaying subjects in ʿIraq al-ʿArab. One is the Bedouin (*ʿurbān*) of the villages. They are occupied with sowing and cultivating. They have camels, cattle, sheep, and buffalo. The other category is the Akhshāmāt, i.e., pastoralists. They live in horsehair tents. They are Bedouin organized into various tribes and clans. They also have camels, horses, sheep, buffalo, and cattle.

Types of people

In the province of Baghdad are Muslims, [M 470] heretical groups, Christians, and Jews. Their languages are Arabic, Persian, Turkish, and Kurdish; also Syriac and Armenian. The Muslims are of the Ḥanafī, Shāfiʿī, and Ḥanbalī rites. There are many religious scholars and shaikhs belonging to these three rites.

Roads and Stages

From Baghdad to Basra: > Madāʾin 6 parasangs > Dayr ʿĀqūl 8 > Jabal 7 > Famm Ṣulḥ 10 > Wāsiṭ 10 > Shahribān [Nahrābān][1] > Qārūth [Fārūt] 8 > Dayr ʿAmmāl 5 Ḥawāthit [Ḥawānīt] 7—the road passes along the Shaṭṭ (Tigris) and through the Marshes, then crosses Nahr Asad and reaches Dijlat al-ʿAwrā > Nahr Maʿqil 10 parasangs > Basra.

From Baghdad to Najaf the road goes through Depe-i[2] [Dīh-i] Ṣarṣar, Depe-i [Dīh-i] Farāsha, Shaṭṭ-ı Nīl, Kūfa, Mashhad ʿAlī which is the barrier (?—*sedd*) [edge[3]—*ser*] of the desert of Najaf.

From Baghdad to Iṣfahān: > Kaylūr 7 [Kangūr 75] parasangs > Bandistān 5 > Nahr-i [Shahr-i] Nihāwand 3 > Depe-i [Dīh-i] Farāmurz 4 > Nahr-i Rūjard [Shahr-i Burūjard] 4 > Janābād 4 > Mayān Rūdān 8 > Miyār 3 > city of Karkh [Karaj] 6 > Dūn Sūn 4 > Sangān 6 > stream of Murgh Kahtarā 6 > Isfarāna 7 > Bahrān [Tīrān] 7 > Mūy-i Kūshk 6 > Iṣfahān 4.

From Baghdad to Damascus: > Tall ʿAqarqūb [ʿAqarqūf], a high mound 3 parasangs > city of Anbār 8 parasangs. (From there) through the desert of Samawāt one reaches Damascus in ten days.

From Baghdad to Mosul: > Yūrān [Bardān] 4 > ʿUkbara 5 > Ḥamnā 3 > Qādisiyya of Mosul 7 > Sāmarrāʾ 7 > Karkh 2 > Ḥalfā [Jabaltā] 7 > Sūdqānat [Sūdqāniya] 5 > Lesser Zāb 5—here it meets the Tigris and there is a bridge > Ḥadītha of Mosul 12 > Banī Ṭamyān 7 > Mosul 7.

From Baghdad to Asadābād in the direction of Hamadān: > Yaʿqūbābād 8 > Sarwīna 7—via this road Shahribān is two parasangs to the right > Ribāṭ Jalūlā 5—it was built by the Seljuk Sultan Malik-Shāh > city of Khāniqīn 5 > Qaṣr-i Shīrīn 5 > Kīl wa Kīlā 5—it is the last place inside the border of ʿIrāq al-ʿArab > Ḥulwān 1, via the pass of Ṭāq-i Kamarābād > Depe-i [Dīh-i] Kawzīd and Chūsmān 8 > Jaqākāwān 6 > Bāḥkārs 5. At Ṣuffa-i Asbanarg there is an image of Khusraw and Shīrīn carved in stone, one parasang to the right; enough water to operate two mills emerges below the ledge > Kīrmānshāhān 6 > city of Jamjāl 6 > one enters Kurdistan. It is then six parasangs to Asadābād via the pass over Mt. Arwāna.

From Baghdad to Hamadān: > Başdolab 4 hours > ford of the Diyāla at Qamariyya, also called ʿUmariyya 8 hours > Bāqūya, also called Alwaniyya, 5 hours > Çubuḳköprü, also called Khāliṣa and ʿAjamiyya 3 hours > village of Ḥāshikiyya, also called Darāfsha 2 hours > Hārūniyya 5 hours > village of Ḳızılribāṭ (and?) Ḥafrābād 6 hours > village of Old Khāniqī 8 parasangs—a large river flows in front of it > Qaṣr-i Shīrīn, also called Depegöz 7—it is a ruined fortress > Sarīra (?) 3 > New Khāniqī 5 > defile of Razgarān 2—it is below Dartang > Basharī, also called Takht-i Kisrā of Dartang, 3 > six to Sarmīl > narrows of the fortress of Karīn 2—it is a fountainhead close to İki İmām > village of Shaikh 6 > Hārūnābād 4 > the defile of Jawādrūz 6 > Nāwar River 3 > Marīq 3—it is near Kūjānī > Nīlūfar, also called the village of Nawrūz, 4 > Kirmānshāh 5 [M 471] > bridge of Shāh 2 > Ṭāq-i Bustān 3—here Khusraw's (horse) Shabdīz is found > Ṣāṭir Anjād 3 > Takht-i Rustam, also called the city of Noah, 3—it is Mt. Bīsutūn, a high place like the mountain of Farhād > village of Ṣaḥna (and?) fortress of Bakhtigān, also called Mt. Lanqāwar (and?) village of Jamjāl 3—it is opposite the city of Noah > village of Bīd-i surkh ("Red Willow"), also called Kaj Ova (Kecova "Crooked Valley"), 3 > Kīngāwar 5—it is somewhat ruined > village of Najm-i Panj Khwārān, also called Asadābād, 5 > village of Kahrīz 3 > village of Göldepe 2 > Saʿdābād 2—it is a delightful city with vineyards and gardens > dam of Ṣūlāq 3 > inn of Parīkhān 5 > village of Yüzbaşı 5 > Salmānābād 5 > Qarhārū 4 > Sārīda 3 > Dervish Ilyās 6 > Darguzīn 3 > Jāmūslu 3 > village of Ṣafīkand 5 >

1 Spellings in brackets here are those in *Nuzha*, 165–172.
2 Place names in M with Depe-i ("Hill of") are a misreading of those in *Nuzha* with Dīh-i ("Village of").
3 Correcting M *sedd* ("barrier") to *ser* ("edge") according to *Nuzha*.

Isimābād 3 > village of Üç Künbed ("Three Tombs"), also called Aḳ Yaylaḳ ("White Summer Pasture"), 6 hours > village of ʿAlī Bey, also called Kandmizāb 5 hours > Hamadān 3 hours.

Status of kings and rulers

Noah had three children, one of whom was Shem. He was intelligent and wise; and the learned and insightful people exalted him above his siblings. Thus, after the Flood, Noah appointed him his successor as prophet. He taught him the secrets of prophethood and the subtleties of wisdom. He entrusted Shem with authority over the noblest portion of the earth. Noah prayed to God that among Shem's descendants there would be most of the prophets and apostles, faithful saints, rulers, sultans, and military commanders. Shem had nine children. One of them was Arfakhshād; most of the prophets were descended from him. Another son was Gayūmarth; the kings were descended from him. All of Gayūmarth's descendants were kings. He put in place the ceremonies and protocols of kingship and made the world orderly through justice. Gayūmarth and his children took possession of Bābil, the two Iraqs (ʿIrāq al-ʿArab and ʿIrāq al-ʿAjam), Fārs, Kirmān, and Khurāsān. They were called the kings of the Persians. Their capital was the city of Madāʾin in Iraq. The first king in the world was Gayūmarth. And the population spread out from that region.

At that time Noah's sacred law was in force and the people followed it. Then, when Bīwarāsb Ḍaḥḥāk became king and Nimrod was his agent in the arable land of Iraq, mankind fell into all kinds of unbelief and heresies. One of their heresies was that they affirmed one god made from light whom they called Yazdan, saying "This is God," and another god made from darkness whom they called Ahriman, saying "This is Satan". Moreover, they worshipped and adored light and avoided and hated darkness. Over time they replaced worshiping light with worshiping fire. With the coming of Zoroaster, who claimed to be a prophet, the previous heresies of the people became stronger. They asserted that good and evil, corruption and purity, were the result of the mixture of light and darkness. They originated the religion of the Magians (i.e., Zoroastrianism).

The first capital of the Persian kings was Madāʾin. The Chaldeans, who were under their control, seized the kings of Bābil. Their first ruler was Nimrod, the agent of Ḍaḥḥāk. This will be discussed in the section on the region of Persia.

In Ḥīra and Anbār, the kings of the Banī Lakhm (text: Laḥm; i.e. Lakhmids) are from the Arab tribe of Āl Mundhir. Their first king was Mālik b. Fahr al-Azdī. When they became aware of the flood of ʿArim (the flood that destroyed the country of Sheba), they migrated from Yemen to Ḥīra [M 472] and settled there. Around that time, Alexander the Greek invaded the Persian empire. The Āl Mundhir came to the banks of the Euphrates, built the cities of Ḥīra and Anbār, and settled there. Mālik was succeeded by his brother ʿAmr, and then by his brother's son Jazīma (Jadhīma) al-Waḍḍāḥ. It is reported that this sultan was the first person to build a siege engine and to domesticate mules. Jazīma assembled a number of slave-boys from the sons of (subject) kings to serve him. One of them, named ʿAdī b. Naṣr, fell in love with Jazīma's sister Raqqāsha and she became pregnant by him. When Jazīma learned of this, ʿAdī fled. Raqqāsha gave birth to a boy whom they named ʿAmr. It happened that Jazīma had no children, so he raised him as his own. One day the jinns abducted this boy. Later, two persons named Mālik and ʿUqayl found him in Wādī Samāwa and brought him to Jazīma. Jazīma then asked them "Is there anything that you desire?" And they replied, "We wish to be your boon companions." For forty years they were his boon companions. This resulted in the proverb *ka-nudamāʾ Jazīma* ("Like the boon companions of Jazīma"). Later Jazīma was murdered and ʿAmr succeeded him.

The cause of Jazīma's death was as follows: When the ruler of Jazīra died, his daughter Nāʾila, who had the nickname Zabbāʾ, took his place and became the governess of Jazīra. Then she plotted to kill Jazīma. She seduced him and when she found an opportunity she killed him. When that happened, ʿAmr b. ʿAdī succeeded Jazīma. Zabbāʾ then hatched a plot to kill ʿAmr. When ʿAmr learned of the plot, he wanted to kill Zabbāʾ. He colluded with Caesar, who was one of the slaves of Jazīma, (to kill her). Caesar cut off his own nose of his own free will. The plan was for Caesar to go to Zabbāʾ and tell her that he had fled from ʿAmr. When he went to Zabbāʾ she treated him kindly and promoted him in her retinue. He began to conduct trade for her and seemed to earn much money, although he was secretly receiving money from ʿAmr, which he gave to Zabbāʾ. He said to Zabbāʾ, "This money is the profit earned in trade." One day he brought 1000 camels loaded with chests. He had put armed men in these chests. When Zabbāʾ saw the camels from a distance, she became suspicious and said: *ʿasā al-Ghuwayr abʾusan* ("It seems there is trouble coming from Ghuwayr"). This expression became proverbial. After the camels entered Zabbāʾs citadel, the armed men emerged from the chests, seized the city, and killed Zabbāʾ.

ʿAmr's son Imruʾ al-Qays succeeded him. After him came his son ʿAmr and after him Aws b. Qalām, who was from the Amalikites. After him there was a second Imruʾ al-Qays b. ʿAmr and then his son Nuʿmān Aʿwar (the one-eyed). This Nuʿmān built the famous Khawarnaq (a palace near Ḥīra and the Euphrates). He was succeeded in turn by his son Mundhir, then his son al-Aswad, then his brother Mundhir, then ʿAlqamat al-Duhaylī, then his son Mundhir, whose mother was called Māʾ al-Samāʾ ("Water of Heaven") because of her great beauty, then al-Ḥārith b. ʿAmr b. Ḥujr al-Kindī. al-Ḥārith adopted the religion of Mazdak, which was the faith of (the Persian ruler) Qubād whose authority he acknowledged. Mundhir, however, did not submit and so was deposed and al-Ḥārith succeeded him. Then Khusraw drove out al-Ḥārith and Mundhir returned to the throne. He was succeeded by his son ʿAmr. Around that time, the Most Noble Apostle (Muḥammad) [M 473] came into the world. Mundhir was succeeded by his brother Qābūs, then his brother al-Mundhir b. al-Mundhir, then his son al-Nuʿmān, who became a Christian. al-Nuʿmān was killed by Khusraw Parwīz. Subsequently there occurred the battle of Dhū Qār between the Arabs and Persians.

The flower called *shaqāʾiq Nuʿmān* (anemones of Nuʿmān, i.e., red anemones) takes its name from him because he went to a place (where they were found) and said *mā aḥsana hādhihiʾl-shaqāʾiq aḥmaruhā* ("How beautiful are these anemones the reddest of all"). And he built a dwelling for himself there. He declared one day unhappy (gloomy, infelicitous) and one day happy (pleasant, felicitous). If someone came to meet him on an unhappy day he put him to death. But if someone came to meet him on a happy day he bestowed favors on him. One day he encountered a Bedouin Arab from the Ṭayy tribe. He was about to kill him when the Arab said, "I have a daughter. I have not recommended her to anyone (as a wife). Give me permission to do so and I promise to return." Nuʿmān said, "Who will be the guarantor for you?" His vizier immediately said "I will be the guarantor." Nuʿmān then said, "If he does not come back, I will kill you in his place." Then the Arab departed. On the day that he had promised, the Arab returned. Nuʿmān then said to him, "I don't know which of you two is more honorable." To his vizier he said, "What compelled you to stand as guarantor for this Arab?" And to the Arab he said, "What compelled you to return?" First the vizier answered, "There is no longer honor among viziers." Then the Arab answered, "People no longer keep their promises." Nuʿmān then said, "Let me not be the third, let it not be said that there is no longer forgiveness among kings." And he abolished the unhappy days.

Afterwards Iyās b. Qabīṣat al-Ṭāʾī became ruler and then Zārūya. Subsequently sovereignty again passed to the Lakhmids and Mundhir b. Nuʿmān became king. The Lakhmids ruled until Khālid (b. al-Walīd) conquered them. With the appearance of the dynasty (*devlet*) of Islam, (Caliph) ʿUmar assigned Khālid to this area and he conquered Ḥīra. The rulers of Ḥīra began to be appointed from among the noble Companions (of the Prophet). Later the Umayyad caliph ʿAbd al-Malik b. Marwān made Yazīd b. Hubayra the governor of Iraq.

Then the Abbasid dynasty arose and the Abbasids conquered Iraq and homage was paid at Kūfa. The capital was then moved from Kūfa to Anbār. The Abbasid dynasty was founded in 132/749. The first of them was Abūʾl-ʿAbbās ʿAbd Allāh whose great grandfather was Ibn ʿAbbās. He was famous as ʿAbd Allāh al-Saffāḥ ("Shedder of Blood"). In 132 (749) homage was paid to him. In 136 (753) he died of smallpox. During his time Abū Muslim rose up. His father was from the village of Marw. When he died in Azerbaijan, Abū Muslim became an orphan and he grew up under the care of ʿĪsā b. ʿAqīl. He became the cynosure of the litterati. One day the Imām Muḥammad b. ʿAlī b. ʿAbd Allāh and others from among the *naqībs* (leaders of the descendants of the Prophet) gathered in Kūfa and when they saw Abū Muslim they were pleased with his words and became his associates. While going to Mecca together, they encountered al-Saffāḥ's brother Ibrāhīm. When he saw Abū Muslim he was pleased and took him into his service. Later when it was necessary to appoint someone to govern Khurāsān, he sent Abū Muslim. Abū Muslim took over Khurāsān and gradually his majesty increased. Meanwhile Saffāḥ was a merciless tyrant and shedder of blood. [M 474] During his time he killed innumerable Muslims. It is an immutable law of God that the lives of rulers who shed blood are short. Thus ʿAbd Allāh died at age 28. He had two viziers, one was Abū Muslim and the other was Khālid b. Jaʿfar al-Barmakī.

After him, allegiance was sworn to his elder brother Abū Jaʿfar ʿAbd Allāh Manṣūr in 137 (754). He was also tyrannical, vain, and a source of corruption. The reason for the enmity between the Abbasids and Alids was that he killed Muḥammad and Ibrāhīm, the two sons of Imām Ḥusayn b. ʿAlī b. Muḥammad b. ʿAbd Allāh b. Ḥasan. And they used them (the sons) as an excuse to abuse many Muslims. For example, he (Manṣūr) imprisoned Imām-i Aʿẓam (The Greatest Imām, i.e., Abū Ḥanīfa), who was the lamp of the (Muslim) people and the exemplar of the religious community, because he refused the position of qadi on grounds of his complete renunciation of the world and his piety. And he died in prison. Because of Manṣūr's utter meanness and baseness, they called him al-Dawāniqī ("Penny Pincher"). He would argue with workers and craftsmen over pennies. Abū Muslim-i Khurāsānī undertook a propaganda campaign to make the people swear allegiance to the Abbasids. But Manṣūr even put him to death. He ordered his companions to wear black clothing and tall hats, and to hang their swords from their arms, and he issued an edict to this effect. Abū Dulāma (Manṣūr's court jester) adhered to this policy. Manṣūr asked him how he was complying with it, and Abū Dulāma replied: "I put my face at my waist with my sword on top of me; I painted my face black; and I fastened God's book to my back." Manṣūr laughed and issued a decree abolishing this policy.

This caliph founded Baghdad on the bank of the Shaṭṭ (Tigris) in 146 (763). He was the first caliph to have books translated into Arabic from Syriac, Persian, and Greek, such as Euclid and *Kalīla wa Dimna* (from Persian), and others. In 158 (775) he set out on the pilgrimage to Mecca and (there) wanted to kill (the mystic) Sufyān al-Thawrī. When he reached the halting place called Bi'r-i Maymūn ("Auspicious Well"), he sent his henchmen to Mecca to seize al-Thawrī and crucify him. But when the people of Mecca learned of this, they warned Sufyān. He then stretched out, placing his blessed head on the knee of Fāḍil b. ʿIyāḍ and his feet on the knee of Sufyān b. ʿUyayna, and being opposite the exalted Kaʿba he went to sleep. Suddenly he stood up and, grasping the covering of the Kaʿba, said, "If Manṣūr comes here, I am quit of you." Then he went back and lay down. Immediately thereafter word came that Manṣūr had left Bi'r Maymūn and that he had mounted his horse and set out on the road to Mecca. However, when he reached the spot exactly midway between the two Juḥfas (i.e., between al-Juḥfa and Rābigh?) he fell from his horse, broke his neck and died. His vizier was Khālid al-Barmakī.

Manṣūr was succeeded by Muḥammad al-Mahdī. He associated with the religious scholars and acquired an excellent education from them. He was esteemed by the cultured and the refined. He hated the heretics and killed a great many of them. He ordered an index of books of disputation (*cedel*) be drawn up in order to refute the claims of the heretics. When he made the pilgrimage to Mecca he performed many wonderful charitable deeds. Men went up to the roof of the Kaʿba and perfumed all God's house with musk, ambergris, and *galiya* (the perfumed compound *Galia moschata*). He also provided three *kiswas* (the cloth covering the Kaʿba) composed of various brocades, one on top of the other. He was the first caliph to establish the tradition of the *ṣurra* (providing funds to the poor of Mecca and Medina). He conquered many provinces from Byzantium. [M 475] He died in 169 (785).

His son Mūsā al-Hādī succeeded him. He was extremely tyrannical. He established the custom of having men bearing swords march in front of the ruler. He died in 170 (786) and his brother Hārūn al-Rashīd took his place on the caliphal throne. This caliph was eloquent and cultured. He and made the pilgrimage to Mecca many times and performed one hundred prostrations every day during prayer. Each year he would alternate between going on pilgrimage and conducting raids against the unbelievers. He was an intimate of the religious scholars and a lover of the Sufi shaikhs and saints. His qadi was Imam Abū Yūsuf, his viziers were Barmakids, his court poet was Marwān b. Abū Ḥafṣa, his chamberlain was al-Faḍl al-Rabīʿ, his singer was Ibrāhīm al-Mawṣilī, and his wife was Zubayda. His reign was a period of charity, good deeds, prosperity, and blessings.

His viziers were primarily Barmakids, who were a large group. The most distinguished of them were Yaḥyā, Muḥammad, Mūsā and their children Khālid, Faḍl, Jaʿfar, and Faḍl b. Yaḥyā b. Khālid b. Satāsif b. Ḥāmāsat al-Barmakī. The great Barmak who was the forefather of this group was one of the kings of Persia. In the province of Fars he built a large fire-temple for the Magians that was very famous. Later he went to Syria and in the presence of the Marwanid (i.e., Umayyad) caliph Hishām b. ʿAbd al-Malik he became a Muslim and was given the name ʿAbd al-ʿAzīz. When the Abbasid dynasty was established, ʿAbd Allāh Saffāḥ brought the Barmakid and made him his vizier. After the death of Saffāḥ, he became the vizier of Manṣūr al-Dawāniqī and became famous. He died during the reign of Mahdī. They say that Hārūn al-Rashīd performed his funeral prayer. Of all his children, none reached a higher level of power than Khālid: neither Yaḥyā for his culture, nor Faḍl for his generosity, nor Jaʿfar for his belles-lettres, nor Muḥammad for his valor, nor Mūsā for his bravery. Hārūn al-Rashīd next made Jaʿfar his vizier and treated him with such great favor that Jaʿfar had authority over his palace, his children and his finances. Indeed Hārūn al-Rashīd addressed Yaḥyā as "father" and Faḍl and Jaʿfar as "brother."

The reason that Hārūn al-Rashīd had such affection for the Barmakids was as follows: When Caliph Hādī died, Hārūn was twenty-two years old and in the city of Ṭūs. Furthermore the people of Baghdad had decided that he should not be made caliph because he occupied himself with pleasure and amusements and thus he would ignore the best interests of the people and matters of state and the caliphate would decline. So they would not give their consent to his succession. It was then that the most prominent of the Barmakids in Baghdad came forward—Yaḥyā, Faḍl, and Jaʿfar—and said, "No one but Hārūn al-Rashīd is suitable for the caliphate." They sent a man to fetch Hārūn and made him caliph. As soon as he became caliph, he summoned Yaḥyā and said, "You are the reason that I was assured the caliphate. Therefore I appoint you my agent and grand vizier in overseeing all matters of state."

Yaḥyā began to administer the state in excellent fashion. After him his son Faḍl became vizier, and then Jaʿfar. After they had served in office eighteen years, Hārūn became estranged from the Barmakids. In that year he made the pilgrimage to Mecca and upon his return executed Jaʿfar. They say that the reason for this was as follows: Whenever Hārūn's sister, a lady named ʿAbbāsa, and Jaʿfar were not present at the council of state, Hārūn took no pleasure in the meeting. Therefore, in order to allow Jaʿfar to be present with his sister and to be able to see his face in the council, Hārūn married him to his sister, but on condiditon that Jaʿfar not go near ʿAbbāsa (i.e., consumate the marriage). [M 476] But Jaʿfar disobeyed Hārūn's orders. He had a child by ʿAbbāsa and for this reason he was executed.

Hārūn died in 193 (808) when he went to Ṭūs. His son Muḥammad Amīn succeeded him. He was a handsome attractive youth. But he mismanaged the state and ignored advice. He was preoccupied with women and amusement. Dhū'l Yamīnayn ("The ambidextrous," i.e., his brother Ma'mūn) instructed Ṭāhir (b. Ḥusayn, Ma'mūn's most important military commander) to kill him. When Amīn learned that Ṭāhir was besieging Baghdad, he scolded the one who brought him the news. This is because he was at that time in the harem with his slave girls catching fish. He addressed the messenger as follows "One of my slave girls has caught two fish while I have caught one." Ṭāhir then took Baghdad by storm and murdered Amīn.

His brother ʿAbd Allāh Ma'mūn took his place on the throne in 198 (813). This caliph was renowned for having mastered the arts of this world and the sciences of hadith, jurisprudence, and medicine. He also had great appreciation of and affection for philosophy and the sciences of the ancients (the Greeks). Informed of the affairs of the world, he sent many raiding parties against Byzantium and conquered forty fortresses. Eventually his judgment went astray. He was the first to use the expression *khuliqa al-qur'ān* ("the Koran is created"). He also ordered that ʿAlī be given primacy of place among all the Companions of the Prophet. He died in Tarsus.

Allegiance was sworn to Muʿtaṣim b. Hārūn al-Rashīd as his successor. They called him Muthamman ("Eightfold") because he conquered eight fortresses, he was caliph for eight years, he was the eighth caliph, and he was born on the eighth day of the eighth month. There were also eight letters on his official seal *al-ḥamdu li-'llāh* ("Praise be to God," only consonants counted). Eight kings stood at his gate. He defeated eight kings. He had eight sons and eight daughters who survived him. He conquered eight cities. He built eight pavilions. He had 18,000 Turkish slaveboys. This caliph was very strong. He could pick up two sheep and hold them up by their legs until they were dressed out. During his reign the Turkish corps was introduced to the council of state of the caliphs for the first time. Following the custom of the Persian rulers, he sent money to Samarqand and Farghāna to purchase Turkish slaveboys. He took 18,000 of them into his service. He adorned them with gold necklaces and clothing made of brocade. The people of Baghdad were disturbed by their behavior. They all went to Muʿtaṣim's gate and said, "Either expel the Turkish soldiers from this city or we will do battle with you." Muʿtaṣim replied, "You are a weak regiment of common people. How can you possibly fight against me?" The crowd answered, "We will fight you with the arrows of magical charms and the sword of the summons to humility in the court of the Creator." Muʿtaṣim responded, "Mercy! I don't have the resources for such a battle. Be patient a while." He then built the city of Sāmarrā' near Baghdad and stationed the Turks there.

It was reported that one day, while Muʿtaṣim was sitting on the throne, news arrived that the Byzantine emperor had raided the lands of Islam and had killed or captured many Muslims. In fact one of the captured women was from the Hashimite clan (i.e., the clan of the Prophet). She cried for help saying, "Ah, Muʿtaṣim, Ah," and an unbeliever struck her saying, "Let Muʿtaṣim come with speckled horses and save you." When the caliph learned of this he immediately dispatched 40,000 speckled horses to Maʿmūriyya (Amorium). When he reached Ankara, he captured it by the grace [M 477] of God and captured the Byzantine emperor. He summoned the Hashimite woman and said to her, "At your service, daughter of my uncle! I have answered your summons."

In 227 (841) this caliph died in Sāmarrā'. Hārūn al-Wāthiq succeeded him. He passed away in 232 (846) and Jaʿfar al-Mutawakkil ʿalā'llāh took his place on the throne. He supported the Sunnis and established the religious rites and ceremonies of Islam. Some strange things occurred during his reign. One of them was the following: One night so many comets appeared in the sky that the stars crashed into each other and shattered in all directions. People thought this was the precursor to the Last Day. And in Egypt, stones rained on the village of Suwaydā. They weighed one of them and it came to 130 dirhams (416.91 gr.). Another unusual phenomenon occurred in Yemen. A stone broke off from its place and kept rolling until it came to a mountain and stopped there. Another phenomenon: A white bird somewhat smaller than an eagle perched on the top of a mountain and cried out forty times, *yā maʿshar al-nās ittaqū 'llāh* ("O assemblage of people, fear God!"). He returned the next day and for the next forty days did the same thing. This phenomenon occurred in Ramadan of 241 (January, 856). It was recorded based on the testimony of 500 people. Afterwards there was a huge earthquake and all the springs in Mecca went dry. When al-Mutawakkil learned of this, he sent 100,000 goldpieces to make water flow from the spring of ʿArafāt.

It happened that in Baghdad and its hinterlands, the number of Turks increased and they became powerful. They became involved in matters of state and appointments and dismissals were in their hands. Finally, their tyranny reached such a point that one night ten Turks entered the caliph's private pavilion and killed him. In 247 (861) the caliph's son Muḥammad al-Muntaṣir bi'llāh succeeded him. One day the caliph's health began to fail. They (the Turks) gave his doctor 30,000 florins. They bled him with a poisoned lancet (i.e., they bribed the doctor). The poison took effect and he died. Abū'l-ʿAbbās Aḥmad al-Mustaʿīn bi'llāh succeeded him. The Turks took control of the country. The caliph was reduced to the status of a shaikh. Each of Mustaʿīn's sleeves was three spans wide. The notables of Mecca today wear clothes of this fashion. In short, the authority of the jerky Turks (*Etrāk-i kemkerde-idrāk*) was so great that when they saw he would not accede to them they dismissed him from the throne and then killed him.

Mustaʿīn's son Muḥammad al-Muʿtazz succeeded him. But the Turks dismissed him and replaced him with Muḥammad al-Muhtadī. They took all the wealth and property of al-Muʿtazz' mother and banished her to Mecca. Her property included 1,000,000 goldpieces, a half *irdabb* (= 76.968 kg) each of pearls and emeralds and one-sixth *irdabb* of red rubies. The Turks also attacked al-Muhtadī and killed him. The next caliph was Aḥmad al-Muʿtamid ʿalā'llāh. He was preoccupied with amusement and pleasure and made his brother Ṭalḥa heir apparent. He was given the title al-Muwaffaq bi'llāh. He served as governor of the provinces of Ḥijāz, Yemen, Fars, Ṭabaristān, and Sind. al-Muwaffaq had a young son named Jaʿfar and gave him the title al-Mufawwiḍ ilā 'llāh. al-Mufawwiḍ was appointed governor of al-Jazīra, Syria, and the Maghreb. al-Muʿtamid gave one a white banner and the other a black banner and stipulated that if he should die while his own son was still little, then the caliphate would belong to al-Muwaffaq. [M 478] They drew up a pact in which they agreed to abide by this stipulation and sent it to Mecca where they stored it in the Noble House (the Kaʿba).

Eventually, the stipulation of this pact came to pass in accordance with the saying *lā yughnī ḥadhar min qadar* ("There is no use in taking precautions against fate"). During the reign of al-Muqtadir the Zanjīs revolted and caused great damage to the Muslims (the Zanj Rebellion, 869–883). Their leader Bahyūlī (?) claimed to be a prophet and to have knowledge of the invisible world. This resulted in a civil war in which more than 100,000 Muslims were killed. Finally al-Muwaffaq bi'llāh, with the help (*tawfīq*) of God, sent a great army against the rebels. After the battle, the wind of victory blew. They crushed so many of the Zanjis that they could not be counted. After this victory the caliph was given the title al-Nāṣir li-Dīn Allāh (The Victorious by the Religion of God). When he died his nephew Aḥmad al-Muʿtaḍid took his place on the throne. Muʿtaḍid was an intelligent and courageous man. He would attack a lion single handed. During his reign he ended all oppression of the people and forbade reprehensible acts (i.e., wine drinking, gambling, and prostitution). After the power of the Abbasids had declined, the caliphate regained its strength. Thus they called al-Muʿtaḍid the second Saffāḥ. During the reigns of pervious caliphs, it was forbidden to leave an inheritance to maternal relatives. Inheritances intended for them were seized and taken into the state treasury. Muʿtaḍid believed in strictly following the rules of Shariah (Muslim sacred law, which allowed maternal relatives to inherit).

He died in 289 (902) and his son al-Muktafī bi'llāh succeeded him. He was handsome and of excellent character. He honored the religious scholars. During his reign, the Carmathians appeared on the scene. This sect believed that what was religiously forbidden was permitted. They believed that the spilling of the blood of Muslims was lawful. They claimed that after the Prophet the leader of the Muslim community should be Muḥammad b. al-Ḥanafiyya. The first of them to revolt was Yaḥyā b. Mihrawayh. This occurred in the district of Hujr.

When al-Muktafī died, his brother Jaʿfar al-Muqtadir bi'llāh succeeded him. He was completely preoccupied with wine and debauchery. He was twice deposed. When he ascended the throne the third time, he mended his ways. He forbade all Jews and Christians from participating in affairs of government. On the day of ʿArafa (i.e., 9 Dhū'l-Hijja), he had 50,000 sheep, in addition to cattle and other animals, slaughtered and distributed to the poor. Every year he spent 7,000 dinars on the hospital (in Baghdad). Once al-Muqtadir held a feast in which he had his five sons circumcised. He spent 600,000 florins on this feast and made everyone rich. One time the envoy of the Byzantine emperor arrived and 160,000 armed troops stood at attention. They stood on both sides of the route to the gate of the caliphal palace. Inside 7,000 eunuchs and 700 chamberlains stood on each side of the caliph. The walls of the palace were hung with 38,000 pieces of cloth and on the floor were spread another 22,000 pieces. In addition to these decorations, they set up an (artificial) tree which moved its branches when the wind blew. On the tree were placed golden and silver birds that sang when the wind blew. Despite such a show of strength, it was during his reign that the Abbasid dynasty began to decline.

al-Muqtadir was killed in battle with the Carmathians in 321 (933) and his brother Abū Manṣūr al-Qāhir bi'llāh succeeded him. [M 479] Because he abused the people, they attacked him and pierced his eyes with a stylus. He was succeeded by Muḥammad al-Rāḍī who was succeeded by Ibrāhīm al-Muttaqī bi'llāh. At that time the Turks seized Baghdad. al-Muttaqī's paternal nephew ʿAbd Allāh al-Mustakfī succeeded him. After him Abū'l-Qāsim Faḍl Allāh became caliph in 334 (945). During his reign the Buwayhids occupied Baghdad and power was in their hands. al-Mustakfī's son ʿAbd al-Karīm al-Ṭā'iʿ bi'llāh succeeded him. During his reign the name of the Abbasid caliphs ceased to be mention in the Friday sermon in Mecca and Medina.

Next Aḥmad al-Qādir became caliph. He was devoted to asceticism, righteousness, and mysticism. He composed a treatise rejecting the claim of those who believed that the Koran was created. His caliphate lasted 41 years. After he died in 422 (1031), his son ʿAbd Allāh al-Qā'im bi-Amr Allāh succeeded him. He was an excellent man. He died in 467 (1074) after 44 years as caliph. His son ʿAbd Allāh al-Muqtadī bi'llāh succeeded him. He was generous to the faith and was one of the most upright of the Abbasid caliphs. One of his blessed pious acts was as follows:

Sultan Malik Shāh, who was of the dynasty of Sabuktegin, wanted to treat the caliph highhandedly. Malik Shāh sent him a message saying, "You should not stay in Baghdad. You should leave. We intend to go there." al-Muqtadī sent an envoy to Malik Shāh and was fairly courteous to him in this matter. Malik Shāh became more insistent and said, "He must definitely leave." al-Muqtadī then implored him to grant a delay. Malik Shāh replied, "I won't grant a single hour." Finally he sent a man to Malik Shāh's vizier asking him to grant ten days and this request was granted. During this period the caliph fasted by day and stayed awake by night, praying to God and imploring Him for assistance and humbling himself. Before ten days were up, the arrow of prayer hit the target of response, and he was saved from that evil.

When al-Muqtadī died in 487 (1094), his son al-Mustaẓhir bi'llāh succeeded him. This caliph had a beautiful calligraphic style. During his reign the Seljuks wrested hegemony (from the Buwayhids). He died in 512 (1118). His son Mustarshid bi'llāh succeeded him. This caliph went to war against the Seljuk Masʿūd b. Muḥammad. But his troops were not successful and, left alone on the battlefield, he was killed. al-Rashīd bi'llāh Manṣūr succeeded him. When he was born, his reproductive organ was plugged; the physicians made an opening, and when he was nine he had a nocturnal emission. He was killed by the Seljuk Masʿūd and succeeded by his paternal uncle Muḥammad al-Muqtafī bi'llāh. al-Muqtafī had an excellent and noble character. On his death his son Yūsuf al-Mustanjid bi'llāh succeeded him. The following event is reported concerning this caliph: Before he became caliph, al-Mustanjid had a dream in which the letter *khā'* was written three times on the palm of his hand. The next morning he described this to the interpreters of dreams. They explained this saying "You will become caliph in the year 555 (1160)". And in fact, this is what happened. His son Ḥasan al-Mustaḍī succeeded him as caliph. He was noble in charcter and an excellent administrator.

When he died his son Aḥmad al-Nāṣir li-Dīn Allāh became caliph. [M 480] He revived the protocols of the caliphate. He ordered that his illustrious name be mentioned in the Friday sermon and be stamped on the coinage. His name was long mentioned after that of al-Mustaḍī in the prayers from the *minbars*. His miraculous graces were manifest. One of them was as follows: The Khwārazm Shāh decided to appropriate to himself the rule over Iraq and set out from Hamadān for this purpose. When (his army) entered a plain, snow fell on them day and night for a long time and they could not find their way. The entire army found itself in desperate straits. Most of it was destroyed and the Khwārazm Shāh was forced to retreat. Another miracle was the following: When the infidel Tatars, who resembled the Gog and Magog of old, burst forth and attacked the lands of Islam, God, Exalted and Sublime, along with the propitious hand of victorious fortune, saved Baghdad from them. (Furthermore) up to that time, the covering of the magnificent Kaʿba had been made of white brocade. al-Nāṣir then ordered that the covering be made of black brocade. The covering of God's Kaʿba has continued to be black to this very day. His caliphate lasted with complete independence for 47 years. In Baghdad "the abode of paradise" he built the Niẓāmiyya madrasa and placed 10,000 volumes of books in it. He died in 622 (1225).

His son al-Ẓāhir bi-Amr Allāh succeeded him, and his son al-Mustanṣir bi'llāh Manṣūr succeeded him in turn. al-Mustanṣir showed honor and respect to the religious scholars of his time. He devoted himself to building madrasas, mosques, and gardens. The Mustanṣiriyya madrasa in Baghdad is his work. Four professors were appointed to this college, each teaching one of the four (Sunni) rites.

When Niẓām al-Mulk became vizier of the victorious Seljuk (Malik Shāh) he worked hard to build the foundations of the state and establish the principles of the institution of the sultanate. He patronized the friends and supporters of religion and state, while not for a moment ignoring the enemies of throne and sultanate. He was generous to the religious scholars and the devout and made large donations to the Holy Cities of Mecca and Medina. Every year, apart from his private purse, he spent more than 600,000 goldpieces from the state treasury on charitable works. Some people, bearing envy and enmity toward the sultanate, denounced the vizier saying, "His huge expenditures on behalf of the caliph (sic) are bankrupting the treasury. If the amount of money that he has consumed had been spent on weapons and soldiers, enemy fortresses would have been conquered and our banners would have been raised over them." They intensified their denunciations of the vizier to the caliph (sic) but he closed his eyes to this. Nevertheless, the enemies continued to conspire and when their accusations reached the sultan's ears he became concerned. One day when the vizier was in his presence he addressed him saying, "Every year you are spending this much money of the public treasury for no good purpose." Niẓām al-Mulk answered as follows:

"Oh, my son! I am a weak old Persian. If I were to go on the market (as a slave), no one would pay a florin to buy me. But you are a youthful Turk. If you were to go on the market, you would fetch thirty goldpieces. God, praise be to Him the Exalted, has chosen me and entrusted the government in my able hands. Responsibilities for matters of state and the interests of the people devolve upon me. I am preoccupied with governing and administration, while you are absorbed with the pleasures and delights of this world. We will both approach the court of the Lord of Glory (and be judged for) the results of our actions. Most of our actions are sin and rebellion rather than gratitude and obedience (to God). Each of the soldiers whom you have trained to repel the enemy has a sword in his hand that is two cubits long [M 481] and the arrows that they shoot go no little distance. Still, they commit all kinds of debauchery and misdeeds. If their corruption is taken into consideration, they are not worthy of achieving conquest and victory; rather, the stones of calamity and misfortune should rain upon them. I (on the other hand) have trained a vast number of troops for you. They are called the army of night and day, for during the long nights when you and your soldiers sleep the sleep of neglect, they are awake and their ranks are drawn up. Their eyes shed tears and their tongues call out in prayer and praise (to God). When they extend their hands (in supplication) to the court of the Qadi of needs (or Fulfiller of needs, i.e., God), they strike their enemies with sword blows that reach as far as the land of Chīn u Māchīn (North and South China). You and your soldiers are sustained by their honor and you conquer and are victorious thanks to the blessings of their prayers."

The caliph (sic) then said, "Bravo, father, bravo! You have done well. Increase the number of these troops. These are the kind of troops I need." Thus, the caliph (sic) accepted the explanation of the vizier and the enviers wallowed in their shame.

During the reign of al-Mustanṣir, the Tatars attacked Iraq several times. By the grace of God they were defeated each time. al-Mustanṣir died in 640 (1242). His son ʿAbd Allāh al-Mustaʿṣim bi'llāh succeeded him. He was the last Abbasid caliph in Baghdad. He was preoccupied with amusement and pleasure. He gave his slaves the title "sultan" and turned all matters of state over to them. Eventually the caliphate lost power and they took control. They opened the gates of bribery. For a few goldpieces every unworthy one could obtain his desire. Ranks and positions were given to people who were unqualified. Muʾayyad al-Dīn ʿAlqamī became vizier. He was a Rāfiḍī (a Shīʿī) and a reviler (*sebbāb*, one who curses the first three caliphs). He was an enemy of al-Mustanṣir and the Sunnis. He would appear friendly but he behaved hypocritically. One day in Baghdad the heretics lorded it over an assemblage of Sunnis and greatly abused them. Even though Ibn ʿAlqamī was a Shīʿī, they raided the homes of his relatives. He ostensibly had no power to take reprisals. Meanwhile he began to work for the downfall of the Sunni dynasty (or state—*devlet*—i.e., the Abbasids). He entered into secret relations and correspondence with Hulagu Khan. His goal was to cut the vein of the Sunnis and to return the caliphate to the Alids. Therefore, using weak excuses, he began to disperse al-Mustaʿṣim's select troops (private guard). He would always say to al-Mustaʿṣim, "We don't need this many troops. They should go. What is needed is a treasury. Once there is a treasury, troops can be found whenever necessary." al-Mustaʿṣim did not have the ability to discern what was happening and to understand ʿAlqamī's plans. Indeed, once he said to 20,000 troops, "I'm giving you leave. Go wherever you wish." He would show al-Mustaʿṣim abundant revenue but the caliph did not understand his intentions. Because of the caliph's great desire for money, he gave him permission to collect it. He did not know that the money that he collected was being amassed for the enemy.

It happened that the Mongol and Chaghatay sultan, Hulagu Khan, attacked the lands of Islam with a vast army. He carried out massacres in most of the countries of Islam. This calamity reached such proportions that while the caliph al-Mustanṣir slept the sleep of pride and neglect and passed his time in luxury and pleasure, Ibn

ʿAlqamī [M 482] cut off all sources of news and information to him right up to the time the Mongols invaded Iraq and began the slaughter of Islam. Finally Hulagu resolved to conquer the city of Baghdad. He sent a man to the caliph and summoned him to his presence. Awakening from the sleep of heedlessness, the caliph began to rub his eyes in bewilderment realizing that there was no hope of escape. His vizier Ibn ʿAlqamī the Rāfiḍī departed with his followers and joined Hulagu. Afterwards he returned to Baghdad and said to the caliph, "This matter can be resolved if you come and show respect (to Hulagu). The hope is that peace can be made by dividing the revenue of Iraq in half." Upon hearing these words of Ibn ʿAlqamī, al-Mustaʿṣim departed Baghdad with 700 state officials and personnel of the caliphate. When they approached Hulagu's tent, the caliph was let in with a retinue of seventeen soldiers. The rest were detained outside. In short, they killed the caliph and all the soldiers who were with him. For forty days they subjected Baghdad to the sword. It is said that they killed two million people. Fearing for their lives, people hid in sewers and cellars. When the Tatar soldiers left they emerged. However, when they emerged they were in such a state that father did not recognize son and brother did not recognize brother.

It happened that one member of the Abbasids, Abū'l-ʿAbbās Aḥmad b. Ṭāhir, escaped Hulagu's sword. He took refuge with the sultan of Egypt, Malik Ṭāhir Sayf al-Din Baybars. Baybars showed him great respect and sent him back to Iraq with a large army. When he approached the Euphrates in 659 (1261), Hulagu's vicegerent Qarabogha set out from Baghdad and there annihilated Aḥmad b. Ṭāhir and his army. Very few escaped alive.

Afterwards another member of the Abbasid family whose name was Aḥmad reached Egypt. They gave him the title "al-Ḥākim bi-Amr Allāh." When his Abbasid lineage was confirmed, they made him caliph and swore allegiance to him. They provided him with an income and residence in Cairo. But he was deprived of the privileges of the caliphate. He was caliph in name only. Whenever a new sultan came to the throne in Egypt, the caliph was brought out for good luck and blessings. The caliph would say to the sultan, "May your sultanate be auspicious." Then he would swear allegiance to the sultan. The last Abbasid caliph descended from him was Yaʿqūb al-Mustamsik bi'llāh. He lived to an old age. He eventually went blind. It was during his time that the late Sultan Selīm Khan wrested the rule of Egypt from the Circassians (Mamluks).

In short, when the Abbasid dynasty was cut off from Iraq, the Mongols invaded the country and remained until 754 (1351). Subsequently the Jalāyirid, that is, the Ilkhanid, dynasty emerged and they took control of Baghdad. The first of their rulers (the Jalāyirids) was Shaikh Ḥasan Kabīr. Their dynasty lasted until 822 (1419). Afterwards Muḥammad Shah b. Qara Yūsuf of the Qaraqoyunlu ruled Baghdad until 973 (error for 873/1468). Then Manṣūr, one of the descendants of Ḥasan-ı Ṭawīl (Uzun Ḥasan) of the Aqqoyunlu dynasty, ruled Baghdad. He remained in power until 914 (1508). Then the Qizilbash (Safavid) Shah Ismāʿīl captured the city. Finally in 941 (1534), Ghazi Sultan Süleymān Khan of the Ottomans conquered Baghdad [M 483] and took control of Iraq.

Chapter 44: Description of the Arabian Peninsula

The 44th chapter: description of the Arabian Peninsula. This is the home of Arab clans and tribes. Around it are the Euphrates River and three seas. However, in two places it is partially connected to land, so they call it a peninsula. On place where it is connected to land is between Suez and ʿArīsh. The other is near Aleppo between the Euphrates and the Sea of Syria (the Mediterranean). There is no other connection apart from these two strips of land.

This peninsula is more distinguished and nobler than those in all other regions because in this region is the sacred house of God and the religion of Islam spread here. The Rightly-Guided Caliphs, the Helpers of the Prophet, and those who Emigrated with him (from Mecca to Medina) lived here. Here the banners of Islam were raised and from here were spread throughout the world. Abraham summoned his people here. It has extensive borders and many districts and includes precious buildings and delightful districts. For within its borders are the lands of the Ḥijāz, Yemen, the region of Sinai, Aḥqāf, Yamāma, Shaḥr, Hijr, ʿUmān, Ṭā'if, Najrān, Ḥijr Ṣāliḥ, the lands of ʿĀd and Thamūd, Bi'r Muʿaṭṭala, Qaṣr Mashīd, Ḥays Shaddād, the tomb of Hūd, the lands of Kinda, Mt. Ṭayy, the houses of Fārihīn, Wādī Ayla, Mt. Sinai, Madā'in Shuʿayb, ʿUyūn Mūsā ("Springs of Moses"), and the Wilderness of the Children of Israel.

The circuit of this region would take six and a half months. Thus, beginning from Ayla, which is on the Sea of Suez and is near Mt. Sinai, and whose fortress is subject to Egypt and is at the frontier: two stages to Madyan, which is now called Mafāyir Shuʿayb; then three stages to the fortress of Azlam, seven to Yanbu, nine to the port of Jār, four to the town of Ḥalī, nine to Jāzān, nine to Ṣalīgha, three to the port of Zabīd, i.e., Buqʿa, three to Mocha, two to Mandab, four to Lajma, one to Aden, nine to Yarūm, which is on the border of the province of Shiḥr, two to Shiḥr, three to the place called Bilād Ḥīrī, the first part of the inhabited area of the province of Mahra, ten to Ẓafār, three to Ḥāsik where the *ḥaṣalbān* grows, five to the port of Ghayya' Ḥashīsh which is the frontier of Shiḥr, two to Ḥamrā' Anfūr which is part of the province of Maqar, three to the city of Qalhāt, two to the city of Maskat (Masqaṭ), five to Ṣuḥār which is the chief town of the province of Oman and is now called Ṣayr Nuʿmān, then three to the port of the province of Bawāṭin—this place is to the west of the strait of the Persian Gulf; in Bawāṭin, the Persian Gulf, on the shore, to Qaṭīf, Bahrain, thirty stages; six from Qaṭīf to Basra, then three to Ḥilla along the bank of the Euphrates and through the province of ʿAliyān Oğlu, two to Hīt, three to ʿĀna, three to Raḥba, one to Dayr, five to Bālis, six to a place opposite Salmiya on the edge of the frontier and the desert, three to Tadmur (Palmyra), five to Azraq, two to Balqā, three to Jabal-i Sharāḥ, and from there three to Ayla for a total of 97 stages. In the region delimited by these places, there is neither lake nor river that is navigable. In the province of Najd there are a few small lakes created by rain. [M Map] [M 484]

Administrative divisions

The Arabian Peninsula has twelve administrative divisions: Tihāma, Yemen, Najd, ʿArūḍ, Aḥqāf, Shiḥr, Oman, Hijr, Ḥijāz, Najd-i Ḥijāz, Tihāma-i Ḥijāz, Yamāma, and Bādiyya.

Description of Yemen

This is a famous and extensive country. It is called Yemen because Yaqṭan son of ʿĀmir son of Ṣāliḥ son of Arfakhshadh son of Shem son of Noah came to this place and settled here. It got its name from the Arab expression *Tayammana Banū Yaqṭan* ("The descendants of Yaqṭan went to the right.")

Borders: The borders of this region are as follows: To the east is the country of Oman, to the south is the Arabian Sea, to the west is part of the Gulf of Suez (Red Sea), and to the north is the border with the Ḥijāz and the country of Hijr. This region is divided into four sections: Tihāma, Yemen proper, Najd, and Aḥqāf. For some time most of this territory has been within the boundary of the Ottoman Empire divided into two provinces (*eyālet*). One is Zabīd, which is the capital of Tihāma, and the other is Ṣanʿāʾ, which is the capital of Yemen. Each consists of a number of sanjaks and qadi districts. They are Ṣanʿāʾ, Taʿizz, Ḥays, Zabīd, Aden, Kawkabān, Ṣabiyyā, Liḥyā, and Shahārā. However, I will divide Yemen and describe it according to my own scheme.

First section: Tihāma

It should be known that the low lying coastal plain is called Tihāma and the places that overlook Tihāma are called Sarawāt ("Hills"). As for its borders, to the west is part of the Gulf of Suez, to the south is the Arabian Sea, to the east is Yemen proper, and to the north is the land of Ḥijāz.

Zabīd: the chief city and the capital of the province. This area is a province of Ashʿarīs. This city is located on level ground next to a river. Its port is Buqʿa. They are one stage apart. Its river comes from the area of Ḥabb, flows to the north of it, and then debouches into the sea. This city has many wells, date groves, and all kinds of fruit. In particular it has a fruit called *ʿanbā* (mango?) which resembles the peach and has a wonderful fragrance. They say that if one of them is placed in a room, its fragrance perfumes the whole room. They eat it as salad (?—*saluta*) when it is unripe. When it is ripe it is very soft. They suck out the juice and throw away the seed. It is extremely tasty.

The wall surrounding this city has eight gates. Ḥasan b. Salāma, the vizier of ʿAbd Allāh of the Zaydī dynasty, built it. Ayyūb b. Tughtigin of the Ayyubid dynasty built two madrasas in this city. One, called the ʿĀṣimiyya, is for the Shāfiʿīs. The other is for the Ḥanafīs. ʿUmar b. Manṣūr of the Rasulid dynasty also built two madrasas, one for the Ḥanafīs and the other for the Shāfiʿīs. This city is to the southwest of Ṣanʿāʾ. There are steep mountains between them, impassable by horses but passable in five days by foot. The town of Ḥays is to the south. It is one stage from Zabīd. There are many streams with habitation between Karābja and Ḥays. In this area there is a valley called Wādīʾl-Ḥusnā because it is a beautiful place for excursions. There are populous villages. On the Zabīd coast is the beautiful town of Ghalāfiqa.

Mocha: This is on the coast of the Gulf of Suez and near Bāb al-Mandab. It is two stages from Zabīd. Its port is a great entrepot. Merchants come to this port from all directions. The city of Mocha has very elaborate and well-adorned buildings. In its gardens are every kind of fruit and tree. It is famous for its beautiful women. Previously the city had no fortification, but the Ottoman Aydınlı Meḥmed Pasha built one. [M 485] Also the water of this city used to be salty. Later Shaikh Shāzilī dug a well and excellent sweet water emerged. Today they call it the well of Shaikh Shāzilī. He is buried there in an elaborate tomb.

Ḥays: A sanjak. Under the Ottomans the pilgrimage caravan started from here. It is one stage south of Zabīd. This is a very prosperous city. It is two stages to Taʿizz which is in front of the Ṣabir Mountains. Between them is located Karābja. All the places around it are very prosperous. It has many flowing streams. There are crops and orchards. The mountains before Ḥays extend to the south, then south of Taʿizz they turn eastward and wind around Cennet Ovası until they reach ʿAmāqiyya.

Bāb al-Mandab: This is the name of a mountain range that stretches 12 miles from east to west. Opposite the western part on the coast of Abyssinia, the mountains of Bīlūla can be seen. From one shore people can be seen on the other. Today Bāb al-Mandab is the name of the strait of the Gulf of Suez.

Mawzaʿ: This is on the road to Aden. It is two stages south of Mocha. From there to Damāwa (?) is five stages. This mountain is extremely steep. In the past the ruler of Yemen had his treasury there. From this fortress to Laḥij is one stage. From there one goes to Aden.

Raqʿ: A small town north of Zabīd.

Bayt al-Faqīh al-Kabīr: They call the town here Zaydiyya. It is the province of Sharīf Muṭahhar on the road of pilgrims from Zabīd. It is five stages from Ṣanʿāʾ.

Laḥya: The town of Rāʿī is also called Laḥya. It is on the coast of the sea. Near it is a place to dive for pearls. Coffee is grown in this area. The coffee tree resembles the cherry tree, as does its fruit.

Bayt Faqīh Ṣaghīr: It is one stage from the sea. Coffee is also grown here.

Ṣalba: Also called ʿAṣāb. Originally coffee was grown only in these three towns. It is now grown in other places, but their coffee is not as good.

Abū ʿArīsh: A prosperous town with a fortress. Its port is called Jāzān. Bitter oranges, lemons, grapes, and other fruit are grown here.

Salāma: There are four villages in the district called Ṣabiyya. The people are *sharīfs* (descendants of the Prophet). Its port is Jāzān. Salt is extracted here, exactly like that of Ūlāh. The water in this area comes from springs. There are very prosperous villages everywhere. There are many bitter orange, lemon, jujube, and *ʿanbā* (mango?) trees. The Ṣabiyyā (?) people who live here are very courageous. They bow to no one. They have 4,000 horsemen clad in armor and cuirasses. They make excellent armor and it is famous everywhere.

Bīsh: A town on the Ḥijāz road. The buildings in most of the cities of Yemen are made of reeds.

Aden: This district has its own ruler. To the north is the district of ʿĪd Raws. This city is very old. It is at the southern extremity of the province of Yemen and on the seacoast. It is surrounded by mountains. It has no flowing streams. Water is brought in from the outside. It has a strong fortress. On three sides of the city are fortress-like mountains. It has two gates. One is on the sea side. The other is on the north side and is called Bāb al-Shāmayn. This gate leads out toward Abyan which is a place now called Ḥū east of Aden and is said to be its city. Because this is the end (?—*āḫir*) of Aden in Yemen it is called Aden Abyan. In short, Aden is the meeting place for ships from Zanj (i.e., Africa), Sind, India, Malabar, Ceylon, and Java. This city takes its name from ʿAdan b. Sinān, who was one of the children of Abraham. [M 486] Near the city is a place to dive for pearls. On the coast raw ambergris washes ashore.

Nūnḥara: An invulnerable fortress on the peak of a mountain near Aden. There is no conceivable way to reach it in order to conquer it because there is only one way to reach it and this route is extremely difficult. There is a large spring in the fortress at the top of the mountain and it irrigates many villages. Most agriculture is devoted to growing *wars* (Yemeni turmeric; saffron in Steingass). Muḥammad b. ʿAbd Allāh of the Zaydīs built this castle. And ʿUmar b. Manṣūr of the Rasulids built a madrasa inside it.

Ribāṭ Mīnā Sulaymān: It is two stages from Aden on the road to Shiḥr.

Bilād Ḥūr: A prosperous district. In the east of it is the stage called Ḥūrā. In the north of it is Abū Maʿbada.

Rayma: A district. It has many villages. Rayma proper is eighteen miles southwest of Aden. Its ruler is someone named ʿAbd Allāh ʿĀl who is from the local populace and is the enemy of Sharīf Muṭahhar.

Second section: Yemen proper

This is divided into two parts. One of them is Yemen proper which is located within the Sarawāt Mountains. The other is Ḥaḍramawt which is east of it.

Description of Yemen proper: An extensive but mountainous country. It has many districts, cities, and towns. The major city is Ṣanʿāʾ, which is the capital of the province of Yemen. Ṣanʿāʾ b. Zāl b. ʿĀmir built it. It is the largest city in Yemen. Next to it is a small river that comes from the north. It flows to Dhamār and then goes to the sea. This city is 76 degrees long and 15 degrees wide. Because of the city's abundant streams and trees, it is likened to Damascus. The air is excellent and healthful. The water is sweet and the soil is fragrant. There are few illnesses or blights. Indeed, flies, vermin, and insects are virtually nonexistent. If one who is ill in another city comes to Ṣanʿāʾ, he is cured. If sick camels graze in the pastures of Ṣanʿāʾ, their illness passes. If meat is left exposed for a week, it does not rot.

When the sun enters Aries, it becomes excessively hot in Ṣanʿāʾ, because Aries, Taurus, and Gemini occur in Yemen in the summer. Cancer, Leo, and Virgo occur in the fall; and Libra, Scorpio, and Sagittarius occur in the winter; and Capricorn, Aquarius, and Pisces occur in the spring. Rainfall in this city occurs in the afternoon. It rains in June,

July, August, and sometimes in September. During this season the people say to each other while they are shopping before noon, "Take care of your business quickly before the rain comes." For the sky may be clear with no sign of rain, but as soon as the sun declines from the zenith, the rain begins. The air of this city is so mild that the people do not have to migrate in winter or summer.

This city has abundant fruit. They say that it is especially famous for its abundant *ʿanbā* (mango?). This fruit increases one's sexual potency. This city manufactures cloaks, turbans, *saḥūlī* robes, and Ṭāʾifī leather which are found nowhere else. There is also a species of fine motley cows patterned in yellow and white. There are many Friday mosques in this city.

Next to Ṣanʿāʾ on the top of a hill is the castle of Ghumdān which is said to date from before Islam. There is today a water-works at this place from which water is brought for the Friday mosque of Ṣanʿāʾ. The Himyarite (king) Yaḥṣub built it. He built Ghumdān towards the four directions, one façade red, one white, one yellow and one green. The interior had seven ceilings, that is, seven stories. [M 487] It was very high. When the sun rose, one could see a distance of three miles. At the top of this castle he built a council chamber of colored marble. The ceiling was made of a single piece of marble. On each pillar he carved a likeness of a lion. Whenever the wind blew, the lions' manes were activated.

It is reported that Caliph ʿUthmān b. ʿAffān ordered Ghumdān to be torn down, but he was told that there was an inscription on it that said, "The one who tears you down will be murdered." ʿUthmān disregarded this and tore down the castle (and, of course, he was later assassinated). It is also reported that when the ruling emir of Yemen was on his way to Abyssinia, Abraha b. Ṣabbāḥ (the Christian king of South Arabia) built a church in Ṣanʿāʾ, the like of which was never seen, and he called it Qalīs (or Qullays). He adorned it with gold, silver, and precious stones. Then he sent a letter to the Negus (king of Abyssinia) saying, "I built a church for you. No king has such a church. I want you to divert the Arab pilgrims to this church." A man from the Banī Kināna heard this and went to Ṣanʿāʾ and defecated and urinated in the church. When Abraha learned of this act he asked who did it. He was told that an Arab from the people of the house which was the object of pilgrimage (i.e., Mecca) did this. Abraha became enraged and wanted to go and destroy the Kaʿba. He immediately prepared a huge army and a white Maḥmūdī elephant and set out for the Kaʿba. Three parasangs from Mecca he called a halt and gave the order to pillage. He carried off 200 camels of ʿAbd al-Muṭṭalib, the grandfather of the Prophet, in a raid. ʿAbd al-Muṭṭalib went to Abraha and asked for the return of his camels. Abraha answered saying, "How strange! Here you are asking for the return of your camels while you abandon the Kaʿba, the greatest pillar of your religion." ʿAbd al-Muṭṭalib then said to Abraha, "I am the owner of the camels, the Kaʿba has its lord. He is the one who will protect it." Abraha then gave ʿAbd al-Muṭṭalib his camels. The next morning he set out to destroy the Kaʿba. However, help and aid came from God to his slaves. And according to the noble verse *wa arsala ʿalayhim ṭayran abābīl* ("And send against them birds in flocks," Koran 105:3), a divine victory occurred.

Taʿizz: The Ayyubid Tughtigin built this city. Its air is mild and it has streams. It is located in the mountains and overlooks the Tihāma. Near it is a place called Ṣahla with tall buildings, gardens, orchards, and places of excursion. The ruler of Yemen brought water to this place of excursion (Taʿizz). Taʿizz is to the south of Zabīd and is northwest of the fortresses called Qurayb and ʿArūs. To the east is Cennet Ovası. It is three stages to Zabīd and there is much habitation along the road. The first stage is called Karābja. It is part of the country of the Banī Ḥujr. Near it is Mt. Ṣabir. This is a very populous and blessed mountain. Karābja is surrounded by a populous area and by blessed mountains. Taʿizz has many streams, gardens, orchards and grain fields. Nearby is Wādīʾl-Jannā (same as Cennet Ovası?), which is populous and has many streams. The Rasulid ʿUmar b. Manṣūr built two madrasas in Taʿizz. (The Rasulid) Malik Manṣūr built a madrasa called the Mujāhidiyya. Malik Afḍal (also a Rasulid) built a Friday mosque and a madrasa called the Afḍaliyya.

ʿUmāqī (?): A place on a mountain. [M 488] It has streams and many villages. Caravans that pass through it pay a toll. It is east of Taʿizz.

Udhn: A fortress on the road. A caravan toll is collected here also.

Dūr: Also called Dūrān, it is a castle on a steep, inaccessible, high place in the mountains of Ṣanʿāʾ. ʿAbd Allāh b. Ḥamza is the possessor of this castle. There are two districts near Ṣanʿāʾ. One is Shahārā and the other is Raw. These two districts are in the mountains and are Zaydī centers.

Shahārā: A very strong fortress on a high hill that would be impossible to capture. There is only one route to it, which is a climb of 1,600 steps. The mountain on which it is located is near Ṣanʿāʾ and to the north of it.

Jabala: It is located between Ṣanʿāʾ and Aden on a mountain and next to two rivers. The Ṣulayḥids built this city after they took control of Yemen. It is one stage from Taʿizz.

Jund: It is north of Taʿizz. Its water is unhealthy. Muʿādh b. Jabal built a Friday mosque in it. Nearby Wādī Samūl is very prosperous and has many villages.

Dhamār: A group of *ahl-i riwāya* (scholars who specialize in the study of the transmission of hadith) trace their origin to this city. On a mountain north of it is the mosque of Muʿādh b. Jabal. Many religious scholars come from there. It is two stages from Ṣanʿāʾ. It has solid walls. It has many villas, orchards, arable fields, and villages. Living is very cheap in this city. It has many streams. One stage distant are the ruins of an ancient building. It was constructed on six marble columns and four additional columns were built on top of four of these columns. Many streams flow below the columns. The people are of the general opinion that these pillars were the throne of Bilqīs (the Queen of Sheba, but see below on her throne).

Majbā: The name means the place where a caravan toll is collected. It is at the foot of a mountain near Ṣanʿāʾ. A toll was collected here also.

Ṣaʿda: The first city of Yemen, east of the Sarawāt Mountains. It is 44 parasangs north of Ṣanʿāʾ and 20 stages from the Kaʿba. The skins of cattle are tanned in this city and transported from there to other cities. This city is agriculturally very productive (fertile).

Khaywān: It has streams, arable fields, and villages. Various tribes have settled here. The Banī al-Ḍaḥḥāk of the Āl Yaghfir, who were from the Tubbaʿs (ancient kings of Yemen), built their palaces there. It is 23 parasangs north of Ṣanʿāʾ. They get their water from rain.

Thalla: A strong fortress on the top of a mountain. It was traditionally the headquarters of Sharīf Muṭahhar.

Kawkabān: It is 20 stages from Ṣanʿāʾ. It has a strong fortress. Near it is the city of Ṭawīla.

Ṭawīla: It is six miles to the east of Kawkabān. This city has a great market.

Jarāf: It is half a stage from the buildings of Ṣanʿāʾ. It has orchards, gardens, and streams and is a place of excursion for people from Ṣanʿāʾ. It is the first station for pilgrims coming from Ṣanʿāʾ.

Mafḥaq: One of the Three Towns, 1 1/2 stages west of Ṣanʿāʾ. The Three Towns are several fortresses belonging to Sharīf Muṭahhar.

Shiyām (? Shibām): It is to the south of ʿĀmira. It is currently under the control of someone named Muḥammad al-Dāʿī. Ṭawīla is also one and a half stages from Mafḥaq. Shiyām is a half stage from ʿĀmira and from Masār al-ʿAmmār. It is one of the Three Towns.

Masār: A strong fortress in the mountains and the seat of Muḥammad al-Dāʿī. He is the enemy of Sharīf Muṭahhar.

Maswār: A strong fortress. It is on the top of a mountain, even steeper than Thalla. It has many streams and grain fields. It is only accessible from one road. One of the marvels of this place is that the wheat, barley, and millet do not rot when left exposed for a long time. [M 489] It is even reported that if one stores wheat here for thirty years and then opens it, he will not find a single rotten grain.

Ẓafīrjaḥa: In this city is the burial ground of the ancestors of the Sharifs. It is also where they keep their wealth and treasure. Its fortress is strong.

Maḥwiyat and Masmāt: These are fortresses belonging to the Sharifs. It has rebellious villages. The people are called Banī al-Khayyāṭ. They are highway robbers. They have an army composed of 4,000 armored archers and spear-throwers and men with muskets and slings. Nearby are the Banī'l-Ahīl. They are also highway robbers and rebellious.

Barʿ (?): One of the valleys (?) of Wādī Laʿsān. It has several villages. The ruler of this city is from the local populace and is the enemy of Sharīf Muṭahhar.

Rayma: It is several villages and is near Ṣanʿāʾ.

Thaqīl: A steep and rocky place on the Thamāra road. The Ottoman pashas who passed through this area have made it level. There are streams all around it. It is in the middle of Sarjī territory. It is to the east of the Ziyāb Mountains and extends to them. This district has two divisions, one in the mountains of ʿAbbās territory and the other at the foot of the mountains of Muḥammad Sarjī territory. The territory of the Banī'l-Ḥārith is in (those) mountains. Thaqīl is to the south of Samāra. They are 12 miles apart.

Around the time that the Prophet appeared, there were two famous soothsayers in Samāra. One was Rabīʿa b. Ḥamdān b. Māzin b. Ghassān who had the byname Saṭīḥ. God created him in a strange way. He had no hands or feet. He only had a head. He had no sinews or bones in his body; only he had sinews in his head and tongue, which were the only parts of him that moved. His eyes could see, his ears could hear, he could speak with his tongue, and he could eat food. They made a throne for him from date palms and palm leaves. They wrapped his body from his feet (!) up to his throat and placed him on the throne. Wherever he wanted to go, they took him on this throne. One day they took this soothsayer to Mecca. Four elders of the Quraysh (tribe) came to see him. They kept their lineage secret and told him that they were from a different tribe. Saṭīḥ said to them, "You are not from the tribe that you said you were from. Rather you are from the Quraysh." Many questions and answers followed. Finally he informed them of the coming of the noblest apostle and his prophethood and his circumstances.

The second soothsayer was Shiqq. God also created him in a strange form. He was half a man. The other half was missing. For example, half of his head was missing. Thus he had one eye, one nostril, half a mouth and neck, one arm and one leg, and half of his chest and back were missing. In sum, these two soothsayers informed people of what was in their hearts.

Ḥabb: A strong fortress. (The Ayyubid) Tughtigin built it. When the Ottomans conquered Yemen in 940 (1533), this fortress was not conquered. Meḥmed Pasha, the *beylerbey* of Yemen, found a way to conquer it in 972 (1564). He found much money and valuable objects there. Ḥabb has places for excursions. The river that flows to Zabīd starts here.

Shalāla: The city of the ʿAmmār tribe. It is at the foot of the mountains.

Āb: A strong fortress in the mountains. It is on the road to Saḥūl and is south of it. It has inns for travelers going from Makhādir to Taʿizz.

Baʿdan and the territory of the Shawāfī: Baʿdan is located at the foot of the mountains in an eastern direction seven miles from Ḥabb. [M 490] East of it at the foot of the mountains is the territory of the Shawāfī.

Shiyam (?): A town at the foot of the mountain of Muḥammad Makrad. The "Mountains of Iram" of Muḥammad Makrad lie to the west of it. This town has 3,000 spear-throwers. It is one stage from Laḥij.

Sarawāt: They call the mountain range that extends within Yemen as far as Ḥijāz the Sarawāt Mountains. It is to the west of Ṣaʿda and Ṣanʿāʾ, extending (south) to the sea coast and to Aden, then (west) to Shiḥr which is called Tihāma. Most of the province of Ṣanʿāʾ and Aden is in these mountains. The air is excellent and there is much running water. It is impossible for the caravans of Ṣanʿāʾ to cross from the towns on the east of this range to those on the west. From Ṣanʿāʾ to Zabīd the road is only wide enough for one pedestrian. It takes five days to go from Ṣanʿāʾ to Zabīd. On the other hand, one can go by horse from Ṣanʿāʾ to the town called Bayt al-Faqīh, which is the major city of Tihāma. The road from Bayt al-Faqīh to the territory of the Banīʾl-Ahīl takes a day and it follows a stream. It is easy to traverse the villages of the Banīʾl-Ahīl. It is also an easy day's journey from there to the territory of the Banīʾl-Khayyāṭ. In order to go from there to the city called Ṭawīl in the territory of (Sharīf) Muṭahhar, one must spend a day climbing mountains. It is a steep climb but the road is wide. From Ṭawīla the road goes to the town called Jabbāba. Kawkabān is to the right. One can go from Jabbāba in one day to Ṣanʿāʾ via a place called Manqib and Ṭayyiba. It is possible for troops to go on this road from Bayt al-Faqīh to Ṣanʿāʾ. If there are no rebels on this road then caravans can travel on it. From Jāzān to Saʿda there is a road that is passable by horse and mule. There is a good road that goes from Jāzān to Nawīdiyya, from there to the fortress of Tarīma, from there to the fortress of Falakī, from there to the place called Khaṭim, and from there to Saʿda. Apart from these two roads, there is no way to go from Tihāma to Najd. But one can go from Ṣanʿāʾ to Zabīd. But caravans often take the round-about way via Taʿizz. The distance is 14 stages. The road goes from Ṣanʿāʾ toward the southwest through the foothills and via Thaqīl, Dhamār, Saḥār, Āb, and ʿAkfar. From there it goes to ʿUmmāq and then via Taʿizz to the west. From Taʿizz it turns northwest to Zabīd and takes three days. From Ṣanʿāʾ it goes south to Aden through the foothills, but goes north to Saʿda. Pilgrims from Ṣanʿāʾ go via Saʿda to Ṭāʾif and then Mecca. Pilgrims from Aden go to Ṣanʿāʾ, or Zabīd, or by boat to Jedda. Pilgrims from Zabīd go to Mecca along the coast on the Tihāma road via Bayt al-Faqīh, Jāzān, and Yalamlam. The area east of Saʿda and Ṣanʿāʾ is called Najd al-Yaman.

Further description of Yemen proper: There is a plant here called *ghalas*, which is a type of wheat. It has two grains on the stalk (? *kelem*). It is found nowhere but in Yemen and is a staple food of Ṣanʿāʾ. Another plant found here is *wars*. It has a pod like sesame and once it is sown it continues to sprout for 20 years. There are also bananas. There is a kind of pear here which, if eaten, causes one to have diarrhea ten times. They extract honey from this pear. They make an electuary from this pear and give it to one who has colic and the illness passes.

Yemen has the ugliest monkeys, but they are the quickest to learn. There is a kind of bearded monkey on the mountainsides of Yemen that is diabolical. If it overtakes someone it throw itself on him and copulates with him. If the local people see someone with this animal they say that he is its spouse and he runs away in despair.

It is reported that the Imām al-Shāfiʿī said, "I once entered a town in Yemen. I saw a person there [M 491] who had a body like a woman from the waist to the feet. From the waist to the head it had two separate torsos, four hands and two heads. Each head had a different face. Sometimes the two upper sections would collide with each other. Sometimes they would stop at one point and not collide. They would both eat and drink. Then for many years I was absent from this town. When I returned I asked the people about this person. They said that one of the torsos had died. They bound the dead one to the living one from below (?—*esfelinden*). When it began to smell, they amputated it. The other body was still wandering about the markets."

Description of Ḥaḍramawt

This district has two important cities, Shibām and Tarīm. Both are old settlements. Qaṣr Mashīd is in Ḥaḍramawt. This is the place mentioned in the Koran along with Biʾr

Muʿaṭṭala (22:45 *wa-biʾrin muʿaṭṭalatin wa-qaṣrin mashīdin* "their wells abandoned and their proud palaces empty"). The people of Ṣāliḥ dispersed after his death. One group went to Aden. This group was thirsty because of a lack of rain and had to bring water from a great distance. God blessed this group with a well (*biʾr*). They were amazed by this. They erected columns over the well according to the number of tribes and placed a bucket for each tribe. They had a just king who distributed the water (to the tribes) fairly. After he died, Satan seduced this group and they began to worship idols. For this reason God sent Ḥanẓala b. Ṣafwān to them (as a prophet). They accused him of falsehood. The prophet said to them, "If you don't abandon the worship of idols, the water in the well will disappear". As soon as the prophet said this they killed him. When they awoke the next morning there was not a drop of water in the well. They went to the idol(s) but Satan did not speak to them. The angels of vengeance appeared with a cry and destroyed them.

Qaṣr Mashīd: The person named Ṣadd b. ʿĀd who built this palace was very strong. He could grasp a tree with one hand and pull it out roots and all. He ate as much food as twenty men. He loved to have sexual intercourse with women. He married more than 700 virgins. He had a son and a daughter by each one. As the number of his children increased, they became tyrants and rebels. Then "a blighting wind" (Koran 51:41) was sent down and he saw that the people of ʿĀd would disappear. So, he built Qaṣr Mashīd. It was very solidly made so that the wind would have no effect. Then he moved with his people to this palace. He killed whoever left the palace. With a single cry (cf. Koran 54:31) from the heavens God destroyed him and his people. And the palace was ruined. The remains of this palace can still be seen. The tomb of (the prophet) Hūd is in this district.

In short, Shibām and Tarīm were the towns of Ḥaḍramawt. They are in steep mountains where carnelian and onyx are mined and where there are villages, arable fields, and streams.

Ḥaḍramawt proper: Ḥaḍramawt is the name of a village. It is four stages from Shiḥr. It is now the home of the Banīʾl-Nimr Bedouin. In the past its ports were Shiḥr and Shīrma. At that time, when affairs in Ḥaḍramawt were unstable, in 610 (1213) Sultan Aḥmad built a channel at Ẓafār and it became the capital. Later in 670 (1271) Sultan Muẓaffar built a fortress in Shiḥr and made it his capital.

Tarīm: A great city. It is fifteen miles from Dhamār. [M 492] It is today very populous.

Biʾr Barhūt: A well near Ḥaḍramawt. It is located in a large valley. The caliph ʿAlī is reported to have remarked, "The place that God hates the most is Biʾr Barhūt." There is a well there with black water.

Shibām: A strong fort located in the mountains.

Maʾrib: This district is the land of the Azd tribe. They are the descendants of Sabā (an ancient king of Arabia Felix). This district has streams which irrigate its fields. It is said that they sow three times a year. The chief city is Sabā. It was built by Sabā b. Yashjub b. Yaʿrub. He was called Sabā because he took many people captive (*saby*). This was the city of Bilqīs (the Queen of Sheba). Today clans of the ʿAmmāra live here. It is reported that in the past the province of Sabā was very prosperous. It was planted with all kinds of trees from the one end to the other. There were so many that a person could travel from one town to another in the shade of these trees without the sun ever striking. It is reported that in ancient times a great flood occurred in this district. It flowed between two mountains and then dissipated in the desert. These two mountains were two parasangs apart. Then one of the Himyarite kings built a dam of stones and tar at the place where the flood occurred. An enormous amount of water that came out (between the mountains) was retained in this dam. At the top and bottom of the dam the people put three spillways so that whenever they needed water they could take it from them. They were built for planting and sowing. It became the most beautiful land of God's lands. However, the people later began to worship the sun, so God sent three prophets to them. The people accused all three of falsehood. So God vanquished them. He sent locusts to attack the dam and they made holes in it with their mandibles. They opened the places that held back the water and thus the water flowed out. It flooded the country. All the gardens, houses, and palaces were flooded. Thus the noble verse confirmed the truth of *fa-arsalnā ʿalayhim saylaʾl-ʿarim* ("So we sent upon them a violent torrent" Koran 34:16). At that time the people of Sabā were from the Banī Ḥimyar and Banī Kahlān (tribes).

To resume, Sabā is three stages from Ṣanʿāʾ. It has a large population, excellent air, many trees, delicious fruit, sweet water, and all kinds of animals. As for vermin, there are no flies, mosquitoes, snakes, scorpions, or similar tormenting insects. God stated in the glorious book, *laqad kāna li-sabāʾ* ("Certainly there was for Sabā …" Koran 34:15). It is reported that there are 12 excellent features (of this city). First, there are no scorpions, snakes, ants, or mosquitoes. Second, if a traveler who has fleas or lice in his clothing enters this city, they immediately disappear. Third, no one ever becomes ill. Fourth, if someone who is not from this city is ill and comes to this city, his illness disappears and be becomes healthy. Fifth, in this city there is no blindness, madness, or paralysis. Sixth, if a madman who is not from this city comes to this city and bathes in its water, his madness disappears. Seventh, when the harvest season is over, by order of God, the wind blows and separates the wheat from the chaff. Eighth, they wear the same clothes winter and summer. Ninth, the weather is mild (in temperature). Tenth, when a man marries a girl, her virginity returns every time he copulates with her. Eleventh, during pregnancy their women suffer no discomfort. Twelfth, their clothing never wears out.

There is a salt mine in Sabā. The Prophet showed it to Ibn Jamāl Lamāznī. He turned it white and distributed it to the poor. The throne of Balqīs (the Queen of Sheba) is in this city and is built on columns. [M 493] The columns are 28 cubits high. This city has a well-built market and a mosque.

Third section: Description of Najd al-Yaman, Bilād Jawf, etc., and Raqīm.

Najd al-Yaman: It extends from the east of the mountains of Ṣanʿāʾ and Ṣaʿda, and passes east of Ḥijāz to Najd al-ʿĀriḍ. Within this area are many villages and caravan stations and various districts.

Bilād Jawf: It has many villages and various Bedouin caravan stations. In southern Jawf are located the areas of Nahm, Bilād Ḥawlān, Bilād Radʿ, and Bilād Raqīm. All are in flat areas. It produces several thousand horsemen.

Radʿ: This town is near Dhamār. It is the town of the Bedouin called Āl Raqīm. It has 2,000 horsemen. They are the enemy of Sharīf Muṭahhar.

Bilād Ḥawlān: It is two stages to the east of Ṣanʿāʾ. It is mountainous and has many villages. It produces 2,000 musketeers. They are the enemy of Sharīf Muṭahhar.

Bilād Nahm: It is one stage to the east of the buildings of Jarāf (?—Jarāf *bunyān*). It is on the top of a mountain. It has many villages. It produces 5,000 archers. They are the enemy of Sharīf Muṭahhar.

Najrān: It is east of Ṣaʿda. This is the place where the monk named Najrān announced the coming of the Prophet. It has many villages. The people who now live there are the Āl-i Muḥammad Bedouin. They have excellent quinces and grapes. They dry them and send them to Yemen and Ḥijāz. They have abundant dates. They tan animal skins beautifully. Their mountains are forested and not barren. This city is considered to be part of the district of Hamdān, a well-watered place with abundant fruit. It is ten stages from Ṣanʿāʾ and twenty stages from Mecca. The renowned event of the killing of the Companions of the Trench (*Aṣḥāb al-Ukhdūd*), which is mentioned in the Koran (Koran 85:4), took place in this district. There was a place in Ukhdūd where a fire burned, brought into being by (the pre-Islamic king of Yemen) Dhū Nuwās to prevent Christians from converting to Judaism (!). Any who did, he threw into the fire.

It is related that during the period between revelations, before the sending (of Muḥammad), there was a Jewish ruler of this area. His title was Dhū Nuwās and his name Yūsuf. He had a vizier named Balniyās who was an unrivaled sorcerer and soothsayer. (One day) he said to the ruler, "My lord, the time of my demise is approaching. Give me a clever and intelligent slaveboy and let me teach him soothsaying." The ruler gave him a boy to instruct. Every morning the boy attended on the vizier and acquired this knowledge. Sometime later, this boy got up in the middle of the night and went to the ruler's palace. While on his way, he heard a mournful and moaning voice coming from underground. The sound disturbed him. In the darkness he found a door leading underground. He opened it and went in. There he saw a monk who said to him, "I came here to get away from people and to live austerely in order to be safe from the wiles of my carnal self and of Satan." In short, the boy had a great desire for the religion of Islam. The monk instructed him in the faith but made him swear an oath that he would not reveal this to anyone. In a short time the boy was honored with serving the old man (or shaikh, spiritual guide) and became one whose prayers are answered.

It happened that one of the chamberlains became blind. The boy said to him, "If you follow me in the faith, I will pray for you." The chamberlain agreed to this and when the boy prayed, his eyes were opened by the grace of God. One day the ruler saw the chamberlain and asked him by what means he was able to see. He insisted on an answer. Out of fear, the chamberlain was compelled to tell him in detail what had happened. Dhū Nuwās summoned the boy to his presence and questioned him. "His eyes were opened with the blessings of my prayer," he replied. [M 494] Dhū Nuwās treated the boy with great contempt. They learned about the monk. The monk, the boy and the chamberlain were all brought to the court. Dhū Nuwās said to the monk, "If you abandon your faith, I will forgive you." When the monk refused to do so, they lit a huge fire and cast into the flames (all) those who held to their faith. They even brought a woman with her baby. Out of fear the woman was about to abandon her faith, (but) by order of God the baby spoke saying "God's fire is more intense than this," so she refused to consent to unbelief. Then without hesitation the boy and his mother threw themselves into the fire of their own accord. God protected them them; they emerged from the fire safe and sound and departed. Afterwards, by the power of God, that fire spread and consumed all the unbelievers who were there.

Wādī'l-Dawāsīr: A valley in the wilderness east of Ṣaʿda. It is a place with streams and date palms. Once a year the people of Yemen, Najd, and Oman gather here and hold a great fair. Then people from India, Yemen, and Ḥijāz come here and sell everything from common goods to precious stones.

Description of al-Aḥqāf (the Sand Dunes): It is located in the eastern part of the province of Shiḥr. It is a place of hills with sandy areas and plains. The sands are shifting and stretch for 350 parasangs. It is a changing landscape. The wind continuously moves (the sand) from place to place. In the Koran the verse (that begins) *wa'dhkur akhā ʿĀd* ("And mention the brother of ʿĀd," 46:21) refers to this place (the name of this Sura is in fact *al-Aḥqāf*). In the past this province was extremely prosperous. It was like a paradise with trees, fruit, and streams. These were the lands of the first ʿĀd, to whom the Prophet Hūd was sent. When they accused him of falsehood, the wrath of God struck them with a "roaring wind" (Koran 41:16, 54:19, 69:6) that created desolation like a wilderness. Hūd was from the tribe of ʿĀd. They were Bedouin who took their name from ʿĀd son of ʿAwṣ son of Arīm son of Shem (son of Noah). They lived in Aḥqāf in tents with columns. The original ʿĀd were thirteen tribes. After Noah sovereignty and rule resided with this people. The verse in the Koran *wa'dhkurū idh jaʿalakum khulafāʾ min baʿdi qawmi Nūḥ* ("And remember when he made you successors of Noah's people," 7:69) refers to them. Also, *wa-zādahum Allāhu basṭatan* (rather, *wa-zādahum fī'l-khalqi basṭatan*, "He [God] increased them in stature," 7:69)—i.e., in height, each one being as tall as a date palm. They possessed (great) strength and ability, lived a long life and had many children. Their ruined monuments can still be seen in Shiḥr. ʿĀd's son Shaddād is the one who built Iram Dhāt al-ʿImād ("the many-columned city of Iram," Koran 89:7). He is called the second ʿĀd. In the Koran the verse *alam tara kayfa faʿala rabbuka bi-ʿĀdin Iram Dhāt al-ʿImād* ("Have you not considered how your Lord dealt with ʿĀd (of) Iram Dhāt al-ʿImād?" 89:6–7) refers to him. In short the country of ʿĀd includes Shiḥr, Ḥaḍramawt, and Aḥqāf. It was very fertile but the wrath of God turned it into a desolate land like a wilderness.

Description of Shiḥr: It is on the coast of Yemen. Its lands connect with those of Ḥaḍramawt. It is the home of the Mahra tribes. It was (part of) the territory of the first ʿĀd. It is currently a large city. There is no stock farming or agriculture. But ambergris is found here. It also has hemp which is exported everywhere. The trees of *ḥaṣalbān* (myrobalan), coconut and gum lac are abundant here. They have very highbred and fast-running camels These camels can be quickly trained and are marvelously intelligent. Each one is given a name and they immediately come when called by their names. Shiḥr has an independent ruler. It has beautiful buildings and many religious scholars and devout people. [M 495] It is seven stages from Aden and eight from Ṣanʿāʾ.

Barūm: A stopping point for boats coming from Aden to Shiḥr. Barūm is on the western border of the province of Shiḥr.

Description of Mahra: This district extends from Ḥīra to the place called Jādhib. It is entirely within the province of Shiḥr. The people of Mahra have an unintelligible language. They speak Himyarite. The people of this country have many camels. Most of their income derives from camels, goats, and fish. They catch the fish in the sea of Shiḥr (Arabian Sea). They dry in the sun one kind of fish that they catch and give it to riding animals as fodder. What the people eat is fish, and what they drink is milk. Some of the people of Mahra become ill if they eat bread and grain. The lac tree is just like the mulberry tree, except that the mulberry tree has leaves while the lac does not. Its fruit, which grows on its branches, is called *kendir* (hemp). They also have myrobalan trees and coconuts.

The province of Ḥīrī: It starts at the border with the province of Mahra and extends along the seacoast.

Qash Ṣaḥūna: A two-day sail to the northeast.

Ẓafār [Dhofar]: A town that is part of the province of Shiḥr. In ancient times it was a capital city. It is at the tip of a spit of land that juts toward the north into the Green Sea (Indian Ocean). By land it is 15 stages from Ṣanʿāʾ. Ships and boats powered by an offshore breeze depart Ẓafār and sail to India. In the plantations of Ẓafār Indian plants like coconuts and hyacinth are grown. Ẓafār was the abode of the kings of Himyar. Thirty-seven of their kings lived there. They reigned for 3,190 years. The last of them was Maʿdikarib. In Ẓafār they produce frankincense which is found nowhere in the world but the mountains of Ẓafār. The trees from which it is taken are three stages from Ẓafār. The people of Ẓafār slash these trees with a knife and the frankincense runs out.

Marbāṭ: It is on the coast of the gulf of Ẓafār. Here too frankincense grows.

Description of the province of Oman: During the time of the Prophet it was conquered for Islam. It was named after ʿUmān b. Naqshān, one of the sons of Abraham. This province borders on the territory of Mahra, which is to the south of it. To the northwest it borders on Yamāma. This province is very prosperous. It has many date palms and fruit including bananas, pomegranates, figs, and grapes. It is related that when God said, *yaʾtīna min kulli fajjin ʿamīq* ("[pilgrims] coming from every remote path," 22:27), he was referring to Oman. It is reported that the Prophet said, *man taʿadhdhara ʿalayhi al-rizq fa-ʿalayhi bi-ʿumān* ("he for whom earning a livelihood is difficult should go to Oman"). It is one of those places that are proverbial for its (fine) air. In 674 (1275) the Ibāḍī rite, which is a branch of the Kharijites, appeared in Oman. They are followers of the Kharijite ʿAbd Allāh b. Ibāḍ. This ʿAbd Allāh appeared during the reign of Marwān b. Muḥammad, the last of the Umayyad caliphs. He was put to death and his evil removed.

Ibn al-Athīr relates in his history: "In 375 (985) a bird that was larger than an elephant emerged from the sea and perched on a hill in Oman. It shouted in a clear voice and a distinct tongue '*Qad qurb, qad qurb.*' Then it dove into the sea. It did the same thing for three days straight and then it disappeared."

There is a kind of snake in this area that always seems to be intoxicated. It only hisses, otherwise is quite harmless. It is reported that [M 496] one of these snakes was once put in container and sent outside Oman. After the container was beyond the borders of the province, they examined it and saw that the snake had disappeared.

In the mountains of this province there are countless monkeys. In large groups they cause a lot of damage. Sometimes, in order to prevent their damage, the people of the province have to take up arms against them. Where this province faces the sea it has a sandy coastal plain. The inland areas far from the sea are mountainous.

Oman proper: A well-fortified town on the coast. There is a mountain next to it. Many streams rises from this mountain and the water for Oman comes from it. There are abundant date groves and gardens in this city. Various kinds of fruit are abundant. Their cereal products are mainly wheat, barley, rice, and millet. A rich Zoroastrian named Abū'l-Faraj brought water to the city. He also built large *inns* in Oman for merchants. It is reported that the revenue from tax farming is 80,000 goldpieces every year. Many ships come to this port of Oman, whether from China, India, Zanj (Africa), or elsewhere. The people of Oman are rich. It is a place where they dive for pearls and Oman is famous for very large pearls.

Ṣuḥār: It is currently called Ṣayr Oman. It was formerly the capital of Oman. It is on the right side (of the country of Oman, looking west). It is an entrepot for India, China, and the Zanj. It has many religious scholars and pious men. They call the district of Oman "the land of Azd."

Maskat (today Masqaṭ): A strong fortress on the edge of the sea. It is an entrepot for India and China. It has many date palms and coconut trees. There are also pepper trees and tamarinds. In the past the Portuguese managed to occupy it. Later, around 1070 (1659), the followers of a Muslim jurist overcame it and took it from the Portuguese. They took all of the infidels prisoners and seized their ships. Today the ships of that jurist are famous. When they encounter a Portuguese ship, they do not fail to capture it.

Qaryāt (?): An anchorage for ships. From here to Qalhāt is one day's sail.

Qalhāt: A small town to the southwest of the city of Maskat. Near it are Ra's al-Ḥadd and Ra's al-Jumjuma.

Description of Hajar, also called Bilād Bahrain

It is called Bahrain (al-Baḥrayn, "The Two Seas") because this district borders on the lake (i.e., the sea between modern Bahrain and Qaṭar) near Laḥsā and the Persian Gulf.

Borders: To the east, the Persian Gulf; to the north, the province of Basra; to the west, Najd al-ʿĀriḍ; to the south, Oman. This province is on the coast of the Persian Gulf. It is currently a sanjak belonging to the ruler of Oman. The people are the ʿAbd Qays. Their territory is al-Aḥsā, Qaṭīf, and Khaṭṭ. The Khaṭṭī lances come from here. It has many rivers and springs. In most places if one digs one or two man's height he will strike water. There is cotton and henna. And the rivers' edge is adorned with lilies. They also raise dates and rice. And they have all kinds of fruit. They have a certain kind of fruit from which wine is made. If one drinks it his clothing turns bright yellow.

The gardens of this country extend for one mile all around the city. When the days are their hottest, the people go (back to the city?) only in the morning and evening.

In the district of Bahrain there are shifting sands. A sand dune appears at one moment and the next moment it has been obliterated. The dunes sometimes overwhelm inhabited areas. In the past Bahrain was connected with Oman by the King's Highway (the Royal Road of the Sasanians?). It is not possible today to walk on this road because of the sand dunes. Today they go from Bahrain to Oman by boat. In Bahrain, there are islands at a distance of two or three (sailing) stages on the way to Basra. They have some buildings and ruins, and hunting grounds.

The Island of Kharg: There is a kind of carrot here that is cut with a hatchet. The inhabitants of this province are afflicted with splenitis. They dive for the finest pearls here, those that are the most perfect and smooth. They say that off the coast of Bahrain, in more than eight fathoms of water, divers go to the bottom with a waterskin in each hand and bring up fresh water in them. Because this water is so sweet and cold, the rulers of this country drink it.

In 275 (888), during the reign of Caliph Muʿtamid b. Mutawakkil, the Carmathians Abū Saʿīd and Abū Ṭāhir who were from this province opposed the Muslim community. They massacred pilgrims and stole the covering of the Kaʿba. They plucked out the Black Stone (from the Kaʿba) and carried it off. The Caliph sent ʿAbbās b. ʿAmr al-ʿAnawī with an army in pursuit of them. The Carmathians virtually annihilated this army and took ʿAbbās prisoner. Then they left him in a fairly deserted place so he could tell the people what had happened. They kept the Black Stone for a long time. Finally Caliph Muṭīʿ biʾllāh bought it back from them for 24,000 goldpieces and put it back in its place.

The Carmathians were a heretical group. They said that it was not necessary to wash after becoming ritually impure (i.e., after sexual intercourse) and that drinking wine was lawful. They said it sufficed to fast only two days a year, *nawrūz* (the Persian new year, the sixth day after the vernal equinox) and *mahrajān* (autumnal equinox). (They claim that) the Prophet's rightful successor was Muḥammad b. al-Ḥanafiyya and that pilgrimage should be made to Jerusalem and that one should face it while praying. The first of the Carmathians to appear was Ḥamdān al-Qarmaṭī in the settled area of Kūfa and Abū Saʿīd al-Jannābī in Bahrain. This person was a grain measurer in Basra and very poor. He joined a man from the Zanj and they spread much evil everywhere in those regions.

al-Aḥsā, today called Laḥsā: This is the province of al-Hajar. In the past it was the capital of the Carmathians. It is two stages west southwest of Qaṭīf. It has many date groves. It resembles the Ghouta region near Damascus. It has streams but the weather is very hot. They grow rice. The port of Laḥsā is a place called Ghufr (?). It is a day's journey away and is now the seat of the *beylerbey*.

Qaṭīf: It is on the coast of the Persian Gulf, two stages from Laḥsā. It is a city with date groves. It has a wall and a moat. There are four gates (into the city). It is the port for this district. In Qaṭīf there are places to dive for pearls. It is six stages from here to Basra. During high tide the water reaches the wall but at low tide some places are exposed. There is a spit that reaches from Qaṭīf into the sea. At high tide, large ships can pass inside it.

Yabrīn: This is on brackish ground. There are two springs on each side, a half day's journey apart. There are many date palms. It has a large plain and its air is bad. There is a famous saying, "He who eats the dates of Yabrīn, drinks its water, and sleeps in its shade will certainly be seized with fever." But its dates are excellent. They resemble the dates of Medina. Yabrīn is three stages from Laḥsā.

Kāẓima: It is between Basra and Qaṭīf. It is a harbor on the Persian Gulf. It is the camping grounds of Bedouin tribes. It has pastures and water. Most of the water comes from wells. At Kāẓima there is a cape that juts into the sea. There is a road from the land that goes out to the cape where there are pastures. [M 498] And there is a gate through which the Bedouin enter. After passing this check point they camp in security.

Bahrain proper: Bahrain is an island in the Persian Gulf. It faces Qaṭīf and is east of it. The city is in the middle of the island, which also has mountains. Its length is only slightly greater than its width. In the middle of the eastern rib (*dilʿ*—i.e., side of triangle?) on a small round island is its fortress. It also has orchards, gardens, and a number of villages.

The Island of Qays: It is four parasangs from shore. It has a place where they dive for pearls.

Chapter: Country of Ḥijāz

This is part of the Arabian Peninsula. It got its name from the fact that it forms a barrier (*ḥājiz*) between Najd and Tihāma (i.e., there is a mountain barrier separating the Ḥijāz coastal plain from the interior). It borders Najd to the east, Tihāma to the west, Yemen to the south, and Syria to the north. Within it are the territory of Mecca and Medina, including the towns of Ḥijr, Ṭāʾif, ʿAraj, Baṭn, Murafraʿ (?), Khaybar, ʿUlā, and Tabūk among others.

Description of its territory

MECCA

First let me describe the location and features of Mecca the Exalted. Because Mecca is square in shape it was named Kaʿba ("Cube"). The city is located along a wadi. Its open square comprises the illustrious cemetery that is called Muʿallā and extends as far as the place called Shubayka, in the direction of Jedda; also near the place where Ḥamza (the Prophet's uncle) was born, in the direction of Yemen. Its latitude crosses over Mt. Jazālī, also called Qayquʿān, and falls a little more than half way above Mt. Abū Qubays. They (also) call this mountain Akhshabān.

Mecca is the location of the Exalted Kaʿba itself, which is in the middle of the Masjid al-Ḥarām. It is described as Bayt al-Ḥarām (The Sacred House) because God glorified and honored it. The Masjid al-Ḥarām is located between two hills (i.e., Ṣafā and Marwa) in the middle of Mecca. The Bayt al-Ḥarām is in the shape of a square. The door of the Kaʿba is the height of a man above the surface of the ground. The leaves of the double doors are sheeted in gold and gilded silver. The length of the doors is six cubits and ten *parmaḳs* (1 *parmaḳ* [finger] = 1.25 inches). The width is three cubits and 18 *parmaḳs*. The length of the Kaʿba is 24 cubits and the width is 23 cubits and one *shibr* (span of the hand). The height of the walls of the Kaʿba is 27 cubits. The circumference of the *ḥijr* (sacred chamber inside the Kaʿba) is 15 cubits. The *ḥijr* is located on the north side of the Kaʿba. Water from the Waterspout (Mīzāb = Altın Oluḳ) pours onto it. The *ḥijr* is paved with marble. The walls are fully draped from top to bottom.

The Black Stone is next to the door of the Kaʿba and in the east corner of the Kaʿba. It is the size of a man's head. It is at a height of three cubits from the ground. The space between the Black Stone and the door of the Kaʿba is called the *multazam* (pledged). The reason that it is called the *multazam* is as follows: In the pre-Islamic period, this is the place where the Arabs swore oaths. If someone came here and called a malediction on an oppressor or if he swore falsely, his punishment would immediately be visited upon him.

Inside the Bayt al-Ḥarām, in the inner courtyard on the side of the west wall, at the sixth cubit is a single piece of onyx. It has streaks of black and white. Its length and width are 12 cubits. Around it is a ring of gold that is three *parmaḳs* wide. It is said that the Prophet rubbed it against his right eyebrow.

The Waterspout is in the middle of the wall of the Kaʿba. It juts out about four cubits from the wall of the Kaʿba. The width of the spout and the height of its sides are each eight *parmaḳs*. The inside of the spout is covered with gold.

[M 499] The Bayt al-Ḥarām is draped inside and out with brocade—especially in our time brocade woven with gold. Every year during the pilgrimage season, the sacred *kiswa* (the cloth covering) is replaced and the Banī Shayba (the family who were keepers of the Kaʿba) take the old one.

Mecca is in the second clime. Its longitude is 70 degrees and its latitude is 21 degrees, 40 minutes. The city's houses are made of black and white stone. Every storey is whitewashed and bright. The people of Mecca originally drank rain water. But then Zubayda Khātūn, the wife of the Abbasid caliph Hārūn al-Rashīd, collected water from wells in a valley in a place called Musallash (?) and brought it (by canal) to a basin in Mecca. Later, during the reign of Caliph al-Muqtadir, it fell into ruin. But he repaired it and water again flowed (to the city). Subsequently, it again fell into ruin but was repaired by Caliph al-Nāṣir. Afterwared it completely fell into ruin once more and Amīr Choban repaired it. Eventually the late Ḫāṣekī Sultan, wife of the late Ottoman Sultan Süleymān Khan, had large stones quarried at a place far from Mecca and used them to build a (new) canal to this safeguarded city and start the water flowing again. This water suffices to meet the needs not only of the city but also of the pilgrims who come each year by caravan from Egypt, Syria, Baghdad, Basra, Laḥsā, Najd, and Yemen. And those who benefit from this water pray for the spirit of the one whose charity provided it.

It is reported that originally the population of Mecca was small. At times the Noble Sanctuary was so deserted that gazelles came down from Mt. Abū Qubays to Ṣafā Hill and entered the Masjid al-Ḥarām from the Ṣafā Gate and wandered about it. Supposedly, at that time the market of Saʿī (for those performing the ceremony of running seven times between Ṣafā and Marwa, a livestock market) was completely empty in midmorning. There was no one doing business. Sometimes a caravan would come from the direction of Bujayla bringing grain to the city. But because they could not find anyone to buy it, they had to sell the grain that they brought cheaply and on credit. But now, praise be to God, under the shade of the Sunni Ottoman sultanate, the population is immense and the blessings are without end.

Mecca is surrounded by mountains. It is possible to cross them on foot but not on horse or camel. If one wishes to cross the mountains with horses, camels or other animals loaded with goods, he must go by one of three routes. They are Muʿallā, Shubayka, and Misfala.

In early times a great wall was erected around Mecca. It extended from Mt. ʿAbd Allāh b. ʿUmar in the direction of Jiʿlān to the mountain across from it. They made the gate of iron. They also put openings here and there for flood waters to pass through. Since then the wall was pulled down and has disappeared. No trace remains; however there are still people in Mecca who remember seeing remnants of the wall in some places.

Sultan Süleymān built a water dispensary next to the place where the canal brings water from Ḥunayn. The lookout at the top of the dispensary commands a fine view. He put window grilles in all four directions. It is a place of excursion for the people in this area. Until recent times, the wall near this dispensary was still standing. [M 500] A large wall was built between the two mountains in the direction of Shubayka. It had gates in two places. It has also been torn down. There was another wall on the road to Yemen from the direction of Misfala. Today it has been torn down. It is reported that there was another wall below this one. It was near a mosque known as Masjid-i Rāya. It was constructed between the mountain known as Laʿlaʿ and the mountain across from it. Some traces of it can be seen today in those mountains.

At the highest place in Mecca, near Biʾr Jubayr, is a noble mosque that was one of the buildings of the Prophet. It is a place of pilgrimage. Originally there were no dwellings for people beyond Biʾr Jubayr. Today many people travel in the direction of Muʿallā. That area is completely built up.

The aforesaid Masjid-i Rāya was built at the place where the Prophet erected his noble standard (*rāya*) when he conquered Mecca. It is reported that the length of Mecca from the Muʿallā Gate to the Mājin Gate is 4,270 cubits. And the distance from the Muʿallā Gate to the Shubayka Gate and then from the Muddaʿāya road to Suwayqa and from Suwayqa to Shubayka is 4,272 cubits.

There are no buildings here higher than the Kaʿba. Indeed, it is related that when Shayba b. ʿUthmān (from the family who were keepers of the Kaʿba) used to go up to a high place and looked over the city, and if he saw a building that was higher than the Kaʿba, he immediately ordered that it be torn down.

Selling houses in Mecca is permitted. As for renting houses, according to Imām-ı Aʿẓam (Abū Ḥanīfa), it is reprehensible to rent houses during the pilgrimage season, but it is not reprehensible to rent houses to those who reside there (i.e., who do not come only for the pilgrimage).

Mecca has several names. Among others it is called al-Balad al-Amīn ("The Safe City"), Umm al-Qurā ("Mother of Villages"), and ʿArūḍ ("Prosody"). It received the last name because Khalīl b. Aḥmad originated the study of prosody there.

It should be known that Imām-ı Aʿẓam (Abū Ḥanīfa), al-Shāfiʿī, and Aḥmad b. Ḥanbal considered Mecca to be more distinguished than Medina. According to Mālik, however, Medina is more distinguished than Mecca. There are many reasons why Mecca should be honored as more distinguished:

- The Noble House (Kaʿba) is there.
- When Muslims go there on the pilgrimage their sins are forgiven and their status increases.
- It was made the qibla for all Muslims, in life and in death.
- Those who have the ability have a religious obligation to make the pilgrimage there once during their lifetime.
- The day that God made the heavens and earth he made Mecca a sanctuary (*ḥarām*); so respecting it (*ḥurmet*) is obligatory; and one is not permitted to enter Mecca without wearing the *iḥrām* (special pilgrimage garments).
- The shrines of the prophets Abraham and Ishmael are there.
- Mecca is the place where Muḥammad was born, where he lived here before his prophethood and for thirteen years after beginning his prophethood.
- Most of the Koran was revealed there.
- Faith and Islam (or, submission to God) began there.
- The Black Stone and the Zamzam Spring are there.

Ibn ʿAbbās reportedly said, "I don't know any city other than Mecca where one good deed is tantamount to one hundred. One prostration during prayer in Mecca is worth a hundred elsewhere. Seeing the buildings of Mecca is worth a lifetime of worship elsewhere. One dirham given as alms in Mecca is worth a thousand anywhere else. (All this is true) only in Mecca."

What is the legal ruling on settling in Mecca? Imām-ı Aʿẓam (Abū Ḥanīfa) and some of his followers said that it was reprehensible. [M 501] This is because they feared that, if people got used to living here, after a long time they would view it as just another city and the Noble House would lose its sanctity. Eventually there would be no reverence for it in their hearts. Therefore the caliph ʿUmar, whip in hand, went among the pilgrims after they had completed the pilgrimage saying, *Yā ahl al-Yaman, Yamanakum; yā ahl al-Shām, Shāmakum; yā ahl al-ʿIrāq, ʿIrāqakum. Fa-innahu ibqāʾ li-ḥurmati bayti rabbikum fī qulūbikum* ("O people of Yemen, to your Yemen; O people of Syria, to your Syria; O people of Iraq, to your Iraq. That will keep the reverance of the house of your Lord in your hearts"). However, Imām Yūsuf (i.e., Abū Yūsuf), Muḥammad (al-Shaybānī), al-Shāfiʿī, and (Aḥmad b.) Ḥanbal were of the opinion that settling in Mecca was desirable.

The meaning of the *ḥaram* ("sanctuary, holy or forbidden area") of Mecca is not like the *ḥaram* (Tk. *harem* "courtyard") of ordinary mosques. Rather, the *ḥaram* of Mecca extends to a three-day journey toward Medina, seven miles toward Yemen or Iraq and ten miles toward Jedda. All the mountains, stones and deserts in the area delimited above are part of the *ḥaram*.

How building the Kaʿba came about: Taqī al-Dīn Fāsī (text: Fārsī) says that the Koran and the Sunna (tradition of the Prophet) confirm that Abraham was the first to build the House of God. There is no evidence of its construction before Abraham. When he built it, he made its height nine cubits. As for its length, he built it 32 cubits from the Black Stone to the Syrian corner and 22 cubits from the Syrian corner to the Iraqi corner. Behind the Noble House where the Waterspout is located, he built it 31 cubits from the Iraqi corner to the Yemeni corner. He built it 24 cubits from the Yemeni corner to the Black Stone. He made the entrance of the Kaʿba level with the ground, not high up. And he did not affix a door. It remained in this state until Tubbaʿ the Himyarite came on the scene. He attached a door and put on it a mechanism to lock it. Abraham had hollowed out a place to the right of where people entered the Noble House. He made it a treasury for donations to the Kaʿba. While Abraham was building the Noble House, Ishmael carried stones on his shoulders and Abraham did the construction. The stone where he stood while building is known as the Shrine of Abraham. When he came to the spot where the Black Stone was to be placed, Ishmael went in search of a stone. The angel Gabriel then appeared and brought the Black Stone to Abraham. God had deposited this stone in trust on Mt. Abū Qubays. After Gabriel brought the stone and put it in its place, so that this should be the sign for people to begin to circumambulate from here.

The Reason Abraham Came (to Mecca): It is related in the revered books that when Abraham was saved from Nimrod, those who had followed him maintained their faith. He married Sarah, the daughter of his paternal uncle and went toward Egypt. At that time one of the first pharaohs was ruler of Egypt. He was informed that someone came to his city and with him was a woman of unrivaled beauty. The tyrant sent a man to demand the woman. Abraham explained the situation to Sarah. She went before the tyrant. When he reached out to touch her, his hand dried up and fell on his chest. The tyrant then realized Sarah's noble status and he said, "If you restore my hand, I will free you in God's name." [M 502] Sarah then prayed saying, "If what he says is true, restore his hand!" And her prayer was answered. Afterward the pharaoh gave Sarah a beautiful Coptic slave girl named Hagar as a gift. Later she returned (with Hagar) to Abraham. She explained to him what had happened and said, "Perhaps God will grant you a child from this slave girl." Abraham had sexual intercourse with her and as a result Ishmael came into the world.

Abraham settled in the land of Palestine between Ramla and Ayla. He was hospitable to all who passed through. God granted him wealth and prosperity. God then brought him the good news (that he would have a son named) Isaac. Sarah became pregnant with Isaac and gave birth to him. The boys grew up. One day Abraham took Ishmael in his lap and sat him on his knee, but had Isaac sit beside him. When Sarah saw this she was hurt and said, "You put the child of a slave girl on your knee but you put my child next to you." She swore an oath saying, "I will cut a piece of flesh from Hagar and disfigure her." But after her anger had subsided, she was astonished at the oath she had made. Finally Abraham said to her, *yā Sāra, iḥfaẓīhā, wa ithqabī udhunayhā* ("Oh Sarah, safeguard her and pierce her ears"). And she did so. This was the beginning of the practices of the Sunna (or circumcision, Turkish *sünnet*). For girls it is called *ḥifāẓa* ("safeguard"), i.e., circumcision (*sünnet*), just as for boys *sünnet* is called *khitān* ("circumcision"). After a while, Ishmael and Isaac began to quarrel with each other. Sarah became irritated over this and swore an oath that she would not live in the same country as Hagar. Thus it was that Abraham sent them (Hagar and Ishmael) to Mecca. At that time the (site of the) Noble House was a red hill. Abraham took them to Ḥijr and had them stay there. He told Hagar to find a shady spot and then he departed. There was a waterskin next to Hagar. After the water was gone, Hagar and Ishmael became thirsty. Hagar looked toward the mountains but saw no one. From there she went to the top of Ṣafā Hill. Again she saw no one. Looking at Ishmael she went back down the hill, but when she returned to the valley, Ishmael was nowhere to be seen. Then she went up to a high place that was nearby and saw Ishmael. But she did not remain there, but went up to Marwa. In this fashion she went up and down one mountain after another. Finally she returned to Ishmael. At that moment the angel Gabriel appeared and struck with his wings the place where the Zamzam spring is found today and water began to bubble up from the ground. When Hagar saw this, she closed up the area around it and made a basin so that the water would not flow into the wilderness and disappear.

One day while Hagar and Ishmael, having put their trust in God, were sitting by the Zamzam Spring, a group of people from the Jurhum tribe of Yemen arrived in the area while on their way to Syria. When they saw a bird flying over Mt. Qubays, they deduced that there was water nearby. In short, they found the water and learned the story behind it. With Hagar's permission they settled there. Thus the first Arab tribe to settle in Mecca was the Jurhum. Subsequently Hagar died and Ishmael remained with the Jurhum and grew up to be a brave young man. He learned their language and married one of their daughters. They call the Arabs descended from Ishmael *ʿarab mutaʿarrib* ("Arabized Arabs"). The language that Abraham spoke was Hebrew. [M 503] The language of the descendants of Ishmael was Arabic.

One day, with Sarah's permission, Abraham came to Mecca. There he learned that Hagar had died. He went to Ishmael's house to ask after him. However, Ishmael had gone hunting, so Abraham received no hospitality. Abraham then said to Ishmael's wife, "Give my regards to your husband and may he change the threshold of his house." He bade her farewell and departed. Later Ishmael returned home and his wife told him what Abraham had said. He immediately understood what his father had meant by these words. He divorced his wife and married another woman.

A while later Abraham came to Mecca again, and again he could not find Ishmael. He asked his wife where he was and she replied that he had gone hunting. She showed honor and hospitality to Abraham and brought him milk and meat. Abraham drank and ate. Then Ishmael's wife said, "You have been travelling. Let me wash your head and comb your hair." She then brought out a stone and sat on it. This was the stone which Abraham stood on when he built the Kaʿba. Afterwards Abraham returned to the place from which he had come. Ishmael then returned and his wife told him what had happened. He kissed the stone on which Abraham's foot had left an impression and saved it as a blessing. Today this stone is known as Maqām Ibrāhīm (the Shrine of Abraham, i.e., the Place where Abraham Stood).

God then ordered Abraham to build the Kaʿba and on His command he built the Noble House. When it was finished, God commanded, "Summon the people to make the pilgrimage to God's house." And Abraham said, "My Lord, who will convey my summons to all of humankind?" And God replied, *minnak al-adhān wa ʿalayya al-balāgh* ("You make the summons; I will take care of conveying it"). Abraham went up to the mountain called Thabīr or the one called Abū Qubays and said to the people, *yā ayyuhā al-nās, ajībū rabbakum labbayka allāhumma labbayka* ("Oh people! Respond to your Lord. (Say) I am certainly at Your service"). Abraham then exalted it (Mecca) in prayer saying, "*My Lord, make this a secure town*" (Koran 2:126). The birthplace of the Prophet became the source of the speech of the One Who is generous by necessity (i.e., of the Koran, which is the speech of God).

Afterwards Gabriel appeared to Abraham and Ishmael and showed them how to circumambulate (the Kaʿba). Then he went with them to Minā and explained how to do the ritual prayer. He stayed with them two days. Then he took them to ʿArafāt and had them camp in the place where pilgrims camp today. He taught them the rites of the pilgrimage. And Abraham said, *ʿaraftu ʿaraftu* ("I understand, I understand"). This is how ʿArafāt got its name.

Later, when Abraham had a dream in which he was instructed to sacrifice his son, he said, "Oh my son! Take some rope and a knife and let us go on this mountain road." His son took some rope and a knife in his hands and followed his father. Satan said to him, "Oh Ishmael! Where are you going with your father? Your father wants to sacrifice you." He replied, "Why would my father want to sacrifice me?" And Satan said, "God commanded him to do so." Ishmael then said, "I hear and obey. If God has commanded it, let him do it immediately." Then Satan said to Abraham, "Why are you going this way?" When Abraham said, "I have business up ahead," Satan responded, "I am afraid that Satan entered your dream [M 504] and deceived you and that you are following his words and are going to sacrifice your son." Abraham realized that the person who was saying this was Satan and said, "Get away from me you cursed one!" Abraham then arrived at his intended destination. That place is called Mt. Thabīr. He informed his son what was going to happen. And Ishmael said, "God has ordered this to be done. God willing, you will find me steadfast." (Cf. Koran, 37:102) Just as Ishmael was firmly bound and about to be slaughtered, Gabriel redeemed him with a handsome horned ram from heaven and it was slaughtered in his place.

It should be known that Ishmael had twelve children by the daughter of Mughāṣ b. ʿAmr of the Jurhum tribe. He lived to the age of 130. God multiplied the Arabs (Bedouin) through the lineages of Thābit and Qaydār. When Thābit died, his grandfather, Mughāṣ b. ʿAmr, became governor of the Noble House. He took Thābit's son to his side and he became king of the Jurhum tribe. He settled above Mecca in a place called Quʿayquʿān. The Amalekites also came and settled below Mecca. They chose someone named Sumaydiʿ (?) as their king. War broke out between Amīr Mughāṣ and Sumaydiʿ and (as a result) Mughāṣ became the uncontested king of Mecca. He expelled the Amalekites from that area. This is the reason why the fortune of the Amalekites declined.

Subsequently the Jurhum tribe did things that they should not have done. They began to commit shameful and heretical acts. Mughāṣ advised people not to do these things, but he was ignored. He realized that it was impossible for the people to mend their ways. Inside the House of God were two golden gazelles. He took them and all the donations to the Kaʿba, no matter what they might be, and at night threw them into the well of Zamzam. (At that time) the water in the well had receded; so he dug it deeper, threw in the donations and buried them. Then he took the descendants of Ishmael who lived there and departed from the Jurhum to some distance away. Later the Khuzāʿa tribe came and took over Mecca and expelled the Jurhum tribe. They became the rulers of Mecca. Later the descendants of Ishmael returned and lived among the Khuzāʿa.

So it was until the time that the key of the Kaʿba came into the possession of Quṣayy b. Kilāb b. Murra. Previously the key of the Kaʿba was in the possession of a certain Abū Ghubshān from the Khuzāʿa. He loved to drink wine. One day there was no wine left to drink. Needing a drink, he took the key of the Kaʿba and sold it for a skin of wine. Quṣayy b. Kilāb gave him the wine and took the key. Ever since that time the Arabs have had a saying for anyone who takes a loss in business dealings: *akhsaru ṣafqatin min Abī Ghubshān* ("A worse bargain than that of Abū Ghubshān").

Afterwards the dynasty and power of Quṣayy b. Kilāb became paramount and one day control of the affairs of Mecca passed into his hands. Because he brought together the scattered clans of the Quraysh tribe, he was called *mujammiʿ* ("the uniter"). And the Quraysh are called the

Quraysh because they gathered around him. People began to flock to him.

It is reported that Quṣayy b. Kilāb was the first to give permission to build houses around the Kaʿba and that the first to be built was Dār al-Nadwa meaning "House of Assembly." When there was important business to attend to, people met there and made decisions. [M 505] It was the council chamber of the Quraysh. It is reported that at that time if one was not 40 years old he was not permitted in Dār al-Nadwa, though there was no obstacle to the entry of descendants of Quṣayy no matter what their age. After this, Dār al-Nadwa was rebuilt several times. It was renovated by the Ottoman Sultan Murād (III) son of Selīm (II), who raised a high ceiling over black and white marble columns, had lofty domes placed on top, and called it Masjid-i Sharīf.

Subsequently Quṣayy build houses around the Kaʿba, leaving only enough room for the circumambulation. The doors of the houses opened toward the Kaʿba. He made an alley between every two houses so those coming to do the circumambulation would have easy access to the Kaʿba. Later Caliph ʿUmar acquired those houses and demolished them and joined (the lots) to the Masjid al-Ḥarām. Later Caliph ʿUthmān and others expanded it.

It is reported that no one in Mecca ever combined in himself as many high offices as Quṣayy b. Kilāb. The responsibilities of *ḥijāba, siqāya, rifāda, nadwa, liwāʾ* and *qiyāda* were all in his hands. *Ḥijāba* meant being custodian of the Noble House. He held the key to God's Kaʿba. *Siqāya* meant providing fresh water to pilgrims. During the pilgrimage it was scarce. *Rifāda* meant supplying pilgrims with provisions at Minā during the pilgrimage. (*Nadwa* meant assembling the Quraysh.) *Liwāʾ* meant banner. When the Quraysh assembled and wanted to go to war, they assembled under his banner. *Qiyāda* meant being commander. Wherever he wished to go, the army would fall in and follow. After passing through several tribes, the responsibilities for *siqāya* and *rifāda* returned to ʿAbd al-Muṭṭalib b. Hāshim.

It is reported that ʿAbd al-Muṭṭalib had a son named Ḥārith. One day ʿAdī b. Nawfal, ʿAbd al-Muṭṭalib's paternal cousin, reproached him saying, "Do you think you can reign over us for a long time? You have only one son. We have many." ʿAbd al-Muṭṭalib replied saying, "Are you reproaching me because I have few sons? May God give me ten sons; then I will sacrifice one at the Kaʿba." God answered his prayer and gave him ten sons. Later he assembled all his sons and informed them of his vow. They said that they would obey their father in the fulfillment of his vow. ʿAbd al-Muṭṭalib ordered each one to take a *qidḥ* in his hand. *Qidḥ* was the name of a piece of wood used for casting lots and for gambling. Each son wrote his name on his *qidḥ*. ʿAbd al-Muṭṭalib then took his sons to the Kaʿba. They went inside where the idol called Hubal was standing. They cast lots according to custom. The lot fell on the youngest son ʿAbd Allāh. Saying, "Such is fate. There is nothing I can do about it," he took the boy by the hand and left the Kaʿba. He took a knife and went to the top of Ṣafā Hill where the idol called Asān stood, intending to sacrifice the boy there. Just as he lay the boy down and was about to bring down the knife, the Quraysh rose to their feet and said, "We do not approve of this. If someone does this, it will immediately become the custom among us. It should not be done. If someone has made (such) a vow, let him redeem it by paying money." At that time there was a soothsayer in Ḥijāz. They went to him and related the matter of ʿAbd al-Muṭṭalib's vow. [M 506] He said, "In your custom how much is the blood money?" They replied, "Ten camels." He said, "Go and bring ten camels and his son and cast the lot again. If the lot falls on him again, add ten more camels. Continue to do this until the lot falls on the camels, then sacrifice them. Your lord will be satisfied with this." In this way their casting of the lot increased the number of camels to 100. On the tenth cast the lot fell on the camels, which were then sacrificed. From that day forward the blood money of a man has been set at 100 camels.

After Abraham, the Quraysh kept the Kaʿba in repair. One day a woman entered the Kaʿba carrying incense. As she burned it, a spark leapt from the censer and reached the ceiling of the Noble House where the wooden beams caught fire. The fire also weakened the walls. Subsequently a great flood swept into the Kaʿba and the weakened walls were breached. The Quraysh then resolved to rebuild it in a much sturdier fashion. They raised the door of the Kaʿba above the ground in order to control access.

At that time, they received word that the Byzantine emperor had sent a ship loaded with marble, iron, and wood for the construction of a church in the province of Abyssinia. There had been a church in that area for a long time. The ship went aground and broke up. At that time Walīd b. Mughīra and some companions came to Jedda and saw that a strong wind had driven the ship ashore and destroyed it. As for the contents of the grounded ship, in conformance with the saying *idhā arāda Allāhu shayʾan hayyaʾa asbābahu* ("When God wills something, He prepares the intermediate causes"), He (i.e., God) prepared it on the Jedda coast for the repair of the Kaʿba. And on the ship was a Copt, and another person named Yāqūm, who were master carpenters.

Now the tribes that were at the Kaʿba divided into several groups. Each group set about building one section of the Kaʿba. The door side of the Kaʿba was given to the

(Banī) Zuhra and the Banī ʿAbd Manāf; the space between the corner of the Black Stone and the Yemeni corner was given to the Banī Makhdūm and some members of the Quraysh; and the *ḥijr* section was given to the Banī ʿAbd al-Wādd, the Banī Asad, and the Banī ʿAdna b. Kaʿb. They collected stones (for this project) and the messenger of God (Muḥammad) was among those who was working alongside them. When it came to the placing of the Black Stone, a disagreement broke out among the tribes. Each tribe wanted to put the noble stone in its place. The dispute became so heated that the tribes were about to go to war.

At that time Abū Umayya b. Mughīra was widely respected and obeyed by the people. He said, "Come and watch the Ṣafā Gate. Let the first person who enters decide where to put the Black Stone. Do whatever he tells you to do." They were pleased with this suggestion and watched the gate. Just then the messenger of God entered. At that time he had not yet begun his prophethood. Among themselves they called him "Muḥammad al-Amīn" (the Trustworthy). Thus they said, "Here comes Muḥammad al-Amīn. We will do whatever he says." When he came, they told him what had happened. He told them to bring a robe. He then placed the Black Stone in the middle of the robe and said, "Have a distinguished member of each tribe come forth, [M 507] grasp a corner of the robe, lift up the blessed stone together and bring it to its place." Following his instruction, they brought the stone to its place. With his own blessed hand he grasped one corner of the robe and put the stone in its place. Everyone was satisfied with this.

When the Quraysh built the Kaʿba, the exterior height was 18 cubits, 9 cubits taller than the one previously erected by Abraham. But they made the breadth on the side of the *ḥijr* a few cubits shorter. This was because the money that they lawfully collected for the expenses of the Kaʿba was only sufficient for that much. They made the door of the Noble House rather in order to control access. The door was raised above the ground the height of a man and had two panels. It faced the east. Inside they erected six columns. They placed three columns on each side from the *ḥijr* to the Yemeni corner. Inside at the Syrian corner, they made a ladder that went to the roof of the Kaʿba. Those two carpenters (i.e., the Copt and Yāqūm mentioned above) built a beautifully crafted ceiling. It is reported that the Prophet was thirty-five years old at that time.

Ḥaṭīm: On the northern side of the Kaʿba, where the Waterspout pours out, is a wall called Ḥaṭīm. It is made of marble. There are six cubits between Ḥaṭīm and the Noble House. The place for the circumambulation is behind it; the circuit of the Ḥaṭīm is 25 cubits and a complete circumambulation is 107 cubits. When the Quraysh built the Noble House, they made it smaller than the one that Abraham had constructed and left that place (i.e., Ḥaṭīm) outside. ʿAbd Allāh b. Zubayr remodeled the building that the Quraysh built and included within it this place. He also put two doors in the noble house. This is because he heard a noble hadith according to which the Prophet said to ʿĀʾisha, "O ʿĀʾisha, if your people—i.e., the Quraysh—had not previously come on the scene and remodeled the Kaʿba (lit., had not been here so recently—*ḳarībü'l-ʿahd olmayalardı*), I would have leveled the Kaʿba and rebuilt it with two doors. I would have opened one door to the east and one to the west. I would have added the six cubits from the *ḥijr* to the Noble House. For when the Quraysh rebuilt the Kaʿba they did not have sufficient funds to incorporate this area in the building. If, after I am gone, your people wish to rebuild the Kaʿba, come and let me show you how much space they left outside." He took ʿĀʾisha and showed her. Later the tyrant known as al-Ḥajjāj b. Yūsuf wrote a letter to ʿAbd al-Malik b. Marwān. He informed him that ʿAbd Allāh b. Zubayr had expanded the Noble House into a place that hadn't been part of it and added a door. ʿAbd al-Malik wrote him a reply saying, "Make it as it was at the time of the Prophet." Afterwards the aforesaid tyrant expanded the wall of the Noble House that was on the Syrian side by six cubits and one span and built a new wall on the foundation that the Quraysh had built. Thus the area called the *ḥijr* is that space that juts out. Later it was rebuilt by ʿAbd Allāh b. Zubayr. After him al-Ḥajjāj b. Yūsuf filled in the wall of the Noble House that was on the side of the Golden Waterspout and made a high wall.

Description of the sublime doors of God's House

After the coming of Islam, ʿAbd Allāh b. Zubayr was the first to gild and adorn (the Kaʿba). [M 508] He covered the Kaʿba and its columns with gold. Later Walīd b. ʿAbd al-Malik sent 36,000 dinars to Khālid b. ʿAbd Allāh, whom he had appointed governor of Mecca, to cover the doors of the Kaʿba, the interior columns, and the Waterspout with gold. The Abbasid caliph al-Mutawakkil had two of the "corners" (*zāwiya* [= *rukn*]) inside the Noble House covered with silver and two covered with gold plating. Later al-Mutawakkil sent gold to have all the "corners" of the House covered with gold. And he made a silver band for the curtain that was suspended inside. The band was three cubits in breadth. He attached a golden collar to the band. The portico of the Kaʿba was made of twenty-three teak logs. Over time it too had to be renovated. The caliph did so and covered the columns with silver. The mother of Caliph al-Muqtadir bi'llāh ordered one of her slaves named Lu'lu' to cover all the columns of the Kaʿba with gold, no matter how many there were. This task was carried out in 310 (922). Again in 549 (1154) Jamāl al-Din (apparently Ṭalā'iʿ b. Ruzzīk), the vizier of the (Fatimid) ruler of Egypt, covered all the interior "corners" (*arkān*) of the Kaʿba with gold.

Finally, in 770 (1368) (the Mamluk sultan) al-Malik al-Ashraf, the grandson of al-Malik al-Nāṣir, adorned the door of the Kaʿba. It had been covered with silver. Later some bandits found an opportunity and began to tear off the silver. Finally this matter was brought before Sultan Süleymān and he ordered that the door be recovered with silver.

Before this, in 958 (1551), some of the wooden beams in the roof of the Kaʿba had deteriorated and when it rained the roof leaked. When the qadi of the Kaʿba, the late Ḫoca Ḳāyinī Meḥmed Çelebi, informed Istanbul of the situation, Sultan Süleymān sent a letter to Ebū's-Suʿūd Ḫoca Çelebi Efendi, the Shaikh al-Islām, asking, "Is it religiously permissible to repair and interfere with (*taʿarruḍ*) the Kaʿba?" He replied, "When it is necessary it is lawful," and he issued a fatwa to this effect. A noble rescript in accordance with this fatwa was sent to Mecca urging the religious scholars of Mecca to deliberate on the matter. On 14 Rabīʿ I, 959 (9 March, 1552), after the Friday prayer, the qadi of the Holy Sanctuary called a meeting in which the distinguished Ḥanafī and Shāfiʿī religious scholars were present. They discussed the issue and questioned an architect. He said, "Two of the beams in the roof of the Kaʿba are broken and the other beams have separated from each other and have dropped 12 inches (*qīrāṭ*). In addition to these beams, the one on the side of the door has dropped 9 inches." He further testified that if the necessary measures were not taken, it was likely that the roof of the Kaʿba would completely collapse. Finally, after much discussion, they made a decision to repair and fortify the roof. Work began on Saturday in the middle of Rabīʿ I in the same year. After the roof was finished, they began to pave the ground of the circuit of the Kaʿba because the old paving stones had come apart. [M 509] Afterwards they repaved the floor of the Masjid al-Ḥarām with gravel. They also covered the door of the Kaʿba with silver nails and placed four silver rings on it. Furthermore they repaired the noble Waterspout, replacing the silver plating and the gilding. (Later) they obtained a spout covered with pure gold from Istanbul and inserted it in the Kaʿba. When it arrived they took out the silver spout and replaced it with the golden one. Afterwards the Ottoman sultan Aḥmed left that golden spout in its place and sent a gilded silver band, also gilded silver windows for the Kaʿba. For the tomb of the Prophet (in Medina) he sent a diamond whose value was estimated at 80,000 goldpieces.

In 1039 (1629) a great flood struck Mecca and damaged some places. Consequently Ghazi Sultan Murād IV issued an edict to renovate the Kaʿba. Two ships loaded with marble and other materials were sent from Suez. However, while en route, they sank by divine providence. When news of this reached Mecca, the Sharifs and the qadi of Medina Ḳoca Naḳīb held a meeting and decided: "If we inform the Imperial Threshold that the ships have sunk, many months will have passed until information (how to proceed) comes back. In the meantime, the best course is to break up the silver band that Sultan Aḥmed had previously sent and mint silver coins from it, because there is no need for the band while the Kaʿba is being rebuilt." And this indeed is what happened. They were said to obtain 36,000 *ġuruş* from the silver and with this money they rebuilt the Kaʿba. In the course of construction, 300 stone cutters were employed. Stones were quarried from Mt. ʿUmar and used for the construction of the building. Also three columns in (i.e., supporting?) the roof were damaged and it was decided that they needed to be replaced. By divine providence twenty years earlier a ship from India came apart at Jedda and three ebony columns were stuck in the sand. At that time no one was interested in them. But now someone in Mecca remembered (this wood) and spread the word. They went (to Jedda), removed the columns from the sand and transported them to Mecca. They proved to be the perfect replacement for the damaged columns which were named Ḥannān, Mannān and Dayyān. They carved up this wood, making prayer beads out of it, and began to sell them to pilgrims. To this day the prayer beads that are made there are said by the locals to be made from that wood. Only the very valuable diamond that Sultan Murād IV sent to the tomb of Muḥammad was put in place.

Objects suspended in the Kaʿba and its covering

For a long time the rulers of Iran used to send money and jewels to the Kaʿba. Sāsān b. Bābāk sent a gazelle inlayed with gold, several superb swords, and a large amount of gold. When ʿUmar b. al-Khaṭṭāb captured Madā'in of the Chosroes (the Sasanian capital), he sent two crescents to the Kaʿba where they were suspended. Caliph Ma'mūn sent a superb ruby. During the pilgrimage season they used to suspend it in front of the Kaʿba with a gold chain. [M 510] Caliph al-Mutawakkil sent a rosette inlayed with rubies, pearls, and chrysolite. They used to suspend it with a gold chain on the front of the Kaʿba during the pilgrimage season. Caliph al-Muʿtaṣim sent a lock fashioned from 1,000 *mithqāls* of gold. But the custodians of the Kaʿba did not do what the caliph instructed (i.e., did not attach it). When this situation was explained to the caliph, he left the old lock in place and donated the gold lock to the custodians. In short, the caliphs and sultans have sent many valuable things to the Kaʿba. Indeed in 984 (1576), the Ottoman Sultan Murād (III) sent three oil lamps inlayed with gold. Two were suspended at the Kaʿba and one at the tomb of the Prophet.

Covering (*kiswa*) of the Kaʿba

Before the coming of Islam, a ruler of Yemen named Asʿad had a dream in which he draped the Kaʿba with a covering. When he awoke he covered the Noble House on all sides with leather mats. Afterwards he had another dream and draped the Kaʿba with a covering. When he awoke he removed the leather mats and draped it with curtains from Yemen. The custom of rulers, kings and emirs draping the Kaʿba with a cover dates from that time. Gradually it became the custom to replace the covering of the Kaʿba twice a year. They first changed it on the day of Tarwiya, which is the day before ʿArafa (eve of the Festival of Sacrifice), and then on the day of ʿĀshūra. During the reign of Caliph al-Maʾmūn it was decreed that the covering should be changed three times a year. On the day of Tarwiyya the Kaʿba was draped with red brocade; on the first of the month of Rajab it was draped with a *qubāṭī* (i.e., Coptic) type of cloth; on the festival of Ramadan they draped it with white brocade.

Later when the status of the Abbasid caliphs had declined and their dynasty had lost its authority, sometimes the sultans of Egypt or the kings of Yemen sent the covering for the Kaʿba. Then the (Mamluk) sultan of Egypt al-Malik al-Manṣūr Qalāwūn purchased two villages in Egypt and established them as a pious endowment (the income from which was used to purchase a new covering each year) for the Kaʿba. Subsequently, the sultans of Egypt sent a (new) covering each year to the Kaʿba. When the Ottoman sultan Selīm (I) freed Egypt from the Circassians (i.e., the Burjī Mamluks), he sent coverings of the same colors and type as before to the Kaʿba and to (the tomb of the Prophet) in Medina. Later, by order of Sultan Süleymān Ḳānūnī, the covering of the Kaʿba continued to be provided in the same fashion. When the two villages (mentioned above) fell into ruin and there was insufficient income (to pay for the coverings), Sultan Süleymān supplemented the endowment with several prosperous villages. (The amount of income generated from this endowment) currently, praise be to God, meets the expenses (for the replacement of the old covering).

As for whether it is (religiously) permissible to (cut up) the old covering after it had been removed and divide it among the people: it is reported that the commander of the faithful ʿUmar b. al-Khaṭṭāb used to remove (the old) covering every year and divide it among the pilgrims.

Addendum: Flooding. It should be known that sometimes great floods come to the Nobel House and enter the Ḥaram al-Sharīf. Indeed, during the reign of Caliph ʿUmar a huge flood occurred. It became famous among the Arabs as the Flood of Nahshal. When this flood inundated Mecca and entered the Masjid al-Ḥarām it swept away the Maqām (Station) of Abraham and left it in the wadi below the city. [M 511] It was later retrieved from there. At that time there was a woman named Umm Nahshal. The flood swept her away and she drowned. Above Mecca was a dam that the Arabs called Rawm Banī Jumaʿ. It collapsed during that flood. Caliph ʿUmar rebuilt it and strengthened it. Today it is called Muddaʿī. From the dam one could see the Kaʿba. When pilgrims came to Mecca, they paused there and said a prayer as soon as they saw the city. It is believed that a prayer made at the spot where one (first) sees the Kaʿba will be answered. However, because there are now so many buildings in the way, one can no longer see it from there.

Another flood that comes is called Flood of the Tyrant. There is a flood that goes through the place called the Wadi of Abraham in the mountains south of Mecca and reaches the quarter of Jiyād. From there it skirts Rukn-i Yamānī (Yemeni corner of the Kaʿba) and then changes direction and spills into the lowest part of the city. This flood occurs about once every ten years.

Note: A great flood struck Mecca at noon on 22 Dhū al-Ḥijja 1093 (22 December 1682) and burst the dikes. It reached the Ḥaram-i Sharīf, going as far as the noble mimbar where the preacher stands to deliver the Friday sermon. Many people drowned in the Ḥaram-i Sharīf. Many camels and other riding animals perished. When word of this situation reached Istanbul, Sultan Mehmet IV issued an order placing his imperial stablemaster, Süleymān Agha, in charge (of relief) and sent him (to Mecca). In the same year, he repaired or rebuilt the dikes, greatly strengthening them, and he cleaned and restored the water channels that brought water from the direction of ʿArafāt. This was a great work.

Public works

After the Masjid al-Ḥarām was (originally) built as a square, it was enlarged twice. The first enlargement was carried out by Caliph al-Mahdī. It is at the highest place of the Masjid al-Ḥarām. And at the lowest place he made additions extending to the Gate of ʿUmra and Gate of Abraham. He also expanded the Masjid al-Ḥarām to what is presently the furthest point of the Syrian side and to Qubbat al-ʿAbbās and the place called Ḥāṣil al-Zayt on the Yemeni side. The distance between the wall of the Kaʿba and the wall of the Masjid-i Sharīf is 49.5 cubits. The area behind it slopes toward the wadi. All this work was done by al-Mahdī.

The second enlargement was carried out by Caliph al-Muqtadir. It occurred on the west side. It is known as the enlargement (*ziyāda*) of the Gate of Ibrāhīm. The Ibrāhīm in question was a tailor who had a shop near the aforesaid gate. He lived to an old age and because he never left that place it was called the Gate of Ibrāhīm. Amīr Sharaf al-Dīn, one of the slaves of Caliph al-Mustanṣir, built a madrasa in Mecca. It is located to the right of the place where one enters the Masjid al-Ḥarām from the Bāb al-Salām (Gate of Peace). This madrasa was endowed with precious books. In 641 (1243) all the books in the madrasa were destroyed. Currently this madrasa is used as a hospice.

On a Friday night at the end of Shawwāl, 802 (June, 1400), during the reign of (the Mamluk sultan) al-Malik al-Nāṣir Faraj, a fire broke out in the Masjid al-Ḥarām. The cause of it was as follows: There was a hospice that adjoined the Gate of Gharwanda, one of the gates of this mosque. They called it Dāhasht. That night, while a dervish was staying there, he lit his lamp (in his room) and then went outside, leaving his room unoccupied. [M 512] Then, by God's wisdom, a rat came into the room, tugged on the wick of the lamp and left it in the room. Everything inside burned and the ceiling caught fire. From there it spread to the Ḥaram-ı Sharīf. Discerning people saw the outbreak of this fire as the portent of a great disaster that would befall humanity in the future. And by God's permission it happened. Around that time Tīmūr came to Syria and Anatolia committing all kinds of evil acts.

Before the fire, a great flood occurred in the same year. It entered the Masjid al-Ḥarām and filled it with water as high as the oil lamps. The water broke through the door of God's House and entered it. It destroyed many houses; and a large number of Muslims drowned. Afterwards, under the supervision of Amīr Bayiq Ṭāhirī, many repairs and reconstruction began. Work on God's House was completed between 803 (1400) and Shaʿbān, 804 (April, 1402). After the fire the Sharīf of Mecca, Ḥasan b.[4] ʿAjlān had another hospice made for the dervishes. He built it just like the one that was destroyed. He used his own money to construct a fine sturdy building for use as a residence for the dervishes. Today this hospice is called Ribāṭ-i Nāẓir al-Khāṣṣ ("Hospice of the Private Overseer"). The reason it is called this is that a person from Egypt rented it and at certain times carried out modifications and repairs. Thus it started to become known by him.

In 807 (1404) Amīr Bayiq put a roof on it (apparently the Masjid-i Sharīf). One had not been previously built (as a result of the flood and fire) because of the delay in obtaining the needed wood and other materials from Anatolia. After this roof was completed, he rebuilt the places set aside in the courtyard of the Masjid-i Sharīf for the four Sunni rites.

In 810 (1407) Sultan Ghiyāth al-Dīn Muʿaẓẓam Shāh, the ruler of Bengal in India, sent a enunuch named Yāqūt to Mecca and Medina with a lot of money. Some of it was to be distributed among the poor as alms and the rest was to be used to build a madrasa and hospice. Yāqūt also brought a letter and gifts for the Sharīf of Mecca, Ḥasan b. ʿAjlān. After accepting them, the Sharīf used one third of the alms funds for the construction of a soup kitchen and other charitable buildings and distributed the rest among the poor at Mecca and Medina. Then Yāqūt built a soup kitchen, madrasa, and hospice and appointed four professors, one for the adherents of each rite, and sixty students. He also established endowments for their support. Sharīf Ḥasan received 12,000 *mithqāls* of gold from Yāqūt for the aforesaid buildings. 30,000 *mithqāls* of the appropriated funds were only for the repair of the spring at ʿArafāt. The Sharīf placed one of his men named Shihāb in charge of this and asked him to examine the condition of the spring of Bāzān. He carried out the renovation of two ponds at Muʿallā that had fallen to ruin. He repaired the water channels from the spring of Bāzān and water again flowed to these ponds.

In 815 (1412) prices became very high in Mecca. One sack of wheat, which was about one camel load, cost twenty florins and one watermelon cost one florin. Subsequently prices fell and there was ease and plenty.

A Marvelous Story: It happened that a camel driver named Fārūnī had a camel. He always loaded the camel with a heavier burden than it could carry. [M 513] One day in Jumādā II, 815 (September, 1412), the camel got loose and ran away. It entered the Masjid al-Ḥarām and began to circumambulate the Nobel House. It attacked whoever tried to catch it and remained free. Finally they left the camel alone. It made the circumambulation three times and then kissed the Black Stone. Then it went to al-Maqām al-Ḥanafī (next to the circumambulation route) and kneeled down in front of the golden Waterspout and began to weep. Then it lay down and died. Later they removed the camel and buried it between Ṣafā and Marwa.

One of the rulers of Sabā (Sheba) repaired many places in Mecca. He completely replaced the marble floor of God's House and the surface of the interior walls of the

4 Correcting M BK to BN as below.

Kaʿba. He also repaired the Bāb al-Janāʾiz, which was one of the doors of the Masjid al-Ḥarām. Because it is near the home of ʿAbbās (b. ʿAbd al-Muṭṭalib, the half-brother of the Prophet's father), it is called Bāb al-ʿAbbās. He carried out major charitable projects on the roads to Mecca. One was the erection of canopies for poor pilgrims; later he arranged for them to be loaded onto camels and set up to shade the poor at the pilgrim stations. He also set up cooking facilities at the stages and stations on the roads and provided water, bread, and biscuits for the poor. Sheep were slaughtered and distributed to the poor pilgrims leaving from and returning to Egypt.

The Nāẓir al-Jaysh (chief quartermaster) ʿAbd al-Bāsiṭ had a well dug on the road to ʿUmra. It was located to the left (of the road) of those going to ʿUmra and is still functioning.

The (Mamluk sultan) al-Malik (al-Ẓāhir) Chaqmaq built many charitable foundations. One of them was a water dispensary and basin that he built in 806 (1403) at Muʿallā. Later, after they had fallen into ruin, the late Ḫātem Sultan, the daughter of Sultan Süleymān, rebuilt them.

In 853 (1449) Bayram Ḫoca, the custodian of the Ḥarām, built several ponds at ʿArafāt.

The great (Mamluk) Sultan Qaytbay established many charitable foundations. He built magnificent Friday mosques and neighborhood mosques in Mecca, Medina, Jerusalem, Damascus, Gaza, and Cairo. These beautiful works are still standing. Around Mecca, he made excellent repairs to the mosques and mihrabs where the Prophet had prayed. In 882 (1477) he built madrasas in Mecca: one madrasa and one hospice for each of the four rites. During his reign the Masjid-i Nabawī burned down. On 12 Ramadan 886 (4 November, 1481) in the last third of the night, while the muezzins were reciting the *tamjīd* (a canticle sung between midnight and dawn), thunder and lightning struck. A bolt struck the crescent on the minaret and broke it in two. Then the minaret caught fire and it spread to the Masjid-i Sharīf. All the furnishing, the Korans, everything in the mosque was consumed by the fire. Only the room where the Prophet had stayed did not burn, and nothing happened to the houses surrounding the Masjid-i Sharīf. When news of the calamity reached Qaytbay, he assembled materials and lumber (needed for reconstruction) and appointed responsible men (to oversee the project). He rebuilt the burned places as they had been before. Then he built a large madrasa and hospice near the Masjid-i Sharīf.

The (Mamluk) Sultan al-Malik (al-Ashraf) al-Ghawrī (r. 1501–1516) (did repairs) on the pilgrimage roads (and) at ʿAqaba-i Īlyā (i.e., ʿAqabat Ayla = Eilat in Jordan), and in Mecca proper he paved the *ḥijr* with marble. He also restored the citadel at Jedda. [M 514] However toward the end of his life he and his army were infected with heedlessness and pride. He began to oppress the poor and to treat the religious scholars with contempt.

Story: A saintly individual and friend of God, one whose prayers are answered, was one day in the Cairo marketplace. He observed a Circassian (Mamluk), one of al-Ghawrī's men, taking merchandise from a broker without paying for it. The broker followed him and demanded his money, invoking the Shariah and saying he would complain to the qadi. The Circassian, saying, "Are you the one who is warning me about the Shariah?" took out his mace and struck the poor broker such a blow on the head that he instantly fell down dead. None of the Muslims who witnessed this incident dared question the Circassian about his action. This was because the oppression in Egypt was so great at that time that anyone who asked about such things was asking for trouble. The saintly individual who was present uttered the following prayer: "Lord Almighty, save the Muslims from these oppressors! May this land be free of such rulers and soldiers." Perfectly timed, the arrow of his appeal struck the target of divine response and his prayer was answered. That night he went to bed and had a dream in which angels descended from heaven, each with broom in hand. They instantly swept Egypt clean of the Circassians and threw them into the Nile. When he awoke from sleep a voice reached his ears. "What is this that I am hearing?" he said. Someone was reading the passage *wa idh nataqnā* … in the Koran ("and when we shook …," 7:171, which concerns the transgressions of the Israelites and how God punished them).

It was the felicitous time of the Ottoman sultanate, when the late Sultan Selīm (I), the Strengthener of the Pillars of the Sultanate of the World, doubled the amount that his father Bāyezīd sent every year to the people of Mecca and Medina as alms. This amount had been 14,000 goldpieces. When the *Ṣadaqāt-i rūmiyya* ("Ottoman alms") arrived at the Ḥaram-i Sharīf in 923 (1517) 500 dinars were recorded at the top of the account books in the name of Mawlānā Sharīf Abū Numayy. The alms have been collected in this manner up to the present. The funds were then distributed to the poor and the pious. And Sultan Selīm continued to send the same amount of alms as the Circassian rulers (the Mamluks) of Egypt (had sent). After the alms were distributed, the religious scholars and dignitaries of Mecca came to the noble Ḥaṭīm and began a complete reading of the Koran. When this was finished, they prayed and presented the registers of alms and the distribution of funds to Sultan Selīm. Amīr Muṣliḥ al-Dīn, who was in charge of the distribution of the alms, appointed 30 people to make a complete reading of the Koran every day, as prescribed by an edict from the sultan, and to donate the religious merit gained from this act to Sultan Selīm. After paying some minor expenses, he assembled some of the poor and gave each one of them three goldpieces. He recorded the name of each person in a ledger. Then he recorded the houses of the jurists in Mecca and three goldpieces to each one under the rubric *ahl buyūt* ("Residents"). This rubric was added to the ledger of the *Ṣadaqāt-i rūmiyya* ("Ottoman alms") and the payments have continued to the present. The number of poor in Mecca was beyond count and they would mob Muṣliḥ al-Dīn. Finally, he assembled them in an open area and put one goldpiece in the hand of each one. [M 515] These alms were recorded in the *Ṣadaqāt* ledger under the rubric *ʿāmma* ("Common People") and the practice has continued to the present.

Sometime later, ships loaded with grain from Egypt arrived at the port of Jedda. An imperial order was issued that Muṣliḥ al-Dīn distribute this grain to the people of Mecca and Medina. It was distributed at that time under the supervision of Sharīf Barakāt. The grain amounted to 7,000 *irdabbs*, of which 2,000 were sent to Medina and 5,000 to Mecca. All the houses of Mecca were recorded and a census was taken of the population. The total number of men, women, and children was 12,000. Afterwards each person was given his share, plus one florin. Furthermore, each of the four qadis (the chief qadi of each rite) was given three *irdabbs* of wheat.

Later, Muṣliḥ al-Dīn remodeled the building at the Maqām al-Ḥanafiyya, which was built in 800 (1397). It had four pillars, a mihrab in the middle, and was covered with a roof. He raised the roof higher than the *ḥijr* and covered it with a dome. Later the emir of Jedda, a Bey named Ḫoşgeldi, changed this structure again. He demolished the dome and made it a square, two-storey *maqām*, assigning the top storey to the muezzins.

Conclusion

According to what the governor of Egypt recorded in the account books during the time of Vizier İbrāhīm Pasha, in 1082 (1672) the monetary gifts and grain supplies sent from Egypt to Mecca and Medina amounted to 376 purses and 24,922 *paras*. Based on this, it was determined that the amount distributed to the Sharīfs of Mecca from the state treasury and from the endowments' income of Mecca and Medina was 38 purses and 5,300 *paras*. The Sharīf of Mecca was given 300 goldpieces, the Sharīf of Ḥammūda 1,000, Sayyid Aḥmad 1,000, and Sayyid Ibn Sulaymān 1,000. As for grain, the Sharīf of Mecca was given 1,000 *irdabbs* of wheat, 500 *irdabbs* of rice, and 500 *irdabbs* of (other) grain. Each of the (other) aforesaid Sharīfs was given 3,000 *irdabbs* of wheat, 100 *irdabbs* of rice, and 200 *irdabbs* of (other) grain. The ration of the qadi of Mecca was 10,188 *paras* worth of grain and that of the qadi of Medina was 11,600 *paras*. Then 190 purses and 19,318 *paras* were given to Mecca and Medina from Egypt. Later, as the Ottoman state increased in prosperity, these amounts were doubled. This happened gradually and it was decided that each year the people of Mecca should be given 3,000 *irdabbs* of grain and the people of Medina should be given 2,000 *irdabbs*.

Zubayda Khātūn (the wife of Hārūn al-Rashīd) started the flow of the spring of ʿArafāt with her own money. Over time the water ceased to flow. Indeed in 939 (1532), because of the lack of water flowing there, one water skin fetched one goldpiece. When the pilgrims were on the verge of dying (from lack of water), God blessed them with rain on ʿArafāt. (The rain was so heavy that) they drank the water flowing below their feet and filled their water skins.

When the spring of ʿArafāt reached this point, Sultan Süleymān issued an imperial edict ordering the springs of Badr Ḥunayn, and ʿArafāt to be repaired and he appointed an overseer, a secretary, and an architect (to carry out this project). The construction was done as the edict required. A dome was built over (the source of) the spring (at ʿArafāt) along with a marble trough that had forty-five pipes. [M 516] When no water flowed through one pipe, it meant that there was that much water lacking in the spring. Mihrimah Sultan, the daughter of Sultan Süleymān, was responsible for many acts of charity and spent a lot of money to bring the water from the spring of ʿArafāt to Mecca. It flowed near the Noble House like the Water of Life.

Süleymān also built four madrasas in the heart of Mecca, one for each of the four rites. He assigned fifteen students of religious studies and one assistant who repeated the lessons to each madrasa. Each professor was paid 50 *ʿuthmānīs* (i.e., *aḳçes*), each assistant four, and each student two (per day).

Of the great (Ottoman) viziers, Meḥmed Pasha carried out charitable works in most of the cities of Islam. He made donations in particular to the people of Mecca and Medina. One of them was to start the flow of the spring of Zarqāʾ in Medina. Because this stream had begun to dry up, he diverted water to it from several well-known wells and he increased the flow substantially.

One of the aforesaid wells was the Well of Idrīs, which was known for its very sweet water, in the district of Qubāʾ. The Prophet graced it with the blessed saliva from his mouth. (One day) the Prophet's seal slipped from Caliph ʿUthmān's fingers into the well. From that day forward ʿUthmān was plagued with civil strife.

Another well was Dhū'l-Khalīfa, which today is known as the well of ʿAlī. When the people of Medina and Damascus visit the wadi where it is located, they put on the pilgrim's garb. It was excavated in excellent fashion and its surface was expanded to ten square cubits. They built a stairway on one side of the well so people could go down to the water.

Another charitable work (of Meḥmed Pasha) was to build at a place near the Masjid al-Ḥarām in Mecca as a shelter where the poor and foreigners could sleep and live if they had no place to stay. (If not for this,) they would crowd into the Masjid al-Ḥarām. He also built a hospital for the sick. Around it he built houses and shops which he placed in an endowment, the income from which was spent on the amenities in that place. (In addition) he built a lofty public bath in the center of Mecca, a hospice, and many other charitable facilities.

During the reign of Sultan Selīm (II), the grain ration was increased by 7,000 *irdabbs*.

The portico of the Masjid-i Sharīf had leaned to the Kaʿba side in such manner that the wooden capitals of some of the pillars had separated from their sockets. In 979 (1571) this was reported and an edict arrived from the Ottoman sultan ordering the structure to be rebuilt solidly, to replace all the wooden pillars with marble pillars, and to erect lofty domes with golden finials. When this edict reached Sinān Pasha, the *beylerbey* of Egypt, he assembled the necessary materials without delay and placed an Egyptian emir named Aḥmed Bey in charge of the project. He set out for Mecca and began work as soon as he arrived. (In the portico) he erected three pillars of marble, which was yellow marble known as "sun stone," and across from them (?) a gate. He erected similar pillars and bases all around the Masjid al-Ḥarām and raised cupolas over them. [M 517] As for the Masjid al-Ḥarām, he completed its east and north sides. After he finished his (most) important work, which was the construction of the Gate of ʿUmra, word arrived of the sultan's death.

As soon as Sultan Murād (III) took the throne, he issued an imperial edict instructing the aforesaid Aḥmed Bey to continue his work on the Masjid. Toward the end of 984 (1576) he finished this project. Afterwards he carried out excavations south of the Noble Sanctuary in order to provide better drainage, because over the years flood waters (sometimes) filled the Masjid entering from the southern doors to a height of three steps and so had reached this point. In earlier times it was the custom once every ten years for the earth to be removed and taken to low places in Mecca. This was the first time in 30 years that such excavations had been carried out. Before this was done, a great flood occurred without warning on the night of Wednesday 10 Jumādā I, 983 (17 August, 1575) and entered the doors of the Masjid al-Ḥarām. The circumambulation area was flooded and the Kaʿba was surrounded with water. It was so high that the Black Stone could not be seen. The water continued to rise, passing over the noble threshold of the raised door of the Kaʿba and reaching almost as high as the lock on the door. The Noble Sanctuary remained under water for a day and a night, which made prayer there impossible. Taking this into consideration, an edict based on an old *ḳānūn* (sultanic law) was issued according to which the drainage area was to be excavated and extended once every two years.

Ḫāṣekī Sulṭān (i.e., Hürrem Sultan or Roxelana, wife of Sultan Süleyman), who was a patroness of charitable works, built a large soup kitchen, hospital and other facilities near the Kaʿba in 1094 (1683). One of them was inside the blessed well into which the Prophet had spat.

After the Sacred House was remodeled, the number of marble pillars that were erected was 311. Of these, 62 were on the east side, 81 were on the north side, and 64 were on the west side. All of the pillars were of marble, except for 6 which were made of syenite. On the south side, opposite the Yemeni corner, were 33 pillars, 11 of which were made of syenite. At the Gate of Ibrāhīm there were 6 pillars and at the Dār al-Nadwa there were 15, one of which was syenite. Of the yellow stone called "sun stone" there were 244 in all. These were made of sections of dressed stone. Some pillars were octagonal and some were hexagonal. They were made of dressed syenite one-third the way up from the bottom. The entire length of the upper part of the pillars was made of the dressed yellow stone called "sun stone". In this way the east and west sides of the Masjid al-Ḥarām each acquired 30 pillars of dressed stone. There were 44 pillars on the north side. While there are 76 on the south side, and four at the "corners" (*erkān*) of the Masjid-i Sharīf.

The number of pillars in the enlargement (*ziyāda*) of the Dār al-Nadwa is 36. In the enlargement of the Gate of Ibrāhīm there are 18. As for the cupolas of the Masjid al-Ḥarām, there are 24 on the east side and 36 on the south side. There is also a dome in the corner of the Masjid al-Ḥarām on the side of the Minaret of Ḥazwara. [M 518] There are 16 cupolas at the enlargement of the Dār al-Nadwa and 15 at the enlargement of the Gate of Ibrāhīm. There are 282 *ṭājins*, that is, *tawās* (small domes that support the main domes).

There are 56 places for public worship. There are three to the east of the Masjid-i Ḥarām, 22 to the north, 16 to the west, and 15 to the south. The Masjid has a total of 1,157 *shurfas* (galleries of minarets from which the call to prayer is made).

The Masjid-i Ḥarām has 19 gates. There are four on the east side. The first is Bāb al-Salām or Bāb Banī Shayba, with one arch. The second is Bāb al-Nabī or Bāb-i Janā'iz, with two arches. The third is Bāb al-ʿAbbās, with three arches; it is also called Bāb-i Janā'iz. The fourth is Bāb-i ʿAlī or Bāb-i Banī Hāshim, with two arches.

There are seven gates on the south side. The first is Bāb-i Bāzān, with two arches. The second is Bāb-i Baghla. The third is Bāb-i Ṣafā or Bāb-i Banī Makhzūm, with five arches. The fourth is Bāb-i Jiyād, with two arches. The fifth is Bāb al-Mujāhidiyya or Bāb al-Raḥma, with two arches. The sixth is Bāb-i Madrasa or Bāb-i Sharīf ʿAjlān. The seventh is Bāb-i Umm Hānī, with two arches.

There are three gates on the west side. The first is Bāb-i Ḥazwara or Bāb al-Wadāʿ, with two arches. The second is Bāb-i Ibrāhīm. The third is Bāb-i ʿUmra—those making the ʿUmra pilgrimage from Tanʿīm (north of Mecca) use this gate—or Bāb-i Banī Sahm, with one arch.

There are five gates on the north side. The first is Bāb-i Sudda (?), previously Bāb-i ʿAmr b. al-ʿĀṣ, with one arch. The second is Bāb al-Bāsiṭiyya. The third is Bāb-i Ziyāda-i Dār al-Nadwa, with three arches. The fourth, north of the Bāb-i Ziyāda, is Bāb al-Zarība, with three arches.

In 1092 (1682), when the Ottoman state sent the aforementioned stablemaster Süleymān Agha to Mecca following the flooding of the Ḥaram al-Sharīf, he made repairs and raised some of the gates by two steps, some by three, and some by four in order to keep out the water.

The Well of Zamzam: It is located opposite the path of circumambulation. In 224 (838) there was little water in the Well of Zamzam. During the reign of al-Manṣūr, the second Abbasid caliph, Muḥammad b. Āl-i Ḍaḥḥāk carried out an excavation in the well making it nine cubits deeper. This resulted in an increased flow of water. There had previously been a pool near the Well of Zamzam where pilgrims were given sweetened sherbets. They called (this place) Qubbat al-Sharāb (Dome of Drink).

The Masjid-i Ḥarām has six minarets. The Abbasid caliph Abū Jaʿfar al-Manṣūr al-Dawāniqī built the first minaret at Bāb-i ʿUmra. Formerly the chief muezzin began the call to prayer from this minaret and the muezzins on the other minarets followed him. Today, however, a muezzin begins the call to prayer from the Dome of Zamzam and then the other muezzins follow him. When the *tamjīd* is recited during the nights of Ramadan, the call to prayer is first made from Bāb al-Salām. Sultan Süleymān rebuilt the aforesaid minaret because it had fallen into disrepair. The second is the minaret at Bāb al-Salām, built by the Abbasid Caliph al-Mahdī. When it fell into ruin, al-Nāṣir Faraj (b.) Barqūq of the Circassian (Mamluk) state rebuilt it. The third is the Minaret of ʿAlī, also built by Caliph al-Mahdī. [M 519] When it became structurally unsafe, Sultan Süleyman rebuilt it as well. The fourth is the Minaret of Ḥazwara, originally built by Caliph al-Mahdī but was later repaired by the ruler of Mosul Ashraf b. Shaʿbān—it was torn down and rebuilt in 771 (1369). The fifth is the Minaret of Bāb-i Ziyāda, built by Caliph al-Muʿtaḍid and later restored by (the Mamluk Sultan) al-Malik al-Ashraf Barsbāy. The sixth is the Minaret of the Qayt Bay Madrasa. It has two balconies. The seventh is the Minaret of Sultan Süleymān. It is very large and tall. It is made of "sun-stone" and has two balconies.

On the upper side of Mecca is the Masjid-i Rāya. This mosque is where the Prophet planted his standard (*rāya*) on the day he conquered Mecca. It has an old minaret. During the nights of Ramadan they light lamps on it and make the evening call to prayer from it.

Jabal Ḥirā, Jabal al-Nūr, Jabal Thawr, Jabal Thabīr, Jabal Ḥafdama, and Jabal Abī Qubays are the blessed mountains around Mecca.

Baṭn Nakhl: A blessed place near Baṭn Murr and one stage from Mecca.

Ṭā'if, also called Wādī al-ʿAbbās: It is two stages from Mecca and behind Mt. Ghazwān. This mountain is the coldest place in Ḥijāz. Sometimes the water on the summit of the mountain freezes. The air of Ṭā'if is truly excellent. It has abundant fruit. Lemons, bitter oranges, and, especially, raisins are very plentiful. The villages of Ṭā'if are Bulaydat al-ʿAbbās, Luqaym[a], Rahaṭ, Tabāla, and ʿAylān. The Banī Ṣaqr, Banī Asʿad, and the tribes of the Hudhayla live here.

Wādī Nuʿmān al-Arāk: It is between Ṭā'if and Mecca.

Bi'r Maʿūna: It is between Mecca and ʿUsfān and a camping ground for the Hudhayla. Near here is

Bi[']r Bāy, also called Rajīʿ al-ʿUrj: A district of Ṭā'if.

Baṭn Murr: A place with many villages and streams, gardens and fruit, including bitter oranges, lemons, dates, and bananas. This wadi is connected to Wādī Nakhla. It is six miles from Mecca and on the road for pilgrims coming from Egypt and Syria.

Jedda: A well-known port two stages west of Mecca. It has well-constructed markets. The tithe is collected from the merchants. Large ships come to the port of Jedda from India, Abyssinia, Aden, Yemen, Zaylaʿ and other places. Every year eight ships arrive from India, six of which are merchantmen and two of these belonged to the ruler of India. A captain who operates these ships for the ruler of India lives in Jedda. He commands the ships. (Every year) 38 galleons arrive from Egypt. Some of them come to Jedda and some to Yanbu. The tithe is collected here from the merchants. Caliph ʿUthmān supervised the construction (of the present harbor). Formerly it had no wall, but because of the plundering of the rebellious Bedouin, the Circassian (Mamluk Sultan al-Malik) al-Ashraf Ghawrī had one built.

There are pools here, but they do not provide sufficient water, and this has been a problem for Jedda. In 1094 (1683) Ḳara Muṣṭafā Pasha, grand vizier of Sultan Meḥmed IV, brought water from the distant mountains. Jedda now has plenty of water. This was truly a great work of charity. In addition he built a magnificent inn, a splendid public bath, and a noble Friday mosque in Jedda.

It is reported that Eve fell (from heaven) to Jedda and that she met Adam at ʿArafāt.

Currently Jedda has two magistrates. [M 520] One is a *bey* (military commander) appointed by the Ottoman sultan and the other is an *amīn* (customs agent) appointed by the Sharif of Mecca. Half of the income from the customs duties goes to the *bey* and half goes to the *amīn* for the Sharif of Mecca.

Wādī Fāṭima: It is one stage from Mecca. When pilgrims come, the noble *maḥmal* stops here. As for the pilgrims, they go on to the Kaʿba to rent lodging. There is much water in this valley and a number of villages. There are gardens and it is a place of excursions. A plant called *kādī* is unique to this place.

Ḥadda: A prosperous town between Jedda and Mecca. It has flowing water and beautiful gardens.

Qarn: It is behind Ṭāʾif on the road to Ṣanʿāʾ.

Dhāt Qarn: It is on the road of the pilgrims from Basra. Here they put on the pilgrimage garb.

Awṭās: The Prophet carried out raids here. It is a wadi between Ṭāʾif and Ḥunayn. There is a small river between them flowing north to south.

Dhāt ʿIrq: A village with several nearby wells. It is four stages from Shuʿrā (?) and two from Mecca. The ground is sandy, but if one digs a bit water emerges.

Tabāla: Two stages southwest of Rabaṭ. There is a stream that flows north to south.

Kāmid: Three stages southwest of Tabāla. There are several villages in the district, which is mountainous. There are a number of streams that flow from north to south.

Ḥudaybiya: One stage from Mecca. There is a tree planted here where the Pledge of Riḍwān took place. The tree took its name from this event.

Zahrān: It is on a mountain southwest of Kāmid and has much water.

Qubā: A village two miles from Medina. It has sweet water. This is where the Taqwā mosque is located. It is the mosque about which the following noble verse was revealed: *la-masjidun ussisa ʿalāʾl-taqwā* … ("Certainly a mosque founded on observance …," Koran, 9:108). The Masjid-i Ḍirār is also located here.

al-Furʿ: It is several villages. It is four stages south of Medina. There are brigands here. Between al-Furʿ and Medina is Wādī al-ʿAqīq.

Qubūr-i Shuhadāʾ ("Tombs of the Martyrs"): Here there is a lake between two mountains formed from rainwater.

Ābār ʿAlī ("Wells of ʿAlī"): They are wells, three fathoms deep, that belonged to ʿAlī. This place is six stages from Medina. The pilgrims from Syria put on the pilgrimage garb here.

Jadīda: A village with abundant flowing water, vineyards, gardens, and date palms. There are watermelon and vegetable gardens. It is on a small hill between two mountains.

Jabal Uḥud (Mt. Uḥud): It is one parasang from Medina. This was the site of the (famous) battle in (early) Islam. This is where the following verse was revealed: *laysa laka min al-amri shayʾun* … ("You have no concern in the matter …," Koran, 3:127). Other verses were also revealed here.

Ḥamr al-Asad: This place is eight miles from Medina.

Baṭn ʿAwna: This is the place called Rayy al-Fiḍḍa ("Irrigation of Silver") near Medina. It is 24 miles from Medina.

al-Ṭaraf: Name of a stream that is 36 miles from Medina.

Medina, the City of the Prophet: Also called Yathrib. Dimashqī in his biography of the Prophet lists 95 names of Medina. He also recounts more than 100 special features of this city. In Spanish it is called Madīnat al-Samā ("City of Heaven"). Medina is the second Haram. It is the noblest place in the world. It is located in a flat area. Jabal Uḥud is two parasangs to the north. Jabal Thabīr is to the south. At a distance of one and a half parasangs to the east is Biʾr ʿAqīq. The road to Mecca goes through here. Most of the land is brackish. There are four wadis here: Wādī Qanāh, Wādī Buṭhān, Wādī ʿAqīq-i Akbar (the Greater ʿAqīq Valley), and Wādī ʿAqīq-i Aṣghar (the Lesser ʿAqīq Valley). [M 521] When it rains and there are floods, water comes to these wadis from Ḥarrat Banī Sulaym. The flood waters pass to Wādī al-Ghāb and Wādī Aḍm and then spread out. The waters flow into two wells, one called Biʾr Rūma and the other called Biʾr ʿUrwa.

During the event of Ḥizb al-Aḥzāb, Salmān al-Fārisī recommended to the Prophet that a trench be dug around Medina and this was done. As a result a famous battle was fought there against 12,000 unbelievers of the Quraysh who had set out for Medina from Mecca. It was for this purpose that the Prophet had ordered the trench to be dug.

(The Buyid ruler) ʿAḍud al-Dawla Fanākhusraw al-Daylamī built a wall around Medina.

Medina is half the size of Mecca. Its air is healthful and it has abundant water. It has agricultural lands and date palms. Famous dates are grown here which are found nowhere else. They are called *tamr bardī* and *tamr ṣayḥānī*.

There is another kind of date called *sulṭānī* that is unique to Medina. *Ḥabb al-bān* (ben nut) is grown here and sent to other cities.

Abyār Ḥamza: A stage between Medina and Wādī al-Qurā. There are wells here. There is a mosque here belonging to Ḥamza (b. ʿAbd al-Muṭṭalib, the paternal uncle of the Prophet). It is two miles from Medina.

Wādī al-Qurā: They call this the district of Quzaḥ. It has a strong citadel, a trench, three gates and Friday mosques. In the mihrab of the Friday mosque is the bone that addressed the Prophet saying *lā taʾkulnī fa-innī masmūm* ("Don't eat me because I am poisoned"). The public bath is outside of town. There is a fortress nearby. It is surrounded by prosperous villages. This wadi is on the road to Mecca and has water and date groves and gardens.

Ḥamā: The name of a stream. It is behind Dhāt al-Qurā.

Ḥijr, also called Qurā Ṣāliḥ ("Villages of Ṣāliḥ"): The mountains here, which are called Athālib, are low and the ground is sandy and dry. The Prophet passed through here on the way to the Battle of Tabūk. He ordered that no one take water from its wells and to pass through quickly. Ḥijr is the land of the people of Thamūd. In the mountains are houses carved out of the rock.

Masjid Ṣāliḥ: A high place, carved out of the rock like a terrace. All around it are strange and marvelous (stone ruins) from the time of the people of Thamūd. It is half a stage from ʿUlā. The mountain in the middle of Ḥijr is called Jabal Inān.

They relate (his genealogy as follows): Ṣāliḥ b. ʿUbayd b. Asaf b. Māshīḥ b. ʿUbayd b. Thamūd b. ʿĀbir b. Aram b. Sām (= Shem son of Noah). The descendants of the paternal uncle of Thamūd were Judays and ʿĀd. They trace their tribes back to Thamūd. They were Bedouins and lived in Ḥijr. God sent Ṣāliḥ to them (as a prophet) because they worshipped idols. He summoned them to monotheism. They demanded that Ṣāliḥ extract a she-camel from stone. Following their demand, a camel stepped from the stone and gave birth to a foal. (As a result) some of these people came to belief in God but others remained in unbelief. They assigned a certain day for the she-camel to drink from their well and it would only drink on that day. The chief of the unbelievers was someone named Qudār. When he killed the she-camel its foal went inside the stone. Ṣāliḥ then said to his people, "Three days from now God will rain down stones on you as punishment and you will be destroyed." And this in fact is what happened. Ṣāliḥ then departed from there with the people of Islam (i.e., who had submitted to God) who had believed in him and went to the town in Palestine called Ramla where he died.

Magharish al-Zīr, also called Aqraʿ: About a half stage (from Masjid Ṣāliḥ) to the south. [M 522] This is the place where the she-camel of Ṣāliḥ was killed. The name of this place, which they call Jabal al-Ṭāq, is Mazḥam. Travelers who reach this place turn east and camp in Ḥijr. They pass through here very quickly, raising a hue and cry, because they don't want to hear the voice of the she-camel coming from inside the rock. If their camels hear her voice they will immediately kneel and never get up.

ʿUlā: A half stage to the south of the lands of Thamūd. It is six stages from Medina. It is a town in the mountains and it is subject to Medina. It has flowing water, gardens, and a fortress. It is a place where everything is very cheap. Sultan Süleymān Ḳānūnī rebuilt the fortress. Before this was done, complaints had been made to ʿĪsā Pasha, the *Beylerbeyi* of Syria, about the marauding of a local Bedouin tribe, so he built a fortress here and stationed guards in it. A tax of one dirham is levied on the harvest of each date tree of this town and this tax is spent on the needs of the troops. This tax has gradually increased to 40 dirhams. The people complained but to no avail. This oppression continues today.

Khaybar: A very prosperous place. It produces a great amount of dates. (In the past) this had been the town of the tribe of the Banī ʿAnaza. In the language of the Jews, *khaybar* means fortress. It is six stages northeast of Medina. There are several forts in this district. The names of the seven most famous of these forts are as follows: Katīb, Nāʿim, Shaqq, Qamūṣ, Naṭāh, Saṭḥ, and Sulālim. In the seventh year of the Hijra the Prophet conquered these fortresses and acquired enormous booty. At the dawn of Islam, through the providence of God, two Jewish tribes residing in Khaybar, the Banī Qurayẓa and the Banī Naḍīr, made a non-aggression treaty with the Prophet. Later, however, they violated the pact and the Prophet went to war against them again. He took possession of their property and women and put the men to death. Shamūr b. ʿĀdiyā was in (?) Wādī Naṭāh and Wādī'l-Shaqq. Near them is a place called Arḍ Sanja. In Wādī'l-Shaqq is a spring that the Prophet named ʿAyn al-Jama. Caliph ʿUthmān built up the area around the spring, reinforcing and restoring it. The Prophet also called this spring Qismat al-Malāʾik ("Lot of the Angels"). Two-thirds of the water from this spring enters a canal and one-third flows away. If one blocks the canal with something the water will overflow and go elsewhere and not a drop will flow through the canal.

In Wādī Naṭāh there is a large spring called Luḥayḥa (?). In year 7 and 19 of the Hijra a fire broke out here and spread to the surrounding area. Caliph ʿUmar ordered the people to give alms and by doing so, by order of God, the fire was extinguished.

Khaybar is extremely feverish. The illness even strikes visitors. The people of Khaybar have the following belief:

If someone who comes to Khaybar puts his hands on the ground and stands on all fours and brays like a donkey, the fever of Khaybar, which is contagious, will not harm him. It has been tested. They call this spot *baʿshīra* which means "Fever is desirous of people" (*ḥumā nāsa ḥarīṣdir*) and it means "sons of the donkey" (*benī ḥimār*).

Ukhayḍir: A fortress that was built early in the reign of Sultan Süleymān. There is a water source inside the fortress. People draw from it and pour it in pools. Ukhayḍir is half way between Damascus and Mecca. Pilgrims usually assemble here. [M 523] It is a wide staging point among the mountains. Twenty soldiers from the Syrian army are stationed here to guard it. They fill three pools from the well. It adjoins the fortress. The fortress was built in 938 (1531) by Ṭarbān b. Farājā, the *bey* of the Ḥāritha Bedouin, by order of Muṣṭafā Pasha, the governor of Syria. This was because the Banī Lām and the Banī ʿUqba would fill in the well and empty the pools when they revolted.

Thaqb Ukhayḍir: A pass near Ukhayḍir. The Bedouin of the Banī Lām sometimes block the entrance with stones.

Birkat al-Muʿaẓẓam: A pool in the desert, built of stones, that fills when rain brings floods. If there is no rain there is no water. It was built by the Ayyubid ruler (of Damascus) al-Malik al-Muʿaẓẓam ʿĪsā.

Tabūk: A staging point for the pilgrims from Syria. It has many date palms. The Prophet himself came to Tabūk in order to carry out religious war. (When he arrived he found) that there was little water in the nearby river, so he took some water in his blessed hand and sprinkled it over the river and immediately it burbled like a large flowing stream.

Sultan Süleymān built the Tabūk fortress from the stones in the ruins of Tabūk. It has a large pool. (The prophet) Shuʿayb came here from Medina to summon the people of al-Ayka, which means thicket (i.e., the "People of the Thicket" mentioned in the Koran, 15:78, 38:13, and 50:14). The people of al-Ayka were unbelievers. They practiced highway robbery and cheated in measuring grain. Shuʿayb summoned them to the way of God. Some of them had faith in Him and some insisted in remaining in unbelief. So, God sent against them an earthquake with a horrifying clamor and darkness and destroyed them. al-Ayka is near Medina.

Chapter: Coastal Plain (*Tihāma*) of Ḥijāz

Borders

To the south, from Baysh in Yemen all the way to Ayla; to the east, Ḥijāz; to the west, the Sea of Suez; to the north, the Syrian desert.

Description of its places and regions

Namāwa: A stage for the pilgrims coming from Zabīd. It is one stage from Baysh.

Ghatūd: A place on the coast of the Sea of Suez where Bedouin tribes gather. A great market/fair is held here.

Dihbān: Name of a wadi. The false bdellium (*muql*) tree grows here. It is a three-hour journey from the coast.

Ḥalī: A city, known as Ḥalī b. Yaʿqūb. It is two stages from the coast. Currently Ḥalī is a city under the authority of the Sharīf of Mecca. It is eight stages from Mecca. The port of Ḥale (sic; = Ḥalī) is Qunfuda (i.e., al-Qunfudha, Kunfuda), a prosperous island. It is called Qunfuda because it is just off shore.

Marsā (Anchorage of) Ibrāhīm: A port between Jedda and Qunfuda.

Sirrayn: It is 19 parasangs north of Ḥale.

Yalamlam: A place among the mountains two stages from Mecca. This is where the people of Yemen put on their pilgrimage garments. Its water comes from wells and springs.

Saʿdiyya: It is close to Yalamlam. This is another place where people put on their pilgrimage garments.

Aydām: It is near Saʿdiyya. People also put on pilgrimage garments here. It has a neighborhood mosque, Friday mosque, market, and public baths. Its market is on the seashore. Most of its buildings are made of wood, including the public baths. Since water cannot be heated by a fire in the baths, it is brought in from outside. The water that they use is rain water. They have livestock and crops. They have abundant honey. Above the city gate is a beautifully built Friday mosque.

Juḥfa: A place near Rābigh where pilgrims from Egypt put on their pilgrimage garments. It is six miles from the sea. It is between Rābigh and Khulayṣ. The Banī ʿĀmir live here. It is five stages from Jedda. The Banī ʿUbayd, who were brothers of ʿĀd, lived here. When the Amalikites drove them out of Yathrib (Medina), they settled here. Later a flood came and swept them away.

Mīl: It is located to the right of Juḥfa. [M 524] This place, which they call Ghadīr Khumm, is in a forest. The word *ghadīr* is derived from "a spring" and is part of the compound Ghadīr Khumm (Khumm being the name of the area where the spring is located). They say that anyone born here cannot live here in peace until adolescence but must move away and find a new residence. There is a mosque of the Prophet (*Masjid-i Nabī*) between Ghadīr and ʿAyn ("the Spring").

al-Abwā: It is eight parasangs to the north of Juḥfa. Amīna, the mother of the Prophet, is buried here.

Safwān: A district belonging to Badr.

Badr: A village with extensive gardens and date palms. There is a fountain in front of it. (And there is) a ruin which is called Ḥunayn. This is not the Ḥunayn where the battle was fought near Ṭā'if (a battle mentioned in the Koran, 9:25–26). There is a spring that bubbles up here. A place here called Ghulayb is where a battle of the Muslims occurred. It is now a date grove. Behind it are the graves of the (Muslim) martyrs (from the aforesaid battle).

Two lakes spread out at Badr. Along their shores are rows of banana trees, grapevines, and date palms. At Badr there is a mountain (or dune) of pure white sand. This place is four stages from Rābigh, but the Syrian pilgrims cover it in two stages. It is four stages from Medina. A clan of the Zabīd tribe lives in Badr. They are called the Muzdār. They levy a fee on caravans for protection.

There is disagreement about the stages on the road between Badr and Mastūra. One route is: Shajarāt al-Amīr; Ḥāṣir al-Bathna; Ḥadr; ʿĪtā—desert, with the sea visible (on the left) and the mountains of Ṣīj on the right where there are rebellious Bedouin; Mastūra. The other route is: Bayḍā'; Barīma; Bathna.

Quṭb (al-Dīn) al-Makkī (al-Nahrawālī) says: At Badr are two high dunes. Sometimes the sound of a drum can be heard there. It is known as "Drum of Victory of the Prophet." That this sound is heard is well known and a matter of public report.

Banī'l-Nakhīl: There is a mosque called Masjid al-Ghamām ("Mosque of the Clouds"). It had a *minbar* and the Friday prayer was given here. The Prophet was shaded here by clouds. The biographies of the Prophet mention this place as Masjid-i ʿArsh ("Mosque of the Throne"). Near here are the ruins of a mosque called Masjid-i Aqṣā ("the Farthermost Mosque", Koran 17:1). The Prophet stopped here when he sought victory against the unbelievers of Badr.

Maḥaṭṭa: Here is a large pool to which one descends by steps. (The Mamluk sultan Qānṣawh) al-Ghawrī built it in 919 (1513). It has a white dome. Also at Maḥaṭṭa is a mosque built by Sharīf Abū Numayy.

The homes of the people of Badr are in the district of Nakhīl. It also has a citadel.

Opposite Badr is a spring called ʿAyn Khaybar which has little water. Its water comes from the slopes the mountain. Where it flows there are some arable fields and date palms. A clan lives there.

Mā'-i Ḥunayn: Some places on Shiʿb Ḥunayn—also called Shiʿb Munādī—where rain water has collected. It is sweeter than the water of Badr.

The people of Badr are from the four noblest clans: the Maḥāsina, Qarābida, Shakra, and ʿItq.

Mastūra ("Covered"—scil. covered wells): This is half way (to Medina). It has wells that can be uncovered.

Ḥarība: It is one stage to the left of the road from Mastūra. It has wells and arable lands. A clan of Bedouin lives here. They are called Rawī Muḥammad and Rawī ʿAmr. They can marshal about 1000 lances. Near the wells of Mastūra, on the left are wells of sweet water. They call this place Mastūra al-ʿUlyā (Upper Mastūra). In the spring, rain provides water. It is one mile from the staging point.

Jār: the (old) port for Medina. It is three stages southwest of Medina. It has a citadel and markets. It is 20 miles from Badr.

[M 525] Yanbu: A fortress. It has date palms, streams, and crops. It has about 170 springs. It is under the control of an independent governor appointed by the Sharif. The residences of the Banī Ḥusayn (Ḥusayn being the grandson of the Prophet) are here. ʿAlī (the son-in-law of the Prophet and father of Ḥusayn) had pious endowments for his children in Yanbu. Mt. Raḍwā is near this city and overlooks it on the east side. Grinding stones are extracted from this mountain and sent to other cities. The mountain has many valleys with many branches. The mountain is not barren but green. There are many streams. The Kaysāniyya sect believes that Muḥammad b. al-Ḥanafiyya (another son of ʿAlī) is still alive and living on this mountain and is expected to reappear. Yanbu is today the port for Medina.

Buwāṭ: A district belonging to Raḍwā. Near it is:

al-ʿUthra: A place in the lowlands of Yanbu.

al-ʿUdayba: A village. In order to get there one must first pass through the town called Dahnā'. This is the town of (the mystic) Sayyidī Aḥmad al-Badawī. The Bedouin call any flat desert area *dahnā'*.

Jabal al-Zubna: It is near Yanbu. Customarily the governor of Yanbu sets out for this place with an imposing regiment and meets the *amīr al-ḥajj* (the leader of the annual pilgrimage from Cairo). They deck out the *maḥmal* (the litter accompanying pilgrims from Cairo to Mecca) and its military escort. The noble *maḥmal* stops here. A prayer rug is spread before the camel (carrying the *maḥmal*) and the governor of Yanbu prays two prostrations on it. When the camel carrying the *maḥmal* is about to move forward, he throws a handkerchief and stands with head bowed. This is an old tradition.

Wāsiṭ: A stage (of the pilgrimage caravan). Here lamps are lit and cannons are fired. It is also customary for everyone to light a candle. They have a belief that if one does not light a candle here his camel will die. They consider this to be a token of the good news of the victory of the Prophet, just as the "Drum of Victory" is heard.

Azlam: A stage. This is a grassless and arid place. The water is salty. It is a desert plain between mountains. Senna (Cassia acutifolia) is grown here. There is a ruined inn here. (The Mamluk Sultan) al-Malik al-Nāṣir Muḥammad b. Qalā'ūn built it. Sultan al-Ghawrī tore it down in the year 916 (1510) and built a fortress. Military guards are stationed there and their stores are kept there.

Madyan: A ruined city on the coast of the Sea of Suez. It is parallel with Tabūk. There are six stages between them. The well from which Moses brought water to the daughters of Shuʿayb (a prophet mentioned in the Koran) is at Madyan. Madyan is the name of a tribe. And Shuʿayb was from this tribe. Later the city took the name of the tribe. This place is mentioned in a verse in the Koran *wa ilā madyana* ... ("And to Madyan ...," Koran 7:85). This city was built by Madyan son of Abraham. Abraham had two sons in addition to Ishmael and Isaac. One was Madyan and the other was Madāyin. There was a rock near the well that Moses removed. They call this place Maghāyir Shuʿayb ("Caverns of Shuʿayb"). There is sweet water in its crevasses. There are date palms, tamarisks, and false bdellium (*muql*) trees. There are tablets here on which kings have inscribed their names. The trace of a city wall is still visible.

Ḥasnā': A staging point in a place called Ḥasnā' in Bi'r ʿArab, about a day's sail from Mt. Sinai.

Tāzān: A rock off Ra's Abū Muḥammad. Ships pass between two sides of it. The passage is harrowing.

Ra's Abū Muḥammad: A promontory that juts into the sea from Bi'r ʿArab. Opposite it in the sea is Tāzān. There is a populated area to the south.

Ayla (Eilat): An old city on the coast of the Sea of Suez. It is where the borders of the Syrian Desert, Ḥijāz, and Egypt meet and the place where Ḥijāz begins. It is on the pilgrimage road from Egypt. Ayla is the place alluded to in the verse in the Koran *wa's'ali'l-qarya* ... ("And ask the town ...," Koran, 12:82). [M 526] There is a tower in Ayla where the governor sent from Cairo resides. Near it is the place where Pharaoh was drowned. Pilgrims from Cairo, Gaza, and Hebron assemble here. During the time of David this was a glorious city. It is six stages from Jerusalem, twelve from Damascus, and five from Suez. There are some crop lands here. Ayla is the place God mentioned in the Koran when He said, ... *ḥāḍirata'l-baḥr* ("[Ask them {the Israelites} about the town that] stood by the sea," Koran 7:163).

The people of Ayla were Jews. God forbade them to fish on Saturday. But so many fish swarmed there on Saturday that one could hardly see the sea. Satan whispered to them and led them astray. He said, "You are forbidden to fish on Saturday, but if you make ponds around the sea and drive the fish that come from the sea into them (at high tide) and then prevent them from escaping, then you can fish for them on Sunday." With this stratagem, they caught so many fish that they had more than enough to eat and salted and sold the rest. The Jewish population of Ayla grew beyond count. They divided into three parties. One third refused to do what God had forbidden. One third said *lima taʿiẓūna* ("Why do you preach [to a people whom God would destroy or whom He would chastise with a severe chastisement]?" Koran 7:164). And one third fell into error by following Satan's command. The first two thirds who refused (i.e., the prohibited and prohibitors) said to those who had fallen into error, "We cannot live in the same place with you." They then divided Ayla into two sections. One section was for those who followed God's prohibition. One morning, when those who had followed God's prohibition got up, they did not see those who had fallen into error. Saying, "Perhaps they are sleeping off drunkenness," they climbed up the walls of their houses and looked around. All (the sinners) were lying on the ground. Within three days death overcame them and they were all destroyed.

Ṭūr (Mt. Sinai), where Moses spoke with God, is one night's journey from Ayla.

At the battle of Tabūk, the ruler of Ayla went to the Prophet and made peace with him. He agreed to pay the poll tax and then gave him a present of a white female mule. A Jewish clan that had lived there a long time claimed to have the cloak of the Prophet which he sent as a token of safe conduct. It was wrapped in a bundle one span in length. They used to remove it from the bundle and display it to the people, and let them make a pilgrimage to it.

The pass (*ʿaqaba*—also a place name, ʿAqaba) near Ayla is difficult. A mounted man can ascend, but it takes an entire day in mid-summer (when the days are longest). Then one comes to the famous Wilderness of Tīh. A man named Fā'iq, who was a slave of (the Egyptian ruler) Khumārawayh b. Aḥmad b. Ṭulūn, improved the road. Later the Circassian (Mamluk Sultan) Qānṣawh al-Ghawrī, toward the end of his life, improved and repaired the road. In 945 (1538), Dāwūd Pasha, the Ottoman governor of Egypt, began to widen and improve the road. One can now cross the pass in seven hours and then camp below it.

Caravans from Jerusalem, Hebron, and Karak meet at Ayla. They bring all kinds of fruit, food, and sheep and goats. The inhabitants of this district are the Ḥawīṭāt Bedouin.

One begins to descend the pass and then goes up and down repeatedly until he reaches a certain place in the mountain. Here the soil is red. For this reason they call it *dār-i ḥamrā* ("Red Abode"). From here there is a long rocky road that goes up hill. Then it goes downhill. Then it enters a narrow pass called Ḥalazūn. Then it makes another long ascent as far as the place called Ḳızıltaşlıḳ ("Red Rock"). [M 527] People rest here for a while, then continue through the pass, traversing great wadis and going down slopes. After glimpsing the sea, they ascend a dark mountain and descend into an open plain, reaching the coast. There are other narrow paths by which the Bedouin travel through the pass. Caravans (after reaching the coast) follow the road with the sea on the right. When they descend to the base of the fortress (of ʿAqaba) they rest for three days.

Chapter: Najd-i Ḥijāz

This clime is located east of Ḥijāz. It is a province composed of several districts, as follows: Janābikh, Khanūqa, Shaʿrā, (Ḥiṣn) Ḍariyya, Tharbā, Qalbān, Jabala, Rass, Rusays, Waḍāḥ, Nafī, Jumrān, Rushā, Wādī al-Sirr, and Shafār.

Khanūqa is also called Balad İblīs ("Land of Satan"). This is because the Prophet called Satan "the Shaikh of Najd." They say that at night one can hear eerie voices and hand claps. It is a place of ruins.

Shaʿrā: It is five stages from Mecca. It has springs and arable fields. It is under the control of the Banī Lām.

Ḥiṣn Ḍariyya: It is to the north of Shaʿrā. A large pond wells up here.

Wahīlak: It is eight stages southeast of Shaʿrā. It is surrounded by many villages. Yamāma is two stages to the northwest (error for northeast). It has many dates and much agriculture.

Rass: A well. It is one stage from Ḥiṣn Ḍariyya. The inhabitants have a lot of livestock and sheep. (In the past) because they worshipped idols, God sent the prophet Shuʿayb to them to warn them. However, they did not heed his warning. Consequently, God made the ground swallow up the unbelievers, their lands, and villages.

Nīr (or Dayr?) Ibn Marand: It is northeast of Mecca.

Karmūsa: It has a stream that flows from west to east. It is south of Ḥūṭa which is Najd al-ʿĀriḍ.

Taman: A village south of Karmūsa. Many dates are grown here.

Bashhar: A village that has a stream that flows from north to south.

Rabṭa: It is southwest of Bashhar and northeast of Tabāla. In olden times it had a fortress.

Chapter: Najd al-ʿĀriḍ ("Middle Najd")

This is a vast area. A mountain crosses it which used to be called Jabal al-ʿĀriḍ and today is called Mt. ʿAmmāriyya. There are only two approaches to it (the mountain), one at ʿAyniyya, the other at Dirʿiyya. They say there are 3,000 villages here. The people do not obey the Sharif. This mountain begins three stages north of Ḥijāz and extends northward beyond Najd al-ʿĀriḍ. The western face of the mountain stands like a white stone wall. The eastern face is sandy. Yamāma and Ḥajr are in the middle (of the area) to the east. From Yamāma to the rocky face (of the mountain) is two stages.

Sirrayn: Also in the middle of (Najd al-)ʿĀriḍ. In this mountain is a wadi called Wādī Banī Ḥanīfa with streams and trees and dates. It is a very beautiful place of excursion. Villages are strung around it. Above it is the town of Dirʿiyya and below it is Ḍubayʿ. The people here are called Banī Tamīm and their shaikhs are called Āl Murīd.

Dirʿiyya: It is on the road of the pilgrims from Laḥsā. Its wadi is very circuitous. The country of ʿĀriḍ is located here. Near it is Mt. Abū ʿAraf.

ʿAyniyya: A beautiful town, northwest of Dirʿiyya. It is famous for its grapes, peaches, and dates. The peaches here grow wild.

Malham: It is six stages from Laḥsā. This town is a very (beautiful) place of excursion. Sugar cane, dates, grapes, and peaches grow here.

Maqran: This town has seven earthen walls. [M 528] It is near Dirʿiyya.

Ḍubayʿ: A town south of Yamāma. It has abundant dates and agriculture and stock farming.

Ḥawḍiyya: It is between Dirʿiyya and Maqran. It has dates and agriculture.

Jazʿa: It is below ʿĀriḍ.

Bilād al-Washm: A district named after the mountain in the Washm wadi. This mountain is not very high and has no trees.

Marʾa: It is two stages from Dirʿiyya. It has water and dates.

Waqf: It is six miles to the northwest of Marʾa. It has spring water. It is two stages from Dirʿiyya.

ʿArama: At the foot of a mountain. It is two stages from the country of Dirʿiyya. It is five stages from Shaqrāfīd. It has its own wadi. Near it is:

Shuqayr: There is fruit here that grows wild.

al-Riyāḍ: This *district* is in the vicinity of Laḥsā and comprises towns and villages. One of the towns is Manfūja. It is six stages from Laḥsā.

Qaṣr: It is near Dirʿiyya.

Miʿkāl: It is east of Dirʿiyya, It is one stage from Yamāma.

Bilād al-Nakhlayn: This is named after a wadi. The upper part is called Jalājil, the lower part Nakhīl, and its mountain Sadīr. The mountain is also called Ṭuwayq. Nakhlayn is also called ʿAwda-i Sadīr. This is an independent district. Its town is called ʿAwda-i Sadīr. It has dates, stock farming and agriculture.

Qārtān: A fortress in a valley a half stage from al-Nakhlayn. It has dates and agriculture. There is a fortress there called Rawḍa which has dates, agriculture, and stock farming.

Tuwaym: A place outside the wadi. It has a fortress, dates, agriculture, and stock farming.

Jalājil: This is located at the uppermost part of the wadi.

Bilād al-Qaṣm: A region north of Najd. It is in a sandy plain below Wādīʾl-Sirr. It has no trees but it has a many date palms.

ʿUnayza: It is on the road of the pilgrims from Basra and is five stages from Dirʿiyya. It has two fortresses and excellent grain crops.

Madhnab (?): It is between Shuqayr and ʿUnayza (text has Shayqar and ʿAnaza). This is the country of al-Ḥamdiyya. It has springs and a fortress. It has many dates.

Wathāl (?): Nearby are springs. There is a place here called Ḍārij. It is one stage from ʿUnayza. It has flowing water and excellent agriculture. It has an earthen fortress.

Ḥamdiyya: It has a large saltpan but it is inactive.

Tanūma: It is north of ʿUnayza. It has an earthen fortress.

District of Mazj (?): It comprises the region of Yamāma. It is about fifteen stages in extent. The land is gentle and lovely. The soil is sand-colored. The crops are irrigated with waterwheels.

Yamāma: In ancient times it was called Jaw. It has a wadi called Kharj where there are many towns and villages. It is famous for its wheat and barley. Indeed there is a proverb: *laysa aṭyabu ṭaʿāman min ḥinṭatiʾl-yamāma wa lā ashaddu ḥalāwatan min tamrihā* ("There is no better food than the wheat of al-Yamāma and nothing sweeter than its dates"). Varieties of dates are grown in Yamāma, the most famous of which are *bardī* and *riqāʾ*. The water of Yamāma is very sweet. Indeed they say in their poetry: *araqqu min māʾ al-Yamāma* ("As pure as the water of al-Yamāma"). The many springs in the wadi of Yamāma are famous; they are ʿAyn Khaḍrāʾ, ʿAyn Hīt, and ʿAyn Rām. And there are three rivers: Jūdat Ḥajr, Shaikh al-Ghamm, and Shaikh al-Niʿām. These rivers rise from Mt. Rām.

Yamāma is a city in the desert. It is part of the country of ʿAwālī. It was named after a woman. The country of ʿAwālī is the country of Banī Ḥanīfa. Musaylima the Liar made his claim to prophethood here.

In Yamāma there is a wadi called Kharj. In ancient times this was the camping grounds of Ṭīm and Judays (Bedouin tribes). They were the descendants of Lāwaz b. Sām (= Shem son of Noah). They settled in Yamāma [M 529] and gradually multiplied. Someone named ʿAmlīq (Amalek) b. Ḥāsha from the Ṭīm became their ruler. He was arrogant and oppressive. He governed as he saw fit.

The following story is related about him: A man and his wife had an argument over their child. They brought their quarrel to ʿAmlīq to resolve it. The man's name was Qābis. He said, "Oh, king! I have fully paid the marriage portion for my wife. Apart from the child that she bore, I have had no benefit from her. How would you judge this?" His wife was called Hazīla. She said, "Oh king! I gave birth to this child after being pregnant for nine months. Then I was content to suckle him for twice that. I received no benefit from this. After the child was weaned my attachment to it was even greater, but my husband took it from me by force, which left me distraught." ʿAmlīq was dismayed (when he heard this) and ordered the child to be detained among his own slaves. When this happened Hazīla spoke angrily of ʿAmlīq. When ʿAmalīq heard her words, he was outraged. He ordered the Judaysites to bring their virgin daughters to him when they were about to get married; and after he took their virginity they would go to their husbands. Thus the daughters of Judays were disgraced and humiliated.

Subsequently, a Judaysite named Aswad b. ʿAffār had a virgin sister named ʿAfra whom he married off to someone. On her wedding night they brought ʿAfra to ʿAmlīq who took her virginity by force. When ʿAfra went outside, her brother Aswad addressed a group of people who had gathered there, saying, "What an injustice that the daughters of Judays are subjected to such despicable acts as brides!" When the tribes heard this their ardor was aroused and they immediately attacked ʿAmlīq and killed him. Then each member of the Judays tribe killed a nobleman from the Ṭīm tribe. Then they began to kill the other members of the Ṭīm tribe. During this time a member of the Ṭīm named Ribāḥ b. Murra fled. He joined Ḥassān b. Tubbaʿ al-Ḥimyarī. He said to Ḥassān, "We are your slaves and your subjects. The Judays tribe is our enemy." And he told him what had happened. When Ḥassān heard his story he attacked the Judays tribe with a huge army. Aswad, the above mentioned leader of the Judays, fled with several persons and hid. Ḥassān killed or took captive everyone else.

Around that time there was a woman from Yamāma who could recognize a man if she saw him a day's journey away. She would spread the alarm if the enemy were coming. Her name was Zarqāʾ. When Ḥassān decided to set out against the Judays, Ribāḥ b. Murra said to him, "Oh, king! I have a sister who is the wife of a member of the Judays. Her name is Zarqā'. She can make out an army a day's journey away. I am afraid that if she sees us she will warn the Judays. I recommend that you order your troops to take branches of trees in their hands and hold them in front of them. When Zarqā' sees this she will be in a quandary." This is what they did. As they neared Yamāma, Zarqā' looked and said, "Oh, people of Judays! The trees are marching toward you and the vanguard of the horsemen of the Himyarites." But the people of Judays accused Zarqā' of falsehood. They said, "Do trees ever march?" Finally the enemy arrived. They plundered and sacked the Judays. They captured Zarqā' and immediately gouged out her eyes. Then they crucified her. [M 530]

It is related that during the time of the Prophet, Musaylima the Liar appeared in this city and the people demanded that he perform a miracle. His miracle was to put an egg inside a bottle. The Banī Ḥanīfa were not very intelligent, so they followed Musaylima. Before Islam, during the time of the Jāhiliyya, the Banī Ḥanīfa took to themselves an idol made of honey and butter. Later some of them were overcome by hunger and ate the idol. The people mocked them and ridiculed their intelligence. A line of poetry was recited on this incident: *akalat ḥanīfatu rabbahā min al-naqmi wa'l-majā'a / laqad yaḥdharū min rabbihim siwā'l-'awāqibi wa atbā'a* ("Ḥanīfa ate its god out of revenge and hunger. / Perhaps they have been warned by their god of a similar end and its consequences"). How did he put an egg inside a bottle? If one soaks an egg for a while in vinegar it gets soft and can be put inside a bottle. Then one pours water over it and it returns to its original state. Musaylima was killed during the caliphate of Abū Bakr by Khālid b. al-Walīd.

The city of Yamāma is six stages from Laḥsā and three stages from Yabrīn. There is a famous villa in Yamāma called Dhāt al-Naw'. There is also a fortress there called Barmakiyya.

Ḥaḍruma: It is northeast of Yamāma. There is a plant here that is useful against the bite of a snake or scorpion.

Yabrīn: It is southwest of Laḥsā and is (in?) the land of Sanja. There are two villages. The air is very unhealthy, but its dates are excellent. They are like the *bardī* dates of Medina. There is always fever here.

Ḍubay'a: It is near Yamāma.

Ḥūṭa: It is south of Yamāma. It has many dates and stock farming. It is three stages from Salmā. It has abundant water, trees, and fruit.

Ḥajr: It is one stage from Yamāma. Yamāma and Ḥajr are the camping grounds of Ḥanīfa and some of Muḍar. The graves of those who died in battle against Musaylima during the caliphate of Abū Bakr are in Ḥajr.

Bār: A tract between the mountains of Yabrīn and Yemen. It was the camping grounds of the people of 'Ād. When God destroyed 'Ād, He gave it over to the jinn. No human can come near this place. Today the *nasnās* (a half-human creature of Arab folklore) lives there. In ancient times it was a very prosperous place with many streams and trees.

Bilād Shammar: A district northwest of Yamāma. It has many towns and villages. In this area are two mountains called Ajā and Salmā which are famous among the Ṭayy people. These two mountains are between Fayd and Ḥā'il. There is also another mountain called 'Awjān. As for Ajā, it was the name of a man. He was the lover of a woman named Salmā. The name of their go-between was a woman named 'Awjā(n). When their husbands learned of this affair they pursued them and killed all three. Each one was crucified on a mountain. Later these mountains were named after them.

Description of Bilād al-Jawf: A district between Syria, Ḥijāz, Iraq, and Najd. It is north of the district of Shammar. Between them is a mountain range called Raml Ḍāḥ. These mountains are of sand and the routes through them are difficult. Its fortress is made of white stone. The people are of the Qarūla tribe. One needs a guide to go to this region.

Subka: It is on a worthless stream at the foot of Manzil Ḍāḥī. It is southwest of Dūmā. There is a fortress here called Dūmat al-Jandal, named after Dūmān son of Ishmael who resided there. Jandal was the name of one of the companions of the Prophet. [M 531] He left his father, entered the service of the Prophet, and became a Muslim. There is a spring here called 'Ayn al-Tamr. There is another spring that rises from a square stone made of marble. Above it is a dome built by Dhū'l-Qarnayn the Himyarite. In 919 (1513) the Bedouin found a treasure in this district. It was composed of drops (i.e., pieces?) of pure silver.

Qarāqir: A watering place between Dūmā and Taymā'.

'Amrī: A watering place east of Azraq.

Shafār: It has many streams. (In order to get here, one goes) first through Wādī'l-Sirr, which is west of Qarāqir.

Qāra: An earthen fortress on a hill east of Dūma.

Sakāla: It is one stage from Qāra. It has an earthen fortress.

Shaqīqa: It is near Raml Ḍāḥ.

Jubba: It is one stage after entering Bilād Shammar. It is on the road from Syria to Najd. It has many streams.

Qalta: It is three stages from Taymā'. Between them is Wādī'l-Ṣawān. It has many streams.

Umm Barqaʿ: Its water is collected from rain. It is north of Azraq and is one of the districts of Lajjāʾ.

Ruwayshdāt: It is opposite Mafraq. Its water is collected from rain.

Ṣaryāt: It is near Qaryatayn. There are many terebinth (*buṭm*) trees here. The ground is rocky.

ʿUdhayb: The name of a stream on the road of the pilgrims from Kufa.

Bānāt: It has water and is one stage from Qaṭana.

Sajj and Kharīq: These two streams are brackish. They are one stage from Ḥāʾil.

Qayn: A halting place with water. It is 24 miles from Qarāqir.

Nabak: A halting place with water. It is 22 miles (from?). Those wishing to go to Najd stop here.

Qanāʾ. It is one stage from Muwaffaq. Wadi Samir is located in its wadi. Their crops are irrigated with water-wheels.

Ḥāʾil: It is north of Muwaffaq. It is on the road of the pilgrims from Baghdad.

Muwaffaq: It is five stages from ʿUlā.

Taymāʾ: A more built-up place than Tabūk. It has abundant dates. The Ṭayy tribe used to live here. This fortress belonged to Samawʾal b. ʿĀdiyā who was from the Ṭayy tribe. This place is also called Ḥiṣn Ablaq ("Piebald Fortress") because it was built of red and white stones. They have a proverbial saying: *awfā min al-Samawʾal* ("More loyal than al-Samawʾal"). The reason for this is as follows: Around the time that the father of Imruʾ al-Qays b. Ḥujr al-Kindī was killed in Anatolia (or Byzantium—Rūm), Imruʾ al-Qays went to the Byzantine emperor to discover who killed his father. En route he stopped at this fortress and entrusted his armor with Samawʾal. When Ḥarth b. Ẓālim al-Ghassānī learned of this he marched on Ḥiṣn Ablaq and demanded the armor from Samawʾal, who refused to give it to him. One day Samawʾal's son went hunting outside the citadel. Ḥarth b. Ẓālim captured him and brought him before the citadel and said, "Give me the armor or I will kill your son!" Samawʾal replied, "I do not go back on my word. Do what you will!" This Ẓālim then killed his son in front of him and departed. As a result of this event Samawʾal became proverbial for being true to one's word.

District of Sāʿida: Its most famous towns are Fayd and Thulayth.

Fayd: It is half way along the road of Iraqi pilgrims going from Kufa to Mecca. It is the place of Mt. Salmā of the Ṭayy Tribe. Pilgrims put some of their goods in safe keeping here. This town has abundant water and dates. Fayd was named after Fayd b. Ḥām (a son of Noah), who was the first person to live here. It is 12 stages from Kufa.

Thaʿlabiyya: A walled town that has much water. It is a third of the way on the road for the pilgrims from Iraq. [M 532]

Ghamr Marzūq: It is two stages from Fayd. Its water belongs to the Banī Asad.

Description of the Islands Located Around the Arabian Peninsula

Islands in the Sea of Suez (Red Sea)

1. Nuʿmān: It is off the coast of the fortress of Alzam.
2. Shaikh Muzīq: It is off the coast of the stage (probably meaning coastal stage for boats) of Ḥawrāʾ.
3. Quwayʿa: It is off the coast between Jār and Rābigh.
4. Qunfuda: It is near the port of Ḥalī. It is a prosperous island.
5. Sharja: It is off the north coast of Luḥayya.
6. Ḥurda: It is south of Sharja. It has an excellent harbor suitable for boats. They dive for pearls here.
7. Muḥāmila: It is opposite Bayt al-Faqīh al-Ṣaghīr.
8. Farasān: It is west of Luḥayya. They dive for pearls here.
9. Ḥudayda: It is off the west coast of Rafʿ.
10. Ṣalgha (?): It is south of Ḥudayda.
11. Kamarān: It is near the coast of Mocha. It is a famous and prosperous island.
12. Khartān and Martān: These two islands are opposite Jawn Ḥashīsh. They are off the coast of Shiḥr. A peculiar Arab people currently live there. They speak the language of the people of ʿĀd. Much fine ambergris washes ashore in this area.

Islands in the Persian Sea (Persian Gulf)

1. Musandam: It is in the strait of the Persian Sea. Ships travelling from Aden to Hormuz stop at this island. They can come from the district of Bawāṭin to this island in one day and can go from here to Hormuz in one day.
2. Bahrain (Baḥrayn): It is opposite Qaṭīf and to the east. Its city is in the middle of the island. There are some mountains in the interior. Its width is less than its length. Off the eastern side of Baḥrayn, off the middle of the island, is a small round island with a fortress.
3. Uwāl: It is off Qaṭīf. It is a day's sail from the coast to this island. There is an excellent fortress on the east coast of this island. There is pearl diving here. The diameter of this island is about two stages. It has three hundred villages. It has vineyards, dates, and bitter oranges. It has a plain and abundant arable land and pasture. There are many springs and much water on this island.

4. Kharg: It is near ʿAbbādān. It has a city and there is pearl diving. East of it is:
5. Kīsh: Also called Qīs. There is pearl diving at Kīsh. There are many dates and mountains and trees. The people irrigate from wells.

Description of the Arabian Peninsula

To summarize: This clime is extremely hot. But the air in the Sarawāt (the mountains above the Tihāma) is mild. Indeed, if someone in Ṣanʿāʾ cooks some meat but doesn't have a chance to eat it and goes on the pilgrimage, when he returns he will find that the meat has not changed. There are all kinds of fruit in the Sarawāt. In particular it has a fruit called *ʿanbā* (mango?) that has a fragrance like musk. If one or two are left somewhere they perfume the area around them.

Apart from Ghalāfiqa (a coastal town in the Tihāma of Yemen, half-way between Ḥudayda and Zabīd), the coasts are dry. The people of this clime wear the same clothes summer and winter. In the summer nights, the air in Mecca is excellent. In the Tihāma the air is oppressive and heavy (humid). During the nights in Oman something like grape molasses falls from the sky. The simoom blows between Mecca and Medina. In the commentary on the Koranic phrase *riḥlata'l-shitāʾi wa'l-ṣayfi* ("their summer and winter journeyings" Koran, 106:2) (it is noted) that the Quraysh spent the winter in Mecca and the summer in Ṭāʾif.

Most of the people in this clime are skinny and wear clothing made of cotton.

Religious groupings

The people of Mecca, Medina, Tihāma, Ṣanʿāʾ, and Fazaj (?) are Sunnis. The villages and towns around Ṣanʿāʾ, Oman, part of the Sarawāt, and the coastal areas of Mecca and Medina are Muʿtazilī. Most of the people in Ṣanʿāʾ and Ḥaḍramawt are Ḥanafīs. In some places there are Shāfiʿīs. In Maʿāfir (an ancient Arabian tribe who lived in Yemen) the rite of Ibn Mundhir[5] has spread. [M 533] In the districts of Najd al-Yaman, the sect of Sufyān (al-Thawrī) is well-known. There are also adherents of the Ḥanbalī rite. The Carmathian sect spread in Ḥajr. The Dāwūdī (Ẓāhirī) rite is found in Oman. The ruler of Yemen adheres to the Zaydī sect. They call him their imam. The Zaydīs trace their line to Zayd b. ʿAlī (the son-in-law of the Prophet)—God forbid that Zayd be associated with that group!

It is reported that before Islam the people of the Arabian Peninsula already adhered to some of the stipulations of the Muhammadan sacred law. For example, it was forbidden to marry one's mother or daughter and to marry two sisters (cf. Koran 4:22) and it was shameful to marry a woman who had been married to one's father. They made a pilgrimage to the Kaʿba like the pilgrimage of the Muslims. They had the same notions of ritual impurity and ablution, rinsing the mouth, snuffing water into the nostrils, cleansing oneself from physical pollution, using a toothpick, clipping the nails, circumcision, wearing the turban with the end under the chin, plucking the armpits, cutting off the right hand of a thief, the reckoning of years and days, and keeping time. Once every three years they had a leap month.

These people worshipped idols for the following reason. When the descendants of Ishmael multiplied they reached a point where they could no longer all fit in the territory of Mecca. So some of them set out to settle in another homeland. They spread throughout Arabia. Each group that left Mecca took a stone from the Holy Precinct with them. In the land where they settled they placed their stone in a clean place and then they visited that stone just as they would have visited the Noble House. Sometime later they abandoned the Scriptures of Abraham (cf. Koran 53:36–37). Thanks to Satan's inducement, they began to worship those stones. The first to order them to do this was ʿAmr b. Ṭayy al-Khuzāʿī. While passing through Balqāʾ he saw that the Amalekites who lived there worshipped idols. And other people had worshipped idols since the time of Noah. Indeed, the Koran records how they had persisted in going astray.

This situation pertained until the time of Abraham. After that the children of Ishmael [multiplied, etc.] as described above. One of them took the idol of Hubal and put it in the Kaʿba. These people began to worship idols from that time. A number of idols were famous among them. They gave each one of them an inappropriate name. These idols were: Aṣāf, which was on the top of Mt. Ṣafā; Nāʾila, on the top of Mt. Marwa; Wadd, in Dūmat al-Jandal and was the idol of Banī Kulayb; Suwāgh, the idol of Banī Huzayl; Yaghūth, the idol of Banī Mudlij; Nasr, the idol of Banī Dhū'l-Kilāʿ; Yaghūq, the idol of the Hamdānīs; Lāt, the idol of Banī Nafīf; Ghayrī, the idol of Banī Kināna; Hubal, the idol of the Aws and Khazraj. The many groups among the Arabs worshipped in different ways. The Sabians worshipped luminous celestial objects and others worshipped angels.

5 Correcting M Banī Munzir. The reference is to the Shāfiʿī scholar Abū Bakr Muḥammad b. Ibrāhīm b. al-Mundhir al-Naysābūrī.

Languages

The language of the people of this clime is Arabic. It is known in Aden, Jedda, and in the deserts of Fārs and Hind. The language of the people of Ṭaraf, Shiḥr, and Mahra is Himyarite. Their speech cannot be understood. The people of Aden pronounce the letter *jīm* as a *kāf*. For example, they say *rakab* instead of *rajab*, and *rakul* instead of *rajul*. Indeed, it is reported that when the Prophet wanted to use small stones to make himself ritually pure (when there was no water available) they brought him a ball of dung (by mistake) and he threw it away saying *hiya rakas*, meaning *hiya rajas* ("it is filth"). The language of Aḥqāf is crude but that of Huzayl is beautiful and it is today the language of Najd. There are all kinds of dialects spoken in Mecca.

Coinage

In this clime, in Yemen and Kawkabān, there is a mint. Their *akçes* are the same as Egyptian money. [M 534] Their silver however is base (i.e., not pure). They also mint copper coins, which have currency in Yemen, Shiḥr, and Najd. In Oman, different coins have currency. But goldpieces (i.e., ducats) and *sivilye* (Sevilla?—Spanish silver coins) are accepted everywhere. In Ḥijāz and Laḥsā, Ottoman coins are accepted.

Trade

This country has four ports: Maskat, Aden, Mocha, and Jedda. Inland the largest trading center is in Minā'. There is no other bazaar in the world as large as this one. The goods of the world come here and are sold here. The trade that is done in this clime is very profitable. Much flour comes from Ethiopia and other places. The most prized comes from Ethiopia. All kinds of aromatics and spices come here. From Oman come musk, saffron, log wood, teak, *sāsan*, ivory, pearls, brocade, onyx, rubies, ebony, coconuts, sugar, gum copal, aloe, iron, lead, cane, wet clay, sandal wood, crystal, pepper, Chinese cinnamon, cloves, indigo, aloes wood, all kinds of cloth and rarities and the like. From Aden come ambergris, syrup, and leopard skins. The ambergris washes ashore along the coast from Aden to Mocha and also at Zaylaʿ. If someone finds a piece of ambergris and takes it to the ruler of that area, he is given a gold coin and a piece of cloth. Ambergris washes ashore when the southern winds blow. From Ethiopia come flour and eunuchs. From Jedda come coffee and marbled carnelian. There are several kinds of marbled carnelian. It comes from carnelian mines. The most prized is *baqirān* marbling. There is also carnelian in ʿArwān, Fārs, and Ethiopia. There are also honeyed and sudorific medicaments, and storax which falls like manna on the Yemeni turmeric trees.

Specialties of Yemen

The coffee of Yemen: According to the report of Hezarfen Ḥüseyin Efendi, coffee is grown in two districts (*nāhiye*). One is the mountains above Zabīd. They call this district, which is opposite Bayt al-Faqīh, Awṣāb. The other one is the districts near the port of Jazān in the district of Nahārī. In both of these places the ground is steep and broken. Coffee trees are planted here in rows. The coffee tree resembles the cherry tree, but its color is darker green and its leaves are somewhat thicker. The height of these trees is about eight cubits. Some trees live for 20–30 years. The flower is white. The stems sprout in twos and threes and are rather long like those of the cherry tree. Its fruit is exactly like the cherry. While green it is somewhat acrid. Later it turns red and becomes a bit sour. After it is ripe it becomes dark red like the sour cherry, large, and sweet. If they are picked with the stems on and mixed with Ḥiṣār cherries (the famous cherries from Rumeli Ḥiṣārı in Istanbul) one cannot tell them apart. While eating them one can only tell the difference by their aroma and double seed. In fact, the coffee bean is even more delicious than the cherry.

They pick the beans while green and spread them over the roofs of the houses which are flat and plastered with mortar. Although they are not ripe, they quickly dry and turn black. Beans at this stage are called "acrid" and "unripe." Then they take the beans to the mill where the husks are broken, winnowed and discarded, leaving the kernels behind. This is the coffee that is taken to Anatolia and other countries and sold there. They do not take ripe beans to the mill but remove the kernels by hand. They dry the decoction like the *bağıryasdı* (?) grape. [M 535] The people of Yemen boil the husks and drink the broth in order to keep cool on hot summer days. It is very tasty. Those who drink it think it is sweetened sherbet. The husks do not go to Anatolia because in Yemen they are sold for more than the kernels.

The best coffee is (when the beans are) plump and greenish. By nature (according to the Galenic theory of medicine) the husks are hot and moist in the first degree. If one drinks the broth of the husks in summer, it acts as a stimulant. If drunk in the morning it eliminates indolence. As for the kernels, they are by nature moist in the first degree like wild chicory and (the broth) also acts as a stimulant. It is very good for fatigue. Coffee is not good if roasted too much; it loses its qualities. It is beneficial if drunk one hour after eating. It eliminates the moistures and excesses in the stomach. It is also beneficial for headaches and colds and good for those who do not want to sleep.

Every year 80,000 bags of coffee are produced, 40,000 of which go to Jedda. The remainder go to Basra and other places. Every bag is three quintals; four quintals being ten batmans or one Syrian quintal.

The first appearance of coffee: Coffee houses appeared in Anatolia and Istanbul in 962 (1554–1555). A man from Syria arrived and opened a coffee house around Taḥtelḳalʿe (Tahtakale in Istanbul). Many of the learned, qadis, refined people seeking pleasure, professors (from madrasas), and mystically inclined dervishes came here. When generous people went to the coffee house they spent as much as they liked in accordance with their status. Gradually this practice spread everywhere. It reached the point that the town wits called the places where the great pleasure seekers gathered *mekteb-i ʿirfān* ("school of culture"). Even the aristocracy, those of high rank and position, began to frequent the coffee houses.

Reason for the appearance of coffee: In 656 (1258) Shaikh Abū'l-Ḥasan Shādhilī made the pilgrimage from the Maghreb. When he was six stages from Mt. Zumrud on the road to Sawākin and between Mt. Abr (or Abrak?) and Mt. ʿAjīn, which was six stages from Abr (or Abrak?), he said to his disciple Shaikh ʿUmar, "I am going to die here. My instructions to you are as follows: (after I die) a man will soon come whose face is covered. Whatever he does or says, you do the same." So, it happened that Shaikh Shādhilī died and the man whose face was covered appeared. They dug a little at that place and, by God's command, water emerged. With this water they washed Shaikh Shādhilī's body and buried him. When the man (whose face was covered) was about to leave, Shaikh ʿUmar grasped the skirt of his robe and said, "Who are you? Tell me!" When he insisted, the man raised the covering from his face. He saw that it was his own shaikh, Shaikh Shādhilī himself. Afterwards Shaikh Shādhilī gave Shaikh ʿUmar a wooden ball with the following instruction, "Wherever this ball stops spinning, you reside there." Shaikh ʿUmar followed it to Sawākin. When he arrived, the ball continued to move. He went by ship to Mocha where he noticed that the ball had stopped. He immediately made a hut from reeds and settled there. [M 536] He dug a well and delightful sweet water emerged. Previously there had been no water in Mocha and it had to be brought in from a distance. Subsequently the people of Mocha were afflicted with an illness. They brought those afflicted to Shaikh ʿUmar who prayed for them and, by God's command, they were cured.

Around the same time the very beautiful daughter of the ruler of Mocha also became ill. They brought her to Shaikh ʿUmar and she remained with him for several days. He recited (the Koran) for her and prayed, and she too recovered her health (text: *afāt*, error for *ifāḳat*). (Because the girl remained with Shaikh ʿUmar for some time) the people began to say unseemly things about him. They said, "Is it possible for the ruler's beloved and beautiful daughter to stay with him this many days and for him not to put a hand on her, not to violate her?" When this gossip reached the ruler he was shamed by it. He ordered that Shaikh ʿUmar be banished to Mt. Awṣāb. So they banished him there with several disciples. When they arrived at this place, the shaikh and his disciples found nothing to eat but coffee. Finally they put some coffee in a pot, boiled it, and drank the broth.

Around that time the people of Mocha were afflicted with scabies. Several of the shaikh's friends in Mocha went to visit him on Mt. Awṣāb. As soon as they drank the coffee broth they were relieved of the scabies. When they returned to Mocha, the people of the city asked them how they were cured. They said, "We went to the shaikh and drank some broth and our illness disappeared." Word of this spread among the people. When the ruler learned of it, he invited Shaikh ʿUmar back to Mocha and showed him great respect. He built the shaikh a magnificent lodge. That lodge still stands and the ball is inside.

The shaikh later married and had a son named Abū'l-Futūḥ. When he grew up and became a young man, the shaikh gave him the following charge: "Go to Sawākin. In the place where Shaikh ʿUmar rested for a few days you should take up residence." Today there is a magnificent lodge in Sawākin. Up to now all the shaikhs there trace their descent from him.

Specialties

- Leather of Zabīd; also its indigo, which is like azure.
- Floor coverings (?—*mişedd*) of Hijr.
- Gunpowder of Saḥūl and Jarīb.
- Leather mats of Ṣaʿda; also its villages (?).
- Leather of Yemen; also its unrivaled Moroccan leather.
- Carnelian of Ṣanʿāʾ.
- Drinking cups of Ḥalī.
- Henna of Yanbu.
- Copper sulfide of Marwa.
- Hemp of Mahra; also its she-camels.
- Turmeric of Aden.
- Aloe of Socotra.
- MQS (?) of Oman.
- Pearls of the island of Kharg. Around this island divers bring up pearl-bearing shells for a fee. Large and small pearls emerge from these shells. The pearls of this

region are famous. Sometimes a solitaire (or unique pearl) emerges. But there is a kind of fish in these waters that attacks the divers, aiming for their eyes. The divers are very wary of them.

Carnelian mines are in Ṣanʿāʾ. Whoever wishes can buy a piece of land in Ṣanʿāʾ and start to dig. If he is lucky he will find something. Sometimes he will find large pieces of carnelian and sometimes nothing.

They say that there are gold mines in Yanbu and in the plain of Marwa.

They say that there are absolutely no fleas in the noble Kaʿba.

Fairs of the Arabs

The Arabs had large fairs at certain places in the time before Islam. [M 537] They were as follows:

- Dūmat al-Jandal: It opened at the beginning of Rabīʿ I and continued until the middle of the month. The people traded in the following manner: If someone saw a garment that he liked, he would toss a small stone on it. By doing so the sale had to be made.
- Mushaqqar: It was set up at the beginning of Jumādā II. At this fair the goods were handled to make sure they were not defrauded.
- Ṣuḥār: It was set up on the tenth of Rajab. At this fair the goods were placed in rows.
- Shaḥr: It was set up in the middle of Shaʿbān. Here too they tossed stones (on the goods for sale).
- ʿIlān: It ran from the beginning of Ramadan until the tenth.
- Ṣanʿāʾ: It was set up in the middle of Ramadan.
- Rābiyya: It was in Ḥaḍramawt.
- ʿUkāẓ: It was set up in ʿUkāẓ, which is near ʿArafāt in the direction of Najd. These two fairs (this one and Rābiyya) were set up on the same day in the middle of Dhū'l-Qaʿda. ʿUkāẓ was the biggest of all the fairs. It ran until the end of the month.
- Naṭāṭ: It was the fair of Baḥīra and Ḥijr which are in Yamāma. This fair ran from the beginning of Muḥarram until the tenth.

Customs

In Yemen, Aden, and Ḥijāz, they perform ritual prayer after a complete reading of the Koran in Ramadan.

They light lamps with fish oil that comes from Mahra.

At the Kaʿba on the 21st of Ramadan the Ḥanbalī imam does the *terāvīḥ* (prayer reformed during the nights of Ramadan) following a complete reading of the Koran. The Shāfiʿī imam does the same thing on the 25th of Ramadan, the Ḥanafī imam does it on the 27th and the Mālikī imam does it on the 29th. At (the end of) each complete reading of the Koran, large lamps are lit and people assemble.

During the last ten days of Ramadan they assemble on a different night for each of the recently deceased Sharifs and light lamps and distribute sweetmeats. For each famous shaikh who dies at Mecca, they assemble each year on the anniversary of his death and the shaikh's successor gives a feast.

In Mecca and in Aden the roofs are decorated two days before the beginning of Ramadan. The people decorate the mountains opposite the Kaʿba (i.e., Ṣafā and Marwa?) and light lamps. They make merriment there with music and singing. At the beginning of Ramadan, people gather in groups and stroll around until *saḥūr* (the meal before daybreak during Ramadan) and recite *qaṣīdas*.

At Mecca they light lamps for six months from the beginning of Rajab until Muḥarram. People go to coffee houses and spend time in conversation until morning. In particular on the night of the breaking of the fast, they decorate the bazaar between Ṣafā and Marwa. They make great merriment and bands play.

From the beginning of Rajab until Muḥarram, the road to ʿUmra is very busy. The people from Mecca and elsewhere go to ʿUmra. The distance to ʿUmra is two hours from Mecca. During the month of Rajab the people of Mecca and from elsewhere go to Medina. They call this caravan the caravan of the Rajabiyīn.

There is a marvelous custom in Mecca. It has 24 Friday preachers. Eight of them are outstanding, four are Ḥanafīs and four are Shāfiʿīs. Once every eight years one of them takes a turn giving the sermon on ʿĪd al-Fiṭr (the feast of the breaking of the fast at the end of Ramadan). The preacher whose turn it is to give the sermon begins to make preparations as ʿĪd al-Fiṭr approaches. They make candy from sugar, almonds, roasted chickpeas, and hazelnuts and put them in plates. When the night of breaking the fast arrives, the Sharifs, religious scholars, and leading men and dignitaries of Mecca [M 538] go in groups to the home of that preacher. The preacher presents them with plates of various candies. He even presents them with candy on which there are tablets of ambergris. In addition to what they eat, those who come fill a handkerchief with a plate-full of candies to take with them. This is an old custom. Even those who come with them (servants?) fill a handkerchief. After the morning prayer, these same persons come again to the preacher's house and line up before him. They form a procession, two by two, each row taking a flag, because the preachers of Mecca place two flags on the *minbar*. After the holiday sermon is read, they again line up in front of the preacher and escort him home.

The preacher then gives them a great banquet. After the meal they all disperse. Then the preacher goes to the Sharif and they exchange holiday wishes. The Sharif gives him a present of 40, 50 or more goldpieces. He goes to the qadi of the Kaʿba, the military commander, the Sharifs, religious scholars and leading men. Each gives him a present according to his status and capacity.

Water

The water in this clime varies. The water of Aden, Mecca, Zabīd, and Yathrib is light. The water from ʿAyn Zarqāʾ is the lightest of all. The water of Ghalāfiqa is sweet but that of Quzaḥ and Yanbu is bad.

Animals

In this clime there are all kinds of animals. Monkeys are mainly in Yemen. *Nasnās* are found in uncultivated areas (reading *arḍ barrīya* for *arḍ wa bār*).

The tithe is collected in the Arabian Peninsula.

In the past there were three rulers of Yemen. One governed Jund and its surrounding area, one governed Ṣanʿāʾ and its surrounding area, and one governed Ḥaḍramawt and its surrounding area. A governor lived in each of these places.

Income

In the past (the expenditure) for Mecca and Medina was 100,000 goldpieces (i.e., ducats?). The income from Yemen was 600,000 goldpieces, from Yamāma and Bahrain 500,000 goldpieces, and from Oman 300,000 goldpieces.

Stages and Roads of the Arabian Peninsula

The Kaʿba is located in the middle of the peninsula. All people are obliged to make a pilgrimage there. According to the Koran *wa-li'-llāhi ʿalā al-nāsi ḥijju'l-bayt* ("pilgrimage to the House [of God] is a duty for mankind," Koran 3:96). Throughout the peninsula are places and residences where prophets and the devout (ancestors) lived. One side of the peninsula faces the Holy Land, which is Syria, where the great apostles and prophets came from and lived. Another side faces Basra and Baghdad, citadel of the saints. Another side faces Yemen, about which there are many Prophetic hadiths.

The stages that one follows to reach Mecca from throughout the peninsula are as follows:

[First road: Damascus to Mecca] We begin with the Syrian pilgrims who set out from Damascus. After assembling in Damascus they go in a great procession with the commander of the pilgrimage to Qubbat al-Ḥāj on the 15th of Shawwāl.

- Kiswa: Here many groups join them from behind and all meet at Muzayrib.
- Inn of Dhū'l-Nūn: Here *tarḫāna* (soup made of dried curds and flour) is cooked and distributed, paid for by the endowment of Ibn al-Ḥuṣnī, and a toll is collected from the Muslim pilgrims.
- Ṣanamayn: On the way they pass through Khān al-Zayt. Ṣanamayn is a village of the Ḳavāsoğlı Turkmen. It has water and all kinds of birds. The creature known as leech is collected from the rushes and taken to Damascus. This is a well known practice.
- Tall Firʿawn.
- Ghabāghib: Here Sultan Selīm built a tower. Descendants of the Ḳavāsoğlı come to the tower and wait for the pilgrims. Their posts stretch from Dhū'l-Nūn to Ṣanamayn.

From Ṣanamayn the pilgrims go to Muzayrib. En route they pass by the stream of Dayla [M 539] which is spring-fed. Here the commander of the pilgrimage makes the pilgrims halt.

- Muzayrib: A spring in the province of Ḥawrān. Sultan Selīm built a fortress here. When the pilgrims arrive they set up a bazaar. Pilgrims halt here 5–10 days or more. Southwest of here is Katība, which has many springs and streams, but is in the wrong direction.
- Azriʿāt: A village half a stage distant (from Muzayrib) in the direction of Ḥawrān. Its water comes from wells.
- Mafraq: A waterless flat place. It is southwest of Ṣalkhad. Out of fear of floods, pilgrims sometimes stop at Azriʿāt instead of here.
- Zarqā: A place with streams.
- Azraq: Name of a ruined fortress. It has dates and water. Zarqā is a day's journey to the northeast.
- ʿUmarī: A place with water. It is on the road to Dūmā toward the east. It has a stream that runs two mills. This water comes from ʿAmmān and flows toward Ghawr (in Palestine).
- Balqā: Also called Maştā and Balāṭ. Pilgrims pass through seven steep mountain roads (passes).
- Qaṭrānī: Sultan Süleymān built a fortress here. The Sultan spent 15,000 florins to clean out its pond which had been filled with dirt.
- Lajjūn: Name of a bridge west of Qaṭrānī. Pilgrims sometimes come here for water.
- Ḥisā: This stage has water. It is northeast of Shawbak, from where the pilgrims must get grain supplies. Shawbak is in the mountains and has streams and pas-

tures. On the mountain to the west it has an inn and a spring, and nearby it has a large bridge.

> Ẓahr ʿUnayza: The road here twists and turns. The orchards of Shawbak fortress can be seen from ʿUna[y]za.

> Fortress of Maʿān: In the past this was an abode of the Umayyads. It belongs to the district of Sharāh. Sultan Süleymān built a fortress here. They discovered water and caused it to flow, but it was not good water.

> Ẓahr al-ʿAqaba: Also called ʿAbbādān. It has no water. It has *ṭabīliyyāt* dates. The saying *laysa warāʾ ʿabbādān qarya* ("There is no village beyond ʿAbbādān") must refer to this place.

> Dhāt al-Ḥajj: Also called Ḥijr. Sultan Süleymān built a fortress here. There are many wild dates. They dig for water here. Now they have started a stream which the people of the fortress use to irrigate their orchards.

> Qāʿ al-Basīṭ: Also called ʿArāʾid. It is a sandy place. There is a high hill in the center. The Bedouin call it Sharūrā.

> Tabūk: Described above.

> Maghāʾir al-Qalandariyya: A small hill with no water.

> Ukhayḍir: Also described above.

> Birka Muʿaẓẓama: Also described above.

> Maghārish al-Zīr: Also called Aqraḥ. It is a half stage (from Birka Muʿaẓẓama).

> Jabal Alṭāf: It is the place where the camel of (the pre-Islamic prophet) Ṣāliḥ was killed. The name of this place is Muzaḥḥam. After passing through this place, pilgrims turn east and continue through Mabrak al-Nāqa, then camp in Ḥijr. On a mountain is Qurā Ṣāliḥ ("Villages of Ṣāliḥ") where the houses are of cut stone. It is reached through sand dunes. There are many wells, but it is forbidden to drink from them. It is one stage from Ḥijr to ʿUlā where there is a mountain called Anān.

> ʿUlā: Described above.

> Qaṭrān: Also called Ṭawāmīr. This is a steep and difficult place to reach. It is located among black rocks. It has no water.

> Shiʿb al-Niʿām: They get their water from rain. Nearby there is water on both sides of the road. They bring water from there.

> Hudayya: They dig for water here. The water gives the people diarrhea. This is because they grow Meccan senna here.

> Faḥlatayn: Two small hills with no water. Nearby is a high mountain. There is (also a place called?) Waqf Ḥiṣār.

> Wādī al-Qurā: There are no streams here, however [M 540] there are many trees growing.

> Abyār Ḥamza: There is a mosque of the Prophet. From here the pilgrims continue to Medina. They rest here two days and then go to Abyār ʿAlī.

> Abyār ʿAlī: There are wells there named after ʿAlī. They are three fathoms deep. Here the pilgrims put on their pilgrimage garments.

> Qubūr al-Shuhadāʾ ("Graves of the Martyrs"): Its water comes from rain. (When it rains) a lake is formed. Sometimes there is no water. It is a place between two mountains.

> Jadīda: A village with streams, orchards, and gardens. There are many dates. It has watermelon and orchards. It is on a small hill between two mountains.

> Badr: A place with streams, orchards, and gardens.

> Wādī Ḥafr: It is on the road going from Jadīda to Badr. There are a number of prosperous villages here.

> Qāʿ al-Buzūh: A sandy plain. There is no water.

> Rābigh: A sandy plain which has wells near the sea. The water emerges from under the sand. If one digs down one cubit he will find water. There are dates and gardens. There are fish, fodder, and sheep. Two clans live here: Mawālī Rawī and Rūma. They call their country Rawī Rawāyā. The pilgrimage caravans pay them a tax. There is another clan called Rawī Jumāʿ.

> Ṭārif: To the left is Bilād al-Ṭārif. It has mountains, villages, and arable lands. It is a prosperous place. It has almond trees. They bring pilgrims butter, yogurt, and grain.

> Khulayṣ: The road is flat until it comes to the place called ʿAqabat al-Suwayq.

> ʿAqabat al-Suwayq: A very steep place. It has gardens and a lake with flowing water. The Zabīd Bedouin live here. They are the most insolent of the Bedouin. Every year they receive 700 goldpieces in subsidies from the Commander of the Hajj. There is a well of sweet water here.

> ʿUsfān: It has a well that has filled with sediment. There are many ruins here. Today it is called Madraj ʿUthmān.

> Baṭn Murr: It is east of ʿUsfān.

> Abū ʿUrwa: A village with streams, orchards, and gardens. It is west of the road at the foot of the mountain. From here one can go to Mecca. Or one can go from ʿUsfān to al-Barqā, from there to Murr al-Ẓahrān, also called Wādīʾl-Murr, and from there to the Kaʿba.

In the past there was another road that went from Damascus to ʿUlā. That road, to the east of the present road, went by stages from Damascus to Buṣra, then by stages to the fortress of Azraq. They are places with water.

> Qarāqir: Places with water, three stages (from Azraq). It has dates.

> Qulta: Two stages (from Qarāqir). From there it is another two stages to a place with water called Ṣubayḥa.
> Tīmāya: Three stages (from Ṣubayḥa).
> Wādī'l-Ṣawān: From here it is four stages to ʿUlā. The road is straight, but there is little water. There are sixteen stages, one every 27 miles.

There is another road from Mecca to Medina. The people from Medina use this road. It goes via Ābār ʿAlī.

> Samḥān: A place between two mountains.
> Mafraḥ: A small mountain. The camel drivers have their own fee which they collect when they get here. From here one passes through Quraysh and then comes to Qubūr al-Shuhadā'.
> Rawḥā: There is a deep well here.
> Shiʿb: It is between two mountains. There is water on the side of the mountain.
> Nāziya: The roads from Medina to Nāziya are hazardous.
> Ḥanīf Banī ʿAmr: A place in a long valley in the middle of two mountains. There are villages, houses, and date groves. It has streams and wells. This is the abode of the Banī ʿAmr and Banī Sālim. Basically Ḥanīf Banī ʿAmr are about 1,000 individuals. They are archers. They are famous among the Bedouin for their wealth.
> Wādī'l-Ṣighar: A place with dates at the foot of a mountain. It has streams and wells. The water from this place is sweeter than ʿAyn Ḥanīf and Badr. [M 541] A group of Zaydī *ashrāf* (descendants of the Prophet) live here. They guard the caravans, from which they collect a fee.

Second Road: Cairo to Mecca. The noble *maḥmal* sets out with the commander of the pilgrimage in a great procession from Cairo and goes to Birkat al-Ḥājj (outside Cairo). Then the road goes to the following:

> Hadfat al-Buwayb: A narrow place between two mountains. There is a pinnacle here and to the right is a long hill.
> Ḥamrā': There are cisterns here and a fountain has been built for pilgrims.
> Nakhīl Ghānim.
> ʿAjrūd Birkesi ("Pond of ʿAjrūd"—i.e., Birkat ʿAjrūd). After this point is the first watering place where there is sweet water that sometimes flows into the wadi. (The Mamluk Sultan) Qānṣawh al-Ghawrī has an inn there with three fountains. The port of Suez is located directly opposite. ʿUyūn Mūsā ("the Springs of Moses") are located in that direction.
> Munṣaraf: One stage (from Birkat ʿAjrūd). There are low lying areas here. Some earlier kings carried out excavations here in order to connect the Sea of Suez with the Mediterranean. They say that traces of those excavations are still to be seen.
> al-Qubaybāt: Here there are sand hills like domes. Before it lies al-Tīh, the wilderness of the Israelites (Sinai desert). To the right is Mt. Sinai and to the left is (the road to) ʿArīsh. This is a vast empty space, 40 parasangs in length and 40 parasanges in breadth. Crossing it is difficult. In the winter it is cold and in the summer it is extremely hot and there is no water. Here the Israelites wandered in confusion for about 40 years in an area covering about two stages.
> Wasaṭ al-Tīh ("Middle of Sinai"): Also called Rawḍ al-Jamal ("Camel Garden").
> Baṭn al-Nakhl ("Date Valley"): Also called Wādī Tajr. It has a spring. (The Mamluk sultan) Ashraf Qānṣawh (al-Ghawrī) built a citadel here. It has a fountain fed from a well. The guards in the citadel protect the water from the Bedouin. ʿAlī Pasha, the *beylerbey* of Egypt, enlarged both the citadel and the fountain.
> Wādī'l-Ghaymā.
> Wādī'l-Qurayḍ.
> Abyār al-ʿUlā': It is reached via a downward slope. It is a vast plain. There are two wells here, one belonging to Yīra and the other ʿUlānī. There is also a basin here that is filled by rain water.
> Malāḥa: It is near a place called ʿArāqib al-Bughla.
> Ra's al-Rakb (al-Rukub?): It is near Jifārāt.
> Saṭḥ al-ʿAqaba: This is the ʿAqaba (or "Pass") of Ayla. In the past this was a great city but today it is in ruins. A mile from here is a well built of masonry. The water is sweet. There are dates. A clan of the Ḥawīṭāt Bedouin lives here.

The next stage is one fourth the way (to Mecca). In this first quarter the water is sweet and plentiful. All (the stages) are along the seashore. To the left lies Mt. Sinai at a distance of several miles. At the end of the road are two descending slopes and a narrow defile and low-lying areas with wells of sweet water.

> Ẓahr al-Ḥimār ("Back of the Donkey"): The ascending road to reach here is rocky.
> Jurfayn ("Two Cliffs").
> Shurfat Banī ʿAṭiyya: there is much firewood here.
> Muṭallāt: It is between two mountains. The Banī Lām guard this place.
> Maghārat Shuʿayb ("Cavern of Shuʿayb"): In the low-lying areas there are streams of sweet water and many date groves, tamarisk trees, and false bdellium trees. There are inscribed panels here like (those at) the pyramids (text: *berāyā*, error for *berābī*) near the Nile. The names of kings are carved on them.

> Qabr al-Ṭawāshī ("Tomb of the Eunuch")
> ʿUyūn al-Qaṣab ("Reed Springs"): A place with water and reeds. It is in an extremely hot wadi. In summer many people drop dead from the heat. The tomb of one of the sons of Abraham is near the sea and a site of visitation.
> Sharm: It is near the sea. To the right is a mountain called Ishāra.
> Muwaylaḥa: It is on the coast of the sea. Its water is somewhat bitter.
> Dār Qaytbay: This place is called Dār Qaytbay ("Abode of Qaytbay") because he spent the night here when he made the pilgrimage. They used to halt at Baṭn Kibrīt ("Sulphur Valley") which is before Dār Qaytbay. This stage [M 542] is in a narrow rocky area.
> Tomb of Shaikh al-Kifāfī Mazrūq: A site of visitation.
> Azlam: This place is at the second quarter of the distance to Mecca (i.e., it is half way between Cairo and Mecca). It is a barren place without pasture. Its water is salty. It is a desert among the mountains. Meccan senna grows here
> Summāq, also called Rakhānīn: A wadi full of thorny plants. After passing through here one comes to a place called ʿAntar's Stable.
> ʿAntar's Stable (Iṣṭabl ʿAntar): A plain in the mountains. The toothbrush tree grows here. There is sweet water around this place.
> Sharnaba: It is the shoulder of a mountain.
> al-Wajh: A wadi with wells of sweet water. (The Ottoman Grand Vizier) İbrāhīm Pasha rebuilt the wells here in 931 (1524). The water comes from rain and floods.
> Biʾr al-Qurwā.
> Ḥarīra.
> al-Ḥawrā: It has brackish water.
> ʿAqīq.
> Ṣaḥn Bayāḍ: A circular (?—*degirmi*) place with many sand dunes and white vipers.
> Nabʿā Fuqāʿ, also called Ḥijār: It has sweet water.
> Ṭarāṭīr Rāʿī.
> Wādīʾl-Nār: A rocky and sandy wadi in the mountains. It is known as Yedi Vaʿr because the road passes seven large rocks.
> Khuḍayrā: A village in the district of Yanbu.
> Jabal Aḥmar.
> Wādī Tamā.
> Jabal al-Zayna: It overlooks Yanbu. The governor of Yanbu comes here. They make the noble *maḥmal* halt here. They spread a carpet before the camel (carrying the *maḥmal*) and the governor of Yanbu performs two prostrations on it.
> Yanbu: There are a number of springs here. It is described above.
> ʿUdaybiyya: A village.
> Dahnā: A town.
> Wāsiṭ: At this stage the pilgrims light lamps and fire cannons.
> Badr: Described above.
> Khabab al-Baza: A wide plain.
> Ghayqa: A place away from the coast.
> ʿAqabat Waddān.
> Rābigh: Here the pilgrims put on their pilgrimage garments.
> Juḥfa: A place nearby. Also called Muhayʿa.

Third Road: Aden to Mecca. Aden > Laḥij > Bukur > Taʿizz > Karābja > Wādīʾl-Ḥusnā.

> Ḥays. In the past, when Yemen was under Ottoman control, the noble *maḥmal* departed from here.

Zabīd: Mocha is to the southwest. There are two stages between them. One is Muwashakh. From there it goes to Zabīd.

Rafʿ: The first stage for the pilgrims of Zabīd. > Bayt al-Faqīh al-Ṣaghīr > Qaṭīʿ > Manṣūriyya > Qalʿat Marāwiʿ > Ghānimiyya.

> Bayt al-Faqīh al-Kabīr. Sometimes the pilgrims from Ṣanʿāʾ come here. It is five stages from Ṣanʿāʾ: > Ḥabāba > Ṭawīla > Bilād Banīʾl-Khayyāt > Bilād Banīʾl-Ahīl > Bayt al-Faqīh.
> Mawr > Maʿūliyya > Dūma > Ḥayrān > ʿĀliyya > Abū ʿArīsh > Salāma > Baysh > Namāwa > Atūd > Shaqīq > Abyār.
> Dahbān: A wadi where false bdellium (*muql*) grows.
> Bark: It was built by the Rasulid ruler ʿUmar b. Manṣūr (apparently al-Malik al-Manṣūr ʿUmar, r. 1229–1250).
> Shafaqa.
> Qanūna: Also called Wādīyayn. It is a stage with water; Layth.
> Haṣum: It has water.
> Saʿdiyya.
> Mīqāt: The people of the Tihāma of Yemen live near here.
> Yalamlam: 18 miles from Mīqāt.
> Aydām: One stage from Bayḍāʾ. It belongs to Mīqāt and has water.
> Bayḍāʾ.
> Umm Qurayn: A stage with water.
> The Kaʿba.

Another Road: The pilgrims from Aden and Ṣanʿāʾ take the road that goes through Najd al-Yaman. It goes north from Aden to Ṣanʿāʾ along the foot of the mountains. The pilgrims from Ṣanʿāʾ go via Ṣaʿda [M 543] to Ṭāʾif and Mecca; or from Aden via Bilād Makr; then Qaʿṭaba, Ḥuqb, Dummat, Shalāla, Yarīma, Zammāra, Zarrāja, Khazafa, Siyyān, and Ṣanʿāʾ.

The pilgrims from Shiḥr who take the land route via Ṣanʿāʾ pass through Ḥaḍramawt which is five stages from Shiḥr and four stages from Ṣanʿāʾ. For pilgrims who travel by land to Ẓafār it is 15 stages to Ṣanʿāʾ. Pilgrims from all these places assemble in Ṣanʿāʾ, then proceed to Zayda, Ayāfath, Khaywān, Aʿmashiyya, Ṣaʿda, ʿUrfa, Hijra, Sarūrāj, Thujja, Kuthba, Abyāt, Jarm, Ḥasana, Nīsha, Niyāla, Wabīh, Kadmī, Ṣughar, Būya, ʿAtq, Jadr, ʿUmar, and finally the Kaʿba.

Fourth Road: Oman to Mecca. (Pilgrims take the road that goes from Oman through) Ḥiṣār, Nazwiyya, ʿAjala, ʿAḍūh, and Biʾr al-Salāḥ. From there it is 21 stages to the Kaʿba. Water is found at only four of these stages, and eight are just sand. For this reason the pilgrims from Oman go to the Kaʿba by sea.

Fifth Road: Laḥsāʾ to Mecca. Laḥsāʾ, Jūda, Ḍān, Dahnā, Daḥl Mayy, Jabal Abū ʿArna, Malham, Jaghr, Rabaḍ, Dirʿiyya, Ḥīsiyya, Mirāh, Shuʿur, Janābikh, Marqab, Munkhanā, ʿAbla, Zakiyya, Dhāt ʿIrq, and finally the Kaʿba.

Sixth Road: Basra to Mecca. Basra.

> Dirhamiyya: Also called Old Basra.
> Ṣughwān.
> Juhra: Two stages, near the shore (?).
> Aṣāfa or Ḥafr: Its wells are more than 30 fathoms deep. In this direction is Ḥisn Banī Mūsā.
> Māwiyya: It has water.
> ʿĀlij Ṣaghīr.

Here the road divides in two, the summer road and the winter road.

– Summer road: > Dujānī 2 stages, with water nearby > Ṣadīr, with water > Washm > Sirr > Jumāna 3 stages, with water > Marrān 4 stages.
– Winter road: > ʿĀlij Kabīr > Tanūma > Quṣam Numayrī 2 stages, with water > Rass, with water > Ḍariyya, with water > Qubā 4 stages > Marrān, with water.

(From Marrān) > Wajra.

> Dhāt ʿIrq: Here pilgrims from Najd put on their pilgrimage garments.
> Bustān Banī ʿĀmir.
> Mecca.

Seventh Road: Baghdad to Mecca. Formerly the noble *maḥmal* set out with the Commander of the Pilgrimage from Baghdad.

> Ṣarṣar Hill.
> Farāshar Hill.
> Euphrates.
> Kufa.
> Mashhad ʿAlī: It is in the desert of Najaf.
> Muʿtaba in Wādīʾl-Sabāʾ: Here there are cisterns.
> The wells of Qurʿa.
> Masjid Saʿd Fazārī, popularly called Saʿd Waqqāṣ.
> Wāqiṣa: Here there are wells dug by (the Seljuk sultan) Malik Shāh, 15 cubits in diameter and 400 cubits deep and made of cut stone.
> ʿAqabat Shayṭān: It has wells.
> Qāʿ Rammāla: It has wells.
> Buyū: It has a pond.
> Bārṭān: [Also called] Qubbat al-ʿIbādī. It has ponds.
> Thaʿlab[iyy]a. Here they are joined by pilgrims from Wāsiṭ. Their route is > Wāsiṭ > Shaʿshaʿa > ʿĪṣ > Dhāt Maʿīn > Shāhiyya > Aḥādiyya > ʿAwja > Sawiyya > Ludd > Thaʿlabiyya.
> Kharīma: It has ponds.
> Ḥuqūqiyyat Fayd: It has a stream.
> Shūr: It has ponds.
> Shamīr: It has wells and ponds. The road here branches off to Medina.
> Maʿdin Banī Sulaym: It has pools and wells.
> Silsila.
> ʿAms: It has ponds and wells.
> Qaʿba: It has a pool and wells.
> al-Mislaḥ: It has a pond and a well. Here some pilgrims put on their pilgrimage garments.
> Bustān Ban[ī] ʿĀmir: It has abundant water.
> Mecca.

The Abbasid Zubayda Khātūn (wife of the Abbasid caliph Hārūn al-Rashīd) and the Seljuk sultan Malik Shāh built many (charitable) facilities at these stages.

Status of rulers of the Arabian Peninsula

The following is a summary account of the status of the kings and rulers before Islam, from the beginning to the present day (sic): After the Flood, Noah had three sons. One of them, Shem, was very intelligent, sagacious, and shrewd. Because he was superior to his brothers, Noah made him his heir. Appointing Shem as his successor, he taught him the secrets of prophethood and the subtle points of divine wisdom. Noah prayed to God that among Shem's descendants there be most of the prophets, apostles, saints, righteous men, wise men, sultans, *amīrs*, and devout men.

Shem had nine children. One was Arfakhshadh—most of the prophets were descended from him. Another was Kayūmarth—his progeny were rulers. Another was Iram—the (people of) ʿĀd were descended from him. Another was Lāwaz, from whom ʿAmlīq was descended—he and his people (the Amalekites) settled in Ṣanʿāʾ. Later they moved to an area near the Kaʿba. Then they settled in Palestine and Syria. Some of them settled with the people of ʿĀd in Ḥaḍramawt and Aḥqāf. The people of ʿĀd worshipped the moon. He married 1,000 women and from his loins came 4,000 children. These people are called the First ʿĀd. Over time they became rebellious and arrogant. (The prophet) Hūd then summoned them to faith in God. But the people of ʿĀd accused him of falsehood. So, God sent flood and wind and they were destroyed in Aḥqāf. At that time 3,044 years had passed since Adam.

When the First ʿĀd were destroyed, Shem's son Arfakhshadh was with his sons in India. When he learned of the destruction of the First ʿĀd, he and his sons went to Yemen and took possession of it. Among the most famous sons of Qaḥṭān (the ancestor of all the South Arabian peoples) were Yaʿrub and Jurhum. Yaʿrub became the ruler of Yemen. (The people of) Jurhum settled in Mecca. They were called ʿArab ʿĀriba ("Arabizing Arabs"). Later Ishmael married one of them. He and his descendants lived in the desert with the Bedouin. They were called ʿArab Mutaʿarriba ("Arabized Arabs") because they learned the Arabic language from them. The language of Ishmael was Hebrew. Afterward Jurhum fought a great battle with the Amalekites and defeated them. They committed disgraceful acts, and so the place was called Faḍiḥ ("Disgrace"). Subsequently Jurhum became extremely rebellious. A flood came and they were all drowned.

Those who remained alive from the ʿArab ʿĀriba were the Qaḥṭān and ʿAdnān (ancestor of the Northern Arabs) peoples. These Arabs became famous for the purity and eloquence of their language. There are certain fields of knowledge (concerning Arabic) which only they know. Compared to other nations they have excelled in certain things. [M 545] One of them is genealogy, another is astronomy, another is history, and another is the interpretation of dreams. (Caliph) Abū Bakr excelled in the interpretation of dreams. There is also the science of physiognomancy. According to this, certain traits and resemblances are passed down from father to son (and thus an expert can tell who is related to whom). The Mudlij tribe is especially skilled in this. Another is the science of traces, which helps find something that is lost.

They are extremely skillful in genealogy. They take six layers of descent into consideration. The first is *shaʿb* ("people"). By this they mean the most distant ancestral line. Thus they trace the genealogy of the Prophet back to ʿAdnān. The second is *qabīla* ("tribe"). (Some) say that *shaʿb* is a division of *qabīla*. They liken the tribes to Rabīʿa and Muḍar. The third is *ʿimāra* ("structure"). Here the lineages branch off. Examples are Quraysh and Kināna, which are carefully determined. The fourth is *baṭn* ("womb"). Here the lineages of the *ʿimāra* branch off. Examples are ʿAbd Manāf and Banī Makhzūm. The fifth is *fakhidh* ("subdivision of a tribe"). The lineage of the *baṭn* branches off from the *fakhidh* (!). Examples are Banī Hāshim and Banī Umayya. The sixth is *faṣīla* ("family"). The *fakhidh* is derived from the *faṣīla* (!). Examples are Banī ʿAbbās and Banī ʿAbd al-Muṭṭalib. There are two further divisions: *ʿashīra* ("tribe") about which they say *ʿashīratu'l-rajuli rahṭuhu al-aqrabūn* ("the tribe of a man is his closest group"); and *ḥayy* ("tribal community") about which they say *ḥayy min al-ʿarab* ("a community of Arabs"). And they say Banī Fulān ("Sons of so-and-so").

The first to speak Arabic was Yaʿrub b. Qaḥṭān. Yaʿrub was the king of Ḥaḍramawt for two years. His son was Yashjub and his son was ʿAbd Shams. By being the first person to seize this land (the verb *sabā* = to capture) he gave it the name Saba' (i.e., Sheba). He built a huge dam in the land of Ma'rib, which became famous as Saba'. They report that the ancestor of most groups of Arabs was descended from a man named Saba'. He was the first person to be taken prisoner. They called him ʿAbd Shams ("Slave of the Sun") because he had a very bright countenance. He became the ruler of Yemen. He assembled the Banī Qaḥṭān. The Banī Hūd encouraged them to go on raids. Together they conquered Babil and killed its inhabitants. (However,) they did keep several children, including Ḥimyar, ʿAmr, and Kahlān.

Ḥimyar got his name from the fact that he always wore red clothing (*aḥmar* = red). When he was king he removed the tribe of Thamūd from Yemen and settled them in Ḥijr. Later Kahlān became king. After him his son Wāyla became ruler, then his son Sāska, then his son Yaʿfar, then Zūrbāsh, then Nuʿmān, then Asmaḥ, then Shaddād b. ʿĀd II who was a descendant of Saba'.

This Shaddād conducted warfare in many climes, even as far as Morocco. He became headstrong and proud. When summoned to faith (in monotheism, or Islam), he said, "If I have faith, what will God give me?" They said, "He will give you palaces, boys and houris in a lofty Garden (of Paradise, cf. Koran 25:10, 52:24, etc.)." He replied, "All of these are easily acquired in this world. I'll build a Garden for myself right here." So Shaddād brought lovely boys and girls from various peoples. He assembled gold and all kinds of jewels, musk and ambergris, aloe wood, pearls and coral and built a Garden as he said. When it was finished he was pleased. He and his retinue set out to tour the Garden. When they were one parasang away, a cry (*ṣayḥa*—cf. Koran 54:31) was heard and Shaddād and his people were destroyed.

Some remains of theirs still stand in this area. They are called the Second ʿĀd. They settled in the areas of Shiḥr and Mahra. [M 546] Even today some remains of theirs are still to be seen.

When Shaddād was destroyed his son Murshid became king. Murshid had previously had faith in Hūd. After him his son ʿAmr became a believer. Then his father's paternal uncle, Luqmān, became king. He became famous as the Second ʿĀd. Then came his brother Dhūshdī and then his son Ḥarith, who was also called Ḥāris Ra'īs and the Greater Tubbaʿ. There were three Tubbaʿs, of which he was the first.

Then came his son Dhū'l-Qarnayn who built the dam of Gog and Magog. Next was his son Dhū'l-Manār Abraha. He built towers (*mināreler*) in the desert so that he could find the way back. After him came his son Ifrīqīs. This king made raids against the Berber people. He went as far as Tangiers. The Berbers had migrated from Palestine and Egypt and settled in the Maghreb. They built the city of Ifrīqiyya (Tunis). That time corresponded with the early reign of David. After Ifrīqīs came his brother Dhū'l-Azhār. He attacked the territory of *nasnās* and took a people captive whose faces were in their chests. Next came Sharjīl and then his son Harhād, who was the father of Bilqīs (Queen of Sheba).

Then Bilqīs became queen. Her people were called Maḥbūs. They worshipped the sun. Then Bilqīs' paternal uncle Nāshir al-Niʿam became king. Then Shamrīr ʿAshr b. Ifrīqis became king. On his way to China he razed Samarqand. Then his son (!) Abū Mālik b. ʿUmrān became king; he was descended from the Kahlān. Next was his brother ʿAmr. Everyday this king would tear to pieces the garment that he had worn, so that no one would wear it again. Then came his brother Aqran and then his son Dhū Jasān. This king unleashed slaughter on the Ṭīm and Judays.

Next was Tubbaʿ b. Aqran and then Kalī Karib and then his son Abū Karib. He was the Middle Tubbaʿ and is mentioned in the Koran (cf. Koran 44:37, 50:14). He was the first to drape the Kaʿba and to devise a door for it. This king followed the Jewish religion. He made war against the Aws and Khazraj. Then his son Ḥassān and then his brother ʿUmar became king. This king found a reference to the prophethood of Muḥammad in the book of his grandfather Ifrīqis. Then came his son ʿAbd Kalāl and then Tubbaʿ b. Ḥassān b. Malik Yakrib, known as the Lesser Tubbaʿ. Next was his brother's son Ḥārith and then Mu'aththir b. Kalāl. Afterwards the kings of Ḥimyar scattered. Then Rabīʿa the son of (Mu'aththir), then Abraha b. Ṣabbāḥ, and then Ibn Daqīqān became king. The last had under his command ʿAmr b. Maʿadī who was renowned for his sharp sword.

Next was Lukhtayʿa Dhū'l-Shanātir. This king practiced pederasty with the sons of kings. From among them he demanded Yūsuf Dhū Nuwās. When he came he secretly brought a knife with him and when no one else was present he drew it and killed him. This is a famous story. Then they made Dhū Nuwās king. He was a Jew and a tyrant and one of the Aṣḥāb Ukhdūd (cf. Koran 85:4) whom we mentioned above. After him Yemen was conquered by Aryāṭ b. Aḍḥam from Ethiopia.

The reason for this conquest was as follows: Dhū Nuwās would burn Christians alive. When the Negus learned of this he sent a great army under the command of someone named Aryāṭ (to conquer his country). Dhū Nuwās drowned while fleeing by sea. The next king was Abraha Ashram, owner of the Elephant (referred to in Koran 105), then his son Yaksūm, and then his brother Masrūq. When the last became king, the Ethiopians intensified their attack on Yemen. At that time Sayf (b.) Dhī Yazan, who had the byname [M 547] Abū Marra, went to the Byzantine emperor and asked him for help but he would not provide any. Consequently, he went to (the Sasanian emperor) Kisrā Anūshirwān and begged him for help. Khusraw sent him back with 800 criminals (condemned men?) in Ethiopia. When they approached the coast of Aden in eight ships they set fire to a number of (enemy) ships. They were eventually victorious in the war against the Ethiopians. Masrūq killed the man who invaded Yemen and Yemen returned once more to the people of Ḥimyar. Subsequently, while Sayf (b.) Dhī Yazan Ḥimyarī and Harz, who had been sent by Khusraw, were drinking wine in a magnificent palace that had been built by Solomon, Sayf (b.) Dhī Yazan was killed by one of his retinue who was an Ethiopian. Thus the line of kings descended from Saba' came to an end.

Anūshirwān appointed Harz to succeed them as ruler. After him came Marzubān, Suḥān, his son Jurjīs, and Bāzān b. Sāsān. Bāzān ruled until the Prophet was sent on his mission (in 610). Afterwards he and all those bound to him became Muslims. The reason for this is as follows: the Prophet sent a letter to King Kisrā who tore it up and ordered Bāzān to bring the Prophet's head. Bāzān sent one of his men to Mecca to gather intelligence. God revealed to the Prophet the purpose for which that man had come. Furthermore, the Prophet was informed that Kisrā had been killed at a certain day and time. As word of this spread, the man who had been sent to Mecca learned of it and he returned in failure. When he informed Bāzān of this and it turned out to be true, Bāzān became a Muslim.

Bāzān died ten years after the Hijra (631) and the Prophet appointed Bāzān's son as his successor. This son of Bāzān was emir of Yemen in the early years of Islam. Afterwards companions and associates of the Prophet began to be governors of Yemen. Later, in 202 (817), a letter reached Caliph Ma'mūn from Yemen concerning the Ashʿarīs who had rebelled. Ma'mūn immediately appointed Muḥammad b. ʿAbd Allah b. Ziyād, who was in his service, emir of Yemen. He ordered him to build a fortress in Wādī Zabīd as soon as he reached the territory of the Ashʿarīs. When the new commander arrived in Yemen he conquered Tihāma and built the city of Zabīd. This was in 204 (819). His power gradually increased. Indeed, his name was mentioned in the Friday prayer in the cities of Ṣanʿāʾ, Ṣaʿda, and Baḥrān. Their dynasty (the Ziyādids) continued until 407 (1016). Ḥusayn b. Salāma, the vizier of the last ruler ʿAbd Allah, built the city of al-Kudr in Wādī Ṣahāma and the city of al-Maʿqir in Wādī Dawāl. He established mile posts, parasang posts, and postal stages along the roads from Ḥaḍramawt to Mecca.

Next the Najāḥid dynasty appeared. The founder was Najāḥ an Ethiopian slave. He had been in the service of Ibrāhīm Ziyād (the last Ziyādid ruler). When Yaʿīsh killed Ibrāhīm, Najāḥ was in great grief. He assembled a great army and sent it against Yaʿīsh who was in Zabīd. Najāḥ captured Zabīd and killed Yaʿīsh. He settled in Zabīd and struck coins in his own name. Around that time there was a city called Ḥarār at the top of Mt. Simār. This city was under the control of a jurist belonging to the Imami (Twelver Shīʿī) rite. His name was ʿAlī b. Muḥammad b. ʿAlī Ṣulayḥī. [M 548] This man's power gradually increased. He sent a slave girl as a gift to Najāḥ. He used her as an agent to poison and kill him. Ṣulayḥī conquered more and more fortresses and began to wage war on the Najāḥids. The Ṣulayḥid dynasty finally died out in 484 (1091). The Najāḥid dynasty was founded before that of the Ṣulayḥid and actually continued beyond it. It died out in 553 (1158).

Then came the Mahdid dynasty. The first of the dynasty was ʿAlī b. Mahdī. At the beginning of his reign he was upstanding but then he went into the mountains and began to plunder caravans and commit highway robbery. He killed Fātik b. Muḥammad, the last Najāḥid ruler and seized Zabīd. He was later succeeded by his son Mahdī who was succeeded in turn by his son ʿAbd al-Nabī. (After the latter came to power) the Ayyubid Tūrān-Shāh came to Yemen from Egypt. He took ʿAbd al-Nabī captive and seized Zabīd. (In this way) Ayyubid control began.

Next came the dynasty of the Rasulids of Ghassa[n]. They were called the Rasulids (*rasūl* = messenger) because a member of the family who had been in the service of some of the Abbasid caliphs in Baghdad was sent to various places (as an envoy). Thus he was called Muḥammad Rasūl. He went to Egypt and entered the service of Tūrān-Shāh. Later Sultan Masʿūd Aqsīs (i.e., al-Malik al-Masʿūd Ṣalāḥ al-Dīn Yūsuf, the son of the Ayyubid sultan al-Malik al-Kāmil) sent him to Yemen. When he (Muḥammad Rasūl) became ill, his son was appointed as his successor; and when Masʿūd died, the Ayyubids confirmed the appointment. Their dynasty continued from generation to generation until the reign of Masʿūd Ṣalāḥ al-Dīn.

Next came the dynasty of the Ṭāhirids. The founder was Ṭāhir b. Muʿawwiḍ b. Tāj al-Dīn al-Qurashī al-Umawī al-ʿUmarī. At that time Masʿūd was in Taʿizz. Leaving Taʿizz, he set out against the Ṭāhirids and was defeated. Thus by his own decision he was removed from power. The Ṭāhirids expanded their rule from Taʿizz. Later ʿĀmir b. ʿAbd al-Wahhāb, who had the byname Ṣalāḥ al-Dīn al-Malik al-Ẓāfir, built a pool in Zabīd. He filled it with sweet sherbet. He sprinkled all the people with musk and rosewater. He distributed 6,000 goldpieces as alms to the poor of Zabīd. He took possession of many places in Yemen. In 918 (1512) he built a water canal (i.e., a dam) in Wādī Zabīd. The length was 66 cubits, the height was 50 cubits, and the width was 15 cubits. It was a great dam. In 922 (1516) hostility arose (between the Rasulids and the Mamluks in Egypt). A great war broke out between al-Ẓāfir and the Egyptian commanders Ḥusayn al-Kurdī and Amīr Sulaymān. The Egyptians were victorious and captured the city of Zabīd. More information was given on this above in the first part of the book while discussing the India campaign.

Next came the Imāmī Zaydī dynasty which began in 953 (1546). The first of the dynasty was Sharīf Aḥmad b. Yaḥyā. His lineage went back to the son of al-Ḥasan al-Sibṭ al-Kabīr (who was the son of ʿAlī and grandson of the Prophet). He was succeeded by his son Sharaf al-Dīn Yaḥyā who has a book entitled *al-Baḥr al-Zakhkhār* which describes the Zaydī rite. The theological principles of the Zaydī rite are found in the book called *Kitāb al-Aḥkām* expounding *ijtihād* (individual reasoning in law) and *daʿwā* (propaganda for the imam who could give mankind good guidance).

During his reign the Ottomans sent Üveys Pasha to Yemen in 953 (1546). He went to Zabīd, Dhamār, and Taʿizz and conquered many fortresses. The reason that he went to Yemen is that when Sultan Selīm took Egypt the Sharīf of Mecca Sayyid Barakāt sent his son, who was Sharīf of Yemen, to Istanbul to congratulate the Ottoman sultan. At that time his son was 13 years old. Sultan Selīm showed him great honor and issued orders granting his wishes. [M 549] He even ordered the execution of Amīr Ḥusayn al-Kurdī in Jedda. The reason for this is that previously Ḥusayn al-Kurdī had inflicted many cruelties upon Sayyid Barakāt and was extremely oppressive. This situation was described to Sultan Selīm who then issued an order concerning it. It later happened that when Sharīf Barakāt went to Mecca, and it was decorated (for Ramadan), he put on the robe of honor that Selīm had sent him and made the circumambulation of the Kaʿba while wearing it. The people of Mecca congratulated him. Afterwards, while he was sitting at the Ḥanafī *maqām*, he called for Ḥusayn al-Kurdī and had the order read that was sent by the sultan. As a result the Sharīf of Mecca sent some men to get Ḥusayn al-Kurdī. As these men approached Jedda by ship and reached the (sailing) stage (called) Bayn al-ʿAmmayn, (their boat sank and) they were drowned. The Circassians (Mamluks) who were saved joined Amīr Barsbāy in the region of Zabīd. After Barsbāy was killed, Amīr Iskandar became ruler in Zabīd. He dressed in Ottoman attire and thus it was that an order arrived from Sultan Selīm appointing him governor of the province of Yemen. After an (Ottoman) deputy was sent from Cairo, the name of Sultan Selīm began to be mentioned in the Friday sermon.

The sermon continued to be read in this fashion for three years until the Ottoman emir, the second Ḥusayn, became regent in Jedda. He was an excellent man and had the confidence of Amīr Khayra Bey, who was the (Ottoman) ruler of Egypt. Khayra Bey gave him the sanjak of Jedda. After Amīr Ḥusayn had arrived in Jedda, (the Mamluk) Sultan al-Ghawrī (sic) sent Amīr Ḥusayn al-Kurdī with a number of ships carrying war materiel to Jedda. When it was learned that Yemen was evacuated, Amīr Ḥusayn requested permission from Amīr Khayra Bey, which he received, and set out for Yemen. He arrived in Yemen in 926 (1520). When Khayra Bey learned that Sultan Selīm had died and was succeeded by Süleymān, he recalled Amīr Ḥusayn to Jedda. Later, because there were many disputes and much fighting among the governors (of Yemen), and because Sultan Süleymān had heard that the Portuguese infidels had invaded the coast of India and that India had not been able to oppose them, his zeal for Islam was aroused and he issued an edict making Süleymān Pasha governor of Egypt and ordering him to Yemen. Süleymān Pasha set out for Yemen and then, while going to India, had a pretext to seize Aden, as discussed in detail above. This occurred in 945 (1547).

Later in 977 (1569), Özdemir Pasha (the first Ottoman governor-general or *beylerbey* of Yemen) was sent to Yemen. He was a *bey* appointed by Üveys Pasha in Yemen. As Üveys Pasha (he was later Ottoman governor of Egypt, 1587–1591) grew in power he was promoted one horsetail. He had criers go about the country with troops announcing his promotion. By this means the troops remained obedient and under his command. Later he killed Pehlivān Ḥasan and seized Ṣanʿāʾ. When Özdemir Pasha reached Yemen he informed Istanbul that he had attained rule (?—*iclās*) over Yemen and requested assistance from Egypt. Sultan Süleymān immediately ordered Dāwūd Pasha to be sent from Egypt with 3,000 musketeers and 1,000 cavalry. After Dāwūd Pasha arrived with these troops in Yemen, he met with Özdemir Pasha (and?) Muṣṭafā Pasha (Ottoman governor of Egypt, 1561–1565, and later Yemen) and settled in the fortress of Qilā. A dispute broke out between them. Muṭahhar (Zaydī leader of resistance against the Ottomans in Yemen) found an opportunity and requested safe conduct from Muṣṭafā Pasha. [M 550] After making peace the Ottoman pashas returned to Egypt.

Özdemir Pasha remained in Yemen. Over time he subjugated and conquered the fortresses of Kahlān, Ḥabash, Sawākin. ʿItra, Makhlāq, and Khanfar. He served as governor in Yemen for seven years. After a while he was dismissed from office and recalled to Istanbul. Sultan Süleymān, realizing his complete knowledge of the affairs in Yemen, reappointed him as commander-in-chief and sent him back accompanied by 3,000 men from Egypt. From Cairo he went to Upper Egypt and from there continued toward Sawākin and immediately built several fortresses in that area. Previously an *emīn* (superintendant) appointed from Cairo had resided in Sawākin. Later that position was elevated to *beylerbey*. Özdemir took several provinces in Ethiopia and in 967 (1559) died in the city of Dawaro and was buried in Massawa. His son ʿOs̱mān Pasha built a lofty dome over his grave.

Özdemir's son then controlled Yemen until he was dismissed as governor and replaced by Muṣṭafā Pasha. (Under the Ottomans) the late Muṣṭafā Pasha was the first to send (from Yemen) a noble *maḥmal* with a commander of the pilgrimage. Previously the pilgrims simply went with caravans. Leading them as commander was a shaikh of the Banī Marzūq. When Muṣṭafā Pasha died his majordomo Yūsuf Pasha succeeded him. Subsequent governors were oppressive and did not provide provender for their troops. Yemen was divided into two provinces each headed by a *beylerbey*. Yemen was not free of strife and discord. In short, when the Ottoman sultan learned of the unstable conditions in Yemen, he sent Sinān Pasha to Yemen as governor. He set out from Egypt and when he reached Jāzān (on the coast of the Tihāma of Yemen) he saw that it was deserted. He learned that the Zaydīs had fortified themselves on Mt. Aghbar. Then he went to Taʿizz and sent ʿOs̱mān Pasha to Mt. Aghbar and he conquered Mt. Aghbar. Indeed the Zaydīs fled, so the fortress fell without fighting.

The fortresses and towns that Sinān Pasha subjugated and conquered were the following: fortress of Taʿkur, Jabala, Khayrān, fortress of Ḥadad, Jabal Ḥubaysh, fortress of Āb, Jabal Baʿdān, Ḥiṣn Ḥabb, Hirrān, Qatdā, Madrarā, Zamān, Wādī Suḥayl, Zayl, Thaqīla, Sumār, Wādī Yarm, fortress of Dawrān, Ṣanʿāʾ, Ḥiṣn Ḥawlān, citadel of Talā, Ḥiṣn Dumarmar, Wādīʾl-Sirr; in Bilād Muṭahhara, fortress of Sām, fortress of al-Muqannab.

Sām is extremely strong, being surrounded on three sides by mountains and having formidable walls. It has a thousand LNBLH (?—perhaps for *gülle* "cannon ball") cast from iron. Its length is 5,000 cubits and the width (of the walls) is 15 cubits. On two sides of it are additional fortresses, one called Qaṣr al-ʿArṣa and the other al-Bāḥa.

Also Kawkabān, which is very high; Wādī Būn; Kahūl; the fortresses of Aʿraj, three fortresses where the Zaydī rulers resided; ʿArūs, Ḥabb al-ʿArūs, citadel of al-Ẓafar, fortress of Bayn al-ʿIzz; Wādī Khayyān, where is the fortress of Rawām; village of al-Luʾluʾ; town of al-Ḥudūb in Bilād Dāʿī; district of Ḥayma, which has many fortresses the most famous of which is Nisyāʿ; Hamdān, al-Ḥarrāz and Jabal Asan.

There are eight tribes that belong to the Zaydi rite. They are the Banīʾl-Azraq, Banīʾl-Shadīd, Banī Muḥammad, Banī Awālīd, Banī ʿAdār, al-Dahārija, Jaʿāfira, and Mujādīl.

(Other places that he conquered were) a strong fortress that they call Qurāb al-Masjid, [M 551] Jabal al-Tays, where Bedouin live, Wādīʾl-Ḥayma, which has four fortresses called Lahnaʿa, Ẓafār, Sawdāʾ, and ʿUshr, and the towns of Sahl Bāqīr. There are many tribes here. They have three fortresses that are loyal to the state. They are Khālid al-Bakr (al-Kabīr?), Khālid al-Ṣaghīr, and al-Kāhil. (Other places were) the fortress of Habtī, Wādī Ṣanʿān, Jabal Sinān, the fortress of Ḥajr al-Rakāḥīz, the citadel of Shamāṭ, the fortress of Barās, Sanām al-Yaʿāfir, and the towns of Ḥarrān. In the last are found the fortresses of Bihār, Daʿla, Banī ʿImrān, Baʿdaʿa, and al-ʿAqaba. (Other places were) the fortresses of Awā, Nanbīh, Ṭaghrān, Qabṣān, Rayḥā, Qibla, al-Qifl, Shaḥab, Qurāna Diyat (?), Sāfa, Rāja, Jaʿr Madmahā, Jabal al-Lawz, the towns of Simāt, Ṭawīla, and Thaqīl Aḥmara.

In short, after Sinān Pasha conquered Yemen, he turned it over to Behrām Pasha and returned to Istanbul. For sometime afterwards the governors of Yemen oppressed the people and were unjust to them and conditions in Yemen gradually deteriorated. In 1038 (1628) Ḳanṣu was appointed governor of Yemen. Later the Zaydīs seized control of Yemen and it is presently in their hands.

Successors (of the Sharifs) of Ḥijāz

After the Companions of the Prophet, Ibn Zubayr. Then the Abbasid ʿAbd Allah al-Saffāḥ appointed his paternal uncle Dāwūd who became governor of Mecca, Medina, and Yamāma. After him the dynasty of the Banī Ukhayḍir came on the scene.

The following are the successors of the Sharifs of Ḥijāz, in four categories, all descended from Ḥasan: 1) Banī Ukhayḍir, 2) Awlād Mūsā, 3) Banī Hāshim—but their genealogy goes back to Ḥusayn, 4) Banī Qatāda.

1) Banī Ukhayḍir. They are the Banī Yūsuf b. Ibrāhīm b. Mūsā ʿAbd Allah al-Muḥaṣṣir b. Muḥsin Muthannā b. Ḥasan b. ʿAlī b. Abī Ṭālib. They are called the Ḥasaniyya. They appeared on the scene in 215 (865). They continued until the Carmathians defeated them in 317 (929). The last (Sharif) of the Ukhayḍir was Ṣāliḥ b. Ismāʿīl. They currently have a dynasty in the city of Ghāna in the Sudan. They later lost supremacy to the Banī Hāshim

2) After Ṣāliḥ the Banī Sulaymān b. Ḥasan al-Muntanī of the Awlād Mūsā became independent in Mecca. At that time an accursed person appeared on the scene. They called him Abū Ṭāhir Sulaymān b. ʿAbd Allah Abī Saʿīd al-Ḥusaynī. Because of the evil of this man, no one was able to make the pilgrimage for a long time. He committed highway robbery. As the number of his followers increased, they began to kill pilgrims. At the end of 310 (922), when all the pilgrims were in the noble Haram on the day of *tarwiya* (8th day of Dhūʾl-Ḥijja, the day on which the pilgrimage begins), they attacked the Kaʿba with many soldiers and killed 30,000 pilgrims at the Kaʿba and around Mecca.

It is reported that in 319 (931) Abū Ṭāhir, who was a dissolute slave, appeared at the beginning of Ramadan and claimed divinity. Many people followed him. This slave ordered them to cut open the bellies of their dead, then to wash them, and fill them with wine. He ordered that anyone who blew out a fire should have his tongue cut off. He ordered them to engage in sodomy with boys, but that they should practice moderation in this, otherwise they should be dragged on the ground face downwards for a distance of 40 cubits; and any boy who refused to participate in sodomy should be forced to do so before the butcher (?). And he ordered them to glorify fire. Eventually the followers of Abū Ṭāhir were exterminated. [M 552]

3) The Hāshimiyya, who were mentioned above, appeared in Medina. They were the descendants of Muhannā b. al-Ḥasan b. Muhannā b. Dāwūd b. Qāsim. Their lineage goes back to Ḥusayn b. ʿAlī b. Abī Ṭālib. Their first (Sharif) was Abū Faliya (?) al-Sharīf Qāsim and the last was Sulaymān.

4) Banī Qatāda b. Idrīs. They also go back to al-Ḥasan b. ʿAlī b. Abī Ṭālib. The Qatāda, who lived in the fortress of Yanbu, triumphed over the Banī Hāshim when they declined. They took over Mecca and established an independent dynasty. The Qatāda line has continued down to the present-day Sharifs of Mecca. In 1094 (1684) the Sharif of Mecca is Sharīf Saʿīd b. Abī Barakāt.

In sum, in the Arabian Peninsula, governments are composed of various emirs, Sharifs, and shaikhs. In Yemen, in parts of the Tihāma and parts of the mountains, the Zaydīs are the rulers. They call their ruler the Imam. Various rulers have established their own independent governments in other places. One example is the independent ruler of the province of Shiḥr. The same is true of Oman. As for Laḥsā', although their rulers are independent, nevertheless they acknowledge Ottoman authority. The rulers of Mecca and Medina are Sharifs. They are subject to the Ottomans. Information on the state of affairs and rulers in the peninsula are found in my book *Jawalān al-Afkār*.

Chapter 45: Region of Shām (Syria)

The clime of Syria is one of the conventional climes. It is also called the Holy Land; the land of Canaan; and, in Greek, Suriyā. After completing my account of the Arabian Peninsula, let me begin my description of Syria.

This is a glorious clime. It is the land of prophets, the center of righteous men, and the mine of great saints. The first qibla was here (i.e., Jerusalem, toward which Muslims prayed before the direction of prayer was changed to Mecca). This is also where the resurrection and assembling for judgment will take place. Here are the Holy Land (Koran 5:21) and worthy convents; glorious frontiers and noble mountains; the sojournments of Abraham and his tomb; the land of Job and his well; the prayer niche of David and his gate; the marvels of Solomon and his cities; the tomb of Isaac and his mother; the birth place of the Messiah and his cradle; the village of Saul and his river (Koran, 2:249); the place where Goliath was killed and his fortress; the pit of Jeremiah and his prison; the rock of Moses (Dome of the Rock); the prayer niche of Zachariah; the battlefield of John; the martyria of the prophets; the villages of Moses; the city of Ascalon; the spring of Silwān; the haunts of Luqmān; the valley of Nuʿmān; the cities of Lot; the Place of Gardens (i.e., Damascus); the endowment of ʿUthmān; the council chamber where the two adversaries were present (reference to the judgment of Solomon?); the tombs of Mary and Rachel.

In short, the land of Syria is the Holy Land. God made this place of descent of prophetic inspiration the abode of prophets and the station of saints. Its water is sweet and its air is wholesome. The manners and opinions of the people are esteemed. One of the features of Syria is that it is never empty of great saints. Through their prayers God has been merciful to His slaves and forgiving of their sins. These great saints are never fewer nor greater than 70. Whenever one of these 70 dies another appears from among the people and takes his place. These saints live nowhere else but on Mt. Lukkām.

ʿAbd Allah (b.) ʿAmr b. al-ʿĀṣ is reported to have said, "Good things were divided into ten parts, of which nine are in Syria and one is in the rest of the world." The food of Syria is famous. The soldiers of Syria are brave and heroic. The obedience of the people to their sultan is proverbial. In short, the people of this clime are mighty in blows and virtuous in conduct. They possess the sciences. In particular their philosophers and religious scholars are incomparable. Indeed, most of the ancient Greek philosophers originated in Tyre and other cities. [M 553]

Shām (Syria) got its name from the fact that a group of people from the Banī Kanʿān (Canaanites) from the land of the Kaʿba "turned left" (*tashāmū*), i.e., went to the left or north. Some say that the name Shām comes from Sām the son of Noah. In Syriac Sām is Shām.

Borders

The western border is along the Mediterranean coast extending from Tarsus, a city of the Armenians, as far as Rafaḥ which is in the Sand Dunes between Egypt and Syria. In the south the border extends from Rafaḥ through the Wilderness of Tīh (Sinai desert) of the Israelites and between Shawbak and Ayla and ends at Balqā'. In the east it extends from Balqā' to Ṣarkhad, encompassing the Ghouta of Damascus, then to Salamiyya and Aleppo, ending at Bālis. In the north it extends from Bālis along the Euphrates to the fortress of Najm, Bīra, Samsat, Ḥiṣn Manṣūr, Bahasnā, Marʿash, Bilād Sīs, and Tarsus, and ends at the Mediterranean.

Administrative subdivisions

In the past Syria was divided into five military districts (*jund*), each ascribed to a region. They are Palestine, Jordan, Damascus, Ḥimṣ, and Qinnasrīn. Within each military district are cities, towns, and villages. They are as follows:

- Palestine: Its main city is Jerusalem. Its other cities are Ramla, Bayt Jibrīn, Gaza, Jaffa, Ascalon, Arsūf, Caesarea, and Arīḥā.
- Urdun (Jordan): Its main city used to be Tiberias but now is Nablus. Its other cities are Qadas, Tyre, Acre, Jenin, al-Lajjūn, Kābil, Baysān, Ṣafad, Ṣalt, and ʿAjlūn.
- Damascus: Its main city is Damascus. Its other cities are Bānyās, Sidon, Beirut, Tripoli, Shaqīf, and Arnūn.
- Ḥimṣ: Its main city is Ḥimṣ. Its other cities are Salamiyya, Tadmur (Palmyra), Ḥayāṣir, Kafarṭāb, Lādhiqiyya (Laodicea), Jabala, Anṭarṭūs (Tortosa) and Ḥiṣn al-Khawābī (text: al-Khawānī).
- Qinnasrīn: Its main city is Aleppo. Its other cities are Anṭākiya (Antioch), Sūwaydiyya, Samsat, Manbij, Payās, Qinnasrīn, Marʿash, Iskandarūn (Alexandretta), Ḥamāh, Shayzar, Maʿarrat al-Nuʿmān, Maʿarrat al-Miṣriyyīn, Maʿarrat Qinnasrīn, Ṣahyūn, Darbasāk (Turbessel), Ḥiṣn Barzara, Shughr, Fāmiya (Apameia),

Rāwandān, Bāb Nirāʿā, ʿAynṭāb, Qalʿat al-Rūm, Bīra, Bālis, Ruṣāfa, and Ḥiṣn Manṣūr.

There are four districts (*ṣuqʿa*) around Damascus—western, eastern, southern and northern—each having its own administrative units. 1) Western district: It has two sections, the coast and the mountains. The coastal section lies along the Mediterranean. Its subdistricts are Gaza, Ramla, Lydda, and Qāqūn. The subdistricts of the mountainous section are Jerusalem, Hebron, and Nablus. 2) Southern district: Its subdistricts are the town of Ḥawrān, Ghawr, Bānyās, Azriʿāt, ʿAjlūn, Balqāʾ, Ṣarkhad, Bosra, and Darāʿ (text: Zaraʿ). 3) Northern district: It has coastal and mountainous sections. The coastal section has the subdistricts of Sidon and Beirut. The mountainous section has the subdistricts of Baalbek, also called Qāʿ al-Buzūh, and Biqāʿ-i ʿAzīzī. The latter was named after (al-Malik) al-ʿAzīz, the son of Saladin. 4) Eastern district: Its subdistricts are Ḥimṣ, Maṣyāf (text: Maṣyāt), Qāra, Tadmur, Raḥba, and Salamiyya.

Previously, during the time of the (Abbasid) caliphate, the people and soldiers of Syria were divided into four classes: 1) Military officers, that is, professional military men. Their top positions were eight in number: *nāʾib Shām, amīr kabīr, ḥājib al-ḥujjāb, nāʾib al-qalʿa, ustādār sulṭān* who was *muqaddam al-qadr*, and *ḥājib dawādār*, who was appointed by the (Ottoman) sultan to be in the service of *nāʾib Shām*. [M 554] These were the emirs. 2) Authorities on Shariah, such as qadis. They were divided into several classes of their own. 3) Government officials. The top three were *kātib al-sirr, nāẓir al-jaysh*, and vizier. Later the title of vizier was replaced by *nāẓir-i niẓām*. Below them were four (sic, error for three?) classes: *nāʾibs*, of which there were seven, one for Jerusalem, Ḥimṣ, Baalbek, Raḥba, ʿAjlūn, Ṣarkhad, Maṣyāf, and Qalʿat Ṣabiyya; *kāshifs*, of which there were one for Qabīla, Ramla, and Nablus; *wālīs*, of which there were one for Sidon, Beirut, and Tadmur. 4) The Bedouin of Syria.

The Bedouin consist of nine tribes.

- Ṭayy. It is the largest of the tribes. Their chiefs come from three consanguineal lines: 1) Āl Faḍl. The Banī Rabīʿa come from them. They are chiefs of all. Their territory goes from Ḥimṣ to Qalʿat Jaʿbar and from there to Raḥba which is on the banks of the Euphrates. It extends into Iraq. They have abundant water all the way to Basra. They are superior to the other chiefs. They call their chief *amīr al-mullā*. But currently the Banī Ṭarpūsh and the Banī Rīsha are the chiefs of that region. They call these tribes ʿArab al-Mawālī. 2) Āl Murād. The Banī Marī b. Rabīʿa b. Faḍl come from them. Their abode is Ḥawrān and Bariyya. Many chiefs have come from them. The Bedouin of the Mafālija are notorious in this region for kidnapping pilgrims. 3) Āl ʿAlī. They are the Banī ʿAlī b. Ḥadīth b. ʿAyniyya b. Faḍl. They too are very prominent. Their territory is the meadow of Damascus and its surrounding Ghouta. It falls between the territories of Faḍl and Murād (text: Umarā).
- Banī Mahdī. They are from the Qaḥṭāniyya. Their abode is Balqāʾ and currently extends to the regions of Jasān and Ṣalt. Because this area is today ruled by the Banī Ṭurābī, they call the Bedouin Arabs Ṭurabiyya. They call Jenin and the provinces that belong to it the province of the Banī Ḥārith. At first it was a hereditary grant and then it became a *beylik*.
- Jurm. The Banī Jurm b. Thaʿlab are from the Ṭayy. Their abode is Gaza and Dārim. This region is currently well known for Sawālim and ʿĀbid Bedouin. The Āl Muḥammad come from them. They meet the Egyptian pilgrims on the pilgrimage road (?—or ʿAqabat al-Ḥajj?) in Hebron. Here are the Bedouin of the Ḥawīṭāt and those of the Banī ʿAṭiyya, who are related to the Sharifs. And in Maghāyir Shuʿayb ("Caverns of Shuʿayb") is the clan of the Banī Lām. Currently the well-know Bedouin around Gaza are the ʿArab al-Waḥīdā. They have many clans.
- Thaʿlaba. They are of the Banī Khālid b. Sallāmān who are from the Ṭayy. There are two lines of the Thaʿlaba descending from Dadirmā (?) and Ruzzīq, the children of ʿAwf b. Thaʿlaba. They are Ḥamdānids (a famous Arab family in Syria). They call them Thaʿlabat al-Shām (the Thaʿlaba of Damascus). They extend from Damascus to ʿAmmān, but they have no chiefs.
- Zubayd. They are from the Kahlān, who are the Banī Muzḥij of (descended from?) Saʿd al-ʿAshīra. Their abode is at Burj Birqā. They take the name of the place where they live, for example, Zubayd of Marj and Zubayd of Ḥawrān. The successors of Zubayd came from them. They live around the Āl-i Faḍl near Raḥba. This is a clan of highwaymen.
- Banī Khālid. They are the Bedouin of Ḥimṣ. They claim descent from Khālid b. al-Walīd (an Arab commander who assisted with the conquest of Syria). Everyone knows, however, that his hereditary line came to an end. They have no chiefs.
- Ḥāritha.
- Banī ʿAqaba. Their abode is Karak and Shawbak. They have chiefs.
- Ghazyat (Ghaziyyat?) Ḥamdānī.

These are the tribal subdivisions (text: *efhaz*, error for *efhāz*). [M 555]

Forests, lakes, rivers, and mountains

Forests: There are many forests in the clime of Syria. The most famous is the great forest of Ascalon. It extends as far as Ramla. The forest of Arsūf is near the ʿAwjā River. It extends as far as Acre. Today they call it the forest of Qalansuwa. It extends from Qāqūn to ʿUyūn al-Tujjār. There are springs of *esāvir* (?) in this forest. There is also the fortress of Qunayṭira and there are several forests around Tripoli and Aleppo. One of them in particular is the forest of Būk, which is called Rūz. Most of the trees in this forest are mulberry trees.

Lakes: The most famous is Lake Zughar (the Dead Sea), also called Stinking Lake and, today, Lake of Lot. It is between Khasīsa and Baysān which are subdistricts of Tiberias. It is ten miles in length and six miles in breadth. Its circuit takes two days. In the middle is a hollowed out rock (?). The Jordan River, also known as the Sharīʿa River, flows into it. There are absolutely no birds or fish in this lake.

However, in some years a substance emerges from the lake, amidst noise and commotion, called *ḥamīr*. It resembles honey and is smeared on grape vines to protect them from a certain kind of insect. It has other uses as well. They derive considerable income from this. There are two administrative authorities over this lake. This *ḥamīr* solidifies in the middle of the lake and the wind blows it to one side. It then belongs to the administrative authority of that side.

In ancient times there were five cities at this lake. The largest was Sodom, the second was Ṣibgha, the third was Gomorra, the fourth was Rūmādir, and the fifth was Ṣaʿda. These were the cities of the people of Lot.

Lot greatly loved Abraham. He emigrated with him and came with him (to Syria) and settled in Sodom. When the people began to practice immorality he forbade it. But they paid no attention to him. As a result, God turned those cities upside down. For that reason they are called al-Muʾtafikāt meaning "Overturned" (Koran 9:70, 53:53, 69:9). Afterwards Lot died in the town of Kafarṭāb which is one parasang from Hebron.

Lake Tiberias (Sea of Galilee): Also called Lake al-Minya. This lake is at the fore of the Ghawr (the Jordan lowland). The Jordan River flows from Lake Bānyās through the Ghawr to Lake Tiberias. The Ghawr is the shore of Lake Tiberias. Stones in the shape of a watermelon come out of this lake. They are called "Jewish stones." Doctors used them (to cure) gallstones.

Lake Bānyās: It is in the vicinity of Damascus and near Bānyās. It is formed from the collected springs that rise on Jabal al-Thalj (Snow Mountain, i.e., Mt. Hermon). The Jordan River rises from this lake and flows into Lake Tiberias. There is a bed of reeds in this lake. It is never winter around this lake (i.e., it never freezes).

Lake Biqāʿ: An accumulation of water flowing from different directions. It is one day's journey west of Baalbek. Several rivers flow into it, including the Ghūrbīl and also the Ḥalazūn (text: Jalzūn), whose water is finest of all. These rivers arise in the Biqāʿ Valley. In short, all these rivers gather in Lake Biqāʿ and then flow into Lake Bānyās.

The Lake of Damascus: This is in northeastern (Syria) in the Ghouta of Damascus. The Baradā and many other rivers flow into it. There is a reed bed here. There are many fish in this lake and a government agent overseeing the fisheries. All kinds of birds flock here.

Lake Qadas: This is the lake of Ḥimṣ. Its length from north to south is three stages. Water from this lake goes to swell the Orontes River. [M 556] A strong stone dam was built on the north side of the lake. It is attributed to Alexander the Great. In the middle of the dam are two towers built of black stone. The length of the dam is 1,287 cubits from east to west, and the width is 18.5 cubits. This dam stores an enormous amount of water. If the dam were destroyed it would cause a flood, the lake (reservoir) would disappear and only a river remain. This lake is one stage to the west of Ḥimṣ. They catch strange kinds of fish in the lake.

Lake Afāmiyya (Apamea): This lake is (actually) composed of a number of shallow bodies of water (marshes). They are among beds of reeds. There are two large lakes (marshes), one to the north and the other to the south. The water for both lakes comes from the Orontes, which flows into the lake on the south forming the marsh. The ʿĀṣī (Orontes) then flows out from the northern end of the marsh. It is the southern of these two lakes which is (properly) Lake Afāmiyya. The width of this lake is about a half parasang. Its depth is shallow, but it is only about the height of a man. The bottom is miry. The southern lake is completely surrounded by reeds and willows. In the middle (reading *ortada* for text: *bir depede* "on a hill") is a thicket (of reeds). In these marshes are countless ducks, geese, and similar kinds of birds. In the spring, Lake Afāmiyya is covered with water lilies. There is a reed bed between Lake Afāmiyya and the second lake to the north. There is a passage through it that boats and skiffs take while going from Lake Afāmiyya in the south to the lake in the north. The northern lake, which belongs to the district of Ḥiṣn Barziyya, is well known as Lake of the Christians because the fishermen at the northern lake are Christians. They have houses (built on piles) in the northern lake. Every year the northern lake becomes four times larger than Lake Afāmiyya. In the middle (of the northern lake) dry land appears. Water lilies grow along its northern and southern shores. There are many birds of all kinds. There is a well-know fish that is caught called eels (*ankālīs*) or snakefish (*yılanbalığı* = eel).

Lake Anṭākiya (Antioch): It is in a flat area between Baghrās and Ḥārim. They call this the land of ʿAmq. It belongs to the district of Aleppo. It is two stages from Aleppo. Three rivers flow into this lake from the north. The easternmost of these is called Nahr ʿAfrīn. The westernmost, which flows below Darbasāk, is called al-Nahr al-Aswad ("Black River"). The third, which flows between these two rivers, is Nahr Yaghrā (text: Baghrā). Yaghrā is the name of a village on its bank. The people are Christians. These three rivers merge to form one river which then enters the lake from the north and then flows out from the south. As one river it joins the ʿĀṣī (Orontes) below al-Jisr al-Ḥadīd ("Iron Bridge") which lies about a mile above Anṭākiya. The circuit of this lake is one day's journey.

Rivers

ʿAwja ("Crooked River"): This river is 12 miles north of Ramla. The source of the river is below a mountain opposite the ruined fortress of Majdal Yāba. It flows from east to west. It flows into the Mediterranean south of the forest (lowlands in Abū'l-Fidā) of Arsūf. They say that if two armies meet on its banks, it is always the western army that wins and the eastern army that is defeated.

Urdun (Jordan): It is known as the Ghawr, the Yūrdān, and also the Sharīʿa River. This river flows from Jabal al-Thalj (Mt. Hermon) to Lake Bānyās. [M 557] It then flows out of it to Lake Tiberias (Sea of Galilee) and from there flows into the Lake of Lot (Dead Sea), also called Stinking Lake.

Yarmuk: It flows past Qaṣrayn and meets the Jordan River near Lake Tiberias.

Nuʿmān: It rises from Mt. Ṣafad and flows past Acre. Then it enters the Mediterranean.

Līṭānī: Its source is in the land of Ḳarak Nūḥ ("Stronghold of Noah"). It rises from Mt. Baalbek where several springs and streams come together to form it. It flows from Mt. Lebanon to Biqāʿ, passes Marj ʿUyūn and Shaqīf and through the apple-growing region and enters the sea between Sidon and Tyre.

Beirut: It rises from Mt. Shūf. There is another river near it which rises from the same place.

Nahr Ibrāhīm ("River of Abraham"): It rises from Mt. Lebanon and Mt. Kasrawān (text: Karwān) and flows into the sea.

Ghaḍbān ("Raging River"): It rises from Ṣāmsiyya and flows past Tripoli to the sea.

al-Bārid ("Cold River"): It rises from a place near Nahr Ghaḍbān.

al-Aʿwaj ("Crooked River"—not to be confused with Nahr ʿAwja above): It rises from Mt. Hermon. Then it flows past Wādī'l-Tīn, Saʿs [aʿ], Kiswa, Ḥūrjūliyya, ʿĀdiliyya, Ghazlāniyya, and then enters a lake (near Damascus).

Baradā: It is formed from a number of springs. Three of these springs arise from the ʿUyūn Tūt ("Mulberry Springs"). They flow from the area above the Zabadānī dervish lodge. They are called the Waters of the Baradā. These waters meet those of the spring of Fīja. Then the Fīja river is raised (by a waterwheel?) about one yard and it waters a village called Hāma. From here the water is channeled into the Yazīd Canal which goes as far as Qāsyūn. It continues through Qābūn and Ḥarasta and ends at Dūmā. In early times it went as far as Māṭirūn (which is near Damascus) and Zanbath. Today it ends at Dūmā and ʿAzar. The Yazīd Canal is ascribed to the Umayyads. Yazīd b. Muʿāwiya widened and deepened this canal and so it was ascribed to him.

Thawrā: It begins near Dummar (Dūmā?) and flows above Būrā as far as Qayṣar. Near Ashrāfiyya it is diverted to the Mizza Canal and from there to the Dārānī Canal. Also from the Būrā Canal (or river?) (the water is taken to the canal called) Nahr Qanawāt. Then it is diverted into the Bānyās Canal and from the Bānyās it is diverted at Rabwā. It passes by the foot of Jabal ʿUrban (?) west of Gök Meydānı. Then it reaches the Mevlevīḫāne (the Mevlevi dervish lodge outside Damascus) and from there enters the walls (of Damascus). Then it reaches the Umayyad Mosque where it is divided into numerous channels, one of them bringing water to all the houses in Damascus.

ʿAqrabāʾ Canal: It is diverted from the Baradā near Baḥṣa. The waters flow past the foot of the western mountains.

After being diverted into the aforesaid canals, the Baradā River passes by Gök Meydānı. The area around Baḥṣa is called Qahwat al-Maqṣif. The ʿAqrabāʾ Canal branches off here. Then both the Baradā and the ʿAqrabāʾ flow through the center of Damascus. (The ʿAqrabāʾ) flows near Shaikh Rasān while the Baradā divides in two near the wadi. One branch is called the Dāghilī Canal, the other the Jarmānī Canal. The Dāghilī flows through Barka, Dāʿa, Kafar Baṭna, and Wathqana where it joins the other branch and becomes the Jarmānī Canal. It flows through Jarmānā (a district of the Ghouta) to the great settled area. After all these canals come together at the village of Manīḥa near the bridge of Ghayḍa, it is once again called the Baradā (river). Then it flows through Ḥadītha, Ḥarzmiyya, Bayta Nā'im and Muḥammadiyya. Then it flows into a lake.

ʿĀṣī ("Rebel River"), also called Urunṭ (Orontes) and Maqlūb ("Inverted River"). Water is raised from the river with waterwheels. This river begins near Baalbek. [M 558] The source is near the village of Ra's which is about one stage from Baalbek. It reaches a place called Qāsim al-Hirmil between Jūsiyya and Ra's. Here, where it passes through a valley, is the main source of the river at a place called Maghārat al-Dhahab ("Gold Cave"; in Abū'l-Fidā', Maghārat al-Rāhib, "Monk's Cave"). When it passes Jūsiyya it flows north. Then it flows into Lake Qadas to the west (of Ḥimṣ). Then it flows out of the lake, passing through Ḥimṣ and then through Rastan and Ḥamā. Then it flows into (Lake) Afāmiyya. Then it flows out of the lake and continues to Darkūsha and then al-Jisr al-Ḥadīd ("Iron Bridge"). Bordering the river thus far to the east has been Mt. Lukkām. When it reaches al-Jisr al-Ḥadīd, Mt. Lukkām comes to an end. Here the river turns and goes to the southwest. It passes the city wall of Anṭākiya and enters the Mediterranean near al-Suwaydiyya.

A number of streams flow into the Orontes. One is al-Nahr al-Kabīr ("Big River"), which rises north of Afāmiyya and flows into Lake Afāmiyya. Another is al-Nahr al-Aswad ("Black River") which flows north passing under Darbassāk. Another is Nahr Yaghrā which rises near (the town of) Yaghrā and flows into the al-Nahr al-Aswad, then flows into the Lake of Anṭākiya. Another is Nahr ʿAfrīn which comes from Anatolia, flows by al-Rāwandān and al-Jūma, then enters (the region of) ʿAmq and there joins al-Nahr al-Aswad.

Quwayq: It rises at a village called Sīnāb. It is seven miles from Aleppo. It joins the Marʿash River and other rivers. Then it flows through Aleppo and (the district of) Qinnasrīn and ends at al-Marj al-Aḥmar ("Red Meadow").

Ceyḥan (Jayḥān, Pyramus): It is colloquially called Jahān Ṣuyı. It rises from the land of Sīs and Maṣṣīṣa near Elbistān and flows into the Sayḥān.

Seyḥan (Sayḥān): It rises from Mt. Ḳuramaz near Kayseri (in Anatolia) and flows through the land of Sīs. Then it flows east of the city wall of Adana and passes under a large bridge at the foot of Adana castle. It continues for a while under the name of Çaḳıd Ṣuyı. Near Maṣṣīṣa it joins the Ayās-Barandī and Jayḥān rivers, then enters the Mediterranean between Ayās and Tarsus.

Samsāṭ (Sumaysāṭ, Samosata): It flows near the city of Samsāṭ (Samosata) and reaches Ḥiṣn Ziyād, i.e., the town of Ḫarpert (Harput). Then it flows past Malatya and into the Euphrates.

Mountains

Jabal Ṭūr Sīnā (Mt. Sinai): This mountain is between Syria and Madyan (Midian). In Hebrew *ṭūr* means mountain. Later this word was applied to several mountains. One of them was Ṭūr Hārūn.

Ṭūr Hārūn (Mt. Aaron, Mt. Hor): This mountain overlooks Jerusalem. The tomb of Aaron is there. They say that when Moses went with Aaron (to this mountain) to converse with God, Aaron saw an excavated tomb and near it was a beautiful couch. He decided to lie down, and when he did so, he died and was buried in that tomb. Some have suspected that the same thing happened to Moses.

Jabal Kalbūdī: It is west of Nablus and stretches along the Jordan River. Here Saul and Goliath fought a great battle and Saul was victorious.

Jabal ʿAwf: It overlooks the Ghawr (Jordan Lowland). The city of ʿAjlūn is found here. It has all kinds of fruit and trees and many streams. Jabal al-Ṣalt is to the east, two stages from Karak.

Jabal al-Sharāh: It is east of (the district of) Balqā'. Currently peasants live here.

Ṭūr Zaytā (Mt. of Olives): A mountain at Jerusalem. There are ancient ruins there today. Seventy thousand of the Banī Jūʿ (The Hungry) perished on Ṭūr Zaytā (an allusion to all the graves there).

Jabal al-Ṭūr: It stretches toward Lake Tiberias.

[M 559] Ṭūr Sīnā (Mt. Sinai): Near this mountain is where Moses spoke with God. When Moses came to this mountain, a cloud (*ghamām*—cf. Koran 2:57, 2:210, 7:16) descended upon it and Moses entered the cloud on this mountain. God conversed with him and when He manifested Himself the mountain crumbled, as stated in the noble Koranic verse (7:143). When the stones broke open, an image of the boxthorn tree appeared. This is the reason the Jews greatly esteem the boxthorn tree.

Jabal ʿĪsā ("Mountain of Jesus"): A mountain three miles from Ṣafad. Jesus usually worshipped on this mountain. He taught his disciples how to pray and also performed some of his miracles on this mountain.

Jabal Tābūr (Mt. Tabor): A high mountain near Tiberias. At the top there is an open space about a parasang long. The air is excellent. There are many vineyards and olive trees. There are all kinds of flowers. In particular, all kinds of birds come here both in winter and summer. Abraham spent a lot of time on this mountain.

Mt. Ṣadīqā: It is between Tyre, Qadas, Bānyās, and Sidon.

Mt. ʿĀmula: It extends south toward the coast. It reaches as far as Tyre. The fortress of Shaqīf is on the west side of this mountain. The mountains of Shūf, Kasrawān, al-Naṣriyya (Nuṣayris, members of a Shīʿī sect), and al-Kalbiyyīn (members of the Kalb or Yemeni faction, see below under Hebron) are also in this area.

Mt. Qāsyūn: It overlooks Damascus. There are ancient ruins here belonging to the prophets. This is one of the most important mountains. On this mountain are caves and grottos and places where the righteous worshipped (temples?). There is a cave near Maghārat Dam ("Cave of Blood") which they say is the cave where Abel was murdered (by Cain). There is another cave called Maghārat Jūʿ ("Cave of Hunger"). They say that forty prophets perished of hunger in this cave.

Jabal al-Thalj ("Snow Mountain"—i.e., Mt. Hermon): It is west of Damascus in the region of Bānyās. It is a lofty mountain, visible from a three days' journey away. It is always covered with snow. There are ruins of buildings where people lived in ancient times. Today there is a prosperous village near the foot of the mountain called ʿArna. This village is covered with snow in the winter. In this season the people live with their animals in caves. They stay in the caves four or five months, until the winter passes. Then they come out and sow their crops.

Jabal al-Rabwa: A lofty mountain one parasang from Damascus. On its summit is a beautiful mosque. The area around it is adorned with gardens, trees, and fragrant herbs. Seeking a course for the Thawrā River, and because the mountain was an obstacle, they carved a channel under the mountain and made it flow through it. The Yazīd River (canal?) is above this mountain and descends below it (?). There is a small cave on this mountain where, they say, Jesus spent some time.

Jabal al-Jawlān (Mt. Golan): It faces Damascus.

Jabal Lubnān (Mt. Lebanon): It is the largest mountain in Syria. It is connected to a chain of mountains that starts with Mt. ʿAraj in Ḥijāz, continues through the mountains of Sinai and Ayla and connects with the mountains of Jerusalem. From there it reaches the Biqāʿ (plain), Ḥimṣ, Aleppo, and *al-thughūr* (the border with Anatolia). From there it extends to Anatolia and to Mt. Lukkām.

Mt. Lebanon has springs of cold water and fruit trees and countless beneficial herbs used in the preparation of theriac (a mixture of various drugs used especially as an antidote to poison). It is reported that the herb of alchemy (?—*ḥaşīş-i kīmyā*) is found on this mountain. The devout and the great mystics worship on this mountain. They have huts made of brush and eat nothing but the local fruit.

[M 560] As part of Mt. Lebanon there is a mountain called Dayr that overlooks the coast of Beirut. They say that this is the mountain about which Kanʿān the son of Noah said, *saʾāwiyā ilā jabalin yaʿṣimuni min al-māʾi* ("I will betake myself for refuge to a mountain that will save me from the water," Koran 11:43). At the foot of this mountain is a village called Ḳarak which, they say, has the tomb of Noah and also the tomb of Shaybān Rāʿī.

Jabal ʿAkkār: It is east of Tripoli. It takes its name from a citadel on this mountain. It extends to the north. Ḥiṣn al-Akrād ("Citadel of the Kurds"—i.e., Krak des Chevaliers) and Ḥimṣ front it on the west. It continues (toward the north) passing Ḥamā and Shayzar as far as Afāmiyya (on the west) and connects with Mt. Lukkām.

Jabal Shaḥshabū: It is named after a village there. This mountain runs from south to north, passing to the west of al-Maʿarra, Sarmīn, and Aleppo. It then [bears west and] joins the mountains of Anatolia.

Jabal Lukkām: This mountain is connected with Mt. Lebanon. It goes as far as Maʿrash and separates Syria from Malatya. This mountain chain extends (south) past Marʿash, ʿAyn Zarba, and Hārūniyya. It is called Mt. Lukkām as far as Lādhiqiyya. From Lādhiqiyya to Ḥimṣ it is called Jabal al-Bahrā. It connects with Mt. Lebanon. This mountain is very populous and has much fruit. It lies "between the two spaces," i.e., between Syria and Mesopotamia.

Jabal Khūshan: It is west of Aleppo. The wife of Ḥusayn b. ʿAlī (the Prophet's grandson) had a miscarriage and the child was buried here. This blessed shrine is known as Mashhad al-Ṭarḥ ("Shrine of the Miscarriage").

Jabal Summāq: A district (*ʿamal*) of Aleppo with many cities and inaccessible fortresses. The mountain gets its name from the fact that much sumac grows here.

Jabal al-Bustān: There is a cleft in this mountain through which a road passes. If those who pass through it do not eat cheese and bread, they become ill from its putrid air. This is a feature for which this province is known.

Eulogy of the clime of Syria

The air is excellent, so most of the people are energetic and healthy. This clime is very prosperous, its climate is moderate, its agricultural production is high, its food is delicious, its homes are attractive, and its people are handsome and skilled craftsmen. In particular traces of buildings of the Barmakids (?) are still standing. It has many neighborhood mosques, Friday mosques, Sufi lodges, and charitable institutions. It has countless madrasas. Above all, its Umayyad Mosque is famous throughout the (Muslim) world.

This clime was so prosperous in ancient times—during the time of the Israelites—that it was divided into 32 separate sultanates each with an independent ruler. The soil of this clime is extremely fertile. Some places in the district of Ḥawrān in particular produce one hundred fold. It has pastures and grasslands and abundant and various medicinal plants. In particular it has all kinds of flowers. Hyacinths, jonquils, tulips, and others grow wild in the mountains and on the plains.

There are all kinds of fruits in this clime: apricots, apples, pears, plums, cherries, sour cherries, pistachios, pine nuts, hazelnuts, bananas, sugarcane, figs, sycamore figs, quinces, peaches, pomegranates, *mersin* (probably the *dağ mersini*, bilberry or whortleberry), *zuʿbūb* (sorb, azerole), lotus fruit, walnuts, almonds, *ʿuqābiyya* (?), wild pistachio, *ḥabb al-ʿazīz* (same as *ḥabb al-zalam*, root of the rush-nut), olives, mulberries, and other fruits beyond count.

There are many kinds of trees, including balsam and oleander. There are all kinds of vegetables, (edible) roots, taro, cauliflower, etc. There is no lack of fruit summer or winter. [M 561] Even manna rains down. The varieties of chickpea grown here are found nowhere else. Fruits include: bitter orange, *raknā* (?), grapefruit, *naffāsh* lemon, citron, peach, lemon, orange.

Different kinds of animals live here. The sheep of the Israelites are especially common. There are also tigers, bears, hyenas, wild asses, leopards, wild boar, foxes, wolves, jackals, various kinds of deer, and rabbits. The animals here are found nowhere else. They hunt gazelles and various birds.

Silk, cotton, sesame, olives, olive oil, alkali, salt, and other products are very abundant. The annual tax revenue from Tripoli alone is 900 purses. (But) it is currently 600. The tax from Aleppo is 800 purses. It had been 600 purses from Ṣafad and Sidon, but now it is 500.

The Israelites relate that at the time of Moses there was created a bird named *ʿanqā*. Its plumage was multicolored. It had wings on all four sides. Its face resembled that of a man. It had a portion and sample of every beautiful thing attached to its limbs. There was a similar bird that lived around Jerusalem. Later it went to the regions of Ḥijāz and Najd. It became used to killing animals and livestock. Khālid b. Sinān al-ʿAbsī was a devout man who some said was a prophet (before Islam). The people complained to him about this bird. He cursed it and the bird became extinct.

Languages

There are various languages in the clime of Syria. Arabic, Turkish, Kurdish, Persian, Hindi, Afghani, and Sulaymani (a dialect of Kurdish) are spoken in Damascus. The Hindi speakers even have their own lodge (*tekke*). In addition, there are those who speak Maghrebi (the Arabic of North Africa), Syriac, and Hebrew. In [I]skandarūn and Tripoli, also in Sidon and Jerusalem there are various groups of Christians who speak Greek, Latin, Italian, French, Spanish, English, German, Polish, Muscovite, Russian, Coptic, Ethiopian, and Armenian.

Religions

The people of this clime are Muslims. But there are also Jews, Christians, and others. There are more Jews than Christians. There are a number of erring sects: the Taymānī Druze in the Shūf Mountains; the Nuṣayrīs in the vicinity of Tripoli; the Kalbiyya (members of the Kalb or Yemeni faction, see below under Hebron) and Ismāʿīlīs in the region of Ḥamā and in Jabal Samʿān in the district of Aleppo. There are Shīʿīs in every city. There are also Yazīdīs who live in several places around Aleppo and Ḥamā; they are Kurds.

Concerning the direction of prayer (*qibla*)

When Moses came on the scene, he instructed the Israelites to face a certain direction (when they prayed). And so it remained until David came on the scene. He instructed them to face Jerusalem. The Christians were instructed to face the equinoctial point of sunrise. They reasoned that paradise was in the direction of the rising sun. The Samaritans faced Jabal al-Burayk. For Magians the direction of prayer was the sun—they prayed to the sun, but at sunrise, sunset and midday they also worshipped fire and water. The Ḥarrāniyya, i.e., the people of Ḥarrān, faced the south pole and the Sabians faced the north pole.

Administrative divisions of Syria

Since I am giving a comprehensive description of Syria, let me now describe its administrative divisions. I will first mention the jurisdictional divisions.

The clime of Syria is divided into a number of provinces (*eyālet*). One is Shām, [M 562] which consists of Damascus. The others are Jerusalem, Tripoli, Sidon, and Aleppo. Syria is then subdivided into ten sanjaks: Damascus, which is the pasha sanjak, Jerusalem, Gaza, Nablus, ʿAj[l]ūn, Lajjūn, Ṣafad, Sidon, Beirut, and Ķarak and Shawbak. However, Jerusalem, Gaza, Nablus, ʿAjlūn and Ṣafad are a separate group. The noblest of the administrative divisions of Syria is Palestine. Let me begin with it because it includes the district of Jerusalem.

Description of Palestine

Palestine is composed of two sanjaks: Gaza and Jerusalem.

Borders: In the southwest the border goes from the Mediterranean and al-ʿArīsh to the Wilderness of the Israelites (Tīh = Sinai desert). In the southeast it is the Dead Sea and the Jordan River. In the north it goes from the Jordan River to the borders of Urdun as far as Caesarea.

In ancient times four tribes of Israelites lived in Palestine: Simon in Hebron; Dan between Ascalon and Lydda; Judah in the environs of Jerusalem; and Benjamin in Jericho where the tomb of Moses is.

Cities

Gaza: A medium-sized town on the coast. In the past it was a hereditary grant of a pasha, but this status was removed and today it is attached administratively to Jerusalem. This city has gardens and vineyards, dates, olives, bitter oranges, lemons, bananas, and other fruits. They have a prized apricot called *mishmish aʿlā* which is very sweet; the sycamore fig, whose fruit resembles a regular fig but has no seeds; and a prized almond with a sweet flavor and a crumbly shell. On the coast there is a beautiful park with all kinds of fruit trees. Between Gaza and the sea are sand dunes where, because of the sand, the fruit trees only grow the height of a man.

The qadi of Gaza is subordinate to the molla of Jerusalem. It has several districts: Majdal, Ramla, and Lydda. The tomb of Hāshim b. ʿAbd Manāf (great grandfather of the Prophet) is in Gaza. In the pre-Islamic period ʿUmar was held prisoner in Gaza. Later, during ʿUmar's caliphate, Muʿāwiya b. Abī Sufyān conquered it. (The jurist) Imām al-Shāfiʿī was born here in 150 (767). He memorized the Koran at age seven and he memorized the *Muwaṭṭa'* (the law book of Mālik b. Anas) at age ten.

During the period of the birthday of the Prophet the *Mawlid-i sharīf* (a panegyric poem in honor of the Prophet) is recited every Friday at blessed places in Gaza. There is a huge crowd. This is a famous fair for the Gazans. They also have the custom of making a promenade to the coast every Saturday; this is called the *sabtiyya* (Saturday stroll).

Ascalon: An old town on an elevated place on the coast, 3 parasangs from Gaza and 4½ from Ramla. The people get sweet water from wells. The town was conquered by Muʿāwiya b. Abī Sufyān. It is reported that in the year 86 (?—perhaps error for 547/1153, date of Baldwin III's assault on Ascalon) the Franks (Crusaders) appeared off the walls of Ascalon in tall ships filled with men and arms. The ships rafted together and stormed the city. Having loaded ships as tall as the walls of Ascalon with men and arms, they steered them as far as the sea walls, then leaped from the ships and took the city by force. It remained in their hands for 35 years until Saladin liberated it from the Franks. When the Franks took Acre, they wanted to recapture Ascalon, but when Saladin learned of this he razed Ascalon.

[M 563] There is a martyrium here for the head of Ḥusayn (the grandson of the Prophet). It is a magnificent martyrium with marble columns. Many gifts come to it from all around. Some marvelous ruins can still be seen in Ascalon.

On the road from Ramla to Gaza is the grave of Abū Hurayra (a companion of the Prophet); it is between Khān Sudūd and Ramla.

Ramla: A famous city, 18 miles from Jerusalem. It was laid out by (the Umayyad caliph) Sulaymān b. ʿAbd al-Malik. The Franks occupied it for a number of years and then Saladin liberated it. ʿAbd al-Malik's brother brought water to the city by an underground channel. Now drinking water is supplied from wells and rain water collected in cisterns. They also have a large pool from which they use to water their riding animals and livestock.

When Syrian merchants return to Ramla from Jerusalem they set up a bazaar here that runs for several days. Many people come from the surrounding districts to shop here. There is a bazaar in Ramla once a week. There is an especially large bazaar on Red Egg Day (Easter).

This city is famous for its apricots and watermelons. Its most important product is cotton. It has beautiful vineyards and many olives. There are workshops that make soap. There is an old temple near Ramla called the White Mosque (*al-jāmiʿ al-abyaḍ*) where many prophets are buried.

Lydda (Ludd): Sulaymān b. ʿAbd al-Malik razed this city and built Ramla. Nearby is a place called Bi'r Zaybak. It is related that Jesus will kill the Antichrist here. This city is one hour from Ramla. There is a bazaar here every week. There is a famous church in this city. The commander of the *ḫāṣekī* regiment is stationed here.

Yāfā (Jaffa): A small city enjoying much affluence. It is a famous coastal town. There used to be a large citadel here. It had thriving markets and commercial agents. It has a harbor for ships. It is 6 miles from Ramla and is the port for Ramla.

Arsūf: A town on the coast with a fortress. It is currently in ruins. Some remains of its walls and markets can be seen. It is 12 miles from Ramla. The tomb of ʿAlī b. ʿAlīm is in the vicinity.

Qaysāriyya (Caesarea): A famous city on the coast built by the Roman emperor. It is 18 miles from Ramla. Its air is very fine. It has many charitable institutions. Its fruit is delicious. When the Franks occupied it they destroyed it and it has remained in this state.

Bayt Jibrīl: A city that is spread over plain and mountain (?—*sahliyya jabaliyya*). Its *rustāq* (cultivated area with its villages) is called Dārūm. It has marble quarries and a fortress.

Bayt Ḥabrūn (Hebron): It is the city of Khalīl al-Raḥmān ("Friend of the Merciful One"—i.e., Abraham; cf. Koran 4:125). Here are the tombs of Abraham, Isaac, Jacob, Sarah, Adam and Eve, Solomon, Zachariah, and Jesus; also the shrines of Ibrāhīm b. Adham and ʿAbd Allah b. Ṣāmit. This city is in a serene place amidst the densely forested mountains. On the plain and in the mountains of this region grow olives, figs, carobs, and other fruit. In particular there are countless vineyards. There are also mines of rock-crystal which is worked here and then sent to the surrounding region for sale. Indeed, crystal is exported from here as far as the *provinces* of Ethiopia and Sudan where it is exchanged for gold dust. The grape molasses and dried grapes of this city are also famous.

Hebron has a formidable citadel in which guards are stationed. [M 564] The people claim that it was built by jinn because the stones used in its construction are huge. Each is a block 6–7 yards in length and 4–5 yards in breadth. It (the citadel?) has a beautiful fountain.

Surrounding the city to a distance of half a stage are vineyards and gardens. Its fruit, including apples, grapes, and quinces, is famous. Banquets are given in this city to celebrate Abraham. There are cooks, bakers, and servants (for this purpose). Everyone, rich and poor, eats lentils cooked in olive oil. There is a marvelous feature to this meal. When one sees this dish he prefers it to all others because it is more delicious. These lentils come every year from Egypt; there is an endowment (for this purpose). And every year there is a great fair, attended especially by pilgrims and merchants from Syria.

The people in this region are divided into two hostile factions: the Yemenis or "Whites" (Aḳlu) and the Qaysis or "Reds" (Ḳızıllu). When they clash, the Reds shout *yā lahū birr* and the Whites shout *yā al-maʿrūf*. These parties have survived from pre-Islamic times and retain the "bigotry of ignorance" (*ḥamiyyatu'l-jāhiliyya*, cf. Koran 48:26).

There is a pool in this city which fills with water when it floods. Near the city is a small mountain called Ṣughar (Zoar) that overlooks the Dead Sea. On the mountain is a mosque built by Abū Bakr Ṣabbāḥī and also a shrine of the Forty which is a place of pilgrimage.

Hebron belongs (administratively) to Jerusalem. Its people are very strict. It is 6 parasangs from Jerusalem. There is a large forest between the two cities.

Arīḥā (Jericho): It is west of the Ghawr and near the Jordan River. Jericho is Qaryat al-Jabbārīn ("Village of the Giants"). It is one day's journey from Jerusalem. Dates, bananas, and sugar cane are grown there. Sulphur is mined nearby. The indigo plant is grown there and indigo is extracted; it is a beautiful dye. The fields are irrigated with water from local springs.

Bayt Maqdis (Jerusalem), also called Īliyā (Aelia Capitolina). This city is now the seat of an Ottoman pasha. He is in charge of the caravan of Syrian pilgrims to and from Mecca. Previously the commander of the pilgrimage was the pasha of Nablus. Later this responsibility was given to the pasha of Jerusalem. Meanwhile, Gaza was attached administratively to Nablus. Its longitude is 68 degrees, 11 minutes; its latitude is 32 degrees.

Originally David built Jerusalem, by God's command, on the Rock (Sakhra). His son Solomon completed it. Afterwards Nebuchadnezzar destroyed it. Some Persian kings rebuilt it. Later, when Constantine and his mother Helena became Christians, they destroyed what had been built on the Rock and constructed the Church of the Holy Sepulcher. When (Caliph) ʿUmar conquered Jerusalem, he cleared away the rubble on the Rock and built a mosque on it. Later, in 721, al-Walīd b. ʿAbd al-Malik built a dome over the Rock. In short, Jerusalem sits on high mountains; one ascends to it from all directions.

The city of Jerusalem is somewhat oblong extending from east to west. It is surrounded by fortifications, built partly on the mountain and partly on the moat. It has eight gates (seven are listed): Zion Gate: When one enters a little way one comes to the Church of Zion, where David is buried and where Jesus and his Apostles ate a meal (the last supper); from here one goes to the Wādī Jahannam (Gehenna) [M 565] where there are Jewish and Christian cemeteries; nearby is the ʿAyn Silwān (Siloam). Gate of Tīh (the Wilderness of Sinai). Gate of Balāṭ (the Palace or Court). Gate of Jubb Urmiyāh (Jeremiah's Pit). Gate of Jericho. Gate of al-ʿAmūd (the Column). Gate of Miḥrāb Dāwūd (David's Oratory).

It has three pools: Pool of the Israelites. Pool of Solomon. Pool of ʿIyāḍ (named after ʿIyāḍ b. Ghanm, a companion of the Prophet).

In the mosque (al-Aqṣā') are twenty cisterns (*jubb*). Every quarter has a well and every house has a cistern. Two large reservoirs have been constructed in a wadi near the city. They are filled when it floods. The water for the al-Aqṣā' Mosque comes from them.

The Rock (Sakhra) is a rock (*ḥajar*) raised above the ground and over it a lofty dome. The Rock is about the height of a man above the ground. One can descend into a hollow below, a plain chamber about the length of the Rock.

There are a number of springs in Jerusalem, but no streams. Nevertheless, it is a fertile area. David's Oratory is there. When al-Walīd b. ʿAbd al-Malik built the dome over the Rock, he built other domes there as well and gave each one a name. They are Qubbat al-Miʿrāj (Dome of the Ascension of the Prophet), Qubbat al-Mīzān (Dome of the Scale), Qubbat al-Silsila (Dome of the Chain), and Qubbat al-Maḥshar (Dome of the Resurrection).

The al-Aqṣā Mosque is mentioned in the Koran (17:2). It is in the eastern part of Jerusalem. David built its foundation. Each one of its stones is ten cubits in length. The courtyard of the mosque is long and wide. The mosque is very beautiful and solidly built. The mosque is built on columns of colored marble. The enclosed mosque has 26 doors. The one opposite the prayer niche is called Bāb al-Nuḥās (the Bronze Door). To the right of it are seven large doors, of which the middle one is made of bronze and is gilded. On the left is the same arrangement (of doors). On the east side of the building are 11 doors and a colonnade with 15 domes supported by marble columns.

The courtyard of the mosque is paved with stones. In the middle is a large platform, five cubits high, with steps on all four sides up to the top. In the middle of that is a large dome set on marble columns over an octagonal space and covered with lead. It has four gates, each one opposite a courtyard (reading *ḳāʿa* for *ḳāh*). They are: Gate of the Qibla (direction of Mecca), Gate of Isrāfīl (the Archangel who will sound the last trumpet), Gate of the Trumpet (al-Ṣūr), and Gate of Praising God (al-Tasābīḥ). There are also gates at the anterooms. The interior space is enclosed by colonnades supported by columns. There is a fourth colonnade on the exterior enclosed by another gallery. Above the columns over the Rock is a circular gallery and above that is a drum that juts into the air on which are large galleries and there is a large dome on the drum. The height from the ground is 100 cubits. The exterior and interior are covered with mosaics, cut stone, and marble. The Rock, which is the object of pilgrimage, is right under the dome. The footprint of the Prophet is found over the row of columns (?). The top of the dome is covered with lead.

The (al-Aqṣā) Mosque has 13 gates (ten are listed):

- Gate of the Prophet (al-Nabī).
- Gate of Mary's Oratory (Miḥrāb Maryam).
- Gate of Mercy (al-Raḥma).
- Gate of the Pool of the Israelites.
- Gate of the Tribes (al-Asbāṭ). When one enters a little way one comes to the Church of Mary. This place is called al-Jasmāniyya (Gethsemane). Mary's tomb is there. This place is connected to the Mt. of Olives. Ascending the mountain, one encounters a church where there are monks and nuns. A little to the east of the mountain, on a flat area, is the tomb of Lazarus whom Jesus brought back to life. On the road to Jericho, [M 566] near al-Kathīb al-Aḥmar (the Red Dune) is the noble grave of Moses. (The Mamluk Sultan) al-Malik al-Ẓāhir Baybars built a dome over it. Every year there is a festival and people make a pilgrimage to it. The city of Jericho is nearby.
- Gate of al-Hāshimī.
- Gate of al-Walīd.
- Gate of Abraham.
- Gate of Umm (Mother of) Khālid.
- Gate of David. The noble tomb of David is here. There are other shrines in this area.

The length of the al-Aqṣā Mosque is 784 cubits and the width is 455 cubits. There are 684 columns. There are 30 columns inside the Dome of the Rock. The Dome of the Rock is sheathed with lead which in turn is sheathed with gilded bronze. In the ceiling of the al-Aqṣā Mosque are 4,000 pieces of wood and stone. The Rock is 33 by 27 cubits. The hollow under the rock has room for 60 people. In the al-Aqṣā Mosque 1,500 lamps were lit and in the Dome of the Rock 464. Every month 100 *qisṭs* (an ancient measure of 20 fluid ounces) of oil were set aside to light these lamps. Every year 100,000 cubits of reed mats came from Egypt for the al-Aqṣā Mosque.

In the past, 200 slaves were assigned to serve the al-Aqṣāʾ Mosque. (The Umayyad caliph) ʿAbd al-Malik b. Marwān appointed them from the prisoners that were his share of the booty, so these servants were called *khumāsī* (the caliph's share of booty was one-fifth = *khums*). Their sustenance came from the state treasury. Currently several thousand lamps are lit in the al-Aqṣāʾ Mosque on the night of 15 Shaʿbān. This is a famous night and people come from all around.

In the middle of the Haram there are many olive trees and there are anterooms on each side. The religious scholars, shaikhs, and devout of Jerusalem sit there. Within the Haram the ritual prayer is done according to the four Sunnī rites. Each rite has its own imam. The first prayer is conducted by the Mālikī imam. The reason for this is that at the last conquest of Jerusalem a group of Maghrebis (Muslims from North Africa who are Mālikīs) were the first to enter the city. In recognition of this they requested precedence in performing the ritual prayer, i.e., that their imam lead the prayer before the others. And so it is.

In Jerusalem is the great church of the Christians known as al-Qumāma (place of the Resurrection—i.e., Church of the Holy Sepulcher). It is in the middle of the city. It is built with great artistry and has beautiful ornamentation. The wealth of this church has no limit. Entering the church from the door on the west side, one is in a domed space. The dome covers the entire church, which is below that door. One cannot go down into the church from any (other) direction. On the north side, however, there is a door—called the Door of St. Mary—from which one goes down 30 steps to the Holy Sepulcher. On the south side is the Door of the Crucifixion (Ṣalūbiyya). Opposite that on the east side is another large church where the Franks perform their rituals. East of here and a bit to the south is the place where Jesus was imprisoned and where he was crucified. Inside the large dome, the upper part of which is open, are images of the prophets, Jesus, Mary, and John the Baptist. In the Holy Sepulcher there are three gold lamps suspended from the ceiling. When one leaves the great church and goes east one comes to Jerusalem (proper). In a certain place in this church is a lamp about which they claim that on a specified day—i.e., on their festivals—a light descends from the sky [M 567] and ignites the lamp. Among Christians this phenomenon is well known. The Greeks and Armenians also believe this but the Franks do not.

When Christian pilgrims enter Jerusalem they are charged two goldpieces and when they enter the Church of the Holy Sepulcher they are charged nine goldpieces. The pasha of Jerusalem takes the Christians to the Dead Sea and to bathe in the Jordan River. They are charged a fee for this. The money that is collected at the Church of the Holy Sepulcher is distributed among the religious scholars, shaikhs, and imams in Jerusalem. This is called *māl-i rubʿa* ("quarter money"). Everyone receives his share.

The Friday preachers go about the city carrying a rod (to encourage attendance at prayer?).

Every year the *ṣurra* (sultanic gifts for the holy cities) comes from Egypt. It amounts to one purse and 11,320 *para*.

There are many madrasas in Jerusalem. Indeed the molla of Jerusalem resides in a madrasa. The pasha has a special palace where he lives.

In Jerusalem there is a spring called ʿAyn Silwān (Spring of Siloam). The people regard it as blessed. This spring irrigates many gardens. There is a quarter in this area that ʿUthmān b. ʿAffān established as a pious endowment for the poor of Jerusalem. Near it is the Valley of Jahannam (Gehenna).

Bethlehem: A village six miles from Jerusalem. This village is the birthplace of Jesus. Christians revere the church of Bethlehem. Between Jerusalem and Bethlehem is the tomb of Rachel, the mother of Benjamin and Joseph. Over this tomb are twelve alcoves and a dome. East of Bethlehem is the Melkite (al-Malikiyya) Church in which are the prayer niche of ʿUmar, the tomb of Solomon, the date palm of Mary, and the cradle of Jesus. The shepherds who announced the birth of Jesus are also (buried) in Bethlehem.

In short, there are many charitable foundations in Jerusalem. From its surroundings come bitter oranges, almonds, dates, walnuts, figs, bananas, and other kinds of fruit. There are two seasonal fairs each year.

The Mt. of Olives is east of the city, overlooking the noble (al-Aqṣā) Mosque. At the beginning of the valley there is the Mosque of ʿUmar. This is where ʿUmar resided when he conquered Jerusalem. And there is a church here (in the valley) at the spot of Jesus' ascension. And there is a place called Sāhira (Wakeful), which they say will be the place of resurrection.

In Jerusalem are workshops that make soap. They cook the soap. The soap made here is highly prized and famous. They also make fine musk-scented soap that is highly prized and found nowhere else. The rosewater of Jerusalem is also famous. Its fragrance wafts to a great distance. There are many kinds of flowers in Jerusalem, but it is the fragrance of the rose that is very strong.

Seasonal fairs

Every year in the spring after Nawrūz (March 22, the vernal equinox), one of the Sufi shaikhs of Damascus and a member of his order conduct the pilgrims of Damascus to Jerusalem. Three months beforehand, the shaikh and a member of his order carrying a banner on his shoulder go to the Umayyad Mosque for Friday prayer. After the prayer they go about the markets shouting "Whoever wants to make a pilgrimage to Jerusalem should get ready. We will depart at such and such a time." In order to drum up enthusiasm among the people they enumerate its pilgrimage sites. One week before they set out, the shaikh and his novitiates go throughout Damascus with drums and tambourines and banners, performing *tevḥīd* (i.e., *ẕikr*). At the end of the week, they proceed to the village called Dāryā and stay there. The next day all the pilgrims depart and go to the quarter called Saʿsaʿ. All the pilgrims and merchants assemble there. This stage is a place of excursion. The Nahr al-Aʿwaj ("Crooked River") passes through here. On the bank of the river is a beautifully designed inn built by Sinān Pasha. There they reside. Villagers arrive and line up in rows along the road [M 568] to view the pilgrims. Each one has a reed or wicker basket in his hand. In the baskets are excellent plump chickens, live or cooked, flaked pastries, soft white bread, strained yogurt, butter, honey in the comb, clarified butter, eggs, various cheeses, clotted cream, and other things all of which are for sale at cheap prices. This stage is on a stony plain.

> Jisr Yaʿqūb (Jacob's Bridge): Here they cross the Sharīʿa (Jordan) River. The Christians call this river the Yūrdān. The banks of this river are covered with so many flower gardens that one does not wish to leave.
> Lake Minya (text: Mīna), also called Lake Ṭabariyya (Tiberias). On the way they pass the Well of Joseph.
> ʿUyūn al-Tujjār (Merchants' Springs). Here is another beautifully designed inn built by Sinān Pasha. Here the road splits in two. One goes to Lajjūn and Ramla and the other goes to Jerusalem. If they take the route to Jerusalem, they leave ʿUyūn al-Tujjār and (first) arrive at Jenin and halt there for one day. Jenin belongs administratively to Lajjūn. At this stage all kinds of fine products are assembled (for sale): dates, rice, henna, sugar cane, and other things. The prices are so low that (one can buy) 30–40 eggs for one *para*, one *kīle* (bushel, 36½ kilos) of barley for one *para*, and one cooked chicken for a *para*.
> Nablus: A beautiful city located between two mountains. It has abundant flowing water and springs. All of its forests are of olive trees. They stay here for three days.
> Bīra
> Jerusalem. The total distance is ten stages, but the distance between them is short. The road is safe and secure and nothing gets lost because there are guards on the road who collect a toll from merchants and dhimmis. Indeed (the road is so safe that) sometimes loads of goods are left on the road and the Bedouin guards take them to the next stages.

As soon as they arrive in Jerusalem, the pilgrims begin to visit pilgrimage sites and the merchants begin to buy and sell. The merchants remain there for more than thirty days. The shaikh and the pilgrims, however, after two weeks proceed to Hebron where they spend three days and then return to Jerusalem. They also go to the tomb of Moses and remain there for a day. The tomb of Moses is beautifully designed. There are no trees at all around it. In place of wood, they light certain stones found there (coal, charcoal) which burn better than wood. Then they go back to Jerusalem. Afterwards the shaikh and the pilgrims return to Damascus.

Chapter: Country of Urdun (Jordan)

It is the second administrative division of Syria (the first being Palestine). The borders are as follows: in the west, part of Palestine and the Mediterranean as far as Tyre; in the south, Palestine and the Sinai desert; in the east, the district of Balqā'; in the north, the borders of Damascus. Previously the capital was Tiberius but now it is Nablus. In ancient times the Israelite tribe of Issachar lived around Nablus, the tribe of Manasseh lived around Jenin, the tribe of Zebulun lived around Ṣafad, the tribe of Ruben lived along part of the Jordan River toward the Ghawr (Jordan lowlands) in the direction of Ṣalṭ, the tribe of Gad lived toward ʿAjlūn, and a clan of the tribe of Manasseh lived around the Yarmuk River.

Nablus: A sanjak. It used to be administered by the commander of the pilgrimage (to Mecca), but now has been attached administratively to Jerusalem and is administered by its qadi on the authority of the molla of Jerusalem. There is a district called Jabal al-Shām which is governed by the deputy qadi of Nablus. Nablus is a beautiful city. It is located between two mountains. It has abundant olives. It has a long market which has a famous mosque in the middle. [M 569] It has much spring water and a river flows through it.

Jenin (Jinīn; text regularly has Jīnīn): It used to be a hereditary grant (hereditary fief) of the Tūrānī-oğlı, but their hereditary grant has been abolished. This city had been the city of the Judah branch of the Israelites. It is located at the foot of Mt. Ephraim. In the past it had been a prosperous city. Its district is a place of abundance. Nearby is:

Town of Baysān: It has no city wall but has streams, springs, and vegetable gardens. The land is very fertile. A small river that rises from a spring flows through the city and divides it in two. It is a half stage from Lajjūn. In this city dates, rice, taro and sugarcane are abundant. It is located south of Tiberius.

Sanjak of Lajjūn: There is a round rock in the middle of (town) with a dome over it and a tomb under it. The people regard it as blessed. They say that when Abraham entered Lajjūn the people of the city complained to him about the lack of water. Abraham struck this rock with his rod and water gushed forth. This water was not only sufficient for the city but also supplied nearby villages and cultivated areas. It is still flowing.

Near Lajjūn is Qāqūn and near it is Jaljūliyya. The ..... (?—'LD'RW) of this area is located in the direction of Ramla.

Sanjak of Ṣafad (Safed): It is now attached administratively to Sidon. It has several districts. They are Tiberias, Acre, Tyre, Jīra, ʿUyūn al-Tujjār, Bashāra, Shaqīf, Marj ʿUyūn, Shuʿayb al-Nabī, and Kafar Kannāh (Cana of Galilee).

Ṣafad: This city, which is the capital of a province, has a strong fortress that overlooks Lake Tiberias. An underground channel provides drinking water to the fortress. It is connected to the gate of the fortress. The vegetable gardens of Ṣafad are on the shores of Lake Tiberias and in the lower (Jordan) valley. The buildings in the outskirts (*rabaḍ*, extramural settlement) of Ṣafad are spread over three mountains. After Malik Ṭāhir (the Mamluk sultan al-Malik al-Ẓāhir Baybars) had freed Ṣafad from the Franks (Crusaders), he made it a military center in order to protect the country along the (Mediterranean) coast. There is a sect of Jews who live in this city. They make excellent felt. Its fame extends in all directions. Near it is the Well of Joseph and nearby is a pool and an inn on the road to Jerusalem.

Shaqīf Bayrūt: A strong fortress near the coast one stage from Ṣafad. Near the territory of Shaqīf is the area of Bashāra.

Acre: A city on the seacoast. Inside is a spring known as Baqara ("Ox"). There is a mosque named after Ṣāliḥ (the pre-Islamic prophet who was sent to convert the tribe of Thamūd). Acre is 24 miles from Tiberius and 12 miles from Tyre. There is a flowing stream called Wādī'l-Nuʿmān. It has an excellent harbor.

Nearby is al-Nawāqir, three high mountains overlooking the sea. There are caves at the foot of these mountains. When the waves of the sea roll into them they make a loud noise that can be heard at a great distance.

Near Acre are Haifa, Ṣughūriyya, and Nazareth. Southeast of Acre is al-Ṣarfan (Sarepta).

Tyre (Ṣūr): A very strong city on the seacoast. It was conquered during the time of ʿUmar b. al-Khaṭṭāb. This city is very old. Most of the philosophers of the ancient Greeks were from Tyre. It has a great citadel and a moat filled with sea water. It is 12 miles from Acre. Tyre has marvelous circumvallation. It has a bridge that is a wonder of the world, spanning the moat with a single arch. [M 570] In no country is there a larger bridge, but there is one similar to it in Toledo in Andalus (Spain).

There is a larger bridge than the one at Tyre over the Neretva (text: Neretra) River in the city of Mostar in Bosnia, recently conquered (by the Ottomans). Its extent is 150 carpenter's cubits. It was described above.

In the past, ships entered the harbor of Tyre by passing under this (text: *bir*, error for *bu*) bridge. A long strong chain was stretched across the harbor to prevent ships from entering.

Sanjak of ʿAjlūn: An Ottoman pasha administers this province. It has a number of districts. They are Ghawr, Banī ʿUlwān, and ʿIrāj. ʿAjlūn is a citadel. It is located opposite Baysān above Mt. Ghawr. It is east of the Sharīʿa (Jordan) River. It has trees and flowing water. This citadel is new (i.e., built after the Muslim conquest?), having been built by ʿIzz al-Dīn, one of the commanders of Saladin the Great. The extramural settlement of this fortress, located on Mt. Ghawr, is called al-Bāʿūtha. This district has beautiful gardens.

Ṣalt: This place is east of Mt. Ghawr and has a fortress. It is south of ʿAjlūn and about one stage distant. It is opposite Jericho. Below the fortress of Ṣalt a large spring emerges and its water flows through the town. Ṣalt has many vegetable gardens. The pomegranates grown here are very famous. The district of Balqā' belongs administratively to this sanjak.

ʿAmmān: An old city and capital of the country. (The region) has a number of towns. The Zarqā' River, which is above the road taken by pilgrims from Damascus, passes below ʿAmmān. ʿAmmān is to the west of the Zarqā' (text: Zawqā) River and north of the Pool of Zabzā', which is one stage from ʿAmmān. On the mountain near ʿAyn al-Zarqā' is Shayb Qaṣr, now in ruins. There are important (ancient) monuments in ʿAmmān. There are arable fields around ʿAmmān and its land is productive and good. The Prophet Lot received a communication in ʿAmmān. Most of the trees of the city are *buṭum* (mastic tree, *pistacia terebinthus*), which are very abundant. Here is Qaṣr Jālūt (Goliath's Castle) with a mountain view; also the tomb of Uriah and the theater of Solomon. The Sardī (?) Arabs live in this area; they conduct raids from Balqā' to ʿUlā.

Ḥusbān: A famous city of Balqā'. This city has valleys, mills, trees, and vegetable gardens. It is one stage from Jericho.

Sanjak of Karak and Shawbak: Karak is a citadel on a lofty place. Below Karak is a valley in which there is a public bath and many vegetable gardens. This valley lies in the direction of Ḥijāz. In Karak are apricots, pears, pomegranates, and other choice fruits. Karak used to be administered by the Arabs, but later, in 1090 (1679), the pasha of Damascus, ʿOṣmān Pasha, found a way to take it from them and he stationed special guards there from the Damascus garrison. Karak is three stages from Shawbak.

Mu'ta: It is one stage from Karak. There are tombs here of Jaʿfar al-Ṭayyār, al-Ḥārith b. Hishām, and other Companions of the Prophet who died in the Battle of the Yarmuk.

Ma'āb: An ancient city. It is now a village called Rabba. It is a half stage from Karak. Near it is Rābiyya. From here the road goes to ʿAmmān via Mawjib.

District of Sharāt: It belongs administratively to ʿAjlūn. It is located between Lake Tiberias and the Dead Sea and (the area of) Balqā' (to the east). Shawbak is the capital of the district. It has a fortress. To the east is Ghawr (the Jordan lowlands), on a high white hill. At the skirt of the fortress are two springs, one to the right and one to the left, [M 571] that provide water to Shawbak and irrigate all of its vegetable gardens. Mt. Sharāt is south of Balqā'.

Rabba: It is behind Sharāt and east of Shawbak. The road is very round about. There are many vineyards and gardens in this town. The ground is made of granite.

Between Shawbak and Balqā' is Wādī Mūsā (Valley of Moses) where is the famous rock from which twelve springs emerged for the Israelites. And towards Mt. Sharāt is Khamīma, where the Abbasids emerged at the time of their caliphate, and where there is a spring giving off sulfurous gases (?)—it is always bubbling and is beneficial for scabies and other illnesses.

al-Raqīm (perhaps Petra): A small town near Balqā'. All the houses are cut from the local rock. They say that "the people of al-Raqīm" were three men who went into a cave to get out of the rain (cf. Koran 18:9: *aṣḥāb al-kahf wa al-Raqīm* "the people of the Cave and the people of al-Raqīm"). Then a huge rock fell at the mouth of the cave and blocked its entrance. The men despaired and discussed among themselves what to do. (They decided that) each one should recall the good deed that he had done and should pray. That is what they did. By God's command, the rock at the mouth of the cave was rolled away and they went out.

Maʿān: A small city. The inhabitants were the Umayyads and their clients. It is on the road of the pilgrims coming from Damascus. Sultan Süleymān built a fortress here and diverted a stream for it. It is one stage from Shawbak.

Chapter: Damascus

Damascus is the capital of the province (of Syria). The post for its molla (chief qadi) pays 500 *akçes* per diem. It has been likened to paradise. (Indeed) it is a city like paradise, a place that increases happiness. Because it is the major city of the Bilād al-Shām (the region of Syria), the name Shām is applied to it.

It is well known that many books have been written on the excellences and beauties of this city. Indeed, many places of this noble clime are mentioned in the Koran, and there are many hadiths confirming the nobility and excellence of this city. In the environs of Damascus and at the foot of Mt. Qāsyūn are the noble tombs of many great prophets—it is related that the tombs of 7,000 prophets are here between Arza and Barza. At the wall of the prayer niche of the Umayyad Mosque are the tombs of many distinguished Companions (of the Prophet) and worthy Successors (*selef*—the generation after the Prophet). Here are the tombs and martyria of many members of the Prophet's family. And the largest Friday mosque is the Umayyad Mosque—a blessed and noble place that gathers so many excellences.

Damascus is today the administrative center and capital of the country. The Ottomans assign a vizier to govern it. Its longitude is 70 degrees and its latitude is 33½ degrees. It has seven districts (*nāhiye*): Marj, Jabat ʿĀl (?), Zabadānī (text: Randānī), Biqāʿ al-ʿAzīz, Qunayṭira, Wādī'l-ʿAjam, and Ḥawrān. And it has nine precincts (*ṣubaşılık*): Ḳaralar, Baalbek, Biqāʿ, Wādī'l-Tamm, Qunayṭira, Banī Kanʿān, Ḥawrān, Ṣāliḥiyya, and the city.

There are differences of opinion about how Damascus got its name. The one with the most veracity is that Dumshāq, one of the sons of Sam (son of Noah), built its city wall and opened seven gates in its circumference. Above each gate he carved an image of one of the seven planets. On the first, the East Gate, where was the White Tower, he carved the sun. On the second, the Gate of Tūmā (St. Thomas), he carved Venus. On the third, the Gate of Janbīq, he carved the moon—later this gate was blocked, and the Gate of Peace (Bāb al-Salāma) was opened. On the fourth, the Gate of Paradise (Bāb al-Farādīs), he carved Mercury. On the fifth, [M 572] the Gate of al-Jābiyya, he carved Jupiter. On the sixth, the Small Gate, he carved Mars. On the seventh, the gate of Kaysān, he carved Saturn. They all still exist. Later Nūr al-Dīn the Martyr (Nūr al-Dīn al-Zangī) opened the Gate of Peace (Bāb al-Salāma) and the Gate of Joy (Bāb al-Faraj). And there is another gate called the Gate of Victory (Bāb al-Naṣr).

The citadel is enclosed within the city wall. It was built by Atsız b. Uvak (a Turkmen chief who conquered Damascus in 1076). It is related that construction was begun on 1 Shawwāl 690/26 September 1292 by order of (the Mamluk Sultan) al-Malik al-Ashraf Khalīl. Inside the citadel are a public bath and a mill, supplied by the Bānyās Canal (from the Baradā River) which passes through the citadel; also a mint and the tomb of Abū al-Dardā' (a Companion of the Prophet). Around the citadel is a moat more than 100 cubits deep. Water enters the moat from several places, and poplars, fruit trees, chestnut trees and medlar trees grow inside it. A wooden bridge goes from the fortress over the moat.

When the noble *maḥmal* and noble banner return from the Kaʿba they are stored in the citadel. Every Friday during the months of Rajab, Shaʿbān, and Ramadan, the noble banner is brought out and taken to the Great Mosque. Just before prayer they place it in a large window in front of the *maqṣūra* (a railed off area). After the prayer the muezzins perform *tevḥīd* (i.e., *ẕikr*) and then take the banner to the Post Gate (Bāb al-Barīd). From there they take it back to the fortress and place it with the *maḥmal*. There they remain until Shawwāl.

In the first week of Shawwāl (in preparation for the annual pilgrimage), they take out the banner and *maḥmal* and take them to the palace of the pasha. They are paraded about the city in grand procession, escorted by the commander of the pilgrimage and soldiers from the Damascus garrison. In the second week of the month, they take them, again in great procession, from the palace of the pasha to the place called Qubbat al-Ḥājj. Here the pasha of Damascus turns them over to the commander of the pilgrimage. The molla of Damascus issues a document certifying the pilgrimage (*ḥujja*). The commander of the pilgrimage then goes to Sahl and camps there. > Kiswa > Ṣanamayn > Muzayrib. They halt at Muzayrib until all the pilgrims are assembled. Then at the beginning of the month of Dhū'l-Qaʿda they set out for the Kaʿba.

Küçük Aḥmed Pasha, when he was governor of Syria, built the aforesaid Qubbat al-Ḥājj. He set aside some charitable income for this purpose. He died in battle against the Qizilbash (Safavids) at Baghdad. His head was brought from Baghdad and buried there.

Outside the city wall, but near the citadel, is the Gate of Sūq al-Arwām ("Market of the Turks"; text: al-Sūq Arwām). Opposite it is the palace called Dār al-Saʿāda (Abode of Felicity). The governors of Syria reside there. Nūr al-Dīn the Martyr (Nūr al-Dīn al-Zangī) built it and named it Dār al-ʿAdl (Abode of Justice), but it is now called Dār al-Saʿāda.

As for the Umayyad Mosque: it is located in the middle of the city wall. No mosque in any Muslim country is more beautiful or has greater expenses and salaries than this one. Its dome over the prayer niche was constructed by the Sabians. They worshipped there for a long time. It was their temple until the Greeks came. Then it became a temple for idolaters. Then it became a Jewish temple. At the time that John son of Zacharias (John the Baptist) became the guardian of the temple ... at the gate called Jayrūn. Then the Christians became were triumphant and turned this mosque into a temple (i.e., church). They greatly venerated it. Then, after Islam appeared and spread, it became a mosque.

It is reported that when the Muslims conquered Damascus, they left half the mosque (which was then a church of course) in the hands of the Christians. The reason for this is as follows: [M 573] During the caliphate of al-Ṣiddīq (Abū Bakr), Abū ʿUbayda was sent to conquer Damascus. When he reached Damascus with the Muslim army, he laid siege on the side of the Gate of al-Jābiyya and attacked the city from there. Meanwhile Khālid b. al-Walīd laid siege on the side of the Eastern Gate and attacked the city from there. He took sword in hand and conquered his half of the city by force. Abū ʿUbayda, however, conquered his half of the city peacefully (i.e., it surrendered). This being the case, half the church was turned into a mosque and the other half remained in the hands of the Christians. This arrangement lasted for 70 years. When al-Walīd b. ʿAbd al-Malik became caliph in 86 (705) the Muslims complained to him saying that it was not proper for Muslims to worship in the same place with idolaters. When the Christians were politely informed of this, they paid no heed. al-Walīd was troubled by this turn of events and he cast around for a solution. It was discovered that the Church of St. Thomas, which was outside the Gate of St. Thomas and at the edge of the city, was not included in the peace treaty for the city. al-Walīd threatened the Christians that he would destroy that church, which they highly venerated, if they would not abandon their half of the other place of worship. They did so; the two halves were joined and the whole became a mosque.

He then began to rebuild the mosque, as follows: He made the roof convex and inside it a *muqarnas* (ceiling that rises in stages) covered with gold. He paved the floor with marble; made the four walls of veined marble and the columns of variegated marble; and covered the places where the capitals rested on the columns with gold. Inside the mosque he placed two rows of large stone columns, alternating green and red, stretching from east to west. He placed long thick wooden beams over the small arches that rest on these large columns. And he built the roof over these beams. The roof rests on the two rows of large columns, 40 in number, every tenth column being a square pier whose sides are covered with marble up to the arch.

In the middle of the mosque is a lofty masonry dome called Qubbat al-Nasr ("Dome of the Eagle"). The exterior of the dome is covered halfway to the top with tiles. The roof is covered end to end with lead slabs. The marble columns in the middle of the mosque are on two terraces. The lower terrace is composed of large columns and the upper one of small columns. The spaces between the columns are decorated with mosaics of gold and other colors, depicting city walls and villages, full of marvels and adorned with all kinds of flowers, trees and branches, and cities abounding with fruit (?—*emṣār eṣmār ile ictinā olunup*), so that one's eyes are filled with pleasure at every moment and the leaves of the fruit trees are never lacking, and the position of the Kaʿba is depicted on top of the chambers (*ḥujurāt*—after Koran 49:4).

The two columns under Qubbat al-Nasr were purchased from Khālid b. Yazīd for 1,500 pieces of gold. And 100 gold-pieces were paid for two slabs of pistachio-green marble that were brought from Alexandria. The noble head of John the Baptist was put at a place that was excavated during construction (reading *ḥafār* as *ḥafar*). The two stone columns in front of one of the doors of the mosque are marvels of the world. They are exceedingly tall and wide. They are outside the Post Gate. [M 574] They resemble the four massive tall chestnut-colored marble columns in the noble Friday mosque of the late Sultan Süleymān in Constantinople (the Süleymaniye).

On the west side of the mosque, on the upper level, are two of the small columns said to be made of malachite. In the wall of the southern courtyard there is a round stone resembling a shield, with red and white speckles. They say that the Franks sent a great amount of money for this stone, but the Muslims refused to part with it. In addition to the gold and silver lamps suspended below the Qubbat al-Nasr and elsewhere, traces of the (original) 600 lamps hung from chains are still to be found. They were replaced with rock-crystal lamps. The chains were made of gilded copper and are still there. And there are two censers over the columns in the mosque courtyard.

Inscribed on the western, southern and eastern walls of the mosque is the twenty-fifth chapter of the Koran ("al-Furqān"). It is in the *thulth muḥaqqaq* style of calligraphy on boards that are two carpenter's cubits above the pavement. The chapter ends west of the *maqṣūra*. The sixty-seventh chapter ("al-Mulk") of the Koran is written on two sides of the wall east of the *maqṣūra*. This inscription is written in *ṭūmār* (scroll) style. The lettering in relief is yellow tin foil and the ground is dark blue.

The aforesaid *maqṣūra* is at the middle of the southern wall. It is a small caged-in area made of lattice. The muezzins' gallery is in the middle of the *maqṣūra*, which is secured by doors on three sides. The Ḥanafī prayer niche and *minbar* are inside this *maqṣūra*. Verses of the Koran are written along the edges of the prayer niche. The copies of the Koran that (the caliphs) ʿUthmān and ʿAlī had made are stored in two cupboards. People make visitations to them.

Adjoining the western side of the *maqṣūra* in the southern wall is a place called Bayt al-Khuṭṭāb (the Preachers' Room) where the preacher waits before the Friday prayer.

The Bānyās Canal flows by this house and goes to Jayrūn.

There are three prayer niches in addition to the Ḥanafī prayer niche. On the western and southern sides of the *maqṣūra* are two niches. The one closest to the *maqṣūra* is the Shāfiʿī niche. The other, on the further side near the ʿAnbarāniyya Gate, is the Ḥanbalī niche. On the eastern side of the *maqṣūra* at the southern wall is the Mālikī prayer niche, also called the niche of the Companions of the Prophet.

This mosque has four (old) gates: ʿAnbarāniyya Gate on the southern side of the building. Gate of al-Shumaysāṭiyya on the northern side, also called the Gate of al-Silsila (the Chain), which opens to the Shumaysāṭiyya madrasa. Post Gate (Bāb al-Barīd) on the western side, with a stone stairway of five or six steps going up to the entrance of the Murād Pasha Market. Jayrūn Gate on the eastern side.

There are also three new gates: Kallāsa Gate that opens to the Kallāsa Madrasa. A small gate nearby. Bāb al-Ṣighar, near the Shumaysāṭiyya Gate, which opens to the Kāmiliyya Madrasa.

The famous head of John son of Zacharias (John the Baptist) is in a small *maqṣūra* made of lattice and located between two large green columns next to one of two rows of columns at a place near the aforesaid *maqṣūra* and between it and the eastern wall. People make visitation to it and recite prayers to his saintly spirit. A special custodian is assigned to it and he keeps the key (to the *maqṣūra*).

Near this *maqṣūra*, to the northeast, is a well with very sweet water. [M 575] The mouth of this large well is made of a single piece of cut marble. A wheel (for a bucket) has been erected (next to it). During the summer, at sunset and evening prayer, those who come to pray in the courtyard of the mosque are given water from this well. There is a place where running water collects under a small dome made of cut stone just inside the eleventh gate opposite the *maqṣūra*. Outside the walls of this water reservoir clean water flows continuously from four protruding spouts.

There is another wall with another roof behind the eastern and western walls of the mosque. This covers two beautiful shrines called the Enclosed (*Muḥāṭ*) and the Roofed (*Musaqqaf*). There are two basins here with continuously flowing water.

On three sides of the courtyard of this great mosque—the eastern, western, and northern—are long terraces, each half a cubit high and covered (with mats or carpets). At the terraces on the north, groups of people from India and Sind used to be continuously instructed in the Koran and its recitation. But today they have been removed from there. At the eastern and western terraces are two martyria where there are two basins of continuously flowing water. Large marble columns have been erected on the large terrace at the eastern side of the courtyard. All four of them have black and chestnut-colored speckles. At the entrance to the Jayrūn Gate, which is between the southern end of this terrace and the northern wall of the mosque, are four large marble columns. Two have black and red speckles and two have green and black speckles. There are nine small columns on these large columns. Two have dark green and white speckles. Five of them are coarse-grained and white and one has red and white speckles. The small column in the middle of this group is light brown and twisted. The three doors here each have an alcove on each side.

Just outside the Jayrūn Gate and down a flight of 27 stone steps one comes to a marvelous basin from which a stream of water as thick as a lance shoots into the air.

Outside the Jayrūn Gate is a beautifully designed stone pavilion called the Jayrūn Pavilion.

Four large marble columns have been erected on the large terrace at the western side of the courtyard. All four have black and chestnut-colored speckles. There are two more large marble columns, with red and black speckles, between the southern edge of this terrace and the northern wall of the mosque. These four large stone columns are in a line. Inside the Post Gate, between it and the two red and black speckled marble columns, are four more large columns with green and black speckles. There are nine thin columns resting on the large columns of the western terrace, two with dark green and white speckles. The small column in the middle of this group is light brown and twisted.

Near the eastern terrace of the courtyard is a dome resting on eight marble columns. The top is covered with lead. Four of these columns are variegated porphyry and four are dark green marble. Today it is called Qubbat al-Bāʿūniyya, because in the past a learned lady used to give lessons here (ʿĀʾisha al-Bāʿūniyya, d. 1517).

Opposite this dome and near the western terrace [M 576] are again eight stone columns. Above the alcove created by them is a masonry dome covered with lead. This alcove is locked. Inside are the registers concerning the endowments of the mosque.

In the middle of the mosque courtyard is a small dome of marble resting on four columns the height of a tall man and made of coarse-grained white marble. The dome is laid out above in the form of eleven compartments. Under this small dome is a small ablution fountain, the size of a large basin, made of a single piece of coarse-grained white marble, with plentiful and continuously flowing water. On either side of this ablution fountain are stone columns ten carpenter's cubits tall. Round iron gratings have been placed over them (the sides of the fountain?). On the gratings are censers the size of a medium-sized tray. During the time of the Umayyads and in the period of some Muslim sultans several *raṭls* (a measure of weight) of aloe wood were burned in these censers every Friday. They were screened off from the rest of the mosque. After the Friday prayer the screen was removed from the censers. A breeze would waft the wonderful fragrance of the aloe throughout Damascus and everyone would smell it.

Near the mosque, at the northwestern corner, is a room made of wood. Here a group of Yemenis sit with their shaikhs.

This mosque has three minarets: 1) One is in front of the eastern martyrium (for the head of John the Baptist) and is connected to the mosque. It is stated in a Prophetic hadith that Jesus will descend (on Judgment Day) at the white minaret in the east of Damascus. It is square, both the length and breadth being four carpenter's cubits, while the height to the top of the finial is about 100 carpenter's cubits. Its stairway has 200 steps. Today it is called the Minaret of Jesus. 2) One is connected to the western wall of the mosque and has two galleries. Today it is called the Western Minaret. 3) One is connected to the Gate of the Chain on the northern side of the mosque courtyard. It is also square and is smaller than the eastern white minaret. It is called Maʾdhanat al-ʿArūs (Bridegroom's Minaret). It was built by (the Umayyad caliph) al-Walīd b. ʿAbd Allāh (rather, ʿAbd al-Malik) who assigned 75 muezzins to recite calls to prayer, in turns of 25 each. After the evening prayer they perform *tevḥīd* (i.e., *ẕikr*) and recite a hymn from the *maqṣūra* before descending. Then they pause in front of the shrine of John the Baptist and say a prayer. In the month of Ramadan they recite the *tamjīd* (a canticle sung from the minarets at night about an hour after the last service of worship) and then assemble at the gate of the Western Minaret. There one of the muezzins, in a loud and beautiful voice, recites *Yā ummat khayr al-anām* ("Oh people of the best of mankind"), referring to Muḥammad. Then all the muezzins proceed with hymns and eulogies and sit under the dome. They eulogize all the great prophets. Then they depart. They do this throughout Ramadan.

The mosque has sixteen imams, four assigned to each of the four Sunnī rites. Thus there are four Ḥanafī imams, four Shāfiʿī imams, four Mālikī imams, and four Ḥanbalī imams. And there are two preachers. This is currently the arrangement.

In this noble mosque the domes, the seals of Solomon, and various lamps are arranged in a line. During the nights of Ramadan 12,000 lamps are lit. After the Friday prayer the shaikhs of the Sufi orders [M 577] come to the mosque and perform *tevḥīd* (i.e., *ẕikr*), each according to his order. They continue until the afternoon prayer. The most famous are the shaikhs of the Rifāʿī, ʿĀrūdakī, Quṭnānī, Burāqī, Samādī, Ḥaddādī, Sīnānī, Qādirī, Naqshbandī, Jalwatī, and Khalwatī brotherhoods. Each shaikh has an assigned place (in the mosque). In addition, there are places where the general public can do religious studies. In these places Ḥanafī, Shāfiʿī, Ḥanbalī, and Mālikī professors give instruction. Shaikhs who are Koran readers instruct in the Seven and the Ten (Koran recitations). Their place in the mosque is opposite (the martyrium of) John the Baptist. They perform complete recitations at the Shāfiʿī prayer niche. In the months of Rajab, Shaʿbān, and Ramadan a shaikh who is a specialist in hadith transmits hadiths of Bukhārī from afternoon to evening prayer under the dome. Previously there were shaikhs transmitting hadiths of Muslim and the *Shifāʾ* from noon to afternoon prayer, but that practice is no longer followed.

This mosque is 548 feet long from east to west and 151 feet wide. It is reported that this mosque cost 5,060,000 goldpieces. Many pious endowments have been established to support this mosque. Daily expenses are 300 goldpieces. Money left over after the payment of these expenses from the endowments goes to the state treasury.

In addition to this large mosque, there is another inside the city wall where Friday prayer is given. It is called the Mosque of the Cotton Merchants (al-Qaṭṭānīn). Apart from these, there are many neighborhood mosques inside the city wall where Friday prayer is not given.

Outside the city wall are many large and small mosques, numbering about 150. Some of these. built in the Ottoman period, are in the style of Turkish (*Rūm*, i.e., Ottoman) mosques. They are:

- The Mosque of the late Sultan Süleymān, *may his earth be sweet*. It is east of the city, between the Upper and Lower Easts (?), on the eastern side of a broad open square called Gök Meydānı. Between the two minarets connected to the eastern and western walls of this mosque is a single small lead-covered masonry dome. The gate of this mosque is opposite its prayer niche. On each side of the gate is a terrace covered with cupolas. The courtyard in front of the terraces is paved with black and white stone similar to marble. At the eastern and western sides of the courtyard is a fine vegetable garden, and beyond that extending from north to south are guest quarters for travelers containing interconnecting rooms—ten rooms on two sides (i.e., five on each side?)—with lead-covered cupolas. Above the courtyard on either side are 12 cupolas over 12 marble columns, so the cupolas number a total of 24. In the middle of the courtyard is a square basin, measuring ten by ten (cubits), with an ablution fountain in the middle made of a single piece of marble. Around the fountain are four brass spouts from which water flows continuously. There is a handrail on the ground on the northern side of the basin and another one opposite it, and the water channel flows between these two handrails. Beyond the handrails is another small garden, and beyond this garden there is a soup kitchen with pantry, cooking area, oven, and other accoutrements. Behind the guest quarters is a madrasa containing 18 rooms and lead-covered masonry cupolas. [M 578] The professor of this madrasa in Damascus who teaches according to the Ḥanafī rite receives a daily salary of 60 *akçes* and the students receive daily (a total of) 200 *akçes*. In addition, each day they are given two meals from the soup kitchen. Travelers who reside in the guest quarters in the courtyard of the mosque have their own table at the soup kitchen.
- The Mosque of Derviş Pasha, a former *beylerbey* of Syria. It is near the Dār al-Saʿāda. It is a beautiful mosque built in Turkish style. In the middle of the courtyard is a well-crafted basin and ablution fountain.
- The Mosque of Sinān Pasha. It is outside the Gate of al-Jābiyya. Its single dome and the top of its vaults are covered with lead. Below these vaults are upper galleries raised on marble columns resting on the floor of the mosque. The stairs to these upper galleries inside the mosque are made of solid stone. The doors are in the middle of the northern wall of the mosque. The open area of the courtyard extends nearly to the end of the terraces on either side of the door of the mosque. In the middle of the courtyard are a basin and an ablution fountain carved from varicolored speckled marble. The courtyard has two gates. The minaret of the mosque is connected to the gate in the western wall of the courtyard. The exterior of the minaret is covered with green tile.
- The Mosque of Ḳara Murād Pasha, a former *beylerbey* of Syria. It is outside the city wall and on the pilgrimage road, in the vicinity of the Gate of al-Muṣallā. This building and its dome are in the Turkish style.

There are madrasas, dervish lodges and clean public baths in Damascus. The most famous baths are the Derviş Pasha Bath near the Umayyad Mosque and the Muṣṭafā Pasha Bath inside the Saddlers' Market (Sarrāj Khāna). In the middle of the Muṣṭafā Pasha Bath is a large round pool that is 20 cubits in circumference and is as deep as the height of a man. Opposite the door of the bath is a pool. Above it is a fountain that they call the "lion's mouth" which is carved from marble. Water flows continuously from it into the pool.

There are many charitable buildings and inns, also flourishing markets, in Damascus. In addition, there are many homes and residential quarters outside the city wall. Nūr al-Dīn the Martyr (Zangid ruler of Syria, r. 1147–1174) built a hospital in this city. It is one of its greatest charitable institutions, with income from many endowments, but today only one in a hundred is still extant. Nevertheless it is still a beneficial institution. There is a chief of physicians and oculists. Compared to the hospital at Ṣāliḥiyya (a large village on the slope of Mt. Qāsyūn), this one is more elaborate.

This city is located on a level plain two hours north of Mt. Qāsyūn. To the south is an endless plain where one edge of the city's gardens reaches to the foot of Mt. Qāsyūn. Together with Ṣāliḥiyya, it has become like a single town. If one views the city and its plain from the Qubbat al-Naṣr at the top of Mt. Qāsyūn, one sees houses and other buildings in the midst of gardens shaped like a ladle.

In short, the city of Damascus with its grand buildings, markets, merchandise and business men, is an excellent place. Above all, the New Murād Pasha Market opposite the Post Gate of the Umayyad Mosque and, directly across from it, a coffeehouse and unrivaled bedestan (covered market) are the pride and ornament of Damascus. The Sipahi Market above the moat, built by Şemsī Aḥmed Pasha, and the Sūq al-Dhirāʿ, which is at the southern wall of the Umayyad Mosque, [M 579] are where most of the cloth and silk woven in Damascus is found. Then there are the Bazūriyya Market, the Market of Chaqmaq (proabaly named after the Mamluk Sultan al-Malik al-Ẓāhir Sayf al-Dīn Chaqmaq, r. 1438–1453) and countless other shops and markets.

On one side of the city is the Qubaybāb Quarter. It is a large new quarter built in a dry area. The houses are large nomadic tents with cone-shaped domes and laid out lengthwise, occupied by Turkmens.

West of the city is Gök Meydānı, which is like an emerald. The area all around is covered with flowers of many colors and all sorts of fruit trees. The gardens of Damascus are places of excursion, the favorite ones being Taḥt al-Qalʿa ("Below the Fortress"), Ṣāliḥiyya, Gök Meydānı, Nayrab, etc. They have clean cooking establishments, offering all kinds of delectable foods. They can prepare a superb banquet in one or two hours and take it to where it is wanted.

The Baradā River flows through the middle of Gök Meydānı, like a ribbon of silver. The lofty villas around Gök Meydānı are renowned. It would be impossible to describe them in the manner they deserve, especially the remains of the buildings of the Barmakids that are still standing.

Another famous place of excursion, known as Jabha, is a square-shaped area of two donums. It has willows, walnuts, poplars and other trees, amidst water channels, pools and basins. This park is located on the banks of the Baradā River. Above it, the Qanawāt Canal flows to the *maqṣaf*. The Bānyās Canal flows to the Mevlevīḫāne, which is a beautiful dervish lodge where two finely crafted waterwheels turn. From there it flows to the Dengiz Mosque, which is located in the place called Sharaf and has 20 large windows overlooking the canals flowing in that direction.

Nearby is the Yelboġa Mosque. In its courtyard is a square pool with a round jet in the middle that shoots water to the height of a man. The windows of this mosque look out over three directions: east toward Taḥt al-Qalʿa—"a watering place for strangers and a pasture for kinfolk" surrounded by houses and villas; west toward the two canals; and south toward the Baradā River. There are trees and flowers in this area. And in the place called Bayn al-Nahrayn ("Between the Canals") is a bridge to a small island. Here the Baradā divides into two branches, one of which takes the name of Shaikh Arslan while the other retains the name of Baradā. In this *maqṣaf*, which is called Jazīra ("The Island"), there are many idlers.

Another place of excursion in this area is called Bahnasiyya. It is a place where trees, fruit, flowers, and springs are found together. The upper part, called Arḍ Nayrabayn, has excellent fruit and abundant flowers. From there one enters Arḍ Rabwa where, at the foot of Jabal-i Sharqī (or, the eastern mountain?) is Mahd ʿĪsā (Cradle of Jesus), at whose gate is a spring called Maltam. In that direction canals branch off from the Baradā. This area is at the foot of Mt. Qāsyūn. Here are the famous *ḥawāqir*, i.e., gardens. Mt. Qāsyūn and Mt. Rabwa are here with the village of Dummar between them and Qubbat Sayyār opposite.

In the boundless plains to the southeast and south of the city are large villages here and there, and arable fields and pastures as far as the eye can see. The inhabited areas appear joined together like one interwoven fabric.

Most of the houses and buildings in this city are made of [M 580] black, white, and red stone and have high arches, with timbers placed on the roofs and over the columns, then plastered with mud. The name for this kind of house is *īwān*. It has a "lake"—i.e., a large basin—in front and a garden on either side. Over against the *īwān* on all four sides is a wall with alcoves (?) and a very high, wide, and richly adorned large vestibule known as *qāʿa*. Between the basin and the *qāʿa* is (a kind of porch called) "Under the Sky" (*taḥt al-samāʾ*). Small houses consisting of (?) *soffa*, *īwān* and *qāʿa* are laid out with square columns 5–6 cubits long and wide and 12 cubits tall. The *īwāns* are open to the north. They are built with the heat of the summer in mind. They are solid buildings. Recently some people have begun to make hearths and rooms in Turkish style and to build an upper floor.

There is no residence or abode which does not have water flowing through it. This city has very clean and richly adorned markets. Most of the roofs are octagonal. There are large workshops where all kinds of cloth are made.

Damascus has all kinds of flowers, fruit, and the like. The most prized flowers grow to the north of the city and the rose is considered the most esteemed of these flowers. There are six kinds of roses, the most prized and esteemed of which is called *ward jawrī*. Aside from this rose there are varieties of jonquils and carnations; also jasmine, wallflower, iris, lily and cowslip; varieties of daisies that turn toward the sun and close up at night, myrtle and sweet basil; tulips, anemones, nightcaps (?—*şebkülāh*); dates, henna, ben trees, rue, wild thyme, costus, savory, marjoram, many kinds of sweet basil, willow, prophet's knot (?—*peyġamber dügmesi*), lavender; and several varieties of water lily, the most prized being blue. The *khaylān* is similar to the willow tree (*ṣafṣāfa*); in spring all its leaves turn red.

The best place for fruit is Arḍ Mizza wa Lawān which is south of the city. There are 21 varieties of apricot, the most prized (fresh apricot) known as *ḥamawī* and the most prized dried apricot known as *baladī*. There are seven kinds of cherries and sour cherries. (In the past) they were not grafted like those in Istanbul, but today they are grafted and have become excellent. There are also many kinds of pears and apples, the most prized apples known as *miskī*, *sukkarī* and *māwardī*. And there are all kinds of peaches, plums and pomegranates, the most prized being the *māwardī*. In the area south and east, from the villages of Yaldā to ʿArbīl, grapes are prized. They say that there are 49 kinds of grapes, the most prized (fresh grape) known as *zaynī* and the most prized dried grape known as *dūrbūlī* (cf. below, *dūrbalī*). There are many kinds of quinces. And there are figs and almonds as well as jujubes, pistachios, hazel nuts and pine nuts; azerole, the most prized of which known as *qamarī*; several kinds of carobs and walnuts, also of blackberries and citrus fruits, including oranges and grapefruits, *naffāsh* lemons and citrons, six kinds of lemons and two kinds of oranges; and all sorts of olives; also several kinds of rhubarbs, bananas, and sugar cane; and wild dates.

Of vegetables there are asparagus, tarragon, cauliflower, and three kinds of cabbage (*kurunb*); also several kinds of cabbage (*laḥana*) and eggplant and one or two kinds of leeks; several kinds of mulberries and of carrots; radishes, garden cress, purslane, spinach, celery, chard; two varieties of wild chicory; garlic, coriander, caraway, cumin, and squash. Furthermore, there are different types of mushrooms, turnips, black-eyed beans, broad beans, rice, [M 581] lentils, and millet; different kinds of safflower, sesame, fleawort seeds, chickpeas, fenugreek, and romaine lettuce; sumac, colocasia, marsh-mallow, okra; two varieties of cucumber (*khiyār*), summer and fall; and many kinds of melon and watermelon.

Cotton and silk are abundant. The most prized silk is the *baladī*. It is more esteemed and more expensive than the other kinds. Salt and potash are produced. Cucumbers (*qiththāʾ*) are especially plentiful. In the village of ʿArbīl the people raise an herb similar to sweet basil that they call clove, and indeed it has the fragrance of cloves and its blossom is like that of sweet basil. They extract the juice from this flower and use it in treating many illnesses, especially shortness of breath and (excessive) phlegm. Production is confined to this village; it is found nowhere else.

Environs of Damascus

Ṣāliḥiyya: It is one hour north of Damascus. It is located slightly up the slope of Mt. Qāsyūn. The foot of the mountain is built up and there are villas and gardens. Its gardens extend to those of Damascus. It is like a quarter of Damascus. It has old and new Friday mosques, madrasas, and public baths. Compared with Damascus its air and water are excellent. It is a charming town. Its old derelict mosques and madrasas, which one still sees here and there but which are in ruins, and their high walls, finely built with varicolored stone, testify that in the past it was a city similar to Damascus.

They say that in the past there was a place with many madrasas called Banī al-Madāris ("Tribe of the Madrasas"). A book entitled *al-Dāris fī (taʾrīkh) al-Madāris* (by al-Nuʿaymī, d. 1521) is devoted to the madrasas of Damascus.

Sultan Selīm built a noble Friday mosque in Arabian style near the mausoleum that he erected over the noble grave of al-Shaikh al-Akbar (i.e., Ibn al-ʿArabī) (on Mt. Qāsyūn). And behind it, on the main road, he built a large soup kitchen with storeroom, oven, kitchen, and other accoutrements. The poor in the Ṣāliḥiyya quarter benefit from this charity. In addition, there are currently twelve other mosques there. Most prominent is the Ḥanbaliyya Mosque. Sultan Saladin built a hospital there as a charitable institution. Its markets are still prosperous.

Two canals flow through Ṣāliḥiyya, the Yazīd and the Thawrān. The Yazīd comes from the direction of Mt. Qāsyūn, while the Thawrān flows a bit below that, near the plain toward Damascus. Ṣāliḥiyya has excellent gardens. Lofty and delightful villas overlook the plain of the city. There are pleasing terraces and meadows along the canal. One would justifiably characterize Ṣāliḥiyya as a splendid place. It is a spiritual place that banishes gloom and gives pleasure to the heart.

Şubaşıs are appointed to Ṣāliḥiyya by the governor of Damascus and the Efendi (i.e., the molla or chief qadi) of Damascus has deputies here. Mulberries and chickpeas, myrtles and peaches are plentiful. On Mt. Qāsyūn grows an herb with a wonderful scent called *khuzām* (lavender). Another herb that grows there is called *shīḥ* (Armenian wormwood). Also grown are sumac, white and red azerole, and carob.

There is a place connected to Ṣāliḥiyya called Saham, which used to be called Sahlam. Most of its trees are mulberries from which they make much silk (from the worms that eat the mulberry leaves).

Drinking glasses are made in Ṣāliḥiyya and sold in Damascus.

Near Ṣāliḥiyya is a place called Bayt Lihyā wa'l-ʿInāya. It was Adam's *bayt al-abyāt* ("house of houses"). They say that Eve lived there a long time. There are many jujubes trees here.

Also near here is Arḍ Saṭrā and its main town. They say that Abel used to live in Saṭrā, [M 582] while Cain lived in Qayniyya. Between these two places is a place of excursion called Bülbülli (Place with Nightingales).

An unusual flower grows in this area. The color is white tending toward red. It covers the ground. They call it *qaranfila*. It has an unusual characteristic whereby it gives off a fragrance from the time of the afternoon prayer until morning but not during the day. There are many quince trees in this area. When they bloom, the social elite of Damascus gather there and channel streams under the trees. On moonless nights they put olive oil in egg shells and light it, then release the shells in the channels. Or they suspend peels of bitter orange from the trees and light them in the same fashion (i.e., with olive oil). This place is near Bahnasa. Also near Bahnasa is Arḍ Mizza wa Lawān which is famous for figs, apricots, cherries and sour cherries.

Behind Mt. Qāsyūn is a famous village called Mīnīn where two revered saints are buried, Shaikh Jandal and Shaikh Abū'l-Rijāl. They say that Shaikh Jandal would not allow anyone to sleep next to him. This village is famous for its walnuts with very thin white shells. This area never lacks for snow. Rhubarb and blackberries grow here.

Northwest of Damascus, in the region of Tall Ḥūrna and ʿAqabat Qābūn al-Rumān, is the district of Jubba ʿAsāl. This district has a deputy of the molla of Damascus assigned to it. Further north are the villages of Ṣaydānāya, Talfītā, and ʿUmra and the more famous ones of Qaṭīfa, Raḥba, Jarūd, Muʿaẓẓamiyya, and ʿAsāl.

Zabadānī: It too has a deputy qadi assigned by the molla of Damascus. It is a delightful place of excursion on the banks of the Baradā. Next come the villages of Ḥāmiyya and Rabwa. This district has many gardens and is famous for its fruit, especially its apples, the like of which are found nowhere else. Its earthenware pots are also famous. Its gardens reach all the way to Damascus. This district is 18 miles from Damascus.

In the past Zabadānī had a city wall. On the road (through this district) Nūr al-Dīn the Martyr (the Zangid ruler) built an inn and established a private endowment for its upkeep. Later the income from these endowments was seized and the inn fell into ruin. The famous villages of Zabadānī are Dummar, Hāma, Ṣabūrā, Kafar, Qūq al-Fistiq, Rīmās, and Wādī Barda. Also here are Khān Maysalūn, Ḥusayniyya, Khān al-ʿArār, and Buqayn. Here (in Zabadānī) is the Friday mosque of al-Dilāʾ. If anyone tells a lie in this mosque he splits in two—this has been tested.

In the district of ʿUyūn al-Tūt (Mulberry Springs) there is a deep well called Hāwiya. Its depth is unknown. People in this region sometimes come just to see the well. They tie up large bundles of wood and brush which they light and drop into the well, watching how far they go. As the bundles fall, the flames get smaller and smaller until they look like fireflies, then disappear. The fact is that fire can be seen from a great distance, as confirmed by the science of optics.

South of Zabadānī is the district of Biqāʿ. A *ṣubaşı* and a deputy qadi of the Damascus Efendi (i.e., the molla or chief qadi) reside here. This is an extensive district with 25 prosperous villages. Its crops are very abundant. Its pastures are famous and its river valley is vast. The Līṭānī River flows through it. Coming from Baalbek, it goes through here, then on to Marj ʿUyūn, Shaqīfa, and Iqlīm al-Tuffāḥ ("Apple Country"), which is a district of Sidon. Near Sidon it flows into the sea. Attached to Sidon are the villages of Rīshiyya, Ḥāṣbiyya, Qalʿat Qabr Ilyās, [M 583] Shūf al-Bayāḍ, Qarʿūn, and Majdal Balhīṣ. The Banī Ḥaymūr, Banī Numayr, and ʿAnaba tribes live here. Also here are ʿAytā, ʿAbbās, Ḥalwā, and Wādī Qurnā. Nearby is where İpşir (Muṣṭafā) Pasha (of Damascus) was defeated (in 1623 by the Druze leader Fakhr al-Dīn II Maʿn).

Wādī'l-Tamīm and, in the same area, Shaqīf Arnūn which is a strong citadel. Wādī'l-Tamīm is administered by a *ṣubaşı* assigned from Damascus. The Banī Shihāb, who are a very fierce people, live here. The Taymānī and Druze also live here. They are divided into two groups, the "whites" and the "reds." The whites belong to the tribe of Amīr ʿAlam al-Dīn and the reds belong to the tribe of Ibn Maʿn (usually Maʿnoğlı in this text). When Sultan Selīm came to Damascus he gave a drum and banner (as symbols of authority) to Ibn Maʿn and established this honor upon its *beys*.

The Druze trace their origin to (the Fatimid caliph) al-Ḥākim bi-Amr Allāh. They are an iniquitous people who believe in *ḥulūl* (indwelling of God in a creature). It is reported that someone composed a book for al-Ḥākim bi-Amr Allāh, stating that the spirit of Adam had been reincarnated in ʿAlī and from him it (eventually) passed to al-Ḥākim bi-Amr Allāh. When the author published this book in the al-Azhar Mosque (in Cairo), they wanted to put him to death. So he fled Egypt and went to Wādī'l-Tamīm where he settled. He convinced the Druze people who lived there to believe him, and he made permissible what is (religiously) forbidden. This people currently live in the Shūf Mountains. They have 600 wise sayings that they attribute to al-Ḥākim bi-Amr Allāh. When they swear an oath they swear "by the 600 wise sayings." When they go to their own provinces, their wives are seen by no one else during the day; but they go out as soon as evening falls. They treat evening as morning. Even their emirs attend their council meeting in the evenings and their subordinates come and kiss their hands and say *ṣabāḥ al-khayr* ("Good morning"). Even a traveler to this region must say the same thing.

In the mountains of Maʿn is a strong fortress called Dayr al-Qamar, mainly inhabited by the Banī Maʿn (usually Maʿnoğulları in this text). The al-Awla River comes from the same mountains. It flows into the sea near Sidon. The al-Nāʿima River also comes from the mountains of Ibn Maʿan. It meets the sea near the al-Awla. The fortresses that the Banī Maʿn hold are the following: Bilād Kafar Kanna, Bilād ʿAkka (Acre) and the coast, Safad, Bilād Bashāra, Bilād al-Shaqīf, Jalāl Jīra, Bilād Ṣafad, Bilād Beyrūt wa Sayda (Beirut and Sidon), Bilād Jabal Karwān, Bilād Jubbat al-Munayṭira, Jubayl, Anṭalyās, al-Batrūn, al-Jarad, al-Gharb, al-Matin, al-Shūf, al-Muqayṭiʿ, al-Shaḥār, al-Biqāʿ, Baalbek, Bilād Ṣūr (Tyre), and al-Maʿshūqa.

Qunayṭira: A *ṣubaşı* and a deputy of the qadi in Damascus reside here. It is two stages from Damascus. In ancient times King Bāzān lived in this region. It has a number of districts, chief among them being Ḥawla. Qunayṭira has a small fortress, an inn, and a fine public bath. But the weather is very cold. One of its famous villages is Khazībī. Coal goes from here to Damascus. Jubā and Khayyām ʿAbs (are attached to it). Ḥawla is the most esteemed district, which they also call Qubalyās. The chief town is Bānyās. This place is very hot. Much cotton is grown here. It has abundant flowers and citrus. Its fortress, called Ṣubaydā, is strong. It is at the foot of Mt. Hermon. Today this district has become an independent tax farm that pays 17 purses (per annum; 1 purse = 50,000 *akçe*) to the government. This district is famous for its water buffalo, which are larger than other water buffalo. The Sharīʿa (Jordan) River rises here. (In spring) the people of Damascus first fetter their horses here [M 584] because these meadows are the first to ripen. In the winter season, Turkmen, Arabs (i.e., Bedouin) and others make this place their winter camp ground. At the time of the Israelites, this district was very prosperous. The Israelite tribes of Naftālim (Napthali) and Īsār (Asher) lived here. The Īsār also lived in the vicinity of Tyre.

Nearby is the large district of the Banī Kanʿān. Its location is to the left (south) of Jisr Yaʿqūb (a bridge over the Jordan), on the ridge of Minya. Here are also the Sharīʿa, Manṭūr, Banī Kināna, the district of Banī Irbid, and the district of Jawlān. In ancient times Ḥāmīm b. ʿAmlīq (of the Amalekites of the Bible) lived in this district.

Wādī al-ʿAjam: A prosperous district southeast of Damascus. It has its own deputy qadi. Its famous villages are Kuswa, Saʿsaʿ, Barāq, and Kanākir. This is where the Prophet's farrier Burqāla is buried. It is reported that if a horse has scrofula and is made to go around the grave of Burqāla, by God's command its illness will disappear—this has been tested.

Bayt Sābir: It is famous for grapes.

Kafar Jawr: The tomb of Nimrod is here.

Bayt Taymā: It is famous for *dūrbali* grapes. There is a strange phenomenon here. Dew never falls. However, near here at a hill called Tall Abū'l-Nadā ("Dew Hill") there is no lack of it.

Qaṭana: A well of Shaikh Ḥasan Rāʿī is here. If a madman is confined to it for three days, by God's command his madness disappears.

Mizza: A village famous for pomegranates and figs.

Ḥawrān: A very famous district with many (subordinate) districts. It has its own *ṣubaşı* and deputy qadi appointed from Damascus who administer the region. In the past it had many towns and villages. No other place has its productivity. In some places it produces crops a hundred fold. The great beauty of the women of Ḥawrān is proverbial. The people are very tall. In ancient times this region was the province of Job. At that time there was a great city here called Ūshīs. Seventy learned men appeared from this city and wrote the Torah and the Gospel. It is reported that Job was a son of Esau. He married Rebecca (text: Rībā, error for Rībakā or Ribqa), the daughter of Jacob, and settled in Bilād Nawā and Jaydūr in the vicinity of Ḥawrān. When Job died, he was succeeded by Ḥūmil and then his son Bishr who was called Dhū'l-Kifl. He was called Dhū'l-Kifl because he always guaranteed (*takafful*) justice in all matters among his people.

Bosra: This city, which is the administrative center of Ḥawrān, is very old. It has a fortress made of black stone, a Friday mosque and a market. At an early time the Arab tribes of Banī Fazāra and Banī Murra lived here. There are beautiful gardens here. Today the Mafālija Arabs live here, as do the Ḥazqiyya Turkmen. When the Companions of the Prophet set out to conquer Syria, this was the first city that they conquered.

Ṣarkhad: A small town in the eastern part of Ḥawrān. It has a fortress, vineyards, and gardens. Water is collected from rain. In the past the Banī Hilāl lived on the mountain of this town. It is on the road to Baghdad; during summer it takes ten days to get to Baghdad.

The districts of Ḥawrān are:

- Jaydūr: Its chief city is Nawā, which was the city of Job. It is 18 miles from Ṣanamayn. Nawā is the city of al-Nawāwī. It is a major source of grain.
- Azriʿāt: Its chief city is Bathīna. It is currently a prosperous village. Its water comes from wells.
- Muzayrīb: A valley with springs. Sultan Süleymān built a fortress above it. [M 585] The grave of Shaikh Saʿd al-Asmar Takrūrī is near Muzayrīb. It has its own endowment.
- Kuthayba: It is located northeast of Muzayrīb. It has many springs. Its famous villages are Ṣanamayn and Najlā. Millstones are produced here.
- Nāmir: It is famous for its wheat. They say that Adam sowed the first wheat here.
- Dhuruʿ and Arḍ Lajā wa Qanawāt in northeastern Ḥawrān. They are rocky areas. They had been prosperous since ancient times. Stone houses are still standing. Water, fruit, and flowers are still present, but there are no inhabitants. Only around it live the Lajā Arabs who are called ʿArab al-Jabal (Mountain Arabs). They are highwaymen and a rebellious good-for-nothing clan who lord it over the area around Damascus. Indeed, they exact guard-money, called *ghafarjī*, from each village. If the village does not pay, they immediately swoop down and plunder it.
- Ṣafā: It is east of Lajā and a rocky place.
- Samāwa Qarāqir: A district between Damascus and Kūfa. In this village there are unusual round stones, like spheres. If one is broken open a kind of soil emerges that is the color of ashes, inclining to red. No one knows what use the soil is or who made these round stones.
- Marj al-Aṣghar: Also in Ḥawrān and is also called Marj Shaqḥat. It is 15 miles from Damascus. This is where Khālid b. Saʿīd b. al-ʿĀṣ was martyred.
- Marj: It is composed of two districts. One is Marj Rāhiṭ, which today is just called Marj. The other is Marj Ghouta, which today is just called Ghouta. This Marj has its own independent deputy qadi. It begins at the end of the lower valley and ends at the lake.

In the year 64 (rather 65/684) at Marj Rāhiṭ there was a famous battle between the Yemenis and the Qaysis. The reason was as follows: Marwān b. al-Ḥakam went to Damascus and demanded the caliphate from Muʿāwiya b. Yazīd. While swearing an oath of allegiance to ʿAbd Allāh b. Zubayr (of the Qaysis), suddenly, by divine wisdom, Muʿāwiya died. In the interim, the people of Damascus agreed to swear allegiance to al-Ḍaḥḥāk b. Qays (leader of the Qaysis and supporter of Ibn Zubayr). Marwān b. al-Ḥakam (leader of the Yemenis) naturally took advantage of this to bring over to his side a number of leading personalities. There were rumors among the people of Damascus about this. At the urging of ʿUbayd (rather ʿAbbād) b. Ziyād (Marwān's commander), al-Ḍaḥḥāk went to Marj Rāhiṭ and assembled those who had sworn allegiance to him. Marwān, with the support of 13,000 troops who had sworn allegiance to him, mounted and marched against al-Ḍaḥḥāk. After 20 days of fighting they killed al-Ḍaḥḥāk b. Qays and the caliphate fell to Marwān.

This district has abundant grain and other (agricultural products). South of Marj are the well-known villages of Ghazlāniyya, Hījāna, Majdalyā, Ḥawrān al-ʿAwāmīd, and Qāsimiyya. North of Marj are the villages of Muʿayṣira, ʿUbbāda, Quṣayr; Ḍumayr—there is a watermelon named for this; Barza; Qābūn—it has an excellent climate and is divided into Upper and Lower Qābūn; ʿAdrā, Dayrānī, ʿAṣrūn; also Qaḍam near Marj Rāhiṭ and Māʾ Ghassān.

District of Ghouta: It is 30 miles long and 15 miles wide. Its villages have so many trees that sunlight never reaches the ground. Ghouta has all kinds of trees, some that bear fruit three or four times (in a year). There are apricots, apples, plums, and pears. Indeed, they say that there are about 130,000 vegetable gardens in the Ghouta of Damascus. [M 586] The pass of Thaniyyat al-ʿIqāb overlooks Ghouta. Its most famous villages are Mizza, Ḥarasta, Kawkabān; Kafar Sūsiyya, which is well known for olive oil; and Dāriyya. The latter two are on the border of Wādī'l-ʿAjam and Ghouta.

Dāriyya is a beautiful village. It is two hours from Damascus, on the road to Jerusalem and Egypt. It has abundant and cheap produce. Nūr al-Dīn the Martyr (the Zangid ruler of Syria) made it an endowment for the poor, and it is still functioning in this manner. Whatever amount of wheat is collected, it is distributed under the supervision of the molla of Damascus to the poor and widows during the month of Muḥarram. However, most of it is swallowed up (i.e., embezzled) and not one-tenth is given to the poor. The agricultural productivity of this village is proverbial. Cotton, grapes, olives, and other grains and fruit are abundant, and it has many streams. Abū Muslim al-Dāranī and Abū Muslim al-Khawlānī (early Muslim ascetic and transmitter of hadith, d. 62/682) (and the) Banī Kharqīl are (from) here. Wonderful watermelons grow here.

South of ʿAqrabāʾ is the noble grave of Sittī Zaynab, the daughter of ʿAlī. It is two hours from Damascus. Nearby is the noble tomb of Sayyidī Mudrik (al-Fazārī, a Companion of the Prophet).

Northeast of ʿAqrabāʾ is the village of Manīhā where Saʿd b. ʿUbāda (early convert to Islam and chief of the clan of Sāʿida at Medina) [is buried]. Near this village is a forest that is a place of excursion called Ghayḍat al-Sulṭān ("The Sultan's Grove"). There are lofty trees here that no one can cut without permission (reading *icāzet* for *icāret*).

They are only cut for mosques or other important things. The forest has a warden appointed by the government. The Baradā River passes through it. There are all kinds of animals in the forest, wild boars being particularly abundant.

Nearby is (another) place of excursion called Sitt al-Shām which has an excellent spring; also a wonderful valley full of trees of all kinds. The Baradā River flows through it and then passes through Ghayḍa(t al-Sulṭān) and toward the sea. There are many quince trees in this valley.

At the lake there are various birds that are hunted and fish that are caught winter and summer. North of this valley is the noble tomb of ʿAyn al-Awliyāʾ Shaikh Arslān b. Yaʿqūb (one of Damascus' most important saints, d. 550/1155), may God sanctify his secret. It is a distance of one bow-shot from Bāb Tūmā, one of the gates in the city wall of Damascus. Shaikh Arslān was a contemporary of the saint Abū ʿĀmir; the Sufi brotherhood of Abū ʿĀmir is traced back to Sarī al-Saqaṭī (a renowned mystic, d. 253/867).

The famous springs of Damascus: They are ʿAyn al-Warākā near Bāb al-Salām, ʿAyn ʿAlī, ʿAyn al-Sukhna, ʿAyn al-Dhahab, ʿAyn Luʾluʾ, ʿAyn al-Jālūd, and ʿAyn al-Zaybiyya, which is the most esteemed. They say that in the vicinity of Dummar on the mountain of Ṣāliḥiyya there are 360 springs.

ʿAdhrā: A famous village near Ḍumayr and Raḥba. To the north are Qaṭīfa and Jarūd. In ʿAdhrā is the grave of Ḥujr b. ʿAdī (Shīʿī agitator of the earliest period of Islam). In the past ʿAdhrā constituted a famous district. It has a fortress. Here and in Ḍumayr saltwort grows. They burn it and obtain potash and earn considerable money from selling it. ʿAdhrā is an independent tax farm.

There is a fortress in Jarūd. Roses grow here and nowhere else in Syria.

Qaṭīfa: It is one stage from Damascus. The (Ottoman) conqueror of Yemen, Sinān Pasha, built a magnificent inn, Friday mosque, and public bath here. Meals are provided to travelers who lodge in the inn. Even their candles are provided for. Such a magnificent inn is found nowhere else.

There is a famous village in the district of Ḥalbūn Ajrū (?) .... North of Qaṭīfa is a stage called Nabak. North of Nabak is al-Qārā and north of it is Ḥasiyya, also called İkiḳapulu.

The district of Ḥasiyya: It has a han and a fortress. There are guards in the fortress. When travelers arrive they send them on to Qaralar/Qārālar which is one stage from Ḥasiyya. [M 587] It is a district of Syria and has a *ṣubaşı* appointed by the pasha of Damascus.

The district of Baalbek: An ancient town and one of the administrative units of Syria. It has a qadi district and a tax farm. It is administered by a *ṣubaşı* appointed by the pasha of Damascus. Previously it was under the control of Ḥarmfūsh-oğlı. It has a famous fortress, an enormous, marvelous, and very strong structure. Each wall of the fortress is made of a single stone. Indeed they say that it was built by demons. It is a fertile area full of trees, streams, and springs (reading *menābiʿ* for *menāfiʿ*). There is an altar here. They say that it is a house belonging to the Sabians and highly revered by them. Inside is the grave of Seth.

ʿAyn al-Jarr: It is south of Baalbek. They are a full stage apart. There is a village here known as Majdal. It is on the road between Baalbek and Wādī al-Taym. A large river rises at ʿAyn al-Jarr and flows to Biqāʿ.

There is a shrine of Abraham in Baalbek. There are also marvelous buildings and monuments and palaces resting on marble columns that have no equal. The prophet Elijah had a monastery in Baalbek. In the past Baalbek was called Bakk. The Israelites worshipped an idol here called Baʿal. The name of the idol was added to the name of the place and it became Baalbek. In the days when its people worshipped that idol, God sent the prophet Elijah to them. They accused him of being a false prophet, so for three years it did not rain in Baalbek.

Sanjak of Sidon (Ṣaydā): It was formerly a sanjak; today a pasha administers it as a tax farm. Safed, Acre, and other places have been attached to it. Sidon has a number of districts: Beirut; Kasrawān; Shūf; Qalʿat Sūlā, which has an excellent quay; Iqlīm al-Tuffāḥ ("Apple Country"); Shūmār; Kharrūb.

[Sidon] fortress is located on the Syrian coast, two stages from Damascus. A number of consuls from the Frankish peoples reside there and trade in silk thread, potash, and other goods.

The Shūf Mountains are located behind the mountains of the Maʿnids (Maʿn-oğlı). Among the administrative units of the Shūf are Bilād Bashān, Qalʿat Tayma, and Qalʿat Rīsha.

Near Sidon and along the coast grows a great amount of sweet basil. The trees on these mountains are all mulberries. Silk is made from (the cocoons of the worms that eat) them. It is a separate tax farm that produces more than 600 purses in annual income.

Also in this region are the mountains of Kasrawān. These mountains are behind Beirut. Here rise the following rivers: Nahr Anṭalyās, Nahr al-Kalb, and Nahr Ibrāhīm. The last enters the sea near the mountains. Also in the mountains of Kasrawān rises the spring of ʿAyn Dārā. The honey, cheese and grape molasses of these mountains are very famous. Many other products come from there, such as silk, cotton, olives, grapes, honey, cheese, sesame, sugarcane, various chickpeas, colocassia, bananas, and *ḥabb al-ʿazīz*.

In this region is the district of Mashgharā, which has a delightful valley. It is crisscrossed with streams and forests. They also call its chief city Mashgharā. It is six miles from the city of Kāmid. The spring known as ʿAyn al-Jard is 18 miles from this city.

In springtime, a kind of finger-length fish appear off the coast of Sidon. The males and females have certain marks. The people catch them when they come to spawn. They are dried and pulverized and then when desired eaten like *ṣufūf* (?). [M 588] They say that it is very potent and increases sexual desire.

From Sidon to Jah (?) is eight miles. In this region is the citadel of Abū Qalamūn, on the coast near the old city of Tripoli. Nearby is Burj al-Buḥṣāṣ where there is an excellent spring. In the vicinity of (Abū) Qalamūn is the forest of Anfā, which extends as far as Beirut. On the coast in this region is a church where blind monks live. Nearby are Wajh al-Ḥajar, then Burj al-Hawā, Bi'r al-ʿAbd, Jasad al-Madfūn which is near Batrūn, then Qalʿat Jubayl. In this region are the administrative districts of Jubbat al-Kūrā and al-Munayṭir.

Beirut: An excellent port. It is a town full of vegetable gardens and flower gardens. It has all kinds of fruit and trees, and especially grapes, *ḥabb al-ʿazīz*, and the like. There is a small stream near the vegetable gardens. There are two places for Friday prayer. There are markets and public baths. There are two towers on the coast. The tomb of the jurist al-Awzāʿī (d. 157/774) is two miles from Beirut. Beirut is 36 miles from the city of Baalbek via the pass of al-Mughīsha. al-Mughīsha and the city of ʿArjamūsh are between Beirut and Baalbek.

From Beirut > ʿArjamūsh 24 miles > Ḥiṣn al-Marādīs 6 miles > Nahr al-Kalb 6 miles > Ḥūbya 4 miles > ʿAṭfa 8 miles > Māḥūr Jubayl, a strong fortress > Nahr Ibrāhīm 3 miles > fortress of Jubayl 5 miles > fortress of Batrūn 10 miles, a strong fortification > Anaf al-Ḥajar 5 miles along the coast > Tripoli 8 miles.

The military district of Ḥimṣ and Qinnasrīn: These are the fourth and fifth military districts of Syria. In the old books they called these two districts the country of Suryā and they called the area from Damascus to Gaza the Holy Land. These are the administrative units where the Canaanites used to live. In more recent times, these two military districts were attached to the land of Syria, just as the clime of Ḥimṣ was attached to Tripoli. It has its own pasha who administers it as a tax farm.

Description of Tripoli: In the past, Tripoli consisted of 21 administrative units. These were Ḥiṣn al-Akrād, ʿAkkār, Balāṭunus, Ṣahyūn, Lādiqiyya, Kahf, Manīfa, ʿAlīqa. Anṭarṭūs, Marqab, Ruṣāfa, Khawābī, Qadmūs, Jubbat al-Munayṭir (probably al-Munayṭira), and a district in the mountains called Sharya. In addition, Batrūn, Ṣāfītā, ʿArafā (?), Safaj—also called Bashrīn, Jabalat Anfā and (?) Kūrā, Jubayl, ʿĀbānī, and Banī Kilāb. Later, when the Ottomans took control, they transformed this region into a province composed of five sanjaks: Syrian Tripoli (Tarābulūs-Shām), Ḥimṣ, Ḥamā, Salamiyya, and Jabala. The tax farms in this province are Jabala, Lādiqiyya, Marqab, Ṣāfītā, Ḥiṣn al-Akrād, ʿAkkār, Waẓīnyā (?), Zāwābā'l-Kawār (?), Jubbat Basharī, Jabal Ṭarṭūs, Minā and Jabala. Batrūn is under the authority of Shaikh Sirḥān. The people (of the Tripoli area) are Qizilbash (Shīʿīs). They are a rebellious people. They only pay taxes at spear point. They pay 40,000 *ġuruş* in taxes annually.

Cities

Tripoli: It is on the edge of the sea and was a Rūmī (Byzantine) city. During the time of Muʿāwiya (in the caliphate of ʿUthmān), Suf[y]ān b. Najīb (rather, Mujīb) conquered it. He settled a group of Jews there. Later the Franks (Crusaders) seized it. In 988 (1580), (the Mamluk Sultan) Qalāwūn conquered it (from the Franks) and then destroyed it. The city of Tripoli is now the administrative center of the province. It has a post for a molla that pays 500 *akçes* a year. [M 589] Its pasha administers it as a tax farm.

It has a number of districts: Ṣāfītā, ʿAkkār, Batrūn, ʿArfa, and ʿArīḍ, This city has many vegetable gardens and trees. Sugarcane, *ḥabb al-ʿazīz*, bananas, and colocasia are grown. The Ghaḍbān river rises in the Masqiyya mountains. (Where it rises) it is called Nahr al-Zāwiya. The Abū ʿAlī river flows from Mt. Shaikh Sirḥān. These two rivers then join and take the name Ghaḍbān. It flows through the middle of Tripoli and into the sea.

In this region there is a famous valley called Wādī Rishfayn which produces abundant mulberries, olives, pomegranates, etc. The soap of Tripoli is world-famous. The silk produced here provides an annual income of 720 purses. It is also famous for its honey, grape molasses, and cheese. It has a fortress and a wall.

Its gates are as follows: Beirut Gate, Bāb al-Aḥmar (Red Gate) near the Mevlevi lodge, Bāb al-Ḥamrāwī, Bāb al-Naṣr (Victory Gate), Bāb al-Asākifa (Bishops' Gate) near the shrine of Shaikh ʿUmar and Birkat al-Badawī, Bāb al-Tabbāna (Straw Vendors' Gate), Bāb al-Maslakh (Gate of the Slaughterhouse) near Jisr al-Jadīd (New Bridge), Bāb al-Dabbāgha (Tanners' Gate), Bāb Ḥān al-Mā' near the Friday mosque of al-Tuffāḥī and [the shrine of] Shaikh

Masʿūd, Bāb al-Zubayriyya, Bāb al-Ḥajjārīn (Stone Masons' Gate), Bāb al-Qill, Bāb al-Raml, etc.

The most famous Friday mosque of this city is the mosque of al-Ṭaylān. It has a number of public baths. It is four stages from Damascus.

Nearby is the fortress of ʿArfa. It has vegetable gardens and is 12 miles from Tripoli and one parasang from the sea. It has a small river called Nahr al-Bārid that comes from the mountains of ʿAkkār. In that region is Mt. Tarbil where ʿAyn al-Ghazāl (Gazelle Spring) rises. Nearby is Bilād al-Ṭīna (Land of Potter's Clay) which is connected to Mt. Lebanon. Near that is Jubba. Sharya is near the mountains of al-Masqiyya. The village of al-Mīna, which is an endowment for the Egyptians (?), is near Mt. Tarbil.

Near Nahr al-Bārid flows the ʿArfa River which comes from Mt. ʿAkkār. It enters the sea near Shahīd al-Baḥr. Near Bilād ʿAkkār is al-Khuzayba. A river rises here and enters the sea near the ʿAkkār River. The river of al-Qunayṭira also comes from Mt. ʿAkkār. al-Nahr al-Kabīr branches off from it and enters the sea nearby. And near Mt. ʿAkkār is a forest called *shaʿrā* where is the shrine of Shaikh ʿAyyāsh.

Near ʿArfa is a bay called the Bay of Tripoli. It is also called Valley of the Jinn. Inside are three fortresses: Qarīʿāt, Ḥiṣn al-Ḥammām and Bāniyya.

Ṭarṭūs: It is on the coast. When the Muslims conquered Ṭarṭūs they destroyed its fortification. Eight miles south of it on the top of a mountain is the fortress of al-Khawābī (text: al-Khawānī), also called Ḥaymūr and Qalʿat Ḥayṣa. The people in this region are Nuṣayrīs.

The fortress of al-Martab is eight miles from here. Martaba is the name of the fortress and also of the city of Bulunyās. This fortress is situated above the sea and is very strong. There is one parasang between the city and the fortress. This city has trees, fruit, various kinds of citrus, and sugarcane. It has many springs. The Muslims built this fortress in 454 (1061). From here numerous streams flow to the Valley of the Jinn. They say that this region supports more than 100,000 water buffalo. Jabala is ten miles from here.

Near the coast opposite Ṭarṭūs is an island called ʿAwrat (rather, Arwād). In the past it was uninhabited, but it has good water [M 590] and Frankish ships used to stop there to take on water. Then Grand Vizier Köprülü Meḥmed Pasha built a fortress on this island and stationed guards there. Since then the Franks can no longer take on water from there.

Near Ṭarṭūs the river called Māʾ al-ʿAmqā flows into the sea. Near it is a forest called Hīshat al-Abrash where Arabs and Turkmen gather in bands. The Abrash River flows through it. This river comes from Mt. Fītā and enters the sea near the Nahr al-Akbar.

North of Ṭarṭūs flows Nahr al-Ḥaṣīn which rises from Mt. Marqab. Near it is the Marqab River and near it is the Bānyās River, which also comes from Mt. Marqab. Sayfiyya-oğlı emerged on this mountain and seized control of Tripoli as a hereditary grant. Nahr al-Malik also flows into the sea near the Bānyās River. In this region are the fortresses of Qadmūs and al-Kāf.

Description of the sanjak of Jabala: The tomb of Ibrāhīm b. Adham (celebrated Sufi, d. 161/777) is in this small city. A large shrine is attached to it. This sanjak has many extensive administrative units. It is 12 miles from Lādhiqiyya (text: Lādiqiyya), as is (?) Qadmūs. Ṣahyūn, Lādiqiyya, and Balāṭunus belong administratively to Jabala. Near Qadmūs is al-Halūliyya where [there is] the road to Shughr; one can also go to Wādīʾl-Qurayshiyya. Near Jabala is a fearful forest mostly of pine trees. There is a river near Jabal[a] and there is also a large river north of Lādhiqiyya. Jonquils grow on a small mountain on the coast of Jabala. There is a type of stone on this mountain: when it is broken open, out comes another stone in the shape of a woman's pudendum.

Lādhiqiyya has an excellent harbor which is preferable to all others. There is a very beautiful monastery here called Fārūs Banāsa. It is inhabited. Lādhiqiyya is 18 miles from Ḥiṣn Harbād. Near it is Jabal al-Aqraʿ (Bald Mountain). In this region is Wādīʾl-Qandīl (Lamp Valley) through which a river flows. The Kalbiyya and Baṣriyya people inhabit these mountains. There is a river that flows toward Quṣayd, then north of Ḍayʿat al-Shaikh and into the sea near the Orontes.

Ṣahyūn: A strong fortress built on a rock cliff. It has many water sources. There are valleys nearby where citrus grows that is found nowhere else. It is one stage from Lādiqiyya. The people who live here follow the false Ismāʿīlī rite.

Notice: The commanders of this fortress were notorious. They were from a sect known among the people as the Fidāʾiyya (Commandoes, i.e., the Assassins). They gained great fame during the time of Malik Ṭāhir (the Mamluk sultan al-Malik al-Ẓāhir Baybars). Members of this sect were very brave. Each one had a fortress (rather, the sect had a number of fortresses). Their fortresses stretched along (the mountains that parallel) the coast from Tripoli and Sidon as far as Aleppo, even as far as Ḥawrān. They numbered 70 fortresses. Ṣahyūn was the most important of these, where their chief Maʿrūf b. Ḥamza lived. He was the shaikh who was notorious for trickery among this sect. There is no basis to the legends that their propagandists recorded about their deceptions. However, as a brave people they frequently engaged in religious war against the

Amirtak (?) sect. They have fabricated stories and lies like the *Ḥamzanāma*.

Description of Ḥimṣ: A notable province, one of the fundamental provinces of Syria. In ancient times it was very prosperous. The Amalekite Ḥimṣ b. Mihr founded the city of Ḥimṣ. Because of the oppression of their rulers and the invasion of the Arabs, most of it is now in ruins. It has a number of districts, most notably, Ḥiṣn al-Akrād and Ḥasiyya (?). [M 591] Stored in the fortress of Ḥimṣ is a copy of the Koran (compiled by caliph) ʿUthmān. It is known as the copy of (the Arab commander) Khālid b. al-Walīd. It is never taken out of the fortress, for when that happens, there are rainstorms and floods—it has been tested!

This city has many vineyards and gardens. The olives are beyond count. Their drinking water comes from the Orontes. The city is located on a flat place. Its soil is very fertile. There are no scorpions or snakes. There is a talisman here against scorpions: over the door of the mosque that is next to the church, the upper half of a man and the lower half of a scorpion.

The people of Ḥimṣ are described as being handsome, spendthrift, and foolish. In the past, 300,000 goldpieces were collected in tax. It has a famous lake near Mt. Ḥarmal.

Ḥiṣn al-Akrād (Crac des Chevaliers): An impregnable fortress. Ḥiṣn al-Akrād was the chief city of the sultanate before Tripoli was conquered (from the Crusaders).

They say that in this area, in Wādī Rābīl, there is a kind of ant the size of a sparrow. The Turkmen people who live here suspend the heads of the ants from the heads of their children to ward off the evil eye.

A river flows through this valley. Trees cover the valley and are lined with nightingale nests. This region is famous for nightingales. Indeed, they are exported from here to Damascus.

Near here, in the direction of Tripoli, is a village called Dayr Ḥamīrā. Across from it is a cave in which is a terrace with a window. Water flows from this cave every Wednesday night. The amount of water increases as it flows. From sunset until the forenoon of the next day it flows like a river. Then it drops off until sunset and remains still until the following Wednesday night.

Sanjak of Tadmur (Palmyra): It used to be one of the administrative districts of Ḥimṣ, but now has its own governor. Most of the ground is a salt marsh. There are date palms and olive trees. It has imposing ancient monuments composed of rocks and columns. It also has a fortress and a city wall. They say that the palace there was built by Solomon. Salt is extracted in this district. It is an independent tax farm.

Sanjak of Salamiyya: The Ottomans have currently placed it in the hands of Arab amirs. Its water comes via an underground channel. It has vegetable gardens. ʿAbd Allāh b. Ṣāliḥ b. ʿAlī b. ʿAbd Allāh b. ʿAbbās founded this city and settled his son here. Most of the inhabitants are from the Banī Hāshim. It is a city of great abundance, situated east of the Orontes River and one stage from Ḥamā. Currently the Mawālī amirs administer this sanjak. They are from the clan of Āl Ḥubār, who originate from two tribes, Āl Ḥamd and Āl Muḥammad. Their authority reaches as far as Aleppo and Raqqa.

It is reported that Ḥamd Abū Nuʿayr demonstrated great skill in his service of the rulers of Egypt while performing religious war. They stuck a plume in his headgear, and so he was called Abū Rīsha ("Father of the Plume"). They gave him several thousand goldpieces so he could buy slaves for himself and triumph over his people. He bought 1,000 slaves with that gold. Today the descendants of those slaves are called the Mawālī. They have no emir other than from them.

Sanjak of Ḥamā: It has an independent government that is currently attached to Tripoli. It has an independent qadi. It has three administrative districts: Ṭāhirī, Bārīn, and Maʿarra. In the past many places belonged administratively to Ḥamā, but today there are few. They are Qadmūs, Kahf, Shayzar, Kilīstī, Būzī, Khān Shaykhūn, Maṣyāf, Marqab, Ḥiṣn al-Khawābī, ʿAllāt Turkmān, Sallūr, Kafarṭāb Raḥba, etc. [M 592]. Near here is the famous Usquf Köprüsü ("Bishop's Bridge").

Ḥamā: An ancient city that is mentioned in the books of the Israelites. The province is a place of excursion. Most of the city is along the Orontes River. It has a fortress of excellent construction located on an elevation. In the city there are water wheels on the river. The most famous of the water wheels turning on the Orontes in Ḥamā is the Muḥammadiyya. It is extremely large and irrigates most of the vegetable gardens.

Mudīq: This city is an endowment for Mecca and Medina and is therefore administered by the Dārüssaʿāde Ağası (chief black eunuch of the sultan's palace in Istanbul).

Shayzar: A strong fortress. The Orontes flows north of it. Trees, vegetable gardens and fruit are abundant. The chief fruit is pomegranates. It is nine miles from Ḥamā. It has an earthen wall with three gates. The first who took this fortress from the Byzantines was a certain Abū al-Ḥasan ʿAlī b. Muqallad Sadīd al-Mulk of the Banī Munqidh.

Nāmiyya: A district of Shayzar. The Orontes flows through the middle of it. It has a famous lake with a kind of black fish called *salūrī* that are caught in large number.

Kafarṭāb: The people of this town came from Yemen. It has little water. The people make pots and pottery for export. It is located between Shayzar and Maʿarra. Its wells are very deep. Some are 300 cubits. Most of their water is collected from rain.

Bārīn: Its fortress is in ruins. It has springs and vegetable gardens. It is one stage southwest of Ḥamā. This city has ancient monuments. The Franks built this fortress in 1400 AD. Later the Muslims tore it down.

Maṣyāf: A mountainous town one parasang north of Bārīn. It has small spring-fed streams and vegetable gardens. It is the center of Ismāʿīlī missionary activity. It is at the eastern foot of Mt. Lukkām. The Ismāʿīlī rite is currently found in ʿAkkār, Ḥiṣn al-Akrād, Ṣāfītā, Maṣyāf, al-ʿUllayqa, Ḥiṣn al-Qarnayn, and elsewhere. It is reported that the founder of the Ismāʿīlī movement was Rashīd al-Dīn Sinān. He was originally from Basra. He died in 588 (1192) (rather, 589/1193). The fortresses that he seized in Syria are called al-Akūb.

Sanjak of Maʿarrat al-Nuʿmān: It is currently an independent sanjak. This is the city of the famous (poet) Abū'l-ʿAlāʾ al-Maʿarrī (d. 449/1058). Many distinguished people came from here. It is proverbial for its excellent climate. In the past it was prosperous and famous. Today most of it is in ruins. It is six hours from Maʿarra to Sarāqiba, and from there, four hours to Khān Tūmān near Aleppo.

Chapter: Military District (*jund*) of Qinnasrīn

It is reported that when the Muslims arrived in this region they first settled in Qinnasrīn. Later Aleppo flourished and Qinnasrīn declined and became almost a village. The Quwayq River passes below Qinnasrīn and to the west. It is one stage from Aleppo.

This place is one of the lands of the Banī Rabīʿa. In the past it had many administrative units. They were Ḍawāḥ, Bahasnā, ʿAyntāb, Marʿash, and al-Rāwandān; the fortresses of Kakhtā, Karkar (Gerger), al-Darbasāk, Baghrās, Abū Qays, and Ḥārim; Kafarṭāb, [A]fāmiyya, Sarmīn, Jabbūl, Jabal Samʿān, and ʿAzāz; the fortress of Tall Bāshir; Manbij, Tazīn, ʿAmal al-Bāb and Buzāʿā, Darkūsh, Anṭākiya (Antioch), Shayzar, Shughr, Ayās, Tarsus, Adana, Sarafandigār, Sīs, Maṣṣīṣa, Malāṭya, Dārende, Divriği, Tall Ḥamdūn, Hārūniyya, Najma, Lu'lu', Bīra, Jaʿbar, Ruhā (Edessa, Urfa), Ḫarpert (Harput; text: Ḥart-bert), Kūmī (Gömi), Qaṣr, Shughr and Bakās, Bārīn, Ḥamūḍ, etc.

At that time the major cities were Anṭākiya [M 593] and the districts (*kūra*) of Qūrs al-Jawm, Manbij, Tazīn, Būqār, Bālis, and Ruṣāfa.

Also at that time there were eight Turkmen clans of Aleppo (seven listed): Dhū'l-Qādiriyya in Elbistān; Özeriye, from the confederation of Dāwūd b. Özer, in Maṣṣīṣa; Ẕikriye, from the confederation of Sālim Ẕikrī, in Çakıt; Ramaẓānoğlı in Adana; Avşariye in Jaʿbar; Varsaḳ in Tarsus; Beyāẓiye and Kebkiye around Aleppo. But today their names have changed. They are: Pehlivānoğlı, Bayāt, Ḳaçar, Avşar who are the Receboğulları, Eymir, Cerīd Silsüper, ʿAbālu, Ördeklü, Qaraqoyunlu, and Aqqoyunlu.

There are two groups of Kurds in Aleppo, one Sunnī and the other Yazīdī. Also the Kurds of Judam, Quṣayr, and Birecik.

There are two Bedouin tribes in Aleppo. One is the Banī Kilāb who are rebellious and do not obey their *beys*. They live near ʿAmq. The other is the Āl Yasār whose lands are in Zūr and near Ḥiṣn Kahlā. Sometimes they obey their *beys* and sometimes they are rebellious.

Description of Aleppo

When the Ottomans conquered Aleppo, they made it into a province (*eyālet*) and attached seven sanjaks to it. They were Adana, Bālis, Birecik, Aleppo, ʿUzayr, Kilīs, and Maʿarra. Later some of the sanjaks, like Adana, became provinces with their own pashas; some, like Maʿarra and Birecik, remained sanjaks; while others, like ʿUzayr and Kilīs, became imperial domains (*ḫāṣ*). Today Aleppo is the capital of the province. It is an old city with a position for a molla that pays 500 *akçes per diem*. It has a strong fortress on a high hill. The construction of this fortress was completed in 690 (1291).

Aleppo has few vegetable gardens. The Quwayq River flows to Aleppo. The houses of Aleppo are beautiful. It has stone walls and a large moat. In the middle of town are water mills. In Aleppo are 74 quarters in which 14,000 houses have been registered. It has many Friday mosques and public baths. A gourd has been erected (as a target) in a field; this is the equestrian arena.

There are two shrines for Abraham in Aleppo, one inside the fortress and the other outside. They are places of pilgrimage. (And there) is a cave where Abraham kept his sheep. It is reported that when Abraham milked his sheep he gave one third of it as alms to the poor every Friday, and for this reason the city was called Aleppo (Ḥalab—*ḥalab* being "milk" in Arabic). But this explanation is highly unlikely, because Aleppo was founded after the time of Abraham, also because its language was Hebrew (not Arabic).

There are also two shrines for Khiḍr in Aleppo, one in the fortress and one at Bāb al-Naṣr.

There is sowing and planting in Aleppo. They irrigate with rainwater. They grow cotton, sesame, watermelon, cucumbers, millet, grapes, pistachios, apricots, apples, and figs. There are madrasas and *martyria*. There is so much commercial activity that Aleppo is known as Little India. The Franks have consuls in Aleppo and their ships drop anchor at the port of Iskandarūn.

On the road outside Bāb al-Faraj, also called Bāb al-Yahūd (Jews' Gate), is a rock where people make vows. Muslims, Jews, and Christians all venerate this rock which, they say, marks the tomb of a prophet.

In the Ḥalāwī Madrasa, on the shore of the lake, is a platform-like stone with some carvings in the middle. The Franks are great believers in this stone. They even tried to buy it from the Muslims, but the Muslims would not sell it.

Outside Bāb al-Ḥadba and Bāb Qūsāran is the Sūq al-Zujāj (Glass Market). [M 594] Those who visit this market are amazed at the abundance of rare and marvelous goods. Aleppo is famous for stirrups and shields.

Fuʿa, Sarmīn, and Maʿarrat al-Miṣriyyīn are a single region (*buqʿa*), one stage south of Aleppo, having olives, figs, and many other trees.

Ḥiṣn Tall Bāshūk: It has water and vegetable gardens. It is in a desert south of Khunāṣra and east of Ḥiyāḍ (?), one stage from Aleppo. The Umayyad caliph ʿUmar b. ʿAbd al-ʿAzīz lived here. Kūrat Ḥiyār, also called Ḥiyāḍ, is in the desert. This place, which was prosperous in the past, is known as Kūrat Ḥiyār and Ḥiyār Banī'l-Qaʿqāʿ. Its people were ʿAbs, Fazāra, and other (tribes), now known as ʿUrayna (?), Āl Faḍl, Āl Murrī, and the Jabbār Arabs.

al-Bāb and Buzāʿa: al-Bāb is a small town with markets, a public bath, a neighborhood mosque, and a Friday mosque. It has many vegetable gardens. Buzāʿa is one of the administrative districts of al-Bāb. It is a small village. Outside it is a famous martyrium, the tomb of ʿAqīl b. Abī Ṭālib (brother of Caliph ʿAlī). It is one stage northeast of Aleppo.

Ruṣāfa: It was founded by Hishām b. ʿAbd al-Malik b. Marwān, who took up residence here when plague broke out in Damascus. It has a wall made of cut stone. It has no (flowing) water. There is a well, some 120 cubits deep, but the water is brackish. For drinking water the people use cisterns inside the city. Their industry is making garments, sacks and nosebags.

Rāwandān: An elevated strong fortress on a high white mountain. It has springs, vegetable gardens, fruit, and a beautiful valley. The ʿAfrīn River passes below the fortress. It is near Nahr al-Dhahab ("Gold River").

Wādī Buṭnān: It is between Aleppo and Manbij. Its excess comes to a salt marsh from which they take salt.

Jabbūl: It is east of Buzāʿa. It is connected to Aleppo west of Bāb Jibrāʾīl.

Atārib: It is three hours west of Aleppo. Its grapes are famous.

Ḥārim: It is two stages northwest of Aleppo. It is a valley between the mountains through which the ʿAfrīn River flows, from north to south, into Lake ʿAmq. There are many olives in this valley. There is a district here called Jūma.

Sarmīn: It has many olive trees but no water source. They get water from cisterns. Its districts are large. It has markets, a neighborhood mosque and a Friday mosque. It is one stage south of Aleppo.

Dar[ba]sāk: It has an elevated fortress, springs, and vegetable gardens. It is a fertile area. It has a neighborhood mosque, a Friday mosque, and a *minbar*. It has an excellent and extensive pasture and many meadows. The Nahr al-Aswad (Black River) flows through it. It is ten miles from Yaghrā. Yaghrā is a village in (the district of) Darbasāk. Its inhabitants are Christians who are fishermen.

Ḥiṣn Barziyya: Its fortress stretches toward Shughr. East of it is a (mountain) known as Khayṭ. Its fortress is at the foot of the mountain and is connected to the lake of Afāmiyya. Streams and lakes come right up to the base of Ḥiṣn Barziyya. A series of lakes stretches for about a stage between [A]fāmiyya and Ḥiṣn Barziyya.

Shughr and Bakās: Today it is called Shughūr. Grand Vizier Köprülü Meḥmed Pasha had an elaborate han built here along with a Friday mosque, a public bath, and a lodge for travelers. Food is also provided for travelers. Shughr and Bakās are two fortresses a bow shot apart. They are located on an elongated mountain. A river flows below both of them. They have many vegetable gardens and vineyards and much fruit. [M 595] There is a neighborhood mosque, a Friday mosque and *minbar*, and settlements (text: *rustāt*, error for *rustāḳ*). These two fortresses lie about halfway between Anṭākya and [A]fāmiyya. East of them is the Kasafhān (?) Bridge, which crosses the river that flows below them. There is a weekly market at this bridge where people gather and conduct business.

Shughr and Bakās are northeast of Ṣahyūn. An extraordinary event occurred in 806/1403 when an earthquake struck Shughr and Bakās and other places and destroyed everything. These two fortresses and their inhabitants were swallowed up. Only 50 people were saved. The ground was split for a distance of one *barīd*, about four parasangs, as far as Balad Qaṣra and Shalfūhama. The town of Shalfūhama is at the top of a mountain. This town, its inhabitants, trees, and animals disappeared. No one knew what happened to them. No one was injured until it reached that place.

Anṭākya (Antioch, Antakya)

It is related that when Alexander the Greek died, the countries (that he ruled) remained in the hands of his subordinates. Among them, King Anṭaqyūs (Antigonus), who was also called Ūrghāmus, took the country of Suryā, i.e., the province of Syria plus Jazīra, and built these cities. Salafqūs (Seleucus I) the heroic general initiated the construction of the city wall of Anṭākya, the fortress of Aleppo, Lādhiqiyya, Silifke (Seleucia), Maṣṣīṣa, Urfa (Edessa), and Ḥamā. This was 21 years after the death of Alexander. He made Anṭākya his capital. Before he built this city, the king assembled his sages and engineers and said, "I want a special city for myself. Its climate should be excellent summer and winter and it should be near the sea." They inspected the region and at a place called Marj al-Dībāj they found an excellent spot that had all the aforementioned specifications. It was a place of excursion or promenade, with both mountains and a plain, and abundant sweet water. On one side flowed the Orontes River and the sea was nearby, so it was suitable both summer and winter. They immediately began to build the city wall, the circuit of which was 12 miles, half in the plain and half in the mountains, the diameter of the circle marking the division between them (?). A lofty fortress was also constructed that could be seen from a great distance.

Because the area around the city was at a high elevation, the sun rose over the city only at the second hour of the day. (i.e., one hour after the regular sunrise). Three hundred and sixty towers were built in this city. A temple was built for the worship of Saturn. It was located east of Qanṭarat al-Samak ("Dike of the Fish"). Every year a three-day festival was held in his honor. Nearby was a public bath that people could enter during the days of the Mihrijān festival (the autumnal equinox in September) without paying a fee. The people of the city made pilgrimages to this temple.

Anṭākya has seven gates: five large and two small. The Orontes passes three of these gates. It flows east, north, and west (as it meanders). In the center of the city was a temple to Mars. Today it is known as Kanīsat al-Sayyida ("Church of the Lady"). There is a hot spring hear the church. The church has 40 doors covered with brass. Its walls are decorated with gold and silver. Its floor is paved with various kinds of marble. Inside the church are 100 gold and silver idols. Five hundred heart-ravishing boys were assigned to minister to the church along with 500 priests. Outside, a golden statue of Mars is over the lofty dome and under his foot are a scorpion and a snake made of copper. An underground channel brings water to the houses and public baths (in the city). [M 596] Today it is known as the Būlaṣ Channel. At the head of the channel are two images, one of a (male) angel and one of his wife.

There are ten mills on the Orontes, each one turning seven millstones. And there are a number of bridges over the Orontes made of white stone. (King Antigonus) had a pavement built from Anṭākya to the river as far as Jisr al-Ḥadīd (Iron Bridge). He also had an excellent harbor built for the quay at Suwaydiyya so ships could dock there securely. He built seven markets in Anṭākya, three are covered with roofs and four are uncovered. In the middle of the city is a place called Balāṭ al-Malik ("King's Palace"). It is paved with various kinds of marble. He had large storerooms erected (in the city) for the storage of grain.

There are seven springs in this city, each with its own medicinal property:

- The spring near the Būlaṣ Gate. If one bathes in it on 8 April at the hour of Mars before sunrise, he will be cured of colic.
- The spring near the al-Janānī Gate is beneficial for scabies and mange. If one bathes in it on Saturday at the hour of Saturn, it will cure the illness.
- The spring near the al-Dabbāgha (Tanners') Gate is beneficial for hemorrhoids and imbalance of the yellow bile. If one washes in it on 8 Kānūn (December or January), at the hour of Mercury he is cured.
- The spring at the Church of Qusiyān which is today called the Great Mosque. If one who has a pain in the abdomen drinks from it at the hour of the sun, the illness disappears.
- The spring called ʿAyn al-Ḥayāt ("Water of Life") near Ghassālāt (Laundresses Gate) is a beneficial sulfurous spring. They say that if one enters this spring on 4 April at the hour of Jupiter, it will cure backache and (illnesses of) the blood vessels.
- The spring next to the mountain, known as Jarna or Martīshā. If one drinks from it on 18 July at the hour of Venus before sunrise, it immediately cures illness of the intestines and pains in the abdomen.
- The spring near Bāb al-Jabal (Mountain Gate), to the south. They say that if one lies down there on 14 July, whatever good or evil will befall him will be revealed (in a dream).

We have previously mentioned the Church of Qusiyān. It was named after the king whose son was brought back to life by the chief of the apostles (i.e., St. Peter). Above the door of the church were placed two alarms for the hours of day and night. Each one announced twelve hours.

Anṭākya subsequently fell into ruin, but today the waters of Anṭākya flow to the homes of the people, to the markets, and to the neighborhood mosques and Friday mosques. Anṭākya consists of five levels. Its public baths, vegetable gardens, and promenades are on the fifth (i.e., highest) level. The reason is that the water of Anṭākya comes down from the mountain overlooking the city, and baths and gardens were maintained where the streams passed.

The churches of Anṭākya were beyond count. They were constructed using stones set in gold and colored glass, and their floors were paved over with mosaics, some of which are still extant.

The public baths of Anṭākya are the finest of baths. This is because their water is fresh and flows continuously. The fuel for the stokeholes of the baths is myrtle wood.

The estates and villages of Anṭākya are very fertile. Its districts are Sūwaydiyya, Altun Ovası, and Qaṣra. The story of Ḥabīb al-Najjār took place in Anṭākya and today it is the site of his grave.

Near Anṭākya in the direction of Maṣṣīṣa [M 597] is a place called Marj al-Qabā'il. There is a lake in the plain of Anṭākya. The Baqarā River, which takes the name of the village through which it passes, flows into it.

Baghrās: It has a lofty fortress, springs, a valley, and vegetable gardens. Zubayda (the wife of Hārūn al-Rashīd) built a guesthouse for travelers here. It is twelve miles from Anṭākya and Iskandarūn. Baghrās is on a mountain overlooking ʿAmq (Depression of) Ḥārim. Ḥārim is east of Baghrās and two stages distant. Baghrās is south of Darbasāk and less than one stage distant.

Baghrās is a fortress on Jabal Mūsā. It is north of Anṭākya and Lake ʿAfrīn, and overlooks Anṭākya, ʿAmq, and Jūma. It is a place in the mountains east of the main road which is called Baghrās Beli (spelled thus; Baghrāṣ Pass). It is a land of many goats.

Sultan Süleymān founded a village here in 959 (1552) and built a Friday mosque and a *han*. The people of the village were exempted from the customary taxes and the village became a large town. The Sultan also built a soup kitchen for travelers in that village and the guesthouse provided food and fodder (for their animals). The village has flowing water and vegetable gardens. Saflān, Darbasāk, and the fortresses of al-ʿAwāṣim are in the mountains of Baghrās.

Baghrās is one stage from Anṭākya, on the eastern side of the Anṭākya road and visible while traveling on this road in the mountains on the right. It is a qadi district with an inn, Friday mosque, and public bath in a delightful place. Sultan Süleymān founded it in 959 (1552). There is a famous hyacinth in this district called the Baghrās (here spelled Baqrāṣ) hyacinth. It is a yellow hyacinth and is mentioned widely.

District of Qaṣra: While going from Anṭākya to Damascus, Darkūsh is to the left of Jisr al-Ḥadīd and the District of Qaṣr[a] is to the right. This district stretches as far as the Village of the Shaikh. The tomb of Shaikh Aḥmad Khalwatī (apparently a shaikh of the Khalwatī brotherhood) is in this village. It is located on one side of Qandīl Valley. Beyond it is the village of Ordu (?), which is attached administratively to Tripoli. Qandīl Valley is almost completely forested. This valley is behind Shuʿūr.

Ḥārim: A small town with a fortress, trees, springs, and a small river. It has excellent seedless juicy pomegranates. It is two stages from Aleppo and one stage from Anṭākya. Near Ḥārim are Ḥalqa and Darkūsh. The pomegranates of the dervish lodge of Darkūsh are famous.

Salqīn: A delightful and prosperous town attached administratively to Ḥārim. Currently the qadi of Ḥārim resides in Salqīn.

Shaqīfa: A place with abundant trees, fruit, and springs.

Kafar Bīn: A famous citadel near Anṭākya belonging to Aleppo.

Iskandarūn (Alexandretta; the text has Bāb Iskandarūna): A city on the Mediterranean coast and near Anṭākya. (Aḥmad) Ibn Abī Dā'ūd al-Iyādī (Muʿtazilite judge, d. 240/854) built it during the caliphate of al-Wāthiq (r. 227–232/842–47). Today it is the port for Aleppo and never lacks for ships of Frankish merchants.

Arīḥā: This is the district of Rūj, also called Rūjīn. It is famous for watermelon, the like of which is found nowhere else.

The district of Idlib Ṣughrā (Lesser Idlib) (and) Idlib Kubrā (Greater Idlib), Tīzīn and Nāmiʿ (?); the district of Jabal Bārīsha, Jabal al-Aʿlā, Darkūsh, Zāwiyat Shuʿūr, Shaikh al-Ḥadīd and ʿAmq.

Nālīs, also called Balīs: A very prosperous sanjak. It is a small town on the west bank of the Euphrates. The Bedouin of the Banī Samak live in this region. They can muster 5–6,000 musketeers. They are subject to the al-Mawālī Bedouin.

Sanjak of Bīra, [M 598] also called Birecik: Formerly it was administratively attached to Aleppo, but today it is a separate unit with its own governor. This city is right just to the northeast of the Euphrates. It has a strong lofty fortress and a valley known as Wādī'l-Zaytūn ("Valley of Olives") with trees and springs. Bīra is also an entrepot. It is one stage from Qalʿat al-Rūm. It is west of Qalʿat Najm and Qalʿat al-Kawākib and southwest of Sarūj. Sarūj is very prosperous.

Qalʿat al-Rūm, known as Rūmḳalʿe: The vegetable gardens and fruit of this place are very plentiful. The river called Marzubān flows through here. It comes from the district of Jabal and meets the Euphrates below the fortress. This fortress is below the Euphrates (sic?). (The Mamlūk) Sultan al-Malik al-Ashraf (Khalīl b.) Qalāwūn (r. 689–693/1290–1293) took this fortress from the Armenians. It is one stage west of Bīra and has a district called Marzubān.

Sanjak of Kilīs and ʿAzāz and (Sanjak of) Manbij: Now Kilīs and ʿAzāz are located northwest of Manbij. It (i.e., ʿAzāz) has a citadel. Its soil is excellent and fertile. It is free of scorpions and vermin—indeed, there are none at all in this district. Manbij was founded by one of the Chosroes (Sasanian rulers of Persia) who had conquered Syria. He named it Manbih (meaning "I am the best," or "Who is better" in Persian) and it was Arabicized to Manbij. He built a fire temple there and placed someone named Ibn Dīnār, who was a son of Ardashīr b. Bābak, in charge of it. Most of the trees of Manbij are mulberry trees cultivated for silk production. The land is fertile. Today these two sanjaks (i.e., of Kilīs wa ʿAzāz and of Manbij) and their districts are an imperial domain belonging to the Valide Sultan (sultan's mother), may her chastity be everlasting. She appoints an overseer (*voyvoda*) for them.

Sanjak of ʿUzayr, also called Maṭakh: It is southwest of Aleppo and one stage distant. It is a separate sanjak and is administratively attached to Aleppo. It consists of prosperous villages. It has water.

In sum, Jūma, Jabal Samʿān, and Jabal Summāq are inhabited by Ismāʿīlīs. Jabal al-Aʿlā, Ghunjura, Jabal Bārīsha, Rūj Dākhil and Rūj Khārij, and Sājūr are all located around Aleppo, as is recorded on its maps (?—*niteki ṣuverinde mestūrdur*).

Provinces of Marʿash and Adana

Marʿash and Adana are two provinces. Some of their territory is considered to be the province of Lesser Armenia and some to be part of the clime of Syria. Because these two provinces are frontier towns of Syria, we have added them to (the section of this book on) Syria.

The province of Marʿash was the territory of the Dhū'l-qadriyya (Zülkadriye, Zülkadiroğulları). The province of Adana was ruled by the family of the Āl Ramaḍān (Ramażānoğulları). Subsequently the Ottomans conquered these two territories and made them into two provinces.

Borders: To the north is the province of Ḳaraman, to the east are the Euphrates and the province of Urfa; to the south, the province of Aleppo; to the west, Adana and the Ce[y]ḥān River. The following are their sanjaks: Marʿash, which is the pasha sanjak; Qārṣ Dhū'l-qadriyya (Kars Zülkadriye); Malatya; ʿAynṭāb; and Sumaysāṭ (Samosata).

Sanjak of Marʿash: This is the administrative center of the province. It is a large city with a post for a molla paying 500 *aḳçes* (per day). It has many small mosques and Friday mosques, Sufi convents, madrasas, public baths, busy markets, and splendid houses. Its fortress is located on a high ridge below (the city) in a flat (text: *dir*, error for *düz*) place. Near the old Friday mosque known as Ulu Cāmiʿ (Great Mosque) is the court of justice and the palace reserved for the pashas. This city has abundant water, agricultural crops, and trees. It is twelve miles from Makhāṣṣat al-ʿAlawī, which is on the Ceyḥān River. It has a number of districts. They are Elbistān, Behisnī, Pazarcıḳ, Hārūniyya, Bürkān, Ḳaraḥasanlı, Andırun, Ẓamantı, Gügercinlik, Cāmūsteli, Peşli (Beşenli), Yeñice Ḳalʿe (text: ḲLV), Kemer, Ḳartā, and İslāmbeyli.

Hārūniyya: A district on the road going from Istanbul to Marʿash. [M 599] This area has some villages, fields (reading *mezāriʿ* for *mażāriʿ*), and the grazing grounds of tribute-paying nomads. The city (of Hārūniyya) was founded by Hārūn al-Rashīd. It was a small city in the direction of Mt. Kān (error for Lukkām?). It is twelve miles from Kenīse (Kilise).

Pazarcıḳ: It is near Marʿash and has abundant flowing water and places of excursion. It has a lake in which there is a village that drifts over the water. It is on an island like a raft composed of the roots of reeds tangled together and which, over time, have been compressed by dust and dirt. Some Armenian mat makers built houses of those reeds and straw and settled there. Gradually they increased in number and created a village. All the people of the village are mat makers and Armenian infidels. This village drifts with the wind in one direction or another, and if it drifts off the people can move it back to shore (?) using thin poles.

Elbistān: A prosperous town on the road from Marʿash to Kayseri. It is a separate/independent qadi district with many villages and fields (reading *mezāriʿ* for *mażāriʿ*). It has a very broad plain. Its districts are Ṣārdas, Ṣadernā, Aḥsender[e], and Orta Niyābet.

Zabtar, also called Zamantı: An Armenian frontier city, nine parasangs from Marʿash, with a city wall built of stone. Abū ʿUbayda (early Muslim commander, d. ca. 18/639) sent Khālid (b. al-Walīd, military commander, d. 21/642) from Manbij to this city. He destroyed it during his conquest. Afterwards (the Umayyad caliph) Muʿāwiya rebuilt it. ʿAbbās b. al-Zabīd carried out additional restoration and built a small mosque and Friday mosque. Subsequently, the Byzantines destroyed it and Ṣāliḥ b. ... (name missing) rebuilt it during the caliphate of al-Manṣūr (Abbasid caliph, r. 136–158/754–75).

Behisnī: A qadi district and one of the districts of Marʿash. Its town is on the road from Marʿash to Kayseri. It has a small river, vegetable gardens, markets, a small mosque and a Friday mosque, and many pious foundations. It is two stages northwest of ʿAynṭāb. Its districts are Dobaz, Gözebaşı, and Somal. The qadi district of Behisnī is Ẓamantı whose districts are Çörsek and Pıñarbaşı.

Sanjak of Qārṣ Dhū'l-qadriyya (Kars Zülkadriye): A town and a qadi district. Miṣṣīṣ (Maṣṣīṣa) is four hours to the east. It has a small fortress. During the winter some nomadic Turkmen have their winter campgrounds nearby. So that region is very lively in winter. In summer, when they go to their summer pastures, no one is left behind.

They are opposite Ḳurdḳulaġı and away from the main road. The qadi districts of Şora and Andırun are administratively attached to it. Its districts are Kānı, Sögüdlü, Sīnās, Mināreli, Maġāra, Gökṣun, Döşek, and Gökṭaşlı.

Sanjak of ʿAynṭāb: ʿAynṭāb is a beautiful city. It has a fortress hewn out of the cliff. It has abundant water and vegetable gardens. It is the administrative center of its districts. It has splendid markets. It is three stages north of Aleppo.

Nearby the fortress of Delük, which is in ruins. It is famous for white grape molasses which is cut with a knife. It is also famous for bows, saddles, and saddle trees; as well as apricots and apples—one apple amounts to one oka. ʿAynṭāb is three stages southeast of Behisnā (Behisnī). Its districts are Tel Yaşar, ʿUrbān, Burc, and Sarūj.

Sanjak of Sumaysāṭ (Samsat, Samosata): It is located on the banks of the Euphrates. The people practice agriculture with and without irrigation. It is west of Qalʿat al-Rūm and north of Ḥiṣn Manṣūr. It is close to both. It is 20 miles to the place where Qalʿat al-Najm is located on the Euphrates and to the Manbij Bridge. In the past Qalʿat al-Najm was called Ḥiṣn Manbij. Then it became known as Ḳalʿe (Qalʿa). One reaches Ḥarrān via this bridge which was built by Sultan Maḥmūd b. Zangī (ruler of Aleppo and then Damascus, 541–569/1147–1174). One stage above this bridge is Ḥiṣn Hudāyā from where [M 600] one proceeds to Sarūj. Its districts are Ḳūrṣ and Reşvān.

Sanjak of Malatya: Malatya is an old town on a broad plain west of the Euphrates. It has a market, public bath, inns, and Friday mosques, and toward the south at the foot of the mountains are countless gardens and streams. In the summer the people usually go to these gardens as their summer pasture.

The city has a wall that overlooks ruins. Muṣṭafā Pasha, the sword-bearer of Ghazi Sultan Murād IV (r. 1623–1640) built an inn there. It is located on one of the major roads going east from Istanbul. If one goes a half parasang east of Malatya on this road, he will reach the Euphrates. Boats wait there to transport people to the other side.

The plain of Malatya is surrounded by mountains. Here walnuts and other fruits grow without cultivation. Malatya is to the north of these mountains and behind them is the province of Sīs. A small river irrigates these ... (word missing?) and it reaches the city wall. Its vegetable gardens are along this river.

The winters in Malatya are very cold. Sivas is three stages to the northwest. Marʿash and Aleppo are reached in eight days via military stages.

There is a river in Malatya called Dayr Masīḥ. Fifteen villages lie among the gardens and vineyards that line the banks of this river from the mountain, which is its source, all the way to the city. During the fruit season, the people spend three or four months in these villages. They stay in the city during the day until afternoon prayer and then depart at the time of the evening prayer. Five or ten watchmen remain behind at night. The village of Asbūzī (Aspuzu), which is on this river, is like a town.

Malatya has another river called Pıñarbaşı which joins the Dayr Masīḥ in Malatya and enters the Euphrates five miles below the city. The mouth of the river and the Euphrates can be seen over the vineyards (?) of Malatya.

Malatya has another river called the Toḥma. Near Malatya is a great bridge called Ḳırḳgöz ("Forty Eyes," i.e., forty arches) that spans this river. The main road that goes through Malatya goes over it.

There is a place in the city which is said to be the birthplace of Baṭṭāl Ghāzī (a champion of the Arabs in the Umayyad period and legendary hero of a Turkish romance) and which, since his time, has been a place of pilgrimage. In 1056 (1646) Melek Aḥmed Pasha erected a stone building with a beautiful dome at this shrine.

The districts of Malatya are Ḥiṣn Manṣūr, Divrigi, Şūramaġ, Bucaḳ, Ṭaşābād, Ṭīneli, Kāḫta, Gerger, and Cūyāş. Malatya has delicious apples that are "inscribed". They are produced as follows: Letters and words are written on a piece of paper. When the apple has ripened, the paper is attached to the side of the apple facing the sun. A few days later when the apple has acquired its full color, the area under the letters (which has seen the sun) turns red and the area under the (rest of the) paper remains yellow. It is as if one wrote in red ink on yellow paper. In particular they write suitable couplets and verses.

Ḥasan Batrīḳ: A prosperous town north of, and immediately adjacent to, Malatya. It is located between Ḥākimḫānı and Malatya at the foot of a large mountain one stage west of the Euphrates and is on the road to the inn. After crossing the Şār ford, those who wish go on to Ḥasan Batrīḳ and those who wish go on to Ḥekīm Ḫānı. This mountain extends from east to west as far as Ḥasan Batrīḳ west of Malatya. The Ḳırḳgöz River rises from this mountain, passes under a stone bridge and above the Şar ford and empties into the main part of the Euphrates. It flows between Malatya and Ḥasan Batrīḳ.

Kākhta: A town on the Euphrates at the spur of the mountain, two stages southeast of Malatya. It has a small river, [M 601] vineyards, and gardens.

Gerg[er]: A fortress in the same mountain spur. Kāḫta is one stage to the southwest.

Ḥiṣn Manṣūr: It is west of the Euphrates and near Sumaysāṭ. It is named after (the Umayyad emir) Manṣūr b. Jaʿwana b. al-Ḥārith al-ʿĀmirī (d. 141/758). During the reign of the Umayyad caliph Marwān b. Muḥammad al-Jaʿdī

(r. 127–132/744–50), Manṣūr undertook the construction of this fortress. Ḥiṣn Manṣūr is north of the Azraḳ River (Gökṣu). West of Ḥiṣn Manṣūr, between it and Malatya, is a mountain where the road to Malatya is reduced to a defile.

Arḳaludya: A strong fortress near Malatya.

The Kurds of Marʿash: The Reşvān Reşī are Yazīdīs who live in Ufacıḳlı, Baḳrāṣlı and Behisnī. Most of the people of Malatya are Kurds. One of their clans in these parts are mischievous rebels and highway robbers.

Province of Adana

This is a small province with one or two sanjaks. Sīs and Tarsus are administratively attached to Cyprus. Sīs is in ruins. The qadi districts of this province are Ayās-Barandī, Ziyāḳar, Ṣarıçam, Sīs, ʿUzeyr-Ḳınıḳ, Ḳaraʿīsālı, Ḳaraḳışla, and Yüregir. Its districts are Dindārlı, Cānib-i Şehr, Cīver, Oflu (?), Esen, Būs (?), Ḳurdḳulağı, Barāḳlu, Ḳaraḳaşlu, and Ḥācılar.

Borders: To the south is the Mediterranean; to the west, the province of Silifke; to the north, Ḳaraman; and to the east, the province of Marʿash.

Adana: This is the seat of government of the province. It has a qadi district. The caliph Hārūn al-Rashīd began the construction of this city and his son Muḥammad completed it. Adana is on the Seyḥan River and is twelve miles from Tarsus. Adana was the westernmost of the cities of the Zāb and the furthest of the districts of the Aghlabids (?).

The Seyḥān River: It is west of Adana and is smaller than the Ceyḥān. There is a stone bridge over the Seyḥān which is a marvelous structure. It is also reported that Adana is on the Çaḳıd River. Adana has a Friday mosque and a madrasa. Pīrī Pasha of the Āl Ramaḍān (Ramażānoğulları) built them.

In summer the air in the city is oppressive, so the people go up to their summer pastures for about six months. Only a few shops stay open in the city for travelers, and the neighborhood watchmen also remain behind. The summer pastures are two stages distant in the mountains behind Adana. The summer pasture of the Ramażānoğulları is on the main road. Their original summer pasture is that of Tekir. The people of Adana have proper homes there as well, and in summer it becomes like a thriving town.

Külek Beli (Külek Pass): The inn at Külek Beli belongs to Grand Vizier Pīrī Pasha—Meḥmed Pasha b. el-Cemālī Lala ʿOṣmānī—who built it in 928 (1522). He built three inns and a Friday mosque near Külek Beli. The summer pasture of the Ramażānoğulları is beyond it and is reached from it.

Külek Ḳalʿe: It is on a mountain near Külek Beli. It is difficult to reach. It is on the easternmost edge of Ḳaraman. For about one or two stages the road passes through mountains, pine forested passes, and rocky areas. The fortress is half way along this road. Two sides of it are carved out of the rock. It is virtually inaccessible and located on the top of a mountain to the right of the main road that passes between two mountains. When crossing the summer pasture of the Ramażānoğulları and to the other side of it, Külek Ḳalʿe is on the right side of the road.

One stage from Uluḳışla one comes to Çifteḫān ("Double Inn"). Nearby is a long slope by which one descends to the inns. Off the main road is a fine mineral spring and a hot spring. Also nearby is a river formed from the streams from surrounding mountains. It is called Ḳırkgeçid ("Forty Crossings") because one frequently encounters and crosses this river. From Çifteḫān to the inn on the bank of the Çaḳıd are mountains covered with dense oak forests. [M 602]

A half stage past Çifteḫān the Ḳırḳgeçid River increases in size and passes under the bridge called Aḳköprü ("White Bridge"). Near this bridge a river called the Ḳaraṣu also rises from the foot of the mountains and joins the Ḳırḳgeçid. There is a bridge at that place. A large spring called Şekerbıñarı also rises from the foot of the mountains, crosses the main road a stone's throw away, and flows into that river. One crosses the bridge, goes up to the mountains where the Ramażānoğulları have their summer pasture, then descends and arrives at Çaḳıd Ḫān. There, in the towering mountains, are many caves and ancient ruins, including gold, silver, and copper mines. Remnants of these places have survived until today. The river passes under the aforesaid bridge and flows through various mountain valleys. It passes above the Çaḳıd River behind the mountains and then joins the Çaḳıd. From Çifteḫān to Çaḳıd and from there to Bayram Paşa takes three or four days in the winter. Several inns have been built along the route.

Sanjak of Sīs: In the past it was prosperous and famous, but today it is in ruins. Sīs is on an elongated mountain. It has a fortress and three walls (i.e., circumvallations). It has vegetable gardens and a small river. In 721 (1321) Sīs was the city and capital of the Armenian king. Originally founded by the Armenian king Lāvī (Leo II, r. 1187–1219), Sīs was one of the Muslim frontier towns (*thughūr*). Every place located near the enemy is called a frontier town. Sīs is 24 miles from ʿAyn Zarba and Maṣṣīṣa. Sīs, which is well known today, was built in 721 (1321) by servants of (Caliph Hārūn) al-Rashīd. Its districts are Ḳāvar (and?) Ḳālī.

Maṣṣīṣa (Miṣṣīṣ, Misis): It is twelve miles from Marj al-Dibāj ("Meadow of Silk Brocade") which acquired its name because of its verdant beauty. Maṣṣīṣa is two cities. One is Maṣṣīṣa and the other is called Kafr Bayyā. They are on opposite sides of the Ceyhān River and are connected by a stone bridge. Maṣṣīṣa is very fertile. It has a mosque and a Friday mosque and overlooks the Mediterranean. It is a half stage from Adana. It has a bridge at the head of which there is only an inn that was built by some merchants in 949 (1542). A toll is collected at the bridge. Near Maṣṣīṣa on a mountain called Jabal al-Nūr grow beautiful hyacinths and all kinds of medicinal herbs, even mandrake. This mountain extends from Maṣṣīṣa almost to the sea.

Yılan Ḳalʿesi ("Serpent Fortress"): An abandoned fortress on a mountain. It is also called Şahmaran Ḳalʿesi. One passes through Maṣṣīṣa and then, when one emerges onto the plain on the way to Ḳurdḳulağı, one sees it in the distance on the left. They say that the King of Serpents (Şāh-ı Mārān) is there.

ʿAyn Zarba or ʿAyn Nāvzer: A town on a mountain, with a fortress and a river. It is one stage southwest of Sīs, between Sīs and Tel Ḥamdūn.

Tel Ḥamdūn: It has a fortress and wall. It stands on a hill. It has an extramural settlement and vegetable gardens. A river flows to the fortress. It is very fertile. The Ceyḥan River is nearby, being one stage distant. It is south of the Ceyḥan. Tel Ḥamdūn is two stages from Sīs.

Ḥiṣn Ḥamūs: It is east of Tel Ḥamdūn and close enough to it to be seen from there.

Bersbert: A citadel on a high mountain. It is the largest stronghold in the land of the Armenians. [M 603] It is about one stage north of Sīs and overlooks it.

Ayās: A new tower in a bay of the Mediterranean. It is two stages from Baghrās and one stage from Tel Ḥamdūn. It has guards. Its harbor is famous. It is an entrepôt for merchants. The Muslims conquered it in 722 (1322), plundered it and left it in ruins. It has saltpans, and the salt exported today along its coasts is from there. Sultan Süleymān built a fortress and inn here. The island of Cyprus is south of Ayās.

Bāyās (Payas): It is one stage (from Ayās?). It has a large soup kitchen, inns, a market, a public bath, a Friday mosque, and a strong fortress. It has some public buildings and some charitable institutions established by ʿAtīḳ Meḥmed Pasha, the vizier of Ghazi Sultan Süleymān. Its air is oppressive. But it is an entrepot. It has abundant lemons and bitter oranges. A picturesque mountain adjoins Payas and stretches a half stage distance along the main road. As the road goes up the mountain it passes the fortress of Merkez and continues on to Baghrās Beli. Payas has summer pastures on this mountain that have no equal. Payas is one stage from Baghrās. Food is given to travelers from the soup kitchen. It is a great public work.

Merkez: A new, specially built fortress at the top of a mountain overlooking Payas and İskenderūn. It has specially appointed guards. The main road passes in front of it.

Serfendigār: A fortress on a pinnacle in a valley. The Ceyḥān River passes near it to the south. Serfendigār is one stage southeast of ʿAyn Zarba, and Derbend pass is one stage east of Serfendigār. Large pine trees grow here the like of which are found nowhere else.

Sanjak of Tarsus: Administratively attached to the province (*eyālet*) of Cyprus. It consists of several qadi districts: Elvānlu, Ulaş, Ḳuṣūn, Gökçelü, and Ḳuştemürbaş. Tarsus is on the Mediterranean and was part of Byzantine territory. Later it passed into the hands of the Armenians and became Armenian territory. It has two city walls (i.e., circumvallations) made of stone. The mountains come right into Tarsus. When Caliph (Hārūn) al-Rashīd came to Tarsus, he rebuilt it and divided its river. There are many pious men and ascetics here. In 350 (961) Tarsus was taken (from) the Byzantine emperor by someone who (became) its governor on behalf of Sayf al-Dawla (emir of Aleppo and northern Syria). It has remained in Muslim hands ever since and is one of the frontier fortresses of the Muslims.

There is a famous place in Tarsus which some believe to be under the protection of jinn. When the Commander of the Faithful Caliph al-Maʾmūn (r. 189–218/813–833) came here to carry out religious war he set up quarters at this place where there is a river of extremely pure water. Al-Maʾmūn sat on the bank looking in the river and saw a fish that was a cubit in length. He ordered it to be caught. The fish was extremely beautiful, like silver. It leapt up and fell back into the river. When it splashed into the river it got al-Maʾmūn's clothes wet. Al-Maʾmūn became angry and ordered that the fish again be caught. Al-Maʾmūn looked at it and said, "Now I'm going to grill you." He gave it to the cook and, shivering, returned to his lodging. The cook grilled the fish and brought it to him. Al-Maʾmūn did not have the strength to eat one bite. He grew weak and his chill intensified and he died.

The story of the Aṣḥāb al-Kahf (Seven Sleepers) was also said to have taken place in Tarsus.

Stages and distances in Syria

[M 604] From Damascus to Cairo: > Dāriyya 2 hours > Saʿsaʿ 7 hours > Qunayṭira 7 > Jisr Yaʿqūb 7 > Minya (text: Mīna) 5 > ʿUyūn al-Tujjār 6. At ʿUyūn al-Tujjār the road divides; one branch goes to Cairo, one to Jerusalem. The Jerusalem branch: > Jenīn 6 hours, Nablus 8 > Bīra 8 > Jerusalem 5. At Jerusalem the road divides again; one branch goes to Gaza, one to Ramla. If one goes to Khalīl al-Raḥmān

(Hebron), the road from Jerusalem takes 8 hours > Bayt Jibrīl 8 hours > either Gaza or Ramla 10 hours. The other branch from ʿUyūn al-Tujjār > Lajjūn > Qāqūn > Jaljūliyya > Ramla > Khān Sudūd > Gaza > Khān Yūnus > Zaʿqa > ʿArīsh > Umm al-Ḥasan > Biʾr al-ʿAbd > Qaṭaba > Biʾr al-Duwaydār > Ṣāliḥiyya > Būlbays (Bilbays) > Khānqāh > Cairo.

From Damascus to Konya: > Qaṭīfa 9 hours > Nīka 8 > İkiḳapulu 12 > Ḥimṣ 10 > Ḥamā 10 > Mudīq 12 > Şuğur Köprüsü 12 > Zanbāqiyya 7 > Anṭākiya 10 > Bīlān 10 > Ḳurdḳulağı 10 > Maṣṣīṣa 6 > Adana 6 > Çifteḥān 7 > Uluḳışla 6 > Eregli 9 > Ḳarapıñar 12 > İsmil 9 > Konya 11.

Or, from Ḥamā: > Khān Shaykhūn > Maʿarra > Khān Tūmān > Aleppo.

From Aleppo to Anṭākiya: > Khān Tūmān > Ḥārim > Jisr Ḥadīd > Anṭākiya. Or, from Aleppo: > Gök Meydānı > Gancara, also called Gancāra > Efrun > bridge of Murād Pasha > Anṭākiya.

Status of kings and rulers of the clime of Syria

(Here will be related a brief account of) the kings and rulers in this country from the time before Muḥammad up to the present.

It is related that after the Flood the Canaanites were the first to build up the province of Syria, where they settled after the change of language (i.e., after Babel). In addition, a group of Amalekites came to Syria and settled in various places after they were defeated in battle by the Jurhum people at the Kaʿba. After these two peoples, the clime of Syria passed into the hands of the Israelites. Many prophets, kings, and rulers appeared among them and the clime of Syria belonged exclusively to them. These Israelites were the descendants of Abraham. Abraham was born in Babylon at the time of Namrūd b. Kanʿān (Nimrod son of Canaan) 1,018 years after the Flood.

Because it would take too long to recount everything (that subsequently happened) in this summary, it will suffice to present an account of the rulers after the appearance of Islam. In sum, during the period of 600 years that passed from Jesus to Muḥammad—which is called the period of *fatra* (interregnum or transition)—Jerusalem and Damascus (or Syria: Shām) were held by the Caesars of Rūm (Byzantine emperors). After the appearance of Islam, the Muslims took possession of them. However, two hundred years after Jesus' ascension, the clan of Banī Ghassān, who were one of the Arab tribes, took control of most of Syria. This clan was from the tribe of Banīʾl-Azd who traced their descent from Sabāʾ.

After the Sayl al-ʿArim (the "Powerful Flood" that destroyed the Maʾrib Dam in Yemen), [M 605] the Ghassānids left the province of Yemen and settled near Nahr Marj in Syria, and the lands of this tribe were called Ghassān. Previously the Qajāʿima Arabs had lived in that region. The Ghassānids expelled them and killed their kings. The first ruler of this clan was Jafna b. ʿAmr who had the nickname al-Muzayyiq ("The Tearer"; text: MZYQYʾ). He was given this name because every day he tore to shreds the clothes that he had worn so no one else could wear them. He was succeeded by his son ʿAmr who was succeeded by his son Thaʿlaba who built Ṣarḥ al-Ghadīr (the Palace of al-Ghadīr) in Ḥawrā[n]. He was succeeded by his son Ḥārith, then by his son Jabala, then by his brother Nuʿmān, then by his brother ʿAmr, then by Jafna al-Aṣghar (the Younger), then by his brother Nuʿmān al-Aṣghar (the Younger), then by Nuʿmān b. ʿAmr who built the palace at Suwaydā and who was succeeded by his son Jabala. This (last) ruler fought against Mundhir b. Māʾ al-Samāʾ. Then came Nuʿmān b. Ayham, then his brother Ḥārith, then his son Nuʿmān, then his son al-Mundhir, then ʿAmr, then Ḥijr, then Ḥārith, then Jabala, then his son Ḥārith who had the patronymic Abū Kurb and the nickname al-Quṭām, then Ayham b. Jabala who became ruler of Tadmur (Palmyra) and erected a great building in the desert, then his brother al-Mundhir, then Shuraḥbil, then ʿAmr, then Jabala, and then Jabala b. Ayham who was the last king of the Ghassānids. At that time ʿUmar was caliph. This Jabala was a Christian, but then he repented. At that time ʿUmar conquered Damascus.

Description of the Muslim dynasties that possessed the clime of Syria

When Abū Bakr was caliph he sent several commanders with Muslim armies to conquer the clime of Syria. Yazīd went to the region of Balqāʾ, Shuraḥbil went to Urdun (Jordan), and Abū ʿUbayda went to Damascus. At that time Khālid (b. al-Walīd) was assigned the region of Iraq. Subsequently when Abū ʿUbayda asked Abū Bakr for help he withdrew Khālid b. al-Walīd from Iraq and sent him to help Abū ʿUbayda. While the Muslims were besieging Damascus news arrived of the death of Abū Bakr. The date was year 13 of the *hijra* (634). ʿUmar then succeeded him as caliph. ʿUmar immediately appointed Abū ʿUbayda commander-in-chief of the Muslim army in Syria. He laid siege to Damascus once more.

For seventy days Abū ʿUbayda besieged Damascus at the Bāb al-Jābiyya and Khālid (b. al-Walīd) besieged it at the Bāb Tūmā. Then Khālid entered at Bāb Tūmā by the sword (i.e., he took that part of the city by force). At the same time, on the other side, the defenders of Damascus capitulated to Abū ʿUbayda and he entered the city by a truce. When the two parts of the army met in the middle of the city, it was in the month of Rajab, year 14 of the *hijra* (August, 635). After conquering Damascus, Abū ʿUbayda went to Ḥimṣ and captured it by capitulation. Then he went to Ḥamā and conquered it and then to Shayzar and took it by capitulation. He took Maʿarra in like manner. Then they conquered Lādhiqiyya by the sword. Then they took Jabala, Anṭarsūs (Anṭarṭūs), and Qinnasrīn, which was then the capital of the province, after a great battle, leaving the fortress in ruins. Then they seized Anṭākiya, Manbij, Sarmīn and all the provinces of that region. Then they went to Marʿash and conquered it and destroyed its fortress. Then they conquered Ḥiṣn Ḥadath (Göynük) and Qaysāriyya (Kayseri). As a result of this, the Caesar (Byzantine emperor) lost all hope of regaining Syria.

[M 606] Then they went to Jerusalem and besieged it. Because of the long siege, the defenders of Jerusalem asked for peace terms. They agreed to surrender but only on condition that Caliph ʿUmar himself would come. When ʿUmar was informed of this he immediately went to Jerusalem and concluded peace.

Subsequently the Umayyad Dynasty appeared in Syria. Its first member was Muʿāwiya b. Abī Sufyān to whom ʿUmar had previously given the emirate of Syria. After this event he became caliph in Syria by force of arms. He died in year 60 of the *hijra* (679). He was buried between Bāb al-Ṣaghīr and Bāb al-Jābiyya (in Damascus). He was succeeded by his son Yazīd, who died of pleurisy and was buried in Ḥawrān. He was succeeded by his son Muʿāwiya, known as Muʿāwiya al-Aṣghar (the Younger), and was buried outside Bāb al-Jābiyya.

He was succeeded by Marwān b. ʿAbd al-Ḥakam. During his reign the people of Syria divided into two parties. The Yemeni party supported Marwān and the Qaysī party supported Ḍaḥḥāk (b. Qays al-Shaybānī). A great battle took place between them at Marj Rāhiṭ (128/746). At the end of it Ḍaḥḥāk was killed and Marwān was independent ruler. One day Marwān reviled his wife but at night, while he was sleeping, she put a pillow over his face, and with the help of her slave girls, held him down until he died.

He was succeeded by his son ʿAbd al-Malik. At first he was very pious and God fearing, but after he became caliph his character changed. He unjustly put many people to death. He was the first of the caliphs to forbid speaking in the council chamber. He is buried at the Bāb al-Ṣaghīr.

ʿAbd al-Malik was succeeded by his son al-Walīd who was of excellent character. He always gave alms to the pious. He also set aside alms for lepers. He forbade begging. He provided servants for the crippled and the blind. He was the first to build a hospital and a hostel in the abode of Islam. He patronized the "bearers" of the Koran (i.e., those who memorized the entire Koran). He paid the debts of debtors. He fame spread to the horizons. He had no equal. He demolished the so-called Church of St. John and built the Umayyad Mosque in its place. They say that he spent (on its construction) 400 chests of gold each containing 20,000 coins, that he hung 600 gold chains for the lamps (suspended in the mosque)—they remained until the reign of ʿUmar b. ʿAbd al-ʿAzīz (99–101/717–720) who removed them and placed them in the treasury. Al-Walīd also built the Dome of the Rock (actually it was ʿAbd al-Malik) and expanded and rebuilt the Masjid al-Nabī (in Medina). During the reign of al-Walīd the clime of Andalusia was conquered as well as most of the lands of the Turks and most of India. He died in 96 (714) and was buried at the Bāb al-Ṣaghīr. His rule lasted nine years and eight months (86–96/705–715).

He was succeeded by Sulaymān b. ʿAbd al-Malik. He was a handsome young man and a *ġāzī*, but he was a glutton. It is reported that when he went on the pilgrimage he stopped en route in Ṭāʾif. There he ate 70 pomegranates, one lamb, and six chickens. Then they brought him raisins and he ate a great amount. *Kunāfa*, which today is called *qaṭāʾif* (a dessert layered with cheese and soaked with honey or syrup), was invented for him. In Syria his residence was at Siqāyat Jayrūn. He sent his brother Maslama on religious war against Istanbul in Rūm (i.e., Constantinople in Byzantium). Maslama reached as far as Galata (on the northern shore of the Golden Horn) where he built a mosque that is still called ʿArab Cāmiʿi ("the Arab Mosque," actually a 13th century Dominican church).

When Sulaymān died, ʿUmar b. ʿAbd al-ʿAzīz became caliph. He immediately forbade the cursing of ʿAlī in the Friday prayer. In place of this, he ordered that the verse *inna Allāha yaʾmuru* ... ("Surely God enjoins ..." Koran, 16:90) be read.

North of the Umayyad Mosque (he built) [M 607] residences for the pious called Khānqāh Shamāṭiyya. He forbade his own relatives to be paid salaries from the state treasury. Indeed, it is reported that one day he became ill. When his emirs and close friends came to visit him they saw that he was wearing a soiled shirt that was not worth four dirhams. They asked his wife why she didn't change his shirt and wash it. She swore that he had no other shirt. He was buried at Dayr Simʿān, which is a district of Ḥimṣ.

He was succeeded by Yazīd b. ʿAbd al-Malik who was thoroughly absorbed in pleasure and died in the vicinity of Urdun (Jordan).

He was succeeded by Hishām b. ʿAbd al-Malik who was very just and, owing to his justice, accumulated much wealth. He built Ruṣāfa in Qinnasrīn and was buried there.

He was succeeded by al-Walīd b. Yazīd who was very dissolute and an adulterer. He even had sex with his own daughter and prayed while not in a state of ritual purity. Ultimately he was murdered and buried outside Bāb al-Farādīs.

He was succeeded by Yazīd b. al-Walīd. He was the first to attend the *bayram namāzı* (morning service on the first day of Ramadan) armed (with a sword) and this became an innovation.

He was succeeded by Ibrāhīm b. al-Walīd.

He was succeeded by Marwān al-Ḥimār ("the Jackass") who was given this title because of his bravery—they have a proverb, *aṣbara min ḥimārin fi'l-ḥarb* (More steadfast than a jackass in battle). During his reign, Abū Muslim (leader of the Abbasid movement) rose up in Khurāsān and (Abū'l-ʿAbbās) al-Saffāḥ (founder of the Abbasid caliphate) rose up in Kūfa. When Marwān advanced against al-Saffāḥ, the latter sent his uncle ʿAbd Allāh (b. ʿAlī) with a large army against him. The two sides met near Mosul. Marwān was defeated and fled. ʿAbd Allāh pursued him and, catching up with him in Egypt, killed him in the Church of [A]bū Ṣīr.

Thus the Umayyad dynasty in Syria came to an end. Their caliphate lasted 92 years, that is, 1,000 months. It ended in year 132 (749) of the *hijra*. The Abbasid Dynasty appeared in Iraq and their supporters captured Syria. After Marwān [al-Ḥimār, the Abbāsid caliph] Abū'l-ʿAbbās al-Saffāḥ gave Syria to his uncle ʿAbd Allāh. In this (the Abbasid) period, Ibn al-Shāṭir observed the stars in Damascus in 215 (830). He was the first to observe the stars in Islam. Later in Syria he observed Banāt (al-Naʿsh) (constellation of the Bear comprising Ursa Major and Ursa Minor).

In 254 (868) the Tulunids appeared in Egypt. They seized Syria. Their rule lasted 38 years. Then the Abbasids again took control of Syria. In 290 (903) the Carmathians raided and plundered Syria. In 292 (905) the (Tulunid) Hārūn b. Khumārawayh was killed and their dynasty came to an end.

The appearance of the Ikhshidids: *Ikhshīd* means "king of kings." It happened that when al-Rāḍī bi'llāh became the Abbasid caliph he appointed someone named Ikhshīd to be governor of Syria and Egypt. This was in year 323 of the *hijra* (935). When the Abbasid caliphate declined, Egypt and Syria remained in the hands of the Ikhshidids. Subsequently, in 330 (941) Syria was seized by the Fatimids. In 414 (1023) the Mirdāsid dynasty appeared (in Northern Syria) and captured Aleppo. Then in 467 (1074) the Seljuk (Sultan) Malik Shāh took possession of Aleppo. When the sun entered the house of Ares, he declared it to be the vernal equinox and he had (astronomers) observe the stars. [M 608] In 461 (1068) a great fire broke out in Damascus and destroyed the beautiful features of the Umayyad Mosque.

In 468 (1075) Malik Shāh gave the province of Syria to his brother Tutush. Then in 499 (1105) (his son) Tughtigin appeared and took the province of Syria. Later in 511 (1117) Il Ghāzī of the Artuqid dynasty, which was a branch of the Seljuks, seized Aleppo. Then in 523 (1129) ʿImād al-Dīn Zangī Ibn Aq Sunqur, who was from another branch of the Seljuks, took Aleppo.

In 549 (1154) his son Nūr al-Dīn Maḥmūd the Martyr became ruler of Syria. He conducted many gazas and took 50 fortresses from the Franks (Crusaders). He built a hospital (*dār al-shifā'*) in Damascus. He used what they called *ḥamāma hawādī* (guided pigeons) to get and send news quickly, from here (i.e., Damascus) as far as Hamadān. He has a Friday mosque in Mosul called the Nuwayrī mosque. He built a Friday mosque in Ḥamā next to the Orontes River and hadith schools in Ruhā (Urfa), Manbij, and Damascus. He assigned a fixed sum to the Bedouin so they would stop plundering pilgrims on the way to the Kaʿba.

Nūr al-Dīn had a dream in which the Prophet said to him three times, "Save me from these two fair-complexioned ones." He immediately went to Medina and apprehended the two men. They had been sent from the infidels to Medina in order to steal the (body of) the Prophet from his noble tomb. When they arrived in Medina they lodged in a house near the Rawḍa Sharīfa (the "Noble Garden" next to the Prophet's tomb). They dug a tunnel from their lodging to the Rawḍa Sharīfa and got very close to the noble tomb. Eventually their tunnel (was discovered and it) and the area around the tomb were filled with lead. Nūr al-Dīn died in 569 (1174).

In 570 (1174) Ṣalāḥ al-Dīn (Saladin) Yūsuf of the Ayyubid Kurds occupied Syria. His gazas are famous. He took many fortresses from the Franks (Crusaders) and even conquered Nubia in the Sudan. It is reported that when he conquered Jerusalem he took 60,000 prisoners. He made peace (on condition that) each man pay ten gold pieces and each woman pay five. He was going to destroy the Church of the Holy Sepulcher [but refrained because] they made peace. He rebuilt the ʿArab Cāmiʿi ("the Arab Mosque") in Galata in Istanbul and sent a *minbar* there from Egypt. He died in 589 (1193) and was buried in Damascus in his own madrasa, the Kāmiliyya. His sword was buried with him—he willed that he be buried with his sword so that he could rest on it on the Day of Judgment.

Afterwards his son al-Afḍal ʿAlī took possession of Damascus and it was later wrested from the Ayyubids by the Turkish dynasty of Egypt (i.e., the Mamlūks) who appeared in 648 (1250). The first of them was al-Malik al-Muʿizz Aybak. In 643 (1245) the Khwārazmians besieged Damascus for about five months. In 658 (1260) Hulagu pillaged Ḥamā, Ḥimṣ, and Aleppo and then fought a great battle with the sultan of Egypt, al-Malik al-Muẓaffar, near Dayr Jālūt (rather ʿAyn Jālūt) in the Ghawr (the Jordan valley). Hulagu's commander was defeated and fled. Afterwards the Ayyubid dynasty (still) had branch states in those countries—there were branches in Egypt, Damascus, Ḥimṣ, [M 609] Ḥamā, Aleppo, Yemen, and Jazīra. Then the Turkish dynasty appeared. In 657 (1259), after al-Malik al-Muẓaffar defeated Hulagu in battle, he took possession of Syria and made (al-Malik al-Mujāhid) Sanjar his deputy in Aleppo and al-Malik al-Saʿīd his deputy in Damascus.

In 740 (1339) fire descended from the sky along the coast of Syria and burned many places.

In 780 (1378) the dynasty of the Dhū'l-qadriyya (Ẕülḳadriyye) appeared in Marʿash and Elbistān. They were Turkmen descended from Ḳaraca b. Ẕū'l-ḳadr, who was the first to appear. The dynasty of the Ramażāniyye appeared at the same time in Adana. They were also Turkmen. They had come against the Byzantine Empire (Bilād-i Rūm) and seized (part of) it.

In 784 (1382) the (Mamlūk) dynasty of the Circassians appeared and took control of Egypt and Syria. Al-Ẓāhir (the text has Ṭāhir as above) Barqūq (r. 784–791/1382–1389) was the first to seize the sultanate. It is reported that they were originally from the seven Israelite tribes who fled (from Palestine). Five of them went to the region of the Caspian Sea and two settled in the province of the Circassians. Indeed even today the Circassians practice some Jewish customs and methods of worship; but they do not claim to be descended from the Israelites. In short, because of their intelligence, beauty and bravery, the people of Egypt have a great desire for Circassians and most of their slave boys and slave girls are Circassian. They gradually increased in number until, conspiring among themselves, they took possession of the sultanate of Egypt. They are (skilled at) using weapons and in horsemanship: they can stand (in the saddle), remove and replace the saddle, and pass under the horse's belly, all while the horse is running.

The last of the Circassians was Abū'l-Naṣr Qānṣawh al-Ghawrī. At first he was a just, circumspect, prudent, and intelligent man, but later greed got the better of him and be began to be oppressive and act unjustly. He began to construct buildings. In the place called Bayn al-Qaṣrayn in Cairo, he built a Friday mosque and a mausoleum. His intention was to be buried there, but fate did not allow this to happen. At the end of his life, heedlessness and pride overcame him and his army, and he became the object of curses because he oppressed the poor.

It is reported that one day al-Ghawrī wanted to kill a poor soul on a trivial pretext. The mufti (chief jurisconsult) of the time learned of this and said, "There is no need to kill him for this. He should not do so," and sent his decision (to the sultan). When al-Ghawrī heard this, he flew into a rage and ordered that the man be hanged at the mufti's door. Less than a month passed (after this event) before word of Sultan Selīm's advance reached Cairo and unsettled al-Ghawrī. He immediately set out for Aleppo. When al-Ghawrī reached Aleppo, Sultan Selīm sent him a letter saying, "We have come to such-and-such a place. Our goal is to make war on the Qizilbash and to make peace with you." Al-Ghawrī had the beard of the man (envoy) who had come cut off and said, "Tell your present master that the Qizilbash are outside the faith and so are you just like them. Be prepared, for your time has come. We will meet at Marj Dābiq." After much vaunting of this sort, al-Ghawrī left Aleppo with 30,000 men and camped at Marj Dābiq. After he remained there for three days, the Ottoman army arrived at noon. When the regimental cannons were fired they frightened the horses and al-Ghawrī fell from his horse. They picked him up [M 610] and put him bareheaded back on his horse, but he fell again. When Sultan Selīm's army attacked, al-Ghawrī and a large part of his army were crushed under the hooves of the victorious horses.

The surviving Circassian troops headed back for Egypt. They installed the Circassian named Ṭūmān Bay, who was the *dawādār* (keeper of the royal inkwell, secretary), as sultan. Sultan Selīm followed right behind (the defeated army), seizing territory as he went, and arrived at the place called Raydāniyya outside Cairo. Ṭūmān Bay set out with the Egyptian army to fight Sultan Selīm, but they did not have the strength to resist for an hour. The hapless Circassians turned, abandoned the field, and fled to Upper Egypt. The commander of Upper Egypt, Shaikh al-ʿArab, seized Ṭūmān Bay and brought him to Sultan Selīm. On 17 Rabīʿ al-Awwal 923/9 April 1517, an edict was issued for the hanging of Ṭūmān Bay, and by imperial order he was hanged at Bāb al-Zuwayla (in Cairo). And so the meaning of the noble verse *fa-quṭiʿa dābiru 'l-qawmi* ("Thus the roots of the people [who did wrong] were cut off," Koran, 6:45) became manifest.

At that time the countries of the Arabs and non-Arabs and the roads to Mecca and Medina came under the protection of the Ottomans.

Military forces of Syria

The garrison troops of Damascus are 2,000 fortress guards and 400 armorers. There are 990 *ḳılıç* (timars) of which 128 are *zeʿāmets* and 868 certified and uncertified timars. According to *ḳānūn* (sultanic law), their holders together with their armed retainers provide 2,600 soldiers.

Province of Ṭarāblus Shām (Tripoli): There are 614 *ḳılıç*, of which 63 are *zeʿāmet*s and 571 certified and uncertified timars. Their holders together with their armed retainers provide 1,400 soldiers.

Province of Aleppo: There are 903 *ḳılıç*, of which 104 are *zeʿāmets* and 799 timars. According to *ḳānūn*, their holders together with their armed retainers provide 2,500 soldiers.

Province of Marʿash: There are 2,169 *ḳılıç*, of which 98 are *zeʿāmet*s and the remainder are certified and uncertified timars. According to *ḳānūn*, their holders together with their armed retainers provide 5,500 soldiers.

In sum, according to *ḳānūn* Damascus, Tripoli, Aleppo, and Marʿash provide a total of 12,000 men. But today not one in a hundred is left.

Chapter: (Province of) İçel

Borders: To the west is Antalya, to the north are Ḳaraman and the province of Adana, to the east is the province of ʿAynṭāb from the *eyālet* of Aleppo, and to the south is the Mediterranean.

Because this province is opposite Cyprus, it has been attached administratively to Cyprus. It is divided into eight sanjaks. Lefkoşa (Nicosia) is the pasha sanjak. The others are Girne, Baf (Paphos), Magosa (Famagusta); and, on the opposite (mainland) coast, İçel, Alanya (text has ʿAlāʾiyya throughout), Tarsus, and Sīs. This province is under the administration of the Ḳapudan Pasha (admiral of the Ottoman fleet). It has its own minister of treasury, minister of timars, and keeper of tax rolls. The four mainland sanjaks are imperial domains (*ḫāṣ*). The maritime sanjaks—Lefkoşa, Baf, and Girne—pay the *sālyāne* (an annual lump sum) and each is responsible for providing one ship. The Pasha has one ship and one auxiliary ship. Like the *beys* of Teke and Menteşe, he sometimes goes to war on land and sometimes on sea with one or two ships. The *beys* of Degirmenlik, Selanik (Thessalonica), and Alexandria and Damietta in Egypt each go to war at sea with one ship. Altogether they provide about 20 ships. The number could be more or less.

There are 138 *zeʿāmets* and 1,434 timars enrolled in the above-mentioned sanjaks, amounting to 1,572 [M 611] *ḳılıç*. Their holders together with their armed retainers provide 4,500 soldiers. Adding the *ʿazab* troops, which number 1,893 men, the total is 6,300. Most of the *zeʿāmet* and timar holders pay *bedel* (a sum in lieu of providing troops).

Chapter: İçel Proper

In the past it was called Bilād-i Lashkarī.

Borders: To the east is Tarsus, to the north is Ḳaraman, to the West is Alanya, and to the south is the Mediterranean.

This emirate is like a natural fortification (reading sūr for ṣūr). It is located in the midst of difficult, rocky, high mountains. It can only be entered by a narrow mountain pass, which serves as the gateway (the Cilician Gates). It is defended by some fortresses which have guards stationed in them. It has eleven fortresses: Silifke, Maʿmūriye, Ermenak, Mut (text has Mud throughout), Badna, Man, Uğadi, İkidam, Māhiyān, Temirreh (?), and Külek. Its towns are Anamur, Bozṭoğan, Zine, Ṣarıḳavaḳ, Selendi, Sinānlı, Ḳarataş-Irğadi, Gülnār, Gülyüzi, and Günyüzi.

Districts of Ermenak

Anamur: This is the city of Maʿmūriye. Today the city is in ruins; it is only a district.

Gülnar: A district and a qadi district. It is located between Silifke and Anamur. To one side is the coast.

Ermenak: A town and a qadi district two stages south of Lārende. It has abundant vineyards, gardens, and flowing water that comes from a cave high in the mountains. The interior of the cave is spacious and it has a pool in the middle. People descend to it bearing torches. There is an outlet for the water (to the exterior) where it flows in different directions, one branch reaching the city. On a high cliff in the middle of Ermenak is a fortress, like the den of a wild beast, where a warden and garrison reside.

Selenti (Selendi): A district between Alanya and Anamur. To one side is the coast.

Sinānlı: A district between Mut and Ermenak.

Mut: A town near Silifke. It has a fortress. It is located between Silifke and Sinānlı.

Ṣarıḳavaḳ: A district located east of Mut in the vicinity of Lārende.

Silifke: It is the pasha sanjak. It is a prosperous town near the coast. The bey of İçel resides there. Mt. Arsuḳ is northeast of Silifke.

Chapter: Sanjak of Alanya

Borders: To the west is Antalya, to the south is the Mediterranean, to the east is the sanjak of İçel, and to the north is Ḥamīd. It is administratively attached to Cyprus. It is an emirate adjacent to Ḳaraman. The people are fruit dealers and Turks. They are characterized by their extreme sternness and ill-temper.

Its qadi districts are (only) Alanya. The districts of Alanya are Manavġat, İbradi, and Señir-Düşenbe.

Description of the city of Alanya: It was founded by the Seljuk Sultan 'Alā' al-Din Kay-Qubād (r. 616–634/1220–1237). It is on a hill next to the Mediterranean. It has a fortress as large as Baghdad. It is very strong and consists of three levels of walls. Its gates open to the sea. West of the fortress along the coast is a great mountain. The fortress is located at the eastern foot of this mountain. Inside the fortress is a cistern; there is no flowing water. Outside the fortress a stream emerges from the rocks and eventually flows into the sea. All the houses and quarters are inside the fortress; there are no buildings outside. In the middle fortress is a small area that is rocky, barren, and covered with oak scrubb. Antalya is two stages west of Alanya. Outside (the fortress) are vineyards and gardens. In the mountains to the north is the garden and palace of (the Seljuk Sultan) Qilich Arslān (the second r. 551–588/1156–1192). Remnants of this building are still to be seen.

Because most of the area around Alanya is mountainous and rocky, the land is not very productive. Mostly they produce cotton, silk, and sesame.

Antalya has several districts. [M 612] One of them is Ovabāzārı, which is about ten miles from Alanya. It is the residence of the *bey*. It has a Friday mosque and three public baths. Today it is in ruins. Seydīşehri is three stages north of Alanya, Silifke is five stages to the east, and Antalya is three stages to the west.

At each level of Alanya fortress there is a Friday mosque. In the outer fortress in front of the gate is a public bath built by Orḫān (r. 724–761/1324–1360).

Berdaniye: Departing Seydīşehri and going west of Külek on the way to Alanya, one comes to the town called Berdāniye. It has a market (*bāzār*). It is one stage north of Alanya.

Rivers

Ḳarġı: It rises from Aḳṭaġ, flows between Alanya and Düşenbe, and enters the sea.

Manavġat: It flows between Düşenbe and Manavġat and meets the sea.

Köprü: It flows between Manavġat and Ḳaraḥiṣār-i Teke and meets the sea.

Alara: A large river. It meets the sea near Alanya. There is not always a place to cross it. Sultan 'Alā' al-Dīn built a bridge over it. Those going from Seydīşehri to Alanya must cross this river.

Mountains

Ṣuṣam Beli: A great mountain. It is uncultivated and very cold. It is impassible in the depth of winter because the wind is so strong that it throws stones for a mile.

Chapter: Island of Cyprus

In the language of the Jews it is called "Çetem" or "Ketem" and in the language of the Greeks it is called "Çıprıs" or "Kipru." This island is in the Mediterranean and off the straits of Ayās. It is 70 miles from the coast of İçel. It is 100 miles from the coast of Tripoli, 550 miles from Alexandria, and 400 miles from the island of Crete. Its longest day is 14 hours. Its circuit is 700 miles. Its length is 220 miles and its width is 60 miles at the widest point.

Borders: To the east is the cape opposite Payas and İskenderūn extending to the north opposite Tarsus; it extends toward this cape as far as the corner that is opposite Selenti. The northern side is longer than the southern side, which faces the open sea toward Egypt.

This island has 11 administrative divisions. They are Baf, Odino, Limosa, Masuto, Memlaḥa, and Olim[p]os, all of which are along the southern coast. The others are in the mountains, in the north, and along the northern coast. They are Ḫırsavi, Bendalya, Girne, Karitaso, and Viçekomita. Salt is produced here (this should refer to Memlaḥa, as below).

The qadi districts of Cyprus are as follows: Ḫırsavi, Baf, Poskipoklan, Urumlimos, Lafkara, Lefḳoşa, Maġosa, Masarya, Karpas, Girne, Omorf, Pendabe, and Lefke. The capital of the island is Lefḳoşa (Nicosia). In the past the city was very prosperous, with 40,000 homes, and was the capital of the Lusignans. It has 11 hills. From its conquest up to the present, it has been the seat of the pasha and governor. It has a position for a molla (chief qadi) paying a salary of 500 *aḳçes* (per diem).

Maġosa (Famagusta): It has a very strong fortress. Its harbor is excellent. It has two gates, one that opens to the sea and one to the land. Inside the seaside gate is the statue of a crouching lion. But one leg is made of a different stone. Behind it is another smaller lion. It is depicted on the other gate and gazing at (the big lion). No one knows the reason it was made in this way. There is a church in Maġosa that is as finely constructed as Ayasofya (in Istanbul).

Çeterya: The meaning of this is the "City of Venus." The Greeks say that Venus resided here in ancient times. It was a city on the coast. Today, however, it is a village known as Ḳondelya, not far from Baf (Paphos). [M 613]

Çira: A city of merchants. Much trading takes place here.

Girne: A separate district, surrounded by mountains.

Kiti: There is a lake near this city. It is about three miles long. Its water is salty. In the summer it dries up leaving salt behind.

Olim[p]os: There is a church of St. Michael here. There is a mountain here with the same name.

Memlaḥa: A famous place. A small river flows nearby. On the other side of the river are two ponds that are close to each other, with something like a bridge passing between them. There is a well-adorned dervish convent with shrine here where a lady who was one of the Companions of the Prophet is buried. The convent has a shaikh who feasts travelers and shows them hospitality.

Mesarya: In a certain season countless quail flock to this. The people hunt them. In flat areas they put out birdlime and snares. They are small yellowish birds. They have a lot of flesh, though their internal organs are as tiny as nutmeg seeds. The people cut open their stomachs and remove their organs. Then they pickle them and preserve them as food.

In the vicinity of Baf, above a large village called İḥtimā, is a mountain covered with brackish soil. The village is at the foot of the mountain. On that mountain is a mine for the Ceylon red stone (garnet), also known as "Cyprus apple."

Description of Cyprus

The agricultural plain of this island resembles that of Egypt and its mountains resemble those of Syria. There is a mountain on this island on which there is a great monastery built of dressed stone. In the monastery is a wooden cross known as *ṣalbūt*. It is suspended in the air between pieces of magnetic stones (lodestones). It is gilded all around and covered with iron. It does not fall from its place because the magnets around it attract the iron. Because foolish people are captivated by it, they are misled.

There is a (certain kind of) stone on this island. The people crush it and extract from it a substance (asbestos?) like cotton which is suitable for spinning and weaving. If they put something woven from this substance into a fire, the flames eliminate the dirt but the textile comes out clean and is not burned.

The air of this island is hot in the summer. In the fall and winter it usually rains. The north wind gives life to the people. Earthquakes are frequent on this island, but the land is fertile. Wheat ripens in April. Fruit is very abundant. There are all kinds of citrus, especially lemons and citrons. There are also abundant apples, plums, apricots, olives, sugar cane, dates, honey, salt both extracted and mined, and capers. The island is also famous for cotton, silk, saffron, emeralds, rock crystal, and alum. It also has iron sulfate, potash, copper, sulphur, gold, silver, a stone similar to diamond. And there is rhubarb, and all kinds of medicinal plants grow in the mountains. The people of the island are warlike.

Animals: There is a type of animal on this island that is a cross between a donkey and an ox. This animal is small in stature but very strong. It can carry as much of a load as a camel. The goats on the island wander about on their own. Absolutely no water buffaloes can live on the island. They were introduced several times, but they did not survive.

Rivers: Two notable rivers rise from Mt. Olinbo (Olympus). They call one the Likos Ḫortina and the other the Labetos. The Likos flows to the south and the Labetos flows to the north. Apart from them, the rivers are like floods that dry up and disappear in the summer.

Mountains: This island has many mountains. The highest, called Olimpo (Olympus), is adorned with all kinds of trees and has a circumference of 44 miles.

Status of rulers

Originally Cyprus and İçel were in the hands of the Greeks (*Yūnāniler*). [M 614] Later the Byzantines (*Rūmīler*) became supreme and took possession. Then the king of Istanbul (the Byzantine emperor) gave the island as a gift to a Spanish nobleman because of a prior claim. At that time the Venetians said to the nobleman, "It will be very difficult to take possession of this island. Come, sell it to us." So he sold it to them for a high price. After Islam appeared and the Muslims conquered the clime of Syria and certain parts of Lesser Armenia, (the Umayyad caliph) Muʿāwiya conquered the island in year 26 (646) of the *hijra* (646). Later, when the Muslims were weak, the Venetians seized the opportunity to retake the island. Subsequently, in 978 (1570), the Ottoman sultan Selīm II conquered it.

As previously mentioned, the Muslims had invaded İçel and had either plundered it or taken certain parts of it. Then the Qutulmushids, who derived from the Seljuks, invaded the province of Rūm and made Konya their capital (referring to Sulaymān b. Qutulmush, d. 479/1086, who founded the Seljuk dynasty of Anatolia.) During the reign of ʿAlāʾ al-Dīn Kay-Qubād, who was one of them, a Sufi named Nūra appeared who was master of the mystical way and ʿAlāʾ al-Dīn believed in him. It is reported that a Turkish clan, fleeing from the Mongols, settled around Ermenak. They sowed and reaped and paid the tithe to the unbelievers. Their chief was called Nūra Ṣūfī. This Nūra had many sons, of whom the eldest was known as Ḳaraman. On a day when the unbelievers were having a fair, the people of fortress of Silifke went out to the fair. At that time the Ḳaramanids had assembled. They deliberated and said, "How long will we obey these unbelievers and pay them tribute? Today is our chance to crush them." So, they put on the clothing of the unbelievers and approached the fortress. Because it was a rainy day, they wrapped and covered themselves and those in the fortress did not recognize them. They entered the fortress and conquered it. Afterwards he (Nūra Ṣūfī) sent word to Sultan ʿAlāʾ al-Dīn requesting that he give the *beylik* of the fortress to his son (Ḳaraman) who would make *gaza* against the unbelievers in that region. When Sultan ʿAlāʾ al-Dīn received this news he was very pleased and gave the *beylik* of that region (to Ḳaraman). Ḳaraman reached Silifke and began to launch religious wars in that region. He conquered certain places. When Sultan ʿAlāʾ al-Dīn learned of this, his affection for Ḳaraman greatly increased. He added Lārende (Ḳaraman) to his holdings and gave him the rank of commander-in-chief. The Ottomans appeared at that time. Afterwards, when the Seljuk dynasty came to an end, the Ḳaramanids took possession of those regions and held them until Sultan Meḥmed the Conqueror came to power (r. 1444–1446, 1451–1481). He became offended by the last of the Ḳaramanids, Pīr Aḥmed, over certain matters. So he wrested those territories from them. This was in 871 (1466).

Distances: Silifke is one stage from Tarsus, two stages from Ermenak, and four stages from Lārende. Konya is one stage from Lārende. Eregli and Anamur are both three stages from Silifke. It is two stages from Anamur to Alanya and from there (an additional) two stages to Antalya. Seydīşehri is two stages from Alanya and Ermenak is three. The island of Cyprus is 60 miles from the coast of Silifke.

Chapter: Province of Ḳaraman

This region was called Ḳaraman because the Ḳaramanids invaded it before the Ottomans. Originally this country extended as far as the sea. After the Ottomans took possession of it, they divided it into two sections, one along the coast, which they called İçel, [M 615] and the other the rest of the country (which continued to be called Ḳaraman). The section other than İçel (i.e., Ḳaraman) is larger and more important. The capital is Konya, which is the pasha sanjak. This section (i.e., Ḳaraman) consists of seven sanjaks: Aḳsarāy, Aḳşehir, Beyşehri, Konya, Ḳırşehri, Kayseri, and Nigde. Ḳaraman has its own minister of treasury, keeper of tax rolls, and minister of timars.

The imperial domain of the *mīrlivā* (= sanjak bey) of Aḳsarāy is 305,000 *aḳçes*, that of Aḳşehir is 190,500, that of Beyşehri is 290,000, that of Ḳırşehri is 267,540, that of Kayseri is 250,000, and that of Nigde is 350,000. This province has 620 *ḳılıç*, of which 116 are *zeʿāmets* and the rest are certified and uncertified timars. According to *ḳānūn*, the emirs (i.e., sanjak beys) and the *zeʿāmet* and timar holders together with their armed retainers provide a total of 4,600 troops.

Fortresses: There are a total of eight fortresses: Konya proper (i.e., the walled city), Lārende, Hindus, Nigde proper, Dedeli, Lüle, Kayseri proper, and Ḳırşehri proper.

Borders: To the east are Adana and Marʿash, to the north are Sivas and part of Anaṭolı, to the west is Anaṭolı, and to the south is İçel.

Chapter: Konya

This city is the capital of the province. It has a position for a molla (chief qadi) paying 500 *aḳçes* per diem. Its qadi districts are Eregli, Eski İl-Aḳçeşehir, Alaṭaġ, İnṣuyı, Bayberd, Birgiri, Birloġand (?—cf. Bulavġand below), Belvīrān, Ḫātūnsarāy, Ṭurġud, Ġaferbād, Ḳırış-Birindi, Lārende, Maḥmūdlar, Lāẕḳiye, and Ḳaraman.

Description of Konya: There are two forked mountains at the western edge of the city. The city is located near the eastern foot of these mountains. It is on a flat area and has streams, vineyards, and gardens. It has a city wall in good condition. South of the city at the foot of the mountains are gardens and a place of excursion called Meram. Rivers flow from the mountain to the city and to Meram. After providing water to the fields, vegetable gardens, and the city, they flow down into the plain near the city and form a lake that surrounds the mountains (sic?).

The Seljuk Sultan Qilich Arslān built the fortress of stone. It was the royal residence and capital. He built a great *eyvān* (chamber or hall open to the outside at one end) in his palace. Later, when the city wall began to fall into ruin, the Seljuk sultan ʿAlāʾ al-Dīn and his emirs rebuilt it. They built the walls in stone starting from the bottom of the moat. The moat was 20 cubits deep and it was 30 cubits to the top of the walls.

The city wall has 12 gates, each one having a tower in the shape of a large palace. Here they built a lofty building. Since the city's water comes from the mountain, there is a large domed structure at the (main) city gate and outside are about 300 pipes of flowing water that is distributed among (the quarters of) the city.

The soil produces cotton, various grains, and much fruit. In this city there is a delicious apricot known as *ḳamerüddīn*. The climate is mild. Most of the orchards are on the mountain side of the city.

In Konya there is a kind of azure flower called tanner's flower (*debbāġ çiçegi*). Its seeds are sown and reaped every year like other agricultural products. Tanners use it to dye leather and Morocco leather sky blue. This leather is exported throughout Turkey (Bilād-i Rūm) and Europe.

Shrine: In Konya is the mausoleum of Mawlānā Jalāl al-Dīn Muḥammed b. Sulṭānu'l-ʿulemā (i.e., Rumi). He died in 672 (1273). His age was 68. Sulṭān Veled, Bahā'eddīn Sulṭānu'l-ʿulemā, Shaikh Kerīmüddīn, Çelebi Ḥüsām, Şems-i Tebrīzī, Shaikh Sadraddīn, Qadı Sirācaddīn, Seyyid [M 616] Burhānaddīn, and Muhaḳḳiḳ-i Tirmiẕī also have shrines and tombs in Konya. The mausoleum of (Seljuk Sultan) ʿAlā' al-Dīn and (the grave of) Eflāṭūn-i Ilāhī (Plato) are in the fortress of Konya.

Lārende (Ḳaraman): It is one stage southeast of Konya. It is on flat land and has a town and a fortress. It has flowing water, vineyards, gardens, a Friday mosque, and a public bath.

Lāẕḳiye-i Ḳaraman: The Yörüks call it Ladik. It is a small town and qadi district on the main road. It has a Friday mosque, a public bath, and numerous inns. It is one stage west of Konya. Between them is a town and qadi district called Zengiçay.

İsmil: It is twelve hours east of Konya. It is (a village) like a town and is located on the main road. It has some inns. The area from Konya as far as the border of Eregli is a plain. İsmil is located on this plain. South of İsmil is an open plain. During times of flooding, water covers the plain like a sea. At the edge of this plain to the north of İsmil is a large mountain range that runs from east to west. The mountains opposite İsmil are reportedly known as the Mountains of Fodul Baba. The people of İsmil say that the wild (reading *vaḥşī* for VSY) sheep there are not to be hunted. They are called "the herd of Fodul Baba" and are under his protection. By a dispensation (*destūr*—i.e., permission of the saint) the people hunt two or three of those sheep as sacrificial animals, but never more than three. It is well known among them that anyone who takes too many sheep suffers severe consequences—indeed, some have died. Once when the pasha of Konya molested those sheep he suffered severe punishment. This herd of sheep numbers about 2,000 or more.

Today there are neither trees nor water on this mountain. However there is a small pool whose water neither increases nor decreases. The animals that live in that area drink this water. Sometimes the lake of Konya overflows and comes close to İsmil and the whole plain becomes like a sea. Because of this, the people of the province say that at one time the Konya Plain was a sea, but Plato by some means made the water disappear.

Alaṭaġ: It is located southwest of Konya and comprises a district and a qadi district. It has red soil and excellent grapes. The hills and valleys have many vineyards.

Ḳarapıñar: A small town with a Friday mosque, a public bath, a small market, and splendid inns. The mosque was built by (the Ottoman) Sultan Süleymān. Its dome is covered with lead. Ḳarapıñar is on the main road, one stage northeast of İsmil and northwest of Eregli. Lārende is south of Ḳarapıñar.

Eregli: Its old name was Heraḳle-i Ḳaraman (i.e., Heraclea). It is a large town with 22 quarters, each with its own neighborhood mosque. There are four Friday mosques. One was built by Ḳaramanoğlı İbrāhīm Bey. Another is the mosque of Şehābeddīn-i Maḳtūl, whose mausoleum is attached to it. (The town also) has two public baths.

In order to prevent the city from being captured, they poured dirt around it like a wall. Today the wall is called Döke Bārū ("Poured Rampart") and the city is inside of it. This rampart is (like) a tall and broad mountain, as though the town were surrounded by some hills. It has a city wall that was built later. One side of it connects to the middle of the rampart.

This town has meadows and streams and is located on a level plain. It is at the foot of Mt. Erdost. In the past, one tenth of its produce was set aside as a pious endowment for the poor of Mecca and Medina. Three hours from town on Mt. Erdost (a spring) emerges from a rock. Although it has little water at the source, [M 617] as the stream approaches the town, there is no place to cross it. 360 channels have been dug along the banks of this river to irrigate fields and vineyards, and many mills have been erected on it.

At the source of the river is a great rock on which the image of a man has been carved that is still visible. It is supposed to represent the *bey* of Abrinos, who was an unbeliever and the emir and great man of the village known as Ābrīz ("Pouring Water"). He holds a sheaf of grain in one hand and two bunches of grapes in the other, implying that the statue watches over the sown fields and the vineyards.

The above-mentioned river does not go very far into those plains, but descends into the valley and then spreads out into the reed beds. Eventually the overflow sinks into a stone hole known as Düdne (?) and disappears. This occurs at the foot of the Bozoğlan and Bulgar mountains, which are parallel to the town of Ḳarapıñar and west of the above-mentioned town (Eregli). They say that this river emerged by miracle of Muḥammad, and for this reason the entire town is today a pious endowment.

In this town is an old Friday mosque of (the Seljuk Sultan) Qilich Arslān and an inn of (the Ottoman grand vizier) Rüstem Pasha. Etmekçioğlı Aḥmed Pasha began the construction of a new inn there, but it remained unfinished. When (Ottoman deputy grand vizier) Bayram Pasha became commander-in-chief in 1047 (1637) and came here, he completed it.

The river water that flows around Eregli becomes petrified. Indeed the inn there was built from stones petrified in this water.

Eregli has abundant grain crops and all kinds of fruit. They say, in fact, that it has 90 varieties of pear alone.

Belvīrān: A town located between Konya and Alaṭaġ.

Pirloġanda: This place consists of a number of villages and a qadi district between Silifke and Alaṭaġ. It is a mountainous and fruit-bearing area, famous for its apples.

Chapter: Sanjak of Nigde

The qadi districts of this sanjak are the following: Andoğu, Ürgüp, Bor, Çamardı, Dedelü, Şücāʿaddin, Ḳay, Ḳaraḥiṣār, Develü, Menend.

Description of Nigde: A castle of three levels surrounds the entire city, the circumvallation consisting of strong stone walls. In one level of the city, inside the fortress, Sultan ʿAlāʾ al-Dīn and Sunḳur Bey have splendid Friday mosques. ʿAlāʾ al-Dīn also has a lofty madrasa called Beyżā built of stone two stories high. There is nothing like it anywhere. The Friday mosque of Ḥasan Çelebi is elegant. There are many other Friday mosques and madrasas, and there are 11 Friday mosques and splendid public baths in the fortress and the castle. It (the mosque of Ḥasan Çelebi?) has three minarets. The fortress, dating to infidel times, is as big as Eski Sarāy (the Old Palace in Istanbul) and is situated on a cliff. From the city it appears as high as several minarets.

Roads were built going from the city to the west. There are lofty houses overlooking the fortress walls. In the middle of the fortress is another fortress the size of three threshing-floors. Inside are the residences of the warden and the guards and a mosque. This second fortress is higher than the first. Somewhat separate from the city is a large quarter that was built later.

Nigde has vineyards, gardens, and much water. Before Nigde is a broad plain. In various places streams flow through it and it is shaded by plane trees. The people enjoy themselves there. Nigde has excellent crops. It has all kinds of fruits, which are abundant. This city is four stages northeast of Konya and two stages north of Eregli.

Aḳsarāy: It is between Nigde and Konya. It is located about three hours north of the stage of Uluḳışla.

Bor: It is four hours from Nigde on the Nigde-Konya road. [M 618] It is a town in a valley with a small earthen circumvallation. Its soil is dry and brackish. There is a large gunpowder mill here belonging to the state, containing 100 mortars. Its river is the same one that flows before Nigde. When it rains, they collect saltpeter for the gunpowder from an old demolished fortress of the unbelievers called Kiliseḥiṣār. This Kiliseḥiṣār is a ruined fortress near Bor, with marble pillars and arches made of large stones. When Sultan ʿAlāʾ al-Dīn was building the fortress of Konya, he transported stones from here (for this purpose). Today it is a flourishing village. They obtain charcoal from vine shoots that they bring from Nigde (and pile up?) like mountains.

Üçḳapulu: A high summer pasture three hours from Nigde and behind the mountains which are to the left of Çifteḫān. It is covered with hyacinths, tulips, and (other) flowers. Mostly Turkmen come here and spend the summer. They make a kind of cheese in a skin. They store the skins in some of the caves at the camp ground. It is extremely good cheese. This summer pasture is a plain situated where the spurs of three mountains come together. For this reason it is called Üçḳapulu ("With Three Gates").

Qadi district of Şücāʿaddin: It consists of a number of villages called Uluḳışla. On the side of a hill overlooking a flat plain in the place called Uluḳışla on the main road between Konya and Syria, (the Ottoman grand vizier) Öküz Meḥmed Pasha built a richly adorned inn, Friday mosque, and market. It is one stage between Ḳarapıñar and Çifteḫān.

Sanjak of Beyşehri

The qadi districts of this sanjak are the following: Bozḳırı, Seydīşehri, Ḳarael, Güci-i Kebīr (Büyükgücü), Güci-i Saġīr (Küçükgücü), and Ḳuşaḳlı.

Description of Beyşehri: A qadi district and town east of a lake. It has a stone fortress on a flat place. Its gate opens to both sides. It was built by Sultan ʿAlāʾ al-Dīn. It has two Friday mosques and a public bath. One of the baths is in the fortress. The marketplace is separate from and some distance outside of the town. On one side are vineyards and gardens. The side bordering on the plain is a pasture known as Yaġan. This qadi district (of Beyşehri) has one district, with a number of villages on the southern side.

Ḳuşaḳlı: It is one stage northwest of Beyşehri. It is a qadi district adjacent to the lake and has five to ten villages. A market is set up once a week. Seven or eight of the villages that are on the lake shore are called Yeñişehir ("New City"). Mt. Anamas is located between this place and Eker. ʿAlāʾ al-Dīn has buildings in Yeñişehir. Three of the villages are called Şehirgölü ("City Lake"). Large pine trees are logged here. A large stream descends from Mt. Anamas and flows into the lake.

Ḳıralı: A qadi district north of the lake. (The people) live in a small town called Beramköy. It has a coffee house, a public bath, and a Friday mosque. A weekly bazaar is set up. All around it are vineyards and gardens. Its water is obtained from wells. This qadi district has 20 villages scattered over the plain a distance of one stage. East of the qadi district are the mountains of Konya and the qadi district of Ṭoġanḥiṣārı. North are the mountains of Aḳşehir and west are the mountains of Ḳaraaġaç. Between these mountains are desolate areas known as Ramażān Beli. There is flowing water throughout the qadi district. Its fruit and sown crops are good. They catch fish in the lake. There are three hot springs here. Two are in the village of Çavuş and one is near the village of Köşk.

Seydīşehri: A town and a qadi district on a plain south of Beyşehri, between it and Bozḳır. Five miles south of the town is a lake, smaller than Lake Beyşehri, [M 619] producing good fish. It is one stage from Bozḳır and one and a half stages from Konya. Behind this town is an oak-covered mountain. From the outlet of Lake Beyşehri water flows below Seydīşehri and into this small lake. A stream descends from the mountain and the people of the town make use of it (for irrigation). There are vineyards and gardens next to the mountain. The gardens of Seydīşehri are more extensive than those of Beyşehri. Water from the outlet of this lake reaches Ṣuġla and irrigates agricultural lands. Then it disappears into the Konya valley (text: *arasında*, error for *ovasında*).

Bozḳır: A qadi district of 32 villages located amidst Alaṭaġ, Alanya, Silifke, and Bulavġand (?—cf. Birloġand above). This qadi district is one and a half stages from Konya and is one stage northwest of Seydīşehri. It has a valley (or fluvial plain, *ova*), a mountain, and a river known as the ʿAyġara that rises from the Ṣorḳūn summer pasture and flows to Ṣuġla. In the village called Moruson and the villages of Farṭ and Seristan there are a number of bridges. This qadi district was the district of Seydīşehri and a weekly bazaar was set up there. Today there are sown fields. It has abundant pears and plums. It has no vineyards.

Lakes

Lake Beyşehri: A fresh-water lake west of Beyşehri. Water flows from it to Seydīşehri and enters another lake. Water flows from its outlet to Seydīşehri. It flows past the east side of town and continues south. The people of the town drink from it. Lake Seydīşehri extends one stage south of the town. Water flows from its outlet to the east. It irrigates the agricultural lands of Ṣuġla. The water disappears in the Konya plain.

Lake Ḳaraeli: It is north of Lake Beyşehri and the same size. One side of it comes close to Ḳaraaġaç. On its shore is an ancient building which prevents the outflow of a great amount of water. Its construction is attributed to the (ancient) sages.

Sanjak of Aḳşehir

The qadi districts of this sanjak are the following: İsḥāklu, Ilġun, and Ṭoġanḥiṣār.

Description of Aḳşehir: A town with flowing water and gardens located near the foot a chain of mountains running east to west and on the southern edge of a broad plain. It is on the road to Konya and three stages distant. Baştekye of Aḳşehir is a delightful place of excursion. On the way (there) are a number of shrines. First one comes to the mausoleum of Buḫārī Dede; then the mausolea of Niʿmetullāh Velī, Aḫi Evren and Kürd Emīr; and finally to Baştekye, a matchless place of excursion. There are a number of dams here. Above them are pine forests with grassy banks and waterspouts on the edge of each bank, each emitting water the width of an arm. The mausoleum of Ḫoca Naṣr al-Dīn is also a famous pilgrimage site.

Ilġun: A town two stages east of Aḳşehir. It has a charming Friday mosque, an inn built by Rüstem Paşa, and a double public bath, which was built in 660 (1262) by (Seljuk) Sultan Ghiyāth al-Dīn Kay-Khusraw b. Qilich Arslān. This town also has an old public bath located about a mile to the west. It is a public bath with a hot springs. The Seljuk ʿAlāʾ al-Dīn had a masonry dome constructed over it. Inside is a pool with two pipes spouting water from the mouths of lions. Bathing in this water is beneficial for curing paralysis and leprosy. East of this town, near the forecourt (?—*fināʾ*) of the city, is a lake, eight miles in circumference, with fresh water and all sorts of fish. The Konya road passes along its shore.

İsḥāḳlı: A town one stage east of Aḳşehir and on the main road. It is located on the road one stage to the west (sic?). There is an inn for travelers. It is a qadi district.

Lake Aḳşehir: This lake is one stage from Ilġun and next to the town (of Aḳşehir). Sometimes it dries up. [M 620] Fish are caught in the lake. When there is a lot of water in it, a stream flows out of it to the west and into Lake Bolvadi. Bolvadi is a small lake nearby and to the west. It has a lot of water.

Sanjak of Aḳsarāy

The qadi districts of this sanjak are the following: Eyübeli and Ḳoçḥiṣār.

Description of Aḳsarāy: A town with flowing water and gardens at the foot of Mt. Ḥasan. A large river flows through the middle of town. It has a fortress. Konya is three stages to the west and Kayseri is the same distance to the east and somewhat to the south. The Seljuk ʿIzz al-Dīn Qilich Arslān b. Masʿūd built its fortress in 599 (1202). It has abundant fruit and grain crops.

Sanjak of Kayseri (Qayṣariyya)

The qadi districts of this sanjak are the following: Develi Ḳaraḥiṣār, Develi, Yaḥyālı, Süleymānlı-i Kebir, Ovanāḥiye, Küstere, and İslāmlı.

Description of Kayseri: In the past it was the capital of the Byzantine emperor. It is a city and fortress at the northern foot of Mt. Erciş (Erciyes). It has flowing water, gardens, vineyards, a Friday mosque, public bath, and markets. Its most famous product is lemon-colored Moroccan leather, extremely fine. The cultivated area is outside the walls. It has a small fortress, one side of which is made of black stone. Its water and air are very good. Most of its villages are at the foot of that mountain.

Snow is never absent from this mountain (Erciyes). From a distance of several stages (the upper) half appears like a white conical hat.

Develi Ḳaraḥiṣār is one stage west of Kayseri. Ḥācı Bektaş is two stages to the north. Sivas is four stages to the east. Develi is to the south. Yaḥyālı is southeast of Develi. Küstere is northeast of Yaḥyālı and İslāmlı is east of Küstere. Ovanāḥiye is southeast of Kayseri.

Among the ancient buildings (of Kayseri) are shrines. There is a Friday mosque belonging to Abū Muḥammad Baṭṭāl. And there is a public bath dating to the founding of Kayseri by the philosopher Apollonius (Belīnās); it is heated with little wood.

There are many snakes on Mt. Erciş, but thanks to a talisman they never leave the mountain. ʿAlāʾ al-Dīn Kay-Qubād rebuilt the city walls. In Kayseri there is a shrine belonging to Muḥammad b. al-Ḥanafiyya (a son of ʿAlī b. Abī Ṭālib who died in Medina in 81 (700–701)). Remnants of the old city wall are still standing. The town is a mile across. Its most famous river is the Ḳuramaz. It rises from Mt. Ḳuramaz and flows to Kayseri.

Mt. Erciş (Erciyes), also called Mt. Ercāsib (Arjāsb). Develi is at the foot of this mountain a little distance from Kayseri. On this mountain are wolves, bears, and lynx. There are all kinds of fruit in this district. There are many Armenians in this district.

Sanjak of Ḳırşehri

The qadi districts of this sanjak are the following: Diñek Keskeni, Süleymānlı-ı Kebir, Ḥācı Bektaş-Nūrda, Ḳonurnuz, and Ciyān.

Description of Ḳırşehri: A town on a plain with excellent air. Above it is a fortress. It has markets and inns. There is a saltpan (great Salt Lake) between Aḳsarāy and Ḳoçḥiṣār, at (?) Hayman[a], and the valley of Ḳarapıñar. It is a huge saltpan, the circuit of which is three or four stages. This saltpan is so large that the Haymana road, which enters it from one side, takes one or two hours to cross it. Signs are set up (along the route). A small river flows into this salt pan. The bed of this lake is pure salt and is farmed out as tax farm (*muḳāṭaʿa*).

Kings and Rulers of Ḳaraman

The first Muslim dynasty that appeared in Ḳaraman was that of the Qutulmushids who were a branch of the Seljuks. [M 621] They arose in the following manner: After the death of Baṭṭāl Ghazi, Dānişmend (Ghazi) the conquerer of Byzantine Anatolia (Diyār-ı Rūm) died a martyr during the conquest of Niksār and his son Meḥmed Ghazi succeeded him. When Meḥmed Ghazi('s lands) were again invaded by the unbelievers, he requested help from the Abbasid caliph al-Muqtadī. The Caliph then asked the Seljuk (ruler) Malik Shāh to send help (to Meḥmed Ghazi). He (Malik Shāh) sent Sulaymān Qutulmush with several thousand troops. When they arrived in Baghdad, the Caliph gave tidings of victory and arranged for the marriage of Sulaymān Shah's sister to Meḥmed Ghazi, thus forming a bond between them. Then they set out in the direction of the unbelievers. Meḥmed Ghazi retook the lands that had been conquered by his father. Then they captured Kayseri, where Sulaymān Shah established himself. Melik Meḥmed Ghazi, who was commander-in-chief, entrusted Kayseri to Sulaymān Shah on a friendly basis, and good relations continued between them for many years.

Later Sulaymān Shah conquered Konya and made it his capital. This occurred in 477 (1084). Then he captured Antakya from the Christians. Later he fought a battle against Muslim b. Quraysh (text: Qūsh), the ruler of Aleppo. Sulaymān was victorious and Muslim was killed in the battle. Sulaymān then took possession of Aleppo. Afterwards, in 479 (1086), he fought a battle against the ruler of Damascus Tutush (son of Alp Arslan), who was the son of his uncle (Alp Arslan was not Sulaymān's uncle) and brother of Malik Shāh. Sulaymān was killed in this battle and his son Qilich Arslān succeeded him. He ruled for nine years. When he died his son Masʿūd succeeded him.

This Masʿūd was a courageous ruler who frequently undertook the jihad. He died after ruling fifty years and his son Qilich Arslān succeeded him in 588 (1192). His reign was replete with conquests. He had eleven sons and each one took possession of a country. ʿIzz al-Dīn (Qilich Arslān II) made his son Ghiyāth al-Dīn Kay-Khusraw heir apparent. Afterwards his brother Sulaymān Shah became sultan, then his son Qilich Arslān III, then Kay Kāwus b. Kay-Khusraw, and then ʿAlāʾ al-Dīn Kay-Qubād b. Kay-Khusraw.

This ʿAlāʾ al-Dīn Kay-Qubād was courageous and generous. He invaded the region of (Lesser) Armenia, Alanya, and other places as far as İskenderūn (Alexandretta). During his reign, the Muslims went (across the Black Sea) from Sinop to (the lands of) the Kipchaks (Crimea), conquered the city of Sughdāq, and took possession of the areas around it. In 618 (1221) he built the walls of Konya and of Sivas. He died in 634 (1236).

During ʿAlāʾ al-Dīn's reign, the forefather of the Ottomans, Erṭuġrul, entered Rūm (Anatolia). He sent one of his sons named Ṣarı Balı to ʿAlāʾ al-Dīn seeking permission to enter his territory and asking that (a certain) land be assigned to him. ʿAlāʾ al-Dīn gave him the region stretching from Ṭomaniç to Ermenak. They went there and settled with 400 tents at Ḳaracaṭaġ, as will be described when we discuss Anaṭolı.

ʿAlāʾ al-Dīn's buildings and charitable works were many. He built 19 cities and many Friday mosques, madrasas, dervish lodges and caravansaries. During his reign the ulema and the pious increased throughout Rūm. Even Mawlānā [M 622] Jalāl al-Dīn Rūmī, may his grave be hallowed, lived during his reign.

He was succeeded by his son Ghiyāth al-Dīn. During his reign the Tatars raided and wreaked havoc in Rūm. He died in 644 (1246) and the Seljuk state came to an end with him. However, he had two sons, Rukn al-Dīn Qilich Arslān and ʿIzz al-Dīn Kay-Kāwus, who quarreled (over the succession). When Rukn al-Dīn emerged triumphant, ʿIzz al-Dīn fled and took refuge in Constantinople. Later, Rukn al-Dīn was put to death by Muʿīn al-Dīn (Sulaymān Parwāna) on behalf of the Tatars and was succeeded by his son Ghiyāth al-Dīn (Kay-Khusraw III). He was killed by the Tatar ruler Arghun (rather Abaqa), who appointed Ghiyāth al-Dīn (Masʿūd) b. Kay-Kāwus (as sultan). During this time there was confusion everywhere and this ruler, who was very poor, poisoned himself.

Later, the second (rather third) ʿAlāʾ al-Dīn Kay-Qubād b. Farāmurz b. Kay-Kāwus was appointed by Ghāzān Khan and ruled under the auspices of the Tatars. At that time Erṭuġrul was very old. He had very close relations with Sultan ʿAlāʾ al-Dīn Kay-Qubād. Sometimes his sons visited ʿAlāʾ al-Dīn bringing gifts. Erṭuġrul had become a frail old man, but his son ʿOs̱mān Ghazi became famous and word of his heroic deeds spread abroad. It even reached Sultan ʿAlāʾ al-Dīn who then gave ʿOs̱mān Ghāzī a drum, banner, sword, and horse and also sent him a splended robe of honor. ʿOs̱mān succeeded his father. Later, when ʿAlāʾ al-Dīn died, the Ḳaramanids occupied his lands, then the Ottomans occupied them in year 700 (1300).

Province of Sivas, also called Province of Rūm

Borders: In the east, part of Erzurum and Diyarbekir; in the south, part of Marʿash and Ḳaraman; in the west, Ḳaraman and Anaṭolı; and in the north, the Black Sea.

This province has seven sanjaks. Sivas is the pasha sanjak. The others are Amasya, Bozoḳ, Canik, Çorum, Divrigi, and Arapgir.

There are seventeen fortresses in this province, some of which are in ruins and some of which are in good repair. There are 3,130 *ḳılıç* (timars) in this province, 109 of which are *zeʿāmets*, the rest certified and uncertified timars. According to *ḳānūn* their holders together with their armed retainers provide 9,000 troops.

Chapter: Sivas

Its qadi districts are as follows: Artuḳābād, Iraḳ, Elbeyli, İnallı, Behrāmşāh, Ballı, Turḫal, Toḳat, Çebni-Çongar, Ḥüseynābād, Ḫān-ı Cedīd, Derbegüşābād, Dünkese, Zile, Ṣonisa, Sivaseli, Şarḳıābād, Ṭaşābād, Ḳazābād, Ḳaraḥiṣār, Ḳaraḳuş, Ḳarayaḳa, Mecidözi, Niksār, Yeñiel, Şarḳpāre, fortress of Ḥācı murād, Ṭopraḳḳalʿe, Düḳanābād, and Tozanlu.

Description of Sivas: It has a small fortress. It has few trees and its winter is severe. The Seljuk ʿAlāʾ al-Dīn Kay-Qubād built the city wall of stone. When Tīmūr destroyed Sivas, the word *ḫarāb* ("destruction") provided the chronogram (= 803/1400–1401). The city is on a hill. The river known as Ḳızılırmaḳ ("Red River") flows in a valley before the city. The water is brackish and is undrinkable. The city has Friday mosques, public baths, and markets. Its crops are grains (and) cotton. It has little fruit. Toḳat is one stage to the north.

Artuḳābād: It is located on a plain between Sivas and Toḳat. It was named after Artuḳ Bey because he had his seat of government here. There is a large village called Pavlos on this plain. Its inhabitants are Muslims and Christians. (Because it is) on the (main) road, some travelers stop there in the winter. There is a great mountain pass on the road that comes from Sivas to this plain. [M 623]

Toḳat: It is located in a valley in the spur of a mountain of red earth. The fortress is on a high cliff at the edge of the mountain, and around the fortress is a dense concentration of houses and quarters. It has abundant vegetable gardens, trees, and fruit. The climate is mild. The town proper is in valleys and depressions amidst the mountains. It has unmatched Friday mosques, public baths, inns, and madrasas. The city has two sections that are connected to each other. Toḳat is an entrepôt for merchants. It faces a beautiful plain.

Turḥal, also called Keşan Ḳalʿesi: A city on the main road five hours west of Toḳat and three hours north of Zile. A level plain stretches from Toḳat to Turḥal. The Ḳoçḥiṣār River flows through the middle of the plain.

Zile: It is located southwest of Toḳat. The town of Zile is very beautiful. There is an open plain to the south.

Niksār: Its fortress is at the foot of the mountains. There are a few gardens on the level area, but there are many gardens up and down the slope opposite (the town). To the west is the Niksār Valley. The region to the east is somewhat mountainous. It is one stage west of Toḳat. It has abundant fruit. Niksār is located on the road to Erzurum.

Ṣonisa: It is located one stage west of Niksār. It is on the main road and at the foot of the Niksār Mountains. Behind the Niksār Mountains is the province of Canik.

Mecidözi, also called Mecidābād: It is near Çorum and to the west. It is a fine town and qadi district with vineyards and gardens. It is located in a valley between two mountains.

Ḥüseynābād: It is one stage southeast of Çorum. A river called the Delıırmaḳ flows into a natural basin and out of it.

Chapter: Sanjak of Canik

Borders: To the east, Trabzon; to the west, Ḳasṭamonı; to the north, the Black Sea; and to the south, Sivas.

Its qadi districts are the following: Erim, İfraz, Aḳçay, Alaçam, Ünye, Ökse, İdecek, Bafra, Terme, Cüreki, Cevizderesi, Ḥiṣārcıḳ, Serkiz, Ṣatılmış, Samsun, Fenaris, Ḳavaḳ, Keşderesi, Meydān, Vone, and Fatsa.

Distinguishing features: It should be known that some villages in these regions contain many quarters (*maḥalle*), each quarter consisting of three or four residences (*ḫāne*). These clusters are rather distant from each other, and for this reason are known as quarters.

The people of Canik are ignorant Turks known for their mean nature and quarrelsome disposition. Nevertheless, the province of Canik is a beautiful and charming place. It has unequalled streams, meadows, and pine-covered mountains. Indeed, the forests on the mountains are almost impenetrable. The people are brutish and rebellious. Canik proper, in the midst of great mountains, is a qadi district composed of several villages, each village consisting of several quarters. These quarters have their own names, but cumulatively they are known as Canik. One quarter in this qadi district abuts the Black Sea. To the northeast is Samsun, to the west is Ḳarayaylak, and to the east is the village of Abdāl.

Bafra: It is located on the seacoast, one stage east of Samsun. The Ḳızılırmaḳ, which comes from ʿOs̱māncıḳ and Boyābād, flows about one parasang west of Bafra and enters the sea just west of Bafra. There is a wooden bridge over this river. Bafra has two Friday mosques and two small public baths. Most of the buildings in Canik are made of wood.

Samsun: [M 624] A famous town on the Black Sea coast opposite Kefe (Caffa in the Crimea). The Amasya River, also called the Çarşamba (Yeşilırmak), enters the sea to the east of it. Samsun is in a lowland area in the midst of mountains—the mountain range to the south extends west and east to the sea. Samsun has a ruined fortress of ancient construction, a Friday mosque, a public bath, and a small market. The air is bad. It has a harbor where ships drop anchor. Sinop is to the east of this town and Trabzon is to the west (sic; actually Trabzon is further to the east). It is on the main road to Ḳasṭamonı. On one side of the city there is a lake. Some houses of the city are located above it. From the outlet of the lake, water flows into the sea (text: *buḥayreye* "into the lake," error for *baḥra* "into the sea").

Alaçam: A qadi district to the east (rather, west) of Bafra and near it on the Black Sea coast. Mostly ship masts are cut down in the mountains near this town. They are tied together in large rafts and brought by sail from the Black Sea to Istanbul.

Chapter: Sanjak of ʿArabkīr (Arapgir)

Its qadi districts are the following: Egin-Şādī.

Description of Arapgir: A sanjak that extends from the extreme northeast of the province of Sivas to the sanjaks of the province of Diyarbekir. The fortress and town of Arapgir are located one or two stages east (rather, southeast) of Divrigi, which is a few miles west of the Euphrates, and one stage south of Egin. Arapkir is a prosperous town. Egin is administratively attached to it.

Zile: A large village, one of the villages of Arapkir east of the Euphrates. It is at a ford where there are a few boats. One crosses here and after travelling a distance of three hours, one reaches a village on the bank of the Murād River where there are also boats. The territory of the *emāret* (= sanjak) is on the west bank and Mt. Mihek is on the east bank.

Egin: A town one stage north of Arapgir and three stages (south) east of Sivas. It is at the foot of a rocky mountain on the west bank of the Euphrates. The water is brackish. Vineyards and gardens stretch from the foot of this mountain to the bank of the Euphrates and there is abundant fruit. Egin is among mountains. It is located at the foot of one of these mountains and from this mountain water emerges as a very sweet spring from under a rock in the town. It then flows into the Euphrates. The town of Egin is located on both sides of the water (that flows from the spring). It consists of houses and quarters built one on top of the other on the side of the mountain. It has Friday mosques and public baths. Opposite the town, on the other side of the Euphrates, are rocky mountains. The town is walled in on the west by very high cliffs. On the other side of the Euphrates is the area of Çemişgezek.

Chapter: Sanjak of Divrigi

The qadi districts of Divrigi are the following: Dārende (followed by a possible lacuna in the text of two or three words).

Description of Divrigi: An important qadi district and center of government located two stages (south) east of Sivas. Its eastern border extends as far as Mt. Çiçek. Its southern border reaches Mt. Ḥasan and the territory of Malatya. It is located at the end of a valley formed by two high, rocky, treeless mountains stretching from east to west. It is a large town in the spur of the mountain on the west. The fortress is on a high place in the spur of the mountain, but most of the houses and markets are (in the valley) on level ground. It takes two hours to reach this valley (from the fortress).

The valley is full of gardens and has all kinds of fruit. On one side of the valley, a river flows south through these gardens. Shortly after emerging from the valley it passes near Mt. Ḥasan. It flows through a plain where that valley comes to an end. Just north of Egin [M 625] it joins another river before flowing into the Euphrates. A little beyond where these two rivers meet, they flow under a bridge and enter the Euphrates. Dams have been built on the river that comes from Mt. Ḥasan and water is brought (from the reservoirs) to the gardens. This river, which reaches Divrigi, comes from the territory of Sivas. Because it flows through a low area, it does not benefit the gardens.

There is an old ruined fortress in the spur of the mountain that is opposite the mountain with the river. It is on the same level as the fortress of the city and is nearby. Behind this mountain and to the east is a village of infidels called Kesme. Pure iron ore is found on that side of the mountain. Wherever one digs he finds abundant iron. The iron that is extracted belongs to the beys. On a nearby mountain, opposite this place to the west and north, is a loadstone mine, producing magnets of high quality.

The Friday Mosque of Aḥmed Pasha (rather, the Mengüjekid ruler Aḥmed Shah) in Divrigi is built in the style of the Ulu Cāmiʿ (Great Mosque) in Bursa.

Dārende (text has Derende throughout): A well-known town. It is two stages south of Divrigi and borders the territory of Malatya. It has its own qadi and a *ṣubaşı* appointed by the *bey* of Divrigi. Overlooking the city of Dārende is a high cliff that has been split in two, seemingly by human artifice. A river called the Aḳṣu (Tohma), which flows between these two high cliffs, passes through the middle of the city of Derende. There is a strong fortress at the top of these rocks.

Chapter: Sanjak of Çorum

It is three stages southwest of Zile and northeast of Ḳalʿecik which belongs to the sanjak of Çankırı. South of it is the sanjak of Bozoḳ.

The qadi districts of Çorum are the following: İskilip, Taḥta'l-ṭarīḳ, Ḥācı Ḥamza, Sazsaḳız, Os̱māncıḳ, Ḳaraḥiṣār, Ḳaraḥiṣār-ı Tīmūrlı, Fevḳa'l-ṭarīḳ, Ḳaṭar, Sarāy, Ḳanar, Mehma, and Selām.

Description of Çorum: The town is a prosperous city. Every week a bazaar is set up. It has splendid inns as well as neighborhood mosques and Friday mosques and flourishing markets. It has its own palace for a pasha. (The city) is located on a wide plain between two mountains. It is not on the main road. One stage east of Çorum is the dervish lodge of Shaikh ʿAlvān (the poet Elvan Çelebi, fl. 14th century). There is a large guesthouse here where travelers are given hospitality.

İskilip: A town and fortress one stage west of Çorum. The mufti Ebū'l-Suʿūd (Koran commentator and shaikh al-Islam, d. 1574) was from this town. He built here a Friday mosque and a school for teachers. The Ḳızılırmaḳ flows between this town and Çorum.

ʿOs̱māncıḳ: It is two stages from Amasya. The Ḳızılırmaḳ flows near it to the west. It is a fortress and town on the main road. One hour from the river is the grave of a saint called Ḳoyun Baba. His tomb is in a shrine that has a dome covered with lead and dervishes live in the attached lodge. Bayram Ketḫüdā (Bayram Pasha, grand vizier, d. 1638) built a Friday mosque here. The Ḳızılırmaḳ flows nearby.

Ḥācı Ḥamza: A village containing a Friday mosque, public bath, inn, and shops.

Chapter: Sanjak of Amasya

Its qadi districts are the following: Baʿẓ-ı Geldigelen, Ḥafṣa also called Ḥavṣa, Ẕū'l-nūnābād, Zeytūn, Gedeġra, Gelikiras, Gümüş, Ladik, Merzifonābād and Merzifon.

Description of Amasya: It is located in a valley above the Toḳat River. It has many quarters, markets, public baths, Friday mosques, madrasas and inns. The most elevated quarter is surrounded by a wall. The Seljuk ʿAlāʾ al-Dīn Kay-Qubād rebuilt its fortress. In the past there was a silver mine here. [M 626]

This city is located between two mountains. The canals that Farhād carved for Shīrīn are in Amasya—in fact he cut them out of local rocks and made large waterways. Sultan Bāyezīd (II, r. 1481–1512) has a magnificent Friday mosque with two minarets there. Also in Amasya is the palace of Isfendiyār (one of the rulers of the Candaroğulları or Isfendiyāroğulları dynasty, 1292–1462). They call this city the Baghdad of Rūm. For some time it was the royal residence.

Merzifon: It is west of Mt. Ṭaşan, which is one stage north of Amasya. It has an old Friday mosque and an old public bath going back to the time of the infidels. Sultan Meḥmed the Wrestler (i.e., Meḥmed I, r. 1413–1421) built a Friday mosque and a madrasa there.

Gümüş: It is just south of Ḥācıköyi, which is three stages southeast of ʿOs̱māncıḳ. Nearby is a large village, also a district, called Ḥamāmözi or Ḥamāmlu, with a fine hot spring covered by a dome.

Bulaḳ: A qadi district near Gümüş.

Gedeḳara (Gedegra), also called Köprü: It is located on a broad plain east of Mt. Ṭaşan. A famous individual named Ṭaşanoğlı conquered the fortress in this place in 812 (1409). It happened this way: The *bey* of the fortress had gone out hunting; Ṭaşanoğlı attacked them, killed everyone, and dressed his followers in their clothes. Then he entered the fortress at night and took it. This fortress is difficult to reach. After living at the foot of it for some time, he moved to the valley. A rich janissary named Ḥācı Yūsuf, who subsequently became the *zāʿim* (*zeʿāmet* holder) in this valley, built a fortress and public bath there. (The Ottoman grand vizier) Köprülü Meḥmed Pasha and Abaza Ḥasan Agha had palaces there. They are places with vineyards and gardens. There is also a place called the Palace of Sultan Muṣṭafā (I, r. 1617–1618, 1622–1623). There is a hot spring between it and Amasya. The water is hot enough to cook eggs. Ṣādī Bey built a dome over this hot spring. Women have one pool and men have two. The water cures certain illnesses.

Tekyeler, also called Aḳyeler: An old and rather prosperous town, off the (main) road and on the other side of the mountain. Here is the dervish lodge called the Tekye of Ḥācı Baba, who is buried inside. They attribute many miracles to him. Indeed they say that he mounted a wall and made it walk. Turḥal is six hours from here. One passes through a place called Derbend.

Gelikiras: It is five hours from Amasya and opposite Merzifon. (It has) neighborhood mosques, Friday mosques and public baths, and its people are pious. It is an endowment for Mecca and Medina.

Ḥācı Köy: It is south of Gümüş. It has a Friday mosque and two inns.

Chapter: Sanjak of Bozoḳ

It is located southeast of Çorum. Its qadi districts are the following: Aḳṭaġ, Emlāk, Budaḳözi, Boğazlıyan, Süleymānlu, Ṣaġir, Ṣorḳun, Gökçubuḳ, Ḳızılḳocalar, Ḥüseyinova, and Ḫān-ı Cedīd.

Special features: This district comprises many villages. Its administrative center is Ḳırşehir. The people are said to be Rāfiḍī (i.e., Shīʿī). They are very hospitable to travelers. Travelers on the Bozoḳ road have no worries about food and drink. Every village is a guesthouse.

Rivers

Nehr-i Sürḫ ("Red River"): It is the Ḳızılırmaḳ that comes from the Çubuḳ plain and the area around Ḳoçḥiṣār and flows from east to west. It reaches Sivas, passes south of Ḳırşehir, and comes to ʿOs̱māncıḳ from Çāşnigīr Köprüsi. It then flows to the qadi district of Ḥācı Ḥamza, passes near the town of Ḳarġı, and comes to the qadi district of Zeytūn. It passes under a wooden bridge near the village of Çaymaḥal, which is between these two qadi districts. [M 627] It flows through the qadi district of Ṭuraġan, between the villages of Gedeḳara and Görende, through a narrows between two rocks, and then enters the Black Sea near Bafra. On a cliff east of this narrows is a small fortress called Boġazḳalʿe.

Amasya (Yeşilırmak): It rises from tributaries around Ḳaraḥiṣār-ı Şarḳī, passes through the valley of Ḳoyuluḥiṣār and Niksār, flows under the bridge north of Toḳat, goes to Amasya where it passes through the city, reaches Canik in the district of Erim, and enters the sea near Çarşamba Bāzārı. At Canik they call this river Çarşamba Ṣuyı.

Çökerek: Near Amasya it meets the river that comes from the direction of Ḳaraḥiṣār-ı Şarḳī. This river flows among the villages of the qadi district of Yüzdepāre and joins the Amasya River.

Mountains

Mt. Yıldız: It stretches from the border of the province of Sivas to Kayseri. It is very high and in the form of one stratum above the other. They are immense mountains with unrivaled summer pastures. The Yıldız River is formed from tributaries coming from these mountains.

Southeast of Sivas, in the region of Divrigi, there are extensive stony tracts and barren summer pastures; but in the northeast there are delightful summer pastures and pine-forested passes. The area to the north (of Sivas) consists mainly of a chain of high mountains along the Black Sea coast (the Pontic range) that are on a level with each other (i.e., at the same elevation?). The area to the south consists mainly of plains, with some mountains.

Mt. Canik: It is connected to the mountains of Trabzon (the Pontic range) and extends to the territory of Amasya. It is a high mountain, difficult to traverse, and replete with an animal called *zerdavā* (the pine marten). The people mostly hunt them and sell their skins. At one or two places at the top of this mountain, holes have been made in boulders and large iron chains attached to them. In many places the rocks, which are the height of a man, have been hewn and made into pillars. It is not known why they were made. The people claim that before Alexander cut the straits to the Black Sea, these pillars were on the seacoast and ships used them as mooring stones.

Every spur of these mountains—whether facing Canik, Amasya, or Niksār—has unrivaled and famous climate. One of them, the Çemen summer pasture, is located in a branch of these mountains. The Turkmen come here in the summer and every spur becomes like a city. There is great bustle.

The inhabitants of this province are mostly Muslims. They are Turkmen, Turks, and Greeks. The Muslims are of the Ḥanafī rite. In some places there are secret Rāfiḍdīs (i.e., Shīʿīs). There are also infidels, most of whom are Armenians.

Routes and stages of the province of Sivas

From Sivas to Erzurum, military stages: > Acıṣubaşı 5 hours > Ḳoçḥiṣār 5 hours > Altı Ḳarye ("Six Villages", Altıköy) 4 hours > Arġanutözi 7½ hours—one goes through a great mountain pass; > Ayaş summer pasture 6 hours—the road goes through pine-forested mountains; > Şaḥneçimen 4 hours > Aḳşār 4½ hours > Barudiḳa 7½ hours > Yārġāzībıñarı 7½ hours > Çimen summer pasture 4 hours—here black tulips grow in the snow; > Yassıçimen 4½ hours > Ḳarabulur 5 hours > Seker plain 3 hours > Canik 5 hours > Toloslar 5½ hours > Aḳdegirmen 4½ hours—here one crosses the Euphrates; > Mamaḫātūn 5½ hours > Penk 4½ hours > Ḫınūs 5 hours > Ilıca 4 hours > Erzurum 4 hours. [M 628]

From Sivas to Ṭosya: > Yıdızṣuyı 4 hours > Meḥmed Paşa Ḫānı 4 hours > Taḫtābād 7 hours; or else Boles—it is a large village on the Artuḳābād plain and one goes through a mountain pass (to reach it); > Toḳat 7 hours > Eynebāzārı 6 hours > Turḫal 4 hours > Eski Eyne 7 hours; or else > Aḳyalar, also called Tekyeler; > Amasya 7 hours > Gelikiras 5 hours > Bulaḳ, near Merzifon, 7 hours > Köşeşaʿbān, at the end of the Merzifon plain, 4½ hours > ʿOsmāncıḳ 8 hours; or else Ḥācıköyi, south of the town of Gümüş—it has two inns and a Friday mosque; > ʿOsmāncıḳ 8 hours > Ḥācı Ḥamza 6 hours > Ṭosya 7 hours. At a place called Kepirbeli there is a mountain pass very difficult to traverse.

From Sivas to Konya: > İskelec 8 hours > Abardi, also called Gedikçayırı or Şārḳışla, 6 hours > Çubuḳçayırı 6 hours > Şarıoğlan 5 hours > Barsin 6 hours > Kayseri 6 hours > İnceṣu 5 hours > Develü Ḳaraḥiṣār 4½ hours > Gölbaşı 4 hours > Nigde 4 hours > Naḳārazen 4 hours—on this road there are large marshes; > Eregli 4 hours > Gölbaşı 7½ hours > Aḳçeşār 7½ hours > Bıñarbaşı 6 hours > Bekāroğlı 6 hours > Konya 6 hours.

From Sivas to Malatya: one goes through the Zelketek mountain pass at the village of Ulaş > Ḳanġal 9 hours; or else > Selçūḳ Ḫānı 6 hours > village of Degirmen, also called Baḥirburnı, 4½ hours > Ḳanġal 4 hours; or else > Alacaḫān > village of Ḥasan Çelebi > Kesikköprü > Ḥasan Batrīḳ; or else > Ḳoçḥiṣār > Ḳazlıgöl > Gölbaşı > Güzelḥiṣār > Öyük > Taḫtalu > Ṭopraḳḳalʿe > Aḳşār > Ḳarayaʿḳūb, also called Arpa > Çatı > Çamurlu > Tercan > Mamaḫātūn > Armudlu > Alacaḫān 5 hours; or else > Ḫātūnçayırı—the inhabitants of this last village are Turkmen and are subjects (i.e., pay taxes) to the pious endowments of the Vālide Sultan; > Aḳçaḳalʿe 5 hours > village of Ḥasan Çelebi—the roads join in the middle of this stage; > village of Ketḫüdā—a small village in the midst of Ḳaraṭaġ; the road is level; > a river called Ḳırḳgeçid ("Forty Fords") 5 hours—the inhabitants are Turkmen who go up to their summer pastures in the spring; > Ḥekīm Ḫānı 5 hours—it was built by Sultan ʿAlāʾ al-Dīn and later restored by someone named Ḥekīm, so it became known by that name; > Ḥasan Batrīḳ—there is a great Friday mosque and inn > Malatya 5 hours.

From Sivas to Ankara: > Ilıcaḫān > Meḥmedpaşa > village of Tolos > plain of Ḫızırlıḳ; or else > Artuḳābād > Sīs > Zile > Üçṭaş > Ḳazanḳaya > Sancar > Budaḳözi > Elbeyli > Deliceırmaḳ > Keskin > Ḳızılırmaḳ > Elmaṭaġı > Ankara.

From Toḳat to Ankara: > Eynebāzārı > Şeyḫnuṣret near Zile > Meşhedābād > Aḳediz > Ḳazanḳaya > Ḥüseynābād > Aḳçeḳoyunlı > Nāzende > Ḳaraḳoġa > Çāşnigīr Köprüsi

("Bridge of the Taster") on the Ḳızılırmaḳ > Çuḳurcuḳ > Lalaçayırı > Ankara.

From ʿOṣmāncıḳ to Erzurum: > Direkliderbend > Merzifon > Ladik > Sepetlübeli > Ṣonisa > Niksār > Telmese > Ḥācımurād > [M 629] Aşḳār (Aḳşār) Ovası > Gercamis (Gercanis) > Kemaẖ > Erzurum.

Status of kings and rulers

After the Persians, Greeks, and Byzantines, and after the rise of Islam, the Abbasid Caliphs carried out campaigns and raids into these regions. At that time Jaʿfar Baṭṭāl Ghazi took Sivas from the infidels and made it part of Dār al-Islām (Islamic territory). When he died the *tek*[*f*]*ur* (Byzantine governor) of Toḳat organized a caravan of merchants and sent it to Sivas. He put warriors inside chests (accompanying the caravan). When they entered Sivas, they took sword in hand, killed the Muslims, and destroyed the city.

Later the dynasty of Melik Danishmend arose. This came about as follows: Baṭṭāl Ghāzī married the daughter of the daughter of his sister to ʿAlī b. Miḍrāb, a notable from the Khwārazm Turkmen. They had a son who was named Aḥmed. When Aḥmed grew up, he acquired much knowledge and so got the title Danishmend ("possessor of knowledge"). Later he learned all about horsemanship as well and joined forces with Ṭursan, the grandson of Baṭṭāl Ghāzī. They received permission from the Abbasid caliphs to fight the infidels. The caliph gave each of them permission in the form of a banner and an edict (*menşūr*) saying that whatever territory they conquered would be theirs. The date was 360 of the *hijra* (970).

Tursan with half the troops went toward Constantinople and made many conquests on the way. When he was on the outskirts of Üsküdar, he built a fortress at ʿAlem Ṭağı and resided there. Danishmend went toward Sivas which he rebuilt and made his seat of government. Later he set out for the city of Sise, i.e., Komenat. It was an ancient city built by Iraj son of Farīdūn and contained 360 churches. After conquering it he took Keşan, i.e., Turḥal, and destroyed it. Then he conquered and destroyed the city called Ḳızḳarye near Ḳazgölü. Then he took Çankırı, Amasya, and Çorum. Then he sent one of his soldiers named ʿOṣmān with 5,000 troops to Ḳasṭamonı. While on the way to conquer it, he conquered Aflanos—it is now called ʿOṣmāncıḳ, after its conqueror—and seized the silver mines in that area. Then he conquered the citadel of Ḳasṭamonı and all the fortresses in the vicinity; then Niksār—they say that in the past this was a very great city, as large as Istanbul; then Canik. Later, while besieging the fortress of Helekne, an arrow struck him in the chest and he died a martyr (d. 497/1104). His son Muḥammad al-Ghāzī, who succeeded him, buried him in Niksār and built a lofty dome over him. Today this martyrium is a place of visitation. They say that Muḥammad al-Ghāzī was the conqueror of the lands of Rūm after Baṭṭāl Ghāzī.

Later the Danishmendid state declined and the infidels invaded it again. The Danishmendids asked (Abbasid) Caliph al-Muqtadī (r. 467–487/1075–1094) for help. He (in turn) informed the Seljuk sultan Malik Shāh (r. 465–485/1073–1092) who sent the Seljuk Sulaymān b. Qutulmush (d. 479/1086) with a few thousand men to help him. The Muslim army assembled and reconquered all the provinces that they had previously. Then the caliph betrothed the sister of Sulaymān Shāh (b. Qutulmush) to Malik Muḥammad Ghāzī and in this way established a bond between them. Sulaymān Shāh settled in Kayseri, in the province of Rūm. [MAP] [M 630] They lived many years in friendship. Little by little the Seljuk state grew in size and they made Konya their seat of government. Meanwhile Muḥammad Ghāzī was succeeded by his son Yağıbasan, then his nephew Ibrāhīm, then his son Ismāʿīl, then Dhū'l-Nūn. During the reign of the last (567–570/1172–1174), the state of the Danishmendids declined. As a result, in 568 (1172–1173) the Seljuk (sultan) Qilich Arslān invaded and put an end to the Danishmendid state.

When the Seljuk state also came to an end, Öljeytü Khan—a descendant of Arghun, who was a descendant of Tolui Khan, who was from a branch of the family of Chingis Khan—invaded the provinces of Rūm. After him Abū Saʿīd took the throne, but because he was young, one of his emirs, Amīr Choban, became his *atabek* (tutor) until he grew up. But Abū Saʿīd became suspicious of Choban and put him to death. At that time Amīr Choban's son Timurtash had been sent by Abū Saʿīd to take control of Sivas and Erzurum. When he learned of his father's murder, he fled to Egypt where he had a falling out with (the Mamluk sultan) al-Malik al-Nāṣir (Muḥammad b. Qalāwūn) who put him to death (728/1328).

Abū Saʿīd died subsequently in 738 (1337). Ertena, the brother-in-law of Timurtash, became ruler of Sivas. He was succeeded by his son Muḥammad who was succeeded by his son (ʿAlī). However, because the latter (rather, the latter's son) was young, ʿAbd Allāh, the governor of Sivas and its qadi, was appointed (to replace him). Then Qadi ʿAbd Allāh's son Burhān al-Dīn Aḥmad, who was an educated and prudent man, became ruler of those regions and made Sivas his seat of government. He established close relations with the Dhū'l-qadriyya ruler and gave his daughter to his son. When Burhān al-Dīn died, he left a young son, so the notables of the province acted prudently and handed the province over to the Ottoman (sultan) Yıldırım Bāyezīd Khan.

Supplement of the Publisher (İbrāhīm Müteferriḳa)

Chapter 46: Description of (Province of) Anaṭolı

The forty-sixth chapter is an account of Anaṭolı. It should be known that the northern border of the conventional clime that the geographers call Asia Minor is the Black Sea stretching from Üsküdar east to Trabzon. Its western border begins at Üsküdar, follows the Aegean coast past the forts on the (Gallipoli) strait, passes beyond Izmir, and (ultimately) reaches Tekirburnı opposite the island of İstanköy (Kos) and the Mediterranean proper. Its southern border is the Mediterranean. The border turns east and goes along the coast from Tekirburnı to Marmaris, Mekre (Fethiye), Finike, Antalya, Alanya, Silifke, and finally reaches Payas and İskenderūn (Alexandretta). From İskenderūn there is an imaginary line on land that goes as far as Bīra (Birecik) on the Euphrates. Its eastern border is the Euphrates from Bīra to Erzincan. From Erzincan there is an imaginary line to Trabzon on the Black Sea.

The lands within these borders are called Asia Minor. Because some areas of these lands were, in the past, under the control of the rulers of Syria, they were described in the chapter on the clime of Syria. And some of them were mentioned subsequently in other contexts. These extensive lands that were fated to be conquered by the present Ottoman state were divided into provinces (*eyālet*). One of them is that of Anaṭolı, described in this chapter. In this section the account of Ebū Bekr b. Behrām ed-Dimaşḳī is cited word for word.

Chapter: Description of the province of Anaṭolı

It is a notable clime in the westernmost part of Asia. It is noteworthy that the Ottomans [M 631] appeared in this clime. In Greek, Anaṭolı means "direction of the rising sun."

Borders: To the east, Ḳaraman and Sivas; to the south, the Mediterranean; to the west, the Gulf of Constantinople (i.e., the Golden Horn) and part of the Aegean; to the north, the Black Sea.

The capital is the city of Kütahya. This province has fourteen sanjaks: Ankara, Aydın, Bolu, Teke, Ḥamīd, Ḫüdāvendigār, Sulṭānöñi, Ṣaruḫān, Ḳasṭamonı, Ḳaraḥiṣār-ı Ṣāḥib, Ḳaresi, Çankırı, Kütahya, and Menteşe.

This province has a *defter ketḫüdāsı* (minister of tax rolls), *timar defterdārı* (administrative director of timars), *ze'āmet* (*defterdārı*), four *müsellem beys* (chiefs who provide military service instead of paying taxes) and eleven *yaya beys* (commanders of regiments of foot soldiers).

The incomes of the imperial domains (*ḫāṣ*) are the following (in *aḳçes*): Ṣaruḫān 40,000, Aydın 613,465, Ḳaraḥiṣār-ı Ṣāḥib 240,299, Ankara 264,380, Ḫüdāvendigār 618,079, Bolu 300,122, Ḳasṭamonı 500,000, Menteşe 400,800, Teke 328,000, Ḥamīd 204,000, Çankırı 258,081, Ḳaresi 300,000, Sulṭānöñi 250,000.

The *ze'āmet* of the *ketḫüdā* (governor) of Anaṭolı generates 100,912 *aḳçes*, the *ze'āmet* of the *defterdār* of Anaṭolı generates 90,596 *aḳçes*, and those of the *müsellems* of Aydın, Ṣaruḫān, Menteşe, Ḳaresi, Bursa, Biġa, Ḳocaeli, and Sulṭānöñi each generate 50,200 *aḳçes*. Altogether the province of Anaṭolı has 7,311 *ḳılıç* (timars) of which 195 are *ze'āmets*; the rest are certified and uncertified timars. They provide 9,700 armed retainers. According to *ḳānūn*, the emirs (i.e., *beylerbeys*, holders of *ḫāṣ*), *zā'ims* (holders of *ze'āmet*), and timar holders provide 17,000 armed retainers.

All the infantry and *müsellems* in Anaṭolı go on campaign. The *beys* of these foot-soldiers and *müsellems* used to send their deputies on campaign. They performed such services as towing artillery, clearing roads and providing provisions. The *müsellem* corps also had farms and paid a tithe and a tax on the grain produced on the farm estates. These were campaign dues for the *yaya* and *müsellem* who were the deputies of the corps. Today the *yaya* and *müsellem* corps have been abolished. The taxpaying subjects have been enrolled (to carry out all their duties) and their farms have been assigned as *ze'āmets* and timars. Currently the *ze'āmet* and timar holders who possess these farms are charged with supporting the admiral and naval operations. In addition there was a corps of *cānbāzān* (special troops employed in dangerous enterprises) and *ġarībān* (cavalry units) that went on campaign, but today these too have been abolished and their farm estates have been turned into timars.

Fortresses of Anaṭolı

- Sanjak of Kütahya: Kütahya, Egrigöz.
- Sanjak of Menteşe: Necde, Bodrum.
- Sanjak of Ṣaruḫān: Manisa, Ḳaraḳoca.
- Sanjak of Ḥamīd: Egridir.
- Sanjak of Ḳaraḥiṣār-ı Ṣāḥib: Ḳaraḥiṣār-ı Ṣāḥib.
- Sanjak of Sulṭānöñi: İnöñi.
- Sanjak of Ḳaresi: Yeñiceḥiṣār.
- Sanjak of Ḫüdāvendigār: Sivriḥiṣār, Bergama.
- Sanjak of Bolu: Amaṣra.
- Sanjak of Ḳasṭamonı: Ḳasṭamonı, Māne, Sinop.

- Sanjak of Ankara: Ankara.
- Sanjak of Çankırı: Çankırı, Ḳılıçcıḳ.
- Sanjak of Teke: Antalya.

There are a total of 20 fortresses.

Chapter: Description of the Sanjak of Germiyan

Around the year 783 (1381), the Sultan of the Germiyanids (Sulaymān Shah), arranging a marriage contract with Ghazi Ḫüdāvendigār (i.e., Murād I), gave the fortress of Kütahya and several of its administrative districts to Ḫüdāvendigār (as dowry). [M 632] For this reason the sanjak of Kütahya was known by the name of its former ruler.

The qadi districts of this sanjak are the following: Ez̲īne, Eşme, Egrigöz, Anay, Baḳlañ (text: Baklami), Banaz, Bozġuş-Ġalçan, Çaḳırca, Çal, Çarşamba, Ḥoma, Ḥotaz, Serge, Selendi, Simav, Şeyḫlü also called Işıḳlu, Ṭazḳırı, Ṭaġardı, Ṭavşanlu, Ṭoyla-ʿOs̲māneli, Uşak, Ḳazıḳlı, Ḳula, Gediz, Kütahya, Ṣıçanlu, Selenti, Göre, Gököyük, Geyikler, Lāz̲ḳiye, Deñizli.

Borders: To the west, Ḫüdāvendigār; to the south, Aydın and Ṣaruḫān; to the east, Ḥamīd and Ḳaraman; to the north, the province of Ḫüdāvendigār.

Special characteristics

Kütahya is the administrative center and has a post for a molla paying 500 *aḳçes* (per diem). The city is located at the foot of a mountain. It has a fortress on a high cliff. It is related that the Germiyanid sultan built this city and its fortification.

Kütahya has seven madrasas; a noble Friday mosque known as Ulu Cāmiʿ (Great Mosque) of the Germiyanid Sultan; the Friday mosques of Emīnzāde, Yorġancızāde, Molla Vācid, Ḳāḍıʿasker Ḫalīl Çelebi and Ḳaraca Pasha; the Ḳāḍıʿasker public bath and the Balıḳlı public bath, which has a pool full of cold water; also six or seven other public baths, a covered market, flourishing markets, inns, neighborhood mosques, vineyards, vegetable gardens, streams and numerous places of excursion.

To the east and north is a verdant and level plain. The Porsuḳ River flows through the middle of it and joins the Ṣaḳarya River near Eskişehir.

Behind the city is a rather high mountain on which is the fortress in two levels, one above the other, and in between the two is the citadel known as Gevhernigīn ("Ringstone"). A stream flows through it and there is a Friday mosque built by the Ottomans. The city proper is at the foot of the mountain. There are 15 town quarters, some reaching the fortress on one side. There the law court is in the (walled) city and the pasha's palace is next to it. (i.e., not in the walled city). The roofs of the houses are covered with dirt.

Places of excursion:

- Ḫıżırlıḳ, near the fortress.
- Aḳṣu, on the side of the mountain. It is a very cold stream. And a small and well-made canal pours into a basin where there is a stream (?).
- Sulṭān Bāġı. It has gardens and vineyards.
- Kebkir, southwest of the city. A spring flows through it.
- The Garden of Sultan Bāyezīd.

In Kütahya apples, pears, and other kinds of fruit are plentiful, but little attention is given to growing grapes.

Kütahya has a natural hot spring fed by a stream of warm water. It is three hours west of the city in a flat plain known as Yoncalı. There is a double public bath (i.e., for men and women) and next to it a Friday mosque. In this plain there are canals and streams with mixed and unmixed hot and cold water. In one of these, hot sand is continually bubbling up from a place the size of a candle tray. This place is evident from the appearance of the water and from the flowing stream that is like a hot spring. Those afflicted with pain in the joints enter the river and witness its benefit. That place is very deep. Next to the pool at the hot spring is a hole full of mud, like a grotto, where ill people take mud baths. Here are located the Friday Mosque of Ahteri and the tombs of Shaikh Germiyanī and Firāḳī Kütahyavī.

[M 633] There is another hot spring with a double (public bath) and next to a small mosque. It is farther from Kütahya but located in (one of) its districts.

This city is three stages southeast of Yeñişehir of Bursa. While going from Bursa to Kütahya one passes Mt. Ṭomaniç and then arrives in Kütahya. Uşak is one stage to the east, Ḳaraḥiṣār-ı Ṣāḥib is two stages to the east, and Eskişehir is two stages to the north.

The districts attached to Kütahya proper are the following: Ṭavşanlu, Gümüş, Altunṭaş, Giregi (Aslanapa), Örencik (Virancık), Nevāḥī-i Şehir, and Emrūdeli. There are two hot springs in Emrūdeli, one of which, Ḳızıl Ilıca, is known for its medicinal properties.

District of Ṭavşanlu: It is eight hours northwest of Kütahya. It is a town with a Friday mosque and a market. Once a week a great bazaar is held here. The grape syrup produced here is very good and in great demand. The district of Ṭavşanlu is two stages from Mudanya and is on the main road. Those who travel to Gelibolu and the (Gallipoli) Straits usually stop at this district.

Altunṭaş: A district northeast of Kütahya.

Gümüş: A district containing a number of villages on the road near Kütahya.

Uşak (Text has ʿUşşāḳ throughout): A town with a fortress one stage east of Kütahya in a valley near Mt. Murād. It is a prosperous qadi district with 150 villages. The town of Uşak is located in the eastern part of a broad plain and its villages are located on that plain. It is famous for its prayer rugs and carpets.

Selendi: A town east of Ḳula and west of Güre. It is a half day's journey to either one. There is a qadi district that is in ruins on a mountain called Serge between this place and Güre. The town of Selendi is on the slope of a valley. There is a rather large village (in this district) called Ḳaraselendi where kilims and prayer rugs are made.

Gediz (Text has Gedüs throughout): A town and qadi district one stage west of Kütahya in a valley surrounded on all four sides by mountains. The Gediz River is south of here. It descends from Mt. Murād and flows through the Gediz plain.

Ḳula: A town in an area of black rocks. It has a ruined fortress. To the west is a partially open plain. There are unrivaled orchards and gardens. In the orchards in the mountainous area are *beglerce* grapes, pears, and other fruits, but pears and grapes are (especially) abundant. The edge of the plain to the west of this town forms the border with Ṣaruḫān and to the south is the qadi district of İnay (text: İnar). Most of the drinking water comes from wells.

Güre: A town in a valley east of Selendi. Banaz is located between it and Kütahya. It trades in carpets. Uşak and Güre are famous for this commodity.

Banaz: A town in a valley west of Mt. Murād. The qadi district of Ṣıçanlu is to the south. Between Ṣıçanlu and Banaz is a mountain called Aḫır. It is a large summer pasture, with excellent air, where Yörüks and Turkmen gather. East of it is Altun Ovası.

Işıḳlu, also called Şeyḫlü: A town two stages west of Ḳaraḥiṣār. It is located at the foot of Aḳ Ṭaġ which is in the eastern part of a broad plain that takes a day to cross. It is a qadi district comprising 84 villages. There are few vineyards and gardens, but grain is grown.

Baḳlañ (Text has Baflañ throughout): A qadi district east of Deñizlü reaching to Şeyḫlü. It has about ten villages.

Çal: A qadi district in the middle of the qadi districts of Uşak, Baḳlañ, and İnar. It contains about 30 villages. [M 634] The Menderez River flows through it. A weekly market known as *ḫālı bāzārı* ("carpet market") is held in a valley on the river.

Ḥoma (Gümüşsu): (It is located) on the Işıḳlu valley at the foot of Aḳṭaġ. Half the valley on the southern side of Işıḳlu is arable land and the other half is a reedbed.

Tazıḳırı (Kocaoluk): A qadi district on the southern side of Baḳlañ. A mountain separates the two. On one side of it is Geyikler and on the other is Sögüt Ṭaġı. It is a place of flat plains and uncultivated land. It has five to ten villages and vineyards and gardens. Their drinking water comes from wells.

Geyikler: A large village and qadi district east of Tazıḳırı. A bazaar is held every week. It has a small river.

Gököyük: A qadi district with several villages adjacent to Aydın. Its inhabitants take offense at the expression *boz yoḳuş* ("grey slope").

Deñizlü (Deñizli), also called Lāẕḳiye: It is called Deñizlü ("With a Sea") because it has many rivers. It is a prosperous town on one side of a plain that is surrounded on four sides by mountains. It has abundant rivers and vineyards and gardens. It has 24 town quarters, 7 Friday mosques, 5 public baths, and numerous inns. It also has a small fortress that belonged to the infidels. A stream that emerges from a pond called Ḥorsoloz turns into a river as it meanders around the town. This is a beautiful place of excursion. West of town is a mountain called Küçük Baba. The eastern part of its plain is Mt. Çöklez where someone named Çöklez is buried; his tomb is a place of visitation. To the south are the Ḳazıḳlubeli Mountains. One crosses them to go to Ḥamīd (and) ʿĀṣī of Ḳaraaġaç.

Eẕīne ("Friday"): A qadi district on the western side of the Deñizlü Valley. It comprises several villages. A bazaar is held every week. It is three hours from Deñizlü.

Çehārşanba (Çarşamba, "Wednesday"): Also in the Deñizlü Valley. A bazaar is held every week in the village of Buladan which is two hours northwest of Eẕīne.

Ḥonaz: A town in the Deñizlü Valley and at the northern foot of a large snow-capped mountain. It has a fortress carved out of a rock cliff that is difficult of access, also vineyards, gardens, and streams. Most of the people are Rūm (i.e., Greek) dhimmis. It is a pious endowment of Sultan Süleymān.

İnay-Eşme: A qadi district east of Gököyük. A bazaar is held here every week.

Rivers

Gediz: It rises from Mt. Murād on the southern side of the town of Gediz, flows through the Gediz Plain, and, instead of going through Gediz, flows near Güre below (south of) Selenti (Selendi), through Aṭala, then passes near Manisa and finally meets the sea at Izmir opposite a place called Sancaḳburnı. At the upper side of Aṭa[la] at the narrows is a great stone bridge that is in ruins. Shaikh Nurullāh Efendi, one of the local notables, built a stone bridge near Güre.

Menderez (Menderes): It rises from a spring known as Pıñarbaşı, one stage from the town of Ḥoma. It flows to

Ḫoma, passes through qadi districts of Baḳlañ and Çal on the Işıḳlu Plain, and reaches the qadi district of Çarşamba.

Murād Ṭaġı: It comes from Banaz, passes through the Işıḳlu Valley (text: *arasından*, error for *ovasından*), and becomes a large river. In the qadi district of Çarşamba it joins the Menderez and together they enter the sea in the qadi district of Balaṭ belonging to Aydın. The Demirṭaş Bridge, which is at the border of the provinces of Aydın and Germiyan, was over this river. A hot spring appeared at the base of the bridge that gradually weakened it so that today it is in ruins.

Chapter: Description of the Sanjak of Ṣaruḫān

Borders: To the east, Kütahya; to the north, Ḫüdāvendigār; to the west, Ṣuġla; to the south, Aydın.

Its qadi districts are the following: Aṭala, Aḳḥiṣār, Ilıca, Perākende, Perākende-i Sālhā, Borlu, Bengi, Terḫāniyāt, Ṭurġudlu, Ḳocalar, Ḳayacıḳ [M 635], Gördek, Gördüs (Gördes), Güzelḥiṣār, Menemen, Mermercik also called Demirci. Other land grants (*muḳāṭaʿa*) of Manisa include Manoḫorya.

Special features of Maġnisa (Manisa)

Because it was the administrative center of the lands of Ṣaruḫān, it became a residence for the Ottoman princes. It is a large city at the foot of a mountain on the edge of a wide plain. The mountain, which is a sheer cliff, is snow-capped and overlooks the city. There is a fortress going back to the time of the infidels. The Gediz River flows in front of the fortress through the plain a distance of two hours and there is a wooden bridge over it.

The city has a covered market, more than ten public baths, and five imposing Friday mosques: 1) Murādiye; 2) Ḫātūniye; 3) Sulṭān; 4) Çāşnigir, where are the Prince's Palace and the summer villa of the Sultan; 5) Ulu Cāmiʿ (Great Mosque), which was built by the Ṣaruḫānid İsḥāḳ Çelebi (d. c. 790/1388), who also built a Mevlevi lodge. (The Murādiye complex includes) the Mosque of Sultan Murād, with its two minarets, a madrasa, a soup kitchen, and a guesthouse. It was completed around the year 1000 (1592). The Mosque of Ḫāṣekī Sulṭān also has two minarets, plus a madrasa on the west side and a soup kitchen and public bath on the north side. It is generally known as the Sulṭān Mosque. Furthermore, Manisa has a public bath in the market, about ten neighborhood mosques, an old han covered with lead, and many other inns.

On a hill south of the city is a fortress that guards the fortification. West and south (of the city) is a high hill called Süsen Dıraz on which ancient monuments are visible. In 994 (1586) (while he was still a prince) Sultan Meḥmed (III r. 1595–1603) did not go up to the Bozṭaġ summer pasture but made Süsen Dıraz his residence. It has abundant water, excellent air, and the ground is covered with grass.

Manisa has many vegetable gardens and vineyards. Most of the gardens are on the north side of the city. In its plains are tulips of many colors which give delight. They are so highly prized that some connoisseurs will buy a single tulip for many *aḳçes*. In the spring, they take a small pick in hand and go out to the tulip fields, dig up the ones they like, and plant them in their gardens. However, they rarely resemble in shape and color the ones (that grow wild) in the plain. The flower of the Kefe tulip is not elongated but rather circular, with a single colorful petal like a red rose and a velvet-like musk-bag (i.e., stamen) within.

The palace of the Ottoman princes is on the east and north sides of the city and completely encloses the north side. The palace's gates open to the north. In front of it is a broad field. In the summer, however, the air is very bad, so the people go up to the summer pastures.

Izmir is six hours distant from Manisa. It is reached by going over the Ṣabuncı mountain pass south of Manisa. Menemen, to the southwest, is very close and is reached in four hours. Nif is four hours to the west.

Ṭurġud[lu]: It is the primary district of Manisa. Palamuṭ and Aḳḥiṣār, which are at the foot of the mountains east of Manisa, are north of it. It is situated at the foot of the mountain east of Manisa. Palamuṭ and Aḳḥiṣār are north of it.

Aḳḥiṣār: A small city northeast of Manisa and two stages distant. It has one or two Friday mosques and neighborhood mosques, two public baths, and a ruined fortress. There are many vineyards. In the summer the air is very unhealthy. It is located east of Gördes and perhaps slightly to the north. Gördes, Aḳḥiṣār, and Palamuṭ form a triangle. The road that goes from Palamuṭ to Manisa also goes between Gördes and Aḳḥiṣār. Palamuṭ is a qadi district containing a number of villages southwest of Başgelemte. It is a district of Manisa. Aḳḥiṣār is to the east of it. Gördes is to the southwest (rather, southeast). [M 636] While going from Palamuṭ to Manisa, Aḳḥiṣār appears on the left. It is connected to Ṣındırġı (to the north).

Marmara: A town with a Friday mosque on the Boz Ṭaġ road between Aḳḥiṣār and Boz Ṭaġ. It is south of Aḳḥiṣār, north of Boz Ṭaġ, and east of Manisa.

Aṭala: A qadi district surrounded by the qadi districts of Simav and Timurci to the east, İnegöz and Aydın to the south, Gördes to the west, and Aḳḥiṣār to the northwest.

Foça: A town and fortress on the seacoast. It has abundant vineyards and gardens. It has a Friday mosque, public bath and market. Most of the inhabitants are Christian Greeks who cultivate the vine; the Ḳaraca Foça wine is famous among them.

Hot spring of Ṣart: This is located one stage southeast of the town of Ṣart on the road between Manisa and Boz Ṭaġ.

Timurci, also called Demirci and Marmaracik: A qadi district near Ṣındırġı. This qadi district has a district called Ḳarataş. Simav is east of Demirci and Mandeḥorya (Kemaliye) is to the north.

Chapter: Description of the Sanjak of Aydın

Borders: To the north, Manisa; to the west, Ṣuġla; to the south, Menteşe; and to the east, Kütahya.

Its qadi districts are the following: Arbaz (Arpaz), Alaşehir, Amasa, Ortaḳçı, Ayaṣluġ attached to Ṣuġla, İnegöl-i Aydın, Balyanbolı, İne, Bayındır, Birgi, Bozṭoġan, Tire, Ṣart, Sulṭānḥiṣārı, Ḳaracaḳoyunlı also called Tirepolması, Nāzilli also called Kestel, Keles, Güzelḥiṣār, Köşk, Vaḳıf, Yeñişehir, Göynük, Keler, Kestere, and Bayramlu-Ḳaraḳoyunlu.

Special features

Description of Tire: It is the capital of Aydın. Between Tire and Manisa is a mountain called Ḳızılca Mūsā Ṭaġı. Tire is located at the foot of another mountain that faces this one. It has a lead-covered Friday mosque, markets and public baths. It is an emporium. The people are prosperous traders. It has an open air prayer-grounds in the shape of a square measuring 200 paces on each side. This open area is surrounded by a stone wall with four gates through which one ascends a stairway of ten steps. In the middle of this area is a fountain from which water flows continuously. The ground of the place of prayer is adorned with a plant called *terfil* (a kind of clover) and there are also some trees. Inside the southern wall, stretching from the easten wall to the western wall are 100 cypress trees. It is a place where people gather to sharpen their wits.

Bayındır: A prosperous town north of Tire. It has markets, public baths, and a Friday mosque. Much cotton is grown. Bayındır is at the southern foot of Ḳızıl Mūsā Ṭaġı (same as Ḳızılca Mūsā Ṭaġı above). Tire is located to the south of it, Birgi to the east, and Ayaṣluġ and Nif to the west. They are all south of Manisa. They are reached by going through the Ḳızıl Mūsā Pass.

Nāzilli: A prosperous town with abundant fruit. A bazaar is held here each week. Deñizlü is one stage to the east and Sulṭānḥiṣārı is to the west. Nāzilli is a charming town famous for its figs. Gardens surround it for a distance of one parasang. Its administrative center is called Kestel. It is located in a level place between Mt. Tire and the mountain south of Mt. Tire. Yeñişehir is south of it. Kestel is located south of Sulṭānḥiṣārı.

Sulṭānḥiṣārı: A qadi district with an old ruined fortress. Köşk is located to the west. Sulṭānḥiṣārı is behind Mt. Kestāne (part of the present Aydın Ṭaġları) and on its southern foothills. Kestel is to the south and between it and Yeñişehir. Sulṭānḥiṣārı is on a level with Balyanbolı. One descends to Güzelḥiṣār via the mountain pass.

Köşk: A town where a bazaar is held each week. Güzelḥiṣār is four hours distant. [M 637] It has abundant vegetable gardens and its figs are excellent.

Güzelḥiṣār (Aydın): A prosperous city surrounded by a fortress. Two of its quarters are outside the walled city. It is located east of a great mountain. Its plain is sandy and therefore not very productive in grain crops. However there are abundant figs. Ez̲īne is four hours distant. Güzelḥiṣār has eight Friday mosques. It has markets and public baths. It is located in the valley in the southern foothills of Mt. Tire. Köşk is east of it. Ortaḳçı is located in the foothills of the mountain opposite this one, on the southern side. There is a place of excursion called Andız on an overlook of the mountain. If one crosses Mt. Tire it is slightly to the west.

Bozṭoġan: A town east of Ortaḳçı, west of Arbaz, and in the northern foothills of Yeñişehir Ravine. Köşk is four hours to the south. The surrounding meadows have many streams.

Arbaz: A qadi district comprising a number of villages between Yeñişehir and Bozṭoġan. It is somewhat removed from the mountain.

Ortaḳçı: A qadi district comprising several villages west of Bozṭoġan. It is located at the foot of the mountain.

Yeñişehir of Aydın: It has a Friday mosque, public bath and market. Many capable people (?—*ḳavābil*) come from here. It is opposite Mt. Tire. South of it is a mountain range running from east to west; Yeñişehir is located at the northern foot of this mountain. By going through the nearby Yeñişehir Ravine (Yeñişehir Boġazı), one reaches Ḥamīd on the other side of the mountain. It is a large, forbidding ravine (*derbend*). The common people say that this is the Ravine of Jālandar (a famous ravine in Punjab). Some old buildings are found there.

Birgi: It is northeast of Tire and is nestled in the mountain. The front of the city and the southern side face open country. There is another mountain to the south of and opposite Mt. Birgi; Tire backs up against it. Tire is southwest of Birgi, at the foot of the mountain that is south of and opposite Balyanbolı of Aydın and Birgi. Marmaracik is east of Birgi and very close to it; like Birgi, is nestled in a mountain.

Birgi has a Friday mosque, public bath, and a market. Of fruits, its pomegranates are especially abundant and are exported abroad. Many excellent people have come from here, including Ḫoca ʿAṭāyī Efendi, Birgili Meḥmed Efendi, and Ḫocazāde. The town has quite a few streams that turn mills. The climate is excellent. The mountain at this place, Boz Ṭaġ, is the summer pasture for the people of Manisa, Tire, Aḳḥiṣār and Marmara.

Marmaracik: A qadi district just east of Birgi. It is nestled in Mt. Birgi.

Balyanbolu of Aydın: A number of villages that are south of Marmaracik and nestled in the mountain opposite it.

Alaşehir of Aydın: A large city and fortress. All the houses are within the city wall. İnegöl of Aydın is to the west, Ḳula is to the east, and Timurcı to the north. Alaṭaġ is not far to the south.

İnegöl of Aydın: Ṣart is not far to the south, Marmaracik is to the west, Alaşehir to the east, and Aṭala to the north. It is a prosperous town, located on flat ground, with a small lake to the north.

Keles: A town. Most of the trees in its gardens are pomegranates and apples; there are also stately poplars. Most of the place is a meadow covered with clover. It is located east of Balyanbolı and backs up against the mountain. To the north is open countryside. It is south of Marmaracik.

Ayaṣluġ (ancient Ephesus): A magnificent city, full of ancient monuments and lofty buildings, but now in ruins. One enters the city at a certain place by a channel at the end of which is a place like a gate. Its market and shops are made of stone. Some [M 638] people reportedly come simply to see this place.

Ṣart (ancient Sardis): It is located southeast of Manisa on the road that goes to the Boz Ṭaġ summer pasture. It is northwest of Boz Ṭaġ. It is one stage from Manisa and one and a half stages from the Boz Ṭaġ summer pasture. Traces of the city wall of Ṣart, which in ancient times was a great city, are still standing. A stream flows through Ṣart. It is in the foothills of some of the mountains connected to the Boz Ṭaġ summer pasture. (From Ṣart) one crosses Boz Ṭaġ via the Aydın Pass and descends to Birgi. İnegöl is northwest of Ṣart and Alaşehir is to the north.

Birgi Ṭaġı (Mt. Birgi) is located between Manisa and Tire. There is a great mountain at the western foot of Birgi Ṭaġı called Boz Ṭaġ. The Ottoman princes used to go up to this summer pasture for about two or three months every year. East of this (text: *bir*, error for *bu*) mountain in these two places (Manisa and Tire?) there are cultivated fields known as Çavdarlıḳ (text: Ḫavdarlıḳ). A lake is there where notables sometimes take a boat out for diversion. On its northern shore there is a great vegetable garden irrigated by a stream from the lake. The circuit of this lake is ten miles. The local people call it Gölcük ("Little Lake"). It has some fish, but the people do not catch them. This lake is in the foothills of Mt. Birgi and more than ten miles away (from Birgi).

They call this the summer pasture of Boz Ṭaġ. When the people who live around Boz Ṭaġ go up to the summer pasture they hold Friday prayer at one or two places. They have set up markets, five public baths, and an open prayer-grounds for Muslim festivals. East and south of this summer pasture is another mountain called Maḳām-i Erbaʿīn ("Shrine of the Forty"). The top of the mountain is very difficult to reach. At the top of Mt. Birgi, on the side that overlooks the summer pasture, is a great rock that reflects the intense heat of midday and ill affects those at the campground. There is another peak right opposite this one. Between them is a dervish lodge built by a Persian descendant of the Prophet named Baba Reşīd. The dervishes of this lodge give a feast for the dervishes who go up to enjoy the summer camp season. This place has a delightful stream.

Mt. Kestāne is a summer pasture south of Tire. It has neighborhood mosques, Friday mosques, and public baths. The people of Tire go up there for most of the summer. They set up markets that provide for all of their needs. After morning prayer, those who go to the city reach it with the rising sun.

Chapter: Description of the Sanjak of Menteşe

Borders: In the north, Tire; in the west and south, the Mediterranean; in the east, Antalya and part of Kütahya.

Its qadi districts are the following: Üzümlü, Eskiḥiṣār, Eşen, Aġartos, Ula, Perākende-i Menteşe, Pernaz, Bozöyük, Peçin, Çine, Düger, Sar, Ulus, Ṣobice, Ṭatya, Ṭaraḫya, Feslegen, Germe, Ḳaraova-Bodrum, Keraniş, Gökābād also called Gökova, Ṭavas, Köycegiz, Mazoz, Gököyük, Mesol, Mekri (Fethiye), Mandalyat, Muġla, and Yerekesigi.

Special features: Muġla is a famous city and the capital of Menteşe; it is four stages south of Ḳuşadası. Deñizlü is two stages to the east; one goes there over a mountain via the Gedelek road. Ṭavas is a fortress and qadi district one stage from Menteşe; a bazaar is set up every week.

Chapter: Description of the Sanjak of Teke

Borders: To the north, Ḥamīd; to the west, Menteşe; to the south, the Mediterranean; to the east, Alanya.

Its qadi districts are the following: Elmalı, Antalya, İgdir, Fenike, Kaş, Ḳalḳanlu, Ḳaraḥiṣār-ı Teke, Ḳaraḳaya, Germegi.

Special features

Antalya: the capital of Teke. Ḍaḥḥāk-i Mārī (Zahhak of the Serpents, a figure from the Persian epic) built its fortress and settled one of his three sons in each of its three levels. [M 639] Currently the Ottoman sultan stations 400 garrison troops there. This city is located on a high place at the seacoast. It has a strong rectangular fortification that forms a high barrier to the sea. It is a port city. The area around it is like a flat plain covered with brush. A river flows to the city on the northern side. They have built aqueducts. The market is outside the fortress on the north side of the city near the port.

The fortress has two gates that open to the sea and one that opens to the land. The city has several Friday mosques and public baths. Its gardens abound in lemons, bitter oranges, and various other fruits, also dates. The fortress looms over the city. Inside the fortress are two Friday mosques and two public baths; another public bath is outside the walled city. The city's vineyards and gardens are located to the east.

At the foot of a mountain north of the city flows a river called the Düden, whose source is Lake Egridir. It flows (out of the lake), then sinks into the ground, travels underground and re-emerges near the gardens of Antalya. It irrigates all the gardens, then goes on to the city and enters the sea. It is a large, navigable river that enters the sea near Köpribāzārı. Köpribāzārı is a district (*nāḥiye*) of the qadi district (*ḳażā*) of Manavġat east of Antalya.

Antalya has many camels that graze on holly bushes. Sesame and strawberries are abundant in this district. It is eleven stages from Bursa. West of Köpribāzārı is a spit called Şerdenburnı that juts into the sea for a distance of two stages. There are many places void of human settlement. It is frequented by the ships of foreign infidel pirates. A fortress should be built in this area. This area is famous for its lemon syrup.

Astanar: A district of Antalya. It is one stage north of Antalya and its summer pasture. The Tefenni qadi district of Ḥamīd is located one stage to the north. Astanar is like a town. The people of Antalya have houses and shops there and migrate there in the summer; only the castle warden and some infidels remain behind. It is located on a high plain. Most of the trees are hazelnuts. There are Friday mosques and public baths. It is under the jurisdiction of a *ṣubaşı* of the *bey* of Teke. It has a large river.

Ḳızılḳaya: A qadi district comprising about ten villages, on a plain between Antalya and Ağlasun. A bazaar is held on a flat place. They set up a booth where the qadi presides.

İgdir: A qadi district comprising five to ten villages, one stage west of Antalya. It has large mountains that are difficult to traverse. It is a rocky country near the sea. It has summer and winter campgrounds. In the summer the people migrate to the villages in the mountains out of fear of the infidels. There are many carob trees in these mountains. There are excellent grapes and figs. They have few sown crops. Their trade is in lumber. Most of the people are weavers.

Finike: A fortress on the border of Ḥamīd.

Rivers

The Aḳṣu flows through Antalya and Ḳaraḥiṣār of Teke and enters the sea. One branch of it flows through Ḳaraḥiṣār Germegi and there enters the sea.

Mountains

Taḫt-ı ʿAlī (Tahtalı) is a great mountain next to İgdir.

Chapter: Description of the Sanjak of Ḥamīd

It is named after Ḥamīd Bey who was governor during the time of the Seljuks.

Borders: To the east, Beyşehri; to the north, Aḳşehir; to the west, Germiyan; to the south, Antalya.

Its qadi districts: Isparta, Ağlasun also called ʿĀṣī of Ḳaraağaç, Ağraş, Avşar, Uluborlu, Egridir, Irla, İncīrbāzārı, Barla, Pavli, Burdur, Lake Tefenni, Ḥoyrān, Seroz, Keçiborlu of Ḥamīd, Kemer, Gölḥiṣār of Ḳaraağaç, Gönen of Ḥamīd, Yalvaç, Yalvaç of Ḳaraağaç, and Yaviçe.

Special features

Isparta: A large city and the capital of the sanjak of Ḥamīd. It has no fortress. It has abundant fruit and a bazaar. [M 640] Its air is excellent, but the winters are severe. It has streams, markets, public baths, and Friday mosques. Someone named Ḫoca Manṣūr ʿAṭṭār brought a canal to the city from a great distance. In the south, in the direction of Ağlasun, is a great mountain. A river flows (from this mountain) and irrigates the vineyards and gardens of Isparta. Then it goes to the city. Grapes and hazelnuts are abundant. The finest twill is made here. There are large dye-houses. The longest day is fourteen and a half hours.

Yalvaç: A town east of Lake Egridir. Its climate is excellent. It has abundant fruit. It is renowned for pears, apples, and cherries. It has many streams.

Pavlu (Pavli): A great mountain near Egridir. It has abundant grapes and white mulberries. Excellent molasses is made here from the mulberries.

Aġrus, also called Aġraş: It is located on a plain. Its inhabitants are craftsmen and merchants.

Barla: It is located between two mountains. It is near the eastern shore of Lake Egridir. It has excellent vineyards and fine grapes.

ʿĀṣī ("Rebel") of Ḳaraaġaç (Acıpayam): A qadi district and town in open country about one stage west of Isparta. The people are obstinate Turks inclined to rebellion. Deñizlü is one stage to the north. After one crosses at Ḳazıḳbeli, Tefenni is one stage to the south.

Tefenni: A border qadi district between Ḥamīd and Antalya. ʿĀṣī of Ḳaraaġaç is located one stage to the south.

Avşar, also called Aḳṣār: It is located one stage north of Isparta. It is one and a half hours from Lake Egridir. It is a town and qadi district on a level place on the road. It has gardens and vineyards. It has a stream, Friday mosque, public bath, and markets. A bazaar is held every week. There are 13 villages, some of which are the size of towns.

One of two prominent villages is Gelendos, whose inhabitants are weavers. They weave twill and have dyeworks. The air is oppressive.

Egridir: A small citadel with strong circumvallation on the western side of the lake and jutting into the lake. It is surrounded by fresh water. It has flourishing markets, Friday mosques, and numerous public baths. The blessed graves of (some) great Naqshbandī saints and other shaikhs are here. A noble Friday mosque and soup kitchen have been built over them. There is also a madrasa in this city. It is three stages from Kütahya. This town has a suburb and gates on both sides.

Opposite Egridir is a small island with vineyards and gardens but no human habitation. North of it is a large island with about 200 houses, half the people Muslims and half infidels (i.e., Christians). The grave of the great saint Shaikh Muṣliḥüddīn Efendi is found there. The people of the island are sailors and the women weave cloth.

Overlooking the aforementioned suburb is a mountain called Sivrināz, on which is a strong fortress built by an infidel named Küleyb. Seydī Baṭṭāl Ghāzī conquered this fortress. It takes six hours to reach the top of this mountain.

Outside the town is a village called Nāzile. In 600 (1203) a holy man and descendant of the Prophet, known as the Shaykh al-Islām, in the province of Bardaʿ (in Yemen) was directed by God to build a Friday mosque and a dervish lodge in this village of Nāzile. (He did so) and resided there. Many of his descendants became saints. The Ottoman government assigned the poll tax paid by seventy people (to the support of the lodge). Until 1070 (1659), food was provided to the dervishes who sojourned there; but in that year the practice was halted.

This town of Egridir has no arable fields. For this reason, at the time of the interregnum, a dervish lodge was built in every town quarter to provide food for the poor. [M 641] Villages were designated as pious endowments for these lodges; but some are discontinued and some are in ruins.

To conclude, this qadi district of Egridir is a mountainous region. It has large mountains, tall trees, and delightful springs. There is one lofty mountain named Isaforos (?) which takes 24 hours to climb. Even when it is extremely hot, the snow on this mountain is 10–15 yards deep. Reportedly, one can find there the *āb-ı zülāl* (snowmelt from the snow worm). And marvelous flowers bloom in the summer. There is a lake of fresh water in a plateau in the middle of the mountain. It is 10 miles long and 5 miles wide and around it are meadows with excellent air in the summer. But there is nothing but snow from the lake to the top of the mountain.

There is another lake of fresh water about six hours from Egridir. The fish in this lake are called "sweet fish" and weigh 10–15 okas. They are caught during *erbaʿīn* (the forty days of midwinter, 22 December–30 January). The lake is 60 miles long and 40 miles wide.

It is reported that Egridir has 36 different kinds of grapes.

Aġlasun: A town and qadi district south of Isparta. It is located on a hill at the foot of the mountain. On the west side are vineyards and gardens. Just as at Ḳaraḥiṣār (text: Ḳarıḥiṣār), there is a single operative gate that opens to the west. Its road ascends. Streams flow through its vineyards. The city has fountains as well as a Friday mosque, public bath, market, and dye-works.

Keçiborlu: A small town and qadi district in a flat place a half stage south of Uluborlu. It has vineyards and gardens. Its inhabitants make twill. Sögüt Ṭaġı faces it. It is north of Isparta.

Ḥoyrān: A fortress (and a qadi district comprising) 36 villages, north of Lake Egridir. To the east is Yalvaç, to the north is Kütahya, and to the south are Avşar and Nāzile. The air is oppressive. Its people originally came from Ulu[borlu] as nomadic Yörüks and they settled in villages. A few of them reside in 30 households dispersed in Ḳaraağaç, Yalvaç, and Avşar.

[Lakes]

Lake Egridir: A fresh-water lake in the middle of Ḥamīdeli. It is four and a half parasangs long from south to north, three parasangs wide, and ten fathoms deep. In the lake, but connected to the western shore, is a citadel called Egridir. And there are two islands in the middle of the lake; the smaller is called Cān Adası, the larger is Nīs Adası. There are five kinds of fish in the lake and they are caught from the beginning of spring until cherry season. It is a marvelous lake. On the shore are white pebbles, some of which, by divine power, spell out the name of God. From the outlet of the lake a stream flows for a distance of two stages. In many places it goes underground. Finally, at a distance of one stage from the city of Antalya, at a place called Düden, it (wells up) like a whirlpool, then goes back underground. Later it emerges at the Customs Gate at Antalya.

Lake Burdur: A lake like Lake Ḥamīd and Lake Egridir. ʿĀṣī of Ḳaraağaç and Gölḥiṣār are located on its shore. Gölḥiṣār was torn down because it had become a nest of brigands.

Chapter: Description of the Sanjak of Ḳaraḥiṣār-ı Ṣāḥib (Afyonkarahisar)

Borders: To the east and north, Aḳşehir; to the west, Kütahya; to the south, Ḥamīd.

Its qadi districts are the following: Barçınlu, Bolavadin, Çay, Çölābād, Sincānlu, Şuhūd, Şandūḳlı, Ḳaramıḳ, the districts of Barçınlu (sic?), Oynaş.

Special features: Ḳaraḥiṣār-ı Ṣāḥib is the capital of the emirate (= sanjak). (The Ottoman grand vizier) Gedik Aḥmed Pasha (d. 1482) has a public bath and dervish lodge here. A number of madrasas have been built. Kütahya is two stages to the west.

Chapter: Description of the Sanjak of Sulṭānöñi

Borders: To the east and south, [M 642] Ḳaraḥiṣār-ı Ṣāḥib; to the west and north, Ḥüdāvendigār.

Its qadi districts are the following: Eskişehir, İnöñi also called Bozöyük, Bilecik, Seydīġāzī, Ḳaracaşehir, Ḳalʿecik, Sulṭānöñi, Aḳbıyıḳ.

Special features: Eskişehir, the capital of Sulṭānöñi, is a town and qadi district located on a broad plain. Friday prayer is held at two places. It has a small market, inns, hot springs, vineyards and gardens. Its market and inns are located in the lower plain near the hot spring. The houses are somewhat removed, on the south side of town. The tombs of Shaikh Edebali (father-in-law of ʿOs̱mān, founder of the Ottoman Empire) and Shihāb (al-Dīn Yaḥyā) al-Suhrawardī (Persian philosopher, d. 587/1191) are here and are places of visitation. To the east of town is Seydīġāzī, to the north is Göynük, to the west is Sögüd, and to the southwest is İnöñi.

Seydīġāzī: A prosperous town on the main road between Eskişehir and Bardaḳlu and Yeñiḫān. There is a magnificent new inn in the town, in the lower plain. The tomb of Seydi Baṭṭāl Ghāzī is on a lofty hill overlooking the city. It is (incorporated in) a large dervish lodge that includes a Friday mosque, cells, a madrasa, and a guesthouse that provides food to travellers. It is a complex of high masonry buildings covered with lead. The public bath and Friday mosque were built by Sultan Süleymān.

This lodge is the residence of Bektashi dervishes. It was built by some emirs of the Miḥaloğulları (Ottoman irregular cavalry in the Balkans). The mausoleum of the Seljuk Sultan ʿAlāʾ al-Dīn was built by his mother, Khātūn.

It is related that when (the Umayyad general) Maslama b. ʿAbd al-Malik besieged Istanbul (in 99–100/717–718), Baṭṭāl Ghazi was in his army. Maslama had to give up the siege, but he swore an oath that he would not leave without seeing Istanbul. (The Byzantines) opened one of the city gates and allowed Maslama to enter alone. He went straight to Ayasofya and entered it on horseback. Then he lifted the siege and returned to Syria. During the time Maslama was in the city, Baṭṭāl Ghāzī stood sentinel at the gate, mounted on his horse and leaning on his lance, saying, "Maslama is alone, may the infidels not harm him."

In the flat plain in the town there is a hot spring where hot water emerges from several places. It has a dome over it, dressing rooms, and two baths, one for men and one for women. There is another hot spring at the end of the vineyards of Eskişehir, two miles from the settled area of the city. A kind of oily substance, resembling boiled olive oil, comes to the surface and is collected by the people, sometimes one or two bowlfulls.

Sögüd: A town on the main road. It is located one stage east of Lefke and one stage west of Eskişehir. It is parallel with Ṭaraḳlı, which is on another road north of this main road. Aḳbıyıḳ is nearby and to the south. Near the main road two miles from town is the grave of Erṭuğrul Bey, the forefather of the Ottoman dynasty. A small dome was built over the grave. It is on the left side of the road while going to Lefke. There are numerous inns in this town as well as a Friday mosque and public bath. It has fine pickled grapes. In 1070 (1659), Grand Vizier Köprülü Meḥmed Pasha built an imposing inn, Friday mosque, public bath and markets at a formerly inhospitable and desolate place near the valleys in the mountains between Lefke and Sögüd that are now cultivated as rice fields.

Lefke (Osmaneli): It is one stage west of Sögüd and one stage east of Iznik. It has an inn and Friday mosque for travellers. Iskender Pasha (Ottoman administrator, fl. 16th century) built the mosque.

İnöñi: It is located at the spur of a high mountain overlooking a plain. It has many caves. Outside of town is a tower that serves as the fortress. [M 643] About nine or ten guards are assigned to it. One of its caves is inhabited by some tax-paying subjects. It can be reached on foot by various paths, but only after a thousand difficulties.

Another cave is east of Aḳbıyıḳ Baba and southwest of Eskişehir. Bozöyük is a qadi district just west of this. It is also called Küçükyöre. The qadi districts of İnöñi and Bozöyük are both attached administratively to this qadi district.

Bilecik: A prosperous town and qadi district in the midst of the qadi districts or Yeñişehir of Bursa, Lefke, İznik, Yalaḳābād (Yalova), and Gemlik. İznik is to the north and Gemlik is to the south. It has workshops where Bursa pillows, mattresses, woolens, velvets and the like are made. But their quality is lower than those made in Bursa.

Ḳaracaşehir: It has an old fortress. ʿOs̱mān Khan Ghazi, the first of the Ottomans, conquered it with an edict from (the Seljuk) Sultan ʿAlāʾ al-Dīn. The Porsuḳ River flows before it. A beautiful town fronted by a plain and with excellent air. It is north of Kütahya and near İnöñi, which is four hours away.

Chapter: Description of the Sanjak of Ankara

It is also called ʿAmmūriyya (Amorium), but the real ʿAmmūriyya is the city known as Anamur in the sanjak of İçel. Ankara is today called Engüri. According to the history books, ʿAmmūriyya is a different place, not far from Ankara and north of Konya. When Hārūn al-Rashīd conquered ʿAmmūriyya he left it in ruins; today there is no trace of it.

(Borders:) To the south, Aḳsarāy; to the east, Ḳırşehir; to the north, Çorum and Ṭosya; to the west, Eskişehir and Ḫüdāvendigār.

Its qadi districts are the following: Ankara, Yörükān, Ayaş, Bacı, Çubuḳābād, Çuḳurcuḳ, Şorba (text: Şorya), Murtażā-ābād, Yabanābād, Haymana, Yerköy and Güdül.

Special features

Ankara is the capital of the province. It has a post for a molla paying 500 *aḳçes* per diem. Sometimes it is given as a stipend to a *ḳāżıʿasker* (chief qadi of the military) who is out of office. The city has an inner and outer fortress, covered market and (other) markets, Friday mosques, public baths, and many quarters (*maḥalle*). Outside the town are some vegetable gardens, vineyards, gardens, and a small river. Most of the inhabitants are Turkmen. Its most famous commodity is mohair and the cloth made from it, produced here as nowhere else.

The inner fortress or citadel, being on an elevated place, is strong, unlike the outer one which is on a flat area. All the houses are inside the city wall. Hārūn al-Rashīd, al-Maʾmūn, and others undertook several religious wars against it. It was built by the Emperor Constans. Ḳasṭamonı is five stages distant. When Hārūn came here he transported the paired door leaves from the fortress gate to Baghdad. Above the gate there was a Greek inscription, saying:

> In the name of God, the Compassionate, the Merciful. O son of Adam! Boldly seize the opportunity when it comes. Do not miss it. Entrust affairs to those qualified to carry them out. Do not let excessive pleasure bring you to grief. Do not bear the cares of the future on your shoulders. Do not imitate the haughty in the accumulation of wealth.

In 762 (1360) Sultan Murād Ghazi, the son of Orḫān, took this city from the usurpers known as the Akhis.

Near Ankara are delightful gardens, with streams and lush vegetation, known as the Ḳayaş Gardens. At a place overlooking the city is the dervish lodge of Ḥüseyin Ghāzī, where some dervishes reside. It has a stream known as Ḳayaş Ṣuyı. Ḥācı Bayram, the founder of the Bayramiye Sufi order, is buried in this city (d. 833/1429–1430). [M 644]

A marvelous story is told in this region of Ankara. There is a ruined church east of the city on this side of the Ḳızılırmaḳ. The church had a dry well called Deliler Ḳuyusı ("Madmen's Well"). They used to take anyone who was mad to that well with his burial shroud and make him look inside it. He would either die while looking in the well or else be cured of his madness. This well is famous among the people of Ankara. Indeed, there is a large cemetery around the well called Deliler Mezārı ("Madmen's Graves"). A person of sound mind who looks in the well sees nothing at all and no harm befalls him; he only smells something like the odor of sulfur.

Murtażā-ābād: A qadi district with many villages, whose town is just west of the villages of the district of Ankara known as Ḳaṣaba Nāḥiyesi ("District of the Town"). It is located on a plain and its inhabitants make mohair thread.

Ayaş: A prosperous town on the main road west of Ankara. Beybāzārı is west of Ayaş, Yerköyi south of it, and Güdül is north of it. Ayaş, Yerköy (sic), and Güdül are close to one another and administratively attached to the qadi district of Ayaş. Between Ayaş and Beybāzārı are a hot spring and a mineral spring whose water has excellent medicinal qualities. Yerköy is a town just to the right of Ayaş. Güdül is south (sic, rather, north) of Ayaş. The Germi River—also called the Germir—flows through it. This river irrigates rice fields. Its tributary approaches Beybāzārı, then flows toward the south and enters the Ṣaḳarya above Geyve.

Yabanābād: A town and qadi district south of Ayaş and north of Şorba. It is located in the mountains.

Bacı: A qadi district composed of a number of villages bordering the territory of Ayaş.

Şorba (text: Şorya): A qadi district without a town located among the summer pastures with good air and water in the pine-forested mountains near Gerede and north of Ayaş. It has several hot springs.

Çubuḳābād: A town and qadi district northeast of Ankara. Çuḳurcuḳ is west of it.

Çuḳurcuḳ: A qadi district without a town one stage east (sic) of Ankara. Elmaṭağı is its summer pasture. Elmaṭağı is a famous high mountain whose summer pastures have good water. It is covered with pine, juniper, and other trees.

Yörükān of Ankara: A qadi district belonging to the Yörüks residing in the sanjak of Ankara. Their qadi attends to their affairs by travelling from city to city; he resides in no particular place.

Haymana: It is in the foothills (south) of Elmaṭağı. Çuḳurcuḳ is north of it. Haymana is a district of Ankara and one of the imperial domains. Its villages are prosperous and its inhabitants are hospitable and friendly to travelers. One border extends as far as Ṭurġudeli. There is a natural hot spring in this district. If one who is ill bathes in the mud of this hot spring he is cured. Most of the people in this district raise horses and camels. The best (horses and camels) are found here.

Ṭurġudeli: Its inhabitants, like those of Haymana, are also hospitable to travelers. No money is demanded of anyone who goes there. They regard it as disgraceful to make any traveler open his baggage. Unmatched horses and camels are raised in Ṭurġudeli, and the livelihood of its inhabitants depends on them. The inhabitants include Turkmen.

To conclude, there are all kinds of fruits in this district of Ankara. There are 36 varities of pear alone. Two of these—the Abbasi pear, and the Bey pear which is spherical—are eaten in the winter. There are various kinds of apples and grapes. And there is a special kind of plum that they shake from the trees on the day of Kasım (8 November). It begins to ripen in *erbaʿīn* (22 December–30 January) and *ḫamsīn* (31 January–20 March) [M 645] and is ripe by the end of *erbaʿīn*. If *paluda* (a kind of jelly or *blanc-mange* prepared from wheat or starch) is cooked with the sweet juice of this plum, it holds up for a long time. There are also abundant sorb-apples.

Rivers

The Çubuḳābād River comes from the west of Ankara and irrigates its arable fields. The İnce River rises from Elmaṭağı near the dervish lodge of Yaʿḳūb.

Chapter: Description of the Sanjak of Kānḳırı (Çankırı)

This sanjak divides Kös Ṭağı in two. More than half the towns are located north of Kös Ṭağı, between it and the mountains of Ilḳas and Budenar.

Borders: To the east, the sanjak of Çorum; to the south, Ankara; to the west, Bolu; to the north, Ḳasṭamonı.

Its qadi districts are the following: Ögez (text: Ögen) Bucorova, Boğaz also called Çerkeş Künbeti, Tuḫt, Çerkeş, Şaʿbānözi, Ṭosya, Ḳarġı, Ḳarıbāzārı, Ḳaracavīrān, Ḳalʿecik, Ḳalʿecik-i Keskin, Keskin (sic), Ḳoçḥiṣār, Ḳurşunlu, Ḳorubāzārı, Milan, and Yeñiceköy.

Special features

Kānḳırı (Çankırı): The capital of the district. It is a fortress and prosperous town on a flat place whose southern side is an open plain, south of Kös Ṭağı, in a spur of its foothills. It has two public baths and two Friday mosques, that of Sultan Süleymān and the Bey Mosque. It also has flourishing markets and inns.

On the side of the town in the plain, on a small hill opposite the fortress, is the Palace of the Sultan. It is the residence of its emirs. All around it are wooden cells and houses where the emirs' entourage reside. Most of the houses and buildings in this town are made of pine and juniper wood and the roofs are covered with tiles.

The fortress, which is in the direction of the mountain from the town, completely dominates the town and is very strong. It goes back to the time of the infidels. It is located on a grooved (text: *yerlü*, error for *yivlü*) cliff as tall as several minarets. One faces a thousand difficulties in going up to the fortress from the town with a loaded animal. Inside are a number of houses and quarters, and a Friday mosque. The fortress has a warden and guards. Those who reside there get their water from the town. There is a large cistern with an iron cover; so if the fortress were besieged, it would not suffer from lack of water. The floor of the cistern is solid rock. A tunnel has been cut in the rock that extends one or two miles down to the level of the town. A stairway like that in a minaret has been carved out of the rock, and in the middle is a basin, also carved from the rock. A channel was made running from one end of the cistern to the other. This created a stream the width of several men's torsoes. The water flows through this rock channel. The people of the town say that this stream is a branch of the river that runs under the town. On hot days, some townsmen and notables go down to the bottom of the well with torches and cook food and kebab.

On one side of the spacious plain that is in front of the town flows the river known as Ḳaraṣu ("Black Water"), which has sweet water. And on the other side flows another river known as Acıṣu ("Bitter Water"). They merge at a place one or two miles below the town and then, after a distance of a few hours, flow into the great river known as Ḳızılırmaḳ ("Red River"). The town is located between these two rivers. At a distance of three hours east of the town, the Acıṣu encounters a saltpan and the water becomes bitter.

Ankara is three stages south of Çankırı. Çerkeş is two stages to the west. Its port, Sinop, is six stages (to the north).

Ṭosya: It is north of Kös Ṭağı. They are some distance apart. Ṭosya is a large town in a broad valley one stage north of Çankırı. [M 646]

Most of its houses and quarters are located on two hillsides. It has a market, inns, and some public baths that are on a flat area in the valley. It has nine public baths, five or six inns, and numerous neighborhood mosques and Friday mosques. The mose famous Friday mosques are the Old Mosque (Eski Cāmiʿ) and those of ʿAbdurraḥmān Pasha, Pıñarbaşı, Alacamescid and Tekke.

In 1070 (1659) Reʾīsülküttāb (Chief Secretary) Şāmīzāde Meḥmed Efendi built an imposing inn near a spring called Aşağıbıñar at the edge of town. It has a public bath and Friday mosque as if it were a walled town.

Ṭosya has a meager earthen fortress in a mountainous area within the town. Some of the inhabitants built it during the time of the Celālīs (companies of brigands who roamed Anatolia between c. 1590 and 1620) but now it is in ruins.

The south side of town faces an open plain that stretches to the Devres (text: Devrek) River nearby and beyond that to the foothills of Kös Ṭağı. It has vineyards, gardens, and abundant fruit.

In the middle of the town of Ṭosya is a spring called Ṭaşbıñarı whose water is excellent. From its outlet the water flows to the tanneries. A dome was built over the spring and there are two or three bowls suspended from chains that one can use to drink the water. The interior of the dome is made of dressed marble. Inside is a square pool about the depth of a man. One side is left open for the water to flow out while the other three sides are enclosed. The dome is laid out just above the spring. The town has other water sources, also excellent and digestive, that do not derive from Ṭaşbıñarı but were brought to fountains from branches of the Ilḳas Mountains behind the town.

Ḳoçḥiṣār-i Gerede: A town on a flat area four hours west of Ṭosya. It is six hours northwest of Çankırı on the main road. The town has markets, public baths, neighborhood mosques, Friday mosques, and inns. Separated from the town, and on a level plain, is a wooden Friday mosque and near it is a court of justice. Every Friday a great bazaar is held at this place. A qadi goes there as well. There is an earthen fortress separate from the town. There are gardens between the town and the fortress. It has abundant fruit. The Devres (text: Devrek) River passes near the town. The Ṭosya road is to one side of the main road and passes through the foothills of Gök Ṭağı. It is a thoroughfare.

Ḳalʿecik: At the edge of the plain in a spur of a mountain facing Gök Ṭağı and to the south of it. It is located at the foot of a lofty grooved (*yivli*) cliff that is somewhat apart (*mun[fa]ṣılca*) from this spur. The agricultural land of the town winds around two sides of the rock and into the corner of the mountain. The aforesaid agricultural land is in the shape of a horseshoe. The fortress of the town is on that rock and is a strong fortress. The fortress has a gate on the mountain side, while on the plain side it overlooks the town. Inside is a Friday mosque, a warden and a garrison. Inside the town are two public baths and three Friday mosques: the Old Friday Mosque (Eski Cāmiʿ) and the ones known as Ḳarabaş Ḫoca and Debbāğḫāne ("Tannery").

Ḳalʿecik has extensive gardens and excellent streams from the mountain behind the town. It is a prosperous town. Most of its buildings are made from juniper trees. Its market is covered with earth. It has abundant fruit. Its apricots in particular are beyond count. Its gardens extend east and a little to the south of Çankırı. Between it and Ḳalʿecik is the Terme Plain, which takes a day to traverse. West of town is an area where forage is cut. The imposing palace of Şehsüvār Pasha overlooks it. [M 647]

Çerkeş: A town located on a broad plain. Sultan Murād IV (r. 1623–1640) built a Friday mosque, inn, and public bath here. It has a squat stone fortress. Inside the town are several quarters and fountains. A small river passes in front of the town and flows to the west. In the middle of the Çerkeş Plain is a low hill with unrivalled pine and juniper trees. On its slope is a large village called Bozoğlı, about two miles from Çerkeş.

Boğaz, also called Çerkeş Künbeti: A low-level qadi district (comprising) 17 villages between Bolu and Çerkeş. It is a land of valleys and hills. Here the main road goes through this narrow pass (*boğaz*). It is a thoroughfare.

Ḳurşunlu: A qadi district comprising 35 villages. The Devres River flows through the middle of it. The qadis reside in a large village called Köprülü. The qadi district of Milan is just north of this. There is a summer pasture most of which is located south of the Devres River. Here there is a summer pasture with dense forest called Mezliva; most of it is located south of the Devres River.

Tuḫt: (A qadi district) toward the southeast of Çankırı between Ṭosya and Ḳoçḥiṣār. Opposite it in Kös Ṭağı is a small earthen fortress surrounded by mountains with a flat area in front and surrounded by vineyards and gardens belonging to Muslims. There is a Friday mosque. Every year a bazaar is held at Bayraḳlubāzārı, a place with unrivalled streams in a rather high summer pasture above the qadi district of the town. The people there mainly deal in *astār* (cloth used for linings).

Milan (Beleni): A qadi district comprising a number of villages in a rocky valley north of Ḳurşunlu. It has vineyards, gardens, and apples beyond number. Its court of justice is in the village of Ṭolab (Dolap). The Milan River passes in front of it and flows to the east.

Bucurava: A qadi district comprising a number of villages between Ḳoçḥiṣār and Ḳaracavīrān. Some of its villages are in the foothills of Ilḳas Ṭağı. The Devres River divides this qadi district in two. Lower down there is a confluence (with the Ḳızılırmaḳ).

Ḳarğu: A town on a flat area with vineyards on all sides surrounded by rocky mountains. It has a Friday mosque and a public bath. It has good quinces and abundant grapes. The Devres River flows near the village of Ḥācı Ḥamza which is south of this town. It flows between the two. Then it empties into the great river called the Ḳızılırmaḳ, which comes from ʿOs̱māncıḳ.

Ögez: A qadi district comprising a number of villages located on both sides of the Devres River. Ṭosya is east of it and Ḳarğu is west of it. It is east of Ṭosya and west of Ḳarğu. It is a district near Ḳarğu.

Ḳarıbāzārı: A qadi district with several villages located on a flat area. It is on the western edge of Kös Ṭağı and south of Boğaz. It holds a Friday market. It comprises a number of villages. There is a Friday mosque in the middle of it. The source of the Devres River is in this qadi district. Şaʿbānözi is to the southeast, Ḳorubāzārı to the northeast.

Şaʿbānözi: A qadi district near Ḳalʿecik. It is located east of Ḳorubāzārı.

Ḳorubāzārı: A qadi district comprising a number of villages.

The area between these three qadi districts in front of Ḳarıbāzārı is full of hills as far as Ḳalʿecik. It is near Ḳalʿecik.

Ḳaracavīrān: A qadi district in an uncultivated area on the main road west of the qadi district of Bucurava. It has a Friday mosque, public bath, and inns. It is surrounded by farms and arable lands.

Ḳalʿecik-i Keskin: A qadi district comprising 30 villages located in a place of valleys and hills south of Ḳalʿecik. Most of the people are Turkmen. The Ḳızılırmaḳ flows between this place and Ḳalʿecik. It has excellent muskmelon and watermelon. [M 648] However, because this is a mountainous place, it is difficult to save the melons from wolves. One of the villages of this qadi district is called Ḳarḳara. Travellers going east from Ankara sometimes lodge here. It has a ruined fortress on this mountain that is difficult to reach.

A few other features: In this sanjak, four hours east of the town of Çankırı, is a village called Ṭuzlu. On a small hill near it there is no trace of any plant. The soil is brackish and the salt prevents (everything from growing). Wherever one digs a saltmine appears. It is an unrivaled salina. Most of the people in this sanjak are Muslim Turks. They earn their living from vineyards and raising mohair goats. Excellent Moroccan leather is produced in this sanjak. There are tanneries in some towns. They make red Moroccan leather. They raise thoroughbred horses and trade in them.

Mountains

Kös Ṭağı divides this sanjak in two. It extends from east to west. It has pine and juniper trees and abundant streams.

Ilḳas Ṭağı (Ilgaz) is along the northern border of the sanjak. It resembles Kös Ṭağı.

Rivers

The Devres River rises at the far end of Kös Ṭağı near Ḳarıbāzārı. It generally flows from west to east. It goes through the Çerkeş plain and then passes near Ḳurşunlu, Ḳoçḥiṣār, and Ṭosya. In the region of ʿOs̱māncıḳ, between Ḳargı and the ruined town of Ḥācı Ḥamza, it enters the Ḳızılırmaḳ, which at that point is flowing from the east.

Chapter: Description of the Sanjak of Ḳasṭamonı

Borders: In the east, Canik; in the north, the Black Sea coast; in the west, Bolu; in the south, Çankırı.

Its qadi districts are the following: Araç, Azṭavay, İstefan, Eflani, Daday, Aḳḳaya, Aḳyörük, Ayandon, Boyalı, Boyābād, Burma-Botan, Çañlıḳ also called Çañlu, Ḥadde, Çeklene, Ḥoşalay, Devrekani, Zarı, Zerdevay, Sāḥil-i Sinop, Şarban, Sarāy, Şart, Şorḳun, Ṭaşköpri also called Aḳyörük (sic?), Yörük, Ṭurağan, Ḳızılağaç, Gerze-i Ḳasṭamonı, Küre-i Ḥadīd, Kūre-i Nuḥās, Kinavul, Gökçeağaç, Göl, Köni, Mergüze, Yortan, and Yörükān-ı Araç.

Description: Ḳasṭamonı is the seat of the province. It has markets, Friday mosques, town quarters, and inns. In Kūre-i Nuḥās, which is the district of Ḳasṭamonı, the people are primarily engaged in making (copper) utensils that are exported in all directions. Sinop is four stages to the north and Ankara is five stages to the southwest.

On the west side of Ḳasṭamonı is a town quarter known as Ḥiṣārardı. The neighborhood mosque of Seyyid Sünnetī Efendi, now known as the Friday Mosque of Şaʿbān Efendi, is there. The shrine of that saint (Seyyid Sünnetī) is a place of visitation. Near that quarter is the neighborhood mosque of Cemāl Agha. Another is the Friday Mosque of Ghāzī Atabeg, which has a madrasa. In the Gökdere quarter is the Friday Mosque of Khwānsālār.

This city is the capital of the Turkmen. It is reported that while Constantinople was in the hands of the infidels the Turkmen used to attack it.

This city has a small fortress on one side of it on a very high cliff. The road to it is difficult.

Ṭaşköpri ("Stone Bridge"): A small town without a wall, and a qadi district. Boyābād is to the northeast. Ṭaşköpri is one stage from Ḳasṭamonı. It has Friday mosques and markets. A small river flows northwest of it. There is a stone bridge over the river; and because it is an old structure, the town is named after it.

Boyābād: This town (and) the district of Eymir are on the Sinop road, also on the road from Canik to Ḳasṭamonı. It has a Friday mosque and a market. [M 649] A bazaar is held there every week. The district of Boyābād is covered with rice paddies. Ṭaşköpri is one stage southwest of here.

Sinop: Off the Black Sea coast is a small square-shaped island connected to the mainland by a slender isthmus of sand. Sinop is a fortress and town on that isthmus. Someone who comes to Sinop enters from the land gate, passes through the town, then enters the island via the gate on the other side. The city is surrounded by sand. A mountain called Bozdepe encompasses the island. It is a wonderful place of excursion and there are streams. At the summit of Bozdepe, at a place where nothing overlooks it, a fine spring bubbles up. It is a marvelous thing. There is also a lake on this mountain. Most of the Black Sea coast at Sinop is composed of rocky precipices.

Sinop is four stages west of Samsun. There is a post for a molla paying 500 *aḳçes* per diem. It has abundant vegetable gardens and fruit.

The Old Friday Mosque has a *minbar* whose floor, ceiling, steps, banisters and door are carved from a single piece of white marble, with hadiths carved in relief. It is a marvelous work of art. On one side of the floor is a crack more than a cubit long, which occurred in the following manner: The late Sultan Süleymān wanted to remove this *minbar* to a Friday mosque that he had built in Constantinople. When he started to pry it from its place, it suddenly cracked, so he decided to leave it where it was.

Sinop is located on a flat place that juts into the Black Sea. It is nearly square in shape. It has twelve quarters, six public baths, markets, and four gates. One of the gates opens to the mainland. The city has the sea on two sides. One reaches it by a sandy road. Its breadth is one mile (text: *mīlden*, error for *mīldir*). The eastern gate opens to the harbor. It is a seaport. About a thousand ships can anchor in the harbor. The northern gate opens to the plain between the fortress and the sea. The aforesaid island is connected to the fortress. From one end of the fortress to the other, the circuit of the island is nine miles. The upper part (of the island) is flat where there are vineyards and gardens. The city's water comes from this side (of the island); it is a wide field. The western gate opens to the sea; but there is no harbor on this side. The citadel is large, high, and difficult of access. It is reached by a drawbridge and is on the mainland side. ʿAlā'eddīn Cāmiʿi is a delightful Friday mosque.

Kūre: A day's travel west (rather, north) of Ḳasṭamonı. It is located on a high place in an area of valleys and hills in the foothills of the mountains of Kūre-i Nuḥās. It has four Friday mosques and numerous public baths. It is four hours east (rather, south) of the port of the town of İnebolı.

İnebolı: A town 100 miles west of Sinop. It is a port and is administratively attached to the qadi district of Kūre. It is located in a valley between two mountains, and a river flows next to it. Behind one of the mountains is a ruined fortress that had previously belonged to the infidels. After 1020 (1611) an Ottoman emir built a fortress and citadel in this town out of fear of the Cossacks. However, the river that flowed by it undermined its walls and destroyed it.

Araç: It is one stage west of Ḳasṭamonı and in a mountainous area. A bazaar is held there every week. It has a Friday mosque and an inn. A large stream called the Araç flows next to it and empties into the Black Sea. On the bank of this river a stream of salt water emerges (from a spring). It is extremely hot.

Boyalu: (A qadi district) to the right of Araç and adjacent to it. A bazaar is held there every week. [M 650] It is composed of a number of villages and there is a Friday mosque. Next to it is a large river known as the Milan. The Mendik, rising from Mt. Külebi, flows into it here. Another river called the Dönerek rises on this side of the mountain and also flows into the Milan; the confluence is where the bazaar is held.

Ṭurağan: A qadi district in the district of Boyābād. This town is west of Boyābād and north of Aymaḥiṣārı.

Ḥiṣār-ı Ayma (Aymaḥiṣārı): A town north of Ṭurağan.

Sarāy: A qadi district in the mountains facing Sinop, with unrivaled summer pastures and streams. It has apples and mountain fruits. It has pine-covered mountains. Most of the pitch collected by the state comes from this qadi district.

Çañlu: A qadi district in the mountains behind Sarāy. It is a place of pine-covered mountains and streams.

Ayandon: A large village on the seacoast west of Sinop. Most of its inhabitants are infidels (i.e., Christians).

Mt. Maʿdin-i Nuḥās (or Baḳır Maʿdeni Ṭağı, "Copper Mine Mountain"): Its circuit is 30 miles. It is many minarets in height. The ground is red soil that looks like it has been burnt. There is vegetation in some places. If one digs into the ground, he extracts copper ore. People collect piles of ore, called columns, and set fire to them. After each has burned by itself for twenty days it turns to liquid sulfur. This product is an independent tax farm with a superintendant assigned to it. The ore is made into ingots and remains in its place. The unrefined metal, known as black copper, is put in furnaces with huge bellows called *zenberek* and smelted for 24 hours. The copper in each *zenberek* is made into ten-batman ingots that must be very highly refined. These are transported to a special large furnace belonging to the state that has a capacity for 3000 batmans. After further smelting and refining, the copper is stirred with green willow poles. Then the molten copper is reheated in that furnace and it finally becomes pure copper. One thousand batmans go to the operators of the furnaces, a thousand to the state, and the remaining thousand to the partners (in the mines). Every year 10,000 batmans of copper are subject to the tax farm—each batman being nine okas—3000 for expenditures and the rest paid to the state treasury. Today the amount going to the treasury has been reduced to 4,000 batmans.

At a certain place on this mountain they dig deeper and deeper into the ground. They say that the best ore is at the bottom of the mine. As they continue digging they bring in special carpenters who install wooden scaffolding to that the tunnel does not collapse. They make a road about 400 fathoms under the ground and in this fashion extract the ore. Since the ore mined in this mountain is found in veins, after it is exhausted in one vein it must be extracted from another. If water emerges where they are digging underground, there are special workers assigned to remove it from the mine. Every fathom there are two men who take the water in large ladles and pour it in a basin one fathom above them. Thus, step by step, the water is brought to the surface. It is a big job. Someone named Meḥmed Pasha, who was the superintendent of the mine, had a tunnel excavated under the mountain as far as the bottom of the shaft and made a drain for the water, thus solving the problem. The water from the shaft flowed through that tunnel and out of the mine. However, [M 651] in 1070 (1659), an earthquake occurred and the shaft completely collapsed. So a new tunnel was begun in another place.

It should be known that the odor (emanating from the mine) is that of a rotting corpse. Most of those who work in it are slaves. Their owners bought them when they were young and left them here. Nevertheless, they look like the dead. They have irons attached to their hands, and if one of them wants to wipe his brow he lacerates his skin. There is a special public bath at the edge of town where they go every day to wash, which eliminates the unpleasant odor.

Sometimes steam issues from the mine. They utter a prayer and it dissipates. Each worker has an oil-burning iron lamp to light his way going in and out the mine. They obtain much sulfur from this mine.

Mt. Sinop: It is on the Black Sea coast, just west of Sinop. The local people refer to it as Bozdepe. It is a mountain with two towers and is devoid of plants. Those coming by sea from Constantinople who want to enter Sinop harbor do not enter the harbor at the foot of this mountain but keep clear of it.

Chapter: Description of the Sanjak of Bolu

It is a sanjak of middle rank in the province of Anaṭolı. Its inhabitants are the best of the Turks and the nearest to humanity.

Borders: To the east, Ḳasṭamonı; to the north, the Black Sea; to the west, Ḳocaeli; to the south, Ḫüdāvendigār.

Its qadi districts are the following: Üskübi, Efteni, Eflani-i Bolu, Aḳlağan, Aḳçeşehir, Aḳṭaş, Amaṣra, Ulaḳderesi, Oniki Dīvān, Ulus, Ovayüzi, Pavlu, Benderkili, Burdur, Perşembe also called Zerzene, Bartın also called Oniki Dīvān, Tefen, Todurġa, Çağa, Çarşamba, Ḥiṣāröñi, Dört Dīvān, Derkene, Devrek, Zerzene, Safranbolu, Sarāy, Şehābeddīn, Ṣamaḳo, Ṭaraḳlıborlu, Ṭaraḳlı Yeñice, Ḳıbrısçıḳ (text: Ḳıbrısçık), Ḳazbel, Ḳoñrapa, Leçnos, Gerede, Gökçeṣu, Nakilbāzārı, Mudurnı, Mengen, Vīrānşehir, Yedidivan, Yılanluca, Yeñice-i Bolu, Yörükān-ı Bolu and Yörükān-ı Ṭaraḳlı.

Description: Bolu is the seat of the province. It has no city wall. It has flourishing markets, numerous Friday mosques, public baths, madrasas, and inns. It is six stages from Istanbul. It comprises 32 villages.

There is a kind of hazelnut here called *festī*. Its tree is like the chestnut tree. It is very fine; it tastes like an almond and looks like a hazelnut.

This city is located on a flat plain. There are mountain barriers surrounding it but with passes in the east and west.

It has three public baths and four Friday mosques, those of Şemsī Pasha, Ḳaraçayır quarter, Debbāğlar ("Tanners"), and Gölboru quarter.

On the southern side of Bolu are two natural hot springs. The one reserved for men has two pools while the one for women has one pool. A small river comes from the summer pastures of Mudurnı, flows between the qadi districts of Gölbāzārı and Ḥiṣārōñi and enters the sea. Behind the Gölboru quarter is a small lake where orrisroot, the root of a reed-like plant, grows and can be harvested.

There are two springs near Bolu. In one the water freezes and turns to stone. In the other the water melts stone and dissolves it. Therefore they put wooden troughs in that fountain. There are no grapes in Bolu. [M 652]

Gerede: It is two stages east of Bolu, between Vīrānşehir and Bolu, at the southern foot of Araḳ Ṭağı. It is on a dry flat plain devoid of trees and plants. It has four quarters, two Friday mosques, a public bath, a market, and numerous inns. (The qadi district of) Gerede comprises 78 villages. The river that comes from Alaṭağ flows through Gerede. It has two small lakes: Ḳara Göl ("Black Lake") is west of it and Ṭuzlu Göl ("Salt Lake") is east of it. Good shoe leather is made there. Most of the comb sellers in Istanbul come from Gerede.

Vīrānşehir: A qadi district on a plain east of Gökçeṣu and Sazaḳ in the district of Bolu. A bazaar is held there each week. It has a summer pasture called Mt. Araḳ and a river that comes from there. It produces saffron and unrivalled honey.

Mudurnı: A qadi district that has frequently held the status of *mevleviyet* (having a molla or chief qadi). It has eleven quarters, two public baths and three Friday mosques. It is five stages from Istanbul and one stage to the south and slightly to the west of Bolu, in a valley between two high bare mountains. In breadth it covers that valley and clings to the skirt of those mountains. In length it stretches about one mile. A river flows through the middle (of the valley). Its mouth is near Ṣapanca where it joins the Ṣaḳarya River.

There is a bridge in the town. The houses in the town are made of timbers and their ceilings are covered with wooden boards. One of their public baths is double (i.e., for men and women) and the other is single. One of its Friday mosques was built by Aṣīl Bey and the other was built by Sultan Yıldırım Bāyezīd (r. 1389–1402). In the middle of town is a madrasa that was also built by Sultan Yıldırım Bāyezīd. There is also a Friday mosque built by Sultan Süleymān called Yeñi Cāmiʿ. Two miles from town on the Istanbul road is an inn known as Dibek Ḫānı, built by Rüstem Pasha (Ottoman grand vizier, d. 1561). And there are two other inns in the town.

Fields and pastures are on the plains in the direction of Istanbul; vineyards and gardens are in the valleys behind the town on its east side. There are numerous fountains for drinking water in the town. Because its river is dirty and the town and its districts are in forested and mountainous places, most of the inhabitants have a malicious disposition. The industry of most of the people is making needles, which is how they earn their living. Most of the needle makers in Istanbul come from Mudurnı.

Two mountains overlook this town. On one of them, northeast of town, there is a fortress that overlooks the town. It is walled on three sides, but on the fourth there is a precipice where no wall is needed. There are houses and a Friday mosque inside. The road from the town to the fortress is difficult to traverse. There are two districts (*nāhiye*) in this qadi district called Doḳurçin and Ḳozyaḳası.

Amaṣra: A strong fortress on a lofty hill on the Black Sea coast. It has a Friday mosque and a warden. On the east and west sides are protected harbors. On the shore opposite the eastern harbor, the town has a small and dirty public bath called Bozḫāne. (This qadi district) comprises 15 villages. Sinop is five stages east of Amaṣra by land, and about 100 miles by sea. Amaṣra is four stages (east) from Eregli by land, and more than 50 miles by sea.

Ḳıbrısçıḳ: A qadi district on Alaṭağ comprising 24 villages. North of it are the summer pasture of Sulṭānoğlı and Bolu. This qadi district is reached by passing through the qadi district of Pavlu one stage east of Mudurnı. Rice is grown in the place called Çeltik Deresi ("Rice Paddy Valley"), which is attached administratively to this qadi district. [M 653]

Pavlu: A low-ranking qadi district at the foot, and southwest, of Alaṭağ. It comprises ten villages. The ground is rocky, so there is little agriculture. However, there is excellent honey. There is a hot spring here. Dāwūd Ḳaramanī, the author of *Gülşen-i Tevḥīd*, is buried near here; his grave is a place of visitation.

Dört Dīvān: A qadi district four hours west of Gerede. It comprises ten villages at the foot of Alaṭağ. It has good pasture. As Sultan Süleymān passed through here on one of his campaigns his sciatica flared up. (He stopped here and) his governing council (*dīvān*) met for four days, so this place was called Dört Dīvān.

Çağa: A small town in a valley north of Alaṭaġ. It has a Friday mosque and a public bath. Between it and the foot of Alaṭaġ is a lake and pasture. This lake is smaller than Lake Ṣapanca. It has a lot of fish. Its people make bows. The bows of Çağa are famous throughout Turkey. Left of Bolu on Mt. Gökçe is Mengen Bāzārı, which is attached administratively to Çağa. From here one crosses the Zevilce summer pasture and reaches its lakes.

Mengen: A qadi district comprising a number of villages north of Çağa. Its bazaar (Mengen Bāzārı) is held next to a stream in a village with five or ten houses. This stream, which descends from Mt. Araḳ, has excellent trout. A kind of plum called *mürdüme* grows here that is very tasty and has a small pit.

Todurğa: A qadi district comprising twenty villages adjacent to Ṭaraḳlı and west of Bolu. Mudurnı is located between it and Bolu. Mountains are north of it and a plain is south of it. It has a famous valley called Nala Tarlası. Its most famous commodities are spoons and combs.

Efnanlu: A qadi district comprising two villages next to a lake. The Milan River flows into this lake, then out the other side and into the sea.

Zaʿferānborlu (Safranbolu): This town has twelve quarters, four Friday mosques and two public baths. It also has 50 villages. Ḫancer Ḫoca (error for [Ḫüseyn] Cinci Ḫoca, d. 1648, exorcist and spiritual advisor of Sultan İbrāhīm)built an imposing inn here.

Ḳızılbeli: A qadi district comprising 12 villages attached administratively to Bolu. It has good vineyards. One side of it is on the Filyas River. It has a summer pasture called Deremahyası.

Ḳoñrapa: A qadi district south of Üskübi and one stage west of Bolu. It comprises 20 villages. It has a number of inns and holds a weekly bazaar. Its (main) products are rice and fine clotted cream. The river called the Uğru, which is south of it, flows into Lake Efnanlu. The place called Düzcebāzārı consists of inns on the main road to Ḳoñrapa and a number of villages in the mountains around it. A bazaar is held next to these inns. Its rice paddies are irrigated by water from the Milan River. The inhabitants raise water buffalo. The rice is red and of low quality. In the mountains in the direction of Bolu and Mudurnı are lynx and beach martin. There are also woodcocks in the mountains. They look like cocks but are larger.

Benderkili (Eregli): A fortress north of Bolu on the Black Sea coast. (Its qadi district) comprises 20 villages. Its harbor is in front of the fortress. It is mostly surrounded by mountains; there are few plains. Its water, a stream that needs a boat to cross, comes from Mt. Çile. North of it is the tomb of Ḥācı Baba, which is a place of visitation. They make good cloth. Their products are timber and fruit.

Ṣamaḳo: A qadi district west of Benderkili, to which it is attached administratively. It comprises 20 villages.

Devrek: A qadi district east of Eregli and inland from the sea. It is north of Mt. Çile. Bolu is south of here. Devrek has a number of inns and shops, and a bazaar is held here every week. This qadi district also has many villages. The Bolu and Mengen Rivers flow nearby and meet the Araç and Vīrānşehir [M 654] Rivers in this qadi district, then enter the sea at Ḥiṣāröni. A bazaar is held here once a year for fifteen days. Yedidivan is in this qadi district.

Derkene: A qadi district south of Devrek comprising eight villages. It is a dry mountainous place. This qadi district is also called Sekiz Dīvān.

Yılanlıca: A qadi district north of Devrek comprising nine villages. It is located in the mountains near the Black Sea. It has excellent fruit. The Filyas River flows through the lower part of this qadi district and enters the sea. There is a place of visitation here named after Abdāl Pasha.

Pencşenbe (Perşembe): A qadi district comprising 20 villages. The Filyas River separates it from Çarşamba.

Gölbāzārı: A qadi district on the Black Sea coast stretching toward the north from Perşembe and comprising 28 villages. The people raise chickens for the state.

Ḥiṣāröni: A qadi district east of Gölbāzārı comprising four villages. The Filyas River flows between the two and enters the sea.

Yeñice: A qadi district two stages east of Bolu comprising twelve villages. It is a dry mountainous place. Mengen is between Yeñice and Bolu. The Filyas River flows through the middle of this qadi district.

Bartın: A qadi district three stages north of Bolu near the Black Sea comprising twenty-four villages. A bazaar is held here every week. A river formed by the confluence of the Derbend, Ova and Ulus Rivers flows through this qadi district and enters the sea. Because it is very deep, ships can sail up its mouth for about two miles and load timber. Mostly galleons are built here. This place is (also called) Oniki Dīvān.

Üskübi: (A qadi district) ten hours north and somewhat west of Bolu. It is south of Aḳçeşār. It has a Friday mosque, public bath, and ten villages. The Debbāğlar Ṣuyı comes from Mt. Çile north of Üskübi and passes through it. The area to the west of it has mountains, forests, and plains. Its arable fields and rice paddies are irrigated with water from the Milan River. It produces low quality rice but excellent cloth. Şemsī Efendi is buried in the village called Çile. His grave site is a place of visitation.

Aḳçeşār: A qadi district north of Bolu and Üskübi. It is six hours from Üskübi and comprises twenty villages. Mt. Çile is located between it and Bolu. This qadi district is located in mountains that have acorn, chestnut, and apple trees. The sea is to the north. Its port is the village of Kerāmeddīn where there is a weekly bazaar on Friday. The port is at the mouth of the Milan River. The qadi districts of Üskübi and Düzce are to the south. There are excellent cherries. Its (main) product is timber.

Gökçeṣu: A district comprising twenty villages in the mountains and plain east of Bolu.

Çile Yaylaḳı: A splendid summer pasture. The Bolu River flows through this district.

Seraḳ: A district east of Bolu and adjacent to Gökçeṣu. It is separated from Gökçeṣu by the Bolu River. It is an area of mountains and plains.

Alaṭağ Yaylaḳı: A district comprising twenty villages. In the village called Tekye is the tomb of a saint known as Emir Sinān. It is a place of visitation.

Aḳṭaş: A qadi district east of Vīrānşehir and composed of six villages. It has areas unfit for cultivation, mountains, and plains. The river of Vīrānşehir flows through here.

Ulaḳ: This (qadi district) is east of Aḳṭaş and comprises ten villages. The river that descends from Alaṭağ irrigates this place.

Zerzene: A qadi district south of Bartın and attached administratively to it. Zerzene comprises seven villages. A bazaar is held every week. Its (main) products are timber and fruit. It is also called Yedidivan.

Ulus: A qadi district south of Amaṣra comprising fifteen villages. There is a market every Friday. The sanjak of Ḳasṭamonı is east of it. It has rocky places that are difficult to traverse.

Ova: A qadi district comprising ten villages in a valley. The border with Ḳasṭamonı is to the east. [M 655] A dry, waterless, rocky place south of Ulus. Most of the people use hand-turned mills.

Eflani: (A qadi district) comprising a number of villages. On the west side it borders on Ḳasṭamonı. There are two Eflanis. The (main) product of this one is grain. There is a weekly bazaar. It has inns.

Tefen: A qadi district near Vīrānşehir. It comprises three villages.

Gücenes: A qadi district adjacent to, and south of, Zerzene. It comprises seven villages. Its (main) products are timber, fruit, and grain. Nearby is a mountain called Ḳırḳ Ṭağı in which is a huge cave known as Güngörmezler Ġārı ("Cavern of those who never see the light of day"). According to the common folk, a certain people live in this cave who die if they see the sun; they only go out at night in order to get provisions. At one time this cave was inhabited. There are remains of habitation—houses, fountains with flowing water, and shops for various crafts. In fact, there are tools and equipment in the shops that show the professions of the craftsmen. All the houses and shops are carved out of stone, and there are corpses in stone coffins. One goes up to this place by ladders. If one goes a certain distance into the cave, one sees paths going off to the right and left. Several times (the villagers) took wind-tapers and followed some of these paths. They emerged after two days and nights without finding an end of the paths. It is estimated that the cave reaches as far as Kūre-i Nuḥās in the mountains of Ḳasṭamonı.

Trees, rivers, and mountains

This province is full of pine, juniper, acorn, plane and hornbeam trees. It has many low and high mountains and extensive forests in some places. There are fruit trees on some of the mountains. Hazelnuts and chestnuts are especially abundant. The Bolu Mountains in Anatolia are proverbial for their height. The most famous of them is Alaṭağ.

Rivers

Milan (Melen): It descends from the Hermerur and Çile mountains and flows toward the east through a rocky river bed in the sanjak of Çankırı, then separating the qadi districts of Düzce and Üskübi and into Lake Efnani. It continues flowing out of the lake, separating the sanjaks of Bursa and İznikmid, then westward to the sanjak of Bursa and Aḳçeşār. It enters the Black Sea branching into several mouths, some of which are navigable for one or two miles and for ships to load timber.

Vīrānşehir: It descends from Mt. Eren, flows westward past Aḳṭaş and Vīrānşehir and, in front of Ḳarabük Tekesi below Borlı (Safranbolu), meets the Araç River coming from Ḳasṭamonı. Both are large rivers. In the qadi district of Tefen, which is one or two stages below it, this river joins the Bolu River and the confluence enters the Black Sea.

Gerede: It descends from Alaṭağ, passes through the qadi districts of Şehābüddīn and Ulaḳ, separates the sanjaks of Ankara, Çankırı, and Ḳasṭamonı from each other, and joins the Araç River. On one side is the sanjak of Bolu and on the other the sanjaks of the others.

Filyas: It flows through the middle of the qadi districts of Ḳızılbel, Yeñice, and Perşembe and then separates the qadi districts of Gölbāzārı and Ḥiṣāröñi and enters the Black Sea.

Derbend, also called Ova or Ulus: It passes through the qadi district of Bartan (Bartın) and enters the Black Sea. Its mouth is deep enough for ships to navigate three miles inland and load timber and other goods.

Mudurnı: It flows into the Ṣaḳarya near the district of Ṣapanca [M 656] called Aṭa (modern Adabazarı), a little below the wooden bridge which is over the Ṣaḳarya.

Those going to Bolu pass through the district of Aṭa and Mudurnı and cross the wooden bridges above them. Lake Çaġa is full of fish like Lake Ṣapanca. Lake Efnanlu (text: Efnaylu) is between Üskübi and Ḳoñrapa (text: Ḳoñraya). The Milan and Uġrı Rivers flow into it, then flow out of it together and enter the Black Sea. There are bridges over these rivers.

Mountains

Alaṭaġ is a great mountain that never lacks snow. From its towers and heights Keşiş Ṭaġı (modern Uluṭaġ) can be seen. Medicinal plants including China-root (*similax china*) are found there. Unrivaled streams are on this mountain. Beybāzārı is south and Bolu is north of it.

Chapter: Sanjak of Ḫüdāvendigār

It received its name as follows: Sultan Orḫān first conquered Iznik and made it his capital. His worthy son Sultan Murād (I) was known as Ghazi Ḫüdāvendigār. Orḫān gave him the sanjak of Bursa. Later, however, Orḫān Ghazi moved to Bursa and made it his capital. However, because it was originally the sanjak of Ḫüdāvendigār it continued to be known as the sanjak of Ḫüdāvendigār.

Borders: To the east, Kütahya; to the south, Manisa and Ḳaresi; to the west, the Gulf of Mudanya and Ḳocaeli; to the north, Bolu and Ḳocaeli.

Its qadi districts are the following: Abgi, Ermeni, Aṭranus, Aḳyazı, Aḳyazı-i Yörük, Ulubad, Edincik, Ilıca, İnegöl, Baḳır, Bergama, Bursa, Beybāzārı, Boyalu, Bayramıç, Tarḫala also called Ṣoma, Harmancıḳ, Sivriḥiṣār, Sögüt, Sincan, Ṣarıçayır, Ṭaġardı, Ṭaraḳlı, Ṭaraḳlı Yeñicesi, Ṭomaniç, Ḳarıbāzārı, Ḳurubāzārı, Ḳaraḥiṣār, Naʿllu, Ḳızılca Ṭuzla, Kebūd, Kite, Kirmasti, Gümüşābād, Gemlik, Gökçeṭaġ, Günbāzārı, Göynük, Günyüzi, Lefke, Manyas, Marmara-Şāmī, Mudanya, Miḫaliç (recte Miḫaliç), Miḫalcıḳ, the districts of Bergama, Yārḥiṣār, Yeñişehir, and Gölbāzārı which is near Lefke.

Description of Bursa

It has a post for a molla paying 500 *aḳçes* per diem. In the early years of the Ottoman conquests, it was the capital of the first sultans. It has a citadel on a lofty hill. Most of the urban space is outside the city wall, extending one parasang lengthwise and about a half parasang wide, north of the foothills of Keşiş Ṭaġı (Uluṭaġ). West of the area around the Friday Mosque of Murād Ḫān, known as the Emir Sultan quarter, as far as the border of Murādiye and Yeñi Ḳapluca is one continuous urban area. Two or three miles beyond this, in a separate low-lying area, are the quarters of Ghazi Ḫüdāvendigār and İki Ḳapluca. The city of Bursa rises level by level from the foot of the mountain to the height of the citadel. Toward the east and north is a broad plateau.

Behind the city wall is a place of excursion known as Pıñarbaşı. At the foot of the mountain behind the citadel a stream emerges from under a rock. It passes through one or two meadows and a flat plain and then reaches the fortress. From the citadel it flows down to some of the houses and town quarters.

Another river that descends from the mountain is Gökdere Ṣuyı. The town quarters east of the citadel get their drinking water from it. Water from this river is distributed throughout the city. It is a delightful place.

Another river is the Aḳçaġlan Ṣuyı. In the city it meets the water needs of the Emir Sultan quarter.

[M 657] The people of this province characterize the sweetness, lightness, or heaviness of these streams in various ways. Thus:

- A dog will never bite the leg of one who drinks the water of Pıñarbaşı.
- A thief will never enter the house of one who habitually drinks the waters of Gökdere Ṣuyı.
- One who drinks from the ʿAlīşīr will not escape stasis; in other words, he will become ill and need a staff in his hand and will suffer from coughing because of an excess of phlegm and will become gluttonous.

These rivers flow to the gardens located below Bursa and during high-water season join the Nilüfer River. In short, the water that enters the city of Bursa (from these streams) is divided and distributed. As in Damascus, water flows to most of the houses and markets.

Bursa has many trees and much fruit. It has abundant Judas-trees (*Cercis Siliquastrum*) in particular.

The pavilion of a certain ʿAbdülmüʾmin is a place of excursion located right at the edge of the city. The Gökdere flows through here. This river runs one or two mills. There are delightful places to relax.

In the Çamlıca and Fıstıḳlu mountains are pine trees.

Silk worms are raised in Bursa and silk is produced of a very high quality. There are shops making a special kind of pillow, mattress, and home-spun woolen cloth (*abai*). Velvet and brocade are also made here.

The chestnuts of Bursa are famous. They are in the foothills of Rāḥib Ṭaġı ("Monk Mountain" = Keşiş Ṭaġı / Uluṭaġ).

Another delightful place of excursion known as Abdāl Murād consists of an extensive open field with a natural dam and terrace at the spur halfway up the mountain overlooking the citadel and the plain (*ova*) of the city. It has a pavilion, shady groves, streams, and the tomb and dervish lodge of Abdāl Murād.

There is another place of excursion called Ḳaranfilli at the spur halfway up the mountain overlooking the city. The ʿAlīşīr River flows through here. They say that it is possible to see all of Bursa from here. Bursa is 61½ miles from Constantinople as the crow flies. By sea, however, Constantinople is about 80 miles from the port of Mudanya, which is about four hours from Bursa.

In 726 (1326) Sultan Orḫān son of ʿO̱smān Khan besieged Bursa and it surrendered. There are guards and others living in the aforesaid citadel. In the middle of the city is an old Friday mosque belonging to the conqueror of the city, Orḫān Ghazi, as well as a soup kitchen and a monastery (remaining) from the Christians that was turned into a madrasa. The dervish lodge and all the Friday mosques above the martyrium of Geyikli Baba (fl. early 14th century) are all attributed to Ghazi Ḫüdāvendigār (Sultan Murād I). Sultan Murād II has a Friday mosque inside the city wall near Sarāy Ḳapusı (Palace Gate) and another Friday mosque, a lofty madrasa, and a soup kitchen near Eski Ḳapluca (Old Hot Spring). Yıldırım Bāyezīd has a Friday mosque, madrasa, soup kitchen, hospital and dervish lodge. Çelebi Sultan Meḥmed (I) has a Friday mosque plus tomb roofed with tiles (Yeşil Türbe), a kitchen known as Yeşil ʿİmāret ("Green Soup Kitchen") and an imperial madrasa. Sultan Murād II has a madrasa and a soup kitchen; also a Friday mosque and another soup kitchen near the martyrium of Shaikh Şemseddīn Seyyid Meḥmed Buḫārī, known as Emīr Sulṭān (patron saint of Bursa, d. 1429).

Hot springs

One is in the western part of Bursa, at one edge of the quarter of the Friday mosque of Ghazi Ḫüdāvendigār. This sultan (i.e., Murād I) had masonry domes erected over this hot spring and also constructed bathing pools. It is called Eski Ḳapluca (Old Hot Spring). Its water is extremely hot: if one puts an egg in the spring the water will cook it. The water is beneficial for mange and itch and the like.

[M 658] There is another hot spring called Yeñi Ḳapluca (New Hot Spring) between Eski Ḳapluca and Bursa. Rüstem Pasha built imposing buildings in the style of a public bath over this hot spring and covered the roofs with lead. The structure has two domes and a dressing room with a fountain. There is another fountain in the tepidarium spouting cold water. Inside, under the (main) dome, is a pool whose circuit is thirty cubits and whose depth is more than the height of a man. Hot water spouts from a lion's mouth made of marble. There are benches here and there around the pool where people sit. Hot water flows continuously. On either side of the door is an alcove with a small pool and privies. Because the design of this building makes for an exhilarating experience, people come from great distances to see it. The water in this spring is very hot.

The molla district (*mevleviyet*) of Bursa comprises two districts (*nāḥiye*): 1) Aḳṣu ("White River"): It has villages in its mountainous areas that are like delightful towns. It has a beautiful town located on its own plain one stage from Bursa on the Kütahya road. It has a river of excellent water called Aḳṣu. 2) Ova ("River Valley"): It also has a number of villages among which flows the Nilüfer River. This district is located on the plain in front of Bursa.

Kite: It is five hours south of Bursa and has a number of villages. Mudanya is opposite it (i.e., to the north) and there are some mountains between the two. Because the qadis (of Kite and Mudanya) have no particular town in which to reside, they live in Bursa. The main road from Mudanya to Bursa is level, but (Mudanya) is somewhat off the main road.

Adranos: A qadi district with a number of villages, one stage southwest of Bursa, on Keşiş Ṭaġı. Gökçeṭaġ is in the middle of it.

Ḫarmancik: A town and qadi district just to the west of Adranos. In the past it was the province of Miḫalbey and was called Ḫarmanḳaya.

Sincan: A qadi district west of Adranos and between the two (Ḫarmancik and Adranos?).

Mudanya: A town about 100 miles southeast of Istanbul, on the southern shore of the Gulf of Gemlik. It is the port for Bursa. It has a number of Friday mosques and public baths. There are special workers here who let themselves out for hire. They find as many packhorses as travelers want and send them to Bursa where their partners receive the packhorses and give them to travelers going to Mudanya. It is six hours from Mudanya to Bursa. Half way one crosses the bridge over the Nilüfer River. Mudanya has 24 villages and an important district called Tirilya. Its most important product is olives. Mudanya has many vineyards, but there are olive trees here and there. In some places there are excellent pomegranates. The soil is very fertile. In fact there are even lemon trees in some places. But the people are bad-tempered and quarrelsome. Half the townsmen are Muslim and half are Christian.

Gemlik: A town at the end of the gulf that is named after it. It has fine pomegranates. The entrance to the Gulf of Gemlik starts at Bozburun where there is a mosque and an inn. One can enter the gulf with a certain wind and then sail east into the gulf. Along the coast opposite Mudanya are Emrūdlı (Armutlu), Fıstıḳlı, and Ḳumla. Gemlik is the easternmost point. From there the shore curves westward (text: eastward) going past Ḳurşunlu, Altunṭaş, Mudanya and finally Tirilya, which is opposite Bozburun. Because the southern [M 659] shore tends toward the southwest, the mouth of the gulf is very wide.

Bāzārköyi: A town east of Gemlik and Engürücük and north of Yeñişehir. It is one stage from Ṣamanlu and is reached by a pass through the Ṣamanlu Mountains. It has many streams and places of excursion. It has Friday mosques, a public bath, and inns for travelers. A bazaar is held there every week.

Exactly between Bursa and Bāzārköyü is the town of Ḳaṭırlı, which has a Friday mosque and an inn. Engürücük is near the shore of the Gulf of Gemlik, between Gemlik and the village of Gencelü, which is near Mudanya. It has excellent grapes. Balabancıḳ is west of Kite on the Nilüfer River. Yarḥiṣār is a town with a Friday mosque and public bath located between Yeñişehir of Bursa and İnegöl. It is one stage from Yeñişehir on the Kütahya road.

Yeñişehir of Bursa: A town northeast of Bursa, southeast of İznik, south of Lefke, west of Aḳbıyıḳ, and in front of Bilecik. It is two stages from Bursa. This place is also called Yeñişehir Ovası (the Plain of Yeñişehir). In the early years of the Ottoman state, ʿOs̱mān Ghazi son of Erṭuğrul built a palace, quarters for soldiers, a Friday mosque, and a public bath in this city. He made it his capital and it was called Yeñişehir ("New City").

Aḳyazı: A qadi district one stage from İzmit and Ṣapanca on the road to Bolu. Its town is Ḥendek. It has a Friday mosque and inns. A bazaar is held there every week.

Miḫaliç: It is on the shore of a large river (Simav Çayı) that flows into the sea. It is at the outlet of Lake Ulubad. It is about forty miles inland from where this river enters the sea, a place called the Strait of Miḫaliç. It is a prosperous city on a rather high place about one mile from its port (on the river) and two hours northwest of the town of Ulubad. Ships and boats can enter the Strait of Miḫaliç and travel in fresh water for about forty miles and reach the quay of Miḫaliç. The area all around is full of willows and other trees.

Ulubad (Ulubat), also called Lake Ayandon: A town and qadi district one stage south of Bursa. The Ulubad River flows through here and enters the lake. The town is located on the lakeshore and has numerous inns. It is two hours from Miḫaliç. Some small boats can operate above Miḫaliç. The river coming from Kirmasti joins the Ulubad River between Miḫaliç and Ulubad. The town has a fortress dating back to Byzantine times; it is now in ruins.

Tarḫala: It is two stages south of Balıkesri. A bazaar is held there every Tuesday. The district of Baḳır is located south of here. Soma is near Tarḫala. It has two Friday mosques, a public bath, and flourishing markets. The qadi of Tarḫala resides there. Baḳır is south of Tarḫala. Bergama is two stages south of it. It has extremely good melons.

Bergama: A town one stage north of Ayazmend. It is located at the foot of a mountain five hours from the coast. It has a strong fortress on the top of a hill that is difficult to reach. The city proper is located below the fortress at the edge of a broad plain that spreads left and right toward the sea and toward the west. It has several Friday mosques, public baths, and markets. It has a town quarter known as Ne Yerde ve Ne Gökde ("Neither on Earth Nor in Heaven"). It received this name because it was built over ancient large masonry water channels, so the ground beneath it is hollow. Its ports are Çandarlu [M 660] and Ayazmend but it is closer to Çandarlu.

Kirmasti: It is near the qadi district of Ṣındırğı and is reached by the passes of Kebsud and Yılancıḳ Ṭağı.

Lake Ulubad: It is between Bursa and Miḫaliç. On the south side of the lake is a steep mountain. On the north side is the outlet of the lake from which its water flows into the Sea of Marmara. There is a daily tide. From dawn to noon this river flows upstream. Afterwards it begins to flow toward the sea again. The water does not flow into the lake, but the waves on the surface of the water move in that direction. The reason for this is that the flow of the water out of the lake bars the current of the river. On the western side in particular, contrary winds blow constantly toward Ulubad fortress. It is a wide lake that extends as far as the foothills of Keşiş Ṭağı to the east. There are bulrushes in the lake and many fish are caught there. Around the lake are Ṣındırğı and Kirmasti (to the south) and the village of Aṭa to the north.

Mountains

Mt. Yund: There are many ancient ruins on this mountain which overlooks Bergama. There are caves in the mountain in which there is something like sun-dried bricks. They are about one cubit square and are stamped in the middle. They are stacked up to one side of the cave. Greedy people who are beguiled by alchemy believe this is some kind of elixir and if they mix it with copper it will become exactly like silver. However, it becomes brittle like enamel but is not malleable.

Ḳaz Ṭağı (Mt. Ida): A famous mountain in the vicinity of Bergama.

Cebel-i Rāhib ("Mountain of the Monk"): This is known as Keşiş Ṭağı ("Mountain of the Monk" = Mt. Olympus,

Uludağ). Bursa is at the foot of this mountain. There are ponds on its summit. They freeze in the winter. Ice is brought from there to Bursa and then is sent to Istanbul. There is a stream at the top of the mountain. Trout are caught in it. These fish are usually found in cold water and in the rivers of summer pastures. They should be eaten as soon as they are caught or they will quickly go stale. This mountain extends east to west for about three stages. All kinds of medicinal plants are found there. There are streams in every spur of the mountain. The climate is excellent. In addition to forests, meadows are scattered here and there.

Remarkable facts:

- Some treasure-seekers say that there is buried treasure on this mountain and have dug holes everywhere.
- At the upper reaches of the mountain is a great plain like a circular pool (a crater?) known as Arḳandı Ḳaya. If you step on one corner the surface moves and the other side rises; if you take your foot away, it returns to the way it was.
- The summit of this mountain is very high and always covered with snow. The skirt of the mountain seems to extend to the sea.
- There are tiny worms known as *zülāl* that reportedly live in some of the old snow beds. People pluck off their heads and sip the delicious cold water inside.
- From the high elevations in Istanbul the towers on the summit of this mountain can be clearly seen like a white cloud at the edge of the horizon. The view is marvelous at sunrise.

From the two skirts of the mountain opposite Cebel-i Rāhib to the foothills of Cebel-i Rāhib is four hours. The area between these two mountains is the area between Bursa and Kite. It is a low mountain, but according to the local people it is part of a chain that extends as far as the mountains of Ḳaraman, Adana, Tarsus and Kelbeyin. The village called Eşkel, which is opposite Eski Ḳapluca and is a pious foundation, is located at a high place on this mountain. It is a village of dhimmis. [M 661] It is on a mountain opposite Bursa. In the past there was a fortress here that was conquered some time after Bursa was conquered.

Cebel-i Ṭomaniç: A great mountain between Kütahya and Ḫüdāvendigār. Originally ʿOs̱mān Ghazi son of Erṭuġrul would go up to this summer pasture with his people. They used to spend the winter between Yarḥiṣār and Bilecik.

The Nilüfer River: It rises from a branch of Keşiş Ṭaġı in the district of Yeñiṭaġ in the qadi district of Adranos. At most places it cannot be crossed. It flows through the middle of the Bursa Plain, passes about two miles north of the city, reaches the Kite Plain, then passes near Balabancıḳ and finally flows into the outlet of Lake Ulubad below Miḫaliç. It flows from east to west. Sultan Orḫān's wife, Nilüfer Ḫātūn, built a stone bridge over this river and the river was named after the bridge. There are trout in the upper reaches of this river. Most of the firewood for Bursa comes from Cebel-i Cedīd ("New Mountain" = Yeni Dağ; text: Cebdīd). The wood that is cut in the district of Adranos is put in this river with a distinctive mark. Then the partners of those who put the wood in the river retrieve the logs with their mark on them near Eski Ḳapluca and take them to the city.

Chapter: Sanjak of Ḳaresi

Borders: To the south, Manisa; to the west, the Mediterranean and the sanjak of Biġa; to the north and east, the province of Ḫüdāvendigār.

Its qadi districts are the following: Edremid, Ayazmend, İvrindi, Balıkesri, Başgelembe, Balya, Biġadiç, Ṣındırġı, Farṭ-Şāmlı, Feslekan-Ḳozaḳ, Ḳaracalar, Kemer-i Edremid.

Description

Balıkesri: This city is the capital of the province. It is four stages south of Bursa. A bazaar is held here every Tuesday. The district of Baḳır is located to the south of Balıkesri and near Bergama. In 737 (1336) Sultan Orḫān took this city by peaceful means from the ʿAjlānids. Balıkesri is located at the foot of a mountain two stages from Miḫaliç. Zaġanos Pasha (Ottoman grand vizier, d. ca. 1464) brought fresh water from this mountain to Balıkesri. The source of this water is the place of excursion for this city. The dervish lodge, Friday mosque, and some other charitable buildings of Shaikh Lüṭfullāh Bayramī are in this city. He is also buried here.

Ayazmend: A town on the (Aegean) seashore.

Edremid: It is located at the far end of a gulf and is one stage north of Ayazmend.

Kemer-i Edremid: A town one stage north of Edremid.

Mendehorya: A large village located on a flat place. It has two inns and a large river. There is a stone bridge over the river. It is three stages from Miḫaliç.

Biġadiç: A town. Most of its people have joined the service of İlyās Pasha.

Gördüs (Gördes): It is next to the qadi districts of Demirci.

Ṣındırġı: A qadi district that detached from Biġadiç. It comprises 50–60 villages.

The *ṣubaşı* of the sanjak of Ḳaresi resides in the village of Ḳāḍıbāzārı, which is three hours east of Biġadiç.

İvrindi: A town west of Balıkesri. Kemer is located south of it.

Başgelembe: A large village located in a region of valleys and hills. It is five stages from Miḫaliç. There are 20 villages around it. It has two inns, a small market, and two Friday mosques. A small river flows in front of it. That river-valley forms the border beween Ḳaresi on this side and Ṣaruḫān on the other side.

Chapter: Description of the Sanjak of Ḳocaeli

This place was conquered by a ghazi named Aḳçeḳoca who was one of the emirs of ʿOs̱mān Ghazi. It was thus named Ḳocaeli ("Province of Ḳoca") after him. The sanjak of Ḳocaeli as well as those of Biġa and Ṣığla are attached administratively to the province of the Ḳapudan Pasha (admiral of the Ottoman fleet). However, [M 662] because it is adjacent to the province of Anaṭolı, we are discussing it in this chapter. In the past the bey of this sanjak used to set out on campaign with one ship. Subsequently, one thousand large masts called *verdenarde* were cut here each year for the imperial arsenal.

Borders of Ḳocaeli: To the east, Bolu; to the north, the Black Sea; to the west, the Bosphorus; to the south, the Gulf of İzmit and the sanjak of Ḫüdāvendigār.

Its qadi districts are the following: Āb-ı Ṣafī, İzmit, İznik, Üsküdar, Aḳyazı, Aḳhiṣār, Bāzārköyi, Ṣarıçayır, Ṣapanca, Geyve, Yalaḳābād, Görele, Yoros, Ḳaramürsel, Ṣamanlu, Ḳaraṣu, Ḳandırı, Şile, and Bāzārṣuyı.

Description

İzmit: A prosperous town located at the end of a gulf one hundred miles east of the Sea of Constantinople (Bosphorus). The town has no wall. Its harbor is well known. It has Friday mosques and inns, an imperial park and palace surrounded by gardens. Because this is the capital of the sanjak of Ḳocaeli, the sanjak bey resides here. The Gulf of İzmit is near Fenar Bāġçesi. It draws in to the east and continues in this fashion.

Pendik and Ḳartal: They are villages on the shore of the Gulf of İzmit.

Ṭuzla: A large village of infidels built on the side of a hill next to the sea. It comes right up to where the shore makes a long wide bend.

Zeytūnburnı: It is located below Gebze.

Dil: At this place (the spit that juts into the gulf from the southern shore) the gulf contracts to a narrow strait from a flat sandy place between two shores. İzmit is at the end of the gulf, which here bends like a bow and then (the shoreline) turns back toward the west. The shores are situated in this manner.

Ṭarlu: An infidel village located at one corner of the promontory known as Bozburun. It is located opposite Fenar Bāġçesi. They are forty miles apart.

As the shore turns westward it comes to Ḳaramürsel. One or two miles inland from the harbor at Dil are the inn, Friday mosque, and public bath of Hersekzāde Aḥmed Pasha (Ottoman grand vizier, d. 1517) and many other inns. This place is called Hersek. Travelers pass Dil and spend the night in this town. This was where the aforesaid Aḥmed Pasha established lodgings.

At Dil there are special small boats that transport travelers. There are frequent storms here. From here to the opposite shore the distance is about five miles. On the opposite side are, in turn, the qadi districts of Yalaḳābād and Görele, then Ṣamanlu and Ḳatırlu. They are about a parasang away.

The area around the gulf is generally mountainous. Some bath keepers cut wood from (the mountains around) these shores for public baths and send it to Istanbul. They are fairly big mountains, known as the Ḳatırlu and Ṣamanlu mountains. The road from Ṣamanlu to Bāzārköyi goes through these mountains.

Görele: A town and qadi district three hours from Bāzārköyi and near Gemlik.

Gegbuza (Gebze): A town between İzmit and Constantinople. It is located on a high flat ridge overlooking the Gulf of İzmit. It is about two parasangs from the sea. Muṣṭafā Pasha, one of the viziers of Sultan Süleymān, built a large Friday mosque, a madrasa, many inns, a public bath, and a soup kitchen here.

İznik: During Byzantine times it was a famous and prosperous walled city; today the city wall overlooks ruins. During the time of the caesars, 318 monks assembled in this city and recorded the Christian creed (the First Council of Nicaea, 325 A.D.). They held many other councils here. Today it has a market, public bath, Friday mosque, madrasas, and soup kitchens. The glazed pottery vessels used in Turkey are made in a number of workshops here. Orḫān Ghazi turned a church (Aya Sofya) in İznik into a Friday mosque and built a soup kitchen. The Eşrefzāde [M 663] Friday mosque, dervish lodge and shrine (of Eşrefzāde-i Rūmī, a poet and mystic, d. 1469) are here and (his tomb) is a place of visitation.

Ṣapanca: A town without a wall located on a flat place on the western (rather, southern) shore of a lake. It is also a qadi district. İzmit is located one stage to the west. Ṣapanca is a well-known stage on the road (to İzmit). It has a Friday mosque, public bath, and market. Rüstem Pasha built an inn and a soup kitchen here. A table is set and candles are provided at each hearth for travelers who sojourn here. Rüstem Pasha also built the Friday mosque in town. There is arable land around the town. However, there is a forest known as Aġaç Deñizi ("Sea of Trees") that comes right up to the edge of town.

Ḳaramürsel: A prosperous town and one of the districts of Üsküdar. It is near İzmit and on the shore of the Gulf of İzmit. It has unrivaled pomegranates. It is at one end of the Gulf of İzmit, south of İzmit.

Yalaḳābād (Yalova): A qadi district at the mid-point of the southern shore of the Gulf of İzmit. It has a mineral spring. Those suffering from diarrhea drink the water. The season for this is August. Ill people come from all around to drink from the spring and many are cured. The place where the spring is found is also called Yalıova.

Ṣamanlu: A town west of Yalıova and on the (southern) shore of the Gulf of İzmit. It has a splendid inn, Friday mosque, and public bath. It is located near the entrance of the gulf. Wood for the public baths of Istanbul is brought from the environs of Ṣamanlu.

Şile: A town on the Black Sea coast thirty miles from Fenar. Wood comes to Istanbul from there.

Kerpe: A district on the outskirts of a small harbor on the Black Sea coast one hundred miles from the Bosphorus. There is another small harbor near this one called Kefke.

The district of Ḳandırı: It is (part of) the district of Üsküdar. It is located east of Aḳbaba and the qadi district of Yoros. It is two stages from Üsküdar. It is located near the Black Sea coast, directly north of İzmit.

Gençeli: Also (part of) the district of Üsküdar. It is located between Ḳandırı and Şile.

Samandıra: Also (part of) the district of Üsküdar. It is three hours east of Üsküdar and one hour west of Mt. Aydos. In the past there was a fortress here. It was conquered during the early years of the Ottomans. Guards were stationed there, but today there is no need for them.

Mt. Aydos is a high mountain four hours east of Üsküdar. A delightful stream emerges from a cave at the summit of this mountain. While the amount of water appears small, it is enough to water all the animals of a village. During the time of the infidels there was an impregnable fortress here.

Ḳāḍıköyi (Kadıköy): Its original name was Chalcedon. It was founded in the year 4861 (of the creation of the world).

Üsküdar: An imperial domain, not included in the emirate (= sanjak) of Ḳocaeli. It is where one crosses the Gulf of Constantinople (i.e., the entrance to the Bosphorus). It is a city east of Istanbul.

Mihrimāh Sulṭān, daughter of the late Ghazi Sultan Süleymān Khan, built a Friday mosque with a dome and two minarets next to the shore, together with a madrasa, kitchen, and guesthouse; and hostels for travelers on each side of the quai. Near them, also next to the shore, Rūm Meḥmed Pasha built a Friday mosque, madrasa, and kitchen. In 985 (1577) the mother of Sultan Murād III built a Friday mosque with a dome and two minarets on a high place at the southern edge of Üsküdar. East of it [M 664] she added a beautiful dervish convent with many rooms; north of it, a lofty madrasa; west of it, a kitchen, guesthouse, and two hostels. In 1053 (1643) the mother of Sultan Ibrāhīm built a Friday mosque, public bath, and school. There are other neighborhood mosques and Friday mosques. And, in particular, the dervish lodge of Maḥmūd Efendi ('Azīz M. Hüdāyī, d. 1628) is a place of visitation.

Üsküdar has a post for a molla paying 500 *akçes* per diem.

On the edge of Üsküdar is Doġancılar Meydānı ("Falconers' Field") where the army is marshaled before setting out on campaign toward Anatolia. Nearby, on the other side of the cemetery of Üsküdar, is a place called Maḥalle-i Mesākīn ("Quarter of the Lepers"). It has a separate mosque and public bath used by those with chronic diseases. There is a servant at the gate where votive offerings are made.

On the eastern shore of Üsküdar is a place called Ḳāḍıköyi Fenar Bāġçesi ("Lighthouse Garden of Ḳāḍıköy"—modern Fenerbahçe). It is south of Üsküdar at the entrance to the Gulf of İzmit. It is a state palace and garden. A tower (Kız Kulesi, Leander's Tower) was built in the sea in front of it to serve as a lighthouse at night. A light is lit on that tower so that ships passing it while going into the Gulf of İzmit do not crash on the reefs.

Üsküdar Bāġçesi: A palace and magnificent garden—the equal of the most noble imperial garden and palace—opposite the new palace. There are splendid imperial gardens and palaces stretching from Üsküdar Bāġçesi to the quay of Üsküdar.

Ḳuzguncuḳ: A town north of Üsküdar on the coast. It has many lofty palaces and splendid homes. Jews live in some of its quarters.

İstavroz: A town north of Ḳuzguncuḳ on the coast. It has many splendid palaces and homes of prominent persons, one of which is Ḫünkār Sarāyı ("Emperor's Palace") and its garden. North of it is Nārlı Bāġçe ("Pomegranate Garden") and Ḳule Bāġçesi ("Tower Garden"). Just above it in the tower (garden) there is flowing water.

Ḳandillü Bāġçe ("Lamp Garden"—modern Kandilli) is a splendid imperial garden near Anaṭolı Ḥiṣārı. The Göksu flows through it. About a mile upstream is a mill. The length of the stream above the mill is short. It is formed from the streams coming from the mountains and hills.

The area up to the Gökṣu is a level field and cypress grove. It is one or two miles in circumference.

Near it is Anaṭolı Ḥiṣārı, also called Güzel Ḥiṣār. It is located on a flat place where the Gökṣu enters the Bosphorus. Fātiḥ Sultan Meḥmed built it when he set about conquering Constantinople. Outside the fortress, on the shore of the Bosphorus and the Gökṣu, are large houses and palaces of prominent people. They continue as far as Ḳanlıca.

Ḳanlıca: A large town. İskender Pasha (Ottoman governor of Egypt in the 16th century) built a Friday mosque, madrasa, public bath, primary school, and a small market here. There are many town quarters and next to the sea are the houses of prominent people.

Above Ḳanlıca is Çubuḳlu Bāġçe ("Sapling Garden") and the imperial gardens known as Sulṭāniye. The villa at Sulṭāniye is in the middle of an inlet from the Bosphorus, like a large pool and harbor. The window shutters, cupboard doors and varnished ceilings are painted with various scenes.

Beyond it is the village of İncīrlü. It is also known for its houses. It is located near Beyḳoz.

Yoros: The qadi of Yoros resides in the town of Beyḳoz, which is located on the shore. It has many town quarters, a Friday mosque, public bath, and market; also gardens and houses belonging to prominent people. At the head of the market near the shore and quay is an unrivaled stream. There is a Friday mosque next to this stream which burbles from a fountain. [M 665] In Beyḳoz there is unrivaled tulum cheese.

Ḳavaḳ: A town located at the northern point of the Bosphorus. It is attached administratively to the qadi district of Yoros. At the end of it is Anaṭolı Boġaz Ḥiṣārı (modern Anadolu Kavağı). In 1033 (1623), when the Cossacks raided Yeñiköy, two small fortresses were built here on opposite sides of the Bosphorus and stationed with guards and cannons. In the middle of the strait here, but closer to Rumeli Ḥiṣārı (modern Rumeli Kavağı), is a large reef the height of a man sticking out from the surface of the water. A lighthouse burns here (to warn ships). Beyond is the open sea. When the infidels controlled Anatolia and the straits, there were apparently forts on opposite sides (of the northern entrance to the Bosphorus). Each one encircled a mountain and extended to the shore. Remains of their walls can still be seen.

Ḥünkār Bāġçesi ("Emperor's Garden") is at a place called Serv Burnı ("Cypress Cape"), between Beyḳoz and the town of Ḳavaḳ. A high dense forest overlooks this garden. According to common belief, the tomb of the prophet Joshua is at the summit (modern Yuşa Türbesi). Extremely good water is found there. From the summit there is a marvelous view of the Bosphorus as it twists and turns like a snake.

Aḳbaba: A town in a pleasant mountainous region two hours from Beyḳoz. It has a Friday mosque, public bath, the grave and dervish lodge of a saint known as Aḳbaba, and town quarters. It also has unrivaled water. Furthermore, it is a place of excursion. The sophisticates of Istanbul go there during two seasons every year. One is the cherry season: when the cherries are ripe, people come to pick a certain kind of black and white Aḳbaba cherry. The other is the chestnut season, because there are wild-growing chestnut trees in the hills of Aḳbaba.

Arva (Aġva): A river in Anatolia beyond where the Bosphorus enters the Black Sea. This river, which rises from the edge of ʿAlem Ṭaġı, flows north and enters the Black Sea. There is a large village at the end of this river which is like the town of the district of Arva. It is four hours from Üsküdar. There are special boats on the Arva River that bring firewood to Istanbul.

Mountains

It is well known that the hills and slopes and valleys on the eastern and western shores of the Bosphorus are all built up. There are vineyards on the mountaintops and gardens, lofty villas and splendid houses in the coastal valleys. There are high mountains in the environs of the Bosphorus. There are large mountains and forests in the environs of Aḳbaba behind Boġaz Ḥiṣārı. The forests are mostly chestnut trees.

Two hours east of Üsküdar are two places of excursion called Çamluca. They are on a large hill in the shape of two circles. One is called Eski Çamluca (Old Çamluca) and the other is called Yeñi Çamluca (New Çamluca). Our felicitous padishah Sultan Meḥmed IV (r. 1648–1687)—may God make his caliphate everlasting!—erected many buildings there. The place was formerly a dervish lodge but now it is a splendid palace. The village of Bulġurlu, which is on the skirt of the road, is the pious endowment of Üsküdari Shaikh Maḥmūd Efendi (Hüdāyi, d. 1628–1629).

Mt. Māldepe (Maltepe) is one military stage east of Üsküdar. It is known as The Hill (*tel*). It extends as far as Ṭuzla on the northern shore of the Gulf of İzmit. One side goes down to the foothills of the shore of the Gulf of İzmit. Based on the name of this mountain (*māl* = wealth, riches), some treasure-seekers became covetous and, hoping to find buried treasure, have dug holes in various places.

ʿAlem Ṭaġı is due east of Istanbul. It is one or two hours north of Māldepe. [M 666] Most of the firewood for Üsküdar comes from there in carts. Some timber is cut there as well.

Ağaç Deñizi ("Sea of Trees") is a large mountain range at the easternmost point of Ḳocaeli and around Lake Ṣapanca. It has abundant forests with pine, oak, hornbeam and acorn trees that reach to the sky. The main road that goes east from Istanbul passes through this mountain. The forest stretches one or two stages.

Lakes

Lake İznik: Its circuit is about two (sic) miles. The water is sweet and different kinds of fish are caught. In particular they catch a kind of fish called *line*, which is about half a span in length. They dry it and take it to Bursa and Istanbul. There are many streams around the lake that flow into it. The outlet of the lake is in the environs of Gemlik where it flows into the sea.

The Ḳırḳgeçid ("Forty Fords") River is formed from the combined streams coming from several mountains in Anatolia. It reaches Ḳırḳgeçid, which is one day's travel east of İznik. It enters the Sea of Rūm (Marmara) near Hersek.

Lake Ṣapanca is oblong shaped and has a circuit of fifteen miles. It is surrounded by what is called Ağaç Deñizi. The road to Bolu is about half a mile from here. One takes a boat across the lake or rides a horse around the shore. This lake is half a stage from the Gulf of İzmit, to the east. Also the Ṣaḳarya River is nearby, three stages distant.

In 909 (rather 999/1591), Sinān Pasha (grand vizier, d. 1596) proposed to excavate canals to connect these three bodies of water. It was the imperial desire for the Ṣaḳarya River to flow into Lake Ṣapanca and Lake Ṣapanca to flow into the Gulf of İzmit. The sultan (Murād III) sent a noble order addressed to the sanjak bey of Ḳocaeli and to the qadis of this sanjak and appointed a team composed of the chief court architect (Sinān), the astrologer and astronomer Molla Fütūḥ, Miʿmār Çavuş, Miʿmār Süleymān, Yūsuf and ʿAlī who were in charge of the aqueducts, the *ṣu nāẓırı* (water inspector) and others. They were to determine if connecting (the three bodies of water) was feasible.

After the matter was clear, the architect and the geometers and the experts reported truthfully that the Ṣaḳarya river is 17½ cubits below the surface level (i.e., the water had to be raised by this amount, or a canal would have to be dug out that deep). When it was investigated if there was a more suitable location further upstream, it was found that to dam the river about one mile upstream was most advantageous. When the level was measured, it was 17 cubits from the water to the surface level, and it was 5½ cubits higher than the place architect Sinān the Elder had determined. The water of the Ṣaḳarya would (have to flow) 9,600 cubits to the lake, but after 1,100 cubits it would join the river Ṣarıdere. And while the Ṣaḳarya would be 12 cubits below surface level, the place where it joined the Ṣarıdere it was (down to) 16 cubits, so that the place where the dam would be built, and the place where the river emptied into the lake, would be at the same level. Since the Ṣarıdere was a ready river-bed flowing towards the lake, which would have to be widened somewhat, and deepened by a few cubits, using this location would reduce the amount of excavation to half of what Sinān the Elder's location would have required.

When they had ascertained that the Ṣaḳarya River could flow into Lake Ṣapanca, they went to the side of the lake facing the Gulf of İzmit. When they measured here very carefully, from both sides, (it was found that) it was 22,000 cubits from the lake to the sea. On this route there were no mountains and rocks, while half of it was covered by forest. Starting from the lake, it would be necessary to dig to the following depths, in turn, for each 1,000 cubits: 7 cubits, 9½ cubits, [M 667] 10 cubits, 6 cubits, 13 cubits, 17 cubits, 18½ cubits, 26 cubits, 30 cubits, 25 cubits, 26 cubits, 17 cubits, 20 cubits, 9 cubits, and for the final 8,000 cubits of distance to the sea 6 cubits (all the way). Thus, after ascending for a while and then gradually descending, after 14,000 cubits the ground surface would be the same level as the lake. After that, the lake (water) would drop 30 cubits to the Gulf of İzmit. This drop would occur in the following manner: after the first 14,000 cubits, to 8 cubits (below the level of the lake) for 1,000 cubits, and for each 1,000 cubits subsequently to 10½, to 11½, 17, 20, 28, and 30 cubits (below the level of the lake).

As the water would drop 30 cubits to the Gulf of İzmit, they would dig 10 cubits down from the lake (or lower the lake by 10 cubits?). In this case the water from the Ṣaḳarya River would drop more than 10 cubits to the lake (whereas previously, it had been at the same level). Thus, the Ṣaḳarya would flow in a stream into the lake like water flowing into a mill chute, because the distance between the Ṣaḳarya and the lake is only one mile.

As a result of these investigations it was agreed that it would be possible to connect the Ṣaḳarya with the lake and the lake with the sea without going through anyone's farm or other property. The members then presented their sealed and signed report to the Court. The sultan immediately issued an order to begin work.

There is a place here abounding in oaks that is desirable to reach. It is Mt. İstıranca. There is enough firewood to last 100,000 years and in general it would be an inexhaustible source of lumber for ships. The galleys to be built at Lake Ṣapanca would be built (from the same kind of wood) as the ships at the dockyard (in Istanbul). Thus it was decided to send more than 50 *yüks* of *aḳçes* (5,000,000 *aḳçes*) to the imperial arsenal. It was thought that there would be many advantages to sending provisions and grain via rafts first to the lake and then from the lake to Istanbul.

However, because of bribery this project was never completed.

Chapter: Description of the Sanjak of Biġa

This is the second sanjak that is attached administratively to the province of the Ḳapudan Pasha (admiral of the fleet). Today it is called the pasha sanjak.

Borders: to the west and south, the Sea of Marmara; to the north and east, Ḳaresi.

Its qadi districts are the following: Eynebāzārı, also called Ez̄īne, Balya, Biġa, Çan, Ḳaz Ṭaġı, Ḳalʿe-i Sulṭāniye (Çanakkale), Güvercinlik, Lapseki, Ḳaputdaġı, and Marmara.

Description: In ancient times the capital of this region was a city called Troy or Truva. It was larger than Istanbul. This city is located on the mainland opposite Bozcaaṭa (Tenedos). It was founded in year 492 after the death of Moses. Its first king was Dardanos. His dynasty lasted 297 years. Then the king of this city was Alexander, son of Alexander son of Priam. While he was the guest of Menelaus, king of the city of Mistra in the Morea (Peloponnesus), this Alexander fell in love with Menelaus's beautiful wife Helen, and because of her this city (Troy) was destroyed.

It is related that [M 668] the wife of Alexander, king of Troy, one day while she was pregnant had a dream in which a fire sprang from her vulva and burnt the city. When she told this dream to the king, he said that the child that she would bear would bring harm to this city and dynasty. The woman in fact gave birth to a son. The king wanted to put him to death, but his wife was unwilling to do so. She wrapped the child in a cloth and left him in a forest in the area. While travelling through those parts a shepherd came across the child and took him to his wife. Because they did not have children, they raised this boy as their own. When he grew up he began to harm the neighborhood boys. The people then complained to the king about him. The king summoned the boy and the shepherd. The king reprimanded (the shepherd) saying, "Why don't you punish your son?" The shepherd then said, "This is not my son. I found him in such-and-such a place wrapped in a cloth and raised him up to now. I didn't know that he would become so bad." The king then realized that this child was his own son. He took him from the shepherd and put him in the palace.

One day the king went out hunting and left the boy in the palace. The king had another son who was the brother of this boy. The two brothers quarreled. The boy struck his brother and killed him. Out of fear, he boarded a small boat and fled. He went to the Morea and became the guest of the king of Mistra. The king had a wife of unmatched beauty and the boy fell in love with her. This king had to go to the island of Crete, which was under his rule. He left his guest in his home and set out for Crete with a number of galleys, leaving one galley in the harbor to bring him news on Crete if necessary. Meanwhile that lord persuaded the king's wife to go with him to his father in Troy. The woman collected all of her jewels and a great amount of gold. They boarded the galley that was left in the harbor and set out for Troy. His father and mother were happy to see him. Because so much time had passed since he had killed his brother and fled, this event had been forgotten. Also they had no other children, so they were very happy.

When the king of Mistra returned and learned that that lord had taken his wife and gone to the city of Troy, he immediately set out after them. Now this king was subject to the king of Troy. He demanded that the king of Troy take his wife from his son and give her back to him. The king of Troy refused this demand and wanted to kill him. The military commander at Troy informed him of this situation and said, "What are you waiting for? The king wants to kill you. Do not delay. Go back to the Morea. I will give you a ship." Out of fear he returned to Mistra.

When he returned to his country he informed the neighboring kings of what had happened and laid everything out to them. In short, they assembled a huge army and, coming by land and sea, besieged the city of Troy. However, the fortress of Troy was strong and its population was large, so they were not able to capture it. In the tenth year of the siege a clever man devised the following plan. They built a large and marvelous, skillfully crafted figure of a horse, put several brave men inside and left it near the fortress, figuring that those inside the fortress would not destroy such a magnificent statue and would take it inside. [M 669] And that was what happened.

They had built a statue in the shape of a horse and put several brave men inside. They put the statue on wheels and left it in the camp, then pretended to flee during the night, leaving one man lying in wait to inform them if they took the statue into the fortress. Those in the fortress saw that the enemy had fled and left the statue behind because they could not take it with them. Some of them said, "It is not empty, let's break it to pieces." But most responded saying, "Wouldn't that be a shame? This is a skillfully crafted object. They were not able to take it with them. Let's take it inside and put it inside the fortress gate." However, the fortress gate was too small, so they demolished it, took the statue inside, and then rebuilt the gate as it was.

The spy who had been lying in wait informed them of the situation and the enemy army immediately returned and besieged the fortress. When they attacked the gate, the men inside the horse emerged and stood before the gate. By this means they were victorious. The army entered, massacred the population, and destroyed the fortress.

Today ruins of this great city are still standing. There are finely-crafted columns and beautiful pieces of marble. Most of the columns and marble in Istanbul have been brought from this city. It is located at the end of the Island of Ḳapuṭağı and opposite Edincik. From Troy one can see the coast of Rumelia and the north coast of Gallipoli. Indeed, Süleymān Pasha had (his father) Orḫān Ghazi look over to Rumelia from here, so that he desired to cross over to it and conquer it. Therefore this place is called Temāşālıḳ ("Lookout"). It is also known as Bilḳīs Sarāyı (Palace of the Queen of Sheba).

Chapter: Description of the Sanjak of Ṣığla

This sanjak is also attached to the province of the Ḳapudan Pasha. It is currently the imperial domain of the steward of the imperial arsenal. Its qadi districts comprise various sanjaks. The most important qadi district and the capital is the city of Izmir.

Its qadi districts are the following: Eẕīne, Ayas̱luġ, Izmir, Aḳçeşehir, Aydın also called Söke, İne also called Met, Baf, Balaṭ, Bayramlu, Çeşme, Urla, Çine, Seferīḥisār, Şahme, Ṣavur, Ṣobıca, Ṭalma, Ḳarpuzlu, Ḳaracaḳoyunlı, Ḳaraburun, Ḳızılḥisār, Ḳuşadası, Mendece, Nif, Menemen.

Description of Izmir

It is the capital of the province of Aydın and Saruḫan. Its port is a great entrepot on the Mediterranean coast. It is a city and fortress. It has two fortifications, one on a high hill south of the city and the other on the seashore northwest of the city. Both have a garrison.

Izmir is an ancient city whose remains and ancient fortifications enclose many buildings. It has three Friday mosques, public baths, markets, and vegetable gardens. Lemons, figs, and raisins are exported in all directions. There are many olives.

This city was conquered by Tīmūr in 805 (1402) when he spent the winter at Tire. At that time it was in the hands of the Franks. Later, in 829 (1425), Sultan Murād II took it from Ḥafīd Bey.

This town is located on the seashore at the northern foot of a low mountain. To the west and north is the sea, to the east are plains and hills and dales and vineyards. There are about twenty town quarters, ten Friday mosques, nine public baths, and many markets that are on the seaside. There are shops and markets that take up about half a town quarter.

The city's water comes from fountains and wells. A small river that rises from the foot of the mountain [M 670] goes around the eastern part of the city, comes into the city as it nears the sea and then flows into the sea. The houses are made of wood and covered with red tiles. All the houses face west and north and back up against the mountain.

The Ulu Cāmiʿ (Great Mosque) is that of Niflizāde, on the seaside near the harbor; it has an upper floor, is made of masonry, has a dome sheathed with lead and one minaret. Another Friday mosque is that of Yaʿḳūb Bey, on the seaside; it has a dome sheathed with lead. Another Friday mosque is that of Ḥācı Ḥüseyn, on the seaside. Elsewhere the city are small Friday mosques.

There are about 60 inns. One, at the entrance to the lower harbor, is triangular in shape and was built by Fātiḥ Sultan Meḥmed. It has one large and one small gate. On one side are houses; on the western and northern sides is the sea; and on the southern is an open square.

There is another small fortress on the summit of the aforesaid mountain. It is oblong in shape and goes back to the time of the infidels. A quarter of it is still standing. Inside is a Friday mosque. In the lower fortress there is a small mosque. The water for these fortresses comes from outside. In one of the ruined fortresses are ancient ruins, forty columns and a huge underground cistern above which is a flat field.

The people are a mixture of all nations and religions. Muslims are the majority. Four *bailos* (European consuls), reside there. The city has many religious scholars (ulema). In front of the gate of the upper fortress is a Greek building called Temāşālıḳ (the ancient Greek theater). It has tiers going up the side of the mountain. Among the vegetable gardens are caves and the remains of buildings.

Its districts are the following: Burunābād, a small town opposite Izmir, to the northeast and on the seaside, with five or six Friday mosques and two or three public baths; Zerpande, southeast of Izmir, comprising several villages; Maḥkeme Ḳaryeleri ("Law Court Villages"), south of Izmir, comprising 24 villages.

Izmir has a delightful place of excursion called Pınarbaşı. A copious spring emerges from under a dome and creates a lakepond about half the size of a threshing-floor. From there it flows into the sea. There are mills at the outlet of the lakepond. It is possible to enter the cave from which this spring emerges. At the end of the cave is a place like a door, but it is shut. Buried treasure is believed to be inside. It is well known to the inhabitants of the province. They say that there is a talisman (guarding the door). They relate that when some of those who wanted (to get the treasure) reached the door a poisonous wind blew from the door and killed them.

Urla: A town southwest of Izmir. It is twelve miles from the sea and a half stage from the mainland. It has one or two Friday mosques, a public bath, and markets. Because there are abundant olive trees, Urla is a source of soap and a center of soap production. The quai and harbor are three miles from the town. Because the town is not at the seaside, porters are always available for hire to carry the loads of travelers from the quai. There is an old inn at the harbor that was built for travelers. Later, at the end of 994 (1586), a small hostel and mosque were built.

(Meḥmed) ʿĀşıḳ (Turkish cosmographer, d. after 1598) says: I saw a female mule with its foal behind it in the possession of these porters. When its owner was asked about it, he said that the female mule had mated with a male donkey and had given birth to this foal and that he had never seen such a thing contrary to nature.

Menemen is near the sea and southwest of Izmir. The qadi districts of Gelembe and Ṭurġudlu are north of it. If one wishes to go from Izmir to Nif he must cross a mountain that is between them. One must also cross a mountain to go to Manisa in the direction of the north and to Ṭurġudlu (in the east).

Just west of Nif is Ḳızılḥiṣār and south of Nif [M 671] is Ayaṣluġ. Nif (today's Kemalpaşa) is near Manisa, to the west and south.

Chapter: Routes, Stages, and Distances of Anaṭolı and Ḳaraman

From Üsküdar: > Pendik 5 hours. The road goes through a defile; travelers sometimes lodge at Māldepe > Gebze 5 hours. There is little water; travelers sometimes lodge at Sulṭānçayırı.

From Gebze there are two roads, one through İzmit and the other crossing (the Gulf of Izmit at) Dil with travelers lodging at Hersek. The travel time (to cross the gulf) is twenty minutes. > Derbend 6 hours > İznik 6 hours > Lefke 12 hours > Sögüt 11 hours. In 1070 (1659) Grand Vizier Köprülü Meḥmed Pasha built a splendid inn, Friday mosque and public bath and endowed them as charitable institutions at an unforgiving and desolate place near the wadis known as Çeltiklik in the mountains between Lefke and Sögüt. > Eskişehir 10 hours.

The road through İzmit. From Gebze > Hereke 4½ hours. There is no place to lodge. There are mills near Ṭuzla that are in ruins. > İzmit 5 hours. Çınarçayırı is between (Hereke and Izmit). At Izmit the road divides, one to the Left Branch (*ṣol ḳol*) and the other to the Middle Branch (*orta ḳol*) which goes to Eskişehir.

The Eskişehir road: > Ḳazıḳlubeli 4½ hours > Dikilüṭaş 10 hours—it is near İznik. Or else, from Ḳazıḳlı Beli > Ḳuşçular 7 hours—a winding mountainous road that is difficult to traverse and stony; > İznik 3 hours > Penbecik 3 hours—it is attached administratively to İznik; > Yeñişehir 3 hours. Or else, from Dikilüṭaş: > Gölbaşı 2½ hours > Yeñişehir 4 hours > Aḳbıyıḳ Baba 4 hours—it is near Sögüt > Pazarcıḳ 5 hours > Bozöyük 4 hours. There are caves in the mountains and a tower. Its fortress has been turned into residences. Formerly nine guards were stationed there. > Eskişehir 6 hours > Aḳvīrān 3 hours > Seydīġāzī 3 hours > Bardaḳlı 3 hours > Ḫüsrev Pasha Ḫānı 5 hours > Bayat 8 hours > Bolvadin 8 hours > İsḥāḳlı 5 hours > Aḳşehir 3 hours > Arḳıd Ḫān 4 hours—it is in ruins; > Ilġın 2½ hours.

From Ilġın: > Cisr-i ʿAtīḳ, also called Balḳam Ṣuyı, 2½ hours > Balisköprüsi > Zengi Ṣuyı > Konya. Or else, from Ilġın: > Ladik 10 hours > Konya 9 hours.

From Konya: > İsmil 11 hours > Ḳarapıñar 9 hours > Eregli 12 hours > Uluḳışla 9 hours > Çifteḫān 6 hours.

Here the road becomes rocky and difficult to traverse, but there is a creek and good drinking water and a hot spring nearby; also a river called Ḳırḳgeçit ("Forty Fords") that travelers cross many times. Between Çifteḫān and Çaḳıd are mountains and oak forests. A half stage from Çifteḫān the Ḳırḳgeçit becomes a large river. It passes under a stone bridge called Aḳköpri and meets the river called Ḳaraṣu. A stream that rises from a large spring called Şekerbıñarı at the foot of a mountain empties into this river. After (the road) crosses this bridge, it goes up the mountains where the summer pastures called Ramażānoğlı are located and then goes down. From Çifteḫān to the summer pastures is 7 hours and from there to Çaḳıd is 11 hours. Next to Çaḳıd is an inn built by Bayram Pasha (grand vizier, d. 1657). [MAP] [M 672] In winter it takes three or four days to reach it. Since then many other inns have been built in between. The late Ḥācı Sulṭān built the inn called Sulṭān Ḫānı on the road from Çifteḫān to Çaḳıd.

From Çifteḫān: > Ṣarı Işıḳ > Çaḳıd > Adana 9 hours. From Çaḳıd to Adana there are two roads. One crosses a mountain of earth called Kargasekmez and descends to the Adana Plain. On this route there is no need to cross the Çaḳıd River. The other goes by the inn next to Çaḳıd and one crosses the Çaḳıd River by foot. On this route one crosses the Çaḳıd River twice before reaching Adana.

Another road goes from Zengi Ṣuyı to Adana. > Zengi Ṣuyı 6 hours > Dike Ḫān, also called Kelmiç Pıñarı, 4 hours—this is an area of plains; > Kārizbaşı, also called Ḥavuz, 3 hours > Konya 4 hours > village of Pıñaroğlı 6 hours—it is in ruins > Pıñarbaşı 6 hours—it is near Lārende; > Fīrūz—near Maʿdenṣārı, also called Eflātun Ṣuyı > Aḳçeşār 8 hours—there is a stream and on one side a fortress built by Murād Pasha (probably Grand Vizier Ḳuyucu Murād Pasha, d. 1611); > Gölbaşı 7½ hours—there is a large river and bridge. Or else > Ḳurdbıñarı > Eregli 4 hours > Çavuş Köprüsi—there is a river > Uluḳışla > Çifteḫān > Tekirbeli and the Ramażānoğlı summer pasture > Ṣarı Işıḳ > Çaḳıd River > Ḫān of Bayram Paşa > Adana.

On the Right Branch (*ṣaġ ḳol*) also the road to Bursa splits, one going through Dil, the other through İzmit. From Bursa it goes to Kütahya. From İzmit: > Yeñişehir of Aydın > Bursa.

From Bursa to Kütahya: > Aḳṣu—it has good flowing water; > İnegöl—it is a broad plain near a lake; or else one lodges at Ḳurşunlu Ḫān > Gümüş—in this stage one goes through the pass of the Ṭomaniç Mountains > Kütahya.

Another route from İznik to Kütahya: > Yeñişehir of Aydın > Aḳbıyıḳ > Aḫircik > Bozöyük > Ṣuluköy > Gencelü—there is a hot spring > Kütahya.

The road from Dil has two branches, one to the Right Branch (*ṣaġ ḳol*), the other going to Bursa and Manisa. From Dil: > Ṣamanlu—one stage to the west > Bursa; or else from Ṣamanlu > Bāzārcıḳ—one crosses the Ṣamanlu and Ḳaṭırlı mountains which are big mountains and the roads are difficult > Balabancıḳ > Ulubad.

The military stages from Bursa to Boğazḥisār (Çanakkale): > Beylik Çayır 3 hours—it is at the head of a lake > the nearby vineyards of Ḳaraağaç 6 hours > Ulubad Köprüsü 2½ hours—it is near Miḫaliç > Ṣarıbey 4 hours—it is attached administratively to Kirmasti > Ṣaldır 4½ hours—it is near Belcekağaç and attached administratively to Manyas > Körpeağaç 4½ hours—it has a bridge and is near Budyan (?) and is attached administratively to Gönen > Güvercinlik Bridge 6½ hours > village of Dimetoka 3 hours by sea and 6½ by land > Ḳurudere 4½ hours > Küreci 6 hours—it is near the sea > Gelibolu opposite Çārṭāḳ 4 hours > Berġos 4 hours > Ḳalʿe-i Sulṭāniye (Çanakkale) 5 hours. [M 673]

From Miḫaliç to Manisa: > Ilıbat > Ṣuṣıġırlıġı—a river flows there, > ravine of Mendehorya > Ḳurıgölcük > Başgelembe > Palamuṭ—it has two inns > Manisa > Izmir.

From Bursa to Bergama: > Ḳaraağac > Ulubad Lake > village of Çeltikli > village of ʿÖmer > Balıkesri > village of Bardaḳçı > village of Tarḫala > village of Cavdır > village of Belücek > Bergama.

Ayazmend > Edremid > İvrindi > Balıkesri.

From Kütahya to Ḥamīd: > Deñizli > Zazḳırı > Keçiborlu > Isparta.

From Deñizlü to Antalya: > ʿĀṣī-Ḳaraağaç—it is south, crossing the pass of Ḳazıḳbeli that lies in between > Tefenni > İstinar > Antalya.

From Deñizlü to Ḳuşadası: > Nāzilli—it is west, passing through Sulṭānḥisārı > Köşk > Güzelḥisār > Ez̠īne 4 hours > Balaṭ > Ḳuşadası.

From İzmit there are several roads to the Left Branch (*ṣol ḳol*) > Ṣapanca 9½ hours; or else > district of Aṭa (Adabazarı)—one crosses the Ṣaḳarya River and the rivers of Mudurnı; > Ḥendek > Düzcebāzārı. The road then goes through Ḳoñrapa and crosses the dam and mountains to Bolu. Or else one crosses the Ṣaḳarya near Ṣapanca, 8 hours > Geyve 9½ hours > Ṭaraḳlu 9 hours Göynük 5½ hours > Mudurnı 9½ hours > Bolu 15 hours > Çaġa 8 hours > Gerede 5½ hours > Gölbaşı 4 hours > Bayındır 5 hours—one heading east can lodge at Ḥamāmlu which is opposite; > Çerkeş 8 hours—the road is narrow > Ḳaracavīrān 8 hours—the people are horse trainers; or else: > Ḳurşunlu > Ḳoçḥisār 10 hours > Ṭosya 10 hours.

There are two roads from Gölbaşı to Ḳasṭamonı: > Vīrānşehir > Borlı > Eflani > Maçillina > Ṣorḳun > Ḳasṭamonı; or else, from Gölbaşı: > Bayındır > Aḳṭaş > Ṣırt > Yörük > Arac > Mergüze > Ḳasṭamonı.

From Ṭosya to Çankırı is one stage.

From Ṭosya to Sivas: > Ḥācıḥamza 8 hours—a road here is a mountain pass called Kepirbeli, difficult to traverse, with four inns > ʿOs̠māncıḳ 6 hours > Köseşaʿbān 8 hours—it is at the end of the Merzifon Plain; or else: > Ḥācıḳöyi > Bulaḳ 4½ hours—one passes near Merzifon, > Kelkiraz 7 hours > Amasya > Eskiigne, also called Aḳyalar, 8 hours > Turḫal 7 hours > İnebāzārı 4 hours > Toḳat 6 hours > Taḫtābād 7 hours; or else: > Bolis—a large village on the Artuḳābād Plain inhabited by Muslims and Christians; > Meḥmed Pasha Ḫānı, today called Yeñişehir, 7 hours—one crosses a great pass to reach it; > Yıldızṣuyı 4 hours > Sivas 4 hours—one crosses a bridge to reach it.

From Ankara to Toḳat: > Lalaçayırı > Çuḳurcuḳ > Çāşnigīr Köprüsi > Ḳaraḳoġa > Narende—it is a qadı district, > Aḳçeḳoyunlı > Toḳat.

From Eskişehir to Toḳat: > Ḳaş > Aḳköpri > Oğlaḳçı—a village opposite Sivriḥisār > Mihare (?)—one crosses the Ṣaḳarya and lodges there > Yüñlü [M 674] > Yılancı > Alacaada > Ankara > Lalaçayırı > Ḥisārlıḳ > bridge at the

Ḳızılırmaḳ > front of the Barani summer pasture > Ṣofılar > Siğa > Gürz > Ḳanıḳ > Çavgar > Yalñızṭam > Ṭarıboğazı > Ḳızıltürbe > Meşhedābād > Ḳazābād > Toḳat.

From Ankara to Sivas: > Elmaṭağı > Ḳızılırmaḳ > Keskin > Deliceırmaḳ > Elbeyli > Budaḳözi > Sancar > Ḳazanḳaya > Üçṭaş > Zile > Sīs > Artuḳābād > Meḥmed Pasha Ḫānı > Ilıca > Sivas.

From Ankara to Kayseri: > Aladağı > Çuḳurcuḳ > Çāş-nigīr > Degirmenli, also called Boldaḳlı > Ṣofılar > Ḳazlıḳ-balḳanı > Ḳabaduraḳ > Kürekçi > Demirciler > Anbārān > Kayseri.

From İzmit to Ankara: > Ṣapanca > bridge near Geyve > Ṭaraḳlı > Ṭorbalı Göynük > Naṣūḥ Pasha Ḫānı > the other Naṣūḥ Pasha Ḫānı > Ṣarılar > Beybāzārı > Ayaş > Istonos > Ankara.

Languages

The language of the people of Anaṭolı is Turkish. Here and there some of the Christian subjects speak Greek or Armenian, but they are very few. There are even some Jews but they speak Spanish. Various Frankish (European) languages are found in Izmir.

Religions

The people of Anaṭolı are all Muslims of the Ḥanafī rite except for some Christian villages. Greeks, Armenians and Jews are found everywhere but they are few in number.

Description of the people

The people are Turks, Turkmen, and Greeks; but there are also a few Armenians and Jews. There are also Franks in Izmir.

Status of kings and rulers who have conquered the province of Anaṭolı

It was Misk son of Yāfith (Japheth son of Noah) who 570 years after the Flood conquered this region from that mountain (Ararat?) all the way to Üsküdar and populated it. After him came Luhrāsb, then Dārā, then Iskandar the Greek (Alexander the Great), and then the Ptolemies, during whose reign Antakya was founded. Then came the Caesars of Rome, during whose time Plato the Divine was in Konya. Then the Persian Nūshirwān ruled again as did the Caesars of Rome (Byzantine emperors) until the dynasty of Islam appeared. The Muslim caliphs conquered some places and raided others. The first Muslim state to arise and be established in Anaṭolı was that of the Ottomans.

It is reported that it came about as follows: When Chingis Khan arose and came to Balkh and took control of Iran, Süleymān Shah was ruling in the city of Māhān near Balkh. In 621 (1224) Süleymān Shah left that place and set out for the region of Aleppo. As he approached Qalʿat Jaʿbar he came to the Euphrates. Süleymān looked for a ford and then spurred his horse into the river and met his fate there. They buried him below Qalʿat Jaʿbar. He was survived by three sons: Sunḳurtekin, Günṭoğdı, and Erṭuğrul. When this happened they turned to Pāsīn Ovası. Erṭuğrul remained there with 400 nomadic households. His brothers returned to their original homelands. Erṭuğrul stayed where he was for a long time. Then he set out for Rūm.

When he was about to enter Rūm, he sent one of his sons [M 675] named Ṣarıbalı to Sultan ʿAlāʾ al-Dīn (Kay-Qubād) to ask permission to enter his province. The Sultan gave him permission and assigned to him the area stretching from Ṭomaniç to Ermenek. Erṭuğrul set out for that area and settled at Ḳaracadağ. Later when Sultan ʿAlāʾ al-Dīn was fighting some enemies, Erṭuğrul and the companions who were with him hastened to his assistance. The Tatars were about the defeat ʿAlāʾ al-Dīn when Erṭuğrul and his forces attacked them and routed them. When ʿAlāʾ al-Dīn saw this he received him honorably and Erṭuğrul kissed his hand. ʿAlāʾ al-Dīn dressed Erṭuğrul in a robe of honor and bestowed gifts on his men. Then Erṭuğrul went to the place called Sögüt for the winter. Ḳaracaḥisār and Bilecik, which were in that area, were subject to ʿAlāʾ al-Dīn. After a while the infidels of Ḳaracaḥisār would not leave Erṭuğrul in peace and began to harass him. Erṭuğrul informed ʿAlāʾ al-Dīn, who gave him permission to attack them. Erṭuğrul besieged Ḳaracaḥisār and conquered it.

Later, in 685 (1286), while ʿAlāʾ al-Dīn was besieging Kütahya, word arrived that the Tatars had broken the truce and made a raid on Eregli. ʿAlāʾ al-Dīn immediately turned the siege of Kütahya over to Erṭuğrul and returned to confront the Tatars. He defeated the Tatar army and killed many of them. He went so far as to order that the scrota of the dead Tatars be sewn together to make a canopy. Today that field of battle is known as Ṭaşaḳ Yazusı ("Plain of Testicles").

Meanwhile Erṭuğrul sallied forth and conquered the fortress (of Kütahya). He captured the *tekfur* (local Christian ruler) and plundered the fortress. He set aside one fifth of the booty and distributed the rest among his warriors. Then he sent the fifth of the booty with news of the conquest to ʿAlāʾ al-Dīn.

Subsequently Erṭuğrul engaged in *gazā* for two years, three months, and four days. Then ʿAlāʾ al-Dīn died and his son Ghiyāth al-Dīn (Kay-Khusraw) ascended the throne in Konya. Erṭuğrul remained in Sögüt and maintained friendly relations with the infidels until the reign of Sultan ʿAlāʾ al-Dīn Kay-Qubād II (rather, III) b. Farāmurz. At that period Erṭuġrul had three sons, ʿOs̱mān, Gündüz, and Ṣarıbalı. But the bravest and most courageous of them was ʿOs̱mān and for this reason the people showed him honor and respect.

At that time the great (Ilkhanid) ruler Ḳāzān (Ghāzān) Khan b. Arghun (r. 694–703/1295–1304) invaded Rūm. Erṭuġrul was then very old, but he had close relations with Sultan ʿAlāʾ al-Dīn. Sometimes ʿOs̱mān Ghazi met with ʿAlāʾ al-Dīn and brought him presents. At that time ʿAlāʾ al-Dīn II had deputies in Eskişehir and İnöñi. ʿOs̱mān Ghazi visited them and cultivated their friendship.

This ʿOs̱mān Ghazi was a devout Muslim and a pious person. Every three days he had food prepared and given to the poor and the righteous whom he had assembled. The extent of his piety is illustrated by the following: One night he was the guest in the house of a village imam. There happened to be a Koran in the window casement, and the owner of the house informed ʿOs̱mān Ghazi of this. [M 676] At night, when he went out to perform the ablution, he turned facing the Koran and stood with his hands folded in respect until morning.

It is reported that ʿOs̱mān Ghazi had a dispute with the *bey* of Eskiḥiṣār (i.e., Eskişehir) over a woman. The *bey* of Eskişehir went to war against ʿOs̱mān Ghazi and the *bey*'s army was defeated. Köse Miḫal, who was the *tekfur* of the infidels of Ḫarmanḳaya and who had come to the aid of the *bey*, was captured. Because Köse Miḫal was a brave warrior, ʿOs̱mān Ghazi could not bring himself to kill him. He forgave his offense and freed him. Köse Miḫal in turn became a Muslim wholeheartedly and together with his followers submitted to ʿOs̱mān Ghazi.

ʿOs̱mān Ghazi and Orḫān did battle with the infidels in their territory and conquered many fortresses. Today their descendants possess their estates. Erṭuġrul died in 687 (1288) at age 93 and was buried in Söğüt. ʿAlāʾ al-Dīn was very sorry when he learned of Erṭuġrul's death. He appointed his son ʿOs̱mān Ghazi to succeed him and sent him a drum and standard, sword and horse, and robe of honor. This is what the author of *Künhü'l-aḫbār* (i.e., Gelibolulu Muṣṭafā ʿĀlī, d. 1600) reports. According to the other histories, however, ʿOs̱mān Ghazi was sent to attack İznik, as will be seen below.

It is reported that the men of the *tekfur* of İnegöl could not refrain from raiding when ʿOs̱mān Ghazi went to his summer pasture. As a result, ʿOs̱mān Ghazi said to the *tekfur* of Bilecik, "They have begun to damage our migration. In order not to break our peaceful relations, let us entrust you with our baggage when we go to the summer pasture." The *tekfur* of Bilecik accepted this but on condition that the baggage be brought and put in the fortress by women and that no Turks or males should enter. So women took the baggage to the fortress of Bilecik. This arrangement occurred several times.

When ʿOs̱mān Ghazi and seventy men were passing through the region of the *tekfur* of İnegöl while returning from Ermenek, the *tekfur* of İnegöl learned of this and waiting in ambush, attacked him with many infidels. By divine favor, ʿOs̱mān Ghazi was victorious and killed most of the infidels. But Balı Ḫoca, the son of his brother Ṣarıbalı, was martyred. He was buried near the village of Ḥamza Bey, which is where the pass of Ermeni Beli ends. Today there is a ruined caravanserai next to his grave. ʿOs̱mān Ghazi then went to the summer pasture and returned.

Among ʿOs̱mān Ghazi's people at that time was a saintly shaikh named Edebali. He had countless worldly goods but lived like a dervish. He built a dervish lodge where he ministered to travelers and where sometimes ʿOs̱mān Ghazi was a guest. One night ʿOs̱mān Ghazi saw in his dream that the moon rose from the breast of this shaikh and entered his own breast. Thereupon a tree grew out of his navel and covered the whole world. In its shade were mountains, at the foot of which springs emerged and flowed, some irrigating gardens and some causing fountains to flow.

The next day ʿOs̱mān Ghazi related his dream to the saint. The shaikh interpreted it saying, "O ʿOs̱mān! Good news to you and your descendants. God has granted the sultanate to you and to them. Now I have a daughter—marry her!" ʿOs̱mān Ghazi became betrothed to her and married her.

As soon as Edebali interpreted his dream, [M 677] ʿOs̱mān Ghazi bound his sword to his waist. One night he rode with his companions and came to İnegöl. They set fire to the nearby fortress of Tulça and killed its infidels. The next morning all the infidels in that area came together and sent a man to the *tekfur* of Ḳaracaḥisār. He said: "Ever since the Turks came and settled here they have begun to reach out in all directions. Unless you deal with the situation they will allow neither you nor us to be here. Now let us drive them from this land."

Then the *tekfur* of Ḳaracaḥisār sent his brother with many troops to the aid of İnegöl. When ʿOs̱mān Ghazi learned of this he assembled his troops. He met the infidel army at the Ṭomaniç Pass. There was a great battle in which ʿOs̱mān Ghazi's brother Ṣarıbalı was martyred. There is a pine tree there known as Ḳandīlli Çam ("Pine with the Lamp") because sometimes a lamp is seen there. The *tekfur's* brother was also killed there. ʿOs̱mān Ghazi ordered that his belly be cut open and that they scratch the ground like a dog and bury him (i.e., in a shallow grave). For this reason that place is called İteşeni ("Dog Scratch"). Ṣarıbalı was brought from there and buried next to Erṭuğrul.

When ʿAlāʾ al-Dīn II learned of the martyrdom of Ṣarıbalı, he said, "Now everyone knows that the *tekfur* of Ḳaracaḥisār has rebelled against us." He gave Eskişehir to ʿOs̱mān Ghazi and gave permission for him to make a raid on Ḳaracaḥisār. So ʿOs̱mān Ghazi assembled his troops and conquered Ḳaracaḥisār. He took the *tekfur* prisoner and distributed his wealth to the Ghazis, setting aside one fifth which he sent with his nephew Aḳtīmūr to ʿAlāʾ al-Dīn II. This conquest occurred in AH 687 (1288).

It is reported that following these conquests the Muslims increased in number and asked for the Friday prayer service and the festival prayer service. When this was mentioned to Ṭursun Faḳīh he said, "Performing the Friday prayer requires the permission of the sultan." When ʿOs̱mān Ghazi heard this, he became angry and said, "These provinces were conquered by my sword. What concern is it to the sultan that we need to consult with him about it?" Thus he appointed Ṭursun Faḳīh to be both qadi and preacher in Ḳaracaḥisār, and the first Friday sermon was recited in the name of ʿOs̱mān Ghazi b. Erṭuğrul.

When ʿOs̱mān Ghazi made Eskişehir his capital, a great bazaar was set up and infidels came from all around to trade at the weekly bazaar. One day it happened that some infidels came from Bilecik bringing a load of cups to sell. Someone from Germiyan took one of the cups and did not pay for it. That infidel went to ʿOs̱mān Ghazi and complained. ʿOs̱mān Ghazi immediately took the money from that Germiyanid and gave it him, thus showing his great justice.

It is also reported that Ḳoca Miḫal (i.e., Köse Miḫal) gave his daughter to the *bey* of Ḳunuzıeli and invited the infidels in the vicinity to the wedding. All the *tekfurs* came bringing gifts, but the gift that ʿOs̱mān Ghazi brought surpassed them all. When the infidels saw it they were amazed and said to one another, "This behavior of ʿOs̱mān shows that soon he will not permit any of us to be here. The best plan is to do something about this immediately." They entrusted the matter to the *tekfur* of Bilecik, because ʿOs̱mān Ghazi had shown great respect in dealing with him. [M 678] The *tekfur* of Bilecik said, "We shall also have a wedding soon and we shall invite ʿOs̱mān Ghazi and the *tekfurs*."

It happened that the *tekfur* of Bilecik was to marry the daughter of the *tekfur* of Yārḥiṣār. ʿOs̱mān Ghazi sent a flock of sheep, saying, "Let my friend the *tekfur* feed these sheep to the wedding guests and for those coming to serve him. I shall also come, God willing." Meanwhile, Köse Miḫal informed him that the infidels were plotting against him. So he sent Miḫal to the *tekfur* with the following instruction: "Convey my greetings and tell him that every year they have borne trouble on our account. He knows of the hostility between us and the Germiyanids. Let them bear trouble for us once again this year and put our baggage in the fortress. We shall soon be going to the summer pasture. Let our baggage once again be entrusted to them and we, secured in this respect, shall render service at the wedding." When Köse Miḫal gave this message to the *Tekfur* he was very pleased, thinking that the Turks would fall into his hands with all their women and children and possessions.

The wedding was held at a place called Çaḳırbıñarı. At the appointed time ʿOs̱mān Ghazi loaded many oxen with bales of felt and, saying they were filled with household effects, put men inside. The womenfolk, who customarily put the baggage in the fortress, brought the trains of oxen inside the fortress walls in the dark of night and left the bales there. Having entered the fortress by this ruse, that very night all the armed men emerged from the bales with drawn swords, rushed to the gate, killed the guards and seized the fortress.

Meanwhile ʿOs̱mān Ghazi dressed several of his braves in women's clothing and sent them to the *tekfur*, saying, "Kindly lodge these ladies in a good place." He himself came in the evening, since his men who seized the fortress had been told that ʿOs̱mān Ghazi would meet the *tekfur* once again after they had entered the fortress. Word was sent that ʿOs̱mān's women were on their way, but before they reached their quarters, ʿOs̱mān Ghazi arrived on horseback, then departed as though he were fleeing. Informed that ʿOs̱mān Ghazi had fled, the *tekfur* immediately set out in pursuit. He caught up with ʿOs̱mān Ghazi near Bilecik. Those dressed as women cut him down from behind and ʿOs̱mān Ghazi cut off his head.

That night he descended upon Yārḥiṣār, seizing its *tekfur* the next morning and taking as prisoners the bride and all those who were coming to the wedding. Then he went to İnegöl, immediately plundered it and killed its *tekfur*. These conquests occurred in 699/1299. That bride's name was Nilüfer. ʿOs̱mān Ghazi gave her in marriage to his son Orḫān. This lady has a dervish lodge (which she endowed) in Bursa at the foot of the fortress near the Ḳapluca Gate. She also built the Nilüfer Bridge. This Nilüfer was the mother of Sultan Murād (I) and Süleymān Pasha.

After conquering Bilecik and Yārḥisār, ʿOs̱mān Ghazi went to İznik and plundered the area around it. Its *tekfur* sent word to the king of Istanbul (the Byzantine emperor) and asked him for help. ʿOs̱mān Ghazi learned of this and informed Sultan ʿAlāʾ al-Dīn of the fortresses that had been conquered. He gave ʿOs̱mān Ghazi the drum and standard, sword and horse and robe of honor. Before the man who had gone to ʿAlāʾ al-Dīn could return, the infidel army set out from Istanbul and started to cross (the Gulf of İzmit) at Dil. ʿOs̱mān Ghazi seized informants from Yaylaḳ Ḥisārı and learned of the infidels' heedlessness. [M 679] When they had crossed from Dil, he attacked them at night and overwhelmed them. Some were killed, some were captured, and some fled to Istanbul.

Meanwhile, the drum and standard, horse and sword and robe of honor arrived from Sultan ʿAlāʾ al-Dīn II at the time of the afternoon prayer. The government officials, statesmen and dignitaries drew up in formation and the Sultan's divan was arranged. ʿOs̱mān Ghazi rose to his feet out of respect and remained standing as the ʿOs̱mānī (or Ottoman) drumbeat was sounded. Thus the Ottoman military band was established on the basis of imperial rule and sultanic protocol. From that time until the reign of Sultan Meḥmed (II) son of Sultan Murād Khan, this was part of Ottoman state ceremony. Whenever the band played as a sign that a military campaign was to begin, the sultan rose to his feet. The same practice was followed during festivals. After Sultan Meḥmed (II) the practice of rising ceased.

A short time later, Sultan ʿAlāʾ al-Dīn passed on to the next world. As he had no children, the vizier who was in charge of his affairs succeeded him. His emirs who governed the provinces became independent. The *beys* of Aydın, Şaruḫān, Germiyan, Ḥamīdeli, Teke, Ḳaresi, Ṭurġud, and Menteşe defended their own provinces and became independent rulers. Afterwards, because they attacked one another and drew their swords against Muslims, the ulema issued fetwas allowing (the Ottomans) to attack them and seize their territories. Some surrendered willingly and were treated respectfully according to their status, while others were overcome by force. Even today those territories bear the names of those *beys*.

It is reported that when ʿAlāʾ al-Dīn died, Ḳasṭamonı and Sinop were in the hands of the clan known as the Ḳızıl Aḥmedlu (i.e., the Candaroġulları or Isfandiyāroġulları) who had ruled those places under the authority of ʿAlāʾ al-Dīn and subsequently became independent. Among those who then ruled was ʿĀdil Bey (r. 1345–1361) and then Bāyezīd-i Zemen ("the crippled" = Kötürüm Bāyezīd, below) b. ʿAlī b. Yaʿḳūb b. Şemseddīn b. Nūr al-Dīn the Martyr. Their lineage goes back to Khālid b. al-Walīd. But the most correct account is that Bāyezīd-i Zemen died in 795 (1392). He was succeeded by his son İsfendiyār (r. 1385–1393, 1402–1440) who was succeeded by his son İbrāhīm Bey (r. 1440–1443). Afterwards, in 865 (rather 866/1462), Ebū'l-fetḥ Ghazi Sultan Meḥmed Khan took the fortresses of Ḳasṭamonı and Sinop from the Ḳızıl Aḥmedlu (i.e. Ḳızıl Aḥmed b. İbrāhīm Bey) by the sword. This Ḳızıl Aḥmed was a descendant of Bāyezīd-i Zemen.

It is reported that around that time someone from the province of Germiyan went to ʿOs̱mān Ghazi and made a request, saying "Sell me the tax concession on this market." ʿOs̱mān Ghazi asked, "What is a tax?" The man said, "I will collect one *aḳçe* from anyone who brings a load of goods to the market." ʿOs̱mān Ghazi replied, "Do you have a debt outstanding against those who come to this market that you wish to exact from them? Or is it a divine command or a saying of the Prophet? Or is it based on the customary law of the rulers of every clime?" The man said, "They are the laws of the sultans." ʿOs̱mān Ghazi became angry and said, "Away from here or I will punish you!"

Several of ʿOs̱mān Ghazi's close companions were present during this exchange. They said, "O Khan, although you do not need the money, it is customary to give a little something to the market guards so that their labor does not go unrewarded." ʿOs̱mān Ghazi said, "Since that is the case, let everyone who sells one load of goods pay one *aḳçe* and if he sells nothing he should pay nothing."

Subsequently he conquered Yeñice and Aḳḥiṣār. [M 680] After these, ʿOs̱mān Ghazi besieged Bursa. Because the siege was protracted, they built two strongholds. ʿOs̱mān Ghazi began to suffer from gout and so turned the siege of the fortress over to his son Orḫān and returned to Yeñişehir. In 726 (1326) Bursa was conquered. Meanwhile ʿOs̱mān Ghazi died in Söğüt (in 724/1324). When Orḫān Ghazi learned of this, he buried ʿOs̱mān Ghazi in the monastery (or, in the Manastır [madrasa], as below) in Bursa, beneath the dome, according to his last will and testament. He was 69 years old and his sultanate had lasted 16 years. His personal effects consisted of a new turban, a saddle blanket, flank armor for his horse, a saltcellar and spoon rack, and a pair of high boots. The turban and saddle blanket were made of Deñizli cloth. He also had a few herds of horses and flocks of sheep. Those sheep were the breeding stock for the *beylik* sheep that now graze in the environs of Bursa.

Then Orḫān Ghazi became sultan and made Bursa his capital. His brother ʿAlāʾ al-Dīn, of his own accord, requested the village of Barḳuṭan, saying, "That is enough for me." And he built a dervish lodge at a place called Kükürtlü. He resided in a mosque in the Bursa citadel just inside the Ḳapluca Gate. Orḫān Ghazi would always consult with him.

Later Orḫān Ghazi besieged İzmit. He assigned the conquest of Aydos to ʿAbdurraḥmān. It happened that the *tekfur* of that fortress had a daughter who was the beauty of the age. One night she had a dream in which she fell into a pit and could not get out. A handsome young man came from across the way, took her out of the pit, removed her old clothing, washed her body and dressed her in silk garments. The girl immediately awoke and marveled at this. While pondering it, word arrived that the Turks had come and were besieging the fortress. The girl came to the edge of the tower and looked down to see how the Turks were fighting. Her gaze happened to fall upon ʿAbdurraḥmān Ghazi and she realized that he was the one she saw in her dream. She immediately wrote a note, attached it to a stone, and threw it down. ʿAbdurraḥmān picked it up and had it read by someone who knew Greek. The note stated: "On such and such night at such and such place, send men whom you trust and I will turn the fortress over to you." ʿAbdurraḥmān Ghazi said: "I will take this upon myself." At midnight he went with a number of *ghazis* to the place the girl had mentioned. The girl saw them. Without delay ʿAbdurraḥmān attached a rope to the rampart, climbed up the fortress and met the girl. Proceeding to the fortress gate, he immediately killed the guard and opened it. The Ghazis entered and took the *tekfur* captive.

When the fortress of Aydos was conquered, ʿAbdurraḥmān brought the *tekfur* and his daughter and all their wealth to Orḫān Ghazi. He gave ʿAbdurraḥmān many gifts from the booty in addition to the girl. Descendants of his are still with us, they are called Ḳara Raḥmān.

Later, in 731 (1331), Sultan Orḫān conquered the cities of Koyunḥiṣārı, İzmit, and İznik. Matters relating to sultanic law—such as coinage, military uniforms, offices and ranks—he arranged based on the instruction of ʿAlāʾ al-Dīn Pasha. He replaced the old Seljuk coinage and struck (dirhams) in his own name. In order to distinguish the military from the tax-paying subjects [M 681] he ordered them to wear red, yellow, and black *börks* (tall felt caps). Later he substituted white for black, and so it remained until the reign of Yıldırım Bāyezīd Khan. He assigned white *börks* to his corps of imperial guards and red *börks* to his notables and attendants. This arrangement remained until the reign of Ebū'l-Fetḥ Sultan Meḥmed who ordered the army (i.e., the *sipāhī* cavalry) to wear white turbans and the footsoldiers (i.e., the janissaries) to wear white *börks*. The latter, embellished with gold, are now called *üsküf*; the red *börks* are reserved for the sultan's attendants.

Creation of the janissaries, by grace of ʿAlā al-Dīn Pasha, brother of Orḫān Ghazi

It is reported that one day Sultan Orḫān consulted his brother ʿAlāʾ al-Dīn Pasha and said, "It is my desire to increase the size of the army and to expand its glory and magnificence." ʿAlāʾ al-Dīn said he should consult the qadi. At that time Ḫayr al-Dīn Pasha, who had been made qadi of Bilecik by ʿOsmān Ghazi, was the qadi in Bursa having been appointed by Orḫān Ghazi—afterward Ghazi Murād (Sultan Murād I) made him *ḳāżıʿasker* (chief qadi of the military) and then he became vizier. In short, they consulted him. He said, "My sultan, let us conscript footsoldiers from the countryside and train them." When the subjects heard of this they showed their enthusiasm saying, "Let us be of service to the Sultan." Many men were conscripted. They donned the white *börk* and marched off. When sufficient footsoldiers had been raised, others came and said, "Now register us as *yamaḳs* (locally hired soldiers at detached forts) and let us go on campaign together and be of service."

According to the scheme devised by Ḳara Ḫalīl, ʿAlāʾ al-Dīn, and Ḫayr al-Dīn, these recruits were to be registered as pages or else assigned as guards to the palace gardens. Those who were registered and went on campaign would each receive one *aḳçe*, equal to a quarter dirham. When they returned from campaign they would go back to agricultural work and no government tax would be collected from them. A leader was assigned to every ten men, every hundred men, and every thousand men. They became known as the infantry (*piyāde*).

Later this corps gradually became corrupt, both in war and in peace time. The sultan decided to conduct a levy of infidel children who were then enrolled in the army. They were to adopt Islam and engage in jihad against the enemies of religion. At that time they levied 1000 boys and each was given an additional dirham, according to his ability. They were called janissaries (*yeñiçeri*, lit. "new army"). The infantry were then assigned agricultural lands and the janissaries were assigned military duties in their stead. But in times of war, those who brought their horses and were ordered to join the regular army were called *müsellems*.

It is reported that during the time of Orḫān Ghazi, many dervishes came to Keşiş Ṭağı in the vicinity of İnegöl and took up residence there. Among them was one dervish who lived in the mountains with the deer (*geyik*). Only Ṭurġud Alp was on close terms with him. This dervish was a disciple of Baba İlyās and a member of the Sufi brotherhood of Seyyid Ebū Elvān (Çelebi). When Orḫān Ghazi summoned him, he would not come. Orḫān Ghazi sent word to him, saying, "God willing, one day I will go there." One day this dervish took a poplar tree on his shoulders and planted it next to the courtyard of the palace in the fortress of Bursa. Sultan Orḫān was informed that the dervish had come and went to meet him, but did not ask why he had planted the tree. Only the dervish said, "As long as our blessing stands this will stand," and he bade him farewell and departed. Vestiges of that tree remain to this day. When that dervish died, Orḫān Ghazi erected a tomb over him according to his last will and testament [M 682] and built a dervish lodge and Friday mosque next to it. The dervish lodge of Geyiklü Baba is still a place of visitation.

Sultan Orḫān went on to conquer the fortresses of Göynük, Ṭaraḳlu, Mudurnı, Gemlik, Balıkesri, Ḳaresi, Bergama, Edremid, Kirmasti, Ulubad, Edincik, Manyas, Miḫalic, Yārḥisār, Ayazmend, Ḳızılca, Ṭuzla, Aydın, and Anaḫor. It then occurred to him that he should cross to the other shore (of the Dardanelles). As he was thinking about this, his son Süleymān Pasha suddenly appeared and he told him his plan. Süleymān Pasha said, "If you command me, your slave, I will cross." Orḫān gave his permission and Süleymān went with the Muslim Ghazis to the province of Ḳaresi. One day while travelling about he came to the promenade of Edincik. Ece Bey and Ghazi Fāżıl said to Süleymān, "My sultan you seem to have something on your mind." Süleymān Pasha said, "I wonder if it is possible to cross this sea?" Ece Bey and Ghazi Fāżıl replied, "If you command us, we two will cross." When Süleymān Pasha asked where they would cross, they answered, "From the nearest point (to the opposite shore) which is a ruined fort."

At night they reconnoitered the area around Çini Ḥiṣārı below Güvercinlik. They got hold of an infidel among its vineyards, put him on a raft, crossed back and took him to Süleymān Pasha. Süleymān Pasha dressed him in a robe of honor and showered him with gifts, then asked, "Is there a place where we can enter your fortress without being noticed by the infidels?" He told them. They immediately built several rafts which Süleymān Pasha boarded with eighty brave men and crossed to the other side. They fell upon the fortress and took it. They unfurled their flags, took out the *ṭabl* and *naḳāre* (two types of drums) and beat them. They did not harm the infidels in the fortress but rather treated them with kindness. They took several ships that were in the vicinity and conveyed them to the army on the other side. So on that day they brought over more than 200 men. Ece Bey, using the horses that were in the fortress, (rode to) Aḳça Liman near Bolayır where he took the boats that were in the harbor and used them to bring over more men continuously. In short, they brought over more than 2000 men and seized several fortresses. These conquests occurred in AH 758 (1356).

Then Süleymān Pasha sent word to his father Orḫān Ghazi saying, "My Sultan, thanks to you Rumelia has been conquered. We need men to station as guards in these fortresses that were conquered with the help of God. Please send men to this side." Sultan Orḫān was pleased with this happy news. There were nomads in the province of Ḳaresi. Orḫān Ghazi drove them out and had them cross over to Rumelia. For some time they settled in the district of Gelibolı.

In 7[5]9 (1357) Süleymān Pasha went on the attack and captured, Gelibolı, Ferecik, Ḥayrabolı, Vize, Tekfurṭağı, Seydiḳavağı, Malġara, İbsala and Çorlu. One day while Süleymān Pasha was hunting, his horse fell. Fate caught up with him and he died. His works were a Friday mosque near the hot spring inside the fortress in Bursa, a soup kitchen in Bolayır and a Friday mosque and madrasa in İzmit.

Sultan Orḫān Ghazi died two months later and was buried in the Manastır Madrasa. [M 683] His works were a soup kitchen in Bursa and another in İznik, a Friday mosque and a madrasa in Bursa, and a dervish lodge named after Geyiklü Baba with a Friday mosque (attached). He also assigned a salary to the men of the religious sciences and the qadis.

His son Sultan Murād Ghazi came to the throne in 761 (1360). He appointed Ḳāḍı Ḫayr al-Dīn Pasha *ḳāżıʿasker* in Bursa. First he punished the evildoers in Anaṭolı and conquered Ankara. Then he crossed over to Rumelia and conquered several fortresses in that region. In 762 (1360) he conquered Edirne and took up residence there. He sent his *lala* (tutor) Şāhīn to Zağra and Filibe and he conquered them.

Around that time a learned man named Ḳara Rüstem came to the province of Ḳaraman. This Rüstem came to Ḫayr al-Dīn Pasha, who was then *ḳāżıʿasker*, and said, "Efendi, you are losing so much sultanic wealth. These ghazis have taken many captives in their raids. By God's command, one-fifth belongs to the sultan." When the *ḳāżıʿasker* informed Sultan Murād of this, he said, "If it is God's command, then from now on take it." Subsequently Ḳara Rüstem himself took up residence there and appointed qadis to take 25 *aḳçes* for (each of) the previously captured prisoners and take one-fifth of the prisoners (captured henceforth). He called them "qadis of the raiders" (*aḳıncı ḳāḍıları*). In this way, many slaveboys were collected and brought to Murād Khan. Ḫayr al-Dīn said, "Let's give them to the Turks, so they will become Muslims and will also learn Turkish; then they should be brought back and made janissaries." Thus the number of janissaries increased day by day. Later, while Sultan Murād was on the way from Edirne to Bursa, he came to Gelibolı where he removed Ḫayr al-Dīn from the post of *ḳāżıʿasker* and made him grand vizier.

It is reported that in AH 774 (1372), Lala Şāhīn Pasha conquered the places called İnecik and Çatalburġaz.

It is also reported that Germiyanoğlı had become very old. He summoned his son Yaʿḳūb Bey and said, "I want this province to remain in your hands. Let me give my daughter to Murād Ghazi's son Yıldırım." He immediately sent İsḥāḳ Faḳīh as an envoy (to Sultan Murād). He took fine horses and gifts. In short, after he arrived a wedding contract was concluded and as a dowry he (Germiyanoğlı) gave Kütahya, Simav, Egrigöz, and Ṭavşanlu. There was a magnificent wedding. The *beys* from all the surrounding areas and even the sultan of Egypt were invited. Envoys came and wedding gifts and rarities began to arrive. They gave precedence to the envoy from Egypt. Evrenoz Ghazi brought gifts. In addition to garments, 100 handsome slaveboys and 100 virgin slavegirls were given as gifts. All were beautifully adorned. Each boy carried in his hands a gold tray and a silver tray. Each tray (respectively) was full of the same amount of silver coins or red (i.e., gold) florins. The beys from the surrounding areas were astonished. Murād Khan distributed the gifts that the Egyptian envoy had brought to the *beys* and the envoys from the surrounding areas.

A remarkable event: After the conquest of Niş (in 1386), Sultan Murād came to a fortress while on the road to İncegiz. It is called Tanrıyaḳduğı ("What God Has Burned"). It had previously been called Polonia (?). The reason it was renamed is as follows. Sultan Murād attacked it and besieged it for several days. [M 684] Blood flowed in streams, but the fortress could not be taken. Finally he lifted the siege and said, "God burn this fortress." They made camp at a place called Ḳabaağaç ("Rough Tree"). Murād Khan sat down with his back against that rough tree. A moment later a man arrived and said, "By the power of God, that fortress has been burned." When Murād Khan heard this, he sent his *lala* (Lala Şāhīn Pasha). He brought back much wealth that was in the fortress, including tall red crowns, each one surmounted with figured gold more than a span in length. The sultan put them on the heads of his imperial guard, and they have been part of their traditional dress ever since. The *sipāhīs* found golden bowls in that fortress and put those on their heads. This was the origin of the *üsküf* (tall headdress of the janissaries). Vestiges of that tree remain to this day and there is a well at its foot. The place is now called Devletlüağaç ("Auspicious Tree").

Murād Khan's greatest religious war: The *tekfur* of Nigbolı (Nikopolis) had sought the help of ʿAlī Pasha (Candarlı), saying, "Please ask the sultan to forgive my offense and I will surrender the fortress of Silistre to him." After peace was made on this basis, the infidel again went back on his word. So Sultan Murād headed for Filibe (Philipopolis) and set up winter quarters there. At this, the infidels from all around communicated with one another and came to an agreement. These infidels—composed of Franks, Hungarians, Wallachians, Moldavians, Croats and Russians, and numbering more than 200,000—assembled on the Plain of Kosovo. There were 500 great *tekfurs*. In short, such an enormous army of cavalry and infantry had never been seen. But God gave victory to the Muslims.

While Sultan Murād Khan Ghazi was viewing these infidels, a murderous infidel rose up among them and headed toward the sultan. His chavushes wanted to stop him, but the sultan said, "Don't harm him, let's see what he wants, let him approach." But that accursed man had hidden a dagger in his sleeve. He approached the sultan as if he wanted to kiss his stirrup and then suddenly struck him with the dagger. [Arabic proverb] "When destiny comes, the eye is blind." The sultan died a martyr and that infidel was cut to pieces. That very hour Yıldırım Bāyezīd was raised to the throne. This event occurred in 792 (1389). Sultan Murād was brought to Bursa and buried near the hot spring.

His works: He built a soup kitchen, Friday mosque, and madrasa near the hot spring in Bursa; a soup kitchen and Friday mosque in Edirne; a Friday mosque in Bilecik; a Friday mosque in Yeñişehir; a mosque in Gökova; and palaces near Bursa and in Edirne. He made several improvements in administration, stipulating that the timar of a deceased holder should be divided among his children, and maintaining a standing army for service in campaigns and in the imperial stables. His first conquest was Alaşehir, which had been left in the hands of the infidels; he then took Sivas from Ḳāḍı Burhān al-Dīn; and he also conquered Toḳat, Amasya, and Ḳasṭamonı. The son of Kötürüm Bāyezīd ("the crippled" = Bāyezīd-i Zemen, above) fled to Yaʿḳūb; later he abased himself before Sultan (Yıldırım) Bāyezīd who granted him Ḳasṭamonı. Meanwhile Menteşeoğlı fled to Tīmūr. In 795 (1392), while besieging Istanbul, he received word that 130,000 infidels had crossed the Danube and were besieging Nigbolı. He immediately broke off the siege of Istanbul and [M 685] went there. He defeated the infidel army and took most of its men captive (798/1396). On his return, he learned that the infidels had built a fortress called Güzelceḥiṣār above Boğazkesen (Rumeli Hisar). He immediately sent a man to the *tekfur* of Istanbul (Byzantine emperor) and said, "This fortress must belong to us. Evacuate it immediately or be prepared for your fate." As soon as he was informed of this, the *tekfur* sent an envoy who said that he would agree to pay 20,000 gold in cash as tribute and provide other things every year. The sultan accepted this offer, on condition that a Muslim quarter and mosque be established in Istanbul and a qadi be appointed. Then the sultan ordered the villagers and castle dwellers of Ṭaraḳlu Yeñicesi (in northwestern Anatolia) to be driven out and settled in Constantinople. So the Muslim quarter was established, a mosque was built, and a qadi was appointed. But later, when Bāyezīd set out to fight Tīmūr, the *tekfur* abolished that quarter and tore down the mosque.

It is reported that, while Sultan Bāyezīd was building a soup kitchen in Edirne, Vılḳoğlu (Vuk Branković) sent an envoy with lavish gifts and a letter, dictated by his mother, promising a uniquely beautiful girl—Driga (Despina), the daughter of Laz (Prince Lazar of Serbia, r. 1373–1389)—in marriage to Bāyezīd. Sultan Bāyezīd accepted; they sent the girl and he got what he wanted. Sultan Bāyezīd learned how to drink wine and dissipate himself from this daughter of Laz; up to that time, no one from the Ottoman family had committed such deeds.

During his reign also there were complaints that the qadis took bribes. The sultan was infuriated and issued an order that they be put to death. Subsequently, on the request of the grand vizier, the *resm-i aḳçe* was allowed to be collected at the rate of 20 *aḳçes* per thousand. The *resm-i ḥüccet* (title-deed tax) was 25 *aḳçes*, the *sicil ücreti* (registration fee) was 7 *aḳçes*, and the *ʿaḳd-ı nikāḥ* (marriage contract fee) was 12 *aḳçes*.

In 800 (1397), the provinces of Morea (Peloponnesus), Tırḥala (Trikkala in western Thessaly), and Athens were conquered. In 804 (1402) occurred the battle (of Ankara) with Tīmūr. Sultan Yıldırım Bāyezīd was taken prisoner and died in Aḳşehir (805/1403). From there he was transported to Bursa and buried in his tomb. First his son Mūsā Çelebi (became sultan) and then his son Sultan Meḥmed (I) emerged from a struggle among brothers for the throne and became sultan.

The works of Sultan Yıldırım Bāyezīd: He built a Friday mosque, a madrasa, two hospitals, a soup kitchen, and a house for Ebū İsḥāk in Bursa. He built a Friday mosque, soup kitchen, and madrasa in Edirne. He built a Friday mosque and madrasa in Balıkesri. He also has Friday mosques in Ḳaraferye (Beroia in Greece), Dimetoka (Didymoteichon in Greece), and Kütahya. He carried out thirteen religious wars (*ġazā*).

It is reported that when Tīmūr learned that Sultan Meḥmed was going to take his father's place on the throne he sent an envoy saying, "Come to me and I will betroth my daughter to you and will again turn over to you the country of Rūm." Sultan Meḥmed found an excuse and did not go.

In 819 (1416), word arrived that the infidels of Wallachia refused to pay the tribute. He immediately set out for the region of the Danube and had fortresses constructed at Yergögi (Giurgiu) and İsaḳçı (Isaccea, near the mouth of the Danube). Afterwards they again submitted.

In 823 (1420) Bedr al-Dīn b. Ḳāḍī Simāvnā appeared in Rumelia and Börklüce appeared in Aydın. In short, Sultan Meḥmed had 42 battles. He died in Edirne in 824 (1421) and was succeeded by his eldest son, Murād II. [M 686]

His works: He built a Friday mosque, a madrasa, a soup kitchen known as Yeşil ʿİmāret, a school, and an inn in Bursa. In Merzifon he built two Friday mosques and two public baths. He built a Friday mosque and a soup kitchen in Filibe. He built a Friday mosque known as Eski Cāmiʿ and a hospital in Edirne. He was the first to assign a sultanic gift for Mecca and Medina from the proceeds of pious endowments.

When Sultan Murād II came to the throne there was disorder throughout the country. In 828 (1424) he took Ḳasṭamonı from the Isfandiyarids (or Candarids). In 830 (1426) he conquered Alacaḥiṣār, which was the Laz capital (i.e., the Serb capital of Kruševać). Then he conquered Thessaloniki, then Albania, and then Semendire (Smederovo in Serbia).

In 846 (rather 848/1444), when his son Sultan Meḥmed Khan ascended the throne, the Ḳaramanid sent an envoy to the king of Hungary, saying "What are you waiting for? The opportunity is yours and mine." So the king of Hungary went on the march, followed by King Yanḳo. Thus they lead on King Yanḳo, and went on the march. Vılḳoğlı, who was the king of Hungary, assembled a huge army composed of men from Hungary, Çeh (Czechia, Bohemia), Leh (Poland), Alman (Germany), Saz (Saxon inhabitants of Transylvania), Bosnia, Wallachia, and Moldavia. With 400 gun carriages they passed through Belgrade and marched on to Niş, Nigbolı and Varna. Meanwhile, the people of the country rose against Sultan Murād, saying, "You have given the country to ruin with your own hands! What are you waiting for?" They tried to detain Sultan Meḥmed in Edirne, but he crossed into Rumelia and attacked the infidels.

In short, there was a great battle in the mountain of Varna and the Muslim army was routed. The sultan, left with only his household, thought of flight. When the sultan's maternal uncle Ḳaraca Bey saw what was happening, he dismounted, held tightly to the sultan's skirt, and said, "What are you doing? If you flee, the infidels will follow us all the way to Edirne." The sultan immediately went up to a high place and took his stance, imploring God and asking for His favor. His banners once again stood firm and his drums were beaten.

At once, a number of ghazis shouted, "Hey ghazis! Where are you going? The infidels have been routed." At that juncture the king of Hungary recklessly attacked the sultan's regiment. By divine favor, they hamstrung his horse and knocked him down. Ḳara Ḫıżır dismounted, cut off the king's head and put it on a lance. At once the cry *Allāhu ekber!* rang out. When the ghazis saw the king's head on the lance, they were overjoyed and began to cut down the infidels right and left. Countless infidels were killed. After the Sultan returned to Edirne he sent gifts and captives to all the surrounding beys, and even sent 25 mounted and armored infidel soldiers to the sultan of Egypt. This conquest occurred in 848 (1444).

In 849 (1445), the Janissary Corps pillaged the viziers' houses. Upon receiving a half-*aḳçe* increase in pay per day, they again placed Sultan Murād on the throne.

In 852 (1448), the new king of Hungary again set out with an army of surrounding infidels—Poles, Czechs, Germans, Latins, the king of Sikult (Siculi, Szeklers in Transylvania) and the king of Lek (?). They passed through Belgrade and assembled at Kosovo. When Sultan Murād got word of this he put his faith in God and headed for Kosovo, reaching there on a Friday. As day dawned and he encountered the infidels, Sultan Murād immediately dismounted, [M 687] performed two prostrations of "need prayer" (*ḥācet namāzı*), rubbed his face on the ground, remounted and attacked. There was a great battle and by evening he had taken many standards of the infidels. Fighting continued through the night and into Saturday. Many of the infidels fell and some fled. The Polish king was killed and the Czech king was captured.

In 855 (1451), Sultan Murād reached God's mercy in Edirne and was buried in Bursa.

His works: He built Friday mosques known as Üç Şerefeli and Murādiye, another Friday mosque, a soup kitchen and two madrasas in Edirne. He built a soup kitchen and a madrasa in Bursa. He erected the Ergene Bridge (in western Thrace) where there was previously a thicket and swamp and nest of brigands. Sultan Murād cleared the thicket and built a bridge with 194 arches. On one side of the bridge he built a town called Ergene and a Friday mosque and a soup kitchen. He established Balıḳ Ḥiṣārı in the district of Ankara as a pious endowment for the poor in Mecca. He distributed 1000 gold pieces every year to the *sayyids* (descendants of the Prophet). He sent 3500 gold pieces every year to Mecca, Medina, Jerusalem, and Khalīl al-Raḥmān (Hebron). He participated in 41 battles. Among the honorable Sufi shaikhs of his reign were Ḥācı Bayram Velī (d. 1429), Shaikh Şemseddīn (Meḥmed) Fenārī (d. 1431), Ḫıżır Pasha, Aḳ Şemseddīn (d. 1459), and Shaikh Emīr Sulṭān (d. 1429).

After him, his son Ebū'l-Feth Ghazi Meḥmed Khan (Fātiḥ Meḥmed II) succeeded independently to the throne. He had a fortress built above Istanbul across from Aḳçeḥiṣār (Anadolu Hisarı) and artillery placed in it, allowing no ships to pass. Then in 857 (1453) he laid siege to Istanbul and stormed the city from fifty places in fifty days. Then he had 400 ships, with sails spread, hoisted from the castle (Rumeli Hisar) overland on carts and lowered (into the Golden Horn) at the base of the city wall. They also made a bridge across the sea (Golden Horn). The ghazis stormed the city and the *tekfur* (Byzantine emperor) was killed in the battle.

Subsequently Sultan Meḥmed was preoccupied with the reconstruction of Istanbul. He brought people from the surrounding areas and settled them in the city. As they took up residence they began to impose rent (lit., to treat their property as land grant or tax farm, *muḳāṭaʿa*). The people hated this. One day a man named Ḳula Şāhīn, who was a veteran in service to the sultan's forebears, said, "My Sultan, your forebears conquered many countries but they never imposed rent. This is not befitting my Sultan." These words impressed the Sultan and he abolished the rent. Later Rūm Meḥmed Pasha (d. 1478) became vizier and led the Sultan astray so that out of greed he again imposed rent.

It is reported that after conquering Enez (Ainos on the Aegean coast of Thrace), Ṭaşözü (the island of Thasos), Memleket-i Las (the country of Ladislas, i.e. Serbia), Serbia and Semendire (Smederevo), he set out again on campaign toward Morea in 862 (1457). During this campaign, he conquered 60 fortresses. And in 864 (1459) he took the fortress of Amaṣra in Anaṭolı from the infidels. He also took Sinop, Küre-i Nüḥās, Trabzon, and the islands of Midillü (Lesbos) and Bozcaaṭa (Tenedos). In the region of Bosnia he took Lofça and Yayiçe. In the region of Morea he took the Island of Ağriboz (Euboea). He took Alanya from Qilich Arslān and Lefke from the Ḳaramanids. He had many households transferred from Aḳsarāy to Istanbul, the quarter that is today called Aḳsarāy. [M 688] He conquered Ḳaraḥiṣār-ı Şarḳī and he conquered Kefe and Menkup (Kaffa and Theodoros in Crimea) from the Franks. He also conquered Azov and Moldavia and most of Albania. In short, many conquests were made during his reign.

His works: First he built the Eski Sarāy ("Old Palace" in Istanbul). Then he preferred another place known as Zeytūnlıḳ ("Olive Grove"). He built another fortress there, where he established his harem and erected lofty palaces, making it the seat of the throne (Topkapı Sarayı). In a noble place in Istanbul he built eight madrasas with a Friday mosque in the middle of them, a soup kitchen opposite, on one side a hospital, and behind the madrasas a preparatory school for theology students where they were provided stew and bread from the soup kitchen. He had a Friday mosque built for Shaikh Vefāzāde (Ebū'l-Vefā', d. 896/1491); a soup kitchen, madrasa, and Friday mosque for Aḳ Şemseddīn for discovering the tomb of Ebū Eyyūb Enṣārī; and a tomb over the grave of Ebū Eyyūb. And he made the Ayasofya into a Friday mosque and established many pious endowments for it.

This sultan had a great interest in history. Indeed, in order to learn about the past, he consulted with Greeks and with monks from Europe, questioning those who knew history and commisioning translations from them.

It was Sultan Meḥmed who established the hierarchy of government offices, beginning with a wage of 20 *aḳçes* (per diem) to teach mathematics and theology in two madrasas; then 30 *aḳçes* to teach the *Miftāḥ* and *Meʿānī* and Ṣadr al-Sharīʿa; 40 *aḳçes* to teach the *Şerḥ-i Mevāḳıf* and *Şerḥ-i Maḳāṣıd*; 50 *aḳçes* to teach the *Hidāya* in the "Outside" madrasas built by viziers; then they would be promoted to the "Inside" madrasas built by the sultan; then to one of his eight madrasas; and finally they would be appointed qadis in Mecca and Medina, Aleppo, Damascus, Cairo, Baghdad, and the Three Towns (Eyyub, Galata, and Üsküdar).

Because the Muslim armies were making many conquests, the tax-paying subjects were required to pay 22 *aḳçes* (per annum) as *çift aḳçesi* (farm tax). When the campaign distance increased, the sultan ordered the *sipāhīs* to exact 30 *aḳçes*. Today 32 *aḳçes* are taken from each of the tax-paying subjects, a practice that goes back to that time.

His character: He held the ulema in great favor and was himself a member of the learned class. Whenever he heard of a capable man he had him brought and given a salary. He even had the Pride of the Scholars (Faḫrü'l-ʿulemā) Mevlānā Shaikh ʿAlī Ḳuşçı (d. 1471) with all his dependants brought from Samarqand and showered him with wealth. Any expert craftsman was definitely brought and appointed a salary.

He built only one lofty palace in Edirne. He died in 886 (1481).

His son Sultan Bāyezīd Khan (II) then came from Amasya and assumed the throne. At that time Cem Çelebi (Bāyezīd's younger brother) was in Bursa. When Sultan Bāyezīd marched there to seize him he fled to Egypt. From there he went back to Rūm (Anatolia) and then to Europe.

During Bāyezīd's reign, Kili, Aḳkirman, Adana, and Tarsus were conquered, as well as the fortresses of İnebaḫtı (Lepanto), Moton and Ḳoron (Methóni and Koróni in Morea), Anavarin (Navarino), and many fortresses in Albania.

His works: He built a soup kitchen and a madrasa in Amasya; a bridge over the Ḳızılırmaḳ at ʿOs̱māncıḳ; a soup kitchen, madrasa, hospital, and Friday mosque in Edirne; a bridge over the Tunca (in Thrace); a large bridge over the Ṣaḳarya; a famous market in Edirne to replace the one that had burned; a soup kitchen, madrasa, Friday mosque, and a school in Istanbul; and a bridge over the Gediz River in the province of Ṣaruḫān. He died in 819 (rather, 918/1512). [M 689]

His son Sultan Selīm Khan (I) came from Trabzon and ascended the throne. This sultan adopted the following excellent practices: most nights were devoted to reading fine books and discussing matters related to the ordering of the world (i.e., affairs of state).

Indeed, one night the sultan and two other people had the same dream. One was Ḥasan Agha, the Ḳapu Ağası (chief white eunuch in the imperial palace), who used to get up in the middle of the night to pray. Sultan Selīm himself had a certain dream, and when he was informed that someone else had the same dream, he ordered the Ḳapu Ağası to relate his dream. The Ḳapu Ağası related as follows:

"This night there was a loud knocking at the door where I was sitting. I went to see what was going on and noticed that the door was slightly ajar. Outside was a group of men with luminous faces and Arab features, wearing hooded cloaks and carrying white banners. When they said to me, 'Do you know why we have come here?' they asked. 'Please tell me,' I replied. 'These men whom you see are the companions of the Prophet. The Prophet sent us here. He greets Sultan Selīm Khan and orders that he set out with his sea-roiling army. He has entrusted him with the service of Mecca and Medina. Go and tell Selīm Khan.'"

Sultan Selīm Khan kept this dream in his heart. After punishing his brother Sultan Aḥmed, he marched against the people of error (i.e., the Safavids). In 920/1514, he went to Seydīġāzī and bestowed favors on the *sipāhīs*, giving to each timar holder 50 *aḳçes* per 1000 (of the value of his timar). Then he went to Sivas and reviewed the troops. He ordered those who were unsuitable for battle to remain between Sivas and Kayseri, so they would not use up the grain supplies in the army, and advanced with 40,000 handpicked soldiers.

The Shah (Ismāʿīl) came to the battle. To ensure that the Sultan would not flee, the Shah sent him (as symbols of derision) now a staff and woolen cloak (appurtenances of a shepherd); now a woman's hoop and headscarf and felt cap; or a pot with *maʿcūn* (aphrodisiac taffy, suggesting the Sultan was impotent).

While en route, the Janissaries more than once rose up en masse against some of the high officials, demanding—now by favor, now by force—to give up the campaign. Indeed, one night, in order to frighten the Sultan, they fired their guns and shot holes in the top of his pavilion. So on the return march the Sultan ordered the creation of the post of Chief of the Janissaries (*yeñiçeri ağası*).

Before that time the *sekbānbaşı* and *zağarcıbaşı* served as chief (*ağa*) and deputy (*ketḫüdā*) of the Janissaries. A great divan was held in Istanbul that lasted four days. The grand vizier Pīrī Pasha was executed as a sower of corruption. The standard-bearer Yaʿḳūb Agha, who had previously been the Sultan's sword-bearer, was placed in charge of the Janissaries. He reorganized the corps by providing for 61 company captains, each in charge of 30 men. While previously the *zağarcıbaşı* served as deputy, now the first company captain was appointed to that office.

According to the old statute, the deputies came every day to the gate of the chief and he resolved issues relating to the corps. Later, when special commanders were appointed to march against the enemies of religion, the deputy marched instead of the chief, and as long as the sultans did not go on campaign, neither did the chiefs; while the deputy's substitute (*ketḫüdā yeri*) stood in for the deputy in Istanbul. Eventually, even when the deputy was in Istanbul, [M 690] the substitute was employed and the deputy (*ketḫüdā bey*) dealt only with major issues. He did not deign to deal with other matters.

The route of promotion to the post of deputy's substitute starts with the *muẓhir* (usher). He graduates to the office of *ʿacemī yayabaşısı* (head of the novice foot soldiers of a province), then *ḳapu yayabaşısı* (head of the foot soldiers of the court), then *deveci* (keeper of the camels, member of one of the first five regiments of the Janissaries), then *ḫāṣekī* (sergeant at arms in the bodyguard of the sultan), then *ṭurnacı* (keeper of the cranes, member of the 73rd regiment of the Janissaries), then *ṣamṣoncı* (keeper of the mastiffs), then *zağarcıbaşı* (commander of the 64th regiment of the Janissaries), and then *ketḫüdā bey*. According to the old *ḳānūn*, there were four *devecis* and three *ḫāṣekīs*. The *ketḫüdā bey* is promoted to sanjak bey (provincial governor) or, if he wishes, to *sekbānbaşı* who in turn is promoted to *yaya bey* (foot-soldier commander) and then sanjak bey. When the sultan goes out on campaign, the *sekbānbaşı* remains in Istanbul to guard the capital. In the Janissary Corps there are posts for 14 *yaya beys*.

To resume: In 920 (1514) Sultan Selīm was camped on high ground at Çaldıran (northwest of Tabriz). Before noon there was an eclipse of the sun. The Sultan and his army then set out and camped on a flat place. The next day, which was Wednesday, the Sultan drew up his regiments and marched into the plain. The battled ensued. Victory came from God. Shah Ismāʿīl fled and one of his beautiful wives was captured. A huge amount of booty was taken. The next day a divan was held. The high state officials kissed the Sultan's hands in congratulations for the victory, and those who had fought heroically or performed excellently were given promotions and cash rewards.

Then the Sultan halted at a place near Tabriz. Molla İdrīs-i Bidlisī came to the Sultan with a surrender document from the people of Tabriz, who then came before the Sultan and spread fabrics beneath his horse's hooves.

After Sultan Selīm returned to Rūm, he ordered the construction of a dockyard in order to build a new navy. He then made preparations for a three-year campaign in order to conquer all the provinces of Persia. He assembled his army and set out in that direction. Arriving in Elbistān, he sent word to the Circassian *beys* (the Circassian Mamluks of Egypt) asking permission to pass through Malatya. Incited by the Qizilbash, and fearing lest Sultan Selīm was *ṣāḥib-ẓuhūr* (i.e., had plans for world conquest), the sultan of Egypt, Qānṣawh al-Ghawrī, marched from Cairo to Aleppo with an army of 50,000 men. When Selīm learned of this, he sent an envoy to al-Ghawrī bearing a letter and gifts and seeking permission to pass through Aleppo in order to take revenge against the enemy of religion, the Qizilbash, and to do battle with them. After sending the envoy, Sultan Selīm Khan went to Elbistān.

When the envoy met with the sultan of Egypt, al-Ghawrī, he became furious at the envoy. As soon as the Sultan heard this he headed for Aleppo. The two armies met at Marj Dābiq and a great battle occurred that lasted until the time between the noon and afternoon prayers. The Circassians were defeated and al-Ghawrī disappeared. Khayra Bay Bey came with all his dependants and submitted to the Sultan.

As Sultan Selīm approached Aleppo, all the inhabitants came out to meet him and subjected themselves to him, surrendering the money, jewels, and precious objects that were in their treasuries. The *sipāhī* troops in general were given cash rewards. The fortresses and towns in that region were taken over. 25 *yüks* (1 *yük* = 100,000) of *aḳçes*, 500 robes of honor and 17 banners were distributed to the Kurdish beys. Then Sultan Selīm headed for Damascus. As he approached Damascus, [M 691] the city notables came out to meet him. He ordered a twelve-day halt. On the fifth of Ramadan he issued an order for the army to spend the winter (in Syria) and Sultan Selīm entered Damascus and took up residence there. When news of al-Ghawrī's death reached Egypt, Ṭūmān Bay succeeded him as ruler.

When Sultan Selīm decided to go to Egypt, his viziers said, "My Sultan, the road to Egypt is desolate and waterless and inhospitable. The army of Islam cannot endure it. For this reason, even Tīmūr, Hulagu, and Chingis Khan abandoned the thought of going there." But the Sultan responded, "I, the rebellious slave of God, will go there for the sake of the guidance of the Lord of the Worlds and the miracles of the Trustworthy Commander (i.e., the

Prophet) and zeal for the religion of Islam." He distributed 200 *yüks* of *aḳçes* to the *sipāhīs* and advanced many *yüks* of *aḳçes* from the treasury to the troops from Rumelia. He sent an envoy to Ṭūmān Bay telling him to submit or be prepared for his fate. Ṭūmān Bay put the envoys to death and sent Janbirdi al-Ghazālī to Syria with 5,000 troops.

The viziers and Sinān Pasha met the Egyptian army at Khān Yūnus, where they had gone before, and defeated them with the help of God. Selīm Khan proceeded to Gaza and (the Ottoman army) halted there. He himself went with several thousand men to visit the holy sites of Jerusalem and Khalīl al-Raḥmān (Hebron). Returning to Gaza, he distributed cash awards to the troops and resumed his march to Egypt. As he approached Cairo, Ṭūmān Bay met him with 30,000 troops. By the grace of God they were routed and Ṭūmān Bāy fled to Upper Egypt. He was later captured, brought to the Sultan, and hanged.

Now the governor of Mecca and Medina, whose office had been passed down from father to son for generations, sent his son Sayyid Barakātzāde Abū Numayy Muḥammad to Cairo with gifts and a letter of congratulations. He arrived and kissed the Sultan's hand and was shown great favor by the Sultan who bestowed money and gifts on his father. He also sent 200,000 florins and a large amount of wheat for distribution to the poor in Mecca and Medina, along with two pious qadis to oversee the distribution. He also bestowed cash awards on the troops, increasing the pay of the *sipāhīs* by two *aḳçes* and that of the Janissaries by one *aḳçe*. The Arab shaikhs throughout Egypt also came with many gifts and tendered their submission. He showed them great favor and dressed them in robes of honor.

Sultan Selīm built a palace at Miqyās al-Nīl (the Nilometer on the island of Roda in Cairo). He issued an edict to send the noble *maḥmal* (annual ceremonial pilgrimage litter) to the Kaʿba. He appointed four (chief) qadis in Egypt—Ḥanafī, Shāfiʿī, Mālikī and Ḥanbalī. He appointed Yūnus Pasha governor of Egypt, but dismissed him when he learned of his greed and replaced him with Khayra Bay because of his veteran service and his understanding of the conditions in Arabia. Then he returned to Damascus where he built a tomb over the grave of the Shaikh al-Akbar, Muḥyī'l-Dīn Ibn al-ʿArabī (famous mystic-philosopher, d. 1240), also a Friday mosque and a soup kitchen.

After returning to Istanbul he ordered supplies for a campaign against Rhodes. When they were prepared, he examined them and said: "How can I go with a three- or four-months supply of gunpowder? It will take eight months to conquer. I cannot go on campaign against such a difficult place with such inadequate provisions. And anyway, the only campaign for me is the journey to the afterlife." [M 692] And indeed, in 926 (1520) they removed a boil from his back and he died.

His conquests: Fortress of Bayburd, Kiġi, Kemaḫ, Āmid (Diyarbekir), Mardin, Ruhā (Urfa), Mosul, Sinjār, Birecik, Antakya, Behisnī, ʿAynṭāb, Dārende, Divrigi, Shayzar (text: Şīrāz), Ḥamā, Ḥimṣ, Damascus, Tripoli, Baalbek, Jerusalem, Ramla, Gaza, Sidon, Ṣafad, fortress of Lajjūn, fortress of Qāqūn (near Ramla in Palestine), Cairo, Alexandria, Damietta, and Mecca and Medina.

His son Sultan Süleymān succeeded him on the throne. He issued an edict for a campaign to Belgrade, another for a fleet to guard the Mediterranean, and a third for 50 boats (*zevraḳ*) and 500 ships (*gemi*) to sail up the Danube toward Belgrade. When he arrived, the *aḳıncı* troops were given leave to attack and, with God's help, Belgrade was conquered in 927 (1520) and the keys to its surrounding fortresses were surrendered.

His second religious war (*ġazā*) was the conquest of Rhodes. He sent his second vizier with 700 ships and himself set out by land for Aydın. The governer of Egypt Khayra Bey sent his brother with 24 *grabs* (a kind of ship). The siege lasted for thirty days and with God's help Rhodes was conquered. Then the keys of İstanköy (Kos), Bodrum, Lindos, and Sönbeki (Syme) were surrendered. At that time an envoy arrived from Shah Ismāʿīl bringing magnificent gifts.

In 932 (1525–1526) Süleymān Reʾīs was sent from Egypt with 20 galleys to the province of Yemen.

In Dhū'l-Qaʿda of the same year (August, 1526), Sultan Süleymān met the Seven Kings (i.e., the Holy Roman Empire) and their innumerable army of infidels on the Plain of Mohács. There was a great battle. By the grace of God, the infidels were routed and their king, Lavoş (Louis II of Hungary and Bohemia) fell from his horse while fleeing, sank into a swamp and drowned. One soldier of the army of Islam cut off the king's head and brought it before the Sultan. He was granted 1000 gold pieces and the governorship of Ḳaraman.

Then the Sultan laid siege to Budin (Buda), the greatest city of Hungary, and the king's wife brought its keys and surrendered them. Many other countries too were brought within the fold of Islam. But Ferdinand, the Kaiser of Austria, seized an opportunity and took back the fortress of Budin. Sultan Süleymān set out to reconquer it on 1 Ramadan 933 (1 June 1527). When he reached the Plain of Mohács, Yanoş (Janós Szapolyai or John Zápolya), who was king of the province of Erdel (Transylvania) came and paid him homage. Because the infidels of the province of Hungary, which had been conquered, were not accustomed to Muslims, Yanoş was made king on condition of paying tribute and was given the standard and horsetail, banner and drums. Thus Budin was reconquered, also Ḳomaran (Komárom) and Yanıḳ (Györ).

In 937 (1530) Süleymān, setting out again against the Shah of Persia, conquered Aḫlāṭ, Adilcevaz and Erciş and subjugated Baghdad. In short, he engaged in 16 religious wars, the last of which was the conquest of the fortress of Szigetvár (in Hungary) after which he was afflicted with dysentery and died there in 974 (1566).

His conquests: (from the Persians,) Rawān, Van, Baghdad, Shahribān, Tabriz, Nakhjiwān, Ganja and Shirwān, and most of the cities of Georgia, including Tortum and Aḳçaḳalʿe, etc., more than 20 fortresses; from the Kurds, Ẓālimḳalʿe and the fortresses around it; from the Armenians, [M 693] Aḫlāṭ, Adilcevaz, Akhtamar (text: Aḥmār) and Erciş; in the province of Mushaʿshaʿ, Jazāʾir, Basra, and Wāsiṭ; and in the other direction (Europe) he took Budin, Belgrade, Egri (Eger, Erlau), Bosnia, Croatia, Ösek (Eszék, Osijek), Sombor, Szegedin, Raçe (Rácz), and Mohács; from the Hungarians, Ṭımışvar (Temesvár), Esterġon (Esztergom), Yanıḳ (Györ), Ḳomaran (Komárom), Üsto[l]nibelġrad (Székesfehérvár), Peçuy (Pécs), Szigetvár, Şıḳlovoş (Siklós) and Pojega (Pozsega); from Spain, Castile; from Germany, Çeh (Czechia, Bohemia) and Venice, many places; the capital of Moldavia, Sicav and Yaş (Suceava and Iaşi); and many fortresses on this side and that side of Vienna; also Jazāʾir (Algiers), Ṭarāblus al-Gharb (Tripoli in Libya), Rhodes, Saḳız (Chios), İstanköy (Kos) and Bodrum. Each of the places I have mentioned had many fortresses.

His works: in Istanbul, a noble Friday mosque with four minarets (Süleymāniye), built in 960 (1553), and around it four madrasas, a madrasa for the study of medicine, a hadith school, a soup kitchen, and a hospital; in memory of the Prince (Şehzāde, his son Meḥmed), a Friday mosque, madrasa, soup kitchen, and their requisite facilities; in memory of Prince Cihāngīr, a small Friday mosque on a hill across (the Golden Horn) from Istanbul. Ḫāṣekī Sulṭān, mother of Sultan Meḥmed Khan, built a Friday mosque, madrasa, soup kitchen, and public bath in Üsküdar; Merḥūme Sulṭān, his second *ḫāṣekī* (i.e., who had also borne him a son), also built a Friday mosque, madrasa, soup kitchen, and school; and his daughter Mihrimāh built a Friday mosque and a madrasa. Among his great charitable works were the Ḳırḳçeşme aqueduct; more than 400 fountains; the Büyükçekmece Bridge; four madrasas in Mecca; a Friday mosque, hostel and soup kitchen in Damascus; and great amounts of provisions sent to Mecca and Medina.

When Sultan Süleymān died at Szigetvár, his son Sultan Selīm II was summoned. He met the army of Islam at Belgrade. Later, when he returned to Istanbul, he bestowed favors on the ulema, the viziers, and the troops.

In 976 (1568), he sent a large number of troops by sea to Kefe (Kaffa in Crimea), ordering them to dig a canal connecting the Volga and Don Rivers where they flowed close to each other. In this way, troops coming by ship from Istanbul could sail from the Black Sea into the Don, from there into the Volga, and from there into the Sea of Jurjān (Caspian). Thus it would be possible to conquer the Persian regions of Shirwān, Gīlān, Ṭabaristān, Khurāsān, and Jurjān. By divine wisdom, the Khan of Kefe and the army of Islam reached as far as the fortress of Ajdarkhān (Astrakhan) and excavated about one third the distance of the canal. But then because of severe winter weather and lack of provisions it was not completed.

In the same year, Malik Muṭahhar, the Sharif of Yemen, killed Murād Pasha, the *beylerbey* of Yemen, and seized the country. The Sultan ordered Sinān Pasha, the governor of Egypt, to proceed there with imperial troops and conquer Yemen. He carried out heroic actions with Özdemüroğlı and reconquered Yemen.

In 978 (1570) he conquered Cyprus.

In 979 (1571), Devlet Giray Khan and a great army of Crimean Tatars marched to the city of Moscow and burnt it, then advanced three days beyond it and returned with a large number of captives.

The Sultan died in 982 (1574) and was succeeded by his son Sultan Murād III. He made his vizier (Lala) Muṣṭafā Pasha commander-in-chief and sent him to Persia.

Works of the late Sultan Selīm II: [M 694] He built a splendid bath in the Imperial Palace that resembled the Bursa hot spring. When it was finished he entered it but he slipped and fell and one side of his body turned dark black (from the bruise). The chief physician at that time was an ignoramus who mistakenly gave the Sultan a phlebotomy. It was his destined hour, and he died. Among his great charitable works were domes constructed over the mausoleum of the Prophet in Medina; a magnificent Friday mosque, madrasa, hadith school, Koran school, hospital and soup kitchen in Edirne; and mosques and madrasas in Cyprus.

To resume: Muṣṭafā Pasha went to Persia. He built the fortress of Kars and placed an armory and garrison inside, then went to Çaldıran and seized its fortress. The governor of Tiflis sent his son Manūchahr to the vizier with keys to his fortresses. They established friendly relations in the noble council. Manūchahr converted (from Shīʿī to Sunni Islam) and the governorship of Akhiskha (in Georgia) was bestowed on him. Özdemüroğlı was appointed *beylerbey* of Erzurum and sent with his sanjak beys to guard Shirwān, while Vizier Muṣṭafā Pasha marched to Rawān (Erivan in Armenia).

In 987 (1579) the observatory of Taḳiyyüddīn was begun in Topḫāne.

In 888 (rather 988/1580) the turbans of the Christians and Jews were changed. The Jews were ordered to wear yellow turbans and black caps while the Christians were ordered to wear blue turbans and calpacs made of broadcloth.

In 990 (1582), on the occasion of the circumcision of prince Meḥmed Khan, an imperial celebration lasting 44 days was held at the At Meydānı (Hippodrome) and 7000 orphaned and indigent boys were also circumcised.

In 991 (1583), Ferhād Pasha rebuilt the fortress of Rawān. ʿOs̱mān Pasha (Özdemiroğlı), assuming the office of grand vizier, became commander of the army facing Persia. He entered Tabriz in 993 (1585) and built the fortress of Heşt Bihişt there.

In 1001 (1592), the Janissaries assembled before the Imperial Gate (Bāb-ı Hümāyūn) and demanded that the *defterdār* (minister of finance) be removed from office. They were warned once or twice to disperse, but they refused. By imperial order the sultan's private guard and the palace guards took up arms and attacked the Janissaries. 110 men were killed and the leaders of the sedition were hanged.

In 1003 (1594), Sultan Murād III died and was succeeded by his son Sultan Meḥmed (III). He added 100 *aḳçes* to the salary of the shaikh al-islam, who was already receiving 600 (per diem).

In 1004 (1595), there was a campaign against the infidels of Hungary. After the fortress of Egri was conquered, the infidels retaliated and a great battle took place. The Muslims were routed and the infidels had the temerity to come before the Imperial Pavilion and plant the standard of disgrace upon the treasury chests. While despair was at its height, the tide of victory turned and the infidel army was defeated. So many infidels were killed that even today their bones are piled in heaps.

In 1009 (1600), İbrāhīm Pasha conquered Ḳanije (Kanizsa in Hungary).

In 1012 (1603), the practice of smoking tobacco appeared and Sultan Meḥmed III died. His son Sultan Aḥmed Khan succeeded him. At that time Shah ʿAbbās invaded. Aḥmed became sultan at age fifteen without a struggle for the throne—the first time this had occurred in the Ottoman dynasty. [M 695] For certain fateful reasons cracks began to appear in the Ottoman state. Brigands caused widespread disorder. In 1016 (1607) Sultan Aḥmed made Ḳoca Murād Pasha grand vizier and ordered him to eradicate that abominable band of brigands. All of them were punished. In 1026 (1617) the Sultan died.

His works: In 1018 (1609) he laid the foundation for the New Friday Mosque (Sultan Ahmed) and it was completed seven years later. He also built a madrasa, soup kitchen, hospital and school. Before this, he had refurbished Medina and had sent a diamond known as Kevkeb-i Dürrī ("Brilliant Star")—it was estimated to be worth 80,000 gold pieces, and in addition was embellished with 227 precious stones—to be placed at the tomb of the Prophet. To Mecca he sent a golden waterspout and screens for the noble Kaʿba made of silver burnished with gold. He also had a latticed screen made for the tomb of the Prophet that was of silver burnished with gold.

Then his brother Sultan Muṣṭafā took the throne. Because of his reclusive character, he gave up the sultanate.

In 1027 (1618) Sultan ʿOs̱mān (II) took the throne. In Rabīʿ II 1029 (March 1620), one night just at the break of dawn, a luminous celestial sign appeared in the sky. It was in the exact shape of a sword and was approximately five lances long and two cubits wide. There was general amazement at this for an entire month. People said it signified that an important event would occur to the Ottomans.

At that time the Ḳapudan Pasha (admiral of the Ottoman fleet) conquered the fortress of Manfredonia in Spain. Then in 1030 (1620), by the power of God, at the end of winter the sea (Bosphorus) froze and remained so for several days. In the spring of that year, the sultan set out on campaign for Poland. He conquered the fortress of Chotin and ordered the Tatar Khan to raid Russia. They took more than 100,000 captives and laid siege to their stockades; but they sued for peace because winter was approaching and when this was accepted they returned to Istanbul.

On 8 Rajab 1031 (19 May 1622), after the afternoon prayer, the Sultan quaffed the sherbet of martyrdom and Sultan Muṣṭafā came to the throne again. The Janissaries again demanded that he abdicate.

Then Sultan Murād IV, son of Sultan Aḥmed, came to the sultanic throne. In 1033 (1623) the inhabitants of Baghdad gave the fortress of Baghdad to Shah ʿAbbās. In 1033 (1623) the Cossacks attacked. They pillaged Yeñiköy near the Bosporus. In Ṣafar 1043 (August 1633) a great fire erupted outside the Cibali Gate (in Istanbul) and much of the city burned. Because of this, the coffeehouses were torn down and forbidden. In 1044 (1634) the Sultan himself marched to the fortress of Rawān and with God's help conquered it in eight days. Then in 1048 (1638) he set out for Baghdad, which he conquered in 39 days. Then he returned to Istanbul and died on 16 Shawwāl 1049 (9 February 1640).

His works: He rebuilt the Kaʿba.

His brother Sultan İbrāhīm Khan [M 696] came to the throne. In 1055 (1645) a campaign was carried out against Crete and the fortress of Ḥanya (Canea) was conquered. Subsequently the sultanate was ceeded to his son Sultan Meḥmed Khan, while his own spirit rose in martyrdom to the upper world (i.e., he was put to death).

Sultan Meḥmed Khan IV ascended the throne in 1058 (1648). In 1066 (1655) Köprülü Meḥmed Pasha became grand vizier. In 1068 (1658) Grand Vizier Köprülü Meḥmed Pasha conquered the fortress of Yanova (Kis Jenö in Transylvania). In 1070 (1659) Ḳapudan Serdār ʿAlī Pasha conquered the fortress of Varat (Várad in Transylvania). At the end of Rajab 1071 (March 1661) there was a total eclipse of the sun and it became so dark that candles had to be lighted. In 1072 (1661) Köprülü died and his son Fāżıl Aḥmed Pasha succeeded him as grand vizier. In 1074 (rather 1073/1663) Fāżıl Aḥmed Pasha conquered the fortress of Uyvar (Újvár in Hungary). In the same year the Sultan went to Edirne and sent off Grand Vizier Fāżıl Aḥmed Pasha to Crete.

In 1077 (1666) Fāżıl Aḥmed Pasha with the imperial fleet crossed over to Crete from the port of Termaş (Monemvasia, at the eastern tip of the Peloponnesus) and dropped anchor in the harbor of the fortress of Ḥanya. They spent the winter at Ḥanya and took on additional supplies, then set out for the fortress of Ḳandiye (Candea, Herakleion). When they arrived, they dug trenches up to Ḳızıl Tabya ("Red Bastion") west of the fortress and began to bombard Ḳandiye. The battle continued between the two sides—winter and summer, day and night—for 29 months. By Divine favor, the fortress surrendered at the end of Rabīʿ II 1080 (2 September 1669).

During that period Sultan Meḥmed Khan Ghazi while hunting came to Yeñişehr-i Fenar (Larissa in Thessaly). The Ṣarıḳamış Cossacks came to seek refuge with the Sultan and offered to become his subjects. Their offer was accepted. A horsetail and standard were bestowed on their hetman, Hetman Doroshenko, and an agreement was made whereby the province of Ukraina, which bordered the Ottoman lands, would remain under their control and no harm would come to them or their territories.

Subsequently the Poles, who had long had an agreement with the Ottomans, intended to harm the Cossacks and to pillage their province, while they were in their province (?). Subsequently, while they were in their province, the sultan sent chaushes once or twice with letters admonishing them and saying, "Restrain your hand from the Cossacks, who are our subjects. If you cause them harm, you will break the treaty." When it was confirmed that they had broken the treaty, he issued an edict for a campaign against the Polish infidels. In 1083 (1672), Ghazi Sultan Meḥmed Khan set out from Edirne for Poland. When he reached İsaḳçı (Isaccea), he ordered a bridge to be built over the Danube and crossed it. Then he came to Chotin, he took over the infidels in the fortress of İzvançe (Zurawno), on the other side of the Dniester, and annexed it to the fortresses of Islam. The Crimean Khan, Selīm Giray Khan, joined forces there and the Sultan bestowed on him a gem-encrusted quiver, a gem-encrusted sword, a sable fur and calpac, and a thoroughbred horse with gem-encrusted saddle. Then a bridge was built over the Danube and the Sultan and army of Islam crossed it.

On Wednesday, 23 Rabīʿ II (18 August 1672), [M 697] the Ottoman army pitched its tents at suitable places facing the fortress of Ḳamaniçe (Kamieniec, Kamenetz), a very strong fortress built of dressed stone. Its defensive ditches were amazingly deep and the rushing river surrounded it. Moreover it had a strong citadel and its outer bastions were difficult to reach. The fortress was immediately surrounded and trenches were dug. The infidels who occupied this fortress were subjected to a terrifying siege. The battle continued in this way for about ten days. The infidels then fled to the citadel. By the grace of God, some of the fortification towers were pierced and undermined and thus collapsed.

Finally, the commander of the infidels surrendered the keys to the fortress and submitted. On 10 Jumādā I (rather 3 Jumādā I/27 August 1672) the fortress and all its dependencies were taken over and its churches were turned into mosques. The fortress of Ḳamaniçe was annexed to the fortresses of Islam and became a province (*eyālet*). It was given to the vizier Ḥalīl Pasha as an emolument.

Now Ḳaplan Pasha and the Tatar Khan were sent to besiege the city of Albu (?). When the siege began, the prominent infidels in the city sent letters of apology to the Tatar Khan, asking him to serve as intermediary with the felicitous Sultan, begging him to forgive their offense and offering, in return for peace, to surrender the fortress of Ḳamaniçe with its province and eight wooden forts, to pay an annual tribute of 20,000 gold pieces, and promising henceforth not to attack those provinces in the possession of the Ṣarıḳamış Cossacks. Selīm Giray Khan served as intermediary and peace was concluded.

Afterwards the Padishah of Islam, Ghazi Sultan Meḥmed Khan, set out from there to Edirne. In Shaʿbān 1087 (November 1676), Grand Vizier Köprülüzade Fāżıl Aḥmed Pasha died. Ḳara Muṣṭafā Pasha succeeded him as grand vizier.

Subsequently, the infidels of Muscovy seized the fortress of Czehrin. As a result and imperial order was issued for a campaign against Muscovy in 1089 (1678). In that year the Ghazi Sultan Meḥmed Khan marched to İsaḳçı and sent Grand Vizier Ḳara Muṣṭafā Pasha to Muscovy. The Grand Vizier reached the fortress of Czehrin and took it by the sword after a siege of 32 days.

In 1093 (1682) Ghazi Meḥmed Khan sent out from Edirne to put the cursed king of Austria, whom the infidels called *çasar* (Kaiser), in his place and in 1094 (1683) reached Belgrade. Then he sent Grand Vizier Muṣṭafā Pasha with a huge army to conquer the fortress of Vienna which was the capital of the king of Austria.

Supplement by the Publisher (İbrāhīm Müteferriḳa): Praise and again praise to God Most High, with whose help and favor this agreeable book has been completed. Trusting in His generosity and benevolence, we have been vouchsafed to carry out what we humbly put forward in the Introduction of the book: to record—in written descriptions and in maps, and to the limit of our knowledge and information—the conditions of the countries and kings of the continent of Asia, beginning with the countries in the Far East and continuing westward as far as Üsküdar, which is across from Istanbul on the eastern side, and the eastern shores of the Black and Mediterranean Seas, which are to the right and left of it. [M 698]

If the imperial desire and sultanic will are manifest, I will endeavor to publish the second volume of this agreeable book, which will begin with Istanbul and continue to the furthest regions of the West, comprising the remaining three divisions of the earth—the continents of Europe, Africa, and America—describing in detail the status of their countries, kings, and peoples. I beseech the court of the Granter of wishes to allow me to exert myself to produce a memorable book that will render admirable and praiseworthy service and will be a mirror revealing the state of the entire world, a work unprecedented in Islamic literature.

This book was completed with the assistance of God, at the hand of this humble one who was permitted to undertake its publication in the imperial printing house in the good city of Constantinople—may it ever continue in splendor and beauty at the hand of its possessor (the Sultan)—on the 10th of the sacred month of Muḥarram in the year 1145 of the *hijra* of the Prophet (3 July 1732), upon him be the most perfect prayers and salutations.

Appendix: Maps and Diagrams

About Maps and Diagrams

Maps and diagrams from R have been inserted into the translation, exactly where Kātib Çelebi indicated it. By contrast, Müteferriḳa, due to the exigencies of the printing process, printed maps and diagrams on separate folios, which were then inserted and bound with the text approximate to their referents. These maps are not paginated, instead, we give here the page numbers before and after the insertion point. Several diagrams are linked to specific points in the text, as will be seen.

Müteferriḳa adopted many of the more elaborate maps and diagrams from R for his edition, or completed maps which exist only as sketches in R. In other instances, he had members of his press produce new maps and diagrams, or he included maps originally produced for different works; for instance, the maps of the Mediterranean and its parts (M 75/76, M 77/78) were originally produced for his edition of Kātib Çelebi's *Tuḥfetü l-kibār fī esfāri l-biḥār*. The world map M 71/72 (except for the surrounding diagrams) is also found in that edition. The illustration of the compass (M 65/66) comes from Müteferriḳa's own treatise, *Füyūżāt-ı Miqnaṭisīye*.

In order to show how both Kātib Çelebi and Müteferriḳa adopted or created maps, we have included in this section not only Müteferriḳa's maps for comparison with the material in R, but also transcriptions and translations. Where Müteferriḳa adopts Kātib Çelebi's maps, his versions are usually more complete, therefore, we have chosen to base our transcriptions on M. We have also transcribed maps from R here that do not have a counterpart in M. Specifically, R's regional maps of Iran are replaced in M with fewer larger maps, which are included here as well. In addition, we provide selected examples of Kātib Çelebi's models, in particular Mercator/Hondius and Cluverius.

There are thousands of place names on the maps of *Cihānnümā*, and most of them are the result of multiple translations, between the local form in a local language through one or more informants to a European map to an Ottoman form. Sometimes a language our mapmakers would be familiar with was involved at one stage, to be overwritten by another later on. The process is complicated by the languages and cultural assumptions of every individual in the chain of transmission, consisting of informants, authors, mapmakers, translators, intermediaries, etc. As Sonja Brentjes has shown, Kātib Çelebi and his successors had choices to make: They could "transliterate every single word in the foreign maps" into Arabic characters (based on Ottoman pronunciation), "they could translate compound names and expressions", they "could replace foreign names and expressions with culturally adequate names and expressions", or they "could create new terms".[1] De facto we find all of these approaches, applied in a mostly eclectic, unsystematic fashion.

Our translation of the maps inevitably adds another layer to this complexity, although paradoxically one of our goals was to increase transparency. Like our mapmakers, we, too, have 'transliterated' (or rather 'transcribed') toponyms. Due to the ambiguities of the Arabic script, many of them contain a good dose of conjecture, but we strove to render them as an Ottoman reader would have read them. We have replaced toponyms with 'culturally adequate names and expressions', that is, we gave the original Arabic and Persian names in the respective regions. For well-known places we substituted either modern English forms (Corinth, Osaka, Lahore, Jerusalem), or current local forms (Durrës, Taranto, Alanya),[2] where it was possible, and as long as there were no reasons (as explained below) to retain the Ottoman form. In many instances, we were able to trace toponyms back to sources in Latin or Italian, and where the correspondence is compelling we have substituted those original forms, even if we where unable to locate a modern equivalent (e.g. M 129 La Redonda, an island off New Guinea). We have used those criteria narrowly, however. On the same map of New Guinea, 'Riūnīrū' suggests an original 'Rio Nero', but because we have not been able to locate such a place on a possible source map we restricted ourselves to the transcription. To be sure, an exhaustive attempt to trace all toponyms back to specific maps and ultimately to actual locations was beyond the scope of our project. Much more work remains to be done in identifying sources based on these toponyms.

Another goal of our treatment was to make those decisions by our mapmakers that Brentjes identified transparent. Where they substituted a 'culturally adequate name' we retained it: Ancient maps of New Guinea show a cape named 'Ancon de la Natividad de nostra Signora', which the Ottoman cartographer translated as 'Mīlād-ı Meryem Burnı' (Cape of the Nativity of Virgin Mary), which we have kept as such, instead of reverting to the original. Explanations included in the map (Insūlī Verdes i.e. Green Islands,

1 Brentjes, Sonja. "Multilingualism in Early Modern Maps." *Archives et Bibliothèques de Belgique: Archief- en Bibliotheekwezen in België* Numéro Spécial 83, (Mélanges offerts à Hossam Elkhadem par ses amis et ses élèves, edited by Frank Daelemans, Jean-Marie Duvosquel, Robert Halleux, David Juste) (2007): 317–328, at 319.

2 Our choices of modern local forms should not be taken to reflect political preferences.

 | DOI:10.1163/9789004441330_006

M 129) are in the original. For the sake of transparency, we have been reluctant to emend even obvious mistakes. The map of the Arabian peninsula (M 483) includes not only several duplications of well-known locations, such as Ma'rib in Yemen, and Qalhāt in Oman; the capital of Oman is spelled as Maskat instead of Masqaṭ (English: Muscat), and we retained that misspelling because it documents that the mapmaker transcribed from a Latin map, and did not connect to the name to its original Arabic. The 'Country of Shajar' in southern Arabia most likely is nothing but a duplicate of the region of Shiḥr, differing from the latter only by one additional diacritical dot, yet the dot is clearly there, indicating that the mapmaker thought of it as a separate name, and thus we show it.

There are many more such instances. Our goal of making the maps legible and transparent inevitably result in competing and conflicting priorities for individual choices, and readers may disagree in specific cases. Brentjes concluded that Ottoman cartographers ended up applying an unsystematic combination of all the strategies available to them. On our part, we have attempted to make the combination as systematic as possible, well aware that perfect consistency is unattainable.

Maps and Diagrams

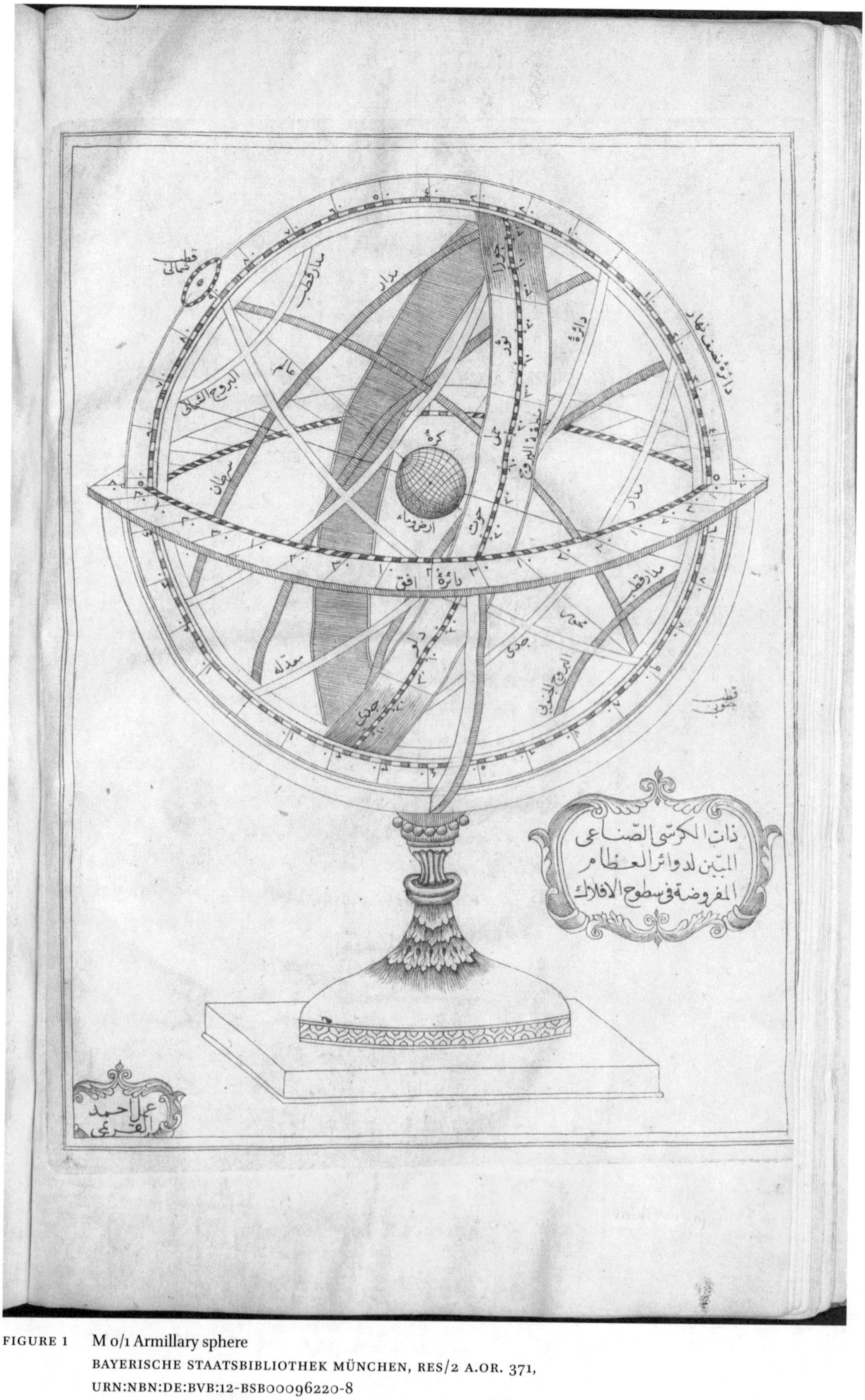

FIGURE 1 M 0/1 Armillary sphere
BAYERISCHE STAATSBIBLIOTHEK MÜNCHEN, RES/2 A.OR. 371,
URN:NBN:DE:BVB:12-BSB00096220-8

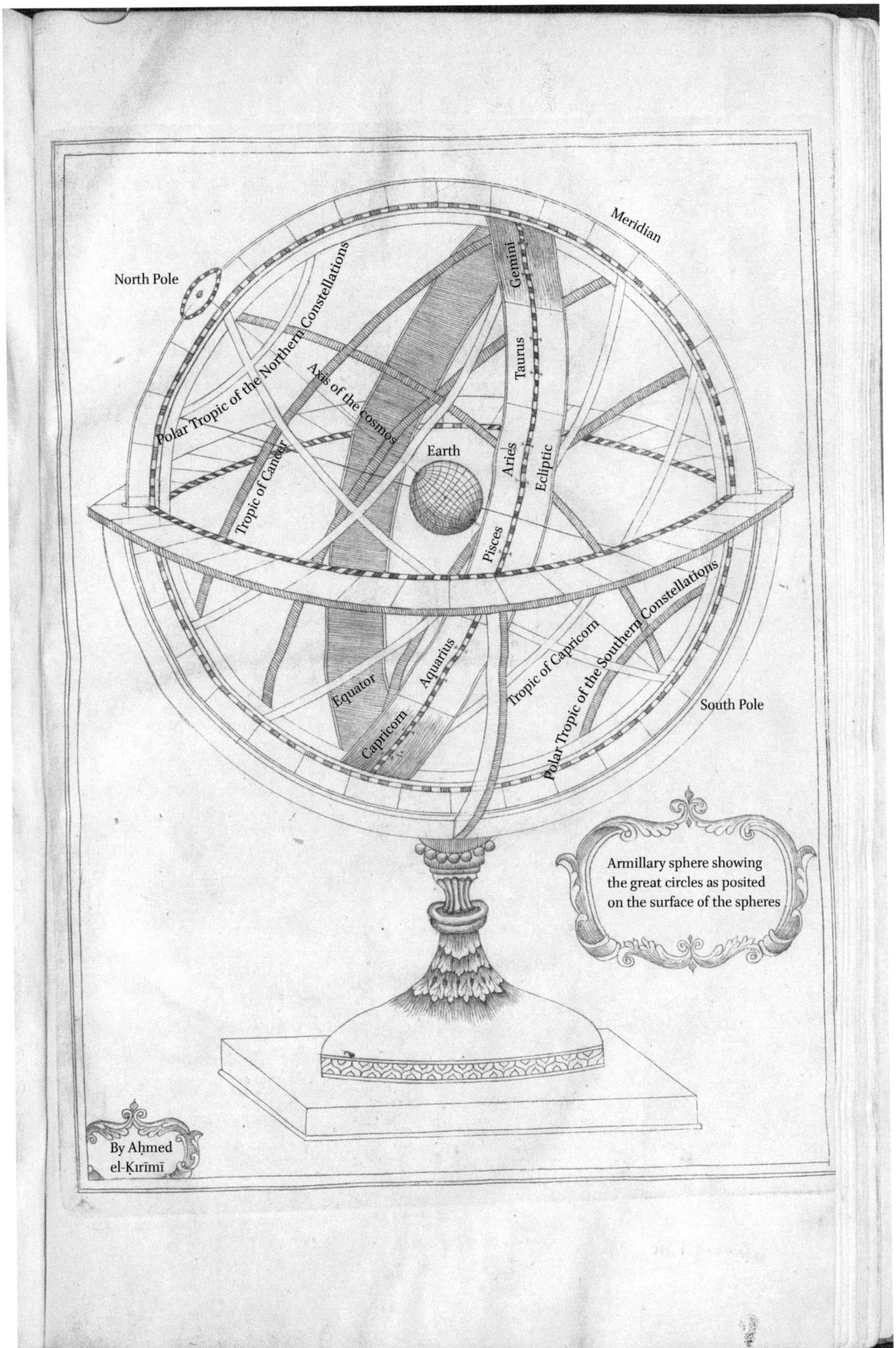

FIGURE 2 M 0/1 Armillary sphere (translation)

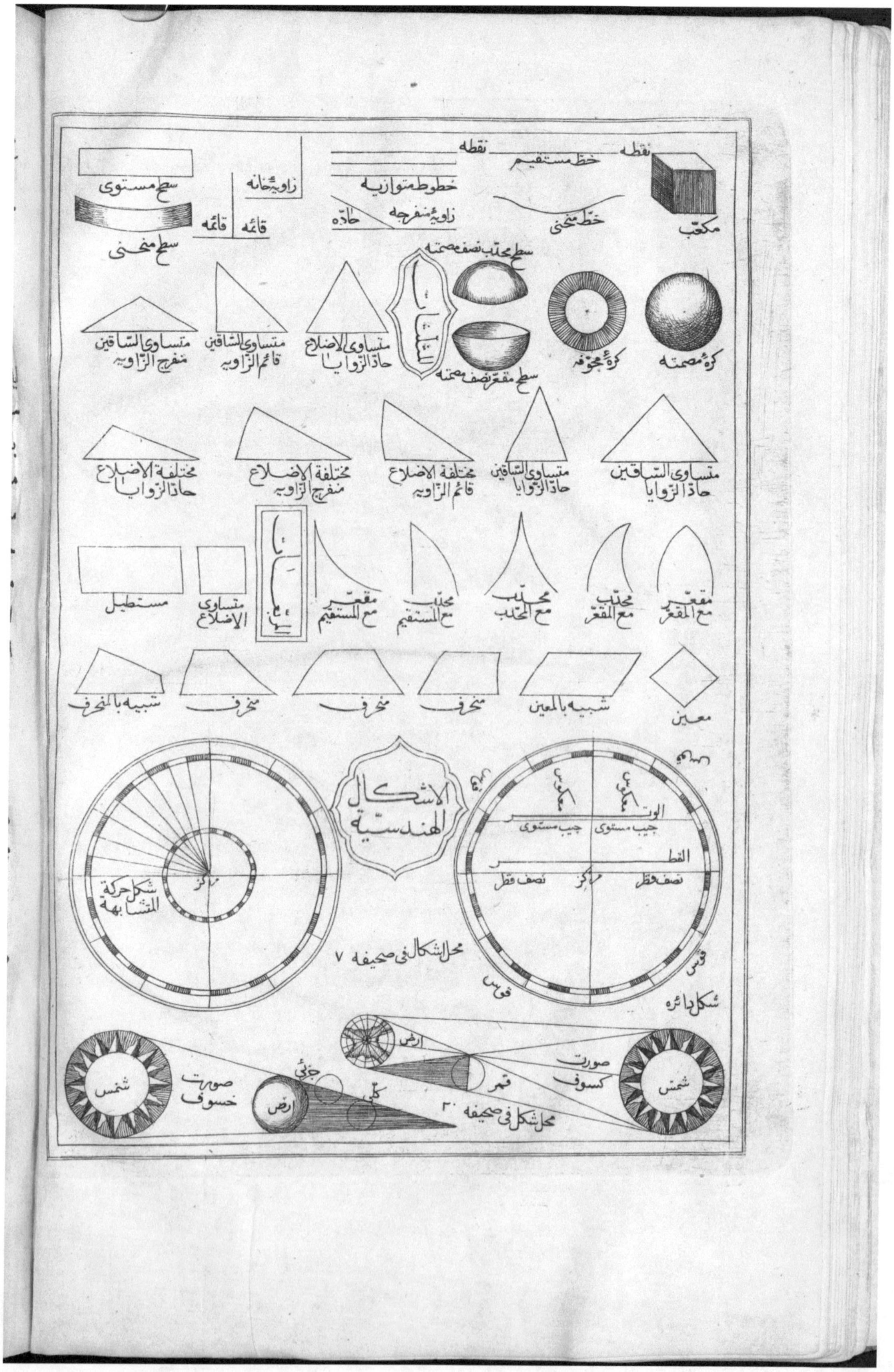

FIGURE 3 M 7/8 Geometrical figures
BAYERISCHE STAATSBIBLIOTHEK MÜNCHEN, RES/2 A.OR. 371, URN:NBN:DE:BVB:12-BSB00096220-8

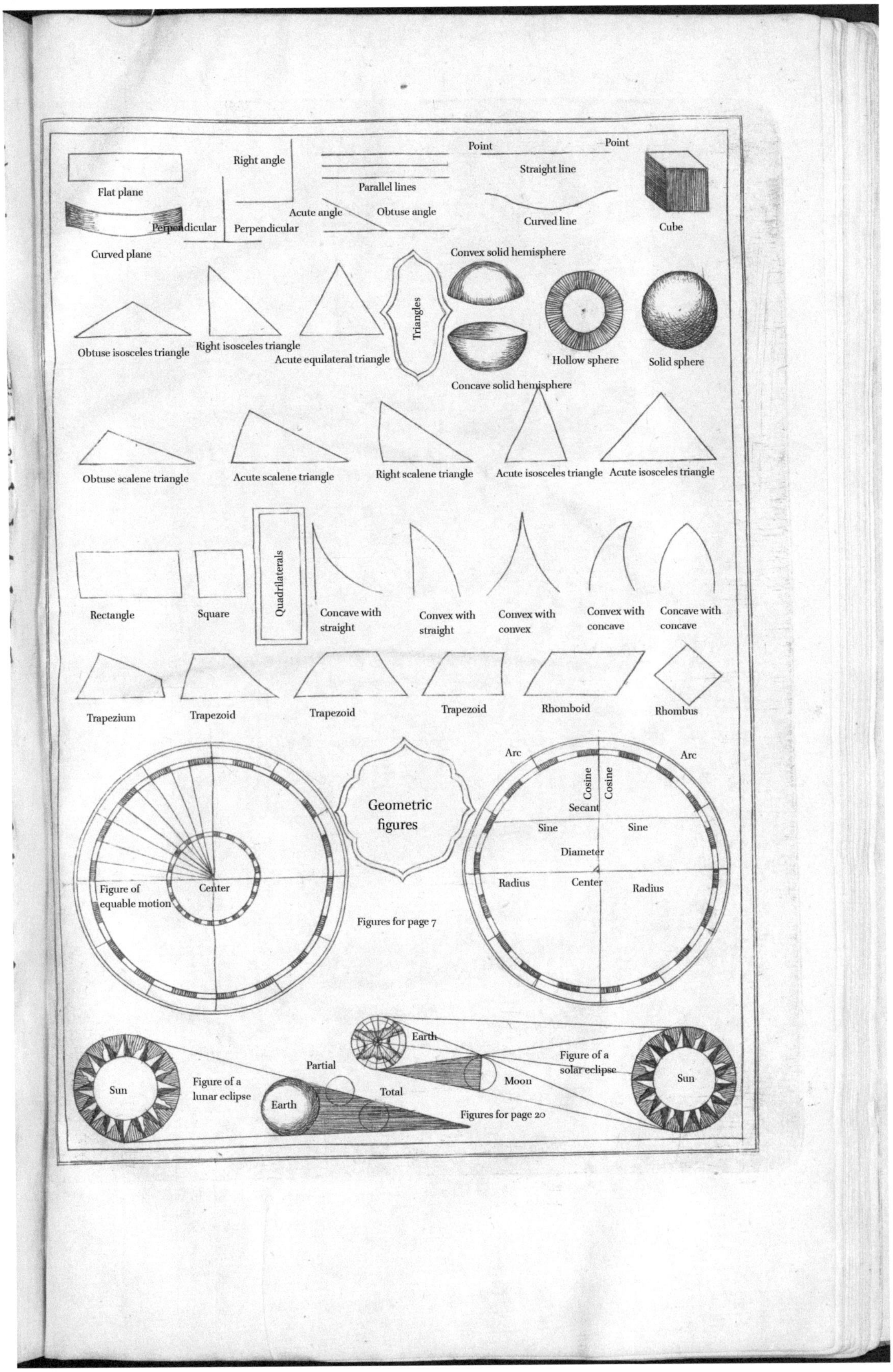

FIGURE 4 M 7/8 Geometrical figures (translation)

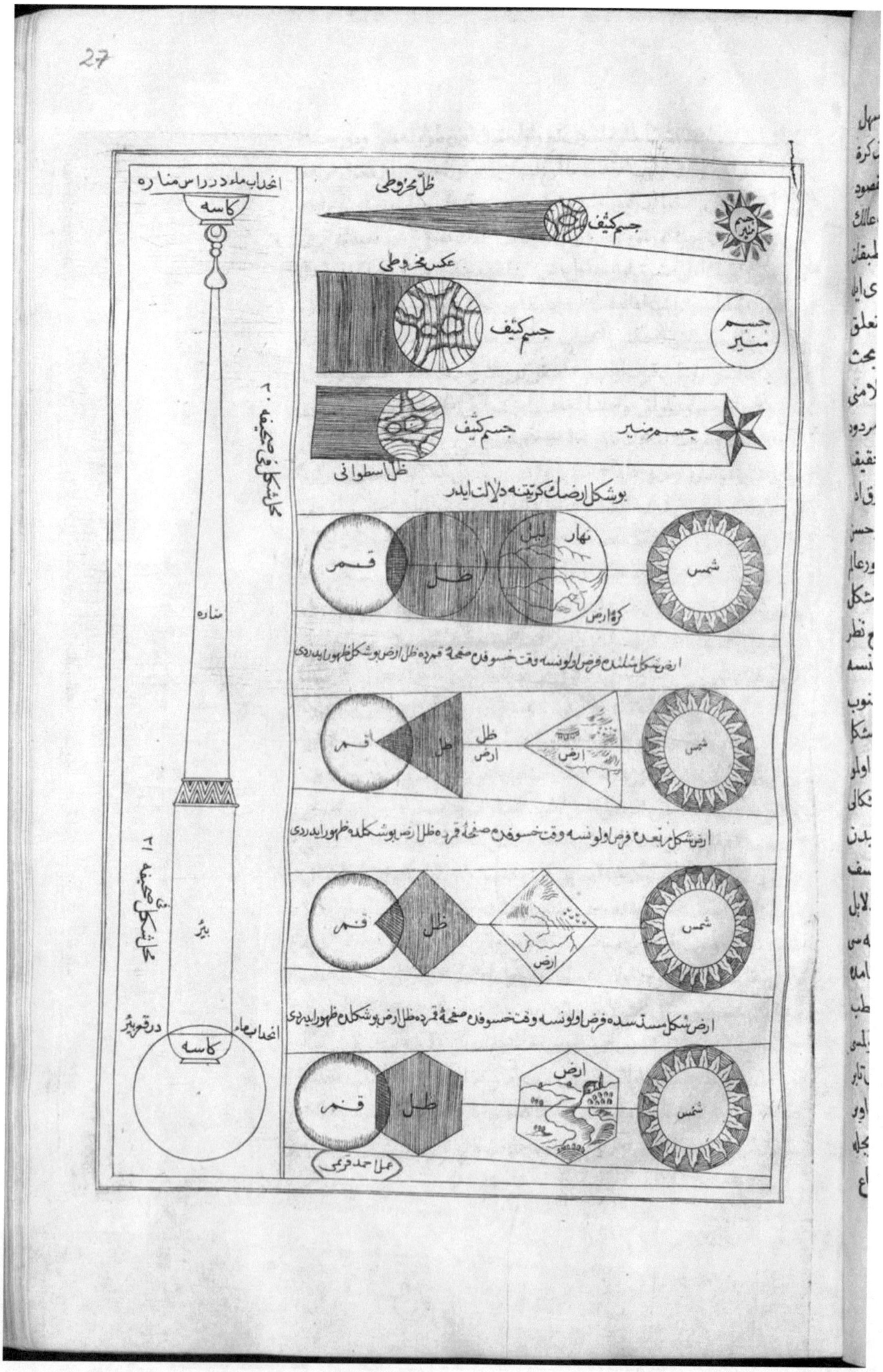

FIGURE 5 M 18/19 Diagram showing the earth's sphericity
BAYERISCHE STAATSBIBLIOTHEK MÜNCHEN, RES/2 A.OR. 371, URN:NBN:DE:BVB:12-BSB00096220-8

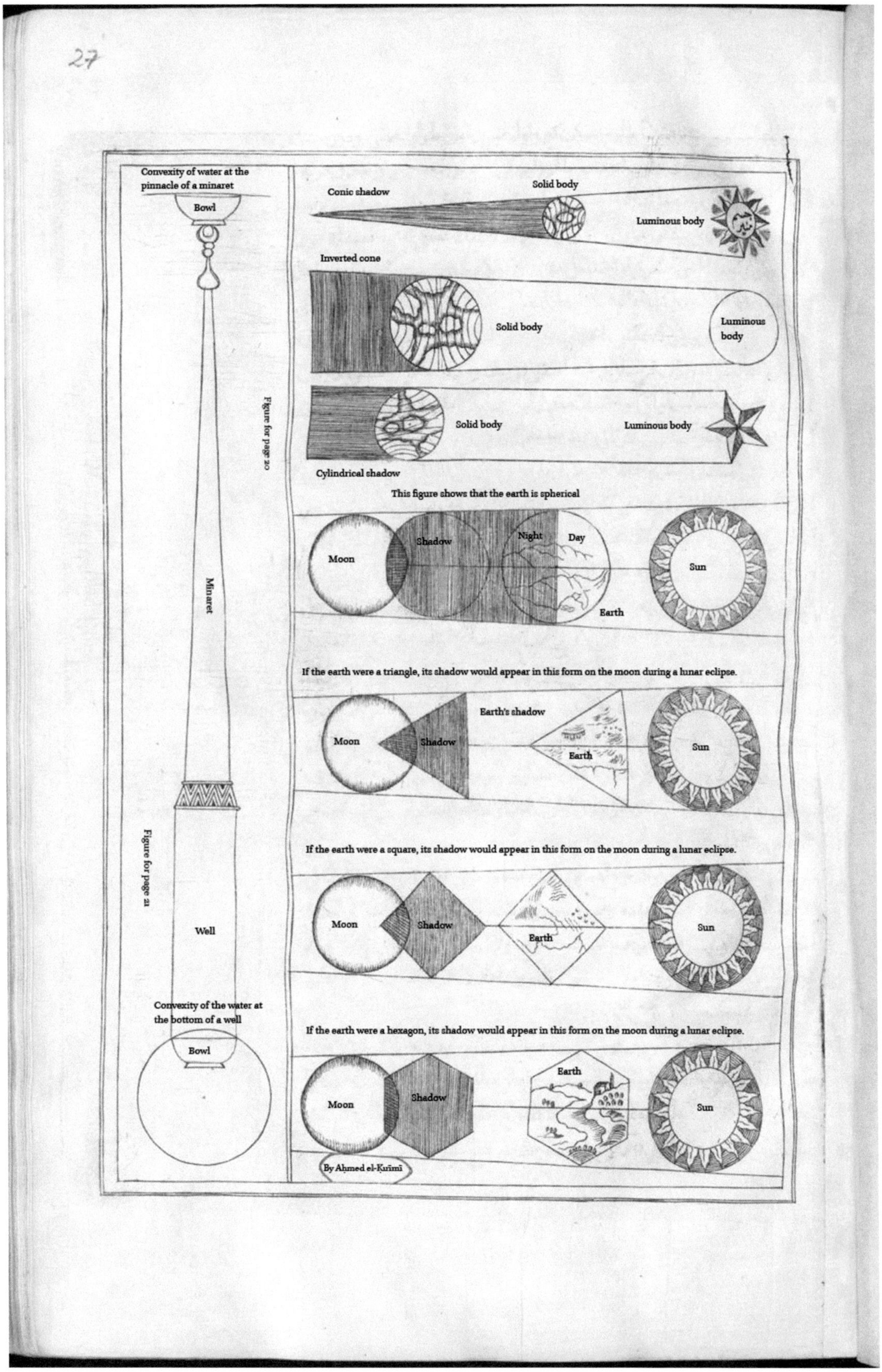

FIGURE 6 M 18/19 Diagram showing the earth's sphericity (translation)

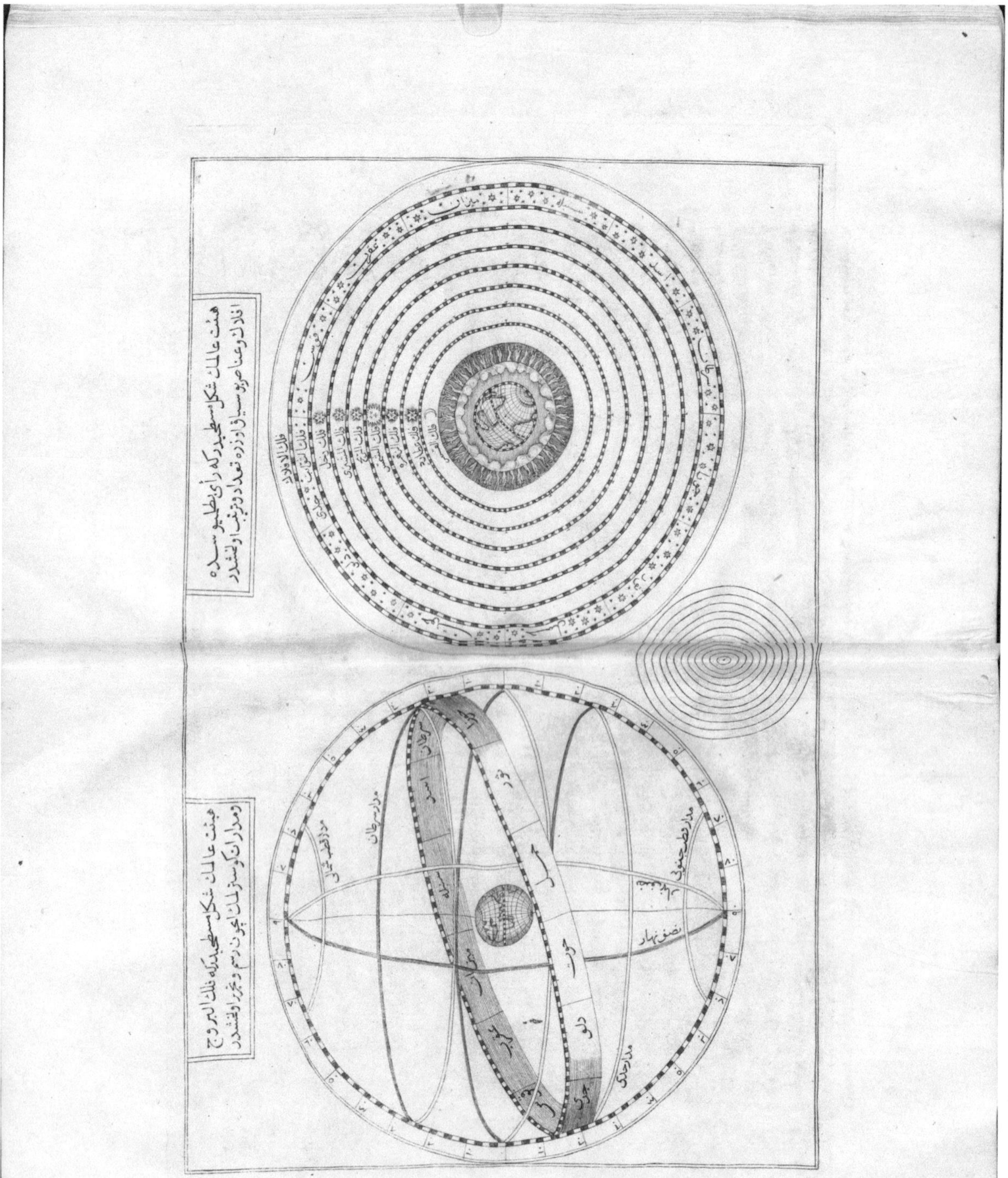

FIGURE 7 M 21/22 Cosmological scheme
BAYERISCHE STAATSBIBLIOTHEK MÜNCHEN, RES/2 A.OR. 371, URN:NBN:DE:BVB:12-BSB00096220-8

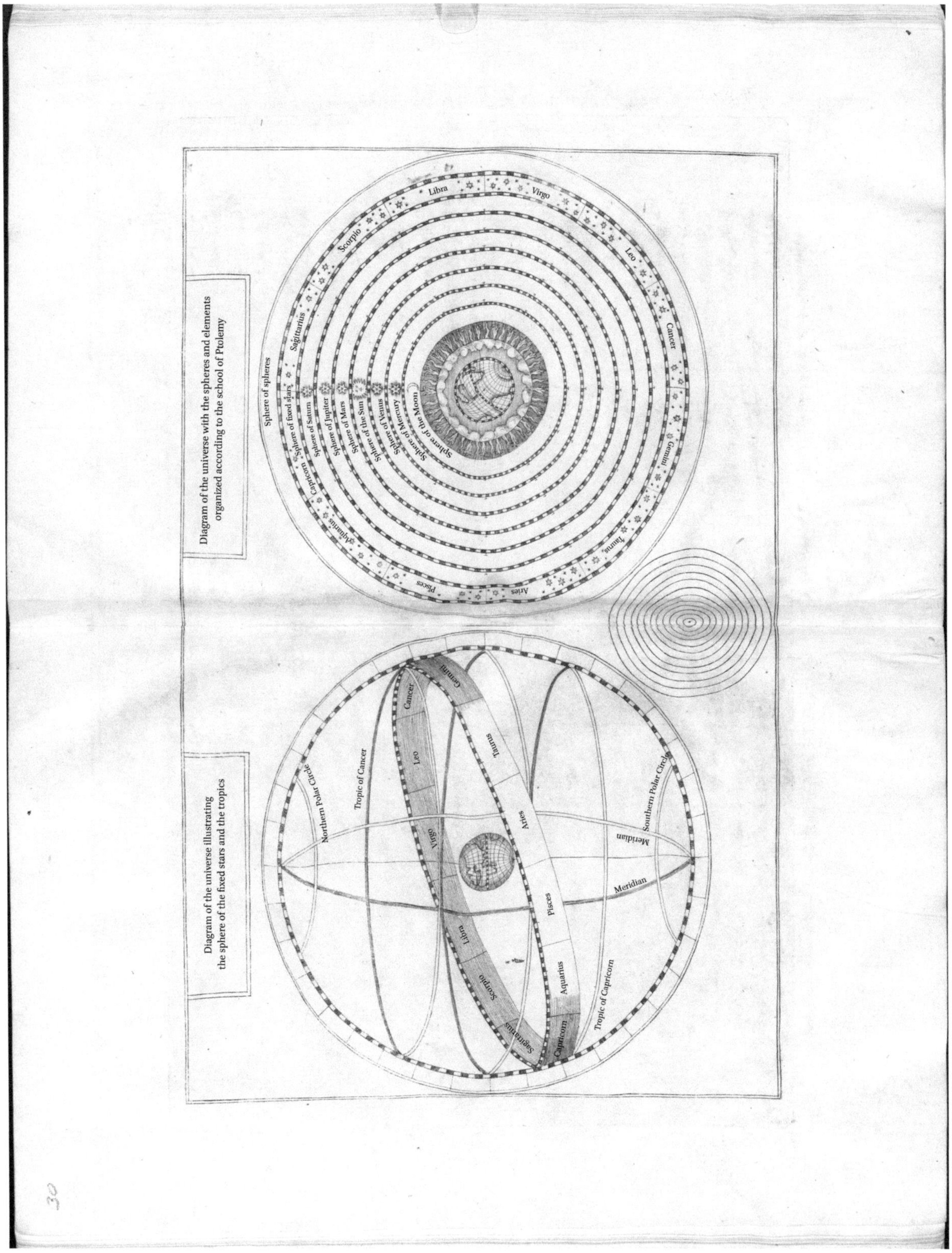

FIGURE 8 M 21/22 Cosmological scheme (translation)

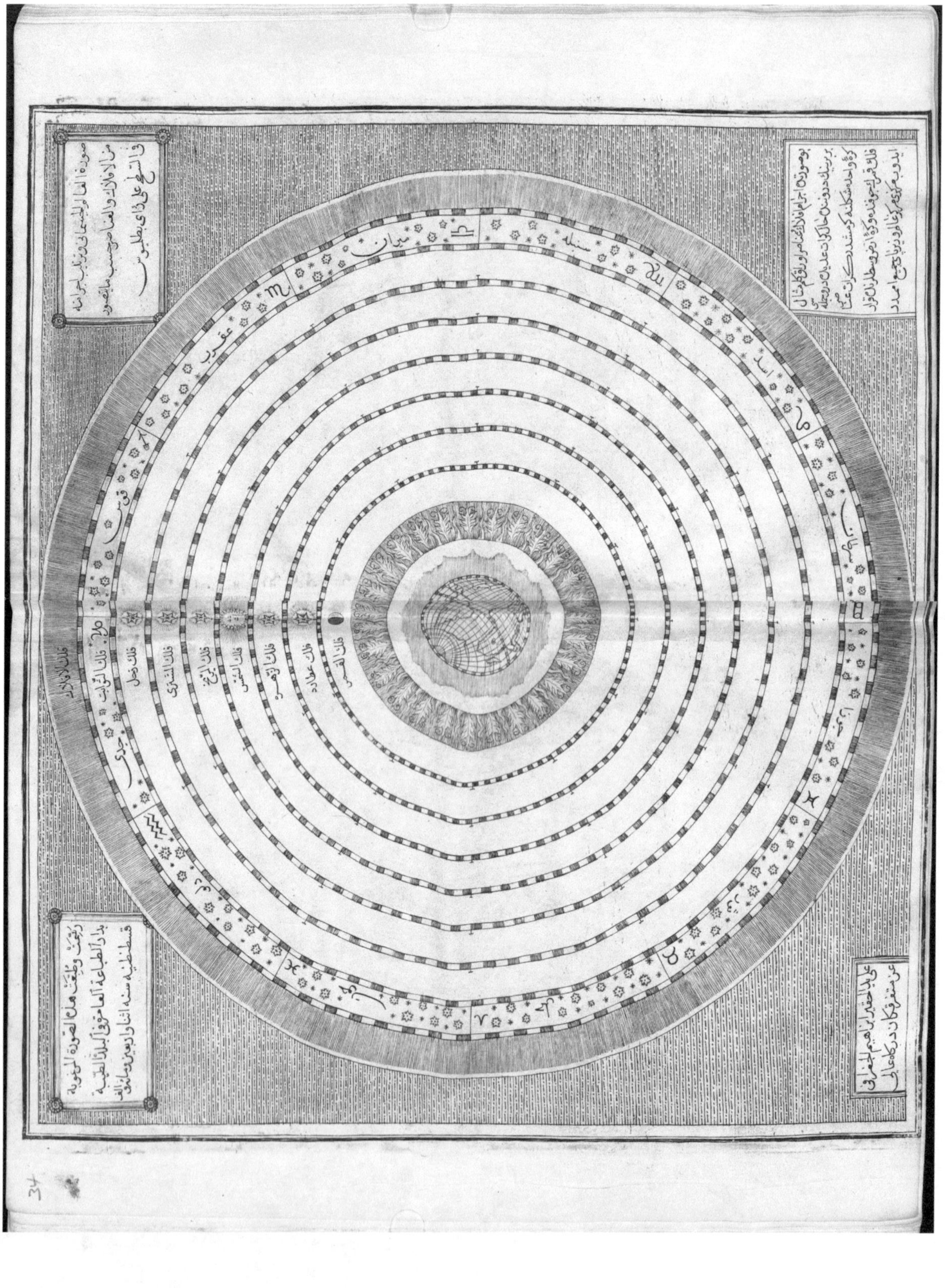

FIGURE 9 M 25/26 Cosmological scheme
BAYERISCHE STAATSBIBLIOTHEK MÜNCHEN, RES/2 A.OR. 371,
URN:NBN:DE:BVB:12-BSB00096220-8

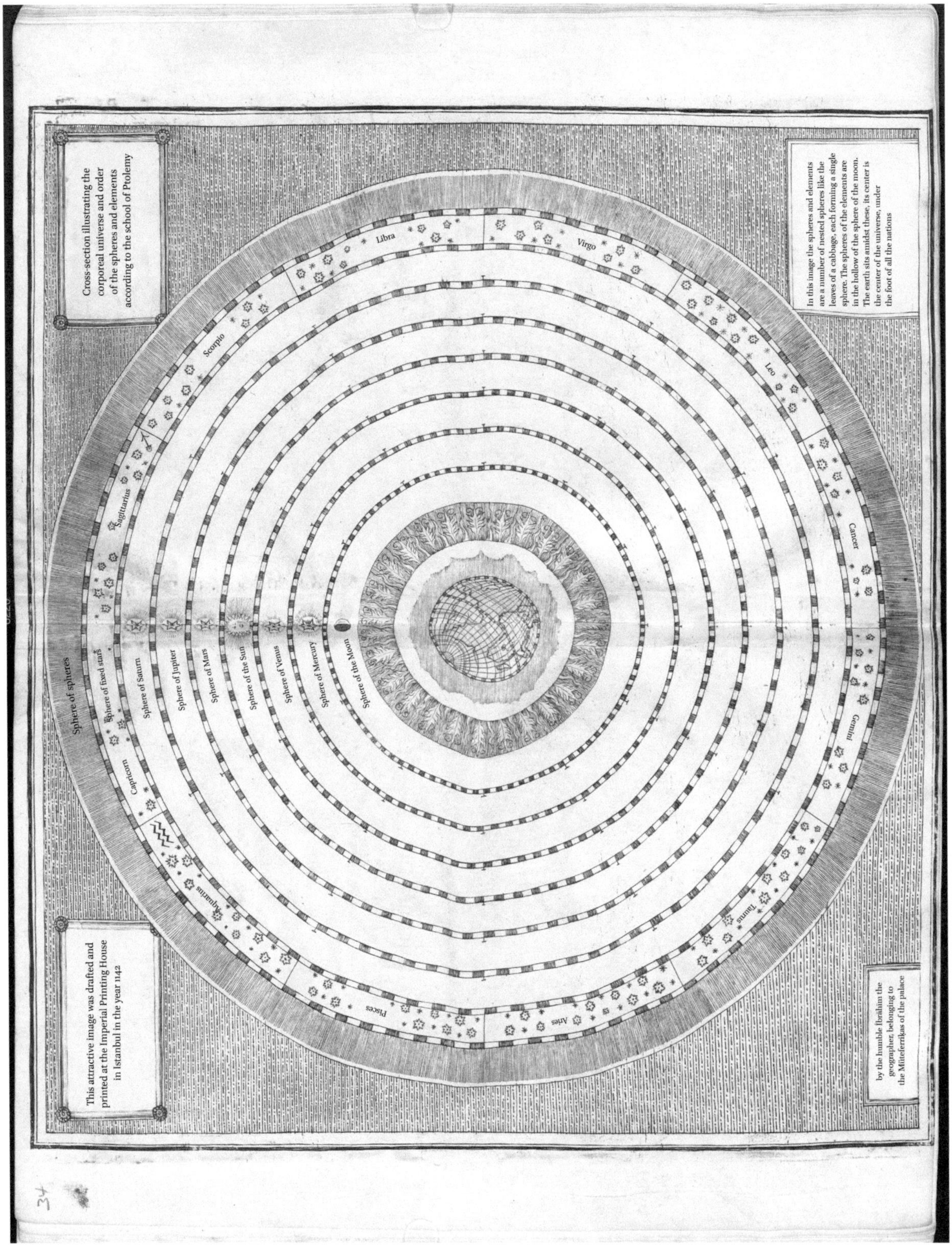

FIGURE 10 M 25/26 Cosmological scheme (translation)

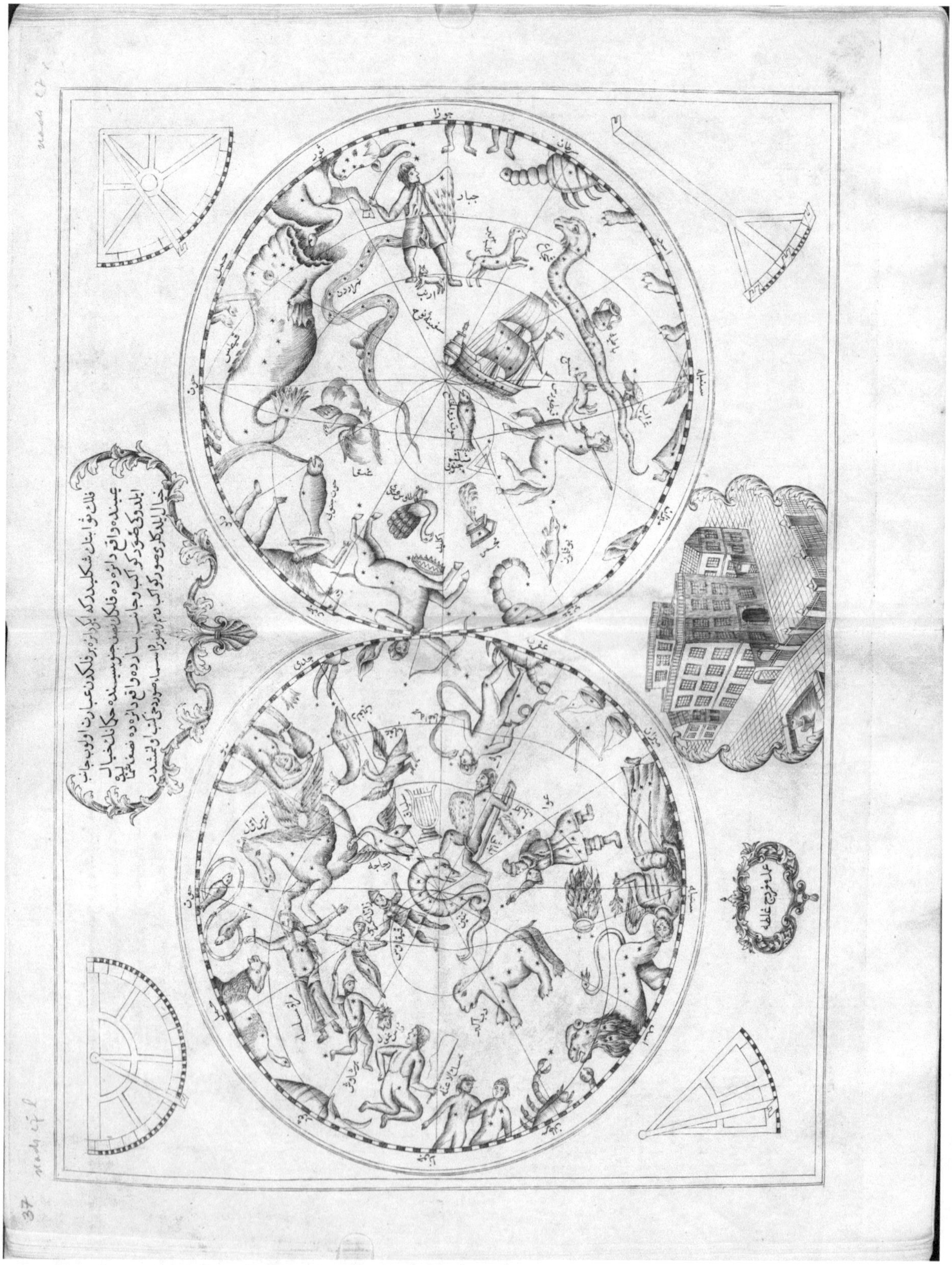

FIGURE 11 M 27/28 Celestial chart
BAYERISCHE STAATSBIBLIOTHEK MÜNCHEN, RES/2 A.OR. 371,
URN:NBN:DE:BVB:12-BSB00096220-8

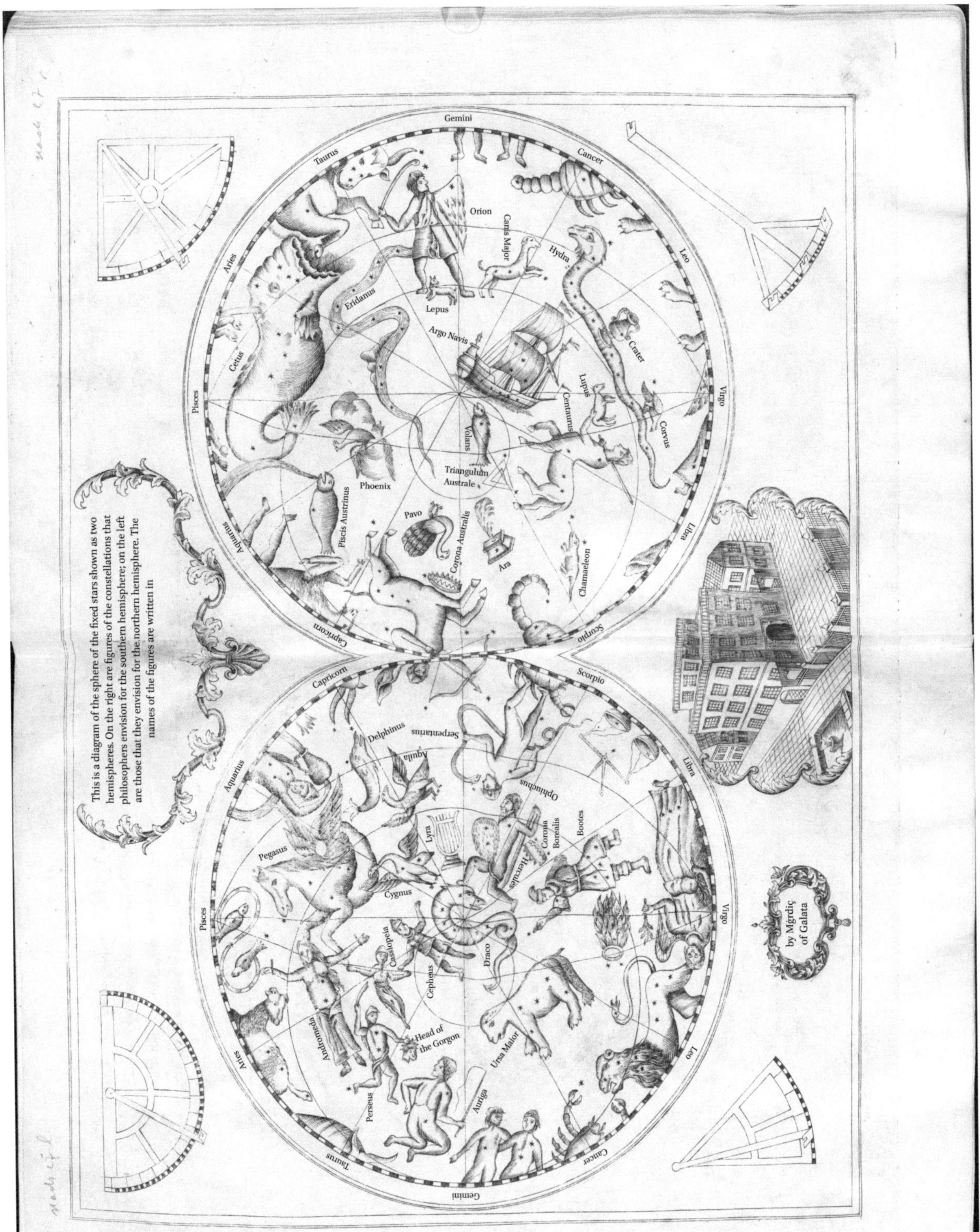

FIGURE 12 M 27/28 Celestial chart (translation)

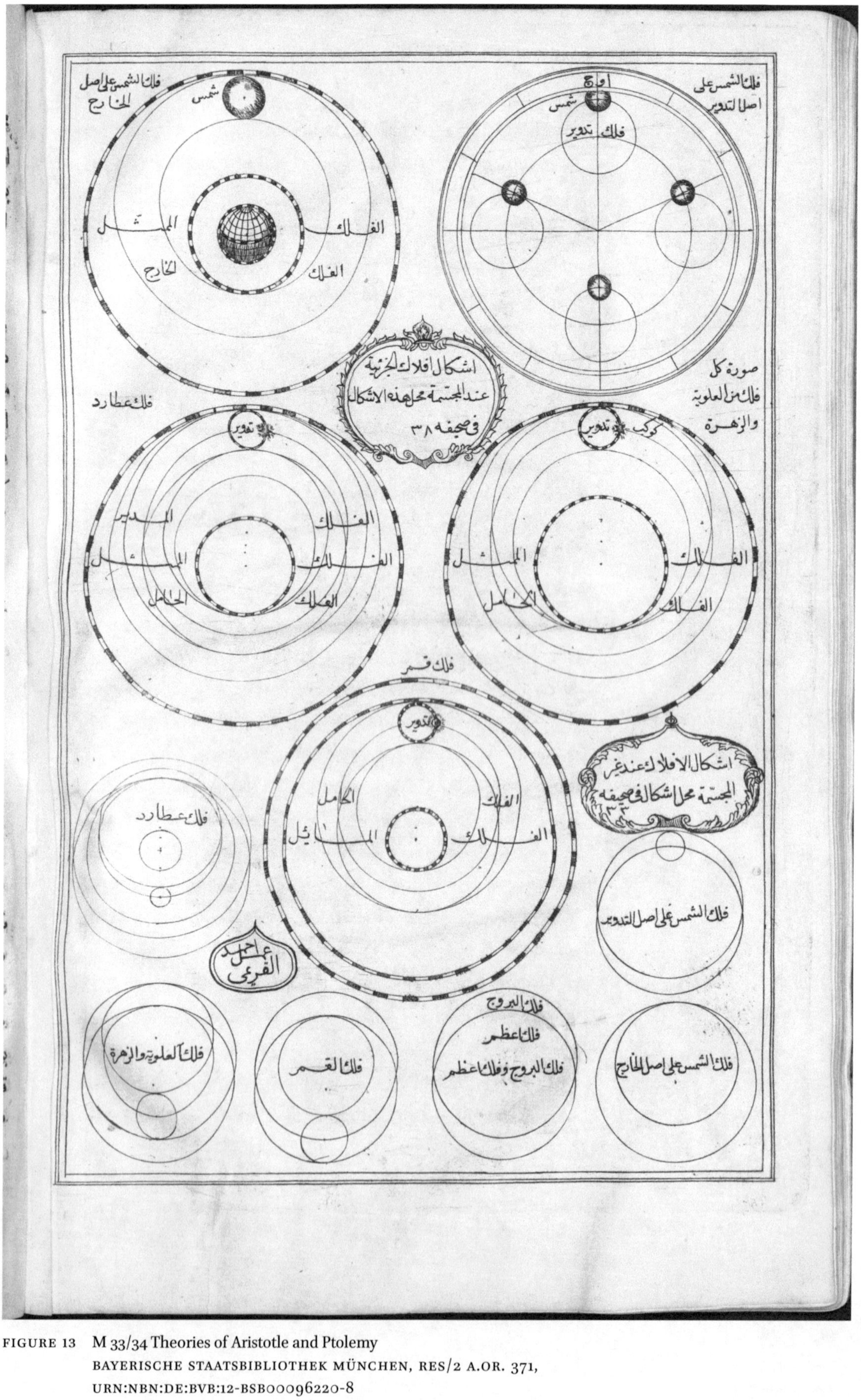

FIGURE 13 M 33/34 Theories of Aristotle and Ptolemy
BAYERISCHE STAATSBIBLIOTHEK MÜNCHEN, RES/2 A.OR. 371,
URN:NBN:DE:BVB:12-BSB00096220-8

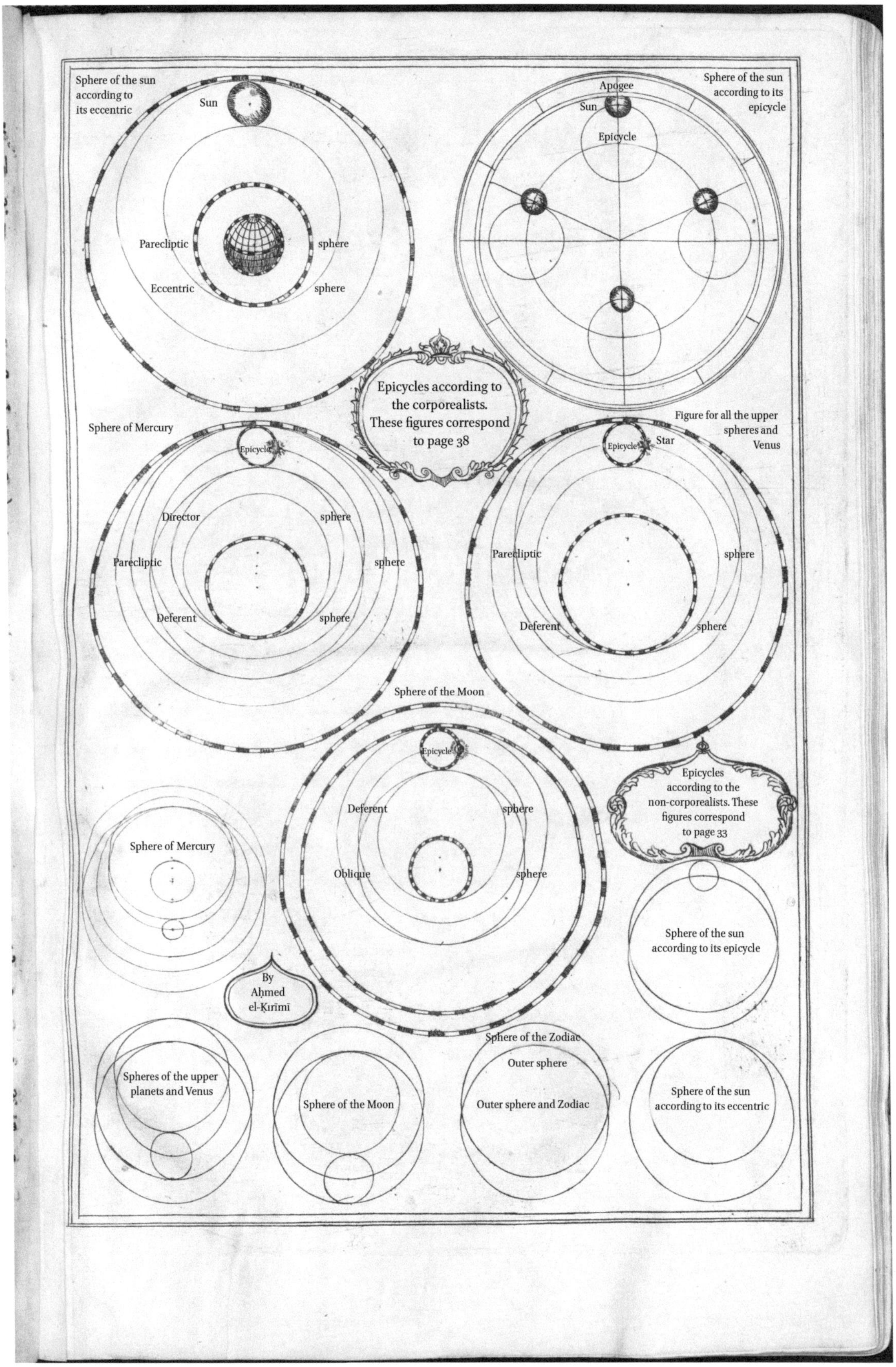

FIGURE 14 M 33/34 Theories of Aristotle and Ptolemy (translation)

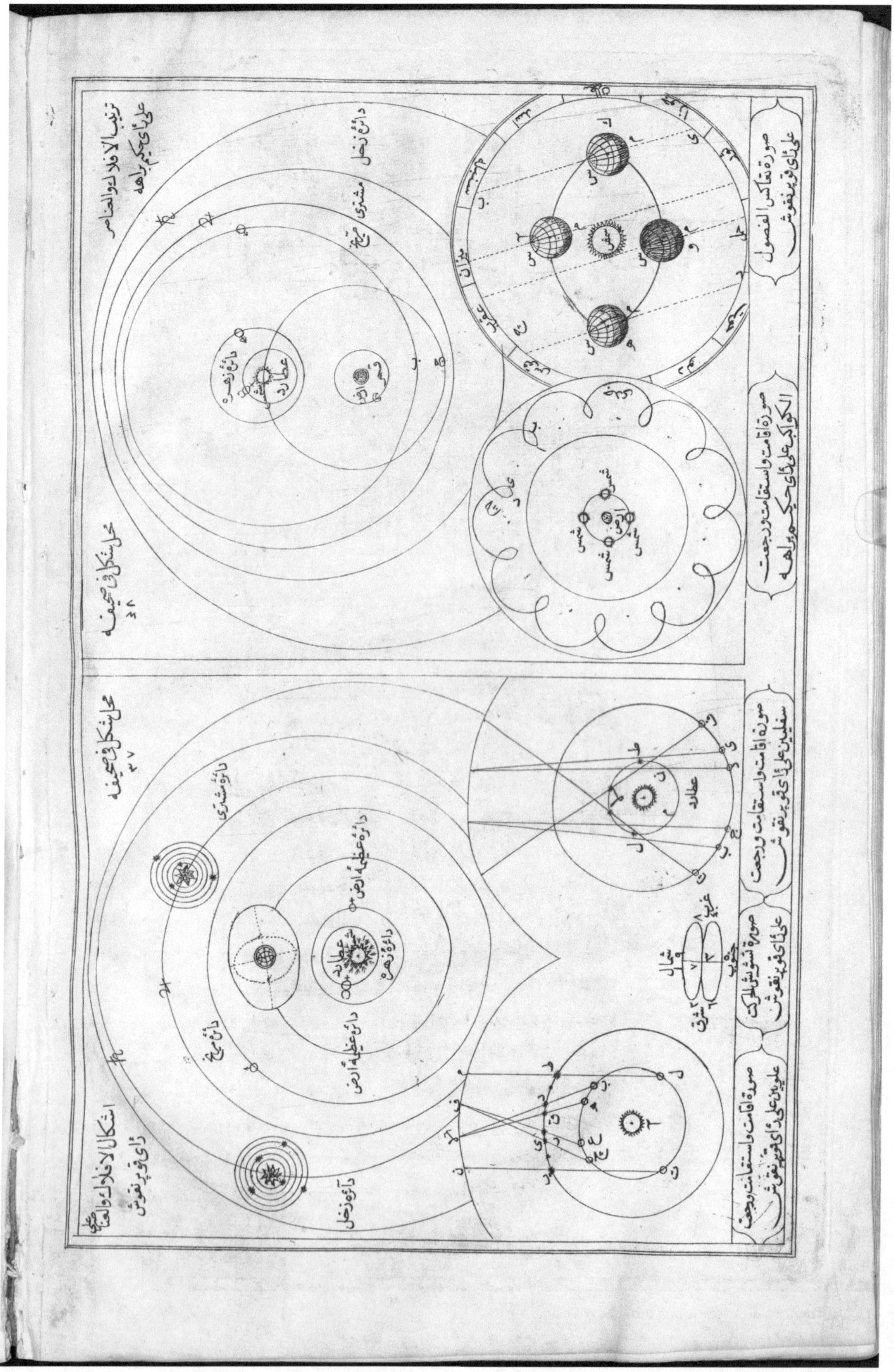

FIGURE 15 M 47/48 Theories of Copernicus and Brahe
BAYERISCHE STAATSBIBLIOTHEK MÜNCHEN, RES/2 A.OR. 371,
URN:NBN:DE:BVB:12-BSB00096220-8

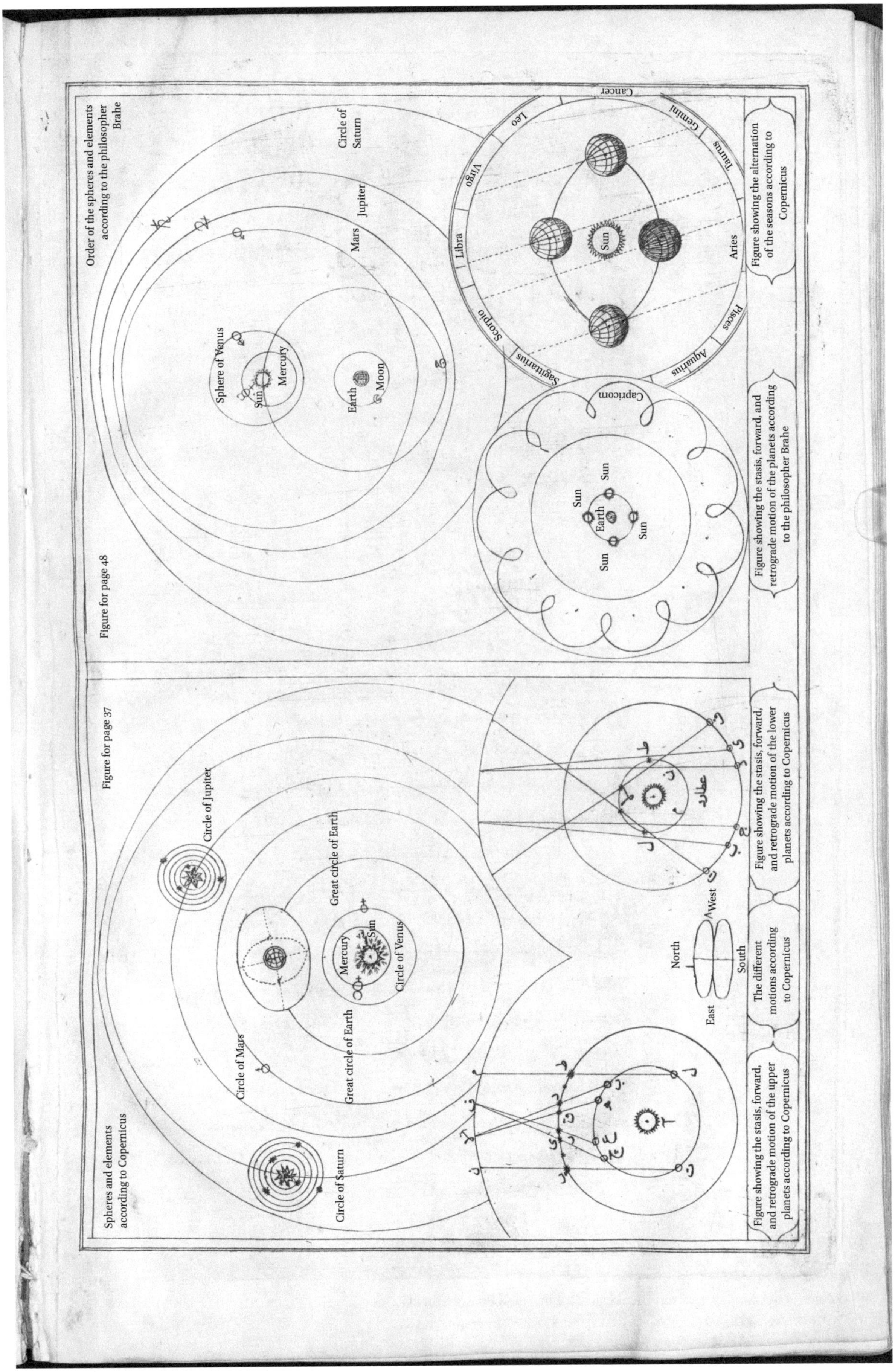

FIGURE 16 M 47/48 Theories of Copernicus and Brahe (translation)

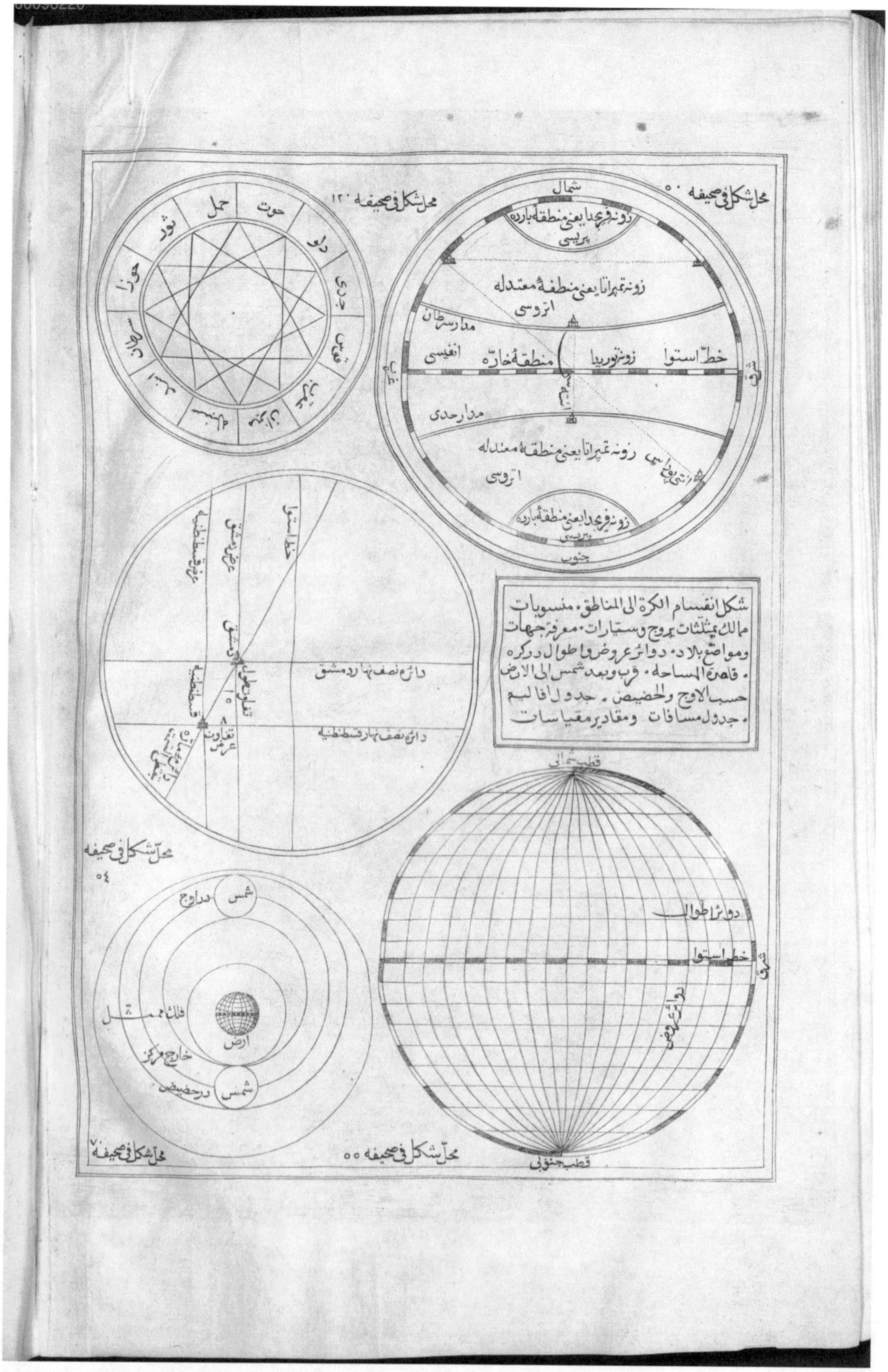

FIGURE 17 M 49/50 Zones, coordinates, astronomical calculations; cf. R 8b, 11a, 11b, 12a, 42b
BAYERISCHE STAATSBIBLIOTHEK MÜNCHEN, RES/2 A.OR. 371,
URN:NBN:DE:BVB:12-BSB00096220-8

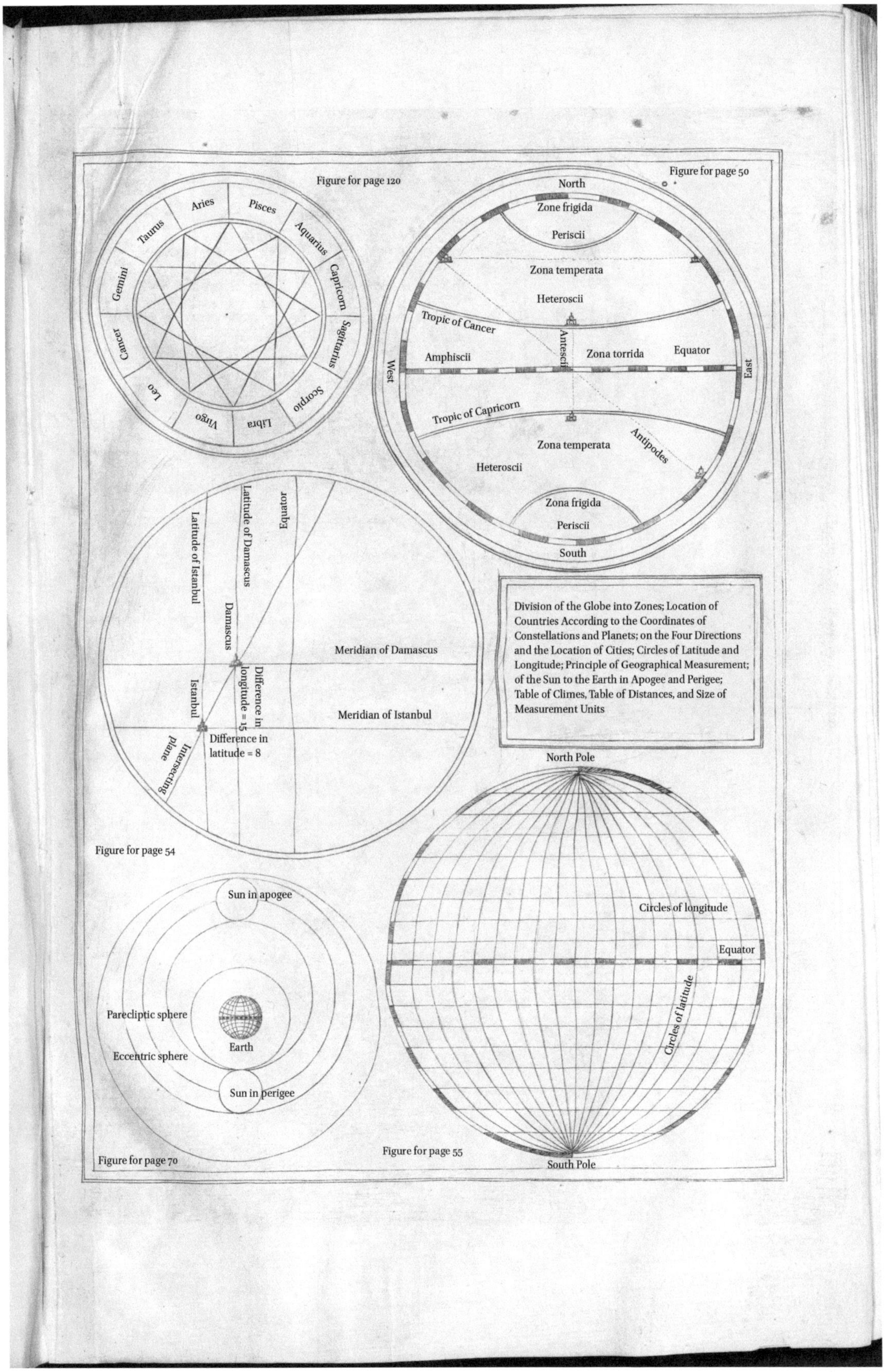

FIGURE 18 M 49/50 Zones, coordinates, astronomical calculations (translation)

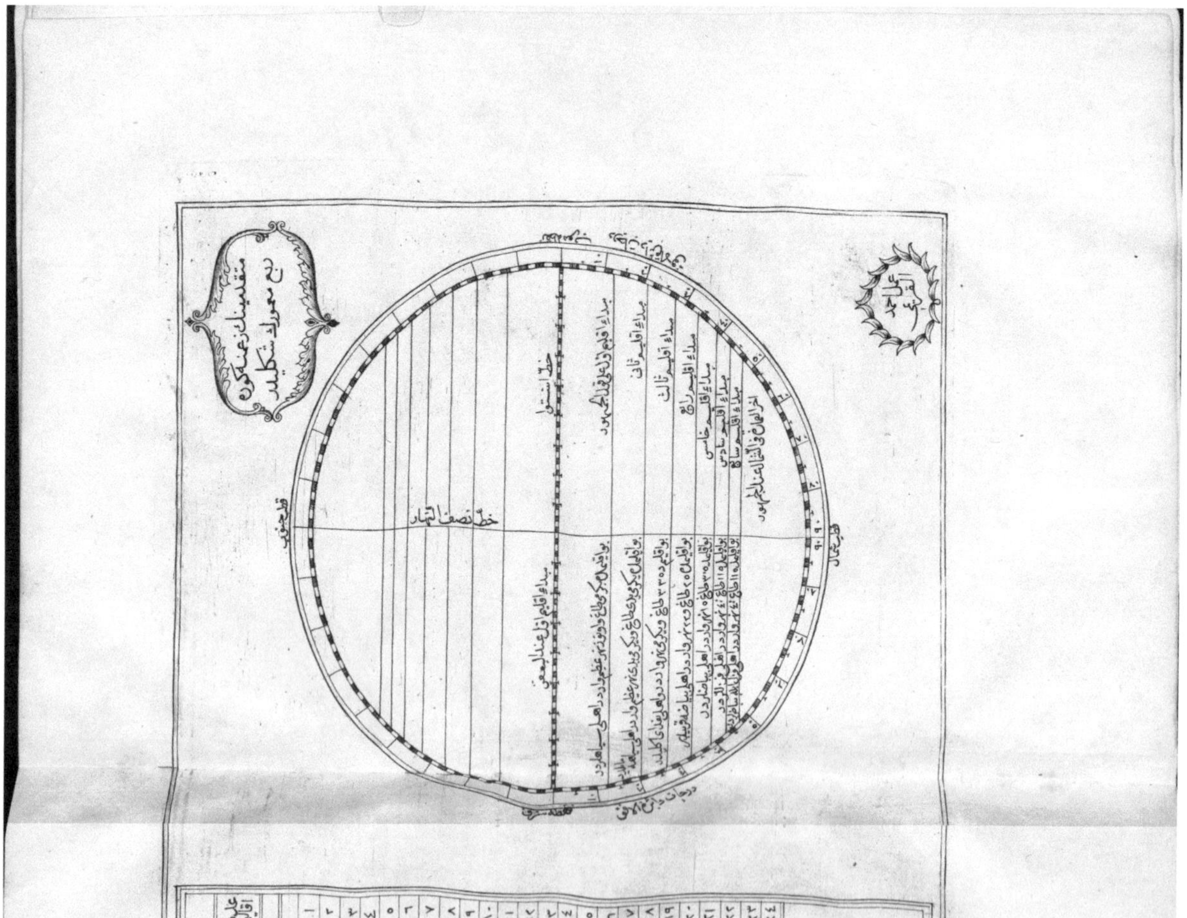

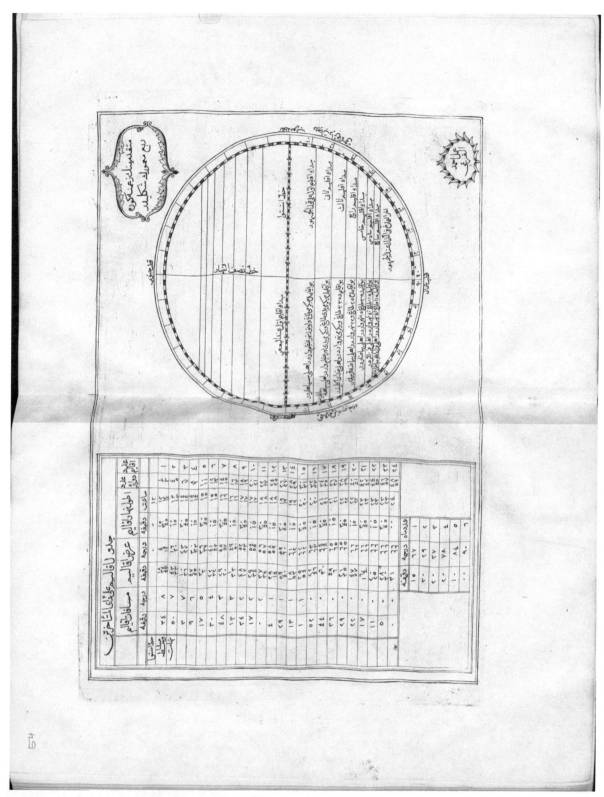

FIGURE 19 M 51/52 The inhabited quarter and the climes according to the ancients; cf. R 10a
BAYERISCHE STAATSBIBLIOTHEK MÜNCHEN, RES/2 A.OR. 371,
URN:NBN:DE:BVB:12-BSB00096220-8

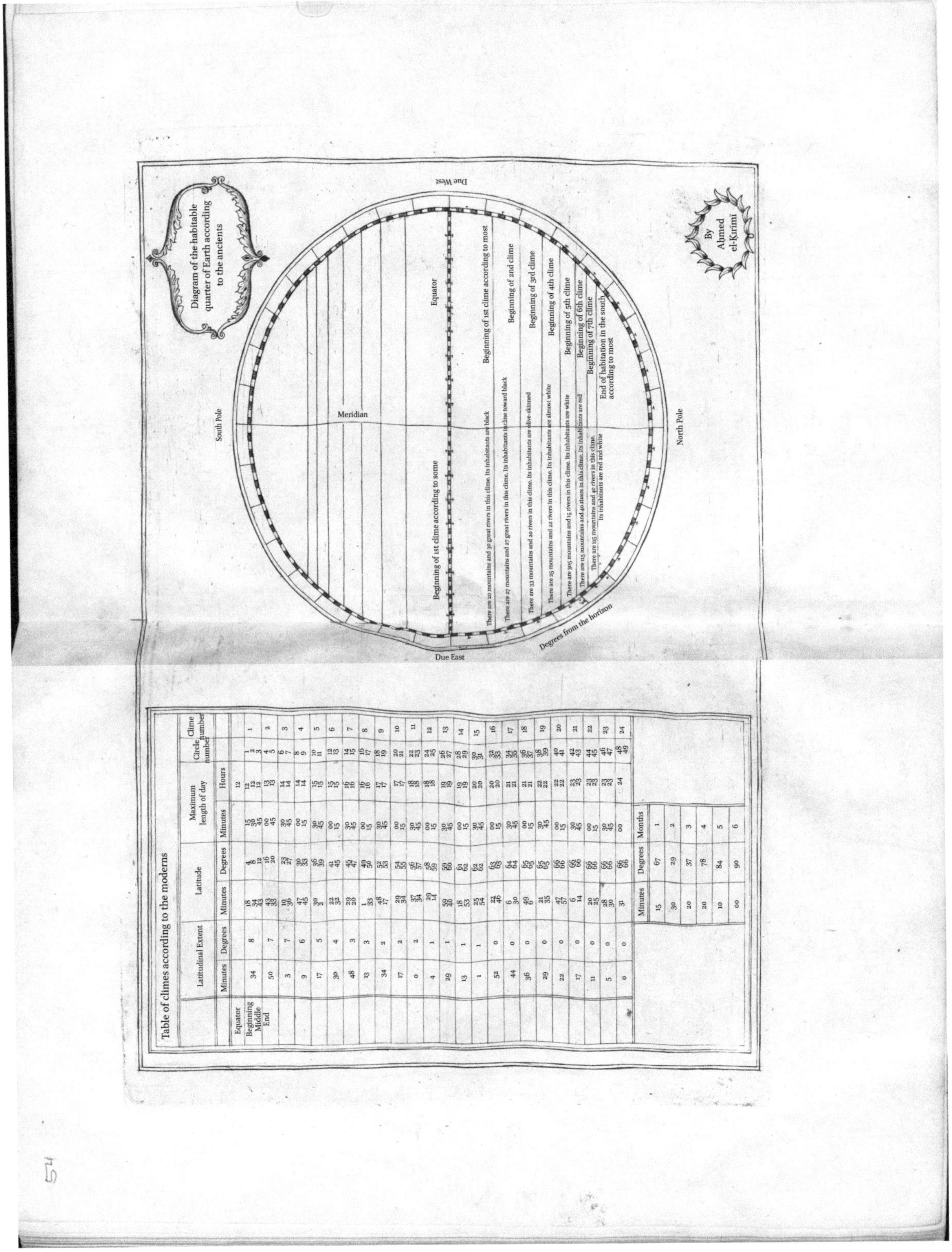

FIGURE 20 M 51/52 The inhabited quarter and the climes according to the ancients (translation)

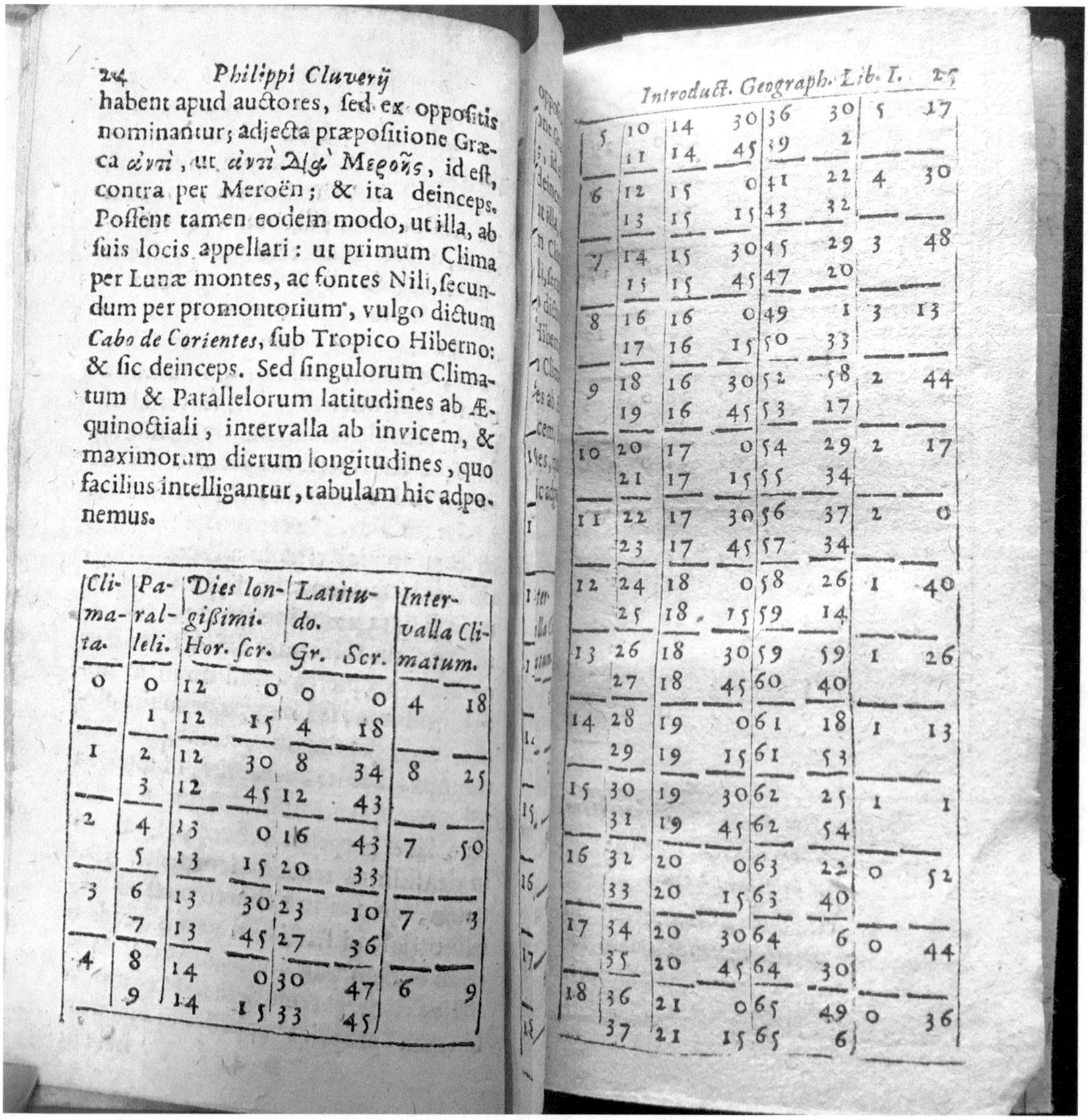

24 *Philippi Cluverij*

habent apud auctores, sed ex oppositis nominantur; adjecta præpositione Græca ἀντὶ, ut ἀντὶ Διὰ Μερόης, id est, contra per Meroën; & ita deinceps. Possent tamen eodem modo, ut illa, ab suis locis appellari: ut primum Clima per Lunæ montes, ac fontes Nili, secundum per promontorium, vulgo dictum *Cabo de Corientes*, sub Tropico Hiberno: & sic deinceps. Sed singulorum Climatum & Parallelorum latitudines ab Æquinoctiali, intervalla ab invicem, & maximorum dierum longitudines, quo facilius intelligantur, tabulam hic adponemus.

Climata.	*Paralleli.*	*Dies longißimi.* *Hor.*	*scr.*	*Latitudo.* *Gr.*	*Scr.*	*Intervalla Climatum.*	
0	0	12	0	0	0	4	18
	1	12	15	4	18		
1	2	12	30	8	34	8	25
	3	12	45	12	43		
2	4	13	0	16	43	7	50
	5	13	15	20	33		
3	6	13	30	23	10	7	3
	7	13	45	27	36		
4	8	14	0	30	47	6	9
	9	14	15	33	45		

Introduct. Geograph. Lib. I. 25

5	10	14	30	36	30	5	17
	11	14	45	39	2		
6	12	15	0	41	22	4	30
	13	15	15	43	32		
7	14	15	30	45	29	3	48
	15	15	45	47	20		
8	16	16	0	49	1	3	13
	17	16	15	50	33		
9	18	16	30	52	58	2	44
	19	16	45	53	17		
10	20	17	0	54	29	2	17
	21	17	15	55	34		
11	22	17	30	56	37	2	0
	23	17	45	57	34		
12	24	18	0	58	26	1	40
	25	18	15	59	14		
13	26	18	30	59	59	1	26
	27	18	45	60	40		
14	28	19	0	61	18	1	13
	29	19	15	61	53		
15	30	19	30	62	25	1	1
	31	19	45	62	54		
16	32	20	0	63	22	0	52
	33	20	15	63	40		
17	34	20	30	64	6	0	44
	35	20	45	64	30		
18	36	21	0	65	49	0	36
	37	21	15	65	6		

FIGURE 21A Cluverius, *Introductio* (1641 edition), pp. 24–25, table of climes
UNIVERSITY OF MICHIGAN LIBRARY, SPECIAL COLLECTIONS, Z 232 .E5 1641EB

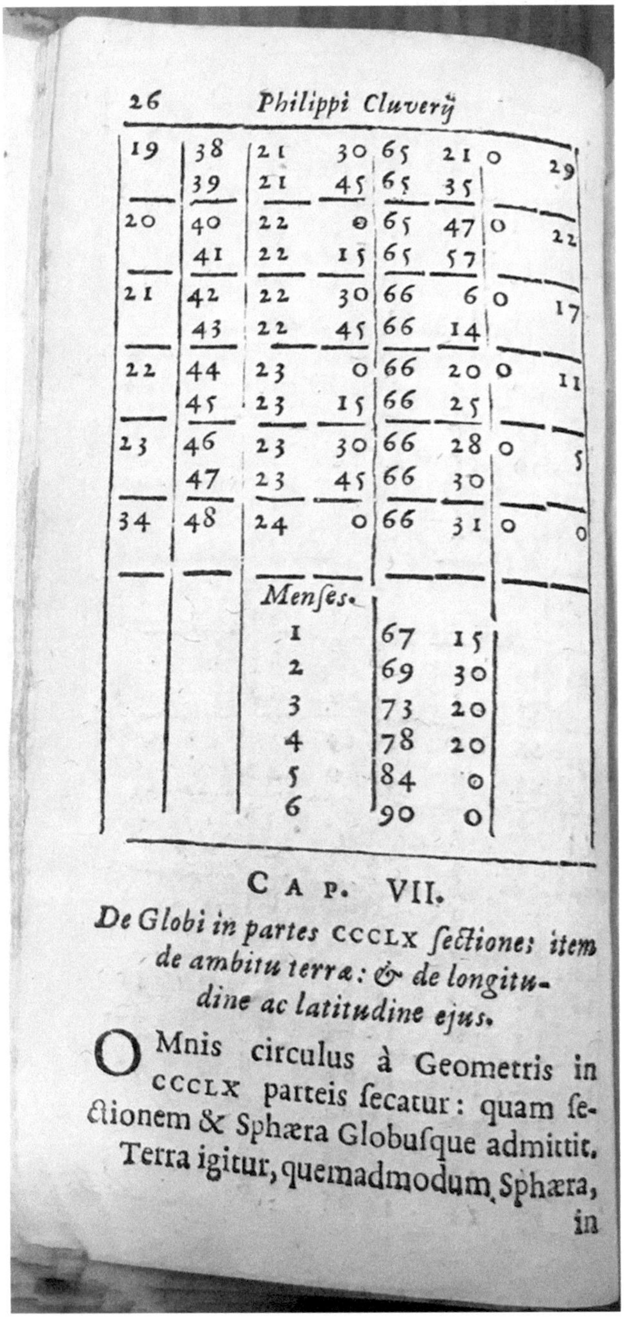

26 *Philippi Cluverij*

19	38	21	30	65	21	0	29
	39	21	45	65	35		
20	40	22	0	65	47	0	22
	41	22	15	65	57		
21	42	22	30	66	6	0	17
	43	22	45	66	14		
22	44	23	0	66	20	0	11
	45	23	15	66	25		
23	46	23	30	66	28	0	5
	47	23	45	66	30		
34	48	24	0	66	31	0	0
		Menses.					
		1		67	15		
		2		69	30		
		3		73	20		
		4		78	20		
		5		84	0		
		6		90	0		

CAP. VII.

De Globi in partes CCCLX *ſectione: item de ambitu terræ: & de longitudine ac latitudine ejus.*

OMnis circulus à Geometris in CCCLX parteis ſecatur: quam ſectionem & Sphæra Globuſque admittit. Terra igitur, quemadmodum Sphæra, in

FIGURE 21B Cluverius, *Introductio*, p. 26, table of climes
UNIVERSITY OF MICHIGAN LIBRARY, SPECIAL COLLECTIONS, Z 232 .E5 1641EB

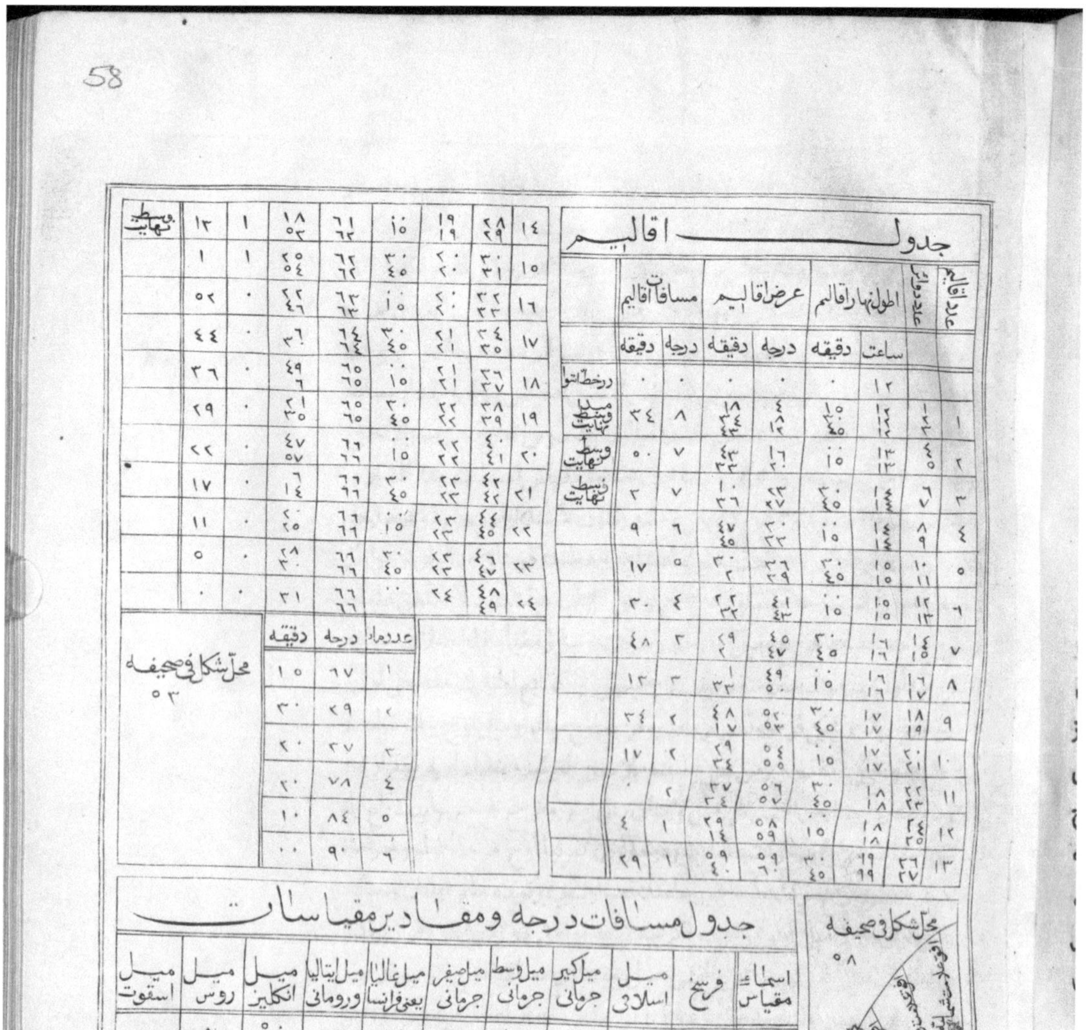

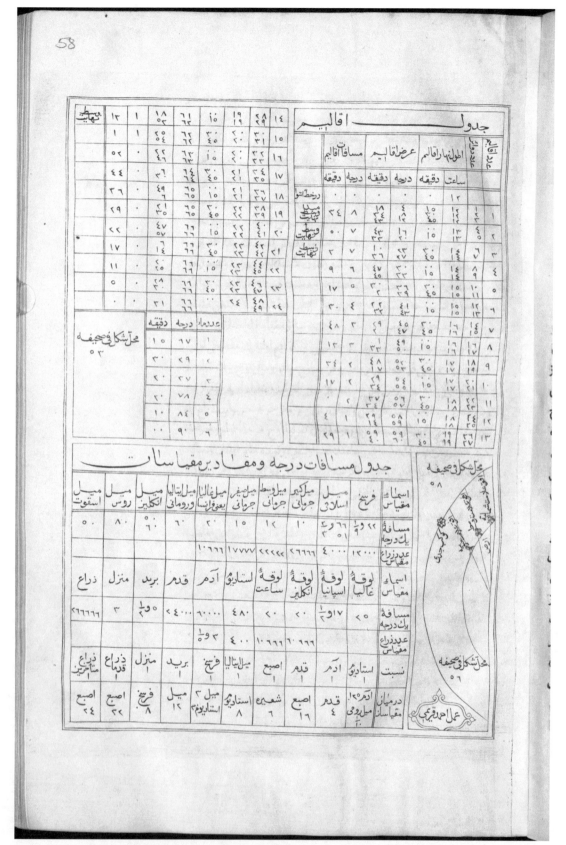

FIGURE 22 M 57/58 Tables of climes and distances; measuring the circumference of the earth; cf. R 12a and 12b

BAYERISCHE STAATSBIBLIOTHEK MÜNCHEN, RES/2 A.OR. 371, URN:NBN:DE:BVB:12-BSB00096220-8

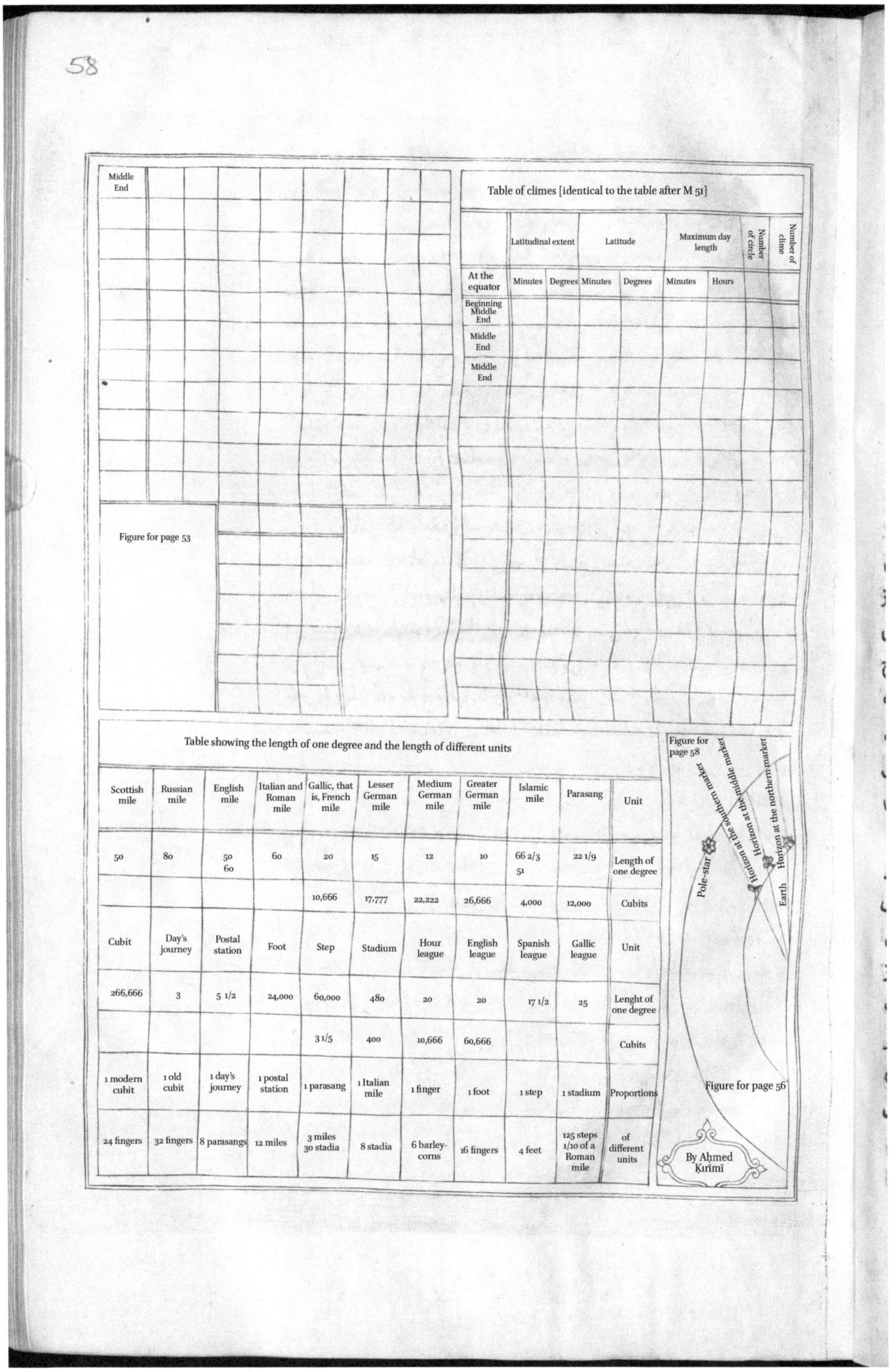

FIGURE 23 M 57/58 Tables of climes and distances; measuring the circumference of the earth (translation)

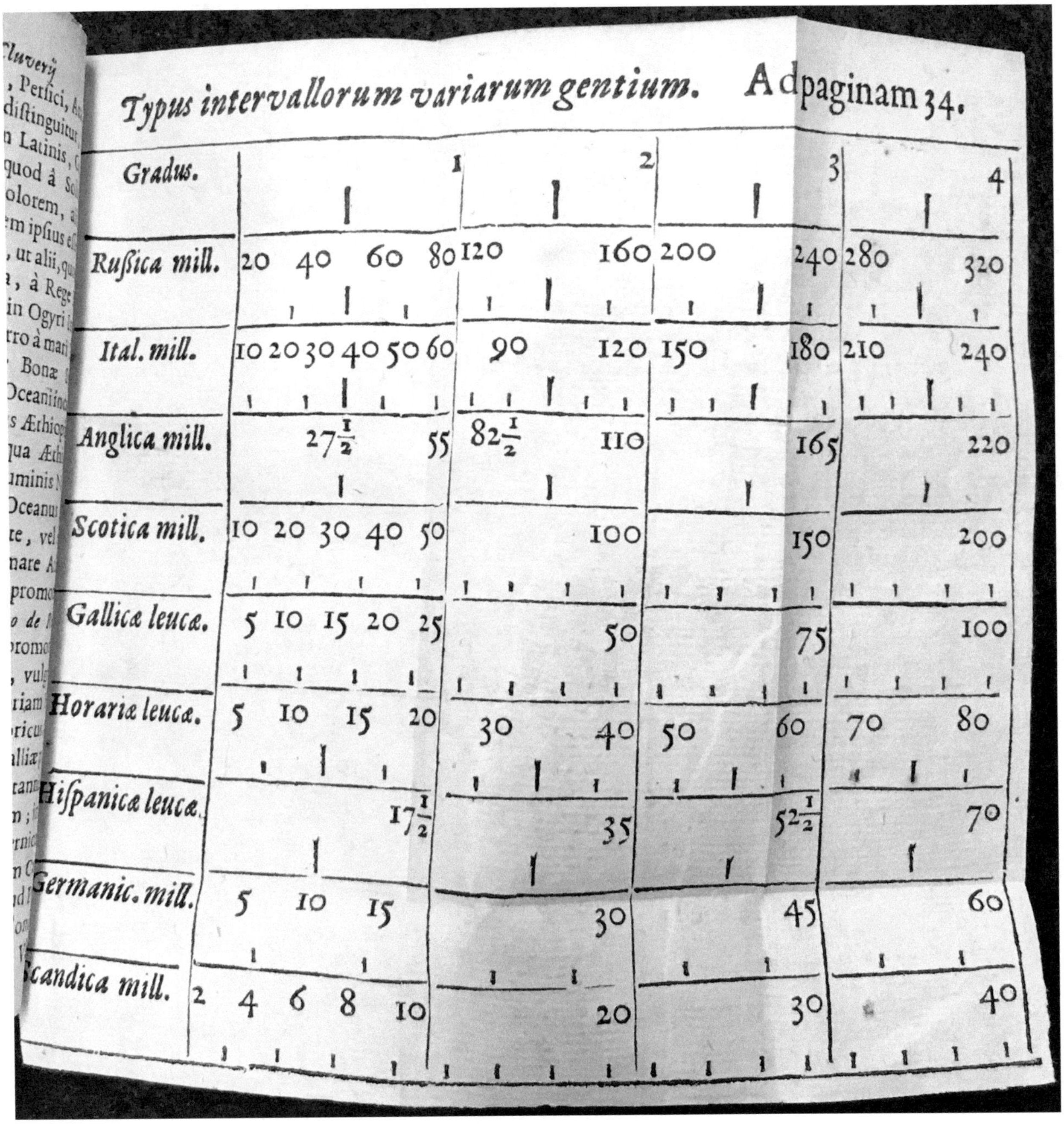

Typus intervallorum variarum gentium. Ad paginam 34.

Gradus.	1	2	3	4
Rußica mill.	20 40 60 80	120 160	200 240	280 320
Ital. mill.	10 20 30 40 50 60	90 120	150 180	210 240
Anglica mill.	$27\frac{1}{2}$ 55	$82\frac{1}{2}$ 110	165	220
Scotica mill.	10 20 30 40 50	100	150	200
Gallicæ leucæ.	5 10 15 20 25	50	75	100
Horariæ leucæ.	5 10 15 20	30 40	50 60	70 80
Hispanicæ leucæ.	$17\frac{1}{2}$	35	$52\frac{1}{2}$	70
Germanic. mill.	5 10 15	30	45	60
Scandica mill.	2 4 6 8 10	20	30	40

FIGURE 24 Cluverius, *Introductio* (1641 edition), after p. 34, table of distances
UNIVERSITY OF MICHIGAN LIBRARY, SPECIAL COLLECTIONS, Z 232 .E5 1641EB

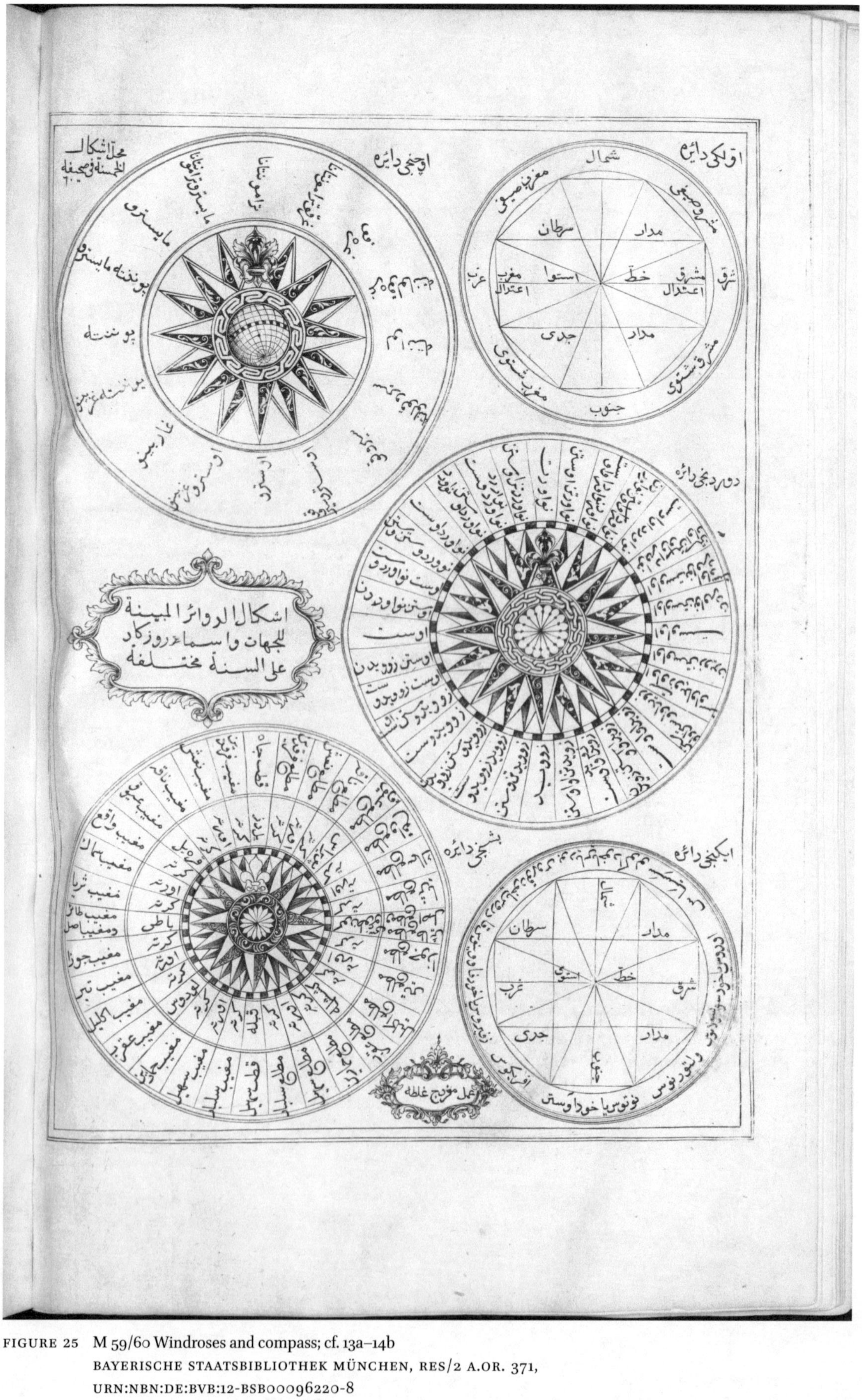

FIGURE 25 M 59/60 Windroses and compass; cf. 13a–14b
BAYERISCHE STAATSBIBLIOTHEK MÜNCHEN, RES/2 A.OR. 371,
URN:NBN:DE:BVB:12-BSB00096220-8

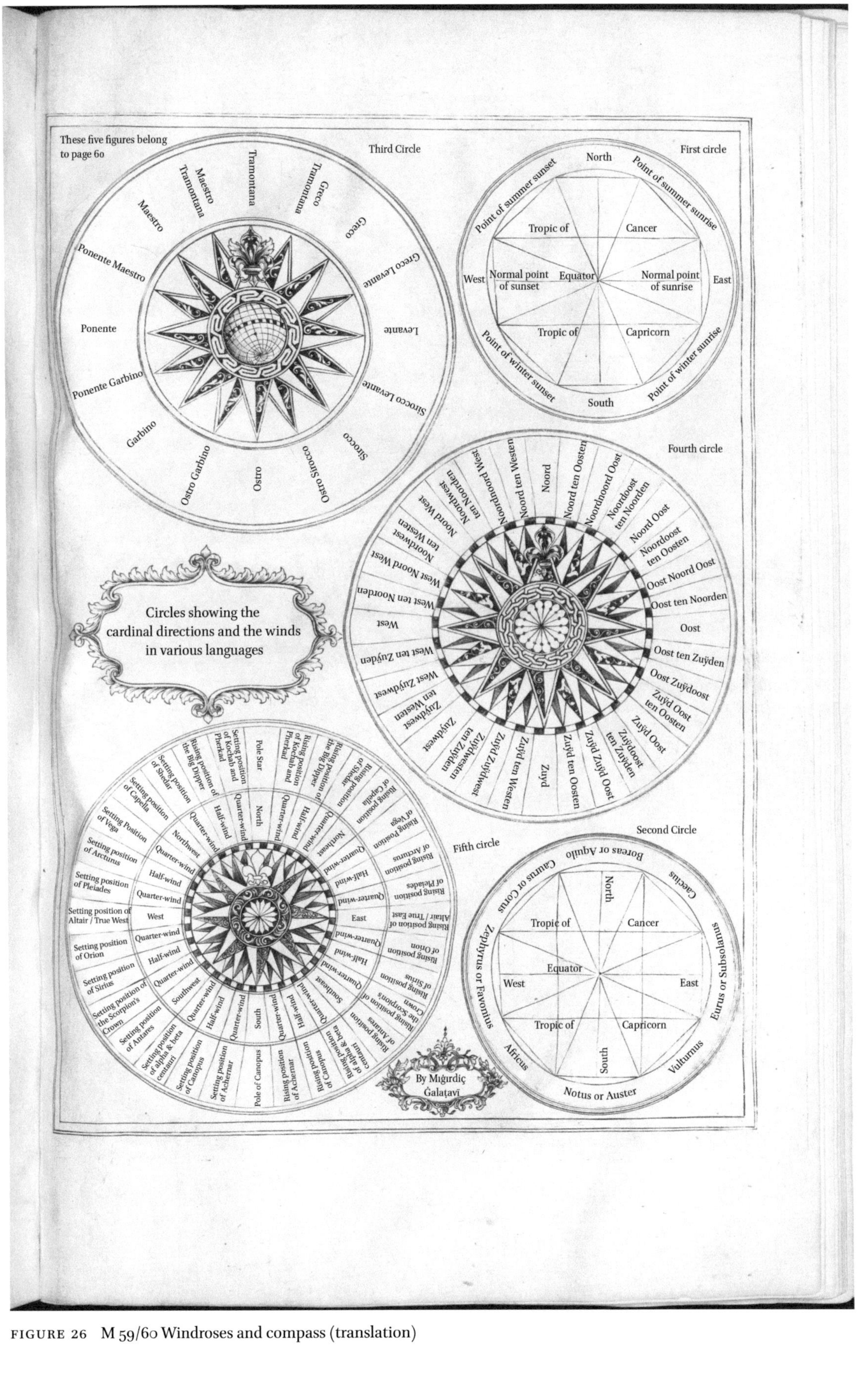

FIGURE 26 M 59/60 Windroses and compass (translation)

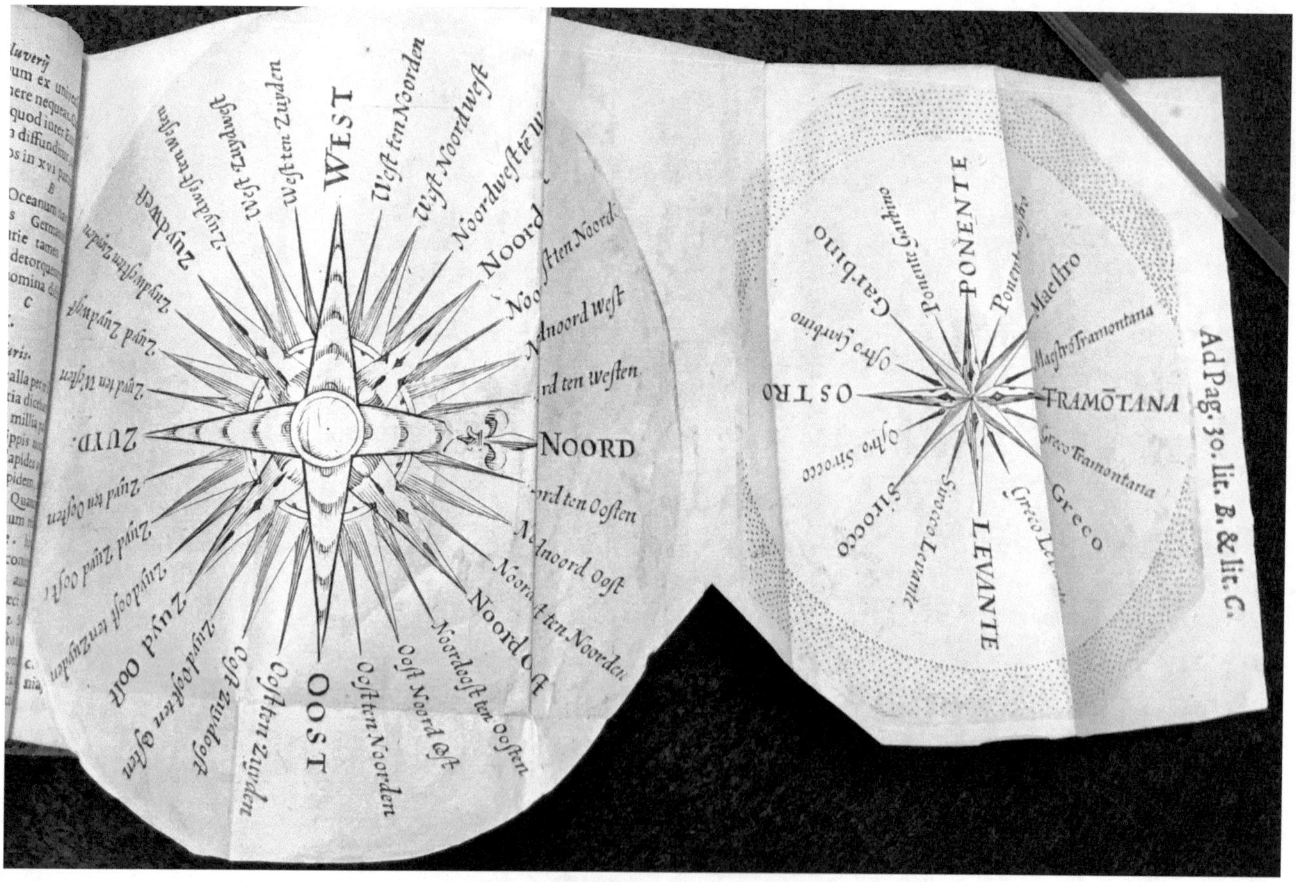

FIGURE 27A Cluverius, *Introductio* (1641 edition), after p. 28, windroses and compass
UNIVERSITY OF MICHIGAN LIBRARY, SPECIAL COLLECTIONS, Z 232 .E5 1641EB

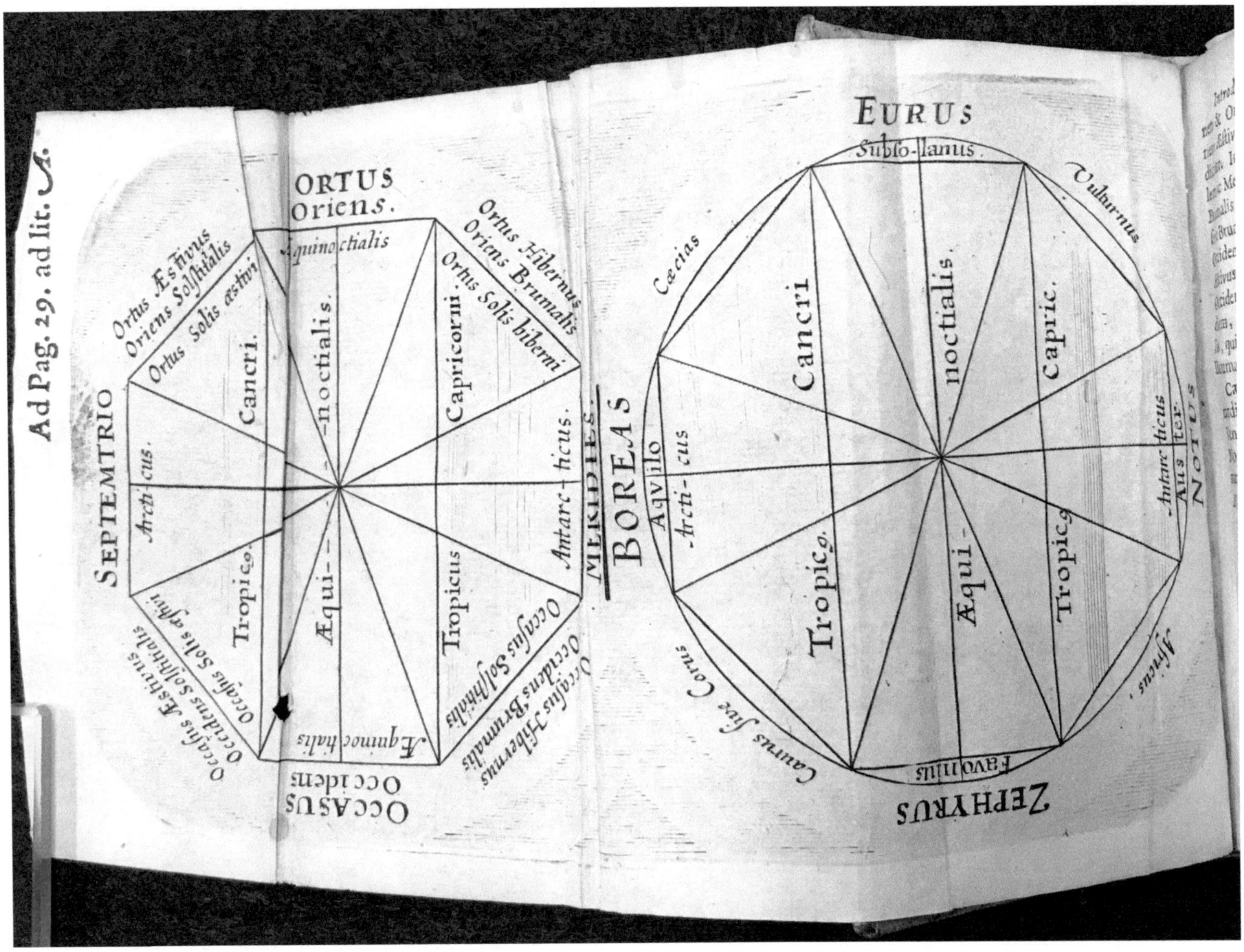

FIGURE 27B Cluverius, *Introductio* (1641 edition), after p. 30, windroses and compass
UNIVERSITY OF MICHIGAN LIBRARY, SPECIAL COLLECTIONS, Z 232 .E5 1641EB

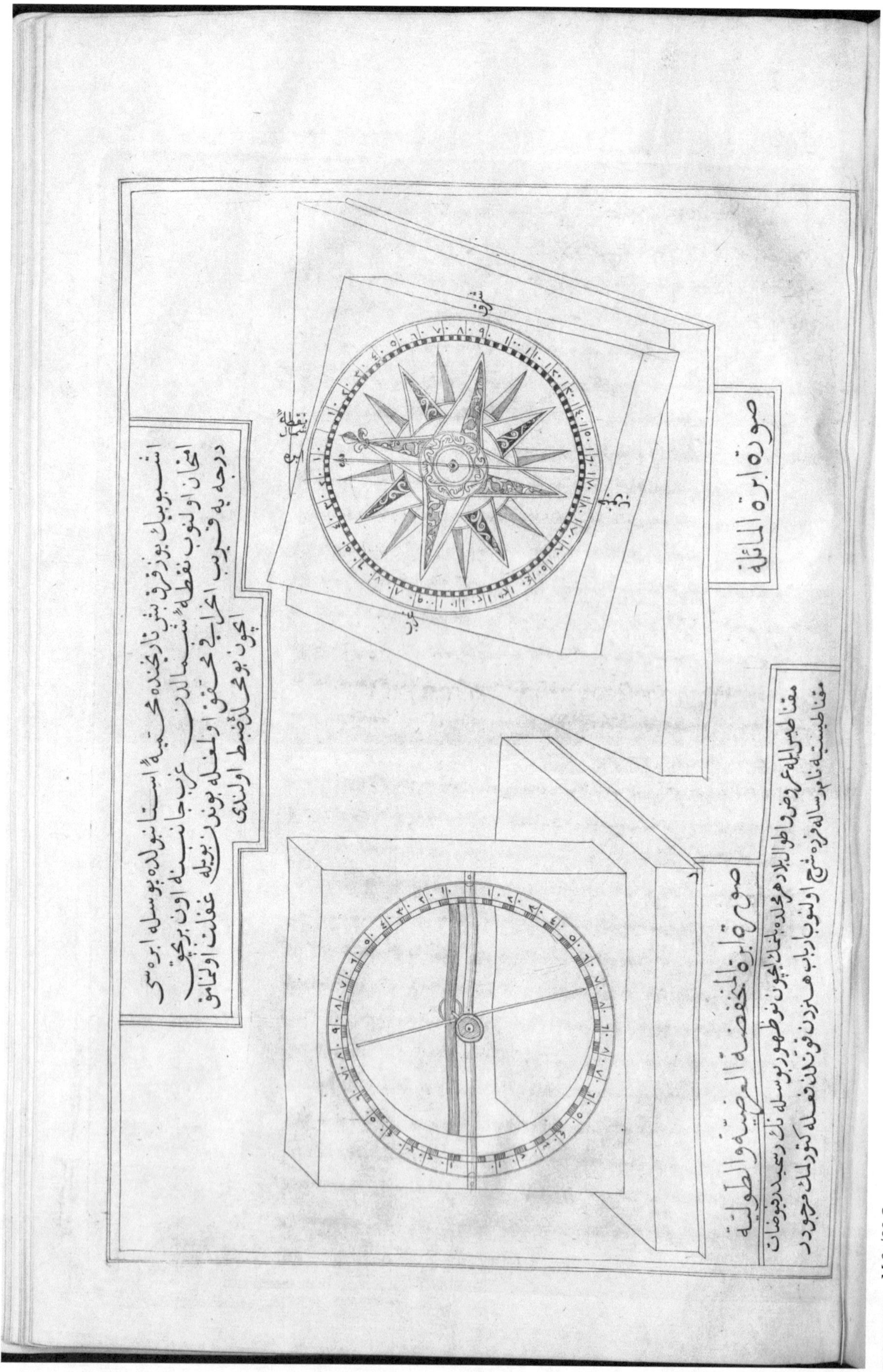

FIGURE 28 M 65/66 Compass
BAYERISCHE STAATSBIBLIOTHEK MÜNCHEN, RES/2 A.OR. 371,
URN:NBN:DE:BVB:12-BSB00096220-8

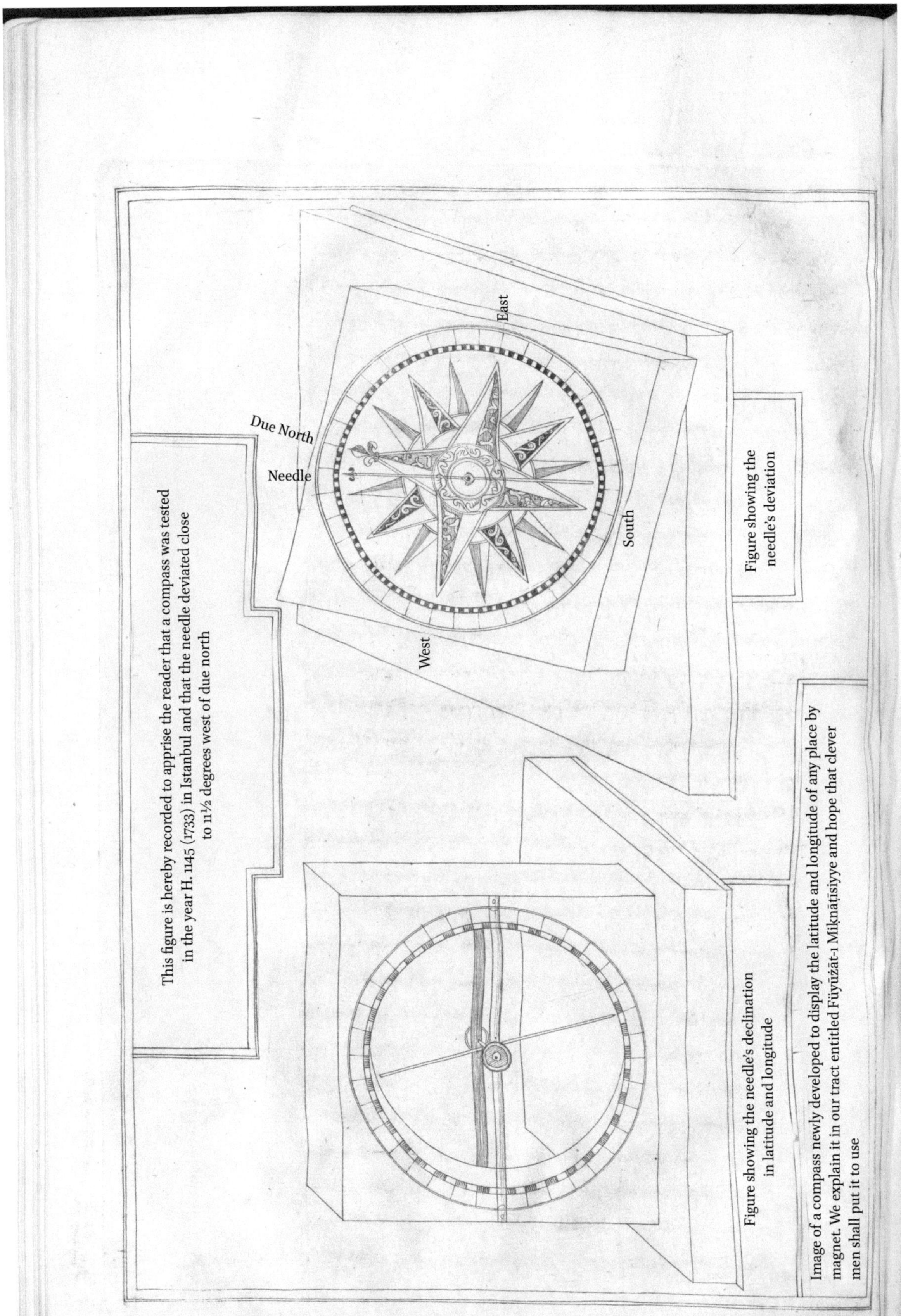

FIGURE 29 M 65/66 Compass (translation)

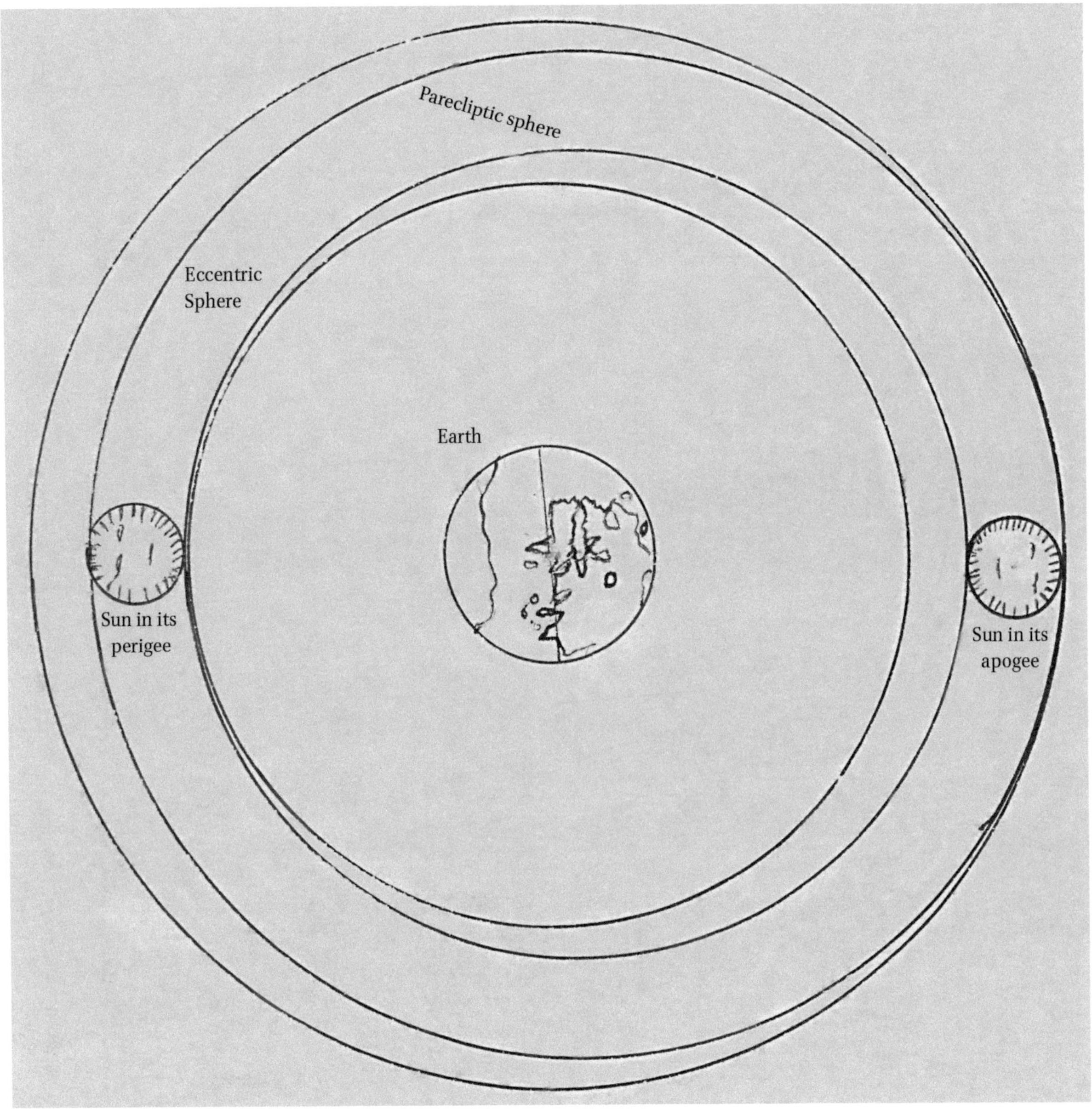

FIGURE 30 R 19b: Scheme showing earth with sun at apogee and perigee (translation)

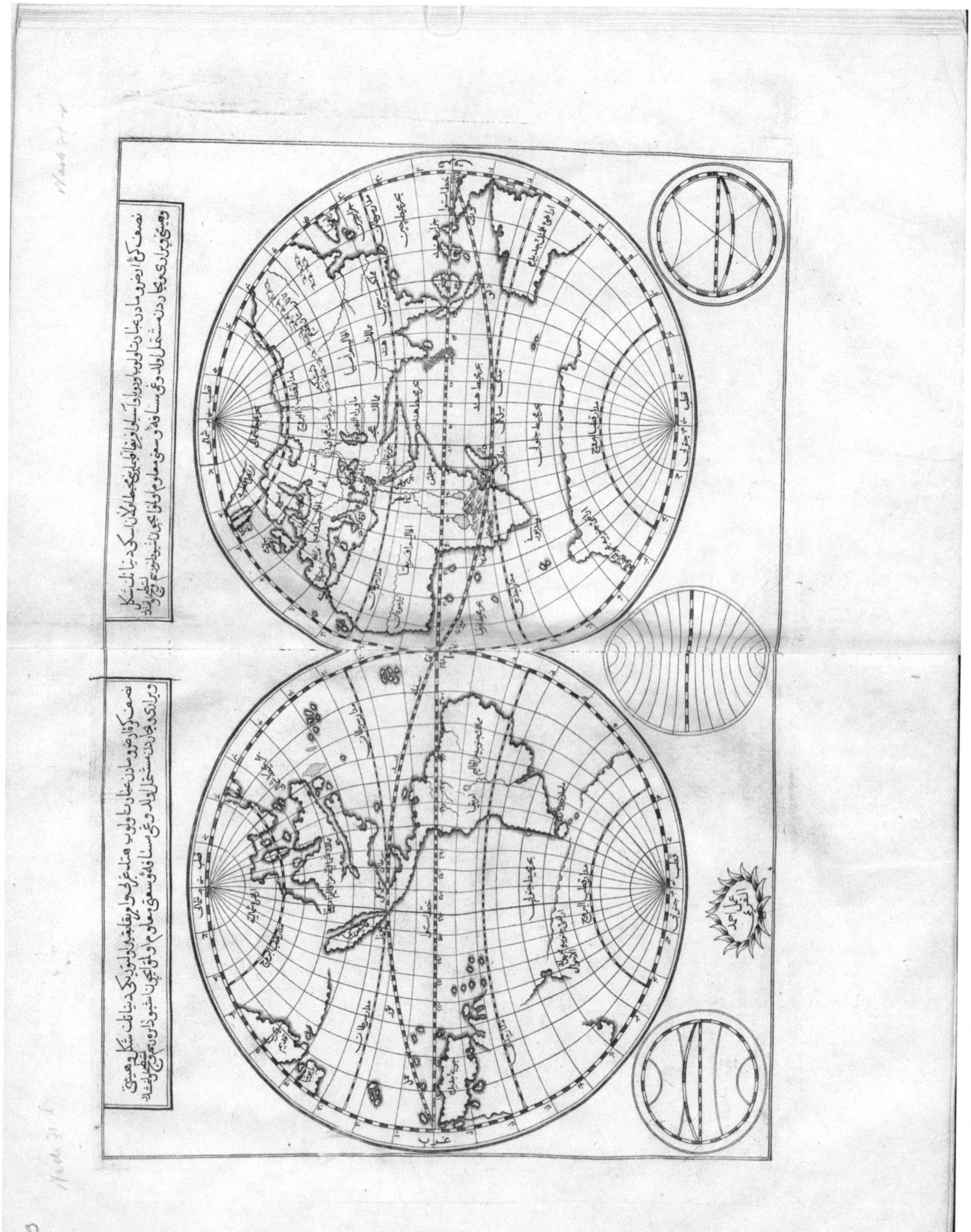

FIGURE 31 M 71/72 World map; cf. R 21a
BAYERISCHE STAATSBIBLIOTHEK MÜNCHEN, RES/2 A.OR. 371,
URN:NBN:DE:BVB:12-BSB00096220-8

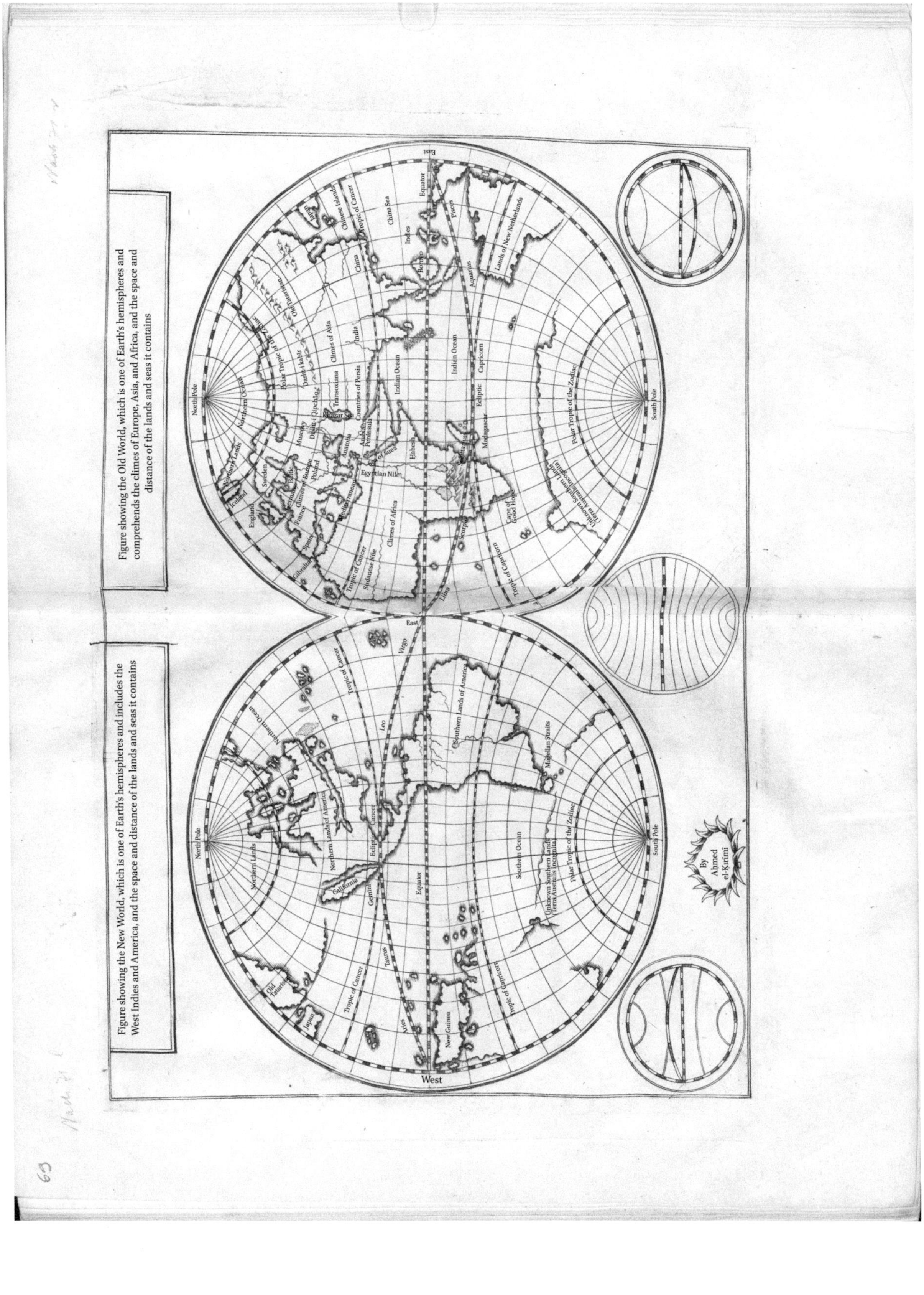

FIGURE 32 M 71/72 World map (translation)

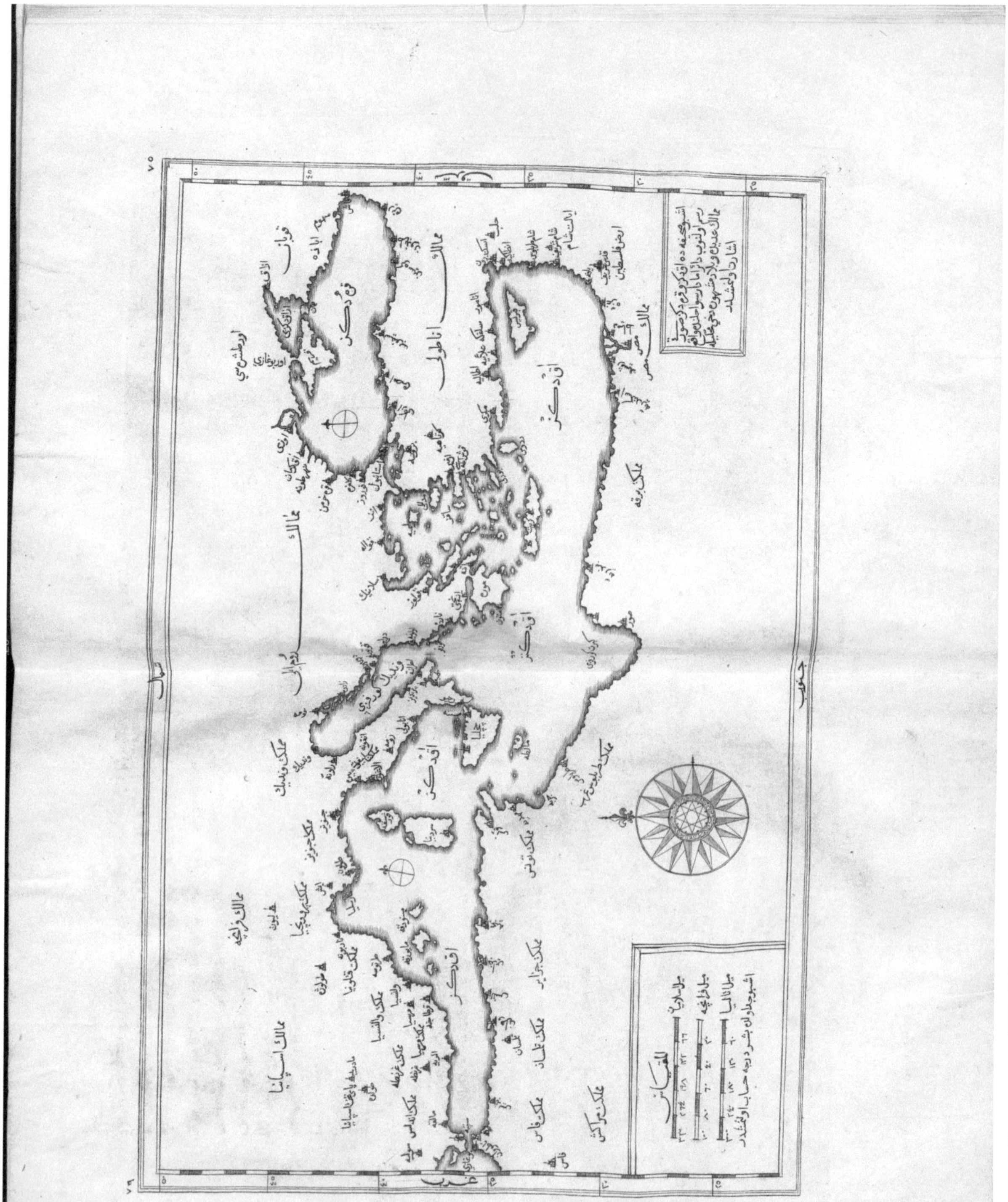

FIGURE 33 M 75/76 Mediterranean and Black Seas
BAYERISCHE STAATSBIBLIOTHEK MÜNCHEN, RES/2 A.OR. 371,
URN:NBN:DE:BVB:12-BSB00096220-8

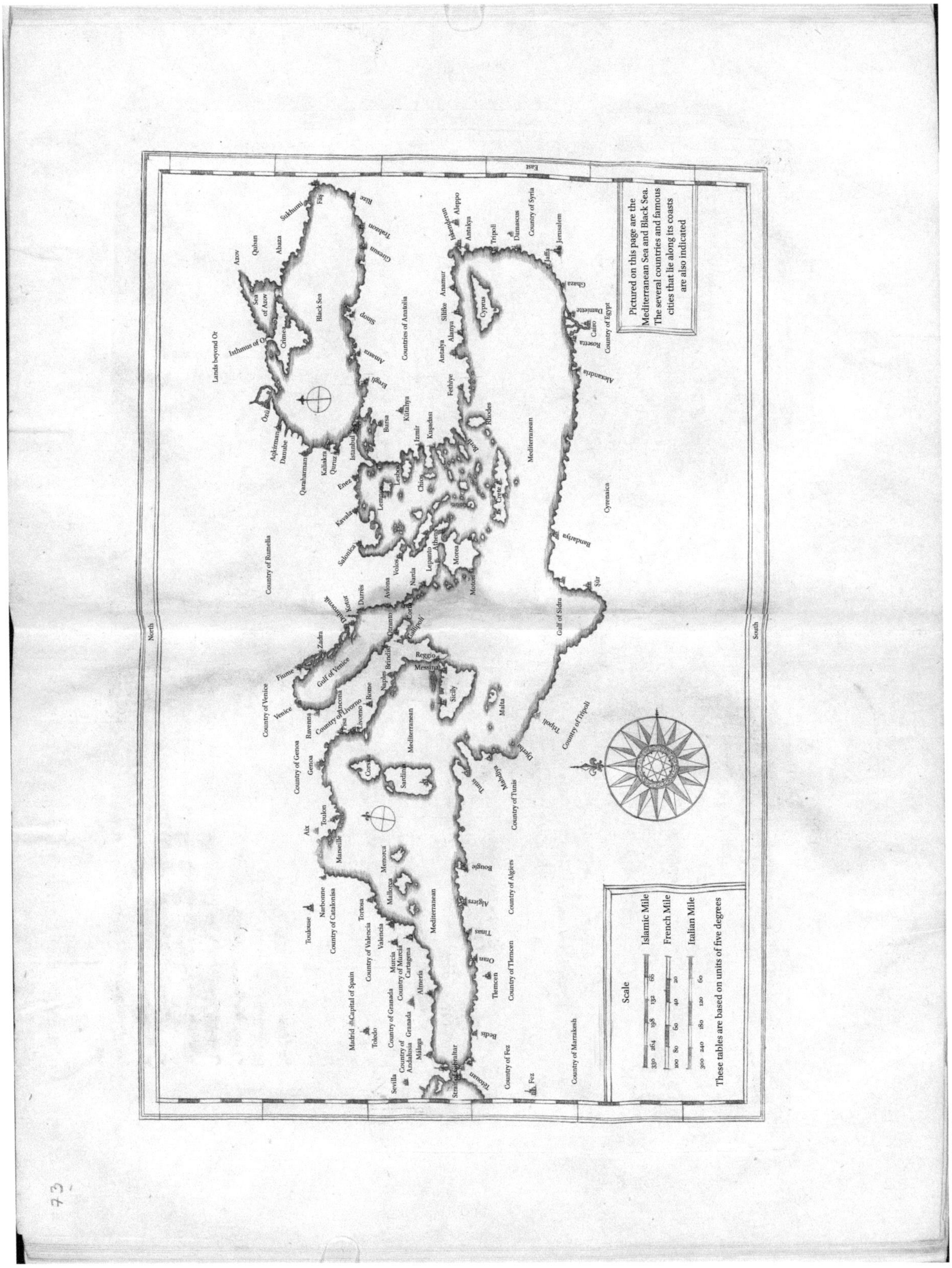

FIGURE 34 M 75/76 Mediterranean and Black Seas (translation)

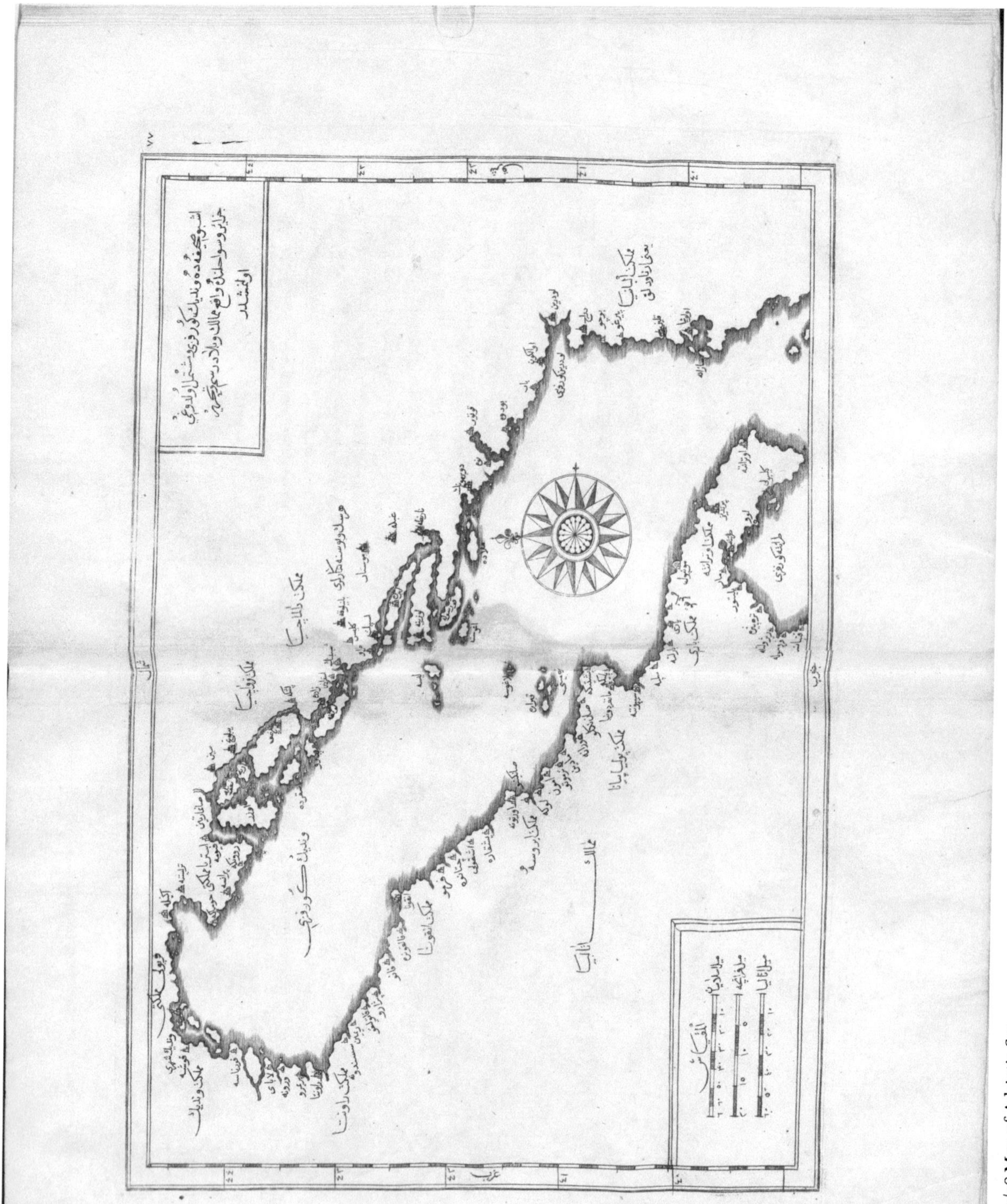

FIGURE 35 M 77/78 Map of Adriatic Sea
BAYERISCHE STAATSBIBLIOTHEK MÜNCHEN, RES/2 A.OR. 371,
URN:NBN:DE:BVB:12-BSB00096220-8

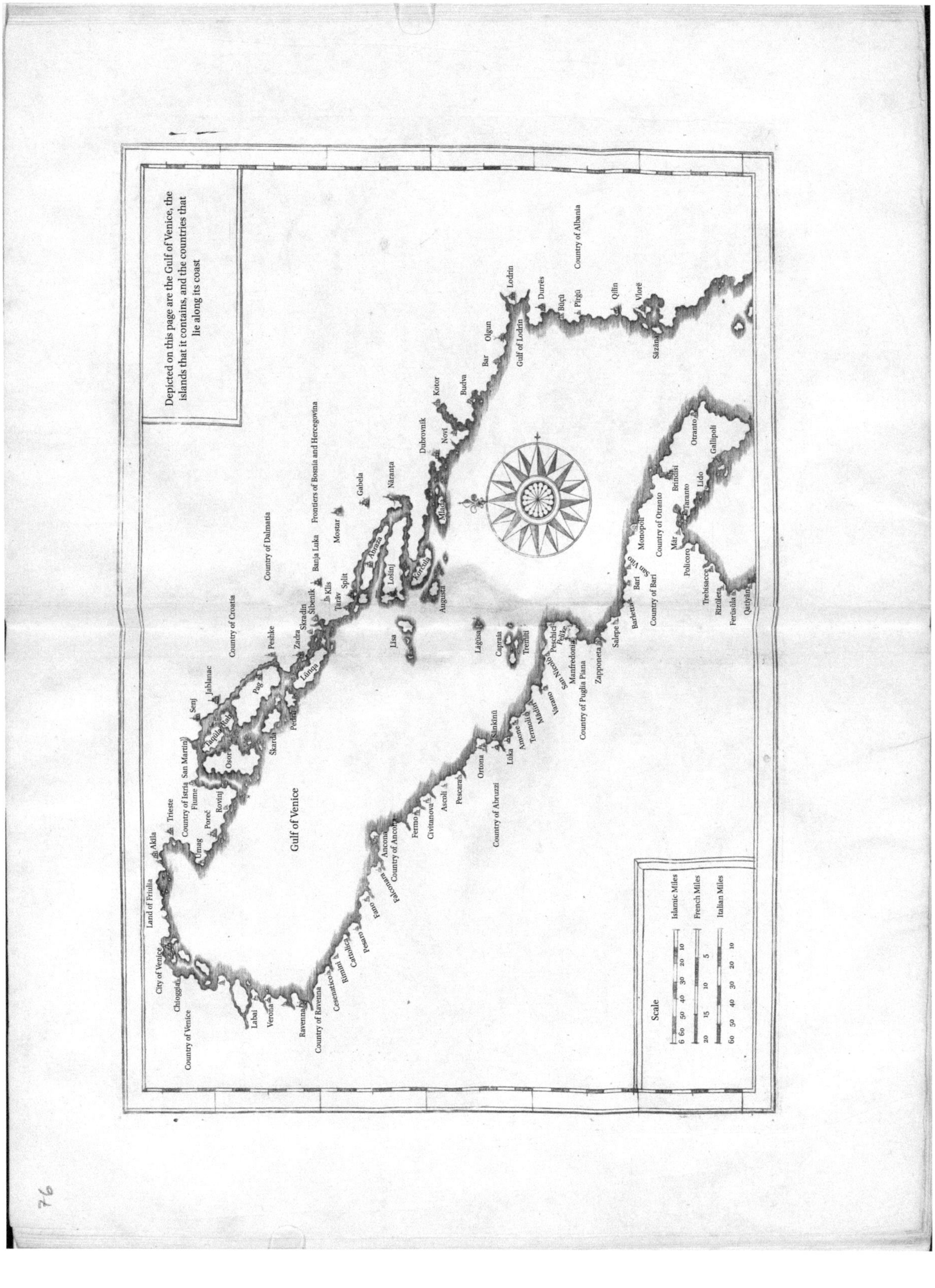

FIGURE 36 M 77/78 Map of Adriatic Sea (translation)

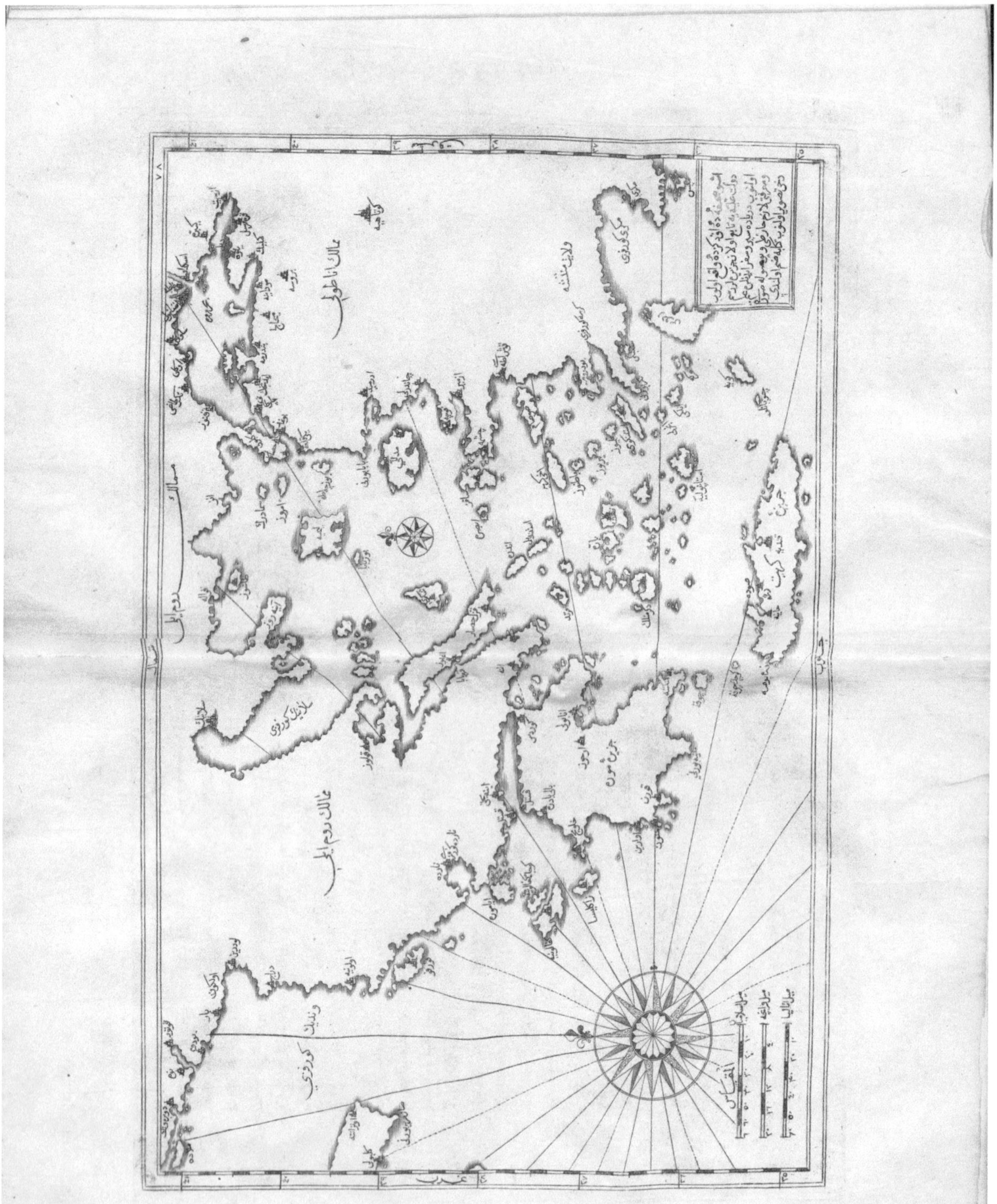

FIGURE 37 M 77/78 Map of Aegean Sea
BAYERISCHE STAATSBIBLIOTHEK MÜNCHEN, RES/2 A.OR. 371, URN:NBN:DE:BVB:12-BSB00096220-8

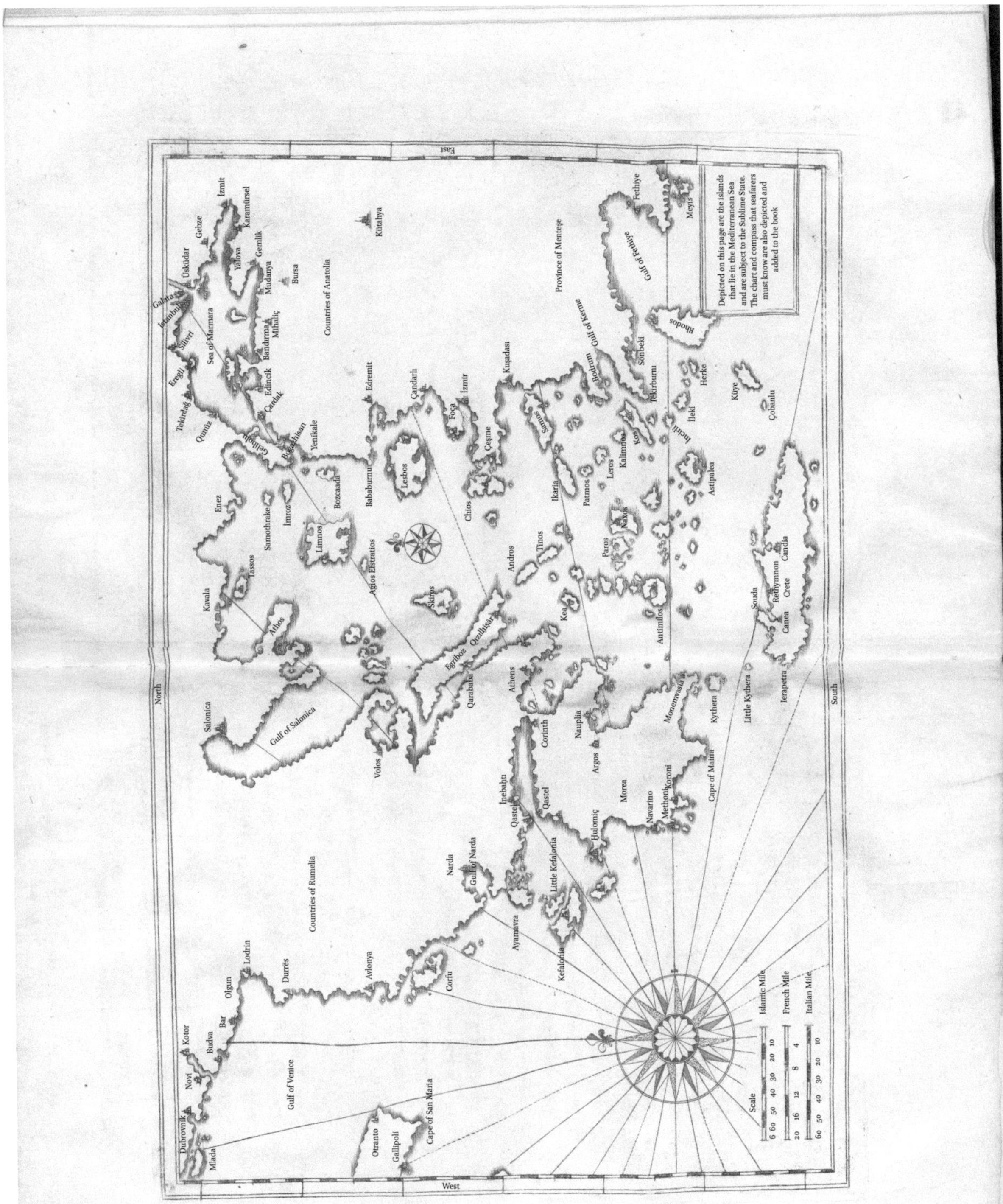

FIGURE 38 M 77/78 Map of Aegean Sea (translation)

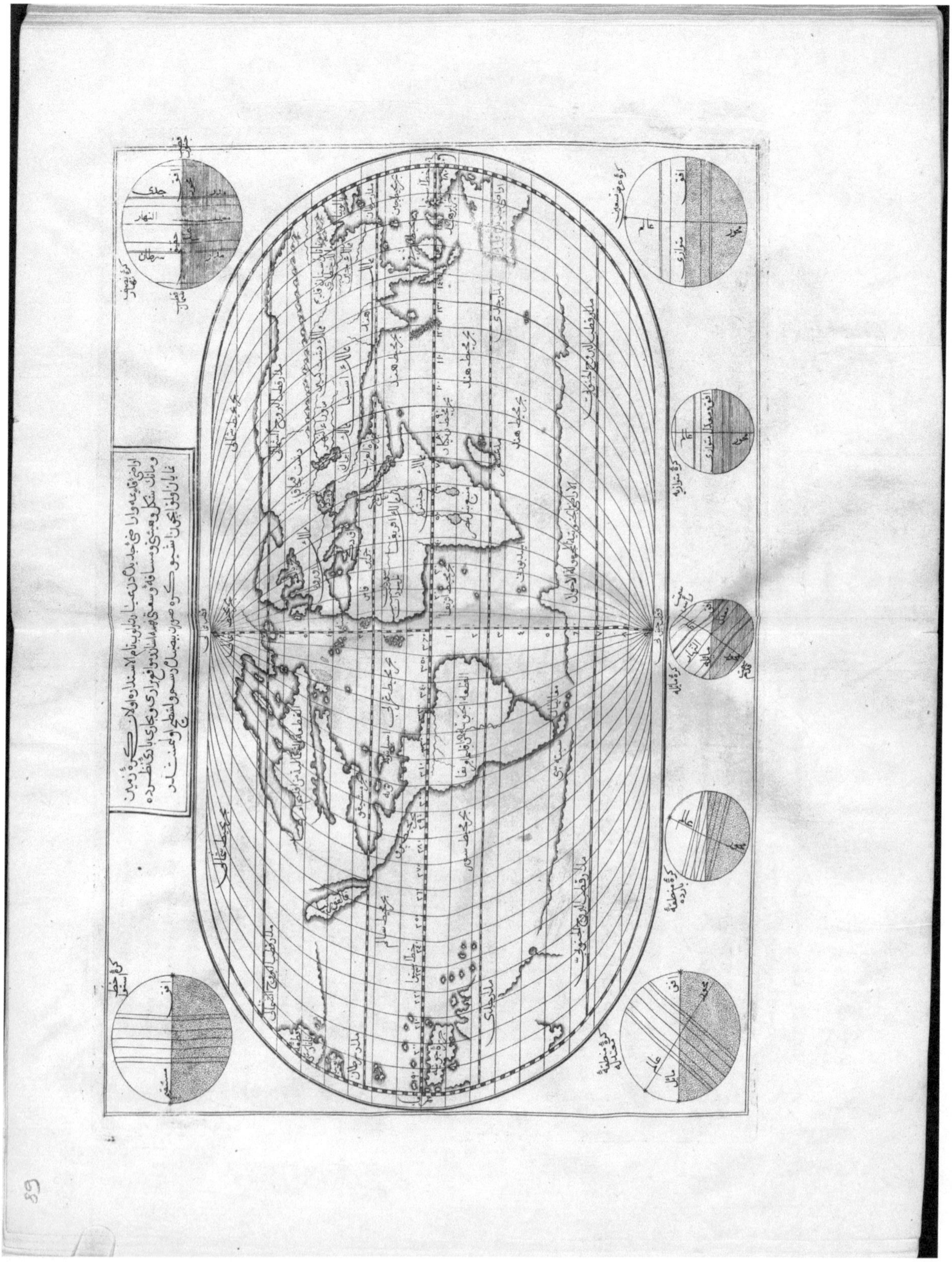

FIGURE 39 M 95/96 World map
BAYERISCHE STAATSBIBLIOTHEK MÜNCHEN, RES/2 A.OR. 371,
URN:NBN:DE:BVB:12-BSB00096220-8

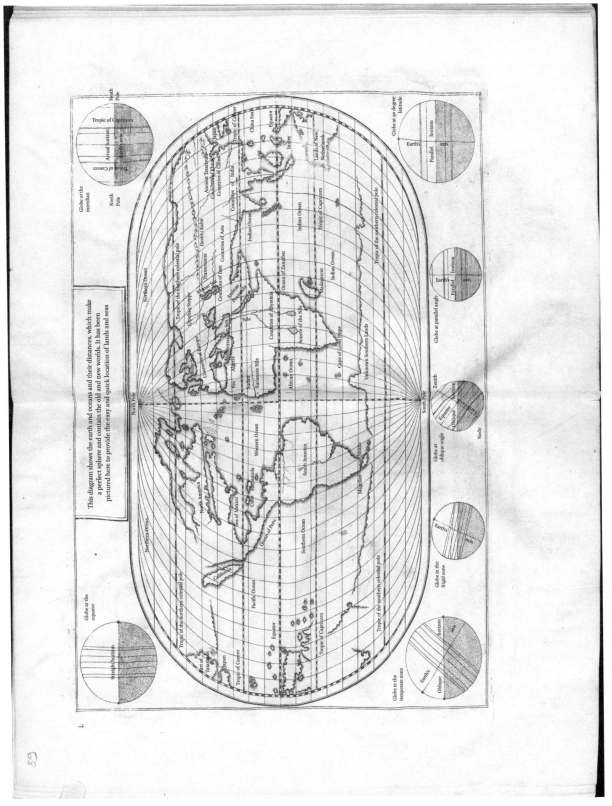

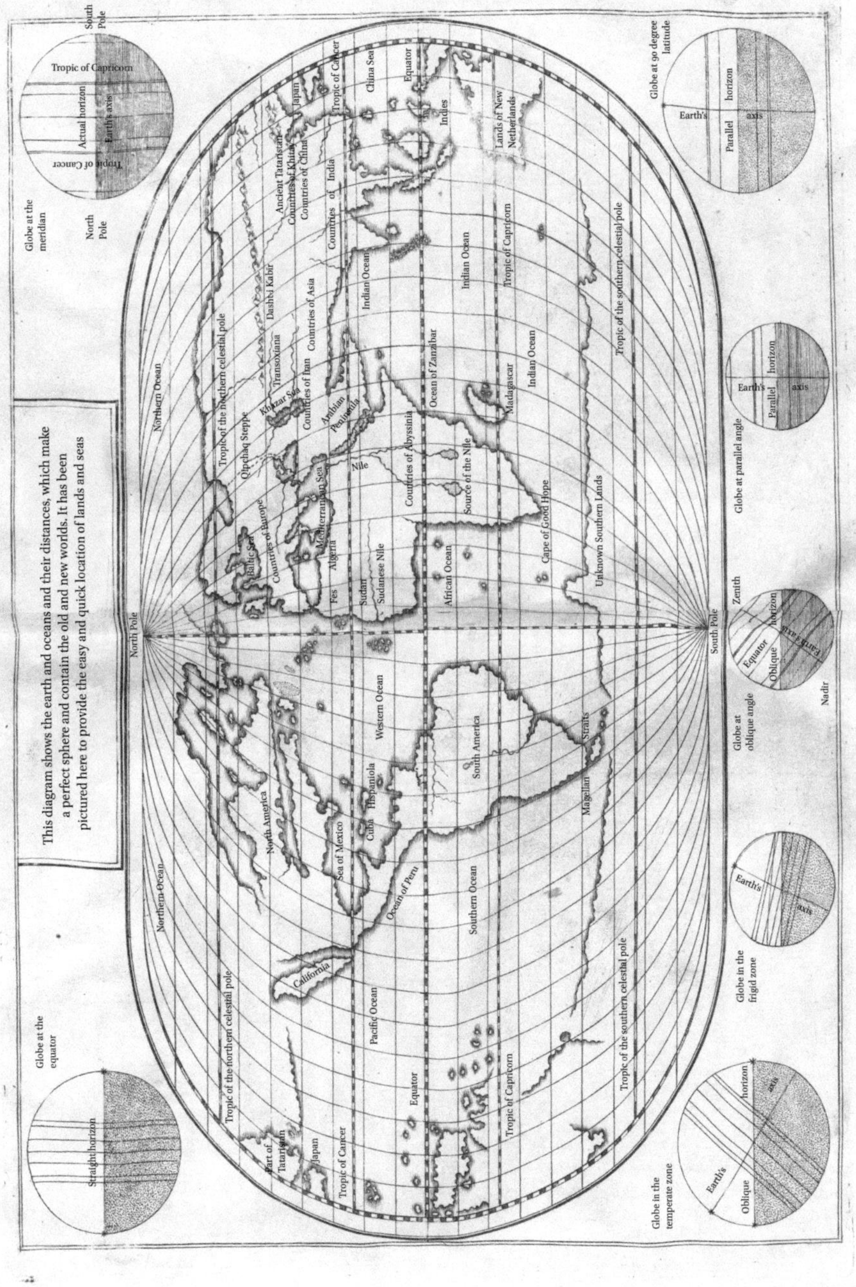

FIGURE 40 M 95/96 World map (translation)

FIGURE 41 M 99/100 Map of Europe; cf. R 32a
BAYERISCHE STAATSBIBLIOTHEK MÜNCHEN, RES/2 A.OR. 371,
URN:NBN:DE:BVB:12-BSB00096220-8

FIGURE 42 M 99/100 Map of Europe (translation)

FIGURE 43 M 101/102 Map of Africa; cf. R 34a
BAYERISCHE STAATSBIBLIOTHEK MÜNCHEN, RES/2 A.OR. 371, URN:NBN:DE:BVB:12-BSB00096220-8

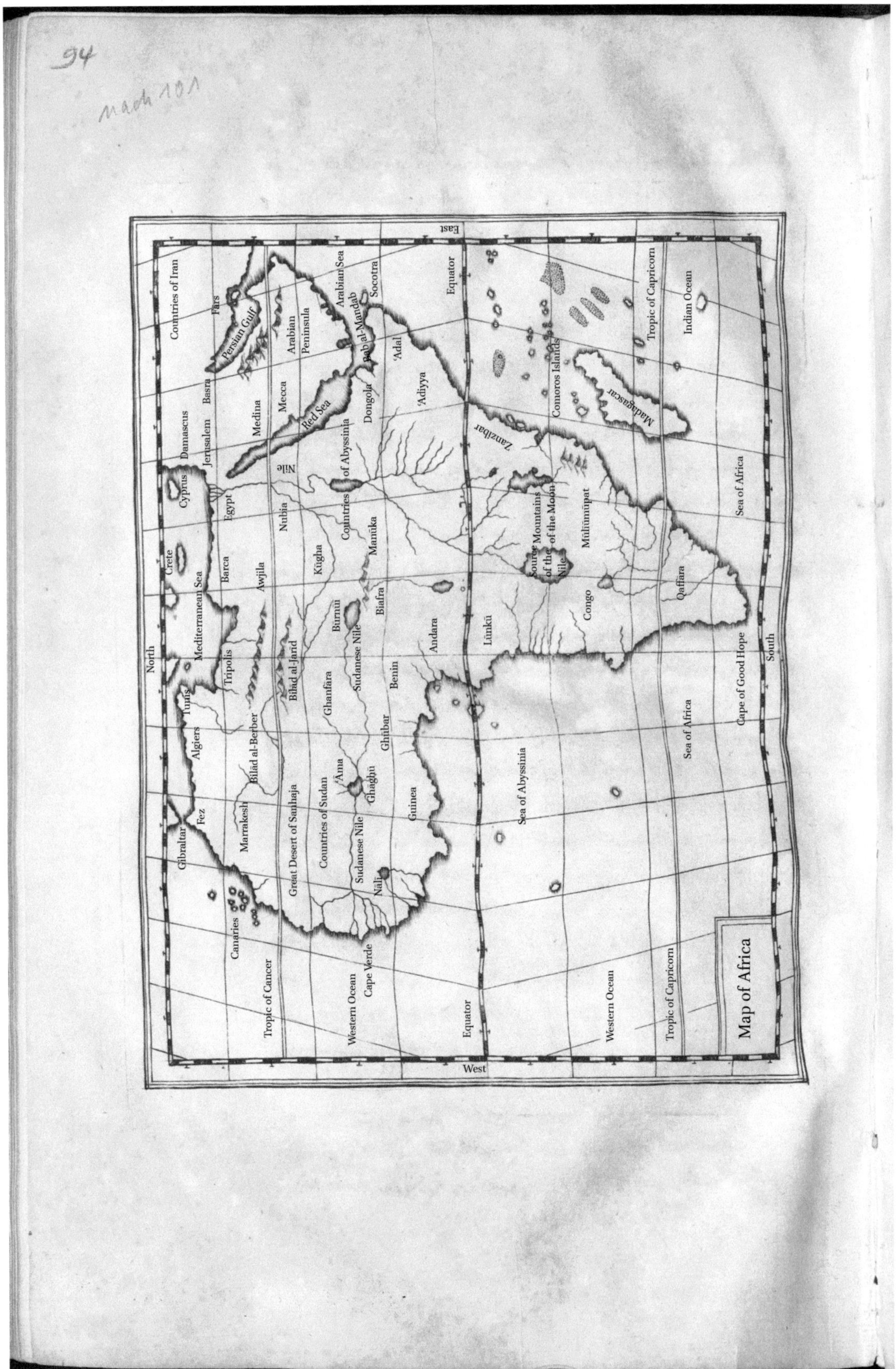

FIGURE 44 M 101/102 Map of Africa (translation)

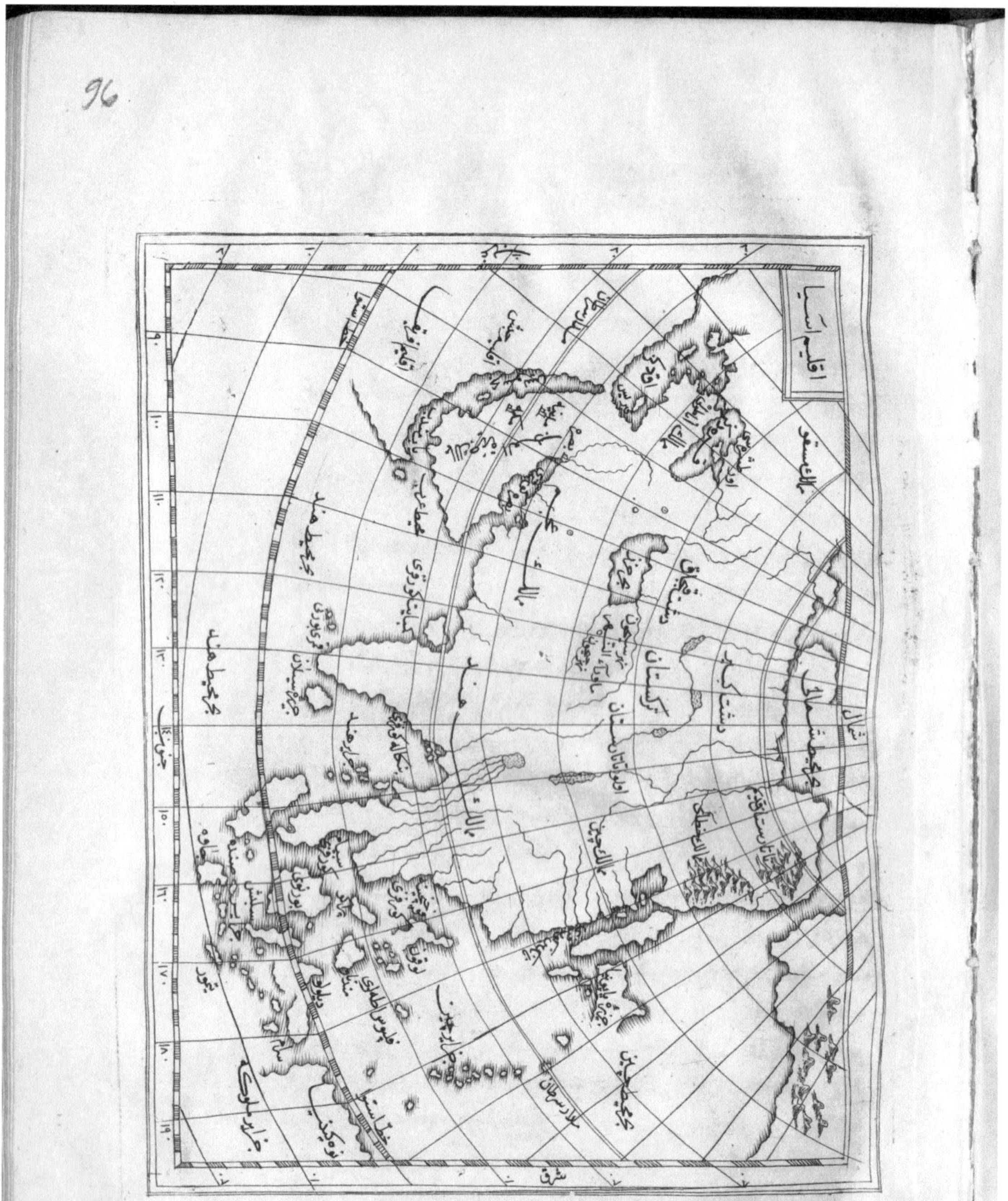

FIGURE 45 M 103/104 Map of Asia; cf. R 35a
BAYERISCHE STAATSBIBLIOTHEK MÜNCHEN, RES/2 A.OR. 371,
URN:NBN:DE:BVB:12-BSB00096220-8

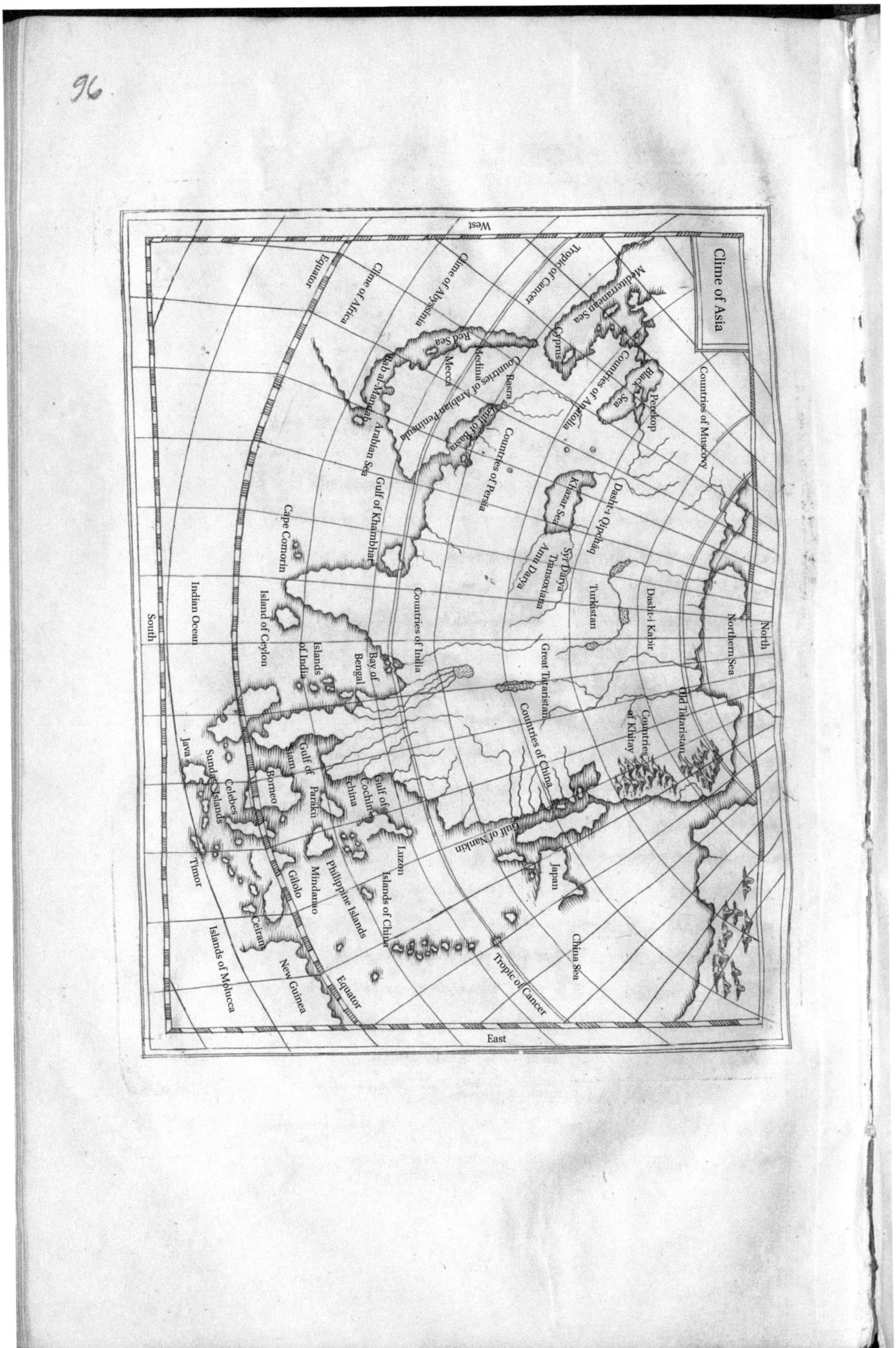

FIGURE 46 M 103/104 Map of Asia (translation)

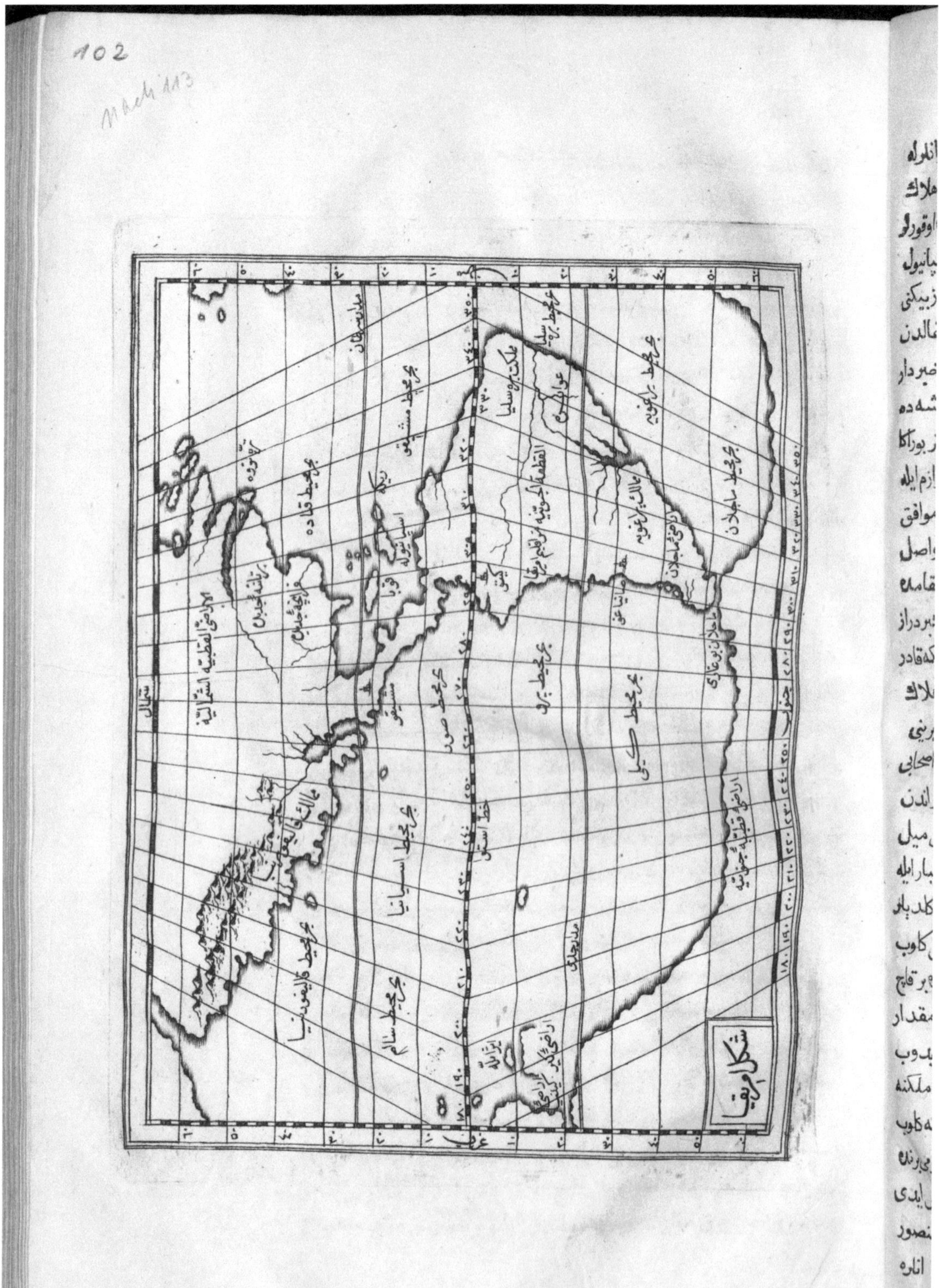

FIGURE 47 M 113/114 Map of America; cf. R 37a
BAYERISCHE STAATSBIBLIOTHEK MÜNCHEN, RES/2 A.OR. 371,
URN:NBN:DE:BVB:12-BSB00096220-8

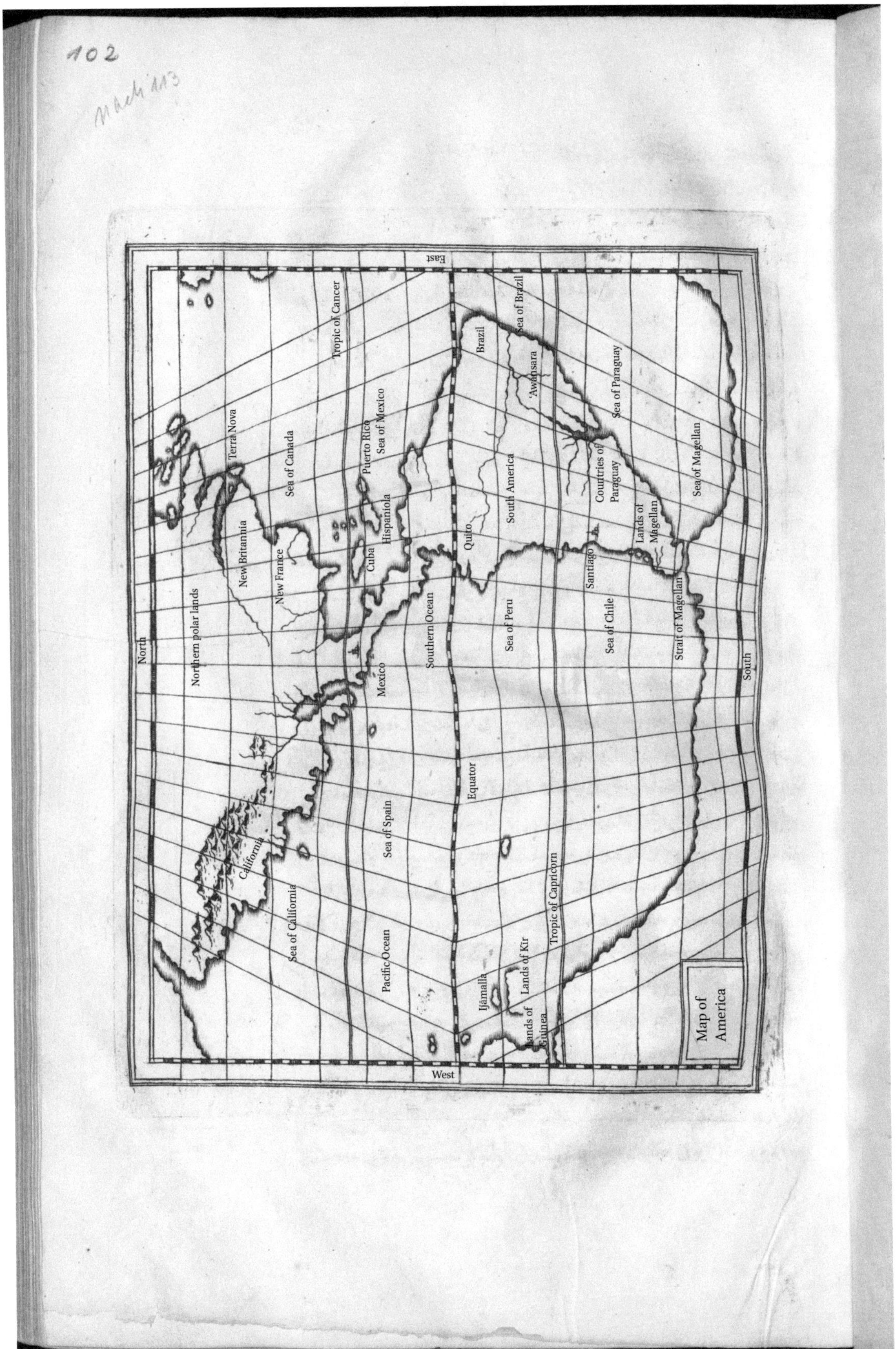

FIGURE 48 M 113/114 Map of America (translation)

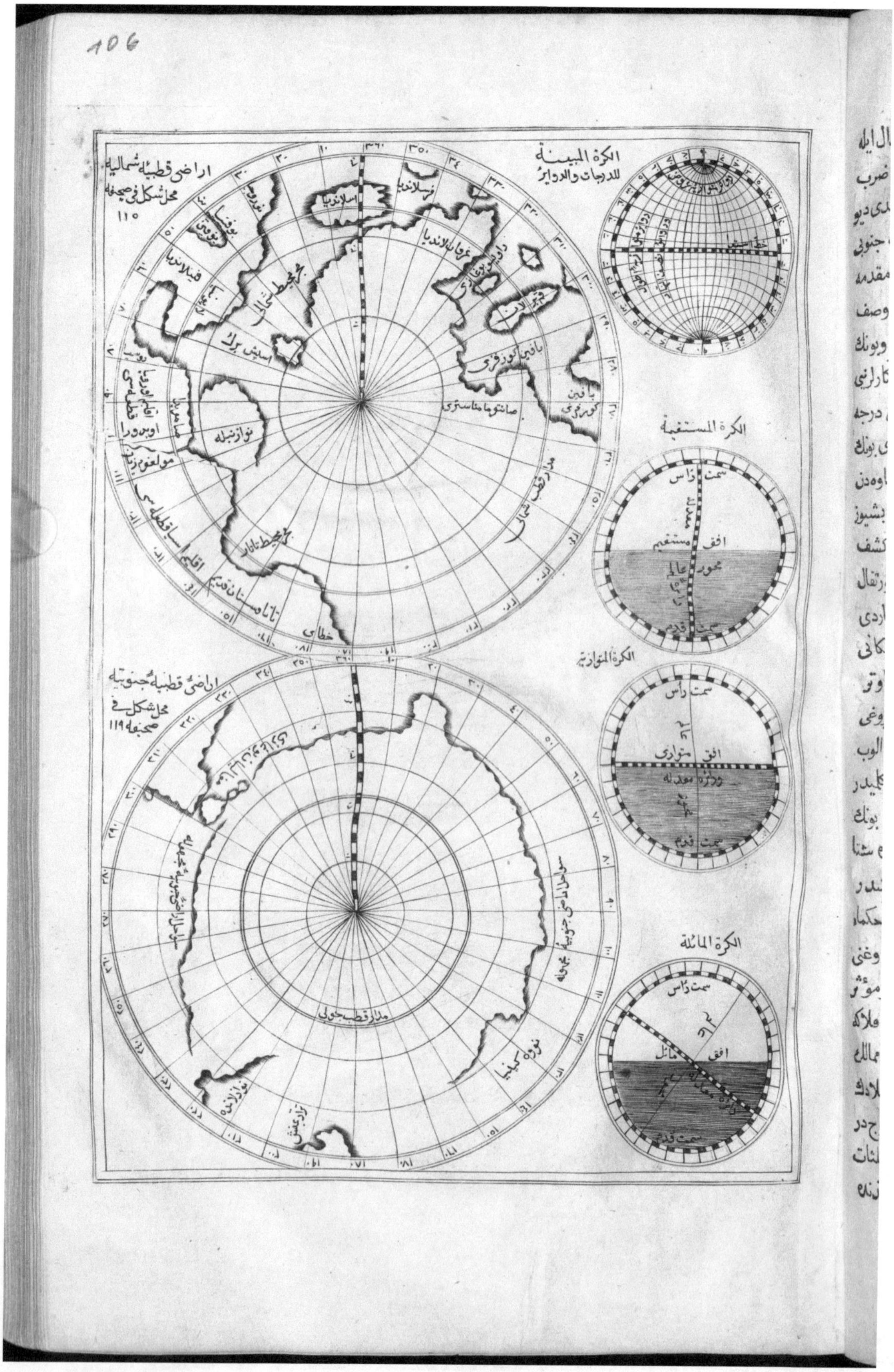

FIGURE 49 M 119/120 Map of Arctic and Antarctic Regions; cf. R 41a, 42a
BAYERISCHE STAATSBIBLIOTHEK MÜNCHEN, RES/2 A.OR. 371,
URN:NBN:DE:BVB:12-BSB00096220-8

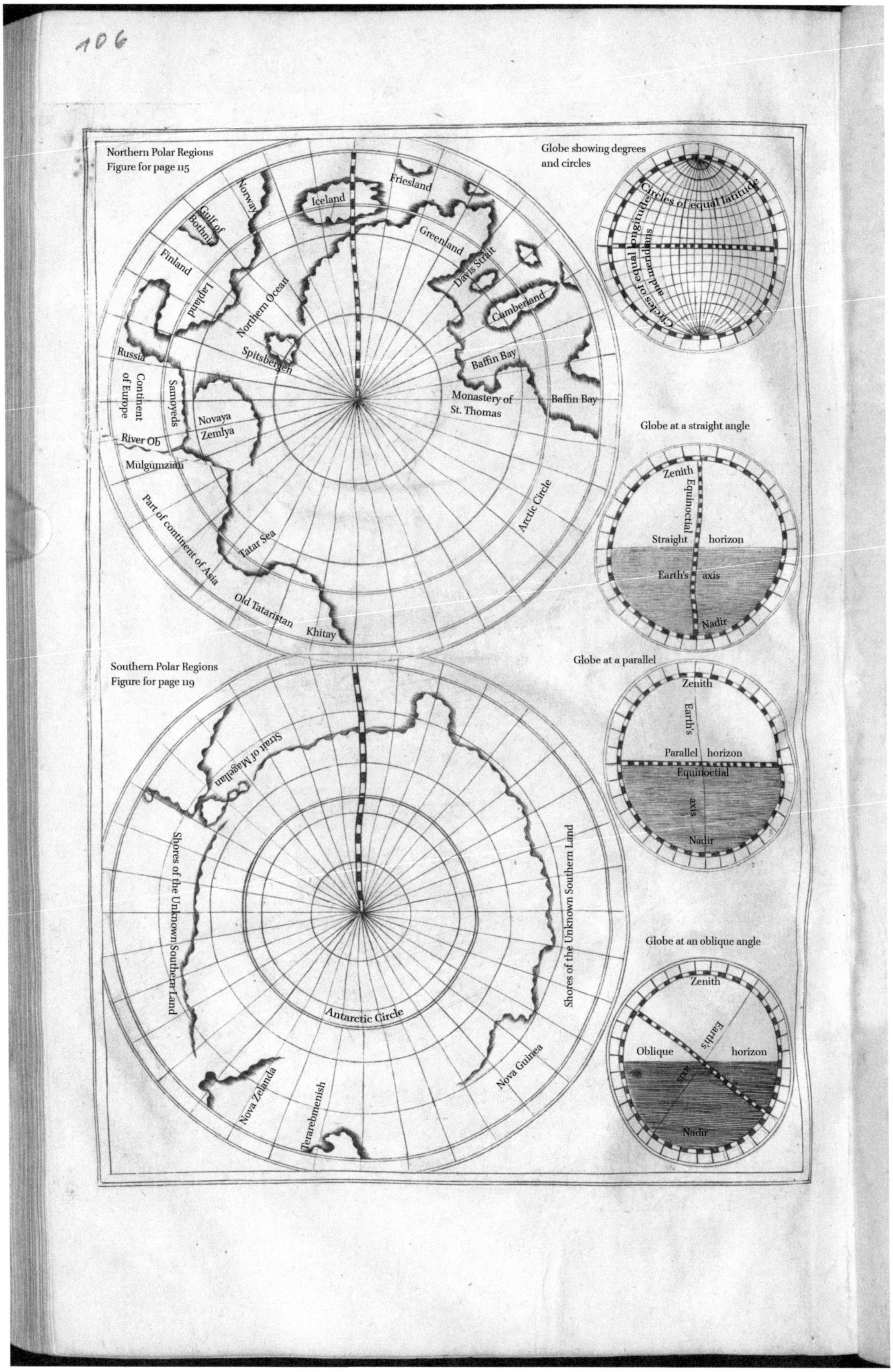

FIGURE 50 M 119/120 Map of Arctic and Antarctic Regions (translation)

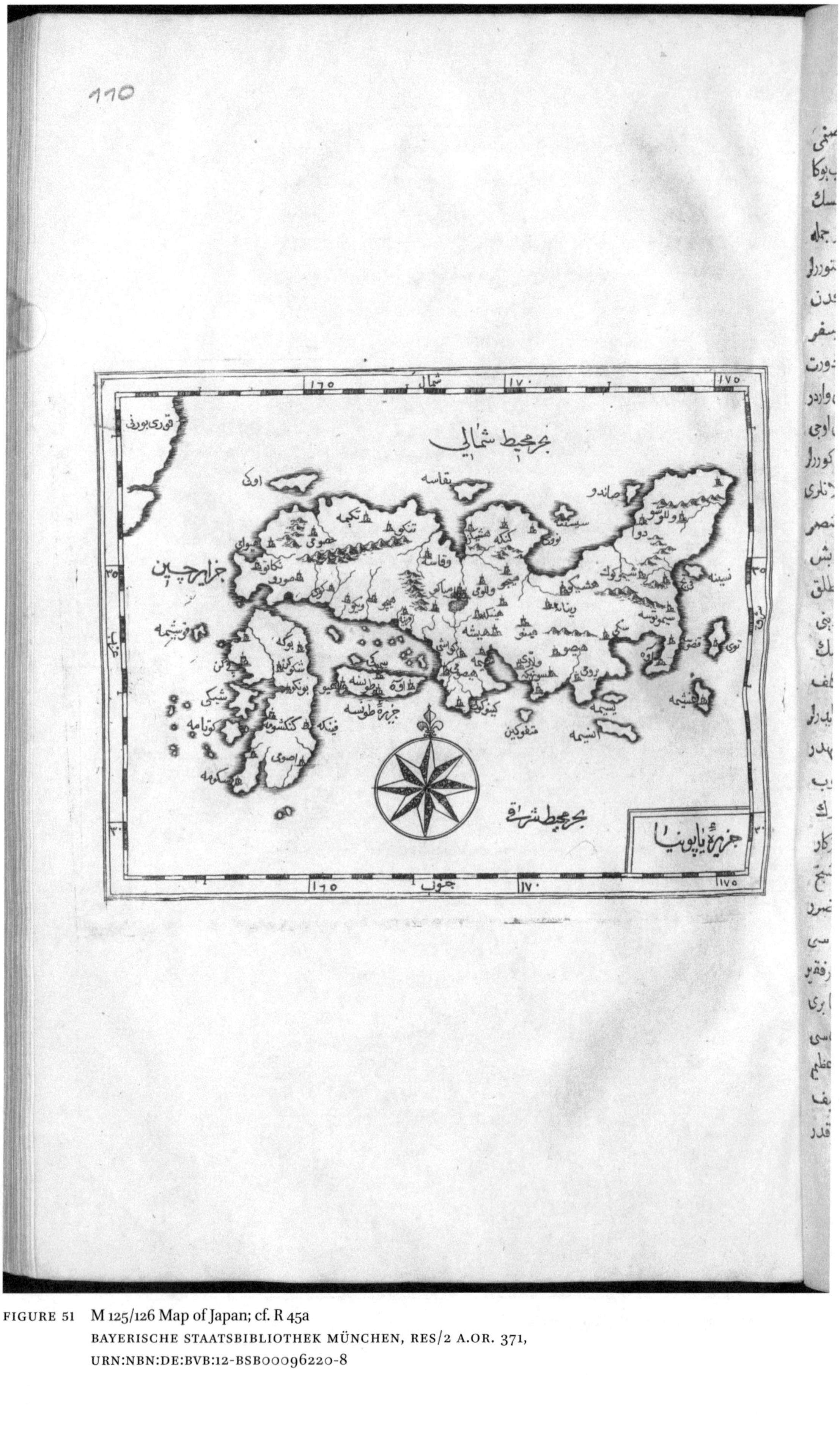

FIGURE 51 M 125/126 Map of Japan; cf. R 45a
BAYERISCHE STAATSBIBLIOTHEK MÜNCHEN, RES/2 A.OR. 371, URN:NBN:DE:BVB:12-BSB00096220-8

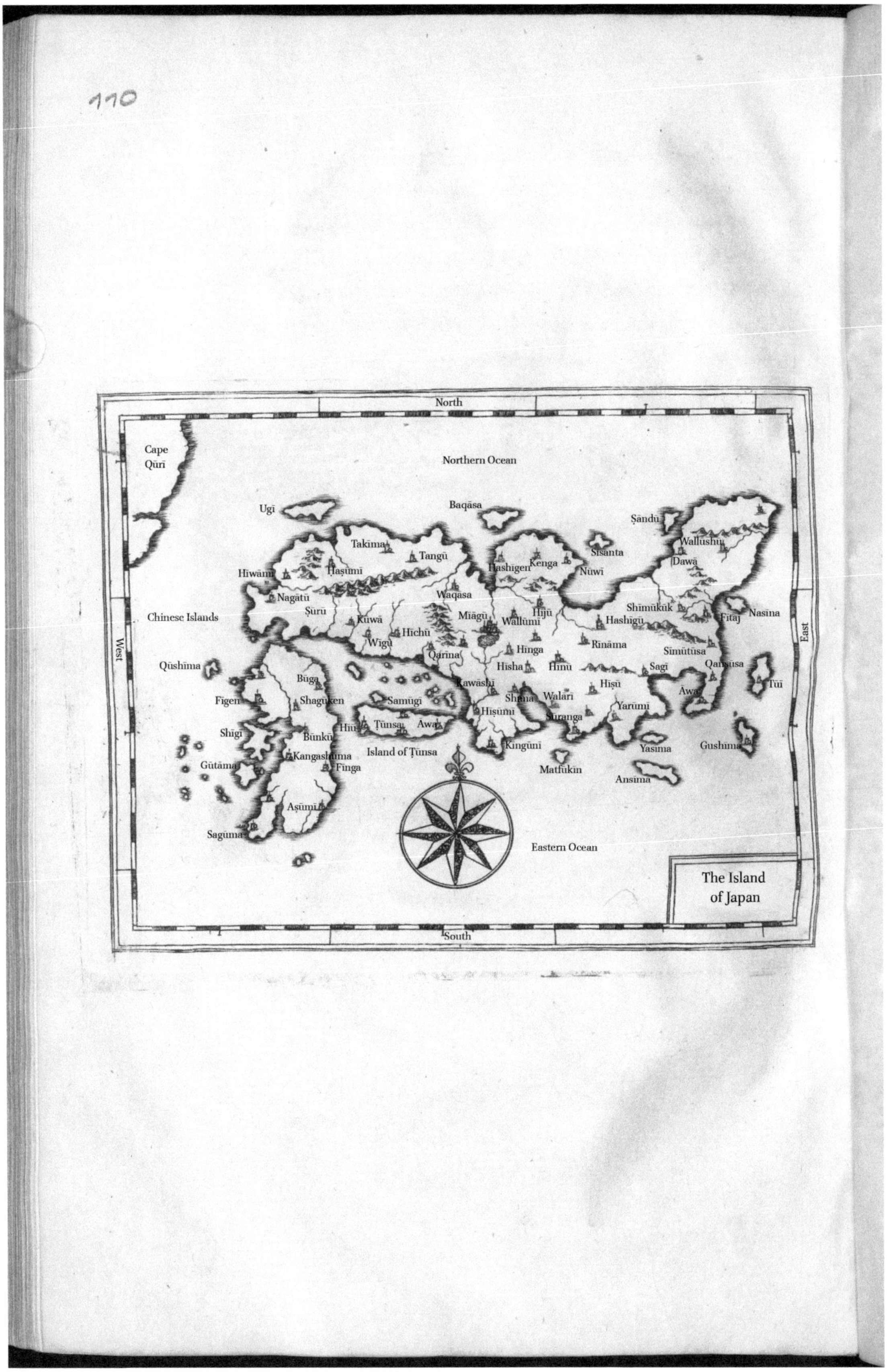

FIGURE 52 M 125/126 Map of Japan (translation)

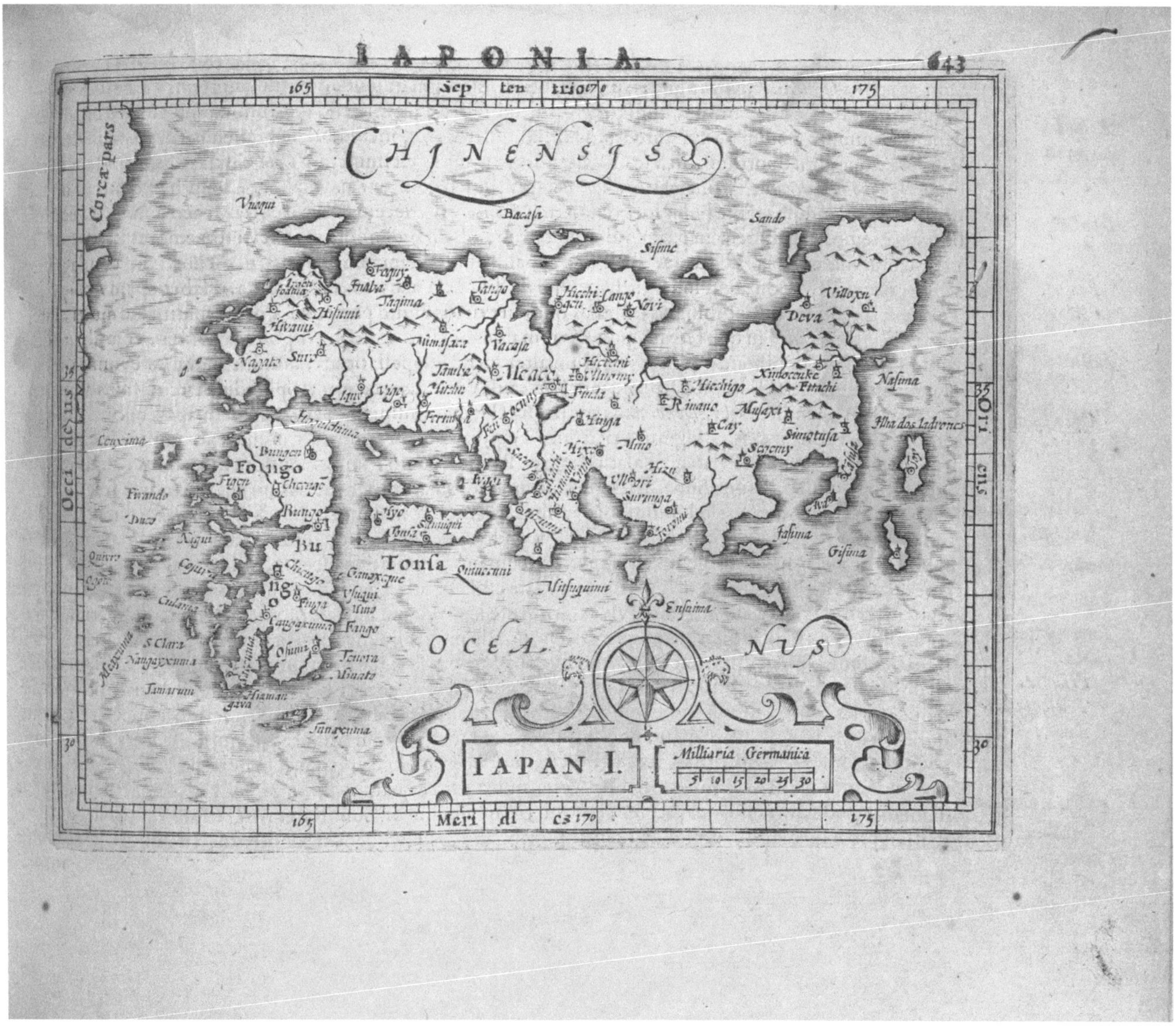

FIGURE 53 Mercator/Hondius, Atlas Minor (1621 edition), p. 627, map of Japan
UNIVERSITY OF MICHIGAN, CLEMENTS LIBRARY, ATL2 1621 ME

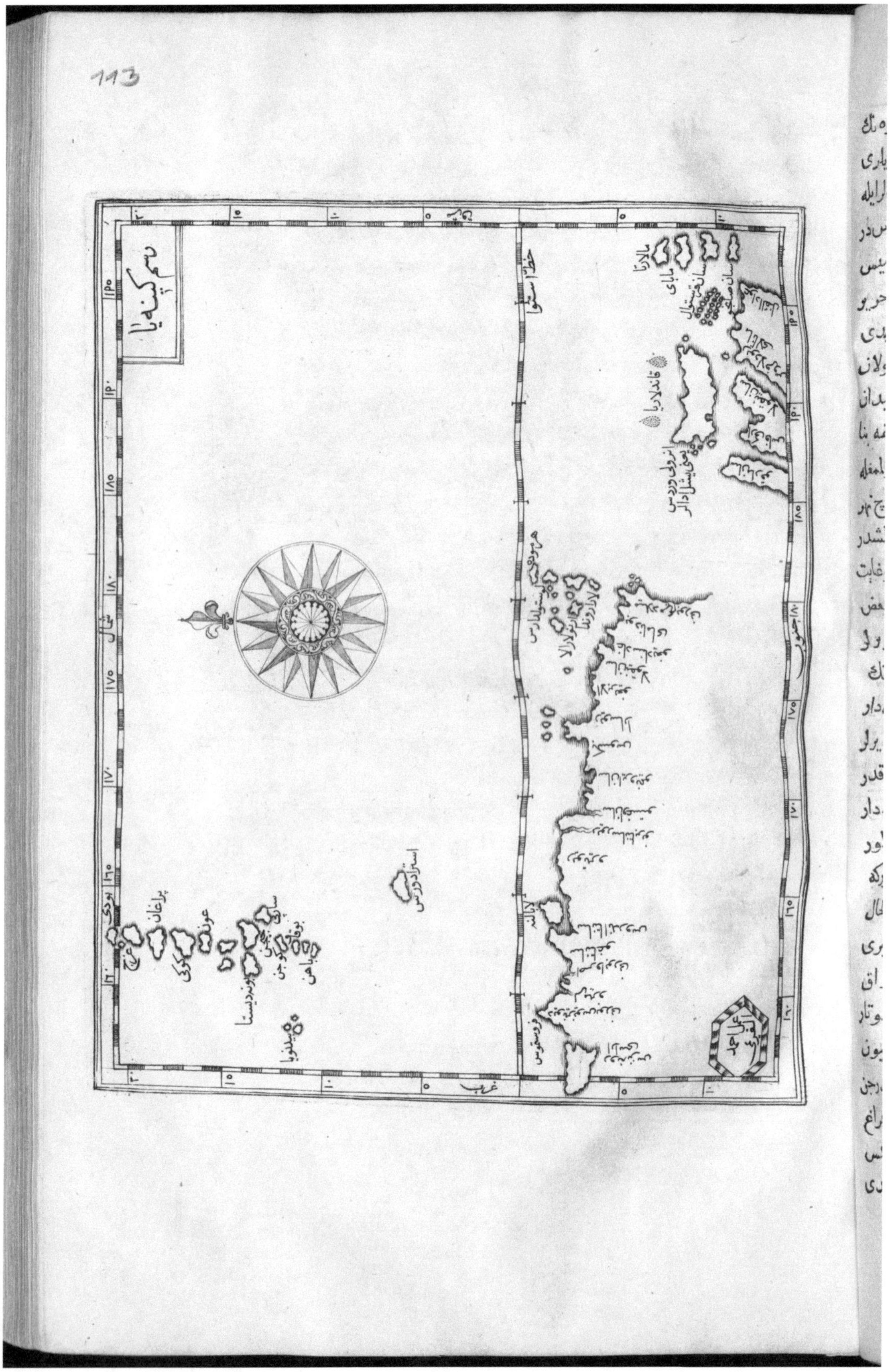

FIGURE 54 M 129/130 Map of New Guinea; cf. R 46b
BAYERISCHE STAATSBIBLIOTHEK MÜNCHEN, RES/2 A.OR. 371,
URN:NBN:DE:BVB:12-BSB00096220-8

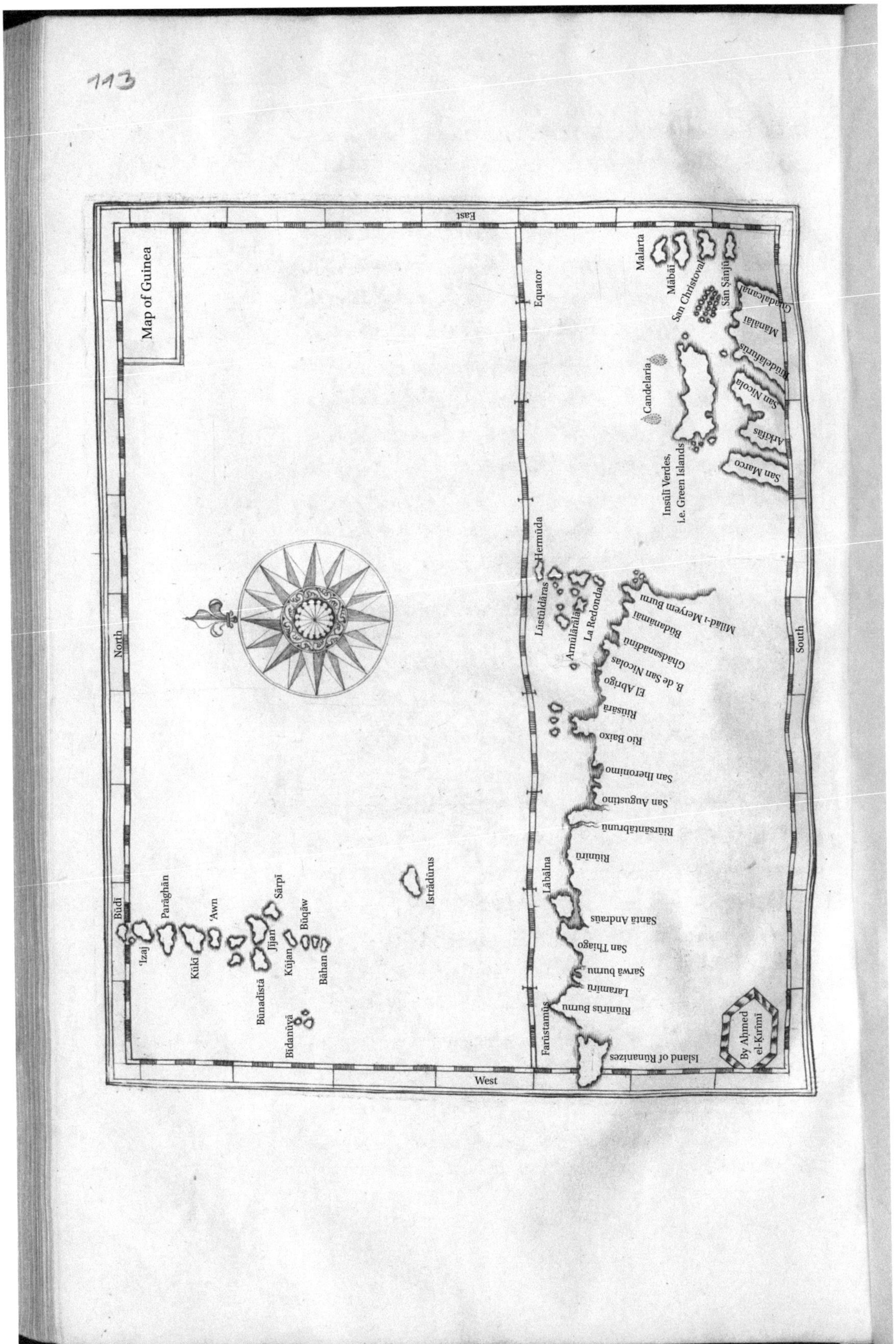

FIGURE 55 M 129/130 Map of New Guinea (translation)

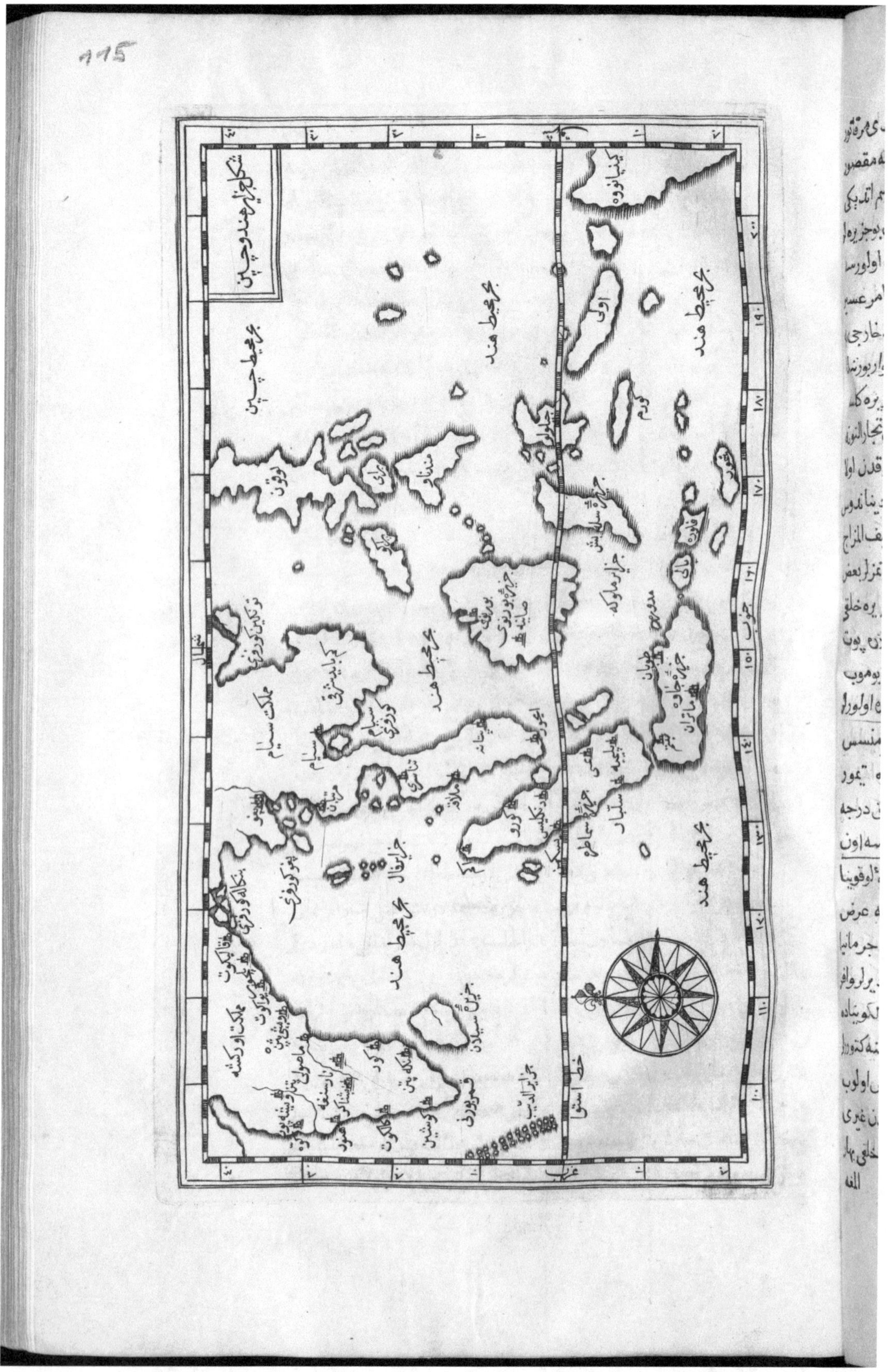

FIGURE 56 M 131/132 Map of East Indies; cf. R 47b
BAYERISCHE STAATSBIBLIOTHEK MÜNCHEN, RES/2 A.OR. 371, URN:NBN:DE:BVB:12-BSB00096220-8

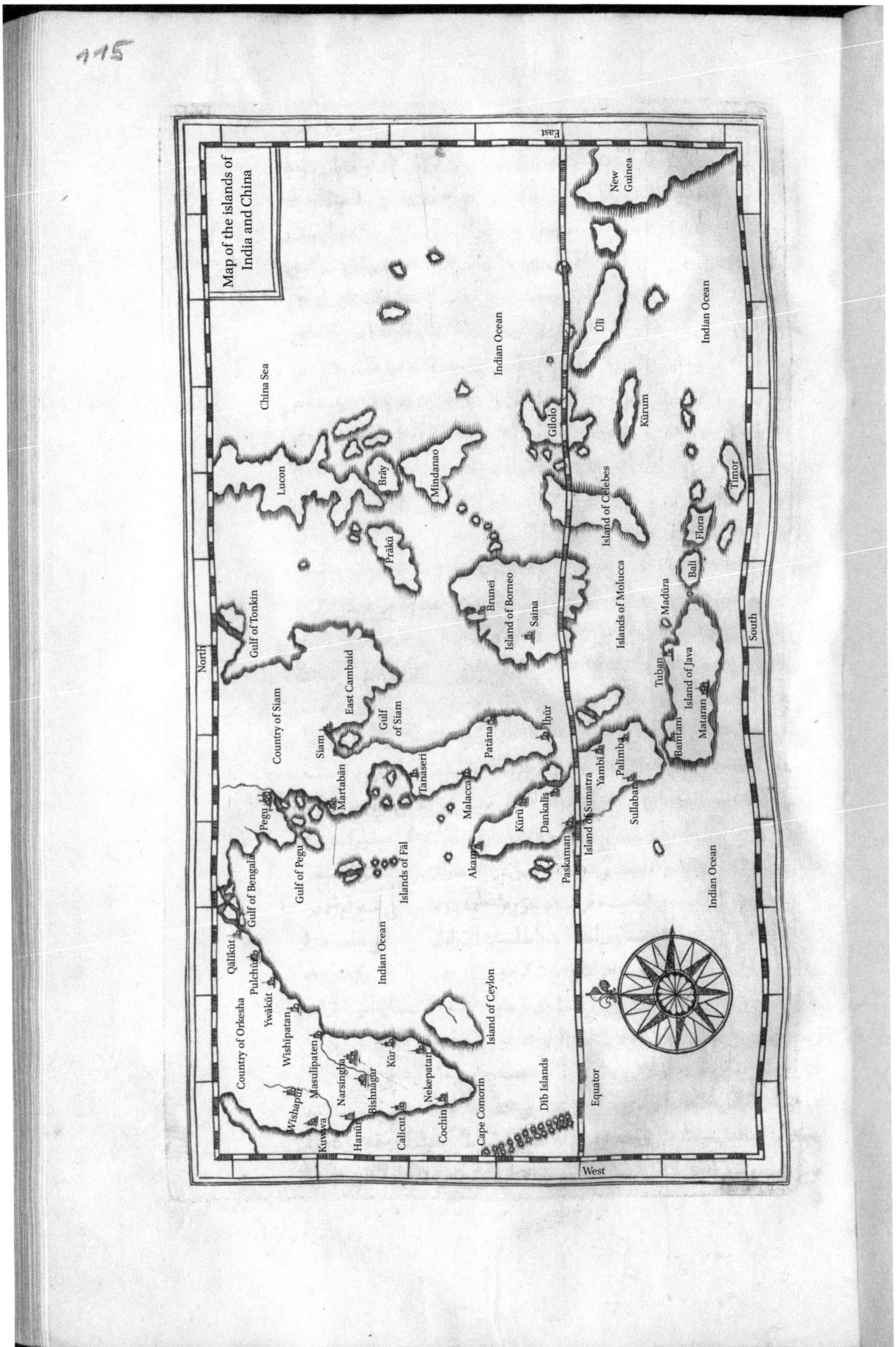

FIGURE 57 M 131/132 Map of East Indies (translation)

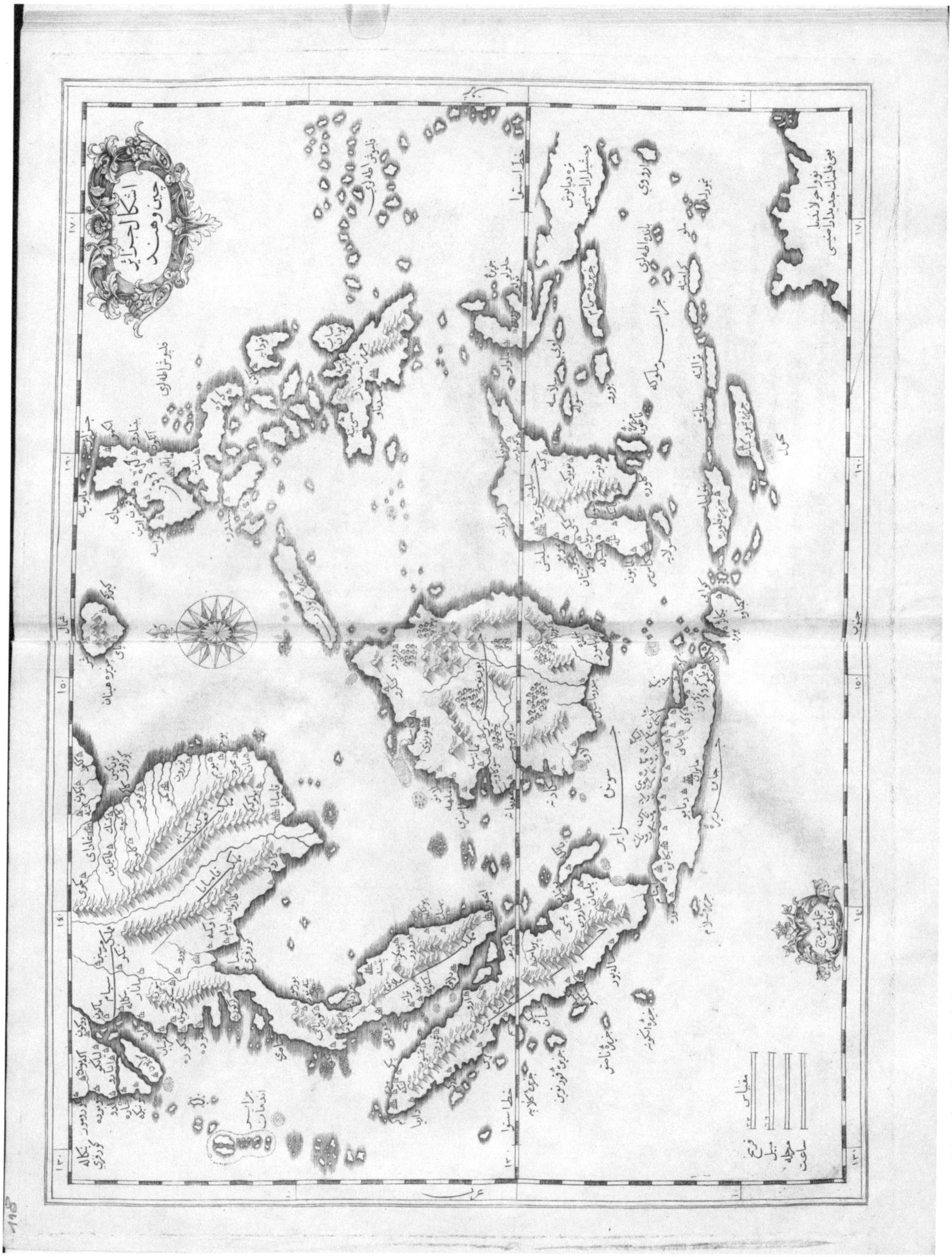

FIGURE 58 M 133/134 Map of Gilolo and Molucca; cf. R 49a
BAYERISCHE STAATSBIBLIOTHEK MÜNCHEN, RES/2 A.OR. 371,
URN:NBN:DE:BVB:12-BSB00096220-8

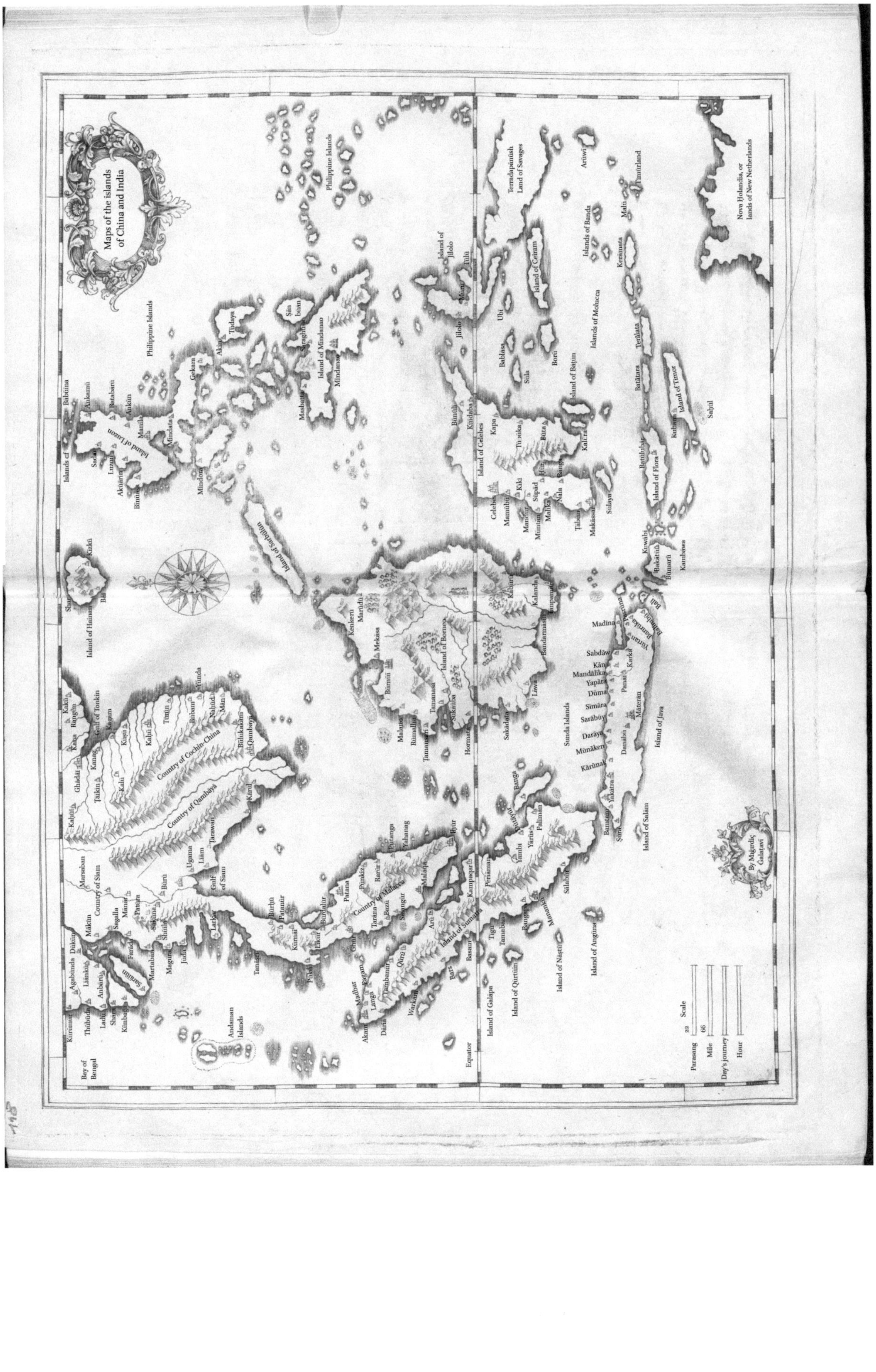

FIGURE 59 M 133/134 Map of Gilolo and Molucca (translation)

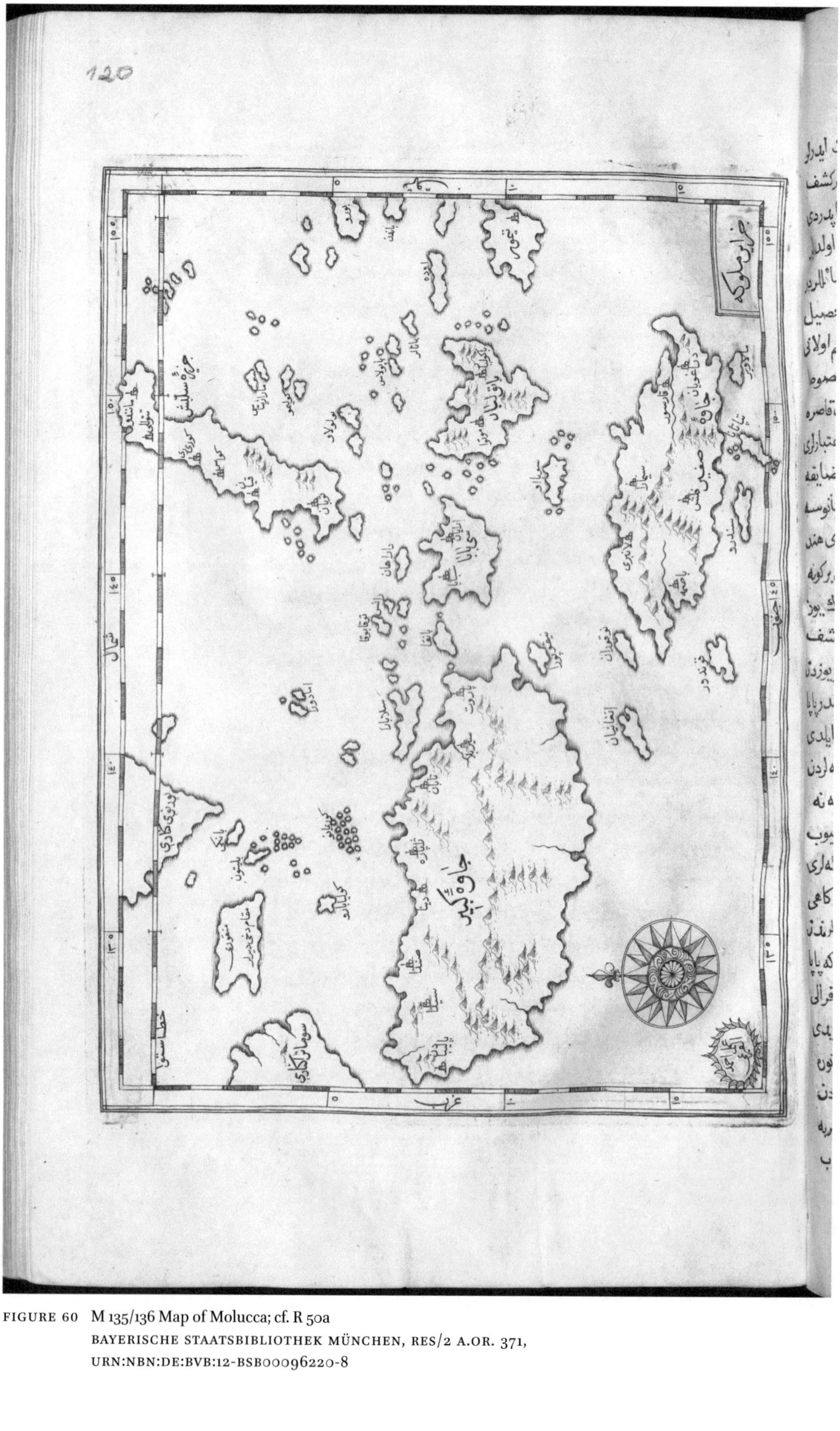

FIGURE 60 M 135/136 Map of Molucca; cf. R 50a
BAYERISCHE STAATSBIBLIOTHEK MÜNCHEN, RES/2 A.OR. 371, URN:NBN:DE:BVB:12-BSB00096220-8

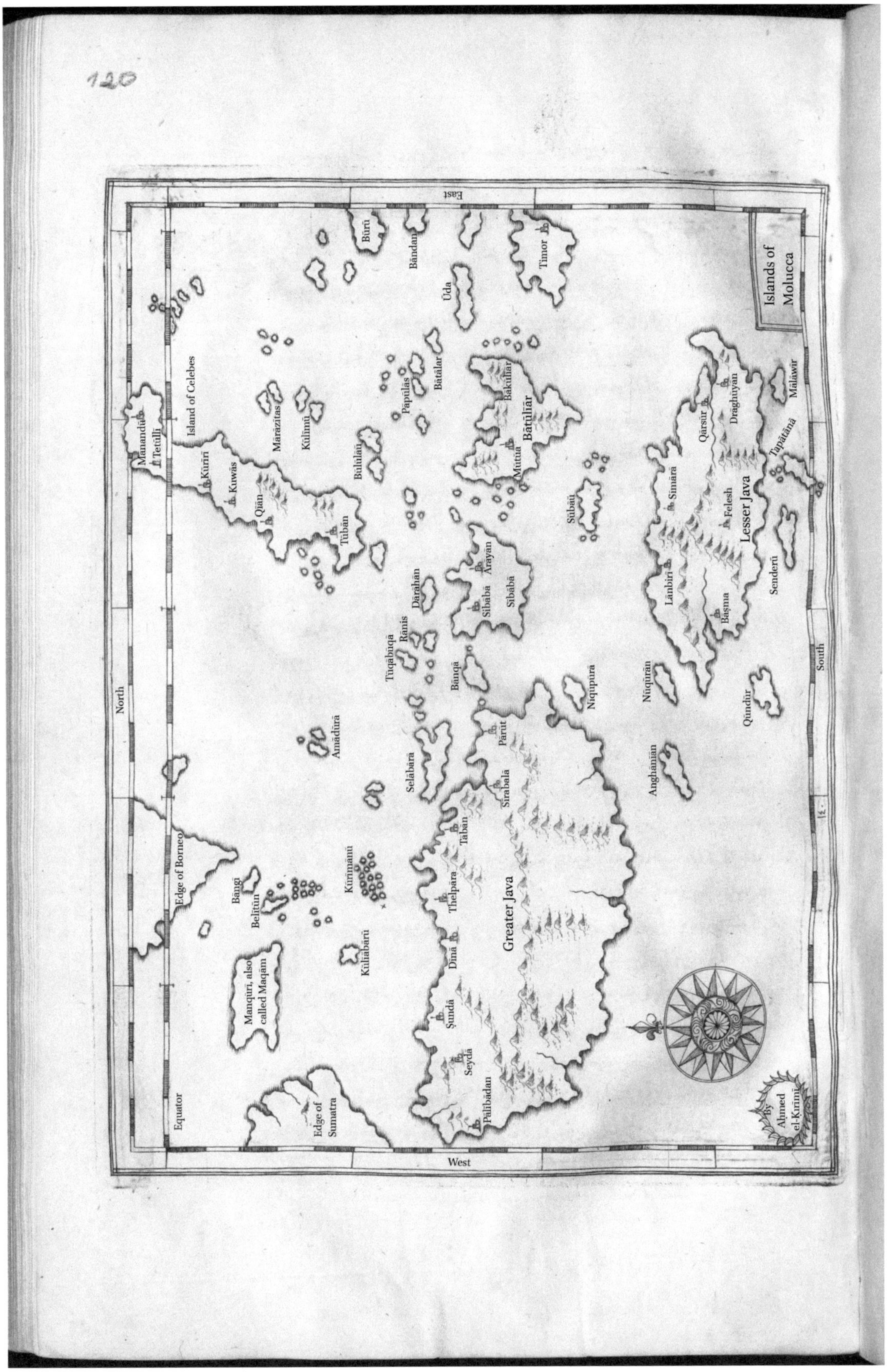

FIGURE 61 M 135/136 Map of Molucca (translation)

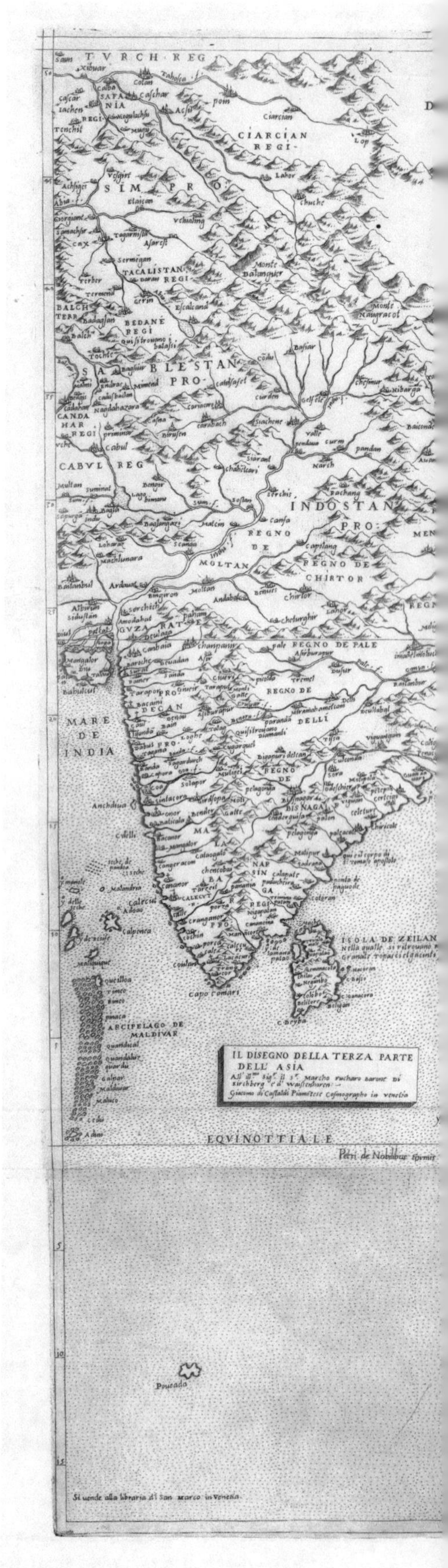
TVRCH REG
CIARCIAN REGI
SIM PRO
TACALISTAN REGI
BEDANE REGI
SABLESTAN PRO
CANDAHAR REGI
CABVL REG
INDOSTAN PRO
REGNO DE MOLTAN
REGNO DE CHIRTOR
GVZARAT
REGNO DE PALE
DECAN
REGNO DE DELLI
MARE DE INDIA
REGNO DE BISNAGA
ARCIPELAGO DE MALDIVAR
ISOLA DE ZEILAN
Capo Comari
IL DISEGNO DELLA TERZA PARTE DELL' ASIA
All' ill.mo sig.r il s.r Marcho fuchard barone di Kirchberg e d' Waissenhoren
Giacomo di Castaldi Piamontese Cosmographo in venetia
EQVINOTTIALE
Petri de Nobilibus formis
Si uende alla libraria di San marco in venetia

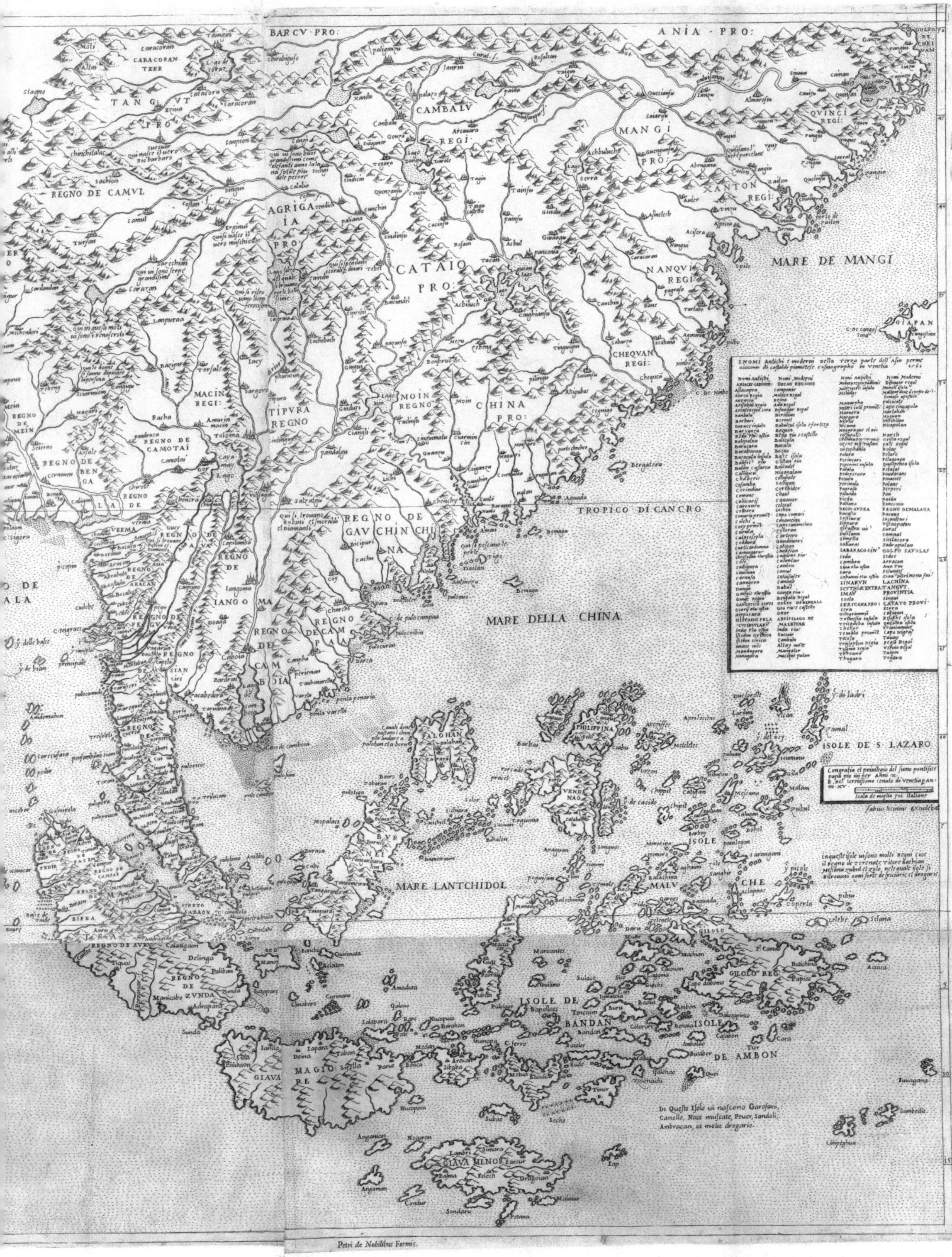

E 62 Giacomo Gastaldi, Il disegno della terza parte dell' Asia, 1580
BARRY LAWRENCE RUDERMAN, ANTIQUE MAPS INC., LA JOLLA, CA, WITH PERMISSION

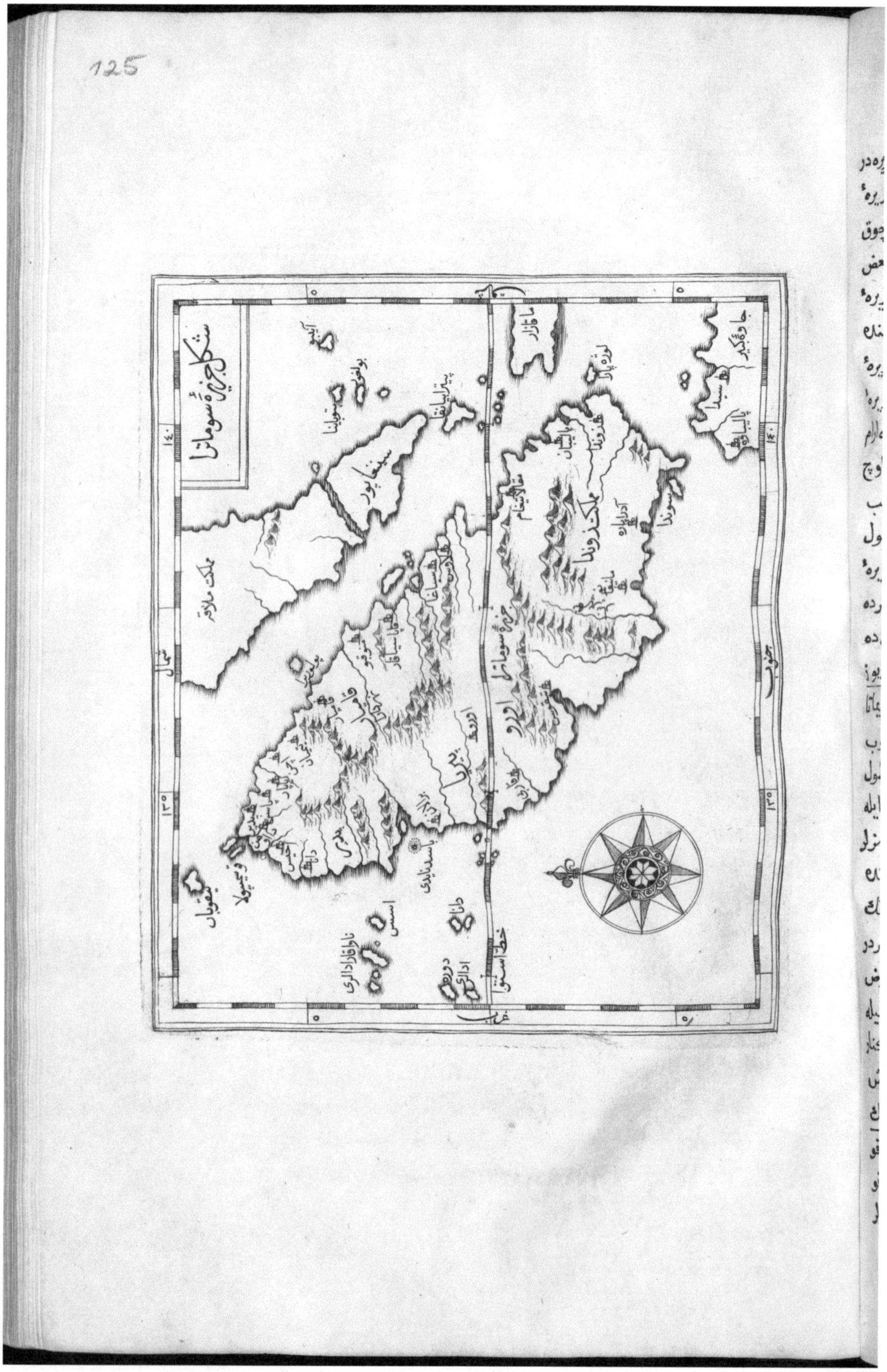

FIGURE 63 M 143/144 Map of Sumatra; R 52a
BAYERISCHE STAATSBIBLIOTHEK MÜNCHEN, RES/2 A.OR. 371, URN:NBN:DE:BVB:12-BSB00096220-8

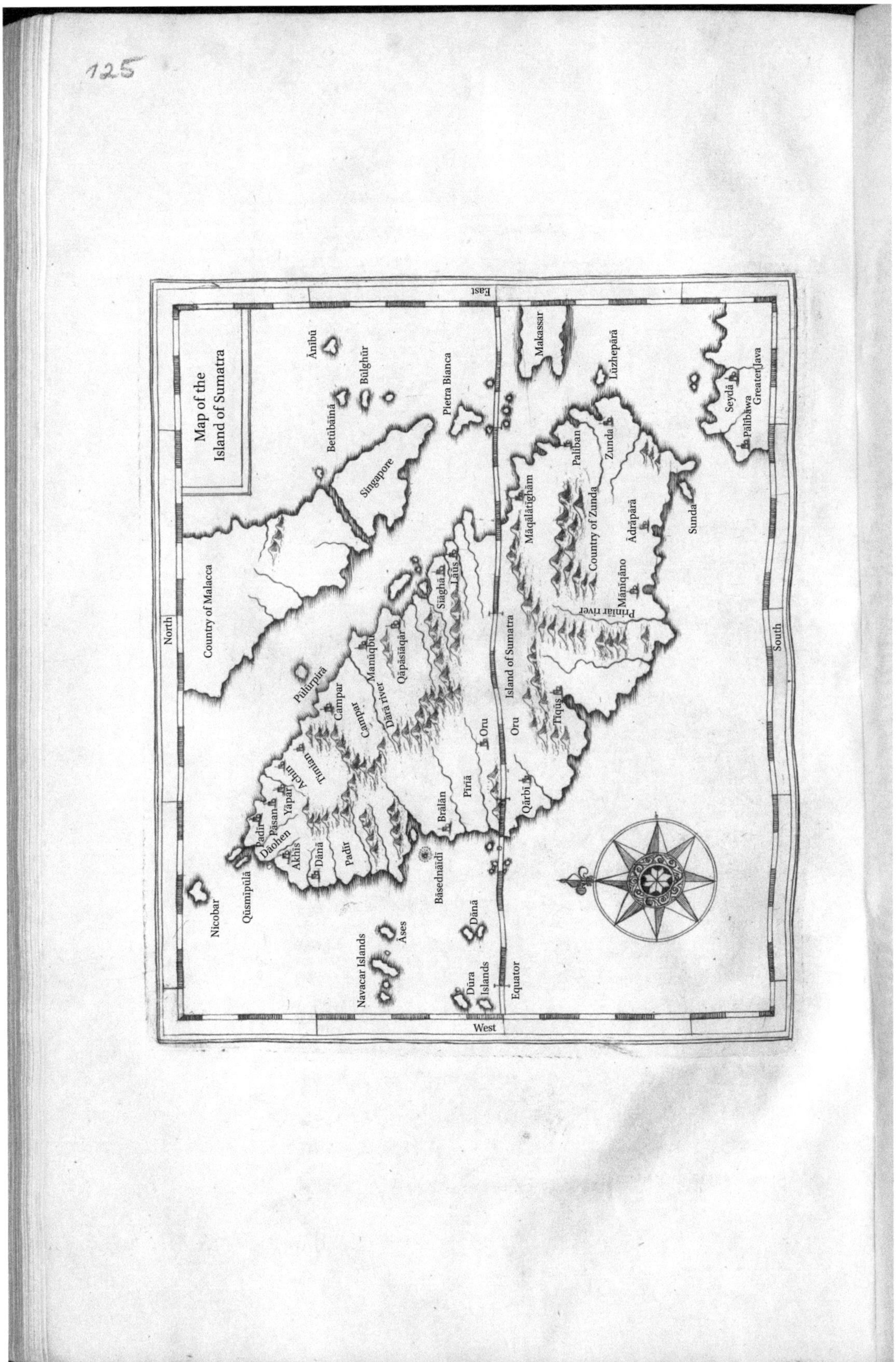

FIGURE 64 M 143/144 Map of Sumatra (translation)

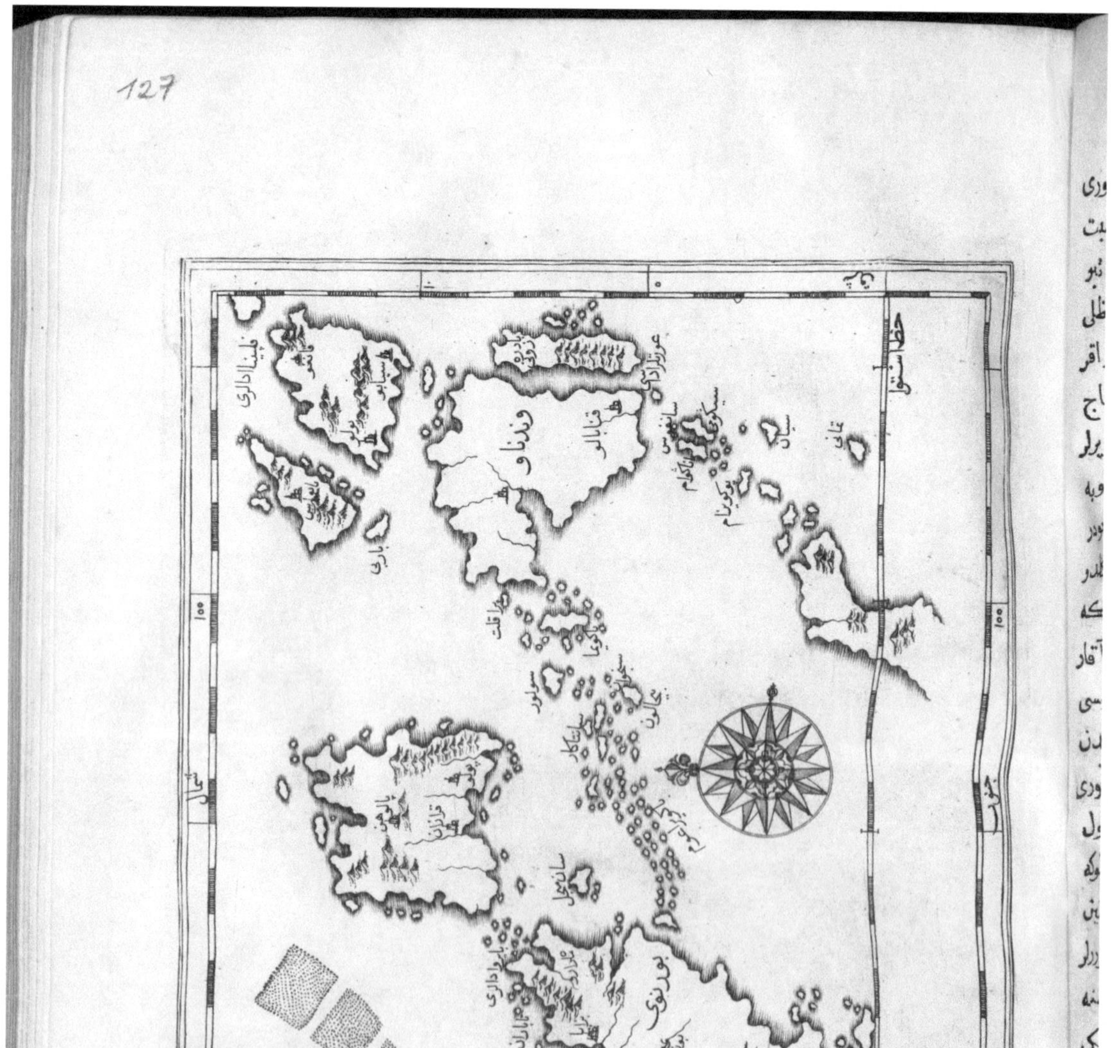

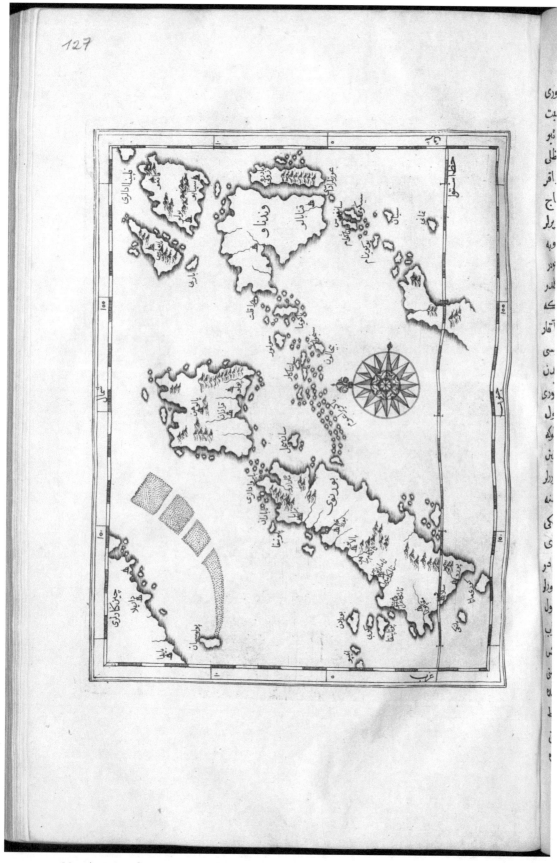

FIGURE 65 M 145/146 Map of Borneo and the Philippines; cf. R 53a
BAYERISCHE STAATSBIBLIOTHEK MÜNCHEN, RES/2 A.OR. 371,
URN:NBN:DE:BVB:12-BSB00096220-8

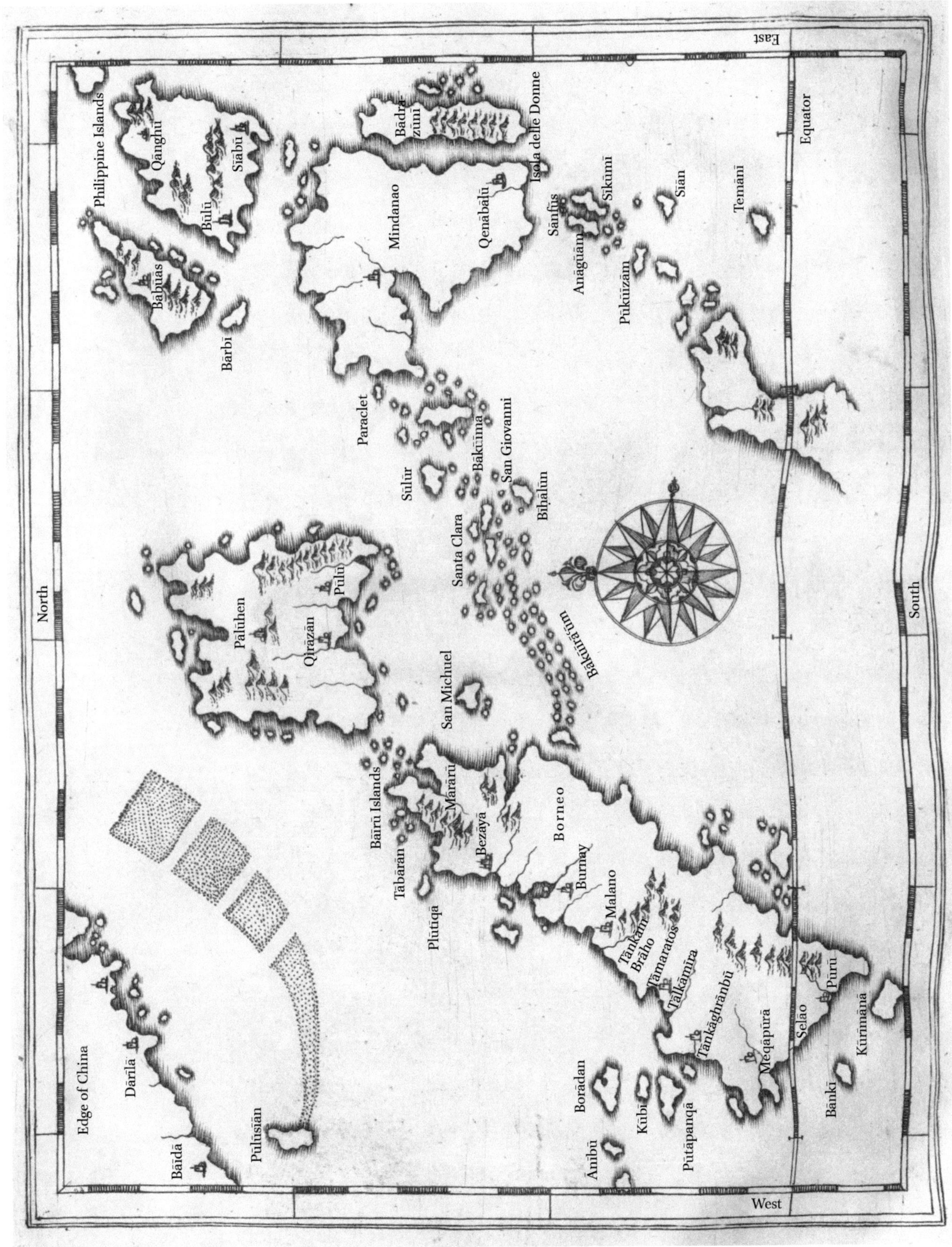

FIGURE 66 M 145/146 Map of Borneo and the Philippines (translation)

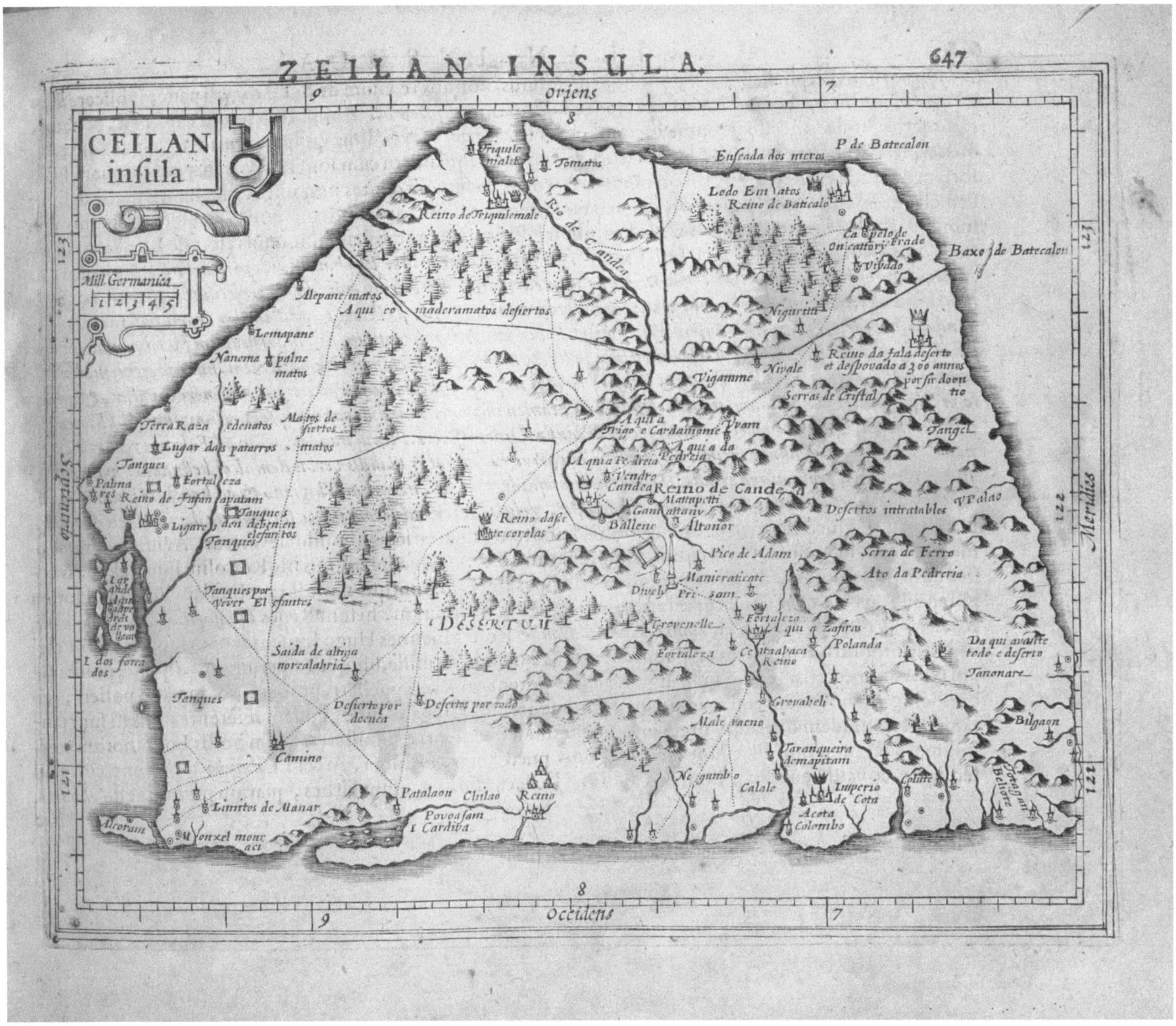

FIGURE 67 Mercator/Hondius, Atlas Minor (1621 edition), p. 647, map of Ceylon
UNIVERSITY OF MICHIGAN, CLEMENTS LIBRARY, ATL2 1621 ME

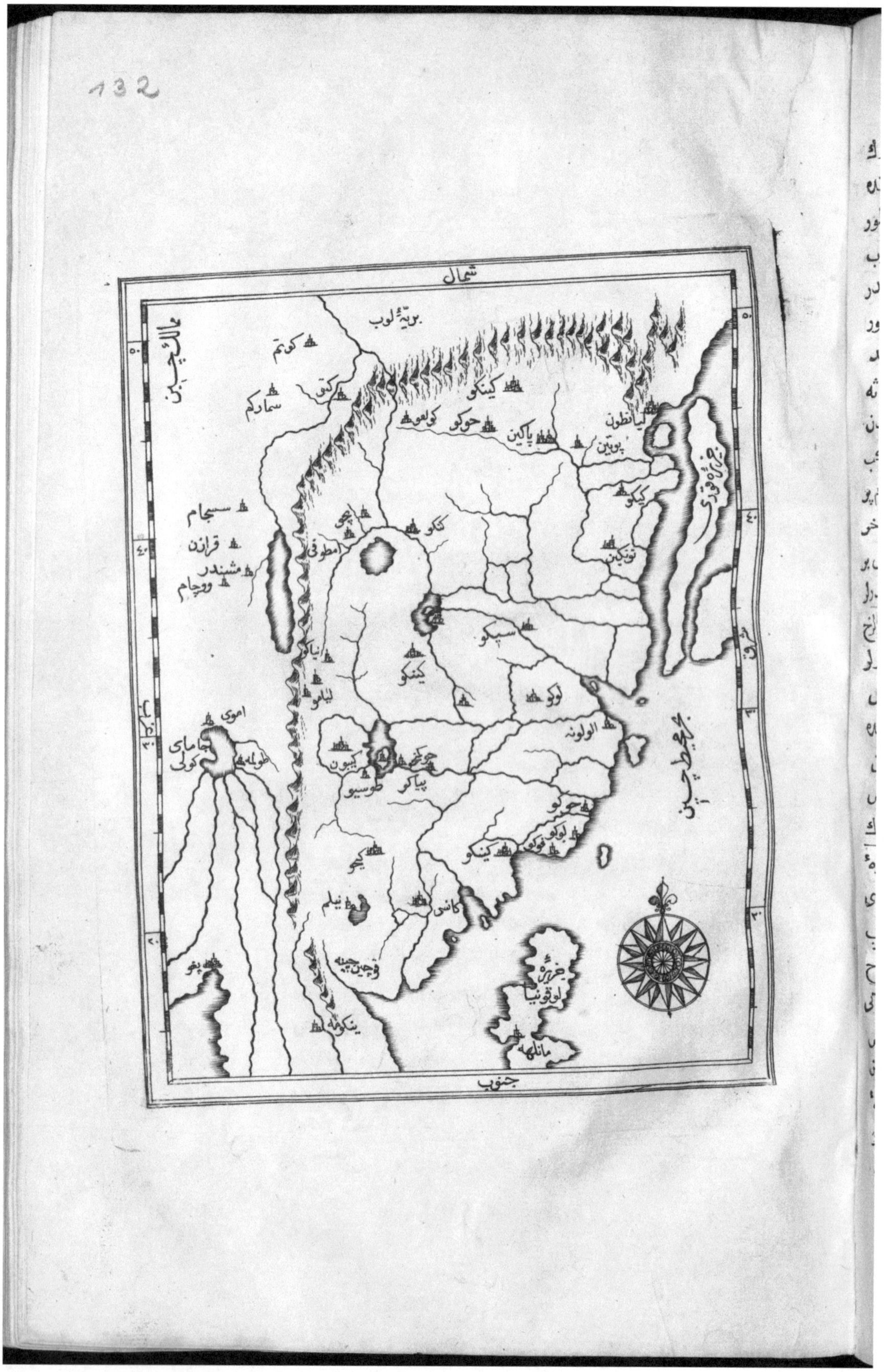

FIGURE 68 M 153/154 Map of [South] China [Çīn]; cf. R 58a
BAYERISCHE STAATSBIBLIOTHEK MÜNCHEN, RES/2 A.OR. 371, URN:NBN:DE:BVB:12-BSB00096220-8

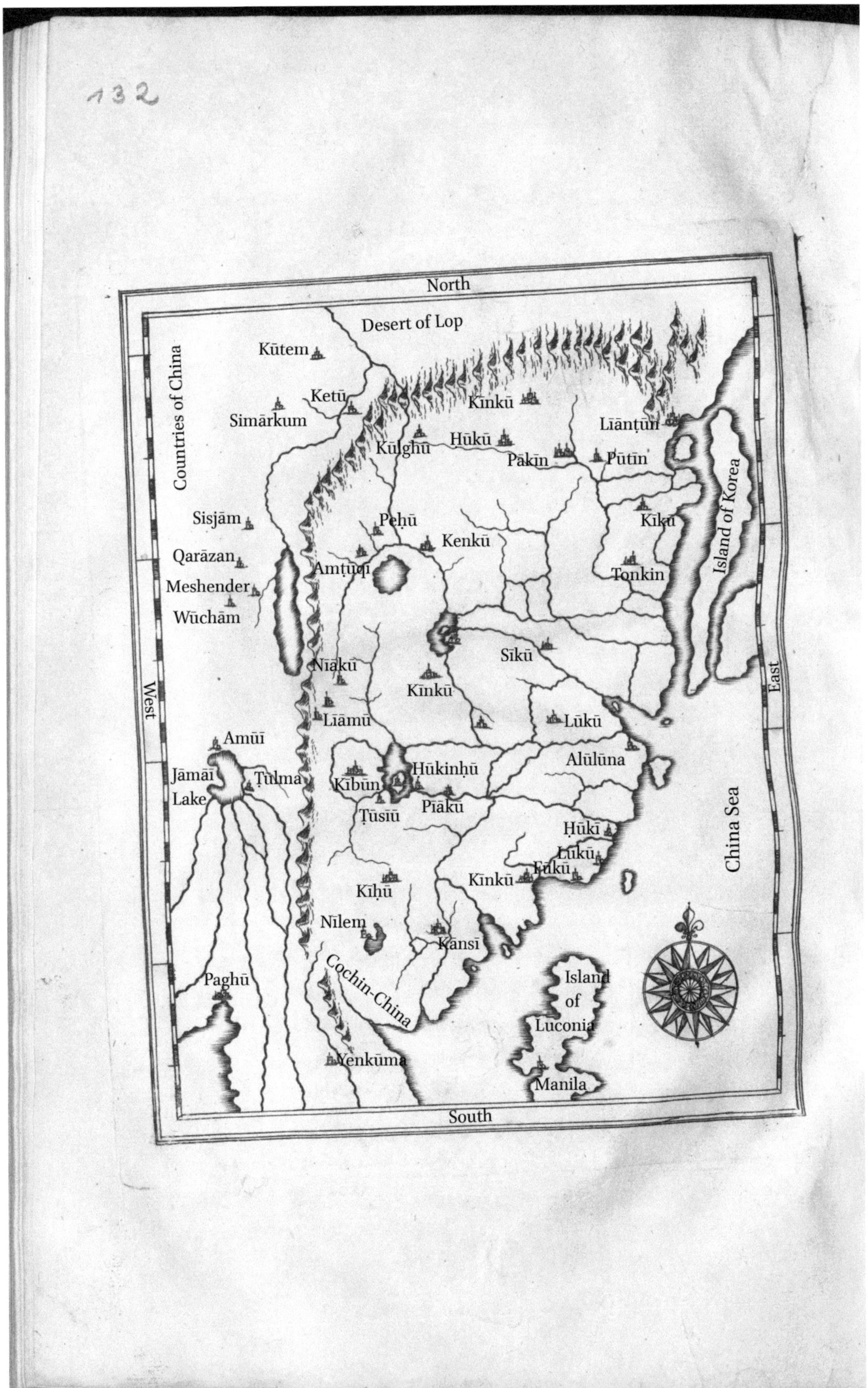

FIGURE 69 M 153/154 Map of [South] China [Çīn] (translation)

FIGURE 70 Mercator/Hondius, Atlas Minor (1621 edition), p. 631, map of China
UNIVERSITY OF MICHIGAN, CLEMENTS LIBRARY, ATL2 1621 ME

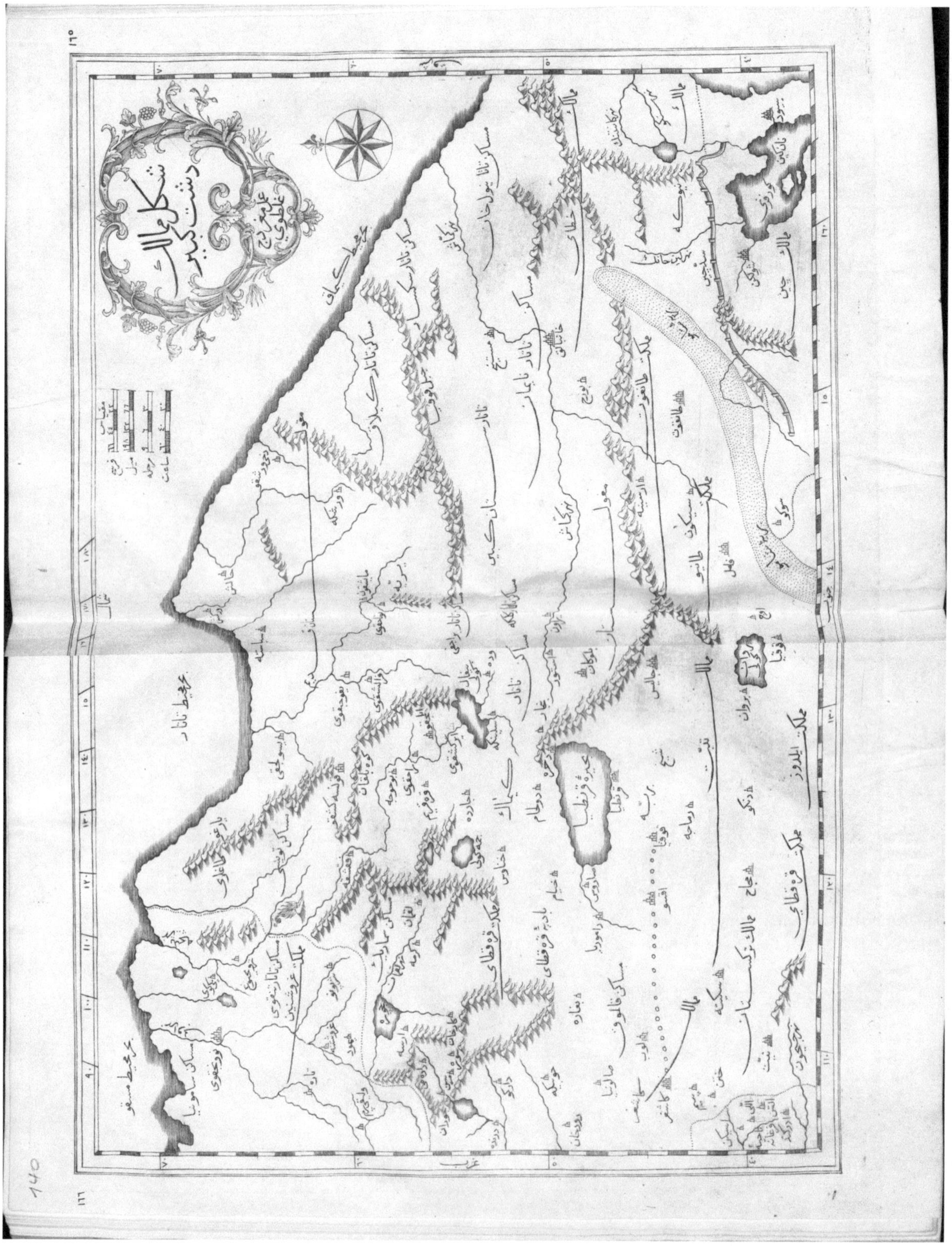

FIGURE 71 M 165/166 Map of [North] China [Cathay]; cf. R 69a
BAYERISCHE STAATSBIBLIOTHEK MÜNCHEN, RES/2 A.OR. 371, URN:NBN:DE:BVB:12-BSB00096220-8

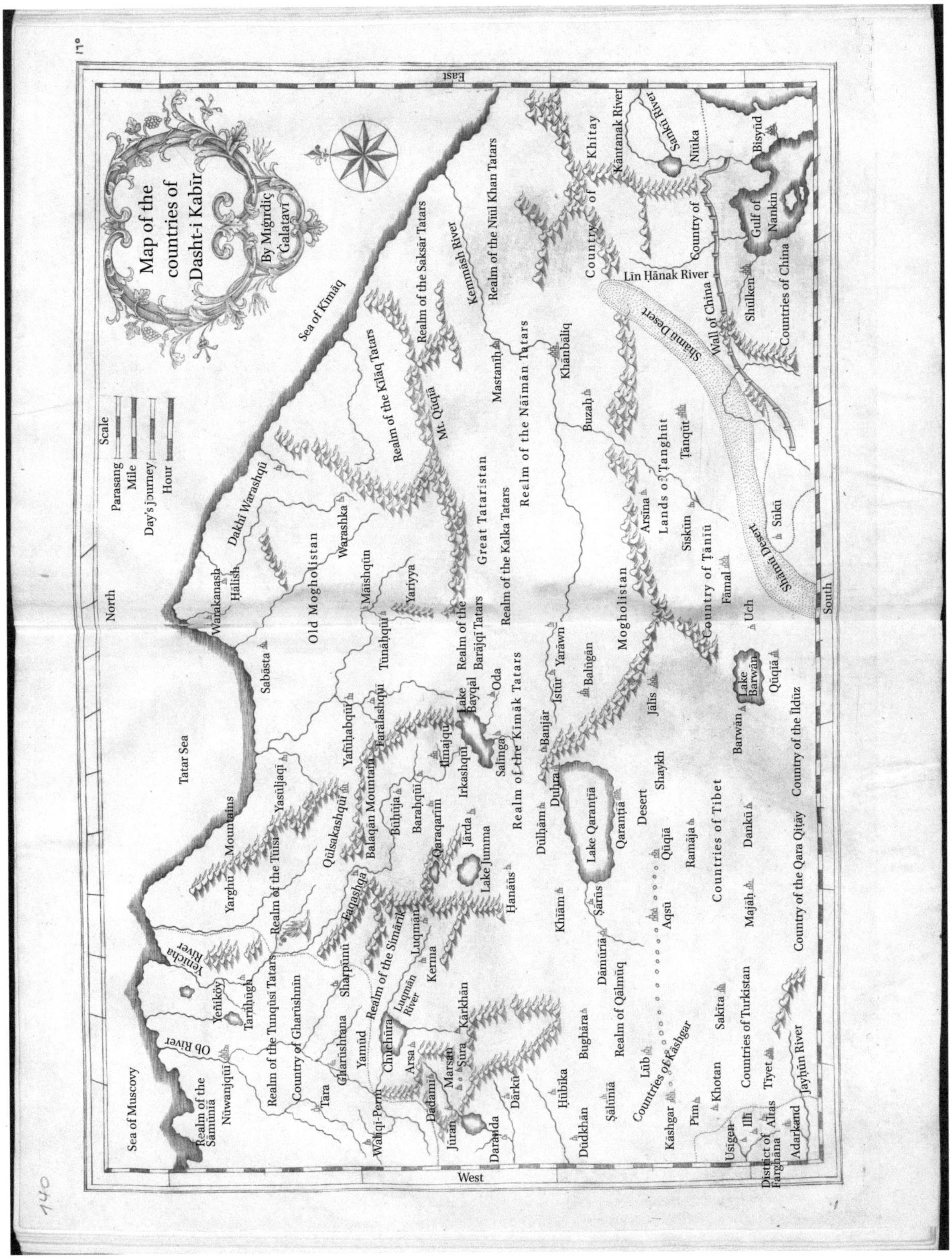

FIGURE 72 M 165/166 Map of [North] China [Cathay] (translation)

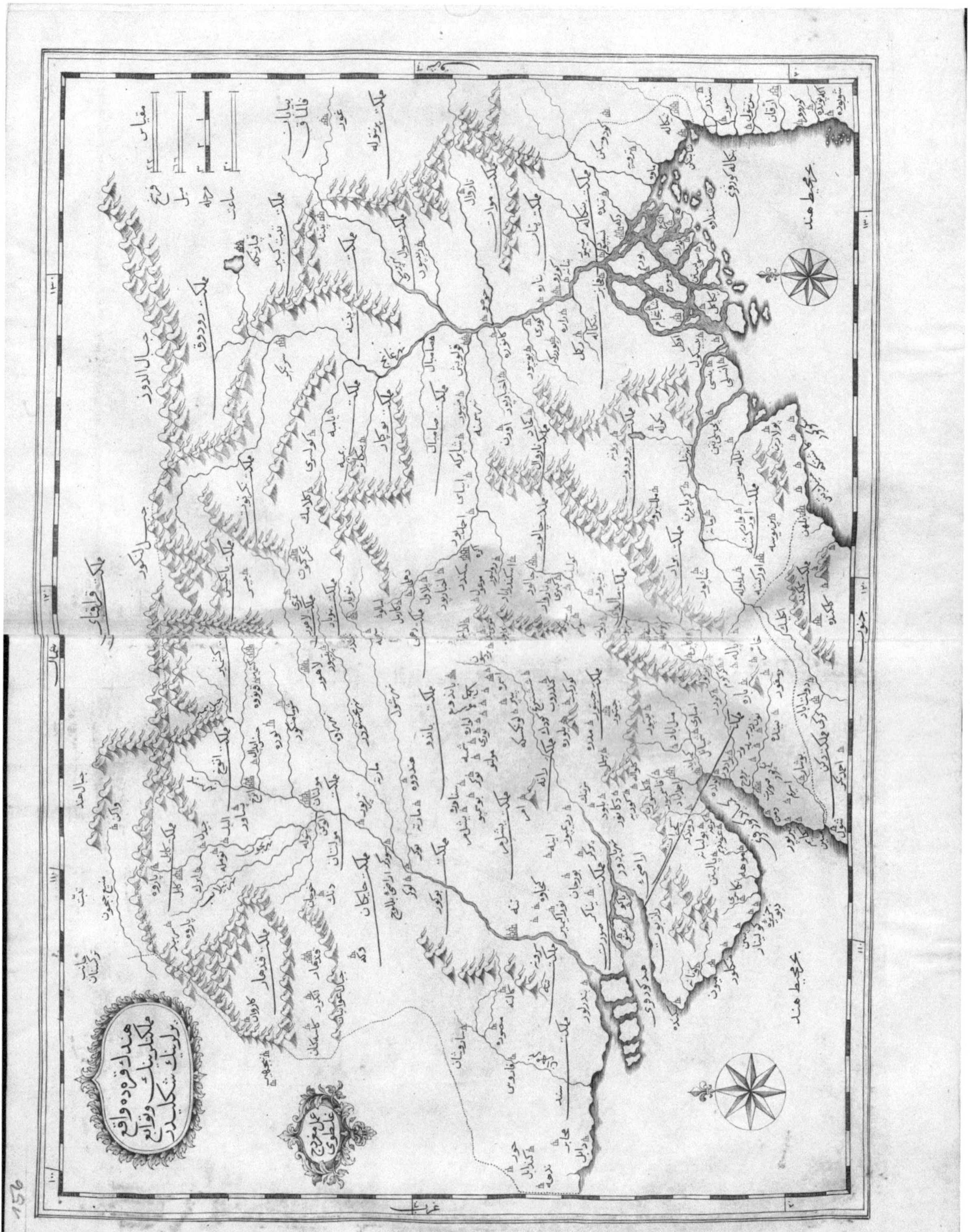

FIGURE 73 M 193/194 Map of India; cf. R 75a
BAYERISCHE STAATSBIBLIOTHEK MÜNCHEN, RES/2 A.OR. 371,
URN:NBN:DE:BVB:12-BSB00096220-8

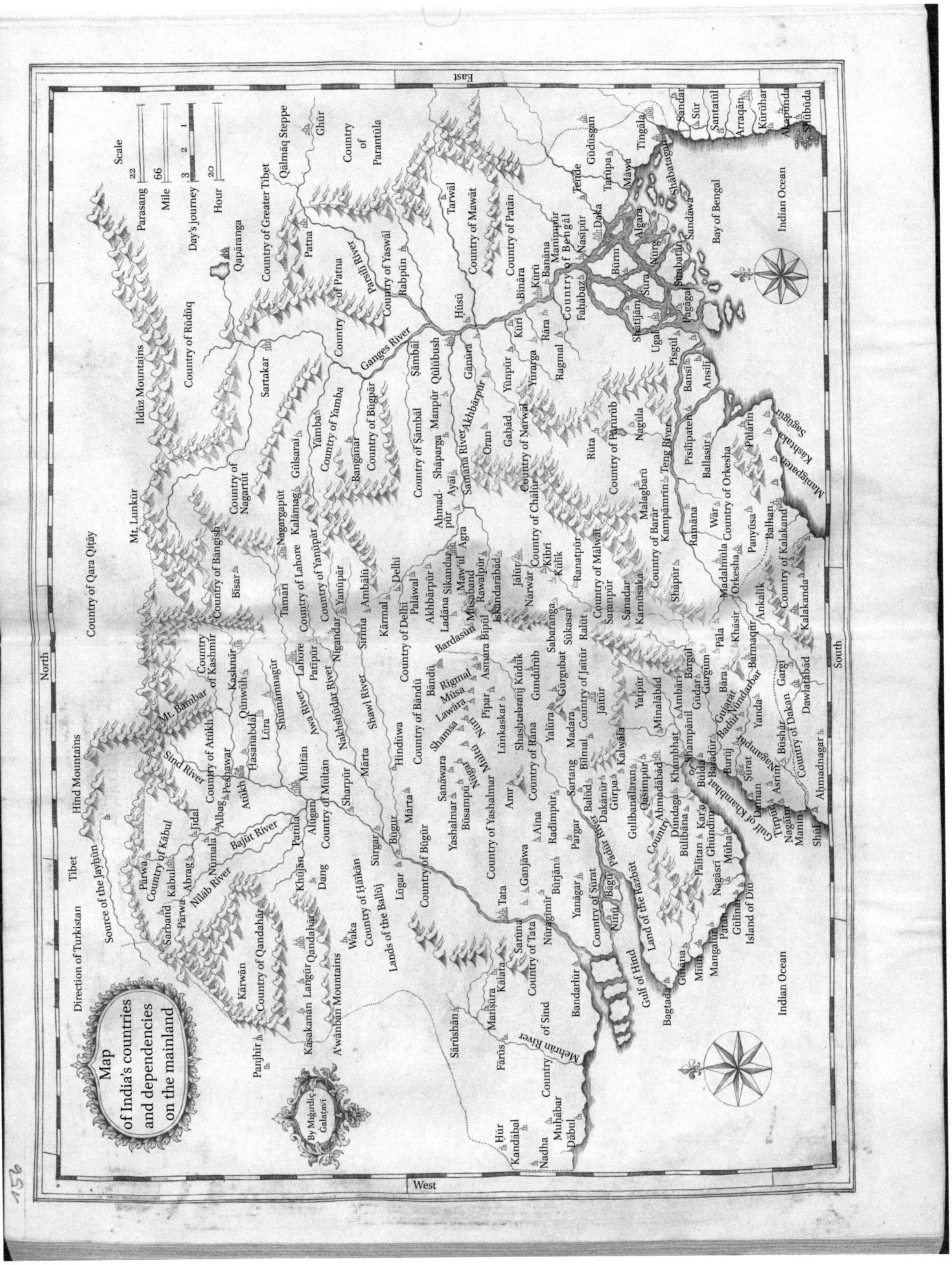

FIGURE 74 M 193/194 Map of India (translation)

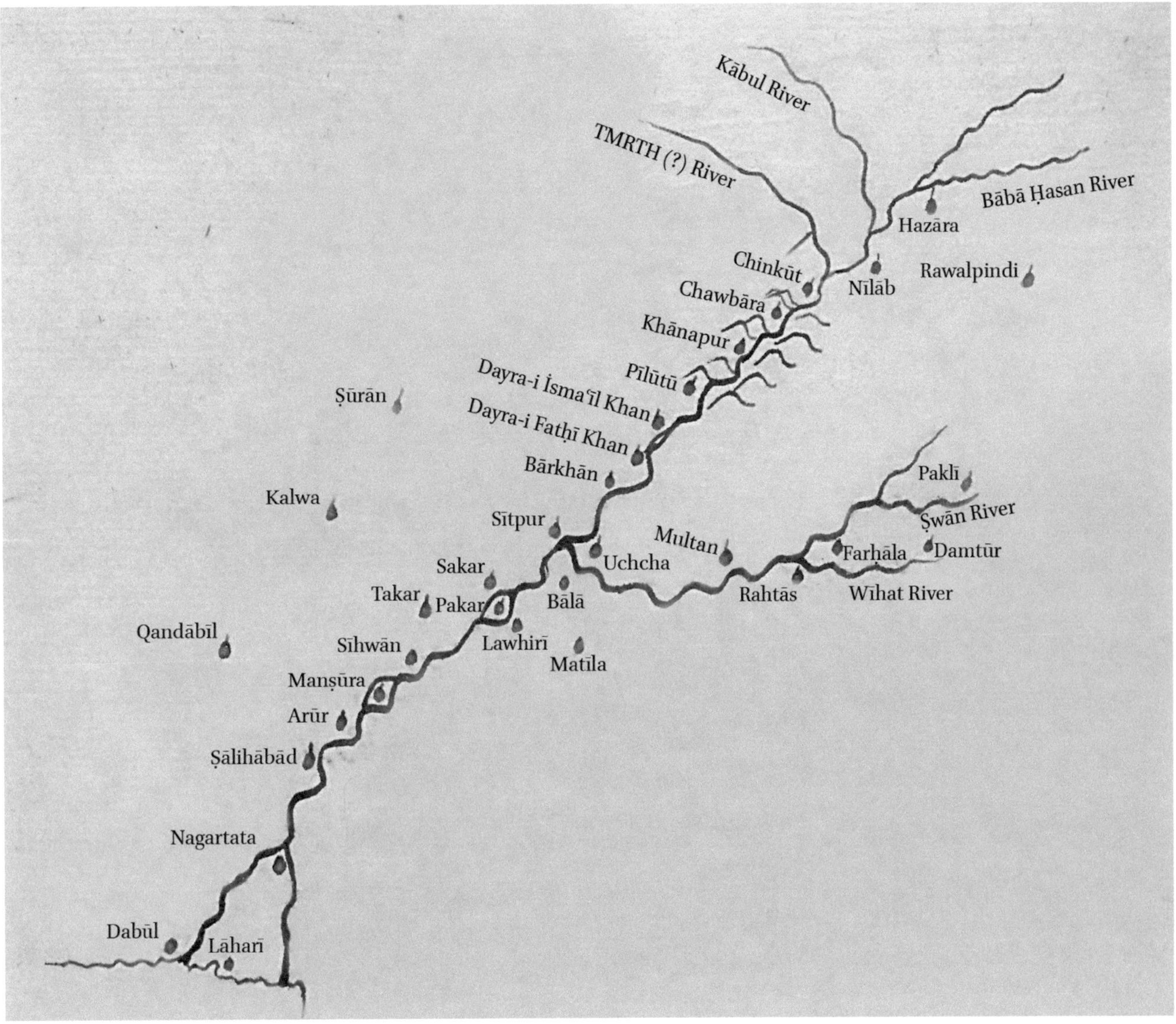

FIGURE 75 R 87a Map of Sind (translation)

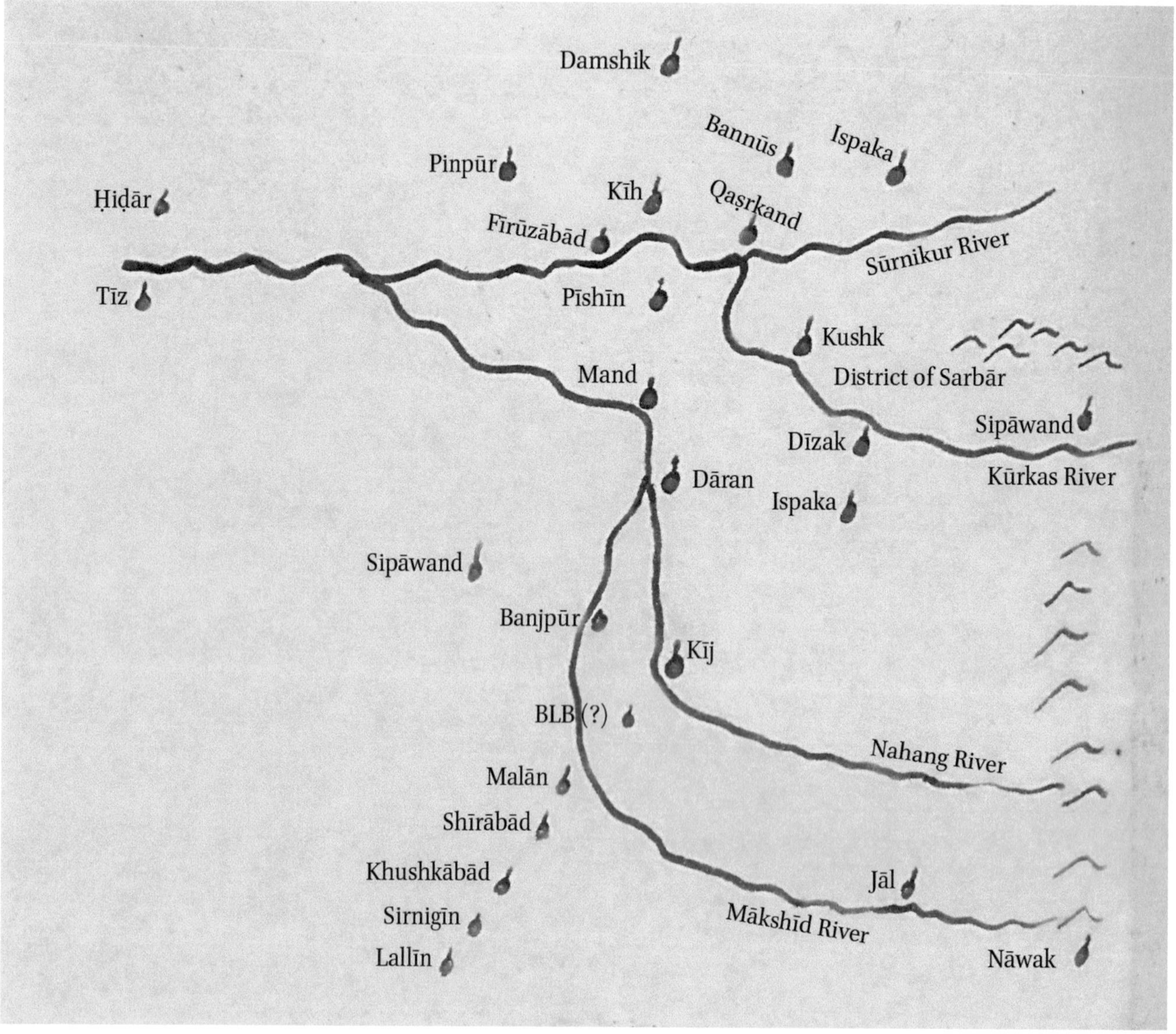

FIGURE 76 R 88a Map of Makran (translation)

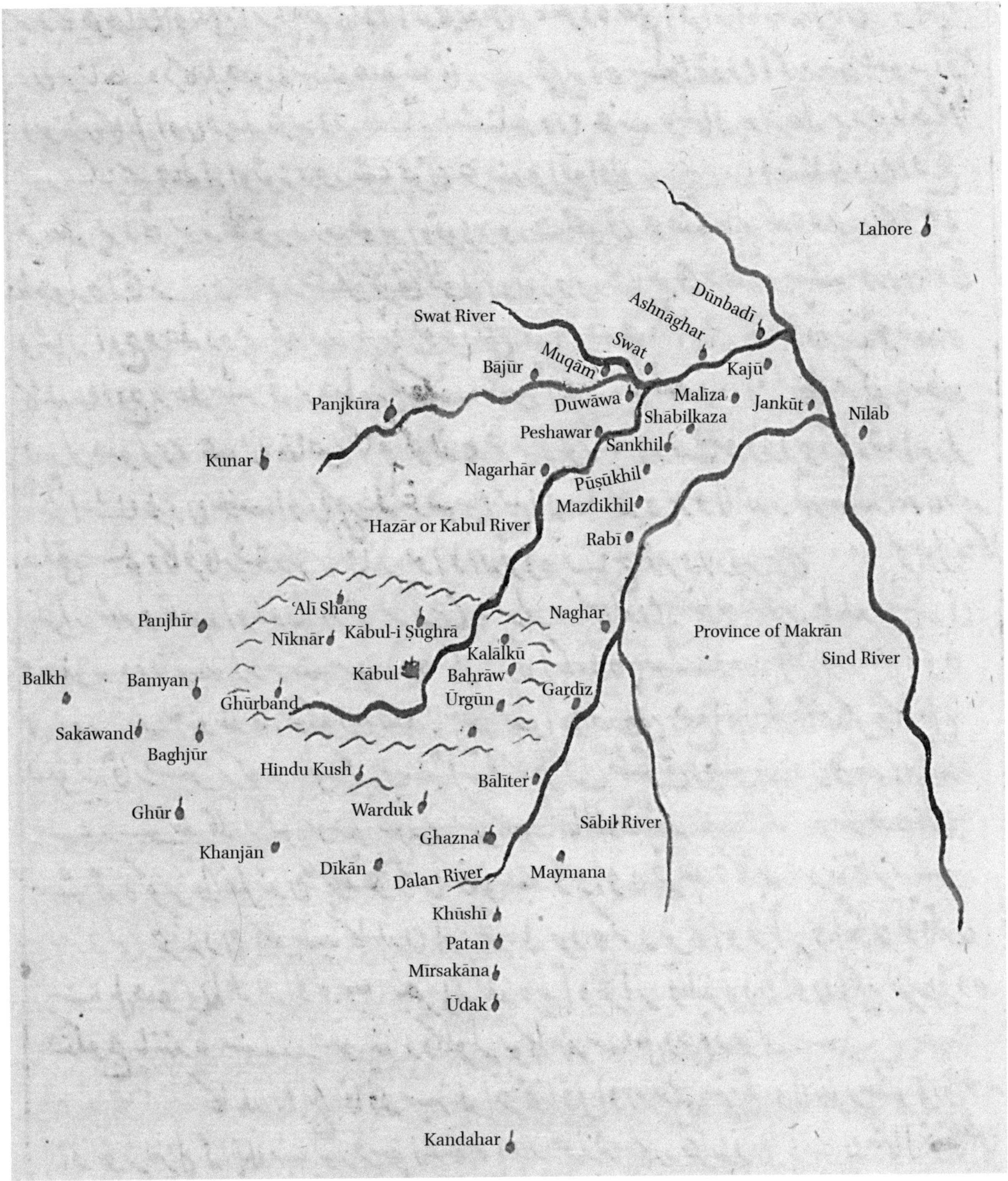

FIGURE 77 R 89b Map of Zābulistān, Khwāst, Ashnāghar (translation)

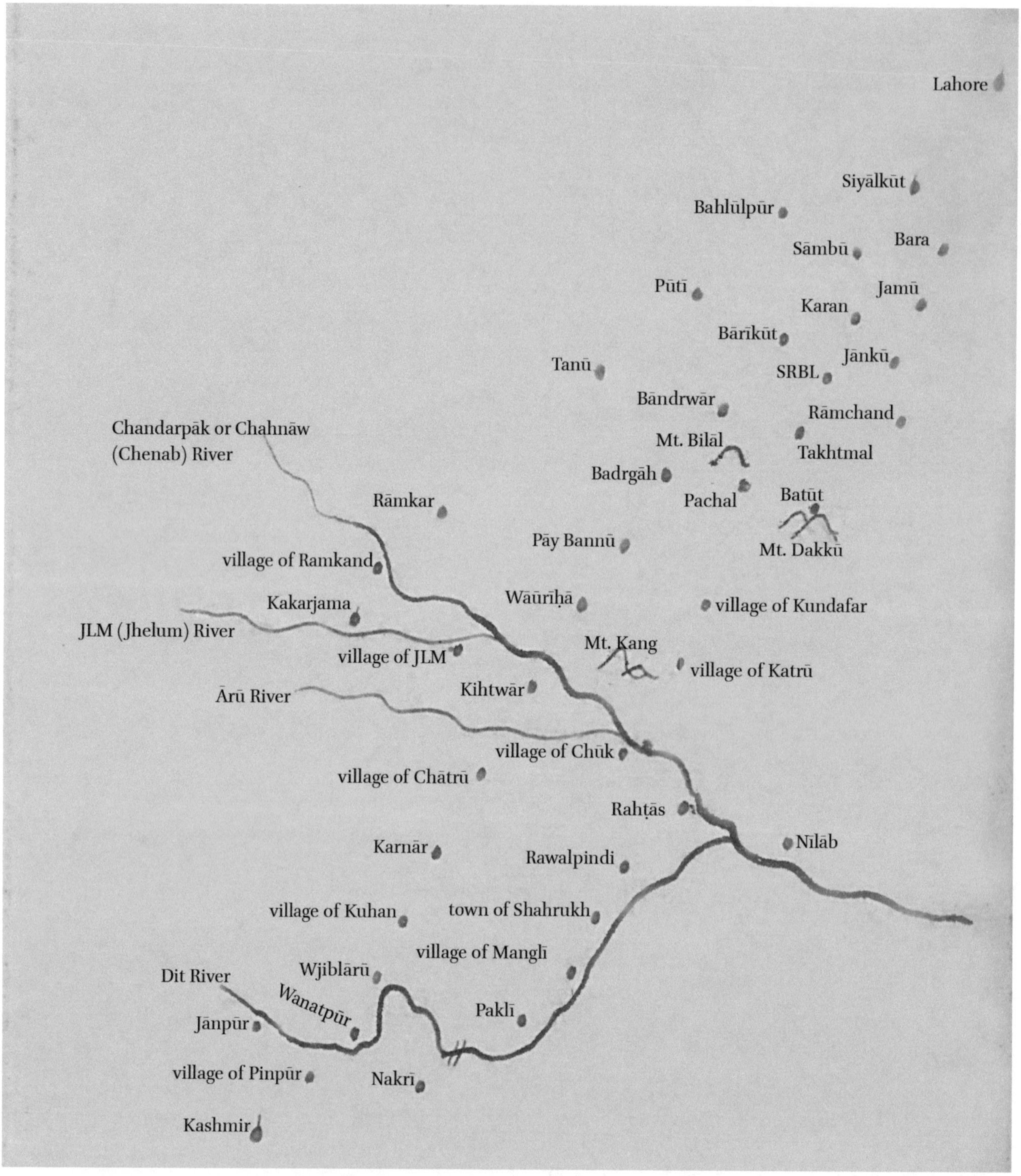

FIGURE 78 R 93a Map of Kashmir and Tibet (translation)

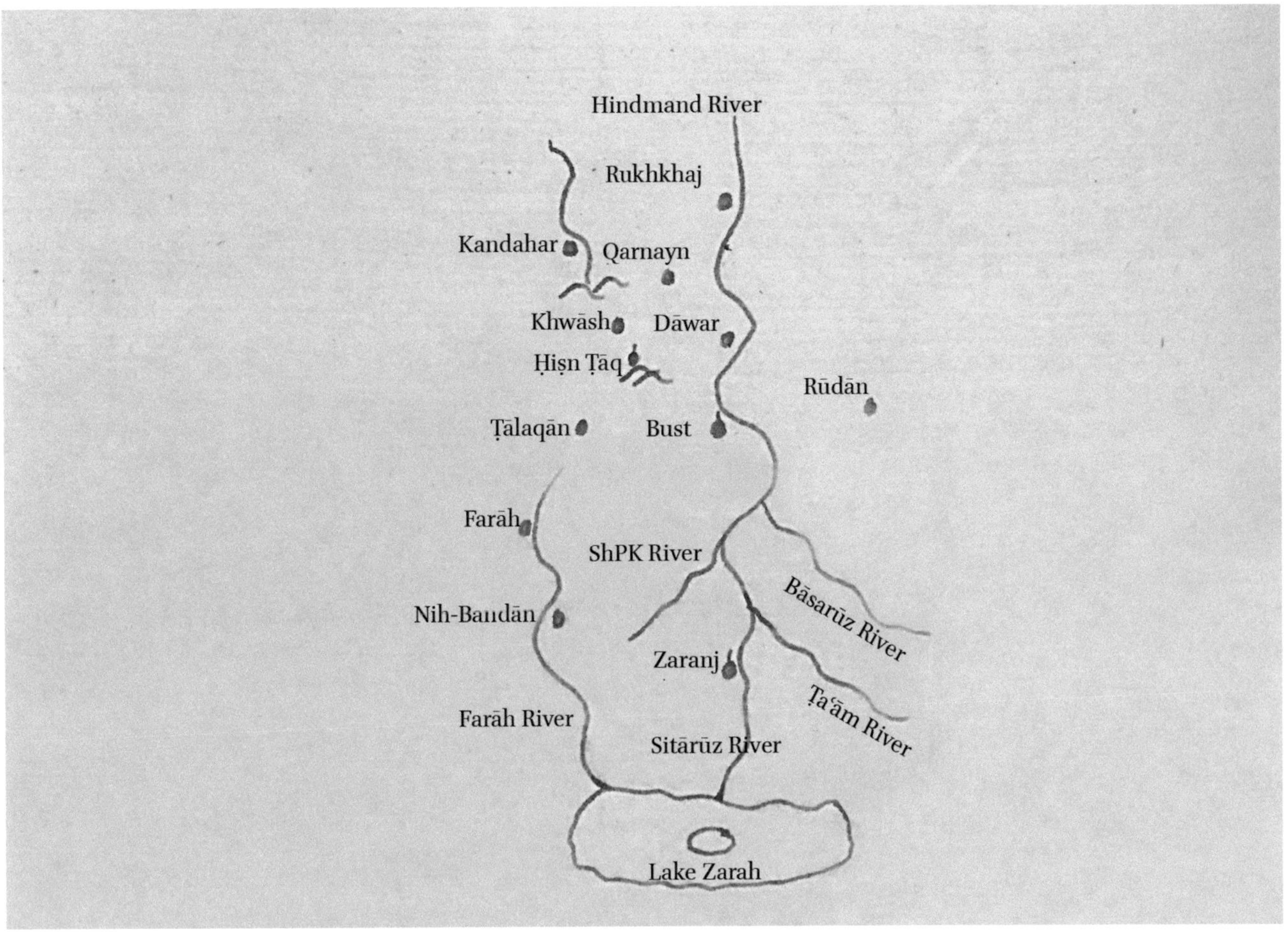

FIGURE 79 R 95a Map of Sijistān (translation)

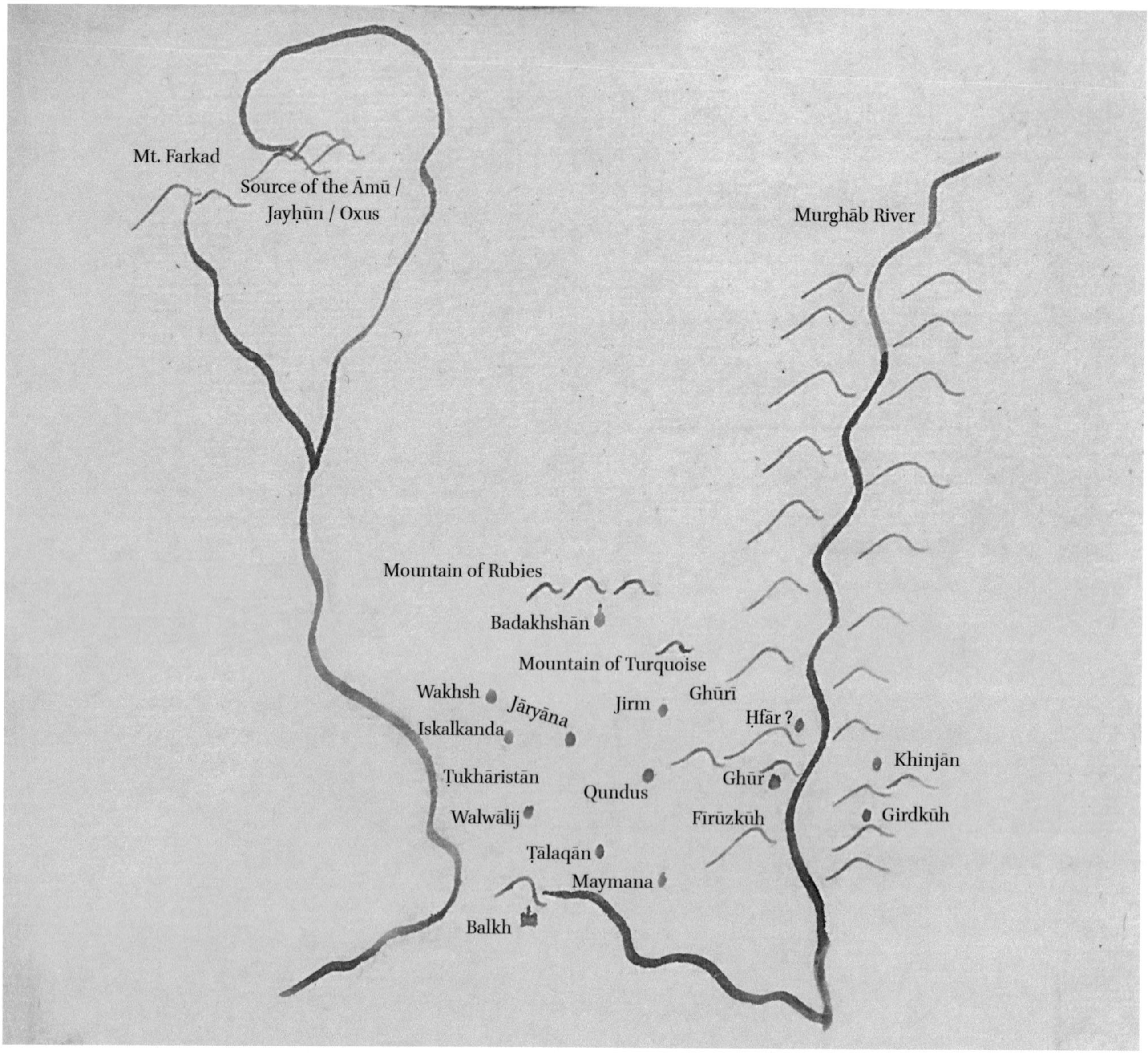

FIGURE 80 R 96b Map of Badakhshān, Ghūr, Ṭukhāristān (translation)

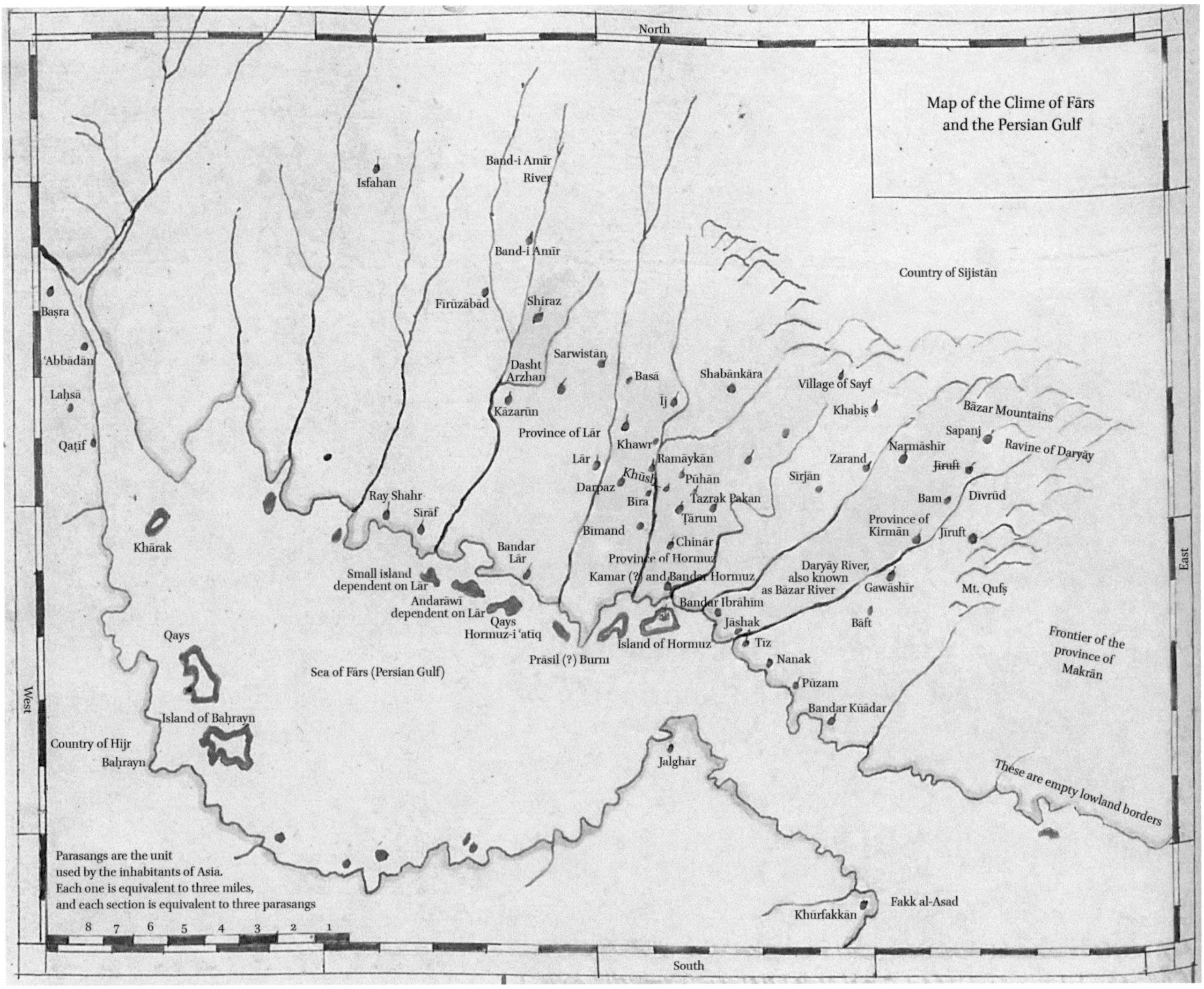

FIGURE 81 R 104a Map of Fārs and Persian Gulf (translation)

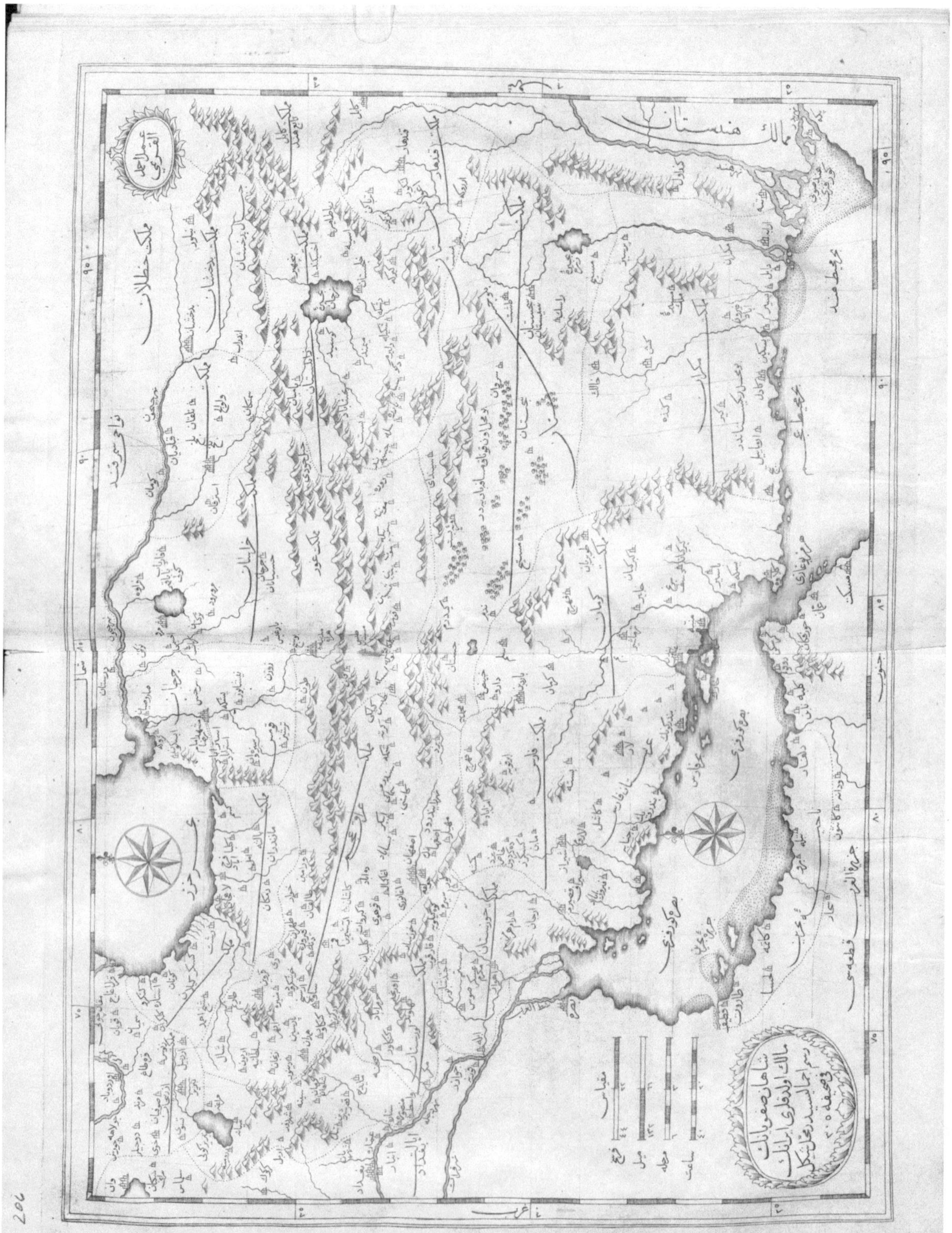

FIGURE 82 M 289/290 Map of Iran
BAYERISCHE STAATSBIBLIOTHEK MÜNCHEN, RES/2 A.OR. 371,
URN:NBN:DE:BVB:12-BSB00096220-8

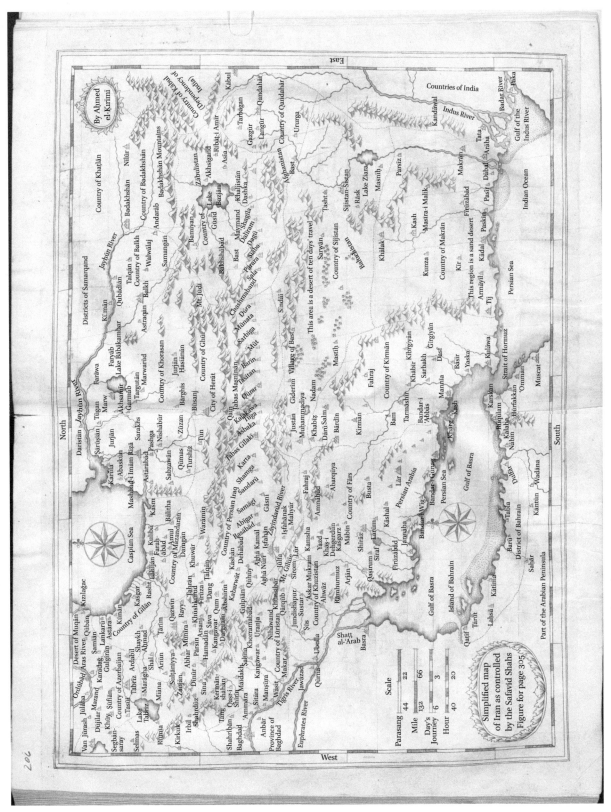

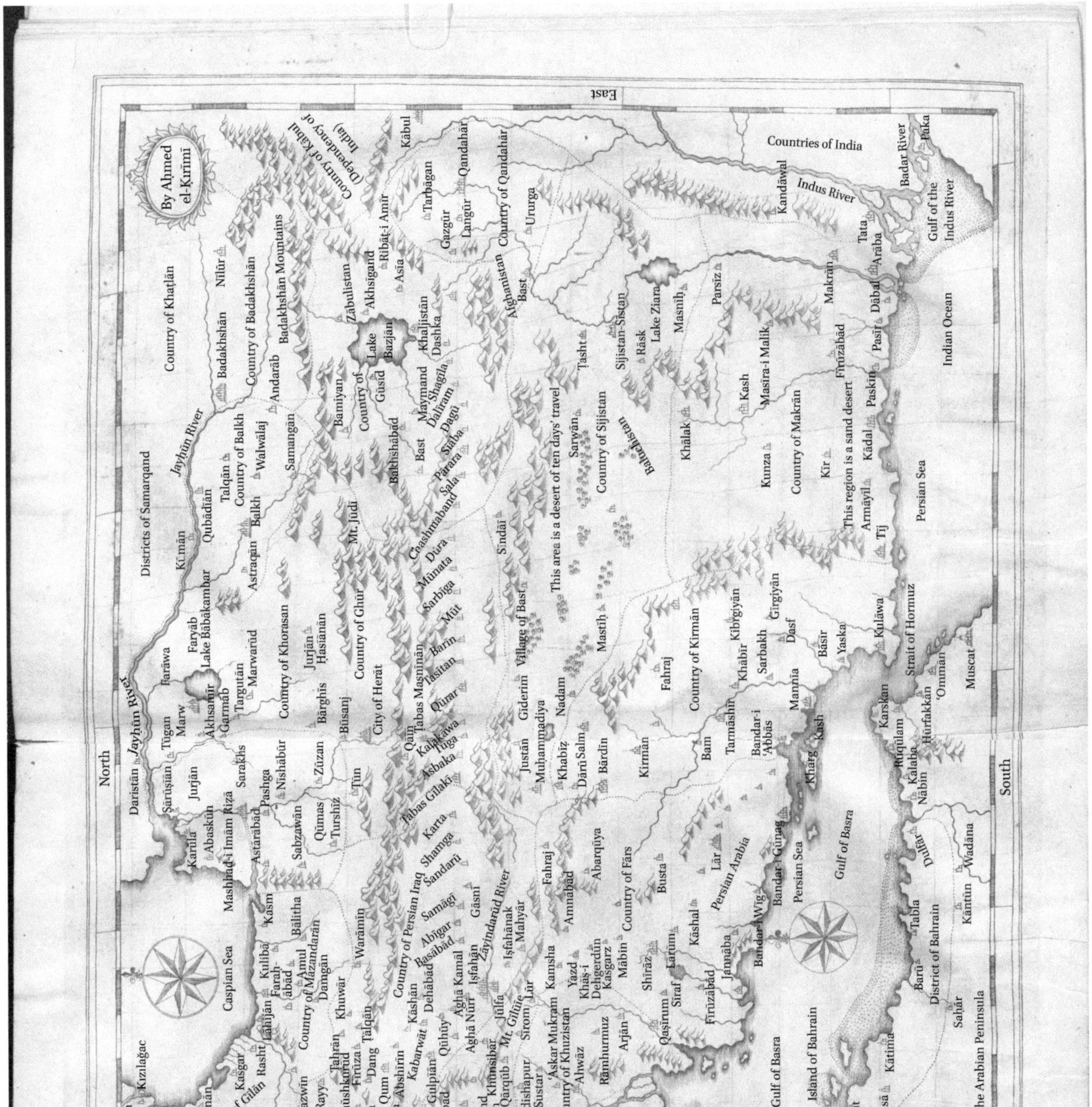

FIGURE 83 M 289/290 Map of Iran (translation)

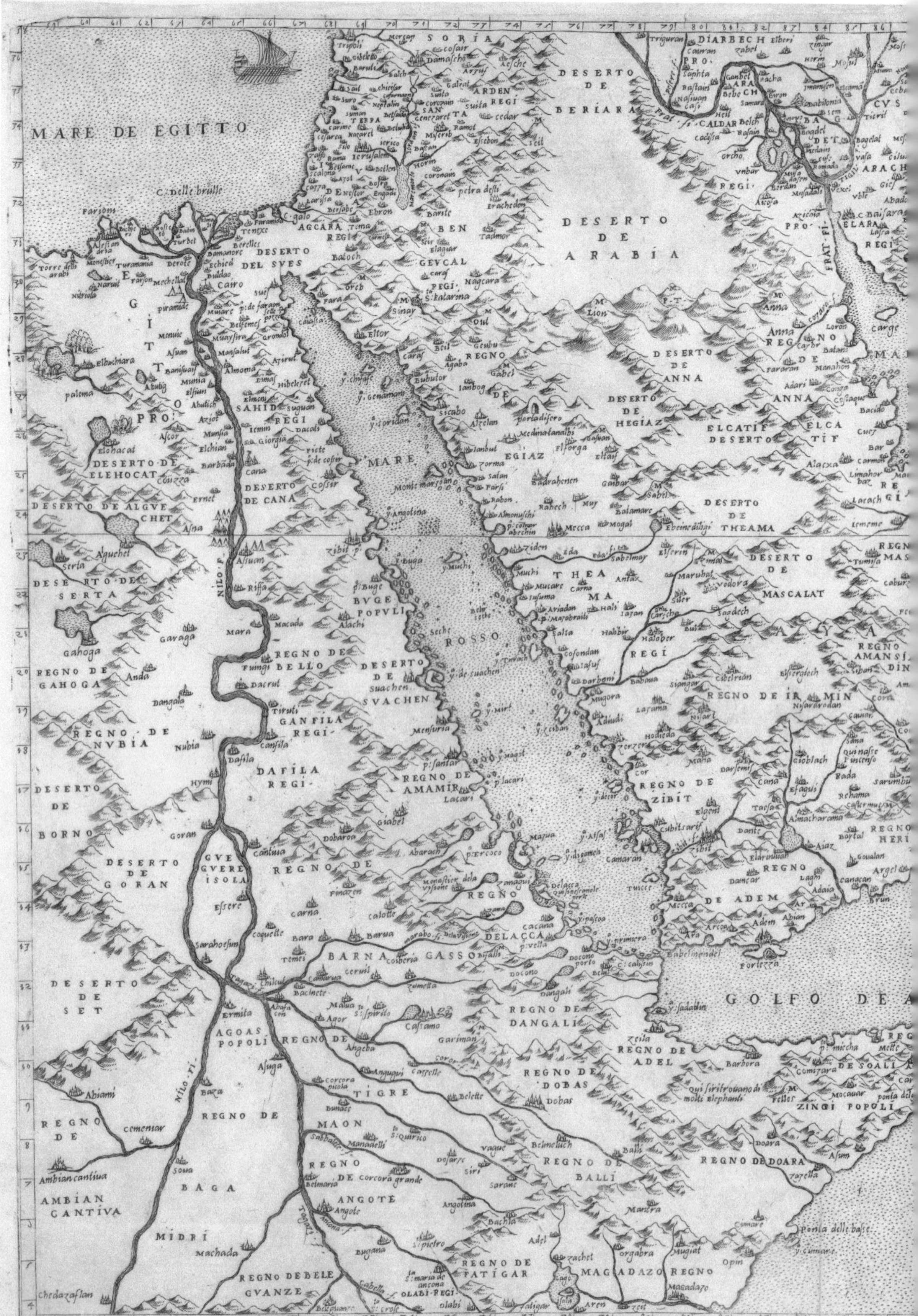

FIGURE 84 [*on this p. 626 and facing p. 627*] Giacomo Gastaldi, Il disegno della seconda parte dell' Asia, 1561

BARRY LAWRENCE RUDERMAN, ANTIQUE MAPS INC., LA JOLLA, CA, WITH PERMISSION

IL DISEGNO DELLA SECONDA PARTE DELL'ASIA
Il quale principia da leuante al fiume indu gia detto indo, et alla prouintia de malabar e da Ponente il fiume Nilo, e parte luochi diserti da settentrione il Parallelo ch' e in 76 gradi di larghezza et dall'Austro il mare d' india, et il fiume di magadazo, et il Regno de Beleguanze: Nel detto disegno si uedeno tutte le scale che fanno le naue, che vano e vengano con le spesciarie dalla citta de calecut, et similmente tutti iluochi per terra nominati al presente, ch' uano e uengono le carouane dal Suachen e Dalacca paese del prete giouani, et similmete da Aden e ormus, et dalla Balsara castello sopra il fiume frat: tutto graduato in Longezza e larghezza, et misurato con miglia italiani.
All' ill^mo sig^r il sig^r Marcho fucharo, Barone di Kirchberg e d' Waissenhoren:-
Giacomo di Castaldi Piamontese cosmographo in venetia:- 1561
Con gratia et priuilegio del Sumo Pontifice Papa Pio iiii p anni X
E del serenissimo senato di venetia per Anni XV
fabio licinio f.
MARE DE INDIA
GOLFO DE ORMVS
GOLFO DI PERSIA
TROPICO DI CANCRO
DESERTO DE MINGIV
DESERTO DE LVT
SIGISTAN PRO:
CIRCAN PRO:
MACRAN REGNO
REGNO DE ERACHAIAN
GVZARATE
REGNO DE CABVL
SABLESTAN PRO:
CORASAN PRO:
ARACH PRO:
FARSI PRO:
REGNO DE ORMVS
REGNO DE MAFFA
DISERTO
ISOLE PALANDVRE
ARCIPELAGO DE MALDIVAR
DECAN
MALABAR PRO:
INDV FI.
Scalla de milia ... italiani

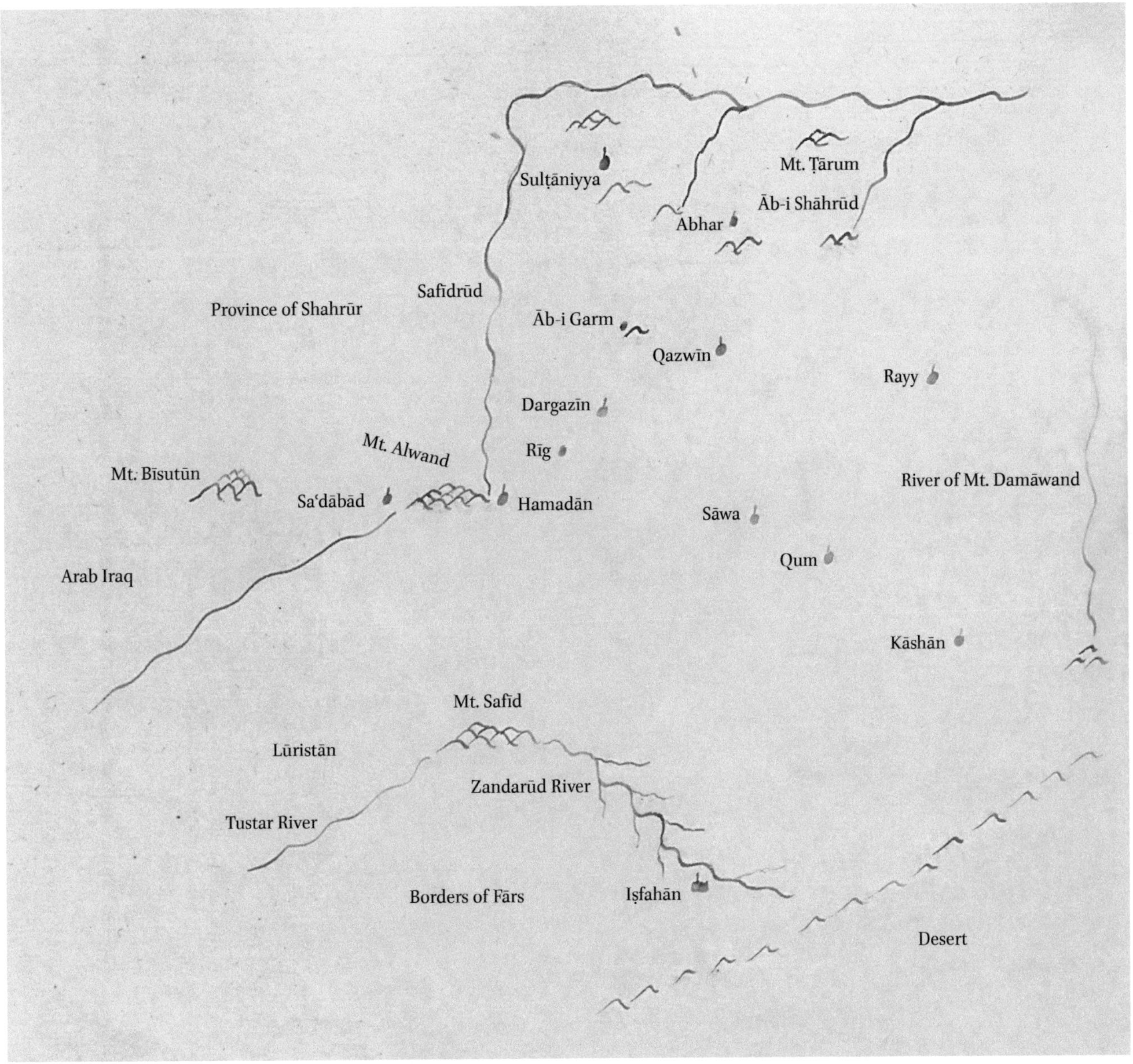

FIGURE 85 R 116a Map of Jibāl or Persian Iraq (translation)

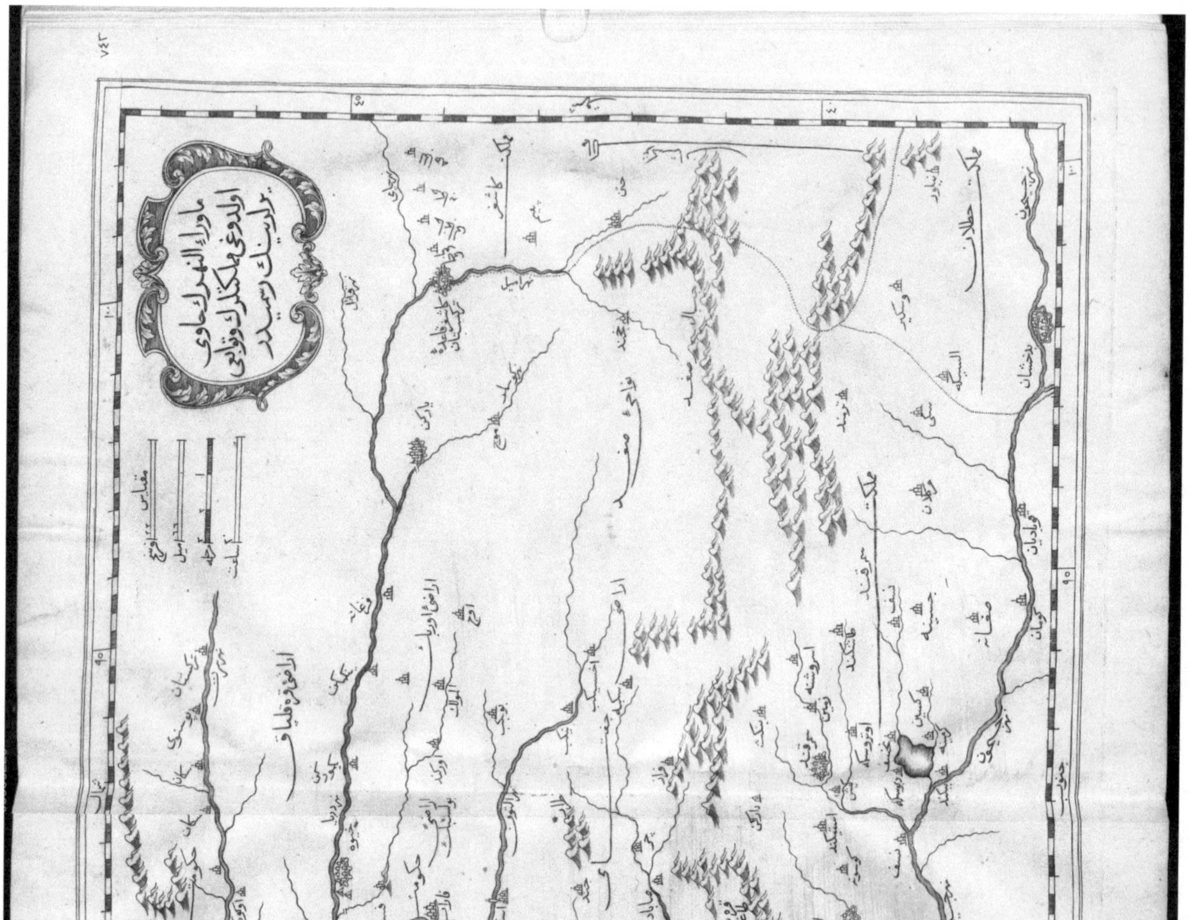

FIGURE 86 M 347/348 Map of Transoxiana
BAYERISCHE STAATSBIBLIOTHEK MÜNCHEN, RES/2 A.OR. 371, URN:NBN:DE:BVB:12-BSB00096220-8

FIGURE 87 M 347/348 Map of Transoxiana (translation)

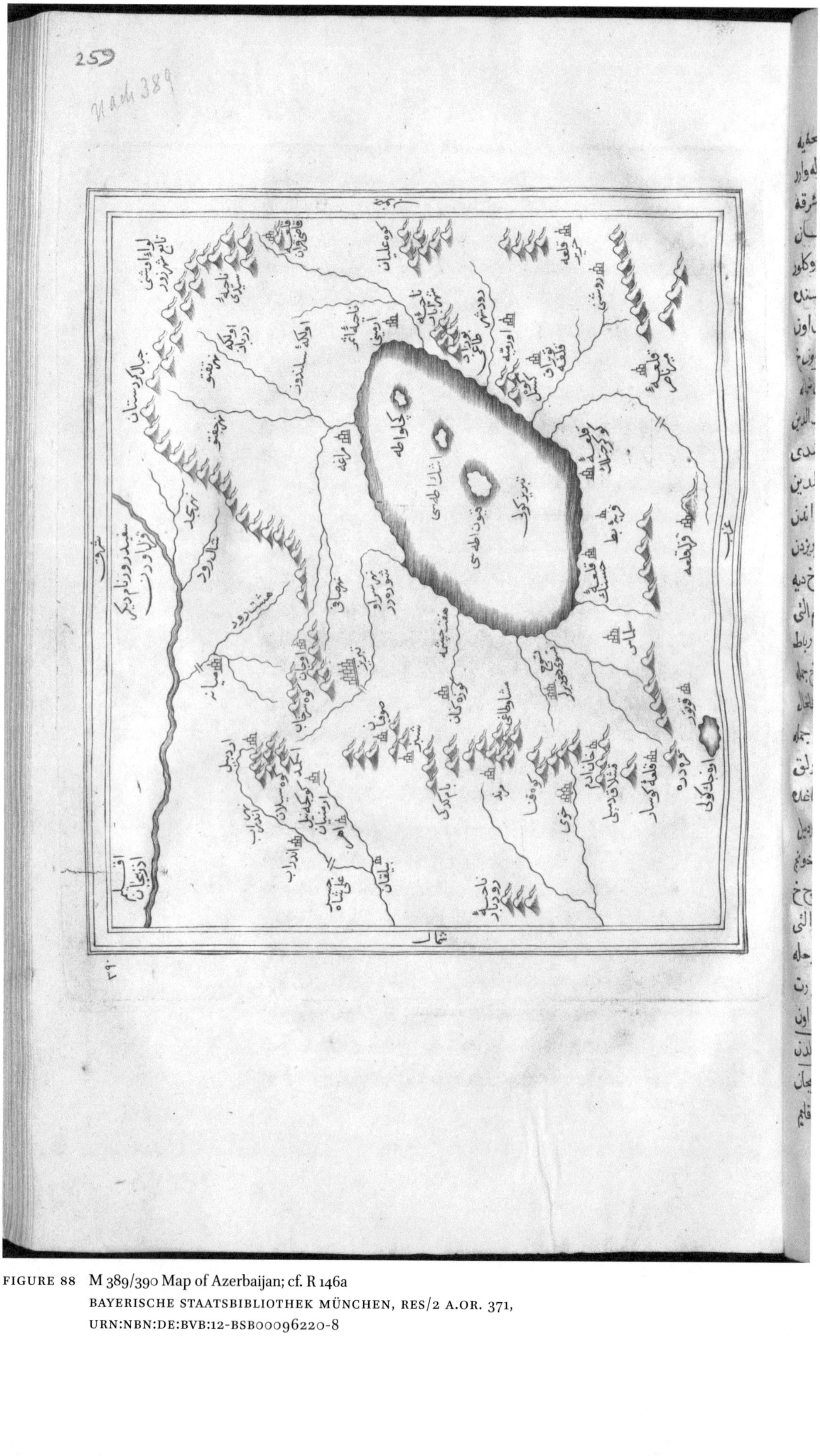

FIGURE 88 M 389/390 Map of Azerbaijan; cf. R 146a
BAYERISCHE STAATSBIBLIOTHEK MÜNCHEN, RES/2 A.OR. 371, URN:NBN:DE:BVB:12-BSB00096220-8

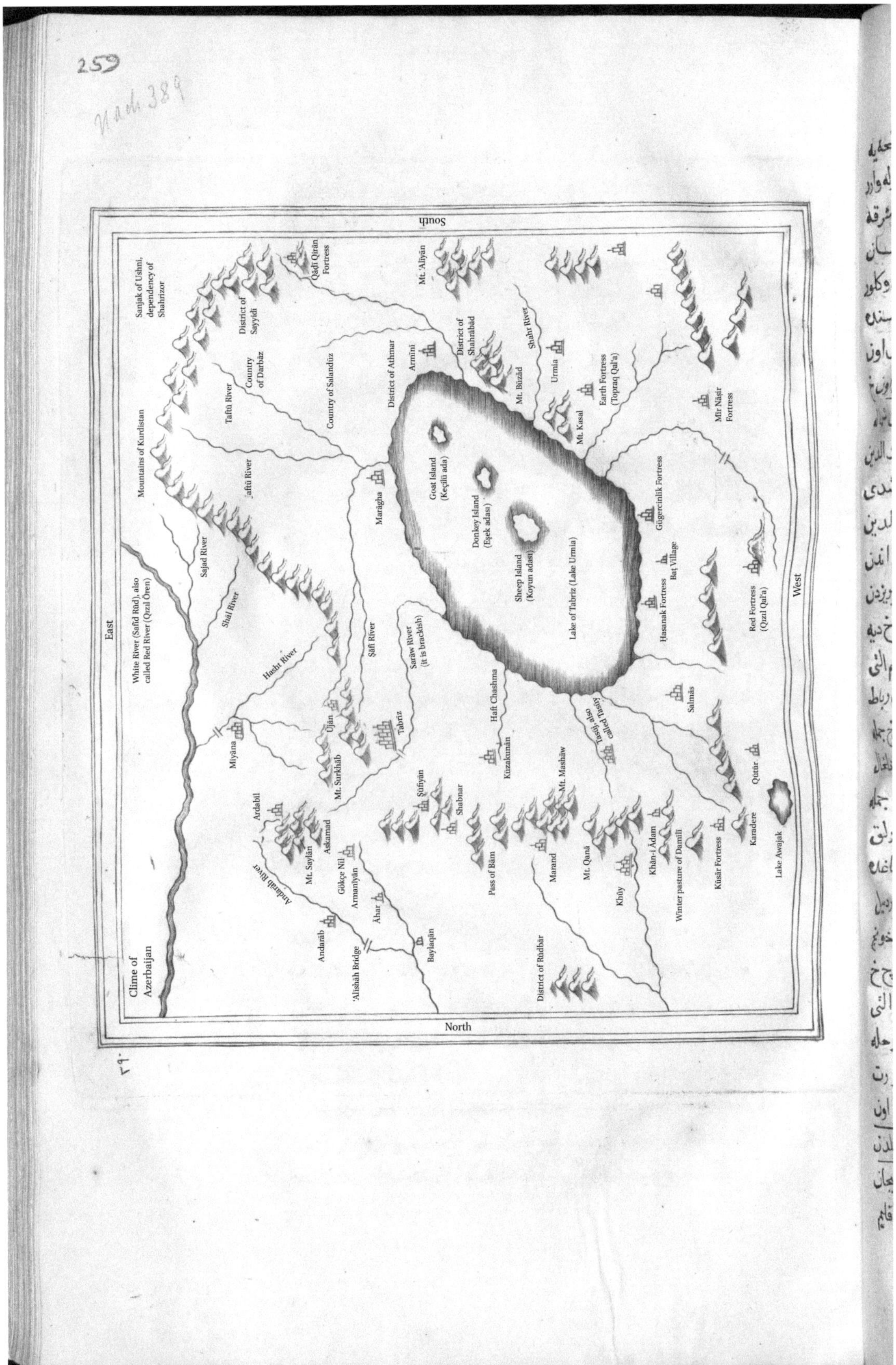

FIGURE 89 M 389/390 Map of Azerbaijan (translation)

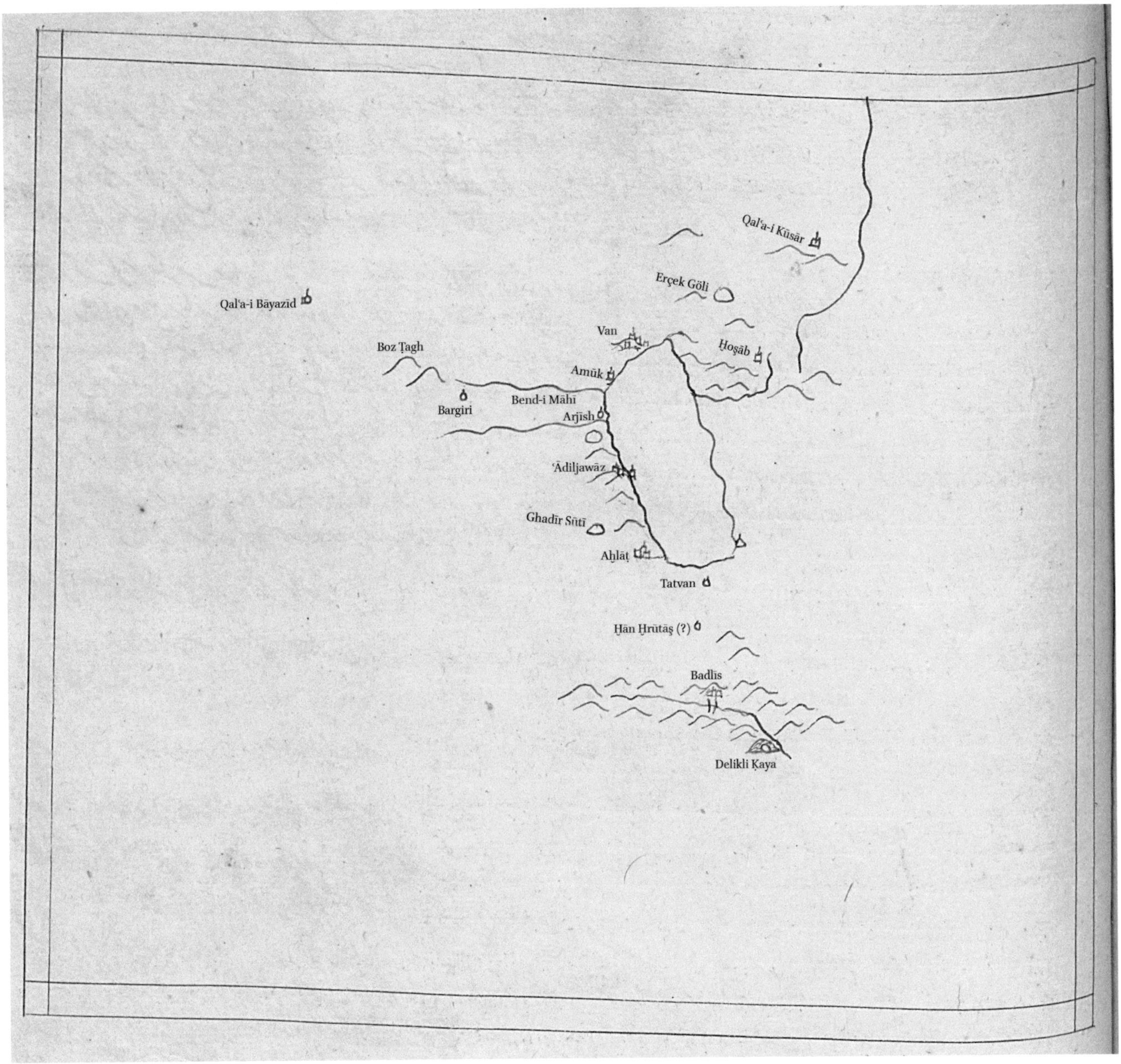

FIGURE 90 R 159a Map of Van Province (translation)

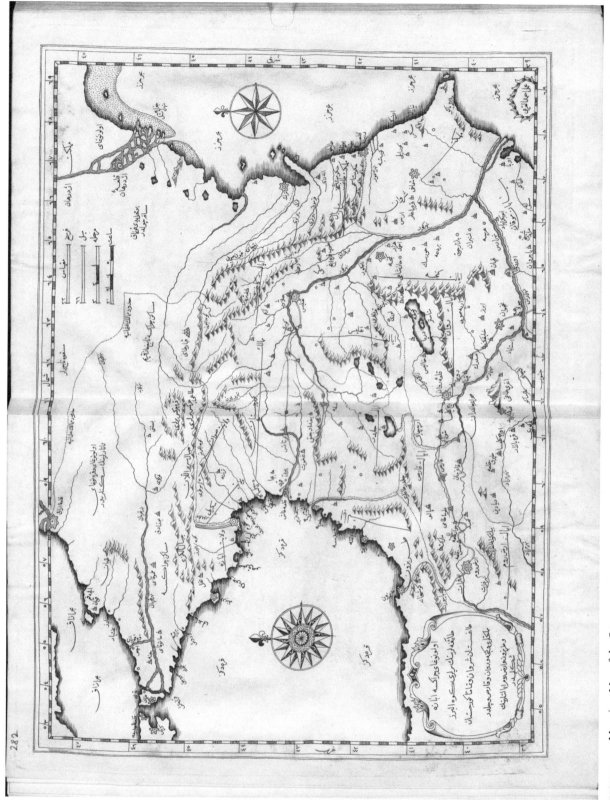

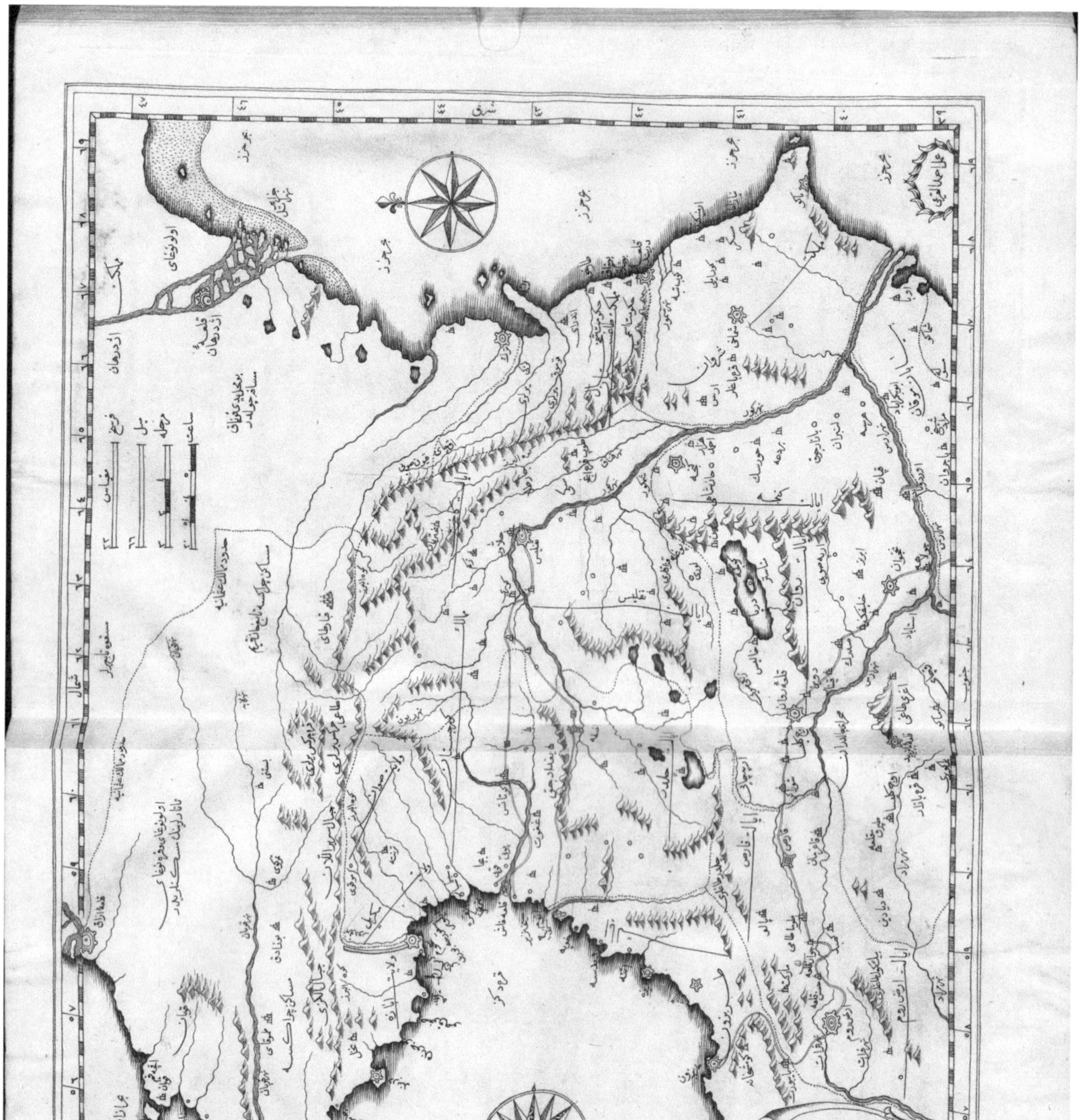

FIGURE 91 M 431/432 Map of the Caucasus
BAYERISCHE STAATSBIBLIOTHEK MÜNCHEN, RES/2 A.OR. 371, URN:NBN:DE:BVB:12-BSB00096220-8

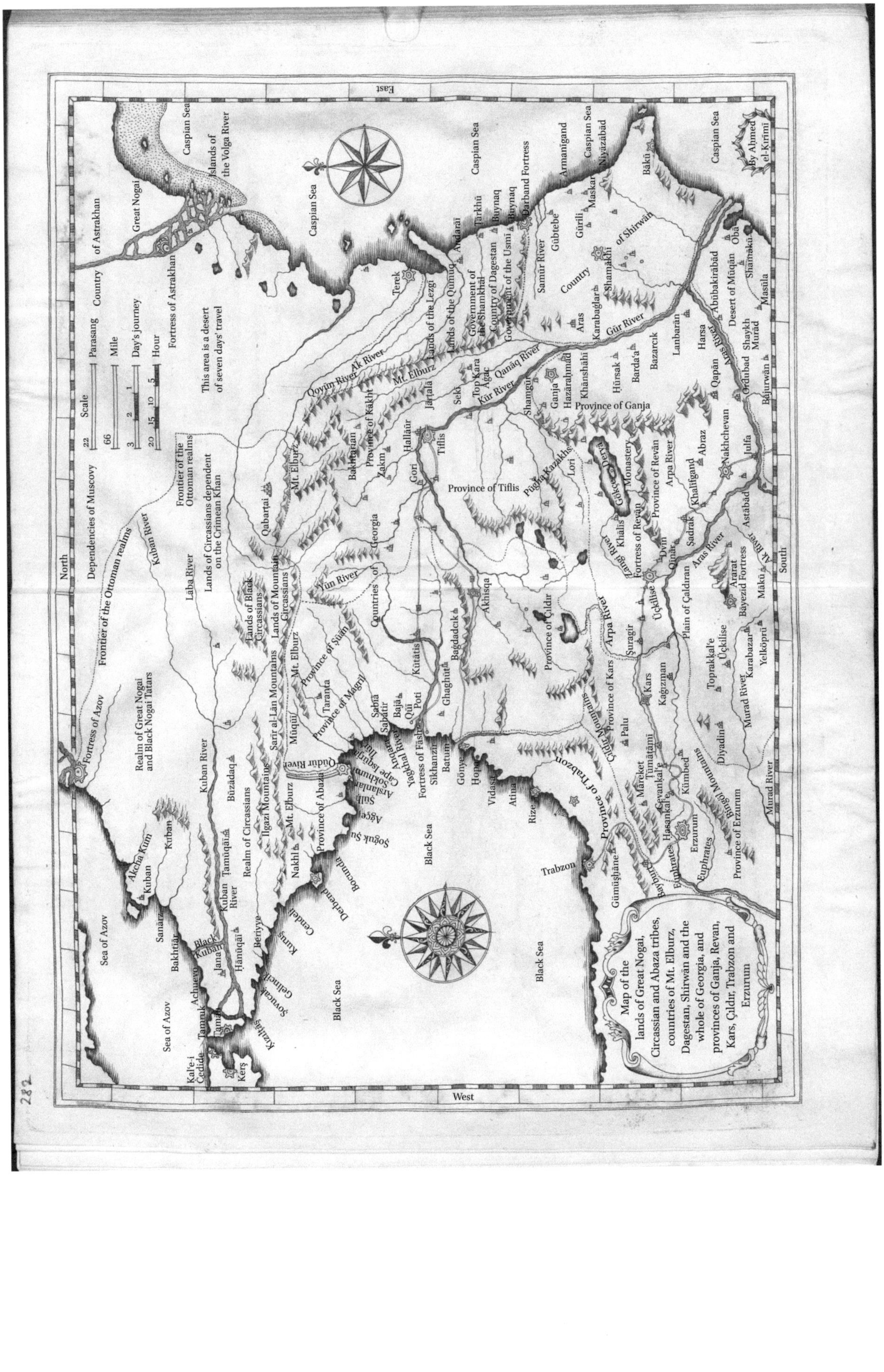

FIGURE 92 M 431/432 Map of the Caucasus (translation)

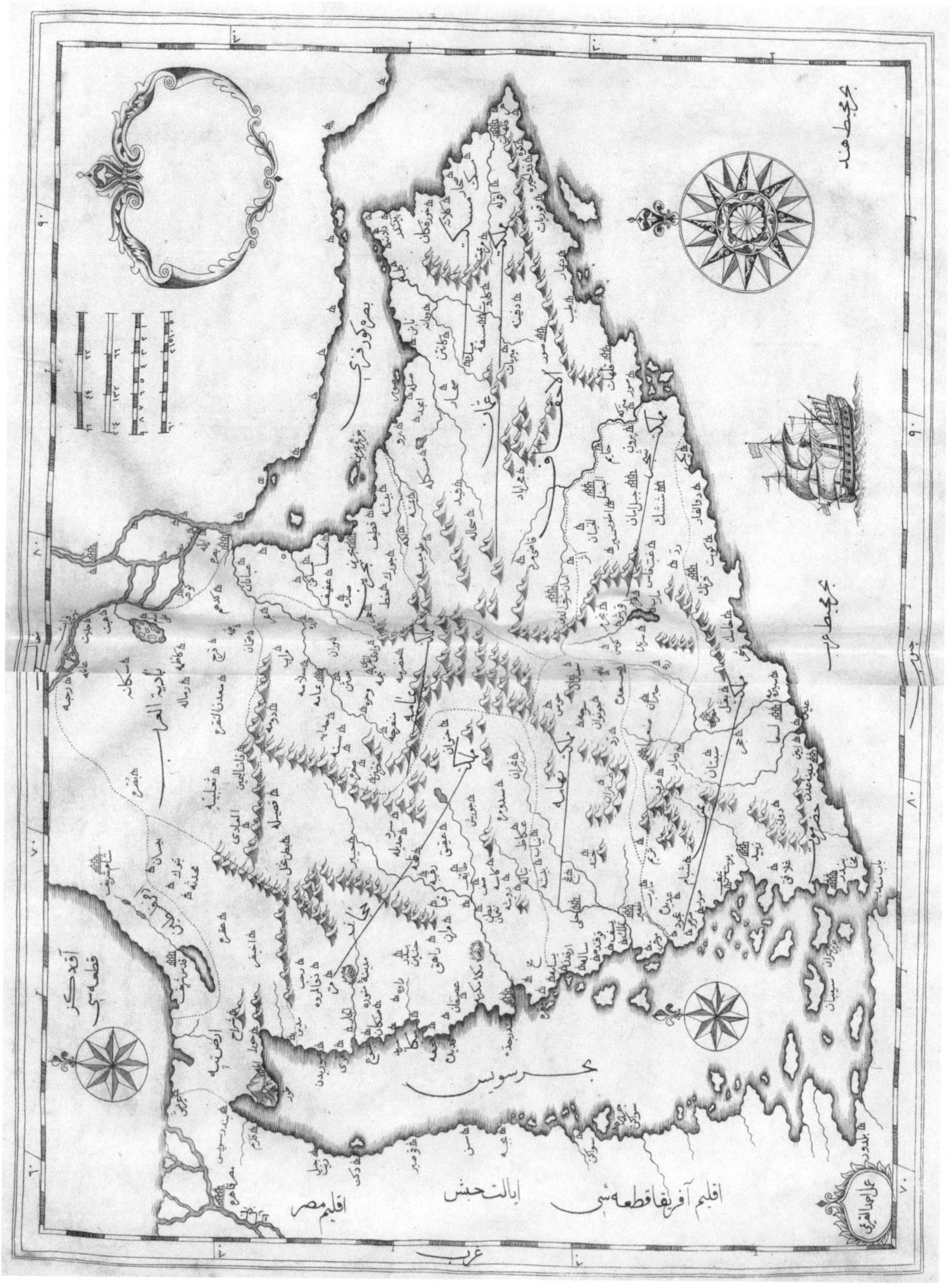

FIGURE 93 M 483/484 Map of Arabian Peninsula
BAYERISCHE STAATSBIBLIOTHEK MÜNCHEN, RES/2 A.OR. 371,
URN:NBN:DE:BVB:12-BSB00096220-8

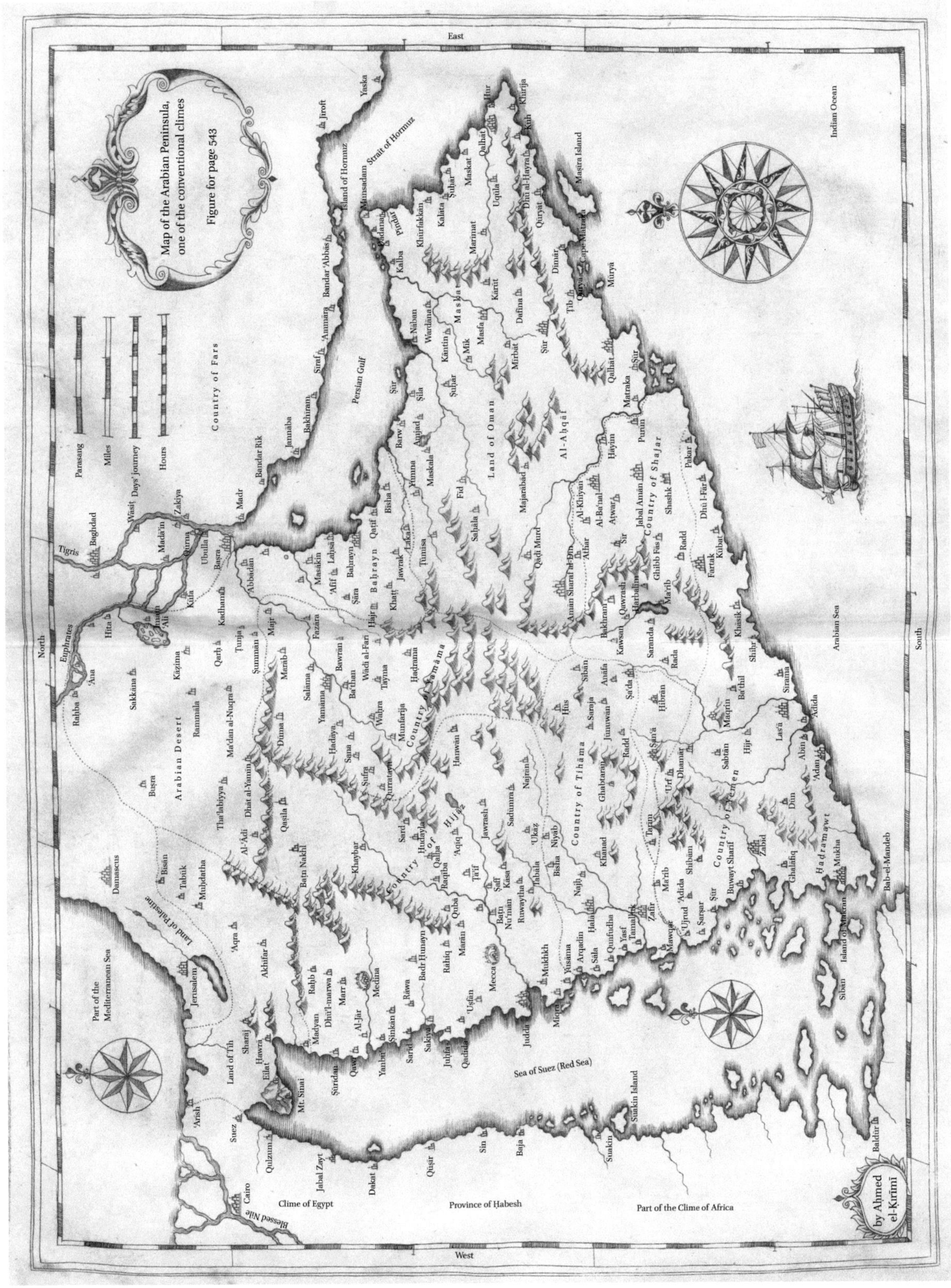

FIGURE 94 M 483/484 Map of Arabian Peninsula (translation)

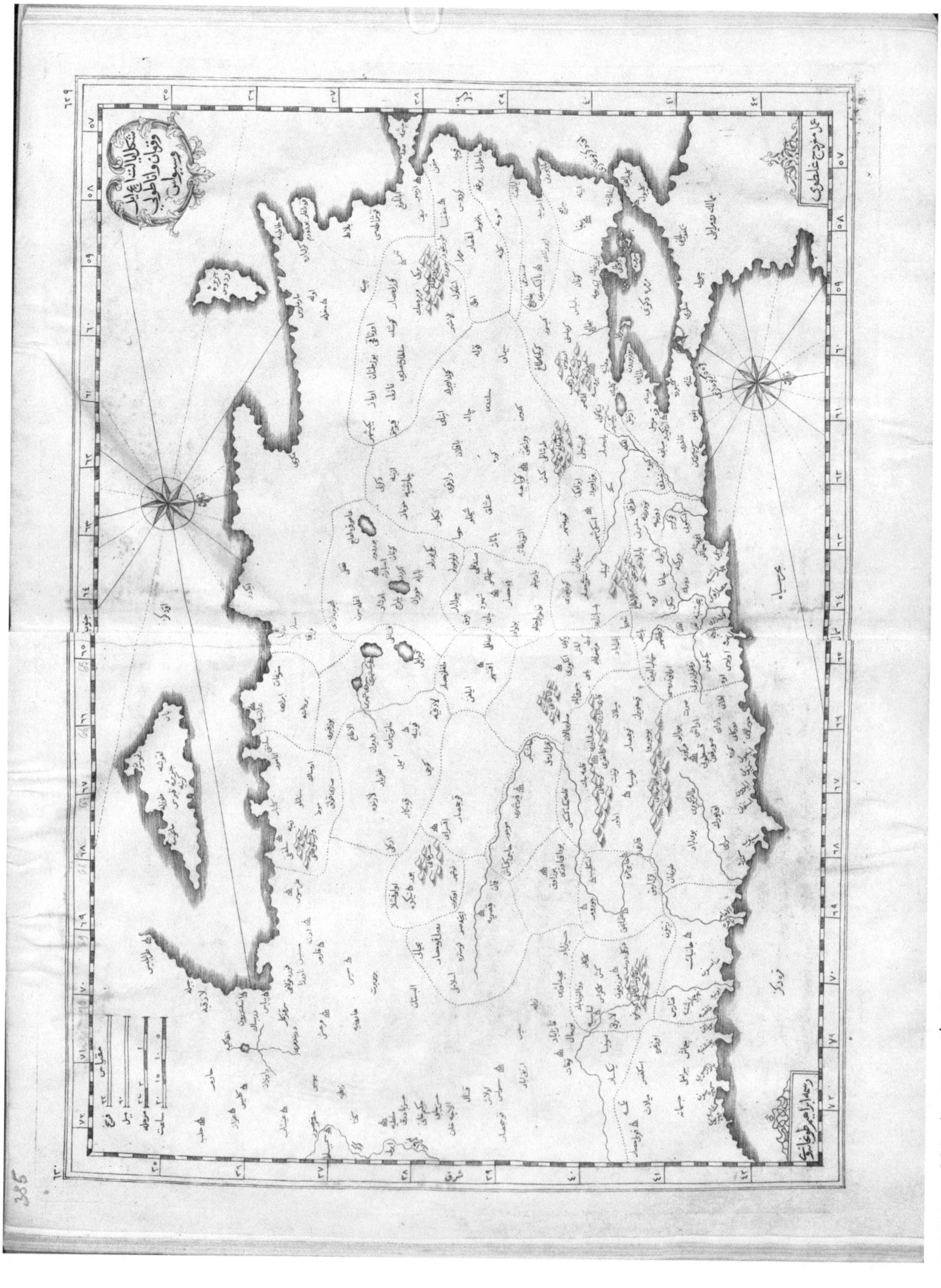

FIGURE 95 M 629/630 Map of İçel, Karaman, Anaṭolı and Sivas
BAYERISCHE STAATSBIBLIOTHEK MÜNCHEN, RES/2 A.OR. 371,
URN:NBN:DE:BVB:12-BSB00096220-8

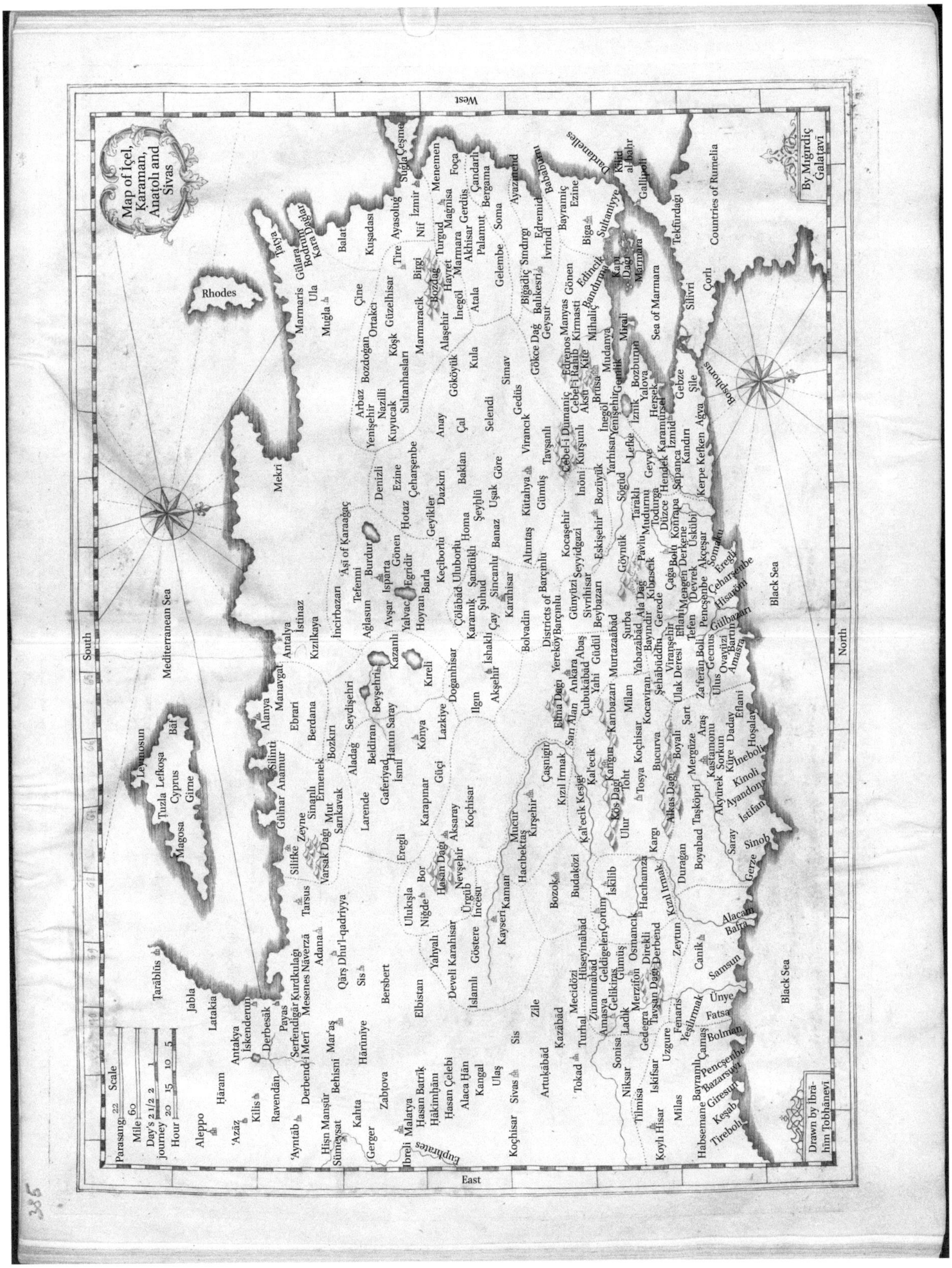

FIGURE 96 M 629/630 Map of İçel, Ḳaraman, Anaṭolı and Sivas (translation)

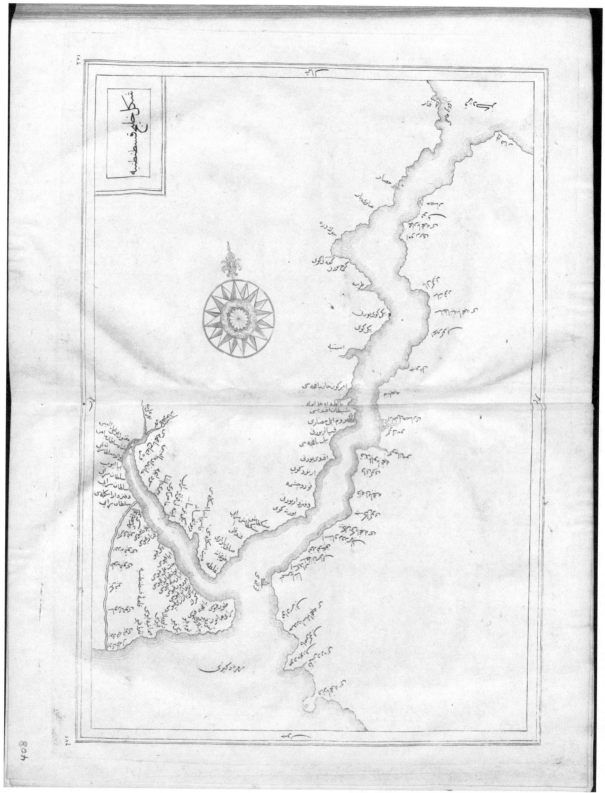

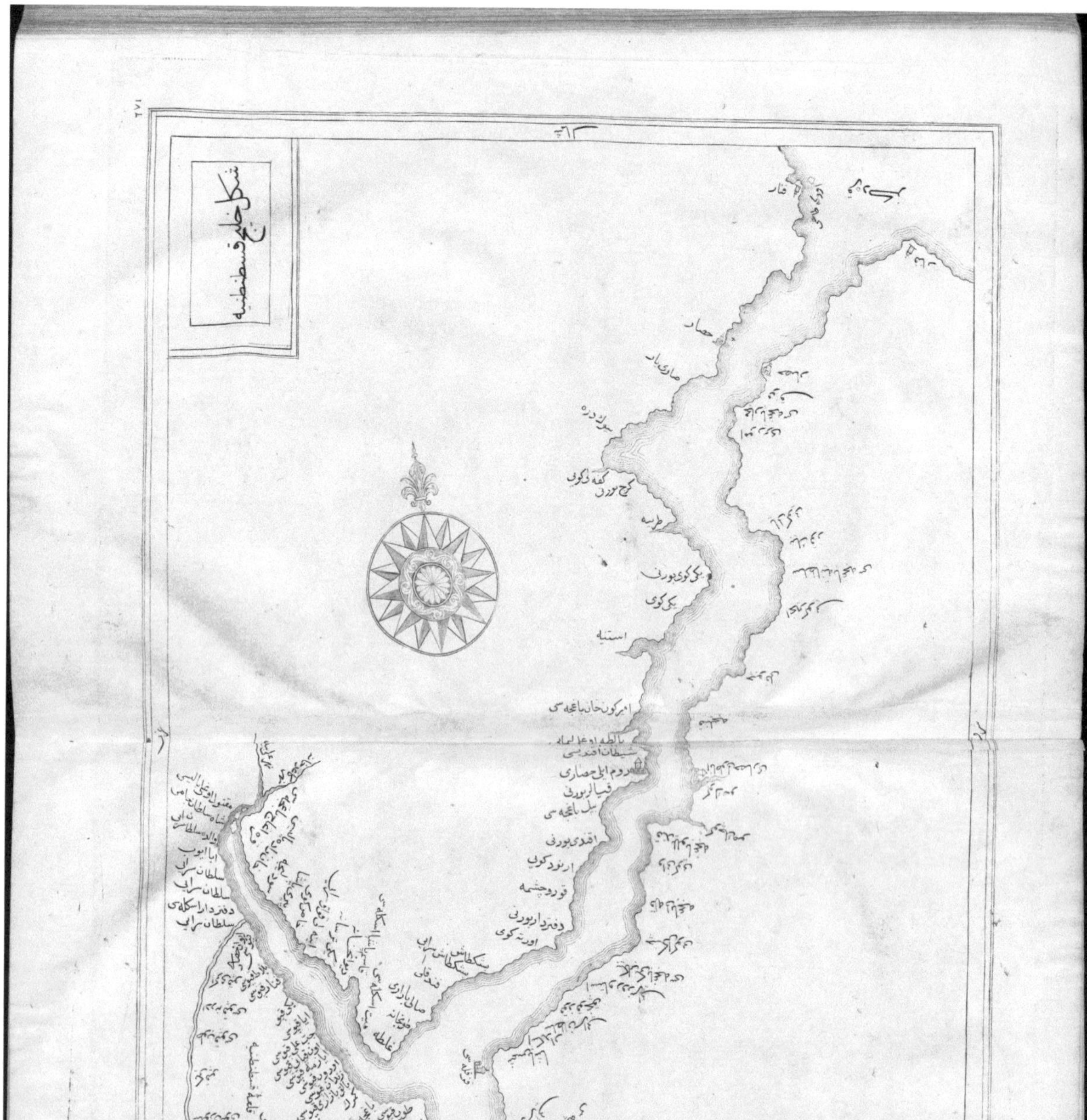

FIGURE 97 M 671/672 Map of the Bosphorus
BAYERISCHE STAATSBIBLIOTHEK MÜNCHEN, RES/2 A.OR. 371,
URN:NBN:DE:BVB:12-BSB00096220-8

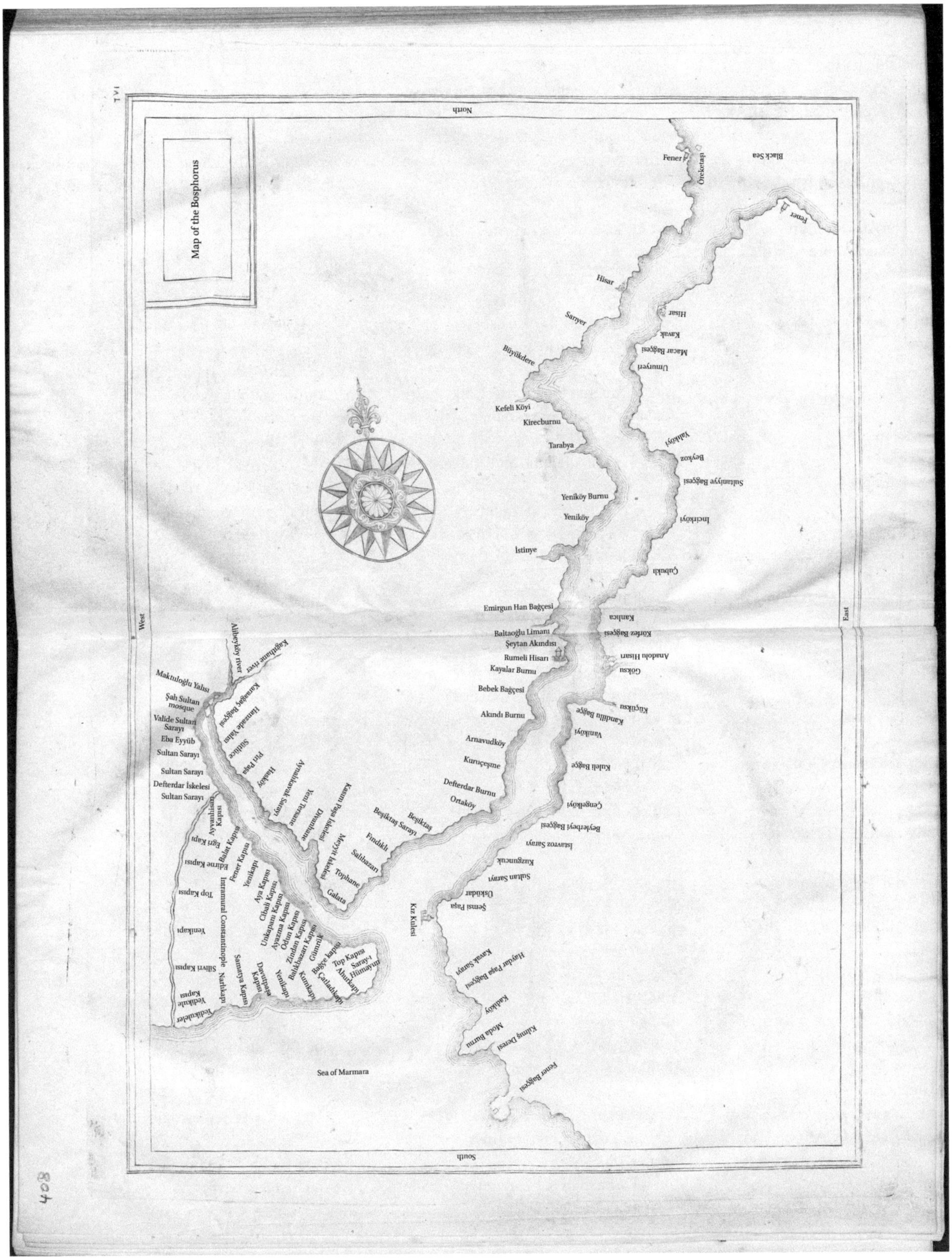

FIGURE 98 M 671/672 Map of the Bosphorus (translation)

Index of *Cihānnümā*

The following are alphabetized together: C-Ç; H-Ḥ-Ḫ; S-Ş-Ṣ; D-Ḍ; T-Ṭ; Z-Ẓ-Ż.

For purposes of alphabetization, ʿayn (ʿ) and hamza (ʾ) are ignored, as is the Arabic definite article (al-, etc.) and the Persian izafet (-i), and b. indicating "son of" in proper names.

Only the text of *Cihānnümā* is indexed. We have not included the introductory chapters, nor the maps.

Only proper nouns are indexed, with a number of further restrictions. We usually have not indexed names of buildings, such as gates, mosques, and markets within cities. We have also omitted many minor items that appear in the text, especially those reflecting Kātib Çelebi's Arabic-script rendering of sources in Latin, etc., much dynastic minutia, and names of heavenly bodies.

Omitted are alternate forms introduced in the text by "also known as," and the like.

Brackets [ ] indicate forms in Kātib Çelebi's sources.

Aaron (brother of Moses) 252, 434
Āb (fortress in Yemen) 388, 428
Āb-i Bayda 277
Āb-i Gard 235
Āb-i Jāsht 233
Āb-i Ṣafī 505
Āb-i Sarw River 196
Āb-i Shūr 278
Āb-i Talkh See Surkhāb
Āba 256
Abaḳa, Abakay 339
Abaqa Khan (Ilkhanid) 249–250, 254, 256, 263, 282, 319, 477
Ābār ʿAlī, Abyār ʿAlī (wells near Medina) 407, 421–422
Abardi 481
Abarkand 300
Abarkandmān 326
Abarqūh 234, 239, 241, 264
Ābaskūn 285–286, 301, 326
Abaza (Abkhaz), Abkhazians, Anjās 329–331, 348
Abaza Ḥasan Agha 480
ʿAbbādān 246, 363–364, 366, 416, 421
Ābbārān 212
ʿAbbās (in Syria) 450
ʿAbbās (in Yemen) 387
ʿAbbās b. ʿAbd al-Muṭṭalib (uncle of the Prophet) 403
ʿAbbās b. ʿAmr al-ʿAnawī 354, 393
ʿAbbās b. al-Zabīd 461
ʿAbbās, Shāh (Safavid) 252, 255, 291, 320–321, 327, 359, 366, 526–527
ʿAbbās Mīrzā (Safavid) 320
ʿAbbāsa, sister of Hārūn al-Rashīd 379
Abbasid(s) 74, 206, 223, 225, 230, 244, 249, 253–254, 263, 265, 269, 280, 286, 292, 301, 306, 308, 312, 319, 327, 347, 352, 357, 364, 366–368, 371, 375, 377–378, 380–383, 394, 399, 427, 429, 443, 467, 476, 482
ʿAbd Allāh (hospice) 222
ʿAbd Allāh b. ʿAbd al-Muṭṭalib (father of the Prophet) 398
ʿAbd Allāh b. ʿAlī (road stage) 365
ʿAbd Allah b. ʿAmr b. al-ʿĀṣ 430
ʿAbd Allāh b. ʿAṭā (road stage) 365
ʿAbd Allāh b. Ibāḍ 392
ʿAbd Allāh b. ʿUmar, Mt. 394
ʿAbd Allāh Yazdī, Mawlānā 234
ʿAbd-i Jān (town in Fārs) 237
ʿAbd al-Malik (Umayyad caliph) 377, 399, 437, 439, 446, 466
ʿAbd al-Masīḥ 97
ʿAbd al-Muʾmin ibn ʿAbd al-Ḥaqq 43
ʿAbd al-Muṭṭalib (grandfather of the Prophet) 386, 398
ʿAbd al-Qādir-i Gīlānī, ʿAbd al-Qādir al-Jīlānī 209, 368
ʿAbd Qays (tribe) 364, 392
ʿAbd al-Raḥmān (stage in Fārs) 241
ʿAbd al-Raḥmān al-Ḥārithī 255
ʿAbd al-Raḥmān b. Rabīʿa 327
ʿAbd al-Raḥīm al-Turkashī, Mawlānā 328, 337
ʿAbd Warqā (road stage) 366
Abdāl (village) 478
Abdāl Murād (place of excursion) 502
Abdāl Pasha 499
ʿAbdallar (road stage) 346
Ābdān-i Ganj 291
ʿAbdī Khāṣ, Lake 213
Ābdīsak Bridge 247
ʿAbdurraḥmān Ghazi 516–517
Abduti 200
Abel son of Adam 435, 450
Abgi 501
Abhar 251, 254–257, 262
Abharrūd River 262
ʿĀbid (tribe) 431
ʿĀbidūn (road stage) 356
ʿAbīrā 229
Abīward 271, 279
Abkhazians See Abaza
ʿAbla 424
Abr (or Abrak?), Mt. 418
ʿAbra-i Amīr al-Muʾminīn (in Iraq) 366
Abraha, Dhūʾl-Manār 426
Abraha b. Ṣabbāḥ (Christian king of South Arabia) 386, 426
Abraham, Prophet Abraham, Ibrāhīm 119, 268, 311, 357–358, 369, 383, 395–399, 401, 411, 416, 430, 432, 434, 438, 441, 453, 457, 465
children / sons of Abraham 385, 391, 423
descendants of Abraham 107, 200
Gate of Abraham (at al-Aqṣāʾ Mosque) 439
Scriptures of Abraham 416
Abrāhistān 236
Abrash River 455
Hīshat al-Abrash 455
Abrinos 473
Abrshahr See Hamshahra
Abrūn (island) 366
ʿAbs (tribe) 458
Abūʾl-ʿAbbās Aḥmad b. Ṭāhir (Abbasid) 383
Abū ʿAbd Allāh Khafīf, Shaikh 315
Abūʾl-ʿAlā-i Ganjī 324
Abū ʿAlī b. Marwān al-Kurdī 357
Abū ʿĀmir (Sufi) 453
Abū ʿAraf, Mt. 412
Abū ʿArīsh 385, 423
Abū ʿArna, Jabal Abū ʿArna (mountain) 363, 424
Abū Bakr b. Wāʾil 350
Abū Bakr Muẓaffar b. Yāqūt (governor of Iṣfahān) 244
Abū Bakr Ṣabbāḥī 438
Abū Bakr Sallabāf Tabrizī 314
Abū Bakr al-Ṣiddīq, Caliph 97, 414, 425, 444, 465
Abū al-Dardāʾ 443
Abū Dharʿa, Sufi shaikh 315
Abū Dulaf 258, 260
Abū Dulāma 378
Abūʾl-Fayḍ See Fayḍī-i Hindī
Abūʾl-Faḍl b. Shaykh Mubārak 192
Abūʾl-Fidāʾ, al-Malik al-Muʾayyad 42, 68, 334
Abū Ghānim (mountainous area in Iran) 229
Abū Ghubshān 397
Abū Ḥāmid al-Gharnāṭī al-Andalusī 255, 290, 315, 327
Abū Ḥanīfa, Imām-ı Aʿẓam 260, 367–368, 377, 395
Abūʾl-Ḥasan al-Kharaqānī, Shaikh 259, 332
Abū Hurayra (a companion of the Prophet) 437
Abūʾl-Ḥusayn Ḥaḍarī 368
Abū Isḥāq 344
Abū Isḥāq al-Ṣābī 373
Abū Karib (ruler of Yemen) 426
Abūʾl-Khaṭṭāb Jabalī 370
Abū Maʿbada (in Yemen) 385
Abūʾl-Majd al-Mawṣilī 42
Abū Mālik b. ʿUmrān (ruler of Yemen) 426
Abū Muḥammad Baṭṭāl See Baṭṭāl
Abū Muḥammad Riʿashī 368

Abū Muslim (leader of the Abbasid movement) 377, 467
Abū Muslim al-Dāranī 452
Abū Muslim al-Khawlānī 452
Abū Naṣr 221
Abū Numayy, Mawlānā Sharīf 403, 410
Abū Qalamūn 454
Abū'l-Qāsim Faḍl Allāh (caliph al-Muṭīʿ) 381
Abū Qays 457
Abū Qubays, Mt. 393–394, 396–397, 406
Abū Ṣadrī (road stage) 365
Abū Saʿīd (Carmathian) 392
Abū Saʿīd, Sultan (Īnjūid) 245, 267
Abū Saʿīd Bahādur Khan (Ilkhanid) 319, 347, 482
Abū Saʿīd al-Bardaʿī 321
Abū Saʿīd b. Mīrānshāh, Guragan (Timurid) 225, 320
Abū Saʿīd al-Jannābī 393
Abū Ṣīr, Church of 467
Abū Ṭāhir (Carmathian) 392
Abū Ṭāhir-i Kurdī, Shaikh 337
Abū Ṭāhir al-Salafī, Ḥāfiẓ 314
Abū Ṭāhir al-Shīrāzī 312
Abū Ṭāhir Sulaymān al-Ḥusaynī 429
Abū ʿUbayda (commander) 444, 461, 465–466
Abū Umayya b. Mughīra 399
Abū ʿUrwa (village) 421
Abū Yaʿqūb al-Buwaiṭī 368
Abū Yūsuf, Imām 368, 378, 395
al-Abwā 409
Abyār, Abyār al-ʿUlā' 422–423
Abyār Ḥamza 408, 421
Abyāt (in Arabia) 424
Abyla (mountain) 91
Abyssinia See Ethiopia
Acharbul 219
Achiq Bash (tribe and region) 331
Acıgöz 356
Acıṣu River 494
Acıṣubaşı 481
Acre, ʿAkka 430, 432–433, 441–442, 453
  Bilād ʿAkka (fortress) 451
ʿĀd (tribe) 383, 391, 408, 414, 415, 424–425
ʿĀd son of ʿAwṣ son of Arīm son of Shem (son of Noah) 391
Ādāb (book) 315
Adam 88, 118, 148, 150, 242, 369, 407, 425, 438, 450, 452, 492
  Adam's Footprint 150
Ādam Khan 216
Adana 434, 457, 461, 463–465, 469, 472, 504, 511, 512, 522
  Adana Plain 512
Aden 102–103, 145, 383–386, 388–389, 391, 406, 417–420, 423, 428
ʿAdhrā 453
ʿĀdil Khan 184
Adilcevaz 334–335, 356, 525
ʿĀdiliyya 433
ʿAdnān (ancestor of the Northern Arabs) 425
ʿAdrā (error for ʿAdhrā?) 452
Adranos 502, 504
Adria 91
Adriatic Sea 93
ʿAḍud al-Dawla Fanā Khusraw al-Daylamī (Buwayhid) 232–234, 244, 246–248, 371, 407
  ʿAḍudī Dam, Band-i ʿAḍudī 240, 262
ʿAḍud al-Dīn al-Ījī 236
ʿAḍūh 424
Aegean Sea 84, 95, 125, 483, 504
Aegyptus See Egypt
Aethiopia See Ethiopia
Afāmiyya, Fāmiya (Apameia) 430, 434–435, 457–458
  Lake Afāmiyya 432, 434, 458
al-Afḍal ʿAlī (Ayyubid) 467
Afghans, Awġāniyān, Afghanistan 187, 195, 207, 211, 215, 220
  Afghani (language) 436
Afghans (i.e., Lodi Dynasty that ruled the Delhi Sultanate, 1451–1526) 198
  Afghan conquest 202
Afghans (i.e., conquerors of Iran in 1722) 330
Aflūn (Apollo, idol) 350
Africa, Āfrīq 46, 69, 77, 84–85, 90–93, 96, 99, 104, 107–109, 112–113, 125–127, 132, 137, 181, 333, 528
  Africa Minor 108
ʿAfrīn 460
  ʿAfrīn River, Nahr ʿAfrīn 433–434, 458
Afrus 107
Afshīn 321
Ağaç Deñizi (forest) 506–508
Ağacbaşı Mountains 349
Ağaçlar (village) 352
Ağagöli 347
Ağakis 334, 341
Agenor 105
Aghbar, Mt. 428
Aghlabids (?) 463
Ağır River 346
Ağlasun 489–490
Agra 184, 187, 192–193, 197–198, 205–206
Ağraş, Ağrus 489–490
Ağrı See Arghī
Ağriboz See Euboea
Aḥādiyya 424
Āhangarān 225
Āhar 318
  Āhar River 318
Aḫi Evren 475
Aḫır summer pasture 485
Aḫircik 512
Aḫlāṭ (Ahlat), Khalāṭ 334–338, 345, 525
  Aḫlāṭ apple 335
  Lake Aḫlāṭ See Van: Lake Van
  quarter of the Aḫlāṭīs (in Cairo) 336
  ruler of Aḫlāṭ 337, 347
Aḥmad (Ilkhanid) 319
Aḥmad al-Badawī, Sayyidī 410
Aḥmad b. Ḥanbal, Aḥmad Ḥanbalī, Imām 368, 395
Aḥmad b. Khalaf, ruler of Sīstān 223
Aḥmad Kahtūb, Shaikh 182
Aḥmad Khan, Khānlar Khānı 199
Aḥmad Shah b. Sultan Muḥammad b. Muẓaffar 182, 199
Aḥmadābād 182–184, 192, 198, 205–206
Aḥmadnagar, Āmadnagar 183–184, 199, 205–206
Aḥmed, Ottoman Sultan 400, 526
Aḥmed (son of Sultan Bāyezīd II) 522
Aḥmed Bey (Ottoman Egyptian emir) 405
Aḥmed Pasha, vizier 331
Aḥnaf b. Qays 269
Aḥnaf (town), Qaṣr-i Aḥnaf 275, 278
Aḥqāf 103–104, 383–384, 391, 417, 425
al-Aḥsā See Laḥsā
Aḥsan al-taqāsīm fī maʿrifat al-aqālīm (book by al-Maqdisī) 39, 206, 232, 236–237, 240, 248, 251, 320
Ahwāz See Khūzistān
Āī (demon) 193
ʿĀ'isha (wife of the Prophet) 366, 399
ʿĀ'isha al-Bāʿūniyya 446
Ajā (mountain) 414
ʿAjā'ib al-makhlūqāt ("Wonders of Creation"—book genre) 152
ʿAjā'ib al-makhlūqāt (book by al-Qazwīnī) 39, 322, 324
ʿAjāyib-i ʿālam (book) 212
ʿAjala 424
Ajdarkhān See Astrakhan
ʿAjīn, Mt. 418
ʿAjlūn 430–431, 434, 436, 441–442
Ajmāsāt (tribe) 367
ʿAjrūd Birkesi 422
Aḳ Şemseddīn 521
Aḳ Ṣu River (in Caucasus) 330
Akarū, Mt. 217
Akbar, Jalāl al-Dīn Muḥammad (Mughal emperor) 188, 192–193, 198
Aḳbaba 506–507
Aḳbıyıḳ, Aḳbıyıḳ Baba 491–492, 503, 511–512
Akça Liman 518
Aḳçaağlan Ṣuyı 501
Aḳçaḳalʿe 340, 346, 348, 353, 481, 525
Aḳçeḥiṣār (Anadolu Hisarı) 521
Aḳçeḳoca (ghazi) 505
Aḳçeḳoyunlı 481, 512
Aḳçeşār, Aḳçeşehir 481, 497, 499–500, 510, 512
Aḳdegirmen 357, 481
Aḳediz 481
ʿAkfar 388
Akhalkalak 332–333
Akhbār al-duwal (book) 160
  Akhalkalak River 324
Aḳḥiṣār 486, 488, 505, 516
Akhiskha, Akhisqa (Akhaltsikhe) 324, 332, 343, 367, 525
Akhshāmāt tribe 367, 375
Akhsīkath 296–297, 299, 301
Akhsū 304

Akhtamar 335, 340, 525
Akhtamar Church 332
Akhūkh (tribe) 329
Mt. Akhūkh 329
Akhzāb mountains 354
ʿAkka See Acre
ʿAkkār, Bilād ʿAkkār 454–455, 457
Mt. ʿAkkār 455
Aḳsarāy 472, 474–476, 492, 522
Aḳkirman 522
Aḳköpri, Aḳköprü bridge 463, 511–512
Aḳşār, Aşḳār, Avşar 481, 490
Aḳşār Ovası, Aşḳār Ovası 344, 482
Aḳşehir 472, 475, 489, 491, 511, 520
Lake Aḳşehir 475
Aḳṣu River 479, 484, 489, 502, 512
Aḳṭaġ 345, 470, 480, 485
Aḳṭaş 497, 500, 512
Aḳtīmūr 515
al-Akūb 457
Ākūtiya 131
Aḳvīrān 511
Aḳyalar 481, 512
Aḳyazı 501, 503, 505
Aḥqāf Āl Abū ʿArīf 367
Āl ʿAlī 431
Āl Amīra 367
Āl Badr 367
Āl Barāq Dynasty (Qutlughkhanids) 231
Āl Dabāb 367
Āl Faḍl 431, 458
Āl Ḥamd 456
Āl Ḥasanayn 367
Āl Ḥawiyya 367
Āl Ḥubār 456
Āl Ḥusayn 366
Āl Khūrshīd (dynasty) 249
Āl Majīr 367
Āl Muḥammad 390, 431, 456
Āl-i Muʾminīn (in Iraq) 366
Āl Mundhir See Lakhmids
Āl Murād 431
Āl Murīd 412
Āl Murrī 458
Āl Muẓaffar (dynasty) 244
Āl Ramaḍān (Ramażānoġulları) 461, 463
Āl Raqīm 390
Āl-i Ṣāliḥ (in Iraq) 367
Āl ʿUmāra, Āl Jalandī 243
Ḥiṣn Ibn ʿUmāra, Jalandī 238, 240–241
ʿAbd Allāh b. Aḥmad al-Jalandī 244
Āl Yaghfir 387
Āl Yaḥyā 367
Āl Yasār 457
ʿAlā al-Dīn (governor of Agra) 197
ʿAlā al-Dīn (ruler of Bengal) 198
ʿAlā al-Dīn (ruler of Kashmir) 219
ʿAlā al-Dīn Aḥmad Shah 199
ʿAlāʾ al-Dīn Atsız See Khwārazmshāh
ʿAlā al-Dīn Iskandar Shah 197
ʿAlā al-Dīn Khaljī 196, 199
ʿAlā al-Dīn Mehran 200
ʿAlā al-Dīn Pasha (son of ʿOṣmān Ghazi) 516–517
ʿAlāʾ al-Dīn Kay-Qubād I (Seljuk) 344, 470–477, 480–481, 491–492, 513
ʿAlāʾ al-Dīn Kay-Qubād II (Seljuk) 477, 514–516
ʿAlāʾ al-Dīn Ṭūr ʿAlī (Aqqoyunlu) 357
Ala Ṭāgh (in Turkistan) 306
Alacaaṭa 512
Alacahān 481
Alacaḥisār 520
Alaçam 478
ʿAlam al-Dīn al-Qamarī, Shaykh 202
Alamūt 256, 261, 264
Alan (tribe) 329
Alān (island) 366
Alanī 362
Alanya 470, 472, 475, 477, 483, 469, 489, 521
Alara River 470
Alaşehir 487–488, 519
Alaṭaġ (district near Konya) 472–475
Alaṭaġ (mountain in eastern Turkey) 339, 345–346
Alaṭaġ (mountain in western Turkey) 488, 498–501
Alaṭaġı 513
ʿAlawīs, Alids (partisans of ʿAlī) 135, 287, 365, 377, 382
Alba 119
Albahar See Liber
Albāḳ 340
Albania, Albanians, Arnavud 91, 106–107, 111, 520–522
Albanian mountains 310
Albertus (Magnus) 92
Alburz, Mt. 323, 325, 328–332, 348
ʿAlem Ṭaġı 482, 507
Aleppo 125, 209, 236, 334, 356, 383, 430, 432–436, 455–462, 465, 467–469, 476, 513, 522–523
Alexander (son of Priam of Troy) 509
Alexander the Great, Alexander the Greek 87, 99, 107, 118–119, 195–196, 221, 225, 233, 237, 243, 252, 266, 270–271, 292, 311, 336, 354, 432, 458–459, 481, 513
Alexander's campaign 203
history of Alexander 204
tomb of Alexander 359
Alexandria 42, 125, 157, 524
Lighthouse of Alexandria 111, 191
Alfonsez (Afonso de Albuquerque) 142, 185, 190
Algeciras See Jazīrat al-Khaḍrā
Algeria See Mauritania
Algiers, Jazāʾir, Jazāyir-i Gharb 125, 525
Alḥarā 212
ʿAlī (b. Abī Ṭālib) 314, 358, 360, 363, 366, 370–371, 379, 389, 405–406, 445, 450, 466
ʿAlī b. Aḥmad al-Maymandī, Abū'l-Hasan 213
ʿAlī b. Ilyās, emir 227
ʿAlī b. ʿĪsā 74
ʿAlī Bey (in Iraq) 376
ʿAlī Ḥammāmı (in Iraq) 373
ʿAlī Pasha (Candarlı) 519
ʿAlī Pasha (beylerbey of Egypt) 422
ʿAlī Pasha, Serdār 527
ʿAlī Ḳuşçı, Mevlānā Shaikh 302, 522
ʿAlī al-Riḍā, Imam 245, 271, 372
ʿĀlī, Gelibolulu Muṣṭafā 44, 84, 514
Alids See ʿAlawīs
Alīghā River 184–185
ʿĀlij Kabīr and ʿĀlij Ṣaghīr 424
Alinjaq 321, 328
Alisas 96
ʿAlīshāh See Khwāja Tāj al-Dīn
ʿAlīşīr River 501–502
ʿAliyan Oġlı (in Iraq) 384
ʿĀliyya 423
Allāh Verdi Khān (of Iṣfahān) 252
Almagest See *Geography*
Alman See Germany
Alona (Alonissos) 96
Alp Arslan (Seljuk) 231, 244–245, 254, 347, 357
Alps 107
ʿAlqamī, Muʾayyad al-Dīn 382
Alqarūd 151
Alqāṣ Mirza 331
Alsin Mountain 348
Altak 334
Altar, Mt. 260
Altay Mountains 307
Altı Ḳarye 481
Altın Köprü River 360, 373
Altun Köprisi (village) 341, 361
Altun Ovası 460, 485
Altunṣuyu Bridge 352, 355
Altunṭaş 484, 503
Alwand See Arwand
Alzam (fortress) 415
Āmadnagar See Aḥmadnagar
Amalek, Amalekite(s) 397, 409, 413, 416, 424–425, 451, 456, 465
Amalfi 83
ʿAmaliyān (oil wells near Baku) 324
Amānghūzākī (Nagasaki) 130
ʿAmāqiyya 385
Amās 229
Amasa 487
Amash 300
Aʿmashiyya 424
Amaṣra 483, 497–498, 500, 521
Amasya 125, 477, 479–482, 512, 519, 522
Amasya River, Çarşamba Ṣuyı (Yeşilırmak) 478, 480
America, New India, New Indies, West Indies, New Spain 35, 42, 46, 87, 90, 96, 99, 101, 104, 112–117, 120, 122, 126–127, 132, 136, 528
historian of New India (i.e., the author of *Tārīḫ-i Hind-i Ġarbī*) 119
Amerigus (Amerigo Vespucci) 112
Āmid See Diyarbekir
Amīn, Muḥammad (Abbasid) 379
Amīn Aḥmad See al-Rāzī

Amīn al-Dīn Muḥammad b. Muẓaffar 315
Amīna (i.e., Āmina, mother of the Prophet) 409
Amīr Dam See Band-i Amīr
Amīr ʿAlam al-Dīn (tribe) 450
ʿĀmir b. ʿAbd al-Wahhāb (Tahirid) 427
ʿĀmira 387
ʿAmlīq b. Ḥāsha (Amalek) 413, 424
ʿAmmān 420, 431, 442
ʿAmmār, ʿAmmāra (tribe) 388–389
Ṣadr ʿAmmār, Ṣadr ʿAmmāra 366–367
ʿAmmāriyya, Mt. 412
ʿAmq, ʿAmq Ḥārim 433–434, 457, 460
Lake ʿAmq 458
ʿAmr b. Layth (Ṣaffārid) 223, 228, 232, 244, 272, 280
ʿAmrī 414
ʿAms 424
Amsterdam 41
Āmū Ṣuyı See Jayḥūn
Āmul 260, 263, 275, 279, 285–287, 289, 300, 325–326
Old Āmul 325
Āmūl-i Shaṭṭ See Jayḥūn
ʿĀmula, Mt. 434
Amūriyya See Ankara
ʿĀna 300, 346, 358, 367, 371, 384
ʿAnaba (tribe) 450
Anaf al-Ḥajar 454
Anafi 96
Anaḫor 518
Anamas, Mt. 474
Anamur, Maʿmūriye 469, 472, 492
Anān (mountain) 421
Anania 43
Anapaya (Anapa) 330
Anār 318
Anāra 241
Anas b. Mālik 364
Anaṭolı (province) 123, 334, 472, 477, 483, 497, 505, 511, 513, 518, 521
Anaṭolı Boġaz Ḥiṣārı (modern Anadolu Kavağı) 507
Anaṭolı Ḥiṣārı (Anadolu Hisarı in Istanbul) 507
Anatolia; and see Rūm 36, 44, 68, 84, 91, 110, 125, 347, 350, 402, 415, 418, 434–435, 494, 500, 506–508
Anavarin (Navarino) 522
Anaxagoras 52, 88
Anay 484
Anbār (in Iraq) 350, 366, 371, 375–377
Anbār (village in Kashmir) 217
Anbārān 513
Anbigana, Anbaghana (Ambon) 118, 138
Anbūn (Ambon) 140
Andalus, al-Andalus (Iberian Peninsula), Andalusia 43, 97, 115, 314
Andāmīsh 262
Andarāb, Andarāba 224–225, 279, 321
Andarāb River 239, 318
Andarāwī, island 229
Andarūd River 318
Andaz 211
Andijān 295–298, 303, 305
Andırun 461, 462
Andkhūr (Andkhūy) 270
Anquman, Anqāmān (Andaman) 142
Greater Andaman and Lesser Andaman 142
Andrānyān 290
Anfā (forest) 454
Anglia See England
Ania, Anian 179
Cape of Anian 161
Gulf of Lonza and Ania 127
Strait of Anian 110, 113, 178
Anjarūd 256
Anjās See Abaza
Ankara, Engüri, Engüriyye, ʿAmūriyya 357, 379, 481–484, 492–496, 500, 512–513, 518, 520–521
battle of Ankara 520
Yörükān of Ankara 493
Anqūja, ʿAnqūja, Almūja 151
Anquman See Andaman
Ansāb (book by al-Samʿānī) 42
Anṣārī, Khwāja ʿAbd Allāh, Shaikh al-Islām 267, 284
Antakya, Anṭākiya (Antioch) 125, 325, 430, 433–434, 457–458, 465–466, 476, 513, 524
Lake Anṭākiya 433–434
Antalya 469, 470, 472, 483–484, 488–491, 512
Anṭalyās (fortress) 451
Nahr Anṭalyās (river) 453
Anṭaqyūs (Antigonus) 458–459
ʿAntar Canal 363, 365
ʿAntar's Stable (Isṭabl ʿAntar) 423
Anṭarṭūs, Ṭarṭūs (Tortosa) 430, 454–455, 466
Antimeroe 70
Antilles 100
Antissa 96
Antonius (Antonio Galvão) 132
Antonius Mora 132
Antonius Debreyu (António de Abreu) 132
Antwerp 112
Anūshirwān, Nūshirwān, Kisrā Anūshīrwān 161, 238, 243, 274, 286, 296, 322–323, 335, 362, 370, 426, 513
Apollonius 476
Aq Band, Ravine 319
Aq Taq 305
ʿAqaba, ʿAqabat Ayla 403, 411
ʿAqabat Qābūn al-Rumān 450
ʿAqabat Shayṭān 424
ʿAqabat al-Suwayq 421
ʿAqabat Waddān 423
ʿAqār, ʿAqāra 363, 365–366
ʿAqāra Canal 365
ʿAqarqūf, Tall ʿAqarqūb 369, 375
ʿAqarqūf Canal 346
ʿAqd 241
ʿAqīl b. Abī Ṭālib 458
ʿAqīq 423
Biʾr ʿAqīq 407
Wādī ʿAqīq 407
Aqqoyunlu dynasty / sultans 314, 320, 337, 347, 352, 361, 366, 383
Aqqoyunlu clan of the Turkmen 357
ʿAqra 373
ʿAqrabāʾ 452ʿ
Aqrabāʾ Canal 433
Aqsunqur Atabegs 352
Aquitane 90, 93
Arab, Arabs 39, 43, 84, 102, 118, 134–135, 143, 146–147, 156, 188, 201, 208, 228–230, 232–233, 250, 354, 358, 361, 373, 377, 386, 394, 397, 401, 415–416, 419, 425, 442, 451–452, 455–456
ʿArab ʿĀriba and ʿArab Mutaʿarriba 425
ʿArab al-Jabal (Mountain Arabs) 452
Arab amirs / chiefs 350, 456
Arab astrologers 103
Arab expression 384
Arab family 353
Arab Iraq See Iraq
Arab kings 357
Arab month 104
Arab pilgrims 386
Arab shaikhs 524
Arab traders 141
Arab tribe(s) / clans 156, 352, 357, 376, 383, 396, 431, 451, 458, 465
Arabia, Arabian Peninsula 36, 69, 90–91, 102–104, 108, 124, 144, 241, 305, 363, 374, 383–384, 393, 416, 420, 424, 429–430, 524
Arabia Troglodytica 108
Arabian horses 172
Arabian islands 103
Arabian Sea, Mare Arabicum 90, 99, 108, 384
Arabian Desert 125
Arabian style (of building) 449
Arabic (language) 39–41, 43, 84, 182, 198, 239, 287, 337, 355–356, 375, 378, 396, 417, 425, 436
Arabic letters, Arabic script 201
Arabic title 311
annals of the Arabs 371
bedouin Arabs 350, 355, 373, 377
Fairs of the Arabs 419
Muslim Arabs 363
Shaṭṭ al-ʿArab See Tigris
ʿArab Cāmiʿi 466–467
ʿArabkīr, Arapgir 477–478
Araç River 499–500
Aracan 181, 189, 194
ʿArafāt 359, 380, 397, 401–404, 407, 419
Aragon 115
ʿAraj 393
Mt. ʿAraj 435
Araḳ, Mt. 498–499
Aral Sea, Lake Khwārazm, Qitay Lake 84, 290, 294, 299, 307, 309
Ārām Shah 196

ʿArama 412
Arapgir see ʿArabkīr
ʿArāqib al-Bughla 422
Aras (town) 322–323
Aras River (Araxes) 316–321, 323–325, 332, 339, 345
Arasbār 324
Arash 326
Arāzīqā 140
Arbījan 295
ʿArbīl 449
Arch of Chosroes 111, 313
ʿArcha 366
Archimedes 59
Archipelago, Archipels, the Islands 91, 105, 125
Archipelagus (island in the Eastern Ocean) 90
Arḍ-ı Surkh 374
Ardabīl 263, 288, 314–317, 319–320, 323, 325–327, 339
Ardabīl campaign 317
Ardabīl River 318
Ardahan, Ḳara Ardahan 324, 332
Ardalān 358
Ardanuç 332–333
Ardashān (castle) 255
Ardashīr son of Darius (Xerxes) 91
Ardashīr b. Bābak, Ardashīr Bābakān (Sasanian) 227–228, 232–234, 240, 243–244, 246, 256, 269, 272, 288, 327, 351, 461
Ardashīr Khūra 232–233, 236
valley of Ardashīr 240
Ardashīr son of Shīrūya 243
Ardistān 257, 262
fire temple in Ardistān 257
ʿArfa 454–455
ʿArfa River 455
Arfakhshād / Arfakhshadh (son of Shem) 376, 424–425
Arġanutözi 481
Argenteus River, Rio de (la) Plata 114
Arghī, Mt., Ağrı Ṭağı (Ararat) 321, 324, 332, 336, 346
Arghun Khan (Ilkhanid) 254, 256, 274, 282–283, 319, 346, 477, 482
Argonauts 122, 137
Arian 110
ʿArīḍ 412, 454
Arīḥā See Jericho
ʿArim 376
ʿArīsh 97, 108, 383, 422, 437, 465
Aristarchus 59
Aristotle 47, 51, 59, 64, 88, 92–96, 98, 105
ʿArj, ʿArjistān (mountains) 292, 325
ʿArja 367
ʿArjamūsh 454
Arjāq 318
Arjīs 319
Arjīsh See Erciş
Arḳaludya (fortress) 463
Arkāmū 215
Arḳandı Ḳaya 504
ʿArkaz 300
Arḳıd Ḫān 511
Arkūb 210
Arman b. Layṭī b. Yūnān 334
Armaniyān, Gökçe Nīl 319
Karīwa-i Armaniyān, Ravine of Gökçe Nīl 318
Armenia, Armenians 36, 110, 302, 307, 338, 343, 347, 349–350, 355, 430, 440, 464, 476–477, 513, 525
Greater Armenia 125, 343
Lesser Armenia 461, 471, 513
Armenian (language) 356, 375, 436
Armenian authors 307
Armenian church 332, 336
Armenian dhimmis, tribute-paying Armenians 337, 355
Armenian infidels 461, 481
king / ruler of Armenia, Armenian king(s) 312, 335, 344, 463
mountains of Armenia 312, 374
Arminiyya (Van province) 312, 319–321, 334, 337, 339
Greater Arminiyya 334
Lesser Arminiyya 334
Armudlu 481
Armūk 339
ʿArna 363, 365, 435
ʿArnaʿāt 367
Arnheim 41
Arnavud See Albania
Arnūn 430
Arotia 69
Arpa Çayı, Arpa River 324, 346, 356
Arpa Khan (Ilkhanid) 245, 319
Arrajān 238–240, 244, 248–249
Arrān son of Japheth 320
Arrān, al-Rān, Rawān (Revān, Erevan) 312, 319–327, 332, 334, 336, 343, 345–346, 361, 367, 525–527
governor of Arrān 327
Rawān campaign 324, 338–339
Arsal, Mt. 277
Arslān b. Yaʿqūb, Shaikh, ʿAyn al-Awliyāʾ 453
Arslan Giray (fort) 330
Arsūf 430, 432–433, 437
Arsuḳ, Mt. 469
Arsūra 222
Artlās 305
Artuḳābād 477, 481
Artuḳābād plain 481, 512–513
Artuqid dynasty 467
Ārū River 216
ʿArūḍ 384
ʿArūs (fortress) 386, 429
ʿĀrūtān River 150
Arva 507
Arva River 507
Arwāna, Mt. 375
Arwand, Alwand, Mt. 248, 258–260, 262, 374
Arwand River 374
Aryāṭ b. Aḍḥam 426
Arza 443
Arzan al-Rūm See Erzurum
Asad Khan 185
Asadābād 259–260, 262–263, 278–279, 375
Aṣāf (idol) 416
Aṣāfa 424
Āṣaf Khan (Mughal vizier) 198
Asal 249
Asān (idol) 398
Asbūzī (village) 462
Ascalon 430, 432, 437
Asfara 297
Asfara, Mt. 300
Asfī (Jīruft) 227
Aṣḥāb See Companions of the Prophet
Aṣḥāb al-Kahf (Seven Sleepers) 464
ʿAshār Canal 365
al-Ashʿarī, Abū Mūsā 247, 249, 263
Ashʿarīs 384, 426–427
Ashkahrān, Mt. 260
Ashkamad Mountains 319
Ashkānīs, Ashkāniyān (Parthians) 243, 352
Ashkmish 224–225
Ashknīr, Mt. 316
Ashnāghar 193, 202, 206, 209, 211–215
al-Ashraf Khalīl (Mamluk Sultan) 400, 443
Ashrāfiyya 433
ʿĀṣī of Ḳaraaġaç (Acıpayam) 485, 489–491, 512
ʿĀṣī River, Orontes 209, 432–434
Asia 36, 46, 84–85, 90, 104, 108, 110–112, 124, 126–127, 132, 137, 152, 161, 176, 181, 231, 307, 310, 328, 333, 343, 483, 528
Asia Major 110
Asia Minor, Anadolu 110, 123, 309, 483
Asia wife of Japheth 110
Asius son of Lidus 110
ʿĀşıḳ, Meḥmed b. ʿÖmer 43, 325, 511
ʿĀshiq and Maʿshūq (near Baghdad) 372, 374
Āsitāna 279
Aşḳāle 346
Aşḳār See Aḳṣār
ʿAskar Mukram, ʿAskar 246–249
Aspās 300
ʿAṣrūn 452
Assur 350, 352
Asṭa 262
Āstalisht 278
Astanar 489
Astarābād 260, 263, 286
Astarghanj 212
Astrakhan, Astraqan, Qara-azhdahān 308–309, 326, 329, 525
Astrakhan road 329
Aswad River 374
Aswad b. ʿAffār (Judaysite) 413
Aswar River 294
Āsyākard 279
ʿAṭā Malik (Juwaynī) 274
Atabeg(s) 231–232, 236–237, 244–245, 248–249, 255, 321, 327, 352

Aṭa (Adabazarı) 501, 503, 512
Ataḫ 207, 209
Ataḳ 353, 355
Aṭala 485–486, 488
Atāq River 373
Atārib 458
Atārkīd, Āthārkīt 300
ʿAṭāyī Efendi, Ḫoca 488
ʿAṭfa 454
Athālib mountains 408
Āthār al-bilād wa akhbār al-ʿibād (book by al-Qazwīnī) 39, 240, 292
Athens, Athenian 112, 520
Atina River 349
ʿAtīḳ Meḥmed Pasha, vizier 464
Atlantis 112
Athos 96
Atil (Volga) 309, 325
Atlantic Ocean See Western Sea
Atlas (mountains) 90
 Atlas coast 99
 Atlas regions 93
 Sea of Atlas See Western Sea
Atlas, Atlas Minor of Hondius 35, 39, 41–43, 47, 69–70, 72–74, 77, 84–85, 98, 100, 104–107, 110–115, 125–127, 132, 134–136, 138–142, 146–149, 152, 154, 156–159, 161–162, 171, 173, 175–176, 179–181, 183, 185–187, 189–191
Atlas Major (book by Mercator and Hondius) 39, 40–41
ʿAtq 424
Aṭrāz See Ṭarāz
Atūd 423
Augustus 99
Auiadat (Lorenzo's source of information) 142
Ausonia 91
Austria 106, 524, 528
 Austrian soldier 107
Avāng 361
Averroes 94
Avicenna 98, 128, 295
Avşar 489–491
Avşar, Avşariye (Turkmen clan) 457
Avżaḥu'l-mesālik ilā maʿrifeti'l-büldān ve'l-memālik (book by Sipāhizāde Meḥmed Efendi) 42
Āwa 261–262
al-Aʿwaj River, Nahr al-Aʿwaj 433, 440
ʿAwālī 413
Awardak 214
al-ʿAwāṣim (fortresses) 460
ʿAwda-i Sadīr 413
ʿAwja 424
 ʿAwjā River 432–433
ʿAwjān mountain 414
al-Awla River 451
ʿAwm 222
Awnī 222
ʿAwrat (island, rather Arwād) 455
Awrīqān 194
Aws (tribe) 416, 426
Awṣāb 417
 Mt. Awṣāb 418
Awsh 296, 298, 300
 Awsh River 296
Awṭās 407
al-Awzāʿī (jurist) 454
Ayāfath 424
Ayandon 495, 497
ʿAyār Çayı 347
ʿAyārān 346
ʿAyārbaşı 347
Ayās 434, 457, 464
 straits of Ayās 470
 Ayās-Barandī River 434
Ayaş 492–493, 513
 Ayaş summer pasture 481
Ayaṣluġ 111, 487–488, 510–511
Ayasofya (in Istanbul) 470, 522
Ayāz (slave of Maḥmūd of Ghazna) 228, 231
Ayazmend 503–504, 512, 518
Aybak, al-Malik al-Muʿizz (Mamluk Sultan) 467
Aydām 409, 423
Aydın 483–488, 510, 512, 516, 518, 520, 524
Aydın Ṭağları 487
Aydınlı Meḥmed Pasha (governor of Yemen) 384
Aydos 516–517
 Mt. Aydos 506
ʿAyġara River 475
al-Ayka 409
Ayla (Eilat) 384, 396, 409, 411, 430, 435
 Wādī Ayla 383
Aymaḥiṣārı, Ḥiṣār-ı Ayma 497
ʿAyn (near Bitlis) 339
ʿAyn ʿAlī Efendi 334
ʿAyn al-Ghazāl (Gazelle Spring) 455
ʿAyn Ḥanīf 422
ʿAyn-i Ḥavz 353
ʿAyn al-Jama 408
ʿAyn al-Jard 454
ʿAyn al-Jarr 453
ʿAyn Silwān (Spring of Siloam) 430, 438, 440
ʿAyn al-Tamr 414
ʿAyn Warda 358
ʿAyn Zarba 435, 463–464
ʿAyniyya 412
ʿAynṭāb 356, 431, 457, 461–462, 469, 524
 ʿAynṭāb River 356
ʿAytā 450
Aytemür (Mughal emir) 196
Ayyubids, Ayyubid Kurds 42, 352–353, 384, 386, 427, 467–468
Āzādmard 236
Āzādwār 274
ʿAzar 433
Azāriqa (Kharijites) 366
ʿAzāz 356, 457, 460–461
Azbūr (castle) 324
Azd (tribe) 389, 392
Azerbaijan, Ādharbāygān 84, 125, 249, 251, 260–261, 284, 287, 302, 312–321, 323–325, 327, 334–335, 337, 359, 361, 377
Azghūr castle 324, 333
al-ʿAzīz billāh (Fatimid ruler) 43
ʿAzīz al-Mulk 199
ʿAzīzī (book by al-Muhallabī) 43, 206
Azkashiya 304
Azlam 383, 410, 423
Azores 70
Azov 309, 329, 521
 Sea of Azov, Lake Maiotis 84, 91–92, 95–96, 105, 110
Azraḳ River (Gökṣu) 463
Azraq (in Arabia) 384, 414–415, 420–421
Azraq (in Khūrasan) 290
Azraq Jāzū, pass 359
Azriʿāt 420, 431, 452

Baal, Baʿal (idol) 125, 453
Baalbek 125, 431, 434, 443, 451, 453–454, 524
 Mt. Baalbek 433
al-Bāb (town) 458
Bāb al-Abwāb See Darband
Bāb al-Farādīs (in Damascus) 443, 467
Bāb Jibrāʾīl 458
Bāb al-Lān 323, 325
Bāb al-Mandab 100, 384–385
Bāb Nirāʿā 431
Bāb al-Zuwayla (in Cairo) 468
Bābā Bayrām 229
Bābā Ḥasan River 191, 193
Baba İlyās 517
Baba Reşīd 488
Bābur Mīrzā, Ẓahīr al-Dīn Muḥammad (Mughal emperor) 198, 212, 284, 292, 296, 302
Babylon, Babylonia, Babylonians, Bābil, Balūkh 110–111, 124–125, 336, 352, 357, 368–370, 376, 425, 465
Baçan 345
Bacchus (commander) 183
Bacı 492–493
Badakhshān 204, 210, 212, 215, 223–225, 269, 279, 291, 296, 298, 300, 303, 306
 kings / shahs of Badakhshān 195, 225
 Badakhshān River 224
 Badakhshān road 300
Badal 218
Bādān 355
Baʿdan 388
Badar 206
Badargāh 216
Badayāl, Mt. 218
Bādeklü 347
Bādghīs 268, 278
 Bādghīs mountains 278
Badīʿ al-Zamān Hamadānī 258
Badīl b. Waraqā 258
Bādir 368
Bādiyya 384

Badjūy 277
Badlīs See Bitlis
Badna 469
Badr 409, 410, 421–423
Badr al-Dīn Luʾluʾ 352
Badr Ḥunayn 404
Badra 372, 366
Badrāy 372
Bādrund 218
Badsangar, Mt. 218
  Badsangar Road 218
Badū 218
Badra 333, 366, 372
  Badra Ravine 324, 333
Badrān 251
Baf (Paphos in Cyprus) 469–470, 510
Bāf (in Iraq) 359
Bāfd, Baft 227, 235
Bafra 478, 480
Bāgh-i Sīrjān 227
Baghashūrā 275
Baghchūr 212
Baghdad 124–125, 223, 244, 247, 250–253, 259–260, 263, 288, 311, 334, 351–352, 359–360, 362–363, 365–375, 378–383, 394, 420, 424, 427, 444, 451, 470, 476, 480, 492, 522, 525, 526
  Baghdad campaigns 250
  Baghdad Patriarchate 187
  conquest of Baghdad 291
  Old Baghdad 372–373
  pilgrims from Baghdad 415
Baghlān 279
Baghrās 433, 457, 460, 464
  Baghrās Beli 460, 464
Bahādur, Amīr Dānishmand 283
Bahādur Khan (vizier) 199
Bahādur Khan, Bahādur Shāh, Sultan Muḥammad 101, 198–199
Bahāʾī Efendi, Shaikh al-Islam 49
Bāḥakath 300
Bahār (in Iran) 229
Bahār (in Kurdistan) 362
Bahār (fire temple of Balkh) 269
Bahasnā, Bahnasa 430, 450, 457
Baḥīra (in Arabia) 419
Bāḥkārs 375
Bahlī (tribe) 373
Bahlūlpūr 218
Bahman (ancient Persian king) 234, 236, 243, 252, 257, 266, 274
Bahmanids, Bahmanid dynasty, Bahmanid Sultans 193, 199–200
Bahnasa See Bahasnā
Bahnasiyya 448
al-Baḥr al-zakhkhār (book by Sharaf al-Dīn Yaḥyā al-Zaydī) 427
Bahrā 203
Baḥrābād 274
Bahrain 240–241, 352, 366, 384, 392–393, 415, 420
Bahraj 230
Bahrām b. Shābūr (Sasanian) 362
Bahrām Chūbīn (Sasanian) 243, 327, 361
Bahrām Shāh (Ghaznavid) 225
Bahrām Shah, son of Iletmish 196
Baḥrān 427
Bahrān 375
Bahrāw 203
Bahrūz 367
Baḥṣa 433
Baḥriye (book by Pīrī Reʾīs) 42, 82, 99, 101, 104, 122
Bāhūrū 219
Bājarwān 319, 321
  Bājarwān River 322
Bajūr 212, 213, 215
Bājwān (tribe), Bājwānlu 350, 352
Bākharz 272, 278
Bakās 457–458
Bakhtāl (tribe; error for Bakhtākin?) 307
Bakhtākin, Bakhtākiyān (tribe) 307
Bākhtarzamīn 211
Bakhtigān 375
  Lake Bakhtigān 240
Bakır 501, 503–504
Bakīr b. ʿAbd Allāh 327
Baḳlañ 484–486
Baknazīl 218
Bakr b. Wāʾil 352, 357
Baḳrāṣlı 463
al-Bakrī, Abū ʿUbayd 43
Baku 142, 282, 322–323, 325–326
  Baku Sea See Caspian
Bālā (in India) 207, 209
Bālā (in Iraq) 367
Balaban 341, 360–361
Balabancıḳ 503–504, 512
Balad İblīs 412
Balad al-Khaṭīb 351
Balad Qaṣra 458
Balāl, Mt. 216
Balar, Mt. 218
Balāsāghūn 304
Balaṭ (near Izmir) 510, 512
Balaṭ (belonging to Aydın) 486
Balāṭunus 454–455
Baleares, Balearics 91, 106
Balı Ḫoca 514
Balıḳ Ḥiṣārı 521
Balıkesri 503, 505, 512, 518, 520
Bālīn 222
Bālis 350, 358, 384, 430–431, 457
Balisköprüsi 511
Balḳās 359
Balkh 125, 212–214, 223–225, 234, 265–266, 269–275, 278–283, 294, 297–301, 306, 513
  Balkh River, Dahās, Shāhī 269–269, 298
  Balkhi crows 179
al-Balkhī, Aḥmad b. Sahl 43
Balniyās (vizier of Dhū Nūwās) 390
Balqāʾ 384, 416, 420, 430–431, 434, 441–443, 465
Baltic, German Sea, Sea of the Goths, Varangian Sea 65, 84, 88, 90, 93, 99
Baluch 208, 229
Balūkh See Babylon
Bālūs (king of Assur) 350, 352
Balya 504, 509
Balyanbolı 487–488
Bām 222, 227, 229
Bāmgān 212
Bamiyan 212, 214, 224, 226, 269, 279
Bamkhū 328
Bamrū 278
Ban 216, 217
Bāna 352
Bānāt (in Arabia) 415
Banaz 484–486
Banbar 218
Band-i ʿAḍudī 262
Band-i Amīr, Amīr Dam 237, 246
Band-i Māhī 319, 325, 335, 347, 356
Bandan 140
Bandar Ibrāhīmī 228, 230
Bandar Mājūr 248
Bandar Ray (Ray Shahr) 229
Bandarjūlī 205
Bāndarwār 216
Bandistān 375
Bāndū 128
Bangāla 187, 190
Banī ʿAbbās 425
Banī ʿAbd al-Muṭṭalib 425
Banī ʿAbd al-Wādd 399
Banī ʿAbd Manāf 399, 425
Banī ʿAdār 429
Banī ʿAdna b. Kaʿb 399
Banīʾl-Ahīl, Bilād Banīʾl-Ahīl 387–388, 423
Banī ʿĀmir 409
Banī ʿAmr 422
Banī ʿAnaza 408
Banī ʿAqaba 431
Banī Artuq b. Eksük 357
Banī Asad 357, 399, 415
Banī Asʿad 406
Banī ʿAṭiyya 431
Banī Awālīd 429
Banīʾl-Azd 465
Banīʾl-Azraq 429
Banī Bakr 357
Banī al-Ḍaḥḥāk 387
Banī Dhūʾl-Kilāʿ 416
Banī Fāris 363
Banī Ghassān 465
Banī Ḥakīm 367
Banī Ḥanīfa 413–414
Banīʾl-Ḥārith 387
Banī Hāshim 425, 429, 456
Banī Ḥaymūr 450
Banī Hilāl 451
Banī Ḥimyar 389
Banī Hūd 425
Banī Ḥujr 386
Banī Ḥusayn 410
Banī Huzayl 416
Banī ʿImrān (fortress) 429
Banī Irbid 451

Banī Jurm b. Thaʿlab 431
Banī Jūʿ 434
Banī Kahlān 389
Banī Kalb See Kalbiyya
Banī Kanʿān (Canaanites) 430, 443, 451
Banī Khālid 431
Banī Kharqīl 452
Banīʾl-Khayyāṭ, Bilād Banīʾl-Khayyāt 387–388, 423
Banī Kilāb 454, 457
Banī Kināna 386, 416, 425, 451
Banī Kulayb 416
Banī Lakhm See Lakhmids
Banī Lām 409, 412, 422, 431
Banī al-Madāris 449
Banī Mahdī 431
Banī Makhdūm 399
Banī Makhzūm 425
Banī Mālik (district) 367
Banī Marī b. Rabīʿa b. Faḍl 431
Banī Marzūq 428
Banī Maʿn See Maʿnoğlı
Banī Mazyad 371
Banī Mudlij 416
Banī Muḥammad 429
Banī Munqidh 456
Banī Murra 451
Banī Muzḥij 431
Banī Naḍīr 408
Banī Nafīf 416
Banīʾl-Nakhīl 410
Banīʾl-Nimr 389
Banī Nizār b. Maʿadd b. ʿAdnān 350
Banī Numayr 450
Banī Qaḥṭān 425
Banī Qatāda 429
Banī Qays 357
Banī Qurayẓa 408
Banī Rabīʿa See Rabīʿa
Banī Rīsha 358, 431
Banī Sālim 422
Banī Samak 358, 460
Banī Ṣaqr 406
Banīʾl-Shadīd 429
Banī Shayba 394
Banī Shihāb 450
Banī Sulaymān b. Ḥasan al-Muntanī 429
Banī Taghlib 352
Banī Tamīm 412
Banī Ṭamyān 375
Banī Ṭarpūsh 431
Banī Thaʿlaba 357
Banī Ṭurābī 431
Banī ʿUbāda 355
Banī ʿUbayd 409
Banī Ukhayḍir 429
Banī ʿUlwān 442
Banī Umayya 425
Banī ʿUqba 409
Banī Zuhra 399
Baniari 200
Bāniyya 455
Bānūāūm [Banoum] 129, 131
Bānyās 430–432, 434–435, 451
  Bānyās Canal 433, 443, 445, 448
  Bānyās River 455, 460
  Lake Bānyās 432–433
Baqrāj 333
Bāqūya 375
Bār (in Arabia) 414
Baradā 432–433
  Baradā River 443, 448, 450, 453
Barakāt (Sharīf of Mecca) 404
Barakchīn 277
Barani summer pasture 513
Barāra River 233
Barās (fortress) 429
Baʿrashūr 278
Barāzrū River 323, 325
Barbadaq 318
Bārbānd River 184
Barca 125
Barcelona 93, 116
Barçınlu 491
Bardaʿ, Bardaʿa 321–322, 326, 334, 490
Bardaḳçı 512
Bardaḳlı, Bardaḳlu 491, 511
Bardasīr 277
Bardāwūd 352
Bardshīr See Gawāshīr
Bardūn 249
Barents, Guillaume (Willem) 120–121
Bargiri 334–335, 340
al-Bārid, Nahr al-Bārid (river) 433, 455
Bārīla 219
Barīma 410
Bārīn 456–457
Bariyya 431
Barjīn Kazīn 161
Bark 423
Barka Dāʿa 433
Barḳuṭan 516
Barla 489–490
Barlās River 321
Barma See Pegu
Barmakids 269, 378, 435, 448
  ʿAbd al-ʿAzīz Barmak 378
  Faḍl b. Yaḥyā al-Barmakī 269, 304, 378
  ʿImrān b. Mūsā al-Barmakī 208
  Jaʿfar al-Barmakī 269, 378–379
  Khālid b. Jaʿfar al-Barmakī 377–378
  Mūsā al-Barmakī 378
  Yaḥyā al-Barmakī 269, 378–379
Barmakiyya (fortress) 414
Bārnahār 217
Barnal 217
  Barnal Road 217
al-Barqā 421
Barqaʿīd 351
Barqūq, al-Ẓāhir (Mamluk sultan) 468
Barra 261
Barsājān 304
Barsarāy 217
Barsbāy, al-Malik al-Ashraf (Mamluk Sultan) 406
Barsbāy, Amīr 428
Bārsīn 262
Bārsinkath 298
Barsīr 278
Barshawar 211
Barsin 481
Bārṭān 424
Barṭās 326
Bartın 497, 499–500
Bārū 230
Barudiḳa 481
Barūm 391
Barut 218
Barwaj (Bharuch, Broach) 101, 182–184, 205–206
Barwand (ruler of Bitlis) 337
Barza 443, 452
Barzān 373
Barzand, Warzand 317, 319, 321, 326
Barzilal 218
Basā 239, 241
Bāsān, Bāsiyān 248–249
Basanū 216
Bāsarūz River 222
Basātara 218
Başdolab 375
Başgelembe, Başgelemte 486, 505, 512
Başḥān 347
Bashāra, Bilād Bashāra 441–442, 451
Basharī 375
  Basharī River 373
Bashghurd (Bashkurt) Tatars 309
Bashhar 412
Bāshqara 373
Basht-ham 241
Bāsīn 268
Baskath 298
Bāskī (fortress) 361
Başnī 248–249
Basra 42, 101, 125, 238, 240–241, 244, 246–247, 249, 334, 346, 363–366, 368, 370, 373–375, 384, 392–394, 407, 413, 420, 424, 431, 457, 525
  Basra Canal(s) 365
Başriyya 455
Bast 277
Bastara 222
Baştekye 475
Bastian (Sebastian), Captain 118
Bāt River 184
Baṭāʾiḥ (Marshes) 368, 365, 373
Bathīna 452
Bathna, Ḥāṣir al-Bathna 410
Bāṭinīs See Ismaʿīlīs
Batman 354
  Batman River 356
Baṭn 393
Baṭn ʿAwna 407
Baṭn Kibrīt 423
Baṭn Murr 406, 421
Baṭn Nakhl 406, 422
Bāṭnā 366
Bātrek (tribe) 345
Batrūn, al-Batrūn 451, 454

Baṭṭāl Ghazi, Jaʿfar, Seydī 462, 476, 482, 490–491
Batum 347, 349
Batūt 216
al-Bāʿūtha 442
Bavaria, Bavarians 41, 107
Bawān b. Īrān b. Aswad b. Sām (eponym of Shiʿb Bawān) 237
Bāward 279
Bawāṭin 384, 415
Bawāzīj 351
Bāwīl 318
Bāwīlrūd River 318
Bāwlā 202
Bāyān 340–341
Bayān (fortress), Bayān Şenbū 361
Bāyās (Payas) 464
Bayat (Turkmen) 457
Bayat (in Turkey) 511
Bayāt (in Iraq) 241, 367, 372, 457
Baybars, al-Malik al-Ẓāhir / Malik Ṭāhir (Mamluk sultan) 383, 403, 439, 442, 448, 455
Bayberd, Bayburd, Bayburt 343–344, 349, 472, 524
Bayḍā 235, 239, 241, 410, 423
al-Bayḍāwī, Qāḍī 312, 327
Baydu (Ilkhanid) 319
Bāyezīd I, Yıldırım, Ottoman Sultan 302, 344, 483, 498, 502, 517, 519, 520
Bāyezīd II, Ottoman Sultan 337, 403, 480, 522
Garden of Sultan Bāyezīd 484
Bāyezīd (fortress and district) 324, 334, 338, 343, 345
Bāyezīd Plain 346
Bāyezīd-i Zemen 516
Bayhaq 274
Bayındır 487, 512
Bayindiriya 347, 357
Baykand 293–294, 299–301
Baylaqān 318–319, 321, 326, 334
Baymand 229
Bayn al-ʿAmmayn 428
Bayn al-ʿIzz, fortress of 429
Bayn al-Nahrayn (near Damascus) 448
Bayn al-Qaṣrayn (in Cairo) 468
Bayraḳlubāzārı 495
Bayram Ḫoca 403
Bayram Paşa (in Adana mountains) 463
Bayram Pasha (grand vizier d. 1638) 473, 479
Bayram Pasha (grand vizier, d. 1657) 512
Ḫān of Bayram Paşa 511
Bayramī cloth 186
Bayramīç 501
Bayramlu (near Izmir) 510
Bayrāmlu 345
Bayramlu-Ḳaraḳoyunlu 487
Baysān 430, 432, 441–442
Baysh (in Yemen) 409, 423
Bayt al-Faqīh (in Arabia) 385, 388, 415, 417, 423
Bayt Jibrīl 438, 465
Bayt Jibrin 430
Bayt Lihyā wa'l-ʿInāya 450
Bayt Maqdis See Jerusalem
Bayt Sābir 451
Bayt Taymā 451
Bayta Nā'im 433
Baʿẓ-ı Geldigelen 480
Bāzān (Sasanian) 426, 451
Bāzān (spring) 402
Bāzar Mountains 229
Bāzār-i Sanlīl 249
Bāzārcıḳ 512
Bāzārköyi 503, 505
Bāzārṣuyı 344, 505
Bazda 294
Bazdān 368
Bāzīrānī 373
Bāzyān 352
Bebek 83
Becanus 105
Bedr al-Dīn b. Ḳāḍī Samāvnā 520
Behbeh 359
Beherdī 353
Behisnī, Behisnā; see also Bahasnā 461–463, 524
Behrām Pasha (governor of Yemen) 429
Behrāmşāh 344, 477
Behremaz 357
Beirut 430–431, 435–436, 453–454
Beirut River 433
Bekāroġlı 481
Bektashi dervishes 491
Belcekaġaç 512
Belgrade 520–521, 524–525, 528
Belücek 512
Belvīrān 472, 474
Bencīn 352
Bend-i Shamāmak 352
Bendalya 470
Benderkili (Eregli) 497, 499
Bengal, Bengalese, Bengalis 102–103, 145, 181, 187–189, 191, 193–195, 198, 203
Bay of Bengal, Ganges Bay 90, 188, 190, 203
governor of Bengal 188, 197
kings / rulers of Bengal, Bengal sultanate 194–195, 198
Bengi 486
Benjamin (Israelite) 437, 440
Benvīd (Kurdish tribe) 355
Beramköy 474
Berāsī clan 355
Berber people 426
Berbera 102
Berberia 108
Berdaniye 470
Berend 359
Bergama 483, 501, 503–504, 512, 518
Bergen 69
Berġos 512
Berke Khan 282, 306
Bersbert 464
Bersī 355
Bertius, Petrus 41
Besbycum 96
Besh Barmaq Mountains 261, 318
Beş Depe, Beştepe 330, 356
Beşīrī 353–354
Besyān, Besyāniyān (tribe) 345–346
Bethlehem 440
Betis (river, Guadalquivir) 93
Beyāẓiye 457
Beybāzārı 493, 501, 513
Beygöl 346
Beyḳoz 507
Beylik Çayır 512
Beyşehri 472, 474–475, 489
Lake Beyşehri 475
Bhat River 218
Bībashar 373
Bibipur 208–209
Bīcān Dādyān 331
Bīd-i surkh 375
Bīdahān 238
Bīdar 184, 194, 199
Biġa 483, 504–505, 509
Biġadiç 504
Bih Dah 238
Bihār (in India) 197
Bihār (in Arabia) 429
Bihrā 191
Bījānagar (Vijayanagar) 183–185, 194–195, 199, 205
sultans of Bījānagar 195
Bijanoġlı Süleymān Bey 336, 341
Bijapur 184
Bījāwar 213
Bīl 359
Bilāl, Mt. 216
Bīlān 465
Bilānī 355
Bilecik 491–492, 503–504, 513–515, 517, 519
Bilqīs (Queen of Sheba) 387, 389–390, 510
Bıñarbaşı 481
Bingāla See Bengal
Bingöl summer pasture 324, 338, 344–346
Binkath 295
Biñlik meadow 324
Bīnī 229–230
Biqāʿ Valley, al-Biqāʿ, Biqāʿ al-ʿAzīz, Biqāʿ-i ʿAzīzī 431–433, 435, 443, 450–451, 453
Lake Biqāʿ 432
Bi'r al-ʿAbd 454, 465
Bi'r ʿAqīq 407
Bi'r ʿArab 411
Bi'r Barhūt 389
Bi'r Bāy 406
Bi'r al-Duwaydār 465
Bi'r Jubayr 394
Bi'r Maʿūna 406
Bi'r Muʿaṭṭala 383, 388, 389
Bi'r al-Qurwā 423
Bi'r Rūma 407
Bi'r al-Salāḥ 424

Bi'r ʿUrwa 407
Bi'r Zaybak 437
Bi'r-i Maymūn 378
Bīra, Birecik 346, 350, 356, 358, 371, 430, 431, 441, 457, 460, 464, 483, 524
Birādūst See Mīr Nāṣır
Bīramkala 218
Birecik See Bīra
Birgi 487–488
  Birgi Ṭaġı / Mt. Birgi 487–488
Birgili Meḥmed Efendi 488
Birgiri 472
Bīrīā [Birea] 144
Birka Muʿaẓẓama 421
Birkat al-Badawī 454
Birkat al-Ḥājj (outside Cairo) 422
Birkat al-Muʿaẓẓam 409
Bīrūn 207–208
Bīrūn Shīrū (strait) 142
al-Bīrūnī, Abū al-Rayḥān 73, 158, 207
Bīsh 385
Bishāwur 237
Bīsnāghār (Visnagar) 185, 204
Bisṭām 261, 284–286
Bīstrūd River 261
Bīsutūn 340–341
  Bīsutūn, Mt. 260–262, 362, 370, 375
Bithynia 96
Bitlis, Badlīs 334, 336–341, 355–356
  Bitlis River 336, 341, 354, 373–374
  emirate of Bitlis 340
Bīwarāsb See Ḍaḥḥāk
Biznī 355
Black River (Niger) 109
Black Sea (also known as Pontus Euxinus) 69, 84, 88, 91–92, 95, 101, 105, 110–111, 123, 125, 137, 312, 325, 329–332, 345–346, 348, 477–478, 480–481, 483, 495–501, 505–507, 525, 528
Blacks, Land of See Nigrita
Blessed Harbor (island, Porto Santo) 109
Bodrum 483, 524–525
Boġaz (of Erzurum) 344, 352
Boġaz (of Kānḳırı) See Çerkeş: Çerkeş Künbeti
Boġazḥiṣār, Boġaz Ḥiṣārı (Çanakkale) 507, 512
Boġazḳalʿe 480
Boġazkesen (Rumeli Hisar) 519
Boġazlıyan 480
Boġrat Lake 347
Bohemia See Czech
Bohus 69
Bolayır 518
Boles 481
Bolu 483, 493, 495, 497–501, 503, 505, 508, 512
  Bolu Mountains 500
  Bolu River 499–500
Bolvadin 491, 511
  Lake Bolvadi 475
Bona Ispiransa See Cape of Good Hope
Bor 96, 474
Borlı, Borlu, Zaʿferānborlu, Safranbolu 486, 497, 499–500, 512
Borneo, Būrnūī 110, 118, 133–134, 139, 143, 145–147, 151
Bosnia, Bosnians, Dalmatia 91, 106–107, 125, 442, 520–521, 525
Bosphorus, Gulf / Sea / Strait of Constantinople / Istanbul 91–92, 105, 110, 238, 505–507, 526
Bosphorus Cimmerius ("Strait of Kerch and Taman") 91
Boṣra 431, 451
Boyābād 478, 495–497
Boyalı, Boyalu 495–496, 501
Boz Ṭaġ 486–488
Bozburun 502–503, 505
Bozcaada (Tenedos) 509, 521
Bozcaṭaġ 346
Bozdepe See Sinop: Mt. Sinop
Bozġuş-Ġalçan 484
Bozḳır 475
Bozḳırı 474
Bozoġlan and Bulgar mountains 473
Bozoġlı 495
Bozoḳ 477, 479–480
Bozöyük See İnöñi
Bozṭaġ summer pasture 486
Bozṭoġan 469, 487
Bögürdelen (fortress) 363
Börklüce 520
Brahmins 128, 183, 186, 193, 195, 200–202, 204
  Brahmin paintings 128
  Brahmin soldiers 195
Brājwīrī 218
Brāshīū (Bizacchium) 108
Brazil 112
  Terra Brazil 100
Britain, Britania, British 90, 93, 106
  British ships 100
Bucaḳ 462
Bucorova 493, 495
Budaḥkath 298
Budaḳözi 480, 481, 513
Būdāsni 350
Budenar 493
Būdī-jān 241
Budin (Buda) 125, 524–525
Bughbūriya 304
Buḫārī Dede 475
Buḥayra 289
Buḫtī (tribe) 354–355
Bujayla 394
Būjen sultanate 128
Būk, forest of 432
Bukhara 183, 275, 279–280, 291, 293–295, 299–302, 305
  Bukhara road 298
  governor of Bukhara 301
Bukhārī (Hadith scholar) 446
Bukht Naṣr (Nebuchadnezzar) 243
Bukur 423
Bula Vista 109
Buladan (village) 485
Bulaḳ 480–481, 512
Bulanın 345
Bulaydat al-ʿAbbās 406
Būlbays (Bilbays) 465
Bulgar(s), Bulghār 42, 106, 125, 304, 308–309, 326, 333
  Bulgar mountains 473
Bulgaria 106
Bulġurlu 507
Bulunyās 455
Būmaḥkath 293, 301
Būnāfād 208–209
Bunam 293
  Bunam Mountains 292–293
Būnqūm, Būnghūm [Bungum] 129–130
Buqʿa See Zabīd
Būqānūr 139
Būqār 457
Būrā 433
Būrā Canal 433
Burāq 234
Burāqī brotherhood 446
Burc 462
Burdur 489, 497
  Lake Burdur 491
Burhān al-Dīn Aḥmad, son of Qadi ʿAbd Allāh 482
Burhān al-Dīn ʿAlī b. Abī Bakr (al-Marghīnānī), Imām 297
Burhanpur 195, 199
Burj 262
Burj Birqā 431
Burj al-Buḥṣāṣ 454
Burj al-Hawā 454
Burma, Burmese See Pegu
Burma-Botan 495
Būrnāy, Būrnīo, Būrnūī See Borneo
Bursa 72, 479, 483–484, 492, 500–504, 508, 512, 516–522, 525
  Bursa Plain 504
Burṭās 304
Burūjird, Burūjard 258, 375
Burunābād 510
Burzal, Mt. 218
Būshanj, Pūshang 268, 279
Būshkāmāt 233
Būsrūd 278
Bust 208–209, 214, 221–222
Bū-ūtūr Mountains 152
Bustādārān 278
Bustān Banī ʿĀmir 424
al-Bustī, Abū'l-Fatḥ 221
Būstīnūs 337
Butam 298
Butar, Mt. 217
Buwāṭ 410
Buwayhids, Būyid dynasty, Daylamites 244–245, 254, 258, 263–264, 291, 366, 381
Būy River 299
Būya (in Arabia) 424
Būyā (in Transoxiana) 300

Buyū 424
Buzā'ā 457–458
Būzjān 275
Bülbülli 450
Bürkān 461
Büyükçekmece 76
  Büyükçekmece Bridge 525
Büyükgöl summer pasture 345
Byzantium, Byzantine(s) 378–379, 456, 461, 471, 482
  Byzantine Anatolia (Diyār-ı Rūm) 476
  Byzantine emperor(s) / Caesar(s) of Rūm / king of Istanbul / *tekfur* of Istanbul 379–380, 398, 415, 426, 464–466, 471, 476, 513, 515, 519, 521
  Byzantine empire / territory 464, 468
  Byzantine times 503, 505

Cádiz, Ghādes 91, 99, 112, 115–116
Caesar, Roman Caesars; and see Julius Caesar 362, 376
  Caesar of Engüriyye 337
  Caesar(s) of Rūm See Byzantium: Byzantine emperor
Caesarea 430, 437
Çağa 497, 499, 501, 512
Cailoco See Terra Galleca
Cairo 97, 125, 182, 184, 193, 266, 336, 383, 403, 411, 422–423, 428, 450, 464–465, 468, 522–524
Çal 484–486
Çaḳıd, Çakıt 457, 511–512
  Çaḳıd Ḫān 463
  Çaḳıd River, Çaḳıd Ṣuyı 434, 463, 512
Çaḳırbıñarı 515
Çaḳırca 484
Calabria 43
Calamianes, Palohan (Palawan) 110, 133–134
Çaldıran 321, 339, 345–346, 523, 525
  defeat at Çaldıran 320
Caledonia 90
Calicut 186–187
Calpe, at the Cape of Spain 91
Çamās 345
Cambaia See Gujarat
Cambodia 181
Cambram River 109
Çamlıca, Çamluca 501, 507
Çamurlu 481
Cāmūsteli 461
Çan 509
Canaan, Canaanites 358, 430, 454, 465
Canada 113
Çanakkale See Ḳal'e-i Sulṭāniye
Canary Islands, Canaria 70, 99, 109, 115–116
Çandarlu 503
Candelara reefs 132
Canfeza 356
Canıgel 356
Canik 477–478, 480–482, 495–496
  Mt. Canik 481
Çankırı, Kānḳırı 479, 482–484, 493–495, 500, 512
Çañlıḳ, Çañlu 496–497
Canta, Canton See Qāntā
Cantabria 90
Çapaḳçur 346, 353, 355
Cape Breton 113
Cape Cimbria 99
Cape Comorin 148, 181, 185–187, 189
Cape Corrientes (Cuba) 70
Cape Fortuna 113
Cape of Good Hope, Cava Bona Ispiransa, Bona Esperansa 90, 92, 100, 108, 181
Cape of Iscandia known as Nort-Kent 106
Cape of Mania (Cape Agrilia) 105–106
Cape of Santagostin (Santo Agostinho) 99, 118
Cape of Vīn-sen, Cape of Tesāwensā (S. Vincente, Cape Finisterre) 105–106
Cape of Wardhuys 105
Cape Parthenium 96
Cape Roca 90
Cape Tabin 110
Cape Verde, Capo Verde, Perīdā 99–100, 108–109, 117
Çārdīvār 360
Caria 91
Caribbean Islands 116
Carmathian(s), Qarmaṭī 281, 380–381, 392–393, 429, 467
  Carmathian sect 416
  Ḥamdān al-Qarmaṭī 393
Çarmur 346
Carolus V, Emperor 137
Çarşamba (of Germiyan), Çehārşanba 484–486
Çarşamba (of Bolu) 497, 499
Çarşamba Bāzārı 480
Çarşamba River, Çarşamba Ṣuyı See Amasya: Amasya River
Çārtāḳ 512
Cartesian astronomy 59
Carthage See Tūnis
Çāşnigīr 512–513
Çāşnigīr Köprüsi 480–481, 513
Caspi and Hyrcani (mountains) 92
Caspian Sea, Baku Sea, Sea of Jurjān, Sea of Khazaz, Sea of Shirwān, Sea of Ṭabaristān 84, 88, 90, 92, 105, 119, 261, 285–288, 290, 299, 304, 307–309, 318, 320–326, 328–329, 331, 468, 525
Castile 112, 132, 525
Çatalburġaz 518
Çatalgedük (mountain) 354
Cathay, Khitāy 87, 90, 120, 124, 152–153, 157–180, 190, 193, 204
  Cathay astrologers 165
  Cathay commanders 166
  Cathay desert 179
  Cathay iron 155
  Cathay language 173
  Cathay notables 181
  Cathayan satins 173
  Cathay steel 163, 179
  Cathayan vaults 177
  jurisconsults of Cathay 170
  Khaqan of Cathay, Khan of Cathay, khans of Cathay, Cathay Khaqan Chīn-Khūār 158, 161–162, 167, 175, 178
  soldiers of Cathay 167
  Sultanate of Cathay 161
  Tatars of Cathay 159
  territories of Cathay 166–167
  Khitāy origins 316
  campaign against Khitāy 302
Çatı 481
Caucasus Mountains 181, 204
Çavdarlıḳ 488
Cavdır 512
Çavgar 513
Çavuş (village) 475
Çavuş Köprüsi 512
Çay 491
Çaymaḥal 480
Cebel-i Cedīd 504
Cebel-i Rāhib See Keşiş Ṭağı
Cebel-i Ṭomaniç See Ṭomaniç
Cebelān 356
Cebu, Sebūt 118, 138, 147
Çeh See Czech
Çehārşanba See Çarşamba
Ceiram 133–134
Çeklene 495
Celālīs, Celālī revolt 320, 494
Celebes (Sulawesi) 110, 133–135, 140, 143
Çelekī (tribe) 353, 355
Cellarius, Daniel 41
Çeltikli, Çeltiklik 511–512
Çeltik Deresi ("Rice Paddy Valley") 498
Cem Çelebi (Cem Sultan) 522
Cem Sultan (road stage in Iraq) 366
Çemen 347
  Çemen summer pasture 481
  Çemen Ṭağı 344
Çemişgezek 347, 353–354, 479
Cenābī (historian) 199, 296
Cenaon 134
Cengūle 359
Cennet Ovası (in Yemen) 385–386
Çepni Mountains 348
Cerīd Silsüper 457
Çerkeş 493–495, 512
  Çerkeş Künbeti 493, 495
  Çerkeş Plain 494–495
Çermik 353, 355
Cerraḥ River 356
Çeşme 84, 510
Çeterya 470
Ceuta 115
  Cape of Ceuta (Gibraltar) 99, 118
  Strait of Ceuta (Strait of Gibraltar), Strait of Hercules 91, 96–97, 105, 108, 112, 115
Cevizderesi 478
Ceyḥan 463
  Ceyḥan River 434, 461, 463–464
Ceylon (Sri Lanka), Seylān, Sarandīb, Sarandīl 103, 110, 133, 147–151, 385

Ceylon cinnamon 138
Ceylon red stone (garnet) 471
Cezāyir-i Ḫālidāt See Canary Islands
Cezīre-i Küfre 354
Chabanlu Tatars 308
Chabdak, Mt. 217
Chābkārī 217
Chāch, Shāsh (Tashkent) 288, 295–301, 303, 305
Rūd-i Chāch, Shāsh River See Sayḥūn
Chāghān 352, 359
Chāghān Pass / defile 360, 367
Chāghān River 352
Chaganat (Jagannath, Juggernaut) 200
Chagatay (son of Chingis Khan) 212, 302, 306
Chagatay (country) 180, 307, 382
Chagatay sultan 382
Chagataid governors 302
Chaghāniyān See Ṣaghāniyān
Chāh-i Bīrūn 291
Chāh-i siyāh 278
Chahār-dīh 284
Chahnī 206
Chak 218
Chaldeans 373, 376
Chalgā-i Yulduz 176, 305
Chalī 216
Chandapūt 203
Chandarpāk 216
Chandarpāk River 216
Chandnī 216
Changalistān 216
Chānpūr 218
Chapa-khāl River 322, 324
Chaprār 217
Chāqdash (ruler of Erzurum) 347
Chaqmaq, al-Malik al-Ẓāhir (Mamluk Sultan) 403, 448
Charal 217
Chārbāna 224
Chardanū 217
Charhārū, Mt. 218
Charhārū Road 218
Charkh-i ẓālim (fortress) 359
Chashma-i Sabz 273
Chashma-i Surāb 278
Chatalwās 217
Chātrū 216
Chawbāra 202, 207
Chenab River, Chanhāw 191, 202–203, 206–207, 215–219
Chendenot (Chiniot) 191
Chigil 304, 307
Chihil Dukhtarān 269
Chil-hazār 272
Chīn u Hind See East Indies
Chīn (island) 366
Chīn son of Yāfith (Japheth son of Noah) 152
China, Chinese, Chīn u Māchīn, Ṣīn 36, 42, 53, 69, 80, 83, 104, 110–111, 120–121, 125, 128–130, 132–134, 144–145, 147–148, 150–160, 163–164, 172, 174–176, 181, 188, 190, 195, 206, 300, 311, 382, 392, 426
China boats 141
China Sea, Sea of China See Eastern Sea
Chinese cities 159
Chinese emperor, Emperor of China, Chinese emperor Terīn-Zū 153, 156–159
Chinese historians 159
Chinese masters (painters) 154
Chinese merchants 145
Chinese porcelain 138, 155
Chinese printing 154
Chinese traders 141
Chinese type (of boat) 141
kings / Faghfur of China 304, 362
mountains of China 219
villages of China 161
wonders of China 161
Chinār 225
Chinārān 284
Chinchāwāt (tribe) 331
Chingis Khan 159, 162, 180, 212, 231, 269–270, 273, 282, 291, 301–302, 336, 482, 513, 523
Chingisids, Chingisid dynasty; and see Ilkhanids, Khaqans 175, 245, 253, 319
Chingisid interregnum 256
Chingisid tradition 162
Chingisids 176, 179–180, 283, 306, 319, 338, 347
Chios, Saḳız 84, 525
Chītawar (Chittor) 192, 206
Choban Bridge 324–325, 332, 345–346
Choban, Amīr 394, 482
Chobanid dynasty 347
Chotin 526–527
Christian(s), Christianity 40–41, 59, 114, 126, 128, 130, 133, 136, 139, 148–149, 151–152, 169, 173, 186–187, 194, 200, 239, 246, 304, 315, 334, 339, 348, 368, 375, 377, 380, 390, 432–433, 436, 438, 440–441, 458, 465, 476–477, 487, 490, 497, 502, 505, 513–514, 526
Christian era 59
Christian nations 58–59, 77
Christian religion 59
Christian (i.e., theological) sciences 185
Christian subjects 513
Christian theologians 52
Christian villages 513
Nestorian Christians See Nestorians
Christopholos, Christophorus See Columbus
Chūhbar 206
Chūk 216
Chulhan 217
Chūn 216
Chuḳur Saʿd 320
Chūtpūr, Udaipur 181, 192–193
Chuyl See Juyl
Çiçek, Mt. 479
Çifteḫān 463, 465, 474, 511–512
Cihānnümā 35, 40–44, 51, 85
Çil (castle and emirate) 340
Çıldır 321, 331–332, 334, 343
Çile 499
Mt. Çile 499–500
Çile Yaylaḳı 500
Cilicia See Silifke
Cimbombom 118
Çınarçayırı 511
Cinci Ḫoca 499
Çine 488, 510
Çini Ḥisārı 518
Çira 470
Circassia, Circassians 125, 310, 328–331, 348
Circassians (Mamluk dynasty in Egypt) 194, 383, 403, 407, 411, 428, 468, 523
Cisr-i ʿAtīḳ 511
Ciyān 476
Cizre 353
Cleanthes of Samos 59
clime-books (i.e., the Muslim cosmographers) 146, 152, 158–159, 290
Cluverius See Philippus
Cochin 149, 187
Khaqan of Cochin 83
Cochin-China 156, 159–160, 190
scholars of Cochin-China 154
Coimbra 88
Colombo (capital of Ceylon) 148–149
Columbus, Christophorus 96, 101, 112, 115–117
Commentator (of the Collegium Conimbricense) 93
Comoros 90, 100, 103
Companions of the Prophet, Aṣḥāb, Ṣaḥāba 87, 263, 318, 364, 366, 377, 379, 398, 414, 426, 429, 443, 445, 451, 523
Four Companions (the first four caliphs) 314
Confluence of the Two Seas (Koran 18:60) 322
Constans, Emperor 492
Constantine 438
Constantinople See Istanbul
Contarenus 94
Copernicus 51, 59–60, 62–64
Copernican astronomy 59, 61, 64–65
Copt, Coptic 396, 398
Coptic (language) 436
Coptic years 61–62
Çoraḳ River, Cū-yı Rūḥ (Çoruḫ) 344, 349
Çorlu 76, 518
Cornelius (Tacitus) 99
Cordoba 125
Coron 91
Corsica 91, 106
Cortes, Captain 138
Çorum 477–479, 482, 492–493
Cossacks 496, 507, 526
Don Cossacks 309
Şarıḳamış Cossacks 526–527
Çökerek 480
Çöklez, Mt. 485
Çölābād 491

Çörsek 461
Crete, Creta 91, 105–106, 123, 125, 470, 509, 527
Crete campaigns 101
Crimea, Lesser Tataria 106, 306, 309–310, 331
Crimean Khan / khans 308, 527
Crimean Tatars 525
Croatia, Croats, Ḫırvat 106, 519, 525
Crustina 309
Ctesiphon See Madāʾin
Cū-yı Rūḥ See Çoraḳ
Cuama River (Zambezi) 109
Cuba 93, 113, 116
Çubuḳ plain 480
Çubuḳābād 492–493
Çubuḳābād River 493
Çubuḳçayırı 481
Çubuḳköprü 375
Çubuḳlu Bāġçe 507
Çuḳurbostān River 352
Çuḳurcuḳ 482, 492–493, 512–513
Cūlāmerg 340–341
Cūlāmerg River 340
Cumani 307–308
Cuzco 113
Cusanus, Nicolaus (Cardinal) 59
Cūyāş 462
Cüdāmen Şāh Bridge 346
Cülāb 356
Çüngüş 353
Cüreki 478
Cyclades 96
Cyprus 91, 96–97, 105, 110, 123, 125, 334, 463–464, 469–472, 525
Cyrenaica 108
Cyrus 196
Czech(s), Çeh, Bohemia 74, 106–107, 520–521, 525
Czehrin 527–528

Dabāl-pur See Dibalpur
Dabīl 207–208, 334, 339
Dābshalīm 196
Dābūl, Dabūl-i Sind 103, 206–207, 209–211, 230
Dabūsiya 294, 300
Dachinpāra 218
Dacia 106
Dād b. Saʿd (fortress) 363, 365
Dādhan 233
Dādūya 278
Dādyān 331
Dāghilī Canal 433
Dagon (idol) 125
Dagestan, Ṭaġistān, Sarīr al-Lān 303–304, 320, 325, 328–331
governor(s) of Dagestan 328
rivers of Dagestan 325
Dahak 222
Dahān-i Shīr mountains 298
Ribāṭ-i Dahān-i Shīr 290
Dahās River See Balkh River
Dahbān 423
Ḍaḥḥāk, Bīwarāsb, Ḍaḥḥāk-i Mārī, Ḍahhāk of the Serpents 225, 242, 260–261, 266, 361, 369, 376, 489
Ḍaḥḥāk (b. Qays al-Shaybānī) 466
Daḥl Mayy 424
Dahnāʾ 410, 423–424
Dahūk 373
Dāʿī, Bilād 429
Dāīmenk Khan 164
Dāīr 131
Daka 366
Dakān 240
Dakān River 240
Dakhna 363
Dakkū (Dakū), Mt. 216–217
Dalan River 213, 214
Dalanguer 307
Dalījān 257
Dalmatia See Bosnia
Daman, Damaon 183
Dāmān 196
Damānisandū 216
Damāwand, Dunbāwand 261, 263, 285–286
Mt. Damāwand 253, 260, 261, 286
Damianus 107
Damascus, Shām 72, 124, 325, 358, 360, 375, 385, 393, 403, 405, 409, 411, 420–421, 430, 432, 435–436, 440–444, 446–450, 452, 455–456, 460, 464–469, 501, 522–525
Ghouta of Damascus 293, 314, 393, 430–433, 452
Lake of Damascus 432
Damāwa 385
Dāmāzī Mountains 152, 181
Dāmghān 261, 263, 265, 278–279, 284, 286
Damietta, Dimyāṭ 126, 237, 469, 524
Lake Damietta, Lake of Tinnīs and Dimyāṭ 97
Damkand 203, 217
Dāmkar 217, 217, 218
Damshik 209, 211
Damtūr 193, 206
Ḍān 424
Dāna 278
Dana-pur 192
Dandān 215
Dandāniqān 275
Dandarhit 216
Dānduwār 217
Dānduwār Road 217
Dania See Denmark
Daniel, tomb of 247
Danishmendids, dynasty of Melik Danishmend 482
Dānişmend (Ghazi), Malik Muḥammad Ghāzī 476, 482
Dankūt 219
Dansāl 216
Danube 520, 524, 527
Dānyāl (village) 217
Dāqūq 352, 372
Dara 276–277, 279
Marghzār-i Dara 278
Dārā b. Dārā, Darius 258, 266, 311, 354, 513
Dārā (in Iraq) 351, 354
ʿAyn Dārā 453
Dārā, Mt. (in India) 216
Darā, Darāʿ (in Syria) 357, 431
Dārāb b. Bahman b. Isfandiyār, eponym of Dārābjird 235
Darabīla 207, 209
Dārābjird (Dārābgird) 232, 235–236, 239, 241
Dārābjird, Mt. 240
Daraj River 255
Dāramī, Mt. 218
Dāran 210
Dārānī Canal 433
Darbād, Mt. 278
Āb-i Darbād River 278
Darband, Bāb al-Abwāb, Demir Ḳapu, Darband-i Khazaz 125, 301, 304, 309, 322–323, 325–327, 329–330
Darband-i Tāj Khātūn 362
Darband-i Zangī 362
Darbasāk 430, 433–434, 457–458, 460
Darbīl 362
Dardanelles See Hellespont
Dardanos 509
Dārende 457, 479, 524
Darghān 290
Darghash 221
Darguzīn 259, 263, 375
Ḍarīḥ 307
Ḍārij 413
Dārim 431
al-Dāris fī taʾrīkh al-madāris (book) 449
Dāriyya 452, 464
Ḍariyya, Ḥiṣn 412, 424
Dārkān 241
Darkūsh, Darkūsha 434, 457, 460
Darna 366–367, 372
Darna River 374
Darpaz 229
Darrāk, Mt. 232–233, 240
Dartang 366–367, 372, 375
Dartang River 374
Dārūm 438
Dārüssaʿāde Aġası 456
Dāryā (village) 440
Daryā Khan, ʿImād al-Mulk b. ʿImād al-Dīn 185, 199
Daryāy, Shiʿb-i Daryāy 228–229
Daryāy River 229
Daskhwāqān 317
Dasht-i Arzan, Lake 240
Dasht-i Bayāḍ 276
Dashtābād 246
Dashtābād River 248
Daspul 241
Daspul River 374
Dastgir 247
Dastkard 229
Dastyārī 210

Davis, Captain John 120
Davis Strait 120
Dawān 233
Dawatabad, Dīwgīr (Devagiri, Deogiri) 184, 192, 199, 205
David (prophet) 356, 411, 426, 430, 436, 438–439
Ḍawāḥ 457
Dawaro 428
Dawlat Khan (governor of Lahore) 198
Dawraq 241, 246, 248–249
Dāwūd Ḳaramanī 498
Dāwud Khan 199
Dāwūd Pasha 411, 428
Dāwūdī (Ẓāhirī) rite, Dawūdis 239, 416
al-Dawwānī, Abū Naṣr 238
al-Dawwānī, Jalāl 315
Daylam, Daylamān, Daylamis 124, 184, 230, 251, 253–254, 285–288, 291, 312, 320, 325
Daylam Mountains 254, 288
Daylamites See Buwayhids
Dayn Dulā 237
Dayr (in Iran) 278
Dayr (in Iraq) 346, 357–358, 384
Dayr, Mt. (in Lebanon) 435
Dayr ʿĀfūl 369
Dayr ʿAmmāl 375
Dayr ʿĀqūl 370, 375
Dayr-i Burqān 277
Dayr Ḥamīrā 456
Dayr Jālūt (rather ʿAyn Jālūt) 468
Dayr-i Kharqān 319
Dayr-i Khwān 319
Dayr Maqlūb 373
Dayr Masīḥ River 462
Dayr al-Qamar (fortress) 451
Dayr Simʿān 466
Dayra 367
Dayrānī 452
Ḍayʿat al-Shaikh 455
Dead Sea, Lake of Lot, Lake Zughar 432–433, 437–438, 440, 442
Debbāğlar Ṣuyı 499
De caelo (book by Aristotle) 64, 92, 94, 96
Deccan, Deccanese 77, 181, 184–185, 193–195, 199–201
Deccan Sultanate 184, 199
Kings of Deccan 199
rulers of Deccan 195
Dedeli, Dedelü 472, 474
Defter-i Ḫāḳānī (book) 338
Degirmen 481
Degirmenli 513
Degirmenlik 469
Delefseme 356
Delük (fortress) 462
Delhi, Dehlī, Dillī 181, 188, 190–192, 196–198, 204, 215, 219
Delhi road 191
Delhi Sultans 199
Old Delhi 192
New Delhi 192, 196
Deliceırmaḳ, Delıırmaḳ 478, 481, 513
Delikli Ḳaya, Deliklüṭaş 337, 341, 355–356
Demir Ḳapu See Darband
Demirci 486, 487, 504
Demirciler 513
Demirkapı 367
Demirṭaş Bridge 486
Deñizli, Deñizlü 484–485, 487–488, 490, 512, 516
Deñizlü Valley 485
Deñizli cloth 517
Denmark, Danes, Dania 65, 74, 90, 106
Derbend (near Iznik) 511
Derbend (near Amasya) 480
Derbend Pasha 352
Derbend pass 464
Derbend River 499–500
Derbend-i Püşt 341
Dere-i Alakis 339
Deremahyası summer pasture 499
Derkene 497, 499
Derviş Pasha 447
Dervish Ilyās 375
Descartes 59–60, 63
Deşt-i Erzen Lake 346
Deveboynı, Deveboynu 346, 357
Develi / Develü Ḳaraḥiṣār 476, 481
Devīn 361
Devlet Giray Khan 525
Devrek 497, 499
Devrekānī 360–361, 495
Devres River 494–495
Deyr-i Ḳıyme 347
Dhamār 385, 387–390, 427
Dharāb 224
Dhāt al-Ḥajj 421
Dhāt ʿIrq 407, 424
Dhāt al-Nawʿ (villa) 414
Dhāt Maʿīn 424
Dhiʾb 103
Dhimmis 337–335, 345, 366, 441, 485, 504
Dhūʾl-Khalīfa (in Arabia) 405
Dhūʾl-Kifl 371, 451
Dhūʾl-Nūn, Inn of 420
Dhū Nuwās 390, 426
Dhūʾl-qadriyya (Zülkadriye, Zülkadiroğulları) 461, 468, 482
Dhū Qār, battle of 377
Dhūʾl-Qarnayn 119, 414, 426
rampart of Dhūʾl-Qarnayn, dam / rampart of Gog 87, 310–311, 426
Dhuruʿ 452
Dhuwālīn 224
Diana, Temple of 111
Dīb islands 151
Dibalpur, Dabāl-pur, Dapār-pur 183, 196–197, 205–206
Dibīl 326
Dih-i Naw 300
Dihbān 409
Dihistān 285–286, 326
Diḥyat al-Kalbī 236
Dijlat al-ʿAwrā See Tigris
Dīkān 213, 214
Dike Ḫān 512
Dikilüṭaş 511
Dil 505, 511–512, 516
Dilimān 288
Dillī (Dili in East Timor) 145
Dillī See Delhi
Dillikār 356
Dīmandal 217
Dimashqī (author of biography of the Prophet) 407
ed-Dimaşḳī See Ebū Bekr b. Behrām
Dimetoka 512, 520
Dimyāṭ See Damietta
Dīn-Ṭāy Khan 173
Dīnār, Mt. 240
Dīnawar 260, 262, 319, 362
Dīnawar River 248
Diodorus 112, 204
Diocletian 97
Diogenes 88
Dīra-i Fatḥī Khan 202
Dīra-i Ismaʿīl Khan 202
Dirakhtistān 277
Dīranda 355
Direkliderbend 482
Dirham b. Ḥusayn (ruler of Sīstān) 223
Dirhamiyya 424
Dirʿiyya 412–413, 424
Dirrikiw 218
Dit River 215
Diu 182–183, 207, 209
Divrigi 457, 462, 477–479, 481, 524
Dīw 219
Dīwasar 217
Dīwgīr See Dawatabad
Dīwkār 218
Dīwnār Valley 217
Dīwrūd (river) 227, 240
Ḍiyā al-Mulk al-Nakhjiwānī 317
Diyādīn 335, 345
Diyāla 370, 374–375
Diyāla River 360, 372–373
Diyarbekir, Diyār Bakr, Āmid 125, 312, 316, 320, 334, 337, 340, 343, 346–347, 350, 352–357, 373–374, 477–478, 524
governor of Diyarbekir 320
Diyār Muḍar See Muḍar
Diyār Rabīʿa See Rabīʿa
Dīzak (in Sind) 209–210
Dīzak (in Transoxiana) 300
Dizfūl 247–248, 251
Dizfūl River 248
Dīzī 340
Dniester River 527
Dome of the Rock 466
Dominica (Santo Domingo) 116
Don River, Tanais 91–92, 105, 110, 303, 307, 309, 525
Don Cossacks See Cossacks
Dönerek River 497
Dört Dīvān 497–498
Dragon's Mouth (harbor) 117

Dromiscum 96
Dujayl River 248
Druze 436, 450
Ḍubay', Ḍubay'a 412, 414
Dujānī 424
Dujayl 367–368
Dujayl River, Nahr Dujayl 248, 351, 368
Dūlū 218
Dūlāb 288
Dūlcevrān 359–360
Dūlcevrān desert 360
Dūma, Dūmā, Dūmā', Dūmat al-Jandal 414, 416, 419–420, 423, 433
Dūmān son of Ishmael 414
Ḍumayr 452–453
Dūmīsh 326
Dummar 433, 448, 450, 453
Dummat 423–424
Dumshāq son of Sam son of Noah 443
Dūn 179
Dūn Sūn 375
Dūnbadī 202, 213
Dunbāwand See Damāwand
Dunbul-i Yaḥyā 339
Dunbulī 334, 339, 355
Dūr (castle) 386
Dūrdān 229
Dūrkāmī 341
Dūrū 216
Duwār 221
Duwārkantal 219
Duwāwa 213, 214, 215
Duwāwa River 213
Duwind 218
Dutch See Holland
Düden 491
Düden River 489
Düzce, Düzcebāzārı 499–500, 512
Dvina River 121
Dwīn 334

East Indies, Chīn u Hind, islands of India 92, 112, 132, 143
Eastern Sea, Eastern Ocean, China Sea, Indian Ocean, Sea of India 43–44, 69, 77, 80, 82, 90, 93, 99–103, 110, 121, 127, 146–147, 151, 161, 181, 183, 185
Eastern astronomers 103
Ebū Bekr b. Behrām ed-Dimaşḳī 43, 329, 332, 343, 349, 358, 483
Ebū Elvān (Çelebi), Seyyid 517
Ebū Sa'īd (in Iraq) 356
Ebū's-Su'ūd Ḥoca Çelebi Efendi, Shaikh al-Islam 400, 479
Ece Bey 518
Edebali, Shaikh 491, 514
Edincik 501, 510, 518
Edirne 84, 518–522, 525, 527–528
Edremid 504, 512, 518
Eflani 495, 497, 500, 512
Eflāṭūn See Plato
Efnanlu 499
Lake Efnani / Efnanlu 499, 501
Efrun 465
Egerlü, Mt. 346
Egil 353, 355
Egin 346, 478–479
Egri (Eger, Erlau) 43, 525–526
Egridir 483, 489–491
Lake Egridir 489–491
Egrigöz 483–484, 519
Egypt, Egyptians, Aegyptus 42, 43, 74, 91, 97, 99, 101, 105, 108–111, 125, 160, 190, 194, 206, 264, 352, 379, 383, 394, 396–406, 409, 411, 426–428, 430, 440, 450, 452, 456, 467–471, 482, 519, 521–525
Upper Egypt 108, 125, 428, 468, 524
Lower Egypt 108
Egyptian lands 125
Egyptian money 417
Egyptian pilgrims 431
Egyptian priests 112
Egyptian pyramid 254
ancient Egypt, ancient Egyptians 125, 129, 202
governor / *beylerbey* of Egypt 354, 404, 422, 524–526
lentils of Egypt 438
reed mats of Egypt 439
endowment for the Egyptians 455
Elbester 362
Elbeyli 477, 481, 513
Elbīnumān 151
Elbistan 334, 434, 457, 461, 468, 523
Elchi, Elchis (official emissaries) 128, 164, 171–173
Elegzi See Lezgi
Eleşkird 324, 343, 345
Eleşkird Plain 321
Elijah (prophet) 453
Elmalı 356, 489
Elmaṭaġı 481, 493, 513
Emīr Dāvūd 340
Emir Sultan, Shaikh 502, 521
Emmanuel, Portuguese King 137
Emrūdeli 484
Emrūdlı 502
Enderay 328–329
Enez (Ainos) 521
England, English, Anglia 40, 74, 90, 93, 99, 106, 112, 115, 120, 123, 125, 185
English (language) 436
English leagues 75
English miles 75
Engüri, Engüriyye See Ankara
Engürücük 503
Enoch son of Seth 98
Epaphus 108
Ephraim, Mt. 441
Epirus 106
Erbil 359–361, 373–374
Erciş, Arjīsh 319, 334–335, 347, 349, 356, 476, 525
Lake Erciş See Van: Lake Van
Mt. Erciş (Erciyes) 476
Erdel See Transylvania
Erdost, Mt. 473
Eren, Mt. 500
Eregli 465, 472–474, 481, 498–499, 511–513
Erġani 353–354, 357
Ergene Bridge 521
Erim 478, 480
Eritra 92
Ermenak, Ermenek 469, 471–472, 477, 514
Ermeni 501
Ermeni Beli pass 514
Ermeşāṭ castle 355
Erṭuġrul 477, 491, 513–515
Erzen 344
Erzen river 355, 373
Erzincan 339, 343–344, 346–347, 483
Erzurum, Arzan al-Rūm 319–320, 325, 331–334, 338, 343–349, 352, 355, 362, 477–478, 481–482, 525
mountains of Erzurum 332
Pasha of Erzurum 344
Esau 451
Esbāber, Esbāberd, Esyāber, Esyākerd 334, 341
Eşek Meydanı 356
Esen Tayshi 167
Eşkel (village) 504
Eski Eyne (near Amasya) 481
Eski Ḳapluca (in Bursa) 502, 504
Eskiḥisār See Eskişehir
Eskiigne 512
Eskişehir, Eskiḥisār 484, 491–492, 511–512, 514–515
Eşrefzāde-i Rūmī 505
Esterġon (Esztergom) 525
Ethiopia, Ethiopian(s), Abyssinia, Ḥabesh 42, 77, 99–100, 108, 125, 178, 194, 385–386, 398, 406, 417, 426, 436
Ethiopean Guinea 93
Euboea, Aġriboz 96, 106, 521
Euclid 315, 378
Eudoxus 99
Euphrates 111, 124–125, 181, 191, 221, 250, 288, 299, 338, 346–348, 350, 354–358, 365–366, 368–374, 424, 430–431, 434, 460–462, 478–479, 481, 483, 513
Euripus Strait 95
Europa 105
Europa Tyria 105
Europe, European, Frengistān 35, 42, 46, 77, 82, 90–91, 99–101, 104–108, 110, 112–113, 123, 126–129, 132, 159, 181, 204, 523, 528
European books 39, 72, 81, 84, 190
European geographies 73
European globes 81
European nations, European countries 41, 59
European painter 148
European printing 154
Eve 150, 407, 438, 450
Evrenoz Ghazi 519
Evṭārī 359
Eymir 457, 496

Eynebāzārı See Ez̲īne
Eyübeli 475
Eyyūb 521–522
Ez̲īne, Eynebāzārı 481, 484–485, 487, 509–510, 512
Ezköy 344
Ezrencik 347
Ez-z̲ikr, Mt. 346

Fabrica del Mondo (book by Lorenzo) 43
Faḍl b. Marwān (ruler of Fārs) 239
Faḥlatayn 421
Fakhrī, Fakhriyya (island and fortress) 364
Fāl Islands 150, 187
Falakī (fortress) 388
Falakī (poet) 323
Fallūja 367
Famm al-Ṣulḥ 370, 375
Fanā Khusraw See ʿAḍud al-Dawla
Fanākath 299
Fārāb 275, 279, 297, 299–300, 304
Farāh 221, 222
  Farāh Mountains 222
  Farāh River 222
Farāhān 257–258
Farāmurz (village) 375
Farasān (island) 415
Farāsha (village) 375
Farāshar Hill 424
Farāwa 224, 275
Farawāt 240
  Farawāt River 240
Farbar 293
Farghān 279
Farghāna 125, 295–300, 379
  Lower and Upper Farghāna 293
  Farghāna road 295–296
  Mt. Farghāna 300
Farhād (legendary figure, lover of Shīrīn) 260–261, 322, 362, 480
  mountain of Farhād 375
Farḥāla 191, 193, 203, 206
Farhāwān 278
Farīd Khan 197
Farīd Shakarganj, Shaikh 183
Farīdūn (ancient Persian king) 236
Fārihīn, houses of 383
Farīwār 262
Fāris b. Nāsūr b. Sām (eponym of Fārs) 231
Farrazīn (castle) 258
Farrukhī 212
Fārs, Pārs 69, 90, 125, 196, 221–223, 227–246, 248–249, 251–252, 262, 264–265, 273, 276–278, 284, 289, 291, 376, 417
  Atabegs of Fārs (see Salghurids) 245
  fire temples of Fārs 241
  kings of 242–243
  Sea of Fārs See Persian Gulf
  tribes of Fārs 232
  Fārsnāma (book) 234, 238
Farshī 278
Fartak 102
Fārūn 235
Farūrmand 222
Fārūs Banāsa (monastery) 455
Farwa 286
Fāryāb 270, 275
Fasā 236, 238, 241
Fāsh 330
Fāsha (Phasis) River 331
Fatḥiyya (fortress) 363, 365–366
Fatimid(s) 43, 467
Fatsa 478
Fayd 414–415
Fayd son of Ham son of Noah 415
Fayṣal (book by Abū'l-Majd al-Mawṣilī) 42
Fayḍī-i Hindī, Shaykh Abū'l-Fayḍ 192
Fazāra (tribe), Banī Fazāra 451, 458
Felek (fortress) 355
Fenar, Fenar Bāğçesi 505–506
Feñarī, Shaikh Şemseddīn 521
Fenāyūra 128
Ferdinand, the Kaiser of Austria 524
Ferecik 518
Ferhād Pasha 321, 327, 526
Ferrando Galleco 132
Ferruḫşād Bey 344
Fersenk 347
Festus 107
Feyek 355
Fez 125
Fez̲leke (book by Kātib Çelebi) 181
Fīānūmā 128
Fidā'iyya (the Assassins) 455
Fīja (spring and river) 433
Filibe 518–520
Fīlsuwār 322
Filyas 500
  Filyas River 499
Fīn (castle) 257
Finike 483, 489
Firdawsī 272, 278, 322
Fire Island (Ateş Adası) 109
Firozabad, Pīrūzābād (in India) 191, 197, 205, 209–211, 225
Fīrūz (Sasanian) 237, 316
Fīrūz (in Anatolia) 512
Fīrūz (in Fārs) 234
Fīrūz Bakht 332
Fīrūz Shah, Fīrūz Khan (Tughluq sultan) 191, 196–199
  Fīrūz Shah's Hunting Ground 192
Fīrūzābād (in Iran) 233, 238, 241, 257, 315
  Fīrūzābād River 240
Fīrūzān 261
Fīrūzkūh 213, 225, 285–286
  Fīrūzkūh mountain 261
Fīrūzmand 222
Fīrūz-Qubād 322
Fıstıḳlı 503
  Fıstıḳlu mountain 501
Fītā, Mt. 455
Flanders, Flemish, Flemings 93, 106, 113, 125, 155
  Flemish bertones 100
  Flemish illustrators 142
Florence 112
Foça 486
Fodul Baba Mountains 473
Fortunata See Canary Islands
France, French, Gallia 40, 45, 69, 74–75, 77, 84, 90–91, 93, 99, 106–107, 112, 123, 125
  French (language) 436
Francisco de Albuquerque, Franciscus 132, 228
Franconia 107
Frank, Franks 42, 141, 228, 437, 440, 442, 445, 454, 457–458, 467, 510, 519, 521
  Frankish books 129
  Frankish geographies and globes 150
  Frankish histories 161
  Frankish languages 513
  Frankish merchants 460
  Frankish ships 455
Frislandia 106, 119–120
Frobisher, Captain Martin 120
  Frobisher Strait 120
Frozen Sea (Arctic Ocean) 112, 161
Fuʿa 458
Fukien 157
  Fukien River 157
Fūmin (or Pūmin) 288–289, 291
Fūr See Por
al-Furʿ 407
Fūshanj 278
  Fūshanj River 271

Gabriel (angel) 351, 396–397
Gaetulia 108
Gagarū 218
Gaikhatu (Ilkhanid) 319
Gajargām 219
Galata 466–467, 522
  Galata Tower 313
Galen 417
Galileo 64
Galita (Marie-Galante) 116
Gallia See France
Gallipoli See Gelibolı
Gancara 465
Gang 217
  Gang, Mt. 217
Gang Diz 70
Gangaridai 196
Ganges River 90, 111, 161, 181, 187–188, 190, 193, 196–197, 202–204
  Ganges Bay See Bengal: Bay of Bengal
  Gulf of Ganges 187
Ganja 318, 320–321, 326, 328, 525
  Ganja River 324
Garamantes 108
Gardīz 213–214
Garm River, Garmrūd 235, 258, 262, 317–318
  Āb-i Garm (stage) 279
Garmīna 300
Garmīniya 294
Garonne River 93
Gashta 294
Gasparus (Varrerius) 112

Gastaldi See Giacomo
Gaur 188, 205
Gawāshīr, Bardshīr 227, 230
Gāwanīsarāy 216
Gawhar Shād Khātūn 265
Gāwkhānī swamp 261
Gāwmāsārūd River 262
Gāwpārī Plain 322
Gayal 217, 218
Gayūmarth (ancient Persian king) 233–234, 242, 286, 376, 424
Gaza 125, 403, 430–431, 436–438, 454, 464–465, 524
Ġāzīḳıran 360
Gāzmār, Mt. 216
Ġazanfer Agha 43
Gebze, Gegbuza 505, 511
Gedeḳara, Gedeġra 480
Gedelek 488
Gedlekşemāḫī 356
Gedik (village) 346
Gedik Aḥmed Pasha 491
Gediz 484–486
Gediz River 485, 522
Geldim (fortress) 360
Gelembe 511
Gelendos 490
Gelibolı, Gelibolu, Gallipoli 483–484, 510, 512, 518
Gelikiras 480–481
Gelincik 331
Gemlik 492, 501–503, 505, 508, 518
Gulf of Gemlik 502–503
Genç 346, 353, 355
Gençeli, Gencelü 503, 506, 512
Genebrardus (Gilbert Génébrard) 98
Genesis See Torah
Genoa 91, 93, 107, 125
Geography, Geographike Hyphegesis, Almagest, geography of Ptolemy 42, 57, 69–70, 73, 110, 119, 146, 332
Geographia Minor (book) 112
Georgia, Georgians 124–125, 320, 322–323, 327–333, 343, 346–349, 525
Georgian (language) 333
Georgian tribes 333
mountains of Georgia, Georgian Mountains 322, 325, 349
Gercanis 343, 482
Gerede 493, 497–498, 500, 512
Gerger 457, 462
Germābe 335
Germany, Germans, Alman 41, 69, 74–75, 77, 90, 98, 106, 123, 125, 155, 175, 520–521, 525
German (language) 156
German miles 66, 75, 106, 108, 110, 113, 127, 133–134, 138, 152, 156, 158, 161, 176, 181–182, 185–186, 190, 203, 436, 521, 525
German Sea, Sea of Germany See Baltic
Germi River 493
Germiyan 484, 489, 516
Germiyanid, Germiyanoğlı 515, 518–519
Gevanis 343–344
Geyikler 484–485
Geyikli Baba, Geyiklü Baba 502, 518
Geylān 347
Ḫān-ı Geylān 346
Geyve 493, 505, 512–513
Ghabāghib 420
Ghaḍbān River 433, 454
Ghādes, Ghades See Cádiz
Ghadīr b. Hārūn (road stage) 365
Ghadīr Khumm 409
Ghaffārī (writer on the Mongols) 306
Ghalāfiqa 384, 416, 420
Ghamr Marzūq 415
Ghāna in the Sudan 429
Ghānimiyya 423
Gharjistān, Gharja 212, 224, 225, 275
Ghassānids, Banī Ghassān 465
Ghataḥ 219
Ghats Mountains 181, 184, 203
Ghatūd 409
Ghawr 420, 431, 432–434, 438, 441–442, 468
Mt. Ghawr 442
al-Ghawrī, Abū al-Naṣr Qānṣawh al-Malik al-Ashraf, Sultan (Mamluk) 356, 403, 407, 410–411, 422, 428, 468, 523
Ghayḍa 433
bridge of Ghayḍa 433
Ghayḍat al-Sulṭān 452–453
Ghayqa 423
Ghayrī (idol) 416
Ghayya' Ḥashīsh 384
Ghazlāniyya 433, 452
al-Ghazālī, Imam 46, 272
Ghāzān Khan, Sultan Maḥmūd Ghāzān (Ilkhanid) 244–245, 250, 253, 283, 313, 315, 319, 322, 371, 477, 514
Ghazi Fāżıl 518
Ghāzī Giray Khan 332
Ghāzī Malik (Tughluq) 197
Ghazna, Ghaznīn (Ghazni) 125, 183, 196, 207–208, 210, 212–214, 220, 222–226, 265, 279, 281, 302, 306
Ghaznavid(s), Ghaznavid dynasty 196, 212–213, 223, 225, 281–282
Sultan Maḥmūd of Ghazna See Maḥmūd
Ghazwān, Mt. 406
Ghazyat Ḥamdānī (tribe) 431
Ghifārī, Qāḍī Aḥmad 251, 327
Ghiyāth al-Dīn, Tughluq Shah 197, 199
Ghiyāth al-Dīn (Masʿūd) b. Kay-Kāwus (Seljuk) 477
Ghiyāth al-Dīn Muʿaẓẓam Shah, Sultan (ruler of Bengal) 402
Ghiyāth al-Dīn Muḥammad al-Rashīdī, Khwāja, Vizier 313, 315, 319
Ghiyāth al-dīn Naqqāsh, Khwāja Ghiyāth 161, 163, 165, 172, 176–179, 186, 189–191, 305
Ghouta 393, 430, 432, 452
Ghumdān 386
Ghunjura 461
Ghūr (region of Afghanistan) 125, 200, 212, 214, 220, 222–226, 265
Ghūrids, Ghūrid kings / sultans 196, 200, 221, 225–226, 281, 284
Ghūr (island south of China) 147
Ghūrband 212
Ghūrbīl River 432
Ghūrī 224–225
Ghuzz See Oghuz
Giacomo, Giacomo from Castile (Jacobo Gastaldi) 132, 134
map of Giacomo 133, 138, 140–144, 147, 157–159
Gibraltar See Ceuta: Strait of Ceuta
Gihindī, Gindī 207, 209
Gīlān, Gīlānāt, Jīlān 125, 251, 261, 285, 287–289, 291, 301, 309, 315, 318, 320, 322, 325, 367–368, 372–373, 525
governor of Gīlān 288
sultans of Gīlān 288
Mt. Gīlān 372
Gilolo 110, 133–134, 139
Gīlūya, Mt. 245
Gīr castle 355
Girdkūh 225, 286
Giresun, Giresin 349
Girne 469–471
Goa 183, 185, 228
captain of Goa 199
Gog and Magog 310–312, 381
dam / rampart of Gog See Dhū'l-Qarnayn
Gold Coast 100
Golden Horn, Gulf of Constantinople 84, 483, 521
Goliath 430, 434
Golkonda 184, 193
Gonave 115
Gori 323, 333
Goths, Sea of See Baltic
Gök Meydan (in Bitlis) 336–337
Gök Meydanı (in Damascus) 443, 447–448, 465
Gök Ṭağı 494
Gökböri, Abū Saʿīd, Muẓaffar al-Dīn (Turkmen) 359–360
Gökçeağaç 496
Gökçe, Lake (Lake Sevan) 324
Mt. Gökçe 499
Gökçe Nīl See Armaniyān
Gökçek, Lake 354
Gökçelu 464
Gökçeṣu 356, 497–498, 500
Gökçeṭağ 501–502
Gökçubuḳ 480
Gökdere (quarter of Ḳasṭamonı) 496
Gökdere Ṣuyı (river) 501
Gökmeydan 433, 465
Gökova, Gökābād 488, 519
Gököyük 484–485, 488
Göksu 356
Gökṣu River 506–507

Gökşun 462
Gökṭaşlı 462
Göktepe 352
Gölbaşı 357, 481, 511–512
Gölbāzārı 498–501
Göldepe (village) 375
Göle 324, 332
Göle Plain 324
Göle Ravine 324
Gölek Baba (road stage) 347
Gölḥiṣār 489, 491
Gönen 489, 512
Gönye 330–331, 347–349
Gördes, Gördüs 486, 504
Görele 348, 505
Görgil 341, 355
Göynük 487, 491, 501, 512, 518
Granada 115, 125
Great Brahmin 97
Greece, Greeks, Graecia 59, 70, 74, 77, 91, 99, 105–106, 108, 126, 137, 154, 181, 201, 243, 307, 440, 442, 444, 470–471, 481–482, 487, 511, 513, 522
Greek (language) 44, 66–67, 69–70, 77, 81, 311, 344, 378, 430, 436, 470, 483, 513, 517
Greek authors 88
Greek books / sources 95, 243, 311, 349, 378
Greek history 119
Greek infidels 345
Greek inscription 492
Greek philosophers 45, 430, 442
ancient Greece / Greeks 152, 200, 202, 309, 379, 442
Groenlandia 119–120
Guinea 93, 132
Gujarat, Gujarati, Cambaia, Kenbāyet 77, 93, 101–102, 181–185, 190–192, 195, 197–198, 200, 204, 206
Bay of Cambaia (or, Gulf of Cambay) 182
Gulf of Kenbāyet 182
musicians of Cambaia 194
rulers of Cambaia, Cambaia sultanate 195
Gujarati king, Sultanate of Gujarat, Sultans of Gujarat, ruler of Gujarat 182, 194, 198
Gujarati Sultan Maḥmūd II 182
Gulābād 278
Gulbarga 199
Gulgūn 326
Gulistān Castle 323
Gulistān, Mt. 278
Gulshan-i rāz (book by Shabistarī) 314
Gūrān (clan) 358, 361
emirs of Gūrān 360
Gurgān See Jurjān
Gurgān (village in Fārs) 237
Gurgānj, Gurganj
Greater Gurgānj (Gurgānj-i Kubrā, Ūrgānj) 289–290
Lesser Gurganj (Gurganj-i Ṣughrā) 289, 298
Gurgīn-i Mīlād (Iranian hero) 361
Guril 332, 349
Gūsfand River See Ḳoyun River
Gushtāsb (Kayānid) 227, 235–236, 243, 266, 292, 322, 327
Gushtāsbī, Gushtāsfī 322, 324, 327
Gutia, Jutia (Jutland) 69
Guwār 239
Guwwa See Goa
Gücenes 500
Güdül 492–493
Gül-ʿanber castle 359
Gülşen-i Tevḥīd (book by Dāwūd Ḳaramanī) 498
Gümsi 352
Gümüş 480–481, 484, 512
Gümüşābād 501
Gümüşḫāne See Urla
Günbāzārı 501
Gündüz 514
Güngörmezler Ġārı 500
Günṭoġdı 513
Günyüzi 469, 501
Güre 485
Gürgür Baba 362
Gürz 513
Güvercinlik, Güvercinlik Bridge 509. 512, 518
Güzelceḥiṣār 519
Güzelḥiṣār 481, 486–487, 512
Gwalior 193

Ḥabāba 423
Ḥabb 384, 388
Ḥabba 367
Ḥabesh See Ethiopia
Sea of Ḥabesh 109
Ḥābilishī 217
Ḥabīb al-Najjār 460
Ḥabīb al-siyar (book) 161, 175, 183–185, 194, 220, 270, 292–293
Ḥābūr 353, 356
Ḥābūse (village) 347
Ḥācı Baba 480, 499
Ḥācı Bayram Velī 492, 521
Ḥācı Bektaş (town) 476
Ḥācı Ḥamza (village) 479, 481, 495, 512
Ḥācıköyi 480–481, 512
Ḥācımurād 477, 482
Ḥadda 407
Hadfat al-Buwayb 422
al-Hādī bi'llāh, Mūsā (Abbasid), Hādī ʿĪsā 254, 368, 378
Ḥadīqa (book by Ḥakīm Sanāʾī) 212
Hadīr 366
Ḥadītha 346, 351, 363, 367, 371, 374–375, 433
Ḥadr 410
Ḥaḍramawt 385, 388–389, 391, 416, 419–420, 424–425, 427
Ḥaḍruma 414
Ḥāfiẓ of Shiraz 65, 233, 335
Ḥafrābād 375
Ḥafrband 290
Ḥafṣa 480
Haft iqlīm (book by al-Rāzī, Amīn Aḥmad) 44, 182–184, 187–188, 191–193, 201, 211, 216, 219, 224, 228, 235–236, 247, 249, 253, 255–256, 267, 269, 271, 289, 292, 294–299, 303, 314–315, 322–323, 333
Hagar mother of Ishmael 396
Ḥāh, Mt. 217
Haifa 442
Ḥāʾil 414–415
al-Hajar 392–393
al-Ḥajjāj b. Yūsuf (Umayyad governor of Iraq) 196, 227, 232, 236, 247, 257, 263, 292, 370, 399
Ḥajr 412, 414, 416
al-Ḥākim bi-Amr Allāh (Fatimid) 450–451
al-Ḥākim bi-Amr Allāh (Abbasid) 383
Ḥakīm Sanāʾī, Shaikh Abū'l-Majd Muḥammad b. Ādam 212
Ḥakīm al-Tirmidhī 298
Ḥakkārī 334, 339–341, 351–352, 359, 361–362, 373
Ḥakkārī clan of the Kurds 340
Hākuwās 218
Hāla 216
Hālan 218
Halāward 298, 300
Ḥalazūn pass 412
Ḥalazūn River 432
Ḥalfā [Jabaltā] 375
Halicarnassus 96
Ḥalī (Red Sea port) 383, 409, 415, 418
Halīk 300
Ḥalīl Pasha (vizier) 527
al-Ḥallāj, Ḥusayn Manṣūr 235, 368
Ḥalqa 460
al-Halūliyya 455
Ḥāluy (tribe) 361
Ham 104
Ḥamā, Ḥamāh (city) 42, 430, 434–436, 454, 456–459, 465–468, 524
Hāma (village) 433, 450
Hamad 229
Hamadān, Hamadānī 125, 200, 231, 245, 251, 255, 257–260, 262–264, 314, 359, 369, 372, 375–376, 381, 467
Ḥamāl (village) 356
Ḥamāma 363
Ḥamāmlu 480, 512
Ḥamās 346
Ḥamd Abū Nuʿayr 456
Ḥamd Allāh Muṣṭawfī (author of *Nuzha*) 44, 232, 234, 237–238, 246, 249, 251, 253–254, 256, 259–261, 265, 272–277, 285–287, 290, 299, 312–317, 319–322, 324, 334–335, 338, 366
Hamdān (district in Arabia), Hamdānīs 390, 416, 429
Ḥamdān River 158–159
Ḥamdānids, Ḥamdānid dynasty 352, 431
al-Ḥamdiyya 413

Ḥamīd, Ḥamīdeli 469, 483–485, 487, 489–491, 512, 516
Lake Ḥamīd 491
Ḥamīdiyya (clan) 351
ʿAqr al-Ḥamīdiyya 351
Ḥāmīm b. ʿAmlīq 451
Ḥāmiyya (village) 450
Ḥammūda 404
Ḥamnā 375
Hampardan, Mt. 216
Ḥamr al-Asad 407
Ḥamr, Ḥamrīn, Mt. 358–359
Ḥamra 372
Mt. Ḥamra 372
Ḥamrāʾ 422
Ḥamrāʾ Anfūr 384
Hamshahra, Abrshahr 322
Ḥamur Mountains 346
Ḥamza (uncle of the Prophet) 408
Ḥamzanāma (book) 456
Ḥamza Bey (village) 514
Ḥanāb 230
Ḥanafī(s) 209, 239, 241, 246, 252–253, 255, 271, 284, 289, 293–295, 297–298, 303, 321, 356, 359, 375, 384, 400, 416, 419, 445–447, 481, 513, 524
Ḥanafī *maqām* (in Kaʿba) 427
Ḥanbalī(s); and see Aḥmad b. Ḥanbal 360, 375, 416, 419, 445–446, 524
Ḥānī 355
Ḥanīf Banī ʿAmr 422
Hanjān 229
Ḫānlūc 347
Hanno 99
Ḫāntepesi 352
Ḥanya (Canea) 527
Ḥanẓala b. Ṣafwān (prophet) 389
Hapāziyas (Hyphasis) River 203
Ḥaqshiyān (fortress) 362
Ḥarādīn 339
Ḥaram al-Sharīf, Holy Precinct / Sanctuary 395, 400–401, 403, 406, 416
Ḥarār 427
Ḥarasta 433, 452
al-Harawī, Shaikh Abu'l-Ḥasan ʿAlī b. Abī Bakr 39
Harawī, Muḥammad Yūsuf 200
Ḥarfān 300
Ḥarība 410
Ḥārim 433, 457–458, 460, 465
Ḥarīr 340–341, 358–361
Ḥarīra 423
Harīrūd See Herat River
al-Ḥārith b. Hishām (companion of the Prophet) 442
al-Ḥārith b. ʿAmr b. Ḥujr al-Kindī 357, 377
Ḥāritha (tribe) 409, 431
Ḥāriz River 352
Ḥarmā 235
Ḥarmal, Mt. 456
Harmancık, Ḥarmanḳaya 501–502, 514
Ḥarmash 293
Ḫarpert, Harput 352–354, 347, 434, 457
Ḥarqal 352
Ḥarrān 125, 357–358, 462
Ḥarrāniyya (sect) 436
Ḥarrat Banī Sulaym 407
Ḥarth b. Ẓālim (Ghassānid) 415
Hārūn al-Rashīd (Abbasid) 227, 253–254, 263, 368, 378–379, 394, 461, 463–464, 492
Hārūnābād 375
Hārūniyya 372, 375, 435, 457, 461
Hārūniyya Bridge 367
Harūr 352
Ḥarūrī 222
Hārūt and Mārūt 368
Harūyāna 350
Ḫarzem 356
Ḥarzmiyya 433
Ḥasan, Mt. 475, 479
Ḥasan b. ʿAjlān (Sharīf of Mecca) 402
Ḥasan b. ʿAlī b. Abī Ṭālib, al-Ḥasan al-Sibṭ al-Kabīr 427, 429
Ḥasan b. Sahl, vizier 370
Ḥasan b. Salāma, vizier 384
Ḥasan Agha, Ḳapu Aġası 522
Ḥasan al-ʿAskarī, Imam 372
Ḥasan Batrīḳ (town) 462, 481
Ḥasan Bey, Ḥasan-ı Ṭawīl (Aqqoyunlu) See Uzun Ḥasan
Ḥasan Bey (Maḥmūdī) 340
Ḥasan Çelebi (village) 481
Ḥasan-i Ṣabbāḥ See Ṣabbāḥ
Ḥasana (in Arabia) 424
Ḥasanābād 229, 359
Ḥasanḳalʿe, Ḥasanḳalʿesi 345–346
Ḫāṣekī Meḥmed Pasha 371
Ḫāṣekī Sulṭān (wife of Sultan Süleyman) 394, 405, 525
Ḥāshikiyya (village) 375
Hāshim b. ʿAbd Manāf 437
Hashimite clan, Hāshimiyya 379, 429
Hasht Bihisht (book by Ḥakīm Idrīs Bidlīsī) 337
Hashtdar 325
Hashtrūd River 318
Ḥāsik 383
Ḥasiyya 453, 456
Ḥasnāʾ 411
Ḥasnigāh 288
Ḥasnūya dynasty 236
Ḥassān b. Tubbaʿ al-Ḥimyarī 413–414, 426
Hastān 222
Hastapālūt 218
Hastawnaz, Mt. 218
Ḥaṣum 423
Ḥatābād 278
Hātapa 219
Ḫātem Sultan (daughter of Sultan Süleymān) 403
Ḫato See Ḫazo
Ḫātūnçayırı 481
Ḥawās 229
Ḥawāthit [Ḥawānīt] 375
Ḥawḍ 277
Ḥawḍiyya 412
Ḥawīja Canal (of Herat) 266
Hawīn 218
Hawīn mountain 249
Ḥawīṭāt (tribe) 411, 422, 431
Ḥawīza 246–248
Ḥawla 451
Ḥawlān: Bilād Ḥawlān 390, 428
Ḥiṣn Ḥawlān 428
Hawmat al-Batt 247
al-Ḥawrā 423
Ḥawrān 420, 431, 435, 443, 451–452, 455, 465–466
Ḥawz al-Sayf 233
Ḥayāṣir 430
Ḥayāt al-ḥayawān (book) 44
Hayʾat (i.e., *al-Mulakhkhaṣ fī'l-hayʾa*, book by Jaghmīnī) 119
Hayʾat (of Erzurum) 346
Hayāṭila (Hephthalites) 224, 268, 296
Ḥaydar, Shaikh (Safavid) 32
Ḥaydar, Bahā al-Dīn 255
Haydarbaġı 335
Ḥaydargöliler pasture 347
Ḥaydariyya (castle and river) 255
Hāyizān 247
Ḥayma, Wādī'l-Ḥayma 429
Haymana 476, 492–493
Ḫayr al-Dīn Pasha 517–518
Ḫayrabolı 518
Ḥayrān 423
Ḥays, Ḥays Shaddād 383–385, 423
Hazara (in Afghanistan) 222, 244
Hazara (in India) 191, 193, 203, 206, 209, 211, 219
Hazara River 193, 202, 213
Hazarasp 289–290, 300
Hazārdirakht 234
Ḫazo, Ḫato 353, 355–356
Ḥazqiyya Turkmen 451
Hebān 352
Hebrew 425, 434, 436, 457
Hebron, Bayt Ḥabrūn, Khalīl al-Raḥmān 411, 431–432, 437–438, 441, 464–465, 521–524
Ḥekīm Ḫānı 346, 462, 481
Helekne (fortress) 482
Helen of Troy 509
Helena mother of Constantine 438
Hellespont, Mediterranean Sea Strait, Dardanelles 91, 96, 105, 110, 518
Helmand River, Hindmand 221–222
Hemnişin (Hemşin) 331
Ḫendek 503, 512
Henli (Turkic clans) 307
Herat, Heratis 161, 198, 213, 222–225, 235, 265–272, 275, 278–284, 298
Herat River, Harīrūd 266, 268, 271–272
Kāzargāh of Herat 235, 267
fire temple of Herat 265
rulers of Herat 279–284
Herbīl (fortress) 361

Hercules 107, 196
Pillars of Hercules 91, 96
Strait of Hercules See Ceuta: Strait of Ceuta
Hereke 511
Hermas River 354
Hermerur (?), Mt. 500
Hermon, Mt. 432, 433, 451
Herodotus 105, 108
Hersek 505, 508, 511
Hersekzāde Aḥmed Pasha (grand vizier) 505
Hesperides 117, 136
Hetman Doroshenko 527
Heyhāt Plain 330
Heyṣem 355
Hezārfen Ḥüseyin Efendi 417
Hezārmerd 359–360
Ḥibān 229
Hībang 219
Hibernia (Ireland) 40, 69, 90, 106, 113
Ḥidār 210
al-Hidāya (book by al-Marghīnānī) 297, 522
Ḥijār 423
Ḥijāz 84, 100, 124–125, 325, 337, 371, 380, 383–385, 388, 390–391, 393, 398, 406, 409, 411–412, 414, 417, 419, 429, 435–436, 442
Hijr, Ḥijr Ṣāliḥ, Qurā Ṣāliḥ, Dhāt al-Ḥajj 383–384, 393, 408, 419, 421, 425
Hijra (in Arabia) 424
Ḥilla 346, 366–369, 371–372, 374, 384
Hīm 285
Ḥimṣ 125, 177, 325, 358, 430–432, 434–435, 454, 456, 465–468, 524
Himyar, Ḥimyar, Himyarite 386, 414, 425
Himyarite (language) 391, 417
Himyarite kings, kings of Ḥimyar 311, 389, 391, 426
Tubbaʿ the Himyarite 395
al-Ḥimyarī, Abū ʿAbd Allāh Muḥammad b. ʿAbd al-Nūr 43
Hind See India; Sind: Sind and Hind
Hind River See Indus
Hindi (language) 185, 436
Hind-i Ġarbī See *Tārīḫ-i Hind-i Ġarbī*
Hindū 279
Hindu(s) 77, 183, 186–188, 191, 193, 200–201, 203–204
Hindu Kush 211, 212, 214
Hindus (fortress, Konya) 472
Hinduwāna (fortress, Balkh), Qalʿa-i Hinduwān 269, 302
Hīnī River 373
Ḥınūs, Ḥınıs 319, 338–339, 343–345, 481
Hipparis Pelagus (island in the Eastern Ocean) 90
Ḥīr 233
Ḥīra 97, 370–371, 376–377, 391
Ḥīra mountains 372
Ḥīrī, Bilād Ḥīrī 383, 391
Hīrpūr 218
Hīrpūr Road 218
Ḥırvat See Croatia
Ḥisā 420
Ḥiṣār (in Arabia) 424
Ḥiṣār (in Transoxiana) 300
Ḥisārlıḳ 512
Ḥiṣāröñi 497, 498–500
Hishām b. ʿAbd al-Malik (Umayyad caliph) 327, 378, 458, 466
Ḥīsiyya 424
Ḥiṣn al-Akrād (Krak des Chevaliers) 435, 454, 456–457
Ḥisn Banī Mūsā 424
Ḥiṣn Barziyya 432, 458
Ḥiṣn Ḍariyya 412
Ḥiṣn Ḥadath (Göynük) 466
Ḥiṣn Ḥamūs 464
Ḥiṣn Harbād 455
Ḥiṣn Hudāyā 462
Ḥiṣn Ibn ʿUmāra See Āl ʿUmāra
Ḥiṣn Kahlā 457
Ḥıṣn Kayfā (Hasankeyf) 353, 355
Ḥiṣn Mahdī 248
Ḥiṣn Manṣūr 431, 462–463
Ḥiṣn al-Marādīs 454
Ḥiṣn al-Qarnayn 457
Ḥiṣn al-Ṭāq 221
Ḥiṣn Tall Bāshūk 458
Ḥiṣn Ziyād 354, 434
Hispaniola 99, 112–113, 115–117
History of Khurāsān (book by Muʿīn al-Dīn Zamjī) 265
History of the New World See *Tārīḫ-i Hind-i Ġarbī*
Hīt 350, 367, 371, 384
ʿAyn Hīt 413
Hīzal River 373
Ḥīzān 334, 341, 353
Ḥıżır Pasha 521
Ḥıżırlıḳ (in Kütahya) 484
Ḥıżırlıḳ plain 481
Hlor 341
Holland, Dutch 39, 41, 70, 75, 99, 121, 132, 156, 181, 185
Dutch language 119
Holy Land 420, 430, 454
Holy Sanctuary, Holy Precinct (in Mecca) See Ḥaram al-Sharīf
Holy Scriptures 145
Holy Sepulcher (in Jerusalem) 438–440, 467
Church of the Holy Sepulcher, al-Qumāma 438–440, 467
Ḥoma 484–486
Ḥonaz 485
Hondius, Ludovicus / Jodocus 39, 41, 132
Hormuz 42, 76, 102–103, 209–211, 227–231, 238, 241, 276, 361, 415
Old Hormuz 228–229
Gulf of Hormuz, Sea of Hormuz 101, 103, 227
Strait of Hormuz 101, 240
sultans of Hormuz 231
Ḥoşāb 339–340
Ḥoşāb River 340
Ḥoşoğlan Mountains; Ḥoşoğlan River 349
Ḥoyrān 489–490
Hubal (idol) 398, 416
Ḥūbya 454
Hūd (prophet) 383, 389, 391, 425–426
Ḥudaybiya 407
Ḥudayda 415–416
Hudayya 421
Ḥudhayfa b. al-Yamān 267
treaty of Ḥudhayfa b. al-Yamān 319
Hudhayla (tribe) 406
Ḥūḥ (village) 357
Ḥujr b. ʿAdī 453
Hulagu Khan 244, 249, 265, 274, 282, 306, 317–319, 321, 341, 352, 366, 382–383, 467–468, 523
Ḥulam Pass 299
Hulbuk 224
Ḥulwān 125, 262–263, 359, 362–363, 366, 369–370, 375
Ḥulwān River 370
Ḥūm 222
Humā, Humāy (Sasanian princess) 234, 243, 257
Ḥūmanjān 262
Humāyūn Mīrzā, Muḥammad (Mughal emperor) 191, 198–199
Ḥunayn 394, 407, 410
Hunbaracı Meḥmed Agha 330
Hungary, Hungarians, Magyar 106–107, 519–521, 524–526
Hungarian campaigns 320
Ḥuqūqiyyat Fayd 424
Ḥūrā, Bilād Ḥūr 385
Ḥūrāb 278
Ḥurda (island) 415
Ḥūrjūliyya 433
Hurmuz (Sasanian) 247
Ḥusām al-Dīn, Mawlānā 337
Ḥusayn, Sultan of Lucknow 197–198
Ḥusayn b. ʿAlī, Imām 371, 377, 429, 435, 437
Ḥusayn b. Salāma (Ziyādid vizier) 427
Ḥusayn Bayqara, Sultan (Timurid) 267–269, 284, 302
Ḥusayn Khalkhālī, Mawlānā 316
Ḥusayn al-Kurdī, Amīr 427–428
Ḥusayn-i Qādirī, Sayyid 209
Ḥusām al-Dīn ʿIwaḍ 200
Ḥusbān 442
Hūshang (ancient Persian king) 234, 368
Ḥūstū mountain pass 347
Ḥūstū Ḫānı 347
Ḥūṭa 412, 414
Hūya (caves) 216
Ḥüdāvendigār (sanjak) 483–484, 486, 491–492, 497, 501, 504–505
Ḥüseyin Ghazi (dervish lodge in Ankara) 492
Ḥüseyin Pasha 366
Ḥüseynābād 477–478, 481
Ḥüsrev Pasha 258, 335, 337
Ḥüsrev Pasha Ḫānı 511
Ḥvāce Aḥmed (near Sivas) 347
Hospice of Ḥvāce Aḥmed 347

Ibāḍī rite 392
Ibeza See Yābisa
Ibn ʿAbbās (traditionalist) 87, 395
Ibn Abī al-Sāj 317
Ibn ʿAlqamī 382–383
Ibn al-ʿArabī, Muḥyī'l-Dīn, al-Shaikh al-Akbar 449, 524
Ibn al-Athīr 151, 287, 392
Ibn al-Furāt (historian) 265, 289
Ibn Ḥawqal 42–43, 84, 209, 224, 240, 288, 295–299, 304
Ibn al-Ḥuṣnī, endowment of 420
Ibn Jazla 42
Ibn al-Kalbī 227
Ibn Khālawayh al-Hamadānī 258
Ibn Khurdād (Abbasid vizier) 312
Ibn Khurradādhbih (geographer) 42, 208
Ibn Maʿn See Maʿnoġlı
Ibn Mundhir (Shāfiʿī scholar) 416
Ibn Saʿīd al-Maghribī 43, 151, 159, 219, 304, 311, 318, 321–322
Ibn al-Shāṭir (astronomer) 467
Ibn al-Wardī, Shaikh Zayn al-Dīn ʿUmar b. Muẓaffar 43, 84, 151
Ibn Zubayr, ʿAbd Allāh 399, 429, 452
Ibrāhīm See Abraham
İbrāhim, Ottoman sultan 35, 527
İbrāhīm Bey 336, 516
Ibrāhīm b. Adham 368, 438, 455
Ibrāhīm b. al-Walīd (Umayyad) 467
Ibrāhīm Khan, Sultan Ibrāhīm (i.e., Ibrāhīm Khan Lodi, Afghan sultan of Delhi) 198–199, 213
Ibrāhīm Khan, Sultan of Lār 231
Ibrāhīm al-Mawṣilī (singer) 378
İbrāhīm Müteferriḳa (referred to in the text as "the Publisher") 35–36, 40, 44, 48, 50, 75, 82, 330, 343, 349, 358, 363, 483, 528
İbrāhīm Pasha, Grand Vizier 334, 404, 423, 526
İbsala 518
Icaria 120
İçel 469–470, 472
Iceland 40
Īdaj 248–249, 262
Idlib Ṣughrā and Idlib Kubrā 460
idol-worshippers 30, 129, 132–133, 136, 143, 148, 150, 162, 173, 182, 186–188, 192–194, 200, 369
Idrīs (prophet) 44, 84
Well of Idrīs 404
İdrīs-i Bidlisī, Ḥakīm, Molla, Mawlānā 337, 523
al-Idrīsī, Sharīf Muḥammad b. Muḥammad 44, 84
Idumea 124
Ifrīqiya (identified with Libya; and see Tūnis, Tunisia) 97
İgdir 489
Īj 236, 244
İḥtimā (village) 471
Ikhshidids 467
Ikhtiyār al-Dīn (castle) 265–266, 283
İki Aḫur 347
İki İmām 375
İkiḳapulu 453, 465
Iklīd 235
Īlāq 300–301
Īlāq River 295
Ilduz (Turkish commander) 226
Iletmish (Iltutmush) Dynasty, Iletmishids 196
Ilġın, Ilġun 475, 511
İlhāmu'l-muḳaddes mine'l-feyżi'l-aḳdes (book by Kātib Çelebi) 49
Ilıbat 512
Ilıca 346–347, 481, 486, 501, 513
Ilıcaḫān 481
Ilḳas (Ilgaz) 493
Ilḳas Mountains, Ilḳas Ṭaġı 494–495
Īlkhān 326
Ilkhanid(s); and see Chingisids 282, 366, 383
Ilkhanid army 321
Illyricum 91
ʿImād al-Mulk See Daryā Khan
ʿImād Yāsir, Shaikh 337
ʿİmādiye, ʿImādiyya 340–341, 352, 359, 361, 367, 373–374
ʿİmādiye plain 341
Imām ʿAlī (in Iraq) 351, 367, 371
Imām-ı Aʿẓam See Abū Ḥanīfa
Imām-ı Aʿẓam (in Iraq) 367–368, 370
Imām Mūsā (in Iraq) 368, 371
Īmān-ı Şāh, defile of 360
Imaus Mountains (i.e., Caucasus) 161, 179, 307, 309–310
Imruʾ al-Qays (poet) 357, 376, 415
Īnānj 305
İnay, İnay-Eşme 485
İncegiz 519
İnceṣu 330, 352, 481
İncīrlü 507
India, Indians, Hind, Hindustan 36, 42, 53, 69, 77, 81, 90, 97, 99–101, 103, 110–111, 117, 124–125, 129, 133, 139, 143, 145, 147–149, 150, 152, 157, 160–161, 163, 175, 181–204, 215–220, 222–225, 251–252, 265, 282, 291, 302, 305, 363, 366, 368, 385, 391–392, 400, 402, 406, 425, 417, 428, 445, 457, 466
India boats 141
India campaign 427
India Orientalis 181
India Pass 215
Indian captives 99
Indian lords 185
Indian merchants 142
Indian Ocean, Sea of India See Eastern Sea; Pacific
Indian plants 391
Indian tutty 234
histories of India 192
History of India (book by Mīrzā Ḥaydar, i.e., *Tārīkh-i Rashīdī*) 219
islands of India See East Indies
kings / rulers / Emperor / Raja of India 190, 193–196, 200, 221, 362
pagans / heathens of India 192, 211
pagan Indian kingdoms 195
Indo-China (*Hind u Chīn*) 102
Indus River, Hind River, Mehran River, Sind River 93, 111, 181–183, 190, 193, 196, 202–203, 206–208, 210, 213–214, 282–283
İne 487, 510
İnebaḫtı (Lepanto) 522
İnebāzārı 509, 512
İnebolı 496
İnecik 518
İnegöl (near Bursa) 501, 503, 512, 514–515, 517
İnegöl (of Aydın) 487–488
İnegöz 486
Inhabited Quarter 123
Injīl River 267, 269
Īnjūids 245
İnöñi, Bozöyük 483, 488, 491–492, 511–512, 514
Introductio (book by Cluverius) 43, 44, 73, 90, 99, 104, 106, 108, 110, 112–113, 122
İpşir Pasha 450
Iqlīm al-Tuffāḥ 450, 453
Īrad 234
Iraj son of Farīdūn 482
Iram Dhāt al-ʿImād, Garden of Iram 184, 391, 424
Iram-like buildings 192
Iran, Iranians 77, 232, 242, 251, 263, 291–292, 298, 302, 306, 313, 316, 320, 334, 336–337, 347, 352, 366, 372, 513
Īrān Shahr 251
cities of Iran 253
conquest of Iran 302
kingdoms of Iran 263
padishahs / ruler(s) of Iran 242, 320, 400
Shah of Iran 348, 367
Iraq, Iraqi, Arab Iraq, ʿIrāq al-ʿArab 84, 221, 231, 243, 246–251, 253, 260–262, 273, 279, 281–284, 291, 296, 301–302, 312, 319–320, 325, 336, 350, 358, 361, 363, 366–367, 369, 371, 374–377, 381–383, 395, 414, 431, 465, 467
Persian Iraq See Jabal
the two Iraqs 254, 334, 376
Iraqi army 281
Iraqi pilgrims 415
Iraqi *raṭl* 177
Iraq road 259
governor(s) of Iraq 263, 279, 369
kings / sultans of Iraq 249, 291
seven cities of Iraq 370
ʿIrāq al-ʿAjam See Jabal
Irbil 341, 351
Īṣ 424
ʿĪsā, al-Malik al-Muʿaẓẓam (Ayyubid) 409
ʿĪsā Beylü (emirs) 339
Isaac 396, 411, 430, 438
Isabella 116
Isaforos (?) mountain 490

İsaḳçı (Isaccea) 520, 527–528
Isbīdbalān 297
Isbījāb See Isfījāb
Iṣfahān 125, 231, 235, 241, 244–249, 251–252, 255, 257–258, 260–265, 276–277, 281, 291, 330, 375
  Iṣfahān collyrium 252
  governor of Iṣfahān 245
  meat and dairy products of Iṣfahān 260
Iṣfahānak 262
Iṣfahānī, Kamāl al-Dīn Ismāʿīl 252
Iṣfahbad 288
Isfandān 235
Isfandiyār (Kayānid) 243, 252, 257, 323
Isfarāyin 274, 279
Isfandiyarids (or Jandarids, Candaroğulları) 516, 520
Isfarāna 375
Isfījāb, Isfijāb 295–296, 298, 300
Isfirār 221
Isfizār 267–268, 275
  Rūd-i Isfizār 267
Isfūrān 275
İsḥāḳ Çelebi 486
İsḥāḳ Faḳīh 518
İsḥāḳlı, İsḥāḳlu 475, 511
Isḥāq b. Ibrāhīm 351
Isḥāqī Canal 351
Ishāra mountain 423
Ishārāt ilā maʿrifat al-ziyārāt (book by al-Harawī) 39
Ishmael 396, 397, 411, 425
Ishtīkhan 294–295
Isidorus 105
Işıḳlu 485
  Işıḳlu Plain / Valley 486
Isimābād 376
Iskalkand 224
Iskandar, Amīr (ruler in Zabīd) 428
Iskandar, Sultan (Sikandar Lodi, Sultan of Delhi) 192, 216
Iskandar, Sultan (of Kashmir) 219
Iskandar b. Qara Yūsuf 347
Iskandarūn See İskenderūn
İskelec 481, 479
İskender Pasha 365, 492, 507
İskenderūn, Iskandarūn (Alexandretta) 430, 436, 457, 460, 464, 470, 477, 483
Islandia See Thule
Islam, Islamic; and see Muslim 41, 87, 125, 141, 173184, 193, 195, 200, 206, 227, 231–232, 234, 243, 249, 251, 255, 263, 269, 271, 281, 287, 299, 304, 308, 327, 329, 333, 337, 371, 373, 379, 383, 386, 390–391, 395, 399, 401, 407, 414, 416, 419, 424, 426, 428, 444, 452–453, 465, 467, 471, 482, 518, 524–525
  Muhammadan sacred law 416
  Islamic authors 88, 99
  Islamic books 35, 106, 118–119, 122, 160, 182, 291
  Islamic conquest 196
  Islamic chronicles 200
  Islamic era / period / times 258, 286, 341
  Islamic geographers 68–73
  Islamic historians 242
  Islamic sources 151
  abode of Islam 466
  army of Islam 243, 260, 524–526
  banner of Islam 219
  cities of Islam 404
  Dār al-Islām (Islamic territory) 482
  "Dome of Islam" 363
  dynasty of Islam 352, 377, 513
  fortresses of Islam 528
  frontier of Islam 300
  history of Islam 36
  land(s) / countries of Islam 42–43, 277, 299, 304, 331, 379, 381–382
  people of Islam 40, 82, 115, 408
  pre-Islamic period / times 394, 437–438
Island of Women (Jazīrat al-Nisā) 147
Ismāʿīl, Shah (Safavid) 250–252, 271, 284, 302, 320, 327, 337, 339, 347, 357, 383, 522–524
Ismāʿīl II (Safavid) 320
Ismāʿīl ʿĀdil Khan 199
Ismāʿīl b. Burhān, Ismāʿīl Shah 200
Ismāʿīl b. Mūsā al-Kāẓim 235
Ismaʿīl-i Sāmānī (Samanid) 223, 272, 301
Ismaʿīlīs, Bāṭinīs, Ismāʿīlī rite 256–257, 264, 276, 436, 455–457, 461
Ismīd 304
İsmil 465, 473, 511
Ispaka 209, 211
Isparta 489–490, 512
İspir 343–345
Israel, Israelite(s) 161, 344, 383, 422, 430, 435–437, 441, 443, 451, 453, 456, 465, 468
Iṣṭahbānān 236
Iṣṭakhr 232, 234–235, 239–241, 320
Istālif 212
Istanbul, Constantinople, Three Towns
  Istanbul 72, 76–77, 83, 123, 328, 344, 366–367, 400–401, 418, 427–429, 449, 461–462, 466–467, 478, 482, 491, 497–498, 503–510, 515–516, 519–526, 528
  Constantinople 36, 41–42, 72–73, 83–84, 91, 107, 123–125, 323, 477, 482–483, 496–498, 502, 507, 528
  Three Towns (Eyyub, Galata and Üsküdar) 522
  Sea / Gulf of Istanbul / Constantinople See Bosphorus; Golden Horn; Sea of Marmara
  Strait(s) of Istanbul / Constantinople See Bosphorus; Hellespont
  king of Istanbul (Byzantine emperor) 471
  public baths of Istanbul 506
  qadi of Istanbul 42
  sultanic mosques of Istanbul 314
İstanköy (Kos) 483, 524–525
İstavroz 506
Iṣṭīfūn, Mt. 151
İstinar 512
İstiranca, Mt. 508
Istonos 513
ʿĪtā 410
Italy, Italian(s) 43, 69, 74, 77, 83, 91, 93, 96, 106–107, 112, 123
  Italian (language) 436
  Italian miles 75, 158, 176
Itarbum (Icarium?) 120
Ithbāt wājib wa tahdhīb (book) 316
Itil 304, 309
ʿItq (clan) 410
ʿItra 428
İvrindi 505, 512
Īwān, Mt. 372
ʿIyāḍ b. Ghanm 337, 347, 350, 352, 438
al-Iyādī, Aḥmad b. Abī Dāʾūd 460
Izmir 483, 485, 510–513
İzmit, İznikmid 500, 503, 505–507, 511–513, 516–519
  Gulf of İzmit 505–509, 511, 516
Iznik 492, 501–503, 505, 508, 511–512, 514–515, 517–518
  Lake İznik 508
İznikmid See İzmit
İzvançe (Zurawno) 527
ʿIzz al-Dīn (Aqqoyunlu) 347
ʿIzz al-Dīn (commander of Saladin) 442
ʿİzzeddīn Şīr Bey 341, 361

Jabal, al-Jibāl, Bilād-i Jabal, Persian Iraq, ʿIrāq al-ʿAjam 69, 84, 235, 244–246, 249, 251, 254, 260, 263–264, 276, 284–286, 297, 302, 318, 359–363
Jabal Aḥmar (in Arabia) 423
Jabal ʿAkkār 435
Jabal al-Aʿlā 460–461
Jabal al-Aqraʿ (in Syria) 455
Jabal ʿAwf 434
Jabal al-Bahrā 435
Jabal Bārīsha 460–461
Jabal al-Burayk 436
Jabal al-Bustān 435
Jabal Ḥafdama (near Mecca) 406
Jabal Ḥamra 374
Jabal Ḥirā (near Mecca) 406
Jabal Inān 408
Jabal ʿĪsā 434
Jabal Kalbūdī 434
Jabal-i Khākhāl 288
Jabal Khūshan 435
Jabal Mūsā 460
Jabal al-Nūr (near Mecca) 406
Jabal al-Nūr (near Misis) 464
Jabal Samʿān 436, 457, 461
Jabal Shaḥshabū 435
Jabal Summāq 435, 461
Jabal Tābūr (Mt. Tabor) 434
Jabal Ṭarṭūs 454
Jabal Thabīr (near Mecca) 406–407
Jabal al-Thalj (Mt. Hermon) 432–433, 435
Jabal Thawr (near Mecca) 406
Jabal al-Ṭūr (in Palestine) 434

Jabal Yūnus (in Turkistan) 307
Jabala 386, 412, 428, 430, 454–455, 466
Jaʿbar, Qalʿat Jaʿbar 358, 431, 457, 513–514
Jabbāba 388
Jabbār Arabs 458
Jabbūl 457–458
Jabha park 448
Jacob from Buscod 120
Jadīda (in Arabia) 407, 421
Jadīda-i ʿAfrād (in Iraq) 366
Jaʿfar al-Ṭayyār (companion of the Prophet) 442
Jaʿfarābād 233
Jaʿfariyya 372
Jaffa 430
al-Jaghmīnī (astronomer) 315
Jaghr 424
Jahānārā (book by Ghifārī) 327
Jahannam (Gehenna), valley of, Wādī Jahannam 438, 440
Jahānshāh (Qaraqoyunlu) 320
Jahīnkūt 202, 206–207, 209, 214
  Jahīnkūt Road 191
Jahrum 236, 239, 241
Jājurm (Jājarm) 274
Jakīnū 230
Jākīr 216, 217
Jāl 209–210
Jalājil 413
Jalāl al-Dawwānī 315
Jalāl al-Dīn ʿAtīqī 315
Jalāl al-Dīn Khaljī, Sultan 192
Jalāl al-Dīn Muḥammad al-Rashīdī 319
Jalandhar, Jālantar 191, 205
  Jālandar, Ravine of 487
Jalandī See Āl ʿUmāra
Jalayirid dynasty 383
Jaljūliyya 351, 441, 465
Jallād Khānī 352
Jalore 193
Jalūlā 369
Jalzūn River 432
Jam, Mt. 233
Jām 272, 279, 315
Jamad 215
Jamaica 113, 116
Jamāl al-Dīn Barka, Sayyid 270
Jamālnagar 218
Jāmas-Fala 145, 151
Jamāsa 373–374, 357–358
Jāmda 373
Jāmāsb 268, 335
Jāmhar 207, 209
Jāmiʿ al-uṣūl (book by Ibn al-Athīr) 287
Jamshīd (ancient Persian king) 234, 251, 260, 370
Jamū 217
Jāmūslu 375
Jān (in India) 217
Jān (tribe, in Iraq) 367
Janābād 375
Janābikh 412, 424
Jānbirdi al-Ghazālī 524
Jand 294
Jandal (companion of the Prophet) 414
Jandanūt 209
Jangband 300
Jangalābād 222
Jangī 222
Jangūla 367, 372
  Mt. Jangūla 372
Jāniknū 216, 217
Jankūj See Jahīnkūt
Jannāba 233, 238, 240
Jānpūr 218
Japan, Japanese 110, 127–132, 150, 153–154, 156–157, 174
  Japanese (language) 129
  ruler of Japan 153
  scholars of LKH (?) and Japan 154
Japheth son of Noah 1–4, 152, 180
Jaqākāwān 375
Jār 383, 410, 415
Jarāf 387, 390
Jarbādiqān 257, 286
Jardūr River 289
Jarīb 418
Jarīr 222
Jarjarāyā 370
Jarlān (tomb in Herat) 269
Jarm 424
Jarma 277
Jarmābān 241
Jarmānā 433
  Jarmānī Canal 433
Jarmaq 277
Jarna mountain 459
Jarūd 450, 453
Jaruh 237
Jarūn 228
Jarwān 230
Jasad al-Madfūn 454
Jasān 366–367, 372, 431
Jāshak 228, 230
Jasht 272
Jaunpur 197–198, 206
Java, Javan, Javanese, the two Javas 110, 122, 133, 141–143, 151, 188, 385
  Greater Java 141, 143
  Lesser Java 142
  Javan frankincense (storax gum) 142
Jawād, Yurtbāzār 324
  Jawād Bridge 324
Jawādrūz, defile of 375
Jawalān al-afkār (book by Ebū Bekr b. Behrām) 367, 430
Jawāzir 346, 365–367, 372–373
Jawf, Bilād Jawf 390, 414
Jawīm Abī Aḥmad 236
Jawlān (Golan) 451
  Jabal al-Jawlān (Golan Heights) 435
Jawlī, Atabeg (Shabānkāraid) 236–237
Jawlī, Fakhr al-Dawla (Seljukid) 240
Jawn Ḥashīsh 415
Jawn River (Jumna, Yamuna) 192, 197
Jawr 240–241
Jayḥān See Ceyḥān
Jayḥūn River, Āmū Ṣuyı, Āmūl-i Shaṭṭ (Amu Darya, Oxus) 84, 111, 177, 180, 212, 223–224, 226, 242, 265, 269, 275, 278–279, 289–291, 293–294, 297–300, 302, 307, 309
Jāyjrūd (Jājrūd) River 261
Jayl 372
Jayrūn 445–446
Jazʿa 412
Jāzān 383, 385, 388, 417, 428
Jazāʾir, Jazāʾir al-Shaṭṭ (in Iraq) 365, 374, 525
Jazāʾir, Jazāyir-i Gharb See Algiers
Jazāʾir Yūsuf 374
Jazalī, Mt. 393
Jazīma (Jadhīma) al-Waḍḍāḥ 376
al-Jazīra, Jazīra (upper Mesopotamia) 124–125, 302, 320, 349–355, 358–359, 362–363, 458, 468
  beys / governor / ruler of Jazīra 355, 376, 380
  governess of Jazīra 376
Jazīrat Ibn ʿUmar 350, 355–356, 373–374
Jedda 100, 102, 388, 393, 395, 398, 400, 403–404, 406–407, 409, 417–418, 427–428
Jengdū 175
Jenin 430–431, 441, 464
Jeremiah 430
Jericho, Arīḥā 430, 437–439, 442, 460
Jerusalem, Bayt Maqdis 111, 124–125, 192, 350–351, 369, 393, 403, 411, 430–431, 434–442, 452, 464–467, 521, 524
Jesuits 129
Jesus, the Messiah 112, 169, 186, 430, 434–435, 437–440, 446, 465
Jews, Jewish, Judaism 107, 158, 169, 173, 194, 200, 211, 239, 246, 252, 251, 255, 304, 329, 375, 380, 388, 390, 411, 434, 436, 442, 454, 458, 506, 513, 526
  Jewish cemetery 438
  Jewish clan 411
  Jewish customs 468
  Jewish homes 373
  Jewish population 411
  Jewish religion 426
  Jewish ruler 390
  "Jewish stones" 432
  Jewish temple 444
  Jewish tribes 408
  historian of the Jews 107
  Jews of Isfahan 252
  language of the Jews 408, 470
Jibāl See Jabal
Jifārāt 422
Jifr-i jāmiʿ (book) 336
Jīghān 240
Jīlān See Gīlān
Jiʿlān 394
Jilum 219
  Jilum (Jhelum) River 219
Jinār 198
Jīr 230

Jīra 441
Jirjīs (St. George) 350
Jirm 224–225
Jīruft 227, 229–230
Jisr al-Ḥadīd 433–434, 459–460, 465
Jisr Sūrān 369
Jisr Yaʿqūb 441, 451, 464
Joannes Barryos (João Barros) 157
Johannes (Jan Huygen van Linschoten) 148
Job 63, 430, 451–452
Book of Job 97
Jochi son of Chingis Khan 306
John the Baptist 430, 440, 444–446
Jonah 350–351
Jordan See Urdun
Joseph 440–442
Josephus 107
Joshua 507
Jū-yi Sard River 249, 261
Jubayl, Qalʿat Jubayl 451, 454
Jubba 414, 455
Jubbā 247
Jubba ʿAsāl 450
Jubbāʾī, Abū ʿAlī 247
Jubbat al-Kūrā 454
Jūda 424
Judam 457
Jūdat Ḥajr River 413
Judays (tribe) 408, 413–414, 426
Mt. Jūdī 354–356
Juḥfa 378, 409, 423
Juhra 424
Juft River, Rūd-i Juft 317–318
Jūkām 217
Julāb 357
Julius Caesar 99, 243
Jūma, al-Jūma 434, 458, 460–461
Jumāna 424
Jūminjān 298
Jumjuma Sultan 369
Junayd, Shaikh (Safavid) 32
Junayd al-Baghdādī 368
Jund 386, 420
Jundīsāpūr 246–249, 262
Junūn River 373–374
Jupiter 105, 443, 459
statue of Jupiter 111
Jūr 232–234
Jurbādiqān 262
Jurfayn 422
Jurjān, Gurgān 125, 237, 264–265, 274–275, 285–288, 323, 525
Jurjān River 286
Sea of Jurjān See Caspian
al-Jurjānī, Ṣāʿid b. ʿAlī 43
Jurhum (tribe) 396–397, 425, 465
Jūsiyya 434
Jutia See Gutia
Juwayn 240, 274
Juyl, Mt., Chuyl 218
Juyl Road 218
Kaʿba 183, 231, 343, 378, 380–381, 386–387, 394–395, 397–403, 405, 407, 416, 419–421, 423–424, 426–427, 429–430, 443–444, 465, 467, 524, 526–527
Ḳabadurak 513
Kābil (in Palestine) 430
Kabrān 277
Kabūd Jāma 286
Kabul 125, 198, 204, 207, 209, 211–215, 224–225, 275, 279, 302
Kabul River 202, 212–214
Kabul road 191, 206
Kābul-i Ṣughrā 212
Kabūtarkhāna 279
Kachahāmū 218
Kadawnī 217
Ḳāḍı Burhān al-Dīn 519
Ḳāḍıbāzārı 504
Ḳāḍıköyi (Kadıköy) 506
Kadīw River 318
Kadmī 424
al-Kāf (fortress) 455
Kafar Baṭna 433
Kafar Bīn 460
Kafar Jawr 451
Kafar Kannāh 441
Bilād Kafar Kanna 451
Kafar Sūsiyya 452
Kafarṭāb 430, 432, 456–457
Kāfir (village) 346
Kāfiristān 212
Kafr Bayyā 464
Kafr Tūthā 351
Ḳağızman 332, 345
Kaghadh-kunān See Khūnaj
Kahf 454, 456
Kahlān (tribe) 425–426, 428, 431
Kahrīz (village) 375
Kāhū 229
Kāhūk (fortress) 355
Kāj 277
Kājdār, Mt. 217
Kajū 213–215
Kajū Khan 213
Kajūr (castle) 287
Kakarjama 218
Kakheti 347
Kakhtā, Kākhta 457, 462
Kakjādū 216
Kākūyids, Kākūyid Dynasty 245, 264
Kalā 216
Kalāhkām 217
Kalālkū 213, 214
Kalamat 230
Kalanahāmū 218
Kalānī 218
Kālār 288
Kalār mountains 240
Kalbān 216
Kalbiyya, al-Kalbiyyīn, Bani Kalb (tribal faction; and see Yemen, Yemeni faction) 372, 435–436, 455
Ḳalʿe-i Cedīde 363
Ḳalʿe-i Sulṭāniye (Çanakkale) 509, 512
Ḳalʿecik 479, 491, 493–495
Kālibī 200
Kālif 275, 298
Kalīla wa Dimna (book) 196, 378
Kalinjar, Kālinjar 193, 198
Kalkā 229
Kalla-khāl River 321, 324
Kalmykia, Qalmāq, Kalmyks, Kalmyk Tatars 163–164, 167, 176, 178–179, 304
Kalmyk invasion 305
Kalmyk khans 180
Kalmyk Steppe 178
Kalmyk territory 303
Kaltin 218
Kalwa 206, 208
Kalwa River 208
Kamāl al-Dīn ʿAbd al-Razzāq 194
Kamāl al-Dīn Chalabī Bey 315
Kamāl al-Dīn Masʿūd, Mawlānā 323
Kamāl al-Dīn Ṭabīb-i Shīrāzī 337
Ḳamaniçe (Kamieniec) 527
Kamarān (island) 415
Kamārukh 237
Kambuja, Kembayed (Cambodia) 77, 190
Cape of Kambuja 190
Gulf of Kambuja (Gulf of Thailand) 190
Kāmfirūz 235, 240
pasture of Kāmfīrūz 239
Kāmid (in Arabia) 407
Kāmid (in Syria) 454
Kamīn 235
Kamīnābād 222
Kamīsha 366
Kāmlanūkūt 218
Kampar 216
Kampar road 203, 218
Kāmrasī tribe 355
Kanʿān son of Noah 435
Kanara 181, 186
Kanawānī 218
Kanbāwa 241
Kand-i Bādām 297
Kand-i Salafān 253
Kanda 300
Kandahar 207, 212, 222–224, 226, 228, 235–237
Kandahar road 205, 228
Ḳandīlli Çam 514
Ḳandillü Bāġçe 506
Ḳandırı 505–506
Ḳandiye (Candea) 527
Kandpār 218
Ḳanġal 481
Ḳanije (Kanizsa) 526
Ḳanıḳ 513
Kanīwa 103
Kānḳırı See Çankırı
Ḳanlıca 507
Kanlu Sevinç River 330
Kannauj, Qinnawj 188, 191, 193, 196–197, 204
Kannauj campaign 204
Kantamal 218

Kanzawal 218
Kanwit 216
Ḳaplan Pasha 527
Ḳapudan Pasha (admiral of the Ottoman fleet) 469, 527
province of the Ḳapudan Pasha 505, 509–510
Ḳapuṭaġı (island) 509–510
Ḳara Çelebizāde Maḥmūd Efendi 40
Ḳara Göl 498
Ḳara Murād Pasha 447
Ḳara Muṣṭafā Pasha, Grand Vizier 327, 332, 366, 407, 527–528
Ḳara Rüstem 518
Ḳara Ṣu 324, 338
Ḳaraaġaç 475, 491, 512
Ḳarabaş (Yazīdīs) 361
Karābja 384, 386, 423
Ḳarabulur 481
Ḳarabük Tekesi 500
Ḳaracaḥisār 513–515
Ḳaracaḳoyunlı 487, 510
Ḳaracalar 504
Ḳaracaşehir 491–492
Ḳaracaṭaġ 347, 352, 354, 360, 477, 513
Ḳaracaṭaġ River 356
Ḳaracavīrān 493, 495, 512
Ḳaradāsni 350
Ḳaradere 356
Ḳaraeli Lake 475
Ḳaraferye 520
Ḳaraḥiṣār (of Nigde) 474
Ḳaraḥiṣār (of Bursa) 501
Ḳaraḥiṣār Germegi 489
Ḳaraḥiṣār-ı Ṣābīn 345
Ḳaraḥiṣār-ı Ṣāḥib (Afyonkarahisar) 483–485, 490–491
Ḳaraḥiṣār-ı Şarḳī 343–346, 480, 522
Ḳaraḥiṣār-i Teke 470, 489
Karaj 257–258, 260–262
Karak, Karak Nūḥ 358, 411, 431, 433–436, 442
Karakh 246
Ḳaraḳoca 483
Ḳaraḳoġa 481, 512
Ḳaraḳovacalar 346
Ḳaraköprü 356
Ḳaraköy 356
Karam (fortress) 330
Ḳaraman 461, 463, 469, 471–472, 476–477, 483–484, 504, 511, 518, 524
Ḳaramanid(s) 472, 477, 520, 521
Ḳaramanoğlı İbrāhīm Bey 473
Karāmjān 222
Ḳaramürsel 505–506
Karan 216
Ḳarand 362, 372
Ḳarapıñar 465, 473–474, 476, 511
Ḳaraselendi 485
Ḳaraseydi 356
Ḳaraṣu 505, 511
Ḳaraṣu River 346, 463, 494
Ḳaraṭaġ 359, 367, 372–373, 481
Ḳaratepe 356
Ḳarayaʿḳūb 481
Ḳarayaylak 478
Karbalāʾ 346, 369, 371
Karbīl 352
Kardānrūd River 262
Kāre mountains 355
Ḳaresi 483, 501, 505, 509, 516, 518
Ḳarġa Çamı 351
Ḳarġabāzārı 347
Kargas 261
Mt. Kargas, Kūh-i Kargas 277
Ḳarġı, Ḳarġu 480, 493, 495
Ḳarġı River 470
Karḫ 244
Ḳarıbāzārı 493, 495, 501
Karīm (fortress) 366
Karīn (fortress) 375
Karinjiyān 288
Ḳarīṣī (clan) 355
Kārisī (tribe) 355
Karīwa-i Armaniyān See Armaniyān
Kārīz 269
Kārizbaşı 512
Kārīzsarī 222
Karj See Kerch
Kārjīkān 339–340
Kārkār 334, 341
Ḳarḳara (village) 495
Karkh (near Herat) 269
Karkh (in Iraq) 375
Karkha 248
Karkha River 248
Karkisa 358
Karkūna 222
Karmalīs 373
Kārmūḥ 356
Karmūsa 412
Kārnakūt 218
Karnū Road 219
Kars, Qārṣ Dhū'l-qadriyya 324, 331–344, 343, 346, 367, 461, 525
Kart, Ghurid kings of 226, 282
Ḳartal 505
Kartan (in India) 217
Kartan (in Iraq) 372
Kartnāma (book by Rabīʿī) 226
Karūd 222
Karun River See Tustar River
Karʿūn 229
Kārzīn 233
Ḳaş 489, 512
Kasal River 307–308
Kasallu Tatars 308
Kāsān 297, 301
Kash 292, 294, 296, 300–301
Kashāb 372
Kāshān, Qāshān 241, 253, 262–263, 276–277
Āb-i Kāshān River 262
Kāshghar 296–297, 303–305
Kashf al-mamālik (book by Khalīl b. Shāhīn al-Ẓāhirī) 43
al-Kāshī, Jamshīd 87
Kashmir 161, 183, 193, 195, 198, 206, 213, 215–219, 234, 303, 305
Kashmir mountains 202–204, 215–220
Kashmir road 180
Kashmir River 215
kings of Kashmir 219
Kashshāf 247
Ḳasımoğlı Çayı 356
Kasīnān 230
Kaskar 288, 291
Kasrawān 453
mountains of Kasrawān 433–434, 453
Ḳasṭamonı 348, 478, 482–483, 492, 493, 495–497, 500, 512, 516, 519–520
Kāt 289–290
Katha 235, 241
Kāthāl, Kātahāl 218–219
Kātib Çelebi, Muṣṭafā Ḫalīfe (referred to in the text as "the poor one," "this humble one," etc.) 35, 39–42, 73, 84–85, 87, 96–97, 101, 107, 110–111, 114, 117, 119, 132, 136, 146, 151, 162, 182, 187, 190, 258, 261, 308–309, 311–312, 322–324, 334, 338, 343
Katība 420
Ḳaṭırbel 347
Ḳaṭırlı, Ḳatırlu 503, 505
Ḳaṭırlı mountain 512
Katrū 216, 217, 218
Katrū Road 216
Ḳavaḳ 478, 507
Ḳavāsoġlı Turkmen 420
Kawādar 210
Kawaj See Kannauj
Kawār 234
Kāwbāy (tribe) 329
Kāwdawān 318
Kawkabān 384, 387–388, 417, 429, 452
Kāwkul 216
Kawkūr 278
Kawn 275
Kawnī 369
Kawzīd Chūsmān (village) 375
Kay-Khusraw (Kayānid) 231, 243, 249, 255, 315
Kay-Khusraw (Seljuk) 475, 477, 513, 514
Kay-Kāwus (Kayānid) 221, 242–243, 269, 292, 369
Kay-Kāwus (Seljuk) 477
Kay-Qubād (Kayānid) 242, 251, 327
Kay-Qubād, ʿAlāʾ al-Dīn (Seljuk) See ʿAlāʾ al-Dīn Kay-Qubād
Kayānīs, Kayānī dynasty, Kayāniyān, Kayānids 242–243, 352
Ḳayaş Gardens (of Ankara) 492
Ḳayaş Ṣuyı 492
Kaydaram 279
Kaylūr [Kangūr] 375
Ḳaynarca 352
Kaysāniyya sect 410
Kayseri, Qaysāriyya, Caesarea 125, 434, 461, 466, 472, 476, 481–482, 513, 522
Kayūmarth 376, 424

Ḳaz Ṭaġı (Mt. Ida) 503, 509
Kaz Valley 217
Ḳazābād 477, 513
Kazak, Kazaks (or Cossacks) 42
Kazan 308–309
  Kazan Horde 307–308
Ḳazanḳaya 481, 513
Kāzargāh See Herat
Kāzarūn 230, 237–239, 241, 244
Ḳazgölü 482
Ḳazıḳbeli, Ḳazıḳlubeli 485, 490, 511–512
Kāẓima 393
Ḳazlıgöl 481
Ḳazlıḳbalḳanı 513
Kebkir 484
Keçiborlu 489–490, 512
Kefe (Kaffa) 309, 330, 478, 521, 525
  Kefe tulip 486
  Khan of Kefe 525
Kefender 354–356
Kefke 506
Kefr-i Zemān (Kafr Zamān) 356
Kekkerān 193
Kelbeyin 504
Kelehrān (clan) 361–362
Keler 487
Keles 487–488
Kelkīd, Kelkīt 343–344, 346
Kelkiraz 512
Kemaḫ 343, 344, 346, 482, 524
  Mt. Kemāḫ 347
Kemāl Re'īs 42
Kemālpaşazāde 291
Kembayed See Kambuja
Kemer, Kemer-i Edremid 461, 489, 505
Kenān Pasha, Vizier 333
Kenbayet See Gujarat
Kenjū (post-station) 172
Kenīse 461
Kenken 102–103
Kense See Quinsai
Kenzik 356
Kepirbeli 481, 512
Kerch, Karj 323
  Strait of Kerch, Taman Strait 91, 329–330
Kerpe 506
Keşāb 348, 360
Kesani 334
Kesikköprü 481
Keşiş Ṭaġı, Cebel-i Rāhib, Rāḥib Ṭaġı (Uludağ) 501–504, 517
Keşīşḫānı 347
Keskin 481, 493, 513
Kesme (village) 479
Kestāne 340–341, 361
  Mt. Kestāne 487–488
Kestel, Nāzilli 487, 512
Keşan See Turḫal
Ketḫüdā (village) 481
Kevelkū River 362
Keysān 341
Khabab al-Baza 423
Khābarān 271
Khabiṣ 223, 227–228, 277
Khabīṣī (author of book on logic) 337
Khabrar 233
Khābsār 222
Khābūr 357–358
  Khābūr River 341, 346, 351, 373
Khabūshān 274
Khadaysar 298
Khaḍir (in Iraq) 351
Khaḍrā', 'Ayn 413
Khajrak (castle) 324
Khalaj, Qālāj 222, 295, 305–306, 333
  Khalaj emirs 196
Khalam 279
Khalāṭ See Aḫlāṭ
Khālid (in Iraq) 366–367
Khālid b. 'Abd Allāh (governor of Mecca) 399
Khālid b. Sa'īd b. al-'Āṣ 452
Khālid b. Sinān al-'Absī 436
Khālid b. Yazīd 444
Khālid b. al-Walīd 97, 377, 414, 431, 444, 456, 461, 465, 516
Khalīl, Shirwānshāh 320
Khalīl b. Aḥmad (philologist) 395
Khalīl al-Raḥmān See Hebron
Khāliṣ (district of Baghdad) 367, 370
Khāliṣ (district of Nahrawān) 370
Khaljī Dynasty 196–197
  Khalj emirs 196
  Khalj kings of Oudh 200
  Khaljis of Ghūr 200
Khalkhāl 299, 312, 315–316
Khalṭ-i Kalām (cave) 359
Khamīma 443
Khamrūd 230
Khān al-'Arār 450
Khān Āzād 239
Khān Lanjān 260
Khān Maysalūn 450
Khān Shaykhūn 456, 465
Khān Sudūd 437, 465
Khān Tūmān 457, 465
Khān Yūnus 465, 524
Khān al-Zayt 420
Khānasitān 234
Khandkhāla River 288
Khānī Beg Khan 347
Khāniqī: Old Khāniqī and New Khāniqī 375
Khāniqīn 263, 369, 372, 375
Khānīsār, Mt. 262
Khanjān 212, 214, 237
Khānpūr 218
Khanbaliq (Peking), Khanbaligh 157, 162–163, 165, 167, 169–170, 172–173, 175–180, 308
Khānqāh (in Egypt) 465
Khānqāh Shamāṭiyya (Damascus) 466
Khansā 180
Khanūqa 412
Khāqān Sa'īd 195
Khāqānī (poet) 323
Khaqans, era of, lineage of, etc.; and see Chingisids 161, 173, 180, 322
Kharār River 224
Kharashkath 298
Kharbuza 219
Kharg (island) 241, 362, 392, 416, 418
Kharīdat al-'ajā'ib, Kharīda (book by Ibn al-Wardī) 42–43, 84
Kharīdat al-'ajā'ib (book by Samarqandī) 119
Khārijites 267, 366, 392
Kharīma 424
Kharīz 235
Kharj wadi 413
Kharjird 315
Kharluj (*recte* Kharlukh) See Qarluq
Kharqān, Kharqānīn 255, 259, 261–262
  Āb-i Kharqān River 262
Kharraqān 286
Khartān and Martān (islands) 415
Khartūs 324, 332
Kharzān Pass 318
Khāsān 222
Khashāb, Qashāw 191, 206, 224
  Khashāb River 224
Khashabāt 364
Khasht 237
Khasīsa 432
Khāṣṣa River 352
Khāst 213
Khaṭim 388
Khaṭṭ 392
  Khaṭṭī lances 392
Khātūniyya lake 354
al-Khawābī, Ḥiṣn al-Khawābī 430, 454–456
Khāwarān 279
Khawarnaq 111, 377
  Khawarnaq canal 371
Khāwas 304
Khawāṣpur 219
Khawr 229–230
  Āb-i Kharw River 278
Khāwrān 222
Khayrī 366
Khayṭ (mountain) 458
Khaywān 387, 424
Khazaz (*recte* Khazar) 304, 307–308, 323, 325–328
  Khazaz language 304
  Khazaz Sea See Caspian
  king of Khazaz 312
Khazībī (village) 451
Khazo 373
Khaṭīb-i Tabrizī, Abū Dhakariyyā Yaḥyā b. 'Alī 315
Khaybar 393, 408–409
Khaymagāh-i Iskandar 336
Khayra Bey, Amīr, Khayra Bay Bey 428, 523–524
Khazraj 416, 426
Khiḍr (al-Khiḍr in the Koran) 243, 457
Khiḍr Khan (Sayyid Khizr Khan) 197
Khīrākhīr 307

Khīrkhīr, Khirkhīr (*recte* Khirkhiz, Qirgiz) 300, 333
Khīsa Kūz 297
Khīwa 290
Khitāy See Cathay
Khinjān 224–225
Khiyābān 267
Khiyār 316, 318, 326
Khoshān (village) 362
Khotan 124, 161, 180, 219, 303, 305
Khotan road 180
Khudābanda, Sultan Muḥammad (Ilkhanid) 283, 319
Khudābanda (Safavid) 320, 327
Khuḍayrā 423
Khujand 295–300, 304
Khujand River 297
Khulayṣ 409, 421
Khūmāqān 235
Khūnaj, Khūna, Kaghadh-kunān 257, 318–319
Khunāṣra 458
Khurāsān, Khurāsāni 69, 158, 161, 193, 204, 212, 215, 219–227, 231, 240–241, 244–245, 251–253, 255, 258, 262, 264–265, 267–285, 289–291, 293–294, 297, 301–302, 306, 319–320, 369–370, 376–377, 467, 525
Khurāsān Desert 276–277
Khurāsān Pass 215
Khurāsān (district of Baghdad) 367
Khurāsānābād 265
Khurdād b. Fāris 247
Khurram Shah See Shah Jahān
Khurramābād River 248, 374
Khursha 236
Khūsaf 276
Khūshāb 366
Khūshanābād 229
Khūshī 213, 214
Khushkābād 210
Khushkrū 288
Khushkrūd River 262
Khusraw, Chosroes, Kisrā (Sasanian kings) 227, 243, 258, 261, 274, 352, 369–370, 426, 461
empire of the Chosroes 242
Arch / Vault of Chosroes (Ṭāq-i Kisrā) 313, 369–370, 373
Madā'in-i Kisrā, Madā'in of the Chosroes See s.v. Madā'in
Khusraw Parvīz 362, 370, 377
Khusraw's Shabdīz See Ṭāq-i Bustān
Khusraw and Shīrīn (poem by Niẓāmī) 260
Khusraw (village) 278
Khūst 278–279
Khusrawjird, Khusrawgird 274, 279
Khuttal, Khuttalān 224–225, 298–299
Khuttalān See Khuttal
Khūy 314, 316–317
Khūy River 319, 339, 347
Khuzā'a (tribe) 397
al-Khuzayba 455
Khūzistān, Ahwāz 125, 231, 238, 240–241, 244–251, 280, 363, 365–366, 374, 379
Khūz (people of Khūzistān) 246
Khūzī language 246
kings of Khūzistān 249–251
Sayr Ahwāz River 374
Khwāf 272, 276
Khwāja Tāj al-Dīn 'Alīshāh al-Tabrīzī 313, 316, 319, 335
Khwāja 'Alīshāh Bridge 318
Khwāja Ghiyāth See Ghiyāth al-Dīn
Khwāja Humām 315
Khwāja Khayrān (village) 270
Khwāja Sihyārān 212
Khwākand 297
Khwān 278
Khwāndamīr 289
Khwār 260
Khwār River 286
Khwārazm, Khwārazmians 206, 225–226, 275, 279, 284–287, 289–291, 298–300, 302
Lake of Khwārazm See Aral Sea
Khwārazm Turkmen 482
Khwārazmshāhs, Khwārazmian emirs 226, 231, 467
Khwārazmshāh 'Alā al-Dīn Atsız 225, 289, 301
Khwārazmshāh Sultan Jalāl al-Dīn 328, 335
Khwārazmshāh Sultan Muḥammad 225, 231, 270, 274, 281–282, 381
Khwāsh 221
Khwāst 193, 209, 211–214, 222, 262
Ḳıbrısçık 497–498
Kīdabarū 219
al-Kifāfī Mazrūq, Shaikh 423
Ḳiğı 343, 345, 524
Kīh 209–211
Kihtwār 216
Kihtwār Road 215
Kīj 207–211, 222
Kīl (in Iran) 230
Kīl wa Kīlā (in Iraq) 375
Kilid al-Baḥr (fortress) 365
Kili 522
Ḳılıçcıḳ (fortress) 484
Kilīs 356, 457, 460–461
Kiliseḥiṣār 474
Kilīstī 456
Kīlū 218
Kilwa 103
Kīmāk, Kimek (Turkic people) 307, 309, 333
Ḳınalızāde 'Alī 292
Kinda 383
kings of Kinda 357
al-Kindī 43
Kīngāwar 375
Kinpin 216
Kipchaks 477
Kipchak Steppe 179, 302–303, 306–307
Ḳıralı 474
Kirih 230
Kīrīl 352
Ḳırḳ Ṭağı mountain 500
Ḳırḳçeşme aqueduct 525
Ḳırḳgeçid River, Ḳırḳgeçit 346, 463, 481, 508, 511
Ḳırḳgöz River and bridge 462
Kirkuk 352, 359–360, 362, 372, 374
Kirkuk desert 360
Kirmān b. Falūj 227
Kirmān, Kirmānī 90, 125, 193, 209, 220–222, 227–231, 235–236, 238, 240–241, 243–245, 248, 260, 265, 272, 276–280, 376
rulers of Kirmān 230–231
Kirmānshāh, Kirmānshāhān, Qirmīsīn 260, 262, 362, 375–376
Kirmasti 501, 503, 512, 518
Kirmiz 217
Kirnār 216
Ḳırşehir, Ḳırşehri 472, 476, 480, 492
Kīsh, island 228, 241, 416
Kīsim 288
Kisrā See Khusraw
Kīsū Pasture 262
Kiswa 420, 433, 444
Kitāb al-aḥkām (book on Zaydī rites) 427
Kītal 205–206
Kite 501–504
Kite Plain 504
Kiti (in Cyprus) 471
Kītī Pasture (in Iraq) 261
Kītīklū, Mt. 217
Kitilpāyī 216
Kītūr 190
Kız Kulesi, Leander's Tower 506
Ḳızderesi 357
Kizhih 286
Ḳızıl Aḥmed Bey 348–349
Ḳızıl Aḥmedlu See Isfandiyarids
Ḳızıl Ilıca 484
Ḳızıl Mūsā Ṭağı, Ḳızılca Mūsā Ṭağı 487
Ḳızıl Mūsā Pass 487
Ḳızılağaç 496
Ḳızılbel, Ḳızılbeli 499, 500
Ḳızılbıñar 347
Ḳızılca 359–360, 501, 518
Ḳızılcaḥān 351
Ḳızılcaḳal'e 360, 367
Ḳızılḥiṣār 510–511
Ḳızılırmaḳ River 477–479, 480–482, 492, 495, 513, 522
Ḳızılcaṭağ, Ḳızılcaṭağı 345, 346
Ḳızılḳaya 489
Ḳızılḳocalar 480
Ḳızılribāṭ 375, 367, 372, 374
Ḳızıltaşlıḳ 412
Ḳızıltepe 356
Ḳızıltürbe 513
Ḳızḳarye 482
Ḳızlar Geçidi 347
Ḳızuçan 343, 345
Koca Meḥmed Pasha 42
Ḳoca Murād Pasha, grand vizier 526

Ḳoca Naḳīb (qadi of Medina) 400
Ḳoca Nişāncı 334
Ḳocaeli 483, 497, 501, 505–508
Ḳocaṭaġ 356
Ḳoçḥiṣār 357–358, 475, 476, 480–481, 493–495, 512, 522
Ḳoçḥiṣār River 478
Ḳolor 344
Ḳomaran (Komárom) 525
Komenat 482
Ḳoñrapa 497, 499, 501, 512
Konya 314, 465, 471–477, 481–482, 492, 511–513
Koran 98, 196, 198, 244, 265, 310, 371, 379, 381, 395, 403, 419, 437, 443, 445, 456, 514
Koran (quoted) 40, 73, 88, 241, 311, 322, 324, 344, 369, 386, 388–391, 397, 403, 407, 409–411, 416, 420, 425–426, 430, 432, 434–435, 438–439, 443–444, 466, 468
Koran "bearers" (memorizers) 466
Koran commentary / exegesis 41, 192, 223, 253
Koran reader(s) 253, 359, 446
Koran school 526
Korea, Zānjīā 132, 158
Ḳoron (Koróni in Morea) 522
Ḳorubāzārı 493, 495
Kosovo 519, 521
Ḳoṭur 338, 345
Ḳoṭur valley 339
Benī Ḳoṭur (Kurdish tribe) 334
Ḳoyuluḥiṣār 480
Ḳoyun Baba 479
Ḳoyun River, Ḳoyun Ṣuyı, Gūsfand River 325, 330
Koyunḥisārı 517
Ḳoyunlıḥiṣār 345
Köpribāzārı 489
Köpriḫān 347
Köprü River 470
Köprülü (village) 495
Köprülü Meḥmed Pasha, Grand Vizier 455, 458, 480, 491, 495, 511, 527
Köprülüzade Fāżıl Aḥmed Pasha, Grand Vizier 527
Körpeaġaç 512
Kös Ṭaġı 345, 493–495
Köse Miḫal 514–515
Köşeşaʿbān 481, 513
Köşk 475, 487, 512
Krīās (Deccan ruler) 185
Kru-Rukhkhaj 222
Kubachi 328
Ḳubād Pasha 42, 101
Kuban 330
Kūbar Mountains 213
Kūbīḥ 365
Kublai Khan 159, 175, 179
Kūbrī 300
Kūchisbān 288
Kūdahlārū Valley 217
Kūfa 239, 252, 360, 363, 365–366, 370–373, 375, 377, 393, 415, 424, 452, 467
Kūfan 271
Kūfdī 202
Kufic script / inscription 265, 336
Kūh-i Mubārak 228, 230
Kūh-i Safīd 213
Kūh-i Siyāh (island) 326
Kūh-i Zarda (mountain) 249
Kūhak mountain 292
Kuhan 215, 216
Kuhbānān 277
Kūhijān 233
Kūhistān 211, 223, 252, 253, 296
Kūjābād 279
Kūjānī 375
Kūjard 228, 230
Kujaz 218
Kūjistān 288
Kūka 288
Kūkchī 218
Ḳula 484–485, 488
Kūlī 192, 205–206
Kūlīkanda 205
Kūlkū River 248
Ḳulp 353
Kūlū 319
Kulwāz 369
Kūmī (Gömi) 457
Kūmīn 230
Ḳumla 503
Kūmrū 228, 230
Kunar 213
Kundafar 203, 216
Kūper Mountains 214
Kūpra See Sūdra and Kūpra
Kupuhra 229
Kur River, Rūd-i Kur 239–240, 320, 322–324, 326, 333
Kur Ṭagh, Kur Taq 305
Kūra-kāt 188, 193
Ḳuramaz, Mt. 434, 476
Ḳuramaz River 476
Kūrash 196
Kūrat Ḥiyār 458
Kurd, Kurds, Kurdistan 125, 208–209, 229, 246, 248, 259–261, 312, 320, 334, 339, 341, 350–351, 358, 361–362, 366, 375, 463, 525
Kurdish (language) 351, 356, 361, 373, 375, 436
Kurdish Ayyubid dynasty, Ayyubid Kurds 353, 467
Kurdish chiefs, beys of Kurdistan 339, 350, 353, 523
Kurdish clans / tribes 232, 338–340, 345–346, 352, 355, 361
mountains of Kurdistan 316, 318, 355–356, 362, 375
Kurdān 279
Ḳurdbıñarı 512
Kurdkān 229
Ḳurdḳulaġı 462–465
Ḳurdyurdı 347
Kūre-i Nuḥās 496, 500, 521
Kūregīl (fortress) 361
Ḳurıgölcük 512
Kūrkang River 210
Kūrkas River 209–210
Kurmānc Kurds 361
Ḳurşunlu, Ḳurşunlu Ḫān 493, 495, 503, 512
Ḳurubāzārı 501
Ḳuruçay 343–344, 346
Ḳurudere 512
Ḳuşadası 488, 510, 512
Kūs, Mt. (Köseṭaġı) 346
Kūsān 222
Ḳuşaḳlı 474
Ḳuşçular 511
Kuşlar Ḳalʿesi 368, 371
Kushānīya 295
Kushk 209–210
Kushk River 210
Kūshk-i Manṣūr 278
Kūshk-i Zar, Kūshk-i Zard 235, 239, 262
Kushmayhan 275
Kuskun-kıran (mountain) 316
Kusna River 219
Ḳuṣūn 464
Kusūr 337, 339
Kusūr River 336
Kūsūya 268
Kūt 366
Kūt Abū Manṣūr 363
Kūt Abū Suwayd 363, 365
Kūt Baḥrān 363, 366
Kūt Banī Manṣūr 365
Kūt Dāwudiyya 365, 363
Kūt Ḥūr 365
Kūt Muʿammar 363, 366
Kūt Sūra 363
Kutal-i Zarmark, Mt. 217
Kūtam 288
Kūtar 217
Kuthayba 452
Kuthba 424
Kūthī 369
Kuve-līnfū 180
Kuwādar 230
Kuwāram, Mt. 219
Kūy 359
Kūyān, Kūyān-shī (lake in China) 157–158
Kūz 229, 340
Ḳuzguncuḳ 506
Küçük Aḥmed Pasha 444
Küçük Baba mountain 485
Külebi, Mt. 496
Külek, Külek Ḳalʿe 463, 469–470
Külek Beli 463
Künbeti 493
Künhü'l-aḫbār (book by ʿĀlī) 44, 84, 514
Kürd Emīr 475
Küre 360, 496
Küreci 512
Kürekçi 513

Küstere 476
Kütahya 483–488, 490–492, 501–504, 512–513, 519–520

Lādhiqiyya (Laodicea) 430, 435, 454–455, 459, 466
Ladik 480, 482, 511
Lāfāq 122
Lafkara 470
Lagzī See Lezgi
Lah (village) 217
Lāharī 182–183, 203, 206–209
Laḥij 385, 388, 423
Lāhijān 288–289, 291
Lahnaʿa 429
Lahore 183–184, 190–193, 196–197, 203, 205–206, 208, 214–216, 225, 305
  Lahore road 184
  governor of Lahore 198
Laḥsā, al-Aḥsā 363, 392–394, 412, 414, 417, 424, 430
Laḥya 385
Lajā, Arḍ Lajā 452
  Lajā Arabs 452
Lajjāʾ 415
Lajjūn 420, 430, 436, 441, 465, 524
Lajma 383
Lakhmids, Banī Lakhm, Āl Mundhir 376–377
Lākhūf 303
Lāl 217
Lala Şāhīn Pasha 518–519
Lalaçayırı 482, 512
Lālar 216
Lāldīw 217
Lallīn 210
Lamghān 212
Lālapūr 206
Land of Fire (Terra del Fuego) 132
Land of Gold 122
Langar, Langarū, Langarūn 288, 325
Langarkunān 323, 325
Lankūtī 200
Laozi 161
Lapseki 509
Lār 227–230, 237, 241
  fortresses of Lār 229
  sultans of Lār 231
Lārende (Ḳaraman) 469, 472–473, 512
Lārijān 286
Larke 355
Lashinshāh 288
Lāshtar 262
Lāshkard 229–230
Lāt (idol) 416
Latin 40, 41, 43, 47, 51, 66–67, 77, 83, 88, 119, 130, 311, 436
  Latin historian 107
  Latin map 532
  Latin poets 59
  Latin sources 349
  Latins (i.e., Latin speakers) 67, 77, 91, 307, 309, 521
Latros (Lathyrus) 99
Lāvī (Leo II, Armenian king) 463
Lavoş (Louis II of Hungary and Bohemia) 524
Lāwakand 298
Lawand (tribe) 329
Lāwaz son of Shem son of Noah 413, 424
Lāyijān 326
Layth b. ʿAlī b. Layth 223
Layth b. Ṣaffār 223
Laz See Lezgi
Laz, Las (Serbia) 520–521
Lazari (island in the Eastern Ocean) 90
Lazarus 439
Lāẓḳiye 472–473, 484
Lebanon, Mt., Jabal Lubnān 325, 433, 435, 455
Lefke (Osmaneli) 470, 491–492, 501, 503, 511, 521
Lefkoşa (Nicosia) 469–470
Leh See Poland
Lek 521
Lekīnhūt 193
Lemnos 96
Leo Africanus 108
Lesbos 96
Levāmiʿü'n-nūr fī ẓulmet Atlas Minor (translation of *Atlas Minor*) 40–41, 125, 127, 136
Lezgi, Lagzī, Lāzkī, Laz (tribe) 329, 348
  Lezgi Mountains, Elegzi, Mt. Legzi 325, 332, 348
Liber Baba (Liber Pater), Albahar the Great 196
Liburnia 91
Libya; and see Ifrīqiya 91, 108, 125
Ligorna (Livorno) 91
Likām, Mt. 325
Līmānqū 179
Limosa 470
Lindos 524
Lisbon 186
Līṭānī 433
  Līṭānī River 450
Lithuania 106
Livonia 90, 308
Lofça 521
Lonza 127
Lopes, Captain Michael 122
Lorenzo, the book of Lorenzo 43, 122, 127, 130, 132–135, 138–143, 145–150, 152–155, 158–159, 161–163, 167, 169–170, 173, 178–190, 192, 194–195, 203–204, 207, 215, 220, 307–310
Lori 323–324, 333
Lot (prophet) 430, 432, 442
  cities of the people of Lot 432
  Lake of Lot See Dead Sea
Lubāb (book) 312, 334
Lucia Church in Rome 108
Luconia, Lekīne islands, Greater and Lesser Lekūnīā 133, 151
Lucknow 192, 196–197
  ruler of Lucknow 198
Lūdarak, Mt. 217
Ludd 424
Lūdū 217
Luḥayya 415
Lūḥ-kān 278
Luhrāsb (Kayānid) 235, 238, 243, 269, 322, 513
Lūkbaūn 216
Lūkbūn 216
Lukkām, Mt. 430, 434–435, 457, 461
Lūlapūr 218
Luqmān 426, 430
Luqmāniyya 366
Lūr, Lūristān, Greater and Lesser Lūr 240, 246, 248, 248–251, 261–262
  Atabegs of Lūr 248
  kings of Lūr 249–250
  mountains of Lūr 246, 261
Lūmīna 298
Lusignans 470
Lūsra 319
Luṭfullāh Nīsābūrī, Mawlānā 201
Luzon 140
Lüle 472
Lydda 431, 437

Māʾ al-ʿAmqā River 455
Māʾ Ghassān 452
Māʾ-i Ḥunayn 410
Maʾāb 442
Maʿāfir 416
Maʿān 421, 443
al-Maʿarra, Maʿarrat al-Nuʿmān 430, 435, 456–457, 465–466
Maʿarrat al-Miṣriyyīn 430, 458
Maʿarrat Qinnasrīn 430
al-Maʿarrī, Abū'l-ʿAlāʾ (poet) 370, 457
Mābadastān 240
Mabādir 248
Mabāḥith mashriqiyya (book by Imam Rāzī) 96
Maʿbar 182, 184
Mabatpal 218
Mābayn 222
Mabker 360
Mabrak al-Nāqa 421
Macedonia 106–107, 119, 123–124, 359
Māchīl, Mt. 217
Machlū 218
Māchīn (Chinese capital; and see China: Chīn u Māchīn) 175, 181
Māchīwān 205
Maçillina 512
Maçka 348
Mactan 118
Madagascar Sea 90
Madāʾin, Ctesiphon 243, 296, 359, 369, 375–376, 400
Madāʾin Shuʿayb 383
Mādarān Castle 262
Madāyin son of Abraham 411
Madeira 100, 109, 115–116
Maʿdikarib 391

Maʿdin Banī Sulaym 424
Maʿdin-i Nuḥās, Mt. 497
Madīna (in Iraq), Madīnat al-Qilāʿ 363, 365
Madīna Mūsā 254
Madīnat Nahr al-Malik 369
Madraka 103
Mādūsang 219
Madyan (Midian) 383, 411, 434
Mafālija (tribe) 431, 451
Mafārḳīn See Miyāfārḳīn
Mafeos (Jean-Pierre Maffei) 132
Mafḥaq 387
Mafraḥ 422
Mafraq 415, 420
Magellan, Captain Ferdinand, Ferdinandos 101, 115, 118, 122, 132, 137–138
  Magellan's people 135
  Magellan's ship 135, 143, 147
  Captain Magellan's expedition 136
  Cape of Magellan 113
  Magellan Strait, Strait(s) of Magellan 90, 99, 112–114, 118, 122
Magellanica, South Polar Region, Terra Incognita of the South 90, 104, 122
Maghārat / Maghāyir Shuʿayb 422, 431
Magharish al-Zīr 408, 421
Maghāʾir al-Qalandariyya 421
Maghreb, Maghrebis 43–44, 80, 97, 105, 115, 125, 380, 426, 439
  Maghrebi (Arabic) 436
al-Maghribī, Ibn Saʿīd 43
Magians; and see Zoroastrians 148, 241, 243, 252, 287, 376, 378, 436
  language of the Magians 293
Maginus, Antonius 41
Magnesia 83
Maġosa (Famagusta) 469–470
Magyar See Hungary
Maḥāsina (clan) 410
Maḥaṭṭa 410
Mahdi (the expected deliverer) 130, 372–373
al-Mahdī (Abbasid) 248, 253, 368, 378, 401–402, 406
Mahdīābād 278
Mahdid dynasty 427
Māhiyān (fortress) 469
Māhiyān, Fountain of 267
Maḥkeme Ḳaryeleri (near Izmir) 510
Maḥmūd, Ottoman Sultan 36, 328
Maḥmūd, Sultan of Ghazna 183, 193, 196, 204, 212, 213, 225, 263–264
Maḥmūd b. ʿIzz al-Dīn (Atabegid) 352
Maḥmūd Efendi, Shaikh (Hüdāyī) 506–507
Maḥmūd Khaljī 196, 199
Maḥmūdābād 188, 319, 321–322, 324, 326
Maḥmūdīs, Maḥmūdī tribe 339–340, 355
Maḥmūdlar 472
Mahra 383, 391, 417–419, 425
Mahraj 241
Maḥraqa 307
al-Mahrī, Sulaymān b. Aḥmad 43–44
Mahrūd (village) 262
Mahrūyān 238–241, 246, 364
Mahū 217
Māḥūr Jubayl (fortress) 454
Māhūrū 218
Maḥwiyat and Masmāt (fortresses) 387
Mahyār 262
Mājarwān 312
Majbā 387
Majdal 437, 453
Majdal Yaba 433
Majmaʿ arbāb al-mamālik (book by Rukn al-Dīn al-Khūyī) 312
Majmūʿ al-ansāb (book) 228
Majmaʿ al-buldān (book) 322
Majorca 91, 106
Mājū 109
Maḳām-i Erbaʿīn mountain 488
Mākān (Buwayhid governor of Khurāsān) 244, 291
Makand 202
Makhādir 388
Makhāṣṣat al-ʿAlawī 461
Makhlaṭ (fortress) 374
Makr, Bilād 423
Makran, Makranis 125, 206–210, 220–221, 227–228, 240, 261, 276
Makrīl (Mingrelia), Mingrelians 329–331, 348–349
  Georgian Mingrelia 330
  khan of Mingrelia 331
Makshafa 365
Mākshīd River 209–210
Māksīn 358
Mākū 312, 340, 367
Malā River 222
Malabar, Manībār 102, 148, 181, 186–187, 204–205, 385
  governor of Malabar 195
Malacca 100, 102–103, 128, 133, 141, 189–190, 195
  Cape of Malacca 110, 181, 190
Malāḥa 422
Malān 210
Malatya, Malaṭiya 125, 334, 346, 350, 354, 356–357, 361, 434–435, 461–463, 479, 481, 523
Malaza 214
Malazgird, Malāzcird, Malādhjird 319, 334, 343, 345–347
  Malazgird River 345–346
Māldepe, Mt. 507, 511
Maldives: islands of Māldīwā and Qāndālūs 187
Malġara 518
Malham 412, 424
Malik Ashraf (Chobanid) 347
Malik Fakhr al-Dīn (governor of Rayy) 253
Malik Shāh, Sultan (Seljuk) 263, 354, 357, 375, 381, 424, 467, 476, 482
al-Malik al-Ashraf (Khalīl, Mamluk sultan) 460
al-Malik al-Muʾayyad See Abū l-Fidāʾ
al-Malik al-Muẓaffar (Quṭuz, Mamluk sultan) 468
al-Malik al-Nāṣir Muḥammad (Mamluk Sultan) 411, 482
al-Malik al-Saʿīd (Mamluk deputy) 468
Malik Quṭb al-Dīn 228
Malik-Patta 222
Mālik b. Ṭawq al-Thaʿlabī 358
Mālik (b. Anas, founder of the Mālikī rite) 395
  Mālikī 419, 439, 445, 446, 524
Mālīn 268, 272, 279
Mālīnar 205
Malīza 213, 214
Malkoç Efendi, village of 357
Malta 106
  lords of Malta (i.e., Knights Templar) 128
Mamaḥātūn 347, 481
Māmaṭīr 285
Mamhāl 217
Mamrevan 343, 345
Māmshārūd, Māmshānrūd 262, 314
al-Maʾmūn (Abbasid) 42–43, 74, 88, 358, 368, 370, 379, 400–401, 426, 464, 492
Maʿmūriye See Anamur
Maʿn-oğlı, Maʿnids, Ibn Maʿn 450, 453
  Maʿn mountains 451
Manaḥ 193
Manak 300
Manastır Madrasa (in Bursa) 516–518
Manāt (idol) 183
Manavġat 470, 489
Manazkām 218
Manazmūh 216
Manbij 430, 457–458, 460–461, 466–467
  Manbij Bridge 462
  Ḥiṣn Manbij 462
Mand 209–211, 213
Mandab 383
Mandalī 372
Mandaljīn 366–367
Māndasht 362
Mandaw 190
Mandeḥorya 487
Māne 483
Manfredonia 526
Manfūja 412
Māngal 219
Mangalore 183, 186, 205, 209
Mani the Painter 154, 243
Manichaean(s), Manicheism 154, 200, 243
Manīḥa (village) 433
Manīhā (village) 452
Manikonko 100
Manila 140
Manisa 483, 485–488, 501, 504, 511–512
Mankanār Road 217
Mankūjiya dynasty (Mengüjekids) 347
Mānkūt 216
Manṣūr (Arab ruler in Molucca) 118, 134
al-Manṣūr, Abū Jaʿfar, al-Dawāniqī (Abbasid) 206, 367–368, 371, 377–378, 406, 461
Manṣūr b. Jaʿwana b. al-Ḥārith al-ʿĀmirī (Umayyad emir) 462–463

Manṣūra (in India) 184, 206–208
Manṣūrābād 286
Manṣūriyya 363, 366, 423
Manuel the King of Portugal 112
Manūchahr son of Ghaza (governor of Tiflis) 333, 525
Manūchihr (ancient Iranian king) 242, 266, 323, 327
Manūkān 228, 230
Manūrza 214
Manyas 501, 512, 518
Mappa Mundi 73, 81
Maqām Ibrāhīm (in Kaʿba) 397, 401
Maqām Imām ʿAlī, Maqām-ı ʿAlī (in Iraq) 346, 365
Maqāmāt (book by al-Ḥarīrī) 334
Maqar 384
al-Maqdisī, Shaikh Shams al-Dīn Muḥammad b. Aḥmad 39
Maʿqil Canal, Nahr Maʿqil 363–365, 375
Maqran 412
Maqrūn 365
Mār-i Ḥasmīn 290
Marʾa 412
Marāgha 234, 263, 312, 317–319, 336, 340–341, 359
Marand 316–317, 319, 347
  Marand Mountains 319
  Marand River 319
Mārandakī, Mt. 221
Maranium River (Amazon?) 114
Maras 326
Marʿash 334, 361, 430, 435, 457, 461–463, 466, 468–469, 472, 477
  Marʿash River 434
Marāṣid al-iṭṭilāʿ (book) 43, 285, 289, 296, 322
Marbāṭ 391
Marcos See Paulus
Mardāwīj (Daylamite) 244, 258, 291
Mardin 336, 353, 357, 524
Mārdū 196
Mardūyīn 300
Marghīnān 291, 296–297
al-Marghīnānī See Burhān al-Dīn ʿAlī b. Abī Bakr
Mare Pigrum 120
Māre Vermīlū [Mare Vermiglio] 127
Maʾrib 389, 425, 465
Marīq 375
Marius 110, 181
Marj, Marj Rāhiṭ 431, 443, 452, 466
  Nahr Marj 465
al-Marj al-Aḥmar 434
Marj Dābiq 356, 468, 523
Marj al-Dībāj 459, 463
Marj al-Qabāʾil 460
Marj ʿUyūn 433, 441, 450
Markh 229
Mārklī 206
Marmara (town) 486, 488, 509
  Sea of Marmara, Propontis, Sea of Istanbul, Sea of Rūm 84, 91, 503, 508–509
Marmaracik 487–488
Marmaris 483
Marmarica 108
Marqab 424, 454, 456
  Mt. Marqab 455
Marrān 424
Mars 443, 459
Marsā Ibrāhīm 409
Marsīn 362
Martab, Martaba (in Syria) 455
Martaban, Marṭaban 102–103, 189–190
  Martaban River 189
Maʿrūf Karkhī 368
Marw, Marw al-Rūd 125, 271, 275, 279
  Marw al-Rūd / Marwrūd River See Murghāb
Marw-i Shāhijān, Marw 270–271, 275, 278, 288, 290–291, 377
Marwa 394, 396, 402, 418–419
  Mt. Marwa 416
Marwān b. Abū Ḥafṣa (poet) 378
Marwān b. al-Ḥakam (Marwān I, Umayyad) 452, 466
Marwān b. Muḥammad, Marwān al-Ḥimār (Marwān II, Umayyad) 317, 369, 392, 462, 467
Marwanid dynasty, Marwāniyya 357, 361, 378
Mary mother of Jesus 430, 439–440
Marzdān 233
Marzghad 300
Marzubān 460
  Marzubān River 460
Mas River 240
Maṣabb Ṭabā 365
Masālik al-mamālik (book; and see "Routes of Countries") 43, 244, 246, 271, 276–277, 285–286, 290, 295–297, 299–300
Masār al-ʿAmmār 387
Mashān 364
Mashgharā 454
Mashhad 222, 245, 271–272, 276
Mashhad ʿAlī (in Iraq) 375, 424
Mashhad al-ʿAshāra 364
Mashhad al-Ṭarḥ 435
al-Mashhadī, ʿAbd al-Raḥīm 43
Mashruqān 248
  Mashruqān River 248
Maṣīra 104
Masjid al-Ḥarām, Masjid-i Sharīf (in Mecca) 393–394, 398, 400–403, 405–406
Masjid-i Rāzān 279
Maskān 235
Maskat 384, 392, 417
Maslama b. ʿAbd al-Malik 327, 466, 491
  sword of Maslama 329
Māspidān 260
Masqiyya mountains 454–455
Massawa 428
Maṣṣīṣa, Misis 434, 457, 459–460, 463–464
Mastūra 410
Masʿūd, Sultan (Seljuk) 196, 327, 381, 476
Masʿūdī, al-Masʿūdī 43, 84, 97, 99, 151, 158, 196, 219, 285, 290, 309, 323
Masʿūdīs 238
Māsūla 288, 325
  Māsūla River 288
Māsūlīpatan 206
Masuto 470
Maswār (fortress) 387
Maṣyāf 431, 456–457
Maṭāliʿ (book on astronomy) 253, 315, 337
Maṭārī 374
Mathura 193
Matīla 207, 209
Māṭirūn 433
Maṭlaʿ al-saʿdayn (book by Kamāl al-Dīn ʿAbd al-Razzāq) 161
Matūth 248–249
Maʿūliyya 423
Maʿūn 222
Mauritania, Mauretania Caesariensis, Algeria 90–91, 108, 125
Mausolus, Tomb of 111
Mawālī (Arab tribes) 358, 421, 431, 456, 460
Mawārid al-kilam (book by Fayḍī-i Hindī) 192
Mawḍiʿpanār 217
Māwiyya 424
Mawjib 442
Mawlānā Jalāl al-Dīn Rūmī 274, 314, 473, 477
Mawlānā Ḥusayn 315
Mawlānā Mūsā 337
Mawlid-i sharīf (poem in honor of the Prophet) 437
Mawr 423
Mawzaʿ 385
Maymana 224
Mayamundar 218
  Mayamundar Road 218
Mayān Rūdān 375
Maybud 234, 241
Maybudī, Kamāl al-Dīn Mīr Ḥusayn 235
Maydān-i Sultan 286
Mayhana See Mihana
Māyīn 235, 262–263, 276–277
  Māyīn River 240
Maymand 213, 214, 234
Māymargh 298
Maymūn 229, 265
  Maymūn Bridge 326
Maynāwar 218
Maywajān 230
Maʿzal (spring) 277
Māzandarān 125, 223, 260, 265, 283–285, 287, 302, 312, 325
  mountains of Māzandarān 286
Mazdak 357, 377
Mazdaqān 257
Mazdikhil 213, 214
Mazgird 353
Mazḥam 408
Mazraʿa 277
Meʿānī (book) 522

Mecca 49, 91, 103, 125, 221, 232, 263, 265, 297, 311, 325, 359–360, 368–369, 377–383, 386–388, 390, 393–410, 412, 415–417, 419–430, 438–439, 446, 468, 473, 480, 520–522, 524–526
Sea of Mecca See Red Sea
Sharif(s) of Mecca 404, 407, 409–410, 420, 431
Mecidözi 477–478
Mecingerd 343–345
Medea (sorceress) 112
Mediterranean, Inland Sea, Ionian Sea, Sea of Rūm, Sea of Rome, Sea of Syria 42, 76–77, 81–82, 84, 88, 90–93, 95–96, 101–102, 105–106, 108, 110–111, 119, 123, 125, 351, 383, 409, 420, 422, 430–431, 433–434, 437, 441, 460, 463–464, 469–470, 483, 488–489, 510, 524, 528
Medina, Yathrib 125, 243–244, 297, 325, 359, 381–382, 393, 395, 400–410, 416, 419, 420–422, 424, 429–430, 452, 467–468, 473, 476, 480, 520–522, 524–526
Meḥmed I, Çelebi, Ottoman Sultan 480, 502, 520
Meḥmed II, Fātiḥ, Ebū'l-fetḥ Ghazi, Ottoman Sultan, Meḥmed the Conqueror 320, 324, 348–349, 507, 511, 516–517, 520–522
Meḥmed III, Ottoman Sultan 43, 486, 526
Meḥmed IV, Ottoman Sultan 35–36, 41, 401, 507, 527–528
Meḥmed (Şehzāde) 525
Meḥmed Cān Efendi, Ḳāḍī 337
Meḥmed İḫlāṣī, Shaikh Meḥmed Efendi 40
Meḥmed Pasha 321, 347
Meḥmed Pasha (Sokullu, Ottoman vizier) 404–405
Meḥmed Pasha, *beylerbey* of Yemen 388
Meḥmed Pasha Ḫānı, Meḥmedpaşa 481, 512–513
Mehran River See Indus
Mehrānī 353
Mekre (Fethiye) 483, 488
Mekrit (tribe) 328
Melek Aḥmed Pasha 462
Melūka See Molucca
Memlaḥa 470–471
Men Lopes 145
Menān River (Chao Phraya) 189–190
Menāẓirü'l-ʿavālim (book by ʿĀşıḳ) 43, 84, 325
Mendehorya 504, 512
Menderez 485–486
Mendik River 496
Menelaus 509
Menemen 486, 510–511
Meneses 133
lord of Meneses (Jorge de Menezes) 133
Mengen 497, 499
Mengen River 499
Mengli Timur 177
Mengu Khan 175
Menībār 182
Menkup 521
Menteşe 469, 483, 487–489, 516
Menzilḫāne 346
Meram park 472
Mercator, Gerardus 39, 41, 104, 120–121, 125, 127, 132, 136, 145, 147–148, 175, 181, 203, 307
Mercator's map 110, 133
Merebek (village) 352
Mergüze 496, 512
Meriç River 84
Merkez (fortress) 464
Merkūh 360
Meroe 69
Meron, Mt. 183
Merula, Paulus 41
Merzifon 480–482, 512, 520
Merzifon Plain 512
Mesarya 471
Meşhedābād 481, 513
Meşhedpıñarı 356
Mesopotamia 181, 349–351, 366, 435
Messiah See Jesus
Messina See Sicily
Messenger of God See Muhammad
Mesālik al-memālik (book by Johannes) 148
Meteora (book, commentary on Aristotle) 88
Meteorology (book by Aristotle) 96
Mevlevi, Mevlevi lodge, Mevlevīḫāne 367, 433, 486
Mexico 93, 113, 133
Meymān 339
Midillü (Lesbos) 521
Miftāḥ, Talkhīṣ Miftāḥ (books, i.e., *Miftāḥ al-ʿulum* by al-Sakkākī and *Talkhīṣ al-Miftāḥ* by al-Qazwīnī) 257, 272, 316, 522
Mihaliç, Miḫaliç 501, 503–505, 512, 518
Strait of Miḫaliç 503
Miḫaloğulları (Ottoman irregular cavalry in the Balkans) 491
Mihana, Mayhana 271
Mihek, Mt. 478
Mihr 217
Mihr u mushtarī (book by Muḥammad ʿAṣṣār) 315
Mihrajān 279
Mihrānrūd River 314
Mihribān 360, 367
Mihrimah Sultan 404, 506, 525
Miʿkāl 413
Mīl (in Arabia) 409
Mīl-i ʿUmarī 278
Milan (in Italy) 107
Milan (in Çankırı) 493, 495
Milan River 495–497, 499–501
Miletus 96
al-Mīna (village) 455
Minā' 398, 417
Mīnāb 228
Mināreliköy 344
Mindanao, Wandanāo; and see Philippina 110, 133–134, 138–140, 147
Mindos 96
Mingrelia See Makrīl
Minhāj al-fākhir fī ʿilm al-baḥr al-zākhir (book by al-Mahrī) 44
Minorca 91, 106
Minya 451, 464
Lake Minya See Lake Tiberias
Mīqāt 423
Miqyās al-Nīl (Nilometer in Cairo) 524
Mīr ʿAlī, calligrapher 315
Mīr Dāmerd (castle) 340
Mīr Khusraw 192, 196–197
Mīr Nāṣır Birādūst (castle) 340
Mīr Sharaf See Muḥammad Barqalaʿī
Mirāh 424
Mīrak, Mawlānā 315
Mirʾat al-shifā (book by Mawlānā Rukn al-Dīn) 324
Mirʾāt al-zamān (book) 44
Mirʾātü'l-memālik (book by Seydī ʿAlīzāde) 241
Mirdāsid dynasty 467
Mirṣād al-ʿibād (book by Najm al-Dīn-i Dāya) 253
Mīrsakāna 213, 214
Mīrzā Ḥaydar 217, 219
Mīrzā Jān 315, 324
Misfala 394
Mishqāṣ 102
Misis See Maṣṣīṣa
Misk son of Japheth son of Noah 513
al-Mislaḥ 424
Mistra 509
Mītla 202–203
Miyāfārkīn, Māfārkīn 353–355
Miyako (Kyoto) 128, 130–131
Miyānaj, Miyāna 317–319
Miyānaj Bridge 318
Miyānaj Plain 318
Miyānaj River, Miyānajrūd 261, 318
Miyār 375
Mizza 451–452
Mizza Canal 433
Arḍ Mizza wa Lawān 449–450
Mocha 100, 383–385, 415, 417–418, 423
Moesia 106
Mogadishu 100, 103
Moghul, Moghulistān; and see Mongols 302, 304–306
Moghul (eponym) 305
Moghul ethnicity 153
Moghul tribe 193
Great Moghul 190
Moghulistan road 180
mountains of Moghulistān 303
Mohács Plain 524–525
Moin 153, 160, 181
Moldavia, Moldavians 519, 520, 525
Molucca, Moluccan Islands 118, 132, 134–137, 140, 143, 147, 155, 160
cloves of Molucca 160
spices of Molucca 220

Mongol(s); and see Moghul 226, 231, 244–245, 249, 252, 257, 265, 268, 273, 291, 302, 346, 366, 371, 382–383, 471
Mongol army / soldiers / troops 268, 270, 273, 282, 289, 335
Mongol conquest / invasion / incursions / onslaught 212, 215, 246, 256, 301
Mongol custom 256
Mongol era / interregnum/ period / time 226, 231, 252, 255–257, 285, 313
Mongol infidels 282
Mongol language 318
Mongol sultan 362, 371, 382
Montanus, Petrus 41
Morea, Mora (Peloponnesus) 91, 105–106, 125, 509, 520–521
Morocco 76, 425
Moscow See also Muscovy 525
Moses 243, 351, 411, 434, 436, 509–510, 520, 522
Moses's Rock (Koran 18:63) 322, 430
tomb / grave of Moses 437, 439, 441
villages of Moses 430
Mostar 442
Mosul 125, 263, 320, 350–361, 366, 373–375, 406, 467, 524
Moton (Methóni) 522
Mount of Olives 439–440
Mountain(s) of the Moon See Nile: source of the Nile
Muʿaddal b. ʿAlī 223
Muʿādh b. Jabal 386–387
Muʿallā (in Mecca) 393, 394, 403
Muʿāwiya b. Abī Sufyān (Muʿāwiya I, Umayyad) 301, 437, 454, 461, 466, 471
Muʿāwiya b. Yazīd, al-Aṣghar (Muʿāwiya II, Umayyad) 366, 452
Muʿayṣira (village) 452
Muʿaẓẓamiyya 450
Mubārakābād 254
Mudanya 484, 501–503
Muḍar, Banī Muḍar 350, 414, 425
Diyār Muḍar 350, 354, 358
Mudéjars 118
Mudīq 456, 465
Mudlij tribe 425
Mudrik, Sayyidī (al-Fazārī, companion of the Prophet) 452
Mudurnı 497–499, 501, 512, 518
al-Mufawwiḍ (governor of al-Jazīra) 380
Mughals; and see Timurids 195
Mughal army 198
Mughal emirs 196
Mughal emperor 195, 223
Mughal tribe 193
Mūghān See Mūqān
Mughāṣ b. ʿAmr (father-in-law of Ishmael) 397
al-Mughīsha 454
Mughūn 229–230
Muğla 488
al-Muhallabī, Ḥusayn b. Aḥmad 43
al-Muhallabī, ʿUmar b. Ḥafṣ known as Hazārmard 206
Muḥāmila (island) 415
Muḥammad, the Prophet, the Messenger of God 40, 173, 362, 377, 395, 399–400, 426, 446, 463, 465, 467, 522–523, 525–526, 528
Companion(s) of the Prophet, Aṣḥāb 87, 263, 270, 318, 364, 377, 379, 398, 405, 414, 417, 426, 429, 443, 445, 451, 523
Muhammadan See Muslim
Muḥammad ʿAṣṣār, Mawlānā 315
Muḥammad Barqalaʿī, Mawlānā, Mīr Sharaf 337
Muḥammad al-Dāʿī (lord in Yemen) 387
Muḥammad Ḥanafī, Mawlānā 315
Muḥammad (al-Shaybānī, jurist) 395
Muḥammad b. al-Ḥanafiyya 241, 380, 393, 410, 476
Muḥammad Jawwād Raḍī, Imām 368
Muḥammad Makrad (mountain) 388
Muḥammad Sharānshī 337
Muḥammad b. Karrām 221
Muḥammad b. Mulkdād 336
Muḥammad b. Qāsim al-Thaqafī 196, 232
Muḥammad b. Yazīd (descendant of Bahrām Chūbīn) 327
Muḥammad, Sultan See Bahādur Khan; Khwārazm Shah
Muḥammadābād 182
Muḥammadiyya (near Damascus) 433
al-Muḥāsibī, Ḥārith 368
Muḥawwal 369
Muḥīṭ (book by Seydī ʿAlīzāde) 43, 104, 141–141, 144–145, 151
al-Muhtadī (Abbasid) 380
Muḥyi'l-dīn, Mawlānā 336
Muʿīn al-Dīn Zamjī 265, 268
Muʿizz al-Dawla (Buwayhid) 366
Muʿjam al-buldān (book by Yāqūt al-Ḥamawī) 43, 232
Mūjawāl 218
Mukān River 234
Mukhtār (in Iraq) 364
Mukhtaṣar-i muʿjam (book) 259
Mukrim b. al-Ghurar 247
al-Muktafī bi'llāh (Abbasid) 380
Mular 216
Multan 181, 183, 190–191, 193–194, 197, 203, 205–208, 215, 279
Multan River 203
Multan road 208
Mu'minābād 276
al-Munayṭir, Jubbat al-Munayṭira 454
Mundhir (tribe) 376
Mundhir b. Imru' al-Qays (Lakhmid) 371, 376–377
Mundhir b. Mā' al-Samā' 477, 465
Mundhir b. Nuʿmān 377
Munjatīn Mountains 299
Munk 224
Munkhanā 424
Munṣaraf 422
al-Muntaṣir bi'llāh (Abbasid) 380
Muqaddima (book by Sharaf-i Yazdī) 306
Muqām 213–215
Mūqān son of Japheth 321
Mūqān, Mūghān 288, 320–322, 324–326
al-Muqtadī (Abbasid) 381, 476, 482
al-Muqtadir (Abbasid) 368, 380–381, 394, 400, 402
al-Muqtafī bi'llāh (Abbasid) 381
Murād I, Ghazi Ḫüdāvendigār, Ottoman Sultan 484, 492, 501–502, 515, 517–519
Murād II, Ottoman Sultan 502, 510, 516, 520–521
Murād III, Ottoman Sultan 42, 291, 333, 398, 401, 405, 506, 508, 525–526
Murād IV, Ottoman Sultan 313, 316, 321, 324, 328, 333, 338–339, 366, 400, 462, 494, 526
Murād Pasha, Ḳuyucu 512
Murād Pasha, bridge of 465
Murād Pasha (*beylerbey* of Yemen) 525
Murād River 345–347, 355, 478
Murād Ṭağı, Mt. Murād 485–486
Murādiye (of Bursa) 501
Murgh Kahtarā (road stage) 375
Murghāb 270, 283
Murghāb mountains 278
Murghāb River, Marw al-Rūd / Marwrūd River 223–224, 271, 278, 318, 325
Murj 240
Mūrqān 230
Murr al-Ẓahrān 421
Murtaḍā Pasha (governor of Baghdad) 368
Murtażā-ābād 492
Murūj al-dhahab (book by al-Masʿūdī) 42, 84, 97, 158, 196, 290, 309, 323
Muş 334, 337–339, 355
Muş Ovası, Muş Valley, Muş plain 338, 346
Mūsā b. Būqā 254
Mūsā Çelebi 520
Mūsā al-Kāẓim, Imam 315, 368, 372
Mūsā River 253
Musandam (island) 415
Musaylima the Liar 413–414
Muscat 102–103, 240
Muscovy, Muscovites, Muscovia, Russia, Russians, Rūs 42, 69, 74, 77, 106–107, 125, 162, 307–310, 328–330, 333, 436, 519, 526
deserts of Muscovy 303
lord / duke / tsar of Muscovy 110, 307–308, 329–330
Russian religion 107
mountains of Russia 309
Mushaʿshaʿ 525
Mushʿashaʿ dynasty 250, 365–366
Mushkīn 316, 318
Mushtarik (book by Yāqūt al-Ḥamawī) 42
Muṣliḥ al-Dīn, Amīr (of Mecca) 403–404
Muslim, Muslims, Muhammadan, people of Islam; and see Islam 35–36, 40, 77, 84, 110, 133, 135, 141, 145–146, 148, 158, 169,

Muslim, Muslims, Muhammadan, people of Islam; and see Islam (*cont.*) 172–173, 176–177, 182, 186, 188, 191, 194, 200, 202, 206, 208, 211, 213, 224–225, 234, 259, 271, 273, 279, 284, 292, 299, 303–305, 310, 315, 322, 327, 329, 333–334, 337, 343, 347–348, 355–356, 366, 375, 377–380, 395, 402–403, 410, 416, 426, 436, 444–445, 455, 457–458, 464–464, 471, 477, 481–482, 490, 495, 502, 511, 513–514, 516, 519, 525–526
Muslim Arabs 363
Muslim army, Muslim armies 301, 317, 444, 465, 482, 521–522
Muslim astronomers 51
Muslim books, books of the Muslims; and see clime-books 110–111, 132, 146, 309
Muslim city 321
Muslim conquest, Muslim conquests 106, 224, 247, 332
Muslim country, Muslim countries 39, 84, 444
Muslim dynasty 476
Muslim eunuchs 165–166
Muslim festivals 488
Muslim frontier towns (*thughūr*) 463
Muslim Ghazis 518
Muslim governors 336
Muslim groups 77, 349
Muslim homes 373
Muslim jurist 171, 392
Muslim philosophers 51–52, 58, 122
Muslim pilgrims 420
Muslim quarter and mosque 520
Muslim rule, Muslim / Muhammadan kings / rulers 75, 195–196, 227, 302
Muslim sailors 91
Muslim scholars 53
Muslim soldiers 227
Muslim surveyors and geometricians 75
Muslim territory 224
Muslim theologians 52
Muslim tribes 295
non-Muslims, non-Muslim subjects (i.e., dhimmis) 300, 334
Muslim (author of *Ṣaḥīḥ*) 363, 447
Muslim b. Quraysh (ruler of Aleppo) 476
al-Mustaḍī (Abbasid) 381
Muṣṭafā, Ottoman Sultan 36, 480, 526
Palace of Sultan Muṣṭafā 480
Muṣṭafā b. ʿAlī 42
Muṣṭafā Pasha, Admiral 83
Muṣṭafā Pasha (governor of Egypt) 428
Muṣṭafā Pasha (governor of Syria) 409
Muṣṭafā Pasha (sword-bearer of Murād IV) 462
Muṣṭafā Pasha (Lala), Vizier 333, 505, 525–526
al-Mustaʿīn (Abbasid) 380
al-Mustakfī (Abbasid) 381
al-Mustamsik bi'llāh, Yaʿqūb (Abbasid) 383
al-Mustanjid bi'llāh, Yūsuf (Abbasid) 381
al-Mustanṣir (Abbasid) 381–382, 402
al-Mustarshid bi'llāh (Abbasid) 381
al-Mutawakkil ʿalā'llāh (Abbasid) 379
Mustawfīs (family) 263
Amīn al-Dīn Naṣr Mustawfī 251
Ḥamd Allāh Mustawfī See Ḥamd Allāh
al-Mustaẓhir (Abbasid) 381
Mut 469
Mu'ta 442
Muʿtaba 424
al-Muʿtaḍid (Abbasid) 223, 380, 406
Muṭahhar, Sharīf, Malik 385, 387–388, 390, 428, 526
Muṭallāt 422
al-Muʿtamid (Abbasid) 364, 392
al-Mutanabbī (poet) 237–238, 315
al-Muʿtaṣim (Abbasid) 321, 368–369, 372, 379, 382–383, 400
al-Mutawakkil (Abbasid) 312, 368, 372, 380, 399–400
Muṭawwal (book by Saʿd al-Dīn) 226, 296
Muʿtazilī(s) 239, 246–247, 258, 290, 416
al-Muʿtazz (Abbasid) 380–381
al-Muṭīʿ bi'llāh (Abbasid) 393
Mūtīqā (island) 141
al-Muttaqī bi'llāh (Abbasid) 381
al-Muwaffaq (Abbasid) 380
Muwaffaq (in Arabia) 415
Muwashakh 423
Muwaṭṭa' (law book of Mālik b. Anas) 437
Muwaylaḥa 423
Mūy-i Kūshk 375
Muẓaffar (sultan of Ḥaḍramawt) 389
Muzūrī (tribe) 373
Muẓaffar al-Dīn See Gökböri; Özbek
Muẓaffarids 246
Muzayrib 420, 444, 452
Muzdaqān 262
Muzdaqān River, Āb-i Muzdaqān 256, 259, 262
Muzīl al-irtiyāb (book by Abū'l-Majd al-Mawṣilī) 42
Muzūrī (tribe) 373
Müküs 334, 340–341, 374
Müküs River 341

Nabʿā Fuqāʿ 423
Nabak 415, 453
Nabateans 368
Nablus 430–431, 434, 436, 438, 441, 464
Nadha 206, 208–209
Nafaḥāt (book) 337
Nafī 412
Nāfiʿ b. Azraq 366
Naftū River 318
Nagarhār 213, 215
Nagarkot, Naugocrot, Naugrocot (mountains) 191, 197, 202, 204, 307
Nagartata See Tata
Naghar 213, 214
Nāghawandī River 185–186, 189, 204
Naghāy (idol) 169
Nah 272
Nahalwāra 222
Nahang River 209–210
Nahārī 417
Nahm, Bilād Nahm 390
Nahr Āftāb 372
Nahr Asad 375
al-Nahr al-Aswad 433–434, 458
Nahr al-Dhahab 458
Nahr al-Ḥaṣīn 455
Nahr Ibrāhīm (in Syria) 433, 440, 453–454
Nahr ʿĪsā (in Baghdad) 367, 369
al-Nahr al-Kabīr, Nahr al-Akbar (river in Syria) 434, 455
Nahr al-Kalb 453–454
Nahr al-Malik 369, 455
Nahr Shāhī 372
Nahr al-Zāwiya 454
Nahrawān 125, 369–370
Nāhūd 196
Nā'ila (idol) 416
al-Nāʿima River 451
Nāīnū 217
Najaf 97, 363, 371, 375, 424
Najāḥid dynasty 427
Najānīkath 298
Najd, Najd-i Ḥijāz 366, 384, 388, 391, 393–394, 412–415, 417, 419, 424, 436
Najd al-ʿĀriḍ 390, 392, 412
Najd al-Yaman 388, 390, 416, 423
Najīram 233, 240
Najm al-Dīn-i Dāya 253
Najm al-Dīn al-Kubrā, Shaikh 290, 337
Najlā (village) 452
Najm (fortress), Qalʿat al-Najm 430, 460, 462
Najrān 383, 390
Nakārazen 481
Nakhīl 363, 410, 413
Nakhīl Ghānim 422
Nakhjiwān, Nakhjiwānis 101, 317, 319, 321, 339–340, 525
Nakhlayn, Bilād al-Nakhlayn 413
Nakhshab 269, 301
Nakīb (fortress) 366
Nakrī 219
Nala Tarlası valley 499
Nālīs 460
Namaklān, Mt. 261
Namāwa 409, 423
Nāmir 452
Nāmiyya 456
Nanak 230
Nandā 193
Nānqī, Nānqīn (Nanking) 157–158
Nānqīn River 157
Naples, Napoli 83, 106–107
al-Naqī, Imam ʿAlī 372
Nāqil 286
Naqra, Mt. 230
Naqsh-i Jahān Park (in Iṣfahān) 251
Naqshbandī 446
Naqshbandī saints 490
Naqshbandi shaikh(s) 294, 303, 316
Nārā (Nara) 131

Narāda (error for Nawāda?): Bidāʿ Narāda 365
Nārbarār 218
Narbonne 91
Narende 512
Narmāshīr 227, 230
Narmāsīr 278
Narsinga 181, 185–187, 190, 204, 206
Narsinga ruler / king / sultanate 149, 194–195, 204
musicans of Narsinga 194
Narthacusa 96
Nartūka castle 268
Narūrū 217
Nasā (in Iran) 271, 279
Nasā, Upper and Lower (in Transoxiana) 297
Nasaf 294, 298
al-Nāṣir Faraj (Mamluk sultan) 402, 406
al-Nāṣir li-Dīn Allāh (Abbasid) 366, 380–381, 394
Nāṣir al-Dīn ʿAbd ʿAllāh b. Muḥammad al-Mufassir, qadi 235
Nāṣir al-Dīn b. Shams al-Dīn, Sultan 219
Naṣīr al-Dīn Ṭūsī See s.v. Ṭūsī
Nāṣir-i Khusraw (shah of Badakhshān) 224
Nāṣir al-Dīn Sabuktagin 196
Naṣīr al-Ṭūsī, Khwāja Naṣīr 317, 321, 336
Nasr (idol) 416
Naṣr al-Dīn, Ḫoca 475
al-Naṣriyya See Nuṣayrīs
Nassau Strait 121
Naṣūḥ Pasha Ḫānı 513
Naṭāh (fort of Khaybar) 408
Wādī Naṭāh 408
Naṭanz 258, 262
Nativita 122
Natural History (book by Pliny) 47
Naugocrot See Nagarkot
Nawā 452
Nawāda (error for Narāda?): Birāgh Nawāda 363
Nawādir-i akhbār (book) 291
Nawādir al-uṣūl (book by Ḥakīm al-Tirmidhī) 298
Nāwak 209–210
al-Nawāqir mountains 442
Nāwar River 375
al-Nawāwī (Shafiʿi jurist) 452
Nawbahār 258
Nawbakht (astrologer) 368
Nawbakht (in Transoxiana) 295
Nawbandjān 237–239, 241
Nawīdiyya 388
Nawkhāla River 288
Nawkhānī 277
Nawrūsī Mountains 161
nawrūz 102
Nawshahr 218, 321, 323–324, 326
Nāyīn 241
Naynusak 219
Nazareth 442
Nāzende 481
Nāzile (village) 490
Nāzilli See Kestel
Nāziya 422
Nazwiyya 424
Nebuchadnezzar 42, 247, 251, 258, 265–266, 438
Necde 483
Negroes 100
Negus (king of Abyssinia) 386, 426
Nerdibānlar 347
Neretva River 442
Nestorians, Nestorian Christians 173, 178, 187
Nevşār ford 346, 355
Nevşehir 347
New Guinea 110, 132–134
Cape of New Guinea 133
New India, New Indies, New Spain See America
New World, *Novus Orbis* 42, 53, 77, 84, 88, 90, 92–93, 104, 112, 119, 121–122
History of the New World See *Tārīḫ-i Hind-i Ġarbī*
Niceas 107
Nicola, Captain 119
Nif 486–487, 510–511
Nigar (Srinagar) 215
Nigbolı 519–520
Nigde 472, 474, 481
Niger River 90, 113
Nigrita, Nigritae, Land of Blacks 108
Nih-Bandān 222, 223
Nihāwand 259, 262–263
Nahr-i [Shahr-i] Nihāwand 375
Nihāyat al-arab (book by al-Nuwayrī) 135, 228, 272
Nīka 465
Niksar 345
Niksār 476–478, 480–482
Niksār Mountains 478
Nīl, Shaṭṭ-i Nīl (in Iraq) 369, 375
Nīlaārū Valley 217
Nīlāb 193, 202, 205–207, 209, 213–215
Nīlān 317
Nile 69, 108, 109–110, 190, 202–203, 210, 299, 309, 403, 422
source of the Nile, Mountain(s) of the Moon 70, 84, 100, 108–109
Nīlūfar, also called the village of Nawrūz 375
Nilüfer River 501–504
Nilüfer Ḫātūn, wife of Sultan Orḫān 504, 515
Nīm Mardān 286
Nīmanda 230
Niʿmat Allāh Walī 227
Niʿmetullāh Velī 475
Nīmpū (Ningbo?) 157, 160
Nimrod, Namrūd b. Kanʿān 338, 357, 369, 376, 396, 451, 465
Namrūd, Mt. (Nemrut Dağı) 338
Nīnavī son of Bālūs 350, 352
Nineveh 350, 352
Niphus 92
Niqāb (fortress) 367
Nīraj 357
Nīrīz 236, 240–241
mountains of Nīrīz 239
Rūd-i Nīriz 240
Niş 519, 520
Nisā 290
Niṣāb-i ṣibyān (book by Abū Naṣr) 221
Nīsha 424
Nishapur, Naysābūr, Nishāwur 222, 271–285
Nishāwur Mountains 240
Nishāwur River 240
Niyābud 276
Niyāla 424
Niyāzābād 323, 325
Nīzā (Nysa) 182
Niẓām (al-Dīn) Awliyā 192
Niẓām al-Mulk (vizier) 244, 381, 382
Niẓām al-tawārīkh (book by Qāḍī Bayḍāwī) 327
Niẓāmī (poet) 257, 260–261
Niẓāmid dynasty 199
Nizib 356
Noah 98, 104, 350, 355, 370, 375–376, 391, 416, 424, 435
Noah's Ark 354, 356
Nogai 125, 308
Noricum 106
North Pole 74, 77, 82–84, 104, 119, 121
Northern Ocean, North Sea, Tatar Sea, Sea of the Tatars 90, 98–99, 105–106, 110, 121, 178
Norway, Norwegians, Norvegia 69, 74, 99, 106, 120
Nova Fransa (Nova Scotia) 113
Nova Zembla 77, 119–122
Nuʿmān / Nuʿaym b. Muqarrin 253, 263
Nūqān 271–272
al-Nuwayrī, author 259
Nubia 42, 467
Nūkām 216
Nukār 229
Nukhaylāt Bākīr 365
Nuʿmān b. Mundhir (Ghassanid) 358, 371, 376–377, 430
Nuʿmān al-Aṣghar (Ghassanid) 465
Nuʿmān Aʿwar (Ghassanid) 376–377
Nuʿmān (island) 415
Nuʿmān River 433
Nuʿmāniyya 370
Numidia 108
Nūnḥara 385
Nūpūr 219
Nūra Ṣūfī 471–472
Nūraza 213
Nuṣaybīn 125, 243, 353–354, 356–359
Nuṣayrīs, al-Naṣriyya 371, 435–436, 455
Nūshirwān See Anūshirwān
Nuzhat al-mushtāq fī ikhtirāq al-āfāq (book by al-Idrīsī) 42, 44, 84

Nuzhat al-qulūb, Nuzha (book by Ḥamd Allāh Mustawfī) 44, 212, 224–225, 232–238, 246–248, 251–254, 258–260, 265, 270–272, 274–276, 284–288, 289, 312–313, 319, 334, 336, 339, 366

Ob River 105–106, 110, 309
Ocean (Circumambient Sea) 88
Ocean Sea See Western Ocean
Odyssia 105
Of 348
Of River 349
Oghuz, Ghuzz (Turkish tribe) 225, 231, 273, 275, 290, 295, 299, 305, 307, 333
Oghuz army 281
Oghuz Khan 305–306, 362
Oğlaḳçı (village) 512
Old Testament See Torah
Old World 88, 90
Oltu 332, 345
Olympus, Mt. Olympus (in Cyprus) 470–471
Oman 125, 240–241, 384, 391–392, 416–418, 420, 424, 430
Sea of Oman 326
Oniki Dīvān 497, 499
Ordeliana River (Orinoco?) 114
Ordūbād, Ordūbādis 317, 340
Orient, Orientals
books of the Orientals 127
philosophers of the Orient 70
Orissa 181, 185, 187, 189, 199–200, 202
Gulf of Orissa 204
ruler of Orissa 199
Orḫān Ghazi, Ottoman sultan 470, 492, 501–502, 504–505, 510, 514–518
Oro 188, 203
Orontes River See ʿĀṣī
Ortaḫān 347, 366
Ortaḳçı 487
Ortelius, Abraham 40–41, 105, 126, 132, 147, 432–434, 455–456, 459, 467
ʿOs̱mān I, Ghazi, Ottoman sultan 477, 492, 503–505, 514–516
ʿOs̱mān II, Ottoman Sultan 526
ʿOs̱mān Pasha, Özdemiroğlı 327, 340, 428, 525–526
ʿOs̱mān Pasha (governor of Damascus) 442
ʿOs̱māncıḳ 478–482, 495, 512, 522
Ottomans, Ottoman Empire / domains / government / state / sultanate 36, 75, 85, 101, 107, 153, 320, 330, 333–334, 338, 358, 361–363, 363–366, 372, 383–384, 388, 403–406, 428, 430, 442, 454, 456–457, 461, 472, 477, 483–484, 489–492, 503, 506, 513, 516, 526–527
The Porte, Sublime Porte, Sublime State 36, 42, 291, 330, 337, 350
Ottoman army / troops, army of Islam 259, 313–314, 331, 339, 346, 468, 524–526, 528
Ottoman attire 428
Ottoman coins 417
Ottoman commanders 327
Ottoman conquest(s) 332, 501
Ottoman emir 496
Ottoman dynasty / family 520, 526
Ottoman governor of Egypt 411, 428
Ottoman pasha(s) 387, 428, 438, 442
Ottoman period 447
Ottoman princes 488
Ottoman provinces 343
Ottoman sanaks 350, 353
Ottoman rulers / sultans 338–340
history of the Ottoman dynasty 337
Otrar 303
Oudh 200
Ova (of Bolu) 500
Ova (of Bursa) 502
Ovabāzārı 470
Ovacuḳ 339
Ovanāḥiye 476
Oxonia (Oxford) 120
Ögetey (son of Chingis Khan) 306
Ögez 493, 495
Öküz Meḥmed Pasha (grand vizier) 474
Öljeytü Sultan / Khan (Mongol) 254, 260, 263, 283, 362, 371, 482
ʿÖmer (road stage) 512
Örencik (Virancık) 484
Ösek (Eszék, Osijek) 525
Öyük 481
Özbek, Muẓaffar al-Dīn 328
Özdemir Pasha (governor of Yemen) 428
Özeriye (clan) 457

Pachal 216
Pacific Ocean, Mare Pacificum, Southern Ocean, Sea of India 90, 110, 112, 127, 132–134, 137, 138, 143, 152
Pahlawī languge 239, 241, 322
Pakar 202, 207, 209, 222
Paklī 193, 206, 219
Paklī Road 219
Paklū 208
Paladium (idol) 110
Palamuṭ 486, 512
Palestine 91, 125, 396, 408, 424, 426, 430, 436–437, 441, 468
Pālītar 213, 214
Palos 115–116
Palohan See Calamianes
Palu 346, 353, 346
Pamphylia 91
Panipat 191, 198, 205–206
Pānhāl 217
Pānhāl Road 216
Panjhīr 212, 224
Panjhīr, Mt. 224
Panjkūra 213, 215
Panjkūra River 213, 214
Panjpūr 209–210, 215
Pannonia 106
Pannūs 209, 211
Parak 236
Paria 112, 117
Parīkhān, inn 375
Paris 43
Paropamisus (mountains) 204
Pārs See Fārs
Parsarām 216
Pashak River 222
Pāshān, Rūd-i (river) 267
Pashkān 279
Pashpishā 288
Pāsīn 319, 324, 332, 339, 343, 345, 513
Pāsīn Ovası 513
Patan 192, 205–206, 213–214, 218, 219
Patmaran 218
Patrū 219
Patta (Afghan tribe) 222
Patyāla 218
Paulus, Marcos (Marco Polo) 131, 158
Pavlos (village) 477
Pavlu, Pavli 489, 497–498
Pavlu mountain 490
Pawunī 218
Payah-pas 291
Payas 430, 470, 483
Pāybasanū 216
Pāyīl 205
Pāynatāl 217
Pazarcıḳ 461, 511
Peçuy (Pécs) 525
Pegu and Barma (Burma) 156, 181, 189–190
Bay of Pegu 189–190
pagans of Pegu 200
Pegu River 189
lords of Barma 153
Pehlivān Ḥasan 428
Penbecik 511
Pencşembe, Perşembe 345, 497, 499–500
Pendik 505, 511
Pentepel Mountains 213
Perekop (Crimea) 106, 308
Peripatetics 50
Perīdā See Cape Verde
Perseus 92
Persia, Persians 74, 80, 84, 98, 110, 124, 126, 147, 181, 186, 196, 211, 231–232, 237, 242–243, 247, 337, 345, 357–358, 366, 370, 372, 377, 382, 482, 513, 523–526
Persian (language) 40, 43–44, 84, 153, 195, 209, 239, 246, 254, 272, 287, 292, 305, 348, 351, 356, 375, 378, 436, 531
Persian book 161
Persian cities 359
Persian dictionaries 151
Persian empire (ancient) 376
Persian fabulists 261
Persian Gulf, Persian Sea, Sea of Fārs, Sea of Hormuz, Sea of Sīrāf 88, 90, 92, 97, 182, 210, 227–229, 231, 240, 246, 248, 346, 363, 365, 374, 384, 392–393, 415
Persian Iraq See Jabal
Persian kings / ruler(s), Persian rule (pre-Islamic) 91, 259, 263, 286, 323, 327–328, 339, 349, 376, 379, 438

Persian manna 295
Persian merchants 145, 344
Persian realm 362
Persian rule 349, 357
Persian Shahs / shah of Persia See Safavids
Persian style (painting, building) 258
Persian Sultanate 110
ancient Persians 76, 200, 272
ancient Persian (language) 272
Pertek 347, 353–354
Peru 122, 132, 133, 139
Pervez 355
Peshawar 213, 214, 215
Pesyān tribe 339
Phāl 217
Pharaoh 99
Philippina islands; and see Mindanao 133–134
Philippos II 134
Philippus (Cluverius) 44, 90, 152, 156, 158, 161, 175, 178, 181–182, 184–187, 189–190, 194, 307
Phoenicians 112
Phrygia 142
Pīdarang River 219
Pīla-suwā (Būyid emir) 322
Pīlūtū 202, 207
Pıñarbaşı 461, 485, 494, 501, 510
Pıñarbaşı River 462
Pıñaroğlı (village) 512
Pinpūr 209–210
Pīr Aḥmed, last of the Ḳaramanids 472
Pīr Ḥüseyin Depesi 347
Pīrchand 276
Pīrī Pasha 463, 523
Pīrī Reʾīs 42, 99–101, 122
Pīristān 216
Pīrkānī 218, 219
Pirloġanda 474
Pīrpanjāl 218
Pīrūzābād See Firozabad
Pīrūzpūr 218
Pīshān 286
Pīshdādīs, Pishdadian dynasty (ancient Persian kings) 242, 374
Pīshīn 209–211
Plato, Plato the Divine, Eflāṭūn-i Ilāhī 51, 58–59, 105, 112, 473, 513
Pliny 47, 90, 99, 107, 110, 178, 196, 203–204
Plutarch 59
Pojega (Pozsega) 525
Poland, Poles, Polish, Polonia, Leh 42, 59, 69, 74, 106–107, 125, 519, 521, 526–527
Polish (language) 436
"a Polish bridge" 107
Polish infidels 527
Polish king 521
Polish monk 59
Polovzi (Polovtsy) 307
Pomerania 90
Pool of Tibet 219
Pope 128, 137
Pope Alexander VI 136
Pope Leo (IV) 108
Por, Fūr (Porus) 196
Porsuḳ River 484, 492
Porte, Sublime Porte See Ottomans
Portugal, Portugese 42, 92–93, 99–100, 110, 112, 123, 125, 132, 137, 142, 149, 181, 184–188, 190, 194–195, 199, 228, 231, 392, 428
Portuguese (language) 130, 149
Portuguese battles 145
Portuguese captain 132, 228
Portuguese deputy 185
Portuguese explorer 133
Portuguese fleet 185
Portuguese invasion 145
Portuguese king, King of Portugal 115, 118, 136–137, 195
Portuguese merchants 146
Portuguese nation 136
Portuguese Ocean See Western Ocean
Portuguese ship 100, 122, 136–137
Portuguese standard 185
Pourchot, Edmond 58, 65
Prague 125
Prester John
sultanate of Prester John 178
Promontory of Women (ʿAvret Burnı) 118
Prophet, the See Muhammad
Propontis See Marmara
Prūnaj 218, 219
Prussia 90
Ptolemy, Ptolemaeus 42–43, 51–53, 61–65, 69–70, 73, 76, 87, 92, 99, 105, 108, 110, 119, 125, 143, 145, 147, 181, 307, 311
Ptolemaic astronomy 61, 64, 85
Ptolemaic Year 53
Ptolemies 513
Pūchī (Pūtī) 217
Puglia 125
Pūhālan 219
Pūhān 229
Pul-i Siyāh and (Pul-i) Safīd 278
Pūlārī 187–188, 205
Pulisangu River, Pūlīs, Pūlī-sānghīs, Pūlīs-Sānkūs 158, 160, 175–176, 179
Pūlūhūn (island of China) 145
Pūlyās 219
Pūmin, Fūmin 288, 325
Purna 193
Purusotam (Puri) 200
Pūshang See Būshanj
Pushna 218
Pūṣūkhil 213, 214
Pūtwāl 193
Pūzam 230
Pūzkān 279
Pyramids of Egypt 111, 190
Pyrenees 107
Pythagoras, Pythagoreans 50–51, 58–61, 200
Qāʿ al-Basīṭ 421
Qāʿ al-Buzūh 421, 431
Qāʿ Rammāla 424
Qabā 328–329
Qaʿba 424
Qabala 322
Qabān 321, 363, 365
Qabān Mountains 317
Qabarṭāy 330
Qabīla (in Syria) 431
Qabr al-Ṭawāshī (in Arabia) 423
Qābūn 433, 452
Qābūs (ruler of Ḥīra) 377
Qaḍam 452
Qadarābād 184, 205
Qadas 430, 434
Lake Qadas 432, 434
al-Qādir (Abbasid) 381
Qādisiyya 363, 366, 370–372, 375
battle of Qādisiyya 243
Qāḍīzāde 119
Qadmūs 454–456
Qāf, Mt. 325
al-Qāhir bi'llāh, Abū Manṣūr (Abbasid) 381
Qahqaha fortress 324
Qahrūd 257
Qaḥṭān, Banī Qaḥṭān, Qaḥṭāniyya 425, 431
Qahwat al-Maqṣif 433
al-Qāʾim bi-Amr Allāh (Abbasid) 381
Qāīm 181
Qalʿa-i ʿAjam 372
Qalʿa-i ʿAjūr 351
Qalʿa-i Jadīda 346, 364, 366
Qalʿa-i Kūh 227
Qalʿa-i Naw 318
Qalʿa-i sapīd 237–238
Qalʿa-i Sadda 365
Qālāj See Khalaj
Qalandar dervish 219
Qalʿat Jaʿbar 513
Qalʿat al-Kawākib 460
Qalʿat Marāwiʿ 423
Qalʿat Rīsha 453
Qalʿat al-Rūm, Rūmḳalʿe 346, 350, 431, 460, 462
Qalʿat Sūlā 453
Qalʿat Tayma 453
Qalāta 373
Qalāwūn (Mamluk Sultan) 401, 454
Qalbān 412
Qalhat 102–103, 384, 392
Qāliʿ (curser of Babel) 369
Qalīqalā (Cilicia) 334
Qalīqalā mountains 325, 346
Qālīqān Mountains 324
Qalta 414
Qāmal 176, 305
Qamariyya 375
Qamarrūd River 262
Qāmpār, Qayṣūr 143–144, 151
Qamjū, Qamchū 173, 180
Qāmrān 255
Qamṭar mountains 262

Qāmūs (book) 246, 291
Qanā' 415
Qanawāt 452
Nahr Qanawāt, Qanawāt Canal 433, 448
Qandābīl 184, 208
Qandīl Valley, Wādī'l-Qandīl 455, 460
Qanjsiyah 216
Qānpīlū 179
Qānpīlū Lake 179
Qānṣawh See al-Ghawrī
Qāntā (Canton) 133, 153, 156–157, 159–160
Qāntā River 156, 159–160
Qāntābaras 109
Qāmpāyā in China 134
Qanqli (Turkic tribe) 305
Qanṭarat al-Nu'mān 262
Qanṭarat al-Samak 459
Qānūn al-adab (book) 324
al-Qanūn al-Mas'ūdī (book by al-Bīrūnī) 73, 158, 207
Qānūnnāme (book) 153, 155, 157, 161–163, 170, 173, 175, 179–180
Qanūna 423
Qāqūn 431–432, 441, 465, 524
Qāra 414, 431
al-Qārā, Qārālar 453
Qara-azhdahān See Astrakhan
Qara Khitāy 231
Qara Ṣu 332
Qara Yūsuf 165
Qara Yülük 'Uthmān (Aqqoyunlu) 347, 357
Qarabāgh 284, 319–322, 325, 332
Qarābida (clan) 410
Qarabogha 383
Qaraja, Mt. 324
Qarakhitay 161, 245
Qarakhoja 176
Qarāmūrān River, Qārūmūrān (Kara-muran) 177, 179
Qarāniyya 367
Qarāqir 414–415, 421–422
Qaraqorum, Qaraqar, Qaraqarim 175, 178, 181, 304, 306
Qaraqoyunlu dynasty 320, 383, 457
Qaraqum, Ulugh Yurt 304–306
Qarawul (fortress) 172
Qarāyān, Qarāīm 178
Qarāyān Lake 179
Qarāzān 178
Qarhārū 375
Qarī'āt (fortress) 455
Qarīnayn 275
Qarluq 305–306, 333
Qarmaṭī See Carmathian(s)
Qarn and Dhāt Qarn 407
Qarnayn 221
Qarqīsā 358
Qārṣ Dhū'l-qadriyya See Kars
Qarṭ b. Ka'b al-Anṣārī 253
Qārtān 413
Qarūla tribe 414
Qārūn (Korah, Croesus) 192
Qārūth [Fārūt] 375
Qaryāt 392
Qaryatayn 415
Qaṣār Dam 240
Qash Ṣaḥūna 391
Qāshān See Kāshān
Qashāw See Khashāb
Qāṣī 222
Qāsim-i Anwār, Sayyid 314
Qāsim al-Hirmil (in Syria) 434
Qāsimiyya 452
al-Qaṣm, Bilād 413
Qaṣr (in Arabia) 412
Qaṣr (in Iran) 222
Qaṣr (in Syria) 457
Qaṣr, Qaṣr Ibn Hubayra (in Iraq) 366, 369
Qaṣr Jālūt 442
Qaṣr Mashīd 383, 388–389
Qaṣra 460
Qaṣrayn 433
Qaṣr-i Aḥnaf 275
Qaṣr-i Ḍaḥak 219
Qaṣr al-Luṣūṣ, Qaṣr-i Duzdān 260, 262–263
Qaṣr-i Shīrīn See Shīrīn
Qaṣrān 253
Qaṣrkand 209–211
Qaṣṣārīn River 294
Qaṣūr 191
Qāsyūn, Mt. 360, 433, 435, 443, 447–449
Qaṭaba 465
Qa'ṭaba 423
Qaṭana 415, 451
Qaṭī' 423
Qaṭīf 240, 363, 366, 384, 392–393, 415
Qaṭīfa 450, 453, 465
Qaṭrabul 368
Qaṭrān (in Arabia) 421
Qaṭrān Valley (in Transoxiana) 300
Qaṭrānī 420
Qā'ūr 194
Qawsān 369
Qaydār (Arab lineage) 397
Qāyin 274, 276, 278–279
Qayn 415
Qayrawān 125
Qays (island) 229, 240, 366, 393
Qays (tribal grouping), Qaysi(s) 438, 452, 466
Qayṣar (near Damascus) 433
Qayṣarān 265
Qaysāriyya (Caesarea in Palestine) 437
Qaysāriyya (in Anatolia) See Kayseri
Qayṣūr See Qāmpār
Qaytaq (people, tribe) 323, 325, 328–329
Qaytaq Mountains, Mt. Qaytaq, Siyāh Kūh 325, 328–329, 348
Qaytbay (Mamluk sultan) 403, 423
Dār Qaytbay 423
Qazaq Tatars 308
Qazmal 211
Qazwīn 125, 251, 253–257, 259–264, 276, 286–289, 301, 319
Āb-i Qazwīn River 262
bows of Qazwīn 260
rulers / governors of Qazwīn 263
al-Qazwīnī, Zakariyyā b. Muḥammad 39, 289
Qīāndū (Shangdu) 179
Qīāntāy 181
Qibchaq (Turkic tribe) 305–306
Qilā (fortress) 428
Qilich Arslān (Seljuk) 470, 472–473, 476–477, 482, 521
Qiniq River 324
Qinnasrīn 430, 434, 454, 457, 466
Qinnawj See Kannauj
Qirān al-sa'dayn (book by Mīr Khusraw) 196
Qirmīsīn See Kirmānshāh
Qitay Lake See Aral Sea
Qizil Aghach 288, 323, 327
Qizilbash See Safavids
Qu'ayba 365
Qu'ayqu'ān 397
Qubā, Qubā' (in Arabia) 404, 407, 424
Qubā (in Transoxiana) 297, 300
Qūbābād 370
Qubād, Qubād-khūra (district of Fārs) 232, 238, 241
Qubād (Sasanian) 237–238, 243, 248, 286, 316, 321–322, 357, 359, 362, 377
Qubādābād 298–299
Qubādhiyān 275
Qubān River 324
al-Qubaybāt 422
Qubays, Mt. 396
Qubbat al-Arḍ See Uzhayn
Qubbat al-Ḥāj 420, 444
Qubbat Sayyār 448
Qubūr al-Shuhadā' 421, 422, 407
Qūdar River 329
Qufṣ, Mt. 229
Qūhad 261
Quhandiz (of Herat) 265–266, 280–281
Quhistān 125, 223, 234, 266, 272–278
Quḥūd 256
Quinsai (Hangzhou), Kense 153, 158, 160, 190
Kense River 159–160
Qūl-Mūkhī 148
Qulta 422
Qulzum 91
Sea of Qulzum See Red Sea
Qum 251, 253, 257, 261–263, 277
Qūmis 125, 262–263, 265, 279, 284–286
Qumistān 235
Qunayṭira 432, 443, 451, 464
river of al-Qunayṭira 455
Qundus, Qunduz 211, 224–225
Qunfuda (island) 409, 415
Qūq al-Fistiq 450
Qurā Ṣāliḥ 408, 421
Qur'a 424
Qūrān Mountains 152
Qurayb (fortress) 386
Quraysh (tribe) 232, 387, 397–399, 407, 416, 425, 455
Quraysh (in Arabia) 422

Qūrna 346, 363, 365–366, 374
Qurqūb 247, 249
Qūrs al-Jawm 457
Quruchāy Valley 319
Quryān 374
Quṣam Numayrī 424
Quṣayd (error for Quṣayr?) 455
Quṣayr 452, 457
Quṣayy b. Kilāb 397–398
Qūsbatīr 181, 194
Qūsh (village) 352
Qūsmīn River 187
Qusṭanṭiniyya-i Gharb (Constantine in Algeria) 125
Qutayba b. Muslim al-Bāhilī 213, 260, 263, 279, 292, 301
Quṭb al-Dīn (ruler of Kashmir) 219
Quṭb al-Dīn ʿAtīqī 315
Quṭb al-Dīn Aybek, Sultan 192, 196
Quṭb al-Dīn Ḥaydarī 272
Quṭb al-Dīn al-Makkī 410
Quṭb al-Mulk 184, 187
Quṭbid dynasty 200
Qutulmush (Seljuk) 234
 Qutulmushids 471, 476
Qūṭūr 367
Quwayʿa (island) 415
Quwayq 434
 Quwayq River 457
Qūyāntū River 179
Quymuq (Kumyk) 328
Quzaḥ 408, 420

Rabaḍ (in Arabia) 424
Rabaṭ (in Arabia) 407
Rabī 213, 214
Rabīʿa, Banī Rabīʿa 350, 357, 425, 431, 457
 Diyār Rabīʿa 351, 353–354, 358
Rābiʿa Khatun 360
Rabīʿa Saṭīḥ 387
Rābigh 378, 409–410, 415, 421, 423
Rabīʿī (poet) 226
Rābiyya 419, 442
Rabṭa 412
Rabwa, Rabwā, Arḍ Rabwa 433, 448, 450
 Mt. Rabwa, Jabal al-Rabwa 435, 448
Raçe (Rácz) 525
Rachel 430, 440
Radʿ, Bilād Radʿ 390
Radaf 225
Rādgānī (tribe) 373
Raḍī al-Dīn ʿAlī Lālā, Shaikh 212
Raḍwā, Mt. 410
Rafʿ 415, 423
Rafaḥ 430
Rāfiʿī, Imam 254–255
Rāfiḍī See Shīʿī; Safavids
Raḥba 346, 350, 357–358, 384, 431, 450, 453, 456
Rāḥib Ṭağı See Keşiş Ṭağı
Raḥmāniyya 363, 365
Rāhmī 207, 209
Rahtās, Rahṭās 193, 206, 215, 219
 Rahtās mountains 198
Rāhūn, Mt. 150
Rāhvā 337
Rajputs 194
 Rajput tribe 184
Rājwīral 218
Rājwīrī 218
Rāk 217
Rām, Mt. 413
 ʿAyn Rām 413
Rāman 262
Rāmand, Mt. 261
Ramāykān 229
Ramażāniyye, Ramażānoğulları, Āl Ramażān 463, 468, 511
Ramażānoğlı summer pasture 512
 Ramażān Beli 475
Rāmchand 216
Rāmhurmuz 246–249
Rāmīn 253, 278–279
Rāmjard, dam 240
Raml Ḍāḥ 414
Ramla 350, 396, 408, 430–433, 437, 441, 464–465, 524
Rāmshahr 221
Rāmū 218
Rana Sanga (country and ruler of Mewar) 182, 192, 198
Raʿnāsh (village) 251
Raʿnāshiyya dynasty 251
Ranbal 219
Rāndsang 216
Ranipur 203
Rantamūr 205
Rantāpūr 192
al-Rāḍī biʾllāh (Abbasid) 381, 467
Raqʿ 385
al-Raqīm, Āl Raqīm, Bilād Raqīm 390, 443
Raqqa 125, 334, 346, 350, 352, 357–358, 456
Raʾs (village) 434
Raʾs Abū Muḥammad 411
Raʾs al-ʿAyn 125, 358
Raʾs al-Ḥadd 392
Raʾs al-Jumjuma 240, 392
Raʾs al-Nāʿūra 351
Rāsad 217
al-Rashīd biʾllāh (Abbasid) 381
Rashīd al-Dīn Sinān 457
Rasht 288–289, 301, 325
 kings of Rasht 291
Rāsib (castle) 298
Rāsin 229
Rāsmand, Mt. 258, 261–262
Rass 412, 424
 Dwellers of Rass (Koran 50:12) 324
Rastan (in Syria) 434
Rastkhāl River 324, 326
Rasulids 385, 427
 Muḥammad Rasūl 427
Rasūq 222
Rawalpindi 193, 203, 206, 219
Rawām (fortress) 429
Rawān See Arrān
Rāwan 277
Rāwān 224
al-Rāwandān 431, 434, 457–458
al-Rawḍ al-miʿṭār fī akhbār al-aqṭār (book by al-Ḥimyarī) 43, 252, 308, 314, 323, 327
Rawḍa Sharīfa (in Medina) 467
Rawḍat al-ṣafā (book) 292, 306
Rawḥā 422
Rāwī River 191, 203
Ray Shahr (Bandar Ray) 238
Rāyagān meadow 272
Raydāniyya 468
Rayma 385, 387
Rāyū [Raju] (a surgeon) 148
Rayy 125, 251, 253, 255–256, 258, 260–264, 276–277, 279, 284, 286–287
al-Rāzī, Amīn Aḥmad 44, 253, 274, 314, 321–322
al-Rāzī, Imām Fakhr al-Dīn Muḥammad b. ʿUmar 96, 226, 253
Rāżiyān 352
Rebecca 451
Red Sea, Sea of Qulzum, Ethiopia Sea, Sea of Ḥabesh, Sea / Gulf of Suez 88, 90–93, 100, 103, 108, 110, 145, 182, 326, 384–385, 411
Reşen (fortress) 374
Reşvān 346, 462–463
Revān See Arrān
Rhaetia 106
Rhodes 84, 91, 96, 110–111, 524–525
 Colossus of Rhodes 111, 131
 conquest of Rhodes 525
Ribāḥ b. Murra 413–414
Ribāṭ River 336
Ribāṭ Azraq 365
Ribāṭ Abūʾl-ʿAbbās 300
Ribāṭ-i ʿAlī b. Rustam 277
Ribāṭ-i Alīwān 319
Ribāṭ-i Arshad 319
Ribāṭ-i Badlaʿī 278
Ribāṭ-i Badr 277
Ribāṭ-i Bū Naʿīmī 278
Ribāṭ-i Būd 291
Ribāṭ-i Chashma 277
Ribāṭ-i Dahān-i Shīr 290
Ribāṭ-i Ḥājib 262
Ribāṭ-i Ḥūrān 277
Ribāṭ-i Jaʿfarī 278
Ribāṭ Jalūlā 375
Ribāṭ-i Kaʿb 279
Ribāṭ-i Kara 278
Ribāṭ-i Kardān 224
Ribāṭ-i Kūrān 278
Ribāṭ Mīnā Sulaymān 385
Ribāṭ-i Muḥammad 278
Ribāṭ-i Mūrcha-yi Khurd 262
Ribāṭ-i Mushrifān 229
Ribāṭ-i Nawshākir 291
Ribāṭ-i Nūrkhā 278
Ribāṭ Saʿd 300
Ribāṭ-i Saddī 278

Ribāṭ-i Sagbā 319
Ribāṭ Sultan 279
Ribāṭ-i Sūrān 291
Ribāṭ-i Wāsiq 262
Ribāṭ-i Zangī 278
Rīg-i Biyābān 278
Rigān River 230
Riphean Mountains 105, 111
Risāla-i Bahā'iyya (book by Fakhr al-Dīn al-Rāzī) 226
Rīshiyya (village) 450
al-Riyāḍ 412
Rize 348–349
Rize River 349
Romania 106
Rome, Romans 73, 77, 99, 106–109, 111–112, 123, 125–126, 243, 347, 349, 357
Roman authors 88
Roman emperor(s) 153, 243, 254, 437
Roman mile 74
Roman rule 112, 349
Roman Caesars, Caesars of Rome 362, 513
Rostock 69
"Routes of Countries" (genre of geographical works; and see *Masālik al-mamālik*) 35, 40, 42–44, 84
Rū'ad 286
Rūbār 230
Rūbār, plain 228
Rūbīn (fortress) 361
Rūd-i Ṣāfī River 317
Rūdān 222, 229, 241
Rūdbār 246, 256, 260, 261, 288
Rūdbār River 261
Rūdgard 260
Rūdrāwar 260, 262
Ruhā, Urfa 125, 356–358, 366, 457, 459, 461, 467, 524
Rūhā River 346
Rūj, Rūjīn, Rūj Dākhil and Rūj Khārij 460–461
Rujān 248
Rukhkhaj 221, 222
Rukn al Dawla Ḥasan (Buwayhid) 232, 251
Rukn al-Dīn, Mawlānā 324
Rukn al-Dīn al-Khūyī, Qāḍī 312
Rukn al-Dīn Sajāsī, Shaikh 314
Ruknābād 232
Rūm, Turkey; and see Anatolia, Byzantine/Byzantium, Rome/Roman 43, 69, 84, 101, 105, 107, 139, 161, 251, 292, 307, 473, 499, 505, 513–514, 522–523
Sea of Rūm See Marmara
Rūm Meḥmed Pasha, vizier 506, 522
Rūmāḥiyya 366–367, 372
Rumāḥiyya River, Nahr Rūmāḥiyya 346, 372
Rumeli Ḥiṣārı (modern Rumeli Kavağı) 507
Rumelia 44, 91, 106, 123, 510, 518–520, 524
Rūmī See Mawlānā
Rūmī (castle) 341
Rūmiya See Urmiya
Rūmḳal'e See Qal'at al-Rūm
Rūs, Russia See Muscovy
Ruṣāfa 368, 431, 454, 457–458, 466
Rūsān 228
Rushā 412
Rūstā-yi Shāh 240
Rustā-yi Zarrīn 296
Rustam son of Zāl (Iranian hero) 221, 242, 258, 361, 370
Rustam (governor of Georgia) 331
Rustam (governor of Sijistān) 243
Rustam (Safavid general) 366
Rustamdār 253, 285–287, 325
Rustān 278
Rustāq 224, 229
Rustāq al-Zuṭṭ 247
Ruwayshdāt 415
Ruwwād, Amīr 312
Rūyān 286–287
Rūydashtīn 261
Rūyīn 230
Rūz 229
Rūy-tāz 193
Rūzkān 262
Rūznāme-i Khitāy (book by Khwāja Ghiyāth) 161, 175
Rüstem Pasha (grand vizier) 335, 473, 475, 498, 502, 506

Sāb 249
al-Sab' Canal 366
Sabā, Sabā' (ancient king of Arabia Felix) 389, 425–426, 465
Sabā, Saba' (Sheba) 389–390, 402, 425
Sabā'īd (village) 278
Ṣa'bānözi 493, 495
Sābāṭ 296, 300
Ṣabbāḥ dynasty, Āl Ṣabbāḥ 264–265
Ḥasan-i Ṣabbāḥ 256
Sabians 358, 373, 416, 436, 444, 453
Sābīkand 301
Ṣabir Mountains 384, 386
Ṣabiyya, Ṣabiyyā, Qal'at Ṣabiyya 384–385, 431
Ṣabrān 299, 303
Sabuktegin, Amir (Ghaznavid) 212, 381
Ṣabuncı mountain pass 486
Sābūr 239–240
Mt. Sābūr 240
Sabzawār 273–274, 283
Sa'd al-'Ashīra 431
Sa'd al-Dīn al-Taftazānī 226, 271
Sa'd b. Abī Waqqāṣ 243, 254, 370, 372, 424
Sa'd b. 'Ubāda 452
Ṣa'da 387–388, 390–391, 418, 423–424, 427, 432
Sa'dābād 340, 375
Ṣadd b. 'Ād 389
Sa'dī, Shaikh 245, 315
Ṣādī Bey 480
Ṣadīqā, Mt. 434
Ṣādır (near Mardin) 353
Ṣadīr (road stage, in Arabia) 424
Sadīr mountain (in Arabia) 413
Sa'diyya 409, 423
Ṣadr al-Dīn, Mawlānā 324
Ṣadr al-Dīn Mūsā 315
Ṣadr al-Sharī'a (jurist) 522
Ṣadr Baḥrayn (in Iraq) 365
Ṣadr-i Āl-i Ṭawīl (road stage) 365
Ṣadr-i Ṭawīl Canal 365
Ṣadrayn 373
Sadw 300
Ṣafā (in Syria) 452
Ṣafā (in Transoxiana) 300
Ṣafā, Mt. (in Mecca) 393–394, 396, 398, 402, 416, 419
Ṣafad 430, 434, 436, 441–442, 451, 524
Bilād Ṣafad (fortress) 451
Mt. Ṣafad 433
Ṣafadī (Khalīl ibn Aybak) 317
Safar Agha, Khudāwand Khan 182
Safavids, Persian shahs, Qizilbash, Rāfiḍīs, Sofis
Safavid dynasty, Safavid shahs 315, 320, 330, 339, 371
Persian shahs, The shahs, Shah of Persia 200, 219, 221, 223, 225, 227–228, 231, 239, 246, 251, 253, 271, 287, 291, 330, 334
Qizilbash, *surḫ ser* ("Redheads") 195, 209, 222, 287, 302, 314, 317, 321, 327, 333, 338, 340, 346, 358, 363, 365–366, 372, 383, 444, 454, 468, 522–523
Rāfiḍīs ("heretics") 258, 284, 291
Sofis 195
Qizilbash khans (i.e., Safavid governors) 223
Qizilbash troops 302
Safavid rulers / shahs 330, 340
Safavid state 340
Sofi Shah of Persia 110
al-Saffāḥ, Abū'l-'Abbās 'Abd Allāh (Abbasid) 366, 371, 377–378, 380, 429, 467
Saffarids 223, 225
Ṣafī al-Dīn Ardabīlī, Shaikh Ṣafī (Safavid) 254, 314–315, 320–321
Ṣafī, son of Shāh 'Abbās (Safavid) 320
Ṣāfī River 318
Safīd Rūbār 288
Safīdrūd River, Sapīdrūd 251, 257, 261–262, 288, 317–319
Ṣafīkand (village) 375
Safkhāy 222
Saflān 460
Safranbolu See Borlı
Ṣafvān River 356
Safwān 409
Sagābād 278
Ṣāghān (village) 275
Ṣaghānī (clan) 355
Ṣaghāniyān, Chaghāniyān 275, 296, 298–301
Chaghāniyān Mountains 292
Ṣaghāniyān River, Rūd-i Chaghāniyān 297

Sagjū 305
Ṣağmān 353–354
Ṣaḥāba See Companions of the Prophet
Sahand, Mt. 312, 314–315, 317–318
Saḥār 384, 388
Saḥar mountains 278
Āb-i Saḥar River 278
Ṣāhil al-Kubrā 241
Sāhira (in Jerusalem) 440
Sahīrand 205
Sahl (pilgrimate stage) 444
Ṣahla 386
Ṣaḥn Bayāḍ 423
Ṣaḥna 262, 375
Ṣaḥne Geçidi 346
Ṣaḥneçimen 481
Ṣāhta 235
Saḥūl 388, 418
Sahwān 202
Ṣahyūn 430, 454–455, 458
Saʿī, market of (in Mecca) 394
Saʿīd (fortress) 363, 365
Saʿīd Khwāja Rashīd (al-Dīn) 313, 315
Saʿīd b. Abī Barakāt (Sharif of Mecca) 429
Sāʿida (district) 415
Sāʿida (clan) 452
Saʿīdābād 278, 286, 319
Sajad River 318
Sajās 262
Ṣakaf Han 356
Ṣaḳaḳ 340
Sakāla 414
Sakān 240
Ṣaḳarya 493, 501, 513, 522
Ṣaḳarya River 310, 484, 498, 508, 512
Sakat 300
Sakāwand 212
Sakirābād 262
Saḳız See Chios
Sāl b. Ṣadra (road stage) 365
Saladin, Masʿūd Ṣalāḥ al-Dīn Yūsuf (Ayyubid) 427, 437, 442, 467
hospice / hospital of Saladin / Ṣalāḥ al-Dīn 239, 449
Salafqūs (Seleucus i) 458
Salāma 385, 423
Salamiyya 358, 430–431, 454, 456
Ṣalba 385
Salbāʾān 279
Ṣaldır 512
Ṣalkhad 420
Salghur (Turkic tribe) 245
Salghurābād 245
Salghurids 231, 245–246
Ṣalīgha 383
Ṣāliḥ (prophet) , 384, 408, 421, 442
people of Ṣāliḥ 389
Ṣāliḥ b. Ismāʿīl (Sharif of the Ukhayḍir) 429
Ṣāliḥiyya (in Syria) 443, 447–450, 453, 465
Ṣāliḥiyya (in Iraq) 363, 365
Salim (village) 278
Salīm, Sultan (Mughal emperor Jahāngīr) 198
Salīm Khan 192
Salmā 414
Mt. Salmā 414–415
Sanām, Mt. 364
Salmān-i Fārisī 247, 370, 407
Salmānābād 375
Salmās 316–319, 334
Salmās Mountains 317
Salmiya 384
Salqīn 460
Ṣalt, Ṣalṭ 430–431, 441–442
Jabal al-Ṣalt 434
Saltpan (great Salt Lake) 476
Saltpan Island (Tuzla Adası) 109
Saltuqids 347
Sālūs 285–286, 288
Sām (fortress) 428
Sām son of Noah 430
Samāḳlu 360–361
Mt. Samāḳlu 360–361
Ṣamaḳo 497, 499
Sāmān 257, 259
Samānā 191, 197, 206
Samandar 325–326
al-Samʿānī 42, 272, 298
Samandıra 506
Samanid(s), Samanid dynasty 272, 280–281, 293, 301
Ṣamanlu 503, 505–506, 512
Ṣamanlu Mountains 503, 505, 512
Samāra (in Arabia) 387
Samaritans 436
Samarqand 125, 176, 212, 237, 250, 280, 284, 290–305, 314, 321, 379, 426, 522
Samarqandī 119
Samarra, Sāmarrāʾ 351–352, 361, 368, 372, 375, 379
Samāwa Qarāqir 452
Samawʾal b. ʿĀdiyā 415
Samāwāt 366–367, 372, 375
Samāwāt River, Nahr Samāwāt 346, 372
Sāmbū 216
Samḥān 422
Sāmira 340–341
Sāmirā wife of Nīnavī 352
Samīrān 236
Samīrān Valley 253
Ṣāmīzāde Meḥmed Efendi 494
Samoyed 120
Samra Māryān 365
Ṣamṣām al-Dawla (Buwayhid) 232, 244
Samsat See Sumaysāṭ
Samsun 349, 478
Samūr River 325, 328
San Luca (Sanlucar) 99, 116, 118
San Toma (church) 186
San Toma Pater See Thomas the Apostle
Ṣanʿāʾ 125, 384–388, 390–391, 407, 416, 418–420, 423–424, 427–428
Sanābād 272
Sanām mountain, Mt. Sanām 363–364
Ṣanamayn 420, 444, 452
Ṣanamgān 233
Sancaḳburnı 485
Sancar 481, 513
Sandabrārī 216
Sandbūr 290
Sangābād 291
Sangalābād 319
Sangān 375
Sang-bar-sang Pass 321
Sānghā 190, 192, 195
Sangīn 222
Sangpūr 218
Sangrū 216
Sanjar, Sultan (Seljuk) 245, 270, 273, 281, 291, 347, 354
Sanjarids 283
Sanjīd-i Khalkhāl 319
Sankhil 213, 214
Sanlīl, Bāzār-i Sanlīl 248–249
Santa Anton 109
Santa Fe 115
Santa Gloria 117
Santa Jacobi 109
Santa Lorensi 109
Santa Lucia 109
Santa Nicola 109
Santa Toma 109, 116
Santa Vīnensī 109
Santoma 119
Ṣār ford 462
Sarbal 207, 216, 218
Ṣapanca 498, 501, 503, 505, 512
Lake Ṣapanca 499, 501, 508–509, 513
Sapanj 228
Sapīdrūd See Safīdrūd
Sapū 216
Saqīna 299
Saqlāb, Ṣaqlāb, Slavs, Slavic 69, 84, 307, 333
Saqrī 291
Sar-i Asad, Mt. 262
Sar-i Maydān 325
Sar-i pul 294
Sarāb 222
Sarah wife of Abraham 396, 438
Sarakhs 268, 271, 275, 278–279
Sarāqiba 457
Sarāw, Sarāh 316, 319
plain of Sarāh 317
Sarāwrūd, Sarāh River, Rūd-i Sar 312, 316, 318–319, 325
Sarāw, Mt. 318
Sarawāt Mountains 384, 385, 387, 388, 416
Sarāy (qadi district in Ḳasṭamonı) 496–497
Sarbal 216, 218
Sarbār 209
Sarbār River 209
Sard 234, 318
Sardrūd River 313, 318
Sardinia 106, 125
al-Ṣarfan (Sarepta) 442
Sārī 285–286, 307
Sarī al-Saqaṭī 368, 453

Ṣarı Işıḳ 512
Ṣarıbalı, brother of ʿOs̱mān 477, 513–515
Ṣarıbey 512
Ṣarıçayır 501, 505
Sārīda 375
Ṣarıdere River 508
Ṣarıḳavaḳ 469
Ṣarılar 513
Ṣarıoğlan 481
Sarīr (in Dagestan) 304, 312, 323, 326, 328, 375
Sarīr al-Lān See Dagestan
Sarīr (in India) 198
Sāriya 285
Sarjahān (castle) 256
Sarjam 319
Sarjī territory (Yemen) 387
Ṣarkhad 430–431, 451
Sarkhīj 183, 205
Sarmatia, Salmatia, Sarmatians 90, 98, 105–106, 110
Sarmaqān 275
hospice of Sarmaqān 241
Sarmatia 307
Sarmīl 367, 375
Sarmīn 366, 435, 457–458, 466
Ṣarṣar 369, 375
Ṣarṣar Hill 424
Ṣart (ancient Sardis) 487–488, 496
Ṣaruḫān 483–486, 505, 510, 516, 522
Ṣaruḫānid 486
Sarūj 357–358, 460, 462
Sarūr 222
Sarūrāj 424
Sarwān 221, 262
Sārwān, Mt. 299
Sarwīna 375
Sarwistān 233, 241
Ṣaryāt 415
Saʿsaʿ 433, 440, 451, 464
Sāsān b. Bābāk 400
Sasanian dynasty, Sasanians 232, 243, 252–254, 263, 269, 285, 316, 352, 359
Sasanian army 251
Sasanian period 270
Ṣāṣūn (fortress) 355
Saṭḥ (fortress) 408
Saṭḥ al-ʿAqaba 422
Ṣāṭīghān (Chittagong) 187–188, 205
Ṣāṭir Anjād 375
Saṭrā, Arḍ Saṭrā 450
Saturn 443, 459
Saul (*sic*) and Goliath 430, 434
Ṣavur 353, 356, 510
Sāwa 253, 255–257, 259, 261–263
Sāwa, Mt. 261
Sawād al-ʿIrāq, Sawād-i Baghdad 350–351, 369
al-Sāwajī, Khwāja Saʿd al-Dīn, Vizier 319
al-Sāwajī, Niẓām al-Dīn Yaḥyā, Amīr 319
Sawākin, Suakin 102, 418, 428
Sawālim (tribe) 431
Sāwghar 301
Sawiyya 424
Sāwkat 300
Ṣawlīna 367
Saxons 107, 520
Sāya Patrū 138
Sāyand 279
Ṣaydānāya (village) 450
Sayf (village) 228
Sayf, Sayf Abī Zubayr, Sayf ʿImāra (seacoast districts of Fārs) 233
Sayf al-Baḥr 233, 240
Sayf b. Dhī Yazan 426
Sayf al-Dawla b. Ḥamdān 353, 464
Sayf al-Dawla Ṣadaqa 371
Sayfiyya-oğlı 455
Sayḥūn River, Rūd-i Chāch, Shāsh River 290–291, 294, 296–299, 304, 309
Sayl al-ʿArim (flood that destroyed the Maʾrib Dam, Koran 34:16) 376, 389, 465
Saylān, Mt. 315–316, 318–319, 321
Ṣaymar 246
Ṣaymara 258, 260, 262
Saynasar 218
Ṣayram 305
Qari Ṣayram 305
Sayyid Barakāt, Sharīf of Mecca 427–428
Sayyid Barakātzāde Abū Numayy Muḥammad, Sharīf Abū Numayy 410, 524
Sayyid Ḥasan River 373
Sayyid Ḥusayn 336
Sayyidpur 192, 205
Saz (Saxons of Transylvania) 520
Sazaḳ 498
Scaliger, Julius 121
Scotland, Scotia, Scots 74, 90, 106
Scythia 90, 124, 307
Şebḫāne 345
Sebūt See Cebu
Sefer Pasha 333
Seferīḥisār 510
Şehirgölü 474
Şehmerān 360
Şehr-i Zūl See Shahrazūr
Seker plain 481
Şekerbıñarı 463, 511
Sekjū 176–177
Sekjū Pass 172, 180
Sekmānābād, Sekmenābād 319, 339–340, 347
Selanik, Thessaloniki 469, 520
Şelbe (village) 356
Selçūḳ Ḫānı 481
Selem 339
Selendi, Selenti 469–470, 484–485
Şelfederesi Çayı 352
Şelīle 347
Selīm I, Ottoman Sultan 161, 320, 337, 339, 344, 347–348, 352, 356–357, 365, 383, 401, 420, 427–428, 449–450, 468, 522–523
Selīm II, Ottoman Sultan 42, 153, 309, 398, 403–405, 471, 525–526
Selīmnāme (book by Şükrī) 337
Selīm Giray Khan 527
Selīūs (Caelius Antipater) 99
Seljuk(s), Seljuk dynasty, Seljukid 223, 227, 230–231, 235–236, 240, 245, 263–264, 270, 281, 304, 306–307, 319, 335, 347, 352, 357, 366, 381, 467, 471–472, 475–477, 480, 482
Seljuk coinage 517
Seljuk governors 301
Seljuk state 477, 482
Seljuk sultan 245, 327, 344, 347, 375, 424, 470, 472, 482, 491
Seljuk rule / time(s) 236, 238, 251, 489
Selmās 334, 339–340
Şemāmek 359
Şemānīn (village) 356
Semāviye 347
Semāvkū 341
Semendire (Smederevo) 520–521
Şemīrāh 359
Semiramis 196
Şemsī Aḥmed Pasha 448
Şemsī Efendi 499
Seneca 98, 112
Senagal River 109
Şenbū, Şenbūner (governors of Ḥakkārī) 341, 361
Señir-Düşenbe 470
Senūr Valley 344
Sepetlübeli 482
Seraḳ 500
Serbeten 347
Serbia, Servia, Sirf 106, 520–521
Serçınar 352
Şerdenburnı 489
Serdī 341
Şerefoğlı Süleymān Bey (Aqqoyunlu emir) 361
Serfendigār 464
Şerḥ-i Maḳāṣıd (i.e., commentary on *Maqāṣid fī ʿilm al-kalām* by al-Taftāzānī) 522
Şerḥ-i Mevāḳıf (i.e., al-Jurjānī's commentary on *al-Mawāqif fī ʿilm al-kalām* by al-Ījī) 522
Seri Yazlulı (castle) 341
Serica See Cathay
Şerīf Efendi 43
Seroz 489
Serūcek 359–360
Seth son of Adam 373, 453
Seven Kings (i.e. Seven Electors of the Holy Roman Empire) 525
Seven Wonders of the World 111
Seville 116, 118
Seybān, Mt. 336
Seydī ʿAlīzāde, Ġalaṭalı 43, 101, 241
Seydīġāzī 491, 511, 522
Seydiḳavaġı 518
Seydīşehri 470, 472, 474–475
Seyḥan River 434, 463
Şeyḫlü 484–485
Şeyḫnuṣret 481
Seyyid Ḥasan River 355

Seyyid Ṣādıḳ (road stage) 352
Shabānkāra 235–237, 241, 244–245
Shabānkāra Dynasty 244
Shābilkaza 213, 214
Shabistarī, Shaikh Maḥmūd 314
Shābūr, Dhū'l-Aktāf (Shābūr II, Sasanian) 237–238, 243, 246–248, 254–255, 272–273, 351
Shābūr Khūra 232, 237, 239, 241, 247, 254
Shaburghān 270, 283
Shādāfarīn 240
Shādaj 218
Shadanū 217
Shaddād son of ʿĀd 391, 425–426
Shādhilī, Shaikh Abū al-Ḥasan, Shaikh Shāzilī 384, 418
Shādī Khwāja 161, 165
Shādirwān (dam) 246, 248
Shādyākh 273
Shafaqa (in Arabia) 423
Shafār 412, 414
al-Shāfiʿī, Imam 388, 395, 437
Shāfiʿīs, Shāfiʿī rite 232, 252–253, 255–256, 274, 286, 295, 304, 324, 328, 337, 348, 354, 356, 359, 361, 375, 400, 416, 419, 445–446, 524
Shaft 288
Shah Jahān, Khurram Shah (Mughal emperor) 198
Shāh-i Shujāʿ Kirmānī (Shāh b. Shujāʿ al-Kirmānī) 227
Shahārā 384, 386
Shāhī River (in Iraq) 346, 374
Shāhī River (in Balkh) See Balkh River
Shahīd al-Baḥr (in Syria) 455
Shāhiyya 424
Shahkūt 218
Shaḥnab 228
Shahnāma (book by Firdawsī) 322
Shaḥr 383, 419
Shahr-i Sabz 284, 294, 300
Shahrābād 286
Shahrabāzār 359–360
Shahrazūr, al-Zūr, Shahr-i Zūr, Şehr-i Zūl 262–263, 312, 334, 340, 350, 352, 358–360, 362
Shahrastān 275–276
Shahristān, Shāristān 229, 237, 252, 255, 259, 267, 294
Shahribān 263, 367–368, 370, 374–375, 525
Shahriyār (Sassanian) 243, 327
Shahriyār, desert of 239, 327
Shāhrūd River 261, 286
Shāhrukh Mīrzā, Sultan Shāhrukh (Timurid) 161, 165, 197, 219, 265, 270, 284, 302, 321
Shahrūqiyā 298
Shahrūqiyā River 296–297
al-Shaikh al-Akbar See Ibn al-ʿArabī
Shaikh ʿAlvān, dervish lodge of 479
Shaikh al-Ghamm River 413
Shaikh al-Ḥadīd (in Syria) 460
Shaikh Muzīq (island) 415
Shaikh al-Niʿām River 413
Shaikh Rasān (near Damascus) 433
Shaikh Ṣādiq (in Transoxiana) 300
Shaikh Zevli (near Diyarbekir) 356
Shajarāt al-Amīr 410
Shākhā 370
Shakra (clan) 410
Shakyamuni 173, 176
Shāl Mountains 318
Shālrūd River 318
Shalāl 217
Shalāla 388, 424
Shalfūhama 458
Shalj 304
Shālūshiyya (fortress) 363, 366
Shālūt 216
Shām See Syria, Damascus
Shām, Shām-i Qazān (in Tabriz) 313
Shamākhī 321–324, 326, 330
Shamāmak 352
Shamīl 230
Shāmil River 213
Shamīr 424
Shamīrān 257, 266, 280–281
Shamīsāṭ See Sumaysāṭ
Shamkhāl 330
mountains of Shamkhāl 325
Shamkūr 322, 326
Shamkūra, Lake 324
Shammar, Bilād 414
Shams al-Dīn (ruler of Kashimir) 219
Shams al-Dīn Muḥammad b. ʿAlī b. Malikdād 314
Shams al-Dīn, Khwāja, Ṣāḥib-dīwān (Juwaynī) 262, 318
Shams al-Dīn ʿUbaydī 315
Shams al-Dīn (road stage) 365
Shankar, Mt. 219
Shaqīf, Shaqīf Bayrūt, Bilād al-Shaqīf 430, 433–434, 441–442, 451
Shaqīf Arnūn, Shaqīfa 450, 460
Shaqīq (road stage) 423
Shaqīq al-Balkhī 269
Shaqīqa 414
Shaqq Ḥusayn 300
Shaqrāfid 412
Shār 217
Shaʿrā 412
Sharaf Khān Badlīsī, Mīr Sharaf 335–341, 373
Sharḥ-i jadīd (book) 315
Sharaf al-Dīn (governor of Nishapur) 273
Sharaf al-Dīn ʿAbd al-Qahhār 268
Sharaf al-Dīn ʿAlī Yazdī, Sharaf-i Yazdī 235, 306
Sharaf al-Dīn al-Dargazīnī, Shaikh al-Islām 259
Sharaf al-Dīn Maḥmūd Shāh (Īnjūid) 232, 245
Sharaf al-Dīn Yaḥyā (Zaydī) 427
Sharaf al-Dīn, Amīr 402
Sharaf al-Dīn, Mt. 338
Sharāh, Sharāt 421, 442
Jabal al-Sharāh, Jabal-i Sharāḥ (sic; error?), Mt. Sharāt 384, 434, 442–443
Sharash (fortress) 363, 365
Sharīʿa (tribe) 451
Sharīʿa River See Urdun
Sharīf Abū Numayy See Sayyid Barakāt-zāde
Sharīf al-Idrīsī 42
Sharīf Muṭahhar (of Yemen) 385, 387, 390, 427
Shāristān See Shahristān
al-Sharīshī (commentator of the *Maqāmāt* of al-Ḥarīrī) 334
Sharja (island) 415
Sharm 423
Sharnaba 423
Sharya 454–455
Shāsh See Chāch
Shaʿshaʿa 424
Shatān, Mt. 278
Shatar 238, 240
Shaṭṭ al-ʿArab See Tigris
Shaṭṭ al-Ḥimār 366
Shawāfi (tribe) 388
Shawbak 420, 421, 430, 431, 436, 442, 443
Shāwkath 296
Shayāh 363
Shayba b. ʿUthmān 395
Shaybān Rāʿī 435
Shayzar 430, 435, 456–457, 466, 524
Shāzilī See Shādhilī
Shem son of Noah 104, 242, 376, 391, 413, 424–425, 430
Shiʿb-i Bawān 237–239, 314
Shiʿb-i Bawān River 240
Shiʿb Ḥunayn 410
Shiʿb al-Niʿām 421–422
Shiʿb-i Sulaymān River 248
Shibām, Shiyam (?) 387–389
Shiblī, Shaikh 298, 368
Shibliya 298
Shifāʾ (book by Avicenna) 128
Shifāʾ (book by Qāḍī ʿIyāḍ al-Yakhṣubī) 447
Shihāb al-Dīn b. Aḥmad, Artuqid 357
Shihāb al-Dīn Muḥammad, Sultan (Ghurid) 196, 200, 225, 228
Shiḥr 102–103, 383–384, 388–389, 391, 415, 417, 424–425, 430
Shīʿī(s), Rāfiḍī(s) 232, 239, 246, 253, 255–258, 274, 284–285, 291, 348, 382–383, 436, 480–481
Imami Shīʿīs / Shīʿīsm, Twelver Shiism 199, 239, 257, 274, 287, 427
Shīr, Mt. 261
Shīr Khan, Shīr Shah 192–193, 198–199
Shīrkhānid dynasty 198
Shīrābād 210
Shiraz 228, 230–241, 244–245, 263, 315
rice of Shiraz 246
al-Shirāzī, Quṭb al-dīn 90
Shīrdār 239

Shīrdūn 241
Shīrīn (beloved of Khusraw Parvīz and Farhād) 243, 362, 370, 375, 480
Qaṣr-i Shīrīn 243, 263, 367, 369–370, 372, 375
Shīrīn River, Sīrīn River 238, 240
Shīrkar 218
Shīrkūh b. Muḥammad 358
Shīrma 389
Shirwān 288, 302, 304, 309, 312, 318, 320–327, 330, 331, 525–526
Shirwānid dynasty, Shirwānids, Shirwānshāh 320
kings of Shirwān 323
Sea of Shirwān See Caspian
Shitābī (island) 364
Shuʿayb (prophet) 409, 411–412, 441
Shubayka 393–395
Shubūrqān 279
Shūf 434, 453
Shūf Mountains 433, 436, 451
Shughr 430, 455, 457–458
Shūman 298
Shūmār 453
Shuqayr 412–413
Shūr 424
Shūra (in Iran) 277
Āb-i Shūra River, Shūrarūd 277–278
Shūra (in India) 203
Shurfat Banī ʿAṭiyya 422
Shurūr 339
emirate of Shurūr 339
Shūsh 351, 373
Shuʿur 424, 460
Sialkot 191, 203, 208–209, 216–218
Siam (Thailand), Siamese, Sian 143, 152, 181, 188–190, 202, 204–205
Gulf of Sian 190
Sea of Siam 127
Siam River 190
Sībān, Mt. 312
Siberia 308
Ṣıçanlu 484–485
Sicav (Suceava) 525
Sicily, Sicilia, Messina 44, 91, 96, 105–106, 119, 123, 125
Sidaw Road 218
Sidon 105, 126, 430–431, 433–434, 436, 441, 450–451, 453–455, 524
Sierra Leone (mountain) 108
Ṣiffīn 358
Ṣığla 505, 510
Sīhrind, Sirhind 191
Sīhwān 207–209
Siʿird, Siʿirt 353, 354, 374
Sijistān, Sīstān 210, 212, 220–224, 227–229, 231, 243, 245, 265, 272, 276–280, 282, 302
Sīstān road 278
Siğa 513
Sīḥ 365
Sikandarī 366
Sikult (Siculi, Szeklers) 521
Sīl 218
Sīlākhūr River 248
Şile 505–506
Silifke, Cilicia 91, 459, 463, 469–470, 472, 474–475, 483
Cilician Gates 469
Silistre 519
Silivri 76
Silsila (road stage) 424
Silwān, spring of See ʿAyn Silwān
Sīmābrūd 229
Simār, Mt. 427
Simav 484, 486–487, 519
Simav Çayı 503
Sīmber, Sīmberī (Cimbri, Cimmerians) 105, 107
Siminjān 224
Simnān 125, 259–260, 263, 284–286
Simnānī, Shaikh ʿAlā al-Dawla 254, 285
Simṭ al-ʿulā (book) 227
Sīn 262
Sīnāb (village) 434
Sinai (region) 383
Mt. Sinai, Ṭūr Sīnā 411, 434–435
Sinān (imperial architect) 508
Sinān Pasha 357, 405, 428, 440–441, 447, 453, 508, 524, 525
Ḥān of Sinān Pasha 357
Sinānlı 469
Sincan 353, 501–502
Sind 69, 80, 103, 124–125, 160, 183–184, 193, 202, 206–209, 215, 220–221, 225, 230, 240, 380, 385, 445
Sind and Hind 97, 100
Sind-hind, Sind wa Hind (book by Great Brahmin) 97, 201
Sind River, river of Sind See Indus
Sind son of Ham 206
Sindān Meadow 237
Sindī, Sindiyān (tribe) 373
Ṣındırğı 486–487, 503–504
Singapore 190
Cape of Singapore 132
Sīnghālīs [Cingales] (Sinhalese), Sīnghālī [Cingalas] 148
Sīnj 222
Sinjān 222
Sinjār 73, 125, 354, 357, 359, 524
Sinjū 305
Sīnk 133
Sinn (town) 351
Sinnī (tribe) 361
Sinop 348, 477, 483, 496–498, 516, 521
Mt. Sinop, Bozdepe 497
Sipāhizāde Meḥmed Efendi 42
Sipanj 277–278
Sipāwand 210
Siperbārik 352
Siqāyat Jayrūn 466
Sīrāf 229, 233, 238–241
Sea of Sīrāf See Persian Gulf
al-Sīrāfī, Abū Saʿīd (grammarian) 233
Sirāj al-Dīn al-Urmawī, Qāḍī, Qadı Sirācaddīn 316, 473
Ṣīrān 345
Sirf See Serbia
Sirḥān, Shaikh 454
Mt. Shaikh Sirḥān 454
Sirhind See Sīhrind
Sīrīn River See Shīrīn River
Sīrjān, Sīrgān 227, 229–230, 241
Sirnigīn 210
Sirohi 193
Ṣīrovası 346
Sirr 422
Sirrayn 409, 412
Ṣırt 512
Sīrūm 340
Sīrvī, Ṣīrvī 334, 340, 361
Sīrzān 275
Sīs, Bilād Sīs 334, 345, 430, 434, 457, 513
Sīstān See Sijistān
Sitāragān River 240
Sitārūz River 222
Sītpur 202–203, 207–209, 222
Sitt al-Shām 453
Sittī Zaynab, daughter of ʿAlī 452
Sivas 302, 334, 343–347, 352, 462, 472, 476–483, 512–513, 519, 523
Siverek 353–356
Sivriḥiṣār 483, 501, 512
Sivrināz mountain 490
Sīwar 191, 203
Siwinj River 325
Siyāh Kūh (in Caucasus) See Qaytaq
Siyāh Kūh (in Sīstān) 222, 277–278
Siyyān 424
Slavs See Ṣaqlāb
Sochi, Ṣūjaq, Ṣūcha 329, 331
Socotra 92, 418
Sodom city of Lot 432
Sofi See Safavids
Sofala 103
Ṣofılar 513
Ṣofyān 347
Ṣoğan Yaylası, Ṣoğanlu Yaylası 332, 346
Soghd 291–295, 299–301, 314
Soghdian (language) 293
Soghd Valley 295
Sohrān (clan) 358, 361
emirate of Sohrān 340
rulers of Sohrān 361
Solinus 92, 204–205
Solomon 36, 132, 145, 152, 234, 239, 309, 350, 426, 438, 440, 442, 446, 456
seals of Solomon 446
Temple of Solomon 111
theater of Solomon 442
Solon the Athenian 112
Sombor 525
Somnath, Sū-manāt 183, 196
Ṣonisa 477–478, 482
Sophists 201
Ṣora 462
Ṣorba 492–493
Ṣorḳun 480, 496, 512
Ṣorḳūn summer pasture 475

South America 70
South Pole 77, 80, 82, 122
Southern Ocean See Pacific
Sögüt, Sögüd 491, 492, 501, 511, 513–514, 516
Sögüt Ṭağı 485, 490
Spain, Spanish, Spaniard; Ispania 45, 69–70, 74, 88, 90–93, 95, 99–100, 105–107, 112, 116–117, 123, 125, 133, 135, 137, 181, 185, 407, 471, 436, 513, 525–526
Spanish leagues 75
Spanish sea captains 70
Cape of Spain See Calpe
emperor / king / ruler of Spain 115–118, 147–148
Spiritus Sancti River 109
Srinagar 218
St. John, Church of 466
Stoic sages 170
Straits See Bosphorus, Hellespont
Strabo 106, 110, 112, 196
Ṣu Şehri 338
Ṣuʿābaşı 347
Suakin See Sawākin
Ṣubaydā (fortress) 451
Ṣubayḥa 422
Subka 414
Suçuk summer pasture 357
Sūd Dawwāra 277
Sudan 438, 467
Sūdqānat 375
Sūdra and Kūpra 191, 209
Suedia See Sweden
Suez 42, 92, 101, 108, 383, 400, 409, 411, 422
Sea / Gulf of Suez See Red Sea
Strait of Suez 110
Ṣuffa-i Asbanarg 375
Sufis, Sufism 83, 227, 284, 315, 337, 359–360
Sufi orders 446
Sufyān al-Thawrī 378, 416
Ṣughar (in Arabia) 424
Ṣughar (Zoar) mountain 438
Sughdāq 477
Ṣughūriyya 442
Ṣughwān 424
Ṣuğla 475, 486–487
Ṣuğni Valley 346
Ṣuğur Köprüsü 465
Ṣuḥār (Sohar in Oman) 384, 392, 419
Sūhqān 276
Suhran (in Kashmir) 218, 219
Suhraward 256
al-Suhrawardī, Abū'l-Najīb 337
al-Suhrawardī, Shaikh Shihāb al-Dīn 301, 368, 491
Sujās 256
Mt. Sujās 256
Ṣūjaq, Ṣūcha See Sochi
Sukra (village) 369
Sūktel 216
Ṣulayḥids, Ṣulayḥid dynasty 386, 427
Sullam al-samāʾ (book by al-Kāshī) 87
Sūkh 297, 300
Sūkh mountains 296–297
Sukhūm, Sukhumi 329–331, 348
Sūkra 193, 206
Sulaymān, Amīr (Qaraqoyunlu commander) 320
Sulaymān b. ʿAbd al-Malik (Umayyad) 437, 466
Sulaymān b. Qutulmush, Sulaymān Shah (Seljuk) 471, 476–477, 482
Sulaymānī (Kurdish tribe and dialect) 373, 436
Sultan Island 109
Sulṭān Bāğı (in Kütahya) 484
Sulṭān Depesi 347
Sulṭān Ḫānı 511
Sulṭān Jamjālābād (town) 362
Sultan Junayd, one of Bābur's emirs 198
Sulṭānçayırı 511
Sulṭānḥiṣārı 487, 512
Sulṭāniyya 125, 253–256, 262, 278–279, 319, 334
Sulṭānöñi 483, 491
Sulṭānpur 191, 205–206, 219
Sulṭānoğlı summer pasture 498
Sulṭānyaylağı Mountains 353
Ṣuluköy 512
Sū-manāt See Somnath
Sumaysāṭ, Şamīsāṭ, Samosata (Samsat) 334, 346, 350, 354, 430, 434
Sumatra 102–103, 133, 139, 141, 143–145, 151
Summāq 423
Sumurkūt 217
Sunbul Mountains 219
Sunda 141, 143, 145
Sundar, Mt. 217
Sundarbār, Mt. 217
Sunni(s) 232, 239, 259, 267, 291, 320, 328, 339–340, 361, 366, 379, 382, 394, 416, 439, 446, 457, 526
Sūnūn, Mt. 216
Supuk 218
Sūq-i Arbaʿā 247–249
Sūq-i Asal 248
Sūr (tribe) 198
Sūr River 307
Ṣūr See Tyre
Surāl 218
Şūramağ 462
Surāqa b. ʿAmr 327
Surat, Bandar Surat 101, 182
Sūrāwān 290
Şūregīlī 346
Sūrghūrmūyān (a prophet) 150
Sūrī (ancestor of Ghūrids) 221
Sūrī (Ghūrid king) 225
Sūrish 218
Sūrnīkūr River 209–210
Surkhāb 259, 263, 285
Surkhāb River See Sarāwrūd
Surkhāb, Mt. 314
Surmaq 235
Sūrqān 230
Suryān (Assyrian kings) 368
Sūrzāde 347
Sūs 247–249
Sūs al-Aqṣā 125
Ṣuṣam Beli mountain 470
Ṣuşehir 344
Ṣuṣığırlığı 512
Suwāgh (idol) 416
Ṣuwān River 193, 203
Ṣuwar-i aqālīm (book) 321
Ṣuwar al-kawākib (book) 53
Suwayb (fortress) 363, 365
Suwaydā (in Egypt) 379
Suwaydā (in Iraq) 465
Sūwaydiyya 430, 434, 459–460
Ṣuwayḥ 365
Suwayqa (in Mecca) 395
Suypūr 219
al-Suyūṭī, Jalāl 43, 84
Sübḥān, Mt. 345
Şücāʿaddin 474
Şühūdpıñarı 356
Şükrī (poet) 337
Süleymān, Kanuni, Ottoman Sultan 42, 101, 291, 331, 334–335, 338, 354, 359, 366, 394, 400–401, 403–406, 408–409, 420–421, 428, 443–444, 447, 452, 460, 464, 473, 485, 491, 493, 496, 498, 505–506, 510, 524–525
Süleymān Agha 401, 406
Süleymān Pasha 101, 428, 510, 515, 518
Süleymān Reʾīs 524
Süleymān Shah (father of Ertuğrul) 513
Süleymānsarāy 339
Şürbḫāne 346
Sürmene 348
Sürmene River 349
Süsen Dıraz (in Manisa) 486
Swabia 99
Swahili coast 102–103
Swat 206, 212–214
Swat River 213–214
Sweden, Swedes, Suedia 69, 74–75, 90, 106–107, 125
Sea of Sweden See Baltic
Syria, Syrians, Shām, Bilād al-Shām 68, 84, 91–92, 96, 105, 124, 249, 251, 293, 325, 334, 337, 339, 350, 352, 361, 380, 393–396, 401–402, 406–409, 414, 420, 424, 430–441, 443–444, 447, 451, 453–454, 456–458, 461, 464–468, 471, 474, 483, 491, 524
Sea of Syria See Mediterranean
Syrian Desert 366, 409, 411
Syrian merchants 437
Syrian pilgrims 410
Syrian quintal 418
ancient Syrians 125
rulers of Syria 483
Syriac (language) 375, 378, 436
Szegedin 525
Szeklers 521
Szigetvár 525

Ṭaʿām River 222
Ṭāb River 238, 240, 246
Tabāla 406–407, 412
Ṭabaqāt (book by Ḳoca Nişāncı) 334
Ṭabaqāt (book by Tiflisī) 324
Ṭabaqāt-i Akbarī (book) 182
Ṭabaqāt-i Nāṣirī (book) 196, 258
Ṭabarak (castle) 253
  Ṭabarak mountain 261
Ṭābarān 271
Ṭabaristān 90, 125, 260–261, 264–265, 284–288, 291, 209, 309, 312, 323, 325, 327, 380, 525
  mountains of Ṭabaristān 286
Ṭabas, Ṭabasīn, Ṭabas-i Gīlakī and Ṭabas-i Mīnān 265, 276–277, 279
Ṭabasarān 329
Tabhandū 219
Tābiʿīn (the second generation of Muslims) 364
Tabriz, Tabrizi 131, 241, 255, 262–263, 282, 312–320, 324, 326–327, 334–335, 338–339, 341, 343, 523, 525–526
  Ghouta of Tabriz 314
  Lake Tabriz and see Lake Urmiya 316–317
Tabūk 393, 408–409, 411, 415, 421
Tadhkira (book) 42
Tadmur (Palmyra) 73, 384, 430, 431, 456, 465
Tadwīn (book by Imam Rāfiʿī) 254
al-Taftazānī See Saʿd al-Dīn
Ṭaġardı 484, 501
Tagharghar See Tughuzghuz
Taghaza 214
Ṭāhān 325
Ṭāhir (ruler of Sīstān) 223
Ṭāhirī (in Syria) 456
Ṭāhirī (in Transoxiana) 291
Ṭāhiriyya 290, 300
Ṭāhir (b. Ḥusayn, commander) 379
Ṭāhirids 223, 263, 273, 281, 427
Tahāfut al-Falāsifa (book by al-Ghazālī) 46–47
Ṭahmāsp, Ṭahmās, Shah (Safavid) 198, 251, 253, 255, 291, 314, 320, 327, 335, 371
Ṭahmāsp (legendary king of Iran) 242, 374
Ṭahmūrath (ancient Persian king) 236–237, 247, 251, 256–257, 370
Ṭahmūrath (governor of Zagam) 331
Ṭahrī (tribe) 355
Taḫt-ı ʿAlī mountain 489
Taḫtābād 481, 512
Taḫtalu 481
Taḫtelḳalʿe (in Istanbul) 418
al-Ṭāʾiʿ biʾllāh, ʿAbd al-Karīm (Abbasid) 381
Ṭāʾif 383, 393, 406, 407, 410, 416, 423, 466
Taʿizz 384–386, 388, 423, 427–428
Tāj al-Dīn ʿAlīshāh, Khwāja 313, 335
Tākht 230
Takht-i Rustam 375
Takhtmal 217
Taḳiyyüddīn (astronomer) 49, 526
Tākmū 217
Takrur (Sudan) 42
Tākwar 192–193
Talā See Urmiya: Lake Urmiya
Ṭālaqān 224–225, 257, 279, 287
  Ṭālaqān Mountains 261–262
Talfītā (village) 450
Ṭalḥa (Companion of the Propet) 364, 366, 380
Talingāna 200
Ṭālish, Tālish 315, 319
  Ṭālish villages 316
Talkhīṣ-i jāmiʿ (book by Muḥammad b. Mulkdād) 336
Talkhīṣ miftāḥ (book) 316
Tall Aʿfar 351
Tall Bashar 356
Tall Firʿawn 420
Tall Ḥūrna 450
Tall al-Tawba 351
Taman (peninsula and castle) 91, 105, 303, 330
  Taman Strait See Strait of Kerch
Taman (in Arabia) 412
Taman River (in Transoxiana) 303
Tanāṣara 102
Tanbīh (book by al-Masʿūdī) 99
Ṭanbūr (fortress) 238
Tandū 304
Tandūk, Mt. 216
Tanduq 178
Tangālīn 288
Tangdār 219
Tangiers See Ṭanja
Tangut 161, 178, 307
Tanīsar 205–206
Tanja (in Iran) 276
Ṭanja, Tangiers 125, 426
Tanka 228
Tankhar 217
Tanrıyaḳduġı (fortress) 519
Tanū 218
Tanūma 413, 424
Ṭanzī 355
Taprobana 110, 143, 147
Ṭāq-i Bustān, Khusraw's Shabdīz 261, 362, 370, 375
Ṭāq-i Kamarābād 375
Taqī al-Dīn Fāsī (historian of Mecca) 395
Taqwīm al-buldān (book by Abū'l-Fidāʾ) 42–43, 68, 73, 135, 151, 158, 177, 182, 186, 191, 193, 206, 210, 213, 219, 221, 227, 229, 233, 235, 247–248, 253, 257, 260, 267, 271–272, 274–276, 285–286, 288–289, 291, 295–299, 303–304, 309, 312, 318, 320–322, 325, 328, 334–335, 337, 339
Taqwīm al-abdān (book by Ibn Jazla) 42
Ṭarablūs See Tripoli
al-Ṭaraf (in Arabia) 407, 417
Tarak summer pasture 321
Ṭaraḳlı, Ṭaraḳlu, Ṭaraḳlıborlu 491, 497, 499, 501, 512–513, 518
Ṭaraḳlı Yeñicesi, Ṭaraḳlu Yeñicesi 497, 501, 519
Tāram 229–230
Ṭarāṭīr Rāʿī 423
Ṭarāz, Aṭrāz 296, 300, 304
Ṭarārak 248
Tarbil, Mt. 455
Tarhāda 206
Tarḫala 501, 503, 512
Ṭarıboġazı 513
Ṭārif 421
Tārīḫ-i Hind-i Ġarbī, Tārīḫ-i Hind-i Cedīd, History of the New World 42, 84, 90, 96, 99, 101, 115, 119, 121–122, 132, 135–136, 138, 147, 151
Tārīkh-i Akbarnāma (book by Abu'l-Faḍl b. Shaykh Mubārak) 192
Tārīkh-i guzīda (book by Ḥamd Allāh Mustawfī) 254, 263
Tārīkh-i Khurāsān (book) 272
Tārīkh-i Kirmān (book) 227
Tārīkh-i Mubārakshāhī (book by Muʿīn al-Dīn Zamjī) 212, 222, 274
Tarīm 388–389
Tarīma 388
Tarjīl River 373
Ṭarkhū 328–329
Ṭarlu (village) 505
Ṭarqān 176, 305
Tarshīz See Turshīz
Tarsus 125, 334, 368, 379, 430, 434, 457, 463–464, 466, 469–470, 504, 522
Ṭarṭūs See Anṭarṭūs
Ṭārum, Ṭārumayn 256–257, 262
  Upper Ṭārum and Lower Ṭārum 257
  Ṭārumayn River, Āb-i Ṭārum 261–262
Tārūza 213, 214
Ṭaryān 238
Ṭaşābād 462, 477
Ṭaşaḳ Yazusı (battleground) 513
Ṭāsānadası 352
Ṭaşan, Mt. 480
Ṭaşanoġlı 480
Ṭaşköpri (of Ḳasṭamonı) 496
Taşköprü River 374
Tāsnī (tribe) 361
Ṭaşözü (Thasos) 521
Ṭasūj 240
Tasūy 317
Tatar, Tatars, Tataria, Tataristan 42, 107, 110, 121, 125, 129, 152–153, 155, 158–159, 161–162, 167, 174–175, 178–180, 194, 203, 228–290, 294, 304–310, 382, 477, 513
  Tatar (eponym) 305
  Lesser Tataria See Crimea
  Tatar Sea, Sea of the Tatars and see Northern Ocean 90, 99
  Tatar army, Tatar soldiers 383, 514
  Tatar attacks 152
  Tatar invasion 296
  Tatar Khan, Tatar khans 306, 317, 526–527
  Tatar merchants 309
  Tatar pagans 309

Tatar River 310
Tatar ruler 477
Tatar tribes 179–180
infidel Tatars 178, 381
Crimean Tatars 526
Kalmyk Tatars 304
Moghul Tatars 310
Nogai Tatars 308
Pitorsi Tatars 310
Qazaq Tatars 308
Tūman Tatars 309
Yashilbash Tatars See Uzbeks
Zāwlaq Tatars 308
Zavolhen Tatars 189
Tatmal 219
Tatvan 336–337, 356
Taurus Mountains 111, 181
Ṭavşanlu 484, 519
Ṭawālish River 261
Ṭawāwīs 294, 300
Ṭawīla, Ṭawīl (in Yemen) 387–388, 423, 429
Ṭāwūs al-Ḥaramayn 234
Tāybād 278
Taymānī 436, 450
Taymā᾿ 414–415
Tays 228
Ṭayy (tribe) 377, 414–415, 431
Mt. Ṭayy 383
Ṭayyiba 388
Tāzān 411
Tazıḳırı 484–485
Tazīn 457
Tazrak 229
Tazrak Pakan 229
Tefen 497, 500
Tefenni 489, 490, 512
Lake Tefenni 489
Teke 469, 483–484, 489, 516
Tekfurṭağı 518
Tekir 463
Tekirbeli 512
Tekirburnı 483
Tekman 343–344
Tekye (village, Bolu) 500
Tekye Çayı 347
Tekyeler, also called Aḳyeler 480
Tel Ḥamdūn 464
Telājīā in Gujarat 145
Telangana 193
Telmese 482
Temāşālıḳ 510
Tendāy 134
Tercan 339, 344, 346–347, 481
Tercil 353, 355
Mt. Tercil 355
Terek 329–330
Terek River 330
Terme 478
Terme Plain 494
Ternāta 136
Terra Brazil See Brazil
Terra Galleca, Cailoco 132
Tesāwensā See Cape of Vīn-sen
Tetrabiblos (*Maqālāt-i Arbaʿa*, book by Ptolemy) 125
Thabīr, Mt. 397
Thābit (Arab lineage) 397
Thaʿlaba (Ghassanid) 465
Thaʿlabat al-Shām, (Banī) Thaʿlaba 431
Thaʿlabiyya 415, 424
Thalla (fortress) 387
Thamāra 387
Thames River 93
Thamūd 383, 408, 425
Thaniyyat al-ʿIqāb pass 452
Thaqīl 387–388
Thawr 366
Thawrā 433
Thawrā River 435
Thawrān (canal) 449
Theatrum (i.e., *Theatrum Orbis Terarum* of Ortelius) 186, 228
Thera (Santorini) 96
Therasia 96
Thessaloniki See Selanik
Thomas the Apostle, San Toma Pater 186, 189
Thrace, Thracia 106, 123
Three Towns See Istanbul
Thujja 424
Thulayth 415
Thule, Islandia 69, 106
Thuringia 107
Ṭīb 247, 249
al-Ṭībī, religious scholar 247
Tiberias 430, 432, 434, 441–442
Lake Tiberias, Lake al-Minya (Sea of Galilee) 432–433, 441–442
Tibet, Tibetan(s) 69, 163–164, 193, 200, 204, 215, 217, 219–220, 300, 303–306
Tibet Pass 215
desert of Tibet and Cathay 179
mountains of Tibet and Tataristan 203
Pool of Tibet 219
Tidore 116
Tiflis 312, 322, 324, 326, 330–334
governor of Tiflis 525
Tiflisī (jurist) 324
Tigris River, Shaṭṭ, Shaṭṭ al-ʿArab 111, 124, 158, 181, 215, 240, 246, 248–249, 336, 340–341, 346, 348, 350–356, 360–361, 363–365, 367–368, 370–375, 378
Dijlat al-ʿAwrā (Tigris below Baṭā᾿iḥ) 365, 375
al-Tīh (Sinai desert) 411, 422, 430, 437
Tihāma 384, 386, 388, 393, 416, 423, 427, 429
Ṭihrān (Tehran) 253
Tikrīt 350–351, 366, 372–373
Ṭīm (tribe) 413, 426
Timaeus (book by Plato) 112
Tīmāya 422
Ṭımışvar (Temesvár) 525
Timocharis of Alexandria 61
Timor Islands 133, 139, 141
Tīmūr, Amīr Tīmūr, Tīmūr Leng 161, 165, 180, 197–198, 226, 244–246, 250, 266, 269–270, 283–284, 291–293, 302, 306, 320–321, 327–328, 331–332, 335, 352, 357, 402, 477, 510, 519–520, 523
Timurids, Timurid dynasty 190, 198, 246, 302
Timurci (Demirci) 486–488
Tīmūrṭaş Pasha 344
Tīmūrtash (Chobanid) 347, 482
Ṭīn, Mt. 238
al-Ṭīna, Bilād 455
Tinnīs See Damietta
Tipora (Tripura) 153, 194
Tīr 234
Tire 487–488, 510
Mt. Tire 487
Tirebolı 348
Tırḥala 520
Tirilya 502–503
Tirişken 356
Tirmidh 275–276, 279, 297–300
Tirshak 277
Tīz 209–210, 230
Todurġa 497, 499
Ṭoġanḥiṣārı 475
Toḥma River 462
Toḳat 345, 477–478, 480, 482, 512–513, 519
Toḳat River 480
Ṭolab (village) 495
Toledo 70, 442
Tolia 84
Tolos, Toloslar 481
Tolui Khan 175, 270, 273, 282, 306, 482
Ṭomaniç 477, 501, 513
Mt. Ṭomaniç 484, 512
Ṭomaniç Pass 514
Ṭopraḳḳalʿe, Ṭopraḳḳalʿesi 351, 373, 477, 481
Torah, Old Testament 63, 451
Book of Genesis in the Torah 98
Ṭorbalı Göynük 513
Ṭoro 355
Tortum 331, 343, 346, 525
Ṭorul (of Erzurum) 343
Ṭorul (of Trabzon) 348
Ṭosya 481, 492–495, 512
Trabzon 125, 329–331, 334, 343–344, 347–349, 478, 483, 521–522
Trabzon Plain 344
Trāhkām 219
Transylvania, Erdel 106, 521
Transoxiana 69, 124–125, 161, 198, 224, 226, 265, 273, 275, 282–283, 289–303, 306, 309, 430, 432, 435–436, 454–456, 460, 469–470, 525
Tripoli, Ṭarāblus Gharb (Tripoli in Libya) 125, 208, 524
Bay of Tripoli 455
Tripoli, Ṭarāblus Shām (Tripoli in Lebanon) 469
Troglodytae 108

Troy, Eski İstanbulluḳ 110, 509–510
Mirror of Troy 186
Tubbaʿ (ancient kings of Yemen) 292, 371, 387, 395, 426
Ṭughā Tīmūr 319
Tughluq Shah dynasty, Tughluqids 197
Tughluqābād 197
Tughtigin (Ayyubid) 386, 388
Tughtigin (Seljuk) 467
Tughuzghuz, Toquz Oghuz 333
Tūḥ 233, 241
Ṭuḥāb 280
al-Tuḥfa (i.e., *al-Tuḥfa al-shāhīya*, book by al-Shirāzī) 90, 320
Tuḥfat al-fuḥūl (book) 43
Tuḥfetü'z-zemān ve ḥarīdetü'l-āvān (book by Muṣṭafā b. ʿAlī) 42
Tūhī River 217
Tuḫt 493, 495
Ṭukhāristān 213, 223–224, 269, 297–298
Tūlam 288
Ṭūlāsī 307
Tulça 514
Tulunids 467
Aḥmad b. Ṭulūn 97
Khumārawayh b. Aḥmad b. Ṭulūn 411
Ṭūmān Bay (Mamluk) 468, 523–524
Ṭumāṭāmı 346
Tūn 276, 278
Tunca River 84, 522
Tūnis, Tunisia, Tunisians, Carthage, Ifrīqiya 105, 108, 112, 125, 426
Tunkat 295
Tūr son of Farīdūn 306
Ṭūr Hārūn (Mt. Hor) 434
Ṭūr, Ṭūr Sīnā See Sinai: Mt. Sinai
Ṭūr Zaytā (Mt. of Olives) 434
Ṭuraġan 480, 496–497
Turan, Tūrān 242–243, 291–292, 298, 302–303, 306
Tūrān Shāh (Ayyubid) 427
Tūrān Shāh (Seljuk) 227, 231
Ṭūrān (in Sind) 206, 208–209
Tūrānī-oġlı 441
Turfan 303
Ṭurġud 472, 516
Ṭurġud Alp 517
Ṭurġudeli 493
Ṭurġudlu 486, 511
Turḫal, Keşan Ḳalʿesi 477–478, 480–482, 512
Turk, Turks, Turkish 42, 145, 197, 215, 261, 266, 273, 290–291, 295, 297, 299–300, 304–305, 307, 310, 317, 323, 362, 379–382, 466, 469, 481, 513–517, 519
best of the Turks 497
ignorant Turks 478
infidel Turks 301
Khan / Khaqan of the Turks 281, 362
King of the Turks 213
land of the Turks, Turkish lands 242, 281, 304
Muslim Turks 305, 495
obstinate Turks 490
pagan Turks 304
tribe(s) of the Turks, Turkish tribes 305, 307, 348
Turkey See Rūm
Turkish (language) 39–41, 43–44, 84, 304, 310, 348, 351, 356, 375, 436, 513, 519
Turkish clime 304
Turkish dynasty (i.e., Mamluks) 467–468
Turkish clan 471
Turkish corps 379
Turkish manner of punishment 310
Turkish method (of referring to the compass directions) 81
Turkish origin 327
Turkish sailors 104
Turkish slaveboys 379
Turkish style (of building) 447–448
Turkish sultans 305
Turkish town 304
Turkish women 289
Turkic people / tribe 245, 333
Turkistan 69, 87, 124–125, 161, 223–224, 265, 295–296, 299–300, 303–307, 316, 325, 333, 344, 474, 481, 485, 492–493, 495–496, 513337, 362
governor of Turkistan 302
ruler of Turkistan 362
Turkistan (village of Kash) 294
Turkmān, Turkmen 207, 305, 333–334, 357, 420, 451, 455–457, 461, 468, 470, 513
Turkmān (village) 319
Ṭursan (grandson of Baṭṭāl Ghāzī) 482
Turshīz, Tarshīz 274, 278
Ṭursun Faḳīh 515
Tuscany, Tuscia 106
Ṭūs 125, 271–272, 278–279, 288, 368, 378
al-Ṭūsī, Naṣīr (al-Dīn) 87, 97, 99
Tūsh River 217
Tustar 125, 232, 241, 246–249
Tustar River (Karun) 246, 248–249
Tutush (Seljuk) 352, 467, 476
Tuwān (castle) 340
Tūsūn clan (Japan) Tuz-ḫurmā 352, 362, 372, 374
Ṭuzla 501, 505, 507, 511, 518
Ṭuzlu (village) 495
Ṭuzlu Göl 498
Türk son of Japheth 180, 305
Türkeri 339
Tycho Brahe 51, 64–65
Tychonic astronomy 64–65
Tyre, Ṣūr 105, 125, 131, 430, 433–434, 441–442, 451
Bilād Ṣūr (fortress) 451

ʿUbaysī (road stage) 365
ʿUbbāda 452
Ubulla 248, 364
Ubulla Canal 237, 363–364
Uchcha 203, 206–207
Ūd 262
Udaipur See Chūtpūr
al-ʿUdayba 410
ʿUdaybiyya 423
ʿUdhayb (stream) 415
Udhn (fortress) 386
al-ʿUdhrī, Aḥmad b. ʿUmar 43
Ufacıḳlı 463
Uġrı River 501
Uḥud, Mt. 407
Ūj 329
Ūjān 313, 315, 318
Ūjān mountains 318
Ūjān River 318–319
ʿUkāẓ 419
ʿUkbarā 368, 375
Ukhayḍir 409, 421
Banī Ukhayḍir (dynasty) 429
Ukhdūd 390, 496
Ukraina (Ukraine) 527
ʿUlā 393, 408, 415, 421–422, 442
Ulaḳ, Ulaḳderesi 497, 500
Ulaş 464, 481
Ultajia (Ultima Thule?) 122
Ulubad 354, 501, 503, 512, 518
Ulubad Köprüsü 512
Ulubad River 503
Lake Ulubad 503–504, 512
Uluborlu 489–490
Ulugh Beg, Mīrzā Ulugh Beg Abū Saʿīd, Sultan Ulugh Beg Muḥammad 165, 212, 292, 302, 305
Ulugh Khan (Khan of Cathay) 162
Ulugh Khan dynasty 196
Ulugh Yurt See Qaraqum
Uluḳışla 463, 465, 474, 511–512
Ulus (qadi district of Bolu) 488, 497, 500
Ulus River 499
Ulus (tribe, clan) 344–345
ʿUmar (in Arabia) 424
Mt. ʿUmar 400
ʿUmar, Shaikh (disciple of Shaikh Shādhilī) 418
ʿUmar b. ʿAbd al-ʿAzīz, Umayyad caliph 227, 231, 279, 355, 448
ʿUmar b. al-Khaṭṭāb, Caliph 230, 243, 249, 253, 255, 259, 263, 319, 327, 337, 350, 352, 355, 357, 363–364, 366, 370, 377, 395, 398, 400–401, 408, 437–438, 440, 442, 465–466
ʿUmar b. Manṣūr (Rasulid) 384–386, 423
ʿUmar b. Sarḥ (on Khurāsān road) 277
ʿUmar Khayyām 273
ʿUmar Shaykh Mīrzā 198
ʿUmarī (in Arabia) 420
Umayyad(s), Umayyad dynasty 135, 255, 263, 269, 301, 319, 366, 377, 421, 433, 443, 435, 446, 466–467
Umayyad Mosque 466–467
ʿUmdat al-mahra (book by al-Mahrī) 43
Umm Barqaʿ (in Arabia) 415
Umm al-Ḥasan (road stage) 465
Umm Qurayn (road stage) 423
ʿUmmāq 388
ʿUmra (near Mecca) 403, 419

ʿUmra (in Syria) 450
ʿUnayza 413, 421
Unj 229
ʿUnṣurī 183
ʿUqaylids, ʿUqaylid dynasty 352
ʿUqbarā 372
Ur (village) 218
Uraniborg 65
ʿUrbān 462
Urdun (Jordan) 430, 437, 441, 465–466
  Urdun River, Jordan River, Sharīʿa River 111, 433, 437, 441–442, 451
Urfa See Ruhā
ʿUrfa (in Arabia) 424
Ūrgūn 213, 214
Urla, Gümüşḫāne 343–344
Urla (near Izmir) 510–511
Urmiya, Rūmiya, Urūmiya 316, 319, 370
  Lake Urmiya, Lake Tabriz 316–318
  Shaikh of Rūmiya, Mahmūd Efendi 316
Ūrṭāgh summer pasture 305
Uşak 484–485
Usāma b. Shurayk 318
Usbānīkath 296
ʿUsfān 406, 421
Ushkaw 218
Ushnūya 316
Usquf Köprüsü 456
Usrūshana 291, 296, 298, 300
  mountains of Usrūshana 292, 295
Ustūrkat 300
Ustuwā 274
Ūştī 359–360
Ustūn 334, 340–341
ʿUtba b. Farqad 319
ʿUtba b. Ghazwān (companion of the Prophet) 363, 366
ʿUtbī 204
ʿUthmān b. Abān 364
ʿUthmān b. ʿAffān (third caliph) 319, 366, 386, 398, 404, 406, 408, 430, 440, 445, 454, 456
ʿUthmān Muḥammad (fortress) 365
al-ʿUthra 410
Uwāl (island) 415
Uyghur (tribe), Uyghurs 305–306
ʿUyūn Mūsā 383, 422
ʿUyūn al-Qaṣab 423
ʿUyūn al-Tujjār 432, 441, 464–465
ʿUyūn al-Tūt 433, 450
Uyvar (Újvár in Hungary) 527
ʿUzayr 457, 461
Uzbek, Uzbeks, Yashilbash Tatars 195, 198, 284, 302
  Uzbek khans 303
  Uzbek armies / troops 284, 304
Uzhayn 70
Ūzkand 298–300
Uzun Ḥasan, Ḥasan Bey, Ḥasan-ı Ṭawīl (Aqqoyunlu) 225, 284, 314, 320, 341, 348–349, 361, 383
Üç Kilise 332
Üç Künbed 376
Üçkapulu 474
Üçpıñar 356
Üçtaş 481, 513
Üçkān 339
Ürgüp 474
Üskübi 497, 499–501
Üsküdar 482–483, 505–507, 511, 513, 522, 525, 528
Üsküdar Bāġçesi 506
Üsküdari Shaikh Maḥmūd Efendi (Hüdāyi) 507
Üstolnibelġrad (Székesfehérvár) 525
Üveys el-Ḳaranī (road stage) 356
Üveys Pasha 427–428

Vālersāyī tribes 355
Valide Sultan (sultan's mother) 461, 481
Van; and see Arminiyya 36, 320, 334–335, 338–343, 349, 352, 356, 361–362, 373, 525
  Lake Van, Lake Aḫlāṭ, Lake Erciş 335, 339–340
  mountains of Van 362
Vān-qūla (Angola) 109
Varangians 90
  Varangian Sea See Baltic Sea
Varat (Várad in Transylvania) 527
Varna 520
Varsak 457
Vasco da Gama 181
Vefāzāde (Ebū'l-Vefā', shaikh) 521
Venice, Venetians 107, 125, 145, 158, 471, 525
Vesṭān 334–335, 339–341
Viborg 69
Vienna 525, 528
Viglia Lopes (Ruy López de Villalobos) 132
Vijayanagar See Bījānagar
Vılḳoġlu (Vuk Branković) 520
Vindelicia 106
Vīrān 249
Vīrānşehir 497–498, 500, 512
  Vīrānşehir River 499
Visayas islands 138, 141
Vize 518
Volga River 105, 304, 307–308, 525

Wabīh 424
Wachūn 279
Waḍāḥ 412
Wadd (idol) 416
Wadhār 298
Wādī al-ʿAjam 443, 451–452
Wādī Banī Ḥanīfa 412
Wādī Barda 450
Wādī Būn 429
Wādī Buṭnān 458
Wādī Dawāl 427
Wādī'l-Dawāsīr 391
Wādī Fāṭima 407
Wādī'l-Ghaymā 422
Wādī Ḥafr 421
Wādī'l-Ḥusnā 384, 423
Wādī'l-Jannā 386
Wādī Khayyān 429
Wādī Laʿsān 387
Wādī'l-Murr 421
Wādī Mūsā 443
Wādī Nakhla 406
Wādī'l-Nār 423
Wādī'l-Nisā' 364
Wādī'l-Nuʿmān 442
Wādī Nuʿmān al-Arāk 406
Wādī'l-Qandīl 455
Wādī'l-Qurayḍ 422
Wādī'l-Qurayshiyya 455
Wādī Qurnā 450
Wādī Rābīl 456
Wādī Rishfayn 454
Wādī'l-Sabā' 424
Wādī Ṣahāma 427
Wādī al-Samā, Wādī Samāwa 372, 376
Wadi Samir 415
Wādī Samūl 386
Wādī'l-Ṣawān 414, 422
Wādī'l-Shaqq 408
Wādī'l-Ṣighar 422
Wādī'l-Sirr 412–414, 428
Wādī Tamā 423
Wādī'l-Tamīm 450
Wādī'l-Tamm 443
Wādī'l-Taym 453
Wādī'l-Tīn 433
Wādī'l-Zaytūn 460
Wafī 293
Wāfī bi'l-wafāyāt (book by Ṣafadī) 317
Wahīj 230
Wahīlak 412
Waḥshī 192
Wahūk 373
Wajān 299
al-Wajh 423
Wajh al-Ḥajar 454
Wajiblārū 215, 216, 217
Wajra 424
Waka 279
Wakhsh 224, 298–299
Wakhshāb 300
  Wakhshāb River 298–299
Wallachia, Walachian(s) 42, 519–520
al-Walīd b. ʿAbd al-Malik (al-Walīd I, Umayyad) 399, 438, 439, 444, 446, 466
Walīd b. Mughīra 398
Walīd b. ʿUqba 319
al-Walīd b. Yazīd (al-Walīd II, Umayyad) 466
Wālīn 279
Walwālij (Walwālish or Walj) 224
Wanatpūr 215
Wandanāo See Mindanao
Wāndaspak, Mt. 217
Waqf, Waqf Ḥiṣār 412, 421
al-Wāqidī 337
Wāqiṣa 424
Waqwāq 151
  Wāqwāqnāme (book) 151
Wār 219

Wārang 222
Wāraq 277
Wardkard 262
Warduk 213, 214
Warīz 300
Warka, Mt. 300
Warnga 217
Warthān 318, 321, 326
Warwaḥ 218
Warzand See Barzand
Wāsang (spring) 217
Wāsangkand 217
Wasaṭ al-Tīh 422
Wāshjird 298
Washm, Bilād al-Washm 412, 424
  Washm wadi 412
Wāsiṭ 244, 246–247, 249, 366–368, 370, 373–375, 410, 423–424, 525
Wasṭām (village) 362
Wat River 217, 218, 219
al-Wāthiq (Abbasid) 368, 372, 379, 460
Wathqana 433
Wāwarīḥājī 216
Wazīrshahrī 290
Wazlū 219
Wene-zārī 194
West Indies See America
western hemisphere 81
Western Sea, Western Ocean, Atlantic Ocean, Sea of Atlas 40, 42, 77, 82, 84, 93, 96, 99, 101, 108–109, 112, 116
Western philosophers 53, 83
Weygats Strait 120
Wīhat River 191, 193, 203
Wīma 286
Wintpūr 217
Wīrta River 203
Wittenberg 69
Wiyāh River 202–203
Wīza 300
Wonders of Creation (book by Solinus) 92

Yabanābād 492–493
Yabhun 218
Yābisa (Ibeza) 91
Yabrīn 393, 414
Yāfā (Jaffa) 437
Yaġan pasture 474
Yaghrā 433–434, 458
  Nahr Yaghrā 433–434
Yaghūq (idol) 416
Yaġmur Deresi 344
Yaḥṣub (Himyarite king) 386
Yahūdiyya 251
Yaḥyā Pasha 366
Yaḥyā b. Mihrawayh 380
Yaḥyā b. Zayd 333
Yaḥyālı 476
Ya'jūj River 304
Ya'ḳūb Agha 523
Ya'ḳūb Bey (Germiyanid) 510, 518–519
Yalaḳābād, Yalıova (Yalova) 492, 505–506
Yalamlam 388, 409, 423
Yaldā (in Syria) 449
Yalñızṭam 513
Yalvaç 489–491
Yamāma 383–384, 391, 412–414, 419–420, 429
Yanbu 383, 406, 410, 418–420, 423, 429
Yanghikand 294, 299
Yañiḥiṣār 303–305
Yanıḳ (Györ) 525
Yanḳo, king 520
Yanoş (king of Erdel) 524–525
Yanova (fortress) 527
Yaqṭan (ancient Yemenite) 384
Ya'qūb (Qaraqoyunlu) 320
Ya'qūb Charkhī 212
Ya'qūb b. Layth 221, 223
Ya'qūbābād 375
Ya'qūbī (historian and geographer) 297
Yāqūm (a carpenter) 398–399
Yāqūt al-Ḥamawī 42–43
Yārġāzībıñarı 481
Yārḥiṣār 501, 503–504, 515, 518
Yarīma 424
Yārkand 303–304
Yārmarak 218
Yarmuk River 433, 441–442
Ya'rub b. Qaḥṭān 425
Yarūm 383
Yaş (Iaşi) 525
Yasht-i Bādām 278
Yasī 296, 303, 305
Yassıçimen 481
Yathrib See Medina
Yayiçe 489, 521
Yayla Mescidi, Mt. 349
Yaylaḳ Ḥiṣārı 516
Yazd 232, 234–235, 237, 239, 241, 244–245, 251, 262–264, 275–276, 278
  Atabegs of Yazd 245
  Yazd road 278
Yazdajird (Sassanian) 102, 243
Yazdajird (fortress) 359
Yazdajird of Lesser Lūr 262
Yazdkhwar 235, 262
Yazīd (commander) 465
Yazīd b. al-Walīd (Yazīd III, Umayyad) 467
Yazīd b. Mu'āwiya (Yazīd I, Umayyad) 433, 466
Yazīd b. 'Abd al-Malik (Yazīd II, Umayyad) 466
Yazīd b. 'Umar b. Hubayra al-Fazārī (governor of Iraq) 369, 377
Yazīd Canal 433, 435, 449
Yazīdī(s), Yazīdī rite / sect, Yazīdī tribes 339–340, 350, 353, 355–356, 361, 436, 457, 463
Yedidivan 497, 499–500
Yelenkān 360
Yemen, Yemeni(s) 100, 124–125, 376, 379–380, 383–388, 390–391, 393–396, 401–402, 406, 409, 414, 416–420, 423, 425–429, 446, 453, 456, 465, 468, 524–525
  kings / rulers / Tubba's of Yemen 264, 292, 311, 371, 401, 420, 425
  Yemeni corner (of Ka'ba) 399, 401, 405
  Yemenite carnelian 349
  Sharif of Yemen 525
Yemen (tribal grouping; and see Kalbiyya),
  Yemeni faction / party, Yemenis 435, 438, 452, 466
Yeñi Ḳapluca (of Bursa) 501–502
Yeñice 516
Yeñice Ḳal'e 461
Yeñice-i Bolu 497, 499–500
Yeñiceḥiṣār 483
Yeñiceköy 493
Yeñiḥān 491
Yeñiköy 507, 526
Yeñişehir of Aydın 487, 511–512
  Yeñişehir Ravine 487
Yeñişehir of Beyşehri 474
Yeñişehir of Bursa 484, 492, 501, 503, 516, 519
  Yeñişehir Ovası, Yeñişehir valley 286, 503
Yeñişehr-i Fenar (Larissa in Thessaly) 527
Yeñiṭaġ 504
Yergögi (Giurgiu) 520
Yerköy 492–493
Yernes 96
Yeşil 'İmāret 502, 520
Yılan Ḳal'esi 464
Yılancı 512
Yılancıḳ Ṭaġı 503
Yılanlıca, Yılanluca 497, 499
Yıldız River, Yıldızṣuyı 481, 512
  Mt. Yıldız 481
Yoghi(s) 215, 219
Yoncalı plain 484
Yorgi (nephew of Rustam Bey) 330
Yoros 505–507
Yörük(s), Yörükān 473, 485, 492–493, 496, 512
  Yörükān of Ankara 493
  Yörükān-ı Araç. 496
  Yörükān-ı Bolu 497
  Yörükān-ı Ṭaraḳlı 497
Yucatan 99
Yund, Mt. 503
Yūnus Pasha 524
Yurtbāzār See Jawād
Yūsuf Khan (governor of Kashmir) 219
Yūsuf Pasha 428
Yūsuf Qāḍī, Mevlānā 165
Yūsuf Qarabāghī, Mawlānā 324
Yūsuf Shāh, Atabeg (governor of Yazd) 244
Yūzkand 303
Yüksekyazı 360
Yüñlü 512
Yüzbaşı (village) 375

Zāb River, Greater and Lesser Zāb 263, 350, 352, 359–360, 362, 373–375, 463
Zabadānī 443, 450
Zabīd 383–386, 388, 409, 416–418, 420, 423, 427–428
  Wādī Zabīd 426

Zabīd (tribe) 410, 421
Zabtar 461
Zābulistān 209, 211–212
Zachariah 430, 438
Zād-i Ākhirat 278
Ẓafar (fortress) 224
Ẓafār (Dhofar) 383, 389, 391, 424, 429
Zaʿferānborlu See Borlu
Ẓafīrjaḥa 387
Zagam 330–331
Zagam Mountains 324
Zağanos Pasha 504
Zağanos quarter (Trabzon) 348
Zağra 518
Zāhid Gīlānī 315
Ẓāhir b. Muḥammad (ruler of Sīstān) 223
al-Ẓāhir bi-Amr Allāh (Abbasid) 381
Ẓahīr al-Dīn ʿAlī, Khwāja 256
al-Ẓāhirī, Khalīl b. Shāhīn 43
Ẓahr al-Ḥimār 422
Ẓahr al-ʿAqaba 421
Ẓahr ʿUnayza 421
Zahrān 407
Zaʿīm İsmāʿīl (road stage) 346
Zaire River 109
Zakho 373
Zakiyya (fortress) 363, 365, 374, 424
Ẓālim, Ẓālim ʿAlī, Ẓālimḳalʿe (fortress) 359, 367, 525
Zamakhshar 289
Zāmīn 295, 297, 300
Zamm 275, 298
Zammāra 424
Zamzam Spring 395, 397, 406
Zanbāqiyya 465
Zandarūd River 249, 251–252, 256, 261–262
Zangābād 367, 372
Zangī River, Zangī Ṣuyı 320, 324
Zangibār, Zanzibar 42, 125, 233
Zangids
ʿImād al-Dīn Zangī 373, 467
Maḥmūd b. Zangī, Sultan 462
Nūr al-Dīn the Martyr (Nūr al-Dīn al-Zangī) 443–444, 447, 450, 452, 467
Saʿd b. Zangī 232, 249
Zangistān 125
Zanj (Africa) 385, 392, 393
Zanjān 251, 256, 262–263, 319
mountains of Zanjān 262
Zanjānrūd River 261–262
Zānjīā See Korea
Zanjīr (fortress) 367
Zanjis 380
Zaʿqa 465
Zar, Mt. 219
Zara, Lake 222
Zard Mountains 248
Zarand 227, 230, 260
Zaranj 221–223, 265
Zard, Mt. 261
Zardusht See Zoroaster
Zarkān 236
Zarnūk (fortress) 365
Zarqāʾ (a false prophetess) 413–414
Zarqāʾ (in Arabia) 420
Zarqāʾ River 442
ʿAyn al-Zarqāʾ, spring of Zarqāʾ 404, 420, 442
Zarrāja 424
Zarrīn Dirakht 247
Zarwadī (village) 367
Zāwul, Zāwulistān 243, 276
Zāwa 272
Zāwān 370
Zawur 218
Zawwāra 258
Zawzan 276, 279
Zāyanda River, Zāyandarūd 261–262
Zayd b. ʿAlī (the son-in-law of the Prophet) 416
Zayda 424
Zaydīs, Zaydī dynasty, Zaydī sect 384–386, 416, 422, 429
Zaylaʿ 102, 194, 406, 417
Cape of Zaylaʿ 108
Zayn al-ʿĀbidīn, Sultan (of Kashmir) 219
Zayn Ḥasan (governor of Kashmir) 219
Zaynī (fortress) 364
Zazḳırı 512
Zaznār Descent 218
Zelketek mountain pass 481
Zenbere, Lake 109
Zenge 359
Zengi Ṣuyı (in Armenia) 511–512
Zengi Ṣuyı (near Konya) 512
Zengiçay 473
Zephyrium 96
Zeres (fortress) 374
Zerīl 340
Zerpande 510
Zerzene 497, 500
Zevilce summer pasture 499
Zevīn 345
Zeytūn (Zaiton, Quanzhou) 157–158, 160
Zeytūn River 158–160
Gulf of Zeytūn 158
Zeytūn (district of Amasya) 480
Zeytūnburnı 505
Zībār, Zībārī 340–341, 361
Zībār River 373
Zīj-i Ilkhānī (book by Naṣīr al-Ṭūsī) 317
Zile 477–478, 481, 513
Zindānbād caves 234
Zinjīrgāh, Mt. 267
Zirḳī (fortress) 354
Zīrkūh 276
Ziyāb Mountains 387
Ziyād b. Samra 366
Ziyād b. Umayya (governor of Fārs) 238
Ziyādābād 241
Ziyādids 427
Zoroaster, Zardusht 243, 376
Zoroastrians 200, 208, 239–240, 243, 246, 269
Zubayd (tribe) 431
Zubayda, daughter of Jaʿfar al-Dawānīqī 223
Zubayda, wife of Hārūn al-Rashīd 257, 312, 378, 394, 404, 424, 460
Zubaydiyya 262
Zubayr (companion of the Prophet) 364, 366
Zughar See Dead Sea
Zulmattū 217
Zumrud, Mt. 418
Zundar 217
Ẓufār 102
Zuqāq-i Qaṣaba 365
Zūr 457
al-Zūr See Shahrazūr
Ẓünnūnābād 480

Printed in the United States
by Baker & Taylor Publisher Services